Y0-BYK-423

AIBS DIRECTORY OF BIOSCIENCE DEPARTMENTS AND FACULTIES

IN THE UNITED STATES AND CANADA

SECOND EDITION

PETER GRAY

Andrey Avinoff Professor
University of Pittsburgh

Distributed by
HALSTED PRESS
A division of
John Wiley & Sons, Inc.

Library of Congress Catalog Card Number: 75–33761
ISBN: 0–470–32272–1

77 76 75 1 2 3 4 5

Manufactured in the United States of America.

LIBRARY OF CONGRESS CATALOGING IN PUBLICATION DATA

American Institute of Biological Sciences.
AIBS directory of bioscience departments and faculties in the United States and Canada.

Published in 1967 under title: Directory of bioscience departments in the United States and Canada.
Includes bibliographical references.
1. Biology--Study and teaching--United States--Directories. 2. Biology--Study and teaching--Canada--Directories. I. Gray, Peter, 1908- II. Title.
QH319.A1A44 1975 574'.07'1173 75-33761
0 470-32272-1

Exclusive distributor: **Halsted Press**
A division of John Wiley & Sons, Inc.

FOREWORD

There are many aspects of the slow and undramatic changes in the world of biology that are not apparent until specific attention is drawn to them. A directory of departments and faculties offers, as one of its purposes, information about the changing structures into which departments are organized. These fluctuations, whether they result from public or political pressure, from competition for funds, or from conflict between the old and the new, inevitably reflect the transition from one generation of scholars to the next. The passing parade, in fact, consists not only of people but of new specialities and new concepts kept in step by recurrent aspirations.

As the biological community gains by such a recording of its history, structure, and participants, the AIBS is especially grateful to those who cooperated in the generation of the required information. All of us owe a particular debt to Dr. Peter Gray, whose perseverance and dedication have made this volume possible.

RICHARD TRUMBULL
Executive Director, AIBS

PREFACE

The preface of a book affords the author an opportunity of speaking to his reader in a comparatively direct and personal manner, and of acquainting the prospective user of the book with the considerations which impelled the author to write it.[1]

I agreed to edit this second edition of the American Institute of Biological Sciences Directory because I felt that public service was not a bad way of occupying one of my declining years. The first edition was supported by a lavish grant, no longer available. However, the AIBS felt that their exiguous budget could provide funds for the printing and distribution of the questionnaires and furnish me with the occasional services of a part-time secretary. A number of people, doubtless in part inflamed by the almost universal dislike of questionnaires, pictured me in a luxurious office surrounded by assiduous secretaries.

"Why," wrote one "respondent" across the front of the questionnaire, "should I use *my* time to make *you* money?" I wish I knew the answer to that one; it would have enhanced my youth and assuaged my old age. Then, there was the Dean who wrote in fury: "You should really overhaul your staff. I called twice yesterday and the phone was not answered." The staff (me) was at the time enjoying hospitality in the original sense of that word. I was far more sympathetic to the Chairman who wrote: "We have 642 majors who are far more important than your [expletive deleted] questionnaire." Another response that I liked was the individual who inscribed, in small and meticulous penmanship: "If you want this information, you should send $1 (with 25 cents handling charge) to the bookstore, who will then send you a *Bulletin*. This department is *not* in the habit of distributing free information." The only one who really puzzled me was the head of a parabiological department in a medical school. "Under no circumstances," he wrote, "may any details of the operation of this department be included in your volume." This points out the difference between an editor, which has been my role, and a compiler. In the former capacity, I am bound by his request. In the latter role, I would have extracted, and published, the necessary information from his bulletin; I cannot believe that it would have appeared as shameful as he feared.

All these people were, of course, a minute minority. As I have also written in another place[2]: "I am confirmed in my opinion that biologists the world over are the most scholarly, the most courteous, and the most cooperative group with which any person could be privileged to associate."

Many people, in addition to returning their questionnaires promptly, asked if there was any way in which they could help, or suggested schools and departments that I might have overlooked. One dean asked for a copy of the questionnaire, copied it, and sent it for action to each of the seven departments under her jurisdiction. This sort of help can only be appreciated by those who have received it.

It is impossible to acknowledge everyone who helped, but particular tribute must be paid to Barbara Hagen, who typed the whole body of the text as it appears. The more than 30,000 cards necessary for the indexes were prepared by Cindy Kane, Lisa Domanski, and Lois Toski. They were then hand-sorted, alphabetized, and filed by me and my wife before being typed by Ms. Toski.

Dowden, Hutchinson & Ross have been a tower of strength, supplying materials and information with generosity and dispatch, as has also the AIBS.

But my greatest debt, which I hope this volume will in part repay, is to the more than 2,400 individuals who filled out and returned questionnaires.

PETER GRAY

[1] *The Bookman's Glossary*, 3rd ed. New York: Bowker [© 1951]. Quoted by permission of the R. R. Bowker Co.

[2] P. Gray, *The Encyclopedia of the Biological Sciences*, 2nd ed. New York: Van Nostrand Reinhold, 1970.

INTRODUCTION

Justification for this volume. The only justification for a reference work is that the information contained in it cannot be obtained from any other single source. The present volume easily meets this criterion. Names of most faculty can be found in *American Men of Science,* but they are not, of course, grouped by department. Names of departments can be found in *American Colleges and Universities* (Washington, D.C.: American Council on Education, 1973), but faculty are not listed, a situation parallelled in *Universities and Colleges of Canada* (Ottawa: Information Canada, 1974). Neither volume lists graduate courses and specialities, although the Canadian work indicates fields in which degrees are offered. Neither lists the field stations attached to many departments, and only the Canadian source indicates the requirements for the various degrees. Not even a complete collection of U.S. college bulletins would wholly replace the present volume, since by no means all of them indicate the specialized areas of degree study within departments with generalized titles.

Organization of this volume. There is no universally accepted definition for the term "bioscience department." It is taken by the editor to mean a department that grants academic, as distinct from, or in addition to, professional degrees. Many schools of the health professions, for example, have divided themselves into divisions of basic sciences and clinical sciences—but others have not, and the many borderline cases can cause confusion. It is obvious, for example, that a department of biochemistry, no matter where administratively located, is a bioscience department often offering baccalaureal, as well as magistral and doctral, degrees. It is equally obvious that a department of neurosurgery, although conducting research of the highest order and giving instruction in superlative skills, is not a bioscience department as this term is usually understood, and specifically applied, in this volume. But what about a department of pathology, either in a medical school or a college of veterinary science? It may be a research-oriented organization offering academic degrees not only in pathology but also, often enough, in parasitology, histology, and cytology. It may also confine itself to offering service courses to those seeking professional degrees, the remainder of its faculty's time being devoted to the pursuit of their professional practices. The editor has endeavored to include the first type and to omit the second, but the establishment of the distinction has not always proved a simple task.

Departmental names, particularly of parabiological departments in professional schools, can frequently be misleading. For example, in the 1950s and 1960s the term "anatomy" fell into disfavor and departments with that title were quick to change either their name, or their habits, or both. The most usual aliases adopted were "biological structure" and "cell biology." Names by no means always followed interests, so that today one may find a "Department of Anatomy" offering a Ph.D. in invertebrate zoology or a "Department of Cytobiology" offering instruction in human gross anatomy.

Contemporary confusion in the biological sciences. The confusion in names, location, and functions is by no means confined to schools of the health professions. In the first place, incredible though it may seem, more than two hundred different names (from Acarology to Zymology) are applied to the bioscience departments here listed. The editor had originally intended to index these, but the task proved hopelessly impractical, since there may be five departments of biology to one page and one department of oenology to one hundred pages.

This appalling confusion derives in large part from the fact that the diseases of "lumping" and "splitting," endemic among taxonomists, have reached epidemic proportions among administrators. One dean, in the hope of quieting the internecine squabbles within his division, may fuse a dozen departments into one. The usual result is to transfer the skirmishes about budgets into vicious arguments about the content of the freshman course. Another dean, with equally pacific intent, may split a feuding group into half a dozen departments, each of which immediately transforms its scholarly differences into financial demands. These results could easily have been forecast by a study of political, as distinct from academic, history. The world is filled with heavily populated areas, the governments of which are seeking, on grounds of common sense, efficiency, and economy, to fuse the numerous villages and townships from which they have grown into metropolitan areas. The statesmen of the world, at Versailles, endeavored to ensure perpetual peace by chopping up the Austro-Hungarian empire into small fragments. Neither group has been any more successful than have deans.

None of this would be of the least importance to the editor of a directory of bioscience departments were it not for the unfortunate fact that the least scholarly division of a university is usually its mailing department. One of these had the courtesy to return to the editor, with the endorsement "no such department in this school," a letter addressed to the "Department of Biology"; this had, in fact, been split into Bacteriology, Botany, and Zoology a scant two years before the letter arrived. Thus raises the question of how many departments, listed in this directory with the comment "no reply received," may never have received the editor's request for information.

There are two more major sources of difficulty in finding and selecting bioscience departments. The first is the fact that names of departments, and this by no means only in schools of the health professions, may be either meaningless or misleading. There was undoubtedly a time when the "Department of Poultry Science" confined its activities to the care, feeding, and marketing of commercially important birds. Today, a department of the same name may be a center of genetic research or, in one classic case, of ethology. There is recorded in this directory a "Department of Horticulture" that offers three graduate-level courses in molecular genetics. The editor may be assumed to have done his best, but there are many places in which he would admit that the assumption appears to be unjustified.

The second source of difficulty referred to is the extraordinary replication of departments in a single institution and their movements within and among schools and divisions. The fact that this situation is confined to the United States does not make it less cogent. In every other area of the world in which institutions of higher learning are found, a university may have a "Department of Bacteriology." This houses all the bacteriologists, each of whom conducts his own particular research and teaching, as well as contributing to such generalized instruction as may be pertinent to students specializing in other fields. A letter addressed to the "Department of Bacteriology" in a university of the United States may be referred to an organization of that name in the College of Arts and Sciences, the Medical School, the Graduate School of Public Health, the College of Agriculture, the School of Pharmacy, and possibly others. Moreover, arcane departments on the cutting edge of contemporary research rarely have much time for undergraduates, since the primary interest of their faculties is in training disciples to follow in their footsteps. In those eras in which esoteric studies are lavishly supported by governmental agencies and private foundations, these departments are welcomed everywhere. When the going gets tough, they tend to congregate in schools of medicine, beloved of wealthy urban donors, or schools of agriculture, beloved of rural legislators. None of this eases the difficulties of an editor in search of bioscience departments.

Method of securing information. The information included in this volume was solicited by the distribution of questionnaires. The original mailing was from the American Institute of Biological Sciences computerized list of bioscience department chairmen. This produced a surprisingly high (50 percent) response, and a later follow-up mailing to nonrespondants from the same list resulted in the collection of a further 10 percent. Unfortunately, the list included a considerable number of two-year colleges, the information from which was not germane to the present volume, and the editor regrets the inconvenience caused to those involved. An unexpected difficulty derived from the use of the chairmen's names. Many were no longer in the same position and, although the questionnaire may have been forwarded to them, many felt justified in ignoring it. An exception, though not a very helpful one, was an individual who received the document while on a field trip in Africa and regretted that the information requested was not, under these circumstances, available to him.

The editor then went through *American Universities and Colleges*, page by page, noting those relevant institutions and departments omitted, for one reason or another, from the AIBS list. This included a considerable number of "Departments of Science and Mathematics" and "Departments of Natural Sciences." These are ambiguous terms, since they cover anything from a two-person department offering a leavening of science in a school devoted primarily to nonscientific topics to a thirty-person department offering majors in a variety of sciences. Moreover, some of these departments had changed their names to Department of Biology, and in a few cases the reverse had occurred. All departments, both biological and general, that had not previously been approached were then sent the original questionnaire and the

NUMBERS OF INSTITUTIONS (AND DEPARTMENTS)
APPROACHED FOR INFORMATION

United States		*Canada*
Alabama 23 (51)	Nebraska 13 (28)	Alberta 2 (10)
Alaska 1 (3)	Nevada 2 (7)	British Columbia 4 (16)
Arizona 5 (22)	New Hampshire 13 (22)	Manitoba 3 (16)
Arkansas 16 (29)	New Jersey 21 (37)	New Brunswick 3 (3)
California 62 (164)	New Mexico 8 (19)	Newfoundland 1 (4)
Colorado 14 (34)	New York 91 (181)	Nova Scotia 6 (10)
Connecticut 19 (37)	North Carolina 42 (79)	Ontario 15 (47)
Delaware 2 (4)	North Dakota 8 (24)	Prince Edward Island 1 (1)
District of Columbia 8 (20)	Ohio 49 (93)	Quebec 11 (35)
Florida 21 (51)	Oklahoma 18 (38)	Saskatchewan 2 (16)
Georgia 29 (49)	Oregon 13 (33)	
Guam 1 (1)	Pennsylvania 85 (139)	*Total* 49 (158)
Hawaii 6 (16)	Puerto Rico 4 (11)	
Idaho 5 (14)	Rhode Island 7 (17)	
Illinois 51 (102)	South Carolina 23 (39)	
Indiana 37 (54)	South Dakota 12 (24)	
Iowa 27 (47)	Tennessee 37 (58)	
Kansas 22 (49)	Texas 60 (118)	
Kentucky 23 (43)	Utah 6 (27)	
Louisiana 19 (24)	Vermont 10 (24)	
Maine 14 (23)	Virgin Islands 1 (1)	
Maryland 20 (48)	Virginia 30 (53)	
Massachusetts 46 (69)	Washington 13 (35)	
Michigan 30 (68)	West Virginia 17 (25)	
Minnesota 23 (43)	Wisconsin 30 (58)	
Mississippi 15 (32)	Wyoming 1 (6)	
Missouri 31 (63)		
Montana 6 (15)	*Total* 1182 (2371)	

follow-up. This raised the responses to about 70 percent. The editor then again went carefully through the pages of *American Universities and Colleges* and *Universities and Colleges of Canada,* eliminating from the final list of departments approached those with general science departments demonstrably too small to offer a bioscience degree and some other departments that, on internal evidence, fell outside the definition given at the commencement of this introduction. This left the 2,371 U.S. and 158 Canadian bioscience departments summarized in the table. A list of the individual colleges, arranged by state and province, is given immediately following the table of contents.

Each nonrespondent on this final list now received a personal letter from the editor accompanied by a stamped and addressed postcard on which the recipient was asked to check either of two boxes, one requesting an additional questionnaire and the other requesting the entry "no information available." (The letter to French Canadian universities and colleges was in the language of their preference, a courtesy immediately reciprocated by the number of responses.) Those who had asked for an additional questionnaire, but had not returned it within two weeks, were then approached on the telephone.

These activities finally resulted in responses being received from 92 percent of the 2,529 departments approached. Of these, approximately 2 percent requested the entry "no information available."

The nature of the questionnaire and the responses. The nature of the questions asked is apparent from the information given in the body of this directory. It is regrettable that considerations of space, in this instance synonymous with economy, made it impossible to follow the example of the first edition in the inclusion of undergraduate courses and of the academic qualifications of the faculty. New to this edition is the list of field stations and the identification of those areas in which a graduate student may specialize. This is reinforced by the listing of courses that may be taken for graduate credit.

The responses to the questionnaire were varied. A scant majority provided the information as and how requested. Many were too pressed for time or short of help to do this and provided tearsheets or copies of relevant portions of their bulletins. This required much additional labor by the editor, since a great deal of the material submitted in this form had to be edited and reduced in bulk.

It must be emphasized that the entry "no reply received" means exactly this and carries no other implication. It may well be that, twice exposed to the double jeopardy of university and federal mails, none of the material may have been received by the department or a response may have been lost in transit. There are also numerous explanations of the request for the entry "no information available." Several respondents explained that their departments were in a state of reorganization; some, in these days of academic depression, were not certain of their survival; others frankly did not agree with the editor that they belonged in a directory of bioscience departments. In any case, the editor is grateful to all of the 92 percent who replied, no matter in what form the reply arrived.

It is unfortunate that it is not possible to make an accurate estimate—an actual count would be impossibly time consuming—of the number of faculty listed. The obvious device of counting the number of names in a column of index, and multiplying this by the number of columns, ignores the fact that numerous individuals share the same name and initials. The name "Williams," for example, occupies eighty lines but requires one hundred and six page references; and "Williams" is by no means the most frequent name. The editor's guess, based largely on the number of index cards consumed, is that about 30,000 individuals are recorded.

In conclusion, the editor would like again to thank all concerned and to apologize for the errors that will inevitably be present and for which he must assume the sole responsibility. He cannot, however, accept all the blame, particularly for the inaccurate spelling of names, and omission of initials, infuriating though these are to the victim. Departmental secretaries, who filled out the questionnaires, may have accounted for some. An editor is, after all, merely a reproducing machine who tries to avoid adding to, but cannot correct, the errors of others.

CONTENTS

INSTITUTIONS GRANTING BIOSCIENCE DEGREES

UNITED STATES

ALABAMA

Alabama A and M University, p.3
Alabama State University, p.3
Athens College, p.3
Auburn University, p.3
 School of Agriculture, p.3
 School of Vetinary Medicine, p.6
Birmingham Southern College, p.6
Florence State University, p.6
Huntingdon College, p.7
Jacksonville State College, p.7
Judson College, p.7
Livingston University, p.7
Miles College, p.7
Mobile College, p.7
Oakwood College, p.7
St. Bernard College, p.7
Samford University, p.8
Spring Hill College, p.8
Stillman College, p.8
Talladega College, p.8
Troy State University, p.8
Tuskegee Institute, p.8
 College of Arts and Sciences, p.8
 School of Applied Science, p.9
 School of Vetinary Medicine, p.9
University of Alabama in Birmingham, p.10
 University College, p.10
 School of Medicine, p.11
University of Montevallo, p.12
University of South Alabama, p.12

ALASKA

University of Alaska, p.14

ARIZONA

Arizona State University, p.15
Grand Canyon College, p.15
Northern Arizona University, p.15
Prescott College, p.16
University of Arizona, p.16
 College of Agriculture, p.16
 College of Liberal Arts, p.18
 Graduate College, p.19
 College of Medicine, p.19

ARKANSAS

Arkansas College, p.20
Arkansas Polytechnic College, p.20
Arkansas State University, p.20
The College of the Ozarks, p.20
Harding College, p.20
Henderson State College, p. 21
Hendrix College, p.21
John Brown University, p.21
Ouachita College, p. 21
Southern State College, p.21
State C ollege of Arkansas, p.21
University of Arkansas, p.22
 College of Arts and Sciences, p.22
 College of Agriculture and Home Economics, p.22
University of Arkansas Medical Center, p.24
University of Arkansas at Little Rock, p.25
University of Arkansas at Monticello, p.25
University of Arkansas at Pine Bluff, p.25

CALIFORNIA

Biola College, p.26
California Baptist College, p.26
California Institute of Technology, p.26
California Lutheran College, p.26
California Polytechnic State University, p.26
 School of Agriculture and Natural Sciences, p.26
 School of Sciences and Mathematics, p.27
California State College, Bakersfield, p.27
California State College, Dominguez Hills, p.27
California State College, San Bernadino, p.28
California State College, Sonoma, p.28
California State College, Stanislaus, p.28
California State Polytechnic University, p.28
 School of Agriculture, p.28
 School of Science, p.29
California State University, Chico, p.29
California State University, Fresno, p.29
 Division of Agricultural Sciences, p.29
 Division of Natural Science, p.30
California State University, Fullerton, p.30
California State University, Hayward, p.30
California State University, Long Beach, p.31
California State University, Los Angeles, p.31
California State University, Northridge, p.31
California State University, Sacramento, p.31
California State University, San Jose, p.32
Chapman College, p.32
Claremont Colleges, p.32
 Claremont Graduate School, p.32
 Pomona College, p. 32
 Claremont Men's, Pitzer and Scripps Colleges, p.33
College of Notre Dame, p.33
Dominican College of San Rafael, p.33
Holy Names College, p.33
Humbolt State University, p.33
 School of Natural Resources, p.33
 School of Science, p.34
Immaculate Heart College, p.34
La Verne College, p.35
Loma Linda University, p.35
 College of Arts and Sciences, p.35
 School of Medicine, p.35
 School of Health, p.36
Lone Mountain College, p.36
Los Angeles Baptist College, p.36
Loyola Marymount University, p.37
Mills College, p.37
Mount St. Mary's College, p.37
Occidental College, p.37
Pacific College, p.37
Pacific Union College, p.38
Pepperdine University, p.38
Point Loma College, p.38
St. Mary's College of California, p.38
San Diego State University, p.38
San Franciso State University, p.40
Southern California College, p.41
Stanford University, p.41
 School of Humanities and Sciences, p.41
 School of Medicine, p.41
University of California Berkeley, p.42
 College of Letters and Sciences, p.42

College of Agricultural Sciences, p.44
School of Public Health, p.46
University of California, Davis, p.46
College of Environmental and Agricultural Sciences, p.46
College of Science and Letters, p.51
School of Vetinary Medicine, p.52
School of Medicine, p.53
University of California, Irvine, p.54
School of Biological Sciences, p.54
College of Medicine, p.55
University of California, Los Angeles, p.55
College of Letters and Science, p.55
School of Medicines, p.56
University of California, Riverside, p.57
University of California, San Diego, p.59
University of California, San Francisco, p.60
University of California, Santa Barbara, p.61
University of California, Santa Cruz, p.61
University of the Pacific, p.61
University of Redlands, p.62
University of San Diego, p.62
University of San Francisco, p.62
University of Santa Clara, p.62
University of Southern California, p.62
College of Arts, Letters and Science, p.62
School of Engineering, p.63
School of Dentistry, p.63
School of Medicine, p.63
Westmont College, p.64
Whittier College, p.64

COLORADO

Adams State College, p.65
The Colorado College, p.65
Colorado State University, p.65
College of Agricultural Sciences, p.65
College of Forestry and Natural Resources, p.66
College of Home Economics, p.67
College of Natural Sciences, p.67
College of Vetinary Medicine and Biomedical Sciences, p.68
Colorado Womens College, p.70
Fort Lewis C ollege, p.70
Loretto Heights College, p.70
Metropolitan State College, p.70
Regis College, p.70
Southern Colorado State College, p.70
United States Air Force Academy, p.70
University of Colorado, p.71
College of Arts and Sciences, p.71
Denver Medical Center, p.72
University of Denver, p.73
University of Northern Colorado, p.73
Western State College, p.73

CONNECTICUT

Albertus Magnus College, p.75
Annhurst College, p.75
Central Connecticut State College, p.75
Connecticut College, p.75
Eastern Connecticut State College, p.75
Fairfield University, p.75
Quinnipiac College, p.76
Sacred Heart University, p.76
Saint Joseph College, p.76
Southern Connecticut State College, p.76
Trinity College, p.76
University of Bridgeport, p.76
University of Connecticut, p.77
College of Liberal Arts and Sciences, p.77
College of Agriculture and Natural Resources, p.77
University of Connecticut Health Center, p.79
University of Hartford, p.79
University of New Haven, p.79
Wesleyan University, p.79
Western Connecticut State College, p.80
Yale University, p.80
Yale College, p.80
School of Medicine, p.80
School of Forestry, p.81

DELAWARE

Delaware State College, p.82
University of Delaware, p.82
College of Agricultural Sciences, p.82
College of Arts and Science, p.82

DISTRICT OF COLUMBIA

The American University, p.83
The Catholic University of America, p.83
Federal City College, p.83
Gallaudet College, p.83
George Washington University, p.83
Columbian College of Arts and Sciences, p.84
School of Medicine, p.84
Georgetown University, p.85
College of Arts and Sciences, p.85
School of Medicine, p.85
Howard University, p.86
College of Liberal Arts, p.86
College of Medicines, p.86
Trinity College, p.86

FLORIDA

Barry College, p.87
Bethune Cookman College, p.87
Eckerd College, p.87
Edward Waters College, p.87
Florida A and M University, p.87
Florida Atlantic University, p.87
Florida Institute of Technology, p.87
Florida Southern College, p.88
Florida State University, p.88
Florida Technological Institute, p.88
Jacksonville University, p.89
New College, p.89
Palm Beach Atlantic College, p.89
Rollins College, p.89
St. Leo College, p.89
Stetson University, p.89
University of Florida, p.89
College of Arts and Sciences, p.90
College of Agriculture, p.91
College of Medicine, p.94
University of Miami, p.95
College of Arts and Sciences, p.95
School of Engineering, p.95
School of Medicine, p.95
Dorothy H. and Lewis Rosentiel School of Marine and Atmospheric Science, p.96
University of Southern Florida, p.97
College of Liberal Arts, p.97
College of Medicine, p.97
University of Tampa, p.98
University of West Florida, p.98

GEORGIA

Agnes Scott College, p.99
Albany State College, p.99
Armstrong State College, p.99
Atlanta University Center, p.99
Atlanta University, p.99
Clark College, p.99
Moorehouse College, p.99
Morris Brown College, p.100
Augusta College, p.100
Berry College, p.100
Brenan College, p.100
Columbus College, p.100
Emory University, p.100
Emory College, p.100

School of Medicine, p.101
Georgia College, p. 102
Georgia Institute of Technology, p. 102
Georgia Southern College, p.103
Georgia Southwestern College, p.103
Georgia State University, p. 103
Medical College of Georgia, p.103
Mercer University, p. 104
North Georgia College, p.104
Oglethorpe University, p.104
Piedmont College, p.104
Savannah State College, p.105
Shorter College, p.105
University of Georgia, p.105
Franklin College of Arts, p.105
College of Agriculture, p.106
College of Vetinary Medicine, p.108
School of Forest Resources, p.108
Valdosta State College, p.109
Wesleyan College, p.109
West Georgia College, p.109

GUAM

University of Guam, p.110

HAWAII

Brigham Young University, p.111
Chammade College, p.111
Hawaii Loa College, p.111
University of Hawaii, p.111
College of Arts and Sciences, p.111
School of Medicine, p.112
College of Tropical Agriculture, p.113
University of Hawaii at Hilo, p.114

IDAHO

Boise State University, p.116
The College of Idaho, p.116
Idaho State University, p.116
Northwest Nazarene College, p.116
University of Idaho, p.116
College of Letters and Science, p.116
College of Agriculture, p.117
College of Forestry, Wildlife, and Range Science, p.118

ILLINOIS

Augustana College, p.119
Aurora College, p.119
Barat College, p.119
Blackburn College, p.119
Bradley University, p.119
Chicago State University, p.119
College of St. Francis, p.120
De Paul University, p.120
Eastern Illinois University, p.120
Elmhurst College, p.120
Eureka College, p.121
George Williams College, p.121
Greenville College, p.121
Illinois Benedictine College, p.121
Illinois College, p.121
Illinois Institute of Technology, p.121
Illinois State University, p.122
Illinois Wesleyan University, p.122
Judson College, p.122
Knox College, p.122
Lake Forest College, p.123
Lewis University, p.123
Loyola University of Chicago, p.123
College of Arts and Sciences, p.123
School of Dentistry, p.123
Stritch School of Medicine, p.124
McKendree College, p.125
MacMurray College, p.125
Milliken University, p.125
Monmouth College, p.125
Mundelein College, p.125
North Central College, p.125
North Park College, p.125
Northern Illinois University, p.126
Northwestern University, p.126
College of Arts and Sciences, p.126
Medical School, p.126
Olivet Nazarene College, p.127
Principia College, p.127
Quincy College, p.127
Rockford College, p.127
Roosevelt University, p.127
Rosary College, p.128
St. Xavier College, p.128
Shimer College, p.128
Southern Illinois University at Carbondale, p.128
College of Liberal Arts and Sciences, p.128
School of Agriculture, p.129
Southern Illinois University at Edwardsville, p.130
Trinity Christian College, p.130
Trinity College, p.131
University of Chicago, p.131
Division of Biological Sciences, p.131
Pritzker School of Medicine, p.131
The University of Health Sciences, p.133
University of Illinois at Chicago Circle, p. 134
University of Illinois at the Medical Center, p.134
College of Dentistry, p.134
College of Medicine, p.134
School of Associated Medical Sciences, p.135
University of Illinois at Urbana Champaign, p.135
College of Liberal Arts and Sciences, p.135
College of Agriculture, p.137
Western Illinois University, p.139
Wheaton College, p.139

INDIANA

Anderson College, p.142
Ball State University, p.142
Bethel College, p.142
Butler University, p.142
Concordia Senior College, p.143
DePauw University, p.143
Earlham College, p.143
Franklin College, p.143
Goshen College, p.143
Hanover College, p.144
Huntington College, p.144
Indiana Central College, p.144
Indiana State University at Evansville, p.144
Indiana University at Bloomington, p.144
Indiana University at Indianapolis, p.145
Indiana University at Kokomo, p.145
Indiana University Northwest, p.145
Indiana University School of Medicine, p.145
Indiana University at South Bend, p.147
Indiana University Southeast, p.147
Manchester College, p.147
Marion College, p.147
Purdue University, p.147
School of Agriculture, p.147
School of Science, p.150
Purdue University - Calumet Campus, p.150
Purdue University at Fort Wayne, p.150
St. Francis College, p.151
Saint Joseph's College, p.151
Saint Joseph's Calumet College, p.151
St. Mary-of-the-Woods College, p.151
Saint Mary's College, p.151
St. Meinrad College, p.151
Taylor University, p.151
Tri-State College, p.151
University of Evansville, p.152
University of Notre Dame, p.152

Valparaiso University, p.153
Wabash College, p.153

IOWA

Briar Cliff College, p.154
Buena Vista College, p.154
Central College, p.154
Clarke College, p.154
Coe College, p.154
Cornell College, p.154
Dordt College, p.154
Drake University, p.155
Graceland College, p.155
Grinnell College, p.155
Iowa State University, p.155
College of Agriculture, p.155
College of Engineering, p.156
College of Science and Humanities, p.156
College of Vetinary Medicine, p.158
Iowa Wesleyan College, p.158
Loras College, p.158
Luther College, p.159
Marycrest College, p.159
Morningside College, p.159
Mount Mercy College, p.159
Northwestern College, p.159
St. Ambrose College, p.159
Simpson College, p.159
University of Dubuque, p.159
University of Iowa, p.160
College of Liberal Arts, p.160
College of Medicine, p.160
University of Northern Iowa, p.162
Upper Iowa College, p.162
Wartburg College, p.162
Westmar College, p.163
William Penn College, p.163

KANSAS

Baker University, p.164
Benedictine College, p.164
Bethany College, p.164
Bethel College, p.164
Emporia Kansas State College, p.164
Fort Hays Kansas State College, p.165
Friends University, p.165
Kansas Newman College, p.165
Kansas State College of Pittsburg, p.165
Kansas State University, p.166
College of Arts and Sciences, p.166
College of Agriculture, p.166
College of Vetinary Medicine, p.168
McPherson College, p.169
Marymount College of Kansas, p.169
Ottowa University, p.169
Saint Mary College, p.169
St. Mary of the Plains, p.169
Southwestern College, p.169
Sterling College, p.170
Tabor College, p.170
University of Kansas, p.170
University of Kansas Medical Center, p.172
Washburn University, p.173
Wichita State University, p.173

KENTUCKY

Asbury College, p.174
Bellarmine College, p.174
Berea College, p.174
Brescia College, p.174
Campbellsville College, p.174
Centre College of Kentucky, p.174
Cumberland College, p.175
Eastern Kentucky College, p.175
Georgetown College, p.175
Kentucky State University, p.175
Kentucky Wesleyan College, p.175
Moorehead State University, p.175
Murray State University, p.175
Northern Kentucky State College, p.175
Pikeville College, p.176
Spalding College, p.176
Thomas More College, p.176
Transylvania University, p.176
Union College, p.176
University of Kentucky, p.176
College of Liberal Arts, p.176
College of Agriculture, p.177
College of Engineering, p.179
University of Kentucky Medical Center, p.179
University of Louisville, p.180
College of Arts and Sciences, p.180
School of Medicine, p.180
Western Kentucky University, p.181

LOUISIANA

Centenary College of Louisiana, p.182
Dillard University, p.182
Grambling College, p.182
Louisiana College, p.182
Louisiana State University and Agricultural and Mechanical College, p.182
College of Arts and Sciences, p.182
College of Agriculture, p.183
College of Chemistry and Physics, p.184
Louisiana State University in New Orleans, p.184
College of Sciences, p.185
Louisiana State Medical Center, p.185
Louisiana Tech University, p.186
Loyola University, p.187
McNeese State University, p.187
Nicholls State University, p.188
Northeast Louisiana University, p.188
Northwestern State University, p.188
St. Marys Dommican College, p.189
Southeastern Louisiana University, p.189
Southern University, p.189
Tulane University, p.190
Liberal Arts College, p.190
Newcomb College, p.190
School of Medicine, p.190
University of Southwestern Louisiana, p.191
Xavier University, p.191

MAINE

Bates College, p.192
Bowdoin College, p.192
Colby College, p.192
Nasson College, p.192
Ricker College, p.192
St. Francis College, p.192
St. Joseph's College, p.192
Unity College, p.192
University of Maine at Farmington, p.193
University of Maine at Fort Kent, p.193
University of Maine at Drono, p.193
College of Arts and Sciences, p.193
College of Life Sciences and Agriculture, p.193
University of Maine at Portland - Gorham, p.195
University of Maine at Presque Isle, p.195
Westbrook College, p.195

MARYLAND

Bowie State College, p.196
College of Notre Dame of Maryland, p.196
Columbia Union College, p.196
Coppen State College, p. 196
Frostburg State College, p.196

Goucher College, p.196
Hood College, p.196
Johns Hopkins University, p.197
Faculty of Arts and Sciences, p.197
School of Medicine, p.197
School of Hygiene and Public Health, p.198
Loyola College, p.198
Morgan State College, p.199
Mount St. Mary's College, p.199
St. Mary's College of Maryland, p.199
Salisbury State College, p.199
Towson State College, p.199
United States Naval Academy, p.199
University of Maryland at College Park, p. 199
College of Arts and Science, p.199
College of Agriculture, p.200
University of Maryland at Baltimore, p.203
School of Medicine, p.203
School of Dentistry, p.204
School of Pharmacy, p.204
University of Maryland, Baltimore County, p.205
Washington College, p.205
Western Maryland College, p.205

MASSACHUSETTS

American International College, p.206
Amherst College, p.206
Anna Maria College, p.206
Assumption College, p.206
Atlantic Union College, p.206
Boston College, p.206
Boston State College, p.207
Boston University, p.207
College of Liberal Arts, p.207
School of Medicine, p.208
Brandeis University, p.208
Bridgewater State College, p.208
Clark University, p.209
College of the Holy Cross, p.209
College of Our Lady of the Elms, p.209
Curry College, p.209
Eastern Nazarene College, p.209
Emmanuel College, p.210
Framingham State College, p.210
Gordon College, p.210
Harvard University, p.210
Harvard College, p.210
Medical School, p.211
School of Public Health, p.212
Lowell State College, p.212
Lowell Technological Institute, p.213
Massachusetts Institute of Technology, p.213
Merimack College, p.213
Mount Holyoke College, p.213
North Adams State College, p.214
Northeastern University, p.214
Regis College, p.215
Salem State College, p.215
Simmons College, p.215
Smith College, p.215
Southeastern Massachusetts University, p.215
Springfield College, p.216
Stonehill College, p.216
Suffolk University, p.216
Tufts University, p.216
College of Liberal Arts, p.216
School of Medicine, p.216
University of Massachusetts, p.217
College of Arts and Sciences, p.217
College of Agriculture, p.218
University of Massachusetts - Boston, p.219
Wellesley College, p.220
Westfield State College, p.220
Wheaton College, p.220
Williams College, p.220
Worcester Foundation for Experimental Biology, p.220
Worcester State College, p.221

MICHIGAN

Adrian College, p.222
Albion College, p.222
Alma College, p.222
Andrews University, p.222
Aquinas College,p .222
Calvin College, p.222
Central Michigan University, p.223
Detroit Institute of Technology, p.223
Eastern Michigan University, p.223
Ferris State College, p.223
Grand Valley State College, p.223
Hillsdale College, p.224
Hope College, p.224
Kalamazoo College, p.224
Lake Superior State College, p.224
Marygrove College, p.224
Mercy College of Detroit, p.224
Michigan State University, p.225
College of Agriculture and Natural Resources, p.225
College of Natural Science, p.226
College of Vetinary Medicine, p.227
Michigan Technical University, p.228
College of Arts and Sciences, p.228
School of Forestry, p.229
Nazareth College, p.229
Northern Michigan University, p.229
Oakland University, p.229
Olivet College, p.230
Saginaw Valley College, p.230
Sienna Heighth College, p.230
University of Detroit, p.230
University of Michigan, p.230
College of Literature, Science and the Arts, p.230
School of Natural Resources, p.231
Medical School, p.232
School of Public Health, p.233
University of Michigan -
Wayne State University, p.234
College of Liberal Arts, p.234
School of Medicine, p.235
Western Michigan University, p.236
College of Arts and Sciences, p.236
College of Applied Science, p.236

MINNESOTA

Augsburg College, p.237
Bemidji State College, p.237
Bethel College, p.237
Carleton College, p.237
College of St. Catherine, p.237
College of Saint Scholastics, p.237
College of Saint Theresa, p.238
College of St. Thomas, p.238
Concordia College, p.238
Gustavus Adolphus College, p.238
Hamline University, p.238
Macalester College, p.238
Mankato State College, p.238
Moorhead State College, p.239
St. Cloud State College, p.239
St. Johns University, p.239
St. Mary's College, p.239
St.Olaf College, p.240
Southwest Minnesota State College, p.240
University of Minnesota, p.240
College of Agriculture, p.240
College of Biological Sciences, p.241
College of Forestry, p.243
Medical School, p.243
University of Minnesota - Duluth, p.245
University of Minnesota - Morris, p.245
Winona State College, p.245

MISSISSIPPI

Alcorn State University, p.246
Belhaven College, p.246
Delta State University, p.246
Jackson State University, p.246
Millsaps College, p.246
Mississippi College, p.247
Mississippi University for Women, p.247
Mississippi State University, p.247
College of Arts and Sciences, p.247
College of Agriculture, p.248
School of Forest Resources, p.250
Mississippi Valley State University, p.250
Rust College, p.250
Tougaloo College, p.250
The University of Mississippi, p.251
The University of Mississippi, Medical Center, p.251
University of Southern Mississippi, p.252
William Carey College, p.252

MISSOURI

Cardinal Glennon College, p.253
Central Methodist College, p.253
Central Missouri State University, p.253
Culver-Stockton College, p.253
Drury College, p.253
Evangel College, p.253
Lincoln University of Missouri, p.253
The Lindenwood College, p.254
Maryville College, p.254
Missouri Southern State College, p.254
Missouri Valley College, p.254
Missouri Western State College, p.254
Northeast Missouri State University, p.254
Northwest Missouri State University, p.254
Notre Dame College, p.255
Park College, p.255
Rockhurst College, p.255
St. Louis University, p.255
College of Arts and Sciences, p.256
Medical School, p.256
The School of the Ozarks, p.257
Southeast Missouri State University, p.257
Southwest Baptist College, p.257
Southwest Missouri State University, p.258
Stephens College, p.258
Tarkio College, p.258
University of Missouri - Columbia, p.258
College of Agriculture, p.258
College of Arts and Sciences, p.260
School of Forestry, p.261
School of Medicine, p.261
School of Vetinary Medicine, p.262
University of Missouri - Kansas City, p.262
University of Missouri - St. Louis, p.262
Washington University, p.262
College of Arts and Sciences, p.262
Division of Biology and Biomedical Science, p.263
Westminster College, p.264
William Jewell College, p.264
William Woods College, p.264

MONTANA

Carroll College, p.265
Eastern Montana College, p.265
Montana State University, p.265
College of Letters and Science, p.265
College of Agriculture, p.266
Rocky Mountain College, p.266
University of Montana, p.266
Western Montana College, p.267

NEBRASKA

Chadron State College, p.268
College of Saint Mary, p.268
Creighton University, p.268
College of Arts and Science, p.268
School of Medicine, p.268
Dana College, p.269
Doane College, p.269
Hastings College, p.269
Kearney State College, p.269
Midland Lutheran College, p.270
Nebraska Wesleyan, p.270
Union College, p.270
University of Nebraska - Lincoln, p.270
School of Life Sciences, p.270
College of Agriculture, p.271
University of Nebraska - Omaha, p.272
College of Arts and Sciences, p.272
College of Medicine, p.273
Wayne State College, p.274

NEVADA

University of Nevada - Las Vegas, p.275
University of Nevada - Reno, p.275
Max C. Fleischmann College of Agriculture, p.275
College of Arts and Sciences, p.276

NEW HAMPSHIRE

Dartmouth College, p.277
Dartmouth College, p.277
Dartmouth Medical School, p.277
Franconia College, p.278
Franklin Pierce College, p.278
Keene State College, p.278
Mount Saint Mary College, p.278
Nathaniel Hawthorne College, p.278
New England College, p.278
New Hampshire College, p.279
Notre Dame College, p.279
Plymouth State College, p.279
Rivier College, p.279
St. Anselm's College, p.279
University of New Hampshire, p.279
College of Liberal Arts, p.279
College of Life Sciences, p.280

NEW JERSEY

Bloomfield College, p.282
College of Medicine and Dentistry of New Jersey, p.282
Graduate School of Biomedical Sciences, p.282
New Jersey Medical School, p.282
Rutgers Medical School, p.283
College of Saint Elizabeth, p.284
Drew University, p.284
Fairleigh Dickinson University, p.284
Maxwell Becton College of Liberal Arts, p.284
School of Dentistry, p.285
Georgian Court College, p.285
Glassboro State College, p.285
Jersey City State College, p.285
Kean College of New Jersey, p.286
Monmouth College, p.286
Montclair State College, p.286
Princeton University, p.286
Rider College, p.287
Rutgers-The State University, Camden, p.287
Rutgers-The State University, Newark, p.288
Rutgers-The State University, New Brunswick, p.288
Saint Peters College, p.289
Seton Hall University, p.290
Trenton State College, p.290
Upsala College, p.290
The William Paterson College of New Jersey, p.290

NEW MEXICO

College of Santa Fe, p.291
Eastern New Mexico University, p.291
New Mexico Highlands University, p.291
New Mexico Institute of Mining and Technology, p.291
New Mexico State University, p.292
 College of Arts and Sciences, p.292
 College of Agriculture and Home Economics, p.292
University of Albuquerque, p.293
University of New Mexico, p.293
 College of Arts and Sciences, p.293
 School of Medicine, p.293
Western New Mexico University, p.294

NEW YORK

Adelphi University, p.295
Alfred University, p.295
Bard College, p.295
Briarcliff College, p.295
City University of New York, p.296
 Bernard M. Baruch College, p.296
 Brooklyn College, p.296
 The City College, p.296
 Graduate School and University Center, p.297
 Lehman College, p.298
 Hunter College, p.298
 Queen's College, p.298
 York College, p.299
Colgate University, p.299
College of Mt. St. Vincent, p.299
College of New Rochelle, p.299
College of Saint Rose, p.299
College of White Plains, p.299
Columbia University, p.299
 Barnard College, p.300
 Columbia College, p.300
 College of Physicians and Surgeons, p.300
Concordia College, p.301
Cornell University, p.301
 New York State college of Agriculture and Life Sciences, p.301
 New York State Vetinary College, p.305
 Division of Nutritional Sciences, p.306
 New York State Agricultural Experiment Station, p.306
 Cornell Medical College, p.307
Dowling College, p.308
D'Youville College, p.308
Eisenhower College, p.308
Elmira College, p.308
Fordham University, p.308
Hamilton College, p.308
Hartwick College, p.309
Hobart and William Smith Colleges, p.309
Hofstra University, p.309
Houghton College, p.309
Iona College, p.309
Ithaca College, p.309
Keuka College, p.309
The Kings College, p.310
Kirkland College, p.310
Le Moyne College, p.310
Long Island University, p.310
 The Brooklyn Center of Long Island University, p.310
 C.W. Post Center, p.310
 Southampton College, p.310
Marist College, p.310
Marymount College, p.311
Mercy College, p.311
Molloy College, p.311
Mt. St. Mary College, p.311
Nazareth College of Rochester, p.311
New York Institute of Technology, p.311
New York University, p.312
 University College of Arts and Sciences, p.312
 College of Dentistry, p.312
 School of Medicine, p.312
Niagara University, p.313
Pace University, p.313
 School of Engineering, p.314
 School of Sciences, p.314
Roberts Wesleyan College, p.314
Rochester Institute of Technology, p.314
Rosary Hill College, p.315
Russel Sage College, p.315
St. Bonaventure College, p.315
St. Francis College, p.315
St. John Fisher College, p.315
St. John's University, p.315
St. Joseph College, p.316
St. Lawrence University, p.316
St. Thomas Aquinas College, p.316
Sarah Lawrence College, p.316
Siena College, p.316
Skidmore College, p.316
State University of New York at Albany, p.317
State University of New York at Binghampton, p.317
State University of New York College at Rockport, p.317
State University of New York College at Buffalo, p.317
State University of New York at Buffalo, p.318
 Faculty of Natural Science and Mathematics, p.318
 School of Health Sciences, p.318
 Rosevell Park Graduate Division, p.319
State University of New York College at Cortland, p.321
State University of New York College of Envioromental Sciences, p.321
State University of New York at Fredonia, p.323
State University of New York at Genesco, p.323
State University of New York College at New Paltz, p.323
State University of New York at Oneonta, p.323
State University of New York College at Oswego, p.324
State University of New York College at Plattsburgh, p.324
State University of New York College at Pottsdam, p.324
State University of New York Downstate Medical Center, p.325
State University of New York at Stony Brook, p.325
State University of New York Upstate Medical Center, p.325
Syracuse University, p.326
 College of Arts and Sciences, p.326
 College of Human Development, p.326
 Utica College, p.327
Union College and University, p.327
 Union College, p.327
 Albany Medical Center, p.327
University of Rochester, p.327
 College of Arts and Sciences, p.327
 School of Medicine and Dentistry, p.328
Vassar College, p.329
Wagner College, p.329
Wells College, p.329
Yeshiva University, p.329
 Yeshiva College, p.329
 Albert Einstein College of Medicine, p.329

NORTH CAROLINA

Appalachian State College, p.331
Barber Scotia College, p.331
Belmont Abbey College, p.331
Bennett College, p.331
Campbell College, p.331
Catawba College, p.331
Davidson College, p.331
Duke University, p.332
 Trinity College of Arts and Sciences, p.332
 School of Medicine, p.332
 School of Forestry, p.334
East Carolina University, p.334
Elizabeth City State University, p.335
Elon College, p.335
Fayetteville State University, p.335
Gardner - Webb College, p.335
Greensboro College, p.335

Guilford College, p.335
High Point College, p.335
Johnson C. Smith University, p.336
Lenoir Rhyne College, p.336
Livingstone College, p.336
Mars Hill College, p.336
Meredith College, p.336
Methodist College, p.336
North Carolina Agricultural and Technical State University, p.336
  School of Arts and Sciences, p.336
  School of Agriculture, p.337
North Carolina Central University, p.337
North Carolina Wesleyan College, p.337
Pembroke State University, p.337
Pfeiffer College, p.337
Queens College, p.337
St. Andrews Presbyterian College, p.337
Saint Augustines College, p.338
Salem College, p.338
Shaw University, p.338
University of North Carolina at Asheville, p.338
University of North Carolina at Chapel Hill, p.338
  College of Arts and Sciences, p.338
  School of Medicine, p.339
  School of Public Health, p.341
University of North Carolina at Charlotte, p.342
University of North Carolina at Greensboro, p.342
University of North Carolina State University at Raleigh, p.343
  School of Agricultural and Life Sciences, p.343
  School of Forestry, p.346
University of North Carolina at Wilmington, p.346
Wake Forest University, p.347
  Wake Forest College, p.347
  Bowman Gray School of Medicine, p.347
Warren Wilson College, p.348
Western Carolina University, p.348
Winston-Salem State University, p.348

NORTH DAKOTA

Dickinson State College, p.349
Jamestown College, p.349
Mary College, p.349
Mayville State College, p.349
Minot State College, p.349
North Dakota State University, p.349
  College of Arts and Sciences, p.349
  College of Agriculture, p.350
  College of Chemistry and Phyics, p.352
University of North Dakota, p.352
  College of Arts and Sciences, p.352
  School of Medicine, p.352
Valley City State College, p.353

OHIO

Antioch College, p.354
Ashland College, p.354
Baldwin-Wallace College, p.354
Bluffton College, p.354
Bowling Green State University, p.354
Capital University, p.354
Case Western Reserve University, p.354
  Adelbert College, p.355
  School of Medicine, p.355
  School of Engineering, p.356
Central State University, p.357
The Cleveland State University, p.357
College of Mount St. Joseph, p.357
College of Steubenville, p.357
College of Wooster, p.357
Defiance College, p.358
Denison University, p.358
Edgecliff College, p.358
Findlay College, p.358
Heidelberg College, p.358
Hiram College, p.358
John Carroll University, p.359
Kent State University, p.359
Kenyon College, p.359
Lake Erie College, p.359
Malone College, p.359
Marietta College, p.360
Miami University, p.360
Mount Union College, p.361
Muskingum College, p.361
Notre Dame College, p.361
Oberlin College, p.361
Ohio Dommion College, p.361
Ohio Northern University, p.361
Ohio State University, p.362
  College of Biological Sciences, p.362
  College of Agriculture and Home Economics, p.364
  College of Medicine, p.367
Ohio University, p.368
Ohio Wesleyan University, p.369
Otterbein College, p.369
Rio Grande College, p.369
University of Akron, p.369
University of Cincinnati, p.369
  McMicken College of Arts and Sciences, p.369
  College of Medicine, p.370
University of Dayton, p.371
University of Toledo, p.371
Urbana College, p.371
Ursuline College, p.371
Walsh College, p.371
Wilberforce University, p.372
Wilmington College, p.372
Wittenberg University, p.372
Wright State University, p.372
Xavier University, p.373
Youngstown State University, p.373

OKLAHOMA

Bethany Nazarene College, p.374
Central State University, p.374
East Central State College, p.374
Langston University, p.374
Northeastern State College, p.374
Northwestern State College, p.374
Oklahoma Baptist College, p.374
Oklahoma Christian College, p.375
Oklahoma City University, p.375
Oklahoma Panhandle State University, p.375
Oklahoma State University, p.375
  College of Arts and Sciences, p.375
  College of Agriculture, p.376
  College of Vetinary Medicine, p.377
Oral Roberts University, p.378
Phillips University, p.378
Southeastern State College, p.378
Southwestern State College, p.378
University of Oklahoma, p.379
  College of Arts and Sciences, p.379
  Health Sciences Center, p.379
University of Science and Arts of Oklahoma, p.380
The University of Tulsa, p.381

OREGON

Eastern Oregon State College, p.382
George Fox College, p.382
Lewis and Clark College, p.382
Linfield College, p.382
Oregon State University, p.382
  School of Agriculture, p.382
  School of Forestry, p.384
  School of Oceanography, p.385
  School of Sciences, p.385
Pacific University, p.387
Portland State University, p.387
Reed College, p.387

Southern Oregon College, p.387
University of Oregon, p.387
College of Liberal Arts, p.387
Medical School, p.388
University of Portland, p.389
Warner Pacific College, p.389
Willamette University, p.389

PENNSYLVANIA

Albright College, p.390
Allegheny College, p.390
Allentown College of Saint Francis de Sales, p.390
Alliance College, p.390
Alvernia College, p.390
Beaver College, p.390
Bloomsburg State College, p.390
Bryn Mawr College, p.391
Bucknell University, p.391
Cabrini College, p.391
California State College, p.391
Carlow College, p.391
Carnegie Mellon University, p.391
Cedar Crest College, p.392
Chatham College, p.392
Chestnut Hill College, p.392
Cheyney State College, p.392
Clarion State College, p.392
College Misericordia, p.392
Delaware Valley College of Science and Agriculture, p.393
Dickinson College, p.393
Drexel University, p.393
College of Science, p.393
College of Engineering, p.394
Duquesne University, p.394
East Stroudsburg State College, p.394
Eastern College, p.394
Edinboro State College, p.394
Elizabethtown College, p.395
Franklin and Marshall College, p.395
Gannon College, p.395
Geneva College, p.395
Gettysburg College, p.395
Grove City College, p.396
Gwynedd-Mercy College, p.396
Haverford College, p.396
Holy Family College, p.396
Immaculata College, p.396
Indiana University of Pennsylvania, p.396
Juniata College, p.397
Kings College, p.397
Kutztown State College, p.397
Lafayette College, p.397
La Roche College, p.397
La Salle College, p.397
Lebanon Valley College,p.398
Lehigh University, p.398
Lincoln University, p.398
Lock Haven State College, p.398
Lycoming College, p.398
Mansfield State College, p.399
The Medical College of Pennsylvania, p.399
Mercyhurst College, p.400
Messiah College, p.400
Millersville State College, p.400
Moravian College, p.400
Muhlenberg College, p.400
Our Lady of Angels College, p.401
Pennsylvania State University, p.401
College of Agriculture, p.401
College of Science, p.403
Milton S. Hershey Medical Center, p.405
Philadelphia College of Pharmacy and Science, p.405
Point Park College, p.406
Rosemont College, p.406
St. Francis College, p.406
Saint Joseph's College, p.406
Saint Vincent College, p.406
Seton Hill College, p.406
Shippensburg State College, p.407
Slippery Rock State College, p.407
Susquehanna University, p.407
Swarthmore College, p.407
Temple University, p.407
College of Liberal Arts, p.407
School of Medicine, p.407
Theil College, p.409
University of Pennsylvania, p.409
College of Arts and Sciences, p.409
School of Medicine, p.410
School Engineering, p.411
School of Dental Medicine, p.411
School of Vetinary Medicine, p.411
University of Pittsburgh, p.412
College of Arts and Sciences, p.412
School of Dental Medicine, p.413
School of Medicine, p.413
Graduate School of Public Health, p.414
University of Pittsburgh at Johnstown, p.414
University of Scranton, p.414
Ursinus College, p.415
Villa Maria College, p.415
Villanova University, p.415
Washington and Jefferson College, p.415
Waynesburg College, p.415
West Chester State College, p.415
Westminster College, p.416
Widener College, p.416
Wilkes College, p.416
Wilson College, p.416
York College of Pennsylvania, p.416

PUERTO RICO

Catholic University of Puerto Rico, p.417
Inter-American University of Puerto Rico, p.417
University of Puerto Rico - Mayaguez, p.417
College of Arts and Science, p.417
College of Agricultural Sciences, p.417
University of Puerto Rico - Rio Pedras, p.418
College of Arts and Sciences, p.418
School of Medicine, p.418

RHODE ISLAND

Barrington College, p.420
Brown University, p.420
Providence College, p.420
Rhode Island College, p.420
Roger Williams College, p.420
Salve Regina College, p.420
University of Rhode Island, p.420
College of Arts and Sciences, p.420
College of Resource Development, p.422

SOUTH CAROLINA

Baptist College at Charleston, p.423
Benedict College, p.423
Central Wesleyan College, p.423
The Citadel, p.423
Clafin College, p.423
Clemson University, p.423
College of Agricultural Sciences, p.423
College of Forest and Recreational Resources, p.425
College of Physical, Mathematical, and Biological Sciences, p.425
Coker College, p.426
College of Charleston, p.426
Columbia College, p.427
Converse College, p.427
Erskine College, p.427
Francis Marion College, p.427
Furman University, p.427
Lander College, p.427
Limestone College, p.427

Medical University of South Carolina, p.428
Newberry College, p.429
Presbyterian College, p.429
South Carolina State College, p.429
University of South Carolina, p.429
Voohees College, p.429
Winthrop College, p.429
Wofford College, p.430

SOUTH DAKOTA

Augustana College, p.431
Black Hills State College, 431
Dakota State College, p.431
Dakota Wesleyan University, p.431
Huron College, p.431
Mount Marty College, p.431
Northern State College, p.431
Sioux Falls College, p.432
South Dakota State University, p.432
University of South Dakota, p.434
College of Arts and Sciences, p.434
School of Medicine, p.434
University of South Dakota at Springfield, p.435
Yanktown College, p.435

TENNESSEE

Austin Peay State University, p.436
Belmont College, p.436
Bethel College, p.436
Bryan College, p.436
Carson Newman College, p.436
Christian Brothers College, p.436
Covenant College, p.437
David Lipscomb College, p.437
East Tennessee State University, p.437
Fisk University, p.437
George Peabody College for Teachers, p.437
King College, p.437
Knoxville College, p.438
Lambuth College, p.438
Lane College, p.438
Lee College, p.438
Lemoyne-Owen College, p.438
Lincoln Memorial University, p.438
Maryville College, p.438
Memphis State University, p.438
Middle Tennessee State University, p.439
Milligan College, p.439
Southern Missionary College, p.439
Southwestern at Memphis, p.439
Tennessee State University, p.440
Tennessee Technological Institute, p.440
College of Arts and Sciences, p.440
College of Agriculture and Home Economics, p.440
Tennessee Temple College, p.440
Tennessee Wesleyan College, p.440
Tusculum College, p.440
Union University, p.440
University of the South, p.440
University of Tennesse at Chattanooga, p.440
University of Tennesse at Knoxville, p.441
College of Liberal Arts, p.441
College of Agriculture, p.442
University of Tennessee at Martin, p.444
University of Tennessee Medical Units, p.445
University of Tennesse at Nashville, p.446
The Vanderbilt University, p.446
College of Arts and Sciences, p.446
School of Engineering, p.446
Medical School, p.446

TEXAS

Abilene Christian College, p.448
Angelo State University, p.448
Austin College, p.448
Baylor College of Medicine, p.448
Baylor University, p.449
College of Arts and Sciences, p.449
College of Dentistry, p.449
Bishop College, p.450
Dallas Baptist College, p.450
East Texas Baptist College, p.450
East Texas State University, p.450
Hardin-Simmons University, p.450
Howard Payne College, p.450
Huston-Tillotson College, p.450
Incarnate Word College, p.450
Jarvis Christian College, p.451
Lamar University, p.451
LeTourneau College, p.451
Lubbock Christian College, p.451
McMurry College, p.451
Mary Hardin-Baylor College, p.451
Midwestern University, p.452
North Texas State University, p.452
Our Lady of the Lake College, p.452
Pan American University, p.452
Prairie View A and M University, p.453
Rice University, p.453
St. Edwards University, p.453
St. Mary's University, p.453
Sam Houston State University, p.454
Southern Methodist University, p.454
Southwest Texas State University, p.454
Southwestern Union College, p.455
Southwestern University, p.455
Stephen F. Austin State University, p.455
Sul Ross State University, p.455
Tarleton State University, p.456
Texas A and I University, p.456
Texas A and M University, p.456
College of Agriculture, p.457
College of Geosciences, p.459
College of Science, p.460
College of Vetinary Medicine, p.460
Texas Christian University, p.460
Texas College, p.461
Texas Lutheran College, p.461
Texas Southern University, p.461
Texas Tech University, p.461
College of Arts and Science, p.461
College of Agricultural Sciences, p.462
Texas Wesleyan College, p.463
Texas Women's University, p.463
Trinity University, p.463
University of Dallas, p.463
University of Houston, p.463
University of St. Thomas, p.464
University of Texas at Arlington, p.464
University of Texas at Austin, p.465
University of Texas Health Science Center at Dallas, p.466
University of Texas Medical Branch at Galveston, p.467
University of Texas at El Paso, p.469
University of Texas at Houston, p.469
University of Texas at Permian Basin, p.469
University of Texas Health Science Center of San Antonio, p.469
Wayland College, p.470
West Texas State University, p.471
College of Arts and Sciences, p.471
College of Agriculture, p.471
Wiley College, p.471

UTAH

Brigham Young University, p.472
Southern Utah State College, p.473
University of Utah, p.473
College of Science, p.473
College of Medicine, p.473
Utah State University, p.474
College of Agriculture, p.474

College of Natural Resources, p.475
College of Science, p.476
Weber State College, p.477
Westminster College, p.478

VERMONT

Castleton State College, p.479
Johnson State College, p.479
Lyndon State College, p.479
Marlboro College, p.479
Middlebury College, p.479
Norwich University, p.479
St. Michael's College, p.479
Trinity College, p.480
University of Vermont, p.480
College of Arts and Sciences, p.480
College of Agriculture and Home Economics, p.480
College of Medicine, p.481
Windham College, p.482

VIRGIN ISLANDS

College of the Virgin Islands, p.483

VIRGINIA

Averett College, p.484
Bridgewater College, p.484
College of William and Mary, p.484
Faculty of Arts and Sciences, p.484
Christopher Newport College, p.484
Eastern Mennonite College, p.484
Emory and Henry College, p.485
George Mason University, p.485
Hampden-Sydney College, p.485
Hollins College, p.485
Longwood College, p.486
Lynchburg College, p.486
Madison College, p.486
Mary Baldwin College, p.486
Mary Washington College, p.486
Old Dominion University,p.487
Radford College, p.487
Randolf-Macon College, p.487
Randolf-Macon Women's College, p.487
Roanoke College, p.487
St. Paul's College, p.488
Sweet Briar College, p.488
University of Richmond, p.488
Richmond College, p.488
Westhampton College, p.488
University of Virginia, p.488
Clinch Valley College, p.488
College of Arts and Sciences, p.488
College of Engineering and Applied Sciences, p.489
School of Medicine, p.489
Virginia Institute of Marine Science, p.489
Virginia Commonwealth University, p.491
School of Arts and Sciences, p.491
School of Graduate Studies, Health Sciences Division, p.491
Virginia Military Institute, p.492
Virginia Polytechnic Institute and State University, p.492
College of Arts and Sciences, p.492
College of Agriculture and Life Sciences, p.493
Virginia State College, p.494
Virginia Union University, p.494
Virginia Wesleyan College, p.495
Washington and Lee University,p.495
Westhampton College, p.495

WASHINGTON

Central Washington State College, p.496
Eastern Washington State College, p.496
Gonzaga College, p.496
Pacific Lutheran University, p.497
St. Martin's College, p.497
Seattle Pacific College, p.497
Seattle University, p.497
University of Puget Sound, p.497
University of Washington, p.497
Center for Bioengineering, p.497
College of Arts and Sciences, p.497
College of Fisheries, p.498
College of Forest Resources, p.499
School of Medicine, p.499
School of Public Health and Community Medicine, p.500
Walla Walla College, p.500
Washington State University, p.501
College of Liberal Arts, p.501
College of Agriculture, p.502
College of Vetinary Medicine, p.504
Western Washington State College, p.504
Whitman College, p.504

WEST VIRGINIA

Alderson-Broaddus College, p.506
Bethany College, p.506
Bluefield State College, p.506
Concord College, p.506
Davis and Elkins College, p.506
Fairmont State College, p.506
Glenville State College, p.506
Marshall College, p.507
Morris Harvey College, p.506
Salem College, p.507
Shepherd College, p.507
West Liberty State College, p.507
West Virginia Institute of Technology, p.507
West Virginia State College, p.507
West Virginia University, p.507
College of Arts and Sciences, p.507
College of Agriculture and Forestry, p.507
School of Medicine, p.509
West Virginia Wesleyan College, p.509
Wheeling College, p.510

WISCONSIN

Alverno College, p.511
Beloit College, p.511
Cardinal Stritch College, p.511
Carroll College, p.511
Carthage College, p.511
Edgewood College, p.511
Lawrence University, p.512
Marian College, p.512
Marquette University, p.512
The Medical College of Wisconsin, p.512
Milton College, p.513
Mount Mary College, p.513
Northland College, p.513
Ripon College, p.513
St. Norbert College, p.514
Silver Lake College, p.514
University of Wisconsin - Eau Clair, p.514
University of Wisconsin - Green Bay, p.514
University of Wisconsin - La Crosse, p.515
University of Wisconsin - Madison, p.515
College of Letters and Science, p.515
College of Agricultural and Life Sciences, p.516
Medical School, p.520
School of Natural Resources, p.521
University of Wisconsin - Milwaukee, p.521
University of Wisconsin - Platteville, p.522
University of Wisconsin - River Falls, p.522
College of Arts and Sciences, p.522
College of Agriculture, p.522
University of Wisconsin - Stevens Point, p.522
University of Wisconsin - Oshkosh, p.523

University of Wisconsin - Parkside, p.523
University of Wisconsin - Stout, p.524
University of Wisconsin - Superior, p.524
University of Wisconsin - White Water, p.524
Viterbo College, p.524

WYOMING

University of Wyoming, p.526
College of Arts and Science, p.526
College of Agriculture, p.526

CANADA

ALBERTA

University of Alberta, p.530
Faculty of Agriculture and Forestry, p.530
Faculty of Science, p.531
University of Calgary, p.531
University of Lethbridge, p.532

BRITISH COLUMBIA

Notre Dame University, p.533
Simon Fraser University p.533
University of British Columbia, p.533
Faculty of Agricultural Science, p.533
Faculty of Science, p.534
Faculty of Medicine, p.535
University of Victoria, p.535

MANITOBA

Brandon University, p.537
University of Manitoba, p.537
Faculty of Agriculture, p.537
Faculty of Medicine, p.537
Faculty of Science, p.538
University of Winnipeg, p.539

NEW BRUNSWICK

Mount Allison University, p.540
Université de Moncton, p.540
University of New Brunswick, p.540

NEWFOUNDLAND

Memorial University of Newfoundland, p.541

NOVA SCOTIA

Acadia University, p.542
Dalhousie University, p.542
Institute of Oceanography, p.542
Faculty of Arts and Sciences, p.542
Faculty of Medicine, p.542
Mount Saint Vincent University, p.543
Nova Scotia Technical College, p.543
St. Francis Xavier University, p.543
St. Mary's University, p.543

ONTARIO

Brock University, p.544
Carleton University, p.544
Lakehead University, p.544
Laurentian University, p.544
McMaster University, p.545
Queen's University, p.545
Faculty of Medicine, p.545
Faculty of Arts and Scienes, p.546
Trent University, p.546
University of Guelph, p.546
College of Biological Sciences, p.546
Ontario Agricultural College, p.547
Ontario Vetinary College, p.548
University of Ottowa, p.548
Faculty of Medicine, p.548
Faculty of Arts and Sciences, p.549
University of Toronto, p.549
Faculty of Medicine, p.549
Faculty of Arts and Science, p.550
University of Waterloo, p.550
University of Western Ontario, p.551
School of Medical Sciences, p.551
School of Science, p.552
University of Windsor, p.552
Wilfred Laurier University, p.552
York University, p.552

PRINCE EDWARD ISLAND

University of Prince Edward Island, p.554

QUEBEC

Bishop's University, p.555
McGill University, p.555
Faculty of Agriculture, p.555
Faculty of Medicine, p.555
Faculty of Science, p.556
Macdonald Campus, p.556
Sir George Williams University, p.557
Université Laval, p.557
Loyola University, p.558
Université de Montréal, p.558
Faculté de Médicine, p.558
Faculté des Arts et Sciences, p.559
Université de Sherbrooke, p.559
Faculté de Médicine, p.559
Faculté des Sciences, p.559
Université de Québec à Chicoutimi, p.560
Université de Québec à Montreal, p.560
Université de Québec à Rimouski, p.560
Université de Québec à Trois - Rivières, p.560

SASKATCHEWAN

University of Regina, p.561
University of Saskatchewan, p.561
College of Agriculture, p.561
College of Medicine, p.562
College of Arts and Science, p.562
College of Vetinary Medicine, p.562

UNITED STATES

ALABAMA

ALABAMA A & M UNIVERSITY

Normal, Alabama 35762 Phone: (205) 859-7011
Dean: B.W. Jones

DEPARTMENT OF BIOLOGY
Phone: Chairman (205) 859-7268

Chairman: A.C. Jenkins
Professors: A.C. Jenkins, B.S. Mangat
Associate Professors: W.L. Gibson, G. Grayson, M. Pinjani
Instructors: E. Golden

Degree Program(s): Biology, Botany, Zoology
Undergraduate Degree: B.S.
Undergraduate Degree Requirements: 32 semester hours of Biology. This must include Invertebrate Zoology, Comparative Vertebrate Anatomy, Genetics and Embryology

ALABAMA COLLEGE

(see University of Montevello)

ALABAMA STATE UNIVERSITY

Montgomery, Alabama 36101 Phone: (205) 262-3581
Dean of Graduate Studies: L. Bell, Jr.
Dean: R. Newman

DEPARTMENT OF BIOLOGY
Phone: Chairman (205) 262-3581 Ext. 319

Chairman: J. Oliviere
Professors: J. Oliviere, U.D. Sharma
Associate Professors: A.O. Glass, B.L. Young
Assistant Professors: A.J. Harris, E.C.K. Igwegbe, J.E. Mayfield
Research Assistants: 3

Undergraduate Degree(s): B.A. (Biology, Medical Microbiology)
Undergraduate Degree Requirements: Invertebrate Zoology, Botany, Vertebrate Embryology, Parasitology, Comparative Vertebrate Anatomy, Genetics, Cell Biology, Animal Histology, Biological Technique, Microbiology, Plant Physiology

Graduate Degree(s): M.S. (Biology)
Graduate Degree Requirements: Radiation Biology, Ecology, Mammalian Physiology, Advanced Plant Physiology, Cytogenetics, Thallophytes, Tracheophytes, Endocrinology, Protozoology, Virology, Advanced Microbiology, Physiological Chemistry

ATHENS COLLEGE

Athens, Alabama 35611 Phone: (205) 232-1802
Dean: T.A. Rodgers

BIOLOGY DEPARTMENT

Chairman: R. Daly
Associate Professors: R. Daly, J. Russell
Instructor: W. Wright

Degree Program(s): Biology, Botany, Medical Technology, Zoology
Undergraduate Degree(s): B.S.
Undergraduate Degree Requirements: 38 hours (semester) major field (Biology, Zoology or Botany) 20 hours (semester) Chemistry

AUBURN UNIVERSITY

Auburn, Alabama 36830 Phone: (205) 826-4650
Dean of Graduate Studies: P. Parks

School of Agriculture

Dean: R.D. Rouse

DEPARTMENT OF AGRONOMY AND SOILS
Phone: Chairman (205) 826-4100

Chairman: L.E. Ensminger
Professors: F. Adams, J.T. Cope, E.D. Donnelly, A.E. Hiltbold, J.T. Hood, C.S. Hoveland, W.C. Johnson, C.E. Scarsbrook, J.I. Wear
Associate Professors: G.A. Buchanan, R. Dickens, C.E. Evans, B.F. Hajek, C.C. King, D.L. Thurlow
Assistant Professors: A.C. Bennett, R.L. Haaland
Research Assistants: 11

Undergraduate Degree(s): B.S. (Agronomy)
Undergraduate Degree Requirements: a total of 210 quarter hours with 33 quarter hours in Agronomy

Graduate Degree(s): M.S., Ph.D. (Crop Science, Soil Science)
Graduate Degree Requirements: Master: 45 quarter hours with a thesis in either crop or soil science. Doctoral: 80 to 120 quarter hours and a dissertation in either crop or soil science

Graduate and Undergraduate Courses Available for Graduate Credit: Commerical Fertilizers (U), Soil Management (U), Soil Resources and Conservation (U), Methods of Plant Breeding (U), Principles of Herbicide Use (U), Soil Morphology (U), Turfgrass Management (U), Soil Physics (U), Soil Microbiology, Experimental Methods, Advanced Plant Breeding, Experimental Evolution, Crop Ecology, Theories in Forage Crop Management, Advanced Soil Fertility, Soil and Plant Analysis, Clay Mineralogy, Soil Chemistry, and Advanced Soil Physics

DEPARTMENT OF ANIMAL AND DAIRY SCIENCES
Phone: Chairman (205) 826-4160

Head: W.M. Warren
Alumni Professors: R.C. Smith, D.R. Strength
Professors: W.B. Anthony, K. M. Autrey, R.Y. Cannon, R.R. Harris, G.E. Hawkins, D.L. Huffman, P.F. Parks, T.B. Patterson, R.C. Smith, D.R. Strength, E.L. Wiggins
Associate Professors: H.H. Daron, T.A. McCaskey, G.H. Rollins, C.D. Squiers, H.F. Tucker
Assistant Professors: D.J. Jones, D.N. Marple, G.B. Meadows, G.L. Zable
Instructors: J.A. Little, J.P. Cunningham

Field Stations/Laboratories: Black Belt Substation - Marion Junction, Piedmont Substation - Camp Hill, Lower Coastal Plains Substation - Camden, Upper Coastal Plains Substation - Winfield, Sand Mountain Substation - Crossville, Tennessee Valley Substation - Belle Mina, Wiregrass Substation - Headland, Gulf Coast Substation - Fairhope

Degree Program(s): Animal Genetics, Animal Physiology, Animal Science, Biochemistry, Dairy Science, Food Science, Meat Science, Nutrition

Undergraduate Degree(s): B.S. (Animal Science, Dairy Science, Food Science)
Undergraduate Degree Requirements: 210 credit hours (quarter basis), 15 cr. chemistry (5 organic), 10 cr.

Mathematics, thru Calculus, 5 cr. Physics, 13 cr. Animal Biochemistry & Nutrition, 10 cr. Animal Physiology, 19 cr. Genetics & Animal Breeding, 24 cr. additional ADS courses

Graduate Degrees: M.S., Ph.D.
Graduate Degree Requirements: Master: 45 cr. graduate courses plus thesis Doctoral: 90 cr. graduate courses plus dissertation

Graduate and Undergraduate Courses Available for Graduate Credit: Swine Production (U), Beef Cattle Production (U), Animal Breeding (U), Dairy Cattle Production (U), Physiology of Lactation (U), Animal Reproduction (U), Applied Animal Nutrition (U), Meat Technology (U), Dairy Chemistry (II), Frozen and Concentrated Dairy Foods (U), Fermented Diary Foods (U), Food Microbiology (U), Food Plant Sanitation (U), Food Plant Sanitation (U), Biochemistry I (U), Biochemistry II (U), Meat Science, Technical Control of Dairy Products, Comparative Animal Nutrition, Advanced Animal Reproduction, Advanced Beef Cattle Production, Advanced Swine Cattle Production, Genetics of Populations, Minerals, Ruminant Nutrition, Microbial Biochemistry, Experimental Methods, Proteins, Lipids, Enzymes, Topics in Biochemistry, Biochemical Research Techniques

DEPARTMENT OF BOTANY AND MICROBIOLOGY
Funchess Hall
Phone: Chairman (205) 826-4830 Others 826-4000

Head: J.A. Lyle
Alumni Associate Professor: R. Rodriguez-Kabana
Professors: E.A. Curl, D.E. Davis, N.D. Davis, U.L. Diener, H.H. Funderburk, Jr., R.T. Gudauskas, J.A. Lyle, N.L. Marshall, R.M. Patterson
Associate Professors: E.M. Clark, R.M. Cody, J.D. Freeman, W.H. Mason (Coordinator, General Biology), G. Morgan-Jones, R. Rodriguez-Kabana, B. Truelove, J.C. Williams, Jr.
Assistant Professors: P.A. Backman, W.T. Blevins, T.C. Davis, W.E. Goslin, V.C. Kelley, A.J. Latham, C.M. Peterson, W.A. Shands, Jr., O.C. Thompson, J.D. Weete, G.R. Wilt
Instructors: G.L. Benson, W.D. Kelley
Teaching Fellows: 13
Research Assistants: 8
NDEA Fellow: 1
University Fellow: 1
Postdoctoral Research Associate: 1

Field Stations/Laboratories: Black Belt Substation - Marion Junction, Brewton Field Station - Brewton, Chilton Area Horticulture Substation - Clanton, Forestry Field Station - Fayette, Gulf Coast Substation - Fairhope, Lower Coastal Plan Substation - Camden, Monroeville Field Station - Monoreville, North Alabama, Horticulture Substation - Cullman, Ornamental Horticulture Field Station - Mobile, Piedmont Substation - Camp Hill, Prattville Experiment Field - Crossville, Tennessee Valley Substation - Belle Mina, Tuskegee Experiment Field Station - Tuskegee, Upper Coastal Plain Substation - Winfield, Wiregrass Substation - Headland

Degree Programs: Botany, Microbiology

Undergraduate Degree: B.S.
Undergraduate Degree Requirements: Botany - 210 quarter hours with 35 in Botany, 25 in Zoology, and 40 in Physical Sciences Microbiology - 210 quarter hours with 45 in Microbiology, 25 in Biology, and 50 in the Physical Sciences

Graduate Degrees: M.S., M.A., Ph.D.
Graduate Degree Requirements: Master: Botany - 30 credit hours in major, 15 in minor, with a final oral examination. Microbiology - 30 credit hours in major, 15 in minor, with a final oral examination Master of Arts for College Teachers: 56 credit hours with 48 in Biological and Physical Sciences, with a final oral examination Doctoral: Botany and Microbiology - 60 to 80 credit hours or more in Botany and Microbiology with one or more minors consisting of about 30 to 40 credit hours, in related fields, a reading knowledge of 2 foreign languages, or a reading knowledge of 1 foreign language and 15 credit hours of courses not in the biological or physical sciences; written and oral preliminary examinations; final examination on dissertation

Graduate and Undergraduate Courses Available for Graduate Credit: Advanced Plant Pathology I (U), Aquatic Plants (U), Biological Statistics (U), Biological Microscopy (U), Microtechnique, and Photography (U), Developmental Plant Anatomy (U), General Plant Ecology (U), General Virology (U), History of Selected Botanical Topics (U), Immunology and Serology (U), Introductory Mycology (U), Marine Botany,(U), Marine Microbiology (U), Microbial Physiology (U), Microbiological Methods,(U), Morphology of the Vascular Plants (U), Paramicrobiology (U), Phycology (U), Physiology of the Fungi (U), Plant Nematology (U), Principles in Plant Disease Control (U), Sanitary Microbiology (U), Systematic Botany (U), Special Problems (15) in Botany and Microbiology, Advanced Medical Microbiology I, II, Advanced Plant Pathology II, Advanced Plant Physiology I, II, III, Advanced Systematic Botany, Biological Processes, Biological Statistics, Chemical Weed Control, Clinical Plant Pathology, Cytology and Cytogenetics, Departmental Forum, Developmental Morphology of the Angiosperms, Ecology of Soil Fungi, Industrial and Applied Microbiology, Least Squares Analysis of Experiments, Nuclear Science in Agriculture, Phytovirology, Plant Ecosystems, Seminar, Special Problems (22) in Biostatistics, Botany and Microbiology, Research and Thesis, Doctoral.

DEPARTMENT OF FISHERIES AND ALLIED AQUACULTURES
Swingle Hall
Phone: Chairman (205) 826-4786

Chairman: E.W. Shell
Professors: J.S. Dendy, J.M. Lawreng, D.D. Moss
Associate Professors: R. Allison, M.M. Damatmat, C.E. Boyd, R.T. Lovell, E.E. Drather, J.S. Ramsey, W.A. Rogers, R.O. Smitherman
Assistant Professors: R.D. Bayne, J.H. Grover, L.L. Loushin, D. Leary, H.R. Schmitton, W.D. Davis, J.A. Plumb, W.L. Shelton
Instructors: J.S. Jensen, D. Hughes, E.W. Scarsbrook
Research Assistants: 20

Undergraduate Degree: B.S. (Fisheries Management)
Undergraduate Degree Requirements: 210 Quarter Hours, 15 hours Biology, 15 hours Mathematics, 20 hours Chemistry, 10 hours Physics, 9 hours English, 35 hours Zoology, 10 hours Botany, 27 hours Fisheries, 69 hours Humanities, Social Sciences and Electives

Graduate Degrees: M.S., Ph.D. (Aquaculture, Fish Pathology, Fisheries Biology)
Graduate Degree Requirements: Master: Minimum of 45 hours plus thesis, 30 hours taken in Department, 15 in other Departments Doctoral: Minimum of 80 hours above B.D. Degree, 50 hours in Department, 30 hours in other Departments. Dissertation, two foreign languages, General exam and Final exam

Graduate and Undergraduate Courses Available for Graduate Credit: Limnology (U), Biological Productivity and Water Quality (U), Hatchery Management (U), Pond Construction (U), Management of Small Impoundments (U), Fisheries Biology (U), General Icthyology (U), Functional Morphology (U), Fish Parasitology (U), Fish Diseases (U), Management of Streams and Large Impoundments (U), Special Problems in Fisheries and Aquacultures (U), Advanced Fisheries Biology, Nutrient Cycles in Aquaculture, Aquaculture, Fish Processing Technology, Fish Nutrition, Advanced Fish Parasitology, Advanced Microbial Fish Diseases

DEPARTMENT OF FORESTRY
M. White Smith Hall
Phone: Chairman (205) 826-4050

Head: W.B. DeVall
Professors: E.J. Biblis, H.E. Christen, W.B. DeVall, G.I. Garin (Emeritus), J.F. Goggans, E.J. Hodgkins, E.W. Johnson
Associate Professors: H.O. Beals, H.S. Larsen, E.S. Lyle, H.G. Posey, S.D. Whipple
Assistant Professors: T.C. Davis, L.E. DeBrunner, K.W. Livingston
Instructors: W. Chiu, G.E. Coleman III, R.H. Crowley, J.R. Frazier, D.R. Hicks, D.M. Hyink, D.J. Janes, K.D. Lynch, R.J. Meier, L.R. Sellmann
Teaching Fellows: 1
Research Assistants: 4
Fellowships: 3

Field Stations/Laboratories: Fayette Experiment Forest - Fayette

Degree Programs: Forest Ecology, Forest Management, Genetics, Physiology, Wood Technology

Undergraduate Degree: B.S. (Forest Management, Wood Technology)
Undergraduate Degree Requirements: Forest Management 227 quarter hours, Recreation option 225 quarter hours, Honors Program 210 hours, Wood Technology 210 hours

Graduate Degrees: M.S., Ph.D.
Graduate Degree Requirements: Master: Minimum of 30 hours in a major subject area and 15 hours in a minor subject area Doctoral: Minimum of 60 hours in a major subject area and 30 hours in a minor subject area, behond the B.S. minimal reading of two foreign languages

Graduate and Undergraduate Courses Available for Graduate Credit: Forest Management (U), Microtechnique of Hard Materials (U), Range Management (U), Photogrammetry (U), Silviculture (U), Forest Research Methods (U), Wood Gluing and Lamination (U), Mechanical Properties of Wood (U), Seasoning and Preservation of Wood (U), Seasoning and Preservation Laboratory (U), Forest Policy and Law (U), Forest Products Marketing (U), Forest Watershed Management (U), Forest Economics I (U), Forest Economics II (U), Small Woodland Management (U), Seminar in Forestry (U), Wood Chemistry, Forest Tree Improvement, Forest Soils, Forest Community Invesitations, Remote Sensing, Directed Study, Special Problems, Research and Thesis, Research and Dissertation

DEPARTMENT OF HORTICULTURE
Funchess Hall
Phone: Chairman (205) 826-4862

Head: D.Y. Perkins
Professors: H.J. Amling, W.H. Greenleaf, J.D. Norton, H.P. Orr
Associate Professors: O.L. Chambliss, H. Harris, F. B. Perry, Jr., K.C. Sanderson
Assistant Professors: W.A. Dozier, W.A. Johnson, K.S. Rymal
Instructors: W.C. Martin, Jr., K. Marcus, J.L. Turner
Research Assistants: 4
Field Superintendant: 1

Field Stations/Laboratories: Chilton Area Horticulture Substation - Clanton, North Alabama Horticulture Substation - Cullman, Spring Hill Station - Mobile

Degree Programs: Food Science, General Horticulture, Landscape and Ornamental Horticulture

Undergraduate Degree: B.S.
Undergraduate Degree Requirements: 210 Quarter hours with a "C" average

Graduate Degree: M.S.
Graduate Degree Requirements: 45 Quarter hours with 30 in the major, Thesis

Graduate and Undergraduate Courses Available for Graduate Credit: Commerical Vegetable Crops (U), Storage Packaging and Marketing of Vegetable Crops (U), Fruit Growing (U), Small Fruits (U), Nut Culture (U), Commerical Vegetable Crops (U), Recent Advances in Small Fruits (U), Care and Maintenance of Ornamental Plants (U), Floricultural Crop Production (U), Nursery Management (U), Planting Design (U), Minor Problems (U), Advanced Plant Propagation (U), Marketing Horticultural Speciality Products (U), Advanced Landscape Gardening (U), Controlled Plant Growth (U), Food Engineering (U), Experimental Methods in Horticulture Seminar, Special Problems in Horticulture, Plant Growth and Development, Nutritional Requirements of Horticultural Plants, Physiology of Horticultural Products Following Harvest, Breeding of Horticultural Crops, Research and Thesis

DEPARTMENT OF POULTRY SCIENCE
Animal Sciences Building
Phone: Chairman (205) 826-4133

Head: C.H. Moore
Professors: G.J. Cottier, S.A. Edgar, C.H. Moore, E.C. Mora
Associate Professors: R.N. Brewer, L.W. Johnson, G.R. McDaniel
Assistant Professors: G.F. Combs, Jr.

Degree Program: Poultry Science (participate in Ph.D. program through interdepartmental program in Physiology, Microbiology and Nutrition)

Undergraduate Degree: B.S.
Undergraduate Degree Requirements: 210 Quarter hours

Graduate Degrees: M.S., Ph.D.
Graduate Degree Requirements: Master: 45 Quarter hours including Thesis Doctoral: 80 Quarter hours including Dissertation

Graduate and Undergraduate Courses Available for Graduate Credit: Poultry Management (U), Poultry Feeding (U), Incubation and Brooding (U), Poultry Diseases and Parasites (U), Poultry Breeding (U), Poultry Marketing (U), Biological Rhythms (U), Advanced Poultry Production, Advanced Poultry Breeding, Advanced Poultry Nutrition, Advanced Poultry Management, Advanced Poultry Diseases I, Advanced Poultry Diseases II, Immunochemsitry, Avian Physiology, Experimental Virology, Transmission and Scanning Electron Microscopy

DEPARTMENT OF ZOOLOGY-ENTOMOLOGY
Funchess Hall
Phone: Chairman (205) 826-4850

Head: F.S. Arant
Professors: F.S. Arant, M.H. Bass, R.S. Berger, G.H. Blake, J.S. Dendy, J.L. Dusi, K.L. Hays, E.E. Jones (Adjunct), R.H. Mount, D.A. Porter (Adjunct)
Associate Professors: H.D. Alexander, M.K. Causey, H.B. Cunningham, C.F. Dixon, J.L. Bodie, G.W. Folkerts, L.C. Frandsen (Adjunct), F.R. Gilliland, L.L. Hyche, W.D. Ivey, C.A. Kouskolekas, W.H. Mason, J.S. Ramsey, D.W. Speake, J.E. Watson
Assistant Professors: P.M. Estes, J.D. Harper, J.E. Kennamer, F.B. Lawrence, M.E. Lisano, T.M. Pullen, J.F. Prichett, J.M. Slack, T.L. Terrel, M.L. Williams, D.W. Young
Instructors: T. Brugh, D. Carroll, N. Willis
Visiting Lecturers: 1

Teaching Fellows: 23
Research Assistants: 20

Field Stations/Laboratories: Cooperative agreement with Gulf Coast Research Laboratory - Ocean Springs, Marine Sciences Consortium - Dauphin Island

Degree Programs: Entomology, Marine Biology, Parasitology, Physiology, Wildlife, Zoology

Undergraduate Degree: B.S.
Undergraduate Degree Requirements: 210 Quarter hours

Graduate Degrees: M.S., Ph.D.
Graduate Degree Requirements: Master: Minimum of 45 Quarter hours of courses, Thesis and Oral Examination Doctoral: 80 to 120 Quarter hours beyond B.S., reading knowledge of two languages, written and oral general examinations, Dissertation, Final Examination

Graduate and Undergraduate Courses Available for Graduate Credit: 30 - 400 level courses; 35 - 600 level courses

School of Vetinary Medicine
Dean: J.E. Greene

DEPARTMENT OF ANATOMY AND HISTOLOGY
Phone: Chairman (205) 826-4427

Head: C.L. Holloway
Professors: J.S. McKibben, C.L. Holloway
Associate Professors: L.M. Krista
Assistant Professors: B.W. Gray, T.M. Reynolds, J. Lafaver
Instructors: H.N. Engel, P.F. Rumph, R.E. Cartee

Degree Programs: Anatomy, Histology, Physiology

Graduate Degrees: M.S., Ph.D.
Graduate Degree Requirements: Master: 30 Quarter hours - Anatomy and Histology; 15 Quarter hours - Microbiology, Parasitology, Pathology, Physiology, Medicine or Surgery Doctoral: Student must major in some aspect of Physiology and Research should be in this area. Minor field or fields may consist of Biochemistry, Physics, Pharmacology, Anatomy and Histology, Nutrition, Psychology, or other supporting fields

Graduate and Undergraduate Courses Available for Graduate Credit: Microbiology, Parasitology, Pathology, Physiology, Medicine and Surgery, Animal Reproduction, Biochemistry, Comparative Animal Nurtition, Advanced Animal Reproduction, Minerals, Ruminant Nutrition, Microbial Biochemistry, Experimental Methods, Biological Statistics, Microbial Physiology, Biological Statistics II, Carbohydrates, Amino Acids and Proteins, Lipids, Enzymes, Intermediate Metabolism, Biochemical Research Techniques, Environmental Physiology and Bioengineering, Advanced Poultry Nutrition, Avian Physiology, Introduction to Biophysics, Physiological Function Tests and Laboratory Diagnosis, Advanced Anatomy, Experimental Neuroanatomy, Advanced Histology of Domestic Animals, Advanced Veterinary Radiation Biology, Electrocardiology and Blood Vascular Physiology, Animal Physiology, Physiology of the Cell, Endocrinology, Animal Behavior

DEPARTMENT OF PHYSIOLOGY AND PHARMACOLOGY
Phone: Chairman (205) 826-4425

Head: C.H. Clark
Professors: S.D. Beckett, M.J. Burns, C.H. Clark, R.W. Redding
Associate Professors: B.T. Robertson, W.M. Pedersoli
Assistant Professor: R.F. Nachreiner
Instructors: C.E. Branch, M.H. Sims, R.L. Boyd
Teaching Fellow: 1
Research Assistants: 4
Graduate Students: 6

Degree Programs: Pharmacology, Physiology

Graduate Degrees: M.S., Ph.D.
Graduate Degree Requirements: Master: 45 Quarter hours plus Thesis Doctoral: 100 hours plus Dissertation

Graduate and Undergraduate Courses Available for Graduate Credit: Intermediate Human Physiology (U), Respiratory Physiology (U), Advanced Renal and Hepatic Physiology, Advanced Endocrinology and Reproduction, Advanced Neurology, Advanced Veterinary Pharmacology, Physiology of Digestion, Small Animal Nutrition, Veterinary Radiation Biology, Electrocardiology and Blood Vascular Physiology

DEPARTMENT OF MICROBIOLOGY
Phone: Chairman (205) 826-4539

Head: T.T. Kramer
Professor: T.T. Kramer
Associate Professors: M.H. Attleberger, P. Klesius (Adjunct), C.R. Rossi, L.J. Swango
Instructors: J.M. Westergaard

Degree Programs: Microbiology, Vetinary Microbiology

Graduate Degrees: M.S., Ph.D.
Graduate Degree Requirements: Master: GRE, Minimum of 45 Quarter Hours, Thesis Doctoral: Advanced GRE, 80 to 120 Quarter Hours, two foreign languages, Thesis

Graduate and Undergraduate Courses Available for Graduate Credit: Advanced Pathogenic Bacteriology, Advanced Immunology, Pathogenesis of Virus Diseases, Clinical Mycology, Bovine Virology, Advanced Epidemiology

BIRMINGHAM-SOUTHERN COLLEGE
Birmingham, Alabama 35204 Phone: (205) 328-5250
Dean: P.C. Bailey

DEPARTMENT OF BIOLOGY
Phone: Chairman (205) 328-5250 Ext. 245

Chairman: T.S. Quarles
Professors: D.C. Holliman
Associate Professors: T.S. Quarles, E.D. Waits
Assistant Professors: L.D. Zettergren

Undergraduate Degree: B.S. (Biology)
Undergraduate Degree Requirements: 9 1/2 Semesters in Biology, 2 Semesters in Chemistry, 2 Semesters in Physics, 2 Semesters in Mathematics

FLORENCE STATE UNIVERSITY
Florence, Alabama 35630 Phone: (205) 766-4100
Dean: F. McArthur

DEPARTMENT OF BIOLOGY
Phone: Chairman (205) 766-4100 Ext. 270

Head: A.L. Hershey
Professors: A.L. Hershey, J.S. Brown, W.B. Hawkins, C.E. Keys, P. Yokey, Jr., W.R. Montgomery
Associate Professors: J.H. Moore, R.W. Williams
Assistant Professors: J.W. Holland, Jr., B.J. Kent

Field Stations/Laboratories: Institute for Freshwater Biology - Florence

Undergraduate Degree: B.S., (Biology)
Undergraduate Degree Requirements: 40 Hours in Biology plus Minor in Chemistry and 8 to 10 Hours in Physics or, 30 Hours in Biology plus 8 Hours of

Chemistry or one year of Marine Biology

HUNTINGDON COLLEGE
Montgomery, Alabama 36105 Phone: (205) 263-1611
Dean: W. Top

BIOLOGY DEPARTMENT
Phone: Chairman (205) 263-1611 Ext. 59

Chairman: M.J.C. Brannon
Professors: H.S. Ward
Assistant Professors: F.S. Cox
Instructors: M. Parker

Undergraduate Degree: A.B., (Biology)
Undergraduate Degree Requirements: 36 Hours in Biology (8 Hours in Plants, 12 Hours in Senior Level), 6 Hours English, 6 Hours Literature, 6 Hours Philosophy, 6 Hours History, 6 Hours Religion, 3 Hours Mathematics

JACKSONVILLE STATE UNIVERSITY
Jacksonville, Alabama 36265 Phone: (205) 435-9820
Dean of Graduate Studies: J.A. Reaves
Dean: R.B. Boozen

BIOLOGY DEPARTMENT
Phone: Chairman (205) 435-9820 Ext. 203
Others: Ext. 241

Head: K.E. Landers
Professors: T. Cochis, K. Landers, L.G. Sanford
Associate Professors: W. Curles, R. Mainland, C.W. Summerour, III.
Assistant Professors: M. Rollins, W.D. Staples, Jr.
Instructors: R. Rollins, F. Woodliff

Degree Programs: Biology, Biology and Education, Medical Technology

Undergraduate Degree: B.S.
Undergraduate Degree Requirements: 32 Semester Hours Biology, 2 Chemistry courses

Graduate Degrees: M.A., M.S. (Biology)
Graduate Degree Requirements: Master: 30 Semester Hours

Graduate and Undergraduate Courses Available for Graduate Credit: Microtechniques, Animal Physiology, Economic Entomology, Plant Anatomy, Plant Taxonomy, Dendrology, Parasitology, Cytology, Techniques in Botany, Economic Botany, Invertebrate Zoology, Vertebrate Zoology, Organic Evolution, Special Problems in Biology, Problems in Biology, Radiation Biology, Plant Ecology, Mycology, Phycology, Mammalogy, Animal Ecology, Protozoology, Advanced Animal Biology, Advanced Plant Biology, Advanced Techniques in Biology, Research and Thesis

JUDSON COLLEGE
Marion, Alabama 36756 Phone: (201) 683-2011
Dean: W.D. Murray

BIOLOGY DEPARTMENT
Phone: Chairman (201) 683-2011 Ext. 58

Head: T.H. Wilson
Professor: T.H. Wilson
Associate Professor: B. Grant (part-time)
Assistant Professor: L. Tarrants
Visiting Lecturer: 1

Undergraduate Degrees: B.S., B.A., (Biology)
Undergraduate Degree Requirements: B.S. - 30 Semester Hours in Biology, Algebra, Organic Chemistry
B.A. - 30 Semester Hours, Biology, Algebra, 2 years Foreign Language

LIVINGSTON UNIVERSITY
Livingston, Alabama 35470 Phone: (205) 652-5241
Dean of Graduate Studies: J. Patrenois
Dean: N. Reed

DIVISION OF NATURAL SCIENCES
Phone: Chairman (205) 652-5241 Ext. 218

Chairman: C.E. Tucker
Professor: C.E. Tucker
Associate Professor: A. Nixon
Assistant Professors: E. Bonds, R. Holland
Instructor: P. Mitchell

Undergraduate Degrees: B.S., B.A., (Biology)
Undergraduate Degree Requirements: 192 Quarter Hours, 50 in Biology, 35 in Minor

Graduate Degree: M.Ed., (Biology)

Graduate and Undergraduate Courses Available for Graduate Credit: Aquatic Biology, Parasitology, Biological Thories, Cell Biology, Anatomy and Morphology of Non-Vascular Plants, Anatomy and Morphology of Vascular Plants, Seminar in Biological Sciences, Directed Studies, Histology, Embryology

MILES COLLEGE
Birmingham, Alabama 35208 Phone: (205) 786-5281
(No reply received)

MOBILE COLLEGE
Mobile, Alabama 36613 Phone: (205) 675-5990
Dean: E. Keebler

BIOLOGY AREA IN DIVISION OF NATURAL SCIENCE

Chairman: C. Eyster
Professor: C. Eyster
Assistant Professor: E. French
Instructor: J. Borom

Field Stations/Laboratories: Marine Science Sea Laboratory - Dauphin Island

Undergraduate Degree: B.S., (Biology)
Undergraduate Degree Requirements: 3 Beginning Courses plus 5 Advanced Courses, of which 3 need to be Junior and Senior Level Courses

OAKWOOD COLLEGE
Huntsville, Alabama 35806 Phone: (205) 837-1630
Dean: E.A. Cooper

BIOLOGY DEPARTMENT
Phone: Chairman Ext. 282 Others Ext. 281

Head: R.C. Smith
Associate Professor: R.C. Smith
Assistant Professors: L. Fish, D. Ekkens

Undergraduate Degree: B.A., (Biology)
Undergraduate Degree Requirements: 45 Hours Biology, 44 Hours in Mathematics, Chemistry and Physics Plus others to equal 192 Hours

ST. BERNARD COLLEGE
St. Bernard, Alabama 35138 Phone: (205) 734-4110
Dean: C. Payne

LIBERAL ARTS DEPARTMENT - LABORATORY SCIENCE SECTION
Phone: Chairman: Ext. 72

Heads: C. Payne - Liberal Arts Department, J. Clark - Laboratory Science Section

Professors: Rev. M. Morgan, N.M. Herrera
Assistant Professors: Rev. V.J. Clark, P.C. Culivan, G. Laux

Degree Programs: Biology, Medical Technology

Undergraduate Degree: B.S.
Undergraduate Degree Requirements: Complete 128 Semester Hours of credit, 45 of which must be of upper level work, that is, of third and fourth year college level. (24 of which must be in Biology. Earn 256 Quality Points, with a "C" average in upper level work

SAMFORD UNIVERSITY

Birmingham, Alabama 35209 Phone: (205) 870-2844
Dean of Graduate Studies: L. Allen
Dean: H. Bailey

DEPARTMENT OF BIOLOGY
Phone: Chairman (205) 870-2844
Others (205) 870-2011

Chairman: H.A. McCullough
Professors: T.E. Denton
Associate Professors: W.M. Howell, E. McLaughlin
Assistant Professors: F.A. Hayse, R. Stiles
Teaching Fellows: 1
Part-time Laboratory Teachers: 2

Field Stations/Laboratories: Dauphin Island Sea Laboratory - Dauphin Island

Undergraduate Degree: B.S., (Biology)
Undergraduate Degree Requirements: 30 Hours in Biology (minimum), 2 years of Chemistry, 1 year of Physics

Graduate Degree: M.S., (Biology)
Graduate Degree Requirements: Master: 30 Hours in graduate Biology courses with 6 of these for Thesis

Graduate and Undergraduate Courses Available for Graduate Credit: Invertebrate Zoology (U), General Physiology (U), Vertebrate Field Zoology (U), Plant Morphology and Anatomy (U), Plant Taxonomy and Local Flora (U), History of Biology (U), Ecology (U), Experimental Embryology, Modern Concepts in Biology, Biosystematics, Speciation, Biogeography, Cytogenetics, Molecular Genetics, Advanced Topics in Genetics

SPRING HILL COLLEGE

Mobile, Alabama 36608 Phone: (205) 460-2325
Dean: C.H. Boyle

DEPARTMENT OF BIOLOGY
Phone: Chairman (205) 460-2325

Chairman: A.F. Hemphill
Professors: A.F. Hemphill, J.P. Macnamara
Assistant Professors: Rev. G.T. Regan,
Instructors: C.T. Wells
Special Lecturers: 2

Field Stations/Laboratories: Dauphin Island Sea Laboratory - Dauphin Island

Undergraduate Degree: B.S.,(Biology)
Undergraduate Degree Requirements: 31 - 32 Semester Hours of Biology, 20 Semester Hours of Chemistry, total of 128 Semester Hours

STILLMAN COLLEGE

Tuscaloosa, Alabama 35401 Phone: (205) 752-2548

MATH-SCIENCE DEPARTMENT
Phone: Ext. 24

Professors: J. Hall, E. Sparks
Associate Professors: M. Collins, E. Sparks
Assistant Professor: M. Collins
Visiting Lecturers: 3

Undergraduate Degrees: B.S., B.A., (Biology)
Undergraduate Degree Requirements: 30 Semester Hours of Biology, 18 Semester Hours of Chemistry, 10 Semester Hours of Mathematics, 8 Semester Hours of Physics

TALLADEGA COLLEGE

Talladega, Alabama 35160 Phone: (205) 362-5152
Dean: R. Braithwaite

BIOLOGY DEPARTMENT

Head: A.L. Bacon
Professors: A.L. Bacon
Associate Professors: W. Hodson, V.S. Rauganathan
Assistant Professor: B. Barlow
Instructors: S.L. Cole
Visiting Lecturers: 1

Undergraduate Degree: B.S., (Biology)

TROY STATE UNIVERSITY

Troy, Alabama 36081 Phone: (205) 566-3000
Dean of Graduate Studies: R.M. Argenti
Dean of College of Arts and Sciences: J.M. Long

DEPARTMENT OF BIOLOGICAL SCIENCES
Phone: Ext. 241

Chairman: J.C. Wilkes
Professors: R. Dietz, R. Kisner, D.C. Widdowson, J.C. Wilkes
Associate Professors: D. Barras
Assistant Professors: W. Adams, D.H. Costes, R. Tucker
Instructor: P.W. Wilkes

Field Stations/Laboratories: Marine Environmental Science Consortium Sea Laboratory - Dauphin Island

Degree Programs: Biology, Botany, Marine Biology, Medical Technology, Microbiology, Zoology

Undergraduate Degree: B.S.
Undergraduate Degree Requirements: 45 Hours in Major, or 25 Hours in Major and 20 Hours in Minor or Second 45 Hour Major, 69 Hours General Studies, 21 Hours Free Electives, 180 Hours Total

Graduate and Undergraduate Courses Available for Graduate Credit: Contemporary Problems in Biology, Teaching Biological Science, Independent Study

TUSKEGEE INSTITUTE

Tuskegee Institute, Alabama 36088 Phone: (205) 727-8217
Dean of Graduate Studies: Z.W. Metcalf, Jr.

College of Arts and Sciences

Dean: H.P. Carter

DEPARTMENT OF BIOLOGY
Phone: Chairman (205) 727-8217

Head: W.J. Sapp
Professors: H.P. Carter, J.H.M. Henderson, R.S. Saini, W.J. Sapp, O.C. Williamson
Associate Professors: J.E. Thomas, J.W. Williams, S. Yamaguchi
Assistant Professors: C.D. Barbour, O. Okelo
Instructors: V.B. Richardson, E.M. White
Research Assistants: 4
Teaching Assistants: 8

Field Stations/Laboratories: Experimental Farm adjacent

to Campus

Undergraduate Degree: B.S., (Biology)
Undergraduate Degree Requirements: 30 Hours of Biology, 20 Hours of Chemistry, 8 Hours of Physics, 8 Hours of College Algebra, 30 Hours of Humanities, 18 Hours of Social Sciences, 12 Hours Foreign Language or Option 4 Hours General Elective

Graduate Degree: M.S., (Biology)
Graduate Degree Requirements: Master: 30 Credit Hours and Thesis

School of Applied Science

Dean: B.D. Mayberry

DEPARTMENT OF AGRICULTURAL SCIENCES
Phone: Chairman: 727-8452

Chairman: M.H. Maloney, Jr.
Professors: P.K. Biswas, C.O. Briles, K.S. Chahal, M.A. Maloney, W. Nelson, B.T. Whatley
Associate Professor: E.T. Miles
Assistant Professors: R.B. Bettis, G.E. Cooper, L.O. Gadt, T.N. Kamalu, C.L. Mannings, J.A. Walls, E. M. Wilson, D. Granberry, I.J. Nelson, K.B. Paul, J.E. Powell, G.S. Rahi
Instructors: D.W. Libby, J. Nichols
Research Assistants: 6

Degree Programs: Agriculture, Agronomy, Animal Genetics, Animal Husbandry, Animal Industries, Animal Science, Crop Science, Dairy Science, Farm Crops, Horticultural Science, Horticulture, Plant Breeding, Plant Industry, Plant Sciences, Poultry Husbandry, Poultry Science, Seed Investigations, Soil Science, Soils and Plant Nurtition, Vegetable Crops, Veterinary Science

Undergraduate Degree: B.S.
Undergraduate Degree Requirements: 129 Semester Credit Hours, maintain a minimum grade-point average of 2.00 out of 4.00, pass the English proficiency Examination and satisfy the reading requirement

Graduate Degree: M.S.
Graduate Degree Requirements: Master: 30 Semester Credit Hours including no more than six credit hours of Research and Thesis; submit an approved Thesis, Maintain a 3.00 GPA

Graduate and Undergraduate Courses Available for Graduate Credit: Physiology of Reproduction (U), Advanced Animal and Poultry Nutrition (U), Immunogenetics (U), Advanced Reproductive Physiology, Population Genetics, Ruminology, Animal and Poultry Seminar, Animal and Poultry Literature Review, Advanced Classical Genetics, Special Problems in Animal and Poultry, Research in Poultry Science, Research in Animal Science, Advanced Crop Production (U), Soil Physics (U), Fertilization and Soil Fertilization (U), Soil Erosion, Silt and Control (U), Growing and Developing of Plants (U), Soil Microbiology (U), Special Problems in Plant and Soil Science (U), Plant and Soil Seminar, Ecology of Crop Geography, Pasture and Meadow Management, Chemistry of Soil and Fertilization, Advanced Fruit Science, Chemistry of Soil Organic Matter, Phytohormones and Vitimins, Physical Chemistry and Mineralogy of Soils, Special Topics in Plant and Soil Science, Research in Plant Science, Research in Soil Science, Bio-Statistics I (U), Bio-Statistics II, Biological Isotope Technique (U), Special Problems in Environmental Sciences, Water Chemistry I (U), Bio-Chemistry Instruction I-II (U), Aquatic Ecology, Pollution Chemistry and Biology I-II, Water Chemistry II, Topics in Environmental Science, Research in Environmental Science

School of Vetinary Medicine

Dean: W.C. Bowie

DEPARTMENT OF MICROBIOLOGY
Phone: Chairman: 727-8475

Head: B.B. Watson
Professors: I.H. Siddique, B.B. Watson
Associate Professors: G.C. Chang, E.M. Jenkins
Assistant Professors: S.W. Highley
Visiting Lecturers: 4
Research Assistants: 4
Technical: 7

Degree Program: Microbiology

Graduate Degree: M.S.
Graduate Degree Requirements: Interdisciplinary are subjects, Admission to Candidacy, Research, Thesis

Graduate and Undergraduate Courses Available for Graduate Credit: Advanced Pathogenic Microbiology, Special Problems in Microbiology I - II, Biochemistry, Molecular Biology, Virology, Immunology, Statistics, I, II Pathology, Graduate Research

DEPARTMENT OF PATHOLOGY AND PARASITOLOTY
Phone: Chairman: 727-8470

Chairman: W.O. Jones
Professors: W.E. Johnson, D.I. Lyles, T.S. Williams
Associate Professor: W.O. Jones
Assistant Professors: Y. Cho, T. Graham, B.E. Mc-Kenzie
Visiting Lecturers: 5

Degree Program: Veterinary Pathology

Graduate Degree: M.S.
Graduate Degree Requirements: Master: Thirty-Two Hours plus Thesis

Graduate and Undergraduate Courses Available for Graduate Credit: Problems in Veterinary Pathology, Clinical Pathology, Special Problems in Parasitology, Veterinary Helminthology, Veterinary Entomology, Veterinary Protozoology, Research

DEPARTMENT OF PHYSIOLOGY AND PHARMACOLOGY

Acting Chairman: W.C. Bowie
Professors: W.C. Bowie, C.A. Walker, R.G. Kammula
Associate Professor: E. Dixon
Assistant Professors: O.D. Verma, K.F. Soliman, K. Dalvi
Instructor: J.O. Owasoyo
Visiting Lecturers: 6

Field Stations/Laboratories: Veterinary Research Center - Moton Field, Carver Research Foundation (on campus)

Degree Programs: Animal Pathology, Animal Physiology, Veterinary Anatomy, Vertèrinary Bacteriology, Veterinary Microbiology, Veterinary Science

Graduate Degree: M.S.
Graduate Degree Requirements: 30 Semester Hours course work, 6 Hours Research, Thesis

Graduate and Undergraduate Courses Available for Graduate Credit: Physiology, Pharmacology, Anatomy, Microbiology, Parasitology, Pathology, Clinical Sciences

DEPARTMENT OF VETERINARY ANATOMY
Phone: Chairman: 727-8471

Head: R.C. Williams
Professors: S. Goldsberry, R.L. Judkins, R.C. Williams
Assistant Professors: G.A. Chibuzo, V.K. Reddy
Student Teaching Assistants: 2

Degree Program: Anatomy

Graduate Degree: M.S.

Graduate Degree Requirements: 32 Semester Hours of Course Credit plus Thesis

Graduate and Undergraduate Courses Available for Graduate Credit: Gross Anatomy I (U), Gross Anatomy II (U), Microscopic Anatomy I & II (U), Anatomy of the Nervous System, Anatomy (Special Problems), Histochemistry and Histochemical Techniques, Anatomy (Research)

UNIVERSITY OF ALABAMA

University, Alabama 35486 Phone: (205) 348-6120
Dean of Graduate Studies: C. Scott

College of Arts and Sciences

Dean: D.E. Jones

DEPARTMENT OF BIOLOGY
Phone: Chairman: 348-5960

Chairman: W.H. Darden, Jr.
Professors: E.L. Bishop, H.T. Boschung, E.A. Cross, W.H. Darden, T.R. Deason, W.F. Hill (Adjunct), J. L. Mego, J.C. O'Kelley, G.A. Rounsefell, C.E. Smith, J.L. Thomas, L.G. Williams
Associate Professors: D.G. Davis, D.C. Holliman (Adjunct) R.D. Hood, B.G. Moore, D.T. Rogers, E.R. Sayers
Assistant Professors: T.R. Bauman, G.F. Crozier (Adjunct) F.C. Gabrielson, G.S. Hand (Adjunct), T.M. Graham, G.R. Ultsch, B.A. Vittor (Adjunct), Z.S. Wochok
Instructor: J.P. Mitchell
Visiting Lecturers: 6
Research Assistants: 3
Graduate Teaching Assistants: 24
Graduate Council Fellows: 2
Graduate Council Research Fellow: 1

Field Stations/Laboratories: Dauphin Island Sea Laboratory-Dauphin Island

Degree Programs: Biology, Marine Science

Undergraduate Degree: B.S.
Undergraduate Degree Requirements: A minimum of 36 Hours in Biology plus 8 Hours General Physics, Chemistry through 8 Hours of Organic Chemistry and Mathematics through One Semester of Analytical Geometry and Calculus

Graduate Degrees: M.S., Ph.D.
Graduate Degree Requirements: Master: Plan 1 - at least 24 Hours of Graduate Credit and a Thesis Plan II - at least 30 Hours of Graduate Credit and No Thesis Doctoral: A minimum of 24 Semester Hours in Major Subject and a minimum of 12 Semester Hours for each of two Minor Subjects. One Minor must be done in the Department of Biology. The second Minor may be done outside the department

Graduate and Undergraduate Courses Available for Graduate Credit: Limnology (U), Physiological Ecology (U), Plant Growth and Development (U), Genetics (U), Dendrology, Plant Geography (U), Economic Botany (U), Cellular Biology (U), Biochemistry I (U), Biochemistry II (U), Phycology (U), Biometry (U), Histology of Vertebrates (U), General Ecology (U), Plant Ecology (U), Seminar in Physiological and Biochemical Ecology (U), Molecular Biology (U), Biology Seminar, Invertebrate Seminar, Invertebrate Morphology, Population Biology, Cytogenetics, Herpetology, Icthyology, Topics in Animal Physiology, Topics in Physiological Reproduction, Topics in Developmental Zoology, Systematics and Evolution of Vascular Plants, Advanced Genetics, Seminar in Genetics, Man and His Natural Environment, Enzymology, Protozoology, Advanced Topics in Morphology, Experimental Phycology, Ecology Seminar, Seminar in Molecular Biology, Cellular Morphogenesis, Invertebrate Embryology, Various Tutorials

DEPARTMENT OF MICROBIOLOGY
(No reply received)

UNIVERSITY OF ALABAMA IN BIRMINGHAM

1919 7th Avenue South Phone: (205) 934-4685
Birmingham, Alabama 35294
Dean of Graduate Studies: S. Barker

University College

Dean: R. Hanson

DEPARTMENT OF BIOLOGY
Phone: (205) 934-4685

Chairman: B. Cline
Professors: C.G. Crispens, C.P. Dagg, W.D. Fattig
Associate Professors: G.B. Cline, W.T. Feary (Adjunct) M.W. Hartley (Adjunct), D.D. Jones, K.R. Marion, G.R. Poirier, T. Shiota (Adjunct), E. Weller (Adjunct)
Assistant Professors: G.F. Crozier (Adjunct), J.J. Gauthier, G.S. Hand (Adjunct), D.T. Jenkins, R. MacGregor, B.A. Vittor (Adjunct)
Instructors: D. Gauthier, V. Nancarrow, L. Naumann, V. Volker
Teaching Fellows: 4
Research Assistants: 4

Field Stations/Laboratories: Marine Research Laboratory - Dauphin Island

Degree Programs: Genetics, Comparative and Cellular Physiology, Reproduction and Development

Undergraduate Degree: B.S., (Biology)
Undergraduate Degree Requirements: A. To qualify for a B.S. in Biology, the student must have at least a 1.00 grade point average overall, a 1.00 average in all Biology courses, and a 1.00 average in all Biology courses taken at UAB B. The student must complete 32 Semester Hours of Biology including: 6 Hours of Introductory Courses, 9 Hours of Intermediate Courses, 16 Hours of Advanced Courses, Twelve Hours of which must be distributed, as shown below, among three tracts: Organismal (6 Hours Minimum) Genetic and Cellular (3 Hours Minimum) Developmental (3 Hours Minimum) C. Chemistry: Inorganic (8 Hours Minimum), Organic (8 Hours Minimum) Physics: 8 Hours Minimum

Graduate Degrees: M.S., Ph.D.
Graduate Degree Requirements: Master: Students in the M.S. program may write a Thesis based on a research project, or, alternatively, may elect to submit a non-research Thesis incorporating a review and analysis of one or more topics of current or historical interest in Biology. M.S. students who write a research-based Thesis must complete a minimum of 24 Hours of committee-approved coursework, whereas students who choose to write a non-research thesis must complete a minimum of 30 Hours of approved coursework. An intellectual tool of research or a foreign language is required for the master's degree. Master's students must enroll in three seminar courses approved by the committee, and one of the seminars must be outside the major area of specialization. Presentation of a departmental seminar is required. To qualify for Candidacy, a student must take either a written or an oral comprehensive examination. The final examination for all candidates will consist of an oral presentation and defense of the research thesis or dissertation, or of the comprehensive review paper. Doctoral: A dissertation embodying the results and analysis of an original experimental investigation is required for Ph.D. candidates. A tool of research and one language or two languages are required for the Ph.D. program. Ph.D. students must enroll in three seminar courses

approved by the committee, and one of the seminars must be outside the major area of specialization. Presentation of a departmental seminar is required. They also may be required to demonstrate proficiency in teaching by delivering formal course lectures or by conducting teaching laboratories. To qualify for candidacy, a Ph.D. student must take both written and oral comprehensive examinations. The final examination for the Ph.D. Degree will consist of an oral presentation and defence of the research thesis or dissertation.

Graduate and Undergraduate Courses Available for Graduate Credit: General Physiology (U), Advanced General Physiology (U), Genetics (U), Embryology (U), Histology (U), Plant Physiology (U), Field Biology, Plants (U), Environmental Monitoring (U), Ecology (U), Mathematics in Physiology (U), Microbial Physiology, Special Topics in Biology, Advanced Biology for Teachers, Cellular Physiology, Reproductive Physiology, Developmental Genetics, Experimental Embryology, Advanced Genetics, Mammalian Embryology, Teratology, Seminars in Genetics, Physiology, Botany, Embryology, and Ecology

School of Medicine

Dean of Medical School: J.A. Pittman, Jr.
Dean of Dental School: C.A. McCallum, Jr.

DEPARTMENT OF ANATOMY
Phone: Chairman: (934-4494

Chairman: G. Hamel, Jr.
Professors: E.C. Crosby (Emeritus), T.E. Hunt (Emeritus), J. Schmidt (Emeritus), J.W. Brown, H.H. Hoffman, C. Klapper
Associate Professor: E.M. Welker
Assistant Professors: R. Mayne, G.S. Hard
Instructors: J.R. Augustine, A. Maier
Visiting Lecturers: 2
Teaching Fellows: 5

Degree Program: Anatomy

Graduate Degrees: M.S., Ph.D.
Graduate Degree Requirements: Course presented by Department and Graduate Committee, Thesis

DEPARTMENT OF BIOCHEMISTRY
Phone: Chairman: 934-4753

Chairman: J.M. McKibbin
Professors: L.M. Hall, C.D. Kochakian, J.M. McKibbin, E.J. Miller, J. Navia, L. Roden, J.R. Smythies, D.W. Urry, K.L. Yielding, C.H. Butterworth, Jr.
Associate Professors: F. Benington, C. Bugg, G.L. Carlson, H.C.Cheung, W.H. Finley, A. Habeeb, B. Johnson, C. Krumdieck, W. Niedermeier, K. Pruitt, W. Reedy, R. Schrohenloher, T. Shiota, K.B. Taylor, W. J. Wingo
Assistant Professors: T. Christian, G.M. Emerson, D. Gaudin, J. Glickson, R.S. Johnson, J.C. Lacey, E.L. Smith, J. Thompson

Degree Program: Biochemistry

Graduate Degrees: M.S., Ph.D.
Graduate Degree Requirements: Master: Introductory Biochemistry, Seminar, a Thesis, other courses requested by the student's Graduate Committee Doctoral: Introductory Biochemistry, Applied Concepts in Biochemistry, Seminar, a Thesis, a facility with one useful foreign language, other courses requested by the student's Graduate Committee including those in two useful minor areas.

Graduate and Undergraduate Courses Available for Graduate Credit: Introductory Biochemistry, Applied Concepts in Biochemistry, Seminar, Protein and Enzyme Chemistry, Biochemistry of Nucleic Acids, Human Genetic Disorders, Seminar in Enzyme Kinetics, Biochemistry of Growth, Aging, and Neoplosia, Nutrients and Intermediary Metabolism, Biochemistry of Ribosomes, Chemical Thermodynamics, Quartum Chemistry, X-Ray Crystallography, Computing Methods in Chemistry

DEPARTMENT OF BIOMATHEMATICS
Phone: Chairman: 934-5431 Others: 934-4011

Chairman: W. Siler
Professors: A.C.L. Barnard, G. Hutchison, J. Macy, Jr., M. Turner
Associate Professors: H.C. Cheung, K.D. Reilly, W. Siler
Assistant Professors: S. Harvey, C. Katholi, A. Segal
Research Assistants: 2

Field Stations/Laboratories: Dauphine Island Sea Laboratory - Dauphine Island

Degree Program: Biomathematics

Graduate Degrees: M.S., Ph.D.
Graduate Degree Requirements: Master: 24 Credit Hours plus Thesis Doctoral: Completion of qualifying Examination and successful defense of Dissertation

Graduate and Undergraduate Courses Available for Graduate Credit: Calculus and Differential Equations, Calculus and Differential Equations with Biomedical Applications, Modelling Biological Systems, Theory of Models, Mathematics for Biomathematicians, Models of Ecological Systems, Selected topics in Microscopic Systems, Selected topics in Organ Systems, Models of the Heart, Circulatory Systems and Hemodynamics, Sensory System Models, Neural Models

DEPARTMENT OF BIOSTATISTICS
Phone: Chairman: 934-4905 Others: 934-4011

Chairman: D.C. Hurst
Professors: D.C. Hurst, M.E. Turner, Jr., A.E. Wolff
Associate Professors: B.J. Trawinski, I. Trawinski
Assistant Professors: E.L. Bradley
Instructors: J.A. Burdeshaw, K.A. Kirk

Degree Program: Biostatistics

Graduate Degrees: M.S., Ph.D.
Graduate Degree Requirements: Master: 24 Semester Hours of credit and a Thesis Doctoral: Course program developed individually for each student

Graduate and Undergraduate Courses Available for Graduate Credit: Modern Mathematics in the Life Sciences, Computer Techniques in Statistical Research, Statistical Analysis and Design of Experiments, Statistical Methods in Biological Assay, Mathematical Theory of Epidemics, Introduction to Probability, Game Theory and Operations Research, Probability, Introduction to Stochastic Processes, Population Analysis, Inference, Introduction to Nonparametric Inference, Least Squares, Sampling, Discrete Data Analysis, Continuous Data Analysis, Biological Models, Advanced Experimental Design, Advanced Analysis of Variance, Stochastic Processes, Advanced Probability Advanced Inference, Non-Parametric Statistics, Multivariate Analysis, Advanced Analysis

DEPARTMENT OF ENGINEERING BIOPHYSICS
Phone: Chairman: 934-5154 Others: 934-4011

Chairman: S. S. West
Professor: S.S. West
Associate Professors: J.E. Lemons, A.A. Lorincz
Assistant Professors: H. Cort, Jr., J.F. Golden, R.E. Hurst, J.M. Menter, T.G. Pretlow, II, J.N. Thompson, E.M. Weller
Instructor: L.D. Love
Visiting Lecturers: 2

Degree Programs: Biophysical Cytochemistry, Engineering Biophysics, Physiology and Biophysics

Graduate Degree: Ph.D.
Graduate Degree Requirements: Engineering Biophysics, Physiology and Biophysics, Biophysical Cytochemistry

Graduate and Undergraduate Courses Available for Graduate Credit: Biophysical Cytochemistry (U), Biomaterials Science (U), Biophysical Anatomy (U), Introductory Engineering Biophysics (U), Measurement Systems (U)

DEPARTMENT OF INFORMATION SCIENCES
Phone: Chairman: 934-2213 Others: 934-4011

Chairman: A.C.L. Barnard
Professors: A.C.L. Barnard, J. Macy, Jr.
Associate Professors: J.M. Fontana, E. Mesel, K.D. Reilly, E.M. Wilson, C.C. Yand
Assistant Professors: E.H. Blackstone, D.A. Klip, L.C. Shappard, S.E. Wixson
Instructors: V.C.P. May, J.K. Murdock, E.M. Strand, M.P. White

Degree Programs: Computer Science

Undergraduate Degree: "Minor" (Computer Science)
Undergraduate Degree Requirements: 6 courses (18 Semester Hours Credit)

Graduate Degrees: M.S., Ph.D.
Graduate Degree Requirements: Master: 12 courses (36 hours credit) plus project Doctoral: Flexible course requirements. Research and Thesis mandatory

Graduate and Undergraduate Courses Available for Graduate Credit: Approximately 30 courses, covering the field of Computer and Information Sciences as defined by ACM Curriculum '68. Applications in medicine and biology are particularly emphasized.

DEPARTMENT OF MICROBIOLOGY
(No reply received)

DEPARTMENT OF PHYSIOLOGY AND BIOPHYSICS
(No information available)

UNIVERSITY OF ALABAMA IN HUNTSVILLE

Huntsville, Alabama 35807 Phone: (205) 895-6260
Dean of Graduate Studies: J.E. Rush
Dean of School of Science and Engineering: J. Hoomani

DEPARTMENT OF BIOLOGY
Phone: Chairman: 895-6360

Chairman: R.C. Leonard
Associate Professors: R.C. Leonard, C.H. Adams, E. Rowland, H.J. Wilson
Assistant Professors: S. Campbell, L.M. Rosing

Degree Program: Biology

Undergraduate Degrees: B.A., B.S.
Undergraduate Degree Requirements: 6 Semester hours English Literature, 6 Semester Hours English Composition, 6 Semester Hours World Civilization, 9 Semester Hours Mathematics (Calculus and Analytical Geometry), 6 Semester Hours Social Behavioral Science (Economics, Sociology, Psychology), 30 Semester Hours Biology above General Biology to include a. General Microbiology b, Genetics, C. A course in Anatomy and Morphology (Zoology or Botany). A course in either 1. Animal Physiology 2. Plant Physiology 3. Cell Physiology. 2 Seminars. Other courses to total 30 Semester Hours - is selected to meet needs of the individual student. Minor - 21 Semester Hours (usually Chemistry is elected by the students)

UNIVERSITY OF MONTEVALLO

Montevallo, Alabama 35115 Phone: (205) 665-2521
Dean of College of Arts and Sciences: J.B. Walters, Jr.

DEPARTMENT OF BIOLOGY
Phone: Chairman: Ext. 235

Chairman: P.G. Beasley
Professors: P.G. Beasley, E.B. Sledge, H.F. Turner
Associate Professor: J. Eagles
Assistant Professor: R.F. McGuire

Field Stations/Laboratories: Dauphin Island Sea Laboratory - Dauphin Island

Degree Programs: Biology, Medical Technology

Undergraduate Degrees: B.S., B.A.
Undergraduate Degree Requirements: (Semester Hours) Biology 36, Chemistry 16, Mathematics 8, English 12, Physics 8, History 6, Social Science 6, Art or Music or Speech 3, Psychology or Philosophy or Religion 3, Electives 25 (Minor from elective hours)

Graduate Degree: M.A.T.
Graduate Degree Requirements: Master: Research, 15 Hours Biology, 9 Hours Minor, 6 Hours Professional Education

Graduate and Undergraduate Courses Available for Graduate Credit: Histology (U), Parasitology (U), Genetics (U), Evolution (U), Field Botany (U), Vertebrate Field Zoology (U), Ornithology (U), Plant Physiology (U), Animal Physiology (U), Cell Physiology, Advanced Anatomy, Current Topics in Biology

UNIVERSITY OF SOUTH ALABAMA

Mobile, Alabama 36688 Phone: (205) 460-6101
Dean of Graduate Studies: F.H. Mitchell
Dean of College: W.W. Kaempfer

DEPARTMENT OF BIOLOGICAL SCIENCES
Phone: Chairman: 460-6331 Others: 460-6101

Chairman: R.J. Beyers
Professors: R.J. Beyers, E.E. Jones, J.M. Rawls, H.M. Phillips
Associate Professors: J.M. Boyles, B.L. Brown, J.F. Fitzpatrick, Jr., M.G. Lelong, D.W. Linzey, M.W. Miller
Assistant Professors: D.M. Dean, J.W. Langdon, R.L. Shipp, L.G. Tate

Degree Programs: Biology, Marine Biology

Undergraduate Degree: B.S., (Biology)
Undergraduate Degree Requirements: Must meet the general requirements for the degrees of Bachelor of Arts or Bachelor of Sciences plus 44 quarter hours of Biology (Biology 121, 122, 123, and 32 quarter hours of Biology electives), Chemistry 226 & 227 or 221, 222, 223; one year Physics, 12 Hours Mathematics

Graduate Degree: M.S.
Graduate Degree Requirements: Master: 48 quarter hours with minimum of B grade, of which 35 hours must be at the 500 level. 24 quarter hours in area of specialization, 12 hours outside specialization, Thesis, one foreign language, 3 quarters in Biology Seminar

Graduate and Undergraduate Courses Available for Graduate Credit: Biology Seminar, Aquatic Microbiology, Phycology, Special Topics, Scientific Data Management, Limnology, Special Research Problems, Physiology of Marine Animals, Advanced Parasitoloty, Aquatic Vascular Plants, Advanced Animal Behavior, Systematic Icthyology, Advanced Molecular Biology, Invertebrate Histological Techniques, Fisheries Economics, Oceanology of the Gulf of Mexico, Benthic Community Structure, Marine Zoogeography, Plankton, Thesis, Biogeography (U), Biological Literature and Terminology (U), Undergraduate Research (U), Marine Botany (U), Plant Physiology (U), Biology of Fungi (U), Taxonomy of Flowering Plants (U), Animal Behavior (U), Marine Vertebrate Zoology (U), Coastal Ornithology (U), Parasitology I & II (U), Medical Entomology (U), Vertebrate Histology (U)

ALASKA

UNIVERSITY OF ALASKA

College, Alaska 99701 Phone: (907) 479-7211

College of Biological Science and Renewable Resources

Dean: C. Behlke (Acting)

DEPARTMENT OF BIOLOGICAL SCIENCES
Phone: Chairman: 479-7542 Others: 479-7211

Head: J.E. Morrow
Professors: H.M. Foder, R.D. Guthrie, J.E. Morrow, B.J'Neiland, R.L. Rausch, L.G. Swartz
Associate Professors: S.F. MacLean Jr., D.F. Murray, R.L. Smith
Assistant Professor: C. Feist
Instructor: A. Scarborough (Lecturer)
Teaching Fellows: 8

Field Stations/Laboratories: NARL- Pt. Barrow, NARL Field Station - Anaktuvuk Pass, Yukon River Field Station - Eagle, Cantwell Reindeer Research Station - Cantwell, Seward Marine Facility, -Seward

Degree Programs: Applied Ecology, Biological Education, Biological Engineering, Biological Sciences, Botany, Ecology, Physiological Sciences, Physiology, Zoology

Undergraduate Degrees: B.A., B.S. (Biological Sciences)
Undergraduate Degree Requirements: 130 Semester Credit Hours

Graduate Degrees: M.S., Ph.D.
Graduate Degree Requirements: Master: 30 Semester Credit Hours beyond Bachelor's Degree Doctoral: Dissertation. No fixed Credit Hour requirements

Graduate and Undergraduate Courses Available for Graduate Credit: Parasitology (U), Systematic Botany (U), Vascular and Non-Vascular Plants (U), Cytogenetics (U), Cell Biology (U), Comparative Physiology (U), Plant Physiology (U), Ichthyology (U), Mammalogy (U), Ornithology (U), Animal Behavior (U), Plant Ecology (U), Animal Ecology (U), Field Ecology (U), History of Biology, Taxonomy, Biogeography, Physiological Ecology, Genetics, Marine Ecology, Advanced Plant Ecology

PROGRAM IN OCEANOGRAPHY AND OCEAN ENGINEERING
Phone: Chairman: 479-7749 Others: 479-7531

Chairman: V. Alexander
Professors: R.J. Barsdate, D.K. Button, R. Elsner, J.J. Goering, R. Neve
Associate Professors: V. Alexander, H.M. Feder,
Assistant Professors: R.T. Cooney, R. Horner, C. McRoy, R.L. Smith
Research Assistants: 20

Field Stations/Laboratories: Seward Marine Station - Seward, Izembek Lagoon Field Station - Cold Bay

Degree Programs: Limnology, Oceanography

Graduate Degrees: M.S., Ph.D.
Graduate Degree Requirements: Master: 30 Graduate Credits, Thesis Doctoral: No fixed course requirements "The degree is awarded for proven ability and scholarly attainment, the exact program is to be determined by the students advisory committee"

Graduate and Undergraduate Courses Available for Graduate Credit: Introduction to Biological Oceanography, Marine Ecology, Cellular Biochemistry, Limnology, Problems in Fisheries Management, Fishery Ecology, Invertebrate Fisheries Biology, Sampling the Marine Environment, Numerous courses in chemical, mathematical and physical oceanography several of which are required

INSTITUTE OF ARCTIC BIOLOGY
Phone: Chairman: 479-7648 Others: 479-7661

Director: P. Morrison
Advisory Scientific Director: L. Irving
Professors: G.C. West, J.R. Luick, R. Rausch (Adjunct) F. Milan (Joint)
Associate Professors: H. Behrisch, J.N. Cameron, R. Dietrich, P.W. Flanagan, C. Genaux (Joint), L.K. Miller, B.A. Philip (Adjunct), R. Van Pelt
Assistant Professors: J.H. Anderson, F.S. Chapin, D. D. Feist, K.J. Kokjer (Joint), M. Rosenmann, K. Van Cleve (Joint), R.G. White
Visiting Lecturers: 8
Research Assistants: 4

Degree Programs: Animal Physiology, Comparative Biochemistry, Ecology, Nutrition, Comparative Environmental Physiology, Plant Sciences, Veterinary Science, Zoology

Graduate Degree: Ph.D. (Arctic Biology)
Graduate Degree Requirements: Doctoral: M.S. or equivalent courses, courses as specified by committee written and oral qualifying examination, Dissertation and oral examination on same

ARIZONA

ARIZONA STATE UNIVERSITY

Tempe, Arizona 25281 Phone: (602) 965-9011
Dean of Graduate School: W.J. Burke

College of Liberal Arts

Dean: C.M. Woolf

DEPARTMENT OF BOTANY AND MICROBIOLOGY
Phone: Chairman: 965-3414 Others: 965-6657

Chairman: H.C. Reeves
Professors: J.M. Aronson, J.E. Canright, R.M. Johnson, W.T. Northey, D.T. Patten, H.C. Reeves
Associate Professors: A.M. Dycus, C.R. Leathers, D. J. Pinkava, J.M. Schmidt, M.R. Sommerfeld
Assistant Professors: S.J. Archer, E.A. Birge, T.H. Nash, R.N. Trelease
Instructors: J.R. Swafford
Visiting Lecturers: 1
Teaching Fellows: 21
Research Assistants: 2
Postdoctoral: 1

Degree Programs: Bacteriology, Bacteriology and Immunology, Biochemistry, Biological Chemistry, Biological Sciences, Biology, Botany, Comparative Biochemistry, Ecology, Farm Crops, Genetics, Health and Biological Sciences, Immunology, Life Sciences, Medical Biophysics, Medical Microbiology, Medical Technology, Microbiology, Physiology, Plant Pathology, Plant Sciences, Science Education, Virology

Undergraduate Degree: B.S.
Undergraduate Degree Requirements: Minimum of 128 Semester Hours including 45 Semester Hours in Major area, of which 18 must be upper division courses. One year of foreign language required

Graduate Degrees: M.S., Ph.D.
Graduate Degree Requirements: Master: Minimum of 30 Semester Hours of course work. Language requirement in Microbiology. No language requirement in Botany. Thesis required. Doctoral: Minimum of 84 Semester Hours of course work. Language required. Dissertation required

Graduate and Undergraduate Courses Available for Graduate Credit: Plant Ecology (U), Plant Geography (U), General Mycology (U), Morphology of the Non-Vascular Plants (U), Experimental Phycology (U), Morphology of the Vascular Plants (U), Algae and Bryophytes (U), Growth and Reproduction (U), Physiology of Lower Plants (U), Physiology of Lower Plants Laboratory (U), Taxonomy of Southwestern Vascular Plants (U), Angiosperm Taxonomy (U), Experimental Plant Systematics (U), Paleobotany (U), Plant Metabolism, Immunology (U), Bacterial Genetics (U), Bacterial Physiology (U), Systematic Bacteriology (U), Virology (U), Immunochemistry, Microbial Enzymology, Pathogenic Bacteriology

ZOOLOGY DEPARTMENT
Phone: Chairman: 965-3571

Chairman: R.H. Alvarado
Maytag Chair: J. Emlin
Professors: R.H. Alvarado, G.L. Bender, E.M. Bertke, G.B. Castle, M. Cazier, G.A. Cole, S.D. Gerking, H. Hanson, E.J. Landers, R.A. Patterson, K. Pike, C.H. Woolf
Associate Professors: J. Alcock, K.K. Church, R. Clothier, M.J. Fouquette, Jr., N. Hadley, F. Hasbrouck, J. Justus, R. McGaughey, W.L. Minckley, R. Ohmart, D.I. Rasmussen
Assistant Professors: D. Belk, J. Foster, E. Goldstein, G. Murdock, R.N. Trelease
Research Assistants: 3
Postdoctoral: 3

Degree Programs: Biology, Entomology, Wildlife Biology, Zoology

Undergraduate Degree: B.S.
Undergraduate Degree Requirements: 126 Semester Hours 35 in Major, 42 in supporting areas

Graduate Degrees: M.S., Ph.D.
Graduate Degree Requirements: Master: With Thesis - 30 Semester hours of which 6 must be thesis and 9 may be research or special studies Without Thesis - 30 Semester Hours of which 6 may be research or special studies Doctoral: A minimum of 84 Semester Hours beyond the bachelor's degree. All candidates must pass a written qualifying examination during their first year of residency. One foreign language is required. A written and oral comprehensive examination is given after completion of course work and language. A dissertation based on original work and defended in an oral examination is required

GRAND CANYON COLLEGE

Phoenix, Arizona 85017 Phone: (602) 249-3300
Dean of College: D. Whitis

DEPARTMENT OF NATURAL SCIENCES
3300 W. Camelback Phone: Chairman: Ext. 263

Chairman: B. Williams
Professors: B. Williams, E. Morris, H. Rush
Associate Professors: P. Youngs
Part-time: 2

Degree Program: Biology

Undergraduate Degree: B.S.

NORTHERN ARIZONA UNIVERSITY

Flagstaff, Arizona 86001 Phone: (602) 523-9011
Dean of Graduate Studies: R.S. Beal
Dean of College: C. Little

DEPARTMENT OF BIOLOGICAL SCIENCE
Phone: Chairman: 523-3538

Chairman: J.R. Wick
Professors: G. Anderson, C. Goin (Adjunct), O. Johnson, L. Mogensen, J. Rominger, T. Vaughan, J. Wick
Associate Professors: R. Balda, K. Derifield, D. English, G. Goslow, R. Hevly, C.D. Johnson, W. Lipke, S. Wilkes
Assistant Professors: G. Bateman, D. Blinn, W. Gaud, M. Glendening, J.N. Grim, G. Pogany, C.N. Slobodchikoff, J. States, R. Tamppari
Teaching Fellows: 19
Research Assistants: 15

Degree Programs: Biology, Botany, Educational Biology

Undergraduate Degree: B.S.
Undergraduate Degree Requirements: 125 Semester Hours credit. 42 General College Courses (Liberal Studies), 36 Biology courses, 17-24 Mathematics and

Chemistry, 23-30 Electives (19 hours Education courses for B.S. in Education

Graduate Degrees: M.A., M.S., Ph.D.
Graduate Degree Requirements: Master: M.A. - 36 Hours course work; oral examination, M.S. - 30 Hours including 6 Hours Thesis; oral examination Doctoral: 60 Hours coursework beyond Bachelor's degree, acceptable Dissertation, 2 languages written and oral preliminary examinations, oral Thesis defence

Graduate and Undergraduate Courses Available for Graduate Credit: Phycology (U), Mycology (U), Agrostology (U), Protozoology (U), Ichthyology (U), Herpetology (U), Mammalogy (U), Ornithology (U), Comparative Animal Physiology (U), Plant Physiology (U), Field Biology (U), Evolution and Man, Advanced Plant Taxonomy, Insect Identification, Insect Morphology, Developmental Biology, Advanced Plant Anatomy, Plant Chemistry, Advanced Genetics, Molecular Biology, Cytology and Cytogenetics, Human Genetics, Organic Evolution, Speciation, Biogeography, Systematic Zoology, Animal Behavior, Advanced Ecology, Paleobotany-Paleoecology, Limnology, Physiological Ecology, Medical Entomology, Economic Botany, Biological Technics, Advanced Studies in Biology, Quantitative Biology, Modern Biology for Teachers

PRESCOTT COLLEGE

Prescott, Arizona 86301 Phone: (602) 445-3254
There are no deans; the vice president (R. Harril) administers academic and personnel policies which are generated by the faculty and various student-faculty committees

There are no departments or departmental chairmen.

Associate Professors: K. Asplund
Assistant Professors: M. Asplund, J. Taylor, C. Tomoff
Instructor: K. Kingsley
Teaching Assistants: 2

Field Stations/Laboratories: While we do much field work, we use mobile laboratories. We work in Mexico,--especially along the perimeter of the Gulf of California, and we sponsor continuing research in all areas of Arizona and specific areas of California, Utah, and Wyoming

Undergraduate Degree: B.A.
Undergraduate Degree Requirements: Breadth, depth and self-direction. 1.) He or she must demonstrate considerable evidence that he or she is capable of slef-directed learning (usually via extended projects or research which is summarized articulately and extensively) and 2.) He or she must demonstrate extensive competence in a specific subject area selected from: Anatomy, Anatomy and Histology, Anatomy and Physiology, Applied Ecology, Biochemistry and Biophysics, Biological Education, Biological Sciences, Biology, Biophysics, Biostatistics, Botany, Conservation, Ecology, Ethology, Genetics, Health and Biological Sciences, Horticultural Science, Horticulture, Human Genetics, Land Resources, Landscape Horticulture, Life Sciences, Limnology, Natural Resources, Natural Science, Nutrition, Physiological Sciences, Physiology, Plant Sciences, Science, Science Education, Soils and Plant Nutrition, Vegetable Crops, Veterinary Anatomy, Watershed Management, Wildlife Management, Zoology, Marine Biology

UNIVERSITY OF ARIZONA

Tucson, Arizona 85721 Phone: (602) 884-2751
Dean of Graduate Studies: H.D. Rhodes

College of Agriculture

Dean of Agriculture: G.R. Stairs

DEPARTMENT OF AGRICULTURAL BIOCHEMISTRY

401 Agricultural Science Building
Phone: Chairman: 884-1798 Others: 884-2751

Head: M.G. Vavich
Professors: J.W. Berry, M. Bier (Visiting), A.J. Deutschman, J.P. Holbrook (Visiting), A.R. Kemmerer (Emeritus), H.W. Kircher, W.F. McCaughey, J.L. Parsons (Visiting), P.B. Pearson (Visiting), J.S. Strong (Visiting)
Associate Professor: A.B. Stanfield
Assistant Professors: D.P. Bourque, W.S. Cozine
Research Associates: 2

Field Stations/Laboratories: University of Arizona Field Station - Mesa, Safford, Tempe, Tucson and Yuma

Degree Programs: Agricultural Biochemistry, Nurtition

Undergraduate Degree: B.S. (Nutritional Sciences)
Undergraduate Degree Requirements: 130 Semester Units including the following: 8 units General Chemistry with laboratory, 8 units Organic Chemistry with laboratory, 3 units Analytical Chemistry with laboratory, 8 units Physics and Geosciences, 8 units Biology or Microbiology, 11 units Mathematics and Statistics

Graduate Degrees: M.S., Ph.D.
Graduate Degree Requirements: Master: A final oral examination and a Thesis based upon research are required in addition to 30 Semester Unite of graduate courses, of which no more than six may be credited to thesis research Doctoral: Minimum admission requirement is a bachelor's degree which includes one year of General Biology, one year of Organic Chemistry with laboratory, one year of Physics and Mathematics, through Calculus.

Graduate and Undergraduate Courses Available for Graduate Credit: Principles of Nutrition (U), Comparative Animal Nutrition (U), Nutritional Biochemistry (U), Sensory Evaluation of Food (U), Biochemistry of the Aging Process (U), Food Chemistry (U), Food Microbiology and Sanitation Laboratory (U), Food Sanitation (U), Food Processing (U), Plant Pathogenic Fungi and Nematodes, Advanced Nutrition, Chemistry and Metabolism of Lipids, Steroid Chemistry and Biochemistry, Plant Pathogenic Bacteria and Viruses, Chemistry and Metabolism of Proteins, Nutritional Biochemistry Techniques, Vitamins and Minerals, Chemistry of Enzymes, Intermediary Metabolism, Chemistry and Metabolism of the Nucleic Acids, Laboratory Methods in Nutrition, Chemistry of Natural Products, Advanced Plant Physiology, Chemistry of Proteins, Mechanism of Enzyme Action, Special Topics in Genetics, Seminar

DEPARTMENT OF AGRICULTURAL CHEMISTRY AND SOILS

"No information available"

DEPARTMENT OF AGRONOMY AND PLANT GENETICS

Phone: Chairman: 884-1945 Others: 884-1977

Head: M.A. Massengale
Professors: M.A. Massengale, R.E. Briggs, A.D. Day, R.E. Dennis, J.E. Endrizzi, C.V. Feaster, W.D. Fisher, K.C. Hamilton, E.B. Jackson, F.R.H. Katterman, W.R. Kneebone, G.M. Loper, D.F. McAlister, D.S. Metcalf, J.R. Mauney, R.T. Ramage, D.D. Rubis, M.H. Schonhorst, L.S. Stith, R.L. Voigt, L.N. Wright, L.H. Zimmerman
Associate Professors: D.R. Buxton, R.G. McDaniel, H. Muramoto, R.G. Sackett, B.B. Taylor
Teaching Fellows: 1
Research Assistants: 16

Field Stations/Laboratories: Mesa Branch Agriculture Experiment Station - Mesa, Yuma Branch Agriculture Experiment Station - Yuma, Safford Branch Agriculture Experiment Station - Safford, Cotton Research Center - Phoenix

Degree Programs: Agronomy, Plant Genetics

Undergraduate Degree: B.S. (Agronomy)
Undergraduate Degree Requirements: A total of 130 Semester units is required for graduation. Thirty-six units are required in Agriculture including 16 in the Major. Twenty-eight units are needed in Biological and Physical Sciences, 12 in Social Sciences and Humanities, 6 in English Composition, 6 in Communications and 2 in Physical Education

Graduate Degrees: M.S., Ph.D.
Graduate Degree Requirements: Master: Minimum of 30 units of graduate work including Thesis if required. Fifteen units or more must be in Major or closely related field. Doctoral: Minimum of 36 units of credit exclusive of Dissertation, must be in area of Major subject. Two semesters of essentially full-time graduate work and at least 30 units of graduate credit must be completed at the University of Arizona

Graduate and Undergraduate Courses Available for Graduate Credit: Forage Production (U), Cotton and Other Fiber Crops (U), Weed Control (U), Crop Ecology (U), Quality Components of Crop Seeds (U), Genetic Principles of Hybrid Seed Production (U), Turfgrass Management (U), Principles of Plant Breeding (U), Plant Breeding Laboratory (U), Field Plot Research (U), Plant Resistance to Insects (U), Advanced Genetics, Advanced Cytogenetics, Theory of Crop Production, Statistical Genetics, Theory of Plant Breeding, Advanced Plant Physiology, Quantitative Genetics and Selection, Seminar, Individual Studies

ANIMAL SCIENCE DEPARTMENT
"No information available"

DEPARTMENT OF DAIRY AND FOOD SCIENCES
Phone: Chairman: (602) 884-1508

Head: G.H. Stott
Professors: J.W. Stull, F.E. Nelson, W.H. Brown, J.D. Schuh
Assistant Professors: F.M. Whiting, T.N. Wegner, R.L. Price, D. Armstrong
Visiting Lecturers: 2
Research Assistants: 12

Field Stations/Laboratories: Dairy Research Center - Tucson

Graduate Programs: Agricultural Biochemistry, Animal Physiology, Dairy Science, Food Science, Microbiology

Undergraduate Degree: B.S., (Dairy Science, Food Science)
Undergraduate Degree Requirements: Basic science background is required with an additional 16 units in the Major subject Dairy Production which includes Cattle Breeding, Management, Nutrition, Physiology and Reproduction

Graduate Degrees: M.S., Ph.D.
Graduate Degree Requirements: Master: Additional 30 Hours above the B.S. degree; 16 units must be in the Major field. Thesis is required Doctoral: The department also cooperates with the Committees on Animal Physiology and Agricultural Biochemistry and Nutrition, and Genetics in both instruction and direction of research leading to the Ph.D. degrees in these fields

DEPARTMENT OF ENTOMOLOGY
Phone: Chairman: (602) 884-1151

Chairman: G.W. Ware
Professors: L.A. Carruth (Emeritus), P.D. Gerhardt, W.L. Nutting, D.M. Tuttle, T.F. Watson, F.G. Werner
Associate Professors: L.A. Crowder, R.E. Fry, L. Moore, M.W. Nielson
Assistant Professors: D.T. Langston, G.L. Lentz, C.E. Mason, G.S. Olton, G.D. Waller
Visiting Lecturers: 1
Research Assistants: 5

Field Stations/Laboratories: Agricultural Experiment Station Field Laboratories - Mesa, Phoenix, Tucson and Yuma

Degree Programs: Acarology, Aprculture, Behavior, Biological Control, Bionomics, Ecology, Host Plant Resistance, Morphology, Pest Management, Physiology, Taxonomy, Toxicology, Veterinary and Public Health Entomology Special Options: Plant Protection, Turf Management

Undergraduate Degrees: B.A., B.S. (Entomology)
Undergraduate Degree Requirements: 16-24 Units of Entomology courses depending on program--the Major is offered in the College of Agriculture (2 programs) and in the College of Liberal Arts

Graduate Degrees: M.S., Ph.D. (Entomology)
Graduate Degree Requirements: Master: 30 Hours of graduate work, including Thesis Doctoral: A minimum equivalent of 3 years of study beyond the bachelor's degree; mastery of statistical procedures, Biochemistry and a foreign language; Dissertation

Graduate and Undergraduate Courses Available for Graduate Credit: Parasites of Domestic Animals (U), Insect Morphology (U), Insect Physiology (U), Insect Toxicology (U), Advanced Economic Entomology (U), Applied Insect Taxonomy (U), Insect Pest Management (U), Public Health Entomology (U), Acarology, Experimental Methods in Pest Management, Plant Resistance to Insects, Advanced Insect Physiology, Insect Ecology, Insect Behavior, Biological Control, Graduate Seminar, Graduate Research, Individual Studies (U), Thesis, Dissertation, Advanced Insect Physiology, Forest Entomology (U)

DEPARTMENT OF HORTICULTURE AND LANDSCAPE AGRICULTURE
Phone: Chairman: (602) 884-2940 Others: 2751

Head: A.E. Thompson
Professors: W.P. Bemis, L. Burkhart, S. Fazio, M.R. Fontes, R.E. Foster, G.S. Greene, R.H. Hilgeman, L. Hogan, W.D. Jones, J.R. Kuykendall, N.F. Oebker, W.D. Pew, D.R. Rodney, G.C. Sharples
Associate Professors: P.M. Bessey, C.M. Sacamand
Assistant Professors: G.W. Charchalis, J.N. Nelson
Research Assistant: 3
Teaching Assistant: 1
Research Associate: 1

Field Stations/Laboratories: Yuma Branch Experimental Station - Yuma, Mesa Branch Experimental Station - Mesa, Salt River Valley Citrus Station - Tempe

Degree Programs: Horticulture, Landscape Horticulture, Genetics

Undergraduate Degree: B.S.
Undergraduate Degree Requirements: 130 Semester Units including 16 in Major

Graduate Degrees: M.S., Ph.D.
Graduate Degree Requirements: Master: At least 30 Unit Units of graduate credit - 6-8 units for Thesis plus 15 additional units in Major. Thesis may be waived. Doctoral: At least 36 units of work exclusive of Dissertation must be in area of Major Subject. Proficiency in one foreign language approved by Major Department. Equivalent of at least 6 Semesters of essentially full-time graduate study required - minimum residence requirement - must spend minimum of 2 regular semesters of essentially full-time academic work

in actual residence and at least 30 units of graduate credit from this institution

Graduate and Undergraduate Courses Available for Graduate Credit: Citriculture (U), Quality Components of Crop Seeds (U), Genetic Principles of Hybrid Seed Production (U), Plant Propagation (U), Technology of Horticultural Products (U), Tropical and Subtropical Horticulture (U), Turfgrass Management (U), Field Plot Research (U), Vegetable Crop Ecology (U), Advanced Genetics, Advanced Cytogenetics, Theory of Plant Breeding, Advanced Plant Physiology, Quantitative Genetics and Selection, Seminar, Special Problems (U), Research, Extended Registration, Thesis, Advanced Landscape Design (U), Perception of Designed Environments: Measurement and Evaluation (U),

DEPARTMENT OF PLANT PATHOLOGY
Phone: Chairman: 884-1828 Others: 884-2511

Chairman: E.L. Nigh
Professors: S.M. Alcorn, R.M. Allen, G. Cummins, R.L. Gilbertson, R.B. Hine, E. Kendrick, M.A. McClure, M.R. Nelson, E.L. Nigh
Associate Professors: H.B. Bloss, R.L. Caldwell, M.E. Stanghellini, J.L. Troutman
Assistant Professors: T.E. Russell
Research Assistants: 5

Field Stations/Laboratories: Yuma Agricultural Experiment Station - Mesa AES, Safford AES, Marana AES

Degree Programs: Botany, Plant Pathology

Undergraduate Degree: B.S.
Undergraduate Degree Requirements: Not Stated

Graduate Degrees: M.S., Ph.D.
Graduate Degree Requirements: Not Stated

Graduate and Undergraduate Courses Available for Graduate Credit: General Mycology, Plant Disease Control, Diagnosis of Plant Pathology, Methods in Plant Pathology

POULTRY SCIENCE DEPARTMENT
"No information available"

DEPARTMENT OF VETERINARY SCIENCE
Phone: Chairman: (602) 884-2355 Others: 884-2211

Head: R.E. Reed
Professors: E.J. Bicknell, L.W. Dewhirst, R.E. Watts
Assistant Professors: T.H. Noon
Research Assistants: 1

Field Stations/Laboratories: Mesa Research and Diagnostic Laboratory - Mesa

Degree Programs: Agricultural Biochemistry, Agriculture, Animal Physiology, Animal Health Science

Undergraduate Degree: B.S.,(Agriculture, Animal Health Science)
Undergraduate Degree Requirements: Not Stated

Graduate Degrees: M.S., Ph.D.
Graduate Degree Requirements: Not Stated

Graduate and Undergraduate Courses Available for Graduate Credit: Parasites of Domestic Animals, Animal Diseases, Laboratory Animal Management, Experimental Surgery

DEPARTMENT OF WATERSHED MANAGEMENT
Phone: Chairman: (602) 884-2313 Others: 884-2314

Head: D.B. Thorud
Professors: M.M. Fogel, H.M. Hull (Adjunct), G.L. Jordan, D.A. King, J.O. Klemmedson, P.N. Knorr, S.C. Martin (Adjunct), P.R. Ogden, E.M. Schmutz, J.L. Thames, R.F. Wagle, M.J. Zwolinski
Associate Professors: S.K. Brickler, P.F. Ffolliott, G.S. Lehman, E.L. Smith
Instructors: R.L. Beschta, B.E. Kynard

Degree Programs: Fisheries Management, Range Management, Watershed Management (including Forestry), Natural Resource Recreation

Undergraduate Degree: B.S.
Undergraduate Degree Requirements: Semester Hours total - 130, 9 Hours English, 3 Hours Speech, 16-30 Hours Professional Courses, 30 Hours Biological and Physical Sciences, 12 Hours Social Sciences, Elective

Graduate Degrees: M.S., Ph.D.
Graduate Degree Requirements: Master: Plan A - minimun of 30 graduate units plus Thesis Plan B - minimum of 36 graduate units plus professional paper Doctoral: minimum 6 semesters of full-time graduate work (20 Semester Hours at University of Arizona), one language plus collateral science (Statistics, Computer Programming, 2nd language), Dissertation

Graduate and Undergraduate Courses Available for Graduate Credit: Resource Protection (U), Silviculture (U), Mensuration (U), Plants and Environment (U), Photogrammetry (U), Photo Interpretation (U), Wood Technology and Utilization (U), Forest Resource Management (U), Forest Resource Economics (U), Watershed Hydrology (U), Watershed Management (U), Forest and Range Policy (U), Management Project in Renewable Natural Resources (U), Snow Hydrology, Modeling of Small Watershed Hydrology, Systems Analysis in Watershed Management, Research Methods, Forest and Range Plant Identification (U), Range Management (U), Ecology of Forest and Range Communities (U), Range Evaluation and Planning (U), Fishery Management (U), Recreation Resource Planning (U), Economics of Outdoor Recreation (U)

College of Liberal Arts
Dean: H.K. Bleibtreu

DEPARTMENT OF BIOLOGICAL SCIENCES
Phone: Chairman: (602) 884-2715

Acting Head: G.A. Dawson
Professors: J.T. Bagnara, W.A. Calder, H.K. Gloyd, R.B. Chiasson, E.L. Cockrum, W.R. Ferris, R.M. Harris, W.B. Heed, J.R. Hendrickson, R.W. Hoshaw, R.R. Humphrey (Emeritus), K. Keck, C.H. Lowe, I. M. Lytle, C.T. Mason, A.R. Mead, J.W. O'Leary, P.E. Pickens, N.A. Younggren
Associate Professors: P.G. Bartels, R.P. Davis, M.E. Hadley, K. Matsuda, R.S. Mellor, R.B. Miller, S.M. Russell, D.A. Thomson, W. Van Asdall
Assistant Professors: A.C. Gibson, H.R. Pulliam, E.A. Stull, W.V. Zucker
Lecturers: C.W. Gaddis, D.B. Sayner, O.G. Ward
Research Associates: 6

Field Stations/Laboratories: Laboratorio de Biologia Marina - Puerto Penasco, Sonora, Mexico, Boyce Thompson Arboretum - Superior, Southwestern Research Station - Portal

Graduate Programs: Animal Behavior, Animal Physiology, Biological Education, Biological Sciences, Biology, Botany, Ecology, Ethology, Fisheries and Wildlife Biology, Genetics, Limnology, Physiology, Plant Sciences, Zoology, Marine Sciences

Undergraduate Degree: B.S.
Undergraduate Degree Requirements: 2 years lower division Biology, 2 years Chemistry, 1 year Mathematics (Fishery Biology 2 years). 12 units upper division Biology

Graduate Degrees: M.S., Ph.D.
Graduate Degree Requirements: Applicants for admission should be prepared in Chemistry, Physics and Mathematics, and must submit both the Aptitude and Advanced Test scores of the Graduate Record Examinations. All applicants must communicate directly with the Department regarding other requirements for admission.

Graduate and Undergraduate Courses Available for Graduate Credit: Biological Materials (U), History of Biology (U), Techniques of Biological Literature (U), Quantitative Biology (U), Cell Biology (U), Microtechniques (U), Scientific Illustration (U), Population Genetics (U), Genetics of Microorganisms (U), General Ecology (U), Plant Ecology (U), Plant Geography (U), Zoogeography (U), Oceanography (U), Oceanography Laboratory (U), Limnology (U), Marine Ecology (U), Wildlife Management (U), Aquatic Resource Biology (U), Wildlife Management Techniques (U), Ecology of Wildlife Reproduction (U), Current Problems in Wildlife Biology (U), Current Problems in Fishery Biology (U), Developmental Plant Anatomy (U), Plant Microtechniques (U), Developmental Biology (U), Comparative Chordate Morphology (U), Plant Physiology (U), Plant-Water Relations (U), An Introduction to ervous System (U), Human Physiology (U), Human Physiology Laboratory (U), Comparative Physiology (U), Endocrine Physiology (U), Environmental Physiology (U), Sensory-Motor Physiology (U), Plant Morphology (U), Systematic Botany (U), Legumues, Grasses and Composites (U), Freshwater Algae (U), Marine Algae, (U), Invertebrate Zoölogy (U), Icthyology (U), Herpetology (U), Ornithology (U), Mammalogy (U), Parasitology (U), Animal Behavior (U), Radioisotopes in Bioloty, Principles of Electron Microscopy, Bioenergetics, Cellular and Molecular Biology: I. Structure, Cellular and Molecular Biology: II. Development, Cellular and Molecular Biology: II. Development Laboratory, Cellular and Molecular Biology: III. Macromolecules, Cellular and Molecular Biology: IV. Biophysics, Cellular and Molecular Biology: IV. Biophysics Laboratory, Applications and Techniques of Human Genetics, Physiological Genetics, Cytogenetics, Speciation, Genetics in Populations, Genetics in Population Laboratory, Laboratory Techniques in Genetics, Topics in Physiological Ecology, Advanced Ecology, Advanced Studies in Marine Biology, Advanced Topics in Morphology, Advanced Topics in Endocrinology, Methods in Plant Physiology, Advanced Plant Physiology, Advanced Vertebrate Physiology, Advanced Topics in Neurophysiology, Comparative Endocrinology, Comparative Endocrinology Laboratory, Special Topics in Genetics, Advanced Systematic Botany, Sèlected Studies of Birds, Selected Studies of Mammals, Seminar, Special Problems, Research, Thesis, Dissertation

Graduate College

Dean: H.D. Rhodes

COMMITTEE ON GENETICS

Phone: Chairman: (602) 884-3584 Others: 884-2751

Chairman: R.M. Harris
Professors: W.P. Bemis, J.E. Endrizzi, R.M. Harris, W.B. Heed, F.S. Hulse, F.R. Katterman, R.T. Ramage, C.B.Roubicek
Associate Professors: J.R. Davis, A.B. Humphrey, R.G. McDaniel, N.H. Mendelson
Assistant Professor: O.G. Ward
Teaching Fellows: 6
Research Assistants: 8
Graduate Fellows: 2

Degree Program: Genetics

Graduate Degrees: M.S., Ph.D.
Graduate Degree Requirements: Master: 30 Units, option with or without Thesis, no language Doctoral: 36 Units in the Major plus 18 Dissertation units, Minor of varying Units, one language

Graduate and Undergraduate Courses Available for Graduate Credit: Depends on area of specialization in genetics, all courses used for degree must be graduate

College of Medicine

Dean: N.A. Vanselow

DEPARTMENT OF ANATOMY

(No information available)

DEPARTMENT OF BIOCHEMISTRY

Phone: Chairman: (602) 882-6025 Others: 882-6024

Chairman: D.J. Hanahan
Professors: D.J. Hanahan, C.K. Mathews
Associate Professors: M.S. Olson, M.A. Wells
Assistant Professors: W.J. Grimes, M.R. Haussler
Teaching Fellows: 5
Research Assistants: 20

Degree Program: Biochemistry

Graduate Degree: Ph.D.
Graduate Degree Requirements: Doctoral: General Biochemistry, Biochemical Literature, Seminar, five advanced courses

DEPARTMENT OF MICROBIOLOGY

(No reply received)

DEPARTMENT OF PHYSIOLOGY

Phone: Chairman: (602) 882-6511

Professors: W.H. Dantzler, P.C. Johnson, D.G. Stuart
Associate Professors: R.W. Gore, R.P. Gruener, G.A. Hedge
Assistant Professors: E.J. Braun, H.D. Kim
Teaching Fellows: 5
Research Assistants: 2
NIH Predoctoral Trainees: 5

Degree Program: Physiology

Graduate Degree: Ph.D.
Graduate Degree Requirements: Doctoral: General requirements as given in University of Arizona graduate catalog

Graduate and Undergraduate Courses Available for Graduate Credit: Human Physiology, Neurosciences, Research Methods in Physiology, Advanced Mammalian Physiology, Special Topics, Seminar

ARKANSAS

ARKANSAS AGRICULTURAL, MECHANICAL AND NORMAL COLLEGE
(See University of Arkansas at Monticello)

ARKANSAS AGRICULTURAL AND MECHANICAL COLLEGE
(See University of Arkansas at Pine Bluff)

ARKANSAS COLLEGE
Batesville, Arkansas 72501
(No Reply Received)

ARKANSAS POLYTECHNIC COLLEGE
Russellville, Arkansas 72801 Phone: (501) 968-0311
Dean of College: H.L. McMillan

DEPARTMENT OF BIOLOGICAL SCIENCES
Phone: Chairman: (501) 968-0293

Head: G.P. Hutchinson
Professor: H.L. McMillan
Associate Professors: H.D. Crawley, R.D. Couser, G.P. Hutchinson, E.E. Hudson
Assistant Professors: T.M. Palko, G.E. Tucker, B.L. Tatum
Instructors: G. Turnipseed

Degree Programs: Biology, Biological Sciences, Fisheries, Wildlife Biology

Undergraduate Degree: B.S.
Undergraduate Degree Requirements: Botany, Vertebrate Zoology, Invertebrate Zoology, Genetics, Plant Taxonomy, Seminar, 16 Hours Chemistry, 8 Hours Physics, 48 Hours General Education

ARKANSAS STATE UNIVERSITY
State University, Arkansas 72467 Phone: (501) 972-2100
Dean of Graduate Studies: E. Smith
Dean of College of Science: J.H. Stevenson

DIVISION OF BIOLOGICAL SCIENCES
Phone: Chairman: (501) 972-3082 Others: 972-2100

Chairman: J.K. Beadles
Professors: J.K. Beadles (Biology, E.L. Hanebrink (Biology), J.A. Hutchinson (Botany), E.L. Richards (Botany), E.G. Wittlake (Botany)
Associate Professors: H.E. Barton,(Zoology), W.W. Byrd (Zoology), G.L. Harp (Zoology, J.R. Hersey (Zoology), L.W. Hinck (Biology), J.M. Hite (Biology), B.D. Johnson (Zoology, D. Timmermann (Botany)
Assistant Professors: R.Z. Gehring (Botany) V.R. McDaniel (Zoology), L.A. Olson (Zoology), P.L. Raines (Botany)
Teaching Fellows: 12

Field Stations/Laboratories: Fish Farming Experiment Station - Walcott

Degree Programs: Biology, Botany, General Science with Emphasis on Biology, Medical Technology, Wildlife Management, Zoology

Undergraduate Degree: B.S.
Undergraduate Degree Requirements: B.S.E. 36 Hours University College Requirements, 37 Hours Divisional, 23 Major (Biology), 23 Hours Professional Education, 5 Hours Electives, B.S., 36 Hours University College, 37 Hours Divisional, 28 Hours Biology Major, 23 Hours Electives; 25 Hours Botany Major, 26 Hours Electives, 22 Medical Technology Major, 33 Hours Clinical, 25 HoursWildlife Management Major, 26 Hours Electives, 23 Hours Zoology Major, 28 Hours Electives, 124 Semester Hours required for each major.

Graduate Degrees: M.S., M.S.E.
Graduate Degree Requirements: Master: M.S. - 24 Hours of course work and 6 Hours of Thesis. M.S.E.- 21 Hours of Major Field courses and nine Hours of professional education courses

Graduate and Undergraduate Courses Available for Graduate Credit: Cytogenetics (U), Physcology (U), Mycology (U), Bryology (U), Biological Seminar, Anatomy of Vascular Plants, Speciation, Advanced Plant Taxonomy, Paleobotany, Medical and Veterinary Entomology, Insect Morphology, Insect Taxonomy, Fishery Biology (U), Cytology (U), Mammalogy (U), Immunology (U), Parasitology (U), Protozoology (U), Ichthyology (U), Herpetology (U), Ornithology (U), Limnology (U), Literature and History of the Biological Sciences, Aquatic Biology, Comparative Ethology, Natural History of the Vertebrates, Endocrinology

THE COLLEGE OF THE OZARKS
Clarksville, Arkansas 72830 Phone: (501) 754-8261
Dean of College: F. Ehren

DIVISION OF SCIENCE AND MATHEMATICS
Phone: Chairman: (501) 754-8261

Chairman: N.D. Rowbotham
Professor: N.D. Rowbotham
Associate Professors: J. Bridgman (Bioloty), S.M. Condren (Chemistry), R.S. Reynolds (Biology, G.W. Shellenberger (Physics)
Assistant Professors: L. MacLaughlin (Physics) M. Moore (Mathematics)

Degree Programs: Medical Technology, Natural Science

Undergraduate Degree: B.S.

HARDING COLLEGE
Box 941 Phone: (501) 268-6161
Searcy, Arkansas 72143
Director of Graduate Studies: N.D. Royse
Dean of College: J.E. Pryor

DEPARTMENT OF BIOLOGICAL SCIENCE
Phone: Chairman: Ext. 362 Others: 362 and 364

Chairman: J.W. Sears
Professor: J.W. Sears
Associate Professors: W.B. Roberson, G.W. Woodruff
Assistant Professor: W.F. Rushton
Instructor: R.H. Doran

Degree Program: Biology

Undergraduate Degrees: B.A., B.S.
Undergraduate Degree Requirements: B.A. - 30 Semester Hours in Biology B.S. - 30 Hours in Biology (minimum) plus 27 Hours in two other sciences plus 6 Hours in a fourth Science and one year of French or German or a reading proficiency in either

Graduate and Undergraduate Courses Available for Graduate Credit: Genetics (U), Biological Techniques (U), Cellular Physiology (U), Workshop in Environmental Studies (U), Science for Elementary School Teachers, Biology for Secondary School Teachers NOTE These courses are offered in connection with our Master's in Teaching Program

HENDERSON STATE COLLEGE

Arkadelphia, Arkansas 71923 Phone: (501) 246-5511
Dean of Graduate Studies: W.A. Dahlstedt
Dean of College: J.C. Wright

DEPARTMENT OF BIOLOGY
Phone: Chairman: Ext. 207 Others: Ext. 217

Chairman: P.R. Dorris
Professors: P.R. Dorris, K.H. Oliver, Jr.
Associate Professor: J.D. Bragg
Assistant Professors: M.L. McBurney, D.W. McMasters, D.L. Marsh

Degree Programs: Science, Science in Education

Undergraduate Degree: B.A.
Undergraduate Degree Requirements: General Biology, General Zoology, Directed Biology Electives 22 Hours, Chemistry

HENDRIX COLLEGE

Conway, Arkansas 72032 Phone: (501) 329-6811
Dean of College: F. Christie

BIOLOGY DEPARTMENT
Phone: Chairman: Ext. 377

Chairman: A.A. Johnson
Professor: A.A. Johnson
Associate Professors: G.T. Clark, A.M. Raymond
Assistant Professor: B. Haggard

Degree Program: Biology

Undergraduate Degree: B.A.
Undergraduate Degree Requirements: Biology -- Concepts in Biology, Botany, General Zoology, Comparative Anatomy, Genetics, Seminar, Four additional Biology Courses, Department Comprehensive Examinations. Non-Biology -- three Humanity courses, three Social Science courses, two Chemistry courses, total of 36 courses including biology

JOHN BROWN UNIVERSITY

Siloam Springs, Arkansas 72761 Phone: (501) 524-3131
Dean of College: R. Cox

DEPARTMENT OF BIOLOGY
Phone: Chairman: Ext. 237

Chairman: G. Griggs
Professors: G. Griggs, I. Wills
Assistant Professor: L. Seward
Visiting Lecturers: 1
Research Assistant: 1

Degree Program: Biology

Undergraduate Degree: B.S.
Undergraduate Degree Requirements: Not Stated

LITTLE ROCK UNIVERSITY

(See University of Arkansas at Little Rock)

ONACHITA COLLEGE

Arkadelphia, Arkansas 71923 Phone: (501) 246-4531
Dean of College: C. Goodson

DEPARTMENT OF BIOLOGY
Phone: Chairman: (501) 246-4531

Chairman: V.L. Oliver
Associate Professors: C.K. Sandifer, R. Brown
Graduate Assistant: 1

Degree Program: Biology

Undergraduate Degrees: B.A., B.S.
Undergraduate Degree Requirements: General Biology 8 Hours, Ecology 4 Hours, Other Biology 20 Hours, Physics, Chemistry, Mathematics to total at least 45 Hours

SOUTHERN STATE COLLEGE

Magnolia, Arkansas 71753 Phone: (501) 234-4120
Dean of College: L.A. Logan

DEPARTMENT OF AGRICULTURE
Phone: Chairman: Ext. 238 Others: 338 and 339

Chairman: O.A. Childs
Professor: O.A. Childs
Associate Professors: J.T. Attebery, W.C. Lee
Assistant Professor: R. Adams
Instructor: D. Higgins

Degree Programs: Agriculture, Agricultural Chemistry

Undergraduate Degree: B.S.
Undergraduate Degree Requirements: 124 Semester Hours. Grade point average of 2.00

DEPARTMENT OF BIOLOGY
Phone: Chairman: Ext. 202 Others: Ext. 212

Chairman: L.A. Logan
Professor: L.A. Logan
Associate Professor: M. King
Assistant Professors: D.R. England, H.A. Johnson, H.W. Robison

Degree Programs: Biological Science

Undergraduate Degrees: B.A., B.S.
Undergraduate Degree Requirements: Minimum of 30 Semester Hours in pure Biological Subjects

STATE COLLEGE OF ARKANSAS

Conway, Arkansas 72032 Phone: (501) 329-2931
Dean of Graduate Studies: H.B. Hardy
Dean of College: O.W. Rook

DEPARTMENT OF BIOLOGY

Chairman: T.J. Burgess
Professors: N.D. Buffaloe, T.J. Burgess, R.A. Collins, J.E. Moore, J.B. Throneberry
Associate Professors: D.E. Culwell, D.D. Smith
Assistant Professors: E.R. Kinser, R.T. Kirkwood, J.R. Nichols, W. Owen
Teaching Fellows: 3

Degree Programs: Developmental Biology, General Physiology, Plant Taxonomy or Plant Morphology, General Microbiology, General Ecology, Seminar. B.S. (Pre-Professional) - Developmental Biology, General Physiology, Histology, General Microbiology, Comparative Vertebrate Anatomy

Undergraduate Degrees: B.S., B.S.E. (Biology)
Undergraduate Degree Requirements: 2 years Chemistry, 1 year Physics, 38 Semester Hours Biology including General Zoology, General Botany, Genetics, Cell Biology as a core for all undergraduate degrees.

B.S.E. - General Physiology, either Plant Morphology or Plant Taxonomy, either Invertebrate Zoology or Vertebrate Zoology, General Ecology, Seminar. B.S. (General Program)

Graduate Degrees: M.S., M.S.E. (Biology)
Graduate Degree Requirements: Master: M.S. - 30 Semester Hours, including 6 Thesis Research, 2 Seminar, Remainder by advisement. Thesis required. M.S.E. - 30 Semester Hours, minimum 12 Hours Education-Psychology, minimum 18 Hours Biology. Course requirements by advisement

Graduate and Undergraduate Courses Available for Graduate Credit: Seminar in Biology (U), Histology (U), Invertebrate Zoology (U), Vertebrate Zoology (U), General Ecology (U), Plant Taxonomy (U), Plant Morphology (U), Developmental Biology (U), General Microbiology (U), Comparative Vertebrate Anatomy (U), General Physiology (U), Graduate Seminar, Special Problems, Thesis Research, History of Biology, Experimental Embryology, Plant Ecology, Aquatic Ecology, Cellular Physiology, Microbial Genetics

UNIVERSITY OF ARKANSAS

Fayetteville, Arkansas 72701 Phone: (501) 575-2000
Dean of Graduate Studies: J. Hudson

College of Arts and Sciences

Dean: R.C. Anderson

DEPARTMENT OF BOTANY AND BACTERIOLOGY
Phone: Chairman: (501) 575-4901

Chairman: L.F. Bailey
Professors: L.F. Bailey, E.E. Dale, R.S. Fairchild, G.T. Johnson, L.J. Paulissen
Associate Professors: R.K. Bower, F.E. Lane, R.L. Meyer, E.B. Smith, D.E. Talburt, J.L. Wickliff
Assistant Professors: W.M. Harris, A.F. Reeves
Research Assistants: 4

Degree Programs: Botany, Bacteriology

Undergraduate Degrees: B.S., B.A.
Undergraduate Degree Requirements: 30 Semester Hours of Botany or Bacteriology, Chemistry through Organic (4 or 8 Hours), Physics 8 Hours

Graduate Degrees: M.A., Ph.D.
Graduate Degree Requirements: Master: 30 Semester Hours, including at least one course in each of the broad areas of either Botany or Bacteriology, Biochemistry, Statistics, and courses in other disciplines. Thesis required for the M.S. Doctoral: Course requirements to meet deficiencies and specialized training needs, reading knowledge of German or French, Candidacy examination, Dissertation Research

Graduate and Undergraduate Courses Available for Graduate Credit: Plant Taxonomy (U), Advanced Taxonomy, Ecology (U), Physiological Ecology, Algology, Mycology, Morphology (U), Physiology (U), Photobiology, Growth Substances, Photosynthesis, Metabolism, Research Methods, Anatomy (U), Cytology (U), Cytogenetics, Pathogenic Bacteriology (U), Physiology (U), Immunology, Virology (U), Public Health and Sanitation (U), Genetics, Industrial, Applied, Metabolism, Research Methods

DEPARTMENT OF ZOOLOGY
Phone: Chairman: (501) 575-3253 Others: 575-3251

Chairman: P.M. Johnston
Professors: F.E. Clayton, W.L. Evans, W.C. Guest, D.A. James, W.L. Money, E.H. Schmitz, J.A. Sealander
Associate Professors: D.A. Becker, R.V. Kilambi, L.R. Kraemer, D.W. Martin, J.M. Walker
Assistant Professor: C.F. Bailey
Instructor: A.F. Posey
Research Assistants: 5
Research Associate: 1

Degree Program: Zoology

Undergraduate Degrees: B.S., B.A.
Undergraduate Degree Requirements: B.S. - 30 Semester Hours exclusive of General Biology or its equivalent to include 1 course in 2000's, 3 in 3000's, at least one with laboratory, 3 in 4000's, at least one with laboratory, and exclusive of 4001. Chemistry 8 Hours, Physics 8 Hours, Algebra and Trignometry 6 Hours B.A. - 30 Semester Hours of Biology and Zoology to include Biology 1004, 1014, 2404, 3323, 4001 and 4544

Graduate Degrees: M.S., Ph.D.
Graduate Degree Requirements: Master: 30 Semester Hours of graduate credit; six hours may be Thesis or 30 semester hours of graduate credit to include nine hours of research credit from three members of the committee directing the student; final oral Doctoral: No specific course requirements except for minimum 18 hours of Dissertation credit. One modern foreign language-reading; 6 weeks of approved marine work. Qualifying examination after one year of full-time graduate work. Candidacy Examination usually after 2 years; final examination primarily covering Dissertation

Graduate and Undergraduate Courses Available for Graduate Credit: Vertebrate Paleontology, Mechansims of Behavior, Research Methods, Cytology, Parasitology, Fresh Water Invertebrates, Principles of Parasitism, Field Studies, Special Topics, Graduate Seminar, Bibliographic Practicum, Comparative Physiology, Experimental Endocrinology, Cellular Physiology, Experimental Genetics, Biochemical Genetics, Physiological Ecology, Experimental Embryology, Comparative Vertebrate Embryology, Microscopic Technique, Protozoology, Invertebrata, Plankton Studies, Fisheries Science, Ichthyology, Herpetology, Ornithology, Mammalogy, Limnology, Animal Behavior, and Ecology. Research course credit in most of the above for 1-6 Hours

College of Agriculture and Home Economics

Dean: G.W. Hardy

DEPARTMENT OF AGRONOMY
Phone: Chairman: (501) 575-2355

Chairman: D.A. Hinkle
Altheimer Chair for Cotton Research: B.A. Waddle
Professors: D.A. Brown, C.E. Caviness, R.E. Frans, J.L. Keogh, M.S. Offutt, A.E. Spooner, R.E. Talbert, L.F. Thompson, J.O. York
Associate Professors: F.C. Collins, E.M. Rutledge, W.E. Sabbe, C.A. Stutte
Assistant Professors: J.T. Cothren, L.H. Hileman, J.W. King, R. Maples, H.D. Scott, H.R. Stoin
Research Assistants: 14

Field Stations/Laboratories: Southwest Branch Station - Hope, Southeast Branch Station - Rohwer, Rice Branch Station - Stuttgart, Cotton Branch Station - Marianna, Northeast Branch Station - Keiser, Livestock and Forestry Branch Station - Batesville

Degree Programs: Agronomy, Crop Science, Genetics, Physiology, Plant Breeding, Soils and Plant Nutrition

Undergraduate Degree: B.S.
Undergraduate Degree Requirements: 132 Semester Hours total, with 32 Hours of Physical and Biological Sciences, 11 Hours Communications, 12 Hours Humanities and Social Sciences, 19-23 Hours of Agricultural Sciences, 24-29 in Agronomy, and 26-23 Hours of Electives

Graduate Degrees: M.S., Ph.D.
Graduate Degree Requirements: Master: 24 Hours of graduate credit, plus 6 Hours for Thesis Doctoral: approximately 60 additional hours of graduate credit, including a Dissertation

Graduate and Undergraduate Courses Available for Graduate Credit: Principles of Experimentation (U), Radiological Applications in Agronomy,(U), Plant Breeding (U), Pasture and Forage Crop Management (U), Principles of Weed Control (U), Fertilizers (U), Soil Physics (U), Soil and Plant Analysis (U), Soil Fertility (U), Soil Chemistry (U), Experimental Designs, Plant Nutrition, Soil Genesis and Morphology, Herbicides and Plant Growth Regulators, Soil Mineralogy

DEPARTMENT OF ANIMAL SCIENCES
Phone: Chairman: (501) 575-4351
Others: 4352, 4353, 4354, 4355

Head: E.L. Stephenson
Professors: J.N. Beasley, C.J. Brown, T.L. Goodwin, N.R. Gyles, G.C. Harris, Jr., P.K. Lewis, T.S. Nelson, P.R. Noland, L.T. Patterson, J.L. Perkins, E.H. Peterson, J.M. Rakes, M.L. Ray, O.T. Stallcup
Associate Professors: L.D. Andrews, J.F. Brown, L.B. Daniels, E.L. Piper, P.W. Waldroup, R. Williams
Assistant Professors: L.L. Barton, J.A. Collins, M.C. Heck, H.P. Peterson
Research Assistants: 19

Degree Programs: Animal Science, Dairy Science, Food Science, Poultry Science, Animal Nutrition

Undergraduate Degree: B.S.
Undergraduate Degree Requirements: Not Stated

Graduate Degree: M.S., Ph.D.
Graduate Degree Requirements: Master: 30 Semester Hours including 6 Hours for Thesis Doctoral: Courses prescribed by Graduate Committee plus an acceptable Thesis

Graduate and Undergraduate Courses Available for Graduate Credit: Due to variety of degrees offered the number of courses that might be used for graduate credit is numerous. These would include college courses in many departments on this campus plus in certain cases courses in the Medical School in Little Rock.

DEPARTMENT OF ENTOMOLOGY
Phone: Chairman: (501) 575-2451

Head: F.D. Miner
Professors: J.L. Lancaster, Jr., C. Lincoln, F.D. Miner, J.R. Phillips, W.D. Wylie, W.C. Yearian
Associate Professors: R.T. Allen, M.V. Meisch, N.P. Tugwell, S.Y. Young, III
Assistant Professors: A.J. Mueller, F.M. Stephen
Research Assistants: 13
Research Associates: 2

Degree Program: Entomology

Undergraduate Degree: B.S.
Undergraduate Degree Requirements: Fifteen semester hours of Entomology, 20 other hours designated by the advisor, 27 to 32 hours of other Physical and Biological Sciences, in a total of 132 hours

Graduate Degrees: M.S., Ph.D.
Graduate Degree Requirements: Master: Thirty semester Hours as designated by the students' Graduate Committee, including a 6-hour Thesis, and a final oral examination Doctoral: At least one year of full-time residence beyond the M.S., completion of a course of study outlined by the students' Graduate Committee including 30 hours in the Major and supporting fields, demonstration of proficiency in one foreign language, completion of an 18-hour Dissertation and passage of qualifying, candidacy and final examinations

Graduate and Undergraduate Courses Available for Graduate Credit: Insect Behavior, Insect Identification, Apiculture, Advanced Applied Entomology, Insects of Man and Animals, Special Problems, Morphology of Insects, Insect Ecology, Special Topics, Immature Insects, Master's Thesis, Seminar, Insect Physiology, Insect Toxicology, Insect Pest Management, Insect Systematics, Doctoral Dissertation

DEPARTMENT OF HORTICULTURAL FOOD SCIENCE
Phone: Chairman: (501) 443-3281

Head: A.A. Kattan
Professors: W.A. Sistrunk, D.R. Tompkins
Associate Professor: J.R. Morris
Assistant Professor: R.W. Buescher
Research Assistants: 3
Graduate Assistants: 9

Degree Program: Food Science

Undergraduate Degree: B.S.
Undergraduate Degree Requirements: 132 Semester Hours including: Communications - 11 Hours, Humanities and Social Sciences - 12 Hours, Agricultural Sciences - 19-23 Hours, Food Science Major - 24-29 Hours, General Electives - 26-33

Graduate Degrees: M.S.
Graduate Degree Requirements: Master: As specified by department of major: 24 semester hours of course work and 6 hours of Thesis Doctoral: Candidates may select Food Science as a minor - requirements are specified by candidate's graduate committee

Graduate and Undergraduate Courses Available for Graduate Credit: Special Problems, Postharvest Physiology, Quality Evaluation and Control, Analytical Food Science, Seminar, Special Problems Research, M.S. Thesis, Special Topics

DEPARTMENT OF HORTICULTURE AND FORESTRY
Phone: Chairman: (501) 575-2604

Head: G.A. Bradley
Professors: J.L. Bowers, G.A. Bradley, J. McFerran, F.M. Meade (Emeritus), J.N. Moore, R.C. Rom, J.E. Vaile (Emeritus), V.M. Watts (Emeritus)
Associate Professor: A.E. Einert
Assistant Professors: H.A. Holt, A.T. McDaniel
Research Assistants: 10

Field Stations/Laboratories: Fruit Sub Experiment Station-Clarksville, Peach Sub Experiment Station - Nashville Strawberry Sub Experiment Station - Bald Knob, Vegetable Sub Experiment Station - Van Buren

Degree Program: Horticulture

Undergraduate Degree: B.S.A.
Undergraduate Degree Requirements: 132 credit hours, 4 optional plans available in which science and mathematic requirements are different. Includes minor options in business or journalism

Graduate Degree: M.S.
Graduate Degree Requirements: Master: 30 hours of graduate level courses, including 6 hours of Thesis credit. An acceptable Thesis must be presented

Graduate and Undergraduate Courses Available for Graduate Credit: Special Problems (U), Tree Fruit Science (U), Small Fruit Production (U), Advanced Vegetable Crops (U), Seminar, Special Research Problems, Advanced Plant Breeding, Master's Thesis

DEPARTMENT OF PLANT PATHOLOGY

Phone: Chairman: (501) 575-2446 Others: 442-8271

Head: D.A. Slack
Professors: D.A. Slack, J.L. Dale, J.P. Fulton, N.D. Fulton, M.J. Goode, J.P. Jones, M. McGuire, R.D. Riggs, H.A. Scott, G.E. Templeton, H.J. Walters
Associate Professor: F.H. Tainter
Research Assistants: 13
Research Associate: 1

Degree Programs: Agriculture, Environmental Science, Nematology, Plant Pathology, Virology, Plant Protection, Environmental Science

Undergraduate Degree: B.S.

Graduate Degrees: M.S., Ph.D. (Plant Pathology)

UNIVERSITY OF ARKANSAS MEDICAL CENTER
4301 West Markham Street Phone: (501) 644-5000
Little Rock, Arkansas 72201
Dean of Medical Studies: W.K. Shorey
Dean of Graduate Studies: J. Hudson

DEPARTMENT OF ANATOMY
Phone: Chairman: (501) 664-5000 Ext. 297

Chairman: J.E. Pauly
Professors: H.N. Marvin, J.E. Pauly, E.W. Powell, L.E. Scheving, J.K. Sherman
Associate Professors: E.R. Burns, M.D. Cave, S.A. Gilmore
Assistant Professors: E.A.Lucas, T.W. Schoultz, R.D. Skinner
Instructor: S. Tsai

Degree Program: Anatomy

Graduate Degrees: M.S., Ph.D.
Graduate Degree Requirements: Master: 30 Semester Hours including Gross Anatomy, Microanatomy, (Histology and Embryology), Neurosciences, Physiology and Biochemistry; research project and Thesis; written and oral preliminary examination on course work, and a final oral examination on Thesis and related material Doctoral: Same basic courses as required for master's followed by First Level Examination (equivalent to preliminary examination for Masters); reading knowledge of at least one foreign language; two day written and oral Doctoral Preliminary Examination; research and Dissertation; final oral examination on defense of Dissertation

Graduate and Undergraduate Courses Available for Graduate Credit: Advanced Gross Anatomy, Advanced Microanatomy, Advanced Neuroanatomy, Microscopic Technique, Cytology, Morphological and Functional Endocirnology, Fetal and Neonatal Anatomy, Low-Temperature Biology, Seminar, Research, Histochemistry, Cytochemistry and Electron Microscopy, History of Anatomy, Research Methods in Anatomy, Experimental Cell Biology, Biology of Neoplasia, Peripherial Nervous System-Anatomy and Physiology, Sleep-Wake Mechanisms, Neuroanatomy-Gross, Three-Dimensional, Neuroscience-Systems Review, Hypothalamus, Limbic Systems, Chronobiology, Principles and Techniques of Electron Microscopy, Master's Thesis, Doctoral Dissertation. Other courses available in Biochemistry, Physiology, Pharmacology, Microbiology, Pathology, etc.

DEPARTMENT OF BIOCHEMISTRY
Phone: Chairman: (501) 664-5000 Ext. 464

Chairman: C.L. Wadkins
Professors: C. Angel, M.F. Cranmer, M. Morris, C.L. Wadkins
Associate Professors: K. Baetcke, C.D. Jackson, W.G. Smith, C.G. Winter, Y.C. Yeh, J.L. York
Assistant Professors: M. Brewster, R.M. Cernosek, Jr., D.L. DeLuca, H. Mohrenweiser, C.A. Nelson, K.D. Straub, P.V. Wagh, G. Wolff
Instructor: C. Bhuvaneswaran
Teaching Fellows: 2
Research Assistants: 12

Degree Program: Biochemistry

Graduate Degrees: M.S., Ph.D.
Graduate Degree Requirements: Master: (1) 24 semester hours and a Thesis (2) A comprehensive examination. (3) A cumulative grade average of 2.85 (individual departments may have higher grade standards). (4) A minimum residence of 30 weeks. Doctoral: The course of studies, seminars and research leading to candidacy for the doctoral degree will be planned by the student and his individual advisory committee. In most cases it is expected that students will enroll in all major courses offered by the department. The student must bass the standard language examination in German, French or Russian. This requirement must be met prior to scheduling of the written comprehensive candidacy examination. Each candidate must perform satisfactorily on written and oral comprehensive examinations in the field of study and complete an acceptable research Thesis with a satisfactory oral defense.

Graduate and Undergraduate Courses Available for Graduate Credit: Chemistry of Biological Compounds, Biochemistry, Biochemical Methods, Biochemistry Seminar, Research in Biochemistry, Master's Thesis, Bio-Organic Reaction Mechanisms, Physical Chemistry of Macromolecules, Metabolism, Chemistry of Enzymes, Biochemical Genetics, Doctoral Dissertation

DIVISION OF BIOMETRY
Phone: Chairman: (501) 664-5000 Ext. 533

Head: J.H. Meade, Jr.
Professors: J.H. Meade, Jr., G.V. Dalrymple, R.A. Dykman, D.W. Gaylor
Associate Professors: J.F. McCoy, R.C. Walls
Assistant Professors: J.G. Farmer, A. Schroeder, C. Thompson
Instructors: J.B. Johnson, J. W. Nance, R.J. Warner
Research Assistants: 2

Degree Program: Biometry

Graduate Degree: M.S.
Graduate Degree Requirements: Not Stated

Graduate and Undergraduate Courses Available for Graduate Credit: Biometrical Methods I (U), Biometrical Methods II (U), Biometrical Methods III, Introduction to Biomedical Computing (U), Theory of Statistics I (U), Theory of Statistics II (U), Theory of Statistics III, Mathematical Biology I, Linear Statistical Models, Sampling, Nonparametric Methods, Special Topics in Biometry, Biometry Seminar, Master's Thesis

DEPARTMENT OF MICROBIOLOGY AND IMMUNOLOGY
Phone: Chairman: (501) 664-5000 Ext. 136

Chairman: A.L. Barron
Professors: A.L. Barron, R.S. Abernathy, J. Bates, C.E. Duffy
Associate Professors: R. Bowling, V. Gordon, H. Hardin, P. Morgan
Assistant Professors: H. Betterton, J. Daly, M. Ito
Instructors: H. Matthews, E. Moses
Visiting Lecturer: 1

Degree Programs: Immunology, Microbiology, Virology

Graduate Degrees: M.S., Ph.D.
Graduate Degree Requirements: Master: A minimum of 24 semester hours and an acceptable Thesis are fundamental requirements. Courses in Microbiology

and Immunology are necessary. Course in related areas can also be taken for credit. A final oral examination must be successfully completed. Doctoral: A student, his major professor and an advisory committee outline a program of study to prepare the student for a fundamental understanding of Microbiology, Immunology and related areas. The student must develop a fundamental knowledge of biomedical computing. A Doctoral Dissertation is required in some area of the student's major interests. Several comprehensive oral and written examinations are administered throughout the course of study

Graduate and Undergraduate Courses Available for Graduate Credit: Medical Microbiology, Advanced Medical Microbiology Immunology, Serology, Advanced Immunological Methods, Virology, Cell and Tissue Culture Methods, Microbial Physiology, Bacterial Genetics, Medical Mycology, Medical Ecology, History of Microbiology, Parasitology, Research in Microbiology, Medical Protozoology, Medical Helmithology, Research in Parasitology, Seminar, Thesis, Dissertation

DEPARTMENT OF PHYSIOLOGY AND BIOPHYSICS
Phone: Chairman: (501) 664-5000 Ext. 207

Head: J.E. Whitney
Professors: A.A. Krum, J.E, Whitney
Associate Professors: T.I. Koike
Assistant Professors: G.C. Bond, H.H. Conaway, J.N. Pasley, W.M. St. John, A.K. Tung

Degree Program: Physiology

Graduate Degrees: M.S., Ph.D.
Graduate Degree Requirements: Master: Course work and Thesis Doctoral: Course work, foreign language and Dissertation

Graduate and Undergraduate Courses Available for Graduate Credit: Medical Physiology, General Physiology, Comparative Physiology I and II, Endocrinology, Research in Physiology, Physiology-Biophysics Seminar, Molecular Biophysics I and II, Selected Readings in Physiology, Advanced Physiology, Special Methods in Biophysics

UNIVERSITY OF ARKANSAS AT LITTLE ROCK
Little Rock, Arkansas 72204 Phone: (501) 568-2200
Dean of Graduate Studies: W. Beard
Dean of College: C.B. Sinclair

DEPARTMENT OF BIOLOGY
Phone: Chairman: (501) 568-2200 Ext. 232

Head: R.L. Watson
Professor: C.B. Sinclair
Associate Professor: D.V. Ferguson, G.A. Heidt, L.F. Morgans, R.L. Watson
Assistant Professors: P.J. Garnett, J.R. Rickett, J.H. Whitesell

Degree Programs: Biology

Undergraduate Degree: B.S.
Undergraduate Degree Requirements: 30 Hours (Semester) in Biology inclusing seminar, 8 Hours in Chemistry (minimum)

UNIVERSITY OF ARKANSAS AT MONTICELLO
Monticello, Arkansas 71655 Phone: (501) 367-6811
Dean: R. Kirchman

DEPARTMENT OF AGRICULTURE
Phone: Chairman: (501) 367-6811 Ext. 70

Chairman: P.E. Grissom
Associate Professors: P.E. Grissom, B.B. Brooks, R.C. Kirst

Degree Programs: Agriculture

Undergraduate Degrees: B.S.
Undergraduate Degree Requirements: 12 Hours English and Humanities, 12 Hours Natural Science and Mathematic Science, 12 Hours Social and Behavioral Science, 6 Hours Physical Education, 34 Hours Agricultural subjects, 24 Hours of a minor, 24 Hours of Electives

DEPARTMENT OF BIOLOGY
Phone: Chairman: Ext. 67

Head: H.C. Steelman
Professors: H.C. Steelman, C.M. Ward
Associate Professors: J.G. Culpepper, A.L. Etheridge
Assistant Professors: J.W.Huey, R.W. Wiley

Degree Program: Biology

Undergraduate Degree: B.S.
Undergraduate Degree Requirements: Total Hours required - 124, Vertebrate and Invertebrate Zoology - 8 Hours General Botany, 4 Hours Comparative Anatomy 5 Hours Taxonomy of Vertebrates 4 Hours, Taxonomy of Flowering Plants 3 Hours, Minor - 24 Hours, General Education - 38 Hours, Senion level Biology Electives - 8 Hours

UNIVERSITY OF ARKANSAS - PINE BLUFF
Pine Bluff, Arkansas 71601 Phone: (501) 535-6700
Dean of College: V. Starlard

DEPARTMENT OF BIOLOGY
Phone: Chairman: Ext. 238

Chairman: R.L. Caine
Professors: C.L. Bentley, R.L. Caine, M.C. Dedrick
Associate Professor: 1
Assistant Professor: 1
Instructor: 1
Research Assistants: 1

Degree Programs: Animal Science, Biological Science, Biology, Crop Science, Dairy Science, Horticultural Science, Poultry Science

Undergraduate Degrees: B.A., B.S.
Undergraduate Degree Requirements: 120 Semester Hours overall

CALIFORNIA

BIOLA COLLEGE
LaMiranda, California 90639 Phone: (213) 941-3224
Dean of College: R.F. Crawford

DEPARTMENT OF BIOLOGICAL SCIENCES
Phone: Chairman: Ext. 215

Chairman: R.R. Payne
Professor: R.L. Stephens
Assistant Professors: L.C. Eddington, P.H. Kuld
Instructor: R.R. Payne

Degree Program: Biological Science

Undergraduate Degree: B.S.
Undergraduate Degree Requirements: Core curriculum of 32 units of Biology, 20 units of Chemistry, 12 units of Physical Science (Mathematics, Physics), courses requires: Experimental Biology, Cellular Biology, Population Biology and Majors Seminar, 30 units Biblical Studies, 44 units General Education, Total 138

CALIFORNIA BAPTIST COLLEGE
Riverside, California 92504 Phone: (714) 689-5771
Dean of College: S. Carleton

DEPARTMENT OF LIFE SCIENCES
Phone: Chairman: 689-5771

Head: L.B. Young
Professor: L.B. Young
Associate Professor: R.P. Roth
Instructor: T.B. Morgan
Visiting Lecturer: 1

Degree Program: Life Science

Undergraduate Degree: B.S.
Undergraduate Degree Requirements: 4 Semester units of Botany, 4 semester units of Zoology, 8 units of freshman Chemistry, plus one of the following: Calculus, Physics, Organic

CALIFORNIA INSTITUTE OF TECHNOLOGY
Pasadena, California 91109 Phone: (213) 795-6811
Dean of Graduate Studies: C.J. Pings

DIVISION OF BIOLOGY
Phone: Chairman: (213) 795-6811 Ext. 1951

Chairman: R.L. Sinsheimer
Albert Billings Ruddock Professor of Biology: M. Delbruck
Thomas Hunt Morgan Professor of Biology: E.B. Lewis
Bing Professor of Behavioral Biology: J. Olds
Hixon Professor of Psychobiology: R.W. Sperry
Professors: G. Attardi, S. Benzer, J.F. Bonner, C.J. Brokaw, E. Davidson, M. Delbruck, W.J. Dreyer, D.H. Fender, N.H. Horowitz, E.B. Lewis, Herschel, K. Mitchell, J. Olds, R.D. Owen, J.P. Revel, R.L. Sinsheimer, R.W. Sperry, F. Strumwasser, J. Vinograd, C.A.G. Wiersma, W.B. Wood
Associate Professors: E.H. Davidson, L.E. Hood
Assistant Professors: H.A. Lester, D. McMahon, J.D. Pettigrew, R.L. Russell, J.A. Strauss
Research Fellows: 55
Research Assistants: 46

Field Stations/Laboratories: Kerckhoff Marine Laboratory - Corona del Mar

Degree Programs: Biochemistry, Biophysics, Genetics, Virology

Undergraduate Degree: B.S. (Biology)
Undergraduate Degree Requirements: Physics - 2 years, Mathematics - 2 years, Chemistry - 2 years, Humanities - 108 units, General Biology, Cell Biology, Organismic Biology, Biochemistry, Genetics, Neurophysiology, 55 units of additional Biology courses, 516 units of total course credit

Graduate Degree: Ph.D.
Graduate Degree Requirements: Doctoral: Satisfaction of requirement in cellular and organismic biology. Passage of candidacy examination in major field. Completion of satisfactory research Thesis.

Graduate and Undergraduate Courses Available for Graduate Credit: Invertebrate Biology, Vertebrate Biology, Developmental Biology of Animals, Biochemistry, Biochemistry Laboratory, Immunology, Immunology Laboratory, Virology, Advanced Cell Biology, Biosystems Analysis, Genetics, Biophysics, Biophysical Chemistry of Macromolecules, Biophysics of Macromolecules Laboratory, Advanced Research Techniques in Molecular Biology, Optical Methods in Biology, Optical Methods in Biology Laboratory, Multicellular Assemblies, Selected Topics in Evolution Theory, Neurophysiology, Behavioral Biology, Brain Studies of Motivated Behavior, Psychobiology, Neurochemistry, Neurophysiology Laboratory, General Biology Seminar, Biochemistry Seminar, Genetics Seminar, Biophysics Seminar, Selected Topics in Neurobiology, Psychobiology Seminar, Advanced Seminar in the Molecular Biology of Development, Advanced Topics in Molecular Biology, Advanced Physiology, Special Topics in Biology, Biological Research

CALIFORNIA LUTHERAN COLLEGE
Thousand Oaks, California 91360 Phone: (805) 492-2411
Dean of College: P. Ristuben

DEPARTMENT OF BIOLOGICAL SCIENCES
Phone: Chairman: Ext. 347

Chairman: C.B. Nelson
Associate Professors: B.J. Collins, C.B. Nelson, P.A. Nickel
Visiting Lecturers: 2

Degree Programs: Biology, Medical Technology

Undergraduate Degrees: B.A., B.S.
Undergraduate Degree Requirements: (Semesters) B.S.- 40 units, 32 U.D., 8 additional hours in Mathematics Statistics or Computer, B.A. - 32 units, 20 U.D.

CALIFORNIA POLYTECHNIC STATE UNIVERSITY
San Luis Obispo, California 93401 Phone: (805) 546-0111
Dean of Graduate Studies: D. Grant

School of Agricultura and Natural Sciences
Dean: J.C. Gibson

ANIMAL SCIENCE DEPARTMENT
(No information available)

CROP SCIENCE DEPARTMENT
Phone: Chairman: (805) 546-2489

Head: C.M. Johnson
Professors: C.M. Johnson, H. Rhoads, F.F. Thrasher, W.T. Troutner
Associate Professors: 4
Assistant Professors: 5

Degree Program: Crop Science

Undergraduate Degree: B.S.
Undergraduate Degree Requirements: 198 Quarter units 60 - General Education courses, 64 - Major course requirements, 30 - in supporting courses, 34 - Free electives

Graduate and Undergraduate Courses Available for Graduate Credit: Crop Physiology, Experimental Techniques, and Analysis, Oil and Fiber Crops, Selected Advanced Topics, Advanced Field Crop Production, Graduate Seminar in Crop Production, Advanced Pomology, Orchard Management, Advanced Fruit and Nut Crop Production, Graduate Seminar in Fruit Production, Vegetable Crop Management, Advanced Vegetable Science

DEPARTMENT OF DAIRY SCIENCE
(No reply received)

NATURAL RESOURCES MANAGEMENT DEPARTMENT
Phone: Chairman: (805) 546-2702

Head: M.J. Whalls
Associate Professors: M.H. Whalls, R.J. Greffenius
Assistant Professors: J.E. Bedwell, J.R. Kitts, W.B. Curtz, A.E. Knable, W.R. Mark, N.H. Pillsbury
Visiting Lecturer: 1

Degree Programs: Fisheries and Wildlife Biology, Forest Resources, Natural Resources, Silviculture, Watershed Management, Wildlife Management, Environmental Services, Environmental Interpretation

Undergraduate Degree: B.S.
Undergraduate Degree Requirements: 198 units and completion of Senior Project

DEPARTMENT OF ORNAMENTAL HORTICULTURE
(No reply received)

School of Science and Mathematics
Dean: C.P. Fisher

BIOLOGICAL SCIENCES DEPARTMENT
Phone: Chairman: (805) 546-2437

Head: R.F. Nelson
Professors: T.G. Call, F.L. Clogston, H.L. Fierstine, H.C. Finch, C.D. Hynes, D.H. Montgomery, R.A. Pimentel, R.J. Rodin, A.E. Roest, S.R. Sparling, W.D. Stansfield, D.H. Thomson, W. Thurmond
Associate Professors: R.J. Brown, D.D. Donaldson, D. F. Frey, D.N. Homan, E.V. Johnson, K.L. Leong, P.C. Pendse, T.L. Richards, J.W. Thomas, D.R. Walters
Assistant Professors: F.P. Andoli, P. Babos, L.A. Barclay, J.S. Booth, L.S. Bowker, R.J. Cano, J.S. Colome A.F. Cooper, A.A. DeJong, R.D. Gambs, D.V. Grady, R.L. Grayson, V.L. Holland, P.T. Jankay, G.N. Knecht, T.R. Koff, M.G. McLeod, M.E. Ortiz, L.R. Parker, E.K. Perryman, R. Riggins-Pimentel, A.M. Waterbury
Lecturers: D.A. Alex, M.E. Bernstein, S.H. Hamada, G.M. Tvrdik

Degree Programs: Biology, Botany, Field Biology, Marine Biology, Medical Laboratory Technology, Microbiology, Plant Pathology and Entomology, Zoology

Undergraduate Degree: B.S. (Biological Science)
Undergraduate Degree Requirements: 58 units Biological Sciences, 20 units Chemistry, 12 units Physics

Graduate Degree: M.S. (Biological Sciences)
Graduate Degree Requirements: Master: 45 units, 27 in Biological Sciences of which 15 are in field of major interest. Thesis. Comprehensive Examination

Graduate and Undergraduate Courses Available for Graduate Credit: Dairy Bacteriology (U), Industrial Microbiology (U), Sanitary Inspection and Control (U), General Virology (U), Food Microbiology (U), Public Health Microbiology (U), Bacterial Cytology and Physiology (U) , Marine Microbiology (U), Advanced Genetics (U), Radiation Biology (U), Evolution (U), Biological Instrumentation(U), Marine Biology (U), Biosystematics (U), Freshwater Ecology (U), Radiation Laboratory Techniques (U), General Cytology (U), Electron Microscopy (U), General Physiology (U), Marine Resources (U), Quantitative Biology (U), Selected Advanced Topics (U), Individual Study, History of Biology, Curriculum and Methods in Biological Sciences, Developmental Biology, Cell Physiology, Multivariate Biometry, Selected Topics in Biology, Introductory Plant Physiology (U), Plant Pathology (U), Plant Nematology (U), Plant Ecology (U), Morphology of Vascular Plants (U), Plant Anatomy (U), Algology (U), Advanced Plant Taxonomy (U), Plant Virology (U), Mycology (U), Advanced Plant Physiology (U), Advanced Plant Pathology (U), Freshwater Fisheries (U), Game Management (U), Aquaculture (U), Economic Entomology (U), Immature Stages of Insects (U), Vertebrate Embryology (U), Mammalogy (U), Biology of Fishes (U), Ornithology (U), Comparative Anatomy of the Vertebrates (U), Vertebrate Field Zoology (U), Invertebrate Zoology, Human Muscle Anatomy (U), Herpetology (U), Vertebrate Embryology Laboratory (U), Introduction to Clinical Pathology (U), Histology (U), Parasitology (U), Serology and Immunology (U), Hematology (U), Comparative Animal Physiology (U), Functional Vertebrate Morphology (U)

CALIFORNIA STATE COLLEGE
BAKERSFIELD
Bakersfield, California 93309 Phone: (805) 2011
Dean of College: J. Coash

DEPARTMENT OF BIOLOGY
Phone: Chairman: (805) 833-2225 Others: 833-2224

Chairman: F. D.Blume
Professor: F.D. Blume
Associate Professors: R.A. Cornesky, T. Murphy, D.S. Hinds, D.H. Ost, B.E. Michals
Assistant Professors: J.C. Amundson, S.R. Seavey
Visiting Lecturer: 1

Degree Program: Biology

Undergraduate Degree: B.S.
Undergraduate Degree Requirements: 55 quarter units in major, 66 quarter units in general education, 65 quarter units in electives (cognate areas - chemistry, physics and mathematics)

CALIFORNIA STATE COLLEGE
DOMINGUEZ HILLS
Dominguez Hills, California 90747 Phone: 532-4300
Dean of College: R.B. Fischer

DEPARTMENT OF BIOLOGICAL SCIENCES
Phone: Chairman: Ext. 581

Chairperson: G.A. Kalland
Professors: L.W. Chi, H.L. Arora
Associate Professors: D. Brest, E. Childress, D. Colvin G. Kalland, R. Kuramoto, C. Lydon
Assistant Professors: F. McCarthy, D. Morafka, L. Phillips
Visiting Lecturers: 2

Degree Program: Biology

Undergraduate Degree: B.S.
Undergraduate Degree Requirements: Biological Science 110-112-114, Chemistry 110-112-114. Physics 120-122, Elements of Physics, General Chemistry, Principles of Biology, Mathematics 110-112. Differential and Integral Calculus, Biology 210-212, Biology 214, Biology-220, Biology 232, Biology 240, Biology 294, Chemistry 216-217 or Chemistry 210

CALIFORNIA STATE COLLEGE SAN BERNADINO

San Bernardino, California 92407 Phone: (714) 887-6311
Dean of Graduate Studies: R.H. Petrucci
Dean of College: J.D. Crum

DEPARTMENT OF BIOLOGY
Phone: Chairman: Ext. 284

Chairman: D. Harrington
Professors: A.S. Egge, G.M. Scherba, A. Sokoloff
Associate Professors: R.E. Goodman, D. Harrington, S.K. Mankau
Assistant Professors: E.L. Taylor, R.C. Wilson

Degree Program: Biology

Undergraduate Degrees: B.A., B.S.
Undergraduate Degree Requirements: B.A. - 4 courses in Chemistry (including Organic Chemistry); two courses in Physics; one course in Mathematics; three lower division core Biology courses; six upper division Biology courses; ten General Education courses; ten free electives. B.S. - one additional course in Chemistry (Quantitative Analysis); for additional upper division Biology courses (Molecular, Genetics, Physiology, Development, Ecology): free electives reduced to five courses. Both degrees require 186 quarter units

Graduate Degree: M.S.
Graduate Degree Requirements: Master: Forty-five quarter units within an approved graduate program; 21,5 units at the 600-level, including Thesis Research and Graduate Seminar; remaining units from 600-, 500-, and 400- levels; maintainence of 3.0 QPA; oral defense of Thesis

Graduate and Undergraduate Courses Available for Graduate Credit: Advanced Topics in Plant Biology, Advanced Topics in Genetics, Advanced Topics in Ecology, Advanced Topics in Evolution; Graduate Seminar, Seminar Chemical and Molecular Biology (G,U), Genetics and Ecology of Populations (G,U), Comparative Endocrinology (G,U), Animal Behavior (G,U), Scanning Electron Microscopy (G,U), Independent Study (research not associated with the Thesis) (G,U), Molecular Biology (U), Genetics (U), Comparative Animal Physiology (U), Foundations in Endocrinology (U), Comparative Plant Physiology (U), Principles of Development (U), Ecology (U), Taxonomy of Vascular Plants (U), Biology of Microorganisms (U), Hematology (U), Immunobiology (U)

CALIFORNIA STATE COLLEGE SONOMA

Darwin Hall Phone: (707) 795-2880
Rohnert Park, California 94928
Dean of Graduate Studies: R.Y. Fuchigami
Provost: J.H. Brumbaugh

DEPARTMENT OF BIOLOGY
Phone: Chairman: (707) 795-2189

Chairman: C.K. Kjeldsen
Professors: J.R. Arnold, R.R. Blitz, J.H. Brumbaugh, R.J. Bushnell, G. Clothier, W.W. Ebert, D.E. Isaac, C.K. Kjeldsen
Associate Professors: R.A. Baker, D.F. Hanes, C.O. Hermans, J.D. Hopkirk, F. R. Lockner, P.T. Northern, T. R. Porter, J.H. Powell, C.F. Quibell, R.J. Sherman
Assistant Professors: P.V. Benko, C.L. Liu
Instructor: K. Fogg
Visiting Lecturers: 2

Degree Programs: Biology

Undergraduate Degree: B.A.
Undergraduate Degree Requirements: Biology 11, 117, 215 = total of 12 units; Upper division units = 28 to include a course from the following: Field Biology, Genetics, Physiology, Developmental/Molecular/ Microbiology; 15-18 units Physical Science

Graduate Degree: M.A.
Graduate Degree Requirements: Master: Thesis; 12 units of 500 level courses and courses selected by candidate's committee appropriate to his M.A.

Graduate and Undergraduate Courses Available for Graduate Credit: Ecology (U), Field Biology (U), General Physiology (U), General Genetics (U), Plant Taxonomy (U), Plant Morphology (U), Bacteriology (U), Natural History - Invertebrates (U), Entomology (U), Natural History of Vertebrates (U), Embryology (U), Nutrition (U), Cont Iss in Biology (U), Community Involvement Project (U), Marine Ecology, Plant Ecology, Evolution, Phycology, Ichthyology, Immunology, Medical Microbiology, Special Studies, Graduate Seminar, Molecular Biology, Marine Ecology, Plant Ecology, Electron Microscopy, Thesis

CALIFORNIA STATE COLLEGE STANISLAUS

800 Monte Vista Avenue Phone: (209) 633-2476
Turlock, California 95380
Dean of College: E.M. Thompson

DEPARTMENT OF BIOLOGICAL SCIENCES
Phone: Chairman: 633-2476

Chairman: J.C. Hanson
Professors: S.J. Grillos, G.A. Hackwell, J.C. Hanson
Associate Professors: J.P. Christofferson, P.S. Mayol
Assistant Professors: J.A. Brown, D.M. Gotelli, W.S. Pierce, P. Roe, W. Tordoff III, D.F. Williams

Degree Program: Biological Sciences

Undergraduate Degree: B.A.
Undergraduate Degree Requirements: 32 credits in Biology, 10 in Chemistry, 4 in Mathematics, Concentration areas require 16 units in either Botany, Entomology or Zoology

CALIFORNIA STATE POLYTECHNIC UNIVERSITY

Pomona, California 97168 Phone: (714) 598-4711
Dean of Graduate School: R. Maurer

School of Agriculture

Dean: C. Christensen (Acting)

ANIMAL SCIENCE DEPARTMENT
Phone: Chairman: 598-4151

Chairman: E.K. Keating
Professors: E.K. Keating, H.D. Fausch, J.T. Gesler, E.A. Nelson, N.K. Dunn, R.H. Packard, M.H. Kinnington
Associate Professors: A. Knight, C.Y. Matsushima, R.J. Schechter, T.W. Westing, J.E. Trei
Assistant Professors: M.T. Savoldi, A.A. Wysocki, K. Jones

Degree Program: Animal Science

Undergraduate Degree: B.S.
Undergraduate Degree Requirements: 198 quarter units, including 10 Basic Biology, 4 Genetics, 5 Bacteriology, 17 Chemistry, 6 Mathematics, 44 Animal Science

School of Science

Dean: V. Parker

DEPARTMENT OF BIOLOGICAL SCIENCES
Phone: Chairman: 598-4444 Others: 598-4141

Chairman: R.W. Ames
Professors: R.W. Ames, H.S. Brown, J.E. Dimitman, J.L. Erspamer, B.L. Firstman, L.M. Knill, H.L. Lint, E.T. Roche, F. Shafia
Associate Professors: R.L. Barlet, L.M. Blakely, R.S. Daniel, W.D. Edmonds, D.C. Force, B.H. Goehler, M.P. Harthill, G.W. Martinek, E.K. Mercer, G.R. Stewart, M.F. Stoner, L.J.Szijj, J.H. Wu
Assistant Professors: J.N. Baskin, T.W. Brown, P. Castro, D. Garcia, J.O. Jackson, R.D. Quinn, R.Z. Riznyk, L.R. Troncale
Instructors: J.L. Bath, J.M. Hogg, D. Steele
Graduate Assistants: 12

Degree Programs: Biology, Botany, Microbiology, Zoology

Undergraduate Degree: B.S. (Biology)
Undergraduate Degree Requirements: 45 quarter units Biological core courses, 8 units Chemistry, 12 units Physics

Graduate Degree: M.S. (Biological Sciences)
Graduate Degree Requirements: Not Stated

Graduate and Undergraduate Courses Available for Graduate Credit: Up to 21 units of 400 level courses may be taken with the approval of the committee

CALIFORNIA STATE UNIVERSITY CHICO

Chico, California 95926 Phone: (916) 895-5356
Dean of Graduate Studies: J. Gregg
Dean of School: F.C. Pennington

DEPARTMENT OF BIOLOGICAL SCIENCES
Phone: Chairman: (916) 895-5656

Chairman: R.I. Ediger
Professors: D.G. Alexander, M.S. Anthony, F.S. Cliff, D.H. Curtis, W.H. Dempsey, W.F. Derr, A.G. Douglas, R.S. King, D.H. Kistner, D.T. Kowalski, V.H. Oswald, F.W. Ramsdell, W.L. Stephens, K.R. Stern, W.M. Struve, D.A. Sutton, R.E. Thomas, D.M. Wootton
Associate Professors: A.F. Baker, G.E. Corson, R.I. Ediger, M.J. Erpino, P.E. Maslin, R.B. McNairn, A.R. Wilhelm
Assistant Professors: M.A. Abruzzo, R.S. Demeree, R.J. Lederer, R.A. Schlising, B. Vasu
Teaching Assistants: 12
Graduate Assistants: 8

Field Stations/Laboratories: Eagle Lake Field Station - Susanville

Degree Programs: Biological Sciences, Botany, Micro-Biology

Undergraduate Degree: B.A.
Undergraduate Degree Requirements: Not Stated

Graduate Degree: M.A.
Graduate Degree Requirements: 30 units approved courses. Thesis. Oral Examination

Graduate and Undergraduate Courses Available for Graduate Credit: Any 200 or 300 level course offered in the Department of Biological Sciences.

CALIFORNIA STATE UNIVERSITY FRESNO

Fresno, California 93740 Phone: (209) 487-9011
Dean of Graduate Studies: P.W. Watts

Division of Agricultural Sciences

Dean: O.J. Burger

DEPARTMENT OF ANIMAL SCIENCE
Phone: Chairman: (219) 487-2971

Chairman: A.S. Hoversland
Professors: J. Bell, F. Hixson, L. Krum, D.M. Nelson, E. Rousek, R. Selkirk, E. Ensminger, G. Lukas, L. Larson
Associate Professor: D.D. Nelson
Assistant Professor: D. Dildey
Teaching Fellows: 3

Degree Programs: Agriculture, Agronomy, Animal Husbandry, Animal Science, Crop Science, Dairy Science, Farm Crops, Fisheries and Wildlife Biology, Horticulture, Natural Resources, Soil Science

Undergraduate Degree: B.S.
Undergraduate Degree Requirements: 128 Semester credits

Graduate Degree: M.S.
Graduate Degree Requirements: Master: 30 credits beyond a B.S., 12 upper division approved courses accepted. Thesis recommended but not mandatory

Graduate and Undergraduate Courses Available for Graduate Credit: Readings in Agriculture, Topics in Animal Science, Endocrine and Reproductive Physiology, Endironmental Physiology of Domestic Animals, Metabolism and Energy Physiology, Vitamin and Mineral Nutrition, Advanced Animal Breeding, Seminar in Animal Science, Independent Study, Thesis

DEPARTMENT OF PLANT SCIENCE
Phone: Chairman: 487-2861 Others: 487-2862

Chairman: H.P. Karle
Professors: W. Biehler, J.R. Brownell, O.J. Burger, A.A. Hewitt, T.T. Ishimoto, V.E. Petrucci, G. Ritenour, M. Van Elswyk, J. Whaley
Associate Professors: S.A. Badr, R.D. Harrison, G. Koch, A.A. Olney
Assistant Professor: H.A. Paul
Research Assistants: 2
Teaching Assistants: 8

Degree Programs: Agronomy, Horticulture, Ornamental Horticulture, Vegetable Crops, Viticulture, Natural Resource Management, Soils, Water Science, Pest Management, Post-harvest Physiology

Undergraduate Degree: B.S.
Undergraduate Degree Requirements: 128 total units 40 upper division of which 20 units must be in major plus 40 units General Education. Specific requirements, Soils, Irrigation, Weeds, Plant Pathology, Chemistry 2A, 2B, 8, Botany 1 or 10, Plant Physiology, Genetics

Graduate Degree: M.S.
Graduate Degree Requirements: 30 total units, 18 units in graduate level in Plant Science, 12 units Electives, in Plant Science or related fields, plus Thesis which may be included in the 30 units

Graduate and Undergraduate Courses Available for Graduate Credit: Soil Classification (U), Agricultural Chemical Application (U), Soil Management (U), Soil Fertility (U), Biometrics in Agriculture, Readings in Agriculture, Topics in Plant Science, Pesticides, Plant Nutrition, Plant Hormones and Regulators, Advanced Plant Breeding, Plant-Water Relationships, Physiology of Cultivated Crops, Plant Disease Control, Seminar in Plant Science, Independent Study, Thesis, and courses in Chemistry or Botany

Division of Natural Science
Dean: B. Kehoe

DEPARTMENT OF BIOLOGY
Phone: Chairman (209) 487-9011

Chairman: B.A. Tribbey
Professors: G. Arce, D. Burdick, J. Carr, E. Daubs, R. Evans, W. Harmon, H. Latimer, J.R. McClintic, R. Meyer, C.J. Pigg, P. Smith, A.Staebler, K. Standing, B. Tribbey, J. Weiler, K. Woodwick
Assiciate Professors: S.F. Cheuk, R. Haas, T. Mallory, J. Mangan, R. Spieler, V. Vidoli
Assistant Professors: R. Brown, D. Chesemore, C. Clay, D. Grubbs, K. Kleeman, F. Schrieber, L. Wiley
Visiting Lecturers: 2
Graduate Teaching Assistants: 19

Field Stations/Laboratories: Moss Landing Marine Biology Laboratory - Monterey Bay

Degree Programs: Biological Sciences, Biology, Botany, Ecology, Medical Technology, Microbiology, Physiology, Zoology

Undergraduate Degree: B.S.
Undergraduate Degree Requirements: General Botany; General Zoology; Introduction to Genetics; Introduction to Cell Biology; Introduction to Ecology (preceding constitute the Biology core required of all majors); Chemistry through Organic; Calculus for Life Science; Statistics; plus the completion of a minimum of 23 units of Biology coursework in one of six degree options.

Graduate Degree: M.S.
Graduate Degree Requirements: Master: Thirty post-baccalaureate units that have been approved by the Department's graduate faculty, of which a minimum of 15 must be graduate work. A research Thesis is required.

Graduate and Undergraduate Courses Available for Graduate Credit: Principles and Great Experiments in Biology, Cytogenetics, Biology of Speciation, Principles of Taxonomy, Field Work in Biology, Insect Toxicology, Insect Taxonomy, Scientific Research Reporting, Zoogeography, Biology Colloquium, Advanced General Microbiology, Immunochemistry, Experimental Virology, Experimental Infectious Pathology, plus a long series of special courses not listed in the catalog but offered under the heading Topics in Biology, Topics in Zoology, Topics in Botany, Topics in Microbiology

CALIFORNIA STATE UNIVERSITY
FULLERTON
Fullerton, California 92634 Phone: (714) 870-2524
Dean of Graduate Studies: G. Brown

DEPARTMENT OF BIOLOGICAL SCIENCE
Phone: Chairman: (714) 870-2440 Others: 870-3614

Chairman: D.B. Bright
Professors: P.A. Adams, N. Barish, L.H. Bradshaw, B.H. Brattstrom, D.B. Bright, C.A. Davenport, T.L. Hanes, M.D. McCarthy, L.L. McClanahan, M.J. Rosenberg, A.H. Rothman, D.D. Sutton, D.L. Walkington
Associate Professors: L.F. Dubin, M.H. Horn, C.E. Jones, C.C. Lambert, K.L. McWilliams, J.D. Smith, J.D. Weintraub, J. Wilson
Assistant Professors: J.H. Burk, P. Dunn, J. Kandel, S.N. Murray, W.F. Presch
Instructors: T. Hoshizaki, R.R. Seapy

Degree Program: Biological Sciences

Undergraduate Degree: B.A.
Undergraduate Degree Requirements: Not Stated

Graduate Degree: M.A.
Graduate Degree Requirements: Master: Thesis, Seminar Undergraduate QP 3.0 in Biology, 2.5 overall. 30 courses of which 15 must be graduate only

Graduate and Undergraduate Courses Available for Graduate Credit: Biogeography (U), Biosystematics (U), Evolution (U), Developmental Biology (U), Biometry (U), General Cell Physiology (U), Population Genetics (U), Molecular Genetics (U), Advanced Human Genetics (U), Neurobiology (U), Limnology-Fresh Water Ecology (U), General Oceanography (U), Biological Oceanography (U), Marine Ecology (U), Biology of Marine Plankton (U), Pathogenic Microbiology (U), Immunology (U), General Virology (U), Microbial Growth and Physiology (U), Microbial Ecology (U), Plant Taxonomy (U), Plant Ecology (U), Plant Physiology (U), Mycology (U), Phycology (U), Advanced Plant Ecology (U), Protozoology (U), Invertebrate Zoology (U), Parasitology (U), Comparative Vertebrate Anatomy (U), Embryology (U), Animal Ecology (U), Animal Behavior (U), Entomology (U), Comparative Animal Physiology (U), Hematology (U), Insect Survey Techniques (U), Natural History of the Vertebrates (U), Ichthyology (U), Herpetology (U), Mammalogy (U), Seminar in Molecular Biology, Seminar in Physiology, Seminar in Genetics, Seminar in Ecology, Seminar in Marine Science, Seminar in Microbiology, Seminar in Immunology, Seminar in Botany, Seminar in Zoology, Advanced Topics in Graduate Biology

CALIFORNIA STATE UNIVERSITY
HAYWARD
Hayward, California 94542 Phone: (415) 881-3000
Dean of Graduate Studies: G. Resnikoff
Dean of School: L.H. Fisher

DEPARTMENT OF BIOLOGICAL SCIENCES
Phone: Chairman: 881-3472

Chairman: D.R. Parnell
Professors: R.J. Baalman, H.L. Cogswell, N. Goldstein, T.C. Groody, P.P. Gross, H.D. Heath, A.E. Heuer, E.B. Lyke, S.M. McGinnis, R.A. Main, J.W. Nybakken, D.R. Parnell, W.K. Schoenholz, H.I. Scudder, A.C. Smith
Associate Professors: L.O. Elkin, J.M. Erickson, M.S. Foster, U.M. Neill, R.A. Symmons, R.E. Tullis

Field Stations/Laboratories: Moss Landing Marine Laboratories, Moss Landing

Degree Program: Biological Sciences

Undergraduate Degree: B.S.
Undergraduate Degree Requirements: 15 hours General Chemistry, 8 or 15 Organic Chemistry, 4 College Mathematics, 4 Statistics, 9 Physics, 3 Physics Laboratory, 15 Foundations of Biological Science, 4 Developmental, 4 Ecology, 4 Genetics, 4 Evolution, 4 Physiology, 15-22 Electives in Biological Science

Graduate Degree: M.A.
Graduate Degree Requirements: Options include Environmental Biology (General Ecology, Vertebrate Ecology Marine Ecology) and Physiological, Biology (Cell Physiology, Plant and Animal Physiology, or Microbial Physiology) Plan A requires 45 units beyond the baccalaureate including 3 units Seminar, 9 Thesis, 8-30 in area of specialization, 3-25 elective graduate courses in area of specialization; 0-18 upper division courses taken as a graduate and approved by the Advisory Committee; comprehensive examination. Plan B requires 45 units beyond the baccalaureate including 3 units Seminar, 4 units Independent study for preparation of a study paper; 12-35 graduate

courses in area of the option; 3-26 elective graduate courses in area of option; 0-18 upper division courses taken as a graduate and approved by Advisory Committee; comprehensive examination

Graduate and Undergraduate Courses Available for Graduate Credit: Selected topics in Physiology, Desert Biology, Marine and Fresh-Water Plankton, Principles and Practices of Vector Control, Cell Biology, Community and Ecosystem Ecology, Population Biology, Selected Topics in Botany, Plant Biosystematics, Mycology, Plant Ecology, Insect Ecology, Advanced Medical Microbiology, Selected Topics in Biochemistry Laboratory, Physiological Ecology, Topics in Invertebrate Zoology, Wildlife ecology, Mammalian Systems Physiology, Analysis of Vertebrate Faunas, Electron Microscopy, Electron Microscopy Laboratory, Graduate Seminar - Ecology, Graduate Seminar Physiology, Graduate Seminar Zoology, Independent Study, Thesis, D.G.S. Also graduate courses are offered in Marine Sciences at the Moss Landing Marine Laboratories. Most upper division courses; up to a maximum of 18 units, can be applied to graduate credit, with approval of Advisory Committee

CALIFORNIA STATE UNIVERSITY HUMBOLT

(see Humbolt State University)

CALIFORNIA STATE UNIVERSITY LONG BEACH

(No reply received)

CALIFORNIA STATE UNIVERSITY LOS ANGELES

5151 State University Drive Phone: (213) 224-0111
Los Angeles, California 90032
Dean of Graduate Studies: G. Stuart
Dean of Letters and Science: D.D. Dewey

DEPARTMENT OF BIOLOGY

Phone: Chairman: (213) 224-3258 Others: 224-0111

Chairman: D.W. Thomas
Professors: R.K. Allen, T.D. Bair, S.M. Caplin, B.Capon, W.O. Griesel, W.R. Hanson, E.K. Oyakawa, M.P. Russell, J.A. Sacher, L.W. Stearns, R.M. Straw, D.W. Thomas, V.J. Vance, R.J. Vogl
Associate Professors: W.P. Alley, M.J. Hartman, J.S. Henrickson, G.E. Jakway
Assistant Professors: R.R. Bowers, D.P. Mahoney, H. Rosen, D.L. Soltz
Graduate Teaching Assistants: 20

Degree Program: Biology

Undergraduate Degrees: B.A., B.S.
Undergraduate Degree Requirements: B.A. - 90 quarter units in major; 186 quarter units total B.S. 109 quarter units in major; 198 quarter units total

Graduate Degree: M.S.
Graduate Degree Requirements: Master: 45 quarter units of which 23 must be graduate level courses and the balance senior level courses. Both thesis and comprehensive examination options

Graduate and Undergraduate Courses Available for Graduate Credit: Approximately 35 senior level courses, fifteen (15) graduate level courses

DEPARTMENT OF MICROBIOLOGY AND PUBLIC HEALTH
Phone: Chairman: (213) 224-3531

Chairman: J.T. Seto
Professor: E. Tamblyn
Associate Professors: K. Anderson, L. Chan, R. Fleming, A. Shum
Visiting Lecturers: 3

Degree Programs: Microbiology, Medical Technology

Undergraduate Degrees: B.A., B.S.
Undergraduate Degree Requirements: B.A. - 192 units, B.S. - 196 units

Graduate Degree: M.S.
Graduate Degree Requirements: Master: 45 units (quarter system)

CALIFORNIA STATE UNIVERSITY NORTHRIDGE

1211 Nordhoff Phone: (213) 885-3356
Northridge, California 91324
Dean of Graduate Studies: E. Lucki
Dean of School of Science and Mathematics: D.E. Bianchi

DEPARTMENT OF BIOLOGY
Phone: Chairman (213) 885-3356

Chairman: G.L. Lefevre
Professors: P. Bellinger, M, Cantor, M. Corcoran, J. Dole, W. Emboden, W. Furumoto, H.R. Highkin, K.C. Jones, J. Kontogiannis, D. Kuhn, G. Lefevre, J. Moore, R. Pohlo, E. Pollock, E. Segal, P. Sheeler, C.R. Spotts, A. Starrett, K. Wilson
Associate Professors: M.L. Barber, K. Daly, R. Potter, R. Swade, J. Swanson, C. Weston
Assistant Professors: A. Gaudin, J. Maxwell, S.B. Oppenheimer

Degree Program: Biology

Undergraduate Degree: B.S.
Undergraduate Degree Requirements: Not Stated

CALIFORNIA STATE UNIVERSITY SACRAMENTO

6000 J. Street Phone: (916) 454-6535
Sacramento, California 95819
Dean of Graduate Studies: E. Thompson
Dean of College: D. Ballesteros

DEPARTMENT OF BIOLOGICAL SCIENCES
Phone: Chairman: (916) 454-6535

Chairman: M. L. Bolar
Professors: E.B. Austin, M.L. Bolar, M.R. Brittan, R.E. Darby, A.L. Delisle, K.H. Eldredge, L.G. Kavaljian, R.L. Livezey, C.M. Love, C.E. Ludwig, R.A. Schinske, M.D.F. Udvardy, H.W. Wiedman
Associate Professors: E.S. Benes, E.A. Christian, Jr., Y.L. Hwang, P.T. Kantz, G.L. Meeker, J.T. Morse, C.R. Moser, R.F. Reichle, C.D. Vanicek, M.L. West
Assistant Professors: M.F. Baad, D.K. Huff, J.M. Langham, G.A. Leidahl, R.M. Metcalf, G.R. Trapp, L.L. Washburn
Instructors: 14

Field Stations/Laboratories: Marine Sciences, Moss Landing

Degree Programs: Animal Biology, Environmental Biology, Human Biology, Microbiology, Plant Biology, Biological Conservation, Clinical Laboratory Technology, Public Health Microbiology, Environmental Health

Undergraduate Degree: B.A.
Undergraduate Degree Requirements: Not Stated

Graduate Degree: M.A. (Biological Sciences)
Graduate Degree Requirements: 30 semester units, 15 units in graduate courses no including Thesis. Thesis

3.0 grade point average

Graduate and Undergraduate Courses Available for Graduate Credit: Advanced Plant Ecology, Cell and Molecular Biology, Cell and Molecular Biology Laboratory, Mammalian Physiology, Radioisotope Techniques in Biological Research, Experimental Morphogenesis, Environmental Physiology, General Immunology, Helminthology, Advanced Animal Ecology, Conservation Policy and Administration, Advanced Fishery Biology and Management, Advanced Wildlife Management, Evolution, Biogeography, Advanced Cytology, Biosystematics, Biological Concepts, Seminar in Botany, Ecology, Genetics, Physiology, Microbiology and Zoology, Problems in Biological Science, Thesis

CALIFORNIA STATE UNIVERSITY
SAN DIEGO
(see San Diego State University)

CALIFORNIA STATE UNIVERSITY
SAN FRANCISCO
(see San Francisco State University)

CALIFORNIA STATE UNIVERSITY
SAN JOSE
San Jose, California 95192 Phone: (408) 277-2526
Dean of Graduate Studies: G. Fullerton
Dean of College: L.H. Lange

DEPARTMENT OF BIOLOGICAL SCIENCES
Phone: Chairman: (408) 277-2355 Others: 277-2526

Chairman: J.H. Young
Professors: J.H. Akiyama, A.G. Applegarth, R.C. Ballard, C.W. Bell, E. Chin, Jr., J.M. Craig, J.G. Edwards, F.A. Ellis, W. Ferguson, W.T. Graf, H.A. Harris, R.J. Hartesveldt, H.T. Harvey, J.J. Hendricks, K.E. Hutton, L.H. Lindeberg, G.A. McCallum, L.R. Mewaldt, G.V. Morejohn, H.R. Patterson, R.G. Pisano, J. Pratley, H.W. Robinson, C.L. Schmidt, H.S. Shellhammer, M.M. Shrewsburg, Jr., W.L. Tidwell, H.G. Weston, J.H. Young
Associate Professors: P.C. Andriese, J.L. Chen, P.L. Grilione, R.J. Hahn, R.D. Haight, R. Hassur, R.L. Ingraham, V.C. Kenk, H.D. Murphy, C.W. Porter, R.E. Richter, W. Savage, R.E. Stecker, T. Thompson, E.C. Weaver
Assistant Professors: R. Carey, L.G. Dorosz, Jr., W.G. Iltis, S.K. Webster
Visiting Lecturers: 1
Research Assistants: 1

Field Stations/Laboratories: Moss Landing Marine Laboratories - Moss Landing, Point Reyes Bird Observatory - Farallon Islands

Degree Programs: Biological Sciences, Marine Biology, Molecular Biology, Park Ranger, Botany, Zoology, Wildlife Zoology, Microbiology, Entomology, Microbiology, Concentrations in Medical Technology, Environmental Health, Public Health Microbiology

Undergraduate Degrees: B.A., B.S,
Undergraduate Degree Requirements: Biology Core (Zoology, Botany, Biology), 12 semester units, Chemistry - 14 units; Physics - 8 units; upper-division Genetics, Physiology, Ecology, Botany, Zoology/ Entomology, Microbiology, Major Electives, 28 units, Several majors and concentrations have minors in Chemistry

Graduate Degree: M.A.
Graduate Degree Requirements: Master: 30 semester units of classes, 15 of which must be graduate level classes. 24 units must be in residence. Candidate must maintain a GPA of 3.0.

CHAPMAN COLLEGE
333 N. Glassel Street Phone: (714) 633-8821
Orange, California 92666
Dean of College: W. Boyer

DEPARTMENT OF BIOLOGY
Phone: Chairman: Ext. 532 Others: Ext. 543

Head: Cheng-Mei Fradkin
Professor: Cheng-Mei Fradkin
Associate Professor: C.A. Westervelt, Jr.
Assistant Professors: T.H. Mortenson, V. Carson

Degree Program: Biology

Undergraduate Degrees: B.A., B.S.
Undergraduate Degree Requirements: Biology - 42 semester units; Chemistry - 14; Mathematics 3-6; Physics 8; Humanities - 12; Social Science - 12; basic communications skills to be met by course or examination. Total units for graduation - 124

CLAREMONT COLLEGES
Claremont, California 91711

Claremont Graduate School
900 North College Avenue Phone: (714) 626-8511
Claremont, California 91711
Dean: P.A. Albrecht

DEPARTMENT OF BOTANY
Phone: Chairman: (714) 626-3489 Others: 626-3922

Chairman: L.W. Lenz
Violetta Horton Chair of Botany: S. Carlquist
Professors: R.K. Benjamin, L. Benson, S. Carlquist, L.W. Lenz, E.A. Phillips, R.F. Thorne
Assistant Professor: R. Scogin

Degree Program: Botany

Graduate Degrees: M.A., Ph.D.
Graduate Degree Requirements: Master: 30 units of graduate credit, one semester of full-time study and Thesis Doctoral: 72 units of graduate credit, one year of full-time study, two foreign languages and a Thesis

Graduate and Undergraduate Courses Available for Graduate Credit: Evolution of Cultivated Plants; Autecology, Principles of Taxonomy, Comparative Plant Anatomy, Chromosomal Cytology, Synecology, Algae and Bryophytes, Evolution of the Plant Kingdom, Phylogeny of Angiosperms, Mycology, Biochemical Systematics, Biochemical Evolution, Cellular Biology, Experimental Taxonomy, Plant Geography, Pteriodophytes and Gymnosperms, Taxonomy, Advanced Field Techniques, Advanced Techniques in Botanical Research, Tutorial Reading, Independent Study, Plant Physiology

Pomona College
6th and College Phone: (714) 626-8511
Claremont, California 91711

DEPARTMENT OF BOTANY
Phone: Chairman: Ext. 2993 Others: 2993

Chairman: E.A. Phillips
Henry Kirke White Bent Professor of Botany: L. Bensen
Professor: I.A. Phillips
Instructor: T. Mulroy
Visiting Lecturers: 8

Field Stations/Laboratories: Pitt Ranch - Cholame

Degree Program: Botany

Undergraduate Degree: A.B.
Undergraduate Degree Requirements: 32 courses, 12

12 courses in concentration, foreign language, (General Education - 3 courses in each of 3 divisions

Graduate and Undergraduate Courses Available for Graduate Credit: Algae and Fungi, Mosses and Ferns and Conifers, Environmental Botany, Principles of Evolution and Taxonomy, Plant Physiology, Plant Ecology, Advanced Classification of Vascular Plants, Plant Anatomy, Plant Microtechniques

DEPARTMENT OF ZOOLOGY
Phone: Chairman: Ext. 2950

Chairman: L.W. Cohen
Willard George Halstead Professor of Zoology: U. Amrein
Professor: Y.U. Amrein
Associate Professor: W.W. Andrus, L.W. Cohen, L.C. Oglesby, W.O. Wirtz II
Assistant Professor: D. Baskin

Degree Program: Zoology

Undergraduate Degree: B.A.
Undergraduate Degree Requirements: General Zoology plus six upper division courses (4 units each) in Zoology and half course (2 units) Senior Thesis. In addition: two years chemistry and one year physics; and general college requirements for graduation

Claremont Men's, Pitzer and Scripps Colleges

11th and Dartmouth Phone: (714) 626-8511
Claremont, California 91711
Dean: A. Fuller - Scripps
Dean: A. Heslop - Claremont Men's College
Dean: A. Schwarts - Pitzer

JOINT SCIENCE DEPARTMENT
Phone: Chairman: Ext. 2679

Chairman: R. Pinnell
Professors: C. Eriksen, M. Mathies
Associate Professors: D. Guthrie, C.R. Feldmeth
Assistant Professors: D. Sasava
Instructors: G. Andrus

Field Stations/Laboratories: Ecological Field Station - Claremont

Degree Programs: Biology with specialization in Psychobiology, Biochemistry, Environmental Studies, Natural History

Undergraduate Degree: B.A.
Undergraduate Degree Requirements: Introductory Biology (1 year), Introductory Chemistry (1 year), Introductory Physics (1 year) 8 advanced courses in Biology (one each with laboratory in Cellular, Organismic and Population Biology, Organic Chemistry may count for one course. Thesis may count for up to two).

COLLEGE OF NOTRE DAME

1500 Ralston Avenue Phone: (415) 593-1601
Belmont, California 94002
Dean of Graduate Studies: E. Zenner
Dean of College: R.J. Gavin

DEPARTMENT OF BIOLOGY AND ENVIRONMENTAL SCIENCE
Phone: Chairman: Ext. 46

Chairman: R. DiGirolamo
Associate Professor: R.DiGrolamo
Assistant Professors: G. Ostarello, C. Pontle, Sr. C. Buhs
Instructors: E. Horvath
Research Assistants: 2

Field Stations/Laboratories: Marine Resources - Belmont

Degree Program: Biology

Undergraduate Degree: B.A., B.S.
Undergraduate Degree Requirements: 120 units - at least 72 units in major area, i.e. 20 units Chemistry, 8 units Physics, 44 units Biology

Graduate Degree: M.S.
Graduate Degree Requirements: Master: 30 units Beyond Bachelors

Graduate and Undergraduate Courses Available for Graduate Credit: Advanced Microbiology (U), Endocrinology (U), Thesis Research, Scientific Writing (U), Graduate Seminar, Virology (U), Parasitology (U)

DOMINICAN COLLEGE OF SAN RAFAEL

San Rafael, California 94901 Phone: (415) 457-4440
Dean of Graduate Studies: H.J. Aigner
Dean of College: H.J. Aigner

Chairman: Sr. M. Aquinas Nimitz
Associate Professors: Sr. M. Aquinas Nimitz, Sr. M. Ward
Assistant Professors: S.L. Volk
Instructors: Sr. J. Engle

Degree Program: Biology

Undergraduate Degree: B.A.
Undergraduate Degree Requirements: General Chemistry (10), Organic Chemistry (8); Physics (8); Mathematics for Biology Majors (3) or more advanced mathematics; General Zoology (4), Plant Biology (4); 24 upper division units in Biology; and a comprehensive, experimental research project or journal research paper

HOLY NAMES COLLEGE

3500 Mountain Boulevard Phone: (415) 436-0111
Oakland, California 94619
Dean of Graduate Studies: Sr. Cabridi Weber
Dean of College: G. Larsen

Chairman: Sr. D.M. Ploux
Professors: Sr. D.M. Ploux
Associate Professor: Sr. M.B. Dean
Instructors: Sr. M. Mitchell, Sr. J. Aquinlivan, Sr. M. Webb
Guest Lecturers: 3-5 each year

Degree Program: Biological Science

Undergraduate Degree: B.S.
Undergraduate Degree Requirements: Program 1 - Emphasis on Physiology: Embryology, Plant Physiology, Comparative Physiology of Vertebrates, Genetics, Cell Physiology, Physiological Chemistry - 6 - 8 units of electives. Program 2 - Emphasis in Microbiology: Parasitology, Medical Bacteriology, Serology, Immunology, Hematology, Physiological Chemistry, Organic Chemistry, Introduction to Biochemistry, College Physics 3-8 units. Program 3 - Teaching Credential Candidates: Molecular Biology, Comparative Anatomy, Embryology, Entomology, Plant Taxonomy, Seminar on Evolution, Physical Sciences

HUMBOLT STATE UNIVERSITY

Arcata, California 95521 Phone: (707) 826-3011
Dean of Graduate Studies: A. Gillespie

School of Natural Resources

Dean: D. Hedrick

DEPARTMENT OF FISHERIES
Phone: Chairman: 826-3448

Chairman: G.H. Allen
Professors: G.H. Allen, J. DeWitt, J.P. Welsh

Associate Professors: T. Roelofs, R. Barnhard (Adjunct)
Assistant Professors: T. Hassler (Adjunct), R. Busch

Field Stations/Laboratories: Marine Laboratory, Trinidad

Degree Programs: Fisheries

Undergraduate Degree: B.S.
Undergraduate Degree Requirements: 1. General Education. 2. Lower Division: One year freshman biology; one year freshman biology; one year freshman chemistry; Conservation; Biometrics; Calculus; Physics 3. Upper Division: Invertebrate Zoology; Aquatic Plants; Physiology; Identification of Fishes; Fish Physiology; Limnology or Oceanography; Fish Ecology, Fisheries Techniques; Fisheries Management; Seminar Senior Project

Graduate Degree: M.S.
Graduate Degree Requirements: Master: Thesis. Total 45-57 Units, with 23 units graduate level, including seminar

Graduate and Undergraduate Courses Available for Graduate Credit: Ichthyology (3 courses); Fish Taxonomy; Limnology (2 courses); Fish Ecology (2 courses); Water Pollution Biology; Fish Population Dynamics (2 courses); Principles Fisheries Management; Fish Culture; Fish Diseases (2 courses); Commerical Fisheries; Economically Important Invertebrates; Reservoir Biology; Advanced Principles of Fisheries Management; Senior and Graduate Seminars; Senior and Graduate Research

DEPARTMENT OF FORESTRY
(No information available)

DEPARTMENT OF NATURAL RESOURCES
Phone: Chairman: (707) 826-4147 Others: 826-3011

Chairman: D.E. Craigie
Professors: R.W. Becking, M.B. Rhea
Associate Professors: D.E. Craigie, D.L. Hauxwell, J.G. Hewston
Assistant Professors: R. VanKirk
Lecturers: 3
Teaching Assistants: 7

Degree Program: Natural Resources

Undergraduate Degree: B.S.
Undergraduate Degree Requirements: Lower division requirements: Biology, Botany, Zoology, Chemistry, Geology, Natural Resources. Upper division requirements: Biology, Economics, Geology, Natural Resources Approved electives: Forty-two units including at least 15 units from the School of Natural Resources and at least 8 units is Psychology and/or Sociology: courses to be approved by the student's advisor and the program leader. Free electives to bring the total units from the B.S. degree to 192.

Graduate Degree: Interdisciplinary M.S. in Natural Recources is offered by the School of Natural Resources, not by the Department

Graduate and Undergraduate Courses Available for Graduate Credit: Fundamentals of Research, Multivariate Biometry

DEPARTMENT OF WILDLIFE MANAGEMENT
Phone: Chairman: (707) 826-3320 Others: 826-3954

Chairman: J.R. Koplin
Professors: R.E. Genelly, S.W. Harris, A.S. Mossman, C.F. Yocom
Associate Professors: R.G. Botzler, J.R. Koplin
Assistant Professors: D.W. Kitchen (Acting Director of Graduate Studies)
Visiting Lecturers: 2
Technical Assistants: 6

Degree Programs: Wildlife Management

Undergraduate Degree: B.S.
Undergraduate Degree Requirements: 192 quarter units of required and approved electives

Graduate Degree: M.S.
Graduate Degree Requirements: 45 quarter units of approved electives

Graduate and Undergraduate Courses Available for Graduate Credit: Advanced Principles of Wildlife Management, Advanced Ornithology, Advanced Topics of Wildlife Disease, Ecology and Control of Wildlife Disease (U), Ornithology II (U), Behavioral Ecology of Wildlife Populations (U), Ecology of Wildlife Populations (U)

School of Science

Dean: R. Barratt

DEPARTMENT OF BIOLOGY
Phone: Chairman: (707) 826-3245

Chairman: T.E. Lawlor
Professors: D.E. Anderson, E.R. Beilfuss, D.H. Brant, J.E. Butler, J.D. DeMartini, W.J. Houck, D.R. Lau Lauck, F.R. Meredith, W.C. Vinyard, J.F. Welsh
Associate Professors: W.V. Allen, W.A. Brueske, G.J. Brusca, R.D. Gilchrist, R.L. Hurley, D.L. Largent, T.E. Lawlor, S.Y. Lee, C.J. Lovelace, R.J. Meyer, D.H. Norris, R.A. Rasmussen, J.O. Sawyer, J.P. Smith, D.K. Walker, J.F. Waters, J.L. Yarnall
Assistant Professors: M.H. Boyd, V.J. Gagliano, M.C. Kaster, T.H. Kerstetter, K.L. Lang, W.L. Lester, C.M. Stuart
Teaching Fellows: 20
Research Assistants: 4

Field Stations/Laboratories: HSU Marine Laboratory - Trinidad, Lanphere-Christensen Dunes - Arcata

Degree Programs: Biology, Botany, Zoology

Undergraduate Degree: B.S.
Undergraduate Degree Requirements: For Biology, Botany, and Zoology majors, 186 total units are required, of which 70 must be general education and 60 are upper division requirements or electives. Specific course requirements may be obtained by examining the Humboldt State University Catalog.

Graduate Degree: M.S. (Biology)
Graduate Degree Requirements: Master: Preliminary wirtten examination; 45 upper division units in Biology or supporting courses, including 23 units of graduate courses; thesis or special problem; final oral examination

Graduate and Undergraduate Courses Available for Graduate Credit: Approximately 85 courses are available to graduate students, 24 of which are strictly graduate courses

IMMACULATE HEART COLLEGE

2021 N. Western Avenue Phone: (213) 462-1301
Los Angeles, California 90027
Dean of Graduate Studies: M.L. Krug
Dean of College: M.G. Shea

DEPARTMENT OF BIOLOGY
Phone: Chairman: Ext. 286 Others: Ext. 289

Chairman: Rev. H.E. Wachowski
Professors: Rev. H.E. Wachowski, M.G. Shea
Associate Professors: D.M. Juge
Visiting Lecturers: 3
Teaching Fellows: 3
Research Assistants: 2

Field Stations/Laboratories: Las Cruces Biological Station - Baja California

Degree Programs: Microbiology, Marine Biology, Pre-Medical, Social Impact Biology, Developmental Biology, Cell Biology

Undergraduate Degree: B.A.
Undergraduate Degree Requirements: Biology 10ab (Molecular, Organismic, Biology), Genetics; Department Seminar, 24 Upper Division units, Chemistry through Organic; Mathematics through Calculus; College Physics

Graduate Degree: M.A.
Graduate Degree Requirements: Master: Undergraduate major in Biology, 30 units upper (graduate) level courses, Dissertation

Graduate and Undergraduate Courses Available for Graduate Credit: Microbial Cytology and Physiology; Comparative Microanatomy; Advanced Genetics, Embryology,, of Invertebrates; Problems in Cellular Biology; Virology; Developmental Biology (U); Medical Mycology (U), Endocrinology (U), Biochemistry (U), Independent Study

LA VERNE COLLEGE

1950 Third Street Phone: (714) 593-3511
La Verne, California 91750
Dean of Graduate Studies: D.W. Clague
Dean of College: M. Preska

DEPARTMENT OF BIOLOGY
Phone: Chairman: Ext. 250

Chairman: R. Neher
Professor: R. Neher
Associate Professors: H. Good, S. Merritt
Research Assistants: 1

Field Stations/Laboratories: Balsa Chica, Pacific Marine Station - Balsa Chica

Degree Programs: Biology, Environmental Management

Undergraduate Degree: B.S.
Undergraduate Degree Requirements: General Biology, Environmental Biology, Developmental Biology, Genetic and Cellular Biology, Molecular Biology, and two or more electives. Competency in mathematics, physics, and organic chemistry. Senior paper and give science seminar

Graduate Degree: M.S.
Graduate Degree Requirements: Master: Systematics of Local Flora; Animal Associations; Graduate Research

Graduate and Undergraduate Courses Available for Graduate Credit: Systematics of Local Flora; Animal Association, Ekistics (Environmental Studies); Graduate Research

LOMA LINDA UNIVERSITY

Loma Linda, California 92354 Phone: (714) 824-0800
Dean of Graduate Studies: J.P. Stauffer

College of Arts and Sciences

Dean: J.P. Stauffer

DEPARTMENT OF BIOLOGY
Phone: Chairman: Ext. 2976

Chairman: L.R. Brand
Professor: A.A. Roth
Associate Professors: L.R. Brand, E.W. Lathrop, H.R. Milliken, N.L. Mitchell
Assistant Professors: G.L. Bradley, A.V. Chadwick, C.D. Clausen, E.S. McCluskey, B.R. Neufeld
Instructors: C. Rosario

Field Stations/Laboratories: Tropical Field Station - Chiapas, Mexico

Degree Program: Biology

Undergraduate Degrees: B.A., B.S.
Undergraduate Degree Requirements: B.A. - 41 quarter units of Biology, including General Biology, seminars, Cell and Molecular Biology, Genetics, Developmental Biology and cognates in Chemistry and Mathematics B.S. - 56 quarter units of Biology including (same as above), and cognates in Chemistry, Physics and Mathematics

Graduate Degrees: M.A., Ph.D.
Graduate Degree Requirements: Master: 48 quarter units of graduate work, 30 units of Biology including: a course or research in a marine or tropical environment (4 quarter units) 1 hour graduate seminar, teaching experience, 15 units of Biology at or above the 500 level (exclusive of research) a course in research techniques (during undergraduate or graduate years) a course in paleontology, or speciation, or history and philosophy of biology Doctoral: 2 languages, Biostatistics, Broad Biology of at least one taxon, Advanced Botany, Advanced Genetics, Animal Physiology or Cell and Molecular Biology, Paleontology, 2 of: Advanced Philosophy, of Biology, Biogeography additional paleontology, Advanced Invertebrate Biology

Graduate and Undergraduate Courses Available for Graduate Credit: Biological Techniques (U), General Ecology (U), Wilderness Ecology (U), Histology (U), Vertebrate Physiology (U), Protozoology,(U), Microbiology (U), Plant Morphology (U), Plant Anatomy (U), Cell and Molecular Biology (U), Introduction to Marine Biology (U), Invertebrate Biology (U), Tropical Plant Ecology (U), Human Ecology (U), Human Genetics (U), Genetics (U), Principles of Development (U), Ornithology (U), Animal Behavior (U), Plant Physiology (U), Systematic Botany (U), Philosophy of Science (U), Biogeography, Readings in Biogeography, Readings in Ecology, Advanced General Ecology, Advanced Invertebrate Biology, Seminar in Animal Behavior, Mammalogy, Biosystematics and Speciation, Developmental Genetics, Advanced Studies in Genetics, History and Philosophy of Biology, Problems in Paleontology, Paleobotany, Physiological and Development of Plants, Advanced Studies in Plant Hormones, Seminar in Biology, College Biology Teaching, Special Problems, Research Techniques in Biology

School of Medicine

Dean: D.B. Hinshaw

DEPARTMENT OF ANATOMY
Phone: Chairman: Ext. 2901

Chairman: W.H. Taylor
Professors: W.H.B. Roberts, H.Shryock, W.M. Taylor,
Associate Professors: C.W. Harrison, G.M. Hunt, J. P. McMillan, R.L. Schultz, H.Smith
Assistant Professors: N.W. Case, A.T. Dalgleish, P. Engen, H.W. Henken, W.Hooker
Instructors: P.B. Nava
Visiting Lecturers: 6

Degree Programs: Anatomy, Anatomy and Histology, Histology

Graduate Degrees: M.S., Ph.D.
Graduate Degree Requirements: Master: 48 quarter units (B average, no grade below C), modern language, written/oral comprehensives, Thesis Doctoral: as Ph.D. but 3 years residence with 12 units per quarter, 2 languages, Thesis, oral Examination

Graduate and Undergraduate Courses Available for Graduate Credit: Gross Anatomy I & II, Histology, Neurosciences, Development, Additional courses selected in conference with the adviser, in hormony with the students particular interests.

DEPARTMENT OF BIOCHEMISTRY
Phone: Chairman: Ext. 2947 Others: 2947

Chairman: R.B. Wilcox
Distinguished Service Professor: R.A. Mortensen
Professors: R.E. Beltz, U.D. Register, R.B. Wilcox, B.H. Ershoff (Research)
Associate Professors: R. Evard, R.W. Hubbard, C.W. Slattery
Assistant Professors: D.J. Gusseck, E.C. Herrmann, G.M. Lessard
Research Assistants: 3
Teaching Assistants: 2

Degree Program: Biochemistry

Graduate Degrees: M.S., Ph.D.
Graduate Degree Requirements: Master: Coursework: Biochemistry, 28 quarter units; Related fields, 11 units; Research 9 units, Thesis or publishable paper; Oral defense; OR Coursework: Biochemistry 28 units, minir sequence, 20 units; written Comprehensive Examination Doctoral: Coursework: Biochemistry 32 units, minor sequence, 20 units GPA: 3.0 Languages 2, Written Comprehensive. Dissertation and Oral Defense

Graduate and Undergraduate Courses Available for Graduate Credit: General Biochemistry, Physical Biochemistry, Metabolism and Regulation, Molecular Biology, Techniques of Biochemistry, Seminar in Biochemistry

DEPARTMENT OF MICROBIOLOGY
Phone: Chairman: Ext. 2971

Chairman: C.E. Winter
Professors: R.L. Nutter, R.E. Ryckman, E.D. Wagner, C.E. Winter
Associate Professors: L.E. Ballas, Y.L. Ho, B.H.S. Lau
Instructors: J.D. Kettering, J.W. Scott

Field Stations/Laboratories: Linda Vista Station - Chiapis, Mexico

Degrèe Program: Microbiology

Graduate Degrees: M.S., Ph.D.
Graduate Degree Requirements: Master: 48 quarter hours course work, Research and Thesis Doctoral: Master's Degree, Additional course work - designated by advisory committee, Research and Dissertation 2 foreign languages, reading and translation

Graduate and Undergraduate Courses Available for Graduate Credit: Medical Microbiology, Microbial Physiology, Molecular Biology of Microorganisms, Advanced Basic Bacteriology, Applied Clinical Microbiology, Advanced Immunology, Bacteriophage Genetics, Microbial Genetics, Bacterial Virology, Animal Virology, Cell Culture, Arthropod Vectors of Infectious Diseases, Field Medical Entomology, Diagnostic Medical Parasitology, Helminthology, Medical Mycology, Seminar in Microbiology, Special Problems in Microbiology

DEPARTMENT OF PHARMACOLOGY, PHYSIOLOGY AND BIOPHYSICS
Phone: Chairman: Ext. 2921

Professors: K.A. Arendt, J. Leonora, L.D. Longo, I. R. Neilsen
Assistant Professors: R.G. Hall, E.S. McCluskey, G. G. Power, T.J. Willey, D.D. Rafuse, W. Zaugg, R. Teel, R.R. Gonzales
Research Assistants: 5

Degree Program: Physiology

Graduate Degrees: M.S., Ph.D.
Graduate Degree Requirements: A minimum of 30 units of coursework in Physiology including 511, 512, is required for the Master's degree. Course 531, 532 is required for the doctorate. Since several programs are offered, it is expected that each student's program will be suited to his requirements and will be subject to the consent of the program faculty. Although a foreign language is not a requirement for the master's degree, students who plan to proceed to a Doctor of Philosophy degree are strongly encouraged to demonstrate, during the course of the master's program, reading ability in at least one of the languages required for the Ph.D.

Graduate and Undergraduate Courses Available for Graduate Credit: Physiology, Lectures in Physiology, Readings in Physiology and Biophysics, Cell and Molecular Biology, Comparative Physiology, Readings in Comparative Physiology, Readings in Circadian Rhythms, Properties of the Nervous System, Regulation in Normal and Cancer Cells, Regulation in Normal and Cancer Cells, Clinical Cardiopulmonary Physiology, Circulatory Physiology, Respiratory Physiology, Gastrointestinal Physiology, Readings in Neurophysiology, Endocrinology, Physiology of Reproduction, Fetal and Neonatal Physiology, Seminar in Physiology, Research, Thesis, Dissertation, Special Problems in Physiology

School of Health
Dean: M.G. Hardinge

DEPARTMENT OF BIOSTATISTICS AND EPIDEMIOLOGY
Phone: Chairman: (714) 796-7311 Ext. 3721

Chairman: J.W. Kuzma
Professors: J.W. Kuzma
Associate Professors: P.Y. Yahiku
Assistant Professors: D.E. Abbey, R.L. Phillips
Research Assistants: 3

LONE MOUNTAIN COLLEGE
San Francisco, California 94118 Phone: (415) 752-2000
Dean of Graduate Studies: E. Klinckmann
Dean of College: T. Walters

DEPARTMENT OF BIOLOGY
Phone: Chairman: (415) 387-6111

Chairman: B. Wislinsky
Associate Professor: B. Wislinsky
Instructor: I. Creps
Visiting Lecturers: 1

Degree Program: Biology

Undergraduate Degrees: B.A., B.S.
Undergraduate Degree Requirements: 64 semester units in General Education, 8 semester units in Chemistry, 8 semester units in Physics, 48 semester units in Biological Sciences

LOS ANGELES BAPTIST COLLEGE
Newhall, California 91321 Phone: (805) 259-3124
Dean of College: L.C. Button

DEPARTMENT OF BIOLOGICAL SCIENCES AND PHYSICAL SCIENCES

Chairman: G.F. Howe
Professor: G. F. Howe
Assistant Professor: D. Englin
Instructors: R. Vandiver

Degree Programs: Biology, Natural Science

Undergraduate Degree: B.S.
Undergraduate Degree Requirements: Principles of Biology, General Botany, General Zoology, Physics - 2 semesters, General Chemistry 2 semesters, Mathematics - 2 semesters, 24 semester hours in upper division biological sciences to include 2 hours of seminar. Natural Sciences is much the same except that is requires less mathematics, physics, and chemistry (1 semester each) and requires 24 hours in upper division sciences geared for teachers primarily.

LOYOLA MARYMOUNT UNIVERSITY
Loyola Boulevard and W. 80th Street Phone: (213) 776-0400
Los Angeles, California 90045
Dean of Graduate Studies: Rev. R. Trame, S.J.
Dean of College: J Foxworthy

DEPARTMENT OF BIOLOGY
Phone: Chairman: Ext. 494

Chairman: C.G. Kadner
Professors: C.G. Kadner, T.D. Pitts
Associate Professors: Rev. F. Jenkins, S.J., R. Schafer
Assistant Professors: P. Haen, V. Merriam, A. Szabo, H. Towner, A. P. Smolders
Research Assistants: 1

Degree Programs: Biological Sciences and Life Sciences

Undergraduate Degree: B.S.
Undergraduate Degree Requirements: Biology minimum total of 26 U.D. units, including Biology, Chemistry, Physics, Mathematics

Graduate Degree: M.A.T.
Graduate Degree Requirements: Master: 6 courses in U.D. or Graduate Biology, 6 Courses in Education (Specified)

Graduate and Undergraduate Courses Available for Graduate Credit: General Physiology, Cellular Physiology, Plant Taxonomy, Microbiology, Genetics, Parasitology

MILLS COLLEGE
Oakland, California 94613 Phone: (415) 632-2700
Dean of Graduate Studies: E. Milowicki
Division Chairman: E.L. Mirmow

Head: D. Bowers
Professors: D.E, Bowers, B. Kasapligil
Assistant Professor: K. Swearingen

Degree Program: Biology

Undergraduate Degree: B.A.
Undergraduate Degree Requirements: 35 semester courses (one course equals 4 semester units)

MOUNT ST. MARY'S COLLEGE
12001 Chalon Road Phone: (213) 272-8791
Los Angeles, California 90049
Dean of Graduate Studies: Sr. R. Clare
Dean of College: Sr. M. Williams

DEPARTMENT OF BIOLOGICAL SCIENCES
Phone: Chairman: Ext. 52

Chairman: Sr. A. Bower
Associate Professor: Sr. M.G. Leahy (Research)
Assistant Professor: Sr. A. Bower (Research), M. Zeuthen
Instructors: G. Anderson, C. Smith, Sr. L. Snow
Research Assistants: 2
Teaching Assistants: 3

Degree Programs: Biological Sciences, Pre-Medical, Psychobiology

Undergraduate Degrees: B.A., B.S.
Undergraduate Degree Requirements: 28-33 credits in Biology, plus courses in Chemistry, Mathematics and Physics

MUIR COLLEGE
(see University of California, San Diego)

OCCIDENTAL COLLEGE
Los Angeles, California 90041 Phone: (213) 255-5151
Dean of Graduate Studies: R. Hallin
Dean of Faculty: W. Gerberding

DEPARTMENT OF BIOLOGY
Phone: Chairman: (213) 255-5151

Chairman: P.H. Wells
Professors: J. Mcmenanmin, J. Stephens, P.H. Wells
Associate Professors: W. Hand, M. Morton
Assistant Professors: L. Baptista, L. Mays, R. Stockhouse
Teaching Fellow: 1
Research Assistants: 3
Undergraduate Teaching Assistants: 12

Field Stations/Laboratories: Oceanographic research and teaching vessel VATUNA, San Pedro. Santa Catalina Island Marine Station Consortium - Santa Catalina

Degree Programs: Biology

Undergraduate Degree: B.S.
Undergraduate Degree Requirements: 35 courses to graduate, including 9 in Biology (minimum), of which 3 are specified and 4 are elected from specified categories. Inorganic and organic chemistry, calculus and physics also are required of the Biology major. Foreign language required

Graduate Degree: M.S.
Graduate Degree Requirements: Master: 6 courses at the graduate level, plus Thesis.

Graduate and Undergraduate Courses Available for Graduate Credit: Microbiology (U), Marine Invertebrate Biology (U), Histology and Histotechniques (U), Cellular Physiology (U), Biology of Insects (U), Ichthyology (U), Environmental Physiology (U), Avian Biology (U), Animal Behavior (U), Biological Oceanography (U), Plant Growth and Development (U), Plant Taxonomy (U), Ecology (U), Special Topics (including all courses offered at Catalina Island Marine Biological Station), Research, Thesis for Master of Arts Degree

PACIFIC COLLEGE
1717 South Chestnut Phone: (209) 251-7194
Fresno, California 93702
Dean of College: D. Reimer

DIVISION OF NATURAL SCIENCE AND MATHEMATICS
Phone: Chairman: 251-7194 Ext. 58

Chairman: D. Isaak
Professors: D.E. Braun (Professor of Chemistry), D.I. Isaak (Professor of Biology)
Assistant Professors: 2
Instructors: 2

Degree Program: Life Sciences

Undergraduate Degree: B.A.
Undergraduate Degree Requirements: 1 year Chemistry, 1 year Physics, 1 year Biology and additional upper division 40 quarter units 30 of these in one depart-

ment, 1 year of Mathematics

PACIFIC UNION COLLEGE
Angwin, California 94508 Phone: (707) 956-6311
Dean of Graduate Studies: J. Scott
Dean of College: J. Christian

DEPARTMENT OF BIOLOGY
Phone: Chairman: (707) 965-6227

Chairman: E.D. Clark
Professors: E.D. Clark, L.E. Eighme, J.G. Fallon, D.V. Hemphill
Associate Professors: S.A. Nagel, T. Trivett
Assistant Professors: G.M. Muth
Teaching Assistants: 3

Field Stations/Laboratories: Mendocino Biological Field Station - Albion, Mendocino County

Degree Programs: Biology, Medical Technology, Life Sciences

Undergraduate Degrees: B.A., B.S.
Undergraduate Degree Requirements: B.A. - Biology 48 hours (27 hours upper division), including 1 summer at the Mendocino Biological Field Station total 192 quarter hours B.S. - Biology 48 hours (31 hours upper division) with 15 hours of specialization in Microbiology--Public Health, Zoology, Ecology, or Human Biology. B.A. - Life Sciences 45 hours (25 hours in upper division. B.S.-- Medical Technology-- 3 years in residence on campus, 1 year clinical (12 months)

Graduate Degree: M.A.
Graduate Degree Requirements: Master: 45 hours (21 hours must be 200-level)

Graduate and Undergraduate Courses Available for Graduate Credit: Embryology (U), Field Nature Study (U), Country Living (U), Flowering Plants (U), Vertebrate Natural History (U), Ornithology (U), Speical Topics in Biology (U), General Entomology (U), General Entomology Laboratory (U), Field Methods in Entomology (U), Sanitary Bacteriology (U), Advanced Bacteriology (U), Communicable diseases (U), Cell Biology (U), Cell Biology, 160 (U), Genetics (U), Histology (U), Physiology (U), Physiology Laboratory (U), Biotic Communities (U), Plant Ecology (U), Animal Ecology (U), Wilderness Ecology (U), Human Parasitology (U), Biological Conservation (U), Marine Biology (U), Marine Biology, 188 (U), Human Ecology (U), Neuroanatomy (U), Philosophy of Biology (U), Seminar (U), Independent Study (U), Reading and Conference, Marine Ecology, Radiation Biology, Medical Entomology, Genetics and Speciation, Field Paleontology and Paleoecology, Graduate Seminar in Biology, Independent Study, Research in Biology, (Thesis)

PASADENA COLLEGE
(see Point Loma College)

PITZER COLLEGE
(see under Clarement Colleges)

PEPPERDINE UNIVERSITY
1121 W. 79th Street Phone: (213) 971-7561
Los Angeles, California 90044
Dean of Graduate Studies: F. Pack
Dean of College: G. Goyne

DEPARTMENT OF SCIENCE AND MATHEMATICS
Phone: Chairman: (213) 971-7641

Chairman: C.E. Wilks
Associate Professors: C.E. Wilks, E. Kinsey
Assistant Professors: G. Anderson, G. Hodge, L.Salter
Instructor: R.R. Zuck

Degree Program: Biology

Undergraduate Degree: B.S.
Undergraduate Degree Requirements: 128 units total (Semester) 36 semester units Biology (must include Zoology,4, Botany 4, Physiology 4, (including 25 upper division) and Genetics 3, plus General Chemistry 8, Organic Chemistry 8, Biochemistry 4, Calculus 8, Physics 8

POINT LOMA COLLEGE
3900 Lomaland Drive Phone: (714) 222-6474
San Diego, California 92106
Dean of College: L.P. Gresham

DEPARTMENT OF BIOLOGY
Phone: Chairman: (714) 222-6474 Ext. 271
Others: Exts. 270,272

Head: D.D. Brown
Associate Professor: D.D. Brown
Assistant Professors: K.M Hyde, J.C. Crandall
Laboratory Assistants: 8

Degree Program: Biology

Undergraduate Degree: B.A.
Undergraduate Degree Requirements: A Biology Major consists of 12 quarter units of Freshman Biology and nine advanced courses (36 units) in Biology as well as Freshman Chemistry and Organic Chemistry

POMONA COLLEGE
(see under Clarement Colleges)

REVELLE COLLEGE
(see University of California, San Diego)

ST. MARY'S COLLEGE OF CALIFORNIA
Moraga, California 94575 Pnone: (415) 376-4411
Dean of College: T.J. Slakey

BIOLOGY DEPARTMENT
Phone: Chairman: Ext. 365

Chairman: P. Leitner
Professor: L.R. Cory
Associate Professors: B.E. Dodd, P. Fjeld, P. Leitner
Assistant Professor: A.K. Hansel

Degree Program: Biology

Undergraduate Degree: B.S.
Undergraduate Degree Requirements: 1 year Calculus, 1 year, General Chemistry, 1 year Introductory Physics, 1 year General Biology, 1 year Organic Chemistry, 7 upper division courses in Biology

SAN DIEGO STATE UNIVERSITY
San Diego, California 92115 Phone: (714) 286-5000
Dean of Graduate Studies: J.W. Cobble

School of Sciences
Dean: A.W. Johnson

DEPARTMENT OF BIOLOGY
Phone: Chairman: 286-6767

Chairman: W.E. Hazen
Professors: A.I. Baer, C.L. Brandt, M.E. Clark, B.D. Collier, D.A. Farris, R.F. Ford, W.E. Hazen, A.W. Johnson, W. McBlair, P.C. Miller, J.W. Neel, J.A.

Parsons, F.J. Ratty, R.R. Rinehart, D.C. Shepard, W.C. Sloan, K.M. Taylor
Associate Professors: F.T. Awbrey, W.F. Daugherty, W.P. Diehl, T.E.Ebert, S.H. Hurlbert, S. Krisans, P.J. Paolini, G P. Sanders, H.C. Schapiro, W.M. Thwaites, P.H. Zedler
Assistant Professors: C.A. Barnett, C.H. Davis, F. Dukepoo, R.L. Hays,D.A. Mauriello, J.B. Zedler
Graduate Assistants:

Field Stations/Laboratories: Santa Margarita Field Station - Fallbrook, Fortuna Mountain - San Diego

Degree Programs: Biology, Genetics, Physiological Sciences

Undergraduate Degree: B.S.
Undergraduate Degree Requirements: Calculus, Physics, Chemistry, Organic Chemistry, Biostatistics, Genetics, Ecology, Physiology, Electives

Graduate Degree: M.S.
Graduate Degree Requirements: Master: Biology - 30 units, including Thesis (Ecology, Physiology, Genetics) Doctoral: Ph.D. offered jointly with University of California, Riverside (Ecology), and University of California, Berkeley (Genetics) requirements: 1 year residency on each campus; foreign language; qualifying examination and Thesis

Graduate and Undergraduate Courses Available for Graduate Credit: Cellular Physiology (U), General Cytology (U), Regional Field Studies in Biology (U), Biology (U), Ecology (U), Aquatic Biology (U), Fisheries Biology (U), Biological Oceanography (U), Advanced Ecology (U), Systems Ecology (U), Environmental Measurement (U), Simulation of Ecological Systems (U), Comparative Animal Physiology (U), Comparative Endocrinology (U), Comparative Endocrinology Laboratory (U), Photophysiology (U), Photophysiology Laboratory (U), Radiation Biology (U), Radiation Biology Laboratory (U), Radioisotope Techniques in Biology (U), Genetics (U), Developmental Biology (U), Cytogenetics (U), Human Genetics (U), Evolution and Population Genetics (U), History of Biology (U), Source Material in the History of Biology (U), Microbial Genetics (U), Ecological Genetics (U), Mutagenesis (U), Behavioral Genetics (U), Advanced Genetics (U), Statistical Methods in Biology (U), Advanced Cellular Physiology (U), Immunochemistry (U), Immunochemistry Laboratory (U), General Botany (U), Phycology (U), Mycology (U), Vascular Plants (U), Advanced Phycology (U), Cultivated Trees and Shrubs (U), Plant Taxonomy (U), Plant Pathology (U), Plant Physiology (U), Plant Metabolism (U), Experimental Plant Metabolism (U), Plant Anatomy (U), Agricultural Botany (U), Palynology (U), Selected Topics in Botany (U), General Microbiology (U), Pathogenic Bacteriology (U), Fundamentals of Immunology and Serology (U), Medical Mycology (U), Microbial Physiology (U), General Virology (U), General Virology Laboratory (U), Hematology (U), Epidemiology (U), Bacterial and Viral Genetics (U), Bacterial and Viral Genetics Laboratory (U), Advanced General Microbiology (U), Marine Microbiology (U), Animal Viruses (U), Experimental Immunology (U), History of Microbiology (U), Electron Microscopy (U), Invertebrate Embryology (U), Embryology (U), Comparative Anatomy of the Vertebrates (U), Histology (U), Marine Invertebrates Zoology (U), Ichthyology (U), Herpetology (U), Ornithology (U), Mammalogy (U), General Entomology (U), Special Topics in Entomology, Immature Insects (U), Insect Ecology (U), Econimic Entomology (U), Medical Entomology (U), Insect Control (U), Parasitology (U), Principles of Pest Management (U), Advanced Invertebrate Zoology (U), Insect Physiology (U), Physiological Zoology (U), Experimental Animal Surgery (U), Principles of Taxonomy, Systematics and Phylogeny (U), Vertebrate Paleontology (U), Mammalian Paleontology (U), Animal Behavior (U)

BOTANY DEPARTMENT
Phone: Chairman: (714) 286-6409

Chairman: J. Alexander
Professors: A.H. Gallup, J. Kummerow, H.L. Wedberg
Associate Professors: J.V. Alexander, D.A. Preston, D.L. Rayle
Assistant Professors: N.M. Carmichael, K.D. Johnson

Degree Program: Botany

Undergraduate Degrees: B.A., B.S.
Undergraduate Degree Requirements: B.A. - 24 upper division units, 19 specified and 5 electives B.S. - 36 upper division units, 27 specified and 9 electives

Graduate Degree: M.S., M.A.
Graduate Degree Requirements: Master: This is a 30 unit program, at least 15 in graduate courses. A research problem and Thesis are required

Graduate and Undergraduate Courses Available for Graduate Credit: General Botany (U), Phycology (U), Mycology (U), Vascular Plants (U), Cultivated Trees and Shrubs (U), Systematic Botany (U), Plant Pathology (U), Plant Physiology (U), Plant Metabolism (U), Experimental Plant Metabolism (U), Plant Anatomy (U), Agricultural Botany (U), Palynology (U), Seminars in: Phycology, Mycology, Vascular Plants, Systematic Botany, Plant Pathology, Plant Physiology, Plant Anatomy, Palynology

DEPARTMENT OF MICROBIOLOGY
Phone: Chairman: (714) 286-6250

Chairman: H.A. Walch
Professors: W.L. Baxter, B.L. Kelly, H.B. Moore, H.A. Walch
Associate Professors: E.A. Anderes, L.N Phelps, J.F. Steenbergen
Instructors: R.M. Heifetz, A.T. Jokela, A.G. Mikolon, R.B. Redmond, B.P. Swinyer, E.A. Watkins
Teaching Assistants: 8

Degree Programs: Microbiology, Microbiology, Environ-Mental Health

Undergraduate Degrees: B.A., B.S.
Undergraduate Degree Requirements: 128 units

Graduate Degree: M.S. (Microbiology)
Graduate Degree Requirements: 30 units

Graduate and Undergraduate Courses Available for Graduate Credit: Pathogenic Bacteriology (U), Medical Mycology (U), General Virology (U), General Virology Laboratory (U), Hematology (U), Epidemiology (U), Bacterial and Viral Genetics (U), Bacterial and Viral Genetics Laboratory (U), Advanced General Microbiology (U), Marine Microbiology (U), Animal Viruses (U), Experimental Immunology (U), Electron Microscopy (U), Cytoplasmic Inheritance, Molecular Biophysics, Seminar in Genetics, Physiological Genetics, Seminar in Phychology, Seminar in Mycology, Seminar, Seminar in Microbial Physiology, Seminar in Pathogenic Bacteriology, Seminar in Bacterial and Viral Genetics, Seminar in Medical Mycology, Seminar in General Microbiology, Seminar in Aquatic Microbiology, Seminar in Virology, Seminar in Immunology and Serology, Bacterial Viruses, Advanced Pathogenic Bacteriology, Bibliography, Research Techniques, Research, Special Study, Thesis or Project

ZOOLOGY DEPARTMENT
Phone: Chairman: (714) 286-6330

Chairman: R.E. Carpenter
Professors: M D. Atkins, K.K. Bohnsack, R.E. Carpenter, T.J. Cohn, D.M. Dexter, R. Estes, R. Etheridge, E.W. Huffman, D. Hunsaker, N. McLean, R.E. Mon-

roe, C.E. Norland, A.C. Olson, W.J. Wilson
Associate Professors: R.H. Catlett, L. Chen, G. Collier, R.W. Cooper, J.A. Lillegraven, H.H. Plymale
Assistant Professors: V.L. Avila, C.O. Krekorian
Visiting Lecturers: 6
Graduate Teaching Assistants: 6

Degree Programs: Faculty participate in Biology Department M.A., M.S., and Ph.D. Programs

Undergraduate Degrees: A.B., B.S.

Graduate Degree: None carrying name of this department on degree
Graduate Degree Requirements: (See response from Biology Department)

Graduate and Undergraduate Courses Available for Graduate Credit: See Biology Department

SAN FRANCISCO COLLEGE FOR WOMEN
(See Lone Mountain College)

SAN FRANCISCO STATE UNIVERSITY
1600 Holloway Avenue Phone: (415) 469-1477
San Francisco, California 94132
Dean of Graduate Studies: D.M. Castleberry

School of Natural Sciences
Dean: J.S. Hensill

DIVISION OF BIOLOGY
Phone: Chairman: (415) 469-1549 Others: 469-1081

Associate Dean: D.W. Fletcher
Professors: C.G. Alexander, G.S. Araki, R.D. Beeman, S.T. Bowen, R.I. Bowman, M.G. Bradbury, C.J. Coppenger, J.T. Duncan, E.F. Estermann, D.W. Fletcher, J.R. Gabel, J.F. Fustafson, J.G. Hall, J. S. Hensill, J.P. Mackey, A.H. Nelson, G.T. Oberlander, D.M. Post, L.W. Swan, J.R. Sweeney, H.D. Thiers, J.T. Tomlinson, A.L. Towle, H.S. Wessenberg, W.G. Wu, H.H. Yonenaka
Associate Professors: J.N. Adams, H.L. Auleb, R.E. Berrend, R.G. Doell, B. Goldstein, R.C. Hunderfund, D.H. Kenyon, J.D. Stubbs, S.C. Williams
Assistant Professors: G.L. Batchelder, A. Catena, J.W. Gerald, J.H. Martin, R. Morelli
Instructors: 49 part-time
Visiting Lecturers: 5
Teaching Fellows: 0
Research Assistants:1-10

Field Stations/Laboratories: Moss Landing Marine Laboratories - Moss Landing, J. Paul Leonard Science Field Campus - Sattley

Degree Programs: Biology, Medical Technology

Undergraduate Degree: B.A. (Biology, Clinical Science)
Undergraduate Degree Requirements: Varies according to program

Graduate Degree: M.A. (Biology)
Graduate Degree Requirements: Master: 30 units of upper division and graduate courses, at least half graduate

Graduate and Undergraduate Courses Available for Graduate Credit: Cell biology (U), Cell Biology Laboratory (U), Genetics (U), Genetics Laboratory (U), Molecular Genetics (U), Cytogenetics (U), Human Genetics (U), Population Genetics (U), Techniques in Molecular Biology (U), Introduction to Biophysics (U), General Animal Development (U), Tissue Culture (U), Microtechnic (U), Microscopy and Photomicrogramph (U), Electron Microscopy (U), General Microbiology (U), Microbial Techniques (U), Sanitary Microbiology (U), Applied Microbiology (U), General Virology (U), General Virology Laboratory (U), Microbial Genetics (U), Medical Microbiology (U), Immunology (U), Medical Mycology (U), Microbial Physiology (U), Microbial Physiology Laboratory (U), Biology of the Protozoa (U), Parasitology (U), Helminthology (U), General Entomology (U), Insect Taxonomy (U), Entomology and Public Health (U), Medical Entomology Laboratory (U), General Arachnology (U), Natural History of Vertebrates (U), Herpetology (U), Systematic Herpetology (U), Ornithology (U), Mammalogy (U), Animal Ecology (U), Zoogeography (U), Marine Zoogeography (U), Vertebrate Paleontology (U), Pleistocene Ecology (U), Evolutionary Processes (U), Comparative Anatomy of Vertebrates (U), Human Anatomy and Evolution (U), Comparative Morphology of the Non Vascular Plants, Algology, Biology of the Fungi, Comparative Morphology of the Vascular Plants, Plant Anatomy, Plant Taxonomy, Ornamental Trees and Shurbs, Plant Physiology, Plant Soil Relationships, Plant Ecology, General Radiation Biology, Radioecology, Marine Invertebrate Zoology, Natural History of Marine Invertebrates, Marine Invertebrate Physiology, Invertebrate Embryology, Introduction to Ichthyology, Introductory Fishery Biology, Limnology, Biological Oceanography, Marine Ecology, Marine Science Diver Training, General Ethology, Invertebrate Behavior, Animal Communication, Principles of Human Physiology, Human Physiology, Human Physiology Laboratory, Neurophysiology, Neuroanatomy, Endocrinology, Reproductive Physiology of Vertebrates, Hematology, Ecological Physiology, Animal Bio mechanics, Comparative Physiology; Circulation, Respiration and Excretion, Comparative Physiology; Metabolic Functions, Comparative Physiology; Neuromuscular Control: Nervous and Endocrine Mechanisms, Biological Literature, Molecular Biology of Development, Developmental Genetics Seminar, Seminar in Advanced Topics in Biogenesis, Photobiology, The Biochemistry of Proteins and Nucleic Acids, The Biochemistry and Physiology of Lipids, Comparative Biochemistry, Analysis of Development, Experimental Embryology, Advanced Tissue Culture, Cell Ultrastructure Seminar, Advanced Electron Microscopy, Microbial Ecology, Marine Microbiology, Immunochemistry, Laboratory Instrumentation and Electronics, Medical Parasitology, Advanced Medical Hematology, Immunofluorescent Methods, Nuclear Radiation Biology, Automated Data Processing and Computer Methods, Advanced Clinical Biochemistry I, Advanced Clinical Biochemistry II, Topics in Medical Laboratory Science, Medical Electrophoresis, Bio-statistics and Quality Control, Acarology, Applied Entomology, Advanced Morphology and Ultrastructure of Marine Invertebrates, Systematic Ichthyology, Plankton, Growth and Development of Marine Algae, The Higher Fungi, Experimental Mycology, Lichenology, Bryology, Palynology, Plant Biosystematics, Ecology of Estuaries and Lagoons, Island Life, Biometry, Population Biology, Quantitative Research Methods in Ecology, Advanced Ethology, Electrophysiology, Advanced Biology, Advances in Cell and Molecular Biology, Advances in Ecology and Systematic Biology, Advances in Marine Biology, Advanced Microbiology, Advances in Physiology and Behavioral Biology

SAN JOSE STATE COLLEGE
(see California State University, San Jose)

SCRIPPS COLLEGE
(see under Claremont Colleges)

SCRIPPS INSTITUTION OF OCEANOGRAPHY
(See University of California, San Diego)

SOUTHERN CALIFORNIA COLLEGE

2525 Newport Boulevard Phone: (714) 545-1178
Costa Mesa, California 92626
Dean of College: R.P. Spittler

BIOLOGY DEPARTMENT
Phone: Chairman: (714) 545-1178

Chairman: L.T. McHargue
Associate Professors: E.D. Lorance, L.T. McHargue

Degree Program: Biology

Undergraduate Degree: B.A.
Undergraduate Degree Requirements: (All units are semester units) 55 Units of General Education, 10 units lower division Chemistry, 8 units of Organic Chemistry with Biochemistry also recommended, 8 units of Physics (lower division), General Botany and General Zoology (both lower division) and at least 24 upper division units in Biology including areas of Cell Biology, Microbiology, Ecological Studies, Anatomy and Physiology, Genetics and Systematics

STANFORD UNIVERSITY

Stanford, California 94305 Phone: (415) 497-1411
Dean of Graduate Division: L. Moses

School of Humanities and Sciences

Dean: H. Royden

DEPARTMENT OF BIOLOGICAL SCIENCES
Phone: Chairman: (415) 497-2413 Others: 497-1411

Chairman: N.K. Wessells
Bing Professor of Human Biology: C.S. Pittendrigh
Herzstein Professor of Biology: C. Yanofsky
Professors: I.A. Abbott, D.P. Abbott, W.R. Briggs, A.M. Campbell, P.R. Ehrlich, P.B. Green, P.C. Hanawalt, R.W. Holm, D. Kennedy, J. Lederberg, D.D. Perkins, J.H. Phillips, Jr., C.S. Pittendrigh, P.M. Ray, D.C. Regnery, R.T. Schimke, N.K. Wessells, D.O. Woodward, C. Yanofsky
Associate Professors: M. Feldman, P. Getting, H.D. Heller, P.K. Hepler, J. Roughgarden, R.D. Simoni, F.E. Stockdale, W.B. Watt
Lecturers: M.K. Allen, C.H. Baxter, E.M. Center, P.A. Sokolove
Research Assistants: 7
Visiting Scholars: 5
Research Associates: 14
Research Fellows: 37

Field Stations/Laboratories: Jasper Ridge Biological Preserve - Portola Valley, Hopkins Marine Station - Pacific Grove

Degree Program: Biological Sciences

Undergraduate Degree: B.S.
Undergraduate Degree Requirements: 1 year of Physics, Mathematics, Organic and Inorganic Chemistry, 3 quarter sequence of Biology Core and 2 quarters of Laboratory, 19 quarter units of Biology electives

Graduate Degrees: M.A., Ph.D.
Graduate Degree Requirements: Master: 3 full quarters, 45 units, 9 units of Research (at least). 36 of the units must be in Biology or related subjects. Doctoral: 3 units required with each of four or more Stanford Faculty. Must complete a minimum of three years of graduate registration (nine full quarters). Must pass qualifying procedure, Orals Committee, Submit Dissertation

Graduate and Undergraduate Courses Available for Graduate Credit: Seminar in Animal Communication (U), Biological Effects of Radiation (U), Bacterial Genetics (U), Advanced Topics in Genetics (U), Membrane Molecular (U), Viruses (U), Advanced Topics in Evolution (U), Advanced Topics in Plant Physiology and Development (U), Biological Oceanography (U), Laboratory in Biological Clocks (U), Protein Synthesis and Degradation in Eukaryotes, Cytogenetics (U), Molecular Biophysics (U), Biophysical Measurements (U), Gene Action (U), Laboratory in Neurophysiology, Drug Interactions with Biological Systems, Molecular Photobiology (U), Physiological Basis of Adaptation (U), Biological Clocks (U), Comparative Biochemistry of Marine Microorganisms, Ecological Physiology (U), Mathematical Analysis of Biological Processes, Mathematical Population Biology (U), Theoretical Population Biology (U)

BIOPHYSICS PROGRAM
Phone: Chairman: (415) 497-2424 Others: 497-2413

Program Director: P.C. Hanawalt
Professors: D. Kennedy, H. McConnell, D. Perkel
Assistant Professor: D. Clayton

Degree Program: Biophysics

Graduate Degree: Ph.D.
Graduate Degree Requirements: Doctoral: Adequate preparation in Biology, Chemistry and Physics, one or more foreign languages, comprehensive examination, Thesis

Graduate and Undergraduate Courses Available for Graduate Credit: Seminar: Physics of Biological Molecules, The Physiological Basis of Behavior, Biological Effects of Radiation, Cytogenetics, Molecular Biophysics, Biophysical Measurements, Gene Action, Molecular Photobiology, Biochemistry Lectures, The Arrangement of Information in Chromosomes, Physical Chemistry of Proteins and Nucleic Acids, Physical Chemistry, Advanced Organic Chemistry, Advanced Physical Chemistry, Chemical Physics, Introduction to a Programming Language, Biological Information Processing, Radio-activation Analysis, Selected Topics in Neurobiology

School of Medicine

Dean: C. Rich

DEPARTMENT OF ANATOMY
(No reply received)

BIOCHEMISTRY DEPARTMENT
Phone: Chairman: (415) 497-6164 Others: 497-6161

Chairman: I.R. Lehman
Willson Professor of Biochemistry: P. Berg
Merner Professor of Biochemistry: A. Rornberg
Professors: R. Baldwin, D. Hogness, A.D. Kaiser, G. Stark, E. Shooter, (Joint with Genetics) A. White (consulting)
Assistant Professors: R. Davis, D. Brutlag
Teaching Fellows: 20
Research Assistants: 8

Degree Program: Biochemistry

Graduate Degree: Ph.D.
Graduate Degree Requirements: Not Stated

Graduate and Undergraduate Courses Available for Graduate Credit: Not Stated

DEPARTMENT OF GENETICS
(no information available)

DEPARTMENT OF MEDICAL MICROBIOLOGY
Phone: Chairman: (415) 497-2004

Head: S. Raffel
Professors: L. Hayflick, S. Raffel, C.E. Schwerdt B.A.D. Stocker

Associate Professors: R.J. Roantree, L.T. Rosenberg, O.A. Soave (Clinical), M.D. Eaton, E.M. Lederberg, J.P. Steward
Instructor: E.J. Stanbridge
Visiting Lecturers: 14
Research Assistants: 8
Laboratory Assistants: 5

Degree Programs: Animal Behavior, Bacteriology, Immunology, Medical Microbiology, Medical Technology, Veterinary Microbiology, Viticulture and Enology, Bacterial Genetics, Cell Biology

Undergraduate Degree: B.S.
Undergraduate Degree Requirements: Biological Sciences, 15 quarter units; Chemistry, 19; Biochemistry, 19; Physics, 12; Medical Microbiology, 25-30

Graduate Degrees: M.S., Ph.D.
Graduate Degree Requirements: Master: 45 quarter units related to microbiology, including 15 in research plus Thesis Doctoral: M.S. requirements, including 45 units in research, plus courses recommended for individual programs; foreign language; University oral examination; Dissertation

Graduate and Undergraduate Courses Available for Graduate Credit: Principles of Immunology (U), General Microbiology (U), Special Problems (U), Immunology, Medical Microbiology, Bacterial Genetics, Virology, Advanced Medical Bacteriology, Mammalian Cell as a Microorganism, Literature Reviews, Current Topics in Immunology

DEPARTMENT OF PHARMACOLOGY
Phone: Chairman: (415) 497-6210 Others: 497-6834

Chairman: L. Aronow (Acting)
Professors: L. Aronow, A. Goldstein, O. Jardetzky, S. Kalman, T. Mansour, R. Schimke
Assistant Professors: H.F. Epstein, L. Wilson
Senior Scientist: D.B. Goldstein

Degree Program: Pharmacology

Graduate Degree: Ph.D.
Graduate Degree Requirements: The Ph.D. program is designed for students with a background in Biology, Chemistry, Physics, or Mathematics who wish to pursue a career of research in a field that lies between Biology and Medicine. A baccalaureate degree in one of these areas is required. The Ph.D. degree in Pharmacology ordinarily requires four years of study. One full year of course credits at Stanford in pharmacology and in basic Medical Sciences related to Pharmacology

DEPARTMENT OF PHYSIOLOGY
(No information available)

UNIVERSITY OF CALIFORNIA BERKELEY

Berkeley, California 94720 Phone: (415) 642-6000
Dean of Graduate Studies: S.S. Alberg

College of Letters and Sciences

Dean and Provost: R. Park

DEPARTMENT OF BIOCHEMISTRY
Phone: Chairman: (415) 642-0550 Others: 642-5252

Chairman: D.E. Koshland, Jr.
Professors: B.N. Ames, C.E. Ballou, H.A. Barker, F.H. Carpenter, M.J. Chamberlin, R.D. Cole, C.A. Dekker, D.E. Koshland, Jr., J.B. Neilands, J.C. Rabinowitz, H.K. Schachman, E.E. Snell, A.C. Wilson, J.A. Bassham (Adjunct), C.A. Knight (Research)
Associate Professors: J.F. Kirsch, S.M. Linn
Assistant Professors: G.Milman, E.E. Penhoet
Teaching Fellows: 21
Research Assistants: 9
NIH Trainees: 41

Degree Program: Biochemistry

Undergraduate Degree: B.A.
Undergraduate Degree Requirements: 1 year each of General Introductory Chemistry, Calculus, Physics and Biology; 2-3 quarters Organic Chemistry; Quantitative Analysis; 2 quarters Physical Chemistry, 3 quarters Biochemistry Lecturers; 2 quarters Biochemistry Laboratory; 1 Biochemistry Proseminar, 6 units related sicence electives. Additional courses to fill degree requirements of the College of Letters and Science

Graduate Degrees: M.A., Ph.D., C. Phil.
Graduate Degree Requirements: Master: 24 units of course work plus a research thesis or 36 units course work and a Comprehensive Examination Doctoral: No specific course work requirement, but 1) demonstrated reading knowledge of a foreign language selected from French, German, Japanese and Russian; 2) an Oral Ph.D. Qualifying Examination of the proposition type, and completion of a significant piece of Original Laboratory Research reported in a Thesis

Graduate and Undergraduate Courses Available for Graduate Credit: Biochemistry 100A-B-C, General Biochemistry Biochemistry 101-A-B, General Biochemistry Laboratory Biochemistry 150 Biochemistry and Society all (U); Biochemistry 201A-B Advanced Laboratory Methods for Graduate Students, 202, Biochemistry of Carbohydrates; 204, Biochemistry of Proteins, 205 Biochemistry of Nucleic Acids, 206, Physical Biochemistry, 207, Comparative Biochemistry, 213, Enzyme Synthesis and Control, 214, Mechanisms of Enzyme Action, 280, Thesis Research, 285, Research Seminar, 290, Graduate Seminar, 299, Special study, reading and Conference, 602, Individual study prepatory to the Qualifying Examination

BOTANY DEPARTMENT
Phone: Chairman: 642-1079

Chairman: L. Machlis
Professors: H.G. Baker, O.R. Collins, L. Constance, R. Emerson, W.A. Jensen, W.M. Laetsch, L. Machlis, R. Ornduff, R.B. Park
Associate Professors: R.L. Jones, D.R. Kaplan, J.A. West
Assistant Professors: R. Schmid, N.J. Vivrette
Teaching Fellows: 27
Research Assistants: 3
Lecturers: 2
Visiting Professor: 1

Field Stations/Laboratories: Sagehen Creek - Truckee, Bodega Marine Laboratory - Bodega

Degree Program: Botany

Undergraduate Degree: B.A.
Undergraduate Degree Requirements: Lower Division Biology 1A-1B; Botany 1; Chemistry 1A-1B, 8A-8B; two additional quarters of Chemistry (from 1C, 5 or 15); or Calculus (Mathematics 1A-1B-1C or 16A-16B or 190A-190B) or Physics 6A-6B-6C. Upper Division Biochemistry 102, two courses of the following Botany 101, 102, 105 or 110, one course of the following: Botany 120, 124, or 154L (Botany 150 is prerequisite); Botany 146, Botany 112 or 130; plus one additional course from any of the preceding groups. Genetics 100 or 150A-150B. Additional courses in Botany or approved courses in related departments to complete a minimum of 36 upper division units.

Graduate Degrees: M.A., Ph.D.
Graduate Degree Requirements: Master: Requires at

least 36 quarter units of upper division and graduate courses, followed by a comprehensive final examination administered by the student's department or group At least 18 of the units must be in graduate courses in the major subject. Doctoral: Student must have a firm foundation in at least one language (German, French or Russian) and demonstrate a reading knowledge of one foreign languate in their first year of graduate work.

Graduate and Undergraduate Courses Available for Graduate Credit: General Botany, Plant Biology, Survey of Mycology, General Phycology, Marine Botany, The Principles of Plant Morphology, Evolutionary Morphology of Vascular Plants, Plants and Man, Taxonomy of Seed Plants, Field Course in Plant Taxonomy and Ecology, The California Flora, Plant Cell Biology, Plant Physiology, Plant Ecology (U), Biology of the Lower Fungi, Biology of the Slime Molds and Higher Fungi, Algology, Pteridology, Advanced Taxonomy, Plant Biosystematics, Evolutionary Ecology, Advanced Plant Physiology, Advanced Plant Ecology, Techniques of Electron Microscopy for Biologists (G)

DEPARTMENT OF MOLECULAR BIOLOGY
Phone: Chairman: (415) 642-0942

Chairman: H.K. Schachman
Professors: M. Calvin, A.J. Clark, P.H. Duesberg, H. Echols, H.L. Fraenkel-Conrat, J.C. Gerhart, D.A. Glaser, C.A. Knight, H. Rubin, H.K. Schachman, G.S. Stent, R.C. Williams
Associate Professors: R. Calendar, J.R. Roth
Assistant Professor: T. Gurney
Research Assistants: 25

Degree Programs: Biochemistry, Biochemistry and Biophysics, Biological Chemistry, Biological Engineering, Biology, Biometry, Genetics, Molecular Biophysics, Virology, Molecular Biology

Graduate Degree: Ph.D.
Graduate Degree Requirements: Doctoral: At least 2 years or 6 quarters of academic residence. Courses Required: Molecular Biology 200A and B plus 5 seminar courses. Teaching Assistant experience for at least two quarters. One foreign language. Qualifying Examination and Thesis

Graduate and Undergraduate Courses Available for Graduate Credit: Introduction to Molecular Biology, Molecular Biology Laboratory, Special Topics in Molecular Biology, Introduction to Research in Molecular Biology, Molecular Biology of Viruses, Microbial Genetics, Techniques in Animal Cell Culture

DEPARTMENT OF PHYSIOLOGY AND ANATOMY
Phone: Chairman: 642-5978

Chairman: J.G. Forte
Professors: W.J. Freeman, H.B. Jones, R.I. Macay, N. Pace, L. Packer, L.L. Rosenberg, H.H. Srebnik, P.S. Tisniras, G. Westhiwer, S.F. Cook (Emeritus)
Associate Professors: M.C. Diamond, J.G. Forte C.S. Nicoll, P. Satir
Visiting Lecturers: 4
Teaching Fellows: 4
Research Assistants: 20

Field Stations/Laboratories: White Mountain Research Station - Bishops

Degree Programs: Physiology, Anatomy, Biophysics, Endocrinology and Neurobiology

Undergraduate Degree: B.S.
Undergraduate Degree Requirements: Not Stated

Graduate Degrees: M.S., Ph.D.
Graduate Degree Requirements: Not Stated

Graduate and Undergraduate Courses Available for Graduate Credit: Not Stated

DEPARTMENT OF ZOOLOGY
Phone: Chairman: (415) 642-4954 Others: 642-3281

Chairperson: R.C. Strohman
Professors: M. Alfert, W. Balamuth, G.W. Barlow, W. E. Berg, H.A. Bern, R.M. Eakin, C.H. Hand,Jr., M. Harris, A.S. Leopold, P. Licht, W.Z. Lidicker, D. Mazia, S. Nandi, D.R. Pitelka (Adjunct), F.A. Pitelka, C.H.F. Rowell, R.I. Smith, R.C. Stebbins, R.C. Strohman, D.B. Wake, F.H. Wilt
Associate Professors: D.R. Bentley, N.K. Johnson, T.E. Rowell (Adjunct), J.E. Simmons, R.A. Steinhardt
Assistant Professors: R.L. Caldwell, R.K. Colwell, M.T. Ghiselin, J.L. Patton, M.L. Pressick, M.H. Wake
Professional Research Zoologists: 6
Research Fellows: 15

Field Stations/Laboratories: Bodega Marine Laboratory - Bodega Bay, Hastings Natural History Reservation - Carmel Valley

Degree Programs: Neurobiology, Zoology, Endocrinology, Parasitoloty

Undergraduate Degree: A.B.
Undergraduate Degree Requirements: Basic Biology, Chemistry, Mathematics and Physics, 36 units of upper division Zoology, Statistics and Biochemistry recommended

Graduate Degrees: M.A., Ph.D.
Graduate Degree Requirements: Master: 30 units graduate Zoology, Modern Language and Thesis or 36 units language, written and oral Examination Doctoral: special program for each student. 5 quarters of foreign language , 2 graduate seminars, Ph.D. oral, Thesis, Teaching Experience.

Graduate and Undergraduate Courses Available for Graduate Credit: Introduction to Physiochemical Biology (U), Vertebrate Embryology (U), Evolutionary and Functional Vertebrate Anatomy (U), Natural History of the Vertebrates, Invertebrate Zoology (U), Animal Evolution (U), Cytology (U), Experimental Embryology (U), Normal and Abnormal Growth (U), Biology of Chemical Mediation (U), Biology of Chemical Mediation, Invertebrate Physiology, Invertebrate Physiology Laboratory (U), Physiological Ecology (U), Animal Behavior (U), The Neurophysiological Basis of Animal Behavior (U), Social Behavior of Animals (U), Ecological Aspects of Behavior (U), Animal Ecology (U), Ecology and Evolution of Biological Communities (U), Marine Ecology (U), Marine Ecology (U), General Protozoology (U), General Animal Parasitology (U), Biology of Marine Invertebrates (U), Experimental Protozoology (U), Evolutionary Cytogenetics of Vertebrates (U), Mammalogy (U), Ornithology (U), Herpetology (U), Ichthyology (U), American Game Birds and Mammals (U), Comparative Histology (U), Biology of Neoplasia (U), Special Topics in the Biology of Neoplasia (U), Molecular and Cellular Aspects of Development, Seminar in Cytology, Seminar in Physiochemical Biology, Somatic Cell Heredity, Seminar on Fine structure, Seminar in Developmental Biology, Special Topics in Biology of Chemical Mediation, Seminar in Comparative Endocrinology, Seminar in Marine Biology, Seminar in Physiological Ecology, Seminar in Comparative Neurophysiology, Seminar in Animal Behavior, Comparative Population Ecology, Seminar in Animal Ecology, Ecological REsearch Reviews, Genetic Ecology, Biology of Parasitic Protozoa, Seminar in Protozoology, Advanced Biology of Marine Invertebrates, Advanced Invertebrate Zoology, Seminar in Invertebrate Zoology, Seminar on Speciation in Vertebrates, Vertebrate Review, Seminar in Wildlife Ecology and Population Dynamics, Chordate Neurology, Tumor Biology Research Review, Seminar on Biology of Neoplasia, Seminar in Comparative Neurochemistry, Principles and Concepts of Modern Zoology, Special Study for Graduate Students, General Biological Microtechniques, Zoology Seminar, Introduction to Aquanautics, Problems in Marine Biology, Comparative Neurophysiology, Advanced Laboratory in Neurophysiology, Animal Behavior Research Reviews, Experimental Helminthology

College of Agricultural Sciences

Dean: Vacant

DEPARTMENT OF CELL BIOLOGY
(No reply received)

DEPARTMENT OF ENTOMOLOGICAL SCIENCES
Phone: Chairman: 642-6660

Chairman: P.S. Messenger
Professors: J.R. Anderson, J.E. Casida, H.V. Daly, R. L. Routt, J.Freitag, D.P. Furman, K.S. Hagen, C.B. Huffaker, C.S. Koehler, E. G. Linsley, W.J. Loher, P.S. Messenger, W.W. Middlekauff, R.L. Pipa, J.A. Powell, E.I. Schlinger, R.F. Smith, E.S. Sylvester, Y. Tanada, R. van den Bosch, D.L. Wood
Associate Professors: D.L. Dahlsten, C.J. Weinmann
Assistant Professors: J.T. Doyen, B. Heinrich, G. Oster
Visiting Lecturers: 1

Field Stations/Laboratories: Blodgett Forest Laboratory - El Dorado County, Deciduous Fruit Field Station - San Jose, Hopland Field Station - Mendocini County, Fresno Mosquito Laboratory - Fresno, Kearney Horticulture Field Station - Parlier

Degree Programs: Entomology, Parasitology

Undergraduate Degree: B.S.
Undergraduate Degree Requirements: Entomological Sciences: 29 units humanities and social sciences, as follows: 10 English Composition, 19 additional units, including foreign language through course 2. Physical sciences, 41 units as follows: 12 Chemistry--Inorganic with Laboratory; 9 Organic with Laboratory, 12 Physics; 8 Mathematics and/or Statistics; 43 units Biological and Agricultural Sciences other than major field as follows: 4 Microbiology/Laboratory, 5 Genetics; 4 Physiology; 4 Pathology; 26 additional biological sciences. Major field, 33 units as follows: 5 general Entomology; 4 Systematic Entomology; 4 insect Ecology; 4 Insect Classification; 6 Anatomy and Physiology of Insects; 5 Field Practice courses; 5 additional Entomology. 35 units additional courses, total 180. Pest Management: 30 units humanities, and social sciences as follows: 10 English Composition; 5 economics, 5 interdepartmental studies; 10 additional courses. Physical Sciences, 29 units as follows: 8 Chemistry -- Inorganic with Laboratory; 9 Organic with Laboratory; 4 Biochemistry; 4 Physics, 4 Mathematics/Statistics. Biological and Agricultural Sciences, 64 units as follows: 12 General Biology; 5 Interdepartmental Studies, 4 Physiology -- animal; 4 plant; 5 Genetics; 4 Ecology; 12 Pathobiology; 4 Data Assessment and Collection; 11 Entomology. Pest Management Practice and Methods, 37 units, including 8 unit Summer Field Course. Additional courses, 20 units, total 180.

Graduate Degrees: M.S., Ph.D.
Graduate Degree Requirements: Master: Plan I requires 30 quarter units and Thesis. At least 12 of these units not granted for Thesis. Plan II : 36 quarter units upper division and graduate courses followed by comprehensive final examination. At least 18 of the units must be graduate courses in major; minimum of 3 quarters of academic residence. Doctoral: Minimum residence 2 years; fundamental courses in chosen field; Dissertation

Graduate and Undergraduate Courses Available for Graduate Credit: General Entomology (U), Insect Classification (U), Functional Insect Anatomy (U), Environmental Physiology of Insects (U), Systematic Entomology (U), Insect Ecology (U), Field Entomology (U), Destructive and Beneficial Arthropods (U), Insect Pest Management (U), Insect Toxicology (U), Pathobiology (U), Control Methods in Pest Management (U), Pest Management Systems (U), Biological Control of Insect Pests and Weeds (U), Insect Pathology (U), Medical and Veterinary Helminthology (U), Medical and Veterinary Entomology (U), Principles and Methods of Entomological Research (U), Biological Deterioration of Wood (U), Principles of Systematic Entomology; Population Ecology, Principles and Problems in Agricultural Entomology, Insect-Crop Relationships, Concepts and Research in Forest Entomology, Advanced Insect Physiology, Biochemistry, and Toxicology, Physiological Mechanisms in Insect Behavior, Biology of Parasitoids, Advanced Insect Pathology, Advanced Medical and Veterinary Entomology, General Nematology, Acarology, Insect Vectors of Plant Pathogens, History of Entomology, Immature Insects, Seminar in Insect Physiology, Seminar in Parasitology, Seminar in Agricultural Entomology, Seminar in Insect Biochemistry and Toxicology, Seminar in Insect Pathology, Seminar in Systematic Entomology, Seminar in Insect Ecology and Biological Control, Seminar in Forest Entomology, Research in Entomology and Parasitology

DEPARTMENT OF FORESTRY AND CONSERVATION
Phone: Chairman: (415) 642-0376

Professors: D.L. Brink, R.A. Cockrell, R.N. Colwell, F.E. Dickinson, R.F. Grah, H.F. Heady, A.S. Leopold, W.J. Libby, A.P. Schniewind, A.M. Schultz, E.C. Stone, H.J. Vaux, J.A. Zivnuska, H.H. Biswell (Emeritus), E. Fritz (Emeritus), M. Krueger (Emeritus)
Associate Professors: J.A. Helms, W.L.M. McKillop, D.E. Teeguarden, P.J. Zinke
Assistant Professors: D.G. Arganbright, D.C. Erman, J.R. McBride, L.C. Wnesel, R.G. Lee (Acting)
Lecturers: A.B. Anderson, P. Casamajor, B.M. Collett, W.G. O'Regan, H.C. Sampert, M. White, W.W. Wilcox, E. Zavarin
Visiting Lecturers: 2
Research Assistants: 50

Field Stations/Laboratories: Whitaker's Forest - Sierra Nevada, Blodgett Forest - Sierra Nevada, Sagehen Creek Wildlife-Fisheries Research Station - Sierra Nevada, Russell Tree Farm - San Francisco Bay Region, Howard Forest - Mendocino County

Degree Programs: Forestry, Wood Science, Wildland Resources

Undergraduate Degree: B.S.
Undergraduate Degree Requirements: Forestry Major -- 195 quarter units, with various specified courses Wood Science Major -- 180 quarter units, with various specified courses

Graduate Degrees: M.S., Ph.D.
Graduate Degree Requirements: Master: For details, consult the department Doctoral: For details, consult the department

Graduate and Undergraduate Courses Available for Graduate Credit: Field Study of Forestry and Wildland Resources (U), Resource Information Systems (U), Forest Photogrammetry and PhotoInterpretation (U), Forest Harvesting Systems (U), Control and Management of Fire (U), Wildland Resource Economics and Planning (U), Forest Regulation and Management (U), Decision-Making in Resource Management (U), Introduction to Natural Resource Policy (U), Recreational Use of Forests and Wildlands (U), Sociology of Natural Resources (U), Dendrology (U), Forest Influences (U), Ecology of Renewable Natural Resources (U), Principles of

Silviculture (U), Advanced Forest Mensuration, Advanced Photographic Interpretation (U), Seminar on Fire as an Ecological Factor, Seminar in Analysis of the Forest Ecology, Seminar in Econimics of Forestry Enterprises, Case Studies in Wildland Resource Management, Seminar in Natural Resource Policy, Seminar in Forest Genetics, Seminar in Forest Influences and Watershed Management, Natural Resource Eco systems, Advanced Silviculture, Remote Sensing of Earth Resources, Assessment of the Environment, Anatomy and Physical Characteristics of Wood (U), Mechanical Processing of Wood (U), Physical Properties of Wood (U), Mechanics of Wood (U), Chemistry and Chemical Processing of Wood (U), Advanced Wood Anatomy, Advanced Wood Physics, Advanced Wood Mechanisms, Chemistry of Polysaccharides, Lignin, and Extractives, Seminar in Wood Science and Technology, Biological Deterioration of Wood, Principles of Range Management (U), Range Animal Nutrition and Management (U), Range Ecology (U), Range Analysis and Planning (U), Wildlife Biology and Management (U), Field Course in Wild life and Fisheries (U), Wildlife Populations (U), Case Histories in Wildlife Management (U), Freshwater Ecology (U), Seminar in Wildlife Biology and Management, Seminar in Freshwater Ecology

NUTRITIONAL SCIENCES DEPARTMENT
Phone: Chairman: 642-5202 Others: 642-6490

Chairman: S.M. Margen
Professors: G.M. Briggs, D.H. Calloway, R.L. Lyman, S.M. Margen, E.L.R. Stokstad
Associate Professors: B.M. Kennedy, R. Ostwald, M. A. Williams
Assistant Professors: L. Bjeldanes, G.W. Chang, J. King, S. Oace
Lecturers: A. Little, J. Richmond
Visiting Lecturers: 2
Teaching Fellows: 6
Research Assistants: 10

Degree Programs: Agricultural Chemistry, Comparative Biochemistry, Fisheries and Wildlife Biology, Nutrition, Dietetics

Undergraduate Degree: B.S.
Undergraduate Degree Requirements: 36 units agricultrual and biological sciences, 48 units physical sciences and Mathematics, 32 units Chemical Sciences

Graduate Degrees: M.S., Ph.D.
Graduate Degree Requirements: Master: 30 upper division units, Thesis Doctoral: 4 years of graduate study, Thesis

Graduate and Undergraduate Courses Available for Graduate Credit: Economics of Food and Nutrients (U), Food Analysis (U), Introduction to Nutritional Sciences (U), Introductory Nutritional Sciences (U), Introductory Nutritional Science Laboratory (U), Food Chemistry (U), Food Chemistry Laboratory (U), Nutrition of Population Groups (U), Experimental Helminthology (U), Introduction to Nutrition (U), Principles of Food Preservation and Processing (U), Introduction to Food Research (U), Food Toxicology (U), Experimental Study of Food Properties (U), Institutional Food Production Service and Organization (U), Nutrition (U), Experimental Nutrition (U), Human Nutrition (U), Human Nutrition Laboratory (U), Therapeutic Nutrition (U), Field Study in Food and Nutritional Sciences (U), Man and His Environment - Crises and Conflicts (U), Seminar in Nutrition, Nutritional Aspects of the Metabolism of Carbohydrates and Lipids, Biochemical Aspects of Protein Nutrition, Innovations in Food Processing, Research Methods in Nutritional Sciences, Instrumentation, Research Methods in Nutritional Sciences, Directed Group Studies, Research in Food and Nutrition, The Profession of Dietetics, Individual Study for Doctoral Students

DEPARTMENT OF PLANT PATHOLOGY
Phone: Chairman: (415) 642-5121 Others: 642-5121

Chairman: D.E. Schlegel
Professors: L.J. Ashworth, K.F. Baker, B.E. Day, A.H. Gold, J.R. Parmeter, R.D. Raabe, D.E. Schlegel, M.N. Schroth, A.R. Weinhold, S. Wilhelm, C.E. Yarwood
Associate Professors: F.W. Cobb, J.G. Hancock
Assistant Professors: O.C. Huisman, S.T. Thomson
Research Assistants: 13

Field Stations/Laboratories: Deciduous Fruit Field Station-San Jose, Hopland Field Station - Hopland, Imperial Valley Field Station, El Centro, Kearney Horticultural Field Station - Parlier, Lindcove Field Station - Exeter, Sierra Foothill Range Field Station - Browns Valley, South Coast Field Station - Santa Ana, Tulelake Field Station - Tulelake, West Side Field Station - Five Points

Degree Program: Plant Pathology

Graduate Degrees: M.S., Ph.D.
Graduate Degree Requirements: Plan I: 30 quarter units Thesis Plan II: 36 quarter units (18 units in major subject); Comprehensive Examination Doctoral: Academic requirements similar to M.S. degree, Comprehensive Oral Examination, Demonstration of original research and Doctoral Thesis

Graduate and Undergraduate Courses Available for Graduate Credit: Diseases of Forest Trees (U), Plant Diseases and the Protection of Plant Resources (U), Plant Diseases (U), Seminar in Plant Pathology, Biology of Plant Pathogenic Fungi, Bacteria in Relation to Plant Diseases, Physiology of Plant Virus Infection, Plant Disease Control, Advanced Plant Pathology, Physiology of Plant Pathogens, Physiology of Plant Pathogens Physiology of Plant Diseases, History and Literature of Plant Pathology, Epidemiology and Diagnosis of Plant Diseases, Physiology of Plant Virus Infection

GENETICS DEPARTMENT
Phone: Chairman: (415) 642-3052 Others: 642-6000

Chairman: S. Fogel
Professors: S.W. Brown, D.R. Cameron, (Emeritus), E.R. Dempster (Emeritus), S. Fogel, I.M. Lerner, W.J. Libby, C. Stern (Emeritus)
Associate Professors: J.W. Fristrom, P. St. Lawrence
Assistant Professors: M.R. Freeling, R.M. Palmour, P.T. Spieth
Teaching Fellows: 3
Research Assistants: 6
Teaching Assistants: 7

Field Stations/Laboratories: Oxford Greenhouse - Berkeley, Gill Tract - Albany

Degree Programs: Genetics, Human Genetics

Undergraduate Degree: B.A.
Undergraduate Degree Requirements: B.A. in Letters and Science. Lower Division Requirements-- In addition to college breadth requirements, major requirements consist of General Biology, Generan and Organic Chemistry, Introductory Physics, Two terms of Calculus. Upper Division Requirements-- Four courses in Genetics and one course in each of Cytology, Biochemistry and Microbiology, plus elective courses in areas close to the field of Genetics

Graduate Degrees: M.S., Ph.D.
Graduate Degree Requirements: Master: Plans I and II, no special requirements Doctoral: Two years resident (4-5 normal). Requirements: 1) No language requirement 2) preliminary biology competence passed by unassembled exam, 3) qualifying examination 4) Dissertation reporting on original research in the field

of Genetics

DEPARTMENT OF SOILS AND PLANT NUTRITION
Phone: Chairman: (415) 642-0341 Others: 642-6000

Chairman: K.L. Babcock
Professors: R.J. Arkley (Lecturer), K.L. Babcock, I. Barshad (Lecturer), P.R. Day, I. Jacobson, C.M. Johnson (Chemist), A.D. McLaren, A. Ulrich (Lecturer), J. Vlamis (Plant Physiologist), D.E. Williams (Lecturer)
Associate Professors: L.J Waldron, R.K. Schulz (Associate Research Soil Chemist)
Assistant Professors: H.E. Doner, P.L. Gersper, N. Terry
Research Assistants: 12

Field Stations/Laboratories: Agricultural Experiment Station - Berkeley

Degree Programs: Soil Science, Plant Physiology

Undergraduate Degree: B.S.
Undergraduate Degree Requirements: Biological and Physical Sciences, Humanities and Social Sciences, Mathematics, Major Field

Graduate Degrees: M.S., Ph.D.
Graduate Degree Requirements: Master: Not Stated Doctoral: Calculus, Biometry, Botany, Chemistry, Physics, Biochemistry, Genetics, Plant Nutrition, amd Soil Science

Graduate and Undergraduate Courses Available for Graduate Credit: Soil Resources Evaluation, Advanced Soil Biochemistry and Soil Biology, Advanced Soil Chemistry, Pedochemistry and Mineralogy of Soils, Soil Summer Field Course (U)

School of Public Health
Dean: W.W. Winkelstein

DEPARTMENT OF BIOMEDICAL AND ENVIRONMENTAL HEALTH SCIENCES
Phone: Chairman: (415) 642-3744

Division Head: C.H. Tempelis
Professors: S.S. Elberg, S.H. Madin, C.H. Tempelis, N.A. Vedros
Associate Professor: J.L. Hardy
Assistant Professor: G. Buchring
Teaching Fellows: 3
Research Assistants: 2
NIH Trainees: 9

Degree Programs: Microbiology, Immunology, Comparative Pathology, Parasitology

Graduate Degrees: M.A., Ph.D.
Graduate Degree Requirements: Master: Depends on Graduate Program Doctoral: Depends on Graduate Program

Graduate and Undergraduate Courses Available for Graduate Credit: Medical Microbiology, Introduction to Animal Viruses, Survey of General Pathology, Advanced Medical Microbiology, Advanced Medical Virology, Advanced Methods in Medical Microbiology

GROUP IN BIOSTATISTICS
Phone: Chairman: (415) 642-3241 Others: 642-3241

Chairmen: C.L. Chiang, E.L. Scott
Professors: D.R. Brillinger, C.L. Chiang, E.R. Dempster, J.L. Hodges, L. LeCam, J. Neyman, W.C. Reeves, E.L. Scott, C. Tobias, W. Winkelstein
Associate Professors: K.A. Doksum, M.E. Tarter
Assistant Professors: R.J. Brand, S. Selvin
Instructor: C.A. Langhauser
Research Assistants: 10
Teaching Assistants: 12

Degree Program: Biostatistics

Graduate Degrees: M.A., M.Ph., Ph.D., Dr, PH.
Graduate Degree Requirements: Master: 36 quarter units and an oral Comprehensive Examination Doctoral: Qualifying Examination, Thesis, Defense of Dissertation (one language examination)

Graduate and Undergraduate Courses Available for Graduate Credit: courses in Statistics, Public Health, Biostatistics, Biology, Mathematics, Physiology, Genetics, Zoology and Computer Sciences are available for graduate credit (They are too numerous to list individually)

UNIVERSITY OF CALIFORNIA
DAVIS
Davis, California 95616 Phone: (916) 752-1011
Dean of Graduate Studies: A.G. Marr

College of Environmental and Agricultural Sciences
Dean: A.F. McCalla

DEPARTMENT OF AGRICULTURAL ENGINEERING
Phone: Chairman: 752-0103 Others: 752-0102

Chairman: R.B. Fridley
Professors: N.B. Akesson, W.J. Chancellor, K.L. Coulson, J.B. Dobie (Lecturer) R.B. Fridley, J.R. Goss, G.F. Hanna (Lecturer), S.M. Henderson, R.A. Kepner, M. O'Brien, H.B. Schultz, W.E. Yates
Associate Professors: R.E. Garrett, S.R. Morrison, L.O. Myrup
Assistant Professors: T.H. Burkhardt, J.J. Carroll, P. Chen (Lecturer), H.E. Studer (Lecturer), A.E. Tangren
Teaching Fellows: 3
Research Assistants: 5

Degree Programs: Engineering -- Major in Agricultural Engineering

Undergraduate Degree: B.S.
Undergraduate Degree Requirements: 180 quarter units including final 45 units in residence. Basic Engineering, Science, and Mathematics, approximately 140 units, including at least 2 courses of agricultural or biological science

Graduate Degrees: M.S., Ph.D.
Graduate Degree Requirements: Master: A minimum of 36 quarter units of courses that includes a minimum of 12 units of graduate engineering courses and a Thesis for the M.S. degree and an engineering project for the M.E. degree Doctoral: Six quarters of academic residence, completion of a qualifying examination, and a Dissertation. A minimum of approximately 55 quarter units compose a program of a major area and one or two minors

Graduate and Undergraduate Courses Available for Graduate Credit: Engineering Economics, Characteristics and Applications of Electric Motors, Engines for Agriculture, Industry and Transportation, Principles of Field Machinery Design, Forest Engineering, Stability and Traction of Off-Raod Vehicles, Testing and Evaluation of Engineering Designs, Hydraulic and Pneumatic Systems, Agricultural Structures: Engironmental Aspects, Agricultural Structures: Construction Aspects, Unit Operations in Agricultural Processing, Engineering Design Projects for Agriculture and Forestry (all U), Soil-Machine Relations in Tillage and Traction, Advanced Unit Operations in Agricultural and Food Processing, Agricultural Waste Management, Design of Mechanical Systems, Environmental Engineering in Agriculture, Design and Analysis of Engineering Experiments, Physical Properties of Agricultural Materials, Seminar

GRADUATE GROUP IN ECOLOGY (Department of Agronomy and Range Science)
Phone: Chairman: (916) 752-2461 Others: 752-1011

Chairman: R.M. Love
Agricultural Economics
Professors: W.E. Johnston, J.H. Snyder
Associate Professor: D.E. Hansen
Agronomy and Range Science
Professors: H.H. Biswell (Emeritus), S.K. Jain, H.M. Laude, R.S. Loomis, R.M Love, D.S. Mikkelsen, D.W. Rains, W.A. Williams
Associate Professors: C.A. Raguse
Animal Science
Professor: R.C. Laben
Associate Professor: D.W. Robinson
Anthropology
Professor: M.A. Baumhoff
Associate Professor: W.G. Davis
Assistant Professors: K.H.K. Chang, M.K. Neville, P. S. Rodman
Applied Behavioral Sciences
Professor: G.R. Hawkes
Avian Sciences
Professor: P. Vohra
Bacteriology
Professor: M.P. Starr
Botany
Professors: D.I. Axelrod, D.W. Kyhos, J.Major, J.M. Tucker, G.L. Webster
Associate Professors: M.G. Barbour
Entomology
Professor: A.A. Grigarick
Associate Professors: W.R. Cothran, R.W. Thorp
Assistant Professor: M.C. Birch
Lecturer: R.K. Washino
Environmental Horticulture
Professors: R.W. Harris, A.M. Kofranek, H.C. Kohl, J.H. Madison
Associate Professors: S.M. Gold, J.A. Harding, A.T. Leiser, J.L. Paul
Assistant Professor: R.W. Hodgson
Environmental Studies
Professors: C.R. Goldman, W.J. Hamilton
Associate Professors: J. McEvoy III, L.O. Myrup,
Assistant Professors: T.E. Dickinson, T.C. Foin, R.A. Johnston, B.S. Orlove, T.M. Powell, P.J. Richerson, P.A. Sabatier, S.I. Schwartz
Food Science and Technology
Professor: H.S. Olcott
Assistant Professors: E.V. Crisan, R.H. Shleser
Lecturer: G.K. York
Genetics
Professors: R.W. Allard, F.J. Ayala
Geography
Associate Professor: S.C. Jett
Lecturer: W.E. Derrenbacher
Geology
Professor: J.W. Valentine
Associate Professors: R. Cowen, J.H. Lipps
Lecturer: R.A. Matthews
Medicine (Veterinary Medicine)
Professor: M.E. Fowler
Nematology
Lecturer: A.R. Maggenti
Nutrition
Professor: F.W. Hill
Wildlife and Fisheries Biology
Professor: W.E. Howard
Associate Professors: D.F. Lott, D.G. Raveling, R.G. Schwab
Assistant Professors: H.W. Li, P.B. Moyle
Lecturer: R.W. Brocksen
Zoology
Professor: R.L. Rudd, G.W. Salt, K.E.F. Watt
Assistant Professor: A.M. Shapiro

Graduate Degrees: M.S., Ph.D. (Ecology)
Graduate Degree Requirements: Master: The Graduate Group in Ecology is an organized program of interdisciplinary nature, involving faculty members in 36 departments in 5 colleges. Doctoral: Required courses for Ecology: Ecology 201A, Ecological Theory; Ecology 201B, Analysis of a Selected Ecosystem; Ecology 201C, The Changing Biosphere, 1 unit (M.S.) or 3 units (Ph.D.) of Ecology 290, Seminar in Ecology

DEPARTMENT OF AGRONOMY AND RANGE SCIENCE
Phone: Chairman: 752-1713 Others: 752-1703

Chairman: P.F. Knowles
Professors: J.P. Conrad (Emeritus), R.C. Huffaker, S.K. Jain, P.F. Knowles, H.M. Laude, R.S. Loomis, R.M. Love, D.S. Mikkelsen, M.L. Peterson, C.O. Qualset, F.L. Smith (Emeritus) E.H. Stanford, W.A. Williams, E.P. Zscheile (Emeritus)
Associate Professors: C.A. Raguse, D.W. Rains
Instructors/Lecturers: B.H. Beard, R.W. Breidenbach, B. Crampton, M.B. Jones, D.E. Seaman, J.N. Rutger, B.D. Webster
Research Assistants: 11

Field Stations/Laboratories: Hopland Field Station - Hopland, Imperial Valley Field Station, El Centra, Rice Experiment Station, Biggs, U.S. Cotton Research Station - Shafter

Degree Programs: Agronomy, Botany, Crop Science, Ecology, Genetics, Plant Physiology, Plant Breeding, Plant Sciences, Range Science, Soils and Plant Nutrition

Undergraduate Degree: B.S.
Undergraduate Degree Requirements: Completion of 180 quarter units with a minimum of a "C" average of courses as listed in the catalog for the designated major

Graduate Degrees: M.S., Ph.D.
Graduate Degree Requirements: Master: Plan I -- completion of 130 units of upper division and graduate courses and a Thesis Plan II -- completion of 36 units of upper division and Graduate courses and a Comprehensive Examination Doctoral: Successful completion of qualifying examination and research problem approved by the Dissertation Committee

Graduate and Undergraduate Courses Available for Graduate Credit: Science and Technology of Field Crop (U), Cereal Crops of the World (U), Forage Crop Ecology (U), Fiber, Oil and Sugar Crops in a Changing World (U), Directed Group Study (U), Special Study for Advanced Undergraduates (U), Design, Analysis and Interpretation of Experiments, Agricultural Research Planning and Management, Advanced Plant Breeding, Quantitative Genetics and Plant Improvement, Selection Theory in Plant Breeding, Advanced Population Biology, Advanced Topics in the Ecology of Crop Plant Communities, Advanced Topics in the Physiology of Crop Plants, Seminar, Group Study, Research, Range Plants (U), Grassland Inventory, Analysis and Planing, Field Course (U), Grassland Ecology (U)

DEPARTMENT OF ANIMAL PHYSIOLOGY
Phone: Chairman: (916) 752-2559 Others: 752-1011

Chairman: V.E. Mendel
Professors: J.M. Boda, R.E. Burger, F.W. Lorenz (Emeritus), A.H. Smith, I.H. Wagman
Associate Professors: H.W. Colvin, Jr., J.M. Horowitz Jr., V.E. Mendel, D.E. Woolley
Assistant Professors: B.A. Horwitz, J.M. Goldberg
Lecturers: M.L. Blair
Teaching Assistants: 11
Readers: 6

Degree Programs: Biological Sciences, Physiology

Undergraduate Degree: B.S.
Undergraduate Degree Requirements: Chemistry (1A-1B-1C, 5, 8A-8B); Mathematics (13, 16A-16B-16C or Physiology 108); Physics (2A-2B-2C); Physiology (100A-100B, 100L, 101L, 110A-110B, 111A-111B). Social science and humanities (including 8 units of English and/or rhetoric): Upper division units which must include either biochemistry and morphology or mathematics, chemistry, physics, and/or engineering. Program should be developed in consultation with major advisor; unrestricted electives

Graduate Degrees: M.S., Ph.D.
Graduate Degree Requirements: Master: Physiology 290 (Seminar) each quarter while in the M.S. program. Physiology 100A,B,L (General Physiology), 7 units, in addition, 11 units of electives Doctoral. Chemistry, Biochemistry, Morphology, Physiology plus 10 courses in selected area of specialization

Graduate and Undergraduate Courses Available for Graduate Credit: General Physiology (U), General Physiology Laboratory (U), Functions of Organ Systems (U), Organ Systems Laboratory (U), Physiology of Growth (U), Physiology of Animal Cells (U), Avian Physiology (U), Avian Physiology Laboratory (U), Mammalian Physiology (U), Mammalian Physiology Laboratory (U), Neurophysiology, Neurophysiology Laboratory, Generla and Comparative Physiol ogy of Reproduction, Physiology and Rhythmicity, Group Study, Comparative Physiology (U), Physiology of Reproduction (U), Physiology of Reproduction Laboratory (U), Physiology of Endocrine Glands (U), Principles of Environmental Physiology (U), Environmental Physiology of Domestic Animals (U), Proseminar (U), Tutoring in Physiology (U), Advanced General Physiology, Advanced General Physiology Laboratory, Advanced Systemic Physiology, Neurophysiology Literature, Physiology of Lactation, Selected Topics in Neuroendocrinology, Seminar, Tutoring in Physiology, Research

DEPARTMENT OF ANIMAL SCIENCE
Phone: Chairman (916) 752-1250 Others: 752-1011

Chairman: G.E. Bradford
Professors: R.L. Baldwin, Jr., G.E. Bradford, F.D. Carroll, P.T. Cupps, W.N. Garrett, I.I. Geschwind, H. Heitman, Jr., R.C. Laben, G.P. Lofgreen, W.C. Rollins, M. Ronning
Associate Professors: C.R. Ashmore, G.A.E. Gall, J. G. Morris, D.W. Robinson
Assistant Professors: G.B. Anderson, J.W. Evans, G.P. Moberg, N.E. Smith
Visiting Lecturers: 2
Teaching Fellows: 6
Research Assistants: 6

Field Stations/Laboratories: Hopland Field Station - Hopland, Sierra Foothill Range Station - Browns Valley, Imperial Valley Field Station, El Centro

Degree Programs: Animal Science, Biochemistry, Genetics, Oceanography, Physiology

Undergraduate Degree: B.S. (Animal Science)
Undergraduate Degree Requirements: 180 quarter units, Lower Division: 1) Biological Sciences 19; 2) Physical Sciences and Mathematics 23. Upper Division: 1) Biochemistry, Genetics, Nutrition and Physiology, one course or course sequence each 24 units; 2) Animal Science specialization - 25 units. Social Sciences and Humanities - 20 units; Unrestricted electives - 69 units

Graduate Degrees: M.S., Ph.D.
Graduate Degree Requirements: Master: 36 units, including 18 in graduate level courses, plus Comprehensive Oral Examination, or 30 units and a Thesis Doctoral: Comprehensive qualifying examination and a Thesis demonstrating ability to do original research

Graduate and Undergraduate Courses Available for Graduate Credit: Several hundred available in a wide variety of biological disciplines

DEPARTMENT OF AVIAN SCIENCES
Phone: Chairman: (916) 752-1300 Others: 752-1300

Chairman: C.R. Grau
Professors: U.K. Abbott, H. Abplanalp, R.E. Burger, R.A. Ernst, C.R. Grau, F.X. Ogassawara, D.W. Peterson, P.N. Vohra, B.W. Wilson, W.O. Wilson
Instructors: L.C. Norris (Emeritus Professor), R. Sawyer (Adjunct Assistant Professor)
Research Assistants: 6
Teaching Assistant: 1
Specialist: D.C. Lowry
Associate Specialist: A.E. Woodard

Degree Programs: Agriculture, Animal Genetics, Animal Physiology, Genetics, Oceanography, Physiology, Poultry Husbandry, Poultry Science

Undergraduate Degree: B.S. (Avian Sciences)
Undergraduate Degree Requirements: 180 units

Graduate Degree: M.S., Ph.D. (Avian Sciences)
Graduate Degree Requirements: Master: Thesis or oral examination, Doctoral: Thesis

Graduate and Undergraduate Courses Available for Graduate Credit: Comparative Avian Microanatomy (U), Avian Physiology (U), Avian Physiology Laboratory (U), Avian Biology (U), Poultry Production (U), Laboratory in Poultry Production (U), Survey of Poultry and Allied Industries (U), Birds, Man, and the Environment (U), Fertility and Hatchability in Birds (U), Environmental Management of Poultry (U), Comparative Nutrition of Avian Species (U), Proseminar in Avian Sciences (U), Directed Group Study (U), Special Study for Advanced Undergraduates (U), Laboratory in Avian Experimental Embryology and Teratology, Seminar, Group Study, Research

DEPARTMENT OF BIOCHEMISTRY AND BIOPHYSICS
Phone: Chairman: 752-3611

Chairman: R.H. Doi
Professors: S. Chaykin, E.E. Conn, R.S. Criddle, R.H. Doi, J.L. Hedrick, L.L. Ingraham, J. Preiss, I.H. Segal, P.K. Stumpf
Associate Professor: G.E. Bruening
Assistant Professors: M.E. Dahmus, M.E. Etzler, A. P. Toliver
Visiting Lecturers: 1
Teaching Fellows: 5
Research Assistants: 3
USPHS NIH Training Grant Trainees: 28

Field Stations/Laboratories: Deciduous Fruit Field Station - San Jose, Hopland Field Station - Hopland, Imperial Valley Field Station - El Centro, Kearney Horticultural Field Station - Parlier, Lindcove Field Station - Exeter, Sierra Foothill Range Field Station - Browns Valley, South Coast Field Station - Santa Ana, Tulelake Field Station - Tulelake, West Side Field Station - Five Points, Richmend Field Station - Berkeley, White Mountain Research Station - Bishop

Degree Program: Biochemistry

Undergraduate Degree: B.S.
Undergraduate Degree Requirements: 50-54 units lower division Chemistry, Mathematics, Physics and Biology, 56 upper division units in Biochemistry, Chemistry, Genetics and restricted electives

Graduate Degrees: M.S., Ph.D.
Graduate Degree Requirements: Master: 36 units (without Thesis), 30 units (with thesis) upper division courses. Comprehensive final for non-thesis degree

Doctoral: 33 units Biochemistry, 5 units electives from Biochemistry, Biological Sciences, Chemistry, Mathematics and Physics. Monthly written Examination, qualifying examination, Thesis, Foreign language advised but not required

Graduate and Undergraduate Courses Available for Graduate Credit: General Biochemistry 101AB (U), General Biochemistry Laboratory (U), Plant Biochemistry (U), Behavior and Analysis of Enzyme Systems (U), Structure and Function of Proteins (U), Biosynthesis of Informational Macromolecules (U), Advanced General Biochemistry 201ABC, Carbohydrates, Nucleic Acids, Biochemical Mechanisms, Physical Biochemistry of Macromolecules, Lipids, Protein Biochemistry, Chemical Modifications Proteins, Principles of Comparative Biochemistry, Kinetics of Biological Systems, Science, the Scientist and Society, Biochemical Aspects of Endocrinology, Selected Topics in Biochemistry, Biochemical Literature, Advanced Research Conference, Current Progress in Biochemistry

DEPARTMENT OF ENTOMOLOGY
Phone: Chairman: 752-0479

Chairman: O.G. Bacon
Professors: O.G. Bacon, R.M. Bohart, N.E. Cary, A. A. Grigarick, H.H. Laidlaw, W.H. Lange, D.L. McLean, T.F. Leigh
Associate Professors: C.L. Judson, G.A.H. McClelland, R.W. Thorp, R.E. Rice, R.K. Washino
Assistant Professors: M.C. Birch, W.R. Cothran, L.E. Ehler, R.W. Bushing
Research Assistants: 10
Teaching Assistant: 1
Postgraduate Researchers: 5

Degree Programs: Agricultural Entomology, Apiculture, Medical Entomology, Nematology, and Systematic Entomology

Undergraduate Degree: B.S.
Undergraduate Degree Requirements: B.S. requires 180 units of formal course work. Preparatory subject matter (73 units) including Biology, Botany, Zoology, Bacteriology, Genetics, Plant Pathology, Plant or Animal Physiology, or Biochemistry, Chemistry, Mathematics, including Statistics, Physics, Earth or Atmospheric Science, plus electives in Biological Science (exclusive of Entomology). Depth subject matter (28 units): Introductory Entomology, Structure and Function in Insects, Insect Physiology, Systematic Entomology, Insect Ecology, Field Taxonomy and Ecology or Insect Classification and another upper division course in entomology which requires a collection of insects. Breadth subject matter (36 units): English and/or rhetoric, Electives in social sciences and humanities plus 43 units of unrestricted electives

Graduate Degrees: M.S., Ph.D.
Graduate Degree Requirements: Master: B.S. in Entomology. Plan I: 30 units of upper division and graduate work of which 12 units must be graduate work in entomology. A Thesis is required. Plan II: 36 units of formal course work of which 18 units in graduate work in entomology. A Comprehensive Oral Examination is required Doctoral: The following requirements must be met: Residence (two years), completion of the minimum preparatory requirements, satisfactory completion of fundamental courses within a chosen field of study, general knowledge of the subject matter of a broad field of study and evidence of scholarly attainments in that field, ability to analyze problems critically as well as to coordinate and correlate data from a number of allied fields, and finally a Dissertation

Graduate and Undergraduate Courses Available for Graduate Credit: Field Entomology (U), Economic Entomology (U), Principles of Agricultural Entomology (U), Biology of Aquatic Insects (U), Chemistry of Insecticides (U), Apiculture (U), Insect Behavior (U), Classification of Immature Insects (U), Insect Vectors of Plant Pathogens (U), Acarology (U), Biological Control (U), Medical Entomology (U), Insect Pest Management (U), Advance Insect Physiology, Principles of Systematic Entomology, Advanced Apiculture, Advanced Medical Entomology, Electronic Principles Related to Entomological Research, Pollination Ecology, Principles and Methods of Entomological Research

DIVISION OF ENVIRONMENTAL STUDIES
Phone: Chairman: (916) 752-2734 Others: 752-3026

Chairman: G.A. Wandesforde-Smith
Professors: C.R. Goldman, W.J. Hamilton III, K.E.F. Watt
Associate Professors: J. McEvoy, L.O. Myrup, G.A. Wandesforde-Smith
Assistant Professors: T.E. Dickinson, T.C. Foin, R. A. Johnston, B. Orlove, T.M. Powell, P.J. Richerson, P.A. Sabatier, S.I. Schwartz
Lecturers: W.B. Goddard, W.A. Harvey
Visiting Lecturers: 2
Research Assistants: 10

Field Stations/Laboratories: Institute of Ecology - Davis

Degree Programs: There is no major or degree program in Environmental Studies. Students can receive a B.A. or B.S. in Environmental Studies through the Individual Major Programs of the several Colleges.

DEPARTMENT OF FOOD SCIENCE AND TECHNOLOGY
Phone: Chairman: 751-1465 Others: 752-1465

Chairman: B.S. Schweigert
Professors: R.A. Bernhard, W.D. Brown, E.B. Collins, W.L. Dunkley, R.E. Feeney, W.F. Jennings, G.L. Marsh (Emeritus), M. Mazelis, M.W. Miller, E.M. Mrak (Emeritus), T.A. Nickerson, H.S. Olcott, R.M. Pangborn, H.J. Phaff, B.S. Schweigert, L.M. Smith, C. Sterling, C.F. Steward, A.L. Tappel, R.H. Vaughn, J.R. Whitaker
Associate Professors: D.W. Gruenwedel, J.M. Henderson, M.J. Lewis
Assistant Professors: E.V. Crisan, R.L. Merson, G. F. Russell, R.A. Shleser
Instructors: A.W. Brant, J.C. Bruhn, S.J. Leonard, B. S. Luh, F.H. Winter, G.K. York
Research Assistants: 30
Teaching Assistants: 2

Degree Program: Food Science

Undergraduate Degree: B.S.
Undergraduate Degree Requirements: 180 quarter units total - 54 quarter units upper division. i.e. Preparatory Biology and Microbiology - 10 quarter units, Chemistry and Biochemistry - 17 quarter units

Graduate Degree: M.S.
Graduate Degree Requirements: Master: Plan I: requires (a) completion of a total of 30 units of upper division (100 series) and graduate (200) courses and (b) a Thesis based on original laboratory research approved by a Thesis Committee of the faculty. Plan II: requires completion of 36 units of upper division and graduate courses and, in lieu of the Thesis, the student must prepare a research report and pass a comprehensive final examination in the major field. Of the 36 units, at least 18 units must be strictly graduate courses in the major field.

Graduate and Undergraduate Courses Available for Graduate Credit: Principles of Food Composition and Preparation (U), Principles of Food Composition and Preparation (U), Principles of Food Composition Preparation Laboratory (U), Principles of Food Composition and Preparation Laboratory (U), Biochemistry and Food

Science (U), Malting and Brewing Technology (U), Physical and Chemical Methods for Food Analysis (U), Food Microbiology, Food Microbiology Laboratory (U), Microbiological Analysis of Foods (U), Industrial Fermentations, 106L - Food and Industrial Microbiology Laboratory (U), Analysis of Variance as Applied to Sensory Evaluation Problems (U), Food Plant Sanitation (U), Food Plant Sanitation (U), Physical Principles in Food Processing (U), Physical Principles in Food Processing (U), Introduction to Food Processing (U), Comparative Aspects of Food Habits and Culture (U), Structure of Food Materials (U), Principles of Dairy Products (U), Packaging Processed Foods (U), Thermal Processing of Foods (U), Proteins: Functional Activities and Interactions, Chemistry of the Food Lipids, Macromolecular Gels, Mycology of Food and Food Products, Mycology of Food and Food Products Laboratory, Isolation and Characterization of Trace Volatiles, Isolation and Identification of Trace Volatiles, Seminar, Group Study, Research

DEPARTMENT OF GENETICS
(No reply received)

DEPARTMENT OF NEMATOLOGY
Phone: Chairman: 752-2121 Others: 752-1403

Chairman: A.R. Maggenti
Professors: B.F. Lownsbery, D.J. Raski, W.H. Hart, A.R. Maggenti, D.R. Viglierchio
Research Assistants: 2

Degree Programs: None - except through Entomology or Plant Pathology

Graduate Degree Requirements: Master: According to requirements of Departments of Entomology or Plant Pathology Doctoral: Same as Master's

DEPARTMENT OF NUTRITION
Phone: Chairman: 752-2089 Others: 752-6650

Chairman: W.C. Weir
Professors: W.C. Weir, F.W. Hill, L.S. Hurley, F.J. Zeman
Assistant Professors: N.L. Canolty, A.J. Clifford, J.S. Stern, R.B. Rucker, J. Vermeersch
Instructors: J. Prophet
Visiting Lecturers: 2
Teaching Fellows: 4
Research Assistants: 9
Readers: 2

Field Stations/Laboratories: Everson Hall - Walker Annex

Degree Programs: Nutrition, Nutrition with a Dietetics Emphasis, Community Nutrition, Dietetics, Food Service Management

Undergraduate Degree: B.S.
Undergraduate Degree Requirements: G.P.A. requirement, residence, requirement, plus individual requirements of different majors

Graduate Degrees: M.S., Ph.D.
Graduate Degree Requirements: Master: 30 units plus Thesis, 36 units plus Comprehensive Examination Doctoral: Satisfy course requirements specified by the Graduate Group in Nutrition - an interdepartmental group from Colleges of Agricultural and Environmental Sciences, Medicine and Veterinary Medicine. Qualifying Examination, residence, Research Thesis

Graduate and Undergraduate Courses Available for Graduate Credit: Advanced Protein and Amino Acid Nutrition, Advanced Animal Energetics and Energy Metabolism, Advanced Vitamin and Mineral Nutrition, Advanced Diet Therapy, Advanced Field Work in Community Nutrition, Concepts of Animal Nutrition, Single Carbon Metabolism in Nutrition, Nutrition and Development, Control of Food Intake, Ruminant Digestion and Metabolism, Natural Toxicants in Foods, Nutrition and Hormonal Control of Animal Metabolic Function

DEPARTMENT OF PLANT PATHOLOGY
Phone: ChairmanL 752-0301

Chairman: T. Kosuge
Professors: E.E. Butler, R.N, Campbell, J.E. DeVay, W.H. English, R.G. Grogan, W.B. Hewitt (Emeritus), L.D. Leach (Emeritus), G. Nyland, J.M. Ogawa, T. A. Shalla, R.J. Shepherd, E.E. Wilson (Emeritus)
Associate Professors: C.I. Kado, R.K. Webster
Research Assistants: 16

Degree Programs: Plant Sciences, Plant Pathology

Undergraduate Degree: B.S. (Plant Sciences)
Undergraduate Degree Requirements: 180 units, Preparatory Subject Matter 36, Depth Subject Matter 34, Brea Breadth Subject Matter 20, Restricted Electives 45, Unrestricted Electives 45

Graduate Degrees: M.S., Ph.D. (Plant Pathology)
Graduate Degree Requirements: Master: Plan I: 30 units of upper division and graduate courses and a Thesis. At least 12 of the 30 units in graduate courses in major field Plan II: 36 units of upper division and graduate courses and a Comprehensive Final Examination; 18 of 36 units in graduate courses in major field Doctoral: Qualifying examination administered by committee of five. Dissertation committee to guide candidate in research and pass upon merits of Dissertation

Graduate and Undergraduate Courses Available for Graduate Credit: Diseases of Crop Plants, Ecology of Plant Pathogens and Epidemiology of Plant Disease, Physiology and Biochemistry of Host-Pathogen Interaction, Genetics of Plant Pathogens, Pathogenic Fungi, Plant Virology, Plant Bacteriology, Advanced Plant Virology, Advanced Plant Pathology, Seminar, Seminar in Host-Parasite Physiology, Seminar in Plant Virology, Seminar in Mycology, Research

DEPARTMENT OF POMOLOGY
(No reply received)

DEPARTMENT OF SOILS AND PLANT NUTRITION
Phone: Chairman: (916) 752-1406

Chairman: L.D. Whittig
Professors: F.E. Broadbent, C.C. Delwiche, E. Epstein, H.M. Reisenauer, V.V. Rendig, P.R. Stout, L.D. Whittig
Associate Professors: R.G. Burau, D.N. Munns
Assistant Professors: D.E. Rolston, M.J. Singer
Instructors: E.L. Begg, A.L. Brown, G.L. Huntington, H.O. Walker
Research Assistants: 6

Field Stations/Laboratories: Deciduous Fruit Field Station-Santa Clara County, Hopland Field Station - Mendocino County, Imperial Valley Field Station - Imperial County, Kearney Horticultural Field Station - Fresno, Lindcove Field Station - Tulare County, Sierra Foothill Range Field Station - Yuba County, South Coast Field Station - Orange County, Tulelake Field Station - Siskiyou County, West Side Field Station - Fresno County

Degree Programs: The Department contributes toward the B.S. in Soil and Water Science, and many M.S. and Ph.D. programs in other Departments.

Undergraduate Degree: B.S. (Soil and Water Science)
Undergraduate Degree Requirements: 180 quarter units of which 57 must be of upper division courses, 37 of soils or related courses, and 25 of physical and biological sciences and mathematics

Graduate and Undergraduate Courses Available for Graduate Credit: Soil Physics, Soil-Plant Interrelationships, Soil Microbiology, Soil Mineralogy, Physical Chemistry of Soils, Special Topics in Soil Science, Current Literature in Plant Nutrition, Group Study, Research, Principles of Plant Nutrition

DEPARTMENT OF VEGETABLE CROPS
Phone: Chairman: (916) 752-1741 Others: 752-0516

Chairman: O.A. Lorenz
Professors: W.J. Flocker, J.F. Harrington, O.A. Lorenz, J.M. Lyons, L.L. Morris, H.K. Pratt, L. Rappaport, C.M. Rick, P.G. Smith, A.R. Spurr, M. Yamaguchi
Associate Professors: B.D. Webster, J.E. Welch, S.F. Yang
Assistant Professors: K.N. Paulson, M.A. Stevens
Instructors: F.D. Howard (Lecturer)
Specialists: J.C. Bishop, H. Timm, M.B. Zahara, F.W. Zink
Teaching Fellows: 2
Research Assistants: 12
Visiting Research Geneticist: M. Holle

Field Stations/Laboratories: Mann Laboratory - Davis

Degree Programs: Vegetable Crops, Soils, Plant Physiology and Genetics, Soil Science, Botany, Agricultural Chemicials

Undergraduate Degree: B.S. (Plant Sciences)
Undergraduate Degree Requirements: 180 quarter units includes 48 in preparatory subject matter, 89 in depth subject matter, and 43 in electives

Graduate Degrees: M.S., Ph.D.
Graduate Degree Requirements: Master: 36 units of upper division and graduate courses in which at least 18 units must be strictly graduate courses in the major subject Doctoral: Depends on the Ph.D. program. All are different.

Graduate and Undergraduate Courses Available for Graduate Credit: Principles of Vegetable Crops (U), Major Vegetable Crops (U), Systematic Olericulture (U), Seed Physiology and Production (U), Vegetables as World Food Crops (U), Field Study of Vegetable Industry (U), Directed Group Study (U), Special Study for Advanced Undergraduates (U), Postharvest Physiology of Vegetables, Vegetable Genetics and Improvement, Vegetable Physiology, Seminar, Seminar in Postharvest Physiology, Group Study, Research

DEPARTMENT OF VITICULTURE AND ENOLOGY
Phone: Chairman: 752-1946 Others: 752-0380

Chairman: A. D. Webb
Professors: M.A. Amerine, H.W. Berg, J.A. Cook, J.F. Guymon, L.A. Lider, K.E. Nelson, H.P. Olmo, V.L. Singleton, R.J. Weaver, A.D. Webb, A.J. Winkler (Emeritus)
Associate Professors: R.E. Kunkee, W.M. Kliewer, C.S. Ough
Assistant Professor: A.C. Noble
Research Assistants: 9
Specialist: 1

Field Stations/Laboratories: Oakville Field Station - Oakville

Degree Programs: Agricultural Chemistry, Bacteriology, Food Science, Genetics, Microbiology, Plant Sciences, Viticulture and Enology

Undergraduate Degree: B.S. (Food Sciences, Plant Sciences)
Undergraduate Degree Requirements: 180 units Food Science, Fermentation Science, Plant Science

Graduate Degree: M.S., Ph.D.
Graduate Degree Requirements: Master: 20 units plus Thesis or Examination, Food Science, Agricultural Chemistry, Horticulture Doctoral: Qualifying Examination and Dissertation, Genetics, Plant Physiology, Agricultural Chemistry, Bacteriology

Graduate and Undergraduate Courses Available for Graduate Credit: Systematic Viticulture including Fruit Maturation and Handling (U), General Viticulture A (U), General Viticulture B (U), Analysis of Musts and Wines (U), Wine Production (U), Wine Types and Sensory Evaluation (U), Wine Processing (U), Disstillation Principles and Brandy (U), Plant Hormones and Regulators, Plant Hormones and Regulators Laboratory, Microbiology of Wine Production, Plant Phenolics, Seminar, Group Study, Research

College of Science and Letters
Dean: L.J. Andrews

DEPARTMENT OF BACTERIOLOGY
Phone: Chairman: 752-0274 Others: 756-0261

Chairman: H.J. Phaff
Professors: R.E. Hungate (Emeritus), J.L. Ingraham, A.G. Marr, H.J. Phaff, D. Pratt, M.P. Starr
Associate Professors: D.M. Reynolds
Assistant Professors: P. Baumann, S. Kustu, J. Maning, M.L. Wheelis
Instructor: M.J. Bowes
Research Assistants: 1
Teaching Assistants: 10

Degree Programs: Bacteriology, Microbiology

Undergraduate Degrees: B.A., B.S.
Undergraduate Degree Requirements: B.A. - 47 lower division courses including Biology, Chemistry, Bacteriology, Mathematics, Physics, 37 upper division courses including Genetics, Biochemistry and Bacteriology. B.S. 51 lower division units and 63 upper division units

Graduate Degrees: M.S., Ph.D.
Graduate Degree Requirements: Master: No information supplied Doctoral: No information supplies

Graduate and Undergraduate Courses Available for Graduate Credit: Microbiology and Society, General Bacteriology (U), Bacterial Diversity: Morphology, Systematics, Habitats (U), Bacterial Diversity: Metabolism, Physiology (U), Bacterial Diversity: Ultrastructure and Morphogenesis (U), Bacterial Physiology and Genetics (U), Bacterial Physiology and Genetics (U), Bacterial Physiology Laboratory (U), Eukaryote Protistology: Yeasts (U), Eukaryote Protistology: Yeasts Laboratory (U), Bacterial Diversity, Ecology, and Systematics, Bacterial Physiology

DEPARTMENT OF BOTANY
Phone: Chairman: 752-0617

Chairman: E.M. Gilfford, Jr.
Professors: F.T. Addicott, F.M. Ashton, D.I. Axelrod, P.A. Castelfranco, H.B. Currier, E.M. Gifford, Jr., H.J. Ketellapper, N.J. Lang, J. Major, C.R. Stocking, J.M. Tucker, G.L. Webster, K. Wells, A.S. Crafts (Emeritus, T.E. Weier (Emeritus)
Associate Professors: M.G. Barbour, B.A. Bonner, D.W. Kyhos
Assistant Professors: R.H. Falk, T.M. Murphy, R.F. Norris, T.L. Rose, R.M. Thornton
Instructors: D.E. Bayer, O.A. Leonard (Lecturers)
Teaching Fellows: 14
Research Assistants: 12

Degree Programs: Botany, Ecology, Genetics, Plant Physiology, Comparative Biochemistry

Undergraduate Degrees: B.A., B.S. (Botany)
Undergraduate Degree Requirements: (B.S.) General

Bacteriology (5), General Zoology (6), Biology (5), Physics (9), Statistics (4), Biochemistry (6), Plant Anatomy (5), Plant Taxonomy (r), Plant Physiology (6), Plant Morphology (5), Genetics (6), Mycology (5) and Phycology (5) or Plant Ecology (4) and 5 upper sivision unit units in botany or related science

Graduate Degrees: M.S., Ph.D.
Graduate Degree Requirements: Master: Plan I: In addition to the undergraduate degree requirements; 30 units of upper division and graduate courses and a Thesis are required. At least 21 of the 30 units must be in graduate courses in the major field Plan II: 36 units and a Comprehensive Final Examination Doctoral: In addition to the undergraduate degree requirements; additional course work beyond the Master's degree requirements will average about 18 units, satisfaction of the foreign language requirement in French, German, or Russian, submission of an acceptable Dissertation based on original and independent investigation

Graduate and Undergraduate Courses Available for Graduate Credit: Survey of Plant Communities of California (U), Weed Control (U), Introductory Plant Physiology Laboratory (U), General Cytology (U), Paleobotany (U), Plant Geography (U), Evolution of Plant Ecosystems (U), Plant Microtechnique (U), Biological Evaluation of Herbicides (U), Ecological Theory, Analysis of a Selected Ecosystem, The Changing Biosphere, Advanced Plant Physiology, Cell Physiology-Protoplasmatics, Plant Cell Metabolism, Physiology of Herbicidal Action, Light and Plant Growth, Advanced Morphology of Vascular Plants, Concept and Measurement of Plant Community, Plant Morphogenesis, Special Topics in Plant Physiology, Biological Electron Microscopy, Pollination Ecology, Principles of Plant Taxonomy, Experimental Plant Taxonomy, Plant Autecology, Plant Synecology

DEPARTMENT OF ZOOLOGY
Phone: Chairman: (916) 752-1272 Others: 752-1273

Chairman: R.J. Baskin
Professors: R.J. Baskin, M. Hildebrand, E.W. Jameson Jr., M.A. Miller (Emeritus), L.E. Rosenberg (Emeritus), R.L. Rudd, G.W. Salt, H.T. spieth (Emeritus), K.E.F. Watt
Associate Professors: P.B. Armstrong, D.W. Deamer, R.D. Grey, S.L. Wolfe
Assistant Professors: J.H. Crowe, W.E. Jacobus, B. Mulloney, D.W. Phillips, A.M. Shapiro, J.A. Stamps, V.D. Vacquier
Teaching Assistants: 30

Field Stations/Laboratories: Bodega Marine Laboratory - Bodega Bay

Degree Program: Zoology

Undergraduate Degrees: A.B., B.S.
Undergraduate Degree Requirements: B.S. Lower Division General Introductory courses in Principles of Biology, Zoology, Botany, Bacteriology, Inorganic Chemistry, Analytic Geometry and Calculus, Statistics, and Physics Upper Division Courses in Organic Chemistry, Biochemistry, Genetics, Evolution, plus 30 quarter units of electives in the Biological Sciences. B.A. Similar to B.S. program, with these exceptions: in lower division requirements, Botany and Bacteriology are not required and there is a choice of courses in Mathematics In upper division work, Biochemistry is not required, and the number of elective units in biological sciences is 27

Graduate Degrees: M.A., Ph.D.
Graduate Degree Requirements: Master: Thesis Plan: Thirty units (quarter system) of upper division or graduate work plus a Thesis Examination Plan: A minimum of 18 units of graduate level courses plus an additional 18 units of upper division or graduate work. In addition, this program requires satisfactory completion of a written examination in General Zoology and in two elective specialized fields of Zoology Doctoral: The doctoral program has no set course requirements. It does require passage of the comprehensive general examination in Zoology (taken at the end of the first year), completion of a course at a biological field station, demonstration of a working knowledge of one foreign language, usually French or German, completion of an Oral qualifying Examination taken no later than after the seventh quarter in residence, and submission of a satisfactory Dissertation to the graduate faculty.

Graduate and Undergraduate Courses Available for Graduate Credit: Embryology (U), Developmental Biology: Cellular Basis of Morphogenesis (U), Developmental Biology: Cell Differentiation (U), Phylogenetic Analysis of Vertebrate Structure (U), Microanatomy (U), Protozoology (U), Invertebrate Zoology (U), Invertebrate Physiological Ecology (U), Principles of Animal Resource Management (U), Principles of Environmental Science (U), Cell Biology (U), Animal Ecology (U), Biology of Cold-Blooded Vertebrates (U), Biology of Cold-Blooded Vertebrates (U), Mammalogy (U), Ornithology (U), Invertebrate Physiology (U), Zoogeography (U), Animal Phylogeny and Evolution (U), Evolution of Ecological Systems (U), Behavior of Animals (U), Advanced Cell Biology (U), Cellular Inheritance (U), Ecological Theory, Analysis of a Selected Ecosystem, The Changing Biosphere, Biomathematics, Global and Regional Modeling, Developmental Biology, Experimental Animal Ecology, Muscle Physiology, Seminar in Advanced Cytology, Seminar in Cell Biology, Seminar in Animal Behavior, Seminar in Developmental, Seminar in Invertebrate Zoology, Seminar in Animal Ecology, Seminar in Systematic Zoology and Evolution

School of Vetinary Medicine
Dean: W.R. Pritchard

DEPARTMENT OF ANATOMY
Phone: Chairman: (916) 752-1177 Others: 752-1011

Chairman: L.J. Faulkin, Jr.
Professors: L.M. Julian, R.L. Kitchell, W.S. Tyler
Associate Professors: L.J. Faulkin, B.L. Hart
Assistant Professor: C.L. Lohse

Degree Program: Anatomy

Graduate Degrees: M.S., Ph.D.
Graduate Degree Requirements: Master: Plan I - Student is required to take a minimum 30 units of upper division and graduate course work and submit a Thesis Plan II - Student is required to take a minimum 36 units of upper division and graduate course work, of which at least 18 units must be in major subject. The candidate must pass a Comprehensive Final Examination in anatomy and physiology Doctoral: Minimum residence requirement of 2 years; a course in Biochemistry; Student is required to specialize by completing a minimum of 10 quarter units in one of the areas of specialization; qualifying oral examination in anatomy, physiology and his specialty area; submittal of a Thesis

Graduate and Undergraduate Courses Available for Graduate Credit: Numerous courses offered in several departments and schools

DEPARTMENT OF PHYSIOLOGICAL SCIENCES
Phone: Chairman: 752-1373

Chairman: A.L. Black
Professors: A.L. Black, V.W. Burns, R.A. Freedland, S.A. Peoples, R.E. Smith (Emeritus)
Associate Professors: J.R. Gillespie, A.A. Heusner,

J.G. Morris, H.R. Parker, Q.R. Rogers
Assistant Professors: G. Conzelman, Jr., D.L. Curry, S.N. Giri, R.J. Hansen, R.M. Joy
Research Assistants: 5

Degree Programs: Biochemistry, Comparative Biochemistry, Nutrition, Comparative Pathology, Veterinary Science

Graduate Degrees: M.S., Ph.D.
Graduate Degree Requirements: Master: Several course requirements and Thesis Doctoral: Several course requirements and Thesis and Qualifying Exam

Graduate and Undergraduate Courses Available for Graduate Credit: Physiological Chemistry, Bioenvironmental Consequences of Nuclear Technology, Special Study (U), Cell Physiology: Biophysical Aspects, Intermediary Metabolism of Animals, Comparative Pharmacology, Use of Isotopes as Tracers in Biological Research, Drug Metabolism, Applied and Clinical Pharmacology, Pharmacogenetics, Medical Toxicology, Pharmacology Literature, Physiochemical Relationships in Drug Action, Comparative Bioenergetics, Experimental Physiology, Seminar, Group Study, Research

DEPARTMENT OF VETERINARY MICROBIOLOGY
Phone: Chairman: (916) 752-1399 Others: 752-1400

Chairman: E.L. Biberstein
Professors: N.F. Baker, E.L. Biberstein, M.M.J. Lavoipierre, D.G. McKercher, J.W. Osebold, M. Shifrine, Y.C. Zee
Assistant Professors: A.M. Buchanan, D.C. Hirsh, J.H. Theis, G.F. Slonka
Visiting Lecturers: 3

Degree Programs: Comparative Pathology, Microbiology, Parasitology (Comparative Pathology, Zoology, Entomology)

Graduate Degrees: M.S., Ph.D.
Graduate Degree Requirements: Master: Vary with program Doctoral: Vary with program

Graduate and Undergraduate Courses Available for Graduate Credit: Immunology (U), Medical Microbiology (U), Animal Virology (U), Parasitology (U), Advanced Immunology, Microbiological Diagnosis, Fundamentals of Immunology (U), Immunology Laboratory (U), Medical Bacteriology and Mycology (U), Biology of Animal Viruses (U), Animal Virology Laboratory (U), Medical Microbiology Laboratory (U), Advanced Immunology, Seminar in Immunology, Seminar in Animal Virology, Seminar in Infectious Diseases, Seminar in Parasitology, Clinical Microbiology, Microbiological Diagnosis, Group Study, Research

DEPARTMENT OF VETERINARY PATHOLOGY
Phone: Chairman: (196) 752-1385

Chairman: D.L. Dungworth
Professors: D.R. Cordy, D.L. Dungworth, P.C. Kennedy, J.E. Moulton
Associate Professors: T.G. Kawakami (Adjunct) B.I. Osburn
Assistant Professors: D.H. Gribble, M.I. Johnson, R.R. Pool, L.T. Pulley, L.W. Schwarts (Adjunct)
Graduate Students: 15

Degree Program: Comparative Pathology

Graduate Degrees: M.S., Ph.D.
Graduate Degree Requirements: Master: D.V.M. or equivalent to enter. Course work and pass oral examination. Doctoral: D.V.M. or equivalent to enter. Course work and research project with Dissertation

School of Medicine
Dean: C.J. Tupper

DEPARTMENT OF BEHAVIORAL BIOLOGY
Phone: Chairman: (916) 752-3361

Chairman: L.F. Chapman
Professor: L.F. Chapman
Associate Professors: V.J. Polidora, G. Mitchell
Assistant Professors: R. Scobey, A. Gabor
Instructor: E. Sassenrath
Visiting Lecturers: 3
Research Assistants: 6

Field Stations/Laboratories: California Center for Primate Biology - Sacramento

Degree Program: Physiology

Graduate Degrees: M.S., Ph.D.
Graduate Degree Requirements: Varies. Doctoral and Master's Degree Programs are "Graduate-Group" basis. Department participates in graduate groups in Physiology, Biomedical Engineering, Child Development and "Individualized" Graduate Degrees

Graduate and Undergraduate Courses Available for Graduate Credit: Group Study Seminars, Research, Neuroendocrine aspects of Stress, Three Dimensional Modeling of the Brain

DEPARTMENT OF BIOLOGICAL CHEMISTRY
Phone: Chairman (916) 752-2929 Others: 752-2925

Chairman: E.G. Krebs
Professor: E.G. Krebs
Associate Professors: R.R. Traut, D.A. Walsh
Assistant Professors: W.F. Benisek, J.W.B. Hershey, J.T. Stull, Jr. (Adjunct), F.A. Troy
Visiting Professors: 2

Degree Programs: Biochemistry, Biophysics, Microbiology

Graduate Degrees: M.S., Ph.D.
Graduate Degree Requirements: Master: 15 units Doctoral: 30 units

Graduate and Undergraduate Courses Available for Graduate Credit: Principles of Comparative Biochemistry, Molecular Biology Laboratory, Current Topics in Biological Chemistry, Current Topics in Protein Synthesis, Group Study, Research

DEPARTMENT OF CLINICAL PATHOLOGY
Phone: Chairman: 752-0153

Chairman: J.J. Kaneko
Professors: D.E. Jasper, J.J. Kaneko, O.W. Schalm
Associate Professor: N.C. Jain
Assistant Professor: K.S. Keeton
Lecturer: E.J. Carroll

Degree Program: Pathology

Graduate Degrees: M.S., Ph.D.
Graduate Degree Requirements: Master: 34 units of course work and Examination or Thesis Doctoral: Thesis

Graduate and Undergraduate Courses Available for Graduate Credit: Hematology, Cytology

DEPARTMENT OF HUMAN ANATOMY
Phone: Chairman: (916) 752-3211 Others: 752-2100

Chairman: A. Barry
Professors: R.H. Brownson, R.L. Hunter, E. Gardner, R. O'Rahilly
Associate Professors: S. Meizel, D.B. Wilson
Assistant Professors: K.J. Chacko, V.K. Vijayan
Research Assistants: 3

Graduate Degrees: Offered through Anatomy Group

DEPARTMENT OF HUMAN PHYSIOLOGY
Phone: Chairman (916) 752-1241 Others: 752-3230

Chairman: E.M. Renkin
Professors: A.G.L. Hsieh, E.M. Renkin
Associate Professors: L. Rabinowitz, R. E. Smith
Assistant Professors: A.M. Goldner, S.D. Gray, J.F. Green, T.C. Lee
Research Assistants: 15
Postgraduate Research Physiologists: 2

Degree Program: Human Physiology is a member of the Physiology Graduate Group on campus. The Physiology Group offers M.S. and Ph.D. degrees

Graduate and Undergraduate Courses Available for Graduate Credit: Information Systems: Design and Analysis of Computerized Information Systems, Advanced General Physiology, Surgical Approaches to Physiology, Renal Physiology, Physiology of the Body Fluids, Physiological Systems Analysis, Respiratory Physiology, Cardiovascular Physiology, Peripheral Circulation, Cardiovascular Research Conference, Pulmonary Function Evaluation

DEPARTMENT OF MEDICAL MICROBIOLOGY
(No reply received)

UNIVERSITY OF CALIFORNIA
IRVINE
Irvine, California 92664 Phone: (714) 833-5011
Dean of Graduate Division: J. Schultz

School of Biological Sciences
Dean: H.A. Schneiderman

DEPARTMENT OF DEVELOPMENTAL AND CELL BIOLOGY
Phone: Chairman: (714) 833-6928

Chairman: M.W. Berns
Professors: E.A. Ball, P.F. Hall, R.K. Josephson, H.M. Lenhoff, H.A. Schniederman, G.C. Stephens
Associate Professors: J. Arditti, M.W. Berns, R.D. Campbell, D.E. Fosket, S.M. Krassner
Assistant Professors: K. Baldwin, H. Bode, P.J. Bryant, S.V. Bryant, M.R. Goldsmith, G. Greenhouse, B. Hamkalo, H. Koopowitz, J.L. Osborne, S.H. White
Teaching Fellows: 6
Research Assistants: 25
Lecturers: 2
Associate: 1

Field Stations/Laboratories: Faculty Research Facilities - Campus and Jamboree Road , Irvine

Degree Programs: Developmental and Cell Biology

Undergraduate Degree: B.S.
Undergraduate Degree Requirements: Biological Science 101 A-B-C-D-E-F-G, Biological Science 101L A-B-C-D-E-F (Core Curriculum), Minimum of 3 Satellite Courses, Physics 3A-B-C, 3LA-B-C or 5A-B-C, Mathematics 2A-B-C Calculus or 2A-B and 7 (Statistics), General Chemistry lA-B-C, lLA-B-C, Organic Chemistry 5lA-B-C, 5lLA-B-C, Humanities lA-B-C (This also covers the English requirements for premedical students)

Graduate Degrees: M.S., Ph.D.
Graduate Degree Requirements: Master: 1. Application for Advancement to Candidacy 2. Course requirements 3. Residence requirement 4. Language Examination 5. Comprehensive Examination 6. Thesis Doctoral: Form I - Nomination for qualifying examination Form II Report on qualifying examination Form III - Report of final examination (Languages - Examinations - Dissertation - Residence requirements)

DEPARTMENT OF MOLECULAR BIOLOGY AND BIOCHEMISTRY
Phone: Chairman: 833-6035 Others: 833-5011

Chairman: R.C. Warner
Professors: E.R. Arquilla, G.A. Granger, C.S. McLaughlin, K. Moldave, R.C. Warner, D.L. Wulff
Associate Professors: K.H. Ibsen, W.M. Stanley,Jr., K.K. Tewari, C.A. Woolfolk
Assistant Professors: S.M. Arfin, J.L. Clark, C.N. Gordon, B.A. Hamkalo, S.J. Hayes, J.E. Manning, D. Pisziewicz, B.M. Sutherland, J. Sutherland, E.K. Wagner
Teaching Fellows: 9
Research Assistants: 12

Degree Programs: The Ph.D. degree is offered in areas included under Molecular Biology and Biochemistry. A separate organized program leading to the Ph.D. in Biophysical Chemistry and Biophysics is operated jointly with the Chemistry Department.

Undergraduate Degrees: Undergraduate degrees and requirements are supervised by the School of Biological Sciences. The separate departments exist only for maintaining separate graduate programs.

Graduate Degrees: M.S., Ph.D.
Graduate Degree Requirements: Master: Plan A: 7 courses and a thesis Plan B: 9 courses and a comprehensive examination Students are not usually accepted for the Masters degree only Doctoral: 1) Completion of graduate core courses and other courses specified for the individual student 2) Comprehensive qaulifying examination 3) Advancement to candidacy examination, chiefly on research progress 4) Thesis and thesis examination. Normal time for completion of degree is four years

Graduate and Undergraduate Courses Available for Graduate Credit: Core: Biochemical Methodology, Biochemistry A and B, Molecular Genetics Other: Biosynthesis of Nucleic Acids, Medical Microbiology, Regulatory Mechanisms and Metabolic Deseases, Biomolecular Structure, Biopolymers in Solution, Biochemical Dynamics, Immunology (U), General Microbiology (U), Biophysical Chemistry (U), Virology (U), Radiation Biology (U), Biogenesis of Cell Organelles (U), Electron Microscopy (U), Structure of the Eukaryotic Chromosome (U)

DEPARTMENT OF POPULATION AND ENVIRONMENTAL BIOLOGY
Phone: Chairman: (714) 833-6007 Others: 833-5011

Chairman: P.R. Atsatt (Acting)
Professors: P.S. Dixon, R.E. MacMillen
Associate Professors: P.R. Atsatt, K.E. Justice, M.M. Littler, P.W. Rundel
Assistant Professors: L.W. Carpenter, G.L. Hunt
Visiting Assistant Professors: 2
Instructor: 2

Degree Programs: Animal Behavior, Animal Physiology, Applied Ecology, Biological Sciences, Ecology, Ethology, Population and Environmental Biology, Marine Ecology, Aquatic Productivity, Terrestrial Ecology, Physiological Ecology

Undergraduate Degree: B.S.
Undergraduate Degree Requirements: Not Stated

Graduate Degrees: M.S., Ph.D.
Graduate Degree Requirements: Master: Maintain a "B" average. Complete 7 upper division or graduate courses and present a thesis, or complete a minimum of 9 upper division and graduate courses with a comprehensive final examination Doctoral: Satisfactory completion of Preliminary Examination (Fall and Spring) during 1st year. Maintain grade point average (B or higher) in all course work. Complete all breadth requirements and doctoral candidacy by end of 3rd year. and approval of thesis project. Each Ph.D. student is required to participate in teaching activities. A

written evaluation of teaching performance becomes a part of the student's permanent record.

Graduate and Undergraduate Courses Available for Graduate Credit: Processes of Evolution (U), Physiological Animal Ecology (U), Behavioral Ecology (U), Aquatic Productivity, Ecology of Terrestrial Communities (U), Evolutionary Ecology, Physiological Animal Ecology, Behavioral Ecology, Productivity Ecology, Ecology and Evolution of Terrestrial Communities

DEPARTMENT OF PSYCHOBIOLOGY
Phone: Chairman: 833-6025

Chairman: J.L. McGaugh
Professors: C.W. Cotman, R.K. Josephson, J.L. McGaugh, E.P. Noble, R.F. Thompson, M. Verzeano, N.M. Weinberger, R.E. Whalen
Associate Professors: R.A. Giolli, G.S. Lynch, A. Starr
Assistant Professors: H.P. Killackey, J.C. Waymire, P.I. Yahr
Instructors: S.A. Deadwyler, P.E. Gold
Visiting Professors: 3
Teaching Fellows: 29
Research Assistants: 7
Postdoctorals: 6

Degree Program: Psychobiology

Graduate Degree: Ph.D.
Graduate Degree Requirements: Doctoral: Preliminary Examination, Candidacy Examination, 2 foreign languages, Thesis

Graduate and Undergraduate Courses Available for Graduate Credit: Graduate Core Program (1 year), An integrated sequence in nucrobiology and behavioral biology. Other courses include: Advanced Analysis of Learning and Memory, Hormones and Behavior, Comparative Neurology, Neurochemistry, Biochemical Neuropharmacology, Attentive Processes, Integrative Neurobiology, Neuroanatomy, Fine Neuroanatomy, Brain and Behavior, Neurological Psychobiology, Animal Behavior, Muscle and Other Effectors. Seminars are offered in all of the above and an Colloquium in Psychobiology which is required.

College of Medicine
Dean: S. van der Noort

DEPARTMENT OF MEDICAL MICROBIOLOGY
Phone: Chairman: (714) 833-5261 Others: 833-5011

Chairman: P.S. Sypherd
Professors: H.S. Moyed, P.S. Sypherd
Associate Professors: D.D. Cunningham, G.W. Hatfield
Assistant Professors: M.T. Kaplan, D.T. Kingsbury, D.J. Raidt
Instructors: 10

Degree Programs: Medical Microbiology, Microbiology

Graduate Degrees: M.S., Ph.D.
Graduate Degree Requirements: Master: Seven graduate quarter courses of which five must be non-research courses. Presentation of a thesis. Doctoral: Sufficient coursework to result in competance in mathematics, chemistry (including biochemistry) and specialized areas of microbiology, as judged by the faculty through examination. Presentation of a thesis on work completed under the direction of a faculty member. Areas include medical microbiology and immunology, enzyme structure and regulation, tumor biology, cell growth, regulation, microbial development and virology

Graduate and Undergraduate Courses Available for Graduate Credit: Medical Virology, Medical Immunology, Tumor Biology, Cell Culture Biology, Microbial Metabolism and Physiology, Regulatory Mechanisms and and Disease, Genetics of Bacteria and Viruses

UNIVERSITY OF CALIFORNIA
LOS ANGELES
405 Hilgard Avenue Phone: (213) 825-4321
Los Angeles, California 90024
Dean of Graduate Studies: R.S. Kinsman

College of Letters and Science
Dean: K. Trueblood

DEPARTMENT OF BACTERIOLOGY
Phone: Chairman: (213) 825-1478

Chairman: S.C. Rittenberg
Professors: C.F. Fox, J. Lascelles, R.J. Martinez, M.J. Pickett, S.C. Rittenberg, W.R. Romig, E.E. Sercarz
Associate Professors: R.J. Collier, F.A. Eiserling, G.J. Jann, D.R. Krieg, D.P. Nierlich
Teaching Fellows: 11
Research Assistants: 27
Postdoctorals: 10

Degree Programs: Microbial Physiology, Microbial Metabolism, Microbial and Phage Genetics, Immunology, Medical Microbiology, Cell and Virus Structure and Assembly

Undergraduate Degree: B.A., (Bacteriology)
Undergraduate Degree Requirements: Inorganic, organic and biochemistry, 7 quarters; physics, 3 quarters; calculus, 3 quarters; biology, 2 quarters; plus 10 upper division courses from a core curriculum in Microbiology

Graduate Degrees: M.A., Ph.D. (Microbiology)
Graduate Degree Requirements: Master: A thesis based on laboratory research; plus a minimum course requirement set by the Graduate Division. This includes a total of 9 upper division or graduate level courses of which at least 5 must be in the 200 series Doctoral: There is neither a standardized curriculum nor a course or unit requirement for the Ph.D. degree. Each program is designed to accomplish the educational goals of the individual student. In general, a program will include a major area of specialization within the department and either a specialized or interdisciplinary minor.

Graduate and Undergraduate Courses Available for Graduate Credit: All of our upper division courses are available for graduate credit.

DEPARTMENT OF BIOLOGY
Phone: Chairman: (213) 825-4373 Others: 825-6155

Chairman: A.A. Barber
Professors: L. Barajas, A.A. Barber, G. Bartholomew, J. Belkin, J. Biale, J. Cascarano, M. Cody, N.E. Eollias, F. Crescitelli, W.T. Ebersold, R. Eckert, E. Edney, F. Engelmann, J.H. Fessler, M. Gordon, A. Grinnell, K. Hamner, T. Howell, T. James, J.L. Kavanau, G. Laties, F.H. Lewis, M.H. Mathias, L. Muscatine, E. Olson, R. Orkand, B.O. Phinney, P. Vaughn, B. Walker, S.G. Wildman, K. Nagy (Adjunct)
Associate Professors: G. Brunk, D. Chapman, A. MacInnis, P. Nobel, J. O'Connor, W. Salser, L. Simpson, J.P. Thornber
Assistant Professors: W.R. Clark, A.E. Gill, G. Gorman, H. Hespenheide, J. Merriam, J.G. Morin, R. Vance
Teaching Assistants: 72

Degree Programs: Biology

Undergraduate Degree: B.A.
Undergraduate Degree Requirements: Mathematics 3-A-

B-C; Physics 6A-B-C; Chemistry 1A-B-C, 21, 22, 24; Biology 1A-1B. Eleven upper division courses in Biology including five from our core list.

Graduate Degrees: M.A., Ph.D.
Graduate Degree Requirements: Master: 9 graduate courses, comprehensive examinations, Thesis
Doctoral: No formal course or languages requirements qualifying examination, Thesis

Graduate and Undergraduate Courses Available for Graduate Credit: All upper division courses are available for graduate credit.

School of Medicine

Dean: S. Mellinkoff

DEPARTMENT OF ANATOMY
(No reply received)

DEPARTMENT OF BIOLOGICAL CHEMISTRY
Phone: Chairman: (213) 825-6494 Phone: 825-6545

Chairman: E.L. Smith
Professors: R.B. Alfin-Slater, S. Eiduson, R.M. Fink, A.N. Glazer, I. Harary, D.R. Howton, R.W. McKee, J.F. Mead, J.R. Nyc, J.G. Pierce, G.J. Popjak, S. Roberts, E.L. Smith, M.E. Swenseid, I. Zabin, S. Zamenhof
Associate Professors: R.J. DeLange, A.J. Fulco, D.G. Glitz, H.R. Herschman, D.S. Sigman, J.E. Snoke, P.J. Zamenhof
Assistant Professors: J.E. Ayling, J.P. Blass, J. Emond, B.D. Howard, W.T. Wickner
Research Assistants: 2
Postgraduate Research Biological Chemists: 16

Degree Programs: Biological Chemistry

Graduate Degrees: M.S., Ph.D.
Graduate Degree Requirements: Master: courses: Bioorganic Catalysis, Proteins and Nucleic Acids, Biological Catalysis, Physical Chemistry of Biological Macromolecules, Cellular Metabolism, Neuclic Acid and Protein Biosynthesis, foreign language and final oral, Thesis Doctoral: Bioorganic Catalysis, Proteins and Nucleic Acids, Biological Catalysis, Physical Chemistry of Biological Macromolecules, Cellular Metabolism, Nucleic Acid and Protein Biosynthesis, Foreign language, oral examination, written examination, Thesis

Graduate and Undergraduate Courses Available for Graduate Credit: Biochemistry (U), Biochemical Methods (U), Biological Chemistry (U), Biological Chemistry (U), Biological Chemistry, Bioorganic Catalysis, Proteins and Nucleic Acids, Biological Catalysis, Physical Chemistry of Biological Macromolecules, Cellular Metabolism, Nucleic Acid and Protein Biosynthesis, Advanced Topics in Biochemistry, Biochemical Preparations, Neurobiochemistry, Seminar in Experimental Neurochemistry, Seminar in Biological Chemistry, The Biochemistry of Lipids, Seminar in the Biochemistry of Proteins, Seminar in Advanced Lipid Biochemistry, Seminar in the Biochemistry of Nucleic Acids, Seminar in the Biochemistry of Differentiation, The Biochemistry of Differentiation

DEPARTMENT OF BIOMATHEMATICS
Phone: Chairman: 825-5349 Others: 825-5547

Chairman: W.J. Dixon
Professors: W.J. Dixon, O.J. Dunn, D.J. Jenden, F.J. Massey, W.S. Yamamoto, M.R. Mickey (Research)
Associate Professors: A.A. Afifi, V.A. Clark, R.I. Jennrich, V.K. Murthy (Adjunct), C.M. Newton
Assistant Professors: P. Bright (Research), M.S. Estes (Adjunct), A.B. Forsythe (Lecturer), J.W. Frane (Adjunct), H. Frey (Adjunct), K. Lange, R.J. Sclabassi (Lecturer), M.A. Spence
Research Assistants: 1
Visiting Research: 3

Field Stations/Laboratories: Health Science Computing Facility - Los Angeles

Degree Program: Biomathematics

Graduate Degrees: M.S., Ph.D.
Graduate Degree Requirements: Master: Required courses: Deterministic Models in Biology, Time Series Analysis, Stochastic Models in Biology, Qualifying Examination, Thesis, Final Examination
Doctoral: Required courses as for M.S with additional courses in mathematics and computer sicence. Teaching preceptorship. Four qualifying examinations, Thesis, Oral Defense

Graduate and Undergraduate Courses Available for Graduate Credit: Elements of Biomathematics (U), Special Studies in Biomathematics (U), Deterministic Models in Biology, Time Series Analysis, Stochastic Models in Biology, Modeling of Cellular Systems, Modeling in Genetic Analysis, Modeling and Analytical Methods in the Neurosciences, Introduction to Biomedical Computation, Biomedical Laboratory Computing, Biomedical Laboratory Computing, Advanced Biomedical Computation, Computer and Biomathematical Applications in Radiological Sciences, Topics in Biological Control Theory, Biomathematics

DEPARTMENT OF MICROBIOLOGY AND IMMUNOLOGY
Phone: Chairman: (213) 825-6568 Others: 825-5661

Chairman: J.L. Fahey
Professors: M.A. Baluda, J.L. Fahey, S. Gorbach, W.H. Hildemann, D.H. Howard, D. Imagawa, J.N. Miller, A.F. Rasmussen, M.I. Sellers, J.G. Stevens, M. Voge, F. Wettstein, T.H. Work, S. Zamenhof
Associate Professors: S. Froman, D.L. McVickar, D. P. Nayak, J.A. Turner, H.E. Weimer
Assistant Professors: R.F. Ashman, B. Bonavida, P.E. Byfield, S. Golub, R.H. Stevens, R. Wall, W.D. Winters
Visiting Lecturers: 19
Teaching Fellows: 10
Research Assistants: 7
Postdoctoral Fellows: 35

Degree Programs: Microbiology and Immunology

Graduate Degrees: M.S. (Exceptional Cases), Ph.D.
Graduate Degree Requirements: Master: Students entering graduate study will be expected to pursue the Ph.D. Degree. Exceptional cases may be considered to pursue the M.S. Degree Doctoral: Written and Oral Qualifying Examinations, Teaching experience, Thesis

Graduate and Undergraduate Courses Available for Graduate Credit: Introduction to Laboratory Research in Immunology, Advanced Immunology Workshop, Seminar in Immunogenetics, Advanced Immunology Co-Seminar, Advanced Immunology, Immunology Forum, Tumor Immunology, Seminar in Immunobiology of Cancer, Molecular Immunology, Immunochemistry, Microbiology and Immunology, Medical Mycology, Seminar in Microbiology and Immunology, Seminar in Medical Virology, Seminar in Medical Parasitology, Seminar in Medical Mycology, Seminar in Host-Parasite Relationships, Animal Virology, Seminar in Viral Oncology, Co-Seminar in Animal Virology, Seminar in Current Topics in Molecular Biology, Directed Individual Study or Research, Preparation for Comprehensive Examination for the M.S. Degree or the Qualifying Examination for the Ph.D. in Microbiology and Immunology, Research for and Preparation of the Doctoral Dissertation in Microbiology and Immunology

DEPARTMENT OF PHYSIOLOGY

Phone: Chairman: (213) 825-5667 Others: 825-5531

Emeritus Professors: V.E. Hall, J. Field
Professors: R.W. Adey, A.J. Brady, M.A.B. Brazier, J.S. Buchwald, J.M. Diamond, G. Eisenmann, J. Field, M.I. Grossman, A.D. Grinnell, S. Hagiwara, V.E. Hall, G.A. Langer, D.B. Lindsley, W.F.H.M. Mommaerts, W.D. Odell, G. Ross, D. Simmons, R.R. Sonnenschein, B.M. Wenzel, F.N. White, E.M. Wright
Associate Professors: C. Baxter (Adjunct), M. Chase, S. Ciani, R.S. Eisenberg, D. Junge, E.H. Rubinstein, J.McD. Tormey, D.O. Walter (Adjunct), C.D. Woody
Assistant Professors: H.L. Batsel (Adjunct), E. Homsher, J. Metzger, G. Szabo, B. Whipp
Lecturers: 1
Teaching Fellows: 3

Field Stations/Laboratories: Los Angeles County Heart Laboratory - Los Angeles

Degree Program: Physiology

Graduate Degrees: M.S., Ph.D.
Graduate Degree Requirements: Master: Students entering graduate study in the Department of Physiology will be expected to pursue the Ph.D. degree Doctoral: Required courses : Physical Chemistry, Neuromuscular and Cardiovascular Physiology, Renal, Respiratory and Gastrointestinal Physiology, Functional Ultrastructure of Cells, Electrical Properties of Cells, Written and Oral Comprehensive, Thesis

Graduate and Undergraduate Courses Available for Graduate Credit: Neuromuscular and Cardiovascular Physiology (U), Renal, Respiratory and Gastrointestinal Physiology (U), Basic Neurology (U), Transport Phenomena in Membranes, Critical Topics in Physiology, Electrical Properties of Cells, Excitation and Contraction, Renal Respiratory and Gastrointestinal Physiology, Physiology of the Nervous System, Physiology of Nerve Cells, Permeability of Biological Membranes to Ions, Cardiovascular Physiology, Neurophysiology, Mathematical Modeling of Physiological Systems, Basic Foundation in Endocrinology, Basic Foundations in Endocrinology, Biological and Artificial Membranes, Bilayer Membranes, Theoretical Problems in Membrane Permeation, Seminar in Physiology, Directed Individual Study or Research,

UNIVERSITY OF CALIFORNIA RIVERSIDE

4045 Canyon Crest Drive Phone: (714) 878-1012
Riverside, California 92502
Dean of Graduate Studies: H. Johnson

College of Agricultural and Biological Sciences

Dean: M. Dugger, Jr.

DEPARTMENT OF BIOLOGY
Phone: Chairman: (714) 787-5902 Others: 787-5904

Chairman: I.W. Sherman
Professors: W.L. Belser, C.R. Bovell, L.H. Carpelan, K.W. Cooper, F.N. David, W.M. Dugger, L.G. Erickson, G.W. Gillett, W.W. Mayhew, J.A. Moore, I.M. Newell, D.R. Parker, E.T. Pengelley, T. Prout, R. Ruibal, I.W. Sherman, H.H. Shorey, W.W. Thomson, I.P Ting, F.C. Vasek
Associate Professors: V.H. Goodman, R.L. Heath, E. C. Pauling, V.H. Shoemaker
Assistant Professors: W.R. Allen, R.D. Farley, R.W. Gill, H.B. Johnson, G.E. Jones, P. Painter, E.G. Platzer, G.K. Snyder, W.J. Vanderwoude

Field Stations/Laboratories: Citrus Experiment Station - Riverside

Degree Programs: Biology, Botany, Ecology, Genetics, Microbiology, Molecular Biophysics, Parasitology, Physiology, Zoology

Undergraduate Degrees: A.B., B.S.
Undergraduate Degree Requirements: A.B. 36 upper division units Biology, 8 units Chemistry, 6 units Physics Foreign language. B.S. 36 upper division units Biology, plus 16 units in field of major, 8 units chemistry, 3 units physics, 2 units mathematics. No language.

Graduate Degrees: M.S., Ph.D.
Graduate Degree Requirements: Master: Thesis degree. 36 units of which at least 24 must be graduate courses in major. Thesis. Oral final. Non-thesis degree. 36 units of which at least 18 must be graduate courses in field of major, including 9 of directed research. Written comprehensive Doctoral: General requirements for graduate division plus one language, thesis, written qualifying and final oral

Graduate and Undergraduate Courses Available for Graduate Credit: Cellular and Developmental Biology, Biochemical Genetics, Experimental Cytology, Population Genetics, Advanced Population Biology, Special Topics in Vertebrate Biology, Special Topics in Physiological Ecology, Special Topics in Physiological Ecology laboratory, Microbial Genetics, The Procaryotic Cell, Microbial Metabolism, Speciation, Morphology and Composition of Fruits, Plant Metabolism, Photobiology, Advanced Plant Ecology, Seminar in Cell and Developmental Biology, Seminar in Invertebrate Zoology, Seminar in Genetics, Seminar in Vertebrate Zoology, Seminar in Evolution, Seminar in Ecology, Seminar in Plant Systematics, Seminar in Plant Ecology, Seminar in Physiology, Seminar in Evolution, Field Course in Evolutionary Biology, Inderdisciplinary Studies: Tropical Biology - An Ecological Approach,

DEPARTMENT OF BIOCHEMISTRY
Phone: Chairman (714) 787-4227 Others: 787-4229

Chairman: R.T. Wedding
Professors: R.T. Wedding, L.C. Erickson, J.B. Mudd, E.A. Noltmann, A.W. Norman, L. Ordin, L.M. Shannon, W.B. Sinclair (Emeritus)
Associate Professors: R.L. Heath, D.D. Holten, N.G. Pon, B.R. Reid
Assistant Professors: M.F. Dunn, C.E. Furlong, R.W. Olsen, J.A. Traugh
Instructors/Lecturers: I.L. Eaks, G.M. Hathaway
Teaching Fellows: 3
Research Assistants: 23
Post Graduate Research Biochemist: 6

Degree Programs: Biochemistry

Undergraduate Degrees: B.A., B.S.
Undergraduate Degree Requirements: Biochemistry, Chemistry, Biology

Graduate Degrees: M.S., Ph.D.
Graduate Degree Requirements: Master: The department offers two plans for the Master's degree (Plan I; Thesis , Plan II: Comprehensive Examination). Both plans require completion of at least 36 course units; for Plan I a maximum of 12 units may be for thesis research Doctoral: A student's course requirements are determined in consultation with a three member advisory committee appointed for each student. An individual course program is suggested for an adequate course preparation is a prerequisite part of the training program the department encourages early involvement of the student in research directed toward his dissertation.

Graduate and Undergraduate Courses Available for Graduate Credit: Elementary Biochemistry Laboratory (U), General Biochemistry (U), Intermediate Biochemistry Laboratory (U), Chemical Dynamics of Biochemical Processes (U), Biochemistry of Macromolecules, Molecular Biology and Biochemical Regulation, Advanced Biochemistry Laboratory, Advanced Topics in Biochemistry, Special Topics in Biochemistry, Graduate Seminar in Biochemistry, Colloquium in Biochemistry, General Seminar in Biochemistry

DEPARTMENT OF ENTOMOLOGY
Phone: Chairman (714) 787-5806

Chairman: F.G. Gunther
Professors: L.D. Anderson, M.M. Barnes, L.R. Brown, G.E. Carman, P.H. DeBach, W. Ebeling, W.H. Ewart, T.R. Fukuto, G.P. Georghiou, F.A. Gunther, I.M. Hall, R.N. Jefferson, E.F. Legner, R.B. March, M.S. Mulla, I.M. Newell, H.T. Reynolds, L.A. Riehl, H. H. Shorey, V.M. Stern, A.M. Boyce, C.P. Clausen, R.C. Dickson, S.E. Flanders, C.A. Fleschner
Associate Professors: R.D. Goeden, T.A. Miller
Assistant Professors: G. Kennedy, R.F. Luck, J.D. Pinto, V. Sevacherian, S.N. Thompson
Lecturers: B.R. Bartlett, L.K. Gaston, D. Gonzalez, L. R. Jeppson, D.L. Lindgren, J.A. McMurty, E.R. Oatman, R.G. Strong
Research Assistants: 50

Field Stations/Laboratories: Citrus Research Center and Agricultural Experimental Station - Riverside, U.C. Meloland Research Station - El Centro, South Coast Research Station - Irvine, U.C. Reedley Field Station - Reedley, Five Points Field Station - Five Points

Degree Programs: Entomology

Undergraduate Degrees: B.A., B.S.
Undergraduate Degree Requirements: 36 units including 6 units Entomology, 5 units Chemistry, 3 units Physics

Graduate Degrees: M.S., Ph.D.
Graduate Degree Requirements: Master: The department offers two plans for the Master's degree (Plan I; Thesis Plan II: Comprehensive Examination). Both plans require completion of at least 36 course units; for Plan I a maximum of 12 units may be for thesis research Doctoral: A student's course requirements are determined in consultation with a three member advisory committee appointed for each student. An individual course program is suggested for each student involving courses in Entomology and courses in a subsidiary field of study. Although an adequate course preparation is a prerequisite part of the training program the department encourages early involvement in the student in research directed toward his dissertation.

Graduate and Undergraduate Courses Available for Graduate Credit: General Entomology (U), Insect Morphology (U), Field Entomology (U), Insect Physiology (U), Insect Systematics (U), Insect Systematics Laboratory (U), Economic Entomology (U), Medical Entomology (U), Laboratory in Medical Entomology (U), Insect Ecology (U), Chemistry and Toxicology of Insecticides (U), Laboratory in Chemistry and Toxicology of Insecticides (U), Introduction to Biological Control (U), Introduction to Biological Control Laboratory (U), Taxonomy of Immature Insects (U), Insect Behavior (U), Insect Biochemistry, Arthropod Resistance to Toxic Agents, Acarology, Advanced Economic Entomology, Insect Vectors of Plant Pathogens, Insect Population Ecology, Advanced Insect Toxicology, Advanced Biological Control, Insect Pathology, Analysis of Pesticide Chemicals, Analysis of Pesticide Chemicals, Seminar in Entomology, Seminar in Economic Entomology, Seminar in Insect Behavior, Seminar in Insect Toxicology, Seminar in Biological Control, Seminar in Medical Entomology, Seminar in Systematic Entomology, Seminar in Insect Pathology

DEPARTMENT OF NEMATOLOGY
Phone: Chairman: 787-4431

Chairman: S.D. Van Gundy
Professors: C.E. Castro, S.A. Sher, I.J. Thomason, S.D. Van Gundy, J.D. Radewald (Adjunct)
Associate Professors: R. Mankau (Research)
Assistant Professors: H. Ferris, M.V. McKenry, E.G. Platzer (Research)
Research Assistants: 4

Degree Programs: Graduate degrees offered in collaboration with Department of Plant Pathology

DEPARTMENT OF PLANT PATHOLOGY
Phone: Chairman: 787-4116 Others: 787-4117

Chairman: L.G. Weathers
Professors: J.G. Bald (Emeritus), S. Bartnicki-Garcia, E.C. Calavan, P.R. Desjardins, J.W. Eckert, R.M. Endo, D.C. Erwin , D.E. Munnecke, P.H. Taso, L. G. Weathers, G.A. Zentmyer, L.J. Klotz, (Emeritus), J.M. Wallace (Emeritus)
Associate Professors: N.T. Keen, J.S. Semancik, J.J. Sims
Assistant Professors: D.J. Gumpf, J.V. Leary, J.A. Menge
Instructors: E.F. Darley, W.O. Dawson, M.J. Kolbezen, A.O. Paulus, H. Schneider
Research Assistants: 16
Postdoctorals: 11

Degree Programs: Nematology, Air Pollution

Graduate Degrees: M.S., Ph.D.
Graduate Degree Requirements: Master: Requires 36 units of upper division and graduate courses, 18 in the 200 series excluding graduate research for thesis or dissertation, and a comprehensive final examination of the major subject Doctoral: Requires completion of the following: Principles of Plant Pathology (lecture and lab), Advanced Plant Pathology, Seminar in Plant Pathology and other subjects in research area as directed by the Graduate Affairs Committee. Successful completion of a comprehensive written and oral exam in this field. A dissertation and a final oral examination

Graduate and Undergraduate Courses Available for Graduate Credit: Mycology (U), Introduction to Plant Pathology, (U), Introduction to Plant Pathology Laboratory (U), Genetics of Fungi (U), Genetics of Fungi Laboratory (U), General Virology (U), General Virology Laboratory (U), Special Studies (U), Research for Undergraduates (U), Principles of Plant Pathology, Principles of Plant Pathology Laboratory, Phytopathogenic Fungi, Control of Plant Diseases, Physiology of Fungi, Physiology of Plant Disease, Advanced Plant Pathology, Field Plant Pathology, The Chemistry of Secondary Metabolites, Seminar in Plant Pathology, Current Research in Plant Pathology, Directed Research, Research for Thesis or Dissertation

DEPARTMENT OF PLANT SCIENCES
Phone: Chairman: (714) 787-4413 Others: 787-1012

Chairman: R.K. Soost
Professors: W.P. Bitters, J.W. Cameron, C.W. Coggins, Jr., T.W. Embleton, B. L. Johnson, W.W. Jones, L.S. Jordan, J. Kumamoto, C.K. Labanauskas, L.N. Lewis, L.F. Lippert, T. Murashige, W. Reuther, R.K. Soost, W.B. Storey, O.C. Taylor, D.M. Yermanos, V.B. Youngner
Associate Professors: M.R. Kaufmann, R.W. Scora, O. E. Smith, R.E. Young
Assistant Professors: A.E. Hall, R.T. Leonard, J.A. Swader
Teaching Fellows: 1
Research Assistants: 12
Visiting Research Scientist: 2
Postdoctoral: 9

Field Stations/Laboratories: U.C. Statewide;

Degree Programs: The Department participates in interdepartmental Ph.D. Programs in Botany, Genetics and Plant Physiology

Undergraduate Degree: B.S. (Plant Sciences)
Undergraduate Degree Requirements: College Mathematics, 1 year each of General Chemistry, Organic Chemistry, Physics and Biology, 2 courses in Plant Physiology (including laboratory), at least 1 course in the following: Genetics, General Botany, Soils or Environmental Science and 1 course in either Plant Pathology, or Entomology plus 12 units of upper division Plant Science. Biochemistry and Statistics recommended.

Graduate Degrees: M.S., Ph.D.
Graduate Degree Requirements: Master: Courses required for the undergraduate major are considered deficiencies. A total of 36 units of course work and research are selected by the student and a guidance committee. Doctoral: Course work requirements are established for each individual student based on his interest and degree objective

Graduate and Undergraduate Courses Available for Graduate Credit: Citrus Fruits and Their Relatives, Citriculture, Tropical and Subtropical Horticulture, Ecology of Crop Plants, Physiology of Crop Plants, Principles of Plant Propagation, Postharvest Physiology of Fruits and Vegetables, Grasses and Grasslands, Plants in the Urban Environment, Science/Technical Writing and Research Repetitive Methods, Design and Interest of Biological Experiments, Principles of Plant Breeding, Instrumental Methods in Plant Physiology, Quantitative Genetics, Physiology and Biochemistry of Herbicides, Physiology and Nutrition of Tree Crops, Methods in Mineral Element Analyses in Plants, Cytogenetics of Crop Plants, Chemotaxonomy, Morphology and Composition of Fruits, Special Topics in Plant Physiology, Plant Cell, Tissue and Organ Culture, Chemistry of Secondary Metabolites, Water Relations and Transport Phenomena, Plant Morphogenesis, Mineral Metabolism of Plants, Photobiology, Plant Growth Substances, Ion Transport in Plants, Advanced Plant Ecology, Seminar in Plant Science, Seminar in Plant Physiology, Directed Studies, Directed Research, Research for Thesis or Dissertation

DEPARTMENT OF STATISTICS
Phone: Chairman: (714) 787-3774

Chairman: F.N. David
Professors: F.N. David, M.J. Garber
Assistant Professors: R.J. Beaver, D.V. Gokhale, R.F. Green, N.S. Johnson, C.A. Robertson, V. Sevacherian, D.J. Strauss, L. Zahn

Degree Programs: Applies Statistics, Population Biology, Statistics, Systems Ecology

Undergraduate Degree: B.S.
Undergraduate Degree Requirements: Not Stated

Graduate Degree: Ph.D.
Graduate Degree Requirements: Not Stated

Graduate and Undergraduate Courses Available for Graduate Credit: Not Stated

UNIVERSITY OF CALIFORNIA
SAN DIEGO
P.O. Box 109 Phone: (714) 453-2000
La Jolla, California 92037
Dean of Graduate Studies: W.A. Nierenberg
Dean of Colleges: R.H. Pierce

DEPARTMENT OF BIOLOGY
Phone: Chairman: 453-2000 Ext. 1722

Chairman: H. Stern
Professors: H. Stern, W.L. Butler, R.W. Dutton, M.E. Friedkin, E.P. Geiduschek, C. Grobstein, D.R. Helinski, J.J. Holland, H.A. Itano, D.L. Lindsley, W. D. McElroy, S.E. Mills, P.D. Saltman, G. Sato, S.J. Singer, S.S. Varon
Associate Professors: S. Brody, M.J. Chrispells, M.H. Green, M. Hayashi, W.F. Loomis, Jr., P.J. Russell, M.I. Simon, M.E. Soule, C.J. Wills, N-Xoung, J. Yguerabide
Assistant Professors: W.C. Borwn, J. Elovson, R.A. Firtel, H. Friedman, P.A.G. Fortes, M. Gilpin, D.K. Hartline, S. Howell, M.N. Nesbitt, R. Pinon, P.A. Price, M. Saier, I. Scheffler, A.I. Selverston, D.W. Smith, N.C. Spitzer
Instructors: B. Mulloney
Visiting Lectures: 5

Degree Program: Biology

Undergraduate Degree: B.A.
Undergraduate Degree Requirementa: All courses necessary for each college in Biology (Revelle, Muir, Third and Fourth Colleges). Each college in Biology has their required core courses for majors

Graduate Degrees: M.S., Ph.D.
Graduate Degree Requirements: Master: 36 credit units other than research. 1 quarter residency from the date of application. Thesis or comprehensive examination Doctoral: Our department does not have a language requirement for the Ph.D.

Graduate and Undergraduate Courses Available for Graduate Credit: Too many courses to list. A General Catalog may be obtained from UCSD Bookstore.

DEPARTMENT OF THE SCRIPPS INSTITUTION OF OCEANOGRAPHY
Phone: Chairman: 453-2116

Professors: E.H. Ahlstrom (Adjunct), A.A. Benson, J.T. Enright, D. Epel, H.T. Hammel, F.T. Haxo, R.R. Hessler, J.D. Isaacs, R. Lasker (Adjunct), R.A. Lewin, J.A. McGowan, R.H. Rosenblatt, B.E. Volcani
Associate Professors: W.F. Heiligenberg, N.D. Holland J.R. Hunter (Associate Adjunct), M.M. Mullin, W.A. Newman
Assistant Professors: P.K. Dayton, K.H. Nealson, G.N. Somero

Degree Programs: Marine Biology, Oceanography

Graduate Degrees: M.S., Ph.D.
Graduate Degree Requirements: Master: Because of limited facilities, the Department does not encourage students who wish to proceed only to the M.S. If circumstances warrant, the degree is normally offered by taking a comprehensive examination after completing course work established by the Department. Doctoral: Appropriate course work, Oral Qualifying Examination, submission of a Dissertation and Final Examination

Graduate and Undergraduate Courses Available for Graduate Credit: Biological Oceanography: Processes and Events, Laboratory in Biological Oceanography, Oceanic Zoogeography, Animal Behavior, Experimental Laboratory in Animal Behavior, Topics in Community Ecology, Applied Statistics, Deep-Sea Biology, Problems in Biological Oceanography, Special Topics in Biological Oceanography, Marine Communities and Environments, Laboratory in Marine Organisms, Environmental Physiology and Biochemistry of Marine Organisms, Physiology of Marine Vertebrates, Laboratory in Physiology, Isotope Tracer Techniques and Related Topics in Physiology, Cell Physiology of Marine Organisms, Marine and Comparative Biochemistry, Methods in the Comparative Biochemistry of Marine Organisms, Cellular Structure and Biochemical Function, Microbial Ecology, Experimental Microbiology, Microbial Ecology, Experimental Microbiology, Microbial Metabolism, Marine Plants, Physiology of Marine Algae, Developmental Biology of Marine Organisms, Laboratory in Developmental Biology, Advanced Invertebrate Zoology, Biology of Fishes, Seminar in Advanced Icthyology, Special Topics in Marine Biology, Marine Biology Seminar, Special Studies in Marine Sciences, Research

UNIVERSITY OF CALIFORNIA SAN FRANCISCO

San Francisco, California 94122 Phone: (415) 666-9000
Dean of Graduate Studies: H. Harper

School of Medicine

Dean: J. Krevens

DEPARTMENT OF ANATOMY
Phone: Chairman: 666-1861

Chairman: H.J. Ralston III
Professors: C.W. Asling, J. deGroot, B.C. Garoutte, L.E. Glass, T. Hayashida, A.L. Jones, M.R. Miller, I.W. Monie, H.J. Ralston, W.O. Reinhardt, S.L. Wissig, S. Wolff
Associate Professors: A.N. Contopoulos, J.J. Elias, J.A. Long, W.R. Mehler (Adjunct)
Assistant Professors: R. Armstrong, P. Calarco, R.L. Hamilton, D.M. McDonald, D. Riley, M.T.T. Wong-Riley
Visiting Lecturers: 16
Teaching Fellows: 6
Research Assistants: 2
Post-doctoral Fellows: 3

Degree Program: Anatomy

Graduate Degrees: M.A., Ph.D.
Graduate Degree Requirements: Master: Plan I: 30 units and Thesis, 3 quarters residence requirement Plan II: 36 units and comprehensive examination, 3 quarters residence requirement Doctoral: Dissertation and 6 quarters residence

Graduate and Undergraduate Courses Available for Graduate Credit: Human Gross Anatomy, Histology and Cell Biology, Neuroanatomy, Developmental Biology, Reproductive Biology, Endocrinology

DEPARTMENT OF BIOCHEMISTRY AND BIOPHYSICS
Phone: Chairman: (415) 666-4324

Chairman: W.J. Rutter
Hertzstein Professor of Biochemistry, W.J. Rutter
Professors: I. Edelman, J.J. Eiler, C. Epstein, D. Greenberg (Emeritus), G. Grodsky, H. Harper, E. Kun, H. Landahl, C.H. Li, B.J. McCarthy, M.F. Morales, W.J. Rutter, T.P. Singer (Adjunct), W. Stoeckenius, H. Tarver, G.M. Tomkins, S. Watanabe, J.T. Yang
Associate Professors: R.A. Fineberg, H. Goodman, H.M. Martinez, L.M. Peller, J. Ramachandran, D.V. Santi.
Assistant Professors: A.R. Cooke, M.J. Dennis, C. Guthrie, R. Kelly, D. Michaeli, J. Supdich, J.A. Watson
Teaching Fellows: 10
Research Assistants: 3

Degree Programs: Biochemistry, Biomathematics and Biophysics

Graduate Degree: Ph.D.
Graduate Degree Requirements: Doctoral: Completion of course requirements. Passing Qualifying Examination. Writing a thesis on original research

Graduate and Undergraduate Courses Available for Graduate Credit: General Biochemistry, Physical Biochemistry, Computation in Biochemistry and Physiology, Introduction to Biomathematics, Biochemistry of Connective Tissues, Current Topics, Biological Transport Systems, Bio-Organic and Enzyme Mechanisms, Preparation for Research in Biochemistry and Biophysics, Seminar, Student Seminar, Research, Special Study

MICROBIOLOGY DEPARTMENT
Phone: Chairman: (415) 666-1211

Chairman: E. Jawetz
Professors: J.M. Bishop, J.W. Goodman, E. Jawetz, L. Levintow, R. Painter, R.S. Speck
Associate Professors: H.W. Boyer, C. Halde, W.D. Linscott, W.E. Levinson, H.E. Varmus
Assistant Professors: P. Coffino
Postdoctoral Fellows: 14

Degree Programs: Microbiology, Virology, Immunology, Microbial Genetics

Graduate Degrees: M.S., Ph.D.
Graduate Degree Requirements: Master: Research, Thesis, Oral Examination Doctoral: M.S., Thesis, Preliminary Examination and Oral Final

Graduate and Undergraduate Courses Available for Graduate Credit: Not Stated

DEPARTMENT OF PATHOLOGY
Phone: Chairman: (415) 666-1701 Others: 666-9000

Chairman: O.N. Rambo (Acting)
Professors: D.G. McKay, O.N. Rambo, W. Rosenau, J.S. Wellington, R.R. Wright
Associate Professors: C.G. Biava, S.H. Choy, D.S. Friend, B. Gondos, E.B. King, W. Margaretten, N.S. McNutt
Assistant Professors: D.F. Bainton, L.M. Friedlander, M.L. Goldberg, E.L. Howes, C.K. Montgomery, S. L. Nielsen, K.H. Woodruff
Visiting Lecturers: 1-2
Teaching Fellows: 2-3
Clinical Faculty: 30

Degree Programs: Experimental Pathology

Graduate Degrees: M.S., Ph.D.
Graduate Degree Requirements: Master: Research Thesis and Oral Examination Doctoral: M.S. research Thesis and preliminary and Final Oral Examination

Graduate and Undergraduate Courses Available for Graduate Credit: General Pathology, Systemic Pathology, Theoretical and Applied Pathology, Seminar, Research, Thesis, Dissertation

DEPARTMENT OF PHYSIOLOGY
Phone: Chairman: 666-1869 Others: 666-1751

Chairman: W.F. Ganong
Morris Herzstein Professor of Physiology: J.H. Comroe
Professors: L.L. Bennett, D.J. Botts, K.T. Brown, J.C. Coleridge, J.H. Comroe, R.P. Durbin, J.M. Felts, W.F. Ganong, R.B. Jaffe, R.H. Kellogg, B. Libet, H.M. Patt, S.S. Rothman, A.M. Rudolph, N.C. Staub
Associate Professors: H. Coleridge, R. Mitchell, R. Steinberg
Assistant Professors: J.W. Adelson (Adjunct), M.F. Dallman, M.J. Dennis, H. Fields, R. Gallo, R.J. Goerke, E. Mayeri, M.M. Merzenich, A.H. Mines, I Reid, S.R. Sampson, J.V. Tyberg, J.A. Williams
Lecturers: C.R. Bainton, D.B. Gordon, D.L. Jewett, N. Panagiotis, G. Searle, D.B. Stone, D.N. Whitten, S. Winston
Visiting Lecturers: 2
Associate Specialists: 4

Degree Programs: Physiology, Endocrinology

Graduate Degrees: M.A., Ph.D.
Graduate Degree Requirements: Master: 30 units of which at least 12 must be graduate physiology. Thesis

Doctoral: Specific courses, language, teaching experience, qualifying examination, Thesis

Graduate and Undergraduate Courses Available for Graduate Credit: Organ System Physiology, Endocrinology, Integrative and Nutritive Systems, Mammalian Physiology 120, Mammalian Physiology 125, Research in Physiology, Research in Endocrinology, Electronic Instrumentation, History of Research on the Circulation, Supervised Study in Physiology, Tutorial in Physiology, Physiology of Vision, Cardiovascular Research Seminar, Cardiopulmonary Research Seminar, Seminar: Topics in Physiology, Science, Health and the Public, Advanced Neurophysiology, Nelroendocrinology, Physiology of the Auditory, Vestibular and Other Sensory Systems, Physiology Seminar, Advanced Cardiovascular Renal and Pulmonary Physiology, Endocrinology Seminar, Graduate Student Seminar, General and Cellular Physiology, Research in Physiology, Research in Endocrinology, Practicum in Teaching Physiology, Scientific Writing, Group Practice in the Art of Lecturing

UNIVERSITY OF CALIFORNIA SANTA BARBARA

Santa Barbara, California 93106 Phone: (805) 961-2311
Dean of Graduate Studies: R. Collins
Dean of College: B. Rickborn

DEPARTMENT OF BIOLOGICAL SCIENCES
Phone: (805) 961-2415

Chairman: G. Taborsky
Professors: J.A. Carbon, J.F. Case, J.H. Connell, J. Cronshaw, J.E. Cushing, D. Davenport, B.B. DeWolfe, A.W. Ebeling, E. Englesberg, A. Gibor, R.W. Holmes, W.N. Holmes, M.F. Moseley, C.H. Muller, W.H. Muller, N. Neushul, I.K, Ross, D.M. Smith, B.M. Sweeney, G. Taborsky, J.L. Walters, A.M. Winner
Associate Professors: J.R. Haller, J.L. King, C. Kung, P.C. Karis, N.L. Lee, W.W. Murdoch, H.I. Nakada, E. Roias, E.L. Triplett
Assistant Professors: M. Cassman, J.J. Childress, S.K. Fisher. R.P. Howmiller, R. Jacobs, D.M. Kohl, A. Oaten, S. Rothstein, C. Samuel, C. Wriaght

Degree Programs: Aquatic Biology, Biochemistry-Molecular Biology, Biological Sciences, Botany, Cellular and Organismal Biology, Environmental and Systematic Biology, Pharmacology, Zoology, Population Biology

Undergraduate Degree: B.S. (Biological Sciences)
Undergraduate Degree Requirements: Common courses requirements for all majors: separate requirements for majors in Aquatic Biology, Biochemistry-Molecular Biology, Biological Sciences, Botany, Cellular and Organismal Biology, Environmental Biology, Pharmacology, Zoology

Graduate Degrees: M.S., Ph.D.
Graduate Degree Requirements: Master: Plan I: Research and Thesis Plan II: Comprehensive Examination
Doctoral: Separate requirements in areas of emphasis in: Biochemistry-Molecular Biology, Cellular and Organismal Biology, Environmental and Systematic Biology, Aquatic and Population Biology

Graduate and Undergraduate Courses Available for Graduate Credit: About fifty graduate courses; some of them offered in alternate years only. In addition most of the faculty offer graduate seminars in their specialty.

UNIVERSITY OF CALIFORNIA SANTA CRUZ

Santa Cruz, California 95064 Phone: (408) 429-0111
Dean of Graduate Studies: J.S. Pearse
Dean of College: G.S. Hammond

Chairman: J.H. Langenheim
Professors: H. Beevers, F.H. Cota-Robles, C.I. Da Davern, W.T. Doyle, R.S. Edgar, J.H. Langenheim, K.S. Norris, S.F. Bailey (Emeritus), L.R. Blinks (Emeritus), K.V. Thimann (Emeritus), R.J. Berger, R.T. Hinegardner
Associate Professors: C.W. Daniel, W.J. Davis, H.R. Hilgard, B.J. LeBoeuf, A.T. Newberry, H.F. Noller, J.S. Pearse
Assistant Professors: J.F. Feldman, A.R. Moldenke, C.L. Ortiz, C.A. Poodry, V.M. Rocha, M.E. Silver, L. Taiz, F.J. Talamantes, Jr., H.H. Wang
Visiting Lecturers: 2
Teaching Fellows: 3
Research Assistants: 15

Degree Programs: Psychobiology, Biology

Undergraduate Degree: B.S.
Undergraduate Degree Requirements: Total of nine (9) biology courses (unspecified) plus one of the following: Senior Thesis, Senior Essay, Advanced Biology Graduate Record Examination (score of 600 and above)

Graduate Degrees: M.S., Ph.D.
Graduate Degree Requirements: Master: Comprehensive Examination (can be satisfied by Ph.D. Oral Examination) Course work above the Bachelor's Degree
Doctoral: Satisfaction of language requirement, Completion of Ph.D. Qualifying Examination, Completion of Doctoral Dissertation

Graduate and Undergraduate Courses Available for Graduate Credit: Topics in Cryptogam Biology, Bryology, Topics in Advanced Plant Science, Experimental Ecology, Seminar on Species Diversity, Seminar on Pollination Ecology, Topics in Invertebrate Morphology, Topics in Ethology, Topics in Marine Vertebrates, Theoretical Biology, Analysis of Development, Cytology and Cytochemistry, Topics in Genetics, Topics in Neurophysiology, Tumor Biology, Structure and Function of Cell Membranes, Plant Metabolism, Neuropsychology Seminar, Topics in Molecular Biology

UNIVERSITY OF THE PACIFIC

3601 Pacific Avenue Phone: (209) 946-2011
Stockton, California 95204
Dean of Graduate Studies: D. Shao

College of the Pacific

Dean: C. Hand

DEPARTMENT OF BIOLOGICAL SCIENCES
Phone: Chairman: 946-2181 Others: 946-2182, 2183

Chairman: F.R. Hunter
Professors: J.D. Carson, A. Funkhouser, F.R. Hunter, F. Nahhas
Associate Professors: L.E. Christianson, A. Hunter, W.M. Kaill
Assistant Professor: D.W. McNeal
Instructors: B. Blum, K. Chaubal
Teaching Fellows: 11

Field Stations/Laboratories: Pacific Marine Station - Dillon Beach

Degree Program: Biological Sciences

Undergraduate Degrees: A.B., B.S.
Undergraduate Degree Requirements: 33 courses, 11-15 biology courses, 2 in each of 4 groups. A.B. - 3 semesters of chemistry, 1 of mathematics B.S. - 6 semesters of chemistry, 2 of physics, 2 of calculus

Graduate Degree: M.A.
Graduate Degree Requirements: Master: 6 graduate courses including Research and Thesis, plus 1 other course plus "Tool" (language, E.M., Computer, etc.)

Graduate and Undergraduate Courses Available for Graduate Credit: Population and Quantification Genetics, Biometry, Advanced Microbiology, Immunology, Cytology andCytogenetics, Biosystematics, Comparative Physiology, Special Topics, Animal Behavior (U), Ecology (U), Vertebrate Biology (U), Embryology (U), Comparative Anatomy (U), Cell Physiology (U), Parasitology (U), Plant Morphology (U), Medical Bacteriology (U), Microbiology (U), Plant Taxonomy (U), Plant Kingdom (U), Histology (U)

UNIVERSITY OF REDLANDS

Redlands, California 92373 Phone: (714) 793-2121

DEPARTMENT OF BIOLOGY
Phone: Chairman: Ext. 292

Chairman: R. Wright
Robertson Professor of Biology: C.D. Howell
Professor: C.D. Howell, R.D. Wright
Associate Professor: G.O. Gates
Assistant Professor: D. Cronkite, R. Sener, L.K. Smith

Degree Programs: Biology

Undergraduate Degree: B.S.
Undergraduate Degree Requirements: 36 units biology, 8 chemistry, 8 physics, 8 mathematics and computer science

UNIVERSITY OF SAN DIEGO

Alcala Park Phone: (714) 291-6480
San Diego, California 92110
Dean of College: E.E. Foster

DEPARTMENT OF SCIENCE AND MATHEMATICS
Phone: Chairman: (714) 291-6480

Chairman: R.H. White
Professors: B. Farrens, C.W. Spanis
Associate Professors: J. Bradshaw, R.E. Dingman, D.G. Reck
Assistant Professors: D.K. Severson

Degree Program: Biology

Undergraduate Degrees: B.S., B.A.
Undergraduate Degree Requirements: General Biology (2 semesters) Embryology, Genetics, Comparative Physiology, Cell Physiology, Upper Division biology electives (8 units), Biology electives (4 units), General Chemistry (2 semesters, Organic Chemistry, General Physics (2 semesters), Mathematics (including Calculus)

UNIVERSITY OF SAN FRANCISCO

San Francisco, California 94117 Phone: (415) 666-0600
Dean of Graduate Studies: Rev. J. Martin, S.J.
Dean of College: L.D. Luckmann

DEPARTMENT OF BIOLOGY
Phone: Chairman: 666-6345 Others: 666-6755

Chairman: G.L. Stevens
Professors: E.K. Kessel, F.P. Filia
Associate Professors: D.A. Mullen, R.A. Schooley, L. Treagan
Assistant Professors: R.J. Brown, P.K. Chien, C.P. Flessed, W.P. Jordan, G.L. Stevens
Research Assistants: 2

Degree Program: Biology

Undergraduate Degree: B.S.
Undergraduate Degree Requirements: 28 units upper division biology, 10 units lower division biology, 13 units chemistry, 8 units physics, 3 units statistics, 6 units English, 3 units Economics, 3 units speech

Graduate Degree: M.S.
Graduate Degree Requirements: Master: 24 units of upper and graduate level courses, organic chemistry, statistics, plus Thesis plan: Completion of Thesis, Examination Plan: Written examination in general biology and an oral examination in two specific areas

Graduate and Undergraduate Courses Available for Graduate Credit: All undergraduate courses in the hundred series are available for graduate credit, plus biochemistry in the Chemistry Department, Graduate Seminar, Lichenology, Seminar in Marine Biology, Comparative Histology, Seminar in Virology

UNIVERSITY OF SANTA CLARA

Santa Clara, California 95053 Phone: (408) 984-4242
Dean of Graduate Studies: W.F.Donnelley
Dean of College: J.F. Drahmann

DEPARTMENT OF BIOLOGY
Phone: Chairman: (408) 984-4496 Others: 984-4242

Chairman: T.N. Fast (Acting)
Professors: F.R. Flaim, J.S. Mooring
Associate Professors: T.N. Fast, G.A. Tomlinson
Assistant Professors: L.J. Homach
Visiting Lecturers: 2

Degree Programs: Biology

Undergraduate Degree: B.S.
Undergraduate Degree Requirements: Biology: lower division (5 courses, upper division (7), Chemistry: Inorganic (2), Organic (2); Physics (2); Mathematics (2); English (2); Humanities (4); Social Science (2); Ethnic Studies (1); Religious Studies (3); and other courses. Total 40 courses or 175 quarter units

UNIVERSITY OF SOUTHERN CALIFORNIA

University Park Phone: (213) 746-2311
Los Angeles, California 90007
Dean of Graduate Studies: C. Mays

College of Arts, Letters and Science

Dean: W. Wagner

DEPARTMENT OF BIOLOGICAL SCIENCES
Phone: Chairman: (213) 746-2929 Others: 746-2928

Chairman: B.C. Abbott
Professors: B.C. Abbott, M.D. Appleman, J.W. Bartholomew, R.F. Bils, R.M. Chew, A. Dunn, J.S. Garth, C.C. Hogue (Adjunct), H.M. Kurtz, J. Katz (Adjunct), W.C. Martin (Adjunct), G.Mead (Adjunct), J.L. Mohr, T.R. Pray, J.M. Savage, K.E. Stager (Adjunct), B.L. Strehler, R.B. Tibby, F.S. Truxal (Adjunct), L.C. Wheeler
Associate Professors: M.M. Appleman, R.F. Baker, G. Bakus, L.G. Bishop, G.F. Jones, J.H. McLean (Adjunct), B.G. Nafpaktitis, D.R. Patten (Adjunct), J.A. Petruska, P.M. Shugarman, R.B. Weg, J.W. Wright (Adjunct), R.L. Zimmer
Assistant Professors: R.L. Bezy (Adjunct), H.F. Cossio C.E. Finch, M.F. Goodman, R.R. Given, K. Fauchald, E.I. Kataja, N.L. Nicholson, E.M. Perkins, R.E. Pieper, D. Straughan (Research, I.R. Straughan, C.C. Swift (Adjunct
Lecturers: A. Siger, H. Starrett, J. Waggoner
Teaching Fellows: 2
Visiting Lecturers: 2

Field Stations/Laboratories: Santa Catalina Marine Biological Laboratory - Santa Catalina Island

Degree Programs: Biological Sciences, General Organismic Biology, Cellular and Molecular Biology.

Undergraduate Degrees: B.S., B.A.
Undergraduate Degree Requirements: 128 units including Foreign Language, Physics, Chemistry (Inorganic and Organic), Calculus, Freshman Biology - 1 year, 3 Core Courses - Genetics, Biochemistry and Physiology, Ecology-Evolution, 5 Upper Division Electives

Graduate Degrees: M.S., Ph.D.
Graduate Degree Requirements: Master: One Graduate-level course from each: Ecology, Evolution and Biosystematics, Genetics and Development, Physiology and Morphology; one or two courses in Directed Research; 2 seminars; minimum 20 units, written Comprehensive and Oral Doctoral: CMB - 28 units during first two years in formal Graduate courses; courses, research and Dissertation to 60 units. Required: Molecular Organisms and Function, Molecular Basis of Cell Reproduction, Diversification and Integration of Cell Function, 4 semesters of Seminar in Cell and Molecular Biology; screening examination after 24 units, Qualifying, 2 semesters teaching, 2 languages or 1 Comprehensive - GPB same as M.S. for 2 years. 2 additional courses, research, dissertation, 60 units assist in 2 laboratory sections of teaching, 2 languages or Comprehensive, Qualifying

Graduate and Undergraduate Courses Available For Graduate Credit: Introduction to Microbiology, Mankind Emerging, Genetics, Ecology, General Embryology, California Plants, Cell Function, Plant Anatomy, Microbial-Host Relationships, Animal Behavior, History of Biology, Principles of Biochemistry, Advanced Biochemistry, Comparative Physiology of Animals, General Animal Parasitology, Principles of Immunology, General Entomology, Microtechnique, Histology, Topics in Advanced Botany, Introduction to Issues in Gerontology, Environmental Microbiology, Special Problems, Microbial Physiology, (All above U) Biology of Aging, Molecular Organisms and Function, Molecular Basis of Cell Reproduction, Diversification and Integration of Cell Function, Laboratory Techniques in CMB, Development and Aging Mechanisms, Marine Invertebrate Zoology, Advanced Invertebrate Zoology, Advanced Molecular Genetics, Biology and Evolution of Tropical Plants, Tropical Forest Ecology, Biology and Evolution of Tropical Insects, Biology of Tropical Epiphytes, Helminthology, Biology of Tropical Vertebrates, Recent Advances in CMB, Recent Advances in Marine Biology, Marine Botany, Advanced Population Biology, Tropical Marine Biology, Tropical Biometerology, Theories of Aging, Human Physiology and Aging, Marine Fouling and Pollution, Marine Ecology, Marine Plankton Ecology, Terrestrial Ecology, Field Ecology in the Southwest, Special Topics in Marine Biology, Current Problems in Marine Sciences, Oceanology, Comparative Physiology of Marine Animals, Concepts and Issues in Gerontology, Biometrics, Developmental Biology of Marine Organisms, Behavior, Brain Function and Aging, Evolutionary Osteology, Directed Research, Research, Dissertation Seminar, Seminar in Marine Invertebrate Zoology, Seminar in Ecology, Seminar in Biosystematics, Seminar in Ultrastructure, Seminar in Physiology, Seminar in Plant Physiology, Seminar in Cellular and Molecular Biology, Seminar in the Biology of Aging, Cytology, Protozoology, Ichthyology, Herpetology, Ornithology, Mammalogy, Crustacean Biology, Malacology, Advanced Neurophysiology, Biosystematics, Physiological Ecology of Marine Organisms, Natural History of Santa Catalina Island, Electrobiology, Biology of Marine Vertebrates, Advanced Marine Invertebrate Biology, Electron Microscopy I, Electron Microscopy II

School of Engineering

Dean: Z.A. Kaprielian

DEPARTMENT OF BIOMEDICAL ENGINEERING
Phone: Chairman: (213) 746-2087 Others: 746-2311

Chairman: F.S. Grodins
Professors: G.A. Bekey, R.E. Kalaba, D.J. Marsh, J.B. Reswick, M.H. Weil (Clinical), F.E. Yates
Associate Professor: G.P. Moore
Assistant Professors: S P. Azen, R.N. Bergman, S.M. Yamashiro
Visiting Lecturers: 2
Research Associates: 4

Degree Program: Biomedical Engineering

Undergraduate Degree: B.S.
Undergraduate Degree Requirements: 128 units with cumulative scholarship average of C (2.00). Required courses in mathematics, physics, biology, chemistry, engineering, humanities and social sciences

Graduate Degrees: M.S., Ph.D.
Graduate Degree Requirements: Master: B.S. degree in Engineering. 27 units beyond B.S., and Comprehensive Examination or Thesis project based on "internship" Doctoral: B.S. in science or engineering, 60 units beyond B.S., Screening Examination, Qualifying Examination, Intensive research and Doctoral Dissertation

Graduate and Undergraduate Courses Available for Graduate Credit: Control and Communication in the Nervous System (U), Physiological Systems (U), Biostatistics (U), Bioinstrumentation, Experimental Projects in Biomedical Engineering, Biomedical Engineering Approaches to Neuromuscular Disability, Measurement and Processing of Biological Signals, Biomedical Engineering Problems in the Hospital Ward, Artificial Organs, Hospital Computer Applications and Health Care Systems, Mathematical Biophysics, Mathematical Analysis of Selected Biological Processes, System Identification in Bioengineering, Introduction to Biomedical Research, Models of Neuronal Systems, Models of Cardiopulmonary Systems, Models of Fluid-Electrolyte and Renal Systems, Models of Humoral Signals and Systems

School of Dentistry

Dean: W. Crawford

(Graduate degrees offered only by joint appointments with Department of Biological Sciences)

School of Medicine

Dean: F.K. Bauer

DEPARTMENT OF ANATOMY
Phone: Chairman: (213) 226-2277 Others: 226-2001

Chairman: D.E. Kelly
Professors: P.R. Patek, S. Bernick, G.F. Hungerford, B.G. Monroe, I. Rehman, D.B. MacCallum (Emeritus)
Associate Professors: R.L. Binggeli, C.K. Haun, G.J. Marshall, B.L. Newman, W.J. Paule, J.E. Schechter, B.G. Slavin, D.E. Rounds (Adjunct)
Assistant Professors: N. Ahmad, K.S. Hung, C.L. Portnoff, V.A. deMignard (Adjunct)
Teaching Assistants: 4

Field Stations/Laboratories: Catalina Marine Biology Laboratory - Catalina Island

Degree Program: Anatomy

Graduate Degrees: M.S., Ph.D.
Graduate Degree Requirements: Master: 24 Units and Thesis Doctoral: 60 Units, One foreign language test (French or German), a Qualifying Examination, Dissertation

Graduate and Undergraduate Courses Available for Graduate Credit: Gross Human Anatomy, Philosophy and Techniques of Teaching in the Life Sciences, Cellular Structure and Function, Neuroanatomy, Histochemistry, Anatomy Seminar, Recent Advances in Anatomy, Regional Dissections, Directed Research, Thesis, Research,

Dissertation

DEPARTMENT OF BIOCHEMISTRY
Phone: Chairman: (213) 226-2151 Others: 226-2158

Chairman: A.F. Brodie
Professors: W.R. Bergren, A.F. Brodie, T.H. Fife, A.L. Fluharty (Adjunct), T. Hall, B. Harding, M.E. Jones, G.E. Lanchantin (Adjunct), W. Marx, M. Nimi, E. Roberts (Adjunct), R. Sterling, D.W. Visser
Associate Professors: D. Levy, F.S. Markland, P. Roy-Burman
Assistant Professors: R.D. Mosteller, R.H. Stellwagen, V. Kalra, S. Oldham, L. Raijman, Z. Tokes
Visiting Lecturers: 1
Research Assistants: 4
Predoctoral Fellows: 3
NIH Predoctoral Trainees: 4

Degree Programs: Biochemistry

Graduate Degrees: Ph.D.
Graduate Degree Requirements: Master: Students are not admitted to prepare for this degree. Under special circumstances a Ph.D. student may terminate his graduate work with this degree Doctoral: Applicant should have completed studies for a Bachelor's degree in one of the natural sciences before entering the graduate program. Courses in Organic Chemistry, Physics, Calculus and at least one Biological Science are required prerequisites. Courses in Physical Chemistry and General Biochemistry are required but may be taken during graduate study

Graduate and Undergraduate Courses Available for Graduate Credit: Biochemistry, Biochemistry Seminar, Chemical Reaction Mechanisms in Biochemistry, Current Topics In Biochemistry, Cofactor Catalyzed Reactions, Enzyme Kinetics and Mechanisms, Lipids, Membrane Structure and Function, Cellular Regulation of Metabolism, Biochemistry of Cellular Differentiation, Oxidative Metabolism, Bioenergetics, Polynucleotide and Polypeptide Biosynthesis, Carbohydrates, Biochemistry of Metabolic Diseases, Protein Chemistry and Structure, Hormones, Advanced Clinical Chemistry Laboratory, Advanced Clinical Biochemistry

DEPARTMENT OF MICROBIOLOGY
Phone: Chairman: (213) 226-2147 Others: 226-2130

Chairman: I. Gordon
Hastings Professor of Microbiology: P.K. Vogt
Professors: F. Aladjem, R.F. Baker, M.O. Biddle (Adjunct), A.P.S. Loyt (Emeritus), D. Ivler, M. Lieb, H. Pearson
Associate Professors: V. Klement, G. Matioli
Assistant Professors: G. Brewer, R. Cross, J.R. Katze, M.M.-C. Lai
Research Assistants: 2
Research Scholars: 3
Fellows or Trainees: 9

Degree Program: Microbiology

Graduate Degree: Ph.D.
Graduate Degree Requirements: Not Stated

Graduate and Undergraduate Courses Available for Graduate Credit: Immunochemistry, Animal Virology, Microbial Regulatory Mechanisms, Micromorphology, Mode of Action of Antibiotics, Methods in Phage Research, Immunochemistry, General Genetics, Differentiation and Hematopoietic Malignancies, Current Topics in Microbiology, Directed Research, Advanced Microbial Genetics, Immunochemistry, Seminar

DEPARTMENT OF PHYSIOLOGY
Phone: Chairman: 226-2241 Others: 226-2242

Chairman: J.P. Meehan
Professors: B.C. Abbott, F.S. Grodins, J.P. Henry, F.E. Russell (Research, P.R. Saunders, S.S. Sobin (Research), I.J. Pincus (Clinical), W.F. Frasher, Jr. (Research)
Associate Professors: M.A. Baker (Research), S.D. Frasier, D.F. Lindsley, G.P. Moore (Research)
Assistant Professors: L.W. Chapman, H.J. Meiselman, G.P. Moore, A.A. Pilmanis, W. Price, W.H. Richmond, C.M. Stevens, D.W. Warren (Research)
Instructor: J.R. Rikel (Teaching)
Research Assistant: 1

Field Stations/Laboratories: University of Southern California Marine Biology Laboratory - Catalina Island

Degree Programs: Physiology

Graduate Degrees: M.D., Ph.D.
Graduate Degree Requirements: Master: B.A. Degree in one of the natural sciences. Undergraduate course work in mathematics, physics, chemistry, organic chemistry and zoology, Physical Chemistry, Comparative Physiology, Cellular Physiology or equivalent experience Doctoral: All of the above, and research, 60 units past the B.A. are required

Graduate and Undergraduate Courses Available for Graduate Credit: Seminar in Physiology, Advanced Physiological Methods, Recent Advances in Physiology, Selected Topics in Physiology, Basic Physiology, Directed Research, Neuronal Signals and Systems, Advanced Circulation, Advanced Endocrinology, Advanced Steroid Methodology, Brain and Behavior, Hyperbaric Physiology, Physiopharmacology of Toxins, Research, Thesis, and Dissertation

WESTMONT COLLEGE
(No reply received)

WHITTIER COLLEGE
Whittier, California 90608 Phone: (213) 693-0771
Dean of College: R. Harvey

BIOLOGY DEPARTMENT
Phone: Chairman: Ext. 261

Chairman: I.M. Hull
James Irvine Professor of Biological Sciences: L.E. James
Professors: J. Arcadi (Adjunct Research), I.M. Hull, L.E. James
Associate Professor: C. Morris
Assistant Professor: S.R. Goldberg, A.W. Hanson
Teaching Assistants: 2

Degree Programs: Biology

Undergraduate Degrees: B.A.
Undergraduate Degree Requirements: Completion of a Field of Concentration with a minimum of 15 modules including both an intensive area and a supportive area Completion of at least one senior Colloquium of three modules: contrast to concentration of 6-9 modules, three Jan. Sessions, four Extended Half Modeuls. Completion of 60 modules or 210 semester credits

Graduate Degree: M.S.
Graduate Degree Requirements: Master: 28 credits in graduate or upper division courses in Biology Thesis and Oral Examination

Graduate and Undergraduate Courses Available for Graduate Credit: Vertebrate Embryology (U), Comparative Anatomy (U), Animal Histology (U), Human Anatomy (U), Plant Anatomy (U), Morphology of Vascular Plants (U), Cell Physiology (U), Plant Physiology (U), Animal Physiology (U), Hematology (U), Immunology (U), Microbiology (U), Parasitology (U), Algology (U),

COLORADO

ADAMS STATE COLLEGE

Alamosa, Colorado 81101 Phone: (303) 589-7011
Dean of Graduate Studies: L. Swenson
Vice-President: J. Turano

DEPARTMENT OF BIOLOGY
Phone: Chairman: 589-7231

Head: V.F. Keen
Professors: J.H. Craft (Division Chairman) H.N. Dixon, V.F. Keen
Graduate Assistant: 1

Degree Program: Biology

Undergraduate Degree: B.S.
Undergraduate Degree Requirements: Minimum of 60 Quarter hours in field including one year of General Chemistry

THE COLORADO COLLEGE

Colorado Springs, Colorado 80903 Phone: (303) 473-2233
Dean of College: R.C. Bradley

DEPARTMENT OF BIOLOGY
Phone: Chairman: (303) 473-2233 Ext. 304
Others: (303) 473-2233 Ext. 315

Chairman: W.G. Heim
Professors: R.G. Beidleman, J.L. Carter, M.A. Hamilton, W.G. Heim
Associate Professor: J.H. Enderson
Assistant Professors: R.L. Capen, R. Hathaway, T.B. Kinraide, A. Vargo
Visiting Lecturer: 1
Assistant In Biology: 1

Degree Program: Biology

Undergraduate Degree: B.A.
Undergraduate Degree Requirements: A student majoring in biology must pass Cellular and Molecular Biology, General Zoology or The Flowering Plant, Environmental Biology or other field course, Cell Biology, Genetics and Evolution and at the least three other approved courses. In addition, he must pass one year of calculus, two years of chemistry, one year of physics and either a comprehensive examination or submit an acceptable thesis. GRE Advanced Test in Biology must be passed

COLORADO STATE UNIVERSITY

Fort Collins, Colorado 80521 Phone: (303) 491-5321
Dean of Graduate Studies: W.H. Bragonier

College of Agricultural Sciences

Dean: D.D. Johnson

DEPARTMENT OF AGRONOMY
Phone: Chairman: (303) 491-6501 Others: 491-1101

Head: R.S. Whitney
Professors: R.E. Danielson, C.J. deMooy, A.D. Dotzenko, T.E. Huas, D.D. Johnson, W.D. Kemper, A. Klute, W.L. Lindsay, J.O. Reuss, B.R. Sabey, W.R. Schmehl, T. Tsuchiya, J.R. Welsh, R.S. Whitney, D.R. Wood
Associate Professors: K.G. Brengle, K.G. Doxtader, J.W. Echols, W.T. Franklin, R.D. Heil, G.O. Hinze, H.O. Mann, C.B. Rumburg, E.G. Siemer, P.N. Soltanpour, W.G. Stewart, V.E. Youngman
Assistant Professors: W.S. Ball, W.A. Berg, R.L. Cuany, H.M. Golus, E.J. Langin, A.E. Ludwick, H.D. Moore, C.W. Robinson, J.G. Walker
Instructors: G.H. Ellis (Research Associate, W.D. Luebbe, T.A. Ruehr
Teaching Fellows: 1
Research Assistants: 35
Postdoctorals: 2

Field Stations/Laboratories: Central Great Plains Field Station - Akron, South Colorado Research Center - Springfield, and Walsh, Arkansas Valley Research Center - Rocky Ford, San Juan Basin Research Center - Cortez, Mountain Meadow Research Center - Gunnison, Fruita Research Center - Grand Junction - San Luis Valley Research Center - Center

Degree Programs: Agronomy, Conservation, Crop Science, Ecology, Crops, Genetics, Land Resources, Crop Physiology, Plant Breeding, Plant Industry, Soil Science, Soils and Plant Nutrition, Soil Microbiology, Soil Physics

Undergraduate Degrees: B.S.
Undergraduate Degree Requirements: 192 quarter credits; minimum of 48 quarter credits in residence

Graduate Degrees: M.S., Ph.D.
Graduate Degree Requirements: Master: 45 quarter credits of which 36 earned at Colorado State University; courses chosen from at least two departments: Thesis and non-thesis plans Doctoral: 108 quarter credits of which 45 credits may be accepted from accredited colleges or universities; minimum of 48 quarter credits in residence

Graduate and Undergraduate Courses Available for Graduate Credit: Seed Production (U), Forage Crops (U), Principles of Genetics (U), Soil Conservation in Land Planning (U), Soil Fertility Management (U), Dryland Soil Management (U), Irrigation Practice (U), Special Crops, Field Crop Breeding (U), Plot Technic (U), Soil Genesis and Survey (U), Forest and Range Soils (U), Disturbed Lands (U), Soil Physics (U), Environmental Agronomy, Cereal Crops, Crop Ecology, Cytogenetics Origin and Evolution of Cultivated Plants, Soil and Water Conservation, Soil Chemistry, Soil and Plant Chemical Analysis, Saline and Sodic Soils, Soil Microbiology, Soil-Plant Relationships, Bio-environmental Relationships, Field Crop Physiology, Advanced Cytogenetics, Mineral Nutrition Plants, Advanced Soil Chemistry, Soil Physics, Advanced Soil Microbiology, Simulation of Soil-Plant Systems, Advanced Plant Breeding, Theory of the Gene, Chemical Equilibria in Soils, Transport Phenomena in Soils

DEPARTMENT OF ANIMAL SCIENCES
Phone: Chairman: (303) 491-6672 Others: 491-1101

Head: T.R. Greathouse
Professors: J.S. Brinks, D.A. Cramer, A.L. Esplin, T.R. Greathouse, M.W. Heeney, J.E. Johnson, E.W. Kienholz, J.K. Matsushima, C.F. Nockels, J.W. Oxley, C.L. Quarles, T.M. Sutherland, R.E. Taylor, G.M. Ward
Associate Professors: T.R. Blackburn, W.R. Culbertson, H.L. Enos, D.E. Johnson, G.R.J. Law, B.F. Miller, V.B. Swanson
Assistant Professors: J.S. Avens, B.W. Berry, J.A. Carpenter, L.M. Slade
Instructor: D.D. Rains

Degree Programs: Animal Disease, Animal Genetics, Animal Physiology, Animal Science, Dairy Science, Meat Sciences, Nutrition, Poultry Science, Radiation Biology, Radiobiology

Undergraduate Degree: B.S. (Animal Science, Avian Science
Undergraduate Degree Requirements: 192 quarter credits 2.0 (C average) All university requirements - 6 cr English Composition, 3 quarters of P.E. Remaining requirements are determined by department

Graduate Degrees: M.S., Ph.S.
Graduate Degree Requirements: Not Stated

DEPARTMENT OF HORTICULTURE
Phone: Chairman: (303) 491-6336 Others: 491-6144

Head: K.M. Brink
Professors: K.M. Brink, J.R. Feucht (Extension Professor) J.J. Hanan, W.D. Holley, M. Workman
Associate Professors: J.D. Butler, D.W. Denna, J.E. Ells, K.L. Goldsberry, C.J. Jorgensen, K.W. Knutson, W.G. Macksam, F.D. Moore, E.A. Rogers, J.K. Stacey, A.H. Hatch, N.S. Leupschen, J.A. Twomey
Assistant Professors: C.W. Basham, B.T. Swanson, Jr. F.D. Schweissing, J.F. Swink
Research and Extension Positions: A.D. Bulla, J.J. Shaughnessy, E.E. Roos
Teaching Fellows: 1
Research Assistants: 8

Field Stations/Laboratories: Austin Rogers Mesa Research Center - Austin, Orchard Mesa Research Center - Grand Junction, San Luis Valley Research Center - Center, Arkansas Valley Research Center - Rocky Ford, Botanical Gardens - Denver

Degree Programs: Horticultural Science, Landscape Horticulture

Undergraduate Degree: B.S.
Undergraduate Degree Requirements: 192 quarter hour credits, including a minimum of 40 in horticulture courses. Students must complete the core curriculum and choose one of the five horticulture options; or if majoring in landscape horticulture, choose one of three options

Graduate Degrees: M.S., Ph.D.
Graduate Degree Requirements: Master: 45 quarter hour creidts in course work and research. Program determined by committee. 23 credits above #500 required Doctoral: 108 quarter hour credits (including 45 credits allowed for Master's degree). Program determined by committee. (See General Catalog for more information).

Graduate and Undergraduate Courses Available for Graduate Credit: Basic Horticulture (U), Indoor Plants (U), Evaluating and Staging Horticultural Materials (U), Suburban Horticulture, Landscape Plants, Landscape Plants, Plant Propagation (U), Plant Propagation II (U), Interaction of Environment and Horticulture (U), Greenhouse Management (U), Floriculture (U), Floriculture (U), Floral Design (U), Nursery Production and Management (U), Elementary Landscape Design and Theory (U), Landscape Management (U), Turf Management (U), Vegetable Crops (U), Fruit Production (U), Horticultural Products Technology (U), Plants in Teaching Biology (U), Carnation Production (U), Landscaping (U), Retail Nursery Management (U), Landscape Planning (U), Turfgrass Science (U), Ecology of Horticultural Plants (U), Breeding of Horticultural Crops (U), Aboriculture (U), Horticulture Seminar (U), Plant Environment Measurements, Research Techniques in Horticulture, Biochemical Genetics, Biochemical Genetics I, Biochemical Genetics II, Biochemical Genetics III, Developmental Genetics, Genetics Seminar

College of Forestry and Natural Resources
Dean: R.E. Dels

DEPARTMENT OF FISHERY AND WILDLIFE BIOLOGY
Phone: Chairman: (303) 491-5020 Others: 491-1101

Head: G.A. Swanson
Centennial Professor of Forestry and Natural Resources: H.W. Steinhoff
Professors: A.T. Crigan, N. French, D.L. Gilbert, W.J. McConnell, R.A. Ryder, H.W. Steinhoff, G.A. Swanson
Associate Professors: R.J. Behnke, C.A. Carlson, M.I. Dyer (Affiliate), R. Gard, H.K. Hagen, D. Hein, J. G. Nagy, G. Post, K. Russell, D.R. Smith
Assistant Professors: J.A. Bailey, E. Bergerson, E. Decker, S.A. Flickinger, J. Gross, J.L. Schmidt (Extension), D.Wade (Extension)
Teaching Fellows: 4
Research Assistants: 40

Field Stations/Laboratories: Pingree Park Campus, - 60 miles n.w. of Fort Collins

Degree Programs: Fishery Biology and Wildlife Biology are listed by Department only, but flexible requirements permit, de facto, a variety of programs.

Undergraduate Degrees: B.S.
Undergraduate Degree Requirements: 192 quarter credits (changes to semesters 1975-76) Fishery Biology: including 28 in Biology, 10 in Calculus, 18 in Communications, 3 in Physical Education, 25 in Chemistry, 5 in Physics, 10 in Statistics, 15 in Social Science, 20 in Fishery Biology, Wildlife Biology: including 32 in Biology, 10 in Mathematics, 16 in Communications, 3 in Physical Education, 15 in Chemistry, 5 in Physics, 10 in Statistics, 15 in Social Sciences, 5 in Soils, 18 in Wildlife Biology, and 12 in Summer Field course

Graduate Degrees: M.S., Ph.D.
Graduate Degree Requirements: Master: Flexible program of study approved by candidate's faculty committee and graduate dean depends upon candidate background and professional goals; 45 quarter credits must include 23 at strictly graduate level. Plan A (Thesis) requires original research. Plan B (Non-Thesis) requires a professional paper based upon less research. Doctoral: Flexible program of study approved by candidate's faculty committee and graduate dean depends upon candidate's background and professional goals. Minimum of 108 credits in course work and research of which about two-thirds in general area of specialization, one-third in supporting areas; includes, the traditional foreign language requirement (reading knowledge of two) may be satisfied by a) greater proficiency (reading and speaking) in one foreign language or b) special preparation in two other "research skills" e.g., statistics

Graduate and Undergraduate Courses Available for Graduate Credit: Ichthyology, Fishery Management, Fishery Management Techniques, Fish Ecology, Fishery Science; Fish Culture, Fish Nutrition, Pond and Reservoir Management, World Fishery Resources, Fish Population Dynamics, Trophic Ecology of Fishes, Seminar, Wildlife Management, Waterfowl, Wildlife Management Techniques, Wildlife Nutrition, Wildlife Values, Large Mammals, Small Game, Nongame Wildlife, Administration, Law and Policy, Wildlife Ecology, Field Studies, Ecological Zoogeography, Design of Wildlife Studies, Population Dynamics, Principles of Research

DEPARTMENT OF FOREST AND WOOD SCIENCES
Phone: Chairman: (303) 491-6637 Others: 491-6911

Chairman: F.F. Wangaard
Professors: C.W. Barney, G.H. Fechner, E.W. Mogren, H.E. Troxell

Assistant Professors: A.A. Dyer, D.V. Sandberg
Instructor: D.R. Betters
Visiting Lecturers: 1
Teaching Fellows: 3
Research Assistants: 24

Field Stations/Laboratories: Pingree Park Field Station - Colorado

Degree Programs: Forest Botany, Forest Chemistry, Forest Entomology, Forest Resources, Forestry, Forestry and Conservation, Wood Science

Undergraduate Degree: B.S.
Undergraduate Degree Requirements: 192 quarter credits satisfying specific requirements of a major in 1) Forest Management Science, 2) Forest Biology, or 3) Wood Science and Technology

Graduate Degrees: M.F., M.S., Ph.D.
Graduate Degree Requirements: Master: M.F. and M.S. degrees: 45 quarter credits with required report or Thesis Doctoral: Ph.D. degree: 108 quarter credits exclusive of language or cultural requirement. One-third of total credits may be earned for research leading to a Dissertation

Graduate and Undergraduate Courses Available for Graduate Credit: Forest Ecology,Silvicultural Systems, Forest Tree Improvement, Forest Biometry, Forest Photogrammetry, Forest Fire Control, Timber Harvesting and the Environment, Wood Anatomy and Properties, Wood and Fiber Identification, Wood-Base Building Materials, Seasoning and Preservation, Forest Products and Man, Wood Chemistry, Pulping Processes, Tree Physiology, Tree Physiology, Regional Silviculture, Forest Growth and Regulation, Natural Resource Sampling Methods, Data Processing in Natural Resource Investories, Decision Making in Forest Resource Management, Forest Management Practices, Forest Photo Interpretation, Principles of Timber Management, Forest Economics, Qualitative Methods in Forest Resource Management, Natural Resource Administration, Product Analysis of Forest Industries, Forest Product Industries, Forest Utilization, Mechanics ofWood and Wood Composites, Wood Mechanics Laboratory, Design of Wood Structures, Bending of Wood, Special Studies in Forestry, World Forestry, Eco-Physiology of Trees, Silvics, Forest Regulation, Forest Policy, Woods of the World, Wood Physics, Wood-Liquid Relations, Tree Growth and Wood Properties, Technology of Wood Adhesion, Genetics Seminar

DEPARTMENT OF RANGE SCIENCE
Phone: Chairman: (303) 491-6620

Head: C.W. Cook
Professors: C.W. Cook, R.M. Hansen, D.A. Jameson, C. Terwilliger, G. Van Dyne, C. Wasser
Associate Professors: C.D. Bonham, R.M. Hyde, P.L. Sims
Assistant Professors: E.T. Bartlett, A.A. Dyer, D.N. Klein, M.J. Trlica
Visiting Lecturers: 7
Research Assistants: 21
Faculty Affiliates: 7

Field Stations/Laboratories: Eastern Colorado Range Research Station - Akron, Southeast Range Research Station - Springfield , San Juan Basin Research Station-Hesperus

Degree Programs: Forest Management, Graduate Range Ecology, Range Soils, Animal Nutrition, Range Ecophysiology, Range Economics and Planning, Range Resources and Decision Making

Undergraduate Degree: B.S. (Range Ecology, Range-Forest Management)
Undergraduate Degree Requirements: Ninety-two quarter hour credits with twenty credits in Range Ecology, and Range Management and Decision Making and Support Courses in Mathematics, Biology and Economics

Graduate Degrees: M.S., Ph.D.
Graduate Degree Requirements: Master: Forty-five quarter credits with a thesis not accounting for more than fifteen and the equivalent of a B.S. in Range Science. Plan B: Fifty-four quarter hours without a Thesis Doctoral: One hundred two quarter hour credits with not more than one third informal courses, two languages or nine quarter hour credits of special courses for each

Graduate and Undergraduate Courses Available for Graduate Credit: Principles of Range Management (U), Range Ecosystem Function I, II, and III (U), Range Ecosystem Measurements (U), Range Economics (U), Range Ecosystem Planning (U), Special Studies in Range Science (U), Range Science Seminar (U), Range Ecosystem Management, Ecology of Brushlands, Range Ecosystem Analysis--Structure, Analysis of Range Ecosystem I, II, and III, Grassland Ecology, Special Studies in Range Science, Seminar in Range Science, Master's Research in Range Science, Doctoral Research in Range Science

College of Home Economics
Dean: E.D. Gifford

DEPARTMENT OF FOOD SCIENCE AND NUTRITION
Phone: Chairman: (303) 491-5093 Others: 491-6535

Chairman: G.R. Jansen
Professors: G. Blaker, J. Dupont, I. Harrill, G.R. Jansen, A. Weis (Emeritus)
Associate Professors: K. Lorenz, J.A. Maga, R. Dowdy (Extension)
Assistant Professors: E.C. Charman, B.L. Henry (Extension) A.M. Kylen, V.A. Lee, J.B. Lough, M. Mc Murry, M. Mathias
Instructors: 3 (Temporary)
Teaching Assistants: 4
Research Assistants: 12

Field Stations/Laboratories: Nutrition Research Laboratory Fort Collins

Degree Programs: Food Science, Nutrition

Undergraduate Degree: B.S.
Undergraduate Degree Requirements: 192 credits

Graduate Degree: M.H.Ec., M.S., Ph.D.
Graduate Degree Requirements: Master: 45 Credits in course work and research (36 to be earned at Colorado State University). Doctoral: 108 credits in course work and research (48 earned at Colorado State University

Graduate and Undergraduate Courses Available for Graduate Credit: Diet and Disease, Infant and Child Nutrition, Geriatric Nutrition, General Nutrition I, Community Nutrition, Laboratory Methods in Nutrition, International Nutrition, Seminar in Food Science and Nutrition Current Concepts of Food Management, Recent Developments in Food Science, Recent Developments in Nutrition, Child Nurtition and Growth, Laboratory Methods in Food Science

College of Natural Sciences
Dean: W.B. Cook

DEPARTMENT OF BIOCHEMISTRY
Phone: Chairman: (303) 491-6443 Others: 491-6441

Chairman: W.S. Caughey
Professors: L.W. Charkey, J.P. Jordan, R.P. Martin, P.G. Squire, A.T. Tu

Associate Professors: P. Azari, D.E. Fahrney
Assistant Professors: J.R. Bamburg, M.R. Paule
Visiting Lecturers: 2
Faculty Affiliate: 2
Joint Appointments: 3

Degree Program: Biochemistry

Graduate Degrees: M.S., Ph.D.
Graduate Degree Requirements: Master: A total of 45 credits of which at least 36 credits, which includes all of the research requirements, the B.C. 500 series (lectures and laboratories) and 1 advanced Biochemistry course, must be taken at Colorado State University Doctoral: A minimum of 48 credits earned at Colorado State University and nine quarters of residence with full time registration (12-15 credits per quarter) in graduate course work and research (108-135 credits total) are required by the Colorado State University Graduate School. Additional information is available from the Department.

Graduate and Undergraduate Courses Available for Graduate Credit: Fundamentals of Biochemistry (U), General Biochemistry, Bipolymers, Biocatalysis, Physical Biochemistry, Biochemical Energetics (U)

DEPARTMENT OF BOTANY AND PLANT PATHOLOGY
Phone: Chairman: (303) 491-6524 Others: 491-1100

Chairman: R.T. Ward
Professors: J. Altman, R. Baker, W.H. Bragonier, L.E. Dickens (Extension), R.L. Dix, J.L. Fults, M.D. Harrison, P.E. Heikes (Extension), C.W. Ross, R.T. Ward
Associate Professors: R.P. Adams, R.G. Hacker, P.J. Hanchey, J.E. Hendrix, A.L. Larsen, C.H. Livingston, N. Oshima, R.V. Parke, F. Reeves, S.M. Stack, R.L. Zimdahl
Assistant Professors: P. Kugrens, M.W. Nabors, R.G. Walter, D.H. Wilken
Visiting Lecturer: 1
Research Assistants: 8
Instructor: 4 (Temporary)
Graduate Teaching Assistants: 11

Field Stations/Laboratories: Pingree Park Campus - 50 miles west of Fort Collins

Degree Programs: Botany, Plant Pathology

Undergraduate Degree: B.S.
Undergraduate Degree Requirements: Botany Major: 44 hours botany and biology; 22 chemistry; 11 mathematics; 15 physics; 6 English, 5 Agronomy, 3 speech, plus electives to total 192 credits Plant Pathology Major: 39 hours botany and biology; 22 chemistry 9 mathematics; 6 english, 10 agronomy, 5 physics, 3 journalism, 5 microbiology, 3 speech, 5 entomology plus electives to total 192 credits

Graduate Degrees: M.S., Ph.D.
Graduate Degree Requirements: Master: Plan A: 45 credits, including approximately 15 for research required Plan B: 48 credits. No Thesis required Doctoral: 108 credits, including research, past the BS, or 63 credits, including research, past the MS

Graduate and Undergraduate Courses Available for Graduate Credit: Biology and Control of Weeds (U), Plant Anatomy (U), Introduction to Phycology (U), Introduction to Mycology (U), Elements of Plant Pathology (U), Forest Pathology (U), Grass Systematics (U), Cytology (U), Botanical Microtechnique (U), Plant Physiology (U), Physiology of Seeds (U), Physiology of Plant Growth and Development (U), Plant Ecology (U), Field Study of Plant Diseases (U), Experimental Plant Taxonomy, Plant Chemo-Systematics, Topics in Mycology, Translocation in Plants, Research Methods in Plant Ecology, Plant Geography, Plant Disease Control, Dendropathology, Dendropathology Laboratory, Phytopathological Technique, Plant Metabolism, Epidemiology of Plant Disease, Physiology of Parasitism, Plant Virology, Physiological Plant Ecology, Special Studies in Botany and Plant Pathology, Botany Seminar, Master's and Doctoral Research

DEPARTMENT OF ZOOLOGY AND ENTOMOLOGY
Phone: Chairman: (303) 491-7011 Others: 491-1101

Chairman: C.L. Ralph
Professors: P.H. Baldwin, H.E. Evans, W.D. Fronk, W.C. Marquardt, D.J. Nash, D. Pettus, C.L. Ralph, C.G. Wilber, T.A. Woolley
Associate Professors: J.W. Brewer, G.M.Happ, R.E. Johnsen, A.S. Kamal, J.F. McClellan, G.C. Packard, R.G. Simpson, W.M. Hantsbarger (Extension)
Assistant Professors: L.E. Jenkins, F.W. Lechleitner, P.N. Lehner, J.F. Ragin, J.V. Ward, B.A. Wunder
Instructors: D. Hoppe
Teaching Fellows: 1
Graduate Teaching Assistants: 21

Field Stations/Laboratories: Pingree Park, - West of Fort Collins

Degree Programs: Entomology, Ecology, Limnology, Zoology

Undergraduate Degree: B.S.
Undergraduate Degree Requirements: 192 total quarter credits; Biology Core Courses Freshman and Sophomore years, 55 credits Physical Science, 20 credits Humanities, 20 credits Humanities, 20 credits Social Sciences, 51 credits Zoological Science, Physical Education, Electives

Graduate Degrees: M.S., Ph.D.
Graduate Degree Requirements: Master: 45-48 credits Thesis or report, foreign language proficiency Doctoral: Credits and course work varies with student. Language and/or ancillary skills. Dissertation

Graduate and Undergraduate Courses Available for Graduate Credit: Comparative Anatomy and Physiology of Vertebrates (U), Invertebrate Zoology (U), Insect Morphology (U), Systematic Entomology (U), Comparative Invertebrate Physiology (U), Animal Behavior (U), Limnology (U), Insect Ecology (U), Aquatic Insects (U), Protozoology (U), Insect Control (U), Animal Ecology (U), Ecology of Parasitism, Advanced Ornithology, Biological Control, Experimental Animal Behavior, Evolution, Acrology, Parasitic Protozoa, Immature Insects, Production Limnology, Toxicology, Helminthology

College of Vetinary Medicine and Biomedical Sciences

Dean: W.J. Tietz

DEPARTMENT OF ANATOMY
Phone: Chairman (303) 491-5847

Chairman: G.P. Epling
Professors: R.W. Davis, G.P. Epling, R.D. Frandson, R.A. Kainer
Associate Professors: Y.Z. Abdelbaki, W.J. Banks, D.C. Billenstein, W.C. Gorthy, G.C. Solomon
Assistant Professors: J.P. Bowman, S.J. Kleinschuster, G.P. Lozlowski, H.O. Nornes
Instructor: L.D. Aldes
Research Assistants: 1
Postdoctorals: 2

Field Stations/Laboratories: Electron Microscopy Training Center - Fort Collins

Degree Programs: Anatomy, Vetinary Anatomy

Graduate Degrees: M.S., Ph.D.

Graduate Degree Requirements: Master: 45 hours after B.S. plus research Doctoral: 45 hours after M.S. plus research

DEPARTMENT OF MICROBIOLOGY
Phone: Chairman: (303) 491-6136

Head: J.E. Ogg
Professors: J.R. Bagby, Jr., W.L. Boyd, G. Cholas, T.L. Chow, J.R. Collier, A.B. Hoerlein, S.M. Morrison, J.E. Ogg, J. Storz, R.P. Tengerdy
Associate Professors: G. Cholas, D.W. Grant, B.K. Joyce, D.A. Klein, K.A. Larson, L.H. Lauerman, Jr., D.C. Lueker, J.G. Osteryoung, G. Post, E.P. Savage, T.G. Tornabene, L.P. Williams
Assistant Professors: J.J. England, D.E. Jensen, M.G. Petit, D.W. Shapiro
Visiting Lecturers: 1
Teaching Fellows: 13
Research Assistants: 28
Lecturer: 1

Field Stations/Laboratories: Colorado Epidemiological Pesticide Study Center - Campus

Degree Programs: Animal Disease, Immunology, Medical Microbiology, Medical Technology, Microbiology, Veterinary Microbiology, Veterinary Science, Environmental Health, Microbiology - Fish Disease Technology Option

Undergraduate Degree: B.S.
Undergraduate Degree Requirements: Microbiology: 1 year of inorlanic chemistry, 1 year of organic chemistry, 1 semester of biochemistry, 1 year of biological sciences, 35 quarter credits of microbiology, 1 semester of calculus, 1 semester of physics. Medical Technology: 1 year of undergraduate chemistry, 1 year of organic chemistry, 1 semester of biological sciences, 30 quarter credits of microbiology, college algebra, college physics (1 quarter. Environmental Health. 1 year of inorganic chemistry, 1 year of organic chemistry, 1 semester of biochemistry, 1 year of biological sciences, 35 quarter credits of environmental health, 1 semester of calculus, 1 semester of physics. Fish Disease Technology: 1 year of inorganic chemistry, 1 semester of biochemistry, 1 year of organic chemistry, 1 year of biological sciences, 35 quarter credits of fish disease technology, 1 semester of calculus, 1 semester of physics

Graduate Degrees: M.S., Ph.D.
Graduate Degree Requirements: Master: 1 year of graduate biochemistry, 1 year of physics, 1 year of calculus computer sciences (Fortran programming, 2 quarters of statistics,* minimum of 15 credits of graduate courses in microbiology, including either medical microbiology, veterinary microbiology, virology or immunology. Research in microbiology and preparation of master's thesis * Environmental health program takes a minimum of 15 credits of environmental health (Public Health Courses) Doctoral: Course requirements as specified for the master of science degree plus proficiency (reading) in one foreign language. Research in Microbiology. Dissertation.

PATHOLOGY DEPARTMENT
Phone: Chairman (303) 491-6144 Others: 491-6145

Head: A.F. Alexander
Professors: A.F. Alexander, M.M. Benjamin, H. Breen, W.W. Brown (Extension), C.P. Hibler, R.L. Jensen, R.D. Phemister, R. Rubin, R.H. Udall, C.E. Whiteman, S. Young
Associate Professors: J.M. Chency, D.W. Hamer, R.S. Jaenke, R.W. Norrdin
Assistant Professors: J.D. DeMartini, G.C. Hill, A.E. McChesney, S.P. Synder
Research Assistants: 1
Residents: 3

Degree Programs: Pathology, Clinical Pathology, Biochemical Pathology and Parasitology

Graduate Degrees: M.S., Ph.D.
Graduate Degree Requirements: Master: Plan A - 30 quarter course credits and Thesis Plan B - 45 quarter course credits. Graduate Committee has broad power of requirements to be met Doctoral: 108 credits including original research and dissertation. No language requirement. Graduate Committee has broad power of requirements to be met.

Graduate and Undergraduate Courses Available for Graduate Credit: Clinical Pathology (U), Veterinary Parasitology (U), Veterinary Parasitology (U), General Pathology (U), General and Specific Pathology (U), Special Pathology (U), Instrumentation in Pathology, Avian Diseases, Special Studies, Advanced Clinical Pathology, Surgical Pathology, Comparative Gross Pathology, General Ultrastructural Pathology, Laboratory Diagnosis and Techniques, Practicum in Gross and Histological Pathology, Supervised Colloquium Teaching, Master's Research in Pathology, Pathology of Nutrient Diseases of Animals, Comparative Gross and Histologic Pathology, Pathology of Exotic Diseases I, II, III, Processes in Laboratory Diagnosis, Advanced Comparative Neuropathology, Pathology of Disease of Nonhuman Primates, Surgery and Pathological Practice, Seminar in Pathology, Doctoral Research in Pathology

DEPARTMENT OF PHYSIOLOGY AND BIOPHYSICS
Phone: Chairman: 491-6187 Others: 491-1101

Chairman: L.C. Faulkner
Professors: L. Ball, J.B. Best, E.J. Carroll, L.E. Davis, L.C. Faulkner, R.A. Herin, M.L. Hopwood, R.W. Phillips, B.W. Pickett, D.H. Will
Associate Professors: J.H. Abel, Jr., C.A. Bonilla, A. B. Goodman, P. Hall, J.F. Masken, R.J. Morgan, G.D. Niswender, E.L. Pautler, A. Pigon, T.N. Solie
Assistant Professors: C.W. Miller, M. Morita, T.N. Nett, G.E. Seidel, Jr.
Instructors: J.B. Grogan, C.L. Schatte
Research Associates: 10
Postdoctoral Fellows: 6
Interns: 3

DEPARTMENT OF VETERINARY RADIOLOGY AND RADIATION BIOLOGY
Phone: (303) 491-5222

Chairman: M.R. Zelle
Professors: W.C. Dewey, E.L. Gillette, J.L. Lebel, J.T. Lett, M.R. Zelle
Associate Professors: G.M. Angleton, J. Barber (Adjunct) L.W. Fraley, J.E. Johnson, F.W. Whicker
Instructor: J.A. Johnson
Research Assistants: 23
Research Associates: 9
Residents: 2

Degree Programs: Veterinary Radiology and Radiation Biology

Graduate Degrees: M.S., Ph.D.
Graduate Degree Requirements: Master: 45 quarter hours credit plus paper (Plan B) or Thesis (Plan A) Doctoral: 108 quarter hours credit plus Thesis

Graduate and Undergraduate Courses Available for Graduate Credit: Radioisotope Techniques (U), Radiology, Nuclear Radiation Physics, Radiobiology, Deterministic Models in Biology, Special Problems, Radiochemical Techniques, Radiation Chemistry, Radioecology, Radiographic Techniques, Ragiographic Interpretation, Radiation Therapy, Radioactive Tracers in General Medicine, Radiological Physics, Radiation Dosimetry, Radiation Public Health, Practicum, Special Studies, Supervised College Teaching, Master's Research,

Radiation Biology I, II, III

COLORADO WOMEN'S COLLEGE

1800 Pontiac Street Phone: (303) 394-6012
Denver, Colorado 80220
Dean of College: N. Slater

DEPARTMENT OF BIOLOGY
Phone: Chairman: (303) 394-6956

Coordinator: E.J. Buecher
Associate Professors: E.P. White, E.J. Buecher, C. Hansman
Assistant Professor: C. Radcliffe
Instructor: J. Kulhanek

Degree Program: Biology

Undergraduate Degree: B.A.
Undergraduate Degree Requirements: 40 hours in Biology: to include a course in Microbiology, Genetics, Zoology, and Botany. 1 year of Introductory Biology, 1 year of Chemistry, 1 year of Physics

DENVER MEDICAL CENTER
(See under University of Colorado)

FORT LEWIS COLLEGE

Durango, Colorado 81301 Phone: (303) 247-7454
Director - School of Arts and Sciences: L. Johnson

DEPARTMENT OF BIOLOGY, AGRICULTURE AND FORESTRY
Phone: Chairman: 247-7454

Chairman: J.G. Erickson
Professors: J.G. Erickson, H.E. Owen
Associate Professors: J.E. Dever, Jr., A.W. Spencer
Assistant Professors: H.H. Hadow

Degree Program: Biology

Undergraduate Degree: B.S.
Undergraduate Degree Requirements: 128 semester hours, 36 hours in biology, several miscellaneous requirements

LORETTO HEIGHTS COLLEGE

3001 S. Federal Phone: (303) 922-4011
Denver, Colorado 80236
Dean of College: A. Parimanath

DEPARTMENT OF NATURAL SCIENCE
Phone: Chairman: (303) 922-4249

Head: M.A. Coyle
Professors: S.M.G. Elsey, S.J.d'Arc Schleicher
Associate Professors: S.M.A. Coyle, J.M. Hayes, D. M. Simmons, G. Sullivan
Instructors: K. Breakstone

Degree Programs: Biology, Environmental Science

Undergraduate Degree: B.S.
Undergraduate Degree Requirements: 16 hours chemistry, 4 hours mathematics, 8 hours physics, 28 hours Biology

METROPOLITAN STATE COLLEGE

250 West 14th Street Phone: (303) 292-5190
Denver, Colorado 80204
Dean of College: S. Sunderwirth

DEPARTMENT OF BIOLOGY
Phone: 292-5190 Ext. 368 Others: Ext. 246 or 302

Chairman: G. Becker
Professors: G. Becker, D.A. Bowles, R. Cohen
Associate Professors: D.K. Alford, J.C. Krenetsky
Assistant Professors: R.R. Hollenbeck, D. Marsh, I. Wesley, D. Voth, C. Steele

Degree Program: Biology

Undergraduate Degree: B.S.
Undergraduate Degree Requirements: 60 quarters of biology courses as directed, 60 hours of upper division credits (any discipline), basic studies requirements in communications, humanities, social studies, mathematics and science, 45 hours in residence

REGIS COLLEGE

W. 50th and Lowell Boulevard Phone: (303) 433-8471
Denver, Colorado 80221
Dean of College: E.J. Lynch

DEPARTMENT OF BIOLOGY
Phone: Chairman: 433-8471 Ext. 311

Chairman: C. Currie
Associate Professor: C. Currie
Assistant Professors: B.A. Finney, B. Pace, G.L. Rank, H.L. Taylor

Field Stations/Laboratories: Ecological Station - Lake City

Degree Programs: Biology, Ecology

Undergraduate Degree: B.S.
Undergraduate Degree Requirements: Basic Courses Requirements: English - 12 semester hours, Modern Languages 6 or 9, Social Science 6, Speech 2 or 3, Philosophy 6 semester hours, Religious Studies 9 semester hours, mathematics or science 6 or 8 Major - 18 upper division hours plus prerequisite courses Minor - 12 upper division hours plus prerequisite courses

SOUTHERN COLORADO STATE COLLEGE

Pueblo, Colorado 81001 Phone (303) 549-0123
President: H. Bowes

DEPARTMENT OF BIOLOGY AND AGRICULTURE
Phone: Chairman: 549-2743

Chairman: J.W. LaVelle
Associate Professors: J. Dorsch, G. Farris, R. Garden, S. Herrmann, J. Linam, H. Murray, N. Osborn, J. Seilheimer
Assistant Professors: L. Thomas

Degree Programs: Agriculture, Biology

Undergraduate Degree: B.S.
Undergraduate Degree Requirements: Approved Biology Electives (Minimum of 25 credits) Approved Electives from Sciences other than Biology (Minimum of 40 credits) in addition to requirements

TEMPLE BUELL COLLEGE
(See Colorado Womens College)

UNITED STATES AIR FORCE ACADEMY

Colorado 80840 Phone: (303) 472-1818
Dean of College: W.T. Woodyard

DEPARTMENT OF LIFE AND BEHAVIORAL SCIENCES
Phone: Chairman: (303) 472-3860 Others: 472-4395

Head: P.B. Carter
Professor: W.E. Ward
Associate Professors: L.F. Wailly, J.W. Williams,

R. Chason, D. Prather, T. Newton, J. O'Connor, M.J. Stansell, R.B. Tebbs, O. Sampson, J. Walters, R. Myers, A. Young
Assistant Professors: D.L. Netzinger, C. Payne, R. Schmitt, C. Thalken, G. Berry, E. Galluscio, D. Brown, D. Logsdon, J. Lynette
Instructors: J.K. Jarboe, W.J. Cairney, G.R. Coulter, R. Coffman, R. Eggleston, R. Harkins, J. Knight, R.F. Lloyd, W. Moe, L. Painter, K. Parker, T. Schmeder, T. Simondi
Research Assistants: 2

Degree Program: Life Sciences

Undergraduate Degree: B.S.
Undergraduate Degree Requirements: 187 semester hours; 46.5 hours in the major

UNIVERSITY OF COLORADO
Boulder, Colorado 80302 Phone: (303) 443-2211
Dean of Graduate Studies: M. Lipetz

College of Arts and Sciences
Dean: W.E. Briggs

DEPARTMENT OF ENVIRONMENTAL, POPULATION AND ORGANISMIC BIOLOGY
Phone: Chairman: (303) 443-2211 Ext. 6422

Chairman: H.M. Smith
Professors: E.K. Bonde, J.H. Bushnell, R.E. Gregg, J.W. Marr, R.W. Pennak, O. Williasm, W.V. Mayer, C.H. Norris, W. Segal, H.M. Smith, J.T. Windell, P.W. Winston, D.W. Crumpacker, E.B. Leopold (Adjunct), D.J. Rogers, S. Shushan, O. Thorne (Adjunct), B. Willard (Adjunct), W.A. Shulls, P. Weiser (Adjunct),
Associate Professors: C.E. Bock, P.J. Webber, J.H. Bock, R.E. Jones, D.O. Norris, H. Nichols (Adjunct)
Assistant Professors: D.H. Horak (Adjunct), A. Cruz, Y. Linhart, N. Richardson, J. Wilson, R. Bernstein
Research Assistants: 10
Teaching Assistants: 40

Field Stations/Laboratories: Mountain Research Station - outside Boulder in foothills, EPO Biology Environmental Natural Area - Fort Collins, The Research Ranch - Elgin, Arizona

Degree Programs: Anatomy, Anatomy and Histology, Anatomy and Physiology, Animal Genetics, Bacteriology, Biochemistry, Biology, Ecology, Entomology, Entomology and Parasitology, Genetics, Histology, Human Genetics, Limnology, Microbiology, Pathology, Parasitology, Physiology, Silviculture, Zoology, Zoology and Entomology

Undergraduate Degree: B.S.
Undergraduate Degree Requirements: Varies with area of emphasis. Total of 30 hours in Biology 16 hours from courses listed 300 or above

Graduate Degrees: M.S., Ph.D.
Graduate Degree Requirements: Master: (only 12 hours may be 400 level) M.A. I (Thesis) 18 hours of course work comprehensives (written and oral) plus 6 hours of Thesis (original reserach) plus defense. Recommended for Ph.D. candidates. M.A. II 30 hours of course work plus comprehensives (written and oral) recommended for teachers Doctoral: Reading ability in one language other than native language. 46 hours: 30 hours of course work (written and oral) comprehensives plus 16 hours of Thesis - defense. All course work must be numbered 500 and above

Graduate and Undergraduate Courses Available for Graduate Credit: Advanced Environmental, Population and Organismic Biology, Avian Communities of Colorado, and New Mexico, Birds of the World, Animal Geography, Population Dynamics, Recent Advances in Animal Ecology, Biological Oceanography, Dynamics of Mountain Ecosystems, Tundra Ecology, Aquatic Botany, Peripatetic Biology, Soil Development and Morphology, Mammalogy, Field Problems in Mountain Plant Ecology, Stream Biology, Limnology, Benthic and Aufwuchs Biology, Advanced Vertebrate Ecology, Ecological Plant Physiology, Quantitative (Plant) Ecology, Plant Biosystematics, Plant Cytology, Taximetrics, Introduction to Biostatistics, Principles and Practices of Biological Taxonomy, Recent Advances in Genetics, Advanced Topics in Population Genetics, Ecological Genetics, Plant Cytotaxonomy, Plant Evolution and Biogeography, Advanced Palynology, Historical Geobotany, Cenozoic Paleobotany and Palynology, General Physiology, Comparative Animal Physiology, Comparative Endocrinology, Experimental Embryology, Ichthyology, Biology of Fish Populations, Advanced Plant Physiology, Modern Plant Embryology, Developmental Plant Anatomy, Lichenology, Range Plants, Bryology, Seminar on Ecophysiology of Alpine and Arctic Plants, Microbial Physiology, Topics in Animal Behavior, Biology of Vertebrate Reproduction, History of Biology, Instrumentation, Comparative Neurophysiology, Mammalian Physiology Research, Endocrinology Seminar, Advanced Experimental Embryology, Reproductive Biology of Flowering Plants, Virology, Applied Microbiology, Ergonomics, Advanced Organismic Biology

DEPARTMENT OF MOLECULAR, CELLULAR AND DEVELOPMENTAL BIOLOGY
Phone: Chairman: 443-2211 Ext. 8059 Others Ext.7743

Chairman: D.M. Prescott
Professors: P. Albersheim, H. Berg, L. Goldstein, D. Kennedy (Adjunct), L.D. Peachey (Adjoint), K.R. Porter, D.M. Prescott, M.N. Runner, N. Sueoka
Associate Professors: M.A. Bonneville, A.S. Flexer, M. Fotino, L. Gold, R.G. Ham, P.L. Keumpel, E.H. McConkey, J.R. McIntosh, J. Pickett-Heaps, L.A. Staehelin, M.J. Yarus
Assistant Professors: M.W. Dubin, D.I. Hirsh, L. Soll
Visiting Lecturers: 6
Research Assistants: 25

Degree Programs: The Division of Biological Sciences offers BA, MA, and Ph.D. degrees in Biology. This department offers specialized programs in molecular, cellular and developmental biology at each level.

Undergraduate Degree: B.A.
Undergraduate Degree Requirements: 30 hours of biological courses including one course in genetics, molecular genetics, cell biology and developmental biology General, organic and biochemistry; two semesters of physics; two semesters of mathematics

Graduate Degrees: M.A., Ph.D.
Graduate Degree Requirements: Master: Bachelor's degree, preliminary examination, comprehensive final examination. A thesis required for Plan I based on original research and 24 hours of course work. Plan II requires 30 hours of course work Doctoral: Bachelor's degree, preliminary examination, comprehensive examination, 30 hours of course work, translating ability in one language, two semesters of teaching experience, thesis and oral defense usually presented in the form of a seminar

Graduate and Undergraduate Courses Available for Graduate Credit: Techniques of Cell Culture, Techniques of Developmental Biology, Electron Microscopy for Biologists, Biological membranes, Cell Electrophysiology, Biology of Cancer Cells, Vertebrate Cell Culture, Principles of Neurobiology, Topics in Plant Cell Biology, Embryology, Advanced topics in Cell Biology, The Plant Cell, Biology of Sensory Phenomena, The Biology of Sex, Experimental Embryology, Mechanisms of Biological Pattern Formation, Cellular Differentiation, Cell and Tissue Interactions in Development,

Mechanisms of Development, Mechanisms of Aging, Biomolecular Organization, Statistical Processes in Molecular Biology, Interactions of Biological Molecules, RNA and Protein Synthesis, Genetics of Mammalian Cells, Bacterial and Viral Genetics, Advanced Topics in Molecular Genetics, Workshop in Electron Microscopy, Special Topics, Graduate Seminar, Independent Study, Doctor's Thesis

Denver Medical Center

4200 East 9th Avenue Phone: (303) 399-1211
Denver, Colorado 80220
Dean of Graduate Studies: S. Katsch
Dean of School of Medicine: H. Ward

DEPARTMENT OF ANATOMY
(No Reply Received)

DEPARTMENT OF BIOCHEMISTRY
Phone: Chairman: 394-7567 Others: 394-7013

Acting Chairman: A. Abrams
Professors: A. Abrams, D. Hagerman, C.G. Mackenzie, J.B. Mackenzie, J.M. Stewart
Associate Professors: J.L. Brown, C. Bublitz, O.K. Reiss
Assistant Professors: G.A. Scarborough
Instructors: G.R. Matsueda

Degree Program: Biochemistry

Graduate Degree: Ph.D.
Graduate Degree Requirements: Not Stated

Graduate and Undergraduate Courses Available for Graduate Credit: Not Stated

DEPARTMENT OF BIOPHYSICS AND GENETICS
Phone: Chairman: (303) 394-7919 Others: 394-8208

Chairman: A. Robinson
American Cancer Society Lifetime Research Professor: T.T. Puck
Professors: J.R. Cann, M.L. Morse, M. Naughton, T.T. Puck, A. Robinson, H.V. Rickenberg, W. Sauerbier, S.M. Ulam
Associate Professors: F.T. Kao, H. Lubs, D.E. Pettijohn, J.R. Sadler
Assistant Professors: C. Abel, C. Jones, N. Pace, D. Patterson, V. Riccardi, N.W. Seeds
Instructors: H.G. Morse, D. Peakman, K.C. Rock, C. Waldren, P. Wuthier
Visiting Lecturers: 5
Teaching Fellows: 5

Degree Programs: Biophysics, Genetics, Cell Biology, Human Genetics

Graduate Degrees: M.S. (Human Genetics), Ph.D.
Graduate Degree Requirements: Master: 1) pass with a grade of B or above a sequence of required courses and accrue 36 quarter hours of graduate credit: 2) do research and submit an acceptable thesis 3) pass a final comprehensive examination. There is no foreign language requirement Doctoral: 1) pass with a grade of B or above a sequence of required courses and accrue 45 or more quarter hours of graduate credit 2) do research and submit an acceptable thesis 3) pass qualifying and comprehensive examinations and an oral defense of the thesis 4) show evidence of proficiency in one foreign language

Graduate and Undergraduate Courses Available for Graduate Credit: Medical Biophysics and Genetics, Basic Biomedical Electronics, Mammalian Cell Seminar, Nucleic Acids, Topics in Molecular Neurobiology, The Immunoglobulins, Environmental Mutagens and their Detection, Microbial and Molecular Genetics, Physical Mechanisms of Enzyme Action, Physics, Mathematics and the Modelling of Biological Systems, Seminar in Biophysics and Genetics, Physicochemistry of Macromolecules in Solution, Selected Topics in Mathematical Biology, Research in Biophysics and Genetics, Master's Thesis, Doctor's Thesis

DEPARTMENT OF BIOMETRICS
Phone: Chairman: (303) 394-7605

Chairman: S.H. Walker
Professors: D.W. Stilson, R.B. Selman, S.H. Walker
Associate Professor: P.G. Archer
Assistant Professor: G.O. Zerbe, F.W. Briese
Instructors: N.P. Dick, J.R. Murphy

Degree Program: Biometrics

Graduate Degrees: M.S., Ph.D.
Graduate Degree Requirements: Master: Applicants for tho program will be expected to have a bachelor's degree in a scientific field, with at least mathematics through intermediate (2nd year) calculus and 2 years of course work in the physical and/or biological sciences No beginning students will be accepted who do not have at least a 3.0 grade-point average as undergraduates Doctoral: Applicants for the program will be expected to have a bachelor's as well as a master's degree in a scientific field

Graduate and Undergraduate Courses Available for Graduate Credit: Biomathematics, Biostatistics Methods, Biomedical Computing-FORTRAN, Models in Human Ecology, Epidemiology, Computer Oriented Statistical Methods, Statistical Theory, Consulting Methods, Bioassay, Sampling and Survey Methods, Nonparametric Statistical Methods, Advanced FORTRAN Programming, Mathematical and Statistical Genetics, Matrix Algebra, Epidemiologic Methods, Bayesian Inference, Survival Curves and Lifetable Analysis, Longitudinal Data Analysis, Complex Variables and Operational Methods, Mathematical Modelling in Medicine

DEPARTMENT OF MICROBIOLOGY
Phone: Chairman: 394-7016 Others: 394-7903

Chairman: L.M. Kozloff
American Cancer Society Professor: S.S. Cohen
Professors: E. Borek, H. Claman, S.S. Cohen, A.J. Crowle, D.J. Cummings, S.G. Dunlop, F. Harold, L.M. Kozloff, D.W. Talmage
Associate Professors: W. Roberts, A.L. Taylor
Assistant Professor: M. Pato
Instructor: C.J. Male
Visiting Lecturers: 10
Research Associates: 7

Degree Program: Microbiology

Graduate Degrees: M.S., Ph.D.
Graduate Degree Requirements: Master: One and one-half years of course work with or without a Thesis Doctoral: Acceptable Thesis in original research

Graduate and Undergraduate Courses Available for Graduate Credit: Properties of Bacteria, Viruses, and Protozoa and Their Interaction with Their Hosts (U), Microbiology Laboratory (U), Immunology (U), Microbial and Molecular Genetics, Topics in Microbial Physiology, Bacterial Viruses, Biochemistry of Animal Viruses, Methods in Microbial Genetics, Delayed Hypersensitivity, Advanced Cellular Immunology, Research in Microbiology

DEPARTMENT OF PHYSIOLOGY
Phone: Chairman: (303) 394-7403 Others: 394-8122

Chairman: A.R. Martin
Professors: A.R. Martin, G. Meschia
Associate Professors: N. Banchero, C. Tucker
Assistant Professors: W. Betz, T. Burke, E. Johnson, M. Neville, W.O. Wickelgren

Degree Programs: Neurophysiology, Physiology

Graduate Degrees: Ph.D.
Graduate Degree Requirements: Master: Not normally given Doctoral: Undergraduate degree in biological or physical sciences with at least one course in calculus and in physical chemistry

Graduate and Undergraduate Courses Available for Graduate Credit: Principles of Physiology, Physiology Seminar, Fetal and Neonatal physiology, Cell Membranes and Transport, Neurobiology, Cardiovascular Physiology, Applied Cardiovascular Physiology

UNIVERSITY OF DENVER

Denver, Colorado 80210 Phone: (303) 753-3661
Dean of Graduate Studies: R.C. Amme
Dean of College of Arts and Sciences: E.A. Lindell

DEPARTMENT OF BIOLOGICAL SCIENCES

Phone: Chairman: 753-3661

Chairman: G.E. Stone
Professors: W.F. Brandom, G.B. David, M.L. Shubert, G.E. Stone, F.N. Zeiner (Emeritus)
Associate Professors: R.A. Anderson, D.A. Belden, Jr.
Assistant Professors: R.W. Angell, T. Ashley (Adjunct), D. Barrett, S.M. Coakley (Visiting) N.L. Couse, P. Feinsinger (Visiting), D.S. May, J.E. Platt (Visiting)
Visiting Lecturers: 5
Teaching Fellows: 1
Research Assistants: 8

Field Stations/Laboratories: Alpine Biological Laboratory - Echo Lake

Degree Programs: Biological Sciences

Undergraduate Degrees: B.A., B.S.
Undergraduate Degree Requirements: B.S. - 45 quarter hours of biological sciences including Concepts in Biology (3 quarter course); Cell Physiology; General Genetics; General Ecology, One year of both general and organic chemistry), Mathematics through Calculus and one year sequence course of Physics; one year equivalent of Foreign Language B.A. - same as above except Mathematics through Calculus, Physics and Modern Language not required

Graduate Degree: M.S.
Graduate Degree Requirements: Master: 45 quarter hours including at least 30 quarter hours in the major field; Thesis based on a research project approved by the chairman and major adviser. Courses may be taken in other academic areas which complement the major field and are approved by the department.

Graduate and Undergraduate Courses Available for Graduate Credit: Microanatomy, Invertebrate Zoology, Protozoology, Entomology, Seminar in Advanced Topics in Invertebrate Zoology, General Ecology Laboratory, Field Ecology, Zoology, Animal Ecology, Plant Ecology, Advanced Topics in Ecology, Advanced Botany, Plant Physiology, Plant Morphology, Lichenology, Biogeography, Molecular Genetics, Advanced Topics in Genetics, Developmental Biology, Advanced Topics in Developmental Biology, Cellular Biology, Introduction to Microtechniques, Advanced Microscopy, Histo- and Cytochemistry, Chromosome Biology, Advanced Topics in Cellular Biology, Principles of Animal Physiology, Endocrinology, Seminar, Independent Study*
*Graduate Only

UNIVERSITY OF NORTHERN COLORADO

Greeley, Colorado 80639 Phone: (303) 351-1890
Dean of Graduate Studies: A. Reynolds
Dean of College: R.O. Schulze

DEPARTMENT OF BIOLOGICAL SCIENCES

Phone: Chairman: 351-2532

Chairman: R.K. Plakke
Professors: J. Gapter, R. Rich, E. Richards, G. Schmidt B. Thomas, B. Thorpe, R.K. Plakke
Associate Professors: W. Buss, I. Lindauer
Assistant Professors: J. Fitzgeral, W. Harmon, M. Heimbrook, E. Peeples
Teaching Fellows: 6

Degree Programs: Biological Sciences, Botany, Zoology

Undergraduate Degree: A.B.
Undergraduate Degree Requirements: 60 quarter hours in Biology, Zoology or Botany, 15 hours of chemistry (minor recomended) 186 quarter hours in total program

Graduate Degrees: M.A., D.A.
Graduate Degree Requirements: Master: 45 quarter hours Doctoral: 90 quarter hours beyond the M.A. Degree

Graduate and Undergraduate Courses Available for Graduate Credit: Evolution (U), Aquatic Biology (U), Cell Physiology (U), Drugs and Human Behavior (U), Biology of Microorganisms (U), Microbiology (U), Techniques of Biological Preparation (U), Topics in Birth Control and Contraception (U), Counseling in Birth and Contraception (U), Seminar in Research I,II (U), Undergraduate Research (U), Conservation of Natural Resources (U), Biological Microtechnique (U), Biological Photography (U), Topics in Field Biology (U), Economic Botany (U), Plant Taxonomy (U), Identification of Trees and Shrubs (U), Morphogenesis of the Nonvascular Plants (U), Morphogenesis of the Vascular Plants (U), Plant Ecology (U), General Plant Physiology (U), Mycology (U), Principles of Plant Culture (U), Ornithology (U), Entomology (U), Comparative Morphogenesis of the Vertebrates (U), Comparative Morphogenesis of the Vertebrates II (U), Animal Ecology (U), General Parasitology (U), Vertebrate Embryology (U), Faunistics (U), Radiation Biology, Population Biology, Human Genetics and Eugenics, Cytology, Evolution and Speciation, Eco system Modification, Analysis and Alteration of Ecosystems, Environmental Pollution, Pathogenic Microbiology, Foundations of Biological Research, Conceptual Schemes of S cis, Immunology and Serology, Virology, Social Implications of Biology Seminar, Problems in Human Genetics, Ecology, Aquatic Ecology and Water Pollution, Taxonomy of Grasses, Plant Anatomy, Physiological Plant Ecology, Plant Geography, Plant Growth and Development, Plant Pathology, Mineral Nutrition of Plants, Plant Water Relations, Comparative Mammalian Anatomy, Advanced Invertebrate Zoology, Helminthology, Medical Entomology, Experimental Vertebrate Embryology, Vertebrate Histology, Mammalian Physiology, Mammalian Physiology II, Endocrinology, The Central Nervous System and Special Senses,

WESTERN STATE COLLEGE

Gunnison, Colorado 81230 Phone: (303) 943-0120
Dean of Graduate Studies: E.H. Randall
Dean of College: C.M. Bjork

DIVISION OF NATURAL SCIENCES AND MATHEMATICS

Western State is not organized on a departmental system. The Natural Science Division includes Chemistry, Physics, Geology, Biology and Mathematics. There is no Biology Departmental organization although classes are divided as to Biology, Botany and Zoology and there is informal cooperation among the staff.
Phone: Chairman: 943-2015 Others: 943-2015

Division Chairman: A.W. Lawrence
Professors: S. Adams Jr., (Biology, H.A. Ferchau (Botany), T.T. Harris (Zoology), H.M. Mobley (Botany)
Associate Professors: K.F. DeBoer, E.K. Longpre

Assistant Professor: R.E. Richards
Graduate Assistants: 2

Degree Programs: Biology, Science

Undergraduate Degree: B.A. (Biology)
Undergraduate Degree Requirements: 187 quarter credits including 60 credits in General education, 45 credits (Minimum) of Biology, Botany or Zoology, courses, some of which must be selected from a required list. 12 credits (Minimum) of Chemistry. More are recommended as is Physics and mathematics.

Graduate Degree: M.A. (Science)
Graduate Degree Requirements: Master: 45 credits (minimum); 27-33 credits in Biology (including Botany or Zoology); 12-18 credits in the minor (usually Chemistry, Mathematics or Geology). Written or oral examination or both. Thesis may or may not be required but credit in research is.

Graduate and Undergraduate Courses Available for Graduate Credit: Biological Techniques (U), Cellular Biology, Biosystematics, Advanced Ecology, Advanced Microbiology, Individual Problems, Rocky Mountain Flora (U), Plant Anatomy (U), Plant Morphology (U), Field Ecology (U), Introduction to Parasitology (U), Rocky Mountain Fauna (U), Histology (U), Vertebrate Morphogenesis (U), Entomology (U), Mammalogy, Ornithology

CONNECTICUT

ALBERTUS MAGUS COLLEGE
(no reply received)

ANNHURST COLLEGE
R.R. 2 Phone: (203) 928-7773
Woodstock, Connecticut 06281
Dean of College: S.H. Bonin

DEPARTMENT OF BIOLOGY
Phone: Chairman: 928-7773 Ext. 149

Chairman: S.A. Celine
Instructors: G. Harrington, T. Wilson

Degree Programs: Biology

Undergraduate Degree: B.A.
Undergraduate Degree Requirements: 120 credits of which 38 in Biology, 20 in Chemistry, 8 in Physics, 6 in Mathematics

CENTRAL CONNECTICUT STATE COLLEGE
1615 Stanley Street Phone: (203) 225-7481
New Britain, Connecticut 06050
Dean of Graduate Studies: A.C. Erickson
Dean of College: T.A. Porter

DEPARTMENT OF BIOLOGICAL SCIENCES
Phone: Chairman: Ext. 294 Others: Ext. 279

Chairman: L.F. Chichester
Professors: H.G. Anderson, O.R. Bissett, L.M. Carluccio, L.F. Chichester, C. Fiore, T.W. Lee, A.H. Tozloski
Associate Professors: R.A. Booth, D.J. NeNuccio, W. Fu, L.J. Gorski, D.C. Newton
Assistant Professors: R.L. Burns, R.L. Davis, G.W. Frost, D.R. Ostrander, T.S. Scheinblum, R.P. Wurst
Instructors: P.M. Kalbach
Graduate Teaching Assistants: 2

Degree Programs: Biology, Environmental Sciences, Plant Sciences

Undergraduate Degree: B.S.
Undergraduate Degree Requirements: 30-37 semester hours in Biology. Specified Chemistry, Physics, Mathematics

Graduate Degree: M.S.
Graduate Degree Requirements: 12-15 graduate courses, Thesis

CONNECTICUT COLLEGE
New London, Connecticut 06320 Phone: (203) 442-5391
Director of Graduate Studies: K. Finney
Dean of College: J. Cobb

DEPARTMENT OF BOTANY

Chairman: B.F. Thomson
Professors: R.H. Goodwin, W.A. Niering
Assistant Professors: S.L. Taylor, R.S. Warren

Degree Programs: Botany

Undergraduate Degree: B.A.
Undergraduate Degree Requirements: Not Stated

Graduate Degree: M.A.
Graduate Degree Requirements: Not Stated

DEPARTMENT OF ZOOLOGY
Phone: Chairman: Ext. 274

Acting Chairman: J.F. Kent (Fall) B. Wheeler (Spring)
Professors: J.P. Cobb, J.F. Kent, B. Wheeler
Associate Professor: P. Fell, J. Prokesch
Assistant Professors: W.B. Hunter, F.C. Roach
Teaching Assistant: 1

Degree Program: Zoology

Undergraduate Degree: A.B.
Undergraduate Degree Requirements: Introduction to Biology, Introduction to Zoology, Comparative Anatomy or Invertebrate Zoology, Mammalian Physiology or Comparative Animal Physiology, Vertebrate Embryology and minimum of three additional courses in department plus minimum of 1 year of Chemistry

Graduate Degrees: M.A., M.A.T.
Graduate Degree Requirements: Not Stated

Graduate and Undergraduate Courses Available for Graduate Credit: Theory and Practice of Electron Microscopy (U), Radiation Biology (U), Cellular Biology (U), Experimental Endocrinology (U), Vertebrate Embryology, (U), Immunology (U), Seminar (U), Marine Biology, Individual Study: Thesis

EASTERN CONNECTICUT STATE COLLEGE
Willimantic, Connecticut 06226 Phone: 423-4581
Dean of Graduate Studies: G. Moore
Dean of College: R. Wickware

DEPARTMENT OF BIOLOGY
Phone: Chairman: Ext. 260

Chairman: N. Shapiro
Professors: H. Roos, G. Rovozzo, N. Shapiro
Assistant Professor: M. Gable, B. Wulff

Degree Program: Biology

Undergraduate Degree: B.A.
Undergraduate Degree Requirements: 41 credits in Biology (minimum of 12 courses) and 1 1/2 years of mathematics, including statistics or calculus, 1 year of physics, 2 years of chemistry, including organic chemistry

FAIRFIELD UNIVERSITY
Fairfield, Connecticut 06430 Phone: (203) 255-5411
Dean of College: Rev. J.H. Coughlin

BIOLOGY DEPARTMENT
Phone: Chairman: (203) 255-5411

Chairman: F.J. Rice
Professors: J.E. Klimas, K.A. Oster (Adjunct), J. Hanks (Adjunct), D.J. Ross
Associate Professors: T.J. Combs, W. Lazaruk, F.J. Rice, B. Rosenberg (Adjunct)
Assistant Professors: M.C. Barone, W. Blogoslawski (Adjunct), S.F. Bongiorno
Visiting Lecturers: 2

Degree Programs: Biology, Pre-Medicine, Pre-Dental,

Education, Naturalist, Health

Undergraduate Degree: B.S.
Undergraduate Degree Requirements: 8 semesters Biology to include Botany, Zoology, Genetics and Ecology. 4 semesters of chemistry, 2 semesters of Physics and 2 semesters of Calculus. Plus university core program

QUINNIPIAC COLLEGE

Mt. Carmel Avenue Phone: (203) 288-5251
Hamden, Connecticut 06518
Dean of School of Allied Health and Natural Sciences: S.S. Katz

DEPARTMENT OF BIOLOGICAL SCIENCES
Phone: Chairman: (203) 288-5251 Ext. 451

Chairman: H.R. Levine
Professors: R.F. Bernard, D.F. Gordon, H.R. Levine, H.A. Rierson, B.B. Ritchie
Associate Professors: I. Beitch, D.C. Borst, A.J. Repak
Assistant Professors: R.M. Martinez, K.R. McGeary, J.W. Streett, D. Van Hemert, J.J. Woods
Part-time Faculty: 6

Degree Programs: Biology, Environmental Health Technology, Laboratory Animal Technology

Undergraduate Degrees: B.A., B.S.
Undergraduate Degree Requirements: Biology - Biology 54, Chemistry 16, Physics 8, Mathematics 6, Liberal Arts 42 total 126 credits, Environmental Health Technology - Biology 66, Chemistry 16, Physics 8, Mathematics 9, Liberal Arts 30 total 129 credits, Laboratory Animal Technology - Biology 56, Chemistry 20, Physics 8, Mathematics 9, Liberal Arts 36 total 129 credits

Graduate and Undergraduate Courses Available for Graduate Credit: Animal Anaesthesiology and Surgery (U), Animal Pathology (U), Human Embryology (U), Epidemiology (U), Environmental Health Practices I and II (U), Environmental Health Administration (U), Environmental Health Education (U), Public Health Microbiology (U), Radiation Biology (U), Biological Techniques (U), Independent Study (U), Research (U), Advanced Topics in Structural Biology, Advanced Topics in Functional Biology, Advanced Topics in Developmental Biology, Advanced Topics in Environmental Biology, Advanced Topics in Systematic Biology, Graduate Seminar, Independent Graduate Study, Methods of Teaching Biology

SACRED HEART UNIVERSITY

Bridgeport, Connecticut 06604 Phone: (203) 374-9441
Dean of College: C.E. Ford

DEPARTMENT OF BIOLOGY
Phone: Chairman: Ext. 213 Others: Ext. 253

Chairman: H.A. Denyes
Professor: H.A. Denyes
Associate Professor: W. Gnewuch
Assistant Professors: R. Green, C. Verses
Instructors: C.M. Schofield
Student Assistants: 5

Degree Programs: Biology, Biological Sciences

Undergraduate Degrees: B.A., B.S.
Undergraduate Degree Requirements: 120 credits from required and elective science courses and from the Core Program (32 credits in biology, 16 in chemistry, 6 in mathematics is the minimum requirement)

SAINT JOSEPH COLLEGE

1678 Asylim Avenue Phone: (203) 232-4571
West Hartford, Connecticut 06117
Dean of Graduate Studies: S.L. Joseph
Dean of College: Rev. C. Shaw

BIOLOGY DEPARTMENT
Phone: Chairman: 232-4571

Chairman: W.G. Stanziale
Professors: W.G. Stanziale
Associate Professor: S.L. Herald
Assistant Professor: C.L. Vigue
Instructor: L. Vigue

Degree Program: Biology

Undergraduate Degree: B.S.
Undergraduate Degree Requirements: C+ or better in major subjects average C or better in other areas. Completion of all college requirements

Graduate Degree: M.S.
Graduate Degree Requirements: Master: Completion of 30 credits good standing in department, must pass a written comprehensive

SOUTHERN CONNECTICUT STATE COLLEGE

New Haven, Connecticut 06515 Phone: (203) 397-2101
Dean of Graduate Studies: P. Rosenstein
Dean of College: L. Kuslan

BIOLOGY DEPARTMENT
Phone: Chairman: 397-2101 Ext. 355

Chairman: J.W. McClymont
Professors: S. Collins, R.L. Hutchinson, J.W. McClymont, E. Shelar, C. Steinmetz, H.Q. Stevenson
Associate Professors: D. Avery, B. Cosenza, B. Dierolf, V. Nelson
Assistant Professors: C. Bosworth, J. Cunningham, I. Leskowitz, P. Pellegrino, D. Smith, P. Tenerowicz
Instructors: N. Proctor, A. Turko
Teaching Fellows: 6

Degree Program: Biology

Undergraduate Degrees: B.A., B.S.
Undergraduate Degree Requirements: B.S. - 10 courses in Biology, 1 year of Chemistry, 1 semester Physics B.S. in Secondary Education - 8 Biology courses, 1 year of Chemistry, 1 semester of Physics B.A. - 8 Biology courses, 1 year of Chemistry

Graduate Degree: M.S.
Graduate Degree Requirements: Master: 30 hours of Biology

TRINITY COLLEGE

300 Summit Street Phone: (203) 527-3151
Hartford, Connecticut 06106
Dean of Faculty: E.P. Ney

DEPARTMENT OF BIOLOGY
Phone: Chairman: 527-3151 Ext: 374 or 408

Chairman: F.M. Child
Professors: J.W. Burger, J.M. Van Stone, F.M. Child, R.B. Crawford
Associate Professors: R.H. Brewer, D.B. Galbraith, J.E. Simmons

Degree Programs: Biochemistry, Biology

Undergraduate Degrees: B.A., B.S.
Undergraduate Degree Requirements: 12 course credits acquired through a combination of departmentan and non-departmental courses

UNIVERSITY OF BRIDGEPORT

Bridgeport, Connecticut 06602 Phone: (203) 576-4767

Dean of College: A. Schmidt

BIOLOGY DEPARTMENT
Phone: Chairman: 576-4767

Chairman: M.E. Somers
Professors: H.A. James, M.E. Somers, C.F. Spiltoir
Associate Professors: M.J. Autuori, W.A. Blogoslowski (Adjunct), A. Calabrese (Adjunct), P.M. Galton, W.E. Pistey (Adjunct), J.J. Poluhowich, N. Raghuvir, R.A. Singletary
Assistant Professors: B. Block, R.A. Busci, D.A. Hoffman, R. LeRud, J. Nelson, M. Rattner (Adjunct), J. VanGundy
Instructor: C. Kaufman
Visiting Lecturers: 2
Teaching Fellows: 6
Research Assistants: 6

Degree Programs: Anatomy and Histology, Animal Physiology, Bacteriology and Immunology, Biological Education, Biological Sciences, Biology, Ecology, Immunology, Medical Technology, Microbiology, Natural Science, Parasitology, Physiological Sciences, Zoology and Entomology, Biotechnology

Undergraduate Degree: B.S.
Undergraduate Degree Requirements: 120 credit hours/ 2.0 QPA. 8 hours Inorganic Chemistry, 6 hours English Comprehension, 4 hours Zoology, 8 hours Organic Chemistry, 8 hours Calculus, 8 hours Physics, 12 hours Foreign Lanugage, 6 Hours Humanities, 6 hours Social Sciences, 3 hours Communication, 4 hours Botany, 4 hours C.V.A., 4 hours Physiology, 2 hours Senior Seminar, 16-24 hours Biology electives

Graduate Degree: M.S.
Graduate Degree Requirements: Master: 32 credits, Thesis, Biology - Biographics 1 credit (course in graphics for publication) Bioglyphics - 1 credit (course in writing for publication)

Graduate and Undergraduate Courses Available for Graduate Credit: Ichthyology (U), Herpetology (U), Ornithology (U), Mammalogy (U), Parasitology (U), Bio-Techniques (U), Endocrinology (U), Biographics (U), Bioglyphics (U), Animal Behavior, (U), Field Biology (U), Human Environment (U), Electron Microscopy (U), Cell Fine Structure (U), Cell Physiology (U), Immunology (U), Invertebrate Paleontology, Evolution, Biogeography, Entomology, Helminthology, Marine Ecology, Limnology, Microbiology Ecology, Estuarine Biology, Comparative Physiology, Cell and Molecular Biology, Neurobiology

UNIVERSITY OF CONNECTICUT

Storrs, Connecticut 06268 Phone: (203) 486-2000
Dean of Graduate Studies: H. Clark

College of Liberal Arts and Sciences

Dean: R.W. Lougee

BIOLOGICAL SCIENCES GROUP
Phone: Chairman: (203) 486-4313 Others: 486-4314

Head: W.K. Purves
Professors: H.N. Andrews, J.A. Cameron, A. Chovnick, H. Clark, R.P. Collins, N.T. Davis, P.F. Goetinck, H. Herrmann, S.M. Heywood, G. Kegeles, C.A. Kind, N.W. Klein, H. Laufer, P.I. Marcus, G.A.L. Mehlquist, L.R. Penner, L.J. Pierro, G.R.J. Pilar, W.K. Purves, J.S, Rankin, C.W. Rettenmeyer, A.H. Romano, J.S. Roth, R.J. Schultz, J.A. Slater, J.F. Spryer, F.R. Trainor, F.D. Vasington, A. Wachtel, D.F. Wetherell, R.M. Wetzel, D.A. Yphantis
Associate Professors: C. Berg, E.H. Braswell, A.H. Brush, J.D. Buck, W.D. Chapple, G.A. Clark, N.B. Clark, A.W.H. Damman, K.A. Doeg, S.Y. Feng, I. M. Greenblatt, T.F. Hopkins, E.A. Khairallah, H.V. Koontz, J. Lucas-Lenard, W.A. Lund, B. Mundkur, H.W. Pfeifer, A.H. Phillips, C.W. Schaefer, T. Schuster, T.L. Schwartz, J.L. Scott, F.A. Streams, T.R. Webster
Assistant Professors: G. Anderson, J. Knox, H.M. Krider, R.S. Norman, P. Rich, I. Schwinck, G.I. Stage, T.M. Terry, R.T. Vinopal, B.L. Welsh
Instructors: A. Indars

Field Stations/Laboratories: Marine Sciences Institute, with biological laboratories at Noank and at Avery Point, Connecticut

Degree Programs: Animal Genetics, Animal Physiology, Biochemistry, Biochemistry and Biophysics, Biology, Biophysics, Botany, Ecology, Entomology, Entomology and Parasitology, Genetics, Microbiology, Parasitology, Physiology, Zoology and Entomology, Cell Biology

Undergraduate Degrees: B.S.
Undergraduate Degree Requirements: Not Stated

Graduate Degrees: M.S., Ph.D.
Graduate Degrees Requirements: Not Stated

SYSTEMATIC AND EVOLUTIONARY BIOLOGY SECTION
Phone: Chairman: (203) 486-4322

Section Head: J.A. Slater
Professors: H. Andrews, G. Mehlquist, L. Penner, C. Rettenmeyer, J. Slater, R. Wetzel
Associate Professors: G. Clark, H. Pfeifer, C. Schaefer, T. Webster
Assistant Professors: G. Anderson, G. Stage

Field Stations/Laboratories: Noank Marine Laboratory - Noank

Degree Programs: Plant Systematics, Paleobotany, Entomology, Parasitology, Systematics and Evolution

Graduate Degrees: M.S., Ph.D.
Graduate Degree Requirements: Master: M.S. (A) - 15 course credits and Thesis (minimum) M.S. (B) - 24 course credits and Thesis (minimum) Doctoral: No formal course requirements - 3-man committee prepares program - Graduate School Approval Needed.

College of Agricultural and Natural Resources

Dean: E.J. Kersting

DEPARTMENT OF ANIMAL INDUSTRIES
Phone: Chairman: (203) 486-2413 Others: 486-2000

Head: W.A. Cowan
Professors: W.A. Alo, R.H. Benson, W.A. Cowan, W.S. Gaunya, N.S. Hale, D.M. Kinsman, A.C. Smith
Associate Professors: L.R. Brown, R.C. Church, L.R. Glazier, J.M. Kays, L.A. Malkus, D.W. Talmadge, C.O. Woody
Assistant Professors: J.W. Riesen
Research Assistants: 6

Degree Programs: Animal Industried, Food Science, Physiology of Reproduction

Undergraduate Degree: B.S. (Animal Industries)
Undergraduate Degree Requirements: 120 credits, quality point requirements, distribution of requirements, requirements for a major

Graduate Degrees: M.S., Ph.D.
Graduate Degree Requirements: Master: Plan A - Thesis, 15 advanced credits, Plan B - 24 advanced credits (Advisory committee may increase) Doctoral: variable

Graduate and Undergraduate Courses Available for Graduate Credit: Advanced Animal Breeding, Milk Secretion,

Physiology of Livestock

DEPARTMENT OF ANIMAL GENETICS
Phone: Chairman: (203) 486-2427 Others: 486-4124

Head: L.J. Pierro
Professors: P.F. Goetinck, N.W. Klein, L.J. Pierro
Research Assistants: 3
Instructor/Laboratory Aide: 1

Degree Programs: Ph.D. and M.S. offered in cooperation with Biological Sciences Group.

PLANT SCIENCE DEPARTMENT
Phone: Chairman: (203) 486-2924 Others: 486-2000

Head: J.R. Guttay
Professors: F.H. Emmert, R.J. Favretti, A.J.R. Guttay, J.S. Koths, J.M. Lent, R.D. McDowell, G.A.L. Melquist, R.A. Peters, E.J. Rubins, M.G. Savos, W. W. Washko, R.W. Wengel
Associate Professors: D.W. Allinson, R.A. Ashley, K.A. Bradley, E.D Carpenter, Jr., E.J. Duda, M.D. Ferrill, G.F. Griffin, W.L. Harper, D.A. Kollas, L.A. Mitterling, R.J. Schramm, Jr., D.B. Schroeder, S. Waxman, W.R. Whitworth
Assistant Professor: D.R. Miller
Visiting Lectures: 1
Research Assistants: 7
Lecturer: J. Scarchuk

Field Stations/Laboratories: Agronomy Research Station - Storrs, Horticulture Research Station - Storrs, Pomology Research Station - Storrs, Vegetable Research Station - Coventry, Bartlett Arboretum - Stamford

Degree Programs: Agronomy, Conservation, Fisheries and Wildlife Biology, Horticultural Science, Horticulture, Landscape Horticulture, Natural Resources, Plant Industry, Plant Sciences, Soil Science, Vegetable Crops, Wildlife Management, Icthyology

Undergraduate Degree: B.S. (Agriculture, Natural Resources)
Undergraduate Degree Requirements: Earned a total of 120 course credits. Have a CQPR of 20 as an Upper Division student. Met the group distribution requirement of the College and earned not less than 30 course credits in a concentration of study.

Graduate Degrees: M.S., Ph.D.
Graduate Degree Requirements: Master: No less than 15 credits of advanced course work and a research Thesis. Doctoral: A competent reading knowledge of at least one foreign language or at least 6 credits of advanced work in a supporting area. Pass a general written and oral examination. Complete an acceptable Dissertation that makes a significant contribution to the candidates field of specialization. Pass an oral final examination dealing mainly with the subject of the dissertation.

Graduate and Undergraduate Courses Available for Graduate Credit: Advanced Plant Breeding, Advanced Study of Economic Plants, Community Landscapes, Post-harvest Physiology, Plant Response to the Environment, Crop Ecology, Soil Microbiology, Soil Chemistry, Soil Physics, Soil Analysis, Ecological Concepts of Open Space Planning, Environmental Measurements, Forest and Field Microclimatology, Wildlife Ecology, Wildlife Field Study, Advanced Wildlife Ecology, Wildlife Management, Ecology of Fishes, Field Studies in Limnology

DEPARTMENT OF NUTRITIONAL SCIENCES
Phone: Chairman: (203) 486-3633

Chairman: K.L. Knox
Professors: J. Czajkowski, H.D. Eaton, R.G. Jensen, K.L. Knox, I. MacKellar, E.P. Singsen, R.G. Somes
Associate Professors: K.N. Hall, W.J. Pudelkiewicz, J. E. Rousseau
Instructors: J. Boelke (The Waterbury Branch), M.W. Wheeler (The Stamford Branch)
Visiting Lecturers: 1
Research Assistants: 4

Degree Programs: Animal Genetics, Biochemistry, Food Science, Nutrition, Nutritional Biochemistry, Nutritional Pathology

Undergraduate Degree: B.S. (Nutrition)
Undergraduate Degree Requirements: a total of 120 credits, at least 20 times as many quality points as the total number of credits for which they have been registered as upper division students. Meet all requirements of the College of Agriculture and Natural Resources

Graduate Degrees: M.S., Ph.D.
Graduate Degree Requirements: Master: The degrees of Master of Science may be earned under either of two alternate plans as determined by the Advisory Committee. The first plan emphasizes research, the second requires a comprehensive understanding of a more general character. Plan A requires not less than 15 credits of advanced work and the writing of a Thesis, while Plan B requires not less than 24 credits of advanced work and no thesis. In either case, Advisory Committees may require more than the minimum number of credits. Doctoral: The equivalent of at least three years of full time study beyond the Bachelor's Degree or two years beyond the Master's Degree is required. When not more than 12 credits of appropriate course work has been completed the student shall have prepared a plan of study with the aid and approval of his Advisory Committee and file the completed form with the Graduate Record's Office for the approval of the Executive Committee. The Plan of work beyond the Master's Degree or its equivalent. Students shall be required to have a complete reading knowledge of at least one foreign language, or at least six credits of advanced work in the supporting area. Students must pass a general examination, conduct independent research, and defend a dissertation in a final examination.

Graduate and Undergraduate Courses Available for Graduate Credit: Advanced Nutrition, Malnutrition, Lipids, Advanced Therapeutic Nutrition, Field Work in Community, Independent Study, Principles of Nutrition (U), Principles of Food Science (U), Readings in Human Nutrition (U), Therapeutic Nutrition (U), Principles of Community Nutrition (U)

DEPARTMENT OF PATHOBIOLOGY
Phone: Chairman: (203) 486-4000

Head: R.W. Leader
Professors: W.H. Daniels, T.N. Fredrickson, A.J. Kenyon, R.W. Leader, R.E. Luginbuhl, S.W. Neilson, M.E. Tourtellotte, H.J. Van Kruiningen, D.S. Wyand, L. Van der Heide
Associate Professors: E.S. Bryant, C.N. Burke, S.N. Kim, W.J. Parizek, L.F. Williams
Instructor: R.P. Pirozok
Research Assistants: 17
Research Associates: 2
Graduate Assistants: 2

Degree Programs: Bacteriology and Immunology, Biochemistry and Biophysics, Pathology, Viticulture and Enology, Wildlife Diseases

Graduate Degrees: M.S., Ph.D.
Graduate Degree Requirements: Master: As prescribed by Committee Doctoral: As prescirbed by Committee

Graduate and Undergraduate Courses Available for Graduate Credit: Advanced Pathobiology, Avian Pathology, Comparative Hematology (U), Comparative Oncology,

Comparative Virology, Diagnostic Veterinary Microbiology, Diseases of Wildlife, Diseases of Domestic Mammals (U), Diseases of Poultry (U), Experimental Hematology, Experimental Virologic Techniques (U), Histologic Structure and Function (U), Pathobiology (U), Research and Independent Study in Animal Diseases, Veterinary Pathology, Seminar, Veterinary Histopathology, Veterinary Chemical Pathology, Ultrastructure I, Ultrastructure II, Ultrastructure IV

UNIVERSITY OF CONNECTICUT HEALTH CENTER

Farmington, Connecticut 06032 Phone: (203) 674-2000
Dean of Graduate School: P. Rice
Dean of School of Medicine: R.U. Massey

DEPARTMENT OF ANATOMY
Phone: Chairman: (203) 674-2617 Others: 674-2000

Head: S.J. Cooperstein
Professor: S.J. Cooperstein
Associate Professor: D.T. Watkins
Assistant Professors: R.L. Church, M.R. Kalt, R.A. Kosher, R.T. Lobl, J.F. McKelvy, C.H. Phelps
Teaching Fellows: 2
Clinical Associate: 1
Advanced Graduate Students: 4

Degree Programs: Anatomy, Cell Biology

Graduate Degrees: Ph.D.
Graduate Degree Requirements: Doctoral: Course work 40 hours, Reading proficiency in 1 language, Qualifying Examination: Written and Oral, Original Laboratory research as basis of thesis

Graduate and Undergraduate Courses Available for Graduate Credit: Biomedical Sciences, Neurobiology, The Biology of Mineralized Tissue, Neurochemistry, Contemporary Research Techniques in Anatomy and Cell Biology, Human Gross Anatomy, Human Histology, Developmental Biology

DEPARTMENT OF BIOCHEMISTRY
(No information available)

DEPARTMENT OF MICROBIOLOGY
Phone: Chairman: (203) 674-2318

Chairman: L.I. Rothfield
Professors: M.J. Osborn, L.I. Rothfield
Associate Professors: S. Pfeiffer, H.C. Wu
Instructors: S. Taube
Postdoctoral Research Associates: 15

Degree Program: Molecular Biology

Graudate Degree: Ph.D.
Graduate Degree Requirements: Doctoral: Programs will usually include at least five of the courses indicated below.

Graduate and Undergraduate Courses Available for Graduate Credit: Biochemistry and Cell Biology, Biochemistry and Molecular Biology, Laboratory Methods, Molecular Biology of Eucaryotic Cells, Supramolecular Organization, Physical Chemistry of Macromolecules, Biochemical Regulatory Mechanisms, Membrane Molecular Biology, Microbial Genetics

DEPARTMENT OF PATHOLOGY
Phone: Chairman: (203) 674-2316 Others: 674-2000

Chairman: P.A. Ward
Professors: E.L. Becker, S. Cohen, P.B. Hukill, P.A. Ward
Associate Professors: P.J. Goldblatt, I. Goldschneider, R.R. Lindquist, R.M. Maenza
Assistant Professors: R.B. Cogen, P.B. Conran, I. Damjanov, E.M. Gross, D. Krutchkoff, J. Oliver, P. Rinaudo, T. Yoshida
Instructor: C.G. Gillies
Visiting Lecturers: 1
Teaching Fellows: 2
Research Assistants: 19

Degree Program: Immunology

Graduate Degree: Ph.D.
Graduate Degree Requirements: Doctoral: Not Stated

Graduate and Undergraduate Courses Available for Graduate Credit: Not Stated

UNIVERSITY OF HARTFORD

200 Bloomfield Avenue Phone: (203) 243-4534
West Hartford, Connecticut 06117
Dean of College: F. Chiarenza

DEPARTMENT OF BIOLOGY
Phone: Chairman: (203) 243-4534 Others: 243-4531

Chairman: E.R. Swain
Professors: S. Pond (Emeritus), E.R. Swain
Associate Professors: W. Duff (Adjunct) R.H. Gwynn, T. Simpson
Assistant Professors: G.T. Brierley (Adjunct), W. Coleman, J. Lylis, T. Maguder
Teaching Fellows: 4

Field Stations/Laboratories: Great Mountain Forest - Norfolk

Degree Programs: Biology, Medical Technology

Undergraduate Degrees: B.A., B.S.
Undergraduate Degree Requirements: Biology - 32 credits Chemistry - 16 credits, Mathematics 8 credits

Graduate Degrees: M.A., M.A.T.
Graduate Degree Requirements: Master: 30 credits including 6 credits thesis research

Graduate and Undergraduate Courses Available for Graduate Credit: Plant Physiology (U), Plant Morphology (U), Vertbrate Anatomy (U), Vertebrate Embryology (U), Microbiology (U), Cytology (U), Cell Physiology (U), Ecology (U), Freshwater Biology (U), Genetics (U), Invertebrate Zoology (U), Mammalian Physiology (U), Field Marine Biology (U), Biochemistry, Histology, Topics in Literature of Biology, Topics in Research Techniques in Biology, Thesis, research

UNIVERSITY OF NEW HAVEN

West Haven, Connecticut 06516 Phone: (203) 934-6321
Dean of Graduate Studies: J. Parker
Dean of Arts and Sciences: D. Robillard

DEPARTMENT OF BIOLOGY AND GENERAL SCIENCE
Phone: Chairman: 934-6321 Ext. 263 Others: 261

Chairman: H.F. Wright
Professor: H.F. Wright
Associate Professors: D.C. Reams, B.T. Staugaard
Assistant Professors: D.L. Kalma, H.E. Voegeli
Instructor: J.W. Blaskey
Adjunct Lecturers: 6
Laboratory Assistants: 3

Degree Programs: Biology, Environmental Studies

Undergraduate Degrees: B.A., B.S.
Undergraduate Degree Requirements: Basic core courses plus electives in field of specialty.

WESLEYAN UNIVERSITY

Hall-Atwater and Shanklin Laboratories

Lawn Avenue Phone: (203) 347-9411
Middletown, Connecticut 06457

DEPARTMENT OF BIOLOGY
Phone: Chairman: (203) 347-9411 Ext. 365 Others: 307

Chairman: E.D. Hanson
Professors: V.W. Cochrane, W. Firshein, R.A. Gortner, E.D. Hanson, L.N. Lukens
Associate Professors: S.J. Berry, A.A. Infante, B.I. Kiefer
Assistant Professors: N.M. Allewell, A. Berlind, J.J. Donady, C.B. Lynch, J.S. Wolfe
Research Assistants: 13
Post-doctoral Fellows: 2

Degree Program: Biology

Undergraduate Degree: B.S.
Undergraduate Degree Requirements: Introduction to Biology (2 semesters) or equivalent; one semester Cell Biology and one semester Genetics, 4 semesters upper level work in Biology (two semesters in cognate areas, by persion, e.g. Biochemistry or Physiological Psychology), Chemistry through Organic (usually 4 semesters) 2 semesters of Physics

Graduate Degrees: M.S., Ph.D.
Graduate Degree Requirements: Master: 2 years of course work (determined individually), research thesis, and comprehensive examinations Doctoral: Course work (determined individually, always includes Biochemistry); Comprehensive (qualifying) examination; reading knowledge of at least one foreign language; thesis and thesis examination

Graduate and Undergraduate Courses Available For Graduate Credit: Microbiology (U), Immunology (U), Evolution (U), Endocrinology (U), Biophysics (U), Cellular Basis of Development (U), Molecular Biology (U), Neurophysiology (U), Advanced Genetics (U), Comparative Animal Behavior (U), Ecology (U), Topics in Developmental Biology, Topics in Cell Biology, Science in Society, Analytical Cytology, Virology, Biochemistry, Biochemistry of Cell Differentiation, Developmental Genetics, Individual tutorials, Selected Topics in Modern Biology, Practicum (Laboratory Techniques), Advanced Research, And selected courses from the other graduate department offerings.

WESTERN CONNECTICUT STATE COLLEGE
181 White Street Phone: (203) 792-1400
Danbury, Connecticut 06810
Dean of Graduate Studies: M.J. Rudner
Academic Dean: G. Braun

DEPARTMENT OF BIOLOGICAL AND ENVIRONMENTAL SCIENCES
Phone: Chairman: (203) 792-1400 Ext. 207

Chairman: R.M. Dole Jr.
Associate Professors: R.M. Dole, Jr., F.J. Dye, W.J. Esposito, J.D. Kreizinger, R. McMahon
Assistant Professors: T.M. Buterworth, K. Kanungo, W.W. Rasor, W.J. Smith
Instructor: P.B. Chipman

Degree Programs: Liberal Arts Biology Program, Secondary Education Biology Program

Undergraduate Degrees: B.A., B.S.
Undergraduate Degree Requirements: General Biology, Ecology, Animal Physiology or Plant Physiology, Genetics, Cell Physiology, Senior Research, General Chemistry, Organic Chemistry, 14 hours Biology Electives; 46 hours General Education, 20 hours Free electives, Secondary Education, Biology, Biology Sequence same;, Prof. Ed. from 34 Elective hours above

Graduate Degree: (College has a M.S.E.D. Graduate Biology courses exist to support this degree)

Graduate and Undergraduate Courses Available for Graduate Credit: Advanced Genetics (U), Evolution (U), Field Biology, Ecolosy and Conservation, Marine Biology, Marine Ecology, Encountering the Environment, Selected Topics in Animal Development, Newer Developments in the Biological Sciences

YALE UNIVERSITY
New Haven, Connecticut 06520 Phone: (203) 436-8320

Yale College
Dean of College: H. Taft
Dean of Graduate School: D.W. Taylor
Director of Graduate Studies: I.M. Sussex

DEPARTMENT OF BIOLOGY
Phone: Chairman: (203) 436-2139

Chairman: T.H. Goldsmith
Ross Granville Harrison Professor of Experimental Zoology: E.J. Boell
Henry Ford II Professor of Biology: C. Markert
Eugene Higgins Professor of Biochemistry: J.S. Fruton
William Robertson Coe Professor of Ornithology: C.S. Sibley
Eaton Professor of Botany: A.W. Galston
Professors: E.J. Boell, F.H. Bormann, M.J. Cohen, M.B. Davis, J.S. Fruton, J.G. Gall, A.W. Galston, T.H. Goldsmith, C.L. Markert, R.S. Miller, D.F. Poulson, L. Provasoli (Adjunct), F.H. Ruddle, C.S. Sibley, B B. Stowe, I.M. Sussex, J.P. Trinkaus, T.H. Waterman
Associate Professors: M.H. Goldsmith, W.D. Hartman, D. Merriman, A. Novick, J.S. Ramus, C.L. Remington, J.L. Rosenbaum, K.S. Thomson, R.C. Williams, R.H. Wyman
Assistant Professors: S. Altman, M.L. Ernst-Fonberg, D.R. Kankel, J.A.W. Kirsch, L.T. Landmesser, L.N. Ornston, J.R. Powell, P.M.M. Rai, J.E. Rodman, R.K. Trench
Visiting Lecturers: 6
Teaching Fellows: 24
Research Assistants: 19
Research Associates: 8
Research Staff Biologists: 10

Field Stations/Laboratories: Yale Biological Station - Guilford, Horse Island - Thimble Islands

Degree Programs: Neurobiology and Comparative Physiology, Molecular Biology and Biochemistry, Cellular, Genetic and Developmental Biology, Evolutionary and Ecological Biology

Undergraduate Degree: B.S. (Biology)
Undergraduate Degree Requirements: Introduction to Biology, 3 of the following with laboratories: Biochem istry, Genetics, Cell Biology, Ecology, Evolutionary Biology, Chemistry and Organic Chemistry with laboratories, Physics with laboratory, Mathematics (Calculus) 2 courses; four additional term courses in biology above the introductory level

Graduate Degrees: M.S., M. Ph., Ph.D.
Graduate Degree Requirements: Master: 5 course gradues (or equivalent) of High Pass, 1 full year of graduate study in residence Doctoral: 1, Minimum of 3 academic years in residence, 2) general oral or written qualifying examination in major subject, 3) foreign language examination, presentation of acceptable dissertation prospectus, submission and defense of a dissertation judged acceptable by the faculty of the department

School of Medicine
Dean: R.W. Berliner

Dean of Graduate School: J. Pelikan

DEPARTMENT OF ANATOMY
Phone: Chairman: (203) 436-3691 Others: 436-4771

Chairman: R.J. Barrnett
Professor: R.J. Barrnett
Associate Professors: J.A. Higgins, T.L. Lentz
Assistant Professors: S.B. Andrews
Postdoctoral Fellows: 6

Degree Program: Anatomy

Graduate Degrees: M.S., Ph.D.
Graduate Degree Requirements: Master: Successful completion of two years of course work and qualifying examinations Doctoral: Successful completion of two years of course work, qualifying examinations, defense of thesis proposal , Dissertation

Graduate and Undergraduate Courses Available for Graduate Credit: Nuerobiology, Cell Biology, Developmental Cytology, Principles and Methods in Electron Microscopy, Histochemistry and Cytochemistry, Applications of Cytochemistry to Electron Microscopy, Neurocytology, Cellular Membranes, Neuro bases of Behavior, Techniques in Developmental Biology

DEPARTMENT OF MOLECULAR BIOPHYSICS AND BIOCHEMISTRY
Phone: Chairman: (203) 436-0648

Henry Ford II Professor of Molecular Biophysics and Biochemistry: F.M. Richards
Eugene Higgin Professor of Biochemistry: J. Fruton
Faculty: Mr. Altman, Dr. Coleman, Mr. Cronan, Mr. Crothers, Mr. Engelman, Mr. Fruton, Mr. Garen, Mr. Howard-Flanders, Mr. Konigsberg, Mr. Lenggel, Mr. Markert, Mr. Moore, Mr. Morowitz, Mr. Paulson, Dr. Radding, Dr. Reid, Mr. Ruddle, Mr. Rupp, Mr. Schmir, Ms. Simmons, Mr. Söll, Ms. Steitz, Dr. Steyer, Dr. Summers, Mr. Ward, Mr. Wyckoff
Lecturer: I. Zelitch
Teaching Fellows: 11

Degree Programs: Biochemistry and Biophysics

Graduate Degrees: M.S., Ph.D.
Graduate Degree Requirements: Master: Successful completion of two years of course work and qualifying examination Doctoral: Successful completion of two years of course work, qualifying examinations, defense of thesis proposal, Dissertation

Graduate and Undergraduate Courses Available for Graduate Credit: Principles of Biochemistry, Introduction to Biochemistry, Molecular Biophysics, Molecular Biophysics - X-ray Diffraction, Molecular Biophysics - Macromolecular Structure and Function, Biochemical Mechanisms of Human Disease, Developmental Genetics, Introduction to Research for First-Year Students, X-Ray Diffraction Analysis of Structure, Special Topics, Theoretical Biology, Enzyme Catalysis and Biochemical Processes, Chemistry of Biological Systems, Molecular Genetics I, Historical Topics in Biochemistry, Advanced Laboratory in Biochemistry and Biophysics, Molecular Genetics II, Chemical Properties of Nucleic Acids, Physical Properties of Nucleic Acids, Metabolic Processes and Their Control, The Mechanism of Enzymatic Reactions, Membrane Structure, Perspectives in Plant Biochemistry, Protein Biosynthesis, Molecular Virology, Animal Viruses, The Structure and Genetics of Immunoglobins, Mechanics, Electromagnetic Theory, Mathematics Methods of Physics, Quantum Mechanics I, Thermodynamics and Statistical Mechanics, Cellular Transport Processes, Molecular Biophysics and Biochemistry, Seminar in Current Research

DEPARTMENT OF PHYSIOLOGY
Phone: Chairman: (203) 436-8935

Chairman: J.F. Hoffman
Sterling Professor of Physiology: G.H. Giebisch
Eugene Higgins Professor of Physiology: J.F. Hoffman
Professors: W.K. Chandler, G.H. Giebisch, J.F. Hoffman
Associate Professors: T.U.L. Biber, E. Boulpaep, L. Cohen, C. Michael, J. Sachs, G. Shepherd, C. Slayman, T. Thach, R. Tsein, F. Wright
Assistant Professors: T. Getchell, S. Long
Instructor: Y. Chan
Research Assistants: 3
Visiting Professors: 2

Degree Program: Physiology

Graduate Degree: Ph.D.
Graduate Degree Requirements: Doctoral: Series of Courses, Written Qualifying Examination, Oral Thesis Proposal, Written Thesis

School of Forestry

Dean: F. Mergen

DEPARTMENT OF FORESTRY AND ENVIRONMENTAL STUDIES
Phone: Chairman: (203) 436-0440

Chairman: F. Mergen
Pinchot Professor of Forestry: F. Mergen
Oastler Professor of Forest Ecology: F.H. Bormann
J.P. Weyerhaeuser, Jr., Professor of Forest Management: G.M. Furnival
Oastler Professor of Wildlife Ecology: R.S. Miller
Morris K. Jesup Professor of Silviculture: D.M. Smith
Margaret K. Musser Professor of Forest Soils: G.K. Voigt
Edwin W. Davis Professor of Forest Policy: A.C. Worrell
Professors: M.B. Davis, A.W. Galston, W.F. Reifsynder, B.B. Stowe
Associate Professors: W.H. Smith, G.P. Berlyn, D.B. Botkin, F.T. Ledig
Assistant Professor: T.C. Siccama
Lecturers: S.H. Berwick, M.L. McManus, P.E. Waggoner, W.T. Wilson, G.M. Woodwell

Field Stations/Laboratories: Union Forest - Union, Keene Forest - Keene, Great Mountain Forest - Norfolk

Degree Program: Forest Science

Graduate Degrees: M.S., Ph.D.
Graduate Degree Requirements: Not Stated

Graduate and Undergraduate Courses Available for Graduate Credit: Forest Vegetation, Mensuration, Surveying, Forest Biology, Physiology of Trees, Forest Genetics, Genecology, Biotic Changes in Ecosystems, Soil Science, Biometerology, Micrometerology, Solar and Ratiation Processes, Air Pollution Meterology, Forest Hydrology, Forest Tree Pathology, Air Pollution and Plants, Pesticides and Ecosystems, Forest Entomology, Wildlife Ecology, Population Ecology, Pattern and Process in Terrestrial Ecosystems, Forest Ecology, Seminar on the Biosphere, Research Methods in Ecosystem Analysis, Theoretical Approaches to Ecology, Ecological Modeling and Simulation, Ecosystem Concepts, Anatomy and Morphogenesis of Vascular Plants, Research Methods in Anatomy and Physiology of Trees, Silviculture

DELAWARE

DELAWARE STATE COLLEGE
(No reply received)

UNIVERSITY OF DELAWARE
Newark, Delaware 19711 Phone: (302) 738-2000
Dean of Graduate School: A.L. Lippert

College of Agricultural Sciences
Dean: W.M. McDaniel

DEPARTMENT OF ENTOMOLOGY AND APPLIED ECOLOGY
Phone: Chairman: (302) 738-2526

Chairman: D.M. Bray
Professors: D.F. Bray, P.P. Burbutis, N.G. Patel (Adjunct)
Associate Professors: E.P. Catts, W.A. Connell, L.P. Kelsey
Assistant Professors: F.J. Murphey, R.R. Roth, R.W. Rust
Research Assistants: 3
Extension Specialist: 1
Agricultural Chemical Specialist: 1
Pest Management Specialist: 1

Field Stations/Laboratory: Georgetown Substation - Georgetown

Degree Programs: Entomology, Entomology - Plant Pathology

Undergraduate Degree: B.S.
Undergraduate Degree Requirements: 30 credit hours in major: Required: Fundamentals of Entomology, Elements of Entomology, Systematic Entomology, Fundamentals of Ecology, Ecology Laboratory, Entomology Seminar, English and Mathematics

Graduate Degree: M.S.
Graduate Degree Requirements: Master: 30 credit hours beyond the B.S., minimum of 12 in Entomology, including 1 credit seminar and 6 in thesis research

Graduate and Undergraduate Courses Available for Graduate Credit: Systematic Entomology, Medical Entomology, Economic Entomology, Research Techniques in Ecology, Ecology Laboratory, Identification and Comparative Ecology of Insects, Principles of Biological Control, Insect Toxicology, Colicidology, Mammalogy, Ornithology, Special Problems (U), Advanced Medical Entomology, Advanced Systematics Entomology, Immature Insects, Insect Morphology, Insect Physiology, Control Systems in Insects, Special Problems, Seminar, Thesis

DEPARTMENT OF PLANT SCIENCES
Phone: Chairman: 738-2531 Others: 2411

Chairman: M.R. Teel
Professors: C.W. Dunham, D.J. Fieldhouse, W.H. Mitchell, M.R. Teel
Associate Professors: 3
Assistant Professors: 1
Instructors: 1
Research Assistant: 1

Field Stations/Laboratories: University Substation - Georgetown

Degree Programs: Agronomy, Plant Pathology, Plant Sciences

Graduate Degrees: M.S., Ph.D.
Graduate Degree Requirements: Master: twenty-four semester hours plus six hours of thesis Doctoral: 1. completed one full year of graduate work 2) passed his foreign language examinations, 3) passed his qualifying examination, 4) shown ability in carrying on research, 5) had a research project accepted by his advisory committee and 6) had a program of study approved

College of Arts and Science
Dean: M. Guttentag

DEPARTMENT OF BIOLOGICAL SCIENCES
Phone: Chairman: (302) 738-2281 Others: 738-2282

Chairman: W.S. Vincent
H. Fletcher Brown Professor: G.F. Somers
Professors: R.W. Bailey, A.M. Clark, F.C. Daiber, G.F. Somers, R.W. Stegner, M.R. Tripp, W.S. Vincent
Associate Professors: R.L. Boord, D.W. Francis, D.S. Herson, J.B. Krause, V.A. Lotrich, P.D. Lunger, J.J. Pene, D.E. Sheppard, S.D. Skopik
Assistant Professors: R.M. Eisenberg, A.B. Gould, R.C. Hodson, L.E. Hurd, H. Ling, S.C. Loken, T.D. Myers, J. Noble-Harvey, L.G. Parchman, M.H. Stetson, R.J. Tasca, M.H. Taylor, R.C. Wagner

Field Stations/Laboratories: Lewes Field Station - Lewes

Degree Program: Biological Sciences

Undergraduate Degree: B.S.
Undergraduate Degree Requirements: Not Stated

Graduate Degrees: M.S., Ph.D.
Graduate Degree Requirements: Not Stated

Graduate and Undergraduate Courses Available for Graduate Credit: Not Stated

DISTRICT OF COLUMBIA

THE AMERICAN UNIVERSITY

Massachusetts and Nebraska Avenues Phone: (202) 686-2000
Washington, D.C. 20016
Dean of Graduate Studies: T. Owen
Dean, College of Arts and Sciences: R. Berendzen

BIOLOGY DEPARTMENT
Phone: Chairman: (202) 686-2177 Others: 686-2000

Chairman: B.R. Griffin
Professors: E.J. Breyere, S.O. Burhoe, G.N. Catravas,
Associate Professors: R.R. Anderson, W.C. Banta, R.H. Fox, R.L. Strautz
Assistant Professors: M.A. Champ, B.J. Clarke, B.R. Griffin, C.R. Wrathall
Teaching Fellows: 6
Research Assistants: 2
Departmental Honor Award: 1

Field Stations/Laboratories: Marine Science Laboratories - Wallops Island, Virginia, Lewes

Degree Programs: Applied Ecology, Bacteriology, Bacteriology and Immunology, Biology, Genetics, Immunology, Medical Technology, Microbiology, Physiology, Plant Sciences, Science Education

Undergraduate Degree: B.S.
Undergraduate Degree Requirements: Ten courses in Biology, to include, General Biology I and II, Morphogenesis of Vertebrates, Genetics, Cell Biology, a course in Plant Science, plus approved electives. Organic Chemistry I and II, Introduction to Physics I and II, Elements of Calculus, and Basic Statistics (Plus electives to equal 32 courses, to satisfy University requirements).

Graduate Degree: M.S.
Graduate Degree Requirements: Master: A total of 32 semester hours of approved coursework, to include 2 semester hours of graduate seminar and 6 semester hours of Master's Thesis research

Graduate and Undergraduate Courses Available for Graduate Credit: Microbial Physiology, Systems Ecology, Cooperative Work Study at the National Aquarium, Cooperative Work Study at the Lightship Chesapeake, Plant Taxonomy, Comparative Mammalian Embryology, Animal Histology, Parasitism and Symbiology, Endocrinology, Immunology, Virology, Scientific Publication, Water Pollution Biology, Water Pollution Biology Laboratory, Human Genetics, Pollution in the Aquatic Environment, Microbial Ecology, Marine Ecology, Marine Ecology laboratory, Graduate Seminar, Transplantation Biology, Advanced Cell Physiology, Radiation Biology, Bacterial Genetics, Experimental Embryology Master's Thesis Research, Special Topics in Ecology (also in Vertebrate Physiology, Microbiology, Genetics, Immunology, Zoology), Limnology, Introduction to Oceanography

THE CATHOLIC UNIVERSITY OF AMERICA

Washington, D.C. 20064 Phone: (202) 635-5267
Dean: E.R. Kennedy

DEPARTMENT OF BIOLOGY
Phone: Chairman: (202) 635-5267 Others: 635-5000

Chairman: B.T. DeCicco
Professors: R.A. Davidson, E.R. Kennedy, R.M. Nardone
Associate Professors: D.C. Braungart, E.C. Cutchins, B.T. DeCicco, J.A. O'Brien, A.M. Peadon
Assistant Professors: M.A. Johnson (Adjunct, S. Polivanov, J.H. Roberts, M. Sochard
Teaching Fellows: 11
Research Assistants: 2
Clinical Associates: 2

Degree Programs: Bacteriology, Biochemistry, Biological Sciences, Biomedical Engineering, Medical Technology, Microbiology, Oceanography, Cell Biology

Undergraduate Degrees: B.A., B.S.
Undergraduate Degree Requirements: B.A. - 40 courses- 6 chemistry, 10 Biology, 2 Physics, 2 Mathematics, 20 Non-Science B.S. 30 courses and one year (12 months) hospital internship

Graduate Degrees: M.S., Ph.D.
Graduate Degree Requirements: Master: M.S. 24 course credits (18 major - 6 minor) and 6 research and Thesis M.T.S. 30 course credits (18 major - 12 minor) No Thesis Doctoral: Ph.D. 35 credits in major - 18 in minor - 2 Foreign languages and Dissertation D.A. 35 credits in major - 22 in minor - 2 foreign languages and Dissertation

Graduate and Undergraduate Courses Available for Graduate Credit: Virology, Pathogenic Bacteriology (U), Immunology (U), Bacterial Genetics, Biological Law and Theories (U), Microbiology, Metabolism I and II, Advanced Virology, General Microbiology (U), Biology Consortium Seminar, Microbiology Seminar, Cell Biology Seminar, Methods in Biological Research, Biomedical Instrumentation (U), Introduction to Radiation Biology, General Cytology (U), Advanced Cytology, Physiology (U), Cell Physiology, Ecology (U), Human Genetics (U), Developmental Genetics (U), Evolution (U), Population Genetics (U), Biostatistics (U), Enzymology, Biosynthesis and Regulation of Macromolecules, Clinical Biochemistry, Plant Taxonomy, Biochemistry I and II

FEDERAL CITY COLLEGE

(No reply received)

GALLLUDET COLLEGE

6th and Florida Avenue N.E. Phone: (202) 447-0740
Washington, D.C.
Dean of Graduate Studies: G.L. Delgado
Dean of College: J.S. Schushmann

DEPARTMENT OF BIOLOGY
Phone: Chairman: (202) 447-0740 Others: 447-0314

Chairman: E.B. Lloyd
Associate Professor: E.B. Lloyd
Assistant Professors: G. Kates, C. Powell, E. Rikuris
Instructor: C. Bateman
Part-time Non-faculty Staff Member: 1

Degree Program: Biology

Undergraduate Degree: B.A.
Undergraduate Degree Requirements: 30 credits in Biology (23 hours required courses and 7 hours electives) 8 hours Chemistry, 4 hours Mathematics

GEORGE WASHINGTON UNIVERSITY

2029 G. Street N.W. Phone: (202) 676-6000

Washington, D.C. 20037
Dean of Graduate Studies: A.E. Burns

Columbian College of Arts and Sciences
Dean: C.D. Linton

DEPARTMENT OF BIOLOGICAL SCIENCES
Phone: Chairman: (202) 676-6091 Others: 676-6090

Chairman: A.H. Desmond
Professors: A.H. Desmond, K.C. Kates (Professorial Lecturer), K. Parker, W.A. Shropshire, Jr. (Professorial Lecturer), R.S. Sigafoos (Professorial Lecturer), R.L. Weintraub
Associate Professors: D.L. Atkins, S.O. Schiff
Assistant Professors: J. Dickens, C.S. Henry, T.L. Hufford, R.E. Knowlton, H. Merchant, W.G. Nash, R.K. Packer, S. Smith-Gill, P.E. Spiegler (Adjunct), B. Timberlake (Assistant Professorial Lecturer)
Teaching Fellows: 8

Degree Programs: Anatomy and Histology, Anatomy, Animal Genetics, Animal Physiology, Biological Sciences, Biology, Entomology, Ecology, Genetics, Histology, Physiology, Radiation Biology, Zoology, Molecular Genetics, Photobiology, Neuroanatomy, Herpetology

Graduate Degrees: M.S., Ph.D.
Graduate Degree Requirements: Master: 24 credits plus Thesis, 36 credits without Thesis Doctoral: 48 credits, General Examination, Dissertation

Graduate and Undergraduate Courses Available for Graduate Credit: Organic Evolution (U), Introductory Microbiology (U), Molecular and Subcellular Biology (U), Cell Biochemistry (U), Genetics (U), Cell Physiology (U), Advanced Genetics (U), Radiation Biology (U), Local Flora (U), Field Botany - lower plants (U), Structure of Seed Plants (U), Lower Plants (U), Flowering Plants (U), Plant Physiology (U), Plant Ecology (U), General Ecology (U), Animal Ecology (U), Aquatic Ecology (U), Histology (U), Protozoa (U), Parasitology (U), Entomology (U), Invertebrate Zoology (U), Advanced Invertebtate Zoology (U), Comparative Vertebrate Anatomy (U), Vertebrate Embryology (U), Vertebrate Embryology (U), Vertebrate Zoology (U), Comparative Endocrinology (U), Insect Physiology (U), Human Physiology (U), Vertebrate Physiology (U), Histophysiology, Histochemistry, Morphogenesis, Comparative Animal Physiology, Seminar: Radiation Biology

School of Medicine
2300 I Street, N.W.
Dean: H. Solomon

DEPARTMENT OF ANATOMY
Phone: Chairman: (202) 331-6511

Chairman: R.S. Snell
Professors: R.S. Snell, F.D. Allan, T.N Johnson, C. M. Goss
Associate Professors: E.N. Albert, M. Koering
Assistant Professors: A. Butler, D.E. Morse, M.D. Olson, I. Leverton
Visiting Lecturers: 7
Teaching Fellows: 4
Research Assistants: 1

Degree Program: Anatomy

Graduate Degree: Ph.D.
Graduate Degree Requirements: Not Stated

Graduate and Undergraduate Courses Available for Graduate Credit: Anatomy, Gross Anatomy, Human Embryology, Neuroanatomy, Human Microscopic Anatomy, Electron Microscopic Histochemistry, Comparative Vertebrate Neurology, The Use of Audio-visual Techniques in Anatomy, The Use of Computers in Anatomy, Cell Biology, Neurobiology, Seminar, Introduction to Anatomical Research, Physical Anthropology, Fetal Anatomy, Human Genetics, Cytology, Electron Microscopy in Cellular Biology, Electron Microscopy in Cellular Biology Laboratory

DEPARTMENT OF BIOCHEMISTRY
Phone: Chairman: (202) 331-6517 Others: 331-0200

Chairman: C.R. Treadwell
Professors: J.M. Bailey, B.S. Smith, C.R. Treadwell, G.V. Vahouny
Associate Professor: G.A. Walker
Assistant Professor: L.L. Gallo (Research)
Visiting Lecturers: 9
Teaching Fellows: 6
Research Assistants: 3

Degree Program: Biochemistry

Graduate Degrees: M.S., Ph.D.
Graduate Degree Requirements: Master: 30 credit hours plus 1 tool subject; research and thesis required Doctoral: 72 credit hours, plus 2 tool subjects; research and thesis required

Graduate and Undergraduate Courses Available for Graduate Credit: General Biochemistry, Enzymology, Biochemistry Procedures, Seminar, Proteins and Amino Acids, Isotopes, Carbohydrate Metabolism, Biochemistry of the Brain, Human Nutrition, Biochemistry of Organ Function, Biochemistry of Steroids, Biochemistry of Lipids, Biochemical Genetics, Biochemistry of Bone and Muscle

MICROBIOLOGY DEPARTMENT
Phone: Chairman (202) 331-6531

Chairman: L.F. Affronti
Professors: L.F. Affronti, R. Hugh, E. Lerner (Research) M.L. Robins
Associate Professors: K-Y. Huang, P. Kind, M. Reich
Assistant Professors: Y-M. Chu, R.D. Donnelly (Research)
Teaching Fellows: 4

Degree Programs: Bacteriology, Bacteriology and Immunology, Genetics, Immunology, Medical Microbiology, Microbiology, Microbial Physiology, Virology

Graduate Degrees: M.S., Ph.D.
Graduate Degree Requirements: Master: 30 hours (24 in course work, 6 in thesis) Doctoral: 72 hours with baccalaureate, 48 with masters, reading knowledge in French and German or one language and Biostatistics, General Examination coveres at least 4 fields

Graduate and Undergraduate Courses Available for Graduate Credit: Microbiology, Pathogenic Microbiology, Tissue Cell Culture and Somatic Variation, Scientific Writing for Graduate Students, Microbial Physiology, Immunology, Immunobiology, Biology of Viruses, Systematic Bacteriology, Clinical Laboratory Bacteriology, Virology, Experimental Immunochemistry, Microbial Genetics, Seminar

DEPARTMENT OF PHYSIOLOGY
Phone: Chairman: (202) 331-6553 Others: 331-6551

Chairman: R.A. Kenney
Henry D. Fry Professor of Physiology: C.S. Tidball
Professors: R.A. Kenney, M.M. Cassidy, C.S. Tidball, M.E. Tidball
Associate Professors: M. McCally, M.J. Jackson
Assistant Professors: R.A. Lavine, D.W. Watkins
Teaching Fellows: 3
Research Assistants: 3

Degree Program: Physiology

Graduate Degrees: M.S., Ph.D.

Graduate Degree Requirements: Master: requires 30 hours of course work, 3 semester hours of research, in addition to the thesis Doctoral: requires 48 semester hours of course work in the field of Physiology, a major field in Physiology, a subfield in Physiology, and a minor field outside of Physiology leading to the Cumulative Examination

Graduate and Undergraduate Courses Available for Graduate Credit: Selected Topics in Human Structure and Function (U)

GEORGETOWN UNIVERSITY

37th and D Streets N.W. Phone: (202) 625-0100
Washington, D.C. 20007
Dean of Graduate School: D.G. Herzberg

College of Arts and Sciences

Dean: Rev. R.B. Davis, S.J.

DEPARTMENT OF BIOLOGY
Phone: Chairman: (202) 625-4706 Others: 625-4126

Chairman: G.B. Chapman
Professors: G.B. Chapman, I. Gray, O.E. Landman, E. Leise (Research), J.A. Panuska (Part-time), D.M. Robinson, R.J. Weber
Associate Professors: M.H. Bauer, R.S. Banquet, P.K. Chen
Assistant Professors: R.C. Baumiller (Joint appointment with Medical School), K.L. Bick, D. Eagles, G.W.M. Ferguson, T.A. O'Keefe, D.M. Spoon
Instructors: A.J. Angerio (Joint appointment with School of Nursing)
Visiting Lecturers: 4
Teaching Fellows: 20

Degree Program: Biology

Undergraduate Degree: B.S.
Undergraduate Degree Requirements: 44 Biology credits including a senior thesis, senior comprehensive examination (currently the undergraduate record examination), one foreign language

Graduate Degrees: M.S., Ph.D.
Graduate Degree Requirements: Master: 24 credits, with a laboratory thesis; or 30 credits with a library thesis, All M.S. candidates serve as teaching assistants Doctoral: 60 credits, with a laboratory thesis, one foreign language, All Ph.D. candidates serve as teaching assistants

Graduate and Undergraduate Courses Available for Graduate Credit: Comparative Anatomy (U), Genetics (U), Microbiology (U), General Physiology (U), Mammalian Physiology (U), Embryology (U), Cytology and Histology (U), Invertebrate Zoology (U), Ecology (U), General Parasitology (U), Topics in Microbial Genetics, Recent Developments in Molecular Biology, Developmental Biology, Radiation Biology, Electron Microscopy, Enzymology, Genetics of Higher Organisms, Comparative Physiology, Population Genetics, Endocrinology, Plant Virology, Thermobiology, Cytochemistry and Histochemistry, Bacterial and Viral Diseases of Fish, Ecology of Aquatic Microcosms, Neurophysiology, Immunochemistry, Biology Research

School of Medicine

4000 Reservoir Road
Dean: J.P. Utz

DEPARTMENT OF ANATOMY
Phone: Chairman: (202) 625-7522 Others: 625-7521

Professors: B.R. Bhussry, O. Solnitzky (Emeritus), I. Telford, B. Vidic
Associate Professors: T. Crisp, G.C. Goeringer, W. Norman
Assistant Professors: M. DeSantis, B. Hamilston, S. Kapur, S. Rao, F. Suarez, L. Warner
Instructors: P. Marlow, D. Rigamonti
Visiting Lecturers: 10
Research Assistants: 4

Degree Programs: Gross Anatomy, Histology, Neurobiology

Graduate Degrees: Ph.D. (Anatomical Sciences)
Graduate Degree Requirements: Master: Coursework - 24 credit hours Thesis - 6 credit hours Doctoral: Coursework and thesis 60 credit hours

Graduate and Undergraduate Courses Available for Graduate Credit: Gross Anatomy (U), Neurobiology (U), Advanced Human Neuroanatomy: Brain Reconstruction, Microscopic Anatomy (U), Advanced Anatomy of the Head and Neck, Dynamic Concepts in Histology, Advanced in Teratology, Generation, Degeneration and Regeneration in the Nervous System (U), Anatomy Seminar, Anatomy Research

DEPARTMENT OF BIOCHEMISTRY
Phone: Chairman (202) 625-7622

Chairman: W.H. Horner
Professors: M. Blecher, O. Gabriel, W. Horner, M. Rubin, R.O. Brady, H.Edelhoch, G.A. Jamieson (Adjunct)
Associate Professor: M.E. Smulson
Assistant Professor: J.C. Cassatt, J.C. Chirikjian, L. Van Lenten
Instructor: R.D. Irwin
Professorial Lecturers: J.J. Canary, F. Eisenberg, Jr., J.C. Houck
Teaching Fellows: 5
Research Assistants: 5
Postdoctoral Fellows: 6

Degree Program: Biochemistry

Graduate Degrees: M.S., Ph.D.
Graduate Degree Requirements: Master: 24 credit hours minimum, Qualifying and Comprehensive Examinations, one language reading requirement, Master's thesis Doctoral: 60 credit hours minimum. Qualifying and Comprehensive examinations, two language reading requirements, Ph.D. thesis and defense

Graduate and Undergraduate Courses Available for Graduate Credit: General Biochemistry, Physical Chemistry of Proteins, Inorganic Biochemistry, Enzymes, Biological Oxidations and Reductions, Structure and Function of Nucleic Acids, Chemistry and Metabolism of Carbohydrates, Chemistry and Metabolism of Lipids, Chemistry and Metabolism of Amino Acids, Chemistry and Metabolism of Hormones, Regulatory Mechanisms in Metabolism, Advanced Biochemistry Laboratory, Seminar

DEPARTMENT OF MICROBIOLOGY
(No reply received)

DEPARTMENT OF PHYSIOLOGY AND BIOPHYSICS
Phone: Chairman: (202) 625-7545

Chairman: L.S. Lilienfield
Professors: L.S. Lilienfield, E.R. Ramey, R.S. Ledley, P.W. Ramwell, J.C. Rose
Associate Professors: M.O. Dayhoff, P.A. Kot, M. Lorber, J.C. Penhos, J.V. Princiotto, L.M. Slotkoff
Assistant Professors: P. Hamosh, D.K. Kasbekar, C.E. McCauley
Visiting Lecturers: 10
Teaching Fellows: 2
Research Assistants: 1

Degree Programs: Physiology, Biophysics

Graduate Degrees: M.S., Ph.D.

Graduate Degree Requirements: Master: B.S. with a major in Biology, Chemistry, Physics, Mathematics or Engineering, Selected students with either and M.D. or a D.D.S. are accepted Doctoral: M.S. with a biology or Physiology major

Graduate and Undergraduate Courses Available for Graduate Credit: Mammalian Physiology, Advanced Cardiovascular Physiology, Advances in Neurobiology, Advanced Neuroendocrinology, Advances in Endocrinology, Advances in Hematology, Survey in Human Physiology, Advanced Renal Physiology, Advanced Pulmonary Physiology, Biological Role of Prostaglandins, Experimental Design and Statistical Methods, Advanced Gastrointestinal Physiology, Biomechanics, Selected Topics in Physiology, Cell and Membrane Physiology, Computers in Medicine, Macromolecular Evolution

HOWARD UNIVERSITY

2400 Sixth Street N.W. Phone: (202) 626-6100
Washington, D.C. 20059

College of Liberal Arts

(No reply received)

College of Medicine

Dean: M. Mann

DEPARTMENT OF ANATOMY
Phone: Chairman: (202) 636-6557

Chairman: L.V. Leak
Professors: L.V. Leak
Associate Professors: R.S. Lloyd, C.G. Santos
Assistant Professors: K.M. Baldwin, F.C. Castro, A.S. Chan, W.Cruce, B.D. Garg, K.C. Gupta, S.T. Hussain, A. Szabo, D. Tanaka, B.H. Turner, D.G. Walker
Instructors: M.N. Flores, Y.C. Hwang
Research Associates: H. Covington, K.S. Rahil
Visiting Lecturer: 1
Research Assistants: 2

Degree Programs: Anatomy, Anatomy and Histology, Anatomy and Physiology, Biochemistry, Genetics, Histology, Microbiology, Pathology, Pharmacy, Physiology

Graduate Degrees: M.S., Ph.S.
Graduate Degree Requirements: Master: 24 semester hours of which 15 must be in anatomy and 15 in graduate level courses, Thesis, Final examination Doctoral: 72 semester hours including anatomy, embryology, histology, neuroanatomy, one foreign language, qualifying examination, Thesis, Final Oral

Graduate and Undergraduate Courses Available for Graduate Credit: Histology-embryology, Advanced Histology, Developmental Biology, Endocrinology and Reproduction, Topics in Cell Structure and Function, Optical Methods in Analytical Cytology, Research in Anatomy, Special Projects in Gross Anatomy, Thesis Writing, Techniques in Electron Microscopy, Gross Anatomy, Evolution and Life History, Topics in Neuroanatomy

BIOCHEMISTRY DEPARTMENT
(No reply received)

MICROBIOLOGY DEPARTMENT
(no reply received)

PHYSIOLOGY DEPARTMENT
(No reply received)

TRINITY COLLEGE

Washington, D.C. 20017 Phone: (202) 269-2000
Dean of College: Sr. M.A. Cook

DEPARTMENT OF BIOLOGY
Phone: Chairman: (202) 269-2264 Others: 269-2000

Chairman: Sr. M.T. Dimond
Professor: Sr. M.T. Dimond
Associate Professor: E.H. Bellmer

Degree Programs: Biology, Medical Technology

Undergraduate Degrees: B.A., B.S.
Undergraduate Degree Requirements: B.A. or B.S. in Biology: at least 7 Biology major courses; General and Organic Chemistry, General Physics (For the B.S. 90 or more semester hours in science and mathematics) B.S. in Medical Technology: three years of college work with emphasis on Biology and Chemistry plus one year at medical technology school

Graduate and Undergraduate Courses Available for Graduate Credit: Comparative Physiology (U), General Microbiology (U)

FLORIDA

BARRY COLLEGE

11300 N.E. 2nd Avenue Phone: (305) 758-3392
Miami, Florida 33161
Dean of College: Sr. R. Schaefer, O.P.

BIOLOGY DEPARTMENT
Phone: Chairman: Ext. 332 Others: Ext. 376

Chairperson: Sr. J.K. Frei, O.P.
Professor: Sr. A.L. Stechschulte
Associate Professor: Sr. J.K. Frei
Assistant Professors: R. Davis, Sr. J.K. Comiskey

Degree Program: Biology

Undergraduate Degree: B.S.
Undergraduate Degree Requirements: 120 hours with 48 of these upper Biology courses 30 of the 129 must fulfill distribution requirements as follows: 9 in Religious studies and/or Philosophy, 6 in written and oral Communication; 6-9 in Humanities; 6-9 in Social Sciences, Students in Biology must minor in either Mathematics or Chemistry. All Biology students follow a core curriculum in biology take calculus, Inorganic and Organic Chemistry and Biochemistry

BETHUNE COOKMAN COLLEGE

(No reply received)

ECKERD COLLEGE

St. Petersburg, Florida 33733 Phone: (873) 867-1166
Dean of College: E.A. Smith

BIOLOGY DISCIPLINE
Phone: Chairman: Ext. 439

Chairman: G.K. Reid
Professors: J.E. Ferguson, G.K. Reid
Associate Professor: W.B. Roess
Assistant Professor: C.A. Jefferson

Degree Program: Biology

Undergraduate Degree: B.S.
Undergraduate Degree Requirements: 32 college courses, 16 courses in natural sciences, 8 courses in Biology

EDWARD WATERS COLLEGE

1658 Kings Road Phone: (904) 355-5411
Jacksonville, Florida 32209
Vice President for Academic Affairs: P.J. Driver

SCIENCE-MATHEMATICS DIVISION
Phone: Chairman: 355-5411 Ext. 70

Chairman: W.P. Kellogg
Professors: S.M. Aijaz, P.J. Driver (Adjunct), W.P. Kellogg
Assistant Professor: W. Fordham

Degree Programs: Biology, Botany, Medical Technology, Science Education

Undergraduate Degree: B.S. in General Science (concentration in Biology)
Undergraduate Degree Requirements: Student must complete 30 semester hours in their area of concentration (Biology). In addition to these courses 8 hours in zoology, 8 in Botany, 8 in General Chemistry, and a combination of 8 hours in qualitative analysis, Organic Chemistry or Biochemistry as well as 8 hours of General Physics; 6 hours of Mathematics, Foreign language

FLORIDA A AND M UNIVERSITY

(No reply received)

FLORIDA ATLANTIC UNIVERSITY

Boca Raton, Florida 33432 Phone: (305) 395-5100
Dean of Graduate Studies: J.T. Kirby
Dean of College: R.M. Iverson

DEPARTMENT OF BIOLOGICAL SCIENCES
Phone: Chairman: Ext. 2706 Others: Ext. 2707

Chairman: S. Dobkin
Professors: M.L. Boss, W.R. Courtenay, Jr., S. Dobkin, H.A. Hoffman, V.R. Saurino, P.L. Sguros, T.T. Sturrock
Associate Professors: R.M. Adams, D.F. Austin
Assistant Professors: B. Ache, R.B. Grimm, J. Hartmann, G.A. Marsh
Research Assistants: 14

Degree Programs: Bacteriology, Bacteriology and Immunology, Biological Sciences, Botany, Ecology, Microbiology, Zoology, Marine Biology, Marine Ecology, Ichthyology, Marine Botany

Undergraduate Degrees: B.A., B.S.
Undergraduate Degree Requirements: 180 quarter hours including: 60-65 quarter hours in Biology, foreign language, 18 credits out of college

Graduate Degree: M.S.
Graduate Degree Requirements: Master: 36 credits of course work minimum of 9 credits of thesis

Graduate and Undergraduate Courses Available for Graduate Credit: Marine Biology (U), Flora of South Florida, Microtechnique, Virology, Biosystematics, Principles of Pathogenesis, Principles of Immunology, Enzymology, Bacterial Physiology I and II, Microbial Physiology in the Sea, Applied Microbiology, Microbial Ecology, Natural History of Fishes, Marine Invertebrate Zoology I and II, Physiological Animal Ecology, Physiological Plant Ecology, Plant Ecology and Phytogeography, General Mycology, Algology, Studies in Algology, Ethology, Ethology Laboratory, Ecological Theory, Freshwater Ecology, Marine Ecology, Experimental Ecology, Seminar in Ichthyology

FLORIDA INSTITUTE OF TECHNOLOGY

Melbourne, Florida 32901 Phone: (305) 723-3701
Dean of College: H.P. Weber

DEPARTMENT OF BIOLOGICAL SCIENCES
Phone: Chairman: Ext. 306, 242 Others: 254, 255

Head: G.C. Webster
Professors: S.R. Defazio (Adjunct), R.H. Jones (Research), T.A. Nevin (Research), J.J. Thomas (Research)
Assistant Professors: K.B. Clark, G.M. Cohen, K.L. Kasweck, J.G. Morris, G.N. Wells
Teaching Fellows: 9
Research Assistants: 2

Degree Programs: Biological Science

Undergraduate Degree: B.S.
Undergraduate Degree Requirements: 192 quarter credits, including one year of Mathematics (including Calculus) one year of Physics, two years of Chemistry, a core program of 72 credits of Biology, and electives in Biology, other Sciences, and Humanities

Graduate Degrees: M.S., Ph.S.
Graduate Degree Requirements: Master: 48 credits plus M.S. Thesis Doctoral: An approved program of study beyond that required for the M.S. degree, plus a doctoral Dissertation

Graduate and Undergraduate Courses Available for Graduate Credit: Ecology (U), Mammalian Physiology (U), Developmental Biology (U), Marine Biology (U), Comprehensive Biochemistry, Biochemical Techniques, Advances in Cell and Molecular Biology, Electron Microscopy, Microbial Genetics, Advances in Immunology, Advanced Microbiology, Marine Microbiology, Marine Ecology, Advanced Invertebrate Zoology, Comparative Animal Physiology, Advanced Plant Physiology, Field Ecology, Current Topics in Ecology, Biological Science Seminar

FLORIDA PRESBYTERIAN COLLEGE
(See Eckard College)

FLORIDA SOUTHERN COLLEGE
Lakeland, Florida 33802 Phone: (813) 683-5531
Academic Dean: B.F. Wade

DEPARTMENT OF BIOLOGICAL SCIENCES
Phone: Chairman: Ext. 447

Chairman: M.L. Gilbert
Professor: M.L. Gilbert
Assistant Professors: L. Campbell, C.F. Erickenberg, J. R. Haldeman, J.R. Tripp
Visiting Lecturers: 1
Research Assistants: 1

Degree Programs: Biology

Undergraduate Degrees: B.A., B.S.
Undergraduate Degree Requirements: Minimum of 30 credits in Biology including Botany and Zoology, General Chemistry, Organic Chemistry, Physics, one semester Calculus. For the B.A., a modern language and Philosophy

FLORIDA STATE UNIVERSITY
Tallahassee, Florida 32306 Phone: (904) 644-2525
Provost, Division VI: R. Johnson
Provost, Division I: R.A. Spivey

DEPARTMENT OF BIOLOGICAL SCIENCE
Phone: Chairman: (904) 644-3700 Others: 644-2525

Chairman: T.Peter
Professors: L. Beidler, T.P. Bennett, A. Collier, A.G. DeBusk, P. Elliott, M. Greenberg, R. Johnson, H. Lipner, M Menzel, R. Short, J.H. Taylor, D.C. White, T. Williams, R. Breen (Emeritus), H. Gaffon (Emeritus), R. Godfrey (Emeritus), M.N. Hood (Emeritus), G. Madsen (Emeritus), J.A. Beech (Adjunct), W. Peters (Adjunct), B. Simpson (Adjunct)
Associate Professors: A. Clewell, S. deKloet, D. Easton, L. Ellias, E. Friedmann, P. Graziadei, W. Heard, W. Herrnkind, P. Homann, R. Mariscal, A. Pates, M. Roeder, D. Simberloff, C. Stasek, H. Stevenson, J. Stuy, L. Wiese
Assistant Professors: J. Byram, J. Elam, L.R. Fox, M. Freeman, K. Hofer, R. Livingston, T. Seale, D. Strong, W. Tschinkel, C.S. Wang, N. Williams
Research Associates: 3

Field Stations/Laboratories: Mission Road Research Station - Tallahassee, Ed Ball Marine Laboratory - Turkey Point, Carrabelle

Degree Programs: Medical Technology, Biological Science

Undergraduate Degree: B.S.
Undergraduate Degree Requirements: 54 quarter hours Biology (including 14 quarter hours Introductory Biology, 5 quarter hours Genetics, 4-5 quarter hours Physiology, 4 quarter hours Evolution or Ecology, 27 quarter hours electives); Mathematics through Calculus; Chemistry through 2 quarter hours of Organic or Physical, 12 quarter hours of Physics

Graduate Degrees: M.S., Ph.D.
Graduate Degree Requirements: Master: 1) at least 45 quarter hours of graduate credit (500 level and above courses and those 400 level courses recommended by student's committee, including 5-9 quarter hours of thesis), 24 quarter hours of which must bear letter grades, and at least 2 quarter hours Supervised Teaching; 2) reading knowledge of one foreign language or at least 6 quarter hours communication approved graduate level credit in Mathematics or Statistics 3) completion of acceptable thesis 4) final comprehensive examination Doctoral: The direction and supervision of graduate work resides primarily with major professor and supervisory committee. Overall requirements are: 1) e quarter hours continuous full time residence 2) communications proficiency 3) satisfaction of teaching requirement 4) preliminary doctoral examination 5) Dissertation 6) defense of dissertation

Graduate and Undergraduate Courses Available for Graduate Credit: Animal Physiology, Plant Biochemistry and Biophysics, Developmental Plant Physiology, Endocrinology, Neurology, Advanced Neurophysiology, Methods of Biophysics, Intermediate Genetics, Contemporary Animal Behavior, Field Course-Marine Biology, Malacology, Biological Animal Parasites, Ichthyology, Physiological Ecology of Fishes, Invertebrate Morphology and Classification, Microbiology, Bacterial Physiology, Comparative Microbiology, Microbial Genetics, Microscopic and Submicroscopic Technique, Py of Sensory Receptors, Behavioral Neurology, Advanced Genetics-Molecular and Biochemical, Compara tive of Chromosomal Organisms, Molecular and Cellular, Bacterial, Advanced Ichthyology, Advanced Ornithology, Topics in Vertebrate Py, Population Biology, Ecology, Comparative Py, Marine Biology, Seminars in Marine Biology, Cell Biology, Genetics, General Physiology, Invertebrate Zoology, Vertebrate Zoology, Experimental Embryology, Parasitology, Microbiology, Current Problems in Psychobiology, Selected Topics in Plant Physiology, Psychobiology Colloquium, Special Research Problems, Supervised Research, Supervised Teaching

FLORIDA TECHNOLOGICAL UNIVERSITY
Orlando, Florida 32816 Phone: (305) 275-9101
Dean of Graduate Studies: L.L. Ellis
Dean of Natural Sciences: O. Ostle

DEPARTMENT OF BIOLOGICAL SCIENCES
Phone: Chairman: (305) 275-2141

Gordon J. Barnett Professor of Environmental Science: R.J. Wodzinski
Professors: J.L. Koevenig, H.A. Miller
Associate Professors: J.F. Charba, L.M. Ehrhart, D.T. Kuhn, D.R. Reynolds, W.K. Taylor, F.F. Snelson, D.H. Vickers, R.S. White, H.O. Whittier
Assistant Professors: R.C. Bullock, R.N. Gennaro, J.C. Mickus, J.A. Osborne, H.J. Price, I.J. Stout, M.J. Sweeney, H.C. Sweet, G.E. VanderMolen
Graduate Assistants Research: 10

Graduate Assistants Teaching: 4

Degree Programs: Options in Biology, Biotechnology, Botany, Freshwater Ecology, Microbiology, Zoology

Undergraduate Degree: B.S. (Biological Science)
Undergraduate Degree Requirements: 55 quarter credits minimum in Biological Sciences, 1 year Mathematics, 1 course Computer Science, 2-3 years Chemistry, 8 credits in Physics

Graduate Degree: M.S. (Biological Science)
Graduate Degree Requirements: Master: 45 quarter credits minimum, required courses in experimental methods, Cytogenetics, Environmental Science, 9 quarter hours Thesis included in 45 Doctoral: Cooperative programs in Biology and Botany with the University of South Florida and University of Florida as degree granting institutions

Graduate and Undergraduate Courses Available for Graduate Credit: Microtechnique (U), Limnology (U), Freshwater Systems (U), Organic Evolution (U), History of Biology (U), Experimental Methods for Organismic Biology, Cell Biology, Population Ecology, Ecology of Running Water, Cytogenetics, Population Genetics, Experimental Ecology, Phycology (U), Mycology (U), Community Ecology (U), Plant Geography (U), Eumycota, Bryology, Plant Biosystematics, Diagnostic Microbiology (U), Microbiology of Water and Waste (U), Microbial Physiology (U), Determinative Microbiology (U), Microbial Ecology (U), Medical Mycology (U), Infectious Process, Virology, Applied Microbiology, Microbial Metabolism, Vertebrate Embryology (U), Invertebrate Zoology (U), Ichthyology (U), Zoogeography (U), Vertebrate Ethology (U), Endocrinology, Ornithology, Herpetology, Mammalogy, Fishery Biology, Zoological Systematics, Aquatic Invertebrates, Comparative Animal Physiology

JACKSONVILLE UNIVERSITY

Jacksonville, Florida 32211 Phone: (904) 744-3950
Dean of College: S. Frank

BIOLOGY DEPARTMENT
Phone: Chairman: Ext. 262

Chairman: K.I. Miller
Associate Professors: T.T. Allen, K.I. Miller, K.G. Relyea, J.S. Robertson
Assistant Professors: J.E. Trainer, T.O. Weitzel
Part-time Lecturers: 2

Degree Program: Biology

Undergraduate Degrees: B.S., B.A.
Undergraduate Degree Requirements: Biology 120 - Introductory Biology, 204 - General Botany, 207 - General Zoology, 401 - Genetics plus 21 elective hours in Biology, numbered 300 and above, 16 hours Chemistry, 8 hours Physics, 8 hours Mathematics, etc.

NEW COLLEGE

Sarasota, Florida 33578 Phone: (813) 355-7131

DIVISION OF NATURAL SCIENCES
Phone: Chairman: 355-8486

Chairman: D. Gorfein
Professors: P. Buri, J.B. Morrill
Assistant Professor: R. Rubin

Degree Program: Biology

Undergraduate Degree: A.B.
Undergraduate Degree Requirements: Too variable to state

PALM BEACH ATLANTIC COLLEGE

1101 South Olive Avenue Phone: (305) 833-8592
West Palm Beach, Florida 33401
Dean of College: N. Robinson

DEPARTMENT OF BIOLOGY
Phone: Chairman: 832-9824

Chairman: J.P. Thomas
Professors: J.P. Thomas, E. Rodriguez

Degree Program: Biology

Undergraduate Degree: B.S.
Undergraduate Degree Requirements: 28 semester hours must include Animal Biology, Plant Biology, Invertebrate Zoology, Botany, Genetics, Microbiology

ROLLINS COLLEGE

A.G. Bush Science Center Phone: (305) 646-2000
Winter Park, Florida 32789
Provost: D.L. Ling

DEPARTMENT OF BIOLOGY
Phone: Chairman: Ext. 2518 Others: 2494, 2379, 2416

Head: D.I. Richard
Associate Professors: D.I. Richard, E.W. Scheer, D.E. Smith
Assistant Professors: M.A. Henderson, J.W. Small
Instructor: C. Sandstrom

Degree Programs: Anatomy, Anatomy and Physiology, Animal Behavior, Applied Ecology, Bacteriology and Immunology, Biochemistry, Biological Chemistry, Biological Sciences, Biology, Botany, Ecology, Entomology, Genetics, Microbiology, Oceanography, Physiology, Science, Science Education, Zoology, Zoology and Entomology

Undergraduate Degree: B.S.
Undergraduate Degree Requirements: Not Stated

ST. LEO COLLEGE

(No Reply Received)

STETSON UNIVERSITY

DeLano, Florida 32720 Phone: (904) 734-4121
Dean of Graduate Studies: R. Fox
Dean of College: R. Chauvin

DEPARTMENT OF BIOLOGY
Phone: Chairman: Ext. 345 Others: 243

Chairman: F.M. Knapp
Professors: D.L. Fuller, K.L. Hansen, F.M. Knapp
Associate Professor: F.E. Clark
Assistant Professors: E.M. Normen, D.A. Stock
Visiting Lecturers: 1

Degree Programs: Biological Sciences, Medical Technology

Undergraduate Degree: B.S.
Undergraduate Degree Requirements: 128 hours (university requirement) Biology: 26 semester hours upper division, 1 year Physics, General Chemistry (1 year), Organic Chemistry (1 year), Calculus, one semester Botanical Science, 1 year of seminar at Junior-Senior levels

Graduate Degree: The M.S. program curtailed in past two years. Now given in limited fashion

UNIVERSITY OF FLORIDA

Gainsville, Florida 32601 Phone: (904) 392-3261
Dean of Graduate Studies: H.H. Sisler

College of Arts and Sciences
Dean: C.A. VanderWerf

DIVISION OF BIOLOGICAL SCIENCES
Phone: Chairman: (904) 392-1175

Director: L. Berner
Professors: L. Berner, F.J.S. Maturo
Assistant Professor: G.A. Olson
Graduate Assistantships: 8

Field Stations/Laboratories: Seahorse Key Marine Laboratory - Cedar Key

Degree Program: The Division itself does not offer a degree program as such. Degrees are offered through the four departments under the Division (Biochemistry, Botany, Microbiology and Zoology)

Graduate and Undergraduate Courses Available for Graduate Credit: Radiation Effects and Radiation Biology (U), Radioisotope Theory and Techniques (U), Radiation Effects on Humans (U), Special Topics in Radiation Biology

DEPARTMENT OF BIOCHEMISTRY
Phone: Chairman: (904) 392-3361 Others: 3362/3363

Chairman: P.A. Cerutti
Professors: R.P. Boyce, M Fried, S. Gurin, R.J. Mans
Associate Professors: P. Chun, T.W. O'Brien, R.M. Roberts, E.G. Sander
Assistant Professors: C.M. Allen, Jr., R.J. Cohen, K.D. Noonan, G. Stein, J.C.M. Tsibris
Instructors: P.V. Hariharan, J. Remsen

Field Stations/Laboratories: Whiteny Marine Laboratory - St. Augustine

Degree Programs: Biochemistry (Program in Medical Biochemistry

Graduate Degrees: M.S., Ph.D.
Graduate Degree Requirements: Master: 35 credit hours and research Dissertation Doctoral: Ph.D. Dissertation, no credit hour requirements

Graduate and Undergraduate Courses Available for Graduate Credit: Biochemistry 1, Biochemistry 2, Biochemistry Laboratory, Introduction to Human Biochemistry, Biochemistry Senior Research, Biochemistry, Biochemistry Laboratory, Current Trends in Biochemistry, Principles of Molecular Biology and Genetics, Recent Advances in Biochemistry, Physical Biochemistry, Research Methods in Biochemistry, Biochemistry Seminar, Special Topics in Biochemistry, Master's Research, Biochemistry of Disease, Molecular Biology 1, 2, 3, Doctoral Research, Current Trends in Biochemistry, Chemistry of Biological Molecules, Principles of Molecular Biology, Biochemical Structure and Function, Metabolism

DEPARTMENT OF BOTANY
Phone: Chairman: (904) 392-1891

Chairman: W.W. Payne
Professors: D.S. Anthony, M.M. Griffith, T.E. Humphryes, J.T. Mullins, W.W. Payne, H.H. Popenoe (Courtesy), L. Shanor, I.K. Vasil
Associate Professors: J.S. Davis, D.G. Griffin III, J.W. Kimbrough, R.M. Roberts (Courtesy), R.C. Smith, D.B. Ward
Assistant Professors: G. Bowes, R.E. Dohrenwend, J. Ewel, T.W. Lucansky, A.E. Lugo
Research Assistants: 5
Postdoctorals: 2
Visiting Associate Professor: 1

Degree Programs: Agricultural Biology, Biological Education, Botany

Undergraduate Degree: B.S.
Undergraduate Degree Requirements: 36 credits (quarter hours) in Botany, plus General requirements for College of Arts and Sciences, including at least one laboratory course in Genetics, Mathematics through Differential Calculus, one year of College Physics, and Chemistry through Organic

Graduate Degrees: M S., Ph.D.
Graduate Degree Requirements: Master: General requirements of Graduate College with specialization in Botany. A Thesis is usually required Doctoral: General requirements of Graduate College. Students must pass written departmental examination in major subject areas prior to qualifying examination. Foreign language ability is required

Graduate and Undergraduate Courses Available for Graduate Credit: Plant Geography (U), Plant Ecology (U), Ecosystems of Florida (U), Intermediate Plant Physiology (U), Introductory Mycology (U), Phycology (U) Mosses and Liverworts (U), Lower Vascular Plants (U), Plant Anatomy (U), Taxonomy of Seed Plants (U), Cytology (U), Ecology of Aquatic Plants, Ecosystems of the Tropics, Tropical Biology: An Ecological Approach Advanced Tropical Botany, Plant Growth and Development, Plant Nutrition, Plant Metabolism, Radiation and Plant Growth, Biology and Taxonomy of the Basidiomycetes, Biology and Taxonomy of Myxomycetes, and Phycomycetes, Biology and Taxonomy of Ascomycetes and their Imperfect Stages, Fungal Physiology, Fungal Genetics, Lichenology, Development and Morphology of Seed Plants, Techniques in Developmental Botany, Advanced Taxonomy, Electron Microscopy of Biological Materials, Plant Cytology, Cytochemistry, Problems, Topics and Thesis Research in Botany

DEPARTMENT OF MICROBIOLOGY
(no reply received)

DEPARTMENT OF ZOOLOGY
Phone: Chairman: (904) 392-1107

Chairman: J.E. Heath
Graduate Research Professor: A.F. Carr
Professors: W. Auffenberg (Adjunct), L. Berner (Adjunct) P. Brodkorb, J.W. Brookbank, E.S. Deevey (Adjunct), J.C. Dickinson, Jr. (Adjunct), J.H. Gregg, J.W. Hardy (Adjunct), F.C. Johnson, E.R. Jones, F.J.S. Maturo (Adjunct), B.K. McNab, M.J. Westfall
Associate Professors: W.E.S. Carr (Adjunct), R.M. DeWitt, T.C. Emmel, C.R. Gilbert (Adjunct), D.W. Johnston, J.H. Kaufmann, C.A. Lanciani, B.B. Leavitt, J.E. Lloyg (Adjunct), F.G. Nordlie, T.H. Patton (Adjunct), J. Reiskind, H.O. Schwassmann, H.M Wallbrunn, S.D. Webb (Adjunct), S.G. Zam
Assistant Professors: J.F. Anderson, F.C. Davis, J.T. Giesel, S.R. Humphrey (Adjunct), G.C. Karp, H.D. Prange
Teaching Fellows: 23
Research Assistants: 9

Field Stations/Laboratories: University of Florida Marine Laboratory - Sea Horse Key, University of Florida Cornelius Vanderbilt Marine Laboratory - Marineland

Degree Program: Zoology

Undergraduate Degree: B.S.
Undergraduate Degree Requirements: 36 credits in Zoology, one year of college Physics, Chemistry through Organic and two courses in Calculus

Graduate Degrees: M.S.T., M.S., Ph.D.
Graduate Degree Requirements: Master: 36 credits of graduate level courses, 9 credits of research and completion of Thesis Doctoral: Course work as required by supervisory committee, reading knowledge of one foreign language and completion of Disserta-

tation

Graduate and Undergraduate Courses Available for Graduate Credit: General Ecology (U), Vertebrate Paleontology (U), Evolution (U), Parasitology (U), Histology (U), Behavioral Ecology (U), History of Biological Sciences (U), Physiological Genetics (U), Cytology (U), Cellular Physiology (U), Animal Physiology (U), Biology of Marine Animals (U), Ancient Vertebrate Faunas, Seminar in Evolution, Zoogeography, Marine Ecology, Helminthology, Protozoology, Community Ecology, Limnology, Seminar in Ecology, Advanced Invertebrate Zoology, Principles of Systematic Zoology, Physiological Ecology, Experimental Embryology, Ichthyology, Herpetology, Mammalogy, Ornithology, Ecological Genetics, Theoretical Population Ecology, Field Population Ecology, Seminar in Animal Behavior

College of Agriculture

Dean: C.B. Browning

DEPARTMENT OF ANIMAL SCIENCE
Phone: Chairman: (904)392-1914 Others: 392-1911

Chairman: T.J. Cunha
Distinguished Service Professor: T.J. Cunha
Professors: C.B. Ammerman, L.R. Arrington, G.E. Combs, J.H. Conrad, J.P. Feaster, J.F. Hentges, Jr., M. Koger, P.E. Loggins, J.E. Moore, A.Z. Palmer, R.J.E. Pace, R.L. Reddish, R.L. Shirley, H.D. Wallace, A. C. Warnick
Associate Professors: F.W. Bazer, B.W. Crawford, K.L. Durrance, D.E. Franke, E.A. Ott, D.L. Wakeman
Assistant Professors: J.F. Easley, M.J. Fields, S. Lieb, L.R. McDowall, R.S. Sand, D.C. Sherp, R.L. West
Visiting Lecturers: 1
Research Assistants: 14

Field Stations/Laboratories: Agrie Research and Education Center - Belle Glade and Quincey, Agricultural Research Center, Ona

Degree Programs: Animal Breeding and Genetics, Animal Nutrition, Physiology, Meats

Undergraduate Degree: B.S. (Animal Science)
Undergraduate Degree Requirements: 96 quarter hours General Education and Science, 96 quarter hours College of Agriculture, 12 hours in major required

Graduate Degrees: M.S., Ph.D.
Graduate Degree Requirements: Master: Master of Agriculture - 48 quarters - 24 hours in 600 level, 18 hours in department Master of Science - 45 quarters - 36 hours course work, 9 hours research and thesis Doctoral: Depends on Committee, usually 3 to 4 years residence in coursework and research

Graduate and Undergraduate Courses Available for Graduate Credit: Animal Nutrition (U), Animal Production in Tropics (U)

DAIRY SCIENCE DEPARTMENT
Phone: Chairman: (904) 392-1981

Chairman: H.H. VanHorn
Professors: S.P. Marshall, L E. Mull, C.J. Wilcox, J.M. Wing
Associate Professors: B. Harris, H.H. Head, W.A. Krienke, K.L. Smith, W.W. Thatcher
Assistant Professors: K.C. Bachman, R.L. Richter, D.W. Webb
Research Assistants: 9
Postdoctoral: 1

Field Stations/Laboratories: Dairy Research Unit - Hague

Degree Programs: Degrees offered in conjunction with Animal Science or Microbiology

Graduate Degrees: M.S., Ph.D.
Graduate Degree Requirements: Same as for admission to Graduate School. Must have taken G.R.E. and supply transcript and letters of recommendation

Graduate and Undergraduate Courses Available for Graduate Study: Physiology of Lactation (U), Diary Science Research Techniques, Graduate Seminar, Endocrinology Advanced Dairy Cattle Management, Energy Metabolism, Advanced Dairy Technology, Advanced Dairy Microbiology, Topics in Genetics

DEPARTMENT OF ENTOMOLOGY AND NEMATOLOGY
Phone: Chairman: 392-1901

Chairman: W.G. Eden
Entomology
Professors: R.M. Baranowski, F.S. Blanton, J.E. Brogdon, R.F. Brooks, H. Cromroy, W.G. Genung, D.H. Habeck, R.B. Johnson, S.H. Kerr, L.C. Kuitert, M. Murphey, J.L. Nation, F.A. Robinson, T.J. Walker, W.H. Whitcomb, R.C. Wilkinson, D.O. Wolfenbarger
Associate Professors: W.C. Adlerz, R.C. Bullock, G.L. Greene, M.J. Janes, L.E. Lloyd, C.W. McCoy, J.L. Strayer, W.B. Tappan, R.E. Waites, R.G. Workman
Assistant Professors: J.F. Butler, S. Fluker, R.A. Hamlen, F.A. Johnson, D.R. Minick, S.L. Poe, J.A. Reinert, D.E. Short, W.E. Denton, J.A. Tsai, J.C. Allen
Nematology
Professors: V.G. Perry, A.J. Overman, G.C. Smart Jr., A.C. Tarjan
Associate Professors: H.L. Rhoades,
Assistant Professors: D.W. Dickson, R.A. Kinloch, J.K. Hoffman
Visiting Lecturers: 14
Research Assistants: 5

Degree Programs: Entomology and Nematology

Undergraduate Degree: B.S.
Undergraduate Degree Requirements: 30 quarter hours of credit in Entomology and Nematology, 72 other quarter hours of credit

Graduate Degrees: M.A., M.S., Ph.D.
Graduate Degree Requirements: Master: M.A. - 18 quarter hour credits in Entomology and Nematology; 30 additional credits of upper division and/or graduate level classes; pass comprehensive written examination M.S. - 18 quarter hour credits in Entomology and Nematology; 18 additional credits of upper division and/or graduate level classes; 9 quarters in research; a thesis; and must pass final examination Doctoral: Ph.D. - 18 quarter hour credits of graduate Entomology and Nematology, 9 full quarters of resident credits; e consecutive quarters of concentrated study; a dissertation; demonstration of proficiency in one foreign language; pass qualifying examination and pass a final examination

Graduate and Undergraduate Courses Available for Graduate Credit: Veterinary Entomology, Tropical Entomology, Insect Pest Management, Tropical Nematology, Problems in Entomology and Nematology, Insect Systematics, Comparative Anatomy of the Hexapoda, Special Topics in Entomology and Nematology, Insect Physiology, Growth and Development in Insects, Insect Behavior, Insect Ecology, Insect Toxicology, Medical Entomology, Mosquitoes, Biological Control of Insects, Immature Insects, Acarology, Techniques in Medical Entomology, Aquatic Insects, Plant Parasitic Nematodes, Morphology and Physiology of Nematodes, Taxonomy of Nematodes, Radiation in Insect Studies, Insect Resistance in Crop Plants 1 and 2

DEPARTMENT OF FOOD SCIENCE
Phone: Chairman: (904) 392-2022 Others: 392-3261

Chairman: R.A. Dennison

Professors: E.M. Ahmed, R.A. Dennison, R.F. Matthews
Associate Professors: H. Appledorf, R.P. Bates, F.W. Knapp, J.A. Koburger, H.A. Moye, R.C. Robins, N. P. Thompson, W.B. Wheeler
Assistant Professors: P.E. Araujo, J. Deng, W.E. McCullough, J.L. Oblinger, R.H. Schmidt, J.G. Surak
Research Assistants: 10

Degree Programs: Food Science, Nutrition

Undergraduate Degree: B.S. (Agriculture)
Undergraduate Degree Requirements: 8 hours College of Agriculture Core Requirements - Communications 38 hours Departmental Core Reeuirements - 5 basic FS courses, Microbiology, Statistics, Organic Chemistry, 50 hours additional requirements and electives - I Food Science, II Food and Consumer Protection, and III Nutrition and Dietetics

Graduate Degree: M.S. (Agriculture)
Graduate Degree Requirements: Master: 45 credits, including no less than 36 credits of regular course work and up to 9 credits of research. At least half of credits must be in Food Sciences

Graduate and Undergraduate Courses Available for Graduate Credit: Research Planning, Advanced Food Microbiology, Nutritional and Toxicological Aspects of Foods, Instrumental Analysis and Separations, Industrial Food Fermentations, Food Processing Systems, Food Chemistry, Psychophysical Aspects of Foods, Food Product Development, Advanced Human Nutrition, Proteins and Amino Acids in Nutrition, Nutritional Aspects of Carnohydrates and Lipids

DEPARTMENT OF FORESTRY
(no reply received)

DEPARTMENT OF FRUIT CROPS
Phone: Chairman: 392-1996, 1997

Chairman: A.H. Krezdorn
Professors: J.A. Attaway, R.H. Biggs, C.W. Campbell, H.W. Ford, J.F. Gerber, W. Grierson, R.C.J. Koo, F.P. Lawrence, C.D. Leonard, H.J. Reitz, R.H. Sharpe, J.W. Sites, J. Soule, I. Stewart, L.W. Ziegler
Associate Professors: C.E. Arnold, J.F. Bartholic, D.W. Buchanan, S.E. Malo, W.B. Sherman, W.J. Wiltbank
Assistant Professors: L.K. Jackson
Research Assistants: 2

Field Stations/Laboratories: Agricultural Research and Education Center - Lake Alfred, Agricultural Research and Education Center - Homestead, Agricultural Research Center Leesburg - Leesburg, Agricultural Research Center Monticello - Monticello

Degree Programs: Horticultural Science (with major in Fruit Crops, Vegetable Crops or Ornamental Horticulture); specializations, Food Science, Fruit Crops or Plant Science (includes Fruit Crops, Vegetable Crops, Ornamental Horticulture and Agronomy)

Undergraduate Degree: B.S. (Agriculture)
Undergraduate Degree Requirements: 198 credits: Lower Division Requirements 96; College of Agriculture Core Requirements 8; Plant Sciences Core (Basic Plant Pathology, Entomology, General Soils, Genetics, Elementary Plant Physiology, Elementary Organic and Biological Chemistry, Fundamentals of Crop Production, Plant Propagation) 37; Departmental Requirements (Introduction to Citrus Culture, Physiology of Fruit Production, Citrus Maturity and Packinghouse Procedures, Citrus Production, Fruit Crops Laboratories 1 and 2) 21: Approved Electives, 36, Specialization in Agricultural Technology, Agricultural Science, Agricultural Business or Agricultural Chemicals

Graduate Degrees: M.S., M.Agr., Ph.D.
Graduate Degree Requirements: Master: 3 quarters full-time residence, final examination M.Agr. - (Non-thesis): .48 credits. At least 24 in courses numbered 600 or above, including 18 in major (Fruit Crops); remainder of credits in courses numbered 300, 400, or 500 in non-major (excludes Vegetable Crops and Ornamental Horticulture) subject areas. Comprehensive written examination prior to admission to candidacy M.S. Agr. (Thesis): 45 credits; at least 36 credits of regular course work and up to 9 credits of master's research (FC 699); 18 credits (exclusive of FC699) in courses 600 or above; 18 credits if a minor is designated Doctoral: Residence: 9 quarters full-time resident graduate study. Course work: No limitation but program should be unified in relation to a clear objective and have approval of student's supervisory committee. If a minor is designated, 18 to 36 credits are taken in the minor field. Language requirement: one foreign language or equivalent, Qualifying examination: written and oral, Dissertation, Final examination

Graduate and Undergraduate Courses Available for Graduate Credit: Major Tropical Fruits, Minor Tropical and Subtropical Fruits, Fruit Crops Seminar, Physiology of Fruit Crops, Morphology of Fruit Crops, Taxonomy of Fruit Crops, Breeding of Fruit Crops, Rootstock-Scion Relationships, Tropical Fruit Production and Research in Florida, Citriculture, Citrus Production Management, Agricultural Meteorology, Environmental Measurements, Topics in Fruit Crops

DEPARTMENT OF ORNAMENTAL HORTICULTURE
Phone: Chairman: 392-1831

Chairman: J.W. Strobel
Professors: J.N. Joiner (Teaching-Research), T.J. Sheehan (Teaching-Research)
Associate Professors: A. Dudeck (Teaching-Research), R. Henley (Extension)
Assistant Professors: R. Bednarz (Extension), C.R. Johnson (Teaching-Research), D.B. McConnell (Teaching-Research), S.E. McFadden (Research), H.G. Meyers (Extension, M R. Sheehan (Teaching), G.S. Smith (Teaching-Research) - 3 vacancies
Postdoctoral: 1

Field Stations/Laboratories: ARC Apopka, AREC - Bradenton, ARC - Ft. Lauderdale, ARC - Monticello

Degree Programs: Ornamental Horticulture

Undergraduate Degree: B.S.
Undergraduate Degree Requirements: 96 credits (quarter system) lower division, 96 credits (quarter system) College of Agriculture, Plant Science and Ornamental Horticulture core requirements

Graduate Degree: M.S.
Graduate Degree Requirements: Master: Master of Agriculture - 48 credits; 24 at 600 level or above - Thesis not required. Master of Science in Agriculture - 45 credits 36 in regular course work and 9 credits of research - Thesis required

Graduate and Undergraduate Courses Available for Graduate Credit: Orchidology (U), Seminar, Nutrition of Ornamental Plants, Nursery Production and Management, Research and Development in Turfgrass Technology, Non-Thesis Research, Thesis Research

PLANT PATHOLOGY DEPARTMENT
Phone: Chairman: 392-1861

Chairman: L.N. Purdy
Professors: A.A. Cook, P. Decker, T.E. Freeman, H.H. Luke, H.N. Miller, R.S. Mullin, D.A. Roberts, N.C. Schenck, R.E. Stall
Associate Professors: D.E. Purcifull, F.W. Zettler
Assistant Professors: J.A. Bartz, R. Charudattan, E. Hiebert, T.A. Kucharek, D.J. Mitchell, D.R. Pring

Teaching Fellows: 1
Research Assistants: 6
Postdoctoral: 1

Degree Program: Plant Pathology

Undergraduate Degree: B.S.
Undergraduate Degree Requirements: Not Stated

Graduate Degrees: M.S., Ph.D.
Graduate Degree Requirements: Not Stated

Graduate and Undergraduate Courses Available for Graduate Credit: Not stated

DEPARTMENT OF POULTRY SCIENCE
Phone: Chairman: 392-1931

Chairman: R.H. Harms
Professors: J.L. Fry
Associate Professors: C.R. Douglas, L.W. Kalch, H.R. Wilson
Assistant Professors: B.L. Damron, D.M. Janky, D.A. Roland, R.A. Voitle
Research Assistants: 6

Degree Program: Poultry Husbandry

Undergraduate Degree: B.S.
Undergraduate Degree Requirements: 192 quarter hours

Graduate Degrees: M.S., Ph.D.
Graduate Degree Requirements: Master: 415 quarter hours Doctoral: 90 quarter hours

Graduate and Undergraduate Courses Available for Graduate Credit: Topics in Poultry Production, Avian Physiology, Advanced Poultry Nutrition, Problems in Egg and Poultry Meat Technology, Advanced Poultry Management, Graduate Seminar in Poultry Science, Topics in Genetics, Advanced Problems in Poultry Production

SOIL SCIENCE DEPARTMENT
Phone: Chairman: (804) 392-1804 Others: 392-1951

Chairman: C.F. Eno
Professors: W.G. Blue, R.E. Caldwell, C.F. Eno, J.G.A. Fiskell, N. Gammon, Jr., L.C. Hammond, H. L. Poponoe, W.L. Pritchett, W.K. Robertson, D.F. Rothwell, G.M Volk, T.L. Yuan
Associate Professors: H.L. Breland, V.W. Carlisle, J. H. Herbert (Extension), D.H. Hubbell, C.C. Hortenstine (Research), R.S. Mansell, J. NeSmith (Extension)
Assistant Professors: F.G. Calhoun, D.A. Graetz, J.B. Sartain, B.G. Volk, L.W. Zelazny
Visiting Lecturers: 1
Research Assistants: 4
Teaching Assistants: 2

Field Stations/Laboratories: Agricultural Research and Education Centers - Belle Glade, Bradenton, Homestead, Lake Aflred, Quincy, Sanford and Tallahassee, Agricultural Research Centers - apopkia, Brooksville, Dover, Ft. Lauderdale, Ft. Pierce, Hastings, Immokalee, Jay, Lakland, Leesburg, Live Oak, Marianna, Monticello and Ona

Degree Programs: Soil Fertility, Soil Science, Soil Chemistry, Soil Physics, Soil Microbiology, Soil Genetics, and Classification

Undergraduate Degree: B.S.
Undergraduate Degree Requirements: 96 quarter credits during Freshman and Sophomore years including 8 credits required by the College of Agriculture, 102 quarter credits in Upper Division (Junior and Senior years) including 22 credits or more in soils and 72 credits in certain required courses and electives total 198 credits

Graduate Degrees: M.Ag, M.S.A., and Ph.D.
Graduate Degree Requirements: Master: M.Ag. - a minimum of 48 quarter credits, 24 of which must be designated strictly for graduate study and 18 must be graduate courses in the Soil Sciences Department M.S.A. - a minimum of 45 quarter credits, including no less than 36 credits of regular course work of which at least 18 must be graduate level courses in Soil Sciences Doctoral: Course work not specified but prescribed by students' supervisory committee. Generally consists of at lea st 45 quarter credits beyond the Masters' Degree, one minor of from 18 to 36 credits or two minors of at least 12 credits each, and at least a years work on a dissertation problem

Graduate and Undergraduate Courses Available for Graduate Credit: Tropical Soils (U), Soils of Florida (U), Topics in Soils, Soil Chemistry, Soil Microbiology, Soil Genesis and Classification, Soil Physics, Soil Fertility, Laboratory Methods of Soil Chemical Analyses, Soil Organic Matter and Organic Soils, Forest Soils, Colloidel and Physical Chemistry of Soils, Advanced Soil Physics, Micronutrients in Soils, Morphology of Florida Soils, Soil Mineralogy, Seminar, Non-Thesis Research, Supervised Research, Supervised Teaching, Master's Research, Doctoral Research

VEGETABLE CROPS DEPARTMENT
Phone: Chairman: 392-1928 Others: 392-1794

Chairman: J.F. Kelly
Professors: H.H. Bryan, D.S. Burgis, H.W. Burdine, V.L. Guzman, C.B. Hall, J.F. Kelley, S.J. Locascio, A.P. Lorz (Emeritus), G.A. Marlowe Jr, M.E. Marvel, J. Montelaro, V.F. Nettles, J.R. Orsenigo, W.T. Scudder, R.K. Showalter, B.D. Thompson, E.A. Wolf
Associate Professors: D.D. Gull, L.H. Halsey, H.Y. Ozaki, J.R. Shumaker
Assistant Professors: M.J. Bassett, D.J. Cantliffe, C.H. Doty, G.W. Elmstrom, L.C. Hannah, J.R. Hicks, S.R. Kostewicz, J.M. Stephens, J.M. White
Temporary: 1

Field Stations/Laboratories: Agricultural Research and Education Centers at Belle Glade, Bradenton, Homestead, Lake Alfred, Quincy, Sanford, Tallahassee, Agricultural Research Centers - Apopka, Brooksville, Dover, Ft. Lauderdale, Ft. Pierce, Hastings, Immo= kalee, Jay, Lakeland, Leesburg, Live Oak, Marianna, Monticello, Ona

Degree Programs: Vegetable Crops

Undergraduate Degree: B.S.
Undergraduate Degree Requirements: 96 quarter hours in University College followed by 102 quarter hours in Agriculture curriculum (44 quarter hours of Plant Science and College of Agriculture core courses, 20 quarter hours of vegetable crop requirements, 38 quarter hours of approved electives) student may elect specialization in Environmental Science or in Agricultural Chemicals

Graduate Degrees: M.A. (Agr.), M.S. (Agr.) Ph.D.
Graduate Degree Requirements: Master: M.A. - 48 quarter hours, 18 in graduate courses in department, 1 or 2 minor fields (8 quarter hours each, No thesis required M.S.A. - 36 quarter hours, 18 in graduate courses in department, 1 or more minor fields (8 quarter hours each) Thesis required Doctoral: No specific course or credit requirement arranged by committee, 9 quarters full time residence, Dissertation required

Graduate and Undergraduate Courses Available for Graduate Credit: Topics in Vegetable Production Principles, Advanced Vegetable Technology, Techniques and Methods of Vegetable Breeding, Postharvest Vegetable Physiology, Practicum in Vegetable Crops, Nonthesis

Research in Vegetable Crops, Seed Physiology (U), Biochemical Genetics

DEPARTMENT OF VETERINARY SCIENCE
Phone: Chairman: (904) 392-2381 Others: 392-1841

Chairman: C.E. Cornelius
Professors: G.T. Edds, C.F. Simpson, F.H. White
Associate Professors: R.E. Bradley, Sr., P.T. Cardeilhac, D.J. Forrester, J.A. Himes, G.W. Meyerholz, F.C. Neal, J.T. Neilson, W.P. Palmore
Assistant Professors: J.M. Gaskin, K.D. Ley
Research Assistants: 3
Teaching Assistant: 1

Degree Program: Veterinary Science

Graduate Degree: M.S.
Graduate Degree Requirements: Master: Good background in zoology, genetics, chemistry, biochemistry, physics, bacteriology, anatomy and physiology, deficiencies in the prerequisites must be satisfied as soon as possible, At least 27 credits of course work are to be selected from the major, including veterinary research techniques, veterinary parasitology, pathology, hematology, microbiology, pharmacology, toxicology and laboratory animal diseases, a related minor subject is to be chosen in consultation with the major adviser

Graduate and Undergraduate Courses Available for Graduate Credit: Animal Parasitology (U), Veterinary Physiology (U), Veterinary Physiology 2 (U), Veterinary Parasitology 1 (U), Veterinary Parasitology 2 (U), Veterinary Pathology, Helminthology, Immunology of Animal Parasites, Veterinary Pharmacology 1, Veterinary Pharmacology 2, Veterinary Toxicology 1, Veterinary Toxicology 2, Veterinary Microbiology, Parasitic Diseases in the Tropics and Subtropics, Veterinary Research Techniques 1, Problems in Veterinary Science 2

College of Medicine
Dean: C.A. Stetson

DEPARTMENT OF IMMUNOLOGY AND MEDICAL MICROBIOLOGY
Phone: Chairman: (904) 392-3311

Chairman: P.A. Small, Jr.
Professors: E.M. Ayoub, L.W. Clem, R.B. Grandall, G.E. Gifford, J.E. McGuigan, C. Moscovici, J.W. Shands, Jr., P.A. Small, Jr.
Associate Professors: H. Baer, A.S. Bleiweis, C.L. Cusumano, Y.M. Centifanto, E.M. Hoffman, R.H. Waldman
Assistant Professors: D.C. Birdsell, D.H. Duckworth, J.M. Gaskin, K.D. Ley
Visiting Lecturers: 1
Research Assistants: 2
Postdoctoral Associates: 4

Field Stations/Laboratories: C.V. Whitney Marine Laboratory - Marineland

Degree Programs: Immunology and Medical Microbiology, Parasitology and Virology

Graduate Degrees: M.S., Ph.D. (Medical Sciences)
Graduate Degree Requirements: Master: Candidates are requested to prepare, present and defend thesis acceptable to their supervisory committee and the Graduate School. The minimum course work is 45 credits including 9 credits of research. A foreign language is not required Doctoral: The minimum residence requirement is none quarters of full-time resident study, or equivalent. Course requirements are not specified but are designed for the needs of the student. A Ph.D. student is required to demonstrate proficiency in a language other than his native tongue, however there are provisions to substitute other academic proficiencies (e.g. competency in statistics and computer language) for the requirement. A written and oral qualifying examination must be successfully passed prior to admission to candidacy. Every candidate for a doctoral degree is required to prepare, present, and defend a scholarly dissertation that shows independend investigation and is acceptable in form and content to his supervisory committee and to the Graduate School.

Graduate and Undergraduate Courses Available for Graduate Credit: Microbiology, Medical Parasitology, Public Health Microbiology, Special Topics in Microbiology, Virology, Virology Laboratory, Research Planning, Experimental Microbiology, The Literature of Microbiology, Microbial Metabolism, Microbial Physiology, Principles of Immunology, Immunology Laboratory, Biology of Uncommon Microorganisms, Microbial Genetics, Parasitic Diseases of the Tropics and Subtripics, Viral Diseases, Microbial Infections, Microbiology I and 2, Regulation in Biological Systems, Seminar, Journal Colloquy, Research Conference, Clinical Immunology

DEPARTMENT OF NEUROSCIENCE
Phone: Chairman: (904) 392-3383

Chairman: F.A. King
Professors: R.L. Isaacson, F.A. King, P.E. Mahan, O.M. Rennert, B.J. Wilder
Associate Professors: J.J. Bernstein, P.E. Mahan, J.B. Munson, C.J. Vierck, Jr.
Assistant Professors: W.E. Brownell, A.J. Dunn, R.L. King, W.G. Littge, G.W. Sypert, F.J. Thompson, C.J. VanHartesveldt, D W. Walker, R.T. Watson, S.F. Zornetzer
Visiting Assistant Professor: 1
Teaching Fellows: 4
Research Assistants: 8
Associate in Neuroscience: 1

Degree Programs: Neurophysiology, Neuroendocrinology, Neurochemistry and Neurobehavior

Graduate Degree: Ph.D. (Neuroscience)
Graduate Degree Requirements: Doctoral: Neuroanatomy, Neurophysiology, Neurochemistry, Neuroendocrinology Neurobehavior

Graduate and Undergraduate Courses Available for Graduate Credit: Neurohumors and Behavior (U), Introduction to the Neurosciences (U), Vision (U), History of the Neurosciences, Pain and Somesthesis, Comparative Neuroanatomy and Neurophysiology, Physiology of the Central Nervous System, Neurophysiology, Research in Medical Sciences, Neurobiology, Neuroendocrinology, Neurohistology, Nerve as a Tissue, Physiology and Pharamcology of Excitable Membranes, Neural-Behavioral-Endocrine Interactions, Neurobehavioral Relations, Information Storage: A Neurobiological Approach, Developmental Neural-Behavioral Endocrine Interactions, Neural Mechanisms of Ingestion and Energy Regulation, Colloquium in Neurobiology, Physiological Basis of Brain Rhythms, Medical Neuroscience, Recent Advances in Neuroscience, Molecular Neurobiology, Motor Systems, Functional Neurochemistry

DEPARTMENT OF PHYSIOLOGY
Phone: Chairman: (904) 392-3791 Others: 392-3261

Chairman: A.B. Otis
Professors: S. Cassin, W.W. Dawson, M.J. Fregly, A.B. Otis, W.N. Stainsby
Associate Professor: M.J. Jaeger
Assistant Professor: P. Posner
Research Assistants: 1
Postdoctorals: 2

Degree Program: Physiology

Graduate Degrees: M.S., Ph.D. (Medical Sciences)
Graduate Degree Requirements: Master: "B" average - original Thesis Doctoral: "B" average - original Dissertation - foreign language

Graduate and Undergraduate Courses Available for Graduate Credit: Principles of Physiology (U), Physiology of Respiration, Physiology of the Circulation of Blood, Renal Physiology, Neurophysiology, Body Temperature Regulation, Recent Advances in Physiology, Research Methods in Physiology, Seminar in Physiology, Neonatal Physiology, A Survey of Sensory Systems, Seminar in Vision, Physiology of the Mammalian Thyroid, Gland, Physiology and Pharamcology of Excitable Membranes

UNIVERSITY OF MIAMI
Coral Gables, Florida 33124 Phone: (305) 284-2211
Dean of Graduate Studies: C.G. Stuckwisch

College of Arts and Sciences
Dean: R.H. Hively

BIOLOGY DEPARTMENT
Phone: Chairman: (305) 284-3973

Chairman: L.J. Greenfield
Maytag Professor of Ornithology: O.T. Owre
Professors: T. Alexander, J. Clegg, L. Gilman, C. Grabowski, L. Greenfield, B. Hunt, W.H. Leigh, M. Mustard, O. Owre, E. Rich, H. Teas, R. Williams
Associate Professors: B. Burkett, D. Evans, W. Evoy, R. Hofstetter, P. Luykx
Assistant Professors: M. Grabowski, T. Herbert, C. Mallery, T. Pliske, J. Prince, A. Smith
Instructor: J. Brach
Research Assistants: 12
Undergraduate Assistants: 20

Field Stations/Laboratories: Pigeon Key - Tropical Field Station in Ecuador

Degree Program: Biology

Undergraduate Degrees: B.A., B.S.
Undergraduate Degree Requirements: 32 credits biology including a core program. 16 credits Chemistry, 1 year of Calculus, 1 year of Physics with laboratory, general arts and sciences requirements

Graduate Degrees: M.S., Ph.D.
Graduate Degree Requirements: Master: 24 credits plus 6 Thesis, comprehensive examination, language examination, defense of thesis Doctoral: 36 credits above masters plus 24 dissertation additional language (one from masters level) qualifying examination, defense of dissertation

Graduate and Undergraduate Courses Available for Graduate Credit: There are 15 of our own courses of this nature and cross list at least that number with other departments

School of Engineering
Dean: H. Harvenstein

DEPARTMENT OF BIOMEDICAL ENGINEERING
Phone: Chairman: (305) 284-2441 Others: 284-2442

Director: J. Kline
Professors: J. Catz, J. Cunio, C. Geeslin (Adjunct, F. Gollan, J. Hirschberg, K.F. Lampe, R. Myerburg, J. Moder, R. Palmer, O.M. Reinmuth, M. Sackner, P. Samet, A. Sarmiento, M. Viamonte, Jr., E.L. Wiener, H.A.B. Wiseman, R. Zeppa
Associate Professors: H. Bolooki, A. Budkin, W. Coulter (Adjunct), J. Elden (Adjunct), L. Fishman, C. Goldsmith, G. Gonzalez, J.W. Keller (Adjunct), S. Lee, G. Light (Adjunct), B. Linn, J.B. Mann, T. Mende, E. Peterson, A. Rogers, W. Rogers, J. Santlucito (Adjunct), E. Schiff, N. Schneiderman, L.S. Somer, B. Spector (Adjunct), P. Tarjan (Adjunct)
Assistant Professors: R. Davis, A. Feingold, R.S. Hosek, D. Hutson, R. Leif (Adjunct), M. Mastandrea, (Adjunct), M. Rosen (Adjunct), L. Rosenstein (Adjunct) C. Stuuon, A. Wanner, R. Winters
Visiting Lecturers: 5
Research Assistants: 14

Degree Program: Biomedical Engineering

Undergraduate Degree: B.S.
Undergraduate Degree Requirements: Not Stated

Graduate Degrees: M.S., Ph.D.
Graduate Degree Requirements: Master: 30 credits with thesis, 36 credits without thesis Doctoral. 48 credits plus thesis plus examination

Graduate and Undergraduate Courses Available for Graduate Credit: Not Stated

School of Medicine
1600 N.W. 10th Avenue
Miami, Florida 33152
Dean: E.W. Papper

DEPARTMENT OF BIOCHEMISTRY
Phone: Chairman: (305) 350-6265

Chairman: W.J. Whelan
Professors: L.S. Dietrich, R. Ho, S.L. Hsia, F. Huijing, G.T. Lewis, D.W. Ribbons, J. Schultz, K.H. Slotta, A.M. Stein, G.A. Tershakovec, W.J. Whelan, J.F. Woessner, A.A. Yunis
Associate Professors: D.G. Anderson, W.M. Awad, K. Brew, S.B. Greer, D.R. Harkness, E.Y.C. Lee, J.M. Marsh, T.J.Mende, K.H. Muench, N.L. Noble, W.V. Shaw, E.E. Smith, A.G. So, A.T. Soldo, R. Werner
Assistant Professors: F. Ahmad, A.H. Brady, K.E. K. Downey, J.J. Marshall, T. Russell, V. Ziboh
Visiting Lecturers: 6
Research Assistants: 9

Field Stations/Laboratories: Field Station - Pigeon Key

Degree Programs: Biochemistry

Undergraduate Degrees: B.S.
Undergraduate Degree Requirements: 24 credits in biochemistry, 28 credits in chemistry, 8 credits in biology, total 120 credits

Graduate Degrees: M.S., Ph.D.
Graduate Degree Requirements: Master: (a) 24 graduate credits, b) research thesis with oral defense Doctoral: a) 36 graduate credits b) 24 credits for research thesis with oral examination by committee and external examiner

Graduate and Undergraduate Courses Available for Graduate Credit: Seminars (U), Physical Biochemistry (U), Proteins and Enzymes (U), Biosynthesis of Macromolecules (U), Metabolic Processes (U), Introductory General Course (U), Laboratory, Introductory Course (U), Research Problems in Molecular Enzymology (U), Research Problems in Biochemistry (U), Advanced Topics in Biochemistry, Molecular Medicine, Molecular Enzymology Instrumentation, Special Work

DEPARTMENT OF BIOLOGICAL STRUCTURE
Phone: Chairman: 350-6691

Chairman: D.E. Kelly
Professors: C. Cabrera (Clinical), W. Copenhaver (Clinical)
Associate Professors: I. Berman, D. Cahill, R. Clark, R. Wood

Assistant Professors: R. Gulley, M. Snow, R. Warren, G. Wise, A. Yakaitis-Surbis
Research Assistant: 1

Degree Program: Biological Structure

Graduate Degrees: M.S., Ph.D.
Graduate Degree Requirements: Master: Thesis
Doctoral: Competence in one foreign language, Written and oral qualifying examinations, Dissertation

Graduate and Undergraduate Courses Available for Graduate Credit: Gross Anatomy, Embryology and Basic Tissues, Organ Histology, Neuroanatomy, Seminar, Special Work Microscopic, Electron Microscope Seminar, Theory of Modern Microscopic Technology, Topics in Ultrasturctural Cytology, Autoradiography, Special Work - Embryology, Advanced Human Embryology, Structural Basis of Cell Motility and Cytomorphogenesis, Special Work - Neuroanatomy, Advanced Neuroanatomy, Neurocytology I and II, Special Work - Gross Anatomy, Advanced Regional Anatomy, Advanced Regional Anatomy - Head and Neck, Advanced Regional Anatomy - Back and Extremities, Gross Anatomical Techniques

DEPARTMENT OF MICROBIOLOGY
Phone: Chairman: 547-6655

Chairman: B. Sallman
Professors: S.B. Greer, J.A. Jensen, F.J. Roth, Jr., B. Sallman, M.M. Sigel
Associate Professors: G.E. Schaiberger, M.M. Streitfeld
Assistant Professors: A.R. Beasley, D. Lopez, I. Fuller (Research), S.M. Gerchakov, (Research), R.C. Leif (Research), W.E. Scott (Research) A. Thorhaug (Research)
Instructors: G.Ortiz (Research), D. Rippe (Research)

Degree Programs: Microbiology

Graduate Degrees: M.S., Ph.D.
Graduate Degree Requirements: Master: 30 graduate credits including specified courses in microbiology, qualifying examination Doctoral: general university requirements for Ph.D. plus physical chemistry, calculus, foreign language

Graduate and Undergraduate Courses Available for Graduate Credit: Medical Microbiology, Microbial Chemistry and Physiology, Microbial Genetics, Immunology and Immunochemistry, Virology and Pathogenic Bacteriology, Mycotic Agents of Disease, Projects in Microbiology, Advanced Microbial Chemistry and Physiology, Advanced Microbial Genetics, Advanced Immunology and Immunochemistry, Advanced Virology, Advanced Mycology, Advanced Pathogenic Bacteriology and Antibiotics, Applied Microbiology

DEPARTMENT OF PHYSIOLOGY AND BIOPHYSICS
Phone: Chairman: (305) 547-6821

Chairman: W.R. Loewenstein
Professors: E.L. Chambers, W.C. Grant, M. Jacobson, M. Kalser, L. Potter, B.C. Pressman, M. Rockstein, M. Sackner
Associate Professors: R.N. DeGasperi, E. Kohen, S.J. Scocolar, C. VanBreeman
Assistant Professors: R. Azarnia (Research), E.F. Barrett, J.N. Barrett, D. Dandowne, W.J. Larsen (Research), K.L. Magleby, D.A. McAffee, R.J. Myerburg, B. Rose (Research), D.L. Wilson
Instructors: J. Chesky (Research), D. Garrison (Research), R. Levine (Research), I. Simpson (Research), J. Zengel-Messer (Research)
Visiting Lecturers: 2
Research Assistants: 3
Pre-doctoral Trainee: 1
Teaching Assistants: 7

Degree Programs: Cellular Aging, Membrane Biology, Neurobiology

Graduate Degree: Ph.D. (Physiology, Biophysics)
Graduate Degree Requirements: Doctoral: Requirements for admission to candidacy include (1) 36 graduate credits (exclusive of dissertation research) (2) demonstration of reading proficiency in French, German, Russian, or another language that the department considers acceptable on scientific grounds; and (3) satisfactory performance on a qualifying examination, consisting of the formulation and defense of a research proposal and a demonstration of mastery of relevant physiological principles and methods. Students working toward the Ph.D. degree may bypass the M.S. degree. The Ph.D. dissertation research must be original work of a quality acceptable for publication in a first-rate scientific journal

Graduate and Undergraduate Courses Available for Graduate Credit: Mammalian Physiology, Physiology of the Arthropoda; Cellular Physiology and Biophysics, Neurophysiology, Systematic Physiology, Membrane Biophysics Seminar, Principles of Membrane Physiology and Biophysics, Topics in Membrane Physiology and Biophysics, Methods in Membrane Research, Cell Organelles and Bioenergetics, Developmental Neurobiology, Molecular Processes and Higher Brain Function, Neuropharmacology, Nerve and Synapse, Principles of Cellular Aging, Special Topics in Cellular Aging

Dorothy H. and Lewis Rosenthiel School of Marine and Atmospheric Science
13 Rickenbacker Causeway Phone: (305) 350-7211
Miami, Florida 33149
Dean: W.W. Wooster

DIVISION OF BIOLOGY AND LIVING RESOURCES
Phone: Chairman: (305) 350-7351 Others: 350-7211

Division Chairman: F. Williams
Maytag Professor of Ichthyology: C.R. Robins
Professors: F. Bayer, J. Bohlke (Adjunct), T. Bowman (Adjunct), H.R. Bullis (Adjunct) J. Bunt, D. Cohen (Adjunct), A. Colwin (Adjunct), L. Colwin (Adjunct), E. Corcoran, L. Holthuis (Adjunct), C. Lane, R. Manning (Adjunct), H. Michel, A. Mitsui, A. Myrberg, C.R. Robins, C. Roper (Adjunct), G. Voss, H.H. Webber (Adjunct), F. Williams
Associate Professors: K. Cooksey, D. DeSylva, J. Fell, E. Iversen, A.C. Jones (Adjunct), G. Krantz, M. Reeve, W. Richards (Adjunct), C. Robins (Adjunct), M.A. Roessler (Adjunct), D. Tabb (Adjunct), D. Taylor, L. Thomas, W. Yang
Assistant Professors: S. Gruber, J. Higman (Research), A. Hine (Research), E. Houde, P. McLaughlin (Research), D. Odell, J. Richard, J. Staiger (Research), R. Stevenson, B. Taylor, N. Voss (Research), J.P. Wise (Adjunct), R. Work (Research)
Instructors: M. Alvarez (Instructor), C. Gordon (Associate), C.C. Lee (Scientist), B. Yokel (Scientist), E. Zillioux (Scientist)

Field Stations/Laboratories: Field Stations at Turkey Point and Pigeon Key

Degree Programs: Physiology, Parasitology, Behavior, Primary Productivity, Taxonomy, Ichthyology, Fisheries and Mariculture

Graduate Degree: M.S., Ph.D.
Graduate Degree Requirements: Master: Minimum of 2 semesters in full time residence or 24 credits. Pass reading test of foreign language if required and prepare thesis (original research 6 credits), pass comprehensive examination Doctoral: 2 semesters beyond the first year of full time graduate study. Prepare dissertation - pass qualifying examination. Pass reading test of 2nd foreign language if required

Graduate and Undergraduate Courses Available for Graduate Credit: Introduction to Marine Biology, Introduction to Marine Biology Laboratory, Advanced Marine Biology, Marine Biochemistry, Fishes and Their Environment, Biological Oceanography Seminar, General Biological Oceanography, General Biological Oceanography Laboratories, Taxonomy of Marine Invertebrates, Ecology of Marine Animals, Invertebrate Embryology, Behavior of Marine Organisms, Advanced Studies in Ethology, Marine Microbiology, Plankton, Phycology, Physiology of Marine Organisms, Systematics of Fishes, Fisher Technology, Introduction to Fishery Science, Saltwater Pollution Technology, Supervised Projects, Fishery Seminar, Fish Stocks and Their Management, Fish Stocks and Their Management Laboratory, Population Enumeration and Dynamics, Ecology of Marine Parasites, Biometrics in Marine Sciences, Economics of Natural Resources, Mariculture

UNIVERSITY OF SOUTH FLORIDA

4202 Fowler Avenue Phone: (813) 974-2011
Tampa, Florida 33620
Dean of Graduate Studies: J. Lawrence

College of Liberal Arts

Dean: J.D. Ray, Jr.

DEPARTMENT OF BIOLOGY

Phone: Chairman: (813) 974-2688

Acting Chairman: S.L. Swihart
Professors: M.R. Alvarez, C.J. Dawes, F.E. Friedl, R. W. Long, N.M. McClung, A.J. Meyerriecks, G.E. Nelson, W.S. Silver, G.E. Woolfenden
Associate Professors: J.V. Betz, L.N. Brown, D.G. Burch, B.C. Cowell, C.E. King, J.M. Lawrence, J.E. Linton, R.L. Mansell, R.W. McDiarmid, D.T. Merner, G. Robinson, J. Simon, S.L. Swihart
Assistant Professors: G.R. Babbel, F.I. Eilers, S. Grove, D.A. Hessinger, K. Stuart, H.C. Tipton
Instructors: C.S. Hendry, A.A. Latina (Lecturer), B. Michaelides
Visiting Lecturers: 1

Degree Programs: Biology, Botany, Microbiology, Zoology

Undergraduate Degree: B.A.
Undergraduate Degree Requirements: Not Stated

Graduate Degrees: M.A., Ph.D.
Graduate Degree Requirements: Not Stated

Graduate and Undergraduate Courses Available for Graduate Credit: Cytology (U), Subcellular Cytology (U), Neurophysiology (U), Molecular Genetics (U), Evolutionary Genetics (U), History of Biology, Chromosome Structure and Chemistry, Ultrastructure Techniques in Electron Microscopy, Biometry, Population Biology, Marine Plankton Systematics, Marine Plankton Ecology, Selected Topics, Graduate Research, Directed Teaching, Graduate Seminar, and Ph.D. Dissertation, Taxonomy of Flowering Plants (U), Physiology of the Fungi (U), Physiology of Plant Growth and Development (U), Phycology (U), Plant Ecology (U), Marine Botany (U), Biosystematics, Biology of Tropical Plants, Laboratory in Tropical Plants, Plant Metabolism Lecture and laboratory, Marine Algal Ecology, Medical Mycology (U), Advanced Bacteriology (U), Determinative Bacteriology (U), Microbial Physiology (U), Virology (U), Bacterial Genetics, Immunology, Parasitology (U), Aquatic Entomology (U), Limnology (U), Ornithology (U), Mammalogy (U), Ichthyology (U), Biology of Echinoderms (U), Comparative Physiology (U), Biology of the Amphibians (U), Biology of the Reptilia (U), Zoogeography (U), Terrestrial Animal Ecology (U), Marine Animal Ecology (U), Mechanisms of Animal Behavior (U), Experimental Embryology, Advanced Mammalogy, Physiological Ecology, Physiology of Marine Animals, Comparative Endocrinology, Invertebrate Reproduction and Development, Advanced Animal Behavior

DEPARTMENT OF MARINE SCIENCE

Phone: Chairman: (813) 898-7411 Ext. 231

Chairman: F.T. Manheim
Professors: H.J. Humm
Associate Professors: R.C. Baird, K.L. Carder, T.L. Hopkins, T.H. Pyle
Assistant Professors: P.R. Betzer, N.J. Blake, L.F. Doyle, K.A. Fanning
Research Assistants: 28
Research Associates: 2

Degree Programs: Marine Science, Oceanography

Graduate Degree: M.S.
Graduate Degree Requirements: 45 credit hours and thesis

Graduate and Undergraduate Courses Available for Graduate Credit: Introduction to Oceanography (U), Chemical Oceanography (U), Geological Oceanography (U), Physical Oceanography (U), Biological Oceanography (U), Selected Topics in Oceanography (U), Scientist-in-the-Sea I, II, III, Methods in Chemical Oceanography, Oceanic Modeling, Marine Algal Ecology, Marine Plankton Systematics, Methods in Biological Oceanography, Marine Plankton Ecology, Dynamics of Marine Benthic Communities, Graduate Research, Selected Topics in Oceanography, Graduate Seminar in Oceanography, M.S. Thesis

College of Medicine

Dean: D.L. Smith

DEPARTMENT OF ANATOMY

Phone: Chairman: (813) 974-2843 Others: 974-2011

Chairman: H.N. Schnitzlein
Professors: H.N Schnitzlein, J.W. Ward (1/2 time)
Associate Professors: J.J. Dwornik, J.M. Thompson (Clinical)
Assistant Professors: R.K. Boler, A.C. Hernandez (Clinical), N.A. Moore, E.G. Salter, Jr.
Instructor: G.C. Morgan, Jr.

Field Stations/Laboratories: Marine Science Department - St. Petersburg

Degree Program: Anatomy

Graduate Degree: Ph.D.
Graduate Degree Requirements: Doctoral: Successful completion of program

Graduate and Undergraduate Courses Available for Graduate Credit: Usual courses offered by the College of Medicine

DEPARTMENT OF PHYSIOLOGY

Phone: Chairman: (813) 974-2590 Others: 974-2527

Chairman: C.H. Baker
Professors: C.H. Baker, D.L. Davis
Assistant Professors: D.K. Anderson, J.A. Boulant, J.M. Downey, R.P. Menninger, G.R. Nicolosi, R. Shannon
Research Assistants: 2
Postdoctoral Research Fellow: 1

Degree Program: Physiology

Graduate Degree: Ph.D.
Graduate Degree Requirements: Not Stated

Graduate and Undergraduate Courses Available for Graduate Credit: Not Stated

UNIVERSITY OF TAMPA

401 West Kennedy Boulevard Phone: (813) 253-8861
Tampa, Florida 33606

DEPARTMENT OF BIOLOGY
Phone: Chairman: Ext.326

Chairman: R.H. Gude
Professors: M.L. Ellison, R.H. Gude
Associate Professors: J. Dinsmore, W.L. Smith
Assistant Professors: S.E. Monaloy, J.G. Waite

Degree Program: Biology

Undergraduate Degree: B.S.
Undergraduate Degree Requirements: 38 semester hours of biology credit to include Biological Diversity, Biological Unity, Plant Morphology or Taxonomy of Flowering Plants, Genetics and 22 hours of biology selected in consultation with the academic advisor, 12 hours of which to be in courses numbered 300 or above. Also required are certain Chemistry, Physics and Mathematics

UNIVERSITY OF WEST FLORIDA

Pensacola, Florida 32504 Phone: (476) 9500-283
Provost: A.B. Chaet

FACULTY OF BIOLOGY
Phone: Chairman: Ext. 283, Others: 284,339,482

Chairman: T.S. Hopkins
Professor: R.D. Reid (Emeritus)
Associate Professors: J.R. Baylis, S.B. Collard, C.N. D'Asaro, T.S. Hopkins, G.A. Moshiri, J.P. Riehm
Assistant Professors: S.A. Bortone, M.I. Cousens, K.R. Rao, P.A. Winter
Visiting Lecturers: 9

Field Stations/Laboratories: EPA Laboratory - Sabine Island

Degree Programs: Biology

Undergraduate Degree: B.S.
Undergraduate Degree Requirements: 8-16 hours Biology Core, 20 additional hours in field of speciality (Biology, Marine Biology)

Graduate Degree: M.S. (Biology, Estuarine Biology)
Graduate Degree Requirements: Master: 45 quarter hours of which at least half must be in courses available only for graduate credit. Thesis

Graduate and Undergraduate Courses Available for Graduate Credit: Medical and Public Health Bacteriology (U), Introduction to Endocrinology (U), Intermediary Metabolism (U), Advanced Genetics (U), Biology of Animal Parasites (U), Enzymology (U), Estuarine Biology (U), Biological Oceanography (U), Plankton Biology (U), Biology of Algae (U), Scientific Illustration (U), Development of Marine Invertebrates (U), Endocrinology of Marine Organisms (U), Introduction to Ethology, Biology of Molluscs, Biology of Crustacea, Biology of Echinoderms, Molecular Genetics, Quantitative Ecology, Ecological Adaptations, Ecological Energetics, Chemistry of Marine Natural Products, Microbial Genetics, Aquatic Microbiology, Biology of Fishes, Biology of Vascular Plants, Aquaculture, Comparative Animal Physiology

GEORGIA

AGNES SCOTT COLLEGE

College Avenue Phone: (404) 373-2571
Decatur, Georgia 30030
Dean of College: J.T. Gary

BIOLOGY DEPARTMENT
Phone: Chairman: Ext. 375 Others: Ext. 376

Chairman: N.P. Groseclose
Professor: N.P. Groseclose
Associate Professor: S. Bowden
Assistant Professor: T. Simpson, H. Wistrand
Instructor: G. Miller

Degree Program: Biology

Undergraduate Degree: A.B.
Undergraduate Degree Requirements: General Course - 12 quarter hours credit; Concepts of Biology and Zoology Surveys, Cytology, Genetics, Cellular Physiology, Plant or Animal Development, Seminar(s) and other biology courses to total 45-60 quarter hours, Organic Chemistry (10 quarter hours); Mathematics, Physics, Languages are strongly recommended

ALBANY STATE COLLEGE

504 Hazard Drive Phone: (912) 439-4234
Albany, Georgia 31705
Chairman - Division of Arts and Sciences: B.C. Black

BIOLOGY DEPARTMENT

Chairman: M.E. Jones, Jr.
Associate Professors: E.W. Benson, B.H. Fort, Jr., E.A. Green, A. Husain, M.E. Jones, Jr.
Assistant Professors: H.B. Bates, Jr., D.C. Robinson
Instructors: O.S. Bailey, O.E. Lockley, E.E. Lyons, M.A. Norman, E.E. Sykes, B.J. Washington

Degree Program: Biology

Undergraduate Degree: B.S.
Undergraduate Degree Requirements: B.S. in biology 188 hours, 1 year of Mathematics through Calculus, 1 year Organic Chemistry, 1 year Inorganic Chemistry, 10 Biology courses, General Education requirements

ARMSTRONG STATE COLLEGE

Savannah, Georgia 31406 Phone: (912)925-4200
Dean of Graduate Studies: J.V. Adams
Dean of College: H.D. Propst

DEPARTMENT OF BIOLOGY
Phone: Chairman: Ext. 241, 242, 243

Chairman: L.B. Davenport, Jr.
Professors: L.B. Davenport, Jr., F.M. Thorne, III
Associate Professors: A.D. Beltz, A.L. Pingel
Assistant Professors: M.S. Brower, L.J. Guillou, Jr.

Degree Programs: Biology, Medical Technology

Undergraduate Degree: B.S.
Undergraduate Degree Requirements: 20 quarter hours lower division (courses specified), 40 quarter hours upper division (5 hours each in Genetics, Ecology and Physiology specified, remainder elective), 25 quarter hours Chemistry (10 general, 15 Organic)

ATLANTA UNIVERSITY CENTER

55 Walnut Street S.W. Phone: (404) 524-5751
Atlanta, Georgia 30309

Atlanta University

223 Chestnut Street, S.W. Phone: (404) 681-0251
Atlanta, Georgia 30134
Dean: B.F. Hudson

DEPARTMENT OF BIOLOGY
Phone: Chairman: Ext. 237

Chairman: L. Frederick
Professors: L. Frederick, R. Hunter, Jr.
Associate Professors: J.M. Brown, E.L. Stevenson
Assistant Professors: J.R. Lumb, J.B. Myers, G.A. Ofosu, J. Ruffin
Instructors: C.W. Clark
Visiting Lecturers: 4

Degree Programs: Biology, Botany, Microbiology, Zoology

Graduate Degrees: M.S., Ph.D.
Graduate Degree Requirements: Master: Completion of 30 semester hours, four of which must be in a research course; thesis, one foreign language Doctoral: Minimum of 72 semester hours in regular courses and research, thesis, two foreign languages or one language and one or more courses in computer science

Graduate and Undergraduate Courses Available for Graduate Credit: Evolution and Origin of Life, General Parasitology, Insect Biology, General Animal Physiology, Modern Genetics, Microbial Genetics, Experimental Biology, Medical Microbiology, Immunology, Virology, Plant Anatomy, Morphology of Non-Vascular Plants, Morphology of Vascular Plants, Experimental Embryology, Systematic Botany, Plant Pathology, Mycology, Cell Biology, Cytogenetics, Microtechnique, Plant Physiology, Tissue Culture Techniques, Advanced Insect Zoology, Protozoology, Comparative Animal Physiology, Microbial Biochemistry, Plant Biochemistry, Growth and Metabolism in Plants, Quantitative Biology, Ultrastructure, Chemistry of Living Systems, General Ctyology, Vertebrate Experimental Physiology, Developmental Physiology, Cytochemistry, Molecular and Cellular Aspects of Development, Special Topics in (Cell Biology, Developmental Biology, Plant Physiology, e.g.) Research in (Microbiology, Parasitology, Developmental Botany, e.g.)

Clark College

(No Reply Received)

Moorehouse College

232 Chestnut Street, S.W.
Atlanta, Georgia
Dean: W.J. Hubert

DEPARTMENT OF BIOLOGY
Phone: Chairman: (404) 681-2800 Ext. 263 Others: 262

Chairman: T.E. Norris
David Packard Professor of Biology: F.E. Mapp
Professors: T.E. Norris, F.E. Mapp, R. Hunter (Adjunct)
Assistant Professors: R.J. Sheehy, J. Bender
Visiting Lecturers: 15
Teaching Fellows: 1
Research Assistants: 2
Student Assistants: 10

Degree Program: Biology

Undergraduate Degree: B.S.
Undergraduate Degree Requirements: Not Stated

Morris Brown College
(No reply received)

AUGUSTA COLLEGE
Augusta, Georgia 30904 Phone: (404) 733-2234
Academic Dean: J.G. Dinwiddie, Jr.

DEPARTMENT OF BIOLOGY

Chairman: D.B. Morris
Professor: D.B. Morris
Associate Professors: J.B. Black, S.L. Wallace, B.B. Webber
Assistant Professors: J. Bickert, G.B. Cook, H.L. Stirewalt, R.E. Stullken

Degree Program: Biology

Undergraduate Degrees: B.S.
Undergraduate Degree Requirements: Completion of the general graduation requirements and thirty credits in the Area IV of the Core relating. This includes five credits in Mathematics 201 or 221, five credits in Chemistry 103, ten credits in Biology 101 and 102, and ten credits from Biology 201, 202, 221 and 222

BERRY COLLEGE
Mount Berry, Georgia 30149 Phone: (404) 232-1732
Dean of College: W.C. Moran

BIOLOGY DEPARTMENT
Phone: Chairman: Ext. 224

Chairman: K.P. Hancock
Dana Professor of Biology: K.F. Hancock
Professors: K.F. Hancock, J.W. McDowell
Associate Professor: R.S. Kiser

Degree Programs: Biology, Medical Technology

Undergraduate Degrees: B.S., B.A.
Undergraduate Degree Requirements: B.S. - Biology, 46 quarter hours, including Principles, Animal Kingdom, Plant Kingdom, 18 hours above 300, B.A. requires an additional 20 quarter hours, language (foreign) B.S. Medical Technology - Highly structured 3 year program of courses plus 12 months course in AMA-approved hospital school of medical technology, plus passing Registry Examination, ASCP. Three-year program includes 64 hours general education, 5 hours Physics, at least 31 hours Biology, at least 28 hours Chemistry

BRENAN COLLEGE
Gainesville, Georgia 30501 Phone: (404) 532-4341
Dean of College: J.E. Sites

DIVISION OF NATURAL SCIENCES
Phone: Chairman: Ext. 22

Chairman: C.L. Andrews
Professor: L.G. Andrews
Assistant Professors: H.M. Langley, W. Nachtmann
Instructor: R. Champion

Degree Program: Biology

Undergraduate Degrees: B.A., B.S.
Undergraduate Degree Requirements: Biology - 10 hours Mathematics (including Statistics), 25 hours Chemistry (including Organic and Biochemistry), 40 hours Biology (including Ecology, Genetics, and Cell Biology) 10 hours Physics

CLARK COLLEGE
(See Atlanta University Center)

COLUMBUS COLLEGE
Columbus, Georgia 31907 Phone: (404) 561-5134
Dean of College: J.E. Anderson

DEPARTMENT OF BIOLOGY
Phone: Chairman: 561-5134 Ext. 392

Chairman: W.C. LeNoir
Professors: W.C. LeNoir, J.B. Lytle, K. Nance
Associate Professors: F.M. Clark, W.Z. Faust, G.E. Stanton
Assistant Professors: S.A. Sonstein

Degree Program: Biology

Undergraduate Degree: B.S.
Undergraduate Degree Requirements: University system of Georgia core requirements, Cytology, Genetics, Physiology, Ecology, Senior Seminars, Biology Electives, Organic Chemistry, Mathematics, Statistics, Electives

EMORY UNIVERSITY
Atlanta, Georgia 30322 Phone: (404) 377-2411
Dean of Graduate Studies: C.T. Lester

Emory College
Dean: J.M. Palms

DEPARTMENT OF BIOLOGY
Phone: Chairman: Ext. 7519 Others: 7516

Chairman: A.A. Humphries, Jr.
Charles Howard Chandler Professor of Biology: C.G. Goodchild
Professors: W.D. Burbanck, A.C. Clement, C.G. Goodchild, A.A. Humphries, Jr. E.L. Hunt, W.H. Murdy, R.B. Platt, C. Ray, Jr.
Associate Professors: W.E. Brillhart, H.L. Ragsdale, D.J. Shure, P.D. Smith
Assistant Professors: W.A. Elmer, D.R. Stokes, C.E. Wickstrom
Teaching Fellows: 13
Research Assistants: 2

Field Stations/Laboratories: Lullwater Field Laboratory

Degree Programs: Biology

Undergraduate Degrees: B.A., B.S.
Undergraduate Degree Requirements: For the B.A. degree a minimum of nine biology courses (45 hours), introductory inorganic chemistry (10 hours), introductory organic chemistry (10 hours, introductory physics (10 hours), 5 hours mathematics (107 or 171), For the B.S. degree a minimum of 9 biology courses (45 hours), introductory inorganic chemistry, (10 hours), introductory organic chemistry (10 hours), introductory physics (10 hours), Mathematics (15 hours)

Graduate Degrees: M.S., Ph.D.
Graduate Degree Requirements: Master: Three quarters of residence, competence in one foreign language, written comprehensive examination, thesis and oral examination Doctoral: Two years' work beyond Master's, competence in second foreign language, written comprehensive examination, dissertation and oral examination

Graduate and Undergraduate Courses Available for Graduate Credit: Biology of the Protozoa, Advanced Plant Ecology, Treatment of Experimental Data, Population Genetics I and II, Problems of Development, Comparative Endocrinology, The Cytology of Development, Developmental Genetics, Topics in Cell Biology, Concepts of the Gene, Problems of Plant Systematics, Analysis of Man-Environment Ecosystems, Ecosystem Dynamics, Mathematical Ecology, Physiology of Animal Parasites, Ecology of Aquatic Communities, Experimental Plant Ecology, Protozoological Problems, Physiology of Development, Biology of Parasitism (U), Comparative Vertebrate Embryology (U), Comparative Vertebrate Anatomy (U), Experimental Embryology (U), Plant Systematics (U), Invertebrate Zoology (U), Biology of Microorganisms (U), Advanced Vertebrate Zoology (U), Ecology (U), Human Ecology (U), Cell Biology (U), Plant Physiology (U), Ecology (U), Human Ecology (U), Cell Biology (U), Plant Physiology (U), Metabolic Biology (U), Genetics (U), Evolution (U), Experimental Genetics (U)

School of Medicine

Dean: J.A. Bain

DEPARTMENT OF ANATOMY

Phone: Chairman: Ext. 7805 Others: 7806

Chairman: J. Sutin
Professors: J.V. Basmanian, R.L. DeHaan, S.W. Gray, R.P. Michael, W.K. O'Steen, J.E. Skandalakis, J. Sutin
Associate Professors: K.V. Anderson, L.J. DeFelice, A. Falek, K. Nandy, J.W. Tigges
Assistant Professors: C.A. Baste, C.M.Dienhart, B.F. Edwards, N.A. Jones, W.D. Letbetter, R. Margeson, B.D. Noe, J.D.Rose, J.W. Scott, C.A. Shear
Instructors: R.P. Olafson, A.W. English
Teaching Fellows: 1
Research Associates: 2

Degree Programs: Anatomy, Anatomy and Histology, Histology, Life Sciences, Medical Biophysics, Neurobiology, Cell Biology

Graduate Degrees: M.S., Ph.D.
Graduate Degree Requirements: Refer to Bulletin, Graduate School of Arts and Sciences, Emory University

Graduate and Undergraduate Courses Available for Graduate Credit: Gross Anatomy, Neurobiology, Advanced Neuroanatomy, Anatomy Seminar, Cell Biology and Biochemistry, Microscopic Anatomy, Topics in Clinical Anatomy, Biology of Aging, Spinal Cord Neurobiology, Nervous System Theory, Developmental Biology, Research, Directed Study, Neural Correlates of Behavior, Electron Microscopy and Cell Fine Structure

DEPARTMENT OF BIOCHEMISTRY

Phone: Chairman: Ext. 7823 Others: 7824

Chairman: A.E. Wilhelmi
Charles Howard Chandler Professor: A.E. Wilhelmi
Professors: F. Binkley, J.B. Lyon Jr., L.E. Reichert Jr., R. Shapira
Associate Professors: F.W. Fales, D.P. Groth, R.F. Kibler (Adjunct), J.B. Mills III, J.R.K. Preedy (Adjunct), A.J. Sophianopoulos, D. Rudman (Adjunct)
Assistant Professors: D. Collins (Adjunct), L. Elsas (Adjunct), J.M. Kinkade, Jr., J.F.R. Kuck Jr (Adjunct), V. Seery, R.C. Shuster
Research Assistants: 15
Seminar Speakers: 8-12/year

Degree Program: Biochemistry

Graduate Degrees: M.S., Ph.D.
Graduate Degree Requirements: Master: This program requires at least four quarters of residence. Requirements include the following courses or equivalents: Biochemistry 332,333,334,336,337 and 491 (three quarters), a reading knowledge of the scientific literature in one foreign language, and the presentation of a satisfactory thesis based upon the student's original research. Doctoral: Admission to advanced standing on basis of M.S. degree or equivalent course work, followed by 6 additional quarters of residence, with advanced course work and thesis research amounting to a minimum of 72 hours of credit

Graduate and Undergraduate Courses Available for Graduate Credit: Cellular Biology and Biochemistry, Basic Radioisotope Principles and Techniques, Fundamentals of Biochemistry I, II, III; Aspects of Biochemical Regulation, Biochemical Methods and Preparations, Clinical Chemistry, Clinical Chemistry Internship, Introduction to Endocrinology, Reading and Composition of Biochemical Papers, Amino Acids and Proteins, Nucleic Acids, Control Mechanisms in Carbohydrate and Lipid Metabolism, Enzymes, Advanced Endocrinology, Physical Behavior of Macromolecules, Neurochemistry

DEPARTMENT OF MICROBIOLOGY

(No reply received)

DEPARTMENT OF PHARMACOLOGY

Phone: Chairman: Ext. 7852

Chairman: N.C. Moran
Professors: J.A. Bain, R.E. Jewett, N.C. Moran, A.P. Richardson, E.L. Frederickson, J.L. McNay
Associate Professors: J.F. Kuo, F.H. Schneider, C.C. Hug
Assistant Professors: R.I. Glazer, H.J. Haigler, J.E. Harris, S.G. Holtzman, D.P.Groth, D.W. Pruitt
Visiting Lecturers: 5
Research Assistants: 2

Degree Program: Pharmacology

Graduate Degrees: M.S., Ph.D.
Graduate Degree Requirements: Master: 30 hours course or seminar, 6 hours research, three quarters residence thesis, Doctoral: M.S., M.A. degree or equivalent, 36 hours course work or directed study - B grade in 30 hours, six quarters residence, general doctoral examination - Dissertation and final oral examination

Graduate and Undergraduate Courses Available for Graduate Credit: Introduction to Pharmacology (lecture), Introduction to Research in Pharmacology, Laboratory exercises in Pharmacology, Special Topics in Pharmacology, Neuropharmacology, Cardiovascular Pharmacology, Molecular Pharmacology, Advanced Chemotherapy

DEPARTMENT OF PHYSIOLOGY

Phone: Chairman: Ext. 7862 Others: 7867

Chairman: J.L. Kostyo
Professors: J.W. Manning, V. Popovic
Associate Professors: D.R. Humphrey, Y. Matsumoto, J.D. Neill
Assistant Professors: E.O. Fuller, S.J. Hersey, J.P. Pooler, G.A. Rinard, L.G. Young, J.F. White
Instructors: C.R. Reagan, M.S. Smith
Visiting Lecturers: 18
Teaching Fellows: 1
Research Associates: 5

Degree Program: Physiology

Graduate Degree: Ph.D.
Graduate Degree Requirements: Doctoral: This program usually requires two to three years after admission to advanced standing (M.S. or equivalent). All students must pass a qualifying examina tion, one foreign language examination and formal training in computer language. A student enters a research program of such quality that the result is a prepared dissertation which must be successfully defended in an oral examination

Graduate and Undergraduate Courses Available for Graduate

Credit: Neurobiology, Mammalian Physiology, Tutorial in Physiology, Analysis of Biophysical Systems, Cellular Excitation and Bioelectricity, Hypothermia, Physiology of Contractile Systems, Cellular Endocrinology, Graduate Neurobiology, Neurobiology Seminar, Physiology of Reproduction, Cellular Transport Processes, Departmental Graduate Seminar, Research

GEORGIA COLLEGE

Milledgeville, Georgia 31061 Phone: (912) 453-4246
Dean of Graduate Studies: E. Hong
Dean of College: W. Simpson

BIOLOGY DEPARTMENT
Phone: Chairman: (912) 453-4246 Others: 453-5290

Chairman: D.J. Cotter
Georgia College Foundation Distinguished Professor: D.J. Cotter
Professors: D.J. Cotter
Associate Professors: J.D. Batson, T.L. Chesnut, C.P. Daniel, H.L. Whipple
Assistant Professors: J.V. Aliff, E.H. Barman, S.D. Caldwell, D.J. Staszak
Instructor: D.P. Moody
Teaching Fellows: 1

Field Stations/Laboratories: Center for Environmental Study and Planning - Lake Sinclair, Milledgeville

Degree Programs: Biology

Undergraduate Degree: B.S.
Undergraduate Degree Requirements: 63 quarter hours of Biology - Introductory Biology sequence (2 courses), Invertebrate Zoology, One Botany Course, 6 Advanced Courses and 3 Seminars

Graduate Degrees: M.S., M.Ed., Ed.S.
Graduate Degree Requirements: Master: M.S. Biology 45 quarter hours of Biology course work and 15 quarter hours of thesis M.Ed. - 25-35 quarter hours of biology and 25 hours of education Ed.S. - Biology - 45 quarter hours beyond M.Ed.

Graduate and Undergraduate Courses Available for Graduate Credit: Field Botany, Field Zoology, Genetics, Evolution, Ecology, Independent Study, Plant Anatomy, and Microtechnique, Freshwater Biology, Limnology, Ichthyology, Recent Advanced in Biological Sciences, Introduction to Scientific Research, Biological Techniques, Biogeography, Population Ecology, Advanced Parasitology, Advanced Physiology, Advanced Microbiology, Advanced Entomology

GEORGIA INSTITUTE OF TECHNOLOGY

Atlanta, Georgia 30332 Phone: (404) 894-2000
Dean of Graduate Studies: S.C. Webb
Dean of College: H.S. Valk

SCHOOL OF BIOLOGY
Phone: Chairman: 894-3735 Others: 3736

Director: J.W. Crenshaw, Jr.
Professors: W.L. Bloom, J.W. Crenshaw, Jr., R.H. Fetner, T.W. Kethley, D.W. Menzel (Adjunct), H.A. Wickoff (Emeritus)
Associate Professors: E.L. Fincher, J.J. Heise, A.W. Hoadley, J.S. Hubbard, N.W. Walls, E.K. Yeargers
Assistant Professors: G.L. Anderson, A.C. Benke, D.B. Dusenbery, D.M. Gillespie, P. Haysman, S.P. Porterfield, J.R. Strange and R.M. Wartell
Instructors: R.M. Smith, R.D. Snyder, G.T. Tibbs
Research Assistants: 7

Degree Programs: Biology, Biophysics, Ecology, Genetics, Microbiology, Physiology, Radiobiology

Undergraduate Degrees: B.S. (Biology)
Undergraduate Degree Requirements: Biology Core Curriculum, 33 hours (quarter), biology hours unspecified, 25; 14 hours introductory Chemistry, 13 hours Organic Chemistry, 15 hours Calculus, 15 hours Physics, 18 hours Social Sciences, or modern language, 23 hours technical electives, 23 hours free electives Total 200 quarter hours

Graduate Degree: M.S.
Graduate Degree Requirements: Master: 33 quarter hours of course work beyond the bachelor's degree plus a thesis. Of the course work at least 18 quarter hours must be in the major field, at least 18 quarter hours must be fully graduate courses

Graduate and Undergraduate Courses Available for Graduate Credit: (Up to 15 quarter hours of junior and senior level undergraduate courses may be included in a graduate program). Ecological Systems, Special Topics in Ecology, Design of Experiments in Quantitative Biology, Selected Topics in Radio Biology, Selected Topics in Experimental Cell Biology, Air Pollution Biology, Instrumental Methods in Biology, Electron Microscopy Laboratory, Mammalian Physiology, Developmental Biology, Mammalian Endocrinology, Neurobiology, Selected Topics in Regulatory Biology, Biological effects of radiation, Seminar in Physiology, and Seminar in Genetics

GEORGIA SOUTHERN COLLEGE

Statesboro, Georgia 30458 Phone: (912) 764-6611
Dean of Graduate Studies: J.N. Averitt
Dean - School of Arts and Sciences: W.F. Jones, Jr.

DEPARTMENT OF BIOLOGY
Phone: Chairman: 764-6611 Ext. 488

Head: E.T. Hibbs
Callaway Professor of Biology: J.H. Oliver, Jr.
Professors: J.A. Boole, E.T. Hibbs, S. McKeever, D.W. Menzell (Adjunct, D.A. Olewine, J.H. Oliver, T. Pennington (Emeritus)
Associate Professors: J.W. Andrews (Adjunct), J.R. Bozeman, D.J. Drapalik, F.E. French, W.K. Hartberg, C.T. Hyde, B.P. Lovejoy, K. Maur, H.L. Windom (Adjunct)
Assistant Professors: S.N. Bennett, H.L. Curtis, R. Marshall, L.V. Sick (Adjunct), R.R. Stickney (Adjunct) P.H. Ankney, Jr.
Research Associates: 2

Degree Program: Biology

Undergraduate Degree: B.S, B.A.
Undergraduate Degree Requirements: B.S, in Biology: Core curriculum 90 hours, upper division biology 30, foreign language thru sophomore level, health and physical education, 190 hours total

Graduate Degree: M.S.
Graduate Degree Requirements: Master: Forty hours graduate courses and thesis based on original research. Reading knowledge of French, German or Russian language.

Graduate and Undergraduate Courses Available for Graduate Credit: Vertebrate Anatomy (U), Parasitology (U), Plant Physiology (U), Plant Anatomy (U), Evolution (U), Cell Biology (U), Genetics (U), Ecology (U), Anatomy (U), Physiology (U), Bacteriology (U), Plant Ecology (U), Acarology (U), Endocrinology (U), Invertebrate Physiology (U), Angiosperms (U), Mycology (U), Plant Taxonomy (U), Plant Pathology (U), Microtechnique (U), Cytogenetics (U), Histology (U), Parasitology, Botany, Genetics, Population Genetics, Physiology, Developmental Physiology

GEORGIA SOUTHWESTERN COLLEGE

Americus, Georgia 31709 Phone: (912) 924-6111
Dean of Graduate Studies: H. Pope
Dean of College: H. Johnson

DEPARTMENT OF BIOLOGICAL SCIENCES
Phone: Chairman: Ext. 215

Chairman: J.W. Russell
Professor: C.K. Ewing
Associate Professors: J.C. Carter, V.N. Powders, J.W. Russell, W.L. Tietjen, R. Westra
Assistant Professor: R. McNeill
Visiting Lecturers: 1
Teaching Fellows: 1

Degree Programs: Biological Education, Biology, Medical Technology

Undergraduate Degree: B.S, A.B., B.S.Ed.
Undergraduate Degree Requirements: B.S. : 180 hours (quarter); including 20 hours of English and Humanities; 20 hours of Social Sciences including Political Science; 20 hours of Science and Mathematics; Genetics: Ecology: Electives in Bioloty to make up a minimum of 45 hours: 25 hours of chemistry through Organic: Mathematics through Calculus

Graduate Degree: M.Ed.
Graduate Degree Requirements: Master: 50 hours including 25 hours of Biology; 25 hours of Professional education; A directed Study

Graduate and Undergraduate Courses Available for Graduate Credit: Biology of the Invertebrates, Field Botany, Ecology, Genetics, Plant Physiology, Natural History of the Vertebrates, Animal Physiology, Aquatic Biology, Biological Chemistry, Special Problems in Biology, Biology for Secondary School Teachers, Biological Techniques for Teachers, History and Philosophy of Natural Sciences, Economic Botany

GEORGIA STATE UNIVERSITY

(No reply received)

MEDICAL COLLEGE OF GEORGIA

Augusta, Georgia 30902 Phone: (404) 724-7111
Dean of School of Medicine: C.H. Carter
Dean of Graduate Studies: S.A. Singal

DEPARTMENT OF ANATOMY
Phone: Chairman: (404) 724-7111 Ext. 8846

Acting Chairman: T.F. McDonald
Professors: L.H. Allen, J.W. McKenzie
Associate Professors: W.A. Wellband, M. Sharawy
Assistant Professors: R.S. Hannah, P.J. Moore, T.H. Rosenquist, H. Troyer, D.A. Welter, E.J. Wheeler, I.K. Hawkins, F.T. Lake
Instructor: J.A. Horner

Degree Program: Anatomy

Graduate Degrees: M.S., Ph.D.
Graduate Degree Requirements: Master: B.S.
Doctoral: B.S.

Graduate and Undergraduate Courses Available for Graduate Credit: Gross Anatomy, Neuroanatomy, Embryology, Histology, High Resolution Microscopy, Histochemistry, Special Topics in Gross Anatomy, Advanced Developmental Anatomy, Special Topics in Neuroanatomy, Special Topics in Cell Biology, Advanced High Resulution Microscopy, Seminar in Anatomy, Investigation of a Problem, Research for Thesis or Dissertation

DEPARTMENT OF CELL AND MOLECULAR BIOLOGY
Phone: Chairman: 828-3361 Others: 828-3271

Chairman: E. Bresnick
Charbonnier Professor of Biochemistry: S.A. Singal
Regents' Professor of Biochemistry: T.H.J. Huisman
Professors: E. Bresnick, S.A. Singal, T.H.J. Huisman, W.K. Hall, A.A. Latif, J.F. Denton, H. Wycoff
Associate Professors: J.C. Howard, G.K. Best, G. Brownell, F. Leibach, L.L. Smith, W.S. Harms, C. Roesel
Assistant Professors: F. Garver, P.M. Prichard, E. Howard, D. Lapp, D. Scott, J.V. Isreal, K. Cutroneo, C.N. Nair, M. Hall, K. Lanclos, J.M. Hill, M. Coryell, R. Wrightstone, E.C. Abraham (Research)
Instructor: J.B. Wilson
Research Assistants: 30
Research Fellows: 4

Field Stations/Laboratories: Gracewood State Hospital - Gracewood

Degree Programs: Biochemistry, Microbiology

Graduate Degrees: M.S., Ph.D.
Graduate Degree Requirements: Master: Acceptable course work and thesis. Doctoral: Acceptable course work, acceptable performance on preliminary examination, and acceptable thesis.

Graduate and Undergraduate Courses Available for Graduate Credit: Biochemistry, Topics in Biochemistry, Experimental Biochemistry, Chemical Kinetics, Physical Methods of Structure Determination, Physical Biochemistry, Biochemical Disorders in Disease, Organic Reaction Mechanisms, Seminar in Biochemistry, Biochemistry Workshop, Investigation of a Problem, Research (for Dissertation and Thesis), General Microbiology: Bacteria and Immunity, Medical Microbiology, Bacteria, Viruses, Rickettsiae and Fungi, Parasitology, Immunology, Bacterial Genetics, Virology, Mycology, Immuno-chemistry, Microbial Physiology, Cell and Tissue Culture, Hematology, Host-Parasite Relationship, Seminar in Microbiology, Biochemistry Workshop Investigation of a Problem

DEPARTMENT OF ENDOCRINOLOGY
Phone: Chairman: (404) 724-7111 Ext. 554

Regents Professorship: V.B. Mahesh
Professor: R.B. Greenblatt (Emeritus)
Associate Professors: J.R. Byrd, T.G. Muldoon
Assistant Professors: T.O. Abney, V.K. Bhalla, A. Costoff, T.M. Mills
Instructors: J.O. Ellegood
Research Assistants: 7
Postdoctoral Fellows: 6
Graduate Teaching Assistants: 10

Degree Programs: Endocrinology, Reproductive Biology

Graduate Degrees: M.S., Ph.D.
Graduate Degree Requirements: Master: Basic graduate courses in Biochemistry, some courses in Physiology, core courses in Endocrinology (see below) are required. An oral or written comprehensive examination is given and a thesis (dissertation) required describing a suitable original research problem. Doctoral: Basic graduate courses in Biochemistry and Physiology are required along with Statistics and Computer Programming. Courses listed below in Endocrinology are required. The student passes a preliminary examination to become a candidate. The degree is awarded after a thesis is presented and defended statisfactorily.

Graduate and Undergraduate Courses Available for Graduate Credit: Introduction to Endocrinology, Clinical Endocrinology, Advanced Clinical Endocrinology, Clinical Aspects in Human Reproductive Physiology, Human Cytogenetics - Laboratory Techniques, Biochemical Methods in Endocrine Research, Laboratory Procedures in Clinical Endocrinology, Chemistry of Steroids,

Endocrinology, Reproductive Physiology, Population Explosion and Contraception, Comparative Endocrinology, Seminar in Endocrinology

DEPARTMENT OF PHARMACOLOGY
Phone: Chairman: 724-7111 Ext. 8811

Chairman: R.P. Ahlquist
Professors: R.P. Ahlquist, A.A. Carr, L.P. Gangarosa, W.F. Geber, J.H.R. Sutherland
Associate Professors: A.M. Karow, Jr., J.M. Kling, M.W. Riley
Assistant Professors: T.S. Chiang, B.W. Fry, D.C. Jerram, M.E. Logan, J.L. Matheny, T.J. Mellinger
Instructors: G.O. Carrier, M.H. Johnson, R.L. Longe, F.W. Warren

Degree Program: Pharmacology

Graduate Degrees: M.S., Ph.D.
Graduate Degree Requirements: Master: 45 credit hours, 3 quarters in residence, comprehensive examination, thesis, final oral examination. This is a terminal degree. Doctoral: 3 academic years, 3 quarters in residence, program of research, foreign languages (2) or research tools (computer, etc.); preliminary examination, dissertation, final oral dissertation defense.

Graduate and Undergraduate Courses Available for Graduate Credit: General Pharmacology, Cryopharmacology, Molecular Pharmacology, Methods in Pharmacology, Neuropharmacology, Advanced Pharmacology, Toxicology, Autonomic Pharmacology, Teratology, Cardiovascular-Renal pharmacology

DEPARTMENT OF PHYSIOLOGY
Phone: Chairman: Ext. 8871 Others: 8809, 8833

Chairman: R.C. Little
Professor: J.M. Ginsburg
Associate Professors: J.F. Delahayes, C.E. Hendrich, S.D. Stoney
Assistant Professors: W.F. Hofman, W.J. Jackson, D.A. Miller, R.A. Vargo, V.T. Wiedmeier
Research Assistants: 4

Degree Program: Physiology

Graduate Degrees: M.S., Ph.D.
Graduate Degree Requirements: Master: 45 quarter hours total plus thesis and examinations. Doctoral: 135 quarter hours total plus thesis and examinations

Graduate and Undergraduate Courses Available for Graduate Credit: Survey of Physiology, Cardiodynamics, Peripheral Circulation, Muscle Physiology, Blood Flow Regulation, Respiration, Body Fluid Regulation, Electrophysiology, Radioisotopes in Biological Research, Motor Systems, Radiobiology, Sensory Systems, Biological Substrates of Learning and Motivation, The Application of Control Theory to Physiological Systems, Physiology of Thyroid and Parathyroid Hormone, Special Topic Seminar Courses in Physiology, Investigation of a Problem

MERCER UNIVERSITY

1400 Coleman Avenue Phone: 743-1511
Macon, Georgia 31207
Dean of Graduate Studies: P. Cable
Dean of College: G.F. Taylor

BIOLOGY DEPARTMENT
Phone: Chairman: Ext. 191 or 192

Chairman: T.R. Haines
Professors: T.R. Haines, J.O. Harrison, G.L. Ware
Associate Professor: R. Slentz
Assistant Professors: B.H. Caminita, L. Morgan
Instructors: V. Baisden
Student Assistants: 5-7

Degree Program: Biology

Undergraduate Degrees: B.A., B.S.
Undergraduate Degree Requirements: B.A. - 8 courses in Biology and 1 seminar course, includes 3 courses Principles Zoology, Botany, other courses tailored to fit interest of student B.S. - 8 courses in Biology and related sciences which degree is personal preference of student

MOOREHOUSE COLLEGE

(see Atlanta University Center)

MORRIS BROWN COLLEGE

(see Atlanta University Center)

NORTH GEORGIA COLLEGE

Dahlonega, Georgia 30533 Phone: (404) 864-2008
Dean of College: H.I. Shott II

BIOLOGY DEPARTMENT
Phone: Chairman: 864-3391 Ext. 18 Others: Ext. 61

Head: M.A. Callaham
Professor: M.A. Callaham
Associate Professor: J.A. Biesbrock
Assistant Professors: A. Hunt, J. Kennedy
Teaching Assistant: 1

Degree Programs: Biology, Science Education

Undergraduate Degree: B.S.
Undergraduate Degree Requirements: 185 quarter hours including: Humanities - 20, Social Science - 20, Mathematics and Natural Science - 20, minor in Chemistry, Physics, Mathematics or Psychology 30, Biology - 40, electives - 55 - including 15 hours of a foreign language

OGLETHORPE UNIVERSITY

Atlanta, Georgia 30319 Phone: (404) 261-7441
Dean of College: M.G. Amerson

BIOLOGY DEPARTMENT
Phone: Chairman: 261-7441 Ext. 43

Chairman: P.P. Zinsmeister
Assistant Professors: H. Henry, P. Zinsmeister

Degree Program: Biology

Undergraduate Degree: B.S.
Undergraduate Degree Requirements: 10 Biology courses, 2 years of Chemistry, 1 year of Physics

PIEDMONT COLLEGE

Demorest, Georgia 30535 Phone: (404) 723-4462
Dean of College: C.A. Carder

DEPARTMENT OF BIOLOGY

Chairman: R.J. Lopez
Professor: R.J. Lopez
Associate Professor: J.M. Knox
Teaching Fellows: 2

Field Stations/Laboratories: Lake Burton, Ralsin County

Degree Programs: Major and minors; Premedical, Dental, Pre-nurse, Medical Technology, Prepharmacy, Pre-Veterinary and Biology - Majors for Secondary Education

Undergraduate Degree: B.S.

Undergraduate Degree Requirements: A major and a minor, fields concentrations plus other subjects to 190 hours

SAVANNAH STATE COLLEGE
Savannah, Georgia 31404 Phone: 354-5717
Dean of Graduate Studies: J.A. Eaton
Dean of College: T.H. Byers

DEPARTMENT OF BIOLOGY
Phone: Chairman: 354-5717 Ext. 227

Chairman Division of Natural Sciences: M.C. Robinson
Head: M.C. Robinson
Professors: M.C. Robinson, F.R. Hunter, G.K. Nambiar
Associate Professors: J.B. Benson, P. Krishnamurti, B.L. Woodhouse
Assistant Professors: I.R. Bacon, C.O. Emeh
Instructor: M.J. Stone

Degree Program: Biology

Undergraduate Degree: B.S.
Undergraduate Degree Requirements: 53 quarters in Biology, 35 quarter hours in Chemistry, 20 quarters hours in Mathematics, 15 quarter hours in Physics

Graduate Degree: M.Ed. in Biology
Graduate Degree Requirements: Master: 35-45 hours - Biological Science, 10-20 hours - Professional Education

Graduate and Undergraduate Courses Available for Graduate Credit: Topics in Molecular and Cellular Biology, Advanced Animal Physiology, Advanced Plant Physiology, Advanced Microbiology, Advanced Genetics, Comparative Morphology of Non-Vascular Plants, Field and Laboratory Botany, The Biological Sciences in the Secondary Schools, Advanced General Ecology, Cellular Physiology

SHORTER COLLEGE
Rome, Georgia 30161 Phone: (404) 232-2463
Dean of College: C. Whitworth

DEPARTMENT OF BIOLOGY AND EARTH SCIENCE
Phone: Chairman: 232-2463 Ext. 50 Others: Ext. 60

Chairman: P.F-C. Greear
Professors: P. F-C. Greear, E.L. Lipps
Assistant Professor: C. Allee
Visiting Lecturers: 2

Field Stations/Laboratories: Fiddlers Point, Panacea, Genesis, Ossabaw Island, Savannah

Degree Programs: Biology, Botany, Ecology, Life Sciences, Medical Technology, Natural Science, Zoology

Undergraduate Degrees: B.S., A.B.
Undergraduate Degree Requirements: 126 S.H. of which at least 42 must be upper division

UNIVERSITY OF GEORGIA
Athens, Georgia 30651 Phone: (404) 302-3030
Dean of Graduate Studies: H. Edwards

Franklin College of Arts
Dean: J.C. Stephens

DIVISION OF BIOLOGICAL SCIENCES
Phone: Chairman: 542-2635 Others: 542-3030

Chairman: J.L. Key
Callaway Professor of Genetics: N.H. Giles
Callaway Professor of Ecology: E.P. Odum
Alumni Foundation Distinguished Professor: E.P. Odum

The Division of Biological Sciences is an administrative "umbrella" organization which encompasses the following departments: Biochemistry, Botany, Entomology, Microbiology and Zoology. Interdisciplinary groups are: Animal Behavior, Developmental and Molecular Biology, Ecology, Genetics, Marine Biology and Limnology and Parasitology.

DEPARTMENT OF BIOCHEMISTRY
Phone: Chairman: (404) 542-1801 Others: 542-1334

Chairman: H.P. Peck, Jr.
Professors: M.J. Cormier (Research), W.L. Williams (Research), R.A. McRorie, L.S. Dure, H.P. Peck, Jr., B.C. Carlton
Associate Professors: J.M. Brewer, R.S. Cole, D.V. DevVartanian, R.J. DeSy, J.W. Lee, L.G. Ljungdahl, R.E. Louius, J.F. Mendicino, M.G. Sausing, J. Travis
Assistant Professors: W.S. Champuey, S.R. Kushner, P.N. Srivastava, J.E. Wampler
Research Assistants: 9
Postdoctorals: 20

Field Stations/Laboratories: Sapelo Island Marine Station-Sapelo Island

Degree Program: Biochemistry

Undergraduate Degree: B.S.
Undergraduate Degree Requirements: Not Stated

Graduate Degrees: M.S., Ph.D.
Graduate Degree Requirements: Master: Thesis - 40 hours course work Doctoral: Thesis, 10 hours of Chemistry, 10 hours of Biology, Enzymology, Nucleic Acid, Macromolecules - one language

DEPARTMENT OF BOTANY
Phone: Chairman: (404) 542-3732

Chairman: C.C. Black, Jr.
Professors: C.C. Black, W.H. Duncan, M.S. Fuller, R.T. Hanlin, C.W. James, J.L. Key, E.S. Luttrell, B.E. Michel, C.D. Monk, G.L. Plummer, D.S. Van Fleet, J.J. Westfall
Associate Professors: D.B. Fisher, S.B. Jones, G.D. Kochert, D. Porter
Assistant Professors: W.M. Darley, E.L. Dunn, E. Franz, A. Hicks, A. Jaworski, G.F. Leeper (Adjunct), J. Rawson, H. Rines, R. Sharitz (Adjunct), D. Walker
Teaching Assistants: 35

Degree Programs: Botany, Ecology

Undergraduate Degree: B.S.
Undergraduate Degree Requirements: See UGA Bulletin pp 54 or 73 plus Mathematics 253,254; Physics 127-128, Chemistry 121-122, 240-241, Biology 101-102 or Botany 121-122, Biology 310, 320, 330 or 340 or 360; Botany 431, 483

Graduate Degrees: M.S., Ph.D.
Graduate Degree Requirements: Meet undergraduate requirements plus requirements of UGA Graduate School pp. 20-22

Graduate and Undergraduate Courses Available for Graduate Credit: Physiology of Woody Plants (U), Biology of Algae I, II (U), Palynology (U), Paleobotany (U), Morphology of Non-Vascular Plants (U), Cytology (U), Morphogenesis (U), Taxonomy of Seed Plants (U), Plant Biosystematics (U), Taxonomy of Grasses (U), Plant Population Ecology (U), Vegetation of North America (U), Community Ecology (U), Nutrition of Green Plants (U), Plant Physiology (U), Physiological Plant Anatomy (U), Biology of Ascomycetes, Morphology of Seed Plants, Biology of Phycomycetes, Cytogenetics, Aquatic Plants, Marine Botany, Autecology, Plant Ecological Periodicities, Advanced Community Ecology, Developmental Mechanisms, Advanced Plant

Physiology, Plant Growth and Development, Plant Water Relations, Histochemistry, Nucleic Acid Metabolism

DEPARTMENT OF ENTOMOLOGY
Phone: Chairman (404) 542-2816

Head: H.O. Lund
Professors: W.T. Atyeo, M.S. Blum, D.A. Crossley, Jr., C.M. Himel (Research), P.E. Hunter, H.O. Lund, H. H. Ross, and A.B. Weathersby
Associate Professors: U.E. Brady, Jr., A. Dietz, R.T. Franklin, H.R. Hermann, A.O. Lea, C.H. Tsao, J.B. Wallace
Assistant Professors: J.N. All, C.W. Berisford, R.W. Matthews
Research Associates: 7

Degree Program: Entomology

Undergraduate Degrees: B.S.A., B.S.
Undergraduate Degree Requirements: B.S.A. 195 hours including 55 in biological subjects, 25 in Entomology, and 55 in other sciences. B.S. 180 hours, including 50 in sciences (20 in Entomology) and 10 in Mathematics

Graduate Degrees: M.S., Ph.D.
Graduate Degree Requirements: Master: 40 quarter hours of graduate level course work of which 15 hours must be in graduate level courses. 5 hours thesis plus qualification, by examination or special course, in one foreign language. 15 hours (graduate or undergraduate level) must be taken in non-Entomology courses Doctoral: 40 quarter hours (undergraduate or graduate level and including 20 hours of non-Entomology courses) beyond the Master's requirements. Language and dissertation requirements as for M.S. degree. A second language or 10 hours of a tool course is required

Graduate and Undergraduate Courses Available for Graduate Credit: Insect Biochemistry, Systematics, Insect Endocrinology, Radiation Entomology, Advanced Medical Entomology I and II, Taxonomy, Advanced General Entomology, Advanced Acarology, Advanced Toxicology, Advanced Insect Ecology, Insect Ecology (U), Introductory Acarology (U), Taxonomy (U), Morphology (U), Insecticides (U), Bionomics of Forest Insects (U), Physiology (U), Field Entomology (U), Behavior (U), Histology (U), Advanced Agricultural Entomology (U), Aquatic Entomology (U), and Insects in Field and Stream (U)

DEPARTMENT OF MICROBIOLOGY
(No reply received)

DEPARTMENT OF ZOOLOGY
Phone: Chairman: (404) 542-8905 Others: 542-3310

Head: P.E. Thompson
Callaway Professor of Genetics: N.H. Giles
Callaway Professor of Ecology: E.P. Odum
Alumni Foundation Professor: W.B. Cosgrove
Alumni Foundation Professor: R.B. McGhee
Professors: M. Agosin (Research), S.I. Auerbach (Research), J.H.D. Bryan, F.B. Golley, W.J. Humphreys, J. Papaconstantinou (Research), B.C. Patten, L.R. Pomeroy, E.E. Provost, H.T.M. Ritter, Jr., D.C. Scott, P.E. Thompson, R.G. Wiegert
Associate Professors: W.W. Anderson, M. Case, S.J. Coward, R.T. Damian, C.W. Helms, R.E. Johannes, H.A. Kent, Jr., D.T. Lindsay, M.E. Mattingly, G.L. Patel, J.J. Paulin, M.H Smith, R.C. Taylor, G.J. Thomas
Assistant Professors: J. Alberts, R.H. Crozier, R.J. Reimold, J.E. Schindler
Research Associates: 9

Field Stations/Laboratories: University of Georgia Marine Institute - Sapelo Island

Degree Programs: Animal Behavior, Biology, Ecology, Genetics, Parasitology, Zoology

Undergraduate Degree: B.S. (Biology, Zoology)
Undergraduate Degree Requirements: Lower Division courses: Animal Structure and Function and Cell Structure and Function or 2 courses in General Biology General and Organic Chemistry (4 courses); Introductory Physics (2 courses); 2 courses in Calculus or 1 course in Calculus and 2 courses in Statistics. Upper division courses: 8 courses including Biochemistry, Genetics, Cell or General Physiology, Ecology and Senior Seminar. Total program of 180 quarter hours

Graduate Degrees: M.S., Ph.D. (Zoology), Ph.D. (Ecology)
Graduate Degree Requirements: Master: Thesis and 40 quarter hours including 15 in courses open only to graduate students. Reading knowledge of one language. Courses equivalent to undergraduate major required for admission. Doctoral: Dissertation and reading knowledge of one language. Courses equivalent to undergraduate major required for admission.

Graduate and Undergraduate Courses Available for Graduate Credit: Advanced Genetics (U), Protozoology (U), Mammalogy (U), Herpetology (U), Ichthyology (U), Ornithology (U), Invertebrate Zoology I (U), Invertebrate Zoology II (U), Evolution (U), Chromosomal Heredity (U), Population Genetics (U), Animal Physiology I (U), Animal Physiology II (U), Marine Ecology (U), Vertebrate Physiology (U), Physiology Seminar (U), Animal Behavior (U), Principles of Ecology (U), Parasitology (U), Analysis of Development I and II (U), Animal Histology (U), Faunistic Zoology, Ecological Energetics, Parasitic Protozoa, Biochemistry of Parasites, Helminthology, Advanced Invertebrate Zoology I and II, Limnology-Oceanography, Marine Biology, Freshwater Biology, Seminar in Hydrobiology, Physiology Seminar, Cellular Physiology, Advanced Physiology, Comparative Neurophysiology, Advanced Endocrinology, Seminar in Parasitology, Cytology, Physiological Ecology, Population Ecology, Ecology Seminar, Pollution Ecology, Systems Ecology I and II, Cell Biology Seminar, Systematic Ecology, Dynamic Analysis I, Genetics Seminar, Seminar in Developmental Biology, Molecular Genetics, Developmental Mechanisms, Cytochemistry, Problems in Zoology, Thesis

College of Agriculture

Dean: H. Garren

AGRONOMY DIVISION
Phone: Chairman (404) 542-2461 Others: 542-3030

Chairman: W.L. Colville
Alumni Foundation Distinguished Professor of Agronomy: G.W. Burton
Terrell Professor of Agronomy: H.D. Morris
Professors: College Station: E.R. Beaty, R.H. Brown, W.L. Colville (Head), A.A. Fleming, J.E. Giddens, H.D. Morris, H.F. Perkins Georgia Experiment Station: O.E. Anderson (Head), F.C. Boswell, H.B. Harris, M.D. Jellum, R.E. Wilkinson Cooperative Extension Service Department of Agronomy: J.B. Jones Jr., J.F. McGill
Associate Professors: College Station: D.A. Ashley, A.R. Brown, R.A. McCreery, K.H. Tan, J.B. Weaver Georgia Experiment Station: R.E. Burns, D.G. Cummins, B.J. Johnson, K. Ohki Cooperative Extension Service: P J. Bergeaux, W.H. Gurley, R.A. Isaac, J.E. Jackson (Head), J.F. Miller
Assistant Professors: College Station: H.R. Boerma, J. E. Elsner, J.S. Schepers, R.L. Todd Georgia Experiment Station: T.C. Keisling, W.H. Marchant, M.B. Parker, M.E. Walker Cooperative Extension Service: G.W. Brown, P.B. Bush, H. Lowery, J.A. McAfee, R.L. Miles, C.R. Roland, W.I. Segars, W.H. Sell, C.W. Swann

Instructor: Coastal Plain Experiment Station: S.H. Baker
Teaching Fellows: 2
Research Assistants: 16

Field Stations/Laboratories: Georgia Mountain Station - Blairsville, Northwest Georgia Station - Calhoun, Southeast Georgia Station - Midville, Southwest Station - Plains, Central Georgia Station, Eatonton, Shade Tobacco Station - Attapulgus, Americus Plant Materials Center - Americus

Degree Programs: Agronomy, Conservation, Crop Science, Genetics, Soil Science

Undergraduate Degree: B.S
Undergraduate Degree Requirements: 2.0 GPA, 195 hours, 90 in residence hours

Graduate Degrees: M.S., Ph.D.
Graduate Degree Requirements: Master: Entrance: B.S. or equivalent, 2.8 GPA and/or 900 GRE. To obtain degree: maintain 3.00 GPA, 45 hours minimum with 15 hours of 800 level courses, satisfactory completion of comprehensive examination and thesis Doctoral: Entrance: M.S. or equivalent, 3.00 GPA in graduate work, 900 GRA. To obtain degree: maintain 3.00 GPA, 45 hours minimum about M.S., one foreign language, satisfactory completion of comprehensive examination and thesis.

Graduate and Undergraduate Courses Available for Graduate Credit: Seed Technology (U), Principles of Experimental Methods (U), Pasture Development and Management (U), Principles of Chemical Weed Control (U), Soil Morphology and Classification (U), Land Use and Soil Conservation (U), Soil Fertility (U), Soil Physics (U), Soil Microbiology (U), Wastes and the Soil (U), Soil Mineralogy (U), Soil and Plant Analysis (U), Soil for Multipurpose Use (U), Advanced Soil Chemistry, Research Methods in Agronomy, Specialized Plant Breeding, Experimental Design, Crop Response to Microclimate, Advanced Soil Fertility, Methodology in Soil Chemistry, Advanced Soil Morphology and Genesis, Soil Physical Factors and Plant Growth, Quantitative Aspects of Plant Breeding

ANIMAL AND DAIRY SCIENCE
Phone: Chairman: 542-6259 Others: 542-1852

Head: L.J. Boyd
Alumni Distinguished Professor: W.J. Miller
Professors: A.E. Cullison, O.T. Fosgate, A.D. Johnson, J W. Lassiter, M.L. Loewenstein, R.S. Lowrey, W. J. Miller, L.J. Boyd
Associate Professors: C.M. Clifron, R.W. Seerley
Assistant Professors: M.B. Neathery, R.D. Scarth, S.J. Speck, C.S. Ward, C.H. White, G. Rampacek
Instructor: H.C. McCampbell
Research Assistants: 10
Postdoctorate: 1

Field Stations/Laboratories: Several Branch Stations for Agricultural Research. Two main stations with Animal Science Departments: Georgia Agricultural Experiment Station - Griffin, Coastal Plain Experiment Station - Tifron

Degree Programs: Animal Science, Dairy Science, Environmental Health Science, Animal Nutrition, Dairy and Food Technology

Undergraduate Degree: B.S.
Undergraduate Degree Requirements: Core curriculum - 90 quarter hours composes of Humanities, 20 hours, Mathematics and Natural Sciences, 20 hours, Social Science, 20 hours and 30 hours of courses related to major. Junior-Senior Years - 45 hours required courses, 34 hours special requirements by selection from prescribed courses, and 20 hours of free electives. 195 hours for graduation

Graduate Degrees: M.S., Ph.D.
Graduate Degree Requirements: Master: 40 quarter hours plus thesis Doctoral: Requirements stipulated by Advisory Committee, one foreign language required Dissertation required.

Graduate and Undergraduate Courses Available for Graduate Credit: Ruminant Nutrition, Monogastric Nutrition, Bioenergetic in Animal Nutrition, Vitamins in Animal Nutrition, Minerals in Animal Nutrition, Proteins and Amino Acids in Animal Nutrition, Carbohydrates and Lipids in Animal Nutrition, Nutrition Seminar, Genetic Improvements of Animals, Physiology of Reproduction in Farm Animals, Advanced Livestock Breeding, Advanced Livestock Feeding, Statistical Methods in Animal Science, Population Genetics, Microbiology for Sanitarians, Microbiology for Fermented Dairy Products, Artificial Insemination, Milk Secretion, Dairy Chemistry, Nutritional Properties of Dairy Products, Chemical Analyses of Dairy Products, Physical Chemistry of Dairy Products, Advanced Dairy Technology

DEPARTMENT OF FOOD SCIENCE
Phone: Chairman: 542-2286

Chairman: C.J.B. Smit
Professors: J.C. Ayres, M.K. Hamdy, J.J. Powers
Associate Professors: J.A. Carpenter, P.E. Koehler, D.A. Lillard, R.T. Toledo
Assistant Professors: R.R. Eitenmiller, J.O. Reagan
Research Assistants: 10
Research Associates: 4

Degree Programs: Food Science, Dairy Manufacturing

Undergraduate Degrees: B.S.A., B.S. (Food Science)
Undergraduate Degree Requirements: 30 hours in major of which t must be in Biological Science, 20 hours Mathematics and Natural Science

Graduate Degrees: M.S., Ph.D.
Graduate Degree Requirements: Master: 3.0 average in courses, foreign language, Thesis, Final Oral Doctoral: 3 years residence, Preliminary Examination, Foreign Language, Thesis, Final Oral

Graduate and Undergraduate Courses Available for Graduate Credit: Too many to list. Determined by students interest.

DEPARTMENT OF HORTICULTURE
Phone: Chairman: (404) 542-2471

Chairman: J.B. Jones, Jr.
Professors: J.B. Jones Jr., C.H. Hendershott, F.E. Johnstone, J H. Tinga, H.M. Vines
Associate Professors: F.A. Pokorny, D. Sparks
Assistant Professors: G.A. Couvillan, N.J. Natarella
Research Assistant: 1

Field Stations/Laboratories: Georgia Experiment Station - Experiment, Coastal Plain Experiment Station - Tifton

Degree Program: Horticulture

Undergraduate Degree: B.S.A.
Undergraduate Degree Requirements: 196 quarter hours of course work

Graduate Degrees: M.S., Ph.D. (Plant Sciences)
Graduate Degree Requirements: Master: Horticulture - 40 quarter hours of course work, Thesis Doctoral: Plant Science - 3 full years of study, reading knowledge of one foreign language, Thesis

Graduate and Undergraduate Courses Available for Graduate Credit: Advanced General Horticulture, Advanced Olericulture, Advanced Pomology, Greenhouse Management I and II, Horticultural Crop Improvement,

Horticultural Research, Horticultural Seminar, Instrumental Analysis Methods, Post Harvest Physiology, Production Practices for Horticultural Crops

DEPARTMENT OF PLANT PATHOLOGY AND GENETICS
Phone: Chairman: (404) 542-2571

Chairman: W.N. Garrett
Professors: W.N Garrett, R.T. Hanlin, F.F. Hendrix, Jr., C.W. Kuhn, E.S. Luttrell, W.M. Powell, J. Taylor
Associate Professors: S.M. McCarter, K.E. Papa, R.W. Roncadori, W.K. Wynn, Jr.
Assistant Professors: R.S. Hussey, G.M. Kozelnicky
Research Assistants: 2

Field Stations/Laboratories: Georgia Experiment Station - Experiment, Coastal Plain Experiment Station - Tifton

Degree Programs: Genetics, Plant Pathology

Undergraduate Degree: B.S.
Undergraduage Degree Requirements: 30 hours in major of which 5 must be in Biological Science, 20 hours Mathematics and Natural Science

Graduate Degrees: M.S., Ph.D.
Graduate Degree Requirements: Master: 3.0 average, in courses, foreign language, Thesis, Final Oral Doctoral: 3 years residence, Preliminary Examination, Foreign Language, Thesis, Final Oral

Graduate and Undergraduate Courses Available for Graduate Credit: Plant Pathology Seminar, Introductory Mycology, Plant Virology, Plant Nematology, Diagnosis and Control of Plant Diseases, Bacterial Plant Pathogens, Advanced Forest Pathology, Research in Plant Pathology, Research Methods in Plant Pathology, Etiology of Plant Diseases, Biology of Ascomycetes, Phytopathology: Principles and Theory, Physiology of Parasitism, Physiology of Fungi, Biology of Plant Parasitic Nematodes, Biology of Phycomycetes, Biology of Basidiomycetes, Research Plant Genetics, Plant Genetics, Plant Breeding, Physiological Genetics, Special Plant Breeding, Advanced Graduate Seminar, Quantitative Genetics

DEPARTMENT OF POULTRY SCIENCE
Phone: Chairman: (404) 542-1351

Chairman: M.G. McCartney
Alumni Foundation Distinguished Professor: W.M. Reed
Professors: H.L. Fuller, T.M. Huston, L.S. Jensen, M.G. McCartney, W.M. Reed
Associate Professors: B. Howarth, K.W. Washburn
Assistant Professors: W.M. Britton, K.K. Hale, Jr., A.P. Raln, M.D. Ruff

Degree Program: Poultry Science

Undergraduate Degree: B.S.A.
Undergraduate Degree Requirements: 195 quarter credit hours

Graduate Degrees: M.S., Ph.D.
Graduate Degree Requirements: Master: 45 quarter hours, thesis Doctoral: Approved Program for each prespective candidate, 1 foreign language, Dissertation

Graduate and Undergraduate Courses Available for Graduate Credit: Avian Physiology, Advanced Poultry Breeding, Ruminant Nutrition, Monogastric Nutrition, Physiology of Avian Reproduction, Parasitic Diseases of Poultry, Poultry Diseases and Parasites, Scientific Writing, Seminar, Parasitology Seminar, Problems in Poultry Science

College of Veterinary Medicine

Dean: L.E. McDonald (Acting)

DEPARTMENT OF PARASITOLOGY

DEPARTMENT OF PHYSIOLOGY AND PHARMACOLOGY
Phone: Chairman: 542-3014

Head: D.D. Goetsch
Professors: N.H. Booth, J.M. Bower, D.D. Goetsch, L.E. McDonald, H.S. Siegel (Adjunct)
Associate Professors: D.B. Coulter, R.C. Hatch, T.L. Huber
Assistant Professors: L.M Crawford, G.W. Horn, F.N. Thompson
Veterinary Medical Residents: 2
Research Assistants: 1

Degree Programs: Physiology, Pharmacology

Graduate Degrees: M.S., Ph.D.
Graduate Degree Requirements: Master: Committee decision Doctoral: Committee decision

Graduate and Undergraduate Courses Available for Graduate Credit: Animal Physiological Chemistry, Mammalian Physiology (3 quarters), Pharmacology, Clinical Pharmacology, Advanced Physiology, Ruminant Physiology, Physiology of Gametes, Comparative Medical Endocrinology, Neurophysiology, Molecular Pharmacology, Chemotherapy, Seminar, Thesis

School of Forest Resources

Dean: A.M. Herrick

SCHOOL OF FOREST RESOURCES
Phone: Chairman: (404) 542-2686 Others: 542-3124

Head: A.M. Herrick (Dean)
Union Camp Professor of Forest Resources and Statistics J.L. Clutter
Professors: C.L. Borwn, J.L. Clutter, L.A. Hargreaves, J.H. Jenkins, J.T. May, A.E. Patterson, M. Reines
Associate Professors: J.E. Bethune, P.E. Dress, P.J. Dyson, J.C. Fortson, R.T. Franklin, J.T. Greene, W.L. Nutter, J.R. Parker, E.E. Provost, J.T. Rice, K. Steinbeck, K. Ware (Adjunct)
Assistant Professors: J.R. Beckwith, G.H. Brister, C.M. Chin, J.P. Clugston (Adjunct), W.L. Cook, R.G. Dudley, C.H. Fitzgerald, A.C. Fox (Adjunct), R.W. Jones, R.L. Marchinton, L.V. Pienaar, R.C. Schultz, S.W. Thacker
Visiting Lecturers: 1
Research Assistants: 36

Field Stations/Laboratories: White Hall Forest - Grant Forest

Degree Programs: Applied Ecology, Avian Disease, Biometry, Biostatistics, Ecology, Fisheries and Wildlife Biology, Forest Botany, Forest Chemistry, Forest Entomology, Natural Resources, Pathology, Wildlife Management

Undergraduate Degree: B.S.
Undergraduate Degree Requirements: 30 hours in major of which 5 must be in Biological Science, 20 hours, Mathematics and Natural Sciences

Graduate Degrees: M.S., Ph.D. (Forest Resources)
Graduate Degree Requirements: Master: 3.0 average in courses, foreign language, Thesis, Final Oral Doctoral: 3 years residence, Preliminary Examination, Foreign Language, Thesis, Final Oral

Graduate and Undergraduate Courses Available for Graduate Credit: Advanced Forest Soils, Physiology of Woody Plants, Botany, Horticulture, Forest Tree Improvement Watershed Hydrology, Forest Range Management, Aquatic Environments, Fish Ecology, Forest Management for Recreation, Timber Characteristics and Utilization, Physical and Mechanical Properties of Wood, Advanced Wood Anatomy, Principles of Micro-Measurements, Biometrics, Forest Mensuration, Timber Management, Advanced Timber Management, Forest Resources History and Policy, Forest Resources Law, Socio-Political Aspects of Forest Resources Management, Forest Land Use Planning, Forest Resources Administration, Informational Methods in Forest Resources, Artificial Regeneration, Practices of Watershed Management, Techniques in Wildlife Population Analysis and Management, Advanced Principles of Wildlife Management, Parasites of Freshwater Fishes, Wildlife Habitat Management, Fishery Management Techniques, Forest Recreation Area Development, Adhesive Properties and Uses, Forest Management Plans, Environmental Manipulation, Forest Resources Economics I, Forest Resources Economics II, Forest Harvesting, Operations Analysis in Forest Resources, Operational Problems in Forest Resources Management, Urban Woodlands, Urban Woodlands Protection, Forest Resources Seminar, Forest Ecosystems, Problems in Forest Resources Seminar, Forest Ecosystems, Problems in Forest Resources, Fisheries and Wildlife Seminar, Quantitative Aspects of Forest Resource Management, Mathematical Programming, Quantitative Methods in Forest Management, Sampling Techniques for Biological Populations, Diseases of Wildlife, Diseases of Wildlife, Introduction to Tropical Forestry, Forest Resource History and Administration

VALDOSTA STATE COLLEGE

Valdosta, Georgia 31601 Phone: (912) 244-6340
Dean of Graduate Studies: F. Pearson
Dean - School of Arts and Sciences: J.D. Daniels

DEPARTMENT OF BIOLOGY
Phone: Chairman: 244-6340 Ext. 352

Head: C.E. Connell
Professors: C.E. Connell, W.R. Faircloth
Associate Professors: S.M. Schmittner, H.H. West
Assistant Professors: E. Bechtel, W.G. Brannen, W.H. Brides, Jr., C.H. Brown, H.K. McIntyre, Jr.
Instructors: W.H. Cribbs, B.S. Purvis, E.E. Sheeley

Degree Programs: Biology, Medical Technology

Undergraduate Degree: B.S.
Undergraduate Degree Requirements: (in quarter hours) Biology: 60 Biology; 15 foreign language; 25-24 Chemistry; 15 Mathematics; 12 Physics; 5 senior English Medical Technology: 40 Biology; 40 Chemistry; 15 Mathematics; 12 Physics; 4 quarters at affiliate hospital

WESLEYAN COLLEGE

(No reply received)

WEST GEORGIA COLLEGE

Carrollton, Georgia 30117 Phone: (404) 834-4411
Dean of Graduate Studies: B. Griffith
Dean of College: R. Dangle

DEPARTMENT OF BIOLOGY
Phone: Chairman: (404) 834-4411 Ext. 321,322

Chairperson: E.E. Gilbert
Professors: E.E. Gilbert, A.W. Gardner, R.K. Lampton, W.P. Maples, R.M. Welch
Associate Professors: R.B. England, F.T. Hickson
Assistant Professors: D. Byrd, M.E. Lieberman, C.O. Manahan, C.J. Quertekmus, J.P. Snow
Teaching Assistants: 2-4

Field Stations/Laboratories: Oak Mountain - Carrollton

Degree Program: Biology

Undergraduate Degree: A.B.
Undergraduate Degree Requirements: 190 hours, 5 quarter sequence in Biology and 4 5-hour courses; Organic Chemistry, 2 quarters of Physics and Mathematics, 20 quarter hours in each of Humanities, and Social Sciences

Graduate Degree: M.S.
Graduate Degree Requirements: Master: 45 hours, thesis - Plan I, 60 hours (40 hours in Biology) - Plan II. Language or statistics or computer science competence written or oral examination

Graduate and Undergraduate Courses Available for Graduate Credit: History of Biology, Population Biology, Paleontology, Marine Biology, General and Comparative Physiology, Advanced Parasitology, Methods in Microscopy, Biological Techniques in Electron Microscopy, Advanced Microbiology, Plant Physiology, Developmental Biology, Advanced Genetics, Endocrinology, Animal Behavior, Aquatic Biology, Advanced Invertebrate Zoology, Cell Biology, Cellular Physiology, Studies in Ecology, The Bryophytes, Angiosperms, Biogeography, Protozoology, Biometry

GUAM

UNIVERSITY OF GUAM

Box EK Phone: 749-2929
Agana, Guam

Dean of Graduate Studies: J.A. McDonough
Dean - College of Arts and Sciences: P. Richardson

BIOLOGY DEPARTMENT
Phone: Chairman: 749-2929 Ext. 263

Chairman: L. Raulerson
Professors: D.R. Smith, L.G. Eldredge (Marine Laboratory), R.S. Jones (Marine Laboratory)
Associate Professors: D.M. Davis, J.A. Marsh (Marine Laboratory, R.T. Tsuda (Marine Laboratory)
Assistant Professors: T.D. Allen, R. Krizman, L. Raulerson, M. Yamaguchi (Marine Laboratory)
Instructor: R. Randall (Marine Laboratory)

Field Stations/Laboratories: Marine Laboratory

Degree Program: Biology

Undergraduate Degree: B.S.
Undergraduate Degree Requirements: Freshman Biology (2 semesters), Plant Diversity, Animal Diversity, Genetics, Cell Physiology, Ecology and 30 hours total College Algebra and Trigonometry (pre-calculus), College Physics (2 semesters), Inorganic Chemistry (2 semesters); Organic Chemistry (2 semesters) (Note: University has additional required "general education courses").

Graduate Degree: M.S.
Graduate Degree Requirements: Master: Meet undergraduate degree requirements biometrics, Scientific Literature and Writing, 1 semester Seminar for every year in residence, 30 semester hours, 18 of which must be graduate level (500) courses (not 400-4006 senior-graduate); 6 of which must be research. A thesis on an experimental research project. An oral qualifying examination. An oral defense of thesis.

Graduate and Undergraduate Courses Available for Graduate Credit: Biosystematics (U), Biochemistry (U), Microtechnique (U), Ichthyology (U), Ornithology (U), Plant Anatomy (U), Plant Geography (U), Bryology (U), General Microbiology (U), Biological Literature and Scientific Writing, Marine Biogeography, Marine Ecology, Biometrics, Animal Behavior, Herpetology Marine Invertebrates , Fisheries Biology, Marine Botany

HAWAII

BRIGHAM YOUNG UNIVERSITY

Laie, Hawaii 96762 Phone: (808) 293-9211
Dean of College: D.W. Anderson

DEPARTMENT OF BIOLOGICAL SCIENCES
Phone: Chairman: (808) 293-9211 Ext. 230

Chairman: D.M. Andersen
Professors: D.M. Andersen, P.D. Dalton
Associate Professors: D.G. Berrett
Assistant Professors: E.C. Devenport

Degree Program: General Biology

Undergraduate Degrees: B.A., B.S.
Undergraduate Degree Requirements: A minimum of 36 semester hours of Biology with 15 hours of physical science - plus general education requirements

CHAMINADE COLLEGE

3140 Waialae Avenue Phone: (808) 732-1471
Honolulu, Hawaii 96816
Dean of Academic Affairs: D. Hiu

DEPARTMENT OF BIOLOGY
Phone: Chairman: (808) 732-1471

Chairman: H.C.F. Chun-Hoon
Professor: H.C.F. Chun-Hoon
Associate Professors: N. Chee, R. Iwamoto
Instructors: 3 (part-time)

Degree Program: Biology

Undergraduate Degree: B.A.
Undergraduate Degree Requirements: Outside of college-wide core requirements; Biology majors must fulfill following: 1 year general chemistry, 1 year Organic Chemistry, 1 year General Biology and 24 credit hours of upper division Biology courses including Genetics, Embryology, Seminar, Senior Research courses which are required (Students may choose among the following electives to complete 24 upper division credits: Microbiology, Histology, Physiology, Structural Biology, Honors Program, Special Topics, Ecology, Field Experience. We also offer 2 other courses which are electives on the lower division level: Oceanography; Man and Nature

CHURCH COLLEGE OF HAWAII

(See Brigham Young University)

HAWAII LOA COLLEGE

Kaneohe, Hawaii 96744 Phone: (808) 235-3641
Dean of College: M.W. Oliphant

DIVISION OF SCIENCE AND MATHEMATICS
Phone: Chairman: 235-3641

Professor: M.W. Oliphant
Associate Professor: C.M. Weinbaum
Assistant Professors: R.W. Frystak, M.A. Ward

Degree Program: Biology

Undergraduate Degree: B.S.
Undergraduate Degree Requirements: 32 courses: 10 distribution courses, 8 major area, 4 sub-major, 10 electives

UNIVERSITY OF HAWAII

244 Dole Street Phone: (808) 944-8111
Honolulu, Hawaii 96822
Dean of Graduate Studies: H.P. McKaughan

College of Arts and Sciences

Dean: D.E. Contois

DEPARTMENT OF BOTANY
Phone: Chairman: (808) 948-8369 Others:948-8111

Chairman: N.P. Kefford
Wilder Professor of Botany - vacant
Professors: E.K. Akamine, B.J. Cooil, M.S. Doty, D.J.C. Friend, N.P. Kefford, C.H. Lamoureux, D. Mueller-Dombois, G.A. Prowse, S.M. Siegel
Associate Professors: M. Awada (Associate Researcher), J.E. Bowen, S. Nakata, E.W. Putman, C.W. Smith, W.L. Theobald
Assistant Professor: K.W. Bridges
Instructors: J.C. Deputy (Junior Researcher), R.A. Gay, K.M. Nagata
Research Assistants: 9

Field Stations/Laboratories: Beaumont Agricultural Research Center - Hilo, Harold L. Lyon Arboretum - Honolulu

Degree Programs: Botany, Botanical Sciences, Plant Pathology

Undergraduate Degree: B.A.
Undergraduate Degree Requirements: Plan A. Major requirements: 32 semester hours. Required courses: a core of three courses, preliminary to 16 credit hours in other courses above 200. Related courses required: Chemistry 243-246, or 241-242 and Agricultural Biochemistry 402-403; Mathematics 205. Plan B. Major requirements: 32 semester hours including not more than 4 credits in courses below 200. Required courses; a core of two Biology, and 3 Botany courses, preliminary to at least 12 credit hours in other courses, above 300. Related courses required: Chemistry 243 Mathematics 205

Graduate Degrees: M.S., Ph.D.
Graduate Degree Requirements: Master: Plan A (Thesis) and Plan B (nonthesis) are separate M.S. programs with distinct purposes. Plan A is the usual program to be taken. Plan B is offered for students who do not intend to make research in botanical sciences their profession. For the M.S. degree--Plan A, 12 credits shall be for thesis and a minimum of an additional 18 credits for courses approved by a candidate's committee are required. For the M.S. degree--Plan B, of the minimum of 30 credits required, 15 credits shall be earned in the major field, or an approved related field, in courses numbered 600-799. Of these credits 6 must be for directed research. Doctoral: Program specified by committee. Language, Thesis. Comprehensive examination

Graduate and Undergraduate Courses Available for Graduate Credit: Plant Anatomy (U), Microtechnique (U), Developmental Biology (U), Mycology (U), Medical Mycology (U), Natural History of the Hawaiian Islands (U), Plant Ecology (U), Vegetation Ecology (U), Systematics of Vascular Plants (U), Principles of Plant Physiology (U), Phycology (U), Botanical Seminar, Advanced Botanical Problems, Morphology Seminar Cytology, Environmental and Space Biology (U), Origin, Evolution and Distribution of Flowering Plants,

Marine Phytoplankton, Physiology of Fungi, Environmental and Space Biology II, Ecology Seminar, Dynamics of Marine Productivity, Advanced Taxonomy, Nomenclature Seminar, Plant Nutrition and Water Relations, Energetics and Biosynthesis in the Plant Kingdom, Techniques in Physiology, Techniques in Physiology-Biochemistry, Physiology Seminar, Phycology-Chlorophyta, Phycology-Phytoplankton, Phycology-Myxophyta and Phaeophyta, Phycology-Rhodophyta, Principles of Plant Pathology (U), Biology and Ecology of Soil-Borne Plant Pathogens (U), Tropical Plant Pathology, Clinical Plant Pathology, Principles of Plant Disease Control, Plant Nematology, Plant Pathology Techniques, Advanced Plant Pathology, Plant Virology, Epidemiology of Plant Diseases, Physiology of Fungi, Plant Pathology Seminar, Host-Parasite Physiology

DEPARTMENT OF MICROBIOLOGY
Phone: Chairman: (808) 948-8603 Others: 948-8111

Chairman: M. Herzberg
Professors: A.A. Benedict, L.R. Berger, C.E. Folsome, K.R. Gundersen, J.B. Hall, M. Herzberg, P.C. Loh
Associate Professors: B.G. Adams, R.D. Allen, B.Z. Siegel
Teaching Fellows: 12
Research Assistants: 2

Degree Program: Microbiology

Undergraduate Degree: B.A.
Undergraduate Degree Requirements: Biology, Organic Chemistry, Physics, Calculus. Electives: (Biochemistry, Genetics, Physical Chemistry, Microbiology 15 hours

Graduate Degrees: M.S., Ph.D.
Graduate Degree Requirements: Master: Directed Research, Thesis and non-thesis plans (Comprehensive examination) 18 hours graduate work. Total of 30 hours Doctoral: No course requirement, Written Comprehensive Examination, Oral Examination, Ph.D. Dissertation

Graduate and Undergraduate Courses Available for Graduate Credit: Immunochemistry, Advanced Microbial Physiology, Marine Microbiology, Virology, Electron Microscopy, Microbial Genetics, Exobiology, Host-Parasite Relationships, Seminar, Directed Research, Special Topics

DEPARTMENT OF ZOOLOGY
Phone: Chairman: (808) 948-8617 Others: 948-8111

Chairman: F.I. Kamemoto
Professors: A.H. Banner, A.J. Berger, I.M. Cooke, F.I. Kamemoto, E.S. Reese, S.A. Reed, A.L. Tester (Senior Professor), S.J. Townsley, P.B. van Weel
Associate Professors: A.N. Popper, E.D. Stevens
Assistant Professors: J.H. Brock, S.R. Haley, R.A. Kinzie III, G.S. Losey, J.S. Stimson
Research Assistants: 3
Teaching Assistants: 16

Field Stations/Laboratories: Hawaii Institute of Marine Biology - Coconut Island

Degree Programs: Animal Behavior, Animal Physiology, Ecology, Ethology, Zoology, Marine Zoology

Undergraduate Degree: B.A. (Zoology)
Undergraduate Degree Requirements: 124 credits with 24 credits in Zoology including 3 laboratory courses in 3 of 5 specified areas. Senior Seminar. A course in another Biological Science. Two years of Chemistry, including Organic Chemistry or Organic Chemistry and Biochemistry, Precalculus.

Graduate Degrees: M.S., Ph.D. (Zoology)
Graduate Degree Requirements: Master: 30 credits including 6 credits in Thesis research or 2-5 credits in Directed Research. Passing score in General Examination. Doctoral: Course in Preparation of Manuscripts. Passing score in one foreign language (ETS). Passing scores in General and Comprehensive Examinations. Doctoral Dissertation.

Graduate and Undergraduate Courses Available for Graduate Credit: Zoology of the Lower Invertebrates (U), Zoology of the Higher Invertebrates (U), Histology (U), Microtechnique (U), Embryology (U), Developmental Biology (U), Animal Physiology (U), Endocrinology (U), Animal Ecology (U), Laboratory in Animal Ecology (U), History of Zoology (U), Natural History of the Hawaiian Islands (U), Avian Biology (U), Animal Evolution (U), Biogeography (U), Comparative Physiology, (U), Comparative Physiology Laboratory (U), Comparative Endocrinology, Comparative Endocrinology Laboratory, Principles of Animal Behavior, Principles of Animal Behavior Laboratory, Growth and Form, Biology of Symbiosis, Topics in Developmental Biology, Seminar in Teaching, Marine Ecology, Isotopic Tracers in Biology, Biometry, Advanced Biometry, Cellular Neurophysiology, Advanced Ichthyology, Seminar in Zoology

School of Medicine

Dean: T.A. Rogers

DEPARTMENT OF ANATOMY AND REPRODUCTIVE BIOLOGY
Phone: Chairman: 948-7131 Others: 948-8111

Chairman: J.C. Hoffmann
Professors: V.J. DeFeo, M. Diamond, J.C. Hoffmann, P.A. Jacobs, R.G. Kleinfeld, R. Yanagimachi
Associate Professors: G.D. Bryant, V.L. Jacobs, M. B. Nelson, R.J. Teichman
Postdoctoral Fellows: 3

Degree Program: Anatomy

Graduate Degree: M.S.
Graduate Degree Requirements: Master: 30 graduate credits, no language, research thesis

Graduate and Undergraduate Courses Available for Graduate Credit: Functional Human Anatomy, Cell Structure and Function, Microanatomy, Human Sexuality, Experimental Methods in the Study of Reproductive Behavior

DEPARTMENT OF BIOCHEMISTRY AND BIOPHYSICS
Phone: Chairman: (808) 948-8490 Others: 948-8111

Chairman: R.J. Guillory
Professors: N.V. Bhagavan, I.R. Gibbons, F.C. Greenwood, R.J. Guillory, M. Mandel, H.F. Mower, L.H. Piette, K.T. Yasunobu
Associate Professors: T. Humphreys, B. McConnell, R. McKay, B. Morton
Instructors: S. Jeng, R. Mathews, W. Morishige, D. Myers, M. Schachet, J. Shieh, M. Tanaka
Visiting Lecturers: 1
Teaching Fellows: 17
Research Assistants: 12

Field Stations/Laboratories: Kewalo Basin Laboratory - Honolulu, Coconut Island Laboratory - Kaneohe, Waikiki Aquarium, Honolulu, Kapiolani Hospital New-Born Psychology Laboratory, Honolulu, Leahi Hospital School of Public Health - Honolulu

Degree Programs: Biochemistry, Biophysics

Graduate Degrees: M.S., Ph.D.
Graduate Degree Requirements: Master: Plan A (thesis) 23 hours course work, 8 hours research Plan B (non-thesis) 29 hours course work, 2 hours research, Clinical Biochemistry (thesis) 22 hours course work, 8 hour research Doctoral: 35 hours course work, plus dir-

rected research and thesis research

Graduate and Undergraduate Courses Available for Graduate Credit: Introduction to Human Endocrinology (U), Introduction to Molecular Biology (U), General Biochemistry, Medical Biochemistry, Advanced Topics in Clinical Biochemistry, Seminar, Special Topics in Biochemistry, Special Topics in Enzymology, Bioenergetics, Nucleic Acids and Viruses, Advanced Protein Chemistry, Survey of Biophysics, Biophysics Laboratory, Molecular Structure and Function of Chromosomes, Special Topics in Biophysics, Molecular Structure and Function, of Cell Organelles, Basic Biochemistry (U),

DEPARTMENT OF GENETICS
Phone: Chairman: 948-7659 Others: 948-8552

Chairman: J.A. Hunt
Professors: G.C. Ashton, C.S. Chung, J.A. Hunt, M. P. Mi, Y.K. Paik
Associate Professors: S.J. Bintliff, M.N. Rashad, D. C. Vann
Assistant Professors: D.T. Arakaki, S.R. Malecha
Research Assistants: 4
Visiting Professors: 2
Postdoctorals: 5

Field Stations/Laboratories: Behavioral Biology Laboratory - Honolulu

Degree Program: Genetics

Graduate Degrees: M.S., Ph.D.
Graduate Degree Requirements: Master: Thesis (Plan A): Molecular Genetics, Population Genetics, Techniques, 4 semesters of Seminar Total of 30 credits, 10 of which are Thesis. Non-Thesis (Plan B): Total of 30 credits, same courses, but no thesis. Doctoral: No formal requirements, but equivalent coursework as for M.S.; Dissertation required.

Graduate and Undergraduate Courses Available for Graduate Credit: Advanced Topics in Genetics, Cytogenetics, Directed Research, Evolutionary Genetics, Genetic Risk Analysis, Genetical Problems, Genetics Clinic, Genetics Seminar, Human Genetics, Human Polymorphisms, Human Population Genetics, Immunogenetics, Molecular Genetics, Population Genetics, Statistical Methodology in Genetics, Techniques in Human Genetics

DEPARTMENT OF PHYSIOLOGY
Phone: Chairman: (808) 948-8652 Others: 948-8640

Chairman: S.K. Hong
Professors: S. Batkin, S.K. Hong, G.C. Whittow
Associate Professors: F.T. Koide, Y.C. Lin, T.O. Moore, M.D. Rayner, R.H. Strauss, R. Tracy, W. Woodard
Assistant Professors: G.A. Gerencser, J.M. Hanne, H. L. Gillary, T.E. Nicholas, R.M. Smith
Visiting Lecturers: 2
Research Assistants: 3

Degree Programs: Physiology

Graduate Degrees: M.S., Ph.D.
Graduate Degree Requirements: Master: Successful completion of 1st year Medical School courses in Physiology, Histology, Biochemistry, Endocrinology and Neuroscience, plus at least 12 credits of advanced graduate courses in Physiology and completion of an acceptable thesis Doctoral: Students are required to pass a qualifying examination following completion of the course work listed ablve; then they become eligible to pursue their thesis research

Graduate and Undergraduate Courses Available for Graduate Credit: Seminar in Physiology, Physiology of Nerve and Muscle, Comparative Physiology of Thermoregulation, Biophysical Concepts in Physiology, Advanced Renal Physiology, Advanced Cardiovascular Physiology, Advanced Respiratory Physiology, Directed Research, Hyperbaric and Diving Physiology, Thesis Research. The department also contributes to the following interdisciplinary courses: Introduction to Organ Systems, Endocrinology and Reproduction, Cardiovascular System, Neuroscience, Respiratory and Renal Systems, Nutrition in Health and Disease, Digestion and Metabolism

College of Tropical Agriculture
Dean: C.P. Wilson

ANIMAL SCIENCES DEPARTMENT
1825 Edmondson Road
Manoa, Hawaii 96822
Phone: Chairman: (808) 948-8356 Others: 948-8217

Chairman: C.C. Brooks
Professors: C.C. Brooks, W.I. High, J.H. Koshi, A. Y. Miyahara, R.M. Nakamura, J.C. Nolan, Jr., E. Ross, O. Wayman, H.R. Donoho, K.D. Reimer
Associate Professors: C.M. Campbell, R.B. Herrick, A.L. Palafox, T. Tanaka, D.W. Vogt
Assistant Professors: S.M. Ishizaki, 2 technicians, 1 temporary appointment
Research Assistants: 3

Field Stations/Laboratories: Waialee Livestock Research Farm - Oahu, Mealani Research Farm, Kamuela, Waiakea Branch Station - Hilo

Degree Programs: Animal Sciences, Animal Genetics, Animal Nutrition or Animal Physiology. Ph.D. programs are conducted in cooperation with other departments.

Undergraduate Degree: B.S. (Animal Sciences)
Undergraduate Degree Requirements: 130 semester hours, 2.0 GPA, complete University and College requirements and courses specific programs

Graduate Degree: M.S.
Graduate Degree Requirements: Master: 30 credits including at least 12 of graduate level, 3.0 GPA, thesis for thesis program; 36 credits including 18 graduate level and 3.0 GPA for non-thesis

Graduate and Undergraduate Courses Available for Graduate Credit: Animal Breeding (U), Physiology of Domestic Animals (U), Animal Diseases and their Control (U), Animal Science Seminar, Ruminant Nutrition, Physiology of Reproduction, Quantitative Genetics

DEPARTMENT OF FOOD SCIENCE AND TECHNOLOGY
Phone: Chairman: (808) 948-8663

Chairman: T. Nakayama
Professors: H.A. Frank, H.Y. Yamamoto
Associate Professors: F.S. Hing, J.H. Moy
Assistant Professor: C.G. Cavaletto
Research Assistants: 1

Degree Programs: Food Technology (including Tropical Fruit and Seafood Processing), Biochemistry, Chemistry, Microbiology, Engineering, Sensory Evaluation, Irradiation, Safety (including detection and metabolism of pesticides and natural toxins), Fermentation, and Waste Product Utilization

Graduate Degree: M.S. (Food Science)
Graduate Degree Requirements: Master: Plan A. Thesis 18 semester hours, 12 hours thesis Plan B. 30 hours final oral

Graduate and Undergraduate Courses Available for Graduate Credit: Food Processing, Microbiology of Foods, Food Engineering, Food Chemistry, Food Safety and Consumer Protection, Special Topics in Food Microbiology, Advanced Food Processing I, II, Seminar, Food Fermentation, Food Safety, Directed Research

DEPARTMENT OF ENTOMOLOGY
2500 Dole Street
Krauss Hall, Room 23
Phone: Chairman: (808) 948-7076

Chairman: W.C. Mitchell
Professors: D.E. Hardy (Senior Professor), J.W. Beardsley, F.H. Haramoto, A.A. LaPlante, W.C. Mitchell, R. Namba, T. Nishida, M. Sherman, M. Tamashiro
Associate Professor: F. Chang
Research Assistants: 2
Teaching Assistants: 2

Field Stations/Laboratories: Hawaii Agricultural Experimental Farms on the islands of Kauai, Hawaii, Maui and Oahu

Degree Program: Entomology

Undergraduate Degree: B.S.
Undergraduate Degree Requirements: University core requirements, College of Tropical Agriculture Core requirements; 1 year foreign language; 1 year Physics, General Entomology, Insect Morphology, Economic Entomology, Systematic Entomology plus electives. Total semester credits 127

Graduate Degrees: M.S., Ph.D.
Graduate Degree Requirements: Master: 30 semester credit hours, 18 credit hours academic courses, 12 credit hours Directed Research and thesis. Program developed by student and his committee. Doctoral: Credit hours variable depending upon committee, pass examination one foreign language, Comprehensive Examination, Final Examination in defense of dissertation

Graduate and Undergraduate Courses Available for Graduate Credit: All 400 courses and above may be given graduate credit.

DEPARTMENT OF HORTICULTURE
3190 Maile Way
Honolulu, Hawaii 96822
Phone: Chairman: (808) 948-8389 Others: 948-8351

Chairman: H. Kamemoto
Professors: J.L. Brewbaker, J.C. Gilbert, R.A. Hamilton, H. Kamemoto, H.Y. Nakasone, P.E Parvin, Y. Sagawa, G.T. Shigeura, R.M. Warner, D.P. Waston, W. Yee
Associate Professors: R.A. Criley, R.W. Hartmann, P. J. Ito, C.L. Murdoch, R.K. Nishimoto, F.D. Rauch
Assistant Professors: J.T. Kunisaki, B.A. Kratky, T.T. Sekioka, J.S. Tanaka
Visiting Lecturers: 1
Teaching Fellows: 2
Research Assistants: 10

Field Stations/Laboratories: Poamoho and Waimanalo Research Stations - Oahu - Waiakea, Volcano, Malama-Ki and Kona Branch Station - Hawaii, Kula and Olinda Branch Station - Maui, Kauai Branch Station - Kauai

Degree Program: Horticulture

Undergraduate Degree: B.S.
Undergraduate Degree Requirements: 2.0 grade point ratio for all registered credits and a minimum of 128 credits. Must complete College course requirements, Agricultural Science course requirements and Horticulture Technology or Horticultural Science curriculum requirements.

Graduate Degrees: M.S., Ph.D.
Graduate Degree Requirements: Master: Minimum of 30 credits and a grade point ratio of 3.0. Must pass a written General Examination and a final oral examination. Can register for either Plan A (Thesis) or Plan B (Non-Thesis). Doctoral; Required to pass a three-hour written Qualifying Examination, Comprehensive Examination and defense of his dissertation

Graduate and Undergraduate Courses Available for Graduate Credit: Plant Propagation and Seed Technology (U), Tropical Horticultural Crop Production (U), Plant Breeding (U), Turfgrass Management (U), Post-Harvest Handling (U), Weed Science (U), Experimental Design Plant Improvement Systems, and Plant Breeding Profession, Advanced Plant Breeding, Cytogenetics, Advanced Vegetable Crops, Advanced Tropical Fruit Science, Orchidology, Biochemical Genetics of Plants, Horticulture Seminar, Growth Regulators in Horticulture, Crop Ecology, Special Topics in Experimental Horticulture

DEPARTMENT OF PLANT PATHOLOGY
3190 Maile Way
Phone: Chairman: (808) 948-7723 Others: 948-8329

Chairman: O.V. Holtzmann
Professors: W.J. Apt (Research), M. Aragaki (Research) I.W. Buddenhagen (Research), O.V. Holtzmann (Research), M. Ishii (Research), A.P. Martinez (Extension), S.S. Patil (Research)
Associate Professors: W.H. Ko (Research), K.G. Rohrbach (Research), E.E. Trujillo (Research)
Assistant Professors: R.R. Bergquist (Research), J.J. Cho (Research associate)
Instructors: A.M. Alvarez (Extension), J B. Pfeiffer (Research associate)
Research Assistants: 4
Postdoctoral: 1

Field Stations/Laboratories: Beaumont Research Center - Hilo, Kauai Branch Station - Kapaa, Kula Branch Station - Kula

Degree Programs: Joint degree program with Botany, all faculty belong to the Botanical Sciences Graduate Faculty.

Graduate Degree: M.S., Ph.D. (Plant Pathology)
Graduate Degree Requirements: Master: Plan A (Thesis) 12 credits shall be for thesis and a minimum of an additional 18 credits of approved courses. Plan B (non-thesis) minimum of 30 credits, 15 of which are earned in major graduate field; 6 credits of directed research Doctoral: Two foreign languages. Comprehensive examination in general botanical sciences and in depth in areas of botanical or related disciplines selected by t he thesis committee and approved by graduate faculty. Dissertation by the graduate faculty and dissertation committee. Dissertation be an original contribution defended at a public seminar followed by an examination by the graduate faculty and dissertation committee.

Graduate and Undergraduate Courses Available for Graduate Credit. Introduction to Plant Pathology (U), Introduction to Plant Pathology Laboratory (U), Diseases of Tropical Crops (U), Clinical Plant Pathology (U), Principles of Plant Disease Control, Plant Nematology, Plant Pathology Techniques, Plant Virology, Physiology of Fungi, Directed Research, Plant Pathology Seminar, Host-parasite Physiology

UNIVERSITY OF HAWAII AT HILO
P.O. Box 1357 Phone: (808) 961-9311
Hilo, Hawaii 96720
Dean of College: C.M. Fullerton (Acting)

DISCIPLINE OF BIOLOGY
Phone: Chairman: (808) 961-9357 Others: 961-9311

Chairman: H.F. Little
Professors: H.F. Little, E.H. Mercer, K. Noda
Associate Professor: R.E. Baldwin
Assistant Professors: J.G. Chan, D.P. Cheny, D.E.

Hemmes
Part-time Lecturer for Laboratories: 1

Degree Program: Biology

Undergraduate Degree: B.A.

Undergraduate Degree Requirements: 32 semester hours in biology. Biology courses required are: General Zoology, General Botany, Ecology, Quantitative Biology, Genetics, Microbiology, General Biochemistry, Cellular Biology. Also: General Chemistry, Organic Chemistry, Statistics, Topics in Calculus

IDAHO

BOISE STATE UNIVERSITY

Boise, Idaho 83725 Phone: (208) 835-1011
Dean of Graduate Studies: G. Maloof
Dean of College: J.B. Spulnik

DEPARTMENT OF BIOLOGY
Phone: Chairman: 385-1411 Others: 385-1526

Chairman: D.J. Obee
Professors: H.K. Fritchman, D.J. Obee
Associate Professors: C. Baker, H.W. Belknap, E. Fuller, H. Papenfuss, G. Wyllie
Assistant Professors: C. Colby, L. Jones, F. Kelley
Visiting Lecturers: 3

Degree Program: Biology

Undergraduate Degrees: B.A., B.S.
Undergraduate Degree Requirements: Not Stated

THE COLLEGE OF IDAHO

East Cleveland Boulevard Phone: (208) 459-5011
Caldwell, Idaho 83605
Dean of Faculty: D.M. Leach

DEPARTMENT OF BIOLOGY
Phone: Chairman: (208) 459-5431 Others: 459-5332

Chairman: J.D. Marshall
Professors: R.D. Bratz, P.L. Packard, L.M. Stanford
Associate Professor: J.D. Marshall
Visiting Lecturers: 1

Field Stations/Laboratories: The College of Idaho Field Biology Expeditions. Alternate summers to Mexico. Each January as part of Winter Session, To Australia Spring Semesters 1971, 1973, next 1976.

Degree Programs: Biology, Zoology, Botany, Medical Technology

Undergraduate Degree: B.S
Undergraduate Degree Requirements: 32 semester hours in Biology, to include Biological Principles, General Botany, General Zoology, Independent Study Experience, and at least 3 courses selected from Genetics, Comparative Anatomy, Physiology, Plant Physiology, General Ecology, and Evolution, and two additional courses in Botanical Sciences. For Zoology, omit botany requirements, etc.

IDAHO STATE UNIVERSITY

Pocatello, Idaho 83209 Phone: (208) 236-0211
Dean of Graduate Studies: L. Rice

College of Liberal Arts

Dean: J.A. Hearst

DEPARTMENT OF BIOLOGY
Phone: Chairman (208) 236-3765 Others: 236-0211

Chairman: A.D. Linder
Professors: E. Fichter, A.D. Linder, W.E. Saul, J.A. White
Associate Professors: R.G. Bowmer, D.E. Bunde, K.E. Holte, E.W. House, G.W. Minshall, J.E. Tullis
Assistant Professors: R.C. Anderson, D.W. Johnson, B.L. Kelley, F.L. Rose, C. Trost, R.R. Seeley
Teaching Fellows: 10
Research Assistants: 7

Degree Programs: Biology, Botany, Conserversation, Zoology

Undergraduate Degrees: B.A., B.S.
Undergraduate Degree Requirements: 1 year Mathematics 1 year Physics, 1 year Foreign Language, 2 years Chemistry, 20 semester hours of core biology and appropriate courses in major.

Graduate Degrees: M.S., Ph.D., D.A. (Biology)
Graduate Degree Requirements: Master: Thirty-two hours beyond B.S. Doctoral: No specified courses or hours, but approximately sixty semester hours beyond Master's Degree.

DEPARTMENT OF MICROBIOLOGY AND BIOCHEMISTRY
Phone: Chairman: (208) 236-2375

Chairman: R.W. McCune
Professor: F.G. Jarvis
Associate Professors: R.W. McCune, L.J. Fontenelle
Assistant Professors: L.D. Farrell, W.R. Fleishchmann, Jr., J.H. McCune (Affiliate), J.T. Ulrich
Teaching Fellows: 3

Degree Programs: Microbiology, Medical Technology

Undergraduate Degree: B.S.
Undergraduate Degree Requirements: General Chemistry, Algebra, Trigonometry, Calculus, General Biology, Cell Biology, Organic Chemistry, General Microbiology, Quantative Analysis, Biochemistry, Media Preparation, Immunology, Pathogenic Microbiology, General Physics, Foreign Language, Microbial Physiology, Virology (Bacterial and Animal), Microbial Genetics, Seminar in Microbiology

Graduate Degree: M.S.
Graduate Degree Requirements: Master: Seminar - 2 semester credits, Thesis and Research - 10 semester credits, Additional courses - 18 semester credits, Reading Knowledge of one foreign language

Graduate and Undergraduate Courses Available for Graduate Credit: Biochemistry (U), Immunology (U), Immunology Laboratory (U), Pathogenic Microbiology (U), Pathogenic Microbiology Laboratory (U), Precambrian Evolution (U), Physiology of Microorganisms (U), Biochemistry (U), Experimental Biochemistry (U), Microbial Genetics (U), Industrial Microbiology (U), Bacterial Virology (U), Bacterial Virology Laboratory (U), Survey of Electron Microscopy (U), Animal Virology (U), Animal Virology Laboratory (U), Special Topics (U), Advanced Methods in Microbiology, Intermediary Metabolism, Advanced Microbial Physiology, Experimental Intermediary Metabolism, Selected Topics in Biochemistry

NORTHWEST NAZARENE COLLEGE

(No reply received)

UNIVERSITY OF IDAHO

Moscow, Idaho 83843 Phone: (208) 855-6111
Dean of Graduate Studies: R. Stark

College of Letters and Science

Dean: E.K. Raunio

DEPARTMENT OF BIOLOGICAL SCIENCES
Phone: Chairman: (208) 885-6280

Head: D.E. Anderegg
Professors: D.E. Anderegg, L.W. Roberts, S.C. Schell
Associate Professors: J.H. Ferguson, O.C. Forbes, D. R. Johnson, E.J. Larrison, J.L. McMullen, R.A. Mead, F.W. Rabe, G.G. Spomer, E.E. Tylutki, R.L. Wallace
Assistant Professors: V.P. Eroschenko, D.M. Henderson, R.J. Naskali, A.W. Rourke
Teaching Fellows: 14
Research Assistants: 4

Degree Programs: Biology, Botany, Zoology

Undergraduate Degrees: B.S.
Undergraduate Degree Requirements: B.S. requires 128 semester hours credit which must include at least 30 hours of biology, 22 hours of chemistry, 8 hours of physics, 9 hours of mathematics, 9 hours of humanities, and 9 hours of social sciences. At least 36 hours must be taken at the upper division level.

Graduate Degrees: M.S., Ph.D.
Graduate Degree Requirements: Master: 30 hours minimum beyond B.S. of which 10 hours may be research and 18 hours must be courses restricted to graduate students. Doctoral: 90 hours beyond B.S. of which 30 hours may be research. One language at the 75 percentile by ETS examination.

Graduate and Undergraduate Courses Available for Graduate Credit: Techniques in Plant Tissue Culture (U), Mineral Nutrition (U), Developing Plant Anatomy (U), Plant Ecology (U), Agrostology (U), Plant Growth Stustitution, Physiological Ecology, Comparative Vertebrate Reproduction (U), Cell Physiology (U), Mammalian Physiology (U), Endocrine Physiology (U), Vertebrate Histology and Organology (U), Limnology (U), Ethology (U), Ichthyology (U), Natural History of Birds (U), Natural History of Mammals (U), Invertebrate Zoology (U), Protozoology (U), Parasitology (U), Herpetology (U), Comparative Animal Physiology, Hydrobiology

College of Agriculture

Dean: A.M. Mullins

DEPARTMENT OF AGRICULTURAL EDUCATION
Phone: Chairman: 885-6358

Head: D.L. Kindschy
Professor: R.C. Haynes
Associate Professor: J.A. Lawrence
Assistant Professor: W. Shane (Extension)

Degree Program: Agricultural Education

Undergraduate Degree: B.S.
Undergraduate Degree Requirements: Not Stated

Graduate Degree: M.S.
Graduate Degree Requirements: Not Stated

DEPARTMENT OF ANIMAL INDUSTRIES
Phone: Chairman: 885-6345

Chairman: J.M. McCroskey
Professors: J.M. McCroskey, R.E. Christian, J. Dahmen, C.F. Peterson, R. Ross, E.A. Santer
Associate Professors: R. Bull, S. Davis, W. Hodgson, G. Sasser
Assistant Professors: D. Howes, J. Jacobs, D. Thalter
Teaching Fellows: 1
Research Assistants: 8

Field Stations/Laboratories: Agricultural Br. Experimental Station - Caldwell

Degree Programs: Animal Genetics, Animal Industries, Animal Physiology, Animal Science, Dairy Science, Meat Science, Nutrition, Poultry Science

Undergraduate Degree: B.S. (Animal Industries)
Undergraduate Degree Requirements: 132 credits

Graduate Degree: M.S.
Graduate Degree Requirements: Master: 30 credits

Graduate and Undergraduate Courses Available for Graduate Credit: Population Genetics, Animal Growth, Animal Adoptation, Endocrine Physiology, Physiology of Reproduction, Animal Nutrition, Energy Metabolism, Ruminant Nutrition, Non-Ruminant Nutrition, Statistical Genetics, Advanced Reproduction, Advanced Endocrine Physiology, Meat Science

DEPARTMENT OF BACTERIOLOGY AND BIOCHEMISTRY
(no reply received)

DEPARTMENT OF ENTOMOLOGY
Phone: Chairman: (208) 885-6595 Others: 885-6595

Chairman: A.R. Gittins
Professors: W.F. Barr, G.W. Bishop, A.R. Gittins, J.A. Schenk, R.W. Stark
Associate Professors: M.A. Brusven, L.E. O'Keeffe, D.R. Scott (Research), H.W. Smith
Assistant Professors: G.P. Carpenter (Research, N.D. Waters (Research)
Research Assistants: 5

Field Stations/Laboratories: Research and Extension Centers at Sandpoint, Parma, Caldwell, Kimberly, Aberdeen

Degree Program: Entomology

Undergraduate Degree: B.S.
Undergraduate Degree Requirements: 20 credits Entomology; 34 Life Sciences (Botany, Zoology, Bacteriology) 14 Physical Sciences (Chemistry, Physics); 4 Mathematics, 19 Social Sciences, 31 electives

Graduate Degrees: M.S., Ph.D.
Graduate Degree Requirements: Master: 30 credits, 10 research and thesis, 20 in major and minor fields including 3 credits of graduate seminar. Doctoral: minimum of approximately 70 credits beyond Bachelor's degree including 6 graduate seminar; 30 research and thesis

Graduate and Undergraduate Courses Available for Graduate Credit: Pesticides in the Environment (U), Immature Insects (U), Plant Resistance to Insects (U), Medical Entomology (U), Biological and Integrated Control (U), Insect Physiology (U), Forest Entomology (U), Aquatic Entomology (U), Insect Anatomy and Physiology (U), Insect Morphogenesis (U), Directed Studies (U), Entomological Research Methods, Entomological Literature, Principles of Control, Insect Ecology, Insect Behavior, Pest Management, Systematic Entomology, Insect Toxicology, Insect Biochemistry, Advanced Forest Entomology, Insect Physiological Ecology, Seminar, Research and Thesis, Dissertation

DEPARTMENT OF PLANT AND SOIL SCIENCES
Phone: Chairman: (208) 885-6274 Others: 885-6276

Head: L. Calpouzos
Professors: L. Calpouzos, L.L. Dean, R.D. Ensign, L.C. Erickson, H.S. Fenwick, A.M. Finley, M.A. Fosberg, D. Franklin, J.G. Garner, J.W. Guthrie, A.W. Helton, W. Kochan, M.J. LeBaron, G.C. Lewis H. McKay, R.E. Ohms, E.E. Owens, W.K. Pope, C. L. Seely, W.R. Simpson, W.C. Sparks, P. Torell, R.D. Watson
Associate Professors: A.A. Boe, J.R. Davis, R.W. Harder, R.E. Higgins, A.S. Horn, J.P. Jones, J.J. Kolar, G.A. Murray, D.V. Naylor, C.G. Painter, R.R. Romanko, H.B. Roylance, G. Stallknecht, A.J. Walz
Assistant Professors: A.R. Campbell, R.H. Callihan,

C. Dallimore, R.D. Johnson, D.J. Makus, R.E. McDole
Visiting Lecturers: 1
Teaching Fellows: 1
Research Assistants: 3
Postdoctoral Fellow: 1
Scientific Aide: 1

Field Stations/Laboratories: Research and Extension Centers at Aberdeen, Kimberly, Parma, Sandpoint and Tetonia

Degree Programs: Crop Science, Horticultural Science, Physiology, Plant Breeding, Plant Pathology, Plant Sciences, Soil Science, Weed Science

Undergraduate Degree: B.S.
Undergraduate Degree Requirements: A total of 132 credits, of which, according to option selected, 60-66 are in field of speciality

Graduate Degrees: M.S., M. Agr., Ph.D.
Graduate Degree Requirements: Master: Master of Science, 18 credits in 500's courses or above, Master of Agriculture, 16 credits in 500's courses or above. M.S. degree without thesis, 30 credits, M.S. degree with thesis, 20 credits of course work. Doctoral: 78 credits beyond a B.S. degree, of those 78 credits, 52 must be in courses numbered 500 or above

Graduate and Undergraduate Courses Available for Graduate Credit: Plant Science - 507 Preparation and Presentation of Scientific Material, Ecology Soil-Borne Plant Pathogenic Organisms, Plant Virology, Physiology of Disease, Environmental Plant Physiology, Tree Physiology, Plant Stress Physiology, Genetics Literature, Advanced Crop Production, Research Methods, Advanced Weed Studies, Cytogenetics, Propatation and Function of Herbicides, Practicum, Internship, Seminar, Directed Study. Soils - Advanced Laboratory Technique, Advanced Forest Soils, Soil Organic Matter, Advanced Soil Chemistry, Chemistry of Plant Nutrients, Advanced Soil Physics, Advanced Soil Fertility, Advanced Soil Genesis and Classification, Practicum

DEPARTMENT OF VETERINARY SCIENCE
Phone: Chairman: 885-7081

Chairman: F.W. Frank
Professor: F.W. Frank
Associate Professor: C.S. Card, A.S. Ward
Assistant Professor: E.H. Stauber, H.W. Vaughn
Teaching Fellows: 1
Research Assistants: 5
Postdoctoral Fellow: 1

Field Stations/Laboratories: Caldwell Veterinary Research Laboratory, Caldwell

Degree Program: Veterinary Science

Undergraduate Degree: B.S.

Graduate Degree: M.S.
Graduate Degree Requirements: Master: 30 credits, minimum of 20 credits course work, remainder being research and thesis credits

Graduate and Undergraduate Courses Available for Graduate Credit: Seminar (U), Workshop (U), Dis. and Care of Laboratory Animals (U), Meat Inspection and Veterinary Hygiene (U), Animal Dissection (U), Virology (U), Virology Laboratory (U), Directed Study (U), Master's Research and Thesis, Seminar, Directed Study, Workshop, Principles of Comparative Pathology, Methods of Animal Experimentation, Practicum, Internship, Research

College of Forestry, Wildlife and Range Science

Dean: J.H. Ehrenreich

Phone: 885-6441 Others: 885-6442

Academic Program Chairmen: D.L. Adams (Forest Resources), K. Hungerford (Wildlife), G.W. Klontz (Fisheries), F.L. Newby (Wildlife Recreation Management), L.A. Sharp (Range), K.M. Sowles (Wood Utilization)
Professors: E.D. Ables, P.D. Dalke (Emeritus), M.E. Deters (Emeritus), J.H. Ehrenreich, R.C. Heller (Re-Search, M. Hironaka, M. Hornocker, J.P. Howe, K. Hungerford, F.D. Johnson, G.W. Klontz, H. Lowenstein, C. MacPhee, F.L. Newby, A.D. Partridge, F. Pitkin, J.A. Schenk, R.H. Seale, L.A. Shapr, R.W. Stark, E.W. Tisdale, C.W. Wang, E.W. Woletz
Associate Professors: D.L. Adams, G.H. Belt, E. Bizeau, T.C. Bjornn, O.M. Falter, F.B. Godfrey, C.R. Hatch, J.M. Peek, K.M. Sowles, J.J. Ulliman,
Assistant Professors: J.R. Fazio, H. Hofstrand, J.E. Houghton, L. Johnson, J.E. Mitchell, S.R. Peterson, E.G. Schuster, R. White
Instructors: G.M. Allen, J.G. King
Teaching Fellows: 10
Research Assistants: 29
Research Associates: 10
Graduate Students: 30

Field Stations/Laboratories: Wilderness Research Field Station - Idaho Primitive Area

Degree Programs: Degrees offered in: Fisheries Resources, Forest Resources, Range Resources, Wildlife Resources, Wildland Recreation Management, Wood Utilization

Undergraduate Degree: B.S.
Undergraduate Degree Requirements: 136 semester hours, 12 of Social Sciences-Humanities, Calculus, two courses in modelling, biometry, computer programming summer camp or work.

Graduate Degrees: M.S., M.F., Ph.D.
Graduate Degree Requirements: Master: 30 semester hours beyond B.S., comprehensive examination, thesis for M.S. Doctoral: 78 semester hours beyond B.S., 52 credits must be graduate level, dissertation, comprehensive examinations, no language

Graduate and Undergraduate Courses Available for Graduate Credit: Too many to list.

ILLINOIS

AUGUSTANA COLLEGE
Rock Island, Illinois 61201 Phone: (309) 794-7311
Dean of College: T. Tredway

DEPARTMENT OF BIOLOGY
Phone: Chairman: (309) 794-7398 Others: 794-7311

Chairman: R. Troll
Professors: F. Neely, R. Troll
Associate Professors: J. Ekblad, I. Larson
Assistant Professors: R.W. Catlin, T. Rennie, R. Turnquist
Undergraduate Laboratory Assistants: 12

Degree Programs: Biology, Medical Technology

Undergraduate Degree: B.S.
Undergraduate Degree Requirements: 42 quarter hours in biology including biological principles, general botany, vertebrate zoology, invertebrate zoology or entomology or parasitology, and plant morphology or plant taxonomy or microbiology. Supporting courses: Chemistry through Organic Chemistry, one year of Physics, Mathematics through Calculus, and one course in computer programming.

AURORA COLLEGE
347 S. Gladstone Phone: (312) 892-6431
Aurora, Illinois 60506
Academic Dean: D. Arthur

DEPARTMENT OF BIOLOGY
Phone: Chairman: (312) 892-6431 Ext. 65

Chairman: J.G. Green
Associate Professor: J.G. Green
Assistant Professors: R. Kimmel, B. Schmidt, D. Hannum
Student Assistants: 10

Degree Programs: Biology, Ecology, Health and Biological Sciences, Histology, Life Sciences, Medical Technology, Physiology, Biophychology, Bioeconomics, Student initiated

Undergraduate Degree: B.A.
Undergraduate Degree Requirements: 40 courses (credits equivalent to 3 school hours) including 4 in enrichment

BARAT COLLEGE
Lake Forest, Illinois 60045 Phone: (312) 234-3000
Dean of College: D. Hollenhorst

DEPARTMENT OF BIOLOGY
Phone: Chairman: (312) 234-3000 Ext. 307

Head: M.R. Seinwill
Professor: M.R. Seinwill
Assistant Professor: L. Bossenga

Degree Program: Biology

Undergraduate Degree: B.S.
Undergraduate Degree Requirements: 120 semester hours

BLACKBURN COLLEGE
Carlinville, Illinois 62626 Phone: (217) 854-3231
Dean of College: J. Dana

BIOLOGY DEPARTMENT
Phone: Chairman: 854-3231 Ext. 230

The faculty is unranked: W.E. Werner, Jr., D. Singh, J. Jackson

Degree Program: Biology

Undergraduate Degree: B.A.
Undergraduate Degree Requirements: General Chemistry, 1 year Algebra, Trigonometry, 1 semester Botany, 1 semester Zoology, 1 semester Cell Biology, 1 semester Ecology, 1 semester Genetics, 1 semester Physiology, 1 semester choice of 2 other courses; seminar 1 year

BRADLEY UNIVERSITY
1501 W. Bradley Avenue Phone: (309) 676-7611
Peoria, Illinois 61606
Dean of Graduate Studies: W. Grimm
Dean of College: R.G. Bjorklund

DEPARTMENT OF BIOLOGY
Phone: Chairman: Ext. 450

Chairman: B.J. Mathis
Professors: B.J. Mathis, R.G. Bjorklund, J.A. DePinto
Associate Professors: E.C. Gasdorf, G.D. Elsoth, R.R. Stephens, H.L. Monoson, A.G. Galsky

Degree Programs: Biology, Medical Technology, Environmental Science

Undergraduate Degree: B.S.
Undergraduate Degree Requirements: Biology - 27-30 semester hours of Biology, 13-17 hours of Chemistry, 7 hours of Physics and Mathematics through Calculus Medical Technology - 16 hours of Biology, 17 hours of Chemistry, and 4 hours of College level Mathematics Environmental Science - 24 hours of Biology, 13 hours of Chemistry, 14 hours of Geology, 7 hours of Physics, 3 hours of computer science and Mathematics through Calculus

CHICAGO STATE UNIVERSITY
95th at King Drive Phone: (312) 995-2000
Chicago, Illinois 60628
Dean of Graduate Studies: R. Prince
Dean of College: E. Washington

DEPARTMENT OF BIOLOGICAL SCIENCES
Phone: Chairman: (312) 995-2183 or 2184

Chairman: A.N. Bond
Professors: A.N. Bond, O.J. Eigsti, J. Fooden, P.W. Titman
Associate Professors: B. Berkson, G.E. Eertmoed, W.H. Hirschfeld
Assistant Professors: R. James, F. Morzlock, J. Rastorfer
Instructor: T. McCague
Faculty Assistants: 2

Degree Program: Biology

Undergraduate Degree: B.S.
Undergraduate Degree Requirements: Two courses in General Biology, Genetics, Botany, Zoology, Physiology, General Chemistry, Organic and Biochemistry, Two courses in Physics, College Mathematics

Graduate Degree: M.S.
Graduate Degree Requirements: Master: Completion of 30 hours of 300 and 400 level courses with a B average. Successful completion of Comprehensive Examination

Graduate and Undergraduate Courses Available for Graduate Credit: Biology of Africa, Biometrics, Evolution, Ecology, Cytogenetics, Local Flora, Economic Botany, Plant Ecology, Histology, Entomology, Embryology

COLLEGE OF ST. FRANCIS

500 Wilcox Street Phone: (815) 726-7311
Joliet, Illinois 60435
Dean of College: H. Blanton

DEPARTMENT OF BIOLOGY
Phone: Chairman: 726-7311 Ext. 277, 262

Chairman: M.V. Kirk
Assistant Professors: S. Miller, D. Resh
Instructors: C. Mayers, J. Bajt
Visiting Lecturers: 2

Degree Programs: Biology

Undergraduate Degree: B.A.
Undergraduate Degree Requirements: 40 hours at Biology including: Microbiology, Genetics, One Physiology course, one Botany course, Biochemistry, Seminar, 16-20 hours of Chemistry, one Mathematics course, one Physics course

DE PAUL UNIVERSITY

1036 W. Belden Avenue Phone: (312) 321-8000
Chicago, Illinois 60614
Dean of Graduate Studies: Rev. M.T. Cortelyou, C.M.
Dean of College: J.P. Masterson

BIOLOGICAL SCIENCES DEPARTMENT
Phone: Chairman: 321-8161 Others: 321-8000

Chairman: R.A. Griesbach
Professors: J.R. Cortelyou, M.A. McWhinnie, R.C. Thommes
Associate Professors: L.E. Fischer, D.V.M. (Adjunct), R.A. Griesbach, M.A. Murray, D.G. Oldfield, D.J. McWhinnie, J.E. Woods
Assistant Professors: D.S. Juras, R.A. Sorensen
Instructors: 1
Visiting Scientist: 1

Degree Programs: Biology, Medical Technology, Regulatory Biology

Undergraduate Degree: B.S.
Undergraduate Degree Requirements: Total Quarter Hours: 180, College courses (Humanities, Behavioral-Social Sciences, etc.) - 72 hours Biology major - 48 hours, Mathematics 20-28 hours, Chemistry - 24 hours, Physics - 12 hours

Graduate Degrees: M.S., Ph.D.
Graduate Degree Requirements: Master: 44 hours of which up to 8 hours may be research, thesis, (or experimental research), successful completion of Candidacy and Final Masters Examinations Doctoral: 108 Beyond the baccalaureate degree - of which up to 36 hours may be applied toward the doctoral dissertation. Successful completion of Preliminary, Candidacy and Final Examinations. Evidence of proficiency in the translation of a modern foreign language

Graduate and Undergraduate Courses Available for Graduate Credit: Plant Anatomy (U), Plant Physiology (U), Vertebrate Physiology (U), Ecology (U), Aquatic Biology (U), Invertebrate Biology (U), Developmental Biology (U), Concepts in Evolution (U), Cell Physiology: Metabolism (U), Immunobiology (U), General Physiology (U), Introduction to Endocrinology (U), Discussions of Selected topics in Biology, Advanced Genetics, Cell Physiology: Interactions, Cellular Events in Immune Response, Comparative Animal Physiology, Physiology of the Endocrine System, Physiology of Reproduction, Comparative Endocrinology, Plant Hormones, Seminars

EASTERN ILLINOIS UNIVERSITY

Charleston, Illinois 61920 Phone: (217) 581-3011
Dean of Arts and Sciences: L.A. Ringenberg

DIVISION OF LIFE SCIENCES, DEPARTMENT OF BOTANY, DEPARTMENT OF ZOOLOGY
Phone: Chairman: (217) 581-3011 Others: 581-2021

Chairmen: Botany: W.W. Scott, Zoology: G.T. Riegel
Professors: Botany: Z.E. Bailey, J.E. Ebinger, W.W. Scott, W.C. Whiteside Zoology: L. Durham, M.B. Ferguson, M.A. Goodrich, W.J. Keppler, V.B. Kniskern, E.B. Krehbiel, H.C. Rawls, G.T. Reigel, B.T. Ridgeway, L.S. Whitley
Associate Professors: Botany: C.B. Arzeni, S.A. Becker, G.G. Gray, R.L. Smith, T.M. Weidner Zoology: R. D. Andrews, P.J. Docter, R.C. Funk, L.B. Hunt, J.A. Maya, E.O. Moll, F.R. Schram
Assistant Professors: Botany: L.E. Crofutt, R.L. Darding, O.F. Lackey, D.H. Murphy, J.M. Speer, W.A. Weiler, U.D. Zimmerman Zoology: K.D. Baumbardner, M.K. Chapman, F.A. Fraembs, F.H. Hedges, W.S. James, B.A. Landes, J.C. Martinez, H.C. Nilsen
Faculty Assistants: 5

Degree Programs: Botany, Zoology, Environmental Biology

Undergraduate Degrees: B.S.
Undergraduate Degree Requirements: 120 semester hours total - all degrees, Botany - 64 semesters in Botany, Life Science and Correlating courses required (Chemistry - 8 hours, Geology - 4 hours, Zoology - 4 hours Electives, 14 hours) Zoology - 64 semester hours in Zoology, Botany, Life Science and correlative courses Environmental Biology - 84-86 hours in Botany, Zoology, Life Science and correlative courses

Graduate Degree: M.S.
Graduate Degree Requirements: Master: 30 semester hours with thesis, 32 semester hours without thesis

Graduate and Undergraduate Courses Available for Graduate Credit: Lichens (U), Systematic Botany (U), Plant Anatomy (U), Phycology (U), Plant Pathology (U), Plant Geography, Bryology, Advanced Bacteriology, Fungi I, II, Ethnobotany, Advanced Plant Physiology I, II, Cytotaxonomy, Advanced Plant Ecology, Independent Study, Seminar, Thesis Cytology (U), Invertebrate Zoology (U), Advanced Entomology (U), Fisheries Management (U), Ichthyology (U), Herpetology (U), Ornithology (U), Mammalogy (U), Terrestrial Ecology (U), Limnology (U), Population Biology (U), History of Biology, Organic Evolution, Systematics, Paleozoology, Protozoology, Invertebrate Field Studies, Arthropodology, Wildlife Management, Animal Behavior, Cell Physiology, Developmental Zoology, Endocrinology, Methods in Biological Research, Independent Study, Seminar, Thesis

ELMHURST COLLEGE

Elmhurst, Illinois 60126 Phone: (312) 279-4100
Dean of College: R.J. Clark

DEPARTMENT OF BIOLOGY
Phone: Chairman: 279-4100 Ext. 272

Chairman: J.A.Jump
Professor: J. A. Jump
Associate Professors: J. Gorsic, J.H. Honour, E.H.

Meseth
Assistant Professors: C.W. Fuller, F.C. Mittermeyer
Visiting Lecturers: 1

Degree Program: Biology

Undergraduate Degrees: A.B., B.S.
Undergraduate Degree Requirements: 33 courses (4 credit hours each) in total, 8 courses in Biology, 4 courses in Chemistry, including two of Organic, 2 courses in Mathematics, 2 courses in Physics, Other required courses in the Humanities, and Social Science are being changed and not yet finalized .

EUREKA COLLEGE
Eureka, Illinois 61530 Phone: (309) 467-3721
Dean of College: C.R. Noe

BIOLOGY DEPARTMENT
Phone: Chairman: 467-3721 Ext. 241

Chairman: J.W. Snyder
Professors: J.W. Synder
Assistant Professor: M.P. Coons
Student Assistants: 4

Degree Programs: Biology, Medical Technology

Undergraduate Degree: B.S.
Undergraduate Degree Requirements: 28 semester hours of Biology, 13 semester hours Chemistry, 8 semester hours Physics, 4 semester hours Mathematics, 6 semester hours English, 9 semester hours Humanities, 9 semester hours Social Sciences, 6 semester hours General Studies and 4 semester hours Physical Education and 33 semester hours electives

GEORGE WILLIAMS COLLEGE
555 - 31st Street Phone: (312) 964-3100
Downers Grove, Illinois 60515
Dean of College: C.H. Rhee

DIVISION OF NATURAL SCIENCES
Phone: Chairman: (312) 964-3100

Division Director: J.E. Norris
Professors: D.E. Misner, G.C. Robinson, H.E. Westerberg
Associate Professor: J.E. Norris
Assistant Professors: J.E. coleman, P.K. Healey, M.E. Langbein, L. Laub, S. Stock
Instructor: N. Merczak
Part-time Professors: 3

Degree Programs: Biology, Ecology, Medical Technology, Natural Science, Physiology, Health Education

Undergraduate Degree: B.S.
Undergraduate Degree Requirements: 48 hours core requirements (including 12 hours of electives), 48 hours distributed general education (Humanities, Social Science, Natural Science), 48 hours supportive advanced education)Specified per degree), 48 hours major course credits

Graduate Degree: M.S. (Health Education)
Graduate Degree Requirements: Master: No information Available

Graduate and Undergraduate Courses Available for Graduate Credit: No information available.

GREENVILLE COLLEGE
Greenville, Illinois 62246 Phone: (618) 664-1840
Dean of College: W.R. Stephens

BIOLOGY DEPARTMENT
Phone: Chairman: 664-1840 Ext. 244

Chairman: M.G. McHenry
Professor: J. Ayers
Associate Professor: W.B. Ahern, M.G. McHenry
Instructor: M. Siefken

Field Stations/Laboratories: Colorado Science Station - Manitou Springs, Colorado

Degree Programs: Biology, Biological Education

Undergraduate Degree: A.B., B.S. In Educ.
Undergraduate Degree Requirements: Not Stated

ILLINOIS BENEDICTINE COLLEGE
Lisle, Illinois 60532 Phone: (312) 968-7270
Dean of College: G. Tysl

DEPARTMENT OF BIOLOGY
Phone: Chairman: 968-7270 Ext. 208 Others: Ext. 321

Chairman: T.D. Suchy, O.S.B.
Professor: L. Bogdanove
Associate Professor: L. Kamin
Assistant Professor: R. Grossberg
Special Lecturer in Ornithology: 1

Degree Program: Biology

Undergraduate Degree: B.S.
Undergraduate Degree Requirements: 4 credit hours in General Zoology, 4 credit hours in General Botany, 4 credit hours in Genetics, total of 38 semester hours in Biology, completed 1 year of Organic Chemistry, completed 1 year Physics, Either Trigonometry or Pre-Calculus, pass Undergraduate Recond Examination, student must have taken 2 advanced courses in animal science and also 2 advanced courses in plant science)

ILLINOIS COLLEGE
Jacksonville, Illinois 62650 Phone: (217) 245-7126
Dean of College: W. Jamison

BIOLOGY DEPARTMENT
Phone: Chairman: 245-7126 Ext. 243

Chairman: L. Rainbolt
Professor: L. Rainbolt
Associate Professor: L. Moehn, B.C. Moulder

Degree Program: Biology

Undergraduate Degrees: B.S., B.A.
Undergraduate Degree Requirements: 16 hours required in (4) Genetics, (4) Developmental Biology (4) Ecology, (4) Cell Biology, 14 hours of electives in Biology, 1 semester of Mathematics, 2 semesters of General Chemistry

ILLINOIS INSTITUTE OF TECHNOLOGY
Chicago, Illinois, 60616 Phone: (312) 225-9600
Dean of Graduate Studies: S.A. Guarlnick
Dean of College: C.M. Grip

DEPARTMENT OF BIOLOGY
Phone: Chairman: Ext. 1281, 1282 Others: 1281,1282

Chairman: T. Hayashi
Professors: W.F. Danforth, F.C.G. Hoskin, K. Kusano, A.H. Roush
Associate Professors: H.W. Bretz, J.A. Erwin, N. Grecz, D.K. Jasper, D.C. Loblick, R.M. Roth, M.G. Tarver (Adjunct), D.A. Webster
Assistant Professors: S.S. Kang, P.K. Sarkar
Instructors: Ph. DuPont, B. Smith

Research Associate: 1

Degree Programs: Biochemistry, Biology, Biomedical Engineering, Biophysics, Genetics, Microbiology, Physiology, Developmental, Neurobiology, Cytology (a B.S.E.S. in Biomedical Engineering can be obtained through the College of Engineering and Physical Science as a special program.)

Undergraduate Degrees: B.S., B.S.L.A. (Biology)
Undergraduate Degree Requirements: 124 credits total. 27 credits Biology, 19 credits Chemistry, 8 credits Mathematics, 7 credits Physics. Additional Mathematics, Chemistry and Physics recommended.

Graduate Degrees: M.S., Ph.D. (Biology)
Graduate Degree Requirements: Master: 32 hours including 6-8 hours research credit and 2 semesters of Graduate Core courses; 2 hours of Instruction in Biology, thesis based on experimental research; Graduate Qualifying Examination; completion with 5 calendar years Doctoral: 96 semester hours including 27-31 hours Graduate Core courses, 8 hours Instruction in Biology, 6-8 hours Colloquium in Biology, 6 hours minimum Pre-Dissertation Research; Language Examination; Ph.D. Comprehensive Examination, Graduate Qualifying Examination, Research and Thesis

Graduate and Undergraduate Courses Available for Graduate Credit: Core III (U), Core IV (U), Basic Neurophysiology (U), General Microbiology (U), Plant Physiology (U), Taxonomy Microorganisms (U), Basic Statistics for the Life Sciences (U), Genetics (U), Seminar in Physiology (U), Seminar in Invertebrate Physiology (U), Topics in Biology (U), Seminar in Biochemistry (U), Topics in Developmental Biochemistry (U), Seminar in Neurosciences (U), Advanced Cell Biology Cytology and Electron Microscopy (U), Core A, B, C, D, Lectures and Laboratory, Selected Topics in Cellular, Developmental, and Molecular Biology, Statistical Design of Experiments for the Life Sciences, Seminar in Statistical Design of Experiments, Seminar in Physiology, Seminar in Microbiology, Seminar in Molecular Biology, Instruction in Biology

ILLINOIS STATE UNIVERSITY

Normal, Illinois 61761 Phone: (309) 438-2111
Dean of Graduate Studies: C.A. White
Dean of College of Arts and Sciences: S. Shuman

DEPARTMENT OF BIOLOGICAL SCIENCES
Phone: Chairman: (309) 438-3669

Chairman: H.R. Hetzel
University Professor: R.O. Rilett
Professors: D.E. Birkenholz, R. Brawn (Adjunct), H.E. Brockman, W.H. Brown, E. Dilks, J.L. Frehn, H.R. Hetzel, A.E. Liberta, L.W. Mentzer, E.L. Mockford, R.M. Reardon (Adjunct), R.O. Rilett, I. Rhymer, J. Verner, R.D. Weigel, E.R. Willis
Associate Professors: L. Brown, R.M. Chasson, T.I. Chuang, K. Fitch, H. Huizinga, D.R. Jensen, O. Mizer, H. Moore, M. Nadakavukaren, J. Tone, J. Tsang (joint with Chemistry), J. Ward, D. Weber
Assistant Professors: L. Cadwell, J.C. Cralley, B. Haines, D. McCracken, M. Miller (Adjunct), M. Neville, B.D. Richards, A. Richardson (joint with Chemistry), F. Schwalm
Instructors: B.M. Parker
Teaching Fellows: 59
Research Assistants: 3
NSF Fellows: 2
NDEA Title IV Fellows: 3

Degree Programs: Botany, Ecology, Genetics, Microbiology, Physiology and Zoology

Undergraduate Degrees: B.A,, B.S., B.S. in Ed.
Undergraduate Degree Requirements: 37 hours in Biology, Chemistry through Organic, 1 year General Physics,

Graduate Degrees: M.S., M.S. in Ed., Ph.D.
Graduate Degree Requirements: Master: 32 hours graduate biology including Seminar and Readings in the Biological Sciences; thesis or Comprehensive Examinations Doctoral: Minimum of 4 semesters in residence after Master's degree; Reading knowledge of two foreign languages or 1 foreign language plus a statistics sequence

Graduate and Undergraduate Courses Available for Graduate Credit: Readings in Biological Sciences (U), Readings in Biological Sciences (U), History of Biology (U), Natural Science for Elementary Teachers, (U), Seminar in Biology (U), Special Problems in Biology (U), Regional and Area Studies (U), Laboratory Techniques (U), Genetics (U), Plant Pathology (U), Taxonomy of Vascular Plants (U), Taxonomy of Non-Vascular Plants (U), Comparative Plant Morphology (U), Introductory Mycology (U), Administration of School Health (U), General Biochemistry (U), Biochemistry Laboratory (U), Sanitation (U), Phycology (U), Applied Human Anatomy (U), The Eye (U), Parasitology (U), Physical Defects--Survey and Rehabilitation (U), Gross Anatomy (U), Evolution (U), Entomology (U), Embryology (U), Protozoology (U), Biology of the Lower Vertebrates (U), Biology of the Higher Vertebrates (U), Advanced Ecology, Aquatic Biology, Seminar in the Teaching of Biology, Cellular Physiology, Cytology, Electron Microscopy, Advanced Genetics, Seminar in Genetics, Cytogenetics, Radiation Biology, Special Topics in Plant Physiology, Enzymology, Plant Anatomy and Histology, Epidemiology, Advanced Studies in Specialized fields, Advanced Mycology, Microbial Physiology, Microbial Genetics, Histology, Sensory Physiology, Mammalian Physiology, Human Development and Behavior, Endocrinology, Ethology, Human Genetics, Internship--Seminar in College Teaching in the Biological Sciences, Developmental Biology, Biophysics, Comparative Animal Physiology

ILLINOIS WESLEYAN UNIVERSITY

Bloomington, Illinois 61701 Phone: (309) 556-3131
Dean of College: J. Clark

DEPARTMENT OF BIOLOGY
Phone: Chairman: (309) 556-2351 Others: 556-3060

Chairman: B.B. Criley
George C. and Ella Beach Lewis Professor of Biology: D.S. Franzen
Professors: B.B. Criley, D.S. Franzen
Associate Professor: W.W. Darlington
Assistant Professors: R.L. Arteman, J. Austin
Teaching Assistants: 9

Degree Program: Biology

Undergraduate Degree: B.S.
Undergraduate Degree Requirements: 9 courses in Biology, 4 courses in Chemistry (including 1 year of Organic); 1 year of Physics, In addition, there are a number of general requirements for the Bachelor's Degree; see our college catalog.

JUDSON COLLEGE

1151 North State Phone: (312) 695-2500
Elgin, Illinois 60120
Dean of College: E. Boss

DEPARTMENT OF SCIENCE - MATHEMATICS
Phone: Chairman: Ext. 563 Others: 560,561,562

Chairman: E.B. Juergensmeyer
Associate Professor: E.B Juergensmeyer
Assistant Professors: F. Averill, R. Myhrman
Instructor: N. Siplock

Degree Program: Biology

Undergraduate Degree: B.A.
Undergraduate Degree Requirements: 25 semester hours in freshman-level courses in three of the following areas: Biology, Chemistry, Physics, Mathematics, 40 semester hours in upper-level cour ses in at least two of the areas above

KNOX COLLEGE

Galesburg, Illinois 61401 Phone: (309) 343-0112
Dean of College: L.S. Salter

BIOLOGY DEPARTMENT
Phone: Chairman: (309) 343-0112

Chairman: B.W. Greer
Professors: R.E. Johnson, G.H. Ward
Associate Professor: P. Schramm
Assistant Professor: E.A. Perry, F. R. Voorhees

Field Stations/Laboratories: Green Oaks Field Biology Station -

Degree Programs: Biology

Undergraduate Degree: B.A.
Undergraduate Degree Requirements: General Genetics, Plant Structure, Animal Structure, General Ecology, Cytology, Human Physiology, and Biochemistry. Also, two courses in General Chemistry are required and one upper division course each must be taken in the areas of Structural Biology, Molecular Biology and Field Biology. Two additional courses must be elected in Mathematics, Physics, Chemistry or Geology.

LAKE FOREST COLLEGE

Sheridan and College Roads Phone: (312) 234-3100
Lake Forest, Illinois 60045
Dean of College: F.K.I. Radandt

DEPARTMENT OF BIOLOGY
Phone: Chairman: Ext. 469

Chairman: F.A. Giere
Professors: F.A. Giere, C.D. Louch
Associate Professors: R.R. Runge, K.L. Weik
Assistant Professor: L.D. Spiess
Instructor: E.E. Powers

Degree Program: Biology

Undergraduate Degree: B.A.
Undergraduate Degree Requirements: 1 course - 4 semester hours, 8 semester hours in Chemistry (credit hours) 32 credit hours in Biology, of which 4 (1 course) must be in Zoology, and 4 (1 course) must be Botany, 4 courses at Junior-Senior level.

LEWIS UNIVERSITY

Lockport, Illinois 60441 Phone: (815) 838-0500

DEPARTMENT OF BIOLOGY
Phone: Chairman: 838-0500 Ext. 409

Professor: K.V. Thiruvathukal
Associate Professor: E. Stedman
Assistant Professors: R. Farrell

Degree Program: Biology

Undergraduate Degree: B.A.
Undergraduate Degree Requirements: 64 hours of Biology, Chemistry, Physics

LOYOLA UNIVERSITY OF CHICAGO

320 North Michigan Avenue Phone: (312) 274-3000
Chicago, Illinois 60611
Dean of Graduate Studies: R.P. Mariella

College of Arts and Sciences

Dean: R.E. Walker

DEPARTMENT OF BIOLOGY
6525 North Sheridan Road
Phone: Chairman: 274-3000 Ext. 765

Chairman: H.W. Manner
Professors: B.J. Jaskoski, H.W. Manner, E.E. Palincsar, W.P. Peters
Associate Professors: A.S. Dhaliwal, R.W. Hamilton, J. Savitz
Assistant Professors: W.C. Cordes, M. Goldie, C.E. Robbins, A.J. Rotermund, B.E.N. Spiroff, K.J. Vener, D.E. Wivagg
Instructors: V.A. Kuta (Laboratory), R.W. Ulbrich (Laboratory)
Teaching Fellows: 9
Research Assistants: 3

Degree Programs: Biology

Undergraduate Degree: B.S.
Undergraduate Degree Requirements: 34-37 Biology with 6 hours Mathematics, 4 courses in Chemistry and 2 courses in Physics

Graduate Degree: M.S.
Graduate Degree Requirements: 24 hours coursework, thesis, one foreign language

Graduate and Undergraduate Courses Available for Graduate Credit: Invertebrate Zoology, General Virology, Evolution, Vertebrate Physiology, General Physiology, Physiological Ecology, Physiological Ecology, Introduction to Nematology, Population Ecology, Developmental Biology II, Plant Growth and Development, Mineral Nutrition of Plants, Plant Cell Physiology, Advanced Genetics, Cell Biology, Radiation Biology, Experimental Morphogenesis, Protozoology, Comparative Animal Physiology, Advanced Parasitology, Limnology, Zoological Taxonomy, History and Philosophy of Biology,Vertebrate Morphology, Photosynthesis in Higher Plants, Molecular Biology and Biochemistry, Research Techniques, Entomology,

School of Dentistry

2160 South First Avenue Phone: (312) 531-3600
Maywood, Illinois 60153
Dean: R.A. Suriano

DEPARTMENT OF ANATOMY
Phone: Chairman: (312) 531-3706 Others: 531-3600

Chairman: J.M. Gowgiel
Professor: A. Leshin (Clinical)
Associate Professors: N. Brescia (Clinical), L. Erikson (Clinical), J.M. Gowgiel
Assistant Professors: M.L. Kiely, W. Schoenheider (Clinical) P.S. Ulinski
Instructors: T. Leischner (Clinical), F. Honig (Clinical) J. Tomasczewski (Clinical)

Degree Program: Oral Biology

Graduate Degree: M.S.
Graduate Degree Requirements: Master: 2 years program - 16 hours

Graduate and Undergraduate Courses Available for Graduate Credit: Advanced General Histology, Anatomy of the Head and Neck, Growth and Physiology of the Masticatory Apparatus, Advanced Oral Pathology, Concepts in Physiology, Bio-statistics and Experimental Design Oral Physiology, Advanced Pharmacology, Radiation

Biology, Oral Microbiology, Genetics, Anthropology, and Paleontology, Neuroanatomy, Dissection of Experimental Animals, Immunobiology, Experimental Pathology

DEPARTMENT OF BIOCHEMISTRY
Phone: Chairman: (312) 531-3577 Others: 531-3578

Chairman: G.W. Rapp
Professor: G.W. Rapp
Assistant Professors: L.L. Yuan, P.D. Oeltgen
Teaching Fellows: 1

Degree Program: Oral Biology

Graduate Degree: M.S.
Graduate Degree Requirements: Not Stated

Graduate and Undergraduate Courses Available for Graduate Credit: Oral Biochemistry, Nutrition

DEPARTMENT OF HISTOLOGY
Phone: Chairman: (312) 531-3705

Chairman: R.J. Pollock, Jr.
Associate Professor: R.J. Pollock, Jr.
Assistant Professors: H.D. McReynolds, C. Siraki
Instructors: G. Dhalinwal (Clinical), L. Erikson (Clinical) R. Nolan (Clinical), W. Walzak (Clinical),

Degree Program: Oral Biology

Graduate Degree: M.S.
Graduate Degree Requirements: Not Stated

Graduate and Undergraduate Courses Available for Graduate Credit: Advanced General Histology, Advanced Oral Histology and Embryology

DEPARTMENT OF MICROBIOLOGY
Phone: Chairman: (312) 531-3520 Others: 531-3000

Chairman: J.V. Madonia
Associate Professor: J.V. Madonia
Assistant Professor: D. Birdsell
Instructors: S. Gelbart, D. Birdsell, J. Hagan, C. Wu, P. VanGorder

Degree Program: M.S.
Graduate Degree Requirements: Not Stated

DEPARTMENT OF PHYSIOLOGY
(No reply received)

Stritch School of Medicine
2160 South First Avenue Phone: (312) 531-3000
Maywood, Illinois 60153
Dean: J.A. Wells

DEPARTMENT OF ANATOMY
(No reply received)

DEPARTMENT OF BIOCHEMISTRY AND BIOPHYSICS
Phone: Chairman: (312) 531-3360 Others: 531-3000

Chairman: H.J. McDonald
Professors: E.W. Bermes, J. Bernsohn, M.V. L'Heureux, H.J. McDonald
Associate Professors: S. Aktipis, A.A.C. Dietz, P.P. Hung, S. Keresztes-Nagy (Adjunct), N.C. Melchior, H.G. Shepherd (Adjunct)
Assistant Professors: M.A. Collins, L.J. Crolla, J.M. Goldberg, M.D. Manteuffel, A. Frankfater, R.M. Schultz
Lecturer: W.H. Holleman
Teaching Fellows: 8
Research Assistants: 12
Graduate Student Assistants: 10

Degree Programs: Biochemistry, selected areas of Biophysics

Graduate Degree: M.S., Ph.D.
Graduate Degree Requirements: Master: Minimum of one academic year in residence; completion of 36 quarter hours of graduate work; Submission of an acceptable thesis Doctoral: Three years of academic work; 108 quarter hours of graduate work beyong the bachelor's degree; must show competence in one pertinent modern foreign language, must sustain written examination for advancement to candidacy for Ph.D. degree; submission of acceptable dissertation

Graduate and Undergraduate Courses Available for Graduate Credit: General and Medical Biochemistry (U), Advanced Biochemistry, Biochemistry and Chemistry of Hormones, Clinical Chemistry, Enzyme Chemistry, Neurochemistry, Biochemistry of Nucleic Acids, Molecular Biology, Radioactive Tracer Techniques, Physical Biochemistry, Biophysics; Bio-Organic Chemistry, Seminar in Biochemistry, Research in Biochemistry

DEPARTMENT OF MICROBIOLOGY
Phone: Chairman (312) 531-3384 Others: 531-3000

Chairman: H.J. Blumenthal
Professors: H.J. Blumenthal, T. Hashimoto, G. Plummer, W. Yotis
Associate Professor: C. Lange
Assistant Professors: D. Birdsell, F. Montiel, I. Stern (Adjunct), M. Stodolsky, J.Vice
Teaching Fellows: 11
Research Assistants: 3

Degree Programs: Microbiology

Graduate Degrees: M.S., Ph.D.
Graduate Degree Requirements: Master: General and Medical Microbiology, Methods and Techniques in Microbiology, Biochemistry plus at least two advanced courses in Microbiology, research, thesis, profiency examination in elementary statistics Doctoral: requirements for M.S. plus at least two additional courses in microbiology, one course outside department, foreign language (one) examination, Comprehensive Written and Oral Examination, and Dissertation

Graduate and Undergraduate Courses Available for Graduate Credit: General and Medical Microbiology, Methods and Techniques in Microbiology, Clinical Microbiology, Virology, Immunology-Immunochemistry, Microbial Physiology, Microbial Metabolism, Microbial-Cytology and Ultrastructure, Microbial Genetics, Special Topics, Current Literature

DEPARTMENT OF PHARMACOLOGY AND EXPERIMENTAL THERAPY
Phone: Chairman: (312) 531-3261 Others: 531-3000

Chairman: A.G. Karczmar
Professors: J. Bernsohn, J R. Davis, A.G. Karczmar, S. Nishi, Y.T. Oester, R.S. Geiger (Adjunct)
Associate Professors: A.H. Friedman, C.L. Scudder, R.S.Schmidt
Assistant Professors: S. Glisson, R. Jacobs
Visiting Lecturers: 3
Teaching Fellows: 6
Research Assistants: 2

Degree Program: Pharmacology

Graduate Degree: M.S., Ph.D.
Graduate Degree Requirements: Master: Not Stated Doctoral: Not Stated

Graduate and Undergraduate Courses Available for Graduate Credit: Medical School Pharmacology, Neuroscience, Toxicology, Qualitative Methods in Pharmacology,

Journal Club, Teaching of Pharmacology, Autonomic Nervous System Pharmacology, Cellular Neurophysiology, Pharmacology of the Central Nervous System, Systematic Approach to Drugs - Invertebrate, Systematic Approach to Drugs - Vertebrate, Somatic Pharmacology

PHYSIOLOGY DEPARTMENT
Phone: Chairman: (312) 531-3330 Others: 531-3330

Chairman: W.C. Randall
Professors: A.R. Dawe (Adjunct), J.P. Filkins, C.N. Peiss
Associate Professors: R.D Wurster
Assistant Professors: B. Braverman, P.B. Dobrin (Adjunct), G.P. Pollock
Research Associates: 3

Degree Program: Physiology

Graduate Degrees: M.S, Ph.D.
Graduate Degree Requirements: Not Stated

Graduate and Undergraduate Courses Available for Graduate Credit: General, Medical Physiology, Neurosciences, Experimental Mammalian Physiology, Experimental Neuroscience, Experimental Neuroscience, Pathophysiology, The Teaching of Physiology, Techniques - Methods and Techniques in Physiological Research, Methods and Techniques in Physiological Research, Statistical Methods and Experimental Design, Electronic Instrumentation, Advanced Physiological Data Processing, Physiological Control Systems - Control Systems Theory, Physiology of Motor Control, Control of Cardiopulmonary Function, Physiology of Temperature Regulation, Regulation of Host-Defense Systl m, Cardiovascular System - Physiology of Circulation, Advanced Physiology of the Heart, Physiology of Circulatory Shock, Cardiovascular Journal Club Metabolism and Endocrine Function - Renal Physiology, Cell and Metabolic Physiology, Endocrine Physiology Nervous System - Physiology of Autonomic Nervous System, Receptor Physiology, Central Sensory Mechanisms, Physiology of Behavior

McKENDREE COLLEGE
(No reply received)

MACMURRAY COLLEGE
Jacksonville, Illinois 62650 Phone: (217) 245-6151

BIOLOGY DEPARTMENT
Phone: Chairman: Ext. 240

Chairman: J. Avery
Professor: R.E. Freiburg
Associate Professor: J. Avery, M.W. Freiburg, J. Husa

Degree Program: Biology

Undergraduate Degrees: B.S., B.A.
Undergraduate Degree Requirements: 7 1/2 courses in Biology (30 semester hours), 2 courses in Chemistry (8 hours), 2 courses in Physics or Mathematics (8 hours)

MILLIKIN UNIVERSITY
1184 West Main Street Phone: (217) 423-3661
Decatur, Illinois 62522
Dean of College: M.H. Forbes

DEPARTMENT OF BIOLOGY
Phone: Chairman: 423-3661 Ext. 230

Chairman: L.C. Shell
Professor: L.C. Shell
Associate Professor: W.P. Anderson
Assistant Professor: N.M. Baird, N.H. Jensen, J.A. Smithson
Instructor: J.W. Fisher
Teaching Fellows: 1

Degree Programs: Biology, Medical Technology

Undergraduate Degrees: B.A., B.S.
Undergraduate Degree Requirements: B.A. in Biology 32 hours General Education, 32 hours Biology, 12 hours Foreign Language, 4 hours Literature

MONMOUTH COLLEGE
700 E, Broadway Phone: (309) 457-2021
Monmouth, Illinois 61462

DEPARTMENT OF BIOLOGY
Phone: Chairman: (309) 457-2021

Chairman: R.H. Buchholz
W.P. Pressly Professor of Natural Sciences: J.J. Ketterer
Professors: D.C. Allison, M. Bowman, J.J. Ketterer

Field Stations/Laboratories: One Ecological Field Station

Degree Program: Biology

Undergraduate Degree: B.A.
Undergraduate Degree Requirements: 8 Biology courses beyond 101-102 College Biology, 2 courses in Physics, Analytical Chemistry, Organic Chemistry

MUNDELEIN COLLEGE
6363 Sheridan Road Phone: (312) 262-8100
Chicago Illinois 60660
Dean of College: Sr. S. Rink

DEPARTMENT OF BIOLOGY
Phone: Chairman: (312) 262-8100

Chairman: Sr. M.N. Murphy
Professor: Sr. C. Bodman
Associate Professor: Sr. M.N. Murphy
Assistant Professors: J.W. Bisbee, D.M. Reisa

Degree Program: Biology

Undergraduate Degrees: B.S., B.A.
Undergraduate Degree Requirements: B.S. requires 33 hours (11 courses). A student must take Principles of Biology, Biology of Organisms, Animal Structure and Function (a two-term sequence), The Plant Kingdom, a course in Physiology, and Genetics. Cognate requirements include Inorganic Chemistry, Organic Chem istry, two courses in Mathematics

NORTH CENTRAL COLLEGE
(No reply received)

NORTH PARK COLLEGE
5125 No. Spaulding Avenue Phone: (312) 583-2700
Chicago, Illinois 60625
Dean of College: C.H. Edgren

DEPARTMENT OF BIOLOGY
Phone: Chairman: (312) 583-2700 Ext. 372

Chairman: E.J. Kennedy
Professors: E.J. Kennedy, R.D. Lowell, C.J. Peterson
Associate Professors: A. Bartha, R. Tofte
Assistant Professors: A.T. Johnson, L. Knipp

Degree Programs: Biology, Medical Technology

Undergraduate Degree: B.S.
Undergraduate Degree Requirements: 9 courses in Bio-

logy (36 semester hours) and 3 courses in Chemistry (12 semester hours)

NORTHERN ILLINOIS UNIVERSITY

DeKalb, Illinois 60178 Phone: (815) 753-1000
Dean of Graduate Studies: J.A. Rutledge
Dean of College: P.S. Burtness

DEPARTMENT OF BIOLOGICAL SCIENCES
Phone: Chairman: (815) 753-1753

Chairman: J.A. McCleary
Professors: W.E. Briles, R. Fry (Adjunct), D. Grahn (Adjunct), W.D. Gray (Adjunct), M. Hall (Adjunct), K. Hardman (Adjunct), B.N. Jaroslow (Adjunct), M.T. Jollie, D.L. Lynch, J.A. McCleary, W.J. McIlrath, S. Mittler, K. Norstog, K.V. Prahlad, C.J. Rohde, O.A. Schjeide, S. Skok, W. Southern, H. Kubitschek (Adjunct), C. Peraino, (Adjunct), Y. Rahman (Adjunct), F. Schlenk (Adjunct), W. Sinclair (Adjunct), B. Tsotsis (Adjunct), B. Webb (Adjunct)
Associate Professors: F. Abdel-Hameed, J.C. Bennett, E. Frampton, D.W. Greenfield, J.H. Grosklags, A. Hampel, K. Harmet, C. Mathers, M.J. Starzyk, B. Von Zellen, J.H. Zar
Assistant Professors: J. Brower, K. Clayton, M. Fenwick, W. Garthe, Q. Haning, L. Hanzely, R. Knauel, J. Mitchell, R. Pearson, P.D. Sorensen, R. Wittrup
Teaching Fellows: 22
Research Assistants: 4

Degree Program: Biology

Undergraduate Degree: B.S.
Undergraduate Degree Requirements: General Education 37 hours, Departmental requirements (27-28 hours), Extradepartmental requirements (26-28 hours), Prerequisites to above (8 hours), Electives (23-26 hours)

Graduate Degree: M.S.
Graduate Degree Requirements: Master: Thesis option, or non-thesis option. Biological Chemistry, Biostatistical Analysis, Seminar (taken each semester in residence)

Graduate and Undergraduate Courses Available for Graduate Credit: Biological Conservation, Genetics and Cytogenetics, Cytogenetics Laboratory, Plant Anatomy, Mycology, Microbial Physiology, Fresh-water Algae, Applied Microbiology, Human Heredity, Plant Pathology, Edaphology, Plant Taxonomy, Immunobiology, Evolution, Histology, Vertebrate Biology, Animal Parasitology, Protozoology, Ornithology, Entomology, Embryology, Comparative Physiology, Icthyology, Mammalogy, Optical and Instrumentation Methods, Comparative Vertebrate Ethology, Wildlife Ecology, Endocrinology, Insect Physiology, Invertebrate Zoology, Comparative Reproductive Physiology, Invertebrate Paleontology

NORTHWESTERN UNIVERSITY

Evanston, Illinois 60201 Phone: (312) 492-3741
Dean of Graduate Studies: R.H. Baker

College of Arts and Sciences

Dean: R.H. Weingartner

DEPARTMENT OF BIOLOGICAL SCIENCES
Phone: Chairman: (312) 492-5521 Others: 492-3741

Chairman: N.E. Welker
Morrison Professor of Biological Sciences: F.A. Brown, Jr.
William Dearing Professorship of Biological Sciences:
Professors: R.F. Acker, F.A. Brown, Jr., A.L. Burnett, H.R. Friederici, I.K. Gilbert, J.W. Kauffman, R.C. King, J.A. Lippincott, R.R. Novales, W.O. Pipes, A. Routtenberg, J.M. Whitten, A. Wolfson
Associate Professors: C. Enroth-Cugell, R.C. Gesteland, E. Goldberg, T.K. Goldstick, S.B. Simpson Jr.,
Assistant Professors: A.J. Beattie, D.C. Culver, E.R. Heithaus, R.C. MacDonald, J. Oschman
Instructor: D. Igelsrud
Research Assistants: 10
Postdoctorals: 9

Degree Program: Biological Sciences

Undergraduate Degrees: B.A.
Undergraduate Degree Requirements: 45 course units including 2 units of calculus, 3 units general chemistry, 2 units organic chemistry, 3 units physics, 9 units biological sciences

Graduate Degrees: M.S., Ph.D.
Graduate Degree Requirements: Master: Nine graduate level courses, Written Master's examination. Doctoral: Minimum of 3 years of full time study. No foreign language requirement. 2 years of 2 quarters each teaching experience as a Teaching Assistant Preliminary Examination, Dissertation Research, Thesis Defense

Graduate and Undergraduate Courses Available for Graduate Credit: Animal Behavior (U), Honors Seminar III (U), Animal Physiology (U), Physical Principals of Biology (U), Comparative Invertebrate Endocrinology (U), Physiology of Reproduction (U), Invertebrate Biology (U), Vertebrate Histology (U), Applied Microbiology (U), Cell Physiology (U), Vertebrate Endocrinology (U), Cellular Endocrinology (U), Submicroscopic Cytology (U), Microbiology: Virology (U), Field Biology (U), Biogeography (U), Special Senses (U), Microphysiology of Excitable Tissues (U), The Biology of Populations: Ecological (U),. The Biology of Populations: Genetic (U), Developmental Biology (U), Independent Study (U), Honors Seminar IV, Biological Rhythms, Orientation and Clocks

Medical School

Dean: J.E. Eckenhoff

DEPARTMENT OF ANATOMY
Phone: Chairman: (312) 649-8249 Others: 649-8250

Chairman: W. Bondareff
Robert L. Rea Professor: W. Bondareff
Professors: B.J. Anson (Emeritus), L.B. Arey (Emeritus), W. Bondareff, H.A. Davenport (Emeritus), A.L. Farbman, W.J.S. Krieg (Emeritus)
Associate Professors: F. Gonzales, M.F. Orr, J.J. Pysh, N.S. Rafferty
Assistant Professors: R. Berry, J. Daniel, J. Disterhoft, J. Dorn, R. Hinkley, L. Miller, R. Nayyar, R. Perkins, J. Story, A. Telser
Instructors: R.S. Baratz, R. Crissman
Visiting Lecturers: 12
Teaching Fellows: 1
Research Assistants: 2

Degree Program: Anatomy

Graduate Degree: Ph.D.
Graduate Degree Requirements: Master: Students are not encouraged to seek admission when the terminal master's degree is the objective: Doctoral: 1. Demonstration of proficiency in three of four subdisciplines of anatomy recognized by the department: Human Anatomy: Cell Biology, Neurobiology: Developmental Biology, 2. Demonstration of knowledge of concepts basic to physiology and biochemistry, 3. Demonstration of comprehensive reading knowledge in French or German. Another language may be substituted, subject to the approval of the sponsor and the graduate committee 4. Satisfactory completion of a written and oral qualifying examination 5. Satisfactory completion of a dissertation based upon original research 6. Oral examination in defense of the dissertation

Graduate and Undergraduate Courses Available for Graduate Credit: Microscopic and Submicroscopic Anatomy, Introduction to Medical Neuroscience, Gross Anatomy, Anatomy of the Head and Neck, Electron Microscopy, Cell Motility, Biology of Cells and Tissue in Vitro, Special Problems Utilizing in Vitro, Neurobiology, Seminar in Recent Concepts in Anatomical Sciences

DEPARTMENT OF BIOCHEMISTRY
Phone: Chairman: (312) 649-8080

Chairman: J.W. Corcoran
Professors: C. Chen, J.W. Corcoran, N. Freinkel (Adjunct), J.E. Garvin, R.A. Jungmann, A. Veis, E.A. Zeller
Associate Professors: E.L. Coe, G.H. Czerlinski, J. Hefferren, T. Inouye, M. Marini, J. Schweppe (Adjunct), W. Wells
Assistant Professors: B. Anderson, S. Ghosh, A. Goldstone (Adjunct), P.F. Hollenberg, H.C. Diefer, R.D. Mavis, J. Majer, J.J. Mieyal (Adjunct), J.H. Pincus (Adjunct)
Instructors: K.L. Arora, P.C. Lee
Research Assistants: 30

Degree Program: Biochemistry

Graduate Degrees: M.S., Ph.D.
Graduate Degree Requirements: Master: Basic core courses Doctoral: Basic core courses, 4 advanced topic courses, Completed research thesis

Graduate and Undergraduate Courses Available for Graduate Credit: Biochemistry of Cellular Systems, Physical Biochemistry, Enzymology, Molecular Mechanisms, Biochemistry and Chemistry of Cellular and Subcellular Systems, Biochemical Techniques, Advanced Topics in Molecular Biology, Advanced Topics in Cell Biology, Advanced Topics in Physical Biochemistry and Biophysics, Advanced Topics in Physiological and Developmental Biochemistry, Seminar, Biochemistry Research Workshop, Special Topics in Bioorganic Chemistry, Seminar in Structure and Function of Biological Membranes, Biomedical Computing, Independent Study, Seminar in Connective Tissue Chemistry and Biology, Biochemical Research, Enzyme Mechanisms

DEPARTMENT OF MICROBIOLOGY
Phone: Chairman: (312) 649-8230 Others: 649-8649

Chairman: G.P. Youmans
Professors: R.D Ekstedt, T.C. Johnson, P.Y. Paterson, H.D. Slade, G.F. Springer, A.S. Youmans, G.P. Youmans
Associate Professors: H. Huramitsu, D. Perry, J.J. Pruzansky, M. Rachmeler
Assistant Professors: J.L. Duncan, B. Halpern
Research Assistants: 5

Degree Probram: Microbiology

Graduate Degree: Ph.D.
Graduate Degree Requirements: Doctoral: Formal courses both required and elective (minimal of eight) Preliminary examination, written and oral. Satisfactory thesis on original work plus a thesis defense examination.

Graduate and Undergraduate Courses Available for Graduate Credit: Graduate Microbiology, Microbial Techniques, Special Topics in Immunology, Special Topics in Microbial Physiology, Special Topics in Virology, Special Topics in Host-Parasite Interaction, Special Topics in Molecular and Microbial Genetics, Special Topics in Cell Physiology, Special topics in Mechanism of Autoimmune Disease, Special Topics in Immediate Hypersensitivity, Biochemistry of Cellular Systems, Biochemistry of Evolution, Molecular Biology of Development

DEPARTMENT OF PHYSIOLOGY
(No reply received)

OLIVET NAZARENE COLLEGE
Kankakee, Illinois 60901 Phone: (815) 939-5011
Dean of Graduate Studies: E. Eustice
Dean of College: W.E. Snowbarger

DEPARTMENT OF BIOLOGICAL SCIENCES
Phone: Chairman: (815) 939-5309 Others: 939-5011

Chairman: W.D. Beaney
Professors: D.J. Strickler
Associate Professors: W.D. Beaney, R.E. Hayes
Assistant Professors: H.F. Fulton, J.P. Marangu, R. W. Wright
Teaching Assistants: 14

Degree Programs: Biological Education, Biology, Botany, Zoology

Undergraduate Degree: B.S.
Undergraduate Degree Requirements: 30 hours and supporting courses in Chemistry and Mathematics

PRINCIPIA COLLEGE
Elsah, Illinois 62028 Phone: (618) 466-2131
Dean of College: K. Johnston

DEPARTMENT OF BIOLOGY
Phone: Chairman: (618) 466-2131 Ext. 357

Chairman: J.F. Wanamaker
Assistant Professor: D. Warren
Instructor: C. Sparwasser
Research Assistants: 2

Degree Program: Biology

Undergraduate Degree: B.S.
Undergraduate Degree Requirements: Not Stated

QUINCY COLLEGE
1831 College Avenue Phone: (217) 222-8020
Quincy, Illinois 62301
Dean of College: M. Crosby

DEPARTMENT OF BIOLOGICAL SCIENCES
Phone: Chairman: (217) 222-8020

Chairman: G.H. Schneider
Professors: F.J.L. Ostdiek
Associate Professors: A. Pogge, G.H. Schneider
Assistant Professors: T. Godish, F.E. Middendorf, J. Natalini
Visiting Lecturers: 1

Degree Programs: Biological Sciences, Medical Technology

Undergraduate Degree: B.S.
Undergraduate Degree Requirements: 35 semester hours in Biological Sciences, 2 semesters of Inorganic Chemistry, 2 semesters of Organic-Biochemistry (recommended), 2 semesters of Physics, Mathematics through Calculus 1, 1 semester of Computer Science, 2 semesters of modern Foreign Language, General Education Courses, Total 127 semester hours

ROCKFORD COLLEGE
3050 East State Street Phone: (815) 226-4133
Rockford, Illinois 61101
Dean of Graduate Studies: G. Thompson
Dean of College: G. Wesner

DEPARTMENT OF BIOLOGY

Phone: Chairman: (815) 226-4133 Others: 226-5050

Chairman: G.L. Forman
Associate Professor: G.L. Forman
Assistant Professor: A.L. Carley, D. Pippitt
Instructor: M. Pippitt (1/2 time)

Degree Programs: Biology, Botany, Zoology

Undergraduate Degree: B.A., B.S.
Undergraduate Degree Requirements: B.A. - General Biology, Botany, Zoology, Genetics, undergraduate research, Senior Seminar, plus a minimum of three additional courses in the department (upper-level courses) B.S. - Same except that additional courses (to be chosen by student) are required within the biology curriculum

ROOSEVELT UNIVERSITY
430 S. Michigan Avenue Phone: (312) 341-3500
Chicago, Illinois 60605
Dean of Graduate Studies: R. Rosen
Dean of College: R. Carnes

DEPARTMENT OF BIOLOGY
Phone: Chairman: (312) 341-3676 Others: 341-3500

Chairman: G.R. Seaman
Professors: J. Corbett, H. Nelson, G. Seaman
Associate Professor: G. Edwin
Assistant Professor: L. Bradford, D. Gorecki, J. Holleman, L. Ichinose
Visiting Lecturers: 1

Degree Programs: Biology (graduate programs in Microbiology and Biochemistry)

Undergraduate Degrees: B.A., B.S.
Undergraduate Degree Requirements: B.A., 32 semester hours Biology, B.S., 32 Semester hours Biology, Minor in Chemistry (1 year Organic, or 1 semester Organic and 1 semester Organic and 1 semester Quantitative and 1 year Physics

Graduate Degree: M.S.
Graduate Degree Requirements: Master: 30 semester hour courses, including 6 semester hours thesis (research and writing)

Graduate and Undergraduate Courses Available for Graduate Credit: Vertebrate Biology (U), Plant and Animal Metabolism (U), Parasitology (U), Biology of the Invertebrates (U), General Entomology (U), Plant Physiology (U), Evolution of Fungi and Primitive Vascular Plants (U), Aquatic Ecosystems (U), Environment and Man (U), Primatology (U), The Biology of Sex and Reproduction (U), General Genetics (U), Microbiology (U), Industrial Microbiology (U), Clinical Microbiology and Immunology (U), Research Techniques in Microbiology (U), Biological Electron Microscopy (U), Problems in Biology (U), Physical Methods in Biology, Physiology of Protozoa, Host-Parasite Interrelationships, General Endocrinology, Biogeography, Bioenergetics, Nucleic Acids, Immunobiology, Microbial Metabolism, Microbial Fermentations, General Virology

ROSARY COLLEGE
7900 W. Division Street Phone: (312) 369-6320
River Forest, Illinois 60305

DEPARTMENT OF BIOLOGY
Phone: Chairman: (312) 369-6320

Chairman: L.K. Gallo
Associate Professor: P. Schulz
Instructor: L.K. Gallo

Degree Program: Biology

Undergraduate Degree: B.A.
Undergraduate Degree Requirements: Biology Major - 31 units including 10 units of Biology, 2-4 units of Chemistry, and 1 unit Physics

ST. XAVIER COLLEGE
103rd and Central Park Avenue Phone: (312) 779-3300
Chicago, Illinois 60655
Dean of College: E.G. Bollinger

DEPARTMENT OF SCIENCE
Phone: Chairman: Ext. 245

Chairman: W.J. Buckley
Associate Professors: W.J. Buckley, M. Johnson
Assistant Professors: S.D Boyer, M.A. Grant, E.L. Philip

Degree Program: Biology

Undergraduate Degree: B.S.
Undergraduate Degree Requirements: Not Stated

SHIMER COLLEGE
Mt. Carroll, Illinois 61053 Phone: (815) 244-2817
Dean of College: D. Hipple

SCIENCE AREA
Phone: Chairman: (815) 244-2817 Ext. 44

Chairman: R. Atkins
Professors: R. Atkins, A. Kirsch, D. Mron, D. Sakuras, J. Myers, T. Burgers, B. Norman, M. Micola, R. Cole
Visiting Lectures: 2
Teaching Fellows: 1
Research Assistants: 2

Degree Program: Biology

Undergraduate Degrees: B.A., B.S
Undergraduate Degree Requirements: 1 Foreign Language, 14 General Education, 1 Physical Education, 1 English, 4 Social Science, 4 Humanities, 4 Natural Science, 2 Philosophy (History) and 8 Biology courses

SOUTHERN ILLINOIS UNIVERSITY AT CARBONDALE
Carbondale, Illinois 62901 Phone: (618) 453-3344
Dean of Graduate Studies: T.O. Mitchell

College of Liberal Arts and Sciences
Dean: J.C. Guyon

DEPARTMENT OF BOTANY
Phone: Chairman: (618) 536-2080 Others: 536-2331

Chairman: R.H. Mohlenbrock
Professors: W.C. Ashby, R.H. Mohlenbrock, A.J. Pappelis, W.E. Schmid, J. Verduin, J.W. Voight,
Associate Professors: L.C. Matten, O. Myers, D. Tindall, D. Ugent, J.H. Yopp
Assistant Professors: P.A. Robertson, B.C. Stotler, R.E. Stotler
Instructor: J.W. Richardson
Teaching Fellows: 14
Research Assistants: 4

Field Stations/Laboratories: Rocky Mountain Field Station-Red Lodge, Montana, Tropical Field Station - Belize, British Honduras

Degree Programs: Biological Sciences, Botany

Undergraduate Degrees: B.A., B.S.
Undergraduate Degree Requirements: 30 semester hours in Botany, including Genetics, Ecology, Morphology, Systematics, Physiology, 1 year of Chemistry with

introduction to Organic, 1 year of Mathematics, 2 years Foreign Language

Graduate Degrees: M.A., M.S., Ph.D.
Graduate Degree Requirements: Master: 30 semester hours beyond the Bachelor's Doctoral: 90 semester hours beyond the Bachelor's 2 foreign languages, or approved substitutes

Graduate and Undergraduate Courses Available for Graduate Credit: Anatomy, Algae, Fungi, Bryology, Taxonomy and Ecology of Bryophytes and Lichens, Morphology of Ferns, Morphology of Seed Plants, Paleobotany, Botanical Microtechnique, Advanced Plant Physiology, Grassland Ecology, Forest Ecology, Analysis and Classification of Vegetation, Tropical Ecology, Field Studies in Latin America, Field Studies in the Western United States, Elements of Taxonomy, Plant Geography, Pathology, Forest Pathology, Science Process and Concepts for Teachers of Grades N-8, Palynology, Advanced Angiosperm Taxonomy, Advanced Plant Genetics, Cytology, Cytogenetics, Plant Growth and Morphogenesis, Energetics of Aquatic Ecosystems, Biosystematics, Experimental Ecology, Upland Flora, Lowland Flora, Advanced Palynology

DEPARTMENT OF PHYSIOLOGY
Phone: Chairman: 453-2583

Chairman: R.W. Stacy
Professors: E.Borkon (Adjunct), T.T. Dunagan, F. Foote, G.H. Gass, H.M. Kaplan, J.E. Moreland, J P. Miranti (Adjunct), A.W. Richardson
Associate Professors: D.M. Miller, R.A. Sollberger, L.E. Strack, E.H. Timmons
Assistant Professors: R.A. Browning, R.E. Falvo, J.H. Myers, L.G. Nequin, G.T. Taylor, W.M. Yaw
Teaching Fellows: 10
Research Assistants: 6

Degree Programs: Anatomy and Physiology, Biochemistry and Biophysics, Bioengineering, Physiological Sciences, Physiology

Undergraduate Degree: B.S.
Undergraduate Degree Requirements: 20 semester hours in Physical Science, 30 semester hours in Physiology and related sciences, General university requirements

Graduate Degrees: M.Sc., Ph.D.
Graduate Degree Requirements: Master: 30 semester hours graduate credit, written and oral examination, Thesis Doctoral: 2 years beyond Master's, written examination (Qualifying), Dissertation, Oral Examination

Graduate and Undergraduate Courses Available for Graduate Credit: Advanced Human Anatomy (U), Mammalian Physiology (U), Experimental Animal Surgery (U), Anatomy and Physiology of Speech and Hearing (U), Pharmacology (U), Comparative Physiology (U), Comparative Endocrinology (U), Cellular Physiology (U), Biophysics (U), Electron Microscopy (U), Biomedical Electronics (U), Advanced Endocrinology, Advanced Cellular Physiology, Advanced Comparative Physiology, Advanced Biophysics, Physiological Techniques, Advanced Physiology Topics, Readings, Thesis, Dissertation

DEPARTMENT OF ZOOLOGY
Phone: Chairman: (618) 536-2314

Professors: R.E. Blackwelder, R.A. Brandon, H.I. Fisher (Emeritus), E. Galbreath, H.J.Haas, W.D. Klimstra, W.M Lewis, J. Martan, H.J. Stains
Associate Professors: J.A. Beatty, W.G. Dyer, D.C. Englert, G. Garoian, W.G. George, E. LeFebvre, J.E. McPherson, B.A. Shepherd, J.B. Stahl, G. Waring
Assistant Professors: T. Anthoney, R.C. Heidinger, A. Paparo
Teaching Fellows: 22
Research Assistants: 10
Special Doctoral Fellowships: 2

Degree Programs: Anatomy and Histology, Animal Behavior Ecology, Entomology, Ethology, Fisheries and Wildlife Biology, Genetics, Histology, Limnology, Parasitology, Wildlife Management, Zoology

Undergraduate Degrees: B.A., B.S.
Undergraduate Degree Requirements: 37 semester hours of Zoology which must include Genetics, Environmental Biology, Invertebrate, Vertebrate, and Embryology as well as senior seminar

Graduate Degrees: M.A., M.S., Ph.D.
Graduate Degree Requirements: Master: To be determined by Committee, Doctoral: To be determined by Committee

Graduate and Undergraduate Courses Available for Graduate Credit: Natural History of Invertebrates (U), Natural History of Vertebrates (U), Protozoology (U), Parasitology (U), Herpetology (U), Vertebrate Histology (U), Vertebrate Paleontology (U), The Invertebrates (U), Freshwater Invertebrates (U), Entomology (U), Limnology (U), Histological Techniques (U), Comparative Endocrinology (U), Emergence of Order in Biological Systems (U), Game Birds (U), Mammalogy (U), Ichthyology (U), Fish Management (U), Ornithology (U), Wildlife Biology (U), Animal Behavior (U), Helminthology, Animal Geography, Advanced Entomology, Advanced Invertebrates, Advanced Limnology, Cytology, Factors in Animal Reproduction, Osteology, Game Mammals, Fish Culture, Physiological Ecology, Population Ecology, Population Genetics, Advanced Taxonomy, Zoological Literature, Zoology Graduate Seminar, Seminar in Animal Behavior, Seminar in Ecosystems, Seminar in Wetland Ecology, Seminar in Wildlife Ecology, Seminar in Parasitology, Seminar in Amphibia, Seminar in Aquaculture, Seminar in Fish Management

School of Agriculture
Dean: G.H. Kroening

DEPARTMENT OF ANIMAL INDUSTRIES
Phone: Chairman (618) 453-2329

Chairman: H.H. Hodson, Jr.
Professors: B.L. Goodman, S.W. Hinners, G.H. Kroening, H.H. Olson
Associate Professors: H.H. Hodson, Jr. W.G. Kammlade, D.D. Lee, Jr., L.E. Strack, G.H. Waring
Assistant Professors: C.L. Hausler, J.R. Males
Researchers: R.D. Carr, R.L. Francis, J.T. Gholson, G.C. McCoy, R.A. Snyder

Field Stations/Laboratories: University Farms, S.I.U. Carbondale

Degree Programs: Animal Behavior, Animal Husbandry, Animal Industries, Animal Physiology, Animal Science, Dairy Science, Poultry Husbandry, Poultry Science

Undergraduate Degree: B.S.
Undergraduate Degree Requirements: Not Stated

Graduate Degree: M.S.
Graduate Degree Requirements: Not Stated

DEPARTMENT OF FORESTRY
Phone: Chairman: (618) 453-3341

Chairman: A.A. Moslemi
Professors: J.W. Andersen, D.R. McCurdy, A.A. Moslemi, P.A. Yambert
Associate Professors: K.C. Chilman, C.K. Losche (Adjunct), P.F. Nowak, C.C. Myers, R.E. Phares (Adjunct), P.L. Roth, D.P. Satchell
Assistant Professors: D. Baumgartner (Adjunct), C.F. Bey (Adjunct), C.A. Budelsky, P.Y.S. Chen (Adjunct),

G.A. Cooper, Jr. (Adjunct), R.S. Ferell, J.S. Fralish, D.T. Funk (Adjunct), G.R. Gaffney, F.H. Kung, R.D. Lindmark (Adjunct), H.N. Rosen (Adjunct), G.T. Weaver
Visiting Lecturers: 4
Teaching Fellows: 3
Research Assistants: 11

Field Stations/Laboratories: Tree Improvement Center - Carbondale

Degree Programs: Forestry, Silviculture, Forest Resources

Undergraduate Degree: B.S. (Forestry)
Undergraduate Degree Requirements: 120-130 semester hours in specified and elective course work, after graduation from high school.

Graduate Degree: M.S.
Graduate Degree Requirements: Master: 30-35 semester hours of graduate course work beyond bachlors. Includes thesis requirement (6 semester hours), Advising Committee supervises work Doctoral: Cooperative interdisciplinary Ph.D. with Botany, Geography, Economics

Graduate and Undergraduate Courses Available for Graduate Credit: Forest Management for Wildlife (U), Forest Resources Decision Making (U), Forest Resources Administration and Policy (U), Forest Resources Economics (U), Forest Genetics I (U), Forest Resources Management (U), Forest Land-use Planning (U), Marketing of Forest Products (U), Park and Wildlands Management (U), Park and Wildlands Management Field Trip (U), Wildlands Watershed Management (U), Regional Silviculture (U), Urban Forestry Management (U), Natural Resources Inventory (U), Forest Soils (U), Principles of Research, Graduate Seminar, Advanced Forest Resources Economics, Forest Genetics II, Advanced Forest Resources, Management, Advanced Park Planning, Recreation Behavior in Wildland Environments, Forest Site Evaluation, Forest Productivity

DEPARTMENT OF PLANT AND SOIL SCIENCE
Phone: Chairman: (618) 543-2496

Chairman: G.D. Coorts
Professors: G.D. Coorts, D.M. Elkins, I.G. Hillyer, U.K. Leasure, J.B. Mowry, H.L. Portz, J.A. Tweedy
Associate Professors: J.H. Jones, O. Myers, Jr., F.J. Olsen, D.J. Satchell
Assistant Professors: R.R. Maleike, D.J. Stucky, E.C. Varsa
Instructor: G. Kapusta
Graduate Assistants: 26

Field Stations/Laboratories: Agronomy Station - Carbondale, Horticulture Station - Carbondale, Plant and Soil Science Research Station - Carbondale, Belleville Research Station - Belleville

Degree Programs: Agronomy, Conservation, Crop Science, Farm Crops, Horticultural Science, Horticulture, Landscape Horticulture, Ornamental Horticulture, Plant Sciences, Soils and Plant Nutrition, Soil Science

Undergraduate Degree: B.S.
Undergraduate Degree Requirements: 45 semester hours of General Studies courses, 45 semester hours of Agriculture courses of which 24 must be in Plant and Soil Science courses, 120 semester hours required for the degree

Graduate Degree: M.S.
Graduate Degree Requirements: Master: 30 semester hours of which 15 must be at the 500 level (no more than 9 semester hours of the 15 may be in unstructured courses). No language required Doctoral: Offered through the Department of Botany as cooperative degree)

Graduate and Undergraduate Courses Available for Graduate Credit: Trends in Agronomy (U), Plant Breeding (U), World Crop Production Problems (U), Crop Physiology and Ecology (U), Forage Crop Management (U), Crop Pest Control (U), Turf Management (U), Greenhouse Management (U), Floriculture (U), Plant Propagation (U), Nursery Management (U), Woody Plant Maintenance (U), Fruit Production (U), Vegetable Production (U), Soil Morphology and Classification (U), Soil Physics (U), Soil Management (U), Fertilizers and Soil Fertility (U), Soil Fertility Evaluation (U), Microbial Process in Soils (U), Radioisotopes, Principles and Practices (U), Weeds--Their Control (U), Principles of Herbicide Action, Growth and Development of Plants, Soil-Plant Relationships, Field Plot Technique

SOUTHERN ILLINOIS UNIVERSITY AT EDWARDSVILLE
Edwardsville, Illinois 62025 Phone: (618) 692-2000
Dean of Graduate Studies: V. Lindsay
Dean of School of Science and Technology: E. Lazerson

DEPARTMENT OF BIOLOGICAL SCIENCES
Phone: Chairman: (618) 692-3928 Others: 692-2000

Chairman: A.C. Zahalsky
Professors: R. Axtell, A. Baich, H. Broadbooks, D. Myer, A. Zahalsky
Associate Professors: N. Davis, R. Keating, F. Kulfinski, M. Kumler, M. Levy, N. Parker, R. Parker, R. Peterson, K. Ratzlaff, J. Thomerson, G. Wittig, S. Nair
Assistant Professor: S. Wood
Visiting Lecturers: 2
Teaching Fellows: 14
Research Assistants: 3

Degree Programs: Animal Physiology, Biochemistry, Biological Education, Biological Sciences, Biology, Medical Technology

Undergraduate Degrees: B.A., B.S.
Undergraduate Degree Requirements: 52 quarter hours in Biology including Cell, Zoology, Botany, Genetics plus at least one course in each of three designated areas: Chemistry through a complete sequence in Organic including some laboratory. College Physics or Mathematics including Statistics and Calculus.

Graduate Degree: M.S.
Graduate Degree Requirements: Minimum of 48 quarter hours as designated by advisory committee. Thesis and non-thesis options available.

Graduate and Undergraduate Courses Available for Graduate Credit: Microbial Physiology (U), Microbial Physiology Laboratory, (U), Cell Organelles and Inclusions (U), Electron Microscopy (U), Advanced Genetics (U), Experimental Embryology (U), Plant Synecology (U), Principles of Parasitism (U), Environmental Microbiology (U), Mammalian Physiology (U), Ethology (U), Integrative Physiology (U), Endocrinology (U), Plant Anatomy (U), Plant Microtechnique (U), Functional Morphology of Vertebrates (U), Field Botany (U), The Algae (U), Field Zoology (U), Principles of Entomology and Insect Pathology (U), Ichthyology (U), Herpetology (U), Ornithology (U), Mammalogy (U), Instructional Regulation, Instructional Innovation in Secondary School Biology, Nucleic Acids, Proteins, Population Genetics, Cytogenetics, Molecular Genetics, Limnology, Plant Geography, Virology, Microbial Metabolism, Physiology of Sense Organs, Mineral Nutrition of Plants, Morphology of the Spermatophytes, Advanced Invertebrates, Helminthology, Advanced Ichthyology

TRINITY CHRISTIAN COLLEGE
Palos Heights, Illinois 60463 Phone: (312) 597-3000

Phone: Chairman: (312) 597-3000 Ext. 57

Assistant Professors: H. Cook, G. VanDyke

Degree Programs: Medical Technology, Biology

Undergraduate Degrees: B.Sc., B.A.
Undergraduate Degree Requirements: 1 year Physics, 2 years Chemistry, 28 semester hours Biology, History, Philosophy, English, Physical Education

TRINITY COLLEGE

2045 Half Day Road Phone: (312) 945-6700
Deerfield, Illinois 60015
Dean of College: J.E. Hakes

DEPARTMENT OF BIOLOGY
Phone: Chairman: (312) 945-6700 Ext. 246

Chairman: G.R. Maple, Jr.
Assistant Professor: G.R. Maple, Jr.
Instructor: S.F. Levi
Laboratory Assistants: 5

Degree Programs: Biology, Medical Technology

Undergraduate Degree: B.A.
Undergraduate Degree Requirements: General Zoology, Botany, Ecology, Genetics, Origins, Comparative Animal Physiology, Biological Science Seminar, General Chemistry, Organic Chemistry, Mathematics (to Calculus), and 12 hours of electives selected in consultation with the department.

UNIVERSITY OF CHICAGO

5801 South Ellis Avenue Phone: (312) 753-1234
Chicago, Illinois 60637
Dean of Graduate Studies: J. Ceithaml

Division of Biological Sciences

(Faculty are shared with Pritzker School of Medicine)

Pritzker School of Medicine

East 58th Street
Chicago, Illinois 60637
Dean: L.O. Jacobson

DEPARTMENT OF ANATOMY
Phone: Chairman: (312) 753-3903 Others: 753-3900

Chairman: R. Singer
The Robert R. Bensley Professor in Biology and Medical Sciences: R. Singer
Professors: S. Altmann, P.P.H. DeBruyn, W L. Doyle, L.M.H. Larramendi, C.E. Oxnard, R. Singer
Associate Professors: W.A. Anderson, C.L. Coulter, D.A. Fischman, B.B. Garber, J.A. Hopson, F. Manasek, L. Radinsky, R. Rhines, L.P. Straus
Assistant Professors: E.S. Kane, R.E. Lombard
Research Assistants: 7

Degree Program: Anatomy

Graduate Degree: Ph.D.
Graduate Degree Requirements: Doctoral: 1) Course work as defined by the Department for the individual student; 2) Passing a preliminary examination emphasizing the field of specialization and cognate fields; 3) Demonstrated competence in a foreign language 4) Completion of a research project forming the basis of a dissertation

Graduate and Undergraduate Courses Available for Graduate Credit: Gross Anatomy, Histology, Neurobiology, Cell Structure and Function, Vertebrate Paleobiology I, II, III, Evolution of the Primates, Neurogenesis, Experimental and Comparative Neurology, Developmental Biology II, The Fine Structure of Developing Systems, Histochemistry, The Origin of Mammals, Molecular Morphology of Biological Systems, Evolution of the Mammals, Bone-Muscle-Joint Biomechanics, The Analysis of Shape and Structure, Morphology of Synapses

DEPARTMENT OF BIOCHEMISTRY
Phone: Chairman: (312) 753-3960 Others: 753-1234

Chairman: D.F. Steiner
Louis Block Professor in the Division of Biological Sciences: J. Fried
Louis Block Professor in the Division of Biological Sciences: M. Rabinowitz
A.J. Carlson Professor in the Division of the Biological Sciences: I.G. Wool
Richard T. Crane Distinguished Service Professor of Pediatrics: A. Dorfman
Maurice Goldblatt Professor in the Division of the Biological Sciences: H.G. Williams-Ashman
A.N. Pritzker Professor of Biochemistry: D.F. Steiner
F.L. Pritzker Professor of Biophysics and Theoretical Biology: R. Haselkorn
Professors: H.S. Anker, J.J. Ceithaml, J.A. Cifonelli (Research Associate), A. Dorfman, E.A. Evans, Jr., C.F. Failey (Visiting) J. Fried, G.S. Getz, E. Goldwasser, R. Haselkorn (Research Associate), E.T. Kaiser, F.J. Kézdy, J.H. Law, S. Liao, M.B. Mathews, M. Rabinowitz, A.M. Scanu (Research Associate) F. Schlenk (Research Associate), D.F. Steiner, S.B. Weiss, J. Westley, H.G. Williams-Ashman, I.G. Wool
Associate Professors: N.R. Cozzarelli, W. Epstein, R.N. Feinstein (Research Associate), H.C. Friedmann, R.L. Heinrikson, R.P. Mackal (Research Associate), T. Nakamoto, T.L. Steck
Assistant Professors: K. Agarwal, G. Dawson (Research Associate), W.T. Hsu (Research Associate), P. Keim (Research Associate), H. Köhler, A. Lernmark (Visiting) R. Lim, A.C. Stoolmiller (Research Associate), H.S. Tager, M. Volini

Degree Program: Biological Sciences with concentration in Biochemistry

Undergraduate Degree: B.A.
Undergraduate Degree Requirements: (a) Twelve common year courses; (b) Six second level non-science courses; (c) Nine non-biology science courses; (d) A common core of six advanced level biology courses; (e) Three quarters of a foreign language and (f) electives as program time permits.

Graduate Degrees: M.S., Ph.D.
Graduate Degree Requirements: Master: (1) A program of at least nine courses at the graduate level in Biochemistry, Chemistry and Biology; (2) A departmental foreign language reading examination with a pass in French, German or Russian; (3) The qualifying examination at the Master's level. Inindividual cases a research problem may be included as one of the requirements for the degree. Doctoral: (1) Courses in Biochemistry and related fields as determined for each candidate by consultation with the faculty adviser. Courses in Fundamentals of Biochemistry (or equivalent), Physical Biochemistry, Protein Biochemistry - Enzymology, Nucleic Acids and Protein Synthesis, Advanced Research Techniques, and participation in Current Literature in Biochemistry are required together with two graduate-level Organic Chemistry courses. (Physical Chemistry may be taken for graduate credit by the candidate who has no experience in the subject at the undergraduate level; (2) Admission to Candidacy which requires (a) performance at the Ph.D. level on the qualifying examination and (b) acceptable performance in the departmental foreign language reading examination in French, German or Russian (3) A research project and presentation of a satisfactory Thesis.

Graduate and Undergraduate Courses Available for Graduate Credit: Introduction to Research, Fundamentals of Biochemistry, Physical Biochemistry, Protein Biochemistry - Enzymology, Nucleic Acids and Protein Synthesis, Advanced Research Techniques, Enzymic Catalysis, Biochemistry of Cancer, Biochemistry of Endocrine Systems, Membrane Biochemistry, Biochemistry of the Nervous System, Seminar, Developmental Biochemistry, Biochemical Genetics, Special Topics in Developmental Biology, Bioorganic Chemistry, Graduate Genetics, Animal Cells and Their Viruses, Bacterial Viruses, Structural Macromolecules of Connective Tissue, Biology and Biochemistry of Complex Carbohydrates, Laboratory and Discussion Teaching, Research in Biochemistry.

DEPARTMENT OF BIOLOGY
Phone: Chairman: (312) 753-2709

Chairman: H. Swift
Distinguished Service Professor: H. Swift
Louis B. Block Professor: A. Moscona
William Wrather Professor: G. Beadle
Professors: S. Altmann, W. Baker, G. Beadle, E. Gerber, J. Hubby, R. Levins, A. Moscona, J. Overton, T. Park, A. Ravin, H. Swift, L. Throckmorton
Associate Professors: M. Esposito, R. Esposito, D. Fischman, H. Gall, B. Garber, M. Lloyd, M. Ruddat, J. Spofford, L. VanValen
Assistant Professors: R. Alderfer, R. Foster, T. Martin, M. Slatkin, J. Terri
Visiting Lecturers: 40 per year
Teaching Fellows: 10 per year
Research Assistants: 10 per year

Degree Programs: Biology, Ecology, Plant Sciences,

Graduate Degrees: M.S., Ph.D.
Graduate Degree Requirements: Master: Not Stated
Doctoral: Not Stated

DEPARTMENT OF BIOPHYSICS AND THEORETICAL BIOLOGY
Phone: Chairman: (312) 753-8366 Others: 753-1234

Chairman: R. Haselkorn
F.L. Pritzker Professor of Biophysics: R. Haselkorn
Louis Block Professor: M.H. Cohen
Professors: M.H. Cohen, J.T. Cowan, A.V. Crewe, J.M. Boldberg, R. Haselkorn, E.J. Jensen, R. Levins, G.N. Ramachandran, S.A. Rice, B. Roizman, P.B. Sigler, C.S. Spyropoulos, E.W. Taylor, R.B. Uretz, E. Zeitler
Associate Professors: D. Agin, K.S. Chiang, N R. Cozzarelli, W. Epstein
Assistant Professors: S. Kauffman, S. Lerner, M.W. Makinen, L. Riccardi, A.D.J. Robertson, L.B. Rothman-Denes, M. Slatkin, H. Wilson, M. Zwick
Research Associates: 6

Degree Programs: Biochemistry, Biochemistry and Biophysics, Biological Chemistry, Biophysics, Ecology, Genetics, Molecular Biophysics, Physiological Sciences, Physiology, Radiation Biology, Radiobiology, Radiation Biophysics, Virology

Graduate Degrees: M.S., Ph.D.
Graduate Degree Requirements: Master: Course requirements; sometimes thesis, or essay Doctoral: Course requirements; preliminary examinations including a creative literature review and a research proposal; dissertation based on original research; final examination on the dissertation.

Graduate and Undergraduate Courses Available for Graduate Credit: Organic and Inorganic Chemistry, Introduction to Linear Network Analysis, Introduction to Stability Theory, Macromolecules, Electric Circuit Theory; X-Ray Diffraction Analysis of the Structure of Proteins, Nucleic Acids, and Viruses; Structure of Biological Macromolecules; Principles of Optics and Electron Optics, Functional Neurophysiology, Membrane Biophysics, Models of Nervous Activity, Population Biology, Advanced Population Biology, Differentiation in Embryonic Systems, Control of Development, Graduate Genetics, Animal Viruses, Bacterial Viruses, Stochastic Processes in Biology, Electrophysiological Techniques, Physiology of Perception, Research in Biophysics and Theoretical Biology.

COMMITTEE ON GENETICS
Phone: Chairman: 753-3948 Others: 753-3943

Chairman: B.S. Strauss
William E. Wrather Distinguished Service Professor of Biology: G.W. Beadle
Richard T. Crane Distinguished Service Professor of Pediatrics and Biochemistry: A. Dorfman
Fanny L. Prizker Professor of Biophysics and Theoretical Biology: R. Haselkorn
Louis Block Professor of Biology: A.A. Moscona
Distinguished Service Professor of Biology: H. Swift
Ralph W. Gerard Professor of Biophysics and Theoretical Biology: R. Uretz
Professors: W. Baker, J.E. Bowman, E.D. Garber, W. H. Kirsten, R. Levins, A. Markovitz, R.W. Ravin, B. Roizman, R. Singer, B.S. Strauss, L. Throckmorton, S.B. Weiss
Associate Professors: K.S. Chiang, W. Epstein, J.L. Hubby, J. Rowley, J. Spofford, L. Van Valen
Assistant Professors: N. Cozzarelli, M. Esposito, R. Esposito, T.E. Martin, M. Slatkin

Degree Program: Genetics

Graduate Degrees: M.S., Ph.D.
Graduate Degree Requirements: Master: 9 course units including Graduate Genetics. Oral examination on reading list. Doctoral: 27 course units including Graduate Genetics and Thesis research. Oral Examination on reading list. Dissertation.

Graduate and Undergraduate Courses Available for Graduate Credit: General Genetics, Cell Genetics, Medical Genetics, Population Biology, Immunogenetics, Experimental Genetics of Drosophila, Biochemical Genetics, Genomic Evolution, Fungal Genetics, Graduate Genetics, The Nucleus, The Cytoplasm, Advanced Genetics

DEPARTMENT OF MICROBIOLOGY
Phone: Chairman: (312) 753-3948 Others: 753-3943

Chairman: B.S. Strauss
Professors: R.M. Lewert, A. Markovitz, J.W. Moulder, A. Ravin, B. Roizman, B.S. Strauss, S.B. Weiss
Associate Professors: E.P. ohen, W.R. Martin
Assistant Professors: J.A. Shapiro, P.G. Spear
Research Associates: 10
Research Associates: (With Rank) 8

Field Stations/Laboratories: Experimental Biology Building Chicago

Degree Program: Microbiology

Graduate Degrees: M.S., Ph.D.
Graduate Degree Requirements: Master: 12 course units including 3 research units. Oral examination on the results of the research. Doctoral: 27 course units including research. Written and Oral Qualifying Examination and thesis defense

Graduate and Undergraduate Courses Available for Graduate Credit: Biology of Microorganisms, Cellular Genetics, Immunity to Parasites, Immunochemistry, Immunobiology, Readings in Immunobiology, Developmental Biology II, Biology of Bacteria, Microbial Metabolism, Biochemical Genetics, Genomic Evolution, Parasitology, Procaryotic Cells and Cellular Biology, Cellular Pathology, Virology, Immunology, Cell Biology

Laboratory, Clinical Patho-Physiology, Clinical Microbiology, Microorganisms in Infectious Disease, Graduate Genetics, Biophysical Instrumentation in Microbiology, Animal Cells and their Viruses, Bacterial Viruses, Seminar on Animal Viruses, Seminar on Bacterial Viruses, Orientation to Microbiological Research

DEPARTMENT OF PATHOLOGY
Phone: Chairman (312) 947-5464

Chairman: W.H. Kirsten
Donald N. Pritzker Professor: R.W. Wissler
Professors: J.E. Bowman, F.W. Fitch, G.S. Getz, S. Glagov, Z. Hruban, H. Rappaport, W.R. Richter, D. A. Rowley, B.H. Spargo, R.W. Wissler
Associate Professors: J.R. Esterly, P.W. Graff, H. Köhler, J.A. Morello, C.E. Platz, H. Rochman, F.H. Straus, M L. Warnock, T.W. Wong
Assistant Professors: J. Borensztajn, G.E. Byrne, Jr., A.C. Daniels, R.L. Hunter, R.H. Kirschner, F.E. Kocka, S.L. Thomsen, D. Variakokis
Visiting Lecturers: 1
Research Associates: 11
Teaching Fellows: 3

Degree Program: Pathology

Graduate Degrees: M.S., Ph.D.
Graduate Degree Requirements: Master: 9 course credits in Pathology, Dissertation, Oral Examination. Doctoral: Courses selected by advisory committee, Quanifying Examination, Dissertation, Final Oral Examination

Graduate and Undergraduate Courses Available for Graduate Credit: Cellular Pathology and Virology, Organ Pathophysiology, Advanced Post-Mortem Histopathology, The Post-Mortem Examination, Current Trends in Experimental Pathology, Developmental Pathology, Biochemistry of Cancer, Oral Diseases: Clin-Path. Diag. Corr. Chemical Pathology, Fundamentals of Anatomic and Clinical Pathology, Problems in Clinical Pathology: Topics in Tropical Disease (U), Experimental Cytopathology, Neoplastic Diseases: Clin-Path. Diag. Corr. ,Clinical Microbiology, Microoorganisms in Infectious Disease, Cell Biology I, II, Current Problems in Surgical Pathology, Advanced Gynecological Pathology, Individual Tutorial Projects, Principles of Immunopathology, Molecular Basis of Nutritional Pathology (U), Principles of Comparative Pathology (U), Liver Diseases: Clin-Path. Diag. Corr., Normal and Abnormal Cell Proliferation, Mechanical Determinants of Tissue Organization(U), Ultrastructural Pathology (U), Experimental Oncology and Comparative Tumor Pathology, Pulmonary Pathology.

DEPARTMENT OF PHARMACOLOGICAL AND PHYSIOLOGICAL SCIENCES
Phone: Chairman: (312) 753-3945 Others: 753-3942

Chairman: A. Heller
Professors: H.A. Fozzard, J.M. Goldberg, L.I. Goldberg, A. Heller, J.O. Hutchens, P. Meier, E. Page, L.J. Roth, J. Rust, C. Schuster, H.G. Williams-Ashmann,
Associate Professors: P.C. Hoffmann, A. Moawad, L. Seiden
Assistant Professor: P. Grobstein
Research Associates: I. Diab, D. Domizi
Visiting Lecturers: about 15 per year

Degree Programs: Pharmacology and Physiology

Graduate Degrees: M.S., Ph.D.
Graduate Degree Requirements: Master: Required courses, Language Requirement, Research Work and Master's Thesis. Doctoral: Required Courses, Language Requirement, Preliminary Examination, Research Work and Doctoral Thesis

Graduate and Undergraduate Courses Available for Graduate Credit: Basic Pharmacology, Clinical Pharmacology, Cellular and Organ Physiology, Organ Physiology and Endocrinology, Neurobiology, Neuropharmacology, Psychopharmacology

THE UNIVERSITY OF HEALTH SCIENCES

The Chicago Medical School

2020 West Ogden Avenue Phone: (312) 226-4100
Dean of Graduate Studies: J.L. Nickerson
Dean of Medical School: M.A. Falk

DEPARTMENT OF ANATOMY
Phone: Chairman: (312) 226-4100 Ext. 242 or 253

Chairman: M. Combs
Professors: J.J. Chiakulas, C.M. Combs
Associate Professors: E.W. Millhouse, J.C. Thaemert, C.H. Thomas
Assistant Professors: 2
Instructors: 1
Visiting Lecturers: 2

Degree Programs: Anatomy, Anatomy and Histology

Graduate Degrees: M.S., Ph.D.
Graduate Degree Requirements: Master: This advanced heneral research degree is conferred on candidates who have met the requirements for any one of the various programs offered at the graduate school. Doctoral: The Ph.D. degree, which is the highest degree conferred by UHS/CMS is a research degree and is awarded in recognition of proficiency in research, breadth and soundness of scholarship, and a thorough understanding of a specific field of knowledge.

Graduate and Undergraduate Courses Available for Graduate Credit: Graduate level courses in Anatomy, Physiology, Biochemistry, etc.

DEPARTMENT OF BIOCHEMISTRY
(No reply received)

DEPARTMENT OF MICROBIOLOGY
Phone: Chairman: Ext. 416

Chairman: E.R. Brown
Professors: R.A. Albach, N.J. Bigley, T.J. Bird (Visiting), E.R. Brown, W.C. Dolowy (Visiting), R. Morrissey (Visiting), G.J. Scheff (Emeritus)
Associate Professors: R.J. Ablin (Adjunct), T. Booden, D.A. Wilson
Assistant Professors: R.A. Zeineh (Adjunct)
Visiting Lecturers: 3
Teaching Fellows: 2
Research Assistants: 2

Field Stations/Laboratories: Fishery - Aurora

Degree Programs: Avian Disease, Bacteriology, Bacteriology and Immunology, Biological Sciences, Biology, Entomology and Parasitology, Fisheries and Wildlife Biology, Genetics, Health and Biological Sciences, Human Genetics, Immunology, Medical Microbiology, Medical Technology, Microbiology, Parasitology, Radiation Biology, Radiobiology, Virology

Undergraduate Degrees: B.A. - In school of allied Health Sciences with major in Microbiology
Undergraduate Degree Requirements: 3 years Chemistry, 1 year Basic Biology, 2 years English, 20 hours Biology

Graduate Degrees: M.S., Ph.D. (Clinical Microbiology, Genetics, Micromolecular Biology, Virology)
Graduate Degree Requirements: Master: 1 language and 30 hours over B.S. with thesis written and orals Doctoral: 2 languages, 60 hours over B.S. or 30

hours over written and orals and 30 hours of thesis

Graduate and Undergraduate Courses Available for Graduate Credit: Medical and Clinical Microbiology (U), Virology, Viral Oncology, Bacterial Physiology, Parasitology, Biochemistry of Intra-and Extra Cellular Parasites, Immunobiology, Molecular Basis of Immunity, Microbial Genetics, Cell Physiology, Seminar and Journal Club, Selected Topics of Microbiology, Virology, Parasitology and Immunology

DEPARTMENT OF PHYSIOLOGY AND BIOPHYSICS
Phone: Chairman: Ext. 205 or 206

Chairman: V.V. Glaviano
Professors: B.B. Blivaiss, V.V. Glaviano, A. Luisada, J.L. Nickerson, J.A. Smith
Associate Professors: B. Coleman, C.E. McCormack, E.J. Sukowski, P.C. Tang
Assistant Professor: E.M. Barr
Visiting Lecturers: 1
Teaching Fellows: 1
Research Assistants: 2
Technician: 1

Degree Programs: Physiology and Biophysics

Graduate Degrees: M.S., Ph.D.
Graduate Degree Requirements: Master: two academic years of graduate work beyond the baccalaureate degree in addition to 45 units of graduate course work in the major field, one academic year in residence, study program and research plan must be approved for the thesis, scientific reading knowledge of one foreign language, a written, oral or both comprehensive examination. Doctoral: three years of full-time graduate study beyond the baccalaureate degree, 90 course units in addition to 45 units of research, two academic years of full-time graduate residence, research plan for a dissertation must be approved, scientific reading knowledge of two foreign languages, preliminary examination

Graduate and Undergraduate Courses Available for Graduate Credit: Medical Physiology, Radiobiology, Physiology Seminar, Physiological Journal Review, Introduction to the Use of Computers in Biomedical Research, Physiology of the Autonomic Nervous System, Physiology of the Liver, Current Problems in Endocrinology, Endicrine System, Progress in Reproductive Physiology, General Hemodynamics of Blood Flow, Cardiac Dynamics, Cardiac Energetics and Metabolism, Dynamics of Cardiac Vibrations, Pathological Physiology of Trauma, Pathological Physiology, Master's Thesis in Physiology, Aerospace Physiology, History of Physiology, Teaching Methods, Doctoral Dissertation in Physiology, Methods in Metabolic Research, Graduate Colloquy, Introduction to Research.

UNIVERSITY OF ILLINOIS AT CHICAGO CIRCLE
(No reply received)

UNIVERSITY OF ILLINOIS AT THE MEDICAL CENTER
1737 West Polk Phone: (312) 996-7000
Chicago, Illinois 60680
Dean of Graduate Studies: A.V. Wolf

College of Dentistry
Dean: S.H. Yale

DEPARTMENT OF HISTOLOGY
(No reply received)

DEPARTMENT OF ORAL ANATOMY
Phone: Chairman: 996-7544 Others: 996-6893

Head: E.L. Du Brul
Professor: E.L. Du Brul
Associate Professors: H.R, Barghusen, R.P. Scapino
Assistant Professors: T.C. Lakars, S.W. Herring, G.S. Throckmorton

Degree Program: Anatomy in conjunction with the Department of Anatomy

Graduate Degrees: M.S., Ph.D.
Graduate Degree Requirements: Master: Minimum of 16 quarter hours formal courses, 10 quarter hours research, 22 quarter hours seminar and elective courses An acceptable thesis, final oral examination
Doctoral: 16 quarter hours formal 400 series courses, 128 quarter hours research, seminars and electives, Qualifying Interview, Oral Preliminary Examination, An acceptable Thesis, Final Thesis Defense

Graduate and Undergraduate Courses Available for Graduate Credit: Physical Anthropology - Biomechanics of the Head and Neck, Physical Anthropology - Evolution of the Hominid Oral Apparatus

College of Medicine
Dean: T. Anderson

DEPARTMENT OF ANATOMY
Phone: Chairman: (312) 996-7595

Head: K.A. Rafferty, Jr.,
Professors: R. Krehbiel, A. LaVelle, H. Monsen, J. Plagge, E. Polley, W.A. Reynelds, A.H. Schmidt, P. Van Alten
Associate Professors: H. Barghusen, L.A. Benevento, N. Johnston, R.E. Kelly, M. Kernis
Assistant Professors: C. Anderson, J.A. Colgan, J.L. Cracraft, M. Lazarus, H.G. Schs, W. Clark
Instructors: G. Stanton, N. Kinderman
Teaching Assistants: 20
Visiting Lecturers: 1
Trainees: 5

Degree Program: Anatomy

Graduate Degrees: M.S., Ph.D.
Graduate Degree Requirements: Master: 48 quarter hours
Doctoral: 144 quarter hours

Graduate and Undergraduate Courses Available for Graduate Credit: Human Neuroanatomy, Gross Anatomy, Biomechanics of Head and Neck, Medical Histology, Cell Structure and Human Histology, Advanced Anatomy, Morphology of Pelvic Structures, Concepts of Morphology, Principles of Teratology, Concepts of Synaptic Function and Morphology, Approaches and Methods in Anatomical Research, Experimental Neurocytogenesis, Experimental Morphogenesis, Developmental Immunobiology, Morphologic and Physiologic Adaptations to Growth, Aging, and Injury, Evolution of Functional Complexes in Vertebrates, Cell Biology of Vertebrate Repair and Regeneration, Fine Structure of Cells and Tissues, Neuroendocrinology, Brain Mechanisms

DEPARTMENT OF BIOLOGICAL CHEMISTRY
(No reply received)

DEPARTMENT OF MICROBIOLOGY
(No reply received)

DEPARTMENT OF PATHOLOGY
Phone: Chairman: 996-7312 Others: 996-7313

Head: E.A. McGrew
Distinguished Research Professor of Pathology: C.A. Krakower
Professors: H.R. Catchpole (Research), L.J. LeBeau, E.P. Leroy, J.R. Manaligod, E.A. McGrew, S.T. Nerenberg, R.L. Wong

Associate Professors: F.A.O. Eckner, K.V. Karachorlu, N. Ressler, U.F. Rowlatt, H.M. Yamashiroya
Assistant Professors: J.R. Manaligod, R.R. McKeil, S. Millner, E. Popescu, H. Rothenberg
Instructors: L.n. Chowdhury, S.G. Ronan
Research Associate: 1

Degree Program: Biology

Graduate Degrees: M.S., Ph.D.
Graduate Degree Requirements: Master: See Graduate College Bulletin Doctoral: See Graduate College Bulletin

Graduate and Undergraduate Courses Available for Graduate Credit: General Pathology (U), Experimental Pathology, Histochemistry, Physiological Basis of Pathology, Ultrastructural Pathology, Trends in Clinical Chemistry, Seminar in Pathology

PHYSIOLOGY DEPARTMENT
Phone: Chairman: 996-7620 Others: 996-7000

Head: A. Omachi (Acting)
Professors: P.O. Bramante, R. Greenberg, S.F. Marotta, A. Omachi
Associate Professors: P.L. Hawley, R.F. Loizzi
Assistant Professors: D.L. Ford, A.D. Hartman, G.L. Humphrey, S.M. Kilen, A.J. Miller, M.S. Millman
Teaching Fellows: 12
Research Assistants: 1
Affiliates: 27
Trainees: 5

Degree Programs: Physiology

Graduate Degrees: M.S., Ph.D.
Graduate Degree Requirements: Master: 5 quarter hours in 400 series courses, 11 quarter hours collateral courses, 10 quarter hours research, 22 quarter hours Seminars and Electives--total 48 quarter hours, Acceptable thesis and a final oral examination, Residence of 3 calender quarters (24 quarter hours in scheduled courses) Doctoral: Pass Qualifying Examination, 10 quarter hours 400 series courses; 6 quarter hours collateral area; 128 quarter hours Research, Seminars and Electives--total 144 quarter hours, pass oral preliminary examination, thesis, residence of 3 consecutive quarters; continuous registration, final examination

Graduate and Undergraduate Courses Available for Graduate Credit: Applied Physiology (Nursing, Others) (U), Physiology of Pregnancy (Nursing) (U), Human Physiology (Dentistry) (U), Physiology of the Heart, Special Topics in Physiology, Cell Physiology, Transport Across Cell Interphases, Physiology of Endocrines, Concepts in Biophysics; Neural Correlates of Behavior, Physiology of Hearing, Vestibular Physiology, Human Physiology (U), Human Physiology (continuation), (U), Methods in Experimental Physiology, Methods in Experimentation (continuation), Methods in Experimental Physiology (continuation), Tactics and Strategy

DEPARTMENT OF MEDICAL RADIOLOGY
Phone: Chairman: (312) 996-7291 Others: 996-6000

Head: V. Capek
Professor: H.C. Dudley

Degree Program: Radiology

Graduate Degree: M.S.
Graduate Degree Requirements: Master: M.D. or D.D.S. plus 6 quarter hours in Radiation Physics plus 20 quarter hours in Radiation Research and Thesis.

Graduate and Undergraduate Courses Available for Graduate Credit: Radiation Physics for Medical applications, Radioisotopes, Research in Radiology

School of Associate Medical Sciences
Dean: T.F. Zimmerman

Phone: Chairman: (312) 996-6695

Professors: H. Goodwin
Associate Professors: G. Brawley, R. Finnegan, B. Loomis, A. Maturen
Assistant Professor: D. Brownold
(Medical Art - 1 Professor, 2 Associate Professors, 2 Assistant Professors; Medical Dietetics - 2 Assistant Professors, 2 Instructors; Medical Laboratory Sciences 3 Associate Professors, 3 Assistant Professors, 2 Instructors; Occupational Therapy - 3 Associate Professors, 5 Assistant Professors, 5 Instructors; Physical Therapy - 1 Associate Professor, 2 Assistant Professors, 2 Instructors)

Degree Programs: Medical Technology, Occupational Therapy, Physical Therapy, Medical Arts, Medical Dietetics, Medical Record Administration

Undergraduate Degree: Each Curriculum culminates in the B.S./Professional Field.
Undergraduate Degree Requirements: Depend on Professional Field, Physical Therapy and Medical Dietetics - 2 years college with specified preprofessional courses. All others, 3 years college with specified preprofessional courses.

Graduate and Undergraduate Courses Available for Graduate Credit: Biological Chemistry (U)

UNIVERSITY OF ILLINOIS
AT URBANA CHAMPAIGN
Urbana, Illinois 61801 Phone: (217) 333-1000
Dean of Graduate Studies: G. Russell

College of Liberal Arts and Sciences
Dean: R.W. Rogers

DEPARTMENT OF BIOCHEMISTRY
Phone: Chairman: 333-3945 Others: 333-2013

Head: L.P. Hager
Professors: L.P. Hager, H.E. Conrad, I.C. Gunsalus, G. Weber, R. Nystrom
Associate Professors: J. Clark, R. Switzer
Assistant Professors: R. Gennis, R. Gumport, W.O. McClure, J. Robinson, D.R. Storm, P. Schmidt, O. Uhlenbeck, M. Glaser
Teaching Fellows: 50
Research Assistants: 20

Degree Program: Biochemistry

Undergraduate Degree: L.A.S. Science and Letters
Undergraduate Degree Requirements: Students in science and letters curriculum take two years of basic work followed by a major in Biochemistry (in this case). Major: Not less than 20 hours in Biochemistry and Chemistry including Biochemistry 350 and 355, Organic Chemistry through Chemistry 336, and one year of Physical Chemistry; Mathematics through 140, 141 or 145, Physics through 102 or108, and two 300-level courses in Life Sciences.

Graduate Degrees: M.S., Ph.D.
Graduate Degree Requirements: Master: M.S. in Biochemistry with or without a thesis. Coursework master's requires two full-time semesters and eight units of formal lecture and laboratory course work. M.S. with thesis requires three semesters of research. No foreign language requirement. Doctoral: Qualification for the Ph.D. program in Biochemistry involves passing a research qualifying examination within the first eighteen months of residency and requires the successful completion of a series of written cumulative examinations. A reading knowledge of one for-

eign language. A thesis based on original research must be presented to a review committee at least two weeks prior to the final examination. The final examination is limited to defense of the thesis research topic. The Ph.D. program usually takes three to five years for completion.

Graduate and Undergraduate Courses Available for Graduate Credit: Introduction to Biochemistry (U), Biochemistry Laboratory (U), General Biochemistry (U), Half of year long sequence, General Biochemistry (U), Other half of year long sequence, Research Topics in Biochemistry, Experimental Techniques in Biochemistry, Biochemistry Seminar, Special Topics in Biochemistry, Chemical Basis of Biological Specificity, The Use of Carbon-14 in Labeling Techniques, Thesis Research

DEPARTMENT OF BOTANY
Phone: Chairman: (217) 333-3260

Head: J.B. Hanson
Professors: J.S. Boyer, J.M.J. deWet, D.B. Dickinson, Govindjee, B.V. Hall (Emeritus), J.B. Hanson, J.F. Nance,(Emeritus), T.L. Phillips, D.P. Rogers, R.W. Tuveson
Associate Professors: C.J. Arntzen, F.A. Bazzaz, Z.B. Carothers, A.W. Haney, L.W. Hoffman, M.L. Sargent, L.N. Vanderhoef
Assistant Professors: T.E. Lockwood, D.S. Seigler, C. A. Shearer
Visiting Lecturers: 3
Teaching Fellows: 24
Research Assistants: 6
Postdoctorals: 6

Degree Program: Botany

Undergraduate Degree: L.A.S. Science and Letters
Undergraduate Degree Requirements: Botany, Chemistry, Mathematics, Physics, Botany must include Plant Taxonomy, Genetics, Plant Physiology, Plant Morphology, Plant Ecology, Individual Study

Graduate Degrees: M.S., Ph.D.
Graduate Degree Requirements: Master: 8 units of at least 3 of which must be in graduate courses, Thesis. Doctoral: Courses as recommended by committee. One preliminary examination, foreign language examination, Thesis, Final Examination

Graduate and Undergraduate Courses Available for Graduate Credit: Individual Topics (U), General Plant Morphology (U), Paleobotany (U), Plant Physiology (U), Experimental Cytology (U), Plant Physiology Laboratory (U), Experimental Cytology Laboratory (U), Field Ecology (U), Plant Anatomy (U), Phycology (U), Viruses I, II (U), Field Botany (U), General Mycology (U), Plant Ecology (U), Aquatic Plant Ecology (U), Angiosperm Phylogeny and Biogeography (U), Molecular Genetics: Chromosome Mechanics, Physiology of Fungi, Molecular Genetics: Gene Action, Botany Discussion, Discussions in Plant Physiology, Discussions in Plant Morphology and Taxonomy, Discussions in Plant Ecology and Plant Geography, Discussions in Photosynthesis and Related Topics, Cytogenetics, Mineral Nutrition of Plants, Discussions in Mycology, Advanced Physiology of Growth, Responses and Reproduction, Advanced Plant Physiology: Photosynthesis, Environmental Plant Physiology, Advanced Taxonomy of Flowering Plants I, II, Origin of Variation in Plants, Plant Products, Advanced Mycology: Special Groups, Plant Geography of North America, Advanced Studies in Botany

DEPARTMENT OF ENTOMOLOGY
Phone: Chairman: (217) 333-2910 Others: 333-1000

Head: J.R. Larsen
Professors: G.S. Fraenkel, S. Friedman, W.R. Horsfall, E.R. Jaycox (Adjunct), W.E. LaBerge (Adjunct), J.R. Larsen, W.H. Luckmann(Adjunct), R.L. Metcalf, L.J. Stannard (Adjunct), J.G. Sternburg, G.P. Waldbauer
Associate Professors: E.G. MacLeod, J.H. Willis
Assistant Professors: F. Delcomyn, P.W. Price
Visiting Lecturers: 1
Teaching Fellows: 15
Research Assistants: 15

Degree Programs: Entomology, Pest Management

Undergraduate Degree: L.A.S. Science and Letters (Entomology Major)
Undergraduate Degree Requirements: Students must complete a major consisting of Biology 110, 111, Entomology 201, 302, and an additional 11 hours of courses at the 200 or 300 level offered within the School of Life Sciences. Students are also required to complete a year of Physics, Chemistry through Organic with Laboratory, a course in Statistics, and Mathematics through Mathematics 120 or equivalent.

Graduate Degrees: M.S., Ph.D.
Graduate Degree Requirements: Master: A candidate must master four of the five general areas of entomology (Bionomics, Control, Morphology, Physiology, and Taxonomy) and complete a research thesis in an area of interest arrived at in consultation with an advisor. Doctoral: A candidate must master the general areas of entomology (Bionomics, Control, Morphology, Physiology, and Taxonomy) and demonstrate professional competence in a specialized area by presenting an acceptable thesis based on original research. Reading proficiency in a foreign language of importance to the area of research, ordinarily German or Russian is also required.

Graduate and Undergraduate Courses Available for Graduate Credit: Introduction to Entomology (U), Classification of Insects (U), Special Problems (U), Entomology for Teachers (U), Insect Ecology (U), Fundamentals of Insect Control (U), Insect Bionomics (U), Individual and Group Behavior of Honey Bees (U), Insect Morphology, Medical and Veterinary Entomology, Chemistry and Toxicology of Insecticides, Insect Physiology, Insect Behavior, Advanced Insect Physiology, Seminar in Entomology

PROVISIONAL DEPARTMENT OF ECOLOGY, ETHOLOGY AND EVOLUTION
Phone: Chairman: 333-7802 Others: 333-7801

Head: E.M. Banks
Professors: E.M. Banks, L.L. Getz, A.W. Ghent, D.F. Hoffmeister, S.C. Kendeigh (Emeritus)
Associate Professors: T.H. Frazzetta, G.H. Keiffer, M. R. Lee, M.R. Matteson, M. Salmon, H.H. Shoemaker, D. Sweeney, M.F. Willson
Assistant Professors: H.W. Ambrose III, G.O. Batzli, R.M. Fagen

Degree Programs: Biometry, Biostatistics, Ecology, Ethology, Limnology, Wildlife Management, Zoology, Functional Morphology

Undergraduate Degree: B.S. (Zoology)
Undergraduate Degree Requirements: Biology 110, 111, Biology 210, 212, 310, and Zoology 346, Mathematics 120 or 135, Chemistry 131 and 134 or Chemistry 136 and 181, Physics 101 and 102 or Physics 106, 107 and 108. 20 additional hours of Life Science courses at 200 level or above.

Graduate Degrees: M.S., Ph.D. (Zoology)
Graduate Degree Requirements: Master: 8 units of Graduate Credit, Thesis. Doctoral: 16 Units of Graduate Credit and competence in one foreign language, preliminary examination, thesis

Graduate and Undergraduate Courses Available for Graduate

Credit: Not Stated

PROVISIONAL DEPARTMENT OF GENETICS AND DEVELOPMENT

Phone: Chairman: (217) 333-7804 Others: 333-1000

Head: E.H. Brown (Acting)
Professors: F.B. Adamstone (Emeritus), L.M. Black, L. Ingle (Emeritus), J.B. Kitzmiller, (Emeritus), F.J. Fruidenier, J.R. Laughnan, W. Luce (Emeritus), D.L. Nanney, R.B. Selander, D.N. Steffensen, R.W. Tuveson, R.L. Watterson (also School of Basic Medical Sciences), C.R. Woese
Associate Professors: N.E. Alger, E.H. Brown, R. Davenport, E. MacLeod (also Entomology), R. MacLeod (Director, Center for Electron Microscopy), M.L. Sargent, D.L. Stocum, G.S. Whitt, J.H. Willis (also Entomology)
Assistant Professors: W.L. Daniel (also School of Basic Medical Sciences), L.E. Maxson, W.W.M. Steiner
Visiting Lecturers: 2
Research Assistants: 11
Research Associates: 7

Degree Programs: Biology, Genetics, Zoology

Undergraduate Degrees: L.A.S. Arts and Sciences (Biology and Zoology)
Undergraduate Degree Requirements: See courses catalog bulletin, University of Illinois at Urbana-Champaign

Graduate Degrees: M.S., Ph.D.
Graduate Degree Requirements: See Graduate Programs bulletin

Graduate and Undergraduate Courses Available for Graduate Credit: See Courses Catalog bulletin, University of Illinois at Urbana-Champaign.

DEPARTMENT OF MICROBIOLOGY

Phone: Chairman: (217) 333-1737 Others: 333-1000

Head: R.D. DeMoss
Professors: M.P. Bryant, R.D. DeMoss, R.E. Kallio, S. Kaplan, R.C. Meyer, Z.J. Ordal, M E. Reichmann, E.W. Voss, C.R. Woese, R.S. Wolfe
Associate Professors: J.T. Wachsmann, D.C. Savage
Assistant Professors: D.D. Burke, P.L. Carl, M.G. Gabridge, A.C. Helm, C.L. Hershberger, J. Koniske, P.R. Starr, M.J. Weber
Research Assistants: 24

Degree Programs: Bacteriology, Genetics, Microbiology, Cell Biology, Molecular Biology

Undergraduate Degrees: B.S.
Undergraduate Degree Requirements: Major: Biology 110, and 111 and twenty hours of microbiology courses including Microbiology 200 and 201. Biology 210 and 211 may be substituted for Microbiology courses, In addition quantitative and Organic Chemistry with Laboratory, Mathematics through Trigonometry and one year of Physics are required. Credit must also be presented in one course to be selected from Calculus, Statistics, or Computer Science, and one course from Physical Chemistry or Biochemistry.

Graduate Degrees: M.S., Ph.D.
Graduate Degree Requirements: Candidates for graduate degrees design their programs of study in consultation with individually assigned faculty committees. Study programs are flexible and are dependent upon the experience, interest and goals of each student.

Graduate and Undergraduate Courses Available for Graduate Credit: Comparative Microbial Chemistry (U), Food and Industrial Microbiology (U), Techniques of Applied Microbiology (U), Genetic Analysis of Microorganisms (U), Pathogenic Bacteriology (U), Immunochemistry (U), Molecular Biology of Microorganisms (U),Microbial Physiology and Anatomy (U), Viruses (U), Molecular Genetics: Chromosome Mechanics, Molecular Genetics: Gene Action, Cultivation and Properties of Microorganisms, Animal Virology, Experimental Virology

DEPARTMENT OF PHYSIOLOGY AND BIOPHYSICS

Phone: Chairman: (217) 333-1734 Others: 333-1735

Head: W.W. Sleator
Professors: H.W. Ades, J.D. Anderson, L. Barr, D.E. Buetow, E. Donchin, H.S. Ducoff, F. Dunn, C. Gianturco (Clinical), Covindjee, J.R. Larsen, A.V. Nalbandov, C.L. Prosser, V.D. Ramirez, W.W. Sleator, A.B. Taylor (Emeritus), H.I. Teigler (Clinical), A.R. Twardock, H. Von Foerster, G. Weber, J.S. Willis, C.R. Woese
Associate Professors: T.G.Ebrey, S.I. Helman, C.H. Hockman, A. Koklo-Cunningham, S.G. Stolpe, D.C. Sweeney
Assistant Professors: D.J. Barker, J.A. Connor, J.A. Harris, P.H. Hartline, E.G. Jakobsson, G.L. Jendrasiak, B.S. Katzenellenbogen, B.W. Kemper, E.E. Kicliter, Jr., W.B. Rhoten, O.D. Sherwood, R.L. Terjung, J.E. Zehr
Visiting Lecturers: 3
Research Assistants: 9
Research Associates: 4
Teaching Assistants: 37
Graduate Students: 90

Degree Programs: Physiology, Biophysics, Neurobiology, Cell Biology

Undergraduate Degree: B.S. (Physiology)
Undergraduate Degree Requirements: 120 hours selected according to field of concentration in consultation with departmental adviser.

Graduate Degrees: M.S., Ph.D. (Physiology, Biophysics)
Graduate Degree Requirements: Note: 1 unit - 4 hours. Master: 8 units of course work and written research report. Doctoral: 24 units of courses at 400 level, one foreign language, and Thesis

Graduate and Undergraduate Courses Available for Graduate Credit: PHYSIOLOGY - General Physiology (U), Animal Physiology (U), General Physiology Laboratory (U), Experimental Physiology Laboratory (U), Endocrinology (U), General Radiobiology (U), Physiology of Systems and Organs, Comparative and Adaptational Physiology, Cellular and Molecular Physiology, Physiological Measurements, Mammalian Physiology Seminar, Structure and Function of the Nervous System Gross Human Anatomy, Experimental Radiobiology, Advanced Comparative Physiology and Laboratory, Advanced Cellular Physiology, Ergonomics (U), Individual Topics, Thesis Research BIOPHYSICS - Introduction to Biophysics (U), Introduction to Radiobiology (U), Advanced Biophysics I and II, Physiological Measurements, Principles of Biophysical Measurements, Radioisotopes in Biological Research, Individual Topics, Thesis Research

DEPARTMENT OF ZOOLOGY
(Currently divided into Provisional Departments of Ethology, Ecology and Evolution, and Genetics and Development.)

College of Agriculture

Dean: O.G. Bentley

DEPARTMENT OF AGRONOMY

Phone: Chairman: (217) 333-3420

Head: R.W. Howell
Professors: S.R. Aldrich, D.E. Alexander, A.H. Beavers, R.L. Bernard, C.M. Brown, A.W. Burger, S. G. Carmer, J.M.J. DeWet, J.W. Dudley, E.B. Ear-

ley, J.B. Fehrenbacher, D.W. Graffis, H.H. Hadley, R.H. Hageman, J.B. Hanson, J.R. Harlan, C.N. Hittle, R.W. Howell, J.A. Kackobs, E.L. Knake, L.T. Kurtz, J.R. Laughnan, E.R. Leng, G.E. McKibben, S.W. Melsted, D.A. Miller, W.R. Oschwald, D.B. Peters, M.B. Russell, W.O. Scott, R.D. Seif, F.W. Slife, F.J. Stevenson, M.D. Thorne, E.H. Tyner, L.F. Welch, C.M. Wilson

Associate Professors: J.D. Alexander, J.S. Boyer, R.L. Cooper, R.C. Hiltibran, T.D. Hinesly, T. Hymowitz, R.L. Jones, C.J. Kaiser, D.E. Keoppe, R.J. Lambert, M.D. McGlamery, W.L. Ogren, E.B. Patterson, T.R. Peck, B.W.Ray, R.W. Rinne, W.E. Stoller, W.M. Walker, L.M. Wax, E.J. Weber, J.M. Widholm, J.T. Wolley

Assistant Professors: C.W. Boast, M.A. Cole, J.E. Harper, T.T. Hassett, R.G. Hoeft, R.R. Johnson, J.J. Faix, D.K. Whigham

Agronomists: L.V. Boone, D.R. Browning, C.G. Chambliss, J.F. Duncan, J.W. Johnson, J.B. Kelly, E.C. Marcuslu, R.L. Nelson, D.E. Millis, D.L. Mulvaney, M.G. Oldham, G.L. Ross, E.L. Ziegler

Visiting Lecturers: 2

Research Assistants: 70

Field Stations/Laboratories: Dixon Springs Research Center (Simpson), Brownstown, Carbondale, DeKalb, Elwood, and Urbana (all research centers; Aledo, Carthage, Dixon, Hartsburg, Kewanee, Newton, Toledo, (All research fields)

Degree Programs: Agriculture, Agronomy, Crop Science, Soil Science, Crop Production, Agronomy Industry

Undergraduate Degree: B.S.

Undergraduate Degree Requirements: 126 hours with a minimum GPA of 3.0 for B.S. in Agriculture, 126 hours with a minimum GPA of 3.5 for B.S. in Agricultural Science, B.S. in Agronomy - a minimum of 40 hours of Agriculture courses, General Education courses, 6 hours Humanities and 9 hours Social Science and 24 hours in Physics and Biological Sciences, B.S. in Ag. Sci - same except 45 hours Biology and Physical Science

Graduate Degrees: M.S., Ph.D.

Graduate Degree Requirements: Master: 8 units (24 hours) - minimum of 6 units of course work Doctoral: 24 units (72 hours) - minimum of 9 units of course work and 2 languages, or as a language substitute, 1 1/2 units of course work for each language

Graduate and Undergraduate Courses Available for Graduate Credit: Not Stated

DEPARTMENT OF ANIMAL SCIENCE

Phone: Chairman: (217) 333-1045

Head: D.E. Becker

Professors: E.M. Banks, D.J. Bray, O.G. Bentley, G.R. Carlisle, J.E. Corbin, P.J. Dziuk, R.M. Forbes, U.S. Carrigus, E.E. Hatfield, A.H. Jensen, S.P. Mistry, A.V. Nalbandov, A.L. Neumann, H.W. Norton, B.A. Rasmusen, S.F. Ridlen

Associate Professors: W.W. Albert, D.H. Baker, W.F. Childers, B.G. Harmon, P.C. Harrison, F.C. Hinds, H.S. Johnson, J.M. Lewis, G.E. Ricketts, J.R. Romans, B.A. Weichenthal

Assistant Professors: G.F. Cmarik, S.E. Curtis, J.R. Diehl, F.N. Owens, G.R. Schmidt

Visiting Lecturers: 1

Field Stations/Laboratories: Dixon Springs Agricultural Center, Simpson

Degree Programs: Agricultural Biochemistry, Animal Behavior, Animal Genetics, Animal Husbandry, Animal Industries, Animal Physiology, Animal Science, Applied Ecology, Biostatistics, Ethology, Genetics, Meat Sciences, Nutrition, Physiology, Poultry Husbandry, Poultry Science

Undergraduate Degrees: B.S. (Animal Science)

Undergraduate Degree Requirements: 126 credit hours including required courses

Graduate Degrees: M.S., Ph.D.

Graduate Degree Requirements: Master: 8 units of course work Doctoral: 8 additional units of course work and 8 units of thesis research

Graduate and Undergraduate Courses Available for Graduate Credit: Beef Production (U), Sheep Production (U), Pork Production (U), Poultry Management (U), Genetics and Animal Improvement (U), Environmental Management (U), Meat Science (U), Nutrition and Physiology of Ruminants (U), Principles of Animal Nutrition (U), Reproduction and Artificial Insemination of Farm Animals (U), Livestock Marketing (U), Introduction to Applied Statistics (U), Human Evolution (U), Ethology (U), Ethology Laboratory (U), World Animal Agriculture (U), Animal Bionomics, Technics and Topics in Animal Research, Concepts in Nonruminant Nutrition, Physiology of Reproduction, Laboratory Methods in Physiology of Reproduction, Muscle Biology, Research Methods in Animal Science, Advanced Endocrinology, Advanced Animal Genetics, Population Genetics and Animal Breeding, Protein and Energy Nutrition, Minerals and Vitamins in Metabolism, Design and Analysis of Biological Experiments, Radioisotopes in Biological Research, Animal Biochemical Laboratory Technics

DEPARTMENT OF DAIRY SCIENCE

(No reply received)

DEPARTMENT OF HORTICULTURE

Phone: Chairman (217) 333-0350

Head: C.J. Birkeland

Professors: C.Y. Arnold, C.J. Birkeland, J.R. Culbert, D.F. Dayton, F. DeVos (Adjunct), D.B. Dickinson, J.B. Gartner, M.T. Hall (Adjunct), E.R. Jaycox, M. B. Linn, J.B. Mowry, W.R. Nelson, R.K. Simons, W.E. Splittstoesser, J.S. Vandemark, C.C.Zych

Associate Professors: M.C. Carbonneau, J.W. Courter, J.H. Hopen, D.B. Meador, M.M. Meyer, C.A. Rebeiz, A.M. Rhodes

Assistant Professors: M.C. Dirr, G.M. Foster, F.A. Giles, T.D. Hughes, J.C. McDaniel, D.C. Saupe, L.A. Spomer, A.J. Turgeon, D.J. Williams

Instructors: C.C. Rebeiz, C.Y. M. Chu, G.J. Gynn, W.A. Kelley, A.G. Ottbacher, P.W. Spencer

Research Assistants: 2

Field Stations/Laboratories: Drug and Horticulture Station - Downes Grove, Illinois Horticulture Experimental Station - Carbondale, Dixon Springs

Degree Programs: Genetics, Horticultural Science, Horticulture, Ornamental Horticulture, Physiology, Plant Breeding, Pomology, Vegetable Crops

Undergraduate Degree: B.S.

Undergraduate Degree Requirements: Not Stated

Graduate Degrees: M.S., Ph.D.

Graduate Degree Requirements: Master: B average 8 units Doctoral: B average, 8 units beyond M.S. plus 8 units of thesis and examination

Graduate and Undergraduate Courses Available for Graduate Credit: Special Problems, Floriculture Physiology, Plant Nutrition, Principles of Plant Breeding, Plant Physiology Laboratory, Economics of Food Production, International Food Production, Introduction to Applied Statistics, Growth and Development of Horticultural Crops, Nutrition of Plants, Post Harvest Physiology, Pigments, Design and Analysis of Biological Experiments, Research Methods, Thesis

DEPARTMENT OF FORESTRY
Phone: Chairman: (217) 333-2770 Others: 333-2771

Head: I.I. Holland (Acting)
Professors: H.R. Gilmore, I.I. Holland, C.S. Walters
Associate Professors: D. Chow, J.K. Guiher, J.J. Jokela, T.R. Yocom
Assistant Professors: D.R. Pelz, J.F. Owyer, T.W. Curtin (Extension), H. Fox (Extension), R.E. Nelson (Extension), G.L. Rolfe
Instructors: M. Bolin, R.A. Young
Research Assistants: 22
Professional: 3

Field Stations/Laboratories: Sinissippi Forest, Oregon, Illinois, Dixon Springs Agricultural Center - Simpson

Degree Program: Forestry

Undergraduate Degree: B.S.
Undergraduate Degree Requirements: 126 Semester hours including 24 semester hours in forestry and related courses in those areas of special topics prescribed by U.S. Civil Service Commission for certification on Agriculture and Biological register.

Graduate Degree: M.S.
Graduate Degree Requirements: Master: 8 graduate units, thesis, optional

Graduate and Undergraduate Courses Available for Graduate Credit: Environment and tree Growth (U), Environment and Plant Ecosystems (U), Introduction to Applied Statistics (U), Mechanical Properties of Wood and Wood-based Materials, Seminar in Forestry, Special Problems in Forest Ecology and Physiology, Design and Analysis of Biological Experiments, Discussions in Forest Policy Thesis Research

DEPARTMENT OF PLANT PATHOLOGY
Phone: Chairman: (217) 333-3170 Others: 333-3171

Head: R.E. Ford
Professors: J.W. Gerdemann, D.Gottlieb, A.L. Hooker, P.D. Shaw, M.C. Shurtleff, J.B. Sinclair, E.B. Himelick, R.D. Neely
Associate Professors: J.D. Paxton, D.W. Chamberlain, D.F. Schoenweiss
Assistant Professors: R.M. Goodman, B.J. Jacobsen, S.M. Lim, R.B. Malek, G.M Milbrath, S.M. Ries, D.G. White, D.I. Edwards, L.E. Gray, W. Harstirn, H. Jedlinski
Research Assistants: 25

Field Stations/Laboratories: Brownstown, South Farm, and too many to list. No outlying departmental field stations, however.

Degree Program: Plant Pathology

Graduate Degrees: M.S., Ph.D.
Graduate Degree Requirements: Master: A candidate for the M.S. degree must complete eight units of credit, including the presentation of an acceptable thesis, and pass an oral examination. Candidate must maintain an average of 3.75 for the program. Doctoral: Eights units of course work including required courses, must be completed beyond the master's to qualify for the preliminary examination. An oral preliminary examination before five faculty members including three from the department and two from outside the department is required A final examination by the same committee follows the presentation of an acceptable dissertation.

Graduate and Undergraduate Courses Available for Graduate Credit: Research Methods in Plant Pathology, Plant Nematology, Forest Tree Diseases and Wood Deterioration, Plant Disease Development and Control, International Food Crops, Plant Disease Diagnosis, Diseases of Forest and Shade Trees, Phytobacteriology, Physiology of Fungi, Plant Virology, Genetics of Plant Pathogen Interactions, Physiology of Plant Parasite Interactions

WESTERN ILLINOIS UNIVERSITY
Macomb, Illinois 61455 Phone: (309) 298-1546
Dean of Graduate Studies: J.H. Sather
Dean of College: W. Olson

DEPARTMENT OF BIOLOGICAL SCIENCES
Phone: Chairman (309) 298-1546

Chairman: E.F. Morris
Professors: L.D. Dove, E.B. Holmes, T.H. Ma, E.F. Morris, R.M. Myers, P.M. Nollen, J.H. Sather, Y. S. Sedman, J.E. Warnock, G.H. Ware (Adjunct)
Associate Professors: H.H. Edwards, E.C. Franks, R. D. Henry, V.K. Howe, J.D. Ives, L.A. Jahn, J.H. Larkin, M.R. Murnik, L.M. O'Flaherty, V.C. Pederson, S. Singer, B.M. Stidd, G.R. Thurow, J.K. Turner, W.M. Walter
Assistant Professors: G.W. Cartwright, T.C. Dunstan, P.J. Nielsen
Research Assistants: 8
Teaching Assistants: 14

Field Stations/Laboratories: The Alice L. Kibbe Life Science Station - Warsaw

Degree Programs: Biology, Botany, Zoology

Undergraduate Degree: B.S.
Undergraduate Degree Requirements: Biology, Botany, Zoology, 54 quarter hours, Chemistry, 16 quarter hours, Physics, 12 quarter hours, Mathematics, 14 quarter hours, Foreign Language, 1 year, 186 quarter hours required for graduation.

Graduate Degree: M.S.
Graduate Degree Requirements: Master: Botany emphasis) = at least 36 quarter hours in Botany with remaining hours in minor or related field. Total = 48 quarter hours. (Zoology emphasis) - at least 36 quarter hours in Zoology with remaining hours in minor or related field Total = 48 quarter hours

Graduate and Undergraduate Courses Available for Graduate Credit: General Microbiology (U), Plant Taxonomy (U), General Mycology (U), Morphology of Angiosperms, Bryology (U), Phycology (U), Metabolism of Plants (U), Radiation Biology (U), Advanced Genetics (U), Plant Ecology (U), Freshwater Biology (U), Economic Botany (U), Principles of Plant Pathology (U), Aquatic Plants, Principles of Taxonomy, Evolution, Cytology, Microbial Physiology, Cellular Physiology, Plant Geography, Ornithology (U), Entomology (U), Mammalogy (U), Herpetology (U), Ichthyology (U), Protozoology (U), Animal Physiology (U), Endocrinology (U), Animal Ecology (U), Parasitology (U), Insect Morphology, Anatomy of Nervous System, Zoogeography, Animal Behavior, Game Management, Experimental Parasitology

WHEATON COLLEGE
Wheaton, Illinois 60187 Phone: (312) 682-5000
Academic Vice President: D. Mitchell

DEPARTMENT OF BIOLOGY
Phone: Chairman: (312) 682-5008 Others: 682-5000

Chairman: R.H. Brand
Professors: R.H. Brand, J.L. Leedy, C.E. Luckman, C.O. Mack, R.L. Mixter
Associate Professors: D.S. Bruce, A.J. Smith
Assistant Professor: P.P. Pun
Instructor: J.A. Arnold (Laboratory Associate)

Field Stations/Laboratories: Black Hills Science Station -

Rapid City, South Dakota

Degree Programs: Biology, Medical Technology

Undergraduate Degrees: A.B., B.S.
Undergraduate Degree Requirements: 45 quarter courses, 15 Liberal Arts courses, in General Education, 13 courses in Biology, 6 courses in cognate Science, Comprehensive Examinations

INDIANA

ANDERSON COLLEGE

Anderson, Indiana 46011 Phone: (317) 644-0951
Dean of the Faculty: D.C. Hoak

DEPARTMENT OF BIOLOGY
Phone: (317) 644-0951 Chairman: Ext. 324

Chairman: J.D. Goodman
Professor: J.D. Goodman
Associate Professors: M.J. Mayo, O.S. Phalora, J.R. Rees
Assistant Professor: J.E. Sipe
Visiting Lecturers: 5

Degree Program: Biology

Undergraduate Degree: B.S.
Undergraduate Degree Requirements: Thirty hours in appropriate biological subjects. Four streams or submajors. Cellular Biology, Medical Biology, Environmental Biology and Teaching Biology

BELL STATE UNIVERSITY

Muncie, Indiana 47306 Phone: (317) 289-1241
Dean of Graduate Studies: R. Koenker
Dean of College: R.L. Carmin

DEPARTMENT OF PHYSIOLOGY AND HEALTH SCIENCES
Phone: Chairman: (317) 285-5961

Chairman: W.E. Schaller
Professors: W. Bock, C.C. Boyer, G. Branam (Adjunct), C.R. Carroll, R.E. Henzlik, H.L. Jones, N. Miller (Adjunct), L. Montgomery (Adjunct), H.D. Paschall, R.E. Siverly
Associate Professor: D.M. Dennison
Assistant Professor: L.R. Ganion, D.D. Gowings, M.S. Jarial, T.A. Lesh, W.A. Payne, J.J. Pelizza, J.H. Shirreffs, R.H. Travis, G.R. Weber, F.A. Wrestler
Education Student Fellows: 7
Graduate Assistants: 12

Degree Programs: Anatomy and Physiology, Health and Biological Sciences, Physiology

Undergraduate Degree: B.S.
Undergraduate Degree Requirements: Teaching major in health and safety--60 quarter hours, Teaching minor in health and safety --36 quarter hours, Endorsement for teaching a subject field: health science and safety--36 quarter hours, Departmental minor in physiology--32 hours, Departmental minor in public health--36 quarter hours

Graduate Degrees: M.S.
Graduate Degree Requirements: Master: Arts: Major area of study including credit for thesis, research paper or creative project...24-45 quarter hours. Minor area of study of 12 hours and/or electives in any area or areas including major...21-0 hours. Total = 45 hours. Master of Science: same as above except thesis is required. Total = 45 hours. Specialist in Education: 1. hold a master's degree 2, two years of teaching experience 3. 45 quarter hours graduate work beyond master's.

Graduate and Undergraduate Courses Available for Graduate Credit: Health Science: Health Science of Distant Areas, School Health Practice (U), Health Problems of School Children, Health in the Family (U), Workshops in Health Science Education, Alcohol Problems (U), Current Progress in Disease Control (U), Drug Dependence and Abuse (U), Health Quackery (U), Health and Aging (U), Health Science Research Techniques, Death and Dying, Internship in Health Science (U), Public Health Practice (U), Current Problems in Public Health, Environmental Health (U), Epidemiology (U), Community Health Program Practices and Techniques, Seminar in Health Science, Special Studies in Health Science (U). Anatomy: Human Gross Anatomy (U), Human Neuroanatomy (U), Medical Neuroanatomy, Human Embryology (U), Histology, Medical Histology-Embryology, Fundamentals of Pathology, Ultrastructural Anatomy and Electron Microscopy, Special Studies in Anatomy Physiology: Mammalian Physiology (U), Endocrinology, Cellular Physiology, Medical Physiology, Special Studies in Physiology Science: Life Science Statistics (U)

BETHEL COLLEGE

1001 W. McKinley Avenue Phone: (219) 259-8511
Mishawaka, Indiana 46544
Dean of College: W. Gerber

DIVISION OF NATURAL SCIENCE
Phone: Chairman: (219) 259-8511 Ext. 63

Chairman: R. Bennett
Assistant Professors: K. Esau, J.P. McLaren

Degree Program: Biology

Undergraduate Degree: B.A.
Undergraduate Degree Requirements: 32 hours Biology, 8 hours General Chemistry, 8 hours Organic Chemistry

BUTLER UNIVERSITY

4600 Sunset Avenue Phone: (317) 283-8000
Indianapolis, Indiana 46208
Dean of Graduate Studies: D.R. Roberts

College of Liberal Arts and Sciences

Dean: D.M. Silver

BOTANY DEPARTMENT
Phone: Chairman: (317) 283-9413

Chairman: J.F. Pelton
Professors: R.N. Webster
Associate Professor: W.F. Yates, Jr.
Visiting Lecturers: 4

Degree Programs: Biology, Botany, Environmental Science

Undergraduate Degree: B.S.
Undergraduate Degree Requirements: 30 semester hours in Botany and 10 in Zoology or 20 semester hours in Botany and 20 in Zoology, plus Chemistry through Organic

Graduate Degree: M.S.
Graduate Degree Requirements: Master: 24 hours and thesis or 36 hours without thesis

Graduate and Undergraduate Courses Available for Graduate Credit: Man and Environment (U), Trees and Shrubs (U), Evolutionary Trends in Higher Plants (U), Plant Anatomy (U), Special Problems (U), Microbiology (U), Biology Seminar (U), Microtechnique (U), Phycology (U), General Genetics (U), Plant Physiology (U), General Ecology (U), Cytology (U), Seminar in Plant

Ecology (U), Plant Taxonomy (U), Biosystematics (U), Mycology (U), Thesis (U)

DEPARTMENT OF ZOOLOGY
Phone: Chairman: (317) 283-9411

Chairman: P.A. St.John
Professors: E. Durflinger, P.A. St.John
Associate Professors: J.W. Berry, D.W. Osgood
Assistant Professors: D.L. Daniell, S.A. Perrill
Instructors: C.E. Russell (part-time)

Degree Program: Zoology

Undergraduate Degree: B.A., B.S.
Undergraduate Degree Requirements: General Zoology, Comparativo Anatomy, plus 20 hours Zoology, 2 years Chemistry, 1 year Physics, 1 semester Mathematics

Graduate Degree: M.S.
Graduate Degree Requirements: Master: 24 hours course work plus 6 hours Research and Thesis

Graduate and Undergraduate Courses Available for Graduate Credit: Vertebrate Histology, Parasitology, Environmental Conservation, Animal Ecology, Animal Behavior, Biology Seminar, Molecular G netics, Laboratory Techniques in Cell Physiology, Invertebrate Zoology, Evolution, Cell Structure and Function, Ornithology, Field Techniques in Biology

CONCORDIA SENIOR COLLEGE
(No Information available)

DePRAUW UNIVERSITY
Greencastle, Indiana 46135 Phone: (317) 653-9721
Dean of Graduate School: D. Anderson
Dean of College: R.H. Farber

DEPARTMENT OF BOTANY AND BACTERIOLOGY
Phone: Chairman: (317) 653-9721 Ext. 496

Head: H.R. Youse
Professor: W.P. Adams, H.R. Youse
Assistant Professor: R.I. Fletcher

Degree Program: Bacteriology, Botany

Undergraduate Degree: B.A.
Undergraduate Degree Requirements: 8 courses plus related subjects

DEPARTMENT OF ZOOLOGY
Phone: Chairman: Ext. 378

Chairman: F.D. Fuller
Professors: F.D. Fuller, J.R. Gammon, A.E. Reynolds (Emeritus)
Associate Professor: C.E. Mays
Assistant Professor: M.D. Johnson

Degree Program: Zoology

Undergraduate Degree: B.A.
Undergraduate Degree Requirements: 8 courses which must include Principles of Animal Biology, Invertebrate Zoology, Vertebrate Zoology, Developmental Biology and Seminar, Additional courses in Physiology, Ecology, and Genetics - evolution are required. The major must include at least four courses from the areas of Chemistry, Physics, Botany, Geology, and/or Mathematics

Graduate Degree: M.A.
Graduate Degree Requirements: Master: 8 courses above the 300 level. A minimum of 3 courses at the 500 level in addition to the thesis. Demonstrated reading knowledge of a foreign language. Oral examination of thesis and field of study.

Graduate and Undergraduate Courses Available for Graduate Credit: Cell Biology (U), Entomology (U), Genetics (U), Evolution (U), Investigational Methods (U), Histology (U), Advanced Ecology (U), Endocrinology (U), Comparative Physiology (U), Parasitology (U), Research Problems (U), Seminar (U), Field Station Zoology (U), Graduate Topics - Morphology, Developmental Zoology, Ethology, Entomology, Invertebrate Zoology, Systematic Zoology, Population Studies, Hydrobiology, Protozoology, Physiological Zoology, Thesis

EARLHAM COLLEGE
Richmond, Indiana 47374
Dean of College: J.E. Elmore

DEPARTMENT OF BIOLOGY

Chairman: W.K. Stephenson
Professor: W.K. Stephenson
Associate Professors: J.H. Woolpy, G.L. Ward (Joint appointment with Joseph Moore Museum)
Assistant Professors: D. Hoyt, W.H. Harvey, O.J. Blanchard, W.H. Buskirk
Lecturer: J.B. Cope (Joint Appointment with Joseph Moore Museum)

Degree Program: Biology

Undergraduate Degree: B.S.
Undergraduate Degree Requirements: Biology - 10 courses (33 1/3 semester hours) including Genetics, Plant Biology, Population Biology, Field Biology, Physiology, Independent Studies, and Senior Seminar. Chemistry: 3 courses (10 semester hours) including one course in Organic Chemistry. Mathematics, Computer Science, and Physics are recommended but not required

FRANKLIN COLLEGE
Franklin, Indiana 46136 Phone: (317) 736-8441
Dean of College: R.M. Park

DEPARTMENT OF BIOLOGY
Phone: Chairman: Ext. 110

Chairman: C.B. Knisley
Professor: R.J. Trankle
Associate Professors: J.R. Curry, C.B. Knisley

Degree Program: Biology

Undergraduate Degree: B.S.
Undergraduate Degree Requirements: 26 hours in Biology (General, Genetics, Development, Ecology, Plant Survey, Seminar, Physiology), 10 hours in Chemistry, 5 hours in Mathematics, 126 hours total plus 3 winter term courses

GOSHEN COLLEGE
1700 South Main Street Phone: (219) 533-3161
Goshen, Indiana 46526
Dean of College: J.A. Lapp

DEPARTMENT OF BIOLOGY
Phone: Chairman: Ext. 250

Chairman: C.F. Bishop
Professors: H.C. Amstutz, C.F. Bishop, M.E. Jacobs, J.N. Roth

Field Stations/Laboratories: Layton Marine Biology Laboratory - Layton (Long Key), Florida

Degree Program: Biology

Undergraduate Degree: B.A.

Undergraduate Degree Requirements: Biology, 24 hours, Chemistry, 11 hours, Plus Additional Natural Science to total 40 hours

HANOVER COLLEGE

Hanover, Indiana 47243 Phone: (812) 866-2151
Dean of College: H.J. Haverkamp

DEPARTMENT OF BIOLOGY
Phone: Chairman: Ext. 310

Chairman: J.D. Webster
Professors: E.G. Prau, J.D. Webster
Associate Professors: J.H. Maysilles
Assistant Professors: P.C. MacMillan, R.H. Sherwin

Degree Program: Biology

Undergraduate Degree: B.A.
Undergraduate Degree Requirements: (one course = about 5 semester hours) 29 courses total, 12 1/2 courses specified outside the natural sciences. For Biology major, 1 course in Physics, 2 courses in Chemistry and 7 courses in Biology. The last must include 2 General Biology, 1 Genetics, and 1 Senior Independent Study

HUNTINGTON COLLEGE

2303 College Avenue Phone: (219) 356-6000
Huntington, Indiana 46750
Dean of College: W. Custer

BIOLOGY DEPARTMENT
Phone: Chairman: Ext. 72

Head: F.D. Morgan
Professor: F.D. Morgan
Assistant Professor: R. Priddy

Field Stations/Laboratories: Thornhill Field Station - 7 miles north of Huntington

Degree Programs: Biology, Biological Education, Medical Technology

Undergraduate Degrees: B.A., B.S.
Undergraduate Degree Requirements: Division of Humanities and Bible -- 7 courses, Division of History and Social Sciences--4 courses, Division of Natural Science and Mathematics--3 courses, Division of Education--1 course

INDIANA CENTRAL COLLEGE

(No reply received)

INDIANA STATE UNIVERSITY AT EVANSVILLE

8600 University Boulevard Phone: (812) 426-1251
Evansville, Indiana 47712
Dean of College: D. Bennett

DEPARTMENT OF LIFE SCIENCES (BIOLOGY)
Phone: Chairman: Ext. 211

Chairman: M. Denner
Associate Professors: M.W. Denner, J.L. Murr
Assistant Professor: M.V. Shaw
Instructors: J.W. Reising, A.G. Denner (Adjunct), J. Schwengel (Adjunct)
Laboratory Assistant: 1

Degree Program: Life Science, Medical Technology

Undergraduate Degree: B.S.
Undergraduate Degree Requirements: 45 semester hours Biology, 18-21 hours Chemistry, 8-10 hours Physics, Mathematics through Calculus I (8-14 hours)

INDIANA UNIVERSITY AT BLOOMINGTON

Bloomington, Indiana 47401 Phone: (812) 337:6284
Dean of Graduate Studies: H. Yamaguchi

College of Arts and Sciences

Dean: V.J. Shiner

MICROBIOLOGY DEPARTMENT
Phone: Chairman: (812)337-6284

Chairman: D. Fraser
Professors: D. Fraser, H. Gest, A. Koch, W. Konetzka, L.McClung, E. Weinberg
Associate Professors: G. Hegeman, G. Sojka, M. Taylor
Assistant Professors: T. Blumenthal, D. White
Instructors: A. Williams (Laboratory Supervisor)
Research Associates: 7
Visiting Assistant Professor: 1

Degree Program: Microbiology

Undergraduate Degree: B.A.
Undergraduate Degree Requirements: General College Regulations.

Graduate Degrees: M.S., Ph.D.
Graduate Degree Requirements: Master: Minimum of four semesters, thesis (or approved refereed publication) Doctoral: All candidates must serve not less than one semester as teaching assistants except in unusual circumstances. Specific program details will depend on the student's background, professional goals, and interests. Language and Research Skill Demonstration of reading proficiency in one language or (preferably) proficiency in a research skill

Graduate and Undergraduate Courses Available for Graduate Credit: Biomedical Sciences Documentation (U), Proseminar (U), Environmental Microbiology (U), Virology (U), Medical Microbiology (U), Viral-Medical Laboratory (U), Molecular and Biochemical Genetics, Biophysics for Biologists, Microbial Development, Bacterial and Viral Genetics, Recent Advances in Microbiology, Physical Techniques in Biology, Regulation of Procaryote Metabolism, Host-Parasite Relationships

DEPARTMENT OF PLANT SCIENCES
Phone: Chairman: (812) 337-4115 Others: 337-5007

Chairman: A. San Pietro
Professors: C.W. Hagen, Jr., C.B. Heiser, Jr., C.O. Miller, M.M. Rhoades, D. Schwartz, A. San Pietro, R.C. Starr
Associate Professor: D.L. Dilcher, P.G. Mahlberg, A.W. Ruesink, B. Shalucha, R.K. Togassaki, G.R. Williams
Assistant Professors: G.J. Gastony, M.R. Tansey
Research Assistants: 7
Associate Instructors: 19
Research Associates: 3

Field Stations/Laboratories: Crooked Lake Biological Station

Degree Programs: Anatomy, Biological Chemistry, Biology Biostatistics, Botany Plant Sciences, Genetics, Pathology Mycology, Plant Physiology, Pomology

Undergraduate Degree: A.B. (Plant Sciences)
Undergraduate Degree Requirements: 6 courses in Biological Sciences

Graduate Degrees: M.S., Ph.D.
Graduate Degree Requirements: Master: 30 advanced hours (beyond A.B. or B.S.), plus language, thesis or tool skill Doctoral: 90 hours beyond A.B. or B.S. plus 2 languages, 1 foreign language and tool skill, qualifying examination passed, thesis

Graduate and Undergraduate Courses Available for Graduate Credit: Plant Anatomy (U), Algae (U), Mosses and Liverworts (U), Fungi (U), Fungi: Laboratory (U), Summer Flowering Plants (U), Cell Physiology (U), Ethnobotany (U), Cell Physiology Laboratory (U), Plant Physiology (U), Ecological Plant Physiology (U), Ecological Plant Physiology Laboratory (U), Science Workshop for Elementary Teachers (U), Introduction to Paleobotany (U), Comparative Morphology of Vascular Plants (U), Angiosperms (U), Biosystematics (U), Advanced Field Biology (U), Photobiology (U), Molecular Aspects of Biology, Molecular Aspects of Biology Laboratory, Problems in Genetics-Higher Organisms, Seminar in Cytogenetics, Bytogenetics, Anatomy and Morphology Seminar, Experimental Mycology, Developmental Plant Physiology, Plant Biochemistry, Algae and Fungi

DEPARTMENT OF ZOOLOGY
Phone: Chairman: 337-7322

Chairman: J.H. Sinclair
Professors: W.R. Breneman (Waterman Professor), R.W. Briggs (Research), P. S. Crowell, D.G. Frey, J.P. Holland, V. Nolan, J. Preer, F.W. Putnam (Distinguished), T.M. Sonneborn (Distinguished), F.N. Young
Associate Professors: R.V. Dippell, J.M. Emlen, G.A. Hudock, A. Mahowald, G. Malacinski, J.H. Sinclair, F. Zeller, C.E. Nelson
Assistant Professors: R. Raff, P. Randolph, R. Richmond, W.J. Rowland, S.L. Tamm
Instructor: S.A. Pollack
Teaching Fellows: 41
Research Assistants: 15

Degree Program: Biology
Undergraduate Degree Requirements: Not Stated

Graduate Degrees: M.A., Ph.D.
Graduate Degree Requirements: Master: 30 advanced hours (beyond A.B. or B.S.) plus language, thesis or tool skill Doctoral: 90 hours beyond A.B. or B.S. plus 2 languages, 1 foreign language and tool skill, qualifying examination passed, thesis

Graduate and Undergraduate Courses Available for Graduate Credit: Not Stated

INDIANA UNIVERSITY AT INDIANAPOLIS

1201 E. 38th Street Phone: (317) 923-1321
Indianapolis, Indiana 46205
Dean of College: W.A. Nevill

DEPARTMENT OF BIOLOGY
Phone: Chairman: Ext. 271 Others: 269

Chairman: R. Samuels
Professors: R. Samuels, R.C. Sanborn
Assistant Professors: W.S. Courtis, R.W. Keck, R.E. Kirk, N.D. Lees, R.G. Pflanzer, R.H. Schaible (Adjunct)
Teaching Fellows: 4
Part-Time Lecturers: 4
Instructor: F.L. Juillerat

Degree Program: Biology

Undergraduate Degree: B.A.
Undergraduate Degree Requirements: 25 credits including core, beyond entering sequence; 1 year Organic Chemistry with Laboratory, 1 year Calculus, 1 year Physics, 1 year language

INDIANA UNIVERSITY AT KOKOMO

2300 S. Washington Phone: (317) 453-2000
Kokomo, Indiana 46901
Dean of Faculties: H.C. Miller

DIVISION OF BIOLOGICAL AND PHYSICAL SCIENCES
Phone: Chairman: (317) 453-2000 Ext. 252

Chairman: P. Haffley
Associate Professors: R.F. Boneham, R.C. Hanig
Assistant Professors: P.G. Haffley, F. Steldt, H.G. Wilhelm, Jr.
Lecturers: G. Dolph
Laboratory Assistants: S. Johnson, J. Kirkwood
Associate Faculty in the Division: 3

Degree Program: Liberal Studies (Biology Major)

Undergraduate Degree: A.B.
Undergraduate Degree Requirements: 120 semester hours

INDIANA UNIVERSITY NORTHWEST

Gary, Indiana 46408 Phone: (219) 887-0111
Dean of College: F.C. Richardson

DEPARTMENT OF BIOLOGY
Phone: Chairman: 887-0111 Ext. 385

Chairman: P.K. Bhattacharya
Associate Professors: J.H. Dustman, G.D. Hanks, F. C. Richardson
Assistant Professors: C.R. LaFrance, S.W. May, T.A. Stabler
Instructors: E. Haller

Degree Program: Biology

Undergraduate Degree: A.B.
Undergraduate Degree Requirements: Total hours 120. Requirements for the major: At least 24 hours of course work above the 100 level. Upper level courses in Physiology (Cellular or Organismal), Genetics, Ecology, and a Seminar course must be included in the 25-hour major. Also, at least 10 hours each of Chemistry through Organic Chemistry, and Physics are required. Additional Electives to meet requirements of 120 hours will include courses in foreign language, Mathematics, Physical and Life Sciences, Social and Behavioral Sciences, Literature, Philosophy and Arts.

INDIANA UNIVERSITY SCHOOL OF MEDICINE

1100 W. Michigan Street Phone: (317) 635-8661
Indianapolis, Indiana 46202
Dean of Graduate Studies: H.G. Yamaguchi
Dean of School of Medicine: S.C. Beering

ANATOMY DEPARTMENT
Phone: Chairman: (317) 264-7494

Chairman: C.E. Blevins
Professors: W. Andrew, C.E. Blevins, M. Ishaq, R. Jersild, C.R. Morgan, R.H. Shellhamer
Associate Professors: J.D. Fix, A.A. Katzberg, L.C. Perkins, J.F. Schmedtje, R. Webster
Assistant Professors: D.L. Felten, B.L. O'Connor
Research Assistants: 2
Associate Instructors: 18

Degree Program: Anatomy

Graduate Degrees: M.S., Ph.D.
Graduate Degree Requirements: Master: Gross Anatomy 10 hours, Histology 5 hours, Neurobiology 4 hours, Minor Field 12 hours (Biological Science or Sciences Basic to Medicine), Thesis Required, Written examination in major and minor field. Doctoral: Courses required: Minimum of 45 hours in courses other than research, including Gross Anatomy, Histology, Neurobiology, and Seminar, Minor: 12 hours in Physiology, biochemistry, Pathology, or Life Sciences. Language and Research Skill: Reading proficiency in Biostatistics. Qualifying Examination - Written Oral Defense of thesis

Graduate and Undergraduate Courses Available for Graduate Credit: Gross Anatomy, Histology, Neurobiology, Human Developmental Anatomy, Histochemistry and Cytochemistry, Comparative Histology, Anatomy of the Organs of Special Sense, Tissue Responses to Hormone Stimulation, Topographic Anatomy, Anatomical Techniques, Autonomic Nervous System, Advanced Anatomy, Electron Microscopy, Histology of the Immune System, Tissue Culture, Primate Histology

DEPARTMENT OF BIOCHEMISTRY
Phone: Chairman: (317) 264-7151

Chairman: D.M. Gibson
Professors: M.H. Aprison, J.F. Bonner, Jr., D.E. Bowman, D.R. Challoner, E.J. Davis, M.D. Gibson, T.K. Li, L.K. Steinrauf
Associate Professors: D.W. Allmann, P.V. Blair, R.T. Blickenstaff, W.P. Bryan, L.T. Graham, E.T. Harper, R.A. Harris, L. Lumeng, R.W. Roeske, A.R. Schulz
Assistant Professors: J.A. Hamilton, J.M. Pinkerton
Visiting Lecturers: 1
Teaching Fellows: 5
Research Assistants: 12
Postdoctoral Research Associates: 6

Degree Program: Biochemistry

Graduate Degrees: M.S., Ph.D.
Graduate Degree Requirements: Master: 24 hours in courses other than research, including 12 in Biochemistry and six in Graduate level Chemistry courses. Thesis required. Doctoral: 45 hours in courses other than research (12 hours in Biochemistry). Two minors are required: 6 hours in graduate Chemistry courses and 12 hours in one of ten other specified areas. Serial qualifying examinations;language and/or research skills. Thesis

Graduate and Undergraduate Courses Available for Graduate Credit: General Biochemistry, Advanced Biochemistry, Enzymology, Advanced Organic Chemistry, Intermediary Metabolism, Advanced Intermediary Metabolism, Bio-Organic Chemistry, Chemistry of Steroids, Biochemical Nutrition, Neurochemistry, Instrumentation and Methods of Analysis I and II, Mathematics for Biochemistry, Problems, Physical Chemistry, Advanced Physical Chemistry, Biophysical and Protein Chemistry, Seminar, Fundamental Molecular Biology

DEPARTMENT OF MEDICAL BIOPHYSICS
Phone: Chairman: (317) 264-7940 Others: 635-8661

Chairman: S. Ochs
Professors: F. Abel, W. Armstrong, J. Friedman, K. Greenspan, R. Jersild, S. Ochs, C. Rothe, L. Steinrauf, W. Zeman
Associate Professors: W. Bryan, J. Norton, R. Paradise
Assistant Professors: H. Besch, R. Bockrath, R. Haak, J. Hamilton

Degree Program: Medical Biophysics

Graduate Degrees: M.S., Ph.D.
Graduate Degree Requirements: Master: 30 semester hours total, 20 hours in Biophysics, including 7 hours in research; remaining hours in related courses, Thesis required. Doctoral: 90 semester hours total, 35 in Biophysics courses other than research, up to 30 hours may be taken on other approved departments. Minor required, qualifying examination and oral thesis defense, research skill- Biostatistics and (computer Science, Biomedical Instrumentation, and/or a foreign language).

Graduate and Undergraduate Courses Available for Graduate Credit: Introduction to Biophysics, Research, Seminar, Special Problems, Introductory Quantum Biophysics, Molecular Structure, Gas and Fluid Flow, Membrane Biophysics, Excitation and Transmission in Neural Systems, Biological Systems, Modeling, Microbial Mutagenesis and Genetics, Fundamental Molecular Biology

DEPARTMENT OF MICROBIOLOGY
Phone: Chairman: (317) 264-7671 Others: 264-7672

Chairman: W.D. Sawyer
Professors: D.C. Bauer, J.S. Ingraham, A.S. Levine, S.A. Minton, Jr., D.J. Neiderpruem, H. Raidt, W.D. Sawyer, E.W. Shrigley, W.A. Summers
Associate Professors: L.C. Olson, W.S. Begener, J.P. Burnett (Adjunct)
Assistant Professor: R.C. Bockrath, A.G. Matthysse, C.H. Miller, R.R. Watson
Research Assistants: 5
Research Fellow: 1

Degree Program: Microbiology

Graduate Degrees: M.S., Ph.D.
Graduate Degree Requirements: Master: 30 credits, Thesis, Doctoral: 90 credits, thesis, skill or language

Graduate and Undergraduate Courses Available for Graduate Credit: Microbiology, Microbiology for Dental Students, Advanced Microbiology, Immunochemistry: Lecture, Immunochemistry: Laboratory, Research in Microbiology, Parasites and Parasitic Diseases: Lecture, Parasites and Parasitic Diseases: Laboratory, Biology and Chemistry of Fungi, Eukaryote-Prokaryote Interaction, Microbial Pathogenicity, General and Medical Microbiology, Virology, Seminar in Microbiology, Microbial Mutagenesis and Genetics: Lecture, Microbial Mutagenesis and Genetics: Laboratory, Microbial Physiology: Lecture, Oral Microbiology, Fundamental Molecular Biology

DEPARTMENT OF MEDICAL GENETICS
Phone: Chairman: (317) 264-7966

Chairman: A.D. Merritt
Professors: D. Bixler, I.K. Brandt, J.C. Christian, W.E. Nance, J.A. Norton, C.G. Palmer
Associate Professors: P.M. Conneally, E. Roach, P.L. Yu
Assistant Professors: R.M. Antley, K.W. Kang, R.H. Schaible
Instructors: T. Reed, P. Bader
Visiting Lecturers: 10
Teaching Fellows: 4
Research Assistants: 8
Graduate Students: 30
Postdoctorals: 10

Degree Program: Medical Genetics

Graduate Degrees: M.S., Ph.D.
Graduate Degree Requirements: Master: Course Requirements: 20 hours in Medical Genetics, including no more than 7 hours in research, remaining in related courses. Thesis and Final Examination Required Doctoral: Course Requirement: 45 hours exclusive of research, including Physiology or Biochemistry and Medical Genetics courses. Up to 30 hours in non-clinical Medical or Dental Courses may apply toward Ph.D. Minors: Fields related to the major, e.g. Antrhopology, Biochemistry, Biology, Clinical Science, Genetics, Microbiology, Pharmacology. Language and research Skills: One of the following: (a) reading proficiency in one foreign language normally selected from French, German or Russian; (b) proficiency in researching skill (Computer Science or Biomedical Instrumentation). Thesis Required.

Graduate and Undergraduate Courses Available for Graduate Credit: Medical Genetics Laboratory I, Medical Genetics Laboratory II, Biochemical Genetics, Medical Genetics Laboratory III, Medical Genetics Laboratory IV, Mammalian Cytogenetics, Mammalian Cytogenetics Laboratory, Population Genetics, Quantitative Genetics, Special Topics in Human Genetics, Introduction to Biostatistics, Methods of Multivariate Analysis, Probability for the Biological Sciences, Medical Genetics Seminar, Methods in Human Genetics, Selected Topics in Biostatistics, Medical Genetics Research.

DEPARTMENT OF PHYSIOLOGY
(No reply received)

INDIANA UNIVERSITY AT SOUTH BEND
1825 Northside Boulevard S. Phone: (219) 282-2341
South Bend, Indiana 46615
Dean of College: D. Snyder

DEPARTMENT OF BIOLOGICAL SCIENCES
Phone: Chairman: (219) 282-2341 Ext. 291

Chairman: E.J. Savage
Associate Professors: J. Thomas, E.J. Savage
Assistant Professors: T. Gieske, R. Mehra, V. Riemenschneider, S. Winicur
Instructor: E.T. Hibbs

Degree Program: Biological Sciences

Undergraduate Degree: B.A.
Undergraduate Degree Requirements: Plant Biology and Animal Biology Core. semester each prerequisites to all upper level biology courses; 25 credit hours of Biology courses beyond the above two courses; one year Inorganic Chemistry, one semester Organic Chemistry, one year Introductory Physics, Mathematics to include one one semester Calculus (Note: 25 hours of Biology must include a course in Ecology, Physiology, and Genetics).

INDIANA UNIVERSITY SOUTHEAST
New Albany, Indiana 47150 Phone: (812) 945-2731
Dean of Faculties: W.B. Hebard

DEPARTMENT OF BIOLOGY
Phone: Chairman: 945-2731

Head: B.J. Forsyth
Professor: W.B. Hebard
Associate Professors: C.M. Christenson, B.J. Forsyth, R.H. Maxwell
Assistant Professor: G.A. Renwick
Instructors: T.R. Forsyth, F.C. Hill, D. Hotchkiss, L.R. Johnson, R. Moody
Laboratory Assistants: 5

Degree Program: Biology, Medical Technology

Undergraduate Degree: B.S.
Undergraduate Degree Requirements: (1) A minimum of 120 hours (2) a minimum cumulative grade-point average of 2.0 (3) a minimum of 30 hours in courses at the 300-400 level. At least 25 hours in major subject. In addition to these general requirements for a B.A. degree, the student must complete 30 hours in Biology above 100 level. One course in Genetics and Evolution, Cell and Organismal Physiology, Developmental Biology, Organismal Biology, Environmental Biology and 2 semesters of Biology Seminar, Chemistry 101 and 102 or 105 and 106 are required.

MANCHESTER COLLEGE
North Manchester, Indiana 46962 Phone: (219) 982-2141
Dean of Graduate Studies: E. Fahs
Dean of College: H. Book

BIOLOGY DEPARTMENT
Phone: Chairman: Ext. 266

Chairman: R.E. Niswander
Edward Kintner Professor of Zoology: R. Niswander
Professors: W. Eberly, P.A. Orpurt, R.E. Niswander
Assistant Professor: D. Kreps
Instructor: J. White

Degree Programs: Biology, Biological Education

Undergraduate Degrees: B.A., B.S.
Undergraduate Degree Requirements: Fundamentals of Botany, Fundamentals of Zoology, Principles of Genetics, Microbiology, Ecology or Limnology, Plant Physiology or Vertebrate Physiology, Special Problems in Biology, Electives--Minimum of 2 courses

Graduate Degrees: M.A. (Sec. Ed.)
Graduate Degree Requirements: Master: Area of Concentration in Biology for Teacher Certification Minimum of 3 courses (from below with at least 2 courses from "graduate only" offerings.

Graduate and Undergraduate Courses Available for Graduate Credit: Principles of Ecology (U), Vertebrate Anatomy (U), Vertebrate Physiology (U), Microbiology (U), Biological Change, Advanced Topics in Biology

MARION COLLEGE
Marion, Indiana 46952 Phone: (217) 674-6901
Dean of College: M. Burns

BIOLOGY DEPARTMENT
Phone: Ext. 275

Chairman: M. Hodson
Associate Professor: M. Hodson
Assistant Professor: D. Chilgreen, M. Hinds

Degree Program: Biology

Undergraduate Degrees: A.B., B.S.
Undergraduate Degree Requirements: A.B. - College General Requirements including 2 years language or equivalent, 30 hours Biology and 10 hours additional Biology or Cognate. B.S. same, language excepted

PURDUE UNIVERSITY
West Lafayette, Indiana 47907 Phone: (317) 749-8111
Dean of Graduate Studies: F.N. Andrews

School of Agriculture
Dean: R.L. Kohls

DEPARTMENT OF AGRONOMY
Phone: Chairman (317) 749-2891

Head: M.W. Phillips
Professors: J.L. Ahlrichs, S.A. Barber, L.F. Bauman, M.F. Baumgardner, B.O. Blair, R.D. Bronson, R.J. Bula, P.L. Crane, W.H. Daniel, H.M. Galloway, D.V. Glover, W.F. Keim, K.J. Lessman, P.F. Low, W.W. McFee, J.V. Mannering, J.E. Newman, W.E. Nyquist, A.J. Ohlrogge, F.L. Patterson, R.C. Pickett, C.L. Rhykerd, D. Swartzendruber, J. L. White, D. Wiersma, J.R. Wilcox
Associate Professors: J.D. Axtell, R.F. Dale, D.P. Franzmeier, M.E. Heath, H.F. Hodges, D.A. Holt, D.W. Nelson, W.D. Reiss, L.H. Smith, C.D. Spies, R.K. Stivers, M.L. Swearingin, J.J. Vorst, J.E. Yahner, A.L. Zachary
Assistant Professors: L.S. Beckman, E.L. Hood, A.W. Kirleis, H.R. Koller, V.L. Lechtenberg, H.W. Ohm, C.B. Roth, R.H. Shaw, L.E. Sommers, E.D. Treese, Sr.
Instructors: M.D. Abel, J.R. Fenwick, H.F. Hodges, J. J. Roberts, R.E. Ronnenkamp, G.L. Steinhardt, W.L. Stirm, J.B. Teigen, G.E. VanScoyoc
Research Assistants: 20
Postdoctorals: 3

Degree Programs: Agronomy, Crop Science, Conservation, Farm Crops, Genetics, Land Resources, Plant Breeding, Plant Industry, Soil Science, Turf, Agricultural Meterology, International Agronomy

Undergraduate Degree: B.S.
Undergraduate Degree Requirements: 131 credit hours including courses in core curriculum

Graduate Degrees: M.S., Ph.D.
Graduate Degree Requirements: Master: Program approved by guidance committee and departmental graduate committee; minimum of 24 credit hours and thesis or 33 credits with no thesis Doctoral: Program approved by guidance committee and departmental graduate committee, minimum of 45 credit hours beyond undergraduate degree, thesis and foreign language

Graduate and Undergraduate Courses Available for Graduate Credit: Pasture Crops and Management (U), Turfgrass Science (U), Field Crops Breeding (U), Crop Ecology (U), Advanced Genetics (U), Microclimatology (U), Biometerology (U), Statistical Climatology (U), Statistics for Experimental Research (U), Advanced Statistics for Experimental Research (U), Soil and Plant Analysis (U), Physical Properties of Soils (U), Soil Classification and Survey (U), American and Foreign Agronomic Problems (U), Intermediate Soil Science (U), Soil Microbiology (U), Soils and Land Use (U), Special Problems (U), Advanced Plant Breeding, Plant Cytogenetics, Environmental Physiology of Crops, Agronomic Research Techniques, Micrometerology, Crystal Structure and Identification of Clay Minerals, Soil Genesis and Classification, Chemistry and Fertility of Soils, Advanced Soil Fertility, Physical Chemistry of Soil, Advanced Soil Physics, Seminar in Atmospheric Sciences, Genetics Seminar, Field Crop Seminar

DEPARTMENT OF ANIMAL SCIENCE
(No reply received)

DEPARTMENT OF BOTANY AND PLANT PATHOLOGY
Phone: (317) 749-2946

Head: M.L. Tomes
Professors: K.L. Athow, C.E. Bracker, R.W. Curtis, R.J. Green, Jr., J.F. Hennen, T.K. Hodges, R.M. Lister, D.J. Morre, M.A. Ross, M.M. Schreiber, M. L. Tomes, J.F. Tuite, E.B. Williams, J.L. Williams
Associate Professors: R.B. Ashman, G.B. Bergeson, A. Dalby, D.M. Huber, C.B. Kenaga, F.A. Laviolette, C.A. Lembi, G.E. Shaner, D.H. Scott, C.Y. Tsai
Assistant Professors: T.S. Abney, A.O. Jackson, R.L. Nicholson, P.C. Pecknold, F.W. Roeth, W.R. Stevenson, H.L. Warren
Extension Assistants: 2
Visiting Lecturers: 1
Teaching Fellows: 3
Research Assistants: 21

Degree Programs: Plant Breeding, Plant Pathology, Plant Sciences, Plant Protection, Weed Science

Undergraduate Degree: B.S. (Plant Protection)
Undergraduate Degree Requirements: 130 credit hours, including 64 hours of Basic Core courses required of aoo graduates, and 66 hours of departmental requirements.

Graduate Degrees: M.S., Ph.D.
Graduate Degree Requirements: The requirements concerning the Master's and the Ph.D. degree are generally the same as those of the University. There is no general foreign language requirement, but any individual student's advisory committee may require a knowledge of one or more foreign languages.

Graduate and Undergraduate Courses Available for Graduate Credit: Advanced Weed Science, Laboratory in Plant Identification, Environmental Botany and Plant Pathology, Diseases in Ornamental Plantings, Plant Disease Control, Aquatic Botany, Methods in Plant Pathology, Fungi and Bacteria, Methods in Plant Pathology - Microtechnique, Photographic Methods in Botanical Research, Diagnosis of Plant Diseases, Principles of Plant Pathology, Plant Virology, Plant Diseases Caused by Nematodes, Plant Parasitic Fungi Physiology of Plant Disease, Advanced Topics in Plant Physiology, Botanical Writing, Physiological Genetics

DEPARTMENT OF BIOCHEMISTRY
Phone: Chairman: (317) 749-2391

Head: B. Axelrod
Professors: B. Axelrod, L.G. Butler, K.H. Kim, G.B. Kohlhaw, D.W. Drogmann, E.T. Mertz, H.E. Parker, F.W. Quackenbush (Emeritus), V.W. Rodwell, E.D. Schall, R.L. Whistler, H. Zalkin
Associate Professors: K.G. Brandt, F.E. Regnier, R.L. Somerville, H. Weiner
Assistant Professors: J.E. Dixon, K. Herrmann

Degree Program: Biochemistry

Undergraduate Degree: B.S.
Undergraduate Degree Requirements: 133 credits total, including 21 credits Chemistry (Freshman Chemistry, Organic, Physical), 8 credits Biology, 8 credits Physics, 18 credits Biochemistry (Analytical, General, Life Processes, Laboratories), 15 credits Mathematics (Algebra, Trigonometry, Calculus), 24 credits Humaniites or Industrial Management, Genetics, Economics, Science Electives

Graduate Degrees: M.S., Ph.D.
Graduate Degree Requirements: Master: Thesis, approximately 15 credits including comprehensive Biochemistry, Seminar. There is no Master's degree program. Doctoral: Thesis, successful completion of qualifying and preliminary examinations, 32 credits including comprehensive biochemistry, and seminar, Advanced Organic Chemistry, 5-7 credits in advanced Biochemistry courses, 12 credits Biology, C Chemistry, Medicinal Chemistry, Pharmacology, etc.

Graduate and Undergraduate Courses Available for Graduate Credit: General Biochemistry (U), Laboratory in General Biochemistry (U), Biochemistry of Carbohydrates, Advanced Topics in Carbohydrate Biochemistry, Biochemistry of Lipids, Polyisoprenoids, Proteins and Amino Acids in Nutrition, Mineral Metabolism, Biochemical Kinetics, Chemistry of Enzyme Action, Biochemical Control Mechanisms, Biochemical Aspects of Gene Action, Analytical Biochemistry, Plant Biochemistry, Biochemistry of Photosynthesis, Biochemistry of Nucleic Acids, Effects of Hormones on Biochemical Processes, Comprehensive Biochemistry (2 semesters), Seminar in Biochemistry, Annual Review of Biochemistry

DEPARTMENT OF ENTOMOLOGY
Phone: Chairman: 493-9185 Others: 749-2405

Head: E.E. Ortman
Professors: L. Chandler, R.C. Dobson, V.R. Ferris (Research), R.L. Gallun (Adjunct), R.L. Giese (Research), D.L. Matthew (Extension), J.V. Osmun (Extension and Research), D.L. Shandland (Research), M.C. Wilson (Research)
Associate Professors: D.B. Broersma, J.M. Ferris (Research), H.F. Goonewardene (Adjunct), R.C. Hall (Research), R.M. Hollingworth (Research, D.F. Sanders
Assistant Professors: G.W. Bennett, W.R. Campbell (Research), F.R. Courtsal (Adjunct), C.R. Edwards (Extension and Research), J.E. Foster (Adjunct), T.L. Harris, W.P. McCafferty, J.L. Overman (International Programs), R.E. Shade (Research), F.T. Turpin

(Research), A.C. York (Extension and Research)
Instructors: T.R. Hintz (Research), R.W. Meyer (Extension and Research), A.V. Provonsha (Research)
Research Assistants: 2

Field Stations/Laboratories: Entomology Field Operations Center, Throckmorton Farm, Andrews Farm - All within Lafayette vicinity

Degree Programs: Entomology, Urban and Industrial Pest Control

Undergraduate Degree: B.S.
Undergraduate Degree Requirements: B.S. Entomology 130 credit hours, B.S. Urban and Industrial Pest Control - 132 credit hours

Graduate Degrees: M.S., Ph.D.
Graduate Degree Requirements: Master: no specified credit hours although must earn two units of residence Doctoral: Ph.D. candidates must earn six units with a minimum of two units earned by continuous residence on the West Lafayette Campus.

Graduate and Undergraduate Courses Available for Graduate Credit: Fundamentals of Entomology, Entomological Information Retrieval, Introductory Insect Physiology and Biochemistry, Immature Insects, Taxonomy of Insects, Fundamentals of Insect Control, Economic Entomology, Principlas of Forest Entomology, Insecticides, Their Formulations and Applications, Urban and Industrial Pests, Medical and Veterinary Entomology, Urban and Industrial Pests II, Entomology for Science Teachers, Insect Natural History, Fundamentals of Nematology, Aquatic Entomology

FORESTRY AND CONSERVATION DEPARTMENT
Phone: Chairman (317) 749-2487 Others: 749-3811

Head: M.C. Carter
Professors: D.L. Allen (Wildlife Ecology, T.W. Beers (Forestry, W.R. Byrnes (Forest Soils), J.C. Callahan (Forest Economics), C.M. Kirkpatrick (Wildlife Management), C. Merritt (Forestry), C.I. Miller (Forestry), R.E. Mumford (Wildlife Management), E.W. Stark (Forestry), S.K. Suddarth (Wood Engineering)
Associate Professors: W.F. Beineke (Forestry), W.R. Chaney (Forestry), C.A. Eckelman (Forestry), R.M. Hoffer (Remote Sensing), M.O. Hunt (Wood Science and Wood Utilization), D.M. Knudson (Forestry), E.J. Lott (Extension Forester), J.W. Moser (Forestry), R.H. Perkins (Forestry), J.F. Senft (Wood Science)
Assistant Professors: H.L. Archibald (Ecology, K.M. Brown (Forestry), W.L. Fix (Forestry), J.L. Hamelink, (Fisheries Biology-Aquatic Ecology), H.C. Krauch (ForestryP, G.R. Parker (Forest Ecology)
Instructors: F.E. Goodrick, R.P. Mroczynski, R.W. Wenger
Teaching Assistants: 7
Research Assistants: 22
Fellowships: 3

Field Stations/Laboratories: Southern Indiana Purdue Agricultural Center, Purdue Wildlife Area, Cunningham Forrstry Farm, Shidler Tract

Degree Programs: Conservation, Forestry and Conservation, Wildlife Management, Wood Utilization

Undergraduate Degrees: B.S. (Resource Conservation, Wildlife Science) B.S.F. (Forest Products, Wood Utilization)
Undergraduate Degree Requirements: 1) Completion of one of several specified plans of Study. 2) Achievement of a minimum graduate index of 4.00.

Graduate Degrees: M.S. (Conservation, Wildlife Management), M.S.F. (Forest Management, Forest Economics, Silviculture, Wood Technology, Wood Utilization) Ph.D. (All topics in which M.S. or M.S.F. are offered)
Graduate Degree Requirements: Master: Completion of satisfactory Plan of Study on thesis and non-thesis options; acceptable research and thesis on the thesis options. Doctoral: Completion of satisfactory Plan of Study and research on doctoral dissertation.

Graduate and Undergraduate Courses Available for Graduate Study: Production Planning and Financial Control of Forestry Operations (U), Principles of Strength Design of Furniture (U), Fluid Transfer in Wood and Other Fibrous Materials (U), Chemical Treatment of Wood and Related Fibrous Materials (U), Plywood and Related Products (U), Physical Properties of Wood (U), Mechanics of Wood (U), Mechanics and Rheology of Fibrous Materials (U), Forest Soil and Water Management (U), Forest Tree Improvement (U), Physiological Ecology of Woody Plants (U), Economics of Natural Resource Systems (U), Ecology and Management of Wildlife (U), Game Management (U), Natural History of Vertebrates I: Mammalogy (U), Natural History of Vertebrates II: Ornithology (U), Ichthyology-Limnology (U), Fisheries Biology and Management (U), Vertebrate Population Dynamics (U), Forest Inventory (U), Research Methods in Forestry (U), Aerial Photo Interpretation (U), Remote Sensing of Natural Resources (U), Dynamics of Forest Populations (U), Natural Resources Institutions for Developing Nations (U), World Forest Resources (U), Remote Sensing Seminar (U), Conservation of Natural Resources I, II, III (U), Outdoor Recreation Administration (U), Recreation Resource Planning (U), Advanced Forest Ecology, Advanced Forest Management, Forest Typology, Advanced Forest Mensuration, Forest Resources Seminar, Topical Problems in Conservation of Natural Resources, Topical Problems in International Forestry, Topical Problems in Forest Production, Topical Problems in Wildlife Biology, Topical Problems in Wood Technology and Utilization

DEPARTMENT OF HORTICULTURE
Phone: Chairman: 749-2261 Ext. 227

Head: H.T. Erickson
Professors: J.H. Cherry, F.H. Emerson, H.T. Erickson, L. Hafen, J.E. Hoff, J. Janick, A.C. Leopold, P.E. Nelson, R.R. Romanowski, G.F. Warren, G.E. Wilcox
Associate Professors: P.L. Carpenter, H.C. Dostal, H. L. Flint, R.A. Hayden, C.M. Jones, S.L. Lam, G.H. Sullivan, E.C. Tigchelaar, T.D. Walker, J.A. Wott
Assistant Professors: R.A. Blakeley, R.H. Dougherty, G.F. Gerlach, P.A. Hammer, C.A. Mitchell, H.A. Robitaille, T.C. Weiler
Instructor: R.S. Grenard
Visiting Lecturers: 10
Teaching Fellows: 3
Research Assistants: 22

Field Stations/Laboratories: Horticulture Farms (2) Lafayette, In., Ornamental Field Laboratory - Lafayette, Southwest Horticulture Farm - Johnson

Degree Programs: Horticulture and Landscape Architecture, Graduate Degrees in Food Science, Plant Physiology, Plant Breeding, Plant Nurtition

Undergraduate Degrees: B.S.
Undergraduate Degree Requirements: Master: No specific credit hours. Course program and thesis followed by written and oral examinations and a thesis defense. Doctoral: No specific credit hours. Comprehensive course program under guidance of a committee. Written and oral examinations and a dessertation, all to be accepted and approved by the committee. No departmental language requirement, but major professor may request proficiency.

Graduate and Undergraduate Courses Available for Graduate Credit: Horticultural Crop Science (U), Landscape Plant Materials (U), Nutrition of Horticultural Crops (U), Plant Protection (U), Postharvest Physiology of

Horticultural Crops (U), Landscape Design and Community Development (U), Breeding of Horticultural Crop Plants (U), Advanced Wood Plant Material (U), Food Processing II (U), Physical and Chemical Principles in Agricultural Research (U), Recreation Resource Planing (U), Herbidical Action, Plant Growth and Development, Physiological and Biochemical Laboratory Techniques in Horticulture, Seminars

School of Science

Dean: F. Haas

DEPARTMENT OF BIOLOGICAL SCIENCES
Phone: Chairman: (317) 749-2665 Others: 749-3150

Head: K. Koffler
Frederick L. Hovde Distinguished Professor: H. Koffler
Wright Distinguished Professor: H.E. Umbarger
Professors: J. Altman, S. Arnott, A.I. Aronson, J.A. Chiscon, F.L. Crane, P.T. Gilham, L.F. Jaffe, J.S. Lovett, D.J. Morre, L.E. Mortenson, M. Moskowitz, R.D. Myers, S. Nakajima, W.L. Pak, S.N. Postlethwait, W.J. Ray, M.G. Rossmann, E.H. Simon, L.D. Smith, I. Tessman
Associate Professors: W.A. Cramer, G.D. Das, R.A. Dilley, D.L. Filmer, M. Forman, E.S. Golub, D.L. Hartl, M. Levinthal, Y. Nakajima, J.B. Olson, S.E. Ostroy, M. Stiller, J.W. Vanable, J.E. Wiebers, L.S. Williams, A.T. Winfree
Assistant Professors: M.O. Chiscon, K. Fujii, D.E. Graham, R.N Hurst, J.B. Kahle, J.R. Karr, M. Levy, J.E. Mittenthal, F.N. Nordland, L.H. Pinto, R.W. Smith, D.S. Woodruff, E.R. Wright
Instructors: C.F. Babbs, Jr., A. Berkovitz, J. Bray, C. R. Carlin, H-Z. Ho, L.R. Hubbard, E. Joern, T.G. Luce, T.M. O'Heron, W.J. Rogers, C.E. Schaier, R.H. Shippe, S.J. Uyeshiro
Visiting Lecturers: 6
Teaching Fellows: 81
Research Assistants: 26
Research Associates: 43

Field Stations/Laboratories: Ross Biological Reserve - West Lafayette

Degree Programs: Biochemistry, Bioengineering, Biological Education, Biology, Biophysics, Biostatistics, Botany, Ecology, Genetics, Land Resources, Microbiology, Medical Technology, Physiology, Zoology, Premedicine, Predentistry, Preventive Medicine, Neurobiology and Behavior

Undergraduate Degree: B.S.
Undergraduate Degree Requirements: Total of 124 hours: 24-28 Biology, 20-22 Chemistry, 8 Physics, 12-13 Mathematics, electives to meet requirements of 124 hours, 6-12 Foreign Language

Graduate Degrees: M.S., Ph.D.
Graduate Degree Requirements: Master: Non-Thesis option; 30 hours (minimum 21 hours, B or better average in primary area, 9 hours in supporting area). Thesis option; no specific course requirements; thesis; course work approved by advisory. Doctoral: No specific course requirements; qualifying examination; thesis; reading proficiency of 1 Foreign Language

Graduate and Undergraduate Courses Available for Graduate Credit: Properties of Biological Systems (U), Biology of Invertebrate Animals (U), Introduction to X-Ray Crystallography (U), Molecular Structure by X-Ray Crystallography (U), X-Ray Diffraction Analysis of Fibrous Structures (U), Molecular Genetics (U), Molecular Biology: Proteins (U), Molecular Biology: Nucleic Acids (U), Cell Biology (U), Cytology (U), Microbiology (U), General Mycology (U), Parasitology (U), Medical Microbiology (U), Immunobiology (U), Genetic Biology (U), Plant Physiology (U), Human Physiology (U), Physiology (U), Endocrinology (U), Plant Structure (U), Tissue Biology (U), Developmental Biology (U), Systematics of Vascular Plants (U), Evolution (U), Ecology (U), Field Ecology (U), Plant Ecology (U), Animal Ecology (U), Ethology (U), Biology of Behavior (U), Special Assignments (U), Bioenergetics, Advanced Crystallography, Cytology and Physiology of Micro organisms, Advanced Mycology, Animal Virology, Microbial Genetics, Developmental Genetics, Advanced Topics in Plant Physiology, Experimental Endocrinology, Neurobiology: Neuroanatomy, Neurobiology: Neurophysiology, Neurobiology: Physiological Psychology, Neurobiology, Neurochemistry, Methods in Biological Research, Laboratory in Molecular Biology

PURDUE UNIVERSITY - CALUMET CAMPUS

2233 171st Street Phone: (302) 844-0520
Hammond, Indiana 46323
Dean of Graduate Studies: J.S. Tuckey
Dean of College: R. Gonzales

DEPARTMENT OF BIOLOGY
Phone: Chairman: (302) 844-0520 Ext. 404

Chairman: K.S. Wilson
Professor: K.S. Wilson
Associate Professor: J.R. Shoup
Assistant Professors: T.J. Dougherty, R.L. Peloquin, J.E. Wermuth, R.J. Werth
Instructors: K.K. Smith
Visiting Lecturers: 2

Degree Program: Biology

Undergraduate Degree: B.S.
Undergraduate Degree Requirements: 2 years Modern Language, 2 1/2 years Chemistry, 2 1/2 years Mathematics, 24 semester hours in Biology, 18 semester hours Social Sciences and Humanities

Graduate Degree: M.S.
Graduate Degree Requirements: 12 hours in Biology - Thesis

Graduate and Undergraduate Courses Available for Graduate Credit: Genetics, Evolution, Microbiology, Virology, Mycology, Plant Morphology, Plant Physiology, Developmental Biology, Cell Biology, Cytology, Comparative Animal Physiology, Plant Geography, Environment al Biology

PURDUE UNIVERSITY AT FORT WAYNE

2101 Coliseum Boulevard East Phone: (219) 482-5271
Fort Wayne, Indiana 46805
Dean of Graduate Studies: F.N. Andrews
Dean of College: J. Gilbert

SECTION OF BIOLOGICAL SCIENCES
Phone: Chairman: (219) 482-5582 Others: 482-5271

Section Chairman: W. Davies
Professor: W. Davies, S. Gottlieb
Associate Professor: B. Becker, J. Jimerez, J. Tobolski
Assistant Professors: P. Gerity, J. Haddock, E. Holt, R.D. Lyung, L. McKane, C. McKinley, N. D. Schmidt, D. Taves

Degree Programs: Bacteriology, Biology, Medical Technology, Pre-Medicine

Undergraduate Degree: B.S.
Undergraduate Degree Requirements: General Education 34 hours, Biology 29 hours, Chemistry 20 hours, Physics 8 hours, Mathematics and Statistics 13 hours, electives (or prof.) 4-21 *including 12 hours Modern-Language

Graduate Degree: See Purdue University

Graduate and Undergraduate Courses Available for Graduate Credit: Biology of Invertebrate Animals (U), Cell Biology (U), Microbiology (U), Genetic Biology (U), Plant Physiology (U), Physiology (U), Developmental Biology (U), Biology of Algae (U), Plant Ecology (U), Animal Ecology (U), Endocrinology (U), Special Assignments (U), Thesis

ST. FRANCIS COLLEGE

2701 Spring Street Phone: (219) 432-3551
Fort Wayne, Indiana 46808
Dean of Graduate Studies: L. Ross
Dean of College: Sr. M.J. Schaeffer

DEPARTMENT OF BIOLOGY
Phone: Chairman: (219) 432-3551 Ext. 292

Chairman: Sr. M.C. Martin
Professors: Sr. M.C. Martin
Associate Professors: Sr. M.C.F. Faulkner,
Assistant Professors: Sr. M.D. Schlaeger, R.T. Hurley
Instructors: G.L. Tieben

Degree Programs: Biology, Biological Education, Medical Technology

Undergraduate Degree: B.S.
Undergraduate Degree Requirements: 30 semester hours of Biology, 14 semester hours, 2 years of Chemistry, 6 semester hours, 1 year of Mathematics, 8 semester of Physics (strongly recommended) B.S. in Medical Technology, 24 semester hours Biology plus 24 semester hours of Clinical Biology, 18 semester hours Chemistry plus 8 semester hours Clinical Chemistry, 8 semester hours Physics, 6 semester hours in Radiation and Instrumentation

Graduate Degree: M.S. in Education
Graduate Degree Requirements: 18 hours of Science, 15 hours of Education

Graduate and Undergraduate Courses Available for Graduate Credit: Principles of Biology, Topics in Biology, Conservation, Taxonomy, Ecology, Entomology, Genetics, Radiation Science, Cell Biology (U), Evolution (U), Introduction to Ornithology (U)

SAINT JOSEPH'S COLLEGE

Rensselaer, Indiana 47978 Phone: (219) 866-7111
Dean of College: R. Garrity

DEPARTMENT OF BIOLOGY
Phone: Chairman: (219) 866-7111

Chairman: A.G. Mehall
Professor: Rev. U.J. Siegrist
Assiciate Professors: A.G. Mehall
Assistant Professor: D.A. Jones

Degree Programs: Biology, Biology/Chemistry, Medical Technology

Undergraduate Degrees: B.S.
Undergraduate Degree Requirements: B.S. in Biology, 36 hours Biology, 16 hours Chemistry, 8 hours Physics, 3 hours Mathematics, B.S. in Biology/Chemistry, 54 hours Biology/Chemistry, 8 hours Physics, 3 hours Mathematics, B.S. Medical Technology, 16 hours Biology, 16 hours Chemistry, 8 hours Physics, 3 hours Mathematics (Senior year 30 hours for in hospital training)

ST. JOSEPH'S CALUMET COLLEGE

(No reply received)

ST. MARY-of-the-WOODS COLLEGE

(No reply received)

SAINT MARY'S COLLEGE

Notre Dame, Indiana 46556 Phone: (219) 232-3031
Vice President for Academic Affairs: W.A. Hickey

DEPARTMENT OF BIOLOGY
Phone: Chairman: (219) 284-4061

Chairman: C.F. Dineen
Professors: C.F. Dineen, G. Bick, Sr. R. Dunleavy
Assistant Professors: A. Susalla, E. Holmes
Instructors: Br. L. Stewart

Degree Programs: Biology, Medical Technology

Undergraduate Degree: B.S.
Undergraduate Degree Requirements: Biology - 34 Biology, 8 Mathematics, 17 Chemistry, Proficiency in English and Modern Language, 3 in either Literature, Philosophy, Theology, Fine Arts, 12 in either Business/ Economics, History, Political Science, Psychology, Sociology, Medical Technology - 28 hours Biology same as for Biology, 30 hours at a medical center or hospital

ST. MEINRAD COLLEGE

St. Meinrad, Indiana 47577 Phone: (812) 357-6520
Dean of College: T. Ostdick

DIVISION OF MATHEMATICS AND NATURAL SCIENCES
Phone: Chairman: (812) 357-6580 Others: 357-6611

Chairman: D. Schmelz
Professor: T. Ostdick
Associate Professor: D. Schmelz
Assistant Professors: G. Carpenter, R. Hindel

Degree Program: Biology

Undergraduate Degree: B.S.
Undergraduate Degree Requirements: total 128 hours with a C cumulation, 28 hours Biology including Seminar, 6 hours each in Mathematics and Chemistry, Research (thesis)

TAYLOR UNIVERSITY

Upland, Indiana 46989 Phone: (317) 998-2751
Dean of College: R.D. Pitts

BIOLOGY DEPARTMENT
Phone: Chairman: Ext. 332

Head: G.W. Harrison
Professors: E. Poe, H.Z. Snyder
Associate Professors: T.J. Burkholder, G.W. Harrison
Assistant Professors: W.R. Slabaugh

Field Stations/Laboratories: Taylor University Field Station, au Sable Trails Camp, Big Twin Lake

Degree Programs: Biology, Botany, Zoology

Undergraduate Degrees: A.B., B.S.
Undergraduate Degree Requirements: 40 semester hours* Biology, 10 hours General Inorganic Chemistry, 2 years foreign language (A.B.), approximately 44 hours General Education, *Includes 6 hours at Taylor University Field Station - or its equivalent (except pre-medical students)

TRI-STATE COLLEGE

Angola, Indiana 46703 Phone: (219) 665-3141
Dean of Arts and Sciences: J. Nortrup

DEPARTMENT OF BIOLOGY
Phone: Chairman: Ext. 276

Chairman: P. Hippensteel
Associate Professor: P. Hippensteel
Assistant Professor: R. Miller

Degree Program: Biology

Undergraduate Degree: B.S.
Undergraduate Degree Requirements: 186 quarter hours total, 47 quarter hours Biology, 13 quarter hours Chemistry, 75 quarter hours General Studies, 6 quarter hours Physical Education, 44 quarter hours electives

UNIVERSITY OF EVANSVILLE
P.O. Box 329 Phone: (812) 477-6241
Evansville, Indiana 47701
Dean of Graduate Studies: E.M. Tapley
Dean of College: G.W. English

DEPARTMENT OF BIOLOGY
Phone: Chairman: (812) 477-2024 Others: 477-6241

Chairman: E.E. Schroeder
Professors: W.P. Mueller
Associate Professor: P.L. Winternheimer
Assistant Professors: J.A. Brenneman, F.S. Grass, K.J. Ott

Degree Program: Biology

Undergraduate Degrees: B.A., B.S.
Undergraduate Degree Requirements: Bachelor of Arts: Introductory Biology, General Zoology and General Botany, Seminar, electives to complete 45 hours quarter total. One year each of General Chemistry, Organic Chemistry and Physics, one course in Mathematics, Two years of reading knowledge of foreign language. Bachelor of Science: Requirements as B.A. except one course in Computing Science is substituted for the Foreign Language

Graduate Degree: M.A.T. offered through the Education Department.

Graduate and Undergraduate Courses Available for Graduate Credit: Plant Taxonomy (U), Environmental Biology (U), Animal Physiology (U), Plant Physiology (U), Bacteriology (U), Genetics (U), Parasitology (U), Cell Biology (U), Evolution (U), Insects of Importance to Man (U), Freshwater Biology, Cytogenetics, Experimental Parasitology, Research

UNIVERSITY OF NOTRE DAME
Notre Dame, Indiana 46556 Phone: (219) 283-6011
Vice President for Advanced Studies: R.E. Gordon

College of Science
Dean: B. Waldman

DEPARTMENT OF BIOLOGY
Phone: Chairman: (219) 283-7496

Chairman: P.P. Weinstein
Clark Professor of Biology: G.B. Craig, Jr.
Professors: H.A. Bender, G.B. Craig, Jr., H.E. Esch, R.E. Gordon, R.P. McIntosh, K.S. Rai, H.J. Saz, R.E. Thorson, J.A. Tihen, K.S. Tweedell, P.P. Weinstein
Associate Professors: J.B. Critz (Adjunct), T.J. Crovello, M.S. Fuchs, J.J. McGrath
Assistant Professors: W.G. Burton, R.W. Green, R.E. Kingsley (Adjunct), D.W. Morgan, J. O'Malley (Adjunct), Q. Ross, T. Troeger (Adjunct), D. Williams
Teaching Fellows: 23
Research Assistants: 1
Postdoctoral Research Fellows: 9

Field Stations/Laboratories: University of NotreDame Environmental Research Center - Land O'Lakes, Wisconsin

Degree Programs: Ecology and Environmental Biology, Genetics, Cell Biology, Developmental Biology, Medical Entomology, Parasitology, Behavior, Interdisciplinary program in Molecular Biology

Undergraduate Degree: B.S. (Biology)
Undergraduate Degree Requirements: Total credit hours, 124, Biology Major: biology 32, Chemistry 16, Mathematics 8, Physics, 8, Science Electives, 6; Free Electives, 15, Other 39. Biology Concentration: similar to major program, but only 24 credit hours in biology and somewhat lesser credits in the other sciences

Graduate Degrees: M.S., Ph.D.
Graduate Degree Requirements: Master: Research Masters: 24 credit hours in course work, 6 in research: Non-research Masters: 30 credit hours in coursework. Doctoral: Total of 72 credits hours in coursework and research.

Graduate and Undergraduate Courses Available for Graduate Credit: General Parasitology (U), Invertebrate Biology (U), Evolution (U), Plants and Human Affairs (U), History of Biology (U), General Ecology (U), Experimental Embryology (U), Molecular Genetics, Aquatic Ecology (U), Developmental Genetics, Plant Anatomy, Protozoology, Medical Entomology, Vector Genetics, Biological Microtechnique, Community Ecology, Comparative Physiology, Comparative Endocrinology, Analysis of Ultrastructure, Topics in Cell Biology, Topics in Botany, Topics in Evolutionary and Systematic Biology, Seminar in Evolution, Seminar in Physiology, Seminar in Parasitology, Experimental Parasitology, Seminar in Biology, Environmental Physiology and Biochemistry, Aquatic Botany, General Entomology (U), Vertebrate Biology (U), Animal Behavior (U), Flowering Plant Taxonomy (U), Biostatistics (U), Cytology (U), Computers in Biology (U), Fundamentals of Human Genetics, Marine Biology, Cytogenetics, Developmental Cytology, Helminthology, Environmental Biology, Physiological Chemistry of Animal Parasites, Causal Growth and Development, Population Ecology, Plant Physiology, Neurobiology, Biological Electron Microscopy, Topics in Physiology, Topics in Ecology, Topics in Developmental Biology, Seminar in Ecology, Seminar in Developmental Biology, Seminar in Genetic, Topics in Molecular Biology, Special Problems, Graduate Research, Limnology-Oceanography, Aquatic Insects, Practicum in Aquatic Biology

DEPARTMENT OF MICROBIOLOGY
Phone: Chairman: (219) 283-7564 Others: 283-6165

Chairman: M. Pollard
Professors: M. Pollard, M Wagner, B.S. Wostmann
Associate Professors: T. Asano, J.R. Pleasants
Assistant Professors: R.J. Erickson, J.M. Jones, C.F. Kulpa, Jr., P.M. Webb
Faculty Fellows: 3
Postdoctorals: 4

Degree Program: Microbiology

Graduate Degrees: M.S., Ph.D.
Graduate Degree Requirements: Master: 30 hours core curriculum Doctoral: 72 hours including core curriculum, One Language, Qualifying Examinations, Thesis and Defense

Graduate and Undergraduate Courses Available for Graduate Credit: Principles of Microbiology (U), Fundamentals of Microbiology (U), Environmental Microbiology (U), Mycology (U), Gnotobiology, Epidemiology, Virology, Immunology, Microbial Physiology, Microbial Genetics, Biophysics, Pathogenic Bacteriology, Pathogenic Microorganisms, Nutrition and Metabolism

VALPARAISO UNIVERSITY

Valparaiso, Indiana 46383 Phone: (219) 464-5000
Dean of College: H. Peters

DEPARTMENT OF BIOLOGY

Chairman: C.H. Krekeler
Professors: W.W. Bloom, W.C. Gunther, R.J. Hanson, C.H. Krekeler, K.E. Nichols
Associate Professors: F.R. Meyer, J.C.H. Tan
Assistant Professors: G.C. Marks

Degree Program: Biology

Undergraduate Degrees: B.S., B.A.
Undergraduate Degree Requirements: 28 semester credits in Biology for B.A., 32 for B.S. including Unity of Life, Diversity of Life, Genetics, Systematic Biology, Proseminar or Problems, and three of the following: Developmental Biology, Ecology, General Physiology, Cytology or Bacteriology. General and Organic Chemistry

WABASH COLLEGE

301 West Wabash Avenue Phone: (317) 362-1400
Crawfordsville, Indiana 47933
Dean of College: V.M. Powell

DEPARTMENT OF BIOLOGY
Phone: Chairman: Ext. 255 or 261 Others: 273

Chairman: T.A. Cole
Norman E. Treves Professor of Biology: W.H. Johnson
Professor: E.C. Williams
Associate Professors: A.E. Brooks, T.A. Cole, L.L. Hearson, R.O. Petty
Assistant Professors: J.K. Crissman, Jr., W.N. Doemel, C.T. Hammond, L.W. Oiler
Laboratory Preparator: 1

Field Stations/Laboratories: W.C. Allee MemorialWoods, Parke County, Indiana

Degree Program: Biology

Undergraduate Degree: B.A.
Undergraduate Degree Requirements: Introductory Biology, Biomolecules (1/2 course), Morphology Course, Genetics, Ecology, Cell Biology, Independent Study (1/2 course), Senior Seminar (1/2 course), Biology Courses to make 9 full courses, Physics (1 course), and Chemistry (2 courses, Inorganic and Organic)

IOWA

BRIAR CLIFF COLLEGE

3303 Rebecca Street Phone: (712) 279-5490
Sioux City, Iowa 51104
Dean of College: Sr. J.A. Wick

BIOLOGY DEPARTMENT
Phone: Chairman: (712) 279-5490 Others: 279-5321

Chairman: Sr. R.M. Collins
Assistant Professors: K.V. Baldwin, R.M. Collins, J. Hey (Research), B.A. Huberty

Degree Programs: Biology, Environmental Science, Medical Technology

Undergraduate Degree: B.S.
Undergraduate Degree Requirements: 10 term courses in Biology, supporting courses in Chemistry, Mathematics and Physics

BUENA VISTA COLLEGE

4th and Grand Phone: (712) 749-2351
Storm Lake, Iowa 50588

DEPARTMENT OF NATURAL SCIENCES
Phone: Chairman: (712) 749-2200 Others: 749-2351

Chairman: R.P. Borgman
Professors: I.H. Bhatti, R.P. Borgman
Associate Professors: G.M. Poff

Degree Program: Biology

Undergraduate Degree: B.A., B.S.
Undergraduate Degree Requirements: (Minimum), 30 hours Biology, 8 hours Chemistry, 8 hours Physics, 3 hours Mathematics

CENTRAL COLLEGE

Pella, Iowa 50219 Phone: (515) 628-4151
Dean of College: J. Graham

DEPARTMENT OF BIOLOGY
Phone: Chairman: Ext. 300

Chairman: J.B. Bowles
Professor: D.M. Huffman
Associate Professors: J.B. Bowles, D.E. Wilson
Assistant Professors: K.E. Tuinstra

Degree Program: Biology

Undergraduate Degree: A.B.
Undergraduate Degree Requirements: Not Stated

CLARKE COLLEGE

1550 Clarke Drive Phone: (319) 588-6300
Dubuque, Iowa 52001
Academic Dean: Sr. H. Thompson

DEPARTMENT OF BIOLOGY
Phone: Chairman: (319) 588-6399 Others: 588-6300

Chairman: D.R. Zusy
Associate Professors: M.F. Guest, Sr. V. Kaeferstein
Assistant Professor: D.R. Zusy

Degree Program: Biology

Undergraduate Degree: B.A.
Undergraduate Degree Requirements: 30 hours in Biology, 16 hours Chemistry

COE COLLEGE

Cedar Rapids, Iowa 52402 Phone: (319) 364-1511
Dean of College: C.W. Veach

DEPARTMENT OF BIOLOGY
Phone: Chairman: (319) 364-1511 Ext. 223

Chairman: R.V. Drexler
Professors: K.M. Cook, E.K. Goellner
Associate Professor: R. Siemer
Assistant Professor: F.R. Sandford

Field Stations/Laboratories: Associated College Midwest Schools - Winderness Field Station - Basswood Lake, Minnesota

Degree Program: Biology

Undergraduate Degree: A.B.
Undergraduate Degree Requirements: 9 courses in Biology, to include Physiology, Genetics, Literature and one Botany, 2 Chemistry, 2 Mathematics

CORNELL COLLEGE

Mt. Vernon, Iowa 52314 Phone: (319) 895-8811
Dean of College: W. Ehrmann

BIOLOGY DEPARTMENT
Phone: Chairman: (319) 895-8811

Chairman: D.L. Lyon
Professors: F. Pray, E. Rogers
Associate Professors: D.L. Lyon, P. Christiansen
Assistant Professor: G. Bedell

Field Stations/Laboratories: Wilderness Field Station - Ely Minn. (in conjunction with other conference schools [Associated Colleges of the Midwest])

Degree Program: Biology

Undergraduate Degree: B.S.S. (Bachelor of Special Studies), P. Phil.
Undergraduate Degree Requirements: 6 courses beyond the General Biology course for a major, 32 "course credits" = 32 courses including 6 above. for B.S.S. and B. Phil. number of formal courses are adjustable

DORDT COLLEGE

Sioux Center, Iowa 51250 Phone: (712) 722-3771
Dean of College: D.C. Ribbens

BIOLOGY DEPARTMENT
Phone: Chairman: (712) 722-3771 Ext. 311 Others. 119

Chairman: A. Mennega
Associate Professors: A. Mennega, G.E. Parker
Assistant Professors: D. VanderZee

Degree Program: Biology

Undergraduate Degree: A.B.
Undergraduate Degree Requirements: 40 courses: 10 Biology, 5 Chemistry

DRAKE UNIVERSITY
27th and Forest Avenue Phone: (515) 271-2011
Des Moines, Iowa 50311
Dean of Graduate Studies: E.L. Canfield
Dean of the College of Liberal Arts: L.P. Johnson

DEPARTMENT OF BIOLOGY
Phone: Chairman: (515) 271-3191 Others: 271-2011

Chairman: R.A. Rogers
Professors: G.C. Huff, L.P. Johnson, P.A. Meglitsch, R.A. Rogers, H.D. Swanson, R.J. VandenBranden
Associate Professors: J.L. Christiansen, P.J. Kingsbury, R.M. Dokma, G.A. Lucas, W.B. Merkley, M. M. Myszewski, F.A. Rogers, F.M. Shawhan (Emeritus), S.L. Wilson
Assistant Professors: D.B. Stratton, R.S. Wacha
Instructor: B.W. Haglan
Teaching Assistants: 15

Degree Programs: Biology

Undergraduate Degree: B.A.
Undergraduate Degree Requirements: 32 semester hours Biology, 8 semester hours General Physics, 8 semester hours General Chemistry, 4 semester hours Organic Chemistry, 4 semester hours Calculus

Graduate Degree: M.A.
Graduate Degree Requirements: Master: 24 semester hours Graduate Biology, 6 semester hours Thesis

Graduate and Undergraduate Courses Available for Graduate Credit: Microbiology (U), Genetics (U), Intermediate Genetics (U), Science and Society (U), Conservation of Man (U), Lower Plants (U), Higher Plants (U), Ecology (U), Parasitology (U), Invertebrate Zoology (U), Embryology (U), Histology (U), Mammalian Physiology (U), Field Botany (U), Fresh-Water Invertebrates (U), Terrestrial Invertebrates (U), Limnology (U), Scientific Literature (U), Plant Physiology (U), Endocrinology (U), Biophysics (U), Neurophysiology (U), Advanced Vertebrate Anatomy (U), Virology (U), Immunology (U), Behavioral Biology (U), Human Genetics (U), Growth and Morphogenesis (U), Cell Physiology (U), Evolution, Growth and Morphogenesis, Advanced Parasitology, Advanced Cell Physiology, Advanced Microbiology, Fundamentals of Scientific Research, Advanced Ecology, Advanced Genetics, Aquatic Biology, Fish Parasitology, Advanced Vertebrate Zoology, Protozoology, Science Teaching in Higher Education, Evaluation in Higher Education, Population Ecology, Advanced Limnology, Immunology Laboratory, Graduate Seminar

GRACELAND COLLEGE
Lamoni, Iowa 50140 Phone: (515) 784-3311
Dean of College: E. Johnson

BIOLOGY DEPARTMENT
Phone: Chairman: (515) 784-3311 Ext. 119

Head: J.A. Edwards
Associate Professors: R.A. DeLong, N. Hartwig, J. Edwards, B. Mortimore
Student Laboratory Assistants: 9

Degree Program: Biology

Undergraduate Degree: B.S.
Undergraduate Degree Requirements: Not Stated

GRINNELL COLLEGE
Grinnell, Iowa 50113 Phone: (515) 236-6181
Dean of College: W. Walker

DEPARTMENT OF BIOLOGY
Phone: Chairman: (515) 236-6181 Ext. 594 Others:300

Chairman: I.Y. Fishman
Norris Professor of Biology: K.A. Christiansen
Stone Professor of Biology: G. Mendoza
Professors: K.A. Christiansen, K. DeLong, L.H. Durkee, I.Y. Fishman, B.F. Graham, G. Mendoza, W.S. Walker
Assistant Professors: J.R. Denbo, I.M. Jones, J.J. Martinek
Lecturers: L.T. Durkee, S.K. DeLong

Field Stations/Laboratories: Henry S. Conard Environmental Research Area - 12 miles S.W. of Grinnell

Degree Program: Biology

Undergraduate Degree: B.A.
Undergraduate Degree Requirements: 124 semester credits for graduation, including 32 credits in the major

IOWA STATE UNIVERSITY
Ames, Iowa 50010 Phone: (515) 294-4111
Dean of Graduate Studies: D.J. Zaffarano

College of Agriculture
Dean: L. Kolmer

DEPARTMENT OF AGRICULTURAL ENGINEERING
(Also in School of Engineering)
Phone: Chairman: (515) 294-1434 Others: 294-2871

Head: C.W. Bockhop
Professors: C.E. Beer, C.W. Bockhop, W.F. Buchele, L. Charity, H. Giese, T.E. Hazen, T.A. Hoerner, H.P. Johnson, S.J. Marley
Associate Professors: H.E. Hansen, D.W. Mongold
Assistant Professors: W. Anderson, J. Baker, C. Bern, D. Bundy, R. Smith
Instructors: C. Anderson, M. Boyd, Cole, G. Olson, P. Preyer, T. Silletto
Research Assistants: 10

Field Stations/Laboratories: Agricultural Engineering - Agronomy Research Center, - Ames Iowa

Degree Programs: Agriculture, Agricultural Engineering

Undergraduate Degree: B.S.
Undergraduate Degree Requirements: 199 credits (quarter)

Graduate Degrees: M.S., M.E., Ph.D.
Graduate Degree Requirements: Master: 45 quarter credits minimum, Doctoral: 110 quarter credits minimum

Graduate and Undergraduate Courses Available for Graduate Credit: Not Stated

DEPARTMENT OF ANIMAL INDUSTRY
(No reply received)

DEPARTMENT OF BIOCHEMISTRY
(see College of Science and Humanities)

DEPARTMENT OF DAIRY SCIENCE
(see Animal Industry)

DEPARTMENT OF ENTOMOLOGY
(see College of Science and Humanities)

DEPARTMENT OF FOOD TECHNOLOGY
Phone: Chairman: (515) 294-3011

Professors: A.F. Carlin, D.E. Goll, E.G. Hammond, P.A. Hartman, C.A. Iverson, E.A. Kline, A.A. Kraft, W.S. LaGrange, W.W. Marion, G.W. Reinhold, H. E. Snyder, D.C. Topel, H.W. Walker
Associate Professors: F.C. Parrish, W.S. Rosenber-

ger, R.E. Rust, M.H. Stromer
Assistant Professor: R.J. Hasiak
Teaching Fellows: 1
Research Assistants: 14
Postdoctoral Associates: 1
Associates: 2

Degree Programs: Dairy Microbiology, Muscle Biology, Water Resources, Food Science

Undergraduate Degree: B.S. (Food Science)
Undergraduate Degree Requirements: 192 credits, including 10 cr Mathematics, 8 cr. Physical Science, 13 cr. Biological Sciences, 18 cr Agricultural Sciences, 36 cr Food Technology

Graduate Degrees: M.S., Ph.D.
Graduate Degree Requirements: Master: Organic Chemistry, Physical Chemistry, Bacteriology, Statistics, Physics, Food Technology, Foreign Language, Thesis, Final Oral Doctoral: As M.S. but with additional Biochemistry, Physical Chemistry and Statistics, 1 foreign language, Preliminary Examination, Thesis, Final Oral

Graduate and Undergraduate Courses Available for Graduate Credit: Food Preservation (U), Food, Milk and Water Sanitation (U), Food Chemistry (U), Dairy Microbiology, Special Problems in Dairy and Food Technology, Introduction to Food Processing Systems, Undergraduate Seminar in Food Technology, Food Technology, Food Industry Regulations, Topics in Food Chemistry, Food Proteins, Advanced Food Microbiology

DEPARTMENT OF FORESTRY
Phone: Head: (515) 294-1166 Others: 294-1458

Head: H.H. Webster
Professors: D.W. Bensend, J.C. Gordon, G.W. Thomson, H.H. Webster
Associate Professors: W.G. Beardsley, F.S. Hopkins, D.R. Prestemon, D.R. Yoesting
Assistant Professors: R.B. Hall, R.B. Heiligmann, L.C. Promnitz, D.W. Rose
Research Assistants: 15
Teaching Assistants: 2
Research Associates: 2

Degree Programs: Forestry and Conservation, Outdoor Recreation Resource Management

Undergraduate Degree: B.S.
Undergraduate Degree Requirements: 19 cr Mathematics (including Statistics and Computer Programming), 14 credits Biology, 14 cr Chemistry, 40 credits Forestry

Graduate Degrees: M.S., Ph.D.
Graduate Degree Requirements: Master: 45 credits selected by committee of which 10 cr in thesis, if thesis degree is selected. 55 course credits without thesis Doctoral: 108 credits selected by committee. Thesis.

Graduate and Undergraduate Courses Available for Graduate Credit: Not Stated

DEPARTMENT OF GENETICS
(No information available)

DEPARTMENT OF HORTICULTURE
Phone: Chairman: 294-2751 Others: 294-1916

Chairman: C.V. Hall
Professors: G.J. Buck, A.E. Cott, E.L. Denisen, C.V. Hall, H.E. Nichols, E.C. Volz, J.L. Weigle
Associate Professors: R.J. Bauske, J.A. Cook, C.F. Hodges, J.D. Kelley
Assistant Professors: P.A. Domoto, M.P. Garber, K.N. Nilsen, H G. Taber
Teaching Fellows: 1
Research Assistants: 4
Associates: 2

Field Stations/Laboratories: Horticulture Station - Ames, Muscatine Island Field Station - Fruitland, Iowa

Degree Program: Horticulture

Undergraduate Degree: B.S.
Undergraduate Degree Requirements: I Communications 12 cr., II, Mathematical Sciences 10 cr., III, Physical Sciences 17 cr., IV Biological Sciences, 37 cr., V. Social Sciences, 12 cr., VI Humanities 6 cr., VII, Horticulture Sciences 36 cr., VIII Agricultural Sciences 8 cr., XI Electives 54 cr. 192 quarter credits

Graduate Degree: M.S., Ph.D.
Graduate Degree Requirements: Master: 45 credits Thesis (or all credits in scheduled classes and with graduate committee approval) successfully completed examination. Doctoral: 108 credits, dissertation, successfully completed Preliminary and Final Examination

Graduate and Undergraduate Courses Available for Graduate Credit: Turfgrass Science (minor only), Orcharding (minor only), Floricultural Science, Preharvest-Postharvest Physiology and Storage (minor only), Marketing Horticultural Products (minor only), Systematic Horticulture (minor only), Horticultural Food Crops (U), Controlled Plant Growth Environments (U), Genetics and Breeding of Horticultural Plants (U), Current Topics in Olericulture, Current Topics in Pomology, Propagation Physiology

DEPARTMENT OF PLANT PATHOLOGY
(see College of Science and Humanities)

College of Engineering
Dean: D.R. Boylan

DEPARTMENT OF AGRICULTURAL ENGINEERING
(see College of Agriculture)

DEPARTMENT OF BIOMEDICAL ENGINEERING
(see College of Vetinary Medicine)

College of Science and Humanities
Dean: W.A. Russell

DEPARTMENT OF BACTERIOLOGY
Phone: Chairman: 294-1630

Acting Chairman: P.A. Hartman
Distinguished Professor: P.A. Hartman
Professors: D.P. Durand, W.R. Lockhart, P.A. Pattee, L.Y. Quinn
Associate Professors: J.G. Holt, F.D. Williams
Research Assistants: 5
Teaching-Research Assistants: 13

Field Stations/Laboratories: Lakeside Laboratories

Degree Programs: Bacteriology, Bacteriology and Immunology, Microbiology

Undergraduate Degrees: B.S.
Undergraduate Degree Requirements: 192 credits including 6 credits Mathematics, 18 credits Biology, 12 credits Physics, 35 credits Bacteriology

Graduate Degrees: M.S., Ph.D.
Graduate Degree Requirements: Master: 45 credits, Thesis Doctoral: Determined by student's Graduate Committee; includes a minor of 18 plus quarter hours

Graduate and Undergraduate Courses Available for Graduate Credit: General Virology (U), Food Preservation (Processing) (U), Microorganisms in Foods (U), Food, Milk, and Water Sanitation (U), Dairy Microbiology (U), Agro-Microbiology (Soils) (U), Applied Microbiology (U), Medical Virology (U), Immunology (U), Soil Microbiology and Biochemistry (U), Advanced Microbiology (U), Advanced Microbiology (3 courses), Special Topics, Systematic Bacteriology, Tissue Cell Culture

DEPARTMENT OF BIOCHEMISTRY AND BIOPHYSICS
(also in College of Agriculture)
Phone: Chairman: (515) 294-6116

Chairman: J. Horowitz
Professors: J.B. Applequist, P.A. Dahm, D. French (Distinguished Professor), H.J. Fromm, D.E. Goll, D.J. Graves, E.G. Hammond, J. Horowitz, J.D. Imsande, D.E. Metzler, H.E. Snyder, S.S. Stone (Adjunct), W.C. Wildman
Associate Professors: A.G. Atherly, D.C. Beitz, J.G. Foss, D.E. Outka, P.A. Rebers (Adjunct), R.M. Robson, J.F. Robyt, M.A. Rougvie, C.L. Tipton, B. White
Assistant Professors: J.A. Thomas, C.M. Warner
Research Assistants: 47
Teaching Assistants: 7

Degree Programs: Biochemistry, Biophysics

Undergraduate Degree: B.S.
Undergraduate Degree Requirements: 35 credits major; optional minor 20 credits. (Chemistry, Mathematics, Physics, Biological Science, Foreign Language)

Graduate Degrees: M.S., Ph.D.
Graduate Degree Requirements: Master: 45 credits, 18 outside major; Thesis, Final Oral Examination
Doctoral: No formal course requirements. Each student has an advisory committee to help develop a program of study. Minimum of 108 graduate credits, 18 outside the major field, ETS foreign language Examination or 1 year foreign Language. Cumulative examinations (written); Preliminary examination (Oral), dissertation, final oral examination

Graduate and Undergraduate Courses Available for Graduate Credit: Physiological Chemistry (U), Principles of Biochemistry (U), Biochemistry (U), Introduction to Biophysics (U), General Biochemistry (U), Laboratory in General Biochemistry (U), Radiobiochemistry (U), Advanced Cell Biology (U), Biophysical Methods (U), Laboratory in Biophysics (U), Microscopy (U), Laboratory in Microscopy (U), Advanced Biochemistry, Carbohydrate Chemistry, Advanced Biophysics, Molecular Biology of Muscle, Advanced Seminar, Seminar in Cell Biology

DEPARTMENT OF BOTANY AND PLANT PATHOLOGY
(Also in College of Agriculture)
Phone: Chairman: (515) 294-3522

Head: F.G. Smith
Professors: C.C. Bowen, J.A. Browning, J.D. Dodd, J. M. Dunleavy, L. Everson, R.G. Franke, H.T. Horner, Jr., D. Isely, G. Knaphus, R.Q. Landers, N.R. Lersten, H.S. McNabb, Jr., D.C. Norton, R.W. Pohl, M.D. Simons, D.W. Staniforth, E.P. Sylwester, L.H. Tiffany, J.R. Wallin
Associate Professors: J.S. Burris, A.H. Epstein, D.C. Foley, C.E. LaMotte, C.A. Martinson, D. Nevins, R.F. Nyvall, C.R. Steward, H. Tachibana, R.B. Wildman
Assistant Professors: C.B. Davis, D.R. Farrar, J.H. Hill, V.M. Jennings, D.C. Lewin, A. van der Valk
Research Assistants: 12
Teaching Assistants: 26

Field Stations/Laboratories: Iowa Lakeside Laboratory, - Milford

Degree Programs: Botany (areas of specialization: Aquatic Plant Biology, Cytology, Ecology, Economic Botany, Morphology, Mycology, Physiology, and Taxonomy), Plant Pathology

Undergraduate Degree: B.S. (Botany, Plant Pathology)
Undergraduate Degree Requirements: Total quarter credits, 192. General Education requirement, 70; Arts and Humanities, 15, Communication, 15, Mathematics and Natural Sciences, 20, Social Sciences, 15, Major requirement, 35, minor and foreign language optional.

Graduate Degrees: M.S., Ph.D.
Graduate Degree Requirements: Master: Equivalent of undergraduate major plus 45 graduate credits. Foreign language and thesis optional Doctoral: Minimum of 108 graduate credits, 54 at Iowa State, including dissertation research credits. Foreign language requirement: 1 for Botany major, optional for Plant Pathology major

Graduate and Undergraduate Courses Available for Graduate Credit: Plant Physiology (U), Plant Anatomy (U), Principles of Plant Pathology (U), General Virology (U), Forest Pathology (U), Wood Deterioration (U), Plant Ecology (U), Cell Biology (U), Seed Biology (U), Biology of Algae, Morphology of the Embryophyta, Principles of Mycology, Plant Nutrition, Plant Growth Regulation, Plant Metabolism, Plant Morphogenesis, Physiological Methods and Techniques, Advanced Cell Biology, Epidemiology and Control of Plant Diseases, Ecology of Aquatic Fungi, Mosses and Liverworts, Pteridology, Sexual Reproduction in Flowering Plants, Paleobotany, Field Biology of Lower Green Plants, Aquatic Plants, Plant Nematology, Field Mycology, Ecology and Systematics of Diatoms, Advanced Plant Ecology, Advanced General Plant Pathology, Host-Parasite Interactions, Epidemiology of Plant Diseases, Phytogeography, Agrostology, Physiology of Fungi, Fine Structure of Plant Cells, General Mycology, Microscopy I and II, Advanced Plant Pathology, Advanced Plant Taxonomy

DEPARTMENT OF ZOOLOGY AND ENTOMOLOGY
(Also in College of Agriculture)
Phone: Chairman: (515) 294-3316 Others: 294-7255

Distinguished Professor: P.A. Dahm, C.F. Curtiss, K.D. Carlander, O.E. Tauber
Professors: R. Bachmann, T.A. Brindley, B.W. Buttrey, K.D. Carlander (Distinguished), P.A. Dahm (Distinguished), J. Dunham, D. Griffith, W.D. Guthrie, D.E. Harding, R. Haupt, E.A. Hicks (Professor in Charge), H.H. Knight, R.E. Lewis, R.B. Moorman, (Wildlife Extension Conservationist), R.J. Muncy, J. A. Mutchmor, J.R. Redmond, H.J. Stockdale (Extension Entomologist), O.E. Tauber (Distinguished), M. J. Ulmer (Professor and Associate Dean of Graduate College), M.W. Weller (Professor in Charge)
Associate Professors: J.R. Baker, G.G. Brown, R.V. Bulkley, R.B. Dahlgren, C.J. Ellis, J.L. Jarvis, E.L. Jeska, J.A. Klun, B.W. Menzel, L.P. Pedigo (Professor in Charge), W. Rowley, L.D. Wing
Assistant Professors: M. Bachmann, C. Beegle, J.R. DeWitt, W.D. Dolphin, H. Fassel, Y.N. Forbes, E. Hart, L.C. Lewis, L.G. Mitchell, J.C. Owens, M. Peterson, E.C. Powell, K.R. Russell (Assistant Leader Wildlife Research Unit), K.C. Shaw, W.B. Showers (Resident Entomologist), J. Viles
Instructors and Associates: G. Bourne, A.F. Conway, C.M. Conway, R.D. Crawford, E.S. Farrar, D.W. Fredericksen, S.D. Hintz (Post-Doctoral Associate), J.G. Laveglia (Research Associate), D. Liesveld, J.C. Magee (Research Associate), R. Rogers (Research Associate), S.O. Ryan (Extension Associate), L.G. Sellers (Post-Doctoral Associate), W.B. Stoltzfus, J. J. Tollefson (Research Associate), J. Witkowski (Research Associate)
Teaching Assistants: 54
Research Assistants: 37

Field Stations/Laboratories: Ankeny USDA-ISU European C Corn Borer Laboratory - Iowa Lakeside Laboratory -

Veterinary Medical Research Institute

Degree Programs: Physiology, Animal Behavior, Ecology, Entomology, Entomology and Parasitology, Fisheries and Wildlife Biology, Limnology, Parasitology, Physiology, Wildlife Management, Zoology, Zoology and Entomology

Undergraduate Degree: B.S.
Undergraduate Degree Requirements: Total quarter credits - 192. General Education requirement, 70, Arts and Humanities, 15, Communication, 15, Mathematics and Natural Science, 20, Social Sciences, 15 Major requirement, 35 , minor and foreign language optional

Graduate Degrees: M.S., Ph.D.
Graduate Degree Requirements: Master: Equivalent of undergraduate major plus 45 graduate credits. Foreign language and thesis optional. Doctoral: Minimum of 108 graduate credits, 54 at Iowa State, including dissertation research credits. Foreign language requirement; 1 for Zoology and optional for Entomology

Graduate and Undergraduate Courses Available for Graduate Credit: Vertebrate Embryology, Ornithology, Principles of Wildlife Conservation, Human Physiology (U), Human Prenatal Development (U), General Entomology (U), Field Entomology (U), Applied Entomology (U), Biological Illustration (U), Fundamentals of Limnology (U), Mammalogy (U), Wildlife Techniques (U), Principles of Physiology (U), Physiology of Reproduction (U), Inland Sport Fisheries Management (U), Ichthyology (U), Fisheries Management (U), Prinicples of Systematic Zoology, Animal Ecology, Primate Evolution (U), Arachnology (U), Nematology (U), Free-Living Protozoa,(U), Ethology (U), Aquatic Ecology (U), Protozoology (U), Parasitic Protozoa (U), Helminthology (U), Cytochemistry (U), Waterfowl Biology and Conservation, Comparative Animal Physiology (U), Advanced Vertebrate Physiology (U), Fishery Aspects of Water Pollution (U), Fish Propagation (U), Insect Resistance in Crop Plants (U), Aquatic Insects (U), Insect Morphology (U), Economic Entomology (U), Medical Entomology,(U), Systematic Entomology (U), Zoological Literature, Ecological Energetics, Population Ecology, Zoogeography, Advanced Limnology, Advanced Parasitology, Microanatomy of Invertebrates, Survey of Developmental Zoology, Wildlife Management, Cell Physiology, Vertebrate Endocrinology, Insect Physiology, Selected Topics in Insect Physiology, Techniques of Fisheries, Fisheries Resources, Insect Ecology and Population Management, Advanced Medical Entomology, Insect Toxicology

College of Veterinary Medicine

Dean: P.T. Pearson

DEPARTMENT OF VETERINARY ANATOMY, PHARMACOLOGY AND PHYSIOLOGY

Phone: Chairman: (515) 294-4373 Others: 294-2440

Chairman: N.R. Cholvin
Professors: R.L. Engen, N.G. Ghoshal, F.B. Hembrough, J.H. Magilton, W.O. Reece, M.J. Swenson, W.C. Wagner
Associate Professors: F.A. Ahrens, H.S. Bal, J.R. Carithers, M.H. Crump, D.D. Draper, D.D. Gillette, B.H. Skold, W.G. VanMeter
Assistant Professors: G.A. Eckhoff, M.H. Greer, E.G. Palmer, D.H. Riedesel, D.R. Adams
Instructors: W.A. Hagemoser, N.L. Defonti, M.E. Shelton
Research Assistants: 3
Associates: 4

Degree Programs: Physiology, Veterinary Anatomy

Graduate Degrees: M.S., Ph.D.
Graduate Degree Requirements: Master: Coursework 45 quarter credits, research dissertation, final examination Doctoral: Coursework 108 quarter credits, research dissertation, final examination

Graduate and Undergraduate Courses Available for Graduate Credit: General Pharmacology (U), General Pharmacology (U), Systematic Anatomy: Ruminant Anatomy, Nonruminant Anatomy, Anatomy for Biomedical Engineering, Avian Anatomy, Neuroanatomy, Anatomy of Laboratory Animals, Physiology of the Autonomic Nervous System, Physiology of the Central Nervous System, Physiology of Endocrinology of Animal Reproduction, Comparative Mammalian Anatomy and Physiology, Comparative Mammalian Anatomy and Microscopy, Alimentary Physiology, Experimental Techniques in Physiology, Qualitative Pharmacology, Isolated Tissues, Quantitative Pharmacology: Bioassay

BIOMEDICAL ENGINEERING PROGRAM

(Also in College of Engineering)
Phone: Chairman: (515) 294-6520

Professor-in-Charge: R.C. Seagrave
Professors: W.H. Brockman, D.L. Carlson, C.P. Burger, D.W. DeYoung, R.L. Engen, D.D. Gillette, R. T. Greer, C.S. Swift
Associate Professors: R.W. Carithers, M.H. Greer
Research Assistants: 11

Degree Program: Biomedical Engineering

Graduate Degrees: M.S., Ph.D.
Graduate Degree Requirements: Master: 45 quarter hours, Thesis Doctoral: 108 quarter hours, Thesis

Graduate and Undergraduate Courses Available for Graduate Credit: Basic Biomedical Electronics, Integrated Circuit Uses in BM Research, Patient Monitoring and Electrical Safety, Electrical Circuits for BME, Electrical Circuits and Systems for BME, Biomedical Applications of Heat and Mass, Transfer, Biomaterials Biomedical Fluid Mechanics, Electrophysiology, Cardiovascular Transport and Control, Theory and Techniques of Biological Instrumentation, Simulation of Biological Systems, Advanced Biological System Simulation, BM Data Processing, Information Processing in Living Systems

DEPARTMENT OF VETERINARY MICROBIOLOGY AND PREVENTIVE MEDICINE

Phone: Chairman (515) 294-5776

Chairman: R.R. Packer
Professors: G.W. Beran, R.E. Dierks, P. Gough, M.S. Hofstad, M.L. Kaeberle, C.J. Mare, A.C. Pier, R.F. Ross, W.P. Switzer
Associate Professors: D.L. Harris, R.M Hogle
Assistant Professors: L.A. Jensen, L.N.D. Potgieter, J.D. Borthelsen
Instructor: M. Abou-Gabal

Degree Programs: Veterinary Microbiology

Graduate Degrees: M.S., Ph.D.
Graduate Degree Requirements: Master: 45 quarter credits, an acceptable thesis, meet foreign language requirement Doctoral: 108 quarter credits, approved dissertation, meet foreign language requirements

Graduate and Undergraduate Courses Available for Graduate Credit: Serology, Immunology (Medical), Pathogenic Bacteriology, Epidemiology, Animal Virology, Mechanisms in Animal Virology, Research, Special Topics, Immunological Disease

IOWA WESLEYAN COLLEGE

(No reply received)

LORAS COLLEGE
Dubuque, Iowa 50001 Phone: (319) 588-7100
Dean of College: Rev. N. Toby

DEPARTMENT OF BIOLOGY
Phone: Chairman: (319) 588-7128 Others: 588-7231

Chairman: J.F. Bamrick
Professors: E.T. Cawley, J.E. Kapler, Rev. W.E. Nye
Associate Professor: G.W. Kaufmann

Degree Programs: Biology, Medical Technology

Undergraduate Degree: B.S.
Undergraduate Degree Requirements: Rhetoric 8 - 12 hours, complete area of concentration, divisional distribution, 36 hours, electives to 120 hours, 2.0 grade point on 4.0 system, thesis or comprehensives

LUTHER COLLEGE
Decorah, Iowa 52101 Phone: (319) 387-2000
Vice President and Dean of the College: G. Nelson

DEPARTMENT OF BIOLOGY
Phone: Chairman: (319) 387-1117

Head: R.R. Rulon
Professors: R.M. Knutson, P.J. Reitan, R.R. Rulon
Associate Professors: D.J. Roslien, J.L. Tojstem
Assistant Professors: J.W. Eckblad
Instructor: W. Stevens
Technician: 1

Degree Programs: Biology (Programs in Nursing, Medical Technology, Psychobiology)

Undergraduate Degree: B.A.
Undergraduate Degree Requirements: Plan I: 22 hours of Biology, 1 year Calculus, 1 year Physics, Organic Chemistry, one year foreign language (2 - high school), Plan II: 28 hours of Biology, 1 year Mathematics, 1 year Chemistry, 1 year Foreign language (2 - high school)

MARYCREST COLLEGE
(No information available)

MORNINGSIDE COLLEGE
Sioux City, Iowa 51106 Phone: (712) 277-5100
Dean of College: R. Nelson

DEPARTMENT OF BIOLOGY
Phone: Chairman: (712) 277-5150

Chairman: H.L. Rundell
Professor: H.L. Rundell
Associate Professor: M.L. Well
Assistant Professor: E. Gardner

Degree Program: Biology

Undergraduate Degree: B.S.
Undergraduate Degree Requirements: Introductory to Botany, Introduction to Zoology, Ecology, General Physiology, Genetics, and electives to total 32

MOUNT MERCY COLLEGE
1330 Elmhurst Drive N.E. Phone: (319) 363-8213
Cedar Rapids, Iowa 52402
Academic Dean: T. Houser

DEPARTMENT OF BIOLOGY
Phone: Chairman: (319) 363-8213 Ext. 68

Chairman: Sr. M.A. McManus
Professors: Z. Lim, E.J. Peters
Instructor: J. Dorman

Degree Program: Biology

Undergraduate Degree: B.S.
Undergraduate Degree Requirements: Not Stated

NORTHWESTERN COLLEGE
Orange City, Iowa 51041 Phone: (712) 737-4821
Academic Dean: E.E. Ericson

BIOLOGY-CHEMISTRY DEPARTMENT
Phone: Chairman: Ext. 60

Chairman: E.A. VanEck
Professors: H.E. Hammerstrom, E.A. VanEck
Associate Professors: P. Hansen
Assistant Professors: G.D. Hegstad, V. Muilenburg

Degree Program: Biology

Undergraduate Degree: B.A.
Undergraduate Degree Requirements: 30 hours, 4 Botany, 4 Zoology, 3 Genetics, 1 Seminar, 8 - General Chemistry, Physics Recommended

ST. AMBROSE COLLEGE
518 E. Locust Street Phone: (319) 324-1681
Davenport, Iowa 52803
Dean of College: D. Moeller

DEPARTMENT OF BIOLOGY
Phone: Chairman: Ext. 213 Others: Ext. 203

Chairman: M.M. Vinje
Professors: C.S. Rice, M.M. Vinje
Associate Professor: R.J. Masat
Assistant Professor: M.J. Dunne

Degree Program: Biology

Undergraduate Degrees: B.A., B.S.
Undergraduate Degree Requirements: B.A. - 32 hours Biology, 8 hours Chemistry, 4 hours Mathematics B.S. - 32 hours Biology - 16 hours Chemistry - 8 hours Physics, Mathematics (including Calculus)

SIMPSON COLLEGE
Indianola, Iowa 50125 Phone: (515) 961-6251
Dean of College: J.W. Walt

DEPARTMENT OF BIOLOGY
Phone: Chairman: Ext. 698

Head: D.G. DeLisle
Professors: M.L. Watson
Associate Professors: D.G. DeLisle, J. Considine
Undergraduate Laboratory Assistants: 5

Degree Program: Biology

Undergraduate Degree: B.A.
Undergraduate Degree Requirements: Eight units of Biology, 3 units of Chemistry, 2 units of Physics, 1 unit of Mathematics. (Total of 36 units required for graduation with B.A.)

UNIVERSITY OF DUBUQUE
2050 University Avenue Phone: (319) 557-2121
Dubuque, Iowa 52001
Dean of College: K. Weeks

DEPARTMENT OF BIOLOGY
Phone: Chairman: (319) 557-2254 Others: 557-2121

Chairman: W.N. Berg

Associate Professor: W.N. Berg
Assistant Professors: R.M. Miller, D. Straley

Degree Programs: Biology, Medical Technology

Undergraduate Degrees: B.A., B.S.
Undergraduate Degree Requirements: Biology: 30 hours in Biology, Chemistry through Organic

UNIVERSITY OF IOWA

Iowa City, Iowa 52242 Phone: (319) 353-2121
Dean of Graduate Studies: D.C. Spriestersbach

College of Liberal Arts

Dean: D.B. Stuit

DEPARTMENT OF BOTANY
Phone: Chairman: (319) 353-5790

Chairman: R.L. Hulbary
Professors: R.L. Hulbary, R.M. Muir
Associate Professors: W.R. Carlson, H.L. Dean, R.W. Embree, T.E. Melchert, J.T. Schabilion, R.D. Sjolund, S.J. Surzycki
Assistant Professor: J. Clay
Teaching Fellows: 14
Research Assistants: 6

Field Stations/Laboratories: Iowa Lakeside Laboratory on West Lake Okoboji - Milford

Degree Programs: Biology, Botany

Undergraduate Degree: B.A.
Undergraduate Degree Requirements: a) One course in each of the following areas: Genetics, Physiology and Cell Biology, Biology of Vascular Plants, Biology of Non-Vascular Plants, Taxonomy, Ecology and Evolution, b) Two course (upper level) in Botany or Zoology or Biochemistry or Microbiology c) Four courses in Chemistry, through Organic Chemistry d) Analytical Geometry or Calculus e) Graduation requirements of the College of Liberal Arts

Graduate Degrees: M.S., Ph.D.
Graduate Degree Requirements: Master: 30 semester hours determined in consultation with a Guidance Committee of three members of the Graduate Faculty. M.S. Examination, written and oral. Doctoral: 72 semester hours determined in consultation with a Guidance Committee of 3 members of the Graduate Faculty, a Ph.D. Comprehensive Examination, wirtten and oral, a Final Examination in defense of dissertation

Graduate and Undergraduate Courses Available for Graduate Credit: Plant Taxonomy (U), Genetics (U), Cytogenetics (U), Phycology (U), Bryology (U), Mycology (U), Plant Physiology (U), Plant Anatomy (U), Ultrastructural Plant Cytology (U), Botanical Microtechnique (U), Experimental Techniques (U), Paleobotany (U), Palynology (U), Evolution (U), Ecology (U), Medical Mycology, Field Botany, Genetics of Cell Organelles, Eukaryotic Cell Biology, Systematics, Seminar: Genetics, Morphogenesis, Advanced Mycology, Advanced Plant Physiology, Seminar: Ecology, Electron Microscopy

DEPARTMENT OF ZOOLOGY
Phone: Chairman: (319) 353-5572 Others: 353-5751

Chairman: J.J. Kollros
Professors: R.V. Bovbjerg, H. Dingle, J. Frankel, R.G. Kessel, J.J. Kollros, R.D. Milkman, J.D. Mohler, E. Spaziani, N.E. Williams, H.W. Beams (Emeritus), G. Marsh (Emeritus)
Associate Professors: G.N. Gussin, J. Hegmann, S. Hubbell, S.B. Kater, J.R. Minninger, M. Solursh, B.A. Stay
Assistant Professors: E. Barrett, G.D. Cain, R. Murphey, C. Newlon, W.J. Platt, D.R. Soll
Teaching Fellows: 36
Research Assistants: 6
Postdoctoral Assistants in Instruction: 4

Field Stations/Laboratories: The Iowa Lakeside Laboratory Milford

Degree Programs: Biology, Ecology, Ethology, Genetics, Histology, Meat Sciences, Parasitology, Physiological Sciences, Physiology, Zoology

Undergraduate Degree: B.A. (Zoology)
Undergraduate Degree Requirements: 16 hours Chemistry, 8 hours Physics, Mathematics to Calculus, Zoology courses required: Introduction to Animal Biology, Genetics, Cell Physiology. Also required are 2 courses chosen from: Ecology, Evolution, Embryology, Invertebrate Zoology, Vertebrate Zoology. Also required are additional courses from the available departmental pool, to add to a total of 33 semester hours in Zoology

Graduate Degrees: M.S., Ph.D. (Zoology)
Graduate Degree Requirements: Master: If with thesis, 30 semester hours, mainly in Zoology, a thesis, plus written examination over one's graduate program, plus an oral defense of the thesis. Without thesis, 34 semester hours, plus written examination over one's graduate program. Doctoral: a) Formal admission to Ph.D. candidacy, b) Comprehensive examination, including written and oral parts. c) Some experience as a teaching assistant d) A satisfactory thesis, and a final oral defense of the thesis e) a formal seminar presentation of the thesis f) at least 72 semester hours of graduate credit

Graduate and Undergraduate Courses Available for Graduate Credit: Embryology, Ecology, Evolution, Advanced Ecology, Systems Ecology, Theoretical Ecology, Behavioral Ecology, Comparative Physiology, Neurophysiology (Fundamentals), Neurophysiology (Comparative Neuroethology), Comparative Animal Behavior, Neuroembryology, Endocrinology, Hormones and Behavior, Parasitology, Helminthology, Advanced Genetics, Developmental Genetics, Behavioral Genetics, Population Genetics, Qualtitative Genetics, Molecular Genetics, Insect Reproduction and Development, Microscopic Anatomy, Cytology, Gametogenesis, Development of Single Cell Systems, Biogenesis of Cell Structure, Sensory Neurophysiology, Integrative Neurophysiology, Electron Microscopic Techniques

College of Medicine

Dean: J.W. Eckstein

DEPARTMENT OF ANATOMY
Phone: Chairman: (319) 353-6963 Others: 353-5905

Head: T.H. Williams
Professors: N.S. Nalmi, W.R. Ingram, W.W. Kaelber, U.L. Karlsson, T.H. Williams
Associate Professors: N.A. Azzam, P.M. Heidger, D.J. Moffatt, J.R. Scranton
Assistant Professors: M. Aydelotte, R. Bhalla, L.E. DeBault, J.Y. Jew, H. Lin, J.A. Oaks, J.C. Searls, R.J. Tomanek
Associates - Assistants in Teaching: M.D. Bishop, A. Black, A. Boyne, T. Chiba, G. Greenwald, H. Murray, S.F. Pang
Visiting Lecturers: 3
Graduate Students: 19
Professional Improvement: 3

Degree Programs: Endocrinology, Neurobiology, Cell Biology with Oncology, Reproduction

Graduate Degrees: M.S., Ph.D.
Graduate Degree Requirements: Master: 30 semester hours graduate coursework, 24 of which are to be non-transfer courses from another University, thesis, and

final examination Doctoral: 72 hours graduate coursework; comprehensive examination, dissertation, and final examination (defense of dissertation)

Graduate and Undergraduate Courses Available for Graduate Credit: Human Gross Anatomy, Human Microscopic Anatomy, Neurobiology and Behavior, Special Microscopic Anatomy, Human Anatomy and Neuroanatomy, Advanced Human Anatomy, Endocrinology for Medical Students, Topics in Basic Endocrinology, Review of Anatomical Neurology, The Visceral Nervous System, Electron Microscopy Theory and Techniques, Advanced Electron Microscopy, Cell Ultrastructure, Problems, Introduction to Research, Anatomy Seminar, Teaching Workshop in Anatomy. Graduate coursework in other Basic Medical Science Departments is encouraged.

DEPARTMENT OF BIOCHEMISTRY
Phone: Chairman: (319) 353-3034 Others: 353-2121

Head: C.S. Vestling
Professors: C.P. Berg (Emeritus), R.L. Blakley, H. Bull (Research Professor Emeritus), R. Chalkley, T.W. Conway, R.L. Dryer, E.M. Gal (Psychiatry), H.P.C. Hogenkamp, G. Kalnitsky, S.G. Koreman (Professor of Biochemistry and Internal Medicine), R. Montgomery (Professor of Biochemistry and Associate Dean, College of Medicine), V.A. Pedrini (Professor of Orthopedic Surgery and Biochemistry), J.I. Routh (Professor of Biochemistry and Pathology), E. Stellwagen, C.A. Swenson, C.S. Vestling
Associate Professors: D.K. Granner (Internal Medicine and Biochemistry), G.F. Lata, B.V. Plapp, A.A. Spector, L.D. Stegink (Biochemistry and Pediatrics)
Assistant Professors: A. Arnone, J.E. Donelson, R. Roskowski, Jr.
Instructor: D.R. Oliver (Instructor in Biochemistry and Associate Director, Physician's Assistant Program)
Visiting Lecturers: 12
Research Assistants: 27
Postdoctoral Fellows: 18

Degree Program: Biochemistry

Undergraduate Degrees: B.A., B.S.
Undergraduate Degree Requirements: (B.S.), Rhetoric, 4-8 hours, Literature Core, 8 hours, History Core, 8 hours, Social Core, 8 hours, Physical Education, 4 hours, Language, 8 hours, Mathematics, 8 hours, Physics, 8 hours, Biological Science, 9-10 hours, Chemistry, 17-20 hours, Biochemistry, up to 8 hours, Snr. Research, 6 hours, Advanced Science Electives at least 17 hours, Physical Biochemistry 4 hours

Graduate Degrees: M.S., Ph.D.
Graduate Degree Requirements: Master: Biochemistry course, Research Techniques, Seminars, Research, plus at least two courses to supplement training. Graduate College requires minimum of 30 hours Doctoral: Graduate College requires minimum of 72 hours. Graduate Biochemistry course, Research Techniques, 5 Seminar Courses, Physical Biochemistry, three graduate level courses with only 2 from any one department outside Biochemistry, Research hours, plus any courses student committee deems necessary, 2-3 semesters teaching experience

Graduate and Undergraduate Courses Available for Graduate Credit: Chemistry of Biological Materials, Metabolism, Molecular Genetics, Physical Biochemistry, Experimental Biochemistry, Physical-Biochemical Techniques, Applied Biochemistry, Senior Research Independent Study (Honors) (U), Biochemistry for (a) Dental Students (b) Medical Students (d) Physician's Assistants; Macromolecules, Biochemistry of Subcellular Structures, Enzyme Catalysis, Bio-Organic, Mechanisms, Clinical Biochemistry, Spectroscopy of Biological Materials, Molecular Endocrinology, Neurobiochemistry, Research Techniques, Biochemistry, Seminar, Research

DEPARTMENT OF MICROBIOLOGY
Phone: Chairman: (319) 353-5596

Chairman: J.R. Porter
Professors: J. Cazin, Jr., L.G. Hoffman, A.J. Markovetz, J.R. Porter, E.W. Six
Associate Professors: G.E. Becker, J.E. Butler, T.L. Feldbush, W. Johnson, R.L. Richardson, J.E. Rodriguez, D.P. Stahly
Assistant Professors: M.G. Feiss, C.D. Cox, Jr., N.A. Crouch, M.F. Stinski, D.H. Walker, Jr.
Instructors: S. Halling
Research Assistants: 16

Degree Program: Microbiology

Undergraduate Degree: B.S.
Undergraduate Degree Requirements: Biology courses; Chemistry, Inorganic, Quantitative Analysis, Organic (1 year), Biochemistry, Mathematics - through Calculus, Statistics, Physics (1 year), Microbiology (15 h hours)

Graduate Degrees: M.S., Ph.D.
Graduate Degree Requirements: Master: Course work, Research and Thesis Doctoral: Course work, Research, and Thesis

Graduate and Undergraduate Courses Available for Graduate Credit: General Microbiology (U), Pathogenic Bacteriology (U), Microbial Physiology (U), Clinical Laboratory Microbiology, Survey of Immunology (U), Experimental Immunochemistry, Problems in Microbiology (U) Animal Virology, Medical Mycology (U), Microbial Genetics (U), Molecular Immunology, Cellular Immunology, Microbial Ecology, Research, Seminar

DEPARTMENT OF PATHOLOGY
Phone: Chairman: (319) 356-2906

Head: G.D. Penick
Professors: P.A. Cancilla, T.H. Kent, J.A. Koepke, E.S. Meek, G.D. Penick, E.F. Rose, F.W. Stamler
Associate Professors: H.C. Bickley, A. Simmons,
Assistant Professors: D.A. Barrett, II, F.R. Dick, C.S. Gleich, D.P. Nicholson, M.L. O'Connor, W.G. Owen, E.E. Pixley, R.S. Shacklett, D.L. Witte
Instructor: C.A. Aschenbrener
Research Assistants: 4
Visiting Lecturers: 3
Postdoctoral Trainees: 2

Degree Programs: Medical Technology, Pathology

Undergraduate Degrees: B.S.
Undergraduate Degree Requirements: 94 semester hours (including basic core requirements) toward Liberal Arts B.S. degree plus 30 semester hours training in professional Medical Technology.

Graduate Degree: M.S. (Pathology)
Graduate Degree Requirements: Master: 30 semester hours plus thesis

Graduate and Undergraduate Courses Available for Graduate Credit: Principles of Pathology (Allied Health Professional Students), Medical Jurisprudence, Clinical Pathology for Physician's Assistants, General Pathology for Medical Students, Systemic Pathology for Medical Students, Principles of Human Pathology (Dental students, Physician's Assistants, other), Research in Pathology, Graduate Instruction in Pathology, Special Topics in Pathology, Clerkship in Pathology, Special Topics

DEPARTMENT OF PHYSIOLOGY AND BIOPHYSICS
Phone: Chairman: (319) 353-3597 Others: 353-2121

Head: F.P.J. Diecke (Acting)
Professors: F.P.J. Diecke, G. E. Fold, Jr. N. Halmi,

C.A.M. Hogben, R. Llinas, B.A. Schottelius, H. Shipton, C.M. Tipton, C. Wunder
Associate Professors: R. Baker, J.N. Diana, D. Hillman, C.J. Imig, J.C. Nicholson, M.I. Phillips, G.W. Searle, J.D. Thompson
Assistant Professors: H.J. Cooke, D.C. Dawson, J.P. Farber, C. Gisolfi, H. Lorkovic
Instructor: F.D. Ingram
Visiting Lecturers: 1
Teaching Fellows: 6
Graduate Trainees: 9

Degree Program: Physiology

Graduate Degrees: M.S., Ph.D.
Graduate Degree Requirements: Master: 38 semester hours including Mammalian Physiology, Endocrinology, General Physiology, Biochemistry, Seminar and Thesis (M.S. in science education does not include thesis). Doctoral: 72 semester hours, including Neurobiology and Behavior, Endocrinology, Medical Physiology, Biochemistry, 2 advanced physiology courses, comprehensive examination and thesis.

Graduate and Undergraduate Courses Available for Graduate Credit: Physiology courses noted above, plus Introduction to Biophysics, Advanced Physiology of Exercise, Topics in Basic Endocrinology, General Physiology, Endocrinology for Medical Students, Advanced Systemic Physiology, Advanced Renal Physiology, Membrane Transport, Advanced Gastrointestinal Physiology, Human Ecology, Advanced Cardiovascular Physiology, Advanced Neurophysiology

RADIATION RESEARCH LABORATORY, DEPARTMENT OF RADIOLOGY
Phone: Chairman: (319) 353-3748 Others: 353-3747

Chairman: T.C. Evans
Professors: J.W. Osborne, E.F. Riley
Associate Professors: R.L. DeGowin, H.F. Cheng
Research Assistants: 13

Degree Program: Radiation Biology

Graduate Degrees: M.S., Ph.D.
Graduate Degree Requirements: Master: As least 30 hours of course work, a comprehensive M.S. Examination and a thesis. Doctoral: At least 72 hours of course work, a Ph.D. Comprehensive Examination and a thesis.

Graduate and Undergraduate Courses Available for Graduate Credit: Introductory Radiation Biology (U), Environmental and Radiological Health Physics (U), Seminar, Physics of Radiobiology I, II, Mammalian Radiobiology, Cellular Radiobiology, Radioisotopes in Biological Research, Radioisotopes in Clinical Investigations, Research Radiobiology, Special Topics, Thesis

UNIVERSITY OF NORTHERN IOWA
Cedar Falls, Iowa 50613 Phone: (319) 273-2311
Dean of Graduate Studies: G.J. Rhum
Dean of College of Natural Sciences: C.G. McCollum

BIOLOGY DEPARTMENT
Phone: Chairman: (319) 273-2456

Head: J.C. Downey
Professors: V.E. Dowell, J.C. Downey, R.C. Goss, C.G. McCollum (Dean, College of Natural Sciences), E.R. Tepaske, L.P. Winier
Associate Professors: C.F. Allegre, B.L. Clausen, L.J. Eilers, A.C. Haman, A.R. Orr, W.E. Picklum, D.L. Riggs, P. Saver, R.J. Simpson, D.D. Smith, N.A. Wilson
Assistant Professors: D.E. Johnson, D.V. McCalley, P.D. Whitson
Teaching Assistants: 5

Field Stations/Laboratories: Lakeside Laboratory - Milford

Degree Program: Biology

Undergraduate Degree: B.S.
Undergraduate Degree Requirements: Biology - several routes on both teaching major and liberal arts major (varying primarily on required courses and optional areas) minima include 12 hours in Chemistry, 2 hours in Seminar, electives from 5 areas in biology, and 55 hours minimum in major

Graduate Degree: M.S.
Graduate Degree Requirements: Master: A) Thesis Option: 30 semester hours including 12 graduate hours (excluding thesis) 6 hours usable research credits, courses in History and Philosophy of Science and Research Methods. b) Non-thesis Option: 37 hours including 23 minimum in Biology, 15 graduate hours, and a comprehensive examination

Graduate and Undergraduate Courses Available for Graduate Credit: Conservation (U), Comparative Anatomy (U), Vertebrate Embryology (U), Invertebrate Zoology (U), Animal Physiology (U), Plant Morphology (U), Plant Physiology (U), Cell Biology (U), Protozoology (U), Genetics (U), Organic Evolution (U), Experimental Embryology (U), General Microbiology (U), Animal Behavior (U), Parasitology (U), Field Zoology of Vertebrates (U), Plant Systematics (U), Ecology (U), Entomology (U), Plant Anatomy (U), Biological Techniques, Aquatic Biology, Special Problems in Biology, Endocrinology, Comparative Physiology, Experimental Microbiology, Research Methods in Biology, Thesis Research, History and Philosophy of Science, Seminar in Science Teaching

UPPER IOWA COLLEGE
Fayette, Iowa 52142 Phone: (319) 425-3311
Academic Dean: C. Clark

DEPARTMENT OF BIOLOGY
Phone: Chairman: Ext. 222

Head: R.W. Coleman
Professors: R.W. Coleman, E.E. Naylor
Instructor: L. Churbuck

Degree Program: Biology

Undergraduate Degree: B.S.
Undergraduate Degree Requirements: A major in Biology consists of a minimum of 28 hours of Biology, 9-20 hours in Chemistry. Hours are in semester hours.

WARTBURG COLLEGE
Waverly, Iowa 50677 Phone: (319) 352-1200
Dean of College: R. Matthias

DEPARTMENT OF BIOLOGY
Phone: Chairman: (386) 352-1200

Chairman: E.W. Hertel
Professors: E.W. Hertel, L.H. Petri
Associate Professors: G.J. Eiben, D.M Wolff
Assistant Professors: S.P. Main, A.E. Ristau

Degree Program: Biology

Undergraduate Degree: B.S.
Undergraduate Degree Requirements: 9 courses in Biology, 3 courses in Chemistry, 2 courses in Mathematics

WESTMAR COLLEGE

Le Mars, Iowa 51031 Phone: (712) 546-7081
Dean of College: J.C. Courter

BIOLOGY DEPARTMENT
Phone: Chairman: Ext. 319

Chairman: J. Divelbiss
Professors: J.E. Divelbiss, W.G. Marty
Associate Professor: M.G. Ulrich

Degree Programs: Biology, Medical Technology

Undergraduate Degree: B.A.
Undergraduate Degree Requirements: Minimum of 35 courses; minimum GPA of 2.00; proficiency in English, Mathematics, Speech, Minimum of 7 courses in general education, minimum of two semesters of Encounter; completion of a major program, residency requirement

WILLIAM PENN COLLEGE

Oskaloosa, Iowa 52577 Phone: (515) 673-8311
Dean of College: E. Redding

DEPARTMENT OF BIOLOGY
Phone: Chairman: Ext. 233

Head: F.R. Gygi (Acting)
Associate Professors: D.D. Gaunt, F.R. Gygi

Degree Program: Biology

Undergraduate Degree: B.A.
Undergraduate Degree Requirements: 30 hours of Biology; 4 hours each of Mathematics, Chemistry and Physics, 30 hours of core requirements, all the rest in electives for a total of 124 hours for the degree

KANSAS

BAKER UNIVERSITY

Baldwin City, Kansas 66006 Phone: (913) 594-6451
Dean of College: N. Malicky

DEPARTMENT OF BIOLOGY
Phone: Chairman: Ext. 418

Chairman: V.E. Nelson
Associate Professor: V.E. Nelson
Assistant Professor: J. Burns, D. Lollis

Field Stations/Laboratories: Baker Wetlands - Lawrence

Degree Program: Biology

Undergraduate Degree: B.S.
Undergraduate Degree Requirements: (In semester hours) Departmental: 16 hours Chemistry, 8 hours Mathematics and Physics, 28 hours Biology (16 must be at Jr.-Sr. level). B.S., A.B. same except that foreign langu age proficiency is required (16 hours or equivalent), and 4 hours Mathematics or Physics

BENEDICTINE COLLEGE

(incorporating Mt. St. Scholastica College)
Atchison, Kansas 66002 Phone: (913) 367-5340
Dean of College: R.C. Henry

DEPARTMENT OF BIOLOGY
Phone: Chairman: Ext. 210

Chairman: E.W. Dehner
Professor: E.W. Dehner
Associate Professor: S.D. Bassi
Instructor: B. Tremmel
Laboratory Assistant: 1

Degree Program: Biology

Undergraduate Degree: B.A.
Undergraduate Degree Requirements: (in semester credit hours) 8 hours general Biology, 20 hours advanced Biology courses plus undergraduate research, 8 hours mathematics, 12 hours Chemistry (minimum), 8 hours Physics, 8 hours Foreign language. Education courses are added for secondary teaching certification.

BETHANY COLLEGE

Lindsborg, Kansas 67456 Phone: (913) 227-3312
Dean of College: L. Foerster

DEPARTMENT OF LIFE AND EARTH SCIENCES
Phone: Chairman: (913) 227-3312

Chairman: L. Lungstrom
Professors: G. Bellan, A. Hahn, L. Lungstrom
Assistant Professor: J. Murphy

Degree Program: Biology

Undergraduate Degree: B.A.
Undergraduate Degree Requirements: 35 courses and other requirements for general requirement, 8 of these Biology courses, 4 of these Chemistry courses, 2 of these Physics courses

BETHEL COLLEGE

North Newton, Kansas 67117 Phone: 283-2500
Dean of College: M. Deckert

DEPARTMENT OF BIOLOGY
Phone: Chairman: Ext. 367

Head: D.R. Platt
Professor: R.W. Schmidt
Assistant Professor: S.G. Schmidt

Field Stations/Laboratories: Sand Prairie Natural History Reservation - Harvey County

Degree Programs: Biology, Environmental Studies

Undergraduate Degrees: B.A., B.S.
Undergraduate Degree Requirements: 25 semester hours in Biology including courses in Cellular Biology, Organismic Biology, Ecology, Genetics, either General Physiology, Biochemistry or Developmental Biology and an upper level seminar, supporting courses must include at least 16 hours in Chemistry, Mathematics and/or Physics

EMPORIA KANSAS STATE COLLEGE

Emporia, Kansas 66801 Phone: (316) 343-1200
Dean of Graduate Studies: H. Durst
Dean of School of Liberal Arts and Sciences: J.E. Peterson

DIVISION OF BIOLOGICAL SCIENCES
Phone: Chairman: Ext. 311, Others: 311,312,313, 236

Chairman: R.F. Clarke
Professors: R.J. Boles, R. Keeling, G.A. Leisman, H. McElree, C. Prophet, E.C. Rowe, D.L. Spencer, J.S. Wilson
Associate Professors: T. Eddy, H.M. LeFever, G. Neufeld, R.L. Parenti, J. Ransom
Assistant Professors: K. Smalley, A.L. Ulrich
Instructors: J. Dick, J. Miller

Field Stations/Laboratories: Ross Natural History Reservation - Emporia

Degree Programs: Biology, Medical Technology

Undergraduate Degrees: B.S., B.A., B.S. Ed.
Undergraduate Degree Requirements: 124 hours overall, 30 hours General Education core, 40/45 hours Biology, supporting courses in Physical Science: Chemistry, Physics, and Mathematics varies, but Inorganic and Organic Chemistry minimum. Biology courses vary with emphasis

Graduate Degree: M.S.
Graduate Degree Requirements: Master: Two programs: 30 hours with thesis (5 hours included), 35 hours with research problem (3 hours included). Baccalaureate degree requirement for entry.

Graduate and Undergraduate Courses Available for Graduate Credit: General Genetics (U), Computer Models (U), Biology Materials (U), Sex Education, Research Design and Analysis, Evolution, Special Topics (varies), Seminar, Plant Taxonomy (U), Plant Kingdom (U), Plant Anatomy, Plant Physiology, Grasses, Workshop/ Conservation (U), Wildlife Management (U), Community Ecology (U), Human Ecology, Limnology, Cell Physiology (U), Diagnostic Bacteriology (U), Immunology (U), Virology, Mycology, Molecular Genetics, Vertebrate Structure and Development (U), Animal Behavior (U), Invertebrate Zoology (U), Natural History of Vertebrates (U), Parasitology (U), Entomology, Animal Physiology, Ichthyology, Advanced Ornithology, Paleontology, Mammalogy

FORT HAYS KANSAS STATE COLLEGE

Hays, Kansas 67601 Phone: (913) 628-4000
Dean of Graduate Studies: J. Rice
Dean of College of Liberal Arts: W.R. Thompson

DEPARTMENT OF BIOLOGICAL SCIENCES
Phone: Chairman: (913) 628-4214

Chairman: G.K. Hulett
Professors: C.A. Ely, E.D. Fleharty, G.K. Hulett, H.C. Reynolds, G.W. Tomanek, N.A. Walker
Associate Professors: D.W. Pierson, E.K. Schroder, T.L. Wenke
Assistant Professors: J.R. Choate, N.J. Herman, R.A. Nicholson, J.L. Watson
Instructors: J.K. Ely, W.D. Harris
Research Assistants: 5

Degree Programs: Biology, Botany, Ecology, Fisheries and Wildlife Biology, Limnology, Natural Resources, Range Science, Wildlife Management, Zoology

Undergraduate Degree: B.S. (Biology, Botany, Zoology)
Undergraduate Degree Requirements: A minimum of 30 hours of courses in Biological Sciences plus selected cognate courses.

Graduate Degree: M.S.
Graduate Degree Requirements: Master: A minimum of 30 hours beyond B.S. degree. Course Programs determined by student's interest.

Graduate and Undergraduate Courses Available for Graduate Credit: Can Man Survive? (U), Heredity (U), Field Biology (U), Cell Biology (U), Secondary School Science (U), Conservation of Natural Resources (U), Microbiology of the Pathogens (U), Problems in Biology (non-research) (U), Conservation Workshop (U), Ecology (U), Cytology (U), Genetics (U), Immunology (U), Apprenticeship in Biology (U), Biometry (U), Evolution (U), Elementary School Science Workshop (U), Problems in Biology (research) (U), History of Biology, Improving Instruction in Science, Biological Scientific Writing, Biogeography, Seminar in Biology, Ecological Field Study and Problem, Readings in Biology, Research in Biology, Thesis, Environmental Botany (U), Problems in Botany (U), Aquatic Biology (U), Taxonomy of Flowering Plants (U), Dendrology (U), Bacteriology (U), Range Management (U), Plant Pathology, Plant Physiology (U), Ecological and Range Techniques (U), Range Condition and Improvement (U), Range Plant Nutrition (U), Field Course in Range Management, Advanced Plant Taxonomy, Plant Metabolism, Plant Population Ecology, Agrostology, Parasitology (U), Entomology (U), Embryology (U), Ornithology (U), Histology (U), Invertebrate Zoology (U), Problems in Zoology (non-research) (U), Principles of Systematic Zoology (U), Mammalogy (U), Animal Population Ecology (U), Ecological and Wildlife Techniques (U), Herpetology (U), Ichthyology (U), Limnology (U), Physiology of Adaptation, Advanced Invertebrate Zoology, Seminar in Zoology, Problems in Zoology (research), Readings in Zoology, Research in Zoology, Thesis

FRIENDS UNIVERSITY

2100 University Phone: (316) 263-9131
Wichita, Kansas 67213
Dean of College: J. Williams

DEPARTMENT OF BIOLOGY
Phone: Chairman: Ext. 282

Chairman: L.L. Ortman
Professor: L.L. Ortman
Associate Professor: G.R. Dove

Degree Program: Biology

Undergraduate Degree: B.S.
Undergraduate Degree Requirements: Zoology, Botany, Genetics (32 quarter hours from Biology Curriculum), General Chemistry, I, II, III, Organic Chemistry I, Analysis I, General Physics I, II

KANSAS NEWMAN COLLEGE

3100 McCormick Avenue Phone: 942-4291
Wichita, Kansas 67213
Academic Dean: Sr. T. Wetta

DEPARTMENT OF BIOLOGICAL SCIENCES
Phone: Chairman: 942-4291 Ext. 71

Chairman: S.P. Singh
Professor: Sr. Claudine Axman
Associate Professor: S.P. Singh
Assistant Professors: C. Bussjaeger, L. Bussjaeger
Instructor: Sr. T. Wetta
Laboratory Assistants: 2

Field Stations/Laboratories: Environmental Biology Research Area - Derby, Kansas

Degree Programs: Biology, Medical Technology

Undergraduate Degrees: B.A., B.S.
Undergraduate Degree Requirements: 120 college semester credit hours including Biology - 36 credit hours, Chemistry - 13 credit hours, Physics - 8 credit hours, Mathematics - 6 credit hours

KANSAS STATE COLLEGE OF PITTSBURG

1701 South Broadway Phone: (316) 231-7000
Pittsburg, Kansas 66762
Dean of Graduate Studies: J.D. Haggard
Dean of College: R.C. Welty

DEPARTMENT OF BIOLOGY
Phone: Chairman: Ext. 253

Chairman: R.W. Kelting
Professors: J.C. Bass, D. Bishop, B. Duncan, H.A. Hays, J.C. Johnson, L.E. Keller, T.M. Sperry
Associate Professors: R.L. Dinkins, E.D. Fairchild
Assistant Professor: R.H. Riches
Teaching Assistants: 3

Degree Programs: Biology, Medical Technology

Undergraduate Degree: B.S.
Undergraduate Degree Requirements: 140 semester hours of Biology to include Botany, Zoology, Genetics, Microbiology and Physiology, Chemistry through Organic

Graduate Degree: M.S. (Biology)
Graduate Degree Requirements: Option I: 30 hours including a 4 to 6 hour thesis. Option II: 32 hours including a 3 hour Research Problem.

Graduate and Undergraduate Courses Available for Graduate Credit: Literature of Biology (U), Advanced Genetics (U), Genetics of Microorganisms (U), Pollution Biology (U), Insect Ecology (U), Limnology (U), Animal Ecology (U), Field Taxonomy (U), Animal Physiology (U), Animal Parasitology (U), Immunology (U), Microbial Physiology (U), Environmental Sanitation (U), Plant Physiology (U), Mycology (U), Seminar, Biometry Recent Literature in Biology, Problems in Genetics, Cytology and Cytogenetics, Community Ecology, Economic and Industrial Entomology, Problems in Zoology, Problems in Microbiology, Problems in Botany, Research and Thesis, Research Problems

KANSAS STATE TEACHERS COLLEGE
(See Emporia Kansas State College)

KANSAS STATE UNIVERSITY
Manhattan, Kansas 66502 Phone: (913) 532-6011
Dean of Graduate School: R.F. Kruh

College of Arts and Sciences
Dean: W.L. Stamey

DIVISION OF BIOLOGY
Phone: Chairman: (913) 532-6615 Others: 532-6011

Head: L.E. Roth
Professors: V.C. Bode, R.A. Consigli, L.R. Fina, H.T. Gier, M.F. Hansen, J.O. Harris, L.C. Hulbert, C.L. Kramer, T.H. Pittenger, R.J. Robel, L.E. Roth, O.W. Tiemeier
Associate Professors: L.C. Anderson, T.M. Barkley, C.A. Buck, B.T. Burlingham, J.A. Goss, J.J. Iandolo, M.P. Johnson, A.E. Kammer, C.H. Lockhart, G.R. Marzolf, C.C. Smith, J.S. Weis, F.E. Wilson , J.L. Zimmerman
Assistant Professors: M.S. Center, G.W. Conrad, R.E. Denell, C.P. Doezema, D.S. Fretwell, H.E. Klaassen, G.L. Marchin, L.S. Rodkey, R.S. Slesinski, B.S. Spooner, J.E. Urban, L.G. Williams
Instructors: M.M. Davis, P.W. Hook, A.M. Jones, A.S. Smith
Research Assistants: 11
Research Associates: 4
Visiting Associate Professors: 2
Temporary Instructors: 3

Field Stations/Laboratories: Konza Prairie Environmental Research - Manhatten, Tuttle Creek Fisheries Laboratory - Tuttle Creek Reservoir

Degree Programs: Biochemistry, Biology, Fisheries and Wildlife Biology, Genetics, Microbiology, Parasitology, Physiology

Undergraduate Degree: B.S.
Undergraduate Degree Requirements: B.S. 120 credit hours

Graduate Degrees: M.S., Ph.D.
Graduate Degree Requirements: Master: (1) a minimum of 30 semester hours of graduate credit including a master's thesis of six to eight semester hours, (2) a minimum of 30 semester hours of graduate credit including a written report of two semester hours either of research or of problem work on a topic in the major field or (3) a minimum of 30 semester hours of graduate credit in course work only but including evidence of scholarly effort such as term papers, production of creative work, and so forth, as determined by the student's supervisory committee. Doctoral: at least three years of two semesters each of graduate study beyond the bachelor's degree equivalent to about 90 semester hours and a dissertation are requires. A full year of residency (a minimum or 24 credit hours of course work) is required.

Graduate and Undergraduate Courses Available for Graduate Credit: Plant Physiology (U), Comparative Embryology (U), Principles of Quantitative Microbiology (U), Bacteriology of Human Diseases (U), Animal Parasitology (U), Invertebrate Zoology (U), Soil Microbiology (U), Introductory Mycology (U), Protozoology (U), Human Heredity and Evolution (U), Molecular Genetics (U), Microbial Metabolism (U), Evolution (U), Developmental Biology II (U), Immunology (U), Wildlife Management Techniques (U), Limnology (U), Limnology Methods (U), Fisheries Management (U), Advanced Mycology (U), Endocrinology (U), Use of Models in Biology (U), General Virology (U), Anatomy of Higher Plants (U), Molecular and Cellular Biology (U), Microorganisms of the Natural Environment (U), Mineral Nutrition of Plants, Advanced Parasitology, Introduction to Research in Biology I, Introduction to Research in Biology II, Light and Temperature Relations of Plants, Plant Physiological Technique, Paleobotany, Advanced Virology, Plant Growth and Development, Immunochemistry, Animal Behavior, The Genetic Analysis of Eukaryotic Organisms, Regulation of Gene Expression, Microbial Genetic Techniques, Advanced Plant Ecology, Cellular and Developmental Biology, Advanced Systematic Botany, Genetics of Microorganisms, Population Ecology, Ecosystems Energetics, Reservoir Limnology

DEPARTMENT OF BIOCHEMISTRY
Phone: Chairman: (913) 532-6125 Others: 532-6121

Head: D.J. Cox
Professors: R.K. Burkhard, R.E. Clegg, D.J. Cox, H.L. Mitchell, P. Nordin, D.B. Parrish, W.S. Ruliffson
Associate Professors: B.A. Cunningham, C. Hedgcoth, Jr., W.E. Klopfenstein
Assistant Professors: K.J. Kramer (Courtesy Appointment) D.D. Mueller, G.R. Reeck, T.E. Roche
Instructor: W.M. Barlow

Degree Program: Biochemistry

Undergraduate Degrees: B.A., B.S.
Undergraduate Degree Requirements: (B.S.) General Chemistry (2 semesters), Chemical Analysis (1 semester), Organic Chemistry (2 semesters), Physical Chemistry (2 semesters), Biochemistry Lecture (2 semesters) Biochemistry Laboratory (2 semesters), Advanced Biochemistry Elective (1 semester), Analytic Geometry (1 semester), Calculus (3 semesters), Engineering Physics (2 semesters), Principles of Biology (1 semester), Organismic Biology (1 semester), Advanced Biology Electives (2 semesters)

Graduate Degrees: M.S., Ph.D. (Biochemistry, Food Science), Ph.D. (Biochemistry, Food Science, Animal Nutrition)
Graduate Degree Requirements: Master: Biochemistry I and II, lecture and laboratory, two of the following three courses: Instrumental Analysis, Systematic Organic Chemistry, Physical Biochemistry, elective courses in Biochemistry, Biology or Chemistry sufficient to complete a minimum of 30 hours, Seminar, Thesis Doctoral: Biochemistry I and II, lecture and laboratory, Instrumental Analysis, Systematic Organic Chemistry, Physical Biochemistry, Advanced Biology (10 hours), Seminar, Dissertation. Cumulative examinations are required for admission to candidacy. Several special topics courses in Biochemistry are available and students normally take 2 or 3 of these as electives.

Graduate and Undergraduate Courses Available for Graduate Credit: Biochemistry I and II (U), Biochemistry laboratory I and II (U), Principles of Animal Nutrition (U), Biochemistry of Toxic Materials, Plant Biochemistry, Vitamins, Animal Nutrition Techniques, Intermediary Metabolism, Hormones, Lipids, Nucleic Acids, Proteins, Chemistry of Carbohydrates, Enzyme Chemistry, Advanced Animal Nutrition, Physical Biochemistry

College of Agriculture
Dean: C.J. Hess

AGRONOMY DEPARTMENT
Phone: Chairman: (913) 532-6101

Head: H.S. Jacobs
Professors: O.W. Bidwell, R. Ellis, Jr., E.G. Heyne, H.S. Jacobs, H.E. Jones, E.L. Mader, L.S. Murphy, L.V. Withee, F.G. Bieberly
Associate Professors: F.L. Barnett, R.H. Follett, G.H. Liang, E.B. Nilson, C.B. Overley, G.M. Paulsen, W.L. Powers, O.G. Russ, I.D. Teare, R.L. Vanderlip, C.E. Wasson, D.A. Whitney, V.H. Peterson,

R.F. Sloan
Assistant Professors: L.A. Burchett, E.T. Kanemasu, C.D. Nickell, C.E. Owensby, L.R. Stone, C. Swallow, T.L. Walter, N.E. Humburg, W.A. Moore, R.J. Raney, M.G. Lundquist
Instructors: D.J. Bonne, D.M. Gronau, C.W. Knight
Research Assistants: 12
Research Associates: 1
Acting Assistant Professor: 1
Assistant Instructor: 1
Assistant Professor, Temporary: 1

Field Stations/Laboratories: Southwest Kansas Experiment Field - Mineola, Newton Experiment Field - Newton, Kansas River Valley Experiment Field - Topeka, East Central Kansas Experiment Field - Ottawa, South Central Kansas Experiment Field - Hutchinson, North Central Kansas Experiment Field - Belleville, Irrigation Experiment Field - Scandia, Cornbelt Experiment Field - Powhattan, Sandyland Experiment Field - St. John, Evaportranspiration Laboratory - Manhattan, Soil Testing Laboratory - Manhattan

Degree Programs: Conservation, Genetics, Land Resources, Plant Breeding, Soil Science, Soils and Plant Nutrition

Undergraduate Degree: B.S.
Undergraduate Degree Requirements: 126 credit hours including Basic Sciences, Social Sciences, Humanities, and 26 credit hours of Agronomy

Graduate Degrees: M.S., Ph.D.
Graduate Degree Requirements: Master: 30 credit hours including 6 to 8 hours or research and thesis or 28 hours and 2 hours on report. Doctoral: 90 credit hours including 30 hours of research and dissertation.

Graduate and Undergraduate Courses Available for Graduate Credit: Soil as a Natural Resource, Plant Improvement, Crop and Soil Management, Soil Fertility,. Crop Ecology, Weed Science, Management of Irrigated Soils, Identification of Range and Pasture Plants, Physical Environmental of Crops and Soils, Chemical Fertilizers, Range Management II, Chemical Properties ofSoils, Principles of Plant Breeding, Agricultural Climatology, Soil and Plant Analysis Applications, Field Course in Range Management, Plant Genetics, Crop Physiology, Agronomic Plant Breeding, Soil Physical Chemistry, Soil Physics, Soil Genesis, Advanced Crop Ecology

DEPARTMENT OF ANIMAL SCIENCE AND INDUSTRY
Phone: Chairman: (913) 532-6131

Chairman: D.L. Good
Professors: D.L. Good, B.A. Koch, D.H. Kropf, W.A. Moyer (Extension), D. Richardson, E.F. Smith, H.J. Tuma, J.D. Wheat
Associate Professors: B.V. Able, D.M. Allen, D.R. Ames, B.E. Brent, L.H. Harbers, R.H. Hines, G.H. Kiracofe, R.M. McKee, J.G. Riley, R.R. Schalles, W.H. Smith, K.O. Zoellner (Extension)
Assistant Professors: G.A. Ahlschwede (Extension), G.L. Allee, K.K. Bolsen, L. Corah, M.E. Dikeman, J.D. Hoover, D.E. Schafer (Extension), H.W. Westmeyer (Extension)
Research Assistants: 5

Field Stations/Laboratories: Fort Hays Experiment Station - Hays, Kansas, Carden City Experiment Station - Garden City, Southeast Experiment Station - Mound City

Degree Programs: Animal Science and Industry, Animal Breeding and Genetics, Animal Nutrition, Food Science, Physiology of Reproduction

Undergraduate Degree: B.S. (Animal Science and Industry)
Undergraduate Degree Requirements: 126 semester hours, at least 18 hours in major field.

Graduate Degrees: M.S., Ph.D.
Graduate Degree Requirements: Master: A. 30 semester hours graduate credit including masters thesis B. 30 semester graduate credit including a written report of two semester hours on a topic in major field. Doctoral: 3 years, 2 semesters each of graduate study beyond B.S. degree equivalent to about 90 semester hours and a dissertation. One full year residency.

Graduate and Undergraduate Courses Available for Graduate Credit: Livestock Feeding (U), Commerical Cattle Feedlot Management (U), Swine Production Unit Operation (U), Beef Cow Herd Unit Operation (U), Animal Nutrition (U), Meat-Packing Plant Operation (U), Environmental Physiology of Farm Animals (U), Population Genetics (U), Animal Science and Industry Problems (U), Meat Technology (U), Meat Processing and Preparation (U), Research Techniques in Animal Reproduction, Analytical Techniques in Animal Science and Industry, Graduate Seminar in Animal Science and Industry, Master's Report, Topics in Ruminant Nutrition, Research in Animal Science and Industry, Topics in Monogastric Nutrition, Animal Breeding Seminar, Advanced Meat Science, Research in Animal Science and Industry

DEPARTMENT OF DAIRY AND POULTRY SCIENCE
(No information available)

DEPARTMENT OF ENTOMOLOGY
Phone: Chairman: (913) 532-6154

Head: H. Knutson
Professors: R.J. Elzinga, D.E. Gates, T.L. Harvey, T.L. Hopkins, E.K. Horber, H. Knutson
Associate Professors: H.D. Blocker, H.L. Brooks, R.B. Mills, C.W. Pitts, H.E. Thompson, G.E. Wilde
Assistant Professors: L.J. DePew, E.L. Eshbaugh, A.H. Kadoum, W.H. McGaughey (Adjunct), G.J. Partida, J.D. Stone
Research Assistants: 4
Research Associate (Post Doctoral): 1

Field Stations/Laboratories: Hays Branch Experiment Station - Hays, Garden City Branch Experiment Station - Garden City, Colby Branch Experiment Station - Colby, Mound Valley Branch Experiment Station - Mound Valley, Tribune Branch Experiment Station - Tribune

Degree Programs: Entomology, Parasitology

Undergraduate Degree: B.S. (Crop Protection with Entomology option)
Undergraduate Degree Requirements: 18 credits Entomology, 18 credits Biology and Agricultural Science, 40 credits Physical Sciences and Mathematics, remainder elective to total 126 credits.

Graduate Degrees: M.S., Ph.D.
Graduate Degree Requirements: Master: Minimum total 30 credits of which 6 to 8 thesis, 6 to 8 supporting (not entomology). Doctoral: Minimum 90 total credits of which 30 research for dissertation and approximately 1/3 to 1/4 supporting (not entomology); very flexible

Graduate and Undergraduate Courses Available for Graduate Credit: Federal and State Regulation of Pesticides (U), External Insect Morphology (U), Advanced Applied Entomology (U), Insect Taxonomy (U), Insects of Stored Products (U), Medical Entomology (U), Medical Entomology Laboratory (U), Topics in General and Systematic Entomology (U), Insect Ecology (U), Entomological Methods (U), Insect Control by Host Plant Resistance (U), Toxicology and Properties of Insecticides (U), Internal Insect Morphology (U), Insect Physiology (U), Advanced Applied Entomology II, III (U), Entomology Seminar (U), Problems in Entomology (U), Integrated Pest Management (U), Report in Entomology, Research In Entomology, Topics in Environmental and Physiological Entomology, Research in Entomology

DEPARTMENT OF HORTICULTURE AND FORESTRY
Phone: Chairman: 532-6170

Head: R.W. Campbell
Professors: R.W. Campbell, J.K. Greig, R.A. Keen, F.D. Morrison
Associate Professors: N W. Miles, R.E. Odom
Assistant Professors: E. Abmeyer (Research), F.J. Dainello (Research), F.J. Deneke, W.A. Geyer, F.B. Hadle (Research), L.D. Leuthold, C.E. Long, B.D. Mahaffey, C.W. Marr, R.H. Mattson, J.C. Pair (Research), S.M. Still, G.A. van der Hoeven, J. Winzer (Research)
Instructors: T. Bilderback
Assistants: 7

Field Stations/Laboratories: North East Kansas Experimental Field - Wathena, South East Kansas Experimental Field - Chetopa, Horticulture Experimental Field - Wichita

Degree Programs: Agriculture, Forestry, Horticulture, Landscape Horticulture, Ornamental Horticulture, Plant Breeding, Pomology, Vegetable Crops, Weed Science, Horticulture Therapy

Undergraduate Degree: B.S. (Agriculture (Horticulture option))
Undergraduate Degree Requirements: Students in the Horticulture Curriculum select courses of study in one of three options. All students in the curriculum are required to take a core of general courses in addition to agriculture and horticultural courses. The horticulture Therapy Curriculum combines seven semesters at the university with one semester at the Menninger Foundation at Topeka. Students receive general college, horticulture, Humanities and Social Sciences, and supporting science courses. Both curriculums require 126 credit hours.

Graduate Degrees: M.S., Ph.D.
Graduate Degree Requirements: Master: Applicants must hold a B.S. or equivalent degree in agriculture or biological sciences, or have equivalent evidence of an appropriate background for undertaking an advanced degree program. An undergraduate average of B or better in the junior and senior year is required. A minimum of 30 semester hours of graduate credit shall be required for the degree. Doctoral: The Ph.D. degree normally requires at least 3 years beyond the Masters degree and is awarded to candidates who demonstrate unique ability or skills as researchers and proficiency in communication. The candidate must display deep familiarity and understanding of the subject matter and make original contributions to knowledge. A minimum of 90 semester hours above the B.S. degree including research is required.

Graduate and Undergraduate Courses Available for Graduate Credit: Park Administration and Management (U), Landscape Horticulture (U), Fruit Production (U), Landscape Contraction and Development (U), Vegetable Crop Ecology (U), Greenhouse Management (U), Park Operations (U), Turf Management (U), Turf Management Laboratory (U), Arboriculture (U), Floriculture (U), Methods of Environmental Interpretation (U), Horticulture Problems (U), Forestry Problems (U), Parks and Recreation Problems (U), Park Management Seminar (U), Concepts of Travel and Tourism (U), Greenhouse Clinical Practicum (U), Garden and Landscape Therapy (U), Plant Protection (U), Handling and Processing Fruits and Vegetables (U), Municipal Forestry (U), Vegetable Crop Physiology (U), Fruit Science (U), Plant Research Methods, Master's Report, Research--M.S., Horticultural Plant Breeding, Topics in Plant Breeding, Advanced Pomology, Horticulture Crop Nutrition, Topics in Plant Genetics, Horticulture and Forest Graduate Seminar, Controlled Plant Environment, Dormancy and Regeneration

DEPARTMENT OF PLANT PATHOLOGY
Phone: Chairman: (913) 532-6176

Chairman: L.R. Faulkner
Professors: O.J. Dickerson, E.D. Hansing, C.L. King (Emeritus), C.L. Kramer (Adjunct), L.E. Melchers (Emeritus)
Associate Professors: L.K. Edmunds (USDA), L.B. Johnson, C.L. Niblett, F.W. Schwenk, D.L. Stuteville, W.G. Willis (Extension)
Assistant Professors: L.E. Browder (USDA), L.E. Claflin (Extension), D.B. Sauer (USDA), R.S. Slesinski (Adjunct), R.W. Tillman
Research Assistants: 7

Field Stations/Laboratories: Colby, Fort Hays, Garden City, Southeast Kansas, Tribune

Degree Programs: Plant Pathology

Undergraduate Degree: B.S. (Crop Protection with Plant Pathology option)
Undergraduate Degree Requirements: 126 semester hours including required and elective courses; courses required vary with option (Pest Management, Business and Industries, Entomology Science, Plant Pathology Science)

Graduate Degrees: M.S., Ph.D.
Graduate Degree Requirements: Master: Minimum: 1 academic year in residence, 30 semester hours graduate credit, including (a) Masters thesis of 6-8 hours or (b) a written report of 2 semester hours, 2.65 (out of 4.0) graduate course work GPA (grades of A or B in 3/4 of KSU credit, excluding research, course grades of A,B, or C. required to receive credit), successful completion of a final oral examination. Doctoral: Minimum: 3 years of 2 semesters each of graduate study beyond the B.S., equivalent to 90 semester hour and a dissertation, one full year of residency, 24 semester hours course work, 2.65 (out of 4.0) graduate course work GPA (grades of A or B in 3/4 of KSU credit, excluding research, course grades of A,B or C required to receive credit; successful completion of preliminary and final examinations.

Graduate and Undergraduate Courses Available for Graduate Credit: Plant Pathology (U), Properties of Pesticides (U), Diseases of Field and Horticultural Crops (U), Integrated Pest Management (U), General Plant Pathology, Problems in Plant Pathology, Seminar in Plant Pathology, Research in Plant Pathology for the M.S. Degree, Plant Virology, Physiology of Plant Disease, and Research In Plant Patholog y for the Ph.D. degree.

College of Veterinary Medicine
Dean: D.M. Trotter

DEPARTMENT OF PHYSIOLOGICAL SCIENCES
Phone: Chairman: (913) 532-5666

Head: E.L. Besch
Professors: E.L. Besch, R. Clarenburg, M.R. Fedde, D.M. Trotter, D.W. Upson, G.K.L. Underbjerg (Emeritus), H.T. Gier (Ancillary), F.W. Oehme
Associate Professors: G.H. Cardinet, C.L. Chen, T.E. Chapman, R.A. Frey, R.R. Gronwall, R.D. Klemm, J. A. Westfall, L.E. Erickson (Ancillary)
Assistant Professors: G.T. Hartke, P.T. Purinton, V.V. E. St. Omer
Instructors: D.F. Erichsen, J.S. Shaw
Visiting Lecturers: 1
Graduate Research Assistants: 1
Graduate Teaching Assistants: 1
Research Associate: 1

Field Stations/Laboratories: Environmental Physiology Research Laboratory - Animal Resource Facility

Degree Program: Physiology

Graduate Degrees: M.S., Ph.D.
Graduate Degree Requirements: Master: 30 semester hours including 6 for thesis requirement (Departmental

Degree Program) Doctoral: 60 additional beyond the M.S. including thesis and core course requirements. (Graduate Group Program -- Interdepartmental)

Graduate and Undergraduate Courses Available for Graduate Credit: Anatomy and Physiology, Introduction to Pharmacology of Farm Animals, Gross Anatomy, Microscopic Anatomy I, Pharmacology, Special Anatomy, Comparative Physiology, Environmental Toxicology, Advanced Physiology, Physiology and Pharmacology of Hormones, Anatomy III, Veterinary Physiology, Veterinary Orientation, Canine Anatomy, Anatomical Techniques, Physiology Constituents of Body Fluids

KANSAS WESLEYAN

Salina, Kansas 67401 Phone: (913) 827-5541
Vice President for Academic Affairs: O.L. Voth

DEPARTMENT OF BIOLOGY
Phone: Chairman: Ext. 254

Head: B.L. Owen
Professor: B.L. Owen
Assistant Professor: A.K. Neuburger

Degree Program: Biology

Undergraduate Degree: B.S.
Undergraduate Degree Requirements: 32 semester hours of Biology

McPHERSON COLLEGE

McPherson, Kansas 67460 Phone: (316) 241-0731
Dean of Academic Affairs: M.L. Frantz

DEPARTMENT OF BIOLOGICAL SCIENCES
Phone: Chairman: (316) 241-0731

Chairman: G.J. Ikenberry
Professors: J.H. Burkholder, G.J. Ikenberry
Assistant Professor: A.N. Dutrow
Visiting Lecturers: 1-2

Degree Programs: Biology, Medical Technology, Agriculture, Environmental Science

Undergraduate Degree: B.A.
Undergraduate Degree Requirements: (Biology, Agriculture, etc.), Thirty two semester hours in a program, with supporting courses required in Chemistry, Physics, and Mathematics

MARYMOUNT COLLEGE OF KANSAS

East Iron and Marymount Road Phone: (913) 823-6317
Salina, Kansas 67401
Academic Dean: Sr. J. Sweat

DEPARTMENT OF BIOLOGICAL SCIENCES
Phone: Chairman: Ext. 26

Chairman: S.J. Zeakes
Professors: Sr. M.C. Giersch, Sr. J. Sweat
Assistant Professor: S.J. Seakes
Instructor: K. Ostrander
Visiting Lecturers: 10

Field Stations/Laboratories: Research programs conducted in conjunction with Kansas State University, and Kansas University at University experiment stations.

Degree Program: Biology

Undergraduate Degrees: B.S., B.A.
Undergraduate Degree Requirements: 8 courses in Biology (total 28 hours, 3 courses in Chemistry (General, Organic, Qualitative Analysis, Biochemistry), 2 courses in Physical Science

MT. ST. SCHOLASTIC

(see Benedictine College)

OTTAWA UNIVERSITY

10th and Cedar Phone: (913) 242-5200
Ottawa, Kansas 66067
Dean of College: H. Germer

DEPARTMENT OF BIOLOGY
Phone: Chairman: Ext. 213 Others: Ext. 297,231

Chairman: J.A. Bacon
Professor: J.A. Bacon
Associate Professors: J.E. Morrissey, E. Roth
Research Assistants: 2

Degree Program: Biology

Undergraduate Degree: B.A.
Undergraduate Degree Requirements: 32 units of College credit with 6 to 10 in the area of the Depth Study, each of our "units" worth four hours standard college credit.

SACRED HEART COLLEGE

(see Kansas Newman College)

ST. BENEDICTS COLLEGE

(see Benedictine College)

SAINT MARY COLLEGE

4100 South Fourth Street Trafficway Phone: (913) 682-5151
Leavenworth, Kansas
Dean of College: Sr. M.C. Miller

BIOLOGY DEPARTMENT
Phone: Ext. 348

Chairman: J.D. Robins
Professors: 2
Assistant Professor: J.D.Robins
Visiting Lecturers: 3

Degree Programs: Biology, Medical Technology

Undergraduate Degree: B.S.
Undergraduate Degree Requirements: 32 hours Biology, 12 hours Chemistry, 8 hours Physics, Mathematics through Calculus recommended

ST. MARY OF THE PLAINS

(No reply received)

SOUTHWESTERN COLLEGE

Winfield, Kansas 67156 Phone: (316) 221-4150
Dean of College: J.H. Barton

DEPARTMENT OF BIOLOGY
Phone: Chairman: Ext. 42

Chairman: R.B. Wimmer
Professor: R.B. Wimmer
Assistant Professors: C. Hunter, L.D. Smith, M.C. Thompson

Degree Program: Biology

Undergraduate Degree: B.A., B.S.
Undergraduate Degree Requirements: B.A. - 30 hours Biology, 5 hours Chemistry, B.S. - 40 hours Biology, 20 hours Chemistry, 10 hours Physics, Precalculus, Statistics

STERLING COLLEGE

Sterling, Kansas 67579 Phone: (316) 278-2173
Dean of College: C.G. Brownlee

DEPARTMENT OF BIOLOGY
Phone: Chairman: Ext. 243 Others: Ext. 272

Head: D.L. Taylor
Professor: D.L. Taylor
Associate Professor: R. Walker

Degree Programs: Biology, Biological Education, Medical Technology

Undergraduate Degree: B.S.
Undergraduate Degree Requirements: Biology 24 hours, Chemistry 15 hours, Mathematics 5 hours of Analytical Geometry and Calculus, Physics 8 hours

TABOR COLLEGE

Hillsboro, Kansas 67063 Phone: (316) 947-3121
Dean of College: W. Prieb

DEPARTMENT OF BIOLOGY
Phone: Chairman: Ext. 38 Others: Ext. 40

Chairman: M.P. Terman
Associate Professor: B. Stockton

Degree Program: Biology

Undergraduate Degree: B.S.
Undergraduate Degree Requirements: Not Stated

UNIVERSITY OF KANSAS

Lawrence, Kansas 66044 Phone: (913) 864-2700
Dean of Graduate Studies: W.J. Argersinger, Jr.

College of Liberal Arts and Sciences
Dean: G.R. Waggoner

DIVISION OF BIOLOGICAL SCIENCES
Chairman: B.R. Buchell

DEPARTMENT OF BIOLOGY
Phone: Chairman: (913) 864-4373

Chairman: K.B. Armitage
Professors: All faculty are drawn from the graduate departments in the Division of Biological Sciences

Field Stations/Laboratories: John H. Nelson Environmental Study Area plus other areas - within 15 miles of campus.

Degree Programs: Anatomy, Anatomy and Histology, Anatomy and Physiology, Animal Behavior, Animal Genetics, Biochemistry, Biological Chemistry, Biological Engineering, Biometry, Biostatistics, Botany, Ecology, Entomology, Entomology and Parasitology, Ethology, Fisheries and Wildlife Biology, Genetics, Health and Biological Sciences, Histology, Life Sciences, Limnology, Parasitology, Physiological Sciences, Physiology, Plant Sciences, Zoology, Population Biology, Environmental Physiology, Cell Biology, Developmental Biology

Undergraduate Degree: B.G.S, B.A, B.S.
Undergraduate Degree Requirements: B.A., 29 semester credits of Biology (which includes 9 hours of required core curriculum and 4 hours introductory course, 1 year Physics, Calculus, and Organic Chemistry. B.S. - 40 semester credits of Biology - choice of three tracts, specific requirements differ for each tract but all tracts require core curriculum and various amounts of Physics, Chemistry and Mathematics. B.G.S. highly flexible, student may individualize his/her own program.

DEPARTMENT OF BOTANY
Phone: Chairman: (913) 864-4305 Others: 864-2700

Chairman: R.W. Lichtwardt
Professors: R.W. Baxter, R. Borchert, J.E. Fox, R.W. Lichtwardt, R.R. McGregor, R.H. Thompson, A.M. Torres, P.V. Wells
Associate Professor: J.D. McChesney
Assistant Professors: W.L. Bloom, J.L. Hamrick, III
Research Assistants: 4
Teaching Assistants: 3

Degree Programs: Plant Anatomy, Plant Biochemistry, Biology, Botany, Ecology, Genetics, Physiology

Graduate Degrees: M.A., Ph.D. (Botany)
Graduate Degree Requirements: Master: Thesis and non-thesis options available. Thirty hours of credit and a final oral examination required with both options. All incoming students required to take written placement examination. Doctoral: A minimum of thirty hours of credit, a comprehensive oral examination, a dissertation, and an oral final examination in defense of the thesis. All incoming students required to take a written placement examination.

Graduate and Undergraduate Courses Available for Graduate Credit: Systematic Botany (U), Plant Ecology (U), Plant Physiology (U), Plant Physiology Laboratory (U), Ecological Plant Physiology (U), Plant Anatomy (U), Plant Kingdom (U), Elements of Plant Geography (U), Botanical Problems (U), History of Botany (U), Phycology (U), Introductory Mycology (U), Plant Pathology (U), Medical Mycology (U), Paleobotany (U), Plant Communities of North America (U), The Vegetation of the Earth (U), Plant Biochemistry (U), Cytology and Cytogenetics (U), Advanced Tropical Botany (U), Field Botany (U), Chemotaxonomy; Chemotaxonomy Laboratory, Plant Genetics, Biosystematics, Physiology of the Fungi, Fungi and Arthropods, Taxonomy of Various Groups, Paleobotany of the Coal Age, Advanced Mycology, Thesis, Advanced Systematic Botany, Modern Concepts in Ecology, Research Topics in Plant Physiology and Biochemistry, Advanced Phycology, Special Problems in Cytology, Special Problems in Cytogenetics, Biogenesis, Seminar in Vegetation Geography

DEPARTMENT OF BIOCHEMISTRY
Phone: Chairman: (913) 864-4021

Chairman: R.T. Hersh
Professors: J.E. Fox, R.T. Hersh, R.H. Himes, P.A. Kitos
Associate Professors: L.L. Houston, R.B. Sanders
Assistant Professors: R. Borchardt, G.M. Maggiora, M.Z. Newmark, R.F. Weaver

Degree Program: Biology (Biochemistry)

Graduate Degrees: M.A., Ph.D.
Graduate Degree Requirements: Master: General Biochemistry, 2 additional courses in Biochemistry, 5-10 hours in Biochemistry or related hours, 2 semesters of seminar Biochemistry, minimum of 30 hours, including thesis, Biochemistry 802. Doctoral: General Biochemistry (year course), Biochemistry Laboratory, Biochemistry Seminar (6 semesters). Other courses designed by student's committee with at least three courses within the Department. Student must pass an oral comprehensive to be admitted to candidacy. The student must submit a dissertation and defend in a public lecture.

Graduate and Undergraduate Courses Available for Graduate Credit: Cellular Physiology (U), General Biochemistry I, II, Biochemistry Laboratory (U), Biological Macromolecules, Plant Biochemistry (U), Biosynthesis of Macromolecules, Biochemical Genetics (U), Cellular Regulatory Mechanisms, Action of Vitamin and Hormones (U), Physical Biochemistry, Quantum Biochemistry, Mechanism of Enzyme Action, Protein

Protein Chemistry, Molecular Asymmetry in Biology

DEPARTMENT OF ENTOMOLOGY
Phone: Chairman: (913) 864-4610 Others: 864-4301

Chairman: C.D. Michener
Elizabeth M. Watkins Professor of Entomology: C.D. Michener
Professors: R.E. Beer, G.W. Byers, J.H. Camin, R. Jander, C.D. Michener
Associate Professors: P.D. Ashlock, W.J. Bell, K.A. Stockhammer, O.R. Taylor
Teaching Fellow: 1
Research Assistants: 7
Museum Assistants: 2

Degree Program: Entomology

Graduate Degrees: M.S., Ph.D.
Graduate Degree Requirements: Master: Graduation in biological sciences, interest in entomology, excellence in transcripts, recommendations, and GRE scores. Doctoral: Graduation in biological sciences, interest in entomology, excellence in transcripts, recommendations, and GRE scores.

Graduate and Undergraduate Courses Available for Graduate Credit: General Entomology (U), Economic Entomology (U), Insects and Public Health (U), Principles of Systematics (U), Comparative Animal Behavior (U), Principles of Ecology (U), Insect Ecology (U), Special Problems (U), External Morphology of Insects (U), Immature Insects (U), Insect Development (U), Classification of Insects (U), Insect Histology (U), Medical Entomology (U), Entomological Survey (U), Insect Community and Population Ecology (U), Insect Physiology and Internal Morphology (U), Insect Biochemistry (U), Principles of Applied Entomology (U), Taxonomy of Acarina (U), Biology of Acarina (U), Entomology Seminar, Graduate Research, Evolutionary Mechanisms, Thesis, Entomology Seminar, Graduate Research, Dissertation

DEPARTMENT OF MICROBIOLOGY
Phone: Chairman: 864-4311

Chairman: D. Paretsky
Professors: J.M. Akagi, R.H. Bussell, L.R. Draper, A.A. Hirata, D. Paretsky, D.M. Shankel
Associate Professor: C.S. Buller
Assistant Professors: D.C. Robertson, H.O. Stone, Jr.
Instructors: E.J. Handley
Visiting Lecturers: 4
Teaching Fellows: 12
Research Assistants: 4

Degree Programs: Microbiology, Medical Technology

Undergraduate Degree: A.B.
Undergraduate Degree Requirements: One year of Physics, Mathematics through Calculus, one year of Organic Chemistry, and 21 hours of Microbiology

Graduate Degrees: M.A., Ph.D. (Microbiology)
Graduate Degree Requirements: Master: 30 hours graduate credit, including 8-10 in Research. Three areas to be selected from Microbial Physiology, Microbial Genetics, Ultrastructure, Immunology, and Virology, 2 other supporting fields, Thesis. Doctoral: Courses selected by committee. Differential and Integral Calculus, Biochemistry and Physics required. Comprehensive Examination. Thesis. Final Oral.

Graduate and Undergraduate Courses Available for Graduate Credit: Man and Microbes, Principles of Microbiology, Principles of Microbiology, Honors, Special Problems in Microbiology, Special Problems in Micro-Biology, Honors, Hematology and Clinical Procedures, Fundamentals of Microbiology, Fundamentals of Microbiology, Honors, Immunology, Pathogenic Microbiology, Microbial Physiology, Microbial Genetics, General Virology, Structure and Function of Microbial Cell Surfaces, Ultrastructure of Biological Material, Advanced Immunology, Physiology of Bacteria, Tumor Virology, Mechanisms of Microbial Pathogenicity, Tumor Immunology, Bacteriophage Biochemistry, Comparative Microbiology, Radiation Genetics, Tissue Culture, Advanced Microbial Genetics, Animal Virology I

DEPARTMENT OF PHYSIOLOGY AND CELL BIOLOGY
Phone: Chairman: (913) 864-3872 Others: 864-4303

Chairman: P.R. Burton
Professors: K.B. Armitage, W.M. Belfour, E.C. Boree, P.R. Burton, F.A. Samson (Adjunct), H.W. Shirer, J.A. Weli, J.M. Yochim, H.C. Wolfe
Associate Professors: W.J. Bell, R. Borcbert, B.R. Burchill, N.A. Dahl, M.J. Maber, W. Osness (Adjunct), D.M. Quadagno, K.A. Stockhammer, C.W. Wyttenbach
Assistant Professors: P.H. Cooke, H.B. Lillywhite
Visiting Lecturers: 1
Teaching Fellows: 10
Research Assistants: 11
Graduate School Fellows: 2

Field Stations/Laboratories: Natural History Reservation, Rockefeller Experimental Tract, Breidentbal Biology Reserve

Degree Programs: Physiology and Cell Biology

Graduate Degrees: M.A., Ph.D.
Graduate Degree Requirements: Master: Nonthesis - 36 hours graduate course work, Thesis - 20 credit hours study in Physiology and Cell Biology and 10 credits outside Physiology and Cell Biology to make total of 30, plus thesis Doctoral: Five-man committee guides graduate study of student. Preliminary Oral examination required, Original research expected with dissertation

Graduate and Undergraduate Courses Available for Graduate Credit: Not Stated

DEPARTMENT OF RADIATION BIOPHYSICS
Phone: Chairman: (913) 864-3867

Chairman: E.T. Shaw
Professors: B.S. Friesen, E.T. Shaw
Associate Professors: R. Riley, J.D. Zimbrick
Assistant Professors: J.L. Beach
Instructors: R.D. Dean, G.V. Oldfield
Teaching Fellows: 1
Research Assistants: 2
Radiation Protection Service: 2-3

Degree Program: Radiation Biophysics

Undergraduate Degree: B.A., B.S.
Undergraduate Degree Requirements: B.A. minimum of 20 hours in radiation biophysics (including General Biology and courses 108, 702,703,704,711 or 760 (2 hours), Mathematics through Differential and Integral Calculus, a course in Organic Chemistry and a course in Modern Physics.

Graduate Degrees: M.S., Ph.D.
Graduate Degree Requirements: Master: 30 hours graduate courses. Thesis. Final Examination. Doctoral: Courses selected by Committee. Comprehensive Examination. Thesis. Oral Examination.

Graduate and Undergraduate Courses Available for Graduate Credit: Physical Foundations of Radiation Biology, Radioisotopes, Radiation Protection, General Biophysics, Radioactive Tracers and Instrumentation, Radiation Detection Electronics, Epidemiology of Radiation, Basic Radiation Biology, Problems in Biophysics, Health Physics Laboratory, Radiation Biology, Radiotracers and Instrumentation, Molecular Biophysics, Radiation Biophysics Seminar, Dosimetry and Shielding, Radiation Genetics, Nuclear Medical Physics, The Physics of Radiation Therapy, The Physics of Diagnostic Radiology, Advanced Health Physics Laboratory

DEPARTMENT OF SYSTEMATICS AND ECOLOGY
Phone: Chairman: 864-4305 Others: 864-2700

Chairman: G. Schlager
Professors: R.E. Beer, G.W. Byers, W.H. Coil, F.B. Cross, W.E. Duellmann, T. Eaton, H.S. Fitch, R.S. Hoffmann, P.S. Humphrey, R. Jander, R. Johnston, R.M. Mengel, C.D. Michener, G. Schlager, P.V. Wells
Associate Professors: P.D. Ashlock, P.W. Hedrich, P. M. Neely, O.R. Taylor
Assistant Professors: J.W. Bee, M.S. Gaines, L.D. Martin, W.J. O'Brien, R.J. Perkins, N.A. Salde

Degree Programs: Systematics and Ecology

Undergraduate Degree: B.S.
Undergraduate Degree Requirements: 1 year Physics, 1 year Mathematics, 1 year Chemistry, Organic Chemistry, Biology Core Courses, 24 semester hours in Systematics and Ecology

Graduate Degrees: M.A., Ph.D.
Graduate Degree Requirements: Master: 1) thesis with 30 semester hours 2) no-thesis with 36 semester hours Doctoral: Equivalent of three years of full-time graduate work beyond B.A., Dissertation, 2 semesters of half-time supervised teaching, reading knowledge of two foreign languages or demonstrated proficiency in two research skills.

Graduate and Undergraduate Courses Available for Graduate Credit: Comparative Anatomy (U), General Invertebrate Zoology (U), Animal Histology and Histotechnique (U), Plant Ecology (U), Classification of Insects (U), Evolution (U), Population Biology (U), Evolution of Vertebrate Communities (U), Invertebrate Paleontology (U), Introduction to Systematics (U), Comparative Animal Behavior (U), Comparative Animal Behavior Laboratory (U), Limnology (U), Animal Natural History (U), Principles of Ecology (U), Natural History Museum Techniques (U), Insect Ecology (U), Insect Community and Population Ecology (U), Introduction to Parasitology (U), Preparatory Techniques in Zoology (U), Fisheries (U), Fisheries Laboratory (U), Fundamentals of Tropical Biology (U), Advanced Tropical Zoology (U), Advanced Population Biology (U), Paleontology of Lower Vertebrates (U), Paleontology of Higher Vertebrates (U), Ichthyology (U), Ornithology (U), Mammalogy (U), Herpetology (U), Field Course in Vertebrate Zoology (U), Field Course in Vertebrate Paleontology (U), Graduate Conference, Graduate Research, Thesis, Special Topics in Systematics and Ecology, Evolutionary Mechanisms, Principles of Parasitism, Biometry Workshop, Correlation and Causation in Biology, Quantitative Ecology, Topics in Statistical Biology, Numerical Taxonomy, Graduate Conference A; GraduateResearch A, Thesis A, Advanced Animal Ecology, Evolutionary Biology of Higher Vertebrates, Animal Distribution

UNIVERSITY OF KANSAS MEDICAL CENTER
39th Street and Rainbow Boulevard Phone: (913) 831-7005
Kansas City, Kansas 66103
Dean of Graduate Studies: A.M. Thompson
Dean of Medical School: W.D. Rieke

DEPARTMENT OF ANATOMY
(No reply received)

DEPARTMENT OF BIOCHEMISTRY AND MOLECULAR BIOLOGY
Phone: Chairman: (913) 831-7007 Others: 831-7005

Chairman: K.E. Ebner
Distinguished Medical Teacher: K.E. Ebner
Sam E. Rogerts Distinguished Professor of Biochemistry: S. Grisolia
Professors: D. Cohn, K.T. Ebner, H. Fisher, S. Grisolia, J. Kimmel, R. Mills, D. Mulford
Associate Professors: W.N. Arnold, D.O. Carr, B.G. Hudson, M.E. Noelken
Assistant Professors: J.O. Grunewald, M.H. Maguire, A. Murdock, R. Silverstein
Visiting Lecturers: 1
Research Associates: 13

Degree Program: Biochemistry

Graduate Degrees: M.A., Ph.D.
Graduate Degree Requirements: Master: 30 hours of creidt, with a minimum of 18 of these hours in Biochemistry, approximately 50% of the total hours should be course work, remainder-research and thesis. Required courses: General Biochemistry, 2 advanced courses in Biochemistry, Seminar, research and Thesis Doctoral: General Biochemistry, Organic Qualitative Analysis, Biochemistry Research Techniques, Physical Chemistry and 4 additional advanced Biochemistry courses, Minimum of 3 courses in related disciplines seminar, research and thesis. The minimum curriculum is 75 semester hours-approximately 60 in Biochemistry . Of these hours 1/3 are course work and seminar with remainder research and thesis.

Graduate and Undergraduate Courses Available for Graduate Credit: General Biochemistry, General Biochemistry Laboratory, Molecular and Physiological Basis of Nutrition, Qualitative Analysis of Bio-Organic Compounds, Nutrition, Research in Biochemistry, Biochemistry Seminar, Intermediate Metabolism I (Nitrogen), Intermediate Metabolism II (Nucleic Acids), Intermediate Metabolism III (Carbohydrates), Readings in Biochemistry, Protein Chemistry, Enzyme Chemistry, Biochemical Genetics, Biochemical Research Techniques, Biochemical Techniques Laboratory, Physical Biochemistry, Advanced Topics, Biosynthetic Mechanisms, Clinical Biochemistry, Clinical Biochemistry Laboratory, Instrumental Methods

DEPARTMENT OF BIOMETRY
(No information available)

DEPARTMENT OF MICROBIOLOGY
(No reply received)

DEPARTMENT OF PHYSIOLOGY
Phone: Chairman: (913) 831-7025

Chairman: A.M. Thompson (Acting)
Professors: E.B. Brown, Jr., D.C. Johnson, F.E. Samson, L.P. Sullivan, A.M Thompson
Associate Professors: R.L. Clancy, N.C. Gonzalez, G.N. Loofbourrow, J.B. Rhodes, C.M. Tarr, J.W. Trank, J.D. Wood
Assistant Professors: P.A. Hensleigh, A.S. Hermreck, W.K. Legler
Student Teaching Assistants: 8

Degree Program: Physiology

Graduate Degrees: M.S., Ph.D.
Graduate Degree Requirements: Master: 24-30 hours course work, original research, and thesis. Doctoral: Neuroanatomy, Biochemistry, Medical Physiology, Endocrinology, Comprehensive Examination, Seminar participation, Research Skills (2 selected from 4), Original research and dissertation, At least 6 graduate courses offered by PHSL Department, Teaching experience

Graduate and Undergraduate Courses Available for Graduate Credit: Reproductive Physiology, Transport, Permeability and Membranes, Gastroenterology, Fluid, Electrolyte and Acid-Base Balance, Neurophysiology, Biology of Sex, Physiological Instrumentation I, II, Cardiovascular Physiology

WASHBURN UNIVERSITY

1700 College Avenue Phone: (913) 235-5341
Topeka, Kansas 66621
Dean of College: C.R. Haywood

BIOLOGY DEPARTMENT
Phone: Chariman: Ext. 270, Others: Ext. 269,270,271

Chairman: P.H. Kopper
Professors: D.R. Boyer, R.E. Johnson
Assistant Professors: G.H. Irwin III, T.M. O'Connor, T.M Wolf
Instructor: A. Hammond

Degree Programs: Biology, Medical Technology

Undergraduate Degree: B.S., B.A.
Undergraduate Degree Requirements: 30 hours for B.S., 24 hours for B.A. General Cellular Biology, Botany, Zoology, Genetics, Microbiology, Seminar, Two laboratory courses in Chemistry, one laboratory course in Physics

WICHITA STATE UNIVERSITY

1845 Fairmount Phone:(316) 689-3110
Wichita, Kansas 67208
Dean of Graduate Studies: L.M. Benningfield
Dean of College: P.J. Magelli

DEPARTMENT OF BIOLOGY
Phone: Chairman: 689-3377 Others: 689-3110

Chairman: H.D. Rounds
Distinguished Professor of Natural Sciences: A. Sarachek
Professors: E. Bubieniec, H. Rounds, A. Sarachek, G. Sweet
Associate Professors: D. Distler, G. Miller
Assistant Professors: J. Bish, B. Craig, V. Eichler, E. Hugh Gerlach (Adjunct), W. Harm, J. Seng, R. Sobieski, J. Watertor, A. Youngman
Instructors: B. Bowman
Assistant Instructors: D. Krokosska, F. Riffel
Research Assistants: 3
Graduate Teaching Assistants: 8
Endowment Fellowship: 1

Field Stations/Laboratories: Prairie Field Station - Cheney Reservoir

Degree Program: Biological Sciences

Undergraduate Degrees: B.A., B.S.
Undergraduate Degree Requirements: B.A. - 30 hours of Biology and must include General Biology, Botany, Zoology, (1 year General Chemistry, 1 year Organic Chemistry), B.S. - 40 hours of Biology and must include General Biology, Botany, Zoology, Cell Biology, Physics, and one of three options (Botany, Zoology, Microbiology).

Graduate Degree: M.S.
Graduate Degree Requirements: Master: Option A - 30 hours graduate work including thesis, proficiency in one research tool Option B - (open to applicants certified in some professional field), 36 hours graduate work, excludes research or research tool.

Graduate and Undergraduate Courses Available for Graduate Credit: Nonvascular Plants (U), Vascular Plants (U), Plant Physiology (U), Invertebrate Zoology (U), Vertebrate Zoology (U), Parasitology (U), Entomology (U), Mammalian Physiology (U), Neurophysiology and Neuroanatomy (U), Comparative Embryology (U), Histology (U), Bacteriology (U), Mycology (U), Ecology and Man (U), Ecology (U), Field Ecology (U), Limnology (U), Cytology (U), Genetics (U), Immunobiology (U), Analytical Methods in Biology (U), Physiological Plant Ecology (U), Animal Behavior (U), Protozoology (U), Pathogenic Microbiology (U), Microbial Physiology (U), Biological Literature (U), Special Problems in Animal Behavior (U), Physiological Basis of Behavior (U), Comparative Animal Physiology (U), Microbial Metabolism (U), Microbial Genetics (U), Physiological Genetics (U), Advanced Immunology (U), Radiation Biology

KENTUCKY

ASBURY COLLEGE

Wilmore, Kentucky 40390 Phone: (606) 858-3511
Dean of College: R. Kusche

DEPARTMENT OF BIOLOGY
Phone: Chairman: (606) 858-3511 Ext. 234

Chairman: C.B. Hamann
Professors: C.B. Hamann, H.H. Howell
Associate Professors: W.B. Smith
Assistant Professor: J.M. Smith
Instructors: K. Houp

Degree Programs: Biology, Medical Technology

Undergraduate Degree: B.S.
Undergraduate Degree Requirements: General Education - 88quarter hours, Major Requirement - 48 quarter hours of Biology, Chemistry 12 quarter hours, Mathematics 16 quarter hours, Electives 28 quarter hours

BELLARMINE COLLEGE

Louisville, Kentucky 40205 Phone: (502) 452-8011
Dean of College: Not Stated

DEPARTMENT OF BIOLOGY
Phone: Chairman: (502) 452-8202 Others: 452-8011

Chairman: J.J. Dyar
Professor: J.J. Dyar
Associate Professor: R. Korn
Assistant Professor: L. Keffler

Field Stations/Laboratories: Bellarmine Biological Field Station - Mt. St. Francis, Indiana

Degree Program: Biology

Undergraduate Degree: B.A.
Undergraduate Degree Requirements: 2 semesters General Chemistry, 2 semesters Physics, 2 semesters Organic Chemistry, 2 semesters Mathematics - at least one semester Calculus, General Biology, 2 semesters Senior Research, General Ecology, Microbiology, Mammalian Physiology, Developmental Biology, Genetics, Cell Physiology

BEREA COLLEGE

Berea, Kentucky 40403 Phone: (606) 896-9341
Dean of College: W.F. Stolte

DEPARTMENT OF AGRICULTURE
Phone: Chairman: Ext. 525 Others: Ext. 526

Chairman: R.L. Johnstone
Professor: N. Stephens
Associate Professors: C.E. Gentry, E.C. Hogg, J.P. Shugars

Degree Program: Agriculture

Undergraduate Degree: A.B., B.S.
Undergraduate Degree Requirements: a. 11 courses in agriculture, 7 of which are specified, b. 2 courses in Zoology and Botany, 2 courses in Chemistry, and 1 course in Mathematics c. Graduate must also have met the general requirements of Berea College

DEPARTMENT OF BIOLOGY
Phone: Chairman: Ext. 592

Chairman: F.B. Gailey
Professors: R.N. Barnes, F.B. Gailey, N. Stephens
Associate Professors: M. Cartledge, C.E. Gentry, E.C. Hogg
Assistant Professors: J.A. Grossman

Degree Program: Biology

Undergraduate Degree: B.A.
Undergraduate Degree Requirements: (in "courses" equivalent to 4 semester hours each), Basic Ideas in Biology, Botany, Genetics, Physiology, Zoology, and a minimum of three electives in Biology, Mathematics, Two courses in Chemistry (General and Organic).

BRESCIA COLLEGE

Owensboro, Kentucky 42301 Phone: (502) 685-3131
Dean of College: J. Weisenbeck

DIVISION OF MATHEMATICS AND NATURAL SCIENCES
Phone: Chairman: Ext. 275

Division Chairman: F. Montgomery
Professor: Sr. C. Czurles (Emeritus)
Assistant Professors: J.M. Baur, Sr. M. Morek
Lecturers: 2

Degree Programs: Biology, Medical Technology

Undergraduate Degree: B.S.
Undergraduate Degree Requirements: 40 hours biology (Botany, Zoology, Anatomy, Physiology, Genetics, Seminar, and 16 hours of Electives), 8 hours Organic Chemistry, 8 hours Calculus, 8 hours General Physics

CAMPBELLSVILLE COLLEGE

Campbellsville, Kentucky 42718 Phone: (502) 465-8158
Dean of College: D. Jester

DEPARTMENT OF NATURAL SCIENCE
Phone: Chairman: Ext. 53

Chairman: R.B. Shiflett
Assistant Professors: M.A. Rogers, B. Evans
Instructor: H.R. Richardson (part-time)

Degree Programs: Biology, Medical Technology

Undergraduate Degree: B.S.
Undergraduate Degree Requirements: 128 semester hours total, B.S. Biology - 30 semester biology including Botany, Zoology, Microbiology, Genetics and Vertebrate Embryology or Comparative Vertebrate Anatomy, 10 hours Chemistry and 10 hours Physics also required.

CENTRE COLLEGE OF KENTUCKY

Danville, Kentucky 40422 Phone: (606) 236-5211
Dean of College: E.C. Reckard

BIOLOGY PROGRAM (DIVISION OF SCIENCE AND MATHEMATICS)
Phone: Chairman: Ext. 304

Chairman: F.W. Leotscher, Jr.
Professor: F.W. Leotscher, Jr.
Associate Professor: B.T. Feese
Assistant Professors: R.K. Hammond, S.M. Moyle

Field Stations/Laboratories: Central Kentucky Wildlife

Refuge - Boyle County, Kentucky

Degree Programs: Biology

Undergraduate Degree: B.A, B.S.
Undergraduate Degree Requirements: Life Science 11, 12, 21 (Genetics, 22 (Ecology), Biology 33 or 34 (Microbiology, Biology 36 (Plant Biology), Biology 50 (Senior Seminar). At least one Biology 30 (Laboratory courses other than those listed above). 2 terms of Mathematics, 4 terms of Physical Sciences.

CUMBERLAND COLLEGE
(No information available)

EASTERN KENTUCKY UNIVERSITY
(No reply received)

GEORGETOWN COLLEGE
Georgetown, Kentucky 40324 Phone: (502) 863-8011
Dean of Graduate Studies: R. Alexander
Dean of College: J. Butler

DEPARTMENT OF BIOLOGICAL SCIENCES
Phone: Chairman: (502) 863-8836

Chairman: D.M. Lindsay
Professors: D.M. Lindsay, M.E. Wharton
Associate Professors: G. Clark, T. Seay
Assistant Professor: W. Jones

Degree Program: Biology

Undergraduate Degrees: B.A., B.S.
Undergraduate Degree Requirements: 8 courses (each of 4 semester hours) including: Biological Principles, Animal Diversity, Environmental Biology and Plants and Man. Plus: Biology Seminar (1/2 course)

KENTUCKY STATE UNIVERSITY
Frankfort, Kentucky 40601 Phone: (502) 564-6260
Dean of College: J. Graves

DEPARTMENT OF BIOLOGY
Phone: Chairman: (502) 564-6066 Others: 564-6070

Chairman: G.C. Ridgel
Professors: W.J. Fleming, G.C. Ridgel
Associate Professor: W.L. Dixon
Assistant Professors: G. Klock, M. Rahman, D.B. Ralin, H.L. Scuvey, A.L. Surratt

Degree Programs: Biology, Biological Sciences, Medical Technology

Undergraduate Degree: B.S.
Undergraduate Degree Requirements: General Education 46 hours, Biology 30 hours, Mathematics 12 hours, Chemistry 20 hours, Physics 6-8 hours, Electives 12 hours

KENTUCKY WESLEYAN COLLEGE
3000 Frederica Street Phone: (502) 684-5261
Owensboro, Kentucky 42301
Dean of College: E.L. Beavin

DEPARTMENT OF BIOLOGY
Phone: Chariman: (502) 684-5261 Ext. 233

Chairman: O.L. Richardson, Jr.
Professors: R.C. Dalzell, Sr., O.L. Richardson, Jr.
Assistant Professor: D.L. Davenport

Degree Program: Biology

Undergraduate Degree: B.S.
Undergraduate Degree Requirements: Biology course work 30 semester hours (minimum), Mathematics 6 semester hours, Physics 8 semester hours, Chemistry 8 semester hours, Gross Total - 128 semester hours

MOREHEAD STATE UNIVERSITY
Morehead, Kentucky 40351 Phone: (606) 783-2221
Dean of Graduate Studies: J.R. Duncan
Dean of College: C.F. Ward

DEPARTMENT OF AGRICULTURE
Phone: Chairman: 783-3305 Others: 783-2221

Head: C.M. Derrickson
Professors: J. Bendixen, C.M. Derrickson
Associate Professors: D. Minion, K. Wade, R. Wolfe
Assistant Professors: R. Kline, R. Lay
Instructors: R. Eckstein, R. Wood
Visiting Lecturers: 1
Research Assistants: 2

Field Stations/Laboratories: University Farm - Morehead (5 miles from Campus).

Degree Programs: Agricultural Education, Agriculture, Agronomy, Animal Science, Crop Science, Horticultural Science, Horticulture

Undergraduate Degree: B.S.
Undergraduate Degree Requirements: 128 semester hours with total 54 semester hours in Agriculture - with options in Agronomy, Animal Science, Agricultural Economics, Horticulture, Agricultural Business and Agricultural Education

MURRAY STATE UNIVERSITY
Murray, Kentucky 42071 Phone: (502) 762-3011
Dean of Graduate Studies: D. Jones
Dean of College: W. Blackburn

AGRICULTURAL DEPARTMENT
Phone: Chairman: (502) 762-3329 Others: 762-3328

Chairman: W.N. Cherry
Professors: C. Chaney, D. Beatty, R. Hudor, J. Mikulick
Associate Professors: L. Jacks, A. Scott, R. Masha, J. Martin, A. Tackett
Assistant Professors: E. Heathcott, V.R. Shelton
Instructors: J. Rudolph

Degree Programs: Agricultural Education, Agriculture, Agronomy, Animal Husbandry, Dairy Science, Horticultural Science, Horticulture, Landscape Horticulture Soil Science

Undergraduate Degree: B.S.
Undergraduate Degree Requirements: Not Stated

DEPARTMENT OF BIOLOGICAL SCIENCES
Phone: Chairman: (502) 762-2786

Chairman: W.J. Pitman (Acting)
Professors: E.M. Cole, H. Eversmeyer, H.M. Hancock (Emeritus)
Associate Professors: R.E. Daniel, M.J. Fuller, M.D. Hassell, R.G. Johnson, W.J. Pitman, C.G. Smith, M.E. Sisk, C.D. Wilder
Assistant Professors: D.O. Abbott, G.W. Kemper
Visiting Lecturers: 4

Field Stations/Laboratories: Hunter Hancock Biological Station - Kentucky Lake at Tennessee River

Degree Programs: Animal Behavior, Bacteriology and Immunology, Biological Education, Biological Sciences, Botany, Ecology, Fisheries and Wildlife Biology,

Limnology, Zoology, Ichthyology

Undergraduate Degree: B.S. (Biology)
Undergraduate Degree Requirements: 32 semester hours Biology (minimum), Chemistry 18 hours (through Organic), Physics 8 hours, Mathematics 5 hours, Minimum of 128 hours for graduation with the B.S. degree

Graduate Degrees: M.A., M.S., M.Ed.
Graduate Degree Requirements: Master: M.S. - 30 semester hours above the B.S. degree plus a thesis. Other options are available for M.A. in Ed.; M.A. in Teaching and the M.A. in College Teaching without a thesis.

Graduate and Undergraduate Courses Available for Graduate Credit: Embryology, Evolution, Protozoology, Ichthyology, Histology, Morphology of Vascular Plants, Morphology of Non-Vascular Plants, Mammalogy, Endocrinology, Advanced Invertebrate Zoology, Limnology, Animal Behavour, Cell Physiology, Herpetology, Ornithology, Pathogenic Microbiology, Immunology --all (U), Fisheries Biology, Plant Morphology, Freshwater Invertebrates, Aquatic Botany, Plant Ecology, Modern Genetics, Advanced Genetics, Plant Anatomy, Advanced Plant Taxonomy, Biophysics, Graduate Seminar, Graduate Special Research, Thesis Research

NORTHERN KENTUCKY STATE COLLEGE

Highland Heights, Kentucky 41076 Phone: (606) 781-2600
Dean of Graduate Studies: N. Melnick
Dean of College: R.A. Tesseneer

DEPARTMENT OF BIOLOGICAL SCIENCES
Phone: Chairman: Ext. 261 Others: Ext. 143

Chairman: J.W. Thieret
Professor: J.W. Thieret
Associate Professors: J.H. Carpenter, C.L. Richards
Assistant Professors: B. Beach, L.A. Ebersole, L.A. Giesmann, T.C. Rambo, G. Ritschel
Instructor: L.J. Giesmann

Degree Programs: Biology, Medical Technology, Pre-Dentistry, Pre-Medicine, Pre-Pharmacy, Pre-Veterinary and Radiologic Technology

Undergraduate Degree: B.S. (Biological Sciences)
Undergraduate Degree Requirements: 36 hours of courses in Biological Sciences; plus General and Organic Chemistry, General Physics, and at least one semester of Mathematics (excluding algebra), plus the courses specified in the "General Studies" requirements of the College.

PIKEVILLE COLLEGE

(No reply recieved)

SPALDING COLLEGE

(No information available)

THOMAS MORE COLLEGE

P.O. Box #85 Phone: (606) 341-5800
Covington, Kentucky 41017
Dean of College: J.A. Ebben

BIOLOGY DEPARTMENT
Phone: Chairman: Ext. 90 Others: Ext. 87

Chairman: W.H. Volker
Professor: W.F. Humphreys
Associate Professors: Sr. M. Laurence, S.N.D. Budde
Assistant Professors: W.S. Bryant, W.H. Volker
Instructor: R.W. Williams

Field Stations/Laboratories: Thomas More College Biology Station - California, Kentucky

Degree Programs: Biology, Medical Technology

Undergraduate Degree: B.A.
Undergraduate Degree Requirements: Total college requirement - 128 credit hours (semester), Minimum: Biology course work 38 semester hours, Chemistry 18 semester hours, Physics 8 semester hours, Mathematics 9 semester hours, General Requirements 21 semester hours, Electives - balance to 128 semester hours

TRANSYLVANIA UNIVERSITY

Lexington, Kentucky 40508 Phone: (606) 255-6861
Dean of College: J.R. Bryden

DEPARTMENT OF NATURAL SCIENCE AND MATHEMATICS
Phone: Chairman: 233-8228

Chairman: M. Moosnick
Distinguished Service Professor of Biology: L.A. Brown
Professor: L. Boyarsky
Associate Professors: J.H. Hamon, R.M. Hays

Degree Programs: Biology

Undergraduate Degree: A.B.
Undergraduate Degree Requirements: Major Courses: 35 hours to include Botany, Genetics, Ecology or a field course, Physiology, and Zoology. Allied Courses 37 hours to include principles of Chemistry (3 quarters) Organic Chemistry (3 quarters) and Principles of Elements of Physics (2 quarters)

UNION COLLEGE

Barbourville, Kentucky 40906 Phone: (606) 546-4151
Dean of College: R. Rose

DIVISION OF NATURAL SCIENCES
Phone: Chairman: Ext. 170

Chairman: D. Myers
Associate Professor: W.B. Kringen
Assistant Professor: A.F. Scott

Field Stations/Laboratories: MACCI Field Biology Teaching and Research Center - LaFollette, Tennessee, Union College Environmental Education Center - Middlesboro, Kentucky

Degree Program: Biology

Undergraduate Degrees: B.A., B.S.
Undergraduate Degree Requirements: (B.A. same as B.S. except 2 years foreign language additional), B.S. Minimum of 128 semester hours with a quality point standing of 2.00 (A=4). General College Requirements of 35 semester hours in humanities, social sciences and natural sciences. 30 semester hours in Biology, 8 in Physics, 7 Mathematics and 12 in Chemistry.

UNIVERSITY OF KENTUCKY

South Limestone Street Phone: (606) 258-9000
Lexington, Kentucky 40506
Dean of Graduate Studies: W.C. Royster

College of Liberal Arts

Dean: A. Gallaher

T.H. MORGAN SCHOOL OF BIOLOGICAL SCIENCES
Phone: Chairman: 258-8641 Others: 257-4711

Director: S.F. Conti
Professors: M.I.H. Aleem, R.W. Barbour, T.C. Barr, J.M. Carpenter, S.F. Conti, W.H. Davis, J.C. Humphries, N.J. Pisacano, R.F. Wiseman
Associate Professors: J.M. Baskin, W.J. Birge, J.C.

Calkins, E.C. Crawford, J.H. Eley, T.C. Gray, D.O. Harris, C.E. Henrickson, J.J. Just, R.A. Keuhne, W. Meijer, G.A. Rosenthal, P.S. Sabharwal
Assistant Professors: M.A. Hafeez, R.S. Hakim, A.D. Hitchins, D.J. Prior, M. Crandall, J. Lesnaw, G. Uglem
Instructors: F. Hakim
Teaching Fellows: 40
Research Assistants: 5

Degree Programs: Zoology, Microbiology and Botany, Plant Physiology, Biological Sciences

Undergraduate Degree: B.S.
Undergraduate Degree Requirements: 60 credits Biological, Mathematics and Physical Science

Graduate Degrees: M.S., Ph.D.
Graduate Degree Requirements: Master: 24 credits of graduate level courses plus thesis, one foreign language, final oral or 30 credits of graduate level courses, report on independent study, one foreign language, written final examination Doctoral: One foreign language requirement. Course work depends upon the requirements set forth by the committee appointed by the Director of Graduate Studies. The Ph.D. degree is not concerned with any particular course requirements although certain courses are generally required, but the ability to do original and note-worthy research.

Graduate and Undergraduate Courses Available for Graduate Research: Genetic Structure of Populations (U), Experimental Techniques in Plant Physiology (U), Statistical Genetics (U), Developmental Genetics (U), Plant Autecology (U), Seminar (U), Anatomy of Vascular Plants (U), Mycology (U), Plant Cytology (U), Introduction to Heredity (U), Plant Pathology (U), Taxonomy of Vascular Plants (U), Plant Microtechnique (U), Plant Morphogenesis, Topics in Modern Botany, Experimental Mycology, Cytogenetics, Plant Metabolism, Mineral Nutrition of Plants, Physiology of Growth and Development, Pathogenic Bacteriology (U), Growth and Death of Bacteria (U), Metabolism of Microorganisms (U), Immunology and Serology (U), Bacteriology of Foods, Bacteriology of Water and Sewage, Electron Microscopy, Viruses and Rickettsiae, History of Bacteriology, Bacterial Anatomy, Taxonomy and Nomenclature, Microbial Genetics, Metabolism of Microorganisms Zoology, Principles of Physiology (U), Heredity (U), Embryology (U), Evolution (U), Zoology Seminar (U), Physiology of Development (U), General Histology (U), Limnology (U), General Entomology (U), Vertebrate Zoology (U), Medical Entomology (U), General Radiation Biology (U), Insect Taxonomy (U), Protozoology (U), Parasitology (U), Helminthology (U), Ichthyology (U), Herpetology (U), Ornithology (U), Comparative Physiology (U), Mammalogy (U), Embryology and Morphology of Vertebrates (U), Embryology and Morphology of Vertebrates (U), Companion Laboratory to Zoology, Invertebrate Zoology (U), Insect Morphology, Advanced Animal Ecology, Insect Ecology and Behavior, Speciation, Principles of Animal Systematics, Seminar in Animal Navigation, Advanced Microtechnique, General Acarology, Insect Physiology, Biological Effects of Radiation, Vertebrate Natural History, Special Topics in Endocrinology

College of Agriculture

Dean: C.E. Barnhart

AGRONOMY DEPARTMENT

Phone: Chairman: (606) 258-2631 Others: 258-9000

Chairman: A.J. Hiatt
Professors: H.H. Bailey, C.E. Bortner, R.C. Buckner W.G. Duncan, V C. Finkner, R.B. Griffith, J.W. Herron, A.J. Hiatt, M.J. Kasperbauer, J.E. Leggett, H.F. Massey, Jr., D.E. Peaslee, R.E. Phillips, J.L. Ragland, R.E. Sigafus, J.H. Smiley, G.W. Stokes, W.G. Survant, N.L. Taylor, T.H. Taylor, W.C. Templeton, Jr., G.W. Thomas, G.T. Webster
Associate Professors: M.K. Anderson, R.A. Andersen, R.I. Barnhisel, R.L. Blevins, H.R. Burton, L.P. Bush G.B. Collins, D.L. Davis, D.B. Egli, C. Grunwald, P.D. Legg, R.H. Lowe, L.W. Murdock, C.G. Poneleit, C.E. Rieck, S.J. Sheen, J.L. Sims, D.M. TeKrony, K.L. Wells
Assistant Professors: M.J. Bitzer, P.L. Cornelius, R. Hayes, J. Herbek, H. Watkins, W. Witt
Research Assistants: 30

Degree Programs: Agronomy, Crop Science, Plant Physiology, Soil Science

Graduate Degrees: M.S., Ph.D.
Graduate Degree Requirements: Not Stated

Graduate and Undergraduate Courses Available for Graduate Credit: Not Stated

DEPARTMENT OF ANIMAL SCIENCES

Phone: Chairman: (606) 258-2686

Chairman: W.P. Garrigus
Professors: N.W. Bradley, C.F. Buck, R.H. Dutt, W.P. Garrigus, N. Gay, (Extension), V.W. Hays, R.W. Hemken, D.R. Jacobson, J.D. Kemp, D.W. MacLaury, G.E. Mitchell, W.G. Moody, D. Olds, W.Y. Varney (Extension), M.D. Whiteker (Extension), P.G. Woolfold
Associate Professors: C. Absher (Extension), J.J. Begin, J.A. Boling, G.L.M. Chappell (Extension), G.L. Cromwell, B. Dean (Extension), D.G. Ely, J.D. Fox, D.D. Kratzer, B.E. Langlois, J.H. Nicholai (Extension), P.A. Thornton, J.W. Tuttle (Extension), F.A. Thrift
Assistant Professors: R.L. BreDahl (Extension), R.L. Edwards (Extension), J.C. Hartley (Extension), G.S. Hess (Extension), R.N. Goodwill, K. Ryen, M.R. Johnson (Extension), D. Liptrap (Extension), R.E. Tucker, R.M. Wendlandt (Extension), E. Wright (Extension)
Instructors: R.D. Cooper, J.L. Krug
Research Assistants: 35

Field Stations/Laboratories: Quicksand and Princeton, plus Owenton, Kentucky

Degree Programs: Animal Nutrition, Animal Foods, Genetics, Physiology of Reproduction

Undergraduate Degree: B.S.
Undergraduate Degree Requirements: 120 hours, 18 in major

Graduate Degrees: M.S., Ph.D.
Graduate Degree Requirements: Master: 24 hours plus thesis Doctoral: 3 years of residence and dissertation or M.S. and 2 years of residence and dissertation

Graduate and Undergraduate Courses Available for Graduate Credit: Reproduction in Dairy Cattle (U), Advanced Food Microbiology (U), Advanced Animal Food Microbiology (U), Chemistry of Animal Products (U), Advanced Genetics (U), Milk Secretion (U), Principles of Animal Nutrition (U), Techniques in Animal Sciences, Advanced Meat Science, Physiology of Reproduction, Population Genetics, Advanced Animal Breeding, Laboratory Methods inAnimal Nutrition, Energy Metabolism, Advanced Ruminant Nutrition, Mineral Metabolism, Advanced Non-ruminant Nutrition, Vitamin Metabolism, Equine Nutrition, Residence Credit for M.S., Residence Credit for Ph.D., Animal Sciences Seminar, Special Problems in Animal Foods, Special Problems in Genetics, Special Problems in Animal Nutrition, Research in Animal Foods, Research in Genetics and Physiology, Research in Animal Nutrition

DEPARTMENT OF ENTOMOLOGY
Phone: Chairman: (606) 258-5638 Others: 258-9000

Chairman: B.C. Pass
Professors: H.W. Dorough (Research and Instruction), F.W. Knapp (Research and Instruction), B.C. Pass (Research Instruction and Extension), J.G. Rodriguez (Research and Instruction), R. Thurston (Research and Instruction)
Associate Professors: D.L. Hahlman (Research and Instruction), P.H. Freytag (Research and Instruction), W.W. Gregory, Jr. (Extension), R.A. Scheibner (Extension)
Assistant Professors: G.L. Nordin (Research and Instruction), H.G. Raney (Extension)
Instructors: H.E. Bryant (Research), G.A. Jones (Research) J.C. Parr (Research)
Research Assistants: 14
Post Doctoral Fellows: (Research) 4

Field Stations/Laboratories: West Kentucky Substation - Princeton, Robinson Substation - Quicksand

Degree Program: Entomology

Undergraduate Degree: B.S.
Undergraduate Degree Requirements: 120 semester credits (University requirements: 34 to 43 hours - English Composition and completion of five general study areas: Pre-professional - 19 or 20 hours - Courses B10, 200, 201, 202, 203 or equivalent and Che 106, 108 and 115 or equivalent; College Requirements: 12 hours - General Agriculture-Social, Animal and Plant; Departmental Requirements: 24 hours, Specialty support outside of major department: 21 hours and electives 10 to 25 hours

Graduate Degrees: M.S., Ph.D.
Graduate Degree Requirements: Master: 24 semester credits (Graduate) plus Thesis - No Modern Foreign Language - GRE - 18 credits earned in Residence on Lexington Campus - Final examination Doctoral: Not less than three years graduate work - courses selected by student, Major Advisor and Special Committee. Minimum of two semesters above M.S. degree in residence at Lexington Campus. Reading knowledge of one Modern Foreign Language - GRE - Written and oral Qualifying Examination - Dissertation

Graduate and Undergraduate Courses Available for Graduate Credit: Economic Entomology, Insect Pest Management, Medical Entomology, Insect Taxonomy, Parasitology, Insect Morphology, Insect Toxicology, Insect Ecology and Behavior, Principles ofAnimal Systematics, General Acarology, Insect Physiology, Advanced Applied Entomology, Immature Insects, Experimental Methods in Entomology, Biological Control of Insects, Residence Credit for the Master's Degree, Residence Credit for the Doctor's Degree, Entomological Seminar, Special Problems in Entomology and Acarology

DEPARTMENT OF FORESTRY
(No reply received)

HORTICULTURE DEPARTMENT
Phone: Chairman: (606) 257-1758 Others: 258-2859

Chairman: G.W. Stokes (Acting)
Professors: R.G. Lockard, H.C. Mohr, C.R. Roberts, Schneider
Associate Professors: C.E. Chaplin, D.E. Knavel, A.M. Hasheen, D.C. Martin, L P. Stoltz
Assistant Professors: J.W. Buxton, T.R. Kemp, H. Schach, B. Tjia
Research Assistants: 2

Field Stations/Laboratories: South Farm Field Laboratory - Lexington

Degree Program: Horticulture

Undergraduate Degree: B.S.
Undergraduate Degree Requirements: 120 hours

Graduate Degree: M.S.
Graduate Degree Requirements: Master: Plan A minimum 24 hours plus thesis (One half these hours at 600 or higher level), Plan B minimum 36 hours course work (One half these hours at 600 or higher level). Doctoral: Not offered. Ph.D. may be done under Aegis of Plant Physiology or Crop Science programs with major work and dissertation in Horticultural area.

Graduate and Undergraduate Courses Available for Graduate Credit: Pomology: Deciduous Fruits I, II, (U), Landscape Construction (U), Vegetable Crops, Greenhouse Vegetable Crops, Turf Management, Landscape Planting Design, Floriculture II, Plant Propagation, Landscape Design of Park and Recreation Areas, Mineral Nutrition of Plants, Physiology of Growth and Development, Phylogeny of Cultivated Plants, Seminar, Special Problems in Horticulture, Research in Horticulture

PLANT PATHOLOGY DEPARTMENT
Phone: Chairman: (606) 258-2759

Chairman: S. Diachun
University of Kentucky Alumni Professorship: S. Diachun
Professors: R.A. Chapman, S. Diachun, J.W. Hendrix, J. Kuc, T.P. Pirone, M.R. Seigel, J. Smiley, G.W. Stokes, H. Wheeler, A.S. Williams
Associate Professors: J.G. Shaw, S. Sheen
Assistant Professors: S.A. Ghabrial, J.R. Hartman, L. Shain
Research Assistants: 3
Fellowship: 1

Field Stations/Laboratories: Princeton Sub-Station - Ribonsin Sub-Station

Degree Program: Plant Pathology

Undergraduate Degree: B.S.
Undergraduate Degree Requirements: 18 hours - departmental major, 102 hours - Preprofessional component, speciality support, electives

Graduate Degrees: M.S., Ph.D.
Graduate Degree Requirements: Master: 24 hours plus a thesis or 30 hours Doctoral: 3 years residence and dissertation

Graduate and Undergraduate Courses Available for Graduate Credit: Forest Pathology (U), Plant Pathology (U), Diseases of Plants (U), Epidemiology and Control of Plant Diseases (U), Nematode Diseases of Plants (U), Plant Pathogenic Fungi, Virus Diseases of Plants, Physiology of Plant Diseases, Residence Credit for the Master's Degree, Residence Credit for the Doctor's Degree, Plant Pathology Seminar, Special Problems in Plant Pathology, Research in Plant Pathology

DEPARTMENT OF VETERINARY SCIENCE
Phone: Chairman: (606) 258-5901 Others: 258-6000

Chairman: J.T. Bryans
Professors: J.T. Bryans, J.H. Drudge, R.W. Darlington, W.H. McCollum
Associate Professors: T.W. Swerczek, M.W. Crowe, E.T. Lyons
Assistant Professor: O.P. Sharma
Instructors: C. Tai (Research Associate), J.C. Wilson (Research Specialist)
Visiting Researchers: 1

Field Stations/Laboratories: Princeton Substation - Princeton

Degree Programs: Veterinary Pathology, Microbiology, Parasitology, Reproductive Physiology

Graduate Degree: Ph.D.
Graduate Degree Requirements: Doctoral: Thesis, Foreign Language, 3 years residence

Graduate and Undergraduate Courses Available for Graduate Credit: Seminar in Veterinary Science, Correlative Pathology, Infectious Diseases of Demostic Animals, Advanced Veterinary Pathology, Advanced Veterinary Parasitology, Advanced Veterinary Microbiology, ReSearch in Veterinary Pathology, Research in Veterinary Parasitology, Research in Veterinary Microbiology

College of Engineering
Dean: J.E. Funk

DEPARTMENT OF AGRICULTURAL ENGINEERING
Phone: Chairman: (606) 258-5658 Others: 258-5687

Chairman: J.N. Walker
Professors: B.F. Parker, I.J. Ross
Associate Professors: J.B. Brooks, B.J. Barfield, C.T. Haan, H.E. Hamilton, G.M. White, E.M. Smith, W. H. Henson, Jr. (Adjunct), J.M. Bunn
Assistant Professors: L.G. Wells, O.J. Loewer
Research Specialists: J.H. Casada
Extension Specialists: G.A. Duncan, G.M. Turner, D.G. Overhults

Degree Program: Agricultural Engineering

Undergraduate Degree: B.S.
Undergraduate Degree Requirements: 128 credits exclusive of freshman Algebra, Trigonometry, Physical Education and R.O.T.

Graduate Degrees: M.S., Ph.D.
Graduate Degree Requirements: Master: 24 hours graduate courses plus thesis Doctoral: courses selected by Committee, one foreign language, Thesis

Graduate and Undergraduate Courses Available for Graduate Credit: Engineering Analysis (U), Plant, Soil and Machinery Relationships (U), Environmental Design for Biological Systems (U), Advanced Soil and Water Conservation Engineering (U), Advanced Agricultural Processing (U), Micrometeorology, Advanced Plant, Soil and Machinery Relationships, Advanced Design of Structures for Biological Systems, Applied Statistical Methods in Water Resources, Electromagnetic Radiation in Agricultural Engineering, Energy and Mass Transfer in Agricultural Processing, Instrumentation in Agricultural Engineering Research, Similitude in Engineering, Special Problems in Agricultural Engineering, Residence Credit for Master's Degree, Residence Credit for Doctor's Degree, Seminar, Thesis

UNIVERSITY OF KENTUCKY MEDICAL CENTER
Dean: W.S. Jordan, Jr.

DEPARTMENT OF ANATOMY
Phone: Chairman: (606) 233-5276 Others: 233-5155

Chairman: H.F. Parks
Professors: R.S. Benton, I. Fowler, L.A. Gillian, H.F. Parks, M.N. Winer
Associate Professors: W.B. Cotter, W.K. Elwood, R.O. Lambson, D. Peck, S.D. Smith, H.H. Traurig
Assistant Professors: D.H. Matulionis, R.E. Papka
Instructor: W.D. Martin
Visiting Lecturers: 3

Degree Program: Anatomy

Graduate Degrees: M.S., Ph.D.
Graduate Degree Requirements: Master: Twenty-four hours of coursework and a thesis, or thirty hours coursework. Doctoral: Three years of residence. One foreign language. Basic coursework in Anatomy plus special coursework needed for thesis research. Thesis.

Graduate and Undergraduate Courses Available for Graduate Credit: Introduction to Anatomy (U), Microscopy and Ultrastructure (U), Developmental Anatomy (U), Anatomy of the Nervous System (U), Independent Work in Anatomy (U), Concepts of Morphology (U), Combined Histology, Embryology, and Special Oral Microanatomy (U), Oral Histology (U), Regional Gross Anatomy, Techniques of Anatomical Research, Advanced Gross Anatomy, Advanced Developmental Anatomy, Advanced Endocrinology, Advanced Neuroanatomy, The Genetic Basis of Human Morphology, Special Projects in Anatomy of Head and Neck Correlated with Neuroanatomy, Ultrastuctural Anatomy

DEPARTMENT OF BIOCHEMISTRY
Phone: Chairman: (606) 233-5549 Others: 233-5546

Chairman: R.L. Lester
Professors: F.J. Bollum, A.S.L. Hu, R.L. Lester, G.W. Schwert
Associate Professors: S.K. Chan, A.D. Winer, J.J. Hutton, Jr.
Assistant Professors: R.E. Rhoads, G.W. Robinson
Instructor: M.S. Coleman
Visiting Lecturers: 12
Research Assistants: 10
Postdoctoral Fellows: 6

Degree Program: Biochemistry

Graduate Degree: Ph.D.
Graduate Degree Requirements: Doctoral: Chemistry through Physical Chemistry, Mathematics through the Calculus, one year Physics and Biological Sciences, Modern Foreign Languages, of which one should be German and the other French or Russian. Programs of study will be fitted to the preparation and interests of individual students; most students will take advanced courses in Chemistry and in Physiology, Cell Biology, Microbiology, Zoology or Botany.

Graduate and Undergraduate Courses Available for Graduate Credit: Fundamentals of Biochemistry, General Biochemsitry, Experimental Methods in Biochemistry, Biochemistry of Lipids and Membranes, Biochemistry and Cell Biology of Nucleic Acids, Structures and Functions of Proteins and Enzymes, Seminar in Biochemistry, Research in Biochemistry, Topics in Biochemistry

DEPARTMENT OF CELL BIOLOGY
Phone: Chairman: (606) 233-5256

Chairman: E.W. Chick (Acting)
Professors: C.T. Ambrose, H.E. Swim
Associate Professors: N.K. Das, J.E. Sisken
Assistant Professors: B.H. Brownstein, T.R. Roszman, D.B. Shah
Instructor: M. Perlin
Teaching Fellows: 2
Research Assistants: 6

Degree Programs: Bacteriology and Immunology, Virology

Graduate Degrees: M.S., Ph.D.
Graduate Degree Requirements: Master: Plan A - 18 semester hours of course work plus research and a dissertation Plan B - 30 semester hours of course work Doctoral: Established by a committee on the basis of individual students' career objectives. Research seminar and most of the courses listed below.

Graduate and Undergraduate Courses Available for Graduate Credit: Cell Biology, Current Topics in Cell Biology, Molecular Biology, Immunology, Microbial Physiology, Microbial Genetics, Animal Virology, Biochemistry - 5 courses available, Cell Physiology

DEPARTMENT OF PHYSIOLOGY AND BIOPHYSICS
Phone: Chairman: (606) 233-5583

Chairman: F.W. Zechman
Professors: J.W. Archdeacon, L.L. Boyarksy, J. Engelberg, F.W. Zechman
Associate Professors: D.T. Frazier, H.R Hirsch, E.P. McCutcheon, B. Peretz, J.D. Pratt, D.R. Wekstein, J. F. Zolman
Assistant Professors: D.A. Lally, D.R. Richardson
Instructor: B.A. Birge
Visiting Lecturers: 2
Post-doctoral Fellow (Research):1

Degree Program: Physiology

Graduate Degrees: M.S., Ph.D.
Graduate Degree Requirements: Master: 24 hours of graduate course work with a B average, thesis and successful defense of thesis Doctoral: Not fewer than 3 years of graduate work;language examination, datisfactory dissertation and promise of scholarly attainment.

Graduate and Undergraduate Courses Available for Graduate Credit: Principles of Physiology, Experimental Methods, Independent Work in Physiology, Introduction to Endocrinology, Medical Physiology, Cellular Physiology, Theoretical Biophysics, Mathematical Biophysics, Physiological Instrumentation, Advanced Neurophysiology, Proseminar in Physiological Psychology, Advanced Topics in Physiology and Biophysics, Systems Physiology, Seminar in Physiological Psychology, Graduate Seminar in Physiology, Research in Physiology.

UNIVERSITY OF LOUISVILLE

Belknap Campus Phone: (502) 636-6556
Louisville, Kentucky 40208
Dean of Graduate Studies: J.A. Dillon

College of Arts and Sciences

Dean: A.J. Slavin

DEPARTMENT OF BIOLOGY
Phone: Chairman: (502) 636-4431 Others: 636-4401

Chairman: B.L. Monroe, Jr.
Tom Wallace Chair of Conservation: S. Neff
Professors: E.O. Beal (Adjunct), W.H. Clay (Emeritus), W.S. Davis, W.F. Furnish, A.T. Hotchkiss, L.A. Krumholz, R.G. Lambert, B.L. Monroe, Jr., S. Neff, I. Poglayen (Adjunct), T.S. Robinson, J.L. Smothers, F.H. Whittaker
Associate Professors: C.V. Covell, G.E. Dillard (Adjunct), L.N. Gleason (Adjunct), R.D. Hoyt (Adjunct), A.J. Karpoff, V.E. Wiedeman
Assistant Professors: R.M. Atlas, L.S. Cronholm

Degree Programs: General Biology, Aquatic Biology, Environmental Biology, Organismic Biology

Undergraduate Degree: B.S., A.B.
Undergraduate Degree Requirements: A.B.: General Education (38 hours), Biology (30 hours, including core of 15 hours), Mathematics (through Calculus), Chemistry (Inorganic and Organic), total 122 hours. B.S.: same except 47 hours Biology, and basic Physics required.

Graduate Degrees: M.S., M.A.T., Ph.D.
Graduate Degree Requirements: Master: 30 hours (12 of which in graduate only courses). Doctoral: No hour requirements; knowledge of 1 language

Graduate and Undergraduate Courses Available for Graduate Credit: Plant Morphogenesis (U), Protozoology (U), Morphology of Vascular Plants (U), Systematic Zoology (U), Zoogeography (U), Endocrinology, (U), Comparative Animal Physiology (U), Microbial Ecology (U), Ornithology (U), Herpetology (U), Ichthyology (U), Mammalogy (U), Molecular Biology (U), Advanced Entomology (U), Limnology (U), Plant Physiology (U), Phycology (U), Aquatic Botany (U), Avian Biology (U), Medical Entomology (U), Ecology (U), Aquatic Entomology (U), Intermediary Metabolism (U), Cytogenetics (U), Special Topics in Molecular Biology, Special Topics in Ecology, Special Topics in Vertebrate Zoology, Special Topics in Evolutionary Biology, Special Topics in Physiology, Special Topics in Invertebrate Zoology, Special Topics in Biosystematics, Special Topics in Conservation, Special Topics in Entomology

School of Medicine

Dean: A.H. Keeney

DEPARTMENT OF ANATOMY
Phone: Chairman: (502) 582-2211 Ext. 385 Others:385

Professors: J.B. Longley, F.J. Swartz, R.H. Swigart, C.E. Wagner
Associate Professors: F.R. Campbell, R.V. Gregg, F.K. Hilton
Assistant Professors: K.P. Bhatnagar, J.P. Friend, G.E. Herbener, J.E. Hubbard, R.D. Rink
Visiting Lecturer: 1

Degree Programs: Anatomy

Graduate Degrees: M.S., Ph.D.
Graduate Degree Requirements: Not Stated

Graduate and Undergraduate Courses Available for Graduate Credit: Not Stated

DEPARTMENT OF BIOCHEMISTRY
Phone: Chairman: (502) 582-2211 Ext. 321

Head: R.L. McGeachin
Professors: R.D. Dallam, C.A. Lang, R.S. Levy, K.P. McConnell (V.A.), R.L. McGeachin, J.R. Taylor, U.F. Westphal (Research)
Associate Professors: M.A. Hilton, J.L. Hoffman, J.A. Yankeelov, Jr.
Assistant Professors: R.W. Benson, M.L. Fonda, R.D. Gray, J.L. Herrman (V.A.)
Visiting Lecturers: 1
Research Assistants: 12
Research Associates: 5

Degree Program: Biochemistry

Graduate Degrees: M.S., Ph.D.
Graduate Degree Requirements: Master: Courses selected by committee, Comprehensive Examination, Thesis. Doctoral: Courses selected by committee, Comprehensive Examination, Thesis

Graduate and Undergraduate Courses Available for Graduate Credit: Graduate Biochemistry, Medical Cell Biology, Biochemistry Seminar, Biochemistry Techniques, Research, Medical Biochemistry, Protein Structure and Function, Lipid Metabolism, Carbohydrate Metabolism, Hemoproteins, Amino Acid Metabolism, Molecular Genetics

DEPARTMENT OF MICROBIOLOGY AND IMMUNOLOGY
Phone: Chairman: Ext. 338

Chairman: J.H. Wallace
Professors: R.D. Higginbotham, P.V. Liu, L.S. Kimsey, J.H. Wallace
Associate Professors: R.J. Doyle, P.B. Johnston
Assistant Professors: D.E. Justus, K.F. Keller, J.M; Mansfield, J.R. McCammon, U.N. Streips
Teaching Assistants: 2
Graduate Assistant: 4
Research Assitants: 4

Degree Programs: Microbial Genetics, Microbial Physiology, Virology, Immunology, Medical Microbiology, Immunoparasitology and Bacteriology

Graduate Degrees: M.S., Ph.D.

Graduate Degree Requirements: Master: Microbiology 601-602 - 8 hours, Cell Biology 604-5 hours, Microbiology 606-2 hours, Microbiology 620-6 hours, and an additional 9 hours of course work to be selected by the advisor (4 hours must be in the Department of Microbiology and Immunology), and an acceptable Thesis and successful completion of the Final Examination. Doctoral: Microbiology 601-2--8 hours, Cell Biology 604--5 hours, Microbiology 606--4 hours, Biochemistry 601--5 hours, an additional 4 courses in Microbiology, satisfactory completion of language requirements, completion of a special laboratory research project in an area other than the one he has chosen, satisfactory passing of a qualifying examination, an acceptable dissertation, satisfactorily passing the final examination and teaching in medicine and dental laboratories in microbiology.

Graduate and Undergraduate Courses Available for Graduate Credit: Medical Microbiology, Immunochemistry, Special Projects in Microbiology, Cell Biology, Seminar, Topics in Advanced Microbiology, Research, Thesis, Cellular Aspects of Immunology, Immunopathology, Immunology, Diagnostic Microbiology, Pathogenesis and Diagnosis of Bacterial Disease, Medical Parasitology, Medical Mycology, Advanced Animal Virology, Fundamentals of Virology, Prokaryotic Cell: Structure and Function, Microbial Genetics, Microbial Physiology

DEPARTMENT OF PATHOLOGY
Phone: Chairman: (502) 582-2211 Ext. 237

Chairman: G.R. Schrodt (Acting)
Professors: W.M. Christopherson, M. Murray, J.E. Parker, W.L. Past, G.R. Schrodt
Associate Professors: W.L. Broghamer, Jr., R.P. Byrd, E. Espinosa, D.R. Kmetz, G. Nedelkoff
Visiting Associate Professors: 6
Assistant Professors: H.B. Carstens, R. Kovachevich, C.L. Songster
Visiting Assistant Professors: 25
Instructors: H. Layman, M. Zady
Visiting Instructor: 1

Degree Program: Pathology

Graduate Degree: M.S.
Graduate Degree Requirements: Master: A student working for the degree of M.D. may secure a Master's Degree in Pathology.

DEPARTMENT OF PHYSIOLOGY AND BIOPHYSICS
Phone: Chairman: Ext. 272

Chairman: T.B. Calhoon
Professors: T.B. Calhoon, J.C. Moore
Assistant Professors: J.R. Meyer, J.C. Passmore, P.F. Pearson, K.H. Reid, B.B. Silver, J.L. Voogt, W.B. Wead

Degree Programs: Biophysics, Physiology

Graduate Degrees: M.S., Ph.D.
Graduate Degree Requirements: Master: 1 year college level Mathematics, 1 year Basic Physics, Biological Sciences, Chemistry, Organic Chemistry, Calculus, Statistics, Physiology 601, Biochemistry 601, Pharmacology, Anatomy, 2 electives in department, Physiology Seminar Doctoral: same, plus 4 electives in department and Physiology Seminar.

Graduate and Undergraduate Courses Available for Graduate Credit: Physiology, Cell Biology, Seminar in Physiology and Biophysics, Selected Topics in Physiology and Biophysics, Clinical and Basic Sciences Correlations in Endocrinology, Neurological Diseases: Brain Mechanisms, Fluid and Electrolyte Physiology, Environmental Physiology, Pain Mechanisms, Topics in Endocrinology, Clinical Pulmonary Physiology, Pathophysiology of Shock, Thyroid Physiology, Cardiac Physiology, Pathologic Physiology, Physiological Control Systems

WESTERN KENTUCKY UNIVERSITY

Bowling Green, Kentucky 42101 Phone: (502) 745-3696
Dean of Graduate Studies: E. Gray (Acting)
Dean of College: M.W. Russell

DEPARTMENT OF BIOLOGY
Phone: Chairman: (502) 745-3696

Chairman: E.O. Beal
Professors: D.W. Bailey, L.B. Lockwood (Adjunct), D. H. Puckett, H.E. Shadowen, E.O. Beal
Associate Professors: G.E. Dillard, L.P. Elliott, S. Ford, R. Loyt, J.H. Jenkins, K.A. Nicely, R. Prins, J.D. Skean, F.R. Toman, J. Winstead, T.A. Yungbluth
Assistant Professors: T.P. Coohill, I. Erskine, L.N. Gleason, M. Houston, P. Pearson
Instructor: J.R. McCurry
Teaching Fellows: 12
Research Assistants: 3

Field Stations/Laboratories: Tech Aqua Biological Station-Center Hill Reservoir, Tennessee

Degree Programs: Aquatic Biology, Biology

Undergraduate Degrees: B.S.
Undergraduate Degree Requirements: Chemistry - 2 semesters, Mathematics - 2 semesters, Physics - 2 semesters, Principles of Biology - 4 credit hours, Protistology - 3 credit hours, Botany - 3 credit hours, Zoology - 3 credit hours, Recommended electives, Ecology - 4 hours, Genetics - 3 hours, Physiology (Cell, plant or animal) - 3-4 hours, Electives in Biology to total at least 30 hours.

Graduate Degrees: M.S., M.A.C.T., M.A. Ed., Ph.D.
Graduate Degree Requirements: Master: M.S. (thesis) 30 semester hours, including thesis; M.S. (non-thesis) 30 semester hours; M.A.C.T., 30 semester hours Doctoral: Thesis, foreign language competency, and approval of graduate committee.

Graduate and Undergraduate Courses Available for Graduate Credit: Biogeography, Limnology, Plant Ecology, Investigations in Biology, Advanced Genetics, Algal Systematics and Ecology, Aquatic Invertebrates, Advanced Parasitology, Intermediate Metabolism, Immunology, Plant Biochemistry, Aquatic Biology, Freshwater Ecology

DEPARTMENT OF AGRICULTURE
(No information available)

LOUISIANA

CENTENARY COLLEGE OF LOUISIANA
2911 Centenary Boulevard Phone: (318) 869-5011
P.O. Box 4188
Shreveport, Louisiana 71104
Dean of College: T. Kauss

DEPARTMENT OF BIOLOGY
Phone: Chairman: (318) 869-5209 Others: 869-5011

Chairman: R.D. Deufel
Professor: O.P. Wilkins
Associate Professor: A.B. McPherson

Degree Program: Biology

Undergraduate Degree: B.S.
Undergraduate Degree Requirements: Biology - 9 courses (36 hours), 4 courses at 300 plus level, Chemistry - 4 courses (16 hours) including Organic, Physics - 2 courses (8 hours), Mathematics 2 courses (6 hours) including Calculus, 2 interim courses (6 hours), English/Fine Arts (9 hours), History/Philosophy or Religion (9 hours), Sociology or Psychology/Economics or Government (9 hours) - 6 courses at 300 plus level, electives, about 25 hours

DILLARD UNIVERSITY
2601 Gentilly Boulevard Phone: (504) 944-8751
New Orleans, Louisiana 70122
Dean of College: D.C. Thompson

DIVISION OF NATURAL SCIENCES (BIOLOGY AREA)
Phone: Chairman: (504) 944-8751 Ext. 262

Chairman: W.W. Sutton
Professors: C.R. Bryan, W.W. Sutton, J.M. Verrett
Associate Professor: M.S. Ivens
Assistant Professors: W.E. Pullin, R. Rogers, N. Roussell, M.E. Smith
Visiting Lecturers: 6
Research Assistants: 1

Degree Programs: Biology, Biology Education

Undergraduate Degree: B.A.
Undergraduate Degree Requirements: All university requirements in the liberal arts, one year of Physics, at least one year of Mathematics, two years of Chemistry, plus a minimum of 32 semester hours in Biology.

GRAMBLING COLLEGE
Grambling, Louisiana 71245 Phone: (318) 247-6941
Dean of College: E.L. Cole

DEPARTMENT OF BIOLOGICAL SCIENCES
Phone: Chairman: (318) 247-6941 Ext. 252

Head: P.L. Young
Professors: V. Henderson, C. Jordan, H.A. Negm, P.L. Young
Associate Professors: J. Capers, J. Smith, K.M.S. Saxena
Assistant Professors: R. Brown, J. Duplantier, I. Harris, A. Hill, M. Marzett, A.M. Mathews, B. Miles
Instructors: L. Bernard, A. Ensminger

Degree Program: Biology

Undergraduate Degrees: B.A., B.S.
Undergraduate Degree Requirements: General Botany 4 hours, Comparative Anatomy 4 hours, Genetics 4 hours, General Zoology 4 hours, Embryology 4 hours, Biology Electives 12 hours, Mathematics 6 hours, English 12 hours, Social Science 15 hours, Languages 12 hours, Physics 8 hours, Chemistry 18 hours, General Education 19 hours, Electives 21 hours.

LOUISIANA COLLEGE
College Station Phone: (318) 487-7776
Pineville, Louisiana 71360
Dean of College: E. Hall

BIOLOGY DEPARTMENT
Phone: Chairman: 487-7611

Chairman: C.J. Cavanaugh
Professor: E.H. Warnhoff
Assistant Professor: D. Martin

Degree Program: Biology

Undergraduate Degrees: B.A., B.S.
Undergraduate Degree Requirements: 30 hours Biology, 20 hours Chemistry or Physics, 10 hours of other (Chemistry or Physics) - Total to 130 hours (semester hours)

LOUISIANA STATE UNIVERSITY AND AGRICULTURAL AND MECHANICAL COLLEGE
Baton Rouge, Louisiana 70803 Phone: (504) 388-6207
Dean of Graduate Studies: J.C. Traynham

College of Arts and Sciences
Dean: I.A. Berg

DEPARTMENT OF BOTANY
Phone: Chairman: (504) 388-8485 Others: 388-8485

Chairman: C.A. Schexnayder
Professors: C.A. Brown (Emeritus), B. Lowy, C.A. Schexnayder
Associate Professors: W.J. Luke, S.C. Tucker
Assistant Professors: R.L. Chapman, M.H. Lieux, R.T. Parrondo, M.A. Piehl
Teaching Fellows: 9

Degree Programs: Botany, Plant Physiology

Undergraduate Degree: B.S. (Botany)
Undergraduate Degree Requirements: Minimum of 128 semester hours, minimum 2.0 grade point average, last year in residence; consisting of 31 hours in liberal arts, including 9 hours of English and 3-13 hours of a foreign language, and 15 hours in the Social Sciences, including 6 hours of History, and 48-85 hours in the Natural Sciences, including a minimum of 24 hours of Botany, 6 hours of Mathematics, 8 hours of Physics and 16 hours of Chemistry, including Organic.

Graduate Degrees: M.N.S., M.S., Ph.D.
Graduate Degree Requirements: Master: Minimum of one academic year in residence, 30 semester hours of graduate work of which 30 hours must be in coursework and 6 semester hours of thesis credit; thesis, minor optional. Doctoral: Minimum of 3 years of graduate work, 60 semester hours of graduate work, including a minor field, reading proficiency in a foreign language, a dissertation, passing of a qualifying general and final examination.

Graduate and Undergraduate Courses Available for Graduate

Credit: Cytology (U), Plant Anatomy (U), Morphology of Vascular Plants (U), Plant Taxonomy (U), Plant Taxonomy (U), Plant Ecology (U), Phycology (U), Mycology (U), Plant Growth and Development (U), Intermediary Metabolism, Mineral Nutrition, Plant Microtechnique, Independent Study, Selected Topics in Plant Physiology

DEPARTMENT OF MICROBIOLOGY
Phone: Chairman: 388-2601 Others: 388-5114

Chairman: M.D. Socolofsky
Professors: H.D. Braymer, A.D. Larson, C.S. McCleskey (Emeritus), V.R. Srinivasan
Associate Professors: R.L. Amborski, J.M. Larkin, R.J. Siebeling
Assistant Professors: R.F. Cooper
Teaching Fellows: 8
Research Assistants: 8
Research Associates: 3

Degree Program: Microbiology

Undergraduate Degree: B.S.
Undergraduate Degree Requirements: 24 hours Biology, 22 Chemistry, 8 Physics

Graduate Degrees: M.S., Ph.D.
Graduate Degree Requirements: Master: 30 hours beyond B.S. Doctoral: 60 hours beyond B.S.

Graduate and Undergraduate Courses Available for Graduate Credit: Introductory Microbial Physiology (U), Microbial Physiology Laboratory (U), Advanced General Microbiology (U), Current Microbiological Literature (U), Immunology and Serology (U), Pathogenic Microbiology (U), Special Problems in Microbiology (U), Genetics of Bacteria and Bacteriophage (U), Soil Microbiology (U), Microbiology of Water, Sewage, and Industrial Wastes (U), Microbiology of the Dairy and Food Industries (U), Industrial Microbiology (U), Cell Culture (U), Introductory Virology (U), Electron Microscopy of Biological Materials, Advanced Seminar, Virology, Microbial Anatomy and Ultrastructure, Higher Bacteria, Molecular Biology,of Microorganisms, Thesis, Research, Microbiology for Teachers, Research Participation, Methods of Research in Microbiology, Dissertation Research

DEPARTMENT OF ZOOLOGY AND PHYSIOLOGY
Phone: Chairman: (504) 388-1132

Chairman: W.J. Harman
Boyd Professor: G.H. Lowery, Jr.
Alumni Professor: G.C. Kent, Jr.
Professors: H.J. Bennett, N.B. Causey, W.J. Harman, B.E. Jackson, G.C. Kent, Jr., W.R. Lee, G.H. Lowery, A.H. Meier, J.H. Roberts, H.J. Werner, J.P. Woodring
Associate Professors: K.C. Corkum, W.L. French, D.A. Rossman
Assistant Professors: R.L. Collins, T.H. Dietz, J.M. Fitzsimons, W.B. Stickle, J.J. Trimble III, E.H. Weidner
Instructors: A.J. Doucette, M.L. Grodner, D.M. Knox, H.R. Mushinsky, E.S. Waldorf
Research Assistants: 8

Degree Programs: Agricultural Education, Agriculture, Agronomy, Wildlife Management, Zoology, Zoology and Entomology

Undergraduate Degree: B.S.
Undergraduate Degree Requirements: 30 semester hours in major, 1 year of Physics, 1 year of Mathematics, 2 years of Chemistry (through Organic), Foreign Language 128 total semester hours

Graduate Degrees: M.S., M.N.S., Ph.D.
Graduate Degree Requirements: Master: 30 semester hours, 24 in course work = M.S., 36 semester hours all coursework in 3 areas of Natural Science - M.N.S.

Graduate and Undergraduate Courses Available for Graduate Credit: Histology (U), Parasitology (U), Ecology (U), Invertebrate Zoology, Experimental Embryology, Physical Genetics, Cytology, Mammalogy (U), Ichthyology, Herpetology (U), Aquatic Invertebrate Zoology, Protozoology, Comparative Physiology (U), Cell Physiology (U), Endocrinology (U), Mammalian Physiology (U), History of Biology, Mutagenesis, Marine Vertebrate Zoology, Marine Invertebrate, Helminthology, Ornithology, Ethology, Physiological Rhythms

College of Agriculture

Dean: R.H. Hanchey

DEPARTMENT OF ANIMAL SCIENCE
Phone: Chairman: (504) 388-3241 Others: 388-4011

Head: G.L. Robertson
Professor: D.M. Thrasher
Associate Professors: T.D. Bidner, R.F. Boulware, F.G. Hembry, P.E. Humes, L.I. Smart
Assistant Professors: A.C. Boston, R.A. Godke, J.L. Kierder
Teaching Fellows: 1
Research Assistants: 6

Degree Programs: Animal Nutrition, Animal Genetics or Animal Breeding, Physiology of Reproduction, and Meat Technology

Undergraduate Degree: B.S. (Animal Science)
Undergraduate Degree Requirements: Total of 134 credit hours - including 35 hours of Animal Science.

Graduate Degrees: M.S., Ph.D.
Graduate Degree Requirements: Master: As determined by committee - 24 hours plus thesis Doctoral: As determined by committee, Dissertation

Graduate and Undergraduate Courses Available for Graduate Credit: Animal Nutrition (U), Animal Breeding (U), Physiology of Reproduction(U), Beef Cattle Production (U), Swine Production (U), Sheep Production (U), Horse Production (U), Meat Technology (U), Advanced Animal Breeding, Advanced Animal Nutrition, Advanced Animal Physiology, Advanced Physiology of Reproduction

DEPARTMENT OF DAIRY SCIENCE
(No reply received)

DEPARTMENT OF ENTOMOLOGY
(No reply received)

SCHOOL OF FORESTRY AND WILDLIFE MANAGEMENT

Director: P.Y. Burns
Professors: P.Y. Burns, E.T. Choong, A.B. Crow, L.L. Glasgon, N.E. Limnartz, R.W. McDermid
Associate Professors: J.W. Avault, R.H. Chabreck, D. D. Culley, P.J. Fogg, C.B. Marlin, R.E. Noble, B. A. Thielges
Assistant Professors: R.B. Hamilton, T.D. Keister, F.M. Truesdale
Instructor: C.W. Brewer
Research Assistants: 44
Temporary Research Professionals: 7

Field Stations/Laboratories: Lee Memorial Forest - Rt. 2 Franklinton, Ben Hur Farm - Baton Rouge, Idlewild Experiment Station - Clinton

Degree Programs: Fisheries and Wildlife Biology, Forestry, Wildlife Management

Undergraduate Degree: B.S.
Undergradua te Degree Requirements: 136 semester hours 3 curricula; 2 in forestry, 1 in wood science

Graduate Degrees: M.S., Ph.D.
Graduate Degree Requirements: Master: 30 semester hours of graduate credit, including a 6 credit thesis. Doctoral: Mastery of subject matter of forestry, 48 semester hours of graduate credit, exclusive of thesis, dissertation and or research problem credit, Reading knowledge of one foreign language, Acceptable Dissertation. Must have a minor subject.

Graduate and Undergraduate Courses Available for Graduate Credit: Recreation in the Forest Environment (U), Seminar in Tropical Forestry (U), Forest Fire Protection and Use (U), Management of Hardwoods (U), Harvesting Timber Crops (U), Game and Range Management in the Forest (U), Forest Management (U), Advanced Forest Management (U), Forest Economics (U), Forest History and Policy (U), Mechanical and Physical Properties of Wood (U), Design and Control of Wood-Using Processes (U), Chemical Properties of Wood (U), Seasoning and Preservation (U), Forest Products (U), Selected or Assigned Forestry Problem (U), Proseminar (U), Research Methods in Forestry, Advanced Silviculture of Southern Forests, Advanced Forest Soils, Advanced Wood Science, Graduate Seminar in Forestry, Fur Animal Management (U), Game Management Techniques (U), Taxonomy and Ecology of Aquatic Plants (U), Limnology (U), Fundamentals of Fish Culture (U), Ichthyology (U), Selected or Assigned Wildlife Problem (U), Game Management, Advanced Game Management--Waterfowl, Wildlife Population Dynamics, Fish Parasites and Diseases, Water Pollution Biology, Fishery Research Techniques, Fisheries Hydrography, Shellfisheries Biology, Mariculture

DEPARTMENT OF HORTICULTURE
Phone: Chairman: 388-2158 Others: 388-2052

Professors: E.P. Barrios, J.F. Fontenot, R.H. Hanchey, P.L. Hawthorne, T.P. Hernandez, T.P. Hernandez, L.G. Jones, J.E. Love, D.W. Newsom, E.N O'Rourke, R.J. Stadtherr, W.A. Young
Associate Professors: R.J. Constantin, L.C. Standifer, W.W. Etzel
Assistant Professors: C.S. Blackwell, J.R. Novak
Associates: 4

Degree Program: Horticulture

Undergraduate Degree: B.S.
Undergraduate Degree Requirements: 134 semester hours with a minimum of C average on all work taken

Graduate Degrees: M.S., Ph.D.
Graduate Degree Requirements: Master: Thirty semester hours including 6 hours credit for thesis research. Thirty-six semester hours formal course work for terminal degree. Doctoral: Minimum of 60 semester hours beyond B.S., with credit for dissertation research Proficiency in one foreign language as demonstrated by minimum score on E.T.S. examination. Minimum of two graduate level courses in Experimental Statistics

Graduate and Undergraduate Courses Available for Graduate Credit: Florist Crop Production (U), Tropical Horticulture (U), Processing of Fruits and Vegetables (U), Nursery Management (U), Advanced Vegetable Crops (U), Advanced Fruit Crops (U), Turf Management (U), Post-Harvest Physiology (U), Breeding of Horticultural Plants, Seminar, Application of Cytogenetics to the Improvement of Crop Plants, Nutrition of Horticultural Crops, Growth and Development of Horticultural Crops, Current Topics in Pomology

DEPARTMENT OF POULTRY SCIENCE
Phone: Chairman (504) 388-4481

Head: A.B. Watts
Professors: W.A. Johnson, R.A. Tukell, A.B. Watts
Associate Professor: B.H. Davis
Assistant Professor: J.R. Daigle
Research Assistants: 4

Field Stations/Laboratories: North Louisiana Agricultural Experiment Station - Calhoun

Degree Program: Poultry Science

Undergraduate Degree: B.S.
Undergraduate Degree Requirements: 16 hours Chemistry, 21 hours Biology, 6 hours Mathematics, 23 hours Poultry Science

Graduate Degrees: M.S., Ph.D.
Graduate Degree Requirements: Master: 30 semester hours minimum graduate courses, Experimental Statistics 8 to 11 hours, Poultry Science 15 semester hours, Animal Science or Dairy Science 4-6 hours, plus thesis Doctoral: Biochemistry 6 hours minimum, experimental Statistics 11-15 hours, Poultry Science 25-28 hours, minor department requirement usually 15-18 hours. Animal Science 3, Dairy Science 3, Dissertation

Graduate and Undergraduate Courses Available for Graduate Credit: Poultry Market Products (U), Poultry Biology (U), Tracer Methodology for Biological Sciences (U), Applied Poultry Nutrition (U), Commerical Poultry Production (U), Advanced Laboratory Technique in Poultry, Advanced Poultry Physiology, Advanced Poultry Nutrition, Advanced Poultry Research, Poultry Science Seminar

College of Chemistry and Physics
Dean: H.B. Williams

DEPARTMENT OF BIOCHEMISTRY
Phone: Chairman: (504) 388-1556

Head: R.S. Allen
Professors: R.S. Allen, J.G. Lee, E.S. Younathan
Associate Professors: S.H. Chang, W.L. Mattice, G.E. Risinger
Assistant Professor: J.A. Bowden
Instructor: E.W. Blakeney
Teaching Fellows: 2
Research Assistants: 9
Graduate Students: 6

Degree Program: Biochemistry

Undergraduate Degree: B.S.
Undergraduate Degree Requirements: 134 semester credits, General Chemistry, Analytical Chemistry, Organic Chemistry, Physical Chemistry, Biochemistry, Physics, Calculus, Foreign Languages, 16 credits Biological Sciences, 18 credits Social Sciences/Humanities, 16 credits Free Electives, Advanced Biochemistry Computer Science

Graduate Degrees: M.S., Ph.D.
Graduate Degree Requirements: Master: Minimum 30 semester credits, Biochemistry major, outside minor Doctoral: Minimum 60 semester credits, Biochemistry major, 1 outside minor

Graduate and Undergraduate Courses Available for Graduate Credit: Physical Chemistry (U), Physiological Chemistry (U), Basic Biochemistry (U), Veterinary Biochemistry (U), Information Retrieval in the Sciences (U), General Biochemistry (U), Biochemistry Laboratory (U), Advanced Experimental Biochemistry (U), Biochemical Reaction Mechanisms (U), Biochemistry of Nucleic Acids, Advanced Biochemistry, Physical Biochemistry, Protein Chemistry, Advanced Enzymology, Special Topics in Biochemistry

LOUISIANA STATE UNIVERSITY
IN NEW ORLEANS
Lakefront Phone: (504) 288-3161

New Orleans, Louisiana 70122
Dean of Graduate Studies: J.R. Bobo

College of Sciences
Dean: B.J. Good

DEPARTMENT OF BIOLOGICAL SCIENCES
Phone: Chairman (504) 288-3161 Ext. 307

Chairman: J.L. Laseter
Professors: D.S. Dundee, A.N.J. Heyn
Associate Professors: C.K. Bartell, S.E. Bryan, M.L. Ibanez, J.L. Laseter, C.A. Olmsted, A.J. Pinter, D. L. Thomas
Assistant Professors: J.C. Francis, S. Githens, J.A. Mayo, M.L. Poirrier, J.S. Rogers, J.G. Shedlarski, Jr.
Instructors: L.E. Baker, M.L. Cabaniss, R.C. Cashner, D.R.G. Holmquist, R.L. Ponthier, N.H. Prevost, C.J. Probst, Jr., C.H. Ross
Teaching Fellows: 20
Research Assistants: 4
Special Lecturers: 2
Research Associates: 4

Degree Programs: Biological Sciences, Medical Technology

Undergraduate Degrees: B.A., B.S.
Undergraduate Degree Requirements: In addition to 41 semester hours of biological science courses a student will be reeuired as a biology major to take courses in Chemistry, Physics, and Mathematics as well as general university requirements appropriate to the path which the student wishes to pursue. Competency in a foreign language (either French, German, or Russian) is also required.

Graduate Degree: M.S.
Graduate Degree Requirements: Master: M.S. in Biological Sciences. The curriculum provides for a balanced program with the areas of Botany, Zoology, Physiology, Molecular Biology and Marine Biology well represented. A total of 30 semester hours are required for a degree, six of which may be thesis research.

Graduate and Undergraduate Courses Available for Graduate Credit: Vertebrate Zoology (U), Biometry (U), Advanced Biochemistry (U), Introduction to Molecular Biology (U), Techniques of Electron Microscopy (U), Cell Physiology (U), Endocrinology (U), Comparative Animal Physiology (U), Plant Physiology (U), Population Genetics and Evolution (U), Aquatic Microbiology (U), Limnology and Oceanography (U), Advanced Bacteriology (U), Plant Morphology (U), Plant Taxonomy (U), Phycology (U), Invertebrate Zoology (U), Advanced Vertebrate Zoology (U), Malacology (U), Entomology (U), Biological Preparations and Instrumentations, Proteins, Lipids, Enzymology, Biophysics, Neurophysiology, Physiological Ecology, Cytogenetics, Systematics and Evolution, Ecology and Systematics of Estuarine Invertebrates, Vriology, Zoogeography

Louisiana State Medical Center
1542 Tulane Avenue Phone: (504) 527-5341
New Orleans, Louisiana 70112
Dean of Graduate Studies: R. Coulson
Dean of School of Medicine: S. O'Quinn

DEPARTMENT OF ANATOMY
Phone: Chairman: (504) 527-8131 Others: 527-8132

Head: M. Hess
Professors: J.C. Finerty, R.F. Gasser, M. Hess, F. Kasten, M.L. Zimny, D. Webster (Adjunct)
Associate Professors: N.J. Adamo, E.R. Allen, J.D. Dunn, R.F. Dyer, C.H. Narayanan, R.D. Peppler, J.R. Ruby, C.D. Enna (Clinical), H. Rothschild (Clinical)
Assistant Professors: J.N. Bagwell, A.J. Castro, T.E. Croley, J.B. Gelderd, J.O.V. Holmstedt, J.J. Jacobs, R. Bruck-Kan, S.G. McClugage, J.R. McClung, M.A. Matthews, R.M. Webster, H.A. Weitsen, I.R. Martinez (Clinical)
Instructor: J.M. Pekarthy
Post-doctoral Trainees: 2

Degree Program: Anatomy, Histology

Graduate Degree: Ph.D.
Graduate Degree Requirements: Doctoral: B.S. degree, GRE adequate performance, letters of recommendation

Graduate and Undergraduate Courses Available for Graduate Credit: Gross Anatomy, Microscopic Anatomy, Neural Sciences, Embryology, Comparative Neurology, Cell Biology, Abnormal Development, Neuroembryology, Research Methods in Anatomy, Special Topics, Seminar, Colloquium

DEPARTMENT OF BIOCHEMISTRY
Phone: Chairman: (504) 527-5044

Head: F.G. Brazda
Professors: F.G. Brazda, R.A. Coulson, H.C. Dessauer, P.M. Hyde, R.E. Reeves (Research Career Awarde NIH)
Associate Professors: S.Q. Alam, L.C. Mokrasch
Assistant Professors: L.C. Gremillion, J.D. Herbert
Instructor: N.J. Nicosia
Visiting Lecturers: 1
Research Assistants: 6
Teaching Assistant: 1
Research Associate: 1
Postdoctoral: 2

Degree Program: Biochemistry

Graduate Degrees: M.S., Ph.D.
Graduate Degree Requirements: Master: 30 semester hours credit; minimum of one year in residence; reading ability in one foreign language;at least 15 hours outside of medical or dental curriculum, and at least six hours in a related field (minor), research and an acceptable thesis. Doctoral: Minimum of 60 hours of credit, with some credit (at least 12 hours) in a minor field(s); minimum residence of one year at Medical Center; two foreign languages (French, German, Russian or Spanish), completed original research project with an acceptable dissertation describing results of same.

Graduate and Undergraduate Courses Available for Graduate Credit: Basic Biochemistry, Introductory Biochemistry Biochemical Analytical Methods, Introduction to Special Methods of Research, Radioisotopes in Biological Research, Biochemical Preparations, Experimental Endocrinology, Biochemistry of the Steroids, Carbohydrate Chemistry, Protein Chemistry, Chemistry of Macromolecules, Intermediary Metabolism, Enzyme Chemistry, Metabolism, The Vitamins and Nutrition, Comparative Biochemistry, Lipid Chemistry, Biochemistry of Nucleic Acids, Neurochemistry,

DEPARTMENT OF BIOMETRY
(No information available)

DEPARTMENT OF MICROBIOLOGY
Phone: Chairman: (504) 527-5949 Others: 527-5947

Head: C. Howe
Professors: C. Howe, D.J. Guidry, G.J. Buddingh (Emeritus)
Associate Professor: W. Pelon
Assistant Professors: B.M. Catchings, W.R. Gallaher, M.L. Murray, R.J. O'Callaghan, J.W. Smith, H.A. Spence, J.J. Thompson, L.A. Wilson
Instructor: L.T. Lee

Degree Program: Microbiology

Graduate Degrees: M.S., Ph.D.

Graduate Degree Requirements: Master: GRE, plus advanced test, Baccalaureate, 30 credit hours, 2 years (1 in residence) Doctoral: Baccalaureate M.S. or equivalent, 60 credit hours, Minimum 3 years, 1 year in residence

Graduate and Undergraduate Courses Available for Graduate Credit: Medical Microbiology, Medical Mycology, Advanced Mycology, General and Molecular Virology, Advanced Virology, Genetics and Molecular Biology of Microorganisms, Advanced Molecular Biology, Techniques in Microbiology, Fundamentals of Immunology, Advanced Immunology, Seminar in Microbiology

DEPARTMENT OF PATHOLOGY
Phone: Chairman (504) 527-5635 Others: 527-5637

Chairman: J.P. Strong
Professors: M.F. Beeler, P. Correa, D.A. Eggen, J.A. Freeman, R.D. Lillie (Research), V. Halperin, N.D. Holmquist, P.A. McGarry, K. Odenheimer, P. Pizzolato, J.P. Strong, R.A. Welsh
Associate Professors: R.F. Carr, C. Garcia, E.O. Hoffmann, B.W. Jarvis, M. Kokatnur, M.C. Oalmann, H.C. Stary, R.E. Tracy
Assistant Professors: P.R. Avet, S. Botero, N. Gagliano Jr., Y-S. Kao, A. Kitiyakara, H.P. Lehmann, M. Marionneaux, W.P. Newman III, W.A. Rock, T. Stewart, A. Suarez
Instructors: M. Boone, R. Druce (Research), B. Farrell, W. Griffin, G.T. Malcom, M.L. Richards, C.C. Vial, L. Vial, C.S. Walters
Visiting Lecturers: 56
Research Associates and Assistants: 9

Degree Programs: Pathology (Clinical Chemistry)

Graduate Degrees: M.S., Ph.D.
Graduate Degree Requirements: Master: A minimum of 16 semester hours of undergraduate chemistry. A minimum of three months will be spent in a clinical Chemistry laboratory of a hospital to gain practical experience in Clinical Chemistry. Doctoral: Same as above. One foreign language is the minimum required for those pursuing work toward a Ph.D. degree. (A minimum of one year will be spent in a clinical Chemistry Laboratory of a hospital to gain practical experience in Clinical Chemistry).

Graduate and Undergraduate Courses Available for Graduate Credit: Physical Biochemistry, Clinical Chemistry Seminar, Introduction to General and Systematic Pathology, General and Systematic Physiology I, General and Systemic Pathology II, Clinical Chemistry (Toxicology), Topics in Clinical Chemistry

DEPARTMENT OF PHYSIOLOGY
Phone: Chairman: (504) 527-5476 Others: 527-5477

Head: J.J. Spitzer
Professors: C.I. Berlin (Conjoint), L. Churney, G.D. Davis, E.R. Hackett (Conjoint), J.B. Heneghan (Conjoint), F.C. Nance (Conjoint), J.J. Spitzer, L.A. Toth
Associate Professors: P.F. Larson (Conjoint), J.R. Little, H.I. Miller, H.M. Randall, R.A. Russell, J.A. Spitzer, A.R. Vella (Conjoint)
Assistant Professors: A.A. Bechtel (Clinical), A.J. Bocage, E.L. Bockman, L.T. Happel (Conjoint), B.G. Jeansonne, H.J. LeBlanc (Conjoint), S.L. Liles, P.A. Mole, T.P. Schilb
Instructors: I. Hikawyj, P.R. Jeffery, M-S. Liu, J.R. Porter
Teaching Fellows: 10

Degree Program: Physiology

Graduate Degrees: M.S., Ph.D.
Graduate Degree Requirements: Master: 30 semester hours of graduate work, not more than six hours being allowed for the thesis. A reading knowledge (450, ets) of one modern foreign language. A thesis, Doctoral: Three full years of graduate study, at least one of them in residence at the Medical Center. Sixty semester hours of course work, at least twelve of them in a minor field. A reading knowledge (450c ets) of two modern foreign languages or a 550 ets score in a single language. A dissertation desenting the applicants original contribution to his field.

Graduate and Undergraduate Courses Available for Graduate Credit: Human Physiology, Neurosciences, Cardiovascular, Physiology, Physiology of the Kidney, Electrophysiology, Neurophysiology, Endocrinology, Auditory Physiology, Digestion, Nutrition and Gnotsbiotics, The Visual System, Electroencephalography, Instrumentation, History and Philosophy of Science

LOUISIANA TECH UNIVERSITY

Ruston, Louisiana 71270 Phone: (318) 257-0211
Dean of Graduate Studies: J.L. Hester

College of Life Sciences

Dean: H.B. Barker

DEPARTMENT OF AGRONOMY AND HORTICULTURE
Phone: Chairman: (318) 257-3275

Head: C. Winstead
Professors: M.J. Howell, J.A. Wright, C.W. Winstead
Associate Professor: B.F. Grafton

Degree Programs: Agronomy, Horticulture

Undergraduate Degree: B.S.
Undergraduate Degree Requirements: 130 semester hours

ANIMAL INDUSTRY DEPARTMENT
Phone: Chairman: (318) 257-2303 Others: 257-0211

Head: C.R. McLellan, Jr.
Professors: R.L. Bailey, H.B. Barker, G.E. Clark, C.R. McLellan, Jr.
Associate Professors: G.R. Stewart
Assistant Professors: H. McClinton, K.E. Sanderlin
Instructors: D.L. Hays, M. Wersham

Degree Programs: Animal Science, Dairy Science

Undergraduate Degree: B.S.
Undergraduate Degree Requirements: 130 hours for B.S.

DEPARTMENT OF BOTANY AND BACTERIOLOGY
Phone: Chairman: 257-4758

Head: D.D. Lutes
Professors: W.P. Hackbarth, J.C. White
Associate Professors: J.A. Christian, H.G. Hedrick, A.W. Lazarys, D.G. Rhodes
Assistant Professors: C.A. Davis, R.E. Jones, S.J. Viator

Degree Programs: Botany, Bacteriology, Wildlife Management

Undergraduate Degree: B.S.
Undergraduate Degree Requirements: 130 semester hours

Graduate Degree: M.S.
Graduate Degree Requirements: Master: Minimum, 30 semester hours, plus thesis

Graduate and Undergraduate Courses Available for Graduate Credit: Sanitary Microbiology (U), Food and Dairy Microbiology (U), Pathogenic Bacteriology (U), Virology (U), Immunology (U), Industrial Microbiology (U), Advanced Microbial Physiology, Genetics of Microorganisms, Advanced Immunology, Microbial Gradation, Advanced Applied Microbiology, Advanced Plant Pathology (U), Plant Microtechnique (U), Economic Botany (U), Statistical Methods (U), Advanced Plant Physiology, Advanced Plant Taxonomy, Field Botany, Advanced Plant Ecology, Advanced Plant Anatomy, History and Literature of Botany

DEPARTMENT OF ZOOLOGY
Phone: Chairman: (318) 257-4573

Chairman: R. Abegg
Professors: R. Abegg, B.J. Davis, R.W. Flournoy, J.W. Geortz, J.L. Murad, S.M. Weathersby
Assistant Professors: J.P. Bogart, L.G. Sellers
Graduate Teaching Assistants: 3

Degree Program: Zoology

Undergraduate Degree: B.S.
Undergraduate Degree Requirements: Botany 7, Microbiology 4, Chemistry 20, Mathematics 6, Physics 8, Zoology 35, Advanced Science electives 6, Orientation 1, English 12, Social Science 12, Foreign Language 6, Electives 3, Speech 3, Total 129 semester hours

Graduate Degree: M.S.
Graduate Degree Requirements: Master: 24 hours course work and 6 hours thesis

Graduate and Undergraduate Courses Available for Graduate Credit: Ecology, Principles Animal Physiology, General Parasitology, Histology, Animal Genetics Laboratory, Vertebrate Embryology, General and Economic Entomology, Animal Genetics, Mammalian Physiology, General Pharmacology, Endocrinology, Principles of Genetics, Mammalian Physiology, General Pharmacology, Endocrinology, Principles of Electron Microscopy, Evolution, Ichthyology, Herpetology, Mammalogy, Ornithology, Limnology, Field Zoology for Teachers, Cell Biology, Biology of Water, History of Zoology, Principles of Zoological Systematics

LOYOLA UNIVERSITY

6363 St. Charles Avenue Phone: (504) 866-5471
New Orleans, Louisiana 70118
Dean of Graduate Studies: J.F. Christman
Dean of College: Fr. W. Byron, S.J.

DEPARTMENT OF BIOLOGICAL SCIENCES
Phone: Chairman: Ext. 211

Chairman: Rev. J.H. Mullahy, S.J.
Professors: E.L. Beard, K.T. Khalaf, W.G. Moore, J.H. Mullahy, S.J.
Associate Professors: J.T. McHale, J.M. Upadhyay
Assistant Professor: J.K. Shull

Degree Programs: Bacteriology, Biochemistry, Biological Sciences, Botany, Entomology, Genetics, Health and Biological Sciences, Life Sciences, Limnology, Medical Technology, Microbiology, Radiation Biology, Zoology, Zoology and Entomology

Undergraduate Degree: B.S.
Undergraduate Degree Requirements: 4 year minor in Chemistry (Organic and Biochemistry included), completion of a 2 year research project; 4 years of biological sciences, 2 years of foreign language, 6 hours Mathematics, 8 hours Physics, 12 hours English, 33 hours Electives

Graduate Degree: M.S.
Graduate Degree Requirements: Master: Not Stated

McNEESE STATE UNIVERSITY

Lake Charles, Louisiana 70601 Phone: (318) 477-2520
Dean of Graduate Studies: R.B. Landers
Dean of College: S.M. Spencer

DEPARTMENT OF BIOLOGY
Phone: Chairman: Ext. 237 Others: Ext. 231, 235

Chairman: G.V.S. White
Professors: J.B. Black, G.W. Cobb, W. Iglinsky, Jr. J.D. Lane, R.G.Robinson, Jr., G.V.S. White
Associate Professors: J.H. Brooks, D.N. Griffin, R.S. Maples, Jr.
Assistant Professors: W.J. Dickson, M.S. Kordisch, J.C. Watson
Instructor: S.S. Dickson

Degree Programs: Botany, Medical Technology, Biology

Undergraduate Degree: B.S.
Undergraduate Degree Requirements: Varies with curriculum (from 28 to 36 semester hours in major) from 128 to 134 semester hours total)

Graduate Degree: M.S.
Graduate Degree Requirements: Master: 30 semester hours total including 6 hours for thesis

Graduate and Undergraduate Courses Available for Graduate Credit: Special Problems in B ology (U), General Plant Pathology (U), Plant Diseases (U), Parasitology (U), Histology (U), Mammalogy (U), Botanical Techniques (U), Plant Morphogenesis (U), Plant Breeding (U), Endocrinology (U), Functional Neuroanatomy (U), Ecological Physiology (U), Advanced Plant Physiology (U), General Ecology (U), Taxonomy of Marsh Plants (U), Ornithology (U), Herpetology (U), Advanced Insect Morphology (U), Aquatic Botany, Phylogeny of Gymnosperms and Angiosperms, Phycology, Speciation, Zoogeography, Seminar, Advanced Plant Taxonomy, Medical Entomology, Advanced Ecology, Systematic Entomology

DEPARTMENT OF MICROBIOLOGY
Phone: Chairman: Ext. 228

Chairman: V. Monsour
Professor: V. Monsour
Associate Professors: B.D. Barridge, G.J. Fister, Jr., R.C. Ritter
Instructor: E.J. Khoury
Visiting Lecturers: 3
Teaching Fellows: 5

Degree Programs: Microbiology, Environmental Science

Undergraduate Degree: B.S.
Undergraduate Degree Requirements: Microbiology - 125 hours, Environmental Sciences - 126 hours

Graduate Degree: M.S.
Graduate Degree Requirements: Microbiology - 30 hours, Environmental Sciences - 30 hours

Graduate and Undergraduate Courses Available for Graduate Credit: Solid Waste Disposal, Ocean Pollution, Air Pollution, Field Study, Economic Poisons in Foods, Solid Waste, Environmental Science, Air Pollution, Stream Pollution - Chemical, Stream Pollution - Biological, Physical Pollution, Food Quality Control, Drug Quality Control, Water Quality Control, Environmental Laws, Public Administration, MICROBIOLOGY-Industrial Microbiology, General Virology, Microbial Physiology, Soil Microbiology, Immunology, Public Health Microbiology, Medical Virology, Microbiology for Teachers Microbiologic,Problems, Marine Microbiology, Molecular Microbiology, Enzymology, Intermediate Virology, Petroleum Microbiology, Microbial Genetics, Recent Advanced in Microbiology, Medical Mycology, Microbiology Seminar

NEWCOMB COLLEGE

(see Tulane University)

NICHOLLS STATE UNIVERSITY
Thibodaux, Louisiana 70301 Phone: (504) 446-8111
Dean of Graduate Studies: V. Pitre
Dean of College: C. Eubanks

DEPARTMENT OF AGRICULTURE
Phone: Chairman: Ext. 414 Others: Ext. 415

Head: C.J. Falcon
Associate Professors: C.J. Falcon, R.N. Falgout, H.A. Heck
Laboratory Technicians: 2

Degree Programs: Animal Science, Plant Science, Agriculture Business

Undergraduate Degree: B.S.Ag.
Undergraduate Degree Requirements: 130 semester hours for the three degrees offered. 24-34 semester hours in major field of interest.

DEPARTMENT OF BIOLOGICAL SCIENCES
Phone: Chairman: 446-8111 Ext. 217

Head: J.H. Green
Professors: M. Handbenger, A. Harris, H. Long, H. Webent
Associate Professors: J. Gann, J. Green, J. Ragan, C. Viator, B. Wilson
Assistant Professors: M. Howell, M. Kilgen, R. Kilgen, P. Templet, A. Templet
Instructor: J. Smith

Field Stations/Laboratories: Nicholls State University Marine Laboratory - Leeville

Degree Programs: Biology, Marine Biology, Medical Technology

Undergraduate Degree: B.S.
Undergraduate Degree Requirements: Not Stated

NORTHEAST LOUISIANA UNIVERSITY
700 University Avenue Phone: (318) 343-2011
Monroe, Louisiana 71201
Dean of Graduate Studies: J.A. McLemore
Dean of College: D.E. Dupree

DEPARTMENT OF BIOLOGY
Phone: Chairman: (318) 343-3103 Others: 343-2011

Head: B.E. Prince
Professors: E.R. Barrett, N.H. Douglas, B.C. Franklin, W.W. Norris, Jr., H.S. Wallace, E.C. Whatley
Associate Professors: L.S. Baum, H.C. Bounds, F.M Boyd, C.E. DePoe, D.T. Kee, W.W. Miller III, B.L. Ricks, R.D. Thomas
Assistant Professor: R.A. Normand
Graduate Teaching Assistants: 20

Field Stations/Laboratories: White River Biology-Geology Field Station - Batesville, Arkansas

Degree Programs: Biology, Botany, Wildlife Management, Zoology

Undergraduate Degree: B.S.
Undergraduate Degree Requirements: 51 semester hours of Biology, Chemistry 14 hours, Mathematics 6 hours, Foreign Language 6 hours, Speech 3 hours, Health and Physical Education 5 hours, English 6 hours, to be chosen from Mathematics, Chemistry, Geology and/or Physics 12 hours, Social Science 6 hours, electives other than science to complete a total of 128 hours for degree

Graduate Degree: M.S.
Graduate Degree Requirements: Master: Undergraduate requirements for a major: 30 semester hours of course in Biology, Required for graduate major: 30 semester hours of Biology and related courses, including Thesis

Graduate and Undergraduate Courses Available for Graduate Credit: Biology: Ecology (U), Problems (U), Genetics (U), Principle of Food and Dairy Bacteriology (U), Pathogenic Bacteriology (U), Advanced General Bacteriology (U), Cellular Physiology (U), General Virology (U), Mountain Field Biology (U), Limnology, Organic Evolution, Research Methods Botany: Mycology (U), Introduction to Marine Botany (U), Plant Physiology (U), Plant Nutrition (U), Field Botany (U), Dendrology (U), Plant Morphology and Anatomy (U), Systematic (U), Plant Pathology (U), Aquatic Plants (U), Phycology, Advanced Mycology, Physiology of the Lower Plants. Zoology: Invertebrates (U), Vertebrates (U), Embryology (U), Herpetology (U), Mammalogy (U), Marine Field Course (U), Ornithology (U), Animal Physiology (U), Insect Taxonomy (U), Water and Marsh Birds (U), Game Mammals (U), Upland Game Birds (U), Fur Bearers (U), Techniques in Wildlife Investigation(U), Wildlife Administration (U), Fisheries Biology (U), Introductory Parasitology (U), Marine Invertebrate (U), Marine Vertebrate Zoology and Ichthyology (U), Ichthyology (U), Comparative Physiology (U), Protozoology (U), Histology (U), Histological Technique (U), Neurology (U), Biology of the Endocrine Glands, Advanced Ichthyology, Field Ornithology

NORTHWESTERN STATE UNIVERSITY
Natchitoches, Louisiana 71457 Phone: (318) 357-6011
Dean of Graduate Studies: L.T. Allbritten
Dean of College: R.J. Bienvenu

DEPARTMENT OF BIOLOGICAL SCIENCES
Phone: Chairman: (318) 357-5323 Others: 357-6011

Chairman: R.K. Baumgardner
Professors: R.K. Baumgardner, B.R. Buckley, R.A. Daspit, D.N. Kruse, R.H. Outland
Associate Professors: A.S. Allen, T.A. Burns, J.C. Lin, D.R. Sanders, D.T. Stalling, J.R. Stothart, K.L. Williams
Assistant Professors: C.E. Viers
Graduate Assistants: 4

Degree Programs: Botany, Medical Technology, Natural Science, Science Education, Wildlife Management, Zoology, Biology Education, Radiologic Technology

Undergraduate Degree: B.S.
Undergraduate Degree Requirements: Biology 101-101L, 102-192L, 124, 125 and at least 1 course selected from each of the following areas: Genetics, Physiology, Morphology and Development and Environmental - total 24 hours.

Graduate Degree: M.S.
Graduate Degree Requirements: Master: 24 hours plus thesis (1) Meet undergraduate requirements (2) Twelve hours of foreign language or reading knowledge of language (3) Three hours of Advanced Statistics

Graduate and Undergraduate Courses Available for Graduate Credit: BIOLOGY: Cell Physiology (U), Microtechnique (U), Genetics (U), Medical Mycology (U), Biology for Elementary Teachers, Principles of Biology, Problems in Secondary School Biology, History of Biology, Literature in Biology, Organic Evolution, Radioisotopes in Biology and Medicine, Plant Ecology, Animal Ecology, Seminar, Problems in Teaching Biology, Thesis BOTANY: Plant Pathology (U), The Non-Vascular Plants (U), the Vascular Plants (U), Taxonomy of Grasses (U), Plant Physiology (U), Mycology, Plant Taxonomy ZOOLOGY: Ornithology (U), Herpetology (U), Histology (U), Wildlife Management (U), Techniques of Wildlife Management (U), Animal

Behavior (U), Mammalogy (U), Embryology (U), Limnology-Aquatic Biology (U), Biology of Fishes (U), Comparative Vertebrate Physiology (U), Migratory Wildlife Resources (U), Experimental Parasitology, Biology of the Vertebrates

DEPARTMENT OF EARTH SCIENCES
Phone: Chairman: (318) 357-5913

Chairman: Z.W. Daughtrey
Professor: H.E. Townsend
Associate Professors: Z.W. Daughtrey, J.D. Waskom
Assistant Professors: D.A. Dobbins, S.A. Misuraca, M.H. Stevens,
Instructors: R.R. Every, J.E. Farrington
Research Assistants: 3

Degree Programs: Agriculture, Agronomy, Animal Science, Plant Science, Soil Science, Geology

Undergraduate Degree: B.S.
Undergraduate Degree Requirements: 120 semester hours minimum; senior year in residence; 30 semester hours in residence; 2.00 QPA; 57-62 semester hours in major field (geology-38 hours)

Graduate Degree: M.S.
Graduate Degree Requirements: Master: 24 semester hours of graduate credit including 18 semester hours in geology and thesis

Graduate and Undergraduate Courses Available for Graduate Credit: Principles of Animal Nutrition (U), Animal Reproduction (U), Animal Breeding (U), Plant Breeding (U), Soil Morphology and Classification (U), Weed Science (U), Turf Management (U), Plant Propagation (U), Non-metalliferous economic Geology (U), Metalliferous Economic Geology (U), Advanced Sedimentation, Introduction to Geophysics (U), Geomorphology (U), Stratigraphy (U), Optical Mineralogy and Petrology (U), Geochemistry (U), Clay Mineralogy (U), Statistical Models in Geology (U), X-Ray Mineralogy (U), Advanced Ingeous and Metabolic Petrology, Geodynamics Applied Geophysics, Seismology, Advanced Geochemistry

ST. MARY'S DOMINICAN COLLEGE
(No reply received)

SOUTHEASTERN LOUISIANA UNIVERSITY
University Station Phone: 345-1400
Hammond, Louisiana 40701
Dean of Graduate Studies: C. Golemon
Dean of College: J.W. Knight

AGRICULTURE DEPARTMENT
Phone: Chairman: Ext. 314

Head: A.D. Owings
Professors: E.C. Bateman, C. Fischer, Jr.
Assistant Professor: A. Baham, C.E. Hyde
Part-time: 1

Field Stations/Laboratories: Southeastern Louisiana University Farm - Hammond

Degree Programs: Animal Science, Plant Science

Undergraduate Degree: B.S.
Undergraduate Degree Requirements: 129 - 130 semester hours of specified and unspecified courses

DEPARTMENT OF BIOLOGICAL SCIENCES
Phone: Chairman: Ext.281

Head: E.R. Wascom
Professor: E.R. Wascom
Associate Professors: H.C. Adelmann, E.B. Carrier, J. L. Crain, A.V. Friedrichs, D.H. Hays, B.T. Kirk, W.R. Wallace
Assistant Professors: D.J. Acosta, L.F. Baehr, W.P. Bond, D. Brown, L. Frederick, G.P. Guidroz, D.P. Shepherd, H.H. Ziller
Graduate Teaching Assistants: 5

Degree Programs: Botany, Medical Technology, Microbiology, Zoology

Undergraduate Degree: B.S.
Undergraduate Degree Requirements: B.S. (Botany, Microbiology, Zoology) 124-126 specified semester hours and elective courses B.S. (Medical Technology) 97-98 semester hours

Graduate Degree: M.S.
Graduate Degree Requirements: Undergraduate preparation: 30 hours of Biology. Masters: (1) Thesis Program--30 semester hours of biology including 6 hours for the thesis. Six to 8 semester hours may be in a related field with the approval of the advisory committee. (2) Non-Thesis Program--30 hours of Biology. Six to 8 semester hours, may be in a related field with the approval of the advisory committee.

Graduate and Undergraduate Courses Available for Graduate Credit: History of Biology (U), Graduate Research Problems, Graduate Seminar, Thesis, Plant Physiology (U), Plant Pathology (U), Economic Botany (U), Systematic Botany of the Flowering Plant (U), Systematic Botany of the Grasses (U), Phycology (U), General Mycology (U), Plant Anatomy (U), Virology (U), Advanced Plant Pathology, Advanced Mycology, Advanced Plant Taxonomy, Pathogenic Bacteria (U), Immunology and Serology (U), Microbiology of the Diary and Food Industries (U), Soil Microbiology (U), Bacterial Physiology (U), Microbiology of Water, Sewage, and Industrial Wastes (U), Industrial Microbiology, Determinative Microbiology, Microbial Physiology, Microbial Genetics, General Entomology (U), Economic Entomology (U), Waterfowl Management (U), Fresh Water and Estuarine Biology (U), Marine Biology and Biological Oceanography (U), Insect Taxonomy (U), Parasitology (U), Ichthyology (U), Radiation Biology (U), Wildlife Biology (U), Advanced Invertebrate Zoology, Herpetology, Mammalogy, Advanced Mammalian Physiology, Experimental Embryology Laboratory, Biology of the Endocrine Glands, Ornithology, Histology

SOUTHERN UNIVERSITY
Baton Rouge, Louisiana 70813 Phone: (504) 771-5210
Dean of Graduate Studies: J. Martin (Acting)
Dean of College: L.L. White

DEPARTMENT OF BIOLOGICAL SCIENCES
Phone: Chairman: (504) 771-5210

Chairman: C.E. Johnson
Professors: R.M. Ampey, J.B. Bryant, Jr., F.A. Christian, C.E. Johnson, L.R. Roddy, G.N. Ross, L. Scott, G. Williams, Jr.
Associate Professors: C.E. Davis, B.E. Frisby, E.H. Higginbotham, R.S. Kakar, L.A. Metevia, M. Singh, D. Thompson, R. Yadav
Assistant Professors: R. Ardoin, R.H. Gobbins, P.L. Hubbard, H. Minis, J.J. Payne, F. Robinson, F. Spencer, B.A. Wade, G.M. Wilson
Instructors: K. Brossette, N. Clarke
Research Assistants: 10

Degree Programs: Biological Science, Medical Technology, Botany, Microbiology, Zoology

Undergraduate Degree: B.S.
Undergraduate Degree Requirements: 128 hours of course work to include at least 36 hours of Biological science courses, 12 hours of Mathematics, 8 hours of Physics and 16 hours of Chemistry

Graduate Degree: M.S.
Graduate Degree Requirements: Master: 24 hours of course work and research and thesis equivalent to 6 hours. Successful completion of a comprehensive examination and defense of thesis.

Graduate and Undergraduate Courses Available for Graduate Credit: Genetics (U), General Physiology (U), Seminar (Biology)(U), Cell and Molecular Biology, Seminar, Research, Mycology (U), Plant Pathology (U), Plant Physiology (U), Advanced Field Botany, Pathogenic Microbiology (U), Immunology (U), Bacterial Physiology (U), Special Problems in Microbiology (U), General Virology, Parasitology (U), Ecology (U), Animal Physiology (U), Invertebrate Physiology (U), Introductory Radiobiology (U), Endocrinology

TULANE UNIVERSITY

6823 St. Charles Avenue Phone: (504) 865-4011
New Orleans, Louisiana 70118
Dean of Graduate Studies: D.R. Deener

Liberal Arts College

Dean: J . Gordon

DEPARTMENT OF BIOLOGY
Phone: Chairman: (504) 865-6229 Others: 865-6226

Chairman: E.P. Volpe
Ida A. Richardson Professor of Biology: J.A. Ewan
Professors: S.S. Bamforth, D.E. Copeland, H.A. Dundee, J.A. Ewan, M. Fingerman, G.E. Gunning, R.D. Lumsden, M. Mizell, S.M. Rose (Adjunct), A. E. Smalley, R.D. Suttkus, E.P. Volpe, A.L. Welden
Associate Professors: J.T. Barber, L.B. Thien
Assistant Professors: J.W. Bennett, E.G. Ellgaard, C.R. Page
Instructors: M. Pelias, K. Roux
Research Assistants: 6

Field Stations/Laboratories: Hebert Riverside Center, housing the F.R. Cagle Center of Environmental Biology

Degree Programs: Biology, Botany, Life Sciences, Zoology

Undergraduate Degree: B.S.
Undergraduate Degree Requirements: Thirty-two (32) units of academic work including two units of English (or acceptable equivalents), four units of foreign language (or acceptable equivalents), two units of Mathematics, Four units from each of two groups outside the major (Humanities, Sciences, and Social Sciences), and a maximum accumulation of 18 units in the major

Graduate Degrees: M.S., Ph.D.
Graduate Degree Requirements: Master: Total of at least 30 semester hours of graduate courses and a satisfactory reading knowledge of one foreign language. A thesis is not required; the M.S. degree is not a requisite for seeking the Ph.D. Doctoral: Total of at least 48 semester hours of semester hours of course work, a written monograph (in lieu of an oral preliminary examination), written dissertation, oral defense, of dissertation, and reading knowledge in two foreign languages (second foreign language may be replaced by specialized courses in Mathematics or computer sciences).

Graduate and Undergraduate Courses Available for Graduate Credit: Mycology, Plant Ecology, Population Genetics, Advanced Cell Biology, Advanced Plant Taxonomy, Cell Biology of Parasitism, Physiological Ecology, Plant Growth and Development, Molecular Genetics, Cytogenetics, Paleobotany, History of Biology, Research Methods in Biology, Plant Geography, Biology of Fishes, Chemical Embryology, Ichthyology, Animal Systematics, Helminthology, Experimental Parasitology, Biology of Mammals, Advanced Comparative Animal Physiology, Biology of Amphibians and Reptiles, Vertebrate Paleontology, Amphibian Oncology, Electron Microscopy, Natural History of Animals, Tumor Biology, Marine Biology, Experimental Embryology, Host Immunity, Developmental Genetics, Phycology, Limnology

Newcomb College

(No reply received)

School of Medicine

Dean: W.J. Thurman

DEPARTMENT OF ANATOMY
(No reply received)

DEPARTMENT OF BIOCHEMISTRY
Phone: Chairman: (504) 588-5291

Chairman: R.L. Stjernholm
Professors: G.A. Adrouny, R.R. Benerito, W. Cohen, Y.T. Li, R.H. Steele
Associate Professors: E. Hamori, J.E. Muldrey, V.B. Philpot
Assistant Professors: W.H. Baricos, M Ehrlich, S.C. Li, J.K. Stanfield, J.S. Tou
Research Assistants: 3

Degree Program: Biochemistry

Graduate Degrees: M.S., Ph.D.
Graduate Degree Requirements: Not Stated

Graduate and Undergraduate Courses Available for Graduate Credit: General Biochemistry, General Biochemistry Laboratory, Physiological Chemistry, Physiological Chemistry Laboratory, Physical Biochemistry, Molecular Biology, Advanced Biochemistry, Selected Topics, Special Problems, Seminar

DEPARTMENT OF BIOSTATISTICS
(No information available)

DEPARTMENT OF MICROBIOLOGY AND IMMUNOLOGY
Phone: Chairman: (504) 588-5150

Chairman: W.A. Pierce (Acting)
Professors: G.J. Domingue, L. Friedman, E.J. Johnson, W.A. Pierce, J.D. Schneidau, Jr.
Associate Professors: M.O. Hornung, M.K. Johnson, J.D. McDonald
Assistant Professor: J.K. Domer
Visiting Lecturers: 1

Degree Programs: Microbiology, Immunology

Graduate Degree: Ph.D.
Graduate Degree Requirements: Doctoral: 48 semester hours of courses, Reading proficiency in one foreign language, Dissertation

Graduate and Undergraduate Courses Available for Graduate Credit: Medical Microbiology, Seminar, Microbial Physiology, Advanced General Microbiology, Principle of Virology, Special Problems, Research, Recent Advances In Microbial Genetics, Immunology, Immunopathology, Fundamentals of Medical Mycology, Clinical Mycology, Advanced Medical Mycology, History of Microbiology, Clinical Bacteriology, Dissertation

DEPARTMENT OF PARASITOLOGY
Phone: Chairman: (504) 588-5448 Others: 588-5449

Chairman: R.G. Yaeger
The William Vincent Professor of Tropical Diseases and Hygiene: P.C. Beaver
Professors: P.C. Beaver, T.C. Orihel, R.G. Yaeger, E.C. Faust (Emeritus)
Associate Professors: J.H. Esslinger, M.D. Little, E.A. Malek, A. Miller
Visiting Lecturers: 2
Research Assistants: 4

Degree Program: Parasitology

Graduate Degrees: M.S., Ph.D.
Graduate Degree Requirements: Master: 24 hours, one language, a thesis Doctoral: 48 hours, two languages, a dissertation

Graduate and Undergraduate Courses Available for Graduate Credit: Medical Helminthology, Medical Protozoology, Medical Entomology, Malacology, Parasitologic Methods, Parasites of Lower Vertebrates, Comparative Microanatomy of Helminths, Special Problems in Parasitology, Seminar, Biochemistry, Medical Microbiology, Immunology, Cytology-Histology, Pathology, Medical Mycology, Biostatistics, Epidemiology, Tropical Ecology, Cellular Biology of Parasitism (U), Microbial Physiology, Clinical Bacteriology, Electron Microscopy

DEPARTMENT OF PHYSIOLOGY
(no reply received)

UNIVERSITY OF SOUTHWESTERN LOUISIANA

E. University Avenue Phone: (318) 233-3850
Lafayette, Louisiana 70501
Dean of Graduate Studies: R.R. Jones
Dean of College: M.E. Dichmann

DEPARTMENT OF BIOLOGY
Phone: Chairman: Ext. 260 Others: 631, 632

Chairman: E.B. Stueben
Professors: J.T. Brierre (Adjunct), M.E. Dakin, M.B. Eyster, L.T. Graham, W.D. Reese, E.B. Stueben
Associate Professors: W.O. Durio, L. Erbe, H.D. Hoese, E.D. Keiser, B.E. Lemmon
Assistant Professors: C.L. Cordes, M.A. Konikoff, R.A. Pecora, B.L. Perkins, V.I. Sullivan, R.W. Walker
Instructors: A.L. Carter, G.L. Eyster
Visiting Lecturer: 1
Teaching Fellows: 13
Research Assistants: 1

Field Stations/Laboratories: Biology Marine Station, - Redfish Point, Vermilion Bay

Degree Programs: Botany, Ecology, Entomology, Entomology and Parasitology, Fisheries, Parasitology, Physiology, Wildlife Management, Zoology, Herpetology, Predental, Premedical

Undergraduate Degree: B.S.
Undergraduate Degree Requirements: Variable based upon various curricula. 1. Biol-Chem. (Pre-dent and Premedical) 2. Botany, 3. Fishery Biology 4. Physical-Therapy (preprofessional), 5. Wildlife Management 6. Zoology

Graduate Degrees: M.S., Ph.D.
Graduate Degree Requirements: Master: Variable, based upon need, 24 credit hours (12 hours at 500 level), plus thesis, research 6 credit hours = 30 hours Doctoral: Variable based upon need plus 1 language plus statistics or computer science), or 2 languages - modern

Graduate and Undergraduate Courses Available for Graduate Credit: Ornithology (U), Economic Plants (U), Wildlife Management - Upland Game (U), Wildlife Management of Fur bearers and Waterfowl (U), Invertebrate Zoology (U), Histology (U), Genetics (U), Parasitology (U), Animal Ecology (U), Plant Ecology (U), Plant Physiology (U), Mammalogy (U), Animal Physiology (U), Evolution (U), Medical Entomology (U), Aquatic Insects, Field Marine Biology (U), Morphology Non-Vascular Plants (U), Bryology (U), Agrostology (U), Limnology (U), Fish Propagation (U), Icthyology (U), Herpetology (U), Fishery Science (U), Parasitic Protozoology (U), Phycology (U), Aquatic and Marsh Plants (U), Dendrology (U), Introductory Oceanography

DEPARTMENT OF MICROBIOLOGY
Phone: Chairman: Ext. 262

Head: W.L. Flannery
Professors: J.M. Sobek, T.E. Wilson
Associate Professor: P.F. Mathemeier
Assistant Professors: F.W. Forney, G. Kelley
Instructor: E. Barry

Field Stations/Laboratories: Marine Biology Laboratory - Red Fish Point, Louisiana

Degree Programs: Biological Sciences, Microbiology

Undergraduate Degree: B.S.
Undergraduate Degree Requirements: 126 credits (minimum) 30 of these for major

Graduate Degrees: M.S., Ph.D.
Graduate Degree Requirements: Master: 30 graduate credits (6 of these for thesis, 24 class credits) Doctoral: varies dependent upon previous preparation, 2 foreign languages, dissertation

Graduate and Undergraduate Courses Available for Graduate Credit: Industrial (U), Parasitology (U), Virology (U), Advanced Microbial Physiology (U), Marine Microbiology (U), Advanced Medical Microbiology (U), Problems (U), Microscopy (U), Advanced Laboratory Methods, Advanced Problems, Special Topics, Colloquim

XAVIER UNIVERSITY

New Orleans, Louisiana 70125 Phone: (504) 486-7411
Dean of Graduate Studies: Sr. J. Lynch
Dean of College: Sr. M.V. Drawe

DEPARTMENT OF MEDICAL TECHNOLOGY
Phone: Chairman: Ext. 253

Chairman: J. Lomasney
Assistant Professors: J. Lomasney
Instructors: P. Boyd, C. Crockett, Sr. M. Loughlin

Degree Program: Medical Technology

Undergraduate Degree: B.S.
Undergraduate Degree Requirements: Biology 24, Chemistry 20, Medical Technology 44 semester hours, Mathematics 6 semester hours, English 6 semester hours, Social Sciences 6 semester hours, the Arts 6 semester hours, Theology 6 semester hours, Philosophy 6 semester hours, Speech 3 semester hours, Free Electives 8 semester hours, 135 semester hours total

MAINE

ARISTOOK STATE COLLEGE
(see University of Maine at Presque Isle)

BATES COLLEGE
(No reply received)

BOWDOIN COLLEGE
Brunswick, Maine 04011 Phone: (207) 725-8731
Dean of College: A.L. Greason

DEPARTMENT OF BIOLOGY
Phone: Chairman: (207) 725-8731 Ext. 584

Chairman: C.E. Huntington
Professors: A.H. Gustafson, J.M. Moulton, C.E. Huntington, J.L. Howland
Associate Professor: C.T. Settlemire
Assistant Professor: R.A. Winston
Teaching Fellows: 2
Research Assistants: 1
Post-doctorals: 2

Field Stations/Laboratories: Bowdoin Scientific Station - Kent Island via Grand Manan, N.B., Canada, Bowdoin Marine Station - Harpswell, Maine

Degree Programs: Biology, Biochemistry

Undergraduate Degree: B.A.
Undergraduate Degree Requirements: Biochemistry of courses put more emphasis on Chemistry. 6 semesters of Biology, Through Organic Chemistry (two years of Chemistry), 1 year of Physics, Through Calculus, 32 courses

COLBY COLLEGE
Waterville, Maine 04901 Phone: (207) 873-1131
Dean of College: P.G. Jenson

DEPARTMENT OF BIOLOGY
Phone: Chairman: Ext. 214, Others: 246

Chairman: M.F. Bennett
Professors: M.F. Bennett, R.L. Terry
Associate Professor: T.W. Easton
Assistant Professors: A.K. Champlin, B. Fowles, W.G. Gilbert, F.M Kestner
Teaching Assistants: 3

Degree Program: Biology

Undergraduate Degree: B.A.
Undergraduate Degree Requirements: 32 credit hours in Biology, 2 semesters of Mathematics, 1 year of Chemistry, 1 additional year of Science, Successful performance on Comprehensive Examination

NASSON COLLEGE
Springvale, Maine 04083 Phone: (207) 324-5340

DIVISION OF SCIENCE AND MATHEMATICS
Phone: Chairman: (207) 324-5340

Chairman: C.R. Gilmore
Professor: R.H. Ciullo
Associate Professors: C.R. Gilmore, G.S. Johnston
Assistant Professors: G.J. Leversee, Jr., J.A. Rollins

Degree Program: Biology

Undergraduate Degrees: B.A., B.S.
Undergraduate Degree Requirements: 12 courses in Biology, 3 courses in Chemistry, 2 courses in Mathematics and 1 in Physics. A candidate for the B.A. degree must complete the language requirement. A biology major will take comprehensive examinations in biology during the second semester of his senior year.

RICKER COLLEGE
Houlton, Maine 04730 Phone: (207) 532-2223
Co-Deans of Instruction: R.E. Burns, H. Bowman

DIVISION OF SCIENCE AND MATHEMATICS
Phone: Chairman: (207) 532-2223 Ext. 70

Division Chairman: J. Elliott
Instructors: J. Clark, T. Williams, S. Rooney

Degree Program: Biology

Undergraduate Degree: B.A.
Undergraduate Degree Requirements: 8 courses in Biology, Mathematics, through Calculus, 2 years Chemistry, 1 year Physics

ST. FRANCIS COLLEGE
(No reply recieved)

ST. JOSEPH'S COLLEGE
N. Windham, Maine 04062 Phone: (207) 892-6766
Dean of College: Sr. Dolores

DEPARTMENT OF BIOLOGY
Phone: Chairman: Ext. 75 Others: Ext. 64

Chairman: J.M. Hancock
Assistant Professors: J.M. Hancock, R.A. Watson
Teaching Assistants: 2

Degree Program: Biology

Undergraduate Degree: B.A.
Undergraduate Degree Requirements: Introduction to Biology, General and Organic Chemistry, Calculus, Physics (recommended), 2 courses Organismic Biology, Genetics, Developmental Biology, Cell Biology, Ecology, Research and Seminar

UNITY COLLEGE
Unity, Maine, 04988 Phone: (207) 948-3131
Dean of College: A.B. Karstetter

CENTER OF ENVIRONMENTAL SCIENCES
Phone: Chairman: Ext. 33

Chairman: M.A. Rosinski
Professors: H. Gray, M.A. Rosinski
Associate Professors: C.G. Cinnamon, D. Knupp, J.F. Sassaman
Assistant Professors: G. Estell, H. Hatch, C. Rabeni
Teaching Fellows: 1

Field Stations/Laboratories: School is field oriented and located such that much field work is done on or adjacent to premises.

Degree Programs: Biological Sciences, Botany, Ecology, Ethology, Forestry, Life Sciences, Natural Resources, Silviculture, Wildlife Management, Zoology

Undergraduate Degree: B.S.
Undergraduate Degree Requirements: B.S. 120 credit hours, no other special requirements.

UNIVERSITY OF MAINE AT FARMINGTON
Farmington, Maine 04938 Phone: (207) 778-3501
Dean of College: T. Emery

BIOLOGY DEPARTMENT
Phone: Chairman: Ext. 241 Others: Ext. 242,260,327

Chairman: J.E. Madge
Professors: R.L. Martin, J.E. Madge
Associate Professors: R.W. Robinson, V. Wells

Degree Program: Biology

Undergraduate Degrees: B.A., B.S.
Undergraduate Degree Requirements: 120-122 credits, 2.o GPA, 30 credits in major (Biology)

UNIVERSITY OF MAINE AT FORT KENT
Pleasant Street Phone: (207) 834-3162
Fort Kent, Main3 04743
Dean of College: G.T. Prigmore

MATHEMATICS-SCIENCE DIVISION

Chairman: J.O. Olson
Associate Professors: B.G. Liles, Jr., S.Z. Peek

Degree Programs: Biology, Environmental Sciences

Undergraduate Degree: B.S.
Undergraduate Degree Requirements: General Education Requirement 33 semester hours, Graduate Requirement 128 semester hours with point average 2.00 (A = 4.00) Programs Above 1 and 3 - Natural Science Major - Biology and Environmental Studies - 30 semester hours, Natural Science Major - Biology - 18 semester hours

UNIVERSITY OF MAINE AT ORONO
Orono, Maine 04473 Phone: (207) 581-1110
Dean of Graduate Studies: F.G. Eggert

College of Arts and Sciences
Dean: K.W. Allen

ZOOLOGY DEPARTMENT
Phone: Chairman: (207) 581-7679 Others: 581-1110

Chairman: J.F. Haynes
Professors: K.W. Allen, A.A. Barden, Jr., J.R. Cook, A.M. Mun, C.W. Major, B.R. Speicher, F.L. Roberts, W.G. Valleau
Associate Professors: R.W. Hatch, B. Sass, J.H. Dearborn, J.F. Haynes, J. McCleave
Assistant Professors: R.W. Gregory, R.L. Lynch, P. Sinnock, R.G. Summers, Jr.
Teaching Fellows: 12
Research Assistants: 24

Field Stations/Laboratories: Darling Center - Walpole, Maine

Degree Program: Zoology

Undergraduate Degree: B.S.
Undergraduate Degree Requirements: General Chemistry 8 credits, Organic Chemistry 10 credits, Physics 8 credits, Calculus 4 credits, Zoology 30 credits

Graduate Degrees: M.S., Ph.D.

Graduate Degree Requirements: Master: 30 credits

Graduate and Undergraduate Courses Available for Graduate Credit: Drug Use and Abuse (U), Vertebrate Biology (U), Comparative Anatomy (U), Biological Ultrastructure (U), Developmental Biology (U), Comparative Embryology (U), Mammalogy (U), Seminar in Quaternary Studies (U), Histology (U), Animal Microtechnique (U), Invertebrate Zoology (U), Animal Ecology (U), Animal Parasitology (U), Ornithology (U), Human Genetics (U), Principles of Genetics (U), Genetics Laboratory (U), Evolution (U), Limnology (U), Limnology Laboratory (U), Introduction to Oceanography (U), Fishery Biology (U), Fishery Biology Laboratory (U), Animal Physiology (U), General Physiology (U), General Physiology Laboratory (U), Problems in Zoology (U), Biological Oceanography, Anatomy and Classification of Fishes, Larval Biology of Marine Invertebrates, Polar Ecology, Animal Distribution, Ichthyology, Genetics of Populations, Advanced Topics in Aquatic Biology, Fisheries Science, Experimental Endocrinology, Cell Mechanisms, Functional Anatomy of Marine Invertebrates, Experimental Embryology, Seminar in Ecology, Cytology and Cytogenetics, Advanced Genetics, Population Dynamics, Estuarine Ecology, Comparative Physiology, Experimental Physiology, Advanced Cell Physiology, Comparative Endocrinology, Problems in Zoology, Problems in Biological Oceanography

College of Life Sciences and Agriculture
Dean: F.E. Hutchinson

DEPARTMENT OF ANIMAL SCIENCE
(No reply received)

DEPARTMENT OF BIOCHEMISTRY
Phone: Chairman: (207) 581-2272 Others: 581-7149

Chairman: F.H. Radke
Professors: H. DeHaas, F.H. Radke,
Associate Professor: J. Lerner
Assistant Professors: R.D. Blake, S.L. Johnson
Instructors: E.L. Brennan, S.C. Jacobs
Visiting Lecturers: 5
Research Assistants: 2
Teaching Assistants: 4
Post-doctoral Research Associate: 1

Degree Programs: Animal Nutrition, Biochemistry

Undergraduate Degree: B.S.
Undergraduate Degree Requirements: 120 hours with 2.0 average (4.0) Mathematics through Calculus, 8 hours of Communications, 15 hours of Humanities and Social Sciences, 22 hours of Chemistry, 20 hours of Biochemistry

Graduate Degrees: M.S., Ph.S.
Graduate Degree Requirements: Master: Minimum of 30 hours including 6 hours thesis (required) Accumulative average of B (3.0). 12 hours in residence; oral examination of thesis Doctoral: Ph.D. in Animal Nutrition; 2 years in residence 1 year plus Master's. Comprehensive examination, 2 languages and/or skills required. Program designed and supervised by committee

Graduate and Undergraduate Courses Available for Graduate Credit: Biochemistry (U), Physical Biochemistry (U), Principles of Biochemistry (U), Advanced Biochemistry (U), Biochemical Laboratory Methods (U), Biochemical Research (U), Molecular Biology, Carbohydrates and Lipids, Proteins and Enzymes, Vitamins and Hormones, Plant Biochemistry, Biochemical Mechanisms, Thesis

DEPARTMENT OF BOTANY AND PLANT PATHOLOGY
Phone: Chairman: (207) 581-7930 Others: 581-7861

Chairman: G.A. McIntyre

Professors: R.J. Campana, J.R. Cook, F.E. Manzer, R.C. McCrum, G.A. McIntyre, C.D. Richards, A.L. Shigo (Associate)
Associate Professors: R.B. Davis, J.A. Frank (Associate), D.A. Gelinas, R.L. Homola, B.F. Neubauer, R.L. Vadas
Assistant Professors: H.J. Dubin, L.J. Laber
Teaching Fellows: 9
Research Assistants: 3

Field Stations/Laboratories: Aroostook State Farm - Presque Isle, Maine, Highmoor Farm - Monmouth, Maine, Blue Berry Hill Farm - Jonesboro, Maine

Degree Programs: Plant Physiology, Botany, Plant Pathology

Undergraduate Degree: B.S.
Undergraduate Degree Requirements: 120 semester hours, 42 hours Botany and related Biological Sciences, 24 hours Physical Sciences, 9 hours Communications, 15 hours Humanities and Social Sciences, 30 hours Electives

Graduate Degrees: M.S., Ph.D.
Graduate Degree Requirements: Master: 30 hours including 6-15 hours thesis Doctoral: Doctoral registration for full program of study and/or research for two consecutive academic years beyond the baccalaureate, or one academic year beyond the Masters degree.

Graduate and Undergraduate Courses Available for Graduate Credit: Plant Ecology (U), Plant Anatomy (U), Seminar in Quaternary Studies (U), Botanical Microtechnique (U), Plant Physiology (U), Plant Physiology Laboratory (U), Intermediate Plant Physiology (U), Plant Pathology (U), Bryology (U), General Mycology (U), Introductory Phycology (U), Taxonomy of Vascular Plants (U), Field Studies in Ecology (U), Taxonomy of Aquatic Flowering Plants (U), Late Wuaternary Paleoecology, Plant Nematology, Advanced Plant Pathology, Advanced Plant Physiology, Comparative Morphology of Vascular Plants, Plant Geography, Marine Benthic Ecology, Photosynthesis and Chloroplast Development, Seminar, Lake Ecology and Productivity, Research Methods in Plant Science, Problems in Botany, Seminar in Ecology

DEPARTMENT OF ENTOMOLOGY
Phone: Chairman: (207) 581-7704

Chairman: J.B. Dimond
Professors: J B. Dimond, R.E. Olson, G.W. Simpson, F.B. Knight
Associate Professors: H.Y. Forsythe, Jr., D.E. Leonard, I.N. McDaniel, R.H. Storch
Assistant Professors: K.E. Gibbs, G.A. Simmons
Teaching Fellows: 3
Research Assistants: 2
Visiting Professor: 1

Field Stations/Laboratories: Blueberry Hill - Jonesboro, Highmoor Farm, Monmouth, Aroostook Farm - Presque Isle, Chapman Farm - Chapman, Rogers Farm - Stillwater

Degree Program: Entomology

Undergraduate Degree: B.S.
Undergraduate Degree Requirements: 120 hours, Entomology 15 hours, Other Biology 40 hours, Physical Science 16 hours, Mathematics 8 hours, Communications 8 hours, Humanities and Social Sciences 15 hours, Electives 17 hours, Physical Education 1 hr

Graduate Degree: M.S.
Graduate Degree Requirements: Master: 24 hours beyond B.S. - 6 hours thesis Doctoral: Available through Zoology, Forest Resources or Plant Science

Graduate and Undergraduate Courses Available for Graduate Credit: The Insect World (U), Insect Biology and Taxonomy (U), Forest Insect Ecology (U), Economic Entomology, Insect Ecology, Medical Entomology, Morphology of Insects, Seminar in Entomology, Biological Control of Insects, Behavior of Arthropods, Seminar in Ecology, Graduate Thesis

SCHOOL OF FOREST RESOURCES
Phone: Chairman: (207) 581-7312

Director: F.B. Knight
Dwight B. Domeritt Professor of Forest Resources: F.B. Knight
Professors: T.J. Corcoran, R.H. Griffin, H.L. Mendall, H.E. Young, M.W. Coulter, F.B. Knight, J.E. Shottafer
Associate Professors: M D. Ashley, E.L. Giddings, R.A. Hale, N.P. Kutscha, R.B. Owen, H.A. Plummer, A.G. Randale, V.B. Richens, W.C. Robbins, S.D. Schemnitz, C.E. Schomaker, J.C. Whittaker
Assistant Professors: C.E. Shuler, M.D. Zagata
Teaching Fellows: 2
Research Assistants: 26

Field Stations/Laboratories: Robert I. Ashman - Summer Camp - Princeton , Maine

Degree Programs: Forest Resources, Forestry, Natural Resources, Wildlife Management, Forest Engineering, Parks and Recreation

Undergraduate Degree: B.S.
Undergraduate Degree Requirements: B.S. (Forestry), B.S. (Wildlife), require 139 credits including summer camp.

Graduate Degrees: M.S., Ph.D.
Graduate Degree Requirements: Master: Generally 2 years - including thesis Doctoral: One foreign language - both comprehensive and Final Examination. Many specializations in forestry and wildlife.

Graduate and Undergraduate Courses Available for Graduate Credit: Courses too numerous to list.

DEPARTMENT OF MICROBIOLOGY
Phone: Chairman: (207) 581-7628

Chairman: D.B. Pratt
Professors: C.E. Buck, D.B. Pratt, A.R. Whitehill
Associate Professor: W.M. Bain, B.L. Nicholson
Assistant Professor: A.J. DeSiervo
Visiting Lecturers: 1

Degree Program: Microbiology

Undergraduate Degree: B.S.
Undergraduate Degree Requirements: General Microbiology 5 hours, Determinative Bacteriology 4 hours, Pathogenic Bacteriology 4 hours, Microbial Physiology 4 hours, Virology 4 hours, Seminar 2 hours, Prerequisite courses in Chemistry, Physics, Mathematic, and Biochemistry and required. Total Program 120 hours

Graduate Degrees: M.S., Ph.D.
Graduate Degree Requirements: 24 credit hours in formal courses. 6 hours of thesis, 12 credit hours of formal courses must be in Graduate School courses (200-level), Exact program of study is at the committee's discretion Doctoral: Two years residence as a full-time student. Exact program at committee's discretion .

Graduate and Undergraduate Courses Available for Graduate Credit: Microbial Physiology (U), Virology (U), Tissue Culture, Microbial Genetics, Marine Microbiology, Immunology

DEPARTMENT OF PLANT AND SOIL SCIENCES
Phone: Chairman: (207) 581-2771

Chairman: R.C. Glenn
Professors: R.C. Glenn, C.S. Brown, W.C. Stiles, R.A. Struchtemeyer, E.G. Lotse
Associate Professors: R.V. Akeley, P.N. Carpenter, P.R. Hepler, A.R. Langille, H.J. Murphy, H.E. Wave, V.H. Holyoke
Assistant Professors: S.M. Goltz, W.H. Erhardt, A.A. Ismail, L. Littlefield, R.V. Rourke, J.E. Swasey
Research Assistants: 7
Teaching Assistants: 3

Degree Programs: Agronomy, Conservation, Crop Science, Horticulture, Natural Resources, Plant Sciences, Soil Science

Undergraduate Degree: B.S.
Undergraduate Degree Requirements: Minimum Degree Hours Required - 120, Basic Sciences 36, Plant or Soil Sequence 28, Life Sciences and Agriculture Electives 12, Humanities, and Soil Sciences 15, Communications 9, Physical Education 1, Free Electives 1

Graduate Degrees: M.S., Ph.D.
Graduate Degree Requirements: Master: Minimum of 30 semester hours, including credit given for the thesis, is required, with the following limitations: (a) the minimum amount of credit given for the thesis is 6 hours and in no case may it exceed 15 hours (b) a minimum grade point average of 3.0 in courses, which contribute to the M.S. Program. Doctoral: The Graduate School does not define a minimum number of total credits required for the Ph.D. in Plant Science. The advisory committee, however has the responsibility of defining the courses and credits essential to the education and development of each candidate. The department suggests 45 credits beyond the B.S. exclusive of research, as a realistic minimum.

Graduate and Undergraduate Courses Available for Graduate Credit: Not Stated

UNIVERSITY OF MAINE AT PORTLAND-GORHAM

96 Falmouth Street Phone: (207) 773-2981
Portland, Maine 04103
Dean of Graduate Studies: R. York
Dean of College: K. Feig

DEPARTMENT OF BIOLOGY
Phone: Chairman: Ext. 396 Others: same

Chairman: H.H. Najarian
Professor: H.H. Najarian
Associate Professors: G.J. Barker, H.L. Greenwood, P.K. Holmes, A.K. Kern, R.H. Riciputi, L. Schwinek
Assistant Professors: H.B. Hartman, M. Mazurkiewicz
Student Assistants: 8

Degree Program: Biology

Undergraduate Degree: B.A.
Undergraduate Degree Requirements: Calculus 4 credits, Chemistry 8 credits, Organic Chemistry 10 credits, General Physics 8 credits, Foreign Language, Physical Education, Humanities 6 credits, Fine and Applied Arts 6 credits, Social Sciences 6 credits and 30 credits in Biology of which General Biology, Genetics, Developmental Biology and Microbiology, Senior Seminar specified, but choice of either Ecology or Limnology and either General Physiology or Comparative Physiology or Plant Physiology

UNIVERSITY OF MAINE AT PRESQUE ISLE

Presque Isle, Maine Phone: (207) 764-0311
Dean of College: R. Koch

DIVISION OF MATHEMATICS AND SCIENCE
Phone: Chairman: Ext. 43

Chairman: J.C. MacLeod
Associate Professors: J. Libby, D. Stimpson
Assistant Professors: C. Loder, G. Moreau

Degree Program: Life Sciences

Undergraduate Degree: B.A.
Undergraduate Degree Requirements: 36 semester hours Life Science, 25 semester hours Chemistry and Mathematics

WESTBROOK COLLEGE

Portland, Maine 04103 Phone: (207) 797-7261
Dean of College: R. Bond

SCIENCE DEPARTMENT
Phone: Chairman: Ext. 66

Chairman: D.A. Vollmer
(Rank System is not used): T. Hoogakier, J. Gordon, C. Lewis, A. Lambert, D. Vollmer
Research Assistant: 1

Degree Program: Medical Technology

Undergraduate Degree: B.S.
Undergraduate Degree Requirements: 120 hours and 12 month internship in ASCP Hostpial School of Medical Technology

MARYLAND

BOWIE STATE COLLEGE

Jericho Park Road Phone: (301) 267-3350
Bowie, Maryland 20715
Dean of College: S.Myers

SCIENCE DEPARTMENT
Phone: Chairman: Ext. 265

Chairman: H. Jones
Professors: S-U-I. Khan, J.H. Gooden, M. Rock, C. Kirksey
Associate Professors: H. Jones, H.S. Tornabene, R.D. Brown, D. Council

Degree Program: Biological Sciences

Undergraduate Degree: B.S.
Undergraduate Degree Requirements: 8 credit hours Zoology, 8 credit hours Botany, 8 credit hours Inorganic Chemistry, 8 credit hours Organic Chemistry, 8 credit hours Physics (General), 4 credit hours General Physiology, 4 credit hours Calculus I, 4 credit hours Elementary Analysis

COLLEGE OF NOTRE DAME OF MARYLAND

(No reply received)

COLUMBIA UNION COLLEGE

7600 Flower Avenue Phone: (301) 270-9200
Takoma Park, Maryland 20012
Dean of College: J. Blanco

DEPARTMENT OF BIOLOGY
Phone: Chairman: Ext. 222

Chairman: R.M. Scorpio
Professor: R.M. Scorpio
Assistant Professors: C.J. Baird, A.G. Futcher
Contact Teacher: 1

Field Stations/Laboratories: Columbia Union College Biological Field Station - Headwater, Virginia

Degree Program: Biology

Undergraduate Degree: B.A.
Undergraduate Degree Requirements: 32 credit hours in Biology courses

COPPIN STATE COLLEGE

(No reply received)

FROSTBURG STATE COLLEGE

Frostburg, Maryland 21532 Phone: (301) 689-6621
Dean of Graduate Studies: P. Hyons

BIOLOGY DEPARTMENT
Phone: Chairman: (301) 689-4355 Others: 689-4166

Head: M.L. Brown
Professors: M.L. Brown, D.A. Emerson, R.H. Gilpin, T.F. Redick, A.E. Schrock, J.R. Snyder
Assistant Professors: W.J. Pegg, R.K. Riley, W.A. Yoder
Instructors: B.J. Clarin, A.S. Harman, M.R. Lane

Degree Program: Biology

Undergraduate Degree: B.S., A.B.
Undergraduate Degree Requirements: 32 semester hours Biology, 6 semester hours Mathematics, 12 semester hours Chemistry, 6 semester hours Physics

Graduate Degree: M.Ed. (Concentration in Biology)
Graduate Degree Requirements: Master: varied according to student need

Graduate and Undergraduate Courses Available for Graduate Credit: Evolution and Systematics (U), Histology (U), General Ecology (U), Ornithology (U), Maryland's Natural Resources, Plant Diseases, Invertebrate Zoology (U), General Parasitology (U), History and Literature of Biology, Special Problems in Biology, Flora of the Alleghenies, Limnology, Plant Ecology, Animal Physiology, Ichthyology, Animal Ecology, Plant Physiology Laboratory, Biological Techniques, BSCS Biology

GOUCHER COLLEGE

Dulaney Valley Road Phone: (301) 825-3300
Towson, Maryland 21204
Dean of College: K. Walker (Acting)

DEPARTMENT OF BIOLOGICAL SCIENCES
Phone: Chairman: Ext. 404 Others: Ext. 410

Chairman: H.M. Webb
Professor of Biological Sciences on the Lilian Welsh Foundation: H.M. Webb
Professors: H.B. Funk, H.M. Habermann, A.M. Lacy, H.M. Webb
Assistant Professors: M. Berlinrood, J.W. Foerster
Teaching Assistant: 1

Degree Program: Biological Sciences

Undergraduate Degree: B.A.
Undergraduate Degree Requirements: 9 courses in Biological Sciences: 25 least 4 at level II, 3 at levelIII (2 of which must have laboratory Integrative exercise), and a 9th course from level II or III. In fulfilling level V requirement, must take, Genetics, Plant Physiology, Embryology or Physiology, Microbiology or Ecology, A course in Organic Chemistry is also required for the major in addition to the 9 courses

HOOD COLLEGE

Frederick, Maryland 21701 Phone: (301) 663-3131
Dean of College: P. Cunnea

DEPARTMENT OF BIOLOGY
Phone: Chairman: Ext. 261 Others: Ext. 263

Chairman: J.H. Gilford
Professor: J.H. Gilford
Assistant Professors: J.D. Helm III, S. Munch, D. Munson
Visiting Lecturers: 2
Research Assistants: 10

Field Stations/Laboratories: Cooperative agreement with Chesapeake Biological Laboratory

Degree Programs: Biology, Medical Technology

Undergraduate Degree: B.S.
Undergraduate Degree Requirements: 124 hours (semester hours), 3 hours of English composition, 1 hour of Health and Physical Education, minimum of 24 hours

in major.

Graduate Degree: M.S.

Graduate Degree Requirements: Master: 24 hours of graduate credit and 6 hours of thesis preparation, 9 hours of 24 required in interdisciplinary work, dealing with Human Sciences (Political Science, Psychology, Sociology, Biology)

Graduate and Undergraduate Courses Available for Graduate Credit: Man and His Environment, Principles of Environmental Biology, Biological Basis for Environmental Standards, Pollution Biology, Environmental Monitoring, Seminar in Environmental Biology, Independent Study in Environmental Biology, Biochemistry

JOHNS HOPKINS UNIVERSITY

Charles and 34th Street Phone: (301) 366-3300
Baltimore, Maryland 21218
Dean of Graduate Studies: S.R. Suskind

Faculty of Arts and Sciences

Dean: G.E. Owen

DEPARTMENT OF BIOLOGY
Phone: Chairman: Ext. 411

Chairman: W.F. Harrington
Henry Walters Professor of Zoology: (Vacant)
William Gill Professor of Botany: (Vacant)
Professors: M. Bessman, L. Brand, J. Cebra, W. Harrington, P. Hartman, Y. Lee, E. Moudrianakis, A. Nason, S. Roseman, H. Seliger (Part-time): D. Brown, J. Ebert, I. Dawid (Members of Department of Embryology, Carnegie Institution of Washington, Baltimore and part-time Professors in Department of Biology). (Joint appointments with JHU School of Medicine): B. Childs, S. Boyer, K. Ishizaka, V. McKusick, B. Cohen (School of Hygiene JHU): S. Suskind (Academic Programs, JHU Homwood Campus)
Associate Professors: R. Ballentine, H. Berger, M. Edidin, R. Huang, S. Roth (Part-time): W. Kundig (Joint Appointment with JHU School of Medicine): E. Murphy
Assistant Professors: V. Pigiet, D. Powers, M. Rhoades, A. Shearn, W. Sofer, E. Weinberg (Part-time): D. Fambrough and R. Reeder (members of Department of Embryology, Carnegie Institution of Washington, Baltimore and part-time assistant professors in Department of Biology)
Research Assistants: 30
Postdoctoral Fellows: 40

Field Stations/Laboratories: Chesapeake Bay Institute - Solomon's Island, Maryland

Degree Programs: Biochemistry, Biology, Genetics, Cell Biology, Developmental Biology, Human Genetics

Undergraduate Degree: B.A. (Biology)

Undergraduate Degree Requirements: Core curriculum of Calculus, General Physics, General and Organic Chemistry, Biological Concepts and Cell and Comparative Biochemistry, Advanced programs in Molecular Biology, Cell Biology, Organismic Biology and Population Biology

Graduate Degree: Ph.D.

Graduate Degree Requirements: Master: Master's degree given if student completes all requirements except thesis; there is no Master's program. Doctoral: Foreign language examination, 4 minor comprehensives, major comprehensive, teaching experience, thesis, oral defense of thesis

Graduate and Undergraduate Courses Available for Graduate Credit: Cellular and Comparative Biochemistry (U), Genetics, Cell Biology (U), Developmental Biology(U), Genetics Seminar I, Biochemical Genetics and Gene Action, Current topics in Genetics, Advanced Genetics, Human Genetics, Advanced Cell and Developmental Seminar I: Immunobiology, NRA and Protein in Animal Cells, Genetic Analysis of Development Neurobiology, Genetic's and Physiology of Somatic Cell Populations, Advanced Biochemistry, Organic Biochemistry and Radiochemistry, Molecular Biology of Proteins and Nucleic Acids, Biochemistry Seminar I: Biochemical Regulation, II: Physical Techniques, III: Fibrous Proteins, Techniques in Biological Research: training in Various Techniques, Photobiology, Radiobiology, Molecular Immunology

DEPARTMENT OF BIOPHYSICS
Phone: Chairman: (301) 955-3077 Others: 955-3012

Chairman: H.M. Dintzis
Professor: R.J. Poljak
Associate Professors: L.M. Amzel, R.Z. Kintzis, B. Zimmerman
Assistant Professors: R.P. Phizackerley
Teaching Fellows: 1
Technicians: 3

Degree Programs: Degree programs are offered in participation with the Biophysics Department at Homewood Campus and the Department of Physiological Chemistry in the School of Medicine.

School of Medicine

725 North Wolfe Street
Baltimore, Maryland 21205
Dean: R.H. Morgan

DEPARTMENT OF ANATOMY
Phone: Chairman: (301) 955-3240

Chairman: D. Bodian
Bayard Halsted Professor of Anatomy: D. Bodian
Professors: D.G. Walker, L. Weiss
Associate Professors: R.A. Bergman, M.E. Molliver, E.A. Ruth, M.A. Schön
Assistant Professors: L.T. Chen, A.M. Cohen, R.Z. Dintzis, J.H. Gross, J.C. Hedreen, P.G. Heltne
Instructor: C.C. LaMotte

Degree Programs: Anatomy, Histology

Graduate Degree: Ph.D.

Graduate Degree Requirements: Not Stated

Graduate and Undergraduate Courses Available for Graduate Credit: Applied Anatomy, Cytology and Ctyodifferentiation, Histology and Histophysiology, Human Anatomy, Human Embryology, Introduction to Neurology, Primate Evolution

DEPARTMENT OF BIOMATHEMATICS
(No reply received)

DEPARTMENT OF BIOMEDICAL ENGINEERING
Phone: Chairman: (301) 955-3131 Others: 955-5000

Chairman: R.J. Johns
Massey Professor of Biomedical Engineering: R.J. Johns
Professors: D.S. Gann, R.E. Gibson, K. Sagawa, R.H. Shepard
Associate Professors: W.C. Ball, W.H. Guier, L.J. Krovetz, J.T. Massey, R.P. Rich, D.A. Robinson
Assistant Professors: G.L. Cryer, W.K. Harrison, J.M. Heinz, J.B. Oakes, M.B. Sachs, L.P. Schramm, D.W. Simborg, A.A. Shoukas, G.N. Webb
Instructors: H.B. Brown, J. Ward
Postdoctoral Fellows: 7

Field Stations/Laboratories: Applied Physics Laboratory - 8621 Georgia Avenue, Silver Spring, Maryland

Degree Programs: Biomedical Engineering, Clinical Engineering

Graduate Degrees: M.S., Ph.D.
Graduate Degree Requirements: Master: 1 1/2 years formal course work plus 1/2 year thesis research and thesis Doctoral: 2 years formal course work devoted to advanced study in engineering science and in Biomedical Science. Life science courses (Biochemistry, Anatomy and Physiology). Oral Examination. Doctoral Dissertation. Foreign Language. Thesis Research

Graduate and Undergraduate Courses Available for Graduate Credit: Introduction to Computers, Quantitative Systems Physiology, Physiological Foundations for Biomedical Engineering, Seminar in Biomedical Engineering, Topics in Biomedical Engineering, Principles of Biomedical Instrumentation, Seminar in Clinical Medicine and Health Care Delivery, Applications of Computers in Clinical Engineering, Biomedical Systems Engineering, Internship in Clinical Engineering, (Thesis Research), Introduction to Computers in Clinical Engineering

DEPARTMENT OF BIOPHYSICS
Phone: Chairman: 366-3300 Ext. 597

Chairman: M. Beer
Professors: M. Beer, F. Carlson, R. Cone, H. Dintzis, M. Larrabee, W.E. Love, R. Poljak
Associate Professor: D. Fambrough
Assistant Professors: L M Amzel, R. Dintzis, S. Lin, W. Wiggins, B. Zimmerman
Instructor: R.P. Phizackerley

Degree Program: Biophysics

Undergraduate Degree: B.A.
Undergraduate Degree Requirements: Specified program including Mathematics, Physics, Chemistry, Biology

Graduate Degree: Ph.D.
Graduate Degree Requirements: Doctoral: 2 terms full time student including specific courses Biophysics, Biology, Chemistry, Mathematics, one Foreign Language, Teaching Experience, Comprehensive Examination, Thesis, Oral Defense.

Graduate and Undergraduate Courses Available for Graduate Credit: Freshman Biophysics Seminar, Selected Topics in Biophysics, Electronics for Research, Advanced Neurophysiology with Laboratory, Advanced Neurophysiology Without Laboratory, Independent Study of Biophysics, Research Problems in Biophysics, X-ray Diffraction: Chemical Crystallography, Principles of Physiology, Structure, Function, and Biosynthesis of Proteins and Nucleic Acids, Biological Electron Microscopy, Hemoglobin, Biophysics Seminar, Laborstory Problems in Biophysics, Molecular Biology Seminar, Motility and Contractility, Photon Correlation and Light-Beating Spectroscopy, Molecular M croscopy, Membrane Physiology Seminar, Topics in Neurochemistry, Biomolecular X-ray Structure Analysis, The Molecular Biology of Proteins and Nucleic Acids

DEPARTMENT OF MICROBIOLOGY
Phone: Chairman: (301) 955-6352/3653 Others: 955-5000

Director: D. Nathans
Boury Professor of Microbiology: D. Nathans
O'Neill Professor of Microbiology: K. Ishizaka
Professor: M.M Mayer, H.O. Smith, T.B. Turner (Emeritus)
Associate Professors: D.H Carver, W.B. Greenough, III, P.H. Hardy, Jr., T. Ishizaka, R.T. Johnson, L.M. Lichtenstein, R.B. Sack, H.S. Shin, B. Weiss
Assistant Professors: L. Aurelian, K.I. Berns, P.C. Charache, C.S. Henney, R.L. Humphrey, H. Kaizer, T.J. Keely, Jr., D.G. Marsh, M.R. Smith, L.A. Sternberger
Visiting Lecturers: 20
Teaching Fellows: 2
Research Assistants: 10

Degree Programs: Bacteriology and Immunology, Biology, Immunology, Microbiology, Virology

Graduate Degree: Ph.D.
Graduate Degree Requirements: Doctoral: Completion of Biochemistry, 5 one-term courses in Biological or Biophysical Sciences (One in advanced Organic Chemistry or equivalent); demonstration of ability to read scientific publications in one foreign language before end of second year, completion of preliminary oral examination end of first year and a second oral examination at the end of the second year; presentation of doctoral dissertation and a public seminar on dissertation work.

Graduate and Undergraduate Courses Available for Graduate Credit: Principles of Medical Microbiology, Topics in Molecular Biology, Topics in Biochemical Virology, Immunology, Bacterial Pathogenicity, and Special Studies and Research

DEPARTMENT OF PHYSIOLOGY
(No reply received)

School of Hygiene and Public Health
615 North Wolfe Street
Baltimore, Maryland 21205
Dean: J.C. Hume

DEPARTMENT OF BIOSTATISTICS
Phone: Chairman: (301) 955-3685

Chairman: A. Ross
Professors: H. Abbey, D.B. Dunca, A.M. Gittelgohn, A.W. Kimball, A.Ross
Associate Professors: C.A. Rohde, R.M. Royall, J.A. Tonascia
Assistant Professors: H.H. Deyal, S.M. S'Souza
Visiting Lecturers: 3
Research Assistants: 4

Field Stations/Laboratories: Centers for Research and Training at Hagerstwon Maryland, Narangwal and Calcutta, India. Other project headquarters in Peru, Afghanistan, Bangladesh, Nigeria, Taiwan

Degree Program: Biostatistics

Graduate Degrees: M.S., Ph.D.
Graduate Degree Requirements: Master: Courses, written examination, Thesis (2 years plus). Doctoral: Courses, Written examination, Preliminary Oral Examination, Dissertation, Final (Thesis) Examination. (4 years)

Graduate and Undergraduate Courses Available for Graduate Credit: 21 different courses in statistical methods, theory, probability, stochastic processes

LOYOLA COLLEGE
4501 North Charles Street Phone: (301) 323-1010
Baltimore, Maryland 21210
Dean of College: F. McGuire

DEPARTMENT OF BIOLOGY
Phone: Chairman: Ext. 238-239-240

Chairman: H.C. Butcher
Professors: Sr. M.B. Cleary
Associate Professors: H.C. Butcher, C.R. Graham, M. A. Lorenzo, S.J.
Assistant Professors: F.E. Giles, J.T. Maier, S.J.

Degree Program: Biology

Undergraduate Degree: B.S.
Undergraduate Degree Requirements: (1 course = 4 credits 8 courses Biology, 4 courses Chemistry, 2 courses

Physics, 2 courses Mathematics, 16 courses Humanities

MORGAN STATE COLLEGE
(No reply received)

MOUNT ST. MARY'S COLLEGE
Emmitsburg, Maryland 21727 Phone: (301) 447-6122
Dean of College: B.S. Kaliss

DEPARTMENT OF SCIENCE AND MATHEMATICS
Phone: Chairman: Ext. 285

Chairman: W.G. Meredith
Professors: D.G. Greco, W.G. Meredith
Associate Professor: P. Gauthier
Visiting Lecturer: 1

Degree Programs: Biology, Medical Technology

Undergraduate Degree: B.S.
Undergraduate Degree Requirements: 32 hours Biology, 18 hours Chemistry, 70 hours Core and Electives.

ST. MARY'S COLLEGE OF MARYLAND
St. Mary's City, Maryland 20686 Phone: (301) 994-1600
Dean of Faculty: F. Honkala

DIVISION OF NATURAL SCIENCE AND MATHEMATICS
Phone: Chairman: Ext. 290

Division Chairman: C.M. Wilson
Professor: R.E. Wilson
Associate Professors: J.J. Eichenmuller, R.W. Serianni, E.J. Willoughby
Assistant Professors: E.A. Bowman, R.T. Novotny
Instructors: C.T. Krebs, P.J. Willoughby (Part-time)
Visiting Lecturer: 1
Laboratory Assistant: 1

Degree Programs: Biology, Natural Science

Undergraduate Degrees: B.A., B.S.
Undergraduate Degree Requirements: Thirty six credits in Biology made up of General Biology, Cell Biology, Non-Vascular Botany, Vascular Botany, Invertebrate Zoology, Vertebrate Zoology, Ecology, Genetics, and a course in Physiology, General Chemistry I and II, Statistics and one other Mathematics course

SALISBURY STATE COLLEGE
Salisbury, Maryland 21801 Phone: (301) 749-7191
Director of Graduate Studies: M. Massuci
Dean of College: T.L. Erskine

DEPARTMENT OF BIOLOGY
Phone: Chairman: Ext. 420 Others: Ext. 369

Chairman: R.A. Hedeen
Professors: R.A. Hedeen, E.E. Estes, V.L. VanBreemen, H.J. Robertson (Adjunct)
Associate Professor: W.F. Standaert
Assistant Professors: A.G. DiGiovanna, J.R. Ransbottom, J.O. Rebach, D.R. Sistrunk, H. Womack, C.O. Wingo
Assistant Instructor: J.B. Kunkle
Undergraduate Student Assistants: 12

Degree Programs: Biology, Biological Education

Undergraduate Degrees: B.S., B.S.Ed.
Undergraduate Degree Requirements: 35 semester hours in Biology, 12 semester hours in related subjects (Chemistry, Physics, Mathematics) (120 semester hours C or better required for degree)

Graduate Degrees: M.S., M.Ed.

Graduate Degree Requirements: Master: 15 hours graduate level Biology courses; 15 hours in graduate education courses

Graduate and Undergraduate Courses Available for Graduate Credit: Contemporary Cell Physiology, Biology and the Environment, Modern Concepts in Biology, Research in Biology (U), Electron Microscopy (U), Cytology

TOWSON STATE COLLEGE
(No reply received)

U.S. NAVAL ACADEMY
Annapolis, Maryland 21402 Phone: (301) 267-7711
Dean of College: B.M. Davidson

BIOLOGY SECTION, DEPARTMENT OF CHEMISTRY
Phone: Chairman: (301) 267-3302

Section Chairman: R.R. Corey
Professor: R.R. Corey
Assistant Professors: D.L. Weingartner, C.V. Gordon
Instructor: M.M. Bundy

Degree Program: Biological Sciences

Undergraduate Degree: B.S.
Undergraduate Degree Requirements: Mathematics 12 hours, Chemistry 22 hours, Humanities 21 hours, Biology 28 hours, Physics 8 hours, Other 54 hours, Total semester hours 145

DEPARTMENT OF ENVIRONMENTAL SCIENCES
Phone: Chairman: (301) 267-3561

Chairman: W.H. Keith
Professors: J. Williams, J.F. Hoffman
Associate Professors: C.N.G. Hendrix
Assistant Professor: D.W. Edsall
Instructors: R.R. Hadfield, V.K. Nield, R.B. Brodehl, D.L. Jones, W.C. Barney, J.D. King, J.T. Welch, C.A. Martinek, H.T. Dantgler

Degree Program: Oceanography

Undergraduate Degree: B.S.
Undergraduate Degree Requirements: General Biology, Physical Geology, General Meterology, General Oceanography, Essentials of Fluid Dynamics, Environmental Dynamics, Transport Phenomena, Naval Oceanographic Applications, Electives

UNIVERSITY OF MARYLAND at COLLEGE PARK
College Park, Maryland 20742 Phone: (301) 454-0100
Dean of Graduate Studies: D. Sparks

College of Arts and Science
Dean: R.F. Davis

DEPARTMENT OF ZOOLOGY
Phone: Chairman: 454-3202 Others: 454-3201

Chairman: J.O. Corliss
Professors: G. Anastos, H.J. Brinkley, J.R.C. Brown, E. Clark, J.O. Corliss, J.F. Eisenberg (Adjunct), S. Grollman, A.J. Haley (Assistant Chairman), R.T. Highton, L.A. Jachowski, G.F. Otto (Adjunct), G.M. Ramm, W.M. Schleidt
Associate Professors: A.J. Barnett, J.F. Contrera, M.D. Goode, R.B. Imberski, H. Levitan, H.J. Linder, D.H. Morse, S.K. Pierce, J.H. Potter, E.B. Small, G.J. Vermeij
Assistant Professors: J.D. Allan, D.A. Bonar, D.E. Gill, W.J. Higgins, E.S. Morton, C.P. Rees, S.A. Woodin
Instructors: L.A. Killeen, V.I. Knox, D. Moore, A.P.

Neidhardt, R.W. Piper, C.A. Spalding
Research Assistants: 13
Graduate Teaching Assistants: 55

Field Stations/Laboratories: Chesapeake Bay Center for Environmental Studies - Edgewater, Maryland, University of Maryland Center for Environmental and Estuarine Studies (Horns Point), Cambridge, Maryland

Degree Programs: Animal Behavior, Ecology, Genetics, Parasitology, Physiology, Zoology, Marine Biology, Endocrinology, Cell Biology, Developmental Biology, Evolutionary Biology

Undergraduate Degree: B.S. (Zoology
Undergraduate Degree Requirements: 1) 26 hours of zoology including at least one course in each of four areas (Cells and Cell Organelles; Tissues, Organs and Organ Systems, Organisms, Populations and Communities of Organisms), 2) 14 of 26 hours in upper division courses 3) supporting courses in Chemistry (1 1/2 years), Physics (1 year), Mathematics (through one year of Calculus) and a course in Biostatistics, and 4) a "C" average in all of the above.

Graduate Degrees: M.S., Ph.D.
Graduate Degree Requirements: Master: 1) Thesis Option - 24 hours of courses with at least 12 at graduate only level; 6 hours thesis research; a "B" average; final oral examination on thesis 2) Non-Thesis Option - 30 hours of course with at least 18 at graduate only level; writing of a scholarly paper, a "B" average, a final written comprehensive examination in three areas of Zoology Doctoral: 1) residence - equivalent of 3 years of full-time study, 2) credit hours - varies among the different areas of specialization, 3) Oral Preliminary Examination, 4) Dissertation seminar, 5) Final oral examination (dissertation defense)

Graduate and Undergraduate Courses Available for Graduate Credit: Cell Biology (U), Biophysics (U), Cell Differentiation (U), Physiology of Excitable Cells (U), Vertebrate Physiology (U), General Endocrinology (U), Vertebrate Embryology (U), Evolution (U), Advanced Evolutionary Biology (U), Molecular Genetics (U), Experimental Genetics (U), Ethology (U), Ethology Laboratory (U), Advanced Animal Ecology (U), Laboratory and Field Ecology (U), Protozoology (U), General Parasitology (U), Aquatic Biology (U), Biology of Estuarine and Marine Invertebrates (U), Marine Vertebrate Zoology (U), Vertebrate Zoology (U), Form and Pattern in Organisms (U), Mammalian Histology (U), Zoology Seminar, Special Problems in Zoology, Cellular Physiology, Electron Microscopy Laboratory, Biological Ultrastructure, Advanced Topics in Cell Biology, Comparative Physiology, Experimental Mammalian Physiology, Comparative Invertebrate Endocrinology, Molecular Neurobiology, Comparative Vertebrate Endocrinology, Electrophysiology, Organogenesis, Biochemical Patterns in Development, Population Genetics, Ecological Genetics, Developmental Genetics, Cellular Genetics, Systematic Zoology, Comparative Behavior, Sociobiology, Analysis of Animal Populations, Cuantitative Zoology, Advanced Aquatic Ecology, Quantitative Field Ecology, Ecological Models, Behavioral Ecology, Ecology of Marine Communities, Physiological Ecology, Ecology of Marine Communities, Physiological Ecology, Ecology of Marine Invertebrates, Marine and Estuarine Protozoa, Lectures in Zoology, Experimental Parasitology, Helminthology, Advanced Topics in Protozoology

College of Agriculture

Dean: G.M. Cairns

DEPARTMENT OF AGRONOMY

Phone: Chairman: (301) 454-3718

Chairman: J.R. Miller
Professors: J.H. Axley, N.A. Clark, A.M. Decker, J.E. Foss, J.H. Hoyert, C.G. McKee, J.R. Miller, E. Strickling
Associate Professors: M.K. Aycock, V.A. Bandel, D.S. Fanning, F.P. Miller, J.V. Parochetti
Assistant Professors: G.W. Burt, J.R. Hall, D.T. Hawes, L. Hofmann, C.L. Mulchi, J.L. Mulchi, J.L. Newcomer, D.C. Wolf
Instructor: C.E. Rivard
Visiting Lecturers: 1
Faculty Research Assistants: 4
Visiting Assistant Professor: 1
Visiting Associate Professor: 1

Field Stations/Laboratories: Agronomy0Dairy Forage Research Farm - Folly Quarter Road, Ellicott City, Maryland, Plant Research Farm - 12000 Cherry Hill Road, Silver Spring, Maryland, Poplar Hill Research Farm - Route 1, Box 61-A, Quantico, Maryland, Tobacco Research Farm, 2005 Largo Road, Upper Marlboro, Maryland, Wye Institute Research Farm, Cheston-on-Wye, Queenstown, Maryland

Degree Programs: Agronomy, Crop Science, Farm Crops, Plant Breeding, Soil Science, Soils and Plant Nutrition, Weed Science

Undergraduate Degree: B.S. (Crop Science, Soil Sciences)
Undergraduate Degree Requirements: Core 23 hours Chemistry, Botany and Agronomy. Crop Science or Soil Science 68 hours specified courses

Graduate Degrees: M.S., Ph.D.
Graduate Degree Requirements: Master: 24 semester hours, not less than 12 in graduate only courses. Doctoral: 50 - 60 semester hours, one foreign language, Preliminary Examination, Thesis, Oral Defense

Graduate and Undergraduate Courses Available for Graduate Credit: Crop Breeding (U), Tobacco Production (U), Turf Management (U), Forage Crop Production (U), Cereal Crop Production (U), Soil Fertility Principles (U), Commercial Fertilizers (U), Soil and Water Conservation (U), Soil Classification and Geography (U), Soil Survey and Land Use (U), Soil Physics (U), Soil Chemistry (U), Soil Biochemistry (U), Soil-Water Pollution (U), Cropping Systems (U), Seed Production and Distribution (U), Weed Control (U), Special Problems in Agronomy (U), Advanced Crop Breeding, Advanced Crop Breeding, Research Methods, Advanced Soil Chemistry, Recent Advances in Agronomy, Agronomy Seminar, Breeding for Resistance to Plant Pests, Technique in Field Crop Research, Advanced Tobacco Production, Herbicide Chemistry and Physiology, Advanced Forage Crops, Advanced Methods of Soil Investigation, Advanced Soil Mineralogy, Advanced Soil Physics

DEPARTMENT OF ANIMAL SCIENCE

Phone: Chairman: (301) 454-4641 or 454-4642

Chairman: E.P. Young
Professors: W.W. Green, E.C. Leffel, E.P. Young
Associate Professors: J. Buric, J.V. DeBarthe, E.E. Goodwin
Assistant Professor: J P. McCall
Instructor: W.A. Curry
Visiting Lecturers: 2

Field Stations/Laboratories: Horse Research Center - Ellicott City, Maryland, Swine Research Unit - Ellicott City, Maryland, Beef Cattle and Sheep Research Farm - Sykesville, Maryland

Degree Program: Animal Science

Undergraduate Degree: B.S.
Undergraduate Degree Requirements: 46 hours specified courses

Graduate Degrees: M.S., Ph.D.
Graduate Degree Requirements: Master: Thesis option -

30 semester hours, 12 in major subject, Thesis, Final Oral Non-Thesis Option - variable program
Doctoral: Courses selected by committee. Preliminary Examination, Thesis, Final Oral

Graduate and Undergraduate Courses Available for Graduate Credit: Fundamentals of Nutrition, Applies Animal Nutrition, Animal Adaptations to the Environment, Advanced Dairy Production, Biology and Management of Shellfish, Introduction to Diseases of Animals, Laboratory Animal Management, Biology and Management of Fish, Wildlife Management, Meats, Livestock Management, Livestock Management, Principles of Breeding, Dairy Cattle Breeding, Analysis of Dairy Production Systems, Physiology of Mammalian Reproduction, Avian Physiology, Physiology of Hatchability, Poultry Hygiene, Avian Anatomy, Poultry Breeding and Feeding, Poultry Products and Marketing, Special Topics in Fish and Wildlife Management, Special Topics in Animal Science, Advanced Ruminant Nutrition, Mineral Metabolism, Vitamins, Electron Microscopy, Advanced Breeding, Experimental Mammalian Surgery, Experimental Mammalian Surgery II, Research Methods, Poultry Literature, Physiology of Reproduction, Physiological Genetics of Domestic Animals, Advanced Animal Adaptations to the Environment, Seminar in Population Genetics of Domestic Animals

DEPARTMENT OF BOTANY
Phone: Chairman: (301) 454-3812

Chairman: H.D.Sisler
Professors: M.K. Corbett, R.A. Galloway, J.G. Kantzes, W.L. Klarman, L.R. Krusberg, D.T. Morgan, Jr., G. W. Patterson, W.L. Stern, L.O. Weaver
Associate Professors: G.A. Bean, C.R. Curtis, E.P. Karlander, J.D. Lockard, O.D. Morgan, R.D. Rappleye
Assistant Professors: N.M. Barnett, P.J. Bottino, C.R. Broome, G.K. Harrison, J.J. Motta, J.L. Reveal, J.C. Stevenson, S.D. Van Valkenburg
Instructors: B.J. Grigg, E.A. Higgins
Visiting Lecturer: 1
Research Assistants: 3

Field Stations/Laboratory: Salisbury, Maryland

Degree Programs: Biological Sciences, Botany, Genetics, Nematology, Pathology, Physiology, Virology

Undergraduate Degree: B.S.
Undergraduate Degree Requirements: General University Requirements 30 credit hours, Departmental Requirements 26 credit hours, Chemistry 8 credit hours, Mathematics 6 credit hours, Physics 8 credit hours, Zoology 4 credit hours, Microbiology 4 credit hours, Electives 34 credit hours

Graduate Degree: M.S., Ph.D.
Graduate Degree Requirements: Master: 24 credit hours plus thesis Doctoral: No specific credit requirements Course program determined by Departmental Student Advisement Committee.

Graduate and Undergraduate Courses Available for Graduate Credit: History and Philosophy of Botany (U), Plant Microtechnique (U), Systematic Botany (U), Teaching Methods in Botany (U), Plant Anatomy (U), Structure of Economic Plants (U), Plant Geography (U), General Plant Genetics (U), Plants and Mankind (U), Principles of Plant Anatomy (U), Field Botany and Taxonomy (U), Natural History of Tropical Plants (U), Research Methods in Plant Pathology (U), Diagnosis and Control of Plant Diseases (U), Diseases of Ornamentals and Turf (U), Mycology (U), Field Plant Pathology (U), Plant Physiology (U), Plant Ecology (U), Ecology of Marsh and Dune Vegetation (U), Algal Systematics (U), Marine Plant Biology (U), Plant Morphology, Plant Cytogenetics, Physiology of Fungi, Physiology of Pathogens and Host-Pathogen Relationships, Plant Virology, Plant Nematology, Advanced Plant Physiology, Plant Biochemistry, Growth and Development, Plant Biophysics, Advanced Plant Ecology, Physiology of Algae

DEPARTMENT OF DAIRY SCIENCE
Phone: Chairman: (301) 454-3926 Others: 454-3925

Chairman: J.F. Mattick
Professors: G.M. Cairns, R.F. Davis, M. Keeney, R.L. King, J.F. Mattick, J H. Vandersall, W.F. Williams
Associate Professor: L.W. Douglass
Assistant Professors: L.S Bull, P.K. Holdaway, D.C. Westhoff
Instructor: D.J. Seely
Visiting Lecturers: 2
Research Assistants: 4
Extension Specialists: 2

Field Stations/Laboratories: Dairy and Agronomy Forage Research Center - Ellicott City, Maryland

Degree Programs: Dairy Science, Food Science, Nutritional Science

Undergraduate Degree: B.S.
Undergraduate Degree Requirements: 120 credits total - General University Requirements 20 credits plus College and department credits 60 plus 30 elective credits

Graduate Degrees: M.S., Ph.D.
Graduate Degree Requirements: Master: Thesis Option - 30 credits total = 6 credits Research and 24 credits in area of specialization 12 of which must be in graduate level courses and 12 in upper division courses Non-Thesis Option - 30 credits total = 18 credits in graduate level courses and 12 in upper division Doctoral: Course credit - based on need of student - No fixed number minimum 1 year residence - oral examination by faculty in discipline area, Thesis and Defense of Thesis

Graduate and Undergraduate Courses Available for Graduate Credit: Applied Animal Nutrition (U), Biology and Management of Shellfish (U), Advanced Dairy Production (U), Biology and Management of Fish (U), Wildlife Management (U), Dairy Cattle Breeding (U), Analysis of Dairy Production Systems (U), Physiology of Mammalian Reproduction (U), Advanced Ruminant Nutrition, Mineral Metabolism, Population Genetics of Domestic Animals, Agricultural Biometrics (U), Advanced Agricultural Biometrics, Principles of Food Processing (U), Food Chemistry (U), Food Products Research and Development (U), Food Microbiology (U), Dairy Products Processing (U), Seafood Products Processing (U), Advanced Food Microbiology, Lipids, Proteins, Carbohydrates, Organoleptic Food Properties, Fermentation, Enzymes and Microorganisms, Flavor Analysis, Food Products Assays, Animal Science Colloquim, Food Science Colloquim, Nutritional Sciences Colloquim

DEPARTMENT OF ENTOMOLOGY
Phone: Chairman: 454-3843 Others: 454-3842

Chairman: E.C. Bay
Professors: E.C. Bay, W.E. Bickley, F.P. Harrison, J.C. Jones, R.E. Menzer, D.H. Messersmith, A.L. Steinhauer
Associate Professors: D.M. Caron, J.A. Davidson, W.C. Wallace, Jr., C.F. Reichelderfer
Assistant Professors: G.P. Dively, J.P. Linduska, F.E. Wood
Instructor: J.L. Hellman
Visiting Lecturers: 2
Research Assistants: 4
Visiting Assistant Professor: 1
Post-Doctorals: 4

Field Stations/Laboratories: U.M. Vegetable Farm - Salisbury, Maryland, Hancock Fruit Station - Hancock, Maryland

Degree Program: Entomology

Undergraduate Degree: B.S.
Undergraduate Degree Requirements: 120 hours, including 30 hours General University Requirements and 18- 23 electives plus required courses in Chemistry, Entomology, Botany, Zoology, Microbiology, Genetics and Mathematics

Graduate Degrees: M.S., Ph.D.
Graduate Degree Requirements: Master: Strong background in Biological or physical sciences. No language requirement. 24 hours including 12 hours graduate level courses. Thesis. Doctoral: One language. Coursework as determined by committee. Thesis.

Graduate and Undergraduate Courses Available for Graduate Credit: Entomology for Science Teachers (U), Advanced Apiculture (U), Insect Taxonomy and Biology (U), Insect Morphology (U), Insect Physiology (U), Economic Entomology (U), Insecticides (U), Insect Pathology (U), Medical and Veterinary Entomology (U), Seminar (U), Insect Ecology, Experimental Honeybee Biology, Advanced in Insect Physiology, Aspects of Insect Biochemistry, Toxicology of Insecticides, Insect Pest Population Management, Culicidology, Entomological Topics, Advanced Entomology

FOOD SCIENCE PROGRAM
Phone: Chairman: 454-3928

Chairman: R.L. King
Professors: R.L. King, A. Kramer, J.F. Mattick, F.C. Stark, B.A. Twigg, R.C. Wiley
Associate Professors: F. Bender, D.E. Bigbee, J. Buric, A.M. Cowan, O.P. Thomas, E. Young
Assistant Professors: J.L. Heath, C.J. Wabeck, D.C. Westhoff
Visiting Lecturers: 2
Research Assistants: 10

Degree Program: Food Science

Undergraduate Degrees: B.S.
Undergraduate Degree Requirements: General University Requirements 30 hours, Division Requirements 11 hours, Curriculum Requirements 50 hours, electives 29 hours

Graduate Degrees: M.S., Ph.D.
Graduate Degree Requirements: Master: Thesis option - 24 hours plus 6 hours thesis, Non-thesis option - 30 hours plus scholarly paper Doctoral: Programs are developed individually including 12 hours dissertation research, no language requirement

Graduate and Undergraduate Courses Available for Graduate Credit: Not Stated

DEPARTMENT OF HORTICULTURE
Phone: Chairman: 454-3614 Others: 454-3606

Chairman: B.A. Twigg (Acting)
Professors: A. Kramer, C.B. Link, C.W. Reynolds, B.L. Rogers, J.B. Shanks, F.C. Stark, A.H. Thompson, B. A. Twigg, R.C. Wiley
Associate Professors: R.L. Baker, J.C. Bouwkamp, F.R. Gouin, F.D. Schales, K.P. Soergel
Assistant Professors: C.E. Beste, R.C. Funt, J.F. Kundt, C.A. McClurg, D.G. Pitt, W.M. Sedovic, H.D. Stiles
Instructors: H.G. Mityga, S.H. Todd
Visiting Lecturers: 1
Faculty Research Assistants: 4
Graduate Research Assistants: 10

Field Stations/Laboratories: Vegetable Research Farm - Salisbury, Maryland, Fruit Research Laboratory - Hancock, Plant Research Farm, Cherry Hill Road, Silver Spring, Maryland

Degree Program: Horticulture

Undergraduate Degree: B.S.
Undergraduate Degree Requirements: Total of 120 credits for graduation. General University Requirements 30 credits. Departmental requirements of all majors students are: 12 credits Botany, 4 credits Soils, 8 credits Chemistry, 4 credits Horticulture, 3 credits Mathematics and fulfillment of requirements of one of 3 options, Floriculture and Ornamental Horticulture, Pomology and Olericulture or Horticulture Education which include 20 to 25 credits of Horticulture and additional credits in Botany, Agronomy or Entomology

Graduate Degrees: M.S., Ph.D.
Graduate Degree Requirements: Master: Thesis option - 24 semester credits and 6 credits Horticulture Research, 12 to 16 credits in major area, 8 to 12 credits in supportive area. Minimum of 12 credits at graduate level; 6 credits of Horticultural Research. Preparation of thesis, examination by graduate committee. Non-Thesis: 30 semester credits, 15 5o 22 credits in major area and 8 to 15 credits in supportive area, minimum of 18 credits at graduate level, preparation of a scholarly paper, examination by graduate committee. Doctoral: No fixed course requirements; minimum of 24 semester credits in ancillary fields with at least half at graduate level, minimum of 12 credits of Horticultural research, Comprehensive Examination by graduate committee previous to admission to candidacy dissertation oforiginal, independent research and final examination, no language requirement

Graduate and Undergraduate Courses Available for Graduate Credit: Technology of Fruits (U), Technology of Vegetables (U), Fundamentals of Greenhouse Crop Production (U), Technology of Ornamentals (U), Woody Plant Materials (U), Production and Maintenance of Woody Plants (U), Systematic Horticulture (U), Physiology of Maturation and Storage of Horticultural Crops (U), Special Topics in Horticulture (U), Methods of Horticultural Research, Special Topics in Horticulture, Edaphic Factors and Horticultural Plants, Chemical Regulation of Growth of Horticultural Plants, Environmental Factors and Horticultural Plants, Current Advances in Plant Breeding, Advanced Seminar, Special Problems in Horticulture

DEPARTMENT OF MICROBIOLOGY
Phone: Chairman: 454-2848, 2849

Chairman: B.G. Young
Professors: R.R. Colwell, R.N. Doetsch, F.M Netrick, N.C. Laffer, B.G. Young, J.E. Faber (Emeritus)
Associate Professors: T.M. Cook, A.M. MacQuillan, B.S. Robertson
Assistant Professors: Z. Vaituzis, M.J. Voll, R.M. Weiner
Instructors: G.F. Howell
Visiting Lecturers: 2
Teaching Fellows: 3
Research Assistants: 8
Graduate Teaching Assistants: 12

Degree Program: Microbiology

Undergraduate Degree: B.S.
Undergraduate Degree Requirements: 120 hours total, 24 hours in Microbiology

Graduate Degrees: M.S., Ph.D.
Graduate Degree Requirements: Master: 24 semester hours (includes 12 hours in Microbiology) 6 hours Research, Dissertation and oral examination Doctoral 24 semester hours in related minor subjects, 12 hours research, Thesis, Preliminary Examinations, Oral Defence.

Graduate and Undergraduate Courses Available for Graduate Credit: Microbiological Literature, Microbiology and the Public, Special Topcs, Microbiological Problems, Systematic Microbiology, History of Microbiology, Epidemiology and Public Health, Pathogenic Microbiology, Immunology, General Virology, Microbial Physiology, Microbial Fermentations, Bacterial Metabolism, Special Topics, Cytology of Bacteria, Advanced Immunology, Virology and Tissue Culture, Advanced Bacterial Metabolism, Genetics of Microorganisms, Microbial Genetics Laboratory, Doctoral Thesis Research

DEPARTMENT OF POULTRY SCIENCE
Phone: Chairman: 454-3835 Others: 454-3836

Chairman: O.P. Thomas
Professors: C.S. Shaffner
Associate Professors: D. Bigbee, J.L. Heath, O.P. Thomas
Assistant Professors: T. Carter, C. Coon, W. Kuenzel, J.L. Nicholson, J. Soares, C. Wabeck
Research Assistant: 1
Research Associate: 1

Field Stations/Laboratories: Broiler Sub-Station - Salisbury, Maryland

Degree Programs: Poultry Nutrition, Poultry Physiology, Poultry Products,Technology

Undergraduate Degree: B.S. (Animal Science)
Undergraduate Degree Requirements: University of Maryland Requirements - 30 credits in Agriculture and Life Sciences, Social Sciences and Arts and Humanities, Department Requirements - 8 credits of Chemistry, 8 credits of Biology, 3 credits of Mathematics, 2 credits of Speech, 23 credits of Basic Animal Science, 9 credits of Poultry Science

Graduate Degrees: M.S., Ph.D.
Graduate Degree Requirements: Master: 24 hours of course work, plus 6 hours of research, a graduate level course in "Statistics" and "Biochemistry" required
Doctoral: In addition to the Master's requirements, an advanced "Statistics" is required.

Graduate and Undergraduate Courses Available for Graduate Credit: Fundamentals of Nutrition, Applied Animal Nutrition, Introduction to Diseases of Animals, Avian Physiology, Physiology of Hatchability, Poultry Hygiene, Avian Anatomy, Poultry Products and Marketing, Mineral Metabolism, Vitamins, Electron Microscopy, Poultry Literature, Physiology of Reproduction, Physiological Genetics of Domestic Animals, Seminar, Special Problems in Animal Science, Master's Thesis Research, Doctoral D ssertation Research

UNIVERSITY OF MARYLAND AT BALTIMORE

660 West Redwood Street Phone: (301) 528-2121
Baltimore, Maryland 21201
Dean of Graduate Studies: J.F. Lambooy

School of Medicine

Dean: J.M. Dennis

ANATOMY DEPARTMENT
Phone: Chairman: (301) 528-7307 Others: 528-2121

Chairman: F.J. Ramsay (Acting)
Professors: O.C. Brantigan (Part-time), V.E. Krahl (Research)
Associate Professors: E.J. Donati, J.M. Masters (part-time), K.F. Mech (part-time), M L. Rennels, G.E. Wadsworth
Assistant Professors: C.P. Barrett, M.H. Bulmash (part-time), A.W. Klein, K W. Petersen, J.P. Petrali (part-time), F.J. Ramsay (Associate Dean, Director of Student Affairs), F.P. Schulter
Visiting Lecturers: 2
Research Assistants: 7

Degree Programs: Anatomy, Anatomy and Histology

Graduate Degrees: M.S., Ph.D.
Graduate Degree Requirements: Master: A minimum of 24 semester hours in approved graduate courses, 6 semester hours of thesis research, at least 2 semesters residency at the University of Maryland, completion of all requirements within a 5 year period.
Doctoral: A minimum of 40 hours in approved graduate courses (30 hours in major field and 10 in supporting fields), 12 semester hours of doctoral research, successful completion of a qualifying examination before being admitted to Candidacy, competence in one foreign language (option: completion of a non-science course), satisfactory performance in thesis examination. Students are also expected to assist in laboratory teaching.

Graduate and Undergraduate Courses Available for Graduate Credit: Not Stated

DEPARTMENT OF BIOLOGICAL CHEMISTRY
Phone: Chairman: (301) 528-7120

Chairman: E. Adams
Professors: E. Adams, E. Bucci, L. Frank, S. Pomerontz
Associate Professors: L. Black, M E. Kintley
Assistant Professors: C.F. Bucci, R.M. Gryder, N.V. Rao, B.P. Rosen, C. Waechter
Instructor: A.V. Brown

Degree Program: Biochemistry

Graduate Degree: Ph.D.
Graduate Degree Requirements: Doctoral: An established program of coursework, listed in the graduate school catalog of this institution, together with thesis work, in the usual way

Graduate and Undergraduate Courses Available for Graduate Credit: Graduate School Catalog should be consulted

DEPARTMENT OF BIOPHYSICS
Phone: Chairman: (301) 528-7940

Chairman: L.J. Mullins
Professors: L.J. Mullins, R.A. Sjodin
Associate Professors: L.A. Beaugé, A. Hybl
Assistant Professors: D.S. Geduldig
Visiting Lecturers: 12
Research Assistants: 3

Degree Program: Biophysics

Graduate Degree: Ph.D.
Graduate Degree Requirements: Doctoral: B.S. plus 2 years Physics and Mathematics plus 12 credits in Life Science at Graduate level

Graduate and Undergraduate Courses Available for Graduate Credit: Not Stated

DEPARTMENT OF CELL BIOLOGY
(No reply received)

DEPARTMENT OF MICROBIOLOGY
Phone: Chairman: (301) 528-7114 Others: 528-7110

Chairman: C.L. Wisseman, Jr.
Professors: R. Traub
Associate Professors: O.R. Eylar, P. Fiset, R.W.I. Kessel
Assistant Professors: D.J. Silverman, W.F. Myers
Visiting Lecturers: 10

Degree Program: Microbiology

Graduate Degrees: M.S., Ph.D.
Graduate Degree Requirements: Master: "B" average as an undergraduate, Completion of accreditec baccalaureate degree, Evidence of academic potential
Doctoral: Same as above but higher requirements

Graduate and Undergraduate Courses Available for Graduate Credit: Medical Microbiology, Fundamentals of Immunobiology, Advanced General Microbiology, Seminar, Special Topics, Microbial Physiology, Clinical Immunology, Advanced Virology and Rickettsiology Lecture, Virology and Rickettsiology Laboratory, Advanced Immunology, Microbiology: Advanced Immunology Laboratory

PHYSIOLOGY DEPARTMENT
Phone: Chairman: (301) 528-7243 Others: 528-7242

Chairman: W.D. Blake
Professors: C.A. Barraclough, A.B. Fajer, E.M. Glaser, G.R. Mason (Part-time), G.G. Pinter.
Associate Professors: C.P. Channing, L. Goldman, S.E. Greisman (Part-time), L.M. Karpeles, D.S. Ruchkin
Assistant Professors: O.R. Blaumanis, O.M. Cramer, A.P. Fertiziger, A.N. Jurf, J.L. Turgeon, B.K. Urbaitis
Research Assistants: 3
Research Fellows: 2

Degree Programs: Physiology, Reproductive Endocrinology

Graduate Degrees: M.S., Ph.D.
Graduate Degree Requirements: Master: 30 hours, 24 course work, 6 research for dissertation Doctoral: 30 hours course work, 12 hours research credits, dissertation

Graduate and Undergraduate Courses Available for Graduate Credit: Mammalian Physiology, Seminars in Cardiovascular, Renal, Neural , General and Endocrine Physiology, Laboratory Techniques, Biological Control Systems, Signal Analysis

School of Dentistry

Dean: J.J. Salley

DEPARTMENT OF ANATOMY
(No reply received)

DEPARTMENT OF BIOCHEMISTRY
Phone: Chairman: 528-7817

Chairman: J. Lambooy
Professor: J. Lambooy
Associate Professor: C.B. Leonard, Jr.
Assistant Professors: N. Bashirelahi, S. Courtade, M. Morris
Research Assistants: 1

Degree Program: Biochemistry

Graduate Degrees: M.S., Ph.D.
Graduate Degree Requirements: Master: A minimum of 30 semester hours including six hours of thesis research credit, of the 24 hours required, not less than 12 must be earned in major subject; minimum of 12 must be selected from courses numbered 600 or above, final oral examination on thesis Doctoral: Specific requirements may be obtained by contacting Department.

Graduate and Undergraduate Courses Available for Graduate Credit: Advanced Biochemistry, Biochemical Endocrinology, Biochemistry of Lipids,Biochemistry Seminar

DEPARTMENT OF MICROBIOLOGY
Phone: Chairman: (301) 528-7538 Others: 528-7539

Chairman: D.E. Shay
Professor: D.E. Shay
Associate Professor: G.N. Krywolap, R.J. Sydiskis
Assistant Professors: Y-F. Chang, A.L. Delisle, W.A. Falkler, Jr., J.M. Joseph , R.K. Nauman
Lecturers: A.H. Jansen, J.P. Libonati, M.J. Snyder
Visiting Lecturers: 3
Graduate Trainees, NIDR: 4

Degree Program: Microbiology

Graduate Degrees: M.S., Ph.D.
Graduate Degree Requirements: Masters: (1) Non-Thesis 30 hours course work, final comprehensive examination (2) Thesis: 24 hours course work, acceptable thesis Doctoral: 54 hours course work, completion of Preliminary Examinations, Acceptable Thesis

Graduate and Undergraduate Courses Available for Graduate Credit: Advanced General Microbiology, Seminar, Serology and Immunology (U), Viral Oncology, Bacterial Genetics, Chemotherapy, Virology (U), Advanced Dental Microbiology and Immunology, Pathogenic Microbiology (U), Microbial Physiology, Special Problems in Microbiology, Mycology (U), Techniques in Microscopy, Methods in Microbial Physiology, Experimental Virology, Public Health, Parasitology (U), Bacterial Fermentations, Microbiology of the Periodontium

DEPARTMENT OF PHYSIOLOGY
Phone: Chairman: (301) 528-7257

Chairman: J.I. White
Professor: J.I. White
Associate Professor: G.W. Kidder III
Assistant Professor: R.B. Bennett, M. Burke
Instructors: B. Nardell, L. Staling
Research Assistant: 1

Degree Program: Physiology

Graduate Degrees: M.S, Ph.D.
Graduate Degree Requirements: Master: Course work and research by individual arrangement Doctoral: Course work and research by individual arrangement

Graduate and Undergraduate Courses Available for Graduate Credit: Principles of Physiology, Principles of Mammalian Physiology, Advanced Physiology

School of Pharmacy

Dean: W.J. Kinnard

DEPARTMENT OF PHARMACOLOGY AND TOXICOLOGY
Phone: Chairman: (301) 528-7509 Others: 528-7510

Chairman: N. Khazan
Emerson Professor of Pharmacology: C.T. Ichniowski
Professors: N. Khazan, W.J. Kinnard
Associate Professors: G.G. Buterbaugh
Assistant Professors: J.S. Adir, R.T. Louis-Ferdinand, D.R. Brown, D.G. Hattan, E.J. Moreton
Teaching Assistantships: 4

Degree Programs: Pharmacokinetics, Toxicology, Neuropharmacology, Pschopharmacology, Biochemical Pharmacology

Graduate Degrees: M.S., Ph.D. (Pharmacology)
Graduate Degree Requirements: Master: Course work 40 credit hours, cumulative written examination, dissertation defense Doctoral: Course work - 40 credit hours, 12 credit hours research, written cumulative examinations, oral comprehensive examination, dissertation defense

Graduate and Undergraduate Courses Available for Graduate Credit: Pharmacodynamics I, II, Toxicology (U), Principles of Drug Action, Seminar, Classical Techniques in Pharmacology, Advanced Biopharmaceutics, Advanced Toxicology, Neuropsychopharmacology, Principles of Biochemical Pharmacology, Physiological Disposition of Drugs

UNIVERSITY OF MARYLAND BALTIMORE COUNTY

5401 Wilkens Avenue Phone: (301) 455-2271
Catonsville, Maryland 21228
Dean of Graduate Studies: J.F. Milligan
Dean of College: C.B.T. Lee

DEPARTMENT OF BIOLOGICAL SCIENCES
Phone: Chairman: (301) 455-2271 Others: 455-1000

Chairman: M. Schwartz
Professor: M. Schwartz
Associate Professors: R.P. Burchard, F.E. Hanson, P.S. Lovett, A.P. Platt, T.F. Roth
Assistant Professors: B.P. Bradley, N.C. Craig, R.C. Gethmann, A.P. Kendal, J.A. Kloetzel, S. Kung, T.V. Marsho, M.C. O'Neill, C.H. Peterson, P.G. Sokolove, W.L. Wardell, C.S. Weber
Instructors: P. Coursey, D. Bramucci, B. Burdick, P. Sokolove
Teaching Assistants: 14

Field Stations/Laboratories: Chesapeake Biological Laboratories - Solomons, Maryland

Degree Program: Biological Sciences

Undergraduate Degree: B.S.
Undergraduate Degree Requirements: The biological sciences major consists of at least 57 credits distributed as follows: 28 hours in Biology, 16 hours in General Chemistry and Organic Chemistry, 6-8 hours in Physics, or Physical Chemistry, 7 hours of Mathematics.

Graduate Degrees: M.S., Ph.D.
Graduate Degree Requirements: Master: 30 hours, including 12 hours of seminars at the graduate level, 6 hours thesis research for the M.S. with thesis or 3 hours laboratory research, and a scholarly paper for non-thesis M.S. Doctoral: 12 hours of graduate seminar, pass written and oral Preliminary Examinations, research thesis, final comprehensive examination with defense of Thesis

Graduate and Undergraduate Courses Available for Graduate Credit: Advanced Developmental Biology (U), Advanced Laboratory Projects in Biological Sciences, Advanced Topics in Cell Biology (U), Advanced Topics in Organismic Biology, Advanced Tutorial Projects in Biological Sciences, Analysis of Development, Biochemistry (U), Biochemistry Laboratory (U), Biology of Photosynthetic Microorganisms (U), Biology of the Bacteria (U), Cell Structure and Function, Dynamics and Statics of Populations, Microbial Genetics (U), Neurobiology (U), Photobiology (U), Processes in Evolution (U), Physiologic Basis of Invertebrate Behavior (U), Physiological Genetics (U), Physiology and Genetics of Bacterial Viruses (U), Physiology of Plant Systems (U), Research Colloquium, Theoretical and Quantitative Biology (U), Topics in Molecular Genetics

WASHINGTON COLLEGE

Chestertown, Maryland 21620 Phone: (301) 778-2800
Dean of College: N. Smith

BIOLOGY DEPARTMENT
Phone: Chairman: Ext. 245 Others: 246

Chairman: E. Gwym
Professors: E. Gwym, K. Yaw
Instructor: J. Goodfellow

Degree Program: Biology

Undergraduate Degree: B.S.
Undergraduate Degree Requirements: 32 semester courses including 8 in Biology, 4 in Chemistry, 2 in Physics, 2 in Mathematics, remainder in distribution and elective courses

WESTERN MARYLAND COLLEGE

Westminster, Maryland 21157 Phone: (301) 848-7000
Dean of College: W. McCormick

BIOLOGY DEPARTMENT
Phone: Chairman: Ext. 80

Chairman: I. Royer
Professor: J. Kerschner
Associate Professor: M. Brown
Assistant Professor: S. Alspach
Instructor: M. Reed

Degree Program: Biology

Undergraduate Degree: A.B.
Undergraduate Degree Requirements: 30 semester hours in Biology, 15 semester hours Chemistry, 6 semester hours Mathematics

MASSACHUSETTS

AMERICAN INTERNATIONAL COLLEGE
Springfield, Massachusetts 01109 Phone: (413) 737-5331
Dean, School of Liberal Arts: M. Birnbaum

BIOLOGY DEPARTMENT
Phone: Chairman: Ext. 354

Chairman: I. Cohen
Professors: I. Cohen, H.W. Aplington, Jr.
Associate Professor: J.P. Murnane
Student Assistants: 6

Degree Program: Biology

Undergraduate Degree: B.S.
Undergraduate Degree Requirements: 2 semesters college Zoology, 1 semester Microbiology, 1 semester Botany plus minimum of 4 semester courses in various areas: Ecology, Oceanography, Botany, Microbiology, etc. plus 2 years Chemistry, 1 year Physics, 1 year Mathematics

AMHERST COLLEGE
Amherst, Massachusetts 01002 Phone: (413) 542-2000
Dean of College: P. Gifford

DEPARTMENT OF BIOLOGY
Phone: Chairman: (413) 542-2063 Others: 542-2000

Chairman: W.M. Hexter
Stone Professor of Biology: G.W. Kidder (Emeritus)
Edward S. Harkness Professor of Biology: H.H. Plough (Emeritus)
Rufus Tyler Lincoln Professor of Biology: O.C. Schotté (Emeritus)
Professors: L.P. Brower, W.M. Hexter, E.R. Leadbetter, H.T. Yost
Associate Professor: W.F. Zimmerman
Assistant Professors: S.G. Fisher, S.A. George, W. Godchaux, P. Karfunkel
Instructor: P.T. Ives

Degree Program: Biology

Undergraduate Degree: B.A.
Undergraduate Degree Requirements: Six courses in Biology, of which four must be laboratory courses. In addition, one semester each of Mathematics, and Physics, and Chemistry through Organic Chemistry are required.

ANNA MARIA COLLEGE
Paxton, Massachusetts 01612 Phone: (617) 757-4586
Dean of Graduate Studies: I. Socquet
Dean of College: B. Madore

DEPARTMENT OF BIOLOGICAL SCIENCES
Phone: Chairman: Ext. 36 Others: 757-4586

Chairman: B. Madore
Professors: E.A. Cole, B. Madore, J. Rini
Associate Professors: A. McMorrow, M.A. Foley
Assistant Professor: P. Miller
Instructors: P. Conrad, L. Sartori

Degree Programs: Biology, Medical Technology, Pre-medicine

Undergraduate Degree: B.S.
Undergraduate Degree Requirements: 6 courses in the History of Ideas and competency in communication, Humanities, Literature, Social and Natural Sciences, Mathematics. Can be displayed through CLEP, Advanced Placement, Achievement tests, or established through formal courses taken at the college.

Graduate Degree: M.A. (Biological Sciences)
Graduate Degree Requirements: 3 seminars, 15 specified courses

Graduate and Undergraduate Courses Available for Graduate Credit: Biochemistry (U), Physiology (U), Microbiology (U), Fundamentals of Radioactivity (U), Science in History (U), Genetics, Parasitology, Histology (U), Human Anatomy and Physiology (U), Biology and the Future of Man (U), Environmental Pollution (U), Organic Chemistry (U), Mechanisms of Radiobiology, Embryology, Biochemistry, Immunology, Teaching of Science, Ecology

ASSUMPTION COLLEGE
Worcester, Massachusetts 01609 Phone: (617) 752-5615
Dean of Graduate Studies: C. Quintal
Dean of College: R.A. Oehling

DIVISION OF NATURAL SCIENCES
Phone: Chairman: Ext. 247

Chairman: E.W. Byrnes
Professors: R.H. MacDonald, R.M. Sutherland
Associate Professors: J.B. Letendre
Assistant Professor: P.J. Mahon
Instructor: R. Levy
Laboratory Instructor: 1

Degree Program: Biology

Undergraduate Degree: B.A.
Undergraduate Degree Requirements: B.A. in Biology: 4 semesters in Chemistry, 2 semesters in Physics, 8 semesters in Biology

ATLANTIC UNION COLLEGE
South Lancaster, Massachusetts 01561 Phone: (617) 365-4561
Dean of College: S. E. Gascay

DEPARTMENT OF BIOLOGY AND HEALTH SCIENCE
Phone: Chairman: Ext. 58 Others: 56

Chairman: F.E. Hauck
Professor: F.E. Hauck
Assistant Professor: S.A. Nyirady
Visiting Lecturers: 3

Degree Programs: Biology, Medical Technology

Undergraduate Degrees: B.A., B.S.
Undergraduate Degree Requirements: General Biology, Development and Morphogenesis, General Physiology, Genetics and Cytogenetics, Philosophy of Biology, Biology Electives - 8 hours. Total for Biology Major 30 hours. Total hours for B.A. degree 128 hours.

BOSTON COLLEGE
Chestnut Hill, Massachusetts 02167 Phone: (617) 969-0100
Dean of Graduate Studies: D.J. White, S.J.
Dean of College: T.P. O'Malley

DEPARTMENT OF BIOLOGY
Phone: Chairman: Ext. 139 Others: Ext: 132, 133

Chairman: Not Stated
Professors: M. Liss, W.D. Sullivan, Y.C. Ting, C.H. Yoon
Associate Professors: M.L. Bade, W.J. Fimian, J.J. Gilroy, J.A. Orlando, D.J. Plocke, A.H. Rule, P. Rieser, J.B. Solomon, C. Stachow
Assistant Professor: J.J. Goldthwaite
Visiting Lecturers: 3
Teaching Fellows: 23
Research Assistants: 3

Degree Programs: Biology

Undergraduate Degree: B.S.
Undergraduate Degree Requirements: Two semesters each of General Chemistry, Organic Chemistry, Introductory Biology and Physics, each with accompanying laboratory course; one semester of Genetics and Bacteriology, each with accompanying laboratory course, two semesters of Calculus, and three additional upper division electives in Biology.

Graduate Degrees: M.S., M.S.T., Ph.D.
Graduate Degree Requirements: Master: Biochemistry and any two of the following: Cell Physiology, Bacterial Physiology and Metabolism, Molecular Basis of Heredity, One Seminar, presentation and oral defense of thesis based on original research. Doctoral: Biochemistry, Cell Physiology, Bacterial Physiology and Metabolism, Molecular Basis of Heredity, four seminars, presentation and oral defense of Thesis based on original research.

Graduate and Undergraduate Courses Available for Graduate Credit: Biochemistry I,II (U) and Laboratories (U), General Endocrinology and Laboratory (U), Plant Growth and Development and Laboratory (U), Seminar in Evolution (U), Seminar in Carcinogenesis (U), Immunochemistry (U), Immunology (U), Human Genetics (U), Cell Physiology, Molecular Basis of Heredity, Radiation and Laboratory, Cell Cycle and Laboratory, Biology of Ultrastructure and Laboratory, Bacterial Physiology and Metabolism, Biology Statistics, Unicellular Development, Current Topics in Endocrinology, Current Topics in Biochemistry

BOSTON STATE COLLEGE

Boston, Massachusetts 02122 Phone: (617) 731-3300
Dean of Graduate Studies: T. Hegarty
Dean of College: J. Jones

DEPARTMENT OF BIOLOGY
Phone: Chairman: Ext. 380,379

Chairman: R.N. McCauley
Professors: L.C. Colt, Jr., E. Fowell, J. Woodland
Associate Professors: A.B. Gesmer, R. Hilton, J. Kunnenker, M. Segelman, D. Shan, F. Veale
Assistant Professors: Armstrong, R. Guimond, C.B. HellQuist, J. Murray, A.M. Olsory, P. Parsons, S. Priest, M. Tierney
Instructor: E.V. Cosgrove
Adjunct Faculty: 9

Degree Program: Biology

Undergraduate Degrees: B.S., B.A.
Undergraduate Degree Requirements: 54-59 Semester hours in Liberal Arts requirements, 30-26 semester hours in Biology including General Botany, Invertebrate Zoology, Vertebrate Morphogenesis, Ecology, Genetics, Cell Biology, also 2 semester hours of Chemistry, Organic Chemistry, 2 semesters of Mathematics, 18-21 semester hours credit in a minor area.

BOSTON UNIVERSITY

755 Commonwealth Avenue Phone: (617) 353-3000
Boston, Massachusetts 02115
Dean of Graduate Studies: P.E. Kubzansky

College of Liberal Arts

Dean: W.J. Ilchman

DEPARTMENT OF BIOLOGY
Phone: Chairman: (617) 353-2432

Chairman: I.A. Macchi (Acting)
Sheilds Warren Professor of Biology: G.P. Fulton
Professors: J.T. Albright, F.A. Belamarich, H.S. Berman, E.H. Blaustein, M.-C. Chang (Adjunct), S. Duncan, G.P. Fulton, A.G. Humes, F.J. Lionetti (Adjunct), I.A. Macchi, R.S. Mackay, G. McLeod (Adjunct), P.J. Morgane (Adjunct), D.I. Patt, I.D. Raache, D. Shepro, R.F. Slechta, C. Terner, N.B. Todd (Adjunct), G. Young
Associate Professors: J. Atema, I.P. Callard, F.C. Ch Chao (Adjunct), T. Foldvari (Adjunct), S. Golubic, H.G. Knuttgen, C.K. Levy, L.A. Margulis, G.R. Patt, K.R.H. Read
Assistant Professors: R.B. Armstrong, U. Banerjee, A. C. Echternacht, D.L. Hall, D. Ianuzzo, R. Jeanne, T.H. Kunz, F. Lang, R. Tamarin, I. Valiela, D.R. Walters
Visiting Lecturers: 0
Teaching Fellows: 37
Research Assistants: 6
Research Associates: 4

Field Stations/Laboratories: Marine Biological Laboratories - Woods Hole, Massachusetts

Degree Programs: Biochemistry, Biology, Comparative Physiology, Cell Biology, Evolution

Undergraduate Degree: A.B.
Undergraduate Degree Requirements: General Biology and Animal Biology; minimum of eight principal courses in Biology, including Plant Biology and Genetics, required related courses in Calculus, General Chemistry, Organic Chemistry, and General Physics, grades not lower than C in principal and related courses

Graduate Degrees: A.M., Ph.D.
Graduate Degree Requirements: Master: Eight graduate-level courses in Biology, reading knowledge of one modern foreign language, thesis based on research, two year residence Doctoral: Eight graduate-level courses beyond A.M. degree, reading knowledge of one modern foreign language, qualifying examination in specialty field, dissertation based on research

Graduate and Undergraduate Courses Available for Graduate Credit: Entomology (U), Parasitology (U), Evolution (U), History of Biology (U), Animal Behavior (U), Ecological Genetics (U), Mammalian Field Ecology (U), Functional Ichthyology (U), Herpetology (U), Ornithology (U), Mammalogy (U), Marine Botany (U), Limnology (U), Marine Mycology (U), Topics in Marine Zoology (U), Zoogeography (U), Population Biology (U), General Endocrinology (U), Principles of Hormone Action and Analysis, Advanced Bacteriology and Immunology (U), Plant Morphology (U), Plant Physiology and Biochemistry (U), Physiology and Biochemistry of Reproduction (U), Comparative Physiology of Muscle and Contractile Systems (U), Neurobiology (U), Fundamental Properties of the Nervous System (U), Vascular Physiology (U), Biophysics (U), Molecular Biology (U), Marine Invertebrates (U), Ecology of the Sea Margin (U), Methods in Marine Biology (U), Ichthyology (U), Coastal Birds and Mammals (U), Marine Protists (U), Marine Algae (U), Comparative Physiology of Marine Organisms (U), Behavior of Marine Animals (U), Developmental Biology of Marine Organisms (U), Tropical Biospeleology (U), Molecular Aspects of Development, Cellular Aspects of Development, Biology of Macromolecules, Cell Genetics, Molecular Genetics, Comparative Physiology, Ichthyology , Marine Invertebrate Zoology, Marine Ecology, Marine Environmental Physiology, Ecology and Taxonomy of Marine Microphytes, Hormones and Reproduction, Seminar in Marine Biology, Experimental Biology, Advanced Biochemistry, Advanced Biochemistry Laboratory, Environmental Biology, Virology, Topics in Molecular Biology, Mechanisms of Evolution, Topics in Vascular Physiology, Quantitative Methods in Vascular Physiology, Topics in Comparative Biochemistry, Advances in Hormone Research, Transducers in Biology and Medicine, Internship in College Biology, Biochemistry Seminar, Research in Marine Biology, Research in Marine Biology

School of Medicine

80 East Concord Street Phone: (617) 282-4200
Dean of School of Medicine: E. Friedman
Graduate Dean: P.E. Kubzansky

DEPARTMENT OF ANATOMY
Phone: Chairman: Ext. 6154

Chairman: A. Peters
Waterhouse Professor of Anatomy: A. Peters
Professors: J. O'Connor, R.E. Stallard
Associate Professors: J.A. Grasso, D.D. Ifft, W.F. Mc Nary, Jr. D.N. Pandya, G.E. Raviola, E.K.F. Ronka
Assistant Professors: J.E. Dittmer, A-W. I.El-Bermani, M.L. Feldman, J.W. Hinds, R. Hoyt, Jr., W.B. Warr, D.W. Vaughan
Post-doctoral Fellows: 1

Degree Program: Anatomy

Graduate Degree: Ph.D.
Graduate Degree Requirements: Master: 1 year fundamental course work, 1 year directly supervised course work Doctoral: 2 years fundamental course work, 2-3 years of directed research

Graduate and Undergraduate Courses Available for Graduate Credit: Microscopic Anatomy, Gross Anatomy, Neurosciences, Cell Biology, Advanced Neuroanatomy, The Orbit, The Ear, EM Techniques, Human Embryology, Seminars, Special Topics in Anatomy

DEPARTMENT OF BIOCHEMISTRY
(No reply received)

DEPARTMENT OF MICROBIOLOGY
Phone: Chairman: Ext. 6135

Chairman: E.E. Baker
Associate Professors: S.A. Broitman, S.R. Cooperband, M.A. Derow, D.R.D. Shaw
Instructors: A. Badger, M. Johns

Degree Program: Microbiology

Graduate Degrees: M.S., Ph.D.
Graduate Degree Requirements: Master: Microbiology and Biochemistry, Seminar, and Thesis Doctoral: Microbiology, Biochemistry, Bacterial Genetics, Bacterial Physiology Immunology, Seminar and others.

Graduate and Undergraduate Courses Available for Graduate Credit: Not Stated

DEPARTMENT OF PHYSIOLOGY
Phone: Chairman: Ext. 6150 Others: Ext.6131

Chairman: B. Kaminer
Professors: A. Essig, A.I.F. Gorman, B. Kaminer, E. MacNichols, W. Ullrick
Associate Professor: H. Kayne
Assistant Professors: A. Fein (Adjunct), R. Fine, W. Hardy, M. Lang, W. Lehman, A. Politoff
Research Associates: 4

Degree Program: Physiology

Graduate Degrees: M.A., Ph.D.
Graduate Degree Requirements: Master: 8 courses, examination and/or minor thesis Doctoral: 16 courses, thesis

Graduate and Undergraduate Courses Available for Graduate Credit: Physiology Seminar, Cellular Physiology, Membrane Transport, Electrical Coupling of Cells, Cell Motility, Mechanics of Muscle

BRANDEIS UNIVERSITY

415 South Street Phone: (617) 647-2727
Waltham, Massachusetts 02154
Dean: J.S. Goldstein

GRADUATE DEPARTMENT OF BIOCHEMISTRY
Phone: Chairman: (617) 647-2731 Others: 647-2727

Chairman: R.H. Abeles
Professors: R.H. Abeles, G.D. Fasman, D.M. Freifelder, L. Grossman, W.P. Jencks, L. Levine, J.M. Lowenstein, A.G. Redfield, S.N. Timasheff, H. Van Vunakis
Associate Professors: T.G. Hollocher, S. Lowey, W.T. Murakami, R.F. Schleif, M. Soodak
Assistant Professors: J-S. Hong
Visiting Professors: 3
Research Assistants and Associates: 18
Post-doctoral Fellows: 18

Degree Program: Biochemistry

Undergraduate Degree: B.S.
Undergraduate Degree Requirements: Either a Chemical or a Biological point of view, each with varying programs.

Graduate Degree: Ph.D.
Graduate Degree Requirements: Doctoral: Fundamental courses in Advanced Biochemistry, Biochemical Techniques, Physical Biochemistry, Biochemical Research Problems, 4 Biochemistry Seminars, German, Qualifying Examination, Thesis and Defense

Graduate and Undergraduate Courses Available for Graduate Credit: Biochemistry: Advanced, Biochemistry: Immuno chemistry, Biochemistry: Molecular Biology, Biochemistry: Introduction to Physical Biochemistry, Seminars, Biochemistry: Current Topics in Molecular Biology, Biochemistry: Metabolic Regulation, Biochemistry: Biosynthesis of More or Less Complex Molecules, Biochemistry: Advanced Physical Techniques, Biochemistry: Biochemical Research Problems

BIOLOGY DEPARTMENT
(No reply received)

BRIDGEWATER STATE COLLEGE

Bridgewater, Massachusetts 02324 Phone: (617) 697-8321
Dean of Graduate Studies: F.J. Hilferty
Dean of College: W.L. Anderson

DEPARTMENT OF BIOLOGICAL SCIENCES
Phone: Chairman: Ext. 315 Others: Ext. 316,317

Chairman: K.J. Howe
Professors: J.R. Brennan, E.F. Cirino, L.B. Mish, W.A. Morin, W.J. Wall
Associate Professor: W.M. Hewitson
Assistant Professors: J.C. Jahoda, F.A. Muckenthaler, H.P. Schaefer
Laboratory Instructors: 4

Degree Program: Biology

Undergraduate Degrees: B.A., B.S.

Undergraduate Degree Requirements: Plan 1: Broad-Based Program leading to Bachelor of Arts or Bachelor of Science, 36 semester hours Biology, 24 or more semester hours including Inorganic Chemistry, Mathematics, Physics, General Education Requirements of the College, 120 semester hours Minimum Plan 2: Career/Profession-Oriented Program leading to Bachelor of Arts or Bachelor of Science, 27 semester hours of courses comprising a Biology core, 12 or more semester hours in a Biological concentration, one year each of Inorganic Chemistry, Organic Chemistry, Mathematics, Physics, General Education requirements of the college, General Electives (may include minor), 120 semester hours Minimum

Graduate Degrees: M.A., M.Ed. (Biology)
Graduate Degree Requirements: Master: M.A. - Preliminary Examination in subject field, Foreign Language Examination, 30 semester hours in Biology including Original research and Thesis, Comprehensive Examination M.Ed. - Teaching experience, Teaching Certification by State, 30 semester hours of course work: 15 or more of biology and 9 or more of professional education, Comprehensive Examination

Graduate and Undergraduate Courses Available for Graduate Credit: Invertebrate Zoology (U), Comparative Chordate Anatomy (U), Field Natural History (U), Plant Anatomy (U), Plant Morphology (U), Biochemistry (U), Heredity and Human Endeavor (U), General and Comparative Physiology (U), Mammalian Physiology (U), Genetics (U), Microbiology (U), Plant Physiology (U), Biological Evolution (U), Marine Biology (U), Ecology (U), Mycology (U), Developmental Biology (U), Sensory Physiology (U), Radiation Biology (U), Cytology (U), Experimental Morphology of Angiosperms, Comparative Morphology of Vascular Plants, Taxonomy of Spermatophytes, Neurophysiology I and II, Mammalogy, Microbial Physiology, Comparative Ethology, Entomology, Developmental Genetics, Biological Electron Microscopy, Research, Directed Study

CLARK UNIVERSITY

950 Main Street Phone: (617) 793-7173
Worcester, Massachusetts 01524
Dean of Graduate Studies: V. Ahmadjian
Dean of College: R.E. Kasperson

DEPARTMENT OF BIOLOGY
Phone: Chairman: (617) 793-7173

Chairman: R.F. Nunnemacher
Professors: V. Ahmadjian, J.J. Brink, J.C. Curtis, R.F. Nunnemacher, J.T. Reynolds
Associate Professors: H.W. Johansen, R.G. Sherman
Assistant Professors: T.A. Lyerla, S.E. Johnson
Adjunct Professors: 2
Teaching Fellows: 10
Affiliates: 8

Degree Program: Biology

Undergraduate Degree: A.B.
Undergraduate Degree Requirements: 8 semester courses in Biology, 4 semester courses in Chemistry, 2 semester courses in Physics, 2 semester courses in Calculus, 16 semester courses in other areas

Graduate Degrees: M.A., Ph.D.
Graduate Degree Requirements: Master: 5 year courses, Thesis or published paper Doctoral: 4 year courses beyond M.A., 1 foreign language, Acceptable Dissertation

Graduate and Undergraduate Courses Available for Graduate Credit: 3/4 of all courses are available to both undergraduates and graduates.

COLLEGE OF THE HOLY CROSS

Worcester, Massachusetts 01610 Phone: (617) 793-2011
Dean of College: Rev. J.R. Fahey, S.J.

DEPARTMENT OF BIOLOGY
Phone: Chairman: (617) 793-2655 Others: 793-2656

Chairman: Rev. J.W. Flavin, S.J.
Professors: Rev. J.W. Flavin, S.J., B.T. Lingappa
Associate Professors: R.S. Crowe, W.R. Healy, J.H. McSweeney
Assistant Professors: M.A. Johnston, J.K. Reynhout
Instructor: J.V. Wyland
Laboratory Assistant: 1

Degree Program: Biology

Undergraduate Degree: B.S.
Undergraduate Degree Requirements: Total number of undergraduate courses: 32 (Biology: 8, Chemistry: 4, Physics: 2, Mathematics: 2)

COLLEGE OF OUR LADY OF THE ELMS

291 Springfield Street Phone: (413) 598-8351
Chicopee, Massachusetts 01013
Dean of College: Sr. M.M. Harrington

DEPARTMENT OF BIOLOGY
Phone: Chairman: Ext. 52

Chairman: Sr. M.L. Wright
Assistant Professors: Sr. M.J. McGrath, Sr. M.L. Wright
Visiting Lecturers: 2

Degree Programs: Biology, Medical Technology

Undergraduate Degrees: B.A., B.S.
Undergraduate Degree Requirements: 120 credits total, 36 credits in Biology, 24 credits minor (e.g. in Chemistry, etc. (at least 2 years Chemistry), 1 year Physics 1 year Mathematics, foreign language - French or German

CURRY COLLEGE

Milton, Massachusetts 02186 Phone: (617) 333-0500
Dean of College: F. Kirschenmann

SCIENCE DIVISION
Phone: Chairman: Ext. 207 Others: Ext. 208

Division Chairman: J. Hovorka,
Professor: H.J. Evans
Assistant Professor: J.R. Holloway

Degree Program: Biology

Undergraduate Degree: B.S.
Undergraduate Degree Requirements: Applied Mathematics - 1 year, Introductory Physics - 1 year, Introductory Chemistry - 1 year, Organic Chemistry - 1 year, Introductory Biology - 1 year, Biology Seminar - 1/2 year, 20 additional semester hours of Biology

EASTERN NAZARENE COLLEGE

23 East Elm Phone: (617) 773-6350
Wollaston, Massachusetts 02170
Dean of College: D.L. Young

DEPARTMENT OF BIOLOGY
Phone: Chairman: Ext. 281 Others: Ext. 282, 285

Head: G.D. Keys
Professor: W.J.V. Babcock
Associate Professors: E. Leland, G.D. Keys

Field Stations/Laboratories: Adirondack Biological Field Station - Wilmington, New York

Degree Program: Biology

Undergraduate Degree: B.S., B.A.
Undergraduate Degree Requirements: 40 semester hours of Biology, 8 semester hours of Inorganic Chemistry, 8 semester hours of Organic Chemistry, 8 semester hours of Physics, 8 semester hours of Calculus

EMMANUEL COLLEGE

400 The Fenway Phone: (617) 277-9340
Boston, Massachusetts 02115
Dean of College: E. Binns

DEPARTMENT OF BIOLOGY
Phone: Chairman: Ext. 375 Others: Ext. 250, 251

Chairman: D.J. Procaccini
Professor: D.J. Procaccini
Assistant Professors: G.D. Crandall, C.E. Dinsmore, B. F. Weiss
Instructor: Sr. F. Donahue
Research Assistants: 6
Department Assistant: 1

Degree Program: Biology

Undergraduate Degree: B.S.
Undergraduate Degree Requirements: Minimum of 10 courses in Biology of which 6 are specified required courses, 4 courses in Chemistry, including 2 courses in General Chemistry and 2 courses in Organic Chemistry, 2 courses in Mathematics/Calculus, 2 courses in Physics

FRAMINGHAM STATE COLLEGE

Framingham, Massachusetts 01701 Phone: (617) 872-3501
Dean of Graduate Studies: C. Jordan
Dean of College: V.J. Mara

DEPARTMENT OF BIOLOGY
Phone: Chairman: Ext. 316

Chairman: D.N. Jost
Professors: D.N. Jost, W.L. Spence, J.P. Previte
Associate Professors: C.T. Roskey, T. Haight
Assistant Professors: P. Stanton, C. Smith, C. Cross, N. Tanner, V. Hodgson, R. LeBlanc

Field Stations/Laboratories: Wild as the Wind Farm - Upton, Massachusetts

Degree Programs: Biology, Medical Technology

Undergraduate Degree: B.S.
Undergraduate Degree Requirements: 32 courses including 12 in liberal arts, and 20 in major and related areas, Chemistry through Biochemistry, Physics, Mathematics, Genetics, Plant and Animal Sciences, Evolution, Ecology, For Medical Technology - one clinical year plus Chemistry and Genetics requirements

Graduate Degree: M.Ed.
Graduate Degree Requirements: Master: Successful teaching experience or practice teaching. 7 courses in subject area including research and/or seminar, 2 graduate courses in education.

Graduate and Undergraduate Courses Available for Graduate Credit: Biochemistry (U), Ichthyology (U), Histology (U), Non-Vascular Plants (U), Immunology (U), Biometrics (U), Plant Taxonomy (U), Ecology (U), Marine Biology Seminar (U), Animal Behaviour (U), Plants in Relation to Man (U), Plant Physiology, Ornithology, Cell Biology, Developmental Biology, Food Microbiology

GORDON COLLEGE

Wenham, Massachusetts 01984 Phone: (617) 927-2300
Dean of College: R. Gross

DEPARTMENT OF BIOLOGY
Phone: Chairman: Ext. 291 Others: Ext. 262, 263

Chairman: T. Dent
Professors: T.C. Dent, R.T. Wright
Assistant Professor: R.R. Camp

Degree Program: Biology

Undergraduate Degree: B.A.
Undergraduate Degree Requirements: 47 quarter courses, Zoology, Botany, Cell Biology, Population Biology, Genetics plus electives for a total of 9 to 13 courses. Introductory Principles, Organic Chemistry, Analytical Geometry, Calculus, Introductory Principles, The remaining course requirements are from the Divisions of Humanities and Social Science

HARVARD UNIVERSITY

Cambridge, Massachusetts 03128 Phone: (617) 495-1000
Dean of Graduate Studies: P.S. McKinney

Harvard College

Dean: H. Rosovsky

DEPARTMENT OF BIOCHEMISTRY AND MOLECULAR BIOLOGY
Phone: Chairman: 495-4795 Others: 495-4107

Chairman: W. Gilbert
Mallinckrodt Professor of Biochemistry: P. Doty
Higgins Professor of Biochemistry:
American Cancer Society Professor of Molecular Biology: W. Gilbert
Professors: K.E. Bloch, P. Doty, J.T. Edsall (Emeritus) W. Gilbert, G. Guidotti, M.S. Meselson, M. Ptashne, J.L. Strominger, J.D. Watson
Associate Professor: R. Losick
Assistant Professors: D. Dressler, S. Elgin, S.C. Harrison, L. Klotz, W. McClure, D.C. Wiley
Visiting Lecturer: 1
Teaching Fellows: 44

Degree Programs: Biochemistry, Molecular Biology

Graduate Degree: A.M., Ph.D.
Graduate Degree Requirements: Master: qualifying examination in Physical and Organic Chemistry, 8 courses at "graduate" or "graduate and undergraduate levels" Doctoral: Principally research in addition to M.S. requirements

Graduate and Undergraduate Courses Available for Graduate Credit: Not Stated

DEPARTMENT OF BIOLOGY
Phone: Chairman: 495-2327 Others: 495-2319, 2305

Chairman: J. Raper
Higgins Professor of Biology: G. Wald
Bussey Professor of Biology: C. Williams
Charles Bullard Professor of Forestry: M.H. Zimmermann, Director of Harvard Forest
Alexander Agassiz Professors of Zoology: A.W. Crompton, H.W. Levi, R.C. Lewontin, E. Mayr, R.C. Taylor, E. Williams
Arnold Professor of Botany and Professor of Dendrology: R.A. Howard
Henry Bryant Bigelow Professor of Ichthology: K.F. Liem
Asa Gray Professor of Systematic Botany: R.C. Rollins
Paul C. Mangelsdorf Professor of Natural Sciences: R.E. Schultes

Cellular and Developmental Biology

Professors: L. Bogorad, D. Branton, J.E. Dowling, J. W. Hastings, F.C. Kafatos, R.P. Levine, A.M. Pappenheimer, J.R. Raper, G. Wald, C.M. Williams
Associate Professors: B.I. Shapiro, T.C. Wegmann

Assistant Professors: M. Brenner, S.A. Counter, D.M. Gill, J.J. Goldthwaite, U. Goodenough, R.M. Losick, A.J. Tobin
Lecturers: C.F. Cleland, R.H. Hubbard, R.O.R. Kaempfer, M.Z. Nicoli, R.C. Ullrich

Organismic and Evolutionary Biology

Professors: E.S. Barghoorn, K.J. Boss, W.H. Bossert, A.W. Crompton, S.J. Gould, B. Holldobler, R.A. Howard, H.W. Levi, R. Lewontin, K.F. Liem, E. Mayr, R.M. Mitchell, R.C. Rollins, R.E. Schultes, O.T. Solbrig, F.E. Smith, C.R. Taylor, P.B. Tomlinson, J.G. Torrey, R.M. Tryon, E.E. Williams, E.O. Wilson, M.H. Zimmerman, C.E. Wood, Z. Kielan-Jaworowska
Associate Professors: J.M. Burns, J.E. Cohen, F.A. Jenkins, C.P. Lyman, T.W. Schoener
Assistant Professors: A.W. Loeblich, T.R. Roberts, R.F. Silberglied, R.L. Trivers, R.M. Woollacott
Lecturers: W.H. Drury, E.M. Gould, J.F. Lawrence, W.H. Lyford, R.A. Paynter, B.G. Schubert, R.D. Turner
Instructor: J.F. White
Associate Members: H.B. Fell, L.A. Garay, B. Kummel, B. Patterson, B.L. Schevill
Teaching Fellows: 59
Research Assistants: 8

Field Stations/Laboratories: Concord Field Station - Concord, Harvard Forest - Petersham, Museum of Comparative Zoology, Arnold's Arboretum, Harvard University Herbaria, Botanical Museum

Degree Programs: Anatomy and Physiology, Animal Behavior, Animal Communication, Animal Physiology, Bacteriology, Bacteriology and Immunology, Biochemistry and Biophysics, Biology (Cell and Developmental), Botany, Ecology, Entomology, Ethology, Forest Botany, Forest Resources, Forestry, Genetics, Histology, Land Resources, Limnology, Microbiology, Molecular Biophysics, Physiological Sciences, Physiology, Plant Pathology, Plant Sciences, Virology, Wildlife Management, Zoology, Zoology and Entomology, Marine Biology, Mathematical Biology, Population Biology, Population Genetics, Neurobiology, Mycology

Undergraduate Degree: B.A. (Biology)
Undergraduate Degree Requirements: Write to departmental office for particulars.

Graduate Degrees: M.A., Ph.D.
Graduate Degree Requirements: Master: 8 half-courses with grade of B- or better (ordinarily should be graduate courses, one of which must be "research course"); one language (foreign, commonly encountered in biological literature) Doctoral: 16 half-courses (grade of B- or better); two foreign languages (see pamphlet enclosed, pg. 5), doctoral thesis on original research.

Graduate and Undergraduate Courses Available for Graduate Credit: Biology of Invertebrates, Biology of the Verte-Brates, Introductory Biochemistry and Molecular Biology, Plants as Organism, Microbiology, Genetics, Cell Biology, Developmental Biology, Topics in Population Biology and Evolutionary Theory, Ecology, Structure and Physiology of the Vertebrates, Ethology, Neurobiology, Introduction to Research in Biology for Undergraduates, Introduction to Aging, The Taxonomy of Vascular Plants, Plants and Human Affairs, Plants of the Tropics, Evolution of Plant Life in Geologic Time, Structure and Physiology of Trees, Human Genetics, Application of Thermodynamics to Biological Systems, Biology of Marine Invertebrates, Biology of the Sea Floor, Comparative Analysis of Invertebrate Development, Neurocytology, Techniques in Neurophysiology, Biology of Insects, Biology and Evolution of Insect Orders, Biology of Fishes, Biology of Amphibians and Reptiles, Biology of Mollusks, Functional Vertebrate Morphology, Evolution of the Vertebrates, Biology of the Algae, Chromosomes, Social Behavior of Vertebrates, Biology of Birds, Plants in Relation to their Environment, Biogeography, Systematic Biology, Structure and Functioning of Plant Communities, Population Models, Po ing of Plant Communities, Population Models, Population Genetics, Predation and Competition in Animals, Environmental Physiology, The Physiology and Biochemistry of Plants, Plant Growth and Development, Plant Form and Structure, General Immunochemistry, General Immunochemistry, Cell Biochemistry and Physiology, Photobiology, Problems and Methods in Plant Systematics, Biochemical Systematics, Plants in the Development of Modern Medicine, Topics in the Evolution, Dispersal, and Paleoecology of Plants, The Phylogeny of the Flowering Plants, Sociobiology, Kinship and Social Behavior, Biology of Ferns, Principles of Evolutionary Biology

Medical School

Dean of Medical School: R.H. Ebert

DEPARTMENT OF ANATOMY
(No reply received)

DEPARTMENT OF BIOLOGICAL CHEMISTRY
(No information available)

COMMITTEE ON HIGHER DEGREES IN BIOPHYSICS
Phone: Chairman: 734-3300 Ext. 491

Chairman: A.K. Solomon
Professors: E.R. Blout, W.H. Bossert, J.E. Dowling, W. Gilbert, J. Woodland Hastings, W.N. Lipscomb, P.C. Martin
Lecturers: E.A. Dawidowicz, M.J. Poznansky, A. Essig, Y. Lange, H. Slayter
Research Fellows: 3

Degree Program: Biophysics

Graduate Degree: Ph.D.
Graduate Degree Requirements: Doctoral: 16 half course (1 semester each), language examination, thesis, qualifying examination and a thesis

Graduate and Undergraduate Courses Available for Graduate Credit: Membrane Biophysics, Electron Microscopy in Molecular Biophysics, Introduction to Laboratory Research, Digital Computer Applications in Biophysics Transport Processes in Biological Membranes, Nucleic Acid Function and the Mechanism of Protein Synthesis, Biological Macromolecules, Synthesis, Structure and Properties, Protein Composition and Structure in Enzyme Action, Electron Microscopy of Macromolecular Structure, Transport Processes in Model Systems and Biological Membranes

DEPARTMENT OF MICROBIOLOGY AND MOLECULAR GENETICS
Phone: Chairman: Ext. 677

Chairman: E.C.C. Lin
Professors: E.C.C. Lin, L. Gorini, J. Becicwith, H. Amos
Associate Professors: D. Fraenkel, A.S. Huang
Assistant Professors: J. Murphy, R. Goldstein
Teaching Fellows: 2
Research Associates: 1
Postdoctoral Fellows: 15

Degree Programs: Microbiology and Molecular Genetics, Virology, Cell Biology, Biochemistry

Graduate Degree: Ph.D.
Graduate Degree Requirements: Doctoral: Eight courses (1 semester each), Preliminary examinations, participation in journal clubs and teaching, doctoral thesis containing original work

Graduate and Undergraduate Courses Available for Graduate Credit: Not Stated

DEPARTMENT OF PHYSIOLOGY
Phone: Chairman: (617) 734-3300 Ext. 485

Chairman: T.H. Wilson
Robert Henry Pfeiffer Professor of Physiology: A.C. Barger
George Higginson Professor of Physiology: J.R. Pappenheimer
Professors: E. Henneman, T.H. Wilson
Associate Professors: 6
Assistant Professors: 2
Instructors: 1
Visiting Lecturers: 2
Teaching Fellows: 14
Research Assistants: 17

Degree Program: Physiology

Graduate Degree: Ph.D.
Graduate Degree Requirements: Doctoral: 3-year residency 1-year-full-time courses, Research leading to a thesis

Graduate and Undergraduate Courses Available for Graduate Credit: General Mammalian Physiology, Respiratory Physiology, Reproductive Biology, Membrane Transport, Circulatory Physiology

School of Public Health
Dean: H.H. Hiatt

DEPARTMENT OF BIOSTATISTICS
Phone: Chairman: (617) 734-3300 Ext. 531

Head: J. Worcester
Professors: J.J. Feldman, R.B. Reed
Associate Professors: Y.M.M. Bishop, M.E. Drolette, T.M. Frazier, O.S. Miettinen
Assistant Professors: J.C. Kleinman, J.H. Warram
Lecturer: E.W. Jones

Degree Program: Biostatistics

Graduate Degrees: M.S., Ph.D.
Graduate Degree Requirements: Master: A two-year program for applicants with a bachelor's degree, 80 credit units, M.P.H.: one-year program for candidates with a doctoral degree Doctoral: S.D.: Master's degree, and a thesis Ph.D. Doctoral degree, and a thesis

Graduate and Undergraduate Courses Available for Graduate Credit: Principles of Biostatistics, Statistical Methods in Research, Mathematical Foundations of Biostatistics, Design and Analysis of Epidemiologic Investigations: Applications, Topics in Epidemiologic Research, Survey Research Methods in Community Health, Introduction to Computing, Health Program Evaluation

DEPARTMENT OF MICROBIOLOGY
Phone: Chairman: (617) 783-0396 Others: 734-3300

Head: R.L. Nichols
Irene Heinz Given Professor of Microbiology: R.L. Nichols
Professors: E.S. Murray, J.C. Snyder
Associate Professor: J.W. Vinson
Assistant Professor: H.R. Buckley, J. Cerny, M.E. Essex, A.B. MacDonald, C.E. O. Fraser, F.Z. Modabber
Research Associates: J.E. Herrmann, L.V. Howard, Jr., N.S. Orenstein
Visiting Lecturers: 1
Research Assistants: 2

Field Stations/Laboratories: Dharan, Saudi Arabia

Degree Programs: Microbiology, Venerial Diseases Research, Chlamydial Research, Rickettsial Research, Mycological Research

Graduate Degrees: M.P.H., D.P.H., D.Sc.
Graduate Degree Requirements: Master: Demonstrated competence in microbiology and immunology, must understand the problems and opportunities in the control of infectious disease in developed and underdeveloped countries. A minimum of 4 approved courses remaining required credits taken as courses, tutorials or research, total 40 credits. Doctoral: Advanced courses in microbiology, immunology and related fields. At least one year in residence beyond completion of the Master's degree. Most of the training beyond the Master's is occupied by completion of a research project and preparation of a thesis. Applied aspects of research are emphasized (40 credits additional).

Graduate and Undergraduate Courses Available for Graduate Credit: Ecology and Epidemiology of Infectious Diseases; Current Research in Infectious Diseases, Clinical Problems in Infectious Diseases, Public Health and Laboratory Aspects of Infectious, Tuberculosis, Fundamentals of Immunology, Immunology of Infectious Diseases, Laboratory in Immunology, Medical Mycology, Intracellular Microorganisms Pathogenic for Man, Case Studies in Epidemiology of Infectious Disease, Problems in Medical Bacteriology, Sexually Transmitted Diseases, Tutorial Programs, Pathogenic Fungi, Rickettsiae, Chlamydia, Viruses, Susceptibility to Infectious Disease, Venereal Diseases, Disease Control in Underdeveloped Countries

PHYSIOLOGY DEPARTMENT
Phone: Chairman: 734-3300 Ext. 765 Others. 547

Chairman: J.L. Whittenberger
Professors: B.G. Ferris, Jr., J. Mead
Associate Professors: M.O. Amdur, J.D. Brain, D.E. Leith, J.B. Little, R.B. McGandy, S.D. Murphy, J.M. Peters
Assistant Professors: S.V. Sawson, F.G. Hoppin, Jr., R.J. Jaeger, R.L.H. Murphy, Jr., (Clinical), S.P. Sorokin, D.W. Underhill, D.H. Wegman, J.R. Williams
Visiting Lecturers: 2
Research Fellows: 12
Research Associates: 5

Degree Program: Physiology

Graduate Degrees: M.Sc., D.Sc.
Graduate Degree Requirements: Master: Appropriate baccalaureate degree: Biostatistics, Epidemiology, 2 year integrated curriculum in special field.
Doctoral: 2 prior years of appropriate graduate work, including Biostatistics and Epidemiology, research Dissertation

Graduage and Undergraduate Courses Available for Graduate Credit: Courses making up an integrated program include the following (it is not feasible to list titles): courses in Physiology Department, courses in other Departments of the School of Public Health, courses in Harvard Medical School, courses in other Harvard faculties and Massachusetts Institute of Technology.

LOWELL STATE COLLEGE
Lowell, Massachusetts 01854 Phone: (617) 454-8011
Dean of Graduate Studies: D. Procopio
Dean of College: P. Goler

DEPARTMENT OF BIOLOGY
Phone: Chairman: Ext. 267

Chairman: E.N. Kamien
Professor: E.N. Kamien
Assistant Professors: D. Eberiel, C. Hinckley, S. Lee, J. Lyon, T. Namm, P. Protopapas, P. Shepherd
Teaching Fellows: 2
Technicians: 2

Degree Program: Biology

Undergraduate Degrees: B.A., B.S.
Undergraduate Degree Requirements: 35-45 semester

hours in Biology, proficiency in Mathematics at level of Introductory Calculus, 2 years of Chemistry, 1 year of Physics

LOWELL TECHNOLOGICAL INSTITUTE
Lowell, Massachusetts 01854 Phone: (617) 454-7811
Dean of Graduate Studies: E.L. Alexander
Dean of College: J.I. Bruce

DEPARTMENT OF BIOLOGICAL SCIENCES
Phone: Chairman: Ext. 515

Chairman: R.M. Coleman
Professors: R.M. Coleman, J.I. Bruce
Associate Professors: T. Macdonald, N.J. Rencricca
Assistant Professors: I. Brunovskis, J.C. Mallett, R.D. Lynch
Visiting Lecturers: 1
Teaching Fellows: 2
Research Assistants: 1

Degree Program: Biological Science

Undergraduate Degree: B.S.
Undergraduate Degree Requirements: 21 credits Humanities, Senior thesis option, 1 semester Radioisotopic Techniques, 40 credits Biology, including Biochemistry, 16 credits Mathematics, 12 credits Physics, 12 credits Chemistry, 16 credits Technical subjects

Graduate Degree: M.S.
Graduate Degree Requirements: Master: Thesis option - 33 semester hours of which 9 may be thesis Project option - 33 semester hours of which 3 shall be project credit

Graduate and Undergraduate Courses Available for Graduate Credit: Endocrinology, Radiation Biology, Virology, Virology Laboratory, Immunobiology, Immunoparasitology, Physiological Ecology, Current Concepts in Biochemistry, Cell Biology, Electron Microscopy, Animal Nutrition, Biochemical Genetics, Graduate Seminar in Biology, Microbial Genetics, Tumor Virology, Erythropoiesis, Environmental Chemistry, Introduction to Bio-Medical Engineering, Biological Control Systems, Biopolymers, Environmental Radiation and Nuclear Site Criteria, Radioisotope Techniques, Environmental Toxicology and Epidemiology, Air Sampling and Analysis

MASSACHUSETTS INSTITUTE OF TECHNOLOGY
77 Massachusetts Avenue Phone: (617) 253-1000
Cambridge, Massachusetts 02139
Dean of Graduate School: I.W. Sizer
Dean of College: R.A. Alberty

DEPARTMENT OF BIOLOGY
(No reply received)

DEPARTMENT OF NUTRITION AND FOOD SCIENCE
Phone: Chairman: (617) 253-5101 Others: 253-3101

Head: N.S. Scrimshaw
Professors: A.L. Demain, S.A. Goldblith (Associate Department Head), J.E. Gordon (Senior Lecturer), M. Karel, R.S. Lees (Director, Arteriosclerosis Center), P.P. Lele, S.A. Miller, H.N. Munro, P.M. Newberne, N.S. Scrimshaw, J.B. Stanbury, S.R. Tannenbaum, D.I.C. Wang, G.N. Wogan, G. Wolf, R.J. Wurtman
Associate Professors: N. Catsimpoolas, E. Pollitt, W.M. Rand, C.K. Rha, R.C. Shank, A.J. Sinskey, V.R. Young
Assistant Professors: M.C. Archer, C.L. Cooney, J.D. Fernstrom, J.M. Flink, C.E. Kimble, F.J. Levinson, M.C. Linder, L.D. Lytle, W.G. Thilly
Visiting Lecturers: 6
Teaching Fellows: 2
Research Assistants: 36
Senior Research Scientist: 2
Visiting Scientist: 3
Postdoctoral Fellows: 6
Predoctoral Fellows: 65
Research Associates: 32

Degree Programs: Animal Pathology, Applied Biology, Food Science, Nutritional Biochemistry and Metabolism, Toxicology, Biochemical Engineering

Undergraduate Degrees: B.S. (Life Sciences)
Undergraduate Degree Requirements: 360 units, 1 year Calculus, 1 year Chemistry, 2 laboratories, 1 year Physics, 1 term Biochemistry Undergraduate Research

Graduate Degrees: A.M., Ph.D., Sc.D.
Graduate Degree Requirements: Master: Core curriculum 66 Units of Credit (42 units of Advanced Graduate Courses) Thesis Doctoral: Core Curriculum plus 54 Elective Units (24 of Minor) Satisfactory Passing of Written and Oral Examination Teaching assignment (1 term) Written research proposal, 3 Progress reports, Satisfactorily defended Thesis

Graduate and Undergraduate Courses Available for Graduate Credit: Laboratory in Applied Biology, Modelling and Analysis of Systems, Human Fertility and the Population, Mammalian Physiology, Food Science and Technology, Food Engineering, Advanced Food Engineering, Food Plant Visits, Physical and Engineering Properties of Food Materials, Physical and Engineering Properties of Food Materials Laboratory, Food Fabrication and Structure, Sensory Qualities of Foods, Physiological and Nutritional Biochemistry, Nutrition, Growth and Development, Human Nutrition, Field Studies of Human Nutrition, Ecology of Malnutrition, The Vitamins, Sleected Topics in Nutrition and Food Science, Industrial Microbiology, Ecology , Food Microbiology, Biochemical Engineering, Food Biochemistry, Advanced Food Technology, Experimental Toxicology and Pharmacology, Toxicology of Food-Borne Substances, Mutagenesis and Teratogenesis, Environmental Toxicology, Pharmacokinetics, Topics in Environmental Chemistry and Toxicology, Pathobiology, Food Marketing, Nutrition, National Development, and Planning, Nutrition Policy and Planning in Selected Countries, Biochemistry of the Neuron and the Synapse, Pharmacology of Behavior, Neuroendocrine Regulation, Clinical and Public Health, Topics in Engineering in Medicine, Human Genetics, Biostatistics, Mammalian Biochemistry and Metabolism, Mammalian Nutrition, Advanced Human Nutrition, The Human Nervous System

MERRIMACK COLLEGE
Turnpike Street Phone: (617) 683-7111
North Andover, Massachusetts 01845
Dean of College: Rev. J.A. Coughlan

DEPARTMENT OF BIOLOGY
Phone: Chairman: 683-7111 Ext. 204 Others: 305,306

Chairman: J.P. McLaughlin
Professor: G.W. Wermers
Associate Professors: J.L. Hart, J.P. McLaughlin, A. S. Thomas
Assistant Professors: M.H. Gregoire, J.R. Jungck, C. W. Kellogg, C.W. Owens
Instructor: J.R. Aprille

Degree Programs: Biology, Biology Education, Medical Technology, Biological Chemistry

Undergraduate Degrees: A.B., B.S.
Undergraduate Degree Requirements: Biology - 8 courses minimum (5 core, 3 selective), Chemistry 2, Physics 2

MOUNT HOLYOKE COLLEGE

South Hadley, Massachusetts 01075 Phone: (413) 538-2000
Dean of Graduate Studies: E.L. Wick
Dean of College: E.L. Wick

DEPARTMENT OF BIOLOGICAL SCIENCES
Phone: Chairman: (413) 538-2149

Chairman: K.M. Eschenberg
Professors: E.A. Beeman, K.M. Eschenberg, C.G. Smith, I.B. Sprague, J.C. Kaltenbach Townsend
Associate Professors: F.J. De Toma, M.Z. Pryor
Assistant Professors: A.E. Comer, P.J. Gruber, S.E. Gruber, T.C. Jones, R.B. Miller
Instructor: K. Holt
Graduate Teaching Assistants: 9
Laboratory Assistants: 2

Degree Programs: Biological Sciences, Biochemistry

Undergraduate Degree: B.A.
Undergraduate Degree Requirements: Biological Sciences-32, including: Genetics, Energetics, and Evolution, three of the following four courses: Genetics and Evolution, Developmental Biology, Organisms and Their Environments, Biochemistry and Biophysics, four advanced courses in Biology, Inorganic and Organic Chemistry Biochemistry - 44, including Fundamental Chemistry, Organic Chemistry I, II, Chemical Thermodynamics, Genetics, Energetics, and Evolution, Genetics and Evolution or Developmental Biology, Biochemistry and Biophysics, Biochemistry and Bioenergetics, three additional advanced courses in Chemistry, Biology, Physics, or Psychology

Graduate Degrees: M.A., M.A.T., Ph.D.
Graduate Degree Requirements: Master: 28, including thesis Doctoral: No definite requirements, this is earned under five-college cooperative program (Amherst, Hampshire, Mount Holyoke, Smith and the University of Massachusetts). Degree is awarded by University of Massachusetts; thesis research may be done at Mount Holyoke

Graduate and Undergraduate Courses Available for Graduate Credit: Histology of Animals (U), Vertebrate Embryology (U), Advanced Genetics (U), Molecular and Cellular Aspects of Development (U), Vertebrate Endocrinology (U), Invertebrate Zoology (U), Evolution of the Vertebrates (U), Biochemistry and Bioenergetics (U), Animal Behavior (U), Comparative Physiology (U), Physiology and Anatomy of Plants: Growth Regulation (U), Physiology and Anatomy of Plants: Assimilation and Transport of Nutrients (U), Evolution of the Plant Kingdom (U), Microbiology (U), Ecology (U), Neurophysiology (U), Electron Microscopy, Special Topics in Biology, Research

NORTH ADAMS STATE COLLEGE

North Adams, Massachusetts 01247 Phone: (413) 664-4511
Dean of College: J. Zaharis

DEPARTMENT OF BIOLOGY
Phone: Chairman: Ext. 247 Others: 664-4511

Chairman: D. MacKenzie
Associate Professor: D. MacKenzie, J. Smosky
Assistant Professors: P. Humora, F. Johns, R. MacMahon
Laboratory Instructor: 1

Degree Programs: Biology, Medical Technology

Undergraduate Degrees: B.A., B.S.
Undergraduate Degree Requirements: College Requirements-Communication skills 9, Social Sciences 9, Humanities 9, Physical Education and Life Sports 2, Biology Department Requirements: Mathematics through Calculus I, Introductory Physics, Chemistry 20 hours, Biology 45 hours, Electives 12 hours

NORTHEASTERN UNIVERSITY

360 Huntington Avenue Phone: (617) 437-2000
Boston, Massachusetts 02115
Dean of Graduate Studies: R. Ketchum
Dean of College: R. Shepard

BIOLOGY DEPARTMENT
Phone: Chairman: (617) 437-2259 Others: 437-2260

Chairman: F.D. Crisley
Professors: F.A. Barkley, F.D. Crisley, C. Gainor, A.K. Khudairi, N.W. Riser (Director, Marine Science Institute)
Associate Professors: C.H. Ellis, Jr. J.Z. Gabliks, C. A. Meszoely, M.P. Morse, J.V. Pearincott, F.A. Rosenberg, E. Ruber, B.I.E. Stuerckow, H.O. Werntz
Assistant Professors: H.D. Ahlberg, H.S. Bialy, H. Lambert, D.F. Levering, Jr., B. Raisbeck, S. Shukri, P.R Strauss
Instructors: S. Brecher, R.A. Cohen, P.A. Montagna
Visiting Lecturers: 7
Teaching Fellows: 33
Research Assistants: 5
Northeastern University Fellows: 6

Field Stations/Laboratories: Marine Science Institute - Nahant, Botanical Research Laboratory - Woburn

Degree Programs: Bacteriology and Immunology, Biochemistry, Biology, Ecology, Health and Biological Sciencies, Microbiology, Physiological Sciences, Physiology, Plant Sciences

Undergraduate Degrees: B.A., B.S. (Biology)
Undergraduate Degree Requirements: B.A.: 6 quarters of Biology Core: General, Animal, Plant, Environmental and Population, Genetics and Developmental, Cell Biology, Freshman Seminar, 4 Biology electives, 2-3 quarters Basic Mathematics or Calculus, 2 quarters Physics, 5 quarters Chemistry, Intermediate Foreign Language B.S.: Same Biology as for B.A., 3 quarters Calculus, 2 quarters Advanced Physics, 5 quarters Chemistry, 2 approved advanced Science electives, Intermediate Foreign Language, Senior Seminar

Graduate Degrees: M.S., M.S. in Health Science, Ph.D.
Graduate Degree Requirements: Master: 36 quarter course credit hours, 4 hours seminar, 6 thesis or special topics credits Doctoral: Departmental and qualifying examinations, course work as recommended by qualifying committee and adviser, dissertation, 2 languages or 1 language and research tool.

Graduate and Undergraduate Courses Available for Graduate Credit: Biometrics, Lower Invertebrates, Coelomate Invertebrates, Malacology, Vertebrate Zoology, Mammalogy, Ichthyology, Dynamics of Aquatic Ecology I, Dynamics of Aquatic Ecology II, Biological Factors in Ocean Engineering, Ecology of Salt Marshes, Principles of Systematics, Human Ecology, Plant Nutrition and Metabolism, Plant Growth and Reproduction, Physiology of Plant Growth and Development, Fossil Plants, Marine Algae, Photosynthesis, Environmental and Population Biology, Mammalian Physiology, Advanced Mammalian Physiology, Cardiovascular Physiology, Animal Nutrition, Comparative Physiology of Regulatory Mechanisms, Vertebrate Endocrinology, Procedures in Endocrinology, Physiological Ecology, Nuclear and Radiobiology, Advanced Developmental Biology, Insect Metabolism, Cell Biophysics and Biochemistry, Neurophysiology, Tropical Field Studies, Evolution, Computers in Biology, Environmental Microbiology, Marine Microbiology, Food Microbiology, Animal Virology, Microbial Genetics, Medical Mycology, Microbial Biochemistry, Seminar, Special Topics in Biology, M.S. Thesis, Special Investigations in Biology, Biological Electron Microscopy, Ph.D. Dissertation, Concepts in Pharmacology, Concepts in Toxicology, Industrial Hygiene, Air Pollution Science, Biochemistry, Mammalian Physiology I, Scientific and

Legal Interactions of Environmental Management, Comparative Vertebrate Anatomy (U), Developmental Anatomy (U), Invertebrate Zoology (U), Parasitology (U), Vertebrate Paleontology (U), General Microbiology (U), Animal Histology (U), Histological Technique (U), Lower Plants (U), Higher Plants (U), Systematic Botany (U), Plant Anatomy (U), Economic Botany (U), Introduction to Plant Physiology (U), Horticulture (U), Microbial Physiology (U), Medical Microbiology (U), Serology Immunology (U), Comparative Animal Physiology (U)

REGIS COLLEGE

Weston, Massachusetts 02193 Phone: (617) 893-1820
Dean of College: S. Williamson

DEPARTMENT OF BIOLOGY
Phone: Chairman: (617) 893-1820 Ext. 303

Chairman: Not Stated
Professor: Sr. C.A. Mulrennan
Associate Professors: Sr. M. Donovan, J. Barnabo
Assistant Professor: Sr. C. Angelli

Degree Program: Biology

Undergraduate Degree: A.B.
Undergraduate Degree Requirements: for Biology major: 16 courses in Natural Sciences, 16 courses in other areas

SALEM STATE COLLEGE

Salem, Massachusetts 01970 Phone: (617) 745-0556
Dean of Graduate Studies: V.L. Howes
Dean of College: M. LaHood

DEPARTMENT OF BIOLOGY
Phone: Chairman: Ext. 267 Others: 745-0556

Chairman: A.L. Borgatti
Professors: A.L. Borgatti, A.E. Harrises, C.M. Paine, T.I. Ryan, M.F. Sak, J.B. Schooley, E.F. Sweeney
Associate Professors: J.K. Moore, F.L. Sullivan
Assistant Professors: A.V. Shea, C.R. Terrell
Instructors: E.J. Cole, Jr.

Degree Program: Biology

Undergraduate Degree: B.A.
Undergraduate Degree Requirements: 30 semester hours of Biology - including Botany, Invertebrate Zoology, Genetics, General Physiology and Seminar, Chemistry Minor.

SIMMONS COLLEGE

300 The Fenway Phone: (617) 738-2191
Boston, Massachusetts 02115
Dean of College: C. Morocco

DEPARTMENT OF BIOLOGY
Phone: Chairman: (617) 738-2191 Others: 738-2193

Chairman: A.E. Coghlan
Professor: A.E. Coghlan
Associate Professors: M.D. Berliner, M.L. Sacks, E.L. Tuttle, E.A. Weiant
Assistant Professors: R.J. Bevmer, B.L. Bowman, N. S. Brown, M.S.Everett, R.P. Nickerson

Degree Program: Biology

Undergraduate Degree: B.S.
Undergraduate Degree Requirements: Language, English, Distribution Requirements, General Biology - 2 semesters, Invertebrate Biology, Chemistry and Biology of Cells, Developmental Biology, Genetics, plus Biology electives, 4 semesters of Chemistry, 1 semester of Calculus (28 semester hours of Biology)

SMITH COLLEGE

Northampton, Massachusetts 01060 Phone: (782) 584-2700
Dean of Graduate Studies: W. Schumann
Dean of College: A.B. Dickinson

THE BIOLOGICAL SCIENCES
Phone: Chairman: (782) 584-2700

Chairman: C.J. Burk
Professors: B.E. Horner, G.W. deVillafranca, C.J. Burk
Associate Professors: D.A. Haskell, E.A. Tyrrell
Assistant Professors: J.A. Powell, J.M. Greene, L. Luckenbill, S.G. Tilley, P.D. Reid, R. Merritt, M. A. Olivo, R.F. Olivo
Visiting Lecturers: 2
Teaching Fellows: 9
Research Assistants: 3

Degree Program: Biology

Undergraduate Degree: A.B.
Undergraduate Degree Requirements: Not Stated

Graduate Degrees: M.A., Ph.D.
Graduate Degree Requirements: Not Stated

Graduate and Undergraduate Courses Available for Graduate Credit: Advanced studies in the following fields: Molecular Biology, Botany, Microbiology, Zoology, and Environmental Biology. Also seminars, research, and thesis.

SOUTHEASTERN MASSACHUSETTS UNIVERSITY

North Dartmouth, Massachusetts 02747 Phone: (617) 997-9321
Dean of Graduate Studies: R. Fontera
Dean of College: J. Sauro

BIOLOGY DEPARTMENT
Phone: Chairman: Ext. 402 Others: Ext.401

Chairman: J.J. Reardon
Professors: J.G. Hoff, J.J. Reardon, N. Sasseville, E. Whitaker
Associate Professors: D.A. Cotter, R.K. Edgar, S.A. Moss
Assistant Professors: Y. Asato, R. Campbell, J.L Cox, F.R. Engelhardt, R. Ibara, P. MacDonald, B. Matsumoto, D. Mulcare, R. Nakamura, F. O'Brien, R. Szal
Graduate Assistants: 5

Degree Programs: Applied Ecology, Biological Control, Biology, Ecology, Fisheries and Wildlife Biology, Medical Technology, Microbiology, Oceanography, Parasitology

Undergraduate Degree: B.S. (Biology, Medical Technology)
Undergraduate Degree Requirements: Biology Core: 16 credits (Biology of Organisms, Biology of Cells, Biology of Populations, Biology electives - 18 credits of Physics, Required, 6 credits of Mathematics Required 6 credits Free electives - 20 credits. The required total of 120 credits include university requirements. A 2.0 average or better in major and overall requirements is necessary.

Graduate Degrees: M.S.
Graduate Degree Requirements: Master: A minimum of 30 semester hours must be completed in approved graduate courses. A thesis may, when approved by the department, be assigned graduate credit. A student must have a grade point average of at least 3.0 in all courses taken towards the degree, exclusive of the thesis.

Graduate and Undergraduate Courses Available for Graduate Credit: Proseminars: Aquaculture (U), Ecology (U), Environmental Health (U), Estuarine Ecology (U),

Estuarine Ecology (U), Water Pollution (U), Graduate Seminars: Marine Mammals, Mammalian Nutrition, Coastal Zone Processes, Entomology, Biological Control, Biology of Fishes (U), Comparative Physiology (U), Developmental Biology (U), Biology of Animal Parasites (U), Design of Experiments in Biology (U), Research Project (U), Biology of Sharks (U), Biological Oceanography, Thesis, Graduate Research Project

SPRINGFIELD COLLEGE

263 Alden Street Phone: (413) 787-2100
Springfield, Massachusetts 01109
Dean of Graduate Studies: E.W. Seymour
Dean of College: P.U. Congdon

BIOLOGY DEPARTMENT
Phone: Chairman: (413) 787-2375 Others: 787-2374

Chairman: J.R. Cohen
Professors: J.W. Brainerd, C.E. Keeney
Associate Professor: J.R. Cohen
Assistant Professors: R.C. Barkman, C.B. Redington, J.F. Ross
Visiting Lecturers: 3
Teaching Fellows: 8

Field Stations/Laboratories: McCabe Field Laboratory - campus.

Degree Programs: Biology, Medical Technology, Environmental Studies

Undergraduate Degrees: B.S., B.A.
Undergraduate Degree Requirements: All-College requirements plus: Minimum of 32 hours Biology, 16 hours Chemistry, 8 hours Physics, 6 hours Mathematics. Requirements for granting degree: Completion of 130 semester hours including Departmental requirements and an Academic Index in Biological and Physical Science courses of at least 2.5 (A=4.0)

STONEHILL COLLEGE

North Easton, Massachusetts 02356 Phone: (617) 238-2052
Dean of College: (Vacant)

DEPARTMENT OF BIOLOGY
Phone: Chairman: Ext. 229 Others: Exts. 203,204,202

Chairman: Rev. F.J. Hurley
Professors: F.J. Hurley
Associate Professor: P.R. Gastonguay
Assistant Professors: S.L. McAlister, P. Paolella
Instructors: D.H. Grindle

Degree Program: Biology, Medical Technology

Undergraduate Degree: B.S.
Undergraduate Degree Requirements: 40 courses (minimum of 14 in major requirement)

SUFFOLK UNIVERSITY

41 Temple Street Phone: (617) 723-4700
Boston, Massachusetts 02114
Dean of College of Liberal Arts and Science: M.Ronayne

DEPARTMENT OF BIOLOGY
Phone: Chairman: Ext. 393 Others: Ext. 391, 397

Chairman: B.L. Snow
Professors: P.F. Mulvey, Jr., B.L. Snow, A.J. West II
Associate Professor: H.C. Lamont
Assistant Professors: J. Fiore, E.H. Romach
Instructor: G. Gillis

Field Stations/Laboratories: The Robert S. Friedman Cobscook Bay Laboratory - Edmunds, Maine

Degree Programs: Life Studies, Medical Technology

Undergraduate Degrees: B.S., A.B.
Undergraduate Degree Requirements: 30 hours in Biology, minimum 16 hours in Chemistry, 6 hours Calculus, 8 hours Physics, 12 hours English, Humanities and Social Science Requirements differ with degree, 12 hours foreign language (for A.B). Total minimum hours are 122 for both degrees.

TUFTS UNIVERSITY

Medford, Massachusetts 02155 Phone: (617) 628-5000
Dean of Graduate Studies: C.G. Nelson

College of Liberal Arts

Dean: G. Mumford

DEPARTMENT OF BIOLOGY
Phone: Chairman: Ext. 422

Chairman: E.S. Hodgson
Professors: R.L. Carpenter (Emeritus), E.S. Hodgson, N.S. Milburn, K.D. Roeder, H.R. Sweet (Emeritus), B.M. Twarog
Associate Professors: B. Dane, M.E. Feinleib, N.H. Nickerson, G.L. Sames, E.C. Siegel, S.A. Slapikoff
Assistant Professors: N.B. Hecht, J. Kimball, E. Maly, J.F. Schaeffer
Postdoctoral Fellows: 2

Degree Program: Biology

Undergraduate Degrees: B.A., B.S.
Undergraduate Degree Requirements: Ten courses including eight courses in Biology and two courses in either Chemistry or Physics.

Graduate Degrees: M.S., Ph.D.
Graduate Degree Requirements: Master: Eight graduate level courses of which four must be in Biology. Reading knowledge of one foreign language. Comprehensive Examination. Doctoral: Research Thesis, presentation of four research proposals and an oral examination, knowledge of two foreign or one foreign and one computer language, variable courses.

Graduate and Undergraduate Courses Available for Graduate Credit: Immunology (U), Biochemistry (U), Physiology (U), Plant Physiology (U), Cell Biology (U), Cellular Neurophysiology (U), Animal Behavior (U), Microbial Genetics (U), Developmental Genetics (U), Ecology (U), Membrane Physiology (U), Physiological Foundation of Animal Behavior (U), Selected Topics in Biochemistry (U), Topics in Developmental Genetics (U)

School of Medicine

136 Harrison Avenue Phone: (617) 423-4600
Boston, Massachusetts 02111
Dean: L.F. Cavazos (Acting)

DEPARTMENT OF ANATOMY
Phone: Chairman: Ext. 241

Chairman: W.D. Belt
Professors: W.D. Belt, L.F. Cavazos
Associate Professors: J. Frommer, S. Jacobson, J. Morehead
Assistant Professors: T. Beringer, J. Borysenko, M. Borysenko, J. Brawer, K. O'Hare, J. Zimmerman, P. Janfaza (Part-time), F. Merk (Part-time)
Instructor: M. Miller (part-time)
Visiting Lecturers: 1
Teaching Fellows: 4
Research Assistants: 4

Degree Programs: Anatomy, Histology, Immunology

Graduate Degree: Ph.D.
Graduate Degree Requirements: Doctoral: Appropriate

coursework, foreign language (2), Preliminary Examination, both written and oral, research, dissertation and dissertation examination

Graduate and Undergraduate Courses Available for Graduate Credit: Gross Anatomy, Applied Anatomy of the Head and Neck, Histological Techniques, Techniques for Electron Microscopy, Histo and Cytochemistry, Fine Structure of Cells and Tissues, Neuroscience, Neurocytology, Somatic Cell Genetics, Graduate Seminar, Special Topics, Thesis, Graduate Research

DEPARTMENT OF BIOCHEMISTRY AND PHARMACOLOGY
Phone: Chairman: Ext. 450

Chairman: H.G. Mautner
Professors: G. Brawerman, M.A. Cynkin, R.L. Kisliuk, N.I.Krinsky, H.G. Mautner, W.S. McNutt, G. Schmidt (Emeritus), L. Shuster
Associate Professors: R.E. Cathou, J.W. Drysdale, M. Randic, B.D. Stollar
Assistant Professors: S. Husain, Y.-H. Kim, N.R. Langerman
Research Assistants: 10

Degree Programs: Biochemistry, Pharmacology (Neurochemistry)

Graduate Degrees: M.S., Ph.D.
Graduate Degree Requirements: Master: A program of research and courses. Doctoral: A program of research and courses.

Graduate and Undergraduate Courses Available for Graduate Credit: Structural and Physical Chemistry of Biologically Important Molecules, Biosynthesis of Macromolecules, Metabolic Pathways and their Regulation, Energy Flow in Biological Systems, Biochemical Techniques, Immunochemistry, Bio-Organic Mechanisms and Enzyme Action, Biosynthesis of Polysaccharides and Glycoproteins, Biochemical Basis for Chemotherapy, Neurochemistry, Membranes, Molecular Basis of Drug Action, Thermodynamics

DEPARTMENT OF MOLECULAR BIOLOGY AND MICROBIOLOGY
Phone: Chairman: 423-4600 Ext. 280 Others: Ext. 281

Chairman: M. Schaechter
American Cancer Society Professor of Molecular Biology: V. Najjar
Professors: V. Najjar, J.T. Park, M. Schaechter
Associate Professors: E.B. Goldberg, M.H. Malamy, A. Wright
Assistant Professors: A. Levy, A.L. Sonenshein
Teaching Fellows: 1
Research Assistants: 15

Degree Program: Molecular Biology

Graduate Degree: Ph.D.
Graduate Degree Requirements: Doctoral: Proficiency in Biochemistry, Physical Biochemistry, Immunology, Molecular Genetics, Thesis.

Graduate and Undergraduate Courses Available for Graduate Credit: Tutorial in Basic Molecular Biology, Experiments in Molecular Biology, The Bacterial Cell Surface, The Temperate Bacteriophages, The Virulent Bacteriophages, Molecular Biology of Episomes and Plasmids, Differentiation in Bacteria, The Regulation of Bacterial Growth, Bacterial Protein Synthesis and its Control

DEPARTMENT OF PHYSIOLOGY
(No reply received)

UNIVERSITY OF MASSACHUSETTS
Amherst, Massachusetts 01002 Phone: (413) 545-2179
Dean of Graduate Studies: E. Piedmont

College of Arts and Sciences
Dean: M.J. Edds, Jr.

DEPARTMENT OF BOTANY
Phone: Chairman: (413) 545-2235 Others: 545-0111

Chairman: S. Shapiro
Torrey Professor of Botany: A.C. Smith
Professors: D.W. Bierhorst, H.E. Bigelow, E.L. Davis, R.B. Livingston, J.A. Lockhart, R.M. Schuster, A. Smith, S. Shapiro, O.L. Stein, C.P. Swanson, O. Tippo
Associate Professors: E.J. Klelowski, Jr., D.L. Mulcahy, A.I. Stern, R.T. Wilce
Assistant Professors: P. Barrett, M.E. Bigelow, S.A. Bultz, P.J. Godfrey, L. Raudzens, B. Rubinstein, W.T. Thompson, J.W. Walker, P. Webster
Visiting Lecturers: 1

Field Stations/Laboratories: Marine Biology - Marlborough

Degree Program: Botany

Undergraduate Degrees: B.A., B.S.
Undergraduate Degree Requirements: 24 credits in Botany, with at least 6 credits from each of the areas: I. Ecology-Evolution, Physiology-Cytology-Genetics and Anatomy-Morphology-Systematics. Supplemental science to include Chemistry, Organic Chemistry, and Biochemistry, year Physics, mathematics, Calculus and Statistics, Competence in German, French, or Russian recommended.

Graduate Degrees: M.A., M.S., Ph.D.
Graduate Degree Requirements: Master: 30 graduate credits, with at least 21 in Botany, For degree with thesis, 6 credits open to graduate students only. For degree without thesis, 12 credits in courses for graduate students only. Competence in one foreign language. Doctoral: One year residence and completion of approved dissertation. Competence in 2 languages (foreign) required, or may be allowed one language and tool or research competence in Computer Science.

Graduate and Undergraduate Courses Available for Graduate Credit: Introduction to Plant Physiology (U), Plant Metabolism (U), Plant Growth (U), Ecological Plant Physiology (U), Plant Ecology (U), Autecology (U), Plant Geography (U), Principles of Evolution (U), General Mycology (U), Phycology (U), Archegoniates (U), Experimental Pteridology (U), Cytogenetics (U), Origin, Evolution and Distribution of Flowering Plants (U), Plant Anatomy and Histological Methods (U), Morphogenesis (U), Plant Morphogenesis (U), Plant Morphogenesis (U), Cytology (U), Special Problems (U), Advanced Plant Physiology, Plant Growth and Regulators, Advanced Plant Ecology, Advanced Phycology, Advanced Phycology, Plant Photosynthesis, Biology of Lower Plants, Advanced Angiosperm Systematics

DEPARTMENT OF BIOCHEMISTRY
Phone: Chairman: (413) 545-2179

Chairman: R.C. Fuller
Professors: R.C. Fuller, H.N. Little, E.W. Westhead
Associate Professors: A.M. Gawienowski, J. Nordin, T. Robinson
Assistant Professors: M. Fischer, M.J. Fournier, T.L. Mason, D. Schneider, L. Slakey, R. Zimmermann
Visiting Lecturers: 1
Teaching Fellows: 7
Research Assistants: 8

Field Stations/Laboratories: Marine Station - Gloucester

Degree Program: Biochemistry

Undergraduate Degrees: B.A., B.S.
Undergraduate Degree Requirements: B.S. General Chemistry (1 year), Organic (1 year) 3 semesters of Calculus, 1 year Physics, 1 year Physical Chemistry, 1 year General Biochemistry, 3 semesters of Biology, 2 Advanced Courses in Biochemistry, Research Option. B.A. 2 semesters Mathematics only, 1 less advanced course.

Graduate Degrees: M.S., Ph.D.
Graduate Degree Requirements: Master: 1 year advanced level Biochemistry with Project Laboratory, 2 semesters of Advanced level Courses in Biochemistry or Related Sciences, Seminar, Research or Library Thesis Doctoral: 1 year Advanced level Biochemistry with Project Laboratory, 1 semester Enzymology, 3 semesters Advanced Level Biochemistry or Allied Courses, Written or Oral Qualifying Examination, 1 semester as Teaching Assistant

Graduate and Undergraduate Courses Available for Graduate Credit: General Biochemistry (U), Enzymes, Plant Biochemistry, Lipid Biochemistry, Biochemistry of Hormones, Nucleic Acid Biochemistry, Structure and Function of Membranes, Protein Physical Chemistry, Protein Synthesis, Biochemistry of Carbohydrates, Specialized Seminars and Journal Club.

DEPARTMENT OF MICROBIOLOGY
Phone: Chairman: 545-1353 Others: 545-2051

Professors: E.Canale-Parola, C.D. Cox, R.P. Mortlock, C.B. Thorne
Associate Professors: C.E. Dowell, S.C. Holt, T.G. Lessie, M.S. Wilder
Assistant Professors: L.C. Norkin, A.M. Reiner
Visiting Lecturers: 10
Teaching Fellows: 9
Research Assistants: 6

Degree Program: Microbiology

Undergraduate Degrees: B.S.
Undergraduate Degree Requirements: General Departmental Requirements

Graduate Degrees: M.S., Ph.D.
Graduate Degree Requirements: Master: General Departmental Requirements Doctoral: General Departmental Requirements

Graduate and Undergraduate Courses Available for Graduate Credit: General Microbiology (U), Microbial Diversity (U), Pathogenic Bacteriology (U), Microbial Genetics (U), Biology of Microorganisms

DEPARTMENT OF ZOOLOGY
Phone: Chairman: (413) 545-2603 Others: 545-2618

Chairman: H. Rauch
Professors: L.M. Bartlett, D. Fairbairn (Commonwealth), B.M Honigberg, J.G. Moner, W.B. Nutting, J.L. Roberts, H.D. Rollason, J.G. Snedecor, A.M. Stuart, H. Rauch
Associate Professors: T.J. Andrews, D.C. Edwards, M.S. Kaulenas, D.J. Klingener, B.R. Levin, S.D. Ludlam, A.P. Mange, W.B. O'Connor, H.E. Postwald, T.D. Sargent, F.M. Scudo, D.P. Synder, G.A. Wyse, L.S. Roberts
Assistant Professors: M.C. Coombs, G.H. Dersham, J. G. Kunkel, D.M. Noden, M.A. Novak, G.S. Rollason, D.G. Searcy, B.J. White, C.L.F. Woodcock, J.R. Ziegler
Teaching Fellows: 40
Research Assistants: 8

Degree Programs: Cell Biology, Developmental Biology, Population Biology, Ecology, Morphology and Systematics, Genetics, Physiology, Paleontology, Neurobiology and Behavior

Undergraduate Degrees: B.S. (Zoology)
Undergraduate Degree Requirements: Botany, Inorganic and Organic Chemistry, Biochemistry, Calculus or Statistics, Physics, Genetics, Cellular Physiology, Vertebrate Morphology, Physiology, Invertebrate Zoology, Ecology

Graduate Degrees: M S., Ph.D.
Graduate Degree Requirements: Master: 30 credits including special problem or thesis, one foreign language Doctoral: Course requirements as determined by program adviser; dissertation and 2 foreign languages

Graduate and Undergraduate Courses Available for Graduate Credit: General Cytology, Electrom Microscopy, Fine Structure and Function of Cells, Experimental Embryology, Advanced Developmental Biology, Physiological Genetics, Advanced Invertebrate Zoology, Metazoan Symbiosis, Selected Topics in Animal Behavior, Biology of Animal Populations, Systematics and Evolutionary Mechanisms, Population and Community Ecology, Comparative Neurophysiology, Physiological Regulatory Mechanisms, Endocrinology, Writing for Life Sciences, Comparative Vertebrate Anatomy (U), Histology (U), Embryology (U), Systems of the Human Body (U), Principles of Genetics (U), Introduction to Population Biology (U), Population Genetics (U), Biology of Protozoa (U), Biology of Lower Invertebrates (U), Biology of Higher Invertebrates (U), General Parasitology (U), Vertebrate Zoology (U), Ichthyology (U), Ornithology (U), Mammalogy (U), Limnology (U), Animal Behavior (U), Cell Physiology (U), Vertebrate Physiology (U), Comparative Physiology (U), Developmental Biology (U)

College of Agriculture
Dean: A.A. Spielman

DEPARTMENT OF ENTOMOLOGY
Phone: Chairman: (413) 545-2285 Others: 545-2283

Head: T.M. Peters
Professors: W.B. Becker, J.F. Hanson, J.H. Lilly
Associate Professors: T.M. Peters
Assistant Professors: P. Barbosa, H.H. Hagedorn, D. W. Hall, G.L. Jensen, J.G. Stoffolano
Teaching Fellows: 3
Research Assistants: 9

Field Stations/Laboratories: Suburban Experimental Station - Waltham, Cranberry Field Station - Wareham

Degree Program: Entomology

Undergraduate Degree: B.S.
Undergraduate Degree Requirements: Animal Physiology, Genetics, Chemistry, Entomology - 5 courses (15 credits)

Graduate Degrees: M.S., Ph.D.
Graduate Degree Requirements: Master: Insect Physiology, Taxonomy, Ecology, Insect Control, Biochemistry, Seminar (every semester) Doctoral: Same as Master plus courses assigned by Guidance Committee

Graduate and Undergraduate Courses Available for Graduate Credit: Insect Societies (U), Forest and Shade Tree Insects (U), Animal Ecology (U), Evolution (U), Insect Behavior (U), Taxonomy of Insects (U), Insect Morphology (U), Medical and Veterinary Entomology (U), Insect Control (U), Insect Physiology (U), Insect Ecology (U), General Entomology (U), Subject-Matter Seminars

DEPARTMENT OF PLANT AND SOIL SCIENCES
Phone: Chairman: (413) 545-2243 Others: 545-0111

Head: F.W. Southwick
Professors: A.W. Boicourt, M. Drake, J.R. Havis, W.

H. Lachman, W.J. Lord, D.N. Maynard, C.L. Thomson, J. Troll, J. Vengris, M.E. Weeks, J.M. Zak
Associate Professors: J.H. Baker, A.V. Barker, W.J. G.B. Goddard, H.V. Marsh, Jr., W.A. Rosenau, G.L. Stewart
Assistant Professors: J.F. Anderson, R.N. Carrow, E.R. Emino, D.W. Greene, P.H. Jennings
Visiting Lecturers: 1
Teaching Fellows: 3
Research Assistants: 15

Field Stations/Laboratories: Horticultural Research Center - Belchertown

Degree Programs: Plant Industry, Plant Sciences, Soil Science

Undergraduate Degree: B.S.
Undergraduate Degree Requirements: 120 Graduation Credits, 2.0 QPA, Communications 6 credits, Humanities 9 credits, Social Sciences 9 credits, Mathematics and Natural Sciences 32 credits, Plant and Soil Sciences 25 credits, Biological Sciences 12 credits, Open Electives - 12 credits, Physical Education 2 semesters

Graduate Degrees: M.S., Ph.D.
Graduate Degree Requirements: 30 graduate credits. If a thesis is offered, at least 6 credits must be earned in 700-900 series courses, if a thesis is not offered at least 12 credits must be earned in 700-900 series courses. No more than 10 credits may be earned by means of a thesis. Doctoral: Preparation of a dissertation satisfactory to Dissertation Committee, successful completion of graduate courses designated by Guidance Committee, pass a Preliminary Comprehensive examination, pass a final at least partly oral examination conducted by the Dissertation Committee, satisfy residence requirement of one academic year full-time graduate work at the University.

Graduate and Undergraduate Courses Available for Graduate Credit: Plant Nutrition, Taxonomy of Economic Crops, Plant Breeding, Post-Harvest Physiology, Forage and Field Crops, Agrostology, Ecology and Control of Weeds, Soil Formation and Classification, Soil Physics, Soil Chemistry, Soil-Plant Mineral Nutrition, Microbiology of the Soil, Morphology of Economic Plants, Plant Growth Regulators, Plant-Water Relationships, Advanced Soil Chemistry, Microbiology Ecology of the Soil, Plant Photosynthesis, Nitrogen Metabolism

DEPARTMENT OF PLANT PATHOLOGY
Phone: Chairman: (413) 545-2280 Others: 545-0111

Head: R.A. Rohde
Professors: C.J. Gilgut, F.W. Holmes, R.A. Rohde
Associate Professor: G.N. Agrios
Assistant Professors: M.S. Mount, T.A. Tattar
Research Assistants: 9

Field Stations/Laboratories: Shade Tree Laboratory - Amherst, Cranberry Experiment Station - East Wareham, Suburban Experiment Station - Waltham

Degree Program: Plant Pathology

Undergraduate Degree: B.S.
Undergraduate Degree Requirements: 120 credits of which 45 must be in the major department.

Graduate Degrees: M.S., Ph.D.
Graduate Degree Requirements: Master: 30 credits of which 10 may be obtained as research. Usually include a thesis, but course work may be substituted with permission of advisor and Graduate Studies Committee. 6 credits in the 700 or above series of course and thesis or 12 credits as above without thesis Doctoral: Must have fulfilled the usual requirements for a bachelor's degree in a related discipline. Must have at least 12 credits in the 700-800 series, must pass Comprehensive Examinations, present and defend dissertation. No set number of credits.

Graduate and Undergraduate Courses Available for Graduate Credit: Plant Pathology, Forest and Shade Tree Pathology, Genetics of Plant-Pathogen Interaction, Plant Virology, Nematology, Biological Transmission of Plant Diseases

DEPARTMENT OF VETERINARY AND ANIMAL SCIENCE
Phone: Chairman: (413) 545-2312 Others: 545-0111

Head: T.W. Fox
Professors: D.L. Anderson, D.L. Black, W.G. Black, B. Colby, R. Damon, Jr., S. Gaunt, R.M. Grover, W.K. Harris, M. Sevoian, R.E. Smith, J.R. Smyth, Jr., G. Snoeyenbos, D. Stern
Associate Professors: A. Borton, G. Howe, H. Fenner, S. Lyford, C. Smyser, O. Weinack
Assistant Professors: R. Duby, J. Marcum, I. Reynolds
Instructors: E. Donohue, L, Jaskiel-Hamilton, B. Mitchell
Teaching Fellows: 2
Research Assistants: 10

Degree Program: Animal Science

Undergraduate Degree: B.S.
Undergraduate Degree Requirements: 120 credits (semester program), University requirements, Departmental requirements

Graduate Degrees: M.S., Ph.D.
Graduate Degree Requirements: Master: 30 semester credits, Thesis research Doctoral: Variable course requirements based on educational background. Thesis research

Graduate and Undergraduate Courses Available for Graduate Credit: Comparative Animal Genetics, Avian Genetics, Genetics of Productive Traits in Poultry, Quantitative Inheritance and Selection, Advanced Animal Genetics, Introductory Animal Physiology, Animal Physiology, Physiology of Reproduction, Advanced Avian Physiology, Mammalian Reproduction, Vertility and Fecundity, Principles of Nutrition, Advanced Avian Nutrition, Ruminant Nutrition, Animal Pathology, Diagnostic Laboratory Techniques, Mammalian Diseases, Avian Diseases, Histopathology, Intermediate Biometry, Advanced Biometry

UNIVERSITY OF MASSACHUSETTS - BOSTON
Boston, Massachusetts 02125 Phone: (617) 287-1900
Dean of College: R. Spaethling (Acting)

DEPARTMENT OF BIOLOGY
Phone: Chairman: Ext. 2611

Chairman: H. Lipke
Professors: H. Lipke, L. Kaplan, N. Weaver, W. Rosen
Associate Professors: R.R. Bennett, J.A. Freeberg, B. H. Harrison, M.R. Matteo, F.M. Safwat, J.H. Schultz, E. Seaman, C.A. VanUmmerson, R.H. White, H.G. Wilkes
Assistant Professors: S.W. Bradford, E.A. Davis, J.J. Hatch, C.J. Kibel, S.G. Krane, G.M. Langford, L.L. Larison, B.A. Menge, D.J. Policansky, M.A. Rex, W. Tiffney
Instructors: R. Stone
Visiting Lecturers: 2-3/year for short courses and seminars
Research Assistants: 2 (Postdoctoral Fellows)

Field Stations/Laboratories: Nantucket Field Station - Gloucester Field Station, (Joint program with Waltham Field Station)

Degree Program: Biology

Undergraduate Degrees: B.A.
Undergraduate Degree Requirements: 1 year course in Introductory Biology, 1 year Introductory Chemistry for science majors with laboratory, 1 year Mathematics, preferably Physics with laboratory, 1 year Mathematics, preferably Calculus, 2 - one year courses in advanced Biology with laboratory, 8 or more additional credits advanced work from: Advanced Biology or Organic Chemistry

Graduate Degree: M.A.
Graduate Degree Requirements: Master: Biology of Cities, Ethnobotany, Archeological Botany, Evolution-Domestic, Taxonomy-Seed Plants, Histology of Marine Invertebrates, Advanced Biology, Biogeography, Developmental Biology, Seminar - Seurobiology, Tissue and Organ Culture

Graduate and Undergraduate Courses Available for Graduate Credit: Biology of Cities, Ethnobotany, Archeological Botany, Evolution-Domestic, Taxonomy-Seed Plants, Advanced Biology, Biogeography, Developmental Biology, Seminar-Neurobiology, Tissue and Organ Culture

WELLESLEY COLLEGE

Wellesley, Massachusetts 02181 Phone: (617) 235-0320
Dean of College: A.S. Ilchman

DEPARTMENT OF BIOLOGICAL SCIENCES
Phone: Chairman: Ext. 631 Others: 632,633

Chairman: V.M. Fiske
Ellen A. Kendall Professor of Biological Sciences: V.M. Fiske
Professors: V.M. Fiske, D.J. Widmayer
Associate Professors: M.D. Coyne, M. Schweber
Assistant Professors: M.M. Allen, S. Banerjee, D. Busch, D.R. Dobbins, N. Machtiger, N.M. Rubenstein, G. Sanford, K. van der Laan
Research Assistants: 2
Laboratory Assistants: 8

Degree Programs: Biological Sciences

Undergraduate Degree: B.A.
Undergraduate Degree Requirements: I. 8 semester courses in Biology of which a) 4 must be laboratory courses, b) 2 must be at the advanced level, c) 2 must be broad in scope at the introductory level, d) 1 must be a second level course in genetics, 3) 1 must be a second level course in cell physiology; II. A minimum of 2 units in chemistry (most majors, 4); III. Reading knowledge of foreign language; IV. Distribution which includes a minimum of 2 courses in the arts and 2 in the social sciences; V. A minimum of 4 advanced courses III, IV, and V are college wide requirements

Graduate Degree: M.A.
Graduate Degree Requirements: Master: 8 units of work at an advanced level in Biology beyond a B.A. degree of which 2 may be in a related field, example Chemistry

Graduate and Undergraduate Courses Available for Graduate Credit: Animal Physiology (U), Plant Physiology (U), Histology-Cytology II: Structure of Organ Systems (U), Seminar. Genetics (U), Embryology (U), Advanced Ecology (U), Plant Morphogenesis (U), Seminar. Endocrinology (U), Microbial Physiology and Cytology (U), Seminar. Topics in Microbiology (U), Terrestrial Vertebrate Zoology (U), Advanced Cytology: Biological Ultrastructure (U), Biochemistry II (U), Seminar, Marine Biology (U), Seminar. Biological Basis of Animal Behavior (U), Research or Individual Study (U), Honors Program (U), Selected Topics in Biochemistry (U)

WESTFIELD STATE COLLEGE

Westfield, Massachusetts 01085 Phone: (413) 568-3311
Dean of Graduate Studies: H. Becker
Dean of College: P. Marrotte

BIOLOGY DEPARTMENT
Phone: Chairman: Ext. 385 Others: Ext. 318

Chairman: J.K. Taylor
Associate Professors: B.H. Harris, I.J. Lepow, J.K. Taylor, S. Majumder
Assistant Professors: G. Chrisanthopolous, A.D. Driscoll, J. Eberlin, D. Lovejoy, D. Pierce, J. Phillips
Instructor: F.W. Bates

Degree Program: Biology

Undergraduate Degree: B.S.
Undergraduate Degree Requirements: 40 hours Biology, 40 hours Electives, 40 hours Core

WHEATON COLLEGE

Norton, Massachusetts 02766 Phone: (617) 285-7722
Dean of College: B.S. Shapiro

BIOLOGY DEPARTMENT
Phone: Chairman: Ext. 411, 415

Chairman: S.L. Beck
Professors: S.L. Beck, E.L. White
Associate Professor: J.C. Kricher
Assistant Professors: G.H. Davidonis, E.Y. Tong, J.C. Urey
Research Assistants: 6 (Undergraduate Students)

Degree Programs: Biology, Biochemistry, Psychobiology

Undergraduate Degree: B.A.
Undergraduate Degree Requirements: Not Stated

WILLIAMS COLLEGE

Williamstown, Massachusetts 01267 Phone: (413) 597-3131
Dean of College: D.W.R. Bahlman

THOMPSON BIOLOGY LABORATORY
Phone: Chairman: (413) 597-2461 Others: 597-3131

Chairman: W.C. Grant, Jr.
Samuel Fessenden Clarke Professor of Biology: W.C. Grant, Jr.
Professor: W.C. Grant, Jr.
Associate Professors: W. DeWitt, G.L. Vankin
Assistant Professors: H.W. Art, L.C. Drickamer, T.L. Koppenheffer, J. Rosinski, A.E.R. Woodcock
Instructors: E.R. Brown, C.A. Paul
Teaching Fellows: 1

Field Stations/Laboratories: Hopkins Memorial Forest - Williamstown

Degree Programs: Biology, Botany, Zoology

Undergraduate Degree: B.S.
Undergraduate Degree Requirements: Two semesters of Introductory Biology, one semester senior tutorial and elective courses to a total of nine

Graduate Degree: M.S.
Graduate Degree Requirements: Master: Eight (8) course credits at the graduate level, four (4) of which may be given in the area of the thesis. Thesis required.

Graduate and Undergraduate Courses Available for Graduate Credit: Cellular Ultrastructure (U), Advanced Physiology (U), Endocrinology (U), Methods in Ecological Research (U), Immunology (U), Topics in Advanced Biology (U)

WORCESTER FOUNDATION FOR EXPERIMENTAL BIOLOGY

Shrewsbury, Massachusetts 01545 Phone: (617) 842-8921

(NO DEPARTMENTS)

The foundation is a basic biomedical research institute of 250 persons, 85 with the doctorate, training primarily postdoctorals. Areas of specialization are reproductive and endocrine biology; neurobiology and biology of normal and abnormal growth.

WORCESTER STATE COLLEGE

Worcester, Massachusetts 01602 Phone: (617) 754-6861
Dean of College: N.J. Reyburn

DEPARTMENT OF BIOLOGY

Phone: Chairman: Ext. 256 Others: 754-6861

Chairman: T.L. Roberts
Professors: C.M. Chauvin, P.A. Holle, T.L. Roberts
Associate Professors: J.F. Eager, M.B. Kreider, W.K. Masterson, S.M. Paracer, B.D. Russell, A.W. Thurston
Assistant Professors: E.A. Boger, T.E. Graham
Departmental Technicians: 2

Degree Program: Biology, Biological Education

Undergraduate Degree: B.A., B.S.
Undergraduate Degree Requirements: Not Stated

Graduate Degree: M.Ed.
Graduate Degree Requirements: Master: 18 semester hours in Education, 15 semester hours in Biology

Graduate and Undergraduate Courses Available for Graduate Credit: Marine Biology (U), Parasitology (U), Reproductive Physiology, Environmental Physiology (U), Mammalian Physiology (U), Social Genetics, Vertebrate Natural History (U), Advanced Genetics (U), Genetic Abnormalities, Entomology (U)

MICHIGAN

ADRIAN COLLEGE

Adrian, Michigan 49221 Phone: (313) 265-5161
Dean of College: D. Pollard

BIOLOGY DEPARTMENT
Phone: Chairman: Ext. 217

Chairman: Not Stated
Professor: R.W. Husband
Associate Professors: C.K. Wu, K.S. Xavier
Assistant Professor: D.H. Nelson

Field Stations/Laboratories: Peter Dawson Arboretum - 10 miles west of campus.

Degree Programs: Biology, Medical Technology

Undergraduate Degrees: B.A., B.S.
Undergraduate Degree Requirements: Minimum of 30 hours of Biology

ALBION COLLEGE

Albion, Michigan 49224 Phone: (517) 629-5511
Dean of Faculty: R. Rosser

DEPARTMENT OF BIOLOGY
Phone: Chairman: Ext. 291 Others: Ext. 292

Chairman: E.A. Stowell
Professors: K.C. Ballou, D.G. Dillery, C.L. Dixon, W.J. Gilbert, J.B. Guyselman, E.A. Stowell
Associate Professors: R. Auito, R.D. Mortensen
Assistant Professors: R.W. Scott

Field Stations/Laboratories: Harvey N. Ott Biological Preserve - Whitehouse Nature Center

Degree Program: Biology

Undergraduate Degree: B.A.
Undergraduate Degree Requirements: Major in Biology: (1 unit = 4 semester hours) Seven units in Biology; 3 units divided between two cognate areas, Comprehensive Examination in Biology

ALMA COLLEGE

Alma, Michigan 48801 Phone: (517) 463-2141
Dean of College: R.O. Kapp

BIOLOGY DEPARTMENT
Phone: Chairman: Ext. 278 Others: Ext. 282,341,397

Chairman: A.L. Edgar
Professors: A.L. Edgar, L.E. Eyer, R.O. Kapp
Assistant Professors: R.A. Roeper, J.H. Wilson, L.W. Wittle

Field Stations/Laboratories: Alma College Ecological Station - located 12 miles west of Alma, Michigan

Degree Program: Biology

Undergraduate Degrees: B.A., B.S.
Undergraduate Degree Requirements: B.A. 136 credits B.S. 136 credits of which 68 must be in science and mathematics, 36 credits in biology, 2100 average (4.0 scale), proficiency in English Composition, Physical Education, 45 credits in advanced level work, certain distributional credits in Humanities and Social Science Divisions.

ANDREWS UNIVERSITY

Berriensprings, Michigan 49104 Phone: 471-7771
Dean of Graduate Studies: G.A. Madgwick
Dean of College: D. Ford

DEPARTMENT OF AGRICULTURE
Phone: Chairman: 471-7771 Ext. 266

Chairman: B. Andersen
Associate Professor: B. Andersen
Assistant Professor: D. Hodge
Instructors: L.V. Rice, W. Zehm
Visiting Lecturers: 2

Degree Program: Agriculture

Undergraduate Degree: B.Sc.
Undergraduate Degree Requirements: 50 quarter credits

DEPARTMENT OF BIOLOGY
Phone: Chairman: (616) 471-7771 Ext. 244

Chairman: A.C. Thoresen
Professors: L.N. Hare, R.M. Ritland, J.F. Stout, A.C. Thoresen
Associate Professors: H.E. Heidtke, B. Chobotar
Assistant Professor: G.E. Snow
Visiting Lecturers: 4
Research Assistants: 2-3

Field Stations/Laboratories: Marine Biological Station - Rosario Beach, Anacortes, Washington

Degree Program: Biology

Undergraduate Degrees: B.A., B.S. (Zoology)
Undergraduate Degree Requirements: B.A. - 45 credits in Biology plus Organic Chemistry and Physics
B.S. - 60 credits in Biology plus Organic Chemistry, Physics, Mathematics

Graduate Degrees: M.A., M.A.T.
Graduate Degree Requirements: Master: M.A. - 44 credits plus thesis M.A.T. - minimum of 16 credits

Graduate and Undergraduate Courses Available for Graduate Credit: Parasitology (U), General Ecology (U), Herpetology (U), Ornithology (U), Mammalogy (U), Entomology (U), Systematic Botany (U), Symbiosis (U), Biogeography (U), Animal Histology (U), Introduction to Paleontology (U), Biology of Bacteria (U), Animal Physiology (U), Plant Physiology (U), Comparative Animal Behavior (U), Animal Behavior (U), Invertebrate Zoology (U), Comparative Physiology (U), Limnology (U), Ichthyology (U), Marine Botany (U), Physiology of the Algae (U), Coastal Flora (U), Biological Oceanography (U), Biophysics (U), Comparative Morphology and Geological History of Vascular Plants (U), Protozoology, Advanced Ornithology, Plant Growth and Development, Physiology of Behavior, Problems in Paleontology, Paleoecology, Issues in Origins and Speciation

AQUINAS COLLEGE

Grand Rapids, Michigan 49506 Phone: (616) 499-8281
Dean of College: Sr. A. Keating

DEPARTMENT OF BIOLOGICAL SCIENCE
Phone: Chairman: (616) 499-8281

Chairman: Sr. B. Hansen

Professor: F.L. Bouwman
Associate Professor: E.W. Smith
Assistant Professors: R.S. Benda, J.T. Teusink, Sr. A. Wittenbach
Instructor: Sr. I. Chrusciel

Degree Program: Biology

Undergraduate Degree: B.S.
Undergraduate Degree Requirements: Zoology, Botany, Genetics, and Chemistry - General, Qualatative and Organic

CALVIN COLLEGE
Grand Rapids, Michigan 49506 Phone: (616) 949-4000
Dean of College: J. VandenBerg

BIOLOGY DEPARTMENT
Phone: Chairman: (616) 949-2847 Others: 949-2833

Chairman: A.L. Bratt
Professors: A.L. Bratt, A. Gebben, B.J. Broek, G.Van Horn
Associate Professors: J. Beebe, M. Karsten, B. Klooster
Assistant Professor: H. Bengelink

Degree Program: Biology

Undergraduate Degrees: A.B., B.S.
Undergraduate Degree Requirements: 8 1/2 courses: Cell Biology, Genetics and Evolution, Plant and Animal Structure and Function, Senior Seminar, Plant and Animal Diversity plus 3 others

CENTRAL MICHIGAN UNIVERSITY
Mt. Pleasant, Michigan 48859 Phone: (517) 774-3151
Dean of Graduate Studies: W.J. Waggoner (Acting)
Dean of College: R.V. Dietrich

DEPARTMENT OF BIOLOGY
Phone: Chairman: 774-3227 Others: 774-3151

Chairman: L.V.L. Curry
Professors: R.F. Burlington, L.D. Caldwell, L.V.L. Curry, N.L. Cuthbert, R.E. Hampton, M.H. Hohn, F.M. Johnston, L.D. Koehler, C.H. Nicholas, C.A. Scheel, M.I. Whiteny, D.E. Wujek
Associate Professors: J.D. Adams, R.G. Bland, E.R. Brockman, J.N. Krull, J.R. Kampky
Assistant Professors: R.E. Bailey, F.A. Eldredge, M.B. Garibanks, A.L. Kontio, H.L. Lenon, G.J. Pellerin, D.A. Valek
Instructor: R.G. Fell
Graduate Teaching Assistants: 9

Field Stations/Laboratories: Central Michigan University Biological Station - Beaver Island, Michigan

Degree Programs: Biology, Aquatic Biology, Limnology

Undergraduate Degree: B.S.
Undergraduate Degree Requirements: 124 semester hours, Preprofessional major - 40 semester hours, ucation major - 35 semester hours

Graduate Degree: M.S.
Graduate Degree Requirements: Master: 30 semester hours (15 hours at 600-700 level), 6 hours cognate area

Graduate and Undergraduate Courses Available for Graduate Credit: Biometrics (U), Teaching Biology (U), Radiation Biophysics (U), Biological Science for Elementary Teachers (U), Fishery Biology (U), Aquatic Insects (U), Embryology (U), Plant Ecology (U), Limnology (U), Limnological Methods (U), Aquatic Vascular Plants (U), Cellular Physiology (U), Wildlife Biology Management (U), Microbial Biology (U), Electron Microscopic Technique (U), Dendrology (U), Research Methods and Scientific Writing (U), Special Topics (U), Biological Colloquium, Biological Practicum, Research in Biology, Thesis

DETROIT INSTITUTE OF TECHNOLOGY
2727 Second Avenue Phone: (313) 962-0830
Detroit, Michigan 48201
Dean of College: F.R. Boos

DEPARTMENT OF BIOLOGICAL SCIENCES
Phone: Chairman: Ext. 208

No designated head
Professors: J. Ehrlich, V. Bailey (Emeritus)
Associate Professor: J. Stevenson
Assistant Professor: S. Schwartz
Instructor: 1 (part-time)
Student Laboratory Course Manager: 1

Degree Programs: Biology, Biological Sciences

Undergraduate Degree: B.S.
Undergraduate Degree Requirements: B.S. - 30 hours in Biological Sciences, 6 hours Mathematics, 8 hours Physics, 15 hours Chemistry, B.A. - 30 hours in Biological Sciences, plus 3-5 hours Mathematics, 9 hours Chemistry, Minor in Social Science or Humanities

EASTERN MICHIGAN UNIVERSITY
(No reply received)

FERRIS STATE COLLEGE
Big Rapids, Michigan 49307 Phone: (616) 796-9971
Dean of Graduate Studies: D.G. Butcher
Dean of College: R.E. Friar

DEPARTMENT OF BIOLOGY
Phone: Chairman: Ext. 397-398

Head: R.E. Friar
Professors: B. Durian, N.O. Levardsen, D.E. McCoy, M. Sandoz, K. Wagner
Associate Professors: T. Colladay, L. Conrad, J. Lehnert, R. McNeill
Assistant Professors: B. Beetley, G. Campbell, W. Heoksema, D. LaBatt, R. Stevenson, S. Zablotney

Degree Programs: Biological Education, Biology, Environmental Engineering, Health and Biological Sciences, Horticulture, Landscape Horticulture, Medical Technology, Ornamental Horticulture, Pharmacy, Science, Science Education

Undergraduate Degree: B.S.
Undergraduate Degree Requirements: Not Stated

GRAND VALLEY STATE COLLEGE
Allendale, Michigan 49401 Phone: (616) 895-6611
Dean: J. Linnell

DEPARTMENT OF BIOLOGY
Phone: Chairman: (616) 895-6611

Chairman: H.J. Stein
Professors: C.J . Bajema, H.J. Stein
Associate Professors: P.A. Huizenga, R.W. Ward
Assistant Professors: F.G. Anders, N. Leeling, W. Redding
Visiting Lecturers: 1

Degree Program: Biology

Undergraduate Degrees: B.A., B.S.
Undergraduate Degree Requirements: Organic Chemistry,

Physics, Statistics, Elementary Analysis (or Calculus), Genetics, Cell Biology, Ecology, a Botany course, a Zoology course, Senior level Plant Physiology or Animal Physiology or Microbiology or Developmental Biology

Thomas Jefferson College

Dean: T.D. Gilmore

(The College is not departmentalized and the faculty are unranked)

Faculty: E. Harrison (Anthropology), W. Harrison (Archeology), J. Warren (Geology), H. C. Wilson (Botany)

Field Stations/Laboratories: Archeological Site - Arizona, Summer Field Project

Degree Program: Natural History

Undergraduate Degree: B.Ph.
Undergraduate Degree Requirements: 180 credits, No required courses, No formal majors or minors

HILLSDALE COLLEGE

Hillsdale, Michigan 49242 Phone: (517) 437-4341
Dean of College: P. Adams

DEPARTMENT OF BIOLOGY
Phone: Chairman: Ext. 365

Chairman: J. Catenhusen
Professor: J. Catenhusen
Associate Professor: D. Toczek
Assistant Professor: T. Platt

Field Stations/Laboratories: Slayton Arboretum

Degree Program: Biology

Undergraduate Degree: B.S.
Undergraduate Degree Requirements: 32 credit hours, specific courses and electives

HOPE COLLEGE

Holland, Michigan 49423 Phone: (616) 392-5111
Provost: D. Marker

DEPARTMENT OF BIOLOGY
Phone: Chairman: Ext. 2345 Others: Ext. 2247

Chairman: E.D. Greij
Professors: A. Brady, R. Ogkerse
Associate Professors: E. Greij, N. Rieck, P. VanFaasen
Assistant Professors: J. Day, J. Dusseau, A. McBride
Visiting Lecturers: 1

Field Stations/Laboratories: Hope College Field Station - Holland, Michigan 49423

Degree Program: Biology

Undergraduate Degree: B.A.
Undergraduate Degree Requirements: 126 semester hours of which approximately 53 constitute a general core. 25 hours biology for major, 1 year Chemistry minimum requirement, most majors take calculus, 1 year Physics, and Chemistry through Organic.

KALAMAZOO COLLEGE

Kalamazoo, Michigan 49001 Phone: (616) 343-1551
Provost: J. Satterfield

DEPARTMENT OF BIOLOGY
Phone: Chairman: Ext. 256 Others: 254,255,256,257

Chairman: A.W. Wiens
Professor: H.L. Batts
Associate Professors: D.A. Evans, A.W. Wiens
Assistant Professors: P.D. Oiexia, M. Sproul
Visiting Lecturers: 2

Degree Program: Biology

Undergraduate Degree: A.B.
Undergraduate Degree Requirements: 8 courses minimum, 5 courses cognate science-mathematics, Senior Research Thesis. Departmental Comprehensive

LAKE SUPERIOR STATE COLLEGE

Sault Ste. Marie, Michigan 49883 Phone: (906) 632-6841
Dean of College: R. Hakala

DEPARTMENT OF BIOLOGICAL SCIENCES
Phone: Chairman: Ext. 267 Others: 382

Head: B.E. Smith
Professors: P. Chandra, A.E. Duwe, G.R. Gleason
Associate Professors: D.J. Behmer, G.J. Gleason, R.E. Reilly, B.E. Smith
Assistant Professors: V. Knudson, R.S. Furr
Temporary Appointments: 1

Degree Programs: Biology, Medical Technology

Undergraduate Degrees: B.A., B.S.
Undergraduate Degree Requirements: (Quarter Hours), B.A. - 4 year, 186 credits including 52 in Biology, 29 Chemistry, 12 Mathematics, B.S. - 4 year, 195 credits including 88 in Biology, 31 Chemistry, 18 Mathematics 12 Physics, 2 year Assoc. 95-99 credits (all required)

MARYGROVE COLLEGE

8425 West McNichols Road Phone: (313) 862-8000
Detroit, Michigan 48221
Dean of Graduate Studies: Sr. A. Ruedisueli
Dean of College: Sr. Y. Denomme

BIOLOGY DEPARTMENT
Phone: Chairman: Ext. 295

Head: D.C. Rizzo
Assistant Professors: D.G. Rizzo, Sr. M.A. Reuter
Instructor: S. Itzkowitz
Adjuncts: 2

Degree Programs: Biology, Natural Science

Undergraduate Degree: B.S.
Undergraduate Degree Requirements: 34 hours Biology, 16 hours Chemistry, 8 hours Mathematics (Calculus), Total of 128 hours to Graduate

MERCY COLLEGE OF DETROIT

8200 West Outer Drive Phone: (313) 531-7820
Detroit, Michigan 48219
Dean of College: Sr. M. Christopher

DEPARTMENT OF BIOLOGY
Phone: Chairman: Ext. 496

Chairman: A.G. Capodilupo
Professor: D.J. Hitchcock
Associate Professor: A.G. Capodilupo
Assistant Professor: N. Wilmes
Instructor: D. Elkins

Degree Programs: Biology, Biological Education

Undergraduate Degree: B.S.
Undergraduate Degree Requirements: 30 credits in Biology, 24 credits in Chemistry and Physics, 12 credits in Mathematics, 58 credits in core curriculum and related areas

MICHIGAN STATE UNIVERSITY
East Lansing, Michigan 48823 Phone: (517) 355-1855
Dean of Graduate School: C.W. Minkel

College of Agriculture and Natural Resources
Dean: L.L. Boger

DEPARTMENT OF ANIMAL HUSBANDRY
Phone: Chairman: (517) 355-8384 Others: 355-1855

Chairman: R.H. Nelson
Professors: B.H. Good, H.A. Henneman, W.T. Magee, R.A. Merkel, E.C. Miller, E.R. Miller, H.D. Ritchie, D.E. Ullrey
Associate Professors: W.G. Bergen, R.J. Deans, W.R. Dukelow, R.J. Dunn, G.D. Riegle
Assistant Professor: D.R. Hawkins

Field Stations/Laboratories: Lake City Experimental Station-Lake City, Michigan, Upper Peninsula Experiment Station - Chatham, Michigan

Degree Program: Animal Husbandry

Undergraduate Degree: B.S.
Undergraduate Degree Requirements: 180 credits of which 58-72 are in specialized field.

Graduate Degrees: M.S., Ph.D. (Animal Genetics, Animal Husbandry, Meat Sciences, Nutrition)
Graduate Degree Requirements: Master: Student must complete minimum of 45 credits beyond a bachelor's degree with a GPA of 3.0 or higher. Plan A - A research program with preparation of a thesis. Research credits must be 8 to 15, Plan B - Work without a thesis. Doctoral: Student must complete a minimum of 36 course credits beyond the master's degree with a GPA of 3.0 or higher. Candidate must register for 36 credits of dissertation research and prepare a thesis. Student must demonstrate competence in a broad area of education outside the usual formal course requirements.

Graduate and Undergraduate Courses Available for Graduate Credit: Meat Animal Breeding (U), The Impact of Animal Resource Management upon the World's Developing Nations (U), Techniques of Nutrition Research, Seminar, Comparative Nutrition-Lipids and Carbohydrates Comparative Nutrition-Protein Metabolism and Developmental Biology, Comparative Nutrition-Minerals, Comparative Nutrition-Vitamins, Genetics of Breed Improvement, Breeding Systems and Plans, Research

DEPARTMENT OF BIOCHEMISTRY
(Also in College of Natural Science)
Phone: Chairman: (517) 353-3257 Others: 355-1855

Chairman: R. Barker
Professors: R.L. Anderson, R. Barker, L.L. Bieber, J.A. Boezi, R.U. Byerrum, W.C. Deal, Jr., R.J. Evans, J.L. Fairley, P. Filner, H. Kitchen, R.W. Luecke, D. McConnell, A.J. Morris, F.M. Rottman, H.M. Sell, J.C. Speck, Jr., C.H. Suelter, C.C. Sweeley, N.E. Tolbert, W.W. Wells, W.A. Wood
Associate Professors: S.D. Aust, P. Kindel, H.A. Lillevik, R.A. Ronzio, J.E. Wilson
Assistant Professors: D.P. Delmer, P.J. Fraker, J.F. Holland
Post-doctoral Trainees: 24
Visiting Lecturers: 1
Research Assistants: 68

Degree Programs: Biochemistry, Agricultural Biochemistry

Undergraduate Degree: B.S.
Undergraduate Degree Requirements: Mathematics through differential equations, physics, a year of Biology, general, analytical, Organic and Physical Chemistry, Biochemistry, General Education courses of the University

Graduate Degrees: M.S., Ph.D.
Graduate Degree Requirements: Master: A minimum of 23 quarter credits of course work; an acceptable thesis. Doctoral: Graduate-level course work totaling 59 quarter credits, including a year of graduate chemistry courses, an acceptable thesis

Graduate and Undergraduate Courses Available for Graduate Credit: Introduction to Biochemical Techniques, Graduate Laboratory in Biochemistry, Graduate Lecture Sequence in Biochemistry, Special Topics in Biochemistry, 1-2 per Term, Advanced Plant Biochemistry and Plant Physiology, Two term sequence in General Biochemistry. Not available to Biochemistry majors for graduate credit.

DEPARTMENT OF FISHERIES AND WILDLIFE
Phone: Chairman: 353-0647 Others: 355-4477

Chairman: N.R. Kevern
Professors: R.H. Baker (Adjunct), R.C. Ball, K.W. Cummins (Adjunct), L.W. Gysel, N.R. Kevern, G.H. Lauff (Adjunct), C.D. McNabb, G.A. Petrides, E.W. Roelofs, P.I. Tack
Associate Professors: T.G. Bahr (Adjunct), F.M. D'Itri, (Adjunct), R.W. George, H.E. Johnson, G.W. Mouser, H.H. Prince, G. Schneider
Assistant Professors: R. Cole, W.H. Conley, C. Liston (Research Associate), R.J. White
Instructors: G. Dudderar, S. Marlatt
Teaching Fellows: 6
Research Assistants: 40

Field Stations/Laboratories: W.K. Kellogg Biological Station - Hickory Corners, Michigan

Degree Programs: Fisheries and Wildlife

Undergraduate Degree: B.S.
Undergraduate Degree Requirements: 180 credits: Communications (6), Biology (12), Chemistry (11), Physics (8), Botany (4), English (9), Social Science (12), Humanities (12), Economics (4), Entomology (5), Geology (4), Soils (5), Mathematics (10), Genetics (5), Ecology (8), remainder in Fisheries and Wildlife courses (15-20) and electives.

Graduate Degrees: M.S., Ph.D.
Graduate Degree Requirements: Master: 45 credits of which 15 may be for the thesis. Program approval by guidance committee and passing oral comprehensive examination and defense of thesis. Doctoral: 36 course credits, 12 credits in non-allied field (e.g., foreign language), dissertation, teaching experience in at least one course. Program approval by guidance committee and passing comprehensive examination and oral defense of dissertation.

Graduate and Undergraduate Courses Available for Graduate Credit: Environmental Conservation Education (U), Wildlife Population Analysis (U), Wildlife Habitat Analysis (U), Ecology of Migratory Birds (U), Wildlife Biology and Management (U), Ichthyology (U), Fishery Biology and Management (U), Fish Culture (U), Limnology (U), Limnological Methods (U), Outdoor Environmental Requirements of Fish, Advanced Biological Limnology, Chemical Limnology and Quantitative Wildlife Ecology

DEPARTMENT OF FORESTRY
(No reply received)

DEPARTMENT OF HORTICULTURE
Phone: Chairman: 355-5191 Others: 355-5191

Chairman: J. Carew
Professors: M.J. Bukovac, J. Carew, R.F. Carlson, R. L. Carolus, W. Carpenter, H. Davidson, A. DeHertogh, F.G. Dennis, D.H. Dewey, D.R. Dilley, S. Honma, J. Hull, Jr., A.L. Kenworthy, J. Moulton,

H.P. Rasmussen, S.K. Ries, J.L. Taylor, S. Wittwer, A. Putnam
Associate Professors: L.R. Baker, W.H. Carlson, R.C. Herner, G.S. Howell, G. Kessler, R. Mecklenburg, H. Price, K.C. Sink, G.Vest
Assistant Professors: C. Kesner, J. Motes, R.L. Spangler, C. Van Den Brink
Research Assistants: 37

Field Stations/Laboratories: Sodus Experimental Farm - Sodus, South Haven Research Station, S. Haven, Graham Experiment Station - Grand Rapids

Degree Program: Horticulture

Undergraduate Degree: B.S.
Undergraduate Degree Requirements: One year English, Social Science and Humanities one year of Botany, 1 course in Entomology, 25 credits (quarter) of Horticulture, one course in Economics, two courses in Communications, one year of Mathematics, one year of Chemistry, and 15 credits of Agriculture.

Graduate Degrees: M.S., Ph.D.
Graduate Degree Requirements: Master: 45 credit hours (Quarter). Plan A with thesis or Plan B without thesis. 23 credit hours of graduate level courses with 8 credits in Horticulture Doctoral: 48 credit hours (Quarter) of courses plus 36 hours of thesis, reading knowledge of a foreign language may substitute for 6 credit hours of course work

Graduate and Undergraduate Courses Available for Graduate Credit: List too extensive. Graduate students in Horticulture may be using courses in as many as 15 other departments, i.e., Botany, Biochemistry, Soil and Crop Science, Statistics, Education, Economics, etc.

MSU/AEC PLANT RESEARCH LABORATORY
(Also in College of Natural Sciences)
Phone: Chairman: (517) 353-2270

Director: A. Lang
Professors: P. Filner, A. Haug, H.J. Kende, D.T.A. Lamport (Research), A. Lang, K. Raschke, J. Varner (Adjunct), C.P. Wolk, J.A.D. Zeevaart
Associate Professors: M. Jost, L. Wilson (Research)
Assistant Professors: D. Delmer, K. Poff
Research Assistants: 22
Research Associates: 8

Graduate Degrees: Biochemistry, Biophysics, Botany and Plant Pathology. The Programs are administered by the respective department; the Plant Research Laboratory provides graduate assistantships and facilities.

DEPARTMENT OF POULTRY SCIENCE
Phone: Chairman: (517) 355-8415

Chairman: H.C. Zindel
Professors: L.R. Champion, T.H. Coleman, D. Polin, R.K. Ringer, C.C. Sheppard, H.C. Zindel
Associate Professors: R.J. Aulerich, T.S. Chang, C.J. Flegal
Assistant Professor: B.J. Marquez
Instructor: S. Varghese
Research Assistants: 9

Degree Program: Poultry Science

Undergraduate Degrees: B.S.
Undergraduate Degree Requirements: 180 credits including specified University, College of Agriculture and Department of Poultry Science requirements. Approximately 55 credits are elective.

Graduate Degrees: M.S., Ph.D.
Graduate Degree Requirements: Master: 45 credits beyond B.S. including research and thesis Doctoral: Minimum of 36 course credits, exclusive of research credits, beyond M.S. (plus 9 credits to demonstrate competence in at least one area of broad education), Research, Thesis

Graduate and Undergraduate Courses Available for Graduate Credit: Avian Nutrition (U), Poultry Breeding and Incubation (U), Poultry Industry, Management and Marketing (U), Avian Physiology (U), Avian Disease Prevention and Treatment (U), The Impact of Animal Resource Management Upon the World's Developing Nations (U), Advanced Poultry - Special Problems, Advanced Poultry Seminar, Research

College of Natural Science
Dean: R.W. Byerumm

DEPARTMENT OF BIOCHEMISTRY
(see under College of Agriculture and Natural Resources)

DEPARTMENT OF BOTANY AND PLANT PATHOLOGY
(No reply received)

DEPARTMENT OF ENTOMOLOGY
Phone: Chairman: (517) 355-4665 Others: 355-4662

Chairman: J.E. Bath
John A. Hannah Professorship: A.W.A. Brown
Professors: A.W.A. Brown, R.L. Fischer, G.E. Guyer, D.L. Haynes, R.A. Hoopingarner, A.J. Howitt, H. King, E.C. Martin, R.F. Ruppel, W.E. Wallner
Associate Professors: R.V. Connin, K.W. Cummins, J. Hoffman, G. Hooper, C.W. Laughlin, H.D. Newson, F.W. Stehr, O. Taboada, J. Webster, S. Wellso, M. Zabik, G.W. Bird, D.C. Cress, R.J. Sauer
Assistant Professors: B.A. Croft, J.A. Knierim, R.W. Merritt, R.L. Tummala
Visiting Lecturers: 1
Research Assistants: 6

Field Stations/Laboratories: Kellogg Biological Station - Hickory Corners, Michigan

Degree Program: Entomology

Undergraduate Degree: B.S.
Undergraduate Degree Requirements: Mathematics, Chemistry, Physics, one Foreign Language, 30 credits in Biology (other than Entomology), 25 credits Entomology plus 15 elective credits either in Entomology or directly involved.

Graduate Degrees: M.S., Ph.D.
Graduate Degree Requirements: Master: College of Natural Science requirements. Department: 3.0 in upper division of undergraduate work. A minimum of 45 credits, of which not more than 12 may be thesis research. Oral examination on thesis and general knowledge before end of final term. Doctoral: M.S. degree or appropriate qualifying examination. No specific number of credits required. No proficiency in foreign language, but does require candidate to acquire significant level of knowledge in an area peripheral to major research area. Doctoral comprehensive examinations in three areas of study in which student should develop competence. College of Natural Science requirements.

Graduate and Undergraduate Courses Available for Graduate Credit: Field Entomology (U), Apiculture and Pollination (U), Seminar (U), Systematic Entomology (U), Aquatic Insects (U), Stream Ecology (U), Economic Entomology (U), Taxonomy of Immature Insects (U), External Morphology of Insects (U), Internal Morphology (U), Insect Physiology (U), Medical Entomology (U), Introductory Nematology (U), Insects in Relation to Plant Diseases (U), Topics in Entomology (U), Advanced Taxonomy, Ecology of Aquatic Insects, Insect Ecology, Advanced Stream Ecology, Insect Biochemistry, Principles of Taxonomy, Insect Toxicology,

Analytical Techniques for Biological Compounds I,II

W.K. KELLOGG BIOLOGICAL STATION
Phone: Chairman: (616) 671-5144 Others: 671-5117

Director: G.H. Lauff
John A. Hannah Professor: A.W.A. Brown
Professors: K.W. Cummins, D.J. Hall, G.H. Lauff, R.G. Wetzel (Year-round faculty and researchers)
3 month summer session faculty varies each year
Assistant Professors: M.J. Klug, E.E. Werner, P.A. Werner
Research Assistants: 16
Post-doctoral: 1

Field Stations/Laboratories: Kellogg Biological Station consists of the following units: Kellogg Bird Sanctuary, Killogg Farm, and Kellogg Gull Lake Laboratories including year-round teaching and research laboratories and other necessary facilities for biological work. The Station is located on Gull Lake in Kalamazoo County and consists of 2,000 acres of farm land, forests, lakes, ponds and streams plus avaiability to other regional habitats.

Degree Programs: Botany, Entomology, Fisheries and Wildlife, Microbiology and Public Health, Zoology

Graduate Degrees: M.S., Ph.D.
Graduate Degree Requirements: Master: Program determined by guidance committee Doctoral: Program determined by guidance committee.

Graduate and Undergraduate Courses Available for Graduate Credit: Courses vary from year to year.

DEPARTMENT OF MICROBIOLOGY
(See under College of Veterinary Medicine)

DEPARTMENT OF PHYSIOLOGY
(See under College of Veterinary Medicine)

MSU/AEC PLANT RESEARCH LABORATORY
(See under College of Agriculture and Natural Resources)

DEPARTMENT OF ZOOLOGY
Phone: Chairman: (517) 355-4640

Animal Behavior and Neurobiology
Professors: M.M. Balaban, J.C. Braddock, J.I. Johnson, J.A. King
Associate Professors: L. Clemens, H.E. Hagerman
Assistant Professor: C.D. Tweedle
Cell Biology
Professor: N.R. Band
Associate Professor: S. Aggarwal
Ecology and Population Biology
Professors: R.C. Ball, W.E. Cooper, D.J. Hall, G.H. Lauff, T.W. Porter
Assistant Professor: R.W. Hill
Embryology and Experimental Morphology
Professors: J.R. Shaver, C.S. Thornton
Associate Professors: S.C. Bromley, H. Ozaki
Genetics
Professors: J.V. Higgins, H.M. Slatis
Assistant Professors: J.H. Asher, Jr., L.G. Robbins
Physiology
Professors: R.A. Pax, E.M. Rivera
Vertebrate Biology and Ecology
Professors: R.H. Baker, M.M. Hensley, J.A. Holman, G.A. Petrides, P.I. Tack
Assistant Professor: D.L. Beaver
Research Assistants: 6
Graduate Teaching Assistants: 21

Field Stations/Laboratories: Kellogg Biological Station - Gull Lake, Michigan

Degree Program: Zoology

Undergraduate Degree: B.S.
Undergraduate Degree Requirements: Not Stated

Graduate Degrees: M.S., Ph.D.
Graduate Degree Requirements: Master: 45 credits plus Thesis, one language, final oral Doctoral: Courses selected by guidance committee. Comprehensive examination, foreign language, thesis, final oral

Graduate and Undergraduate Courses Available for Graduate Credit: Comparative Physiology I and II (U), Biological Concepts for Engineers (U), Animal Behavior (U), Vertebrate Paleontology (U), Genetics (U), Experimental Ecology (U), Experimental Embryology (U), Vertebrate Biology (U), Invertebrate Zoology (U), Limnology (U), Protozoology (U), Cytochemistry (U), Zoogeography (U), Endocrinology (U), Malacology, Animal Behavior, Ecology, Tropical Biology, Genetics, Vertebrate Zoology, Ultrastructure, Endocrinology, Neuroembryology, Embryology, Cellular Biology, Zoology Seminar

College of Veterinary Medicine

Dean: N.W. Armstead

DEPARTMENT OF ANATOMY
Phone: Chairman: (517) 353-6380 Others: 355-1855

Chairman: B.E. Walker
Professors: M.L. Calhoun (Emeritus), J.L. Conklin, A.L. Foley, T.W. Jenkins, A.W. Stinson, B.E. Walker
Associate Professors: R.E. Carrow, R. Echt, N.S. Henderson, J.D. Manges, L.M. Ross, C.W. Welsch
Assistant Professors: D.R. Adams, M.H.L. Gibson, A.W. Jacobs, G.M. Lew, M.H. Ratzlaff, C.W. Smith, J.F. Taylor, M.B. Zaleski
Electron Microscopy Specialist: 1

Degree Program: Anatomy

Graduate Degrees: M.S., Ph.D.
Graduate Degree Requirements: Not Stated

Graduate and Undergraduate Courses Available for Graduate Credit: Microscopic Anatomy (U), Veterinary Anatomy (Microscopic and Gross), Microscopic Anatomy (Human), Functional Medical Cytology and Histology, Introduction to Human Gross Anatomy, Problems in Anatomy, Anatomy of the Nervous System

DEPARTMENT OF PHARMACOLOGY
Phone: Chairman: (517) 353-7147 Others: 353-7145

Chairman: T.M. Brody
Professors: T. Akera, T.M. Brody, A.M. Michelakis, K.E. Moore, R.H. Rech
Associate Professors: G.L. Gebber, J.E. Gibson, J.B. Hood, D.A. Reinke, T. Tobin
Assistant Professors: J.L. Bennett, R.K. Ferguson, J.I. Goodman, D.E. Rickert, J.L. Stickney, S. Stolman, F. Welsch
Instructor: J.E. Thornburg
Visiting Lecturers: 4
Teaching Fellows: 1
Research Assistants: 1

Degree Program: Pharmacology

Graduate Degrees: M.S., Ph.D.
Graduate Degree Requirements: Master: Histology, Biochemistry, Physiology, Pharmacology, Research, Thesis Doctoral: Histology, Biochemistry, Physiology, Pathology, Pharmacology, Biometry, Synaptic Transmission, Organ Pharmacology, Advanced Principles of Pharmacology, Elective Courses, Dissertation

Graduate and Undergraduate Courses Available for Graduate Credit: Principles of Pharmacology, Pharmacodynamics, Toxicology, Biometry, Synaptic Transmission, Organ System Pharmacology, Advanced Principles of of Pharmacology and Toxicology, Seminar, Problems and Thesis Research, Drug Abuse (U)

DEPARTMENT OF MICROBIOLOGY AND PUBLIC HEALTH
(Also in College of Natural Science)
Phone: Chairman: (517) 355-6465 Others: 355-1855

Chairman: P. Gerhardt
Associate Chairman: D.E. Schoenhard
Professors: E.S. Beneke, S.H. Black, B.R. Burmester, G.R. Carter, R.N. Costilow, C.H. Cunningham, A. Herrer-Alva (Adjunct), W.N. Mack, N.B. McCullough, J.P. Newman, H.L. Sadoff, C.L. San Clemente
Associate Professors: R.C. Belding, R.R. Brubaker, H. W. Cox, V.H. Mallmann, H.D. Newson, F.R. Peabody, E. Sanders,J.M. Tiedje, B. Wentworth (Adjunct)
Assistant Professors: J.A. Breznak, T.R. Corner, G.F. Dardas (Clinical), T.J. Dardas (Clinical), J.W. Dyke (Adjunct), R.C. Gordon, M.J. Klug, H.C. Miller, R. J. Moon, R.A. Patrick, R.J. Patterson, C.A. Reddy, L.R. Snyder, D W. Twohy, R.L. Uffen, L.F. Velicer, J.F. Williams
Instructor: T.C. Beaman
Visiting Lecturers: 5
Teaching Fellows: 2
Research Assistants: 5
Specialists: 2

Field Stations/Laboratories: Kellogg Biological Station - Gull Lake, Hickory Corners, Michigan

Degree Programs: Medical Microbiology, Microbiology

Undergraduate Degree: B.S.
Undergraduate Degree Requirements: General Education courses of the University College, completion of 180 credits with at least a 2.0, 15 credits - Mathematics, 15 credits - Foreign Language, 12 credits, Physics, 25 credits - Chemistry, 12 credits - Biological Science, 8 credits, Biochemistry, and 28 credits of Microbiology plus any electives

Graduate Degrees: M.S., Ph.D.
Graduate Degree Requirements: Master: minimum of 45 credits; for Plan A, 12 credits are taken in independent laboratory research, for Plan B, 8 credits of independent laboratory of library study. Majority of students elect Plan A. Each program individually planned. Doctoral: Generally 3 or more academic years of study and research including 36 credits of research. Each program individually planned.

Graduate and Undergraduate Courses Available for Graduate Credit: General Microbiology (U), General Microbiology Laboratory (U), General Virology (U), General Virology Laboratory (U), Microbial Physiology (U), Microbial Physiology Laboratory (U), Microbial Genetics (U), Microbial Genetics Laboratory (U), Immunobiology and Laboratory (U), Microbiology of Infectious Diseases (U), Introductory Medical Parasitology (U), Environmental Microbiology (U), Medical Mycology (U), Infectious Diseases, Biochemistry (U), Molecular Virology, Immunochemistry, Special Problems, Medical Microbiology Clerkship, Topics in Microbiology, Seminar, Introductory Mycology (U), Food Microbiology (U), Special Problems in Cytology, Instructional Methods Analysis, Laboratory Animal Medicine, Statistics (U), Quantitative Biology (U), General Pathology (U), Microscopic Anatomy (U)

DEPARTMENT OF PHYSIOLOGY
(Also in College of Natural Science)
Phone: Chairman: (517) 355-6475

Chairman: F.J. Haddy
Professors: T. Adams, D.K. Anderson, C. Chou, L.A. Cohen, E.D. Collings, J.M. Dabney, R.M. Daugherty Jr., W.R. Dukelow, T.E. Emerson, Jr., W.L. Franz, P. O. Fromm, F.J. Haddy, H.D. Hafs, S.R. Heisey, J.R. Hoffert, J. Meites, J.E. Nellor, H.W. Overbeck, E.P. Reineke, R.K. Ringer, J.B. Scott, L.F. Wolterink
Associate Professors: M.D. Bailie, R.A. Bernard, E. M. Convey, J.G. Cunningham, R.F. Johnston, G.M. Purcell, G.D. Riegle, N.E. Ribinson, J.M. Schwinghamer, B.H. Selleck, H.A. Tucker
Assistant Professors: G.J. Grega, R.P. Pittman
Research Fellows: 5 (Post-doctorals)

Degree Programs: Physiology, Neuroscience

Undergraduate Degree: B.S. (Physiology)
Undergraduate Degree Requirements: Chemistry through Organic with Physical Chemistry strongly recommended year of Physics, Mathematics through Differential Equations, year of Language, Humanities, English, Social Science, General Biology, Comparative Anatomy Embryology, 15 term credits in Physiology including: Cellular, Comparative, Mammalian Systemic, and electives.

Graduate Degrees: M.S., Ph.D.
Graduate Degree Requirements: Master: Entry: a B.S. degree in Biology with Mathematics, Physics and Chemistry as above. Forty-five term credits in graduate physiology, Biochemistry and Biology, courses as set-up by guidance committee including a research thesis of not more than 12 credits, Final Oral Examination Doctoral: Guidance committee determines program and monitors research progress. No foreign language required, G.P.A. required: 3.0, general examina tion and a final oral thesis defense required.

Graduate and Undergraduate Courses Available for Graduate Credit: Cell Physiology, Comparative Physiology, Advanced Mammalian Physiology, Neuroendocrinology, Physiology of Respiration, Cardiovascular Physiology, Kidney Physiology and Electrolyte Metabolism, Physical Principles of Biological Systems, Radiobiology, Reproduction, Lactation, Neurophysiology, Sensory Physiology, Endocrinology, Neurobiology, Special Problems, Seminar

MICHIGAN TECHNICAL UNIVERSITY

Houghton, Michigan 49931 Phone: (906) 487-1885
Dean of Graduate Studies: D. Yerg

College of Arts and Sciences

Dean: W. Powers

DEPARTMENT OF BIOLOGICAL SCIENCES
Phone: Chairman: (906) 487-2025 Others: 487-1885

Head: R.C. Stones
Professors: R.T. Brown, F.H. Erbisch, R.M. Linn (Adjunct), J.D. Spain, R.C. Stones, B.K. Whitten, R.F. Ziegler (Adjunct)
Associate Professors: B.M. Allison, J.C. Holland, R.A. Janke, K.J. Kraft, K.R. Kramm, R. Krear, D.J. Remondini, T.D. Wright
Assistant Professors: J. Glime, C. Gianoulakis, E. Kahan, D. Nevalainen
Visiting Lecturers: 1
Research Assistants: 6
Graduate Teaching Assistants: 12
Laboratory Associates: 2
Laboratory Coordinator: 1

Field Stations/Laboratories: "Herbert Dow Wilderness Area" - Keweenaw Peninsula of Upper Michigan

Degree Programs: Biological Sciences, Ecology/Environmental Biology, Limnology, Medical Technology, Clinical Chemistry

Undergraduate Degree: B.S.
Undergraduate Degree Requirements: B.S. in Biological Sciences: 196 credits (quarter system) in four groups Biology Core Subjects (47 credits), supporting courses

in Chemistry, Physics, Mathematics (52-56 credits, University Requirements (34 credits) Electives (59-63 credits) B.S. in Medical Technology: 196 credits in each of 4 options, 1) Medical Technology, Medical Technology and Teacher Certification, Medical Technology with Business Administration, Medical Technology with Biological Science

Graduate Degrees: M.S., Ph.D.
Graduate Degree Requirements: Master: minimum GPA of 3.0 curing last 2 years of undergraduate work. 45 total credits (quarter), minimum of 30 hours of coursework (9-15 hours thesis) oral defense of thesis Doctoral: Demonstrated mastery of subject matter in chosen field of environmental biology, language requirement flexible, dissertation research of 13-30 credit hours. Preliminary and final defense of dissertation examinations.

Graduate and Undergraduate Courses Available for Graduate Credit: Fungi and Lechens (U), Genetics (U), Endocrinology (U), Biochemical Techniques (U), Limnology (U), Immunochematology and Serology (U), Ichthyology (U), Cardiovascular Physiology (U), Epidemiology (U), Bryology (U), Ecosystems of the World (U), Organic Evolution (U), Animal Behavior (U), Histology (U), Advanced Technical Simulation (U), Medical Laboratory Instrumentation (U), Clinical Hematology (U), Fishery Biology (U), Cardiac Hemodrynamics (U), Special Problems in Biology (U), Radioisotope Techniques in Biology (U), Virology (U), Cell Biology (U), Biometerology (U), Freshwater Invertebrates (U), Advanced Techniques in Microscopy (U), Physiological Chemistry (U), Clinical Chemistry (U), Aquatic Plants (U), Stress Aquatic Ecosystems (U), Medical Parasitology (U), Clinical-Laboratory Cardiology (U), Steroid Biochemistry, Comparative Biochemistry and Physiology, Special Topics in Biology, Special Topics in Physiology, Advanced Plant Ecology, Biogeography, Enzyme Biochemistry, Taxonomy of Adult Insects, Systems Ecology, Primary and Secondary Productivity, Research Problems in Ecology of Natural Areas, Advanced Plant Physiology, Immunology, Physical Ecology, Advanced Animal Ecology, Physical-Chemical Limnology

School of Forestry

Dean: E.A. Bourdo, Jr.

DEPARTMENT OF FORESTRY
Phone: Chairman: (906) 487-2454 Others: 487-1885

Head: G.A. Hesterberg
Professors: C.R. Crowther, V.W. Johnson, H.M. Steinhilb
Associate Professors: R.K. Miller, N.F. Sloan
Assistant Professors: M.S. Coffman, D.J. Frederick, M.F. Jurgensen, R.L. Sajdak, F.A. Stormer, B.C. Sun
Instructor: C.H. Hein
Visiting Lecturers: 4

Field Stations/Laboratories: Ford Forestry Center - Alberta, Michigan

Degree Programs: Forest Chemistry, Forestry, Forestry and Conservation, Genetics, Natural Resources, Plant Pathology, Silviculture, Soil Science, Wildlife Management

Undergraduate Degree: B.S.
Undergraduate Degree Requirements: 189 credits, plus 19 credits of Forestry Summer School. At least 15 credits of approved Humanities or Social Sciences must be scheduled.

Graduate Degree: M.S.
Graduate Degree Requirements: Master: At least 30 credits must be earned in course work other than the thesis. At least 18 of these 30 credits must be earned in courses numbered the 500 series or 600 series and not more than 12 credits in the 400 series may be applied toward the M.S. degree. Of the total of 45 creidts, 9 to 15 must be assigned to the thesis.

Graduate And Undergraduate Courses Available for Graduate Credit: Graduate Research in Forestry, Current Topics in Resource Economics and Policy, Marketing Forest Products, Forest Synecology, Forest Production Ecology, Forest Influences, Advanced Forest Pathology, Forest Genetics, Advanced Forest Management, Animal Population Dynamics, Wildlife Investigational Techniques, Advanced Soils, Soil Properties and Plant Growth, Advanced Microbial Ecology, Data Processing, Advanced Computer Application in Forest Management Problems, Recreation Area Planning and Design, Insect Morphology, Graduate Seminar in Forestry

NAZARETH COLLEGE

(No reply received)

NORTHERN MICHIGAN UNIVERSITY

Marquette, Michigan 49855 Phone: 227-1000
Dean of Graduate Studies: R.S. Strolle
Dean of College: R.B. Glenn

DEPARTMENT OF BIOLOGY
Phone: Chairman: 227-2311 Others: 227-2310

Head: L.E. Peters
Professors: G.D. Gill, W.J. Merry, L.E. Peters, W.L. Robinson, R.K.E.Thoren, F.A. Verley, L.S. West (Emeritus)
Associate Professors: M.C. Bowers, P.A. Doepke, M.L. Kopenski, R.A. Parejko, O.B. Reynolds, N.A. Snitgen
Assistant Professors: T.G. Froiland, W.J. Vande Berg, J.K. Werner
Lecturer: L.A. Contois
Research Assistants: 3
Graduate Assistants (Teaching): 11

Field Stations/Laboratories: Northern Michigan University Field Station - Melstrand, Michigan

Degree Programs: Biology, Biological Education, Medical Technology

Undergraduate Degree: B.A., B.S.
Undergraduate Degree Requirements: 32 semester hour credits in Biology, including Genetics, Physiology, and Ecology, Physical Science minor: 16 semester hour credits in Chemistry, 8 semester hour credits in Physics

Graduate Degree: M.A.
Graduate Degree Requirements: Master: Minimum of 20 semester credit hours in Biology, minimum of 12 semester hour credits in approved cogantes or additional Biology

Graduate and Undergraduate Courses Available for Graduate Credit: Advanced Microbiology (U), Limnology (U), Biometrics (U), Biochemistry of Development (U), Invertebrate Zoology (U), Vertebrate Zoology (U), General Parasitology (U), General Entomology (U), Plant Anatomy (U), Plant Physiology (U), Algology (U), Fisheries Management (U), Wildlife Management (U), Problems in Biology (U), Special Topics in Biology (U), Advanced Physiology, Advanced Ecology, Population Genetics, Biochemical Genetics, Taxonomy of Insects, Plant Biosystematics, Bryology, Mycology, Research in Biology, Thesis in Biology

OAKLAND UNIVERSITY

Rochester, Michigan 48063 Phone: (313) 377-2100
Dean of Graduate Studies: G.P. Johnson
Dean of College: R. Torch

DEPARTMENT OF BIOLOGICAL SCIENCES
Phone: Chairman: (313) 377-3550 Others: 377-2100

Chairman: N.J. Unakar
Professors: F.M. Butterworth, W.C. Forbes, C.V. Harding (Adjunct), V.E. Kinsey, V.N Reddy, R. Torch (Dean), N.J. Unakar, W.L. Wilson
Associate Professors: J.D. Cowlishaw, M.J. Pak, J.R. Reddan, M.V. Riley, A.K. Roy
Assistant Professors: T.B. Friedman, E.M. Goudsmit, E.W. Henry, R.D. Hunter, P.A. Ketchum, C.B. Lindemann, J T. Romeo, B.S. Winkler
Visiting Lecturers: 4
Research Assistants: 3
Research Associates: 3

Degree Program: Biological Sciences

Undergraduate Degrees: B.A., B.S.
Undergraduate Degree Requirements: B.A. - 40 credits Biology, 14 credits Chemistry, 10 credits Physics, and Mathematics through a standard pre-calculus course. B.S. - 40 credits Biology, 2 years Chemistry (1 year General and 1 year Organic), Mathematics through Integral Calculus and a one-year Calculus-requiring General Physics course. In addition, 1 of 3 alternatives: a) senior paper based upon research, b) senior paper based upon a literature search on a research-oriented topic taken, c) a comprehensive examination. Also, all related laboratories for above.

Graduate Degree: M.S.
Graduate Degree Requirements: Master: 36 credits in Biology (16 in courses numbered 500 and above) including 4 credits in each of the 4 areas, Development Biology-Morphology, Biochemistry-Biophysics, Cell Physiology and Genetics. 8 credits must be a combination of credits received in graduate laboratory courses and credits received in graduate research. 1 credit in the Biology Seminar course and an acceptable thesis are also required.

Graduate and Undergraduate Courses Available for Graduate Credit: Ecology and Laboratory (U), Parasitology and Laboratory (U), Developmental Biology and Laboratory (U), Microbiology (U), Genetics and Laboratory (U), Aquatic Biology and Laboratory (U), Physiology of the Central Nervous System (U), Protozoology, Cellular Biochemistry and Laboratory (U), Experimental Embryology and Laboratory (U), Differentiation (U), Biophysics and Laboratory (U), Cytochemistry and Laboratory (U), Developmental Genetics and Laboratory (U), Virology and Laboratory (U), Microbial Genetics and Laboratory (U), Ultrastructure and Laboratory (U), Cell Biology and Laboratory (U), Advanced Physiology-Nerve and Laboratory (U), Advanced Physiology-Muscle and Laboratory (U), Advanced Topics in Cellular Biochemistry and Biophysics and Laboratory, Advanced Topics in Cell Physiology and Laboratory, Advanced Topics in Developmental Biology and Laboratory, Advanced Topics in Genetics and Laboratory, Graduate Biology Seminar

OLIVET COLLEGE
(No reply received)

SAGINAW VALLEY COLLEGE
2250 Pierce Road Phone: (517) 793-9800
University Center, Michigan 48710
Dean of Graduate Studies: H. Peterson
Dean of College: C. McCray

DEPARTMENT OF BIOLOGY
Phone: Chairman: Ext. 323 Others: Ext. 314,307,310

Chairman: W. Chase
Professor: W. Owsley
Associate Professors: P. DeJong, C. Pelzer
Assistant Professors: W. Chase, W. Rathkamp
Undergraduate Teaching Assistant: 1

Degree Program: Biology

Undergraduate Degree: B.S.
Undergraduate Degree Requirements: Principles of Biology (8 hours), Genetics (4 hours), Cell Physiology (4 hours), Developmental Biology (4 hours), Ecology (4 hours), one course in Natural Science, 30 hours of Biology, General Chemistry (10 hours), Mathematics (8 hours), and Physics (10 hours)

SIENNA HEIGHTS COLLEGE
(No reply received)

THOMAS JEFFERSON COLLEGE
(see under Grand Valley State College)

UNIVERSITY OF DETROIT
Detroit, Michigan 48221 Phone: (313) 927-1000
Dean of Graduate Studies: T.W. Walters
Dean of College of Arts and Sciences: T.E. Porter

DEPARTMENT OF BIOLOGY
Phone: Chairman: (313) 927-1182 Others: 927-1000

Chairman: W.J. Nunez
Professors: P.F. Forsthoefel, A.J. Haggis, P.J. Wood
Associate Professors: T.S. Acker, R.G. Albright, R.W. Balek, J.D. LaCroix, W.J. Nunez, R.J. Smith
Research Assistant: 1
Tuition Remission Scholars: 7

Degree Programs: Biology, Medical Technology

Undergraduate Degree: B.S.
Undergraduate Degree Requirements: 32 semester hours in Biology, Chemistry through Organic, year of Physics. Other semester hours (96 other than Biology) General College.

Graduate Degree: M.S. (Biology)
Graduate Degree Requirements: Master: 30 semester hours, including minimum of 24 non-thesis. Research project (original) for thesis required.

Graduate and Undergraduate Courses Available for Graduate Credit: Biology of Vertebrates (U), Developmental Biology (U), Cell Biology (U), Genetics (U), Genetics Laboratory(U), Evolution (U), Microbiology (U), Biology of Plants (U), Biology of Invertebrates (U), Ecology (U), Advanced Ecology and Animal Behavior (U), Plant Physiology (U), Histo-Physiology (U), Organogenesis, Cells of the Reticulo-Endothelial System, Developmental Genetics, Principles of Immunology, Biochemical Genetics, Biochemistry of Development, Pathogenic Microbiology, Molecular Biology, Special Problems

UNIVERSITY OF MICHIGAN
Ann Arbor, Michigan 48104 Phone: (313) 764-1817
Dean of Graduate Studies: D.E. Stokes

College of Literature, Science and the Arts
Dean: F.H.T. Rhodes

DEPARTMENT OF BOTANY
Phone: Chairman: (313) 764-1440 Others: 764-1441

Chairman: C.B. Beck
Harley Harris Bartlett Professor of Botany: R. McVaugh
Professors: C.B. Beck, S.L. Allen, W.S. Benninghoff, H.A. Crum, R.H. Davis, D.M. Gates, K.L. Jones, P.B. Kaufman, R.J. Lowry, R. McVaugh, G.A. Norman, R.L. Shaffer, A.H. Smith, E.E. Steiner, A.S.

Sussman, E.G. Voss, W.H. Wagner, C.S. Yocum
Associate Professors: H.A. Douthit, R. Ford, R.B. Helling, H. Ikuma, L.D. Nooden
Assistant Professors: J.P. Adams, J.A. Doyle, G.F. Estabrook, G.E. McBride, C.F. Yocum
Teaching Fellows: 44
Research Assistants: 4
Research Associates: 3

Field Stations/Laboratories: Biological Station - Pellston, Matthaei Botanical Gradens - Ann Arbor, University Herbarium - Ann Arbor

Degree Programs: Biology, Botany, Microbiology (Interdepartmental Program in Biology, A.M. or M.S.; Biophysics M.S., Ph.D.)

Undergraduate Degree: B.S.
Undergraduate Degree Requirements: Botany, Chemistry, Physics and cognate fields

Graduate Degrees: M.S., Ph.D.
Graduate Degree Requirements: Master: 34 credits, of which 12 must be graduate credits in 4 areas of Botany, foreign language Doctoral: Courses chosen by advisory committee, area comprehensive examination, thesis, 2 modern languages

Graduate and Undergraduate Courses Available for Graduate Credit: Economic Botany, Ethnobotany, Introduction to Genetics, Introductory Biochemistry, Biophysical Ecology, Immunobiology, Lectures in Cell and Molecular Biology, Laboratory in Cell and Molecular Biology, Ecology and Genetics of Populations, Systematic Botany, Field Botany, Phytogeography, Anatomy of Vascular Plants, Biology of Algae, Morphology and Evolution of Vascular Plants, Paleobotany, Plant Cytology, Introduction to Mycology, General Plant Ecology, Numerical Taxonomy and Estimation of Evolutionary Histroy, Introductory Plant Physiology, Physiology of Plant Development, Principles and Mechanisms of Organic Evolution, Quaternary Paleoecology, Paleoecology Laboratory, Natural History for Teachers, Advanced Instruction, Nuclear Cytology and Cytogenetics, Biophysical Chemistry, Mathematical Biology, Aquatic Flowering Plants, Biological Electron-Microscopy, Advanced Systematic Botany, Analysis of Development, Molecular Aspects of Gene Action, Tropical Biology: An Ecological Approach, Tropical Botany, Photobiology, The Physiology and Biochemistry of Mitochondria, Seminar in Evolutionary Biology, Seminar in Developmental Biology and Genetics, Plant Physiology Seminar, Ecology Seminar, Seminar on Quaternary Environment and Chronology

DEPARTMENT OF ZOOLOGY
Phone: Chairman: (313) 764-1467 Others: 764-1442

Chairman: C. Gans
Professors: R.D. Alexander, J.M. Allen, S.L. Allen, R.M. Bailey, R.E. Beyer, J.B. Burch, I.J. Cantrall, J.N. Cather, W.R. Dawson, F.C. Evans, M. Foster, B.E. Frye, C. Gans, H. Gay, K.F. Futhe, N.C. Hairston, E.T. Hooper, N.E. Kemp, L.J. Kleinsmith, M.M. Martin, R.R. Miller, T.E. Moore, G.W. Nace, T.M. Rizki, D.G. Shappirio, R.W. Storer, F.H. Test, D.W. Tinkle, H. van der Schalie
Associate Professors: S.S. Easter, B.A. Hazlett, D.H. Janzen, A.G. Kluge, L.A. Loewenthal, R.G. Northcutt, B. Oakley, R.B. Payne, G.R. Smith, J.H. Vandermeer
Assistant Professors: D.M. Allen, H.D. Blankespoor, J.P. Chamberlain, G.H. Jones, P.Kilham, R.A. Nussbaum, K.G Porter, R.R. Schafer
Lecturers: D. Baic, P.J. Scott
Teaching Fellows: 98
Research Assistants: 15
Postdoctoral Scholars: 6

Field Stations/Laboratories: University of Michigan Biological Station - Pellston, Camp Filibert Roth - Iron Mountain, Camp Davis - Jackson Hole, Wyoming

Degree Programs: Animal Behavior, Animal Genetics, Animal Physiology, Animal Science, Biochemistry, Biological Sciences, Biology, Ecology, Entomology, Ethology, Genetics, Histology, Immunology, Limnology, Medical Technology, Physiological Sciences, Physiology, Zoology

Undergraduate Degrees: B.S. (Biology, Medical Technology)
Undergraduate Degree Requirements: Introductory Biology, 16 hours Chemistry, 8 Physics, 8 Mathematics, 16 Zoology, including Genetics, Physiology, and two other areas, 6 advanced Natural Science, 2 years foreign language proficiency.

Graduate Degrees: M.S., Ph.D.
Graduate Degree Requirements: Master: 24 hours graduate credit, 6 of which must be in a cognate field, 12 in Zoology Doctoral: Area examination (written and oral), oral preliminary examination, two foreign language examinations (or mathematics substitute for second foreign language), 2 cognate courses, course requirements as determined by individual doctoral committee, thesis, final examination

Graduate and Undergraduate Courses Available for Graduate Credit: Genetics, Biochemistry, Biophysical Ecology, Immunobiology, Cell and Molecular Biology, Population Genetics, Regulation of Internal Environment, Physiology of Nervous Systems, Neurobiology, Cell and Developmental Genetics, Physiological Genetics, Endocrinology, Biological Effects of Radiation, Comparative Vertebrate Neurolnatomy, Biology of Invertebrates, Caribbean Marine Environments, Entomology, Biology of Insects, Limnology, Biology of Mammals, Natural History of Vertebrates, Habitats and Organisms, Ecology, Quantitative Ecology Laboratory, Ornithology, Animal Geography, Numerical Taxonomy, Animal Behavior and Evolution, Ethology, Biological Ultrastructure and Histogenesis, Vertebrate Developmental Biology, Evolutionary Ecology, Analysis of Fish Adaptation, Comparative Biochemistry, Nuclear Cytology and Cytogenetics, Evolutionary Aspects of Protein Biosynthesis, Biophysical Chemistry, Evolution and Systematics of Vertebrates, Physiology and the Environmental, Experimental Invertebrate Embryology, Mathematical Analogies in Evolutionary Biology, Experimental Limnology, Population and Community Ecology, Ecology of Animal Communities, Systematics and Evolution, Seminar in Herpetology (Reptiles); Herpetology (Amphibians), Animal Cytotaxonomy, Malacology, Birds of the World, Biological Electron Microscopy, Mammalian ReproductiveEndocrinology, Sensory Physiology, Developmental Neurobiology, Invertebrate Neurobiology, Analysis of Development, Neurochemistry, Molecular Aspects of Gene Action, Tropical Biology, Photobiology, Cell Physiology, Physiology and Biochemistry of Mitochondria, Animal Energetics, Endocrine Control of Development, Neuroscience

School of Natural Resources
Dean: C.E. Olsen, Jr.

SCHOOL OF NATURAL RESOURCES
Phone: Chairman: 764-6453

Chairman: (Vacant)
George Willis Pack Professor of Resource Economics: G. R. Gregory
Samuel Trask Dana Professor of Outdoor Recreation: S.R. Tocher
Professors: B.V. Barnes, W.R. Bentley, C.W. Cares, Jr., J. Carow, L.E.Craine, W.D. Drake, R.D. Duke, A.G. Feldt, F.F. Hooper, C.D. Johnson, W.J. Johnson, K.F. Lagler, J.W. Leonard, J.T. McFadden, D. N. Michael, R.L. Patterson, K.J. Polakowski, S.B. Preston, W.B. Stapp
Associate Professors: J.R. Bassett, J.R. Boyle, G.P.

Bruneau, J.W. Bulkley, A.B. Cowan, G. Fowler, R. Kaplan, S.D. Marquis, Jr., D.R. McCullough, H.L. Morton, G. Schramm
Assistant Professors: R.N.L. Andrews, B. Bryant, J.E. Crowfoot, A.L.Jensen, B.S. Low, D.N. McEwen, P. M. Pollack, J.W. Porter, C.J. Richardson, P.M. Sandman, R.F. Shceele, K. Shapiro, P. Webb, J. Witter, E.A.H. Woodman
Teaching Fellows: 35
Research Assistants: 20
Staff Assistants: 30

Degree Programs: Natural Resources, Forestry, Landscape Architecture, Regional Planning, Fisheries, Wildlife Management, Forestry

Undergraduate Degrees: B.S., B.S.F.
Undergraduate Degree Requirements: 120 hours for B.S. 130 hours for B.S.F.

Graduate Degrees: M.S., Ph.D.
Graduate Degree Requirements: Master: 30 hours. Six to twelve of these hours are thesis Doctoral: Not Stated

Graduate and Undergraduate Courses Available for Graduate Credit: Not Stated

Medical School
Dean: J.A. Gronvall

DEPARTMENT OF ANATOMY
(No information available)

DEPARTMENT OF BIOLOGICAL CHEMISTRY
Phone: Chairman: (313) 763-0203 Others: 764-8154

Chairman: M.J. Coon
Professors: B.W. Agranoff, I.A. Bernstein, H.N. Christensen, M.J. Coon, E.E. Dekker, D. Dziewiatkowski, I.J. Goldstein, G.R. Greenberg, W.E.M. Lands, L.H. Louis, V. Massey, J.L. Oncley, N. Radin
Associate Professors: D. Aminoff, W.D. Block, P.K. Datta, J.A. Fee, F.L. Hoch, D.E. Hultquist, M.J. Hunter, G.W. Jourdian, M. Ludwig, M. Mason, F. Medzihradsky, E.A. Napier, G.L. Nordby, D.L. Oxender, T.R. Riggs, J.A. Shafer, P.A. Weinhold, C.H. Williams, C. Wu, R. Zand
Assistant Professors: R.L. Armstrong, E.A. Duell, W.J. Ferrell, W.R. Fold, A.K. Hajra, V.C. Hascall, R.S. Kowalczyk, D. McCann, J. Memon, A. Payne, A.R. Price, J.H. Schacht, H.J. Whitfield
Instructors: D. Ballou, J. Vanderhoek, T. Vander Hoeven
Teaching Fellows: 30
Research Assistants: 40

Degree Programs: Biological Chemistry, Biophysics, Neurosciences, Medicinal Chemistry, Cellular and Molecular Biology

Graduate Degrees: M.S., Ph.D.
Graduate Degree Requirements: Master: enroll 2 terms, 24 credit hours of courses, no language requirement, no thesis requirement Doctoral: enroll 6 terms, 32 credit hours of courses, thesis required, no language requirement

Graduate and Undergraduate Courses Available for Graduate Credit: Introductory Biological Chemistry I and II Lecture and Laboratory, Biochemical Techniques I, II, and III Laboratory, Seminar I and II, Advanced Biological Chemistry I, II Lectures

DEPARTMENT OF HUMAN GENETICS
Phone: Chairman: 764-5490 Others: 764-5490

Chairman: J. Van Gundia Neel
Lee R. Dice University Professor: J.V. Neel
Professors: G.J. Brewer, E.H.Y. Chu, H. Gershowitz, M. Levine, J.V. Neel, D.L. Rucknagel, D.C. Shreffler, R.E. Tashian
Associate Professors: T.D. Gelehrter, C.F. Sing, R.D. Schmickel
Assistant Professor: P.E. Smouse
Teaching Fellows: 3
Research Assistants: 5
Research Associates: 16
Research Scientists: 11

Degree Program: Human Genetics

Graduate Degrees: M.S., Ph.D.
Graduate Degree Requirements: Master: Courses in two related fields with a minimum of 30 credit hours, 12 of which must be in human genetics. Usually only persons who already hold and advanced degree are admitted for this degree (M.D. or D.D.S.) Doctoral: Completion of specific course work to be determined individually by student and advisor. Completion of preliminary written examination in general genetics, biology, cytogenetics and population genetics, a research proposal, publication of dissertation and final oral examination.

Graduate and Undergraduate Courses Available for Graduate Credit: Principles of Human Genetics, Introduction to Human Population Genetics, Molecular and Cellular Genetics, Cytogenetics, Immunogenetics, Molecular Aspects of Gene Action, Methodology of Human Genetics, Seminar, Predissertation Research, Postdissertation Research

DEPARTMENT OF MICROBIOLOGY
Phone: Chairman: (313) 763-3531 Others: 763-3531

Chairman: F.C. Neidhardt
Professors: R. Freter, A.G. Johnson, E. Juni, L.L. Kempe, W.J. Loesche, W.H. Murphy, F.C. Neidhardt
Associate Professors: D.B. Clewell, S. Cooper, D.I. Friedman, R.H. Olsen, M.A. Savageau, F. Whitehouse
Assistant Professors: E.M Britt, A.J. Faras, D.A. Jackson, E.N. Jackson, N.R. Harvie, J.E. Nierderhuber, C. Shipman, Jr.
Visiting Lecturers: 2
Teaching Fellows: 15
Research Assistants: 15

Degree Program: Microbiology

Graduate Degrees: M.S., Ph.D.
Graduate Degree Requirements: Master: Courses will be selected in cooperation with advisory committee. A. thorough knowledge of Biochemistry is required. 28 credits (8 in microbiology) Doctoral: Course selection in co-operation with committee. Comprehensive examination. Thesis. Defense

Graduate and Undergraduate Courses Available for Graduate Credit: Microbiology for Dental Students, Microbiology for Engineers, The Microbial Cell (U), The Microbial Cell Laboratory (U), Microbial Interactions (U), Microbial Interactions Laboratory (U), Pathogenic Bacteriology, Pathogenesis of Infectious Diseases, Systems Analysis of the Microbial Cell, Industrial Microbiology, Molecular Aspects of Gene Action, Bacterial Viruses, Molecular Biology of Animal Viruses, Immunology, Microbial Metabolism and its Regulation, Advanced Immunology, Independent non-Dissertation, Research for Graduate Students

DEPARTMENT OF PHYSIOLOGY
Phone: Chairman: 764-4352

Chairman: H.W. Davenport
Professors: P.H. Abbrecht, M. Alpern, J. Bean, D.F. Bohr, H.W. Davenport, J.F. Faulkner, J. Jacquez, K. Jochim, R. Malvin, J.B. Ranck, Jr., L.T. Rutledge, A.J. Vander, W.S. Wilde

Associate Professors: K.L. Casey, B.J. Cohen, J.H. Sherman, H.V. Sparks
Assistant Professors: L.G. D'Alecy, M. Kluger, D. Mouw

Degree Program: Physiology

Graduate Degrees: M.S., Ph.D.
Graduate Degree Requirements: Master: The general requirements of the University of Michigan, School of Graduate Studies, Advanced Physiology and Biochemistry Doctoral: M.S. requirements plus statistics, one foreign language, preliminary examination, thesis

Graduate and Undergraduate Courses Available for Graduate Credit: Human Physiology, Mammalian Reproductive Endocrinology, Neurosciences Laboratory, Neurosciences Seminar, Directed Reading in Physiology I, II, III, IIIa, IIIb, The Use and Care of Laboratory Animals, Mechanism of Vision, Cardiovascular Regulation, Methods in Research Physiology, Hearing, Elementary Electrophysiology of Cells, Renal Physiology, Principles of Animal Surgery, Theoretical Physiology, Techniques in Physiology Instruction, Transport of Amino Acids and Sugars, Compartmental Analysis, Bioengineering Physiology, Bioengineering Physiology Seminar, Animal Models for Biomedical Research, Advanced Electrophysiology of Cells, Vertebrate Temperature Regulation, Seminar in Respiration, Cellular Physiology

School of Public Health

Dean: R. Remington

DEPARTMENT OF BIOSTATISTICS

Phone: Chairman: 764-5450 Others: 764-1817

Chairman: R.G. Cornell
Professors: J.A. Jacquez, F.E. Moore, M.E. Patno, M. A. Schork
Associate Professors: H.L. Johnson, B.M. Ullman (Research), R.B. Zemach (Adjunct)
Assistant Professors: A.M. Feldstein, J.D. Flora, Jr., J.W. McGuire, G.W. Williams
Lecturer: 1

Degree Program: Biostatistics

Graduate Degrees: M.P.H., Ph.D.
Graduate Degree Requirements: Master: 60 semester hours, including a core curriculum in public health and basic courses in statistical theory and methodology, more advanced courses in selected specialized areas and experience in the design and analysis of research projects. Doctoral: A year of advanced course work after the M.P.H. degree in Biostatistics or equivalent graduate study, successful completion of a preliminary examination on biostatistics and a cognate area and a dissertation involving the development of new methodology or the innovative application of known techniques to an important biomedical problem.

Graduate and Undergraduate Courses Available for Graduate Credit: Elements of Biostatistics I, Statistical Methods in Public Health, Data Processing, Elements of Statistical Decision Making, Statistics for Health Research I, II, III, Applications of Nonparametric Statistics in the Biomedical Sciences, Biostatistics I, II, Demographic Methods I, II, Statistical Methods in Biological Assay, Matrix Algebra and Analysis for Statistics, Regression Analysis, Analysis of Variance for Experimental Designs in the Health Sciences, Statistical Methods for Epidemiological Studies, Statistics in Health Service Programs, Health Applications of Statistical Decision Making, Applications of Stochastic Processes I, II, Health Applications of Nonparametric Statistics, Health Applications of Multivariate Analysis, Analysis of Categorical Data, Field Experience in Health Services Statistics, Seminar in Biostatistics, Large Sample Theory, Supervised Consulting, Compartmental Analysis, Readings in Biostatistics, Research in Biostatistics, Introduction to Linear Models, Nonparametric Statistics, Multivariate Linear Statistical Models

DEPARTMENT OF ENVIRONMENTAL AND INDUSTRIAL HEALTH

Phone: Chairman: (313) 764-3188

Chairman: M.S Hilbert
Professors: I.A. Bernstein, E.A. Boettner, D H. Byers, H.H. Cornish, R.A Deininger, J.J. Gannon, R. Hartung, I.T. Higgins, M.S Hilbert, H.J. Magnuson, K. H. Mancy, R.G. Smith, G.H. Whipple
Associate Professors: A.P. Jacobson, P.A. Plato, E.J. Siegenthaler
Assistant Professors: D.G. Brown, B. Chin, D.P. Chynoweth, R.H. Gray, L.H. Hecker, W.W. Joy, P.G. Meier, F.L. Vaughan
Instructors: H.E. Allen, G.L. Ball, C.P. McCord, G.W. Rose, H.B. Russelmann
Visiting Lecturers: 25
Research Assistants: 8
Post Doctoral Scholars: 2
Assistant Research Scientists: 2
Research Associates: 1
Research Associates II: 2
Laboratory Assistants: 1
Animal Technicians I: 1

Degree Programs: Environmental Health Science, Industrial Health

Graduate Degrees: M.S., M.P.H., D.P.H., Ph.D.
Graduate Degree Requirements: Master: 36 hours of credit (3 terms in 12-month period) If no previous experience, 60 hours of credit in 5 term period. Doctoral: Usually two-three years post-Master's.

Graduate and Undergraduate Courses Available for Graduate Credit: Environment and Health, Environmental Physiology, Principles of Environmental Health, Computer Applications in Environmental Health, Elements of Environmental Biology, Environmental Aspects of Cellular Chemistry and Function, Radiation in the Environment, Equipment Sanitation Standards, Design, and Evaluation, Environmental Health Administrative Practice, Housing Seminar, Environmental Health in Developing Areas, Economic Problems in Resources Administration Seminar, Environmental Planning, Principles of Microbiologic Decontamination, Environmental Chemistry, Fundamentals of Instrumental Methods of Chemical Analysis, Fundamentals of Electron Microscopy, Essentials of Toxicology, Industrial Toxicology, Ecological Toxicology, Methods in Toxicology, Principles and Methods of Industrial Health, Principles of Community Air Pollution, Health Factors in Air Pollution, Health Factors in Air Pollution, Water Quality Management, Water Quality Management Practices, Physics of Fluids and Plumbing Principles, Water Pollution Biology, Water Pollution Biology Field Survey and Laboratory Procedures, Water Bacteriology, Microbial Ecology, Water Resources Development, Water Quality, Field and Laboratory Procedures, Systems Analysis of Water Resource Systems, Environmental Systems Engineering, Readings in Environmental and Industrial Health, Research in Environmental and Industrial Health, Environmental Health Seminar, Special Problems in Solid Waste Engineering, Chemical Analysis of Water, Applied Electrochemistry, Instrumental Methods of Chemical Analysis, Physiochemical and Biochemical Methods of Separation, Chemical and Functional Mechanisms of Accommodation in Biological Systems, Tracer Techniques in Biological Research, Industrial Hygiene, Industrial Hygiene Control, Air Sampling and Analysis, Industrial Hygiene Sampling and Analysis, Ventilation Control of Contaminants, Industrial Safety, Industrial Health Seminar, Industrial Health Problems, Sampling Methods in Air Pollution, Air Pollution Seminar, Air Pollution Problems Analysis of Air Pollutants, Air Pollution Chemistry, Advanced Studies in Water Science and Engineering, Water Supply and Quality Control Seminar, Milk Sanitation, Advanced Milk and Food Laboratory, Advanced Food Sanitation and Technology, Radiological Health Seminar, Radiation Physics, Radiation Dosimetry, Radiation Biology, Sources and Control of Radioactive Waste, Applied Radiation Control, Field Experience, Seminar in Environmental Health Planning, Advanced Seminar in Air Pollution

DEPARTMENT OF EPIDEMIOLOGY
Phone: Chairman: (313) 764-5435 Others: 764-5453

Chairman: F.M. Davenport
Professors: W.W. Ackermann, G.C. Brown, K.W. Cochran, F.M. Davenport, E.A. Eckert, W.C. Eveland, I.T.T. Higgins, H.F. Maassab, F.E. Payne, R.J. Porter
Associate Professors: A.V. Hennessy, M.W. Higgins, E. Minuse, A.S. Monto
Assistant Professors: E.M. Dusseau, N.H. Maverakis, G.E. Moss (Adjunct)
Research Scientists: 2
Assistant Research Scientists: 2
Research Investigators: 2
Resident Lecturers: 3
Non-resident Lecturers: 2

Degree Programs: General Epidemiology, Epidemiologic Science

Graduate Degrees: M.P.H., D.P.H., M.S., Ph.D.
Graduate Degree Requirements: Master: M.P.H. (General Epidemiology) 36-60 credit hours, M.P.H. (Laboratory Practicum), 36-60 credit hours, M.S. (Epidemiologic Science), 24 credit hours Doctoral: Dr.P.H. (General Epidemiology), Dr.P.H. (Laboratory Practicum) Ph.D. (Epidemiologic Sciences) No set number of credit hours, program geared to student's needs.

Graduate and Undergraduate Courses Available for Graduate Credit: Strategy and Uses of Epidemiology, Principles and Methods of Epidemiology, Epidemiologic Techniques, Epidemiology of Infectious Deseases, Social Epidemiology, Concepts and Methods, Concepts of Parasitism, Parasitic Diseases IIIa, Laboratory in Parasitology, Medical Entomology, Virus Diseases, Virus Laboratory Methods, Advanced Virology, Methods in Experimental Virology, Epidemiology of Chronic Diseases, Epidemiological Methods Applied to Community Problems, Public Health Laboratory Practice, Advanced Public Health Laboratory Practice, Public Health Laboratory Methods, Fluorescent Antibody Laboratory Methods, Research in Public Health Laboratory Practice, Seminar in Public Health Laboratory Organization and Administration, Molecular Interpretations in Virology, Field Experience in Epidemiology, Reading Course in Tropical Disease, Advanced Studies in Epidemiology, Applications of Epidemiology, Readings in Epidemiologic Science I, II, Research in Tropical Diseases, I, II, III, IIIa, and IIIb, Research in Epidemiology, Research in Epidemiologic Science, Dissertation/Pre-Candidate, Dissertation/Candidate, Doctoral Thesis

DEPARTMENT OF POPULATION PLANNING
Phone: Chairman: (313) 764-7516

Chairman: L. Corsa
Chairman: S.B. Kar (Acting)
Professors: S.J. Behrman, H. Meyer, F.C. Munson, G.D. Ness, G.P. Olsson
Associate Professors: C. Chilman, J.D. Clarkson, J.W. Eliot, J.L. Finkle, T. Poffenberger, J.Y. Takeshita, A.A. Yengoyan
Assistant Professors: J.M. Fields, J.W. McGuire, J.Y. Peng, G. Simmons
Lecturers: L.W. Hoffman, D. Oakley, E.M. Weiss
Research Associates: 6
Visiting Lecturers: 4

Field Stations/Laboratories: Field Partnerships - Michigan, Malaysia

Degree Program: Population Plannint

Graduate Degrees: M.S., M.P.H., D.P.H.
Graduate Degree Requirements: Master: 36 credits
Doctoral: Varies according to background of student

Graduate and Undergraduate Courses Available for Graduate Credit: Foundations of Population Planning, Human Reproductive Biology, Population Problems, Administration Factors in Population Planning Programs, Communications in Population Planning, Methods of Evaluation and Research in Population Planning, Population Programs, Proseminar in Population Planning, Population and Human Affairs, Topics in Population Planning, Economics of Population Growth, Field Experience in Population Planning, Readings in Population Planning

UNIVERSITY OF MICHIGAN - FLINT

1321 E. Court Street Phone: (313) 767-4000
Flint, Michigan 48503
Chancellor: W.E. Moran

BIOLOGY DEPARTMENT
Phone: Chairman: (313) 767-4000 Ext. 305

Chairman: R.W. Dapson
Professors: R.W. Dapson, J.G. Otero, J. Taylor
Associate Professors: P.A. Adams, G.L. Pace, E.H. Studier
Lecturers: 2

Degree Program: Biology

Undergraduate Degree: B.S.
Undergraduate Degree Requirements: Departmental: 30 semester hours in Biology including a 4-course "core" program (Organismal Biology, Population Biology, Cell Biology and Genetics). College: English Composition (6 hours), Humanities (9 hours), Social Sciences (12 hours), specified sequence of courses in foreign language or mathematics.

WAYNE STATE UNIVERSITY

Detroit, Michigan 48202 Phone: (313) 577-2424
Dean of Graduate Studies: T.C. Rumble

College of Liberal Arts

Dean: M. Stearns

DEPARTMENT OF BIOLOGY
Phone: Chairman: (313) 577-2876

Chairman: A. Siegel
Professors: W. Chavin, D.R. Cook, T.J. Curtin (Adjunct), D.L. DeGiusti, A.G. Edward (Adjunct), S.K. Gangwere, S.B. Horowitz (Adjunct), J.M. Jay, S.T. Kitai (Adjunct), L. Levine, L.H. Mattman, K. Mayeda, W. Prychodko, F.L. Rights (Adjunct), C.M. Rogers, H.W. Rossmoore, A. Siegel, W.L. Thompson
Associate Professors: K.C. Chen, J.W. Cosgriff, Jr., W.E. Foor, W.W. Mathews, H. Mizukami, J.D. Taylor, R.R. Teodoro
Assistant Professors: K. Cost, D.J. Donaher, C.T. Duda, R.B. Goldberg, R.W. Harkaway, R.A. Hough, R.H. Monheimer, W.S. Moore, C.J. Swanson, J.S. VandeBerg, T. Waring (Adjunct), M. Weisbart, M.A. Yund
Research Associates: 3
Teaching Fellows: 56
Research Assistants: 5
Post Doctoral: 1

Degree Program: Biology

Undergraduate Degree: B.S.
Undergraduate Degree Requirements: (in quarter hours) Bachelors: 17 quarter credits in introductory courses;

General Chemistry, an additional 36 credits in Biology courses including 1 course in Genetics, 1 Morphology, 1 Physiology. 18 credits must be taken in residence. Foreign Language. B.S. requires 23 hours in Chemistry, 12 hours Mathematics, 12 hours Physics, French, German, Russian or Spanish.

Graduate Degrees: M.S., Ph.D.

Graduate Degree Requirements: Master: Plan C without a thesis; 45 hours of course work including certain core courses. Plan A with thesis, 36 hours course work plus 12 hours thesis, final oral examination. Graduate Record Examination and Advanced Test in Biology. Doctoral: A minimum of 135 quarter credits for completion of the Ph.D. program. 90 hours of course work should include the fields of Genetics, Morphology, Physiology, and supporting courses in Physics, Chemistry, and Mathematics. Graduate Record Examination and Advanced Test in Biology . Qualifying examinations and final oral defense.

Graduate and Undergraduate Courses Available for Graduate Credit: Radiobiology (U), Methods in Radiobiology (U), Biological Literature (U), History of Biology (U), Biometry (U), Microtechnique (U), General Cytology (U), Genetics (U), Methods in Genetics (U), Evolution (U), Limnology I (U), Biogeography (U), Bioecology Lecture (U), Biological Fine Structure (U), Bioecology Laboratory (U), Biology of Laboratory Animals (U), Laboratory: Biology of Laboratory Animals (U), Systematic Biology (U), Field Investigations in Biology (U), Principles of Natural Resource Management (U), Sanitary Bacteriology (U), Food Microbiology (U), Pathogenic Bacteriology I (U), Aquatic Microbiology (U), Immunology (U), Serology (U), Mycology (U), Plant Physiology (U), Plant Development (U), Systematic Botany (U), Invertebrate Zoology (U), Vertebrate Embryology (U), Analysis of Development (U), Vertebrate Histology (U), Natural History of Invertebrates (U), Vertebrate Physiology (U), Endocrinology (U), Methods in Endocrinology (U), Animal Behavior (U), Natural History of Vertebrates (U), Paleontology of Vertebrates (U), Ornithology (U), Mammalogy (U), Insect Biology I: Systematics and Morphology (U), Ichthyology (U), General Protozoology (U), Parasitic Protozoa (U), Biology of Parasitism I (U), Animal Behavior Laboratory (U), Cell Physiology I (U), Biological Instrumentation (U), Methods in Microbial Genetics (U), Techniques in Electron Microscopy I (U), Human Genetics (U), Genetics of Microorganisms (U), Population Genetics (U), Biosynthesis and Metabolism (U), Limnology II (U), Introductory Biophysics (U), General Bacteriology (U), Pathogenic Bacteriology II (U), Microbial Ecology (U), Aquatic Plants (U), Physiological Ecology (U), Neurophysiology (U), Comparative Physiology (U), Insect Biology II: Distribution and Behavior (U), Insect Biology III: Insects of Medical Importance (U), Biology of Parasitism II (U), Cell Physiology II, Gene Structure and Function, Methods in Microbial Genetics, Advanced Electron Microscopy: Physiological Genetics, Genetics of Microorganisms, Population Genetics, Evolution, Biological Fine Structure, Histogenesis, Limnology II, Biophysics I, Biophysics II, Physiology of Bacteria, Comparative Immunology, Advanced Pathogenic Bacteriology, Microbial Ecology, Microbial Ecology Laboratory, Genetics and Development of Fungi, Molecular Plant Physiology, Advanced Plant Physiology, Analysis of Development, Neurophysiology, Comparative Physiology, Endocrinology, Animal Behavior, Animal Behavior Laboratory, Ornithology, Mammalogy, Insect Biology II: Distribution and Behavior, Biology of Parasitism II, Methods in Parasitology, Helminthology, Research Problems in Biology, Special Topics in Biology, Recent Advanced in Parasitism

School of Medicine

40 E. Canfield Avenue
Detroit, Michigan 48201
Dean: R.D. Coye

DEPARTMENT OF ANATOMY

Phone: Chairman: (313) 577-1961

Chairman: C.A. Fox
Professors: M.H. Bernstein, B. Boving, C.A. Fox, M. Goodman, S.T. Kitai, G.W. Lasker, H. Maisel, D.B. Meyer, N.J. Mizeres, R. O'Rahilly
Associate Professors: C.R. Dutta, J. Rafols, R. Ramlau, M. Rodin, W. Schneider
Assistant Professors: J. Alcala, J. Beal, M. Cooper, J. DeFrance, A. Hamparian, J. Hazlett, L. Hazlett, D. Kennedy, I. LuQui, J.A. Mitchell, J. Plant (Associate), R. Pourcho, E. Tracy, P. Waggoner, A. Weinsieder
Instructor: J. Wood
Research Assistants: 2

Degree Program: Anatomy

Graduate Degree: Ph.D.

Graduate Degree Requirements: Doctoral: The requirements for the Ph.D. degree include 135 credits, of which 45 credits are in Dissertation Research and Direction and 33 credits are in the fundamental courses, Gross Anatomy, Neuroanatomy and Microscopic Anatomy. Additional course work in Physiology, Biochemistry and other basic medical sciences are strongly encouraged.

Graduate and Undergraduate Courses Available for Graduate Credit: Neuroanatomy, Cell and Tissue Ultrastructure, Gross Anatomy (Head and Neck; Thorax, Abdomen and Pelvis, Back and Limbs), Special Dissection, Immunobiology and Primate Evolution, Experimental Morphology, Histology, Principles of Neuroanatomy, Special Projects in Anatomy, Living Anatomy, Fetal and Neonatal Anatomy, Histological and Histochemical Techniques, Human Reproduction, Anatomy of the Visual System, Human Biology, Advanced Neuroanatomy, Autonomic Nervous System, Historical Aspects of Anatomy, Neurophysiology, Experimental Neuroanatomy and Neurophysiology, Comparative Neuroanatomy, Human Microscopic Anatomy, Human Developmental Anatomy, Experimental Embryology, The Fine Structure of the Nervous System, Directed Study in Physical Anthropology, Research, Seminar, Seminar in Neurophysiology, Special Topics in Anatomy, Master's Thesis Research, and Direction, Doctoral Dissertation Research and Direction

DEPARTMENT OF BIOCHEMISTRY

Phone: Chairman: (313) 577-1511 Others: 577-2218

Chairman: R.K. Brown
Professors: S.C. Brooks, Jr., R.K. Brown, A.C. Kuyper, J.M. Orten, S.N. Vinogradov
Associate Professors: D. Dabich, M.S. Doscher, A.R. Goldfarb, R.A. Hudson, L.I. Malkin, R.A. Mitchell, C.J. Parker, Jr., J.D. Shore (Adjunct), D. Tsernoglou
Assistant Professors: J.C. Bagshaw, A.S. Barrett, L. I. Grossman, P.H. Johnson, R.M. Johnson, J.J. Lightbody, Jr.
Instructors: G.A. Petsko, J. Rozhin
Teaching Fellows: 12
Research Assistants: 3
Research Associates: 7

Degree Program: Biochemistry

Graduate Degrees: M.S., Ph.D.

Graduate Degree Requirements: Master: 45 quarter hours including graduate level organic chemistry and thesis required (12 hours credit). Doctoral: 135 quarter hours including 30 hours of graduate level course work in biochemistry. Graduate level physical and organic chemistry and proficiency in one foreign language. Minors available in Organic Chemistry, Physical Chemistry, Physical-Organic Chemistry, Microbiology and Immunology, Physiology or Biology. Dissertation required (45 hours credit)

Graduate and Undergraduate Courses Available for Graduate Credit: General Biochemistry Lecture and Laboratory (U), General Biochemistry Lecture and Laboratory, Advanced Intermediary Metabolism, Biological Macromolecules, Enzymology, Bioenergetics, Advanced Biochemistry Laboratory, Nucleic Acids, Biochemistry of Disease, Seminar in Biochemistry

DEPARTMENT OF IMMUNOLOGY AND MICROBIOLOGY
Phone: Chairman: (313) 577-1594 Others: 577-1591

Chairman: N.R. Rose
Professors: R.S. Berk, C.D. Jeffries, M.A. Leon, M.D. Poulik (Adjunct), M.A. Rich (full time affiliate), F.L. Rights, C.C. Stulberg (full-time affiliate), A.R. Taylor, L.M. Weiner
Associate Professors: D.L. Boros, Y.M. Kong, S. Levine, R.H. Swanborg
Assistant Professors: L.D. Bacon, W.J. Brown (full time affiliate), B. Choe, P. Frost, J.D. Jollick, L.A. Jones, W.D. Peterson, Jr. (full time affiliate), V.F. Righthand, R.S. Sundick
Instructor: L. Carrick, Jr.
Visiting Lecturers: 2-4/month
Research Assistants: 3
Research Associates: 4
Associates of the Department: 3

Field Stations/Laboratories: Medical Research Building, 540 E. Canfield, (1st floor)

Degree Programs: Bacteriology, Bacteriology and Immunology, Genetics, Immunology, Medical Microbiology, Microbiology, Parasitology, Virology

Graduate Degrees: M.S., Ph.D.
Graduate Degree Requirements: Master: The minimum requirement is 45 credits, at least 36 of which must be taken at the University. Students must fiel a plan of work, including a thesis (a total of 12 credits), and complete all requirements within a six year time limit. (A credit is a quarter hour of work, four quarters are an academic year - one optional). Doctoral: The meeting of requirement is tested primarily by examinations and the presentations of the dissertation rather than a summation of courses, grades, and credits. Students have a seven year time limit, must complete a minimum of 136 credits and must be proficient in at least one foreign language.

Graduate and Undergraduate Courses Available for Graduate Credit: Bacteriology and Immunology, Pathogenic Fungi and Parasites, Pathogenic Microorganisms, Medical Mycology, Medical Microbiology I, Medical Microbiology II, Advanced Bacteriology, Descriptive Bacteriology, Clinical Microbiology Practice, Immunology, Immunology Laboratory, Recent Advances in Microbiology, Virology, Virology Laboratory, Bacterial Metabolism, Bacterial Metabolism Laboratory, Microbial Genetics, Techniques in Microbial Genetics

DEPARTMENT OF PHYSIOLOGY
(No reply received)

WESTERN MICHIGAN UNIVERSITY

Kalamazoo, Michigan 49001 Phone: (616) 383-1600
Dean of Graduate Studies: G.G. Mallinson

College of Arts and Sciences

Dean: C. Loew

BIOLOGY DEPARTMENT
Phone: Chairman: (616) 383-1674 Others: 383-1600

Chairman: C.J. Goodnight
Professors: A. Barbiers (Adjunct), W.G. Birch (Adjunct), R. Brewer, W.E. Dulin (Adjunct), K.B. Haas (Adjunct), I.V. Holt, K.T. Kirton (Adjunct), A.Robert (Adjunct), B. Schultz, P.B. Stott (Adjunct), L.C. VanderBeek, W.C. VanDeventer
Associate Professors: D.A. Buthala, R. Eisenberg, J.G. Engemann, G. Ficsor, D. Fowler, S.B. Friedman, E. Inselberg, J.J. Josten, J. Lawrence, R.W. Pippen, J.R. Schultz (Adjunct), G.C. Sud, J.S. Wood
Assistant Professors: L.J. Beuving, A.Y. Chang (Adjunct), R.C. Deur, A.R. Diani, D.W. DuCharme (Adjunct), W.E. Johnson, R.W. Olsen, P. Rutherford, D. Schumann, M. Spradling
Teaching Assistants: 29
Research Assistants: 1
Summe Research Assistants: 7

Degree Program: Biology

Undergraduate Degree: B.S.
Undergraduate Degree Requirements: Total hours: 30 including General Biology (3 hours), Zoology (3 hours), Botany (3 hours), Ecology (3 hours), General Physiology (3 hours), Genetics (3 hours), 12 hours in Chemistry including Organic, 8 hours in Physics, 8 hours in Mathematics

Graduate Degree: M.S.
Graduate Degree Requirements: Master: 30 hours of course work including either thesis or research paper.

Graduate and Undergraduate Courses Available for Graduate Credit: Selected Experiences in Biology (U), Human Ecology (U), Genetics of Eukaryotes (U), Heredity (U), Microbial Genetics (U), Recent Advances in Biology (U), Evolution (U), Virology (U), Physiology of Reproduction (U), Health Problems (U), Pathogenic Microbiology (U), Bacterial Physiology (U), Experimental Microbial Physiology (U), Cellular Physiology (U), Integrative Physiology (U), Comparative Animal Physiology (U), Systematic Botany (U), Phycology (U), Phytogeography (U), Paleobotany (U), Economic Botany (U), Biological Constituents (U), Mycology (U), Plant Physiology (U), Biology of Non-Vascular Plants (U), Biology of Vascular Plants (U), Environmental Education (U), Experimental Animal Physiology (U), Field Natural History (U), Animal Behavior (U), Cell and Organ Culture Techniques (U), Invertebrate Zoology (U), Entomology (U), Protozoology (U), Developmental Biology (U), Histology (U), General Cytology (U), Ornithology (U), Animal Ecology (U), Ecology of Southwestern Michigan (U), Parasitology (U), Plant Ecology (U), Limnology (U), Physiological Ecology (U), Immunology (U), Tropical Marine Ecology (U), Tropical Terrestrial Ecology (U), Radiation Biology (U), Biology of Lower Vertebrates (U), Biology of Higher Vertebrates (U), Readings in Biology (U), Independent Studies in Biology (U), Special Investigations, Seminar (various areas), Master's Thesis, Independent Research, Professional Field Experience, Specialist Project

College of Applied Science

Dean: W.C. Fitch

DEPARTMENT OF AGRICULTURE
Phone: Chairman: (616) 383-1986

Head: L.O. Baker
Professor: L.O. Baker
Assistant Professor: C. Stuewer
Instructor: L. Harris

Degree Program: Agriculture

Undergraduate Degree: B.S.
Undergraduate Degree Requirements: 122 semester hours graduation, 30 semester hours Agriculture

MINNESOTA

AUGSBURG COLLEGE
7th Street and 21st Avenue Phone: (612) 332-5181
Minneapolis, Minnesota 55404
Dean of the College: K.C. Bailey

BIOLOGY DEPARTMENT
Phone: Chairman: (612) 332-5181 Ext. 548

Chairman: R.L. Sulerud
Professor: R.L. Sulerud
Associate Professors: E.D. Mickelberg, N.O. Thorpe
Assistant Professors: R.S. Herforth
Teaching Assistant: 1

Degree Program: Biology

Undergraduate Degree: B.A.
Undergraduate Degree Requirements: Plan 1 - 7 courses Biology, 4 courses Chemistry, 2 courses Physics, Mathematics: Introductory Calculus Plan 2 - 9 courses Biology, 3 courses Chemistry

BEMIDJI STATE COLLEGE
14th and Birchmont Phone: (218) 755-2920
Bemidji, Minnesota 56601
Dean of Graduate Studies: B. Sellon
Dean of College: R. Beitzel

DEPARTMENT OF BIOLOGY
Phone: Chairman: (218) 755-2920 Others: 755-2921

Chairman: F.M. Saccoman
Professors: E. Hazard, E. Nordheim, H.T. Peters
Associate Professors: R. Baker, H. Borchers, C. Holt D. Kraft, R. Melchlor, W. Wanek
Assistant Professors: A. Lindgren, P. Trihey

Field Stations/Laboratories: Bald Eagle Center - Cass Lake, Minnesota

Degree Programs: Biology, Medical Technology

Undergraduate Degree: B.A., B.S.
Undergraduate Degree Requirements: 56 quarter hours

Graduate Degree: M.A. (Biology)
Graduate Degree Requirements: Master: 45 quarter hours

Graduate and Undergraduate Courses Available for Graduate Credit: Radiation Biology, Organic Evolution, Biological Techniques, Conservation of Natural Resources, Plant Ecology, Nuclear Science, Pollution Ecology, Molecular Biology, Biophysics, Field Zoology for Elementary Teachers, Field Botany for Elementary Teachers

BETHEL COLLEGE
3900 Bethel Drive Phone: (612) 641-6400
St. Paul, Minnesota 55112
Dean of College: V.A. Olson

DEPARTMENT OF BIOLOGICAL SCIENCES
Phone: Chairman: (612) 641-6313 Others: 641-6400

Chairman: P.J. Christian
Professor: P.J. Christian
Associate Professor: R.W. Johnson
Assistant Professor: T.L. Goff

Degree Programs: Biology, Botany, Medical Technology, Zoology

Undergraduate Degree: B.A.
Undergraduate Degree Requirements: 34 courses: 17 courses in General Arts (Including 3 interim courses) 17 courses in Biology and Cognate Areas.

CARLETON COLLEGE
Northfield, Minnesota 55057 Phone: (507) 645-4431
Dean of College: B. Morgan

BIOLOGY DEPARTMENT
Phone: Chairman: Ext. 422

Chairman: P. Jensen
Professors: P. Jensen, W.H. Muir, R.L. Shoger
Assistant Professors: S.M. Dendinger, G.J. Hill, M.C. McCutchan, G.E. Wagenbach
Instructor: R. Lindner
Technician in Biology: 1

Field Stations/Laboratories: Is member of consortium - Associates Colleges of the Midwest - which operates Wilderness Field Station at Ely, Minnesota

Degree Program: Biology

Undergraduate Degree: B.A.
Undergraduate Degree Requirements: Total of 204 credits including Rhetoric, Language proficiency and 4 area distribution requirements: Major: minimum of 7 courses in Biology representing Organismic Biology, Population Biology and Biochemistry/Physiology. Introductory Physics, Introductory Chemistry and either Quantitative or Organic

COLLEGE OF SAINT BENEDICT
(see St. John University)

COLLEGE OF ST. CATHERINE
2004 Randolph Avenue Phone: (612) 698-5571
St. Paul, Minnesota 55105
Dean of College: Sr. K. Kennelly

BIOLOGY DEPARTMENT
Phone: Chairman: Ext. 324

Chairman: Sr. E. Wittry
Professors: Sr. T. Judd, Sr. E. Wittry
Associate Professors: C. Lennon, Sr. H.J. Sanschagrin
Assistant Professor: F. Vukmonich

Degree Program: Biology

Undergraduate Degree: B.S.
Undergraduate Degree Requirements: Based on semesters, Biological Science I, II, Cell Physiology, Genetics, 6 other upper division courses (4 credits each), 2 courses in chemistry required (4-5 courses recommended)

COLLEGE OF SAINT SCHOLASTICA
Duluth, Minnesota 55811 Phone: (218) 728-3631
Dean of College: Sr. M.O. Cahoon

DEPARTMENT OF BIOLOGY
Phone: Chairman: Ext.495

Chairman: J.L. McLaughlin
Professor: Sr. M.O. Cahoon
Associate Professor: J.L. McLaughlin
Assistant Professor: M. Stiedemann, Sr. D. Schroeder
Advanced Undergraduates to teach some of the Laboratory Sections.

Degree Program: Biology

Undergraduate Degree: B.A.
Undergraduate Degree Requirements: 12 courses in Biology, 6 courses in Chemistry, 3 courses in Physics, 19 elective courses, Directed Action Project (equal to 4 courses)

COLLEGE OF SAINT THERESA
(No reply received)

COLLEGE OF ST. THOMAS
2115 Summit Avenue Phone: (612) 647-5000
St. Paul, Minnesota 55105
Dean of Graduate Studies: J.A. Byrne
Dean of College: C.J. Keffer

DEPARTMENT OF BIOLOGY
Phone: Chairman: (612) 647-5343 Others: 647-5000

Chairman: R. Bland
Professors: P.J. Germann, L.J. McCann, J.F. McMillan
Associate Professor: W.B. Silverman
Assistant Professor: R. Bland, R. Meierotto
Instructor: J. Pavlek

Degree Program: Biology

Undergraduate Degree: B.A.
Undergraduate Degree Requirements: 32 courses, 2 English, 3 Theology, 2 Philosophy, 3 Foreign Language, 1 History, 1 Psychology, 1 Fine Arts, 2 Mathematics, 2 Physics, 4 Chemistry, 8 Biology, Physical Education-no credit, 3 electives

CONCORDIA COLLEGE
Moorhead, Minnesota 56560 Phone: (218) 299-4321
Dean of College: P. Dovre

BIOLOGY DEPARTMENT
Phone: Chairman: (218) 299-3085

Chairman: H H. Osborn
Professors: C. Paulson, O. Torstveit
Associate Professors: H.H. Osborn, J.R. Powers, N. Sundet
Assistant Professors: I. Johnson, T. McCune, R. Nellermoe, E. Torstveit, G. Van Amburg
Instructor: J. Curtiss
Technician: 1

Degree Program: Biology

Undergraduate Degree: B.A.
Undergraduate Degree Requirements: 32 semester credits of Biology, 122 credits in all

CONCORDIA COLLEGE, SAINT PAUL
275 N. Syndicate Phone: (612) 646-6157
St. Paul, Minnesota 55104
Dean of College: G. Meyer

DEPARTMENT OF BIOLOGY
Phone: Chairman: Ext. 287

Chairman: E. Warnke
Professor: O.B. Overn
Associate Professor: J.E. Buegel
Assistant Professors: R.E. Holtz, J.F. Surridge

Field Stations/Laboratories: Carver Nature Center - Carver County, Minnesota

Degree Program: Biology

Undergraduate Degree: B.S.
Undergraduate Degree Requirements: Not Stated

GUSTAVUS ADOLPHUS COLLEGE
(No reply received)

HAMLINE UNIVERSITY
1536 Hewitt Phone: (612) 641-2800
St. Paul, Minnesota 55104
Dean of College: K.L. Janzen

DEPARTMENT OF BIOLOGY
Phone: Chairman: (612) 641-2413 Others: 641-2800

Chairman: E.J. Wyatt
Professor: W.L. Downing
Associate Professor: E.J. Wyatt
Assistant Professors: J.J. Brennan, M.A. Crayton

Degree Programs: Biology, Environmental Studies, Medical Technology

Undergraduate Degree: B.A.
Undergraduate Degree Requirements: Graduate Preparatory Major: Includes pre-Medicine, 10 Biology courses (6 Core - A Electives), General Analytical and Organic Chemistry, College Physics - Advise Mathematics through Calculus, Non-Graduate Preparatory Majors: 8 Biology courses (6 plus - 2 electives), 2 additional Biology electives, Approved sequence in Chemistry, Physics or Mathematics

MACALESTER COLLEGE
Saint Paul, Minnesota 55105 Phone: (612) 647-6100

DEPARTMENT OF BIOLOGY
Phone: Chairman: (612) 647-6372 Others: 647-6100

Chairman: C.A. Welch
O.T. Walter Professor of Biology: C.A. Welch
Professors: J.A. Jones, E.J. Robinson, Jr. C.A. Welch
Associate Professors: E.P. Hill, J.R. Smail, R.A. Whitehead
Assistant Professor: G.V. Dahling

Field Stations/Laboratories: Katharine Ordway Natural History Study Area - Inver Grove Heights, Minnesota

Degree Program: Biology

Undergraduate Degree: B.S.
Undergraduate Degree Requirements: The biology major consists of eight courses of biology, two courses of Chemistry, and two elected courses in any department of the science division or the philosophy department.

MANKATO STATE COLLEGE
Mankato, Minnesota 56001 Phone: (507) 389-1111
Dean of Graduate Studies: W. Benson
Dean of College: E. Ehrle

DEPARTMENT OF BIOLOGICAL SCIENCES
Phone: Chairman: (507) 389-2786 Others: 389-2787

Chairman: L.W. Zell
Professors: V. Burton, H.T. Choe, D. Gordon, R. C Coomes, K. Krabbenhoft, A. Lund, W. McEnery, C. Sehe, L. Zell
Associate Professors; N. Ballard, W.G. Bessler, H. Engh, M. Frydendall, R. Hybertson, H.W. Quade
Assistant Professors: J. Frey, T.B. Johnson, D.

Nielsen
Graduate Teaching Assistants: 12

Degree Programs: Biological Education, Biology, Ecology, Medical Technology, Microbiology

Undergraduate Degree: B.S.
Undergraduate Degree Requirements: B.S. 60 credit hours, plus an appropriate supporting minor, B.S. (Teaching) 52 credit hours, B.A., 41 credit hours, Science B.S. (Teaching), 64 credit hours

Graduate Degrees: M.S., M.A.
Graduate Degree Requirements: Master: M.S., 45 credit hours, with thesis, 52 credit hours without thesis, M.A., 45 credit hours with thesis

Graduate and Undergraduate Courses Available for Graduate Credit: Stream Limnology, Terrestrial Field Ecology, Cytology, Organic Evolution, Principles of Systematic Biology, Cell Physiology, Histological Technique, Genetics, Experimental Genetic, Limnology, Parasitology, Entomology, Reproductive Physiology, Ornithology, Comparative Chordate Anatomy, Vertebrate Embryology, Animal Behavior, Comparative Neurology, Comparative Endocrinology, Mammalogy, Plant Physiology, Plant Taxonomy, Developmental Plant Anatomy, Mycology, Advanced Microbiology, Applied Microbiology, General Virology, Immunology, Pathogenic Microorganism, Determinative Microbiology (U), Laboratory Experiences in Biology, Teaching Methods and Materials (U), Aquatic Ecology, Advanced Cytology, Cytogenetics, Speciation, Advanced Entomology, Teratology, Neuroendocrinology, Morphology of Algae, Morphology of Embryophytes, Modern Trends in Biology Teaching, Modern Biology and Society

MOORHEAD STATE COLLEGE

Moorhead, Minnesota 56560 Phone: (218) 236-2011
Dean of Graduate Studies: B. McCashland
Dean of College: W.B. Treumann

DEPARTMENT OF BIOLOGY
Phone: Chairman: (218) 236-2576 Others: 236-2572

Chairman: R.J. Tolbert
Professors: M.H. Bartel, T.W. Collins, O.W. Johnson, J.L. Parsons, K.R. Skjegstad, R.J. Tolbert
Associate Professors: Y.C. Condell, R.H. Pemble, M. A. Shimabukuro
Assistant Professors: P.A. Harber, R.S. Weibust

Degree Programs: Biology, Biology (Teaching), Medical Technology

Undergraduate Degrees: B.S.
Undergraduate Degree Requirements: Biology: Cell Biology, General Zoology, Invertebrate, General Zoology: Vertebrate, General Botany I, General Botany II, Principles of Ecology, Genetics, Organic Evolution, Microbiology, Physiology, Seminar, General Chemistry I, II, III, Organic Chemistry I, II, III, Elementary Physics I, II, III, 12 credits Mathematics. Medical Technology: Cell Biology, General Zoology: Invertebrate, General Zoology: Vertebrate, Histology, Animal Physiology, Microtechnique, Microbiology, Parasitology, Immunology, General Chemistry I, II, III, Organic Chemistry I, II, III, Biochemistry I, II, 4 credits Mathematics, 52 credits Internship

ST. CLOUD STATE COLLEGE

St. Cloud, Minnesota 56301 Phone: (612) 255-0121
Dean of Graduate Studies: L. Gillett
Dean of College: W. Armstrong

DEPARTMENT OF BIOLOGICAL SCIENCES
Phone: Chairman: (612) 255-2037 Others: 255-2036

Chairman: C. Rehwaldt
Professors: H. Barker, C. Bruton, J. Coulter, A. Grewe, H. Hopkins, A. Hopwood, V. Johnson, M. Partch, D. Peterson
Associate Professors: T. Clapp, W. Ezell, D. Grether, R. Gundersen, S. Lewis, L. Lindstrom, J. McCue
Assistant Professors: D. Barker, K. Knutson, D. Kramer, D. Mork, J. Peck, C. Pou, N. Gonzalez
Research Assistants: 8

Degree Program: Biology

Undergraduate Degree: B.S.
Undergraduate Degree Requirements: Not Stated

Graduate Degree: M.S.
Graduate Degree Requirements: Not Stated

ST. JOHN'S UNIVERSITY (COLLEGE OF ST. BENETICT)

Collegeville, Minnesota 56321 Phone: (612) 363-2011
Dean of College: O.W. Perlmutter

DEPARTMENT OF BIOLOGY
Phone: Chairman: (612) 363-3492

Chairman: N.L. Ford
Professors: Fr. B. Niggemann, Sr. M. Grell
Associate Professors: N.L. Ford, G. Rolfson, N.K. Zaczkowski, Sr. D. Plantenberg
Assistant Professor: R. Henry
Instructor: B. Reaney
Undergraduate Assistants: 18

Degree Program: Biology

Undergraduate Degrees: B.A., B.S.
Undergraduate Degree Requirements: 9 courses in Biology (5 upper division), 2 years of Chemistry, 1 year of Physics, 1 year of Mathematics

ST. MARY'S COLLEGE

Winona, Minnesota 55987 Phone: (507) 452-4430
Dean of Graduate Studies: P. McClean
Dean of College: U. Scott

DEPARTMENT OF BIOLOGY
Phone: Chairman: Ext. 228

Chairman: R. Kowles
Professor: Br. C. Severin
Associate Professor: R. Kowles
Assistant Professors: D. McConville, R. Vose, A. Azluha, Br. V. Sieben
Research Assistants: 3

Field Stations/Laboratories: St. Mary's Hydrobiology Station - Homer, Minnesota

Degree Program: Biology

Undergraduate Degree: B.A.
Undergraduate Degree Requirements: 124 semester hours 52 semester hours in Biology, and related sciences, Research and Bachelor's thesis

Graduate Degrees: M.S., M.A.T.
Graduate Degree Requirements: Master: 30 semester hours Biology, Research and Master's thesis - Comprehensive Written Examinations

Graduate and Undergraduate Courses Available for Graduate Credit: Cell Biology (U), Molecular Biology (U), Developmental Biology (U), Human Ecology (U), Microbiology (U), Field Taxonomy (U), Aquatic Biology (U), Radiation Biology (U), Summers Only: (Not all the same Summer): Microbiology, Biochemistry, Mammalian Physiology, Plant Physiology, Cell Physiology,

Advanced Genetics, Evolution, Terrestrial Ecology, Aquatic Ecology

ST. OLAF COLLEGE
Northfield, Minnesota 55057 Phone: (507) 663-3100
Dean of College: W. Nelson

DEPARTMENT OF BIOLOGY
Phone: Chairman: (507) 663-3100

Chairman: H.D. Orr
Professors: H.D. Orr, H.W. Hansen, A.J. Peterson
Associate Professors: J. Zischke, A. Burton, M. Madson, D.J. Palm
Assistant Professors: E. Bakko, R. Goering

Degree Program: Biology

Undergraduate Degree: B.A.
Undergraduate Degree Requirements: General Biology, and 7 additional courses in Biology (one of which must be a level III Interim (one month course in January), a year of Chemistry is required

SOUTHWEST MINNESOTA STATE COLLEGE
Marshall, Minnesota 56258 Phone: (507) 537-6194
Dean of College: R. Frazier

BIOLOGY DEPARTMENT
Phone: Chairman: (507) 537-6194

Chairman: L.A. Halgren
Professors: E. Hsi, T. Surdy
Associate Professors: L. Barker, L. Halgren, A. Holmes

Degree Programs: Biology, Medical Technology

Undergraduate Degree: B.S.
Undergraduate Degree Requirements: Minimum 41 quarter credits Biology, minimum 15 quarter credits Chemistry, minimum 12 quarter credits Physics, minimum 8 quarter credits Mathematics, minimum 56 quarter credits General Studies, Minimum Institution graduation requirement 180 quarter credits

UNIVERSITY OF MINNESOTA
Minneapolis/St. Paul, Minnesota 55455/55101
Dean of Graduate School: M. Brodbeck

College of Agriculture
St. Paul, Minnesota 55105 Phone: (612) 373-2851
Dean: A.J. Linck

DEPARTMENT OF ANIMAL SCIENCE
(No reply received)

DEPARTMENT OF AGRONOMY AND PLANT GENETICS
Phone: Chairman: (612) 373-0866 Others: 373-0855

Head: H.W. Johnson
Professors: R.N. Andersen, D.K. Barnes, R. Behrens, L.W. Briggle, V.E. Comstock, L.J. Elling, A.W. Hovin, J.W. Lambert, G.C. Marten, G.R. Miller, D.N. Moss, H.J. Otto, D.C. Rasmusson, R.G. Robinson, A.R. Schmid, L.H. Smith
Associate Professors: W.A. Brun, V.B. Cardwell, R.E. Heiner, D.R. Hicks, E.A. Oelke, R.L. Phillips, O.E. Strand, R.E. Stucker, D.D. Stuthman, R.L. Thompson
Assistant Professors: W.A. Elliott, J.L. Geadelmann, B.G. Gengenbach, C.E. Green, N.P. Martin, D.L. Wyse
Teaching Fellows: 1
Research Assistants: 25
Post-Doctorals: 2

Degree Programs: Agronomy, Plant Breeding, Plant Physiology, Genetics

Undergraduate Degree: B.S. (Agronomy)
Undergraduate Degree Requirements: 192 quarter credits 21 credits in Communication, Language and Symbolic Systems, 45 credits in Physical Sciences, 14 credits in Man and Society, 8 credits in Artistic Expression, and 36 credits in Major.

Graduate Degrees: M.S., Ph.D.
Graduate Degree Requirements: Master: Residence requirement of three quarters, 20 credits in major field and 8 in one or more related fields, minimum grade point average of 2.8, no language requirement, thesis Doctoral: Residence requirement of 9 quarters, minimum grade point average of 3.0, no minimum credit requirement, degree granted on basis of mastery of subject as indicated in a written and an oral comprehensive examination and thesis.

Graduate and Undergraduate Courses Available for Graduate Credit: Problems in Agronomy for Advanced Students (U), Pasture and Grassland Crops (U), Weed Control (U), Adaptation, Distribution and Production of Field Crops (U), Growth, Development and Culture of Field Crops (U), Maturation, Harvest and Storage of Field Crops (U), Research in Agronomy, Agronomy Seminar, Advanced Weed Science, Physiology of Field Crops, Pasture and Forage Research Techniques, Applied Statistics, Introduction to Plant Breeding (U), Principles of Plant Breeding I, Principles of Plant Breeding II, Application of Quantitative Genetics to Plant Breeding, Cytogenetics, Plant Genetics in Relation to Plant Improvement, Plant Breeding Seminar, Current Topics in Plant Breeding, Orientation to Field Crop Breeding, Research in Plant Genetics

DEPARTMENT OF ENTOMOLOGY, FISHERIES, AND WILDLIFE
Phone: Chairman: (507) 373-1701

Head: A.C. Hodson
Professors: M.A. Brooks, H.C. Chiang, E.F. Cook, L.K. Cutkomp, D.L. Frenzel, B. Furgata, P.K. Harein (Entension), A.C. Hodson, H.M. Kulman, J.A. Lofgren (Extension), W.H. Marshall, A.G. Peterson, R.D. Price, A.G. Richards, L.L. Smith, T.F. Waters
Associate Professors: E.B. Radcliffe, P. Jordan
Assistant Professor: J.H. Cooper
Instructor: D.M. Noetzel (Extension)
Research Associate: S. Broderius
Teaching Fellows: 7
Research Assistants: 32

Field Stations/Laboratories: Itasca Forestry and Biological Station - Cedar Creek, Natural History Area

Degree Programs: Entomology, Wildlife, Fisheries

Undergraduate Degree: B.S.
Undergraduate Degree Requirements: Not Stated

Graduate Degrees: M.A., Ph.D.
Graduate Degree Requirements: Master: Residence requirement of three quarters, 20 credits in major field and 8 in one or more related fields, minimum grade point average of 2.8, no language requirement, thesis Doctoral: Residence requirement of 9 quarters, minimum grade point average of 3.0, no minimum credit requirement, degree granted on basis of mastery of subject as indicated in a written and an oral comprehensive examination and thesis

Graduate and Undergraduate Courses Available for Graduate Credit: Entomology: Field Entomology (U), House and Garden Insects (U), Insect Morphology (U), Embryology and Development of Insects (U), Insect Metabolism and Coordination (U), Forest Entomology (U), Aquatic Entomology (U), Aquatic Entomology (U), Insect Taxonomy (U), Principles of Systematic Entomology (U), Apiculture (U), Integrated Control (U), Insects in Relation to Plant Diseases, (U), Principles of Economic Entomology (U), Medical Entomology (U), Experimental Ecology (U), Special Lectures in Entomology (U), Problems in Microtechnique (U), Biological Microscopy (U), Current Topics in Forest Entomology, Experimental Ecology Laboratory, Insect Ecology, Biology of Immature Insects, Topics in Insect Physiology, Insect Microbiology, Insecticides and their Action, Insecticides Laboratory Fisheries: Basic Fishery Biology, Aquatic Entomology (U), Experimental Ecology (U), Ecology of Fishery Populations (U), Fishery Management (U), Techniques of Fishery Biology (U), Fishery Ecology in Polluted Waters (U), Fisheries and Wildlife Administration (U), Fishery Biology, Production Biology of Fishery Environments Wildlife: Basic Wildlife Biology (U), Wildlife Ecology and Management I (U), Wildlife Ecology and Management II (U), Fisheries and Wildlife Administration (U), Wildlife Management: Upland Game, Wildlife Management: Waterfowl, Wildlife Management: Big Game

DEPARTMENT OF HORTICULTURE SCIENCE AND LANDSCAPE ARCHITECTURE
Phone: Chairman: (612) 373-1028 Others: 373-1026

Chairman: A.A. Duncan
Professors: D.W. Davis, A.A. Duncan, C.G. Hard, (Extension), A. Kallio (Extension), F.I. Lauer (Research), R.E. Nylund, L.C. Snyder (Extension and Director of Landscape Arboretum), E.J. Stadelmann, O.C. Turnquist (Extension), D.B. White, R.E. Widmer
Associate Professors: P.D. Ascher, L.B. Hertz (Extension), P.H. Li (Research), J.P. McKinnon (Extension), R. Mullin, H.M. Pellett, P.E. Read, C. Stushnoff, H.F. Wilkins (Research and Extension)
Assistant Professors: M.L. Brenner, M.J. Burke, S.L. Desboroligh (Research), R.H. Forsyth, M. Eisel (Extension), S.T. Munson (Research), P.J. Olin, L.R. Parsons
Research Associates: A.G. Johnson, M.H. Smithberg
Research Assistants: 40
Research Fellow: 1

Field Stations/Laboratories: Landscape Arboretum - Chaska, Minnesota, Horticulture Research Center - Excelsior, Minnesota, Horticulture Center - Duluth, Minnesota

Degree Programs: Horticulture, Landscape Horticulture

Undergraduate Degree: B.S.
Undergraduate Degree Requirements: Horticulture - 4 years, Landscape Architecture - 4 years - accredited program

Graduate Degrees: M.S., Ph.D.
Graduate Degree Requirements: Not Stated

Graduate and Undergraduate Courses Available for Graduate Credit: Not Stated

DEPARTMENT OF PLANT PATHOLOGY AND PHYSIOLOGY
(No reply received)

DEPARTMENT OF SOIL SCIENCE
Phone: Chairman: (612) 373-1063 Others: 373-1062

Head: W.P. Martin
Professors: R.S. Adams, H.F. Arneman, D.G. Baker, G.R. Blake, A.C. Caldwell, R.S. Farnham, R.G. Gast, J. Grave, L.D. Hanson, W.E. Larson, J.M. MacGregor, C.J. Overdahl, R.H. Rust, E.L. Schmidt, C.A. Simkins
Associate Professors: C.E. Clapp, R.H. Dowdy, S.D. Evans, E E. Fenster, G.E. Ham, J.A. Molina, J.P. Swan
Assistant Professors: P.P. Antoine, H.R. Finney, D.F. Grigal, C.F. Halsey
Research Assistants: 12

Field Stations/Laboratories: Southern Experiment Station - Waseca, Minnesota, North Central School and Experimental Station - Grand Rapids, Minnesota, Southwest Experiment Station - Lamberton, Minnesota, West Central School and Experiment Station - Morris, Minnesota, Northwest Experiment Station - Crookston, Minnesota

Degree Program: Soil Science

Undergraduate Degree: B.S.
Undergraduate Degree Requirements: Not Stated

Graduate Degrees: M.S., Ph.D.
Graduate Degree Requirements: Not Stated

Graduate and Undergraduate Courses Available for Graduate Credit: Not Stated

College of Biological Sciences
Dean: S.R. Caldecott

DEPARTMENT OF BIOCHEMISTRY
Phone: Chairman: (612) 373-1302

Head: F. Wold
Professors: V.A. Bloomfield, S. Dagley, J.E. Gander, R.L. Glass, L.M. Henserson, R. Jenness, S. Kirkwood, I.E. Liener, H. Schlenk, U.S. Seal, H.R. Warner, F. Wold, P. Rogers (Microbiology), J.M. Wood (Freshwater Biology Institute)
Associate Professors: J.S. Anderson, P.J. Chapman, R. E. Lovrien, E. Muenck (Freshwater Biology Institute), R.E. Barnett (Chemistry
Assistant Professors: J.A. Fuchs, K.G. Mann, G.L. Nelsestuen, C.K. Woodward, G.R. Gray (Chemistry)
Visiting Lecturers: 1
Research Assistants: 15
Teaching Assistants: 5
Postdoctoral Fellows: 12

Field Stations/Laboratories: The Department has close associations with the Freshwater Biological Institute, an independent research unit in the College of Biological Sciences.

Degree Program: Biochemistry

Undergraduate Degree: B.S.
Undergraduate Degree Requirements: Completion of a total of not less than 180 credits with grade of A, B, C, or S.

Graduate Degrees: M.S., Ph.D.
Graduate Degree Requirements: Master: Thesis research, advanced level proficiency in chemistry and quantitative biology Doctoral: Thesis research, advanced level proficiency in chemistry and quantitative biology.

Graduate and Undergraduate Courses Available for Graduate Credit: Biophysical Chemistry: Structure, Energetics, Dynamics, General Biochemistry, General Biochemistry Laboratory, Advanced Biochemical Techniques, Tracer Techniques, Special Topics in Biochemistry: Carbohydrates, Enzymes, Lipids, Nucleic Acids, Proteins, Vitamins, Graduate Research, Graduate Seminar, Graduate Courses in Chemistry, Biophysics, Genetics and Cell Biology, Microbiology, Plant and Animal Sciences, Health Sciences

DEPARTMENT OF BOTANY
Phone: Chairman: (612) 373-2211 Others: 373-2851

Head: A.W. Frenkel
Professors: E.C. Abbe (Emeritus), E.J. Cushing, A.W. Frenkel, E. Gorham, J.W. Hall, H. Jonas, D.B. Lawrence, T. Morley, G.B. Ownbey, D C. Pratt
Associate Professors: W.L. Koukkari, D.J. McLaughlin, T.K. Soulen, C.M. Wetmore
Assistant Professors: I.D. Charvat, A.M. Hirsch

Field Stations/Laboratories: Lake Itasca Biological Station - Cedar Creek Natural History Area

Degree Programs: Biology, Botany, Plant Physiology

Undergraduate Degree: B.S.
Undergraduate Degree Requirements: 180 credits total, including 29 credits in Biology, 15 credits Physics, 24 credits Chemistry, 15 credits English, and 25 credits additional in Biology, Physical Science, or Mathematics

Graduate Degrees: M.S., Ph.D.
Graduate Degree Requirements: Master: Plan A: with thesis, 20 credits Botany, 8 credits related fields Plan B: without thesis, 44 credits including 20 credits Botany, 8 credits related fields Doctoral: No special course requirements beyond Master's.

Graduate and Undergraduate Courses Available for Graduate Credit: Biology of Nonvascular Plants, Morphology of Vascular Plants, Developmental Plant Anatomy, Survey of Plant Physiology and Laboratory, Plant Metabolism, Water Minerals and Translocation, Plant Growth and Development, Flora of Minnesota, Survey of Angiosperm Families, Principles of Antiosperm Phylogeny, Introduction to the Study of Algae, Summer Flora of Minnesota, Aquatic Flowering Plants, Freshwater Algae, Bryophytes, Lichens, Seminar, Special Topics, Research Problems

DEPARTMENT OF ECOLOGY AND BEHAVIORAL BIOLOGY
Phone: Chairman: (612) 373-5177

Head: J.R. Tester
Professors: E.V. Bakuzis, H.C. Chiang, E. Gorham, H. M. Kulman, D.B. Lawrence, W.H. Marshall, D.F. McKinney, D.F. Parmelee, R.E. Phillips, E.L. Schmidt, J. Shapiro, J.R. Tester, H.B. Tordoff, J.M. Wood, H.E. Wright
Associate Professors: F.H. Barnwell, E.C. Birney, R C. Bright, K.W. Corbin, E.J. Cushing, R.O. Megard, D.B. Siniff
Assistant Professors: C.D. Hopkins, P.J. Regal, R.J. Taylor
Research Assistants: 15

Field Stations/Laboratories: Cedar Creek Natural History Area and Lake Itasca Forestry and Biological Station

Degree Program: Ecology

Graduate Degrees: M.S., Ph.D.
Graduate Degree Requirements: Master: Plan A (with a thesis) 20 quarter credits in major, 9 quarter credits in minor Plan B (additional course work and special projects) (substitute for a thesis) 44 quarter credits Doctoral: no specific credit requirement in major minimum of 18 quarter credits in minor

Graduate and Undergraduate Courses Available for Graduate Credit: Ecology of Plant Communities, Nutrients and Energy in Terresterial Ecosystems, Ecological Plant Geography, Predators, Predation Laboratory, Animal Behavior, Behavioral Adaptations, Physical Aspects of Field Biology I, II, III, Advanced Ecysystem Analysis, Population Ecology, Evolutionary Ecology, Limnology, Case Studies in Weather in the Biosphere, Aquatic Ecology, Topics in Limnology, Community Ecology, Field Ethology, Ecology of Fresh Water Algae, Vertebrae Ecology, Quantitative Ecology, Soils and the Ecosystem, Wetland Ecology, Quaternary Paleoecology, Quantitative Aspects of Ecological Systems, Analysis and Modeling of Ecological Systems, Advanced Work in Ecology and Behavioral Biology, Advanced Limnology, Methods for Analysis of Natural Waters

DEPARTMENT OF GENETICS AND CELL BIOLOGY
Phone: Chairman: (612) 373-0966 Others: 373-2851

Head: F. Forro, Jr.
Regents Professor: R.E. Comstock
Professors: V.E. Anderson, F.E. Enfield, R.K. Herman, S.C. Reed, M.D. Rosenberg, J. Rubenstein, D.P. Sunstad, L.A. Snyder, V.W. Woodward
Associate Professors: W.P. Cunningham, A.B. Hooper
Assistant Professors: C.S. Deppe, J.R. Sheppard
Visiting Lecturers: 1
Post-Doctorals: 4

Field Stations/Laboratories: Lake Itasca Field Biology Station - Cedar Creek Biology Station

Degree Programs: Biology, Cell Biology, Genetics

Undergraduate Degree: B.S. (Biology)
Undergraduate Degree Requirements: Calculus, Physics, General Chemistry, Organic Chemistry, Core Biology Courses in Organismic Biology, Biochemistry, Genetics and Cell, Biology, Ecology

Graduate Degrees: M.S., Ph.D.
Graduate Degree Requirements: Master: Genetics - 2 of 3 genetics core courses, 44 credits, oral examination Cell Biology - 2 cell biology core courses, 44 credits, oral examination Doctoral: Genetics - 3 genetics core courses, preliminary oral and written examinations, dissertation, Cell Biology - 2 cell biology core courses, preliminary oral and written examinations, dissertation

Graduate and Undergraduate Courses Available for Graduate Credit: Intermediate Genetics I, II, III (U), Population, Quantitative Genetics (U), Human Genetics (U), Methods in Human Genetics (U), Intermediate Cell Biology (U), Quantitative Techniques in Cell Biology (U), Membranes and Interfaces (U), Biochemistry of Behavioral Genetics, Cellular Regulation, Current Topics, Cell Biology Laboratory (U), Laboratory: Genetics (U), Electron Microscopy

DEPARTMENT OF ZOOLOGY
Phone: Chairman: (612) 373-3649 Others: 373-2851

Head: M. Olson
Professors: H. Chiang (Adjunct), A. Hodson (Adjunct), N. Kerr, R.G. McKinnell, D.J. Merrell, A.G. Richards (Adjunct), M. Rosenberg (Adjunct), W.D. Schmid, O. H. Schmitt (Adjunct), N.T. Spratt, H.B. Tordoff (Adjunct), J.C. Underhill, F.G. Wallace, D.W. Warner
Associate Professors: F.H. Barnwell, M. Brooks (Adjunct), D.E. Gilbertson, W.S. Herman, A.B. Hooper (Adjunct), C.W. Huver, R.G. Johnson, D.F. McKinney (Adjunct), J.D. Sheridan, A. Sinha
Assistant Professors: E.C. Birney (Adjunct), S.F. Goldstein, P.J. Regal
Teaching Fellows: 23
Research Assistants: 6

Field Stations/Laboratories: Lake Itasca, Cedar Creek

Degree Program: Zoology

Undergraduate Degree: B.S.
Undergraduate Degree Requirements: 180 credits total, including 29 credits Biology, 15 credits Physics, 24 credits Chemistry, 15 credits Mathematics, 30 credits Liberal Arts, 15 credits Foreign Language, 10 credits English, and 25 credits additional in Biology, Physical Science or Mathematics

Graduate Degrees: M.A., Ph.D.
Graduate Degree Requirements: Master: Plan A (with a thesis) 20 quarter credits in major, 9 quarter credits in minor Plan B (additional course and special projects) (substitute for a thesis) 44 quarter credits Doctoral: no specific credit requirement in major minimum of 18 quarter credits in minor

Graduate and Undergraduate Courses Available for Graduate

Credit: General and Comparative Embryology, Histology, Invertebrate Biology, Introductory Ornithology, Introduction to Animal Parasitology, Comparative Animal Physiology, Protozoology, Physiology of Excitable Cells, Ichthyology, Vertebrate Biology, Vertebrate Fauna Laboratory, Herpetology, Advanced Mammalogy, General and Comparative Endocrinology, General and Comparative Endocrinology Laboratory, General and Comparative Endocrinology, Parasitic Protozoa, Experimental Parasitology, Fine Structure of Animal Cells, Advanced Cytology Laboratory, Physiological Ecology, Genetics and Speciation, Natural History of Invertebrates, Natural History of Vertebrates, Field Ornithology, Animal Parasites, Special Topics in Comparative Endocrinology, Winter Ecology, Topics in Ecological Genetics

College of Forestry

St. Paul, Minnesota 55108
Dean: F.H. Kanfert

DEPARTMENT OF FOREST BIOLOGY

Phone: Chairman: (612) 373-0840 Others: 373-0825

Head: A.C. Mace, Jr.
Professors: E.V. Bakuzis, B.A. Brown, D.W. French (Associate Member), H.L. Hansen, H.M. Kulman (Associate Member), E.I. Sucoff
Associate Professors: D.F. Grigal (Associate Member), A.C. Mace, Jr., C.A. Mohn, H. Scholten
Assistant Professors: A.A. Alm, V. Kurmis
Instructor: A.P. O'Hayre
Research Assistants: 26

Field Stations/Laboratories: Lake Itasca Forestry and Biological Station and Cloquet Forestry Center

Degree Program: Forest Biology

Undergraduate Degree: B.S.
Undergraduate Degree Requirements: Completion of 192 quarter credits of which 38 credits are elective with an overall average of C or better.

Graduate Degrees: M.F., M.S., Ph.D.
Graduate Degree Requirements: Master: M.S. Plan A - 28 credits, 20 major, 8 related area, thesis, plus oral examination, Plan B - 44 credits, 20 major, 8 related area, plus oral examination M.F. 44 credits, 20 major, 8 related area plus oral examination, completion of core forestry sequence Doctoral: Course work is developed to meet the educational objectives of the student. May select a minor or supporting fields of 18-24 credits and a collateral field in lieu of a language. Completion of a dissertation and written and oral examination and required.

Graduate and Undergraduate Courses Available for Graduate Credit: Silviculture (U), Forest Biology Seminar, Field Silviculture (U), Principles of Silvics (U), Regional Silviculture (U), Forest-Tree Physiology (U), Multiple Use Silviculture (U), Forest Meterology and Hydrology (U), Advanced Forest Hydrology (U), Forest Genetics (U), Forest Ecosystems, Research Problems: Silviculture, Forest-Tree Physiology, Forest-Tree Genetics, Forest Hydrology

DEPARTMENT OF FOREST RESOURCES

Phone: Chairman: (612) 373-0840 Others: 373-0825

Head: A.C. Mace, Jr.
Professors: E.V. Bakuzis, B.A. Brown, H.L. Hansen, F.D. Irving, A.C. Mace, L.C. Merrian, M.P. Myer, W.R. Miles, R.A. Skok, E.I. Sucoff, K.A. Winsness
Associate Professors: P.V. Ellefson, D.J. Gerrard, H.M. Gregersen, A.R. Hallgren, C.A. Mohn, H. Scholten, M.E. Smith, E.H. White
Assistant Professors: A.A. Alm, T.B. Knopp, V. Kurmis, D.W. Rose, D.W. VanOrmer
Instructors: A.P. O'Hayre, K.N. Olson, R.W. Sando
Research Assistants: 65

Field Stations/Laboratories: Cloquet Forestry Center, Lake Itsaca Forestry and Biological Station

Degree Programs: Forest Resources, Forestry, Forestry and Conservation

Undergraduate Degree: B.S.
Undergraduate Degree Requirements: 192 quarter credits

Graduate Degrees: M.F., M.S., Ph.D.
Graduate Degree Requirements: Master: M.S. Plan A - 28 credits, 20 major, 8 related area, thesis, plus oral examination, Plan B - 44 credits, 20 major, 8 related area, plus oral examination M.F. - 44 crediot, 30 major, 0 related area plus oral examination, completion of core forestry sequence Doctoral: Course work is developed to meet the educational objectives of the student. May select a minor or supporitng fields of 18-24 credits and a collateral field in lieu of a language. Completion of a dissertation and written and oral examination are required.

Graduate and Undergraduate Courses Available for Graduate Credit: Fundamental Wood Properties I: Wood Fluid Relations, Fundamental Wood Properties II: Mechanical Properties, Fundamental Wood Properties III: Wood Chemistry, Fundamental Wood Properties IV: Wood Deterioration and Preservation, Wood Processing I: Drying and Impregnation Technology, Wood Processing II: Pulp and Paper Technology, Wood Processing III: Manufacturing Process, Design and Analsysi, Wood Processing IV: Fiberboard and Particleboard Technology, Advanced Pulp and Paper Technology, Manufactured Housing Systems, Pulp and Paper Process Calculations, Pulp and Paper Process Operations, Woody Tissue Microtechnique, Moisture Relations in Wood, Advanced Wood Deterioration and Preservation, Advanced Wood Chemistry, Mechanics and Structural Design with Wood Products, Advanced Forest Products Marketing, Pulp and Paper Technology: Special Topics, Interfacial Phenomena, Research Methods in Forestry, Aerial Photo Interpretation, Forest Inventory and Photographic Interpretation, Multiple Use, Forest Economics, Forest Management and Administration, Forest Mensuration, Forest Fire, Range Management, Management of Recreational Lands Principles of Outdoor Recreation Design and Planning, Forest Protection, Senior Seminar, Timber Harvesting, Interpretation of Aerial Photographs, Forest Policy, Advanced Forest Economics, Remote Sensing of Natural Resources, Advanced Forest Mensuration, Advanced Forest Management and Administration, Planning, Control in Forestry, Recreational Land Policy, Outdoor Recreation Economics, Recreation Land Amenities and the User, Advanced Management of Recreational Lands Forestry and Economic Development

Medical School

Minneapolis, Minnesota 55455
Dean: N.L. Gault, Jr.

DEPARTMENT OF ANATOMY

Phone: Chairman: (612) 373-2603

Head: A. Lazarow
Professors: A.M. Carpenter, P. Dixit, C. Heggestad, M. Smithberg, D. Sundberg
Associate Professors: E Bauer, O. Hegre, D. Robertson, R. Sorenson
Assistant Professors: D. Coulter, M. Fenstad, K. Goodman, L. Hoilund, P. Kuipers, I. Ram, A. Shemesh, R. Younoszai, E. Spedel (Research Associate), W. Schoener (Research Associate)
Instructors: G. Herbst, J. Baker (Research Fellow), R. Leonard (Research Fellow)
Teaching Assistants: 28

Field Stations/Laboratories: Field Station at Itasca State

Park

Degree Program: Anatomy

Graduate Degrees: M.S., Ph.D.
Graduate Degree Requirements: Master: 18 credits - major, 9 credits - minor, Thesis Doctoral: 35-40 credits - major, 18-20 credits - minor or supporting program, Thesis, 1 Foreign Language

Graduate and Undergraduate Courses Available for Graduate Credit: Not Stated

DEPARTMENT OF BIOCHEMISTRY
Phone: Chairman: (612) 373-3336

Head: C.W. Carr (Acting)
Professors: C.W. Carr, I.D. Frantz, Jr., H.R. Gutmann, R.T. Holman, J.F. Koerner, A. Rosenberg, L. Singer, Q.T. Smith (Adjunct), F. Ungar, J.F. Van Pilsum, D.B. Wetlaufer
Associate Professors: J.W. Bodley, M.E. Dempsey, R.D. Edstrom, E.D. Gray
Assistant Professors: J.B. Howard, R.J. Roon
Visiting Lecturers: 2
Teaching Fellows: 23
Research Assistants: 8

Degree Program: Biochemistry

Graduate Degrees: M.S., Ph.D.
Graduate Degree Requirements: Master: 18 credits - major, 9 credits - minor, Thesis Doctoral: 35-40 credits - major, 18-20 credits - minor or supporting program, Thesis, 1 Foreign Language

Graduate and Undergraduate Courses Available for Graduate Credit: Biochemistry (Medical Students) (U), General Biochemistry, Biochemistry Laboratory, Problems in Biochemistry, Advanced Endocrinology and Steroid Chemistry, Nucleic Acid Structure and Function, Topics in Lipid Metabolism, Protein Chemistry, Biochemistry of Specialized Tissues, Current Topics in Basic and Applied Enzymology, Carbohydrate Metabolism

DEPARTMENT OF MICROBIOLOGY
Phone: Chairman: (612) 373-8070

Head: D.W. Watson
Professors: K.G. Brand, F. Busta, M. Dworkin, V.W. Greene, W.H. Hall, H.M. Jenkin, R.K. Lindorfer, C.F. McKhann, P. Rogers, E.L. Schmidt, H.M. Tsuchiya, L.W. Wannamaker, D.W. Watson, H. Zinnerman
Associate Professors: D.L. Anderson, P. Chapman, A.Y. Elliott, B.H. Gray, J.M. Matsen, G. Needham, P.G. W. Plagemann, J.T. Prince
Assistant Professors: P.P. Cleary, A. Dennis, H.R. Gaumer, D. Klein, W. Liljemark, L. McKay, D. Peterson, J. Schmidtke, S. Sudo, J.F. Zissler
Instructors: H. Bauer, D. Blazevic, G.M. Ederer, A.B. Hooper, A.G. Karlson, R.L. Simmons
Research Associates: 8

Degree Program: Microbiology, Medical Microbiology

Undergraduate Degrees: B.S., B.A. (Microbiology
Undergraduate Degree Requirements: Not Stated

Graduate Degrees: M.S., Ph.D.
Graduate Degree Requirements: Master: Microbiology is offered only under the Plan A program which includes major and minor course work, satisfactory performance on a written qualifying examination and a research thesis. A final oral examination is required. Doctoral: Major course work: No minimal program in terms of credit hours can be specified for the Ph.D., the sole criterion being a high degree of competence in microbiology as evidenced by the academic record, the written examination, the thesis, and the oral examinations. The standard of proficiency in the general field of Microbiology for any Ph.D. candidate irrespective of his special field, is at least that required for the M.S. degree. The number and nature of the courses to comprise the major on the three year program will be determined by the advisor, as indicated by the interests and competence of the advisee. Microbiology courses or closely allied courses may make up the major.

Graduate and Undergraduate Courses Available for Graduate Credit: Biology of Microorganisms (U), Microbiology for Medical Students, Microbiology for Dental Students, Immunology (U), Medical Microbiology (U), Microorganisms and Disease (U), Physiology of Bacteria (U), Physiology of Bacteria Laboratory (U), Biology of Viruses (U), Ecology of Soil Microorganisms (U), Biology of Microorganisms, Microbial Genetics, Microbiology Laboratory, Advanced Immunology Laboratory, Oral Microbiology, Immunochemistry and Immunobiology, Advanced Medical Microbiology, Methods in Clinical Microbiology Part I, Diagnostic Bacteriology and Serology, Methods in Clinical Microbiology Part II. Medical Mycology, Parasitology, and Diagnostic Virology, Preceptorship in Medical Microbiology, Bacteriology, Fluorescent-Antibody, Antibiotics, Parasitology, Instrumentation, Preceptorship in Medical Microbiology, Mycobacteriology and Mycology, Diagnostic Microbiology, Laboratory Methods, Applied Animal Cell Culture and Virology, Regulation of Metabolism, Colloquium in Microbiology, Advanced Immunology, Research in Microbiology

DEPARTMENT OF PHYSIOLOGY
Phone: Chairman: (612) 373-2908

Head: Not Stated
Regents Professor Emeritus: M.B. Visscher
Professors: M. Bacaner, H.M. Cavert, I.J. Fox, E. Grim, R.B. Harvey, J.A. Johnson, N. Lifson, V. Lorber, A.E. Seeds, C. Terzuolo
Associate Professors: J. Bloedel, C. Lee, J. Lee, D. Levitt, M. Meyer, R. Poppele, R. Purple, A. Rescigno
Assistant Professors: G. Kepner, C. Knox, R. Kronenberg, J. Love, I. Martinson, R. Stish, O.D. Wangensteen
Instructors: G. Bloom, N. Paradise
Visiting Lecturers: 1
Teaching Fellows: 9
Research Assistants: 4
Research Fellows: 5

Degree Program: Physiology

Undergraduate Degree: B.A.
Undergraduate Degree Requirements: Required preparatory courses: one year each of college chemistry, Physics, and Mathematics (through integral Calculus), Major requirements: 3055, 3056 plus a total of 20 upper division credits in approved courses from related fields such as Mathematics, Physics and Chemistry.

Graduate Degrees: M.S., Ph.D.
Graduate Degree Requirements: Master: For a major or minor in Physiology, background in Mathematics, Physics, Chemistry and Morphology acceptable to the graduate faculty is required. In addition to transcript of prior course work, applicants are encouraged to take the Graduate Record Examination (verbal and quantitative sections). Programs are highly individualized and are developed to meet the needs of each student. Doctoral: The doctoral candidates, the 6-quarter sequence 8103 to 8108 is strongly recommended There is no language requirement for the M.S. degree For the Ph.D. degree, students, in consultation with their adviser, will elect to demonstrate a reading knowledge in one foreign language or to complete a collateral field of knowledge. Master's degree students will be given a final oral examination.

Graduate and Undergraduate Courses Available for Graduate Credit: Human Physiology, General Physiology, Neurophysiology, Cardiovascular Physiology, Respiratory Physiology, Alimentary Physiology, Nehprology, Systems Analysis for Biologists, Problems in Physiology, Biophysics of Nerve Cells, Mathematical Neurophysiology, Biophysical Approaches to Physiology, Chronophysiology, Physiology of Intrauterine Development, Literature Seminar, Readings in Physiology, Research in Physiology, History of Physiology, Selected Topics In Alimentary Physiology, Selected Topics in Permeability, Selected Topics in Heart and Circulation, Selected Topics in Respiration, Selected Topics in Nephrology, Selected Topics in Neurophysiology, Properties of Receptor Systems, Physiology of Visual Systems, Spinal Cord Physiology and Motor Control, Methods of Analysis, Methods in Physiology, Transport Process in Biology, Respiration, Acid-Base Chemistry, and Electrolyte Metabolism, Bioenergetics of Cardiac Contraction, Hemodynamic Measurements, Neural and Humoral Control of Circulation, Physiology of Lymphatic System and Microcirculation

UNIVERSITY OF MINNESOTA - DULUTH

Duluth, Minnesota 55812 Phone: (218) 726-8000
Dean of Graduate Studies: M.H. Lease
Dean of College: W.R. McEwen

DEPARTMENT OF BIOLOGY
Phone: Chairman: (218) 726-7263 Others: 726-7264

Head: T.O. Odlaug
Professors: J.B. Carlson, P.B. Hofslund, B.O. Krogstad, P.H. Monson, D.I. Mount (Adjunct), T.O. Odlaug
Associate Professors: G.E. Ahlgren, K.E. Biesinger (Adjunct), H.L. Collins, S.C. Hedman, K.E. Hokanson (Adjunct)
Assistant Professors: R.L. Anderson (Adjunct), C.E. Firling, W. Fluegel, D.Z. Gerhart, L.L. Holmstrand, J.H. McCormick (Adjunct)
Instructors: N.E. Fontaine, H.B. Hanten
Teaching Fellows: 6

Field Stations/Laboratories: Lake Superior Limnological Research Station - 6008 London Road, Duluth, Minnesota

Degree Programs: Biology, Botany, Zoology

Undergraduate Degree: B.S. (Biology)
Undergraduate Degree Requirements: 50 quarter credits in Biology plus 10 credits in General Chemistry, 5 credits in Organic Chemistry, 5 credits in College Mathematics, 28-32 credits minor, 48 credits in liberal education courses, Total 180 credits

Graduate Degree: M.S.
Graduate Degree Requirements: Master: Plan A - 18 credits in major, 9 in minor, thesis written and oral examination, Plan B - 45 credits in course work, 21 of which must be in major field, 18 credits in related fields. An in-depth paper in each of three advanced courses, Final written or oral examination

Graduate and Undergraduate Courses Available for Graduate Credit: Human Genetics (U), Organic Evolution (U), Plant Anatomy (U), Animal Physiology (U), Cell Metabolism (U), Biochemical Genetics (U), General Microbiology (U), Morphology of Non-Vascular Plants (U), Morphology of Vascular Plants (U), Plant Physiology (U), Plant Taxonomy (U), Physiology of Development (U), Advanced Plant Taxonomy (U), Comparative Anatomy of Invertebrates (U), Comparative Anatomy of Vertebrates (U), Ichthyology (U), Mammalogy (U), Ornithology (U), Entomology (U), Advanced Insect Biology (U), Animal Parasitology (U), Helminthology (U), Fish Biology (U), Animal Behavior (U), Limnology (U), Ecology of Animal Populations (U)

UNIVERSITY OF MINNESOTA - MORRIS

Morris, Minnesota 56267 Phone: (612) 589-1812
Dean of College: G. Bopp

DEPARTMENT OF BIOLOGY (DIVISION OF SCIENCE AND MATHEMATICS)
Phone: Chairman: (612) 589-1644 Others: 589-2211

Discipline Coordinator: R.S. Abbott
Professors: R.S. Abbott, J.Y. Roshal
Associate Professors: E. Ordway, T. Straw
Assistant Professor: K. Horton
Temporary Instructors: 2

Field Stations/Laboratories: University of Minnesota (See listing under Minneapolis Campus)

Degree Program: Biology

Undergraduate Degree: B.A.
Undergraduate Degree Requirements: Minimum degree requirements, all courses require laboratories, Introduction Biology: 10 quarter credits, Genetics 5 quarter credits, Microbiology 5 quarter credits, Cell Physiology 5 quarter credits, Ecology 5 quarter credits, Senior Seminar 1 credit, Elective in Comparative Biology 5 credits, Other Biology Electives 10 credits, General Chemistry 10 credits, Organic Chemistry 5 credits, Humanities 10 credits, Social Sciences 10 credits, 180 credits total

WINONA STATE COLLEGE

(No reply received)

MISSISSIPPI

ALCORN STATE UNIVERSITY

P.O. Box 870 Phone: (601) 877-3711
Lorman, Mississippi 39096
Dean of Graduate Studies: N.A. Edney

DEPARTMENT OF BIOLOGY
Phone: Chairman: Ext. 249

Chairman: N.A. Edney
Professors: N.A. Edney, H.L. Parker, M.P. Sharma
Associate Professors: A. Latif, V.R. Lawson, A. Russell
Assistant Professors: O.N. Jones, S. Naqvi
Instructors: H.C. Banks, W.D. Humphrey, V.B. Hayden, V. Spinks
Research Assistants: 1

Field Stations/Laboratories: USDA Microbial Conversion Laboratory and Cyclic AMP Growth Laboratory both located at Alcorn State University

Degree Programs: Biology, Biological Educat ion, Medical Technology

Undergraduate Degrees: B.S., B. Ed.
Undergraduate Degree Requirements: Students must maintain a 2.0 average. Students majoring in Biology must arrange their programs in consultation with the chairman of the department or other departmental representatives. A student must accumulate approximately 139 hours with which approximately 48 of the 139 hours are in your major.

BELHAVEN COLLEGE

1500 Peachtree Street Phone: (601) 355-1281
Jackson, Mississippi 39202
Dean of College: S.D. Buckley, Jr.

DEPARTMENT OF BIOLOGY
Phone: Chairman: (601) 355-1281 Others: Same

Chairman: W.W. Walley
Professor: W.W. Walley
Associate Professor: M.R. Reid
Assistant Professor: E.N. Peeler
Laboratory Assistants: 6

Field Stations/Laboratories: Affiliated with Gulf Coast Research Laboratory - Ocean Springs, Mississippi

Degree Program: Biology

Undergraduate Degree: B.S.
Undergraduate Degree Requirements: (1) 32 semester hours in Biology to include: General Biology, Botany, Invertebrate Zoology, and Selected topics in Modern Biology (2) Inorganic and Organic Chemistry (3) General College Physics (4) Biology majors should elect German or French to meet language requirements for graduation (5) 6 semester hours of approved Mathematics (6) 50 hours of approved electives

DELTA STATE UNIVERSITY

Cleveland, Mississippi 38732 Phone: (601) 843-2606
Dean of Graduate Studies: R.N. Walters
Dean of College: W.F. LaForge

DEPARTMENT OF BIOLOGICAL SCIENCES
Phone: Chairman: (601) 843-5521

Head: J.D. Ouzts
Professors: J.D.Ouzts, J.S. White
Associate Professors: M. Raspet, J.S. Steen
Assistant Professors: T. Branning, M. Miller, T. Millican, R. Stewart
Instructor: T. Ferretti
Visiting Lecturers: 2

Degree Programs: Biological Science, Medical Technology

Undergraduate Degree: B.S.
Undergraduate Degree Requirements: Zoology, Botany, Cell Biology, General Chemistry

Graduate Degree: M.S. (Biological Sciences)
Graduate Degree Requirements: Master: Independent Research, Scientific Literature, 9 hours Chemistry, 18 hours Biological Science

Graduate and Undergraduate Courses Available for Graduate Credit: Ecology (U), Advanced Plant Physiology (U), Greenhouse Techniques (U), Advanced Microbiology (U), Histology (U), Vertebrate Embryology (U), Medical Entomology (U), Fishery Biology (U), Entomology (U), Parasitology (U), Biophotography (U), Marine Invertebrate Zoology (U), Introduction to Marine Zoology (U), Current Principles in Biological Sciences, Scientific Literature, Plant Life, Problems in Teaching Science in Elementary Schools, Limnology, Microbiological Techniques, Advanced Genetics, Parasites and Man, Economic Entomology

JACKSON STATE UNIVERSITY

Jackson, Mississippi 39217 Phone: (601) 948-8533
Dean of Graduate Studies: O.A. Rodgers
Dean of College: R.H. Smith

DEPARTMENT OF BIOLOGY
Phone: Chairman: Exts. 286-288

Head: J.E. Uzodinma
Professors: J.E. Uzodinma
Associate Professors: E.J. Harvey, F.S. Nelson, B. Raj, B.S. Sekhon
Assistant Professors: R. Anthony, S. Badger, G. Barues, R. Chertok, G. Hardy, H. Nixon, R. Powell, G. Washington
Instructors: B. Henderson, C. Spann

Degree Programs: Biology, Medical Technology

Undergraduate Degree: B.S.
Undergraduate Degree Requirements: General Zoology, General Botany, Introductory Genetics, Microbiology, Comparative Anatomy

MILLSAPS COLLEGE

Jackson, Mississippi 39210 Phone: (601) 354-5201
Dean of the Faculty: J.H. Saunders

DEPARTMENT OF BIOLOGY
Phone: Ext. 286 Others: Exts. 286, 213

Chairman: R.E. Bell
Professor: R.E. Bell
Associate Professors: J.P. McKeewn, R.B. Nevins
Assistant Professor: A.E. Yensen

Field Stations/Laboratories: Affiliated with Gulf Coast Research Laboratories - Ocean Springs, Mississippi

Degree Program: Biology

Undergraduate Degrees: B.A., B.S.
Undergraduate Degree Requirements: Zoology, Botany, Genetics, Taxonomy, Ecology, Physiology, and Seminar

MISSISSIPPI COLLEGE
Clinton, Mississippi 39058 Phone: (601) 924-5131
Dean of Graduate Studies: E.L. McMillan
Dean of College: C.E. Martin

DEPARTMENT OF BIOLOGICAL SCIENCES
Phone: Chairman: Ext. 222

Head: L.C. Temple
Associate Professor: P.G. Cox
Assistant Professors: E.S. Pearson, L.C. Temple
Instructors: S. Gibson, B.E. Bandy, S. Gibson, W.H. Turcotte
Visiting Lecturer: 1

Degree Program: Biological Sciences

Undergraduate Degrees: B.A., B.S.
Undergraduate Degree Requirements: 32 semester hours Biology, 16-18 semester hours Chemistry, Mathematics through Calculus II, 120 total semester academic hours

Graduate Degrees: M.S., M.S.Ed.
Graduate Degree Requirements: Master: 21 semester hours Biology plus 9 semester hours Education for the M.S.Ed. degree, 24 semester hours Biology plus 6 hours thesis for the M.S.

Graduate and Undergraduate Courses Available for Graduate Credit: Vertebrate Histology (U), Histological Technique (U), Anatomy of Seed Plants (U), Cellular Biology (U), Developmental Biology (U), Ornithology (U), Biology Seminar (U), Independent Studies and Research (U), Marine Botany (U), Marine Invertebrate Zoology I (U), Marine Invertebrate Zoology II (U), Marine Vertebrate Zoology and Ichthyology (U), Estuarine and Marsh Ecology (U), Basic Principles of Modern Biology I, Basic Principles of Modern Biology II, Comparative Embryology, Field Animal Biology, Taxonomy of Plants, Problems in Zoology (Marine)

MISSISSIPPI UNIVERSITY FOR WOMEN
Columbus, Mississippi 39701 Phone: (601) 328-6841
Dean of Graduate Studies: H. Cromwell
Dean of College of Arts and Sciences: D. King

DEPARTMENT OF BIOLOGICAL SCIENCES
Phone: Chairman: (601) 328-0812

Chairman: H.L. Sherman
Professors: M. Fulton (Coordinator of Medical Technology), H.L. Sherman
Associate Professor: J.R. Fortman
Assistant Professors: J.D. Davis, N.L. Howell, R.B. Lacey, W.S. Parker
Instructor: D. Keys

Degree Programs: Biology, Microbiology, Medical Technology

Undergraduate Degrees: B.A., B.S.
Undergraduate Degree Requirements: Major in Biology: 32 hours of Biological Sciences, 8 hours General Chemistry and 6 hours Mathematics, Organic Chemistry and Physics highly recommended. Major in Microbiology: 28 hours Microbiology, General and Organic Chemistry.

Graduate Degree: M.S. (Biological Sciences)
Graduate Degree Requirements: Master: 30 hours, including 6-hour thesis or 4-hour special problem

Graduate and Undergraduate Courses Available for Graduate Credit: Parasitology (U), Cell Biology (U), Comparative Physiology (U), Plant Physiology (U), Adaptations (U), The Protista (U), The Metaphyta (U), The Invertebrates (U), Ecology (U), Microbial Physiology (U), Hematology and Immunology (U), Virology (U), Microbial Genetics (U), Medical Mycology (U), Biological Literature and Research Design, Seminars (Behavior, Ecology, Genetics, Animal Physiology, Plant Physiology, Radiation Biology, History of Biology, Zoogeography, Phytogeography, Microbiology), Special Problems, Thesis

MISSISSIPPI STATE UNIVERSITY
Mississippi State, Mississippi 39762 Phone: (603) 325-3211
Dean of Graduate Studies: J.C. McKee

College of Arts and Sciences
Dean: L.C. Behr

DEPARTMENT OF BOTANY
Phone: Chairman: (601) 325-4623

Head: G.W. Johnston
Professors: G.W. Johnston, J.F. Locke
Associate Professors: M.L. Hare, H.C. Lane (Adjunct), J.R. Watson
Assistant Professors: S.T. McDaniel, J.A. Price
Teaching Fellows: 4

Field Stations/Laboratories: Gulf Coast Research Laboratory - Ocean Springs, Mississippi

Degree Program: Botany

Undergraduate Degree: B.S.
Undergraduate Degree Requirements: Completion of 144 semester hours, carrying 288 quality points, 24-30 hours in Botany

Graduate Degrees: M.S., Ph.D.
Graduate Degree Requirements: Master: Minimum of 30 semester hours of graduate study, presentation of a thesis, reading knowledge of one foreign language Doctoral: At least 3 years of course work beyond the bachelor's degree; presentation of a dissertation, reading knowledge of 2 foreign languages or one language and special study in a related field.

Graduate and Undergraduate Courses Available for Graduate Credit: Taxonomy of Spermatophytes (U), General Plant Physiology (U), Methods in Botanical Techniques (U), Plant Ecology (U), Plant Anatomy (U), Cytology (U), Morphology of Lower Vascular Plants (U), Morphology of Fungi (U), Physiology of Wood Plants (U), Bryology of Angiosperms, Cytogenetics, Advanced Taxonomy, Advanced Plant Physiology I, Advanced Plant Physiology II, Seminar

MICROBIOLOGY DEPARTMENT
Phone: Chairman: (601) 325-4816

Chairman: R.G. Tischer
Professors: L.R. Brown, J.C. Mickelson, B.J. Stojanovic (Adjunct), R.G. Tischer
Associate Professors: J. McCamish, A.W. Wang
Assistant Professors: D. Cook, J.L. Mahloch (Adjunct), B.H. Turner
Instructors: P.K. Brasher, F.S. Tucker, J.F. Wyman
Research Assistants: 12

Field Stations/Laboratories: Mississippi Test Flight Facility - Bay Street, St. Louis, Mississippi, Gulf Coast Research Laboratory - Ocean Springs, Mississippi

Degree Programs: Microbiology, Medical Technology

Undergraduate Degree: B.S.
Undergraduate Degree Requirements: 60 credit hours in Physical, Biological and Mathematical Science (at least 30 in Microbiology). 12 credit hours French or German. Total of 144 credits

Graduate Degrees: M.S., Ph.D.
Graduate Degree Requirements: Master: 18-21 hours major field, 9-12 hours in minor, 1 foreign language, thesis, final examination Doctoral: 3 academic years beyond B.S. Courses vary according to student needs. Qualifying examination, Preliminary Examination, 2 languages, thesis, final examination

Graduate and Undergraduate Courses Available for Graduate Credit: Quantitative Methods I (U), Quantitative Methods II (U), Pathogenic Microbiology (U), Advanced General Microbiology (U), Microbiological Literature (U), Seminar (U), Soil Microbiology (U), Microbiology of Foods (U), Microbiology for Sanitary Engineering (U), Microbial Physiology I (U), Determinative Microbiology (U), Immunology (U), Principles of Virology (U), Advanced Pathogenic Microbiology (U), Radioisotope Techniques I (U), Biological Application of Electron Microbiology Genetics, Applied Microbiology, Radioisotopes Techniques II, Seminar

DEPARTMENT OF ZOOLOGY
Phone: Chairman: (601) 325-5722

Chairman: J.T. Morrow
Professor: J.D. Yarbrough
Associate Professors: G.H. Clemmer, A.A. de la Cruz, J.A. Jackson, J.L. Wolfe
Assistant Professor: R.G. Altig
Teaching Fellows: 7
Research Assistants: 3

Degree Programs: Animal Physiology, Zoology

Undergraduate Degree: B.S.
Undergraduate Degree Requirements: 3 hours Political Science, 12 hours English, 12 hours Foreign Language 12 hours Inorganic Chemistry, 8 hours Organic Chemistry, 3 hours History, 9 hours Physics, 6 hours Mathematics, 13 hours General Electives, 13 hours Zoology Electives, 44 hours Required Biological Sciences, 9 suggested Electives, 144 Total Semester Hours

Graduate Degrees: M.S., Ph.D.
Graduate Degree Requirements: Master: 30 hours beyond B.S., 6 Research, 9-12 hours in Minor, 12-15 hours in Major, 1 Foreign Language Doctoral: 60 hours beyond M.S., 2 Foreign Languages, 20 hours above Research, 40 hours of above Course Work

Graduate and Undergraduate Courses Available for Graduate Credit: Cellular Physiology (U), Animal Physiology (U), History of Biology (U), Mammalogy (U), Ornithology (U), Aquatic Invertebrates (U), Evolutionary Biology (U), Advanced Animal Biology (U), Comparative Physiology, Advanced Invertebrate Zoology, Ichthyology, Herpetology, Vertebrate Ethology, Advanced Ecology, Scientific Writing, Zoogeography, Special Topics in Zoology

College of Agriculture

Dean: C.E. Lindley

DEPARTMENT OF AGRONOMY
Phone: Chairman: (601) 325-4181

Chairman: R.G. Creech
Professors: D.N. Baker (Adjunct), R.G. Creech, J.C. Delouche, A.G. Douglas, J.N. Jenkins (Adjunct, Research), W.E. Knight (Adjunct, Research), J.D. Lancaster, N.C. Merwine, V.E. Nash, L.E. Nelson, B.J. Stojanovic, H.B. Vanderford, C.Y. Ward, F.D. Whisler
Associate Professors: C.H. Andrews, B.L. Burson, D.L. Myhre (Adjunct, Research), R.H. Pluenneke, H.C. Potts, C.E. Vaughan, V.H. Watson
Assistant Professors: A.H. Boyd, G.M. Dougherty, L.M. Gourley, W.F. Jones, E.L. McWhirter (Research), B.C Murphy (Research), J.O. Sandford (Research)
Research Assistants: 3

Degree Programs: Agronomy, Conservation, Crop Science, Farm Crops, Genetics, Physiology, Plant Breeding, Soil Science, Soils and Plant Nutrition, Turfgrass

Undergraduate Degree: B.S.
Undergraduate Degree Requirements: 128 hours

Graduate Degrees: M.S., Ph.D.
Graduate Degree Requirements: Master: Minimum of 30 semester hours with 1/2 or more of the program exclusive of thesis credits must be made up of courses at the 8000 and 9000 levels Doctoral: Minimum of 60 hours beyond the masters degree which includes 20 hours of dissertation and research. Demonstrate reading knowledge in 2 languages or collateral training in a supporting area (6 hours). One full year of residency

Graduate and Undergraduate Courses Available for Graduate Credit: Forage and Pasture Crops (U), Cotton Production (U), Grain Crops (U), Plant Breeding (U), Turf Management (U), Golf Course Operations (U), Pasture Development, Advanced Plant Breeding, Crop Ecology, Genetic Variation, Cytogenetics in Plant Breeding, Crop Plant Metabolism, Methods in Agronomy, Seminar, Seed Production (U), Seed Analysis and Laboratory Practices (U), Seed Technology (U), Seed Physiology (U), Fertilizers (U), Soil Fertility (U), Soil Classification (U), Soil Conservation (U), Soil Chemistry (U), Physical Edaphology (U), Clay Mineralogy, Advances Soil Chemistry, Advanced Soil Fertility, Soil Physics, Soil Microbiology, Advanced Fertilizers

DEPARTMENT OF ANIMAL SCIENCE
(No reply received)

DEPARTMENT OF BIOCHEMISTRY
Phone: Chairman: 325-2640

Chairman: B.F. Barrentine
Professors: B.F. Barrentine, P.A. Hedine (Adjunct), R.B. Koch, S.D. Upham (Adjunct)
Associate Professors: J.R. Heitz, T.F. Kellogg, R.H. Pluenneke, R.P. Wilson
Assistant Professors: R.D. Arthur, A.C. Thompson (Adjunct)

Degree Program: Biochemistry

Undergraduate Degree: B.S.
Undergraduate Degree Requirements: Minimum 128 semester hours. 51 hours of Chemistry, including 17 hours of Biochemistry

Graduate Degree: M.S.
Graduate Degree Requirements: Master: Minimum 30 semester hours. Maximum 6 hours for thesis and thesis research. Major: 18-21 semester hours in Biochemistry, Minor: 9-12 semester hours in minor subject. Thesis and one foreign language required.

Graduate and Undergraduate Courses Available for Graduate Credit: Biochemistry Laboratory I (U), Biochemistry Laboratory II (U), General Biochemistry I (U), Biochemistry of Specialized Tissues (U), Seminar, Phytochemistry, Analytical Biochemistry, Biochemistry of Toxic Materials, Special Topics in Biochemistry, Advanced Carbohydrate Metabolism and Biological Oxidation, Advanced Protein and Amino Acid Metabolism, Advanced Lipid Metabolism, Enzymes, Special Problems, Thesis Research/Thesis

DEPARTMENT OF DAIRY SCIENCE
Phone: Chairman: (601) 325-2330

Head: H.J. Bearden
Professors: J.T. Cardwell, E.W. Custer, D.S. Ramsey
Associate Professors: K.R. Cummings, J.W. Fuquay, J. W. Lusk, W.H. McGee
Assistant Professor: J.T. Marshall
Research Assistants: 6

Field Stations/Laboratories: Northeast Mississippi, Branch Experiment Station - Verona, Mississippi, North Mississippi Branch Experiment Station - Holly Springs, Mississippi, Coastal Plains Branch Experiment Station - Newton, Mississippi

Degree Programs: Dairy Science (with options in Dairy Production and Dairy Manufacturing), Food Science, Environment: Dairy, Food and Public Health Sanitation, Animal Physiology, Nutrition and Genetics

Undergraduate Degree: B.S.
Undergraduate Degree Requirements: 128 semester hours total credit to include 12-16 hours Chemistry, 6 hours Mathematics, 6 hours English Composition, 16 hours Social Sciences, 16 hours Basic Sciences

Graduate Degrees: M.S., Ph.D.
Graduate Degree Requirements: Master: 24 semester hours of formal course work plus 6 hours credit for thesis research and thesis. No language is required. Doctoral: 40 semester hours of formal course work plus 20 semester hours credit for dissertation research and dissertation. One foreign language is required.

Graduate and Undergraduate Courses Available for Graduate Credit: Analysis of Dairy and Food Products (U), Dairy and Food Plant Mechanics (U), Fluid Milk and Food Processing (U), Public Health Regulations (U), Fermented Dairy and Food Processing (U), Concentrated and Frozen Dairy and Food Processing (U), Dairy and Food Plant Management (U), Quality Control of Dairy and Food Products (U), Dairy and Food Packaging and Sales (U), Physiology of Lactation (U), Nutrition (U), Physiology of Reproduction (U), Dairy Herd Systems (U), Dairy Herd Management (U), Dairy and Food Science Seminar, Epidemiology of Food Born Infections, Dairy and Food By-products, Dairy Literature, Colloidal Systems of Foods, High Energy Components of Foods, Flavor and Food Acceptance, Dairy and Food Plants, Homeostatic Regulation and Physiological Stress, Advanced Seminar, Advanced Dairy Production, Dairy Farm Management, Ruminant Nutrition, Specialty Nutrition, Comparative Nutrition, Monogastric Nutrition, Micro-Nutrient Nutrition, Genetics of Population, Physiological Genetics, Physiology of Digestion and Metabolism, Avian Nutrition, Advanced Physiology of Reproduction, Statistical Genetics, Methods in Nutrition and Food Science Research, Endocrine Secretion, Environment and Animal Productivity, Advanced Breeding

DEPARTMENT OF ENTOMOLOGY
Phone: Chairman: (601) 325-4541

Chairman: F.G. Maxwell
Professors: F.G. Maxwell, T.B. Davich (Adjunct), L.W. Hepner, D.F. Martin (Adjunct), H.N Pitre, C.A. Wilson
Associate Professors: T.S. Brook, H.W. Chambers, R.L. Combs, W.H. Cross (Adjunct), H.B. Green, D.D. Hardee (Adjunct), F.A. Harris, E.P. Lloyd (Adjunct), R.E. McLaughlin (Adjunct), M.E. Merkl (Adjunct), N. Mitlin (Adjunct), R.C. Morris (Adjunct), W.W. Neel, W.L. Parrott (Adjunct), R.H. Roberts (Adjunct), P.P. Sikorowski, D.F. Young (Adjunct)
Assistant Professors: J.C. Bailey (Adjunct), F.D. Brewer (Adjunct), F.M Davis (Adjunct), J.L. Frazier, J.H. Hatchett (Adjunct), M.L. Laster, B.R. Norment, C.F. Sartor (Adjunct), J.D. Solomon (Adjunct)
Research Assistants: 1
Post-Doctorals: 3

Field Stations/Laboratories: Gulf Coast Research Laboratory - Ocean Springs, Mississippi, (Several universities and colleges have a connection with this laboratory). The Mississippi Agricultural and Forestry Experiment Station at various locations throughout Mississippi

Degree Program: Entomology

Undergraduate Degree: B.S.
Undergraduate Degree Requirements: 128 semester hours within the approved curriculum in Entomology

Graduate Degrees: M.S., Ph.D.
Graduate Degree Requirements: Master: 30 semester hours of graduate study. One-half or more of the program, exlucsive of thesis credits, must be made up of courses at the 8000 and 9000 levels Doctoral: Course requirements: At least three academic years beyond the bachelor's degree are necessary to meet course requirements. The number of course hours will vary according to the specific requirements of the department concerned and the student's needs.

Graduate and Undergraduate Courses Available for Graduate Credit: Field Crop Insects (U), Insect Physiology and Biochemistry (U), Medical and Veterinary Entomology (U), Toxicology and Insecticide Chemistry (U), Forest Entomology (U), Principles of Insect Systematic Taxonomy of Immature Insects, Advanced Insect Physiology and Biochemistry, Insect Behavior, Advanced Toxicology, Advanced Toxicological Methods, Insect Pathology, Diagnostic Insect Pathology, Insects in Relation to Plant Diseases, Cotton Pests, Structural and Stored Products Pests, Topics in Applied Entomology, Arachnology, Principles of Pest Management, Insect Population Dynamics, Biological Control, Insect Control by Host Plant Resistance, Biological Photography

DEPARTMENT OF HORTICULTURE
Phone: Chairman: (601) 325-3223

Head: C.C. Singletary
Professors: C.R. Ammerman, C.O. Box, E.L. Moore, J.P. Overcash, C.E. Parks, B.C. Stojanovic
Associate Professors: L.A. Bartlett, H.L. Hammett, E.C. Martin, C.L. Taylor
Assistant Professors: H.C. Landphair, J.C. Harris, R. Null
Instructors: L.A. Estes, R.A. Calloway, J A. Perry, C.E. Scoggins
Graduate Assistants: 12

Field Stations/Laboratories: 12 Experiment Branch Stations located strategically through the state of Mississippi

Degree Programs: Food Science, Ornamental Horticulture, Horticulture, Vegetable Crops

Undergraduate Degree: B.S.
Undergraduate Degree Requirements: Not Stated

Graduate Degree Requirements: M.S., Ph.D.
Graduate Degree Requirements: Not Stated

Graduate and Undergraduate Courses Available for Graduate Credit: Not Stated

DEPARTMENT OF PLANT PATHOLOGY AND WEED SCIENCE
Phone: Chairman: 325-3138

Head: W.W. Hare
Professors: A.W. Cole, M.C. Futrell (Adjunct), C.H. Graves, Jr., A.B. Wiles (Adjunct)
Associate Professors: E.E. Rosenkranz (Adjunct), J.A. Spencer
Assistant Professors: V.D. Ammon, W.E. Batson, Jr., G.E. Coats, L.L. Singleton

Research Assistants: 9

Field Stations/Laboratories: 9 Branch Stations

Degree Programs: Plant Pathology and Weed Science

Undergraduate Degree: B.S.
Undergraduate Degree Requirements: Science Option - 128 hours semester, 12 hours Humanities, 12 hours Mathematics, 6 hours Foreign Language, 18 hours Chemistry, 21 hours Botany, 26 Other Sciences, 14 our Courses, 14 Electives, Plant Protection Option - 128 hours, 27 hours less Mathematics, Botany, Chemistry, no foreign language, 27 hours Economics, Speech, Agronomy, Entomology and Horticulture

Graduate Degrees: M.S., Ph.D.
Graduate Degree Requirements: Master: M.A. - 10 hours Weed Science, 10 hours Pathology, 10 hours Entomology, plus 1 hour Internship M.S. 24 hours courses, 6 hour thesis Doctoral: About 90 hours, 2 languages, 20 hours thesis

Graduate and Undergraduate Courses Available for Graduate Credit: Plant Pathology, Mycology (U), Control of Plant Diseases (U), Special Problems, Virology, Diseases of Field Crops, Advanced Plant Pathology I, II, Seminar, Methods in Plant Pathology and Weed Science, Advanced Plant Physiology I, Growth and Development, Weed Science - Principles of Weed Control (U), Special Problems, Methods in Plant Pathology and Weed Science, Advanced Weed Science, Mode of Action, General Biochemistry I, Statistical Methods, Experimental Design

DEPARTMENT OF POULTRY SCIENCE
(No reply received)

School of Forest Resources
Dean: R.R. Foil

DEPARTMENT OF FORESTRY
Phone: Chairman: (601) 325-2946 Others: 325-2131

Head: E.F. Thompson
Professors: J.E. Moak, E.G. Roberts, J.W. Starr, G.L. Switzer
Associate Professors: W.W. Elam, W.F. Miller, R.D. Ross, A.D. Sullivan
Assistant Professors: S.B. Land, W.F. Watson
Instructors: M.E. Mitchell
Research Assistants: 5

Degree Program: Forestry

Undergraduate Degree: B.S., B.F.
Undergraduate Degree Requirements: 136 semester hours

Graduate Degrees: M.S., M.F.
Graduate Degree Requirements: Master: 30 semester hours

Graduate and Undergraduate Courses Available for Graduate Credit: Forest Site Evaluation (U), Silviculture I, II (U), Advanced Forest Mensuration (U), Tember Harvesting (U), Forest Policy Seminar (U), Physiology of Woody Plants (U), Forest Photogrammetry (U), Forest Management (U), Forest Management Plans (U), Forest Valuation (U), Forestry Economics (U), Special Problems, Thesis Research/Thesis, Seminar in Forest and Wild Land Resources and Use, Advanced Forest Inventory, Forest Management Seminar, Advanced Forest Management, Managerial Economics of Forestry

DEPARTMENT OF WILDLIFE AND FISHERIES
Phone: Chairman: 325-3133 Others: 325-3134

Head: D.H. Arner
Professor: D.H. Arner
Associate Professors: G.A. Hurst, D.E. Wesley
Assistant Professors: R.E. Reagan, H.R. Robinette
Adjunct Professors: G.H. Clemmer, D.T. Gardner, W.J. Lorio, E.G. Roberts, L.R. Shelton, B.R. Tramel, T.L. Wellborn, J.L. Wolfe
Research Assistants: 18

Degree Programs: Fisheries Management, Wildlife Ecology

Undergraduate Degree: B.S.
Undergraduate Degree Requirements: Minimum credit for degree: 128 semester hours, Normal length of time: 8 semesters, Required courses: 104 semester hours, Electives: 24 semester hours

Graduate Degree: M.S.
Graduate Degree Requirements: Master: Minimum credit for degree: 30 semester hours, Minimum for thesis, 6 semester hours, Residence requirement: 30 weeks, Oral examination

Graduate and Undergraduate Courses Available for Graduate Credit: Fish Culture (U), Limnology (U), Diagnosis and Treatment of Diseases of Warm Water Fish (U), Environmental Problems Dealing with Forests (U), Water (U), and Wildlife (U), Game Conservation and Management (U), Wildlife Techniques (U), Parasites of Game and Fish (U), Seminar in Resource Management for Recreation (U), Seminar in Fisheries and Wildlife (U), Fishery Biology (U), Fish Physiology (U), Wildlife Techniques II (U), Special Problems, Thesis Research and Thesis, Seminar in Fisheries and Wildlife, Waterfowl Management, Upland Game Management, Pond and Stream Management, Fish Pathology, Comparative Physiology of Aquatic and Marine Organisms, Biological and Limnological Aspects ofWater Pollution

MISSISSIPPI VALLEY STATE UNIVERSITY
Itta Bena, Mississippi 38441 Phone: (610) 254-2321
Dean of College: D.F. Blake

DEPARTMENT OF BIOLOGICAL SCIENCE
Phone: Chairman: 254-2321 Ext. 206

Chairman: S. Peyton
Professors: R. Bahadur, L.W. Coats, W.A. Fingal, S.L. Sethi, J. Singh, R.E. Young
Instructor: M.P. Davis

Degree Program: Biology

Undergraduate Degrees: B.S., B.A.
Undergraduate Degree Requirements: Biology 32, Chemistry 16, Physics 8, Mathematics 9, General Education 65

RUST COLLEGE
Holly Springs, Mississippi 38635 Phone: (601) 252-4661
Dean of College: T. Rihan

DEPARTMENT OF BIOLOGY
Phone: Chairman: Ext. 246

Chairman: E. Igbokwe
Associate Professors: E. Igbokwe, M. Gunasekaran
Assistant Professors: L. Debro, C. Caldwell, H. Bronson, F. Osuji
Visiting Lecturer: 1

Degree Program: Biology

Undergraduate Degree: B.S.
Undergraduate Degree Requirements: 16 hours of Chemistry, 8 hours of Physics, 8 hours of Mathematics, 32 hours in Biology

TOUGALOO COLLEGE
Tougaloo, Mississippi 39174 Phone: (601) 956-4941

Dean of College: N.J. Townsend

DEPARTMENT OF BIOLOGY
Phone: Chairman: Ext. 16

Chairman: G.S. Asokasrinivasan
Associate Professor: G.S. Asokasrinivasan
Assistant Professor: B. Mehrotra, R. Costello, C. Haire
Research Assistants: 3
Laboratory Technician: 1

Degree Program: Biology

Undergraduate Degree: B.S.
Undergraduate Degree Requirements. 30 courses in all for B.S. in Biology, 2 Introductory Biology courses, 1/2 Biochemical Techniques, 1 Genetics, 1 Ecology, 1 Cell Biology, plus 2 electives, plus 2 General Chemistry, plus 2 Organic Chemistry, plus 2 College Physics

THE UNIVERSITY OF MISSISSIPPI

Oxford, Mississippi 38677 Phone: (601) 232-7203
Dean of Graduate Studies: J. Sonn

College of Liberal Arts

Dean: M.B. Hunnicutt

BIOLOGY DEPARTMENT
Phone: Chairman: (601) 237-7203

Chairman: L.A. Magee
Professors: F.M. Hull, Sr. (Emeritus), I.C. Kitchin, W. D. Longest, L.A. Magee, M.E. Morrison (part-time), Y.J. McGaha, W.H. Norman, T.M. Pullon, W. St. Amand
Associate Professors: W. Keith, L. Knight, I. McClurkin, D. McClurkin (part-time), D. Russell
Assistant Professors: N.T. Barden, P.H. Darst, D. Cook, (part-time), R. Overstreet, S. Smith

Degree Programs: Bacteriology, Biological Sciences, Biology, Botany, Microbiology, Physiological Sciences, Physiology, Zoology, Zoology and Entomology

Undergraduate Degrees: B.A., B.S.
Undergraduate Degree Requirements: B.A.: 24 semester hours Biology (or sub-specialty), 1 year Inorganic Chemistry, 1 year Mathematics B.S.: 42 semester hours Biology, 2 years Chemistry, 1 year Mathematics

Graduate Degrees: M.A., M.C.S., Ph.D.
Graduate Degree Requirements: Master: 30 semester hours credit, including thesis (6 hours) Doctoral: 54 semester hours, credit beyond the bachelor's degree, of which 36 hours must have been completed on this campus

Graduate and Undergraduate Courses Available for Graduate Credit: History of Biology, Electron Microscopy, Essentials of Biology, Laboratory Methods in Biological Science I, II, Advanced General Botany, Advanced General Zoology, Problems in Biological Photography and Photomicrography, Problems in Instrumentation and Materials for Teaching Modern Biology Botany - Plant Morphology, The Algae, Survey of the Fungi, Algal Physiology, Electron Microscopy, Advanced Plant Taxonomy, Field Botany, Agrostology, Research Methods in Plant Taxonomy, Photophysiology of Plants, Problems in Plant Taxonomy, Embryology of Angiosperms, Problems in Mycology Marine Biology - Marine Science, Marine Fisheries Biology, Esutarine and Marsh Ecology, Marine Botany, Marine Invertebrate Zoology, Marine Invertebrate Zoology, Marine Vertebrate Zoology and Ichthyology, Parasites of Marine Animals, Marine Microbiology, Biological Electron Microscopy I, II, Microbiology/Immunology - History of Microbiology, Aquatic Microbiology, Food Microbiology, Microbial Genetics, Problems in Microbiology, Advanced General Microbiology, Tissue Culture, Virology, Immunology and Immunochemistry, Medical Microbiology and Immunology, General Pathology, Systemic Pathology, Leukocytic Function, Microbiology Seminar, History of Immunology, Problems in Microbiology, Thesis, Dissertation Zoology - Cell Physiology, Comparative Physiology, Endocrinology, Experimental Embryology, Entomology, Medical and Veterinary Entomology, Advanced Genetics, Advanced Histology, Protozoology, Radiation Biology, Zoological Bibliography, Electron Microscopy, Animal Ecology, Limnological Methods, Limnology, Ichthyology, Fishery Biology, Field Zoology, Biosynthetic Pathways in Metabolism, Cytogenetics, Advanced Cytology, Insect Taxonomy, Insect Morphology, Invertebrate Zoology, Advanced Parasitology, Instrumental Methods in Zoological Research, Tissue Culture, Seminar in Mouse Genetics, Thesis

THE UNIVERSITY OF MISSISSIPPI MEDICAL CENTER

2500 North State Street Phone: (601) 362-4411
Jackson, Mississippi 39216
Chairman of Graduate Council: W.L. Williams
Dean of Medical School: N.N. Nelson

DEPARTMENT OF ANATOMY
Phone: Chairman: Ext. 2352

Chairman: W.L. Williams
Professors: C.R. Ball, W.L. Williams
Associate Professors: A.D. Ashburn, B.R. Clower
Assistant Professors: A.D. Collins, M.W. Kendall, J.L. McNair, F.C. Salter

Degree Programs: Anatomy and Histology, Anatomy

Graduate Degrees: M.S., Ph.D.
Graduate Degree Requirements: Master: 45 quarter hours which may be limited to anatomy courses plus thesis research Doctoral: 82 quarter hours consisting usually of 4 major anatomy courses and dissertation plus physiology and pathology. The minor must consist of at least 20 quarter hours.

Graduate and Undergraduate Courses Available for Graduate Credit: Gross Human Anatomy, Developmental Anatomy, Microscopic Anatomy, Neuroanatomy, Advanced Anatomy, Embryology (Advanced), History of Medicine, Research in Anatomy, Experimental Methods Related to Research, Dissertation and Dissertation Research, Thesis and Thesis Research

DEPARTMENT OF BIOCHEMISTRY
Phone: Chairman: (601) 362-4411 Ext. 2358

Chairman: L.L. Sulya
Professors: C.S. McCaa, H.B. White, Jr.
Associate Professors: C.L. Dodgen, V.H. Read
Assistant Professors: O.E. Bell, Jr., J.N. Brown, C.L. Woodley
Research Assistants: 3

Degree Program: Biochemistry

Graduate Degrees: M.S., Ph.D.
Graduate Degree Requirements: Not Stated

Graduate and Undergraduate Courses Available for Graduate Credit: Not Stated

DEPARTMENT OF MICROBIOLOGY
Phone: Chairman: (601) 362-2541

Chairman: C.C. Randall
Professors: B.R. Byers, G.A. Gentry, C.C. Randall
Associate Professors: L.G. Gafford, J.B. Grogan, L.A. Lawson, L.A. Magee, D.J. O'Callaghan
Assistant Professors: J.L. Arceneaus, J.L. Arceneaux,

M.B. Batson, G.M. Butchko, R.J. Christie, J.M. Cruse, W.R. Lockwood, R.R. Watson
Instructors: P.F. Hinton, V.G. Lockard
Research Assistants: 5
Technicians: 3
Laboratory Assistants: 4
Student Assistants: 5
Trainees: 5

Degree Programs: Bacterial Genetics, Immunology, Microbiology, Bacterial Physiology, Virology

Graduate Degrees: M.Sc., M.S., Ph.D.
Graduate Degree Requirements: Master: M.C.S. - 45 quarter hours of which 9 quarter hours is outside the major area, comprehensive examination, no thesis required. M.S., 45 quarter hours, comprehensive examination, thesis required Doctoral: 82 quarter hours of which 20 quarter hours are in a minor field, one language, preliminary examination, dissertation and final oral

Graduate and Undergraduate Courses Available for Graduate Credit: Medical Microbiology, Viruses, Seminar, Research in Microbiology, Radioisotopes in Experimental Biology and Medicine, Special Problems in Virology and Tissue Culture, Special Problems in Cell Physiology, Special Problems in Electron Microscopy, Special Problems in Immunology, Special Problems in Microbial Genetics, Special Problems in Oncology, Special Problems in Mycology, An Introduction to Animal Experimentation, Microbial Genetics, Molecular Genetics, Biology of Membranes, Mycology, Immunology and Immunochemistry, Cellular Immunology

DEPARTMENT OF PHARMACOLOGY AND TOXICOLOGY
Phone: Chairman: (601) 326-2206

Chairman: Not Stated
Professors: R.L. Klein, W.O. Berndt
Associate Professors: A.S. Hume, A. Thureson-Klein
Assistant Professors: G.B. Gatipon, M.A. Pfaffman, B. Hoskins, F. McEwen (part-time), B. Parks (part-time)
Teaching Fellows: 2
Research Assistants: 5

Degree Programs: Pharmacology, Toxicology

Graduate Degrees: M.S., Ph.D.
Graduate Degree Requirements: Master: 45 credit hours - Research Thesis Doctoral: 82 credit hours Preliminary Examination required, Reading of one Foreign Language, Research Dissertation and Publication of this work

Graduate and Undergraduate Courses Available for Graduate Credit: Evolution of Pharmacological Concepts, Pharmacology, Quantitative Pharmacology, Isotopic Techniques, General Preparative Techniques for Electron Microscopy, Special Preparative Techniques for Electron Microscopy, Electron Microscopy, Research in Pharmacology, Neuropharmacology, Cardiovascular Pharmacology, Renal Pharmacology, Membrane Pharmacology, Toxicology, Analytical Toxicology, Analytical Toxicology, Forensic Toxicology, Techniques of Forensic Science and Psychopharmacology

DEPARTMENT OF PHYSIOLOGY
(No reply received)

UNIVERSITY OF SOUTHERN MISSISSIPPI
Hattiesburg, Mississippi 39401 Phone: (601) 266-7011
Dean of Graduate Studies: R.T. van Aller
Dean of College: S.F. Thames

DEPARTMENT OF BIOLOGY
Phone: Chairman: (601) 266-7185 Others: 266-7011

Chairman: B.J. Grantham
Professors: J.W. Cliburn, B.C. Smith
Associate Professors: A.G. Fish, B.J. Grantham, G.F. Pessoney
Assistant Professors: F.G. Howell, J.M. Kelley, J.B. Larsen, B.J. Martin, D E. Norris, S.T. Ross, R.W. Scheetz, L.E. Story, J.W. Wooten, S.W. Rosso

Degree Programs: Biology, Botany, Zoology, Marine Biology

Undergraduate Degree: B.S. (Biology)
Undergraduate Degree Requirements: 32 hours minimum, Emphasis areas - Botany, Zoology, Marine Biology, Wildlife Biology, Fisheries Biology

Graduate Degrees: M.S., Ph.D.
Graduate Degree Requirements: Master: 30 semester hours beyond B.S. with thesis Doctoral: Minimum of 78 semester hours beyond B.S. or minimum of 48 semester hours beyond M.S. with 2 foreign languages or one foreign language and one additional minor

Graduate and Undergraduate Courses Available for Graduate Credit: History of Biology (U), Electron Microscopy (U), Histology (U), Embryology I, II (U), Experimental Embryology (U), Protozoology (U), Human Protozoology (U), Medical Entomology (U), Mammalian Physiology I, II (U), Comparative Animal Physiology I,II (U), Endocrinology (U), Invertebrate Zoology I, II (U), Entomology (U), Ichthyology (U), Ornithology (U), Mammalogy (U), Principles of Nomenclature (U), Introductory Mycology (U), Introductory Phycology (U), Plant Anatomy (U), Morphology of Non-Vascular Plants (U), Morphology of Vascular Plants (U), Aquatic and Marsh Plants (U), Plant Physiology (U), Economic Botany (U), Plant Pathology (U), Taxonomy of High Plants (U), Biogeography (U), Ecology I, II (U), Freshwater Biology (U), Limnology (U), Introduction to Biological Oceanography (U), Ecology of Parasitic and Vector-Borne Diseases (U), Plant Ecology (U), Field Biology (U), Biological Problems, Advanced Botany, Biological Photography, Analytical Methods in Biological Research, Cytology, Cell Ultrastructure, Advanced Systematic Herpetology, Advanced Systematic Ichthyology, Speciation, Planktology, Invertebrate Physiology, Helminthology, Physiology of Marine Animals, Seminar in Animal Behavior, Seminar in Physiology, Experimental Endocrinology, Invertebrate Embryology, Advanced Taxonomy of Seed Plants, Experimental Phycology, Experimental Mycology, Topics in Marine Biology, Fisheries Biology, Biological Oceanography, Marine Ecology, Aquaculture, Biology Seminar

WILLIAM CAREY COLLEGE
(No reply received)

MISSOURI

CARDINAL GLENNON COLLEGE

(No reply received)

CENTRAL METHODIST COLLEGE

Fayette, Missouri 65248 Phone: (812) 248-3391
Dean of College: J.D. McBrayer

DEPARTMENT OF BIOLOGY AND GEOLOGY
Phone: Chairman: Ext. 334

Chairman: B.J. Leach
F.H. Dearing Professor of Biology: B.J. Leach
Professors: B.J. Leach, F.H. Woods (Emeritus)
Associate Professors: E.H. Li, H.L. Momberg, G.A. Vaughan
Undergraduate Student Laboratory Teaching Assistants: 20

Degree Programs: Biology, Human Biology, Wildlife Biology, Zoology

Undergraduate Degree: A.B.
Undergraduate Degree Requirements: 67 semester hours in area of distribution requirements, 57 semester hours in Principal and Related Work that includes: General Biology, Environmental Botany, Genetics, Comparative and Cellular Physiology, Special Problems, Major Readings, General Chemistry, Organic Chemistry, Physics, Statistics and Calculus

CENTRAL MISSOURI STATE UNIVERSITY

Warrensburg, Missouri 64093 Phone: (816) 429-4933
Dean of Graduate Studies: H. Sampson
Dean of College: S. Hewitt

DEPARTMENT OF BIOLOGY
Phone: Chairman: (816) 429-4934 Others: 429-4933

Head: H.P. Savery
Professors: J. Belshe, D. Castaner, A. Elliott, O. Hawksley, H. Savery
Associate Professors: B. Peck, C. Hinerman, G. Oshima
Assistant Professors: J. Hess, E. Keppner, S. Mills, J. Smith
Assistantships: 3

Degree Programs: Biology, Medical Technology

Undergraduate Degrees: B.S.Ed., B.S., B.A.
Undergraduate Degree Requirements: Total 124 semester hours, 30 semester hours minimum (major), 20 semester hours minimum (minor), 45 semester hours minimum (General Studies), 25 semester hours minimum Electives

Graduate Degree: M.S.
Graduate Degree Requirements: Master: 32 semester hours of approved courses beyond the B.S. degree requirement

Graduate and Undergraduate Courses Available for Graduate Credit: Contemporary Issues, Biology (U), Special Problems, Biology (U), Introduction to Cytology (U), Microtechnique (U), Spectal Topics Biology, Introduction to Experimental Biology, History and Literature, Algae and Fungi (U), Mammalogy (U), Animal Histology (U), Parasitology (U), Plant Physiology (U), Biology of Endocrine Glands (U), Genetics and Evolution (U), Field Ecology (U), Comparative Ethology (U), Plant Associations (U), Teaching Biological Science (U)

CULVER-STOCKTON COLLEGE

Canton, Missouri 63435 Phone: (314) 288-5221
Dean of College: H. Harris

DEPARTMENT OF BIOLOGY
Phone: Chairman: Ext. 74 Others: Ext. 40

Chairman: J.T. Haldiman
Associate Professors: J. Bursewicz, J.T. Haldiman, C.T. Wiltshire

Degree Programs: Biology, Medical Technology

Undergraduate Degrees: A.B., B.S.
Undergraduate Degree Requirements: A.B. - 30 semester hours (15 specified) in Biology plus General Chemistry B.S. - 36 semester hours (27 specified) in Biology plus General Chemistry, General Physics, and 20 hours of Education

DRURY COLLEGE

(No reply received)

EVANGEL COLLEGE

Springfield, Missouri 65802 Phone: (417) 865-2811
Dean of College: Z. Bicket

BIOLOGY DEPARTMENT
Phone: Chairman: Ext. 41

Chairman: S.E. Davidson
Professor: S.E. Davidson
Assistant Professors: T. Collins, J. Blizzard

Degree Program: Biology

Undergraduate Degree: B.S.
Undergraduate Degree Requirements: 24 hours Chemistry, 31 hours Biology, 10 hours Physics, 7 hours Mathematics

FONTBONNE COLLEGE

6800 Wydown Boulevard Phone: (314) 862-3456
St. Louis, Missouri 63105
Dean of College: D. Ziemke

DEPARTMENT OF LIFE SCIENCES
Phone: Chairman: Ext. 266 Others: Ext. 266

Chairman: R. Cook
Professor: E. Lissant
Associate Professor: Sr. M.E. Tucker
Assistant Professor: R. Cook
Instructors: P.T. Loughlin, M. Peters

Degree Programs: Life Science, Medical Technology, Environmental Studies

Undergraduate Degree: B.S.
Undergraduate Degree Requirements: Environmental and Structural Biology, Genetics, Developmental Physiology, Cell Biology, Physiology, General Chemistry, Organic Chemistry

LINCOLN UNIVERSITY OF MISSOURI

(No reply received)

THE LINDENWOOD COLLEGE
(No reply received)

MARYVILLE COLLEGE
13550 Conway Road Phone: (314) 434-4100
St. Louis, Missouri 63141
Dean of College: J.C. Stam

DIVISION OF MATHEMATICS AND NATURAL SCIENCES (BIOLOGY)
Phone: Chairman: Ext. 356 Others: Exts. 277,314

Chairman: Sr. P. Thro
Associate Professor: Sr. R. Connell
Assistant Professor: N.L. Woldow
Instructor: E. Hirce

Degree Programs: Biology, Medical Technology

Undergraduate Degree: B.S.
Undergraduate Degree Requirements: Introduction to Botany, Introduction to Zoology or Human Anatomy/ Physiology, Cell Physiology, Vertebrate Biology, Embryology, Genetics, Microbiology, Organic Chemistry, Biochemistry, Physics

MISSOURI SOUTHERN STATE COLLEGE
Joplin, Missouri 68801 Phone: (417) 624-8100
Dean of College: F. Belk

BIOLOGY DEPARTMENT
Phone: Chairman: Ext. 226

Head: E.S. Gibson
Professor: O.E. Orr
Associate Professors: G. Elick, S. Gibson
Assistant Professors: W. Ferron, D. Kirkham, V. Prentice, W. Stebbins

Degree Programs: Biology, Biological Education

Undergraduate Degree: B.S.
Undergraduate Degree Requirements: B.S. Biology - 124 hours, 40 hours in Biology, 13 hours in Chemistry, 5 hours in Mathematics, 12 upper division in Biology

MISSOURI VALLEY COLLEGE
Marshall, Missouri 65340 Phone: (816) 886-6183
Dean of College: A. Brown

DEPARTMENT OF BIOLOGY
Phone: None

Chairman: R.A. Kepner
Associate Professor: R.A. Kepner
Undergraduate Assistants: 8

Degree Program: Biology

Undergraduate Degree: B.S., B.A.
Undergraduate Degree Requirements: 32-36 hours Biology, 15 hours Chemistry, 10 hours Physics, 6-12 hours Language

MISSOURI WESTERN STATE COLLEGE
St. Joseph, Missouri 64507 Phone: (816) 233-7192
Dean of College: Not Stated

DEPARTMENT OF BIOLOGY
Phone: Chairman: (716) 233-7192

Chairman: H.F. Force
Associate Professors: R. Crumley, H. Force, D. Robbins
Assistant Professors: T.E. Rachon, L. Galloway, W. Andersen
Instructor: R. Boutwell

Degree Programs: Biology, Biological Education, Medical Technology

Undergraduate Degrees: B.S., B.A., B.S.Ed.
Undergraduate Degree Requirements: Biology: 36-40 hours dependent on Degree, Chemistry: 13 hours including, Organic and Biochemistry, Physics: 5 hours, Mathematics: 6 hours, including Statistics

NORTHEAST MISSOURI STATE UNIVERSITY
Kirksville, Missouri 63501 Phone: (816) 665-5121
Dean of Graduate Studies: E.F. Mittler
Dean of Instruction: D.W. Krueger

DIVISION OF SCIENCE
Phone: Chairman: Ext. 7204

Chairman: Not Stated
Professors: M.E. Bell (Botany), M.Q. Freeland (Chemistry), W.J. Magruder (Chemistry), R.R. Nothdurft (Physics), R.J. Peavler (Physics, D.A. Rosebery (Biology), G.D. Sells (Biology), W.L. Selser (Science), J.H. Settlage (Science)
Associate Professors: M.E. Eichor (Chemistry), D.L. Hanks (Mycology), R.S. Mason (Radiation Science), O.B. Mock (Zoology), D.E. Walker (Chemistry)
Assistant Professors: V.H. Cochran (Science, M.L. Conrad (Biology), A.J. Copley (Earth Science), R.P. Cornell (Physiology), J.E. Dimit (Biology), D.C. Evans (Earth Science), K.R. Fountain (Chemistry), G.R. Franke (Chemistry), P.C. Goldman (Zoology), A.E. Jay (Genetics), D.A. Kangas (Ecology), J.H. Shaddy (Ecology), E.W. Smith (Physics), J.H. Wells (Organic Chemistry)
Temporary Instructors: 4
Teaching Assistants: 2

Degree Programs: Biology, Botany, Science Education, Zoology

Undergraduate Degree: B.S. (Biology, Botany, Zoology)
Undergraduate Degree Requirements: General Botany, General Zoology, Genetics, Bioecology, Physiology, 12 hours of Biology Electives, required support: General Chemistry, Organic Chemistry, College Physics I, Statistics, 5 hours of Electives in Physics, Chemistry or Earth Science, and 27 hours of unrestricted elective as approved by advisor

Graduate Degree: M.S. (Science Education)
Graduate Degree Requirements: Master: Methods of Research, Problems in Teaching Science, Education Elective, minimum of 4 hours from the following: Fundamental Processes in Chemistry I, II, Physics I, II, History and Philosophy of Science, Methods of Scientific Research, Readings in Physical Science, General Science I, II, and 17 hours of electives in Biology as approved by advisor.

Graduate and Undergraduate Courses Available for Graduate Credit: Human Genetics (U), Limnology (U), Biometry (U), Evolution (U), Herpetology (U), Cytology (U), Ornithology (U), Problems of Environmental Education (U), Advanced Plant Taxonomy (U), Comparative Plant Morphology (U), Ecology (U), Ethology (U), Ichthyology (U), Mammalogy (U), Fisheries Biology, Advanced Invertebrate Zoology, Fundamental Processes in Biology I, Biology II, Speciation, Advanced Plant Physiology, Advanced Plant Anatomy, Microbial Physiology, Advanced Ecology

NORTHWEST MISSOURI STATE UNIVERSITY
Maryville, Missouri 64468 Phone: (816) 582-4851
Dean of Graduate Studies: L. Miller

DEPARTMENT OF AGRICULTURE
Phone: Chairman: (816) 582-2721

Chairman: J.C. Beeks
Professors: J.C. Beeks, D.D. Padgitt
Associate Professors: F.B. Houghton, F.W. Oomens, W.W. Treese
Assistant Professors: H.J. Brown, G.L. Gille, J.B. Kliebenstein
Instructor: R.J. Knudsen

Degree Programs: Agricultural Education, Agriculture, Agronomy, Animal Science, Dairy Science, Horticulture, Agricultural Business

Undergraduate Degrees: B.S., B.S.Ed., B.T.
Undergraduate Degree Requirements: General Requirements: B.S. 41, B.S.Ed., 41-44, B.T. 30, Area of Specialization, B.S. 54-60, B.S.Ed. 54-58, B.T. 75, Education, B.S.Ed. 22

Graduate Degree: M.S. (Agriculture, Agricultural Education
Graduate Degree Requirements: Master: Courses in Agriculture M.S. 20, M.S.Ed. 20, Courses in Education, M.S.Ed. 6, Unspecified M.S. 12, M.S.Ed. 7, Total M.S. (32), M.S.Ed) 32

Graduate and Undergraduate Courses Available for Graduate Credit: Agricultural Prices (U), Futures Marketing (U), Futures Marketing Laboratory (U), Problems in Agricultural Economics (U), Soil and Water Management (U), Soil and Water Management Laboratory (U), Materials Handling (U), Materials Handling Laboratory (U), Problems in Agricultural Mechanization (U), Soil Surveying and Land Appraisal (U), Soil Surveying and Land Appraisal Laboratory (U), Soil Conservation (U), Soil Conservation Laboratory (U), Plant Breeding (U), Plant Breeding Laboratory (U), Soil Amendments (U), Corn and Soybean Production (U), Grain Crops (U), Problems in Agronomy (U), Advanced Livestock Selection and Evaluation (U), Advanced Livestock Selection and Evaluation Laboratory (U), Applied Animal Nutrition II (U), Environmental Physiology (U), Breeding and Improvement of Livestock (U), Problems inA nimal Science (U), Physiology of Milk Secretion (U), Endocrinology of Domestic Animals (U), Problems in Dairy Science (U), Problems in Horticulture (U)

DEPARTMENT OF BIOLOGY
Phone: Chairman: (816) 582-3584 Others: 582-5914

Chairman: K.W. Minter
Professors: R.A. Hart, K.W. Minter, I.M. Mueller, B.D. Scott
Associate Professors: D.A. Easterla, D.M. Smith
Assistant Professors: M.C. Grabau, P.J. Lucido, P.F. Wynne
Graduate Assistant: 1

Degree Programs: Biology, Biological Education, Botany, Medical Technology

Undergraduate Degree: B.S.
Undergraduate Degree Requirements: 13 hours Chemistry, 8 hours Physics, 4 hours College Algebra and Trigonometry, 4 hours Geology, 34-40 hours Biology

Graduate Degree: M.S.
Graduate Degree Requirements: Master: M.S. in Education - 21 hours Biology, 6 hours Education, 5 hours Electives, M.S. - 22 hours Biology including Thesis, 1 hour Elective

Graduate and Undergraduate Courses Available for Graduate Credit: Conservation of Biological Resources (U), Cytology (U), Plant Anatomy (U), Gross Anatomy (U), Vertebrate Embryology (U), Vertebrate Histology (U), Plant Physiology (U), Animal Physiology (U), Animal Physiology II (U), Marine Botany Gulf Coast (U), Advanced Microbiology (U), Mycology, Pathogenic Bacteriology (U), Marine Microbiology Gulf Coast (U), Herpetology (U), Advanced Ornithology (U), Ichthyology (U), Phycology (U), Plant Ecology (U), Animal Ecology (U), Biology Materials and Methods for Teachers (U), Field Biology for Elementary Teachers (U), Methods of Information Retrieval in Science, History of Biology, Aquatic Invertebrates, Plant Pathology, Evolution, Population Genetics, Principles of Biosystematics, Immature Insects, Readings in Ecology, Limnology

NOTRE DAME COLLEGE

320 E. Ripa Avenue Phone: (314) 544-0455
St. Louis, Missouri 63125
Dean of College: Sr. C.A. Collins

UNIT OF NATURAL SCIENCES
Phone: Chairman: Ext. 37

Head: Sr. R. Bast
Professors: R. Bast, E.M. Winking

Degree Programs: Biology, Medical Technology

Undergraduate Degree: B.S.
Undergraduate Degree Requirements: Biology - Biology I (5 hours), Chemistry 1 and 2, 7 upper division Biology courses, 3 upper division Biology, Mathematics, Chemistry, Psychology, Sociology or Philosophy of Science Medical Technology - 28 hours Chemistry, 16 hours Biology, 8 hours Physics, 3 hours Mathematics

PARK COLLEGE

Parkville, Missouri 64152 Phone: (816) 741-2000
Dean of College: W.C. Pivonka

DEPARTMENT OF BIOLOGY
Phone: Chairman: Ext. 141 Others: 140

Chairman: J.M. Hamilton
Findlay Professor of Science: J.M. Hamilton
Professor: J.M. Hamilton
Associate Professor: A. Dusing

Degree Program: Biology

Undergraduate Degree: B.A.
Undergraduate Degree Requirements: Terminal: 34 hours Biology, 1 year Chemistry, Statistics, 1 year Modern Language. Preprofessional: 30 hours Biology, Chemistry through Organic, 1 year Physics, Mathematics through Calculus, three semesters of French or German

ROCKHURST COLLEGE

Kansas City, Missouri 64110 Phone: (816) 363-4010
Dean of College: Fr. A.J. Blumeyer, S.J.

DEPARTMENT OF BIOLOGY
Phone: Chairman: Ext. 268

Chairman: R.E. Wilson
Professor: E.S. Kos
Associate Professor: G.M. O'Connor
Assistant Professors: F.V. Rooney, S.J., N.E. Walker, R.E. Wilson

Degree Program: Biology

Undergraduate Degree: B.S.
Undergraduate Degree Requirements: 18 credit hours in Biology, 12 credit hours in related area, 1 Calculus course Recommended

ST. LOUIS UNIVERSITY

221 North Grand Boulevard Phone: (314) 535-3500
St. Louis, Missouri 63103
Dean of Graduate Studies: W. Stauber

College of Arts and Sciences
Dean: T. Knipp

BIOLOGY DEPARTMENT
Phone: Chairman: Ext. 641

Chairman: R.R. Walsh
Professors: J. Dwyer, D. Feir, P. Raven (Adjunct), R. Walsh
Associate Professors: N. Aspinwall, M. Bakula, J. Mulligan, D. Rooney
Assistant Professors: R. Aloridge, S. Dina, C. Hoessle (Adjunct), J. Medoff, J. Severson
Teaching Fellows: 12
Research Assistants: 1

Field Stations/Laboratories: Fr. Reis Biology Station - Cuba, Missouri

Degree Program: Biology

Undergraduate Degree: A.B.
Undergraduate Degree Requirements: 25 semester hours (Minimum) Biology, 7 semester hours (Minimum), Advanced Chemistry

Graduate Degrees: M.S., M.S.(R.),Ph.D.
Graduate Degree Requirements: Master: 30 semester hours Advanced Biology or 24 semester hours Advanced Biology plus Research thesis Doctoral: 60 semester hours Advanced Biology and/or Related subjects plus Doctoral Research and Dissertation

Graduate and Undergraduate Courses Available for Graduate Credit: Not Stated

Medical School
1402 South Grand Avenue Phone: (314) 664-9800
St. Louis, Missouri 63104
Dean: R. Felix

DEPARTMENT OF ANATOMY
Phone: Chairman: Ext. 404 Others: Ext. 401, 402, 403

Chairman: P.A. Young
Professors: K. Christensen (Emeritus), C.A. Richins, P.A. Young
Associate Professors: U.K. Hwang, L.C. Massopust, J.J. Taylor, V.L. Yeager
Assistant Professors: B.J. Briggs, N.A. Connors, V.W. Fischer, M.G. Murphy, M.W. Rana
Instructor: J.F. Shea
Graduate Fellows: 12
Assistants: 2
Research Associates: 1

Degree Program: Anatomy

Graduate Degree: M.S., Ph.D.
Graduate Degree Requirements: Master: 30 hours including thesis research Doctoral: 72 hours including dissertation research, language requirement in French or German

Graduate and Undergraduate Courses Available for Graduate Credit: Human Gross Anatomy, Human Neuroanatomy, Anatomy of the Head and Neck, Human Gross Anatomy, Special Problems in Gross Anatomy, Human Histology, Histological Techniques, Histological Research, Special Problems in Histology, Human Embryology, Embryological Research, Special Problems in Embryology, Human Neuroembryology, Human Neuroanatomy, Neurohistological Techniques, Neurohistological Research, Special Problems in Neuroanatomy, Advanced Human Neuroanatomy, Organs of Special Sense, Basic Research Techniques in Anatomy, Autonomic Nervous System, Autonomic Nervous System Research, Special Problems in Autonomic Nervous System, Histochemical Techniques, Histochemical Research, History of Anatomy, Introduction to Anatomical Research, Ultrastructural Techniques, Ultrastructural Research

DEPARTMENT OF BIOCHEMISTRY
Phone: Chairman: Ext. 406

Chairman: R.E. Olson
Alice A. Doisy Professor of Biochemistry and Chairman of the Edward A. Doisy Department of Biochemistry: R.E. Olson
Professors: C. Coscia, E.A. Doisy, Jr., W. Elliott, M. Horwitt, P. Katzman, W. Longmore, A. Martonosi, H. Nicholas
Associate Professors: R. Cockrell, W. Davis, K. Dus, C. Fitch
Assistant Professors: G. Bazzano, R. Dorner, F. Francis, R. Hawkins, C. Hirschberg, J. Hunsley, M. Jellinek, H. Lozeron, B. Marrs, C. Tan, H. Van Kley
Instructors: M. Johnston, T. Munns, M. Searcey
Visiting Lecturer: 1
Teaching Fellows: 14
Research Assistants: 12

Degree Program: Biochemistry

Graduate Degrees: M.S., Ph.D.
Graduate Degree Requirements: Master: Courses as determined by Chairman, but at least half in Biochemistry, Thesis Doctoral: 48 credits at least 12 outside the field of Biochemistry, one foreign language Thesis

Graduate and Undergraduate Courses Available for Graduate Credit: Fundamentals of Biochemistry, Biochemistry for Medical Technologists (U), Biochemical Techniques in Laboratory Medicine (U), Advanced Nutrition (U), General Biochemistry, Experimental Methods in Biochemistry, Advanced Biochemical Techniques, Special Topics in Molecular Biology, Advanced Topics in Biochemistry, Biochemistry of Muscle Contraction, Statistical Methods in Biochemical Research

DEPARTMENT OF MICROBIOLOGY
Phone: Chairman: Ext. 425 Others: Ext. 426

Chairman: M.M. Weber
Professors: M. Green (Director, Institute for Molecular Virology), N.E. Melechen, B.M. Pogell, M.M. Weber
Associate Professors: S.K. Bose, L.W. Hedgecock (Adjunct), A. Kaplan, E.L. Minard, I.T. Schulze
Assistant Professors: C.J. Bellone, C.A. Morris
Visiting Lecturers: 5
Teaching Fellows: 10
Research Assistants: 8
Assistant: 1

Degree Program: Microbiology

Graduate Degrees: M.S., Ph.D.
Graduate Degree Requirements: Master: Microbiology and Biochemistry (see below) Doctoral: Microbiology and Biochemistry (see below)

Graduate and Undergraduate Courses Available for Graduate Credit: Molecular Biology of Microorganisms, Experimental Methods in Microbiology, Medical Microbiology and Immunology, Introduction to Microbiological Research, Immunology, Topics in Advanced Microbiology, Cellular Regulatory Mechanisms, Biochemical and Molecular Basis of Cancer, Genetic Functions in Bacteriophage

DEPARTMENT OF PHYSIOLOGY
Phone: Chairman: Ext. 415 Others: Ext. 412

Chairman: A.R. Lind
Professors: B.D. Bhagat, A.R. Lind, L.S. Senay, Jr.
Associate Professors: L.S. D'Agrosa, F. Hertelendy, P.S. Rao, R.H. Secker-Walker, K.W. van Beaumont
Assistant Professors: E.M. Burns, T.E. Dahms,

M.S. Ellert, W.S. Hunter, F.M Ruh
Instructors: L.A. Solberg, G.L. Totel
Visiting Lecturers: 3
Research Assistants: 5
Post Doctoral Fellow: 1

Degree Program: Physiology

Graduate Degrees: M.S., Ph.D.
Graduate Degree Requirements: Master: The courses must include General Physiology or equivalent, Introduction of Physiological Research and Physiology Journal Club. The program usually requires at least two years of full-time work, with a maximum of six semester hours of research. Doctoral: The course program of all candidates for the doctorate in physiology will include both basic and elective courses in Physiology as well as selected courses from other departments. The selection of all courses is designed to give every student a broad base of understanding of physiology and to give each student the opportunity to develop maximally in individual areas of interests. The selection of courses is made by the student and his adviser, and must be approved by the Chairman of the Department. Two thirds of the course program will be selected from the course offerings in physiology and general comparative physiology. Dissertations for the doctorate in physiology will be written in the field of physiology under the direction of the faculty of the Department of Physiology.

Graduate and Undergraduate Courses Available for Graduate Credit: General Physiology, Advanced Physiology Laboratory, Human Physiology, Experimental Physiology of the Central Circulation, Peripheral Circulation, Body Fluids, Introduction to Physiological Research, Nerve-Muscle Physiology, Physiology of the Autonomic Nervous System, The Catecholamines, Advanced Techniques in Experimental Physiology, Seminar in Physiology, Respiratory Physiology, Exercise Physiology, Physiology of Thermoregulation, Recent Advances in the Mechanism of Action of Steroid Hormones, Reproductive and Fetal Physiology, Gastrointestinal Physiology, Basic and Clinical Endocrinology, Physiology Journal Club, Research Topics

THE SCHOOL OF THE OZARKS

Point Lookout, Missouri 65726 Phone: (417) 334-3101
Dean of Academic Affairs: W. Huddleston

DEPARTMENT OF BIOLOGY
Phone: Chairman: (417) 334-4000 Others: 334-3101

Chairman: W.S. White
Professors: W.S White, W. Davis
Associate Professor: K.C. Olson
Visiting Lecturers: 10-15
Undergraduate Student Assistants: 8

Field Stations/Laboratories: We have two Nature Preserves on our own campus.

Degree Program: Biology

Undergraduate Degrees: B.A., B.S.
Undergraduate Degree Requirements: General Education - Contemporary Issues - 2 hours, English - 6 hours, Speech - 3 hours, Physical Education, Health, Psychology 7 hours Natural Science - 7 hours from fields of Biology, Mathematics, Chemistry and Physics with at least 2 fields represented, Economics, Political Science Sociology and History (2 fields) - 9 hours, Non-Western World (Art, History, Religion, Social Science, etc. - 3 hours), Humanities (Music, Literature, Drama, Art - 6 hours, Christian Philosophy of Life (Religion and Philosophy) - 6 hours, Total General Education Hours Requirement - 49 hours, Biology Major - 50 hours, Additional electives - 21 hours, Total hours to graduate 120

SOUTHEAST MISSOURI STATE UNIVERSITY

Cape Girardeau, Missouri 63701 Phone: (314) 334-8211
Dean of Graduate Studies: R. Foster
Dean of College: G. Loftis

DEPARTMENT OF AGRICULTURE
Phone: Chairman: Ext. 264

Head: W.E. Meyer
Associate Professor: W.E. Meyer
Assistant Professor: C. Korns
Instructor: G. Bieber
Part-time Instructors from University of Missouri Extension

Degree Programs: Agriculture, Agriculture Business

Undergraduate Degree: B.S.
Undergraduate Degree Requirements: 124 hours - minimum of 25 hours in Agriculture, minimum of 30 senior college hours, minimum of 60 hours General Education

BIOLOGY DEPARTMENT
Phone: Chairman: Ext. 284 Others: Exts. 228, 288

Chairman: J.P. Huckabay
Professors: N.L. Braasch, S.G. Diehl, E. Dudgeon, J.B. Hinni, R.J. Kuster
Associate Professors: R.E Cook, P.L. Leye, D.D. Jewel, R.G. Kullberg, C.T. Train
Assistant Professors: O. Ohmart, D.E. Phillips, C.G. Twitchell
Instructors: L.M. Cordonnier, B.W. Rader

Field Stations/Laboratories: Affiliated with Gulf Coast Research Laboratory - Ocean Springs, Mississippi

Degree Programs: Biology, Botany

Undergraduate Degrees: B.S., B.S.Ed. (Biology), B.S. (Botany)
Undergraduate Degree Requirements: 40 hours for B.S. including core courses plus electives, 30 hours for B.S.Ed., including core courses plus electives

Graduate Degrees: M.A.T., M.N.S.
Graduate Degree Requirements: Master: (Master of Arts in Teaching), 32 hours (half in 600 level courses or above), 6-10 hours "professional education" courses. M.N.S. (Master of Natural Science) degree program in planning stage.

Graduate and Undergraduate Courses Available for Graduate Credit: Cytology (U), Ecology (U), Advanced Genetic (U), Evolution (U), Topics in Biology (U), Biology Seminar (U), Plant Anatomy (U), Freshwater Algae (U), Mosses and Ferns (U), Protozoology (U), Principles of Physiology, Biological Literature, Experimental Design, Native Aquatic and Terresterial Flora I, II, Native Aquatic and Terresterial Invertebrates I, II, Natural History of Vertebrates, Courses offered for credit through affiliation with Gulf Coast Research Laboratory, Ocean Springs, Mississippi - Special Problems in Advanced Histology (U), Marine Microbiology (U), Marine Botany (U), Marine Invertebrate Zoology (U), Marine Vertebrate Zoology and Ichthyology

SOUTHWEST BAPTIST COLLEGE

623 S. Pike Phone: (417) 326-5281
Bolivar, Missouri 65613
Dean of College: G.H. Surrette

DEPARTMENT OF BIOLOGY
Phone: Chairman: Ext. 342 Others: Ext. 341

Chairman: J.A. Clark
Professors: J.A. Clark, M.A. Kort
Assistant Professor: C.F. Huser

Degree Programs: Biology, Medical Technology

Undergraduate Degrees: B.A., B.S.
Undergraduate Degree Requirements: B.A. - General Education 43 hours, Biological Science 30 hours, Foreign Language 6-8 hours, Electives 43-45 hours, B.S. - General Education - 43 hours, Biological Sciences 30 hours, Additional Physical Sciences 8 hours, Electives - 43 hours

SOUTHWEST MISSOURI STATE UNIVERSITY

901 South National Avenue Phone: (417) 831-1561
Springfield, Missouri 65802
Dean of Graduate Studies: I.D. Bartley
Dean of School of Science and Technology: C.C. Thompson

DEPARTMENT OF LIFE SCIENCES
Phone: Chairman: Ext. 223

Chairman: R.T. Stevenson
Professors: R.L. Philibert, P.L. Redfearn, Jr., T.A. Stombaugh, R.F. Wilkinson
Associate Professors: L.L. Denney, R.L. Irgens, G.L. Pyrah, W.R. Weber
Assistant Professors: A.R. Gordon, S.L. Jensen, G.A. Kahler, R.L. Myers, C.A. Taber, M.S. Topping
Instructor: B.G. Taber
Teaching Fellows: 17

Degree Programs: Education-Biology, Biology, Wildlife Conservation and Management, Medical Technology

Undergraduate Degrees: B.S., B.A. (Education-Biology)
Undergraduate Degree Requirements: General Botany, General Zoology, Genetics and at least one course from four of the following areas: Ecology, Molecular and Cellular Biology, Morphology, Physiology, Systematics, a minimum of 10 hours of Chemistry, including one course in Organic and Pre-Calculus Mathematics

Graduate Degree: M.S. (Biology)
Graduate Degree Requirements: Master: 32 hours with a minimum of 16 semester hours of graduate courses, Thesis or Seminar Option

Graduate and Undergraduate Courses Available for Graduate Credit: History of Biology,(U), Microbiology (U), Immunology and Serology (U), Histological Techniques (U), Methods in Electron Microscopy I (U), Pathogenic Microbiology (U), Cell Biology (U), Limnology (U), Estuarine and Marsh Ecology (U), Field Biology (U), Microbial Nutrition (U), Ecolomic Botany (U), Identification of Aquatic Flowering Plants (U), Phycology (U), Introduction to Plant Ecology (U), Plant Geography (U), Marine Botany (U), Agrostology (U), Bryology (U), Marine Microbiology (U), Animal Physiology (U), Comparative Animal Physiology (U), Aquatic Entomology (U), Ichthyology (U), Herpetology (U), Mammalogy (U), Marine Vertebrate Zoology (U), Parasitology (U), Marine Invertebrate Zoology I (U), Marine Invertebrate Zoology II (U), Zoogeography (U), Histology (U), Research Methods and Information Retrieval (U), Principles of Experimental Microbiology, Processes of Organic Evolution, Methods in Electron Microscopy II, Seminar in Biological Ultrastructure, Advanced Limnology, Advanced Topics in Biology, Advanced Plant Taxonomy, Parasited of Marine Animals, Advanced Vertebrate Zoology, Comparative Endocrinology, Reproductive Physiology, Seminar in Biology, Colloquium in Biology, Organization and Writing of Scientific Papers and Theses, Research, Thesis

STEPHENS COLLEGE

Columbia, Missouri 65201 Phone: (314) 422-2211
Dean of College: R.N. Funk

DEPARTMENT OF NATURAL SCIENCE
Phone: Chairman: Ext. 500

Chairman: L. Howell
Professors: H. Hansen, A. Harvey, L. Howell, C. Laun, A. Novak, D. Otto
Assistant Professors: C. Ansbacher, R. Saunders
Research Associate: 1

Field Stations/Laboratories: Stephens College Mountain Science Station - Steamboat Springs, Colorado, Stephens College Marine Science Station - Key Largo Florida

Degree Programs: Environmental Science, Marine Science, Anthropology, Allied Health

Undergraduate Degree: B.A.
Undergraduate Degree Requirements: Skills in English and Physical Education, three special studies breadth: Equivalent of three courses in two of the three areas oustide the concentration area (science and mathematics) Concentration: at least five courses above the introductory level within the department, 40 courses are required for graduation, with a 2.0 grade point average

TARKIO COLLEGE

Tarkio, Missouri 64491 Phone: (816) 736-4143
Dean of College: T. Bundenthal

BIOLOGY DEPARTMENT
Phone: Chairman: Ext. 222

Head: H.W. McGehee
Associate Professors: G. Jensen, H W. McGehee

Degree Program: Biology

Undergraduate Degree: B.A.
Undergraduate Degree Requirements: 28 hours Biology, 8 hours Chemistry, 4 hours Physics, 8 hours Mathematics (Minimum), 124 hours (total)

UNIVERSITY OF MISSOURI - COLUMBIA

Columbia, Missouri 65201 Phone: (314) 882-2121
Dean of Graduate Studies: L. Berry

College of Agriculture

Dean: E.R. Kiehl

DEPARTMENT OF AGRONOMY
Phone: Chairman: (314) 882-8239 Others: 882-6534

Chairman: E.C.A. Runge
Professors: L.E. Anderson, J.D. Baldridge, O.H. Fletchall, E.R. Graham, G. Kimber, G.F. Krause, K.L. Larson, A.G. Matches, W.J. Murphy, E.J. Peters, J.M. Poehlman, G.P. Redei, E.C.A. Rungle, W.P. Sappenfield, C.L. Scrivner, E.R. Sears, G.E. Smith, G.H. Wagner, H.N. Wheaton, G.M. Woodruff, M.S. Zuber
Associate Professors: K.H. Asay, R.W. Blancher, J.R. Brown, L.E. Cavanah, C.M. Christy, G.G. Doyle, T.R. Fisher, R.D. Horrocks, C.J. Johannsen, H.D. Kerr, E.M. Kroth, R.L. Larson, C.J. Nelson, A.L. Preston, J.A. Roth, J.H. Scott, D.T. Sechler, W.J. Upchurch
Assistant Professors: J.M. Bradford, D.R. Johnson, J. G. Shannon, J.B. Beckett
Instructors: J.C. Baker, J.C. Doll, V.D. Luedders, R.E. Mattas
Visiting Lecturers: 3 (for brief Periods)
Research Assistants: 46
Post Doctorals: 4

Field Stations/Laboratories: Delta Center - Portageville, Missouri, Southwest Center - Mt. Vernon, Missouri, North Center - Spickard, Missouri, Agronomy Research Center - Columbia, Missouri

Degree Programs: Agronomy, Crop Science, Cytogenetics, Farm Crops, Genetics, Land Resources, Plant Breeding, Plant Sciences, Seed Investigations, Soil Science, Soils and Plant Nutrition.

Undergraduate Degree: B.S. (Agronomy)
Undergraduate Degree Requirements: Varies by program.

Graduate Degrees: M.S., Ph.D. (Agronomy)
Graduate Degree Requirements: Varies by Program.

Graduate and Undergraduate Courses Available for Graduate Credit: International Agronomy (U), Crops and Soil Management (U), Problems (U), Fertilizers (U), Forage Crops (U), Grain Crops (U), Advances in Crop Science (U), Weed Control (U), Physical Properties of Soil (U), Soil Conservation (U), Cotton and Other Fiber Crops (U), Soil Microbiology (U), Soil Fertility and Plant Nutrition (U), Soil Fertility and Plant Nutrition Laboratory (U), Crop Physiology (U), Soil Chemistry (U), Soil Genesis Mapping and Classification (U), Field Crop Breeding (U), Rural Real Appraisal (U), Evolution of Genetic Concepts (U), Soil Management Problems (U), Cytogenetics (U), Cytogenetics Laboratory (U), Problems Isotopes in Soil Studies, Advanced Soil Chemistry, Soil Physics, Seminar, Advanced Soil Fertility, Advanced Crop Physiology, Topics in Agronomy, Development of Plant Breeding Concepts, Applied Quantitative and Statistical Genetics, Cytogenetics in Crop Breeding, Research

DEPARTMENT OF ANIMAL HUSBANDRY
Phone: Chairman: (314) 882-8336 Others: 882-7266

Chairman: Not Stated
Professors: C.M. Bradley, B.N. Day, A.J. Dyer, J.F. Lasley, J.W. Massey, W.H. Pfander, C.V. Ross, H.B. Sewell, G.B. Thompson
Associate Professors: J.M. Asplund, J.C. Rea, J.E. Ross, T.L. Veum
Assistant Professor: J. L. Clark
Instructors: M.A. Alexander, G.W. Jesse, W.E. Loch, G.L. Martin
Research Assistants: 35

Field Stations/Laboratories: Nutrition Laboratory, Physiology Reproduction Laboratory, Metabolism Laboratory - all located at the Animal Science Research Center, Beef Farm, Swine Farm, Swine Testing Station, Beef Test Station, Sheep Farm, - all located on Route 1, Columbia Missouri, Pig Mama, located in the Old Livestock Pavilion, North Missouri Research Center - Spickard, Missouri

Degree Programs: Animal Genetics, Animal Husbandry, Nutrition, Physiology

Undergraduate Degrees: B.S. (Animal Husbandry)
Undergraduate Degree Requirements: Communications - 12 credit hours, Natural Science and Mathematics - 16 credit hours, Social Sciences and Humanistic Studies - 14 credit hours, Business and Economics - 8 credit hours, Departmental Requirements - 48 credit hours, Additional Requirements - 30 credit hours, Total - 128 credit hours (Minimum of 2.0 average based on 4.0 grading scale)

Graduate Degrees: M.S., Ph.D.
Graduate Degree Requirements: Master: Minimum of 32 credits (at least 75% of which must be passed with A or B), Thesis required except in special cases Doctoral: Two foreign languages (or proficiency in 1 foreign language plus 1 collateral field for in 2 collateral fields). Doctoral dissertation is required.

Graduate and Undergraduate Courses Available for Graduate Credit: Physiology of Reproduction (U), Genetics of Livestock Improvement (U), Beef Production and Management (U), Applied Animal Genetics (U), Sheep Production and Management (U), Pork Production and Management (U), Field Experience in Animal Science and Technology (U), Problems (Animal Husbandry), Livestock Production and Management, Research Methods, Animal Nutrition, Animal Breeding Investigation, Livestock Feeding Investigation, Ruminant Nutrition, Topics in Animal Husbandry

DEPARTMENT OF DAIRY HUSBANDRY
Phone: Chairman: (314) 882-4553 Others: 882-4454

Chairman: H.D. Johnson
Professors: J.R. Campbell, H.D. Johnson, F.A. Martz, F. Meinershagen, C.P. Merilan, J D. Sikes
Associate Professors: R.R. Anderson, R.E. Ricketts
Assistant Professors: H.A. Garverick, L. Hedlund

Field Stations/Laboratories: Animal Science Research Center, Missouri Climatic Laboratory - both on UMC campus

Degree Program: Dairy Husbandry

Undergraduate Degree: B.S.
Undergraduate Degree Requirements: 25 credit hours in Chemistry, Physics, Zoology, Bacteriology, Mathematics and Physiology, in addition to required courses in major

Graduate Degrees: M.S., Ph.D.
Graduate Degree Requirements: Master: 30 credit hours not less than 16 at graduate level, no language, preparation of paper for publication Doctoral: 2 years beyond M.S. Qualifying examination, courses selected by committee. Comprehensive examination, 2 foreign languages, preparation of research project for publication, final oral

Graduate and Undergraduate Courses Available for Graduate Credit: Not Stated

DEPARTMENT OF ENTOMOLOGY
Phone: Chairman: (314) 882-4445 Others: 882-7384

Chairman: Not Stated
Professors: H.E. Brown (Emeritus), W.S. Craig, W.R. Enns, M.L. Fairchild, C.M. Ignoffo, A.L. Jenkins (Emeritus), C.O. Knowles, C.W. Wingo
Associate Professors: G.M. Chippendale, D.J. Farish, J.L. Huggans, A.J. Keaster, W.H. Kearby, R.E. Munson, G.W. Thomas, T.R. Yonke
Assistant Professors: K.E. Brown, K. Harrendorf
Research Associates: K.D. Biever, J.D. Hoffman, D.L. Hostetter, N.L. Marston, B. Puttler, G.T. Schmidt, G.D. Thomas
Graduate Research Assistants: 15
Post-Doctoral Fellows: 4
Research Specialists: 7

Field Stations/Laboratories: Delta Center - Portageville, Missouri, Southwest Center, Route 3 - Mt. Vernon, Missouri, North Missouri Center - Route 1, Spickard, Missouri

Degree Programs: Entomology, Pest Management

Undergraduate Degree: B.S.
Undergraduate Degree Requirements: 50 hours, Program Requirements: 48 hours with emphasis on entomology, Plant Pathology, and Weed Science and supportive courses in Biochemistry, Ecology, and Statistics, Additional Requirements: 30 hours of which 12 will be outside of theCollege of Agriculture

Graduate Degrees: M.S., Ph.D.
Graduate Degree Requirements: Master: The course work phase of the program will be arranged in committee with the candidate, his or her advisor (as Chairman), and two other faculty members participating. Course work will be designed according to the student's interests. A minimum of 30 hours of appropriate graduate

level course work must be taken. A thesis is usually required. Doctoral: A committee consisting of three departmental faculty members and two professors from other departments will meet with the student and plan the program keeping in mind the desires and needs of the student. A dissertation is required of all Ph.D. candidates.

Graduate and Undergraduate Courses Available for Graduate Credit: Comparative Morphology of Insects (U), Systematic Entomology (U), Aquatic Entomology (U), Bionomics of Insect Pests (U), Field Crop Insects (U), Medical and Veterinary Entomology, Physiology of Insects, Insect Ecology, Entomological Literature and History of Entomology, Biological Control of Insects, Insects in Relation to Plant Diseases, Taxonomy of Immature Insects, Acarology--Mites and Ticks, Insect Toxicology, Advanced Systematic Entomology, Topics in Entomology, Research, Seminar

DEPARTMENT OF HORTICULTURE
Phone: Chairman: (314) 882-2336 Others:882-2646

Chairman: R.A. Schroeder
Professors: A.E. Guas, D.D. Hemphill, A.D. Hibbard, V.N. Lambeth, C.W. Lobenstein, M.N. Rogers, R.A. Schroeder, J.E. Smith, Jr., R.E. Taven
Associate Professors: D.E. Hartley, R.R. Rothenberger, L.C. Snyder, Jr.
Assistant Professors: J.H. Dunn, G.G. Long
Research Assistants: 5

Field Stations/Laboratories: Natural Environment Laboratory-New Franklin, Missouri

Degree Programs: Horticultural Science, Horticulture, Landscape Horticulture, Ornamental Horticulture, Pomology, Vegetable Crops, Floriculture

Undergraduate Degree: B.S. (Agriculture)
Undergraduate Degree Requirements: 128 hours with certain specific requirements

Graduate Degrees: M.S., Ph.D.
Graduate Degree Requirements: Master: 32 hours plus completion of committee approved program. Doctoral: Completion of committee approved program.

Graduate and Undergraduate Courses Available for Graduate Credit: Problems, Post-Harvest Physiology, Fruit Production, Commerical Vegetable and Truck Crop Growing, Vegetable Forcing, Advanced Landscape Design, Advanced Landscape Construction, Turf, Nursery Crop Production and Management, Fall Greenhouse Crops, Srping Greenhouse Crops (U), Chemistry and Physics of Spraying, Topics in Horticulture, Physiology of Woody Plants, Plant Growth Regulating Substances, Breeding of Horticultural Plants, Nutrition of Horticultural Plants, Seminar, Plant Virology, Methods of Horticultural Research, Fungus Physiology, Experimental Pomology, Advanced Olericulture, Non-Thesis Research and Research

DEPARTMENT OF PLANT PATHOLOGY
Phone: Chairman: (314) 882-2418 Others: 882-2643

Chairman: R.N. Goodman
Professors: V.H. Dropkin, R.N. Goodman, W.Q. Loegering, D.F. Millikan, T.D. Wyllie
Associate Professors: M.F. Brown, O.H. Calvert, A. Novacky, E.W. Palm, O. Sehgal
Assistant Professors: C.H. Baldwin, A.L. Karr
Postdoctoral Fellows: 5
Visiting Lecturer: 1
Research Assistants: 10
Research Associates: 2

Field Stations/Laboratories: Field Station/Laboratory - Portageville, Missouri

Degree Program: Plant Pathology

GraduateDegrees: M.S., Ph.D.
Graduate Degree Requirements: Master: 30 hours Doctoral: Minimum Requirements by Completion of Degree: 1) Pathology - Introduction to Comparative Pathology and Seminar; 2) Microbiology - 1 basic and 1 advanced course; 3) Biochemistry - 1 introductory plus 2 advanced courses; 4) Genetics - 1 introductory plus 1 advanced; 5) Statistics - 1 introductory course; 6) Scientific Instrumentation course; 7) Mathematics-Calculus or equivalent

Graduate and Undergraduate Courses Available for Graduate Credit: Introduction to Plant Pathology (U), Mycology (U), Insects in Relation to Plant Disease (U), Clinical Plant Pathology (U), Problems, Diseases of Field Crops, Diseases of Plants (Viral), Diseases of Plants (Bacterial), Diseases of Plants (Fungal), Diseases of Plants (Nematode), Seminar, Biochemistry and Physiology of Plant Diseases, Fungus Physiology, Comparative Pathology, Research, Non-thesis, Introduction to Electron Microscopy, Electron Microscopy Laboratory, Research - Thesis, Genetics of Plant Disease (U)

DEPARTMENT OF POULTRY SCIENCE
(No reply received)

College of Arts and Sciences
Dean: A.F. Yanders

DIVISION OF BIOLOGICAL SCIENCES
Phone: Chairman: (314) 882-6650 Others: 882-4068

Director: A. Eisenstark
Professors: R.P. Breitenbach, A.B. Burdick, E.H. Coe, Jr., B.G. Cumbie, R.M. deRoos, D.B. Dunn, A. Eisenstark, J.N. Farmer, W.R. Fleming, C.S. Gowans, A.P. Harrison, Jr., C.L. Kucera, J. Levitt (on leave), W.Q. Loergering, D. Mertz, Y.G. Neuffer, J.M. Wood
Associate Professors: R.V. Brown, D.H. Hazelwood, G. Y. Kikudome, D.E. Detter, M.W. Sorenson, J.W. Twente
Assistant Professors: P.F. Agris, J.E. Carrel, L.F. Chapman, J.D. David, D.J. Farish, H.C. Gerhardt, J.A. Grunau, C,D. Miles, W.F. Sheridan, L.A. Sherman, R.J. Wang
Instructor: F. Landa
Teaching Fellows: 32
Research Assistants: 5

Field Stations/Laboratories: Research Greenhouse - Columbia, Tucker Prairie - Columbia

Degree Programs: Biological Sciences, Microbiology

Undergraduate Degree: B.A.
Undergraduate Degree Requirements: 30 hours Biology, 18 hours Chemistry, 8 hours Physics, 3 hours Mathematics

Graduate Degrees: M.A., Ph.D.
Graduate Degree Requirements: Master: 32 hours, 15 hours 400 level courses Doctoral: No specific course requirements, candidates must complete minimum of 2 semesters with 12 hours graduate level work in residence

Graduate and Undergraduate Courses Available for Graduate Credit: General Biology - Introductory Radiation Biology (U), Biometry (U), General Ecology (U), Genetics-Mammalian Cell Genetics (U), Genetic Techniques (U), Evolution of Genetic Concepts (U), Genetics of Microorganisms (U), Physiological Genetics (U), Genetics of Plant Disease (U), Cytology (U), Cytogenetics (U), Genetics of Populations (U), Gene Structure and Function, Ultrastructural Basis of Cell Function Animal Biology - Evolution (U), Systematic Entomomology (U), Parasitology (U), Oceanography (U), Herpetology (U), Comparative Animal Physiology (U), Histology of Vertebrates (U), Comparative Animal Ethology (U), Animal Communication (U), Biology of Animal Populations (U), Animal Ecology Laboratory (U), Cellular Physiology (U), Endocrinology, Physiological Ecology, Comparative Endocrinology, Advanced Animal Ethology, Comparative Vertebrate Reproduction, Avian Physiology Microbiology - General Phycology (U), Mycology (U), Taxonomy of Bacteria (U), Advanced Bacteriology (U), The Biology of Nucleic Acids Plant Biology - Comparative Morphology of Vascular Plants (U), Plant Anatomy (U), Plant Physiology (U), Agrostology (U), Paleobotany (U), Palynology (U), Micro-Paleobotany (U), Physiological Responses to Environment, Cell Metabolism, Physiology of the Mineral Elements, Organization and Function of Terrestrial Ecosystems, Physiology of Growth and Development, Plant Morphogenesis, Photosynthesis, Plant Geography, Speciation, Advanced Taxonomy

School of Forestry

Dean: E.R. Kiehl

SCHOOL OF FORESTRY, FISHERIES AND WILDLIFE

Phone: Chairman: (314) 882-6446 Others: 882-7242

Director: D.P. Duncan
Rucker Professor of Wildlife: W.H. Elder
Professors: K.T. Adair, T.H. Baskett, R.S. Campbell, G.S. Cox, D.P. Duncan, W.H. Elder, F.A. McGinnes, A.J. Nash, R.C. Smith, A. Wise, Jr.
Associate Professors: G.N. Brown, M.T. Brown, L.H. Fredrickson, W.H. Kearby, K.E. Moore, J.M. Nichols, L.K. Paulsell, R.B. Polic, C.D. Steeergren, J.P. Slusher
Assistant Professors: R.O. Anderson, H.S. Bhullar, M. C. Brown, T.M. Hinculey, R.A. Musbach, J.P. Pastoret, J.B. Reynolds, R.D. Sparrowe
Research Associates: 18
Instructor: P.F. Grat, L.C. Tennyson
Research Assistants: 29
Postdoctoral Fellows: 2

Field Stations/Laboratories: University Forest - Williamsville, Missouri, Ashland Wildlife Area - Ashland, Missouri, Gaylord Research Laboratory - Puxico, Missouri, Schuabel Woods - Columbia, Missouri

Degree Programs: Forestry, Forest Ecology, Fisheries and Wildlife

Undergraduate Degree: B.S.

Undergraduate Degree Requirements: 125 to 135 hours depending upon curriculum - C average, Fisheries and Wildlife - 125 semester hours, Forest Management - 135 semester hours, Forest Science and Specialization 130 semester hours, Urban and Recreational Forestry - 130 semester hours, Wood Products - 130 semester hours

Graduate Degrees: M.S., Ph.D.

Graduate Degree Requirements: Master: 30 semester hours - B average - thesis or special problem Doctoral: Satisfactory Completion of qualifying, comprehensive, and final examinations, Dissertation, languages or collateral fields

Graduate and Undergraduate Courses Available for Graduate Credit: Resource Measurements (U), Environmental Quality in Forest Systems (U), Forest Inventory (U), Wood Technology (U), Wood Engineering (U), Forest Fire Control and Use (U), Professional Integration (U), Forest Entomology (U), Weed Science (U), Light Construction (U), Wood Processes (U), Problems (U), Forest Photogrammery (U), Mammalogy (U), Watershed Management (U), Forest Hydrology (U), Ichthyology (U), Fish Husbandry (U), Timber Management (U), Waterfowl Biology (U), Forest Economics (U), Advanced Forest Management (U), Recreational Land Management (U), Tree Genetics and Improvement (U), Range and Wildlife Habitat Management (U), Wildlife Management Laboratory (U), Limnology (U), Recreational Land Management and Planning (U), Special Readings (U), Public Resource Policy (U), Wood Industries (U), Land-Use Planning (U), Decision-Making in Natural Resources (U), Science Seminar in Natural Resources (U), Research Methods in Hydrobiology, Forest-Soil Site Relations, Applied Silvidulture, Seminar, Quantitative Fisher Science, Wood Anatomy - Wood Chemistry Relationships, Advanced Ichthyology, Research Methods, Economic Analysis in Forestry, Advanced Forest Mensuration, Fishery Management, Wildlife Idology, Advanced Forest Photogrammery, Physiology of Trees, Plant Environmental and Biochemical Relationships, Ecology of Aquatic Ecosystems

School of Medicine

Dean: W.D. Mayer

DEPARTMENT OF ANATOMY
(No reply received)

DEPARTMENT OF BIOCHEMISTRY

Phone: Chairman: (314) 882-8795 Others: 882-8795

Chairman: B.J. Campbell
Professors: B.J. Campbell, M.S. Feather, G.B. Garner, C.W.Gehrke, O.J. Koeppe, T.D. Luckey, A.P. Martin, M. Muhrer, B.L. O'Dell, E.E. Pickett, M. Vorbeck, R.L. Wixom
Associate Professors: J.M. Franz, C.A. Ghiron, S.R. Koirtyohann, E. Moscatelli, W.D. Noteboom, B. Ortwerth, D.B. Shear, A.A. White, R. Wood
Assistant Professors: C.N. Cornell, R. Larson, D.D. Randall, J.H. Wyche, W.L. Zahler
Instructor: P. Rexroad, D. Quissell, J. Burnett
Research Assistant: 1
Research Associate: 1
Post Doctoral Fellows: 3
Research Specialists: 3
Research Analytical Chemist: 1

Field Stations/Laboratories: Experiment Station Chemical Laboratories, Research Reactor - Dalton Research Center, etc. - on campus, also Sanborn Field - Sinclair Research Farm etc. - near campus

Degree Program: Biochemistry

Graduate Degrees: M.S., Ph.D.

Graduate Degree Requirements: Master: General Biochemistry, Biochemistry Seminar (1), Research (4), one graduate level course beyond the prerequisites in an area outside of the Department, teaching experience and a thesis based on original research. Doctoral: General Biochemistry, Biophysics, Advanced Biochemistry - 6 hours from upper level Biochemistry courses as specified by the Department, Seminar (4 hours minimum), Biology courses in 2 areas beyond prerequisite requirements, 6 hours of 300 or 400 level course work in one field outside of the Department, teaching experience, thesis based on original research.

Graduate and Undergraduate Courses Available for Graduate Credit: Problems, Biophysics, Techniques in Nutritional Biochemistry, General Biochemistry Lectures, Biochemistry Laboratory, Trace Analysis, Interpretation of Molecular Spectra, Instrumental Methods of Analysis, Biochemistry 320 and 322, Chromatography, Topics in Biochemistry, Plant Biochemistry, Advanced Physiological Chemistry of Domestic Animals, Comparative Biochemistry, Comparative Nutrition and Metabolism, Seminar, Reproductive Biology Seminar, Biochemistry of Hormones, Chemistry of Enzyme Cofactors, Analytical Biochemistry - Chromatography, Analytical Biochemistry - Multiple Automatic Microanalysis, Analytical Biochemistry - Mass Spectrometry, Biophysics Topics, Hormones and Metabolism, Research, Advanced Carbohydrate Metabolism and Biological Oxidations, Advanced Metabolism: Proteins and Nucleic Acids, Advanced Metabolism: Amino Acids, Regulation of Energy Metabolism

DEPARTMENT OF MICROBIOLOGY
(No reply received)

DEPARTMENT OF PHYSIOLOGY
(No reply received)

School of Veterinary Medicine
Dean: K. Weide

DEPARTMENT OF VETERINARY ANATOMY-PHYSIOLOGY
Phone: Chairman: (314) 882-7011 Others: 882-2121

Chairman: G.A. Van Gelder
Professors: H.D. Dellmann, J.E. Breazile, E.M. Brown, H.E. Dale, C.W. Foley, R.C. McClure, R.H. Schiffman, F.E. South, G.A. Van Gelder
Associate Professors: D.P. Hutcheson, G.D. Osweiler, V.V. St. Omer, M.E. Tumbleson
Assistant Professors: V.S. Cox, G.R. Kirk, R.E. Doyle
Instructor: C.T. Boyd
Research Assistants: 5

Field Stations/Laboratories: Veterinary Medical Research Farm, Sinclair Farm - Equine Center

Degree Programs: Multi-departmental area programs in Anatomy, Physiology, Pharmacology, and Nutrition leading to the Ph.D. The Department offers a Master's degree in the above areas.

Graduate Degrees: M.S., Ph.D.
Graduate Degree Requirements: Master: Physiology - 15 semester hours physiology, 6 Physiological Chemistry, and 3 hours Statistics Pharmacology - 12 semester hours of Pharmacology, 3 hours of Biochemistry, 3 hours of Mathematics, 3 hours Statistics Anatomy - 30 semester hours and ability to demonstrate proficiency in neuromicroscopic developmental gross anatomy Biochemistry and Nutrition - 14 semester hours in biological chemistry, Statistics and 10 hours of electives. Doctoral: Anatomy - Courses in Gross Developmental Microscopic and Neuroanatomy with supporting work in Biochemistry or Physiology Nutrition Area Program - Specific requirements available from Director of Area Program, at the University of Missouri Physiology Area Program - The graduate program is individually arranged but will include training in Mammalian Physiology, Statistics, Biochemistry, and Physical Chemistry. A dissertation is required Pharmacology - The Degree is offered jointly with the Department of Pharmacology in the School of Medicine.

Graduate and Undergraduate Courses Available for Graduate Credit: Cytology, Histology and Microscopic Anatomy of Domestic Animals, Histological and Anatomical Techniques, Embryology andDevelopment of Domestic Animals, Canine Dissection, Anatomy of Common Domestic Animals, Veterinary Pharmacology, Principles of Physiologic Adaptation, Adaptation to Xenobiotics, Advanced Microscopic Anatomy, Correlative Neuroanatomy, Veterinary Physiology I, Veterinary Physiology II, Fate ofDrugs in the Animal Body

DEPARTMENT OF VETERINARY MICROBIOLOGY
Phone: Chairman: (314) 882-6550 Others: 882-2121

Chairman: Not Stated
Professors: D.C. Blenden, C.R. Dorn, R.W. Loan, B.D. Rosenquist
Associate Professors: H.K. Adldinger, O.R.Brown, G.M. Corwin, E.L. McCune, L.A. Selby, R.F. Solorano, R. F. Sprouse
Assistant Professors: J.N. Berg
Research Associates: 3

Degree Program: Veterinary Microbiology

Graduate Degrees: M.S., Ph.D.
Graduate Degree Requirements: Master: 30 courses selected by committee to include 9 credit hours outside the department. Biochemistry or Biostatistics required. Thesis, Final Oral Doctoral: Courses selected by advisory committee will include Calculus, Analytical Geometry, Physical Chemistry and Biostatistics and 18-20 hours graduate Microbiology, one foreign language, thesis, final examination

UNIVERSITY OF MISSOURI - KANSAS CITY
(No reply received)

UNIVERSITY OF MISSOURI - ST. LOUIS
8001 Natural Bridge Road Phone: (314) 453-5811
St. Louis, Missouri 63121
Dean of Graduate Studies: E. Walters
Dean of College: R.S. Bader

DEPARTMENT OF BIOLOGY
Phone: Chairman: (314) 453-5811

Chairman: L.D. Friedman
Professors: F.H. Moyer, M.W. Strickberger
Associate Professors: H.P. Friedman, L.D. Friedman, G.T. Heberlein, M. Sage
Assistant Professors: J.E. Averett, M.Bekoff, A. Derby, T.H. Fleming, C.R. Granger, D.E. Grogan, B.R. Holt, J.E. Ridgway
Instructors: D. Judd, A. Wilke
Visiting Lecturers: 2
Research Assistants: 2
Graduate Teaching Assistants: 7

Field Stations/Laboratories: Weldon Springs Laboratory - Weldon Springs, Missouri

Degree Program: Biology

Undergraduate Degree: B.S.
UndergraduateDegree Requirements: Each Biology major must complete at least 33 hours of Biology including Biology 10, 224, 276, 289 and a minimum of one course to be taken from each of the following areas: Cellular, and Molecular Biology, Organismal Biology and Population and Ecology Biology

Graduate Degree: M.S.
Graduate Degree Requirements: 30 hours total: Master: 15 or more credits at the 400 level, no more than 10 credits for research (Biology 490), and four, but no more than 8 credits of graduate seminar, (Biology 489)

Graduate and Undergraduate Courses Available for Graduate Credit: Morphology of Nonvascular Plants and Laboratory (U), Morphology of Vascular Plants and Laboratory (U), Cellular Biology (U), Virology and Laboratory (U), Immunobiology and Laboratory (U), Advanced Genetics and Laboratory (U), Microbial Genetics and Laboratory (U), Advanced Development and Laboratory (U), Plant Physiology and Development and Laboratory (U), Population and Community Ecology and Laboratory (U), Techniques in Electron Microscopy and Laboratory (U), Advanced Biological Chemistry and Laboratory (U), Advanced Animal Behavior (U), Biosystematics (U), Advanced Animal Behavior Laboratory (U), Biosystematics Laboratory (U), Behavioral Genetics (U), Special Topics (U), Special Topics (U), Field Biology (U), Marine Biology and Laboratory (U), Graduate Seminar, Graduate Research in Biology, and Topics in Biology

WASHINGTON UNIVERSITY
St. Louis, Missouri 63130 Phone: (314) 863-0100
Dean of Graduate Studies: R.E. Morrow

College of Arts and Sciences

Dean: M. Kling

DEPARTMENT OF BIOLOGY
Phone: Chairman: (314) 863-0100 Ext. 4386 Others: Ext. 4365

Chairman: J.W. Hopkins
University Professor: T.S. Hall
Edward Mallinckrodt Distinguished Professor Emeritus: V. Hamburger
Professors: B. Commoner, W. Maxwell Cowan, H. Cutler (Adjunct), R. Levi-Montalcini, W.H. Lewis, F. Moog, P.H. Raven, O.J. Sexton, H.D. Stalker, J.E. Varner
Associate Professors: G.E. Allen III, O.P. Chilson, E.R. Eisendrath (Emerita), D.L. Kirk, D.H. Kohl, J. Maniotis, H. W. Nichols, B.G. Pickard, S.D. Silver, N. Suga
Assistant Professors: R.W. Coles (Adjunct), A.P. Covich, G.B. Johnson, M. Krukowski, J.E. Morhardt, P.S.G. Stein, V. Walbot

Field Stations/Laboratories: Tyson Research Center - Eureka, Missouri

Degree Programs: The Department of Biology, as a member of the Division of Biology and Biomedical Sciences, participates with eight basic science departments of the School of Medicine in five graduate programs: Molecular Biology; Cellular and Developmental Biology; Neurology; Plant Biology; and Ecology and Population Biology

Undergraduate Degrees: B.S. (Biology)
Undergraduate Degree Requirements: 2 semesters each of Calculus, Physics, Inorganic and Organic Chemistry, 18 upper level units of Biology

Graduate Degrees: M.S., Ph.D. (Biology)
Graduate Degree Requirements: Master: 24 semester hours of graduate level biology and thesis or, in unusual cases, 30 semester hours of graduate level biology with no thesis Doctoral: 72 semester hours of graduate level biology and thesis

Graduate and Undergraduate Courses Available for Graduate Credit: Vertebrate Physiology (U), Phycology (U), Structure and Functionof Higher Plants: Physiology (U), Physiology of Development (U), Introduction to Cell Biology (U), History of Scientific Thought to Newton (U), General Biochemistry (U), Somatosensory System (U), Vision (U), Electrobiology (U), Seminar in Advanced Biology(U), Study for Honors (U), Independent Work; Human Anatomy, Biology of Cultured Cells, Intrauterine Development, Pharmacology, Current Topics in Pharmacology, Fundamentals of Circulation, Cell Biology, General Pathology, Mechanism of Disease, Pathology Research Seminar, Advanced Biochemistry, Medical Microbiology, Physical-Chemical Basis of Techniques in Molecular Biology, Protein Chemistry and Enzyme Mechanisms, Molecular Biology of Animal Viruses, Seminar in Molecular and Cellular Biology, Topics in Neurobiology, Seminar in the Cellular Basis of Behavior, Neurological Pharmacology, Striated Muscle, Nerve Muscle and Synapse, Seminar in Population Biology, Research, Evolution of Man and Culture (U), Experimental Aquatic Biology (U), Structure and Function of Higher Plants: Development (U), Evolutionary Biology (U), Developmental Neurobiology (U), Cell Physiology (U), Advanced Genetics (U), Biochemistry Laboratory (U), Structural and Functional Organization of the Nervous System (U), Techniques in Field Biology (U), General Physiology, Environmental Pathology, Microscopic Anatomy, Human Growth and Development, Cellular Immunology and Immunopathology, Methods in Experimental Pathology, Seminar in Developmental Biology, Mechanism and Regulation of Protein Biosynthesis, Topics in Animal Viruses, Physical Chemistry of Macromolecules, Biochemistry of the Nervous System, Long Term Changes in the Nervous System, Seminar in Plant Biology

Division of Biology and Biomedical Sciences
660 S. Euclid Avenue
St. Louis, Missouri 63110
Dean: P.R. Vagelos

Phone: Chairman: (314) 454-2422 Others: 454-2000

Director: P.R. Vagelos
Professors: D.H. Alpers, R.A. Bradshaw, B.I.Brown, D.H. Brown, R.P. Bunge, A.I. Cohen, B. Commoner, J.R. Cox, W.M. Cowan, H.C. Cutler (Adjunct), G. R. Drysdale, A.C. Enders, C. Frieden, L. Glaser, S. Goldring, T.S. Hall, V. Hamburger (Emeritus), A. M. Holtzer, J.W. Hopkins, C.C. Hunt, F.E. Hunter, D.E. Kennell, S.C. Kinsky, P.E. Lacy, R. Levi-Montalcini, W.H. Lewis, O.H. Lowry, F. Matschinsky, D.M. McDougal, C.E. Molnar, F. Moog, B.W. Moore, R.R. Peterson, R.R. Pfeiffer, P.H. Raven, A. Roos, M.J. Schlesinger, D. Schlessinger, O.J. Sexton, J.R. Smith, H.D. Stalker, R.E. Thach, L.J. Tolmach, R. Torack, M. Trotter (Emeritus), J.E. Varner, D.F Wann (Research), J.C. Warren
Associate Professors: A.W. Alberts, G.E. Allen, D. Apirion, L.J. Banaszak, M.P. Blaustein, S. Boyarsky, M.B. Bunge, H.B. Burch, R.M. Burton, O.P. Chilson, L.L. Costantin, J.M. Davie, N. Daw, P.J. DeWeer, E.R. Eisendrath (Emerita), J.A. Ferrendelli, D.M. Geller, M.N. Goldstein, P.M. Hartroft, W.F. Holmes, C.M. Jackson, L. Jarett, J. Jeffrey (Research), E.G. Jones, D.L. Kirk, G.S. Kobayashi, D.H. Kohl, C. Kuhn, III, S. Lang, J.R. Little, J. Maniotis, G.R. Marshall, F.S. Mathews, J.G. Miller, P. Needleman, H.W. Nichols, C.W. Parker, A. Pearlman, B.G. Pickard, H.J. Raskas, C.M Rovainen, S. Schlesinger, W.R. Sherman, D.F. Silbert, S. Silver, E. Simms, A.C. Sonnenwirth, N. Suga, L.J. Thomas, T.W. Tillack, F.A. Valeriote, J.R. Williamson
Assistant Professors: C.D. Barry, E.R. Bischoff, I. Boime, H. Burton, R.W. Coles, A.P. Covich, J. Fleischman, L.D. Gelb, R.J. Graff, J. Hanaway, J.E. Harvey, D.C. Hellam, G.B. Johnson, B.F. King, S. Kornfield, M. Krukowski, L.F. Lake, H. Liebhaber, R.G. Lynch, P.W. Majerus, J.J. Marr, G. Medoff, D.N. Menton, R.N. Miller, J.E. Morhardt, J.L. Price, D. Purves, R.G. Roeder, S.J. Schlafke (Research), P.D. Stahl, P.S.G. Stein, V. Walbot, S.W. Weidman, M. Willard, T.A. Woolsey
Visiting Lecturers: 122

Field Stations/Laboratories: Tyson Research Center - Eureka, Missouri

Degree Programs: Ph.D. in Anatomy, Biological Chemistry, Biology, Microbiology, Genetics, Pathology, Pharmacology and Physiology and Biophysics, Ph.D. Programs in Cellular and Developmental Biology, Evolutionary Biology and Ecology, Molecular Biology, Neural Sciences and Plant Biology

Undergraduate Degree: B.S. (Biology)
Undergraduate Degree Requirements: Eighteen units in advanced Biology courses, including at least one semester of genetics, one year of College Physics, one year of General College Chemistry, one semester of Organic Chemistry Laboratory, two semesters of Organic Chemistry lectures or one semester of Organic Chemistry lectures plus one semester of Physical Chemistry

Graduate Degrees: Ph.D.
Graduate Degree Requirements: Doctoral: Seventy-two hours of course and research credits, successful completion of a preliminary examination and an original dissertation and defense of same.

Graduate and Undergraduate Courses Available for Graduate Credit: Phycology (U), Experimental Aquatic Biology, Laboratory and Field (U), Structure and Function of Higher Plants-Physiology (U), Structure and Function of Higher Plants-Development (U), Evolutionary Biology (U), Physiology of Development (U), Developmental

Neurobiology (U), Advanced Genetics (U), Neurocytology (U), Somatosensory System (U), Vision (U), Techniques in Field Biology (U), Human Anatomy, General Physiology, Biology of Cultured Cells, Environmental Pathology, Intrauterine Development, Microscopic Anatomy, Pharmacology, Current Topics in Pharmacology, Fundamentals of Circulation, Human Growth and Development, Cell Biology, General Pathology, Mechanisms of Disease, Cellular Immunology and Immunopathology, Pathology Research Seminar, Methods in Experimental Pathology, Properties of Cell Membranes, Seminar in Developmental Biology, Advanced Biochemistry, Mechanism and Regulation of Protein Biosynthesis, Medical Microbiology, Physical-Chemical Basis of Techniques in Molecular Biology, Physical Chemistry of Macromolecules, Protein Chemistry and Enzyme Mechanisms, Molecular Biology of Animal Viruses, Seminar in Molecular and Cellular Biology, Protein Crystallography, Biochemistry and Physiology of Peptide Hormones, Topics in Neurobiology, Seminar in the Cellular Basis of Behavior, Neural Sciences, Biochemistry of the Nervous System, Nerve Muscle and Synapse, Seminar in Plant Biology, Seminar in Population Biology, Seminar in Techniques in Field Biology

WESTMINSTER COLLEGE

Fulton, Missouri 65251 Phone: (314) 642-3361
Chairman: Division of Natural Sciences: H.W. Williams

DEPARTMENT OF BIOLOGY
Phone: Chairman: Ext. 246

Chairman: J.E. McClary
Cameron Day Professor of Biology: H.P. Hinde
Professors: D.R. Fichess, H.W. Williams
Teaching Assistants (Undergraduate Only): 5-10

Degree Program: Biology

Undergraduate Degree: B.A.
Undergraduate Degree Requirements: 20 hours in Biology in addition to the Introductory level course of 8 hours, Chemistry through the level of Organic Chemistry (minimum of 16 hours including Laboratory), Fulfill College requirements of minimum of 124 hours, 28 of which must be within two divisions outside the Science division

WILLIAM JEWELL COLLEGE

Liberty, Missouri 64068 Phone: (816) 781-3806
Dean of College: B. Thomson

DEPARTMENT OF BIOLOGY
Phone: Chairman: Ext. 249

Head: B.L. Wagenknecht
Professor: B.L. Wagenknecht
Associate Professors: J.T. Buss, C. Newlon
Visiting Lecturer: 1

Degree Programs: Biology, Medical Technology

Undergraduate Degrees: B.S. (Medical Technology), B.A. (Biology)
Undergraduate Degree Requirements: 32 hours in Biology, including 1 course in Botany, Genetics, Physiology and Ecology plus seminar and an independent study project, College requires 124 hours

WILLIAM WOODS COLLEGE

Fulton, Missouri 65251 Phone: (314) 642-2251
Dean of College: C.M. Shipp

DEPARTMENT OF BIOLOGY
Phone: Chairman: Ext. 275

Chairman: A.L. Hinde
Professor: J.S. Summers
Associate Professors: A.L. Hinde, N. Nagle

Degree Programs: Animal Science, Biology, Medical Technology

Undergraduate Degree: B.S.
Undergraduate Degree Requirements: Total - 122 hours, Major - 30 hours, 2 Minors - 18 hours each, No grade lower than C accepted in major

MONTANA

CARROLL COLLEGE
(No reply received)

EASTERN MONTANA COLLEGE
Billings, Montana 59101 Phone: (406) 657-2011
Dean of College: R. Rodney

DEPARTMENT OF BIOLOGICAL SCIENCES
Phone: Chairman: (406) 657-2341 Others: 627-2011

Head: G.L. Bintz
Professors: W.L. Milstead, N.D. Schoenthal
Associate Professors: G.L. Bintz, G.A. Davidson, J.F. Kirpatrick
Assistant Professor: R.G. Beaver

Field Stations/Laboratories: Beartooth Mountains - 70 miles from campus

Degree Program: Biology

Undergraduate Degree: B.A., B.S.
Undergraduate Degree Requirements: B.A. - 24 credits, foreign language, 49 credits biology, 18 of which must be upper division, 35 credits in Mathematics and Chemistry, sufficient additional credits in general education and electives to meet 192 credit graduate requirement. B.S. - 64 credits Biology, 10 Mathematics, 35 credits minor in Chemistry/Mathematics, 2nd Education - 52 credits Biology and Chemistry emphasized but not required

MONTANA STATE UNIVERSITY
Bozeman, Montana 59715 Phone: (406) 994-0211
Dean of Graduate Studies: K. Goering

College of Letters and Science
Dean: R.M. McBee

DEPARTMENT OF BIOLOGY
Phone: Chairman: (406) 994-4548

Head: J.M. Pickett
Professors: N. Anderson, L. Drew, R. Eng, K. Mills, R. Moore, D. Quimby, G. Roemhild, J. Rumely, D. Scharff, P.D. Skaar, S. Visscher, H. Watling, D. Worley, J. Wright
Associate Professors: D. Cameron, T. Carroll, D. Collins, W.J. Dorgan, B. Hahn, R. Mackie, H. Picton
Assistant Professors: C. Kaya, J. McMillan, H. Mecklenburg, D. Phillips, E. Vyse, B. Weaver
Instructor: S. Eversman
Research Assistants: 15
Teaching Assistants: 11

Degree Programs: Applied Ecology, Botany, Ecology, Entomology, Fisheries and Wildlife Biology, Forest Botany, Genetics, Limnology, Parasitology, Science Education, Watershed Management, Wildlife Management, Zoology

Undergraduate Degree: B.S.
Undergraduate Degree Requirements: 192 credits including electives in appropriate curriculum (Botany with Biology Teaching Option, Fish and Wildlife Management, Premedicine and Zoology.

Graduate Degrees: M.S., Ph.D.
Graduate Degree Requirements: Master: Option A: 30 course credits and 15 credits thesis Option B: 45 course credits Doctoral: Botany - acceptable pro gram and thesis Entomology, Fisheries and Wildlife Management - Zoology - Acceptable program, thesis and one language in depth

Graduate and Undergraduate Courses Available for Graduate Credit: Animal Physiology (U), Parasitology (U), Cytology (U), Genetics Laboratory (U), Physiological Genetics (U), Limnology (U), Human Genetics (U), Fresh Water Algae (U), Advanced Plant Taxonomy (U), Icthyology (U), Mammalogy (U), Ornithology (U), Herpetology (U), Invertebrate Zoology (U), Advanced Entomology (U), General Development (U), Basic Wildlife Physiology (U), Human Environment Physiology (U), Functional Biology of Fishes (U), Invertebrate Physiology (U), Evolution (U), Independent Problems (U), Special Topics (U), Seminar, Animal Ecology, Advanced Parasitology, Experimental Embryology, Cell Physiology, Advanced Plant Physiology, Fish Management, Wildlife Management - Mammals, Wildlife Management - Birds, Wildlife Management - Habitat, Algal Physiology, Advanced Physiological Genetics, Functional Anatomy of Vertebrates, Insect Morphology, Insect Physiology, Plant Geography, Plant Geography Laboratory, Advanced Animal Physiology, Population Genetics, Plant Autecology, Plant Autecology Laboratory, Plant Synecology, Plant Synecology Laboratory, Fresh-Water Invertebrates, Environmental Physiology, Principles of Aquatic Ecology, Biology of Water Pollution, Individual Problems, Special Topics, Graduate Consultation, Master's Thesis, Doctoral Thesis

DEPARTMENT OF MICROBIOLOGY
Phone: Chairman: (406) 994-2903 Others: 994-0211

Head: W.G. Walter
Professors: J.W. Jutila, R.H. McBee, N.M. Nelson, K.L. Temple, W.G. Walter
Associate Professors: T.W. Carroll, A.G. Fiscus, W. D. Hill, G.A. McFeters, F. Newman, N.D. Reed, D.G. Stuart
Visiting Lecturers: 4
Teaching Fellows: 4
Research Assistants: 4
Post Doctoral Fellow: 1

Degree Programs: Microbiology, Medical Technology, Environmental Health

Undergraduate Degree: B.S.
Undergraduate Degree Requirements: 192 quarter credits with required courses in Mathematics, Chemistry, Biology, Physics, Microbiology, required electives in Humanities and Social Sciences, and General Electives

Graduate Degrees: M.S., Ph.D.
Graduate Degree Requirements: Master: Minimum of 45 quarter credits in B option (No thesis), or 30 plus 15 thesis credits in A option. GRE scores in verbal and quantitative required when considering applicants with B average in U.G. program. Doctoral: 70 quarter credits of classwork beyond the B.S., one foreign language satisfied in one of several ways, comprehensive examinations, satisfactory dissertation and defense

Graduate and Undergraduate Courses Available for Graduate Credit: Immunobiology (U), Medical Bacteriology (U), Virology (U), Medical Mycology (U), Environmental Microbiology (U), Principles of Epidemiology (U), Systematic Bacteriology (U), Physiology of Microorganisms (U), Anaerobic Techniques (U), Geomicrobiology (U), Advanced Clinical Microbiology (U), Transplantation and Transfusion (U), Advanced Microbial Physiology, Biological Electron Microscopy, Basic Mammalian Pathology, Advanced Immunobiology, Seminar

College of Agriculture

Dean: J.A. Asleson

DEPARTMENT OF ANIMAL AND RANGE SCIENCES
Phone: Chairman: (406) 994-0211

Head: R.L. Blackwell
Professors: R.L. Blackwell, J.C. Boyd, J.Drummond, A.M. El-Negoumy, C.W. Newman, G.F.Payne, D.E. Ryerson, D.D. Thomas, J.L. VanHorn
Associate Professors: P.J. Burfening, J.R. Dynes, E.L. Moody, B.R. Moss
Assistant Professors: A.F. Buckler, K.L. Colman, D.D. Kress, J.E. Taylor
Instructor: L.C. Gagnon
Research Assistants: 3
Research Associates: 7

Degree Programs: Animal Science, Range Science

Undergraduate Degree: B.S.
Undergraduate Degree Requirements: 192 credits (Quarter basis)

Graduate Degree: M.S.
Graduate Degree Requirements: Master: 30 credits plus thesis or 45 credits

Graduate and Undergraduate Courses Available for Graduate Credit: Endocrine Physiology (U), Commerical Feeds and Feeding (U), Advanced Animal Breeding, Advanced Physiology of Reproduction, Endocrine and Reproductive Physiology, Nutrition, Research Techniques, Advanced Sheep and Wool Technology, Advanced Animal Nutrition, Range Ecosystems Measurements, Range Watershed Management Problems, Range Nutrition, Range Animal Influences, Harsh Site Development, Applied Range Production

DEPARTMENT OF PLANT AND SOIL SCIENCE
(No reply received)

DEPARTMENT OF VETERINARY SCIENCE
Phone: Chairman: (406) 994-4705 Others: 994-0211

Head: R.E. Dierks
Professors: R.E. Dierks, B.D. Firehammer, M.W. Hull, D.E. Worley
Associate Professors: C.K. Anderson, J.E. Catlin, D.H. Fritts, E.A. Lozano, L.L. Myers, F.S. Newman
Assistant Professor: M. Reilly
Research Assistants: 3
Post Doctoral Research Fellow: 1

Field Stations/Laboratories: 12 Agriculture Experiment Stations - and other stations throughout the state.

Degree Program: Veterinary Science

Graduate Degrees: M.S., Ph.D.
Graduate Degree Requirements: Master: Minimum 30 course credits and 15 credits thesis work. Major areas are defined as Microbiology, Biochemistry, Parasitology and Virology . Doctoral: Minimum 70 course credits and 25 credits thesis work. Major areas are defined as Microbiology, Biochemistry, Parasitology and Virology

Graduate and Undergraduate Courses Available for Graduate Credit: Seminar, Techniques for Research with Experimental Animals, Individual Problems, Special Topics, Graduate Consultation

ROCKY MOUNTAIN COLLEGE

(No reply received)

UNIVERSITY OF MONTANA

Missoula, Montana 58901 Phone: (406) 243-0211
Dean of Graduate Studies: J.M. Stewart

College of Arts and Sciences

Dean: R.A. Solberg

DEPARTMENT OF BOTANY
Phone: Chairman: (406) 243-5222 Others: 243-0211

Chairman: S.J. Preece, Jr.
Professors: M.J. Behan, M. Chessin, C.C. Gordon, J. Habeck, L.H. Harvey, S.J. Preece, Jr., G.W. Prescott, R.A. Solberg
Associate Professor: G.N. Miller, Jr.
Assistant Professors: R.P. Sheridan, M.L. Thornton, T.J. Watson
Teaching Fellows: 13
Research Assistants: 3
Research Associates: 1
Faculty Affiliates: 2

Field Stations/Laboratories: University of Montana Biological Station - Yellow Bay

Degree Programs: Botany, Environmental Studies, Wildlife Biology, Biological Sciences, Biology

Undergraduate Degrees: B.A., B.S.
Undergraduate Degree Requirements: Botany - 45 credits in Botany and Biology including: Ecosystem Biology, Ecosystem Biology Laboratory, Cell and Molecular Biology and Laboratory, Genetics, Evolution, Developmental Biology and Molecular Biology and Laboratory, General Botany, Botanical Literature and 15 credits of upper-division Botany courses. Also required are 50 credits outside ofBotany and Biology in allied sciences including Animal Kingdom.

Graduate Degrees: M.A., M.S., M.S.T., Ph.D.
Graduate Degree Requirements: Master: A minimum of 45 credits of graduate work with at least 30 in Botany and an acceptable thesis Doctoral: A minimum of 90 credits of graduate work with usually two-thirds in Botany and an acceptable thesis.

Graduate and Undergraduate Courses Available for Graduate Credit: Plant Physiology (U), Cellular Physiology (U), Principles of Plant Ecology (U), Fundamentals of Plant Systematics (U), Agrostology (U), Aquatic Flowering Plants (U), Forest Pathology (U), Introduction to Plant Disease (U), Phytochemistry I and II (U), Mineral Nutrition (U), Algal Physiology (U), Morphogenesis (U), Cytology (U), Phycology (U), Bryology (U), Pteridiology (U), Non-Vascular Plants (U), Organography of Vascular Plants (U), Reproduction in Vascular Plants (U), Problems in Plant Science (U), General Ecology (U), Ecological Systems Analysis (U), Biological Effects of Air Pollution (U), Botanical Effects of Water Pollution (U), Individual Problems in Pollution Studies (U), Limnology (U), Mycology (U), Paleobotany (U), Palynology (U), Cytogenetics (U), Population and Ecological Genetics (U), Seminar in Biology (U), Botanical Literature (U), History of Biology, Problems in Plant Physiology, Problems in Plant Anatomy and Morphology, Problems in Plant Cytology and Genetics, Problems in Plant Ecology, Advanced Systematics, Phytogeography, Problems in Plant Taxonomy, Problems in Mycology and Forest Pathology, Molecular Biology Seminar, Problems in Paleobotany, Graduate Seminar, Research, Advanced Molecular Biology Laboratory and Thesis

DEPARTMENT OF MICROBIOLOGY
Phone: Chairman: (406) 243-4582 Others: 243-0211

Chairman: M.J. Kakamura
Professors: C.L. Larson (Director, Stella Duncan Memorial Research Institute), M.J. Nakamura, J.J. Taylor, F. Sogandares, R.N. Ushijima
Associate Professors: G.L. Card, R.A. Faust, J.A. Rudbach
Assistant Professors: W.L. Koostra
Instructor: K.B. Read
Visiting Lecturers: 15
Teaching Fellows: 9
Research Assistants: 5

Field Stations/Laboratories: Biological Station - Flathead Lake, Yellow Bay, Montana, Lubrecht Experimental Station - Greenough, Montana

Degree Programs: Microbiology, Medical Technology

Undergraduate Degrees: A.B. (Microbiology), B.S. (Medical Technology
Undergraduate Degree Requirements: 196 credits, 45 credits in Microbiology, Biochemistry, Parasitology, English Composition, English Literature, 1 year each of Physics, Mathematics, Two years of Chemistry (Inorganic, Qualitative Analysis, Organic, Quantitative Analysis), 1 year of Biology Core Curriculum (Cell Physiology, Genetics, Developmental Biology)

Graduate Degrees: M.S., Ph.D.
Graduate Degree Requirements: Master: 45 credits, thesis Doctoral: 90 credits, Dissertation (or M.A. degree plus 60 credits, Dissertation)

Graduate and Undergraduate Courses Available for Graduate Credit: Medical Microbiology (U), Applied Microbiology (U), Environmental Health (U), Immunology and Serology (U), Microbial Physiology (U), Molecular Genetics (U), Clinical Microbiology (U), Hematology (U), Epidemiology (U), Medical Mycology (U), Yeasts (U), Mycoplasma and L-Forms (U), Virology (U), Population Genetics (U), General Parasitology (U), Helminthology (U), Medical Protozoology (U), Host-Parasite Relationships (U), Strictly Graduate Courses: Advanced Topics in Microbiology, Advanced Immunology, Seminar, Medical Parasitology, Microbiology Literature, Pathology, Microbial Cytology, Advanced Virology, Advanced Microbial Physiology, Immunochemistry, Molecular Biology Seminar, Advanced Molecular Biology Laboratory

DEPARTMENT OF ZOOLOGY
Phone: Chairman: (406) 243-5122 Others: 243-0211

Chairman: D.A. Jenni
Professors: R.B. Brunson, J.J. Craighead, E.W. Pfeiffer, G.F. Weisel, P.L. Wright
Associate Professors: J.H. Lowe, Jr., L.H. Metzgar, A. L. Sheldon, J.F. Tibbs
Assistant Professors: R.P. Canham, D.L. Kilgore, G. Patent
Teaching Fellows: 14
Research Assistants: 5

Field Stations/Laboratories: Biological Station - Flathead Lake, Bigfork, Montana

Degree Programs: Zoology (Interdepartmental Programs in Biology, Wildlife Biology, Environmental Studies, Master of Science in Teaching)

Undergraduate Degree: B.A.
Undergraduate Degree Requirements: 195 Total quarter credits including: 25 Credits of Biology, 51 credits of Zoology, 15 credits of Mathematics, 5 credits of Botany, 3 credits of English, 9 Credits of Chemistry, 15 credits in non-biological sciences of which 9 must be in one science, 23 credits of a foreign language

Graduate Degrees: M.A., Ph.D.
Graduate Degree Requirements: Master: Qualifying examination during first year. 45 quarter credits, 15 credits thesis maximum, 15 credits outside of Department maximum. Reading ability, one foreign language, Thesis, final oral examination Doctoral: Qualifying examination during first year. 90 quarter credits beyond Bachelors. Reading ability one foreign language, Comprehensive examination in student's area. Dissertation and final oral defense of Dissertation.

Graduate and Undergraduate Courses Available for Graduate Credit: Comparative Vertebrate Anatomy and Embryology (U), General Parasitology (U), Animal Microtechnique (U), Herpetology (U), Aquatic Biology (U), Ornithology (U), Mammalogy (U), Ichthyology (U), Vertebrate H stology (U), Protozoology (U), Invertebrate Zoology (U), Invertebrate Zoology II (U), Biological Literature (U), Cellular Physiology (U), Comparative Physiology (U), Mammalian Physiology (U), Mammalian Physiology II (U), Water Quality (U), Population Ecology (U), Community Ecology (U), Entomology (U), Aquatic Insects (U), Human Genetics and Evolution (U), History and Development of Biological Conservation (U), Animal Behavior (U), Fishery Science (U), Zoogeography (U), Endocrinology (U), Cytology (U), Marine Invertebrates (U), Limnology (U), Cytogenetics (U), Population and Ecological Generalizations (U), Helminthology (U), Host-Parasite Relationships (U), Problems in Zoology (U), Special Topics, Areas and Concepts of Zoology, Advanced Animal Behavior, Behavioral Ecology Seminar, Ethology Seminar, Ecological Theory, Ecology Seminar, Evolution Seminar, Advanced Topics in Physiology, Molecular Biology Seminar, Speciation, Advanced Zoological Problems, Advanced Molecular Biology Laboratory I, II, III, Thesis

WESTERN MONTANA COLLEGE

Dillon, Montana 59725 Phone: (406) 683-7011
Dean of Graduate Studies: D. Tash
Dean of College: J.E. Short

DEPARTMENT OF BIOLOGY
Phone: Chairman: (406) 683-7321 Others: 683-7011

Head: D.G. Block
Professors: K.J. Bandelier, D.G. Block
Associate Professors: R.L. Timken

Degree Program: Biology

Undergraduate Degree: B.S.
Undergraduate Degree Requirements: 54 credits in Science including 10 credits of Chemistry.

NEBRASKA

CHADRON STATE COLLEGE

Chadron, Nebraska 69337 Phone: (308) 432-4451
Dean of Graduate Studies: G.B. Barteli
Dean of College: J. Swanson

BIOLOGY DEPARTMENT
Phone: Chairman: Ext. 295

Chairman: (Rotating Chairmanship)
Associate Professors: J.D. Druecker
Assistant Professors: J.C. Gibson, H.R. Lawson, R. Weedon

Degree Programs: Biology, Biological Education, Medical Technology

Undergraduate Degrees: B.A., B.S.
Undergraduate Degree Requirements: Minimum - 10 hours Chemistry, Requirement - Chemistry minor

Graduate Degree: M.S. (Biological Education)
Graduate Degree Requirements: Master: Minimum 18 hours Science, minimum 12 hours Education

COLLEGE OF SAINT MARY

1901 South 72nd Street Phone: (402) 393-8800
Omaha, Nebraska 68124
Dean of College: Sr. M.M. Hill

DEPARTMENT OF BIOLOGY
Phone: Chairman: Ext. 58

Chairman: Sr. M.C. Macaluso
Professor: Sr. M.C. Macaluso
Assistant Professor: C. Oswald
Instructor: Sr. M.C. Snegoski
Visiting Lecturer: 1

Degree Programs: Biology, Medical Technology

Undergraduate Degree: B.S.
Undergraduate Degree Requirements: 32 hours of Biology, 16 hours of Chemistry (Inorganic and Organic), 8 hours of Physics, 6 hours of Mathematics, 37 hours of General Education, 29 hours of Electives

CREIGHTON UNIVERSITY

2500 California Street Phone: (402) 536-2700
Omaha, Nebraska 68133
Dean of Graduate Studies: Rev. R. Shanahan, S.J.

College of Arts and Sciences

Dean: R. Passon

DEPARTMENT OF BIOLOGY
Phone: Chairman: 536-2811

Chairman: R.W. Belknap
Professors: R.W. Belknap, A.B. Schlesinger
Associate Professors: C.B. Curtin
Assistant Professors: H. Nickla, J. Platz, J. Roberts
Teaching Fellows: 8

Degree Programs: Biology, Science Education

Undergraduate Degree: B.S.
Undergraduate Degree Requirements: A minimum of 25 semester hours of 100-level courses including Biometry, Genetics and Physiology. These courses must be supported by one year of Physics and Organic Chemistry. Recommended but not required are Mathematics through Calculus, and 2 years of a foreign language

Graduate Degree: M.S.
Graduate Degree Requirements: Master: Must complete a minimum of 30 semester hours of graduate credit of which 24 must be course credit hours and 6 thesis credit hours.

Graduate and Undergraduate Courses Available for Graduate Credit: Not Stated

School of Medicine

Dean: J.M. Holthaus

ANATOMY DEPARTMENT
Phone: Chairman: (402) 536-2914

Chairman: W.E. Dossel
Professors: J.J. Baumel, W.E. Dossel, C.C. Turbes
Assistant Professors: D.C. Cress, A.F. Dalley, R.J. Thomas, G.L. Todd
Instructor: R.M. Sorenson
Research Assistant: 1
Pre-doctoral Assistant: 1
Post-doctoral Assistant: 1

Degree Program: Anatomy

Graduate Degrees: M.S., Ph.D.
Graduate Degree Requirements: Plan A: 30 semester hours of graduate credit: 24 hours of course credit and 6 hours of thesis credit. Reading knowledge of French or German, comprehensive examinations, written and oral. Plan B: 33 hours of graduate credit, no thesis, or language requirement. Written comprehensive examination, no oral. Doctoral: Ninety semester hours of credit beyond the bachelor's or sixty semester hours beyond the Master's degree. Minimum grade-point average of 3.00 (B) over total program. Proficiency in one or more foreign languages or alternate research tools. Comprehensive qualifying examination. Acceptable dissertation

Graduate and Undergraduate Courses Available for Graduate Credit: Neuroscience, Human Gross Anatomy, Advanced Human Gross Anatomy, Human Histology, Advanced Histology, Fundamentals of Electron Microscopy, Interpretation of Fine Structure, Human Embryology, Human Neuroanatomy, Advanced Neuroanatomy, Brain Research, Readings in Neurology, Neurological Basis of Behavior, Comparative Anatomy of Selected Laboratory Animals, Avian Anatomy, Anatomical Methods, Anatomy Seminar, Research in Anatomy -- Master' s Research, Research in Anatomy -- Doctoral Research Master's Thesis (A-B), Doctoral Dissertation (A-B-C-D)

DEPARTMENT OF BIOLOGICAL CHEMISTRY
Phone: Chairman: (402) 536-2917

Chairman: I.C. Wells
Professors: D. Gambal, E.L. Rongone, I.C. Wells, D.D. Watt
Associate Professors: J.S. Baumstark, E.A. Carusi, A.P. Fishkin, R. Fried, H.C. Lankford
Assistant Professors: D.R. Babin
Research Fellow: 1

Degree Program: Biological Chemistry

Graduate Degrees: M.S., Ph.D.

Graduate Degree Requirements: Master: 30 semester hours, 24 of course work, 6 of thesis, French or German or research tool, Comprehensive examination, Acceptable thesis Doctoral: 1. Ninety semester hours, 2. Grade point of 3.00, 3. Language or research tool, 4. Qualifying examination 5. Publishable dissertation

Graduate and Undergraduate Courses Available for Graduate Credit: Principles of Biochemistry, General Biological Chemistry, Experimental Biochemistry, Advanced Biological Chemistry, Clinical Biochemistry, Biochemical Preparations, Seminar

DEPARTMENT OF MEDICAL MICROBIOLOGY
Phone: Chairman: (402) 536-2921 Others: 536-2922

Chairman: W.E. Sanders, Jr.
Professors: F.M. Ferraro, W.E. Sanders
Associate Professors: E.A. Chaperon, M.J Severin, R.G. Townley
Assistant Professors: S.M. Polly, C.C. Sanders, D. Seshachalam, C.A. Walker

Degree Program: Medical Microbiology

Graduate Degrees: M.S., Ph.D.
Graduate Degree Requirements: Not Stated

Graduate and Undergraduate Courses Available for Graduate Credit: Microbiological Survey, Medical Microbiology (U), Mechanisms of Microbial Pathogenicity, Diagnostic Microbiology, Medical Mycology, Instrumentation and Methodology in Current Microbiology, Immunopharmacology, Recent Developments in Immunopharmacology, Microbial Physiology, Immunobiology and Immunochemistry, Methods in Microbial Physiology, Techniques in Immunobiology and Immunochemistry, Selected Topics in Advanced Immunology, General Virology, Epidemiology and Public Health, Antimicrobial Agents and Chemotherapy, Advances in Antimicrobial and Chemotherapeutic Research, Cell Biology, Microscopy, Departmental Seminar, Research in Microbiology, Master's Thesis

DEPARTMENT OF PHYSIOLOGY-PHARMACOLOGY
Phone: Chairman: (402) 536-2925

Head: D.F. Magee
Professors: H.S. Badeer, R.O. Creek, J.T. Elder, E.H. Grinnell, N.M. Huffman, J.R. Johnson, D.F. Magee, H.J. Phillips, P. Prioreschi, A. Czerwinski, J. Crampton
Associate Professor: N.W. Scholes
Assistant Professor: H.H. Gale
Visiting Lecturers: 1

Degree Programs: Physiology, Pharmacology

Graduate Degrees: M.S., Ph.D.
Graduate Degree Requirements: Master: Plan A - 24 hours plus 6 hours, thesis, minimum QPA of 3.0 Plan B - 33 hours, thesis, minimum QPA 3.0 Doctoral: 90 credit hours, Thesis, QPA minimum of 3.0

Graduate and Undergraduate Courses Available for Graduate Credit: Endocrine Glands, Laboratory in Endocrinology, Cell Adaptation in Culture, Biophysics and Statistics, Medical Physiology, Physiological Processes in Respiration, Advanced Topics in Endocrinology, Medical Pharmacology, Principles of Pharmacology, Steroids, Structure and Pharmacological Activity, Autonomic Pharmacology, Pharmacological Techniques in the Study of Bioelectric Physiology, G.I. Physiology and Pharmacology, Methods of G.I. Research, Seminar in Cardiorespiratory Physiology, Electrophysiology of the Heart, Pharmacology of Myocardial Infarction and Agina Pectoris, Mammalian Electrocardiography, Cardiovascular Drugs, Cardiodynamics, Muscle Physiology and Biophysics, Kidney and Electrolytes, Physiology-Pharmacology Seminar, Research Physiology-Pharmacology, Research in G.I. Physiology

DANA COLLEGE

Blair, Nebraska 68008 Phone: (402) 426-4101
Dean of College: R.J. Glass

DEPARTMENT OF BIOLOGY
Phone: Chairman: Ext. 229 Others: Ext. 328

Chairman: L.E. Stone
Professor: L.E. Stone
Associate Professor: G.E. Grube (Director of Environmental Studies)

Degree Program: Biology

Undergraduate Degree: B.S.
Undergraduate Degree Requirements: (a) 30 hours of Biology (including 4 hours Biological Principles, 4 hours Vertebrate Anatomy and 1 hour Research) (b) 8 hours of Chemistry

DOANE COLLEGE

Crete, Nebraska 68333 Phone: (402) 826-2161

NATURAL SCIENCE DIVISION
Phone: Chairman: Ext. 263 Others: Ext. 217

Chairman: D. King
Professor: K.M. Buell
Assistant Professors: R.D. Muckil, J.A. Walker

Degree Program: Biology

Undergraduate Degree: B.A.
Undergraduate Degree Requirements: Seven units of Biology including Biology 101. Completion of two terms of major colloquium (Activity 117,118). Two cognate units in Chemistry: Chemistry 105-106 or 125-126

HASTINGS COLLEGE

Hastings, Nebraska 68901 Phone: (402) 463-2402
Dean of College: A.L. Langvardt

DEPARTMENT OF BIOLOGY
Phone: Chairman: Ext. 64

Chairman: D.C. Aylward
Associate Professors: D.C. Aylward, G.L. Adrian
Assistant Professors: C. Springer

Degree Program: Biology

Undergraduate Degree: A.B.
Undergraduate Degree Requirements: 1) 35 units (1 unit equals 4 semester hours) to receive A.B. 2) 8-10 units for Biology major, plus 5 units allied fields Chemistry and Mathematics, 3) 12 units upper division (Junior-Senior) courses

KEARNEY STATE COLLEGE

Kearney, Nebraska 68847 Phone: (308) 236-4141
Dean of Graduate Studies: L.J. Bicak
Dean of College: J.M. McFadden

DEPARTMENT OF BIOLOGY
Phone: Chairman: (308) 236-4539 Others: 236-4141

Chairman: O.A. Kolstad
Professors: L.J. Bicak, J.C.W. Bliese, R.W. Ikenberry, D.L. Lund
Associate Professors: H.E. Cole, M.G. Fougeron, H.N. Nagel, A.E. Poorman, C.E. True, M.L. Williams
Assistant Professors: J.P. Farney, S.N. Longfellow

Degree Program: Biology

Undergraduate Degree: B.S.
Undergraduate Degree Requirements: Biology courses, 37 hours, Chemistry courses, 16 hours, Mathematics courses, 5 hours

Graduate Degree: M.S.
Graduate Degree Requirements: Master: 36 hours of Biology

Graduate and Undergraduate Courses Available for Graduate Credit: Microbiology (U), Plant Physiology (U), Plant Ecology (U), Limnology (U), General Plant Morphology (U), Plant Taxonomy (U), Taxonomy of Grasses (U), Special Topics in Biology (U), History of Biology (U), Genetics (U), Human Genetics and Societal Problems (U), Animal Behavior (U), Physiology (U), Radiation Biology (U), Entomology (U), Methods in Secondary Science Teaching (U), Ichthyology (U), Ornithology (U), Mammalogy (U), Protozoology (U), Biology Seminar (U), Organic Evolution, Advanced Cell Evolution, Advanced Cell Biology, Introduction to Graduate Study and Research, Biological Research, Plant Pathology, Developmental Genetics, Biostatistics, Endocrinology, Thesis

DEPARTMENT OF ENVIRONMENTAL STUDIES
Phone: Chairman: 236-4536

Chairman: H.G. Nagel
Professors: E. Grundy, J. McFadden
Associate Professors: H.G. Nagel, P. Blickensderfer, J.L. Roark
Assistant Professors: M.E. Glasser, C.E. True, J. Enns, M. Stone
Instructor: E.A. Sechtem

Degree Programs: Environmental Studies Minor only is offered.

Undergraduate Degree Requirements: 20 hours required courses, 4 hours electives

MIDLAND LUTHERAN COLLEGE

Fremont, Nebraska 68025 Phone: (402) 721-5480
Dean of College: Not Stated

DEPARTMENT OF BIOLOGY
Phone: Chairman: (402) 721-5480

Chairman: D.A. Becker
Associate Professors: D.A. Becker, J.R. Johnson

Degree Program: Biology

Undergraduate Degree: B.A.
Undergraduate Degree Requirements: nine courses - Biology, two courses - Chemistry, two courses - Mathematics

NEBRASKA WESLEYAN

Lincoln, Nebraska 68504 Phone: (402) 466-2371
Dean of College: F. Blumer

DEPARTMENT OF BIOLOGY
Phone: Chairman: Ext. 266

Chairman: W.L. Staudinger
Professor: M. Bichel
Assistant Professors: W. Boernke, V. Carver, G. Dappen

Degree Program: Biology

Undergraduate Degrees: B.A., B.S.
Undergraduate Degree Requirements: B.S. - 35 hours in Biology and 30 hours support, B.A. - 25 hours

UNION COLLEGE

(No reply received)

UNIVERSITY OF NEBRASKA - LINCOLN

Lincoln, Nebraska 68508 Phone: (402) 472-2711
Dean of Graduate Studies: J.V. Drew (Acting)

School of Life Sciences

Dean: M. Larsen (Acting)

SCHOOL OF LIFE SCIENCES
Phone: Chairman: (402) 472-2720 Others: 472-2721

Professors: E. Ball, M. Banerjee, M. Boosalis, R. Borchers, M. Brakke, J. Brumbaugh, J.M. Daly, J. Davidson, W. Gauger, C. Georgi, C. Gugler, H. Gunderson, N. Gupta, P. Johnsgard, R. Johnston, R. Kaul, R. Lommason, W. Militzer, D. Miller, W. Ray, M. Schuster, T. Thompson, T. Thorson, J. Weihing
Associate Professors: E. Barnawell, R. Boohar, E. Davies, B. Doupnik, G. Hergenrader, R. Hill, J. Janovy, Jr., E. Kerr, H. Knoche, P. Landolt, J. Lynch, J. McClendon, R. Meints, B. Nickol, G. Peterson, G. Riffle, G. Tharp, J. Van Etten, G. Vidaver, F. Wagner, D. Wysong
Assistant Professors: R. Dam. V. Day, L. Dunkle, J. Hazel, R. Klucas, W. Langenberg, E. Martin, L. Palmer, L. Parkhurst, M.L. Pritchard, P. Rand, J. Rosowski, J. Steadman, J. Tribble, A. Vidaver, A. Welch

Degree Programs: No formal statement has been developed yet; the School was formed on July 1, 1973.

Graduate Degrees: M.S., Ph.D.
Graduate Degree Requirements: Master: 36 semester credit hours which may include a minimum of 6 hours of thesis research final oral examination Doctoral: 90 semester credit hours of which no more than 60 may be course work. One language or research tool, qualifying examination, written comprehensives, oral defense of thesis

Graduate and Undergraduate Courses Available for Graduate Credit: Cell Biology and Genetics Section - Cytology, Yeasts, Molds and Actinomycetes, Activities of Fungi, Cellular Physiology, History and Literature of Zoology, Developmental Biology, Advanced Genetics, Advanced Cell Biology, Plant Hormones, Comparative Hematology, Protozoan Physiology, Developmental Genetics, Genetic Mechanisms of Evolution, Human Cytogenetics, Transmission Electron Microscopy, Physiology and Biochemistry Section - Endocrinology, Physiology of Domestic Animals, Physiology of Domestic Animals, Environmental Physiology, Introduction to Physical Laws in Biology, Comparative Physiology, Vertebrate Physiology, Experimental Physiology, Comparative Endocrinology, Analytical and Biochemical Techniques, Radioisotopic Methods, Advanced Vertebrate Physiology, Advanced Comparative Physiology, Plant Biochemistry, Physiology of Exercise, Metabolism of Carbohydrates, Metabolism of Lipids and Steroids, Mechanisms of Biochemical Regulation, Plant Biophysics, Plant Metabolism, Physical Methods in Plant Physiology, Chemical Methods in Plant Physiology, Modern Topics in Plant Physiology, Biochemistry of Nutrition Organismic Biology Section—Plant Microtechnique, Microscopical Techniques, Scanning Electron Microscopy, History and Literature of Biology, Dynamics of Wildlife Populations, Experimental Behavior, Physiological Animal Ecology, Zoogeography, Current Problems in Plant Ecology, Advanced Plant Taxonomy, Principles of Taxonomy and Zoological Nomenclature, Local Flora, Vegetation of North America, Microclimate: The Biological Environment, Environmental Quality and Control - Land, Air and Water, Limnology, Advanced Limnology, Marine Ecology and Paleoecology, Animal Behavior, Fungi, Algae, Procaryotes, Experimental Phycology, Paleobotany, Agrostology, Aquatic Plants, Plant Anatomy, Plant Morphogenesis, Bryophytes and Pteridophytes, Phylogeny of Seed Plants, Protozoology, Helminthology, Invertebrate Fauna of the Great Plains, Invertebrate Embryology, Aquatic Insects, Invertebrate Phylogeny, Advanced Invertebrate Zoology, The Elasmobranches, Ichthyology, Fisheries Biology, Herpetology, Ornithology, Mammalogy, Phylogeny of the lower Vertebrates, Vertebrate Morphology Microbiology, Immunology and Plant Pathology - Microbial Physiology, Advanced Biochemistry and Physiology of Microorganisms, Advanced Biochemistry and Physiology of Microorganisms, Microbial Genetics, Microbial Cytology and Anatomy, Pathogenic Bacteriology, Immunology, Viruses, Mechanisms of Bacterial Pathogenicity, Food, Sanitary and Environmental Microbiology, Food Microbiology II, Soil Microbiology, Industrial Microbiology, General Plant Pathology, Insect Transmission of Plant Pathogens, Nematode Disease of Plants, Plant Pathogenic Bacteria, Plant Pathology-Physiology, Plant Pathology-Virus Diseases, Plant Pathology-Epidemiology, Research Methods in Plant Pathology, Seminar in Plant Pathology

College of Agriculture

Dean: D. Acker

DEPARTMENT OF AGRONOMY
Phone: Chairman: (402) 472-2811

Chairman: D.G. Hanway
Meyer Katzman Professor of Agronomy: C.L. Gardner
Bert Rodgers Professor of Agronomy: F.A. Haskins
Professors: O.C. Burnside, L.A. Daigger, A.F. Dreier, J.D. Eastin, C.R. Fenster, A.D. Flowerday, J.D. Furrer, C.O. Gardner, H.J. Gorz, D.G. Hanway, F.A. Haskins, V.A. Johnson, W.R. Kehr, P.J. Mattern, A.P. Mazurak, T.M. McCalla, M.K. McCarty, D.P. McGill, M.R. Morris, L.C. Newell, J.T. Nichols, R.A. Olson, G.A. Peterson, W.M. Ross, D.H. Sander, J.W. Schmidt, W.M. Schultz, J.H. Williams
Associate Professors: L. Chesnin, M.D. Clegg, W.A. Compton, E.C. Conard, L.F. Elliott, D.W. Lancaster (courtesy), T.L. Lavy, J.W. Maranville, R.S. Moomaw, L.E. Moser, R.F. Mumm, G.W. Rehm, R.C. Sorensen, G.A. Wicks, R.A. Wiese
Assistant Professors: G.M. Dornhoff, K.D. Frank, P.H. Grabouski, L.A. Kelpper, D. Knudsen, D.T. Lewis, A.R. Martin, L.A. Nelson, P.T. Nordquist, L.J. Perry, E.J. Penas, J.E. Stroike, J.S. Webster
Instructors: W.A. Anderson, J.R. Ellis, J.M. Hart, J.R. Martineau, L.N. Mielke, L.A. Morrow, G.E. Schuman, L. Svajgr (Courtesy)
Research Assistants: 8

Field Stations/Laboratories: University of Nebraska Northeast Station - Concord, University of Nebraska North Platte Station - North Platte, University of Nebraska Panhandle Station - Scottsbluff, University of Nebraska Sandhills Agricultural Laboratory - Tryon, University of Nebraska Northwest Agricultural Laboratory - Alliance, University of Nebraska South Central Station - Clay Center, University of Nebraska Field Laboratory - Wahoo,

Degree Programs: Plant Breeding and Genetics, Soil Science, Crop Physiology and Production, Forages and Range Management, Weed Science

Undergraduate Degree: B.S. (Soil Science)
Undergraduate Degree Requirements: 128 hours required for graduation. A minimum of 25 hours of Agriculture including Introductory Crop Science, Soil Science, Feeds and Feeding, and Agricultural Economics. Biological Sciences 18-19 hours including 8 hours Chemistry and 4 hours Mathematics. Humanities and Social Sciences 24 hours. Remaining hours to be used for option requirements, agricultural courses or free electives.

Graduate Degrees: M.S., Ph.D. (Agronomy)
Graduate Degree Requirements: Master: Completion of 30 hours, including 12-16 hours at 800 level or above, 8-12 hours at 900 level and 6-10 hours of thesis. If a thesis is not prepared, student must complete a total of 36 hours, including either a minor or an additional 10 hours at the 900 level. Completion of a written and an oral examination required. Doctoral: A program of study and thesis subjects selected by supervisory committee. Must show proficiency in (a) a foreign language, (2) a special research technique or (3) a collateral field including at least 9 hours of study. He must pass a written and an oral examination and present an acceptable thesis.

Graduate and Undergraduate Courses Available for Graduate Credit: Crop Ecology (U), The Biological Environment (U), Agricultural Climatology, Crop Physiology (U), Physiological Genetics (U), Plant Breeding (U), Turfgrass Management (U), Statistical Methods in Research (U), The Range Ecosystem (U), Forage and Weed Physiology (U), Range Plants (U), Radioisotopic Methods, Soil Chemistry (U), Soil Chemical Measurements (U), Soil Microbiology (U), Physical Properties of Soil (U), Soil Morphology, Classification and Survey (U), Independent Study in Agronomy (U), Seminar, Advanced Plant Breeding, Laboratory Experiments in Genetics, Plant Cytogenetics, Plant Genetics, Pesticide Dissipation in Soils and Plants, Experimental Design and Statistical Interpretation, Population Genetics, Biometrical Genetics, and Plant Breeding, Digital Computer Methods for Statistical Data Processing, Forage Evaluation, Research Methods in Soils, Theoretical Aspects of Physical Chemistry of Soils, Soil Physics, Soil Fertility, Soil Genesis and Classification, Research in Crops, Research in Soils, Masters Thesis, Doctoral Dissertation, Livestock Management on Range and Pasture

DEPARTMENT OF ENTOMOLOGY
Phone: Chairman: (402) 472-2125

Chairman: E.A. Dickason
Professors: H.J. Ball, E.A. Dickason, R.E. Hill, G.R. Manglitz, E.S. Raun, R.E. Roselle, R. Staples
Associate Professors: L.W. Andersen, J.B. Campbell, T.J. Helms, D.L. Keith, L.L. Peters, K.P. Pruess
Assistant Professors: J.H.L. Bell, A.F. Hagen, S.D. Kindler, Z.B. Mayo
Instructor: W.J. Gary
Teaching Fellow: 1
Research Assistants: 8
Research Associate: 1

Fiels Stations/Laboratories: University of Nebraska Northeast Station - Concord, University of Nebraska South Central Station - Clay Center, University of Nebraska North Platte Station - North Platte, University of Nebraska Panhandle Station - Scottsbluff

Degree Program: Entomology

Undergraduate Degree: B.S.
Undergraduate Degree Requirements: 128 semester hours, which includes 31-33 agriculture (including 13-15 Entomology), 31 biological sciences, 16 Physical Sciences, 26-28 Humanities and Social Sciences, and 20-24 open electives

Graduate Degrees: M.S., Ph.D.
Graduate Degree Requirements: Master: 24 semester hours, including thesis credits, and at least one-half of the work in major. Doctoral: 96 semester hours, including a minimum of 20 semester hours for Dissertation, minimum residence of one academic year.

Graduate and Undergraduate Courses Available for Graduate Credit: Insect Physiology (U), Aquatic Insects (U), Comparative Insect Anatomy and Histology (U), History and Literature of Entomology (U), Insect Ecology

(U), Insects Affecting Plants and Animals (U), Insect Control by Host Plant Resistance (U), Field Entomology (U), Recognition of Adult Insects (U), Current Problems in Economic Entomology (U), Advanced Insect Physiology, Principles of Systematic Entomology, Independent Study, Research, M.S. Thesis, Ph.D. Dissertation.

DEPARTMENT OF HORTICULTURE AND FORESTRY
Phone: Chairman: (402) 472-2854 Others: 472-7211

Chairman: J.O. Young
Professors: D.P. Coyne, E.J. Kinbacher, R.B. O'Keefe, N.J. Rosenberg and J.O. Young
Associate Professors: W.T. Bagley, R.E. Neild and C.Y. Sullivan
Assistant Professors: B.L. Blac, S.S. Salac

Field Stations/Laboratories: University Field Laboratory - Mead, North Platte Station - North Platte, Panhandle Station - Scottsbluff, Northwest Agricultural Laboratory - Alliance, South Central Station - Clay Center, Northeast Station - Concord, High Plains Agricultural Laboratory - Sidney, Snadhills Agronomy Laboratory - Tryon, Horning State Forest - Plattsmouth

Degree Programs: Agricultural Climatology, Physiological, Genetic, Morphological and Environmental aspects of the production and utilization of horticultural and forestry crops.

Undergraduate Degree: B.S. (Horticulture)
Undergraduate Degree Requirements: A Total of 128 credit hours in Agricultural courses, 15-16 credit hours in Biological Science courses, 16 hours in Physical Science courses, 24 hours in Humanities and Social Sciences, and 45-47 credit hours of optional requirements and free electives.

Graduate Degrees: M.S., Ph.D. (Horticulture)
Graduate Degree Requirements: Master: A course of study which involves preparation of thesis. Course work constituting a major in Horticulture may include: Agronomy 203,205,307,310,311. Doctoral: Candidates for the degree of Doctor of Philosophy must present a dissertation which contains results of original research and which is directed by a member of the Department. The requirements for admission for candidacy and for the dissertation are those of the Graduate College.

Graduate and Undergraduate Courses Available for Graduate Credit: Crop Ecology, Microclimate: The Biological Environment, Crop Physiology, Turfgrass Management, Soil Chemistry, Physical Properties of Soil, Independent Research, Agricultural Climatology, Micrometeorology of the Biological Environment -- Advanced Topics, Crop Responses to Environment, Horticulture Crop Improvement in Breeding, Population Genetics, Seminar in Horticulture, Soil Fertility

LABORATORY OF AGRICULTURAL BIOCHEMISTRY
Phone: Chairman: (402) 472-2932 Others: 472-2711

Head: H.W. Knoche
Petrus Peterson Professor of Biochemistry: J.M. Daly
Professors: R.L. Borchers, H.W. Knoche
Associate Professors: R.M. Hill, F.W. Wagner
Assistant Professors: R. Dam, R.V. Klucas, R.L. Ogden
Teaching Fellows: 3
Research Assistants: 5

Degree Programs: Biochemistry, Physiology and Biochemistry

Graduate Degrees: M.S., Ph.D.
Graduate Degree Requirements: Master: 30 semester hours, thesis Doctoral: 90 semester hours, thesis

Graduate and Undergraduate Courses Available for Graduate Credit: Courses are offered in the School of Life Sciences

DEPARTMENT OF POULTRY AND WILDLIFE SCIENCES
Phone: Chairman: (402) 472-2052

Chairman: G.W. Froning
Professors: G.W. Froning, E.D. Cleaves, T.W. Sullivan
Associate Professors: F.B. Mather, H.L. Wiegers
Assistant Professors: R.M. Case
Research Assistants: 5

Field Stations/Laboratories: Rogers Farm (Turkey Research) 20 miles north of Lincoln

Degree Program: Poultry Science

Undergraduate Degree: B.S.
Undergraduate Degree Requirements: 128 credit hours

Graduate Degree: M.S.
Graduate Degree Requirements: Master: Option 1 - 30 credit hours including thesis Option 2 - 36 credit hours (no thesis)

Graduate and Undergraduate Courses Available for Graduate Credit: Seminar, Poultry Problems, Poultry Nutrition, Computer Feed Formulation, Poultry Breeding and Reproduction, Avian Biology, Independent Study, Animal Industry Seminar, Research, Animal Nutrition Seminar, Advanced Poultry Nutrition, Advanced Science of Poultry Products, Advanced Poultry Physiology, Masters thesis, Doctoral Thesis

DEPARTMENT OF PLANT PATHOLOGY
Phone: Chairman: (402) 472-2858 Others: 472-2859

Chairman: M.G. Boosalis
Professors: M.G. Boosalis, M.K. Brakke, G.W. Peterson, M.L. Schuster, J.L. Van Etten, D.S. Mysong
Associate Professors: E.M. Ball, W.G. Langenberg, J.W. Riffle, A.K. Vidaver
Assistant Professors: L.D. Dunkle, L.T. Palmer, J.R. Steadman
Research Assistants: 6

Field Stations/Laboratories: Panhandle Station - Scottsbluff, North Platte Station - North Platte, South Central Station - Clay Center, North East Station - Concord

Degree Program: Plant Pathology

Undergraduate Degree: B.S.
Undergraduate Degree Requirements: Not Stated

Graduate Degrees: M.S., Ph.D.
Graduate Degree Requirements: Not Stated

Graduate and Undergraduate Courses Available for Graduate Credit: Not Stated

UNIVERSITY OF NEBRASKA - OMAHA

60th and Dodge Street Phone: (402) 544-2998
Omaha, Nebraska 68101
Dean of Graduate Studies: E. Carter

College of Arts and Sciences

Dean: J.V. Blackswell

BIOLOGY DEPARTMENT
Phone: Chairman: 554-2641 Others: 554-2998

Chairman: P.V. Prior
Professors: M. Brooks, K.H.D. Busch, S.R. Lunt, P.V. Prior
Associate Professors: W. deGraw, C. Ingham, R. Sharpe, D. Sutherland
Assistant Professors: C. Nordahl, W. O'Dell, D. Patach, R. Stasiak, A.T. Weber
Instructors: J.D. Fawcett, R. Todd

Graduate Assistants: 17

Degree Program: Biology

Undergraduate Degrees: B.A., B.S.
Undergraduate Degree Requirements: 34-36 credits hours Biology, 18 hours Chemistry, 5-10 hours Physics, Geology recommended

Graduate Degrees: M.A., M.S.
Graduate Degree Requirements: M.A., 30 hours graduate Biology, Final Oral, M.S. 36 hours graduate Biology, Final Comprehensive

Graduate and Undergraduate Courses Available for Graduate Credit: Microtechnique (U), Ecology (U), Morphology of Lower Plants (U), Flora of Great Plains (U), Morphology of Higher Plants (U), Plant Anatomy (U), Fauna of Great Plains (U), Histology (U), Embryology (U), Cell Biology (U), Limnology (U), Organic Evolution (U), Animal Behavior (U), Bryology (U), Ichthyology (U), Mycology(U), Plant Physiology (U), Taxonomy of Vascular Plants (U), General Bacteriology (U), Experimental Endocrinology (U), Endocrinology (U), Animal Physiology (U), Vertebrate Zoology (U), Herpetology (U), Invertebrate Zoology (U), Parasitology (U), Entomology (U), Ornithology (U), Seminar in Biology, Problems in Biology, Current Topics in Biology, Current Topics in Microbiology, Biosystematics, Biomorphology, Advanced Topics in General Physiology, Environmental Physiology, Thesis

College of Medicine
42nd and Dewey Street Phone: (402) 541-4400
Dean: P. Rigby

DEPARTMENT OF ANATOMY
Phone: Chairman: (402) 541-4030

Chairman: Not Stated
Professors: W.L. Hard, E.A. Holyoke, R.D. Meader, W.K. Metcalf
Associate Professors: A.M. Earle, P.J. Gardner
Assistant Professors: R.H. Jensen, N. Metcalf, G.C. Moriarty, J.G. Sharp, W.W. Stinson
Instructors: P. Bunger, C. Spaur
Research Associate: 1

Degree Programs: Anatomy, Anatomy and Histology

Graduate Degrees: M.S., Ph.D.
Graduate Degree Requirements: Master: 45 quarter hour credits of which 9 to 15 hours in thesis research and minimum 24 quarter hour credits in Anatomy Doctoral: Minimum 135 quarter hour credits of which 35 to 60 hours may be thesis research

Graduate and Undergraduate Courses Available for Graduate Credit: Gross Anatomy I, Histology I, Neuroanatomy, Human Embryology, Gross Anatomy II, Histology II, Human Embryology II, Special Topics in Anatomy, Advanced Human Embryology, Advanced Human and Comparative Neuroanatomy, Advanced Gross Anatomy, Fundamentals of Electron Microscopy, Selected Problems in Electron Microscopy, History of Anatomy

DEPARTMENT OF BIOCHEMISTRY
Phone: Chairman: (402) 541-4417

Chairman: W.R. Ruegamer
Professors: K.L. Barker, M.J. Carver, D. Harman, W. Himwich, J.T. Matschiner, W.L. Ryan, A. Schaefer, R.B. Tobin
Associate Professors: A.J. Barak, E. Cavalieri, J.H. Copenhaver, D.P.J. Goldsmith, G.T. Haven, P. Issenberg, J. Johnson, K.Y. Lee, T.A. Mahowald, S. Sirvish, C. Raha, R. Ramaley
Assistant Professors: D.E. Cook, M.L. Heidrick, C.K. Phares, R. Langenbach, E. Smith
Instructors: A.K. Willingham, G.L. Curtis, J. McClurg, D. Tuma

Degree Program: Biochemistry

Graduate Degrees: M.S., Ph.D.
Graduate Degree Requirements: Master: Required to complete Cellular and Systems Biochemistry, complete a minimum of 45 quarter hours credits consisting of 30-36 quarter hour credits of regular course work and a thesis equivalent to 9-15 quarter hour credits, participation in the department's seminar series. Doctoral: 135 quarter hours credit is required to obtain a Ph.D. in Biochemistry completion of Cellular Biochemistry and Systems Biochemistry, completion of required advanced topics courses, participation in the Seminar Series.

Graduate and Undergraduate Courses Available for Graduate Credit: Cellular Biochemistry, Systems Biochemistry, Enzymes, Carbohydrates, Proteins, Nuclear Acids, Hormones, Lipids, Special Topics, Series of Advanced Techniques in Biochemistry, Instrumental and Physical Procedures, Microbiological and Animal Procedures, Radioisotope Procedures, Seminar in Biochemistry

DEPARTMENT OF MEDICAL MICROBIOLOGY
Phone: Chairman: 541-4040 Others: 541-4041

Chairman: H.W. McFadden, Jr.
Professors: H.W. McFadden, Jr., N.G. Miller, V.L. von Riesen
Associate Professors: W.E. Dye, G.R. Dubes, J. Jones, R.E. McCarthy, M.M. Tremaine, R.J. White
Assistant Professors: M.I. Al-Moslih, K. Phares
Instructors: J. Fervasi, P. Yam
Visiting Lecturers: 10
Research Assistants: 2
Teaching Assistants: 3
Assistant Instructors: 2
Teaching Supervisors: 2

Field Stations/Laboratories: Bacteriology Laboratory - University of Nebraska Hospital, Virology Laboratory, University of Nebraska Medical Center, Antibiotic Laboratory, University of Nebraska Medical Center

Degree Programs: Medical Microbiology

Graduate Degrees: M.S., Ph.D.
Graduate Degree Requirements: Master: 45 quarter hours of which at least 30 including research and thesis must be in Medical Microbiology and a research thesis. Satisfactory performance in a comprehensive written examination and defense of the thesis Doctoral: 135 quarter hours of which at least 90 including research and thesis must be in Medical Microbiology and a research thesis. Satisfactory performance on a comprehensive written examination. Defense of the thesis.

Graduate and Undergraduate Courses Available for Graduate Credit: Medical Microbiology I (U), Medical Microbiology II (U), Physiology of Microorganisms, Metabolism of Microorganisms, Principles of Immunology, Medical Bacteriology I, Systematic Microbiology, Medical Mycology, Biology of Animal Viruses, Medical Virology, Medical Parasitology and Tropical Medicine

DEPARTMENT OF PHYSIOLOGY AND BIOPHYSICS
Phone: Chairman: (402) 541-4173 Others: 541-4426

Chairman: J.P. Gilmore
Professors: J.P. Gilmore, F. Paustian, F. Ware
Associate Professors: F.J. Clark, E.D. Gerlings, T.P. K. Lim, C.M. Moriarty, G.G. Myers, J.A. Ramaley
Assistant Professors: D. Haack, W.L. Joyner, W.T. Lipscomb, M.D. Mann, I.H. Zucker

Degree Programs: Physiology and Biophysics

Graduate Degrees: M.S., Ph.D.

Graduate Degree Requirements: Master: Medical Physiology, Medical Biochemistry, Statistics, Biomedical Instrumentation, a minimum of two advanced courses in Physiology, one research paper outside of major field first year, thesis suitable for publication, oral defense of thesis (additional course work to fulfill hour requirements). Doctoral: Medical Physiology, Medical Biochemistry, Statistics, Biomedical Instrumentation, a minimum of four advanced courses in Physiology, two literature papers during first two years, thesis, research, thesis dissertation, one or more manuscripts, (additional course work to round out his program and fulfill hour requirements).

Graduate and Undergraduate Courses Available for Graduate Credit: Medical Physiology, Statistics, Biomedical Instrumentation, Techniques in Experimental Physiology, Advanced Electrophysiology, Advanced Cardiovascular Physiology, Advanced Respiratory Physiology, Advanced Renal Physiology, Advanced Gastrointestinal Physiology, Advanced Neurophysiology, Biophysics of Membrane Transport, Advanced Neuroendocrinology, Comparative Physiology, Computer Techniques

WAYNE STATE COLLEGE

Wayne, Nebraska 68787 Phone: (402) 375-2200

Dean of Graduate Studies: E. Elliott

Dean of College: N. Edmunds

DEPARTMENT OF BIOLOGY

Phone: Chairman: Ext. 247, Others: Ext. 260

Chairman: R.C. Sutherland

Professor: R.C. Sutherland

Associate Professors: B.J. Hirt, C.R. Maier

Assistant Professors: H.V. Pankratz, A.J. Schock

Degree Programs: Biology, Medical Technology

Undergraduate Degree: B.S.

Undergraduate Degree Requirements: Thirty-two hours to include Animal Biology, Plant Biology, Environmental and twenty hours of Electives. Or fifty hours to include Animal, Plant, Environmental and 12 hours of Chemistry.

NEVADA

UNIVERSITY OF NEVADA - LOS VAGAS
(No reply received)

UNIVERSITY OF NEVADA - RENO
North Center Street Phone: (702) 784-1110
Dean of Graduate Studies: T.D. O'Brien

Max C. Fleischmann College of Agriculture
Dean: D.W. Bohmont

ANIMAL SCIENCE DIVISION
Phone: Chairman: (702) 784-6644 Others: 784-1110

Chairman: V.R. Bohman
Professors: C.M Bailey, W.D. Foote, D.W. Marble, R.L. Taylor, H.J. Weeth , W.C. Behrens (Extension Specialist), E.L. Drake (Veterinary Extension Specialist)
Associate Professors: A.L. Lesperance, T.P. Ringkob, C.R. Torell
Assistant Professors: D.A. Reynolds, C.F. Speth
Instructor: M.B. Radmall (Dairy Specialist)

Field Stations/Laboratories: Central Nevada Field Laboratory, Austin, Knoll Creek Field Laboratory, Main Station Field Laboratory - Reno, Newlands Field Laboratory - Fallon, S-S Field Laboratory - Wadsworth, Valley Road Field Laboratory - Reno

Degree Program: Animal Science

Undergraduate Degree: B.S.
Undergraduate Degree Requirements: Social Sciences, 18 credits, Mathematics, etc., 8 credits, Biology, Chemistry, 11 credits, Basic Agricultural Resources, 12 credits, Animal Science, 24 credits, Plant, Soil and Water Sciences, 3 credits, Biology, Zoology, 9 credits, Chemistry, Biochemistry, 6-12 credits, Electives 24-31 credits, 72-73 Credits Total

Graduate Degree: M.S.
Graduate Degree Requirements: Master: 30 Credits as outlined by Graduate Committee

Graduate and Undergraduate Courses Available for Graduate Credit: Livestock Selection (U), Physiology of Domestic Animals (U), Infectious Diseases (U), Feeds adn Feeding (U), Seminar (U), Water Metabolism (U), Animal Genetics (U), Animal Nutrition (U), Physiology of Reproduction (U), Techniques in Livestock Reproduction (U), Physiology of Lactation (U), Endocrinology (U), Independent Study (U), Special Topics (U), Graduate Seminar, Arid Land Animal Nutrition, Graduate Topics, Physiological Surgery, Individual Study, Professional Paper, Thesis

DIVISION OF BIOCHEMISTRY
Phone: Chairman: (702) 784-6031

Chairman: C.R. Heisler
Professors: W.H. Arnett, C.R. Blincoe, R.J. Morris
Associate Professors: R.W. Lauderdale, R.S. Pardini, H.G. Smith
Assistant Professors: R.A. Lewis, B.R. Payne, W.H. Welch
Teaching Fellows: 2
Research Assistants: 5

Field Stations/Laboratories: Valley Road Field Laboratory - Reno

Degree Programs: Agriculture, Biochemistry, Pest Control

Undergraduate Degree: B.S. (Agriculture, Pest Control)
Undergraduate Degree Requirements: 128 credit hours, with 72 hours designed to give student a broad educational basis for identifying and solving problems of pests affecting man, his aniamls and crops.

Graduate Degrees: M.S. (Pest Control, Biochemistry), Ph.D. (Biochemistry)
Graduate Degree Requirements: Master:30 graduate credit hours, with 6 in thesis research. Doctoral: 72 graduate credit hours (after B.S.) with 24 in Dissertation research.

Graduate and Undergraduate Courses Available for Graduate Research: General Biochemistry (U), Plant Biochemistry (U), Radiotracer Techniques (U), Biochemical Techniques, Structural Biochemistry, Metabolism, Physical Biochemistry, Nucleic Acids, Mitochondrial Structure and Function, Mineral Metabolism, Independent Study (U), Enzymology, Seminar (U), General Economic Entomology (U), Insect Pests of Plants (U), Insect Pests of Animals (U), Insect Ecology, Pesticide Residue Analysis Techniques

RENEWABLE NATURAL RESOURCES DIVISION
Phone: Chairman: (702) 784-6763 Others: 784-6763

Chairman: E.L. Miller
Professors: R.E. Eckert (Adjunct), R.A. Evans (Adjunct), J.H. Robertson (Emeritus), C.M. Skau, P.T. Tueller, J.A. Young (Adjunct)
Associate Professors: D.A. Klebenow, F.A. Groves (Adjunct), E.L. Miller, R.O. Meeuwig (Adjunct)
Assistant Professors: W.H. Blackburn, C.F. Tiernan (Adjunct), T. Trelease (Adjunct)
Instructors: R.C. Beall, R.A. Shanks, R.F. Masse
Research Assistants: 2
Extension Range Specialist: 1

Field Stations/Laboratories: Central Vevada Field Laboratory - Austin, Knoll Creek Field Laboratory - Jackpot, Main Station Field Laboratory - Sparks, Newlands Field Laboratory - Fallon, Pahrump Field Laboratory - Pahrump, S. Bar S Field Laboratory - Wadsworth, South Nevada Field Laboratory - Logandale, Valley Road Field Laboratory - Reno

Degree Programs: Applied Ecology, Conservation, Forest Resources, Forestry, Forestry and Conservation, Land Resources, Natural Resources, Range Science, Silviculture, Watershed Management, Wildlife Management, Recreation Area Management (The division participates in an interdisciplinary Hydrology Ph.D. program, but does not offer an independent Ph.D.)

Undergraduate Degrees: B.S.
Undergraduate Degree Requirements: 36-40 core credits 16-20 credits in area option

Graduate Degree: M.S.
Graduate Degree Requirements: Master: Plan A. 12 of 24 graduate credits in major field, 6 in minor field. Thesis, one foreign language Plan B. 15 of 32 graduate credits in major field, 8 in minor field Doctoral: (See note above)

Graduate and Undergraduate Courses Available for Graduate Credit: Logging Systems (U), Forest Management (U), Introduction to Remote Sensing (U), Integrated Natural

Resource Management (U), Fisheries Management (U), Big Game Management (U), Fish and Wildlife Habitat Management (U), Range Agrostology (U), Natural Resources Interpretation and Communication (U), Thesis, Recreation Resource Seminar (U), Recreational Land Use Planning (U), Watershed Management (U), Special Topics (U), Seminar on Environmental Issues (U), Range and Forest Ecology (U), Administration and Policy (U), Perspectives in Renewable Natural Resources, Individual Study, Advanced Research Concepts, Special Topics in Graduate Studies, Professional Paper

DIVISION OF PLANT, SOIL AND WATER SCIENCES
Phone: Chairman: (702) 784-6947 Others: 784-6981

Chairman: R.A. Young
Professors: H.P. Cords, R.O. Gifford, O.J. Hunt (Adjunct), E.H. Jensen, F.F. Peterson, R.A. Young
Associate Professors: J.C. Cuitjens, H.R. Guenthner, C.N. Mahannah, J.E. Maxfield, R.H. Ruf
Assistant Professors: W.W. Miller, R.L. Post
Instructors: J. Gallian, B.J. Hartman
Visiting Lecturers: 1
Research Assistants: 8

Field Stations/Laboratories: Valley Road Field Laboratory - Reno, Main Station Field Laboratory - Reno, Newlands Field Laboratory - Fallon, S Bar S Field Laboratory - Wadsworth, Central Nevada Field Laboratory - Austin, Knoll Creek Field Laboratory - Contact, Southern Nevada Field Laboratory - Logandale, Pahrump Field Laboratory - Pahrump

Degree Programs: Plant Science, Soil Science, Water Science, Crop and Soil Management, Horticulture, Plant Pathology, Weed Science, Bioclimatology

Undergraduate Degree: B.S.
Undergraduate Degree Requirements: 128 semester credits. At least 40 credits in upper division courses. Six credits in English, 15 in Social Sciences, Arts and Humanities, satisfy a Military requirement (0-1 credit), and pass a test on Constitutions of the U.S. and Nevada. Depending on option, 8-20 in Mathematics and Physics, 12-23 in Biology, 4-16 in Chemistry, 3-9 in Agricultural Economics, 9 in basic Agricultural resource courses, 15-30 in Plant, Soil and Water, 20-23 in electives

Graduate Degrees: M.S.
Graduate Degree Requirements: Master: 30 credits including Thesis for which 6 credits are allowed or 32 credits (non-thesis) including a 2-credit professional paper. Half the formal course work must be 900-level (for graduate credit only). Specific requirements determined by the Graduate Advisory Committee.

Graduate and Undergraduate Courses Available for Graduate Credit: Principles of Plant Production (U), Plant Breeding (U), Environmental Quality and Agriculture (U), Special Topics (U), Graduate Seminar, Selected Topics, Research Methodology, Environment and Plant Response, Bioclimatology (U), Advanced Bioclimatology, Forage Crops (U), Herbicides, Plant Pathology (U), Plant Virology (U), Physiology of Plant Pathogenic Organisms, Soil Morphology and Classification (U), Soil Fertility and Management (U), Soil Chemistry (U), Soil Physics (U), Advanced Soil Physics, Soil Chemistry and Fertility, Irrigated Soil Management, Irrigation Principles and Practices (U), Hydrology for Resource Management (U), Irrigation System Management (U), Farm Irrigation System Design (U), Drainage of Agricultural Lands (U)

College of Arts and Sciences
Dean: R.M. Gorrell

DEPARTMENT OF BIOLOGY
Phone: Chairman: (702) 784-6188 Others: 784-1110

Chairman: H.N. Mozingo
Professors: D.G. Cooney, I.L. Rivers, H.N. Mozingo, F. Ryser, F.D. Tibbitts
Associate Professors: P.L. Comanor, J. Knoll, R.W. Mead, D.C. Prusso, B.K. Vig
Assistant Professors: C.E. Dreiling, A.A. Gubanich, E.F. Kleiner, A.C. Risser
Teaching Fellows: 8-1/2

Field Stations/Laboratories: George Whittell Forest and Wildlife Area,- Washoe Valley (15 miles South of Reno)

Degree Programs: Botany, Zoology, Ecology, Cellular Biology and Microbiology

Undergraduate Degrees: B.S.
Undergraduate Degree Requirements: 38 credits in biology with a core curriculum including General Biology, Genetics or Evolution, Cell Biology or Physiology and Ecology, 128 credits required for graduation.

Graduate Degrees: M.S., Ph.D.
Graduate Degree Requirements: Master: Plan A - 30 credits and thesis. Plan B substitutes course work and examination. Residence Requirement: 21 credits
Doctoral: 72 credits including those obtained for the Masters; two-thirds of the total credits must be taken in the major field.

Graduate and Undergraduate Courses Available for Graduate Credit: Biology Courses - Principles of Genetics, Genetics Laboratory, Discussion in Genetics, Human Genetics (U), Bacteriology (U), Organic Evolution (U), Cellular Biology (U), Cellular Biology II (U), Principles of Animal Behavior (U), Biology Hournal Seminar (U), History of Biology (U), Ecology of Pollution (U), Microbial Physiology (U), Limnology (U), Electron Microscopy (U), Electron Microscopy Laboratory (U), Biology Colloquium (U), Cytogenetics (Chromosomal Mechanisms) (U), Biological Survey Techniques (U), Tropical Ecology (U), Topics in Pollution Ecology, Study in Electron Microscopy, Supervised Teaching in College Biology, Genetics of Microorganisms, Advanced Microbiology, Advanced Cytogenetics, Cellular Physiology, Systems Modeling in Ecology, Topics in Ecology, Ecology of Decomposition, Botany Courses - Systematic Botany of Flowering Plants (U), Plant Anatomy (U), Ecology of Xerophytes (U), Plant Physiology (U), Introductory Mycology (U), Systematics and Ecology of Fungi (U), Advanced Systematic Botany, Problems in the Physiology of Growth, Physiological Ecology, Vegetation Analysis, Zoology Courses - Comparative Vertebrate Anatomy (U), Parasitology (U), Ichthyology (U), Herpetology (U), Ornithology (U), Mammalogy (U), Adaptations for Desert and Montane Life (U), Survey of Invertebrates (U), Survey of Invertebrates II (U), Mammalian Physiology, (U), Mammalian Physiology II (U), General Endocrinology (U), General Entomology (U), General Entomology Collection (U), Embryology (U), Comparative Histology (U), Histological Techniques (U), Animal Ecology (U), Comparative Physiology (U), Comparative Physiology Laboratory (U), Reproductive Endocrinology (U), Neuroendocrinology (U), Fish Hatchery Management (U), Comparative Population Ecology (U), Seminar in Zoology (U), Vertebrate Reproductive Biology, Zoological Symbiosis, Advanced Ornithology, Invertebrate Physiology, Special Topics in Endocrinology Experimental Endocrinology, Current Research in Developmental Biology, Uterus, Placenta and Fetus, Advanced Animal Ecology, Advanced Wildlife Ecology, Advanced Wildlife Ecology, Advanced Vertebrate Population Ecology,

NEW HAMPSHIRE

DARTMOUTH COLLEGE
Hanover, New Hampshire 03226 Phone: (603) 646-1110
Dean of Graduate Studies: J.F. Horning

Darthmouth College
Dean: L.M. Rieser

DEPARTMENT OF BIOLOGICAL SCIENCES
Phone: Chairman: (603) 646-2364 Others: 646-2378

Chairman: M. Spiegel
Ira Allen Eastman Professor of Biology: R.P. Forster
Gross Taylor and Cornelia Pierce Williams Assistant Professor of Biology: H.L. Robinson
Professors: R.P. Forster, J.H. Copenhaver, A.E. De Maggio, D.S. Dennison, W.T. Jackson, M. Spiegel, T.B. Roos, C.T. Gray (Adjunct), E.R. Pfefferkorn, Jr. (Adjunct), M. Lubin (Adjunct), W.W. Ballard (Adjunct Research)
Associate Professors: J.J. Gilbert, R.T. Holmes, W. Reiners, N.J. Jacobs (Adjunct), W.M. Layton (Adjunct)
Assistant Professors: H.L. Robinson, T. Gale (Adjunct), J.W. Inselburg (Adjunct), K. Merritt (Adjunct), D.S. Gephart (Adjunct)
Instructor: G.E. Land (Research)
Teaching Fellows: 12
Research Assistants: 3
Research Associate: 1
Curator of Jesup Herbarium: 1

Degree Program: Biology

Undergraduate Degree: B.S.
Undergraduate Degree Requirements: General Chemistry, Introductory Biology, and eight other Biology courses comprise the Biology major. College requires 33 course credits and 6 terms in residence, one of which is a summer term.

Graduate Degrees: M.S., Ph.D.
Graduate Degree Requirements: Master: 8 courses carrying graduate credit (4 may be graduate research) Doctoral: 8 courses carrying graduate credit

Graduate and Undergraduate Courses Available for Graduate Credit: All courses offered by the Department of Biological Sciences numbered 25 to 99 may be taken for graduate credit (see Officers, Regulations and Courses Bulletin for course names and descriptions) (excluding Biology 87). Courses designed for graduate students are Advanced Topics in Developmental Biology, Genetics, Regulatory Biology, Ecology, Systematic Biology, Structure and Function of Cellular Membranes, Cytology and Biochemistry of Cell Organelles, Graduate Research I and Graduate Research II.

Dartmouth Medical School
Hanover, New Hampshire 03755 Phone: (603) 646-2980
Dean: J.C. Strickler

DEPARTMENT OF ANATOMY/CYTOLOGY
Phone: Chairman: (603) 646-2286 Others: 646-1110

Chairman: V.H. Ferm
Professors: V.H. Ferm, W.F. Chambers, W.M. Layton
Associate Professor: S.J. Carpenter
Assistant Professor: T.F. Gale, J. Estavillo
Research Assistants: 3

Degree Programs: No graduate degrees are offered by this Department.

DEPARTMENT OF BIOCHEMISTRY
Phone: Chairman: (603) 646-2605 Others: 646-1110

Chairman: H.A. Harbury
Professors: J.H. Copenhaver, Jr., (Adjunct), H.A. Harbury, L.H. Noda, E.L. Smith
Associate Professor: G.E. Linenhard
Assistant Professors: W.J. Culp, S. Cushman (Adjunct), P.M. Horowitz (Adjunct), J.W. Patrick, O.A. Scornik, J.J. Sharp, B.L. Trumpower
Research Assistants: 10
Research Associates: 9

Degree Program: Biochemistry

Graduate Degree: Ph.D.
Graduate Degree Requirements: Doctoral: Programs of study are arranged individually. Requirements: Two terms general biochemistry, three terms research project rotation, minimum of three other graduate-level courses in biochemistry or related discipline, thesis research. Examinations: Qualifying examination: thesis defense.

Graduate and Undergraduate Courses Available for Graduate Credit: General Biochemistry, Structure and Function of Proteins, Regulation of Gene Expression, Biochemistry of Membranes, Bioenergetics, Mechanisms of Enzymatic Reactions, Regulation of Metabolism, Neuro biochemistry, Spectroscopy in the Study of Biological Systems, Biochemistry of Human Disease, Research Rotation in Biochemistry

DEPARTMENT OF MICROBIOLOGY
Phone: Chairman: (603) 2704 Others: 646-2702

Chairman: C.T.Gray
Professors: C.T. Gray, L. Kilham, M. Lubin, E.R. Pfefferkorn
Associate Professors: N.J. Jacobs, J.W. Inselburg (All members of department above rank of Assistant Professor have adjunct appointments in Biology Department)
Assistant Professor: K.M. Brown (Adjunct)
Instructors: P. Phillips, F. Cahn
Visiting Lecturers: 1
Research Assistants: 8

Degree Programs: Degrees are offered through other Departments.

Graduate and Undergraduate Courses Available for Graduate Credit: General Microbiology - Advanced courses for undergraduates and graduates, Viruses and the Immune Response, Microbiology 121, Medical students and graduate students from this program, Elective courses, Project Research # 127 , medical students from other programs

DEPARTMENT OF PHARMACOLOGY AND TOXICOLOGY
Phone: Chairman: (603) 646-2612 Others: 646-2612

Chairman: R.E. Gosselin
Irene Heinz Given Professor of Pharmacology: R.E. Gosselin
Professors: W.O. Berndt, H.L. Borison, R.E. Gosselin, G.H. Mudge (Adjunct), R.P. Smith
Assistant Professors: R. Rosenstein, J.M. Stolk (Adjunct), C.D. Thron
Visiting Lecturers: 2
Research Associates: 3

Degree Program: Pharmacology

Graduate Degree: Ph.D.
Graduate Degree Requirements: Doctoral: 1. Successful completion of courses in Physiology, Biochemistry, Pharmacology, or their equivalents, and 5 elective courses. 2. Proficiency in reading one foreign language. 3. Three research rotations. 4. Successful completion of a qualifying examination. 5. Original laboratory research leading to the preparation of a thesis of publishable quality. 6. Public defense of the research thesis.

Graduate and Undergraduate Courses Available for Graduate Credit: Pharmacology, General Pharmacology, Neuropharmacology, Permeability of Biological Membranes, Chemical Fallout: Toxic Hazards in the Environment, Behavioral Pharamcology, Catecholamines and Indoleamines: Cellular, Biochemistry, Physiology and Pharmacology, Advanced Topics in Biochemistry, Advanced Topics in Respiratory Physiology, Advanced Endocrinology, Advanced Renal Physiology, Biophysics, Cellular Physiology, Comparative Physiology, Cytology, Microbiology, Nervous System, Pathology

DEPARTMENT OF PHYSIOLOGY
Phone: Chairman: (603) 646-2207

Chairman: S.M. Tenney
Professors: F.V. McCann, A.U. Munck. R.E. Nye, Jr., H.A. Schroeder (Emeritus), G.R. Stibitz (Emeritus)
Associate Professor: V.A. Galton
Assistant Professors: D. Bartlett, Jr., J.A. Daubenspeck, R.L. Detar, B.R. Edwards, J.E. Remmers, H.W. Sokol,
Research Associates: 2
Teaching Fellows: 4

Field Stations/Laboratories: Trace Element Laboratory - Brattleboro, Vermont

Degree Program: Physiology

Graduate Degree: Ph.D.
Graduate Degree Requirements: Not Stated

Graduate and Undergraduate Courses Available for Graduate Credit: Medical Physiology (U), Advanced Mammalian Physiology (U), Respiratory Physiology, Endocrine Physiology, Cardiac Physiology, Renal Physiology, Topics in Applied Physiology, Developmental Physiology, Gastrointestinal Physiology, Topics in Mathematical Physiology, Vascular Physiology

FRANCONIA COLLEGE
Franconia, New Hampshire 03580 Phone: (603) 823-8086
Dean of College: L. Botstein

DEPARTMENT OF NATURAL SCIENCES
Phone: Chairman: (603) 823-8086

Chairman: J. Cassista
Note: There is no formal rank assignment
A.J. Cassista (Biologist), T. Jervis (Physicist), P. Knapp (Chemist)

Degree Program: Biology

Undergraduate Degrees: B.A., B.S.
Undergraduate Degree Requirements: Class and independend research-study through submission of a thesis. Some classwork often taken at other institutions.

FRANKLIN PIERCE COLLEGE
Rindge, New Hampshire 03461 Phone: (603) 899-5111
Dean of College: J.L. Maes

DEPARTMENT OF BIOLOGY
Phone: Chairman: Ext. 343, Others: Ext. 341

Chairman: E.C. Preston
Professor: L.T. Evans
Associate Professor: R.L. Burns, W.C. Preston
Assistant Professor: E.M. Vollerston

Degree Program: Biology

Undergraduate Degree: B.S.
Undergraduate Degree Requirements: Not Stated

KEENE STATE COLLEGE
Keene, New Hampshire 03431 Phone: (603) 352-1909
Dean of Graduate Studies: J. Stewart
Dean of College: C. Davis

DEPARTMENT OF SCIENCE (DIVISION OF BIOLOGY)
Phone: Chairman: Ext. 240 Others: Ext. 247

Coordinator: J.D. Cunningham
Professors: J.D. Cunningham, H.A. Goder
Associate Professors: A.E. Cohen (Research, E.A. Gianferrari, D.P. Gregory, H.M. Oliver (Research)
Instructor: B. Bobes

Field Stations/Laboratories: Lake Nubanusit Field Station (20 minutes from campus)

Degree Program: Biology

Undergraduate Degrees: B.A., B.S., B.S. in Education
Undergraduate Degree Requirements: varies with degree but 32-36 Biology credits plus 1 1/2 - 2 years Chemistry plus 0-1 year Physics plus Mathematics through Calculus

MOUNT SAINT MARY COLLEGE
Hooksett, New Hampshire 03106 Phone: (603) 485-9536
Dean of College: J. Mara

DEPARTMENT OF NATURAL SCIENCE
Phone: Chairman: (603) 485-9536 Ext. 21

Chairman: Sr. G. Quinn
Professor: Sr. G. Quinn
Associate Professor: Sr. M.R. Griffin
Instructors: Sr. J. Godzyk, Sr. E. Kurtz

Degree Program: Biology

Undergraduate Degree: B.A.
Undergraduate Degree Requirements: 30 Courses, 8 courses in Humanities, 2 courses in Social Science, 15 courses in Biology and supportive areas (Chemistry, Physics, Mathematics)

NATHANIEL HAWTHORNE COLLEGE
Antrim, New Hampshire 03440 Phone: (603) 588-6341
Dean of College: R.P. Smith

DEPARTMENT OF BIOLOGY

Chairman: Not Stated
Assistant Professor: A.H. Marin

Degree Programs: Biology, Botany, Zoology

Undergraduate Degrees: B.A., B.S.
Undergraduate Degree Requirements: Total 124 hours, 30 Credit hours, General and Organic Chemistry, Physics

NEW ENGLAND COLLEGE
Henniker, New Hampshire 03242 Phone: (630) 428-2211
Dean of College: C.R. Puglia

DIVISION OF NATURAL SCIENCES

Phone: Chairman: (603) 428-2231 or 2236

Dean: C.R. Puglia
Professor: C.R. Puglia
Associate Professors: E.B. Allison, M. Wirth
Assistant Professors: K.R. Hall, R. Niemeck

Degree Programs: Biology, Natural Science

Undergraduate Degrees: B.S.
Undergraduate Degree Requirements: Biology: 9 courses, 7 beyond courses number 299 (Chemistry 517 will be considered as a Biology courses, Chemistry 201-202, Chemistry 411-412. Two courses in Physics recommended and 1 in Calculus.

NEW HAMPSHIRE COLLEGE
(No reply received)

NOTRE DAME COLLEGE
(No reply received)

PLYMOUTH STATE COLLEGE
Plymouth, New Hampshire 03264 Phone: (603) 536-1550
Dean of Graduate Studies: L.H. Douglas
Dean of College: J.C. Foley

DEPARTMENT OF NATURAL SCIENCE
Phone: Chairman: Ext. 325

Chairman: M.T. Sylvestre
Professors: M.G. Bilheimer, R.H. Frey, G.B. Salmons, M.T. Sylvestre
Associate Professors: A.H. Davis, W.C. Neikam, L.T. Spencer
Assistant Professors: L.K. Cushman, R.A. Fralick, S.G. Murray, W.J. Taffe
Instructor: D.A. Marocco

Degree Programs: Biology, Biological Education

Undergraduate Degree: B.S.
Undergraduate Degree Requirements: 122 semester hours, with 45 semester hours in the Natural Sciences

RIVIER COLLEGE
Clement Street Phone: (603) 888-1311
Nashua, New Hampshire 03060
Dean of Graduate Studies: Sr. M.J. Beroit
Dean of College: Sr. R. Creteau

BIOLOGY DEPARTMENT
Phone: Chairman: (603) 888-9803

Chairman: Sr. C. Bileau
Professor: Sr. C. Bileau
Assistant Professors: B.C. Dufeur, R.D. Harrington
Instructor: J. Kotopoulis

Degree Programs: Biology, Biological Education, Medical Technology

Undergraduate Degrees: B.A., B.S.
Undergraduate Degree Requirements: Interpretive Literature - 2 courses, Social Sciences - 2 courses, Western Man in Historical Perspective - 2 courses, Mathematics and or Natoral Sciences - 2 courses, Major Field: No fewer than 8 courses and no more than 10, Modern Language and Literature - 2 courses, Philosophy - 2 courses, Religious Studies - 2 courses, Physical Education (1 hour per week) freshman, Electives: 14 or 15 courses.

Graduate Degree: M.S.
Graduate Degree Requirements: Master: 1. Admission to candidacy, 2. 30 semester hours of credit (24 hours in graduate work and 6 for a dissertation) OR: 33 hours of credit without a dissertation. 3. Comprehensive written examination in the major field

Graduate and Undergraduate Courses Available for Graduate Credit: Invertebrate Zoology, Genetics (U), Microbiology (U), Ecology (U), Modern Trends in the Teaching of Biology, History and Biological Theories, Field Biology, Modular Botany (U), Animal Physiology (U), Cellular Physiology (U), Histology (U), Histological Techniques (U), Medical Microbiology (U), Endocrinology (U), Parasitology (U), Isotopes in Biology and Medicine (U), Immunology and Serology (U), Virology, Biochemistry

ST. ANSELM'S COLLEGE
St. Anselm's Drive Phone: (603) 669-1030
Manchester, New Hampshire 03102
Dean of College: P.H. Riley

DEPARTMENT OF BIOLOGY
Phone: Chairman: Ext. 357 Others: Ext. 242

Chairman: B.J. Stahl
Professors: A.A. Bibeau, B.J. Stahl
Associate Professors: B.S. Fozdar, R.W. Lawrence, T. F. Lee, R.F. Normandin
Assistant Professors: N. Berning, J.M. Doherty, J.R. Feick, A.M. Sullivan

Degree Programs: Biology, Natural Science

Undergraduate Degree: B.A.
Undergraduate Degree Requirements: 10 semester courses in Biology (Biology) 6 (Natural Science), 2 semester courses in Physics, and distribution requirements in the Humanities

UNIVERSITY OF NEW HAMPSHIRE
Durham, New Hampshire 03824 Phone: (603) 862-1234
Dean of Graduate Studies: R.L. Erickson

College of Liberal Arts
Dean: A. Spitz

DEPARTMENT OF MICROBIOLOGY
Phone: Chairman: (603) 862-2250, 2251

Chairman: T.G. Metcalf
Professors: W.R. Chesbro, G.E. Jones, T.G. Metcalf, L.W. Slanetz
Assistant Professors: T.G. Pistole, R.M. Zsigray
Teaching Fellows: 5
Research Assistants: 6
Department Technicians: 7

Field Stations/Laboratories: Jackson Estuarine Laboratory-Adams Point, - Durham, New Hampshire

Degree Program: Microbiology

Undergraduate Degree: B.S.
Undergraduate Degree Requirements: 128 credit hours - College requirement of which 32 credit hours must be taken in Microbiology.

Graduate Degrees: M.S., Ph.D.
Graduate Degree Requirements: Master: Satisfactory background in Microbiology and related subject matter plus completion of a thesis in the laboratory. Doctoral: Evidence of a command of Microbiologic Theory and practice with emphasis directed to the Doctoral Dissertation.

Graduate and Undergraduate Courses Available for Graduate Credit: Immunology and Serology, Virology, Pathogenic, Systematic, Advanced Microbiology, Marine Micro biology, Microbial Biogeochemistry, Microbial Physiology, Microbial Cytology, Microbial Genetics, Cell Culture, Seminar

DEPARTMENT OF ZOOLOGY
(No reply received)

College of Life Sciences and Agriculture
Dean: H. Keener

DEPARTMENT OF BIOCHEMISTRY
Phone: Chairman: (603) 862-2473 Others: 862-2474

Chairman: E.J. Herbst
Professors: D.M. Green, E.J. Herbst, M. Ikawa, D.G. Routley, S.C. Smith, A.E. Teeri
Associate Professors: G.L. Kleppenstein, J.A. Stewart
Visiting Lecturers: 1
Teaching Fellows: 4
Research Assistants: 12

Field Stations/Laboratories: Jackson Estuarine Laboratory - Adams Point, Great Bay Estuary

Degree Program: Biochemistry

Undergraduate Degree: B.S.
Undergraduate Degree Requirements: 128 credits

Graduate Degrees: M.S., Ph.D.
GraduateDegree Requirements: Master: 30 credits, M.S. thesis, Doctoral: Ph.D. thesis, language (1), written preliminary, oral qualifying, final oral defense of thesis

Graduate and Undergraduate Courses Available for Graduate Credit: Physiological Chemistry and Nutrition (U), Principles of Biochemistry (U), Plant Metabolism (U), Biochemical Genetics (U), Enzyme Chemistry (U), The Nucleic Acids (U), Investigations (U), Biochemistry of Lipids, Biochemical Regulatory Mechanisms, Physical Biochemistry, Advanced Biochemistry Laboratory, Biochemistry Seminar

DEPARTMENT OF BOTANY AND PLANT PATHOLOGY
Phone: Chairman: (603) 862-2060

Chairman: R.O. Blanchard
Professors: A. Hodgdon, A. Mathieson, A. Rich, R. Schrieber, A. Shigo (Adjunct)
Associate Professor: A.L. Bogle
Assistant Professors: A. Baker, R. Blanchard, M. Havgstad, R. Kinerson, W. MacHardy, T. Tattar (Adjunct)
Research Assistants: 2

Degree Programs: Botany, Plant Pathology

Undergraduate Degrees: B.S., B.A.
Undergraduate Degree Requirements: General Botany, The Plant World, Systematic Botany, Plant Physiology, Plant Morphology or Plant Anatomy, General Zoology, 2 semesters Inorganic Chemistry, 2 Botany Electives, 1 year language for B.A. , 128 semester hours

Graduate Degrees: M.S., Ph.D.
Graduate Degree Requirements: Master: 30 semester hours to be determined by committee; Research Thesis Doctoral: Dissertation, courses to be determined by committee; 1 language (Reading knowledge)

Graduate and Undergraduate Courses Available for Graduate Credit: Plant Physiology (U), Freshwater Phycology (U), Marine Phycology (U), Marine Algal Ecology (U), Freshwater Algal Ecology (U), Cell Biology (U), Cell Physiology (U), Ecosystem Analysis (U), Physiological Ecology (U), Aquatic Higher Plants (U), Plant Pathology (U), Mycology (U), Forest Pathology (U), Principles of Plant Disease Control (U), Plant Anatomy (U), Morphology of Vascular Plants (U), Microtechnique (U), Advanced Systematic Botany (U), Advanced Plant Physiology, Advanced Marine Phycology, Morphogenesis, Cell Culture, Advanced Plant Pathology, Plant Geography, Methods in cology, the Plant and the Microclimate

ENTOMOLOGY DEPARTMENT
Phone: Chairman: (603) 862-1707

Chairman: G.T. Fisher
Professor: R. Blickle
Associate Professors: G.T. Fisher, M. Roovos
Assistant Professors: J. Bowman, A. Mason (Adjunct)

Degree Program: Entomology

Undergraduate Degree: B.S.
Undergraduate Degree Requirements: 32 hours of Entomology, 12 hours of Chemistry, 4 hours Mathematics

Graduate Degrees: M.S.
Graduate Degree Requirements: Master: 30 hours plus Thesis, Entomology Doctoral: offered in Zoology with Major Professors and thesis Problem in Entomology

Graduate and Undergraduate Courses Available for Graduate Credit: Insect Taxonomy, Insect Morphology, Insect Physiology, Aquatic Insects, Immature Insects, Chemical Control of Insects, Biological Control of Insects, Structural Insect Control, Agricultural Entomology, Medical Entomology, Taxonomy of Arachnida, Regulatory Entomology, Advanced Graduate Entomology

INSTITUTE OF NATURAL AND ENVIRONMENTAL RESOURCES
Phone: Chairman: (603) 868-1020

Chairman: D.P. Olson
Professors: R.A. Andrews, J.R. Barrett, J.R. Bowring, P.E. Burns, G.L. Byers, W.H. Drew, F.R. Hall, O. F. Hall, W.F. Henry, J.L. Hill, W.H. Hocker, Jr., A.B.Prince, G.E. Frick (Adjunct)
Associate Professors: O.B. Durgin, N. Engalichev, B.B. Foster, E.F. Jansen, D.P. Olson, N.K. Peterson, M.M. Reeves, O.P. Wallace, S.B. Weeks, R.R. Weyrick, C.A. Federer (Adjunct), B.W. Leak (Adjunct), N.L. LeRay (Adjunct), R.S. Pierce (Adjunct)
Assistant Professors: R.D. Harter, W.W. Mautz, D.E. Morris, R.P. Sloan, P.W. Garrett (Adjunct)
Research Assistants: 10
Research Associates: 2

Degree Programs: Forest Resources, Soil Science, Wildlife Management, Resource Economics, Hydrology

Undergraduate Degree: B.S.
Undergraduate Degree Requirements: Not Stated

Graduate Degrees: M.S.
Graduate Degree Requirements: Master: Satisfactory graduate training, thesis, final oral

Graduate and Undergraduate Courses Available for Graduate Credit: Environmental Resources - Natural Resources Policy, Soils and Community Planning, Statistical Methods II, Sampling Techniques, Pollution of Water: Causes and Control, Remote Sensing, Forest Recreation Seminar, Approach to Research, Operations Control Seminar, Forest Resources - Forest Tree Improvement, Game Management I, Game Management II, Forest Management, Operations Control and Analysis, Wood Products Manufacture and Marketing, Forest Industry Economics, Forest Resources Management Seminar, Forest Management Seminar, Utilization Seminar, Operations Control Seminar, Wildlife Management Seminar, Advanced Mensuration, Forest Protection Seminar Resource Economics - Applied Statistics I, Structure and Planned Change in Non-Urban Communities, Economics of Resource Development, Research Methods inSocial Science, Linear Programming Methods, Law of Community and Regional Planning, Regional Economic Analysis, Economics of Production and Resource Use, Statistical Analysis, Agricultural Economics, Introduction to the Location of Economic

Activity, Investigations in Resource Economics

PLANT SCIENCE DEPARTMENT
Phone: Chairman: (603) 862-1205 Others: 862-1234

Chairman: L.C. Peirce
Professors: G.M. Dunn, L.C. Peirce, O.M. Rogers, D.G. Routley
Associate Professors: G.O. Estes, J.B. Loy, J.R. Mitchell, O.S. Wells
Assistant Professors: D.A. Hopfer, M.B. Loyle (Adjunct), Y.T. Kiang, D.W. Koch, J.E. Pollard
Research Assistants: 7
Teaching Assistant: 1

Degree Program: Plant Sciences

Undergraduate Degree: B.S.
Undergraduate Degree Requirements: 128 total credits, including Organic and Inorganic Chemistry, Botany, Entomology, Plant Pathology, Genetics, Soils, 4 Plant Science courses, including "Concepts in Plant Growth", "Environment and Plant Response", and Plant Physiology (plus 1 elective)

Graduate Degrees: M.S., Ph.D.
Graduate Degree Requirements: Master: Determined by committee, but must include 30 credits of which 8 must be 800 level courses (graduate students only), teaching participation or equivalent required, thesis required Doctoral: Determined by Guidance Committee, Teaching participation required, Dissertation required

Graduate and Undergraduate Courses Available for Graduate Credit: Plant Nutrition (U), Plant Metabolism (U), Plant Growth Regulations (U), Population Genetics (U), Developmental Genetics (U), Evolutionary Biology (U), Methods and Theory of Plant Breeding(U), Radioisotopes Techniques for Life Sciences (U), Advanced Topics (U), Plant Growth and Development, Plant Genetics, Cytogenetics, Supervised Teaching, Research in Plant Science, Graduate Seminar, Master's Thesis, Doctoral Thesis

NEW JERSEY

BLOOMFIELD COLLEGE
(No reply received)

COLLEGE OF MEDICINE AND DENTISTRY OF NEW JERSEY
100 Bergen Street Phone: (201) 456-2000
Newark, New Jersey 07103

Graduate School of Biomedical Sciences
Dean: M.F. Shaffer

GRADUATE SCHOOL OF BIOMEDICAL SCIENCES
Phone: Dean: (201) 456-5332 Others: 456-4511

Faculty are drawn from the New Jersey College of Medicine and Dentistry

Degree Programs: Anatomy, Anatomy and Histology, Bacteriology and Immunology, Biochemistry, Biological Chemistry, Immunology, Medical Microbiology, Microbiology, Pharmacology, Physiology, Virology

Graduate Degrees: M.S., Ph.D.
Graduate Degree Requirements: Master: See under Departments of New Jersey College of Medicine and Dentistry Doctoral: See under Departments of New Jersey College of Medicine and Dentistry

Graduate and Undergraduate Courses Available for Graduate Credit: See under Departments of New Jersey College of Medicine and Dentistry

New Jersey Medical School
Dean: S.S. Bergen (Acting)

DEPARTMENT OF ANATOMY
Phone: Chairman: (201) 456-4414

Chairman: A.V. Boccabella
Professors: A.V. Boccabella, L.E. House, G. Kozam
Associate Professors: A. Gona, M. Hollinshead, P. Weis, A. Siegel, S. Gilani, P. Mirouti (Clinical Associate)
Assistant Professors: P. Berendsen, D. DeForw, A. Fasano, L. Laemle, M. Yu, J. DeVoy (Clinical Assistant)
Instructor: N. Pontilena
Teaching Fellows: 10

Degree Program: Anatomy

Graduate Degrees: M.S., Ph.D.
Graduate Degree Requirements: Master: 24 credits course work, thesis acceptable to the committee on Graduates Doctoral: 40 credits course work, Qualifying Examination, thesis based upon independent research work

Graduate and Undergraduate Courses Available for Graduate Credit: Introduction to Anatomy, Human Gross Anatomy, Anatomy of the Back and Extremities, Anatomy of the Head and Neck, Anatomy of the Body Cavities and Viscera, Microscopic Anatomy of Cells and Tissues, Microscopic Anatomy of Organ Systems, Electron Microscopy, Ultrastructural Research, Histochemistry, Neuroanatomy, Neuroscience, Neuroanatomical Research, Human Development, Teratology, Fetal Anatomy, Reproductive Biology, Selected Techniques in Anatomical Research, History of Anatomy, Anatomy Seminar

DEPARTMENT OF BIOCHEMISTRY
Phone: Chairman: (201) 456-4411 Others: 456-4000

Chairman: W.R. Frisell
Professors: W.R. Frisell, S.C.J. Fu, E. Hirschberg
Associate Professors: N. Brot (Adjunct), M.A. Lea, K. Lewis, H.S. Shapiro, S.I. Sherr, R.G. Wilson
Assistant Professors: S. Kumar, Hsiang-fu Kung (Adjunct), N. Lasser, R.H.L. Marks, C. Neurath (Adjunct), W.R. Redwood, C.S. Yang
Visiting Lecturers: 1
Teaching Fellows: 8
Research Assistants: 6
Post-doctorate Fellows: 7

Degree Program: Biochemistry

Graduate Degrees: M.S., Ph.D.
Graduate Degree Requirements: Master: A total of 24 credits for course work above the baccalaureate level. Seven of these credits must be for work done in a department other than that of the candidate; An examination in an acceptable foreign language. A thesis proposal acceptable to the candidate's Thesis Advisory Committee and to the Committee on Graduate Studies; A thesis acceptable to the Thesis Examination Committee and to the Committee on Graduate Studies. Doctoral: A total of at least 40 credits for course work above the baccalaureate level. A minimum of 12 credits must be completed in departments other than that of the candidate; An examination in an acceptable foreign language. A Qualifying Examination; A thesis proposal acceptable to Thesis Advisory Committee and the Committee on Graduate Studies; A thesis based upon independent research shall be prepared by the candidate and shall be evaluated by an Examination Committee which will advise the Committee on Graduate Studies of its acceptability; A defense of the thesis before the Examination Committee.

Graduate and Undergraduate Courses Available for Graduate Credit: Biochemistry, Biochemical Techniques, Biochemistry of the Proteins, Basic Enzymology, Enzyme Mechanisms and Kinetics, Enzyme Mechanisms and Kinetics, Chemistry of Metabolism of Harbohydrates, Chemistry and Metabolism of Lipids, Structure and Function of Nucleic Acids, Current Topics in Biochemistry, Seminar in Biochemistry, Biochemistry of Cancer, Membrane Biochemistry, and Membranes and Transport.

DEPARTMENT OF MICROBIOLOGY
Phone: Chairman: (201) 456-4483 Others: 456-4484

Chairman: B.A. Briody
Professors: P.F. Bartell, G. Furness, R.E. Gillis, M.F. Shaffer
Associate Professors: E.W. Bassett, L.A. Feldman, I. Imaeda, Z.C. Kaminski, A. Saha, M.N. Schwalb
Assistant Professors: R.C. Coles, V.K. Jansons, A.E. Kriksgens, S. Sambury
Visiting Lecturer: 1
Teaching Fellows: 1
Research Assistants: 12

Degree Program: Microbiology

Graduate Degrees: M.S., Ph.D.
Graduate Degree Requirements: Master: Microbiology (with specialization in Bacteriology, Microbial Genetics, Medical Microbiology, Immunology, Immunochemistry, Virology.) Doctoral: Microbiology (with

specialization in Bacteriology, Microbial Genetics, Medical Microbiology, Immunology, Immunochemistry, Virology)

Graduate and Undergraduate Courses Available for Graduate Credit: Microbiology for Graduate Students, Medical Microbiology, Microbiology Seminar, Principles of Chemotherapy, Oral Microbiology, Microbial Physiology, Research in Microbiology, Immunochemistry, Immunology, Experimental Virology, Electron Microscopy, Clinical Microbiology and Parasitology, Frontiers of Immunology, Analytical Techniques in Microbiology, Microbial Genetics, Microbial Genetics Laboratory, Genetic Markers, Evolution and Protein Structure, Infectious Disease Seminar, Experimental Immunology

DEPARTMENT OF PHARMACOLOGY
Phone: Chairman: 456-4444

Chairman: G.A. Condouris
Professors: G.A. Condouris, S.B. Gertner, D.E. Hutcheon
Associate Professors: H.E. Brezenoff, M.J. Mycek
Assistant Professors: E.T. Eckhardt, E.J. Flynn, R.D. Howland, H.E. Lowndes, J.J. McArdle, S. Von Hagen, R.S. Sandhu (Adjunct)
Visiting Lecturers: 10

Degree Program: Pharmacology

Graduate Degree: Ph.D.
Graduate Degree Requirements: Doctoral: Course work, comprehensive examinations (qualifying), thesis research and thesis defense.

Graduate and Undergraduate Courses Available for Graduate Credit: General Pharmacology, Receptor Theory of Drug Action, Pharmacokinetics, Biometrics, Advanced Biometrics and Experimental Design, Biochemorphology, Pharmacogenetics, Neuropharmacology, Cardiorenal Pharmacology, Topics in Biochemical Pharmacology, Smooth Muscle Pharmacology, Clinical Pharmacology, Toxicology, Methods in Pharmacology, Radioisotopic Theory and Techniques, Research in Pharmacology, Pharmacology Seminar, Pharmacology Journal Club, Electronics for Biomedical Instrumentation

DEPARTMENT OF PHYSIOLOGY
Phone: Chairman: (201) 456-4463 Others: 456-4460

Chairman: J.B. Nolasco (Acting)
Professors: L. Horn, A.J. Kahn, J.B. Nolasco, D.F. Opdyke
Associate Professors: J.W. Bauman, J. Boyle III, P. Farnsworth, F.L. Ferrante, W. Perl
Assistant Professors: J. Bullock, H. Edinger, N. Ingoglia
Teaching Fellows: 4
Research Assistants: 1
Visiting Professor: 1
Visiting Associate Professor: 1

Degree Program: Physiology

Graduate Degrees: M.S., Ph.D.
Graduate Degree Requirements: Master: 1) 25 credits course work above baccalaureate level 7 of which must be in another department. 2) Foreign language proficiency, 3) Thesis Doctoral: 1) Minimum of 40 credits course work of which 12 must be in another department; 2) Foreign language proficiency 3) Qualifying examination 4) Thesis based on independent research 5) Defense of Thesis

Graduate and Undergraduate Courses Available for Graduate Credit: Human Physiology (U), General Physiology and Biophysics, General Endocrinology, Neuroscience, Cardiovascular Physiology, Advanced Biophysics, Environmental Physiology, Special Topics in Physiology, Seminar in Physiology, Physiological Methods and Techniques, Research in Physiology, Membranes and Transport

Rutgers Medical School
Piscataway, New Jersey 08854 Phone: (201) 564-4567
Dean of Graduate Studies: J.W. Green (Acting)
Dean of Medical School: J.W. Mackenzie

DEPARTMENT OF ANATOMY
Phone: Chairman: (201) 564-4785 Others: 564-4580

Chairman: A. Hess, S. Malamed
Associate Professors: G. Krauthamer, G.J. MacDonald,
Assistant Professors: D. Seiden, F. Wilson
Instructor: G.L. Trupin
Visiting Lecturers: 2
Teaching Fellows: 3
Research Assistants: 3

Degree Program: Anatomy

Graduate Degree: Ph.D.
Graduate Degree Requirements: Doctoral: Courses, thesis

Graduate and Undergraduate Courses Available for Graduate Credit: Cytology, Investigative Methods

DEPARTMENT OF BIOCHEMISTRY
Phone: Chairman: (201) 564-4541

Chairman: D.J. Prockop
Professor: D.J. Prockop
Associate Professors: G. Avigad, R.A. Harvey, B.R. Olsen
Assistant Professors: R.A. Berg, J.J. Uitto
Instructors: H.P. Hoffmann
Research Assistants: 3 1/2
Graduate Fellow: 1
Post-Doctoral Fellow: 1
Teaching Assistants: 4

Degree Programs: Biological Chemistry

Graduate Degrees: M.S., Ph.D.
Graduate Degree Requirements: Master: 23 semester hours of approved course work; 7 semester hours of research credit culminating in a written thesis. Doctoral: 36 semester hours of approved course work, 36 hours of research credits, a written dissertation, facility in one language

Graduate and Undergraduate Courses Available for Graduate Credit: Biological Chemistry, Biochemistry, Advanced Biochemistry Laboratory, Biophysical Chemistry of Organized Systems, Proteins and Enzymes, Regulatory Enzyme Mechanism, Plant Biochemistry, Biological Regulation, Metabolic Pathways, Nucleic Acids, Organelles, Bioenergetics, Lipids and Carbohydrates, Physical Biochemistry, Advanced Studies in Biochemistry, Seminar in Biochemistry, Special Topics in Biochemistry

DEPARTMENT OF MICROBIOLOGY
Phone: Chairman: (201) 564-4567

Chairman: R.W. Schlesinger
Professor: D.T. Dubin
Associate Professors: J.A. Holowczak, K. Raska, Jr. H. Rouse, W.A. Strohl, V. Stollar,
Assistant Professor: W.T. McAllister
Instructors: G.R. Cleaves, M. Esteban (Research Associate), K.A. Harrap (Research Associate), A. Igarashi (Research Associate), V. Rubio (Research Associate)
Visiting Lecturers: 1
Predoctoral Trainees: 15

Degree Programs: Microbial Biochemistry, Physiology, Molecular Biology, Microbial Genetics, General Microbiology, Chemistry of Microbial Products, Large-scale Fermentation, Virology, Immunochemistry, Cellular Biology, Microbial Ecology, Marine Microbiology, Mycology, Medical Bacteriology

Graduate Degrees: M.S., Ph.D.
Graduate Degree Requirements: Master: 24 credits, thesis Doctoral: 38-45 credits, thesis, teaching experience

Graduate and Undergraduate Courses Available for Graduate Credit: General Microbiology, History of Microbiology, Basic Research Methods in Bacteriology, Analytical Cytology, Practical Microscopy, Microbial Biochemistry, Nutrition and Metabolism of Microorganisms, Chemistry of Microbial Products, Cellular Dynamics, Industrial Microbiology, Fundamentals of Large Scale Fermentation, Radiobiology of Microorganisms, Genetics of Bacteria, Their Viruses - Laboratory, Seminar in Physiological Genetics, Developmental Genetics, Organelle Genetic Systems, Mutation, Immunochemistry, Cell Biology, Medical Microbiology, Tissue Culture and Elementary Virology, Virology, Bacterial Viruses, Genetics of Bacteria and their Viruses, Advanced Virology, Bacterial Physiology, Pathogenic and Diagnostic Bacteriology, Yeasts, Pathogenic Mycology, Microbial Ecology, Leukemia-Sarcoma and Related Neoplasms, Molecular Interactions in Organized Systems, Analysis of Development, Seminar in Microbial Genetics

DEPARTMENT OF PHYSIOLOGY
Phone: Chairman: (201) 427-4550 Others: 564-4551

Chairman: R.K. Crane
Professor: R.K. Crane
Associate Professors: A. Eichholz, B.K. Ghosh, J. Lenard, N. Edelman (Adjunct)
Assistant Professors: K. Briden, A.K. Sinha, N. Stevenson, W.S. Stirewalt, M. Takahashi, H. Weiss, V. Murthy (Adjunct), M. Karetzky, R. Reinhart (Adjunct)

Degree Programs: (M.S. and Ph.D. programs in conjunction with Rutgers University Physiology Section in Physiology)

COLLEGE OF SAINT ELIZABETH

Convent Station, New Jersey 07961 Phone: (201) 539-1600
Dean of College: Sr. J. Glazewski

DEPARTMENT OF BIOLOGY
Phone: Chairman: Ext. 312 Others: Ext. 313

Chairman: Sr. T.A. Madden
Professors: D.R. Brebbia (Adjunct), Sr. A.C. Lawlor (Emeritus)
Assistant Professors: S. Kapica, K. Liebhauser, Sr. T.A. Madden

Degree Program: Biology

Undergraduate Degrees: B.A., B.S.
Undergraduate Degree Requirements: B.A. - 8 courses in Biology, 32 credits - 2 courses in Chemistry, 8 credits, B.S. - 8 courses in Biology, 32 credits - 4 courses in Chemistry, 16 credits, 2 courses in Calculus, 8 credits

COOK COLLEGE

(see Rutgers - The State University - New Brunswick)

DOUGLASS COLLEGE

(see Rutgers - The State University - New Brunswick)

DREW UNIVERSITY

36 Madison Avenue Phone: (201) 377-3000
Madison, New Jersey 07940
Dean of Graduate Studies: B. Thompson

College of Liberal Arts
Dean: I.G. Nelbach

DEPARTMENT OF BOTANY
Phone: Chairman: Ext. 360 Others: Ext. 362

Chairman: R.K. Zuck
Professor: R.K. Zuck
Associate Professors: G.N. Bistis
Instructor: F.M. Zuck
Research Assistants: 2

Degree Program: Botany

Undergraduate Degree: B.S.
Undergraduate Degree Requirements: Plant Biology, Molecules and Life, Principles of Animal Biology, Microbiology, Seminar in Botany, Five upperlevel courses in Botany, Introduction to Genetics, Additional science courses are recommended, depending on the needs and plans of the student, Comprehensive Senior Project (Botany 199), 3 credits

DEPARTMENT OF ZOOLOGY
Phone: Chairman: Ext. 368 Others: Ext. 210

Chairman: J.B. Phillips
Professors: E.G. S. Baker, J.B. Phillips
Associate Professors: J.J. Nagle, H.C. Rohrs
Assistant Professors: M. Kozak, L.W. Pollock
Instructor: M. Christie

Degree Program: Zoology

Undergraduate Degree: B.S.
Undergraduate Degree Requirements: Organismic Biology, Molecules and Life, Chemical Energy and Life, Reactions and Mechanisms, Genetics, Organismic Biology (Vertebrate or Invertebrate), Cellular Biology or Immunology, Population Biology (Evolution or Ecology), 12 additional credits - elective

FAIRLEIGH DICKINSON UNIVERSITY

West Passaie and Montross Avenues Phone: (201) 933-5000
Rutherford, New Jersey 07070

Maxwell Becton College of Liberal Arts
Dean: A.A Anastasia

DEPARTMENT OF BIOLOGICAL SCIENCES
Phone: Chairman: Ext. 345,346 Others: 274,279

Chairman: E. Szebenyi
Professors: A.A. Anastasia, C.B. Dugdale, V.B. Kaczor, E. Szebenyi
Associate Professors: S.C. Catovic, G. Dyer, R.C. Klosek, G. Schreckenberg
Assistant Professors: R.L. Jacques, R.W. LoPinto, M.P. McGinn
Visiting Lecturers: 5

Field Stations/Laboratories: West Indies Laboratory - St. Croix, Virgin Islands, Hackensack Meadowlands

Degree Programs: Biology, Marine Biology, Medical Technology

Undergraduate Degree: B.S.
Undergraduate Degree Requirements: 128 total credits, 8 credits in Biology plus the following courses: Genetics, Comparative Anatomy, Vertebrate Embryology, Biology Seminar, Also required are 8 credits of general Chemistry, 8 credits of Organic Chemistry, 8 credits of Physics, 8 credits of Mathematics, and the remainder liberal arts courses

Graduate Degree: M.S.
Graduate Degree Requirements: Master: 32 credits of classwork, thesis optional

Graduate and Undergraduate Courses Available for Graduate Credit: Evolution, Cell Physiology, Advanced Microbiology, Microbial Genetics, Radiation Biology, Introductory Virology, Advanced Virology, Endocrinology, Experimental Morphogenesis, Invertebrate Zoology, Protozoology, Plant Anatomy, Mycology, Organic Molecular Biology, Biology Seminar, Introductory Immunology, Advanced Immunology, Hematology, Developmental Neurology, Human Embryology, Experimental Parasitology, Micro-ecology, Advanced Entomology, Medical Entomology

School of Dentistry
110 Fuller Place Phone: (201) 836-6300
Hackensack, New Jersey 07601
Dean: L.J. Boucher

DEPARTMENT OF HUMAN ANATOMY AND HISTOLOGY
(No information available)

DEPARTMENT OF BIOCHEMISTRY
Phone: Chairman: Ext. 574

Chairman: D.J. Smith
Professor: D.J. Smith
Assistant Professors: J.W. Coffey (Adjunct), J.G. Hamilton (Adjunct), W. Kreutner (Adjunct), C. Lin (Adjunct), S. Harayanan (Adjunct)

Degree Program: Biochemistry

Graduate Degree: M.S.
Graduate Degree Requirements: Master: 32 credits including research and thesis

Graduate and Undergraduate Courses Available for Graduate Credit: Human Biochemistry, Enzymology, Protein Biochemistry, Lipid Biochemistry, Carbohydrate Biochemistry, Human Physiology

DEPARTMENT OF HUMAN PHYSIOLOGY
Phone: Chairman: Ext. 584

Chairman: L.J. Ramazzotto
Professors: R. Neri (Adjunct), L.J. Ramazzotto
Associate Professors: A. Barnett (Adjunct), J. Long (Adjunct), J. Malick (Adjunct)
Assistant Professors: R. Carlin (Adjunct), G. Warchalowski
Instructors: R. Engstrom, J. Galat (Adjunct)
Visiting Lecturers: 23
Research Associate: 1

Degree Program: Human Physiology

Graduate Degree: M.S.
Graduate Degree Requirements: Not Stated

Graduate and Undergraduate Courses Available for Graduate Credit: Endocrine Physiology, Pulmonary Physiology, Developmental Physiology, Cardiovascular Physiology, Kidney and Gastrointestinal Physiology, Radiotracer Methodology, Environmental Physiology, Advanced General Physiology (Modern Instrumentation), Human Physiology, Neurophysiology, Research Seminar, Research and Thesis, Advanced Special Projects

GEORGIAN COURT COLLEGE
Lakewood, New Jersey 08701 Phone: (201) 364-2200
Dean of College: Sr. B. Williams

DEPARTMENT OF BIOLOGY
Phone: Chairman: Ext. 45

Head: Sr. M.M. Cook
Professor: N.H.C. Shen
Assostant Professor: Sr. M.M. Cook
Instructor: M. Bedell
Visiting Lecturer: 1

Degree Programs: Biology, Biochemistry, Medical Technology

GLASSBORO STATE COLLEGE
Glassboro, New Jersey 08028 Phone: (609) 445-5000
Vice-President for Academic Affairs: L. Brown
Dean of Liberal Arts and Sciences: A.B. Donovan

LIFE SCIENCE DEPARTMENT
Phone: Chairman: (609) 445-6181 Others: 445-6182

Chairman: R.N. Renlund
Professors: G. Gershenowitz, C. Green, S. Husain, R. Renlund, M. Sawyer, H. Stoudt, V. Vivian
Associate Professors: S.H. Crim, E.W. Nichols
Assistant Professors: G. Bisazza, N. Hornstein, E. Landecker, R. Meagher, G. Patterson, A. Prieto, R. Raimist, D. Riblet, P. Sparks

Field Stations/Laboratories: New Jersey Marine Sciences Consortium

Degree Programs: Biology, Biological Education

Undergraduate Degree: B.S.
Undergraduate Degree Requirements: Basic Education, 48 semester hours, Specialization 30 semester hours, Free Electives, 42 semester hours, total 120 semester hours, Required: Biology I and II, Inorganic Chemistry I and II, Organic Chemistry I and II, Physics I and II, Mathematics II, Calculus I

Graduate Degree: M.A.
Graduate Degree Requirements: Master: Basic Professional 6 - 9 semester hours, Teaching Specialty 12-21 semester hours, Curriculum and Instruction 3-9 semester hours, Seminar and Research, 4 semester hours, Electives 6 semester hours, 32 semester hours total

Graduate and Undergraduate Courses Available for Graduate Credit: Biological Techniques, General Taxonomy, Evolutionary Theory, History and Philosophy of Science, Modern Biology I and II, Comparative Morphology of Vascular Plants, Plant Physiology, Animal Physiology, Limnology, Techniques in Plankton Sampling and Analysis, Biology of Marine Plankton, Marine Botany, Primary Production: Techniques and Measurement, Synecology, Pine Barrens Ecology, Ecology of the Estuary, Entomology, Molecular Genetics, General Embryology of Animals, Human Genetics, Introduction to Environmental Education, Trends in Environmental Education, Curriculum Guides and Materials for Environmental Education Programs, Conservation Workshop, Environmental Land Use - Resources and Recreation, Animal Ethology, Cell Biology

JERSEY CITY STATE COLLEGE
3039 Kennedy Boulevard Phone: (201) 547-3054
Jersey City, New Jersey 07305
Dean of Graduate Studies: J. Burks
Dean of College: D. Kahn (Acting)

BIOLOGY DEPARTMENT
Phone: Chairman: (201) 547-3055 Others: 547-3054

Chairman: J.E. Garono
Professors: J.E. Barone, C.E. O'Neill, E.K. Rosenberg
Associate Professors: L.N. Galia, C.H. Miller, B.J. Pettit, M.X. Turner
Assistant Professors: A. Chisholm, O. Donovan, H. Kish, M. Finstein, G. McCoy, F. Mele, G.A. Schultz C.J. Scott, J.R. Siniscalchi, W.R. Stimson
Instructors: G. Cannon, C. Egli, H. Kotsonis
Research Assistants: 2
Laboratory Technicians: 2

Degree Program: Biology

Undergraduate Degree: B.A.
Undergraduate Degree Requirements: The requirement for a major in Biology is a minimum of 36 credits in Biology. The required courses are: Principles of Biology I and II, Principles of Physiology or Comparative Anatomy or Plant Physiology, Microbiology or Cytology, Principles of Ecology or Principles of Urban Ecology, Embryology or Genetics. The major biology may elect up to a maximum of 54 credits in Biology.

KEAN COLLEGE OF NEW JERSEY

Union, New Jersey 07083 Phone: (201) 527-2000
Dean of Graduate Studies: M.M. Rosenberg
Dean of College: S. Haselton

BIOLOGY DEPARTMENT
Phone: Chairman: 527-2460 Others: 527-2000

Chairman: D.B. Linden
Professors: M.A. Hayat, G. Hennings, D.B. Linden, C.R. Madison, R.A. Virkar
Associate Professors: H.B. Reid, Jr., R.W. Schuhmacher, A.L. Smith, D.K. Ward
Assistant Professors: E.L. Boly, F.H. Osborne, J.J. Mahoney, J. Rosenthal, C.K. Shannon
Instructors: M.D. Barber (Adjunct), V.M. Jani (Adjunct), J. Lysko, Jr., (Adjunct), P.R. Maynard, R.T. More (Adjunct), R.B. Smittle (Adjunct)

Degree Programs: Biology, Medical Technology

Undergraduate Degrees: B.A., B.S.
Undergraduate Degree Requirements: 2 years Mathematics, 2 years Chemistry, 30 credits in Biology of which 18 are specified.

LIVINGSTON COLLEGE

(see Rutgers - The State University - New Brunswick)

MONMOUTH COLLEGE

West Long Beach, New Jersey 07764 Phone: (201) 222-6600
Dean of Graduate Studies: N. Coe
Dean of College: R. Rouse

DEPARTMENT OF BIOLOGY
Phone: Chairman: Ext. 224 or 215 Others: Ext. 224

Chairman: L.E. Spiegel (Acting)
Professor: L.E. Spiegel
Associate Professors: B. Shidlovsky
Assistant Professors: V. Churchill, D. Dorfman, I. Gepner, H. MacAllister, D. Volkert
Visiting Lecturers: 2

Degree Programs: Biology, Medical Technology

Undergraduate Degree: B.S.
Undergraduate Degree Requirements: 45 semester hours Biology, 2 semesters College Mathematics including Statistics, 2 semesters College Physics, 4 semesters Chemistry including Organic, 2 semesters English, 2 semesters History, Electives, total 128 semester hours

MONTCLAIR STATE COLLEGE

Upper Montclair, New Jersey Phone: (201) 893-4000
Dean of Graduate Studies: M.H. Freeman
Dean of College: T.L. Wilson

DEPARTMENT OF BIOLOGY
Phone: Chairman: (201) 223-4397 Others: 893-4397

Chairman: S.M. Kuhnen
Professors: G.L. Daniels, L.K. Koditschek, S.M. Kuhnen
Associate Professors: M. Arny, H.L. Asterita, R.T. Kane, P.P. Shubeck
Assistant Professors: L.D. Cribben, O.D. Gona, A.D. Hoadley, S.J. Koepp, J.M. McCormick, A.C. Pai, M.S. Sawits, J.A. Shillcock, G. Sichuk
Teaching Fellows: 3

Degree Program: Biology

Undergraduate Degree: B.A.
Undergraduate Degree Requirements: I. Biology Requirements - Cell Biology, Plant Kingdom, Invertebrate Zoology, Vertebrate Zoology, Genetics II. Collateral Requirement - 30 credit hours from Chemistry, Physics, Mathematics, III. Electives - 14-15 credit hours of Biology courses on the Junior and Senior level (300 and 400 number courses).

Graduate Degree: M.A.
Graduate Degree Requirements: Master: I. Specialization - 22-26 semester hours, A. Research requirement-either: I. Master's Thesis (Credit by arrangement) or 2.514 Biology and Problems of Society - 2 semester hours, B. Minimum of 16 semester hours must be completed within Biology Department at M.S.C. II. Free Electives - 6-10 semester hours III. Comprehensive Examination

Graduate and Undergraduate Courses Available for Graduate Credit: Research Seminar in Biological Literature (U), Biology Independent Study (U), History and Philosophy of the Life Sciences (U), Economic Botany (U), Morphology of Flowering Plants (U), Elementary Plant Physiology (U), New Jersey Flora (U), Field Ornithology (U), Entomology (U), Medical Entomology (U), Mammalian Anatomy and Histology (U), Comparative Vertebrate Embryology (U), Cell Physiology (U), Immunology (U), Advanced Bacteriology (U), Biological Oceanography (U), Limnology (U), Marine Biology (U), Marine Invertebrate Zoology (U), Marine Botany (U), Biology of Marine Plankton (U), Marine Ichthyology (U), Techniques, Field Methods in Oceanography and Marine Biology, Plankton Sampling and Analysis, Primary Production in the Marine Environment: Techniques and Measurement, Plant and Animal Histological Techniques, The Teaching of Biology in Secondary Schools, Histology and Histological Techniques, Advanced Genetics, Instrumentation and Techniques for Biological Sciences, Biological Problems of Society, Radiation Biology, Biogeography, Plant Physiology, Field Studies of Flowering Plants, Plant Pathology, Mycology, Parasitology, Insect Ecology and Behavior, Comparative Human Anatomy, Experimental Embryology and Cellular Differentiation I, II, Topics in Microbiology, Intermediary Metabolism I, Intermediary Metabolism II, Ecology, Physiological Plant Ecology, Benthic Ecology, Ecology of the Estuary, Research in Biology

NEWARK STATE COLLEGE

(see Kean College)

PRINCETON UNIVERSITY

Princeton, New Jersey 08540 Phone: (609) 452-3000
Dean of Graduate Studies: A.B. Kerman

Departments of Arts and Sciences

Dean: N.L. Rudenstine

DEPARTMENT OF BIOCHEMICAL SCIENCES
Phone: Chairman: (609) 452-3658 Others: 452-3000

Chairman: J.R. Fresco
The Donner Chair of Science: A.B. Pardee
Professors: B.M. Alberts, J.R. Fresco, C. Gilvarg, R. Langridge, A.B. Pardee
Associate Professors: A.J. Levine, A. Worcel
Assistant Professors: M.L. Applebury, M.W. Kirchner, U.K. Laemmli, H.M. Weintraub
Instructor: J.A. Littlechild
Teaching Fellows: 7

Research Assistants: 21
Research Associates and Fellows: 33

Degree Program: Biochemistry

Undergraduate Degree: B.S.
Undergraduate Degree Requirements: 30 courses, Junior Independent Work (paper), Senior Independent Work (thesis based on research), Departmental Examination

Graduate Degrees: M.S., Ph.D.
Graduate Degree Requirements: Master: Incidental only, Pass general examination, language requirement
Doctoral: Language requirement, General Examination, Dissertation based on original research, Final Public Oral Examination (no course requirements)

Graduate and Undergraduate Courses Available for Graduate Credit: Physical Biochemistry, Intermediary Metabolism, Biological Macromolecules, Molecular Physiology, Experimental Methods in Biochemistry, Virology, Diffraction Methods in Molecular Biology, Biochemical Genetics, Immunology, Nucleic Acids and Polynucleotides, Laboratory in the Structure of Biological Macromolecules, Special Topics in Biochemistry. (Graduate school does not operate on the credit system. Students may take any undergraduate or graduate course offered by any department which may be required to prepare them for for research and examinations.)

DEPARTMENT OF BIOLOGY
Phone: Chairman: (609) 452-3841 Others: 452-3850

Chairman: J.T. Bonner
The Class of 1877 Professorship of Zoology: V.G. Dethier
The Edwin Grant Conklin Professorship of Biology: F.H. Johnson
The George M. Moffett Professorship of Biology: J.T. Bonner
Professors: J.T. Bonner, V.G. Dethier, W.P. Jacobs, F.H. Johnson, R.D. Lisk, R.M. May, M.S. Steinberg
Associate Professors: E.C. Cox, A. Gelperin, H.S. Horn, M. Konishi, W.A. Newton, J.W. Terborgh
Assistant Professor: E.H. Cohen, J.A. Endler, F. Meins, Jr., W.G. Quinn, T.G. Sanders, R. Tompkins, G.B. Witman
Instructors: R.G. Fowler, D.H. Much, J.F. Novak, L.A. Reuter
Visiting Lecturers: 4
Teaching Fellows: 3
Research Assistants: 7
Research Biologist: 1
Senior Research Biochemist: 1

Degree Program: Biology

Undergraduate Degree: B.S.
Undergraduate Degree Requirements: 4 years

Graduate Degree: Ph.D.
Graduate Degree Requirements: Not stated

Graduate and Undergraduate Courses Available for Graduate Credit: Development in Lower Organisms, Mechanisms, of Animal Morphogenesis, Developmental Physiology of Higher Plants, Developmental Genetics, Current Topics in Animal Behavior, Physiological Control Mechanisms, Biological Reaction Rates, Developmental Biochemistry, Neuroembryology, Principles and Problems in Neuroendocrinology, Theoretical Ecology, Population Biology, Tropical Ecology, Topics in Neural Science and Behavior, plus Colloquia in specialized areas of Development, Neurobiology and Population Biology

RIDER COLLEGE

2083 Lawrenceville Road Phone: (609) 896-0800
Dean of School of Liberal Arts and Science: D. Iorio

BIOLOGY DEPARTMENT
Phone: Chairman: Ext. 297 Others: Ext. 292

Chairman: R.L. Simpson
Professors: T.C. Mayer, J.H. Carlson
Associate Professors: M.A. Leck, L. Oddis
Assistant Professors: R.L. Simpson, M.B. Talmadge, D.F. Whigham

Degree Program: Biology

Undergraduate Degree: B.A.
Undergraduate Degree Requirements: 36 hours Biology, General Zoology, Botany, Genetic Biology, Developmental Biology, General Ecology, Invertebrate Zoology, Cellular Biology, Vertebrate Physiology and one elective, 16 hours Chemistry: Principles of Chemistry, Organic Chemistry I and II, Quantitative Methods in Chemistry, 8 hours Physics: Principles of Physics I and II, 8 hours Mathematics: From Algebra and Trigonometry and higher

RUTGERS COLLEGE

(see Rutgers - The State University - New Brunswick)

RUTGERS - THE STATE UNIVERSITY CAMDEN

311 North 5th Street Phone: (609) 964-1766
Camden, New Jersey 08102
Dean of Graduate Studies: H. Torrey
Dean of College: W.K. Gordon

BIOLOGY DEPARTMENT

Chairman: W.J. Bacha, Jr.
Professors: W.J. Bacha, Jr., J.B. Durand, G.S. Weissman
Associate Professors: R. Good, H. Lee, H. Stempen
Assistant Professors: R. Hastings, J. Kerr, L. Loercher
Instructors: H. Jones
Visiting Lecturer: 1
Teaching Assistants: 3
Research Assistants: 4

Field Stations/Laboratories: Marine Laboratory at Little Egg Harbor Inlet, New Jersey

Degree Program: Biology

Undergraduate Degrees: B.S.
Undergraduate Degree Requirements: 128 credits C or better cumulative, must satisfy course requirements of major

Graduate Degree: M.S.
Graduate Degree Requirements: Master: 30 credits, with or without thesis, B or better grades in course work is normally expected; no more than 9 credits bearing the grade of C may be used in meeting the requirements of the degree

Graduate and Undergraduate Courses Available for Graduate Credit: Marine Biology, Plant Geography, Cell Physiology, Plant Growth and Development, Plant Physiological Chemistry, Helminthology, Immunology, Cell and Tissue Culture, Cell Ultrastructure and Function, Ichthyology, Limnology, General Bacteriology (U), Advanced Bacteriology (U), General Biochemistry (U), Genetics (U), Evolution (U), Cell Biology (U), General Ecology (U), Special Problems in Biology (U), Developmental Botany (U), Systematics and Ecology (U), Plant Ecology (U), Field Ecology (U), General Physiology (U), Animal Physiology (U), Plant Physiology (U), Comparative Anatomy of Vertebrates (U), Vertebrate Embryology (U), Invertebrate Zoology (U), Vertebrate Zoology (U), Animal Histology (U), Animal Parasites (U), (In no case may the candidate for the M.D. degree receive toward that degree more than 12 units of credit for undergraduate courses of study).

Rutgers Medical School
(see College of Medicine and Dentistry of New Jersey)

RUTGERS - THE STATE UNIVERSITY
NEWARK
Neward, New Jersey 07102 Phone: (201) 648-1766
(See also New Jersey College of Medicine and Dentistry)

College of Arts and Sciences
Dean: R.C. Robey

DEPARTMENT OF BOTANY
Phone: Chairman: (201) 648-5131 Others: 648-1766

Chairman: J.H. Crow
Professor: S.S. Greenfield
Associate Professor: J.H. Crow, R.F. Davis, C.F. Meyer
Assistant Professors: A.E. Kasper, J.D. Koppen, J.M. Maiello, T.J. Monahan, G.H. Morton
Teaching Assistants: 11

Degree Program: Botany, Biology (Secondary Education)

Undergraduate Degree: B.A.
Undergraduate Degree Requirements: 32 credits selected from lists of courses in the Botany Department and some electives from other departments, in addition to general education requirements of the college

Graduate Degrees: (Graduate work in conjunction with and the same as Department of Botany at the New Brunswick Campus)

Graduate and Undergraduate Courses Available for Graduate Credit: Mycology (U), Phycology (U), Field Ecology (U), Plant Growth and Development (U), Plant Evolution (U), Seminar in Botany (U), Topics in Botany (U), Problems in Botany (U), Plant Biochemistry

DEPARTMENT OF ZOOLOGY
(No reply received)

RUTGERS - THE STATE UNIVERSITY
NEW BRUNSWICK
New Brunswick, New Jersey 08903 Phone: (201) 932-1766
Dean of Graduate Studies: J.W. Green (Acting)
Deans of Constituent Colleges:
Cook College: C.E. Hess
Douglass College: M. Foster
Routgers College: R. McCormick
University College: N. Pallone
Livingston College: E. Nesthene

The Bioscience Departments of these colleges have been combined as:
NEW BRUNSWICK DEPARTMENT OF BIOLOGICAL SCIENCES
Phone: Chairman: (201) 932-7142

Chairman: B.B. Stout
Professors: C.J Avers, C.H. Bailey (Research), R. Bartha, W.R. Battle, J.W.C. Bird, M.L. Brown, V. Bryson, D.J. Burnes, R.A. Cappellini, S.P. Champe, N.F. Childers, H.E. Clark, R.H. Daines (Research), F.F. Davis, S.H. Davis, R.F. Dawson, R.H. De Boer, G.R. Di Marco, W.O. Drinkwater, D.J. Durkin (Research), P. Eck, J.H Ellison (Research), R.E. Engel, J.L. Evans, D.E. Fairbrothers, A. Farmanfarmaian, H. Fisher, K.W. Fisher, J.A. Forgash (Research), W. Foster, H.M Frankel, C.R. Funk, B.K. Ghosh, S. Gilbert, A.A. Gottlieb, P. Granett (Research), P. Griminger, J.W. Green, J. Grun, J.E. Gunckel, L.E. Hagemann (Research), E.J Hansens (Research), H.H. Haskin, C.E. Hess, A.F. Hopper, L.F. Hough (Research), P.H. Hsu, R.D. Ilnicki, W.R. Jenkins, D.M. Jobbins (Research), S.E. Katz, B.W. Koft, P. Lachance, J.O. Lampen, J.H. Leathem, H.A. Lechevalier, J. Leeder, G.D. Lewis, C. Litchfield, S. Lund, J.D. Macmillan, P.E Marucci (Research), R.E. Mather, L.G. Merrill (Research), R.H. Merritt, R.G. Mitchell, J.P. Mixner, R.E. Morse, W.J. Nickerson, E.R. Orton (Research), N.C. Palczuk, B.F. Palser, J.W. Paterson, P.G. Pearson, J.L. Peterson, O.J. Plescia, D. Pramer, C. A. Price, B.L. Pollack, S.R. Race (Research), R.P. Reece, C.P. Schaffner, J.B.Schmitt, E. Seltzer, M. E. Singley, O. Shifriss (Research), C.R. Smith, W.E. Snyder, M. Solberg, M. Solotorovsky, M.A. Sprague, J.E. Steckel, E.F. Steir, B.B. Stout, P.D. Sturkie, H.T. Streu (Research), D.J. Sutherland (Research), F.C. Swift (Research), G.A. Taylor, D.C. Tudor, W. W. Umbreit, E.H. Varney, G.W. Vander Noot, W.W. Wainio, W.V. Welker, R.F. West, B.R. Wilson, E. M. Witkin
Associate Professors: M. Abou-Sabe', E.V. Adams, M. J. Babcock, R.J. Barfield, D.L. Bishop, E.G. Brennan (Research), T. Chase, J.C.W. Chen, T. Chen (Research), E.A. Cook, R.J. Cousins, B.D. Davis, J.J. Dowling, R.W. Duell, D.E. Eveleigh, R.T.T. Forman, S.A. Garrison, N. Gerber (Research), R.E. Gordon, A.P. Gupta, N. Haard, J.L. Hall, B.A. Hamilton, S.M. Hardy, N.H. Hart, R. Herman, G. Jelenkovic, F.C. Jenifer, J.R. Justin, E. Karmas, D. Kleyn R.J. Kuchler, I. Kujdych, I.A. Leone (Research), R. E. Loveland, M.F. MacDonnell, L.E. McDaniel, J.J. McGrath, L. Miller, B.C. Moser, R.F. Myers (Research), J. Quinn, C.H. Ramage, J.R. Reed (Research) H.C. Reilly, J. Rosen, E. Santamarina, O. Schwabe, R.W. Simpson, C.C. Still, A.W. Stretch (Research), D.H. Strumeyer, R.P. Tewari, F.B. Trama, T. van Es, L.M. Vasvary, N.A. Walensky, G.F. Walton, W.W. Weathers, G. Winnett (Research), R.R. Wolfe
Assistant Professors: A.D. Antoine, J.E. Applegate, V. G. Archer, D.E. Axelrod, J. Burger, S.W.C. Chan (Research), J.Chase, W.J. Crans, H. Daun, S.A. Decter, E.A. Dennis, P. Edwards, F.B. Essien, G.L Floyd, C. Frenkel, R.L. Gilbreath, J.A. Grande, A. A. Greene, M.R. Henninger, C.F. Leck, C.D. Litchfield, L.R. Mendiola-Morgenthaler, J.K. Mitchell, R.A. Niederman, K.R. Olson, C. Page, H.C. Passmore, T.P. Perper, R.P. Petriello, R. Pietruszko, R.D. Poretz, M.L. Rivas, D. Rosenberg, W. Rouslin, J.N. Sacalis, L. Sciorra, A.C. Vasconcelos, A.J. Verlangieri, J.F. Webster, R.W. Willemsen, L.J. Wolgast
Instructors: E.B. Boshko, G. Roberts, W.G. Smith
Visiting Lecturers: 9
Teaching Fellows: 120
Research Assistants: 130

Field Stations/Laboratories: Cape Shore Oyster Laboratory-Cape May Courthouse, New Jersey, Shellfish Research Laboratory - Monmouth Beach, New Jersey, Cream Ridge Research Laboratory, Cream Ridge, New Jersey, Adelphia Research Laboratory - Freehold, New Jersey, Blueberry Research Laboratory - Pemberton, New Jersey, South Jersey Research and Development Center - Bridgeton, New Jersey, Willow Wood Research Laboratory - Gladstone, New Jersey

Degree Programs: Areas of Undergraduate Specialization - Agricultural Biology, Agricultural Education, Agriculture, Animal Science, Bacteriology, Biochemistry, Biological Sciences, Biology, Botany, Conservation, Crop Science, Entomology, Fisheries and Wildlife Biology, Food Science, Forestry, Genetics, Health and Biological Sciences, Horticultural Science, Human Genetics, Landscape Horticulture, Medical Technology, Natural Resources, Ornamental Horticulture, Plant Sciences, Wildlife Management, Zoology
Areas of Graduate Specialization: Agronomy, Animal Science, Biochemistry, Biology, Botany, Ecology, Entomology, Horticulture, Microbiology, Nutrition, Physiology, Plant Pathology, Soil Science, Wildlife Management, Zoology, Environmental Science, Genetic Counsel, Plant Physiology

Undergraduate Degree: B.S.
Undergraduate Degree Requirements: Minimum of 120

semester credit hours with a minimum of 24 credits in particular area of Biology. Usual distribution requirements that vary slightly from college to college.

Graduate Degrees: M.S., M.S.T., Ph.D.

Graduate Degree Requirements: Master: Plan A: 30 semester credit hours of which 6 may be for thesis work. Plan B: All course work - 30 semester credit hours. Doctoral: Minimum of 48 non-research semester credit hours. Residency and language requirements vary between programs. Thesis.

Graduate and Undergraduate Courses Available for Graduate Credit:

Biochemistry: Biological Chemistry, Biochemistry, Advanced Biochemistry Laboratory, Biophysical Chemistry of Organized Systems, Proteins and Enzymes, Regulatory Enzyme Mechanisms, Plant Biochemistry, Biological Regulation, Metabolic Pathways, Nucleic Acids, Organelles, Bioenergetics, Lipids and Carbohydrates, Physical Biochemistry, Advanced Studies in Biochemistry, Seminar in Biochemistry, Special Topics in Biochemistry, Research in Biochemistry

Botany: Cytogenetics, Biogeography and Comparison Ecosystems, Angiosperm Reproduction and Embryology, Comparative Morphology, Biosystematics and Advanced Taxonomy, Experimental Algal Physiology and Ecology, Phycology, Physiological Ecology, Ecosystem and Community Dynamics, Botany Seminar, Developmental Physiology, Research in Botany

Entomology and Economic Zoology: Insect Embryology, Internal and Microscopic Anatomy of Insects, Metamorphosis and Development of Insects, Mosquito Biology and Control, Economic Entomology, Biological Control of Insect Pests, Histological Techniques for Insect Tissues, Advanced Insect Taxonomy, Principles of Systematic Entomology, History of Entomology, Immature Insects, Insect Physiology, Toxicology of Economic Poisons, Advanced Studies in Insecticides, Seminar in Entomology, Research in Entomology

Environmental Science: Waste Treatment I - Sewage Treatment, Stream Sanitation, Environmental Chemistry and Analyses, Waste Treatment II - Water Treatment, Water Treatment III - Industrial Wastes, Principles of Aquatic Chemistry, Application of Principles to Treatment Problems, Pollution Microbiology, Ichthyology, Ichthyology and Fishery Management, Applications of Aquatic Chemistry, Principles of Solid Waste Management and Treatment, Chemistry of the Troposphere, Introduction to Radiation Chemistry, Radiation Biology, Topics in Radiation Biology, Seminar in Radiation Science, Radiation and Radioactivity, Fundamentals of Radiation and Instrumentation, Radiation Instrumentation and Dosimetry, Special Topics in Radiological Health, Experimental Problems in Radiological Health, Seminar in Environmental Sciences, Advanced Special Problems, Experimental Problems in Environmental Science, Environmental Science Research

Horticulture: Advanced in Horticulture Research, Evolution under Domestication, Plant Science Techniques, Genetics of Sexuality, Breeding of Asexually Propagated Crops, Applied Physiology of Horticulture Crops, Plant Senescence, Plant Senescence Laboratory, Physiology and Mineral Nutrition of Fruit Crops, Systematic Pomology, Advanced Vegetable Crops, Seminar in Horticulture, Special Problems in Horticulture, Research in Horticulture

Microbiology: General Microbiology, History of Microbiology, Basic Research Methods in Microbiology, Analytical Cytology, Practical M croscopy, Microbial Biochemistry, Nutrition and Metabolism of Microorganisms, Chemistry of Microbial Products, Cellular Dynamics, Industrial Microbiology, Fundamentals of Large Scale Fermentation, Radiobiology of Microorganisms, Genetics of Bacteria, their Viruses - Laboratory, Seminar in Physiological Genetics, Developmental Genetics, Organelle Genetic Systems, Mutation, Immunochemistry, Cell Biology, Medical Microbiology, Tissue Culture and Elementary Virology, Virology, Bacterial Viruses, Genetics of Bacteria and their Viruses, Advanced Virology, Bacterial Physiology, Pathogenic and Diagnostic Bacteriology, Yeasts, Pathogenic Mycology, Microbial Ecology, Cancer: Leukemia-sarcoma and Related Neoplasms, Molecular Interaction in Organized Systems, Analysis of Development, Seminar in Microbial Genetics, Independent Studies, Special Topics in Microbiology, Research in Microbiology

Nutrition: Minerals in Animal Nutrition, Principles of Nutrition, Vitamins in Nutrition, Developmental Nutrition, Comparative Nutrition, Analytical Techniques in Nutrition, Nutrition Seminar, Proteins in Biology, Research in Nutrition

Physiology: Mammalian Physiology, Mammalian Physiology Laboratory, Comparative Physiology, Cell Physiology, Physiology of Heart and Circulation, Medical Physiology, Environmental Physiology, Respiratory Physiology, Advanced Nerve Muscle Physiology, Electrophysiology and Biophysics, Avian Physiology, Vertebrate Adaptation, Seminar in Physiology, Advanced Studies in Physiology, Research in Physiology

Plant Pathology: Principles of Plant Pathology, Advanced Plant Pathology, Advanced Mycology, Physiology of Fungi, Plant Virology, Pathologican and Mycological Problems, Seminar in Plant Pathology, Seminar in Air Pollution Related to Vegetation, Plant Disease Clinic, Research in Plant Pathology

Plant Physiology: Deductive Plant Physiology, Plant Physiology, Plant Growth and Reproduction, Inorganic Plant Nutrition, Problems in Plant Physiology, Seminar in Plant Physiology, Research in Plant Physiology

Soils and Crops: Advanced Plant Breeding, Ecology of Crop Adaptation and Distribution, Properties and Functions of Herbicides, Crop Science Seminar,

Zoology: Experimental Endocrinology, Elements of Oceanography, Malacology, Ecology of the Estuary, Immunity to Animal Parasites, Advanced Problems in Zoology, Experimental Embryology and Morphology, Population Genetics, Neuroanatomy, Experimental Embryology Laboratory, Advanced Invertebrate Zoology, Host-parasite Relationships, Protozoology, Ecology of Freshwater Organisms, Helminthology, The Behavior of Animal Populations, Comparative Sensory Physiology and Behavior, Histology, Neural Science, Cytology, Investigative Methods, Gross and Developmental Anatomy, Physiological Bases of Behavior, Radioecology, Quantitative Ecology, Human Biology, Topics Advanced Ecology, Population Ecology, Serology and Immunology, Topics in Immunology, Arthropods and Human Disease, Research in Zoology

SAINT PETER'S COLLEGE

Jersey City, New Jersey 07306 Phone: (201) 333-4400

Dean of College: Rev. A.C. McMullen

DEPARTMENT OF BIOLOGY

Phone: Chairman: Ext. 253 Others: 333-4400

Chairman: Rev. J.E. Schuh

Professors: J.J. McGill, Rev. J.E. Schuh

Associate Professors: Rev. R.G. Belmonte, Rev. J.H. Gruszczyk, R.P. Kelly, S.L. O'Malley

Assistant Professors: P. Alexander, Jr., M.A. Fingerhut

Instructors: V.E. Cassaro (Adjunct), A.J. Robbins

Degree Programs: Biology, Biological Chemistry (Interdisciplinary with Chemistry Department)

Undergraduate Degrees: A.B., B.S.

Undergraduate Degree Requirements: 30 credits in Biology (12 beyond 24 credits required core courses (General Biology, Comparative Morphogenesis, Cell Physiology and Molecular Biology); Cognate Requirements: General Chemistry, Organic Chemistry, and Analytical Chemistry, General Physics and Differential and Integral Calculus.

SETON HALL UNIVERSITY

South Orange, New Jersey 07079 Phone: (201) 762-9000
Dean of Graduate Studies: P. Buonaguro (Acting)
Dean of College: N.D. DeProspo

DEPARTMENT OF BIOLOGY
Phone: Chairman: Ext. 253 or 254

Chairman: F.F. Katz
Professors: N.D. DeProspo, F.F. Katz, E.V. Orsi, A. Stauble
Associate Professors: J.R. Keller, S.Z. Kramer
Assistant Professors: J.W. Batey, E. Doman, E.C. Enslee, R.A. Garrick, M. Ghayasuddin, E. Krause
Graduate Assistants: 10

Degree Program: Biology

Undergraduate Degree: B.S.
Undergraduate Degree Requirements: General Biology 8 credits, Biology Electives 24 credits, General Chemistry 8 credits, Physical Chemistry 3 credits, Analytical Chemistry 2 credits, Calculus 8 credits, Physics 8 credits, Organic and Biochemistry 8 credits, College Core and electives to total 61 credits.

Graduate Degree: M.S.
Graduate Degree Requirements: Master: Three plans, Plan A requires 6 credits of thesis research. Otherwise, all require 34 credits total. Only courses specific are Biostatistics and Seminar

Graduate and Undergraduate Courses Available for Graduate Credit: Parasitology, Invertebrate Zoology, Vertebrate Physiology, Evolution, Radiation Biology, Biological Research and Instrumentation

TRENTON STATE COLLEGE

Pennington Road Phone: (609) 771-2371
Trenton, New Jersey 08625
Dean of Graduate Studies: D. Petersen
Dean of College: W. Curry

DEPARTMENT OF BIOLOGY
Phone: Chairman: 771-2371 Others: 771-2356

Chairman: W.S. Klug
Professors: A.Eble, J. Vena
Associate Professors: S. Klug, G. Lipton, E. Rockel, A. Star, H. Treuting
Assistant Professors: O. Heck, E. Hager, R. Fangboner, K. Lee, S. Haworth
Instructor: D. Rose

Field Stations/Laboratories: N.J. Marine Sciences Consortium - Sandy Hook, New Jersey

Degree Programs: Biology, Biological Sciences, Biological Education

Undergraduate Degree: B.S.
Undergraduate Degree Requirements: Not Stated

Graduate Degrees: M.Ed., M.A.T.
Graduate Degree Requirements: Master: Teaching Certification

UNIVERSITY COLLEGE

(see Rutgers - The State University - New Brunswick)

UPSALA COLLEGE

East Orange, New Jersey 07019 Phone: (201) 266-7000
Dean of College: M.D. Schneider

DEPARTMENT OF BIOLOGY
Phone: Chairman: 266-7207 Others: 266-7000

Chairman: G.P. Sellmer
Professors: R.R. Hein, G.P. Sellmer
Associate Professor: L.P. Fanale
Assistant Professor: E.S. Kubersky
Instructor: S.Y. Choi
Lecturer: 1

Degree Program: Biology

Undergraduate Degrees: A.B., B.S.
Undergraduate Degree Requirements: 28 hours Biology, 16 hours Chemistry, 8 hours Physics, 8 hours Mathematics

THE WILLIAM PATERSON COLLEGE OF NEW JERSEY

300 Pompton Road Phone: (201) 881-2000
Wayne, New Jersey 07470
Dean of Graduate Studies: H.L. Burstyn
Dean of College: J.F. Ludwig

DEPARTMENT OF BIOLOGICAL SCIENCES
Phone: Chairman: (201) 881-2245

Chairman: A.F. Shin
Professors: R.F. Callahan, R.O. Capella, L.S. Emrich, J.H. Rosengren, M.L. Spivak, J.R. Voos
Associate Professors: C.Y. Hu, A. Isaacson, D.M. Levine, O. Newton, D. Weisbrot
Assistant Professors: F.E. DiBenedetto, M.E. Hahn, S.L. Hanks, R. Lovell, E. Wallace, J.M. Werth
Instructors: E.G. Atkins, A.C. Barry, C.T. Lyons
Research Assistants: 1

Degree Programs: Biological Sciences, Biology

Undergraduate Degree: B.A.
Undergraduate Degree Requirements: 2 years Core - Botany, Zoology, Cell Biology, Genetics, 30 credits Biology Total, 2 years Chemistry, 1 year Physics, 1 year Mathematics

Graduate Degree: M.A.
Graduate Degree Requirements: Master: 36 credits in Biology, Chemistry and/or related fields, Options include a thesis, Library Research and/or Comprehensive Examination.

Graduate and Undergraduate Courses Available for Graduate Credit: General Ecology, Biometry, Electron Microscopy, Parasitology, Fungi of Woodlands, Limnology, Protozoology, Insect Physiology, Entomology, Environmental Microbiology, Mycology, Phycology, Plant Associations, Biological Conservation, Environmental Management, Human Physiology, Topics in Genetics, Biophysics, Current Topics in Biology, Cold-Blooded Vertebrates, Warm-Blooded Vertebrates, Radiobiology, Ecology of the Estuary, Research and Thesis, Independent Study

NEW MEXICO

COLLEGE OF SANTA FE

Santa Fe, New Mexico 87501 Phone: (505) 982-6411
Dean of College: Br. G. Wright

DEPARTMENT OF SCIENCES AND MATHEMATICS
Phone: Chairman: Ext. 6411

Chairman: Br. J. Walsh
Professor: Br. J. Walsh
Associate Professors: W.B. Grabowski, Br. P. Quin
Assistant Professors: R. Berry, A.P. Butler
Instructors: A. Cano, Br. R. Langlinais

Degree Programs: Biology, Biological Education

Undergraduate Degree: B.S.
Undergraduate Degree Requirements: Biology: 16 hours Chemistry, 8 hours Physics, 32 hours Biology, 8 hours Mathematics, General Science: 54 hours in composite of Physics, Chemistry and Biology with minimum of 12 hours in each area.

EASTERN NEW MEXICO UNIVERSITY

Portales, New Mexico 88130 Phone: (505) 562-2121
Dean of Graduate Studies: J.E. Sublette
Dean of College: D. Whisenhunt

DEPARTMENT OF BIOLOGICAL SCIENCES
Phone: Chairman: 562-2241 Others: 562-1011

Chairman: P.A. Buscemi
Professors: P.A. Buscemi, D.A. Yos
Associate Professors: A.L. Gennaro, O.M. Hofstad, N. Jorgensen, J.B. Secor, R.G. Taylor
Teaching Fellows: 3

Degree Programs: Biology, Wildlife Management, Environmental Health

Undergraduate Degrees: B.A., B.S.
Undergraduate Degree Requirements: Biology 101-102, General Biology or equivalent- Biology, Genetics, Botany Plant Physiology, Zoology, Animal Physiology, Electives 20 hours, Chemistry 101-102 or 103, General Chemistry, Chemistry 341-342 Organic Chemistry (A total of 128 hours are required for graduation)

Graduate Degree: M.S.
Graduate Degree Requirements: Master: 32 hours non-thesis, 30 hours with thesis

Graduate and Undergraduate Courses Available for Graduate Credit: Biology - Molecular Biology (U), Teaching of Natural Sciences (U), Limnology (U), Evolution (U), Biological Literature (U), Biology Field Trip (U), Topics in Biology (U), Workshop in Biology (Summer Session), Individual Research, Graduate Seminar Botany - Plant Physiology (U), Plant Ecology (U), Advanced Plant Physiology, Master's Thesis Microbiology - Principles of Immunology (U), Introduction to Plant Pathology (U), Soil Microbiology (U), Microbial Physiology, Master's Thesis Zoology - Animal Ecology (U), Herpetology (U), Ichthyology (U), Warm-blooded Vertebrates (U), Animal Physiology (U), Advanced Histology and Cytology, Comparative Animal Physiology, Master's Thesis

NEW MEXICO HIGHLANDS UNIVERSITY

Las Vegas, New Mexico 87701 Phone: (505) 425-7511
Dean of Graduate Studies: R.C. Smith
Dean of College: S. Maestas

DEPARTMENT OF BIOLOGY, EARTH SCIENCE, AND ENVIRONMENTAL HEALTH
Phone: Chairman: Ext. 258-259

Chairman: M.W. McGahan
Professors: W. Bejnar, R.G. Lindeborg, M.W. McGahan, L.M. Shields
Associate Professors: F.A. Hopper, R.H. Lessard
Assistant Professors: M. Romine, J.W. Spencer
Instructor: T. Pierce
Visiting Lecturers: 3
Graduate Assistants: 6

Degree Program: Biology

Undergraduate Degree: B.S.
Undergraduate Degree Requirements: Total: 192 quarter hours Major: 48 quarter hours, minor in another science plus 12 quarter hours in a third science and in Mathematics

Graduate Degree: M.S.
Graduate Degree Requirements: Master: 48 quarter hours of graduate work plus a thesis.

Graduate and Undergraduate Courses Available for Graduate Credit: Vertebrate Physiology (U), Bacterial Physiology (U), Biology of the Vertebrates (U), Pathogenic Bacteriology (U), Field Zoology (U), Field Botany (U), Mycology (U), Bioecology (U), Parasitology (U), Vertebrate Histology (U), Quantitative Methods in Biology, Microbial Genetics, Microbial Ecology, Environmental Biology

NEW MEXICO INSTITUTE OF MINING AND TECHNOLOGY

Socorro, New Mexico 87801 Phone: (505) 835-5011
Dean of Graduate Studies: M.H. Wilkening
Dean of College: F.J. Kuellmer

BIOLOGY DEPARTMENT
Phone: Chairman: (505) 835-5321 Others: 835-5011

Head: G. Sanchez
Associate Professors: J.A. Brierley, G. Sanchez, D.K. Shortess
Assistant Professor: J.A. Smoake
Teaching Fellows: 1

Degree Programs: Biochemistry, Biology, Medical Technology

Undergraduate Degree: B.S.
Undergraduate Degree Requirements: General Biology 8 credits, Cell Biology 5 credits, Microbiology 4 credits, Animal Physiology 4 credits, Parasitology 4 credits, Genetics 4 credits, Seminar 1 credit, Plus 8 additional Biology credits, General Chemistry 8 credits Physical Chemistry 3 credits, Organic Chemistry 3 credits, Plus 3 additional Chemistry credits, General Physics 10 credits, Calculus 9 credits, Computer Science 2 credits, Foreign Language 9 credits, Humanities 15 credits, Social Science 9 credits, Electives 21 credits

Graduate Degrees: M.S.
Graduate Degree Requirements: Master: 30 hours of course work above the 400 level approved by the Advisory Committee, 6 hours of thesis credit included

Graduate and Undergraduate Courses Available for Graduate

Credit: Genetics (U), Comparative Physiology (U), Immunology (U), Special Problems (U), Advanced Genetics, Advanced Microbiology, Chemolithotropic Bacteriology, Physiology of Parasitism, Comparative Neurophysiology

NEW MEXICO STATE UNIVERSITY
Las Cruces, New Mexico 88001 Phone: (505) 646-0111
Dean of Graduate Studies: M. Thomson

College of Arts and Sciences
Dean: T. Gale

DEPARTMENT OF BIOLOGY
Phone: Chairman: 646-3611

Chairman: W.A. Dick-Peddie
Professors: W.A. Dick-Peddie, R.T. O'Brien, R.J. Raitt, E. Staffeldt, C.S. Thaeler, W. Whitford, J. Zimmerman
Associate Professors: G. Cunningham, M P. Dunford, J. LaPointe, J. Ludwig, R. Spellenberg
Assistant Professors: M. Bernstein, J. Botsford, C. McCarthy, V. Villa, N. Zucker

Degree Programs: Biology, Medical Technology

Undergraduate Degree: B.S.
Undergraduate Degree Requirements: B.S. (Biology) 2 years Mathematics, 1 year Physics, 7 specified Biology courses, Total credits required 132, B.S. (Medical Technology). As Biology but Biology electives replaced by 32 credits of hospital internship

Graduate Degrees: M.S., Ph.D.
Graduate Degree Requirements: Not Stated

Graduate and Undergraduate Courses Available for Graduate Credit: Special Topics (U), Medical and Veterinary Entomology (U), Molecular Biology for Non-Biologists (U), Microorganisms in Human Affairs (U), Biology in the Elementary School (U), Human Ecology (U), Endocrinology (U), Entomology (U), Natural History of the Vertebrates (U), Aquatic Entomology (U), Evolution (U), Genetics of Microorganisms (U), Cytology (U), Genetics of Higher Organisms (U), Plant Ecology (U), Plant Anatomy (U), Immunology (U), Virology (U), Soil and Water Microbiology (U), Applied Microbiology (U), Microbial Genetics and Physiology (U), Pathogenic Microbiology (U), Animal Histology (U), Herpetology (U), Parasitology (U), Insect Physiology (U), Laboratory Methods in Insect Physiology (U), Principles of Systematic Biology, Principles of Cytogenetics, Cytogenetics Techniques, Seminar in Population Genetics, Quantitative Ecology, Environmental Measurements, Systems Ecology, Plant Biosystematics, Advanced Plant Ecology, Plant Geography, Southwestern Plants, Microbial Ecology, Survey of Microbial Physiology, Microbial Physiology, Seminar in Plant Pathology, Mycology, Field Mycology, Eucaryotic Cell Development, Plant Physiology: Metabolism, Plant Physiology Growth and Development, Plant Water Relations and Mineral Nutrition, Environmental Physiology of Plants, Parasitism in Animals, Animal Speciation, Seminar in Animal Ecology, Seminar in Zoogeography, Environmental Physiology, Techniques in Comparative and Environmental Physiology, Mammalogy, Ornithology, Radiation Biology and Radioactive Tracer Techniques, Seminar in Comparative Endocrinology, Advanced Endocrinology

College of Agriculture and Home Economics
Dean: P.J. Leyendecker

DEPARTMENT OF DAIRY SCIENCE
(No reply received)

DEPARTMENT OF ANIMAL, RANGE AND WILDLIFE SCIENCES
Phone: Chairman: (505) 646-2514 Others: 646-0111

Head: A.B. Nelson
Professors: D.W. Francis, C.H. Herbel (Adjunct), L.A. Holland, D.D. Miller, A.B. Nelson, E.E. Ray, R.H. Robinson, J.L. Ruttle, G.S. Smith
Associate Professors: C.A. Davis, G.B. Donert, W.W. Howard, Jr., D.W. Kellogg, W.D. McFadden, R.D. Pieper, B.J. Rankin, J.D. Wallace, D.L. Zartman
Assistant Professors: R.F. Beck, L.N. Burcham, P.R. Turner
Instructor: E.E. Parker
Research Assistants: 17
Teaching Assistants: 8

Field Stations/Laboratories: F.F. Stanton Experimental Ranch - Capitan, New Mexico, College Ranch, Las Cruces, New Mexico

Degree Programs: Animal Science, Fisheries and Wildlife Biology, Range Science, Wildlife Management

Undergraduate Degree: B.S.
Undergraduate Degree Requirements: 128 credits, G.P.A. 2.000 (2.0 = C), 40 credits in courses numbered 300 and above

Graduate Degrees: M.S., Ph.D.
Graduate Degree Requirements: Master: 30 credits minimum, 15 credits 500 level courses, 1/2 of credits in major Doctoral: 90 credits minimum

Graduate and Undergraduate Courses Available for Graduate Credit: Physiology of Reproduction (U), Animal Nutrition (U), Sheep and Wood Production (U), Dairy Production (U), Animal Breeding (U), Poultry Production (U) Wool Qualities and Techniques (U), Animal Products Sanitation (U), Non-ruminant Nutrition (U), Ruminant Nutrition (U), Laboratory Animal Management (U), Advanced Animal Nutrition, Science of Meat and Meat Products, Genetics of Animal Improvement, Laboratory Techniques in Nutrition, Endocrinology of Domestic Animals, Range Nutrition Techniques, Research Methods in Animal Science, Graduate Seminar, Problem in University Instruction, Special Research Programs, Master's Thesis, Research, Topics in Animal Science, Doctoral Dissertation, Range Ecology (U), Range Analysis (U), Advanced Range Management (U), Methods in Range Research, Mexico's Range Plants, Advanced Range Ecology, Range Plant Ecophysiology, Principles and Evolution of Range Improvements, Management of Game Mammals (U), Wildlife Administration and Law (U), Management of Game Birds (U), Game Mammals Populations (U), Limnology (U), Management of Wildlife Enterprises (U), Advanced Fishery Science (U), Population Ecology, Wildlife Research Methods, Reservoir Ecology

DEPARTMENT OF BOTANY AND ENTOMOLOGY
Phone: Chairman: (505) 646-3225 Others: 646-3226

Head: J.G. Watts
Professors: G.O. Throneberry, J.G. Watts
Associate Professors: J.A. Booth, J.J. Ellington, H.G. Kinzer, D.L. Lindsey, R.B.Turner
Research Assistants: 6

Field Stations/Laboratories: Experiment Branch Stations: Espanola Valley - Alcalde, New Mexico, Middle Rio Grande - Los Lunas, New Mexico, Northeastern - Tucumcari, New Mexico, Plains - Clovis, New Mexico, San Juan - Farmington, New Mexico, Southeastern - Artesia, New Mexico

Degree Program: Agricultural Biology

Undergraduate Degree: B.S.
Undergraduate Degree Requirements: A balance of agricultural and biology courses supported by the basic sciences, humanities and communications arts.

DEPARTMENT OF HORTICULTURE
Phone: Chairman: 646-1521

Head: F.B. Widmoyer
Professors: M.D. Bryant, J.N. Corgan, D.J. Cotter, D.T. Sullivan
Associate Professor: R.M. Nakayama
Instructor: B. Corley
Teaching Fellows: 1.5
Technicians: 3
Research Assistants: 1

Field Stations/Laboratories: Espanola Valley Branch Station-Alcalde, New Mexico, Middle Rio Grande Branch Station - Los Lunas, New Mexico, Northeastern Branch Station - Tucumcari - Nex Mexico, Plains Branch Station - Clovis, New Mexico, Southeastern Branch Station - Artesia, New Mexico, San Juan Branch Station - Farmington, New Mexico, Mora Agricultural Research Station - Mora, New Mexico

Degree Programs: Horticultural Science, Horticulture, Ornamental Horticulture, Pomology, Vegetable Crops

Undergraduate Degree: B.S.
Undergraduate Degree Requirements: Not Stated

Graduate Degree: M.S.
Graduate Degree Requirements: Not Stated

DEPARTMENT OF POULTRY SCIENCE
Phone: Chairman: (505) 646-3435

Head: D.W. Francis
Professor: R.H. Roberson

Field Stations/Laboratories: There are 5 Branch Colleges 6 Branch Experiment Stations, Extension Division and Others.

Degree Program: (Ph.D. is offered in conjunction with Animal Science)

Graduate and Undergraduate Courses Available for Graduate Credit: Not Stated

UNIVERSITY OF ALBUQUERQUE

Albuquerque, New Mexico 87140 Phone: (505) 831-1111
Dean: Not Stated

DEPARTMENT OF BIOLOGY
Phone: Chairman: Ext. 212

Chairman: C.H. Pfeifer
Professors: C.C.C. YuSun
Associate Professors: H.E. Milford
Assistant Professors: J.C. Cook, C. Pantle, E. Pantle, E. Schoenfeld
Instructors: S. Bryan, J. Green

Degree Program: Biology

Undergraduate Degree: B.S.
Undergraduate Degree Requirements: 32 hours Biology, 8 hours Physics and 12-16 hours Chemistry

UNIVERSITY OF NEW MEXICO

Albuquerque, New Mexico 87106 Phone: (505) 277-0111
Dean of Graduate Studies: D. Benedetti

College of Arts and Sciences

Dean: N. Wollman

DEPARTMENT OF BIOLOGY
Phone: Chairman: (505) 277-3411

Chairman: C.S. Crawford (Acting)
Professors: C. Bogert (Adjunct), R. Conant (Adjunct), C.S. Crawford, H.J. Dittmer, J.J. Findley, D.E. Kidd, W.J. Koster, W.C. Martin, R. McClellan (Adjunct), L.D. Potter, M.L. Riedesel, G. Rypke (Adjunct), P.H. Silverman
Associate Professors: W.G. Degenhardt, D.W. Duszynski, G,V. Johnson, W.W. Johnson, P.R. Kerkoff, J.D. Ligon, D. Lundgren (Associate), M.L. Rosenzweig, J.R. Gosz
Assistant Professors: J.J. Altenbach, L. Bardon, E.W. Bourne, R.F. Cooper, L.S. Demski (Joint Appointment) D. Landan, D. Jennings (Adjunct)
Instructors: J.J. Bruner, M.T. Dilley, M.E. Seidel
Teaching Assistants: 35

Degree Programs: Animal Behavior, Bacteriology, Biology, Botany, Ecology, Genetics, Limnology, Parasitology, Physiology, Zoology

Undergraduate Degree: B.S. (Biology)
Undergraduate Degree Requirements: 1 year undergraduate Mathematics, Chemistry through Organic, 1 year undergraduate Physics, 37 hours in Biology as indicated in catalog.

Graduate Degrees: M.S., Ph.D.
Graduate Degree Requirements: Master: Varies beyond Graduate School requirements, Doctoral: Varies beyond Graduate School requirements.

Graduate and Undergraduate Courses Available for Graduate Credit: Biochemistry, Flora of New Mexico, Invertebrate Zoology, Plant Morphogenesis, General Vertebrate Zoology, Biometrics, Concepts of Ecology, Genetics, Genetics Laboratory, Arid Land Invertebrate Population Biology, Comparative Embryology of the Vertebrates, General Entomology, Insect Ecology, Histology, Cytology, Comparative Vertebrate Anatomy, Cellular Physiology, Vertebrate Physiology, Teaching of Biology, Comparative Physiology, Pathogenic Bacteriology, Ethology: Animal Behavior, Immunology, Ethology Laboratory: Animal Behavior, Physiology of Bacteria, Mycology and Plant Pathology, Plant Anatomy, Pharmacology, Economic Botany, Plant Physiology, Environmental Conservation, Medical Entomology, Parasitic Protozoa and Helminths, Limnology, Ornithology, Ichthyology, Herpetology, Mammalogy, Histological Technique, Radiobiology, Current Topics in Biology, Research Procedures, Environmental Physiology, Advanced Invertebrate Zoology, Advanced Genetics, Genetics of Speciation, Current Concepts of Biology, Problems, Advanced Parasitic Protozoology, Advanced Mammalogy, Theoretical Ecology, Advanced Plant Taxonomy, Physiological Plant Ecology, Ecology of North American Vegetation, Plant Mineral and Water Relations, Plant Metabolism andGrowth, Advanced Field Biology

School of Medicine

Albuquerque, New Mexico 87131 Phone: (505) 277-3333
Dean: L.M. Napolitans

DEPARTMENT OF ANATOMY
Phone: Chairman: 277-2221

Chairman: A.J. Ladman
Professors: A.J. Ladman, L.M. Napolitano
Assistant Professors: W.H. Dail, L.S. Demski, S.E. Dietert, A.P. Evan, R.O. Kelley, R.E. Waterman
Research Assistants: 1

Degree Program: Medical Sciences (Anatomy)

Graduate Degree: Ph.D.
Graduate Degree Requirements: Doctoral: Variable

Graduate and Undergraduate Courses Available for Graduate Credit: Seminar in Anatomy, Selected Topics in Developmental Biology, Current Topics in Morphology, Neurobiology and Behavior of Marine Animals

DEPARTMENT OF BIOCHEMISTRY
Phone: Chairman: 277-4928 Others: 277-3333

Chairman: F.N. Lebaron
Professor: R.B. Loftfield
Associate Professors: T.J. Scallen, L. Smith
Assistant Professors: A.C. Atencio (Assistant Dean of Student Affairs), J.L. Omdahl, P. Reyes, D.L. Vander Jagt, G.C. Wild, B.M. Woodfin
Research Assistants: 8

Degree Programs: Biochemistry, Medical Sciences (Biochemistry)

Graduate Degree: Ph.D.
Graduate Degree Requirements: Doctoral: Ph.D. (Chemistry, Biochemistry), Ph.D. Medical Sciences (Biochemistry)

Graduate and Undergraduate Courses Available for Graduate Credit: Introductory Biochemistry, Steroids, Proteins, Vitamins, Lipids, Regulation

DEPARTMENT OF MICROBIOLOGY
Phone: Chairman: (505) 277-2609 Others: 277-3333

Chairman: L.C. McLaren
Professors: L.C. McLaren, J.V. Scaletti, S. Tokuda, J.A. Ulrich
Associate Professor: T.I. Baker
Assistant Professor: C.E. Cords, D.M. Enger (Adjunct), E.H. Goldberg, R.J. Radloff
Visiting Lecturers: 2
Research Assistants: 4

Degree Programs: Microbiology, Medical Science (Microbiology)

Graduate Degrees: M.S., Ph.D.
Graduate Degree Requirements: Master: minimum of 24 semester hours of course work, minimum of 6 semester hours of graduate level courses, minimum of 18 semester hours in residence at University, minimum of 6 semester hours of thesis credit Doctoral: minimum of 48 semester hours of course work, minimum of 18 semester hours of graduate level courses, minimum of 24 semester hours in residence at University, either one foreign language or a collateral field of study

Graduate and Undergraduate Courses Available for Graduate Credit: Microbiology (U), Clinical Laboratory Microbiology (U), Medical Virology (U), Medical Mycology (U), Pathobiology, Research Techniques, Advanced Microbiology, Microbial Physiology, Biochemical Genetics, Immunology, Advanced Virology, Seminar

DEPARTMENT OF PATHOLOGY
(No information available)

DEPARTMENT OF PHYSIOLOGY
Phone: Chairman: (505) 277-4730 Others: 277-3333

Chairman: None
Professor: S. Solomon
Associate Professors: D. Priola, A. Ratner, J. Weiss
Assistant Professors: W. Galey, K. Kastella, R. Shannon
Instructor: H. Spurgpon
Visiting Lecturers: 6
Research Assistants: 7
Postdoctorals: 2

Degree Program: Physiology

Graduate Degrees: M.S., Ph.D.
Graduate Degree Requirements: Master: 24 semester hours of graduate credit, comprehensive examination, thesis and Public Defense of thesis. Doctoral: 48 semester hours of graduate credit, qualifying examination, comprehensive examination (written and oral), dissertation and Public Defense of Dissertation

Graduate and Undergraduate Courses Available for Graduate Credit: Advanced Physiology, Biological Membrane Structure and Function, Integrative Functions of the Endocrine System, Advanced Cardiovascular Physiology, Renal Water and Electrolyte Metabolism, Hormonal Control of Sex and Reproduction, Control Mechanisms in BiologicalSystems, Advanced Neurobiology, Special Topics in Physiology, Physiological Techniques, Seminar in Physiology, Advanced Respiratory Physiology

WESTERN NEW MEXICO UNIVERSITY
(no reply received)

NEW YORK

ADELPHI UNIVERSITY

Graden City, New York 11530 Phone: (516) 294-8700
Dean of Graduate Studies: M A. Iverson
Dean of Graduate School of Arts and Sciences: M.A. Iverson

BIOLOGY DEPARTMENT
Phone: Chairman: Ext. 7501, 7502 Others: Ext. 7503

Chairman: J.K. Hampton, Jr.
Professors: A.H. Brenowitz, W. Warren, B. Eickelberg, I. Fand (Adjunct), R.S. Grillo, J.K. Hampton, Jr., R.J. Lacey, J.J. Napolitano
Associate Professors: J.M Bassin, S.E. Gochenaur, H.S. Grob, R.C. Johnsen, H.G. Kalicki, K.M. Smart
Assistant Professors: K. Bagdonas, E.D. Brodie, Jr., D. Cooperstein, C. Diakow, J.K. Dooley, V.A. Fischette (Adjunct), K.R. Harris, R.D. Jones, R. Lund, J. Macri (Adjunct), L. Sokoloff
Teaching Fellows: 17
Research Assistant: 1

Field Stations/Laboratories: Adelphi Institute of Marine Science - Oakdale, Long Island, New York

Degree Programs: Animal Physiology, Bacteriology, Bacteriology and Immunology, Biology, Physiology, Zoology, Marine Science

Undergraduate Degrees: B.A., B.S. (Biology)
Undergraduate Degree Requirements: 38 credits in Biology, 12 credits in Chemistry

Graduate Degree: M.S. (Biology)
Graduate Degree Requirements: Master: Thesis - 33 Biology credits (includes 6 credits for thesis), Presentation of thesis. Non-thesis - 36 credits (24 in Biology)

Graduate and Undergraduate Courses Available for Graduate Credit: Mammalian Physiology, Human Physiology, Evolutionary Ecology, Immunobiology, Biochemical Control of Cellular Processes, Marine Seminar, Graduate Genetics, Special Research Problems, Thesis Research, Graduate Seminar, Parasitology (U), Cell Physiology (U), Herpetology (U), Entomology (U), Biogeography (U), Ecological Systems, Endocrinology, Virology, Advanced Microbiology, Protozoology, Marine Botany, Marine Ecology, Marine Microbiology, Cytology, Cell and Tissue Culture, History of Biology

ALBANY COLLEGE OF PHARMACY

(see under Union College and University)

ALBANY MEDICAL COLLEGE

(see under Union College and University)

ALFRED UNIVERSITY

Alfred, New York 14802 Phone: (607) 871-2111
Dean of Graduate Studies: L. Butler
Dean of College: J. Taylor

DEPARTMENT OF BIOLOGY
Phone: Chairman: (607) 871-2205

Chairman: G.E. Rough
Professors: P.S. Finlay, G.E. Rough
Associate Professors: C. Shively
Assistant Professors: B. Bowden, J.P. Rousch, B.N. Rock
Instructor: H.H. Tucker, Jr.
Teaching Fellows: 1-2

Field Stations/Laboratories: Cooperative field station through membership in College Center of the Finger Lakes. Finger Lakes Institute Laboratory located at Watkins Glen, New York, onSeneca Lake

Degree Program: Biology

Undergraduate Degree: B.A.
Undergraduate Degree Requirements: 6 credits Biology, 4 credits Chemistry, 20 cognate credits

Graduate Degree: M.S. Ed.--Biology
Graduate Degree Requirements: Master: Minimum 16 hours 400-500 level courses in Biology plus 14-16 hours in Psychology-Education. Comprehensive examination in Biological Subjects.

Graduate and Undergraduate Courses Available for Graduate Credit: Freshwater Vertebrates (U), Parasites of Aquatic Animals (U), Systematic Botany (U), Problems in Ecology (U), General Biochemistry (U), Independent Study (U), Fundamentals of Cytogenetics (U), Bacteriology (U), Physiology (U), Research, Modern Evolutionary Theory

BARD COLLEGE

Annandale-on-Hudson, New York 12504 Phone: (914) 758-6822
Dean of Faculty: C. Selinger

BIOLOGY DEPARTMENT
Phone: Chairman: Ext. 150 Others: Ext. 150

Head: B.L. Josephson
Assistant Professors: B.L. Josephson, W.T. Maple
Instructor: E. Kiviat
Teaching Fellows: 2

Field Stations/Laboratories: Ecology Field Station - Bard College

Degree Programs: Biology

Undergraduate Degree: B.A.
Undergraduate Degree Requirements: General Biology (2 semesters), Physiology (1 semester), Genetics (1 semester), Senior Project (2 semesters), General Chemistry (2 semesters), Organic Chemistry (2 semesters), Language and Literature electives (2 semesters) Social Studies Electives (2 semesters), Art, Music, Drama, Dance Electives (2 semesters)

BERNARD M. BARUCH COLLEGE

(see under City University of New York)

BARNARD COLLEGE

(see under Columbia University)

BRIARCLIFF COLLEGE

Briarcliff Manor, New York 10510 Phone: (914) 941-6400
Dean of College: W. Chizinsky

DEPARTMENT OF SCIENCE AND MATHEMATICS
Phone: Chairman: Ext. 731 Others: Ext. 730

Chairman: R. Weiner
Professors: W. Chizinsky
Associate Professor: J.A. Swatek

Degree Program: Biology

Undergraduate Degree: B.A.
Undergraduate Degree Requirements: 36 courses for graduation. 1 year General Biology, Vertebrate and Invertebrate Biology, 2 years Chemistry, 4 advanced courses (Physiology, Genetics, Evolution, Developmental Biology, Environmental Biology, Parasitology), Biology Seminar

BROOKLYN COLLEGE
(see under City University of New York)

BROOKLYN COLLEGE OF PHARMACY
(see under Long Island University)

CANISIUS COLLEGE
Buffalo, New York 14208 Phone: (716) 883-7000
Dean of College: J. Bieron

BIOLOGY DEPARTMENT
Phone: Chairman: Exts. 870, 871

Chairman: A.A. Alexander
Professors: V.P. Stouter
Associate Professors: A.A. Alexander, K.R. Barker, R.F. Blasdell, K.P. Treanor, W.F. Zapisek
Assistant Professors: S. Chidambaram, I.J. Lorch, J. A. Tomasulo

Degree Program: Biology

Undergraduate Degrees: B.A.
Undergraduate Degree Requirements: College Requirements = 40 courses, Biology Science Requirements, Mathematics - 2 semesters, Physics - 2 semesters, Chemistry - 4 semesters, Biology: 6 required courses, 2 required 1 hour seminars, 3 electives

CITY COLLEGE
(see under City University of New York)

CITY UNIVERSITY OF NEW YORK
55 East 804 Street Phone: (212) 360-2141
New York, New York 10021

BERNARD M. BARUCH COLLEGE
OF THE CITY UNIVERSITY OF NEW YORK

BROOKLYN COLLEGE
OF THE CITY UNIVERSITY OF NEW YORK
Bedford Avenue and Avenue H Phone: (212) 780-5485
Brooklyn, New York
Deputy Chairman of Graduate Studies: R.A. Eckhardt
Dean of the School of Science: D.M. Gabriel

DEPARTMENT OF BIOLOGY
Phone: Chairman: (212) 780-5397 Others: 780-5485

Chairman: D.D. Hurst
Professors: M.M. Belsky, J.R. Collies, N.R. Eaton, G.H. Fried, S. Goldstein, R. Guttman, D. Hurst, P. F.A. Maderson, L.G. Moriber, S.N. Salthe, M.P. Schreibman, F.L. Schuster, M.I. Selsky, I.A. Tittler, G.E. Wheeler, C.L. Withner
Associate Professors: C.A. Beam, C.J. Burdick, D.R. Franz, M. Himes, N.L. Levin, P.K. Nelson, P.K.C. Pang
Assistant Professors: J. Blamire, R.A. Eckhardt, J.J. Eppig, R.H. Gavin, E.S. Hawley, N.A. Khan, R.E. McGowan
Teaching Fellows: 40
Research Assistants: 14
Post-doctoral Fellows: 2

Degree Programs: Biochemistry, Biology

Undergraduate Degrees: B.S., B.A., or B.A./M.A.
Undergraduate Degree Requirements: General Biology, General Chemistry, Organic Chemistry, Physics, Calculus, Botany Elective, Zoology Elective, Physiology, Genetics

Graduate Degrees: M.A., Ph.D.
Graduate Degree Requirements: Master: 30 credits, Comprehensive Examination, Thesis and one language Doctoral: 60 credits, Comprehensive Examinations, Thesis and two languages

Graduate and Undergraduate Courses Available for Graduate Credit: Advanced Biochemistry Lectures, Basic Laboratory Techniques for Research in Biochemistry, Research Toward the Doctoral Dissertation, Biochemistry of the Lipids, Enzymology, Nucleic Acid Metabolism and Function, Metabolic Pathways and their Control Mechanisms, Biopolymers, Quantum Biochemistry, Genetics, Evolution and Systematics - Lectures in Genetics, Microbial Genetics, Genetics of Multicellular Organisms, Cell Heredity, Evolution, Plant Systematics, Animal Systematics, Taxonomy of Vascular Plants, Numerical Systematics, Zoology and Phylogeny of the Chordata, Biochemical Evolution and Systematics, Developmental Genetics, Population Genetics, Comparative Morphology of Vascular Plants, Paleobotany, Microevolutionary Processes, Molecular and Cell Biology - Problems in Nuclear Cytology, Cellular Physiology, Molecular Biology, Experimental Microbiology, Comparative Biochemistry, Cell Biology, Cell Biology Internship, B ological Effects of Chemical Agents, Bacteriophage, Cytology, Fine Structure of Cells, Biological Electron Microscopy, Cytogenetics, Animal Behavior - Neurobiology, Biological Basis of Animal Behavior, Current Issues in Behavioral Ontogeny, Sensory Physiology, Animal Communication, Field Studies in Animal Behavior Physiology - Animal Physiology, Endocrinology, Comparative Neuroendocrine Mechanisms, Comparative Animal Physiology, Endocrine Cytology, Plant Physiology Biochemistry and Biophysics - Introduction to Biophysics, Biophysical Techniques in Physiology, Radiation Biology, Radioisotopes in Biology, Photobiology, Electrobiology, Phytochemistry, Structure and Metabolism of Macromolecules Developmental Biology - Developmental Biology, Special Problems in Developmental Biology, Plant Morphogenesis, Molecular Basis of Development Ecology and Biological Oceanography - Microbial Ecology, Ecology, Community Ecology, Population Ecology, Limnology, Marine Plankton Dynamics, Marine Benthos, Fishes and Fisheries Biology, Marine Microbiology, Marine Ecology, Principles and Practices of Aquaculture, Physiological Ecology, Experimental Parasitology, Plant Ecology, Plankton Zoogeography and Paleoecology Mathematical Biology - Mathematical Biology, Advanced Mathematical Biology

THE CITY COLLEGE
OF THE CITY UNIVERSITY OF NEW YORK
Convent Avenue and 138th Street Phone: (212) 621-7501
New York, New York 10031
Dean of College: H. Lustig

DEPARTMENT OF BIOLOGY
Phone: Chairman: (212) 621-7501

Chairman: J.A. Organ
Professors: L. Aronson (Adjunct), D. Bliss (Adjunct), R.G. Busrel (Adjunct), L.J. Crockett, R.R. Feiner,

M. Hamburgh, K. Krishna, J.J. Lee, L. Levine, J.A. Organ, G.S. Posner, J. Rozen (Adjunct), M. Sacks, C. L. Smith (Adjunct), W.N. Tavolga, E. Tobach (Adjunct), A.O. Wasserman

Associate Professors: A.W.H. Bé (Adjunct), D.M. Cooper, G.W. Cooper, R.P. Goode, J. Grossfield, P.L. Krupa, D.C. Miller, R.A. Ortman, J. Osinchak, J. Rozé, N. M. Saks, R.J. Shields, M. Sideman (Adjunct), J.H. Tietjen, S. Wecker

Assistant Professors: M.A Benjaminson (Adjunct), C. J. Berg, Jr., N.G. Grant, J.G. Griswold, J.P. Hanks, E. LeBourhis (Adjunct), D. Loskutoff (Adjunct), T.C. Malone, L.H. Mantel, N.M. Schwartz, R. Zuzolo

Lecturer: 1

Teaching Fellows: 19

Research Assistants: 2

Adjunct Lecturers-Part Time: 16

Field Stations/Laboratories: Affiliations with: Lamont-Doherty Geological Observatory, Animal Behavior Department of the American Museum of Natural History, and Greenbrook Sanctuary

Degree Programs: Animal Behavior, Biological Sciences, Biology, Ecology, Ethology, Genetics, Microbiology, Oceanography

Undergraduate Degrees: B.S., B.A. (Biology

Undergraduate Degree Requirements: Every student majoring in Biology must complete specialization of 32 credits. This must include the following: a) The Biology Core (7 courses: Vertebrate Biology, Invertebrate Biology, Botany, Principles of Genetics, Principles of Ecology, Cell Physiology and Developmental Biology) b) 8 credits of Biology courses beyond the core requirements as elective work. Other: 7 mathematics, 14 Chemistry, 4 Physics

Graduate Degrees: M.A., Ph.D.

Graduate Degree Requirements: Master: A minimum of 30 crédits of which eleven are required as follows: 1) Colloquium Biology - 1 credit each semester 2) Seminar - 3 credits, Thesis Research 3 credits Remaining 12 credits approved course work in Biology Doctoral: 60 credit of course work; First Examination, two foreign languages, normally French and German (Computer techniques in Lieu of second foreign language. Second Examination. Dissertation. A minimum of two semesters of teaching experience.

Graduate and Undergraduate Courses Available for Graduate Credit: Cell Heredity (U), Animal Systematics (U), Population Genetics (U), Seminar in Evolution (U), Cellular Physiology (U), Comparative Neuroendocrine Mechanisms (U), Radiation Biology (U), Neurobiology (U), Biological Basis of Animal Behavior (U), Current Issues in Behavioral Ontogeny (U), Evolution of Behavior (U), Sensory Physiology (U), Animal Communication (U), Field Studies in Animal Behavior (U), Seminar in Animal Behavior (U), Seminar in Behavioral Aspects of Ecology (U), Selected Topics in Animal Behavior (U), Seminar in Genetics, Seminar in Acoustic Communication in Animals (U), Microbial Genetics (U), Developmental Genetics (U), Developmental Endocrine Cytology (U), Developmental Biology (U), Special Problems in Developmental Biology (U), Plant Morphogenesis (U), Molecular Basis of Development (U), Seminar in Molecular Genetics (U), Seminar in Developmental Biology (U), Community Ecology (U), Population Ecology (U), Limnology (U), Marine Plankton Dynamics (U), Marine Benthos (U), Fishes and Fisheries Biology (U), Marine Microbiology (U), Marine Ecology (U), Principles and Practices of Aquaculture (U), Physiological Ecology (U), Experimental Parasitology (U), Plant Ecology (U), Seminar in Ecology (U), Seminar in Biological Oceanography (U), Mathematical Biology (U), Seminar in Special Topics (U), Colloquim (U), Advanced Study (U), Tutorial (U), Basic Laboratory Techniques for Research (U). A maximum of nine credits (3-4 courses) in the following advanced undergraduate courses may be taken toward the M.A. and Ph.D. Degree. Principles of Ecology, Cell Biology, Developmental Biology, Vertebrate Histology, Parasitology, Entomology, Comparative Animal Physiology, Vertebrate Endocrinology, Microbiology, Microbial Physiology, Microbial Ecology, Animal Behavior, Biological Oceanography, Organic Evolution

GRADUATE SCHOOL AND UNIVERSITY CENTER OF THE CITY UNIVERSITY OF NEW YORK

33 West 42nd Street Phone: (212) 790-4395
New York, New York 10036

Dean of Graduate Studies: H. Hillerbrand

DOCTORAL PROGRAMS IN BIOLOGY

Phone: Chairman: (212) 780-5406

Executive Officer: L.G. Moriber

Professors: S. Aaronson, H.G. Albaum, S. Anderson (Adjunct), L.R. Aronson, A. Bé (Adjunct), W. Bertsch, D. Bliss (Adjunct), M. Brody, R.G. Busnel, J.R. Collier, A. Cronquist (Adjunct), N. Eldredge (Adjunct), R.R. Feiner, G.H. Fried, M. Gabriel, S. Goldstein, M. Hamburgh, M.K. Hecht, I.H. Herskowitz, D.D. Hurst, T.E. Jensen, J.S. Krakow, K. Kirshna, J.J. Lee, L. Levine, B. Maguire (Adjunct), L.F. Marcus, D. Marien, R.C. Mawe, J. Mickel (Adjunct), L.G. Moriber, J.A. Organ, G.S. Posner, O.A. Roels, C. Rogerson (Adjunct), D.E. Rosen (Adjunct), J.G. Rozen (Adjunct), M. Sacks, R. Sager, N.M. Saks, S.N. Salthe, M.P. Schreibman, F.L. Schuster, M.I. Selsky, L. Short (Adjunct), C.L. Smith (Adjunct), W. C. Steere (Adjunct), M. Wasserman, G.E. Wheeler, C.L. Withner, D. Wilkie (Adjunct), P.S. Woods

Associate Professors: E. Balboni, G. Bard, D. Basile, C.A. Beam, J. Berech, I. Blei, T.A. Borgese, C.J. Burdick, P.C. Chabora, W.D. Cohen, G.W. Cooper, L.J. Crockett, D. Franz, S.M. Friedman, J. Golubow, R. Goode, R.J. Grant, M. Green, J. Grossfield, H. Guthwin, E.S. Handler, E.E. Handler, A.E. Haschemeyer, M.H. Himes, S. Koulish, P.L. Krupa, N.L. Levin, K.M. Lyser, G.T. MacIntyre, C.R. Martin, D.C. Miller, R. Ortman, J. Osinchak, G. Prance (Adjunct), J.W. Rachlin, J. Roze, R. Rudner, E. Shahn, Y. Shecter, R.J. Shields, L.J. Smith, S. Taub, J. Tietjen, J.G. Valdovinos, S.S. Wallace, A.O. Wasserman, S. Wecker, E. Worley

Assistant Professors: G. Barry, C. Berg, J. Blamire, C. Ceccarini, R. Chappell, L.A. Ciaccio, K.P. Dumont (Adjunct), J. Eppig, N.G. Grant, J. Griswold, J.P. Hanks, E. Hawley, A. Held, J.L. Hill, P. Kashin, N.A. Kahn, A. Lall, E.H. Leiter, T.C. Malone, L.H. Mantel, R.E. McGowan, C.V.A. Michels, B.P. Moffitt, M.E. Nathanson, G.C.Offutt, N.M. Schwartz, J. Szalay, E.M. Tarjan

Research Associates (Postdoctorals): 2

Field Stations/Laboratories: Kalbfleisch

Degree Program: Biology

Undergraduate Degrees: See individual colleges of the City University of New York

Graduate Degrees: M.A., Ph.D.

Graduate Degree Requirements: Master: 45 credits (one foreign language, paper), for "en route" Master's degree Doctoral: 60 credits (2 foreign languages, thesis)

Graduate and Undergraduate Courses Available for Graduate Credit: Genetics, Evolution and Systematics - 37 courses (Graduate), Molecular and Cell Biology - 26 courses (Graduate), Animal Behavior - 9 courses (Graduate) plus Biopsychology courses, Physiology -11 courses (Graduate), Biochemistry and Biophysics - 10 courses (Graduate), Ecology and Biological Oceanography - 31 courses (Graduate), Mathematical Biology

4 courses (Graduate), Seminars: 29 courses (Graduate), Independent Doctoral Research

DEPARTMENT OF BIOCHEMISTRY
Phone: Chairman: 790-4273

Chairman: P. Lukton
Professors: S. Aaronson, H.G. Albaum, A. App (Adjunct), W. Bertsch, M. Brody, B. Bulkin, N. Eaton, M. Eidinoff, W.B. Essman, M.M. Fishman, G.H. Fried, S. Garattini (Adjunct), M. Glantz, D. Goldberg, P. Haberfield, T.H. Haines, A.E. Haschmeyer, J.F. Hogg, H. Hoyer, H. Jacobson, S. Kortiz, J. Krakow, A. Kukton, A. Mazur, G. Oster, I. Oreskes, O.A. Roels, A.F. Rosenthal (Adjunct), C. Russell, R. Sager, S. Udenfriend (Adjunct), A.E. Woodward
Associate Professors: E. Abbott, D.S. Beattie, J. Berech, D. Beceridge, R. Bittman, R. Engel, M.S. Friedman, J. Golubow, M. Green, H.J. Li, F. Margolis (Adjunct), R. Rudner, S.N. Salthe, M. Tomasz, B.E. Tropp
Assistant Professors: F. Downs, F.R. Naider, Y.S. Papir, H. Schulz, E. Tarjan
Visiting Lecturers: 2

Degree Program: Biochemistry

Graduate Degree: Ph.D.
Graduate Degree Requirements: Doctoral: 60 credits of course work--6 credits of which must be in Biology and 3 credits of which must be in Physical Chemistry. The student is required to participate in Biochemistry seminars, colloquim programs, and research work. Included are: First Examination, Foreign Language, Field Experience, Second Examination and Dissertation

Graduate and Undergraduate Courses Available for Graduate Credit: Advanced Biochemistry Lectures, Basic Laboratory Techniques for Research in Biochemistry, Advanced Organic Chemistry I (Structure and Mechanism), Introduction to Quantum Chemistry, Physical Biochemistry, Seminar in Biochemistry, Biochemistry of the Lipids, Enzymology, Nucleic Acid Metabolism and Function, Metabolic Pathways and their control Mechanisms, Biopolymers, Quantum Biochemistry

LEHMAN COLLEGE OF THE CITY UNIVERSITY OF NEW YORK

Bedford Park Boulevard West Phone: (212) 960-8235
Bronx, New York 18543
Dean of College: G. Wheeler

DEPARTMENT OF BIOLOGICAL SCIENCES
Phone: Chairman: 960-8235

Chairman: C.R. Jones
Professors: M. Anchel (Adjunct), A. Barksdale (Adjunct), H. Becker (Adjunct), A. Cronquist (Adjunct), A. Hervey (Adjunct), H. Irwin (Adjunct), T. Jensen, R. Jones, B. Maquire (Adjunct), J. Mickel (Adjunct), G. Prance (Adjunct), C. Rogerson (Adjunct), W. Steere (Adjunct), J. Valdovinos
Associate Professors: G. Bard, D. Basile, T. Boyese, J. Golubow, H. Guthwin, A. Held, T. Koyama (Adjunct), S. Leone (Adjunct), J. Rachlin, Y. Shechter
Assistant Professors: W. Anderson (Adjunct), G. Barry, S. Canham (Adjunct), K. Dumont (Adjunct), D. Giannasi (Adjunct), E. Henry, N. Holmgren (Adjunct), P. Holmgren (Adjunct), J. Keithly, B. Moffitt, G. Smith (Adjunct), S. Wallace
Instructors: S. D'Onofrio, P. Szabo
Teaching Fellows: 25
Research Assistants: 1

Degree Program: Biological Sciences

Undergraduate Degree: B.S.
Undergraduate Degree Requirements: 24 credits Biology, 19 credits Chemistry (1 year General Chemistry and 1 year Organic Chemistry), 8 credits Physics, 1 year Mathematics, 1 semester Calculus plus another semester of Calculus or Biostatistics

Graduate Degree: M.S.
Graduate Degree Requirements: Master: 30 credits of course work or 24 credits of course work, thesis, Comprehensive

HUNTER COLLEGE OF THE CITY UNIVERSITY OF NEW YORK

695 Park Avenue Phone: (212) 360-2384
New York City, New York 10021
Dean of Graduate Studies: W. Eisenberg
Dean of College: F.J. Weyl

DEPARTMENT OF BIOLOGICAL SCIENCES
Phone: Chairman: 360-2384

Chairman: R.C. Mawe
Professors: W. Bertsch, M. Brody, A. Haschemeyer, I. Herskowitz, J. Krakow, R.C. Mawe, R. Sager
Associate Professors: E. Balboni, W.D. Cohen, S.M. Friedman, R. Grant, M. Green, E. Handler, E. Handler, K. Lyser, C. Martin, R. Rudner, E. Shahn
Assistant Professors: C. Ceccarini, R. Chappell, S. Raps
Instructors: R. Hirschberg, A. Kumar, L. Margulies, H. Miller, N. Scherzer
Visiting Lecturers: 6
Teaching Fellows: 40
Research Assistants: 10

Degree Program: Biology

Undergraduate Degree: B.A.
Undergraduate Degree Requirements: 26 credits Biology, General and Organic Chemistry, 1 year Physics (8 credits), 1 year (6 credits) Calculus

Graduate Degrees: M.A., Ph.D.
Graduate Degree Requirements: Master: Comprehensive Examination, 30 graduate course credits, one language or computer Doctoral: 60 credits (approximately 20 credits can be tutorial), 2 languages, written examination after 1st year, thesis, oral examination on thesis proposal

Graduate and Undergraduate Courses Available for Graduate Credit: Molecular and Cell Biology - Molecular Biology, Cell Biology, Fine Structure of Cells, Basic Laboratory Techniques for Research Genetics and Developmental Biology - Problems in Microbial Genetics, Genetics of Multicellular Organisms, Cell Heredity, Bacteriophage, Developmental Biology Physiology and Biophysics - Comparative Animal Physiology, Endocrinology, Plant Physiology, Introduction to Biophysics, Biophysical Techniques in Physiology, Radiation Biology, Radioisotopes in Biology, Photobiology, Electrobiology, Mathematical Biology, Anatomy and Physiology of the Nervous System Population Biology - Evolution, Ecology

QUEENS COLLEGE OF THE CITY UNIVERSITY OF NEW YORK

Flushing, New York 11367 Phone: (212) 520-7418
Dean of Graduate Studies: S. Axelrad
Dean of College: D.H. Speidel

BIOLOGY DEPARTMENT
Phone: Chairman: 720-7418

Chairman: M.K. Hecht
Professors: S.A. Aaronson, M.K. Hecht, T.S.K. Johansson, M.L. Kaplan, L.M. Marcus, D. Marien, M. Wasserman, P. Woods
Associate Professors: J. Berech, Jr., P.C. Chabora, G.T. MacIntyre, S. Pierce, L.J. Smith
Assistant Professors: D.W. Alsop, E. Boylan,

A.M. Greller, C. Hirsch, P. Kashin, C. Michels, F. Minutoli, M. Nathanson, J. Rifkin, U. Roze, J.A. Sperling, J. Szalay
Instructors: R.A. Calhoon, P. Hollander, A. Pace, M. Santos
Teaching Fellows: 14

Degree Programs: Cellular Biology, Developmental Biology, Evolutionary Biology

Undergraduate Degree: B.S.
Undergraduate Degree Requirements: 36 credits in Biology, courses in Physics, Chemistry and Mathematics (see instructions in catalog).

Graduate Degrees: M.S., Ph.D.
Graduate Degree Requirements: Master: 30 credits in Biology and a thesis Doctoral: Not Stated

Graduate and Undergraduate Courses Available for Graduate Credit: Not Stated

YORK COLLEGE
OF THE CITY UNIVERSITY OF NEW YORK
150-14 Jamaica Avenue Phone: (212) 969-4311
Jamaica, New York City, New York 11432
Dean of College: R.C. King

DEPARTMENT OF BIOLOGY
Phone: Chairman: (212) 969-4404 Others: 969-4387

Chairman: J.M. Schlein
Professor: A.F. D'Adamo
Associate Professor: L. Lewis
Assistant Professors: K. Cooper, M.S. Giannini, S. Hampton, M. Levandovsky, W. Moyer, D. Patton (Adjunct), J.M. Schlein, C. Suerth
Instructors: A. Beulig (Adjunct), B. Borowsky (Adjunct), D. Sarot (Adjunct)
Technicians: 4

Degree Program: Biology

Undergraduate Degree: B.S.
Undergraduate Degree Requirements: 1 year General Chemistry, 1 semester Organic Chemistry, 1 year General Physics, 1 year General Biology, 1 semester Genetics, 1 of Chordate, Botany and Invertebrate, and 10 additional credits in Biology

COLGATE UNIVERSITY
Hamilton, New York 13346 Phone: (315) 824-2000
Dean of College: J.S. Morris

BIOLOGY DEPARTMENT
Phone: Chairman: (315) 824-4344 Others: 824-2000

Chairman: P.G. Crook
Chas. A. Dana Professor: P.G. Crook
Professors: P.G. Crook, R.E. Goodwin, R.A. Hoffman
Associate Professors: W.J. Oostenink, F.W. Weyter
Assistant Professors: R.W. Hoham, J.A. Novak, T.L. Pearce

Degree Program: Biology

Undergraduate Degree: B.A.
Undergraduate Degree Requirements: 32 courses plus 3 January Projects (or 31 and 4), G.P. of 1.8 overall, 2.0 in concentration (including 8 courses)

COLLEGE OF MT. ST. VINCENT
(Includes Biology Department of Manhattan College)
Bronx, New York 10471 Phone: (212) 549-8000
Dean of College: Sr. B. Kennedy

BIOLOGY DEPARTMENT
Phone: Chairman: Ext. 247 Others: Exts. 245,246

Chairman: R.E. Beardsley
Professors: R.E. Beardsley, J. Kouba, U. Naf, T. Stonier, Br. J. Walton
Associate Professors: M. Lynch, Br. E. Quinn, Sr. K. Tracey
Assistant Professors: H. Koritz, Sr. M.E. Zipf
Instructors: G. Brick, S. Bruno, D. Duszinski
Visiting Lecturers: 3
Research Assistants: 1
Teaching Assistants: 4

Field Stations/Laboratories: Marine Biology Field Station-Nassau, Bahamas

Degree Programs: Biology, Microbiology

Undergraduate Degrees: B.A., B.S.
Undergraduate Degree Requirements: General Biology, General Chemistry, Organic Chemistry, Physics, Calculus, 24 credits in Advanced Biology courses including requirement in Genetics (4 credits) and either Physiology (4 credits) or Ecology (4 credits) plus 2 credits Senior Seminar

COLLEGE OF NEW ROCHELLE
New Rochelle, New York 10801 Phone: (914) 632-5300
Dean of College: K. Henderson

DEPARTMENT OF BIOLOGY
Phone: Chairman: Ext. 288 Others: 632-5300

Chairman: R.B. Reggio
Professors: M.G. Connell, Sr. Estelle Ghidoni
Associate Professors: R.B. Reggio
Assistant Professor: L.C. Rudolph

Degree Program: Biology

Undergraduate Degree: B.A.
Undergraduate Degree Requirements: A total of 120 credits to include: a minimum of 38 credits in Biology, one year of Mathematics, two years of Chemistry and one year of a modern foreign language.

COLLEGE OF SAINT ROSE
(No reply received)

COLLEGE OF WHITE PLAINS
White Plains, New York 10603 Phone: (914) 949-2950
Dean of College: J. O'Brien

DIVISION OF SCIENCE AND MATHEMATICS
Phone: Chairman: Ext. 38

Divisional Chairperson: Sr. M.C. Casciano
Professor: Sr. M.L. Cutler
Associate Professor: Sr. M. Eckert
Assistant Professors: Sr. M.C. Casciano, L. Haber, W. Jordan, Sr. M.C. Kelley, G.B. Villanueva, H.K. Wurf
Instructor: S. Merritt

Degree Programs: Biological Science (with majors in Botany, Human Life Studies, Ecology, Environmental Science, General Biology, Pre Medical, Zoology)

Undergraduate Degree: B.S.
Undergraduate Degree Requirements: Biology: 1 year of Introductory Biology, General Chemistry, Organic Chemistry, Physics, Mathematics, 30 credits of major level courses

COLUMBIA UNIVERSITY
116th Street and Broadway Phone: (212) 280-1754

New York, New York 10027
Dean of Graduate Studies: G.K. Fraenkel

Barnard College
606 West 120th Street Phone: (212) 280-1754
New York, New York 10027
Dean: DeR. Brewnig

DEPARTMENT OF BIOLOGICAL SCIENCES
Phone: Chairman: (212) 280-5103 Others: 280-3628

Chairman: D.D. Ritchie
Professors: P.L. Dudley, W.A. Corpe, D.D. Ritchie
Associate Professors: F.E. Warburton
Assistant Professors: M.D. Menon
Laboratory Director: E.L. Noback

Degree Programs: Biological Sciences

Undergraduate Degree: B.A.
Undergraduate Degree Requirements: 8 courses in Biology, including 5 laboratories, 2 courses in Physics and/or Geology, Chemistry through Organic - plus General Degree Requirements

Graduate Degree: Ph.D.

Columbia College
Dean: P. Pouncey

DEPARTMENT OF BIOLOGICAL SCIENCES
Phone: Chairman: (212) 280-2495 Others: 280-4581

Chairman: S. Beychok
William R. Kenan, Jr., Professor of Biology: C. Levinthal
Professors: S. Beychok, W.J. Bock, C.R. Cantor, W.A. Corpe, A. Conquist (Adjunct), J.E. Darnell, Jr. H.S. Irwin, Jr. (Adjunct), H. Levene, C. Levinthal, F.G. Lier, B. Maguire (Adjunct), C.T. Rogerson (Adjunct), G. Zubay
Associate Professors: E. Holtzman, A.L. Mancinelli
Assistant Professors: C. Bancroft, L.A. Chasin, B. Filner, J. Greer, E. Macagno, D.B. Mowshowitz, R.R. Sederoff, J.A. Wechsler
Visiting Lecturer: 1
Faculty Fellows: 36
Research Assistants: 20
Part-time Lecturers: 9
Postdoctorals and Research Associates: 18

Degree Programs: Biological Sciences, Biochemistry, Biophysics

Undergraduate Degree: B.A.
Undergraduate Degree Requirements: Standard Columbia College B.A. degree requirements plus 24 points of Biology courses including core program of General Biology, Genetics, Biochemistry, Cell Biology, one project laboratory, and additional courses at advanced undergraduate or at graduate level. Two years of Chemistry with laboratory, through Organic. One year each of Calculus and Physics.

Graduate Degrees: M.A., M. Phil., Ph.D. (Biological Sciences
Graduate Degree Requirements: Master: 2 Residence Units. Courses to ensure background in Genetics, Cell Biology, Biochemistry and Biophysics, Physiology and Neurobiology, and Population Biology, with proficiency at advanced level in 2 of them. Part I Qualifying Examination. One year college Mathematics, 8 hours teaching. Doctoral: 6 Residence Units. Reading knowledge of one modern foreign language (French or German), 16 hours teaching, 2 years College Mathematics, 3 Qualifying Examinations and course work to ensure satisfactory performance on them. Present and defend dissertation.

Graduate and Undergraduate Courses Available for Graduate Credit: Physiology, Plant Physiology, Plant Photobiology, Cell Biology-Neurons, Genetic Regulation in Cultured Cells, Biology of Microorganisms, Microbial Ecology, Biological and Biophysical Chemistry, Theory and Principles of Evolution, Introduction to Theoretical Population Biology, Comparative Morphology and Phylogeny of Vertebrates, Mathematical Methods in Biology, Biometrics, Biochemistry, Seminar in Cellular Physiology, Seminar in Nucleic Acids and Related Topics

College of Physicians and Surgeons
630 West 168th Street Phone: (212) 579-4011
New York, New York 10032
Dean: D.F. Tapley (Acting)

DEPARTMENT OF ANATOMY
Phone: Chairman: (201) 579-3447 Others: 579-3597

Chairman: C.R. Noback (Acting)
Professors: M.B. Carpenter, E.W. Dempsey, M.L. Moss, C.R. Noback
Associate Professors: F.J. Agate, P.W. Brandt, C.A. Ely, R.M. Hoar (Adjunct), L. M. Saletijn
Assistant Professors: E.W. April, W.G. Dilley, M.D. Felix (Adjunct), R.C. Henrikson, M.M.L. Lee, W.P. Luckett, G.P. Pereira, J.J. Rasweiler, S.C. Shen (Adjunct)
Visiting Lecturers: 2
Teaching Fellows: 6

Degree Programs: Anatomy, Anatomy and Histology, Anatomy and Physiology

Graduate Degree: Ph.D.
Graduate Degree Requirements: Doctoral: six residence units - adequate knowledge of one foreign language - oral defense of an original research project.

Graduate and Undergraduate Courses Available for Graduate Credit: Microscopic Anatomy, Systematic Human Anatomy, Structure and Function of the Nervous System, Developmental Biology and Related Specialized Courses (such as Cellular Biology, Oral Histology, et cetera).

DEPARTMENT OF BIOCHEMISTRY
Phone: Chairman: (212) 579-3669 Others: 579-4011

Chairman: P.R. Srinivasan (Acting)
Professors: R. Benesch, M. Eisenberg, P. Feigelson, A. Krasna, S. Lieberman, B. Low, M. Rapport, D. Sprinson, P. Srinivasan
Associate Professors: R.E. Benesch, A. Gold, D. Grunberger, J. Karkas
Assistant Professors: L. Abell (Adjunct), G. Alexander, E. Brunngraber, M. Feigelson, E. Collub (Adjunct), I.Goodman, R. Hanson, H. Meltzer, R. Moyer, T. Peters, T. Rosenberry, L. Skogerson, B. Weiss
Research Associates: M. Adlersberg, J. Dayan, H. Kaufman, S. Mahadik, U. Maitra, L. Ponticorvo, H. Preston, I. Radichevich, L. Rosen, A. Rudko, D. Srinivasan, J. Stavrianopoulos, H. Tamir, R. Weisner

Degree Program: Biochemistry

Graduate Degree: Ph.D.
Graduate Degree Requirements: Doctoral: six residence units - adequate knowledge of one foreign language - oral defense of an original research project - comprehensive examination in Biochemistry must be taken at end of third term.

Graduate and Undergraduate Courses Available for Graduate Credit: Chemistry and Metabolism of Cell Constituents, Biosynthesis of Nucleic Acids and Proteins, X-Ray Cyrstallography: Structural Chemistry and Biochemistry, Special Methods of Biochemistry, Biochemistry of Supramolecular Structure, Seminar in selected Topics in Biochemistry, Seminar in Biochemistry of the Hormones, General Biochemistry (U), Biochemistry of Nucleic Acids and Protein Synthesis (U), Introductory Biochemistry (Summer)

DEPARTMENT OF HUMAN GENETICS AND DEVELOPMENT
Phone: Chairman: (212) 579-3746 Others: 579-4011

Chairman: R.S. Krooth
Professors: K.C. Atwood, G. Jagiello, E.A. Kabat, R. S. Krooth, H. Levene, P.A. Marks, O.J. Miller, R.A. Rifkind, J.D. Rainer (Clinical Psychiatry), S. Spiegelman, S. Udenfriend (Adjunct), A. Weissbach (Adjunct), H. Weissbach (Adjunct)
Associate Professors: A. Bank, L. Erlenmeyer-Kimling
Assistant Professors: M. Bartalos (Adjunct), V. Dev, D. Kacian, F. Kramer, D. Dufe, G. Maniatis, D.A. Miller, D. Mills, M. Terada, D. Warburton
Instructor: H. Calvin
Visiting Lecturers: 5
Research Associates: 9
Staff Associates: 4

Degree Programs: Human Genetics and Development

Graduate Degrees: M.A., M. Phil., Ph.D.
Graduate Degree Requirements: Master: Only doctoral candidates are admitted; the M.A. must, however, be earned as a pre-requisite for the Ph.D. Successful completion of Introductory Clinical Genetics, Mammalian Genetics and Development, Methods in Human Genetics and Development, Seminar in Human Genetics and Development, 2 residence units. Doctoral: Successful completion of Departmental Qualifying Examination; attainment of M.A. and M. Phil. degrees, 6 residence units, preparation and defense of doctoral dissertation .

Graduate and Undergraduate Courses Available for Graduate Credit: Introductory Clinical Genetics, Mammalian Genetics and Development, Genetics of Nervous and Mental Disorders, Mammalian Meiosis, Abnormal Human Development, Methods in Human Genetics and Development, Advanced Immunochemistry, Techniques in Immunochemistry, Seminar in Genetics and Development, Clinical Genetics, Special Research

DEPARTMENT OF MICROBIOLOGY
Phone: Chairman: (212) 579-3647 Others: 579-2500

Chairman: H.S. Ginsberg
John E. Borne Professor of Medical and Surgical Research: H.S. Ginsberg
Professors: P.D. Ellner, G.F. Erlanger, C.L. Fox, Jr., H.S. Ginsberg, G.C. Godman, D.H. Harter, E.A. Kabat, W. Manski, C. Morgan, H. Rosenkranz, H.J. Vogel
Assistant Professors: J.E. Coward, D.L. Engelhardt, S.L. Morrison, A.C. Sampath, A. Shahidi (Adjunct), S.J. Silverstein, T. Tokumaru, S.M. Vratsanos, C.S. H. Young
Research Associates: 4
Senior Staff Associate: 1
Staff Associates: 3
Special Lecturers: 2

Degree Programs: Bacteriology and Immunology, Bacteriology, Immunology, Microbiology, Virology

Graduate Degrees: M.A., M Phil., Ph.D.
Graduate Degree Requirements: Master: Prerequisite to the M. Phil. and Ph.D. degrees and it will be awarded to candidates for the Ph.D. degree who have completed one academic year of advanced study and passed in preliminary examination. Master of Philosophy - courses listed below. Doctoral: After completion of all requirements for the M. Phil.degree, the student must prepare and successfully defend a dissertation.

Graduate and Undergraduate Courses Available for Graduate Credit: Courses required for M. Phil prerequisite to Ph.D. Microbiology of Bacterial and Mammalian Cells, Techniques of Microbiology of Bacterial and Mammalian Cells, Advanced Immunochemistry, Virology, Microbial Chemistry and Metabolism, Techniques of Clinical Bacteriology and Serology, Mycology, Special Topics in Cell Biology, Techniques in Cell Biology, Infectious Diseases, Seminar in Selected Topics in Microbiology, Research in Microbiology

DEPARTMENT OF PHYSIOLOGY
Phone: Chairman: (212) 579-3546 Others: 579-4011

Chairman: J.V. Taggart
Dalton Professor of Physiology: J.V. Taggart
Professors: S. Chien, E.R. Kandel, W.L. Nastuk, D. Schachter, J.H. Schwartz, W.A. Spencer, J.F. Taggart
Associate Professors: M. Blank, J.S. Britten, L.J. Cizek, R. Emmers, D. Halmagyi, A. Karlin, M.R. Nocenti, J.P. Reuben, P. Witkovsky
Assistant Professors: E.R. Batt, M. Ferin, J. Fischbarg, K. Jan, J.D. Koester

Degree Program: Physiology

Graduate Degrees: M.A., M. Phil., Ph.D.
Graduate Degree Requirements: Master: Not Stated
Doctoral: six residence units - adequate knowledge of one foreign language - oral defense of an original research project

Graduate and Undergraduate Courses Available for Graduate Credit: Introduction to Basic Principles, Human Physiology, Cell Physiology and Biophysics, Control Mechanisms in Physiology, Structure and Function of Sensory Systems

CONCORDIA COLLEGE

171 White Plains Road Phone: (914) 337-9300
Bronxville, New York 10708
Dean of College: Not Stated

DEPARTMENT OF BIOLOGY
Phone: Chairman: (914) 337-9300 Ext. 254

Chairman: E.M. Walle
Professor: C. Peterson
Associate Professors: S. Farelli (Adjunct), E.M. Walle
Assistant Professors: G. Fuhrman, S. Rosohke
Visiting Lecturer: 1

Degree Program: Biology

Undergraduate Degree: B.A.
Undergraduate Degree Requirements: 42 semester hours including Chemistry, 8 hours General Biology, 6 hours General Ecology, 3 hours, General Physiology 4 hours Comparative Vertebrate Morphology 4 hours History and Philosophy ofBiology 3 hours Electives to Total 42 hours

CORNELL UNIVERSITY

Ithaca, New York 14850 Phone: (607) 256-1000
Dean of Graduate Studies: W.W. Lambert

New York State College of Agriculture and Life Sciences

Dean (College of Agriculture and Life Sciences): W.K. Kennedy
Dean (College of Arts and Sciences): H.E. Levin

DEPARTMENT OF AGRONOMY
Phone: Chairman: (607) 256-5459 Others: 256-1000

Chairman: M.J. Wright
Charles Lathrop Pack Professor of Forest Soils: E.L. Stone
Professors: M. Alexander, W.H. Allaway,

D.R. Bouldin, M.G. Cline, B.E. Dethier, M. Drosdoff, D.J. Lathwell, E.R. Lemon, R.F. Lucey, H.A. MacDonald, R.D. Miller, R.B. Musgrave, W.D. Pardee, M. Peech, T.W. Scott, R.R. Seaney, E.L. Stone, M.J. Wright, M.T. Vittum, P.J. Zwerman
Associate Professors: R.W. Arnold, W.B. Duke, D.L. Grunes, J. Kubota, D.L. Linscott, R.L. Obendorf, G.W. Olson, W.S. Reid, F.N. Swader
Assistant Professors: J.M. Duxbury, G.W. Fick, R.H. Fox, W.R. Kanpp, W.W. Knapp, D.A. Paine, J.H. Peverly, R.M. Weaver
Research Assistants: 36
Teaching Assistants: 6
Research Associates: 6

Field Stations/Laboratories: Branch Agricultural Experiment Stations: 6 , Isle of Shoals

Degree Programs: Agronomy, Crop Science, Soil Science

Undergraduate Degree: B.S.
Undergraduate Degree Requirements: College requirements, plus 15 semester hours in Agronomy courses, plus one course in Plant Physiology is Basic; for "Science" emphasis, additional Mathematics, Chemistry and Physics. Atmospheric Science requirements: differ, not listed here.

Graduate Degrees: M.S., M.P.S., Ph.D.
Graduate Degree Requirements: Master: M.S. - 2 semesters residence; all others set by Special Committee M.P.S. - 2 semesters residence; 30 semester hours, Problem Analysis Doctoral: Ph.D. - 6 semesters residence; all others set by Special Committee

Graduate and Undergraduate Courses Available for Graduate Credit: Not Stated

SECTION OF BIOCHEMISTRY, MOLECULAR AND CELL BIOLOGY
Phone: Chairman: (607) 256-7757 Others: 256-2203

Chairman: J.L. Gaylor
Albert Einstein Chair: E. Racker
Greater Philadelphia Area Professor: Q.H. Gibson
Professors: G.G. Hammes (Joint with Chemistry), L.A. Heppel, G.P. Hess, D.E. McCormick (Joint with Nutrition), G. Schatz, L.D. Wright (Joint with Nutrition), R. Wu, D.B. Zilversmit (Joint with Nutrition)
Associate Professors: J.M Calvo, S.J. Edelstein, J.M. Fessenden-Raden, A.J. Gibson, E B. Keller, R.E. MacDonald, R.E. McCarty, A.L. Neal, D.B. Wilson
Assistant Professors: G.W. Feigenson, P.C. Hickle, V. Utermohlen-Lovelace, J.K. Moffat, J.I. Roberts
Instructors: R.M. Alexander, J.M. Griffiths, S. Jacobson
Research Assistants: 48
Postdoctoral Fellows: 34

Degree Programs: Agricultural Biochemistry, Biochemistry, Biochemistry and Biophysics, Biological Chemistry, Biophysics

Undergraduate Degree: B.S.
Undergraduate Degree Requirements: A grounding in Basic Science courses in Chemistry, Physics, and Mathematics. Breadth within the Biological Sciences is achieved by progressing from more introductory to specialized courses. Chemistry through Physical, Organic, and Biological are required.

Graduate Degrees: M.S., Ph.D.
Graduate Degree Requirements: Master: Not encouraged generally Doctoral: Major,Biological orChemical specialization. Additional courses in other areas for balance. No formal language requirement. Oral Admission to Ph.D. Degree Candidacy and thesis examinations are required, as well as seminars throughout course of study.

Graduate and Undergraduate Courses Available for Graduate Credit: Basic Biochemical Methods (U), Principles of Biochemistry, Lectures (U), Principles of Biochemistry, Individualized Instruction (U), Cell Biology, Ultrastructure (U), Cell Biology, Dynamics (U), Intermediate Biochemical Methods (U), Intermediate Biochemistry (U), Biochemistry of the Vitamins and Coenzymes (U), Advanced Biochemical Methods I, Advanced Biochemical Methods II, Basic and Applied Science Coordination Course in Biochemistry, Research Seminar in Biochemistry, Advanced Biochemistry, Biochemistry Seminar, Aspects of Metabolism (Plant Physiology), Vertebrate Biochemistry (Veterinary and Medicine)

DIVISION OF BIOLOGICAL SCIENCES
Chairman: 256-5042

Director: R.D. O'Brien
New York State Einstein Professor: E. Racker
Greater Philadelphia Area Professor: Q. Gibson
Professors: J.M. Anderson, H.P. Banks, A W. Blackler, T.J. Cade, R.T. Clausen, R.K. Clayton, L.C. Cole, T. Eisner, E.L. Gasteiger, J.L. Gaylor, Q.H. Gibson, P.W. Gilbert, L. Heppel, G.P. Hess, J.W. Hudson, A.T. Jagendorf, W.T. Keeton, J.M. Kingsbury, E.H. Lenneberg, G.E. Likens, W. McFarland, D.E. McCormick, W. Nelson, R.D. O'Brien, D. Paolillo, E. Racker, M. Salpeter, G. Schatz, A.M. Srb, H. Stinson, L. Uhler, B. Wallace, R. Whittaker, W. Wimsatt, R. Wu
Associate Professors: K. Adler, J.P. Barlow, J.M. Camhi, R.R. Capranica, W.C. Dilger, M. Eldefrawi, S. Emlen, J.F. Raden, G.R. Fink, A.J. Gibson, B.P. Halpern, E. Keller, S.A. Levin, R. MacDonald, R.J. MacIntyre, R. McCarty, A.L. Neal, T. Podelski, F. H. Pough, R. Spanswick, C. Uhl, D. Wilson, S. Zahler
Assistant Professors: K. Arms, P.J. Bruns, B. Chabot, P.J. Davies, R. Grossfeld, R. Hallberg, P.C. Hinkle, W.M. Howell, H. Howland, R. Hoy, P. Marks, J.K. Moffat, M. Parthasarathy
Lectures and Instructors: 15
Visiting Lecturers: 1
Teaching Fellows: 3
Research Assistants, Research Associates and Postdoctorals: 50

Field Stations/Laboratories: Appledore Island Field Station/Laboratory - Maine

Degree Programs: Biology, Ecology, Neurology, Zoology

Undergraduate Degrees: A B., B.S. (Biological Sciences)
Undergraduate Degree Requirements: 1 year Introductory Biology, 2 semesters Mathematics (Calculus), Organic Chemistry, Physics, Genetics, Biochemistry, plus about 24 hours more

Graduate Degrees: M.S., Ph.D. (Biological Sciences)
Graduate Degree Requirements: Master: Minimum of 2 residence units, thesis, oral examination Doctoral: 6 residence units, thesis, two oral examinations.

Graduate and Undergraduate Courses Available for Graduate Credit: Not Stated

BIOMETRICS UNIT IN THE DEPARTMENT OF PLANT BREEDING AND BIOMETRY
Phone: Chairman: (607) 256-3036

Sub-Chairman: W.T. Federer
Professors: W.T. Federer, D.S. Robson, S.R. Searle, F.B. Cady
Associate Professor: D.L. Solomon
Assistant Professor: C.L. Wood
Visiting Lecturers: 2-5
Research Assistants: 2-3
Visiting Research Professors: 3-4

Degree Programs: Biometry, Biostatistics

Undergraduate Degree: B.S.
Undergraduate Degree Requirements: 33-36 semester hours in Mathematics, Statistics and Computer Science, B.S. Thesis, one summer of practice related to Statistics

Graduate Degrees: M.S., Ph.D.
Graduate Degree Requirements: Master: 2 semesters of residence, M.S. Thesis Doctoral: 4 (beyond M.S.) semesters of residence, Ph.D. Thesis, Foreign language examination for Field of Statistics, English language examination for Field of Biometry

Graduate and Undergraduate Courses Available for Graduate Credit: Statistics and the World we Live In (U), Probability and Statistics I and II (U), Matrix Algebra (U), Statistical Methods I and II (U), Linear Models, Design and Analysis of Experiments I, II and III, Biological Sampling and Bioassay, Multivariate Analysis, Statistical Modelling, Statistical Genetics

DEPARTMENT OF ECOLOGY AND SYSTEMATICS
Phone: Chairman: (607) 256-4522

Chairman: S.A. Levin
Professors: T.J. Cade, L.C. Cole, J.W. Hudson, G.E. Likens, W.N. McFarland, L.D. Uhler, R.H. Whittaker
Associate Professors: J.P. Barlow, S.A. Levin, F.H. Pough
Assistant Professors: P.F. Brussard, B. Chabot, W.M. Howell, P.L. Marks
Research Assistants: 3
Visiting Assistant Professor: 1
Visiting Associate Professor: 1

Field Stations/Laboratories: Cayuga Lake Station - Meyers, New York

Degree Programs: Ecology, Systematics

Undergraduate Degree: B.S. (Biology)
Undergraduate Degree Requirements: a year of Biology, General Chemistry, College Mathematics and Physics, Organic Chemistry, Genetics, Biochemistry, 2 breadth courses, Concentration area courses (6 areas), Modern Language Qualification

DEPARTMENT OF ENTOMOLOGY
Phone: Chairman: (607) 256-3057

Chairman: E.H. Smith
Professors: C.O. Berg, J.L. Brann, Jr., W.L. Brown, Jr., J.E. Dewey, P.P. Feeny, J.G. Franclemont, G.G. Gyrisco, W.T. Johnson, J.G. Matthysse, R.A. Morse, A.A. Muka, R.L. Patton, L.L. Pechuman, D. Pimentel, E.H. Smith
Associate Professors: G.C. Eickwort, J.P. Kramer, E.M. Raffensperger, R.B. Root, J.L. Saunders, M. Semel, M.J. Tauber, C.F. Wilkinson, R.G. Young
Assistant Professors: E.W. Cupp, R.G. Helgesen, W.M. Tingey

Field Stations/Laboratories: Geneva Experiment Station - Geneva

Degree Program: Entomology

Undergraduate Degree: B.S.
Undergraduate Degree Requirements: 14 hours Chemistry, 8 hours Physics, 6 hours Foreign Language, 8 hours Biological Sciences, 5 hours Genetics, 3 hours Ecology, 3 hours Physiology, 4 hours Botany, 17 hours Entomology

Graduate Degrees: M.S., Ph.D.
Graduate Degree Requirements: Master: Final Examination, thesis, 2 units residence, program individually planned Doctoral: Thesis, language, 6 units residence, program individually planned

Graduate and Undergraduate Courses Available for Graduate Credit: General - Insect Biology, Techniques of Biological Literature Apiculture - Introductory Beekeeping, Biology of the Honey Bee, Environmental Entomology - Bionomics of Freshwater Invertebrates, Biological Control, Environmental Biology, Insect Ecology Field Course, Insect Behavior Seminar, Seminar in Aquatic Ecology Economic Entomology - Introductory Applied Entomology, Insect Pest Management, Arthropod Pests of World Importance, Special Topics in Economic Entomology Medical Entomology and Insect Pathology - Medical Entomology, Insect Pathology Taxonomy, Morphology and Acarology - Insect Morphology, Introductory Insect Taxonomy, Acarology, Taxonomy of the Smaller Orders of Insects, Taxonomy of the Immature Stages of Holometabola, Taxonomy of the Coleoptera and Lepidoptera, Taxonomy of the Diptera and Hymenoptera, Physiology, Biochemistry, and Insecticidal Chemistry, Insect Physiology, Insect Biochemistry, Insect Toxicology and Insecticidal Chemistry

DEPARTMENT OF FLORICULTURE AND ORNAMENTAL HORTICULTURE
Phone: Chairman: (607) 256-3044

Chairman: J.W. Boodley
Professors: A. Bing, J.W. Boodley, R.W. Langhans, J.G. Seeley, H.B. Tukey, Jr.
Associate Professors: M.I. Adleman, C.C. Fischer, R.T. Fox, G.L. Good, C.F. Gortzig, R.J. Lambert, A.S. Lieberman, R.G. Mower, R.J. Scannell (quarter time), E.F. Schaufler, P.L. Steponkus
Assistant Professors: J.E. Kaufmann, P.S. Tresch
Lecturers: R.L. Dwelle, A.M. Elliot
Teaching Fellow: 1
Research Assistants: 10

Field Stations/Laboratories: Ornamentals Research Laboratory - Farmingdale, New York

Degree Programs: Horticulture, Landscape Architecture, Ornamental Horticulture

Undergraduate Degree: B.S.
Undergraduate Degree Requirements: One hundred twenty credit hours, 4 semesters of Physical Education

Graduate Degrees: M.S., Ph.D.
Graduate Degree Requirements: Master: Minimum of 2 residence units. Full time study for one semester with satisfactory accomplishment constitutes one residence unit. Completion of all requirements is four years. Doctoral: Six residence units, at least four of which must be earned as full time student. Two of the last four in successive terms of full-time study on the Cornell campus. Completion of all requirements in seven years from date of first registration.

Graduate and Undergraduate Courses Available for Graduate Credit: Physiology of Horticultural Plants (U), Florist Crop Production (U), Woody Plant Materials (U), Greenhouse Production Management (U), Special Problems (U), Seminar

DEPARTMENT OF NATURAL RESOURCES
Phone: Chairman: (607) 256-2298 Others: 256-3257

Chairman: W.H. Everhart
Professors: W.H. Everhart, L.S. Hamilton, R.R. Morrow, D.A. Webster, F.E. Winch, Jr.
Associate Professors: R. Baer, Jr. (Courtesy), H.B. Brumsted, A. Dickson, A.W. Eipper (Courtesy), R.J. McNeil, A.N. Moen, R.T. Oglesby, M.E. Richmond (Courtesy), D.Q. Thompson (Courtesy), B.T. Wilkins
Assistant Professors: J.W. Kelley, J.B. Nickum (Courtesy), G.R. Reetz, W.D. Youngs
Visiting Lecturers: 1

Teaching Fellows: 23
Research Assistants: 28
Senior Research Associates: 4
Research Associates: 4

Field Stations/Laboratories: Biological Field Station at Oneida Lake, Bridgeport, New York, Arnot Teaching and Research Forest, Cayuga Lake, Ithaca, New York

Degree Programs: Wildlife Science, Fishery Science, Outdoor Recreation, Aquatic Science, Environmental Conservation

Undergraduate Degree: B.S.
Undergraduate Degree Requirements: 120 credits

Graduate Degrees: M.S., Ph.D.
Graduate Degree Requirements: Master: Candidates for the M.A., or M.S. normally take between one and two years of satisfactory full-time study to complete the degree requirements (the minimum is two units, semesters), Doctoral: Candidates for the Ph.D. degree normally take four or five years of satisfactory full-time study to complete the degree requirements (the minimum residence is six units, six semesters)

Graduate and Undergraduate Courses Available for Graduate Credit: Courses are taken as required by the students Graduate Committee.

SECTION OF NEUROBIOLOGY AND BEHAVIOR
Phone: Chairman: (607) 256-4517

Chairman: W.T. Keeton
Richard J. Schwartz Professor of Science and Society: R.S. Morison
Professors: T. Eisner, E. Gasteiger, P. Gilbert (Director Mote Marine Laboratory, Sarasota, Florida), W.T. Keeton, E. Lennebert, R. Morison, R.D. O'Brien (Director, Division of Biological Sciences), M. Salpeter, D. Tapper
Associate Professors: K. Adler, J. Camhi, R. Capranica, W. Dilger, M. Eldefrawi, S. Emlen, B. Halpern, H. Howland, T. Podleski
Assistant Professors: K. Arms, R. Buskirk, R. Hoy
Visiting Professors: 2

Field Stations/Laboratories: Isle of Shoals, New Hampshire and Mote Marine Laboratory - Sarasota, Florida

Degree Programs: Neurobiology and Behavior

Graduate Degrees: M.S., Ph.D.
Graduate Degree Requirements: Master: Final examination Thesis Doctoral: 2 semesters teaching, high proficiency in language other than English, Thesis, seminar on thesis, final examination

Graduate and Undergraduate Courses Available for Graduate Credit: Neurobiology and Behavior (U), Physiological Psychology Laboratory (U), Cellular Organization of the Nervous System (U), Elementary Neurophysiology (U), Behavioral Maturation (U), Principles of Neurobiology Laboratory (U), Comparative Vertebrate Ethology (U), Animal Communication (U), Brain and Behavior (U), Neuropharmacology (U), Sensory Function (U), Neurochemistry (U), Research in Neurobiology and Behavior, The Evolution of Social Systems (U), Behavioral Neurophysiology (U), Behavioral Neurophysical Laboratory (U), Bioelectrical Systems (U), Seminar in Neurology and Behavior (U), Biological Sciences, Biological Sciences Laboratory, Introductory Biology, Biological Discovery, Interactive Computing for Students of Biological Sciences, Biology for Mon-Majors, Biology and Society

DEPARTMENT OF PLANT BREEDING AND BIOMETRY
Phone: Chairman: (607) 256-2180 Others: 256-1000

Head: R.L. Plaisted
Professors: F.B. Cady, L.V. Crowder, W.T. Federer, C.O. Grogan, N.F. Jensen, C.C. Lowe, H.M. Munger, R.P. Murphy, W.D. Pardee, R.L. Plaisted, D.S. Robson, S.R. Searle, D.H. Wallace
Associate Professors: R.E. Anderson, D.L. Solomon
Assistant Professors: V.E. Gracen, C.L. Wood
Teaching Fellows: 6
Research Assistants: 17
Trainees: 9

Degree Programs: Agriculture, Plant Breeding, Biometrics,

Undergraduate Degree: B.S. (Agriculture)
Undergraduate Degree Requirements: 120 hours credit, cumulative grade average C- or better, Final term grade average C- or better, 45 hours distributed between Physical, Biological, and Social Sciences, 4 terms of Physical Education, Minimum of 45 hours of electives in College of Agriculture and Life Sciences

Graduate Degrees: M.S., Ph.D.
Graduate Degree Requirements: Master: 2 residence units, thesis, one oral examination Doctoral: 6 residence units, thesis, two oral examinations

Graduate and Undergraduate Courses Available for Graduate Credit: Methods of Plant Breeding, Physiological Genetics of Crop Plants, Research Orientation, Experimental Methods, Advanced Topics in Plant Genetics and Breeding, Special Problems in Research and Teaching, Probability and Statistics (U), Stochastic Models in Biology (U), Deterministic Models in Biology (U), Matrix Algebra (U), Biometry Seminar, Statistical Methods I, II, Design ofExperiment, Linear Models

DEPARTMENT OF PLANT PATHOLOGY
Phone: Chairman: (607) 256-3245

Chairman: D.R. Bateman
Professors: D.F. Bateman, C.W. Boothroyd, R.C. Cetas, R.S. Dickey, R.P. Korf, J.W. Lorbeer, W.F. Mai, R.L. Miller, A.F. Sherf, H.D. Thurston, M. Zaitlin
Associate Professors: M.B. Harrison, R.K. Horst, E.D. Jones, O.E. Schultz, W.A. Sinclair, R.E. Wilkinson, C.E. Williamson
Assistant Professors: J.R. Aist, P.A. Arneson, S.V. Beer, W.E. Fry, H.D. Van Etten, O.C. Yoder
Research Associates: 4

Field Stations/Laboratories: Nematode Research Laboratory Farmingdale, New York, Long Island Vegetable Research Farm, Riverhead, New York, Uihlein Potato Research Farm - Lake Placid, New York, Golden Nematode Research Farm - Prattsburgh, New York, Orange County Research Laboratory - Florida, New York

Degree Programs: Plant Pathology, Mycology

Undergraduate Degree: B.S. (Agriculture)
Undergraduate Degree Requirements: Eight terms of residence plus 120 credit hours with an accumulated grade of C- (1.7) or above.

Graduate Degrees: M.P.S. (Agriculture), M.S., Ph.D.
Graduate Degree Requirements: Master: Two terms of residence, an approved thesis, and satisfaction of the student's graduate committee consisting of a minimum of two faculty members, one representing a major and the other a minor field of study. Doctoral: Six terms of residence, an approved thesis, and satisfaction of the student's graduate committee consisting of a minimem of three faculty members, one representing a major and the other two representing minor fields of study

Graduate and Undergraduate Courses Available for Graduate Credit: General Plant Pathology (U), Comparative Morphology of Fungi (U), Advanced Plant Pathology, Principles of Plant Disease Control, Plant Virology, Plant Nematology, Bacterial Plant Pathology, Disease and Pathogen Physiology, Experimental Methods in Plant Pathology, Special Problems in Mycology or Plant Pathology, Philosophy of Plant Pathology, Advanced Plant Nematology, Advanced Mycology, Taxonomy of Fungi, Plant Pathology Colloquim

DEPARTMENT OF POMOLOGY
Phone: Chairman: (607) 256-5438

Head: L.J. Edgerton
Professors: L.J. Edgerton, L.E. Powell, F.B. Sands, R. M. Smock
Associate Professors: G.D. Blanpied, L.L. Creasy, G. H. Oberly, J.P. Tomkins
Assistant Professor: D.C. Elfving
Research Assistants: 4
Visiting Professor: 1

Degree Program: Pomology

Undergraduate Degree: B.S.
Undergraduate Degree Requirements: Candidates for the degree of Bachelor of Science normally must be in residence for eight terms and earn 120 credits with a cumulative grade average of C- (1.7) or above, and a grade average of C- or above in the last term.

Graduate Degrees: M.S., Ph.D.
Graduate Degree Requirements: Master: A minimum of two terms of residence beyond the Bachelor's degree, with course and thesis requirements determined by the student's special committee. Language requirements generally determined by the Field. Doctoral: A minimum of six terms of residence beyond the Bachelor's degree, with course and thesis requirements determined by the student's special committee. Language requirements generally determined by the Field.

Graduate and Undergraduate Courses Available for Graduate Credit: Postharvest Physiology, Handling and Storage of Fruits (U), Advanced Laboratory Course (U), Economic Fruits of the World (U), Advanced Pomology, Special Topics in Experimental Pomology, Research, Growth and Development of Woody Plants

DEPARTMENT OF POULTRY SCIENCE
Phone: Chairman: (607) 256-3115

Chairman: Not Stated
Professors: R.C. Baker (Adjunct), M.L. Scott, A. van Tienhoven, R.J. Young (Teaching and Research), C.E. Ostrander (Extension), E.A. Schano (Extension), R.K. Cole, F.B. Hutt, A.L. Romanoff (Emeritus)
Associate Professors: A. Bensadoun (Adjunct), S.E. Bloom (Teaching and Research), O.F. Johndrew, Jr. G.H. Thacker (Extension)
Assistant Professors: R.E. Austic, H.F. Brotman, J.M. Regenstein (Teaching and Research)
Research Assistants; 20
Post-Doctorals: 2

Field Stations/Laboratories: Cornell University Duck Research Laboratory - Box 217, Old Country Road, Eastport, Long Island, New York 11941

Degree Programs: Nutrition, Physiology, Genetics and Animal Breeding, Zoology, Food Science

Undergraduate Degree: B.S.A. (Animal Science)
Undergraduate Degree Requirements: Minimum of 120 hours including - 12 hours Physical Science, 12 hours Biological Sciences, 15 hours Humanities, 55 hours Agricultural and Statutory Electives, 26 hours Electives, 120 hours total

Graduate Degrees: M.S., Ph.D.
Graduate Degree Requirements: Master: 2 resident units and submission of an approved thesis. Doctoral: 6 resident units and submission of an approved thesis

Graduate and Undergraduate Courses Available for Graduate Credit: The Graduate School does not require specific courses or course credit. Graduate students can take any course, both undergraduate and graduate are advised by their committee as to the recommended courses based on their background and experience.

DEPARTMENT OF VEGETABLE CROPS
Phone: Chairman: (607) 256-5401

Chairman: D.H. Wallace (Acting)
Professors: S.L. Dallyn, E.E. Ewing, F.M.R. Isenberg, E.C. Kelly, P.A. Minges, H.M. Munger, J.L. Ozbun, R. Sheldrake, Jr., R.D. Sweet, D.H. Wallace, B.T. Whatley (Adjunct)
Associate Professors: P.L. Minotti, R.F. Sandsted, L.D. Topoleski
Assistant Professor: P.E. Brecht
Research Assistants: 17
Research Associates: 5
Extension Specialists: 1

Field Stations/Laboratories: Long Island Vegetable Research Farm - Riverhead, New York

New York State Veterinary College
Dean: G.C. Poppensiek

DEPARTMENT OF ANATOMY
Phone: Chairman: (607) 256-3161

Head: R.E. Habel
Professors: H.E. Evans, A. de Lahunta, R.E. Habel, W.O. Sack
Associate Professor: J.F. Cummings
Teaching Assistants: 3
Visiting Lecturers: 2

Field Stations/Laboratories: Snyder Hill and Physical Biology, Ithaca, Poultry and Mastitis Laboratories - Kingston, Eastport, Canton, Earlville

Degree Programs: Anatomy and Histology, Veterinary Anatomy

Graduate Degrees: M.S., Ph.D.
Graduate Degree Requirements: Master: Minimum residence: two semesters; courses as required by the Special Committee; thesis Doctoral: minimum residence, six semesters, courses as required by the Special Committee, thesis

Graduate and Undergraduate Courses Available for Graduate Credit: (All courses pertain to domestic animals.) Gross Anatomy (U), Developmental and Microscopic Anatomy (U), Neruoanatomy (U), Applied Anatomy (U), Advanced Anatomy, Advanced Clinical Neurology, Comparative Anatomy of the Digestive System, Vertebrate Morphology (Last two courses include many other species).

DEPARTMENT OF AVIAN DISEASES
Phone: Chairman: (607) 256-5449

Chairman: S.P. Hitchner
Professors: B.W. Calnek, J. Fabricant, M.C. Peckham
Associate Professor: L. Leibovitz
Research Associate: B. Cowen

Field Stations/Laboratories: Regional Poultry Disease Laboratory, Kingston, New York, Duck Research Laboratory - Easport, Long Island, New York

Degree Programs: No degrees are offered by the Department; M.S. and Ph.D. degrees are granted in cooperation with the Cornell Graduate School

DEPARTMENT OF VETERINARY MICROBIOLOGY
(No reply received)

DEPARTMENT OF VETERINARY PATHOLOGY
Phone: Chairman: (607) 256-5409 Others: 256-5454

Chairman: L. Krook (Acting)
Professors: C.I. Boyer, Jr., L. Coggins, J.R. Georgi, L. Krook, F. Noronha, C.G. Rickard, L. Saunders (Adjunct), J.H. Whitlock, K.E. Wolf (Adjunct)
Associate Professors: E.J. Andrews, J.M. King, J.N. Shively
Assistant Professor: G.V. Lesser (Adjunct)
Senior Research Associates: E. Dougherty III, C.L. Gries, M.J. Kemen, J.E. Post, F.E. Waterman
Visiting Professors: 2

Degree Programs: Veterinary Pathology, Parasitology, Physiology

Graduate Degrees: M.S., Ph.D.
Graduate Degree Requirements: Master: minimum residence, two semesters, courses as required by the Special Committee, thesis Doctoral: minimum residence, six semesters, courses as required by the Special Committee, thesis

Graduate and Undergraduate Courses Available for Graduate Credit: Introductory Parasitology and Symbiology, General Pathology, Special Pathology, Basic Parasitology, Applied Parasitology, Wildlife Pathology, Postmortem Pathology, Fish Pathology, Pathology of Nutritional Diseases, Advanced Work in Animal Parasitology, Laboratory Methods of Diagnosis, Advanced Work in Pathology, Reproductive Pathology, Ultrastructural Pathology

DEPARTMENT OF VETERINARY PHYSIOLOGY, BIOCHEMISTRY AND PHARMACOLOGY
Phone: Chairman: (607) 256-2121

Chairman: C.E. Stevens
Professors: A.L. Aronson, E.N. Bergman, A. Dobson, T.R. Houpt, A.F. Sellers, C.E. Stevens, J.F. Wootton
Associate Professor: W.J. Arion
Assistant Professor: W.S. Schwark
Research Assistants: 5

Field Stations/Laboratories: Veterinary Virus Institutė - Ithaca, Poultry Disease Research - Ithaca, New York State Mastitus Control Program - Ithaca, Canton, Erlville and Kingston

Degree Programs: Physiology, Pharmacology, Biochemistry

Graduate Degrees: M.Sc., Ph.D.
Graduate Degree Requirements: Master: Dependent upon Graduate Committee Doctoral: Dependent upon Graduate Committee

Graduate and Undergraduate Courses Available for Graduate Credit: Vertebrate Biochemistry (U), Physiology for Veterinary Students (2 semesters), (U), Principles of Pharmacology, Toxicology and Anesthesiology (U), Clinical Pharmacology (U), Veterinary Animal Behavior (U), Methods in Gastroenterological Research, Comparative Gastroenterology, Physiological Disposition of Drugs and Poisons (U), Special Projects in Physiology

Division of Nutritional Sciences
Director: M.C. Nesheim

DIVISION OF NUTRITIONAL SCIENCES
Phone: Chairman: (607) 256-2229

Head: M.C. Nesheim
James Jamison Professor of Nutrition: R.H. Barnes
Babcock Chair of Food Economics: Vacant
American Heart Association Career Investigator Award: D.B. Zilversmit
Professors: R.H. Barnes, J.L. Gaylor, M.C. Latham, D.B. McCormick, M.A. Morrison, M.C. Nesheim, J.M. Rivers, M. Washbon, L.D. Wright, D.B. Zilversmit
Associate Professors: G. Armbruster, A. Bensadoun, M. Devine, R. Klippstein, B. Lewis, N. Mondy, D.A. Roe, D. Sanjur, R. Schwartz, K. Visnyei
Assistant Professors: J. Bowering, K. Clancy-Hepburn D.A. Levitsky
Instructor: M. Pimentel
Extension Associates: 8

Degree Programs: Nutrition, Joint Faculty supervise Ph.D. in Biochemistry, Physiology, Food Economics, Food Science, Pathology, Animal Science, Poultry Science

Undergraduate Degree: B.S.
Undergraduate Degree Requirements: Basic science preparation required in Chemistry, Biology, Mathematics, Organic Chemistry, Physiology and Biochemistry and a concentration of professional courses in Nutrition or Food or Community Nutrition

Graduate Degrees: M.S., Ph.D.
Graduate Degree Requirements: Master: Completion of work in one major and one minor subject Doctoral: A major and two minors must be satisfied.

Graduate and Undergraduate Courses Available for Graduate Credit: Ecology of Human Nutrition and Food, Introductory Foods, Maternal and Child Nutrition, Physiological Basis of Human Nutrition, Introduction to Physiochemical Aspects of Food, Human Nutrition, Nutrition and Disease, Community Nutrition and Health, Physiochemical Aspects of Food, Experimental Food Methods, Proteins and Amino Acids, Lipids and Carbohydrates, Nutritional Energetics, Minerals and Vitamins, Carbohydrate Chemistry, Nutrition and Growth, Food Supply and Human Nutrition, Nutrition and the Chemical Environment, International Nutrition Problems, Policy, and Programs.

New York State Agricultural Experiment Station
Geneva, New York 14456 Phone: (315) 787-2211
Dean: W.K. Kennedy

DEPARTMENT OF ENTOMOLOGY
Phone: Chairman: 787-2321 Others: 787-2323

Head: E.H. Glass
Professors: P.J. Chapman (Emeritus), E.H. Glass, S.E. Lienk, G.A. Schaefers, E.P. Taschenberg, H. Tashiro
Associate Professors: W.S.Bowers, A.C. Davis, R.J. Kuhr, W.L. Roelofs, K. Trammel
Assistant Professors: C.J. Eckenrode, B.J. Fiori, W.H. Reissig
Research Associates: C.M. Splittstoesser, R.W. Straub R.W. Weires, Jr.
Research Assistants: 26
Postdoctoral Research Associates: 9

Field Stations/Laboratories: Hudson Valley Laboratory - Highland, New York, Vineyard Laboratory - Fredonia, New York

Degree Program: Entomology

Graduate Degrees: M.S., Ph.D.
Graduate Degree Requirements: Master: Established by Cornell University Graduate School and Field of Entomology Doctoral: Established by Cornell University Graduate School and Field of Entomology

Graduate and Undergraduate Courses Available for Graduate Credit: No formal courses offered, graduate students obtain course instruction at main Cornell University campus at Ithaca. Our faculty provide thesis research direction for a limited number of students.

DEPARTMENT OF FOOD SCIENCE AND TECHNOLOGY

Phone: Chairman: (315) 787-2254 Others: 787-2255

Chairman: W.B. Robinson
Professors: M.C. Bourne, R.L. LaBelle, L.H. Massey, L.R. Mattick, J.C. Moyer, W.B.Robinson, R.S. Shallenberger, O.F. Splittstoesser, K.K. Steinkraus, J.P. Van Buren
Associate Professors: T.E. Acree, J.B. Bourke, D.L. Downing, L.R. Kackler, G. Hrazdina, C.Y. Lee, J.R. Stamer, G.S. Stoewsand
Assistant Professors: M.A. Rao, R.H. Walter
Research Associate: Y.D. Hang

Degree Programs: Food Chemistry, Food Microbiology, Food Science (General), International Food Development

Graduate Degrees: M.S., Ph.D.
Graduate Degree Requirements: Master: Minimum: 2 units (2 semesters) and thesis Doctoral: Minimum: 6 units (6 semesters) and thesis

Graduate and Undergraduate Courses Available for Graduate Credit: Thesis only (courses offered on Cornell Campus at Ithaca).

DEPARTMENT OF POMOLOGY AND VITICULTURE
Phone: Chairman: (315) 787-2231 Others: 787-2011

Head: W.J. Kender
Professors: C.F. Forshey, N.J. Shaulis, R.D. Way
Associate Professors: J.N. Cummins, O.F. Curtis, Jr., W.J. Kender, R.C. Lamb, D.K. Ourecky
Assistant Professor: A.N. Lakso
Research Associates: 4
Visiting Scientists: 2

Field Stations/Laboratories: Vineyard Laboratory - Fredonia, New York, Hudson Valley Laboratory - Highland, New York

Degree Program: Pomology

Graduate Degrees: M.S., Ph.D.
Graduate Degree Requirements: Master: Thesis and final oral examination. Doctoral: Thesis and 2 examinations administered by a Special Committee (comprehensive admission to candidacy examination and final examination in defense of the thesis). There is no specific foreign language requirement.

Graduate and Undergraduate Courses Available for Graduate Credit: Courses offered on Ithaca Campus.

DEPARTMENT OF PLANT PATHOLOGY
Phone: Chairman: (315) 787-2326

Head: J.E. Hunter
Professors: A.J. Bram, R.M. Gilmer, M. Szkolnik
Associate Professors: J.D. Gilpartick, J.K. Uzemoto, R. Prouvidenti
Assistant Professors: G.S. Aborvi, H.S. Aldurickle, F.J. Polach, R.C. Larson
Research Assistants: 13

Field Stations/Laboratories: Hudson Valley Research Laboratories - Highland, New York, Vineyard Laboratory - Fredonia, New York

Degree Program: Plant Pathology

Graduate Degrees: M.S., Ph.D.
Graduate Degree Requirements: Master: Thesis and final oral examination. Doctoral: Thesis and 2 examinations administered by a Special Committee (Comprehensive admission to candidacy examination and final examination in defense of the thesis.) There is no specific foreign language requirement.

Graduate and Undergraduate Courses Available for Graduate Credit: All courses taken in the sister department at at the Ithaca campus. We are a research department and students may only do their thesis research at this location.

DEPARTMENT OF SEED AND VEGETABLE SCIENCES
Phone: Chairman: (315) 787-2217

Chairman: M.T. Vittum
Professors: B.E. Clark, G.A. Marx, L.W. Nittler, M.T. Vittum
Associate Professors: M.H. Dickson, A.A. Khan, N.H. Peck, R.W. Robinson, S. Shannon
Assistant Professor: G.E. Harman

Degree Programs: This is a research department not offering degrees

Cornell Medical College
1300 York Avenue Phone: (212) 472-5670
New York, New York 10021
Dean of Graduate Studies: T.H. Meikle, Jr.
Dean of Medical College: J.R. Buchanan

DEPARTMENT OF BIOCHEMISTRY
Phone: Chairman: (212) 472-6212, 6313

Chairman: A. Meister
Israel Rogosin Professor of Biochemistry: A. Meister
Professors: B.L. Horecker (Adjunct), A.S. Posner, J.R. Rachele, A.L. Rubin (Surgery)
Associate Professors: R.W. Bonsnes, E.M. Breslow, J. Goldstein, R.H. Haschemeyer, C. Ressler, K.H. Stenzel (Surgery), D. Wellner, K.R. Woods
Assistant Professors: J.S. Cheigh (Surgery), G.W. Dietz, Jr., G.F. Fairclough, Jr., H Gilder (Surgery), R.R. Riggio (Surgery), W.B. Rowe, E.T. Schubert, J.F. Sullivan (Surgery), S.S. Tate
Instructor: L. Tapia (Surgery)

Degree Program: Biochemistry

Graduate Degrees: M.S., Ph.D.
Graduate Degree Requirements: Master: Same as in Doctoral except minimal residence of one year and one modern foreign landuage Doctoral: For Admission: Comprehensive background in Chemistry, evidence of knowledge of Biology, general experimental Physics and Mathematics, Graduate Record Examinations. For Award of Degree: Minimal residence of three years admission-to-candidacy examinations, thesis, final examination, (defense of thesis), proficiency in two modern foreign languages, or one modern foreign language and proficiency in computer science language

Graduate and Undergraduate Courses Available for Graduate Credit: Basic Medical Biochemistry, General Biochemistry (Biochemistry 2A), General Biochemistry (Biochemistry 2B), Advanced Biochemistry, Biochemical Preparations and Techniques, Research in Biochemistry and Elective Options in Biochemical Research during fourth year

DEPARTMENT OF BIOLOGICAL STRUCTURE AND CELL BIOLOGY
(No reply received)

DEPARTMENT OF BIOMATHEMATICS
Phone: Chairman (212) 472-6959 Others: 472-5670

Field Representative: S.I.Rubinow
Professors: S.I. Rubinow
Visiting Professors: 2
Research Assistant: 1

Degree Program: Biomathematics

Graduate Degrees: M.S., Ph.D.
Graduate Degree Requirements: Master: Register for study in one major and one or more minor field. Courses determined by committee. Residence - 2

units, Final Examination: oral or oral and written, Thesis Doctoral: Register for study in one major and one or two minor fields. Residence: 6 units, Final Examination: oral and written, Thesis

Graduate and Undergraduate Courses Available for Graduate Credit: Introductory Biomathematics, Biomathematics Seminar, Special Topics

DEPARTMENT OF GENETICS
(No reply received)

DEPARTMENT OF MICROBIOLOGY
Phone: Chairman: (212) 472-6540

Chairman: W.F. Scherer
Professors: W.M. O'Leary, W.F. Scherer
Associate Professors: R.W. Dickerman, L.B. Senterfit, D.H. Sussdorf
Assistant Professors: J.L. Beebe, Z.P. Harsanyi, M.J. Lyons, M.E. Wiebe
Teaching Assistants: 13

Degree Programs: Bacteriology, Microbial Chemistry and Physiology, Microbial Genetics, Immunology, Virology

Graduate Degree: Ph.D.
Graduate Degree Requirements: Doctoral: Individual program selected by Committee, 2 foreign languages, admission to candidacy examination, Thesis, oral examination

Graduate and Undergraduate Courses Available for Graduate Credit: General Microbiology, Microbiology and an Introduction to Infectious Diseases, Advanced Diagnostic Microbiology, Microbial Chemistry and Physiology, Advanced Microbial Genetics, Advanced Immunology, Advanced Virology, Microbial Ecology, Methods and Materials of Research, History and Philosophy of Science

DEPARTMENT OF PHYSIOLOGY
Phone: Chairman: (212) 472-5229

Chairman: E.E. Windhager
Professors: R.F. Pitts, E.E. Windhager, R.L. Greif, B. Grafstein
Associate Professors: T.M. Maack, S.B. Baruch, C. Fell
Assistant Professors: O. Andersen, C. Liebow, D. Gardner
Visiting Lecturers: 20
Teaching Fellows: 5
Research Assistants: 7

Degree Program: Physiology

Graduate Degrees: M.S., Ph.D.
Graduate Degree Requirements: Master: Thesis Doctoral: Thesis

Graduate and Undergraduate Courses Available for Graduate Credit: Courses in Medical Physiology, Electives in Physiology

DOWLING COLLEGE
Oakdale, New York 11769 Phone: (516) 589-6100
Dean of College: R.H. Krupp

BIOLOGY DEPARTMENT
Phone: Chairman: (516) 589-6100

Coordinator: M.A. Kamran
Professor: R.Z. Brown
Associate Professor: M.A. Kamran, H.W. Moeller
Assistant Professor: S.J. Shafer

Degree Program: Biology

Undergraduate Degree: B.A.
Undergraduate Degree Requirements: 32-38 hours Biology, 18 hours Chemistry, 10 hours Physics, 7 hours Mathematics, 120 hours all subjects

D'YOUVILLE COLLEGE
320 Porter Avenue Phone: (716) 886-8100
Buffalo, New York 14201
Dean of College: Sr. R. P. Smith

BIOLOGY DEPARTMENT
Phone: Chairman: Ext. 207

Chairman: M. Cohen
Associate Professors: H. Dandit, P. Costisick
Assistant Professor: M. Cohen
Instructor: F. Griffin
Visiting Lecturers: 2

Degree Programs: Biology, Medical Technology

Undergraduate Degrees: B.A., B.S.
Undergraduate Degree Requirements: Not Stated

EISENHOWER COLLEGE
Seneca Falls, New York 13148 Phone: (315) 568-7361
Dean of College: W. Hickman

DIVISION OF SCIENCE AND MATHEMATICS
Phone: Chairman: (315) 568-7361

Division Director: G.L. Miller
Professor: A.A. McAuley
Associate Professors: P.C. Curtis, G.L. Miller

Degree Programs: Biology, Environmental Studies

Undergraduate Degree: B.A.
Undergraduate Degree Requirements: nine semester courses in Biology, two semesters Organic Chemistry, two semesters Physics, two semesters Calculus

ELMIRA COLLEGE
Elmira, New York 14901 Phone: (607) 734-3911
Dean of College: G.H. Spremulli

(Elmira College does not have departments or divisions)

Professors: L. Potter, R. Whitney
Associate Professors: D. Foster, W. Lindsay
Assistant Professor: R. Neumann

Degree Programs: Biochemistry, Biology, Medical Technology

Undergraduate Degree: B.S.
Undergraduate Degree Requirements: Not Stated

FORDHAM UNIVERSITY
(No reply received)

HAMILTON COLLEGE
Clinton, New York 13323 Phone: (315) 859-4221
Dean of College: D. Lindley (Acting)

DEPARTMENT OF BIOLOGY
Phone: Chairman: (315) 859-4221

Chairman: N.J. Gerold
Stone Professor of Biology: N.J. Gerold
Professors: A.D. Chiquoine, J. Ellis, N.J. Gerold, L. McManus
Assistant Professor: J. Bland

Degree Program: Biology

Undergraduate Degree: B.A.
Undergraduate Degree Requirements: 8 courses in Biology, 2 courses in Physics, 4 courses in Chemistry

HARTWICK COLLEGE

Oneonta, New York 13820 Phone: (607) 432-4200

DEPARTMENT OF BIOLOGY
Phone: Chairman: Ext. 236 Others: Ext. 231

Chairman: E.E. Deubler, Jr.
Professors: E.E. Deubler, Jr., R.M. Stagg
Associate Professors: C. Bocher, D. Hutchinson, B. Masters, M.R. Segina, R. Smith
Assistant Professors: W. Elliott, W. Kruczynski

Field Stations/Laboratories: Pine Lakes Field Station at Pine Lake Campus of Hartwick College, Through participation in the Cellege Center of the Finger Lakes we also have use of the Finger Lakes Institute on Seneca Lake, Watkins Glen, New York

Degree Programs: Biology, Medical Technology

Undergraduate Degrees: B.A., B.S.
Undergraduate Degree Requirements: 38 credits in Biology which must include: Cell Physiology, Genetics, Embryology, Ecology, Plant Science (General Botany or other), Senior Seminar, Independent Study, Chemistry plus 16 hours Physics, 8 hours Mathematics, 8 hours each major must also take the GRE and Department Comprehensive Examination their senior year.

HERBERT H. LEHMAN COLLEGE

(see under City University of New York)

HOBART AND WILLIAM SMITH COLLEGES

Geneva, New York 14456 Phone: (315) 789-5500
Dean of College: R.A. Skotheim

DEPARTMENT OF BIOLOGY
Phone: Chairman: Ext. 261

Chairman: R.A. Ryan
Professors: R.A. Ryan, L.F. Nellis
Associate Professors: D.A. Sleeper, E.T.Wolff
Assistant Professors: T.J. Glover, J.T. Kerlan

Degree Program: Biology

Undergraduate Degrees: B.A., B.S.
Undergraduate Degree Requirements: 10 courses in Biology, 1 year Organic Chemistry, 1 year Mathematics (Must include Calculus)

HOFSTRA UNIVERSITY

Hempstead Turnpike Phone: (516) 560-3261
Hempstead, New York 11550
Dean of Graduate Studies: H. Lichtenstein
Dean of College: Adkinson (Acting)

DEPARTMENT OF BIOLOGY
Phone: Chairman: (516) 560-3264 Others: 560-3261

Chairman: I. Galinsky
Professor: C.T. Wemyss
Associate Professors: R. McCourt, E. Kaplan, J.W. Schneiweiss, C. Phillips
Assistant Professors: R. Johnson, K. Erb, G. Grimes, C. Prover, E. Rosen, C. Saladino, E. Snoek, R. Stillwell, R. Belkin
Instructor: B. Wilder
Teaching Internes: 2

Degree Programs: Biology, Natural Science

Undergraduate Degree: B.A.
Undergraduate Degree Requirements: 124 semester hours with cum of 2.0, 94 semester hours must be in liberal arts. The last 30 semester hours including at least 15 in the field of specialization, must oridiarily be completed in residence at Hofstra.

Graduate Degree: M.A.
Graduate Degree Requirements: Master: B.A. in Biology, Any undergraduate deficiencies (including Physics Mathematics and Organic Chemistry) must be make up before the completion of 15 semester hours of graduate work. Candidates must complete 30 semester hours of graduate work if thesis course 301-302 is taken, otherwise 33 semester hours.

Graduate and Undergraduate Courses Available for Graduate Credit: Not Stated

HOUGHTON COLLEGE

Houghton, New York 14744 Phone: (716) 567-2211
Dean of College: F. Shannon

DEPARTMENT OF BIOLOGY
Phone: Chairman: Ext. 239

Head: D.W. Munro
Professors: A.M. Whiting
Associate Professor: J.K. Moody
Assistant Professors: J.K. Boon, E.E. Cook

Degree Program: Biology

Undergraduate Degrees: B.S., B.A.
Undergraduate Degree Requirements: B.S. - General Biology, Genetics, Seminar, Advanced Biology, General Chemistry, Organic Chemistry, General Physics and Calculus B.A. - Same as above except Organic Chemistry, Physics, and Calculus are not required.

HUNTER COLLEGE

(see Under City University of New York)

IONA COLLEGE

(No reply received)

ITHACA COLLEGE

Ithaca, New York 14850 Phone: (607& 274-3011
Dean of Graduate Studies: S. Schneeweis
Dean of Humanities and Sciences: T.S. Baker

DEPARTMENT OF BIOLOGY
Phone: Chairman: (607& 274-3161 Others: 274-3011

Chairman: J.M. Bernard
Professors: J.M. Bernard, L.E. DeLanney, H.C. Yingling, Jr.
Associate Professors: M.I. Brammer, I.A. Tamas, S.R. Thompson, R.S. Wodzinski
Assistant Professors: J.L. Confer, L.A. Schmieder, G. Swenson
Part-time Assistant Professor: 1
Part-time Instructor: 1

Degree Program: Biology

Undergraduate Degree: B.A.
Undergraduate Degree Requirements: 120 credit hours: 37 hours of Biology, 17 hours of Chemistry, 8 hours of Physics, 41-44 hours outside the natural sciences

KEUKA COLLEGE

Keuka Park, New York 14478 Phone: (315) 536-4411
Dean of College: W. Odom

DEPARTMENT OF BIOLOGY
Phone: Chairman: Ext. 268 Others: Ext. 268, 288

Chairman: J.E. White
Professors: E.E. Webber, J.E. White
Associate Professor: D.B. Prest
Assistant Professor: R.D. Everson

Degree Program: Biology

Undergraduate Degree: B.A.
Undergraduate Degree Requirements: Animal Form and Function, Plant Form and Function, Cell Biology, Physiology, Chemistry, Eight elective units in Biology

THE KING'S COLLEGE

Briarcliff Manor, New York 10510 Phone: (914) 941-7200
Dean of College: S.J Barkat

DEPARTMENT OF BIOLOGY
Phone: Chairman: Ext. 282 Others: 278, 286

Chairman: W. Frair
Professors: A.L. Bleecker, W. Friar
Assistant Professors: D. Abb, R. Paisley

Degree Program: Biology

Undergraduate Degree: B.A.
Undergraduate Degree Requirements: A minimum total of 130 hours including 53 hours of specified nonscience and mathematics core. Minimum required courses: 24 hours in Biology, 26 hours in Chemistry, Mathematics, Physics, and 12 elective hours in natural sciences or Mathematics.

KIRKLAND COLLEGE

(No information available)

LE MOYNE COLLEGE

Syracuse, New York 13214 Phone: (315) 446-2882
Dean of College: Rev. J.A. Dinneen, S.J.

BIOLOGY DEPARTMENT
Phone: Chairman: Ext. 270

Chairman: L.D. DeGennaro
Professor: L.D. DeGennaro
Associate Professors: G.J. Lugthart, J.F. O'Brien
Assistant Professors: A. Szebenyi, R. Nentwig
Visiting Lecturers: 1

Degree Program: Biology

Undergraduate Degrees: B.S., A.B.
Undergraduate Degree Requirements: B.S. - 4 years Biology, 2 years Chemistry, 1 year Physics, 1 year Mathematics, A.B. - 3 years Biology, 1 year Chemistry, 1 year Mathematics, 2 years Humanities

LEHMAN COLLEGE

(see under City University of New York)

LONG ISLAND UNIVERSITY

University Center Phone: (516) 299-2518
Dean of Graduate Studies: L. Jacobson

The Brooklyn Center of Long Island University

385 Flatbush Avenue Ext., Phone: (212) 834-6000
Brooklyn, New York 11201
Dean: E.A. Clark

BIOLOGY DEPARTMENT
Phone: Chairman: (212) 834-6121 Others: 834-6120

Chairman: D.M. Curley
Professors: D. Amsterdam (Adjunct), S. Carito, J. Cribbins, D. Curley, N. Firriolo, L. Grosso, D. Hammerman, A. Iovino, C. Rand (Adjunct), N. Rothwell, W. Smith
Associate Professors: R. Ballweg, J. Hirshon, R. Lewis, G. Wendt
Assistant Professor: F. Dowd
Graduate Assistants: 3

Degree Programs: Biology, Medical Technology, Cytotechnology, Microbiology, Cell Biology, Physiology and Medical Microbiology

Undergraduate Degree: B.S. (Biology
Undergraduate Degree Requirements: 128 credits total, 32 credits in Biology, 16 credits in Chemistry, 8 credits in Physics, 8 credits in Mathematics

Graduate Degree: M.S. (Biology)
Graduate Degree Requirements: Master: 30 credits plus a research thesis, 36 credits, without thesis, in Medical Microbiology only.

Graduate and Undergraduate Courses Available for Graduate Credit: Evolution, Protozoology, Radiation Biology, Biochemistry, Cytology, Sensory Physiology, Genetics, Morphogenesis, Microbial Genetics, Comparative Histology, Advanced Microbiology, Mycology, Immunology, Vertebrate Physiology, Neuroendocrinology, Hematology, Marine Microbiology, Reproductive Physiology, Virology, Medical Microbiology

C.W. Post Center of Long Island University

(No reply received)

Southampton College of Long Island University

Southampton, New York 11968 Phone: 283-4000
Dean of College: W. Burke

DIVISION OF NATURAL SCIENCES
Phone: Chairman: 283-4000

Division Chairman: T. Haresign
Professors: E.I. Coher, T. Haresign
Associate Professors: D. Duberman, E. Hehre, J. Price (Adjunct), R. Welker
Assistant Professor: L. Penny

Field Stations/Laboratories: Marine Science Laboratory - Southampton, New York

Degree Programs: Biology, Marine Biology

Undergraduate Degrees: B.A., B.S.
Undergraduate Degree Requirements: B.A. - Molecular Biology, (choice of Botany and Field B ology, Human Biology, Entomology), Genetics, 2 semesters Animal Science, 1 semester Botanical Science, Biochemistry, Physiology, Ecology, 1 semester (Quantitative or Organic), to Calculuc for B.S. - Calciphysics, 2 semesters Chemistry, Evolution and Mathematics

MANHATTAN COLLEGE

(see College of Mt. St. Vincent)

MARIST COLLEGE

Poughkeepsie, New York 12601 Phone: (914) 471-3240
Dean of College: R.A. LaPietra

DIVISION OF NATURAL SCIENCES
Phone: Chairman: Ext. 228 Others: Ext. 243, 297

Division Chairman: G.B. Hooper
Professor: G.B. Hooper
Associate Professor: H.P. Turley

Assistant Professors: J.S. Bettencourt, W.T. Perrotte
Part-time Lecturers: 3

Degree Program: Biology

Undergraduate Degree: B.A.
Undergraduate Degree Requirements: 32 credits in Biology courses, 10 credits in Chemistry courses (required), 20 credits in related fields (Chemistry, Mathematics, Language, Physics, etc.), 58 credits in electives

MARYMOUNT COLLEGE
Tarrytown, New York 10591 Phone: (914) 631-3200
Dean of College: R.E. O'Brien

SCIENCE DEPARTMENT
Phone: Chairman: Ext. 418 Others: 631-3200

Chairman: R.J. Stojda
Professors: M.J. McGuinness
Associate Professor: R.T. O'Brien
Assistant Professors: B. Becker, C. Heogler, R. Stojda
Instructor: R. Madden
Adjunct Lecturers: 2

Degree Programs: Biology, Medical Technology, Pre-Medicine

Undergraduate Degree: B.S.
Undergraduate Degree Requirements: Biology Major: 2 courses in Physics, four courses in Chemistry, two courses in Mathematics, eight courses in Biology

MERCY COLLEGE
Dobbs Ferry, New York 10522 Phone: (914) 693-4500
Dean of College: J. Melville

DEPARTMENT OF NATURAL SCIENCES
Phone: Chairman: (914) 693-4500

Chairman: J.M. Neillis
Professor: J.M. Neillis
Associate Professor: A. Rice, J. McClure, J. Melville
Assistant Professor: E. Duffy
Instructors: S. McGlinchy, M. Keiner
Visiting Lecturers: 3

Degree Program: Biology

Undergraduate Degree: B.S.
Undergraduate Degree Requirements: 120 credits, including General Chemistry, General Biology and 27 credits in Biology, Organic Chemistry, Physics recommended

MOLLOY COLLEGE
1000 Hempstead Avenue Phone: (516) 678-5000
Rockville Centre, New York 11570
Dean of College: Sr. M. Halpin

BIOLOGY DEPARTMENT
Phone: Ext. 51

Chairman: Sr. J. Jones
Professors: Sr. J. Buettner, Sr. J. Jones
Assistant Professors: D.G. Kelly, Sr. M.L.F. Monaghan, Sr. V. Morley
Instructor: G. Gentile, (Adjunct)

Degree Program: Biology

Undergraduate Degree: B.S.
Undergraduate Degree Requirements: 30 semester hours in Biology and Senior Seminar, 1 year General Physics with laboratory, 1 year Inorganic/Organic Chemistry with laboratory, 1 year Mathematics, (9 credits lower level courses) (21 credits upper level courses)

MT. ST. MARY COLLEGE
Newburgh, New York 12550 Phone: (914) 561-0800
Dean of College: Sr. A. Boyle

DEPARTMENT OF BIOLOGY
Phone: Chairman: Ext. 261

Chairman: R.S. Peckham
Professor: Sr. M.J. McGivern
Assistant Professor: P. DeLuca
Lecturer: D. Gawyrs

Degree Program: Biology

Undergraduate Degree: B.A.
Undergraduate Degree Requirements: 1 year Calculus, 1 year Physics, 2 years Chemistry (Inorganic, Organic), 6 upper division Biology

NAZARETH COLLEGE OF ROCHESTER
4245 East Avenue Phone: (716) 586-2525
Rochester, New York 14610
Dean of Graduate Studies: Sr. J. Slattery
Dean of College: Sr. M. Hoctor

DEPARTMENT OF BIOLOGY
Phone: Chairman: Ext. 226 Others: Exts. 353,354

Chairman: Sr. G. Geisler
Assistant Professors: Sr. K. Maloney, F.W. Peek
Visiting Lecturer: 1

Degree Program: Biology

Undergraduate Degree: B.S.
Undergraduate Degree Requirements: 32 hours Biology, 16 hours Chemistry, 8 hours Physics, 6-8 hours Mathematics, (Calculus is recommended)

NEW YORK INSTITUTE OF TECHNOLOGY
Wheatley Road Phone: (516) 626-3400
Old Westbury, New York 11568
Dean of College: H. Fox

DEPARTMENT OF LIFE SCIENCES
Phone: Chairman: Ext. 273

Chairman: C. Bogin
Professors: C. Bogin, M. Halpern, B. Spector (Research)
Associate Professor: S. Polansky
Assistant Professors: E. Mitacek, W. Muller
Instructor: B. Kasper, M. Kossove

Field Stations/Laboratories: Chemistry and Biology Laboratories on campus.

Degree Program: Life Sciences

Undergraduate Degree: B.S.
Undergraduate Degree Requirements: 128 credits - 39 credits in Life Sciences, 25 credits Liberal Arts, 23-29 credits in Life Science option, 30 credits in Humanities

NEW YORK STATE AGRICULTURAL EXPERIMENT STATION
(see under Cornell University)

NEW YORK STATE COLLEGE OF AGRICULTURE AND SCIENCES
(see Under Cornell University)

NEW YORK STATE VETERINARY COLLEGE
(see under Cornell University)

NEW YORK UNIVERSITY
Washington Square Phone: (212) 598-1212
New York, New York 10003
Dean of Graduate Studies: R. Raymo

University College of Arts and Sciences
Dean: P. Mayerson

DEPARTMENT OF BIOLOGY
Phone: Chairman: 598-3436 Others: 598-3435

Chairman: G. Stozky
Professors: I. Brick, S.S. Brody, W.J. Crotty, A.S. Gordon, C.J. Heusser, H.I. Hirschfield, J. LoBue, J. Mitra, A. Perlmutter, C. Siegel, G. Stozky, F. Strand
Associate Professors: H.G. Dowling, J.F. Gennaro, M. P. Kambysellis, E. Karp, J.S. Mellett
Assistant Professors: R.L. Borowsky, P.S. Coleman, V. Likhite, L. Katz
Instructor: H. Babich, M. Roche
Visiting Lecturers: 2
Teaching Fellows: 29
Research Assistants: 10
Adjuncts: 9

Field Stations/Laboratories: New York University Center - Sterling Forest, Tuxedo, New York

Degree Programs: Biology, Medical Technology, Psychobiology

Undergraduate Degree: B.S.
Undergraduate Degree Requirements: 120 points with not less than 2.0 index, to include six four-point courses in Biology beyond the year of Introductory Biology (General Chemistry, Organic Chemistry, Physical Chemistry, Physics and required Mathematics are strongly recommended)

Graduate Degrees: M.S., Ph.D.
Graduate Degree Requirements: Master: 32 course credits with a grade of B or better in each course, proficiency in one foreign language, and a laboratory research thesis. Some students may substitute a library thesis for the research thesis, but this is for a termainal degree in this department. Doctoral: 40 or more course credits beyond the M.S. degree depending on area of specialization, written qualifying examination, preliminary oral examination on research proposed and defense of thesis.

Graduate and Undergraduate Courses Available for Graduate Credit: Microbiology, Special Topics in Physiology, Ultrastructural Techniques, Experimental Biochemistry, Experimental Microbiology, Developmental Biology, Quantitative Biology, Introduction to Ecology, Behavioral Ecology, Current Topics in Developmental Biology, Natural History of Vertebrates

College of Dentistry
342 East 26th Street Phone: (212) 598-7554
New York, New York 10010
Dean of Graduate Studies: A.W. Bernheimer
Dean of College of Dentistry: H. Blechman

DEPARTMENT OF ANATOMY
Phone: Chairman: (212) 598-7094 Others: 598-7522

Chairman: O.G. Mitchell
Associate Professors: O.G. Mitchell, I.J. Singh
Assistant Professors: A. Mercurio, N. Terebey, M. Silber (Adjunct, Clinical)
Instructor: A. DeCicco (Adjunct, Clinical)
Visiting Lecturers: 2
Teaching Fellows: 4

Degree Program: Anatomical Sciences

Graduate Degrees: M.S., Ph.D.
Graduate Degree Requirements: Master: 32 **credits**, thesis Doctoral: 72 credits, thesis

Graduate and Undergraduate Courses Available for Graduate Credit: Gross Anatomy - Human, Head, Neck, Thorax and Introduction to Neuroanatomy - 6 credits, Gross Anatomy - Abdomen, Trunk and Extremities - 6 credits, Seminar in Anatomy - 1 credit, Special Topics in Anatomy 1-2 credits, Research in Anatomy - 1-4 credits

DEPARTMENT OF BIOCHEMISTRY
Phone: Chairman: (212) 598-7075

Chairman: L. Fishman (Acting)
Associate Professors: L. Fishman, I.T. Weiss, M. Oratz (Adjunct)
Assistant Professors: R. Boylan, C. H. Singh (Adjunct)
Instructors: C. Johnson, J. Utting

Degree Program: Biochemistry

Graduate Degrees: M.S., Ph.D.
Graduate Degree Requirements: Master: 32 credits, a research thesis, Language Proficiency Doctoral: 72 credits, written and oral qualifying examinations, thesis, defense of thesis, language proficiency

Graduate and Undergraduate Courses Available for Graduate Credit: Biological Chemistry, Biological Chemistry Laboratory, Proteins, Enzymes, Seminar in Biochemistry and Tutorials in Biochemistry

DEPARTMENT OF MICROBIOLOGY
(No reply received)

DEPARTMENT OF PHYSIOLOGY-PHARMACOLOGY
Phone: Chairman: (212) 598-7571 Others: 598-7572

Chairman: E.A. Neidle
Professor: E.A. Neidle
Assistant Professors: N. Overweg, F. Rakowitz, J. Schiff, S. Turetsky (Adjunct)
Instructors: A. Eichen, S. Wolfe
Teaching Fellows: 2

Degree Program: Physiology

Graduate Degrees: M.S., Ph.D.
Graduate Degree Requirements: Master: 32 credits, a Research Thesis, Language Proficiency Doctoral: 72 credits, written and oral qualifying examinations, thesis, defense of thesis, language proficiency

Graduate and Undergraduate Courses Available for Graduate Credit: Mammalian Physiology, Research in Physiology, Seminar in Endocrine Physiology

School of Medicine
550 First Avenue Phone: (212) 679-3200
New York, New York 10016
Dean of Graduate Studies: R.B. Winder
Dean of School of Medicine: I.L. Bennett, Jr.

DEPARTMENT OF CELL BIOLOGY
Phone: (212) 679-2092 Others: 679-3200

Chairman: D.D. Sabatini
Professors: D.D. Sabatini, M.P. Rabinovitch
Associate Professors: V.H. Black, B.I. Bogart, J.A. Lake, L. Prutkin, K. Rubinson, J.L. Shafland, M. Cunningham-Rundles (Adjunct), R.E. Kessler (Clinical)
Assistant Professors: M.B. Adesnik, G. Kreibich, T. Morimoto, E.S. Robbins (Adjunct), V.R.Zerbino (Clinical)
Instructors: G. Conroy, C.L. DeLemos, M.A. Lande
Visiting Lecturers: 12
Teaching Fellows: 2
Teaching Assistants: 2

Degree Program: Cell Biology

Graduate Degree: Ph.D.
Graduate Degree Requirements: Doctoral: 1) 36 points of course work including Biochemistry and Cell Biology and 36 points of thesis research 2) Qualify examination covering areas of Molecular and Cell Biology, Genetics and Immunology 3) Thesis research and Dissertation

Graduate and Undergraduate Courses Available for Graduate Credit: Biology of Cells, Tissues and Organs (Cell Biology), Electron Microscopy, Spread Topics in Cellular and Molecular Biology: each year on a different subject such as Regulation of Gene Expression, Membrane Structure and Function, Biogenesis, Structure and Function of Subcellular Organelles, Gross Human Anatomy

DEPARTMENT OF MICROBIOLOGY
Phone: Chairman: (212) 679-2317 Others: 679-3200

Chairman: M.R.J. Salton
Professors: W.L. Barksdale, A.E. Bernheimer, E. McFall, J.H. Schwarts, J. Vilcek, S. Dales (Research), K.G. Hirst (Research), B. Mandel (Research), P. Margolin (Research), A.G. Osler (Research)
Associate Professors: J.C. King, J.E. Winter, D.A. Dubnau (Research), R.P. Novick (Research), M. Oishi (Research)
Assistant Professors: T.F.R. Celis, K.S. Kim, M.S. Nachbar, J.D. Oppenheim
Visiting Lecturers: 2
Research Assistants: 5
Visiting Professors: 3

Degree Programs: Microbiology, Virology

Graduate Degrees: M.S., Ph.D.
Graduate Degree Requirements: Master: Completion of graduate studies totalling at least 32 points and a thesis Doctoral: Completion of graduate studies totalling at least 72 points, pass a qualifying examination, and present an acceptable dissertation.

Graduate and Undergraduate Courses Available for Graduate Credit: Microbiology (Medical Microbiology, Immunology), Microbial Genetics, Molecular Biology of Animal Virus - Cell Interactions

DEPARTMENT OF PHYSIOLOGY AND BIOPHYSICS
Phone: Chairman: (212) 679-2814 Others: 679-3200

Chairman: W.J. Sullivan (Acting)
Professors: J. Rosenbluth, W.J. Sullivan, J. Zadunaisky
Associate Professor: F. Aull
Assistant Professors: J. Barker, V. Castellucci, E. Gardner, I.R. Moss, W.Z. Snyder
Instructors: F. Haas, L.L. Vacca

Degree Program: Physiological Sciences

Graduate Degree: Ph.D.
Graduate Degree Requirements: Doctoral: 72 points, at least 36 points in courses and tutorials; written and oral qualifying examinations, written thesis and oral defense

Graduate and Undergraduate Courses Available for Graduate Credit: Mammalian Physiology, Cellular Physiology, Neural Science, Tutorials in Physiology and Biophysics, Biological Chemistry, Biology of Cells, Tissues and Organs, Pharmacology, Gross Anatomy, Other

NIAGARA UNIVERSITY

Niagara University, New York 14109 Phone: (716) 285-1212
Dean of College: Rev. C.J. Leonard

BIOLOGY DEPARTMENT
Phone: Chairman: Ext. 214

Chairman: T.H. Morton
Professors: L.J. Kiely, T.H. Morton, J.J. Reedy
Associate Professors: B.T. Britten, E.J. McKeegan

Degree Programs: Biology, Life Sciences

Undergraduate Degrees: B.A., B.S.
Undergraduate Degree Requirements: 10 units Biology, 3 units Chemistry, 2 units Mathematics, 2 units Physics

Graduate Degree: M.S.
Graduate Degree Requirements: Master: 24 credits, thesis, comprehensive examination

Graduate and Undergraduate Courses Available for Graduate Credit: Comparative Vertebrate Anatomy and Laboratory, Vertebrate Embryology and Laboratory (U), Microscopic Anatomy and Laboratory (U), Physiology and Laboratory (U), Parasitology and Laboratory (U), Genetics and Laboratory (U), History of Biology and Evolution, Cell Biology (RP), Biometrics and Laboratory, Biomedical Applications of Biostatistics (RP), Cell Physiology (RP), Geographical and Tropical Medicine (RP), Endocrine Physiology (RP), Human Physiology, Biochemistry and Laboratory, Biochemistry and Chemistry of Hormones (RP), Chemical Aspects of Biology (RP), Bioengineering (RP), Chemical Pathology (RP), Pathophysiology (RP), Pathology of Common Laboratory Animals (RP), Invertebrate Zoology and Laboratory, Vertebrate Zoology and Laboratory, Human Genetics, Molecular Genetics (RP), Species and Community Ecology and Laboratory, Conservation of Natural Resources, Population Ecology and Laboratory, Radiation Biology and Laboratory, Biophysical and Radiobiological Aspects of the Cell Cycle (RP), Scientific Photography and Laboratory, Bacteriology and Laboratory, Fundamentals of Virology (RP), Viral Oncology (RP), Biostatistics Seminar (RP), Microbiology Seminar (RP), Immunochemistry Seminar (RP)
Note: (RP) - courses taught at Roswell Park Memorial Institute

PACE UNIVERSITY

Pace Plaza Phone: (212) 285-3000
New York, New York 10038
Dean of Graduate Studies: T. Bonaparte (Acting)
Dean of College: J.E. Houle

DEPARTMENT OF BIOLOGY
Phone: Chairman: (212) 285-3504 Others: 285-3000

Chairman: D. Cox
Professors: S.H. Hutner (Haskins Adjunct Research Professor), D.E. Keller
Associate Professors: L. Chunosoff, D. Cox, H.M. Dembitzer (Adjunct), M. Yankow (Adjunct)
Assistant Professors: C. Bacchi, E. Benoit (Adjunct), H. Betros (Adjunct), M.G. Browne (Adjunct), T. Cogen (Adjunct), S. Cohen, S. Goppel (Adjunct), S.L. Marcus (Adjunct), G. Mauriello (Adjunct)
Instructors: R.L. Kahn (Adjunct), D. Kender, M.E. McEnery (Adjunct), M. Schiffenbauer

Field Stations/Laboratories: Haskins Laboratories (Microbiological Section), 41 Park Row, New York, New York 10038

Degree Programs: Biology, Medical Technology

Undergraduate Degree: B.S.
Undergraduate Degree Requirements: 4 year Liberal Arts courses including Mathematics, Chemistry, Biology, English, Speech, Social Science, Language and several electives

QUEEN'S COLLEGE
(see under City University of New York)

RENSSELAER POLYTECHNIC INSTITUTE
Troy, New York 12181 Phone: (518) 270-6000
Dean of Graduate Studies: S.E. Wiberley

School of Engineering
Dean: A.A. Burr

BIO-ENVIRONMENTAL ENGINEERING DIVISION
Phone: Chairman: (518) 270-6363 Others: 270-6000

Division Chairman: W.W. Shuster
Professors: M.T. Amirana (Adjunct), D.B. Aulenbach, P.B. Daitch, R.E. Dutton, Jr., A.M. Hakim (Adjunct), G. Moss (Research), S.R. Powers (Adjunct), W.W. Shuster, A.A. Stein (Adjunct), M.H. Thompson (Adjunct), M.J. Tsapogas (Adjunct)
Associate Professors: E.R. Altwicker, N.L. Clesceri, H.M. El-Baroudi, L.J. Hetling (Adjunct), R.P. Leather (Adjunct), R.W. Shade
Assistant Professors: T.C. Gabriele (Adjunct), A.R. Hinman (Adjunct), G.J. Stensland, L.K. Wang
Teaching Fellows: 10
Research Assistants: 11

Field Stations/Laboratories: Rensselaer Fresh Water Institute on Lake George - Putnam Station - New York

Degree Programs: Biomedical Engineering, Environmental Engineering

Undergraduate Degree: B.S.
Undergraduate Degree Requirements: Students are considered for admission if they have high level secondary school records in subject areas which include Physics, Chemistry, Elementary and Intermediate Algebra, Plane Geometry, Trigonometry and four years of English. The requirements for completing the undergraduate degrees are specified for each curriculum in the Rensselaer Catalog. Approximately 132 credits of prescribed courses are required.

Graduate Degrees: M.S., Ph.D.
Graduate Degree Requirements: Master: Proper admission to the Graduate School with an accredited undergraduate degree in either Science or Engineering, completion of an approved program of study in Environment Engineering with completion of an Engineering Project or thesis and fulfillment of the general requirements for graduate study described in the Rensselaer Catalog. Doctoral: Proper admission to the Graduate School, completion of an accredited master's program in either Science or Engineering, completion of sixty credits beyond the Master's degree which includes 30 course credits and at least 30 credits of original research culminating in a formally written and presented thesis. Preliminary and candidacy examinations are required as well as a thesis examination and the fulfillment of General Requirements for graduate study described in the Rensselaer Catalog.

Graduate and Undergraduate Courses Available for Graduate Credit: Unit Operations in Envrionmental Engineering (U), Unit Processes in Environmental Engineering (U), Environmental Acoustics (U), Water and Wastewater Engineering (U), Environmental Engineering Laboratory and Design (U), Atmospheric Pollution I (U), Solid Waste Engineering (U), Industrial Waste Treatment and Disposal (U), Systems Physiology (U), Engineering Physiology Laboratory (U), Public Health Chemistry for Environmental Engineers, Management and Planning for Pollution Abatement, Mathematics Modeling of Environmental Systems, Water Resources, Stream Pollution Control Atmospheric Pollution II, Solid Waste Laboratory, Environmental Radiation Safety Controls, Environmental Engineering Seminar, Biomedical Engineering Seminar, Clinical Engineering, Air Pollution Meteorology, Environmental Law in Engineering Practice, Advanced Waste Treatment Processes, Atmospheric Chemistry, Environmental, Impact Assessment, Air Pollution Laboratory

School of Science
Dean: G.H. Handelman

DEPARTMENT OF BIOLOGY
Phone: Chairman: (518) 270-6375 Others: 270-6000

Chairman: J.V. Landau
Professors: H.L. Ehrlich, C. Hurwitz (Adjunct), W.H. Johnson, J.V. Landau, D.E. Wilson, Jr.
Associate Professors: W. Auclair, L.S. Clesceri, C.J. Pfau
Assistant Professors: J.J. Diwan, J.E. Estes, C.W. Boylen, T.D. O'Neal, R.H. Parsons, D.H. Pope
Teaching Fellows: 18
Research Assistants: 10
Post Doctoral Research Fellow: 1

Field Stations/Laboratories: Fresh Water Institute - Lake George, New York

Degree Programs: Biochemistry, Biochemistry and Biophysics, Biology, Biophysics, Microbiology, Molecular Biophysics, Physiology, Cellular, Developmental

Undergraduate Degree: B.S. (Biology)
Undergraduate Degree Requirements: 3 semesters Mathematics, 3 semesters Physics, 6 semesters Chemistry, 10 semesters Biology, 8 semesters Humanities and Social Science, Electives to a minimum total of 124 credit hours

Graduate Degrees: M.S., Ph.D.
Graduate Degree Requirements: Master: 30 credit hours beyond B.S., Thesis presentation Doctoral: 90 credits beyond B.S., Candidacy examination, presentation and defense of thesis.

Graduate and Undergraduate Courses Available for Graduate Credit: Plant Physiology (U), Radiation Biology (U), Comparative Physiology (U), Cell Biology (U), Human Physiology (U), Advanced Microbiology (U), Geomicrobiology (U), Introductory Virology (U), Molecular Biology (U), Biochemistry (U), Biophysics (U), Radiation Biology and Toxicology, Microbiology, Bacterial Chemistry, Systematic Microbiology, Microbial Genetics, General Virology, Microbiology Seminar, Cell and Developmental Biology I, II, Developmental Genetics, Advanced Physiology, Topics in Biochemistry, Topics in Biophysics, Seminar in Biology, Selected Readings in Biology

ROBERTS WESLEYAN COLLEGE
2301 Westside Drive Phone: (716) 594-9471
Rochester, New York 14624
Dean of College: D. Keolee

DEPARTMENT OF BIOLOGY
Phone: Chairman: Ext. 337 Others: Ext. 231

Head: D. Barnes
Professors: P.H. Harden
Associate Professors: D. Barnes, D. Scott

Degree Program: Biology

Undergraduate Degrees: A.B., B.S.
Undergraduate Degree Requirements: 1 course - 5 quarter hours, 14 courses, basic general education, 9-10 courses in Biology plus 3 to 8 supporting courses plus electives for a total of 37 courses

ROCHESTER INSTITUTE OF TECHNOLOGY
One Lomb Memorial Drive Phone: (716) 464-2496

Rochester, New York 14623
Dean of College: T.P. Wallace

DEPARTMENT OF BIOLOGY
Phone: Chairman: (716) 464-2496

Head: W.A. Burns (Acting)
Professors: D.M Baldwin, E. Stark
Associate Professors: E.A. Arthur, J.L. Baird, W.A. Burns, J. Klingensmith, R. Sowinski
Assistant Professors: M. D'Ambruso, C. Sack, F. Seischab

Degree Program: Biology

Undergraduate Degree: B.S.
Undergraduate Degree Requirements: B.S. degree in Biology: 60 hours in Biology must be distributed as follows: 3 Quarter courses in General Biology, 2 quarter courses in each area of Molecular and Cellular Biology, Developmental Biology, Genetics and Ecology, Organismal Biology, and 1 quarter course in Biological Techniques

ROSARY HILL COLLEGE
4380 Main Street Phone: 839-3600
Buffalo, New York 14226
Dean of College: Sr. M. Lannan

DEPARTMENT OF NATURAL SCIENCES CONCENTRATION
Phone: Chairman: Ext. 235 Others: Ext. 279,242

Chairman: A.C.S. Wang
Associate Professors: A.C.S. Wang, A. Pleshkewych
Assistant Professor: J. Muller
Instructor: S. Benoit
Part-time Lecturers and Faculty: 5

Degree Programs: Biology, Biochemistry, Premedicine, Medical Technology

Undergraduate Degree: B.S.
Undergraduate Degree Requirements: Minimum total of college credits is 122 for graduation, details vary according to program

RUSSELL SAGE COLLEGE
Troy, New York 12180 Phone: (518) 270-2000
Dean of Graduate Studies: J.R. O'Connell
Dean of College: L.C. Smith

DEPARTMENT OF BIOLOGY
Phone: Chairman: (518) 270-2236 Others: 270-2244

Chairman: E.G. Horn
Professors: R.F. Gehrig
Associate Professors: B.D. Bhatt, W.P. Rockwood
Assistant Professors: E.D. Hickey-Weber, E.G. Horn, N.G. Slack, J. Westin

Degree Programs: Biology, Medical Technology

Undergraduate Degree: B.A.
Undergraduate Degree Requirements: 7 hours Chemistry, General Biology, plus 8 courses advanced Biology, Physical Chemistry and Calculus recommended.

ST. BERNADINE OF SIENA COLLEGE
(see Siena College)

ST. BONAVENTURE UNIVERSITY
St. Bonaventure, New York 14778 Phone: (716) 372-0300
Dean of Graduate Studies: K.E. Anderson
Dean of College: B. Litzinger

DEPARTMENT OF BIOLOGY
Phone: Chairman: Ext. 378 or 379

Chairman: A.F. Finocchio
Professors: K.E. Anderson, R.C. Bothner, S.W. Eaton, R.E. Hartman
Associate Professors: F.A. Jacques, J.P. White
Assistant Professor: R.J. Wordinger
Teaching Assistants: 8

Degree Programs: Biology, Botany, Microbiology, Zoology

Undergraduate Degree: B.S. (Biology)
Undergraduate Degree Requirements: Biology 31 credit hours, Calculus 8 credit hours, Physics 8 credit hours, Chemistry 16 credit hours

Graduate Degrees: M.S., Ph.D.
Graduate Degree Requirements: Master: Course work 24 credit hours, Thesis 6 credit hours, one foreign language Doctoral: Course work 30 credit hours, Dissertation 20 credit hours, Two foreign languages

Graduate and Undergraduate Courses Available for Graduate Credit: Not Stated

ST. FRANCIS COLLEGE
180 Remsen Street Phone: (215) 526-2300
Brooklyn, New York 11201
Dean of College: Vacant

BIOLOGY DEPARTMENT
Phone: Chairman: Ext. 288

Chairman: J.J. Martorano
Professors: Sr. M.V. Cotter, J.J Martorano
Associate Professor: C. Taschdjian
Assistant Professor: J. Corrigan

Degree Program: Biology

Undergraduate Degree: B.S.
Undergraduate Degree Requirements: Core Curriculum: English 6, Fine Arts 6, Speech 3, Philosophy 9, Liberal Arts 6, History 3, Sociology 3, Biology Major: Mathematics - Calculus 6, Chemistry 18, Physics 8, Language 6-12, Biology 32

ST. JOHN FISHER COLLEGE
Rochester, New York 14618 Phone: (716) 586-4140
Dean of College: E.B. Schick

BIOLOGY DEPARTMENT
Phone: Chairman: Ext. 385

Chairman: M.J. Wentland
Associate Professor: M.J. Wentland
Assistant Professors: J.M. Rowan, V.J. Palese, G.T. Crombach, S.R. Gawlik

Degree Program: Biology

Undergraduate Degrees: B.A., B.S.
Undergraduate Degree Requirements: B.A.: 33 hours Biology, 16 hours Chemistry, 8 hours Physics, 4 hours Calculus B.S. - 41 hours Biology, 16 hours Chemistry, 8 hours Physics, 4 hours Calculus

ST. JOHN'S UNIVERSITY
Grand Central and Utopia Parkways Phone: (212) 969-8000
Jamaica, New York 11439
Dean of Graduate Studies: P.T. Medici
Dean of College: Rev. R.J. Devine, C.M.

BIOLOGY DEPARTMENT
Phone: Chairman: Ext. 269 Others: 287-288

Chairman: J.N. Concannon

Professors: W.H. Beckert, D.M Lilly, M.A. Pisano, A.V. Liberti
Associate Professors: C. Efthymiou, W.J. Hartig, R.C. Jack
Assistant Professors: G. Rio, R. Stalter, J.D. Campbell, A. Mignone
Instructors: J. Hanshe, K. Dunker
Visiting Lecturers: 3
Teaching Fellows: 4
Research Assistants: 1

Field Stations/Laboratories: Long Island (New York Ocean Science Laboratory

Degree Programs: Biology, Cell Biology, Microbiology, Physiology

Undergraduate Degree: B.S. (Biology)
Undergraduate Degree Requirements: 32 semester hours, Including Biology 21, 22, and 24 electives hours in Biology. Biology Majors are required to take Chemistry 21, 22 and 23, 24, Mathematics 69, 70 and Physics 1, 2 or 3, 4

Graduate Degrees: M.S., Ph.D.
Graduate Degree Requirements: Master: 3 years Undergraduate study in Biology (at least), including Genetics, 2 years of Chemistry including Organic Chemistry, 1 year Physics and 1 year of College Mathematics
Doctoral: B+ or better average in previous graduate work, evidence of completion of all requirements for M.S. degree at accredited institution, letters of recommendation from major professor or department appraising academic standing and potential.

Graduate and Undergraduate Courses Available for Graduate Credit: A new program is in course of development

ST. JOSEPH COLLEGE

245 Clinton Avenue Phone: (212) 622-4696
Brooklyn, New York 11205
Dean of College: Sr. M.F. Burns

DEPARTMENT OF BIOLOGY
Phone: Chairman: (212) 622-4696

Chairman: Sr. M.B. Schneller
Assistant Professors: C.J. Haves, Sr. R.C. Stevens

Degree Program: Biology

Undergraduate Degree: B.S.
Undergraduate Degree Requirements: 1 year General Biology, 24 additional credits in Biology including Research Seminar, and Research (Independent), 1 year General Chemistry, 1 year Organic Chemistry, 1 year General Physics, Mathematics through Calculus

ST. LAWRENCE UNIVERSITY

Canton, New York 13617 Phone: (315) 379-5294
Dean of College: D.K. Baker

DEPARTMENT OF BIOLOGY
Phone: Chairman: (315) 379-5294

Chairman: L.E. Bowers
Professors: L.E. Bowers, R. Crowell
Associate Professors: W.J. Ash, K. Crowell, J. Green, R.F. Wells
Assistant Professors: T. Budd
Instructors: D. Hornung, J. Westerling
Visiting Lecturer: 1

Degree Program: Biology

Undergraduate Degree: B.S.
Undergraduate Degree Requirements: Eight units within the department

ST. THOMAS AQUINAS COLLEGE

Sparkill, New York 10976 Phone: (914) 359-6400
Dean of College: Vacant

DEPARTMENT OF NATURAL SCIENCES
Phone: Chairman: Ext. 297

Chairman: J.A. Keane
Associate Professors: Sr. M.M. McPhillips
Assistant Professors: D.W. Fenbert, J.A. Keane

Degree Programs: Medical Technology, Natural Science

Undergraduate Degree: B.S.
Undergraduate Degree Requirements: 54 credits in Sciences and Mathematics by approval of advisor

SARAH LAWRENCE COLLEGE

Bronxville, New York 10708 Phone: (914) 337-0700
Dean of Faculty: R. Wagner

DEPARTMENT OF BIOLOGY

Faculty are not ranked: M.H. Bailin, R. Clarke, R. Beck, A. Gressin
Instructors: N. Einstein, Passo, Feirstais, A. Bain, King
Visiting Lecturers: 5

Degree Programs: Biology, Human Genetics

Undergraduate Degree: B.A. (Biology)
Undergraduate Degree Requirements: Not Stated

Graduate Degree: M.S. (Human Genetics)
Graduate Degree Requirements: Master: 30 credits

Graduate and Undergraduate Courses Available for Graduate Credit: Mendelian Molecular Genetics (U), Laboratory Techniques in Human Genetics, Cytogenetics and Biochemical Genetics, Advanced Human Physiology, Genetic Counseling, Human Genetics Clinic Field Work

SIENA COLLEGE

Loudonville, New York 12211 Phone: (518) 783-2300
Dean of College: D. Horgan

DEPARTMENT OF BIOLOGY
Phone: Chairman: (518) 783-2457 Others: 783-2300

Head: E.J. Larow
Professors: R.H. Arnett, Jr., G.F. Bazinet, T.A. Whalen
Associate Professor: E.J. Larow
Assistant Professors: P.S. Brown, K.P. Wittig, D.F. Fraser

Degree Program: Biology

Undergraduate Degree: B.S.
Undergraduate Degree Requirements: 30 semester hours Biology, 20 semester hours Chemistry, 8 semester hours Mathematics, 8 semester hours Physics, 36 semester hours Humanities Core, 18 semester hours electives

SKIDMORE COLLEGE

Saratoga Springs, New York 12866 Phone: (518) 584-5000
Dean of Faculty: E. Moseby

DEPARTMENT OF BIOLOGY
Phone: Chairman: Ext. 317

Chairman: D.W. Crocker
Professor: D.W. Crocker
Associate Professors: H.H. Howard, R.P. Mahoney

Assistant Professors: W.S. Brown, C. DeSha, B.R. Golden, R.S. Meyers
Laboratory Coordinator: 1
Teaching Assistants: 2

Field Stations/Laboratories: Associates with E.N. Huyck Preserve - Rensselaerville, New York

Degree Program: Biology

Undergraduate Degrees: B.S.
Undergraduate Degree Requirements: 10 course units in Biology and 2 in Chemistry, Interdepartmental major programs are offered in conjunction with the departments of Chemistry, Philosophy-Religion, and Psychology

SOUTHAMPTON COLLEGE

(see under Long Island University)

STATE UNIVERSITY OF NEW YORK AT ALBANY

(No reply received)

STATE UNIVERSITY OF NEW YORK AT BINGHAMPTON

Binghampton, New York 13901 Phone: (607) 798-2000
Dean of Graduate Studies: W.C. Hall
Dean of College: P. Vukasin

DEPARTMENT OF BIOLOGICAL SCIENCES

Phone: Chairman: (607) 798-2755 Others: 798-2438

Chairman: A.H. Haber
Professors: W. Battin, J. Christian, G. Fattal (Adjunct), J. Fischthal, A. Haber, W. Hall, S. Landry, G. Schumacher, A. Shrift, J. Wilmoth
Associate Professors: P. Bonamo, J. Duquella (Adjunct), A. Frankel, J. Grierson, N. Lazaroff, J. Lee (Adjunct), A. Mueller, H. Posner, R. Trumbore
Assistant Professors: J. Hall, J. Haugh, F. Kull, D. Mann, S. Murr, C. Paas (Adjunct)
Visiting Lecturers: 4
Teaching Fellows: 11
Research Assistants: 4
NSF: 1
NIH: 4

Degree Program: Biological Sciences

Undergraduate Degrees: B.A., B.S.
Undergraduate Degree Requirements: Introductory Biology, Biology of the Cell, plus two of the following three: Biology of Organisms, Animals, Biology of Organisms, Plants, Biology of Organisms, Microbes, Chemistry 111, 112 and 112L. Four advanced courses with department. Four courses from the following cognate sciences: Mathematics, Physics, Chemistry, Computer Science

Graduate Degrees: M.A., M S., Ph.D.
Graduate Degree Requirements: Master: 30 semester credit hours beyond the Bachelor's degree. Demonstration of ability to read scientific writing in German, French or Russian. Completion of thesis, Passing of oral examination on subject matter of thesis plus related biological knowledge Doctoral: 72 semester credit hours beyond the Bachelor's degree. Passing examinations in two languages. Completion of required course work with a minimum grade point average of 3.0 Passing of the comprehensive examination

Graduate and Undergraduate Courses Available for Graduate Credit: Serology and Immunology, Limnology, Pathology, Pathology Laboratory, Enzymes, Theory and Use of Radioisotopes, Endocrinology, Cellular Control Mechanisms, Cellular Control Mechanisms Laboratory, Biochemistry of Nucleic Acids, Microbial Systems, Paleobotany, Advanced Functional Morphology, Mammalogy, Techniques in Electron Microscopy, Cellular Physiology, Neuroendocrinology, The Teaching of College Biology, Fishery Biology, Radiation Biology, Seminar, Special Studies, Independent Studies in Biology, Seminar for Secondary School Teachers

STATE UNIVERSITY OF NEW YORK COLLEGE AT BROCKPORT

Brockport, New York 14420 Phone: (716) 395-2193
Dean of Graduate Studies: J.C. Crandall
Dean of College: K.T. Finley

DEPARTMENT OF BIOLOGICAL SCIENCES

Phone. Chairman: (716) 395 2193

Chairman: T.J. Starr
Faculty Exchange Scholar of SUNY: T.J. Starr
Professors: C. Barr, J. Bobear, F. Cloutier, K. Damann, R. Dilcher, G. Gehris, R. McLean, J. Mosher, T. Starr, J. Syrocki, C. Thomas, R. Thompson, R. Wallin
Associate Professors: T. Bonner, V. Farris, M, Fox, D. Hammond, R. Hellmann, K. Kline, P. Prichard, D. Smith
Assistant Professors: M. Appley, D. Brannigan, R. Ellis, A. Gianfagna, T. Haines, J. Kowalski, J. Makarewicz
Teaching Fellows: 6
Research Assistants: 4

Field Stations/Laboratories: Fancher Biology Laboratory - Fancher, New York

Degree Programs: Biology, Biological Sciences, Botany, Zoology

Undergraduate Degree: B.S. (Biological Sciences)
Undergraduate Degree Requirements: 30 hours selected from Botany, Zoology, Microbiology, Ecology, Developmental Biology, Genetics, Physiology, 1 year Chemistry, 1 semester Organic, Mathematics

Graduate Degree: M.S.
Graduate Degree Requirements: Master: 30 semester hours with thesis, 34 semester hours without thesis

Graduate and Undergraduate Courses Available for Graduate Credit: Biochemistry, Instrumentation Methods, Limnology, Eugenics, Conservation, Field Biology, Pollution, Water Analysis, Tissue Culture, Biometrics, Microbial Physiology, Topics, Cell Physiology, Evolution, Function, Productivity, Ultrasturcture, Growth, Seminar, Advanced Physiology, Local Flora Taxonomy, Algal Evolution, Plant Ecology, Vascular Plants, Phycology, Entomology, Ornithology, Ichthyology, Ethology, Mammalogy, Chromosome, Limnology, Developmental Biology, Endocrinology, Neurobiology, Independent Study

STATE UNIVERSITY OF NEW YORK COLLEGE AT BUFFALO

1300 Elmwood Avenue Phone: (716) 862-4000
Dean of Graduate Studies: R.W. Williams
Dean of College: J.K. Hichar

DEPARTMENT OF BIOLOGY

Phone: Chairman: (716) 862-5203 Others: 862-4000

Chairman: A.E. Smith
Professors: T.E. Eckert, J.D. Haynes, J.K. Hichar, G.M. Laug, V.J. Nadolinski, W.C. Schefler, H.G. Sengbusch, R.C. Stein, R.A. Sweeney, J. Urban, M.L. Wilson
Associate Professors: H.M. Collins, H. Isseroff, R.E. Moisand, F.W. Price, R.M. Reuss, A.E. Smith, C.R. Sweeney (Adjunct)

Assistant Professors: W. Ainsworth, N.J. LoCascio, E. Randall, J.R. Spotila, W.B. Wickland
Research Assistants: 8
Teaching Assistants: 5

Field Stations/Laboratories: College Camp. Franklinville New York, Great Lakes Laboratory Campus and 1 Porter Avenue, Buffalo, New York

Degree Program: Biology

Undergraduate Degrees: B.A., B.S. (Education)
Undergraduate Degree Requirements: B.A.: 36 hours Biology including 4 hours courses in Cell Biology, Botany, and Zoology, 12 hours Chemistry, 8 hours Calculus 8 hours Physics B.S. in Education: 30 hours Biology, including above 3 beginning courses, 12 hours Chemistry, 6 hours Physics, 6 hours Geosciences, 3-6 hours Mathematics, 24 hours Education

Graduate Degrees: M.A., M.S.
Graduate Degree Requirements: Master: M.A.: 30 graduate hours Biology, including 6 thesis, proficiency in one foreign language. M.S. (Biology Education): 18 graduate hours, Biology, 3-6 hours project or graduate thesis, 3 graduate hours Education, 6 graduate hours electives

Graduate and Undergraduate Courses Available for Graduate Credit: General Ecology, Human Genetics, Basic Human Anatomy and Physiology, Problems in Environmental Biology, Man and Evolution, Field Techniques in Biology, Field Studies in the Conservation of Natural Resources, Institute in Biology, Special Project, Morphogenesis in Vascular Plants, Mycology, Biometrics II, Molecular Genetics, Protozoology, Topics in Ecology, Enzymology, Taxonomy of Vascular Plants, Advanced Bacteriology, Topics in Animal Physiology, Seminar, Ethology, Plant Ecology, Symbiology, Physical Ecology, Algology

STATE UNIVERSITY OF NEW YORK AT BUFFALO

Dean of Graduate Studies: McA. H. Hull, Jr.

Faculty of Natural Science and Mathematics

Provost: C.H. Nancollas

BIOLOGY DEPARTMENT
Phone: Chairman: (716) 831-2635 Others: 831-2634

Chairman: P.G. Miles
Professors: R.A. Flickinger, D.A. Larson (Associate Vice-President for Health Sciences), F.A. Loewus, P.G Miles, C.M. Osborn, C.A. Privitera, M. Rothstein, V. Santilli, H.L. Segal, G.E. Swartz, T.Y. Wang
Associate Professors: L. Berlowitz, A.K. Bruce, M.W. Fransworth, C.F. Herreid, C.E Smith, K.M. Stewart, J.F. Storr, N. Strauss
Assistant Professors: C.R Fourtner, P.S. Gold, W.F. Hadley, A.G. Harford, J.R LaFountain, J.E. Tavares, T.C. Williams
Visiting Assistant Professor: 1

Degree Program: Biology

Undergraduate Degree: B.A.
Undergraduate Degree Requirements: 32 semester hours in Biology plus one year each of Inorganic Chemistry, Calculus and Physics, Organic Chemistry recommended.

Graduate Degrees: M.A., Ph.D.
Graduate Degree Requirements: Master: 18 credits with thesis or 36 credits without thesis Doctoral: Proficiency in a foreign language-either French, Russian, or German recommended. Courses covering broad areas of Biology are expected to be completed. Normally a program should include a minimum of 45 graduate course credits in addition to research for the thesis.

Graduate and Undergraduate Courses Available for Graduate Credit: Graduate courses offered in other Departments may be taken for graduate credit when approved by the student's advisory committee. According to regulations of the State University undergraduate courses may not be used for graduate credit with the exception that upon petition two such courses may be taken by the student providing that extra work is required to insure that the performance is at the graduate level.

School of Health Sciences

Deans: W. Feagans - Dentistry, C. Randall - Medicine, M.A. Schwartz - Pharmacy

DEPARTMENT OF ANATOMY
(No information available)

DEPARTMENT OF BIOCHEMISTRY
Phone: Chairman: (716) 831-4611 Others: 831-2727

Chairman: E.A. Barnard
Einstein Professor: J.H. Wang
Professors: O.P. Bahl, E.A. Barnard, A.C Brownie, M.E. Chilcote (Clinical), W.B. Elliott, R H. McMenamy, M. Reichlin (Research), B.E. Sanders, W.R. Slaunwhite, Jr., D.M. Surgenor, J.H. Wang
Associate Professors: C. Bishop (Clinical), M. Derechin, W.E. Farnsworth (Research), J.V. Fopeano, J. D. Klingman, E.J. Massaro, J.F. Moran, R.W. Noble, D. Papahadjopoulos (Research), D.A. Pragay (Clinical), M. Tunis (Research)
Assistant Professors: Z.F Chmielewicz (Research), L. Edwards (Clinical), M.J. Ettinger, D.J Kosman, M. Kreimer-Birnbaum (Research), R.S. Lane, M.H. Meisler, M.D. Garrick
Research Assistants: 40
Postdoctoral Fellows: 12

Degree Program: Biochemistry

Undergraduate Degree: B.S.
Undergraduate Degree Requirements: Chemistry, General Organic andvanced Organic, Physical, Biology, Mathematics, Physics, Biochemistry: General, Laboratory, Senior Research, plus electives

Graduate Degrees: M,A., Ph.D.
Graduate Degree Requirements: Master: Chemical Basis of Biochemistry, Laboratory, Physical Biochemistry, Metabolic Biochemistry, and 1 600 level (advanced) course plus 9-15 hours research, 6 credit hours in collateral area, thesis/oral defense or project defined by student's research committee. Doctoral: Chemical, Physical, Metabolic Biochemistry, Laboratory, Seminar, 1 600-level course, collateral area, preliminary examinations, research and presentation of research progress, dissertation and oral defense.

Graduate and Undergraduate Courses Available for Graduate Credit: Medical Biochemistry, Dental Biochemistry, Biochemistry Seminar, Advanced Laboratory, Clinical Biochemistry Laboratory, Chemical Basis of Biochemistry, Physical Biochemistry, Metabolic Biochemistry General Biochemistry, Independent Studies, Supervised Teaching, Blood, Lipids, Energy Metabolism, Neurochemistry, Protein Conformation, Analysis and Simulation of Biochemical Systems, Biochemistry of Human Disease, Environmental Biochemistry, Biochemical Evolution, Advanced Amino Acid Metabolism Carbohydrate Structure and Metabolism, Enzymes, Molecular Interrelationships of Nucleic Acids and Proteins, Biochemical Endocrinology, Molecular Basis of Chemical Transmission Processes, Membrane Studies in Human Critical Ilness, Clinical Biochemical Laboratory Procedures, Research

CENTER FOR THEORETICAL BIOLOGY

Phone: Chairman: (716) 831-1322 Others: 831-1323

Director: J.F. Danielli

Faculty, at present numbering 57, are drawn from other Departments, including Biochemistry, Biophysics, Chemistry, Computer Science, Mathematics, Physics, etc. The Ph.D. degrees of students working in the Center for Theoretical Biology are granted in these other Departments.

DEPARTMENT OF MICROBIOLOGY
Phone: Chairman: (716) 831-2907 Others: 831-2907

Chairman: F. Milgrom
Professors: J.G. Allen, G.A. Andres, C.E. Arbesman (Clinical), E.H. Beutner, G. Cudkowicz, A.L. Grossberg (Research), R Guthrie (Research), J.H. Kite, Jr., J.M Merrick, F. Milgrom, J.F. Mohn, E. Neter, P.L. Ogra, C J. van Oss, D. Pressman (Research), K. Wicher
Associate Professors: C.J. Abeyounis, W.R. Bartholomew (Clinical), P.E. Bigazzi, E. Cohen (Research), T.D. Flanagan, E.A. Gorzynski, T.S. Grafton (Research), K. Kano, R.M. Lambert, C.J. O'Connell (Clinical), J. Puleo (Clinical), H.Y. Whang (Research)
Assistant Professors: A.R. Collins, R.K. Cunningham, R.T. Evans, C. Gillman (Research), W.L. Hale (Clinical), L.L. Matz, W.L. Miethaner, D.T. Mount (Clinical), R.J. Nisengard, H.G. Rosamilia (Research), M. W. Stinson, S.P. Targowski, H.R. Thacore, J.I. Wypych (Research)
Instructors: P. Lal (Research), U. Loza (Research, R.W. Plunkett, K. Zelenski (Clinical)
Teaching Fellows: 1
Recipients of local endowments: 3

Field Stations/Laboratories: Virology Laboratory - Buffalo, Blood Group Research Unit - Buffalo, Immunochemistry Laboratory - Buffalo, Immunofluorescence Laboratory-Buffalo, Tissue Immunology Laboratory - Buffalo, Cell Immunology Laboratory - Buffalo, Microbial Physiology Laboratory - Buffalo

Degree Program: Microbiology

Graduate Degrees: M.A., Ph.D.
Graduate Degree Requirements: Master: 36 hours courses, research report or publication of paper. Doctoral: Courses as selected by committee, thesis, oral examination.

Graduate and Undergraduate Courses Available for Graduate Credit: Medical Microbiology and Immunology, Medical and Oral Microbiology, Clinical Diagnostic Bacteriology, Virology and Serology, Supervised Teaching, Critical Analysis of Microbiological Literature, Fundamentals of Virology, Advanced Medical Microbiology, Human Blood Group and Transplantation Immunology, Special Instruction in Microbiology, Principles and Techniques of Cell and Organ Culture, Principles of Immunology, Fundamentals of Immunochemistry, Clinical Immunology Seminars, Microbial Physiology, Microbiology Seminar

Research Courses: Immunopathology, Microbial Physiology, Cellular Immunology, Immunology of Blood Groups, Oral Microbiology, Tissue Immunology and Transplantation, Immunohistology, Bacterial Immunology, Tumor and Transplantation Immunology, Immunohematology, Immunochemistry, Tissue and Transplantation Immunology, Cell Biology of Humoral and Cell-Mediated Immunity, Animal Virology, Heterogenetic Bacterial Antigens, Clinical Microbiology, Virology

DEPARTMENT OF ORAL BIOLOGY
Phone: Chairman: (716) 831-2844 Others: 831-9000

Chairman: S.A. Ellison
Professors: S.A. Ellison, J.A. English, E. Hausmann, P.H. Staple
Associate Professors: C. DeLuca, R.J. Genco, J.K. Gong, P.A. Mashimo, W.A. Miller
Assistant Professors: F.G. Emmings, R.T. Evans, G. Krygier, M.J. Reed

Degree Program: Oral Biology

GraduateDegree: Ph.D.
Graduate Degree Requirements: Doctoral: Admission generally requires D.D.S. or equivalent, Preliminary examination, Defense of thesis

Graduate and Undergraduate Courses Available for Graduate Credit:Cell Biology, Histochemical Techniques, Oral Microbiology, Immunologic Techniques, Radiation Biology Techniques, Dynamics of Bone, Oral Histology, Seminars in Periodontal Biology, Research

DEPARTMENT OF PATHOLOGY
(No information available)

DEPARTMENT OF PHYSIOLOGY
Phone: Chairman: (716) 831-2738 Others: 831-2738

Head: D.W. Rennie
Professors: J.C. Eccles, H. Rahn, J.W. Boylan, L.E. Farhi, C.V. Paganelli, W.K. Noell, D.W. Rennie, H.D. Van Liew
Associate Professors: B. Bishop, P.M. Hogan, B.J. Howell, E. Kownig, R.A. Nicoll (Research), E.A. Ohr, A.J. Olszowka, R.B. Reeves, R. Srebro
Assistant Professors: B. DeW. Erasmus (Clinical), G.I. Allen, R.F. Miller (Research), S.A. Nunneley, D. Organisciak (Research), D.R. Pendergast (in Health Education), J.B. VanLiew (Research)
Instructors: D.H. Nielsen (Part-time), N.L. Urbscheit
Research Associates: 2

Degree Programs: Physiology

Graduate Degree: Ph.D.
Graduate Degree Requirements: Doctoral: Minimum of 3 years graduate study, Minimum of 72 semester hours, Thesis

Graduate and Undergraduate Courses Available for Graduate Credit: Neurophysiology, Human Physiology, Mammalian Physiology, Cellular Physiology, Physiological Basis of Nervous and Sensory Diseases, Experiments in Physiology, Neurobiology: Cellular Mechanisms, Application to Computers to Physiological Problems, Clinical Applications of Acid-Base Physiology, Mammalian Physiology, Research Seminars in Neurophysiology, Research Seminars in Respiratory Physiology, The Electrophysiological Basis of Cardiac Arrhythmias, Neurosensory Colloquium, Advanced Topics in Neurophysiology, Experimental Research: Neurophysiology, Experimental Research: Cell Membrane Phenomena, Experimental Research: Cardiovascular Physiology, Experimental Research: Gas Transport,Mechanisms, Experimental Research: Renal Physiology, Experimental Research: Environmental Physiology

Roswell Park Graduate Division
Director: E.A. Mirand

DEPARTMENT FOR BIOLOGY

Professors: T. Hauschka (Research), C. Helmstetter (Research), W. McLimans (Research), E.A. Mirand (Research), K. Paigen (Research)
Associate Professors: D. Doyle (Research), O. Piercucci (Research), C. Saltarelli (Research)
Assistant Professors: V.M. Chapman (Research), E. Cohen (Research), R. Elliott (Research), B. Paigen (Research), T. Shows (Research), R.T. Swank (Re-

search)

Graduate Degrees: M.S., Ph.D.
Graduate Degree Requirements: Master: 30 hours course work, comprehensive test, thesis or completion of special project Doctoral: Normally 3 academic years of graduate study (minimum 2 years). Work supervised by committee, thesis (one reader from outside department)

Graduate and Undergraduate Courses Available for Graduate Credit: Problems in Zoology for Graduate Students, Problems in Cell Physiology for Graduate Students, Problems in Genetics for Graduate Students, Bioengineering, Cell Biology, Molecular Biology of Disease, Molecular Genetics, Molecular Genetics, Regulation Mechanisms in Eukaryotic Cells

DEPARTMENT FOR BIOPHYSICS
Phone: (716) 854-3063

Professors: H. Box (Research), D. Harker (Research), R. Rein (Research), L. Weiss (Research)
Associate Professors: J. Berger (Research), G. Kartha (Research), P.D. Papahadjopoulos (Research), R. Parthasarathy (Research)
Assistant Professors: R.E. Baier (Research), E. Mayhew (Research), S. Nir (Research), G. Osman (Research)

Graduate Degrees: M.S., Ph.D.
Graduate Degree Requirements: Master: 30 hours course work, comprehensive test, thesis or completion of special project Doctoral: Normally 3 academic years of graduate study (minimum 2 years). Work supervised by committee, thesis (one reader from outside department)

Graduate and Undergraduate Courses Available for Graduate Credit: Surface Phenomena Relevant to Biology and Pathology, Magnetic Resonance Spectroscopy, Chemical Physics of Crystals, X-Ray Diffraction and Crystal Structure, Experimental X-Ray Diffraction, Quantum Biophysics, Advanced Photography Theory and Practice

DEPARTMENT OF BIOSTATISTICS
Phone: Chairman: (716) 854-3063

Professor: I. Bross (Research)
Assistant Professors: L. Blumenson (Research), N. Slack (Research)

Graduate Degrees: M.S., Ph.D.
Graduate Degree Requirements: Master: 30 hours course work, comprehensive test, thesis or completion of special project Doctoral: Normally 3 academic years of graduate study (minimum 2 years). Work supervised by Committee. Thesis (one reader from outside department)

Graduate and Undergraduate Courses Available for Graduate Credit: Biomedical Applications of Biostatistics, Introduction to Biostatistics, Graduate Research

DEPARTMENT OF PHYSIOLOGY

Phone: Chairman: (716) 854-3063

Professors: T. Dao (Research), A. Sandberg (Research) J. Sokal (Research)
Associate Professors: J. Cairns (Research), R. Fiel (Research), E. Sarcione (Research), K. Shimaoka (Research)
Assistant Professors: C. Huang (Research), P.R. Libby (Research), M.P. McGarry (Research), B. Munson (Research), D. Sinha (Research)

Graduate Degrees: M.S., Ph.D.
Graduate Degree Requirements: Master: 30 hours course work, comprehensive test, thesis or completion of special project Doctoral: Normally 3 academic years of graduate study (minimum 2 years). Work supervised by Committee, Thesis (one reader from outside department)

Graduate and Undergraduate Courses Available for Graduate Credit: Endocrine Physiology, Pathophysiology, Physiology Seminar, Problems in Physiology for Graduate Students, Cell Physiology

DEPARTMENT OF BIOCHEMISTRY
Phone: Chairman: (716) 854-3063

Professors: M. Laskowski (Research), G. Markus (Research), F. Rosen (Research), H. Weinfeld (Research), C. Wenner (Research)
Associate Professors: C. Carruthers (Research), C. Coutsogeorgopoulos (Research), M Hakala (Research) A. Mittelman (Research), M. Pine (Research), G. Tritsch (Research)
Assistant Professor: R. Kirdani (Research)

Graduate Degrees: M.S., Ph.D.
Graduate Degree Requirements: Master: 30 hours course work, Comprehensive test, thesis or completion of special project Doctoral: Normally 3 academic years of graduate study (minimum 2 years). Work supervised by Committee, Thesis (one reader from outside department)

Graduate and Undergraduate Courses Available for Graduate Credit: Biochemical Oncology, Biochemistry and Chemistry of Hormones

DEPARTMENT OF EXPERIMENTAL PATHOLOGY
Phone: Chairman: (716) 854-3063

Professors: J.L. Ambrus (Research), G.P. Murphy (Research), E. Sproul (Research)
Associate Professors: C.W. Aungst (Research), E. Holyoke (Research), E. Klein (Research), J. Pickren (Research), J.E. Plager (Research), L. Sinks (Research), L. Stutzman (Research)
Assistant Professors: R. Bourke (Research), A. Bremer (Research), T.M. Chu (Research), I.B. Mink (Research), S. Piver (Research), J.S. Wang (Research), J. Webster (Research)

Graduate Degrees: M.S., Ph.D.
Graduate Degree Requirements: Master: 30 hours course work, Comprehensive test, thesis or completion of special project Doctoral: Normally 3 academic years of graduate study (minimum 2years). Work supervised by Committee, Thesis (one reader from outside department)

Graduate and Undergraduate Courses Available for Graduate Credit: Chemical Pathology, Problems in Pathology, Seminar: Recent Advances in Pathology, Pathophysiology, Techniques in Pathology, Introduction to Histology, Functional Anatomy, and Neuroanatomy, Graduate Research, Geographic Pathology, Introduction to Histology, Functional Anatomy and Neuroanatomy

DEPARTMENT OF MICROBIOLOGY
Phone: Chairman: (716) 854-3063

Professors: W.A. Carter (Research), W.H. Munyon (Research)
Associate Professors: R.F. Buffett (Research), A. Fj Fjelde (Research), J.S. Horoszewicz (Research), J. Minowada (Research), R.F. Zeigel (Research)
Assistant Professors: S.K. Arya (Research), M.J. Evans (Research), O.A. Holtermann (Research), K.F. Manly (Research), W.F. Noyes (Research), J.A. O'Malley (Research), L. Pothier (Research), E. Sulkowski (Research), M.J. Surgalla (Research), R.L. Taber (Research)

Graduate Degrees: M.S., Ph.D.
Graduate Degree Requirements: Master: 30 hours

course work, Comprehensive test, thesis or completion of special project Doctoral: Normally 3 academic years of graduate study (minimum 2 years). Work supervised by Committee, Thesis (one reader from outside department)

Graduate and Undergraduate Courses Available for Graduate Credit: Problems in Microbiology, Viral Oncology, Microbiology Seminar, Fundamentals of Virology, Graduate Research

DEPARTMENT OF PHARMACOLOGY
Phone: Chairman: (716) 854-3063

Professors: C. Ambrus (Research), R. Ellison (Research), F. Mihich (Research)
Associate Professors: A. Bloch (Research), G. Chheda (Research), S. Gailani (Research), T. Gessner (Research), S.F. Zakrzewski (Research)
Assistant Professors: C. Dave (Research), G. Grindey (Research), H. Gutoo (Research)

Graduate Degrees: M.S., Ph.D.
Graduate Degree Requirements: Master: 30 hours course work, Comprehensive test, thesis or completion of special project Doctoral: Normally 3 academic years of graduate study (minimum 2 years). Work supervised by Committee, Thesis (one reader from outside department)

Graduate and Undergraduate Courses Available for Graduate Credit: Principles of Drug Action, Vitamins and Coenzymes in Pharmacology, Advances in Cancer Therapy, Principles of Methods of Toxicology

STATE UNIVERSITY OF NEW YORK COLLEGE AT CORTLAND

Cortland, New York 13045 Phone: (607) 763-2715
Dean of Graduate Studies: A.M. Banse
Dean of College: H.K. Reynolds

DEPARTMENT OF BIOLOGICAL SCIENCES
Phone: Chairman: (607) 753-2715

Chairman: J.A. Gustafson
Professors: L.I. Cohen, A.H Cook, J.A. Gustafson, K. Horak, E.C. Waldbauer
Associate Professors: R.C. Doney, T.D. Fetzgerald, D.J. Houck, M. Nasrallah, W.R. Newman, A.P. Spence, C.R. Wilson
Assistant Professors: B.L. Batzing, E.B. Mason, N.B. Reynolds
Teaching Fellows: 2

Field Stations/Laboratories: Outdoor Education Center - Raquette Lake, New York, Hoxie Gorge Campus - Cortlandville, New York

Degree Program: Biology

Undergraduate Degree: B.S.
Undergraduate Degree Requirements: Liberal Arts - 36 hours, Biology, 6 hours Mathematics (Calculus A and B), 8 hours Physics, 12 hours Chemistry, 24 hours Basic studies outside Mathematics and Science, 4 hours Physical Education, English Composition, For B.A. - Intermediate proficiency in one foreign language, Secondary Biology - 30 hours Biology, 6 Mathematics (Calculus A and B), 8 Physics, 12 Chemistry, 6 Geology, 20 hours in Education including Practice Teaching, 24 hours Basic Studies outside Mathematics and Science, 4 hours Physical Education, English Compositition, Total 124 hours in each.

Graduate Degree: M.S.Ed.
Graduate Degree Requirements: Master: 6 hours Education, 15-18 in sciences (recommend 15 in Biology), 3 hours in seminar, 6 hours electives, Comprehensive examination

Graduate and Undergraduate Courses Available for Graduate Credit: Invertebrate Zoology (U), Vertebrate Zoology (U), Microbiology (U), Cellular Physiology (U), Conservation of Natural Resources (U), Freshwater Algae (U), Field Biology (U), Entomology (U), Animal Behavior.(U), Plant Physiology (U), Ornithology (U), General Ecology (U), Taxonomy of Vascular Plants (U), Vertebrate Physiology (U), Radiation Biology (U), Comparative Anatomy (U), Embryology (U), Plant Morphology (U), Histology, Parasitology, Comparative Physiology, Fungi, Limnology, Endocrinology, Plant Anatomy, Ichthyology, Physiological Genetics, Biological Techniques and Materials, Special Problems in Biology

STATE UNIVERSITY OF NEW YORK COLLEGE OF ENVIRONMENTAL SCIENCES

Syracuse, New York 13210 Phone: (315) 473-8611
Dean of Graduate Studies: W.E. Graves
Dean of College: S.W. Tanenbaum

DEPARTMENT OF FOREST BOTANY AND PATHOLOGY
Phone: Chairman: (315) 473-8801 Others: 473-8611

Chairman: H.B. Tepper
Professor: E.H. Ketchledge, J.L. Lowe, S.B. Silverborg, H.B. Tepper, F.A. Valentine, C.J.K. Wang, H.E. Wilcox, R.A. Zabel, L.L. McDowell (Research)
Associate Professors: D.H. Griffin, P.D. Manion, M. Schaedle
Assistant Professors: J.W. Geis, M. Flashner
Research Assistants: 13

Field Stations/Laboratories: Cranberry Lake Biological Station - Cranberry Lake, New York

Degree Programs: Biochemistry, Biological Control, Biological Sciences, Botany, Ecology, Forest Botany, Plant Pathology, Physiology

Undergraduate Degree: B.S.
Undergraduate Degree Requirements: A year of Biology, Genetics, General Ecology, General Physiology, Statistics, Calculus through Integral, General Physics, General Chemistry, Organic Chemistry, 9 hours of Humanities and Social Sciences, 12 hours in a major area of Biology, 6 hours in a minor area of Biology, 30 credits of electives

Graduate Degrees: M.S., Ph.D.
Graduate Degree Requirements: Master: 30 credit hours of course work and a thesis, Doctoral: course work and thesis

Graduate and Undergraduate Courses Available for Graduate Credit: Advanced Limnology (U), Ecology of Freshwaters (U), Chemical Ecology (U), Fundamentals of Genetics (U), Fundamentals of Genetics Laboratory (U), Evolutionary Genetics (U), Laboratory in Evolutionary Genetics (U), Cytogenetics, Physiological Ecology, Population Dynamics, Bioclimatology, Histochemical Techniques, Membranes and Biological Transport, Biology Seminar, Mycology (U), Phycology (U), Systematic Botany (U), Adirondack Flora (U), Ecology of Forest Communities (U), Forest Ecology (U), Bryoecology (U), Plant Physiology (U), Plant Physiology Laboratory (U), Principles of Forest Pathology (U), Wood Deterioration by Microorganisms (U), Plant Anatomy (U), Fungus Physiology, Phytopathology, Advanced Systematic Botany, Techniques in Plant Physiology, Topics in Phytopathology, Botany Seminar, Research in Forest Botany, Advanced Mycology, Mycomycetes, Phycomycetes, Fungi Imperfecti, Physiology of Growth and Development, Population Genetics, Thesis, Elements of Forest Entomology (U), Forest and Shade Tree Entomology (U), Insects Affecting Forest Products (U), Survey of Entomological Literature and History (U), Forest and Aquatic Insects (U), Insect Morphology (U), Medical

Entomology (U), General Insect Taxonomy, Aquatic Entomology, Insect Physiology, Toxicology of Insecticides, Population Dynamics of Forest Insects, Special Topics in Forest Entomology, Seminar, Research Problems in Forest Entomology, Advanced Insect Taxonomy, Invertebrate Zoology (U), Vertebrate Taxonomy (U), Ichthyology (U), Terrestrial Community Ecology (U), Invertebrate Ecology (U), Vertebrate Ecology (U) Limnology (U), Ecology of Adirondack Fishes (U), Vertebrate Population Ecology (U), Animal Physiology (U), Fishery Biology (U), Wildlife Ecology (U), Wildlife Methods (U), Principles of Animal Behavior (U), Behavioral Ecology (U), Histology (U), Vertebrate Anatomy (U), Invertebrate Symbiosis, Comparative Endocrinology, Vertebrate Behavior, Forest Zoology Trip, Topics in Soil Invertebrate Ecology, Zoogeography, Advanced Wildlife Management, Forest Zoology, Seminar, Problems in Forest Zoology, Physiological Ecology, Invertebrate Physiology, Topics in Wildlife Biology, Topics in Animal Behavior

DEPARTMENT OF FOREST CHEMISTRY
Phone: Chairman: (315) 473-8824

Chairman: K.J. Smith
SUNY Distinguished Professor: M. Szwarc
Professors: R. Lalonde, J Meyer, C. Schuerch, M. Silverstein, J Smid, K. Smith, M Szwarc, T. Timell
Associate Professors: A. Sarko, D. Walton
Assistant Professors: P. Caluwe, B. Levin

Degree Programs: Biochemistry, Forest Chemistry

Undergraduate Degree: B.S.
Undergraduate Degree Requirements: Master: 60 hours thesis Doctoral: Thesis

Graduate and Undergraduate Courses Available for Graduate Credit: Nuclear and Radiation Chemistry (U), Biochemistry (U), Biochemistry Laboratory (U), Principles of Biochemistry (U), Topics in Plant Biochemistry, Polymer Techniques (U), Polymer Chemistry, Polymer Physical Chemistry, Physical Chemistry of Polymers, Organic Natural Products Chemistry

DEPARTMENT OF FOREST ENTOMOLOGY
Phone: Chairman: (315) 473-8754 Others: 473-8611

Chairman: J.B. Simeone
Professors: J. Brezner, H.C. Miller, J.B. Simedne
Associate Professors: D.C. Allen, R.W. Campbell (Adjunct), H. Jamnback (Adjunct, Research), F.E. Kurczewski, G.N. Lanier, T. Nakatsugawa
Assistant Professors: C.D. Morris (Adjunct)
Teaching Fellows: 2
Research Assistants: 12

Field Stations/Laboratories: Cranberry Lake Biological Station - Cranberry Lake, New York, Pack Demonstration Forest - Warrensburg, New York, Adirondack Ecological Center - Newcomb, New York

Degree Program: Biology, Entomology, Forest Entomology

Undergraduate Degree: B.S. (Biology)
Undergraduate Degree Requirements: Not Stated

Graduate Degrees: M.S., Ph.D.
Graduate Degree Requirements: Master: Thirty semester hours of courses and thesis combined, courses selected on basis of student background and goals Doctoral: Dissertation guided by major professor and selected faculty committee, courses or self-study desirable to research goal, Tools: statistics to analysis of variance and a foreign language

Graduate and Undergraduate Courses Available for Graduate Credit: Survey of Entomological Literature and History, Insect Morphology, Medical Entomology, Aquatic Entomology, Insect Physiology, Toxicology of Insecticides, Population Dynamics of Forest Insects, Special Topics in Forest Entomology, Research Problems in Forest Entomology, Advanced Insect Taxonomy

DEPARTMENT OF FOREST ZOOLOGY
Phone: Chairman: (315) 473-8841 Others: 473-8611

Chairman: M.M. Alexander
Professors: M.M Alexander, R.C. Hartenstein, H. H. Payne, W.L. Webb
Associate Professors: R.E. Chambers, D.L. Dindal, W.E. Graves, D. Muller-Schwarze, E.O. Price, R.G. Werner, W.D. Youngs
Senior Research Associates: D.F. Behrend, R.H. Brocke
Research Associate: G.F. Mattfeld
Visiting Lecturer: 1
Teaching Fellows: 7
Research Assistant: 1

Field Stations/Laboratories: Huntington Wildlife Forest Station - Newcomb, New York

Degree Programs: Forest Biology, Forest Zoology, Wildlife Biology, Fishery, Soil Invertebrate, Physiology and Behavior

Undergraduate Degree: B.S. (Forest Biology)
Undergraduate Degree Requirements: 131 credit hours, 1 year English, Mathematics through Integral Calculus, two semesters Organic Chemistry, 4 semesters Social Sciences and Humanities, General Biology, electives in Taxonomy, Anatomy, Physiology, Ecology, Genetics, 12 credits in concentration areas.

Graduate Degrees: M.S., Ph.D. (Forest Zoology)
Graduate Degree Requirements: Master: 30 credit hours including 6-18 credits for thesis project Doctoral: No set requirements - program developed individually

Graduate and Undergraduate Courses Available for Graduate Credit: Terrestrial Community Ecology (U), Limnology (U), Ecology of Adirondack Fishes, Vertebrate Population Ecology, Invertebrate Symbiosis, Comparative Endocrinology, Vertebrate Behavior, Forest Zoology Trip, Topics in Soil Invertebrate Ecology, Zoogeography, Advanced Wildlife Management , Forest Zoology Seminar, Problems in Forest Zoology, Physiological Ecology, Invertebrate Physiology, Topics in Wildlife Biology, Topics in Animal Behavior

DEPARTMENT OF SILVICULTURE AND FOREST INFLUENCES
Phone: Chairman: (315) 473-8646 Others: 473-8611

Chairman: J.W. Johnson
Professors: A.R. Eschner, J.W. Johnson, A.L. Leaf, L. S. Minckler (Adjunct)
Associate Professors: J.V. Berglund, P.E. Black, P.J. Craul, L.P. Herrington, R.V. Lea, N.R. Richards, R.D. Westfall (Research)
Assistant Professor: G.M. Heisler (Adjunct)
Temporary Appointment Instructor: 1

Field Stations/Laboratories: Heiberg Forest - Tully, New York, Huntington Forest - Newcomb, New York, Pack Memorial Forest - Warrensburg, New York, Experiment Station - Syracuse, New York

Degree Program: Silviculture

Undergraduate Degree: B.S.
Undergraduate Degree Requirements: 137 credit hours, (99 core or required courses, 38 electives)

Graduate Degrees: M.S., Ph.D.
Graduate Degree Requirements: Master: 30 credit hours including thesis, Statistics, no foreign language Doctoral: Completion of committee-prescribed course program; dissertation, statistics, one foreign language

STATE UNIVERSITY OF NEW YORK
AT FREDONIA

Fredonia, New York 14063 Phone: (716) 673-3111
Dean of Graduate Studies: M.D. Dowd
Dean of College: R. Butwell

Chairman: K.A. Fox
Distinguished Teaching Professor: A.H. Benton
Professors: R.W. Boenig, W.F. Stanley (Emeritus), K.G. Wood, G.P. Zimmer
Associate Professors: S.J. Cudia, K.A. Fox, B. Polacek (Adjunct), M.L. Sharma
Assistant Professors: K. Bernstein, V. Dunham, A. Lisanti, K. Mantal, R. Mitchell, S. Nicholson, W. Yunghans
Teaching Fellows: 5
Research Assistants: 2

Field Stations/Laboratories: Lakeside Laboratory - Lake Erie, Chautauqua Laboratory - Chautauqua Lake

Degree Programs: Biochemistry, Biological Education, Biology, Ecology, Medical Technology, Physiology

Undergraduate Degrees: B.A., B.S.
Undergraduate Degree Requirements: Students in Biology, Secondary Education must complete a semester of Professional Education; B.A. candidates, a year of language. All must take 36 semester hours in Biology which must include one semester of each of the following: General Biology, General Botany, General Zoology, Environmental Biology, Biochemistry, Genetics or Developmental Biology, and Microbiology or Cell Biology. Also: A year of General Chemistry, a year of Organic Chemistry, a year of College Physics, and one semester of Calculus

Graduate Degree: M.S.
Graduate Degree Requirements: Master: All candidates must take one year of Biochemistry. Candidates for M.S. Ed. take 30 hours which may include 6 hours Education, Candidates for M.S., Biology complete 6 hour thesis research and write and defend thesis. All candidates must successfully pass comprehensive examination.

Graduate and Undergraduate Courses Available for Graduate Credit: Biochemistry of Life Processes I and II, Enzymes, Intermediary Metabolism, Molecular Genetics, Cytology and Cytochemistry, Advanced Microbiology, Plant Morphogenesis, Experimental Embryology, Radiation Biology, Comparative Endocrinology, Methods in Biological Research, General Ecology, Limnology, Mammalogy, Physiology of Behavior, Environmental Science, Current Concepts in Biology, Cell Regulation, Reproductive Physiology, Advanced Ecology

STATE UNIVERSITY OF NEW YORK
AT GENESEO

Geneseo, New York 14454 Phone: (716) 245-5211
Dean of Graduate Studies: B. Ristow
Dean of College: T. Colahan

DEPARTMENT OF BIOLOGY
Phone: Chairman: (716) 245-5301 Others: 245-5211

Chairman: A.F. Reid
Professors: H.S. Forest, L.J. King, A.F. Reid, A. Reid, E. Ritter, R.M. Roecker
Associate Professors: H.L. Huddle, D.L. Meyer, R.W. Reilly, D H.B. Ulmer, Jr.
Assistant Professors: A. Barnitt, J.G. Baust, W.J. Graham, A.H. Latorella, J.V. Logomarisino, C.E. Vasey

Degree Programs: Biology, Medical Technology

Undergraduate Degrees: B.A., B.S.
Undergraduate Degree Requirements: 32 semester hours Biology, including at least one plant science and one animal science course. Chemistry through Organic Mathematics through Calculus, one year Geology or Physics

Graduate Degrees: M.A., M.S.
Graduate Degree Requirements: Master: 32 credit hours of which at least 24 must be at graduate level. One foreign language, thesis, final oral

Graduate and Undergraduate Courses Available for Graduate Credit: Modern Applications of Biological Science, Economic Biology, The Origin and Implications of Evolution, Advanced Plant Taxonomy, Phycology, Bryology and Lichens, Mycology, Mechanics of Inheritance, Ecological Microbiology, Parasitology, Protozoology, Ichthyology, Herpetology, Principles of Mammalogy, Developmental Plant Physiology, Comparative Animal Physiology, Physiological Basis of Animal Behavior, Environmental Physiology, Cytology, Plant Ecology, Animal Ecology, Limnology, Horticultural Science, Plants of New York State

STATE UNIVERSITY OF NEW YORK
COLLEGE AT NEW PALTZ

New Paltz, New York 12561 Phone: (914) 257-2552
Dean of College: P. Vukasin

DEPARTMENT OF BIOLOGY
Phone: Chairman: (914) 257-2552

Chairman: S.J. Spencer
Professors: H. Meng, L. Nydegger, J. Slater
Associate Professors: D. Baker, A. Bregman, H.H. Ho, D. Krieg, A. Nemerofsky, H. Osburg, T. Santoro, P. Stein
Assistant Professors: D. Moran, B.L. Redmond
Instructor: C. Rietsma

Degree Program: Biology

Undergraduate Degrees: B.A., B.S.
Undergraduate Degree Requirements: Not Stated

Graduate Degrees: M.A., M.S.
Graduate Degree Requirements: Not Stated

Graduate and Undergraduate Courses Available for Graduate Credit: Not Stated

STATE UNIVERSITY OF NEW YORK
AT ONEONTA

Oneonta, New York 13820 Phone: (607) 431-3500
Dean of Graduate Studies: J. Sanik
Dean of College: A. Acholonu

DEPARTMENT OF BIOLOGY
Phone: Chairman: (607) 431-3344 Others: 431-3703

Chairman: J.G. New
Professors: J.K. Ang, W. Bukovsan, W. Butts, J.G. Holway, J.G. New, C.A. Ryder, W.D. Wilson
Associate Professors: B.R. Dayton, W.N. Harman, C. D. Marr, M. Singh, L.P. Sohacki
Assistant Professor: W.J. Settle
Instructor: R.R. Phillips
Graduate Assistants: 2

Field Stations/Laboratories: Biological Field Station - Cooperstown, New York

Degree Program: Biology

Undergraduate Degrees: B.A., B.S.
Undergraduate Degree Requirements: 122 hours MAJOR FIELD - General Botany, General Zoology, Genetics, Microbiology, General Physiology, General Ecology, and 18 hours in additional biology courses RELATED WORK: General Chemistry, Organic Chemistry, Physics and Mathematics. Additional Hours. 12 to ful-

fill distribution requirements, 51 electives.

Graduate Degree: M.A.
Graduate Degree Requirements: 1. At least 32 semester hours of approved courses beyond the baccalaureate degree consisting of: a. at least 15 semester hours in 300 and 400 level courses, b. at least 18 semester hours in Biology courses other than seminar, thesis, and independent study, Biochemistry and Statistics c. Biochemistry (3 semester hours), and Statistics (3 semester hours), unless completed on the undergraduate level d. No more than 6 semester hours credit each for Independent Study and/or Thesis. 2. Research Thesis and oral defense. 3. Comprehensive examination.

Graduate and Undergraduate Courses Available for Graduate Credit: Population Biology (U), Developmental Biology (U), Evolution (U), Quantitative Biology (U), Computer Methods in Biology (U), Economic Botany (U), Generally Mycology (U), Morphological Survey of the Plant Kingdom (U), Plant Geography (U), Vascular Plants (U), Trees and Shrubs (U), Invertebrate Zoology (U), Entomology (U), Parasitology (U), Histology (U), Endocrinology (U), Natural History of Vertebrates (U), Ornithology (U), Animal Behavior (U), Physiology of Plants (U), Comparative Anatomy of Vertebrates (U), Cell Biology (U), Human Ecology (U), Aquatic Biology (U), Limnology (U), Population Genetics, Advanced Developmental Biology, Cytology, Morphology of Vascular Plants, Plant Anatomy, Plant Taxonomy, Protozoology, Malacology, Medical Entomology, Endocrinology of Reproduction, Ichthyology, Mammalogy, Advanced Plant Physiology, Environmental Pollution, Plant Ecology, Behavioral Ecology

STATE UNIVERSITY OF NEW YORK COLLEGE AT OSWEGO

Oswego, New York 13126 Phone: (315) 341-2500
Dean of Graduate Studies: R.A Terry
Dean of Division of Arts and Sciences: R.P. Soter

DEPARTMENT OF BIOLOGY
(No reply received)

DEPARTMENT OF ZOOLOGY
Phone: Chairman: (315) 341-3031 Others: 341-2500

Chairman: J.R. Harrison
Professors: J.R. Harrison, G.R. Maxwell, J.A. Miller H.O. Powers, E.G. Wise
Associate Professors: R.A. Engel
Assistant Professors: J.T. Brunson, J.A. Lackey, K.H Martin, S.O. Nelson, Jr.
Instructors: R. DeNeve

Field Stations/Laboratories: Rice Creek Biological Field Station - Oswego, New York, Lake Ontario Environmental Laboratory - Oswego, New York

Degree Programs: Biology, Zoology

Undergraduate Degree: B.A.
Undergraduate Degree Requirements: 4 hours Heredity and Reproduction, 4 hours Animal Function and Adaptation, 23 hours elected in areas of specialization with 6 hours possible in another academic department with approval of the advisor, 15-16 hours Chemistry, 6 hours Mathematics, 8 hours Physics

Graduate Degree: M.S.
Graduate Degree Requirements: Not Stated

Graduate and Undergraduate Courses Available for Graduate Credit: Advanced Ornithology, Vertebrate Populations, Advanced Invertebrate Zoology, Advanced Endocrinology, Advanced Mammalogy, Advanced Field Zoology, Advanced Animal Ecology, Independent Study

STATE UNIVERSITY OF NEW YORK COLLEGE AT PLATTSBURGH

Plattsburgh, New York 12901 Phone: (518) 564-2000
Dean of Faculty of Science and Mathematics: H.J. Perkins

DEPARTMENT OF BIOLOGICAL SCIENCES

Chairman: J.S. Waterhouse
Professors: R.M. Clark, H.Z. Liu, R.D. Moore, V. Munk, F.R. Nevin (Emeritus), R.D. Sudds, P.C. Walker, J.S. Waterhouse
Associate Professors: G.K. Gruendling, H. Klein, D. Lee, J. Nolan, K.S. Schin
Assistant Professors: W.D. Graziadei, C.L. Harris, J.C. McGraw
Visiting Lecturers: 6-12/year
Graduate Teaching/Research Assistants: 7

Degree Programs: Biochemistry, Biochemistry and Biophysics, Biological Sciences, Biophysics, Ecology, Medical Technology, Microbiology

Undergraduate Degrees: B.A., B.S.
Undergraduate Degree Requirements: 122 credit-hours required for graduation, including at least 48 hours of upper division courses.

Graduate Degrees: M.A., M.S.
Graduate Degree Requirements: Master: M.A. - 30 graduate credits, of which up to 9 hours in thesis research, a M.A. Thesis is required M.S. - 32 credits, and a thesis or research report is required.

Graduate and Undergraduate Courses Available for Graduate Credit: Advanced Genetics, Microbial Physiology, Evolution of Biological Concepts, Biological Aspects of Environmental Conservation, Marine Ecology, Plant Ecology, Mammalogy, Ornithology, Parasitology, Protozoology, Animal Cells and Their Viruses, Advanced Invertebrate Zoology, Biophysics, Molecular Biophysics, Bioenergetics, Insect Ecology, Concepts of Developmental Biology, Neuroscience, Graduate Seminar, Recent Advances in Biology

STATE UNIVERSITY OF NEW YORK COLLEGE AT POTSDAM

Potsdam, New York 13676 Phone: (315) 268-2700
Dean of Graduate Studies: K. Gant
Dean of School of Liberal Arts: R.E. Hutcheson

BIOLOGY DEPARTMENT
Phone: Chairman: (315) 268-2985 Others: 268-2986

Chairman: W.G. Hamilton
Professors: R.H. Cerwonka, P.E. Hafer, W.G. Hamilton, G.R. Isenberg, P. Juo
Associate Professors: A.D. Robinson, L.D. Simone
Assistant Professors: P.A. LaHaye, M.S. Rutley
Instructor: C.E. Foster, D.M. Osterberg
Graduate Assistant: 1

Degree Program: Biology

Undergraduate Degree: B.A.
Undergraduate Degree Requirements: 30 semester hours of Biology, including a Genetics course and a Physiology course based on one year of Organic Chemistry.

Graduate Degree: M.A.
Graduate Degree Requirements: Master: 30 semester hours of Biology, including at least 6 to 8 credits in research and a research thesis.

Graduate and Undergraduate Courses Available for Graduate Credit: Advanced Cell Physiology, Animal Reproduction, Graduate Genetics, Cell and Tissue Culture, Advanced Microbiology, Limnology

STATE UNIVERSITY OF NEW YORK
DOWNSTATE MEDICAL CENTER
450 Clarkson Avenue Phone: (212) 270-1000
Dean of Graduate Studies: P. Dreizen

College of Medicine
Dean: L. Laster

DEPARTMENT OF ANATOMY
Phone: Chairman: (212) 270-1014 Others: 270-1015

Chairman: J.B. Hamilton
Professors: D.H. Ford, I.M. Murray, W. Riss, R.S. Speirs, G.B. Talbert
Associate Professors: B.S. Dornfest, M. Halpern, A. Hirschman, F.R. Scalia, B.S. Sherman, W.T. West
Assistant Professors: R.M. Carriere, J.E. Jakway, J.G. Stempak, R.T. Ward
Instructor: C.A. Cardasis, K.W. Chung, E.B. Cramer, M.S. Joshi, C.B. Ware
Teaching Fellows: 4
Research Assistants: 10

Degree Program: Anatomy

Graduate Degree: Ph.D.
Graduate Degree Requirements: Doctoral: Students must take specified courses, totalling 48 credits, after completion of which they must pass comprehensive written and oral examinations. Reading knowledge of one foreign language and teaching experience as a teaching assistant is required. Finally, a dissertation (awarding 24 credits) must be prepared and defended.

Graduate and Undergraduate Courses Available for Graduate Credit: Not Stated

DEPARTMENT OF BIOCHEMISTRY
(No information available)

DEPARTMENT OF MICROBIOLOGY AND IMMUNOLOGY
Phone: Chairman: (212) 270-1238 Others:270-1000

Chairman: S.I. Morse
Professors: J.P. Craig, G. Schiffman
Associate Professors: R. Bablanian, M. Howe, H. Neimark, B.M. Sultzer, J.J. Woodruff
Assistant Professors: D. Anshel, L.K. Eveland, J. Quigley
Visiting Lecturers: 10
Research Assistants: 6

Degree Programs: Microbiology, Immunology, Parasitology

Graduate Degree: Ph.D.
Graduate Degree Requirements: Not Stated

Graduate and Undergraduate Courses Available for Graduate Credit: Microbiology and Immunology, Immunochemistry, Immunobiology, Virology, Microbial Genetics

DEPARTMENT OF PHYSIOLOGY
(No reply received)

STATE UNIVERSITY OF NEW YORK
AT STONY BROOK
Stony Brook, New York 11790 Phone: (516) 246-5000
Dean of Graduate Studies: H. Weisinger

College of Arts and Sciences
Dean: A. Upton

DEPARTMENT OF CELLULAR AND COMPARATIVE BIOLOGY
Phone: Chairman: (516) 246-5000

Chairman: C. Walcott
Distinguished Professor: H.B. Glass
Professors: E. Carlson, F. Erk, H.B. Glass
Associate Professors: E. Battley, A. Carlson, L. Edmunds A. Krikariam, C. Lent, H. Lyman, R. Merriam, B. Tunik, C. Walcott
Assistant Professors: G. Katz, R. Knott, D.G. Smith

Field Stations/Laboratories: Marine Laboratory - Jamaica, Field Laboratory - Lincoln, Massachusetts

Degree Program: Biology

Undergraduate Degree: B.S.
Undergraduate Degree Requirements: Not Stated

Graduate Degrees: M.S., Ph.D.
Graduate Degree Requirements: Master: 24 credits in approved subjects, thesis and comprehensive examination Doctoral: Completion of approved courses of study, thesis, comprehensive examination and thesis defense, 2 years residence

Graduate and Undergraduate Courses Available for Graduate Credit: Not Stated

DEPARTMENT OF PHYSIOLOGY AND BIOPHYSICS
Phone: Chairman: (516) 444-2265 Others: 444-2287

Chairman: W.G. Van der Kloot
Professors: P. LeFevre, H. Levy
Associate Professor: M. Mendelson
Assistant Professors: S. Masiak, J. Fara, S. McLaughlin
Visiting Lecturers: 1
Teaching Fellows: 3
Research Assistants: 7

Degree Programs: Biophysics, Physiology

Graduate Degree: Ph.D.
Graduate Degree Requirements: Not Stated

Graduate and Undergraduate Courses Available for Graduate Credit: Not Stated

STATE UNIVERSITY OF NEW YORK
UPSTATE MEDICAL CENTER
766 Irving Avenue Phone: (315) 473-4570
Syracuse, New York 13210
Dean of Graduate Studies: D.C. Goodman
Dean of College: R.P. Schmidt

DEPARTMENT OF ANATOMY
Phone: Chairman: (315) 473-5145 Others: 473-5120

Chairman: D.C. Goodman
Professors: H.D. Stefano, D.C. Goodman, A.M. Garcia, F.A. Horel
Associate Professors: I. Ames, D. Robertson
Assistant Professors: C. Benzo, B. Berg, J. McKauna, J. Mitchell, D. Packard, D. Stelzner, M. Tseng
Instructor: B.P. Austin
Teaching Fellows: 2
Predoctoral Trainees: 8
Postdoctoral Fellow: 1

Degree Program: Anatomy

Graduate Degrees: M.S., Ph.D.
Graduate Degree Requirements: Master: Permission of Department since M.S. given only in special circumstances, Thesis, 30 hours credit Doctoral: Language requirement, Departmental Examination, Graduate School Qualifying examination, Dissertation, 90 credit hours

Graduate and Undergraduate Courses Available for Graduate Credit: Gross Anatomy, Microscopic Anatomy, Neurosciences (with Department of Physiology), Cellular Biology (with Department of Microbiology), Embryology, Anatomy Seminar, Methods in Anatomy, Special Topics in Anatomy, Comparative Neurology, Topics in Neurobehavior and Neurobiology

DEPARTMENT OF BIOCHEMISTRY
Phone: Chairman: (315) 473-5128 Others: 473-4570

Chairman: W.W. Westerfeld
Professors: R.J. Doisy, M.J. Kronman, W.W. Westerfeld
Associate Professors: C.L. Burger, J.C. Elwood, R.Y. Hsu
Assistant Professors: R.L. Cross, R.D. Jacobs
Visiting Lecturers: 1
Research Assistants: 4

Degree Program: Biochemistry

Graduate Degrees: M.S., Ph.D.
Graduate Degree Requirements: Master: Satisfactory completion of at least 20 semester hours of coursework and 10 semester hours of thesis work for a minimum total of 30 semester hours Doctoral: Requires a minimum of 90 credit hours of which 30 - 45 credit hours are usually devoted to the thesis

Graduate and Undergraduate Courses Available for Graduate Credit: General Biochemistry, Biochemistry in Clinical Pathology, Biochemistry Seminar, Advanced Biochemistry, Special Topics in Biochemistry, Enzymology, Isotopes in Biochemistry, Characterization of Biological Macromolecules, Research Techniques in Biochemistry, Research in Biochemistry

DEPARTMENT OF MICROBIOLOGY
Phone: Chairman: (315) 473-5453 Others: 473-4570

Chairman: G.G. Holz, Jr.
Professors: R.M. Dougherty, J.F. Mueller, M. Ycas
Associate Professors: B.F. Argyris, A.A. Marucci, F.D. Meyer, M.B. Rotheim
Assistant Professors: A.R. Bassel, B.A. Bassel, Jr.
Research Assistants: 10
Research Associate: 4

Degree Program: Microbiology

Graduate Degrees: M.S., Ph.D.
Graduate Degree Requirements: Master: 20 semester hours course credit, 10 semester hours Thesis credit, Research Thesis Doctoral: 30-60 semester hours course credit, 30-60 semester hours Thesis credit Research Thesis, Total 90

Graduate and Undergraduate Courses Available for Graduate Credit: Oncogenic Viruses, Immunochemistry, Immunobiology, Molecular and Cell Biology, Electron Microscopy, Medical Microbiology, Topics in Theoretical Biology, Microbial Biochemistry, Animal Parasite Biochemistry and Physiology, Microbial Genetics, Human Genetics

DEPARTMENT OF PHYSIOLOGY
(No reply received)

SYRACUSE UNIVERSITY

108 College Place Phone: (315) 428-1870
Syracuse, New York 13210
Dean of Graduate Studies: D.E. Kibbey

College of Arts and Sciences

Dean: K. Goodrich

DEPARTMENT OF BIOLOGY
Phone: Chairman: (315) 423-3186 Others: 423-1870

Chairman: D.G. Lundgren
Professors: E. Balbinder, L.E. Davis, M. Druger, P.B. Dunham, J.R. Florini, H.R. Levy, D.G. Lundgren, S.J McNaughton, J.H. Miller, A.W. Phillips, W.D. Russell-Hunter, F.G. Sherman, R.A. Slepecky
Associate Professors: J.K. Bryan, T.P. Fondy, J.S. Garvey, F.R. Hainsworth, C.C. Kuehnert, L. Lebowitz, J.E. Smith, D.T. Sullivan, J.T. Tupper, L.L. Wolf
Assistant Professors: R.L. Bratcher, S.H.P. Chan, P.A. DeBenedictis, H.E. Hemphill, M.S. Kerr, J.N. Vournakis, F.D. Warner
Visiting Lecturers: 1

Degree Programs: Biochemistry, Biology, Botany, Entomology, Genetics, Immunology, Microbiology, Physiology, Zoology

Undergraduate Degree: B.A., B.S. (Biology)
Undergraduate Degree Requirements: General Biology, plus 24 upper level courses

Graduate Degrees: M.A., M.S., Ph.D.
Graduate Degree Requirements: Master: 24 hours of course work, 6 hours of thesis credits Doctoral: Minimum of 48 hours of course work and dissertation credits, successful completion of a qualifying examination, and an original research problem upon which doctoral dissertation is based.

Graduate and Undergraduate Courses Available for Graduate Credit: General Microbiology (U), Pathogenic Microbiology (U), Plant Physiology (U), Protobiology (U), Virology (U), General Ecology (U), General Ecology Laboratory (U), Biology of Invertebrates II (U), Quantitative Ecology (U), Cellular Plant Physiology (U), Geology of Algae (U), Advanced Developmental Biology (U), General Biochemistry I (U), Biochemistry Laboratory (U), Special Problems in Research, Enzyme Regulation, Developmental Genetics, Biological Literature, Advanced Immunology, Topics in Animal Ecology, Topics in Molecular Biology, Aspects of Invertebrate Physiological Ecology, Determinative and Physiological Microbiology (U), Laboratory in Physiological Bacteriology (U), Photobiology (U), Parasitology (U), Mechanisms of Infection (U), Microbiology of Algae (U), Neuroanatomy and Physiology (U), Evolutionary Genetics (U), Ecology of Algae (U), General Biochemistry II (U), Immunology (U), Immunology Laboratory (U), Isolation and Chemistry of Proteins, Biophysical Characterization of Proteins, Enzyme Kinetics and Mechanism, Biochemistry Laboratory for Graduates, Biological Literature, Aquatic Productivity, Topics in Animal Ecology and Behavior, Advanced Topics in Plant Ecology, Cell Differentiation, Membrane Transport

College of Human Development

Dean: M. Marge

DEPARTMENT OF NUTRITION AND FOOD SCIENCE
Phone: Chairman: (315) 423-2396

Chairman: M.V. Dibble
Professors: V.F. Thiele
Associate Professors: S.H. Short, P. Turkki
Instructors: M. Yuan
Lecturers: 2
Teaching Assistants: 3

Degree Programs: Nutrition, Food Sciences

Undergraduate Degree: B.S.
Undergraduate Degree Requirements: General Chemistry, Organic Chemistry, Biochemistry, Bacteriology, 26-28 credits in Food Science and Nutrition

Graduate Degrees: M.S., Ph.D. (Nutrition)
Graduate Degree Requirements: Master: 12-15 credit hours in major field, 9-12 hours in supporting fields, Thesis Doctoral: Program (generally 90 hours), selected by advisory committee, qualifying examination, comprehensive examination, thesis

Graduate and Undergraduate Courses Available for Graduate Credit: Experimental Food Study, Seminar in Foods, Problems in Nutrition and Food Science, Problems and

Programs in Community Nutrition, Man, Environment, and Nutrition, Metabolism, Advanced Nutrition, Special Topics, Analytical Methods, Readings in Foods, Food Investigation, The Nutrition of Growth and Development, Nutrition Problems in Disease, Field Experience in Community Nutrition, Readings in Nutrition, Vitamins and Minerals, Seminar in Nutrition Science, Research in Foods, Problems in Human Metabolism,

Utica College

Burrstone Road Phone: (315) 792-3111
Utica, New York 13502

Division of Science and Mathematics

Chairman: W.F. Pfeiffer

DEPARTMENT OF BIOLOGY

Phone: Coordinator: (315) 792-3130 Others: 792-3111

Coordinator: W.H. Gotwald, Jr.
Professors: J.L. Chamberlain
Associate Professors: R.S. Connor, E.B. Cutler, W.H. Gotwald, Jr.
Assistant Professors: A.C. Checchi, L. Kimler, R. Lucchino

Field Stations/Laboratories: Willowvale Natural Area - Chadwicks, New York

Degree Program: Biology

Undergraduate Degree: B.S.
Undergraduate Degree Requirements: Total hours required: 128, major course requirements: 32 hours, major related course requirements: 18 hours core courses 27-36 hours, free electives: 42-51 hours

UNION COLLEGE AND UNIVERSITY

Schenectady, New York 12308 Phone: (518) 346-8751

Union College
(No reply received)

Albany Medical College

47 New Scotland Avenue
Dean of Graduate Studies: J. Glenn
Dean of Medical College: S. Bondurant

DEPARTMENT OF ANATOMY

Phone: Chairman: (518) 445-5380 Others: 445-5379

Chairman: R.R. Cowden
Professors: R.A. Miller, N.L Strominger
Associate Professors: E. Burack-Cohn, E.V. Crabill R.A. Edmunds, G.L. Haibak (Adjunct), R.G. Skalko, J.L. Zamrernard
Assistant Professors: S.K. Curtis, F.W. Harrison, F. Nagy
Instructors: P. Azer (Adjunct), P. Sadag (Adjunct), O. Stern (Adjunct)
Visiting Lecturers: 2
Research Assistants: 3

Degree Program: Anatomy

Graduate Degrees: M.S., Ph.D.
Graduate Degree Requirements: Master: 45 credit hours (25 of which must be from formal courses, thesis required, oral candidacy and thesis defense, minimum 1 year full-time residency Doctoral: 90 credit hours (40 of which must be from formal courses, dissertation, oral candidacy and a dissertation defense, minimum 1 year residency

Graduate and Undergraduate Courses Available for Graduate Credit: Human Gross Anatomy, Human Neuroanatomy, Histology, Special Dissection, Special Histology, Advanced Human Neuroanatomy, Electives

DEPARTMENT OF BIOCHEMISTRY

Phone: Chairman: 445-5364

Chairman: J.L. Glenn
Professors: J.L. Glenn, F. Maley (Adjunct), J.M. Reiner
Associate Professors: S.H.G. Allen, D.A. Beeler, D.S. Berns, H.H. Brown, B. Hacker (Research), H.I. Jacobson, S.G. Joshi (Research), H.T. Narahara, D.H. Treble, M. Vanko, K.W. Lam (Research
Assistant Professors: L.B Hof, D.H. Jones, A.W. Koeppen, P.B. Weber, C.R. Savage (Research), A. H. Richards (Adjunct)
Research Assistants: 4

Degree Program: Biochemistry

Graduate Degrees: M.S., Ph.D.
Graduate Degree Requirements: Master: 30 hours, 24 course work, 6 research Doctoral: 90 hours, 45 course work, 45 research

Graduate and Undergraduate Courses Available for Graduate Credit: Mammalian Biochemistry, Nuclic Acids, Enzymology, Lipids, Proteins, Biochemical Control, Neurochemistry, Glycoproteins

DEPARTMENT OF MICROBIOLOGY

Phone: Chairman: (518) 445-5166 Others: 445-5165

Chairman: L.A. Caliguiri
Professors: D.A. Berberian (Emeritus), L.A. Caliguiri, S.V. Covert (Emeritus), J.E. Hotchin (Research), R.J. Pickering
Associate Professors: R. Deibel (Adjunct), M.A. Gordon (Adjunct), C. Hurwitz (Adjunct), R.J. Laffin, T.F.D. Oram (Adjunct), K.F. Soike
Assistant Professors: J.L. Barlow (Adjunct), L.D. Beckman, S.F. Bocchieri, H. Gaafar (Adjunct), J.J. McSharry, L.S. Sturman (Adjunct), D.K. Vedder (Adjunct)
Instructors: S.G. Thrasher, S.M. Chen (Research Associate)

Degree Programs: Immunology, Microbiology

Graduate Degree: Ph.D.
Graduate Degree Requirements: Doctoral: 90 credits, qualifying examination, thesis defense

Graduate and Undergraduate Courses Available for Graduate Credit: Medical Microbiology, General Virology, Immunology, Bacterial Physiology and Genetics, Colloquium Microbiology

DEPARTMENT OF PHYSIOLOGY

(No reply received)

UNIVERSITY OF ROCHESTER

Wilson Boulevard Phone: (716) 275-2121
Rochester, New York 14627
Dean of Graduate Studies: P.R. Gross

College of Arts and Sciences

Dean: K. Clark

DEPARTMENT OF BIOLOGY

Phone: Chairman: (717) 275-3835 Others: 275-3835

Chairman: P.R. Gross
Tracy L. Harris Professor of Biology: J.H. Hofreter
Professors: T.T. Bannister, J. Brown, E. Caspari, P. Gross, G. Hoch, W. Muchmore, R. Selander
Associate Professors: S. Hattman, C. Istock, J. Kaye, U. Nur, S. Prakash
Assistant Professors: R. Cafferata, M. Gorvosky, D. Hinkle, C. Moore, J. Olmsted, R. Simon
Visiting Lecturers: 8
Teaching Fellows: 6

Degree Programs: Genetics, Physical/Chemical Biology, Cellular/Developmental Biology, Organismal Biology, Evolutionary and Population Biology

Undergraduate Degree: B.A. (Biology)
Undergraduate Degree Requirements: 2 years of Chemistry, 1 year of Physics, 3 semesters of Mathematics, 6 regular courses of Biology (excluding service introductory courses) One course normally carries 4 credits; normal semester load is 16 credits.

Graduate Degrees: M.S., Ph.D. (Biology)
Graduate Degree Requirements: Master: Plan A or Plan B: Plan A requires a research thesis, Plan B a comprehensive examination. Under either one, students normally register for at least 30 credit hours of course work. Doctoral: 90 credit hours of course work, satisfactory performance in a two-part qualifying examination, part one in three minor areas, part two on the research area defense of a scholarly dissertation

Graduate and Undergraduate Courses Available for Graduate Credit: Introduction to Cell Biology, Genetics, Evolution, Animal Development, Cell Biology of the Protozoa, Seminar in Developmental Neurobiology, Animal Behavior, Ecology, Ecological Analysis, Advanced Animal Behavior, Immunology, The Algae, Laboratory in Biochemical Genetics, Mechanisms and Regulation of Biological Energetics, Structure and Chemistry of the Nucleus, Photobiology, Evolutionary Biology, Developmental Biochemistry, Developmental Genetics, Genetic Recombination, Cytogenetics, Advanced Topics in Experimental Biology, Biology Colloquium

School of Medicine and Dentistry
Dean: J.L. Orbison

DEPARTMENT OF ANATOMY
Phone: Chairman: (716) 275-2591

Chairman: K.M. Knigge
Professors: K.M. Knigge, P.D. Coleman, V.M Emmel, R.J. Loynt, R.S. Snider, P.L. Townes, K.E. Mason
Associate Professors: W.G. Aldridge, K. Goh, D.P. Penney, P.T. Rowley, D.E. Scott, M.N. Sheridan, L.R. Weitkamp, H.R. Ziegler
Assistant Professors: M.C. Ching, R.A. Doherty, A.H. Gates, F.P. Gibbs, M.N. Goldstein, S.A. Joseph, J.H. Morton, J.R. Sladek, S. Sorrentino
Teaching Fellows: 3

Degree Program: Anatomy

Graduate Degrees: M.S., Ph.D.
Graduate Degree Requirements: Master: 30 hours plus thesis Doctoral: 90 credit hours plus thesis

Graduate and Undergraduate Courses Available for Graduate Credit: Not Stated

DEPARTMENT OF BIOCHEMISTRY
Phone: Chairman: (716) 275-2766

Chairman: E.H. Stotz
Professors: A.L. Dounce, G.V. Marinetti, E.H. Stotz
Associate Professors: R. Hilf, B. Love
Assistant Professors: H.E. Auer, A. Nahas, A.E. Senior, A.M. Tometsko, J.L. Wittliff
Visiting Lecturers: 3
Research Assistants: 6
Post-doctoral Fellows: 2

Degree Program: Biochemistry

Graduate Degrees: M.S., Ph.D.
Graduate Degree Requirements: Master: General Biochemistry, Biochemistry Seminar, Physical Biochemistry, Molecular Genetics, Enzymes-Structure and Function Doctoral: Above courses plus Tracer Chemistry, Protein Structure, Basic Endocrinology, Cell Biology

Graduate and Undergraduate Courses Available for Graduate Credit: Not Stated

DEPARTMENT OF MICROBIOLOGY
Phone: Chairman: (716) 275-3402 Others: 275-2121

Chairman: F.E. Young
Professors: P.Z. Allen, J.R. Christensen, J.D. Hare, H.R. Morgan, F.E. Young
Associate Professors: P.C. Balduzzi, N. Cohen, R.G. Douglas, J. Leddy, J. Maniloff, R.E. Marquis, A.L. Ritterson, H.W. Taber
Assistant Professors: G. Abraham, A.N. Chatterjee, J.W. Kappler, A. Meisler, B. Ohlsson-Wilhelm, R.G. Robertson, J.M. Wilhelm, G.A. Wilson
Postdoctoral Fellows: 9

Degree Program: Microbiology

Graduate Degrees: M.S., Ph.D.
Graduate Degree Requirements: Master: Undergraduate major in either Microbiology, General Biology, or Chemistry. One year each of General Biology, General Chemistry, Analytical Chemistry, and Organic Chemistry. The course program includes Medical Microbiology Plan A Master's degree (for those who career goal is research) requires a thesis based on research of a scientific nature. In most cases, the candidate must spend approximately two years to complete the program. Plan B Master's is available for those whose career goals are other than research. This program consists of approximately 27 hours of course work in addition to a written essay critically reviewing an area of literature. A final oral examination is also required . Doctoral: Same as above plus one year Mathematics and Physics. Physical Chemistry is also desirable. Core curriculum includes Graduate Microbiology (or Medical Microbiology), Biochemistry, Immunology, Microbial Physiology or Cell Biology, and Virology or Molecular Genetics. Ninety semester hours of study beyond the Bachelor's degree, or 60 hours beyond an acceptable Master's degree are required. An oral qualifying examination must be passed at least 7 months before the final examination may be taken. A thesis is required on each candidate for the Ph.D. degree.

Graduate and Undergraduate Courses Available for Graduate Credit: Microbial Physiology, Introduction to Basic and Clinical Immunology, Medical Microbiology and Immunology, Animal Parasitology, Graduate Microbiology, Diagnostic Laboratory Methods, Advanced Topics in Immunology, Advanced Topics in Immunochemistry, Special Topics in Immunology, Virology, Experimental Virology, Speical Seminars

DEPARTMENT OF PHYSIOLOGY
(No reply received)

DEPARTMENT OF RADIATION BIOLOGY AND BIOPHYSICS
Phone: Chairman: (716) 275-3723 Others: 275-3905

Chairman: W.F. Neuman
Wilson Professor: W.F. Neuman
Professors: W.F. Bale, T.T. Bannister, G.W. Casarett, T.W. Clarkson, V.F. DiStefano, I. Feldman, J. Ferin, G.B. Forbes, J.W. Howland, R.W. Hyde, V.G. Gaties, T.T. Mercer, S.M. Michaelson, L.L. Miller, P.E. Morrow, W.F. Neuman, F. Sherman, H.M. Sobell, I.L. Spar, J.N. Stannard, T.Y. Toribara, R.I. Weed, B. Weiss, T.F. Williams, F.E. Young
Associate Professors: W.G. Aldridge, K.I. Altman, G.G. Berg, J.R. Coleman, R.L. Collin, D.A. Goldstein, P.L. LaCelle, C.W. Lawrence, M.A. Lichtman, H.D. Maillie, J. Maniloff, D.A. Morken, F.A. Smith, H.W. Taber, D.R. Taves, A.R. Terepka, J.S. Wiberg, D.A. Young
Assistant Professors: H.E. Auer, W.A. Bernard,

J.S. Brand, B.E. Dahneke, R.A. Doherty, H.I. Eberle, H.L. Evans, A.H. Gates, K. Gunter, T.E. Gunter, G.A. Kimmich, R.W. Kilpper, C.S. Lange, L.J. Leach, M.W. Miller, M.W. Neuman, L. Prakash, M. A. Resnick, A. Shamoo, J.W. Stewart, D.A. Weber, R.H. Wilson
Instructor: R.L. Goodland
Visiting Lecturers: 24
Research Assistants: 43

Degree Programs: Radiation Biology, Biophysics, Toxicology

Graduate Degrees: M.S., Ph.D.
Graduate Degree Requirements: Master: 30 credit hours plus essay or thesis Doctoral: 90 credit hours, qualifying examination and thesis

Graduate and Undergraduate Courses Available for Graduate Credit: Introduction to Graduate B ology for Students in the Physical Sciences, Introduction to Radiation Biology and Biophysics, Introduction to Research, Membrane Biology, Radiation Biochemistry, Elementary Quantum Biophysics, Intermediate Quantum Biophysics, Calcium, the Regulator, and its Regulation, Mutagenesis and DNA Repair, X-Ray Crystallography, Replication of Cellular DNA, The Structure and Properties of Biological Macromolecules, Health Physics, Basic Radiological Physics, Basic Radiological Physics, Basic Radiological Physics Laboratory, Cellular and Molecular Radiobiology, Mammalian Radiological Physics Laboratory, Cellular and Molecular Radiobiology, Mammalian Radiobiology, Radiobiology Laboratory, Physical Chemistry of Solutions, Photobiology, Evaluation of Radiation Hazards, Radiation Cytology, Tracer Chemistry

VASSAR COLLEGE

Poughkeepsie, New York 12601 Phone: (914) 452-7000
Dean of College: B. Wells

BIOLOGY DEPARTMENT
Phone: Chairman: Ext. 39 Others: Ext. 104,103,233

Chairman: P. Johnson
John Guy Vassar Chair of Natural History: A. Zorzoli
Professors: F.V. Ranzoni, E. Tokay, M. Wright, A. Zorzoli
Associate Professors: P. Johnson, E.S. Lumb, D.B. Williams
Assistant Professors: C. Fahlund, R.B. Hemmes
Visiting Assistant Professor: S.M. Sward, L. Mehaffey

Degree Program: Biology

Undergraduate Degree: B.A.
Undergraduate Degree Requirements: For concentration in Biology: 12 units (approximately 1/3 of units needed for graduation) including Chemistry 115, 145 and 245 (Inorganic and Organic Chemistry), Biology 105 (Introductory Biology - or its equivalent): 2 units from 202 (Botany), 204 (lower Plants and Animals), 205 (Microbiology), 220 (Mammalian Physiology), and 240 (Comparative Morphology of Vertebrates): 4 units of Grade III courses including 320 (Biochemistry). (Only 1 unit of ungraded work at the 300-level may be credited toward the minimum). Senior Year Requirements: 2 units of Grade III courses other than 301 (Senior Seminar), or 399 (Independent Study), except by permission of advisor.

WAGNER COLLEGE

Stanten Island, New York 10301 Phone: (212) 390-3000
Dean of Graduate Studies: J.J. Bois
Dean of College: E.O. Wendel

DEPARTMENT OF BIOLOGY
Phone: Chairman: (212) 390-3197 Others: 390-3000

Chairman: W.W. Kanzler
Professors: M.E. Annan
Associate Professors: R.B. Priddy, A. Ruark, D. Yarns
Assistant Professors: J.D. Christianson, W.W. Kanzler, C. Kiley

Field Stations/Laboratories: Field Station - Camp Herrlich, Brewster, New York

Degree Program: Biology

Undergraduate Degree: B.S.
Undergraduate Degree Requirements: 28 credits in Biology, Inorganic, Organic Chemistry

WELLS COLLEGE

Aurora, New York 13026 Phone: (315) 364-7161

DEPARTMENT OF BIOLOGY
Phone: Chairman: (315) 364-7161

Chairman: T. Zorach
Professor: C. Burch
Associate Professor: T. Zorach
Assistant Professors: P.A. Sullivan, B. Byrne, B. Byrne

Degree Programs: Basic Biology, Environmental Studies

Undergraduate Degrees: B.A.
Undergraduate Degree Requirements: Introductory Biology - 2 semesters, Ecology, Genetics, Senior Seminar, 5 additional Biology courses, Inorganic Chemistry, 2 semesters, Organic Chemistry - 2 semesters

YESHIVA UNIVERSITY

500 West 158th Street
New York, New York 10033
Dean of Graduate Studies: J.R. Warner

Yeshiva College

(No reply received)

Albert Einstein College of Medicine

1300 Morris Park Avenue Phone: (212) 430-2000
Bronx, New York 10461
Dean: E. Friedman

DEPARTMENT OF ANATOMY
Phone: Chairman: (212) 430-2836 Others: 430-2000

Chairman: B.V. Scharrer (Acting)
Professors: W. Etkin, R. Ger, H. Nathan (Emeritus), J. Padawer, G.D. Pappas, B. Scharrer
Associate Professors: G.J. Fruhman, I. Pesetsky, M.C. Weitzman
Assistant Professors: F. Baker-Cohen
Visiting Lecturers: 2
Research Associate : 1

Field Stations/Laboratories: Croton-on-Hudson, New York

Degree Program: Anatomy

Graduate Degree: Ph.D.
Graduate Degree Requirements: Not Stated

Graduate and Undergraduate Courses Available for Graduate Credit: Not Stated

DEPARTMENT OF CELL BIOLOGY
Phone: Chairman: (212) 430-2815 Others: 430-2000

Chairman: M.D. Scharff
Professors: B. Bloom, H. Eagle, S. Nathenson, E. Robbins, M.D. Scharff, D. Summers, J. Warner
Associate Professors: S. Baum, E. Ehrenfeld, B. Fields, M. Horwitz, J. Maio, C. Schildkraut, R. Soeiro
Assistant Professors: B. Birshtein, S. Horwitz,

A. Rubinstein, D. Shafritz, A. Skoultchi
Visiting Lecturers: 42
Research Assistants: 20

Degree Program: Cell Biology

Graduate Degree: Ph.D.
Graduate Degree Requirements: Not Stated

Graduate and Undergraduate Courses Available for Graduate Credit: Genetics, Molecular Pharmacology, Membranes, Molecular Immunology, Biochemical Genetics, Bacterial Physiology, Virology, Regulation of Gene Expression, Experimental Endocrinology, Biophysics

DEPARTMENT OF BIOCHEMISTRY
Phone: Chairman: (212) 430-3024 Others: 430-2235

Chairman: J. Marmur (Acting)
Professors: O. Blumenfeld, R. Briehl, S. Englard, H. Hoberman, B. Lowy, J. Marmur, S. Seifter, H. Strecker, J. Warner
Associate Professors: J. Betheil, E. Bloch, M. Daly, G. Fujimoto, L Gidez, R. Ledeen, I. Listowsky, M. Makman, C. Moore, E. Seifter
Assistant Professors: B. Kaplan, H. Kern, S. Nakagawa, T. Takahashi, B. Thysen
Instructors: S. Bellhorn, B. Oppenheim, J. Swaney
Visiting Lecturers: 3
Principal Associates: 3
Associates: 2

Degree Program: Biochemistry

Graduate Degree: Ph.D.
Graduate Degree Requirements: Doctoral: Course requirement determined individually by Student Committee from those listed below and others, Doctoral Dissertation

Graduate and Undergraduate Courses Available for Graduate Credit: Intermediate Biochemistry, Apprenticeship in Biochemistry, Bio-Organic Reaction Mechanism, Biochemistry Seminars, Biochemical Regulation of Enzyme Activity, Physical Chemistry of Marcomolecules

DEPARTMENT OF BIOPHYSICS
Phone: Chairman: (212) 430-2248 Others: 430-2593

Head: P. Aisen
Professor: P. Aisen
Associate Professor: A. Kowalsky
Assistant Professor: C.W. Wu
Instructor: Y.H. Wu
Research Assistants: 3

Degree Program: Biophysics

Graduate Degree: Ph.D.
Graduate Degree Requirements: Doctoral: Coursework in Biophysics, Biochemistry, Organic reaction mechanisms, Physical Chemistry and a selection of advanced courses, dissertation based on original research, proficiency in foreign language.

Graduate and Undergraduate Courses Available for Graduate Credit: Animal Virology, Biochemical Genetics, Bioorganic Reaction Mechanisms, Biophysics and Biophysical Techniques, Developmental Biology, Gene Expression in Animal Cells, Genetics, Membranes, Spectroscopy in Biological Systems, Physical Chemistry of Macromolecules, Microbial Physiology

DEPARTMENT OF MICROBIOLOGY AND IMMUNOLOGY
Phone: Chairman: (212) 430-2812 Others: 430-2811

Chairman: E.J. Hehre
Samuel H. Holding Professor of Microbiology: E.J. Hehre
Danciger Distinguished Scholar in Microbiology and Immunology: B.R. Bloom
Professors: J.R. Battisto, B.R. Bloom, E.J. Hehre, S.G. Nathenson, D.F. Summers
Associate Professors: S.G. Baum, B.N. Fields, D.S. Genghof, M.S. Horwitz, M. Lev, D.M. Marcus, R. Soeiro
Assistant Professors: C.F. Brewer, S.C. Edberg, E.B. Robbins, G. Szilagyi
Instructor: S.L. Nehlsen
Visiting Lecturers: Variable
Research Assistants: 8
Post-doctoral Research Associates: 10

Field Stations/Laboratories: Scientific Park and Biological Field Station - Croton, New York

Degree Program: Microbiology, Virology

Graduate Degree: Ph.D.
Graduate Degree Requirements: Doctoral: Formal background in Biological Sciences equivalent to 1 1/2 - 2 years of basic science course work. Initial training in research laboratories of one or more staff members, participation in departmental research seminars and in current topics seminars throughout training period, qualifying examination followed by thesis research.

Graduate and Undergraduate Courses Available for Graduate Credit: Research Seminar in Microbiology and Immunology, Current Topics in Microbiology and Immunology, Research Methods in Microbiology and Immunology, Bacterial Physiology, Molecular Immunology, Virology, Cellular Membranes

DEPARTMENT OF PHYSIOLOGY
Phone: Chairman: (212) 430-2917 Others: 430-2000

Chairman: H.D. Lauson
Professors: S. Baez, R.W. Briehl, M.I. Cohen, S.M. Crain, A. Finkelstein, H.D. Lauson, P.S. Roheim, J. B. Wittenberg
Associate Professors: B.M Altura, B.G. Bass, N.M. Buckley, A.B. Kostellow, G.A. Morrill, E.M. Scarpelli, J. Scheuer, M.H. Williams, E.L. Yellin
Assistant Professors: D.R. Giblin, B.A. Wittenberg, T. Yipintsoi
Postdoctoral Research Fellows: 2

Degree Programs: Animal Physiology, Physiology

Graduate Degrees: Ph.D.
Graduate Degree Requirements: Thesis, language, one of French, German, or Russian, appropriate courses in Physiology, Biochemistry, Anatomy, Neurobiology, Pharmacology, Pathophysiology

Graduate and Undergraduate Courses Available for Graduate Credit: Mammalian Physiology for Medical and Graduate Students, Advanced Cardiovascular Physiology, Developmental Physiology (Early Embryology), Laboratory Apprenticeship, Departmental Graduate Student Seminar

YORK COLLEGE
(see under City University of New York)

NORTH CAROLINA

APPALACHIAN STATE UNIVERSITY
Boone, North Carolina 28608 Phone: (704) 262-3025
Dean of Graduate Studies: C. Williams
Dean of Arts and Sciences: W. Strickland

BIOLOGY DEPARTMENT
Phone: Chairman: (704) 262-3025 Others: 262-3026

Chairman: I.W. Carpenter, Jr.
Professors: I.W. Carpenter, Jr., F.R. Derrick, H. Hurley, J.F. Randall, K.R. Robinson
Associate Professors: W.R. Hubbard
Assistant Professors: J.J. Bond, W.C. Dewel, R.A. Dewel (Adjunct), S. Glover, E. Greene, F. Helseth, R. Henson, F. Montaldi

Degree Programs: Not Stated

Undergraduate Degree: Not Stated

ASHVILLE BILTMORE COLLEGE
(see under University of North Carolina at Ashville)

BARBER-SCOTIA COLLEGE
145 Cabarrus Avenue West Phone: (704) 786-5171
Concord, North Carolina 28025
Dean of College: M.P. McLean

CENTER FOR NATURAL SCIENCES AND ALLIED HEALTH PROFESSIONS
Phone: Chairman: Ext. 239 Others: 786-5171

Chairman: K. Rajasekhara
Professor: K. Rajasekhara
Associate Professor: M. Boyd (Mathematics)
Assistant Professors: G.H. Painter (Mathematics, D.S. M. Prasad (Chemistry), D.D. Sharma (Physics and Mathematics), E.C. Witherspoon (Biology)
Instructors: A. Nowsu (Biology, S. Smalls (Mathematics)
Laboratory Assistant: 1
Teaching Assistant: 1

Degree Programs: Biology, Medical Technology, Biology Education

Undergraduate Degree: B.S.
Undergraduate Degree Requirements: B.S. (Biology): General Education - 45 hours, Biology - 30 hours, Chemistry - 20 hours, Physics - 8 Hours, Mathematics - 12 hours, Electives - 10 hours, 125 hours, B.S. (Biology Education): General Education - 45 hours, Biology - 24 hours, Chemistry 16 hours, Earth Science 3 hours, Physical Science - 3 hours, Mathematics - 12 hours, Professional Education - 22 hours, 125 hours, B.S. (Medical Technology): General Education - 39 hours, Biology - 22 hours, Chemistry - 20 hours, Mathematics - 6 hours, Electives - 8 hours, Clinical Training - 32 hours, 127 hours

BELMONT ABBEY COLLEGE
Belmont, North Carolina 28012 Phone: (704) 825-3711
Dean of College: J. Solari

DEPARTMENT OF BIOLOGY
Phone: Chairman: Ext. 327

Chairman: J. Stuart
Professors: J. Stuart
Instructor: B. Kowalczyk
Visiting Lecturers: 4

Degree Program: Biology

Undergraduate Degrees: B.A,, B.S.
Undergraduate Degree Requirements: 120 semester hours (Total)

BENNETT COLLEGE
Greensboro, North Carolina 27420 Phone: (919) 275-9791
Dean of College: J.H. Sayles

DEPARTMENT OF BIOLOGY
Phone: Chairman: Ext. 27

(No further information available)

CAMPBELL COLLEGE
Buies Creek, North Carolina 27506 Phone: (919) 893-4111
Dean of College: L.B. Ledgerwood

DEPARTMENT OF BIOLOGY
Phone: Chairman: Ext. 257

Chairman: L.S. Beard
Professor: L.S. Beard
Associate Professor: R.L. McIntyre
Assistant Professors: H. Matthews, P.K. McCall, R.F. Soots, Jr., C.G. Yarbrough

Degree Program: Biology

Undergraduate Degree: B.S.
Undergraduate Degree Requirements: Not Stated

BOWMAN GRAY SCHOOL OF MEDICINE
(see under Wake Forest University)

CATAWBA COLLEGE
Salisbury, North Carolina 28144 Phone: (704) 636-5311
Dean of College: C. Turney

BIOLOGY DEPARTMENT
Phone: Chairman: Ext. 78

Chairman: J.A. Buxton
Professor: D.E. Kirk
Associate Professor: E.B. Newell
Assistant Professor: M. Bayanski
Student Laboratory Assistants: 3

Degree Program: Biology (Cooperative program in Forestry with Duke University and in Medical Technology with Bowman Gray College of Medicine)

Undergraduate Degree: B.A.
Undergraduate Degree Requirements: 36 units, each one quarter of ten weeks length, 16 of the total must be from Biology, Chemistry and Physics, Balance from Social Sciences, Fine Arts and Humanities

DAVIDSON COLLEGE
Davidson, North Carolina 28036 Phone: (704) 892-8021
Dean of College: J.M. Berau

DEPARTMENT OF BIOLOGY

Phone: Ext. 325 Others: Ext. 320,330

Chairman: D.L. Kimmel, Jr.
Professor: T. Daggy
Associate Professor: W.T. Lammers
Assistant Professors: C.T. Grant, D.C. Grant, J.L. Putnam

Degree Programs: Biology, Premedicine

Undergraduate Degrees: B.S.
Undergraduate Degree Requirements: 2 general courses plus 8 advanced courses plus 2 independent study courses in major plus college degree requirements

DUKE UNIVERSITY
Durham, North Carolina 27706 Phone: (919) 684-8111
Dean of Graduate Studies: J.C. McKinney

Trinity College of Arts and Sciences
Dean: A. Flowers

DEPARTMENT OF BOTANY
Phone: Chairman: (919) 684-3056

Chairman: R.L. Wilbur
James B. Duke Professors of Botany: W.D. Billings, A.W. Naylor, P.J. Kramer
Professors: L.E. Anderson, W.D. Billings, W.L. Culberson, H. Hellmers, T.W. Johnson, Jr., P.J. Kramer, A.W. Naylor, J. Philpott, D.E. Stone, R.A. White, R.L. Wilbur
Associate Professors: J. Antonovic, J.E. Boynton, R.B. Searles, B.R. Strain
Assistant Professors: W.F. Blankley, N. Christensen
Senior Lecturer: C.F. Culberson
Teaching Assistants: 19
Research Assistants: 11

Field Stations/Laboratories: D.U. Marine Laboratory - Highlands Biological Station

Degree Program: Botany

Undergraduate Degrees: A.B., B.S.
Undergraduate Degree Requirements: 8 courses beyond the introductory level, Up to 3 of the 8 can be substituted for by Physical Chemistry and Advanced Mathematics and Physics courses.

Graduate Degrees: M.A., Ph.D.
Graduate Degree Requirements: Master: 30 units - 1 year Doctoral: 60 units - 2 years

Graduate and Undergraduate Courses Available for Graduate Credit: Cytogenetics (U), Anatomy of Woody Plants (U), Lichenology (U), Bryology (U), Phycology (U), Mycology (U), Major Global Ecosystems (U), Plant Biosystematics (U), Plant Metabolism (U), Plant-Water Relations (U), Principles of Plant Distribution (U), Physiology of Growth and Development (U), The Environment,(U), Physiological Plant Ecology (U), Analysis and Classification of Vegetation (U), Population Genetics (U), Introductory Marine Microbiology, Anatomy, Mycology, Marine Phycology, Systematics, Plant Diversity, Ecology, Physiology, Plant Metabolism, Plant Systematics, The Environment, Analysis and Classification of Vegetation, Principles of Genetics, The University Program in Genetics, Program in Tropical Biology, The University Program in Marine Sciences,

DEPARTMENT OF ZOOLOGY
Phone: Chairman: (919) 684-3583 Others: 684-2507

Chairman: D.J. Fluke
Chairman: J.R. Bailey (Acting)
James B. Duke Professor of Physiology: K. Schmidt-Neilson
James B. Duke Professor: K.M. Wilbur
Professors: J.R. Bailey, C.G. Bookhout, J D. Costlow Jr., D.J. Fluke, J.R. Gregg, P.H. Klopfer, D.A. Livingstone, R.B. Nichlas, K. Schmidt-Koenig (Adjunct), K. Schmidt-Nielsen, V.A. Tucker, K.M. Wilbur
Associate Professors: R.T. Barber, N.W. Gillham, S. Vogel, S.A. Wainwright, C.L. Ward
Assistant Professors: J. Bergeron, R.B. Forward, J.G. Lundberg, D.R. McClay, J.P. Sutherland, H.M. Wilbur
Instructors: 1
Research Assistants: 3
Visiting Scholar: 1
Hargitt Research Fellow: 1

Field Stations/Laboratories: Field Station for Animal Behavior, Primate Facility, Duke University, Marine Laboratory

Degree Program: Zoology

Undergraduate Degree: B.S.
Undergraduate Degree Requirements: 8 courses beyond the introductory level, up to 3 of the 8 can be substituted for by Physical Chemistry and Advanced Mathematics and Physics courses.

Graduate Degrees: M.A., Ph.D.
Graduate Degree Requirements: Master: 30 units - 1 year Doctoral: 60 units - 2 years

Graduate and Undergraduate Courses Available for Graduate Credit: Animal Behavior (U), Introduction to Comparative Behavior (U), Marine Ecology (U), Population and Community Ecology (U), Elements of Theoretical Biology (U), Biological Oceanography (U), Limnology (U), Paleobiology (U), Vertebrate Zoology (U), Morphogenetic Systems (U), Systematic Zoology (U), Biogeography (U), Radiation Biology (U), Physical Biology (U), Introductory Biochemistry (U), Physiological Ecology of Marine Animals (U), Comparative Physiology (U), Fluid Flow and Living Systems (U), Laboratory Research Methods (U), Advanced Cell Biology (U), Cytological Materials and Methods (U), Topics in Cell Structure and Function (U), Marine Invertebrate Zoology (U), Invertebrate Zoology (U), Endocrinology of Marine Animals (U), Invertebrate Embryology (U), Principles of Genetics (U), Evolution (U), The Cell in Development and Heredity (U)

School of Medicine
Vice President for Health Affairs: W.G. Anlyan

DEPARTMENT OF ANATOMY
Phone: Chairman: (919) 684-4124

Chairman: J.D. Robertson
Professors: J.W. Everett, M.J Moses, T. Peele, J.D. Robertson
Associate Professors: S. Counce, K.L. Duke, W. Longley, M.K. Reedy
Assistant Professors: M.R. Adelman, F.H. Bassett, J. A. Bergeron, C.A. Blake, M. Cartmill, H.P. Erickson, W.C. Hall, W.L. Hylander, K.E. Johnson, R. Kay, T. Strickler, L. Tyrey
Research Assistants: 11

Field Stations/Laboratories: Duke University Primate Facility, Duke University Animal Field Station - both in Duke Forest

Degree Program: Anatomy

Graduate Degrees: M.A., M S., Ph.D.
Graduate Degree Requirements: Master: Basic Core Anatomy course, other courses, GSFLT, written Master's thesis Doctoral: Basic core Anatomy course, other courses, GSFLT, written doctoral dissertation

Graduate and Undergraduate Courses Available for Graduate

Credit: Human Anatomy of the Trunk, Contractile Processes, Molecular and Cellular Basis of Development, Human Evolution, Functional and Evolutionary Morphology of Primates, Mechanisms of Biological Motility, The Primate Fossil Record, History of Generation and Mammalian Reproduction, History of Anatomy, Mammalian Embryology and Developmental Anatomy, Mammalian Embryology and Developmental Anatomy, Topics in Cell Structure and Function, Comparative Neurology and Psychology, Molecular Basis of Anatomy, The Light Microscope, the Electron Microscope, and X-Ray Diffraction in Biology, The Cell in Development and Heredity, Membrane Structure, Special Topics in Nerve Ultrastructure, Gross Anatomy, Gross Human Anatomy, Microscopic Anatomy, and Neuroanatomy, Neuroanatomical Basis of Behavior, Advanced Neuroanatomy of Sensory and Motor Mechanisms, Research Techniques in Anatomy, Reproductive Biology

DEPARTMENT OF BIOCHEMISTRY
Phone: Chairman: (919) 684-5326 Others: 684-8111

James B. Duke Professors of Biochemistry: P.H. Handler, R.L. Hill, C. Tanford
Professors: M.L.C. Bernheim (Emeritus), I. Fridovich, S.R. Gross (Director of the Genetics Division), W.R. Guild, J.S. Harris, H. Kamin, N. Kirshner, K.S. McCarty
Associate Professors: S.H. Appel, R.C. Greene, B. Kaufman, S.H. Kim, W.S. Lynn, K.V. Rajagopalan, J.A. Reynolds, H.J. Sage, L. Siegel, R. Webster
Assistant Professors: R.M. Bell, R.L. Habig, D.H. Hall, P.O. Hagen (Adjunct), P.H. Harriman, W.N. Kelley, N.M. Kredich, R. Lefkowitz, P. McKee, D.C. Richardson, J.B. Sullivan, R.W. Wheat
Instructors: J. Bittikofer, J. Bonaventura, J. McCord, Y. Nozaki, H. Steinman
Visiting Lecturer: 1
Research Associates: 30

Field Stations/Laboratories: Duke University Marine Laboratory - Beaufort, North Carolina

Degree Programs: Protein Structure and Function, Crystallography of Macromolecules, Nucleic Acid Structure and Function, Lipid Biochemistry, Membrane Structure and Function, Molecular Genetics, Enzyme Chemistry, Neurochemistry

Graduate Degree: Ph.D. (Biochemistry)
Graduate Degree Requirements: Doctoral: 15 credit hours per semester until completion of the preliminary examinations (three written examinations and one oral research proposition) and the language requirements.

Graduate and Undergraduate Courses Available for Graduate Credit: Introductory Genetics, Laboratory Methods in Biochemistry, Independent Study, Molecular Genetics, Molecular and Cellular Basis of Development, Structure of Biological Macromolecules, General Biochemistry, Experimental Genetics, Current Topics in Immunochemistry, The Carbohydrates and Lipids of Biological Systems, Macromolecules, Nucleic Acids and Macromolecular Synthesis, Enzyme Mechanisms, Biological Oxidations, Intermediary Metabolism, Regulation of Cellular Metabolism, Nutrition, Neurochemistry, Biochemistry of Membranes, Biochemical Pharmacology, Introductory Biochemistry (U)

DEPARTMENT OF MICROBIOLOGY AND IMMUNOLOGY
Phone: Chairman: (919) 684-5138 Others: 684-8111

Chairman: W.K. Joklik
James B. Duke Professor of Microbiology and Immunology: W.K. Joklik
James B. Duke Professor of Microbiology: N.F. Conant
James B. Duke Professor of Immunology: D.B. Amos
Professors: R.O. Burns, E.D. Day, J.E. Larsh, R.S. Metzgar, S. Osterhout, H.P. Willett
Associate Professors: C.E. Buckley, R.H. Buckley, J.J. Burchall (Adjunct), W.F. Rosse, H.F. Seigler, F.E. Ward, R.W. Wheat, H.J. Zweerink
Assistant Professors: D.W. Bigner, D.P. Bolognesi, G.B. Hill, D.J. Lang, P.K. Lauf, N.L. Levy, J.L. Nichols, D.W. Scott, R.E. Smith, T.C. Vanaman, J.L. Wagner, C.M. Wilfert, P.J Zwadyk
Associates: P. Cresswell, J. Dawson, S. Hsia, S.E. Miller, E. Reisner, W.K. Smith
Research Assistants: 15

Field Stations/Laboratories: Oak Ridge Institute of Nuclear Studies - Oak Ridge, Tennessee, Duke University Marine Laboratory - Beaufort, North Carolina, Animal Behavior Station and Primate Facility - Durham, North Carolina, Triangle Universities Computation Center - Research Triangle Park, North Carolina, Duke Forest Durham, North Carolina

Degree Programs: Microbiology, Immunology, Virology, Molecular Microbiology, Immunology

Graduate Degree: Ph.D. (Microbiology, Immunology)
Graduate Degree Requirements: Master: 24 semester hours in approved course work, 6 hours thesis research Doctoral: 60 semester hours graduate credit, Preliminary examination: Oral and Written, Required courses: Immunology, Virology, Molecular Microbiology, Enzyme Mechanisms, Dissertation, Final Examination

Graduate and Undergraduate Courses Available for Graduate Credit: Molecular and Cellular Basis of Development, Medical Microbiology, Microbiology (U), General Animal Virology and Viral Oncology, Molecular Microbiology, Immunology I and II, Immunochemistry, Immunohematology, Medical Mycology, Medical Immunology, Immunogenetics, Cellular Immunophysiology

DEPARTMENT OF PHYSIOLOGY AND PHARMACOLOGY
Phone: Chairman: (919) 684-3049 Others: 684-8111

Chairman: D.C. Tosteson
James B. Duke Professor of Physiology and Pharmacology: D.C. Tosteson, F. Bernheim
Professors: F. Bernheim, J.J. Blum, D.L. Coffin (Adjunct), I. Diamond, W.F. Durham (Adjunct), G.H. Hitchings (Adjunct), F. Jobsis, E.A. Johnson, L. Lack, R.A. Maxwell (Adjunct), J.W. Moore, T. Narahashi, C.A. Nichol (Adjunct), S. Schanberg, G. Somjen, D.C. Tosteson
Associate Professors: N. Anderson, J.R. Clapp, G.B. Elion (Adjunct), R. Erickson, R.E. Fellows, P.K. Lauf, M. Lieberman, T.J. McManus, D. Menzel, E. Mills, A. Ottolenghi, G. Padilla, H.S. Posner (Adjunct), J. Salzano, M. Spach, R.M Welch (Adjunct), M. Wolbarsht
Assistant Professors: M.B. Abou-Donia, R. Baron, D. Blenkarn, R.Y. Chuang, W.A. Cook, W. Duran, H. Elford, E.H. Ellinwood, A. Escueta, B.E. Gisin (Adjunct), J.E. Greenfield, L.E. Gutman, R.B. Gunn, J. Gutknecht, J.E. Hall, F. Hempel, R. S. Jones, R.G. Kirk, J.M. Lootsey, H.E. Lebovitz, L. Mandel, G.M. McKenzie (Adjunct), L. Mendell, D.H. Namm (Adjunct) P. Prinz, F. Ramon, G. Rosen, M. Rosenthal, A.P. Sanders, D.W. Schomberg, J.M Schooler, B. Shrivastav, T. Slotkin, H.C. Strauss, H.P. Ting-Beall, M.T. Tosteson, A.G. Wallace, C.H. Wu, W. Yarger, J.Z. Yeh
Instructors: P.A.W. Anderson, R.W. Anderson, R.J. Bach, F. Cobb, D. Dreyer, J.C. Fuchs, P. McHale, S. Simon, W. Wilson
Visiting Lecturers: 3
Postdoctoral Research Associates: 28

Field Stations/Laboratories: The Duke University Marine Laboratory - Beaufort, North Carolina, Animal Behavior Station - Durham, North Carolina, Primate Facility - Durham, North Carolina, Duke Forest - Durham, North Carolina

Degree Programs: Physiology, Pharmacology

Graduate Degrees: M.S., Ph.D.
Graduate Degree Requirements: Master: The graduate student must (1) have made passing grades in the first 12 units of course work, (2) have made a grade of G or E on at least 3 units of this work, and (3) have received the approval of the major department. Doctoral: (1) major or related courses, (2) supervisory committee for program of study (3) residence (4) preliminary examination, (5) dissertation and (6) final examination. In order to be considered for candidacy for the Ph.D. degree, the student must have passing grades in all of his course work, and on at least 9 units of this course work he must have made a grade of G or better.

Graduate and Undergraduate Courses Available for Graduate Credit: Introduction to the Physiology of Man, Introduction to Physiology, Advanced Physiology, Respiratory System in Health and Disease, Individual Study and Research, Membrane Physiology and Osmoregulation, Topics in Developmental Physiology, Contractile Processes, Membrane Transport, An Introduction to Neuronal Physiology and Pharmacology, Molecular and Cellular Basis of Development, Pharmacology: Mode of Action of Drugs, Cellular and Chemical Pharmacology, Mammalian Toxicology, Human Nutrition, Student Seminar in Physiology and Pharmacology, Physiological Basis of Medicine, Gastrointestinal and Renal Physiology, Pharmacological Basis of Clinical Medicine, Laboratory Methods in Pharmacology, Pharmacodynamics, Current Topics in Cardiac Muscle Physiology, Research in Physiology and Pharmacology, Physiological Instrumentation, Laboratory Methods in Electrophysiology, Integrative and Clinical Neurophysiology, Metabolic and Developmental Physiology, Analysis of Physiological Systems, Biophysics of Excitable Membranes, Cellular Endocrinology, Reproductive Biology, Topics in Mathematical Physiology, Cellular Immunophysiology, Advanced Seminar in Endocrinology and Reproductive Physiology I and II.

School of Forestry

Dean: C.W. Ralston

SCHOOL OF FORESTRY
Phone: Chairman: (919) 684-2135 Others: 684-2421

Dean: C.W. Ralston
Professors: R.F. Anderson, R.L. Barnes, L.E. Chaiken, M.S. Heath (Adjunct), H. Hellmers, F.C. Joerg, K.R. Knoerr, C.W. Ralston, W.J Stambaugh
Associate Professors: E.W. Clark (Adjunct), C.S. Hodges (Adjunct), L.J. Metz (Adjunct), F.M. Vukovich (Adjunct), D.O. Yandle
Assistant Professors: F.J. Convery, A.L. Sullivan, F.J. White, J E. Wuevscher
Visiting Lecturers: 1
Research Assistants: 3
Associate Faculty: 2

Field Stations/Laboratories: Marine Laboratory - Beaufort, North Carolina

Degree Programs: Forest Management, Forest Protection, Forest Business Management, Environmental Management, Forest Sciences, Forest Economics, Environmental Science

Graduate Degrees: M.P., M.E.M., M.S., Ph.D.
Graduate Degree Requirements: Master: M.F. - 30 - 70 units depending on undergraduate preparation, M.E.M. - 60-70 units depending on undergraduate preparation, M.S. - 30 units (minimum) beyond baccalaureate Doctoral: Ph.D. - 60 units (minimum) beyond baccalaureate, 30 units (minimum) beyond Master's, one foreign language, waiver may be granted in some areas, dissertation

Graduate and Undergraduate Courses Available for Graduate Credit: Tree Physiology, Tree Growth and Development, Anatomy of Woody Plants, Chemistry of Woody Plants, Physiology of Wood Formation, Biology of Forest Insects and Diseases, Forest Pathology, Chemical Aspects of Forest Protection, Forest Entomology, General Entomology, Natural Resource Ecology, Seminar in Natural Resource Allocation and Efficiency, Mi Microtechnique of Woody Tissue, Forest Tree Biochemistry, Phytopathological Technique in Forestry, Microbiology of Forest Soils, Toxicology of Insecticides, Ecology of Forest Insects, Entomological Research Techniques, Ecological Principles in Environmental Management, Natural Resource Ecology - Environmental Management Seminar, Natural Resource Ecology - Environmental Management Seminar, Quantitative Analysis of Ecological Environmental Systems, Seminar in Forest Protection

EAST CAROLINA UNIVERSITY

Greenville, North Carolina 27834 Phone: (919) 758-6131
Dean of Graduate Studies: J.G. Boyette
Dean of Arts and Sciences: R.L. Capwell

DEPARTMENT OF BIOLOGY
Phone: Chairman: (919) 758-6718 Others: 758-6725

Chairman: J.S. McDaniel (Acting)
Professors: G.J. Davis, T. Ito, C.B. Knight, E.C. Simpson
Associate Professors: V.J. Bellis, C.E. Bland, P. Daugherty, C. Heckrotte, D.B. Jeffreys, J.S. Laurie, E.P. Ryan, P.P. Sehgal
Assistant Professors: W.E. Allen, M.M. Brinson, C. W. O'Rear, W.J. Smith
Instructor: F.P. Belcik
Teaching Fellows: 16
Research Assistants: 5

Field Stations/Laboratories: Coastal and Marine Resources Center - Roanoke Island, Manteo, North Carolina

Degree Programs: Biology, Biochemistry

Undergraduate Degrees: B.A., B.S.
Undergraduate Degree Requirements: B.S. (Professional) 50 quarter hours Biology, 21 Chemistry, 15 Physics, 25 Mathematics, 0-20 Foreign Language or cognate option. B.A. - 50 quarter hours Biology, 17 Chemistry, 12 Physics, 15 Mathematics B.S. (Teaching): 50 quarter hours Biology, 17 Chemistry, 12 Physics, 10 Mathematics, 31 Education and Psychology B.S. Biochemistry: 31 quarter hours Biology, 37 Chemistry, 15 Physics, 25 Mathematics, 12 science cognates

Graduate Degrees: M.A., M.A.Ed.
Graduate Degree Requirements: Master: M.A.: 45 quarter hours Biology, 23 graduate hours, Chemistry through Organic, several specified Biology courses in background, Foreign Language Examination, 9 minor area (usually Biology). M.A. Biology Education: 45 quarter hours total 23 graduate hours, 33-36 Biology, Chemistry through Organic, several specified Biology courses in background, 9-12 Education and Psychology.

Graduate and Undergraduate Courses Available for Graduate Credit: Radiotracer Techniques in Biology (U), Comparative Endocrinology (U), Animal Parasitology (U), Laboratory in Comparative Endocrinology (U), Introductory Mycology (U), Ornithology (U), Plant Anatomy and Mcrphology (U), Plant Growth and Development (U), Immunology (U), Invertebrate Zoology (U), Phycology (U), Limnology (U), Ecology (U), Heredity (U), Biological Processes and the Chemistry of Natural Waters (U), Plant Biochemistry (U), Cytology (U), Histology (U), Biological Electron Microscopy (U), General Ichthyology (U), Comparative Animal Physi ology (U), Principles of Biology III (U), Intermediary Metabolism (U), Proteins and Nucleic Acids (U), Biometry (U), Biological Applications of Digital Computers (U), Molecular Genetics (U), Microbial Physiology (U), Virology (U), Internship (U), Plant Systematics and Ecology (U), Vertebrate Systematics and Ecology (U), Seminar, Marine Biology, Topics in Cellular Biology, Animal Behavior, Distribution of Organisms, Vertebrate Endocrinology, Experimental Embryology, Organic Evolution: Topics for Advanced Students, Neurophysiology III: Information Processing, Special Problems in Biology, Introduction to Research, Thesis

ELIZABETH CITY STATE UNIVERSITY

Parkview Street Phone: (919) 335-0551
Elizabeth City, North Carolina 27909
Dean of College: F.B. Holley

BIOLOGY DEPARTMENT
Phone: Chairman: Ext. 357

Chairman: S.A. Khan
Professors: H.G. Cooke, S.A. Khan
Associate Professors: J.R. Jenkins, D.E. Thomas
Assistant Professors: H.H. Muldrow, C.D. Turnage

Degree Programs: Biology, Botany, Medical Technology, Zoology

Undergraduate Degree: B.S.
Undergraduate Degree Requirements: Not Stated

ELON COLLEGE

Elon College, North Carolina 27244 Phone: (919) 584-9711
Dean of College: T. Strum

DEPARTMENT OF BIOLOGY
Phone: Chairman: Ext. 223

Chairman: P.S. Reddish
Professor: P.S. Reddish
Associate Professors: V.F. Morgan, R. Rao
Assistant Professors: H.M. Fields, G.L. Ryals, Jr.

Degree Program: Biology

Undergraduate Degree: A.B.
Undergraduate Degree Requirements: Biology - 31 semester hours, General Physics, General Chemistry, Organic Chemistry

FAYETTEVILLE STATE UNIVERSITY

Fayetteville, North Carolina 28301 Phone: (919) 483-6144
Dean: Not Stated

DEPARTMENT OF BIOLOGICAL AND PHYSICAL SCIENCE
Phone: Chairman: Ext. 343

Chairman: J.L. Knuckles
Professors: T.T. Chao, J.L. Knuckles
Associate Professors: P.V.N. Murthy
Assistant Professors: V. Dix, C.B. Huff, F. Klotz, R. Robinson, F. Waddle
Instructor: M.B. Pitts
Laboratory Assistants: 2

Degree Program: Biology

Undergraduate Degree: B.S.
Undergraduate Degree Requirements: Varied but usually a minimum of 120 semester hours of credits, including 30 semester hours in major are required with a "C" or better average.

GARDNER-WEBB COLLEGE

Boiling Springs, North Carolina 28017 Phone: 434-2361

BIOLOGY, CHEMISTRY, GEOLOGY DEPARTMENT
Phone: Chairman: Ext. 343

Chairman: Not Stated
Professors: M. Harrelson, M. Moseley
Associate Professors: L. Brown, J. Fite, S. Partish, R. Chalcraft
Assistant Professors: C. Cash, P. Stacy

Field Stations/Laboratories: College Forest

Degree Programs: Biology

Undergraduate Degree: B.S.
Undergraduate Degree Requirements: 30 hours major, 15 supportive, 15 complementary

GREENSBORO COLLEGE

Greensboro, North Carolina 27420 Phone: (919) 272-7102
Dean of College: R. Hites

DEPARTMENT OF BIOLOGY
Phone: Chairman: Ext. 266

Chairman: A.F. Van Pelt
Professors: K. Callahan, A.F. Van Pelt
Research Professor: 1

Degree Programs: Biology, Medical Technology

Undergraduate Degrees: A.B., B.S.
Undergraduate Degree Requirements: B.S. 31 semester hours Humanities, 9 hours Mathematics, 30 hours Biology, 20 hours Chemistry, 8 hours Physics A.B. - 55 hours (semester) Humanities, 9 hours Mathematics, 20 hours Chemistry, 8 hours Physics. In each case to total 124 semester hours

GUILFORD COLLEGE

Greensboro, North Carolina 27410 Phone: (919) 292-5511
Dean of College: C. Harvey

DEPARTMENT OF BIOLOGY
Phone: Chairman: (919) 292-5511

Head: R.R. Bryden
Professor: R.R. Bryden
Associate Professors: W. Fulcher
Assistant Professor: J. Carver
Instructors: J. Parker
Part-time: 3
Assistants: 6

Degree Program: Biology

Undergraduate Degree: B.S.
Undergraduate Degree Requirements: 8 courses Biology, 1 year Chemistry, 1 year Physics, 1 year Mathematics

HIGH POINT COLLEGE

High Point, North Carolina 27262 Phone: 885-5101
Dean of College: C. Cole

DEPARTMENT OF BIOLOGY
Phone: Chairman: Ext. 52 Others: Ext. 53

Head: L. Weeks
Professors: L. Weeks
Assistant Professors: J.E. Ward, Jr., F. Yeats

Degree Programs: Biology, Medical Technology

Undergraduate Degree: B.S.
Undergraduate Degree Requirements: Semester Hours - General Biology (8), Embryology (4), Cell Physiology (4), Advanced Physiology (4), Genetics (4), Seminar (2), Research (2), Electives (8) - minimum, Inorganic

Chemistry (8), Organic Chemistry (8), Algebra and Trigonometry (3)

JOHNSON C. SMITH UNIVERSITY

100 Beatties Ford Road Phone: (704) 372-2370
Charlotte, North Carolina 28216
Vice President for Academic Affairs: L.C. Collins

DEPARTMENT OF BIOLOGY
Phone: Chairman: Ext. 294 Others: Ext. 207

Head: W.A. Keith
Associate Professors: W.A. Keith, B.K. Chopra (Adjunct)
Instructors: V. Washington, G.F. Henry (Adjunct)
Visiting Lecturer: 1

Degree Program: Biology, Medical Technology

Undergraduate Degree: B.S.
Undergraduate Degree Requirements: A major in Biology requires 32 semester hours, student entering in Medical Technology are required to take 9 hours in the Humanities and Fine Arts and 9 hours in Social Sciences during their Senior year.

LENOIR RHYNE COLLEGE

Hickory, North Carolina 28601 Phone: (704) 328-1741
Dean of College: J.M. Unglaube

DEPARTMENT OF BIOLOGY
Phone: Chairman: (704) 328-1741 Ext. 304 Others: Ext. 204

Chairman: R.L. Spuller
Associate Professors: R.L. Spuller, C.V. Wells
Assistant Professors: M. Fanning
Instructor: T. Huss
Assistant in Instruction: 1

Degree Programs: Biology, Medical Technology, Pre-Medicine

Undergraduate Degree: B.S.
Undergraduate Degree Requirements: 128 total hours, 32 hours in Biology, 17 hours in Chemistry, 6-10 hours in Mathematics, 46 hours in Core courses, 4 Interim Courses (4 hours each)

LIVINGSTONE COLLEGE

Salisbury, North Carolina 28144 Phone: (204) 633-7960
Dean of College: B.J. Verbal

DEPARTMENT OF BIOLOGY
Phone: Chairman: Ext. 81

Chairman: E.M. Harrington
Professors: E.M. Harrington, M.T. Langerbeck, S.N. Munavalli
Associate Professors: R. Boyd, J. Doubles, A. Weierich, G. Nelson, L. Walker
Assistant Professors: R.C. Janu, C. Anderson, B. Chakrapani

Degree Program: Biology

Undergraduate Degree: B.S.
Undergraduate Degree Requirements: Not Stated

MARS HILL COLLEGE

Mars Hill, North Carolian 28754 Phone: (704) 689-1011
Dean of College: R.L. Hoffman

DEPARTMENT OF BIOLOGY
Phone: Chairman: (704) 689-1249 Others: 689-1144

Chairman: F.W. Quick
Chair for Ecological Research: L.M. Outten
Professor: L.M. Outten
Associate Professor: F.H. Diercks
Assistant Professors: W. Hutt, F.W. Quick

Degree Program: Biology

Undergraduate Degree: B.S.
Undergraduate Degree Requirements: Eight or more courses in Biology to include: (a) one year freshman Biology; (b) two or more botany courses, (c) two or more Zoology courses, and (d) two or more combined plant and animal courses. One year freshman Chemistry, intuitive calculus.

MEREDITH COLLEGE

Raleigh, North Carolina 27611 Phone: (919) 833-6461
Dean of College: C.A. Burris

DEPARTMENT OF BIOLOGY

Chairman: C. Bunn
Professors: J.A. Yarbrough, (Emeritus)
Assistant Professors: J.H. Eads, Jr., G. Hoffman

Degree Program: Biology

Undergraduate Degree: A.B., B.S.
Undergraduate Degree Requirements: Not Stated

METHODIST COLLEGE

Fayetteville, North Carolina 28301 Phone: (919) 488-7110
Dean of College: S.J. Womack

DIVISION OF MATHEMATICS AND SCIENCE

Chairman: P. Longest
Assistant Professors: M. Falsom, P. Longest

Degree Program: Biology

Undergraduate Degree: B.S.
Undergraduate Degree Requirements: 30 semester hours of Biology, 16 semester hours Chemistry, and the general college requirements

NORTH CAROLINA AGRICULTURAL AND TECHNICAL STATE UNIVERSITY

312 North Dudley Street Phone: (919) 379-7500
Greensboro, North Carolina 27411
Dean of Graduate Studies: A. Spruill

School of Arts and Sciences

Dean: F. White

DEPARTMENT OF BIOLOGY
Phone: Chairman: (919) 379-7909

Chairman: A.P. Graves
Professors: A.P. Graves, A. Hill, E. Marrow, A.R. Vick, A. Webb, J. White, J.A. Williams, Jr.
Assistant Professors: E. Clark, T.E. McFadden, W. Mitchell

Degree Programs: Biology

Undergraduate Degree: B.S.
Undergraduate Degree Requirements: 4 semester hours Botany, 4 semester hours Zoology, 3 semester hours Genetics, 4 semester hours Embryology, 4 semester hours Physiology, 2 semester hours Seminar, 6 semester hours Electives, 18 semester hours Chemistry, 8 semester hours Physics, 8 semester hours Mathematic Calculus, 6 semester hours foreign language 3 semester hours psychology

Graduate Degree: M.S.
Graduate Degree Requirements: Master: 18 semester

hours Graduate Biology, 6 semester hours Education, 6 semester hours Electives

Graduate and Undergraduate Courses Available for Graduate Credit: Not Stated

School of Agriculture
Dean: B.C. Webb

DEPARTMENT OF ANIMAL SCIENCE
(No reply received)

DEPARTMENT OF PLANT SCIENCE AND TECHNOLOGY
Phone: Chairman: (919) 379-7543

Chairman: Not Stated
Professors: S.J. Dunn, C.A. Fauntain, B.C. Webb
Associate Professors: H.P. Hermanson, R. Harrison, I. Ruffa
Assistant Professor: E.S. Carr, L.A. Yates,
Instructor: M.P. McCleone
Research Assistant: B.J. McCallum
Visiting Lecturers: 10
Research Assistants: 5

Degree Programs: Agronomy, Crop Science, Environmental Engineering, Horticultural Science, Plant Science, Soil Science

Undergraduate Degree: B.S.
Undergraduate Degree Requirements: 124 semester hours 30 semester hours in major

NORTH CAROLINA CENTRAL UNIVERSITY
(No reply received)

NORTH CAROLINA STATE UNIVERSITY
(see University of North Carolina State University at Raleigh)

NORTH CAROLINA WESLEYAN COLLEGE
Rocky Mount, North Carolina 27801 Phone: (919) 442-7121
Dean of College: R.E. Bauer

DEPARTMENT OF BIOLOGY
Phone: Chairman: (919) 422-7121 Ext. 22

Chairman: A.W. Sharer
Professor: A.W. Sharer
Associate Professor: E.E. Brandt
Assistant Professor: B.E. Kane, Jr.

Degree Programs: Biology, Fisheries and Wildlife Biology, Environmental Science

Undergraduate Degrees: B.A., B S.
Undergraduate Degree Requirements: B.S. Principles of Biology, 3, Methods in Biology 1, Animal Physiology 3, Seminar 1, one chosen from 3 others. B.A. Principles of Biology 3, Methods in Biology 1, Developmental Anatomy and Laboratory 5, either Heredity 3 and Investigations in Genetics 2 or Animal Physiology 3 and Investigations in Physiology 2, Ecology 4, and Seminar plus one course chosen from 3 others.

PEMBROKE STATE UNIVERSITY
Pembroke, North Carolina 28372 Phone: (919) 521-4214
Dean of College: C.M. Fisher

DEPARTMENT OF BIOLOGY
Phone: Chairman: (919) 521-4214

Chairman: R.F. Britt
Professor: D.K. Kuo
Associate Professor: J.B. Ebert
Assistant Professors: J.A. McGirt, R.L. Mason, H.D. Maxwell
Instructor: L.B. Oxendine

Degree Programs: Biology

Undergraduate Degree: B.S.
Undergraduate Degree Requirements: Principles of Biology, General Botany, General Zoology, Principles of Ecology, Cell Biology, Genetics plus 10-12 semester hours of electives

PFEIFFER COLLEGE
Misenheimer, North Carolian 28109 Phone: (704) 463-3111
Dean of College: J.G. Haesloop

BIOLOGY DEPARTMENT
Phone: Chairman: (704) 463-3111

Head: R.H. Crowl
Professors: R.H. Crowl, J.G. Haesloop, J.O. Manly, C.H. Robertson
Associate Professor: S.C. Dial

Degree Programs: Biology (Interdepartmental majors are available in: Environmental Studies, Allied Health, Pre-Medical)

Undergraduate Degree: A.B.
Undergraduate Degree Requirements: 1 year General Zoology, 1 year General Botany, 1 semester Genetics, 3 to 5 other courses in Biology, General Chemistry required

QUEENS COLLEGE
Charlotte, North Carolina 28207 Phone: (704) 332-7121
Dean of College: A.O. Canon

DEPARTMENT OF BIOLOGY
Phone: Chairman: Ext. 260

Chairman: J H. Fehon
Dana Professor of Biology: J.H. Fehon
Associate Professors: V.L. Martin, J.C. Coffey
Assistant Professors: J.R. Diebolt

Degree Program: Biology

Undergraduate Degrees: B.A., B.S.
Undergraduate Degree Requirements: B.A. - Twenty-nine credit hours from Biology courses approved by major advisor. Students concentrating inBiology are required to take eight credit hours of Chemistry and Mathematics. Students concentrating in Biology are advised to take Physics and Organic Chemistry. B.S. - 17 courses in Biology, as approved by the major advisor, two years of Chemistry including a term of Organic Chemistry, Physics, Mathematics, In addition intermediate-level proficiency is required in a foreign language.

ST. ANDREWS PRESBYTERIAN COLLEGE
Laurinburg, North Carolina 28352 Phone: (919) 276-3652
Dean of College: V. Arnold

DIVISION OF MATHEMATICAL, NATURAL AND HEALTH SCIENCES
Phone: Chairman: 276-3652

Program Chairman: J.C. Clausz
Associate Professor: A.L. Applegate
Assistant Professors: J.C. Clausz , C.E. Styron

Degree Programs: Biology, Medical Technology

Undergraduate Degrees: B.A., B.S.
Undergraduate Degree Requirements: Contracted major generally including 1 year Mathematics, 1 year Chem-

istry and Biology courses (half of which are junior-senior level courses).

SAINT AUGUSTINE'S COLLEGE
(No reply received)

SALEM COLLEGE
Winston-Salem, North Carolina 27108 Phone: (919) 723-7961
Dean of College: J. Somerville

DEPARTMENT OF BIOLOGY
Phone: Chairman: Ext. 243

Chairman: J.W. Edwards
Professor: J.W. Edwards
Associate Professor: D.E. McLeod
Assistant Professors: C.E. Warnes, S.R. Nohlgren
Laboratory Instructor: J. Perlmutter

Degree Programs: Biology, Medical Technology

Undergraduate Degrees: B.S., B.A.
Undergraduate Degree Requirements: B.S.: Biology, 10 courses, Physics, 2 courses, Chemistry 4 courses, Mathematics 2 courses, B.S. (Medical Technology): Biology 6 courses, Physics 2 courses, Chemistry 4 courses, Mathematics 2 courses, B.A.: Biology 8 courses, Chemistry 1 course

SHAW UNIVERSITY
106 E. South Street Phone: (919) 755-4800
Raleigh, North Carolina 27602
Dean of College: K.K. Ghosh

DEPARTMENT OF BIOLOGY
Phone: Chairman: 775-4992 Others: 755-4800

Chairman: M.R. Jones
Professors: R.K. De, S.R. Toue
Assistant Professor: M.R. Jones
Instructor: D. Kwasikpui

Degree Program: Biology

Undergraduate Degree: B.S.
Undergraduate Degree Requirements: General Biology, Zoology, Botany, Anatomy, Physiology, Embryology, Genetics, Biochemistry, Bacteriology, Molecular Biology, Ecology, Seminar

TRINITY COLLEGE
(see under Duke University)

UNIVERSITY OF NORTH CAROLINA AT ASHEVILLE
University Heights Phone: (704) 254-7415
Asheville, North Carolina 28804
Dean of College: R.A. Riggs

DEPARTMENT OF BIOLOGY
Phone: Chairman: Ext. 225 Others: Ext. 305

Chairman: J.D. Perry
Professor: H.H. Johnston
Associate Professors: J.J McCoy, J D. Perry
Assistant Professor: J.C. Bernhardt
Visiting Lecturer: 1

Degree Programs: Biology, Pre-Medical

Undergraduate Degree: B.A.
Undergraduate Degree Requirements: 8 hours Introductory Biology and Laboratory, 12 hours Genetics, Biology of Lower Plants, Invertebrate Zoology, 12 hours Elective courses in Biology, 8 hours General Chemistry, 6 hours Statistics, Computer Program, one additional course from Organic Chemistry, Physics, or Mathematics

UNIVERSITY OF NORTH CAROLINA AT CHAPEL HILL
Chapel Hill, North Carolina 27514 Phone: (919) 966-1100
Dean of Graduate Studies: L.V. Jones

College of Arts and Sciences
Dean: J.R. Gaskin

BIOLOGY CURRICULUM
Phone: Chairman: (919) 933-3775 Others: 933-3776

Chairman: T.K. Scott
All Professors are connected with other Departments they teach Biology courses under the Biology Curriculum.

Degree Program: Biology

Undergraduate Degrees: A.B., B.S.
Undergraduate Degree Requirements: A.B. - General College requirements and six additional Biology courses taken in up to six departments. B.S. - General College requirements and seven additional Biology courses which must include Genetics, Ecology, Cell Biology, Language, two advanced Chemistry and Mathematics requirements.

CURRICULUM IN GENETICS
Phone: Chairman: (919) 966-4318 Others: 966-4265

Chairman: J.B. Graham
Alumni Distinguished Professor: J.B. Graham
All Professors are connected with other Departments they teach Genetics courses under the Genetics Curriculum.

Degree Program: Genetics

Graduate Degrees: M.A., M.S., Ph.D.
Graduate Degree Requirements: Master: Written and oral examination, Research, Thesis, Doctoral: Written and oral examination, Research, Dissertation

Graduate and Undergraduate Courses Available for Graduate Credit: Biochemistry and Molecular Biology (U), Computer Programming (U), Genetics and Human Evolution (U), Human Genetics (U), Genetics of Bacteria and Viruses, Developmental Genetics, Molecular Genetics, Plant Genetics, Induced Recombination, Mutagenesis, Regulatory Mechanisms, Cytogenetics, Genetic Systems, Statistical Methods in Human Genetics, Genetics for Medical Students

DEPARTMENT OF BOTANY
Phone: Chairman: (919) 933-3775 Others: 933-3776

Chairman: T.K. Scott
University Distinguished Professor of Botany: L.S. Olive
Professors: C.R. Bell, R.M. Brown, M.H. Hommersand, W.J. Koch, E.J. Kuenzler, H.H. Lieth, J.F. McCormick, A.E. Radford, T.K. Scott
Associate Professors: E G. Barry, A.J. Domnas, J.J. Kohlmeyer, C.R. Parks, W.J. Woods
Assistant Professors: W.C. Dickison, N.G. Miller
Teaching Fellows: 2
Research Assistants: 8
Teaching Assistants: 21
Fellows: 6

Field Stations/Laboratories: Institute of Marine Sciences Morehead City, North Carolina

Degree Programs: Botany, Ecology, Genetics

Undergraduate Degree: A.B.
Undergraduate Degree Requirements: A.B. in Botany, General University requirements plus General Botany and Zoology (or General Biology), Plant Diversity, Field Botany, Plant Physiology and Morphogenesis, and 3 additional Botany courses, General Chemistry, General Physics or General Geology, and preferably Organic Chemistry.

Graduate Degrees: M.A., M S., Ph.D.
Graduate Degree Requirements: Master: M.A.: At least 30 semester hours of graduate courses in Botany and related fields and a Thesis. M.S.: Same, but 6 semester hours of courses may be substituted for the Thesis. Doctoral: Ph.D.: At least 2 years residence beyond Master's degree, graduate courses selected by students and their Committees, Dissertation, at least one foreign language.

Graduate and Undergraduate Courses Available for Graduate Credit: U and G: Algae, Fungi, Bryophytes, Plant Physiology, Taxonomy of Vascular Plants, Evolutionary Mechanisms, Ecology, Geobotany, Comparative Morphology of Vascular Plants, Plant Anatomy, Cell Biology, Cytology, Genetics, Plant Genetics, Paleobotany, G: Mycetozoa, Phycomycetes, Ascomycetes, Basidiomycetes, Marine Mycology, Algal Physiology, Plant Biochemistry, Plant Growth and Development, Floristics, Variation and Evolution in Plants, Phylogeny and Classification of Vascular Plants, Chemotaxonomy, Systems Ecology, Ecology of Phytoplankton, Cytological Methods, Cytogenetics, Genetics Systems, Seminar Courses, Speical Topics in Botany Courses, Master's Thesis, Doctoral Dissertation

DEPARTMENT OF ZOOLOGY
Phone: Chairman: (919) 933-2077 Others: 933-2266

Chairman: I.R. Hagadorn
Kenan Professor of Zoology: D.P. Costello
Professors: C.D. Beers (Emeritus), A.F. Chestnut, D.P. Costello, W.L. Engels (Emeritus), I.R.Hagadorn, D.G. Humm, C.E. Jenner, C.S. Jones (Vice Chancellor for Business and Finance), H.E. Lehman, J.C. Lucchesi (Director of Graduate Studies), E.A. McMahan, A.E. Stiven (Chairman of Ecology Curriculum), M. Whittinghill (Emeritus)
Associate Professors: J.A. Feduccia, D.W. Misch, H.C. Mueller, D.W. Stafford
Assistant Professors: B.S. Baker, M.A. Bleyman, A.K. Harris, T.C. Long, S.R. Reice, R.M. Rieger, N.A. Smith, J. White, R.H. Wiley
Instructor: L.S. DeSaix, M.B. Vogel
Visiting Professor: 1
Teaching Fellow: 1
Research Assistants: 3
Teaching Assistants: 36
Unversity Research Assistants: 3

Field Stations/Laboratories: Bermuda Biological Station-Institute of Marine Sciences - Morehead City, North Carolina, Wrightsville Biomedical Marine Laboratory, North Carolina

Degree Programs: Animal Behavior, Animal Physiology, Biochemistry, Biology, Ecology, Genetics, Human Genetics, Microbiology (Cell Biology), Physiology, Zoology, Embryology, Marine Biology

Undergraduate Degrees: A.B., B.S. (Biology, Zoology)
Undergraduate Degree Requirements: For the A.B.: In addition to Zoology 11, 11L (or Biology 21, 21L), seven approved Zoology courses are required for the A.B., of which five must be taken with Laboratory. The following allied sciences are required: Chemistry 21, 21L, 61, 62, Physics, 24, 25./5 to 8 non-divisional electives must also be taken./ For the B.S.: This degree requires, in addition to the requirements for the A.B., an additional approved zoology course, Chemistry 41, 41L, Computer Science 16, Mathematics 32, French, German or Russian 1,2,3,4, and three free Science Electives

Graduate Degrees: M.A., M.S., Ph.D.
Graduate Degree Requirements: Master: (a) A minimum of 30 semester hours, including courses recommended at the time of the Preliminary Oral Evaluation, (b) at least one Zoology course in an approved area of specialization; (c) at least one Zoology Seminar; (d) research in a field numbered 300 or higher not to exceed 6 hours credit; (e) a written examination to be taken no later than the 4th semester of residence, (f) an approved thesis prepared in acceptable form presenting the results of original research (M.A.) or an approved library report (M.S.), (g) final oral examination Doctoral: (a) A minimum of 60 hours, including Zoology courses recommended at the time of the Preliminary Oral Evaluation (b) at least one Zoology course in an approved area of specialization (c) at least one Zoology Seminar (d) research in a field numbered 300 or higher not to exceed 6 hours credit (e) an written examination to be taken no later than the 6th semester of residence (f) a defense of dissertation oral examination, and finally (g) laboratory teaching equivalent to 6 contact hours per week for at least two semesters

Graduate and Undergraduate Courses Available for Graduate Credit: Topics in Cell Biology (U), Topics in Organismal and Environmental Biology (U), Structure and Evolution of Vertebrates (U), Genetics Laboratory (U), C Comparative Physiology and Laboratory (U), Introduction to Cell Physiology and Laboratory (U), Human Genetics (U), Introduction to Neurophysiology (U), Oceanography (U), Biology of Insects and Laboratory (U), Invertebrate Development, Larvae and Plankton (U), Biological Oceanography (U), Special Problems in Marine Biology (U), Marine Ecology (U), Experimental Vertebrate Embryology (U), Developmental Genetics (U), Molecular Biology (U), Molecular Genetics (U), Electron Microscopy (U), Comparative Endocrinology (U), Experimental Methods in Vertebrate Embryology, Functional Cytology, Methods in Cytological Analysis, Introduction to Neurobiology, Population Ecology, Methods in Population Analysis, Experimental Invertebrate Embryology, Advanced Marine Ecology, Experimental Advanced Marine Ecology, Experimental Endocrinology, Advanced Cellular Physiology, Experimental Methods in Cellular Physiology, Special Topics in Physiology, Characterization of Biological Macromolecules, Experimental Neurophysiology, Systems Ecology and Laboratory, Ichthyology, Techniques for Sampling Marine Fishes

School of Medicine
Dean: C.C. Fordham, III

DEPARTMENT OF ANATOMY
(No reply received)

DEPARTMENT OF BACTERIOLOGY AND IMMUNOLOGY
Phone: Chairman: (919) 966-1191

Chairman: G.P. Manire
Kenan, Professor of Bacteriology and Immunology: G.P. Manire
Professors: W.J. Cromartie, H. Gooder, G. Haughton, G.P. Manire, J.S. Pagano (joint appointment), J.H. Schwab, D.G. Sharp, M.S. Silverman (joint appointment), J.K. Spitznagel, W.R. Straughn, M.L. Tyan (joint appointment)
Associate Professors: K.F. Bott, M.H. Edgell, J. Fischer (joint appointment), C.A. Hutchison, P.F. Sparling (joint appointment), R. Twarog, W.J. Yount (joint appointment)
Assistant Professors: J.B. Baseman, J.D. Folds, J.E. Newbold, G. Wertz, P. Wyrick
Instructors: B.F. Maxwell, W.R. Rumpp
Visiting Lecturers: 2
Teaching Fellows: 8

Research Assistants: 3
Research Associates: 3

Degree Programs: Bacteriology and Immunology

Graduate Degrees: M.S., Ph.D.
Graduate Degree Requirements: Master: Minimum of 30 hours of credit Doctoral: Four semesters with full credit, two of which must be on this campus

Graduate and Undergraduate Courses Available for Graduate Credit: Biology and Biochemistry of Microorganisms (U), Infection and Immunity (U), Introduction to Microbiology (U), Special Topics in Bacteriology or Immunoloty (U), General Bacteriology (U), Advanced M crobiology and Immunology, Seminar in Microbiology, Seminar/Tutorial in Microbiology Chemistry and Genetics, Seminar/Tutorial in the Biology of Cancer, Seminar/Tutorial in Animal Virology, Seminar/Tutorial in Immunology, Viruses and the Immune Response, Genetics Systems,

DEPARTMENT OF BIOCHEMISTRY AND NUTRITION
Phone: Chairman: (919) 966-1236/4302

Chairman: J.L. Irvin
Kenan Professor of Biochemistry: J.L. Irvin
Professors: C.E. Anderson, J.C. Andrews, M. Caplow, I. Clark, H.J. Fallon, E. Glassman, J. Hermans, J. L. Irvin, H. Kingdon, M. Lipton, G.L. Mechanic, J.C. Parker, R. Penniall, H.A. Schneider, G.K. Summer, R. H. Wagner, J.R. White, J. E. Wilson, R.V. Wolfdenen
Associate Professors: F.E. Bell, M.K. Berkut, D.J. Holbrook, R.L. Lundblad, P. Morell, S.N Nayfeh, G. Piantadosi, J. Savory
Assistant Professors: J. Benjamins, M. Bleyman, C.B. Chae, S.G. Chaney, K.A. Koehler, K. Marushige, E.O. Oswald (Adjunct), R.C. Reitz, D.W. Stafford, B.R. Switzer
Instructors: N.M. Davidian, T.D. Entingh, D.J. Entingh, D.M Frazier, R. Hartman, A. Hassid, M. Hershkowitz, K.K. Kumaroo, G. Mitchell, L. Shen, D. Wentworth
Research Assistants: 4

Degree Program: Biochemistry

Graduate Degrees: M.S., Ph.D.
Graduate Degree Requirements: Master: 30 semester hours of graduate work, M.S. core curriculum, M.S. written comprehensive examination, thesis, final oral examination Doctoral: 4 semesters of graduate residency, Ph.D. oral comprehensive examination, research, dissertation, final oral examination

Graduate and Undergraduate Courses Available for Graduate Credit: Biochemistry for Students of Biology and Chemistry, Biochemistry Laboratory, Biochemistry for Dental Students, Introduction to Comparative and Evolutionary Biochemistry, Introduction to Biochemistry and Molecular Biology, Introduction to Neurobiology, Protein Chemistry, Nucleic Acid Chemistry and Enzymology, Enzyme Mechanisms and Kinetics, Physical Biochemistry, Structure of Biological Macromolecules, Membrane Chemistry, Clinical Chemistry, Techniques in Clinical Chemistry, Molecular Genetics, Biochemical Preparations, Advanced Biochemistry Laboratory, Biochemistry of Lipids, Bioenergetics, Regulatory Mechanisms, Neurochemistry, Seminar in Biological Macromolecules, Seminar in Biological Macromolecules, II, Seminar in Biochemical Regulation, Seminar in Chemical Neurobiology, Seminar in Enzyme Mechanisms, Seminar in Biopolymers, Seminar in Antibiotics and Antimetabolites, Seminar in Lipid Metabolism, Genetics Systems, Seminar in Neurobiology, Advanced Chemical Neurobiology, Research in Biochemistry, Research in Neurobiology

DEPARTMENT OF PATHOLOGY
Phone: Chairman: (919) 966-4676

Chairman: J.W. Grisham
Professors: W.R. Benson, K.M Brinkhous, F.G. Dalldorf, J.D. Geratz, R.A. Boyer, J.B. Graham, J.W. Grisham, R.P. Hudson, W.D Huffines, M.R. Krigman, R.D. Langdell, R.G. Mason, A.J. McBay, W.W. McLendon, H.R. Roberts, M.C. Swanton, R.H. Wagner, W.P. Webster
Associate Professors: E.M. Barrow, R.L. Lundblad, J.R. Pick, J. Savory
Assistant Professors: C.N. Carney, H. Y-K. Chuang, H.A. Cooper, N.A. Hoffman, L.M. Killingsworth, K.A. Hoehler, J.M. McDonagh, R.P. McDonagh, P. Mushak, W. Nopanitaya, R.W. Shermer, R.F. Turk
Visiting Lecturers: 2
Research Assistants: 20
Visiting Professors: 2
Visiting Assistant Professors: 4

Field Stations/Laboratories: Francis Owen Blood Research Laboratory - University Lake, Chapel Hill, North Carolina

Degree Program: Pathology

Graduate Degrees: M.S., Ph.D.
Graduate Degree Requirements: 2 semesters or 1 academic year residency, 30 hours successful completion of coursework, successful completion of a modern foreign language or 2 courses in Biostatistics or Computer Science, written comprehensive examination, thesis, oral examination Doctoral: 4 semester of residency two of which must be consecutive, satisfactory demonstration of adequate background in Pathology, written comprehensive examination, preliminary oral examination, dissertation, oral examination, same modern foreign language requirement as for Master degree

Graduate and Undergraduate Courses Available for Graduate Credit: Surgical Pathology, Autopsy Pathology, Neuropathology, Hematopathology, Human Genetics and Constitutional Pathology, Speical Methods in Pathology Seminar in Neurobiology, Seminar in Pathology, Research in Pathology, Research in Neurobiology, Introduction to Neurobiology (U), Genetics Systems (U), Pathology (U), Experimental Pathology (U), General, Systemic and Clinical Pathology (U)

DEPARTMENT OF PHYSIOLOGY
Phone: Chairman: (919) 966-3127

Chairman: E.R. Perl
Professors: N.A. Coulter (Joint), F.L. Eldridge (Joint), C.W. Gottschalk, E.B. Kokas (Part-time), M. Kuno, A.T. Miller, Jr., R.L. Ney (Joint), E.R. Perl, J.H. Perlmutt
Associate Professors: M.L. Berkut (Joint), R.G. Faust, R.L. Glasser, E.R. Kafer (Joint), R.A. King (Joint), A. Rustioni (Joint), B.L. Whitsel, R.L. Yonce, G.S. Malindzak, Jr. (Adjunct)
Assistant Professors: W.J. Arendshorst, J.C. Daw, P.B. Farel, R.P. McDonagh (Joint), D.L. McIlwain G.W.D. Meissner (Joint), V. Neelon (Joint), D.L. Trevino, J.N Weakly
Instructors: M.A. Cook, P.R. Loe
Research Fellows: 1
Research Associates: 3

Degree Programs: Physiology, Neurobiology

Graduate Degree: Ph.D.
Graduate Degree Requirements: Doctoral: Courses of study to be specified by Graduate Advisor, Written and oral preliminary examinations, Written Dissertation on area of research, Final written and oral examinations

Graduate and Undergraduate Courses Available for Graduate Credit: Introduction to Biomedical Engineering, Introduction to Neurobiology, Biomedical Instrumentation, Introduction to Biomedical Data Processing, Introduction to Biomathematics, Cell and Organ System Physiology, Mammalian Structure and Function, Current Topics in Physiology, Advanced Physiology, Special Topics in Physiology, Biological Control Systems, Seminar in Physiology, Neural Information Processing, Seminar in Neurobiology, Advanced Neurophysiology

School of Public Health

Dean: B.G. Greenberg

DEPARTMENT OF BIOSTATISTICS

Phone: Chairman: (919) 966-1107 Others: 996-1108,1109

Head: J.E. Grizzle
Kenan Professor of Biostatistics: B.G. Greenberg
Professors: E.J. Coulter, R.C. Elandt-Johnson, R.C. Elston, B.G. Greenberg, J.E. Grizzle, L. Guralnick, D.H. Horvitz (Adjunct), R.R. Kuebler, F.E. Linder, A.S. Lunde (Adjunct), D.E.A. Quade, G. Sabagh (Adjunct), P.K. Sen, H.B. Wells
Associate Professors: J.R. Abernathy, D.G. Hoel (Adjunct), G.G. Koch, P.A. Lachenbruch, W.B. Riggan, B.V. Shah (Adjunct)
Assistant Professors: R.E. Bilsborrow, C.E. Davis, M.E. Francis, D.B. Gillings, R.W. Helms, C.R. Hogue, P. B. Imrey, D.G. Kleinbaum, L.L. Kupper, C.H. Langley (Adjunct), J.W. Lingner, J.J. Palmersheim (Adjunct), W.C. Nelson (Adjunct), A.W. Rademaker, R.H. Shachtman, W.C. Smith, J.R. Stewart, C.H. Suchindran, M. J. Symons, C.D. Turnbull, O.D. Williams
Instructors: J.K. Dias, P.A. Guild (Adjunct), E.B. Kaplan, A.M Sorant
Research Assistants: 20
Research Associate: 5

Degree Program: Biostatistics

Graduate Degree: M.S., Ph.D.
Graduate Degree Requirements: Master: Advanced Calculus, Linear Algebra, Computer Techniques, 13 hours Statistics, 9 hours Cognate subjects, Thesis, Final examination, Teaching Experience Doctoral: M.S. requirements plus 18 hours of advanced Statistics, and 15 hours cognate subjects, Doctoral written and oral examinations, thesis, Final Oral, Teaching or consulting experience

Graduate and Undergraduate Courses Available for Graduate Credit: Public Health Statistics, Management of Public Health Data, Principles of Statistical Inference, Mathematical Methods in Biostatistics, Introduction to Statistical Data Processing, Introduction to Statistical Data Processing and Computer Programming, Probability and Statistics, Problems in Biostatistics, Inference, Elements of Statistical Analysis, Sample Survey Methodology, Analysis of Categorical Data, Applied Stochastic Processes, (U), Data of Health Statistics in Administration, Statistical Methods in Health Services Research, Information Systems in Mental Health, Evaluation and Programming Elements in Mental Health, Specialized Methods in Health Statistics, Nonparametric Procedures in Biometric Research, Large Sample Theory, Linear Models in Categorical Data Analysis, Linear Models I, Linear Models II, Demographic Techniques I, Demographic Techniques II, Statistics in Population Programs, Mathematical Models in Demography, Statistical Methods in Human Genetics, Statistical Methods in Epidemiology, Field Observation in National Health Statistics, Field Training in Public Health Statistics, Seminar in Biological Models, Statistical Consulting in the Health Sciences, Training in Statistical Teaching in the Health Sciences, Seminar in Population Statistics, Research in Biostatistics, Master Thesis, Doctoral Dissertation

ENVIRONMENTAL SCIENCES AND ENGINEERING

Phone: Chairman: (919) 966-1171

Head: R.F. Christman
Kenan Professor of Environmental Engineering: D.A. Okun
Professors: A. Altshuller (Adjunct), E.T. Chanlett, R.F. Christman, W.A. Cook (Adjunct), D.A. Fraser, M.S. Health, Jr. (Joint), D.H. Howells (Joint), L.J. Goldwater (Adjunct), M.M. Hufschmidt (Joint), J.D. Johnson, E.J. Kuenzler, J.C. Lamb III, D.A. Ikun, C. R. O'Melia, P.C. Reist, L.A. Ripperton (Adjunct), M.A. Shiffman, A.C. Stern, A.W. Waltner (Joint), C.M. Weiss
Associate Professors: M.C. Battigelli (Joint), J.C. Brown, T.D. English (Adjunct), R.L. Harris, D.T. Lauria, L.W. Little, F.E. McJunkin, F.O. Mixon, (Adjunct), D.H. Moreau (Joint), J.K. Sherwani, P.C. Singer, D.G. Willhoit, W.E. Wilson, Jr. (Adjunct)
Assistant Professors: D.L. Fox, H.E. Jeffries, F.K. Pfaender, M.S. Shuman, A.G. Turner, Jr., P.J. Walsh (Adjunct)
Research Associates and Assistants: 7
Lecturer: 1

Field Stations/Laboratories: UNC Wastewater Research Center - Chapel Hill, Ambient Air Research Facility - Chatham County, Radiation Biophysics Laboratory - Chapel Hill, Institute of Marine Sciences - Morehead City

Degree Programs: Water Resources Engineering, Environmental Chemistry and Biology, Radiological Hygiene, Environmental Management and Protection, Air and Industrial Hygiene

Graduate Degrees: M.S., Ph.D.
Graduate Degree Requirements: Master: Minimum residence 12 months, minimum 30 semester hours course work ind uding core courses, thesis or technical report required, no language requirement, oral comprehensive examination Doctoral: Master's degree, minimum residence 4 semesters with 2 semesters in continuous enrollment, minor or supporting area of at least 15 semester hours, doctoral written and oral examinations, 1 research skill or language required. dissertation, final oral examination is defense of dissertation

Graduate and Undergraduate Courses Available for Graduate Credit: Elements of Environmental Health (U), Man and His Environment (U), Applied Electron Microscopy (U), Quantitative Studies for Environmental Sciences (U), Water Chemistry (U), Soils and Surfaces (U), Biology in Environmental Science (U), Limnology and Water Pollution (U), Environmental Biology (U), Environmental Microbiology (U), Elements of Air Hygiene (U), Elements of Industrial Hygiene (U), Applied Physiology and Toxicology (U), Air Pollution Measuring, Monitoring and Survey (U), Instrumentation and Data Acquisition (U), Industrial Hygiene Engineering Control Design (U), Microbiology of the Institutional Environment (U), Management of the Institutional Environment (U), Elements of Radiological Hygiene (U), Modern Physics for Environmental Sciences (U), Radiation Instrumentation (U), Field Observations in Radiological Hygiene, Principles of Water Quality Management (U), Water and Wastes Treatment Processes (U), Ground Water Hydrology (U), Special Topics in Water Resources (U), Environmental Management, Planning and Development of Environmental Hygiene Programs, Environmental Issues and Decisions, Environmental Assessment, Systems Analysis in Environmental Planning, Environmental Systems Analysis I: Deterministic Models, Instrumental Methods of Analysis, Special Topics in Aquatic Chemistry, Trace Analysis, Special Topics in Aquatic Biology, Microbial Ecology, Ecology of Phytoplankton, Introduction to Aerosol Science, Industrial Hygiene Practices, Air and Its Contaminants, Industrial Hygiene Laboratory, Air Pollution Control, Biological Effects of Air Pollution, Chemistry of the Troposphere, Industrial Medicine, Air Pollution Meterology, Environmental Protection I, II, III, Solid Wastes Management, Radiation Biophysics, Health Physics, Radiation Hazards Evaluation I, II, Modeling in Natural

Aquatic Systems, Technology of Engineered Water Systems, Water and Wastewater Treatment Plant Design, Advanced Water and Wastes Treatment Processes I, II, Industrial Water Quality Management, Engineering Project Design, Topics in Advanced Hydrology, Natural Resource Law and Policy, Special Project in Water Quality Planning, Environmental Health Seminar, Environmental Health Problems in Developing Countries, reading and research courses

DEPARTMENT OF EPIDEMIOLOGY
Phone: (919) 966-2241

Chairman: Not Stated
Professors: J.C. Cassel, M.A. Ibrahim, B.H. Kaplan, A. Omran, C. Slome, H.A. Tyroler
Associate Professors: B.S. Hulka, R. Patrick
Assistant Professors: D. Andjelkovic, C. Becker, J. Cornoni, S.A. James, A. McMichael
Lecturer: 1

Graduate Degrees: M.S., Ph.D.
Graduate Degree Requirements: Master: Bachelor of Arts from recognized institution, minimum residence for 2 semesters, all work must be completed within five years, application for admission to candidacy, application for degree, thirty semester hours, in department of major, 6 of which master's thesis may be included, written comprehensive examination, thesis, oral presentation of thesis Doctoral: Bachelor of Arts from recognized institution, minimum of 2 years graduate studies, one of which has to be continious at University, major covering field of major interest, Doctoral oral examination, written examination in field of major interest, application for admission to candidacy, application for a degree, dissertation, final oral examination

Graduate and Undergraduate Courses Available for Graduate Credit: Problems in Epidemiology, Principles of Epidemiology, Epidemiology in Population Dynamics and Family Planning Program, Epidemiology in Environmental Health, Behavioral Science Measurement in Health Programs and Research, Determinants of Communicable Disease, Applied Methods in Epidemiology and Health Services, Alcohol Behavior and Health, Research Methodology in Alcohol Usage, Epidemiological Foundations for Disease Control Programs, Dental Epidemiology, Epidemiology of Program Acceptance, Culture and Health, History of Epidemiology, Epidemiological Investigation, Statistical Methods in Epidemiology, Field Training in Epidemiology, Research in Epidemiology, Environmental Epidemiology,Seminar, Evaluative Research Methods

DEPARTMENT OF PARASITOLOGY AND LABORATORY PRACTICE
Phone: Chairman: (919) 966-1067 Others: 966-3285

Chairman: J.E. Larsh
Professors: H.T. Goulson, J.E. Larsh, R.B. Watson, N.F. Weatherly
Associate Professors: E.F. Chaffee, J.R. Hendricks, J.K. Read
Assistant Professor: J.R. Clem
Visiting Lecturer: 1
Research Assistants: 2

Degree Program: Parasitology

Graduate Degrees: M.S., Ph.D.
Graduate Degree Requirements: Master: One academic year residence, at least 30 semester hours course work, written report on relevant subject, final written overall departmental courses, five-year limit. Doctoral: Minimum two academic years, minimum 18 semester hours, Doctoral Oral examination, Doctoral Written examination, Dissertation on independent research, Final doctoral oral examination in defense of dissertation, must be completed in eight years.

Graduate and Undergraduate Courses Available for Graduate Credit: Parasitism and Human Disease, Human Parasitology, Problems in Parasitology, Problems in Public Health LaboratoryPractice, Public Health Bacteriology, Public Health Virology, The Nature of Parasitism, Parasitological Methods, Malariology, Medical Entomology, Problems in Public Health Laboratory, Methodology, Public Health Laboratory Methods, Public Health Laboratory Management, Seminar in Parasitology, Seminar in Public Health Laboratory Practice, Research in Parasitology, Research in Public Health, Laboratory Methodology

UNIVERSITY OF NORTH CAROLINA AT CHARLOTTE

Charlotte, North Carolina 28223 Phone: (704) 597-2285
Dean of Graduate Studies: D. Turner
Dean of College: S. Burson (Acting)

DEPARTMENT OF BIOLOGY
Phone: Chairman: (704) 597-2315 Others: 597-2316

Chairman: M D. Arvey
Distinguished Professor: P. Hildreth (Acting Vice Chancellor for Academic Affairs)
Professors: M.D. Arvey, H.H. Hechenbleikner, J. Matthews
Associate Professors: R. Hogan, E. Menhinick
Assistant Professors: D. Bashor, J. Butts, J. Darner, N. Edwards, W. Graul, J. Haviland, P. Johnson, R. Ostrowski, J. Travis
Teaching Fellows: 6

Degree Programs: Animal Behavior, Animal Genetics, Animal Physiology, Bacteriology and Immunology, Biological Sciences, Biology, Botany, Ecology, Genetics, Histology, Medical Microbiology, Medical Technology, Microbiology, Parasitology, Physiology, Plant Sciences, Radiation Biology, Wildlife Management, Zoology

Undergraduate Degree: B.A.
Undergraduate Degree Requirements: B.A. - 32 semester hours of Biology, including Biology 211, 221, 231, 491, and one course in three areas chosen from development, Ecology, Genetics and Physiology, a minimum of 5 laboratory courses. (Also Chemistry and Mathematics)

Graduate Degrees: M.A., M.S.
Graduate Degree Requirements: M.S. - 32 hours in approved courses in Biology and related work, 16 hours must be from courses open only to graduates, Comprehensive examination and thesis required.

Graduate and Undergraduate Courses Available for Graduate Credit: Microbiology (U), Parasitology (U), Genetics (U), Embryology (U), Ecology (U), Plant Physiology (U), Animal Physiology (U), Plant Morphogenesis (U), Cytogenetics (U), Immunology (U), Endocrinology (U), Cell Physiology (U), Radioisotope Techniques (U)

UNIVERSITY OF NORTH CAROLINA AT GREENSBORO

Greensboro, North Carolina 27412 Phone: (919) 379-5000
Vice Chancellor for Graduate Studies: J.W. Kennedy
Dean of Arts and Sciences: R.L. Miller

DEPARTMENT OF BIOLOGY
Phone: Chairman: 379-5391

Chairman: B.M. Eberhart
Professors: L.G. Anderton, B.M. Eberhart, P.E. Lutz, J.F. Wilson
Associate Professors: W.K. Bates, V. Gangstad, H.T. Hendrickson, E. McCrady, III, R.M. Morrison, H.J. Rogers
Assistant Professors: R.E. Cannon, L.J. Cutter,

S. Sands, R.C. Schauer, R.H. Stavn
Instructors: J.S. Curtis, B.S. Madden, O.R. Patrick, L. Shepard
Visiting Lecturers: 3
Teaching Fellows: 3
Research Assistants: 1

Degree Program: Biology

Undergraduate Degree: B.A.
Undergraduate Degree Requirements: 24-36 semester hours Biology, Chemistry, Physics and Mathematics recommended, thesis

Graduate Degree: M.A.
Graduate Degree Requirements: Master: 30 semester hours Biology beyond undergraduate major

Graduate and Undergraduate Courses Available for Graduate Credit: Microscopy and Photomicrography: Theory and Technique (U), The Development of Modern Concepts in Biology (U), Local Flora (U), Plant Histology and Anatomy (U), Terrestrial Ecology (U), Microbial Ecology (U), Aquatic Ecology (U), General Biochemistry (U), Radiation Biology and Radiotracer Methods (U), Experimental Embryology (U), Natural History of Vertebrates (U), Physiology of Activity (U), Physiology of Vertebrates (U), Cellular Physiology (U), General Microbiology (U), Pathogenic Bacteriology (U), Immunology (U), Cytogenetics, Genetics (U), Advanced Genetics (U), Seminar in Ecology, Developmental Physiology of Insects, Biochemical Genetics, Biochemistry, Mammalian Cytogenetics, Evolution and Systematics, Morphogenetic Processes in Development, Techniques in Biological Research

UNIVERSITY OF NORTH CAROLINA STATE UNIVERSITY AT RALEIGH

Raleigh, North Carolina 27607 Phone: (919) 755-2011
Dean of Graduate Studies: W.J. Peterson

School of Agricultural and Life Sciences

Dean: J.E. Legates

ANIMAL SCIENCE DEPARTMENT

Phone: Chairman: (919) 755-2755 Others: 755-2011

Head: I.D. Porterfield
Reynolds Professor of Animal Science: G.H. Wise, J.E. Legates, (Dean)
Professors: E.R. Barrick, A.J. Clawson, L. Goode, E.J. Eisen, J.M. Leatherwood, J.G. Lecce, J.E. Legates, B.T. McDaniel, R.D. Mochrie, A.H. Rakes, H.A. Ramsey, O.W. Robisnon, J.A. Santolucito (Adjunct), H.A. Schneider, L.C. Ulberg, G.H. Wise
Associate Professors: E.V. Caruolo, D.G. Davenport, E.U. Dillard, R.W. Harvey, W.L. Johnson, E.E. Jones, J.J. McNeill, R.M Myers, J C. Wilk
Assistant Professors: B.D. Harrington (Adjunct), B.H. Johnson, R.B. Lediy (Adjunct)
Visiting Lecturers: 3
Research Assistants: 27

Field Stations/Laboratories: Reidsville, Laurel Springs, Waynesville, Salisbury, Plymouth, North Carolina

Degree Programs: Animal Science, Biochemistry, Genetics, Microbiology, Nutrition, Physiology

Undergraduate Degree: B.S.
Undergraduate Degree Requirements: 130 semester hours

Graduate Degrees: M.S., Ph.D.
Graduate Degree Requirements: Master: M.Agr. - 36 semester hours, M.S. - 30 semester hours and thesis Doctoral: Not Stated

Graduate and Undergraduate Courses Available for Graduate Credit: Reproductive Physiology of Vertebrates (U), Diseases of Farm Animals (U), Genetics of Animal Improvement (U), Tropical Livestock Production (U), Topical Problems in Animal Science (U), Population Genetics in Animal Improvement, Experimental Animal Physiology, Principles of Biological Assays, Mineral Metabolism

DEPARTMENT OF BIOCHEMISTRY

Phone: Chairman: (919) 737-2581 Others: 737-2581

Head: G. Matrone
William Neal Reynolds Professor of Biochemistry: G. Matrone
University Professor of Biochemistry: F.B. Armstrong
Professors: F.B. Armstrong, H.R. Horton, J.S. Kahn, I.S. Longmuir, A.R. Main, G. Matrone, S.B. Tove
Associate Professors: J.A. Knopp, E.C. Sisler
Assistant Professor: E.C. Theil
Visiting Lecturers: 2
Teaching Fellows: 3
Research Assistants: 15
Traineeships and Fellowships: 6
Research Associates: 3
Self-Supported: 4
Postdoctoral Fellows: 3

Degree Program: Biochemistry

Undergraduate Degree: B.S. (Biological Sciences)
Undergraduate Degree Requirements: 29-30 credits Chemistry, 18 credits Mathematics, 8 credits Physics, 20 credits Biology

Graduate Degrees: M.S., Ph.D.
Graduate Degree Requirements: Master: 9 credit hours, seminars, thesis Doctoral: 21 credit hours, seminar, thesis, one foreign language, teaching experience

Graduate and Undergraduate Courses Available for Graduate Credit: Experimental Biochemistry, General Biochemistry, Physiological Biochemistry, Radioisotope Techniques in Biology, Introductory Enzyme Kinetics, Biochemical and Microbial Genetics, Special Topics in Biochemistry, Physical Biochemistry, Biochemical Research Techniques, Mineral Metabolism, Intermediary Metabolism I, Intermediary Metabolism II, Natural Products, Seminar in Biochemistry, Speical Topics in Biochemistry

DEPARTMENT OF BOTANY

Phone: Chairman: (919) 727-2727 Others: 737-2011

Head: G.R. Noggle
Professors: A.W. Cooper, R.J. Downs, J.W. Hardin, W.W. Heck, H. Seltmann, J.R. Troyer
Associate Professors: C.E. Anderson, U. Blum, D.W. DeJong (Adjunct), R.C. Fites, R.L. Mott, H.E. Pattee E.D. Seneca
Assistant Professors: G.M. Jividen (Adjunct), C.G. Van Dyke, A.M. Witherspoon, T.E. Wynn
Instructor: L.M. Stroud
Visiting Lecturers: 2
Teaching Fellows: 11
Research Assistants: 13
Research Associates (Postdoctorals): 2

Field Stations/Laboratories: 16 agricultural research stations

Degree Programs: Sciences, Ecology, Air Conservation, Botany

Undergraduate Degree: B.S. (Botany)
Undergraduate Degree Requirements: 27 hours Botany, Genetics, Microbiology, 12-16 Chemistry, 5-9 Physics 10-12 Mathematics, electives to total 130 hours

Graduate Degrees: M.S., Ph.D.
Graduate Degree Requirements: Master: (non-thesis) 36 hours, 15-18 Botany, 3 independent study, 15-18

minor M.S. - 30 hours, 15-18 Botany, 6 research, 12-15 minor Doctoral: Each candidates' program of study is formulated by a graduate advisory committee

Graduate and Undergraduate Courses Available for Graduate Credit: Economic Botany (U), Systematic Botany (U), Cell Biology (U), Plant Anatomy, Plant Physiology (U), Advanced Morphology and Phylogeny of Seed Plants, Plant Diversity (U), Grasses, Sedges and Rushes, Plant Geography, Advanced Plant Physiology I and II, Principles of Ecology, Physiological Ecology, Phycology, Fungi, Plant Morphogenesis, Advanced Taxonomy, Advanced Mycology, Water Relations of Plants, Plant Growth and Development, Introduction to Thermodynamics of Biological Systems, Advanced Topics in Ecology I and II, Seminar

DEPARTMENT OF CROP SCIENCE
Phone: Chairman: (919) 737-2647 Others: 737-2011

Head: Not Stated
William Neal Reynolds Professor of Crop Science: P.H. Harvey, D.U. Gerstel, W.C. Gregory, J.A. Weybrew
Professors: C.T. Blake, C.A. Brim, D.S. Chamblee, J.F. Chaplin, W.K. Collins, W.A. Cope, S.H. Dobson, D.A. Emery, H.D. Gross, S.N. Hawks, Jr., G.L. Jones, J.A. Lee, W.M Lewis, F.M. McLaughlin, P.A. Miller, R.P. Moore, D.E. Moreland, A. Petty, L.L. Phillips, J.C. Rice, D.L. Thompson, D.H. Timothy, J.B. Weber, E.A. Wernsman, A.D. Worsham, W.H. Wessling (Adjunct), F. Yoshikawa (Research)
Associate Professors: J.C. Burns, T.H. Busbice, J.G. Clapp, Jr., H D. Coble, F.T. Corbin, W.T. Fike, W.B. Gilbert, G.R. Gwynn, R.C. Long, C.F. Murphy, R.P. Patterson, W.G. Toomey
Assistant Professors: E.L. Kimbrough, E.G. Krenzer, Jr., J.W. Schrader, G.A. Sullivan, C.F. Tester, W.W. Weeks, J.C. Wynne
Instructor: C.E. Collins
Research Graduate Assistants: 11
Teaching (Graduate) Assistants: 3

Field Stations/Laboratories: Border Belt Tobacco Research Station, Central Crops Research Station, Harbor House Marine Science Center, Hatteras Marine Research Station, Horticultural Crops Research Station, Horticultural Crops Research Station, IES Eastern Area Office, Lower Coastal Plain Tobacco Research Station, Minerals Research Laboratory, Mountain Horticultural Crops Research Station, Mountain Research Station, Oxford Tobacco Research Station, Pamlico Marine Laboratory, Peanut Belt Research Station, Piedmont Research Station, Sandhills Research Station, Seafood Laboratory, Tidewater Research Station, Upper Coastal Plain Research Station, Upper Mountain Research Station, Upper Piedmont Research Station

Degree Programs: Pest Management for Crop Protection, Weed Science (Control), Crop Production and Physiology, Forage Crops Ecology, Plant Chemistry

Undergraduate Degree: B.S.
Undergraduate Degree Requirements: Graduate of an accredited high school, College Entrance Examination Board Scholastic Aptitude Test (SAT)

Graduate Degrees: M.S., Ph.D.
Graduate Degree Requirements: Master: Bachelor's degree from a recognized college or university and at least a "B" grade average in his undergraduate major for full standing Doctoral: M.S. degree from a recognized college or university

Graduate and Undergraduate Courses Available for Graduate Credit: Weed Science (U), Tobacco Technology, Physiological Aspects of Crop Production, Principles and Methods in Weed Science, Plant Breeding Methods, Plant Breeding Field Procedures, Origin and Evolution of Cultivated Plants, Plant Breeding Theory, Herbicide Behavior in Plants and Soils, Special Problems, Graduate Seminar, Research

DEPARTMENT OF ENTOMOLOGY

Head: K.L. Knight
Professors: R.C. Axtell, C.H. Brett, W.V. Campbell, W.C. Dauterman, M H. Farrier, J.R. Fouts (Adjunct), D.S. Grosch (Adjunct), F.E. Guthrie, E. Hodgson, W.J. Mistric, Jr., H.H. Neunzig, R.L. Rabb, R.L. Robertson (Extension), L.M. Russell (Adjunct), H.E. Scott (Extension), T.J. Sheets, C.F. Smith, G.T. Weekman (Extension), D.A. Young
Associate Professors: J.R. Bradley, Jr., W.M. Brooks, A.L. Chasson, J.M. Falter (Extension), H.B. Moore, G.C. Rock, C.G. Wright, R.T. Yamamoto
Assistant Professors: J.R. Baker (Extension), R.C. Hillmann (Extension), K.A. Sorensen (Extension), R.E. Stinner
Research Associates: 9
Extension Specialists: 4

Field Stations/Laboratories: Highlands Field Research Laboratory - Highlands, North Carolina

Degree Programs: Entomology, Pest Management

Undergraduate Degree: B.S.
Undergraduate Degree Requirements: 130 semester hours total (27 hours Entomology, 28-32 hours Physical and Biological Sciences, 22-24 additional science, 34 hours Humanities, 12 hours free electives).

Graduate Degrees: M.S., Ph.D.
Graduate Degree Requirements: Master: 30 hours plus thesis (research) Doctoral: Program as designed by Advisory Committee plus research thesis

Graduate and Undergraduate Courses Available for Graduate Credit: Insect Diversity (U), Functional Systems of Insects, Bibliographic Research in Biology (U), numerous graduate courses

DEPARTMENT OF FOOD SCIENCE
Phone: Chairman: (919) 737-2951 Others: 737-2953

Head: W.M. Roberts
William Neal Reynolds Distinguished Professor: M.L. Speck
Professors: L.W. Aurand, T.A. Bell, T.N. Blumer, J.A. Christian, E. Cofer, H.B. Craig, J.L. Etchells, M.E. Gregory, M.W. Hoover, I.D. Jones (Emeritus), N.C. Miller, A.E. Purcell, M. L. Speck, H.E. Swaisgood
Associate Professors: D.E. Carroll, H.P. Fleming, S.E. Gilliland, D.D. Hamann, A.P. Hansen, M.K. Head, V.A. Jones, F.R. Tarver, W.M. Walter, N B. Webb
Assistant Professors: D.M. Adams, H.R. Ball, W.Y. Cobb (Adjunct), G.G. Giddings, M.K. Hill, B.R. Johnson, B. Ray (Visiting)
Instructor: L.G. Turner
Research Assistants: 5
Research Technicians: 18

Field Stations/Laboratories: Seafood Laboratory - Morehead City, North Carolina

Degree Program: Food Science

Undergraduate Degree: B.S.
Undergraduate Degree Requirements: Mathematics 11 semester hours, Chemistry, 16 semester hours, Physics 8 semester hours, Biological Sciences 8 semester hours electives, 12 semester hours, language 12 semester hours, Food Sciences 26 semester hours, Total 130 semester hours

Graduate Degrees: M.S., Ph.D.
Graduate Degree Requirements: Master: A minimum of 30 semester hours, minimum of 20 semester hours in 500 and 600 level courses, Dissertation Doctoral:

Student's advisory committee sets specific requirements on an individual basis. Dissertation

Graduate and Undergraduate Courses Available for Graduate Credit: Food Analysis (U), Advanced Food Microbiology (U), Food Research and Development (U), Quality Control of Food Products (U), Food Preservation, Post-Harvest Physiology (U), Special Problems in Food Science (U), Theory of Physical Measurements of Biopolymers, Seminar in Food Science, Special Research Problems in Food Science, Research in Food Science

DEPARTMENT OF GENETICS
Phone: Chairman: (919) 727-2291

Head: D.F. Matzinger (Acting)
William Neal Reynolds Professor: S.G. Stephens
Professors: D.S. Grosch, W.D. Hanson, C.S. Levings, H.V. Malling (Adjunct), T.J. Mann, D.F. Matzinger, L.E. Mettler, R.H. Moll, T. Mukai, G. Namkoong, B.W. Smith, A.C. Triantaphyllou
Associate Professors: L.G. Burk, W.E. Kloos, H.E. Schaffer, C.W. Stuber
Assistant Professors: F.M. Johnson, W.H. McKenzie
Visiting Lecturer: 1
Teaching Fellows: 1
Research Assistants: 5
Research Fellows: 7
Associate Geneticist: 1

Field Stations/Laboratories: 15 field stations located throughout state operated cooperatively with the North Carolina Department of Agriculture, Genetics Nursery - Raleigh

Degree Program: Genetics

Graduate Degrees: M.S., Ph.D.
Graduate Degree Requirements: Master: 30 semester hours, thesis Doctoral: Each candidate's program of study is formulated by an advisory committee, thesis

Graduate and Undergraduate Courses Available for Graduate Credit: Human Genetics, Genetics I, Genetics II, Genetics of Animal Improvement, Special Topics in Cytogenetics, Poultry Breeding, Biological Effects of Radiations, Evolution, Plant Breeding Methods, Plant Breeding Field Procedures, Origin and Evolution of Cultivated Plants, Experimental Evolution, Biochemical and Microbial Genetics, Population Genetics, Plant Breeding Theory, Statistical Concepts in Genetics, Mathematical Genetics, Physiological Genetics, Colloquium in Genetics, Seminar, Special Problems in Genetics

DEPARTMENT OF HORTICULTURAL SCIENCE
Phone: Chairman: (919) 737-3131 Others: 737-2011

Head: J.W. Strobel
Professors: W.E. Ballinger (Research), A.A. Banadyga (Extension), F.D. Cochran, F.E. Correll, G.J. Galletta, F.L. Haynes, Jr. (Research), M.H. Kolbe (Extension), R.L. Lower, C.H. Miller, P.V. Nelson, D.T. Pope (Research), R.L. Sawyer (Adjunct), W.A. Skroch (Research)
Associate Professors: J.F. Brooks (Extension), F.T. Cannon, W.R. Henderson (Research), G.R. Hughes (Extension), T.R. Konsler (Research), C.M. Mainland (Extension), J.F. Monaco, W.B. Nesbitt (Research), W.W. Reid, H.J. Smith (Extension), C.R. Unrath, D.C. Zeiger (Research)
Assistant Professors: M.A. Cohen (Extension), L.K. Hammett, D.M Pharr (Research), D.C. Sanders (Extension), R.M. Southalll, J.H. Wilson (Extension)
Instructor: V.H. Underwood (Research)
Visiting Lecturers: 2
Graduate Research Assistants:

Field Stations/Laboratories: Central Crops Research Station, Horticultural Crops Research Station, Horticultural Crops Research Station, Mountain Horticultural Crops Research Station, Sandhills Research Station, Tidewater Research Station

Degree Program: Horticultural Science

Undergraduate Degree: B.S.
Undergraduate Degree Requirements: 130 semester hours

Graduate Degrees: M.S., Ph.D.
Graduate Degree Requirements: Master: 30 semester hours of approved study with a "B" average. A thesis based on research conducted by the student. Doctoral: At least 6 semesters beyond the bachelor's degree at some accredited graduate school, at least two residence credits (registration for a full load in 2 consecutive semesters), passing preliminary examination, satisfactorily completing a doctoral dissertation and passing the final oral and written examination

Graduate and Undergraduate Courses Available for Graduate Credit: Nursery, Residential Landscaping, Fruit Production, Vegetable Production, Floriculture I, Floriculture II, Arboriculture, Senior Seminar in Horticultural Science (U), Principles and Methods in Weed Science, Food Preservation, Plant Breeding Methods, Plant Breeding Field Procedures, Growth of Horticultural Plants, Postharvest Physiology, Research Principles, Plant Breeding Theory, Herbicide Behavior in Plants and Soils, Methods and Evaluation of Horticultural Research, Mineral Nutrition in Plants

DEPARTMENT OF MICROBIOLOGY
Phone: Chairman: (919) 737-2391 Others: 737-2011

Head: J.B. Evans
Professors: F.B. Armstrong (Associate), W.J. Dobrogosz, G.H. Elkan, J.B. Evans, J.L. Etchells (Associate), P.B. Hamilton (Joint Appointment), J.E. Lecce (Associate), J.J. Perry, M.L. Speck (Associate)
Associate Professors: R.E. Kanich (Adjunct), W.E. Kloos (Associate), J.J. McNeill (Associate), J.J. Tulis (Adjunct), A.G. Wollum (Associate)
Assistant Professors: E.C. Hayes III, P.H. Ray
Teaching Fellows: 3
Research Assistants: 8
Research Associates: 2

Degree Program: Microbiology

Undergraduate Degree: B.S.
Undergraduate Degree Requirements: Not Stated

Graduate Degrees: M.S., Ph.D.
Graduate Degree Requirements: Not Stated

Graduate and Undergraduate Courses Available for Graduate Credit: Not Stated

DEPARTMENT OF PLANT PATHOLOGY
Phone: Chairman: (919) 737-2730 Others: 737-2011

Head: R. Aycock
Professors: J.L. Apple, R. Aycock, K.R. Barker, C.N. Clayton, E.B. Cowling, E. Echandi, G.V. Gooding, Jr., T.T. Hebert, G.H Hepting (Adjunct), C.S. Hodges, Jr., G.B. Lucas, R.D. Milholland, L.W. Nielsen, N.T. Powell, J.P. Ross, J.N. Sasser, H.W. Spurr, Jr., D.L. Strider, F.A. Todd, H.H. Triantaphyllou, J.C. Wells, N.N. Winstead
Associate Professors: C.W. Averre, III, M.K. Beute, H.E. Duncan, E.R. French (Adjunct), L.F. Grand, A.S. Heagle, D. Huisingh, S.F. Jenkins, Jr., J.W. Koenigs, (Adjunct), E.G. Kuhlman (Adjunct), K.J. Leonard, M.P. Levi, L.T. Lucas, C.E. Main, R.A. Reinert, R.E. Welty
Assistant Professors: N.S. Henderson, R.K. Jones, N.A. Lapp (Adjunct), P.B. Shoemaker, J.H. Wilson, C.G. Van Dyke
Research (Graduate) Assistants: 23
Post-Doctorals: 3

Field Stations/Laboratories: Central Crops Research Station, Horticultural Crops Research Station, Lower Coastal Plain Tobacco Research Station, Mountain Horticultural Crops Research Station, Sandhills Research Station, Seafood Laboratory, Upper Piedmont Research Station

Degree Program: Plant Pathology

Graduate Degrees: M.S., Ph.D.
Graduate Degree Requirements: Master: 10 hours courses Plant Pathology, 6 hours research Doctoral: 18 hours Plant Pathology, plus Mycology, Nematology, Virology, thesis

Graduate and Undergraduate Courses Available for Graduate Credit: Plant Disease Control (U), Phytopathology I (U), Identification of Plant Pathogenic Fungi (U), Pathogenic Microbiology (U), The Fungi, Special Problems in Plant Pathology (U), Morphology and Taxonomy of Nematodes (U), Plant Virology (U), History of Plant Pathology (U), Current Phytopathological Research (U), Advanced Plant Nematology (U), Plant Pathogenesis, Nematode Development, Cytology and Genetics (U), Advanced Mycology (U)

DEPARTMENT OF POULTRY SCIENCE
Phone: Chairman: (919) 737-2626 Others: 737-2628

Head: R.E. Cook
Professors: R.E. Cook, H.L. Bumgardner, W.E. Donaldson, P.B. Hamilton (Extension), J.R. Harris (Extension), C.H. Hill, W.C. Mills, Jr. (Extension)
Associate Professors: T.B. Dameron (Adjunct), T.B. Morris (Extension), J.P. Thaxton, J.B. Ward, G.A. Martin (Extension)
Assistant Professors: D.M. Briggs, F.W. Edens, J.D. Garlich, C.R. Parkhurst, W.R. Prince
Instructor: J.R. West (Extension)

Field Stations/Laboratories: Piedmont Research Station - Salisbury, North Carolina

Degree Programs: Genetics, Nutrition, Physiology, Poultry Science

Undergraduate Degree: B.S.
Undergraduate Degree Requirements: 130 semester hours

Graduate Degrees: M.S., Ph.D.
Graduate Degree Requirements: Master: 36 hours including thesis Doctoral: No specified hour requirement - Dissertation

Graduate and Undergraduate Courses Available for Graduate Credit: Not Stated

DEPARTMENT OF ZOOLOGY
Phone: Chairman: (919) 737-2741 Others: 737-2011

Chairman: D.E. Davis
Professors: F.S. Barkalow, Jr., B.J. Copeland, R. Harkema, W.W. Hassler, D.W. Hayne, J.E. Hobbie, C.F. Lytle, B.S. Martof, G.C. Miller, T.L. Quay, J.R. Roberts, D.E. Smith
Associate Professors: P.C. Bradbury, M.T. Huish
Assistant Professors: G.T. Barthalmus, D.S. de Calesta, P.D. Doerr, J.M. Miller, K.E. Muse, G.B. Pardue, G.G. Shaw, J.M Whitsett, T.G. Wolcott
Teaching Fellows: 19
Research Assistants: 30

Field Stations/Laboratories: Hatteras Marine Research Laboratory - Pamlico Marine Laboratory

Degree Programs: Zoology, Wildlife, Conservation, Medical Technology, Ecology

Undergraduate Degree: B.S.
Undergraduate Degree Requirements: Total of 130 hours, including specific courses

Graduate Degrees: M.S., Ph.D.
Graduate Degree Requirements: Master: 30 hours course work, thesis Doctoral: 1 year of residence, 1 language, thesis

Graduate and Undergraduate Courses Available for Graduate Credit: Biological Basis of Man's Environment (U), Bibliographic Research inBiology (U), Cell Biology (U), Cellular and Animal Physiology Laboratory (U), Fishery Science (U), Vertebrate Physiology (U), Ichthyology (U), Ornithology, Comparative Pschology, Adaptive Behavior of Animals, Comparative Physiology, Growth and Reproduction of Fishes, Population Ecology, Limnology, Comparative Endocrinology, Biological Oceanography, Biological Effects of Radiations, Evolution, Herpetology, Mammalogy, Experimental Evolution, Principles of Wildlife Science, Protozoology Principles of Ecology, Physiology of Invertebrates, Helminthology, Medical and Veterinary Entomology, Speical Studies, Topical Problems, Advanced Parasitology, Current Aspects of Animal Behavior, Advanced Cell Biology, Advanced Limnology, Fishery Science, Advanced Topics in Ecology I, Advanced Topics in Ecology II

School of Forestry

Dean: E.L. Ellwood

DEPARTMENT OF FORESTRY
Phone: Chairman: (919) 737-2893

Head: C.B. Davey
Carl Alwin Schenck Professor of Forestry: T.E. Maki
Edward F. Conger Professor of Forestry: B.J. Zobel
Professors: F.S. Barkalow, A.W. Cooper, E.B. Cowling, J.W. Duffield, A.B. Davey, M.H. Farrier, J.W. Hardin, C.S. Hodges, J.O. Lammi, T.E. Maki, G. Namkoong, T.O. Perry, L.C. Saylor, R.R. Wilkinson, B. J. Zobel, W.D. Miller (Emeritus), R.J. Preston (Emeritus), G.H. Hepting,(Adjunct), N.E. Johnson (Adjunct), L.J. Metz (Adjunct), C.G. Wells
Associate Professors: L.F. Grand, W.L. Hafley, J.W. Koenigs (Adjunct), E.G. Kuhlman (Adjunct), D.H.J. Steensen, B.F. Swindel
Assistant Professors: D.L. Holley, R.C. Kellison
Visiting Lecturer: 1
Teaching Fellow: 1
Research Assistants: 2

Field Stations/Laboratories: Hill Forest - Bahama, North Carolina, Scheuck Memorial Forest - Raleigh, North Carolina, Hoffman Forest - Deppe, North Carolina

Degree Programs: Biometry, Forest Resources, Forestry, Forestry and Conservation, Silviculture, Watershed Management, Forest Genetics, Forest Soils, Forest Pathology

Undergraduate Degree: B.S.
Undergraduate Degree Requirements: 139 credit hours

Graduate Degrees: M.S., Ph.D.
Graduate Degree Requirements: Master: 30 credit hours plus thesis (M.S.), (M.F.) 30 credit hours plus Problem Doctoral: Committee Decision plus 2 residence credits (24 credit hours), plus dissertation

Graduate and Undergraduate Courses Available for Graduate Credit: Not Stated

UNIVERSITY OF NORTH CAROLINA AT WILMINGTON

Wilmington, North Carolina 28401 Phone: (919) 791-4330
Vice Chancellor for Adacemic Affairs: C.L. Cahill

DEPARTMENT OF BIOLOGY
Phone: Chairman: (919) 791-4330

Chairman: D.J. Sieren

Professors: J.F. Parnell, D.B. Plyler
Associate Professors: F.H. Allen, W.C. Biggs, J.F. Dermid, C.M. Fugler, D.J. Sieren
Assistant Professors: P.E. Hosier, D.F. Kapraum, C.V. Lundeen, A.B. McCrary, J.F. Merritt, D.B. Roye
Instructors: L. Cockerham, A.B. Pittman

Degree Programs: Biology, Marine Biology

Undergraduate Degrees: B.A.B.S.
Undergraduate Degree Requirements: Not Stated

WAKE FOREST UNIVERSITY

Winston-Salem, North Carolina 27109 Phone: (919) 725-9711
Dean of Graduate School: H.S. Stroupe

Wake Forest College

Dean: T.E. Mullen

DEPARTMENT OF BIOLOGY
Phone: Chairman: Ext. 450

Chairman: J.C. McDonald
Babcock Professor of Botany: W.S. Flory
Professor: C.M. Allen
Associate Professors: R.D. Amen, J.F. Dimmick, G.W. Esch, R.E. Kuhn, J.C. McDonald, A.T. Olive, R.L. Sullivan, P.D. Weigl, R.L. Wyatt
Assistant Professors: V.E. Becker, R.V. Dimock, H.E. Eure, N. Gengozian (Adjunct), J.W. Gibbons (Adjunct), H.C. Lane, S.H. Richardson (Adjunct), M.B. Thomas
Research Assistants: 1

Field Stations/Laboratories: Belews Lake Biological Station

Degree Program: Biology

Undergraduate Degree: B.A.
Undergraduate Degree Requirements: a minimum of 8 courses (some required) beyond the Introductory courses which include both plant and animal areas and a minimum grade average of "C" on all full courses in Biology attempted, four courses in Physical Sciences

Graduate Degrees: M.A., Ph.D.
Graduate Degree Requirements: Master: a minimum of 30 hours (7-8 courses) beyond the courses required for the B.A., a thesis, a foreign language or special skill as determined by the department, examinations Doctoral: coursework as needed, foreign languages or special skills as determined by the department, a dissertation, written and oral examinations

Graduate and Undergraduate Courses Available for Graduate Credit: Genetics (U), Evolution (U), Economic Botany (U), Chordates (U), Parasitology (U), Plant Anatomy (U), Microorganisms (U), Non-Vascular Plants (U), Vascular Plants (U), Invertebrates (U), Vertebrates (U), Entomology (U), Plant Taxonomy (U), Ecology (U), Marine Biology (U), Physiology (U), Developmental Physiology (U), Development (U), Biochemistry (U), Cytology-Histology (U), Philosophy of Biology (U), Seminar (U), Topics in Biology, Cytogenetics, Advanced Invertebrates, Physiological Ecology, Cellular Physiology, Comparative Physiology, Biosystematics

Bowman Gray School of Medicine

Dean: R. Janeway

DEPARTMENT OF ANATOMY
Phone: Chairman: (919) 727-4369 Others: 727-4368

Chairman: N.M. Sulkin
Professors: W.J. Bo, N.M. Sulkin
Associate Professor: C.E. McCreight
Assistant Professors: D M. Biddulph, R.A. Rinch, W.A. Krueger, I.J. Miller, Jr., J.E. Turner
Research Associates: 1
Associates: 1

Field Stations/Laboratories: Wrightsville Marine Biomedical Laboratory - Wrightsville Beach, North Carolina

Degree Program: Anatomy

Graduate Degrees: M.S., Ph.D.
Graduate Degree Requirements: Master: Knowledge of gross anatomy, histology, cell biology, neuroanatomy, developmental biology and specialty areas. Original research thesis in an area of anatomical science. Doctoral: Knowledge of major fields of anatomy with teaching experience in at least two areas Reading ability in a foreign language. In depth knowledge of one or more specialty areas, Dissertation

Graduate and Undergraduate Courses Available for Graduate Credit: Special Topics in Gross Anatomy, Special Topics in Histology, Special Topics in Neuroanatomy, Methods in Biological Research, Cell Morphology, Endocrinology, Human Developmental Anatomy, Experimental Embryology, Gross Anatomy (U), Cellular Basis of Medicine (U), Organology (U)

DEPARTMENT OF BIOCHEMISTRY
Phone: Chairman: (919) 727-4688 Others: 727-4689

Chairman: C.F. Strittmatter
Odus M. Mull Professor of Biochemistry: C.F. Strittmatter
Professors: R.W. Cowgill, C.N. Remy, C.F. Strittmatter
Associate Professors: L.R. DeChatelet, F.H. Hulcher, B.M. Waite
Assistant Professor: C.C. Cunningham
Research Associates: 2
Associates: 2

Degree Program: Biochemistry

Graduate Degrees: M.S., Ph.D.
Graduate Degree Requirements: Master: Minimum of 24 semester hours of appropriate didactic course work, in addition to a research thesis, approximately two years required Doctoral: Individually determined series of formal course work in Biochemistry, at least 18 hours of appropriate graduate course work outside the department, research thesis, approcimately four years required.

Graduate and Undergraduate Courses Available for Graduate Credit: General Biochemistry, Cellular Basis of Medicine, Biochemistry Literature Seminar, Introduction to Biochemical Research, Biochemical Techniques, Advanced Topics in Biochemistry (cyclical course of five sections extending over two years), Physical Biochemistry, Enzymology

MICROBIOLOGY DEPARTMENT
Phone: Chairman: (919) 727-4471 Others: 727-4472

Chairman: Q.N. Myrvik
Professors: Q.N. Myrvik, S.H. Richardson
Associate Professors: J.D. Acton, H. Drexler, S.H. Love, E.S. Leake
Assistant Professors: D.L. Groves, E.R. Heise, A.S. Kreger, L.S. Kucera

Degree Programs: Genetics, Immunology, Medical Microbiology, Microbiology, Virology

Graduate Degrees: M.S., Ph.D.
Graduate Degree Requirements: Not Stated

Graduate and Undergraduate Courses Available for Graduate Credit: General Microbiology, Medical Microbiology, Cellular Basis of Medicine, Advanced Virology, Patho genesis of Infectious Diseases, Bacterial Physiology, Advanced Immunology, Microbial Genetics, Ultrastructure of Microbial and Mammalian Cells, Seminar in Microbiology, Teacher Training

DEPARTMENT OF PHYSIOLOGY
(No information available)

WARREN WILSON COLLEGE

Swannanoa, North Carolina 28778 Phone: (704) 298-3325
Dean of College: J. Godard

BIOLOGY DEPARTMENT

Chairman: W.T. Penfound
Professors: H.W. Jensen, W.T. Penfound
Associate Professor: W.A. Eggler
Assistant Professors: C. Riddick, D. Stockdale

Degree Programs: Biology

Undergraduate Degree: B.A.
Undergraduate Degree Requirements: 128 semester hours total to include 30 hours in Biology, plus at least a year of Chemistry and a Semester of Mathematics.

WESTERN CAROLINA UNIVERSITY

Cullowhee, North Carolina 28723 Phone: (704) 293-7244
Dean of Graduate Studies: M.B. Morrill
Dean of College: J.E. Dooley (Acting)

DEPARTMENT OF BIOLOGY
Phone: Chairman: (704) 293-7244

Head: J.L. West
Professors: R C. Bruce, J.G. Eller, H. Horton, R.H. Lumb
Associate Professors: F.D. Hinson, H.R. Mainwaring, J.D. Pittillo, J.W. Wallace, J.L West
Assistant Professors: F.A. Coyle, A.M. Moore, L.J. Perry, C.P. Wright
Teaching Fellows: 6
Research Assistants: 1

Degree Program: Biology

Undergraduate Degree: B.S.
Undergraduate Degree Requirements: Quarter Hours: Humanities - 30, Social Sciences - 15, Health and Physical Education - 6, Biology 51, Chemistry 25, Physics - 12, Mathematics - 10, General Electives - 43, total 192

Graduate Degrees: M.S.
Graduate Degree Requirements: Master: Quarter hours: 45 in Biology including course work and thesis, must pass language examination and written and oral examinations in Biology and thesis defense.

Graduate and Undergraduate Courses Available for Graduate Credit: Ultrastructure of Cells, Secondary Compounds of Biological Systems, Microbiology, Plant Physiology, Biochemistry, Cytology, Comparative Physiology, Microbial Ecology, Biogeography, Modern Environment, Problems, General Ecology, Plant Ecology, Limnology, Human Genetics, Developmental Genetics, Molecular Genetics, Mycology, Plant Anatomy, Local Flora, Dendrology, Phycology, Animal Behavior, Ichthyology, Invertebrate Zoology, Vertebrate Morphology I and II, Entomology, Vertebrate Natural History, Histology, All the preceding are open to Advanced Undergraduates, Special Topics, Thesis Research, General Virology, Cell Physiology, Insect Physiology, Population Ecology, Energy Transfer in Ecosystems, Graduate Students only

WINSTON-SALEM STATE UNIVERSITY

Winston-Salem, North Carolina 27102 Phone: (919) 725-3563
Dean: L. Parker

NATURAL SCIENCE DEPARTMENT
Phone: Chairman: Ext. 69

Chairman: J.R. Shepperson
R.J. Reynolds Professor of Biology: J.R. Shepperson
Professors: B. Sidhu, M. Singh
Associate Professors: W. Atkinson, L. Oliver, F. Sadek
Assistant Professors: J. Fountain, A. Lipkin
Instructors: A. Terrell, A. Weigl

Degree Programs: Biology, Medical Technology

Undergraduate Degree: B.S.
Undergraduate Degree Requirements: 127 semester hours Biology 30 semester hours, Chemistry 16 semester hours, Physics 8 semester hours, Mathematics 6 semester hours

NORTH DAKOTA

DICKINSON STATE COLLEGE

Dickinson, North Dakota 58601 Phone: (701) 227-2330
Dean of College: P.C. Larsen

BIOLOGY DEPARTMENT
Phone: Chairman: (701) 227-2111 Others: 227-2112

Chairman: J.H. MacDonald
Professor: J.H. MacDonald
Associate Professor: M.H. Freeman

Degree Program: Biology

Undergraduate Degrees: B.A., B.S.
Undergraduate Degree Requirements: 54-59 quarter credits in Biology, Statistics, Algebra or higher Mathematics, 1 year college Chemistry with Laboratory

JAMESTOWN COLLEGE

Jamestown, North Dakota 58401 Phone: 252-4331
Dean of College: M.W. Andersen

DEPARTMENT OF BIOLOGY
Phone: Ext. 365

Chairman: W.J. Claflin
Associate Professors: W.J. Claflin, K.S. Cherian

Degree Program: Biology

Undergraduate Degree: B.A.
Undergraduate Degree Requirements: Biology - 9 1/4 course credits in Biology including Genetics and Senior Seminar (37 semester hours), 5 courses in Chemistry (20 semester hours), 2 courses in Mathematics (at least 1 semester of Calculus), 1 course in Physics (Total Graduate Requirements = 36 course credits)

MARY COLLEGE

Apple Creek Road Phone: (701) 255-4681
Bismarck, North Dakota 58501
Dean of College: Not Stated

DEPARTMENT OF BIOLOGY
Phone: Chairman: Ext. 355

Chairman: None
Professor: D.A. Nix
Associate Professor: T.A. Glum
Assistant Professors: N. Theisen, C. Kalberer

Degree Programs: Medical and Radiological Technology

Undergraduate Degree: B.S.
Undergraduate Degree Requirements: 128 semester hours, 32 hours in Medical Technology internship, 46 hours in Radiological Technology internship

MAYVILLE STATE COLLEGE

Mayville, North Dakota 58257 Phone: (701) 786-2301
Dean of College: G. Leno

DEPARTMENT OF BIOLOGY
Phone: Chairman: (701) 786-2634

Chairman: R.D. Ralston
Assistant Professor: K. Wortham
Visiting Lecturers: 2

Degree Programs: Biology, Biological Education

Undergraduate Degree: B.S.
Undergraduate Degree Requirements: General Biology 15 hours, Chemistry (General), 15 hours, Organic Chemistry 15 hours, Humanities, Literature, Mathematics, Psychology, General Education, Genetics, Cell Biology, General Botany, General Zoology, Microbiology, Ecology (Plant), Geology, Ecology (Animal), Developmental Biology, 72 hours total in major, quarter hours

MINOT STATE COLLEGE

Minot, North Dakota 58701 Phone: (701) 838-6101
Dean of College: J. Davy

BIOLOGY DEPARTMENT
Phone: Chairman: (701) 838-6101 Ext. 271

Head: M.B. Thompson
Professors: D.T. Disrud, P.D. Leiby, R.S. Lipe, O.E. Madhok
Assistant Professors: A. Haskins, M.B. Thompson, J.A. Ward

Degree Program: Biology

Undergraduate Degree: B.S.
Undergraduate Degree Requirements: 52 quarter hours of Biology in seven basic areas, Principles, Structural, Environmental, Developmental, Systematics, Functional, Techniques, Two years of Chemistry through Organic and Biochemistry, one year of Physics and at least College Algebra

Graduate Degree: M.S.
Graduate Degree Requirements: Master: Minot State College is a graduate outreach center for work leading to a masters through the two state universities, University of North Dakota and North Dakota State Univeristy.

Graduate and Undergraduate Courses Available for Graduate Credit: Plant Physiology (U), Parasitology (U), Histology (U)

NORTH DAKOTA STATE UNIVERSITY

State University Station Phone: (701) 237-8011
Fargo, North Dakota 58102
Dean of Graduate Studies: J. Sugihara

College of Arts and Science

Dean: L.W. Hill

DEPARTMENT OF BOTANY
Phone: Chairman: (701) 237-7224 Others: 237-7411

Chairman: W.C. Whitman
Professor: W.C. Whitman
Associate Professors: W.T. Barker, M.E. Duysen, T. P. Freeman, D.S. Galitz, H. Goetz, D.R. Scoby
Assistant Professor: G.K. Clambey
Research Assistants: 9
Teaching Assistants: 3

Field Stations/Laboratories: Institution maintains 7 branch Agricultural Experiment Stations, but no other specific field laboratories or stations.

Degree Programs: Biological Sciences, Biology, Botany, Range Science, Science Education

Undergraduate Degree: B.S.
Undergraduate Degree Requirements: 36 quarter hours in Botany, 60 quarter hours in related subjects, 27 quarter hours in Social Sciences, Humanities, Language and Speech, 9 quarter hours in English, 8 quarter hours in Physics

Graduate Degrees: M.S., Ph.D.
Graduate Degree Requirements: Master: Approval of admission, 45 quarter hours, with at least 30 of these in major, 3 quarters of residence Doctoral: Approval of admission, approved program of study, Normally 3 years of study beyond bachelor's degree, at least 1 year residence, foreign language required

Graduate and Undergraduate Courses Available for Graduate Credit: Biogeography (U), Anatomy of Seed Plants (U), Agrostology (U), Plant Microtechnique (U), Range Management and Improvement (U), Autecology (U), Range Plants (U), Techniques in Range Evaluation (U), Western Range Lands (U), Morphology of Vascular Plants (U), Paleobotany (U), Aquatic Plants (U), Field Botany (U), History of Botany (U), Plant Metabolism (U), Water Relations and Mineral Nutrition of Plants (U), Plant Growth (U), Physiology of Fungi (U), Intermediate Botany for Teachers (U), Advanced Botany for Teachers, Field Experience, Crop Ecology, Advanced Plant Ecology, Range Improvement, Range Plant Communities, Advanced Systematic Botany, Plant Evolution, Physiology of Weeds, Photosynthesis, Nitrogen Metabolism in Plants, Techniques in Electron Microscopy, Cell Ultrastructure, Ecological Methods, Range Survey and Mapping

ZOOLOGY DEPARTMENT
Phone: Chairman: (701) 237-8436 Others: 237-7412

Chairman: J.F. Cassel
Professors: J.F. Cassel, G.W. Comita, R. Goforth (Adjunct), K.W. Harmon (Adjunct), D.W. Johnson (Adjunct), R.A. Leopold (Adjunct)
Associate Professors: J.W. Gerst, J.W Grier, R.E. Molnar
Teaching Fellows: 2
Research Assistants: 8
Postdoctoral Fellow: 1

Degree Programs: Animal Behavior, Fisheries and Wildlife Biology, Medical Technology, Zoology

Undergraduate Degrees: B.A., B.S.
Undergraduate Degree Requirements: B.S. 3 credits Physical Education, 9 credits English, 27 credits Humanities and Social Science - B.A. - B.S. language plus 9 more Humanities and Social Sciences - Both require 38 credits of Zoology and 57 credits in other fields

Graduate Degrees: M.S., M.A., Ph.D.
Graduate Degree Requirements: Master: 45 graduate courses normally required at majors, A plan - includes 10-15 credits in thesis, B plan - includes course work and paper Doctoral: Determined by Advisory Committee during a minimum of three years of fulltime study following the baccalaureate degree, 24 credits in minor fields

Graduate and Undergraduate Courses Available for Graduate Credit: Protozoology, Comparative Embryology and Morphology of the Chordates, Vertebrate Histology, Animal Behavior, Mammalian Physiology, Cytology, Invertebrate Zoology, Ichthyology, Herpetology, Avian Biology, Mammalogy, Limnology, Ethology, Wildlife Ecology, Principles of Zoological Research, Animal Population Dynamics, Fisheries Biology, Fisheries Management, Water Pollution Biology, Cellular Physiology, Endocrinology, Environmental Physiology, History of Biology, Principles of Systematics - All (U) - can be taken by both graduates and advanced undergraduates

College of Agriculture
Dean: A.G. Hazen

DEPARTMENT OF AGRONOMY
Phone: Chairman: (701) 237-7971

Chairman: J.F. Carter
Professors: A.E. Foster, R.C. Frohberg, K.A. Lucken, H.R. Lund, S.S. Maan, L.W. Mitich, J.D. Nalewaja, P.C. Sandal, G.S. Smith, N.D. Williams
Associate Professors: D.C. Ebeltoft, J.R. Erickson, J.J. Hammond, R.H. Hodgson, L.R. Joppa, J S. Quick, A.B. Schooler, R.H. Shimabukuro, D.A. Whited, H.D. Wilkins
Assistant Professors: H.Z. Cross, E.L. Deckard, A.G. Dexter, G.N. Fick, C.G. Messersmith, D.W. Meyer, A.A. Schneiter
Instructors: J. Miller, S. Miller
Visiting Lecturers: 1-2/year
Research Assistants: 16

Field Stations/Laboratories: North Dakota Agricultural Experiment Station - Fargo, Branch Stations at Casselton, Carrington, Langdon, Minot, Dickinson, Williston, Hettinger, Oakes

Degree Programs: Agronomy, Genetics within Agronomy, Weed Science within Agronomy

Undergraduate Degree: B.S.
Undergraduate Degree Requirements: 184 quarter credits with 2.0 average (A equals 4), 55 credits must be in 300 or 400 series, 36 credits in major.

Graduate Degrees: M.S., Ph.D.
Graduate Degree Requirements: Master: 45 quarter credits including required thesis Doctoral: 135 quarter credits approximately including required Ph.D. thesis, written preliminary and oral qualifying examinations required plus final examination on thesis

Graduate and Undergraduate Courses Available for Graduate Credit: Weed Control in Field Crops, Principles of Plant Breeding, Intermediate Genetics, Crop Production, Introduction to Cytogenetics, Reports on Crop Production, Speical Topics, Individual Study, Seminar, Advanced Weed Science, Laboratory Methods in Week Science, Advanced Genetics I, Experimental Designs I and II, Advanced Genetics II, Crop Breeding, Techniques, Cytogenetics, Cytogenetics of Aneuploids, Advanced Breeding--Small Grains, Advanced Breeding--Corn and Forage Crops, Biometrical Genetics in Plant Breeding, Population Genetics

ANIMAL SCIENCE DEPARTMENT
Phone: Chairman: (701) 237-7641

Chairman: M.L. Buchanan
Professors: W.E. Dinusson, V.K. Johnson, M. Light, J. Sell
Associate Professors: C. Edgerly, D.O. Erickson, R. Harrold, C. Haugse, J.N. Johnson
Assistant Professors: R. Danielson, R. Johnson
Instructors: R. Amundson, B. Moore, R. Wollmuth, R. Zimprich
Research Assistants: 8

Degree Programs: Animal Genetics, Animal Husbandry, Animal Science, Genetics, Nutrition

Undergraduate Degree: B.S. (Animal Science)
Undergraduate Degree Requirements: 183 quarter credits, minimum of C average with 55 credits in junior or senior courses, 36 credits required in major department.

Graduate Degrees: M.S., Ph.D.
Graduate Degree Requirements: Master: Minimum of 45 credits in addition to B.S. with not less than 30 credits in one department or field of study. Doctoral: Minimum of three years full time study beyond baccalaureate degree with appropriate thesis, oral and written examinations.

Graduate and Undergraduate Courses Available for Graduate Credit: 121 course credits in Animal Science in the areas of Animal Breeding, Animal Nutrition, Physiology of Reproduction, Animal Reproduction and Animal Production plus supporting courses in Biochemistry, Zoology, Botany, Agronomy and Economics

DEPARTMENT OF BACTERIOLOGY
Phone: Chairman: (701) 237-7667

Chairman: K.J. McMahon
Professors: M.C. Bromel, K.J. McMahon, E.R. Renner, (Adjunct), B.P. Sleeper, P.P. Williams (Adjunct)
Associate Professor: B.R. Funke
Assistant Professors: D.L. Berryhill, J.A. Doubly
Instructor: B.A. Baldwin
Research Assistants: 5

Degree Program: Bacteriology

Undergraduate Degree: B.S.
Undergraduate Degree Requirements: Total quarter hours credits = 183, This includes Mathematics 9, Physics 12, Chemistry including Biochemistry 27, Biology, Botany and Zoology 8, Genetics 3, Bacteriology (general 10, pathogenic 5, immunology 4, physiology 3, seminar 1) plus 15 selected from other bacteriology courses and related fields. Remainder 86 (electives, English Social Sciences and Humanities and Physical Education)

Graduate Degree: M.S.
Graduate Degree Requirements: Master: Total quarter credits 45 with 30 in major field. A minor with up to 15 credits is advised. A B average in courses for graduate credits is required. The thesis carries from 8 to 15 credits. The student's Advisory Committee helps him plan his specific program.

Graduate and Undergraduate Courses Available for Graduate Credit: General Bacteriology (U), Microbial Ecology (U), Soil Microbiology (U), Food and Diary Microbiology (U), Bacterial Physiology (U), Immunology and Serology (U), Pathogenic Bacteriology (U), Advanced Bacterial Physiology, Bacterial Viruses, Microbial and Molecular Genetics, and Graduate Seminar.

DEPARTMENT OF CEREAL CHEMISTRY AND TECHNOLOGY
Phone: Chairman: (701) 237-7711

Chairman: O.J. Banasik
Professors: O.J. Banasik, K.A. Gilles, Vice President for Agricultural
Associate Professors: L.D. Sibbitt (Research), C.E. McDonald, W.C. Shuey (Adjunct), D.E. Walsh, B.L. D'Appolonia
Assistant Professors: C. Baker, M. Breen
Visiting Lecturers: 1
Research Assistants: 13

Field Stations/Laboratories: Langdon, Hettinger, Dickinson, Williston, Carrington, Minot and Oakes

Degree Program: Plant Science

Graduate Degrees: M.S., Ph.D.
Graduate Degree Requirements: 45 quarter credit hours in Graduate School courses; 30 quarter credit hours in Cereal Chemistry and Technology (major), 15 quarter credit hours in a related minor. Thesis and oral examination required. Doctoral: 120-140 quarter credit hours in Graduate School courses; 24 quarter credit hours to minor fields. Comprehensive oral examination on academic subjects. Dissertation and final oral examination on research required. No language required.

Graduate and Undergraduate Courses Available for Graduate Credit: Cereal Technology, Industrial Food Processing, Cereal Chemistry, Advanced Cereal Technology, Cereal Technology Methods, Advanced Cereal Chemistry, Fundamentals of Milling, Other Special Problems,

DEPARTMENT OF ENTOMOLOGY
Phone: Chairman: (701) 237-7581

Chairman: J.T. Schulz
Professors: J.A. Callenbach, L.E. LaChance (Adjunct), E.P. Marks, G.B. Mulkern, R.L. Post, M.S. Quraishi
Associate Professors: R.A. Bell (Adjunct), R.B. Carlson, R.D. Frey, D. North (Adjunct), D.E. Wagoner (Adjunct)
Assistant Professors: T.S. Adams, (Adjunct), A.W. Anderson
Research Assistants: 10

Field Stations/Laboratories: Branch Experiment Stations at Carrington, Dickinson, Hettinger, Langdon, Minot and Williston, North Dakota, Potato Research Station, Grand Forks, North Dakota

Degree Programs: Entomology

Undergraduate Degree: B.S.
Undergraduate Degree Requirements: 183 quarter credits required for graduation, B.S. degree granted in either College of Agriculture or Science and Mathematics, Choice of Business, Production or Science options.

Graduate Degrees: M.S., Ph.D.
Graduate Degree Requirements: Master: A. 45 quarter credits required, 30 of which in a cohesive field of study (major), B. Thesis or comprehensive study options available Doctoral: Major field of study a group of cohesive and related courses, credits required variable and dependent on background and interest of student, minor field of study - 24 quarter credits.

Graduate and Undergraduate Courses Available for Graduate Credit: Systematic Entomology, Immature Insects, Insect Morphology, Insect Physiology, Principles of Insect Control, Biological Control of Insects, Biophysical Control of Insects, Entomological History, Entomological Literature and Scientific Writing, Principles of Systematics and Nomenclature, Insect Vectors of Plant Disease, Medical Entomology, Acarology, Insect Ecology, Insect Toxicology, Biometrics, Quantitative Biology

DEPARTMENT OF HORTICULTURE AND FORESTRY
Phone: Chairman: (701) 237-8161 Others: 237-8162

Chairman: E.P. Lana
Professors: N.S. Holland, D.C. Nelson, R.H. Johansen
Associate Professors: D.C. Herman, E.W. Scholz

Degree Programs: Agriculture, Horticultural Science

Undergraduate Degrees: B.S.
Undergraduate Degree Requirements: 36 credits in major field, 24 credits in minor field, 183 credits total

Graduate Degrees: M.S.
Graduate Degree Requirements: Master: 45 credits minimum, 30 major, 15 minor

Graduate and Undergraduate Courses Available for Graduate Credit: Woody Plant Materials, Herbaceous Plants, Greenhouse Floriculture, Potatoes, Vegetable Crops, Small Fruits, Principles of Landscaping, Plant Propagation, Tree Fruits, Turf Management, Arboriculture (U), Nursery Management, Landscape Design I (U), Landscape Design II (U), Horticultural Crop Production (U), Breeding Horticultural Crops (U), Seminar (U)

DEPARTMENT OF PLANT PATHOLOGY

Phone: Chairman: (701) 237-7561 Others: 237-7562

Chairman: R.L. Kiesling
Professors: J. Hugelet, R. Kiesling, V. Pederson, R. Timian (Adjunct), D. Zimmer (Adjunct)
Associate Professors: W. Bugbee (Adjunct), R. Hosford L. Litterfield, E. Lloyd, G. Statler
Assistant Professors: J. Miller
Research Assistants: 8

Field Stations/Laboratories: Branch Experiment Stations at Williston, Minot, Hettinger, Langdon, Carrington, Dickinson, Oakes, University Branch at Bottineau (Forestry

Degree Program: Plant Pathology

Undergraduate Degree: B.S.
Undergraduate Degree Requirements: Not Stated

Graduate Degrees: M.S., Ph.D.
Graduate Degree Requirements: Not Stated

Graduate and Undergraduate Courses Available for Graduate Credit: Diseases of Field and Forage Crops (U), Diseases of Horticultural Crops (U), Plant Pathology Principles and Practices (U), Special Topics (U), Fungal Biology (U), Advanced Mycology, Fungal Genetics, Plant Virology, Bacterial and Fungal Diseases of Plants, Nematode Diseases of Plants, Special Topics, Individual Study, Seminar, Graduate Thesis

College of Chemistry and Physics
Dean: L.W. Hill

DEPARTMENT OF BIOCHEMISTRY
Phone: Chairman: (701) 237-7678 Others: 237-8011

Chairman: H.J. Klosterman
Professors: H.J. Klosterman, G. Graf, D.C. Zimmerman, J.E. Bakke (Adjunct), G.L. Lamoureux (Adjunct), D.R. Nelson (Adjunct), G.D. Paulson (Adjunct), G.G. Still (Adjunct)
Associate Professors: D.W. Bristol, A.G. Fischer, J.R. Fleeker, A.E. Oleson
Teaching Fellows: 2
Research Assistants: 8

Degree Program: Biochemistry

Graduate Degrees: M.S., Ph.D.
Graduate Degree Requirements: Master: Biochemistry-15 credits, Life Sciences plus Chemistry - 15 credits, Thesis Doctoral: Biophysics and Biochemistry - 30 credits, Life Sciences and Chemistry - 40 credits, Thesis

Graduate and Undergraduate Courses Available for Graduate Credit: Biophysics (U), Biochemistry, Biochemistry Laboratory, Carbohydrates, Proteins, Nucleic Acids, Lipids, Physical Biochemistry, Enzymes, Metabolism

UNIVERSITY OF NORTH DAKOTA
University Station Phone: (701) 777-2731
Grand Forks, North Dakota 58201
Dean of Graduate Studies: A.W. Johnson

College of Arts and Sciences
Dean: B. O'Kelly

BIOLOGY DEPARTMENT
Phone: Chairman: (701) 777-2621

Chairman: J.K. Neel
Professors: M. Behringer, V. Facey, H.L. Holloway, Jr., P.B. Kannowski, J.K. Neel
Associate Professors: F. Duerr, S.M. Jalal, O.R. Larson, L.W. Oring, J.B. Owen, R.W. Seabloom, M.K. Wali
Assistant Professors: R.T. Pollock, L.E. Shubert, W.J. Wrenn
Visiting Lecturers: 2
Teaching Fellows: 18
Research Assistants: 8

Field Stations/Laboratories: University of North Dakota Biological Station - Devils Lake, North Dakota, Forest River Field Station - Inkster, North Dakota, Oakville Prairie - Emerado, North Dakota

Degree Programs: Biology, Botany, Zoology

Undergraduate Degrees: B.S.
Undergraduate Degree Requirements: 36 hours Biology, plus Mathematics, Chemistry, Physics

Graduate Degrees: M.S., Ph.D., D.A. (D.A.T.)
Graduate Degree Requirements: Master: 20 credit major: 10 credits minor, thesis Doctoral: D.A. - 90 credits, 60 discipline, 30 area of concentration, 15 credits in teaching, research problem comprehensive examination Ph.D. - 90 credits, dissertation, 2 scholarly tools, including 1 language, comprehensive examination, original research

Graduate and Undergraduate Courses Available for Graduate Credit: Literature and Scientific Writing, Seminars, Principles of Taxonomy, Cytotaxonomy, Phycology, Aquatic Plants, Aquatic Invertebrates, Helminthology, Parasitology, Forest Ecology, Grasslands, Ecology, Biogeography, Physiological Ecology, Animal Population Ecology, Animal Communication, Insect Societies, Cytogenetics, Biochemistry Genetics, Population Genetics, Developmental Morphology Plants, Sampling Theory, Fishery Biology, Wildlife Disease, Evolution, General Ecology, Systematic Botany, Ichthyology, Ornithology, Mammalogy, Limnology, Plant Morphology, Biometry, Wildlife Management, General Physiology

INSTITUTE FOR ECOLOGICAL STUDIES
Phone: Chairman: (701) 777-2851 Others: 777-2011

Director: P.B. Kannowski
Professors: R.E. Beck (Law), A. Cvancara (Geology), G.O. Fossum (Civil Engineering), E.A. Noble (Geology), J.R. Reid (Geology), J.R. Reilly (Biology)
Associate Professors: J.E. Bowes, II (Journalism), R.E. Frank (Chemistry), S.M. Jalal (Biology, K.J. Klabunde (Chemistry), R.D. Ludtke (Sociology), E.S. Mason (Civil Engineering), L.W. Oring (Biology), D.F. Paulsen (Political Science), R.W. Seabloom (Biology), F.T.C. Ting (Geology), M.K. Wali (Biology), M. Winger (Mathematics)
Assistant Professors: G.E. Johnson (Geography), R.E. Lewis (Geography), L.L. Loendorf (Anthropology and Archaeology), F.E. Schneider (Anthropology and Archaeology), M.H. Somerville (Mechanical Engineering)
Instructor: R.D. Kingsbury (Economics)
Research Assistants: 12
Research Associates: 2
Technician: 1
Librarian: 1

Degree Programs: The Institute does not offer degrees or courses except through other departments.

School of Medicine
Dean: J.W. Vennes

DEPARTMENT OF ANATOMY
Phone: Chairman: (701) 777-2101

Chairman: D.A. Ollerich
Professors: F.N. Low (Research), T. Snook
Associate Professors: D.L. Matthies, J.O. Oberpriller, D.A. Ollerich
Assistant Professors: K.E. Nicolls, J.C. Oberpriller

Degree Program: Anatomy

Graduate Degrees: M.S., Ph.D.
Graduate Degree Requirements: Master: 30 semester hours, roughly divided 2/3 in Anatomy, 1/3 in minor, plus thesis on original research Doctoral: 90 semester hours, roughly divided 2/3 in Anatomy, 1/3 in minor, plus dissertation on original research, two languages required for scholarly tools.

Graduate and Undergraduate Courses Available for Graduate Credit: Introduction to Research in Anatomy, Seminar in Anatomy, Techniques for Histological Research, Electron Microscope Techniques, Problems in Development, Cellular and Extracellular Fine structure, Gross Anatomy, Histology and Organology, Developmental Anatomy, Neuroanatomy, Readings in Special Problems in Anatomy, Advanced Anatomy, Research in Anatomy Prosection

BIOCHEMISTRY DEPARTMENT
Phone: Chairman: (701) 777-3937

Chairman: W.E. Cornatzer
Professors: J.L. Connelly, F.A. Jacobs, R.C. Nordlie, Y.P. Lee, P.D. Ray
Assistant Professor: M. Luper
Teaching Fellows: 2
Research Assistants: 12

Field Stations/Laboratories: Human Nutrition Laboratory - U.S. Department of Agriculture

Degree Program: Biological Chemistry

Graduate Degrees: M.S., Ph.D.
Graduate Degree Requirements: Master: 30 hours (semester), thesis Doctoral: 90 semester hours, Dissertation

Graduate and Undergraduate Courses Available for Graduate Credit: Biochemistry, Advanced Biochemistry (lipids, carbohydrates, hormones, macromolecules, trace elements, physical properties of proteins), Enzyme Chemistry, Instrumentation in Biochemistry, Biochemical Literature, Seminar in Biochemistry, Radioactive Tracers in Biochemistry, Research in Biochemistry

DEPARTMENT OF MICROBIOLOGY
Phone: Chairman: (701) 777-2214

Chairman: R.G. Fischer
Professors: J.A. Duerre, R.G. Fischer, R.M. Marwin
Associate Professors: J.J. Kelleher, J.R. Waller

Degree Programs: Medical Microbiology, Microbiology, Virology

Graduate Degrees: M.S., Ph.D.
Graduate Degree Requirements: Master: For M.S. degree 20 credits in major and 10 credits minor, GPA of 3.0 and thesis Doctoral: 90 semester credits beyond bachelors degree, reading knowledge of one foreign language, thesis

Graduate and Undergraduate Courses Available for Graduate Credit: Not Stated

DEPARTMENT OF PHYSIOLOGY AND PHARMACOLOGY
(No reply received)

VALLEY CITY STATE COLLEGE
(No reply received)

OHIO

ADELBERT COLLEGE
(see Case Western Reserve University)

ANTIOCH COLLEGE
(No reply received)

ASHLAND COLLEGE
Ashland, Ohio 44805 Phone: (419) 289-4008
Vice President for Academic Affairs: E.J. Kazma

BIOLOGY DEPARTMENT
Phone: Chairman: (419) 289-4008 Others: 289-5145

Chairman: R. Rhoades
Professors: J.C. Hadder, W.E. Meredith, M.R. Newkirk, R. Rhoades
Assistant Professors: A.M. Goff, J.L. Sledge, M.M. Lroxel

Degree Programs: Teacher Education, Pre-medicine, Pre-veterinary, Medical Technology, Pre-nursing

Undergraduate Degree: B.S.
Undergraduate Degree Requirements: Biology 32 hours in Biology and Mathematics, Chemistry, Physics amounting to 26 hours

BALDWIN-WALLACE COLLEGE
Berea, Ohio 44017 Phone: (216) 826-2262
Dean of College: D.E. Meyer

DEPARTMENT OF BIOLOGICAL SCIENCES
Phone: Chairman: (216) 826-2262

Chairman: D.S. Dean
Professors: D.S. Dean, O.R. Schneider, G.W. Peterjohn
Associate Professors: C.A. Smith, J.W. Miller
Assistant Professor: S.D. Hilliard

Degree Program: Science

Undergraduate Degree: B.S.
Undergraduate Degree Requirements: 186 hours College credit, 40 hours Biology, 10 hours Chemistry

BLUFFTON COLLEGE
Bluffton, Ohio 45817 Phone: (419) 358-8015
Dean of College: E. Neufeld

SCIENCE DIVISION
Phone: Chairman: Ext. 231

Chairman: R. Suter
Professors: M. Kaufmann, R. Pannabecker

Field Stations/Laboratories: Swinging Bridge Nature Preserve - Bluffton, Ohio

Degree Programs: Biology, Medical Technology

Undergraduate Degree: B.A.
Undergraduate Degree Requirements: 35 course units, total, 12 courses required for major (1 Mathematics, 2 Chemistry, 9 Biology), 10 general education courses are chosen from a group

BOWLING GREEN STATE UNIVERSITY
Bowling Green, Ohio 43402 Phone: (419) 372-2332
Dean of Graduate Studies: C. Leone
Dean of College: J. Eriksen

DEPARTMENT OF BIOLOGICAL SCIENCES
Phone: Chairman: (419) 372-2332

Professors: G. Acker, M.M. Brent, T.R. Fisher, R.C. Graves, C.W. Hallberg, W.B. Jackson, I.I. Oster, K.M. Schurr
Associate Professors: W.D. Baxter, R.E. Crang, N.W. Easterly, E.S. Hamilton, H.T. Hamre, W.D. Hann, E.W. Martin, F.C. Rabalais
Assistant Professors: J.D. Graham, C.S. Groat, V. Hiatt, R. Horvath, R.L. Lowe, L.A. Meserve, R.D. Noble, C.L. Rockett, R.C. Romans, K.W. Thornton, S.H. Vessey
Instructors: S. Conner, A. Graves
Research Assistants: 3
Teaching Assistants: 26
Teaching Fellows: 1
Non-Teaching Fellows: 10

Degree Programs: Animal Behavior, Animal Genetics, Applied Ecology, Biology, Botany, Ecology, Entomology and Parasitology, Genetics, Immunology, Limnology, Medical Technology, Microbiology, Parasitology, Physiology, Virology, Zoology

Undergraduate Degree: B.S. (Biology
Undergraduate Degree Requirements: 48 hours Biology, 10 hours Mathematics (Calculus), 25 hours Chemistry, (Organic and Biochemistry), 10 hours Physics, College and University Group Requirements.

Graduate Degrees: M.S., Ph.D.
Graduate Degree Requirements: Master: 45 hours graduate courses including Biostatistics and remedial work to satisfy undergraduate requirements
Doctoral: 45 hours beyond M.S.

Graduate and Undergraduate Courses Available for Graduate Credit: Not Stated

CAPITAL UNIVERSITY
2199 East Main Street Phone: (614) 236-6011
Columbus, Ohio 43209
Dean - College of Arts and Sciences: T.S. Ludlum

BIOLOGY DEPARTMENT
Phone: Chairman: (614) 236-6900 Others: 236-6011

Chairman: P.E. Zimpfer
Professor: P.E. Zimpfer
Associate Professors: L.F. DeWein, R.M. Jordan, T.J. Long
Instructors: J.J. Gaunt, S. Gaunt

Degree Program: Biology

Undergraduate Degree: B.A.
Undergraduate Degree Requirements: 35 full courses (3.5 semester hours per course), plus two quarter courses in Physical Education

CASE WESTERN RESERVE UNIVERSITY
2040 Adelbert Road Phone: (216) 368-2000
Cleveland, Ohio 44106
Dean of Graduate School: J.G. Taaffe

Adelbert College

Dean: H.B. Willard

DEPARTMENT OF BIOLOGY

Phone: Chairman: (216) 368-3556 Others: 368-2000

Chairman: M.J. Rosenberg
Francis Hobart Herrick Professor of Biology: A.G. Steinberg
Professors: H.D. Mahan (Adjunct), B. Schmidt-Nielsen (Adjunct), M. Singer, N. Rushforth, A. Steinberg
Associate Professors: N A. Alldridge, R.P. Davis, D.L. Foreman, R. Kuerti (Emeritus), G. Lesh-Laurie, R. Rustad, T. Voneida, J Zull
Assistant Professors: A. Caplan, V. Chen, J. Koonce, D. Munish, P. Stambrook, M Teraguchi, J. Tischfield
Instructors: L. Dickerman, V. Flechtner, M. Rosenberg
Teaching Fellows: 22
Research Assistants: 3
Postdoctoral Fellows: 2

Field Stations/Laboratories: Ecosystems Research Center - Squire Valleevue Farm, Hunting Valley

Degree Programs: Animal Behavior, Animal Physiology, Biology, Biostatistics, Ecology, Genetics, Human Genetics, Physiology, Plant Sciences, Zoology

Undergraduate Degree: B.S. (Biology)
Undergraduate Degree Requirements: 120 credit hours, 27 hours in Biology, including courses in Genetics, Cell and Molecular Biology, and 3 upper level laboratory courses, One year of introductory Biology also required

Graduate Degrees: M.S., Ph.D.
Graduate Degree Requirements: Master: 30 hours of upper level courses; 18 of these hours in "graduate student" courses, overall B average, a written thesis, if "degree with thesis" program is selected, written and oral examination Doctoral: Coursework is emphasized only during the first year, beyond first year, original research is emphasized, dissertation must be written and defended, qualifying examination required, B average must be maintained, 5 semesters of teaching is required

Graduate and Undergraduate Courses Available for Graduate Credit: Graduate courses open to Undergraduates: Fundamentals of Molecular Biology, Physiological Adaptation to the Environment, Cellular Regulatory Mechanisms, Selected Topics in Cellular and Molecular Biology, Advanced Developmental Biology, Molecular Biology of Development, Invertebrate Developmental Biology, Neuroembryology, The Mitotic Cycle, Statistical Methods in Biological and Medical Sciences, Molecular Aspects of Developmental Biology, Radiation Biology, Population Genetics, Problems and Methods in Human Genetics, Genetics and Evolution, Physiological Genetics, Physiological Ecology, Cellular Neurobiology Courses for Graduate Students only: Seminar in Current Problems in Ecology, Speical Problems in Mammalian Reproduction, Seminar in Cell Biology, Seminar in Comparative Physiology, Seminar in Experimental Botany, Seminar in Molecular Biology, Seminar in Developmental Biology, Seminar in Experimental Biology

School of Medicine

2119 Abingdon Road Phone: (216) 368-2000
Cleveland, Ohio 44106
Dean: F.C. Robbins

DEPARTMENT OF ANATOMY

Phone: Chairman: (216) 368-3430 Others: 368-2000

Chairman: M. Singer
Henry Willson Payne Professor of Anatomy: M. Singer
Professors: E.H. Bloch, A.F.W. Hughes, M.N. Macintyre, D.B. Scott (Joint appointment - School of Dentistry)
Associate Professors: J.D. Caston, A.L. Hopkins, J. Ilan, I.R. Kaiserman-Abramof, R.J. Lasek, H.I. Perlmutter (Joint-appointment - Radiology), R.J. Przybylski, N. Robbins, R. Rustad (Joint appointment Radiation Biology, B. Tandler (Joint appointment - Dentistry), T.J. Voneida
Assistant Professors: C.H.U. Chu, J. Ilan, S. Krohn (Adjunct), D.S. Love, R. Poritsky (Adjunct), N. Taslitz, D. Verne (Adjunct)
Instructors: J.S. Froelich (Adjunct), R. Pawlowski (Adjunct Clinical Instructor)
Lecturers: M. Egar, N. Krishnan, R. Nordlander
Teaching Fellows: 5
Senior Research Associates: 3

Degree Program: Anatomy

Graduate Degree: Ph.D.
Graduate Degree Requirements: Not Stated

Graduate and Undergraduate Courses Available for Graduate Credit: Principles of Developmental Biology (U), Gross Anatomy (U), Histology and Ultrastructure (U), Laboratory in General Histology (U), Developmental Biology Center Lecture Series, Advanced Developmental Biology, Molecular Biology of Development (U), Special Topics Seminar in Developmental Biology, Invertebrate Developmental Biology (U), Neuroembryology, The Mitotic Cycle (U), Molecular Aspects of Developmental Biology, Radiation Biology, Functional Organization of the Central Nervous System, Interhemispheric Mechanisms, Cellular Neurobiology, Applied Electron Microscopy (U), Seminar: Topics in Neurobiology, Etiology of Neuromuscular Diseases, Human Embryology and Teratology, Seminar in Cell Biology, Regulation of Gene Expression in Eukaryotes

DEPARTMENT OF BIOCHEMISTRY

Phone: Chairman: (216) 368-3344

Director: M.F. Utter
University Professor: H.G. Wood
Professors: C. Cooper, D.A. Goldthwait, H. Hirschmann, H.Z. Sable, W. Sakami, L.T. Skeggs
Associate Professors: H.B. Bensusan, D.M. Carlson, J.A. Harpst, J.R. Leonards, H.H. Evans, E.L. Kean, K.E. Neet, O.F. Nygaard, A.E. Powell, R.A. Meigs, D. Neiderheiser, J.C. Occino, N. Oleinick, K.E. Lentz
Assistant Professors: J. Biaglow, R.B. Billiar, D. Dearborn, F. Dorer, H. Gershman, D. Kerr, I. Kline, M. Levine
Instructors: A. Gallo, S. Polmar, D. Rynbrandt
Visiting Lecturers: 6
Teaching Fellows: 10
Research Assistants: 10

Degree Program: Biochemistry

Graduate Degrees: M.S. (Biochemical Research Technology), Ph.D.
Graduate Degree Requirements: Master: 3 years in residence--30 semester hours Doctoral: one year's residency, major requirements -- as arranged, minor requirements, 15 semester hours

Graduate and Undergraduate Courses Available for Graduate Credit: General Biochemistry, Enzymes and Their Regulation, Special Topics in Biochemistry

DEPARTMENT OF BIOMETRY

Phone: Chairman: (216) 791-7300/2903

Chairman: H.B. Houser (Acting)
Professors: H.B. Houser, J. Rosenblatt
Associate Professor: N. Rushforth
Assistant Professors: R. Fuller, P. Jones, J. Knoke, R. Lake, G. Saidel
Instructors: P. Clarke, W. MacKay, A. Martin (Adjunct) J. Neill (Adjunct)

Teaching Fellows: 1
Research Assistants: 3

Degree Programs: Biostatistics, Biometry

Graduate Degrees: M.S., Ph.D.
Graduate Degree Requirements: Not Stated

Graduate and Undergraduate Courses Available for Graduate Credit: Epidemiology (U), Computers in Health Sciences, Health Science Information Systems, Statistical Methods I, Applied Stochastic Models, Population Dynamics, Biostatistics I, Multivariate Biostatistics, Special Topics in Biostatistics, Special Topics: Computers in Health Science, Supervised Practicum in Computer Applications, Integrated Biological Science

DEPARTMENT OF MICROBIOLOGY
Phone: Chairman: (216) 368-3420

Chairman: L.O. Krampitz
Professor: A.B. Stavitsky
Associate Professors: L. Astrachan, R.W. Hogg, C.W. Shuster
Assistant Professors: S. Badger, S.D. Barbour, L.A. Culp, M.W. Fanger, A.H Evans, C.G. Miller
Teaching Fellows: 2
Research Assistants: 13

Degree Program: Microbiology

Graduate Degree: Ph.D.
Graduate Degree Requirements: Doctoral: 4 semesters of core courses, 3 advanced intradepartmental courses, 4 advanced extradepartmental courses, Qualifying examinations, Foreign language examinations, Thesis

Graduate and Undergraduate Courses Available for Graduate Credit: None

DEPARTMENT OF PATHOLOGY
Phone: Chairman: (216) 368-2480 Others: 368-2000

Chairman: J.R. Carter
Professors: L. Adelson, M. Aikawa, B.Q. Banker, J.R. Carter, R.L. Friede, J. Kleinerman, R.R. Kohn, S. Soletsky, J.W. Reagan, J.D. Reid,A.L. Robertson, Jr., I. Sunshine
Associate Professors: R.C. Graham, Jr., B.J. Helyer, J.R. Kahn, M A. Leon, O.P. Malhotra, A B.P. Ng, J.J. Opplt, E.V. Perrin, G.B. Reed, Jr., E Wolinsky
Assistant Professors: M. Abellera, E. Bechtold, R.L. Cechner, R.W. Chen, R.T. Cook, M.E. Cowan, C.R. Cowdrey, K.C. Feller, C.R. Hamlin, A.D. Heggie, C. R. Hirsch, B.L. Horvat, K.H. Hu, L.E. Lee, Jr., T.R. Mabini, R.L. Martine, J.T. Makley, R.P. Misra, W. Morningstar, M. Petrelli, J.R. Pomeranz, M.A. Radivoyevitch, C.A. Rasch, E.L. Robbins, U. Roessmann, E.P. Rossi, A J. Segal, W.K. Sterin, O. Sudilovsky, P.L. Tang
Instructors: C.R. Abramowsky, D P. Agamanolis, E. Allen, E.T. Audrick, E.K. Balraj, J.J. Barklow, A.S. Fernandez, R.B. Foreny, Jr., W.H. Harrington, J. Hutchinson, J. Levy, D.L. Martin, D.E. Niewoehner, M C. Park, J.R. Rabbege, J. Sobonya, J. Suarez-Hoyos, P. Sullivan, P. Sumpter, J.C. Valentour, J. Wingenfeld
Visiting Lecturers: 50
Teaching Fellows: 21

Field Stations/Laboratories: Squire Valleevue Farm - Chagrin Falls

Degree Program: Experimental Pathology

Graduate Degrees: M.S., Ph.D.
Graduate Degree Requirements: Master: Minimum GPA 2. 2.50 in all courses. Plan A: 18-21 semester hours course work plus thesis plus written or oral examination Plan B: 27 semester hours course work plus Comprehensive Examination Doctoral: Minimum cumulative GPA of 2.75 in all courses. Semester courses in computer programming or equivalent. Appropriate body of graduate course work. General examination. Dissertation research. Written Comprehensive dissertation. Defense of dissertation and final oral examination, Teaching Experience required.

Graduate and Undergraduate Courses Available for Graduate Credit: Integrated Biological Sciences, Instrumental Techniques, Computer Applications in Pathology, Experimental Pathology Seminar, Experimental Pathobiology, Molecular Pathology and Pharmacology of Neoplasia, Systemic Pathology Tutorial, Special Problems

DEPARTMENT OF PHYSIOLOGY
(No information available)

DEPARTMENT OF REPRODUCTIVE BIOLOGY
Phone: Chairman: (216) 791-7300 Ext. 451

Chairman: A.F.W. Hughes
Arthur H. Bill Professor of Obstetrics-Gynecology: A.B. Little
Professors: A.F.W. Hughes, A.B Little, I. Merkatz, M. Rosen, I. Rothchild, A.H. Steinberg, W.B. Wentz
Associate Professor: R.B. Billiar
Assistant Professors: R.A. Meigs, P.T. Schnatz, R.J. Sokol
Instructor: S. Rahman
Postdoctoral Fellows: 5

Degree Program: Reproductive Biology

Graduate Degrees: M.S., Ph.D.
Graduate Degree Requirements: Master: Completion of a minimum number of courses and credits (32). Passing a comprehensive examination. Writing an acceptalbe thesis Doctoral: Completion of a minimum number of courses and credits (60), passing the general (qualifying) examination, writing a dissertation, acceptable defense of dissertation (final examination)

Graduate and Undergraduate Courses Available for Graduate Credit: Reproductive Biology (U), Steroid Hormones (U), Special Topics in Reproductive Biology (U)

School of Engineering

Dean: D.K. Wright, Jr.

DEPARTMENT OF BIOMEDICAL ENGINEERING
Phone: Chairman: (216) 368-4093 Others: 368-2000

Head: L.D. Harmon
Professors: V. Frankel (Adjunct), D.F. Gibbons, L.D. Harmon, J.I. Kleinerman (Adjunct), F.D. Miraldi (Adjunct), M.N. Levy (Adjunct), J. Liebman (Adjunct), Y. Nosé (Adjunct), R. Plonsey
Associate Professors: E. Bahniuk (Adjunct), J.S. Brodkey (Adjunct), A. Burstein (Adjunct), E.H. Chester (Adjunct), B. Friedman (Adjunct), P.G. Katona, M. Macklin, J.T. Mortimer, M R. Neuman (Adjunct), L.E. Ostrander, G.M Saidel, W.S. Topham
Assistant Professors: H.P. Apple (Adjunct), K.L. Barnes (Adjunct), R. Cechner (Adjunct), P.W. Cheung, L. Chik (Adjunct), M D. Graham, R.J. Kiraly (Adjunct), R.B. Lake (Adjunct), R.J. Lorig (Adjunct), P. Martin (Adjunct), F.P. Primiano, Jr. (Adjunct), C.W. Thomas
Teaching Assistants: 2
Research Assistants: 12
Research Associate: 1
Technicians: 5

Degree Programs: Biomaterials, Biomechanical Control, Clinical Engineering, Biomedical Instrumentation, Physiological Systems

Undergraduate Degree: B.S.
Undergraduate Degree Requirements: 130 credit hours in-

cluding 21 in Biomedical Engineering and 13 in a related engineering field. Undergraduates must also satisfy Case Institute of Technology and School of Engineering core requirements.

Graduate Degrees: M.S., Ph.D.
Graduate Degree Requirements: Master: 27-30 credit hours. Thesis usual but not necessary. Students take courses in engineering and the physical and life sciences or concentrate in the representative areas shown above. Doctoral: Minimum of 2 years and 16-20 courses beyond the B.S. level. Course work must include the 10-course BME core, 6 semester hours in Basic Science, 12 semester hours in an approved discipline other than Biomedical Engineering. In addition students must satisfy a life-science concentration requirement. Candidates must pass the Biomedical Comprehensive and Thesis Qualifying Examinations and an oral defense of a thesis based on original research.

Graduate and Undergraduate Courses Available for Graduate Credit: Physiology-Biophysics I and II (U), Bioelectrical Phenomena, Introductory Biomedical Instrumentation, Biomedical Instrumentation, Materials for Prosthetic and Orthotic Use, Polymers in Medicine, Applied Neural Control, Clinical Instrumentation, Clinical Engineering, Clinical Engineering Laboratory, Systems and Signal Analysis in Life Science, Biomechanics of Musculoskeletal System, Digital Processing of Biomedical Signals, Functional Anatomy, Introduction to Biomedical Computation, Population Dynamics, Electro-Physiology and Nervous System Function I and II, Isotope Methodology in Life Sciences, Physiological Systems Analysis Laboratory, Physiological Systems Analysis, Diagnostic Imaging, Artificial Organs, Optimization Techniques in Health Care Delivery, Inferential Processes in Data Analysis, Topics in Cardiovascular Science

CENTRAL STATE UNIVERSITY

Wilberforce, Ohio 45384 Phone: (513) 376-6011
Dean of College: D.W. Hazel

BIOLOGY DEPARTMENT
Phone: Chairman: (513) 376-3626 Others: 376-6624

Chairman: M.A. Johnson, Jr.
Professors: T.J. Craft, M.A. Johnson, Jr.
Associate Professors: J.H. Cooper
Assistant Professors: M.H. Fine, D.C. Rubin, W.J. Washington
Instructors: M.K. Neulieb
Student Research Trainees

Degree Programs: Biology, Education

Undergraduate Degree: B.S.
Undergraduate Degree Requirements: Total Quarter hours required for biology majors - 186 (Biology 45, Chemistry 33, Physics 12, Mathematics 15 Total Quarter hours required for Biology Education Majors - 186 (Biology 45, Chemistry 20, Physics 12, Mathematics 15, Education 29)

THE CLEVELAND STATE UNIVERSITY

1983 Euclid Avenue Phone: (216) 687-2440
Cleveland, Ohio 44115
Dean of Graduate Studies: R.G. Schultz
Dean of College: J.A. Soules

DEPARTMENT OF BIOLOGY AND HEALTH SCIENCES
Phone: Chairman: (216) 687-2241 Others: 687-2440

Chairman: J.H. Morrison
Professors: P.C. Baker, L.J. Brenner, M.F. Bumpus (Adjunct), R.L. Clise, F. DeMarinis, S.S. Deodhar (Adjunct), P.A. Khairallah (Adjunct), J.W. King (Adjunct), L. Messineo, J.H. Morrison, R.R. Smeby (Adjunct)
Associate Professors: R.C. Dickerman, C.M. Ferrario (Adjunct), T.L. Gavan (Adjunct), R.J. Gee, C.A. Goodrich, K.M. Hoff, J. Kerkay, S. Lewis, J.B. Senturia, D.S. Weis
Assistant Professors: A.L. Ehrhart (Adjunct), B.W. Freeman, M. Hall, A.E. Ramm, A. Faller
Teaching Fellows: 9
Research Assistants: 3

Degree Programs: Biology, Regulatory Biology

Undergraduate Degree: B.S.
Undergraduate Degree Requirements: General Biology, Developmental Biology, Cell Physiology, Vertebrate or Plant Physiology, Microbiology, Genetics, Ecology, Senior Seminar, General Chemistry, Organic Chemistry, Biochemistry, Physics, Calculus and Statistics

Graduate Degrees: M.S. (Biology), Ph.D. Regulatory Biology)
Graduate Degree Requirements: Master: Non-Thesis Option - Biochemistry, three seminars, 38 additional hours, Qualifying Examinations Thesis Option - Biochemistry, three seminars, thesis, 22 additional hours Doctoral: 60-82 credit hours - courses, 48-70 credit hours - Research and Dissertation, Research Proposal Defense, Comprehensive Examination

Graduate and Undergraduate Courses Available for Graduate Credit: Microbial Analysis (U), Immunohematology (U), Applied Microbiology (U), Blood Bank Science (U), Vertebrate Physiology (U), Plant Physiology (U), Ecology (U), Biogeography (U), Parasitology (U), Comparative Embryology and Evolution (U), Local Flora (U), Biological Chemistry, Macromolecular and Membrane Biology, Mathematical Biological Control Systems, Cellular Control Systems, Physiological Control Mechanisms I, Physiological Genetics, Population Biology and Evolution, Ecosystem Biology, Advances in Biology, Seminar in Physiology, Seminar in Genetics, Seminar in Developmental Biology, Seminar in Immunology, Seminar in Biological Chemistry, Seminar in Ecology, Seminar in Cell Biology, Animal Models in Research, Biological Chemistry II, Radiobiology, Pharmacodynamics, Toxicology, Bioenergetics, Immunology, Microbial Physiology, Immunobiology, Immunohematology, Blood Bank Science, Histo-Cyto-Chemistry, Protozoology, Physiological Control Mechanisms II, Pathobiology, Endocrinology, Comparative Animal Physiology, Functional Neuroanatomy, Human Genetics, Population Genetics, Biochemical Genetics, Limnology Great Lakes, Limnology Great Lakes Laboratory, Environmental Chemistry, Population Ecology, Physiological Ecology, Host-Parasite Interactions, Developmental Biology, Advanced Comparative Plant Morphology

COLLEGE OF MOUNT ST. JOSEPH

(No reply received)

COLLEGE OF STEUBENVILLE

Steubenville, Ohio 43952 Phone: (614) 283-3771
Dean of College: Rev. J.P. Long

DEPARTMENT OF BIOLOGY
Phone: Chairman: Ext. 272

Chairman: R.E. Cerroni
Professors: T.H. Campbell, R.E. Cerroni
Assistant Professor: E.J. Bessler, Jr.

Degree Programs: Biology, Medical Technology

Undergraduate Degrees: B.A., B.S.
Undergraduate Degree Requirements: 24 hours in Bio-

logy beyond General Botany and General Zoology including Comparative and Developmental Anatomy, Genetics, Cell Physiology and Coordinating Seminar, 16 hours in Chemistry, Introductory Chemistry, Organic Chemistry, 12 hours Foreign Language, 7 hours Mathematics, 8 hours General Physics, 48 hours General Education Requirements

COLLEGE OF WOOSTER

College and Pine   Phone: (216) 264-1234
Wooster, Ohio 44691
Dean of College: F.W. Cropp

DEPARTMENT OF BIOLOGY
Phone: Chairman: Ext. 379 Others: Ext. 379

Chairman: D.L. Wise
Danforth Professor: D.L. Wise
Mateer Professor: C.W. Hinton
Professors: C.W. Hinton, A.A. Weaver, D.L. Wise
Associate Professor: F.L. Downs
Assistant Professors: R.E. Gatten, Jr., J.E. Perley, J.A. Robertson

Degree Program: Biology

Undergraduate Degree: B.A.
Undergraduate Degree Requirements: A minimum of 8 - 300 level courses plus 2 term research thesis.

DEFIANCE COLLEGE

Defiance, Ohio 43512   Phone: (419) 784-4010
Dean of College: R. Ryan

DEPARTMENT OF NATURAL SYSTEMS
Phone: Chairman: Ext. 136

Chairman: J.G. Yuhas
Professors: G.C. deRoth, J.R. Frey, B.C. Mikula
Assistant Professors: J. Birk, R. Jagger, J.G. Yuhas

Degree Programs: Natural Systems with emphasis in Biology, Chemistry or Earth Sciences

Undergraduate Degrees: B.S., A.B.
Undergraduate Degree Requirements: 1 year (2 semesters) General Chemistry - 1 semester Botany, 1 semester Zoology, minimum of 6 other semesters approved courses in consultation with Natural Systems advisory board

DENISON UNIVERSITY

Granville, Ohio 43023   Phone: (614) 587-0810
Dean of College: A. Sterrett

DEPARTMENT OF BIOLOGY
Phone: Chairman: Ext. 334 Others: Ext. 261

Chairman: P. Stukus
Professors: R.W. Alrutz, R. Haubrich, G.R. Norris
Associate Professors: A. Rebuck
Assistant Professors: K.P. Klatt, K.V. Loats, R.K. Pettegrew
Senior Teaching Fellows: 3

Degree Program: Biology

Undergraduate Degrees: B.A., B.S.
Undergraduate Degree Requirements: B.A. Biology - 32 credits Biology (from 4 group areas) 1 year of Chemistry Physics or Geology, B.S. Biology - 32 credits Biology (from 4 group areas), 2 years Chemistry, 1 year Physics, 1 semester Geology and 1 year Intermediate language

EDGECLIFF COLLEGE

2220 Victory Parkway   Phone: (513) 961-3770
Dean of College: W.C. Wester

BIOLOGY DEPARTMENT
Phone: Chairman: Ext. 252 Others: 961-3770

Chairman: Sr. E.M. Charters
Professor: Sr. E.M. Charters
Assistant Professor: G.T. McDuffie
Instructors: Sr. M.J. Wethington, Sr. M E. Stockelman

Degree Programs: Biology, Medical Technology

Undergraduate Degrees: B.A., B.S.
Undergraduate Degree Requirements: 36 semester hours Biology, 20 semester hours Chemistry, 6 semester hours Mathematics, 48-52 hours "Core" courses (Humanities, Social Sciences, Fine Arts), electives to total 128

FINDLAY COLLEGE

1000 North Main Street   Phone: (419) 422-8313
Findlay, Ohio 45840
Dean of College: W.J. McBride

DIVISION OF NATURAL SCIENCES
Phone: Chairman: 275 Others: 272

Chairman: A.J. Wilfong
Associate Professors: J. Joseph, A.J. Wilfong

Field Stations/Laboratories: Findlay College Research Station - Findlay City Reservoir

Degree Program: Science

Undergraduate Degree: B.S.
Undergraduate Degree Requirements: 3 courses in Biology, 3 courses in Chemistry, 3 courses in Physics, 3 courses in Mathematics

HEIDELBERG COLLEGE

Tiffin, Ohio 44883   Phone: (419) 448-2693
Dean of College: A. Porter

BIOLOGY DEPARTMENT
Phone: Chairman: 448-2693

Chairman: G. Barlow
E.R. Kuck Professor of Microbiology: P.L. Lilly
Professors: G. Barlow, H. Hintz, P. Lilly
Associate Professor: D. Baker
Assistant Professors: R. Murray, G. Joyce (Visiting)

Degree Program: Biology

Undergraduate Degree: B.S.
Undergraduate Degree Requirements: 9 hours Social Studies, 20 hours Humanities, 27 hours Biology, 8 hours Physics, 8 hours Chemistry, 6 hours Mathematics

HIRAM COLLEGE

Hiram, Ohio 44234   Phone: (216) 569-3211
Dean of College: R. MacDowell

BIOLOGY DEPARTMENT
Phone: Chairman: Ext. 274

Chairman: D.H. Berg
Professors: J.H. Barrow, D.H. Berg
Assistant Professors: W.S. Cool, W.J. Laughner, D. Brewbaker
Instructor: K. Herndon
Visiting Lecturers: 30 per year
Research Assistants: 1

Field Stations/Laboratories: Hiram College Biology Station

Degree Programs: Biology

Undergraduate Degree: B.A.
Undergraduate Degree Requirements: 186 hours, 60 hours in Biology

JOHN CARROLL UNIVERSITY

20700 North Park Boulevard Phone: (216) 491-4911
University Heights, Ohio 44118
Dean of Graduate Studies: D. Gavin
Dean of College: L.V. Britt, S.J.

DEPARTMENT OF BIOLOGY
Phone: Chairman: (216) 491-4251 Others: 491-4252

Chairman: J.G. Allen
Professor: J. Cummings
Associate Professors: J.G. Allen, E.B. McLean, F.D. Moore, E.J. Skoch, A.M. White, C.H. Wideman
Assistant Professor: P.A. Khairallah
Visiting Lecturer: 1

Degree Program: Biology

Undergraduate Degree: B.S.
Undergraduate Degree Requirements: At least 28 hours of Biology including the following courses: Biology 103,104,206,213 and four or more upper-division courses chosen with advisory approval.

Graduate Degree: M.S.
Graduate Degree Requirements: Master: Plan A - (at least eight courses consisting of a minimum of twenty four hours together with a thesis. Plan B - ten courses and research essay

Graduate and Undergraduate Courses Available for Graduate Credit: Invertebrate Zoology, Vertebrate Zoology, Human Genetics, Biological Technique, Parasitology, General Entomology, Ecology, Mycology, Radioisotope Methodology, Endocrinology, Seminar, Principles of Mammology, Advanced Bacteriology, Current Problems in Cell Physiology, Current Problems in Mammalian Physiology, Advanced Genetics, Herpetology, Limnology, Behavior, Ornithology, Cytology, Cytogenetics

KENT STATE UNIVERSITY

Kent, Ohio 44242 Phone: (216) 672-2121
Dean of Graduate Studies: J. McGrath
Dean of College: R. Buttlar

DEPARTMENT OF BIOLOGICAL SCIENCES
Phone: Chairman: (216) 672-2266 Others: 672-2121

Chairman: C.V. Riley
Professors: D. Anderson, T. Cooperrider, R. Dexter, K. Ewing, M. Ferguson, B.A. Foote, V. Gallicchio, A. Graham, C. Hobbs, L.P. Orr, C. Riley
Associate Professors: W. Adams, A. Cibula, D. Cooke, F. Dela, G. Larkin, R. Rhodes, R. Stokes
Assistant Professors: R. Heath, R. Mack, S. Mazzer, D. Waller
Teaching Fellows: 3

Degree Programs: Biology, Botany, Conservation, Zoology, Biological Sciences

Undergraduate Degrees: B.A., B.S.
Undergraduate Degree Requirements: Not Stated

Graduate Degrees: M.A., M.S., Ph.D.
Graduate Degree Requirements: Master: The M.A. and M.S. programs are open only to qualified individuals who can enroll as full-time students. There is no universal language requirement for the master's degree in the department. A final oral examination is required for the M.A. and M.S. degree Doctoral: The Ph.D. program is open only to qualified individuals who can enroll as full-time students. The applicant must possess the master's degree or its equivalent. The department does not have a universal language requirement for the Ph.D. Degree.

Graduate and Undergraduate Courses Available for Graduate Credit: Cytology (U), Cell Biology (U), Cell Biology II (U), Developmental Systems (U), Human Heredity (U), Advanced Genetics (U), Organic Evolution (U), History of Biology (U), Advanced General Bacteriology (U), Immunology (U), Special Topics in Biology (U), Phycology (U), Plant Anatomy (U), Systematic Botany (U), Plant Taxonomy (U), Morphology of Lower Plants (U), Morphology of Angiosperms (U), Palynology (U), Limnology (U), Plant Ecology (U), Cellular Physiology (U), Endocrinology (U), Vertebrate Physiology I (U), Vertebrate Physiology II (U), Comparative Animal Physiology I (U), Comparative Animal Physiology II (U), Animal Behavior (U), Animal Histology (U), Vertebrate Embryology (U), Neuroanatomy (U), Wildlife Research (U), Fisheries Management (U), Vertebrate Zoology (U), Animal Parasites and Advanced Parasites (U)

KENYON COLLEGE

Gambier, Ohio 43022 Phone: (614) 427-2244
Provost: B. Haywood

Chairman: A. Wohlpart
Professors: R.D. Burns, F.W. Yow
Associate Professors: T.C. Jegla
Assistant Professors: S.H. Anderson, D. Jegla (Adjunct) J.R. Stallard, A. Wohlpart

Degree Program: Biology

Undergraduate Degree: B.A.
Undergraduate Degree Requirements: Introductory Biology and Laboratory in Introductory Biology. Six or more advanced courses in the Department, laboratory experience in at least four areas. Supportive course work in Chemistry, Physics, Psychology and Mathematics

LAKE ERIE COLLEGE

Painesville, Ohio 44077 Phone: (216) 352-3361
Dean of College: C.W.R. Larson

DEPARTMENT OF BIOLOGY
Phone: Chairman: Ext. 300

Instructional Coordinator: B.N. Wise
Associate Professor: B.N. Wise
Assistant Professors: K.M. Foos

Degree Program: Biology

Undergraduate Degrees: B.A., B.S.
Undergraduate Degree Requirements: Minimum of 9 courses in "Area of Concentration" (unique for each student), including at least 5 in Biology.

MALONE COLLEGE

515 25th Street N.W. Phone: (216) 454-3011
Dean of College: R. Chambers

DIVISION OF SCIENCE AND MATHEMATICS
Phone: Chairman: Ext. 282 Others: Ext. 281

Chairman: D.S. Thomson
Professor: D.S. Thomson
Associate Professors: A.W. Fritz, M.W. Stephens, R.D. Ritter
Assistant Professors: R.G. Johnson, R. Nisbet, M B. Niver, G.E. Lipely

Degree Program: Biology

Undergraduate Degree: B.A.
Undergraduate Degree Requirements: Biology 25 hours Biology, 9 hours Chemistry, 9 hours Physics, 6 hours Mathematics

MARIETTA COLLEGE
Marietta, Ohio 45750 Phone: (614) 373-4643
Dean of College: A.W. Bosch

HARLA RAY EGGLESTON DEPARTMENT OF BIOLOGY
Phone: Chairman: Ext. 240

Chairman: D.F. Young
Professors: R.L. Walp, D.F. Young
Associate Professor: W.P. Brown
Assistant Professors: P.L. Anderson, B.P. Clark, F.E. Farley, J.W. Perry

Degree Program: Biology

Undergraduate Degree: B.S.
Undergraduate Degree Requirements: Biology 101, 102, and 26 additional hours in Biology. Minimum 12 hours in Chemistry or 8 hours in Chemistry and 8 hours in Geology. Mathematics: minimum 6 hours credit chosen from courses numbered 120 and higher but not including 213 and 214. General Physics 127 and 128. As many as 8 hours in Chemistry, Geology, Mathematics, or Physics that are in excess of the minimum requirements listed above may be substituted for hours in Biology, College graduation requirements must be completed.

MIAMI UNIVERSITY
Oxford, Ohio 45056 Phone: (513) 529-2161
Dean of Graduate School: S. Peterson

College of Arts and Sciences
Dean: C.K. Williamson

DEPARTMENT OF BOTANY
Phone: Chairman: (513) 529-5031 Others: 529-5321

Chairman: C. Heimsch
Professors: J.W. McClure, D.W. Newman, T.K. Wilson, W.E. Wilson
Associate Professors: W.H. Blackwell, Jr., T.J. Cobbe, W.H. Eshbaugh, K.R. Mattox, K.D. Stewart, K.G. Wilson, R.E. Wilson
Assistant Professors: H.K. Goree, D.M. Travis, J.L. Vankat
Instructor: M.C. Dalgarn
Teaching Fellows: 7
Research Assistants: 4
Dissertation Fellow: 1
Graduate Assistants: 1

Degree Program: Botany

Undergraduate Degrees: B.A., B.S.
Undergraduate Degree Requirements: B.A. - 36 quarter credit hours in Botany plus 24 credit hours in related courses B.S. - 54 quarter credit hours in Botany plus 59-83 credit hours in related courses. B.S. (Environmental Management) - 54 quarter credit hours in Botany plus 77 credit hours in related courses

Graduate Degrees: M.A., M.S., M.A.T., Ph.D.
Graduate Degree Requirements: Master: M.A. and M.S. 45 quarter credit hours, including 6-12 credit hours in research and thesis M.A.T. - 45 quarter credit hours along or including 6-12 credit hours in research and thesis Doctoral: Minimum of 90 quarter credit hours beyond the master's degree or its equivalent plus reading knowledge of one foreign language or approved substitute research tool.

Graduate and Undergraduate Courses Available for Graduate Credit: Plant Anatomy (U), Plant Histology (U), Mycology (U), Plant Pathology (U), Plant Geography (U), Physiological Ecology (U), Biometry, Genetics Seminar, Advanced Plant Taxonomy, Organic Constituents of Higher Plants, Plant Metabolism, Physiology of Plant Growth and Development, Morphology of Vascular Plants, Phycology, Plant Ultrastructure, Techniques of Electron Microscopy, Synecology, Plant Genetics, Plant Biosystematics, Graduate Seminar, Advanced Topics in Botany, Research in Botany

DEPARTMENT OF MICROBIOLOGY
Phone: Chairman: (513) 529-5422

Chairman: R.J. Brady
Professors: J.K. Bhattacharjee, R.J. Brady, I. Kochan, S.W. Rockwood, C.K. Williamson
Associate Professor: R.W. Treick
Assistant Professors: D.D. Barnhart, G. Breidenbach
Instructor: D.B. Stroupe
Teaching Fellows: 3
Research Assistants: 3
Teaching Assistants: 6

Degree Programs: Microbiology, Medical Technology

Undergraduate Degrees: A.B., B.S.
Undergraduate Degree Requirements: Microbiology 44 quarter hours, Chemistry 30 quarter hours, Physics 15 quarter hours, Mathematics - Calculus

Graduate Degrees: M.S., Ph.D.
Graduate Degree Requirements: Master: 45 quarter hours, thesis Doctoral: 90 quarter hours beyond M.S. dissertation

Graduate and Undergraduate Courses Available for Graduate Credit: Microbial Physiology (U), Advanced Microbial Physiology, Applied Microbiology (U), Genetics of Bacteria and Viruses (U), Molecular GEnetics, Pathogenic Microbiology (U), Advanced Pathogenic Microbiology, Immunology and Serology (U), Advanced Immunology and Serology, Virology (U), Advanced Virology, Cell and Tissue Culture Techniques (U), Advanced Molecular Physiology, Special Topics: Instrumentation, Mycology, Immune Mechanisms

DEPARTMENT OF ZOOLOGY
Phone: Chairman: (513) 529-4918 Others: 529-4593

Chairman: C.M. Vaughn
Professors: M.W. Boesel (Curator, Entomology), E.J. DeVillez, T.G. Gregg, R.A. Hefner (Curator, Museum) E.M. Ingersoll, C.M. Vaughn, R W. Winner
Associate Professors: A.L. Allenspach, G.W. Barrett, D.W. Bergstrom, P.M. Daniel, D.L. Leonier, J.A. Goldey (Adjunct), R.E. Griffith, S.I. Guttman, R.E. Hayes, H.E. Klaaren (Adjunct), W.J. MacMasters, R.R. Nielson, D.H. Taylor, J.C. Vaughn, H. Weller, R.L. Wiley, D.F. Wilson, T.E. Wissing
Assistant Professors: D.L. Claussen, M P. Farrell (Adjunct), R.A. Grassmick, D.G. Kaufman, D.R. Osborne, R.J. Pfohl, R.E. Rayle
Instructors: M.A.M. Addington, L.Y. Becker, D.S. Klaaren
Post Doctoral Teaching Fellow: C.F. Thompson
Teaching Fellows: 4
Research Assistants: 3
Teaching Assistants: 20

Degree Programs: Ecology, Entomology, Ethology, Genetics, Limnology, Parasitology, Physiology, Zoology, Developmental Biology, Cell Biology

Undergraduate Degree: B.A. (Zoology)
Undergraduate Degree Requirements: Requirements (in quarter hours) 40 hours Zoology including an elementary course in department, courses selected in consultation with advisor. Related hours 27 including

at least year of Chemistry, plus courses in Botany, Chemistry, Geography, Geology, Mathematics, Microbiology or Physics.

Graduate Degrees: M.A., M.S., M.A.T., Ph.D. (Zoology)
Graduate Degree Requirements: Master: (in quarter hours) M A./M.S. 45 hours including at least 24 hours of formal courses in Zoology plus 6-15 hours of thesis credit. Remaining hours from related fields with advisor's permission. Course work must include at least one course in each subarea: Ecology and Systematics, Developmental Biology and Physiology. M.A.T. 45 hours including 24 hours of formal courses in Zoology, remainder from related fields with advisor's permission. Doctoral: Ph.D. Master's degree or equivalent, advanced course in 3 subareas, proficiency in a foreign language, written and oral comprehensive examination in 3 subareas, dissertation and its defense

Graduate and Undergraduate Courses Available for Graduate Credit: Advanced Entomology (U), Analytical Cytology, Animal Biosystematics, Advanced Vertebrate Physiology, Biochemical Development, Biochemical Genetics (U), Community and Ecosystem Ecology, Comparative Animal Physiology (U), Comparative Endocrinology (U), Current Topics in Ecology and Systematics, Developmental Biology, Genetics, Physiology, Cytology (U), Environmental Physiology, Ethology, Experimental Animal Physiology, Experimental Embryology, Genetics Seminar, Herpetology, History of Zoology (U), Insect Ecology, Limnology (U), Population Ecology, Techniques of Electron Microscopy, Zoological Literature

MOUNT UNION COLLEGE

Alliance, Ohio 44601 Phone: (216) 821-5320
Dean of College: T. Turnquist (Acting)

DEPARTMENT OF BIOLOGY
Phone: Chairman: Ext. 293

Head: J.L. Blount
Lichty Professor of Biology: J.L. Blount
Professor: J.L. Blount
Associate Professors: C.H. Brueske, G.O. Osterman
Assistant Professor: L. Epp

Field Stations/Laboratories: Mt. Union Field Station - 6 miles South of Alliance

Degree Program: Biology

Undergraduate Degree: B.S.
Undergraduate Degree Requirements: 1) thirty-six courses (total), 2) at least 8 courses within the department 3) two courses Physics 4) four courses Chemistry 5) two courses Mathematics

MUSKINGUM COLLEGE

New Concord, Ohio 43762 Phone: (614) 826-8211
Dean of College: W. Fisk

DEPARTMENT OF BIOLOGY
Phone: Chairman: (614) 826-8221 Others: 826-8220

Chairman: C.E. Dasch
Professor: C.E. Dasch
Associate Professors: W. Adams, Jr., D. Quinn, V. Saksena
Assistant Professor: W. Reist

Degree Program: Biology

Undergraduate Degree: B.S.
Undergraduate Degree Requirements: 35 hours major of Biology courses, 8-16 hours of Chemistry, general degree requirements

NOTRE DAME COLLEGE

4545 College Road Phone: (216) 381-1680
South Euclid, Ohio 44121

BIOLOGY DEPARTMENT
Phone: Chairman: (216) 381-1680

Chairman: Sr. M. Lisanne
Professors: Sr. M. Lisanne, Sr. M. St. Damian

Degree Programs: Biology, Medical Technology

Undergraduate Degree: B.S.
Undergraduate Degree Requirements: B.S. - 128 hours (credit hours), Associates in Sciences (64 credit hours), Distributed among major field and liberal arts

OBERLIN COLLEGE

Oberlin, Ohio 44074 Phone: (216) 774-1221
Dean of College of Arts and Sciences: R.M. Longsworth (Acting)

DEPARTMENT OF BIOLOGY
Phone: Chairman: Ext. 4150

Chairman: A.R. Brummett
Professors: A.R. Brummett, G.T. Scott, W.F. Walker, Jr.
Associate Professors: D.H. Benzing, D.A. Egloff, T.F. Sherman, R.A. Levin
Assistant Professor: D.N. Luck, D.H. Miller
Visiting Lecturer: 1
Teaching Fellow: 1
Research Assistant: 1
Technical Assistants: 2

Degree Program: Biology

Undergraduate Degree: B.A.
Undergraduate Degree Requirements: (112 semester hours plus 3 winter terms), Biology Major - Introduction to Cell and Molecular Biology (4 hours), 20 additional hours in Biology to include: 3 laboratory courses, 1 Botany course, 1 Zoology course, Organic Chemistry (8 hours)

OHIO DOMINICAN COLLEGE

Columbus, Ohio 43219 Phone: 253-2741
Dean of College: Sr. M.A. Matesich

BIOLOGY DEPARTMENT
Phone: Chairman: Ext. 52

Chairman: J.E. Kennedy
Assistant Professors: G. Beery, J.E. Kennedy, W.G. Smith
Part-time: 1

Degree Programs: Biology, Medical Technology

Undergraduate Degrees: B.A., B.S.
Undergraduate Degree Requirements: 1) eight Biology courses including 2 in general Biology - 1 in Genetics 1 in General Microbiology, and 4 Biology elective courses 2) minimum of 2 Chemistry and and 1 Mathematics courses, 3) senior comprehensives and 4) General College Requirements

OHIO NORTHERN UNIVERSITY

Ada, Ohio 45810 Phone: 634-9921
Dean of College of Liberal Arts: B.L. Linger

DEPARTMENT OF BIOLOGY
Phone: Chairman: Ext. 224 Others: Ext. 492

Chairman: R.L. Bowden

Professors: R.L. Bowden, D.R. Butler, J.E. Dawson, S.L. Meyer
Associate Professors: C.C. Laing, E.V. Nelson, R.E. Tipple
Assistant Professors: T.D. Keiser, N.J. Moore
Lecturers: 2

Degree Program: Biology

Undergraduate Degree: B.A.
Undergraduate Degree Requirements: 182 quarter hours including 70 hours general education, 50 hours Biology, 25 hours Chemistry, Mathematics, Physics, electives

OHIO STATE UNIVERSITY

190 North Oval Drive Phone: (614) 422-6446
Columbus, Ohio 43210
Dean of Graduate Studies: A.L. Roaden

College of Biological Sciences

Dean: R.H. Bohning

ACAROLOGY LABORATORY
Phone: Chairman: (614) 422-7180 Others: 422-7186

Chairman: G.W. Wharton
Professors: D.E. Johnston, R.D. Mitchell, E.F. Paddock, R.M. Pfister, S.S.Y. Young
Associate Professor: W.B. Parrish
Visiting Lecturers: 2
Trainees: 4

Degree Programs: Graduate degrees in Acarology are offered through the Departments of Entomology and Zoology

Graduate and Undergraduate Courses Available for Graduate Credit: General Acarology, Medical-Veterinary Acarology, Agricultural Acarology

DEPARTMENT OF BIOCHEMISTRY
Phone: Chairman: (614) 422-6771

Chairman: G.S. Serif
Professors: G.A. Barber, E.J. Behrman, W.A. Bulen, F.E. Deatherage, R.W. Doskotch, W.J. Harper, D.H. Ives, R.O. Moore, A.L. Moxon, R.A. Scott III, G.S. Serif, J.F. Snell, Q. Van Winkle
Associate Professors: E.L. Gross, G.A. Marzluf, C.B. Meleca, G.P. Royer
Assistant Professors: G.E. Means
Research Assistants: 5
Postdoctoral Training: 5

Degree Program: Biochemistry

Undergraduate Degrees: B.S., B.A.
Undergraduate Degree Requirements: 196 credit hours of which 65 must be in one department. Those wishing to be admitted to graduate work in Biochemistry must take proficiency examinations in Organic and Physical Chemistry.

Graduate Degrees: M.S., Ph.D.
Graduate Degree Requirements: Master: Six specified courses in Biochemistry plus 6 graduate level credit hours in Chemistry, Mathematics, Physics or Biology. Thesis Doctoral: Six specified courses in Biochemistry plus 15 graduate level credit hours in Chemistry, Mathematics, Physics or Biology

Graduate and Undergraduate Courses Available for Graduate Credit: Introduction to Biological Chemistry (U), Biochemistry of Physiological Processes (U), Biochemistry and Molecular Biology (U), Biochemistry and Molecular Biology (U), Chemistry of Foods and Food Processing (U), Molecular Genetics (U), Individual Studies (U), General Biological Chemistry (U), Physical Biochemistry (U), Molecular Photobiology (U), Research Principles and Techniques (U), Advanced Topics in Molecular Genetics, Enzymes, Carbohydrates, Seminar in Biological Chemistry, Special Topics in Food Chemistry

DEPARTMENT OF BOTANY
Phone: Chairman: (614) 422-8952 Others: 422-6446

Chairman: T.N. Taylor
Professors: R.A. Popham, E.D. Rudolph, J.A. Schmitt, C.A. Swanson, C.E. Taft, T.N. Taylor
Associate Professors: M.G. Cline, M.L. Evans, R.M. Giesy, G.E. Gilbert, T.J. Johnson, V. Raghavan, R.L. Seymour, R.L. Stuckey, T.F. Stuessy
Assistant Professors: D.G. Fratianne, H.P. Hostetter, R.S. Platt
Teaching Fellows: 27
Research Assistants: 3
Postdoctoral Fellows: 2

Field Stations/Laboratories: F.T. Stone Laboratory - Put-in-Bay, Ohio

Degree Programs: Botany, Developmental Biology, Environmental Biology

Undergraduate Degrees: B.A., B.Sc.
Undergraduate Degree Requirements: 40 quarter credit chour major program, liberal arts courses to total 180 hours for graduation

Graduate Degrees: M.A., M.Sc., Ph.D.
Graduate Degree Requirements: Master: 45 quarter credit hours beyond bachelor's degree, coursework in four areas of Botany, acceptable thesis Doctoral: 135 quarter credit hours beyond bachelor's degree, coursework in 5 areas of Botany, acceptable dissertation

Graduate and Undergraduate Courses Available for Graduate Credit: Basic Concepts in Botany,(U), Introduction to Ecological General Systems Theory (U), Field Botany (U), History of Biology (U), Higher Aquatic Plants (U), Taxonomy of Vascular Plants (U), Basic Principles of Plant Ecology (U), Field Plant Ecology (U), Plant Physiology (U), Physiological Ecology (U), Plant Physiology Laboratory (U), Bryophytes, Pteridophytes, and Gymnosperms (U), Morphology of the Angiosperms (U), Plant Microtechnic (U), Developmental Plant Anatomy (U), Algae (U), Experimental Phycology (U), Diatom Ecology and Systematics (U), Mycology (U), Medical Mycology (U), Aquatic Mycology (U), Individual Studies (U), Group Studies in Botany, Plant Morphogenesis (U), [Botany] Honors Course (U), Experimental Taxonomy, Seminar in Plant Ecology, Advanced Plant Physiology: Metabolism, Advanced Plant Physiology: Growth, Advanced Plant Physiology: Water and Solute Relations, Seminar in Plant Physiology, Seminar in Plant Anatomy and Morphology, Seminar in Cryptogamic Botany, Interdepartmental Seminar in Developmental Biology, Environmental Biology Seminar, Interdepartmental Seminar in Polar and Alpine Studies, Interdepartmental Seminar in Natural Resources

DEPARTMENT OF BIOPHYSICS
Phone: Chairman: (614) 422-2733

Chairman: L.E. Lipetz
Professors: H.R. Blackwell, S.A. Corson, R.M. Hill, P.B. Hollander, L.E. Lipetz, J. Rothstein, S.W. Smith, R.W. Stow, Q.Van Winkle
Associate Professors: W.R. Biersdorf, J.Y. Cassim, G.E. Gilbert, C.R. Ingling, K. Kornacker, R.T. Ross, J. Snell
Assistant Professors: R.W. Hart
University Fellows: 1
Research Associates: 3
Teaching Associates: 3

Field Stations/Laboratories: Institute for Research in Vision - 1314 Kinnear Road - Columbus, Ohio

Degree Programs: Biophysics, Radiation Biophysics

Undergraduate Degrees: B.Sc., B.A.
Undergraduate Degree Requirements: Specified courses in Mathematics, Physics, Chemistry, Biology and Biophysics, totalling a minimum of 49 credit hours.

Graduate Degrees: M.Sci., Ph.D.
Graduate Degree Requirements: Master: A, with thesis, B, without. B also requires a critical review of a topic in one minor area and an original research proposal in the second. Dcotoral: A major in one area of Biophysics and a minor in a second; or an interdisciplinary integration of minors in 3 areas. a critical review of the literature in one area, an original research proposal in another, written and oral examinations, a dissertation in one of the areas.

Graduate and Undergraduate Courses Available for Graduate Credit: Psychophysical Measurement (U), Sensory Psychophysics (U), Biophysics of Cell Membranes (U), Sensory Neurophysiology (U), Introduction to Molecular Biophysics (U), Bioenergetics (U), Principles and Techniques of Molecular Biophysics: Applications to Contractility (U), Physical Analysis of Organized Systems in Biology (U), Seminar in Biophysics Research (U), Advanced Experimental Methods in Biophysics (U), Neural Integration of Multiple Sensory Inputs (U), Molecular Biophysics I and II (U), Statistical Thermodynamics in Biology (U), Mechanisms of Psychobiological Intergration (U), Sensory Biophysics I and II, Principles of Nervous System Integration, Functional Study of Sensory Abnormality, Quantum Biology, Information Processing in Sensory Systems (U)

DEPARTMENT OF ENTOMOLOGY
Phone: Chairman: (614) 422-8209

Chairman: L. Goleman
Professors: B.D. Blair, D.J. Borror, J D. Briggs, N.W. Britt, F.W. Fisk, R.P. Holdsworth, D.E. Johnston, H. Niemczyk, R.W. Rings, W.C. Rothenbuhler, G.F. Shambaugh, J.P. Sleesman, G.R. Stairs, R.E. Treece, C.A. Triplehorn, G.W. Wharton
Associate Professors: B.D. Barry, W.J. Collins, W.F. Hink, D.J Horn, C.C. King (Adjunct), J.K. Knoke (Adjunct), H.R. Krueger, T.L. Ladd (Adjunct), R.L. Miller, G.J. Musick, L.R. Nault, A.C. Waldron, R.N. Williams
Assistant Professors: L.J. Connor, W.A. Foster, F.R. Hall, R.K. Lindquist, W.F. Lyon, D.G.Neilsen
Visiting Lecturers: 2
Teaching Fellows: 14
Research Assistants: 4

Field Stations/Laboratories: Franz T. Stone Laboratories - Put-In-Bay, Ohio

Degree Program: Entomology

Undergraduate Degrees: B.S., B.A.
Undergraduate Degree Requirements: Not Stated

Graduate Degrees: M.S., Ph.D.
Graduate Degree Requirements: Master: Completion of 45 hours course work, achievement of cumulative point-hour ratio of 3.0, registration in final quarter of degree program, completion of final examination. Doctoral: Completion of 135 quarter hours of course work, completion of general examination at least 2 quarters before expected degree, completion of acceptable dissertation, passing of a final oral examination on the dissertation.

Graduate and Undergraduate Courses Available for Graduate Credit: Biology of the Honey Bee (U), Field Entomology (U), Aquatic Entomology (U), External Morphology of Insects (U), Insect Physiology (U), Advanced Economic Entomology (U), Medical Entomology (U), Insect Toxicology (U), General Acarology (U), Individual Studies (U), Group Studies (U), Insect Pathology (U), Systematic Entomology (U), Immature Insects (U), Honors (U), Entomology Seminar, Research, Living Insects, Internal Morphology of Insects,Advanced Insect Physiology, Biological Control, Medical Veterinary Acarology, Agricultural Acarology, Interdepartmental Seminar, Environmental Biology, Interdepartmental Seminar, Polar and Alpine Studies, Interdepartmental Seminar in Natural Resources

GENETICS DEPARTMENT
Phone: Chairman: (614) 422-7542 Others: 422-8084

Chairman: B. Griffing
Mershon Professor of Genetics: B. Griffing
Professors: B. Griffing, S.S.Y. Young, V.H. House, E.F. Paddock
Associate Professors: C.W. Birky, Jr., R.V. Skavaril
Assistant Professors: P.S. Perlman, R.L. Scholl, R.H. Essman
Graduate Teaching Associates: 7

Degree Program: Genetics

Graduate Degrees: M.S., Ph.D.
Graduate Degree Requirements: Not Stated

Graduate and Undergraduate Courses Available for Graduate Credit: Interdepartmental Seminar in Development, Cytogenetics (U), General Genetics (U), Genetics Laboratory (U), Molecular Genetics (U), The Cytological Basis of Genetics (U), Analysis and Interpretation of Biological Data I (U), Analysis and Interpretation of Biological Data II (U), Computer Applications in Genetics (U), Individual Studies (U), Group Studies (U), Genetics and Biogenesis of Cell Organelles (U), Research in Genetics, Genetics Seminar, Advanced Topics in Molecular Genetics, The Nature of Gene Action, Mathematical Genetics, Theoretical and Experimental Population Genetics, Transmission Genetics Theory, Quantitative Genetics and Selection Theory

DEPARTMENT OF MICROBIOLOGY
Phone: Chairman: (614) 422-2301

Chairman: R.M. Pfister
Professors: G.J. Banwart, E.H. Bohl, F.W. Chorpenning, M.C. Dodd, P.R. Dugan, L.C. Ferguson, J.I. Frea, J.P. Kreier, R.H. Miller, R.M. Pfister, C.I. Randles, M.S. Rheins, S. Rosen, R.L. St. Pierre, S. Saslaw, J.F. Snell, T. Suie, G.P. Willson III, D.S. Yohn
Associate Professors: T.J. Byers, J.C. Copeland, B.J. Kolodziej, D.A. Wolff
Assistant Professors: A.D. Barker (Adjunct), G.D. Cagle, J.R. Chipley, W.B. Krueger, E.M. Mote, W. R. Sharp, B.S. Zwilling
Instructor: A.M. Ackermann

Degree Program: Microbiology

Undergraduate Degrees: B.S.
Undergraduate Degree Requirements: 30 hours of Microbiology courses, 15 specifically designated. 10 additional hours of courses 300 level or above in the College of Biological Sciences or Department of Microbiology.

Graduate Degrees: M.S., Ph.D.
Graduate Degree Requirements: Master: Plan A - 45 credit hours of graduate course work. Successful completion of a research problem and oral examination Approved thesis Plan B - Non-thesis - 60 credit hours of graduate course work. A departmental comprehensive examination, written and oral. Doctoral: 90 additional credit hours of graduate course work beyond the 45 or 60 credit hours required for the M.S. degree. Successful completion of a general examination, written and oral. Completion of acceptable re-

search. Approved dissertation and final examination.

Graduate and Undergraduate Courses Available for Graduate Credit: Microbiology in Relation to Man (U), General Microbiology (U), General Microbiology 609 (U), Principles of Infection and Resistance (U), Microbial Parasitism (U), Pathogenic Protozoology (U), Cellular Aspects of the Immune Response (U), Water Microbiology (U), Food Microbiology (U), Aquatic Microbiology (U), General Cellular Biology (U), Cell Differentiation (U), Cytologic Preparations in Electron Microscopy (U), Individual Studies in Microbiology (U), Group Studies (U), History of Microbiology and Allied Fields (U), Immunology and Immunochemistry (U), Bacterial Pathogens (U), Advanced Food Microbiology (U), Basic Virology (U), Physiology of Bacteria (U), Microbial Cytology (U), Protozoan Growth and Reproduction (U), Microbial Genetics (U), Special Groups of Microorganisms (U), Microbiology Colloquium (U), Advanced Virology, Advanced Immunology, Isoantigens of Man and Animals, Advanced Topics of Bacterial Physiology, Seminar in Microbiology, Interdepartmental Seminar in Developmental Biology, Interdepartmental Seminar in Environmental Biology, Interdepartmental Seminar in Polar and Alpine Studies, Interdepartmental Seminar in Natural Resources, Interdepartmental Seminar in Nutrition and Food Technology, Group Studies, Research inMicrobiology

DEPARTMENT OF ZOOLOGY
Phone: Chairman: (614) 422-8088

Chairman: T.J. Peterle
Professors: T.A. Bookhout (Adjunct), W.J. Clench (Adjunct), P.A. Colinvaux, J.L. Crites, R.S. Davidson (Adjunct), M.L. Giltz, W.J. Kostir (Emeritus), D.F. Miller (Emeritus), J.A. Miller (Emeritus), R.D. Mitchell, T.B. Myers, W.C. Myser, T.J. Peterle, J.W. Price (Emeritus), L.S. Putnam, C.R. Reese (Emeritus), W.C. Rothenbuhler, D.H. Stansbery, M.A. Trautman (Adjunct), M.B. Trautman (Emeritus), R.A. Tubb (Adjunct)
Associate Professors: W.E. Carey, T.M. Cavender, W. L. Hartman (Adjunct), C.E. Herdendorf III, F.W. Kessler, S.I. Lustick, M. Miskimen, W.T. Momot, W.B. Parrish, V.C. Stevens, R.A. Tassava, B.D. Valentine
Assistant Professors: T. Berra, R.D. Curnow (Adjunct), J.F. Downhower, A.S. Gaunt, L. Greenwald, B. Griswold (Adjunct), T. Grubb, J.D. Harder, L. Hillis-Colinvaux, D. Martin, P.W. Pappas, F.L. St. John, K.R. Smith, E.L. Troutman
Instructors: J.T. Addis, R.K. Burnard, J.M. Condit (Adjunct), R. Jezerinac
Research Assistants: 25

Field Stations/Laboratories: Stone Laboratory, Cooperative Program with NASA Plum Brook Station, Ohio Division of Wildlife, Olentangy Wildlife Experimental Station, Waterloo Wildlife Experimental Station, Crane Creek Wildlife Experimental Station Cooperative Research Program, Winous Point Club, Pt. Clinton, Ohio

Degree Programs: Environmental Biology, Developmental Biology

Undergraduate Degree: B.S.
Undergraduate Degree Requirements: 40-45 hours for major, 20-25 hours in Zoology

Graduate Degrees: M.S., Ph.D.
Graduate Degree Requirements: Master: 45 Academic hours, Plan A thesis-examination, Plan B Examination Doctoral: 90 Academic hours, General Examination, oral and written Doctoral thesis

Graduate and Undergraduate Courses Available for Graduate Credit: Animal Parasites, Animal Parasitology, Invertebrate Zoology, Zoology of Vertebrates, Ichthyology, Herpetology, Biology of Birds, Advanced Ornithology, Mammalogy, Biology of Fishes, Comparative Embryology, Vertebrate Physiology, Animal Behavior, Principles of Animal Ecology, Field Zoology, Limnology, Fish Ecology, Fisheries Biology, Wildlife Biology, Wildlife Biological Techniques, Radiation Biology, Principles of Biogeography, Fish and Wildlife Parasitology, Individual Studies, Advanced Zoology of Invertebrates, Biological Effects of Ionizing Radiation, Helminthology, Behavior Genetics, Bioacoustics, Population Ecology, Seminar on Historical Ecology, Principles of Systematics, Interdepartmental Seminar in Developmental Biology, Interdepartmental Seminar in Environmental Biology in Polar and Alpine Studies, and in Natural Resources, Research in Zoology

College of Agriculture and Home Economics
Dean: R.M. Kottman

DEPARTMENT OF DAIRY SCIENCE
(No information available)

DEPARTMENT OF AGRONOMY
Phone: Chairman: (614) 422-2002 Others: 422-2591

Chairman: G.W. Volk
Professors: T.G. Arscott, L.D. Baver (Emeritus), L.E. Bendixen, R.B. Clark (Adjunct), R.L. Clements (Adjunct), R.R. Davis, E.J. Dollinger, W.R. Findley (Adjunct), R.E. Franklin, D.T. Friday, G.R. Gist, Jr., F. Haghiri, P.R. Henderlong, D.E. Herr, F.L. Himes, D.J. Hoff, N. Holowaychuk, C.A. Lamb (Emeritus), H.L. Lafever, E.O. McLean, H.J. Mederski, R.H. Miller, R.W. Miller, O.L. Musgrave, M.H. Niehaus, J.L. Parsons, D.A. Ray, G.J. Ryder, J.D. Sayre (Emeritus), B.L. Schmidt, L.N. Shepherd, P.E. Smith, E.W. Stroube, P. Sutton, G.S. Taylor, R.W. Teater, G.B. Triplett, D.M. Van Doren, R.W. Van Keuren, G.W. Volk, L.D. Wilding, C.J. Willard, W.T. Yamazaki, R.E. Yoder (Emeritus)
Associate Professors: A.J. Baxter, D.W. Bone, K.R. Everett, G.F. Hall, F.E. Heft, W.H. Schmidt, J.R. Vimmerstedt, A.C. Waldron, J.D. Wells, J.H. Brown
Assistant Professors: A.L. Barta, J.E. Beuerlein, F.W. Chichester (Adjunct), E.H. Derickson (Adjunct), W.M. Edwards, R.H. Follett, R. Geottemoeller (Adjunct), N W. Hopper, D.L. Jeffers, M.L. Kroetz, D.L. Lindell, T.J. Logan, D.P. Martin, N.E. Smeck, J.G. Streeter, J.F. Trierweiler, J.T. Underwood
Instructors: D.G. Alsdorf, J.R. Donelson (Adjunct), D. H. Donelson (Adjunct), N.R. Fausey (Adjunct), M. M. Gahn (Adjunct), J.L. Felton, D.K. Myers, M.H. Warner, J.H. Wilson
Teaching Fellows: 6
Research Assistants: 15
Graduate Students: 45

Field Stations/Laboratories: North Western, Hoytville, Ohio, Western - Springfield, E.O.R.D.C. Coldwell, Ohio, Mahoning, Canfield, Ohio, Muckcrops - Willard, Ohio, Southern Branch, Ripley, Ohio

Degree Programs: Agronomy, Bacteriology, Conservation, Crop Science, Ecology, Farm Crops, Microbiology, Plant Breeding, Plant Industry, Plant Sciences, Seed Investigations, Soil Science, Soils and Plant Nutrition, Weed Control

Undergraduate Degree: B.S.
Undergraduate Degree Requirements: 208 credits

Graduate Degrees: M.S., Ph.D.
Graduate Degree Requirements: Master: 45 to 75 credits Doctoral: 90 to 120 credits

Graduate and Undergraduate Courses Available for Graduate Credit: Crop Production in Developing Countries, Field Crop Breeding, Pedology and Edaphology, Fieldwork, Individual Studies, Group Studies, Field Crop Ecology, Principles of Grassland Management, Crop Physiology and Production, Principles of Turfgrass Management, Agroclimatology, Tropical and Subtripical Soils, Advanced Soil Classification, Morphology and Genesis, Soil Microbiology, Soil Fertility, Soil Physics, Chemistry of Soils and Fertilizers, Organic Soils, Soil Mineralogy, Radioactive Tracers (all U and G), Physiological and Biochemical aspects of Herbicides, Advanced Crop Breeding, Soils of the Cold Region, Chemistry of Soil Organic Matter, Soil Plant Relationships, Advanced Soil Physics, Physical Chemistry of Soils, Research Principles and Techniques, Experimental Design, Seminars, Research

ANIMAL SCIENCE DEPARTMENT
Phone: Chairman: (614) 422-6401 Others: 422-6402

Chairman: G.R. Johnson
Professors: D.S. Bell (Emeritus), V.R. Cahill, J.H. Cline, B.A. Dehority, R.H. Grimshaw, W.R. Harvey, G.R. Johnson, E.W. Klosterman (Associate Chairman), R.M. Kittman (Dean), L.E. Kunkle (Emeritus), T.L. Ludwick, C.M Martin, A.L. Moxon, H.W. Newland, H.W. Ockerman, C.F. Parker, R.L. Preston, R.R. Reed, T.S. Sutton (Dean Emeritus), L.A. Swiger, W.J. Tyznik, B.D. VanStavern, W.G. Venzke, G.R. Wilson, R.F. Wilson
Associate Professors: P.G. Althouse, H.M. Barines, W. H. Bruner (Emeritus), H.E. Goldstein (Adjunct), J.K. Judy, R.F. Plimpton, Jr.
Assistant Professors: C.R. Boyles, R.J. Borton, D.B. Gerber, A.P. Grifo, Jr., C.A. Hutton, G.A. Isler, D. C. Mahan, K.E. McClure, N A. Parrett, D.L. Reed, L.G. Sanford, R.O. Smith, W.W. Wharton, E.E. Zorn
Research Assistants: 12
Technical Assistants: 6

Field Stations/Laboratories: Ohio Agricultural Research and Development Center-Wooster, Ohio, Western Branch - S. Charleston, Ohio, Eastern Ohio Research and Development Center, Caldwell, Ohio

Degree Programs: Agriculture, Animal Science, Agriculture, Agricultural Science, Agricultural Industry, Food Technology, Nutrition, Journalism, International Agriculture, Nutrition, Animal Breeding, Physiology, Meat Science

Undergraduate Degree: B.S. (Agriculture)
Undergraduate Degree Requirements: B.S. 196 quarter hours - each program has specific requirements

Graduate Degrees: M.S., Ph.D. (Animal Science)
Graduate Degree Requirements: Master: 45 hours graduate courses cummulative point-hour ratio of at least 3.0 (B), Comprehensive final examination Doctoral: 90 Graduate Credit hours, cumulative point hour ratio of at least 3.0 (B); General Examination, Dissertation

Graduate and Undergraduate Courses Available for Graduate Credit: Physiology of Lactation (U), Physiology of Reproduction and Growth (U), Laboratory in Reproductive Physiology and Artificial Insemination (U), Nutrition and Feeding of Monogastric Animals (U), Nutrition and Feeding of Ruminant Animals (U), Advanced Meat Technology (U), Laboratory Analysis of Meat Products (U), Individual Studies (U), Group Studies (U), Advanced Reproductive Physiology (U), Genetics of Animal Populations (U), Advances in Physiology of Domestic Animals , Current Topics in Animal Genetics, Advanced Studies in Nutrition, Interdepartmental Seminar in Nutrition and Food Technology, Research

DEPARTMENT OF HORTICULTURE
Phone: Chairman: (614) 422-1800 Others: 422-6446

Chairman: H.A. Rollins, Jr.
Professors: E.K. Alban, G.A. Cahoon, J.L. Caldwell, J.F. Gallander, J.R. Geisman, W.A. Gould, F.O. Hartman, R.G. Hill, Jr., M. Kawase, D.C. Kiplinger, D.W. Kretchman, H.A. Rollins, Jr., H.K. Tayama, E.C. Wittmeyer
Associate Professors: S.Z. Berry, W.M. Brooks, W.L. George, Jr., P.C. Kozel, A.C. Peng, E.M. Smith, Jr., G.L. Staby, J.D. Utzinger
Assistant Professors: W.L. Bauerle, D.E. Crean, D.C. Ferree, T.A. Fretz, R. Blake, A.R. Mosley, W.A. Oitto, R C. Smith, E.J. Stang, T.D. Sydnor
Research Assistants: 15

Field Stations/Laboratories: Ohio Agricultural Research and Development Center - Wooster, Ohio

Degree Program: Horticulture

Undergraduate Degree: B.S.
Undergraduate Degree Requirements: Common Requirements: Physical Education - 3 hours, University College 100 - 1 hours, English 100 - 1 hour, Humanities Option - 15 hours, Social Science Option - 15 hours, Communications Option - 6-10 hours, National Defense Option - 12 hours, Other Basic Requirements - 30-51 hours, Core - 30-43 hours, Specific - 40 hours, Electives - 1-30 hours, Total credit hours required for graduation - 196

Graduate Degrees: M.S., Ph.D.
Graduate Degree Requirements: Master: Plan A - 45 hours, Graduate course work plus thesis Plan B - 50 hours Graduate course work including "special problems" Doctoral: Minimum of 135 quarter hours of graduate course work beyond the baccalaureate degree - dissertation embodying results of an original investigation - general comprehension examination - final oral examination.

Graduate and Undergraduate Courses Available for Graduate Credit: Horticultural Plant Breeding (U), Post-Harvest Physiology of Horticultural Crops (U), Weed Control in Horticultural Crops (U), Tropical and Subtropical Fruit and Vegetable Production (U), Greenhouse Environmental Control (U), Commercial Floriculture (U), Commerical Floriculture II (U), Commercial Floriculture III (U), Unit Operations in Processing Fruits (U), Unit Operations in Processing Fruits II (U), Unit Operations in Processing Fruits III (U), Analysis of Fruits Vegetables and Related Products (U), Greenhouse Vegetable Crops (U), Advanced Vegetable Crops (U), Arboriculture (U), Management of Nursery and Garden Store Operations (U), Physiology of Ornamental Plants (U), Food Regulation and Product Examination (U), Group Studies in Processing (U), Advanced Plant Nutrition I (U), Advanced Plant Nutrition II (U)

FOOD SCIENCE AND NUTRITION DEPARTMENT
Phone: Chairman: (614) 422-6281

Chairman: T. Kristoffersen
Professors: J.B. Allred, W.J. Harper, T. Kristoffersen, W.L. Slatter
Associate Professors: J.L. Blaisdell, P.M T. Hansen, R.E. Jenkins (Adjunct), E.M. Mikolajcik, R.B. Watts (Adjunct)
Assistant Professors: J.R. Chipley, R.B. Holtz, R.V. Josephson, J.P. Kenyon, J.B. Lindamood
Teaching Fellows: 2
Research Assistants: 14
Research Associate: 1
Technical Assistant: 1

Field Stations/Laboratories: Ohio Agricultural Research and Development Center - Wooster, Ohio

Degree Programs: Agriculture, Food Science, Nutrition

Undergraduate Degree: B.S. (Agriculture)
Undergraduate Degree Requirements: 196 hours of specified courses according to program

Graduate Degrees: M.S., Ph.D.
Graduate Degree Requirements: Master: Plan A: Thesis Minimum of 45 credit hours of which 10-15 hours in research Plan B. Non-Thesis Minimum of 50 credit hours Doctoral: Minimum of 90 credit hours of which approximately 30 hours in Dissertation research

Graduate and Undergraduate Courses Available for Graduate Credit: Food Additives (U), Food Sanitation (U), Food Systems (U), Food Contaminants and Toxicants (U), Food Structure and Interaction (U), Food Processing Wastes (U), Food Thermodynamics (U), Advanced Nutrient Utilization (U), Sporeforming Bacteria in Foods (U), Food Fermentations, Food Lipids, Food Proteins, Research Methods, Advanced Topics in Nutrition

SCHOOL OF NATURAL RESOURCES
Phone: Chairman: (614) 422-1279 Others: 422-2265

School Director: R.W. Teater
Professors: R.D. Cowen, G.E. Gatherum, E.E. Good, C.S. Johnson, H. Schick, R.D. Touse, H.B. Kriebel, M.M. Larson
Associate Professors: W.T. Momot, R.E. Roth, T.M. Stockdale, J.H. Brown, J.P. Vimmerstedt, A.R. Vogt, F.W. Whitmore
Assistant Professors: M.L. Bowman, J.F. Disinger, P.A. Hackney, T.W. Townsend, K.A. Wenner, J.H Wheatley, D.B. Houston
Instructors: N.J. Andrew, T.B. Flickinger, T.R. Mitchell, J.M. Pierce
Visiting Lecturers: 2
Teaching Fellows: 12
Research Assistants: 6

Field Stations/Laboratories: Barnebey Center - Lancaster, Ohio, OARDC, Wooster, Ohio

Degree Programs: Forest Industrial Management, Forest Resource Management, Forest Science, Wildlife Management, Fisheries Management, Interpretive Work, Parks and Recreation Administration, Natural Resources Communication, Resource Development

Undergraduate Degree: B.S.
Undergraduate Degree Requirements: 146 credit hour s with from 40 to 60 specified credits according to selected major.

Graduate Degrees: M.S., Ph.D.
Graduate Degree Requirements: Master: Thesis option - 45 quarter hours of which 15 may be thesis credit. Non-thesis option - 55 quarter hours, two papers produces as individual studies. Comprehensive examination
Doctoral: Courses as specified by major advisor, thesis

Graduate and Undergraduate Courses Available for Graduate Credit: Introduction to Conservation of Natural Resources (U), Conservation Agencies (U), Outdoor Recreation in the U.S.A. (U), Work Experience in Natural Resources (U), Natural History of Ohio (U), Land Economics (U), Principles of Park and Recreation Management (U), Natural Resources Problems, Programs, and Policies (U), Interactions in Natural Resources Management (U), Field Course in Conservation and Outdoor Education (U), Management of Fisheries (U), Principles of Wildlife Management (U), Field Laboratory in Renewable Natural Resources Management (U), Urban Parks and Recreation Management (U), Outdoor Recreation by Private Enterprise (U), Park Design (U), Internships in Natural Resources Professions (U), Workshop in Environmental Education(U), Individual Studies in Natural Resources (U), Simulation in Natural Resources Management (U), Research Methods in Natural Resources Management (U), Program Development in Environmental Education, Policies Relating to Governmental Recreation Areas, Coniferous Dendrology (U), Hardwood Dendrology (U), Silvics (U), Principles of Forestry (U), Silviculture (U), Forest Mensuration (U) Forest Management (U), Wood Structure and Properties (U), Manufacturing Forest Products (U), Analysis of Forest Industry Management (U)

DEPARTMENT OF PARKS AND RECREATION
Phone: Chairman: (614) 422-1600

Chairman: H. Schick
Assistant Professor: K. Wenner
Instructor: N. Andrew, T. Flickinger, J.M. Pierce
Teaching Fellows: 2
Research Assistants: 2

Field Stations/Laboratories: Barnebey Center - Lancaster, Ohio

Degree Program: Natural Science

Undergraduate Degree: B.S.
Undergraduate Degree Requirements: 196 quarter hours minimum, 2.0 cumulative point average (based on 4.0 point system), minimum 40 hours in major specialization, 30-36 core courses in natural resources, 89-91 Basic University Courses, 29-37 electives

Graduate Degrees: M.S., Ph.D.
Graduate Degree Requirements: Master: 45 hours Plan A (Thesis), 55 hours Plan B (Non-thesis), minimum of 2.7 cumulative point to enter, graduate school plus acceptance by graduate committee
Doctoral: (only offered on interdisciplinary basis)

Graduate and Undergraduate Courses Available for Graduate Credit: Natural Resources (U), Governmental Policies and Natural Resources, Natural Resources 897, Seminar in Natural Resources, Natural Resources 642 (U), Urban Parks and Recreation, Natural Resources 643 (U), Private Recreation, Natural Resources 644 (U), Park Design Natural Resources 540 (U), Parks and Recreation Management, Natural Resources 840, Parks and Recreation Policies Natural Resources 610 (U), Interpretive Work, City and Regional Planning, Open Space Recreation, Geography (U), Urban Geography, Business Administration (U), Agricultural Engineering (U), Agriculture Forest Hydrology (U), Public Relations Law, Conservation Law, Work and Leisure, Political Science, Politics and Policy Making

DEPARTMENT OF PLANT PATHOLOGY
Phone: Chairman: (614) 422-1865 Others: 422-1375

Chairman: A.F. Schmitthenner
Professors: I.W. Deep, C.W. Ellett, B.F. Janson, C. Leben, A.F. Schmitthenner, L.S. Dochinger, C.L. Wilson (Adjunct)
Associate Professors: O.E. Bradfute, J.D. Farley, M.O. Garraway, D.T. Gordon, L.J. Herr, H.A.J. Hortink, P.O. Larsen, R.M. Riedel, A.W. Troxel, T.C. Weidensaul, R. Louie, H.L. Porter (Adjunct)
Assistant Professors: B.M. Jones, C.C. Powell Jr., R.E. Gingery (Adjunct)
Teaching Fellows: 1

Degree Programs: Plant Pathology

Undergraduate Degree: B.S.
Undergraduate Degree Requirements: 196 credit hours

Graduate Degrees: M.S., Ph.D.
Graduate Degree Requirements: Master: 45 credit hours including thesis. Doctoral: 135 credit hours including dissertation.

Graduate and Undergraduate Courses Available for Graduate Credit: Advanced Plant Pathology (U), Ornamentals Plant Pathology (U), Fruit and Vegetable Crop Diseases (U), Field Crop Diseases (U), Field Plant Pathology (U), Physiology of Parasitism, Bacterial Pathogens, Plant Virology, Plant Nematology, Principles of Plant Pathology

DEPARTMENT OF POULTRY SCIENCE
Phone: Chairman: (614) 422-4821

Chairman: E.C. Naber
Professors: R.L. Baker, E.H. Bohl, K.I. Brown, P.C. Clayton, W.R. Harvey, R.G. Jaap, G.A. Marsh,

E.C. Naber, J.F. Stephens
Associate Professors: J.B. Allred, W.L. Bacon, K.E. Nestor, Y.M. Saif
Assistant Professors: J.R. Chipley, J D. Latshaw, A.H. Cantor
Research Assistants: 5

Field Stations/Laboratories: The Ohio Agricultural Research and Development Center, Wooster, Ohio

Degree Programs: Poultry Science

Undergraduate Degree: B.S.
Undergraduate Degree Requirements: The minimum course requirements for a major in poultry science consist of 25 credit hours in Poultry Science 610,611,630,640, and 650, taken as a sequence during the junior and senior years. These courses provide the scientific and technical basis for employment in the poulty industry, as well as a base for graduate study in the Department

Graduate Degrees: M.S., Ph.D.
Graduate Degree Requirements: Master: Defined by the Graduate Committee of the Department. Doctoral: Defined by the Graduate Committee of the Department

Graduate and Undergraduate Courses Available for Graduate Credit: Avian Growth and Meat Production, Avian Reproduction and Egg Production, Nutrition and Feeding of Monogastric Animals, Prevention and Control of Avian Diseases, Egg and Poultry Products Technology Genetics of Animal Populations, Advances in Physiology of Domestic Animals, Current Topics in Animal Genetics, Advanced Studies in Nutrition

DEPARTMENT OF VETERINARY SCIENCE
Phone: Chairman: (216) 264:1021

Chairman: L.C. Ferguson
Professors: E.H. Bohl, R.F. Cross, C.K. Smith
Associate Professors: E.M. Kohler, P.D. Moorhead, Y. M. Saif
Assistant Professors: D.R. Redman, K.L. Smith
Instructors: J.E. Jones
Research Assistants: 10

Graduate Degree: (Thesis Research done in Department in conjunction with Ohio State University degree programs)

College of Medicine
Dean: H.G. Cramblett

DEPARTMENT OF ANATOMY
Phone: Chairman: (614) 422-4761 Others: 422-4831

Chairman: R.S. St. Pierre
Professors: G.A. Ackerman, R.C. Baker (Emeritus), I. Eglitis, J A. Eglitis (Emeritus), G.R.L. Gaughran, G. F. Martin, R.L. St. Pierre
Associate Professors: R.L. Beran, J.M Delphia, J. Gersten (Emeritus), T.G. Hayes, A.O. Humbertson, J.S. King, M.E. Sucheston, D.G. Vernall, B.L. Wismar
Assistant Professors: G.D. Boston, D.L. Clark, R.M. Folk (Adjunct), J.R. Hostetler, R.K. Hostetter, J.A. Negulesco
Instructors: M.H. Hines
Visiting Lecturers: 4
Research Assistants: 6

Degree Programs: Anatomy

Graduate Degrees: M.S., Ph.D.
Graduate Degree Requirements: Master: Plan A - 45 hours course work plus thesis; Plan B - 50 hours course work plus comprehensive examination. Doctoral: 135 hours course work plus acceptable dissertation and passing of final oral examination

Graduate and Undergraduate Courses Available for Graduate Credit: Clinical Anatomy, Human Anatomy, Histology, Applied Anatomy, Mammalian Histology, Human Gross Anatomy, Human Developmental Anatomy, Human Neuroanatomy, Anatomy of the Visual System, Medical Education, Anatomical Techniques, Seminar in Anatomy, Individual Studies in Anatomy, Advanced Studies in Anatomy: Blood and Hemopoiesis, Connective Tissue and Bone, Embryology, Microscopic Anatomy, Neuroanatomy, Epithelium, Ultrastructure of Central Nervous System, Electron Microscopy, Principles of Human Cytogenetics, Design of Computer Teaching Programs, Anatomy of Newborn

DEPARTMENT OF MEDICAL MICROBIOLOGY
Phone: Chairman: (614) 422-5525

Chairman: F.A. Kapral
Professors: B.U. Bowman, H.G. Cramblett, V.V. Hamparian, R.E. Haynes, F.A. Kapral, R.W. Lang, C.R. Macpherson, A.C. Ottolenghi, R.L. Perkins, S. Saslaw, N.L. Somerson
Associate Professors: P.H. Azimi, R.J. Fass, J.D. Pollack
Assistant Professors: J.H. Hughes, D.C. Thomas
Instructors: A.M Durham
Research Assistants: 13

Degree Programs: Medical Microbiology

Graduate Degrees: M.S., Ph.D.
Graduate Degree Requirements: Master: Medical Microbiology, Immunology, Virology, seminar, Physiological Chemistry research, (Clinal Microbiology curriculum, also mycology, diagnosis of bacteria, parasitology); Doctoral: Above plus Mycoplasma Richettsia Chlamycia, Dynamics of bacteria infection, (Clinal Microbiology, curriculum above master requirement, plus diagnosis of virology, antibiotics, dynamics of bacterial infection)

Graduate and Undergraduate Courses Available for Graduate Credit: Medical Microbiology, Medical Immunology, Clinical Medical Mycology, Medical Virology, Bacteriophagy, Individual Studies, Seminar, Medical Parasitology, Experimental Medical Microbiology, Mycoplasma Richettsia Chlamydia, Molecular Basis Antibiotic Action, Dynamic Aspects of Bacterial Infection, Group Studies, Research

DEPARTMENT OF PHYSIOLOGY
(No reply received)

Division of Environmental Education
124 West 17th Avenue Phone: (614) 422-5589
Columbus, Ohio 43210
Dean: R.W. Teater
Dean of Graduate Studies: G. Gatherum

DEPARTMENT OF FORESTRY
Phone: Chairman: 422-2816 Others: 422-2817

Chairman: G.E. Gatherum
Professors: J.H. Brown, W.F. Cowen, Jr., G.E. Gatherum, H.B. Kriebel, M.M. Larson, R.D. Touse
Associate Professors: J.D. Kasile, J.P. Vimmerstedt, A.R. Vogt (Associate Chairman), F.W. Whitmore
Assistant Professors: D.B. Houston, J.R. McClenahen
Instructor: T.R. Mitchell
Research and Teaching Assistants: 10

Field Stations/Laboratories: Armstrong Memorial Forest - Wayne County, Ohio, Pomerene Forest Laboratory - Coshocton, Ohio, Malabar Farm (Richland County), Ohio, Barnebey Center for Environmental Studies - Fairfield County, Ohio

Degree Programs: Forest Resource Management, Forest Industries Management, Forest Science, Forest Ecology, Forest Genetics, Tree Physiology, Forest Soils

Tree Nutrition, Forest Biometry

Undergraduate Degree: B.S. (Natural Resources)
Undergraduate Degree Requirements: 196 hours formal course work comprised of: a. 98-102 hours - Basic Requirements, b. 32-33 hours - Core Requirements, c. 40-Major Specialization d. 21-26 hours - Free electives

Graduate Degrees: M.S., Ph.D. (Natural Resources)
Graduate Degree Requirements: Master: 45 hours formal course work and Thesis Doctoral: Interdisciplinary with Departments of Botany, Genetics, or Agronomy. 135 hours formal course work and Dissertation

Graduate and Undergraduate Courses Available for Graduate Credit: Forestry 593, Individual Studies - Special Problems in Field of Forestry and Forest Products (U), Natural Resources, Forestry Internships in Natural Resources Professions, Natural Resources, Workshop in Environmental Education, Natural Resources, Individual Studies in Natural Resources, Group Studies on the Nature and Management of Natural Resources Encompassed in Forestry, Natural Resources, Research for thesis or dissertation purposes only.

DIVISION OF ENVIRONMENTAL EDUCATION
Phone: Chairman: (614) 422-5589 Others: 422-0959

Chairman: C.S. Johnson
Professor: C.S. Johnson
Associate Professor: R.E. Roth
Assistant Professor: M.L. Bowman, J.F. Disinger, J.F. Wheatley
Instructor: S.L. Frost
Visiting Lecturer: 1
Teaching Fellows: 5

Field Stations/Laboratories: Barnebey Center for Environmental Studies - Lancaster, Ohio

Degree Programs: Interpretive work, Environmental Education, Environmental Communities

Undergraduate Degree: B.S. (Natural Resources)
Undergraduate Degree Requirements: 196 credits of which 48 must be in specified core courses and 50 in the field of major specialization.

Graduate Degree: M.S. (Natural Resources)
Graduate Degree Requirements: Master: Thesis option - 45 hours of which 15 may be in thesis. Non thesis option 55 hours with two papers, one an individual study comprehensive examination

Graduate and Undergraduate Courses Available for Graduate Credit: Natural History of Ohio (U), Natural Resources Problems, Programs, and Policies (U), Interactions in Resource Management (U), Workshop in Environmental Education (U), Internship in Environmental Education (U), Workshop in Environmental Education (U), Group Studies in Conservation and Outdoor Education (U), Simulation in Resource Management (U), Program Development in Environmental Education.

OHIO UNIVERSITY
Athens, Ohio 45701 Phone: (614) 594-7761
Dean of Graduate Studies: N.S. Cohn
Dean of College: J.G. Jewett

DEPARTMENT OF BIOLOGY
Phone: Chairman: (614) 594-3677 Others: 594-7761

Chairman: C.E. Miller
Professors: N.S. Cohn (Distinguished), C.E. Miller, I.A. Ungar, W.A. Wistendahl
Associate Professors: A.H. Blickle, J.C. Cavender, J. H. Graffius, M.J. Jaffee, L.A. Larson
Assistant Professors: J.P. Braselton, R.M Lloyd, J.P. Mitchell, I.K. Smith
Research Assistants: 2
Teaching Assistants: 21

Degree Programs: Biochemistry, Cell Biology, Ecology, Morphology, Mycology, Paleobotany, Physiology, Pteridology and Systematics

Undergraduate Degree: B.S. (Botany)
Undergraduate Degree Requirements: A.B. minimum 45 hours, B.S. minimum 55 hours; B.S. and A.B. candidates must complete Biology Principles, Plant Biology, Algae and Mosses, Vascular Plant Morphology, Ohio Flora, Biology of Fungi, Biology Discussions, Plant Physiology, Plant Ecology, Cytology, Speciation and Evolution, General Zoology, Genetics, Inorganic and Organic Chemistry, Physics, Mathematics through Introduction to Calculus, 10 hours English, Arts and Sciences College Requirements.

Graduate Degrees: M.S., Ph.D.
Graduate Degree Requirements: Master: M.S. degree minimum 45 hours (30 hours courses, 15 hours research). A research thesis (M S.) resulting from original research is required. A non thesis terminal M.S. degree is also available. All graduate students are required to teach a minimum of two quarters during their tenure in the department. Doctoral: M.S. or M.A. Degree or Masters requirements as above plus Biochemistry and proficiency in one foreign language (French, German, Russian or other, depending on research needs). All graduate students in Botany are required to teach a minimum of two quarters during their tenure in the department. A dissertation resulting from original research is required.

Graduate and Undergraduate Courses Available for Graduate Credit: Plant Physiology, Plant Ecology, Vegetation Analysis, Paleobotany, Molecular Genetics, Instrumentation and Techniques, Phycology, Biology of Fungi, Thesis, Cytology, Plant Biochemistry, Topics in Cell Biology, Plant Cytology, Cell Division and Differentiation, Ultrastructure, E.M. Techniques, Developmental Physiology, Experimental Ecology, Topics in Plant Ecology, Plant Systematics, Topics in Fungi, Aquatic Phycomycetes, Ascomycetes, Cellular Slime Molds, Advanced Topics in Botany, Histochemistry, Advanced Biochemistry, Vascular Morphology, Radiation Biology, Botanical Pedagogy, Supervised Study, Pteridology, Soil Microbiology, Seminar, Research, Dissertation

DEPARTMENT OF ZOOLOGY AND MICROBIOLOGY
Phone: Chairman: (614) 595-5344

Chairman: R.J. Downey
Professors: R.J. Downey, J.B. Lawrence, J.M. McQuate, W.J Peterson, H.J. Seibert
Associate Professors: C.G. Atkins, F.C. Hagerman, O.B. Heck, R.S. Hikida, W.D. Hummon, P. Jones, S. Maier, W.S. Romoser, J S. Rovner, R.V. Walker, J.A. Wilson, W. Witters
Assistant Professors: J. Svendsen, C.H. Page
Instructors: B. Allen, J. Gault
Visiting Lecturers: 1

Field Stations/Laboratories: Dysart Woods and Dow Lake Field Station

Degree Programs: Microbiology, Medical Technology, Zoology

Undergraduate Degrees: A.B., B.S.
Undergraduate Degree Requirements: The major requirement for the A.B. and B.S. Degrees is a minimum of 36 and 45 quarter hours respectively in approved departmental courses. At least 15 hours must be numbered 400 or above.

Graduate Degrees: M.S., Ph.D.

Graduate Degree Requirements: Master: 45 quarter hours A non-thesis masters program is open to secondary and junior college teachers. Doctoral: 135 quarter hours in Zoology or Microbiology and approved related courses. Proficiency with two foreign languages (Computer may substitute for one)

Graduate and Undergraduate Courses Available for Graduate Credit: Comparative Vertebrate Anatomy (U), Histology (U), Human Genetics (U), Invertebrate Zoology (U), Protozoology (U), Entomology (U), Medical Entomology (U), Parasitology (U), Helminthology, Physiology of Exercise, Cell Physiology (U), Endocrinology Ethology, Population Biology, Evolution (U), Comparative Physiology, Comparative Neurophysiology, Neurophysiological Techniques, Microbial Genetics, Virology, Immunology, Electron Microscopy, Biological Ultrastructure, Microbial Physiology, Pathogenic Bacteriology, Immunochemistry, Molecular Genetics, Advanced Topics, Research, Dissertation

OHIO WESLEYAN UNIVERSITY

Delaware, Ohio 43015 Phone: (614) 369-4431
Dean of College: J. Chase

DEPARTMENT OF BOTANY AND BACTERIOLOGY
Phone: Chairman: Ext. 251

Chairman: A.A. Ichida
Professors: G.W. Burns, A.A. Ichida, B.R. Roberts (Adjunct), C.E. Seliskar (Adjunct), E.B. Shirling, D. Worley (Adjunct)
Associate Professors: J.E. Sanger
Assistant Professors: J.M. Decker,
Instructor: J.M. Ichida
Visiting Lecturer: 1
Research Assistants: 1
Department Assistant: 1

Degree Programs: Botany, Bacteriology

Undergraduate Degree: B.A.
Undergraduate Degree Requirements: a sequence of distribution requirements and a major in a department or interdepartmental program.

DEPARTMENT OF ZOOLOGY
Phone: Chairman: Exts: 253, 254 Others: Exts. 253,254

Chairman: W.K. Patton
Professors: A.S. Bradshaw, W.K. Patton, W.D. Stull
Associate Professor: A.E. Fry
Assistant Professors: J.M. Freed, D.C. Radabaugh

Degree Program: Zoology

Undergraduate Degree: B.A.
Undergraduate Degree Requirements: General Zoology (2 terms), Cellular Physiology, Comparative Anatomy or Devel opmental Biology, one additional Zoology course, Calculus (2 terms), Physics (2 terms), Chemistry through Organic (5 terms), OR General Zoology (2 terms), eight additional Zoology courses with one from each of five groups, Plant Kingdom and an additional Botany or Geology course, Chemistry (2 terms)

OTTERBEIN COLLEGE

Westerville, Ohio 43081 Phone: (614) 891-3117
Academic Dean: R. Turley

DEPARTMENT OF LIFE SCIENCE
Phone: Chairman: (614) 891-3117

Chairman: J. Willis
Professor: J. Willis
Associate Professors: M. Herschler, A. Leonard, G. Phinney, T. Tegenkamp
Instructor: N. LeVora

Field Stations/Laboratories: Small Outdoor Laboratory on Flood Plain of Alum Creek, Westerville, Ohio

Degree Program: Biology

Undergraduate Degree: B.A.
Undergraduate Degree Requirements: Biosphere I and II, Genetics, Cellular Physiology, Systems Physiology, Ecology, Microbiology and Morphogenesis I and II

RIO GRANDE COLLEGE

(No reply received)

UNIVERSITY OF AKRON

Akron, Ohio 44325 Phone: (216) 375-7155
Dean of Graduate Studies: D. Griffin
Dean of College: R.A. Oetjen

DEPARTMENT OF BIOLOGY
Phone: Chairman: (216) 375-7155

Chairman: D. Jackson
Professors: D. Jackson, R. Keller, N. Ledinko, L.W. Macior
Associate Professors: E. Flaumenhaft, R. Nokes, D. Nunn, J. Olive, W. Sheppe, W.P. Stoutamire
Assistant Professors: H. Dollwet, J. Frola, J. Gwinn, R. Mostardi, S. Orcutt, D. Ott, B. Richardson, L. Watson
Graduate Teaching Assistants: 22

Degree Programs: Biology, Medical Technology

Undergraduate Degree: B.S.
Undergraduate Degree Requirements: 192 quarter hours with 54 hours of Biology included in that; 2 years of Chemistry, and 2 years of a foreign language

Graduate Degree: M.S.
Graduate Degree Requirements: Satisfactory completion of a total of at least 45 graduate quarter hours, with an overall grade point average of a minimum of 3.0 Completion of a comprehensive Qualifying Examination Committee approval of the Master's Thesis. Reading proficiency in a second language appropriate to the major field of study.

Graduate and Undergraduate Courses Available for Graduate Credit: Plant Development (U), Plant Physiology (U), Plant Anatomy (U), Mycology (U), Phycology (U), Plant Morphology (U), Plant Morphology (U), Population Ecology (U), Limnology (U), Applied Aquatic Ecology (U), Physiology of the Fungi (U), Comparative Physiology (U), Bacterial Physiology (U), Pathogenic Bacteriology (U), Immunology (U), Virology (U), Developmental Anatomy (U), Vertebrate Zoology (U), Advanced Genetics (U), Biological Problems (U), Biology of Behavior (U), Radiation Biology (U), Plant Biosystematics (U), Human Physiology (U), Endocrinology (U), Laboratory Animal Management (U), Experimental Microbial Physiology, Cytology, Experimental Embryology, Environmental Physiology

UNIVERSITY OF CINCINNATI

Clifton Avenue Phone: (513) 475-8000
Cincinnate, Ohio 45221
Dean of Graduate School: G. Stern

McMicken College of Arts and Sciences
Dean: C. Crockett

DEPARTMENT OF BIOLOGICAL SCIENCES
Phone: Chairman: (513) 475-4231 Others: 475-8000

Head: B.L. Umminger (Acting)
Dieckmann Professor of Forestry: J.L. Caruso
Professors: F.J. Etges, A.S. Fraser, J.L. Gottschang

Associate Professors: J. Beatley, A. Butz, J.L. Caruso, L.C. Erway, D.H. Gist, C.A.Huether, B.L. Umminger, C. Whitney, G.D. Winget
Assistant Professors: W.F. Duggleby, T.C. Kane, E. Kaneshiro, S.J. Keller, R.R. Meyer, M. Miller, A.J. Mukkada, J.A. Snider, J. Trela, J.R. Vestal
Instructor: R. Hehman
Professors Emeriti: W.A. Dreyer, M. Fulford
Research Assistants: 2

Degree Programs: General Biology, Cell Biology, Developmental Biology, Physiology, Ecology, Systematics-Evolution

Undergraduate Degrees: B.S. (Biological Sciences)
Undergraduate Degree Requirements: 33 quarter credits in Chemistry (General, Organic, Chemistry Analysis), 15 quarter credits in Physics (Introductory), 6 quarter credits in Introductory Calculus, 46 quarter credits in Biology

Graduate Degrees: M.S., Ph.D.
Graduate Degree Requirements: Requirements for M.S. and Ph.D. vary according to program. Biochemistry (9 credits) common to most programs. Minimum credits: M.S. --- 45 , Ph.D. --- 135

Graduate and Undergraduate Courses Available for Graduate Credit: Cell Biology (U), General Bacteriology (U), Genetics (U), Human Genetics (U), Physiology (U), Developmental Biology (U), Evolution (U), Ecology (U), General Entomology (U), Insect Physiology (U), Insect Ecology (U), Vertebrate Zoology (U), Limnology (U), Experimental Plant Morphology (U), Environmental Physiology (U), Microbial Ecology (U), Comparative Endocrinology (U), Parasitology (U), Helminthology (U), Molecular Biology (U), Plant Ecology (U), Ecological Genetics, Population Dynamics, Advanced Plant Physiology, Molecular Aspects of Developmental Biology, Morphology Aspects of Developmental Biology, Cell Biology Seminar, Physiology Seminar, Ecology Seminar, Metabolic Regulation

College of Medicine

Dean: R.S. Daniels (Interim)

DEPARTMENT OF ANATOMY

Phone: Chairman: (513) 872-5612 Others: 872-3000

Chairman: R.C. Crafts
Francis Brunning Professor of Anatomy: R.C. Crafts
Professors: R.T. Binhemmer, R.C. Crafts, F.M. Deuschle, J.L. Hall, H.A. Meineke, A.R. Vanderahe (Emeritus), J.L. Wilson
Associate Professors: G.C. Blaha, M.G. Menefee, R.S. McCuskey, R.L. Slemmer (Adjunct)
Assistant Professors: C.R. Bason (Adjunct), R.J. Niewenhuis, P. Tornheim, S.F. Townsend
Instructor: A.A. Lamperti
Teaching Fellows: 7
Research Assistants: 6

Degree Programs: Anatomy, Anatomy and Histology, Gross Anatomy, Microscopic Anatomy, Neuroanatomy, Developmental Anatomy

Graduate Degrees: M.S., Ph.D. (Anatomy)
Graduate Degree Requirements: Not Stated

Graduate and Undergraduate Courses Available for Graduate Credit: Microscopic Anatomy, Research Techniques, Gross Anatomy of the Body Wall and Cavities, Gross Anatomy of the Head and Neck, Gross Anatomy of the Upper and Lower Extremities, Neural Sciences, Advanced Anatomy, Research, Seminar, Applied Anatomy, Anatomy of the Nervous System, Correlative Microscopic Anatomy, Teratology, Applied Neuroanatomy, The Autonomic Nervous System, Reproductive Neuroendocrinology

DEPARTMENT OF BIOLOGICAL CHEMISTRY

Phone: Chairman: (513) 872-5643

Chairman: H. Rudney
Andrew Carnegie Professor of Biological Chemistry: H. Rudney
Professors: D.I. Crandall, R.A. Day (Adjunct), J.B. Liugrel, F.H. Mattson (Adjunct), H.G. Petering (Adjunct), H. Rudney, J.W. Vester
Associate Professors: N.C. Davis (Adjunct), J.H. Freisheim, R.C. Krueger, J. MacGee (Adjunct), C.L. Mendenhall (Adjunct), J.D. Ogle, A.J. Pesce (Adjunct), C.C. Smith (Adjunct)
Assistant Professors: R.W. Armentrout, W.D. Behnke, R.C. Boziam (Adjunct), I.W. Chen (Adjunct), R.E. Ganschow (Adjunct), B.C. Moulton (Adjunct), P.T. Russell (Adjunct)
Research Assistants: 16
Post Doctoral Fellows (Research): 8

Degree Program: Biological Chemistry

Graduate Degree: Ph.D.
Graduate Degree Requirements: Doctoral: 51 credits of Graduate course work, 84 credits of Ph.D. thesis research, 2 research proposals must be prepared, defended and accepted for admission to candidacy

Graduate and Undergraduate Courses Available for Graduate Credit: General Biochemistry, Molecular Genetics I, II, and III, Enzyme Catalysis and Kinetics, Protein Structure and Function Relationships I and II, Bioenergetics, and Control of Intermediary Metabolism

DEPARTMENT OF MICROBIOLOGY

Phone: Chairman: (513) 872-5664 Others: 872-3100

Head: H.C. Lichstein
Professors: P.F. Bonventre, H.C. Bubel, J.G. Michael, H.C. Lichstein
Associate Professors: J.C. Loper, M. Carsiotis
Assistant Professor: D.R. Lang
Instructor: J.F. Cicmanec
Research Assistants: 18

Degree Programs: Bacteriology, Immunology, Microbiology

Graduate Degrees: M.S., Ph.D.
Graduate Degree Requirements: Master: Courses plus thesis Doctoral: Courses plus dissertation

Graduate and Undergraduate Courses Available for Graduate Credit: Microbiology (U), Advanced Medical Microbiology, Advanced Immunology, Molecular Genetics, Microbial Physiology, Virology, Seminar, Special Problems, Research

DEPARTMENT OF PHYSIOLOGY

Phone: Chairman: (513) 872-5636 Others: 872-5637

Chairman: D.L. Kline
Joseph Eichberg Professor of Physiology: D.L. Kline
Professors: G. Eckstein (Emeritus), A. Brodish
Associate Professors: J. DiSalvo, W.W. Leavitt, P. Nathan, K.N.N. Reddy (Research Associate)
Assistant Professors: R.O. Banks, D.S. Faber, R.W. Lowenhaupt, R.R. Manalis
Instructors: H.C. Cheng

Degree Programs: Physiological Sciences

Graduate Degrees: M.S., Ph.D.
Graduate Degree Requirements: Master: If the student cannot proceed to the Ph.D. degree, he may be permitted to complete satisfactory research project and take required courses to achieve M.S. Degree. Doctoral: Needs B.A., M.S. or M.A., or M.D. to enter. After first 2 years of basic science, must pass comprehensive examination. Last 2 years are spent on thesis research and some teaching. Thesis

must be defended at open seminar.

Graduate and Undergraduate Courses Available for Graduate Credit: General and Mammalian Physiology, Cardiovascular Physiology and Pharmacology, Endocrinology, Renal Physiology, Fetal and Neonatal Physiology, Reproductive Physiology, Advanced Neurophysiology, Synaptic Physiology, Neoruendocrinology, Physiology for Graduate Students, Physiology Seminars, Physiology Research, Physiology of Tissue Transplantation, Menbrane Transport and Metabolism.

UNIVERSITY OF DAYTON

300 College Park Phone: (513) 227-2500
Dayton, Ohio 45469
Dean of Graduate Studies: J.A. Stander
Dean of College: L.A. Mann

DEPARTMENT OF BIOLOGY
Phone: Chairman: (513) 229-2525 Others: 229-2500

Chairman: G.B. Noland
Professors: J.J. Cooney, P.J. Faso, D.R. Geiger, O.C. Jaffee, R.A. Joly (Emeritus), G.B. Noland
Associate Professors: P.K. Bajpai, C.J. Chantell, D.A. Fleischman (Adjunct), R.C. Lachapelle, K.J. McDougall, J.M. Ramsey, G.D. Shay, G.L. Willis
Assistant Professors: A.J. Burky, J.D. Laufersweiler, F.D. Schwelitz, D.A. Taylor (Clinical), P.K. Williams
Visiting Lecturers: 2
Teaching Fellows: 20
Research Assistants: 4

Degree Programs: Biology, Medical Microbiology

Undergraduate Degree: B.S.
Undergraduate Degree Requirements: 1 year Mathematics (usually Calculus), 1 year Physics, Chemistry through Organic, Biology major requirements - usually 35 or so semester hours, total at least 70 hours Science, 50 hours Humanities, Arts, Etc.

Graduate Degrees: M.S., Ph.D.
Graduate Degree Requirements: Master: 30 semester hours minimum, must include 2 semesters Instrumentation, 2-4 special topic seminars, 2-4 departmental seminars, thesis, preliminary examination thesis defense Doctoral: As above plus more courses to fill in gaps, special topics seminar and department seminar every semester, dissertation, preliminary, candidacy and yearly examinations to level of competence

Graduate and Undergraduate Courses Available for Graduate Credit: Vertebrate Zoology, Radiation Biology, Biochemistry, Bacterial Physiology, Endocrinology, Cytology, Biochemical Genetics, Immunology, Advanced Microbiology, Cell Physiology, Comparative Animal Physiology, Experimental Embryology, Vertebrate Morphology, Community Ecology, Vertebrate Paleontology, Field Biology, Evolutionary Ecology, Biosystematics, Plant Physiology, Biometrics, Biological Instrumentation, Electron Microscopy, Clinical Topics, Physiological Advanced Genetics, Bacteriology, Pathogenic Bacteriology, Lower Plants, Higher Plants, Cell Biology, Embryology

UNIVERSITY OF TOLEDO

2801 W. Bancroft Street Phone: (419) 537-2065
Toledo, Ohio 43606
Dean of Graduate Studies: (Vacant)
Dean of College: A. Cave

DEPARTMENT OF BIOLOGY
Phone: Chairman: (419) 537-2065

Chairman: C.J. Smith
Professors: W.H. Jyung, M. Masatir
Associate Professors: H. Lee, D. Pribor, H.C. Shaffer, C.J. Smith, R. Shore, E.J. Tramer
Assistant Professors: W.L. Bischoff, Jr., C. Creutz, L. Glatzer, P. Fraleigh, S. Goldman, L.A. Jones
Teaching Fellows: 18
Research Assistants: 3

Degree Programs: Biology, Medical Technology

Undergraduate Degree: B.S.
Undergraduate Degree Requirements: The major requires 45 hours and consists of: (a) the two-year Biology core of 39 hours with an emphasis on the areas of Physiology, Ecology, Development and Genetics (b) 15 hours from the upper division courses, Additional requirements include one year of calculus, 1 year of Engineering Physics, and Chemistry through Organic Chemistry.

Graduate Degrees: M.S., Ph.D.
Graduate Degree Requirements: Master: Student must complete 45 quarter hours beyond the requirements for the bachelor's degree, 9 to 15 hours of research and thesis, 3 hours of graduate seminar, at least 4 courses in the Current Problems series. Doctoral: Normally, 135 quarter hours of study beyond the bachelor's degree are required; (a) 3 hours of graduate seminar, (b) the graduate core program (Current Topics series in Cell Biology, Genetics, Development, Physiology, Ecology and Molecular Biology), (c) 2 foreign languages (d) 1 year of teaching.

Graduate and Undergraduate Courses Available for Graduate Credit: Biophysics, Biostatistics, Advanced Genetics, Subcellular Aspects of Development, Environment and Plant Responses, Comparative Animal Physiology, Field Botany, Advanced Aq uatic Ecology, Principles of Immunology, Ecology of Eastern North America, Extramural Studies in Biology, Graduate Research. Current Problems in Cell Biology, Current Problems in Organismal Biology, Current Problems in Developmental Biology, Current Problems in Genetics, Seminars in Biology, Special Problems in Biology, Graduate Readings in Biology, Current Problems in Environmental Biology, Advanced Cell Biology

URBANA COLLEGE

College Way Phone: (513) 652-1301
Urbana, Ohio 43078
Dean of College: B.L. Cooper

BIOLOGY DEPARTMENT, INTEGRATED SCIENCE AREA
Phone: Chairman: Ext. 334

Chairman: C.M. Frederick
Associate Professors: J.K. Detling, C.M. Frederick
Instructor: D.E. Fulton
Research Assistants: 5

Degree Program: Biology

Undergraduate Degrees: B.A., B.S.
Undergraduate Degree Requirements: Biology major, 60 quarter hours, with at least 35 upper level credit hours, Chemistry, Physics, and Mathematics

URSULINE COLLEGE

(No reply received)

WALSH COLLEGE

2020 Easton Street Phone: (216) 499-7090
North Canton, Ohio 44720
Dean of College: Br. A. Cote

DEPARTMENT OF BIOLOGY
Phone: Chairman: Ext. 22

Chairman: Not Stated

Assistant Professors: C. Gutermuth, Jr., K. Schwenk
Instructor: J. Benedetto

Degree Program: Biology

Undergraduate Degree: B.S.
Undergraduate Degree Requirements: 8 credits Biology, 6 credits Mathematics, 8 credits Physics, 16 credits Major elective

WILBERFORCE UNIVERSITY

Wilberforce, Ohio 45384 Phone: (513) 376-2911
Dean of College: Y. Taylor

NATURAL SCIENCE DIVISION
Phone: Chairman: Ext. 315

Chairman: S.K. Saini
Professor: S.K. Saini
Associate Professors: W. Ball, G. Mitchell
Assistant Professor: P.K. Saini

Degree Program: Biology

Undergraduate Degree: B.S.
Undergraduate Degree Requirements: Biology - 37 credit hours, Chemistry - 14 credit hours, Mathematics - 12 credit hours, Physics - 9 credit hours, General Studies-52 credit hours (Trimester), (Total 124 trimester hours)

WILMINGTON COLLEGE

Wilmington, Ohio 45177 Phone: (513) 382-6661
Dean of College: S.P. Olmsted

AGRICULTURE DEPARTMENT
Phone: Chairman: Ext. 280

Chairman: G.L. Karr
Associate Professors: G.L. Karr, G.S. Zimmerman
Instructors: J. Cleveland, J. Niver (Part-time), R. Truesdale (Part-time), W. Penquite (Part-time), L. Cholak (Part-time)

Field Stations/Laboratories: Four College Farms

Degree Program: Agriculture

Undergraduate Degree: B.S.
Undergraduate Degree Requirements: Introductory Animal Science, Introduction to Economics, Essentials of Public Speaking, Algebra-Trigonometry, General Biology - 3 terms, Grain Crop Production, Agricultural Economics, Animal Nutrition, Soils, Farm Management, Molecular Structure, Elemental Analysis

BIOLOGY DEPARTMENT
Phone: Chairman: Ext. 238 Others: Ext. 237

Chairman: T.K. Wood
Associate Professor: T.K. Wood
Assistant Professors: F.S. Amliot, A. Vinegar, G. Williams

Degree Program: Biology

Undergraduate Degree: B.S.
Undergraduate Degree Requirements: 28 hours Biology - only required series is 12 hours General Biology, 15 hours Chemistry and Physics

WITTENBERG UNIVERSITY

Springfield, Ohio 45501 Phone: (513) 327-6231
Dean of College: R. Ballard

DEPARTMENT OF BIOLOGY
Phone: Chairman: (513) 327-7021 Others: 327-6231

Chairman: R.A. de Langlade
Geenawalt Chair of Biology: E.E. Powelson
Professor: E.E Powelson
Associate Professors: N.J. Bolls, Jr., J.B. Hitt, R.A. deLanglade, L.J. Laux
Assistant Professors: D.L. Mason, C.F. Shaffer, Jr.

Degree Program: Biology, Radiation Medicine, Laboratory Medicine

Undergraduate Degree: B.A. (Biology)
Undergraduate Degree Requirements: Biology Require ments: 101 Continuity of Life, Principles of Biology, Cell Biology, Genetics, Other Biology courses (4), Chemistry - 3 courses including Organic Chemistry, 4 other courses selected from the departments of Mathematics, Chemistry, Physics and/or Geology.

WRIGHT STATE UNIVERSITY

Dayton, Ohio 45431 Phone: (513) 426-6550
Dean of Graduate Studies: A.C. MacKinney
Dean of College: B.L. Hutchings (Acting)

DEPARTMENT OF BIOLOGICAL SCIENCES
Phone: Chairman: Ext. 803/531

Chairman: B.L. Hutchings
Professors: P.P. Batra, S. Honda, J.H. Hubschman, B.L. Hutchings, E. Kmetec
Associate Professors: H.I. Fritiz, R.J. Hay, G.J. Kantor, J.D. Lucas, C.R. McFarland, N.S. Nussbaum, J.D. Rossmiller, M.B. Seiger
Assistant Professors: L.G. Arlian, J.B. Conway, A. Foley, R.M. Glasser, H.W. Keller, A.J. Kuntzman, R.A. Morgan, T.S. Wood
Instructors: P.E. Galvas, K.A. Mechlin, R. Webb

Degree Programs: Biology, Medical Technology

Undergraduate Degrees: B.A., B.S.
Undergraduate Degree Requirements: Student must fulfill: College of Science and Engineering Requirements: General Education requirements of the University (35 credits), Departmental unit of Biology courses (65.5-71.5 credits), Required supporting courses in Chemistry, Physics, Mathematics (65.5-74.0 credits), Electives (15.5-30 credits), complete 196 credit hours, 2.0 cumulative average (and in major also)

Graduate Degree: M.S. (Biology)
Graduate Degree Requirements: Student must fulfill: Master: 45 quarter credits (at least 30 credits in graduate level courses only), maximum of 10 credits may be transferred, quarterly participation in Graduate Seminar, 3.0 cumulative average with only 9 hours of C applicable, submit and orally defend a thesis, register for 3 consecutive quarters in final academic year, must satisfy the recommendations of language proficiency made by advisory committee.

Graduate and Undergraduate Courses Available for Graduate Credit: Aquatic Environment, Aquatic Communities, Biology Problems - Water Pollution, Evolution, Biochemistry, Pathogenic Microbiology, General Microbiology, Advanced Genetics, The Invertebrates, Marine Field Trip, Radiation Biology, Parasitology, Economic Physiology of Aquatic Animals, Animal Behavior, Comparative Vertebrate Physiology, Special Problems in Biology (U), Selected Topics in Biology (U), Enzymes (U), Cell Physiology (U), Photobiology (U), Microbiology Physiology (U), Molecular Genetics (U), Radioisotope Principles (U), Microinstrumentation (U), Protein and Vitamin Nutrition (U), General Endocrinology (U), Invertebrate Development (U), Experimental Morphology (U), Tissue Regression in Development (U), Graduate Seminar (U), Metabolic of Control Processes (U), Biochemistry of Natural Products (U)

XAVIER UNIVERSITY
(No reply received)

YOUNGSTOWN STATE UNIVERSITY

410 Wick Avenue Phone: (216) 746-1851
Youngstown, Ohio 44503
Dean of Graduate Studies: L. Rand (Acting)
Dean of College: B.J. Yozwiak

DEPARTMENT OF BIOLOGICAL SCIENCES
Phone: Chairman: Ext. 370 Others: Ext. 371

Chairman: P.D. Van Zandt
Professor: G. Kelley
Associate Professors: G. Karas, P. Peterson, L. Schneeder A. Sobota
Assistant Professors: D. Cannon, D. Fishbeck, R. Kreutzer, D. Machean, C. Rufh, N. Sturm, J. Toepfer, J. Yemma
Instructors: J. Brennan, C. Chucy, A. Sebastiani, E. Staudt

Degree Program: Biological Sciences

Undergraduate Degrees: A.B., B.S.
Undergraduate Degree Requirements: B.S. - 53 quarter hours in Biology, A.B. 45 quarter hours in Biology

Graduate Degree: M.S.
Graduate Degree Requirements: Master: 45 quarter hours, written examination and oral review of the candidate

Graduate and Undergraduate Courses Available for Graduate Credit: Ecology (U), Aquatic Biology, Aquatic Biology Laboratory (U), Mycology (U), Taxonomy of Flowering Plants (U), Plant Anatomy (U), Plant Physiology (U), Advanced Genetics (U), Bacterial Physiology (U), Radioisotopes (U), Biological Seminar (U), Cell Biology (U), Vertebrate Physiology (U), Cytology and Techniques (U), Animal Parasitology (U), Biometry (U), Growth and Differentiation (U), Protozoology (U), Mammalogy (U), Helminthology (U), Comparative Animal Physiology I, Experimental Design, Evolution, Advanced Ecology, Ecosystem Analysis, Physiological Ecology, Advanced Molecular Biology, Marine Biology, Analytical Histochemistry, Plant Growth and Development, Pathogenic Bacteria, Medical Mycology, Virology, Experimental Parasitology, Acarology, Systematic Zoology, Master's Thesis Research, Botany Topics, Vertebrate Zoology Topics, Genetics and Evolution Topics, Parasitology Topics, Environmental Biological Topics, Molecular Biology Topics, Vertebrate Physiology Topics

OKLAHOMA

BETHANY NAZARENE COLLEGE

Bethany, Oklahoma 73008 Phone: (405) 789-6400
Dean of Graduate Studies: D. Beaver
Dean of College: R. Griffin

DEPARTMENT OF BIOLOGY
Phone: Chairman: Ext. 5187

Chairman: R.W. Judd
Professor: R.W. Judd
Associate Professors: S.C. Young, L.F. Finkenbinder
Instructors: S.R. Gould, T.S. Sprenger, G.L. Lehman
Research Assistants: 4
Teaching Assistants: 4

Degree Programs: Biology, Zoology

Undergraduate Degree: B.S.
Undergraduate Degree Requirements: 30-32 semester hours to include: Zoology, Botany, Cell Biology, Genetics, Seminar, and Electives

CENTRAL STATE UNIVERSITY

1000 South University Drive Phone: (405) 341-2980
Edmond, Oklahoma 73034
Dean of Graduate Studies: B. Fisher
Dean of College: N. Russell

DEPARTMENT OF BIOLOGY
Phone: Chairman: (405) 341-2924

Chairman: W. Smith
Professors: B. Cox, L. Hornuff, W. Smith
Associate Professors: R. Anderson, R. Bogenschutz, F. Frazier, D. Frosch, R. Hocker
Assistant Professors: H. Callaway, C. Drabek, P. Guthrie, M. Hamilton, T. Harrison, M Mays, J. Vaughan

Degree Programs: Biology, Biological Education

Undergraduate Degrees: B.S., B.S. Ed.
Undergraduate Degree Requirements: B.S. Degree - General Education - 50 hours, Biology 35 hours, Chemistry 15 hours, Electives to total 124 hours (At least 40 hours of 3000 and 4000 level courses), B.S. Ed. - General Education - 50 hours, Biology 30 hours, Chemistry 10 hours, Physics 10 hours, Professional Education 23 hours

EAST CENTRAL STATE COLLEGE

Ada, Oklahoma 74820 Phone: (405) 332-8000
Dean of Graduate Studies: J. Danley

DEPARTMENT OF BIOLOGY
Phone: Chairman: (405) 332-3653 Others: 332-8000

Chairman: W.A. Carter
Professors: W.A. Carter, T.J. McKnight
Associate Professors: C.E. Butler, H.S. Love
Assistant Professors: E.R. Brown, D.J. Noble

Degree Programs: Biology, Medical Technology

Undergraduate Degrees: B.S., B. Ed.
Undergraduate Degree Requirements: Biology - Biology 36 hours, Chemistry 16 hours, Medical Technology - Biology, 26 hours, Chemistry 17 hours, Mathematics 3 hours

LANGSTON UNIVERSITY

(No reply received)

NORTHEASTERN STATE COLLEGE

Tahlequah, Oklahoma 74464 Phone: (918) 456-5511
Dean of College: L.E. Wallen

DEPARTMENT OF BIOLOGY
Phone: Chairman: (918) 456-2608

Chairman: G. Clarke
Professors: J.M. Anderson, J.D. Reeves
Associate Professors: G.R. Clarke
Assistant Professors: C. C. Smith, J.B. Taylor, E.M. Grigsby

Degree Programs: Biology, Zoology

Undergraduate Degree: B.S.
Undergraduate Degree Requirements: Biology 28 hours plus 16 hours minor plus 37 hours lower division Zoology - 35 hours plus 16 hours minor plus 37 hours lower division

NORTHWESTERN STATE COLLEGE

Alva, Oklahoma 73717 Phone: (405) 327-1700
Dean of Graduate Studies: F.R. Lawson
Dean of College: F.R. Lawson

DEPARTMENT OF AGRICULTURE
Phone: Chairman: Ext. 2300 Others: Ext. 2300

Head: L.S. Brandt
Instructors: L.V. Hill, 1 temporary appointment

Degree Program: Agriculture (Ecology), Agriculture (Business)

Undergraduate Degrees: B.A., B.S.
Undergraduate Degree Requirements: Not Stated

DEPARTMENT OF BIOLOGY
Phone: Chairman: (405) 327-1700 Ext. 2211

Head: S.R. Bigham
Professor: L.J. Bouchard
Associate Professor: S.R. Bigham
Assistant Professors: S. Hensley, P. Nighswonger

Degree Programs: Biology, Botany, Medical Technology, Zoology

Undergraduate Degree: B.S.
Undergraduate Degree Requirements: Not Stated

OKLAHOMA BAPTIST UNIVERSITY

Shawnee, Oklahoma 74801 Phone: (405) 275-2850
Dean of College: W. Mitchell

BIOLOGY DEPARTMENT
Phone: Chairman: (405) 275-2850

Chairman: J.E. Hurley
Professor: J.E. Hurley
Assistant Professors: J. Black

Degree Program: Biology

Undergraduate Degree: B.S.

Undergraduate Degree Requirements: 34 courses, 11 courses in Area of Concentration

OKLAHOMA CHRISTIAN COLLEGE
(No reply received)

OKLAHOMA CITY UNIVERSITY
Northwest 23rd and Blackwelder Phone: (405) 525-5411
Oklahoma City, Oklahoma 73120
Dean of Graduate Studies: W. Coffia
Dean of College: J. White, Jr.

DEPARTMENT OF BIOLOGY
Phone: Chairman: Ext. 2506

Chairman: J.C. Branch
Associate Professors: J.C. Branch, J.R. Hampton, M.H. Brooks
Assistant Professor: L. Kruschwitz

Degree Program: Biology

Undergraduate Degrees: B.A., B.S.
Undergraduate Degree Requirements: B.A. - 30 hours in Biology, 16 hours of Chemistry, 8 hours of Physics, 6 hours of Calculus, B.S. - 32 hours in Biology including 4 hours of Independent Research, 4 hours of Chemistry, 4 hours of Physics, 3 hours College Algebra

OKLAHOMA PANHANDLE STATE UNIVERSITY
Goodwell, Oklahoma 73939 Phone: (405) 349-2611
Dean of College: J. Martin

AGRONOMY DEPARTMENT
(No reply received)

ANIMAL SCIENCE DEPARTMENT
Phone: Chairman: Ext. 228 Others: Ext. 262

Head: M.W. England
Professors: M.W. England, E.E. Firestone, J. Martin
Associate Professors: T.H. Montgomery

Degree Program: Animal Science

Undergraduate Degree: B.S.
Undergraduate Degree Requirements: Not Stated

DEPARTMENT OF BIOLOGY
Phone: Chairman: Ext. 248

Head: S. Ramon
Professors: J. Martin, S. Ramon, E. Reeves, K. Schafer
Instructor: G. Schafer

Degree Programs: Biology, Medical Technology

Undergraduate Degree: B.S.
Undergraduate Degree Requirements: Total hours 124, Biology 36 hours, Chemistry 12-20 hours, Physics 2 courses (Biology Core: General Biology 8 hours, Physiology 4 hours, Genetics 3 hours, Ecology 4 hours, Independent Study 2 hours)

OKLAHOMA STATE UNIVERSITY
Stillwater, Oklahoma 74074 Phone: (405) 372-6211
Dean of Graduate Studies: N. Durham

College of Arts and Sciences
Dean: G. Gries

DEPARTMENT OF BOTANY AND PLANT PATHOLOGY
Phone: Chairman: (405) 372-6475 Others: 372-6471

Head: J.E. Thomas
Professors: G.L. Barnes, L.A. Brinkerhoff, G.W. Todd, J.E. Thomas, E. Basler, A.G. Carroll, H.C. Young,Jr
Associate Professors: C.C. Russell, R.V. Sturgeon, Jr., D.F. Wadsworth
Assistant Professors: J.K. McPherson, L.S. Morrison, P.E. Richardson, R.J. Tyrl
Instructor: J.A. Steinle
Teaching Fellows: 10
Research Assistants: 9
Research Associate: 1

Degree Programs: Botany, Plant Pathology

Undergraduate Degree: B.S.
Undergraduate Degree Requirements: Mathematics to Calculus, General Chemistry, Plant Taxonomy, Anatomy, Pathology, Physiology, Ecology, Genetics, Organic Chemistry

Graduate Degrees: M.S., Ph.D.
Graduate Degree Requirements: Master: B.S. Degree plus 30 hours and thesis Doctoral: M.S. Degree and 60 hours and Thesis

Graduate and Undergraduate Courses Available for Graduate Credit: Botanical Microtechnique (U), Principles of Plant Identification (U), Plant Anatomy (U), Plant Pathology (U), Plant Physiology (U), Forest Pathology (U), Plant Geography (U), Plant Ecology (U), Range Ecology (U), Range Grasses (U), Honors work in Botany (U), M.S. Research, Non-Vascular Aquatic Plants, Mycology, Problems in Plant Pathology, VascularAquatic Plants, Cytology, Plant Physiological Laboratory Techniques, Phylogeny and Classification of Flowering Plants, Physiological Action of Herbicides and Plant Growth Regulators, Physiology of Ion Metabolism, Advanced Ecology, Physiology of Growth and Development, Problems in Plant Physiology, Plant Morphology, Botany Seminar, Environmental Plant Physiology, Ph.D. Research, Virus Diseases of Plants, Plant Disease Genetics, Physiology of Fungi, Bacterial Diseases of Plants, Plant Nematology, Principles of Plant Pathology, Fungus Disease of Plants

DEPARTMENT OF MICROBIOLOGY
Phone: Chairman: (405) 372-7179

Head: L.L. Gee
Professors: L.L. Gee, E.A. Grula, E. Gaudy, N.N. Durham
Associate Professors: L. Richardson, O. Barla
Research Assistants: 5
Graduate Teaching Assistants: 5

Degree Programs: Microbiology, Medical Technology

Undergraduate Degree: B.S.
Undergraduate Degree Requirements: 124 semester credit hours

Graduate Degrees: M.S., Ph.D.
Graduate Degree Requirements: Master: 30 semester credit hours, Doctoral: 90 semester credit hours

Graduate and Undergraduate Courses Available for Graduate Credit: Graviation and Properties of Microorganisms (U), Immunology (U), Immunology (U), Immunology Laboratory (U), Serology (U), Food and Industrial Microbiology (U), Microbiological Laboratory (U), Soil Microbiology (U), Virology (U), Laboratory Techniques (U), Bioenergetics, Advanced Immunology, Research Techniques, Medical Mycology, Genetics of Microorganisms, Advanced Virology, Recent Advances in Microbiology, Microbial Anatomy

DEPARTMENT OF ZOOLOGY
Phone: Chairman: (405) 372-6050

Head: J. Wilhm
Professors: L.H. Bruneau, T.C. Dorris, B.P. Glass,

R.J. Miller, J.A. Morrison, R.J. Summerfelt
Associate Professors: M.R. Curd, J.W. Thornton, D.W. Toetz, J.L. Wilhm
Assistant Professors: A.K. Andrews, J.S. Barclay, S.L. Burks, J.C. Lewis, W.C. Sanford, J.H. Shaw

Degree Programs: Limnology, Zoology, Wildlife Ecology

Undergraduate Degree: B.S.
Undergraduate Degree Requirements: Zoology - 40 hours in Field of Concentration which includes 20 hours in Zoology, 5 hours in Organic Chemistry, 15 hours in Related Fields. Wildlife - 40 hours in Field of Concentration which includes Wildlife Ecology, Wildlife Management, Environmental Biology and Field Botany or Forest Ecology and related electives Zoology (Ecology) - 40 hours in field of Concentration which includes selection of courses from five subject areas.

Graduate Degrees: M.S., Ph.D.
Graduate Degree Requirements: Master: 30 semester hours of graduate credit which includes 6 hours in thesis and 2 hours in seminar, research proposal and thesis and final oral examination required Doctoral: 90 semester hours of graduate credit beyond baccaluareate which includes 30 hours in thesis and dissertation research and 4 hours in seminar, research proposal and dissertation, written and oral qualifying examination, public seminar, and final oral examination.

Graduate and Undergraduate Courses Available for Graduate Credit: Crisis in the Environment, Invertebrate Zoology, Vertebrate Morphology, Field Problems in Wildlife Ecology, Principles of Wildlife Ecology, Natural History of the Vertebrates, Regional Analysis and Planning, Evolution, Ichthyology, Ornithology, Mammalogy, Limnology, Embryology, History, Microtechnique, Wildlife Management, Fisheries Management, Research Problems (U), Teaching Zoology, Special Problems, Wildlife Management Techniques, Aquaculture, Biology of Fishes, Advanced Fishery Science, Water Pollution, Ecology, Analysis of Environmental Contaminants, Population Dynamics, Advanced Wildlife Ecology, Wetland Wildlife Ecology, Environmental Cytology, Ethology, Biology for Teachers

College of Agriculture

Dean: J.A. Whatley

DEPARTMENT OF AGRONOMY
Phone: Chairman: Ext. 278 Others: 372-6211

Head: R.S. Matlock
Professors: J.M. Davidson, F. Gray, W.W. Huffine, J.Q. Lynd, J.C. Murray, L.W. Reed, R.M. Reed, D. A. Sander, P.W. Santelmann, E.L. Smith, J.F. Stone, B.B. Tucker, D.E. Weibel
Associate Professors: D.J. Banks, L.I. Croy, C.E. Denman, L.H. Edwards, E.L. Granstaff, H.A.L. Greer, F.E. LeGrand, W.E. McMurphy, L.G. Morrill, E.E. Sebesta, C.M. Taliaferro, L.D. Tripp, L M. Verhalen
Assistant Professors: R. Ahring, J.M. Baker, R. Foraker, J S. Kirby, E.S. Oswalt, H. Pass, J. Powell, W.L. Richardson, L.M. Rommann, J.H. Stiegler, J.F. Stritzke, B.B. Webb
Instructor: C. Galeotti, D.W. Ringwald, J. Trybom
Research Assistants: 28
Teaching Assistants: 4
Research Associate: 1
Staff Assistant Research: 1

Field Stations/Laboratories: Agronomy Research Station, Agronomy Research Station, Sandy Land Research Station, Caddo Research Station, Eastern Pasture Research Station, Fort Reno Livestock Research Station, Irrigation Research Station, Southwest Agronomy Research Station, North Central Research Station, South Central Research Station, Panhandle Research Station, Agronomy Research Station, Southern Great Plains Field Station

Degree Programs: Agronomy, Crop Science, Range Science, Soil Science, Turf Business

Undergraduate Degree: B.S.
Undergraduate Degree Requirements: 120 hours of specified courses

Graduate Degrees: M.S., Ph.D.
Graduate Degree Requirements: Master: 30 credit hours (including 6 hours research) Doctoral: 90 credit hours, including specific course requirements

Graduate and Undergraduate Courses Available for Graduate Credit: Pasture Management and Forage Production (U), Soil Morphology, Genesis and Classification (U), Physical Properties of Soils (U), Seed Technology (U), Market Grain Technology (U), Soil Chemistry (U), Soil Survey (U), Weed Control (U), Crop Cultures and Growth (U), Soil Fertility and Management (U), Plant Breeding (U), Problems and Special Study (U), Fiber and Special Crops (U), Range Techniques (U), Soils of the World (U), Advanced Crop Culture and Growth, Physiological Genetics, Advanced Genetics, Cytogenetics, Advanced Soil Morphology, Advanced Plant Breeding, Advanced Soil Fertility, Soil Physics, Advanced Soil Biology, Advanced Soil Chemistry, Evaportranspiration, Advanced Range Management, Physical Chemistry of Soil Colloids, Classical Evolution, Soil Mineralogy and Crystallography

DEPARTMENT OF ANIMAL SCIENCES AND INDUSTRY
Phone: Chairman: Ext. 7201

Head: J.C. Hillier
Professors: R.L. Henrickson, J.B. Mickle, R.L. Noble, R. Totusek, E.J. Turman, L. Walters, J.V. Whiteman
Associate Professors: L. Bush, R.R. Frahm, J.J. Guenther, C. Maxwell, F.N. Owens, D.G. Wagner, M. Wells
Assistant Professors: R.K. Johnson, R. Von Gunten, R. Wettemann
Instructors: G. Adams (full-time), B. Kropp (3/4 time), D. Stiffler (1/2 time)
Teaching Assistants: 5
Research Assistants: 20

Field Stations/Laboratories: Fort Reno Livestock Research Station - El Reno, Oklahoma 74034

Degree Programs: Animal Breeding, Animal Science, Dairy Science, Food Science, Nutrition (Animal), Poultry Science

Undergraduate Degree: B.S. (Animal Breeding)
Undergraduate Degree Requirements: 130 semester credit hours required. A Program of study typical of a land grant college Bachelor of Science in Agriculture with a strong major in the animal sciences.

Graduate Degrees: M.S., Ph.D.
Graduate Degree Requirements: Master: A minimum of 30 hours plus a thesis or 32 hours. Advanced training in animal breeding, nutrition, reproduction, statistics, and methods and principles of biochemistry. (Food Science requirements are equivalent but not the same) Doctoral: A minimum of 90 hours beyond a baccalaureate with a thesis required. Strong supporting courses from biochemistry, physiological sciences agronomy, statistics in addition to selected courses from the departmental offerings.

Graduate and Undergraduate Courses Available for Graduate Credit: M.S. Research and Thesis, Special Problems, Seminar, Research Techniques in Food Science, Advances in Meat Science, Advanced Animal Breeding, Experimental Methods in Animal Science, Advanced Animal Nutrition, Rumenology, Rumenology Laboratory, Carbohydrate and Lipid Nutrition, Protein Nutrition, Vitamin and Mineral Nutrition, Ph.D. Research and Thesis, Population Genetics I, Population Genetics II,

DEPARTMENT OF BIOCHEMISTRY
Phone: Chairman: (405) 372-7104 Others: 372-7105

Head: R.E. Koeppe
Regents Professor: K.E. Ebner
Professors: D.C. Abbott, R.K. Gholson, R.E. Koeppe, F.R. Leach, G.V. Odell, G.R. Waller, J.E. Webster (Emeritus)
Associate Professors: K.L. Carraway, B.G. Hudson, E. D. Mitchell, E.C. Nelson, H.O. Spivey
Assistant Professors: R.C. Essenberg
Instructor: J. Hall
Visiting Lecturers: 16
Teaching Fellow: 1
Research Assistants: 26

Degree Program: Biochemistry

Undergraduate Degree: B.S.
Undergraduate Degree Requirements: 124 hours of total course work, Mathematics through Calculus, one year of foreign language, 24 hours of upper division Chemistry and Biochemistry and 16-23 hours of upper division electives including undergraduate research. An honors degree is offered.

Graduate Degrees: M.S., Ph.D.
Graduate Degree Requirements: Master: 24 hours of course work, 6 hours of research, acceptable thesis, successful defense of thesis in an oral examination. Doctoral: 40 hours of course work, 50 hours of research, satisfactory performance on a comprehensive qualifying examination, 1 year of language (undergraduate) or equivalent, acceptable thesis, defense of thesis in an oral examination

Graduate and Undergraduate Courses Available for Graduate Credit: Survey of Biochemistry (U), Biochemical Laboratory (U), Biochemistry, Biochemistry Laboratory, Special Problems, Research, Biochemical Principles, Biochemical Laboratory Methods, Metabolism, Advanced Biochemical Techniques, Biochemical Regulation, Enzymes and Cofactors, Physical Biochemistry, Selected Topics in Biochemistry

DEPARTMENT OF ENTOMOLOGY
Phone: Chairman: (405) 372-7055

Head: D.C. Peters
Professors: W.A. Drew, S. Coppock, R.D. Eikenbary, D.E. Howell, D. C. Peters, J. Young
Associate Professors: J.A. Hair, R.G. Price, J.R. Sauer, K.J. Starks
Assistant Professors: J. Coakley, R. Berberet, R. Burton
Instructor: D. Arnold
Teaching Fellows: 2
Research Assistants: 16

Degree Programs: Agriculture, Entomology

Undergraduate Degree: B.S. (Agriculture)
Undergraduate Degree Requirements: 12 credits Biological Sciences, 16 credits Physical Sciences, 9 credits Mathematics, 12 credits Agricultural Sciences, 44 controlled electives

Graduate Degrees: M.S., Ph.D.
Graduate Degree Requirements: Master: Prerequisites of 30 semester hours of Entomology and Biological Sciences. Applicants must be approved by departmental graduate committee. 30 credit hours including not more than 6 credit hours for required thesis. A minimum of 22 hours in residence. Doctoral: Approved by the departmental graduate committee. Individually developed plan of study including a minimum of 90 semester credit hours. (1) pass a qualifying examination (2) prepare an acceptable dissertation (3) demonstrate ability to do independent study , (4) show qualities of leadership in his chosen field, (5) pass a final examination, and (6) comply with departmental requirements of selected readings, foreign language or 12 credits of broadening course work.

Graduate and Undergraduate Courses Available for Graduate Credit: Not Stated

DEPARTMENT OF FORESTRY
Phone: Chairman: (405) 372-6211 Ext. 363

Head: J.E. Langwig (Acting)
Professors: N. Walker (Emeritus), E.E. Sturgeon
Associate Professors: C.W. Lantz, D.W. Robinson, J.L. Teate
Assistant Professors: R.L. Miller, T.H. Silker
Instructor: D.C. Ketchum
Teaching Fellow: 1
Research Assistants: 6

Field Stations/Laboratories: Kiamichi Experiment Station - Idabel, Oklahoma

Degree Program: Resource Management

Undergraduate Degree: B.S.
Undergraduate Degree Requirements: Not Stated

Graduate Degree: M.S.
Graduate Degree Requirements: Master: Thirty credits hours including six hours on Thesis

Graduate and Undergraduate Courses Available for Graduate Credit: Forest Enivornment and Related Resources (U), Timber Harvesting (U), Aerial Photogrammetry (U), Forest Products (U), Forest Management (U), Forest Administration (U), Forest Problems (U), Forest Recreation (U), Interpretive Services in Recreation (U), Seminar (U), Regional Silviculture, Forest Protection (U), Forest Genetics and Regeneration (U), Wood Properties (U)

DEPARTMENT OF HORTICULTURE
Phone: Chairman: (405) 372-6211 Ext. 302

Head: W.R. Kays
Professor: W.R. Kays
Associate Professors: R.N Payne, C.E. Whitcomb, G. Taylor, J.S. Ownby, H.A. Hinrichs
Assistant Professors: R. Campbell, P.J Mitchell
Instructor: J. Maxson
Teaching Fellows: 2
Research Assistants: 3

Field Stations/Laboratories: Pecan Research Station - Sparks, Oklahoma, Vegetable Research Station - Bixby Perkins Horticulture Station - Perkins

Degree Programs: Horticultural Science, Horticulture, Landscape Horticulture, Ornamental Horticulture, Phomology, Vegetable Crops

Undergraduate Degree: B.S.
Undergraduate Degree Requirements: 130 semester credits

Graduate Degree: M.S.
Graduate Degree Requirements: Master: 30 semester credits with a thesis, 32 semester credits without a thesis

Graduate and Undergraduate Courses Available for Graduate Credit: Not Stated

College of Veterinary Medicine

Dean: W.E. Brock

DEPARTMENT OF VETERINARY PARASITOLOGY AND PUBLIC HEALTH
Phone: Chairman: (405) 372-6325

Chairman: M.L. Frey
Professors: P.B. Barto, R.E. Corstvet, M.L. Frey,

H.E. Jordan
Assistant Professors: J.T. Homer, A.A. Kocan, D.W. MacVean
Instructors: J.A. Jackson, J.G. Williams
Visiting Lecturer: 1
Research Assistant: 1

Field Stations/Laboratories: Pawhuska Research Station - Pawhuska, Oklahoma

Degree Programs: Veterinary Parasitology and Public Health, Veterinary Microbiology, Veterinary Parasitology

Graduate Degrees: M.S., Ph.D.
Graduate Degree Requirements: Master: 30 credit hours including not more than 6 credit hours for the thesis Dcotoral: 90 semester credit hours beyond the Bachelor's degree, 60 credits beyond Master's degree (including thesis credits)

Graduate and Undergraduate Courses Available for Graduate Credit: Veterinary Parasitology, Introduction to Veterinary Microbiology and Immunology, Advanced Veterinary Microbiology, Thesis, Special Problems, Thesis, Seminar, Biology of Parasites, Techniques in Parasitology, Advanced Helminthology, Parasitic Protozoa

ORAL ROBERTS UNIVERSITY

7777 S. Lewis Phone: (918) 743-6161
Tulsa, Oklahoma 74105
Dean of College: C. Hamilton

DEPARTMENT OF NATURAL SCIENCES
Phone: Chairman: Ext. 263 Others: Exts. 260,261

Chairman: R.D. Hartman
Professor: L.D. Thurman
Associate Professors: R.W. Couch, J.M. Nelson
Assistant Professor: E.N Nelson
Instructors: P. Wayne (Adjunct)

Degree Programs: Biology, Biology Education, Medical Technology

Undergraduate Degrees: B.S., B.A.
Undergraduate Degree Requirements: Biology - 128 hours total, 42 hours General Education (non-science), 32 hours in Biology, including Introductory Biology I and II, Microbiology, Ecology, Seminar, and Senior Research, 16 hours Chemistry (1 year General, 1 year Organic), 8 hours Physics, 7 hours Mathematics including Statistics

PHILLIPS UNIVERSITY

Enid, Oklahoma 73701 Phone: (405) 237-4433
Dean of College: N. Jacobs

DEPARTMENT OF BIOLOGY
Phone: Chairman: Ext. 313

Chairman: C.R. Williams
Professors: B. Murphy, C.R. Williams
Associate Professor: L.T. Hall

Field Stations/Laboratories: Phillips University Colorado Campus - Monte Vista, Colorado 81144

Degree Program: Biology

Undergraduate Degree: A.B., B.S.
Undergraduate Degree Requirements: A.B. 24 semester hours of Biology, 14 hours upper division, B.S. - 40 semester hours of Biology, 20 hours upper division, 10 hours of Organic Chemistry

SOUTHEASTERN STATE COLLEGE

Durant, Oklahoma 74701 Phone: 924-0121
Dean of Graduate Studies: E. Boynton
Dean of College: E. Sturch

DEPARTMENT OF BIOLOGICAL SCIENCES
Phone: Chairman: (405) 924-0121 Ext. 2405

Head: E.B. Kilpatrick
Professors: D.B. Hazell, E.B. Kilpatrick, R.J. Taylor
Associate Professor: F. Wade
Assistant Professor: J.E. Lester
Temporary: 2

Degree Programs: Biology, Biological Education

Undergraduate Degree: B.S.
Undergraduate Degree Requirements: 50 hours General Education, 30 hours major, 16 hours minor, 16 hours minor, 12 hours electives, 124 hours total, including 40 hours Junior and Senior Courses

Graduate Degree: M.S.
Graduate Degree Requirements: Master: I. Professional Education minimum 8 hours, including Introduction to Research and Learning Theory, II, Specialized Education minor 8 hours, III General Education minimum 2 hours, IV, Elective 14 hours, 32 hours total

Graduate and Undergraduate Courses Available for Graduate Credit: Heredity, History of Biology, Directed Reading Special Studies, Seminar, Research, Plant Physiology, Plant Morpnology, Bacteriology, Systematic Botany, Ecology, Conservation of Wildlife Resources, Soil Management and Conservation, Range Management, Wildlife Management Techniques, Field Zoology, Entomology, Mammalogy, Ornithology, Human Anatomy and Physiology, Invertebrate Zoology, Ichthyology, Limnology (U), Problems in Biology, Advanced General Biology, Directed Reading in Biology, Special Studies in Biology, Seminar in Biology, Problems in Botany, Directed Reading in Botany, Special Studies in Botany, Seminar in Botany, Problems in Zoology, Directed Readings in Zoology

SOUTHWESTERN STATE COLLEGE

Weatherford, Oklahoma 73096 Phone: (405) 772-6611
Dean of Graduate Studies: H. Massey
Dean of College: L. Morris

DEPARTMENT OF BIOLOGICAL SCIENCES
Phone: Chairman: (405) 772-4302

Chairman: J.F. Lovell
Professors: B. Ballard, R. Dick, T. Gray, J. Lovell
Associate Professors: H. Henson, R. Lynn, D. Messmer, R. Segal, R. Seibert, G. Wolgamott
Assistant Professors: A. Badgett, M. Kerley
Instructors: M. Cox, L. Dickerson, H. Kirkland

Field Stations/Laboratories: Biological Science Field Station - Five miles north of Weatherford on Caddo road

Degree Programs: Biology, Biological Education, Environmental Health, Medical Technology

Undergraduate Degree: B.S.
Undergraduate Degree Requirements: 1 course in each Biological Concepts, Botany, Zoology and Genetics - 16 credit hours, 1 course in each Anatomy, Physiology and Field Study - 12 credit hours, and remaining courses as electives with 40 hours Biology, credit as minimum for Biological Science degree, other programs vary slightly, plus 18 hours of Chemistry and Mathematics through Trigonometry are ancillary requirements.

Graduate Degree: M.S. (Environmental Health)
Graduate Degree Requirements: Master: Prerequisite re-

quirements - 20 undergraduate semester hours in Biological Sciences and a 3.0 grade average.

UNIVERSITY OF OKLAHOMA
600 Parrington Oval Phone: (405) 325-4115
Norman, Oklahoma 73069
Dean of Graduate Studies: A. Gentile

College of Arts and Sciences
Dean: P. Mulhollan

DEPARTMENT OF BOTANY AND MICROBIOLOGY
Phone: Chairman: (405) 325-4321 Others: 325-0311

Chairman: H.W. Larsh
Professors: L. Beevers, N.H. Boke, T.B. Clark, G.C. Gozad, G.J. Goodman, H.W. Larsh, E.L. Rice, D.C. Cox, J.R. Estes, J.H. Lancaster, P.G. Risser, J.J. Skvarla
Associate Professors: W.O. Felkner, J.S. Fletcher, R.W. Leu, T.H. Milby, J.W. Murphy, L. Pfiester, F.A. Rinehart, W.G. Sorenson
Research Assistants: 24

Field Stations/Laboratories: Biological Station and Biological Survey - Willis, Oklahoma

Degree Programs: Botany, Microbiology

Undergraduate Degree: B.S.
Undergraduate Degree Requirements: Botany - 24 hours of major work are required for the degree of Bachelor of Science with one course from each of the following areas: Anatomy-Morphology, Ecology, Genetics, Physiology, Taxonomy, Mathematics, Physica and Organic Chemistry Microbiology-Students must take courses in Basic Microbiology, Introduction to Molecular Biology, Pathogenic Microbiology and Immunology, Introduction to Mycology, Physiology of Microorganisms

Graduate Degrees: M.S., Ph.D.
Graduate Degree Requirements: Not Stated

Graduate and Undergraduate Courses Available for Graduate Credit: Botany - Plant Anatomy, Genetics, Plant Pathology, Principles of Plant Ecology, Phycology, Literature in Botany and Microbiology, Optical Methods in Biology, Instrumental Methods in Biology, Cellular Physiology, Principles of Plant Physiology, Advanced Cell Biology, Introductory Mycology, Methods of Teaching Biology as Inquiry, Systematic Botany, Readings in Botany, Selected Studies in Botany, Advanced Plant Physiology, Morphology of Vascular Plants, Cytology Ultrastructure, Cytogenetics, Agrostology, Investigations in Botany, Problems in Natural Sciences, Ecology and Taxonomy of Aquatic Plants, Plant Community Ecology, Physiological Plant Ecology, Advanced Plant Taxonomy Microbiology - Basic Microbiology, Literature in Botany and Microbiology, Introduction to Molecular Biology, Pathogenic Microbiology and Immunology, Microbial Ecology, Optical Methods in Biology, Instrumental Methods in Biology, Cellular Physiology, Advanced Cell Biology, Introductory Mycology, Hematology, Determinative Microbiology, Industrial Microbiology, Immunology, Physiology of Microorganisms, Microbial Genetics, Virology, Investigations in Microbiology, Directed Readings in Microbiology, Selected Studies in Microbiology, Medical Mycology, Experimental Medical Mycology, Recent Advances in Microbial Physiology, Problems in Natural Sciences

DEPARTMENT OF ZOOLOGY
Phone: Chairman: (405) 325-4821 Others: 325-4822

Chairman: V.H. Hutchison
David Ross Boyd Professor of Zoology: H.W. Frings
George Lynn Cross Research Professor of Zoology: C.F. Hopla
Regents Professor of Zoology: J.T. Self
Professors: G. Braver, H.P. Brown, C.C. Carpenter, H.P. Clemens, W. Friedburg (Adjunct), R.A. Goff, V.H. Hutchison, C.E. Melton (Adjunct)
Associate Professors: J.K. Greer, H.B. Haines, L.G. Hill, C. Lent, F. Seto, F.J. Sonleitner
Assistant Professors: M.A. Chartock, F. deNoyelles, Jr., W.L. Dillard, A.A. Echelle (Adjunct), D.L. Perkins, G.D. Schnell, M.R. Whitmore
Instructor: D. Lollis
Teaching Fellows: 3
Research Assistants: 10
Research Associate: 1
Teaching Assistants: 39

Field Stations/Laboratories: University of Oklahoma Biological Station - Kingston, Oklahoma, Aquatic Biology and Fisheries Research Laboratory - Noble, Oklahoma

Degree Programs: Animal Behavior, Biometry, Biostatistics, Ecology, Entomology, Ethology, Fisheries and Wildlife Biology, Genetics, Histology, Limnology, Medical Technology, Natural Science, Parasitology, Physiological Sciences, Physiology, Zoology

Undergraduate Degree: B.S. (Zoology, Medical Technology)
Undergraduate Degree Requirements: At least 24 semester hours in Zoology to include Invertebrate Zoology, Comparative Vertebrate Anatomy and four of the following: Genetics, Evolution, Ecology, Principles of Physiology, and Embryology, one year of college Physics, Organic Chemistry, and Calculus. Intermediate level in one foreign language. General Botany and Microbiology are recommended.

Graduate Degrees: M.S., Ph.D.
Graduate Degree Requirements: Master: At least 30 semester credit hours, including a course in Biostatistics and one in professional aspects of Biology, Completion and acceptance of a thesis, reading knowledge of a foreign language. Satisfactory completion of qualifying examinations. Doctoral: At least 90 semester credit hours, including credit for a master's degree, a course in biostatistics, and one in professional aspects of Biology. Reading knowledge of two foreign languages, or one foreign language and computer science. Satisfactory completion of general examinations and a dissertation. Program is planned by an advisory committee.

Graduate and Undergraduate Courses Available for Graduate Credit: Evolution (U), Principles of Physiology (U), Vertebrate Embryology (U), Genetics (U), Ecology (U), Natural History of Vertebrates and Invertebrates (U), Microtechnique, Zoogeography, Mammalogy, Ornithology, Protozoology, Entomology, Herpetology, Cellular Physiology, Vertebrate Physiology, Comparative Physiology, Physiological Measurements, Neurophysiology, Advanced Cell Biology, Ecology of Host-parasite relationships, Hematology, Experimental Genetics, Oceanography, Limnology, Field Entomology, Ichthyology, Analysis of Development, Parasitology, Quantitative Biology, Symbiology, Animal Behavior, Endocrinology, Birds of the World, Aquatic Invertebrates, Physiological Ecology, Fish Culture, Medical Entomology, Physiological Bases of Animal Behavior, Population Ecology, Ethoecology, Fisheries Management, Seminars, and Special Study courses.

Health Sciences Center
P.O. Box 26901 Phone: (405) 271-4000
Oklahoma City, Oklahoma 73190
Dean of College of Medicine: R.M. Bird
Dean of College of Dentistry: W.E. Brown

DEPARTMENT OF ANATOMICAL SCIENCES
Phone: Chairman: (405) 271-5470

Chairman: W.J.L. Felts
Professors: J.E. Allison, R.E. Coalson, K.K. Faulkner, W.J.L. Felts, J.F. Lhotka, R.S. Nanda, K.M. Richter

Associate Professors: L.G. Gumbreck, T.D. McClure, P.A. Roberts
Assistant Professors: K.H. Dugan, R. Grubb, R.M. Harkins, R.I. Howes
Teaching Fellows: 5
Research Assistants: 2

Field Stations/Laboratories: University of Oklahoma Biological Station - Willis, Oklahoma

Degree Program: Anatomical Sciences

Graduate Degrees: M.S., Ph.D.
Graduate Degree Requirements: Master: General requirements as stated in Graduate College Bulletin, 30 hours at graduate level, original thesis, participation in departmental seminar, courses of gross anatomy, histology and organology. Doctoral: 90 hours at graduate level, original dissertation for which a total of 30 hours may be allowed, participation in departmental seminar, reading knowledge of one foreign language, completion of courses in gross anatomy, histology and organology, human embryology, neuroanatomy.

Graduate and Undergraduate Courses Available for Graduate Credit: Gross Anatomy, Radiologic Anatomy, Neuroanatomy, Histology and Organology, Human Embryology, Anatomical Techniques, Optical Methods in Anatomical Research, Seminar, Research for Master's Thesis, Advanced Histology, Advanced Embryology, Clinical Anatomy, Advanced Cytology, Advanced Histotechnique, Histochemistry, Tissue and Organ Culture, Ultrastructure, Experimental Embryology, Advanced Gross Anatomy, Anatomy of the Fetus and Infant, Electron Microscopy, Advanced Neuroanatomy, Guided Reading in Classical and Modern Anatomical Literature, Anatomical Basis of Physical Anthropology, Comparative Morphology, Research for Doctor's Dissertation, Specialized Studies in Anatomy

DEPARTMENT OF BIOCHEMISTRY AND MOLECULAR BIOLOGY
Phone: Chairman: (405) 271-4221

Chairman: B.C. Johnson
Professors: P. Alaupovic, R.H. Bradford, M. Carpenter, R. Carubelli, K. Dubowski, W. Friedberg, A.C. Kurtz, P.B. McCay, J. Metcoff (Adjunct), B. Rabinovitch J. Sokatch, J. Tang
Associate Professors: J.H. Anglin, R. Bottomley, T. Briggs, A.M. Chandler, R. Coleman, A.C. Cox, R. Delaney, M.J. Griffin, D.S. Hodgins, E.G. Larsen, J.S. Mayes, J. Ontko, J.R. Seely, L. Unger
Assistant Professors: J.J. Ferretti, P. Gray, J. Hartsuck, D. Lee, M. Mameesh (Research), T. Whayne (Research)
Teaching Fellows: 6
Research Assistants: 20
Research Associates: 12

Degree Programs: Biochemistry and Molecular Biology, Biochemistry, Biochemistry and Biophysics, Biological Chemistry, Nutrition, Genetics

Graduate Degrees: M.S., Ph.D.
Graduate Degree Requirements: Master: 30 hours with thesis (4 hours), 32 hours no thesis Doctoral: 90 hours including at least 30 hours advanced courses, 30 hours dissertation research

Graduate and Undergraduate Courses Available for Graduate Credit: General Biochemistry, General Biochemistry Part I, Biochemistry Laboratory, General Biochemistry Part II, Biochemistry Laboratory, Macromolecules and Genetics, Physical Biochemistry, Biochemistry Seminar, Molecular Oncology, Special Problems

DEPARTMENT OF BIOSTATISTICS
(No reply received)

DEPARTMENT OF MICROBIOLOGY AND IMMUNOLOGY
Phone: Chairman: (405) 271-5054 Others: 271-5051

Chairman: L.V. Scott
Professors: F.G. Felton, R.M. Hyde, M.H. Ivey, F.C. Kelly (Emeritus), H.G. Muchmore, R.A. Patnode, L.V. Scott, J.T. Self (Adjunct), P.E. Smith (Adjunct), J.R. Sokatch
Associate Professors: G.S. Bulmer, W.A. Cain, J.J. Ferretti, J.L.W. Jackson, S.R. Oleinick, E.R. Rhoades, W.R. Schmieding, V.S. Smith (Adjunct), L. Unger
Assistant Professors: R.L. Carpenter (Research), M.P. Lerner, R.E. McCallum
Graduate Assistants: 3

Degree Programs: Immunology, Medical Microbiology

Graduate Degrees: M.S., Ph.D.
Graduate Degree Requirements: Master: 30 hours beyond the B.S. degree, with a "B" average and a thesis based on original research. A no-thesis M.S. degree program may be allowed for certain students. Doctoral: 90 hours beyond the B.S. degree, with a "B" average and a dissertation based on original research.

Graduate and Undergraduate Courses Available for Graduate Credit: Basic Medical Microbiology, Microorganisms as Infectious Agents, Microorganisms as Infectious Agents Laboratory, Principles of Immunology, Topics in Infectious Pathogenic Bacteriology, Medical Mycology, Virology, Immunobiology, Advanced Immunology Laboratory, Immunologically Mediated Diseases, Transplantation Immunology, Bacterial Metabolism, Tumor Viruses, Microbial and Molecular Genetics, Research for Doctor's Dissertation, Medical Parasitology, Diagnostic Microbiology, Environmental Microbiology, Physiologic and Immunologic Nature of Parasitism, Advanced Protozoology and Helminthology, Diagnostic Virology

DEPARTMENT OF PHYSIOLOGY AND BIOPHYSICS
Phone: Chairman: (405) 271-4281

Chairman: J.M. Keyl
Distinguished Professor: G.A. Brecher
Professors: T.K. Chowhury, L.B. Hinshaw (Research), C.E. Melton (Research, A.J. Stanley (Emeritus), J. W. Woods (Research)
Associate Professors: E.D. Frohlich (Research), W.H. Massion (Research), R.E. Thies (Adjunct)
Assistant Professors: R.D. Bell (Research), A. Higgins (Research), O.R. Kling (Research), J. McKenzie (Research), R.J. Person, R.D. Stith
Instructor: R.J. Sinclair (Adjunct)
Research Assistant: 1

Degree Programs: Physiology, Biophysics

Graduate Degrees: M.S., Ph.D.
Graduate Degree Requirements: Master: At least 30 hours graduate work including thesis Doctoral: A minimum of ninety semester hours of which at least forty-five hours must be in formal course work, the remaining hours constitute credit for research leading to the writing of a dissertation.

Graduate and Undergraduate Courses Available for Graduate Credit: None

UNIVERSITY OF SCIENCE AND ARTS OF OKLAHOMA
Chickasha, Oklahoma 73018 Phone: (405) 224-3140
Dean of College: C.C. Ferree

DISCIPLINE OF BIOLOGY
Phone: Chairman: Ext. 316 Others: Ext. 315

Head: N.S. Wirt
Assistant Professors: L.K. Magrath, N.S. Wirt

Degree Program: Biology

UndergraduateDegree: B.S.
Undergraduate Degree Requirements: 33 hours of Biology, 16 hours of Chemistry, 10 hours of Physics, 8 hours of Mathematics (through Calculus), 8 hours of language (foreign, or equivalent research tool such as computer science, etc.)

THE UNIVERSITY OF TULSA

600 South College Phone: (918) 939-6351
Tulsa, Oklahoma 74104
Dean of Graduate Studies: T.F. Staley
Dean of College: E.B. Strong, Jr.

DEPARTMENT OF LIFE SCIENCES

Phone: Chairman: Ext. 204, 205

Head: C.A. Levengood
Professors: A.P. Blair, C.A. Levengood
Associate Professors: P. Buck, H.L. Lindsay, Jr., B. Shirley
Assistant Professors: N. Carpenter, E. Levetin, S.H. Rogers, M. Woolsey
Teaching Fellows: 3
Graduate Assistants: 2

Degree Programs: Biology, Botany, Medical Technology, Microbiology, Zoology

Undergraduate Degrees: B.A., B.S.
Undergraduate Degree Requirements: Total hours required 124 (semester hours) to include: 27 hours of General Education (Humanities, Social Sciences and Science), 12 hours Mathematics and Quantitative Subject for B.S., Equivalent of 16 hours of a foreign language for B.A., 50=58 hours in Area of Concentration (30-38 major plus 20 supportive)

Graduate Degree: M.S.
Graduate Degree Requirements: Master: 30 semester hours, Thesis required-maximum credit allowed is six hours

Graduate and Undergraduate Courses Available for Graduate Credit: Introduction to Virology (U), Genetics (U), Pathogenic Microbiology and Immunology (U), Advanced Microbiology (U), Advanced in Immunochemistry, Ecology (U), Molecular Biology (U), Cellular Biology (U), Genetic Biology, Developmental Biology, Organismic Biology, Environmental Biology, Algae-Fungi (U), Morphology Vascular Plants (U), Plant Physiology (U), Advanced Plant Morphology, Advanced Plant T xonomy, Mammal Physiology (U), Parasitology (U), Invertebrate Zoology (U), Histology (U), Endocrinology (U), Independent Study - Selected Topics

OREGON

EASTERN OREGON STATE COLLEGE

La Grande, Oregon 97850 Phone: (503) 963-2171
Dean of Graduate Studies: W.D. Spear
Dean of College: G.E. Young

DIVISION OF SCIENCE AND MATHEMATICS
Phone: Chairman: Ext. 328 Others: 963-2171

The College is not departmentalized.
Professors: E.C. Anderson, S.C. Head, D.E. Kerley
Associate Professor: A.E. Anderson

Field Stations/Laboratories: Lily White Environmental Field Station

Degree Program: Biology

Undergraduate Degrees: B.A., B.S.
Undergraduate Degree Requirements: 45 hours Biology, Mathematics, through Calculus, 24 hours Chemistry

GEORGE FOX COLLEGE

Newberg, Oregon 97132 Phone: (503) 538-2101
Dean of College: W. Green

DIVISION OF NATURAL SCIENCE
Phone: Chairman: Ext. 211

Chairman: D.E. Chittick
Professor: E. Voth
Associate Professor: G.D. Orkney

Degree Programs: Biology, Biological Education

Undergraduate Degree: B.S.
Undergraduate Degree Requirements: 51 term hours, 39 term hours 300 or more to include 10 hours Plant Morphology, 3 hours senior seminar. Additionally 12 hours General Chemistry, 8 hours Organic, 4 hours Quantitative Analysis and 4 hours Mathematics. A research paper is required.

LEWIS AND CLARK COLLEGE

Portland, Oregon 97219 Phone: (503) 244-6161
Dean of College: J.E. Brown

DEPARTMENT OF BIOLOGY
Phone: Chairman: Ext. 316 Others: Exts. 317, 318

Chairman: T.D. Darrow
Associate Professors: T.D. Darrow, D.S. McKenzie, P.L. Stallcup
Assistant Professors: W. Ryan, S. Gould

Field Stations/Laboratories: Part of a consortium maintaining Malheur Environmental Field Station in Oregon desert near Burns, Oregon.

Degree Program: Biology

Undergraduate Degree: B.S.
Undergraduate Degree Requirements: 10 courses in Biology (50 term hours), plus 3 courses in Chemistry (15 hours) and two additional courses (10 hours) in Mathematics, Physics or Chemistry.

LINFIELD COLLEGE

McMinnville, Oregon 97128 Phone: (503) 472-5215
Dean of College: C.H. Hinrichs

BIOLOGY DEPARTMENT
Phone: Chairman: (503) 472-4121

Chairman: R. Jones
Professor: J.R. Crook
Associate Professor: D. Martinsen
Assistant Professors: C.R. Ault

Field Stations/Laboratories: Malheur Environmental Field Station - Burns , Oregon, Good Samaritan Hospital - Portland, Oregon (Both consortiums)

Degree Programs: Biology, General Science, Pre-Medicine

Undergraduate Degrees: B.S.
Undergraduate Degree Requirements: Satisfactory completion of 24 full courses, at least 23 of which must be taken from the regular curriculum. Requirements for a major in Biology include six full courses in the department. In addition to fulfilling departmental major requirements, all students are required to satisfy certain course requirements from a core of general education courses.

Graduate Degree: M.S. (Neurophysiology)
Graduate Degree Requirements: Master: Requires 30 hours of approved graduate credit, in approximately one year of full-time study. Minimum prerequisite is the successful completion of the first two years (pre-clinical) at a standard medical school.

Graduate and Undergraduate Courses Available for Graduate Credit: Lectures and discussion in Neurophysiology, Lectures and discussion in Medical Electronics, Lectures and discussion in Statistics, Reading and conference in Neurophysiology, Reading and conference in Neuroanatomy, Lectures, reading and laboratory in Neuropathology, Reading and conferences in Experimental Psychology, Research, Thesis

OREGON STATE UNIVERSITY

Corvallis, Oregon 97331 Phone: (503) 754-0123
Dean of Graduate Studies: E.N. Castle

School of Agriculture

Dean: W.T. Cooney

DEPARTMENT OF ANIMAL SCIENCE
Phone: Chairman: (503) 754-3431

Head: J.E. Oldfield
Professors: R. Bogart, D.C. Church, D.C. England, W.D. Frischknecht, D.H. Gates, J.H. Landers, Jr., R.J. Raleigh, A.T. Ralston, J.E. Oldfield
Associate Professors: H.P. Adams, W.H. Kennick, F. Stormshak, F.M. Stout, A.S.H. Wu
Assistant Professors: P.R. Cheeke, D.W. Claypool, W.D. Hohenboken, W.C. Krueger, R.L. Phillips, L.R. Rittenhouse, L.V. Swanson, M. Vavra, A.H. Winward, J.O. Reagan, R.E. Pulse, G.D. Savelle
Research Associate: H.R. Burkhart
Research Assistants: S.K. Martin, L.T. McDaniel, L.R. Shull, M. Wing
Instructors: J. Adair, S.T. Gashler, P.M. Rutland
Graduate Research Assistants: 43

Field Stations/Laboratories: Columbia Basin Research Center, Squaw Butte Experiment Station, Eastern Oregon Experiment Station, Klamath Experiment Station.

Degree Programs: Animal Science, Animal Nutrition, Animal Physiology, Animal Genetics, Meat Science, Rangeland Resources

Undergraduate Degree: B.S.
Undergraduate Degree Requirements: 192 credit hours in a planned program meeting institutional requirements, school requirements and departmental requirements.

Graduate Degrees: M.S., M. Agric., Ph.D.
Graduate Degree Requirements: Master: The student must complete a program of study totalling not less than 45 term hours including thesis and courses approved for graduate credit. Approximately 2/3 of the work (30 term hours) must be in the major and 1/3 (15 term hours in the minor). Master of Agriculture Degree: Forty-five hours are required with a minimum of 9 hours in each of at least three agricultural or agriculturally related fields with not more than 21 hours in any one field. An advisory committee will consist of representatives in these fields. Doctoral: The equivalent of at least three years of full-time graduate work beyond the bachelor's degree is required (approximately 120 hours beyond the bachelor's degree).

Graduate and Undergraduate Courses Available for Graduate Credit: Ruminant Nutrition, Comparative Nutrition, Dairy Production (U), Sheep Production (U), Swine Production (U), Beef Production (U), Physiology of Lactation, Animal Improvement, Graduate Seminar, Topics in Animal Nutrition - 9 hours, Topics in Animal Breeding - 9 hours, Rangeland Analysis (U), Physiology of Reproduction in Domestic Animals, Livestock Genetics, Research Projects, Thesis, Reading and Conference, Range Topics (U), Range Management Planning, Perspectives in Range Research, Rangeland Ecology, Rangeland Management

DEPARTMENT OF FISHERIES AND WILDLIFE
Phone: Chairman: (503) 754-1531

Head: C.E. Warren (Acting)
Professors: C.E. Bond, P. Doudoroff, H.F. Horton, L.W. Kuhn, R.E. Millemann, C.E. Warren, H.M. Wright (Adjunct)
Associate Professors: W.P. Breese, G.E. Davis, J.R. Donaldson, J.D. Hall, B.J. Verts, P.A. Vohs Jr., L.J. Weber
Assistant Professors: R.S. Caldwell, H.J. Campbell (Adjunct), R.L. Garrison, R.L. Jarvis, J.E. Lannan, W.C. Lightfoot (Adjunct), E.C. Meslow (Adjunct), J.D. Mc Intyre (Adjunct), C.D. Snow (Adjunct), H.E. Wagner (Adjunct)
Instructors: R.L. Lantz (Adjunct), W.K. Seim, E.T. Juntunen
Research Associates: J.L. Hedtke, G.L. Larson, J.R. Sedell, F.J. Triska
Research Assistants: 10
Graduate Research/Teaching Assistants: 35

Field Stations/Laboratories: Oak Creek Laboratory, Pacific Fisheries Laboratory, Swanson Aquaculture Laboratory

Degree Programs: Fisheries Science, Wildlife Science

Undergraduate Degree: B.S.
Undergraduate Degree Requirements: Wildlife Science: 44 hours Wildlife Science courses, 18 hours Communications, 18 hours Humanities and Social Sciences, 62 hours Science, 5 hours Physical Education and Health, 45 hours Electives Fisheries Science: 38 hours Fisheries Science courses, 18 hours Communications, 18 hours Humanities and Social Sciences, 6 hours Science, 3 hours Physical Education and Health, 3 hours Environmental Engineering Fundamentals completion of 192 term hours with a minimum 2.00 GPA for both Wildlife and Fisheries Sciences.

Graduate Degrees: M.S., Ph.D.
Graduate Degree Requirements: Master: Must complete a program of study totaling not less than 45 term hours including thesis and courses approved for graduate credit. Program requires preparation of a thesis. Doctoral: No rigid credit requirement - equivalent of at least 3 years of full-time graduate work beyond the bachelor's degree is required. At least 1 academic year must be spent in continuous residence at Oregon State University with a minimum of 36 hours of graduate work. Program requires preparation of a thesis.

Graduate and Undergraduate Courses Available for Graduate Credit: Vertebrate Pest Control (U), Biology of Game Birds (U), Fishery Biology (U), Principles of Symbiosis (U), Fish Culture (U), Fishery Limnology (U), Fishery Limnology Laboratory (U), Management of Big Game (U), Commerical Fisheries (U), Invertebrate Fisheries (U), Water Pollution Biology (U), Parasites and Diseases of Fish (U), Wildlife Field Trip (U), Fish Genetics, Wildlife Investigational Techniques, Research Perspectives, Population Dynamics, Pollution Problems in Fisheries, Functional Ichthyology, Systematics of Fishes, Special Topics in Ichthyology, Wildlife Ecology (U)

DEPARTMENT OF FOOD SCIENCE AND TECHNOLOGY
Phone: Chairman: (503) 754-3131

Head: P.E. Kifer
Professors: A.F. Anglemier, R.F. Cain, P.E. Kifer, L.A. McGill, M.E. Morgan, H.W. Schultz, R.O. Sinnhuber
Associate Professors: D.V. Beavers, D.D. Bills, F.W. Bodyfelt, D.L. Crawford, W.D. Davidson, P.H. Krumperman, K.D. Law, H.P. Milleville, M.W. Montgomery, R.A. Scanlan, J.H. Wales, R.E. Wrolstad, H. Y. Yang, T.C. Yu
Assistant Professor: G.W. Varseveld
Instructors: G.B. Putnam, M.R. Soderquist
Research Assistants: 2
Research Associates: 7

Field Stations/Laboratories: Seafoods Laboratory - Astoria, Oregon, Food Toxicology and Nutrition Laboratory - Corvallis, Oregon

Degree Programs: Food Science

Undergraduate Degree: B.S.
Undergraduate Degree Requirements: Physical Science - 37 credits, Mathematics 18 credits, Biological Science 18 credits, Food Science and Technology - 53 credits, Communications - 18 credits, Humanities and Social Science - 18 credits, Electives - 30 credits

Graduate Degrees: M.S., Ph.D.
Graduate Degree Requirements: Master: 46 term hours including thesis and courses approved for graduate credit. 30 hours. Must be in major and 15 hours in the minor. Residence requirement, one academic year. Doctoral: No rigid credit requirements, however, the equivalent of at least three years of full-time graduate work beyond the bachelor's degree is required. A thesis embodying results of research giving evidence of originality and ability in independent investigation is required. Nine credits of coursework not usually included in the major or minor fields for special skills development.

Graduate and Undergraduate Courses Available for Graduate Credit: Food Science (U), Federal and State Food Regulations (U), Food Analysis (U), Quality Control Systems, Food Packaging, Current Topics in Food Science, Research, Thesis, Reading and Conference Seminar, Carbohydrates in Foods, Food Flavors and Evaluation, Lipids in Foods, Pigments and Color Evaluation, Proteins in Foods, Enzymes of Foods

DEPARTMENT OF HORTICULTURE
Phone: Chairman: (503) 754-3695

Head: C.J. Weiser

Professors: S.B. Apple, J.R. Baggett, R. Garren, Jr., R.H. Groder, H.J Mack, A.N. Roberts, C.J. Weiser, M.N. Westwood
Associate Professors: M.H. Chaplin, G. Crabtree, R.L. Stebbins, M.M. Thompson, S.E. Wadsworth
Assistant Professors: L.H. Fuchigami, N.S. Mansour, D.G. Richardson
Visiting Lecturers: 1
Research Assistants: 6
Acting Assistant Professor: 1

Field Stations/Laboratories: North Willamette Experiment Station, Mid-Columbia Experiment Station, Southern Oregon Experiment Station

Degree Programs: Horticulture, Genetics

Undergraduate Degree: B.S.
Undergraduate Degree Requirements: Institutional Requirements, School of Agriculture Requirements, Department of Horticulture Requirements

Graduate Degrees: M.S., Ph.D.
Graduate Degree Requirements: Master: Institutional Requirements - 45 term hours of credit (30 in major) with 3.00 grade point - 1 academic year residence Doctoral: Institutional Requirements - 3 years full time work (1 in residence with 36 term hours of credit - a thesis Department of Horticulture Requirements 1 term teaching experience

Graduate and Undergraduate Courses Available for Graduate Credit: Spraying, Dusting and Fumigation (U), Horticultural Plant Nutrition (U), Fruit Handling and Distribution (U), Systematic Pomology (U), Systematic Vegetable Crops (U), Research, Thesis, Reading and Conference, Seminar, Plant Genetics, Horticultural Plant Growth and Development, Post-Harvest Physiology, Selected Topics in Horticulture

DEPARTMENT OF POULTRY SCIENCE
Phone: Chairman: (503) 754-2301 Others: 754-2301

Head: G.H. Arscott
Professors: P.E. Bernier, J.A. Harper, J.E. Parker
Associate Professor: C.J. Fischer
Assistant Professor: R.W. Dorminey
Graduate Research Assistants: 2

Degree Program: Poultry Science

Undergraduate Degrees: B.S., B. Agr.
Undergraduate Degree Requirements: Institutional Requirements, School of Agriculture Requirements, Department of Poultry Science Requirements

Graduate Degrees: M.S., M.Agr., Ph.D.
Graduate Degree Requirements: Master: Institutional Requirements - 45 term hours of credit (30 in major) with 3.00 grade point - 1 academic year residence Doctoral: Institutional Requirements - 3 years of full time work (1 in residence with 36 term hours of credit a thesis Department of Poultry Science Requirements - 1 term teaching experience

Graduate and Undergraduate Courses Available for Graduate Credit: Feeds and Feeding (U), Poultry Nutrition, Poultry Meat Production (U), Egg Production (U), Poultry Breeding (U), Population Genetics and Breeding Improvement, Avian Environmental Physiology and Reproduction

DEPARTMENT OF SOIL SCIENCE
Phone: Chairman: (503) 754-2441

Head: H.B. Cheney
Professors: L. Boersma, M.D. Dawson, E.H. Gardner, M.E. Harward, T.L. Jackson, D.P. Moore, G.H. Simonson, J.A. Vomocil, C.T. Youngberg
Associate Professors: J.L. Young, V.V. Volk
Assistant Professors: C.H. Ullery, B.L. Harris, G.F. Kling
Research Assistants: 19

Degree Program: Soil Science

Undergraduate Degree: B.S.
Undergraduate Degree Requirements: Not Stated

Graduate Degrees: M.S., Ph.D.
Graduate Degree Requirements: Not Stated

Graduate and Undergraduate Courses Available for Graduate Credit: Not Stated

School of Forestry

Dean: C. Stoltenberg

DEPARTMENT OF FOREST ENGINEERING
Phone: Chairman: (503) 754-1952 Others: 754-0123

Head: G.W. Brown
Associate Professor: H.A. Froehlich
Assistant Professors: R. Beschta, E. Berglund
Research Assistants: 4

Degree Program: Watershed Management

Graduate Degree: M.S., Ph.D.
Graduate Degree Requirements: Master: 45 quarter credits, thesis, completion of basic courses in soils atmospheric science, plant physiology, hydrology, with specialization in watershed management. Doctoral: 90 quarter credits beyond M.S., same general requirements as above.

Graduate and Undergraduate Courses Available for Graduate Credit: Watershed Management (U), Forest Hydrology, Water Quality and Forest Land Use, Environmental Measurement Techniques

FOREST PRODUCTS DEPARTMENT
Phone: Chairman: (503) 754-2017

Head: H. Resch
Professors: H. Resch, G.H. Atherton, M.D. McKimmy
Associate Professors: W.J. Bublitz, S.E. Corder, R.A. Currier, R.D. Graham, J.W. Johnson, R.L. Krahmer, M.L. Laver, C. Maxey, R.O. McMahon, A.C. Van Vliet, J.D. Wellons, C.J. Kozlik
Assistant Professors: D.J. Miller, A. Polensek, J.B. Wilson
Teaching Fellow: 1
Research Assistants: 12

Field Stations/Laboratories: Forest Research Laboratory

Degree Programs: Wood Science, Wood Industry Management

Undergraduate Degree: B.S.
Undergraduate Degree Requirements: 204 hours of University level courses including: Written communication 9 hours, oral communication 6 hours, humanities arts and social sciences 17 hours, physical and biological sciences 24 hours, forest products subject courses 43 hours, electives.

Graduate Degrees: M.S., Ph.D.
Graduate Degree Requirements: Master: Master of Science: Requirements in the major 9 credits of specified forest products courses, 6 to 12 credits thesis, 9 to 15 courses in forest products and/or departments related to major field of study, 30 credits total in major. Requirements in the minor 15 credits. Total in M.S. program 45 credits. Doctoral: The Ph.D. program is written to include all courses beyond the Bachelor's degree. There is no total credit requirement. A major and a minor must be specified. Three forest products courses at the graduate level must be

taken if this has not been done before. 1 graduate seminar.

Graduate and Undergraduate Courses Available for Graduate Credit: Mechanical Properties II (U), Mechanical Conversion I (U), Mechanical Conversion II (U), Pulp and Paper Processes (U), Wood Industry Problems (U), Forest Products Merchandising (U), Wood Microtechnique, Wood Anatomy, Wood Growth-Quality Relationships, Advanced Wood Physics, Selected Topics in Wood Physics, Wood Chemistry, Wood Industry Management, Wood Technology, Selected Topics in Wood Chemistry, Advanced Pulp and Paper.

School of Oceanography

Dean: J.V. Byrne

SCHOOL OF OCEANOGRAPHY
Phone: Chairman: (503) 754-3504

Professors: H. Curl, Jr., H.F. Frolander, J.W. Hedgpeth, R.Y. Morita, W.G. Pearcy
Associate Professors: A.G. Carey, Jr., J.E. McCauley, L.F. Small
Assistant Professors: W.O. Forster, J.J. Gonor, D.R. Hancock, C.B. Miller, W.C. Renfro
Research Associate: N. Cutshall, R.L. Holton, S. Richardson
Visiting Lecturers: 2
Research Assistants: 70

Field Stations/Laboratories: Marine Science Center - Newport

Degree Program: Oceanography

Graduate Degrees: M.S., Ph.D.
Graduate Degree Requirements: Master: 45 credit hours selected by committee. Thesis, Final Examination Doctoral: 80 credit hours (35 in thesis). Language requirement as determined by committee. Thesis, Final Examination.

Graduate and Undergraduate Courses Available for Graduate Credit: (Only courses of primarily biological content are listed) Marine Zooplankton (U), Principles of Biological Oceanography (U), Marine Radioecology, Special Topics in Marine Radioecology, Biological Oceanography, Marine Nekton, Marine Nekton Laboratory, Marine Phytoplankton Ecology, Marine Phytoplankton Physiology, Marine Primary Production, Marine Benthic Ecology, Speical Topics in Biological Oceanography, Ecology of Foraminifera

School of Sciences

Dean: R.W. Krauss

DEPARTMENT OF BOTANY AND PLANT PATHOLOGY
Phone: Chairman: (503) 754-3451 Others: 754-0123

Chairman: T.C. Moore
Professors: T.C. Allen, A.F. Bartsch (Courtesy), N.I. Bishop, H.R. Cameron, K.L. Chambers, W.W. Chilcote, R.H. Converse (Courtesy), M.E. Corden, H.J. Evans, J.R. Hardison (Courtesy), C.E. Horner (Courtesy), H.J. Jensen, C.M. Leach, I.C. MacSwan, H.K. Phinney, L.F. Roth, J.R. Shay, B. Zak (Courtesy)
Associate Professors: W.H. Brandt, D.L. Coyier (Courtesy), A.J. Culver Jr., (Courtesy), W.C. Denison, R.O. Hampton (Courtesy), R.G. Linderman (Courtesy), T.E. Maloney (Courtesy), C.D. McIntire, R.L. Powelson, R.S. Quatrano, F.R. Rickson, J.M. Trappe (Courtesy), E.J. Trione (Courtesy)
Assistant Professors: R.L. Dennis, H. Dooley (Courtesy), D.M. Knutson (Courtesy), P.A. Keopsell, L.W. Moore, E.E. Nelson (Courtesy), D.B. Zobel
Instructors: J.R. Dilworth, Jr., L.D. Johnston, B.J. Moore, R.F. Obermire
Visiting Lecturers: 12
Teaching Fellows: 18
Research Graduate Assistants: 13
Research Assistants: 10
Research Associates: 6

Field Stations/Laboratories: Botany and Plant Pathology Field Station Laboratory - Corvallis, Oregon

Degree Programs: Botany, Plant Pathology

Undergraduate Degree: B.S.
Undergraduate Degree Requirements: 192 credit hours, including 36 hours in Botany-Biology, two years of Chemistry, one year of Mathematics, one year of Communication skills, and 18 hours of Humanities and Social Sciences.

Graduate Degrees: M.S., Ph.D.
Graduate Degree Requirements: Master: For the Master of Science degree, the student must complete a program of study totaling not less than 45 term hours including thesis and courses approved for graduate credit. Approximately two-thirds of the work (30 term hours) must be in the major and one-third (15 term hours) in the minor. Doctoral: The degree of Doctor of Philosophy is granted primarily for attainments and proved ability. There is no rigid credit requirement; however, the equivalent of at least three years of full-time graduate work beyond the bachelor's degree is required.

Graduate and Undergraduate Courses Available for Graduate Credit: Morphology of Nonvascular Plants (U), Morphology of Vascular Plants (U), Forest Pathology (U), Agrostology (U), Advanced Systematic Botany (U), Plant Taxonomy (U), Bioenergetics of Plants (U), Photobiology of Plants (U), Plant Growth and Development (U), Advanced Plant Ecology (U), Plant Pathology (U), Plant Disease Diagnosis (U), Epidemiology and Disease Control (U), Mycology(U), Microtechnique (U), Plant Anatomy (U), Paleobotany (U), Research, Thesis, Reading and Conference, Seminar, Fresh-Water Algae, Marine Algae, Forest Pathology, Research Methods in Bioenergetics and Photosynthesis, Research Methods in Plant Growth and Development, Mineral Metabolism, Physiology and Biochemistry of Plant Development, Plant Geography, Plant Virology, Nematode Diseases of Plants, Physiology of Fungi, Physiology of Parasitism, Electron Microscopy Laboratory in Botany, Cytological Microtechnique, Plant Cytogenetics, Biological Micrography

DEPARTMENT OF BIOCHEMISTRY AND BIOPHYSICS
Phone: Chairman: (503) 754-1511 Others: 745-0123

Chairman: R.W. Newburgh
Professors: R.R. Becker, I. Isenberg, W.D. Loomis, D.L. MacDonald, D.J. Reed, K.E. Van Holde
Associate Professors: S.R. Anderson, D.J. Baisted, W. Gamble, W.C. Johnson
Assistant Professors: J. Cardenas, G.D. Pearson, H. W. Schaup
Visiting Lecturers: 12
Teaching Fellows: 2
Research Assistants: 26

Field Stations/Laboratories: Oregon State University Marine Science Center - Newport, Oregon

Degree Programs: Biochemistry, Biochemistry and Biophysics, Biophysics

Undergraduate Degree: B.S.
Undergraduate Degree Requirements: 192 credits

Graduate Degrees: M.S., Ph.D.
Graduate Degree Requirements: Master: 45 credits for non-thesis, 36 credits for thesis and 9 credits thesis Doctoral: No formal course requirements (usually 36-45 credits), cumulative examination system, oral preliminary examination, 1 language, thesis

Graduate and Undergraduate Courses Available for Graduate Credit: General Biochemistry (U), General Biophysics (U), Biophysics (U), Biochemistry (U), Biochemistry Laboratory (U), Research, Thesis, Reading and Conference, Seminar, Selected Topics in Biochemistry, Plant Biochemistry, Plant Biochemistry Laboratory, Physical Methods in Biophysics and Biochemistry, Selected Topics in Biophysics, Biochemistry for Science Teachers

DEPARTMENT OF ENTOMOLOGY
Phone: Chairman: (503) 754-1733 Others: 754-0123

Head: P. Oman
Professors: V.J. Brookes, J. Capizzi, H.H. Crowell, G. Ferguson, R.L. Goulding, G.W. Krantz, J.D. Lattin (sabbatical), P.O. Ritcher, J.A. Rudinsky, H.A. Scullen (Emeritus), W.P. Stephen, K.G. Swenson, L.C. Terriere
Associate Professors: N.H. Anderson, W.P. Nagel, R.R. Robinson, R.G. Rosenstiel
Assistant Professors: M.T. AliNiazee, R.E. Berry, D.M. Burgett
Research Associates: R.D. Schonbrod, S. Yu
Research Entomologist: J.A. Kamm
Teaching Fellows: 2
Research Assistants: 12

Field Stations/Laboratories: Aquatic Insect Laboratory - Corvallis, Fresh Insect Laboratory - Corvallis

Degree Programs: Pest Management, Applied Entomology, Forest Entomology, Systematic Entomology, Acarology, Aquatic Entomology, Veterinary Entomology, Insect Physiology, Insect Toxicology, Plant Virus Transmission, Insect Pathology

Undergraduate Degrees: A.B., B.S. (Entomology, Pest Management)
Undergraduate Degree Requirements: One year approved biological science, one year approved physical science, 18 term hours humanities or social sciences, 9 term hours in communication skills, statistical methods, 27 term hours of Entomology

Graduate Degrees: M.A., M.S., Ph.D. (Entomology)
Graduate Degree Requirements: Master: 45 term hours of approved courses including thesis, two thirds in a major field, one third in a minor field. Doctoral: Equivalent of 3 years of full-time graduate work, one years residence required at Oregon State. Thesis required.

Graduate and Undergraduate Courses Available for Graduate Credit: Entomology, Anatomy and Physiology of Insects (U), Forest Entomology (U), Forest Insect Dynamics (U), Aquatic Entomology (U), Principles of Insecticide Usage (U), Plant Protection Entomology (U), Medical and Veterinary Entomology (U), Systematics and Adaptations of Insects (U), Principles of Symbiosis (U), Acarology (U), Comparative Animal Behavior, Insect Ecology and Biological Control (U), Research Methods of Insect Population, Selected Topics in Entomology, Insect Transmission of Plant Viruses, Immature Insects, Developmental Physiology of Insects, Principles of Systematics, Speciation and Distribution

DEPARTMENT OF MICROBIOLOGY
Phone: Chairman: (503) 754-1441

Chairman: P.R. Elliker
Professors: A.W. Anderson, P.R. Elliker, R.Y. Morita, L.W. Parks, K.S. Pilcher, W.E. Sandine, J.L. Fryer
Assistant Professors: L.R. Brown, R.J. Seidler
Instructors: D.D. Curran, C.W. Roth
Teaching Fellows: 6
Research Assistants: 35

Field Stations/Laboratories: Fish Disease Laboratory

Degree Programs: Microbiology

Undergraduate Degree: B.S.
Undergraduate Degree Requirements: 192 term hours minimum, 45 hours minimum in upper division, 36 hour minimum in major, 3 hours English Composition, 3 hours Physical Education, 1 term Personal Health

Graduate Degrees: M.S., Ph.D.
Graduate Degree Requirements: Master: 45 term hours minimum (approximately 30 term hours in major and approximately 15 term hours in minor. Thesis and oral examination Doctoral: At least three years full-time graduate work beyond bachelor's degree, foreign language, thesis, preliminary oral examination and final oral examination.

Graduate and Undergraduate Courses Available for Graduate Credit: Systematic Microbiology, Pathogenic Microbiology, Immunology and Serology, Virology, Food Microbiology, Dairy and Industrial Biotransformations, Microbial Ecology, Marine Microbiology, Microbial Genetics, Bacterial Viruses, Microbial Physiology, Selected Topics in Microbiology, Selected Topics in Soil Microbiology

DEPARTMENT OF ZOOLOGY
Phone: Chairman: (503) 754-3705 Others: 754-0123

Chairman: E.J. Dornfeld
Professors: R.H. Alvarado, F.P. Conte, H.H. Hillemann, A.W. Pritchard, P.A. Roberts, R.M. Storm
Associate Professors: P.S. Dawson, F.L. Hisaw, A. Owczarzak, J.A. Wiens
Assistant Professors: C.J. Bayne, J.E. Morris
Instructors: T.L. Mullen, W.E. Sype
Teaching Fellows: 31
Research Assistants: 7
Research Associates: 3

Field Stations/Laboratories: Oregon State University Marine Science Center - Newport, Oregon

Degree Programs: Ecology, Genetics, Parasitology, Physiology, Zoology, Cell Biology, Invertebrate Zoology, Vertebrate Zoology, Developmental Biology

Undergraduate Degree: B.A., B.S. (Zoology)
Undergraduate Degree Requirements: Zoology/Biology 55, Chemistry 27, Physics 12, Mathematics 12, (in quarter hours)

Graduate Degrees: M.A., M.S., Ph.D.
Graduate Degree Requirements: Master: Equivalent of undergraduate major; 30 hours graduate work in Zoology (including thesis and seminar), 15 in supporting minor field Doctoral: At least 3 years full-time graduate work beyond bachelor's level, approximately 60 percent of program in major (including thesis) and 40 percent in minors.

Graduate and Undergraduate Courses Available for Graduate Credit: Comparative Vertebrate Embryology (U), Comparative Vertebrate Anatomy (U), Developmental Biology (U), Invertebrate Embryology (U), Vertebrate Physiology (U), General Physiology (U), Comparative Physiology (U), Biochemical Adaptations (U), Invertebrate Zoology (U), Integrative Mechanisms in Invertebrates (U), Principles of Symbiosis, Parasitology (U), Comparative Vertebrate Histology (U), Microtechnique, Ornithology (U), Mammalogy (U), Herpetology (U), History of Zoology, Fetal Physiology, Differentiation and Growth, Mammalian Physiology, Endocrinology, Selected Topics in Physiology, Theoretical Genetics, Protozoology, Selected Topics in Invertebrate Zoology, Biology of the Cell, Selected Topics in Cellular Biology, Electron Microscopy, Zoogeography, Population Biology, Selected Topics in Vertebrate Ecology

PACIFIC UNIVERSITY

2043 College Way Phone: (503) 357-6151
Forest Grove, Oregon 97116
Dean of Arts and Sciences: P.J. Cole

DEPARTMENT OF BIOLOGY
Phone: Chairman: Ext. 229

Chairman: R.T. Carter
Professors: D.R. Malcolm, J. Baver
Assistant Professors: A. Mozejko, R.T. Carter

Field Stations/Laboratories: Member of consortium for Malheur Environmental Field Station

Degree Program: Biology

Undergraduate Degree: B.S.
Undergraduate Degree Requirements: Biology Electives 8 semester hours, General Chemistry, Organic Chemistry, Chemistry Electives 4 hours, Mathematics through Calculus, College Physics 8 hours

PORTLAND STATE UNIVERSITY

P.O. Box 751 Phone: (503) 229-3000
Portland, Oregon 97207
Dean of Graduate Studies: D.T. Clark
Dean of College: K. Dittmer

DEPARTMENT OF BIOLOGY
Phone: Chairman: (503) 229-3851

Head: E. Fisher, Jr.
Professors: D. Ferguson (Adjunct), E. Fisher, T.N. Fisher, J. McNab (Emeritus), R. Macy (Emeritus), D. Malcolm (Adjunct), P. Olilvie (Adjunct), W.H. Taylor
Associate Professors: D.W. Boody, C.L. Calvin, R. Forbes, B.E. Lippert, J.W. Myers, L. Newman, K.E. Payne, V. Reierson, L. Simpson, R. Tocher, J. Wirtz, M.L. Taylor (Research)
Assistant Professors: M.S. Lea, R. Peterson, E. Rosenwinlsel, T. Steen, R. Tinnin
Research Assistants: 2
Graduate Teaching Assistantships: 21

Field Stations/Laboratories: Member of Malheur Field Station - Malheur, Oregon

Degree Programs: Environmental Science, Biology

Undergraduate Degrees: A.B., B.S.
Undergraduate Degree Requirements: 45 units Biology, This must include Genetics and Physiology, Chemistry through Organic, 12 units Mathematics, year college level Physics (for Science majors)

Graduate Degrees: M.A.T., M.S.T., M.A., M.S., Ph.D.
Graduate Degree Requirements: Master: Student must complete at least 30 credits in Biology of which at least 22 credits must be in 500 level Biology courses. Maximum 15 credits as electives in fields related to Biology, thesis required, final oral examination required Doctoral: Environmental Sciences Biology Requirements: Environmental Science courses 9 units, Environmental Science Seminar 2 years, one term Statistics, one term computer, Foreign Language, Comprehensive Examination in Biology, student must submit prospectus of proposed research

Graduate and Undergraduate Courses Available for Graduate Credit: Mammalogy, Viruses, Algae, Plant Anatomy, Cell Physiology, Plant Physiology, Invertebrate Zoology, Plant Ecology, Limnology and Aquatic Ecology, Microbiology, Selected Topics in Pathogenic Bacteria, Selected Topics in Immunology, Selected Topics in Immunochemistry, Comparative Vertebrate Endocrinology, Comparative Invertebrate Endocrinology, Microbial Genetics, Molecular Genetics, Problems in Phycology

REED COLLEGE

Woodstock Boulevard Phone: (503) 771-1112
Portland, Oregon 97202

DEPARTMENT OF BIOLOGY
Phone: Chairman: Ext. 337

Chairman: L.N. Ruben
Professors: B. Brehm, G.F. Gwilliam, L.H. Kleinholz (Research), L.N. Ruben, H.A. Stafford
Assistant Professors: S.A. Arch, J. Freedman, P.J. Russell
Research Fellows: 2
Research Assistants: 8

Degree Program: Biology

Undergraduate Degree: B.A.
Undergraduate Degree Requirements: 1 year Mathematics (including Calculus), 1 year Physics, 2 years Chemistry (including Organic), 1 year Introductory Biology plus 5 elective semester courses in Biology, 1 year Thesis Research

SOUTHERN OREGON COLLEGE

1250 Siskiyou Boulevard Phone: (503) 482-3311
Ashland, Oregon 97520
Director of Graduate Studies: B.L. Dunlop
Dean of College: A. Kreisman

DEPARTMENT OF BIOLOGY
Phone: Chairman: (503) 482-6341

Chairman: M.J. Flower
Professors: M. Coffey, D.W. Linn, R. Welton
Associate Professors: S. Cross, R. Lamb, F. Lang, D. Mitchell, C. Skrepetos, W. Sorsoli
Assistant Professors: M. Flower, R. Nitsos, F. Smith, J. Sullivan

Degree Program: Biology

Undergraduate Degree: B.S.
Undergraduate Degree Requirements: 36 hours "core curriculum": Principles of Biology, Plants as Organisms, Animals as Organisms, Environmental Biology, Genetics, Cell Biology, Developmental Biology, Senior Seminar, plus 14 elective hours, Also: General Chemistry (15 hours), Mathematics (12 hours), Physics (12 hours), and Organic Chemistry (8 hours)

UNIVERSITY OF OREGON

Eugene, Oregon 97403 Phone: (503) 686-3111
Dean of Graduate Studies: A. Novick

College of Liberal Arts

Dean: (Not Stated)

DEPARTMENT OF BIOLOGY
Phone: Chairman: (503) 686-4526 Others: 686-3111

Head: S.S. Tepfer
Professors: A.S. Bajer, R.W. Castenholz, C.W. Clancy, P.W. Frank, P. Grant, J. Gray, G. Hoyle, J. Kezer, E.A. Maynard (Adjunct), B.H. McConnaughey, R.W. Morris, F.W. Munz, A. Novick, E. Novitski, B.T. Scheer, W.R. Sistrom, A.L. Soderwall, F.W. Stahl, G. Streisinger, D.E. Wimber
Associate Professors: H.T. Bonnett, Jr., G.C. Carroll, S.A. Cook, J.C. Fentress, P.P. Rudy, F.P. Sipe, J.A. Weston
Assistant Professors: D.L. Barker, W.E. Bradshaw, D. R. Hague, I. Herskowitz, C.B. Kimmel, J.H. Postlethwait, R.C. Terwilliger, W.A. Wiitanen
Instructors: M C. Heimbigner, H.M. Howard, G.J. Murphy, E. Schabtach, H.P. Wisner
Research Associates: 10
Visiting Lecturers: 1

Teaching Fellows: 25
Research Assistants: 13
Trainees: 17

Field Stations/Laboratories: Oregon Institute of Marine Biology - Charleston, Oregon 97420

Degree Programs: Animal Behavior, Animal Physiology, Biochemistry, Biology, Botany, Ecology, Genetics, Limnology, Molecular Biology, Physiology, Zoology, Developmental Biology, Neurophysiology

Undergraduate Degrees: B.A., B.S. (Biology)
Undergraduate Degree Requirements: 1 year Mathematics, 2 years Chemistry, 1 year Physics, 3 years Biology to include a 7 term Core Curriculum plus 2 senior level terms of electives.

Graduate Degrees: M.A., M.S., Ph.D. (Biology)
Graduate Degree Requirements: Master: 45 credit hours, including writing a "critical essay" and defending it. Doctoral: Individually designed programs. Minimum 3 years of work, usually 4-5, years after B.S. degree or 3-4 years after M.S.

Graduate and Undergraduate Courses Available for Graduate Credit: Courses Offered Only at Institute of Marine Biology, Summer: Invertebrate Zoology (U), Comparative Biochemistry (U), Invertebrate Embryology (U), Biology of Marine Organisms (U), Planktonology (U), Marine Ecology (U), Laboratory and Field Methods in Biology (U) Upper-Division Courses Carrying Graduate Credit: General and Comparative Physiology (U), Comparative Neurobiology (U), Environmental Physiology (U), Genetics (U), Genetics Laboratory (U), Human Genetics (U), Evolution (U), Cell Organelles (U), Nuclear Cytology (U), Mycology (U), Algae (U), Methods of Pollen Analysis (U), Cenozoic Paleobotany (U), Systematic Botany (U), Biology of Vascular Plants (U), Developmental Biology (U), Developmental Biology Laboratory (U), Histology (U), Invertebrate Zoology (U), Biochemistry (U), Parasitology (U), Principles of Ecology (U), Marine Environment (U), Limnology (U), Animal Behavior (U), Biology of Prokaryotic Organisms (U), Biology of Prokaryotic Organisms Laboratory (U), Microbial Ecology (U), Microbial Ecology Laboratory (U), Molecular Biology of Phage (U), Molecular Biology of Bacteria (U), Membrane Structure and Function (U), Historical Biogeography (U), History of Biological Ideas (U), Supervised Tutoring Practicum, Vertebrate Endocrinology, Physiology of Reproduction, Endocrinology Laboratory, Advanced Mammalian Neurobiology, Neurochemistry, Neurobiological Basis of Behavior, Advanced Genetics, Principles of Microscopic Techniques, Developmental Genetics, Advanced Systematic Botany, Advanced Plant Physiology, Biology of Fishes

Medical School

3181 S W. Sam Jackson Park Road Phone: (503) 225-8311
Portland, Oregon 97201
Dean: C.N. Holman

DEPARTMENT OF ANATOMY

Phone: Chairman: (503) 225-7811 Others: 225-7811

Chairman: V. Critchlow
Professors: R.L. Bacon, V. Critchlow, E. Jump, A.A. Pearson (Emeritus), H. Spies, W.A. Stotler
Associate Professors: R.S. Connell, H.W. Davis, R. Quinton-Cox, A.R. Tunturi, H. Weitlauf
Assistant Professors: R. Norman, T.C. Richards
Instructors: M. Tseng
Research Assistants: 5
Research Associate: 1
Affiliate: 1

Degree Program: Anatomy

Graduate Degrees: M.S., Ph.D.
Graduate Degree Requirements: Master: 3 academic terms in residence. 30 hours in major field, 15 hours in minor field, thesis. Doctoral: 6 academic terms in residence. 100 hours in major field, 35 hours in minor field, one foreign language, dissertation.

Graduate and Undergraduate Courses Available for Graduate Credit: Gross Anatomy, Histology, Neuroanatomy, Microscopic Techniques, Research, Reading and Conference, Seminar, Neuroendocrinology, Analytical Embryology, Analytical Histology, Special Dissections, Advanced Neuroanatomy and Computer Techniques, Computers in Medicine, Embryology, Mammalian and Human Cytogenetics

DEPARTMENT OF BIOCHEMISTRY

Phone: Chairman: (503) 225-7782 Others: 255-8311

Chairman: R.T. Jones
Professors: J.H. Fellman, R.T. Jones, M. Litt, H.S. Mason, D.A. Rigas, J T. Van Bruggen
Associate Professors: J.P. Bentley, J.A. Black, D. Kabat
Instructors: T.S. Fugita, R.J. Jolley, Jr.
Research Assistants: 17
Postdoctoral Fellows: 7

Degree Programs: Biochemistry, Biochemistry and Biophysics

Graduate Degrees: M.S., Ph.D.
Graduate Degree Requirements: Master: B.S. degree, MCA Test Doctoral: B.S. degree, MCA Test

Graduate and Undergraduate Courses Available for Graduate Credit: General Biochemistry, Cell Organization and Function, Introduction to Biochemical Research, Nucleic Acids and Protein Synthesis, Bioenergetics, Advanced Intermediary Metabolism, Enzymology, Biophysical Chemistry, Biochemical and Biophysical Techniques, Topics in Advanced Biophysical Chemistry, Biochemical Properties of Proteins, Biochemical Mechanisms, of Disease, Structure and Function of Biological Membranes, Biophysical Properties of Proteins

DIVISION OF MEDICAL GENETICS

Phone: Chairman: (503) 225-7703 Others: 225-8311

Head: R.D. Koler
Professors: M. Litt, D.A. Rigas
Associate Professors: R.H. Bigley, N.R.M. Buist, F. Hecht, E. Lovrien, G. Prescott (Dental School)
Assistant Professors: J. Macfarlane, J. Templeton
Instructors: N. Kennaway, J. Vidgoff, H. Wyandt
Visiting Lecturers: 15
Teaching Fellows: 2
Research Assistants: 12

Degree Programs: Degrees and course work are offered through other departments. Research is conducted in this department.

DEPARTMENT OF MICROBIOLOGY AND IMMUNOLOGY

Phone: Chairman: (503) 225-7768 Others: 225-7798

Chairman: J.V. Hallum
Professors: A.W. Frisch, J.V. Hallum, A. Malley, E.A. Meyer, R.B. Parker, B. Pirofsky, M.B. Rittenberg
Associate Professors: H.R. Creamer, B.H. Iglewski, G.A. Leslie, A.L. Rashad
Assistant Professors: E.J. Bardana, D.R. Burger, S.A. Morse
Instructors: N.B Gerhardt
Teaching Fellow: 1
Research Assistants: 8

Degree Programs: Immunology, Microbiology

Graduate Degrees: M.S., M.D., Ph.D.
Graduate Degree Requirements: Master: 30 terms hours

graduate credit in major department plus 15 hours in minor department. Doctoral: 135 term hours of graduate credit. 100 hours in major department.

Graduate and Undergraduate Courses Available for Graduate Credit: Advanced Virology, Bacterial Physiology, Microbial Genetics, Pathogenesis, Immunochemistry, Advanced Microbiology and Immunology, Introduction to Medical Microbiology

DEPARTMENT OF PHYSIOLOGY
Phone: Chairman: (503) 225-8262

Chairman: J.M. Brookhart
Professors: J.M. Brookhart, J.J. Faber, A.J. Rampone, B.B. Ross, R.E. Swanson
Associate Professor: J.A. Resko
Assistant Professors: J.L. Keyes, R.E. Talbott
Research Assistants: 3
Research Associates: 2
Clinical Research Associates: 4

Degree Program: Physiology

Graduate Degrees: Ph.D.
Graduate Degree Requirements: Doctoral: For the degree of Doctor of Philosophy, 135 term hours of graduate credit are required. Of this total, a minimum of 100 hours must be credit in the major department. The student must also complete a minimum of 20 hours of credit in a minor field if he wishes to graduate with a single minor.

Graduate and Undergraduate Courses Available for Graduate Credit: Physiological Instrumentation and Techniques, Advanced Cardiovascular Physiology, Biological Transport Processes., Advanced Neurophysiology, Digestion and Absorption, Respiratory Gas Transport, Fetal and Neonatal Physiology

UNIVERSITY OF PORTLAND
5000 N. Willamette Phone: (503) 283-7129
Portland, Oregon 97203
Dean of College: J. Powers

DEPARTMENT OF PHYSICAL AND LIFE SCIENCES
Phone: Chairman: (503) 283-7122

Head: J.W. McCoy
Professors: J.G. Anderson, C. Bonhorst, B.H. Carleton (Emeritus), J.E. Nohlgren, G. Vassallo (Emeritus), P.E. Wack
Associate Professors: E.H. Collins, J.W. McCoy, J.C. Neeley, M A. Starr, K. Wetzel
Assistant Professor: M.D. Snow

Degree Programs: Life Sciences, Biochemistry, Medical Technology

Undergraduate Degree: B.S.
Undergraduate Degree Requirements: 25 credits Biology, 18 credits Chemistry, 8 credits Physics, 4 credits Mathematics

WARNER PACIFIC COLLEGE
2219 S.E. 68 Phone: (503) 775-4368
Portland, Oregon 97215
Dean of College: C. Loewen

DEPARTMENT OF SCIENCE AND MATHEMATICS
Phone: Chairman: (503) 775-4368

Chairman: L. Corbin
Professors: L. Corbin, W. Davis, S.L. Espey
Assistant Professors: G. Moore

Field Stations/Laboratories: Malheur Field Station - Burns, Oregon (a consortium)

Degree Program: Biology

Undergraduate Degrees: A.B., B.S.
Undergraduate Degree Requirements: 60 hours to include: 12 hours Biology, 15 hours Chemistry, 3 hours Seminar, 30 hours Advanced Biology

WILLAMETTE UNIVERSITY
Salem, Oregon 97301 Phone: (503) 370-6303
Dean of College of Liberal Arts: P. Duell

BIOLOGY DEPARTMENT
Phone: Chairman: 370-6333

Chairman: D.R. Breakey
Professors: D.R. Breakey, M.E. Springer
Associate Professors: G.O. Thoresett
Assistant Professor: S.D. Hawke
Instructors: E.A. Yocom (Part-time)

Field Stations/Laboratories: Member of the Consortium of the Malheur Environmental Field Station near Burns Oregon

Degree Programs: Biology

Undergraduate Degrees: B.A., B.S.
Undergraduate Degree Requirements: B.A. (University requires foreign language competancy). University requirements. Biology Major requires Principles of Biology, 1 Zoology course, 1 Botany course, Senior Seminar courses and 3 other courses in Biology plus two courses in college Chemistry, B.S. does not require language. Senior evaluation required (oral and written examinations, senior research).

PENNSYLVANIA

ALBRIGHT COLLEGE

Reading, Pennsylvania 19604 Phone: (215) 921-2381
Dean of College: R. McBride

BIOLOGY DEPARTMENT
Phone: Chairman: Ext. 232 Others: Ext. 296

Chairman: E.L. Bell
Henry Pfeiffer Chair of Biology: J.S. Hall
Professors: E.L. Bell, J.S. Hall
Associate Professors: D.L. Daniel
Assistant Professors: J.L. Gehres, R.C. Heller
Lecturer: M.A. Dougherty (Part-time)

Degree Programs: Biology, Medical Technology

Undergraduate Degree: B.S.
Undergraduate Degree Requirements: 32 credit hours in Biology, including 8 hours in General Biology, 6 hours - Analytical Geometry and Calculus, 8 hours Physics, 18 hours - General Analytical Chemistry and Organic Chemistry Cooperative Forestry Curriculum (with Duke University) - 2 years at Albright, 1 year at Duke, B.S. degree awarded and end of year at Duke. Program General Studies with 6 hours Economics or Social Science, 12 hours Biology including Botany plus 7 hours in Geology or Biology, 8 hours in Chemistry, 8 hours in Physics, 1 year at Duke University with 14 hours of Biological Sciences

ALLEGHENY COLLEGE

Meadville, Pennsylvania 16335 Phone: (814) 724-5360
Dean of College: J.E. Helmreich

BIOLOGY DEPARTMENT

Chairman: J.R. Wohler
Professor: G.S. Reisner
Associate Professors: W.E. Curtis, J.R. Wohler
Assistant Professors: E.S. Chapman, G.R. Finni

Field Stations/Laboratories: Bousson Researve - Guys Mills, Pennsylvania

Degree Programs: Biology

Undergraduate Degrees: B.S., B.A.
Undergraduate Degree Requirements: General Biology - 1 term, Botany - 1 term, Zoology - 1 term, Microbiology - 1 term, Ecology - 1 term, Genetics - 1 term, Physiology - 1 term, Junior Seminar, Senior Thesis

ALLENTOWN COLLEGE OF SAINT FRANCIS DE SALES

Center Valley, Pennsylvania 18034 Phone: (215) 282-1100
Dean of College: Rev. AIT. Pocetto, OSFS

BIOLOGY DEPARTMENT
Phone: Chairman: 282-1100

Chairman: A.F. Answini
Associate Professors: Rev. F. W. Spaeth, OSFS
Assistant Professors: A.F. Answini, S.J. Grabowski, E.J. Hallinan, OSFS

Degree Program: Biology

Undergraduate Degrees: B.A., B.S.
Undergraduate Degree Requirements: Eight courses in Biology, four course units in Chemistry, two course units in Physics, and, two course units in Mathematics

ALLIANCE COLLEGE

Cambridge Springs, Pennsylvania 16403 Phone: (814) 398-4611
Dean of College: N. Wagner

DEPARTMENT OF BIOLOGY

Chairman of Division, N. Wagner
Assistant Professors: D.L. Graybill, J. Makarski, J.A. Wilson

Degree Programs: Biology

Undergraduate Degree: B.S.
Undergraduate Degree Requirements: General Biology, Genetics, plus 21 additional hours in Biology, General Chemistry, Organic Chemistry, Statistics

ALVERNIA COLLEGE

(No information available)

BEAVER COLLEGE

(No reply received)

BLOOMSBURG STATE COLLEGE

Bloomsburg, Pennsylvania 17815 Phone: (717) 489-2400
Dean of Graduate Studies: H.F. Heller (Acting)
Dean of College: E.J. Drake

DEPARTMENT OF BIOLOGY
Phone: Chairman: Ext. 2409 Others: Ext. 0111

Chairman: C.L. Himes
Professors: J.E. Cole, P.A. Farber, M. Herbert, C.L. Himes, J.J. Klenner, J.R. Kroschewsky, T.R. Manley, D.D. Rabb, J.P. Vaughan
Associate Professors: G.J. Gellos, L.V. Mingrone, S. A. Rhodes, R.G. Sagar
Assistant Professor: J.R. Fletcher
Graduate Assistants: 2

Field Stations/Laboratories: Marine Science Consortium Wallops Island, Virginia and Lewes, Delaware

Degree Programs: Biology, Biology Education, Medical Technology

Undergraduate Degrees: B.A.,B.S., B.S.Ed.
Undergraduate Degree Requirements: Not Stated

Graduate Degrees: M.S., M.Ed.
Graduate Degree Requirements: Master: 30 credits Post Baccalaureate including Thesis (M.S.) or 30 credits Post Baccalaureate including Directed Study (M.Ed.)

Graduate and Undergraduate Courses Available for Graduate Credit: Radiation Biology (U), Evolution (U), Studies in Speciation (U), Southeast States, Southwest State Rocky Mountain States, Cytology and Cytogenetics (U), Ethology, Environmental Microbiology (U), Biological Photographic Techniques (U), Cell Physiology (U), Systematic Zoology, Systematic Botany, Developmental Biology, Biochemical Genetics, Conservation of Biological Resources, Limnology,

Animal Ecology, Plant Ecology, Parasitology, Biology of Arthropods, Entomology, Ichthyology, Herpetology, Vertebrate Morphology, Endocrinology, Comparative Animal Physiology, Thesis

BRYN MAWR COLLEGE
(No reply received)

BUCKNELL UNIVERSITY
Lewisburg, Pennsylvania 17837 Phone: (717) 524-1124
Dean of Graduate Studies: P.H. DeHoff
Dean of College: Vacant

BIOLOGY DEPARTMENT
Phone: Chairman: (717) 524-1124

Chairman: J.E. Harclerode
Herbert Spencer Associate Professor of Biology: D. Pearson
Professors: J.E. Harclerode, H. Magalhaes
Associate Professors: D. Pearson, W. McDiffett, R. Ellis
Assistant Professors: W. Abrahamson, D. Hoffman, J. Lonski, S. Nyquist, J. Tonzetich
Instructor: B. Powers
Research Assistants: 1

Degree Program: Biology

Undergraduate Degrees: B.A., B.S.
Undergraduate Degree Requirements: 8 Biology courses including: Principles of Biology, Organismal Physiology, Molecular and Cellular Biology, and Population Biology, 2 semesters Organic Chemistry, 2 semesters Calculus, 2 semesters Physics, 2 major-related courses

Graduate Degrees: M.A., M.S., M.A.T.
Graduate Degree Requirements: Master: 7 graduate courses plus thesis

Graduate and Undergraduate Courses Available for Graduate Credit: Microbial Physiology (U), Microbiology (U), Teaching of Biology (U), Botany (U), Chordate Morphology (U), Developmental Morphology (U), Seminar (U), Advanced Physiology (U), Teratology (U), Principles of Genetics (U), Cytogenetics (U), Comparative Endocrinology (U), Radiation Biology (U), Limnology (U), Metabolism (U) Advanced Plant Physiology (U), Recent Advances in Biology (U), Advanced Developmental Biology (U), The Natural Community (U), Marine Biology (U), Immunobiology (U), Biology of Cellular Organelles (U), Advanced Cell Biology (U), Invertebrate Zoology (U), Graduate Seminars, Graduate Research, Thesis

CABRINI COLLEGE
Eagle and King of Prussia Roads Phone: (215) 687-2100
Radnor, Pennsylvania 19087
Dean of College: D.M. Brown

BIOLOGY DEPARTMENT
Phone: Chairman: Ext. 67

Chairman: A.C. Kruse
Associate Professors: J.E. DeTurck
Assistant Professors: A.C. Kruse, R.N. Verdile
Instructor: A. Curry

Degree Programs: Biology, Medical Technology

Undergraduate Degree: B.S.
Undergraduate Degree Requirements: 130 credits with 28 credits in Biology, 26 credits in related fields and 26 credits from upper division courses. The remainder are general education requirement credits.

CALIFORNIA STATE COLLEGE
California, Pennsylvania 15419 Phone: (412) 938-4200
Dean of Graduate Studies: Pavlak
Dean of College: Coleman

BIOLOGICAL SCIENCES DEPARTMENT
Phone: Chairman: (412) 938-4200

Chairman: B.B. Hunter
Professors: J.W. Balling, S.C. Bausor, F.E. Billheimer, W.L. Black, T.P. Buckelew, W.E. Gabor, S. Hood, B.B. Hunter, W.R. Lister, T.C. Moon, R.J. Serinko, E. Zadorozny
Associate Professors: M.M. Bailey, W.H. Buell, R.A. Catalano, E.C. Krueger, W.E. Slosky, M A. Sylvester
Assistant Professor: P. Chang
Research Assistant: 1

Degree Programs: Biology, Environmental Science, Medical Technology

Undergraduate Degree: B.S.
Undergraduate Degree Requirements: Not Stated

Graduate Degree: M.S., M.Ed.
Graduate Degree Requirements: Master: M.S. - Research Thesis plus 26 credits, M.Ed. - research project or thesis 28 or 26 credits

Graduate and Undergraduate Courses Available for Graduate Credit: Not Stated

CARLOW COLLEGE
Pittsburgh, Pennsylvania 15213 Phone: (412) 683-4800
Dean of College: Sr. E. McMillan

DEPARTMENT OF BIOLOGY
Phone: Chairman: Ext. 282

Chairman: W.A. Uricchio
Professors: H.M. Uhlrich, W.A. Uricchio
Assistant Professors: R. McCloskey, J. McGivern

Degree Programs: Biology, Medical Technology

Undergraduate Degrees: B.S., B.A.
Undergraduate Degree Requirements: Seven to ten Biology courses, Four to six courses in Chemistry, Two courses in Physics, two to three courses in Mathematics

CARNEGIE MELLON UNIVERSITY
Pittsburgh, Pennsylvania 15213 Phone: (412) 621-1100
Dean of Graduate Studies: A.F. Strehler
Dean of College: A.A. Bothner-By

DEPARTMENT OF BIOLOGICAL SCIENCES
Phone: Chairman: Ext. 608

Head: R.V. Rice
Professors: C. R. Rice, R.V. Davies, T. Harrison, C.R. Worthington
Associate Professors: R.D. Goldman, D.J. Hartshorne, E.W. Jones, P.A. Lemke (Adjunct), J.F. Nagle
Assistant Professors: W.E. Brown, J.R. Ellison, R.S. Fager, W.S. Kelley, J.E. Mayfield, H. Morimoto
Research Staff: 16

Degree Programs: Biochemistry, Cell Biology, Molecular Biology

Undergraduate Degrees: B.A., B.S.
Undergraduate Degree Requirements: Total minimum units required: 385. For B.S.: 9 Mathematics, 9 Physics, 54 Chemistry, 72 Biology for standard B.S., or specialized Biology Research Program. B.S.: 9 Mathematics, 9 Physics, 54 Chemistry, 63 Biology

Graduate Degree: Ph.D.
Graduate Degree Requirements: No standard require-

ments; each student is individually directed.

Graduate and Undergraduate Courses Available for Graduate Credit: Not Stated

CEDAR CREST COLLEGE

Allentown, Pennsylvania 18104 Phone: (215) 437-4471
Dean of College: C.E. Peterson

DEPARTMENT OF BIOLOGY
Phone: Chairman: Ext. 272 Others: Ext. 273

Chairman: M. Kayhart
Professor: M. Kayhart
Associate Professor: M.S. Ubben
Assistant Professors: R. Halma, R. Scott
Instructors: B. Benson, S. Dierolf

Degree Programs: Biology, Medical Technology

Undergraduate Degrees: B.A., B.S.
Undergraduate Degree Requirements: Biology: Biology, Microbiology, Ecology, Genetics, Senior Seminar, plus two additional upper-level courses. Chemistry: Inorganic, Organic, Mathematics: one year

CHATHAM COLLEGE
(No reply received)

CHESTNUT HILL COLLEGE
(No reply received)

CHEYNEY STATE COLLEGE

Cheyney, Pennsylvania 19319 Phone: (215) 399-6880
Dean of Graduate Studies: C. Coleman
Dean of Arts and Sciences: J.A. Jones
Vice President for Academic Affairs: B.S. Proctor

BIOLOGY DEPARTMENT
Phone: Chairman: Ext. 323 Others: 324

Chairman: C.J. Lane
Professors: L. Eisenstat, R.J. Wallace
Associate Professors: T. Anderson, C.J. Lane, K.A. Onyia, M.L. Overton, G. Samuel
Assistant Professors: J.L. Robinson, R. Sutcliffe

Degree Program: Biology

Undergraduate Degree: B.S.
Undergraduate Degree Requirements: A. Curriculum: Liberal Arts (1) Core - General Education Requirements: 54 semester hours, (2) Major Requirements - Zoology I and II, Botany I and II, General Physiology I and II, Biology Seminar, Biology Electives, (3) Inorganic Chemistry I and II and General Electives B. Curriculum: Secondary Education-Biology (1) Core - General Education Requirement - 54 semester hours (2) Zoology I and II, Botany I and II, Biology Seminar and Biology Electives (3) Inorganic Chemistry I and II, Physics I and II, Earth Science, and Electives (4) Professional Education Course (5) Thesis

Graduate Degrees: M.S.
Graduate Degree Requirements: Master: General Science: Liberal Arts I. Core in Area of Concentration (Biology) - Special Problems and Biology Seminar (3 semester hours) II. Biology Concentration (16 semester hours) III. Related Science (Chemistry, Physics, Earth and Space Science (15 semester hours) Master of Education in General Science: Secondary Education - I. Professional Core (9 semester hours) - II. Biology Core - (3 semester hours) - III. Biology Concentration (12 semester hours) - IV. Related areas (6-10 semester hours) Thesis (4 semester hours)

Graduate and Undergraduate Courses Available for Graduate Credit: Histological and Histochemical Techniques, Current Advances in Biological Sciences, Current Advances in Biological Techniques, Plant Systematics, Animal Systematics, Microbial Genetics, Comparative Animal Physiology, Plant Physiology Animal Ecology, Plant Ecology, Contemporary Ecology and Ecological Problems, History of Biology, Evolution, Biological Literature, Seminar, and Thesis

CLARION STATE COLLEGE

Clarion, Pennsylvania 16214 Phone: (814) 226-6000
Dean of Graduate Studies: W. McCauley
Dean of College: J. Bodoh

DEPARTMENT OF BIOLOGICAL SCIENCES
Phone: Chairman: Ext. 530

Chairman: B.H. Dinsmore
Professors: W.D. Chamberlain, B.H. Dinsmore, G.A. Harmon, W.R. Kodrich, K.R. Mechling, J.R. Moore, J.E. Williams
Associate Professors: E.C. Aharrah, J.A. Donachy, N. R. Donachy, K.J. Linton, G.L. Twiest
Assistant Professors: D.O. Cook, W. Jetkiewicz, G.A. McCaslin

Degree Programs: Biology, Biological Education, Medical Technology

Undergraduate Degrees: B.A., B.S.
Undergraduate Degree Requirements: B.S. in Biology - A minimum of 128 credits for graduation of which 64 are in the field of Biology and other sciences. B.S. in Medical Technology. A minimum of 128 credits for graduation, 96 on the campus and 32 in an approved hospital laboratory. B.A. in Biology. A minimum of 128 credits for graduation, including a foreign language, and at least 53 credits ofBiology and other science courses. B.S. in Secondary Education (Biology Major): A minimum of 128 credits for graduation including 50 credits of Biology and other Sciences, and one semester student teaching.

Graduate Degrees: M.S., M.Ed.
Graduate Degree Requirements: Master: M.S. Biology - A minimum of 30 credits including, 6 credits of research and thesis. M.Ed. - Biology - A minimum of 30 credits

Graduate and Undergraduate Courses Available for Graduate Credit: Radiation Biology (U), Microbial Physiology (U), Immunology (U), Animal Physiology (U), Plant Physiology (U), Comparative Vertebrate Morphology (U) Vertebrate Embryology (U), Histology (U), Animal Ecology (U), Plant Ecology (U), Parasitology (U), Evolution (U), Graduate Seminar, Advanced Radiation Techniques, Experimental Designs in Biology, Special Topics in Biology, Identification and Quantification of Terrestrial Plants, Identification and Quantification of Invertebrates, Indentification and Quantification of Aquatic Plants, Identification and Quantification of Vertebrates, Fisheries Biology, Microbial Physiology, Virology, Immunology and Epidemiology, Cell Physiology, Advanced Animal Physiology, Advanced Plant Physiology, Advanced Developmental Biology, Experimental Embryology, Plant Morphology, Behavioral Ecology, Microbial Ecology, Aquatic Community Ecology, Terrestrial Community Ecology, Limnology, Ecology of Aquatic Insects, Experimental Ecology, Fungal Ecology, Forest Ecology, Biome Studies, Alpine Ecology, Advanced Topics in Genetics, Microbial Genetics, Population Genetics

COLLEGE MISERICORDIA

Dallas, Pennsylvania 18612 Phone: (717) 675-2181

DEPARTMENT OF BIOLOGY

Phone: Chairman: Ext. 251 Others: Ext. 291

Chairman: S. Knapich
Associate Professors: T. Petrychenko, S. Knapich
Assistant Professors: C. Konecke, Sr. M.E. Convery

Degree Programs: Biology, Medical Technology

Undergraduate Degrees: B.S.
Undergraduate Degree Requirements: 32 hours Biology, 23 hours Chemistry, 8 chours Calculus, 4 hours Physics

DELAWARE VALLEY COLLEGE OF SCIENCE AND AGRICULTURE
Route 202 and New Britain Road Phone: (215) 545-1500
Doylestown, Pennsylvania 18901
Dean of College: J. Feldstein

DEPARTMENT OF AGRONOMY
(No reply received)

DEPARTMENT OF ANIMAL HUSBANDRY
Phone: Chairman: Ext. 323 Others: Exts. 231,322,324

Chairman: T. Pelle
Professor: T. Pelle
Assistant Professors: J. Plummer, F. Hofsaess, G. Brubaker, C. Hill
Instructors: R. Gilbert, H. Thirey, J. McCaffree
Visiting Lecturers: 5-6

Degree Programs: Animal Husbandry

Undergraduate Degree: B.S.
Undergraduate Degree Requirements: 4 years, 136 credits and off campus supervised program in own field for 24 weeks

DEPARTMENT OF BIOLOGY
Phone: Chairman: Ext. 261

Chairman: E. French
Professors: L.M. Adelson, E. French
Associate Professors: R. Deering, J. Mertz
Assistant Professors: W. Allison, R. Berthold, J. Miller
Instructor: R. Johnson, J. Standing

Degree Program: Biology

Undergraduate Degree: B.S.
Undergraduate Degree Requirements: 32 credits in general studies. Total of 133-136 credits are required

HORTICULTURE DEPARTMENT
Phone: Chairman: Ext. 334

Chairman: C.R. Blackmon
Professors: C.R. Blackmon, D. Blumenfield
Assistant Professors: L. Polites, N.J. Vincent, B. Muse
Instructors: D.A. Claycomb, F.T. Wolford, L.R. Zehnder

Degree Programs: Horticulture, Ornamental Horticulture

Undergraduate Degree: B.S.
Undergraduate Degree Requirements: 136 credit hours

DEPARTMENT OF DAIRY HUSBANDRY
Phone: Chairman: Ext. 320

Chairman: J.R. Plummer
Professors: G. Brubakery, F. Itofsaess, G. Hill H.E. Thirey

Degree Program: Dairy Science

Undergraduate Degree: B.S.
Undergraduate Degree Requirements: Not Stated

DICKINSON COLLEGE
Carlisle, Pennsylvania 17013 Phone: (717) 243-5121
Dean of College: G. Allan

DEPARTMENT OF BIOLOGY
Phone: Chairman: Ext. 329

Chairman: W.B. Jeffries
Professors: P. Biebel, W.B. Jeffries, B.B. McDonald, D.J. McDonald
Associate Professors: R.M. Lane
Assistant Professors: F.J. Shay
Instructor: C.J. Ralph
Research Assistants: 6
Teaching Assistants: 22

Field Stations/Laboratories: The Florence Jones Reineman Wildlife Sanctuary - Perry County, Pennsylvania

Degree Program: Biology

Undergraduate Degree: B.S.
Undergraduate Degree Requirements: Major: seven courses including one of the following upper-level courses in Botany: Biology 222, 223, or 225. Chemistry 131, 132, 251, 252, Mathematics 121, 122 or 141,142; and Physics 111, 112 or 131, 132 are required The seven courses required for the major may not include more than one course in independent study. Minor: six courses. In addition, Chemistry 131, 132 are required.

DREXEL UNIVERSITY
32nd and Chestnut Street Phone: (215) 895-2000
Pheladelphia, Pennsylvania 19104
Dean of Graduate Studies: O. Witzell

College of Science
Dean: F.K. Davis

DEPARTMENT OF BIOLOGICAL SCIENCES
Phone: Chairman: (215) 895-2624 Others: 895-2000

Head: J.W.A. Burley
Betz Professor of Ecology: W.O. Pipes
Obold Professor of Biology: W.R. Nes
Professors: J.W.A. Burley, W.R. Nes, W.O. Pipes, K.P. West
Associate Professors: E. Fromm, J. Lipetz, A. List, I.R. Moore, S. Segall
Assistant Professors: S.E. Dubin, S.L. Rotenberg
Visiting Lecturers: 2
Research Assistants: 3

Degree Programs: Biology, Environmental Science, Biochemistry, Environmental Engineering and Science, Biomedical Engineering and Science

Undergraduate Degrees: B.S. (Biology)
Undergraduate Degree Requirements: 16 quarter hours of Calculus, 13 quarter hours of Physics, 22 quarter hours of Chemistry, 9 quarter hours of Biochemistry, 74 quarter hours of Biology

Graduate Degrees: M.S. (Biology, Environmental Science) Ph.D.
Graduate Degree Requirements: Master: 45 credit hours including 9 credits maximum for thesis research. Doctoral: 1 year, full-time, residency requirement, proficiency in 1 foreign language, written and oral preliminary examination, dissertation including a defense, 3 years of full time study beyond the bachelor's degree is normally required to complete the requirements for the degree.

Graduate and Undergraduate Courses Available for Graduate Credit: Biological Sciences: Life Science, Physiology of Growth and Development, Cell Physiology (U), Endocrinology (U), Radioisotope Methodology (U),

Biochemistry I (U), Biochemistry II (U), Special Topics in Biological Sciences. Biomedical Engineering and Science: Anatomy I, Anatomy II, Physics of Living Systems, Medical Sciences I, Medical Sciences II' Medical Sciences III, Medical Sciences IV, Principles of Bioengineering and Instrumentation I, II, III, Biosimulation I, II, Pharmacology, Biological Control Systems I, II, III, Advanced Physiology Seminar Environmental Engineering and Science: Ecology, Environmental Physiology, Sanitary Microbiology, Aquatic Ecology, Environmental Health, Epidemiology, Toxicology, Radiological Health, Radiobiology

College of Engineering

Dean: R. Woodring

DEPARTMENT OF BIOMEDICAL ENGINEERING AND SCIENCE
Phone: Chairman: 895-2240 Ext. 41

Director: H.H. Sun
Professors: R. Beard, W. Burley
Associate Professors: E. Fromm, F. DiMeo, T. Moore, G. Moskowitz, O. Tretiak
Assistant Professors: S. Dubin, I. Kamel, R. Klafter, E. Kresch
Visiting Lecturers: 4
Teaching Fellows: 2
Research Assistants: 15

Degree Programs: Biomedical Engineering, Biomedical Sciences

Graduate Degrees: M.S., Ph.D.
Graduate Degree Requirements: Master: 45 credits with research thesis Doctoral: 3 years minimum residence credits, Candidate Examination, Research Thesis and Oral Examination

Graduate And Undergraduate Courses Available for Graduate Credit: Medical Sciences, Pharmacology, Anatomy, Physics of Living Systems, Analog Computer, Principles of Bioengineering and Instrumentation, Biosimulation I and II, Biological Control Systems

DUQUESNE UNIVERSITY

Pittsburgh, Pennsylvania 15219 Phone: (412) 434-6332
Dean of Graduate Studies: F.B. Gross
Dean of College: J.A. McCulloch

DEPARTMENT OF BIOLOGICAL SCIENCES
Phone: Chairman: (412) 434-6333 Others: 434-6332

Chairman: H.G. Ehrlich
Professors: H.G. Ehrlich, P.T. Liu, E.I. Sillman
Associate Professors: F. Baron, P.A. Castric, E.C. Raizen, T. Subhas
Assistant Professor: S. Thomas
Research Associate: 1

Degree Program: Biology

Undergraduate Degree: B.S.
Undergraduate Degree Requirements: Not Stated

Graduate Degree: M.S.
Graduate Degree Requirements: Not Stated

Graduate and Undergraduate Courses Available for Graduate Credit: Not Stated

EAST STROUDSBURG STATE COLLEGE

East Stroudsburg, Pennsylvania 18301 Phone: (717) 424-3211
Dean of Graduate Studies: J. Reed
Dean of Faculty of Sciences: H. Shwe

DEPARTMENT OF BIOLOGY
Phone: Chairman: (717) 424-3343 Others: 424-3211

Chairman: H.N. Fremount
Professors: H.N. Fremount, B.L. Haase, R.C. Kelsey, B.K. Rao, L M. Rymon, R.K. Salch
Associate Professors: N.O. Anderson, F.B. Buser, W.E. Eden, B.K. Knapp
Graduate Assistants: 4

Field Stations/Laboratories: Member of Marine Science Consortium which has stations at Lewes, Delaware and Wallops Island, Virginia

Degree Programs: Biology, Medical Technology

Undergraduate Degrees: B.A., B.S., B.S. (Sec. Ed.)
Undergraduate Degree Requirements: B.A. and B.S. Biology 30 (minimum); Chemistry 16 (including Organic Chemistry I and II, Mathematics 6, Physics 8
B.S. in Ed. - Biology 30, Chemistry 16 (including Organic Chemistry I and II), Mathematics 6, Physics 4 (All in semester hours)

Graduate Degrees: M.Ed., M.S.
Graduate Degree Requirements: (In semester hours), Master: equivalent of undergraduate major M.Ed. (non-thesis) - Biology 22 (minimum), related courses 6, general and professional education 6, M.Ed. (Thesis), - Biology 15 (minimum), Thesis 3, related courses 6, general and professional education 6, M.S. - Biology 18 (minimum), Thesis 6, related areas 6 (maximum)

Graduate and Undergraduate Courses Available for Graduate Credit: Man and His Environment, Zoogeography, Plant Ecology, Animal Ecology, Graminology, Bryology Plant Anatomy, Advanced Mycology, Insect Morphology, Insect Physiology, Medical Entomology, Pathogenic Micro-organisms, Protozoology, Helminthology Molecular Genetics, Developmental Genetics, Organic Evolution, Developmental Biology, Endocrinology, Endocrinology of Reproduction, Comparative Physiology, Cell Physiology, Physiological Biochemistry, Cytology, Histochemistry, Ecology of Fishes, Limnology, Biology of the Plankton, Biology of Aquatic Insects, Radiation Biology, Biological Instrumentation, Introduction to Research, History of Marine Science Consortium, General Entomology, Ecology of Water Pollution

EASTERN BAPTIST COLLEGE

(see Eastern College)

EASTERN COLLEGE

St. Davids, Pennsylvania 19087 Phone: (215) Mu8-3300
Dean of College: H.H. Howard

BIOLOGY DEPARTMENT
Phone: Chairman: Ext. 347

Chairman: J. Moore
Associate Professors: J. Moore, J.K. Sheldon
Assistant Professors: C.A. Padgett, M W. Meyers

Degree Programs: Biology

Undergraduate Degree: A.B.
Undergraduate Degree Requirements: Introduction to Cell Biology (4) General Botany (4), Invertebrate (3) or Vertebrate Zoology (4), Ecology (U), Genetics (4), General Physiology (4) and Biological Problems and Literature (3), plus electives for minimum 18 hours above Cell Biology andBotany, 24 hours other Sciences, Liberal Arts Requirement Total 126 credits

EDINBORO STATE COLLEGE

Edinboro, Pennsylvania 16412 Phone: (814) 732-2000
Dean of Graduate Studies: J. Williams
Dean of College: R.I. Weller

DEPARTMENT OF BIOLOGY
Phone: Chairman: (814) 732-2445 Others: 732-2000

Chairman: F.T. Bayliss
Professors: S. Bowne, R. Carls, D. DeFigio, T. Legge, B. Lowenhaupt, J. Paxson, D. Snyder, R. Tammariello, P. Thomas, D. Wheeler, E. Williams
Associate Professors: F. Bayliss, E. Bernice, S. Fertig, E. Friend, H. Harpst, R. Mitchell, C. Robinson, P. Zarenko
Assistant Professors: F. Fisher, E. Kline, J. Spaulding, D. Wilcox, K. Wilz
Research Assistants: 2

Field Stations/Laboratories: Cooperative program with University of Pittsburgh "Pymatuning Laboratory of Ecology

Degree Programs: Biology, Medical Technology, Secondary Education in Biology, Botany, Zoology, Microbiology - Molecular Biology, Developmental Biology - Genetics

Undergraduate Degrees: B.A., B.S.
Undergraduate Degree Requirements: B.S. Biology - 36 hours in Biology (Required: Primate Biology, Zoology, Botany, Genetics, Physiology and Ecology), 21 hours Chemistry, 8 hours Physics, 8 hours Calculus B.S. M.T., B.A., B.A. Sec. Ed.) - 30 hours Biology and 20 hours Chemistry plus 8 hours Physics plus 8 hours Mathematics

Graduate Degree: M.S.
Graduate Degree Requirements: Not Stated

Graduate and Undergraduate Courses Available for Graduate Credit: Ecology (U), Mycology (U), Plant Anatomy (U), Algology (U), Entomology (U), Parasitology (U), Ornithology (U), Embryology (U), Field Zoology and Systematics (U), Radiation Biology (U), Marine Science Courses (Approximately 15), Behavioral Ecology, Forest Ecology, Terrestrial Ecology, Limnology, Modern Genetics, Human Genetics, Organic Evolution, Basic Concepts in Biology for Elementary Teachers, Developmental Genetics, Bacterial Physiology, Ecology of Bacteria, Plant Morphogenesis, Plant Pathology, Ichthyology, Immunology, Mammalogy, Biophysics, Independent Study, Seminar in Biology, Thesis

ELIZABETHTOWN COLLEGE

Elizabethtown, Pennsylvania 17022 Phone: (717) 367-1151
Dean of College: R.L. Hanle

DEPARTMENT OF BIOLOGY
Phone: Chairman: (717) 367-1151

Chairman: R.E. Pepper
Professor: R.E. Pepper
Associate Professors: J.R. Heckman, F. Hoffman, M.J. Kenney, R.L. Laughlin
Assistant Professor: J.L. Dively

Degree Programs: Biology, Forestry

Undergraduate Degree: B.S.
Undergraduate Degree Requirements: College requirements: 128 semester hours including core requirements. Department: 30 semester hours Biology, 20 semester hours Chemistry, 8 semester hours Physics, 6 semester hours Mathematics, language B.S. Forestry - cooperative program with Duke University

FRANKLIN AND MARSHALL COLLEGE

Lancaster, Pennsylvania 17604 Phone: (717) 393-3621
Dean of College: R. Traina

DEPARTMENT OF BIOLOGY
Phone: Chairman: Ext. 285

Chairman: Not Stated
B.F. Fackenthal Chair of Biology: J.J. McDermott
E.P. and F.H. Reiff Chair of Biology: K.R. John
Professors: K.R. John, J.J. McDermott
Associate Professors: D B. King, J.L. Richardson
Assistant Professors: I.N. Feit, J.W. Ferner, D.W. Hosier, C.S. Pike, W.F. Ward

Degree Program: Biology

Undergraduate Degree: B.S.
Undergraduate Degree Requirements: Eight courses in Biology

GANNON COLLEGE

Erie, Pennsylvania 16501 Phone: (814) 456-7523
Dean of Graduate Studies: J. Scottino
Dean of College: Vacant

DEPARTMENT OF BIOLOGY
Phone: Chairman: Ext. 295

Chairman: Rev. A. O'Toole
Professors: R.A. Gammon, E.F. Kohlmiller
Associate Professors: A.J. O'Toole, S.J. Zagroski
Assistant Professors: Rev. J.C. Gregorek
Visiting Lecturer: 1

Degree Programs: Biology, Biological Education, Environmental Science, Medical Technology

Undergraduate Degree: B.S.
Undergraduate Degree Requirements: 40 credits in Biology, 16 credits Chemistry, 8 credits Physics, 6-8 credits Mathematics, 60 credits Liberal Studies and Electives

Graduate Degree: M.Ed.
Graduate Degree Requirements: Master: 6 credits Education, 12-18 credits Science, 6 credits Thesis

Graduate and Undergraduate Courses Available for Graduate Credit: Limnology (U), Modern Genetics (U), Microbial Physiology, Environmental Systems, Environmental Problems, Cell Biology

GENEVA COLLEGE

Beaver Falls, Pennsylvania 15010 Phone: (412) 846-5100

DEPARTMENT OF BIOLOGY
Phone: Chairman: Ext. 335

Chairman: T.M. McMillion
Professors: T.M. McMillion, M.W. Wing
Associate Professors: J. Cruzan, R.P. Fatula, C.B. Freeman
Assistant Professors: J.A. Brushaber

Degree Program: Biology

Undergraduate Degree: B.S.
Undergraduate Degree Requirements: 30 semester hours in Biology, two years of Chemistry, one year of Physics, one year of Mathematics

GETTYSBURG COLLEGE

Gettysburg, Pennsylvania 17325 Phone: (717) 334-3131
Dean of College: J.D. Pickering

DEPARTMENT OF BIOLOGY
Phone: Chairman: Ext. 244

Chairman: N W. Beach
Charles H. Graff Professor of Biology: R.D. Barnes
Professors: R.D. Barnes, W.C. Darrah
Associate Professors: N.W. Beach, A.R. Cavaliere
Assistant Professors: H.H. Darrah, S.S. Hendrix,

R.E. Logan, A.C. Schroeder, J.R. Winkelmann
Instructors: J.E. Mikesell
Laboratory Instructors: 7 (Part-time)

Degree Program: Biology

Undergraduate Degree: B.A.
Undergraduate Degree Requirements: 8 courses in Biology and one year Inorganic Chemistry

GROVE CITY COLLEGE
Grove City, Pennsylvania 16127 Phone: (412) 458-6600
Dean of College: E. Groesbeck

DEPARTMENT OF BIOLOGY
Phone: Chairman: Ext. 281

Chairman: M.W. Fabian
Professor: M.W. Fabian
Associate Professors: A.D. Brower, F.J Brenner
Assistant Professor: G.G. Hartman

Degree Program: Biology

Undergraduate Degree: B.S.
Undergraduate Degree Requirements: 36 semester hours Biology, 8 semester hours Chemistry, 4 semester hours Calculus

GWYNEDD-MERCY COLLEGE
Gwyneed Valley, Pennsylvania 19437 Phone: (215) Mi6-7300
Academic Dean: Sr. M. Colman

NATURAL SCIENCE DIVISION
Phone: Chairman: Ext. 247

Chairman: Sr. M. Paul
Professors: G. de la Vega,
Associate Professors: Sr. E. Marie, Sr. M. Colman, Sr. M. Paul
Assistant Professors: Sr. B. Mary, F. Corsaro, H. McDevitt,
Instructors: E. Gillis, Sr. M Jean
Visiting Lecturers: 7

Degree Programs: Biology, Medical Technology

Undergraduate Degree: B.S.
Undergraduate Degree Requirements: 125 semester hours of credit: Humanities Division 24 semester hours, Language, Literature, Fine Arts Division 18 semester hours, Natural Science 6 - 8 semester hours, Field of Concentration - 30 semester hours, Electives - complete 125 semester hours. Satisfy specific requirements in field of concentration. Minimum quality point average, 2 - earn a grade of C or better in major areas.

HAVERFORD COLLEGE
Haverford, Pennsylvania 19041 Phone: (215) Mi9-9600
Dean ofCollege: D. Potter

DEPARTMENT OF BIOLOGY
Phone: Chairman: Ext. 304

Chairman: A.G. Loewy
Professors: I. Finger, M. Santer
Associate Professors: D. Kessler
Assistant Professors: M. Showe
Instructor: S. Mataric
Research Assistants: 7
Research Associate: 1

Degree Programs: Molecular and Cell Biology

Undergraduate Degree: B.S.
Undergraduate Degree Requirements: 2 sophomore level courses in Molecular and Cell Biology, 2 junior level laboratory courses in Molecular and Cell Biology, 2 Advanced seminars in Molecular and Cell Biology, 2 senior research courses in Molecular and Cell Biology, 1 senior seminar

HOLY FAMILY COLLEGE
Philadelphia, Pennsylvania 19114 Phone: (215) 637-7700
Dean of College: Sr. M. Immaculata

BIOLOGY DEPARTMENT
Phone: Chairman: Ext. 35

Chairman: Sr. R. Ann
Professors: Sr. M. Lillian, J.F. Lontz (Adjunct), Sr. R.A.
Instructors: J. DeStefano, Sr. J. Marie

Degree Programs: Biology, Medical Technology

Undergraduate Degrees: B.A., B.S.
Undergraduate Degree Requirements: 30 hours in Biology excluding General Biology, 16 hours (minimum) Chemistry, 3-6 Mathematics, 4-8 Physics

IMMACULATA COLLEGE
Immaculata, Pennsylvania 19345 Phone: (215) 647-4400
Dean of College: Sr. M. Roseanne

DEPARTMENT OF BIOLOGY
Phone: Chairman: Ext. 264

Chairman: B. Piatka
Associate Professors: B. Piatka, Sr. R. Maria
Assistant Professors: Sr. M. Bernard

Degree Program: Biology

Undergraduate Degree: A.B.
Undergraduate Degree Requirements: Calculus, General Physics, Inroganic and Organic Chemistry, 20-30 credits in Biology, depending on concentration.

INDIANA UNIVERSITY OF PENNSYLVANIA
Indiana, Pennsylvania 15701 Phone: (412) 357-2352
Dean of Graduate Studies: J. Gallanar
Dean of College: F. McGovern

BIOLOGY DEPARTMENT
Phone: Chairman: (412) 357-2352 Others: 357-2100

Chairman: F.W. Liegey
Professors: G. Ferrence, W.W. Gallati, L.L. Gold, F.W. Leigey, G.F. Schrock, M.L. Stapleton, R. Strawcutter, H.H. Vallowe, R.F. Waechter, C. Zenisek
Associate Professors: F.T. Baker, T.E. Conway, W. Forbes, L.L. Hue, J. Humphreys, R.E. Merritt, J.H. Miller, R.N. Moore, J.L. Pickering, W. Waskoskie
Assistant Professors: M. Charnego, W.E. Dietrich, Jr., W.S. Greaves

Field Stations/Laboratories: College Lodge Station (300 acres), Indiana, Pennsylvania, Marine Science Consortium, Lewes, Delaware, Wallops Isle, Virginia, Ivan McKeaver Resource Education Center - Clarion County, Pennsylvania, Mahoning River Station - Yellow Creek State Park

Degree Programs: Biology, Biological Education

Undergraduate Degrees: B.A., B.S., B.S.Ed.
Undergraduate Degree Requirements: 124 credit hours 33 Biology, 20 Chemistry, 8 Physics, 8 Mathematics

Graduate Degrees: M.S., M.S.Ed.
Graduate Degree Requirements: 16-22 semester hours, Thesis

Graduate and Undergraduate Courses Available for Graduate Credit: Biometry, Molecular Genetics, Instrumentation, Comparative Plant Morphology, Protozoology, Microtechnique, Dendrology, Plant Ecology, Taxonomy of Plants, Principles of Animal Taxonomy, Advanced Ornithology, Animal Ecology, Animal Morphology, Endocrinology, Advanced Entomology, Herpetology, General Physiology, Animal Physiology, Physiology of Plants, Biology Practicum, Mammalogy, Radiation Biology, Mycology, Parasitology, Micrology Physiology, Pathogenic Microbiology, Immunology, Taxonomy and Ecology of Bacteria, Methods of Research in Biology

JUNIATA COLLEGE

Huntingdon, Pennsylvania 16652 Phone: (814) 643-4310

BIOLOGY DEPARTMENT
Phone: Chairman: Ext. 71

Chairman: R.P. Zimmerer
Professors: K.H. Rockwell, R.P. Zimmerer
Associate Professors: R.L. Fisher, J.P. Senft, J.L. Gooch
Assistant Professor: J.W. Moser

Field Stations/Laboratories: Lake Raystown - Huntingdon, Pennsylvania

Degree Programs: Biology, Medical Technology

Undergraduate Degree: B.S.
Undergraduate Degree Requirements: 15 units in a program of emphasis

KING'S COLLEGE

133 N. River Street Phone: (717) 824-9931
Wilkes-Barre, Pennsylvania 18711
Dean of College: R. Schleich

DEPARTMENT OF BIOLOGY
Phone: Chairman: Ext. 218 Others: Ext. 236

Chairman: R.A. Paoletti
Professors: W.H. Donahue, E. Jenkins (Adjunct, Research), E.J. Minsavage
Associate Professors: C.C. Lee, R.A. Paoletti, T.J. Tobin
Assistant Professor: H.J. Wehman

Degree Programs: Biology

Undergraduate Degree: B.S.
Undergraduate Degree Requirements: 120 credits total, 32 biology, 6 Mathematics, 18 Chemistry, 8 Physics

KUTZTOWN STATE COLLEGE

Kutztown, Pennsylvania 19530 Phone: (215) 683-3511
Dean of Graduate Studies: P. Drumm
Dean of College: W. Warzeski

DEPARTMENT OF BIOLOGY
Phone: Chairman: Ext. 331 or 431

Chairman: W.A. Green
Professors: W.A. Green, F.P. Muzopappa, C.W. Yarrison
Associate Professors: J.W. Bahorik, P.J. Duddy, D.C. Evans, N.P. Laufer, J. Piscitelli, R.R. Rhein, R.H. Seewald, G.R. Webb
Assistant Professors: R.G. Gray, S.C. Gundy
Teaching Fellows: 1

Field Stations/Laboratories: Lewes, Delaware

Degree Programs: Biology, Medical Technology

Undergraduate Degrees: A B., B.S.
Undergraduate Degree Requirements: 128 semester hours Total, 60 semester hours General Education, 8 semester hours Organic Chemistry, 1 semester hour Calculus or Statistics, 3 semester hours Scientific Writing, 30 semester hours Biology, 8 semester hours Physics, 18 semester hours Free Electives

Graduate Degrees: M.S., M.S. (Sec. Ed.)
Graduate Degree Requirements: Master: 30 semester hours of Biology - Thesis optional

Graduate and Undergraduate Courses Available for Graduate Credit: Radiation Biology, Evolution, Techniques in Environmental Education

LAFAYETTE COLLEGE

Easton, Pennsylvania 18042 Phone: (215) 253-6281
Dean of College: W. Jeffers

DEPARTMENT OF BIOLOGY
Phone: Chairman: Ext. 287 Others: 286

Head: L.T. Stableford
Professor: L.T. Stableford
Associate Professors: R.S. Chase, B. Fried
Assistant Professors: G. Hoskin, S. Majumdar
Laboratory Associates: 2
Student Laboratory Assistants: 20

Degree Program: Biology

Undergraduate Degree: A.B.
Undergraduate Degree Requirements: 8 courses in the Major Field, Physics, Organic Chemistry

LaROCHE COLLEGE

Pittsburgh, Pennsylvania 15237 Phone: (412) 931-4312
Dean of College: M.D. Henry

DIVISION OF NATURAL SCIENCE
Phone: Chairman: Ext. 163 Others: Ext. 166

Chairman: Sr. K. Angel
Professor: Sr. M. M. Hauser (Emeritus)
Assistant Professors: C. Cooper, Sr. M. Thoma
Instructor: M. Bergt
Visiting Lecturers: 3

Degree Programs: Biology, Medical Technology

Undergraduate Degrees: B.A., B.S.
Undergraduate Degree Requirements: Biology: 32 credits in Biology and supporting studies in Chemistry, Mathematics, and Physics. Medical Technology: 24 credits in Biology and supporting studies in Chemistry Mathematics, and Physics - year internship in Hospital Program. Natural Science: 32 credits in Biology, Chemistry, Mathematics, and Physical Sciences

LA SALLE COLLEGE

20th Street and Olney Avenue Phone: (215) 848-8300
Philadelphia, Pennsylvania 19141
Dean of College: H.N. Albright

DEPARTMENT OF BIOLOGY
Phone: Chairman: Ext. 373

Chairman: J.J. Muldoon
Professors: R. Holroyd (Emeritus), J.S. Penny
Associate Professors: J. Bogacz, T.J. Lowery, J.J. Muldoon, N. Sullivan
Assistant Professors: N.F. Belzer, R. Hawley, R. Ksiazek

Field Stations/Laboratories: Biostation - Penllyn Natural Area, Penllyn, Pennsylvania

Degree Programs: Biology

Undergraduate Degree: B.A.
Undergraduate Degree Requirements: Student must: a) Satisfy College Liberal Arts Requirements, b) Biology Requirements: Chemistry through Organic, Principles of Physics, Mathematics to Calculus, 33 credit hours in Biology (Principles, Physiology, Biochemistry and Others at election).

LEBANON VALLEY COLLEGE

Annville, Pennsylvania 17003 Phone: (717) 867-3561
Dean of College: C.Y. Ehrhart

DEPARTMENT OF BIOLOGY
Phone: Chairman: Ext. 281

Chairman: P.L. Wolf
Associate Professors: O.P. Bollinger (Emeritus), P.L. Wolf
Assistant Professors: J. Argot, D.M. Gring, S. Williams, S. Williams, A.F. Wolfe, A. Henninger-Trax

Degree Program: Biology

Undergraduate Degree: B.S.
Undergraduate Degree Requirements: A total of 120 hours are required which must include the following: 33 hours in Biology, 16 hours in Chemistry, 8 hours in Physics, 3 hours in Calculus, foreign language at the intermediate level (3-6 hours), 6 hours in English Composition, 9 hours in Social Sciences, 9 hours in Humanities, 6 hours in Religion

LEHIGH UNIVERSITY

Bethlehem, Pennsylvania 18015 Phone: (215) 691-7000
Dean of Graduate Studies: R. Stout
Dean of College: J. Hunt

DEPARTMENT OF BIOLOGY
Phone: Chairman: Ext. 581 Others: Exts. 582,583

Chairman: R.G. Malsberger
Professors: S.B. Barber, T.C. Cheng, S.S. Herman, R. G. Malsberger, B.B. Owen
Associate Professors: H.N. Prichard
Assistant Professors: B. Bean, D.M Bell, S. Krawiec
Teaching Fellows: 7
Research Assistants: 3
Endowed Fellowship: 1

Field Stations/Laboratories: Field Laboratory - Wetlands Institute, Stone Harbor, New Jersey 08247

Degree Program: Biology

Undergraduate Degrees: B.A., B.S.
Undergraduate Degree Requirements: B.A. - 28 hours Biology, one year Physics, one year Organic Chemistry, one-half year Analytical or Physical Chemistry. B.S. - 34 hours Biology, one year Physics, one year Organic Chemistry, one year Physical Chemistry

Graduate Degrees: M.S., Ph.D.
Graduate Degree Requirements: Master: 30 hours including one year of Biochemistry and one year of Statistics Doctoral: Approximately 20 hours beyond M.S., one foreign language, and Dissertation

Graduate and Undergraduate Courses Available for Graduate Credit: Advanced Invertebrate Zoology (U), Ecology (U), Aquatic Biology (U), General Biology (U), Vertebrate Embryology (U), Evolution (U), Cell Physiology (U), Animal Physiology (U), Animal Behavior (U), Nonvascular Plants (U), Evolution of Vascular Plants (U), Symbiosis (U), Evolution of Land Plants (U), Biology of Marine Animals (U), Virology (U), Sanitary Microbiology (U), Elements of Biochemistry (U), Advanced Biochemistry (U), Comparative Animal Physiology, Special Topics in Biology, Biological Seminar, Biological Research, Advanced Morphology, General Cytology, Biological Membranes, Advanced Ecology, Cytochemistry, Immunology, Marine Ecology, Biological Oceanography, Morphogenesis of the Lower Invertebrates, The Biology of Transplantation, Biological Electron Microscopy, Growth and Development in Plants, Ethology, Marine Botany, Marine Zooplankton, Marine Science Seminar

LINCOLN UNIVERSITY

Lincoln University, Pennsylvania 19352 Phone: (215) 932-8300
Dean of College: L.D. Johnson

DEPARTMENT OF BIOLOGY
Phone: Chairman: Ext. 272 Others: Exts. 204,205

Chairman: J.L. Harrison
Professors: J.L. Harrison, E.D. Heuser
Associate Professors: D.O. Farny, S.K. Reddy
Assistant Professors: W.G. Bush

Degree Program: Biology

Undergraduate Degree: A.B.
Undergraduate Degree Requirements: 8 courses in Biology, 4 courses in Chemistry (2 in Organic), 2 courses in Physics, 2 courses in College Mathematics, 4 courses in English, 2-4 courses in a Modern Language, 3 courses in Social Sciences (3 different departments) 5 to 7 elective courses of student's own choice, 4 January periods

LOCK HAVEN STATE COLLEGE

Lock Haven, Pennsylvania 17745 Phone: (717) 748-5351
Dean of College: H. Williamson (Acting)

DEPARTMENT OF BIOLOGY
Phone: Chairman: (717) 748-5351

Professors: B.T. Carbaugh, P.F. Klens, R. Scherer, P. Schwalbe, K. Settlemyer

Field Stations/Laboratories: Fresh Water Biological Station - Siege Conference Center, Lamar, Pennsylvania

Degree Programs: Biology, Medical Technology

Undergraduate Degrees: B.A., B.S., B.S.Ed.
Undergraduate Degree Requirements: All programs require 60 semester hours in General Education. Bachelor of Arts degree requires 19 semester hours supporting courses to major and 28 semester in major, plus 12 semester hours foreign language and 9 semester seminars in L.A. B.S. 28 semester hours supporting courses, 40 semester hours major, B.S. in Ed., 28 Major, 21 semester hours supporting

LYCOMING COLLEGE

Williamsport, Pennsylvania 17701 Phone: (717) 326-1951
Dean of College: J. Jose

DEPARTMENT OF BIOLOGY
Phone: Chairman: Ext. 254 Others: Ext. 258

Chairman: R.B. Angstadt
Associate Professor: A.G. Kelley
Assistant Professors: J, Diehl, W. Green, L. Mayers, K.B. Sherbine, R.A. Zaccaria

Degree Program: Biology

Undergraduate Degree: B.A.
Undergraduate Degree Requirements: 1 year Mathematics, 1 year Chemistry, 1 year Introductory Biology plus Microbiology, Genetics, Cell Physiology, Animal Physiology, Ecology and one other Elective

MANSFIELD STATE COLLEGE
Mansfield, Pennsylvania 16933 Phone: (717) 662-2114
Dean of Graduate Studies: D. Peltier
Dean of College: M. Pineus

DEPARTMENT OF BIOLOGY
Phone: Chairman: Exts. 364/365 Others: 364/365

Chairman: C. Weed
Professors: E.B. Gasslier, V.P. Smichowski, C.E. Weed
Associate Professors: R.C. Goff, R.J. Hall, L.R. Honeyell, G.B. Young
Assistant Professor: K.A. Meyer

Degree Programs: Biology, Medical Technology

Undergraduate Degrees: B.A., B.S.
Undergraduate Degree Requirements: B.A. Biology - 64 semester hours General Education, 40 semester hours Biology, B.S. Medical Technology, 64 semester hours, 31 semester hours Biology, 32 semester hours, Clinical Training

MARYWOOD COLLEGE
(No reply received)

THE MEDICAL COLLEGE OF PENNSYLVANIA
3300 Henry Avenue Phone: (215) 842-6000
Philadelphia, Pennsylvania 19129
Dean of Graduate Studies: D. Goldman
Dean of College: R. Slater (Acting)

DEPARTMENT OF ANATOMY
Phone: Chairman: (215) 842-7032 Others: 842-6000

Chairman: L.L. Ross
Professors: M.A. DiBerardino, D.E. Goldman, L.L. Ross
Associate Professors: A.U. Barnes, A.B. Beasley, M.E. Goldberger, A.C. Kulangara, R.J.C. Levine, M. Murray, W. Rubin
Assistant Professors: T.J. Cunningham, M.J. Elfvin, L. M. Pubols, J.D. Smith
Visiting Lecturers: 12
Teaching Fellows: 2
Research Assistants: 2

Degree Program: Anatomy

Graduate Degree: Ph.D.
Graduate Degree Requirements: Doctoral: 32 Residence credits, Thesis

Graduate and Undergraduate Courses Available for Graduate Credit: Gross Anatomy, Microanatomy, Neuroanatomy

DEPARTMENT OF BIOCHEMISTRY
Phone: Chairman: (215) 842-7048 Others: 842-7054

Chairman: F.D. DeMartinis (Acting)
Assistant Professors: A.P. Berg, H.A. Bertrand, T.I. Jiamondstone
Instructor: K. Wagner
Teaching Fellows: 1

Degree Program: Biochemistry

Graduate Degree: Ph.D.
Graduate Degree Requirements: Doctoral: Course work, written and oral comprehensive examinations, one foreign language, Thesis

Graduate and Undergraduate Courses Available for Graduate Credit: Biochemistry-Physiology I, II, III, Journal Club, Biochemistry Seminar, Enzymes and Proteins, Metabolism, Macromolecules, Neurochemistry, Molecular Endocrinology, Techniques in Lipoprotein Chemistry

MICROBIOLOGY DEPARTMENT
Phone: Chairman: (215) 842-7064 Others: 842-7060

Chairman: K. Paucker
Professor: K. Paucker
Associate Professor: C.A. Ogburn
Assistant Professors: B.A. Zajac, R.W. Gilpin, A.S. Sideropoulos, J. Erickson (Research)
Visiting Lecturers: 6

Degree Program: Microbiology

Graduate Degree: Ph.D.
Graduate Degree Requirements: Doctoral: Students enrolled in the Ph.D. program avail themselves of courses offered at participating universities and colleges in the area.

DEPARTMENT OF PATHOLOGY
Phone: Chairman: (215) 842-7083 Others: 842-6000

Chairman: J. Leighton
Professors: I.N. Dubin, G.J. Kaldor, J. Leighton, R. Mark, M.M. Porter, R. Shuman
Associate Professors: R. Knight, S. Mansukhani, D. Sawhill
Assistant Professors: N.A. Abaza, D.G. Bartuska, D. Dunn, J J. Godleski, G. Justh, J.R. Krause, T.B. Krouse, J. Mobini, H.W. Schmidt, R. Tchao
Instructors: P. Agarwal, R.T. Gustafson, H. Malhotra
Visiting Lecturers: 20
Research Assistant: 1

Degree Programs: Clinical Pathology, Pathology

Graduate Degrees: M.S., Ph.D.
Graduate Degree Requirements: Master: B.S. or equivalent Doctoral: Pre-medical requirements

Graduate and Undergraduate Courses Available for Graduate Credit: Investigative Pathology Seminars, Investigative Pathology Techniques, Investigative Pathology Research Projects, Cancer Biology Journal Club

DEPARTMENT OF PHARMACOLOGY
Phone: Chairman: (215) 842-7175 Others: 842-7170

Chairman: J. Roberts
Professors: J. Roberts, B. Weiss
Associate Professors: E. Foltz, K. Gabriel, J. Jepson, G.J. Kelliher
Assistant Professors: S.I. Baskin, N. Guiha, E. Johnson
Instructors: P.B. Goldberg, C.M. Lathers
Visiting Lecturers: 2
Teaching Fellows: 1
Research Assistants: 4

Degree Program: Pharmacology

Graduate Degree: Ph.D.
Graduate Degree Requirements: Doctoral: 72 credit hours excluding research, one Foreign Language, Preliminary Examination, Dissertation

Graduate and Undergraduate Courses Available for Graduate Credit: Biochemistry, Physiology, Biostatistics, Microbiology, Gross Anatomy, Neuroanatomy, Microscopic Anatomy, Computer Science, Pharmacology Seminar, Research Methods in Pharmacology, Pharmacology and Experimental Therapeutics, Biomedical Instrumentation and Techniques, Advanced Topics in Pharmacology, Cyclic Nucleotide Journal Club, Teaching Methods in Pharmacology

DEPARTMENT OF PHYSIOLOGY AND BIOPHYSICS
Phone: Chairman: (215) 842-7048 Others: 842-7045

Chairman: F.D. DeMartinis (Acting)
Professors: D.E. Goldman, G.J. Kaldor

Associate Professors: G.K. Chacko (Research), F.D. DeMartinis, I.D. Zimmermann
Instructors: W.J. DiBattista (Research), K. Wagner
Teaching Fellows: 1

Degree Programs: Physiology and Biophysics

Graduate Degree: Ph.D.
Graduate Degree Requirements: Doctoral: course work, written and oral comprehensive examinations, one Foreign Language, Thesis

Graduate and Undergraduate Courses Available for Graduate Credit: Biochemistry-Physiology I, II, III, Journal Club, Special Topics in Physiology, Behavior and Models of Neurophysiological Systems, Molecular Endocrinology, Membrane Biophysics, Molecular Basis for Muscular Action, Systems Analysis, Techniques in Lipoprotein Chemistry

MERCYHURST COLLEGE

501 E. 38 Street Phone: (814) 864-0681
Erie, Pennsylvania 16501
Dean of College: W. Garvey

DEPARTMENT OF BIOLOGY
Phone: Chairman: Ext. 266

Chairman: Sr. E. Poydock
Professor: Sr. E. Poydock
Associate Professor: Sr. M Smith
Assistant Professor: E.M. Lignowski
Instructor: R.L. Tipton
Visiting Lecturer: 1
Research Assistants: 2

Field Stations/Laboratories: Member of a Marine Biology Consortium, Palm Beach, Florida

Degree Program: Biology

Undergraduate Degree: B.S.
Undergraduate Degree Requirements: Not Stated

MESSIAH COLLEGE

Grantham, Pennsylvania 17027 Phone: (717) 766-2511
Dean of College: D. Chambelzand

DEPARTMENT OF BIOLOGY
Phone: Chairman: Ext. 263

Chairman: G.D. Hess
Professor: K.B. Hoover
Assistant Professors: G.D. Hess, B.W. Myers
Instructor: N W. Falk

Degree Program: Biology

Undergraduate Degree: A.B.
Undergraduate Degree Requirements: Minimum total 126 semester hours for A.B. in Biology - core courses Physiology, Genetics, Ecology, Developmental Biology (each semester) - Five additional 1 semester courses, Organic Chemistry, 1 semester of Calculus-Statistics

MILLERSVILLE STATE COLLEGE

Millersville, Pennsylvania 17551 Phone: (717) 872-5411
Dean of Graduate Studies: E.R. Thomas
Dean of College: R. Sasin

BIOLOGY DEPARTMENT
Phone: Chairman: Ext. 208 Others: Ext. 278

Chairman: A.K. Fontes
Professors: A.K. Fontes, A. Henderson, S. Radinovsky, W. Ratzlaff, H.R. Weirich, W.J. Yurkiewicz
Associate Professors: M Z. Bierly (Adjunct), S.J. Ha, A.C. Hoffman, K.G. Miller, J.C. Parks, J. Rorabaugh, G.L. Steucek
Assistant Professors: T.R. Klei, D. Ostrovsky
Instructors: J. Stephan

Field Stations/Laboratories: Lewes Marine Science Station - Lewes, Delaware, Wallops Island Marine Science Station - Wallops Island, Virigina

Degree Programs: Biological Sciences, Biological Education, Medical Technology

Undergraduate Degrees: B.S., B.S. Ed.
Undergraduate Degree Requirements: 120 semester hours 60 hours in General Education, 32 hours of Biology, 16 hours of Chemistry, 8 hours of Physics, 7 hours of Mathematics, Foreign Language through Intermediate level

Graduate Degrees: M.A., M.S., M.Ed.
Graduate Degree Requirements: 30 semester hour s including thesis and graduate seminar

Graduate and Undergraduate Courses Available for Graduate Credit: Methods of Biological Research (U), Plant Ecology (U), Biological Techniques (U), Cytogenetics (U), Field Zoology (U), Advanced Genetics (U), Aquatic Entomology (U), Animal Behavior (U), Ichthyology (U), Arthropod Behavior (U), Plant Evolution (U), Biochemistry, Thesis, Plant Anatomy and Morphology (U), Advanced Plant Physiology, Comparative Animal Physiology (U), Plant Geography, Physical Biology (U), Population Dynamics, Seminar in Biology, Radioiostopes in Biological Research (U), Vertebrate Morphology, Limnology (U), Experimental Embryology, Problems in Biology

MILTON S. HERSHEY MEDICAL CENTER

(see Pennsylvania State University)

MORAVIAN COLLEGE

Main and Elizabeth Phone: (215) 865-0741
Bethlehem, Pennsylvania 18018
Dean of College: J.J. Heller

DEPARTMENT OF BIOLOGY
Phone: Chairman: (215) 865-0741

Head: A.E.H. Gaumer
Professors: K.A. Bergstresser, A.E.H. Gaumer
Associate Professor: J.B. Mitchell
Assistant Professor: J.M. Bevington
Instructor: T. Stanglein

Degree Program: Biology

Undergraduate Degree: B.S.
Undergraduate Degree Requirements: 31 courses total, 12 Biology Courses, 4 Chemistry

MOUNT MERCY COLLEGE

(see Carlow College)

MUHLENBERG COLLEGE

Allentown, Pennsylvania 18104 Phone: (215) 433-3191
Dean of College: C. McClain

DEPARTMENT OF BIOLOGY
Phone: Chairman: Ext. 351 Others: Ext. 350

Chairman: J.R. Vaughan
Professors: C.S. Oplinger, R.L. Schaeffer, J.E. Trainer, J.R. Vaughan, J.C. Weston
Assistant Professor: P.W.H. Weaver, Jr.
Part-time Lecturer: 1

Field Stations/Laboratories: Owl Hollow Wildlife Preserve - Kempton, Pennsylvania

Degree Program: Biology

Undergraduate Degree: B.S.
Undergraduate Degree Requirements: 30 credits Biology, 16 credits Chemistry, 8 credits Physics, 6 credits Calculus, 12 credits in Humanities, 15 credits in Social Sciences, 7 credits in Religion, Foreign Language, through Intermediate Level, 8 quarters Physical Education. Total credits for degree 120.

OUR LADY OF ANGELS COLLEGE

Aston, Pennsylvania 19014 Phone: (215) 459-0905
Dean of College: Sr. M. Marietta

DEPARTMENT OF BIOLOGY
Phone: Chairman: Ext. 40

Chairman: L.G. Shaulis
Associate Professor: L.G. Shaulis
Assistant Professor: A. Curry
Instructor: Sr. J. Marie
Visiting Lecturers: 2

Degree Program: Biology

Undergraduate Degree: B.S.
Undergraduate Degree Requirements: CORE, 34 credits Major: Botany, Zoology, Genetics, Seminar, plus General Chemistry (1 year), Organic Chemistry (1 year) Calculus, Statistics plus 18 hours of Biology 300 level or higher. Electives: 33 hours

PENNYSLVANIA STATE UNIVERSITY

University Park, Pennsylvania 16802 Phone: (814) 865-7517
Dean of Graduate Studies: J. Barton

College of Agriculture

Dean: J. Beattie

DEPARTMENT OF AGRONOMY
Phone: Chairman: (814) 865-6541 Others: 865-6542

Head: J.L. Starling
Professors: D.E. Baker, R.F. Barnes (Adjunct), R.W. Cleveland, J.M. Duich, J.D. Harrington, G.A. Jung (Adjunct), L.T. Kardos, W.A. Kendall (Adjunct), H.G. Marshall (Adjunct), R.P. Matelski, G.W. McKee, J. B. Washko
Associate Professors: J.M. Bollag, E.J. Ciolkosz, R.H. Cole, R.L. Cunningham, J.K. Hall, R.R. Hill, Jr. (Adjunct), L.J. Johnson, M.W. Johnson, L.F. Marriott, R. Pennock, Jr., G.W. Peterson, R.P. Pfeifer, M.L. Risius, A.S. Rogowski (Adjunct), J.S. Shenk, D.V. Waddington, A.C. Wilton (Adjunct)
Assistant Professors: C.C. Berg (Adjunct), D.D. Fritton, D.L. Gustine (Adjunct), N.L. Hartwig, L.D. Hoffman, D.P. Knievel, T.L. Watschke
Instructors: G.W. Fissel (Adjunct), C.F. Gross (Adjunct), R.F. Guyton, J.E Hook, S.A.B. Hornick, R.C. Sidle, W.L. Stout, R.L. Williams
Teaching Fellows: 3
Research Assistants: 20

Field Stations/Laboratories: Southeast Field Research Laboratory - Landisville, Pennsylvania, Southwest Field Research Laboratory - Rector, Pennsylvania

Degree Programs: Crop Science, Soil Science, Plant Breeding, Remote Sensing

Undergraduate Degree: B.S. (Agronomy)
Undergraduate Degree Requirements: 126 semester credits, 12 credits Communications, 9 credits Quantification, 39 credits Natural Science, 6 credits Arts and Humanities, 6 credits Social and Behavioral, 28 credits Agronomy and Related Areas, 23 credits Electives

Graduate Degrees: M.S., M. Agr.
Graduate Degree Requirements: Master: M.S. 30 credits plus thesis, M. Agr. 30 credits plus thesis

Graduate and Undergraduate Courses Available for Graduate Credit: Soil Composition and Physical Properties (U), Chemistry of Soils and Fertilizers (U), Crop Science (U), Breeding of Field Crops (U), Soil Morphology, Mapping and Land Use (U), Soil Genesis and Classification (U), Forest Soils (U), Case Studies in Soil, Plant and Water Management (U), Forage Crop Management (U), Field Crop Management (U), Weed Science (U), Advanced Soil Fertility, Soil Chemistry, Soil Physics, Genetics of Crop Plants, Applications of Cytogenetics to Plant Breeding, Field Plot Technique, Farm Crops Ecology, Growth and Management of Forage Crops, Nature of Soil Minerals, Application of Statistics to Field Experiments

DEPARTMENT OF ANIMAL SCIENCE
Phone: Chairman: (814) 865-1362 Others: 865-1364

Head: B.R. Baumgardt
Professors: B.R. Baumgardt, R.L. Cowan, E.W. Hartsook, G.R. Kean, T.A. Long, T.L. Merritt, J. D. Sink, L.L. Wilson, D.E. Younkin
Associate Professors: C.A. Baile (Adjunct), L.A. Burdette, J.P. Gallagher, T.V. Hershberger, R.J. Martin, B. Morgan, G.W. Sherritt, J.H. Ziegler
Assistant Professors: E.H. Cash, W.R. Jones, P.J. Wangsness, J.L. Watkins
Instructors: E. Keck, R. Kimble (Adjunct)

Degree Programs: Animal Industry, Animal Science, Agriculture, Interdepartmental Graduate Program in Animal Nutrition

Undergraduate Degrees: B.S.
Undergraduate Degree Requirements: Credits for Animal Industry and Animal Science respectively are: Communications 12, 12, Quantification 6,6, Natural Science 32, 50-56, Arts and Humanities 6,6, Social and Behavioral Science 6, 6, Major field and related courses 49-50, 29-30, Health Science and Physical Education 4,4, Electives,15, 11-18

Graduate Degrees: M.S., Ph.D.
Graduate Degree Requirements: Master: No specific requirements, program determined by committee
Doctoral: No specific requirements, program determined by committee

Graduate and Undergraduate Courses Available for Graduate Credit: Courses in the Animal Industry Program (only) Special Topics (U), Meat, Animal Management (U), Science of Meat (U), Advanced Livestock Selection (U), Animal Industry Seminar (U), Advanced Meat Selection and Grading (U), Research in Meats, Advanced Animal Breeding, Animal Science Research Methods, Animal Growth and Development, Colloquium, Research. Other courses available in related disciplines.

DEPARTMENT OF DAIRY SCIENCE
Phone: Chairman: (814) 865-6152 Others: 865-5444

Head: D.V. Josephson
Evan Pugh Research Professor of Agriculture: S. Patton
Professors: D.L. Ace, R.S. Adams (Extension), D.O. Almquist, R.P. Amann, R.J. Flipse, H.C. Gilmore (Extension), P.G. Keeney, E.M. Kesler, M.E. Mason (Adjunct), R.D. McCarthy, D.N. Putnam (Extension), H.E. Shaffer (Extension), L.W. Specht (Extension), M.P. Thompson (Adjunct), G.H. Watrous
Associate Professors: W.P. Anderson (Extension), J.E. Barnard (Extension), D.M. Buckalew, P.S. Dimick, E.D. Glass, M. Kroger, P.R. Shellenberger,

S.B. Spencer, T.Y. Tanabe, M.W. Thoele
Assistant Professors: A.E. Branding, G.L. Hargrove, K. Ostovar
Instructors: W.W. Coleman (Assistant)
Research Assistants: 2

Field Stations/Laboratories: Dairy Breeding Research Center, Dairy Production Center, Centra Milk Testing Laboratory

Degree Programs: Animal Industries, Animal Science, Dairy Science

Undergraduate Degree: B.S. (Animal Industries, Animal Sciences)
Undergraduate Degree Requirements: 131 credits

Graduate Degrees: M.S., Ph.D.
Graduate Degree Requirements: Master: Applicants with 2.70 junior-senior grade point average will be considered, 30 credits for graduation, plus M.S. Thesis, Average of 3.00 or better Doctoral: Ph.D. Dissertation, grade point minimum of 3.00, No required minimum of credits but minimum residence requirement of 3 terms, One foreign language (Intermediate), Candidacy Examination, Comprehensive Examination, final oral examination.

Graduate and Undergraduate Courses Available for Graduate Credit: Dairy Cattle Management (U), Dairy Survey (U), Milk Secretion (U), Dairy Production Problems (U), Physiology of Reproduction in Farm Animals (U), Dairy Cattle Management, Dairy Seminar, Dairy Cattle Nutrition, Advanced Studies in Milk Secretion, Dairy Cattle Nutrition, Advanced Studies in Milk Secretion, Dairy Cattle Breeding, Advanced Physiology of Reproduction in Farm Animals, Artificial Breeding of Farm Animals, Dairy Science Literature, Dairy Radiophysiology

DEPARTMENT OF ENTOMOLOGY
Phone: Chairman: (814) 865-1895

Head: B.F. Coon
Professors: L.E. Adams, D. Asquith, B.F. Coon, S.G. Gesell, R.O. Mumma, C.W. Rutschky, T. Smyth, Jr., R.J. Snetsinger, W.G. Yendol
Associate Professors: A.W. Benton, W.W. Clarke, C.D. Ercegovich, A.A. Hower, K.C. Kim, A. Mallis, R.C. Tetrault
Assistant Professors: W. Bode, R.A. Byers (Adjunct), E.A. Cameron, R.B. Colburn (Adjunct), E.D. Eckess (Adjunct), S. Green, G. Jubb, Z. Smilowitz, A.G. Wheeler (Adjunct)
Instructor: D. Schneider
Research Assistants: 5

Field Stations/Laboratories: Fruit Research Laboratories at Biglerville and North East, Pennsylvania, Field Research Laboratories at Landisville, Rector and Rock Springs, Pennsylvania

Degree Programs: Ecology, Entomology, Physiology

Undergraduate Degree: B.S.
Undergraduate Degree Requirements: Entomology option available under the Animal Science Major.

Graduate Degrees: M.S., Ph.D.
Graduate Degree Requirements: Master: 24 graduate credits in Entomology and related Biological and Physical Sciences, Chemistry through Organic; Physics, Mathematics and Statistics recommended. Degree requirements: 30 graduate credits; seminar participation, Minimum 6 credits research, thesis, oral examination Doctoral: Three consecutive terms in residence; 1 foreign language, competency in English, Seminar Participation, Candidacy, Comprehensive and Final Oral Examinations, Thesis, Department determines degree credit requirements.

Graduate and Undergraduate Courses Available for Graduate Credit: Medical and Veterinary Entomology, Insect Biology and Management, Field Entomology, Insect Morphology and Phylogeny, Insect Physiology and Biochemistry, Insect Ecology and Control, Forest Entomology, Arachnology, Morphology and Systematics, Insect Physiology and Biochemistry (Advanced), Insect Behavior (Advanced), Pest Management, Insect Toxicology, Insect Behavior, Biological Control, Insect Pathology, Systematics, Individual Studies, Colloquium, Research

DIVISION OF FOOD SCIENCE AND INDUSTRY
Phone: Chairman: (814) 865-6152 Others: 865-5444

Division Chairman: D.V. Josephson
Professors: B.R. Baumgardt, K. Goodwin, R.W. Hepler, G.R. Kean, P.G. Keeney, R.D. McCarthy, M.E. Mason (Adjunct), J.D. Sink, M.P. Thompson (Adjunct), G.H. Watrous
Associate Professors: F.J. McArdle, P.S. Dimick, E.D. Glass, M. Kroger, J.E. Barnard (Extension), G.D. Kuhn (Extension), J.H. Ziegler
Assistant Professors: R.B. Beelman, W.R. Jones (Extension), M.G. Mast (Extension), K. Ostovar, J.L. Watkins
Instructor: W.W. Coleman (Assistant)
Research Assistants: 2

Field Stations/Laboratories: Horticulture Processing Laboratory, Meat Laboratory, University Creamery, Poultry Processing Laboratory

SCHOOL OF FOREST RESOURCES
Phone: Chairman: (814) 865-7541 Others: 865-3281

Director: W.W. Ward
Professors: P.W. Fletcher, H.D. Gerhold, R.J. Hutnik, W.K. Murphey, W.E. Sopper, W.W. Ward
Associate Professors: F.C. Beall, F.Y. Borden, J.L. George, J.S. Lindzey (Adjunct), R.E. Melton, R.D. Shipman
Assistant Professors: R.G. Anthony, R.D. Baldwin, D.R. DeWalle, P.C. Kersavage, G.L. Storm (Adjunct), B.J. Turner
Instructors: B.M. Kent, L.H. McCormick, C.H. Strauss
Teaching Fellows: 4
Research Assistants: 24

Field Stations/Laboratories: Stone Valley Experimental Forest - Huntingdon, Pennsylvania

Degree Programs: Forest Resources, Wildlife Management

Undergraduate Degree: B.S.
Undergraduate Degree Requirements: 127 credits

Graduate Degrees: M.F., M.S., Ph.D.
Graduate Degree Requirements: Master: 30 to 40 credits including Thesis, supervised by committee
Doctoral: Thesis, Foreign Language, Oral Examination and courses approved by committee

Graduate and Undergraduate Courses Available for Graduate Credit: Forestry Graduate Courses: Dendrology, Forest Resources Seminar, Forest Recreation, Silviculture, Forest Range Management, Forest Economics and Finance, Introduction to Operations Research, Aerial Photos in Forestry, Forest Resources Management, Policy and Administration, Problems in Forestry, Advanced Forest Ecology, Forest Genetics, Forest Microclimatology, Forest Hydrology, Advanced Silviculture, Forest Land Use Alternatives, Design and Analysis of Experiments, Timber Management, Applications of Forest Economics and Finance, Colloquium, Individual Studies, Research Wildlife Graduate Courses: Mammalogy, Wildlife Ecology, Wildlife Management, Field Research Techniques, Problems In Wildlife, Laboratory Techniques, Advanced Wildlife Management, Colloquium, Individual Studies,

Research Wood Science Graduate Courses: Physical Properties of Wood, Mechanical Properties of Wood, Chemical Properties of Wood, Pulp and Fiber Technology, Processing and Machining of Wood, Adhesives and Finishes for Wood Products, Drying and Preservation, Forest Products Manufacturing Systems and Standards, Problems in Wood Science, Logging and Lumbering, Forest Products Production Management, Small Sawmills, Wood Fibers, Physical Properties, Wood and Fibers, Wood Chemistry, Mechanical Behavior of Wood, Theory ofAdhesion, Forest Products Industrial Operation Analysis, Colloquium, Individual Studies

DEPARTMENT OF HORTICULTURE
(No reply received)

DEPARTMENT OF PLANT PATHOLOGY
Phone: Chairman: (814) 865-7448 Others: 865-3761

Head: J. Tammen
Evan Pugh Professor: R.R. Nelson
Professors: J.R. Bloom, J.S. Boyle, H. Cole, R.S. Dickey (Adjunct), C.H. Kingsolver (Adjunct), L.R. Kneebone, F.H. Lewis, P.E Nelson, R.R. Nelson, L.P. Nichols, J.W. Oswald, D.H. Petersen, L.C. Schisler, R.T. Sherwood (Adjunct), S H. Smith, J. Tammen, T.A. Toussoun (Adjunct), and P.J. Wuest
Associate Professors: J.E. Ayers, K.T. Leath (Adjunct) F.L. Kukezic, W. Merrill, R.F. Stouffer
Assistant Professors: D.D. Davis, R.E. Hite (Adjunct), R.A. Krause, D.R. MacKenzie, A.A. MacNab, E.J. Pell, S.P. Pennypacker
Research Assistants: 6
Visiting Professors: 3

Field Stations/Laboratories: Fruit Research Laboratory, Biglerville, Pennsylvania

Degree Program: Plant Pathology

Graduate Degrees: M.S., Ph.D.
Graduate Degree Requirements: Master: Minimum of 30 credits to include: (1) Minimum of 6 credits thesis research, (2) Minimum of 18 credits in graduate level courses, (3) Minimum of 12 credits in courses in major field, and (4) Final Examination and Thesis Doctoral: M.S. Degree plus minimum of 70 credits to include: (1) 28 credits in Plant Pathology courses, (2) 15 credits in-depth study as determined by Committee, (3) courses in statistics and Plant Physiology, (4) Comprehensive Examination, and (5) Thesis and Final Examination

Graduate and Undergraduate Courses Available for Graduate Credit: Theory and Concepts of Plant Pathology (U), Diseases of Economic Plants (U), Plant Pathological Techniques (U), Plant Pathogenic Bacteria (U), Introduction to Plant Virology (U), Environmental Pathology (U), Phytonematology (U), History of Plant Pathology (U), Clinical Plant Pathology, Plant Disease Control, Physiology of Plant Disease, Epidemiology of Plant Diseases, Pathogen Variation and Host Resistance, Pathological Plant Anatomy and Principles of Plant Pathology

POULTRY SCIENCE DEPARTMENT
Phone: Chairman: (814) 865-3411

Head: K. Goodwin
Professors: G.O. Bressler, E G. Buss, E.B. Hale, F.W. Hicks, R.M. Leach, J.H. MacNeil, W.J. Mueller
Associate Professors: H.B. Graves, H.C. Jordan
Assistant Professors: O.D. Keene
Instructor: T.W. Burr
Research Assistants:

Degree Programs: Animal Science, Animal Industry, Poultry Science (Department participates in M.S. and Ph.D. programs in Genetics, Physiology, Animal Nutrition, Food Science, and Animal Behavior)

Undergraduate Degree: B.S. (Animal Science, Animal Industry
Undergraduate Degree Requirements: 12 hours Communications, 6 hours Mathematics, 50-56 hours Natural Sciences, 6 hours Arts and Humanities, 6 hours Social and Behavioral Sciences, 29-30 hours Major Field, 4 hours Health and Physical Education, 8-15 hours Electives

Graduate Degrees: M.S., Ph.D.
Graduate Degree Requirements: Master: 30 hours of graduate credit plus thesis or acceptable paper Doctoral: Satisfactory completion of courses, one language, thesis, 3 terms of residency

Graduate and Undergraduate Courses Available for Graduate Credit: Special Topics, Poultry Production Technology Animal Bahavior - Ethology, Animal Behavior Laboratory, Advanced Poultry Nutrition, Poultry Farm Management, Poultry Meat and Egg Technology, Research in Animal Behavior

DEPARTMENT OF VETERINARY SCIENCE
Phone: Chairman: (814) 865-7696

Head: A.L. Bortree
Professors: H.W. Dunne, R.F. Gentry, P.J. Glantz, J.F. Hokanson
Associate Professors: R.J. Eberhart, S. Gordeuk, Jr., J.F. Kauanaugh, D.C. Kradel, N. Rothenbacher, R. W. Scholz, A. Zarkower
Assistant Professors: M.O. Braune, F.G. Ferguson, L.C. Griel, Jr., R.H. Latt, W.H. Patton
Instructors: G.R. Bubash, G.L. Cusanno

Degree Programs: Animal Pathology, Animal Science, Physiology, Veterinary Microbiology, Veterinary Science

Undergraduate Degree: B.S. (Animal Science)
Undergraduate Degree Requirements: 131 semester credits

Graduate Degrees: M.S., Ph.D.
Graduate Degree Requirements: Master: 30 credits including research and thesis Doctoral: one language, Candidacy Examination, Comprehensive Examination, Research and Defense of Thesis

Graduate and Undergraduate Courses Available for Graduate Credit: Ecology of Animal Disease, Laboratory Animal Management, Animal Cell Culture, General Animal Pathology, Metabolic Disorders, Experimental Surgery

College of Science

Dean: T. Wartik

DEPARTMENT OF BIOLOGY
Phone: Chairman: (814) 865-4562

Head: J.G. O'Mara
Professors: A. Anthony, E.D. Bellis, R.L. Butler, E.L. Cooper, H.C. Dalton, C.L. Fergus, A.R. Grove, P. Grun, E.B. Hale, R.H. Hamilton, C.J. Hillson, W. C. Hymer, A.J. Kovar, J.G. O'Mara, R.D. Schein, W. Spackman, A. Traverse, J.E. Wright
Associate Professors: S.L. Dachtler, W.A. Dunson, E. Hibbard, C.S. Keener, W.H. Neff, R.A. Pursell, C.D. Therrien, E.W. Wickersham, F.M. Williams, F.H. Witham
Assistant Professors: P.L. Abplanalp, F.S. Adams, D.E. Arnold, E.V. Gaffney, T.M. Hollis, J.W. MacCluer, R.B. Mitchell, A.A. Reimer
Teaching Fellows: 22
Research Assistants: 15

Degree Programs: Biology, Botany, Ecology, Genetics, Physiology, and Zoology

Undergraduate Degree: B.S. (Biology, Zoology)
Undergraduate Degree Requirements: The program requires 35 credits in Biology, of which 16 must be in senior-level courses, 9 credits in Mathematics and Statistics, 14-16 credits in Chemistry and 8 credits in Physics. The total number of credits required for the degree is 124.

Graduate Degrees: M.S., Ph.D.
Graduate Degree Requirements: Master: No specified number of credits is required. Generally, approximately 30 credits are secured, of which 18 are in graduate-level courses. A thesis is usually required, but some students are permitted to substitute additional courses in place of the thesis. Doctoral: No specified number of credits or courses is required. The Degree is awarded after acceptable performance in Candidacy, Comprehensive and Final Oral Examinations and upon the presentation of an approved thesis. The master's degree is not invariably a prerequisite for the doctoral degree.

Graduate and Undergraduate Courses Available for Graduate Credit: Vertebrate Neuroanatomy, Practicum in Experimental Neuroanatomy, Seminar in Neuroanatomy, Plant Anatomy, Biology of Aging, General Limnology, Advanced Systematic Botany, Invertebrate Zoology, Mycology, General Ecology (U), Comparative Anatomy of Vertebrates (U), Advanced Genetics (U), Introductory Palynology, Introductory Cytogenetics (U), Evolution (U), Population Genetics (U), Molecular Biology of the Gene (U), Comparative Plant Morphology (U), Terrestrial Ecology (U), Ecology of Lakes and Streams (U), Laboratory in Freshwater Ecology (U), Histology (U), Embryology (U), Plant Physiology (U), Phytohormones (U), Ornithology (U), Plant Autecology (U), Plant Synecology (U), Ichthyology (U), Animal Behavior - Ethology (U), Cytology (U), Vertebrate Physiology (U), Cellular Physiology (U), General Endocrinology (U), Marine Biology (U), Aquatic Botany (U), Plant Physiology Seminar, The Physiology of the Fungi, Comparative Anatomy of Vascular Plants, Advanced Plant Physiology, Biology Seminar, Ecological Plant Geography, Fish Behavior as Related to Aquatic Ecology, Zoogeography, Myxomycetes, Phycomycetes, and Ascomycetes, Basidiomycetes and Fungi Imperfecti, Seminar in Genetics, Bryology, Morphology of the Tracheophyta Exclusive of Angiosperms, Morphology of Angiosperms, Principles of Microscopic Histochemistry, Analytical Histochemistry Laboratory, Phycology, Comparative Physiology, Systematics, Ecosystem Dynamics, Ecology of Populations and Communities, Neurogenesis, Animal Physiology

DEPARTMENT OF BIOCHEMISTRY
Phone: Chairman: (814) 865-6992

Head: J.H. Pazur
Professors: C.O. Clagett, M.F. Mallette, R.L. McCarl, J.H. Pazur, A.T. Phillips, J.W. Shigley
Associate Professors: W.W. Karakawa, W.G. Niehaus, R.S. Schraer
Assistant Professors: N.N. Aronson, Jr., S.R. Fahnestock, R.H. Hammerstedt, C.R. Hartzell, P.K. Warme, H.C. Winter
Instructors: J.O. Kane
Research Assistants: 3
Postdoctoral Fellow: 1

Degree Program: Biochemistry

Undergraduate Degree: B.S.
Undergraduate Degree Requirements: Master: Placement and cumulative Examinations, advanced Biochemistry courses, Seminars, six credits in related areas, Research presented as Thesis and defended orally. Doctoral: Placement and cumulative examinations, advanced Biochemistry courses, Seminars, fifteen credits in related areas, Candidacy and Comprehensive Examinations, paper published in refereed journal, significant research presented as thesis and defended orally.

Graduate and Undergraduate Courses Available for Graduate Credit: General Biochemistry (U), General Biochemistry (U), Experimental Biochemistry (U), Biochemical Methods (U), Physiological Chemistry (U), Physiological Methods (U), Enzymes and Biocatalysts, Biochemical Problems, Seminar, Proteins, Carbohydrates, Lipids, Biochemistry of Nucleic Acids, Biosynthesis of Macromolecules, Topics in Metabolism, Contemporary Topics in Biochemistry and Molecular Biology

BIOPHYSICS DEPARTMENT
Phone: Chairman: (814) 865-2538

Head: W.D. Taylor
Professors: H. Schraer, W. Ginoza, S. Person, R.A. Deering, W. Snipes, G.K. Strother, T. Smyth (Joint), W.D. Taylor
Associate Professors: R.S. Morgan, P.W. Todd, A. Keith
Assistant Professor: A. Rake
Visiting Lecturers: 1
Teaching Fellows: 0
Research Assistants: 6
Postdoctoral Fellows: 2

Degree Programs: Biophysics

Undergraduate Degree: B.S.
Undergraduate Degree Requirements: Not Stated

Graduate Degrees: M.S., Ph.D.
Graduate Degree Requirements: Not Stated

Graduate and Undergraduate Courses Available for Graduate Credit: Molecular Biology of Genes (U), Introductory Biophysics (U), Molecular Biophysics I, II (U), Cell Biology Seminar, Biophysics of Nucleic Acids, Radiation Biophysics, Biological Ultrastructure, Photobiology, Biological Macromolecules, Ultracentrifugation, Neurophysiology (U), Physiology of Nerves, Muscles and Sense Organs, Mammalian Cell Culture, Evolution Resonance Spectroscopy, Microspectrophotometry

DEPARTMENT OF MICROBIOLOGY
Phone: Chairman: (814) 865-3072

Chairman: Not Stated
Professors: R.W. Bernlohr, L.E. Casida, Jr., E.S. Lindstrom, E.H. Ludwig, R.W. Stone, L.N. Zimmermann
Associate Professors: W.S. Ceglowski, T.M. Joys, C.F. Pootjes, D.R. Tershak, R.F. Unz
Assistant Professors: J.E. Brenchley, J.J. Docherty, D.Y.C. Fung, P.E. Kolenbrander
Instructors: D.Z. Grubbs, M. Luger, T.E. Rucinsky
Degree Programs: Microbiology, Medical Technology

Undergraduate Degree: B.S. (Microbiology, Medical Technology
Undergraduate Degree Requirements: 124 credits total including 9 credits Biology, 18 credits Microbiology, 25 credits Chemistry, 8 credits Physics, 9 credits Mathematics

Graduate Degrees: M.S., Ph.D.
Graduate Degree Requirements: Master: 30 credits total including 8 credits Microbiology, 2 years Chemistry, 1 year Physics, 1 semester Calculus, Thesis Doctoral: 15 credits in minor field, 15 in major field, preliminary examination, comprehensive examination, thesis, final oral

Graduate and Undergraduate Courses Available for Graduate Credit: Introductory Environmental Microbiology (U), Advanced Bacteriology (U), Immunology and Serology (U), Microbial Soil Ecology (U), Food Microbiology (U), Introduction to Animal Viruses (U), Industrial

Microbiology, Epidemiology (U), Bacterial Viruses (U), Seminar, Bacterial Physiology, Experimental Immunology, Bacterial Genetics, Biochemical Virology, Aquatic Microbiology

Milton S. Hershey Medical Center
500 University Drive Phone: (717) 534-8521
Hershey, Pennsylvania 17033
Dean of Graduate Studies: J. Bartos
Dean of Medical College: H. Prystowsky

DEPARTMENT OF ANATOMY
Phone: Chairman: (717) 534-8651 Others: 534-8521

Chairman: B.L. Munger
Professor: I.L. Baird
Associate Professor: B.H. Pubols
Assistant Professors: A.E. Leure-duPree, L.P. McCallister, L.M. Pubols
Instructor: S.L. Quattropani
Visiting Lecturers: 2
Research Assistants: 3

Degree Program: Anatomy

Graduate Degrees: M.S., Ph.D.
Graduate Degree Requirements: Master: Basic courses with B minimum average, plus an acceptable published paper, or thesis Doctoral: Basic course requirement with B minimum average plus qualifying examination, Comprehensive Examination, Dissertation, and Final Examination

Graduate and Undergraduate Courses Available for Graduate Credit: Gross Human Anatomy, Microscopic Anatomy, Human Embryology, Dissection, Submicroscopic Anatomy, Comparative Neurology, Comparative Vertebrate Otology, The Peripheral Nervous System, Seminar in Quantitative Opitcs

DEPARTMENT OF BIOLOGICAL CHEMISTRY
Phone: Chairman: (717) 534-8586 Others: 534-8585

Chairman: E.A. Davidson
Professors: E.A. Davidson, A. Rosenberg
Associate Professors: L.F. Hass, C.W. Hill
Assistant Professors: A McPherson, M. Miljkovic, J.I. H. Patterson, D.E. Roark, C.L. Schengrund, R. Shiman, J. Taylor
Research Associates: V.P. Bhavanandan, H. Den, P. Feil, J. Tkacz
Research Assistants: 2
Graduate Assistants: 15

Degree Programs: Biological Chemistry, Genetics

Graduate Degrees: M.S., Ph.D.
Graduate Degree Requirements: Master: Basic courses with B minimum average, plus an acceptable published paper, or thesis Doctoral: Basic course requirements with B minimum average plus qualifying examination, comprehensive examination, dissertation, and final examination

Graduate and Undergraduate Courses Available for Graduate Credit: Biological Chemistry, Molecular Genetics, Biological Chemistry Laboratory, Biological Chemistry of Macromolecules, Metabolism, Seminar, Kinetics and Mechanism of Enzyme Action, Biochemical Techniques, Special Topics and Research Problems

DEPARTMENT OF MICROBIOLOGY
Phone: Chairman: (717) 534-8253

Chairman: F. Rapp
Professor: F. Rapp
Associate Professors: R.M. Glaser, J. Kreider, A. Lipton
Assistant Professors: R. Christensen, F. Funk, L. Geder, R. Hyman, R.N. Lausch, E.D. Marquez, S.C. de St. Jeor
Teaching Fellows: 6

Degree Programs: Microbiology, Genetics

Graduate Degrees: M.S., Ph.D.
Graduate Degree Requirements: Master: Grade Point Average 3.0 Minimum, Thesis Doctoral: Grade Point Average 3.0 Minimum, Dissertation

Graduate and Undergraduate Courses Available for Graduate Credit: Medical Microbiology, Medical Microbiology Laboratory, Science of Virology, Immunology of Infection, Microbial Physiology and Metabolism, Microbial Genetics, Electron Microscopic Techniques, Topics In Microbiology, Seminar in Microbiology, Literature Reports, Selected Readings, Laboratory Experiments In Microbiology, Special Projects in Microbiology

DEPARTMENT OF PHYSIOLOGY
Phone: Chairman: (717) 534-8567 Others: 534-8521

Chairman: H.E. Morgan
Evan Pugh Professor of Physiology: H.E. Morgan
Professors: H.E. Morgan, G.E. Mortimore, R.F. Zellis
Associate Professors: L.S. Jefferson, J.R. Neely, R.C. Rose
Assistant Professors: M.J. Hernandez-Perez, J.B. Li, S.H. Nellis, D.E. Rannels, W.F. Ward, J.I. Wenger, C.F. Whitfield, K.H. Woodside
Research Associates: 3

Degree Program: Physiology

Graduate Degrees: M.S., Ph.D.
Graduate Degree Requirements: Master: 30 credit hours of which 18 are in major and 6 in minor field. Must pass a candidacy examination covering the areas of Physiology and Biochemistry. Overall grade point average in major must be 3.0 Doctoral: No credit requirement. Must pass a candidacy examination in the areas of Physiology and Biochemistry, be judged competent in 1 foreign language, must pass a comprehensive examination, maintain a 3.0 grade point average, write an acceptable thesis, and pass a Final Oral Examination.

Graduate and Undergraduate Courses Available for Graduate Credit: Medical Physiology, General Physiology, Metabolic and Endocrine Physiology, Heart and Skeletal Muscle, Gastrointestinal Physiology, Seminar, Special Topics, Research Problems

PHILADELPHIA COLLEGE OF PHARMACY AND SCIENCE
43rd Street and Kingsessing Avenue Phone: (215) 386-5800
Philadelphia, Pennsylvania 19104
Director of Graduate Studies: L.A. Reber
Dean of College: L.F. Tice

DEPARTMENT OF BIOLOGICAL SCIENCES
Phone: Chairman: Ext. 244

Chairman: G.V. Rossi
Professors: A. DerMarderosian, F.M. White, B. Witlin, E.W. Packman, G.V. Rossi
Associate Professors: S.H. Joshi, F.C. Roia, P.J. Goldstein, R.F. Orzechowski
Assistant Professors: M.E. Melius, K.W. Thomulka
Instructors: L.B. Jordan, R.C. Kent, R.A. Smith, G.C. Johnson
Visiting Lecturers: 5
Research Assistants: 2
Graduate Assistants: 10

Field Stations/Laboratories: Lacawac Santuary (cooperative with Academy of Natural Sciences - Philadelphia

Degree Programs: Biology, Pharmacology, Pharmacognosy

Undergraduate Degree: B.S. (Biology)
Undergraduate Degree Requirements: 132 semester hours of credit of which 48 credit hours must be in biological subjects and 10 credit hours in "Humanities".

Graduate Degrees: M.S., Ph.D.
Graduate Degree Requirements: Master: 26 semester hour credits minimum, Research Project required
Doctoral: 52 semester hours credit; thesis, required

Graduate and Undergraduate Courses Available for Graduate Credit: Medical Microbiology (U), Microbial Physiology (U), Comparative Anatomy (U), Histology (U), Sanitary and Industrial Microbiology (U), Virology (U), Plant Developmental Biology (U), Animal Developmental Biology (U), Immunology and Serology (U), Cellular Ummunity (U), Mycology (U), Human Parasitology (U), Cell Biology (U), Genetics (U), Toxic Natural Products (U), Marine Pharmacognosy (U), Economic Botany (U), Tissue Culture Techniques (U), Analysis of Natural Products (U), Pharmacology (U), Toxicology (U), General Pathology (U), Advanced Physiology

POINT PARK COLLEGE

Wood Street and Boulevard of the Allies Phone: (412) 391-4100
Pittsburgh, Pennsylvania 15222
Dean of College: A.F. McLean

DEPARTMENT OF NATURAL SCIENCES AND TECHNOLOGY
Phone: Chairman: (412) 391-8374 Others: 391-8255

Chairman: J.M. Simon
Professors: S. Banerjee, C.F. Decker, L.E. Decker
Associate Professor: J . Sunder

Degree Programs: Biology, Medical Technology

Undergraduate Degree: B.S.
Undergraduate Degree Requirements: Biology: Minimum credits for B.S. is 125 credits; 39 of them are Department Requirements and 22 credits are Natural Science Prerequisites (Human Biology, Chemistry and Calculus), Medical Technology: 3 years general college requirements (94 credits) plus 1 year Internship at Hospital (34 credits)

ROSEMONT COLLEGE

Rosemont, Pennsylvania 19010 Phone: (215) 527-0200
Dean of College: M. Healy

DIVISION OF NATURAL SCIENCES AND MATHEMATICS
Phone: Chairman: Ext. 310 Others: Ext. 271

Division Chairman: J. Manning
Associate Professors: W.F. Ward
Assistant Professors: R.J. Kroll

Degree Program: Biology

Undergraduate Degree: B.A.
Undergraduate Degree Requirements: 2 semesters Life Science, 1 semester of each of the following: Genetics, Microbiology, Vertebrate Anatomy, Vertebrate Embryology, Plant Physiology, 2 semesters Senior Seminar, 2 Biology electives from Ecology, Parasitology, Biochemistry, Histology, 2 semesters of Inorganic Chemistry

ST. FRANCIS COLLEGE

Loretto, Pennsylvania 15940 Phone: (814) 472-7000
Dean of College: Q.L. Hartwig

DEPARTMENT OF BIOLOGY
Phone: Chairman: Ext. 296 Others: Ext. 295

Chairman: L.V. Pion
Professors: E. Reseonich, W. Durvea
Associate Professor: M. Kirsch
Assistant Professor: W. Takacs
Instructor: J. Yelenc

Degree Programs: Biology, Medical Technology

Undergraduate Degree: B.S.
Undergraduate Degree Requirements: 132 hours including Zoology, Botany, Comparative Anatomy, Embryology, Genetics, Microbiology, Evolution and Senior Seminar Physiology (Cell or Vertebrate), Inorganic and Organic Chemistry, 1 year Physics, Calculus (1 year), and Philosophy, English, History, Social Sciences etc.

SAINT JOSEPH'S COLLEGE

54th Street and City Avenue Phone: (215) 879-1000
Philadelphia, Pennsylvania 19131
Dean of College: Rev. T.F. Gleeson

DEPARTMENT OF BIOLOGY
Phone: Chairman: Ext. 219 Others: Ext. 458

Chairman: L.S. Marks
Professors: R.W. Fredrickson, L.S. Marks, C.B. Nash
Associate Professor: C.S. Nash
Assistant Professors: F.C. Monson, J.J. Watrous
Visiting Lecturers: 1
Undergraduate Laboratory Assistants: 16

Degree Program: Biology

Undergraduate Degree: B.S.
Undergraduate Degree Requirements: Fulfilment of college general education requirement, plus 24 hours of Biology, 18 of Chemistry, 10 of Physics, 8 of Mathematics and 7 of Modern Language

SAINT VINCENT COLLEGE

Latrobe, Pennsylvania 15650 Phone: (412) 539-9761
Dean of College: R.R. Gorka

DEPARTMENT OF BIOLOGY
Phone: Chairman: (412) 539-9761

Chairman: O.H. Roth
Professors: M.G. Duman, J.R. Lieb, O.H. Roth, J.H. Taubler
Instructor: M.A. Polechko
Research Assistants: 4

Degree Programs: Biology, Medical Technology

Undergraduate Degrees: B.A., B.S.
Undergraduate Degree Requirements: 124 credits, C average, 30 credits in Biology, 16 in Chemistry, 8 in Physics and 8 in Mathematics. In addition 48 credits in a cor-curriculum. All these are included in the 124 total.

SETON HILL COLLEGE

Greensburg, Pennsylvania 15601 Phone: (412) 834-2200
Dean of College: Sr. C. Toler

DEPARTMENT OF BIOLOGY
Phone: Chairman: Ext. 374

Chairman: W.H. Walker
Professor: Sr. A. Infanger
Associate Professor: W.H. Walker
Instructor: Sr. B. Fondy

Degree Program: Biology

Undergraduate Degree: B.S.
Undergraduate Degree Requirements: Program in Biology is completely coordinated with Saint Vincent College program one set of courses is shared by majors of

both institutions.

SHIPPENSBURG STATE COLLEGE
Shippensburg, Pennsylvania 17257 Phone: (717) 532-9121
Dean of Graduate Studies: W.E. Kerr
Dean of College: D. Morningstar

DEPARTMENT OF BIOLOGY
Phone: Chairman: (717) 532-9121

Chairman: W.E. Peightel
Professors: R.A. Barr, W.R. Kelley, L.W. Kreger, W.E. Peightel, A.R. Slysh
Associate Professors: P.J. Buhan, A.V. Gunter, H.E. Hays, Jr. , G.L. Kirkland, Jr., B.O. Ochs, W.E. Rogers
Assistant Professors: J.F. Davidson, W.J. Morrison, R.D. Reed, R.W. Wahl
Instructor: E.L. Nollenberger
Visiting Lecturers: 8
Research Assistants: 3

Field Stations/Laboratories: Marine Science Consortium Stations at Lewes, Delaware and Wallops Island, Virginia

Degree Programs: Biology, Medical Technology

Undergraduate Degrees: B.A., B.S.
Undergraduate Degree Requirements: 128 semester hours

Graduate Degrees: M.S., M.Ed.
Graduate Degree Requirements: Master: 30 graduate credits

Graduate and Undergraduate Courses Available for Graduate Credit: Plant Ecology, Hydrobiology, Plant Anatomy, Plant Taxonomy, Plant Pathology, Acoelomates, Coelomates, Problems in Ornithology, Radioisotope Techniques in Biology, Endocrinology, Ichthyology, Evolution, The Origin and Growth of Modern Biology, Paleontology, Microtechniques, Cytogenetics, Problems in Plant Ecology, Comparative Plant Morphology, Mycology, Problems in Plant Physiology, Cellular Physiology, Comparative Animal Physiology, Physiological Mechanisms of Animal Behavior, Problems in Animal Ecology, Vertebrate Ecology and Taxonomy, Helminthology, Modern Genetics, Topics in Mammalian Biology, Physiological Chemistry, Elements of Research,

SLIPPERY ROCK STATE COLLEGE
Slippery Rock, Pennsylvania 16057 Phone: (412) 794-2510
Dean of Graduate Studies: R.A. Lowry
Dean of College: C.F. Dresden

DEPARTMENT OF BIOLOGY
Phone: Chairman: (412) 794-7295 Others: 794-2510

Chairman: F.A. Pugliese
Professors: G.L. Dryden, K.S. Erdman, R.G. Hart, R J. Medve, K.E. Michel
Associate Professors: P.A. Archibald, T.W. Gaither, T.H. John, H.Y. McAllister, C.F. Mueller, M.A. Shellgren, R E. Taylor
Assistant Professors: F.M. Hoffman, W.S. Morrison, F.A. Pugliese
Graduate Assistants: 3

Field Stations/Laboratories: Moraine State Park

Degree Programs: Biology, Biological Education

Undergraduate Degrees: B.A., B.S. B.S.Ed.
Undergraduate Degree Requirements: (In semester hours) 31 hours Biology, 16 Chemistry, 3 Mathematics, Foreign Language, B.A. 39 hours Biology, 19 Chemistry, 7 Mathematics, 7 Physics B.S., (Certification in Biology, 31 Biology, 16 Chemistry, 3 Mathematics, 28 Education. For all of the above 60 hours general studies. Total hours for graduation 128.

Graduate Degrees: M.S., M.Ed.
Graduate Degree Requirements: (In semester hours) M.S., 30 hours approved graduate coursework to include 3-6 hours thesis and 3 hours biometry. M.Ed. (Biology): 12-15 hours Biology, 6 research skills, 6-9 supportive electives; 3-6 free electives, 3-6 thesis or non-thesis program (0 credit)

Graduate and Undergraduate Courses Available for Graduate Credit: Taxonomy of Vascular Plants (U), Plant Morphology (U), Evolution (U), Biometry (U), Histology (U), Endocrinology (U), Topics in Plant Physiology, Advanced Bacteriology, Cytogenetics, Terrestrial Ecology, Aquatic Ecology, Comparative Animal Physiology, Herpetology, Mammalogy, Invertebrate Zoology, Graduate Special Problems, Laboratory Practicum in Biology, New Curriculum Developments in Biology

SUSQUEHANNA UNIVERSITY
Selinsgrove, Pennsylvania 17870 Phone: (717) 374-2345
Dean of College: W. Reuning

BIOLOGY DEPARTMENT
Phone: Chairman: Ext. 322 Others: Ext. 322,366

Chairman: H.E. DeMott
Professor: H.E. DeMott
Associate Professor: B.D. Presser
Assistant Professor: G.C. Boone, R.P. Harrison

Degree Program: Biology

Undergraduate Degree: B.S.
Undergraduate Degree Requirements: 9 courses in Biology, Chemistry through Organic, Introductory Physics Mathematics through Calculus I

SWARTHMORE COLLEGE
Swarthmore, Pennsylvania 19081 Phone: (215) 544-7900
Dean of College: H.E. Pagliaro

DEPARTMENT OF BIOLOGY
Phone: Chairman: Ext. 220

Chairman: N.A. Meinkoth
Professors: L.J.Flemister, L.G. Livingston, N.A. Meinkoth, K.S. Rawson
Associate Professors: J.B. Jenkins, R.E. Savage
Assistant Professors: J.C. Hickman, M.L. Miovic, B.W. Snyder
Teaching Assistants: 3

Degree Program: Biology

Undergraduate Degree: B.A.
Undergraduate Degree Requirements: 8 semesters of Biology, General Chemistry, Organic Chemistry, 2 semesters of Mathematics, including Calculus.

TEMPLE UNIVERSITY
Broad Street and Montgomery Avenue Phone: (215) 787-7000
Philadelphia, Pennsylvania 19122
Dean of Graduate Studies: M.C. Ebersole

College of Liberal Arts

DEPARTMENT OF BIOLOGY
(No reply received)

School of Medicine
3400 North Broad Street
Philadelphia, Pennsylvania 19140

Dean: R.W. Sevey

DEPARTMENT OF ANATOMY
Phone: Chairman: (215) 221-3161 Others: 221-3160

Chairman: J.R. Troyer
Professors: A.K. Christensen, L.G. Rodriguez-Peralta, C.D. Schneck, J.R. Troyer, R.C. Truex
Associate Professors: M.N. Bates, G.S. Crouse, S.J. Phillips, N.E. Pratt, M. Sodicoff, J.S. Way
Assistant Professors: S.A. Ernst, S.C. Ernst, L.G. Paavola
Instructor: J.W. Mills

Degree Program: Anatomy

Graduate Degree: Ph.D.
Graduate Degree Requirements: Doctoral: 1) Reading Proficiency - one foreign language 2) Teaching--one semester for each medical anatomy course 3) Preliminary examination 4) Research project resulting in acceptable Dissertation 5) Defense of thesis

Graduate and Undergraduate Courses Available for Graduate Credit: Human Gross Anatomy, Advanced Embryology, Histology, Neurosensory Sciences, Topics in Cell Biology, Topics in Reproductive Biology, Current Anatomical Literature, Anatomical Methods I: Gross, Histological and Histochemical, Anatomical Methods II: Basic Electron Microscopy, Anatomical Methods III: Advanced Electron Microscopy and Cytochemistry, Functional and Surgical Anatomy of the Head and Neck, Functional and Surgical Anatomy of Back and Extremities

DEPARTMENT OF BIOCHEMISTRY
Phone: Chairman: (215) 221-3263

Chairman: G.W.E. Plaut
Professors: E. Farber, R. Hamilton (Emeritus), G. Litwack, G.W.E. Plaut, R. Suhadolnik, S. Weinhouse
Associate Professors: J. Cilley, R. Hanson, K.J. Hoober, R. Knauff, L. Norcia, W.K. Paik, R. Pieringer, M.C. Scrutton, S. Winsten
Assistant Professors: R. Adelman, T. Borun, A. Buszynski (Research), R. Davies (Research), E. Kirby, P. Lotlikar, D. Marks, M. Patel (Research), S. Rajalakshmi (Research), R. Roxby, V.L. Schramm, C.M. Smith
Instructors: T. Aogaichi, D.S. Deshmukh (Research), H. Siplet
Teaching Fellows: 4
Research Assistants: 8

Degree Program: Biochemistry

Graduate Degree: Ph.D.
Graduate Degree Requirements: Doctoral: Students must complete at least ten graduate courses (approximately 30 credits) which are to be chosen by the student in consultation with his committee. Also, each student is required to demonstrate a reading knowledge of one Foreign Language. Moreover, the student may, at the discretion of his committee, be required to pass a written Qualifying Examination. All students must pass an oral Preliminary Examination, and must submit an acceptable dissertation.

Graduate and Undergraduate Courses Available for Graduate Credit: Biochemistry for Medicine, Introduction to Research Methodology, Biochemistry Seminar, Cell Biology, Advanced Biochemistry Lecture, Enzymes and Proteins, Metabolism, Structure, Synthesis and Degradation of Macromolecules, Biology of Human and Experimental Cancer, Advanced Topic Courses (a series of six courses: Metabolic Regulation, Mechanisms of Blood Coagulation and Fibrinolysis, Special Topics in Molecular Biology, Mechanism of Action of Hormones, Special Topics in the Molecular Biology, Mechanism of Action of Hormones, Special Topics in the Mechanism of Enzymic Catalysis, and Special Topics in Lipids, Membranes and Transport Processes), Cellular and Molecular Aspects of Aging, Developmental Biochemistry, Research in Biochemistry and Doctoral Dissertation

DEPARTMENT OF BIOMETRICS
Phone: Chairman: (215) 221-4661

Chairman: S. Schor
Professors: S. Schor, J. Norton, S.M Free (Adjunct)
Associate Professors: S. Brauerman, W. Westlake (Adjunct), E. Wollin,(Adjunct), J. Pauls (Adjunct), M. Ball, R. Sharma
Assistant Professors: M. Leyton, M. Rosenzweig, W. Wilkes (Adjunct)
Research Fellows: 10

Degree Program: Biometry

Graduate Degrees: M.S., Ph.D.
Graduate Degree Requirements: Master: 30 semester hour credits, oral and written comprehensives, Masters project or thesis Doctoral: 54 semester hours credits, Masters examinations, Preliminary Examination, Oral and Written, acceptable Ph.D. Dissertation, Computer Programming as a Foreign Language

Graduate and Undergraduate Courses Available for Graduate Credit: Fundamentals of Biostatistics, Applied Biostatistics, Sample Survey, Non-Parametric Statistics, Multivariate Analysis, Analysis of Variance, Program Evaluation, Mathematics and Statistics, Stochastic Models, Intermediate Biostatistics, Design of Experiment, Statistical Methods in Biological Assay, Decision Theory as Applied to Medical Problems, Simulation and Other Computer Applications, Systems Analysis and Operations Research, Demography, Computers in Health Sciences, Categorical Data, Regression Theory, Seminar in Sample Survey Methods

DEPARTMENT OF MICROBIOLOGY AND IMMUNOLOGY
Phone: Chairman: (215) 221-3207

Chairman: G.D. Shockman
Professors: F. Blank, H. Friedman (Research), H.F. Havas, M. Klein, A.H. Nowotny, J. Prier (Adjunct), J. Smolens, E.H. Spaulding, R. Suhadolnik, N. Yamamoto
Associate Professors: W. Ceglowski, K. Cundy, H. Goldner, M.L. Higgins, T.J. Linna, H. Lischner, L. Daneo-Moore, H.C. Oels, V. Pidcoe, W. Pollack, (Research), R. Swenson, G. Weinbaum, L. Zubrzycki
Assistant Professors: H. Beilstein (Adjunct), C.L.H. Chen, B.H. Cooper, A. Deforest, C. Dietz, T.K. Eisenstein, S. Grappel, K.M. Lam, A.J. Lamberti, B. Lorber, A. Schwartz, J. Thompson, K. Thompson, D. Tripodi
Visiting Lecturers: 10/year
Research Assistants: 30
Postdoctoral Fellows: 4
Predoctoral Fellows: 32

Degree Programs: Immunology, Medical Microbiology, Medical Technology, Microbiology

Graduate Degrees: M.S., Ph.D.
Graduate Degree Requirements: (In semester hours)
Master: 24 hours, including Medical Microbiology, 4 Microbiology Graduate Student Seminar, Thesis
Doctoral: 48 hours, including Medical M crobiology, 6 Microbiology Graduate Student Seminar, 12 from the following: 3 Genetics, 3 Immunobiology, 3 Immunochemistry, 2-1 Microbial Physiology, 3-4 1/2 Mycology, 3 Virology, Dissertation

Graduate and Undergraduate Courses Available for Graduate Credit: Medical Microbiology and Immunology for Medical Students, Chemical and Physical Methods in Microbiology and Immunology, Presentation of Scientific Information, Microbiology and Immunology

Graduate Student Seminar, Medical Microbiology and Immunology for Graduate Students, General Microbiology and Microbial Genetics, Immunology and Immunochemistry, Medical Microbiology, Medical Laboratory Microbiology, Clinical Microbiology Seminar, Molecular Immunobiology, Cellular Immunobiology, Medical Virology, Immunochemistry, Microbial Genetics, Medical Mycology, Microbial Metabolism, Growth and Control, Infectious Process, Procaryotic Cell, Immunopathology, Biophysical Aspects of Microbiology, Cellular Action of Antibiotics, Synthesis of Macromolecules, Modern Clinical Microbiology, Biology of the Fungi, Pathogenic Fungi, Medical Mycology Seminar, Infectious Process Seminar, Special Topics in Microbial Biochemistry and Macromolecular Biosynthesis, Immunology Seminar, Current Topics in Microbial Genetics, Tumor Immunology, Microbiology and or Thesis Research

DEPARTMENT OF PHARMACOLOGY
Phone: Chairman: (215) 221-3237 Others: 221-3236

Chairman: G.D.V. vanRossum (Acting)
Professors: M.W. Adler, C.T. Bello, S.C. Glauser, C.A. Papacostas
Associate Professors: M. Black, E. Glauser, C. Harakal, J.G. McElligott, M.M. Reidenberg, B.F. Rusy, L. Salganicoff, R.J. Tallarida, D. Mills (Research)
Assistant Professors: D. Drayer, M.H. Loughane
Research Assistants: 2

Degree Program: Pharmacology

Graduate Degrees: M.S., Ph.D.
Graduate Degree Requirements: Master: For Admission: Applicants must have a B.S. in Biology, Chemistry, Pharmacy or related area, Graduate Record Examination and 2 letters of recommendation Degree Requirements: 30 credits total, 6 credits of which are research. Language requirements, Thesis or non-thesis option Doctoral: For admission: B.S. in Pharmacy, Chemistry, Biology or health realted area, Graduate Record Examination, 2letters of recommendation Degree Requirements: 16 courses, 2 foreign languages, Qualifying Examination, book written and oral comprehensive examination, Research thesis submitted must be defended orally.

Graduate and Undergraduate Courses Available for Graduate Credit: Pharmacology, Mathematical Biology, Physical Chemistry of Biological Reactions, Biomedical Instrumentation, Structure and Optical Properties of Biological Compounds, Introductory Pharmacology, Renal Pharmacology, Cardiovascular Pharmacodynamics, Respiratory Pharmacology, Neuropharmacology, Pharmacology of the Contractile Process, Subcellular Pharmacology, Theory and Techniques in Experimental Neurophysiology, Seminar in Neurophysiology and Neuropharmacology, Pharmacological Principles, Pharmacology of the Cell Membrane

DEPARTMENT OF PHYSIOLOGY
Phone: Chairman: (215) 221-3272 Others: 221-3272

Chairman: A.R. Freeman
Professors: A.R. Freeman, G.C. Henny (Emeritus), P.R. Lynch, M.P. Wiedeman
Associate Professors: G. Ascanio, J.H. Gault
Assistant Professors: J.L. Heckman, C.R. Michie, M.B. Wang
Instructors: E.K. Caldwell, H.N. Mayrovitz, T.N. Tulenko
Visiting Lecturers: 4
Teaching Fellows: 2
Research Assistants: 6

Degree Program: Physiology

Graduate Degree: Ph.D.
Graduate Degree Requirements: Doctoral: No specific credit hours are required, but the following courses are required (Gross Anatomy, Histology, Cell Biology, Biochemistry, Physiology and Pharmacology). A preliminary examination must be passed and a thesis project of original work must be defended.

Graduate and Undergraduate Courses Available for Graduate Credit: Theory and Techniques in Experimental Neurophysiology, Seminar in Neurophysiology and Neuropharmacology, Advanced Cardiovascular, Pulmonary and Neurophysiology, Microcirculation, Radiology-Physiology Seminar, History of Physiology, Electronics for Biological Scientists

THIEL COLLEGE

Greenville, Pennsylvania 16125 Phone: (412) 588-7700
Dean of College: O.M. Todo

DEPARTMENT OF BIOLOGY
Phone: Chairman: Ext. 406

Chairman: K R. Gordon
Associate Professors: R.J. Ingersoll, P.E. Ode
Assistant Professors: K.R. Gordon, W.H. Mason

Degree Program: Biology

Undergraduate Degrees: B.A.
Undergraduate Degree Requirements: 44 hours in the following courses: Fundamentals of Biology (2 courses), The Plant Kingdom or the Animal Kingdom, Cell Biology, General Physiology, Genetics, Ecology, Energy Flow or Population Ecology, Experimental Animal Laboratory or Field Biology or Independent Study, Physical Biology, Biology Seminar (3 courses) Chemistry of the Covalent Bond, Calculus I

UNIVERSITY OF PENNSYLVANIA

Philadelphia, Pennsylvania 19104 Phone: (215) 594-5000
Dean of Graduate Studies: V. Gregorian

College of Arts and Sciences

Dean: V. Gregorian

DEPARTMENT OF BIOLOGY
Phone: Chairman: (215) 594-7121 Others: 594-5000

Chairman: W.H. Telfer
University Professor of Biology: D.R. Goddard
Professors: H.G. Borei, A.H. Brown, A.O. Dahl, A. N. Epstein, R.O. Erickson, W.J. Ewens, P. George, D.R Goddard, S. Inoue, N.R. Kallenbach, H.L. Li, D.J O'Kane, R. Patrick (Adjunct), L.D. Peachey, W.H. Telfer
Associate Professors: J. Bryan, G.W. Ellis, G.Epple (Adjunct), C.E. Goulden (Adjunct), W.W. Moss (Adjunct), R.E. Ricklefs, H. Sata, W.J. Smith, Y. Suyama, L.G. Tilney, T. Uzzell (Adjunct), I.L. Waldron
Assistant Professors: T.L. Bott (Adjunct), F.B. Gill (Adjunct), J.H. Gillespie, J Mears (Adjunct), A.P. Smith, R.I. Yeaton, B.R. Gerber
Teaching Fellows: 24
Lecturers: 5

Degree Programs: Cellular and Molecular Biology, Developmental Biology, Ecology and Evolution, Animal Behavior and Neurophysiology, Genetics, Limnology

Undergraduate Degree: B.S. (Biology
Undergraduate Degree Requirements: 15 course units

Graduate Degree: Ph.D. (Biology)
Graduate Degree Requirements: Doctoral: 20 course units (of which 12 must be completed at the University of Pennsylvania) One Modern language other than English, Preliminary Examination, Dissertation

Graduate and Undergraduate Courses Available for Graduate Credit: Palynology (U), Plant Physiology (U), The Chemistry of Living Cells (U), Cellular Chemistry

Laboratory (U), Biology of the Fungi (U), Evolution (U), Brain and Behavior (U), Population and Community Ecology (U), Topics in Animal Behavior and Brain Function (U), Laboratory in Social Ethology (U), Animal Communication (U), Genetics (U), The Cell (U), Bioenergetics (U), Cell Structure and Function Laboratory (U), Biology of the Invertebrates (U), Mendelian Genetics (U), Human Genetics (U), Biochemical Systematics (U), Genetics Laboratory I (U), Advanced Developmental Biology (U), Neural Bases of Behavior (U), Mathematical Biology (U), Principles of Bioenergetics (U), Marine Life (U), Developmental Plant Morphology (U), Physical and Chemical Limnology (U), Biological Limnology (U), Aquatic Communities (U), Ecological Methods (U), Systematic Biology (U), Population Genetics (U), Ecosystem Structure and Function (U), Current Topics in Genetics (U), Physical Methods in Cell Research (U), Electron Microscope Techniques (U), Biophysical Cytology (U), Mathematical Ecology (U), Molecular Genetics (U), Biometry (U), Mathematics for Biologists (U), Statistics for Biologists (U), Ecology of Social Behavior (U), Adaptation (U), Marine Ecology Field Course (U), Taxonomy and Ecology of Aquatic Insects (U), Current Topics in Developmental Biology, Seminar in Environmental and Evolutionary Biology, Environmental Physiology and Ecology of Plants (Seminar)

School of Medicine

Dean: E. Stemmler

DEPARTMENT OF ANATOMY

Phone: Chairman: (215) 594-8041 Others: 594-8046

Chairman: J.M. Sprague
Joseph Leidy Professor of Anatomy: J.M. Sprague
Professors: W.W. Chambers, L.B. Flexner, H. Holtzer, R.J Johnson, J. Lash, C.N. Liu, A.M Nemeth, F.A. Pepe, J. Piatt
Associate Professors: G. de la Haba, M. Harty, P.A. Liebman, V.T. Nachmias, P. Sterling
Assistant Professors: M.B. Burnside, R. Minor, D. Roberts, A.C. Rosenquist, J. Sanger

Degree Program: Anatomy

Graduate Degree: Ph.D.
Graduate Degree Requirements: Doctoral: 20 course credits, 1 year residence University of Pennsylvania, 1-2 foreign languages, preliminary examination, thesis examination, thesis

Graduate and Undergraduate Courses Available for Graduate Credit: Gross Human Anatomy, Histology, Neurobiology, Structural Adaptions of Vertebrate Body, Comparative Microscopic Anatomy, Introduction to Nervous System, Developmental Anatomy, Advanced Neruobiology, Cell Differentiation, Tissue Interactions and Metabolic Responses, Molecular Organism of Striated Myofibrils, Histochemistry and Cytochemistry, Topics in Embryology and Development, Advanced Human Dissection, Form and Function in Central Nervous System, Advanced Dissection, Cytochemistry, Comparative Mammalian Cytogenetics

DEPARTMENT OF BIOCHEMISTRY

Phone: Chairman: (215) 594-8025 Others: 594-8022

Chairman: J.J. Ferguson, Jr.
Professors: H.J. Bright, F. Charalampous, S.W. Englander, J.J. Ferguson, Jr., J.G. Flaks, H. Rasmussen, A. Weber
Associate Professors: R.G. Kallen, E.K. Pye, N. Kefalides
Assistant Professors: R.L. Barchi, D. Berkowitz, C. Clark, R.O. Viale, M.K. Weibel
Research Associates: 4
Post-doctoral Fellows: 15

Degree Program: Biochemistry

Graduate Degree: Ph.D.
Graduate Degree Requirements: Doctoral: 20 credit units of course work and supervised research satisfactory completion of Preliminary Examination satisfactory defense of dissertation

Graduate and Undergraduate Courses Available for Graduate Credit: Introductory Biochemistry, Intermediate Biochemistry, Advanced Biochemistry, Enzyme Technology, Laboratory Rotation, Electron Microscope Techniques, Biochemistry Computer Workshop, Biochemical Endocrinology, Molecular Genetics, Mathematics for Biochemists, Macromolecules, Neurochemistry, Biochemistry of Inherited Diseases, Mechanism of Enzymatic Reactions, Biochemical Aspects of Metabolism and Human Disease, Introduction to Bioenergetics, Biochemistry of Collagen

DEPARTMENT OF BIOPHYSICS AND PHYSICAL BIOCHEMISTRY

Phone: Chairman: (215) 243-7159 Others: 243-5000

Chairman: B. Chance
Eldridge Reeves Johnson Professor of Medical Physics: B. Chance
Biophysics
Professors: T.F. Anderson, D.W. Bronk, G.L. Gerstein, R.K. Perry, W.D. Phillips
Associate Professors: J.K. Blasie, J.J. Higgins, L. Peachey
Assistant Professors: D.C. DeVault, A. Horwitz, J.S. Leigh, J.A. McCray
Physical Biochemistry
Professors: W.D. Bonner, M. Cohn, A. Kovach, A. Mildvan, K.G. Paul, I. Rose, J.R. Williamson, T. Yonetani
Associate Professors: T. Asakura, L. Cohen, H. Davies, J Glusker, F. Kayne, C. Lee, B.O.N. Rao, B.T. Storey, M.K F. Wikström, D.F. Wilson, E. Wilson
Assistant Professors: H.R. Drott, P.L. Dutton, M. Erecinska, J. Klinman, K. LaNoue, G. MacDonald, L. Mela, T. Ohnishi, N. Oshino, G.R. Reed, A. Scarpa, R.G. Thurman
Research Associates: 4
Postdoctoral Fellows: 24

Degree Program: Biophysics

Undergraduate Degree: B.S.
Undergraduate Degree Requirements: 13 credit units in basic Mathematics (3), Physics (3), Chemistry (5), and Biology (2), 2 credit units in Biophysics

Graduate Degree: Ph.D.
Graduate Degree Requirements: Doctoral: 2 courses graduate Physics or Physical Chemistry, 1 course graduate Biochemistry, 2 courses graduate Physiology and/or Microbiology, 3 courses graduate Biophysics, 4 courses Advanced graduate Physics or Chemistry, and 4 courses Independent study in Biophysics

Graduate and Undergraduate Courses Available for Graduate Credit: Bioenergetics (U), Problems of Molecular Structure: x-ray Neutron diffraction and magnetic resonance (U), Problems of Molecular Structure: Kinetics (U), Problems of Molecular Structure: Spectroscopy (U), Biochemistry: Macromolecules (U), Biochemistry: Biological Membranes (U), Independent Study (U)

DEPARTMENT OF PHYSIOLOGY

Phone: Chairman: (215) 594-8725

Chairman: R.E. Forster
Isaac Ott Professor of Physiology: R.E. Forster
Professors: S.Y. Botelho, J.R. Brobeck, F.P. Brooks, A.B. DuBois, R.E. Forster, G.L. Gerstein, G. Karreman, A.K. Kleinzeller, L.H. Peterson,

A.P. Somlyo, S. Winegrad
Associate Professors: M.M. Civan, R.F. Coburn, R.H. Cox, M. Delivoria-Papadopoulos, G.M. Fischer, C. L. Hamilton, S. Lahiri, M. Morad, D.G. Moulton, M. Pring, D. Scott, Jr., B.T. Storey
Assistant Professors: R.J. Bagshaw, A.L. Beckman, G.M. Carlson, E.D. Crandall, J.J. Deren, M.W. Edwards, A.B. Fisher
Research Assistants: 9
Associates: 8
Research Specialists: 1

Degree Program: Physiology

Graduate Degrees: Ph.D.
Graduate Degree Requirements: Not Stated

Graduate and Undergraduate Courses Available for Graduate Credit: Advanced Neurobiology, Animal Physiology, Biological Control System Analysis, Biological Membranes, Biomathematics, Cell Physiology, Current Topics in Cardiac Physiology, General Principles of Physiology, Medical Physiology, Selected Readings in Physiology in German Language, Selected Readings in Physiology in Russian Language and Special Problems in Physiology

School of Engineering
Dean: A.E. Humphrey

DEPARTMENT OF BIOENGINEERING
Phone: Chairman: (215) 243-5881

Chairman: A. Noodergraaf
Professors: C.T. Brighton, J. Hale, A.E. Humphrey, D.E. Kuhl, H. Kwart (Adjunct), W.G. Custead (Adjunct), M. Litt, A. Noordergraaf, J.A. Quinn, H.P. Schwan, S.R. Warren, C. Weygandt
Associate Professor: S.C. Batterman, D. Garfinkel, J. Higgins, F.D. Ketterer, E. Korostoff, H.N. Kritikos, S.R. Pollack, S. Takashima, H.W. Wallace
Assistant Professors: P.H. Bloch, K.B. Cambpell, D. Graves, G.R. Neufeld
Teaching Fellow: 1
Research Assistants: 10

Degree Program: Bioengineering

Undergraduate Degree: B.S.
Undergraduate Degree Requirements: 4 CU Mathematics, 6 CU in Sciences, 10 Bioengineering and related areas, 9 CU electives in Bioengineering and related areas, 9 electives in Engineering and Physical and Life Sciences, to constitute a minor, 11 CU electives in areas of Mathematics (1), Social Sciences and Humanities (7) and free (3)

Graduate Degrees: M.S., Ph.D.
Graduate Degree Requirements: Master: 8 CU and thesis Doctoral: 20 CU and Written and Oral Preliminary and Thesis

Graduate and Undergraduate Courses Available for Graduate Credit: Not Stated

School of Dental Medicine
Dean: W. Cohen

DEPARTMENT OF BIOCHEMISTRY
Phone: Chairman: (215) 594-8936

Chairman: J.B. Marsh
Professors: J.B. Marsh, J.L. Rabinowitz (Partially Affiliated), J. Rosenbloom
Associate Professors: P.S. Leboy, I.M. Shapiro
Assistant Professors: B. Howard, R. Alper
Instructor: J. Cotmore
Research Assistant: 1
Post-doctoral Fellows: 2

Degree Program: Biochemistry

Graduate Degrees: Ph.D.
Graduate Degree Requirements: Doctoral: Ph.D. through University of Pennsylvania School of Arts and Sciences

Graduate and Undergraduate Courses Available for Graduate Credit: Biochemistry

DEPARTMENT OF HISTOLOGY, EMBRYOLOGY AND GENETICS
Phone: Chairman: (215) 594-8994

Chairman: C.E. Wilde, Jr.
Professors: C.E. Wilde, Jr.,
Associate Professors: R.G. Herold, G.E. Levenson, R. Piddington
Assistant Professors: H.T. Graver
Instructors: E. Macarak,
Teaching Fellows: 1
Research Assistants: 5

Degree Program: Histology, Embryology, Genetics

Graduate Degree: Ph.D.
Graduate Degree Requirements: Doctoral: Ph.D. through University of Pennsylvania's Graduate School of Arts and Sciences

Graduate and Undergraduate Courses Available for Graduate Credit: Dental Medicine, Histology and Embryology, Biology, Current Problems in Developmental Biology

DEPARTMENT OF MICROBIOLOGY
Phone: Chairman: (215) 594-8987 Others: 594-8961

Chairman: B.F. Hammond
Professors: B.F. Hammond, N.B. Williams (Emeritus)
Associate Professors: G.H. Cohen, P.C. Montgomery, B. Rosan
Associate Professor: R.J. Eisenberg
Visiting Lecturers: 4
Teaching Fellow: 1
Research Assistants: 4

Degree Program: Microbiology

Graduate Degree: Ph.D.
Graduate Degree Requirements: Doctoral: Ph.D. through University of Pennsylvania's Graduate School of Arts and Sciences

Graduate and Undergraduate Courses Available for Graduate Credit: Microbiology

School of Veterinary Medicine
Dean: R.R. Marshak

DEPARTMENT OF ANIMAL BIOLOGY
Phone: Chairman: (215) 594-8586

Chairman: D.K. Detweiler
Professors: D.A. Abt, R.L. Brinster, R.E. Davies, D.K. Detweiler, D.C. Dodd, D.E. Johnson, D. Kritchevsky, D.S. Kronfeld, R.R. Marshak, J.T. McGrath, W. Medway, J. Melbin, E.N. Moore, D.F. Patterson, M. Reynolds, W.H. Rhodes, L.F. Rubin, R.M. Schwartzman, L.R. Soma, E.J.L. Soulsby, B. Wolf
Associate Professors: K. Bovee, R.O. Davies, A.M. Kelly, R.M. Kenney, R.A. McFeely, J.S. Reif, S.A. Steinberg, W.T. Weber
Assistant Professor: V.K. Ganjam

Field Stations/Laboratories: New Bolton Center - Kennet Square, Pennsylvania

Degree Programs: Comparative Medical Sciences, Cardiology, Ophthalmology, Internal Medicine, Surgery, Dermatology, Radiology

Graduate Degrees: M.Sc., Ph.D. (Comparative Medical Sciences)
Graduate Degree Requirements: Master: 8 course units. Preliminary examination at the end of year one. A student cannot go on for a degree program unless he has performed satisfactorily in this examination, and obtained approval for a research proposal leading to a thesis. Doctoral: The Ph.D. program ordinarily requires 4 years. Total of 20 course units, 8 of these in basic science subjects. Reading knowledge of one approved foreign language.

Graduate and Undergraduate Courses Available for Graduate Credit: Graduate courses in basic medical sciences, Bioengineering, and Material Science

DEPARTMENT OF PATHOBIOLOGY
Phone: Chairman: (215) 594-8867 Others: 594-8842

Chairman: E.L. Soulsby
Professors: D.C. Dodd, A.J. Girardi, I. Live, H. Martin (Emeritus), J.T. McGrath, J.R. Rooney, E.L. Soulsby, E.L. Stubbs (Emeritus), W.C. Wilcox, B. Wolf
Associate Professors: L.J. Bello, S. Chacko, J. Hwang, A.M Kelly, W. Lawrence, J. Palm, G.A. Schad, W.T. Weber
Assistant Professors: O.O. Barriga, P.B. Khoury, R. Minor, W.H. Riser (Research), G.B. Solomon, B.E. Stromberg
Instructor: H.A.R. Cimprica
Visiting Lecturers: 5
Research Associate: 1

Field Stations/Laboratories: New Bolton Center - Kennett Square, Pennsylvania

Degree Programs: Parasitology, Microbiology

Graduate Degree: Ph.D. (a very few M.S. degrees)
Graduate Degree Requirements: Master: 8 credit units plus a thesis Doctoral: 20 credit units plus doctoral thesis

Graduate and Undergraduate Courses Available for Graduate Credit: Pathology - Introduction to Pathology, Neuropathology, Comparative Pathology, Veterinary Pathology, Surgical Pathology, Selected Topics in Animal Membrane Research Microbiology - Basic Microbiology, Basic Microbiology (Cell Biology), Immune Mechanisms, Biology and Biochemistry of Microbial Cells, Biology and Biochemistry of Mammalian Cells, Microbiology Laboratory, Mechanisms of Infection, Biological Membranes, Specific Topics in Virology, Specific Topics in Microbial Metabolism and Physiology, Specific Topics in Genetics, Veterinary Micro-Biology, Independent Study and Research Parasitology-General Parasitology, Seminars in Parasitology, Research in Parasitology, Immunology of Host Parasite Relationships, Parasitic Zoonoses I - Protozoa, Parasitic Zoonoses II - Helminths, Biology of Parasites I - Adaptation to Parasitism, Biology of Parasites II - Natural Immunity and Host Specificity, Independent Study and Research

UNIVERSITY OF PITTSBURGH

Pittsburgh, Pennsylvania 15260 Phone: (412) 624-4141
Dean of Graduate Studies: E. Baranger

College of Arts and Sciences

Dean: J.L. Rosenberg

DEPARTMENT OF BIOLOGY
Phone: Chairman: (412) 624-4675 Others: 624-4266

Chairman: R.T. Hartman
Andrey Avinoff Professor of Biology: P. Gray
Professors: E.F. Carell, U.M Cowgill, R.T. Hartman, S. Lesher (Adjunct), C.R. Partanen
Associate Professors: W.P. Coffman, H.O. Corwin, F.J. Gottlieb, J. Gottlieb (Adjunct), R. Moore (Adjunct), K.C. Parkes (Adjunct), S. Shostak, E.M. Stricker
Assistant Professors: F.P. Doerder, S. McClure, M. Mares, R.J. Raikow, S.J Roux, A. Sillman, C. Walsh, M.A. Walsh, D.T. Wicklow, M. Zigmond
Lecturers: C.McCoy (Adjunct), J. Parodiz (Adjunct)

Field Stations/Laboratories: Pymatuning Laboratory of Ecology - Linesville, Pennsylvania 16424

Degree Programs: Biology, Human Genetics (Graduate Degrees in Interdisciplinary Programs)

Undergraduate Degree: B.S. (Biology
Undergraduate Degree Requirements: Organismic Biology II, Population and Evolutionary Biology, Molecular Biology, 14 credits in 100 level biology or approved related courses, 8 credits General Chemistry,8 credits Organic Chemistry, Algebra and Trigonometry (if necessary), Analytical Geometry and Calculus I, Introductory Physics

Graduate Degrees: M.S., Ph.D.
Graduate Degree Requirements: Master: Course work will be determined by the student's directing committee. M.S. Comprehensive Examination in four areas of biology. M.S. Thesis, Final Oral Examination (thesis defense). (one seminar per academic year for 1 credit) Doctoral: Course work will be determined by the student's respective committee. M.S. Comprehensive Examination in four areas of Biology. Ph.D. Preliminary Examination in the area of specialty. Ph.D. Comprehensive Examination for research proposal. Ph.D. thesis. Final Oral Examination (thesis defense). Presentation of a seminar.

Graduate and Undergraduate Courses Available for Graduate Credit: Genetics (U), Biology of Agression (U), Primate Behavior (U), Environmental (U), Vertebrate Morphology and Laboratory (U), Biology of the Vertebrates (U), Animal Physiology (U), Community Ecology (U), Population Ecology (U), Identification and Quantification of Algae and Vascular Plants in Field Communities (U), Identification and Quantification of Bacteria and Fungi in Field Communities (U), Mycology (U), Biogeography (U), Ecosystems and Man (U), Identification and Quantification of Insects in Field Communities (U), Identification and Quantification of Invertebrate and Vertebrate Animals in Field Communities (U), Biology of the Invertebrates (U), Neurophysiology (U), Comparative Animal Physiology (U), Photobiology (U), Embryology and Laboratory (U), Cytology(U), Cytology Laboratory (U), Developmental Mechanisms: Molecular to Cell (U), Developmental Mechanism: Cell to Organism (U), Plant Physiology (U), Cell and Organelle Physiology (U), Neuropharmacology (U), Psychopharmacology (U), Psychobiology of Homeostasis: Water and Electrolyte Balance (U), Psychobiology of Homeostatis: Energy Balance (U), Physiological Genetics (U) (U), Developmental Genetics (U), Mutation (U), Social Implications of Biology Seminar, Seminar in Systematics and Evolution, Seminar in Developmental Cytology, Microscopy, Seminar in Ultrastructure, Developmental Seminar I, II, and III, Biogeochemistry Laboratory, Biogeochemistry, Physical and Chemical Limnology, Seminar in Ecology, Seminar in Plant Biochemistry, Seminar in Plant Physiology, Comparative Endocrinology, Seminar in Neuroendocrinology, Seminar in Biochemistry and Behavior, Experimental Genetics, Seminar in Genetics, Experimental Designs in Ecology, Ecological Field Trip, Habitat Ecology: Terresterial, Habitat Ecology: Aquatic, Ecology of Aquatic Insects

DEPARTMENT OF BIOCHEMISTRY
Phone: Chairman: (412) 624-3076 Others: 624-4141

Chairman: R. Bentley
University Professor of Experimental Medicine and Biochemistry: K. Hofmann
Professors: R. Abrams, R. Bentley, A.E. Chung, C. Ho (Joint), R.H. McCoy, I.R. McManus
Associate Professors: M. Edmonds, F. Finn (Joint), J.S. Franzen, J.R. Gilbertson (Joint), M.H. Vaughan
Assistant Professors: I.M. Campbell, T.G. Cooper, D.J. Edwards (Joint)
Research Assistants: 15
Teaching Assistants: 8

Field Stations/Laboratories: Allegheny Observatory, Computer, RIDC, Primate Research Laboratory, Pymatuning Laboratory of Ecology

Degree Programs: Biochemistry (interdisciplinary program in Crystallography)

Undergraduate Degree: B.S.
Undergraduate Degree Requirements: Core program in Biology, Chemistry, Physics, Mathematics in first 2 years. Last two years, 13-15 credits in approved 100-level courses. Extra credits required for honors.

Graduate Degrees: M.S., Ph.D. (Joint program with Medical School Department)
Graduate Degree Requirements: Master: 6 credits of electives, successful completion of comprehensives and final oral and thesis Doctoral: 4 electives, 6 credits of advanced Topics, successful completion of Preliminary, Comprehensive and Final oral Examinations and Thesis

Graduate and Undergraduate Courses Available for Graduate Credit: Methods in Biochemical Research (U), Principles of Biochemistry (U), Enzymology (U), Integration of Metabolism, Physical Biochemistry, Molecular Aspects of Gene Function, Enzymology, Biochemistry, Seminar, Advanced Topics

DEPARTMENT OF BIOPHYSICS AND MICROBIOLOGY
Phone: Chairman (412) 624-4646 Others: 624-4645

Chairman: M.A. Lauffer
Andrew Mellon Professor of Biophysics: M.A. Lauffer
Professors: I.J. Bendet, C.C. Brinton, Jr., C. Ho, M. A. Lauffer, R.A. McConnell (Research)
Associate Professors: T.E. Cartwright, L. Jacobson, C.L. Stevens
Assistant Professors: T.R. Chay, R. Hendrix, K. Ihler
Research Assistants: 8
Lecturer: 1
Research Associates: 11

Degree Programs: Biophysics, Microbiology

Undergraduate Degree: B.S.
Undergraduate Degree Requirements: 2 years of Chemistry and of Life Sciences Core Curriculum, 1 year of Mathematics and one of Physics plus a minimum of 13 credits in upper division courses in the area of specialization

Graduate Degrees: M.S., Ph.D.
Graduate Degree Requirements: Master: Graduate Record Examination in Chemistry and Physics, Required courses one year each of Biochemistry, Microbiology, Physical Chemistry and Biophysical Methods, Written report of research required, formal thesis optional Doctoral: Courses: same as Master's Degree, In Biophysics GRE are required in Chemistry, Physics and Biology, In Microbiology GRE are required in Chemistry and Physics, GRE scores higher than Masters are required Proficiency in one foreign language, Oral Comprehensive Examination, Thesis

Graduate and Undergraduate Courses Available for Graduate Credit: Introductory Biophysics (U), Molecular Biophysics (U), Special Topics (U), Introduction to Microbial and Molecular Biology 1 (U), Bacterial Genetics and Physiology Laboratory (U), Virology (U), Virology Laboratory (U), Seminar, Theory of Biophysical Methods, Advanced Laboratory

School of Dental Medicine

Dean: E.J. Forrest

DEPARTMENT OF ANATOMY
Phone: Chairman: (412) 624-3181 Others: 624-4141

Chairman: R.D. Mundell
Professor: S.R. Bononi
Associate Professors: J.D. Paltan, J.T. Wallace, N.M. Kanczak, D.R. Dickson (Adjunct)
Assistant Professors: G.C. Gaik, H.L. Langdon, O.K. Andrius, W.M. Dickson (Part-time), C.G. Saracco (Part-time), G. Sotereanos (Part time)

Degree Program: Anatomy

Graduate Degrees: M.S., Ph.D.
Graduate Degree Requirements: Master: Minimum 24 hours courses, 6 hours thesis research, Comprehensive examination, Thesis Doctoral: Minimum 90 hours courses; 30 hours dissertation research, preliminary and comprehensive examinations, Dissertation

Graduate and Undergraduate Courses Available for Graduate Credit: Gross Anatomy, Anatomy of the Head and Neck, Histology, Histotechnique, Connective Tissue Biology, Neuroanatomy, Tools of Research, Anatomy Seminars, Special Topics in Anatomy, Oral Histology, Surgical Anatomy

DEPARTMENT OF MICROBIOLOGY
(No reply received)

School of Medicine

Dean of Graduate Studies: R.E. Basford
Dean of School of Medicine: D.N. Medearis, Jr.

DEPARTMENT OF ANATOMY AND CELL BIOLOGY
(No reply received)

DEPARTMENT OF BIOCHEMISTRY
Phone: Chairman: (412) 624-2494

Chairman: E.C. Heath
The William S. McEllroy Professor in Biochemistry: E.C. Heath
Professors: A.E. Axelrod, R.E. Basford, E.C. Heath, K. Hofmann, D. Nakada
Associate Professors: G.M. Ihler, W.F. Diven
Assistant Professors: C.J. Coffee, N.P. Curthoys, R.H. Glew, S.L. Johnson, S.L. Phillips, W.F. Prouty
Postdoctoral Fellows: 12

Degree Programs: Biochemistry

Graduate Degrees: M.S., Ph.D.
Graduate Degree Requirements: Not Stated

Graduate and Undergraduate Courses Available for Graduate Credit: Not stated

DEPARTMENT OF MICROBIOLOGY
Phone: Chairman: (412) 624-2618 Others: 624-2617

Chairman: J.S. Youngner
Professors: D.S. Feingold, S.B. Salvin
Associate Professors: P.A. Hoffee, B.A. Phillips
Assistant Professors: G. Keleti, H.R. Thacore (Research)
Visiting Lecturers: 1
Research Assistants: 10

Degree Programs: Bacteriology, Bacteriology and Immunology, Biochemistry, Genetics, Human Genetics, Immunology, Medical Microbiology, Microbiology,

Parasitology, Virology

Graduate Degrees: M.S., Ph.D. (Microbiology
Graduate Degree Requirements: Master: Successful completion of required courses; preliminary examination, completion of dissertation Doctoral: Successful completion of required courses, preliminary examination, comprehensive examination, completion of dissertation

Graduate and Undergraduate Courses Available for Graduate Credit: General Microbiology, Microbial Genetics, Microbiology Laboratory I and II, Immunology, Microbial Genetics, Advanced Topics, Bacterial Physiology (Advanced), Advanced Topics in Immunology

DEPARTMENT OF PHYSIOLOGY
Phone: Chairman: (412) 624-2446 Others: 624-4141

Chairman: E. Knobil
Richard Beatty Mellon Professor of Physiology: E. Knobil
Professors: E. Knbil, S.G. Schultz, W. Tong
Associate Professors: A.B. Borie, F. Fuchs, V.L. Gay, J. Hotchkiss (Adjunct), E.S. Redgate
Assistant Professors: E.H. Blaine, W.R. Butler (Research), G.R. Fritz (Research), R.A. Frizzell, J.N. Howell, L.C. Krey (Research), R.J. Schimmel
Research Associate: 4
Postdoctoral Fellows: 7
Graduate Students: 2

Field Stations/Laboratories: Primate Research Center - 709 New Texas Road, Pittsburgh, Pennsylvania 15239

Degree Programs: Physiological Sciences

Graduate Degree: Ph.D.
Graduate Degree Requirements: The candidate for the Ph.D. degree must satisfactorily complete courses in Physical Chemistry, Mammalian Physiology, Neurosciences and Biochemistry and electives as decided in conjunction with advisor. A comprehensive examination is given, usually in the third year. A defense of the thesis constitutes the final oral examination.

Graduate and Undergraduate Courses Available for Graduate Credit: Mammalian Physiology, Neurophysiology, Neuroendocrinology, Reproductive Endocrinology, Regulation of Metabolism, Thyroid Function, Parathyroid Function and Calcium Metabolism, Endocrine Regulation of Growth, Membrane Transport, Muscle Physiology, Electrophysiology, Renal Physiology

Graduate School of Public Health

Dean: H.E. Griffin

DEPARTMENT OF BIOSTATISTICS
Phone: Chairman: (412) 624-3020 Others: 624-3035

Head: C.C. Li
Professors: I. Altman (Adjunct), J. Cornfield (Adjunct), S. Cutler (Adjunct), P.E. Enterline, W. Haenzel (Adjunct), C.C. Li, N. Mantel (Adjunct), B.R. Rao (Research), M.A. Schneiderman (Adjunct), G.K. Tokuhata (Adjunct), J.H. Turner
Associate Professors: C.K. Redmond
Assistant Professors: P.P. Breslin (Research), A.A. LeGasse (Research), S. Mazumdar (Research), H.E. Rockette (Research), J.H. Waller (Research)
Teaching Fellows: 2
Research Assistants: 3

Degree Programs: Animal Genetics, Biometry, Biostatistics, Genetics, Health and Biological Sciences, Human Genetics, Cytogenetics

Graduate Degrees: M.S., Ph.D. Sc.D. (Biostatistics, Human Genetics)
Graduate Degree Requirements: Master: 1 - 1/2 to 2 years, dissertation Doctoral: 3 to 5 years, dissertation

Graduate and Undergraduate Courses Available for Graduate Credit: Principles of Statistical Reasoning, Vital and Administrative Statistics, Design and Analysis of Experiments, Analysis of Nonorthogonal Experiments, Sampling Methods for Community Studies, Community Survey Methods, Problems in Statistical Analysis, Introduction to Statistical Methods I, Introduction to Stochastic Processes, Problems in Statistical Analysis, Introduction to Medical Care Statistics, Introduction to Statistical Methods II, Statistical Methods in Public and Industrial Health I, Statistical Methods in Public and Industrial Health II, Introduction to Biol ogical Assay, Statistics in Biomedical and Public Health Research, Applied Regression Analysis, Statistical Estimation, Multivariate Analysis, Nonparametric Methods in Statistics, Recent Developments in Statistical Inference, Mathematical Methods for Statistics I, Mathematical Methods for Statistics II, Linear Models, Data Processing, Biostatistics Research and Dissertation for the Doctoral Degree, Human Genetics, Human Population Genetics, Problems in Human Genetics Analysis, Genetics Seminar, Chromosomes and Human Diseases, Cytogenetic Techniques, Quantitative Genetics, Principles of Genetic Counseling, Behavioral Genetics

DEPARTMENT OF MICROBIOLOGY
Phone: Chairman: (412) 624-2692

Chairman: M. Ho
Professor: M. Ho
Associate Professors: J.A. Armstrong, R.W. Atchinson, R.B. Yu, R. Cypess
Assistant Professor: B. Singh
Teaching Fellows: 2
Research Assistants: 3

Degree Programs: Microbiology, Medical Microbiology

Graduate Degrees: M.S., Sc.D.
Graduate Degree Requirements: Master: graduate courses in Biochemistry, Bacteriology, Virology, Eipdemiology, Biostatistics Doctoral: same as above plus foreign language, and dissertation

Graduate and Undergraduate Courses Available for Graduate Credit: Graduate courses in Bacteriology, Virology, Advanced Topics in Microbiology, Infectious Diseases, Parasitology

UNIVERSITY OF PITTSBURGH AT JOHNSTOWN

Johnstown, Pennsylvania 15904 Phone: (814) 266-5841
Dean of College: R. Mead

BIOLOGY DEPARTMENT
Phone: Chairman: (814) 266-5841, Ext. 328

Chairman: E.A. Vizzini
Professors: H.J. Idzkowsky (Emeritus)
Associate Professor: D.L. Brown
Assistant Professors: H. Mackey, C. Thompson, J. Harris, C. Jones

Degree Programs: Biological Education, Biological Sciences, Biology, Ecology, Natural Science

Undergraduate Degree: B.S. (Biology)
Undergraduate Degree Requirements: 28 credits in Biology, 2 terms General, 2 terms Organic Chemistry, Biostatistics, 1 term Calculus, 2 terms Physics

UNIVERSITY OF SCRANTON

Scranton, Pennsylvania 18510 Phone: (215) 347-3321
Dean of Graduate Studies: H. Strickland
Dean of College: W. Parente

BIOLOGY DEPARTMENT
Phone: Chairman: Ext. 260

Chairman: J.T. Evans
Professors: P.R. Beining, J.J. Callaghan, F. MacEntee
Associate Professors: V. Delvecchio
Assistant Professors: B. Anderson, P. Langer

Degree Program: Biology

Undergraduate Degree: B.S.
Undergraduate Degree Requirements: 32 credits in Biology, 22 credits in Chemistry

URSINUS COLLEGE
Collegeville, Pennsylvania 19426 Phone: (215) 489-4111
Dean of College: R.G. Bozorth

DEPARTMENT OF BIOLOGY
Phone: Chairman: Ext. 215

Chairman: A.C. Allen
Associate Professors: A.C. Allen, C.E. Kruse, R.S. Howard
Assistant Professor: P.F. Small
Instructor: E.J. Shinehouse

Degree Program: Biology

Undergraduate Degree: B.S.
Undergraduate Degree Requirements: General Biology, Embryology, Comparative Anatomy, Genetics, Microbiology, Organic Chemistry

VILLA MARIA COLLEGE
Erie, Pennsylvania 16505 Phone: (814) 838-1966
Dean of College: Sr. M.M. Doubet

DEPARTMENT OF SCIENCE AND MATHEMATICS
Phone: Chairman: Ext. 255, Biology Program Director: Ext. 243

Chairman: J. Zaranek
Biology Program Director: Sr. R.M. Bohrer
Professor: Sr. R.M. Bohrer
Associate Professor: M. Harsch
Instructor: D. Gustafson

Degree Programs: Biology, Medical Technology, Biology Education

Undergraduate Degree: B.S.
Undergraduate Degree Requirements: 34 courses

VILLANOVA UNIVERSITY
(No reply received)

WASHINGTON AND JEFFERSON COLLEGE
Washington, Pennsylvania 15301 Phone: (412) 222-4400
Dean of Graduate Studies: W. Leake
Dean of College: W. McGill

DEPARTMENT OF BIOLOGY
Phone: Chairman: Ext. 259

Chairman: E.E. Sweet, Jr.
Professors: N.W. Vogel, E.E. Sweet, Jr.
Associate Professors: V.M Lawrence
Assistant Professors: R.A. Calderone, T.W. Hart, R.A. Ickes, D.G. Trelka
Visiting Lecturers: 3

Degree Program: Biology

Undergraduate Degree: B.A.
Undergraduate Degree Requirements: Eight courses in Biology beyond Biology 101.

Graduate Degree: M.A. (Education)
Graduate Degree Requirements: Master: Ten courses, 3 or 4 in education, the rest in Biology

Graduate and Undergraduate Courses Available for Graduate Credit: These have not yet been firmly established

WAYNESBURG COLLEGE
51 College Street Phone: (412) 627-8191
Waynesburg, Pennsylvania 15370
Dean of College: G.W. Smith (Academic Dean)

BIOLOGY DEPARTMENT
Phone: Chairman: Ext. 281 Others: Ext. 283

Chairman: R.R. Williams
Professors: O.L. Dryner, R.R. Williams
Associate Professors: L.B. Barnett, J.C. Cummings

Degree Programs: Biology, Medical Technology

Undergraduate Degree: B.S.
Undergraduate Degree Requirements: (Biology), Complete basic and distributive requirements of course core, complete college total program - 124 semester hours with 2.00 or better, 35 semester hours of Biology (21 hours specific, remaining hours elected), Chemistry through 2 semesters of Organic, 1 year of Physics, Mathematics through Calculus I (Medical Technology) Complete college core, complete total of 96 semester hours on campus, minimum of 16 semester hours Biology, (12 semester hours special remaining electives) Chemistry through 2 semesters Organic, 1 year Physics Mathematics through Calculus I, 12 months at AMA approved hospital school of medical technology

WEST CHESTER STATE COLLEGE
West Chester, Pennsylvania 19380 Phone: (215) 436-1000
Dean of Graduate Studies: W.J. Trezise
Dean of College: R.K. Rickert

DEPARTMENT OF BIOLOGY
Phone: Chairman: (215) 436-2638 Others: 436-1000

Chairman: R.W. Bernhardt
Professors: R.B. Brown, C.L. Cinquina, T. DeMott, J.E.C. Dorchester, W.R. Overlease, R.F. Romig, R.I. Woodruff
Associate Professors: R.W. Bernhardt, M.G. Cullen, M.Y. Martinez, J.M. McDonnell
Assistant Professors: W.C. Brown, M.K. Eleuterio, H.G. Jones, T.P. Sullivan, S.W. Webster
Temporary Instructor: 1
Graduate Teaching Assistants: 1

Field Stations/Laboratories: Robert B. Gordon Natural Area for Environmental Studies - Chester County

Degree Program: Biology

Undergraduate Degrees: A.B., B.S. in Ed.
Undergraduate Degree Requirements: A.B. - Biology 34, Chemistry 19, Mathematics 6, Physics 8, B.S. in Ed. Biology 35, Chemistry 15, Mathematics 6

Graduate Degrees: M A., M.Ed.
Graduate Degree Requirements: M.A.: 30 semester hours of Biology, M.Ed.: 34 semester hours, including 20-22 in Biology and 10-12 in Education

Graduate and Undergraduate Courses Available for Graduate Credit: Methods and Materials in Biological Research, Modern Techniques of Teaching Biology, Introductory Biochemistry, Mammalian Physiology, Ecological Concepts, Field Ecology and Natural History, Functional Anatomy, The Living Forest, Common Trees of Chester County, Summer Wild Flowers of Chester County, Summer Birds of Chester County, History of Biology, Seminar in Cellular Biology, Animal Histology, GeneticTheory, Morphology, Human Heredity, Experimental Embryology, Developmental Anatomy, Seminar in Organismic Biology, Comparative Parasitology, Advanced Plant Anatomy and Morphology, Insect Morphology, Comparative Mammalogy, Advanced Human Anatomy, Seminar in Molecular Biology, Biochemistry I, II, Advanced Bacteriology, Physiology of Plants, Endocrinology, Nutrition, Seminar in Populational Biology, Applied Ecology, Economic Entomology, Limnology, Animal Ecology, Freshwater Invertebrates, Human Ecology, Systematic Botany, Systematics Zoology, Medical Entomology, Seminar in Biological Principles, Experimental Biology, Microtechnique Laboratory, Ecological Techniques

WESTMINISTER COLLEGE

New Wilmington, Pennsylvania 16142 Phone: (412) 946-6710
Dean of College: P.A. Lewis

DEPARTMENT OF BIOLOGY
Phone: Chairman: Ext. 44 Others: Exts. 44 and 45

Chairman: C.E. Harms
Professors: E.C. Gese, C.E. Harms
Associate Professor: R.V. Travis
Assistant Professor: J.P. Fawley, P.C. McCarthy, V.D. Rhoton
Undergraduate Teaching Assistants: 10 to 15

Degree Program: Biology

Undergraduate Degree: B.S.
Undergraduate Degree Requirements: Biology, Principles of Biology I and II, Organismic Biology, Genetics, General Physiology, Ecology, 3 to 6 electives (Independent Study is encouraged) Collateral fields required: Statistics, Calculus I (Calculus II recommended), Chemical Principles I and II, Organic Chemistry I and II, Principles of Physics I and II

WIDENER COLLEGE

17th and Chestnut Streets Phone: (215) 876-5551
Chester, Pennsylvania 19013
Dean of College: J.M. Rodney

SCIENCE GROUP
Phone: Chairman: (215) 872-0795 Others: 876-5551

Chairman: J.S. Conroy
Professor: J.S. Conroy
Associate Professors: R.P. Boekenkamp, T.F. DeCaro, H.V.R Rao
Assistant Professors: T.J. O'Tanyi, A.A. Smith

Degree Program: Biology

Undergraduate Degree: B.S.
Undergraduate Degree Requirements: Principles of Biological Systems: Phytology, Zoology, Ecology, Cell Physiology, Genetics, Animal Physiology, Senior Project, plus three Biology electives. 4 semesters of Chemistry, 2 semesters of Physics, 2 semesters of Calculus, 5 semesters of Humanities.

WILKES COLLEGE

Wilkes-Barre, Pennsylvania 18703 Phone: (717) 824-4651
Dean of Graduate Studies: A. Shaw
Dean of College: R. Capin

DEPARTMENT OF BIOLOGY
Phone: Chairman: Ext. 240

Chairman: C.B. Reif
Professor: C.B. Reif
Associate Professors: R. Achason, W. Hayes, G. Kimball, R. Ogren, D. Tappa
Assistant Professors: C. Hoseknecht, W. Richkus, L. Turoczi

Degree Programs: Biology, Medical Technology

Undergraduate Degrees: B.S. ,B.A.
Undergraduate Degree Requirements: 34 semester hours Biology, 20 semester hours Chemistry, 8 semester hours Mathematics, 8 semester hours Physics

Graduate Degree: M.S. (Biology)
Graduate Degree Requirements: Master: 24 graduate credits, one foreign language, thesis, final oral

Graduate and Undergraduate Courses Available for Graduate Credit: Endocrinology(U), Bacteriology (U), Genetics (U), Ecology (U), Physiology (U), Evolutionary Mechanisms, Cell Biology and Differentiation, Physiology of Bacteria, Current Concepts in Genetics, Topics in Ecology, Reproductive Physiology, Advanced Physiology, Immunology and Immunochemistry, Invertebrate Biology, Selected Topics in Biology

WILSON COLLEGE

Chambersburg, Pennsylvania 17201 Phone: 264-4141
Dean of College: M.A. Burns

DEPARTMENT OF BIOLOGY

Chairman: D.G. Grove
Professor: D.G. Grove
Associate Professor: E.L. MacDonald
Assistant Professors: H.E. Holzman, J.J. Dropp

Degree Programs: Biology, Medical Technology

Undergraduate Degree: B.S.
Undergraduate Degree Requirements: 36 courses in the college; at least 8 courses in Biology and 2 in Chemistry, 3 courses in an allied field, fulfillment of college distribution requirements; 2.0 overall-all grade point average

YORK COLLEGE OF PENNSYLVANIA

Country Club Road Phone: (717) 843-8891
York, Pennsylvania 17405
Dean of College: W.A. DeMeester

DEPARTMENT OF BIOLOGY
Phone: Chairman: Ext. 263

Chairman: R.J. Clark
Professor: R.F. Denoncourt
Associate Professor: B.B. Smith
Assistant Professors: I. Austin, R.J. Clark
Instructor: C.L. Carlisle

Degree Program: Biology

Undergraduate Degree: B.S.
Undergraduate Degree Requirements: 30 semester hours in Biology plus college Mathematics, Physics, General Chemistry, Organic Chemistry and Statistics

PUERTO RICO

CATHOLIC UNIVERSITY OF PUERTO RICO
(No reply received)

INTER AMERICAN UNIVERSITY OF PUERTO RICO
San Germain, Puerto Rico 00756 Phone: (809) 892-1095
Dean of College: L.D. Bender

DEPARTMENT OF BIOLOGY
Phone: Chairman: Ext. 260

Chairman: V.A. Carpiles
Professors: C. Ancclay, J.B. Villella
Associate Professors: V.A. Capriles, N.E. de Rios
Instructors: M. Acosta, S.L. Kaplan
Visiting Lecturers: 2
Laboratory Technicians: 4

Degree Programs: Biology, Medical Technology

Undergraduate Degrees: B.A., B.S., B.S.C.
Undergraduate Degree Requirements: For B.A. - 30 hours including Freshman Biology, Zoology and Botany, For B.S. - 40 hours including Freshman Biology, Zoology, Botany, Genetics, General Physiology, Seminar and a Taxonomy course. Related courses in Inorganic Chemistry, Mathematics, College Physics, Foreign Language and 12 hours of English For B.S. in Medical Technology - 30 hours in Medical Technology courses, 16 hours in Biology including Microbiology, 22 hours in Chemistry and 8 hours in College Physics

UNIVERSITY OF PUERTO RICO - MAYAGUEZ
Mayaguez, Puerto Rico 00708 Phone: (809) 832-4040
Dean of Graduate Studies: J.A. Ramos

College of Arts and Science
Dean: W. Ocasio - Cabanãs

DEPARTMENT OF BIOLOGY

Chairman: Not Stated
Professors: J.A. Rsmos, J.F. Morge, G.W. Miskimen, G.R. de Miskimen, R.S. del Toro, A.G. Mas, L.A. Rovie, D. Colon, J. Rivero, V. Bisggi
Associate Professors: F.Cofresis, F. Padovarii, C. Moore, G. Kuno, L. Rodriguez, L. Martinez, E. Saavada, M. del Socorro Torres, A. Berrios, D. Walker
Assistant Professors: L. Rodriguez, Y. Aymath
Instructor: C. Betbncourt

Field Stations/Laboratories: Entomological Research Laboratory, Tropical Mycology Laboratory

Degree Program: Biology

Undergraduate Degree: B.S.
Undergraduate Degree Requirements: 141 credits

Graduate Degree: M.S.
Graduate Degree Requirements: Master: 30 credits including a thesis

Graduate and Undergraduate Courses Available for Graduate Credit: (All courses in Spanish)

College of Agricultural Sciences
Dean: F. del Río

DEPARTMENT OF AGRONOMY
(No reply received)

DEPARTMENT OF ANIMAL INDUSTRY
Phone: Chairman: (809) 832-4040 Ext. 257

Chairman: J.H. Sanfiorenzo
Professors: I. Carlo, B. Rodríguez López
Associate Professors: J. M. Quintana, R.F. Perez Sosa, P.F. Randol, G.R. Carbó, J.H. Sanfiorenzo, F. Suarez
Instructors: A.A. Custodio Gonzalez

Field Stations/Laboratories: Lajas and Isabela Substation

Degree Program: Animal Science

Undergraduate Degree: B.S.A.
Undergraduate Degree Requirements: a) A total of 144 credits with a minimum grade index of 2.00, b) A minimum of 14 credits in professional departmental electives, including a 3-credit Summer Practicum at the end of the Junior year, is required for majors in this field c) An additional 14 credits are also required in general electives

Graduate Degree: M.S.
Graduate Degree Requirements: Master: A minimum of 30 semester hours of credit in approved graduate courses of which at least 24 must be earned at the University of Puerto Rico, b) Not more than six credit of "Courses for Advanced Undergraduates" are accepted toward the degree, c) at least 15 credit hours shall be earned in the major subject and six hours shall be taken in graduate courses in related fields, d) all candidates must present a thesis representing investigation or research. (6 credit hours)

Graduate and Undergraduate Courses Available for Graduate Credit: Dairy By-products (U), Veterinary Surgery (U), Dairy Plant Operation (U), Veterinary Parasitology (U), Advanced Poultry Production (U), Dairy Cattle Management, Advanced Animal Breeding, Milk Secretion, Meat Animal Production, Animal Nutrition, Special Problems Graduate Seminar, Advanced Dairy Bacteriology, Dairy Chemistry

DEPARTMENT OF HORTICULTURE
Phone: Chairman: Ext. 3004

Chairman: F.L. Jordán Molero
Professors: F.L. Jordán Molero, E. Alvarez Rossario, J. Toro Rosario
Associate Professors: A. Perez, López, M. Rico Ballester
Assistant Professors: C.A. Fierro Berwart, J. Cuevas Ruíz, A. Cedno Maldonado
Instructor: L. Flores Flores

Field Stations/Laboratories: Adjuntas Substation, Isabela Substation, Fortuna Substation

Degree Program: Horticulture

Undergraduate Degree: B.S.
Undergraduate Degree Requirements: 147 credit hours

Graduate Degrees: M.S.
Graduate Degree Requirements: Master: 30 credit hours

Graduate and Undergraduate Courses Available for Graduate Credit: Advanced Vegetable Gardening, Advanced Floriculture (U), Seminar in Horticulture (U), Advanced Plant Breeding, Food Processing Plant Manatement and

Sanitation, Nuclear Techniques in Agricultural Research, Advanced Plant Propagation, Advanced Tropical Fruits, Post Harvest Physiology and Manipulation of Horticultural Corps, Physiology of Vegetable Crops, Physiology of Fruit Production, Growth Regulators in Horticulture

UNIVERSITY OF PUERTO RICO - RIO PEDRAS
Rio Pedras, Puerto Rico 00931 Phone: (809) 764-0000
Dean of Graduate Studies: E. Ortiz

College of Arts and Sciences
Dean: J.S. Curet

DEPARTMENT OF BIOLOGY
Phone: Chairman: Ext. 636, 482, 201

Chairman: C.R.T. Acosta
Professors: C.R.T. Acosta, M.I. Camuras de Vazquez, G. Candelas, G. Candelas, H. Lugo, Lugo, E. Ortiz, J. Rosado Alberio, M.J. Velez, R. Woodbury
Associate Professors: O. Baerga de Rodriguez, D.L. Bruck, M.D. Byer, B. Cestero de Rivera, A.F. El Koury, L.B. Liddle, E. Mendez de Ortiz, F.L. Renaud, C. Rossy Valderrama
Assistant Professors: C. Buso de Mendez, Y. Garica-Castro, W.P. Hall III, R. Hann, G.V. Hillyer, Y. Perez-Chiesa, E. Preston, L Preston,
Instructors: J. Bobonis, G. Escalona de Motta, M. Hernandez, L. Otero de Ortia, R. Ramos, K. Rodriguez, S. Saavedra de Sánchez, R. Saylor, V. Toro de Suarez,
Teaching Assistants: 25

Field Stations/Laboratories: El Yungue Biological Station

Degree Programs: Biology

Undergraduate Degree: B.S.
Undergraduate Degree Requirements: 30 credit courses in Biology, 2 years of Chemistry (1 year Organic and Inorganic), Mathematics to Calculus 1, 1 year of Physics, Basic Courses: 2 years - Spanish, 2 years English, 2 years Humanities, 1 year Social Sciences, 1 year Biological Sciences, 1 year Physics Science, 13 credits in elective courses

Graduate Degree: M.S.
Graduate Degree Requirements: Master: 24 credits courses at Graduate level, 6 credits of thesis research (Includes thesis dissertation), Graduate Comprehensive examinations

Graduate and Undergraduate Courses Available for Graduate Credit: Animal Behavior, Topics in Animal Behavior, Experimental Ecology, Marine Ecology, Bioconservation, Animal Morphogenesis, Plant Morphogenesis, Molecular Aspects of Development, Development in Algae, Genetic Analysis, Population Biology, Cytogenetics, Biology of Terrestial and Freshwater Invertebrates of Puerto Rico, Morphology of Thallophytes, Morphology of Vascular Plants, Cellular Physiology, Comparative Physiology, Cellular Biochemistry, Laboratory of Cell Biochemistry, Experimental Protozoology, Immunology, Hormones and the Environment, Marine Invertebrates, Marine Invertebrates, Marine Algae and Fungi, Cytology Biogeography, Biometrics, Research, Microbial Genetics, Special Topics in Modern Biology, Biochemical Techniques and Instrumentation, Thesis Research, Seminars: Tropical Ecological Formations, Analysis of Development, Genetics, Systematics, Plant Physiology, Neurophysiology, Biochemistry, Perspectives on Marine Biology a) Protozoology, Photobiology, Chemical Ecology, Insect Physiology

School of Medicine
Director of Graduate Study: J.I. Colón
Dean of School of Medicine: C. Girod

DEPARTMENT OF ANATOMY
Phone: Chairman: Ext. 482

Chairman: W.L. Stiehl
Professors: R.E. Alegría, C.H. Conaway, F. Ervin, J. Frontera, C.A. Pfeiffer, W.L. Stiehl, J. Szepsenwol
Associate Professors: S.O.E. Ebesson (Ad Honorem), I. Lastra, L.R. Otero, M.A. Suarez
Assistant Professors: J. Bayona, F.J. Llera
Instructors: P. Crowle, H. Lopez de Serrano, R. McClish, J. Muriente

Degree Program: Anatomy

Graduate Degree: M.S., Ph.D.
Graduate Degree Requirements: Master: 30 credit hours no more than six of which are research, Thesis, general Master's Examination and Thesis defense
Doctoral: 60 credit hours no more than fifteen of which are research, Comprehensive examination, Thesis and Thesis defense

Graduate and Undergraduate Courses Available for Graduate Credit: Anatomy, Human Gross Anatomy, Embryology, Histology, Neuroanatomy, Functional Head and Neck Anatomy, Functional Head and Neck Anatomy, Morphology, Biochemistry, Physiology and Pharmacology of the Eye and Related Structures, History of Anatomy, Seminar in Physical Anthropology, Methods of Research in Anatomy, Endocrinology of Reproduction, Advanced Neuroanatomy, Cytology, Histochemistry, Principles of Morphology, Comparative Vertebrate Neuroanatomy, Practice in Teaching

DEPARTMENT OF BIOCHEMISTRY AND NUTRITION
Phone: Chairman: 767-9626 Ext. 235

Chairman: E. Toro-Goyco
Professors: E. Toro-Goyco, J. Chiriboga (Ad Honorem)
Associate Professors: M. Cancio de Toro (Ad Honorem), J.J. Corcino (Ad Honorem), S. El-Khatib, A.F. Rosenthal (Ad Honorem)
Assistant Professors: N.A. Fernandez, M. Rosas del Valle, E. Sanchez
Instructors: A.M. Preston, L.D. Rodriguez, N. Rodriguez de Pérez, ,W. Torres de Rivera
Visiting Lecturers: 3
Research Assistants: 3

Degree Programs: Biochemistry and Nutrition

Graduate Degrees: M.S., Ph.D.
Graduate Degree Requirements: Master: 30 credit hours and thesis Doctoral: 60 credit hours and thesis

Graduate and Undergraduate Courses Available for Graduate Credit: An 8 credit-hours, 2semester general Biochemistry course, A course on the Physical Chemistry of Biomolecules, Introduction to Computer Programming and Data Evaluation, A two credit-hours seminar course, Special Topics in Biochemistry. A 3 credit-hours course offered by outstanding visiting investigators in any given area of Biochemistry.

DEPARTMENT OF MICROBIOLOGY
(No reply received)

DEPARTMENT OF PHYSIOLOGY AND BIOPHYSICS
Phone: Chairman: (809) 765-7550

Chairman: M. Martinez-Maldonado (Acting)
Professors: F. Alvarado, M. Martinez-Maldonado, R. Reinecke, R. Torres-Pinedo
Associate Professor: A. Bonnet
Assistant Professor: J.J. Keene
Associate in Physiology: O.R. Rendon
Instructors: M. Chaves, A. Firpi, P. Van Loon
Visiting Lecturers: 8
Research Assistants: 1

Degree Programs: Physiology, Biophysics

Graduate Degrees: M.S., Ph.D.
Graduate Degree Requirements: Master: 30 credits, 24 credits in course work and 6 credits in Thesis research
Doctoral: 60 credits, 40 credits in course work and 15 in thesis research

Graduate and Undergraduate Courses Available for Graduate Credit: Human Physiology, Laboratory Course for Human Physiology, Seminar in Physiology, Physical Instrumentation for Biologists, Cytogenetics, Physiology of the Kidney and Body Fluids, Gastrointestinal Physiology, Introduction to Computer Programming, Fortran Programming for IBM 1130, Cell Membrane, Introduction to Neurosciences, Brain and Behavior Seminar

RHODE ISLAND

BARRINGTON COLLEGE
Barrington, Rhode Island 02806 Phone: (401) 246-1200
Dean of College: R. Goodnow

NATURAL SCIENCES AND MATHEMATICS DEPARTMENT
Phone: Chairman: (401) 246-1200 Ext. 269

Chairman: H.E. Snyder
Professor: H.E. Snyder
Associate Professor: W.H. Chrovser
Visiting Lecturers: 2

Field Stations/Laboratories: Barrington College Marine Biology Laboratories

Degree Programs: Biology, Marine Biology, Medical Technology

Undergraduate Degree: B.S.
Undergraduate Degree Requirements: Core Courses outside major, 16 semester hours Chemistry, 8 semester hours Mathematics, 8 semester hours Physics, 40 semester hours Biology courses selected to meet major requirements, Total 72 hours science

BROWN UNIVERSITY
(No reply received)

PROVIDENCE COLLEGE
Providence,Rhode Island 02918 Phone: (401) 865-1000
Dean of Graduate Studies: C.P. Forster
Dean of College: T.H. McBrien

BIOLOGY DEPARTMENT
Phone: Chairman: (401) 865-2200 Others: 865-1000

Chairman: D.E. Leary
Professors: W.A. Fish, R.I. Krasner, C.V. Reichart
Associate Professors: E.H. Donahue, D.E. Leary, G.P. O'Leary, R.M Zarcaro
Assistant Professor: C.B. Crafts

Degree Program: Biology

Undergraduate Degrees: A.B., B.S.
Undergraduate Degree Requirements: Requirements for the B.A. degree: Biology Core Curriculum, plus 8 elective hours of Biology courses; Chemistry, Physics, Mathematics, plus the collegiate core. Requirements for the B.S. degree: In addition to the courses listed above (for the B.A. degree) 8 additional hours of science credits must be elected. The additional credits may be elected in Mathematics and approved Psychology courses

RHODE ISLAND COLLEGE
Providence, Rhode Island 02908 Phone: (401) 831-6600
Dean of Graduate Studies: L. Weber
Dean of College: R.F. Shinn, Jr.

DEPARTMENT OF BIOLOGY
Phone: Chairman: Exts. 355, 354

Chairman: G.C. Hartmann
Professors: F. Dolyak, G Hartmann, M. Keeffe, P. Pearson
Associate Professors: R.N. Keogh, T. Lemeshka, H. McCutcheon, A. Silver
Assistant Professors: C. Foltz, N. Gonsalves, M.J. Haagens, K. Kinsey, I. Lough, F. Pearson, S.S. Wasti, R. Young
Instructor: J. Guisti
Teaching Fellows: 6

Degree Programs: Biological Education, Biological Sciences

Undergraduate Degree: A.B.
Undergraduate Degree Requirements: 1 - 30 semester credits Biology to include 4 core courses: Cell and Molecular Biology, Genetics, Developmental Biology, Ecology and Electives in Biology, 2 Organic Chemistry (1 year), and 1 year Physics, 3. General Studies requirements

Graduate Degree: M.A.T. (Master of Arts in Teaching)
Graduate Degree Requirements: Master: M.A.T. - C: Biology 12 credits, Education 23 credits, M.A.T. - I: Biology 21 credits Education 9 credits

Graduate and Undergraduate Courses Available for Graduate Credit: Developmental Biology (U), Nonvascular Plants (U), Vascular Plants (U), Plant Physiology (U), Ecology (U), Invertebrate Zoology (U), Vertebrate Zoology (U), Entomology (U), Comparative Vertebrate Anatomy (U), Human Anatomy (U), Vertebrate Physiology (U), Physiology of Exercise (U), Microbiology (U), Biological Techniques (U), Evolution (U), Mycology, Biogeography, Mammalogy, Biochemistry, Cytology, Cellular Physiology, Advanced Microbiology, Immunobiology

ROGER WILLIAMS COLLEGE
Bristol, Rhode Island 02809 Phone: (401) 255-1000
Dean of College: B. Vehling

BIOLOGY AREA
Phone: Chairman: (401) 255-2226 Others: 255-1000

Area Coordinator: W.R. Mershon
(Faculty not ranked at this institution): G.A. Ficorilli, M.D. Gould, T.J. Holstein, C.R. Jungwirth, G.P. Murphy, M. Villalard-Bohnsack

Field Stations/Laboratories: The college has recently acquired a portion of Gould Island in center of Narragansett Bay

Degree Programs: Biology, Marine Biology

Undergraduate Degree: B.A.
Undergraduate Degree Requirements: 2 semesters Introductory Biology, 2 semesters General Chemistry, 2 semesters Organic Chemistry, 2 semesters Physics, 6 upper level Biology courses, 1 course in Communications in the Biological Sciences.

SALVE REGINA COLLEGE
(No reply received)

UNIVERSITY OF RHODE ISLAND
Kingston, Rhode Island 02881 Phone: (401) 792-1000
Dean of Graduate Studies: N.A. Potter (Acting)

College of Arts and Sciences
Dean: R. Lepper, Jr. (Interim)

DEPARTMENT OF BOTANY

Phone: Chairman: (401) 792-2161 Others: 792-1000

Chairman: R.D. Goos
Professors: L.S. Albert, C.E. Beckman (Joint), N.E. R.L. Hauke, R. Lepper, Jr., E.A. Palmatier, T.J. Smayda (Joint), R.D. Wood
Associate Professors: J.P. Mottinger, E. Swift (Joint)
Assistant Professors: W.L. Halvorson, M.M Harlin, P.E. Hargraves (Joint)
Research Assistants: 11

Field Stations/Laboratories: Alton Jones Campus - Exeter

Degree Programs: Biology, Botany

Undergraduate Degrees: B.A., B.S.
Undergraduate Degree Requirements: Not Stated

Graduate Degrees: M.S., (Botany), Ph.D. (Biological Science)
Graduate Degree Requirements: Master: M.S. Botany 0 24 hours course credits, 6 hours of thesis research. 30 hours plus thesis Doctoral: 72 credits beyond baccalareate, proficiency in a foreign language or research tool, and dissertation.

Graduate and Undergraduate Courses Available for Graduate Credit: Developmental Plant Anatomy, Morphology of Vascular Plants, Methods in Plant Ecology, Plant Geography, Physiology of Fungi, Phytopathological Technique, Experimental Mycology, Medical Mycology, Cytogenetics, Physiological Ecology of Marine Macroalgae, Biogeography of Marine Algae, Environmental Plant Physiology, Systematic Botany (U), Field Aquatic Plant Ecology (U), Marine Botany (U), Freshwater Botany (U), Advanced Practicum in Aquatic Plant Ecology (U), Plant Ecology (U), Mycology (U), Advanced Practicum in Aquatic Plant Ecology (U), Cytology (U), Marine Ecology (U)

DEPARTMENT OF MICROBIOLOGY AND BIOPHYSICS
Phone: Chairman: (401) 792-2205

Chairman: N.P. Wood
Professors: V.J. Cabelli (Adjunct), P.L. Carpenter, H.W. Fisher, C.W. Houston, J. M. Sieburth (Joint), R.W. Traxler, N.P. Wood (Joint), R.D. Goos
Associate Professors: P.S. Cohen, K.A. Hartman, J.C. Prager
Assistant Professors: D.C. Laux, D.W. Shivvers
Instructors: L. Hufnagel
Teaching Fellows: 10
Research Assistants: 6
Special Assistants: 2

Degree Programs: Microbiology, Biological Sciences, Biophysics, Medical Technology

Undergraduate Degree: B.S.
Undergraduate Degree Requirements: Minimum of 30 credits in Microbiology, two semesters each of Biology, Inorganic and Organic Chemistry, Calculus, Mathematics, Physics, Language, one semester each of quantitative analysis and Biochemistry

Graduate Degrees: M.S., Ph.D.
Graduate Degree Requirements: Master: 15-21 credits in Microbiology, Biochemistry, thesis Doctoral: Total of 72 hours credit and proficiency in Calculus, Physical Chemistry, Biochemistry, Biophysics, Statistics or Biometry, Genetics, and one foreign language, courses in related fields such as Virology, Mycology, and Phycology, Dissertation

Graduate and Undergraduate Courses Available for Graduate Credit: Food Microbiology (U), Pathogenic Bacteriology (U), Immunity and Serology, Physiology of Bacteria, Microbial Genetics, Marine Bacteriology, Systematic Bacteriology, Advanced Microbiology (U), Virology, Mycology, Literature of Bacteriology

INSTITUTE OF ENVIRONMENTAL BIOLOGY
Phone: Chairman: (401) 792-2161 Others: 792-2372

Director: J.S. Cobb
The institute has no faculty - it is a coordinating institute for Interdisciplinary Activities.
Research Assistants: 2

Field Stations/Laboratories: W. Alton Jones Campus, Saunderstown Research Area

DEPARTMENT OF ZOOLOGY
Phone: Chairman: (401) 792-2372

Chairman: D.J. Zinn (Acting)
Professors: D.E. Bass (Adjunct), M.E. Carriker (Adjunct), R.K. Chipman, H.G. Dowling (Adjunct), R.H. Gibbs (Adjunct), C.S. Hammen, R.W. Harrison, K.E. Hyland, Jr., S.B. Saila, K.E. Schaefer (Adjunct), H.E. Winn, D.J. Zinn
Associate Professors: R.F. Costantino, C.C. Goertemiller, Jr., F. Heppner, R.B. Hill, W.H. Krueger, J.A. Mathewson, J.M. Mottinger, C.R. Shoop
Instructor: M. McConnell (Special)
Research Assistants: 1

Field Stations/Laboratories: W. Alton Jones campus, Federal Water Quality Laboratory, Saunderstown, Narragansett Marine Laboratory

Degree Programs: Zoology, Biology

Undergraduate Degrees: B.A. (Biology), B.S. (Zoology)
Undergraduate Degree Requirements: B.S. - Zoology, A minimum of 30 credits in Zoology including Animal Physiology, Introductory Ecology, Chordate Anatomy and Morphogenesis, Invertebrate Zoology, Seminar in Zoology and Genetics. General Botany, one year of Inorganic Chemistry, one year Organic Chemistry, Intermediate Calculus, with Analytic Geometry, General Physics, and a modern language through the intermediate level. B.A. Biology A minimum of 28 credits in Biological Sciences including General Biology (Animals), or General Botany and General Zoology, General Microbiology, 6 credits in Botany (exclusive of General Botany), and 6 credits in Zoology (exclusive of General Zoology plus 4-6 credits one or all of the areas in biology.

Graduate Degrees: M.S., Ph.D.
Graduate Degree Requirements: Master: 30 credits (including 2 semesters of Graduate Seminar in Zoology--1 credit each semester--and 6 credits of M.S. thesis researcy), Written Comprehensive Examination, Thesis Doctoral: 72 credits, 42 taken at University of Rhode Island (including 2 semesters of Graduate Seminar in Zoology --1 credit each semester-- and usually 24 credits of Ph.D. Dissertation research). Written and Oral Comprehensive Examination, 2 foreign languages, dissertation, oral defense of dissertation

Graduate and Undergraduate Courses Available for Graduate Credit: Principles of Taxonomy (U), General (Cellular) Physiology (U), Mammalian Physiology (U), Marine Ecology (U), Marine Ecology Laboratory (U), Animal Ecology (U), Limnology (U), Vertebrate Biology (U), Animal Behavior (U), Mammalogy (U), Causes of Evolution (U), Human Genetics (U), Systematic Entomology (U), Modeling of Physiological Systems (U), Modeling and Analysis of Dynamic Systems (U), Biological Photography, Fine Structure, Seminar in Zoological Literature, Mechanisms of Development, Advanced Parasitology Seminar, Comparative Physiology (2 semesters), Biology of Reproduction in Animals, Endocrinology, Neurophysiology, Seminar in Morphogenetic Theory, Seminar in Behavioral Ecology, Ichthyology, Oceanic Ichthyology, Herpetology, Ornithology, Developmental Genetics, Ecological Genetics, Advanced Genetics Seminar, General Acarology,

Medical and Veterinary Entomology, Graduate Seminar in Zoology (2 semesters), Seminar in Physiology, Advanced Mammalian Physiology, Seminar in Environmental Physiology, Seminar in Icthyology, Physiological Ecology, Laboratory in Physiological Ecology, Advanced Ecology Seminar, Animal Communication, Biological Clocks and Orientation, Assigned Work, Zoological Problems, Masters Thesis Research, Doctoral Dissertation Research

College of Resource Development

Dean: G. Donovan

DEPARTMENT OF ANIMAL PATHOLOGY
(No information available)

DEPARTMENT OF ANIMAL SCIENCE
Phone: Chairman: (401) 792-2477

Chairman: L.T. Smith
Professors: J.W. Cobble, C.J. Cosgrove, R.L. Coduri (Adjunct), L.T. Smith
Associate Professors: W.K. Durfee, B. Henderson, Jr., R.S. Hinkson, Jr., T.L. Meade, R.I. Millar, A.G. Rand
Assistant Professor: H.G. Gray
Instructor: M. Nippo
Research Assistants: 4

Degree Programs: Animal Science, Natural Resources, Wildlife Management, Agraculture

Undergraduate Degree: B.S.
Undergraduate Degree Requirements: 130 hours

Graduate Degree: M.S.
Graduate Degree Requirements: Master: 30 hours

Graduate and Undergraduate Courses Available for Graduate Credit: Physiology of Lactation (U), Food Analysis (U), Laboratory Animal Techniques (U), Special Projects (U), Animal Science Seminar, Research Problems, Master's Thesis Research , Animal Breeding (U), Food Quality (U), Population Genetics (U), Physiology of Reproduction (U), Special Projects (U), Advanced Animal Nutrition, Experimental Design, Research Problems

DEPARTMENT OF FOOD AND RESOURCE CHEMISTRY
(No information available)

DEPARTMENT OF FOREST AND WILDLIFE MANAGEMENT
Phone: Chairman: (491) 792-2370

Chairman: W.P. Gould
Professors: E.F. Patric (Research)
Associate Professors: J.H. Brown, Jr., W.P. Gould, J.J. Kupa
Assistant Professor: F.C. Golet
Research Assistants: 4

Field Stations/Laboratories: Alton W. Jones Campus, West Greenwich, Rhode Island

Degree Programs: Natural Resources

Undergraduate Degree: B.S.
Undergraduate Degree Requirements: Basic Core (Required courses (6)) Biological Sciences, Physical Sciences, Mathematics, Social Sciences, Humanities, Communications, Major Area of Concentration - 24 credits, Directed Electives - 21 credits, Free Electives 11-17 credits, 130 credits total

DEPARTMENT OF PLANT PATHOLOGY-ENTOMOLOGY
Phone: Chairman: (401) 792-2481

Chairman: R.W. Traxler
Professors: C.H. Beckman, A.M. Kaplan (Adjunct), T.W. Kerr, W C. Mueller, C.M. Tarzwell (Adjunct), R.W. Traxler
Associate Professor: N. Jackson
Assistant Professors: L. Englander
Research Assistants: 2
Research Associate: 1
Graduate Assistants: 4

Field Stations/Laboratories: Turfgrass Field House Laboratory

Degree Programs: Entomology, Plant Protection, Plant Pathology

Undergraduate Degree: B.S. (Plant Protection)
Undergraduate Degree Requirements: 130 hours, Basic Core 75-83 credits, Major Area Concentration - 24 Credits, Directed Elective - 12 credits, Free Electives 11-19 credits

Graduate Degrees: M.S., Ph.D.
Graduate Degree Requirements: Master: 30 hours credit, Thesis and seminar participation Doctoral: Dissertation

Graduate and Undergraduate Courses Available for Graduate Credit: Diseases of Turfgrass, Trees and Ornamental Shrubs (U), Nematology (U), Plant Virology (U), The Nature of Plant Diseases, Research Problems, Experimental Mycology, Phytopathological Technique, Physiology of Fungi

DEPARTMENT OF PLANT AND SOIL SCIENCE
Phone: Chairman: (401) 792-2791 Others: 792-2494

Chairman: W.E. Larmie
Professors: W.E. Larmie, E.C. Roberts, V.G. Shutak, C.R. Skogley, I H. Stuckey, R.C. Wakefield
Associate Professors: J.D. Dunnington, A.E. Griffiths, R.J. Hindle, R.J Hull, J.J. McGuire, C.G. McKiel, J.E. Sheehan, P.H Wilson
Assistant Professors: H.P. Conlon, D.T. Duff, J.A. Jagschitz, J.L. Pearson, R.J. Shaw, W.R. Wright
Research Assistants: 4

Degree Programs: Plant Sciences, Soil Sciences

Undergraduate Degree: B.S.
Undergraduate Degree Requirements: 130 credits

Graduate Degree: M.S.
Graduate Degree Requirements: 30 credits (6 thesis)

Graduate and Undergraduate Courses Available for Graduate Credit: Plant Propagation (U), Commercial Floriculture (U), Turfgrass Management (U), Environmental Aspects of Landscape Design (U), Soil Conservation and Land Use (U), Identification of Ornamental Plants (U), Weed Science (U), Soil Genesis and Classification (U), Plant Breeding (U), Plant Nutrition (U), Growth and Development of Economic Plants, Post Harvest Physiology, Physiology of Plant Productivity

SOUTH CAROLINA

BAPTIST COLLEGE AT CHARLESTON

P.O. Box 10087 Phone: (803) 797-4133
Charleston, South Carolina 29411
Dean of College: J A. Barry

DEPARTMENT OF BIOLOGY
Phone: Chairman: (803) 797-4205

Chairman: C. Chesnutt
Professor: C. Chesnutt
Assistant Professor: S. Best

Degree Program: Biology

Undergraduate Degree: B.S.
Undergraduate Degree Requirements: Zoology, Botany, Microbiology, Cell Physiology, Ecology, Genetics, Chemistry, Mathematics, 8 hours electives

BENEDICT COLLEGE

Harden and Blanding Streets Phone: (803) 779-4930
Columbia, South Carolina 29204
Dean of Graduate Studies: S.C. Davis

BIOLOGY STUDY PROGRAM
Phone: Chairman: Ext. 359

Director: J.S. Scott
Professor: R. Singh
Associate Professors: A. Davis, M.F. Finlay
Assistant Professor: J. Scott

Degree Program: Biology

Undergraduate Degree: B.S.
Undergraduate Degree Requirements: General Chemistry - 2 semesters, Organic Chemistry 2 semesters, Mathematics 3 semesters, Physics 2 semesters, Biology 4 semesters, General College requirements and these Science Courses must total 115 semester hours

CENTRAL WESLEYAN COLLEGE

Central, South Carolina 29630 Phone: (803) 639-2453
Dean of College: P.B. Wood

DEPARTMENT OF BIOLOGY
Phone: Chairman: (803) 639-2453

Chairman: M. LaBar
Professors: M. LaBar, R.R. Nash

Degree Programs: Biology, Biological Education, Medical Technology

Undergraduate Degree: B.A.
Undergraduate Degree Requirements: Biology, Botany, Zoology, Ecology, Microbiology, Genetics, Anatomy-Physiology, Seminar, 2 semesters Chemistry, 1 semester Organic Chemistry or Calculus

THE CITADEL

Charleston, South Carolina 29409 Phone: (803) 723-0611
Dean of Graduate Studies: D.L. Bowman
Vice President for Academic Affairs: W.E. Anderson

DEPARTMENT OF BIOLOGY
Phone: Chairman: (803) 723-0611 Ext. 416

Head: J.K. Reed
Associate Professors: R.E. Baldwin, E.S. Crosby, G.L. Runey
Assistant Professors: T.C. Bowman, W.B. Ezell, B.J. Kelley, D.M. Forsythe, F.S. Seabury, F.L. Wallace
Visiting Lecturers: 1

Degree Programs: Biology, Biological Education

Undergraduate Degree: B.S.
Undergraduate Degree Requirements: Biology 30 hours, Chemistry 16 hours, Physics 12 hours, Mathematics 6 hours, Foreign Language 12 hours, English 15 hours, American History 6 hours, Elective - 24 hours, ROTC 14 hours, 135 semester hours total

Graduate Degree: M.A.T.
Graduate Degree Requirements: Master: Biology - 18 21 hours, Education 9-12 hours

Graduate and Undergraduate Courses Available for Graduate Credit: Biological Science, Morphology and Evolution of Animals, Morphology and Evolution of Plants, General Physiology, Advanced Microbiology, Pollution Ecology, Marine Invertebrates, Methods and Procedures for Teaching High School Biology, Seminar in Environmental Studies, Genetics (U), The Vascular Flora of South Carolina (U), Plant Morphology, Mycology (U), General Entomology (U), Mammalian Physiology (U), Plant Physiology (U), Ecology (U), Ornithology (U), Marine Biology (U), Vertebrate Natural History (U), Comparative Physiology (U), Plant Anatomy (U)

CLAFLIN COLLEGE

(No information available)

CLEMSON UNIVERSITY

Clemon, South Carolina 29631 Phone: (803) 656-3311
Dean of Graduate Studies: E.A. Schwartz

College of Agricultural Sciences

Dean: L.P. Anderson

AGRONOMY AND SOILS DEPARTMENT
Phone: Chairman: (803) 656-3104 Others: 656-3104

Head: G.R. Craddock
Professors: G.R. Craddock, C.M. Jones, C.N. Nolan B.J. Gossett, U.S. Jones , T.C. Peele, L.H. Harvey, H.L. Musen, J.B. Pitner
Associate Professors: L.R. Allen, W.D. Graham, A. Manwiller, E.A. Rupert, R.E. Currin, M.W. Jutras, J.D. Maxwell, R.F. Suman, E.B. Eskew, K.S. La Fleur, C.L. Parks, J.R. Woodruff
Assistant Professors: D.A. Benton, J.S. Rice, E.F. McClain, B.R. Smith, J.H. Palmer, H.D. Yonce
Instructors: R.P. Alston, R.L. Stephens
Research Assistants: 5

Field Stations/Laboratories: Simpson Experiment Station, Edisto Experiment Station, Pee Dee Experiment, Sandhill Experiment Station

Degree Programs: Agronomy, Plant Physiology

Undergraduate Degree: B.S. (Agronomy)
Undergraduate Degree Requirements: 134 total semester hours, GPR 2.0

Graduate Degrees: M.A., M S., Ph.D.

Graduate Degree Requirements: 30 semester hours: minimum of 12 semester hours with 800 listing, thesis, GPR average 3.0 Doctoral: Approximately 60 semester hours beyond B.S. Dissertation, no language requirement, GPR average 3.0

Graduate and Undergraduate Courses Available for Graduate Credit: Fertilizers, LandPollution Control, Soil Classification, Plant Breeding, Principles of Weed Control, Soil and Plant Analysis, Cotton and Other Fiber Crops, Grain Crops, Tobacco and Special Use Crops, Forace and Pasture Crops, Forace Crops Laboratory, Soil Fertility and Management, Seminar, Crop Physiology and Nutrition, Pedology and Soil Classification, Theory and Methods of Plant Breeding, Soil Fertility, Special Problems , Soil Physics, Soil Chemistry, Crop Ecology and Land Use, Pesticide Residues in the Environment

DEPARTMENT OF ENTOMOLOGY AND ECONOMIC ZOOLOGY
Phone: Chairman· (803) 656-3112 Others: 656-3111

Head: S.B. Hays
Professors: T.R. Adkins, M.D. Farrar (Emeritus), R.C. Fox, S.B. Hays, E.W. King, J.B. Kissam, C.A. Thomas, S G. Turnipseed
Associate Professors: J.A. DuRant, W.C. Nettles (Emeritus), T.E. Skelton, L.M. Sparks, L.G. Webb
Assistant Professors: G.R. Carner, J.D. Hair, R.L. Holloway, A.W. Johnson, H.A. Loyacano, G P. Noblet, R. Noblet, D.K. Pollet, B.M. Shepard, M.J. Sullivan
Instructors: R.P. Griffin, F.J. Howard, Jr., D.C. Smith
Research Assistants: 24
Teaching Assistants: 3

Field Stations/Laboratories: Pee Dee Experiment, Edisto Experiment Station

Degree Programs: Agriculture, Entomology, Wildlife Biology, Economic Zoology

Undergraduate Degree: B.S.
Undergraduate Degree Requirements: 134 Semester Hours

Graduate Degrees: M.S., Ph.D.
Graduate Degree Requirements: Master: Established by Committee Doctoral: Established by Committee

Graduate and Undergraduate Courses Available for Graduate Credit: Field Crop and Stored Grain Insects, Fruit, Nut and Vegetable Insects, Insect Morphology, Insect Taxonomy, Medical and Veterinary Entomology, Pest Control, Introduction to Research, Insect Physiology, Insect Pathology, Entomology, Insect Ecology, Insect Toxicology, Insect Pest Management, Special Problems in Entomology, Wildlife Management, Fish Culture, Seminar, Principles of Wildlife Biology, Applied Wildlife Biology, Special Problems in Wildlife Biology

DEPARTMENT OF FOOD SCIENCE
Phone: Chairman: (803) 656-3397

Head: W.P. Williams
Professors: R.F. Borgman, J.H. Mutchell, W.P. Williams
Associate Professor: D.M Henricks
Assistant Professors: J.C. Acton, R.G. Bursey, I.M. Imbrahim, J.J. Jen, M.G. Johnson, T.C. Titus

Degree Program: Food Science

Undergraduate Degree: B.S.
Undergraduate Degree Requirements: Must complete 134 semester hours with 2.0 GPR out of possible 4.0

Graduate Degree: The Food Science Department does not offer a graduate program per se but is participating Department for interdepartmental programs in Nutrition and Physiology.

Graduate and Undergraduate Courses Available for Graduate Credit: Food Chemistry I (U), Food Chemistry II (U), Nutrition and Dietetics (U), Quality Assurance and Sensory Evaluation (U), and Quality Assurance and Sorsory Evaluation Laboratory (U)

DEPARTMENT OF HORTICULTURE
Phone: Chairman: (803) 656-3404

Head: T.L. Senn
Professors: J.R. Haun, M.B. Hughes, W.L. Ogle, T.L. Senn, E.T. Sims, Jr., G.E. Steinbridge, L.O. Van Blaricom
Associate Professors: J.B. Aitken, R.A. Baumgardner, J.A. Brittain, D O. Ezell, R.J. Ferree, J.P. Fulmer, R.G. Halfacre, M.G. Hamilton, W.S. Jordan, M.L. Robbins, H.J. Sefick, B.J. Skelton, F.W. Thode
Assistant Professors: W.P. Cook, C.E. Gambrell, Jr., E.V. Jones, A.R.Mazur, L.C. Miller, A.J. Pertuit
Instructors: A.R. Kingman, A.J. Lewis III, J.D. Ridley
Research Assistants: 8
Graduate Teaching Assistants: 4

Field Stations/Laboratories: Sandhill Experimental Station-Columbia, Truck Experimental Station - Charleston, PeeDee Station,- Florence, Edistor Experimental Station - Blackville

Degree Programs: Horticulture

Undergraduate Degree: B.S.
Undergraduate Degree Requirements: As specified in Undergraduate Catalog of Clemson University.

Graduate Degrees: M.S., M.Agr.
Graduate Degree Requirements: As specified by Clemson University Graduate School

Graduate and Undergraduate Courses Available for Graduate Credit: Floriculture (U), Commercial Pomology (U), Nut Tree Culture (U), Nursery Technology (U), Landscape Design (U), Turf Management (U), Advanced Turfgrass Culture, Small Fruit Culture (U), Vegetable Crops (U), Problems in Landscape Design (U), Landscape Design Implimentation (U), Postharvest Horticulture (U), Introduction to Research (U), Horticultural Therapy (U), Problems in Small Fruit Production, Research Systems in Horticulture, Experimental Olericulture, Scientific Advances in Ornamental Horticulture, Physiochemical Procedures for Determining Quality in Horticultural Crops, Postharvest Physiology and Handling of Horticultural Crops, Pomology, Special Investigations in Horticulture, Seminar I, Seminar II, Quantitative Exposition of Plant Development

NUTRITION PROGRAM
Phone: Chairman (803) 656-3230 Others: 656-3311

Chairman: W.A. King
Professors: B.D. Barnett, R.F. Borgman, W.A. King, J.H. Mitchell, Jr., G.C. Skelley, R.F. Wheeler, J. N. Williams II, W.P. Williams, Jr.
Associate Professors: R.L. Edwards, D.L.Handlin, D. M. Henricks, J.E. Jones, G.D. O'Dell, F.E. Pardue, D.E. Turk
Assistant Professors: J.C. Acton, R.G. Bursey, D.L. Cross, L W. Hudson, B.F. Jenny, J.C. McConnell
Research Assistants: 10
Associate Chemist: J.T. Gillingham

Field Stations/Laboratories: Uses facilities of S.C. Agricultural Experiment Station

Degree Program: Nutrition

Graduate Degrees: M.S., M.N.S., Ph.D.
Graduate Degree Requirements: Master: M.S. - A minimum of 24 semester hours of graduate credit, plus 6 semester hours of research and an acceptable thesis. For Master of Nutritional Science - a minimum of 30

semester hours of graduate credit. Doctoral: A minimum of 54 semester hours of graduate credit beyond the baccalaureate degree, plus 18 semester hours of research an an acceptable thesis.

Graduate and Undergraduate Courses Available for Graduate Credit: Fundamentals of Nutrition (U), Human Nutrition (U), Therapeutic Nutrition, Public Health Nutrition, Nutrition Education, Feeding Methods, Field Training in Nutrition, Topical Problems in Nutrition, Monogastric Nutrition, Polygastric Nutrition, Metabolism of Nutrients, Nutrition Techniques with Large Animals, Nutrition Techniques with Laboratory Animals, Amino Acids and Protein Nutrition, Vitamins and Minerals, Nutrition Seminar

DEPARTMENT OF PLANT PATHOLOGY AND PHYSIOLOGY
Phone: Chairman: (803) 656-3450

Head: W.M. Epps
Professors: L.W. Baxter, D.F. Cohoon, W.M. Epps, W.R. Sitterly, F.H. Smith (Extension), W. Witcher
Associate Professors: N.D. Camper, G.C. Kingsland
Assistant Professors: J.D. Arnett (Extension), O.W. Barnett, C.W. Blackman, G.E. Carter, J.C. LaPrade, R.W. Miller, S.A. Lewis, E.I.Zehr
Research Assistants: 2
Lecturer: 1

Field Stations/Laboratories: Sandhill Experimental Station - Columbia, Pee Dee Experimental Station - Blackville, Truck Experimental Station - Charleston

Degree Programs: Plant Physiology, Plant Pathology

Undergraduate Degrees: B.S. (Plant Pathology)
Undergraduate Degree Requirements: 134 semester credits, 210 cumulative GPR, Specific curriculum

Graduate Degrees: M.S., M. Ag., Ph.D.
Graduate Degree Requirements: Master: M.S. Major 12 credits, minor 6 credits, Research 6 credits, Other 6 credits M.Ag. Major 15 credits, minor 6 credits, Research 3 credits, Other 6 credits Doctoral: Major and minor - no specified hours, Dissertation, Languages - None

Graduate and Undergraduate Courses Available for Graduate Credit: Plant Pathology, Forest Pathology, Bacterial Plant Pathogens, Plant Virology, Plant Parasitic Nematodes, Advanced Plant Pathology, Physiological Plant Pathology, Techniques and Methods in Plant Pathology, Plant Disease Diagnosis

DEPARTMENT OF POULTRY SCIENCE
Phone: Chairman: (803) 656-3162

Head: B.D. Barnett
Professors: B.D. Barnett, B.W. Bierer, M.A. Boone, W.H. Wiley
Associate Professors: J.B. Cooper, K.A. Holleman, J.E. Jones, D.E. Turk, W.S. Walker
Assistant Professors: D.L. Cross, J.W. Dick, B.L. Hughes, C.F. Risher, J.F. Welter
Research Assistants: 9

Field Stations/Laboratories: Sandhill, PeeDee, Coast, Edisto

Degree Program: Poultry Science

Undergraduate Degree: B.S.
Undergraduate Degree Requirements: 134 credits

Graduate Degrees: M.S., M.Agr., Ph.D.
Graduate Degree Requirements: Master: 30 credits, Thesis Doctoral: (Interdepartmental doctorates in Nutrition and Animal Physiology)

Graduate and Undergraduate Courses Available for Graduate Credit: Animal Environment Technology, Poultry Nutrition, Breeder Flock and Hatchery Management, Poultry Products Technology, Avian Microbiology and Parasitology, Management of Egg, Broiler and Turkey Enterprises, Poultry Pathology, Digestive and Excretory Physiology, Cardiovascular and Respiratory Physiology

College of Forest and Recreational Resources
Dean: W.H.D. McGregor

DEPARTMENT OF FORESTRY
Phone: Chairman: (803) 656-3302 Others: 656-3303

Chairman: R.M. Allen
Professors: R.M. Allen, B.M. Cool, D.D. Hook, J.R. Warner, T.E. Wooten
Associate Professors: N.B. Goebel, G.D. Kessler, C.L. Lane, R.E. Schoenike, W.A. Shain, D.H. Van Lear
Assistant Professors: B.A. Dunn, P. Labosky, S.A. Marbut, A.E. Miller, L.E. Nix, L.D. Reamer, G.E. Sabin
Instructor: A.T. Shearin
Research Assistants: 5

Field Stations/Laboratories: Hobcaw Barony - Georgetown, Sandhill Experiment Station - Columbia

Degree Programs: Forestry, Wood Utilization

Undergraduate Degree: B.S.
Undergraduate Degree Requirements: Forestry - 148 semester hours, 7 hours Chemistry, 10 hours Biology, 8 hours Mathematics, 75 hours area specialization Wood Utilization - 140 semester hours, 4 hours Biology, 15 hours Chemistry, 11 hours Mathematics, 4 hours Physics, 65 hours area specialization

Graduate Degrees: M.S., M F.
Graduate Degree Requirements: Master: M.S. - Minimum of 30 semester hours, M.F. Minimum of 36 semester hours

Graduate and Undergraduate Courses Available for Graduate Credit: Silvics (U), Forest Plants (U), Forest Engineering (U), Forest Mensuration (U), Forest Products (U), Forest Economics (U), Elements of Forestry (U), Wood and Wood Fiber Identification, Aerial Photographs in Forestry (U), Silviculture, Forest Ecology, Logging and Milling, Forest Soils Seminar, Multiple-Use Forestry, Harvesting Forest Products, Forest Protection, Management Plans, Forest Policy and Administration, Forest Management and Regulation, Forest Evaluation, Forest Products, Wood Properties Lectures in Forestry, Forest Genetics and Tree Breeding, Wood Chemistry, A Survey of Forest Policy, Wood Processing, Wood Design, Data Processing in Forestry Problems, Advanced Mensuration, Photo Interpretation, Advanced Forest Economics, Cost Studies in Harvesting and Processing

College of Physical, Mathematical and Biological Sciences
Dean: H.E. Vogel

BOTANY DEPARTMENT
Phone: Chairman: (803) 656-3452 Others: 656-3456

Head: C.J. Umphlett
Professors: R.P. Ashworth, J.B. Whitney, Jr., C.J. Umphlett
Associate Professors: N.D. Camper (Joint)
Assistant Professors: C.R. Dillon, L.A. Dyck, J.E. Fairey III, T.M. McInnis, Jr.
Research Assistants: 1

Field Stations/Laboratories: Member of Highlands Biological Station - Highlands

Degree Programs: Botany, Plant Physiology

Undergraduate Degree: B.S. (Botany)
Undergraduate Degree Requirements: 24 credits in Botany, Chemistry through Organic, Mathematics through Calculus, 12 credits in Zoology, 2 courses in Physics, Several Social Science - Humanities electives, Total 134 credits

Graduate Degrees: M.S. (Botany), Ph.D. (Plant Physiology)
Graduate Degree Requirements: Master: 30 credits, six of which are research, 6 credits in minor, others in Botany - Thesis required Doctoral: 3 courses in Plant Physiology, others as desirable for student program. Dissertation, No languages

Graduate and Undergraduate Courses Available for Graduate Credit: Cytology (U), Plant Anatomy (U), Taxonomy (U), B ological Oceanology (U), Mycology (U), Plant Physiology (U), Phycology (U), Ecology (U), Advanced Mycology, Plant Metabolism, Plant Growth and Development, Growth Substances, Advanced Taxonomy, Advanced Phycology, Fungus Physiology

DEPARTMENT OF MICROBIOLOGY
Phone: Chairman: (803) 656-3057

Head: M.J.B. Paynter
Professor: R.K. Guthrie
Associate Professors: A.W. Baxter, J.H. Bond, B.V. Bronk (Joint), M.J.B. Paynter
Assistant Professors: O.W. Barnett (Joint), M.G. Johnson, L.L. Larcom (Joint)
Graduate Assistants: 13

Degree Program: Microbiology

Undergraduate Degree: B.S.
Undergraduate Degree Requirements: Calculus - 4 credits Chemistry 8 credits, Organic Chemistry 8 credits, Physics 8 credits, Genetics 4 credits, social Sciences 9 credits, English Language plus literature 12 credits, Botany 4 credits, Zoology 4 credits, Mathematics or Science electives 7 credits, Biochemistry 3 credits, Public Speaking 3 credits, Microbiology 31 credits, Electives 23 credits (Semester hour credits in all cases)

Graduate Degrees: Ph.D., M.S. (Emphasis in Microbiology)
Graduate Degree Requirements: Master: 24 credits of formal course work, 6 credits of Research, A thesis is required with must be defended Doctoral: Work depends on a committee. No specified requirements other than 18 hours Research, course, thesis. Qualifying examination. Final Oral and Defense of Thesis

Graduate and Undergraduate Courses Available for Graduate Credit: General Microbiology (U), Public Health Microbiology (U), Advanced Bacteriology (U), Dairy Microbiology (U), Food Microbiology (U), Soil Microbiology (U), Pathogenic Bacteriology (U), Bacterial Physiology (U), Industrial Microbiology (U), Basic Immunology (U), Microbial Genetics (U), Introductory Virology (U), Bacterial Taxonomy, Bacteriological Techniques, Special Problems in Microbiology, Seminar, Soil Microbiology, Bacterial Cytology and Physiology, Bacterial Metabolism, Advanced Microbial Genetics

ZOOLOGY DEPARTMENT
Phone: Chairman: (803) 656-3247

Head: S.A. Gauthreaux, Jr.
Professors: H.S. Min, A.S. Tombes
Associate Professors: S.A. Gauthreaux, Jr., R.L. Hays, W.K. Willard
Assistant Professors: D.R. Helms, B.R. Ingram, H.A. Loyacano, Jr., R.R. Montanucci, G.P. Noblet, S.L. Pimm, E.B. Pivorun, R.F. Walker
Visiting Lecturers: 2
Research Assistants: 7
Graduate Assistants: 24

Degree Programs: Animal Behavior, Animal Physiology, Ecology, Histology, Limnology, Parasitology, Zoology, Developmental Biology

Undergraduate Degree: B.S. (Zoology)
Undergraduate Degree Requirements: Required Zoology courses, 28 semester hours, Required Science Course 28 hours, Required non-science courses 15 hours, Humanities and Social Sciences 9 hours, Zoology electiv es 21 hours, Free Electives, 24 hours (135 hours total)

Graduate Degrees: M.S., Ph.D.
Graduate Degree Requirements: Master: 29 semester hours of course work and research, and a minimum of 6 semester hours thesis research Doctoral: A minimum of 18 semester hours of dissertation research and 12 hours of advanced course work.

Graduate and Undergraduate Courses Available for Graduate Credit: Advanced Invertebrate Zoology (U), Anatomy (U), Animal Behavior (U), Animal Ecology (U), Animal Histology (U), Cell Physiology (U), Developmental Biology (U), General Endocrinology (U), General Physiology (U), Herpetology (U), Ichthyology (U), Invertebrate Endocrinology (U), Limnology (U), Mammalogy (U), Ornithology (U), Parasitology (U), Protozoology (U), Advanced Animal Histology, Advanced Cell Physiology, Behavioral Ecology, Comparative Animal Physiology I and II, Evolution, Histochemistry-Cytochemistry, Histological Techniques, Physiological Ecology, Population Dynamics, Principles and Methods of Systematic Zoology, Radiology, Special Problems in Zoology, Use of Radioisotopes in Biological Research, Zoology Seminar

COKER COLLEGE

Hartsville, South Carolina 29550 Phone: (803) 322-1381
Dean of College: Not Stated

BIOLOGY DEPARTMENT
Phone: Chairman: Ext. 410 Others: Ext. 413

(There is no chairman - administrative duties are shared)
Associate Professor: R.L. Swallow
Assistant Professor: G.P. Sawyer, Jr.

Degree Program: Biology

Undergraduate Degree: A.B.
Undergraduate Degree Requirements: 9 quarter courses in major, 3 quarter courses in each of four (Choice from 6) academic areas, electives to total 36 quarter courses

COLLEGE OF CHARLESTON

Charleston, South Carolina 29401 Phone: (803) 722-0181
Dean of Graduate Studies: R. Crosby
Dean of College: C.H. Womble

BIOLOGY DEPARTMENT
Phone: Chairman: Ext. 300, Others: Exts. 201,360

Chairman: H.W. Freeman
Professors: N. Chamberlain, H.W. Freeman, J. Harrison, M.T. Pennington
Associate Professors: W.D. Anderson, Jr., J. Smiley
Assistant Professors: C.K. Biernbaum, T.T. Ellis, D.L. Johnson, K. Kelly, J.J. Manzi, M.W. Runey

Field Stations/Laboratories: Grice Marine Biological Laboratory - Charleston

Degree Programs: Marine Biology, Marine Botany, Oceanography, Zoology

Undergraduate Degrees: B.S.
Undergraduate Degree Requirements: Not Stated

Graduate Degree: M.S. (Marine Biology)
Graduate Degree Requirements: Master: 30 credit hours, Foreign Language, Thesis, Oral

Graduate and Undergraduate Courses Available for Graduate Credit: Ichthyology, Histology, Oceanography I, II, Marine Botany, Marine Invertebrate Zoology, Advanced Ichthyology, Marine Physiological Ecology, Seminar in Marine Biology

COLUMBIA COLLEGE

Columbia, South Carolina 29203 Phone: (803) 786-3012
Dean of College: W. Butler

DEPARTMENT OF MATHEMATICS AND SCIENCE
Phone: Chairman: (803) 786-3750

Chairman: P.E. Graef
J.M. Reeves Professor of Biology: P.E. Graef
Professors: P.E. Graef, D.N. Mercer

Degree Program: Biology

Undergraduate Degree: A.B.
Undergraduate Degree Requirements: 28 semester hours in Biology, 8 semester hours in Chemistry

CONVERSE COLLEGE

Spartanburg, South Carolina 29301 Phone: (803) 585-6423
Dean of Graduate Studies: T.R. McDaniel
Dean of College: C.D. Ashmore

DEPARTMENT OF BIOLOGY
Phone: Chairman: Ext. 342

Chairman: R.W. Powell, Jr.
Associate Professors: J.H. Cromer, R.W. Powell, Jr.
Assistant Professors: J.A. Lever

Degree Program: Biology

Undergraduate Degree: A.B.
Undergraduate Degree Requirements: 32 semester hours Biology beyond Freshman course, Chemistry through Organic, Basic Physics

Graduate Degree: M.A.T.
Graduate Degree Requirements: Master: 18 hours science for major of 30 (no thesis or language)

Graduate and Undergraduate Courses Available for Graduate Credit: Biology for Teachers (U), Environmental Biology (U), Advanced Microbiology (U), Advanced Plant Taxonomy (U), Plant Morphology (U), Horticultural Botany (U), Developmental Biology (U), General and Cellular Physiology (U)

ERSKINE COLLEGE

Due West, South Carolina 29634 Phone: (803) 379-2131
Dean of College: R.K. Ackerman

DEPARTMENT OF BIOLOGY
Phone: Chairman: (803) 379-8817 Others: 379-2131

Chairman: R.C. Clark
Associate Professor: R.C. Clark
Assistant Professors: J.H. Haldeman, F.E. Nussbaum, G.W. Shiflet

Degree Program: Biology

Undergraduate Degree: B.A.
Undergraduate Degree Requirements: 32 semester hours in Biology, including Introductory Botany, Introductory Zoology, Genetics, Ecology Molecular/Cellular Biology, Organismal Physiology, 8 semester hours of Chemistry, 6 semester hours of Mathematics, other general requirements of the college (120 semester hours total)

FRANCIS MARION COLLEGE

Florence, South Carolina 29501 Phone: (803) 669-4121
Dean of College: J.W. Baker

DEPARTMENT OF BIOLOGY
Phone: Chairman: Ext.318

Chairman: J.S. Boyce, Jr.
Professors: J.P. Boyce, Jr.
Associate Professors: L.F. Swails, F.F. Welbourne
Assistant Professors: L.M. Croshaw, T. Roop
Research Assistants: 1

Degree Program: Biology

Undergraduate Degree: B.S. (Biology)
Undergraduate Degree Requirements: 24 semester hours in Biology above freshman level, 12 semester hours Chemistry (General - 8, Organic - 4), 8 semester hours Physics

FURMAN UNIVERSITY

Greenville, South Carolina 29613 Phone: (803) 246-3550
Dean of College: S. Patterson

DEPARTMENT OF BIOLOGY
Phone: Chairman: Ext. 344, 293

Chairman: C.L. Rodgers
Professors: P.L. Fisher, R.W. Kelly, C.L. Rodgers
Associate Professors: G.W. Fairbanks, W.P. Pielou
Assistant Professors: R.E. Kerstetter, G.C. Smith, J.A. Snyder, L.P. Stratton

Degree Program: Biology

Undergraduate Degrees: B.A., B.S.
Undergraduate Degree Requirements: Generally 8 courses in Biology above the Introductory level plus three courses in Chemistry

LANDER COLLEGE

Greenwood, South Carolina 29646 Phone: (803) 229-5521
Dean of College: Not Stated

DEPARTMENT OF SCIENCE
Phone: Chairman: (803) 229-5521 Ext. 141

Chairman: S.E. Stewart
Professor: S.E. Stewart
Associate Professor: L. Verean
Assistant Professors: L. Lundquist, H. House
Instructor: M. Runyan

Degree Program: Biology

Undergraduate Degrees: B.S., A.B.
Undergraduate Degree Requirements: 124 semester hours, 36 semester hours in Biology, 16 semester hours in Chemistry

LIMESTONE COLLEGE

College Drive Phone: (803) 489-7151
Gaffney, South Carolina 29340
Dean of College: W. Taylor

BIOLOGY DEPARTMENT
Phone: Ext. 167

Vice-Chairman: K.N. Mehra
Assistant Professor: C. Waters

Degree Program: Biology

Undergraduate Degree: B.S.
Undergraduate Degree Requirements: 32 hours of Biology, 16 hours of Chemistry, 4 hours Comparative Anatomy, 4 hours of Human Anatomy, 4 hours of Human Physiology, 8 hours of General Chemistry, 8 hours of Organic Chemistry

MEDICAL UNIVERSITY OF SOUTH CAROLINA

80 Barre Street Phone: (803) 792-0211
Charleston, South Carolina 29401
Dean of College of Medicine: J.W. Zemp
Dean of College of Dental Medicine: J.J. Sharry
Dean of Graduate Studies: F.W. Kinard

DEPARTMENT OF ANATOMY

Phone: Chairman: (803) 792-3521 Others: 792-3522

Chairman: W.C. Worthington
Professors: M.H. Knisely (Research), I. Lackard, I.S.H. Metcalf, E. Tabor, W.C. Worthington
Associate Professors: B.S. Barrington, T.C. Davies, R. Dom, H.S. Debacker, W.J. Dougherty, S.H. Hardin, W.K. Mylon, R.C. Pennington
Assistant Professors: S. Canaday, T.P. Fitzharrin, R.R. Markwald, R.W. Ogilvie
Associates: B. Barzansky, F.M. Kneuss
Research Fellows: 2

Degree Program: Anatomy

Graduate Degrees: M.S., Ph.D.
Graduate Degree Requirements: Master: 36 course credit hours plus thesis Doctoral: 67 course credits plus Dissertation

Graduate and Undergraduate Courses Available for Graduate Credit: Gross Anatomy, Human Histology and Development, Neuroanatomy, Developmental, Normal and Abnormal Anatomy, Advanced Developmental Anatomy, Biology of Sex Reproduction, CNS lesions, Advanced Cell Biology, Optical Theory, E.M. Methods, Anatomy of Hypothalamus, Advanced Histology of Head and Neck, Microacrulation, Neuroendocrinology

DEPARTMENT OF BIOCHEMISTRY

Phone: Chairman: (803) 792-2359 Others: 792-2331

Chairman: B. Baggett
Professors: B. Baggett, R.H. Gadsden, E.L. Hogan, A. R. Krall, M.E. Yannone, J.W. Zemp
Associate Professors: W.O. Boggan, S. Brostoff, E.B. Cunningham, W.W. Fish, J.B. Hynes, H.T. Jonsson, Jr., J.D. Karam, J.W. Ledbetter, Jr., A.A. Swanson, C. Schwabe
Assistant Professors: N.S Buckholtz, L.J. Heere, B.E. Ledford, L.D. Middaugh, R L Miller, R. Nair, D.G. Priest, L.W. Stillway, T.A. Grover (Associate), G.M Thorne (Associate)
Instructor: F.B. Culp
Visiting Lecturers: 40
Teaching Fellows: 7
Research Assistants: 20

Degree Program: Biochemistry

Graduate Degrees: M.S., Ph.D.
Graduate Degree Requirements: Master: In addition to courses covering general biochemistry, all students are required to complete one quarter of laboratory apprenticeship, two advanced topic courses in Biochemistry, and 6 quarter hours of coursework outside biochemistry. A suitable research plan must be prepared, executed and defended. Doctoral: In addition to courses covering general biochemistry, all students are required to complete two quarters of laboratory apprenticeship, three additional advanced topic courses, and 24 quarter hours of course work outside biochemistry. Both written and oral qualifying examinations must be completed. A research plan requiring an original investigation must be proposed, executed and defended to the satisfaction of student's research advisory committee.

Graduate and Undergraduate Courses Available for Graduate Credit: Introductory Biochemistry, Comprehensive Biochemistry, Biochemical Methodology, Biochemical Laboratory Apprenticeship, Molecular Basis of Behavior, Clinical Biochemistry, Molecular Basis of Disease, Nutrition, Plant Biochemistry, Special Topic in Biochemistry, Seminar in Biochemistry, Neurochemistry, Biochemistry of Environmental Problems, Special Projects in Biochemistry, Advanced Physical Biochemistry, Enzyme Kinetics and Mechanisms, Endocrinology and Biochemical Regulation, Lipids and Steroids, Bioenergetics, Molecular Genetics

DEPARTMENT OF BIOMETRY

Phone: Chairman: (803) 792-2261

Chairman: M.C. Miller, III
Professors: M.C. Miller, III
Associate Professors: R.C. Duncan, C.B. Loadholt, W.E. Groves (Adjunct)
Assistant Professors: L.M. Chansky (Adjunct), C.S. Donley, H.H. Hunt, C.F. Lam, D. Lurie, R.L. Mason I.S. Metts (Adjunct), R.D. Small, I.S. Tolins, C.L. Gionet, C. Smith
Teaching Fellows: 2
Research Assistants: 2

Degree Programs: Biometry, Biostatistics, Biomedical Computing

Graduate Degree: M.S.
Graduate Degree Requirements: Master: 36 Quarter hours

Graduate and Undergraduate Courses Available for Graduate Credit: Introduction to Biostatistics, Introduction to Experimental Design, Intermediate Experimental Design, Nonparametric Methods in Biology and Medicine, Multivariate Methods in Biology and Medicine, Sampling Methods in Biology and Medicine, Regression Methods in Biology and Medicine, Stochastic Biological Processes, Medical Economics, Health Care Delivery Systems, Epidemiological Foundations, Special Problems in Epidemiology, Principles of Digital Computing and Programming, Medical Information Systems, Clinical Laboratory Computing, Analog and Hybrid Computation, Digital Signal Analysis, Biomedical Instrumentation, Biological Systems Analysis I - Linear System, Biological Systems Analysis II - Nonlinear Systems, Biological Systems Analysis III - System Identification, Modeling and Simulation I - Continuous Systems, Operations Research in Biology and Medicine I

DEPARTMENT OF BASIC AND CLINICAL IMMUNOLOGY AND MICROBIOLOGY

Phone: Chairman: (803) 792-2564 Others: 792-4421

Chairman: C.D. Graber
Distinguished Professor of Biomedicine: A. Sabin
Professors: N. Anderson, E.H. Eylar, H.H. Fudenberg, C.D. Graber
Associate Professors: J S. Haskill, T. Mahvi, G. Virella, A.C. Wang
Assistant Professors: J.M. Goust, T.B. Higerd, R.C. Johnson, G.E. Newlin, I. Wang
Associate: J.L. Caldwell
Instructor: G. Wilson
Teaching Fellows: 6
Research Assistants: 12

Degree Programs: Microbiology, Immunology, Virology

Graduate Degrees: M.S., Ph.D.
Graduate Degree Requirements: Master: 22 hours in major department, 6 hours outside the major department 9 hours in courses above the 600 level which have been approved for graduate study Doctoral: 36 hours

in major department, 24 hours outside major department, 27 hours in courses above the 600 level which have been approved for graduate credit

Graduate and Undergraduate Courses Available for Graduate Credit: Medical Bacteriology, Medical Mycology, Medical Virology, Microbial Physiology and Genetics, Microbiology for Dental Students, Methods in Mycology, Diagnostic Virology, Advanced Immunology, Advanced Microbial Physiology, Advanced Microbial Genetics, Advanced Virology, Viral Oncogenesis, Interferon and Anti-Viral Agents, Slow Viruses and Neurological Diseases, Methods in Diagnostic Bacteriology, Methods in Diagnostic Mycology, Methods in Diagnostic Cutaneous, Subcutaneous and Systemic Mycoses, Special Research Projects in Bacteriology, Immunology, Mycology and Virology

DEPARTMENT OF PHYSIOLOGY
Phone: Chairman: (803) 792-3648 Others: 792-0211

Chairman: H.G. Hempling
Professors: H.G. Hempling, A.D. Horres, F.W. Kinard
Associate Professors: S. Katz, W.C. Wise
Assistant Professors: J.G Blackburn, R.L. Green, J.G. Ondo, J.B. Pritchard, D.D. Wheeler, G.B. Whitfield

Degree Program: Physiology

Graduate Degrees: M.S., Ph.D.
Graduate Degree Requirements: Master: 36 Quarter hours (22 hours in major department, 6 hours outside major, 9 hours in courses above 600 level, 9 hours in research and preparation of thesis, thesis Doctoral: 67 Quarter hours (36 hours in major department, 24 hours outside major department, 27 hours in courses above 600 level 23 hours research and preparation of dissertation, dissertation

Graduate and Undergraduate Courses Available for Graduate Credit: Physiology for Medical Students, Physiology for Dental Students, Neurophysiology of Motor Systems-Principles and Perspectives, Pulmonary Physiology, Special Projects in Physiology, Selected Topics in Neurophysiology, Circulatory Physiology, Cardiac Physiology, Topics in Membrane Physiology, Physiological Control and Regulation, Physiological Regulation of Fluid and Electrolytes, Neurophysiology, Central Reflex Control of Breathing, Medical Electronics

NEWBERRY COLLEGE
(No reply received)

PRESBYTERIAN COLLEGE
Clinton, South Carolina 29325 Phone: (803) 833-2820
Dean of College: W.F. Chapman

DEPARTMENT OF BIOLOGY
Phone: Chairman: Ext. 316 Others: 833-2820

Chairman: J.D. Stidham
Professors: J.D. Stidham
Associate Professor: F.C. James
Assistant Professors: A.E. Hilger, J. Holt, P.R. Ramsey
Instructor: J. Burns

Field Stations/Laboratories: Affiliations with Highlands Biological Station, Gulf Coast Research Laboratory

Degree Programs: Biology

Undergraduate Degree: B.S.
Undergraduate Degree Requirements: Twenty-eight hours of Biology above the general biology (8 hours) which muct include core courses ofBotany, Genetics, Ecology, and Physiology, and a senior seminar, 3 semesters of Mathematics, 4 semesters of Chemistry (through Organic Chemistry, 1 year Physics

SOUTH CAROLINA STATE COLLEGE
(No reply received)

UNIVERSITY OF SOUTH CAROLINA
Columbia, South Carolina 29208 Phone: (803) 777-4142
Dean of Graduate Studies: G. Reeves

College of Arts and Sciences
Dean: J.R. Durig

DEPARTMENT OF BIOLOGY
Phone: Chairman: (803) 777-2116 Others: 777-4141

Head: W.D. Dawson
Professors: W.T. Batson, B.T. Cole, W.D. Dawson, J.M. Horr, J.G. Scandalios, F.J. Vernberg, W.B. Vernberg (Public Health Professor), N. Watabe
Associate Professors: D.L. Claybrook, B. Coull, G.T. Cowley, S.P. Craig, J.M. Dean, P.J. DeCoursey, F. H. Lauter, N C. Mishra, D.H. Rembert, D.L. Rohlfing, H.L. Stevenson, L.T. Wimer
Assistant Professors: B E. Ely, M.R. Felder, A H. Huang, D D. Husband, E.H. Liu, P.E. Mirkes, E.F. Thompson, M.H. Vodkin, R.G. Zingmark
Teaching Assistants: 36

Field Stations/Laboratories: Hobcaw Plantation - Georgetown

Degree Programs: Biology, Botany, Ecology, Genetics, Zoology, Marine Science, Developmental Biology

Undergraduate Degrees: B.S. (Biology)
Undergraduate Degree Requirements: 24 hours credit in Biology courses above the 200 level, including four semesters of 300 level courses, two of which must be taken in the same core and include laboratory. 6 hours of Biology courses at the 500 level or 600 level, Chemistry 111, 114, 231 and 232, Physics 201 and 202 are recommended, Mathematics through Calculus is highly recommended

Graduate Degrees: M.A.T., M.S., Ph.D.
Graduate Degree Requirements: Master: Twenty four semester hours formal course work, plus thesis. Reading knowledge of one modern foreign language, Final oral Comprehensive Examination Doctoral: Formal course work determined by advisory committee no minimal hour requirement, reading knowledge of one modern foreign language, Dissertation acceptable for publication. Written qualifying examination, written and oral comprehensive examination, and oral defense of dissertation

Graduate and Undergraduate Courses Available for Graduate Credit: Plant Morphology (U), Plant Anatomy (U), Mycology (U), Local Flora (U), Microscopic Anatomy (U), Parasitology (U), Comparative Physiology (U), Plant Physiology (U), Principles of Ecology (U), Developmental Biology (U), Phycology (U), Biological Chemistry (U), Cell Physiology (U), Advanced Microbiology (U), Biological Oceanography (U), Limnology (U), Evolutionary Genetics (U), Molecular Biology (U), Advanced Genetics, Aquatic Bacteriology, Bryology, Phylogeny of Angiosperms, Angiosperm Embryology, Plant-Soil Relations, Marine Phytoplankton, Advanced Phycology, Biology of Fish,,Advanced Invertebrate Zoology, Experimental Parasitology, Vertebrates, Biochemical Genetics, Developmental Genetics, Electron Microscopy

VOORHEES COLLEGE
(No reply received)

WINTHROP COLLEGE
Rock Hill, South Carolina 29730 Phone: (803) 323-2111
Dean of Graduate Studies: H.B. Gilbreth

Dean of College: J.B. Olson

BIOLOGY DEPARTMENT
Phone: Chairman: (803) 323-2111

Chairman: J.A. Freeman
Professors: L.V. Davis, J.A. Freeman, R.D. Houk, J.B. Olson
Associate Professor: E N. King
Assistant Professors: J.E. Dille, K W. Gregg

Degree Programs: Biology, Medical Technology

Undergraduate Degree: B.A., B S.
Undergraduate Degree Requirements: 30 semester hours including general botany, general Zoology, Cell Biology and Genetics

Graduate Degree: M.A.T.
Graduate Degree Requirements: Master: 30 semester hours including 15 semester hours Biology and 6 and 9 in two other fields, one of which must be education

Graduate and Undergraduate Courses Available for Graduate Credit: Comparative Animal Physiology (U), Selected Topics inMicrobiology (U), Advanced Cell Biology (U), Organic Evolution (U), Natural Environments and the Air Pollution Problems, Biological Methods and Materials, Developmental Biology

WOFFORD COLLEGE

Spartanburg, South Carolina 29301 Phone: (803) 585-4821
Dean of College: B.G. Stephens

DEPARTMENT OF BIOLOGY
Phone: Chairman: Ext. 218

Chairman: W.R. Leonard
Reeves Professor of Biology: W.R. Leonard
Professors: H.D. Dobbs
Associate Professor: E G. Patton
Assistant Professor: W.B. Hubbard

Degree Program: Biology

Undergraduate Degree: B.S.
Undergraduate Degree Requirements: All must be completed to the satisfaction of the dpeartmental staff, 4 basic biology courses, 4 electives selected from advanced offerings. Senior Seminar, Departmental reading Program in the junior year. Comprehensive examination in the senior year.

SOUTH DAKOTA

AUGUSTANA COLLEGE

Sioux Falls, South Dakota 57102 Phone: (605) 336-0770
Dean of College: W.R. Matthews

DEPARTMENT OF BIOLOGY
Phone: Chairman: (605) 336-4719 Others: 336-0770

Chairman: L.M. Prescott
Professors: S.G. Froiland, L.G. Johnson, D.J. Rogers, L.L. Tieszen
Associate Professor: L.M. Prescott
Assistant Professor: G.B. Blankespoor
Instructor: B.H. Haglund
Research Assistant: 1
Technician: 1

Field Stations/Laboratories: Black Hills Natural Sciences Field Station

Degree Program: Biology

Undergraduate Degree: B.S.
Undergraduate Degree Requirements: Four biology core courses (Man in the Ecosystem, Organic Biology, Genetics, Cell and Molecular Biology), plus four additional courses, including one field-oriented and one experimental offering. At least one course each in Mathematics and Organic Chemistry are also required.

BLACK HILLS STATE COLLEGE

Spearfish, South Dakota 57783 Phone: (605) 642-6133
Dean of Graduate Studies: M. Fitzgerald

DIVISION OF SCIENCE AND MATHEMATICS
Phone: Chairman: (605) 642-6133

Chairman: G.D. Shryock
Professors: V.W. Backens, J.B. Coacher, E. Follette, C.P. Haight, M. Hilpert, G.D. Shryock
Associate Professor: C.E. Berry
Assistant Professor: M. Durgin, E. Grieb, J. Lozier, L. Reuppel

Degree Programs: Biology, Medical Technology

Undergraduate Degree: B.S.
Undergraduate Degree Requirements: Not Stated

DAKOTA STATE COLLEGE

Madison, South Dakota 57042 Phone: (605) 256-3551
Dean of College: C.K. Brashier

DIVISION OF NATURAL SCIENCES/HEALTH SERVICES
Phone: Chairman: Ext. 267 Others: Exts. 264 or 281

Chairman: C L. Churchill
Professor: C K. Brashier
Associate Professor: R. Buckman
Assistant Professor: V A. Hall

Field Stations/Laboratories: Black Hills Natural Science Field Station near Rapid City

Degree Programs: Biology, Biology (Education)

Undergraduate Degree: B.S.
Undergraduate Degree Requirements: Major: 33 hours including Biological Principles, General Botany, General Zoology and 1 course from each of the following groupings - I Cellular Biology, II. Environmental Biology, III Morphology, IV Taxonomy

DAKOTA WESLEYAN UNIVERSITY

(No reply received)

HURON COLLEGE

Huron, South Dakota 57350 Phone: 352-8721
Dean of College: W. Sutterfield

DEPARTMENT OF BIOLOGY
Phone: Chairman: Ext. 281

Chairman: A. Okonkwo
Professors: A. Okonkwo
Associate Professor: D. Dunlay

Degree Program: Biology

Undergraduate Degree: B.S.
Undergraduate Degree Requirements: Thirty two semester hours of Biology, five course chosen from Chemistry, Physics or Mathematics

MOUNT MARTY COLLEGE

Yankton, South Dakota 57078 Phone: (605) 668-1011
Dean of College: Sr. J. Klimisch

BIOLOGY DEPARTMENT
Phone: Chairman: (605) 668-1512 Others: 668-1011

Chairman: Not Stated
Professor: Sr. V. Fasbender
Associate Professor: Sr. M. Diggins
Assistant Professor: J. Rasmussen

Degree Program: Biology

Undergraduate Degree: B.A.
Undergraduate Degree Requirements: 7 courses in Biology including Ecology, 2 courses in Chemistry, Demonstrated proficiency in college algebra and trigonometry

NORTHERN STATE COLLEGE

Aberdeen, South Dakota 57401 Phone: (605) 622-2456
Dean of Graduate Studies: H. Wollman
Academic Dean of College: L.A. Clarke
Dean of Arts and Sciences Division: R.O. Brock

DEPARTMENT OF MATHEMATICS, NATURAL SCIENCES AND HEALTH PROFESSIONS
Phone: Chairman: (605) 622-2456

Chairman: G.N. Miller
Professors: J.K. Saunders, J A. Fries
Associate Professors: W.W. Hein, M.R. Karim
Assistant Professors: E. Roberts, G. Williams, A.L.W. Woo

Degree Programs: Biology, Medical Technology

Undergraduate Degrees: B.S., B.A.
Undergraduate Degree Requirements: B.S. in Biology - General Biology, Plant Morphology, Invertebrate Zoology

Graduate Degree: M.Ed.

Graduate Degree Requirements: Master: Master in Education - with emphasis in Natural Science

Graduate and Undergraduate Courses Available for Graduate Credit: Principles of Biology, Field Botany, Field Biology, Genetics Seminar, New Concepts in the Teaching of Biology

SIOUX FALLS COLLEGE

Sioux Falls, South Dakota 57101 Phone: (605) 336-2850
Dean of College: J.L. Butler

DEPARTMENT OF BIOLOGY

Phone: Chairman: Ext. 147

Chairman: J.E. Christensen
Professors: Not Stated
Associate Professors: Not Stated
Assistant Professors: Not Stated

Degree Programs: Biology

Undergraduate Degrees: B.A., B S.
Undergraduate Degree Requirements: Completion of 30 semester hours in department including Core courses and electives plus two semesters of college Mathematics and two semesters college Chemistry (Organic Chemistry strongly recommended).

SOUTH DAKOTA STATE UNIVERSITY

Brookings, South Dakota 57006 Phone:(605) 688-4151
Dean of Graduate Studies: H.S. Bailey

College of Agriculture and Biological Sciences

Dean: D. Dearborn

DEPARTMENT OF ANIMAL SCIENCE

Phone: Chairman: (605) 688-5165

Head: J.W. McCarty (Acting)
Professors: C W. Carlson, C.A. Dinkel, L.B. Embry, L.D. Kamstra, P.H. Kohler, W.C. Morgan, R.C. Wahlstrom
Associate Professors: F.R. Gartner, J.K. Lewis, J.W. McCarty, W.C. McCone
Assistant Professors: G.H. Deutscher, D.H. Gee, E. Guenthner, P.E. Plumart, A.L. Slyter
Instructor: G.W. Libal
Research Assistants: 11
Teaching Assistants: 2

Field Stations/Laboratories: Antelope Range Field Station, Buffalo, Central Substation, Highmore, Cornbelt Agricultural Research and Extension Center, Beresford, James Valley Agricultural Research and Extension Center - Redfield, North Soiux Valley Crop and Soils Research Station, Watertown; Pasture Research Center, Norbeck; Range and Livestock Research Station, Cottonwood; West River Agricultural Research and Extension Center, Rapid City

Degree Programs: Animal Science, Range Science

Undergraduate Degree: B.S.
Undergraduate Degree Requirements: Animal Science - Communications 11, Physical Education 2, Animal or Range Science 27-30, Social Science/Humanities 17, Natural Science 15, Biological Science 17, Electives 17, Range Science, Communications 11, Physical Education 2, Animal or Range Science 12 (12 AS), Social Science/Humanities 22-26, Natural Science 39, Biological Science 17-21, Electives 17-21

Graduate Degrees: M.S., Ph.D.
Graduate Degree Requirements: Master: A minimum of 30 graduate credits beyond the B.S. degree with 20 graduate credits in residence. A minimum of 19 credits out of the 30 must be earned in the major, thesis must account for 5 to 7 of these; remaining credits in minor and supporting courses Doctoral: At least 60 of 90 required credits must be earned in the major., At least 15 credits of the 90 required must be earned in a minor or in supporting courses.

Graduate and Undergraduate Courses Available for Graduate Credit: Population Genetics, Animal Nutrition, Animal Nutrition Laboratory, Avian Nutrition, Meat Technology, Wild Lands Seminar, Research Problems, Special Topics, Ruminology, Experimental Procedure, Advanced Physiology of Reproduction, Nutritional Interrelationships, Graduate Seminar, Nutrition Seminar, Thesis, Master of Science Thesis, Ph.D., Principles of Animal Breeding (U), Feed Technology (U), Horse Production (U), Poultry Management (U), Advanced Livestock Judging (U), Livestock Reproduction (U), Beef Cattle Production (U), Sheep and Wool Production (U), Swine Production (U), Animal Science Seminar (U), Principles of Range Management (U), Range Ecosystems (U), Range Analysis (U), Range Improvement (U), Field Studies in Range Management (U), Range Management Planning (U)

BOTANY AND BIOLOGY DEPARTMENT

Phone: Chairman: (605) 688-6141 Others: 688-6141

Head: G.A. Myers
Professors: W.L. Miller (Emeritus), D.J. Holder, W. Morgan, C.A. Taylor
Associate Professors: J.R obinson, V. Fasbander, K. Gaggins,C. Brasier, A. Okonkuro (Adjunct), R.H. Whalen, C.H. Chen, E.S. Olson, N. Granholm, H.L. Hutcheson
Assistant Professors: R. Steinberg, G. Peterson, L. Haertel, K. Morril
Instructors: J. Jackson (Adjunct), J. Larsen, C.R. McMullen
Research Assistants: 2
Teaching Assistants: 5

Field Stations/Laboratories: Natural Sciences Field Station - Rapid City

Degree Programs: Biology, Botany

Undergraduate Degree: B.S.
Undergraduate Degree Requirements: 136 hours toward graduation. 2.0 GPA (2.2 for Education block)/4.0, 24 hour minimum for major, Chemistry minor, 1 year Physics, 1 year Mathematics, 20 hours Social Science, Humanities

Graduate Degrees: M.S. (Biology)
Graduate Degree Requirements: Master: Consent of thesis adviser, GPA 2.4 in graduate school, provisional, 2165 conditional, 3.0 unconditional admission

Graduate and Undergraduate Courses Available for Graduate Credit: Plant Ecology (U), Plant Anatomy (U), Plant Physiology (U), Advanced Plant Ecology, Advanced Plant Anatomy, Advanced Plant Physiology, Plant Morphology (Nonvascular), Plant Morphology (Vascular), Plant Morphogenesis, Plant Growth and Development, Advanced Plant Taxonomy, Aquatic Seed Plants, Thesis, Genetics (U), Genetics Laboratory (U), Seminar (U), Problems in Biology (U), Principles of Techniques in Electron Microscopy, Biology of the Algae, Introduction to Oceanography, Strategies in Science Teaching

DAIRY SCIENCE DEPARTMENT

Phone: Chairman: (605) 688-4116

Head: J.H. Martin
Professors: J.H. Martin, R.J. Baker, K.R. Spurgeon, H.H. Voelker
Associate Professors: E. Bartle (Emeritus), J.G. Parsons

D.J. Schingoethe, S.W. Seas
Assistant Professors: E.U. Kurtz, M. Owens
Research Assistants: 9

Degree Programs: Dairy Science

Undergraduate Degree: B.S.
Undergraduate Degree Requirements: 136 semester hours

Graduate Degree: M.S.
Graduate Degree Requirements: Master: 30 graduate credits beyond B.S. degree

Graduate and Undergraduate Courses Available for Graduate Credit: Dairy Plant Management, Technological Control Dairy Products II, Dairy Seminar, Dairy Farm Management, Dairy Breeds and Breeding, Physiology of Lactation, Advanced Dairy Microbiology, Advanced Laboratory Techniques in Dairy Science, Ruminology

ENTOMOLOGY-ZOOLOGY DEPARTMENT
Phone: Chairman: (605) 688-6176

Head: R.J. Walstrom
Professors: W.L. Berndt (Extension), N.A. Hartwig, E.J. Hugghins, B.H. Kantack (Extension), V.M. Kirk (Research), B. McDaniel, A. Greichus, M.H. Roller, W.N. Stoner (Research), R.N. Swanson
Associate Professors: E.U. Balsbaugh, Jr., R.W. Kieckhefer (Research), J.L. Krysan (Research), G.R. Sutter (Research), D.D. Walgenbach (Research), J.D. Haertel
Assistant Professors: G.A. Thibodeau, T. Branson (Research), R.D. Gustin (Research)
Assistants in Zoology: 2

Field Stations/Laboratories: Southeast Experiment Farm - Centerville, Black Hills Biological Field Station

Degree Programs: Entomology, Zoology

Undergraduate Degree: B.S.
Undergraduate Degree Requirements: A total of 136 semester credits completed with a grade point of at least 2.72 Specific courses required in each curriculum.

Graduate Degree: M.S.
Graduate Degree Requirements: Master: Thirty semester credits of graduate level courses to include 5 to 7 thesis credits

Graduate and Undergraduate Courses Available for Graduate Credit: Insects, Insect Control Methods, Insect Ecology, Insectary Methods, Insect Anatomy, Insect Physiology, Insect Toxicology, Special Topics in Entomology, Taxonomy of Insect Groups, Graduate Seminar in Entomology, Thesis in Entomology, Mammalian Physiology, Mammalogy, Invertebrate Zoology, Vertebrate Zoology, Embryology, Vertebrate Histology, Histological Techniques, Comparative Vertebrate Anatomy, General Parasitology, Mammalian Anatomy, Advanced Systemic Physiology, Endocrinology, Special Topics in Zoology

DEPARTMENT OF HORTICULTURE-FORESTRY
Phone: Chairman: (605) 688-5136 Others: 688-5136

Head: R.M. Peterson
Professor: P.E. Collins
Associate Professors: L.C. Johnson, P. Prashar
Assistant Professors: P.E. Nordstrom, J.E. Klett
Instructor: J.R. Waples
Research Assistants: 3

Degree Programs: Horticulture, Landscape Design, Park Management

Undergraduate Degree: B.S.
Undergraduate Degree Requirements: 136 credits, Horticulture and Park Management require 2 summers of work experience in addition to the 136 credits

MICROBIOLOGY DEPARTMENT
Phone: Chairman: (605) 688-4116 Others: same

Head: R.M. Pengra
Professors: R.J. Baker, P.R. Middaugh, G.C. Parikh, G. Semeniuk
Associate Professors: D E. Reed, T.R. Wilkinson
Assistant Professors: R.P. Ellis, C.A. Kirkbride, J.P. McAdaragh, C.A. Westby
Instructors: R.S. Shave, R.J. Stangeland
Research Assistants: 2

Field Stations/Laboratories: Experiment Station farms and ranches

Degree Programs: Microbiology

Undergraduate Degree: B.S.
Undergraduate Degree Requirements: 136 semester credits total, 24 semester credits in Microbiology

Graduate Degree: M.S.
Graduate Degree Requirements: Master: 30 semester hours total, 19 of which in Microbiology balance in minor or supporting fields

Graduate and Undergraduate Courses Available for Graduate Credit: Advanced Dairy Microbiology, Virology, Molecular and Microbial Genetics, Systematic Bacteriology, Microbiology Problem, Graduate Seminar, Industrial Microbiology, Microbial Metabolism

DEPARTMENT OF PLANT SCIENCE
Phone: Chairman: (605) 688-5121

Head: C.R. Krueger
Professors: B.L. Brage (Adjunct), P.L. Carson, C.D. Dybing (Adjunct), L.O. Fine, W.S. Gardner, M.L. Horton (Adjunct), D.G. Kenefick, R.C. Kinch, C.J. Mankin, C.M. Nagel, L.H. Penny (Adjunct), P.B. Price (Adjunct), J.G. Ross, M.D. Rumbaugh, G. Semeniuk, D.B. Shank, F.E. Shubeck, D.G. Wells, F.C. Westin, E.M White
Associate Professors: W.E. Arnold, G. Buchenau, J.D. Colburn, E.D. Gerloff (Adjunct), S.G. Jensen (Adjunct), R.A. Kohn, A.O. Lunden, R.W. Pylman, Jr.
Assistant Professors: J.J. Bonnemann, G.W. Erion, P.D. Evenson, J.T. Green, J.R. Johnson, Q.S. Kingsley, C.L. Lay, D.D. Malo, J.D. Otta, D.L. Reeves, J.R. Thysell (Adjunct)
Instructor: H.A. Geise
Teaching Assistantships: 2
Research Assistants: 10
Assistants in: 9
Research Associates: 3

Field Stations/Laboratories: Pasture Research Center - Norbeck, Central Crops and Soils Research Station - Highmore, West River Agricultural Research and Extension Center - Rapid City, James Valley Agricultural Research and Extension Center - Redfield, Corn Belt Agricultureal Research and Extension Center - Centerville, North Sioux Valley Crops and Soils Research Station - Watertown; additional small acreages leased for special purposes.

Degree Programs: Agronomy, Crop Science, Plant Pathology, Soil Science

Undergraduate Degree: B.S.
Undergraduate Degree Requirements: 26 credits required

Graduate Degrees: M.S., Ph.D. (Agronomy)
Graduate Degree Requirements: Master: 30 graduate credits beyond bachelors. Minimum resident requirement is 20 graduate credits. Minimum of 1 semester or 2 summer sessions of full-time graduate

work must be spent on campus. A minimum of 19 credits must be earned in the major. Thesis must account for 5-7 of these. Doctoral: 3 academic years of full-time work beyond the bachelor's degree. Minimum resident requirement is 50 credits. At least 60 of the 90 credits for the degree must be earned in the major. Research tools include foreign language, statistics, computer programming. Credits earned in obtaining these not included in degree program. Thesis should represent 1 academic year of full-time research.

Graduate and Undergraduate Courses Available for Graduate Credit: Seed and Grain Technology (U), Soil Geography and Land Use Interpretation (U), Grain and Seed Production and Processing (U), Forage Crops and Pasture Management (U), Soil Fertility and Fertilizers (U), Principles of Plant Pathology II (U), Physical Environment of Soils and Plants (U), Conservation and Management of Soils (U), Soil Chemistry (U), Plant Breeding (U), Mycology (U), Irrigation-Crop and Soil Practices (U), Special Problems (U), Undergraduate Seminar (U), Advanced Plant Pathology I, Advanced Plant Pathology II, Physical Properties of Soils, Chemical Properties of Soils, Environmental and Physiological Aspects of Crop Production, Advanced Genetics and Cytogenetics, Special Topics, Advanced Special Problems

VETERINARY SCIENCE DEPARTMENT
Phone: Chairman: (605) 688-5171

Professors: G.S. Harshfield, J Taylor, M.E. Bergeland
Associate Professors: M.W. Vorhies, C.A. Kirkbride, D. E. Reed, G.R. Ruth, H. Black, J. McAdaragh
Assistant Professors: C. Daley, R.P. Ellis
Instructors: W. Knudtson, R. Pierce, H. Shave, K. Wohlgemuth

Degree Programs: (The department assists with graduate training programs of students in the department of microbiology, wildlife, botany-biology, entomology and zoology and animal science.)

DEPARTMENT OF WILDLIFE AND FISHERIES SCIENCES
Phone: Chairman: (605) 688-6121 Others: same

Professors: R.L. Linder (Research), P.A. Vohs, Jr.
Associate Professor: D.C. Hales (Research)
Assistant Professors: L.D Flake, C.G. Scalet
Instructors: D.C. Hamm, R.L. Applegate
Research Assistants: 16

Degree Programs: Fisheries Science, Wildlife Science

Undergraduate Degree: B.S.
Undergraduate Degree Requirements: Not Stated

Graduate Degree: M.S.
Graduate Degree Requirements: Not Stated
Doctoral: (Program in cooperation with other departments)

Graduate and Undergraduate Courses Available for Graduate Credit: Limnology (U), Fisheries Science (U), Upland Game Management (U), Big Game Management (U), Waterfowl Management (U), Wildlife Research Problems (U), Aquatic Ecology, Animal Population Dynamics

UNIVERSITY OF SOUTH DAKOTA

Vermilion, South Dakota 57069 Phone: (605) 677-5011
Dean of Graduate Studies: H. Cobb

College of Arts and Sciences

Dean: D. Habbe

BIOLOGY DEPARTMENT
Phone: Chairman: (605) 677-5211 Others: 677-5210

Chairman: W.H. Sill, Jr.
Professors: R.D. Dillon, D.G. Dunlap, B.E. Harrell, G.R. Hoffman, J C. Schmulbach, T. Van Bruggen, E.P. Churchill (Emeritus), W.H. Sill, Jr.
Associate Professors: F.A. Einhellig, M. Goldman, J.F. Heisinger, A.D. Johnson, M.H. Peaslee, H.L. Smith
Instructors: P. World (Laboratory Coordinator)
Research Associates: D. Martin, N. Benson
Research Assistants: 2

Field Stations/Laboratories: Black Hills Natural Sciences Field Station- Rapid City

Degree Program: Biology

Undergraduate Degree: B.A., B.S.
Undergraduate Degree Requirements: Not Stated

Graduate Degrees: M.A., M.N.S., Ph.D.
Graduate Degree Requirements: Not Stated

Graduate and Undergraduate Courses Available for Graduate Credit: Not Stated

School of Medicine

Dean of Graduate Studies: N.W. Heimstra
Dean of School of Medicine: K.H. Wegner

DEPARTMENT OF ANATOMY
Phone: Chairman: (605) 677-5321

Chairman: G.C. Rinker (Acting)
Professors: G.C. Rinker, E.B. Scott
Associate Professor: J.C. Moore
Assistant Professors: J.A. Burbach, F. Chan, T.B. Cole Jr.
Visiting Lecturers: 6
Teaching Assistants: 2

Degree Program: Anatomy

Graduate Degrees: M.M.S., Ph.D.
Graduate Degree Requirements: Master: Master of Medical Science (open to M.D. only) Doctoral: Ph.D

Graduate and Undergraduate Courses Available for Graduate Credit: Microscopic Technique, Human Gross Anatomy Regional Human Gross Anatomy, Advanced Gross Human Anatomy, Microscopic Anatomy, Histochemistry, Cytology, Autoradiography, Neuroanatomy, Introduction to Electromyography, Special Problems

DEPARTMENT OF BIOCHEMISTRY
Phone: Chairman: (605) 677-5237

Chairman: O.W. Neuhaus
Professors: O.W. Neuhaus, R.J. Peanasky
Associate Professors: F.D. Marshall, Jr., G.D. Small
Assistant Professors: F.O. Brady, J.A. Thomas
Research Assistants: 8
Postdoctoral Fellow: 1

Degree Program: Biochemistry

Graduate Degrees: M.A., Ph.D.
Graduate Degree Requirements: Master: 30-32 credit hours minimum (Biochemistry 18, Ancillary 6-8, Thesis 4-6). Doctoral: 84 semester hours (Biochemistry 34, Minor, 20, Thesis, 30)

Graduate and Undergraduate Courses Available for Graduate Credit: Biological Chemistry, Biophysics, Enzymes, Neurochemistry, Vitamins and Hormones, Nucleic Acids, Radioisotope Techniques, Carbohydrates, Lipids, Proteins and Amino Acids, Biochemistry Seminar

DEPARTMENT OF MICROBIOLOGY
Phone: Chairman: (605) 677-5253

Chairman: P.F. Smith
Professors: J.N. Adams, R.J. Lynn, P.F. Smith

Associate Professor: C.R. Gaush
Assistant Professors: T.A. Langworthy, K.J. Carson-Mayberry, W.R. Mayberry
Teaching Fellows: 2
Research Assistants: 3

Degree Program: Microbiology

Graduate Degrees: M.A., Ph.D.
Graduate Degree Requirements: Master: 30 semester hours graduate credits, cumulative GPA of 3.0 out of maximum of 4.0, research thesis, comprehensive examination. Doctoral: 84 semester hours of graduate credit, cumulative grade point average of 3.0 out of maximum of 4.0, two foreign languages, preliminary examination, research thesis, final oral examination

Graduate and Undergraduate Courses Available for Graduate Credit: Fundamental Microbiology (U), Infectious Diseases (U), Microbiological Techniques (U), Radioisotopes, Mycology, Parasitology, Protozoology, Immunology, Virology, Cell Culture Techniques, Microbial Genetics, Microbial Physiology, Seminar

DEPARTMENT OF PHYSIOLOGY AND PHARMACOLOGY
Phone: Chairman: (605) 677-5479 Others: 677-5470

Chairman: W.O. Read
Professors: W.O. Read, J D. Welty
Associate Professors: W.L. Jones, J.W. Kakolewski
Assistant Professors: K.L. Bailey, J.L. Johnson, M.J. McBroom, R.L. Reinke, M C. Rost
Instructors: R.M. Henderson, M. Linkenheil, R.L. Schreiber
Visiting Lecturers: 10
Teaching Fellows: 2
Research Assistants: 2

Degree Program: Physiology

Undergraduate Degree: B.S.
Undergraduate Degree Requirements: 60 hours in a school of nursing and R.N certification. 64 hours in anesthesia program.

Graduate Degrees: M S., Ph.D.
Graduate Degree Requirements: Master: Plan B - 32 didactic hours of which 8 hours must be outside the department. Plan A - 24 didactic hours of which 6 hours outside department and 6 hours of thesis research. Doctoral: 54 hours didactic of which 34 hours in physiology and 30 hours of dissertation research

Graduate and Undergraduate Courses Available for Graduate Credit: None

UNIVERSITY OF SOUTH DAKOTA AT SPRINGFIELD

(No reply received)

YANKTON COLLEGE

12th and Douglas Streets Phone: (605) 665-3661
Yankton, South Dakota 57078
Dean of College: A.J. Catana

DEPARTMENT OF BIOLOGY
Phone: Chairman: Ext. 110

Chairman: L.M. Duffey
Professors: L.M. Duffey, A.J Catana, Jr.

Degree Program: Biology

Undergraduate Degree: B.A.
Undergraduate Degree Requirements: B.A. Degree in Biology: About 69 hours in the Division of Natural Science including 40 hours of Biology, 8 hours of Chemistry, 11 hours of Mathematics and 10 hours of of Physics, Also 8 hours in each of the following Divisions - Fine Arts, Humanities, and Social Science 124 hours Total required.

TENNESSEE

AUSTIN PEAY STATE UNIVERSITY

Clarksville, Tennessee 37040 Phone: (615) 648-7781
Dean of Graduate Studies: W. Stamper
Dean of Arts and Sciences: H. Stallworth

DEPARTMENT OF BIOLOGY
Phone: Chairman: (615) 648-7781

Chairman: H.C. Phillips
Professors: C.N. Boekms, W.H. Ellis, F.M. Ford, M. M. Provie, B.P. Stone
Associate Professors: F.L. Brown, E.W. Chester, D.H. Snyder
Assistant Professors: L.C. Bousman, B.D. Cole, D.L. Findley, D.I. Friendley

Degree Programs: Biology, Environmental Science

Undergraduate Degrees: B.S.
Undergraduate Degree Requirements: 44 quarter major, 12 Quarter hours Chemistry

Graduate Degrees: M.S.
Graduate Degree Requirements: Master: 48 quarter hour including a research paper, 45 quarter hours including a thesis

Graduate and Undergraduate Courses Available for Graduate Credit: Plant Anatomy (U), Histology (U), Field Botany (U), Plant Ecology (U), Animal Ecology (U), Entomology (U), Animal Physiology (U), Plant Physiology (U), Genetics (U), Experimental Plant Morphology (U), Biological Photography (U), General Embryology (U), Comparative Anatomy (U), Bacteriology (U), Field Zoology (U), Wildlife Management (U), Plant Taxonomy, Paleobotany, Ornithology, Herpetology, Cytogenetics, Parasitology, Phycology, Water Bacteriology and Pollution Control, Laboratory Techniques in Biological Research, Advanced Topics in Plant Physiology, Mammalogy, Higher Cryptogams, Plant Geography, Advanced Invertebrate Zoology, Aquatic Biology, Thesis

BELMONT COLLEGE

Nashville, Tennessee 37203 Phone: (615) 383-7001
Dean of College: Not Stated

DEPARTMENT OF BIOLOGY
Phone: Chairman: Ext. 293

Chairman: D.R. Ramage
Professors: D.R. Ramage
Associate Professors: D.R. Hill, R. Barrett

Field Stations/Laboratories: Tech Aqua - Cookville

Degree Programs: Biology, Botany, Medical Technology, Zoology

Undergraduate Degree: B.S.
Undergraduate Degree Requirements: 128 total hours 30 Biology

BETHEL COLLEGE

McKenzie, Tennessee 38201 Phone: (901) 352-5761
Dean of College: R. Burroughs

DEPARTMENT OF BIOLOGY - DIVISION OF SCIENCE AND MATHEMATICS
Phone: Chairman: (901) 352-5761

Chairman: L. Black
Professor: L. Black
Assistant Professor: D. Smith

Degree Program: Biology

Undergraduate Degrees: B.S., B.A.
Undergraduate Degree Requirements: 12 quarter hours Zoology, 12 quarter hours Botany, 3 hours Genetics, 2-6 hours Independent Study, 16 hours Junior-Senior courses, 12 hours Chemistry, 9 hours Mathematics

BRYAN COLLEGE

Dayton, Tennessee 37321 Phone: (615) 775-2041
Dean of College: J. Bartlett

DEPARTMENT OF BIOLOGY
Phone: (615) 775-2041

Head: R.B. Paisley
Professors: W.L. Henning
Associate Professor: R.B. Paisley

Degree Programs: Biology, Natural Science

Undergraduate Degree: B.A., B.S
Undergraduate Degree Requirements: 29 semester hours Biology, 8-12 semester hours Chemistry

CARSON-NEWMAN COLLEGE

Jefferson City, Tennessee 37760 Phone: (615) 475-9061
Dean of College: W.R. Guyton

BIOLOGY DEPARTMENT
Phone: Chairman: Ext. 253 Others: 255, 254

Departmental Coordinator: J.A. Chapman
Professors: J.A. Chapman, G.W. Naylor, B.L. Sloan
Associate Professors: W.J. Cloyd, H. Dickenson

Field Stations/Laboratories: Field Biology Teaching and Research Center - Norris Lake, LaFollette

Degree Program: Biology

Undergraduate Degrees: B.S., B.A.
Undergraduate Degree Requirements: 27 hours major, 9 semester hour requirement in English, 12 in foreign language, 6 in social studies, 6 in religion, 3 in fine arts. Comprehensive examination in Biology required of all graduates.

CHRISTIAN BROTHERS COLLEGE

650 E. Parkway So. Phone: (901) 278-0100
Memphis, Tennessee 38104
Dean of College: E J Doody

DEPARTMENT OF BIOLOGY
Phone: Chairman: Ext. 240

Chairman: R.J. Staub
Professor: R.J. Staub
Associate Professor: D. Dunn
Assistant Professor: J.K. Tuech

Degree Programs: Biology, Natural Science, Medical Technology

Undergraduate Degree: B.S.
Undergraduate Degree Requirements: Biology: 138 semester hours including 36 in Biology, 20 in Chemistry, 8 in Physics, 8 in Calculus and the General Education Requirements (GER), Natural Science: 138 semester hours including 40 hours in 3 fields of Science plus GER, Medical Technology: 138 semester hours including 20 hours Biology, 16 Chemistry, 3 Mathematics, plus 12 months in approved hospital and GER, Medical Records: 138 semester hours including 19 hours Biology, 24 hours in Business, plus 9 months in approved hospital and GER

COVENANT COLLEGE

Lookout Mountain, Tennessee 37350 Phone: (404) 831-6531
Dean of College: N.P. Barker

BIOLOGY DEPARTMENT
Phone: Chairman: Ext. 251

Chairman: J.E. Lothers, Jr.
Associate Professor: J.E. Lothers, Jr.
Assistant Professor: J. Wenger

Degree Program: Biology

Undergraduate Degree: B.S.
Undergraduate Degree Requirements: 23 semester hours of biology beyond general Biology including Ecology Genetics, General Physics, Chemistry through Organic Calculus strongly recommended

DAVID LIPSCOMB COLLEGE

Nashville, Tennessee 37203 Phone: (615) 269-5661
Dean of College: M.W. Craig

DEPARTMENT OF BIOLOGY
Phone: Chairman: Ext. 314

Chairman: H.O. Yates
Professors: R.C. Artist, J.E. Breeden, W.C. Owens
Assistant Professors: J.T. Arnett, D.F. Haslam
Teaching and Laboratory Assistant: 1

Degree Programs: Biology, Biological Education

Undergraduate Degrees: B.A., B.S.
Undergraduate Degree Requirements: (in quarter hours) Biology Major: 40 hours Biology, 27 Chemistry, 10 Mathematics, 10-15 hours science electives, Teaching Major in Biology: 45 hours Biology, 15 Chemistry, 38 hours professional education. B.A. degree requires 15 hours foreign language

EAST TENNESSEE STATE UNIVERSITY

Johnson City, Tennessee 37601 Phone: (615) 929-4112
Dean of Graduate Studies: E. McMahan
Dean of College: G.N. Dove

DEPARTMENT OF BIOLOGY
Phone: Chairman: (615) 929-4329 Others: 929-4112

Chairman: T.P. Copeland
Professors: T.P. Copeland, J.D. Moore
Associate Professors: W.A. Tarpley, J.E. Lawson, T.M. Johnson
Assistant Professors: F.J. Alsop, D.B. Benner, G.J. Gonsoulin, R.D. Ikenberry, J.W. Nagel, D.N. Pau, J.C. Warden, R.E. Widdows
Instructors: G. Hicks, M. Davis
Teaching Fellows: 18
Research Assistants: 4

Degree Program: Biology

Undergraduate Degrees: B.S., B.A.
Undergraduate Degree Requirements: 45 quarter hours in Biology

Graduate Degree: M.S.
Graduate Degree Requirements: 45 quarter hours in Biology

FISK UNIVERSITY

Nashville, Tennessee 37203 Phone: (615) 329-9111
Dean of Graduate Studies: I.W. Elliott
Dean of College: E.O. Woolfolk

DEPARTMENT OF BIOLOGY
Phone: Chairman: Ext. 237

Chairman: G. Hull, Jr.
Professors: G. Hull, Jr., M.E. Williams
Associate Professors: M.E. McKelvey, S.R. Whitmon
Assistant Professors: J.C. Sauer, D.L. Sauer
Instructor: W. Butler

Degree Program: Biology

Undergraduate Degree: B.A.
Undergraduate Degree Requirements: A minimum of 32 semester hours in Biology courses including: General Zoology, General Botany, Cell Biology, Senior Seminars, and eight semester hours selected from a core curriculum.

GEORGE PEABODY COLLEGE FOR TEACHERS

Nashville, Tennessee 37203 Phone: (615) 327-8227
Dean of Graduate Studies: C.B. Hunt

DIVISION OF NATURAL SCIENCES
Phone: Chairman: (615) 327-8227 Others: same

Chairman: G. Tomlinson
Professor: G. Tomlinson
Associate Professors: P. Ward, J. Holloway, M. Warren
Assistant Professors: J. Murrell, R. Hoyle, C. Hofwolt, D. Albaugh
Instructors: R. Shuffett, T. Richardson
Teaching Fellows: 5
Research Assistants: 2

Field Stations/Laboratories: Hill Camp

Degree Programs: Biology, Natural Science

Undergraduate Degree: B.S.
Undergraduate Degree Requirements: 124 semester hours

Graduate Degree: M.S.
Graduate Degree Requirements: Master: Thesis or Research

Graduate and Undergraduate Courses Available for Graduate Credit: Normal complement of Biology, Chemistry, Physics, and Natural Science courses plus courses available at Vanderbilt University whose campus joins Peabody's campus.

KING COLLEGE

Bristol, Tennessee 37620 Phone: (615) 968-1187
Dean of College: W. Wade

DEPARTMENT OF BIOLOGY
Phone: Chairman: Ext. 58

Chairman: D.L. Mac Fadden
Professor: D.L. MacFadden
Associate Professor: C.A. Owens
Research Assistants: 3

Field Stations/Laboratories: Biological Field Station on Norris Lake in LaFolette

Degree Program: Biology

Undergraduate Degree: B.S.
Undergraduate Degree Requirements: Biology 28 semester hours, Chemistry 16 semester hours (General and Organic), Mathematics 8 semester hours (Calculus), Physics 8 semester hours (General), English 12 semester hours (6 semester hours Literature), French or German 12 semester hours

KNOXVILLE COLLEGE
Knoxville, Tennessee 37921 Phone: (615) 546-0751
Dean of College: R. Harvey

BIOLOGY DEPARTMENT
Phone: Chairman: Ext. 271 Others: (615) 546-0751

Chairman: M.M. Brown
Professor: W. McArthur
Instructor: D. Bonds
Visiting Lecturers: 2

Field Stations/Laboratories: A member of MACCI Biological Field Station

Degree Programs: Biology, Food Science, Medical Technology

Undergraduate Degree: B.S.
Undergraduate Degree Requirements: 151 credit hours including 10 major courses and 10 related area courses

LAMBUTH COLLEGE
Jackson, Tennessee 38301 Phone: (901) 427-6743
Dean of College: W. Whybrew

DEPARTMENT OF BIOLOGY
Phone: Chairman: Ext. 43

Head: R.A. Carlton
Professors: R.A. Carlton, L.P. Lord
Assistant Professors: J. Booth, W.J. Davis

Degree Program: Biology

Undergraduate Degrees: B.S., B.A.
Undergraduate Degree Requirements: 40 semester hours in Biology

LANE COLLEGE
Jackson, Tennessee 38301 Phone: (901) 424-4600
Dean of College: C.O. Epps

DEPARTMENT OF BIOLOGY
Phone: Chairman: (901) 424-4600

Chairman: S.C. Mahagan
Professor: S.C. Mahagan
Associate Professor: S.E. Braxter
Assistant Professor: J.T. Martin

Degree Programs: Biology

Undergraduate Degree: B.S.
Undergraduate Degree Requirements: Not Stated

LEE COLLEGE
Cleveland, Tennessee 37311 Phone: (615) 472-2111
Dean of College: R.H. Gause

DEPARTMENT OF NATURAL SCIENCES
Phone: Chairman: Ext. 213

Chairman: M. Fleming
Professors: M. Fleming
Associate Professors: C. Dennison, R. O'Bannon, M.P. Riggs

Degree Programs: Biology, Medical Technology

Undergraduate Degrees: B.S.
Undergraduate Degree Requirements: Not Stated

LEMOYNE-OWEN COLLEGE
807 Walker Avenue Phone: (901) 948-6626
Memphis, Tennessee 38126
Dean of College: V.F. Sheppard

DEPARTMENT OF BIOLOGY
Phone: Chairman: Ext. 36

Chairman: M.K. Mohanty
Professor: N.K. Lachhmanan
Assistant Professor: M. Rice
Part-time: M. Richmond

Degree Program: Biology

Undergraduate Degree: B.S.
Undergraduate Degree Requirements: 36 hours in Biology total hours, 120 hours (semester hours)

LINCOLN MEMORIAL UNIVERSITY
(No information available)

MARYVILLE COLLEGE
Maryville, Tennessee 37801 Phone: (615) 983-6412
Dean of College: C. Blair

DEPARTMENT OF BIOLOGY
Phone: Chairman: (615) 982-5181

Chairman: A.R. Shields
Professor: A.R. Shields
Associate Professor: R.C. Ramger
Instructor: T. Taylor

Field Stations/Laboratories: Macci Field Biology Teaching and Research Center

Degree Programs: Biology, Medical Technology

Undergraduate Degree: A.B.
Undergraduate Degree Requirements: Variable with needs of student

MEMPHIS STATE UNIVERSITY
Memphis, Tennessee 38152 Phone: (901) 321-1101
Dean of Graduate Studies: G. Peterson
Dean of College: W.R. Smith

DEPARTMENT OF BIOLOGY
Phone: Chairman: (901) 321-1955 Others: 321-0111

Chairman: C.D. Brown
Professors: C.D. Brown, E.T. Browne, V.E. Feisal, C.G. Hollis, G.L. Howell, R.W. McGowan, G.E. Peterson, O.E. Smith, H. Vogel (Adjunct)
Associate Professors: H.R. Bancroft, C.J. Biggers, S.F. Boyd, K.B. Davis, H. Edwards (Adjunct), D.B. Folden, M.J. Harvey, J.S. Layne, N.A. Miller, G. W. Parchman, J.F. Payne, P.S. Rushton, B.A. Simco, P.R. Simonton, W.E. Wilhelm, B. Wise
Assistant Professors: H.D. Black, H.D. Brown, W.D. Forrest, M. Kennedy, P.M. Ridgway, H. Nishimura (Adjunct)
Instructors: C.A. Balzen, M.P. Brown
Research Assistants: 6
Graduate Teaching Assistants: 32

Field Stations/Laboratories: Meeman Biological Field Station

Degree Programs: Biology, Botany, Invertebrate Zoology, Microbiology, Vertebrate Zoology

Undergraduate Degree: B.S. (Biology
Undergraduate Degree Requirements: 36 semester hours in Biology, including General Botany, General Zoology, Principles of Biology and Genetics, Organic Chemistry

Graduate Degrees: M.S., Ph.D.
Graduate Degree Requirements: Master: Courses as required by committee, including 12 credits of seminar. One foreign language, written examination each semester. Thesis, Final Oral Doctoral: 3 academic years beyond B.S. 30 semester hours in residence beyond M.S. Two foreign languages. Preliminary examination, Thesis, Final examination

Graduate and Undergraduate Courses Available for Graduate Credit: Cellular Physiology, Limnology, Radiation Biology, Organic Development, Histological Techniques, Marine Botany, Mycology, Phycology, Plant Physiology, Plant Taxonomy, Ecology of Forests and Arable Lands, Pathogenic Bacteriology, Immunology, Sanitary Bacteriology, Microbiology of Foods, Marine Vertebrate Zoology and Ichthyology, Ethology, Vertebrate Histology, General Endocrinology, Field Zoology, Ichthyology, Comparative Neurology, Mammalogy, Herpetology, Marine Invertebrate Zoology, Invertebrate Zoology, Insect Morphology, Insect Physiology, Aquatic Entomology, Seminar in Biological Literature, Unifying Principles of Biology, Cytology, Cytogenetics, Instrumentation, Statistics in Biology, Advanced Morphology of the Flowering Plants, Advanced Plant Ecology, Advanced Mycology, Intermediary Plant Metabolism, Photosynthesis, Microbial Ecology, Virology, Bacterial Physiology, Industrial Microbiology, Seminar in Biology, Mammalian Anatomy, Mammalian Physiology, Animal Ecology, Protozoology, Advanced Parasitology, Fresh-Water Invertebrate Zoology, Advanced Insect Morphology, Advanced Insect Physiology, Advanced Systematic Entomology, Classification of Insect Larvae, Applied Entomology

MIDDLE TENNESSEE STATE UNIVERSITY

Murfreesboro, Tennessee 37130 Phone: (615) 898-2300
Dean of Graduate Studies: R.C. Aden
Dean of College: E.S. Voorhies

DEPARTMENT OF BIOLOGY
Phone: Chairman: (615) 989-2847 Others: 989-2300

Chairman: J.A. Patten
Professors: C.M. Chandler, M.C. Dunn, J.L. Fletcher, J.G. Parchment, J.A. Patten, E.S. Rucker, C.W. Wiser
Associate Professors: K.E. Blum, T.E. Memmerly, C.R. McGhee, G.G. Murphy, H.B. Reed, R.E. Sharp, E.F. Strobel, M.R Wells
Assistant Professors: P.J. Doyle, P.M. Mathis, D.A. Pierce
Instructors: J.R. Kemp, S H Swain
Teaching Fellows: 15

Field Stations/Laboratories: Member of a consortium of Regional Universities, Tech Aq ua Biological Station - Cookeville, Tennessee

Degree Program: Biology

Undergraduate Degrees: B.S., B.A.
Undergraduate Degree Requirements: 30 semester hours in Biology, 2 minors (18 semester hours each), 1 year of Chemistry, College Algebra or equivalent

Graduate Degrees: M.S.T. and M.S.
Graduate Degree Requirements: M.S.T. - 30 semester hours. May all be in Biology of certified to teach or may have Biology Major and allied minor. If not certified, minor must be in education M.S. - 30 hours in Biology or major in Biology and minor in allied field. Modern foreign language, research and thesis

Graduate and Undergraduate Courses Available for Graduate Credit: Embryology (U), Comparative Anatomy (U), Entomology (U), Parasitology (U), History and Philosophy of Biology (U), General Physiology (U), Botany of Flowering Plants (U), Histology (U), Invertebrate Zoology (U), Microtechnique (U), Endocrinology (U), Vertebrate Zoology (U), Advanced Bacteriology and Virology (U), Principles and Concepts of Taxonomy (U), Cell Physiology (U), Ichthyology (U), Topics in Microbial Ecology (U), Ecology (U), Limnology (U), Nature Study (U), Radiation Biology (U), Advanced Botany (U), Advanced Genetics (U), Advanced Invertebrate Zoology (U), Terrestrial Ecology, Aquatic Ecology, Ornithology, Economic Biology, Principles of Microbiology, Mammalogy, Population Biology, Speciation, Protozoology, Herpetology, Comparative Animal Physiology, Introduction to Graduate Study, Biological Literature

MILLIGAN COLLEGE

Milligan College, Tennessee 37682 Phone: (615) 928-1165
Dean of College: R. Wetzel

BIOLOGY DEPARTMENT
Phone: Chairman: Ext. 42

Chairman: E.D. Leach
Associate Professors: C. Gee, E. Leach
Assistant Professor: G. Wallace
Visiting Lecturers: 2

Degree Programs: Biology, Biological Education

Undergraduate Degrees: B.A., B.S.
Undergraduate Degree Requirements: B.S. - 128 hours total, 36 hours in Biology with 20 hours specified courses, 2 years Chemistry, 1 year Physics, 1 year Mathematics, plus College requirements: B.A. - 128 hours total: 24 hours in Biology, 6 courses specified, 14 hours Chemistry, 1 year Mathematics, Foreign Language through Intermediate level, plus College requirements

SOUTHERN MISSIONARY COLLEGE

Collegedale, Tennessee 37315 Phone: (615) 394-4366
Dean of College: C. Futcher

DEPARTMENT OF BIOLOGY
Phone: Chairman: (615) 396-4366

Chairman: H.H. Kuhlman
Associate Professors: D. Houck
Assistant Professors: E Grundset

Degree Program: Biology

Undergraduate Degrees: B.A., B.S.
Undergraduate Degree Requirements: 128 hours 30 of which in Biology

SOUTHWESTERN AT MEMPHIS

Memphis, Tennessee 38112 Phone: (901) 274-1800
Dean of College: R.G. Patterson

BIOLOGY DEPARTMENT
Phone: Chairman: Ext. 301

Chairman: R.L. Amy
Professors: A.I. Smith, R.L. Amy, J.T. Darlington
Associate Professors: C.O. Warren, J.D. Witherspoon B.R. Jones
Assistant Professors: C.K. Wagner

TENNESSEE STATE UNIVERSITY
(No reply received)

TENNESSEE TECHNOLOGICAL INSTITUTE
North Dixie Avenue, Phone: (615) 528-3134
Cookerville, Tennessee 38501
Dean of Graduate Studies: M Peters

College of Arts and Sciences
Dean: R.H. Frazer

DEPARTMENT OF BIOLOGY
Phone: Chairman: (615) 528-3134

Chairman: G.E. Hunter
Professors: E.S. Dooley, G.E. Hunter, R.E. Martin, B.L. Ridley
Associate Professors: S.K. Ballal, W.H. Farley, N.A Harrison
Assistant Professors: H.T. Andrews, F.J. Bulow, C.B. Coburn, R.D Estes, C P. Goodyear, J.W. Harris, O. R. Jordan, E.L. Morgan
Research Assistant: 1

Field Stations/Laboratories: Tech Aqua, Center Hill Lake near Smithville, Tennessee

Degree Programs: Biology, Wildlife Management

Undergraduate Degree: B.S.
Undergraduate Degree Requirements: 47 hours of Biology, 12 hours of Physics, 15 hours of Chemistry

Graduate Degrees: M.S.
Graduate Degree Requirements: 45 hours in graduate courses and a thesis

Graduate and Undergraduate Courses Available for Graduate Credit: Cell Physiology (U), Bioinstrumentation and Research Technology (U), Molecular Genetics (U), Field Botany (U), Plant Anatomy (U), Plant Physiology (U), Plant Pathology (U), Invertebrate Zoology I, II (U), Ornithology (U), Fisheries Management (U), Game Management (U), Bacteriology (U), Phycology (U), Ichthyology (U), Herpetology (U), Limnology (U), Comparative Physiology, Pond Management, Fishery Science, Fish Culture, Cytogenetics, Mammalogy (U), Cytology, Virology, Technical Reports, Environmental Sciences Seminar, Water Resources Management Seminar, Graduate Problems in Biology, Plant Ecology, Principles of Taxonomy, Animal Ecology, Speciation, Wildlife Field Technology

College of Agriculture and Home Economics
Dean: S.A. Griffen

DEPARTMENT OF ANIMAL SCIENCE
Phone: Chairman: (615) 528-3155

Chairman: W.C. Hyder
Associate Professor: B.T. Parham
Assistant Professors: J.L. Bohannon, W. Easterly, R. Johnson

Degree Program: Agriculture (Animal Science)

Undergraduate Degree: B.S.
Undergraduate Degree Requirements: 198 quarters of work.

DEPARTMENT OF PLANT AND SOIL SCIENCE
Phone: Chairman: (615) 528-3373 Others: 528-3149

Chairman: W.W. Frye
Professor: W.W. Frye
Associate Professor: H.C. Funk
Assistant Professors: B.R. Fleming, C.C. Pangle

Field Stations/Laboratories: Tennessee Tech Shipley Farm

Degree Programs: Agronomy, Crop Science, Horticultural Science, Natural Resources, Soil Science

Undergraduate Degree: B.S.
Undergraduate Degree Requirements: 198 quarter hours, Plant and Soil Science - 56 quarter hours, other Agriculture 25 quarter hours, Natural Science 57 quarter hours, Social Sciences 15 quarter hours, Communications 18 quarter hours, Electives 22 quarter hours, Other 7 quarter hours

TENNESSEE TEMPLE COLLEGE
Chattanooga, Tennessee 37404 Phone: 693-6021
Dean of College: W. Porter

DEPARTMENT OF MATHEMATICS AND SCIENCE
Phone: Chairman: Ext. 261

Chairman: L.E. Eimers
Professor: J. Prikle

Degree Program: Biology (minor only)

Undergraduate Degree: B.S.
Undergraduate Degree Requirements: 20 hours Botany 4 semester hours, Zoology 4 semester hours, Ecology 3 semester hours, Microbiology 3 semester hours, Parasitology 3 semester hours, Genetics 3 semester hours

TENNESSEE WESLEYAN COLLEGE
(No reply received)

TUSCULUM COLLEGE
(No reply received)

UNION UNIVERSITY
Jackson, Tennessee 38301 Phone: (901) 422-2576
Dean of College: G.W. Brown

DEPARTMENT OF BIOLOGY
Phone: Chairman: Ext. 62 Others: Exts. 61 or 62

Chairman: H.A. Sierk
Professor: H.A. Sierk
Assistant Professor: E.Y. Smith

Degree Program: Biology

Undergraduate Degree: B.S.
Undergraduate Degree Requirements: Total of 128 semester hours of which at least 50 must be at the junior or senior level, Biology major - 1 course in each Microbiology, Zoology, Botany, Genetics, Development or Physiology, Environment, Independent Study and Seminar, Total of 30 semester hours in Biology

UNIVERSITY OF THE SOUTH
(No reply received)

UNIVERSITY OF TENNESSEE AT CHATTANOOGA
Chattanooga, Tennessee 37401 Phone: (615) 755-4011
Dean of Professional Studies and Graduate School: D. Hyder
Dean of Arts and Sciences: D. Harbaugh

BIOLOGY DEPARTMENT
Phone: Chairman: (615) 755-4341

Chairman: R.E. Garth
Professors: J.R. Freeman, R.E. Garth
Associate Professors: R.M. Durham, R.G. Litchford, C.H. Nelson, P.A. Perfetti, G.N. Vredeveld
Assistant Professors: M.E. Edwards, S.R. Kupor,

G.S. Van Horn, B.A. Walton

Degree Programs: Biology, Environmental Science

Undergraduate Degrees: B.A., B.S.
Undergraduate Degree Requirements: 8 hours Mathematics, 16 hours Chemistry, 8 hours Physics, 32 hours Biology

UNIVERSITY OF TENNESSEE AT KNOXVILLE

Knoxville, Tennessee 37916 Phone: (615) 974-2591
Vice Chancellor for Graduate Studies: H.A. Smith

College of Liberal Arts

Dean: A.H. Nielson

DEPARTMENT OF BIOCHEMISTRY
Phone: Chairman: (615) 974-5148

Head: K.J. Monty
Associate Professors: J.E. Churchich, J.G. Joshi, T. P. Salo
Assistant Professors: R.H. Feinberg, S.W. Hawkinson, B.C. Kline
Graduate Assistants: 18

Degree Program: Biochemistry

Graduate Degrees: M.S., Ph.D.
Graduate Degree Requirements: Master: Four years Chemistry including 1 year each of Organic, Analytical and Physical, Calculus to Differential Equations, at least 12 hours Advanced Biology, 25 Biochemistry, Thesis Doctoral: Five years Chemistry including 1 year each of organic, analytical, Physical and Reaction Mechanisms, Calculus to Differential Equations, 1 year Physics, 18 hours advanced Biology, 35 Biochemistry, Research Thesis, M.S. not required

Graduate and Undergraduate Courses Available for Graduate Credit: Cellular and Comparative Biochemistry, Introduction to Physical Biochemistry, Membranes, Compartments, and the Regulation of Energy Metabolism, Structures and Functions of the Nucleic Acids, Graduate Research Participation, Current Topics in Biochemistry, Enzyme Kinetics and Mechanisms of Enzyme Action, Cellular and Comparative Biochemistry Laboratory, Metabolism of Nitrogen Containing Compounds, Protein Structure and Enzyme Function, Protein Synthesis and its Role in Metabolic Regulation, Experimental Techniques, Special Topics, Functions of the Vitamins, Functions of the Trace Elements, Structure and Function of Macromolecules, Biochemical Genetics, Metabolic Regulation, Biological Energy Transformations, Antigen-Antobody Interactions, Biochemistry of Specialized Physiological Processes

DEPARTMENT OF BOTANY
Phone: Chairman: (615) 974-2258 Others: 974-2256

Head: R.W. Holton
Distinguished Service Professor: A.J. Sharp
Professors: H.R. De Selm, W.R. Herndon (Vice Chancellor for Academic Affairs), R.W. Holton, G.E. Hunt, L W. Jones, F.H. Norris, J.S. Olson,(Adjunct), R.H. Petersen, A.J. Sharp, P.L. Walne
Associate Professors: S.L. Bell, J.D. Caponetti, E.E.C. Clebsch, A.M Evans, A.S. Heilman
Assistant Professors: C C. Amundsen, M.W. Bierner, K.W. Hughes, O.J. Schwarz
Instructors: M.A. Dickey, L.E. Edwards, H.H. Shugart
Teaching Fellows: 27
Research Assistants: 2
Professional Assistants: 3

Degree Program: Botany

Undergraduate Degree: B.S
Undergraduate Degree Requirements: 4 courses Biology, 23 upper division hours ofBotany, and 4 hours of upper division courses from related Biological Science (Zoology, Microbiology, Biochemistry, Agricultural Biology, Forestry, or Plant and Soil Science). 12 hours Chemistry, 6 hours Physics

Graduate Degrees: M.S., M.A.C.T. (Biology), Ph.D.
Graduate Degree Requirements: Master: Graduate School requirements plus: Thesis problem, written formulation and oral defense, successful performance on language examination or A or B in French or German 3030, 2 credit hours 6000 level, 30 minute departmental seminar, teaching and/or ancillary service. Doctoral: Dissertation problem, written formulation and oral defense, written comprehensive preliminary examination, presentation of one or more cognate areas outside of department, 9 credit hours, B average Foreign language examination or A or B in French or German 3030, 9 credit hours at 6000 level, one-hour departmental seminar, educational service, teaching and/or ancillary services

Graduate and Undergraduate Courses Available for Graduate Credit: Plants in Evolution, Environmental Botany, Socio-Economic Impact of Plants, Genetics and Society, Biology and Human Affairs, Introductory Plant Pathology, Introductory Plant Physiology, Principles of Plant Evolution, Plant Anatomy, Paleobotany, Plant Ecology, Morphology and Evolution of the Ascomycetes Field Mycology, Bryology, Lichenology, Morphology and Evolution of Basidiomycetes, Agrostology, Advanced Morphology of Flowering Plants, Biosystematics, Advanced Plant Physiology, Radioactive Isotopes in Physiological Investigations, Developmental Plant Physiology, Quaternary Problems, Plant Geography, Special Problems in Botany, Analysis of Plant Communities, Seminar in the Teaching of College Botany, Systems Ecology, Plant Cytology, Experimental Nuclear Cytology, Experimental Cell Biology, Cytogenetics, Methods and Instrumentation in Laboratory Investigations, Field Methods in Plant Ecology, Microbes in Ecosystems, Methods in Instrumentation in Field Investigations, Developmental Plant Morphology, Phylogenetic Plant Taxonomy, Doctoral Research and Dissertation, Advanced Topics in Morphology of Vascular Plants, Advanced Topics in Cryptogamic Botany, Photobiology, Advanced Topics in Cytology and Cell Biology, Ecosystems of the World, Advanced Topics in Genetics, Seminar in the History of Botany, Radiation Ecology, Advanced Topics in Plant Physiology, Advanced Topics in Ecology, Advanced Topics in Systematic Botany

GRADUATE PROGRAM IN ECOLOGY
Phone: Chairman: (615) 974-3065

Director: J.F. McCormick
(This is an interdepartmental program in which all faculty are basically members of departments (Botany, Forestry, Zoology, etc.). Except for the Director there are no faculty staff members.)

Degree Program: Ecology

Graduate Degrees: M.S., Ph.D.
Graduate Degree Requirements: Master: 45 quarter hours of graduate credit, including certain specified courses and 9 hours of thesis. Final examination. Doctoral: Course requirements determined by the student's faculty committee. Reading knowledge of one foreign language. Dissertation. Preliminary and final examinations.

Graduate and Undergraduate Courses Available for Graduate Credit: Principles of Ecology, Ecology for Planners and Engineers, Implementation of Environmental Policy, Special Topics in Ecology. (Most courses used in this interdepartmental program are offered by departments such as Botany, Forestry, Plant and soil Science, Zoology, etc.)

DEPARTMENT OF MICROBIOLOGY
Phone: Chairman: (615) 974-3441

Head: A. Brown
Professors: R.W. Beck, A. Brown, J.H. Coggin, Jr., J.O. Mundt, J.M. Woodward
Associate Professors: T.C. Montie, W.S. Riggsby, C.J. Wust
Assistant Professor: J. Becker
Visiting Lecturers: 3

Degree Programs: Microbiology

Undergraduate Degree: B.S.
Undergraduate Degree Requirements: Organic Chemistry, Analytical Chemistry, Genetics, Cell Biology, Introductory Microbiology, Bacterial Physiology, plus 25 hours in Microbiology

Graduate Degrees: M.S., Ph.D.
Graduate Degree Requirements: Master: Statistics, meet all deficiencies relative to undergraduate program. Total 45 credit hours at least 1/2 all graduate level, 9 to 18 quarter hours, research, Acceptable Thesis with defense Doctoral: Statistics, Calculus, Physical Chemistry sequence, meet all deficiencies relative to undergraduate program. Total more than 90 hours for graduate credit, Independent research project leading to acceptable dissertation and publication, Preliminary Examination

Graduate and Undergraduate Courses Available for Graduate Credit: Pathogenic Bacteriology (U), Immunology (U), Microbial Physiology (U), Virology (U), Microbiology Genetics (U), Yeasts and Molds (U), Medical Mycology (U), Food Bacteriology (U), Taxonomy, Seminar in Microbiology, Journal Clubs, Immunochemistry, Concepts in Immunology, Microbial Physiology, The Oncogenic Viruses, The Bacterial Viruses, Advanced Topics in Microbial Physiology, Advanced Topics in Microbial Pathogenesis, Advances in Virology, Advanced Microbial Genetics

INSTITUTE OF RADIATION BIOLOGY
Phone: Chairman: (615) 974-2371

Director: J.G. Carlson
Alumni Distinguished Service Professor: J.G. Carlson
(This is an interdepartmental program that utilizes the services of instructors from a number of units, Departments of Biochemistry, Botany, Microbiology, Zoology, Chemistry, Physics, UT. Memorial Research Center of the University of Tennessee, the Biology, Environmental Sciences, and Health Physics Divisions or Oak Ridge National Laboratory; the Medical Division of Oak Ridge Associated Universities; the Comparative Animal Research Laboratory in Oak Ridge. In summary there are 33 staff members in Molecular Radiation Biology, 18 in Cellular Radiation Biology, 14 in Organismal Radiation Biology and 8 in Radiation Ecology)

Degree Program: Radiation Biology

Graduate Degrees: M.S., Ph.D.
Graduate Degree Requirements: Master: A minimum of 45 hours of quarter hours credit, including Foundations of Radiation Biology (8 hours), Biometry (3 hours), Radioactivity and its Applications or Radioactive Isotopes in Physiological Investigations (3 hours), Biochemistry (6 hours and Thesis (9 hours) Doctoral: In addition to the M.S. course requirements and those specified by the student's Faculty Committee: Atomic and Nuclear Physics (9 hours), Physical Biochemistry (9 hours), Dissertation, and reading knowledge of 1 foreign language

DEPARTMENT OF ZOOLOGY
Phone: Chairman: (615) 974-2371

Head: J.C. Daniel, Jr.
Professors: J.C. Daniel, Jr., J.G. Carlson, D.L. Bunting, II, A.C. Cole, R.C. Fraser, N. Gengozian, R.F. Grell, B. Hochman, J.C. Howell, A.W. Jones, M.M. Ketchel, J.N. Liles, C.A. Shivers, J.T. Tanner S.R. Tipton (Emeritus)
Associate Professors: R M Bagby, K D. Burnham, D. A. Etnier, K.W. Jeon, J.R. Kennedy, M.C. Whiteside, G.L. Whitson
Assistant Professor: P.B. Coulson, D.J. Fox, M.A. Handel, A.M. Jungreis, J A. MacCabe, S.E. Riechert, G.L. Vaughan
Teaching Assistants: 36

Degree Program: Zoology

Undergraduate Degree: B.A.
Undergraduate Degree Requirements: (General Genetics Cell Biology, General Ecology) 12 quarter hours Chemistry at sophomore level or above. Corequisites are Mathematics (6 quarter hours), and a 1-year sequence in Physics

Graduate Degrees: M.S., Ph.D.
Graduate Degree Requirements: Master: 45 quarter hours which may be entirely in major or may be partially in minor. Thesis. Oral examination Doctoral: Minimum registration of 36 quarter hours of course 6000 required of all doctoral candidates before dissertation will be accepted. Preliminary examination. Reading knowledge of at least 1 foreign language relevant to major field of study and course requirements as determined by candidate's faculty committee.

Graduate and Undergraduate Courses Available for Graduate Credit: Natural History of the Vertebrates, Comparative Vertebrate Embryology, Comparative Vertebrate Anatomy, Immunology, Principles of Animal Physiology, General Entomology, Invertebrate Zoology, Herpetology, Physiology of Reproduction, Histology Study of Animal Tissues, Minicourse in Zoology (selected advanced topics in Zoology), Plant Parasitic Nematodes, Bioethics, Developmental Biology, Practicum in Zoology, Mammalogy, Ichthyology, Animal Ecology, Comparative Animal Physiology, Comparative Endocrinology Herpetology, Ornithology, Animal Cytology, Microtechnique, General Genetics Laboratory, Organic Evolution, Human Genetics, General Parasitology, Medical Entomology, Protozoology, Freshwater Fishery Biology, Comparative Animal Pathology, Comparative Animal Pathology Laboratory, Limnology, Arachnology, Comparative Animal Behavior, Comparative Animal Behavior Laboratory, Insect Morphology and Taxonomy, Physiology of Exercise, Thesis, Graduate Research Participation, Special Problems, Zoological Bibliography, Fresh Water Invertebrate Zoology, Plant Parasitic Nematodes, Physiology of Hormones, Advanced Neuromuscular Physiology, Insect Physiology, Quaternary Problems, Seminar in the Teaching of Advanced Ornithology, Animal Populations, Foundations of Radiation Biology, Methods of Experimentation with Laboratory Mammals, Physiology of Development, Cellular Immunology, Advanced Genetics, General Vertebrate Neuroanatomy, Radiation Cytology, Radiation Physiology, Insect Morphology, Method of Taxonomy, Taxonomy of Immature Insects, Aquatic Insects, Insect Autecology, Geographic Distribution of Animals, Insect Synecology, Social Insects, Doctoral Research and Dissertation, Seminar in Cellular Biology, Seminar in Immunobiology, Seminar in Parasitology, Seminar in Genetics, Seminar in Ornithology, Seminar in Aquatic Biology, Seminar in Ecology, Seminar in Entomology, Seminar in Radiation Biology

College of Agriculture
Dean: O.G. Hall

DEPARTMENT OF AGRICULTURAL BIOLOGY
Phone: Chairman: (615) 974-7135

Head: C.J Southards

Professors: L. Johnson, N. Hall
Associate Professors: C. Hadden, J. Hilty, R. Mullet, C. Pless, H. Reed
Assistant Professors: R. Gerhardt, L. Wilson, H. Williams
Research Assistants: 8

Field Stations/Laboratories: West Tennessee Experiment Station - Jackson, Highland Rim Experiment Station - Springfield, Greeneville Tobacco Station - Greeneville

Degree Program: Agricultural Biology

Graduate Degree: M.S.
Graduate Degree Requirements: Master: Background in Biological or Agricultural Sciences, Meet prerequisites for courses, Complete with "B" average or better 45 quarter hours minimum.

Graduate and Undergraduate Courses Available for Graduate Credit: Microbiology of Soils, Plant Disease Control, Plant Nematology, Forest Entomology, Insects of Shade Trees and Ornamentals, Insects of Man and Animal, Lake and Farm Pond Management, Fish Biology

DEPARTMENT OF ANIMAL SCIENCE
Phone: Chairman: (615) 974-7286

Head: R.R. Johnson
Buford Ellington Distinguished Professor: G.M. Merriman
Professors: M.C. Bell, J.K. Bletner, C.C. Chamberlain, H.M. Jamison,(Extension), R.R. Johnson, G.M. Merriman, R.L. Murphree, R.R. Shrode, E.W. Swanson, R.L. Tugwell (Extension)
Associate Professors: W.R. Backus, K.M. Barth, E.R. Lidvall, F.B. Masincupp (Extension), J.B. McLaren, M.J. Montgomery (Extension), D.O. Richardson (Extension), H.V. Shirley
Assistant Professors: J.A. Corrick, H.C. Goan, (Extension), J.D. Smalling
Instructors: J. Bacon (Research), H.M. Crowder (Research), M.E. Fryer (Research), W.A. Lyke (Research)
(The following people are considered full-time members of the Animal Science Department but perform their duties at sub-stations located away from the University:
Assistant Professors: B.J. Bearden (Research), H.H. Dowlen (Research)
Instructors: G.B. Boyd (Research)

Field Stations/Laboratories: Field Stations: Tobacco Experiment Station - Greeneville, U.T.-CARL, Oak Ridge, Alcoa Farms - Alcoa, Highland Rim Experiment Station - Springfield, Dairy Experiment Station - Lewisburg, Middle Tennessee Experiment Station - Spring Hill, West Tennessee Experimental Station - Jackson, Ames Plantation - Grand Junction

Degree Programs: Animal Science

Undergraduate Degree: B.S.
Undergraduate Degree Requirements: 198 hours total, 20 hours General Agriculture, 39 hours Animal Science, 16 Agricultural Electives, 12 hours Biological Sciences, 20 hours Physical Sciences, 12 hours Mathematics

Graduate Degrees: M.S., Ph.D.
Graduate Degree Requirements: Master: Minimum 45 hours past the B.S. degree Doctoral: Minimum 90 hours beyond the Master's degree

Graduate and Undergraduate Courses Available for Graduate Credit: Anatomy and Physiology of Farm Animals (U), Physiology of Reproduction (U), Animal Nutrition (U), Feeds and Ration Formulation (U), Heredity in Animals (U), Principles of Animal Breeding (U), Animal Hygiene and Sanitation (U), Avian Diseases (U), Nutrition and Management of Laboratory Animals (U), Physiology of Lactation (U), Avian Physiology (U), Applied Reproduction in Farm Animals (U), Feeding Systems for Ruminants and Horses (U), Feeding Systems for Poultry and Swine (U), Applied Animal Breeding (U), Beef Cattle Production and Management (U), Dairy Production and Management (U), Pork Production and Management (U), Poultry Production and Management (U), Light Horse Production and Management (U), Lamb and Wool Production and Management (U), Meat Science (U), Thesis, Problems in Lieu of Thesis, Special Problems in Animal Science, Endocrine Relations in Animal Production, Advances in Mammalian Reproduction, Advanced Studies of the Secretion of Milk, Analytical Technology in Animal Nutrition, Energy in Animal Nutrition, Proteins in Animal Nutrition, Vitamins and Minerals in Animal Nutrition, Genetics of Animal Populations, Methods of Evaluation Experimental Data in Animal Science, Design and Interpretation of Experiments in Animal Science, Seminar, Doctoral Research and Dissertation, Topics in Milk Constituents - Food Technology, Topics in Dairy Microbiology-Food Telchnology, Advanced Topics in Animal Physiology, Environmental Physiology of Farm Animals, Animal Growth and Development, Advanced Topics in Animal Nutrition, Advanced Topics In Animal Breeding, Animal Breeding Research Methods and Interpretation, Advanced Topics in Animal Products - Food Technology, Seminar

DEPARTMENT OF FORESTRY
Phone: Chairman: (615) 974-7126 Others: 974-7126

Chairman: J.W. Barrett
Professors: J.W. Barrett, H.A. Core, T.R. Ripley (Adjunct), E. Thor, and F.W. Woods
Associate Professors: E.R. Buckner, R.W. Dimmick, M.R. Pelton, K.F. Schell
Assistant Professors: R.L. Hay, D.M. Ostermeier, J. C. Rennie, J.G. Warmbrod, G.R. Wells
Instructor: W.G. Minser, W.M. Moschler
Research Assistants: 14
Graduate Assistants: (Teaching) 2

Field Stations/Laboratories: Forestry stations at four locations

Degree Programs: Forestry, Wildlife Management

Undergraduate Degrees: B.S.F., B.S.
Undergraduate Degree Requirements: B.S.F. 213 quarter hours including 10-233k summer session, B.S. with major in wildlife and Fisheries Science 198 quarter hours

Graduate Degree: M.S.
Graduate Degree Requirements: Master: M.S. with major in forestry 45 quarter hours including 9 hours of thesis; M.S. with major in wildlife and fisheries science, 45 quarter hours including 9 hours of thesis

Graduate and Undergraduate Courses Available for Graduate Credit: Dendrology (U), Mensuration (U), Wood Technology (U), Forest Economics (U), Forest Product and Utilization (U), Wildlife Management (U), Silviculture (U), Conservation (U), Utilization (U), Forest Inventory (U), Forest Practice (U), Forest Ecosystems (U), Silvicultural Methods (U), Forest-Resource Management (U), Forest Policy (U), Forest Tree Improvement (U), Regional Silviculture of the United States (U), Forest Recreation (U), Game Mammals (U), Game Birds (U), Seminar (U), Thesis, Special Problems in Forestry, Seminar in Wildlife Conservation, Seminar in Forest Tree Biology, Seminar in Forest Management, Seminar in Forest Genetics, Recreation Planning for Forests, and Associates Lands, Seminar, Wildlife Diseases

DEPARTMENT OF ORNAMENTAL HORTICULTURE AND LANDSCAPE DESIGN
Phone: Chairman: (615) 974-7324

Chairman: D.B. Williams
Professor: D.B. Williams
Associate Professors: L.M. Callahan, E.T. Graham, H. van de Werken
Assistant Professors: J.S. Alexander, G.L. McDaniel, J.L. Pointer, D.F. Wagner
Research Assistant: 1

Degree Program: Agriculture (with major in Ornamental Horticulture and Landscape Design)

Undergraduate Degree: B.S.
Undergraduate Degree Requirements: 198 quarter hours including 37 in Ornamental Horticulture and Landscape Design

Graduate Degree: M.S.
Graduate Degree Requirements: Master: Minimum 45 quarter hours

Graduate and Undergraduate Courses Available for Graduate Credit: Plant Propagation (U), Floral Design (U), Greenhouse Management (U), Plant Materials (U), Landscape Design (U), Nursery Management (U), Park Design (U), Principles of Turf Management (U), Floriculture (U), Individual Problems Study (U), Special Problems in Lieu of Thesis, Special Problems in Ornamental and Landscape Design, Golf Course Design, Development and Management, Park and Public Grounds Management Systems

DEPARTMENT OF FOOD TECHNOLOGY AND SCIENCE
Phone: Chairman: (615) 974-7147

Head: J.T. Miles
Professors: J.O. Mundt, W.W. Overcast
Associate Professors: J.L. Collins, B.J. Demott
Assistant Professors: E.A. Childs, H.C. Holt, H.O. Jaynes, I.E. McCarty, C.C. Melton, S.L. Melton
Research Assistants: 8

Degree Programs: Food Technology and Science

Undergraduate Degrees: B.S.
Undergraduate Degree Requirements: 198 hours, General Agriculture 20, Food Technology and Science 36, Englsih and Communications 18, Mathematics 9, Physics 12, Chemistry 24, Social Science 18, Microbiology 9, Agricultural Mechanisms 4, Statistics 3, Nutrition 5 and Food Science-Home Economics 3 hours

Graduate Degrees: M.S., Ph.D. (in cooperation with Animal Science)
Graduate Degree Requirements: Master: 45 quarter hours, Thesis 9 hours, total 23 hours, courses numbered 5000 or above Doctoral: 108 hours beyond the B.S. degree. 36 hours Doctoral Dissertation, 36 hours courses 5000 and above, 24 hours outside major area.

Graduate and Undergraduate Courses Available for Graduate Credit: DairyProducts I, Food Composition, Food Preservation, Evaluation and Grading Dairy Products, Meat Evaluation and Grading, Meat Science, Problems in Food Technology, Food Technology and Science Seminar, Dairy Products II, Advanced Food Composition, Food Plant Sanitation, Food Quality Assurance, Food Additives, Food Packaging, Food Crop Products, Microbiology in Food Manufacturing, Fermented Foods, Meat Products Manufacturing, Physical Phenomena of Foods, Food Color, Food Enzymology, Food Flavors, Fats and Oils, Research, Food Production Development, Food Thermobiology, Advanced Food Quality Assurance, Meat Technology, Microorganisms Common in Food Products, Microbial Cultures in Foods

DEPARTMENT OF PLANT AND SOIL SCIENCE
Phone: Chairman: (615) 974-7101 Others: 974-0111

Head: L.F. Seatz
Clyde B. Austin Distinguished Professor of Agriculture: L.F. Seatz
Professors: F.F. Bell, H.A. Fribourg, L.M. Josephson, W.L. Parks, L.N. Skold, M.E. Springer, H.D. Swingle
Associate Professors: G.J. Buntley, D.L. Coffey, R.J. Lewis, J.H. Reynolds, H.C. Smith
Assistant Professors: W.D. Barber, L.S. Jeffery, W.A. Krueger, V.H. Reich

Field Stations/Laboratories: Several Branch Agricultural Experiment Stations

Degree Programs: Soils, Plant Breeding and Genetics, Crop Physiology and Ecology

Undergraduate Degree: B.S. (Plant and Soil Sciences)
Undergraduate Degree Requirements: 198 quarter hours credit, 20 quarter hours basic Agriculture, 18 quarter hours English and Communications, 9 quarter hours Mathematics, 12 quarter hours Biological Science, 16 quarter hours Physical Science, 18 quarter hours Social Sciences and Humanities, 30 quarter hours Departmental courses

Graduate Degrees: M.S., Ph.D. (Plant and Soil Science)
Graduate Degree Requirements: Master: 45 quarter hours graduate credit, 9 quarter hours thesis, At least half of credit hours in graduate courses. Doctoral: 108 quarter hours graduate credit, 36 quarter hours dissertation and research, 30 quarter hours in graduate courses, 9 quarter hours of doctoral courses

Graduate and Undergraduate Courses Available for Graduate Credit: Crop Ecology (U), Crop Physiology (U), Soil Fertility and Fertilizers (U), Grain and Oil Crops (U), Forage Crops (U), Cotton and Tabacco (U), Fruit Crops Management (U), Soil Management (U), Soils in Forestry (U), Commercial Vegetable Production (U), Interpretation of Agricultural Research (U), Soil Productivity and Management, Pedology, Design and Interpretation of Experiments, Soil Physics, Advanced Soil Fertility, Soil Physical Chemistry, Seminar (M.S.) Advanced Plant Genetics, Quantitative Genetics, Advanced Plant Breeding, Crop Climatoloty, Principles of Weed Science (U), Soil Chemistry (U), Principles of Crop Breeding (U), Soil Analysis (U), Agricultural Chemicals and the Environment (U), Soil Formation, Morphology and Classification (U), Problems in Plant and Soil Science (U), Special Problems in Lieu of Thesis, Special Problems in Plant and Soil Science, Soil Crop Relationships, Advanced Crop Physiology and Ecology, Mechanisms of Herbicide Action, Special Topics in Soil Science, Special Topics in Plant Breeding, Special Topics in Crop Physiology and Ecology, Advanced Soil Physics, Advanced Soil Physical Chemistry, Experimental Designs, Growth Control with Chemicals

UNIVERSITY OF TENNESSEE AT MARTIN
Martin, Tennessee 38238 Phone: (901) 587-7111
Dean of Graduate Studies: N. Campbell
Dean of College: M Simmons

DEPARTMENT OF BIOLOGICAL SCIENCES
Phone: Chairman: (901) 587-7002 Others: 587-7003

Chairman: T.R. James
Professor: J.M. Moore
Associate Professors: B.L. Berry, J.W. Henson, T.R. James, W.F. Nelson, C. Sharma, G.D. Taylor
Assistant Professors: J.E. Deck, W. Dillon, H. Kittilson, H.G. Morris, T.D. Pitts, W.A. Sliger, W.L. Smith, C.E. Slack
Laboratory Instructors: 5

Field Stations/Laboratories: Reelfoot Biological Station - Reelfoot Lake, Tennessee

Degree Program: Biology

Undergraduate Degrees: B.A., B.S.
Undergraduate Degree Requirements: Freshman sequence of Biology and Chemistry plus 27 hours of upper division biology courses

Graduate Degree: M.S. in Education with certification in Biology
Graduate Degree Requirements: Not Stated

UNIVERSITY OF TENNESSEE MEDICAL UNITS
Memphis, Tennessee 38103 Phone: (901) 528-5500

College of Basic Medical Sciences
Dean: R.H. Alden

DEPARTMENT OF ANATOMY
Phone: Chairman: (901) 528-5965

Chairman: G.G. Robertson
Goodman Professor of Anatomy: S.R. Bruesch
Goodman Professor of Anatomy: H.H. Wilcox
Professors: R.H. Alden, S.R. Bruesch, S.A. Cohn, A.A. Fedinec, J.F. Roger, G.G. Robertson, R.L. Summitt, H.H. Wilcox
Associate Professors: R.L. Atnip, C.E. Corliss, L.R. Fitzgerald, J.R. Holbrook, K. Hashimoto, F.J. Longo, L.R. Murrell, C. Sebelius
Assistant Professors: E.S. Craig, D.J. Donaldson, J.S. Evans, M.H. Garrett, B.E. Magun, B.J McLaughlin, C.N. Smith, C.E. White, J.L. Wilson, F.B. Wild, J. G. Wood
Instructors: M.D. Gardner, E.F. Johnson, L.E. King, G.M Reed
Teaching Fellows: 3
Predoctoral Trainees: 6
Postdoctoral Trainees: 4

Degree Program: Anatomy

Graduate Degrees: M.S., Ph.D.
Graduate Degree Requirements: Master: Average of 3.0 or higher in coursework, minimum credit of 45 quarter hours, minimum residence of 3 quarters, faculty committee thesis and defense Doctoral: Average of 3.0 or higher in coursework, minimum of 9 quarter sof graduate study, reading knowledge of one foreign language, pertinent to field of research, faculty committee, comprehensive preliminary examination; dissertation on original research

Graduate and Undergraduate Courses Available for Graduate Credit: Gross Anatomy, Microscopic Anatomy, Developmental Anatomy, Neuroanatomy, Medical Genetics, History of Anatomy, Histochemistry and Cytochemistry, Cytology, Cell and Organ Culture, Cellular Aspects of Developmental Biology

DEPARTMENT OF MICROBIOLOGY
Phone: Chairman: (901) 528-6175 Others: 528-6177

Chairman: B.A. Freeman
Professors: B.A. Freeman, A.N, Roberts, W M. Todd
Associate Professors: N. Incardona, K.L. Smiley, Jr.
Assistant Professors: P.H. Dorsett, J.F. Kane, W.R. Phillips, C.H. Yang
Instructors: D.M. Kenney, T. Wainscott
Visiting Lecturers: 11
Teaching Fellows: 7

Degree Program: Microbiology

Graduate Degrees: M.S., Ph.D.
Graduate Degree Requirements: Master: 45 quarter hours credit, thesis Doctoral: credit requirements variable, thesis, one foreign language

Graduate and Undergraduate Courses Available for Graduate Credit: Microbiology, Elements of Microbiology, Master's Thesis and Research, Applied Microbiology, Microbiology Seminar, Special Topics in Microbiology Principles of Laboratory Instruction, Microbiology Research, Doctoral Dissertation and Research, Immunology and Immunochemistry, Virology, Microbial Physiology and Metabolism, Advanced Pathology Bacteriology and Advanced Microbial Genetics and Regulation

DEPARTMENT OF MOLECUIAR BIOLOGY
Phone: Chairman: (901) 528-6054 Others: 528-6055

Chairman: J. Autian
Professors: J. Autian, K.J. Goldner, W.H. Lawrence, W.P. Purcell
Associate Professors: G.E. Bass, Jr., E.O. Dillingham, J.W. Lawson, L.J. Powers, N P. Rivers, G.C. Wood
Assistant Professors: S.J. Jackson, L.J. Nunez, P.L. Whyatt, N. Wojciechowski, G.W. Hung, N.E. Webb (Research)
Visiting Lecturers: 20
Research Assistants: 6

Degree Programs: Molecular Biology, Pharmacy

Graduate Degrees: M.S., Ph.D.
Graduate Degree Requirements: Master: 30 credits plus a thesis Doctoral: No set number of credits - dissertation, Ph.D. Committee sets minimum requirements, Note: Students can secure a Ph.D. specializing in toxicology and molecular biology.

Graduate and Undergradua te Courses Available for Graduate Credit: Toxicology, Materials Science Toxicology, Computer Techniques, Natural Products, Research Techniques, Radioactive Isotope Techniques, Biosynthesis of Natural Drug Products, Pedagogic Techniques, Relationships Among Plants, Insects, and Insecticides, Pharmacological Screening Techniques and Classification of Natural Products, Antibiotics, Physicochemical Interpretations in Molecular Biology

DEPARTMENT OF PHYSIOLOGY AND BIOPHYSICS
Phone: Chairman: (901) 528-5822 Others: 528-5500

Chairman: L. Share
Professors: D.A. Brody, C.E. Grosvenor, R.R. Overman, L.D. Partridge, C.W. Sheppard, L. Van Middlesworth, H.H. Vogel, R.E. Taylor, Jr.
Associate Professors: C.M. Blatteis, E. Bowen, J.M. Ginski, L.B. Reynolds, H.W. Smith, E.G. Schneider, P. Zee
Assistant Professors: T.W. Barrett, J.D. Beard, L.H. Blackwell, A.A. Manthey, D.F. Nutting, K.W. Scheel, R.E. Shade, R.N. Stiles, H. Nishimura, N. Whitworth, J.G. Yager, Y.Y. Yeh
Instructors: M. Yamamoto, P.H. Burns, W.Q. Sargent, J.C. Grubbs
Graduate Teaching Assistants: 5
Research Assistants: 1

Degree Programs: Physiology, Biophysics

Graduate Degrees: M.S., Ph.D.
Graduate Degree Requirements: Master: Minimum residence requirement; 3 quarters of full-time study, time limit, 6 calendar years, course requirements: total minimum credit of 45 hours of approved graduate courses, a major research project reported in the form of of a thesis, oral examination covering general field of study and thesis Doctoral: Minimum residence requirements: 9 quarters full-time study, time limit; 5 years after admission to candidacy, excellence in an established program of courses, reading knowledge of one foreign language, preliminary oral or written examination, substantial evidence of independently achieved and original results in research, final examination covering dissertation, special field, etc.

Graduate and Undergraduate Courses Available for Graduate

Credit: Physiology, Fundamentals of Physiological Instruction, Applications of Statistics to Biological Problems, Physiology Seminar, Physiological Research, Master's Thesis and Research, Doctoral Dissertation and Research, Recent Advances in Physiology, Introduction to Computer Methods, History of Physiology, Principles of Mathematical Physiology, Advances in Cardiovascular Physiology, Recent Advances in Neuroendocrinology, Recent Advances in Nerve-Muscle Physiology, Mechanical Behavior of the Lung

UNIVERSITY OF TENNESSEE AT NASHVILLE
323 McLemore Street Phone: (615) 254-5681
Nashville, Tennessee 37203
Dean of College: Not Stated

DEPARTMENT OF BIOLOGICAL SCIENCES (UNDER: NATURAL SCIENCES)
Phone: Chairman: Ext. 266, 267, 268

Corrdinator: J.L. Wilson
Professors: J.L. Wilson, J. Mallette
Assistant Professor: T. Snazelle
Instructor: F. Lackey
Part-time Instructors: 7

Degree Programs: Areas of Concentration: Humanities, Natural Sciences, Social Sciences

Undergraduate Degrees: B.A., B.S. (Natural Sciences)
Undergraduate Degree Requirements: 194 quarter hours for B.A. or B.S. 63 hours in one of 3 areas of concentration - minimum 36 upper division, minimum 18 hours upper division in single subject within concentration area, 12 hours upper division in second subject within arts and sciences

THE VANDERBILT UNIVERSITY
Nashville, Tennessee 37235 Phone: (615) 322-7311
Dean of Graduate Studies: J.F. Kilroy (Acting)

College of Arts and Sciences
Dean: W.G. Holladay

DEPARTMENT OF MOLECULAR BIOLOGY
Phone: Chairman: (615) 322-2008 Others: 322-7311

Chairman: O. Touster
Professors: S. Fleischer, L.S. Lerman, G. Mosig, O. Touster
Associate Professors: T.Bibring, G. Di Sabato, D.L. Friedman, R.M. Litman (Research), R.J.Neff, J.H. Venable, Jr., J.D. Weil, H. Weismeyer
Assistant Professors: J. Bibring (Research), B. Fleischer (Research),W.M. LeStourgeon, G.Meissner (Research)
Research Assistants: 7
Research Associates: 9

Degree Programs: Molecular Biology

Undergraduate Degree: B.S.
Undergraduate Degree Requirements: 27 hours of molecular biology with required courses in Cell Physiology, Genetics and Biochemistry, Physical Chemistry and Developmental Anatomy may be counted in the 27 hours although they are offered by other departments

Graduate Degrees: M.S., Ph.D.
Graduate Degree Requirements: Master: 24 semester hours credit plus master's thesis Doctoral: 72 semester hours credit plus research dissertation

Graduate and Undergraduate Courses Available for Graduate Credit: Cellular Physiology (U), General Bacteriology (U), Biologyof Bacterial Viruses (Bacteriophages) (U), Microbial Metabolism (U), Genetics Laboratory (U), Principles of Genetics (U), Microbial Genetics (U), Physical and Quantitative Approaches in Molecular Biology (U), Fundamentals of Biochemistry (U), Advanced Cellular Physiology, Seminar in Molecular Biology, Bio-Organic Chemistry, Physico-Chemical Properties of Biological Molecules, Focal Topics in Genetics, Lectures on Research Progress in Molecular Biological Mechanisms, Methods in Molecular Biology

School of Engineering
Dean: H.L. Hartman

DEPARTMENT OF BIOMEDICAL ENGINEERING
Phone: Chairman: (615) 322-3521 Others: 322-7311

Chairman: P.H. King
Professors: C.E. Goshen, L.E. Johnson, G.W. Malaney, J.J. Wert
Associate Professors: R.J. Bayuzick, J.R. Bourne, P.H. King, P.D. Krolak, R.L. Lott
Assistant Professors: R.G. Shiavi, D.C. Schmidt
Visiting Lecturers: 1
Teaching Fellows: 3

Degree Programs: B.S. in Biomedical Engineering only, advanced work through other departments at present)

Undergraduate Degree: B.S.
Undergraduate Degree Requirements: 125 semester hours

Medical School
Dean: A.D. Bass (Acting)

DEPARTMENT OF ANATOMY
Phone: Chairman: (615) 322-2134

Chairman: J. Davies
Professors: J. Davies, J.E. Brown, A.M. Burt III, V.S. LeQuire, J.W. Ward (Emeritus)
Associate Professors: R.B. Adkins, Jr., G.R. Davenport J.A. Freeman, W.H. Olson
Assistant Professors: L.H. Aulsebrook, L.H. Hoffman
Research Assistants: 1
Electron Microscopic Technician: 1

Degree Programs: Anatomy, Anatomy and Histology, Anatomy and Physiology

Graduate Degree: Ph.D.
Graduate Degree Requirements: Not Stated

Graduate and Undergraduate Courses Available for Graduate Credit: Those courses which are basically Medical School courses, but may be taken for graduate credit are indicated by (M), Gross Anatomy (M), Experimental Methods in Neurology, Experimental Methods in Histology, Morphological Basis for Biochemical Activity, Histology (M), The Nervous System (M), Neurophysiology Seminar, Experimental Methods in Electrophysiology - Central Nervous System, Reproductive Biology

DEPARTMENT OF BIOCHEMISTRY
Phone: Chairman: (615) 322-3318 Others: 322-3315

Chairman: L.W. Cunningham
Professors: H.P. Broquist, J. Brown, S. Cohen, J.G. Coniglio, W.J. Darby, W.J. Hayes, O. Touster
Associate Professors: J.P. Carter, F. Chytil, W.R. Faulkner, T. Inagami, G.V. Mann, R.A. Neal, J.D. Puett, C. Wagner, B.J. Wilson
Assistant Professors: R.N. Brady, P.W. Felts, D.W. Frederiksen, H.L. Greene, R.D. Harbison, C.G. Hellerqvist, K.J. Lembach, P.G. Lenhert, G. Nichoalds, L.G. Warnock
Instructor: L.A. Holladay
Visiting Lecturers: 2
Teaching Fellow: 1
Research Assistants: 10

Degree Program: Biochemistry

Graduate Degrees: M.S., Ph.D.

Graduate Degree Requirements: Master: 24 hours of graduate course work one-half of which must be taken in department, thesis describing piece of original work Doctoral: Pass qualifying examination after 36 hours of graduate work, 72 hours of graduate work--36 hours of formal course work including a minor of 12 hours in s separate or related fields, Dissertation, Final Examination

Graduate and Undergraduate Courses Available for Graduate Credit: Fundamentals of Biochemistry, Biochemistry, Biochemistry Laboratory, Special Problems and Experimental Techniques, Bio-Organic Chemistry, Seminars in Biochemistry Literature, Fundamentals of Human Nutrition, Lipid Chemistry, Metabolism and Transport, Advanced Biochemistry I, II, Chemistry of Biopolymers, Chemical Mechanisms of Enzyme Catalysis, Toxicology, Clinical Biochemistry I, II, Nutritional Biochemistry Laboratory, Special Topics in Neurochemistry, Reproductive Biology

DEPARTMENT OF MICROBIOLOGY

Phone: Chairman: (615) 322-2087 Others: 322-7311

Chairman: A.S. Kaplan
American Cancer Society-Charles Hayden Foundation Professor of Microbiology: S.P. Colowick
Professors: J.H. Hash
Associate Professors: T.B. Porat, S.Harshman, J. Robinson
Assistant Professors: D.G. Colley, M.A. Melly, C.M. Stoltzfus, F.C. Womack
Instructors: M. Colowick
Visiting Lecturers: 2
Research Assistants: 7
Research Associates: 6

Degree Program: Microbiology

Graduate Degrees: M.S., Ph.D.

Graduate Degree Requirements: Master: A research thesis is required for the master's degree. The department requires each candidate for the master's degree to present a minor of at least six hours in one or more spearate departments giving work related to the major. Doctoral: Each Ph.D. candidate must present at least 12 semester hours of work in one or more outside departments. The department's policy is to permit variations in the number of formal course hours required above the basic minimum of 24 for the Ph.D. degree, with each continuing committee determining the selection of courses and the total number of course hours on the basis of the student's background and need. Proficiency in one foreign language.

Graduate and Undergraduate Courses Available for Graduate Credit: An Introduction to Immunochemistry, Microbiology, Introductory Microbiology, Pathogenic Bacteria Fungi, Introductory Immunology, Introductory Virology, Experimental Methods in Microbiology, Advanced Topics in Microbiology, Microbial Toxins and Enzymes, Antibiotics and Microbial Metabolism, Immunochemistry, Basic Virology, Advanced Immunochemistry Laboratory, Special Seminars, Lectures on Research Progress in Molecular Biological Mechanisms, Master's Thesis Research, Ph.D Dissertation Research

TEXAS

ABILENE CHRISTIAN COLLEGE

ACC Station Phone: (915) 677-1911
Abilene, Texas 79601
Dean of Graduate Studies: F. Dunn
Dean of College: B.J. Humble

DEPARTMENT OF AGRICULTURE
Phone: Chairman: Ext. 711 Others: Ext. 706

Head: J.K. Justice
Professors: J.K. Justice, F.M. Churchill, L.E. DuBose
Associate Professor: T.W. Colby
Research Assistants: 2
Laboratory Assistants: 4

Field Stations/Laboratories: School Farm Near Campus

Degree Program: Agriculture

Undergraduate Degree: B.S.
Undergraduate Degree Requirements: 128 semester hours Total, 33-40 hours in major area, 12 hours Chemistry, 11 hours Biology, 54 hours General Education

DEPARTMENT OF BIOLOGY
Phone: Chairman: Ext. 705 Others: Ext. 709

Chairman: W.C. Stevens
Professors: F.M. Churchill, W.C. Stevens
Associate Professors: J.C. Little, R. Shake, K.B. Williams, J.E. Womack
Assistant Professor: W.M. Kemp
Teaching Fellows: 3
Research Assistants: 3

Degree Programs: Biology, Wildlife Management

Undergraduate Degree: B.S.
Undergraduate Degree Requirements: Language (8 hours, French or German), Mathematics (4 hours Pre-Calculus or Calculus), Chemistry (16 hours), Physics (8 hours), Biology (8 hours General Biology, Cell Biology, Environmental Biology, Genetics, Seminar plus 15 additional hours in advanced biology)

Graduate Degree: M.S.
Graduate Degree Requirements: Master: Thirty hours of graduate work including 6 hours of thesis credit. Major field must include 18-24 hours and the minor field, to be taken in Chemistry, Physics, or Mathematics, must include 6-12 hours. At least 15 hours must be taken in courses open only to graduate students.

Graduate and Undergraduate Courses Available for Graduate Credit: Genetics (U), Microbiology (U), Immunology (U), Plant Physiology (U), Environmental Biology (U), General Entomology (U), Plant Taxonomy (U), Field Zoology (U), Wildlife Management (U), Histology (U), Histological Technique (U), Advanced Wildlife Biology (U), Biometry (U), Embryology (U), Parasitology (U), Radiation Biology (U), Human Anatomy (U), Human Physiology (U), Agrostology (U), Problems in Biology (U), Animal Ecology, Limnology, Topics in Vertebrate Biology, Topics in Physiology, Cell Physiology, Endocrinology, Cytology, Environmental Conservation Workship

ANGELO STATE UNIVERSITY

San Angelo, Texas 76901 Phone: (915) 942-2131
Dean of Graduate Studies: B. Young
Dean of College: G. Welch

DEPARTMENT OF BIOLOGY
Phone: Chairman: (915) 942-2189

Head: C.M. Rowell, Jr.
Professors: G.C. Creel, C. Jones, G.G. Raun, W.A. Thornton
Associate Professor: E.H. Todd
Assistant Professors: R. Ballinger, A. Bloebaum, A. Flury
Instructor: J.R. Capeheart
Research Assistants: 2
Graduate Assistants: 6

Degree Program: Biology

Undergraduate Degree: B.S.
Undergraduate Degree Requirements: 34 hours Biology, 20 hours Chemistry, 8 hours Physics, 6 hours Mathematics

Graduate Degrees: M.A.T., M.S.
Graduate Degree Requirements: Master: M.S. 30 hours plus thesis, M.A.T. - 30 - 36 hours no thesis

Graduate and Undergraduate Courses Available for Graduate Credit: Molecular Biology (U), Evolution (U), General Physiology (U), Biometrics and Experimental Design, Advanced General (M.A.T.), Population Genetics, Population Dynamics, Plant Systematics, Animal Systematics, Terrestrial Ecology, Limnology, Field Biology, Desert Plant Biology

AUSTIN COLLEGE

(No reply received)

BAYLOR COLLEGE OF MEDICINE

1200 Moorsund Avenue Phone: (713) 790-4951
Houston, Texas 77025
Dean of Graduate Studies: J.L. Melnik
Dean of College of Medicine: E.C. Lynch

MARRS McLEAN DEPARTMENT OF BIOCHEMISTRY
Phone: Chairman: (713) 790-4528 Others: 790-4521

Chairman: S.J. Wakil
Professors: C.T. Caskey, N.M. DiFerrante, A.M. Gotto, S.J. Wakil
Associate Professors: D.M. Desiderio, M W. Noall, L.C. Smith, J.E. Stouffer
Assistant Professors: K.C. Aune, E.M. Barnes, S.H. Bishop, M H. Dresden, V.C. Joshi, R.E. Moses, Y.H. Oh, D.J. Roufa, H. Shizuya, J.H. Wilson
Instructors: M.J. Arslanian, J.K. Stoops
Visiting Lecturers: 12
Research Assistants: 16

Degree Programs: Biochemistry, Genetics

Graduate Degree: Ph.D.
Graduate Degree Requirements: Doctoral: Perform successfully on series of courses; show ability to perform original research and write a dissertation from it.

Graduate and Undergraduate Courses Available for Graduate Credit: Introductory Biochemistry I, Introductory Biochemistry II, Analytical Biochemistry, Enzyme Mechanisms, Biochemistry of Macromolecules, Biochemical Genetics, Regulation of Cellular Metabolism, Biochemistry of Disease, Biomembranes,

DEPARTMENT OF BIOCHEMICAL VIROLOGY
Phone: Chairman: (713) 790-4507 Others: 790-4508

Head: S. Kit
Professors: D.R. Dubbs, S. Kit
Assistant Professors: D. Trula, K. Somers
Research Associates: G.N. Jorgensen, W.C. Leung
Research Assistant: 1

Degree Program: Virology

Graduate Degree: Ph.D.
Graduate Degree Requirements: Doctoral: A total of 108 quarter hours of credit of which 20 must be outside the major field. Qualifying examination, Final Dissertation and defense of dissertation.

Graduate and Undergraduate Courses Available for Graduate Credit: General Virology, Biochemical Virology, Seminars, Journal Club, Introduction to Epidemiology, Methods in Epidemiology, Viral Oncology, Biophysics of Viruses, General Virology,Laboratory, Biochemical Virology Laboratory, Methods in Molecular Virology, Viral Genetics, Special Projects, Selected Reading, Basic andViral Immunology

DEPARTMENT OF CELL BIOLOGY
Phone: Chairman: (713) 521-4701 Others: 521-4703

Chairman: B.W. O'Malley
Professors: S. Burstein, P.O. Kohler, J.H. Perry
Associate Professors: J.H. Clark, S.G. Glasser, C. Huckins, R.L. Jackson, A.R. Means, D. Medina, C. Teng
Assistant Professors: C.B. Bordelon, L.C.B. Chan, J.P. Comstock, R.L. Deter, S.C. Harvey, M.E. McClure, E.J.Peck, J.M. Rosen, W.T. Schrader, S.H. Socher, C.T. Tend, R. H. Thalmann, R. Vitale, G.W. Wray, V.P. Wray
Instructors: R.D. Colligan, S.E. Harris, J. Norris, R. G. Smith, M.J. Tsai, N.T. Van
Visiting Lecturers: 1
Teaching Fellows: 2

Degree Program: Cell Biology

Graduate Degree: Ph.D.
Graduate Degree Requirements: Doctoral: The degree of Doctor of Philosophy is awarded upon demonstration that the student has the quality of intellect and the basic scholarly and technical tools necessary for a career in biomedical research and teaching. The Ph.D. program must be pursued on a full-time basis. The Ph. D. Degree is based upon satisfactory achievement in studies amounting to 108 quarter hours, of which a minimum of 40 quarter hours must be devoted to the Major field, 20 quarter hours to other fields (or to the Minor if one is chosen by the student). The remaining 48 hours may be applied to either area of study and may be course or research work depending upon Department requirements. A minimum of 3 quarters of academic work, in residence, must be completed before the Qualifying Examination may be taken. This regulation applies to students with and without transfer of credit from another institution.

Graduate and Undergraduate Courses Available for Graduate Credit: Cell Structure and Function, Cell Biology, Regulation Biology of the Eucaryotic Cell, Gross Anatomy I and II, Reproductive Biology, Experimental Endocrinology, Introduction to Experimental Oncology, Developmental Biology, Cytology, Neurohistology, Seminar in Experimental Oncology, Special Regional Dissections, Special Projects, Neuroscience

DEPARTMENT OF MICROBIOLOGY AND IMMUNOLOGY

Chairman: V. Knight
Professors: W.T. Butler, R.B. Couch, V. Knight, R.P. Williams
Associate Professors: W.J. Fahlberg, R.A. Jensen, J. A. Kasel, R.R. Martin, H.D. Mayor, R.D. Rossen, T.W. Williams
Assistant Professors: B.E. Gilbert, E.M. Hersh (Adjunct), M W. Bradshaw, J.R. Davis, D.M. Musher, R.R. Rich, D.M. Mumford (Adjunct), R.D. Wende
Instructors: B.S. Criswell, S. Farrow, D.H. Kearns, S. Solliday, S.Z. Wilson
Research Assistants: 4
Research Associates: 2

Degree Program: Microbiology

Graduate Degree: Ph.D.
Graduate Degree Requirements: Doctoral: Oral Qualifying Examination, Dissertation, defense of Dissertation

Graduate and Undergraduate Courses Available for Graduate Credit: Medical Microbiology (includes Immunology) (U), General Microbiology, Infection and Immunity, Microbial Physiology, Molecular Biophysics, Immunology, Laboratory Preceptorships

DEPARTMENT OF PHYSIOLOGY
(No reply received)

BAYLOR UNIVERSITY
Waco, Texas 76707 Phone: (817) 755-1011

College of Arts and Sciences

DEPARTMENT OF BIOLOGY
(No reply received)

College of Dentistry
800 Hall Street Phone: (214) 824-6321
Dallas, Texas 75226
Dean of Graduate Studies: J. Bishop
Dean of College of Dentistry: K. Randolf

DEPARTMENT OF BIOCHEMISTRY
(No reply received)

DEPARTMENT OF MICROSCOPIC ANATOMY
Phone: Chairman: Ext. 265

Chairman: R.E. Dill
Professors: J.L. Matthews, N.L. Biggs
Associate Professors: P.L. Blanton, R.E. Dill, J.H. Martin
Assistant Professors: J.E. McIntosh, W.L. Davis, H.W. Sampson, D.R. Fowler
Visiting Lecturers: 2
Teaching Fellows: 2
Research Assistants: 3

Degree Program: Anatomy

Graduate Degrees: M.S., Ph.D.
Graduate Degree Requirements: Master: Thirty semester hours, at least 18 hours in Anatomy and 6 hours in the minor field. A thesis is required. Doctoral: Seventy-eight semester hours, one foreign language proficiency, one or two minor fields of study, and a dissertation.

Graduate and Undergraduate Courses Available for Graduate Credit: Gross Anatomy, Fetal Anatomy, General Histology, Oral Histology and Embryology, Advanced Histology and Microscopic Anatomy, General Embryology, Cytology and Microscopic Technics, Histochemistry, Growth and Development, Head and Neck Anatomy, Neuroanatomy, Research for the Master's Thesis, Thesis, Neuroanatomy Seminar, Special Problems in Microscopic Anatomy, Cell Mechanisms, Special Dissections in Anatomy, Special Problems in Neuroanatomy, Doctoral Research, Dissertation.

DEPARTMENT OF PHYSIOLOGY
Phone: Chairman: Ext. 301 Others: Ext. 255

Chairman: H.L. Dorman
Professors: E.R. Cox, J.Bishop
Associate Professors: H. Varon, J. Finney
Assistant Professors: L. Fraizer, F. Williams
Teaching Fellows: 1
Research Assistants: 1

Degree Programs: (Temporary suspension on new student for Master's and Ph.D. programs).

BISHOP COLLEGE
(No reply received)

DALLAS BAPTIST COLLEGE
P.O. Box 21206 Phone: (214) 331-8311
Dallas, Texas 75211
Dean of College: C. Harris

BIOLOGY DEPARTMENT
Phone: Chairman: Ext. 302

Chairman: C.S. Jordan
Professors: C.S. Jordan, H. Robbins
Associate Professors: L.G. Adair, R. Hogue

Degree Program: Biology

Undergraduate Degrees: B.A., B.S.
Undergraduate Degree Requirements: 30 hours major in Biology

EAST TEXAS BAPTIST COLLEGE
Marshall, Texas 75670 Phone: (214) 934-1905
Dean of College: J.Q. Mason

DEPARTMENT OF BIOLOGY
Phone: Chairman: (214) 938-1905

Chairman: S. Handler
Professor: S. Handler
Instructor: R. Johnston

Degree Programs: Biology, Medical Technology

Undergraduate Degree: B.S.
Undergraduate Degree Requirements: Zoology, Botany, Comparative Anatomy, Human Anatomy and Physiology, Histology, Microbiology, Embryology, Heredity

EAST TEXAS STATE UNIVERSITY
East Texas Station Phone: (214) 468-2224
Commerce, Texas 75428
Dean of Graduate Studies: J.D. Morris
Dean of College: J.E. Thomas

DEPARTMENT OF BIOLOGY
Phone: Chairman: (214) 468-2224 Others: 468-2212

Head: A.M. Pullen
Professors: E.P. Fox, A.M. Pullen, J.S. Norwood, E. P. Roberts, R.K. Williams, B.E. Wilson
Associate Professors: D.A. Ingold, E.F. Klaus
Assistant Professors: D.R. Lee, J. McFeeley
Teaching Fellows: 6
Research Assistant: 1

Degree Programs: Biology, Botany, Zoology

Undergraduate Degrees: B.A., B.S.
Undergraduate Degree Requirements: Total of Forty Courses, 15 courses in General Education, 8 courses in Major, 8-6 courses in 2nd major or minor, Electives to complete forty courses

Graduate Degree: M.S.
Graduate Degree Requirements: Master: 10 course program, can all be taken in major or have a six course major and a four course minor, 12 course program; (non-thesis) program, minimum of six course major, additional courses can be taken in related science areas.

Graduate and Undergraduate Courses Available for Graduate Credit: Seminar, Advanced Genetics, Thesis, Cell Biology, Microbiology, Ecology, Science Workshop, Aquatic Biology, Histo-cytological Techniques, Radiation Biology, Research, Literature and Technique Advanced Botany, Plant Physiology, Taxonomy of Flowering Plants, Plant Ecology, Non-Flowering Plants, Dendrology, Aquatic Invertebrates, Physiology Ornithology, Vertebrate Ecology, Field Zoology, Invertebrate Zoology, Endocrinology, Agricultural Entomology

HARDIN-SIMMONS UNIVERSITY
(No reply received)

HOWARD PAYNE COLLEGE
Brownwood, Texas 76801 Phone: 646-2502
Dean of College: D.L. Stephenson

DEPARTMENT OF BIOLOGICAL SCIENCE
Phone: Chairman: Ext. 243

Head: J.W. Stanford
Professor: J.W. Stanford
Associate Professors: D. Armentrout, G. DeViney
Assistant Professors: E.C. Morgan, D. Jackson (Part-time)
Temporary Appointment: 1

Degree Program: Biology

Undergraduate Degrees: B.A., B.S.
Undergraduate Degree Requirements: 32-38 hours Biology, Chemistry through Organic, Mathematics through Calculus, Physics - Recommended - not required of all majors.

HUSTON-TILLETSEN COLLEGE
Austin, Texas 78702 Phone: (512) 476-7421
Dean of College: E.A. Delco, Jr.

DEPARTMENT OF BIOLOGY
Phone: Chairman: Ext. 306

Head: J.Q. Heplar
Professors: J.Q. Heplar, C.S. Lin, E.A. Delco, Jr.
Associate Professors: M.A.R. Hammond

Degree Program: Biology

Undergraduate Degree: B.S.
Undergraduate Degree Requirements: Total credits - 128 B.S. (Biology Major) - 30

INCARNATE WORD COLLEGE
4301 Broadway Phone: 828-1261
San Antonio, Texas 78209
Dean of Graduate Studies: B. O'Halloran
Dean of College: Sr. G. Corbin

DEPARTMENT OF BIOLOGY
Phone: Chairman: Ext. 222

Chairman: Sr. J.M. Armer
Moody Professor of Biology: D. McLain
Professors: Sr. Armer, Sr. M.L. Corcoran, Dr. McLain
Assistant Professor: D. Hester
Visiting Lecturers: 2

Research Assistants: 4

Degree Program: Biology

Undergraduate Degree: B.S.
Undergraduate Degree Requirements: 30 semester hours in Biology - 18 of which are advanced, 18 semester hours minor (preferably Chemistry), One course in Chemistry required

Graduate Degrees: M.A., M.S.
Graduate Degree Requirements: Master:36 semester hours including Thesis

Graduate and Undergraduate Courses Available for Graduate Credit: Cytology, Biological Systematics, Cytogenetics, Advanced Immunology, Bacterial Physiology, Medical Microbiology, Hematology (U), Parasitology (U), Animal Physiology (U), Animal Behavior (U), Ecology (U), Embryology (U), Invertebrate Zoology (U), Bacterial Systematics (U), Immunology (U), Cellular Physiology (U), Ecology, Environment and Man (U), Biochemical Genetics (U), Mycology (U), History of Biology (U)

JARVIS CHRISTIAN COLLEGE

Hawkins, Texas 75765 Phone: (214) 769-2174
Dean of College: A.R. Lewis

DEPARTMENT OF BIOLOGY
Phone: Chairman: Ext. 168

Chairman: D.D. Malik
Professor: D.D. Malik
Associate Professor: S. Yaden
Instructor: J.F. Johnson

Degree Programs: Biology, Biological Education

Undergraduate Degree: B.S.
Undergraduate Degree Requirements: (Semester hours) 63 hours general Education Requirements, 32 hours in Biology, 8 hours Chemistry, 8 hours Physics, 6 hours Mathematics, Total = 124 hours

LAMAR UNIVERSITY

P.O. Box 10037 Phone: (713) 838-7329
Beaumont, Texas 77710
Dean of College: E.B. Blackburn

DEPARTMENT OF BIOLOGY
Phone: Chairman: (713) 838-7329

Head: M.E. Warren
Professors: E.S. Hayes, R.J. Long, W.R. Smith, H. Waddell, J.J. Ramsey
Associate Professor: W.T. Fitzgerald, J.L. McGraw, Jr., R.C. Harvel, M.E. Warren
Assistant Professors: G.A. Bryan, W.C. Runnels, G.W. Gatlin, P. Malnassy, P.B. Robertson
Instructor: M. D. Hunt
Teaching Fellows: 3

Field Stations/Laboratories: Pleasure Island at Port Arthur, Texas

Degree Programs: Biology, Medical Technology

Undergraduate Degree: B.S.
Undergraduate Degree Requirements: 42 hours Biology, 24 hours Chemistry, 8 hours Physics, 6 hours Mathematics

Graduate Degree: M.S. (Biology)
Graduate Degree Requirements: Master: 30 hours beyond B.S., 24 hours coursework, 16 hours Thesis

Graduate and Undergraduate Courses Available for Graduate Credit: Ornithology (U), Parasitology (U), Entomology (U), Limnology(U), Terresterial Ecology (U), Marine Biology (U), Protistology (U), Cell Biology, Cell Physiology (U), Vertebrate Natural History (U), Graduate Seminar, Materials and Techniques Research, Advanced Ornithology, Helminthology, Mammalogy, Ichthyology, Herpetology, Comparative Physiology, Marine Invertebrate Zoology

LE TOURNEAN COLLEGE

Longview, Texas 75601 Phone: (214) P13-0231

(Biology Department in course of organization)

LUBBOCK CHRISTIAN COLLEGE

5601 West 19th Street Phone: (806) 792-3221
Lubbock, Texas 79407
Dean of College: N. Keener

AGRICULTURE DEPARTMENT
Phone: Chairman: Ext. 281 Others: Exts. 282, 283

Chairman: E.D. Cook
Professor: E.D. Cook
Associate Professor: D.L. Smith
Instructor: R. Blackwood, E. Moudy

Degree Program: Agriculture

Undergraduate Degree: B.S.
Undergraduate Degree Requirements: Not Stated

DEPARTMENT OF BIOLOGY
Phone: Chairman: Ext. 281

Chairman: L.L. Sherrod
Associate Professors: L.L. Sherrod, D.L. Smith, G. Estep
Assistant Professor: J.F. Hay, Jr.
Teaching Fellows: 5

Field Stations/Laboratories: Chama, New Mexico

Degree Programs: Biology, Medical Technology

Undergraduate Degree: B.S.
Undergraduate Degree Requirements: 36 hours General Education, 30 hours Biology, 22 hours Chemistry, 8 hours Physics, 10 hours Mathematics, 14 hours Biblical Studies, Electives to 132 hours

McMURRY COLLEGE

Abilene, Texas 79605 Phone: (915) 692-4130
Dean of College: A.F. Cordts

DEPARTMENT OF BIOLOGY
Phone: Chairman: Ext. 252

Chairman: C.W. Beasley
Professor: R.D. Moore
Associate Professors: C.W. Beasley, B.L. Pilcher
Instructor: B. Beasley (Part-time Laboratory)

Degree Programs: Biology, Biological Education, Medical Technology

Undergraduate Degrees: B.A., B.S.
Undergraduate Degree Requirements: B.A. in Biology - 32 semester hours, plus 6 semester hours Mathematics, 16 hours Chemistry, 8 hours Physics, B.S. with Biology Teaching Field: 8 hours Chemistry, 25 hours Biology, B.S. in Medical Technology - 22 hours Biology, 20 hours Chemistry, 8 hours Physics

MARY HARDIN-BAYLOR COLLEGE

Belton, Texas 76513 Phone: (817) 939-5811

Dean of College: D. Johnson (Interim)

BIOLOGY DEPARTMENT
Phone: Chairman: Ext. 59

Chairman: Dr. Johnson
Associate Professor: Dr. Johnson
Assistant Professor: Dr. Casto
Laboratory Instructors: 2

Degree Program: Biology

Undergraduate Degree: B.S.
Undergraduate Degree Requirements: Not Stated

MIDWESTERN UNIVERSITY

3400 Taft Boulevard Phone: (817) 692-6611
Wichita Falls, Texas 76308
Dean of Graduate Studies: E.L. Dickerson
Dean of College: J.W. Meux

DEPARTMENT OF BIOLOGY
Phone: Chairman: Ext. 253 Others: Exts. 254, 297

Chairman: A.F. Beyer
Professors: A.F. Beyer, W.W. Dalquest
Associate Professors: J.L Boswell, J.V Grimes, N.V. Horner
Assistant Professor: N. Scott
Teaching Assistants (Graduate): 4

Degree Programs: Biology, Medical Technology

Undergraduate Degrees: B.S., B.S. (Medical Technology)
Undergraduate Degree Requirements: 29-40 hours of Biology, Chemistry through Organic, Physics - 1 year, Computing Science, Foreign Language - 1 year, Algebra and Trigonometry or Calculus

Graduate Degree: M.S.
Graduate Degree Requirements: Master: 30 hours of Biology including Thesis

Graduate and Undergraduate Courses Available for Graduate Credit: Field Zoology (U), Bacteriology (U), Wildlife Management (U), General Physiology (U), Genetics (U), Genetics Laboratory (U), Vertebrate Embryology (U), Entomology (U), Systematic Botany (U), Microtechnique (U), Invertebrate Zoology (U), Plant Physiology, Limnology (U), Seminar in Biology (U), Vertebrate Paleontology (U), Human Genetics (U), Ornithology (U), Histology (U), Cytology (U), Plant Anatomy (U), Animal Parasitology (U), Ecology (U), Plant Morphology (U), Cell Physiology (U), Literature and History of Biological Sciences, Herpetology, Ichthyology, Mammalogy, Wood Technology, Biology of Algae and Fungi, Protozoology, Paleobotany, Radiation Biology, Physiological Genetics, Electron Microscopy, Advanced Topics in Molecular Genetics

NORTH TEXAS STATE UNIVERSITY

Box 5218, N.T. Station Phone: 788-2521
Denton, Texas 76201
Dean of Graduate Studies: R.B. Toulouse
Dean of College of Arts and Sciences: J.B. Pearson

DEPARTMENT OF BIOLOGICAL SCIENCES
Phone: Chairman: 788-2011

Chairman: K.R. Johansson
Distinguished Professor: J.K.G. Silvey
Professors: J.R. Lott, D.R. Redden, A.W. Roach, E.A. Schlueter, K.W. Stewart, G. R. Vela, K.R. Johansson, C. Mankinen (Adjunct), R.D. Baumann, J.P. Schmidt
Associate Professors: B.G. Harris, A.S. Kester, B.D. Vance
Assistant Professors: D.L. Busbee, G. Crawford, L.C. Fitzpatrick, B.A. Hatten, W.D. Pearson, D. W. Smith, E.G. Zimmerman, J.A. Besso, J. Stanford
Instructor: V. Rudick (Lecturer)
Teaching Fellow: 1
Research Assistants: 15
Teaching Assistants: 35

Field Stations/Laboratories: Nike Missile Site (20 acres) for aquatic biology program. Shared with Physics and Institute for Environmental Studies.

Degree Programs: Biological Sciences, Botany, Ecology, Genetics, Medical Technology, Microbiology, Physiology, Zoology, Cell Biology, Plant Physiology

Undergraduate Degrees: B.A., B.S.
Undergraduate Degree Requirements: B.A.: Biology (26 semester hours), Chemistry (16), Physics (8), Mathematics (6), B.S.: Biology (36), Chemistry (21, Physics (8), Mathematics (6), plus one additional upper level course from a Science Department

Graduate Degrees: M.S., Ph.D.
Graduate Degree Requirements: Master: 30 semester hours Doctoral: 90 semester hours (minus hours applied to master's degree

Graduate and Undergraduate Courses Available for Graduate Credit: Not Stated

OUR LADY OF THE LAKE COLLEGE

411 S.W. 24th Street Phone: (512) 434-6711
San Antonio, Texas 78285
Dean of Graduate Studies: Sr. M. Malloy
Director of Mathematics and Science Division: Sr. I. Ball

Head: Sr. M.H. Christopher
Associate Professor: Sr. M.H. Christopher
Assistant Professors: P.J. Brown, D. Weniger
Instructor: M.B. Clark

Degree Programs: Biology, Medical Technology

Undergraduate Degrees: B.A., B.S.
Undergraduate Degree Requirements: 36 semester hours in Biology (20 semester hours advanced), 20 hours Chemistry (4 semester hours advanced), Mathematics through First semester of Calculus, General Physics

PAN AMERICAN UNIVERSITY

Edinburg, Texas 78539 Phone: (512) 381-3161
Dean of Graduate Studies: B.E. Reeves
Dean of College: L.O. Sorensen

BIOLOGY DEPARTMENT
Phone: Chairman: (512) 381-3161

Head: S.I. Sides
Professors: P. James, W. Ware, S.L. Sides, L.O. Sorensen
Associate Professors: T. Allison, V. Foltz, N.L. Savage
Assistant Professors: F. Judd, R. Lonard, D. Lyles, J. Ortega
Instructors: B. Cisneros, O.D. Cockrum, T. de la Pena, R. Ross

Degree Program: Biology

Undergraduate Degree: B.S.
Undergraduate Degree Requirements: 30 semester hours in Biology, of which 15 semester hours must be advanced, must include Biology 1401, 1402, 4100, at least eight semester hours from Biology 2401, 2402, 2403, 3402, 3404, 3407, 4402 and at least six semester hours from Biology 3309, 3401, 3408, 4305, 4406, 4410

Graduate Degree: M.S.
Graduate Degree Requirements: Master: (SPECIAL REQUIREMENTS): a. Biological Problems: At least one Biological problem will be required of all, and for those choosing the thesis plan a Biological Problem should be successfully pursued prior to starting the thesis. b. Biology 6101, 6102, 6103 will be required.

Graduate and Undergraduate Courses Available for Graduate Credit: Marine Plant Science, Ecological Concepts, Concepts of Biological Literature, Advanced Plant Physiology, Advanced Mycology, Concepts of Biological History, Advanced Marine Zoology, Biology Graduate Seminar, Cell Biology, Marine Ecology, Advanced Ecology, Host-Parasite Relationships, Biometry, Field Botany, Biological Techniques, Freshwater Phycology, Marine Phycology, Immunology, Ecological Physiology, Advanced Ornithology, Advanced Plant Systematics, Advanced Field Zoology, Biological Problems for Graduate Students: Botanical, Marine, Microbial, Zoological, Thesis

PRAIRIE VIEW A AND M UNIVERSITY

Prairie View, Texas 77445 Phone: (713) 857-3911
Dean of Graduate Studies: W. Webster
Dean of College: Vacant

BIOLOGY DEPARTMENT
Phone: Chairman: (713) 857-3911

Head: E.W. Martin
Professors: J.E. Berry, E.W. Martin
Associate Professors: R.R. Calhoun, R.D. Humphrey
Assistant Professors: A. Washington
Instructor: L.E. Henderson, S.A. Smith
Teaching Fellows: 5
Research Assistant: 1

Degree Programs: Biology, Biological Education

Undergraduate Degree: B.S.
Undergraduate Degree Requirements: 34 hours of Biology, 8 hours of Physics, 8 hours of Chemistry

Graduate Degree: M.S.
Graduate Degree Requirements: Master: A minimum of 20 hours in Biology, plus a minimum of 10 hours in a minor area. The total 30 hours may be done in biology if the candidate so desires, that is, if the candidate does not choose to select a minor.

Graduate and Undergraduate Courses Available for Graduate Credit: Invertebrate Zoology, Parasitology, Experimental Embryology, Histology, Vertebrate Zoology, Advanced Physiology, Genetics, Microbiology, Evolution, Ecology

RICE UNIVERSITY

6100 South Main Phone: (713) 528-4141
Houston, Texas 77001
Dean of Graduate Studies: J.L. Margrave
Dean of Science and Engineering: W.E. Gordon

DEPARTMENT OF BIOLOGY
Phone: Chairman: Ext. 491 Others: Ext. 631

Chairman: J.W. Campbell
Professors: J.W. Campbell, F.M Fisher, Jr., C.W. Philpott, R. Storck, S. Subtelny, C H. Ward
Associate Professors: K.D. Ansevin, R.M. Glantz, C.R. Stewart
Assistant Professors: H. Bultmann, A. Eskin, B.R. Hammond, P. Harcombe, D.M. Johnson
Research Assistants: 4
Instructor: 1
Research Associate: 1
Postdoctoral Fellows: 3

Degree Programs: Genetics, Parasitology, Physiology

Undergraduate Degree: B.A.
Undergraduate Degree Requirements: At least 40 courses with associated laboratory, computation, or tutorial sessions as appropriate, must be passed. At least 14 courses must be on an advanced level, eight of which must be in Biology.

Graduate Degrees: M.A., Ph.D.
Graduate Degree Requirements: Master: The degree of Master of Arts may be obtained after the completion of 30 hours of graduate study, six hours of which must be earned by the completion and public defense of a thesis embodying the results of an original investigation Doctoral: Complete 3 or more years of graduate study, with at least the last 2 years at Rice University, complete an original publication, and submit a doctoral thesis, perform satisfactorily as a teaching assistant for at least 4 semesters pass a comprehensive preliminary examination, defend publicly his thesis, and present a seminar on his research.

Graduate and Undergraduate Courses Available for Graduate Credit: Advanced Cell Physiology, Advanced Comparative Biochemistry, Concepts of Nervous Systems Functions, Physiological Clocks and Their Uses, Cell Biology, Topics in Cell Biology, Special Projects in Developmental Biology, Cell and Tissue Interactions in Development, Advanced Developmental Biology, Topics in Microbiology, Topics in Plant Biology, Topics in Population Biology, Physiology of Parasitism, Topics in Symbiology, Marshland and Estuarine Biology, Topics in Advanced Genetics, Topics in Ecosystem Biology, Topics in Invertebrate Physiology, Advanced Environmental Physiology, Topics in Microbial Genetics

ST. EDWARD'S UNIVERSITY

Austin, Texas 78704 Phone: (512) 444-2621
Dean of College: Br. H. Altmiller

BIOLOGY DEPARTMENT
Phone: Chairman: Ext. 225 Others: Ext. 223

Chairman: Br. D. Lynch
Professor: Br. D. Lynch
Assistant Professors: J.T. Mills, R.W. Reese
Instructor: J. Schafer

Degree Program: Biology

Undergraduate Degrees: B.A., B.S.
Undergraduate Degree Requirements: Total semester hours B S. or B.A. - 120, B.S. Biology 30 hours, Chemistry 16 hours, Physics 8 hours, Mathematics 9 hours, B.A. - Biology 30 hours, Chemistry 8 hours, Physics 8 hours, Mathematics 4 hours

ST. MARY'S UNIVERSITY

2700 Cincinnati Avenue Phone: (512) 433-2311
San Antonio, Texas 78284
Dean of Graduate Studies: L. Brown
Dean of College: J.W. Langlinais

DEPARTMENT OF BIOLOGY
Phone: Chairman: Exts. 389, 395

Chairman: J.T. Donohoo
Professors: J.T. Donohoo, A.J. Kaufmann, B.M. Machia
Assistant Professors: L.W. Goerner, R.M. Moerchen

Degree Program: Biology

Undergraduate Degrees: B.A., B.S.
Undergraduate Degree Requirements: 8 semesters

Biology (32 credit hours), 6 semesters in related minor (24 hours), 128 credit hours total

SAM HOUSTON STATE UNIVERSITY

Huntsville, Texas 77340 Phone: (713) 295-6211
Dean of Graduate Studies: B.M. Hayes
Dean of College: E.L. Dye

AGRICULTURE DEPARTMENT
Phone: Chairman: Ext. 2902

Head: D.G. Moorman
Professors: V.A. Amato, R.J. Agan, M.A. Brown, J.A. Nance, D.G. Moorman
Associate Professors: B.R. Harrell, P.L. Little, E.J. Simpson
Assistant Professors: J.A. Chandler, W.E. Jowell, Y.C. Moseley, L.N. Sikes, H. Schümann, T.J. Honeycutt, V. Stewart
Instructor: J.J. Snelgrove
Teaching Fellows: 3

Field Stations/Laboratories: Agriculture Center - Country Campus, Huntsville, Texas

Degree Programs: Agricultural Education, Agriculture, Agronomy, Animal Husbandry, Crop Science, Horticultural Science

Undergraduate Degree: B.S.
Undergraduate Degree Requirements: 128 semester hours

Graduate Degree: M.S.
Graduate Degree Requirements: Master: 36 semester hours

Graduate and Undergraduate Courses Available for Graduate Credit: Not Stated

BIOLOGY DEPARTMENT
Phone: Chairman: Ext. 2292

Chairman: M.E. Yoes
Professors: Not Stated
Associate Professors: Not Stated
Assistant Professors: Not Stated
Instructors: Not Stated

Field Stations/Laboratories: One is in the planning stages with ground presently secured.

Degree Program: Biology

Undergraduate Degrees: B.S., B.A., B.A.T.
Undergraduate Degree Requirements: Core requirement 19 hours Biology, B.S. 17 hours Biology, 16 hours Chemistry, 8 hours Physics, 8 hours Mathematics, B.A., 11 hours Biology, 8 hours Chemistry, 6-8 hours Mathematics, B.A.T. 11 hours Biology, 8 hours Chemistry, 8 hours Physics, 6 hours Mathematics

Graduate Degrees: M.A., M,S.
Graduate Degree Requirements: Master: 36 hours - selected by a committee after taking an introductory proficiency examination

Graduate and Undergraduate Courses Available for Graduate Credit: Classification of Vertebrates, Entomology, Aquatic Biology, Electron Microscopy Techniques, Plant Physiology, Plant Ecology, Animal Ecology, Parasitology, Comparative Animal Physiology, Animal Behavior, Economic Botany, Cytology, Phycology, Endocrinology, Plant Anatomy, Animal Invertebrate, Evolution, Mycology, Biometry, Virology, Microphysiology, Ichthyology, Herpetology, Avian Biology, Mammalogy, Limnology, Cellular Physiology, Paleontology, Population Genetics, Electron Microscopy

SOUTHERN METHODIST UNIVERSITY

Dallas, Texas 75275 Phone: (214) 692-2000
Dean of College: A.L. McAlester

DEPARTMENT OF BIOLOGY
Phone: Chairman: (214) 692-2730 Others: 692-2000

Chairman: J.L. McCarthy
Professors: J.P. Harris, J.L. McCarthy, W.B. Stallcup (Associate Provost of University)
Associate Professors: R.S. Sohal, J.E. Ubelaker
Assistant Professors: V.F. Allison, N.M Hall, W.F. Mahler, R.O. McAlister, J.O. Mecom, C. Nations
Instructors: L. Laury, J. Flook (Herbarium Botanist)
Teaching Fellows: 7
Research Assistants: 3
Technician: 1

Field Stations/Laboratories: Fort Burgwin, New Mexico

Degree Program: Biology

Undergraduate Degrees: B.A., B.S.
Undergraduate Degree Requirements: B.A. 25 semester hours Biology with 16 semester hours selected from core courses,), plus 12 semester hours of Chemistry, B.S. - At least 33 semester hours Biology (16 semester hours selected from core courses) plus 16 semester hours Chemistry, 8 semester hours Physics, and 6 semester hours Calculus

Graduate Degree: M.S.
Graduate Degree Requirements: Master: Course work and 6 semester hours thesis work to total 30 semester hours

Graduate and Undergraduate Courses Available for Graduate Credit: Cell Biology (U), Biological Chemistry (U), Metabolic Pathways (U), Modern Concepts of Genetics (U), Ecology of Parasites, Histology and Histochemistry (U), Principles of Plant Physiology (U), Laboratory Methods in Plant Physiology (U), Bryology (U), Entomology (U), Evolution and Systematics, Comparative Physiology, Literature of Biological Sciences, Seminar, Selected Topics in Systematic Biology, Selected Topics in Cytology and Physiology, Techniques in Electron Microscopy, Biologic Electron Microscopy, Biological Ultrastructure, Aquatic Vascular Plants, Experimental Parasitology, Helminthology, Aquatic Biology, Biology of Mammals, Biology of Birds, Endocrinology Physiology, Experimental Endocrine Physiology

SOUTHWEST TEXAS STATE UNIVERSITY

San Marcos, Texas 78666 Phone: (512) 245-2111
Dean of Graduate Studies: J.F. Dawson
Dean of College: O.L. Dorsey

DEPARTMENT OF AGRICULTURE
Phone: Chairman: (512) 245-2130 Others: 245-2154

Chairman: R.V. Miller, Jr.
Professors: T.R. Buie, J.D. Elliott, C.M. Gregg, R.V. Miller, Jr.
Associate Professors: G.B. Champagne, G.M. Rydl
Assistant Professors: R.B. Harvey, R.E. Helm, R.R. Shell

Field Stations/Laboratories: Southwest Texas State University Farm, San Marcos

Degree Program: Agriculture

Undergraduate Degree: B.S.
Undergraduate Degree Requirements: 128 semester hours

BIOLOGY DEPARTMENT
Phone: Chairman: (512) 245-2171

Chairman: Not Stated
Professors: M. Alexander, S. Edwards, W. Emery, R.T. Gary, D. Tuff
Associate Professors: W.K. Davis, D. Green, S. Sissom, D. Whitenberg, B.G. Whiteside, W.C. Young
Assistant Professors: C. Benjamin, R.F. Brown, D. Huffman, R. Koehn, D.B. Lambert, G. Meyers, R.C. Mills, E. Schneider
Instructors: G. Engeling, G. Farr, E.A. Morgan
Teaching Fellows: 4
Laboratory Assistants: 40

Field Stations/Laboratories: Aquatic Station - San Marcos Texas

Degree Programs: Botany, Zoology, Aquatic Biology, Microbiology, Physiology, General Biology

Undergraduate Degree: B.S.
Undergraduate Degree Requirements: Core Courses: Botany, Zoology, Human Physiology, Genetics, 10 courses in Biology, 4 courses in Chemistry, 3 courses in Mathematics, 2 courses in Physics

Graduate Degree: M.S.
Graduate Degree Requirements: Master: 8 courses, 2 hours seminar, 6 hours research

Graduate and Undergraduate Courses Available for Graduate Credit: Evolution, Field Biology of Animals, Field Biology of Plants, Parasitology, Ichthyology, Biometry, Comparative Physiology, Experimental Techniques, Cell Physiology, Advanced Genetics, Entomology, Limnology, Cytology and Microtechnique, Natural History Expedition, Seminar, Radiation Biology, Topics in Ecology, Fisheries Management, Parasites and Diseases of Fishes, Developmental Biology

SOUTHWESTERN UNION COLLEGE

Keene, Texas 76059 Phone: (817) 645-3921
Dean of College: L. Larson

BIOLOGY DEPARTMENT
Phone: Chairman: Ext. 274

Chairman: D.F. Beary
Associate Professors: R.L. McCluskey, J. Irvine

Field Stations/Laboratories: Acapulco, Mexico

Degree Program: Biology

Undergraduate Degrees: B.A., B.S.
Undergraduate Degree Requirements: B.A. 44 quarter hours, B.S. 52 quarter hours

SOUTHWESTERN UNIVERSITY

Georgetown, Texas 78626 Phone: (512) 863-6511
Dean of College: F.B. Clifford

DEPARTMENT OF BIOLOGY
Phone: Chairman: Ext. 234

Chairman: E.C. Girvin
Professors: E.C. Girvin, E.M. Lansford
Assistant Professors: D. Deeds, H.S. Jacob

Degree Programs: Biology, Medical Technology

Undergraduate Degree: B.S.
Undergraduate Degree Requirements: Core of 37 semester hours, 32 hours Biology (most specified courses), 18 hours minor, 3 hours Calculus, 8 hours Physics and 16 hours Chemistry if not in minor, plus electives to 124 hours

STEPHEN F. AUSTIN STATE UNIVERSITY

Nacogdoches, Texas 76961 Phone: (713) 569-2011
Dean of Graduate Studies: J.N. Gerber
Dean of School of Science and Mathematics: G.T. Clayton

DEPARTMENT OF BIOLOGY
Phone: Chairman: (713) 569-3601

Chairman: R.C. Faulkner
Professors: J.H. Burr, Jr., W.W. Gibson, V.J. Hoff, K.D. Mace, H.S. McDonald, E.S. Nixon, W.V. Robertson, H.T. Russell, Jr.
Associate Professors: J.D. McCullough, F.L. Rainwater, H.B. Weyland
Assistant Professors: C.D. Fisher, C.W. Mims, W.G. Slagle, B. VanDover
Instructors: G.A. Cook, N.F. McCord
Graduate Teaching Assistants: 20

Field Stations/Laboratories: S.F.A. Biology Department Field Station - Lake Sam Rayburn Reservoir

Degree Programs: Botany, Zoology

Undergraduate Degree: B.S.
Undergraduate Degree Requirements: 30 semester hours of Biology at least 15 of these hours must be at the advanced level (300 or 400 level)

Graduate Degree: M.S.
Graduate Degree Requirements: Master: 30 semester hours of Biology (Thesis), 36 semester hours of Biology (Non-thesis)

Graduate and Undergraduate Courses Available for Graduate Credit: Plant Anatomy, Cellular Biology, Plant Physiology, Comparative Animal Physiology, Pathogenic Microbiology, General Cytology, Advanced Microbiology, General Ecology (U), Virology, Plant Pathology, Invertebrate Natural History, Ornithology, Plant Taxonomy, Mammalogy, Herpetology, Ichthyology, Parasitology, Animal Histology, Limnology, Plant Ecology, Animal Ecology, Advanced Plant Taxonomy, Protozoology, Advanced Vertebrate Zoology, Cytogenetics, Organic Evolution, Special Topics in Animal Physiology, Cryptogamic Botany, Morphology of the Seed Plant, Research Techniques, Biological Literature, Aquatic Ecology, Modern Biology for Teachers

SUL ROSS STATE UNIVERSITY

Alpine, Texas 79830 Phone: (915) 837-3461
Dean of Graduate Studies: J. Houston
Dean of College: C. Lamb

BIOLOGY DEPARTMENT
Phone: Chairman: Ext. 241

Chairman: B.H. Warnock
Professors: A. Hoefling, A.M. Powell
Associate Professor: J. Scudday
Assistant Professors: R. Welker
Instructor: K. Whitley

Field Stations/Laboratories: Boquillas Research Center (Big Bend Natural Park)

Degree Program: Biology

Undergraduate Degree: B.S.
Undergraduate Degree Requirements: 32 semester hours in major and 18 semester hours in minor field or 24 semester hours in 2nd teaching field

Graduate Degree: M.S.
Graduate Degree Requirements: Master: 18 semester hours in Biology, 6 for thesis and 6 for minor

Graduate and Undergraduate Courses Available for Graduate Credit: (No undergraduate courses are available for graduate credit). Cacti and Succulents of Southwest, Cytogenetics, Advanced Plant Systematics, Principles and Methods of Systematic Zoology, Paleobotany, Field Botany, Field Zoology

DEPARTMENT OF RANGE ANIMAL SCIENCE
Phone: Chairman: Ext. 235

Chairman: E.E. Turner
Professors: P.R. Weyerts, E.E. Turner
Associate Professor: S.N. Little
Assistant Professor: C.W. Hanselka
Graduate Teaching and Laboratory Assistants: 6-8

Degree Programs: Animal Disease, Animal Science, Meat Sciences, Range Science, Wildlife Management

Undergraduate Degree: B.S.
Undergraduate Degree Requirements: (1). At least thirty semester hours of work completed in residence at Sul Ross State University, which requirement cannot be satisfied in less than two semesters of a long session, one semester of a long session and two six-week summer terms, or five six-week summer terms. Forty-five hours of general cultural education are required for every bachelor's degree, exclusive of four semester hours of physical education. (2) A major of forty-eight hours in range animal science, twenty-one of which must be advanced. (3). A minor of twenty hours, six of which must be advanced. (4). A minimum of forty semester hours of advanced Courses, including twenty-four which must be completed in residence at Sul Ross State University, (5). 128 semester hours to graduate

Graduate Degree: M.S.
Graduate Degree Requirements: Master: 30 hours Program: (1). Thirty hours of graduate level courses (2). The student must complete at least eighteen semester hours in the major field. (3). The student must take six hours of thesis which may be used in eighteen hours in the major field. 36-Hour Short Thesis Program: (1). Thirty-six hours of graduate level courses. (2) The student must complete at least eighteen hours in the major field. (3) The student may with the approval of the major department, include one minor of six to twelve hours, or two minors of six to nine hours each. (4) The student must complete a short thesis in a 5000+ level course.

Graduate and Undergraduate Courses Available for Graduate Credit: Interpretations of Research Techniques, Advanced Livestock Marketing, Ranch Foreman, Advanced Animal Nutrition, Physiology of Animal Breeding, Principles of Animal Breeding, The Control of Animal Diseases, Advanced Livestock Management, Land-use Economics, Wildlife Population Dynamics, Farm and Ranch Management, Principles of Range Management, Range Improvement Techniques, Advanced Animal Science, Problems in Range Animal Science, Problems in Range Animal Science

TARLETON STATE UNIVERSITY

P.O. Box T219, TSU Phone: (817) 965-4411
Stephenville, Texas 76401
Dean of Graduate Studies: R.H. Walker
Dean of Arts and Sciences: R.C. Fain

DEPARTMENT OF BIOLOGICAL SCIENCES
Phone: Chairman: (817) 968-2715

Head: L. Johanson
Professors: L. Johanson, A.B. Medlen, E.O. Morrison
Associate Professors: W.H. Gehrmann, D. Pittman
Assistant Professors: H.W. Garner, J. Knovicka, R.J. Tafanelli
Instructor: G.W. Luker
Teaching Fellows: 6

Field Stations/Laboratories: Agricultural Experiment Station

Degree Programs: Biological Sciences, Botany, Fisheries and Wildlife Biology, Medical Technology, Wildlife Management, Zoology

Undergraduate Degree: B.S.
Undergraduate Degree Requirements: 4 semesters of English, 2 semesters of Mathematics, 3 semesters of Chemistry (4 in classical), 2 semesters Physics, 9 semesters of Biology - some stipulated

Graduate Degree: M.S.T.
Graduate Degree Requirements: Master: 36 semester hours with 24 in department with Statistics

Graduate and Undergraduate Courses Available for Graduate Credit: Vertebrate Histology (U), Plant Physiology (U), Animal Physiology (U), Ichthyology (U), Mammalogy (U), Ornithology (U), Immunity (U), Parasitology (U), Biochemistry (U), Methods of Systematic Zoology Cellular Biology, Epidemiology of Zoonoses, Vertebrate Endocrinology, Biology of Fishes, Aquatic Environment, Development Biological Concepts, Quantitative Biology, Practicum, Field Problem, Internship

TEXAS A AND I UNIVERSITY

Kingsville, Texas 78363 Phone: (512) 595-3803
Dean of Graduate Studies: J.C. Rayburn
Dean of College: R.D. Rhode (Acting)

DEPARTMENT OF BIOLOGY
Phone: Chairman: (512) 595-3803

Chairman: J.T. Peacock
Professors: E.R. Boqusch, A.H. Chaney, R.B. Davis, J.T. Peacock, D.R. Pratt
Associate Professors: G.B. Fink, J E. Gillaspy, H.H. Hildebrand, G.G. Williges, C.E. Wood
Assistant Professor: J.C. Perez

Field Stations/Laboratories: Texas A and I University Biological Station

Degree Program: Biology

Undergraduate Degrees: B.A., B.S.
Undergraduate Degree Requirements: B.S. 34 credits Biology, 16 credits Chemistry, 6 credits Mathematics, 8 credits Physics, B.A. (Physics not required)

Graduate Degree: M.S.
Graduate Degree Requirements: Master: Thesis option 30 semester hours, thesis, comprehensive examination, Non-Thesis 36 semester hours, research report, comprehensive examination

Graduate and Undergraduate Courses Available for Graduate Credit: Research Methods in Biology, Advanced Topics in Biology, Advanced Biological Principles and History, Advanced Biological Materials and Preparations, Marine Biology, Biology of Estuarine Organisms, Wildlife Biology, Wildlife Resource Management, Modern Concepts in Biology, Advanced Materials and Preparations for Teaching Biology, Investigations in Biology, Research Problems, Vertebrate Embryology (U), Genetics (U), Population Genetics (U), Bacteriology (U), Immunology (U), Plant Physiology (U), Vertebrate Zoology (U), Physiology (U), Plant Pathology (U), Ornithology (U), Parasitology (U), Ichthyology (U), Seminar (U), Taxonomy of the Higher Plants (U), Ecology (U), Biological Problems (U), Plant Anatomy (U)

TEXAS A AND M UNIVERSITY

College Station, Texas 77843 Phone: (713) 845-1011
Dean of Graduate School: G.W. Kunze

College of Agriculture

Dean: H.O. Kunkel

ANIMAL SCIENCE DEPARTMENT

Phone: Chairman: (713) 845-1541 Others: 845-1011

Head: O.D Butler
Professors: J W. Bassett, O.D. Butler, Z.L. Carpenter, T.C. Cartwright, C.W. Dill, W.C. Ellis, F. Hale (Emeritus), R.E. Leighton, H.E. Randolph, J.K. Riggs, A.M. Sorensen, T.D. Tanksley, C. Vanderzant
Associate Professors: B.F. Cobb, H.W. Franke, N.M. Kieffer, G.T. King, D.C. Kraemer, G.T. Lane, G.D. Potter, L.M. Schake, G.C. Smith, D.G. Weseli, L. D. Wythe
Assistant Professors: C.A. Brodoriok, J. Caldwell, T R Dutson, P.G. Harms, J.H. Hesby, R.L. Hostetler, R. E. Lichtenwainer, C.R. Long, A.A. Melton
Instructors: E.H. Bird, L.L. Boleman, R.A. Bowling, G. Davis, R.C. George, G.N. Robinson, T.K. Scribner, D.L. Zink
Visiting Lecturer: 1
Teaching Fellows: 12
Research Assistants: 18
Non-Teaching: 8

Field Stations/Laboratories: USDA Southwestern Great Plains Research Center - Bushland, TAMU Agricultural Research and Extension Center - Vernon, TAMU Agricultural Research Center - McGregor, TAMU Agricultural Research and Extension Center - Overton, TAMU Agricultural Research and Extension Center - San Angelo, TAMU Agricultural Research and Extension Center - Uvalde, TAMU Agricultural Research and Extension Center - Weslaco, TAMU Agricultural Research Station - Angleton, TAMU Agricultural Research Station - Beeville, TAMU Agricultural Research Station - Sonora, TAMU Agricultural Research Station - Spur, TAMU Ag riculture Research and Extension Center - Corpus Christi

Degree Programs: Animal Science, Dairy Science, Food Science, Physiology of Reproduction, Animal Nutrition, Animal Breeding, Methods of Agriculture

Undergraduate Degree: B.S.
Undergraduate Degree Requirements: Total hours - 132 History 6 hours, Political Science 6 hours, Chemistry 13-20 hours, Mathematics 3-9 hours, Biology 12 hours, English 9 hours, Genetics 4 hours, Electives 16-20 hours

Graduate Degrees: M.S., Ph.D.
Graduate Degree Requirements: Master: Minimum GPR 3.0, minimum GRE 800, Minimum total hours 32-36 Doctoral: Minimum GPR 3.0, Minimum GRE 800, Minimum total hours 62 plus M S., or 96 beyond B.S.

Graduate and Undergraduate Courses Available for Graduate Credit: Beef Cattle Production (U), Meat Science and Technology (U), Growth and Finishing of Beef Cattle (U), Swine Production (U), Sheep and Angora Goat Production (U), Principles of Equine Production and Utilization (U), Reproduction in Farm Animals (U), Artificial Breeding of Livestock (U), Marketing and Grading of Livestock and Meats (U), Marketing of Livestock (U), Livestock and Meats Evaluation (U), Animal Nutrition (U), Livestock Practicum (U), Food Plant Production (U), Milk Secretion and Milking (U), Dairy Herd Improvement (U), Feeding and Management of Dairy Cattle (U), Seminar (U), Dairy Herd Improvement (U), Problems (U), Protein and Energy Nutrition, Experimental Nutrition, Ruminant Nutrition, Advancements in Beef Cattle Production, Meat Science and Research Techniques, Feedlot Management, Behavior and Training of Domestic Animals, Quantitative Genetics, Experimental Techniques in Meat Science, Advancements in Sheep and Angora Goat Production, Advancements in Swine Production, Animal Breeding, Physiology of Reproduction, Concepts in Reproduction, Technology of Meat Processing and Distribution, Seminar, Professional Internship, Problems, Research, Dairy Production, Chemistry of Foods, Microbiology of Foods, Seafood Preservation and Processing

DEPARTMENT OF BIOCHEMISTRY AND BIOPHYSICS

Phone: Chairman: (713) 845-5032 Others: 845-1011

Chairman: W.A. Landmann
Professors: B.J. Camp, J.R. Couch (Emeritus), J.W. Dieckert, H.O. Kunkel, W.A. Landmann, R.W. Lewis, J. Nagyvary, J.M. Prescott, R. Reiser (Distinguished) H.A. Röller
Associate Professors: G.W. Bates, N.R. Bottino, C.M. Cater, G.A. Donovan, R.D. Grigsby, E.F. Meyer, R.D. Neff, C.N. Pace, E.A. Tsutsui
Assistant Professors: E.D. Harris, A.R. Hanks, J.M. Magill
Instructor: P.W. Moeller
Visiting Lecturers: 6
Teaching Fellows: 8
Research Assistants: 26
University Health Fellow: 1

Field Stations/Laboratories: Biochemistry and Biophysics Field Laboratory - College Station, Texas

Degree Programs: Biochemistry, Nutrition, Biophysics, Food Technology

Undergraduate Degree: B.S. (Biochemistry)
Undergraduate Degree Requirements: 128 semester credit hours in curriculum, includes 4 credit hours Physical Education, 6 credit hours Political Science, 6 credit hours American History

Graduate Degrees: M.S., Ph.D.
Graduate Degree Requirements: Master: 32 hours plus thesis or 36 hours (Non-thesis degree) Doctoral: 96 hours beyond baccalaureate, 64 credit hours beyond Master's

Graduate and Undergraduate Courses Available for Graduate Credit: Biochemistry - Comprehensive Biochemistry I (U), Comprehensive Biochemistry II (U), Biochemistry Laboratory I (U), Biochemistry Laboratory II (U), General Biochemistry I, General Biochemistry II, Experimental Techniques in Biochemistry, Chemistry and Metabolism of Lipids, Proteins and Enzymes, Nutritional Mechanisms and Metabolic Regulation in Man, Chemistry and Metabolism of Nucleic Acids, Seminar Biophysics - Interpretation of Organic Mass Spectra, Thermodynamics of Biochemical Equilibria, Structural Biochemistry, Radioisotopes Techniques, Seminar

DEPARTMENT OF ENTOMOLOGY

Phone: Chairman: (713) 845-2516 Others: 845-1011

Chairman: P.L. Adkisson
Professors: P.L. Adkisson, H.R. Burke, F.W. Plapp, Jr., N.M. Randolph
Associate Professors: D.G. Bottrell, G.W. Frankie, R.L. Hanna, L.L. Keeley, T.L. Payne, M.A. Price, J.C. Schafner, J.W. Smith, Jr., W.L. Sterling, H.W. Van Cleave, S.B. Vinson, J.K. Walker
Assistant Professors: D.E. Bay, R.N. Coulson, M.K. Harris, J.K. Olson
Instructor: C.L. Meek
Research Assistants: 20
Teaching Assistants: 3

Degree Program: Entomology

Undergraduate Degree : B.S.
Undergraduate Degree Requirements: 129 semester hours, minimum of 24 semester hours in Entomology

Graduate Degrees: M.S., M. Agr., Ph.D.
Graduate Degree Requirements: Master: M.S. - 32 semester hours, including research M.Agr. - 36 semester hours Doctoral: Minimum of 96 semester

hours beyond bachelors degree, including research

Graduate and Undergraduate Courses Available for Graduate Credit: Principles of Insect Pest Management (U), General Economic Entomology (U), Medical Entomology (U), Insect Ecology (U), Principles of Systematic Entomology, Phylogeny and Classification of Insects, Aquatic Entomology, Economic Entomology, Insect Physiology, Acarology, Medical and Veterinary Entomology, Insect Toxicology, Insect Biochemistry and Endocrinology, Special Topics in Entomology

DEPARTMENT OF FOREST SCIENCE
Phone: Chairman: (713) 845-5033 Others: 845-3211

Head: R.G. Merrifield
Professors: D.F. Durso, P.R. Kramer, R.G. Merrifield, J.P. Van Buijtenen
Associate Professors: R.L. Bury, D.M. Moehring, R.R. Rhodes, E.P. VanArsdel, T.L. Payne
Assistant Professors: J.A. Emery, C.A. Hickman, R.A. Woessner
Instructor: B.R. Miles
Teaching Fellows: 2
Research Assistants: 7

Field Stations/Laboratories: Texas A and M University Agriculture Research and Extension Centers at Beaumont, and Eagle Lake, USDA Southwestern Great Plains Research Center at Bushland, Texas A and M University Agricultural Research and Extension Center at Chillicothe-Vernon, Texas A and M University Research and Extension Center at Dallas, Texas A and M University Agricultural Research Center at El Paso, Texas A and M University Agricultural Research and Extension Center at Lubbock, Texas A and M University Agricultural Research Center at Mc Gregor, Texas A and M University Agricultural Research and Extension Center at Overton, Texas A and M University - Prairie View A and M University Cooperative Research Center at Prairie View, Texas A and M University Agricultural Research and Extension Center at San Angelo, Blackland Research Center at Temple, Texas A and M University Agricultural Research and Extension Center at Uvalde, Texas A and M University Agricultural Research and Extension Center at Weslaco, Texas A and M University Agricultural Research Stations at Angleton and Beeville, Texas A and M University Poultry Disease Laboratory at Center, Texas A and M University Rio Grande Plains Research-Demonstration Station at Crystal City, Texas A and M University Poultry Disease Laboratory at Gonzales, Texas A and M University Agricultural Research Station at Iowa Park, Texas A and M University-Texas Tech University Cooperative Research Unit, Lubbock, Texas A and M University Fruit Research Demonstration Station at Montague, Texas A and M University Vegetable Research Station at Munday, Texas A and M University Agricultural Research Station at Pecos, Texas A and M University Agricultural Research Station at Sonora, Texas A and M University Agricultural Research Station at Spur, Texas A and M University Tarleton Experiment Station at Stephenville, Texas Experimental Ranch, Throckmorton, Texas A and M University-Texas A and I University Cooperative Research Unit, Weslaco, Texas A and M University Plant Disease Research Station at Yoakum

Degree Programs: Forest Management, Urban Forestry, Natural Research Conservation

Undergraduate Degree: B.S.
Undergraduate Degree Requirements: Completion of 135 semester hours of coursework with minimum GPA of 2.00, The state has certain requirements in history, political science and physical education.

Graduate Degrees: M.S., M of Agr., Ph.D.
Graduate Degree Requirements: Master: M.S. - completion of 32 hours of coursework with a minimum GPA of 3.00. Successful completion and defense of a thesis M. of Agr. - Completion of 36 hours of coursework with minimum GPA of 3.00. Preparation and defense of written report. Doctoral: Completion of 64 hours of coursework beyond the Masters degree with a minimum of GOA of 3.000. Successful qualifying examination. Successful completion and defense of dissertation.

Graduate and Undergraduate Courses Available for Graduate Credit: Forest Products (U), Forest Management (U), Forest Policy, Harvesting Systems, Quantifications Methods in Forest Management (U), Arboriculture (U), Urban Forestry, Forest Ecology, Advanced Silviculture, Forest Practices, The Research Process, Seminar, Professional Internship

DEPARTMENT OF PLANT SCIENCE
(No reply received)

DEPARTMENT OF POULTRY SCIENCE
Phone: Chairman: (713) 845-1931 Others: 845-1011

Head: W.F. Krueger
Professors: C.R. Creger, T.M. Ferguson, W.F. Krueger
Associate Professors: R L. Atkinson, R.C. Fanguy, F. A. Gardner, D.B. Mellor, C.B. Ryan
Assistant Professors: J.W. Bradley, J.R. Cain
Research Assistants: 8
Fellowship: 1
Teaching Assistants: 2

Degree Programs: Poultry Science, Poultry Breeding and Genetics, Nutrition, Physiology, Products Technology and Environmental Physiology

Undergraduate Degree: B.S.
Undergraduate Degree Requirements: Science option and Inststrial option -both requiring 132 student credit hours for graduation

Graduate Degrees: M. Agr., M.S., Ph.D.
Graduate Degree Requirements: Master: Master of Science: Minimum of 32 student credit hours including an acceptable thesis in Avian genetics, Nutrition, Physiology, Products Technology, Environmental Physiology or Management, M. of Agr.: Minimum of 36 student credit hours and one semester of inservice training with emphasis on production and personnel Management. Doctoral: Minimum of 64 student credit hours and an acceptable doctoral dissertation. A satisfactory Graduate Record Examination score, a preliminary examination and defense of the doctoral dissertation are required.

Graduate and Undergraduate Courses Available for Graduate Credit: Marketing of Poultry Products, Nutrition and Feed Formulation, Avian Genetics and Biometrics, Environmental Physiology, Commercial Egg Production, Avian Incubation and Embryology, Environmental and Developmental Relationships, Avian Physiology, Poultry Products Technology, Breeding Systems Concept, Avian Nutrition

DEPARTMENT OF RANGE SCIENCE
Phone: Chairman: (713) 857-6531 Others: 857-6533

Chairman: J.L. Schuster
Distinguished Professor: F.W. Gould
Professors: J D. Dodd, C.L. Leinweber
Associate Professors: R.H. Haas, M.M. Kothmann, C.J Scifres, F.E. Smeins
Assistant Professors: A.J Dye, S. Hatch, R.E. Whitson
Research Assistants: 11

Degree Program: Range Science

Undergraduate Degree: B.S.
Undergraduate Degree Requirements: 132 credit hours

Graduate Degrees: M.S., M.A., Ph.D.

Graduate Degree Requirements: Master: 32 hours
Doctoral: 66 hours over M.S.

Graduate and Undergraduate Courses Available for Graduate Credit: Range Resource Use, Ecology and Land Uses, Range Research Methods, Range Economics, Vegetation Influences, Plant and Range Ecology, Range Grasses and Grasslands, Control of Noxious Range Plants, Range Management Practices, Policies and Administration, Advances in Range Improvements

DEPARTMENT OF SOIL AND CROP SCIENCES
Phone: Chairman: (713) 845-3041

Head: M.E. Bloodworth
Professors: H.T. Blackhurst, M.E. Bloodworth, M.S. Brown, E.E. Burns, J.W. Collier, J.B. Dixon, C.L. Godfrey, E.C. Holt, G.W. Kunze, K.F. Mattil, M.G. Merkle, M.H. Milford, G.A. Niles, R.C. Potts, R.H. Rucker, J.R. Runkles, O.D. Smith, J.B. Storey, C.H.M. Van Bavel
Associate Professors: W.B. Anderson, A.J. Bockholt, H. H. Bowen, K.W. Brown, C.M. Cater, R.L. Duble, E.C. Gilmore, W.L. Hoover, L.R. Hossner, J.E. Larsen, K.J. McCree, M.E. McDaniel, E.L. McWilliams, J.R. Melton, J.F. Mills, A.E. Nightingale, L.M. Pike, L.W. Rooney, D.T. Smith, R.D. Staten, A.R. Swoboda, N.A. Tuleen, E.L. Whitely
Assistant Professors: R.W. Weaver
Geneticists: E.C. Bashaw, P.A. Fryxell, R.J. Kohel, O.G. Merkle, K.F. Schertz
Research Assistants: 28

Degree Programs: Agronomy, Fisheries and Wildlife Biology, Horticultural Science, Horticulture, Oceanography, Silviculture, Plant and Soil Science

Undergraduate Degree: B.S.
Undergraduate Degree Requirements: 130 credit hours, 6 hours Botany, 8 hours Chemistry, 9 hours English, 3 hours Mathematics, 6 hours History, 6 hours Political Science, 3 hours Physics, 4 hours Genetics, 3 hours Nutrition, 3 hours Plant Pathology, 3 hours Plant Physiology, 3 hours Economics, 3 hours Social Science, 3 hours Microbiology, 15 hours of Agronomy or Horticulture

Graduate Degrees: M.S., Ph.D.
Graduate Degree Requirements: Master: 32 credit hours, 3 hours statistics, 1 hour seminar, 4-8 hours research, other requirements set by committee Doctoral: 64 credit hours, 1 hour seminar, 8-16 hours research, other requirements set by committee

Graduate and Undergraduate Courses Available for Graduate Credit: Grain and Cereal Crops, Forage Crops, Cytological and Histological Principles in Plant Breeding, Pedology, Soils, Conservation and Man, Forage Crops Management, Advanced Soil Physics, Advanced Soil Analysis, Saline and Sodic Soils, Principles of Crop Ecology, Environmental Aspects of Crop-Water Relations, Experimental Crop Ecology, Physical Chemistry of Soils, Colloidal Chemistry of Soils and Clays, Soil Mineralogy, Soil Fertility Relationships, Cereal Grains for Human Food, Chemical and Physical Characteristics of Cerelas, Oilseed Proteins for Foods, Oil and Fat Food Products, Plant Breeding I, Plant Breeding II, Chemical Weed Control, Growth and Development of Turfgrass, Seminar, Professional Internship, Soil Microbiology (U), Soil Genesis and Classification (U), Pasture Management (U), Chemistry and Fertility of Soils (U), Fertilizer Technology (U), Turf Management (U), Soil Chemistry and Fertility Laboratory (U), Soil-Plant-Water Relations (U), Soil Physics (U), Chemical Weed Control (U), Agronomy Seminar (U), Problems (U)

DEPARTMENT OF WILDLIFE AND FISHERIES SCIENCES
Phone: Chairman: (713) 845-6751

Head: J.G. Teer
Caesar Kleberg Chair in Wildlife Ecology: J.C. Teer
Professors: J.R. Dixon, R.K. Strawn, J.G. Teer, F.R. Walther
Associate Professors: K.A. Arnold, W.J. Clark, J.M. Inglis, W.H. Neill
Assistant Professors: S.L. Beasom, B.W. Cain, M.E. Chittenden, C.R. Field (Research), F.S. Handricks, H.D. Irby (Research), J.R. Kelley, J.D. McEachran, R.L. Noble, R.E. Quinn, R.D. Reimer, D.J. Schmidly, N.J. Silvy, C D. Simpson
Instructors: B.A. Brown, Jr., V.F. Cogar
Research Assistants: 40
Teaching Assistants: 13
Research Associates: 9

Field Stations/Laboratories: TAMU Research and Extension at Uvalde and the TAMU Research and Extension Center at Corpus Christi

Degree Programs: Wildlife Ecology, Natural History, Fisheries Ecology, Aquaculture, Museum Science, and Environmental Education

Undergraduate Degree: B.S.
Undergraduate Degree Requirements: (1) complete with at least a C average one of the regular courses of study leading to a degree, (2) maintain at least a C average in major field, (3) complete at least 132 academic hours, 60 of the last 66 of which must be earned in residence at the Texas A and M campus.

Graduate Degrees: M. Agr., M.S., Ph.D.
Graduate Degree Requirements: Master: 32 hours beyond B.S. plus thesis (M.S.), 36 hours beyond the B.S. (M.Agr.), 36 hours beyond the B.S., 24 in major (M.S. non-thesis) Doctoral: 96 hours beyond the B.S. or 64 hours beyond the Masters, plus a dissertation

Graduate and Undergraduate Courses Available for Graduate Credit: General Mammalogy (U), General Ornithology (U), Animal Ecology (U), Techniques of Wildlife Management (U), Conservation and Management of Fishes (U), Limnology (U), Animal Population Dynamic (U), Nature Centers for Learning (U), Ecology for Teachers (U), Museums and their Functions (U), Ethology (U), Aquaculture (U), Field and Laboratory Methods (for teachers), Vertebrate Systematics, Vertebrate Ecology, Systematic Wildlife Research Methods, Estuarine Ecology, Marine Ichthyology, Shore and Estuarine Fishes, Analytical Procedures in Limnology, Mariculture, Vertebrate Ethology

College of Geosciences

Dean: E. Cook

DEPARTMENT OF OCEANOGRAPHY
Phone: Chairman: (713) 845-7211

Chairman: R.A. Geyer
Professors: R.A Geyer, L. Berner, Jr., A.H. Bouma, W.R. Bryant, R.M. Darnell, S.Z. El-Sayed, T. Ichiye, W. Pequegnat, R.O. Reid, R. Rezak, W. Sackett
Associate Professors: T.J. Bright, J. Cochrane, D. Fahlquist, L. Jeffrey, W. Nowlin, Jr., D.R. Schink, A.C. Vastano
Assistant Professors: C.W. Poag, B.J. Presley, T.W. Spence, J. Wormuth
Research Assistants: 11

Degree Program: Oceanography

Graduate Degrees: M.S., Ph.D.
Graduate Degree Requirements: Master: B.S. or B.A. degree, GRE examination, Prescirbed course work in all scientific areas of oceanography, thesis directed by the student's committee. Doctoral: M.S. degree, GRE examination, dissertation directed by the student's committee

Graduate and Undergraduate Courses Available for Graduate

Credit: Sea LaboratoryTechniques, Physical Oceanography I and II, Theoretical Physical Oceanography, Elements of Ocean Wave Theory, Dynamics of the Ocean and Atmosphere, Long Waves and Tides, Theory of Ocean Waves, Theories of Ocean Circulation, Underwater Sound, Marine Boring and Fouling Problems, Biological Oceanography, Analysis of Benthic Communities, Marine Zooplankton, Marine Phytoplankton, Deep-Sea Pelagic and Demersal Fishes, Organic Cycles of the Sea, Ecology of the Continental Shelf, Biology of Coral Reefs, Field Studies on Atlantic Coral Reefs, Geological Oceanography I and II, Benthic Marine Microfauna, Carbonate Sediments I and II, Techniques in Geological Oceanography, Marine Biostratigraphy I and II, Lithophycology, Chemical Oceanography I and II, Geochemistry of the Ocean, Isotope Geochemistry, Marine Organic Geochemistry, Meterological Oceanography, Ocean Boundary Layer Problems, Synoptic Physical Oceanography, Quatenary History of the Oceans

College of Science

Dean: Not Stated

DEPARTMENT OF BIOLOGY

Phone: Chairman: (713) 845-6131 Others: 845-1011

Head: W.P. Fife
Professors: S.O. Brown, L.S. Dillon, W.J. Dobson, H.L. Gravett, H. Kleerekoper, W.R. Kelmm, G.M. Krise, H.A. Röller, I.V. Sarkissian, J.H. Sperry, W.A. Taber, V. Bavel, V. Overbeek
Associate Professors: W.J. Clark, E.R. Cox, B.G. Foster, H.H. Harry, M H. Sweet, U.G. Whitehouse
Assistant Professors: J.W. Anderson, J.D. Geoschl, K.P. Kuchnow, D.M. Mueller, E.L. Thurston, B.F. Watson
Instructors: W.E. Finn, J.M Neff, R.J. Newton

Degree Programs: Animal Physiology, Biology, Botany, Immunology, Medical Microbiology, Microbiology, Physiology, Zoology

Undergraduate Degree: B.S.
Undergraduate Degree Requirements: 128 semester hours with approximately 34 hours in B ological Science

Graduate Degrees: M.S., Ph.D.
Graduate Degree Requirements: Master: Not less than 32 semester hours with 8 semester hours thesis Doctoral: Not less than 64 hours above the Masters with reading competence in one foreign language demonstrated

Graduate and Undergraduate Courses Available for Graduate Credit: Any upper division course may be counted for graduate credit as long as it is part of an approved course program. At the Master's level only 8 hours of undergraduate courses may be used.

College of Veterinary Medicine

Dean: G.C. Shelton

DEPARTMENT OF VETERINARY ANATOMY

Phone: (713) 845-2828

Head: R.F. Sis
Professors: R.G. Greeley, W.E. Haensly, A.G. Kemler
Associate Professors: J.E. Martin, N H. McArthur, G.G. Stott, M.E. Tatum
Assistant Professors: B.V. Beaver, G.R. Bratton, M.A. Herron, J.E. Smallwood

Degree Program: Veterinary Anatomy

Undergraduate Degree: B.S. (Biomedical Science Program)
Undergraduate Degree Requirements: Not Stated

Graduate Degrees: M.S., Ph.D.
Graduate Degree Requirements: Master: Acceptance by Graduate College, 32 credit hours, thesis
Doctoral: Acceptance by Graduate College, 96 hours Dissertation

Graduate and Undergraduate Courses Available for Graduate Credit: Anatomy (U), Histology (U), Embryology (U), Neuroanatomy (U), Applied Anatomy (U), Systems Anatomy (U), Anatomy, Histology, Neuroanatomy, Neuroendocrinology Anatomy, Anatomy of Laboratory Animals, Avian Anatomy, Anatomy of Reproductive Systems, Microscopic Histochemistry

DEPARTMENT OF VETERINARY MICROBIOLOGY
(No reply received)

DEPARTMENT OF VETERINARY PARASITOLOGY
(No information available)

DEPARTMENT OF VETERINARY PHYSIOLOGY AND PHARMACOLOGY

Phone: Chairman: (713) 845-7261 Ext. 23

Head: J.D. McCrady
Professors: B.J. Camp, R.H. Davis, J.W. Dollahite, D. Hightower
Associate Professors: J.G. Anderson, E.M. Bailey, L.D. Claborn, D.R. Clark, D.O. Wiersig
Assistant Professors: E.L. Akins, S.M. Hartsfield, L.D. Rowe
Research Associates: L.G. Gayle, (Associate), H.L. Kim (Associate)
Visiting Lecturers: 1
Teaching Fellows: 1
Research Assistants: 2
Fellow: 1
Graduate Assistants: 3

Degree Programs: Animal Physiology, Radiation Biophysics Toxicology

Graduate Degrees: M.S., Ph.D.
Graduate Degree Requirements: Master: 32 semester hours and thesis Doctoral: 96 semester hours and thesis

Graduate and Undergraduate Courses Available for Graduate Credit: Not Stated

DEPARTMENT OF VETERINARY PUBLIC HEALTH

Phone: Chairman: (713) 845-5261 Others: 845-3211

Head: A.I. Flowers
Professors: L.H. Russell, Visiting Professors: 12
Associate Professors: A.B. Childers, R.P. Crawford
Assistant Professor: G.N. Joiner
Visiting Lecturers: 2
Teaching Fellows: 1
Research Assistants: 1

Degree Programs: Epidemiology, Laboratory Animal Medicine, Veterinary Medical Science

Graduate Degrees: M.S., Ph.D.
Graduate Degree Requirements: Master: Ordinarily 32 semester hours of coursework beyond the bachelor's degree, including up to 8 hours for research and thesis (required) Doctoral: A minimum of 96 credit hours beyond the bachelor or 64 credit hours beyond the master's research and dissertation required. Language requirement determined by committee.

Graduate and Undergraduate Courses Available for Graduate Credit: Food Hygiene, Epidemiology (Applied), Food Toxicology, Public Health Concepts, Animal Diseases in Comparative Medicine

TEXAS CHRISTIAN UNIVERSITY

Fort Worth, Texas 76129 Phone: (817) 926-2461
Dean of Graduate Studies: F. Reuter
Dean of College: W.M. Wiebenga

DEPARTMENT OF BIOLOGY

Phone: Chairman: Exts. 461, 462, 463

Chairman: J.D. Smith
Professors: J.W. Forsyth, E.W. Gardner, S.T. Lyles, C.E. Murphy
Associate Professors: E.F. Couch, J.D. Smith
Assistant Professors: J.C. Britton, D.E. Keith, M.D. McCracken, L.W. Newland, W.J. Barcellona, W.L. Chaffin, G.W. Ferguson
Teaching Fellows: 7
Research Assistants: 2
Graduate Assistants: 5

Degree Programs: Biology, Medical Technology

Undergraduate Degrees: B.A., B.S.
Undergraduate Degree Requirements: B.A., Total of 124 hours including: Major - 30 semester hours, Biology, Minor - any approved by College of Arts and Sciences, B.S., Total of 132 hours including: 36 semester hours Biology, 16 hours Chemistry, 6 hours Mathematics including Calculus, 4 hours Physics

Graduate Degrees: M.A., M.S., M.A.T.
Graduate Degree Requirements: Master: Total of 30 approved hours, including at least 12 in B ology, 6 in thesis, and 3 in Biology seminars

Graduate and Undergraduate Courses Available for Graduate Credit: Biochemistry, Physiological Psychology, Advanced Physiological Psychology, Biochemistry Laboratory, Radioisotope Techniques, Physiology of Bacteria, Limnology, Marine Ecology, Aquatic Biology, Parasitology, Teaching of Biology, Invertebrate Morphology and Physiology, Cellular Physiology, Advanced Cell Biology, Microbial Ecology , Medical Bacteriology, Cellular Physiology, Advanced Cell Biology, Microbial Ecology, Medical Bacteriology, Immunology, Systematics, Advanced Invertebrate Zoology, Fisheries Biology, Field Techniques in Environmental Biology, Virology, Advanced Genetics, Biology Seminar, Assigned Problems in Various Fields of Biology. Other courses not on a regular schedule: History of Biology, Mathematical Biology

TEXAS COLLEGE

2404 North Grand Avenue Phone: (214) 593-8311
Tyler, Texas 75701
Dean of College: A. VanWright, Jr.

DEPARTMENT OF BIOLOGY
Phone: Chairman: Ext. 75

Head: S.P. Mouftah
Associate Professors: S.P. Mouftah
Assistant Professors: E.C. Coleman, I.C. Dugas

Degree Programs: Biology, Biological Education

Undergraduate Degree: B.S.
Undergraduate Degree Requirements: A minimum of 128 semester hours including 67 hours of General Freshman and Sophomore courses and the following: 36 hours of Biology, 20 hours of Chemistry, 11 hours in Mathematics, and 8 hours of Physics. 12 hours of electives are included

TEXAS LUTHERAN COLLEGE

Seguin, Texas 78155 Phone: (512) 379-4161
Dean of College: C.H. Oestreich

BIOLOGY DEPARTMENT
Phone: Chairman: (512) 379-4161

Chairman: H.W. Bischoff
Professor: H.W. Bischoff
Assistant Professors: I.G. Patterson, R.L. Torgerson

Degree Program: Biology

Undergraduate Degree: B.S.
Undergraduate Degree Requirements: Not Stated

TEXAS SOUTHERN UNIVERSITY

3201 Wheeler Street Phone: (713) 528-0611
Houston, Texas 77004
Dean of Graduate Studies: J. Jones
Dean of College: L.L. Clarkson

DEPARTMENT OF BIOLOGY
Phone: Chairman: Ext. 251

Head: J.J. Session
Professors: T.D. Cotton, J. Jones, J.J. Session, R.J. Terry, J. Race
Assistant Professors: O.O. Fadulu, D. Ghosh, L.L. Henderson, M. Hillar, Y. Hogan, R. McGee, W. Williams, L. Johnson
Instructors: C. Bennett, D. Colbert
Research Assistant: 1
Research Associate: 1

Degree Programs: Biology, B.ology for Teaching, Medical Technology, Premedicine, Biology for Teaching

Undergraduate Degree: B.S.
Undergraduate Degree Requirements: At least 30 semester hours Major courses and 94 or more University requirements, At least 30 semester hours Biology course and 94 or more University requirements.

Graduate Degree: M.S. (Biology for Teaching)
Graduate Degree Requirements: Master: A minimum of 30 semester hours, 24 of which must be taken in biology courses and six (6) semester hours of research. A minimum of 36 semester hours in Biology with six (6) semester hours of research.

Graduate and Undergraduate Courses Available for Graduate Credit: Biology Research (U), Cytology I and II Cytogenetics and Immunogenetics, Radiation Biology, Parasitology, Genetics, Plant Physiology, Plant Anatomy, Heredity and Evolution, Life Science, Plant Biology, Biology for Teachers I and II, Histology, Laboratory Methods in Animal Biology, Histological Techniques, Microbiology, Cellular Physiology, General Physiology, Biochemical Preparations, Enzymology, Intermediary and Cellular Metabolism, Biosynthetic Mechanisms, Biostatistics

TEXAS TECH UNIVERSITY

Lubbock, Texas 79409 Phone: (806) 742-0111
Dean of Graduate Studies: J.K. Jones , Jr.

College of Arts and Science

Dean: L.L. Graves

DEPARTMENT OF BIOLOGICAL SCIENCES
Phone: Chairman: (806) 742-7238 Others: 742-0111

Chairman: R.C. Jackson
Horn Professor: R.W. Strandtmann
Professors: R.J. Baker, F.J. Behal (Adjunct), J.D. Berlin, C.C. Black (Adjunct), E.D. Camp, R.C. Jackson, J.K. Jones, J.S. Mecham, J.M. McKenna (Adjunct), R.W. Mitchell, R.L. Packard, V.W. Proctor, F.L. Rose, R.W. Strandtmann, S.P. Yand (Adjunct)
Associate Professors: A.C. Allen, W.R. Atchley, C.L. Baugh, J.M. Burns, D.C. Carter, M.W. Coulter, A.M. Elliot, I.C. Felkner, J.E. George, J.R. Goodin, L.C. Kuhnley, S.S. Lefkowitz (Adjunct), P.R. Morey, J. Morrow (Adjunct), M.K. Rylander, D.W. Thayer
Assistant Professors: D.K. Northington, P.I. Tilton
Research Assistants: 15
Teaching Assistants: 30
Part-time Instructors: 20

Field Stations/Laboratories: Kermit Field Station - Kermit,

Junction Center - Junction

Degree Programs: Biology, Microbiology, Medical Technology, Microbiology, Zoology

Undergraduate Degrees: B.A., B.S.
Undergraduate Degree Requirements: 36 credit hours in courses in the Department of Biological Sciences in addition to physics and chemistry; plus courses in the humanities and social sciences for a total of 128 to 134 credit hours.

Graduate Degrees: M.S., Ph.D.
Graduate Degree Requirements: Master: 18 hours of course work, plus 6 hours of thesis in the major, with 6 hours in a minor for a total of 30 credit hours Doctoral: 3 years graduate study beyond Bachelors Degree (usually 60 credit hours), Thesis

Graduate and Undergraduate Courses Available for Graduate Credit: Microbiology - Instrumental Methods of Microbiology, Research in Microbiology, Selected Topics in Microbiology, General Virology, Microbial Genetics, Immunochemistry, Microbial Metabolism, Advanced General Microbiology, Immunobiology, Advanced Bacterial Physiology Zoology - Selected Topics in Invertebrate Physiology, Problems in Zoology, Principles and Methods of Systematic Zoology, Herpetology, Advanced Invertebrate Zoology, Field Zoology, Advanced Studies in Mammalogy, Physiological Ecology of the Vertebrates, The Arachnids, Biology of the Acarina, Advanced Ornithology, Zoogeography, Experimental Embryology, Comparative Endocrinology, Ichthyology, Advanced Vertebrate Anatomy, Vertebrate Zoology for Advanced Students, Comparative Physiology for Advanced Students, Biology - Advanced Experimental Heredity, Special Problems in Biometry, Advanced Experimental Laboratory in Developmental Biology, Biological Electron Microscopy, Instructional Methods in Biology, Advanced Developmental Biology, Population Genetics, Selected Topics in Radiation Biology, Application of Radioactive Tracers in Biology, Biological Fine Structure, Techniques in Biological Electron Microscopy, Special Problems in Genetics, Advanced Population Biology, Ecology of Inland Waters, Advanced Cell Biology, Cytogenetics, Biometry, Biochemical Genetics Botany - Problems in Botany, Vector Relationships in Plant Diseases for Advanced Students, Advanced Plant Anatomy, Field Botany, Taxonomy of Lower Green Plants, Morphology of the Vascular Plants, Advanced Taxonomy of the Vascular Plants, Plant Speciation, Morphogenesis and Plant Growth Regulators, Plant Growth and Development, Plant Pathology for Advanced Students, Plant-Water Relationships, Advanced Structure and Physiology of Woody Plants, Readings in Plant Geography, Morphology of Fungi for Advanced Students, Experimental Plant Anatomy, Plant Chemosystematics

College of Agricultural Sciences

Dean: A.R. Bertrand

DEPARTMENT OF AGRONOMY

Phone: Chairman: (806) 742-7291

Chairman: H.E. Dregne
Horn Professor of Agronomy: H.E. Dregne
Professors: B.L. Allen, C.I. Ayers, W.F. Bennett, A.R. Bertrand, J.D. Downes, G.O. Elle, L.H. Gile (Adjunct), C. Harvey, J.W. Hawley (Adjunct)
Associate Professors: E.A. Coleman, C.C. Jaynes, D. R. Kreig, R.E. Meyer
Assistant Professor: R.G. Stevens
Teaching Fellows: 1
Research Assistants: 12

Field Stations/Laboratories: Pan Tech (Amarillo, Texas), Junction, Texas

Degree Programs: Agronomy, Crop Science, Soil Science

Undergraduate Degree: B.S.
Undergraduate Degree Requirements: 128 hours plus 4 hours of band, Physical Education or ROTC, C average in all course work; 6 hours of history and 6 hours of government; minimum of 24 hours and maximum of 42 hours in agronomy courses.

Graduate Degrees: M.S., Ph.D.
Graduate Degree Requirements: Master: 24 hours of graduate course work plus 6 hours of thesis work, 18 hours in major field, B average in all course work, acceptable GRE score, residence for 1 full academic year. Doctoral: 60 or more hours of work beyond the bachelor's degree, normally, minimum of three years of graduate study beyond the bachelor's degree, minor of at least 18 hours, competence in foreign language or 12 hours of "tool" subjects, B. average in major and for all graduate work outside the major, 4 year time limit, acceptable GRE score.

Graduate and Undergraduate Courses Available for Graduate Credit: Seminar, Problems in Field Crops, Problems in Soils, Instrumental Analysis for Plants and Soils, Pasture Management, Soil Fertility and Fertilizers, Pedology, Research, Inorganic Plant Metabolism, Soil and Plant Relationships, Methods in Plant Breeding, Laboratory Methods in Plant Breeding, Soil Physics, Organic Plant Metabolism, Environmental Crop Physiology, Soil Mineralogy, Advanced Soil Classification, Herbicidal Action in Plants, Selected Topics in Crop Science, Selected Topics in Soil Science, Master's Thesis, Advanced Genetics, Advanced Soil Chemistry, Advanced Soil Physics, Metabolism of Crop Plants, Fruit and Vegetable Research, Fruit and Vegetable Crop Behavior, Selected Topics in Vegetable Production, Selected Topics in Fruit Production, Post Harvest Physiology of Fruit and Vegetable Crops, Breeding of Fruit and Vegetable Crops, Horticultural Research Methodology

ANIMAL SCIENCE DEPARTMENT

Phone: Chairman: (806) 742-1155

Chairman: F.A. Hudson (Acting)
Professors: R.C. Albin, J.H. Baumgardner, S.E. Curl, R.M. Durham, F.A. Hudson, L.F. Tribble, C.B. Ramsey, D.W. Zinn, L. Sherrod, G.F. Ellis (Adjunct), R.D. Furr (Adjunct)
Associate Professor: C.A. O'Brien
Assistant Professors: C.T. Gaskins, C.E. Sasse, L.H. Thompson
Research Assistants: 8

Field Stations/Laboratories: Texas Tech University Center Amarillo

Degree Programs: Animal Genetics, Animal Husbandry, Animal Physiology, Animal Science, Meat Sciences, Nutrition, Animal Business

Undergraduate Degree: B.S.
Undergraduate Degree Requirements: 132 credit hours plus 2 hours organized activities

Graduate Degrees: M.S., Ph.D.
Graduate Degree Requirements: Master: 24 hours of graduate course work plus 6 hours thesis Doctoral: 60 hours of work beyond bachelor's degree exclusive of dissertation credit, 18 hours in minor field, Qualifying, Preliminary and Final Examinations, Reading knowledge in one foreign language or completion of 12 hours outside of major and minor fields.

Graduate and Undergraduate Courses Available for Graduate Credit: Seminar, Animal Protein Nutrition, Animal Mineral and Vitamin Nutrition, Animal Energy Utilization, Ruminant Nutrition, Developmental Growth and Fattening, Environmental Physiology of Domestic Animals, Techniques in Animal Research, Research in Animal Science, Endocrinology, Biometry, Advanced

Animal Breeding, Physiology of Reproduction, Science of Meat and Meat Products, Computer Processing of Biological Data, Advanced Analysis of Biological Data, Advanced Studies in Specialized Areas of Animal Science, Advanced Beef Production, Advanced Swine Production, Advanced Meat Science

DEPARTMENT OF HORTICULTURE
(No reply received)

DEPARTMENT OF RANGE AND WILDLIFE MANAGEMENT
Phone: Chairman: (806) 742-7295 Others: 742-7296

Chairman: D.F. Burzlaff
Professors: B.E. Dahl, H.A. Wright
Associate Professors: J.R. Hunter, R.E. Sosebee, D.N. Ueckert
Assistant Professors: J.T. Flinders, J D. Garcia, R.D. Pettit, D.A. Quinton
Research Assistants: 14

Degree Programs: Range Science, Wildlife Management

Undergraduate Degree: B.S.
Undergraduate Degree Requirements: 132 semester hours, Basic core of 73 credit hours plus area of emphasis plus electives

Graduate Degrees: M.S., Ph.D.
Graduate Degree Requirements: Master: Thesis Program-24 credit hours of graduate level courses and 6 credit hours of thesis Doctoral: 60 hours beyond B.S. degree exclusive of dissertation credit. A minor of 18 credit hours. One full year of residence. Language optional

Graduate and Undergraduate Courses Available for Graduate Credit: Range Seminar, Wildlife Seminar, Fire Behavior and Ecology, Synecology, The Physiological Basis for Grazing Management, Range Research, Wildlife Research, Ecology of Arid Lands, Range Research Methods, Contemporary Resource Affairs, Experimental Design and Analysis, Advanced Range Management Planning (U), Wildlife Epizootiology and Pathobiology, Advanced Studies in Wildlife Habitat (U), Waterfowl Ecology (U), Advanced Upland Game Ecology and Management (U), Plant Ecophysiology, Advanced Range Ecology (U), Wildlife Conservation and Management (U), Ecology of Renewable Natural Recourses (U), Advanced Big Game Ecology and Management (U), Wildlife Ethology (U)

TEXAS WESLEYAN COLLEGE

P.O. Box 3277 Phone: (817) 534-0251
Fort Worth, Texas 76105
Dean of College: J.B. Gross

DEPARTMENT OF BIOLOGY
Phone: Chairman: Ext. 270

Chairman: J.C. Streett
Professor: J.C. Streett
Associate Professor: W.G. Blanton
Assistant Professors: A.G. Cleveland, G.L. Vertrees, S. Wilcox

Degree Program: Biology

Undergraduate Degree: B.S.
Undergraduate Degree Requirements: Biology 36 hours, Chemistry 16 hours, Mathematics 6 hours

TEXAS WOMAN'S UNIVERSITY

Denton, Texas 76204 Phone: (817) 387-6266
Dean of Graduate Studies: M.E. Huey
Dean of College: E. Davis

DEPARTMENT OF BIOLOGY
Phone: Chairman: (817) 387-6266

Chairman: K.A. Fry
Professors: R. Fuerst, K.A. Fry, E.W. Hupp
Associate Professors: A.W. Cockerline, J Aune, H. Erdman, R.A. Sims
Assistant Professors: J. Hines, M. Rudick
Teaching Fellows: 18
Research Assistants: 10

Degree Programs: Biology, Radiation Biology, Molecular Biology

Undergraduate Degrees: A.B., B.S.
Undergraduate Degree Requirements: At least 30 hours in Biology and a minor (18 hours), 18 hours must be upper class

Graduate Degrees: M.S., Ph.D.
Graduate Degree Requirements: Master: At least B average entrance, maintain B average, thesis, at least 30 hours to complete qualifying examination Doctoral: Satisfactory masters for entrance, two foreign languages (one may be substituted), 60 hours courses, total 90 hours, oral and written candidacy examination, dissertation

Graduate and Undergraduate Courses Available for Graduate Credit: Cell Biology (U), Industrial Microbiology (U), Immunology, Microbial Ecology (U), Mycology (U), Entomology (U), Comparative Physiology (U), Biostatistics, Advanced Genetics, Evolution, Modern Ecology, Cytology and Cytogenetics, Molecular Bio-Physics, Cellular Metabolism, Biophysics/Biochemistry Instrumentation, Advanced Histotechniques, Radiation Protection and Dosimetry, Radiation Effects in Biological Systems, Radioecology, Mammalian Radiation Biology, Radioisotope Techniques, Virology, Bacteriology Physiology, Microbial Genetics, Molecular Genetics, Radiation Effects in Microbiology, Plant Physiology, Plant Morphology, Radiation Botany, Advanced Animal Physiology, Experimental Endocrinology

TRINITY UNIVERSITY

(No reply received)

UNIVERSITY OF CORPUS CHRISTIE

(see Texas A. and I. University)

UNIVERSITY OF DALLAS

(No reply received)

UNIVERSITY OF HOUSTON

3801 Cullen Boulevard Phone: (713) 749-1011
Houston, Texas 77006
Dean of Graduate Studies: R.F. Bunn

College of Arts and Sciences

Dean: R.H. Walker

DEPARTMENT OF BIOLOGY
Phone: Chairman: (713) 749-4661

Chairman: G.D. Aumann
Professors: G.D. Aumann, A.H. Bartel, N.C. Cominsky, J. Evans, R.L. Hazelwood, S.E. Huggins, D.L. Jameson, A.L. Lawrence
Associate Professors: E. Bryant, W.H. Clark, L.E. Franklin, H.T. Freebairn, E.P. Goldschmidt, H. Henney, P. Jurtshuk, D. Mailman, P. Snider, S. Venketeswaran
Assistant Professors: G. Cameron, J. Cowles, N. Fotheringham, M. Harry, R. Jones, B. Schall
Instructor: P. Hacker
Visiting Lecturers: 2
Teaching Fellows: 43
Research Assistants: 4

Field Stations/Laboratories: Houston Coastal Center - Hitchkock, Texas

Degree Programs: Bacteriology, Biology, Botany, Ecology, Genetics, Microbiology, Physiology, Plant Sciences, Zoology, Population Biology

Undergraduate Degree: B.S . (Biology)
Undergraduate Degree Requirements: Chemistry through Organic, Mathematics through Integral Calculus, One year Physics,Thirty-four semester hours of Biology

Graduate Degrees: M.S., Ph.D.
Graduate Degree Requirements: Master: Plan I. 30 semester hours and thesis, Plan II - 36 semester hours and library paper Doctoral: M.S. I or equivalent, one foreign language, course work as determined by committee, defended thesis

Graduate and Undergraduate Courses Available for Graduate Credit: Population Biology Laboratory, Selected Topics in Biology, Special Problems, Field Biology, Ecological Methods, Cellular Physiology Laboratory, Population Biology, Animal Behavior, Endocrinology, Ecology, Cellular Physiology, Comparative Animal Physiology, Radioisotopes I, Introductory Biometrics, Pathogenic Microbiology, Immunology, Biology of the Myxomycetes, Invertebrate Zoology, Marine Biology, Histology, Comparative Histology, Applied Microbiology, Microbial Physiology, Developmental Biology, Cytology, Instrumental Biology, Advanced Bacterial Physiology, Evolution, Bacterial Genetics, Membrane Physiology, Advanced General Microbiology, Comparative Biochemistry, Advanced Plant Physiology, Microbial Ecology, Tissue Culture Techniques, Marine Plants, Radioisotopes II, Graduate Electron Microscopy, Biometrics, Advanced Population Biology, Community Ecology, Theoretical Biology, Photobiology, Advanced Cellular Physiology, Mammalian Physiology

DEPARTMENT OF BIOPHYSICAL SCIENCES
Phone: Chairman: (713) 749-2801 Others: 749-1011

Chairman: A.H. Bartel
Professors: J.C. Allred, A.H. Bartel, A.P. Kimball, J. Oro
Associate Professors: J. Eichberg, J.E. Evans, E.P. Goldschmidt, H.B. Gray, Jr., D.S. Mailman, P.G. Snider
Assistant Professors: W.E. Jeffery, T.G. Spring
Visiting Lecturers: 1
Teaching Fellows: 2
Research Assistants: 10
Research Associates: 2
Postdoctoral: 1

Field Stations/Laboratories: Camp Wallace - Hitchcock

Degree Program: Biophysical Sciences

Undergraduate Degree: B.S.
Undergraduate Degree Requirements: 24 advanced hours in biophysical sciences, 12 hours Mathematics, 8 hours Biology, 15 hours Chemistry, 8 hours Physics, 18 hours basic science electives (Advanced), 9 hours English, 6 hours History, 6 hours Political Science, 18 hours core distribution electives, 2 hours core distribution electives, 2 hours physical education

Graduate Degrees: M.S., Ph.D.
Graduate Degree Requirements: Master: 30 hours of course work, includes 6 hours of thesis work, 24 hours selected with the help of the student's graduate committee. Student must score at least 1100 on GRE and have a g.p.a. of 2.6 or better on last 60 hours for entrance to program. Doctoral: M.S. degree or 30 hours of approved course work with minimum grade point average of 3.0 for entrance. 24 hours of course-work and research over and above requirements for admission to Ph.D. program. dissertation evidencing significant independent research, competence in one foreign language, qualifying, diagnostic and oral proposition examinations required.

Graduate and Undergraduate Courses Available for Graduate Credit: General Biophysics (U), General Biochemistry (U), Nucleic Acids and Proteins (U), Biophysical Sciences Seminar, Enzymes, Biological Radiochemistry, Biophysical Chemistry, Molecular Control Systems Biochemical Evolution, Advanced Nucleic Acids, Seminar in Protein Structure and Function, Biophysical Instrumentation, Neurobiochemistry, Biochemical Oncology, Modern Techniques in Bio and Medical Physics Modern Techniques in Lipid and Neurobiochemistry, Modern Techniques in Enzyme and Protein Biochemistry, Modern Techniques in Nucleic Acid Biochemistry, Modern Techniques in Radiochemical and Biomolecular Analysis, Modern Techniques in Biochemical Oncology A maximum of 9 semester hours of undergraduate courses may be accepted for graduate credit.

UNIVERSITY OF ST. THOMAS
3812 Montrose Phone: (713) 522-7911
Houston, Texas 77006
Dean of College: J.T. Sullivan

DEPARTMENT OF BIOLOGY
Phone: Chairman: Ext. 256 Others: 522-7911

Chairman: J.L. Meyers
Professor: J.L. Meyers
Associate Professor: A.C. Avenoso
Assistant Professor: L. Nordyke
Instructors: H.C. Browning, A.C. Pinkerton

Degree Program: Biology

Undergraduate Degree: B.A.
Undergraduate Degree Requirements: Not Stated

UNIVERSITY OF TEXAS AT ARLINGTON
Arlington, Texas 76019 Phone: (817) 273-2011
Dean of Graduate Studies: L.L. Schkade (Acting)
Dean of College: W.R. Meacham (Acting)

DEPARTMENT OF BIOLOGY
Phone: Chairman: (817) 273-2871

Chairman: W.C. McDonald
Professors: C.C. Hall, T.E. Hellier, T.E. Kennerly, W.B. McCrady, W.R. Meacham, W.F. Pyburn
Associate Professors: R.B. Boley, L.H. Bragg, J.K. Butler, R.R. Eller, B.L. Frye, F.G. Gladden, R.L. Neill
Assistant Professors: M.A. Clark, A. Hopkins, R. Knaus, R.F. McMahon, M.S. Sharp, D.H. Whitmore
Instructor: J. Perryman
Teaching Fellows: 12
Research Assistants: 1

Degree Program: Biology

Undergraduate Degree: B.S.
Undergraduate Degree Requirements: Requirements vary depending on which of 6 options student chooses. Options include General Biology, Pre-Health Profession, Botany, Ecology, Molecular and Cell Biology, Microbiology

Graduate Degree: M.S.
Graduate Degree Requirements: Master: 30 hours of Biology including a thesis, 1 foreign language, or 36 hours of Biology (non-thesis option)

Graduate and Undergraduate Courses Available for Graduate Credit: Selected Topics in Biology, Evolution, History of Biology, Bacterial Metabolism, Biogeography, Physiological Ecology, Animal Ecology, Plant Ecology, Ichthyology, Herpetology, Mammalogy, Ornithology, Aquatic Microbiology, Palynology, Biological Electron Microscopy, Radioecology, Medical Mycology, Advanced Genetics, Developmental Genetics, General Physiology, Advanced Bacteriology, Acarology, Microbial Genetics, Protozoology, Phycology, Plant Physiology (U), Bacterial Ecology (U), Aquatic Biology (U), Radiation Biology (U), Histology, Research Methods in Cell Biology (U), Physiology II (U), Bacterial Physiology (U), Field Biology (U), Introduction to Virology (U)

UNIVERSITY OF TEXAS AT AUSTIN

Austin, Texas 78712 Phone: (512) 741-1233
Dean of Graduate School: G. Lundzey

College of Arts and Sciences

Dean: P. Olum

DEPARTMENT OF BOTANY
Phone: Chairman: (512) 471-5858

Chairman: T. Delevoryas
C.L. Lundell Chair of Systematic Botany: (vacant)
Professors: C.J. Alexopoulos, D.P Bloch, H.C. Bold, W.V. Brown, T. Delevoryas, V.E. Grant, M.C. Johnston, D A. Levin, C.L. Lundell (Adjunct), T.J. Mabry, C. McMillan, C.H. Muller (Adjunct), I. Spear, S. Spurr, G.A. Thompson, Jr., B.L. Turner, W.G. Whaley
Associate Professors: M.D. Summers, M.J. Wynne
Assistant Professors: J.J. Brand, G.T. Cole, S.J. Kirchanski
Teaching Assistants: 9
Research Assistants: 15

Field Stations/Laboratories: Brackenridge Field Station - Austin, Institute of Marine Sciences - Pt. Arausas

Degree Programs: Botany, Biological Science

Undergraduate Degree: B.A.
Undergraduate Degree Requirements: 24 hours of upper division (Junior/Senior) courses beyond the under-division courses

Graduate Degrees: M.A., Ph.D.
Graduate Degree Requirements: Master: 30 hours (21-24 in major plus 6-9 minor) plus thesis Doctoral: Program of courses plus dissertation, acceptable to Graduate Studies Committee

Graduate and Undergraduate Courses Available for Graduate Credit: Cytology, Methods in Virology: Plant and Insect Viruses, Advanced Study and Research, Selected Topics in Plant Physiology, Plant Speciation, Plant Physiology, General Phycology, Paleobotany, Mycology, Plant Genetics, Evolution, Population Dynamics of Plants, Physiological Phycology, Chemistry and Biology of Membranes, Fungal Morphogenesis and Ultrastructure, Morphology of Bryophytes and Vascular Plants, Experimental Mycology, Advanced Topics in Plant Systematics, Biochemical Studies of Evolutionary Relationships, Marine Phycology, Plant Ecology, Selected Topics in Plant Science

DEPARTMENT OF MICROBIOLOGY
Phone: Chairman: (512) 471-5105

Chairman: L.J. Berry
Professors: L.J. Berry, C.E. Lankford, W.J. Mandy, V.T. Schuhardt, O. Wyss
Associate Professors: H.R. Bose, D.T. Gibson, L.J. Rode, J.R. Walker
Assistant Professors: C.F. Earhart, C. Parker, P.J. Szaniszlo, R.F. Tabita, M R.F. Waite
Instructors: S. Archer, R.Gustafson, K. Todar, M. Wells
Research Assistants: 16
Postdoctoral Fellows: 9

Field Stations/Laboratories: Balcones Research Center

Degree Programs: Microbiology, Medical Technology

Undergraduate Degrees: B.A., B.S. (Medical Technology)
Undergraduate Degree Requirements: 8 hours Mathematics, 8 hours physics, 16 hours Chemistry, 3 hours Biology, 24 hours Microbiology 1. B.A. Degree with a major in Microbiology, 2. Bachelor of Science in Medical Technology, 6 hours Mathematics, 8 hours Physics, 20 hours Chemistry, 14 hours Biology, 15 hours Microbiology

Graduate Degrees: M.A., Ph.D.
Graduate Degree Requirements: Master: Qualifying examination, courses selected by advisor, thesis Doctoral: qualifying examination, courses selected by committee, foreign language, comprehensive examination, thesis

Graduate and Undergraduate Courses Available for Graduate Credit: Microbiology and Man (U), Introductory Virology (U), Industrial Microbiology (U), Bacterial, Rickettsial and Viral Diseases of Man (U), Immunology (U), Public Health Bacteriology (U), Metabolism and Biochemistry of Microorganisms, Microbial Physiology, Metabolsim and Biochemistry of Microorganisms Laboratory, Microbial Ecology, Mammalian Cell Culture, Microbial Genetics, Biology of Fungi, Biostatistics, Advanced Mycology, Advanced Microbial Genetics, General Microbiology, Immunochemistry, Problems in Host-Parasite Biology, Applied Public Health and Medical Microbiology, Perspectives in Microbiology, Host-Microbial Interactions, Advanced Microbial Biochemistry, Microbial Anatomy, Cellular Immunology, Problems in Microbial Physiology, Advanced Virology

DEPARTMENT OF ZOOLOGY
Phone: Chairman: (512) 471-7131

Chairman: H.S. Forrest
Professors: J.J. Biesele, W.F. Blair, O.P. Breland, F.H. Bronson, H.S. Forrest, T.H. Hamilton, Y. Hiraizumi, C. Hubbs, A.G. Jacobson, B.H. Judd, J.L. Karimer, M. Menaker, J.E. Myers, J.A.C. Nicol, C.P. Oliver (Emeritus), C. Pavan, E.L. Powers, A.F. Riggs, B.G. Sanders, A.R. Schrank, H.E. Sutton, R.P. Wagner, M.R. Wheeler, D.E. Wohlschlag
Associate Professors: R.H. Barth, G.D. Bittner, G.L. Bush, C. Desjardins, G. Freeman, J.L. Fox, J. Kagowski, B. Maguire, E.R. Pianka, R.H. Richardson
Assistant Professors: L. Gilbert, L. Lawlor, C.S. Lee, D. Otte, R.Perez, D.G. Pollak, M.A. Rankin, A.R. Templeton
Lecturers: A.C. Faberge, W.K. Long
Visiting Lecturers: 1
Temporary Instructors: 2
Temporary Assistant Professors: 4

Field Stations/Laboratories: Brackenridge Field Laboratory W.F. Blair, Director,

Degree Program: Zoology

Undergraduate Degree: B.A.
Undergraduate Degree Requirements: 6 hours Mathematics, 6 hours Organic Chemistry, 30 hours Biological Sciences including 18 hours upper division Zoology

Graduate Degrees: M.A., Ph.D.
Graduate Degree Requirements: At least 30 semester hours of work beyond the Bachelor's degree are required. A thesis is required but foreign language proficiency is not. Doctoral: No specific number of hours are required for the Ph.D. Degree, but the students program must be approved by the Zoology Graduate Staff, his Supervising Committee and the Graduate Dean. Doctoral students are required to pass both a written and an oral qualifying examination and must display "practical reading competence" in one foreign language.

Graduate and Undergraduate Courses Available for Graduate Credit: Seminar: Cytology, Advanced Readings in Zoology, Basic Biological Concepts and Techniques of Analysis, Graduate Cytology and Genetics, Current Concepts in Zoology (Cytology, Genetics, and Vertebrate Zoology), Current Concepts in Zoology (Physiology, Invertebrate Zoology, and Embryology), Graduate Systematic and Environmental Zoology, Graduate Physiology and Biophysics, Comparative Physiology, Biological Aspects of Ionizing Radiation, General Physiology and Biophysics Seminar, Physiology of Marine Animals, Graduate Studies in Development and Reproduction, Seminar in Development and Reproduction, Seminar: Biology of Animal Populations, Topics in Population Biology, Symposium on Concepts and Practices in Taxonomy, Seminar in Genetics

UNIVERSITY OF TEXAS HEALTH SCIENCE CENTER AT DALLAS

5233 Harry Hines Boulevard Phone: (214) 631-3220
Dallas, Texas 75235
Dean of School of Biomedical Sciences: R.W. Estabrook
Dean of Medical School: F.J. Bonte

DEPARTMENT OF BIOCHEMISTRY

Phone: Chairman: Ext. 456

Chairman: Not Stated
Virginia Lazerby O'Hara Chair in Biochemistry: R.W. Estabrook
Professors: J J. Banewicz, E.R. Biehl, H.A. Jeskey (Adjunct), W.B. Dempsey, R.W. Estabrook, J.M. Johnston, J. LoSpalluto, M.D. Prager, P.A. Srere, D.S. Wiggans, H.C. Tidwell (Emeritus)
Associate Professors: E. Bellion (Adjunct), R.A. Butow, G.L. Cottam, R. Frenkel, L.B. Hersh, B.S. Masters, J.D. McGarry, J A. Peterson, T.E. Smith, K. Uyeda
Assistant Professors: A. Bollon, G.R. Faloona, W.T. Garrard, T.C. Linn, G.E. Mize, R.A. Prough, M.R. Waterman
Instructors: T. Delahunty, B. Griffin
Teaching Fellows: 4
Research Assistants: 17
Research Scientist: 1
Postdoctoral Fellows: 6
Research Fellows: 11
Predoctoral Fellows: 4

Degree Programs: Biochemistry

Undergraduate Degree: B.S.
Undergraduate Degree Requirements: 1) a report of scores on the Graduate Record Examination, 2) two letters of recommendation, with at least one from a Chemistry Professor, 3) complete transcript of all academic work, 4) a completed application for admission to graduate school.

Graduate Degrees: M.A., Ph.D.
Graduate Degree Requirements: Master: 1) at least 30 semester hours of course credit, 2) pass qualifying examination, 3) submit a thesis Doctoral: 1) pass a qualifying examination, written and/or oral, 2) to prepare a dissertation which must be defended before a specially appointed committee

Graduate and Undergraduate Courses Available for Graduate Credit: Fundamentals of Biochemistry, Enzymology, Special Topics in Biochemistry, Recent Advances in Nutrition, Nutrition Seminar, Biochemistry Seminar, Biochemical Research, Biochemical Research Techniques, Biochemical Genetics, Physical Chemistry, Advanced Organic Chemistry, Human Genetics, Immunology, Advanced Physical Chemistry, Intermediary Metabolism and Its Regulation, Biophysical Chemistry, Biological Oxidations, the Nature of Enzyme Catalysis

DEPARTMENT OF BIOMATHEMATICS

(No reply received)

DEPARTMENT OF BIOPHYSICS

Phone: Chairman: (214) 688-2593

Chairman: R.M Dowben
Professors: R.M. Dowben, J.Y. Lettvin (Adjunct)
Associate Professors: L.G. Cottam
Assistant Professors: J.R. Bunting, M.M. Judy, E.G. Richards, R. Roe, Jr., W.R. Romans, A.D. Sherry (Adjunct)
Instructor: W.L. Schaar
Teaching Fellows: 2
Research Assistants: 3

Degree Program: Biophysics

Graduate Degrees: M.S., Ph.D
Graduate Degree Requirements: Master: 30 semester hours course credit, research thesis Doctoral: research thesis, two semester residence requirement, language proficiency or Fortran, preliminary examination covers Biochemistry and Physiology in addition

Graduate and Undergraduate Courses Available for Graduate Credit: Physical Chemistry, Biomedical Instrumentation, Introduction to Molecular Spectroscopy, Optical and Magnetic Instrumentation and Measurements, Biophysical Chemistry, Advanced Biophysical Chemistry, Cellular Biophysics, Advanced Spectroscopic Techniques, Seminars in Biophysics, Special Topics in Biophysics

GRADUATE PROGRAM IN CELL BIOLOGY

Phone: Chairman: Ext. 2347 Others: Ext. 2231

Program Chairman: W.B. Neaves
Professors: R.E. Billingham, C.R. Hackenbrock, J.W. Streilein, H.T. Weathersby
Associate Professors: A E. Beer, W.B. Neaves
Assistant Professors: R.G.W. Anderson, P.M. Andrews, R.S. Decker, G.C. Ericson, W.J. Ganyea, F. Grinnell, V.K. Miyamoto, E.J. Moticka, H.R. Toben
Visiting Lecturer: 1
Teaching Fellows: 3

Degree Program: Cell Biology

Graduate Degree: Ph.D.
Graduate Degree Requirements: Doctoral: 2 years' residence, 45 semester hours of prescribed courses, laboratory teaching experience, general qualifying examination, approved dissertation, final oral defense

Graduate and Undergraduate Courses Available for Graduate Credit: Biology of Cells and Tissues, Cell Biology Seminar, Genetics, Gross Anatomy, Human Visceral Gross Anatomy, Immunology, Introduction to Histological and Histochemical Techniques, Modern Concepts in Development, Neurobiology, Readings in Transplantation

DEPARTMENT OF MICROBIOLOGY

Phone: Chairman: (214) 631-3220 Ext. 2160

Chairman: J.W. Uhr
Professors: J D. Capra, R.A. Finkelstein, R. Pike (Emeritus), E. Rosenblum, J.W. Uhr
Associate Professors: J. Klein, E. Vitetta
Assistant Professors: M. Boesman, L Eidels, J. Forman, D.A. Hart, K.V. Holmes, C.-Y. Kang, J.R. Kettman, U. Melcher, D.V. Moore
Instructor: D. Klapper
Teaching Fellows: 2
Research Assistants: 4

Degree Program: Microbiology

Graduate Degrees: M.S., Ph.D.
Graduate Degree Requirements: Master: Thirty semester hours graduate credit, satisfactory thesis, courses in medical microbiology and other supporting work.

Doctoral: Satisfactory dissertation, courses in medical microbiology, biochemistry, statistics, and other supporting work

Graduate and Undergraduate Courses Available for Graduate Credit: Medical Microbiology, Parasitology, Seminar, Research, Immunology, Current Topics in Immunology, Virology, Microbial Genetics, Bacterial Physiology, Parasitology, Pathogenic Mechanisms, Special Topics

DEPARTMENT OF PHARMACOLOGY
Phone: Chairman: Ext. 2268

Chairman: A. Goth
Professors: E.G. Erdos, A. Goth, W.A. Pettinger, S.I. Said, P.A. Shore, A. Tauroq, G.B. Weiss
Associate Professors: W.G. Clark
Assistant Professors: H.R. Adams, A. Giachetti, F.R. Goodman, A.R. Johnson
Instructor: J.C. Garriett
Research Assistants: 2

Degree Program: Pharmacology

Graduate Degree: M.A., Ph.D.
Graduate Degree Requirements: Master: Medical Pharmacology course Doctoral: Medical Physiology, Neurobiology, Fundamentals of Biochemistry, Endocrinology, and Human Reproduction, Medical Pharmacology, Statistics of Medicine and Biology

Graduate and Undergraduate Courses Available for Graduate Credit: Medical Pharmacology, Pharmacology Seminar, Advanced Pharmacology, Special Topics in Pharmacology (special courses in specific areas of pharmacology)

DEPARTMENT OF PHYSIOLOGY
(No reply received)

UNIVERSITY OF TEXAS MEDICAL BRANCH AT GALVESTON

Galveston, Texas 77550 Phone: (713) 765-1011
Dean of Graduate Studies: J.P. Saunders
Dean of Medical Branch: E.N. Brandt, Jr.

DEPARTMENT OF ANATOMY
Phone: Chairman: (713) 765-1293

Chairman: W.J. Hild
Professors: R.E. Coggeshall, G.V. Russell, D. Duncan, E.K. Sanland, W.D. Willis
Associate Professors: C.W. Kischev
Assistant Professors: G. Callan, M.S. Cannon, M.B. Hancock, A.F. Payne, M.C. Young, R.C. Yu

Degree Program: Anatomy

Graduate Degrees: M.A., Ph.D.
Graduate Degree Requirements: Master: Courses as required, thesis, final comprehensive Doctoral: Courses as required, one foreign language, qualifying examination, thesis, oral

Graduate and Undergraduate Courses Available for Graduate Credit: Not Stated

DEPARTMENT OF HUMAN BIOLOGICAL CHEMISTRY AND GENETICS
Phone: Chairman: (713) 765-1961 Others: 1962, 1963

Chairman: B.H. Bowman
Professors: C. Abell, B.H. Bowman, B.R. Brinkley, J. Chang, A. Goldstein, J. Kittredge, V. Koenig, G. Mills, A. Ormsby, B. Papermaster, R.G. Schneider, L. Smith, A. Suttle
Associate Professors: J. Alperin, S. Barrance, III, B. Haber, D. Hodgins, V Holoubek, F. Houston, L. Lockhart, P. Poffenbarger, G. Powell, L. Sordahl
Assistant Professors: K. Arai, D. Barnett, D. Baur, J. Chen, B. Daggett, G. Fuller, J. Gan, N.S. Harris, G. Harrison, J. Hooper, H.T. Hutchison, R. Klebe, D. McAdee, P. Rayford
Instructors: G. Cohen, A. Kurosky
Lecturer: J. Lindsey, T. Monahan

Degree Programs: Biological Chemistry, Human Genetics

Graduate Degrees: M.S., Ph.D.
Graduate Degree Requirements: Master: 30 semester hours, thesis Doctoral: M.S. or equivalent, one foreign language, preliminary examination, qualifying examination, thesis, final oral

Graduate and Undergraduate Courses Available for Graduate Credit: Biochemical Genetics, Advanced Human Genetics, Medical Genetics, Cellular Genetics, Recent Advances in Genetics, Current Concepts in Genetics, Inborn Errors of Metabolism, Problems in Genetics, Evolution and Protein Molecules, Organization and Function of the Eucaryotic Genome, Immunogenetics, Genetics Seminar, Research in Human Genetics, Special Topics in Genetics, Biochemistry, Physiology and Genetics of Hemostasis, Advanced Study of Multifactorial Genetic Diseases, Developmental Genetics, Cellular Radiobiology, Principles of Histochemistry, Biological Ultrastructure I. Methods Biological Ultrastructure II. Application and Interpretation, Cell Proliferation Kinetics of Normal and Neoplastic Cells, Biochemistry of Nucleotides and Nucleic Acids, Clinical Biochemistry, Biochemistry of Amino Acids and Proteins, Special Topics in Biochemistry, Immunology, Calculus for Biomedical Courses, Biometry

DEPARTMENT OF MICROBIOLOGY
Phone: Chairman: (713) 765-2325 Others: 765-2321

Chairman: W.F. Verwey
Professors: W.F. Verwey, L.J. Olson, A. Packchanian
Associate Professors: E.D. Box, Q.T. Box, A. Ewert, J.C. Guckian, E.M. Macdonald, R.C. Wood
Assistant Professors: J.M. Bray, J.W. Peterson
Visiting Lecturers: 8
Research Assistants: 3
Faculty Associates: G.J. Stanton, J.C Reitmeyer

Degree Programs: Medical Microbiology

Graduate Degrees: M.A., Ph.D.
Graduate Degree Requirements: Master: 30 semester hours graduate instruction with 18-24 hours in the major and a minimum of 6 hours in a minor, A thesis based on independent research. Doctoral: Instruction in major and minor areas as determined by the supervisory committee with written and oral examinations, proficiency in an approved foreign language, teaching experience, a dissertation based on independent research, public defense of dissertation.

Graduate and Undergraduate Courses Available for Graduate Credit: Medical Microbiology, Medical Microbiology Tutorial, General Virology, Pathogenic Bacteriology, Mycology, Parasitology, Immunology, Bacterial Chemistry and Physiology, Chemotherapy, Seminar, Advanced Work

DEPARTMENT OF PATHOLOGY
Phone: Chairman: (713) 765-2889 Others: 765-2856

Chairman: F.L. Jennings
Professors: E. Baird, J. Fuller, F.L. Jennings, R. Marshall, W. McCormick, R. Rigdon, A. Rodin, L.C. Stout
Associate Professors: G. Beathard, E. Dahl (Research), S. Schochet
Assistant Professors: D. Folse, R. Gillum, P. Gilmer, S. Moore, E. Mueller, P. Zaharopoulos
Instructor: R. Passey, F. Urry
Visiting Lecturers: 5

Degree Program: Pathology

Graduate Degree: M.A.
Graduate Degree Requirements: Master: 30 semester hours - graduate instruction, thesis, final examination

Graduate and Undergraduate Courses Available for Graduate Credit: General Pathology, Systemic Pathology, Advanced Pathology, Research in Pathology

DEPARTMENT OF PHYSIOLOGY
Phone: Chairman; (&13) 765-2224 Others: 765-1826

Chairman: M.M. Guest
Ashbel Smith Professor of Physiology: M.M. Guest
Professors: A.M. Brown, M.M Guest, C.E. Hall, S.N. Kolmen, L.B. Nanniga (Research)
Associate Professors: R.D. Baker, T.P. Bond (Research), H.M. Fishman, C.E. Hall (Research), F.H. Rudenberg, D.W. Stubbs, D.L. Traber, C.H. Wells
Assistant Professors: R.E. Barrow, D.C. Eaton, H.A. Germer, D.L. Kunze, J.M. Russell, J.R. Walker
Visiting Lecturers: 2
Research Assistants: 9

Degree Program: Physiology

Graduate Degrees: M.A., Ph.D.
Graduate Degree Requirements: Master: 30 semester hours of graduate instruction, 18-24 semester hours, including the thesis, must be in the major area of instruction. Six hours in supporting subject or subjects outside the major area. The written thesis is subject to the approval of the supervisory committee appointed by the Dean of the Graduate School and ultimatoly to thc approval of the Dean of the Graduate School. In many areas a candidate may, in addition to preparing a thesis, be required to pass an oral or written examination (or both) conducted by his supervisory committee Doctoral: There are five basic requirements for admission: 1) a Bachelor's degree from an accredited institution in the U.S. or proof of equivalent training; 2) a satisfactory grade point average; 3) a combined score of 1000 on the verbal and quantitative sections of the Graduate Record Examination Aptitude Test; 4) adequate subject preparation for the proposed major area; and 5) acceptance by the Committee on Graduate Studies of the proposed major area.

Graduate and Undergraduate Courses Available for Graduate Credit: Experimental Endocrinology, Aviation and Space Physiology, Physiology of the Kidney, Neurophysiology, Introductory Cellular Physiology, Advanced Cellular Physiology, Biophysics, Comparative Physiology, Bioinstrumentation, Physiology Research, Weekly Seminar in Advanced Physiology, Calculus for Biomedical Courses, Physical Chemistry, Cardiac Physiology, Gastrointestinal Physiology, Membrane Transport, Special Topics, Rheology and Microcirculation, Cardiovascular Physiology, Basic Neuroscience, Cell Biology, Endocrinology, Seminar in Marine Biomedicine, Integrated Functional Laboratory I, Orientation in Physiology, Integrated Functional Laboratory II, Comparative Neurophysiology

DEPARTMENT OF PREVENTIVE MEDICINE AND COMMUNITY HEALTH
Phone: Chairman: (713) 765-1128 Others: 765-1011

Chairman: D.W. Micks
Professors: E.N. Brandt, Jr., J.G. Bruhn, W.F. Dodge, F.L. Duff (Adjunct), J.E. Overall, M. Patterson
Associate Professors: D.L. Creson, W.W. Kemmerer, (Adjunct), D.J. Kilian (Clinical), J E. Perry (Clinical), M.L. Ross, N.M. Trieff, E.B. Whorton, Jr.
Assistant Professors: D.J Barlow (Adjunct), J.C. DeWitt, C.A. Jernigan (Clinical), H.G. Levine, A.A. O'Donell, J.T. Phillips, D.W. Rowden, W.M. Thompson
Instructors: D.A. Bosshart, J.M. Mahan, B.U. Philips, G.H. Smith
Visiting Lecturers: 20
Research Assistants: 2

Degree Programs: Preventive Medicine and Community Health

Graduate Degrees: M.A., Ph.D.
Graduate Degree Requirements: Master: B average in last two undergraduate years, and combined score of 1000 on the verbal and quantitative parts of the GRE for admission, 30 hours of credit, including a thesis for degree Doctoral: Awarding of the Ph.D. degree is based upon attainment of an adequate level of knowledge and research competence rather than upon the completion of a specified number of hours, also a reading knowledge of a foreign language is required.

Graduate and Undergraduate Courses Available for Graduate Credit: Preventive Medicine and Community Health, Introduction to Epidemiology, Community Health Research, History and Philosophy of Medicine and Community Health, Medical Sociology, Survey Research in the Health Sciences, Medical Care Organization, Medical Education, Water Supply and Pollution Control, Chemical Ecology, Environmental Chemistry, Environmental Health, Environmental Toxicology, Occupational Health, Air Sampling and Analysis, Principles and Procedures of Statistical Reasoning, Statistical Design of Sample Surveys, Statistical Methodology I, II, Fundamentals of Data Processing, Special Topics in Biostatistics

UNIVERSITY OF TEXAS AT EL PASO

El Paso, Texas 79968 Phone: (915) 747-5000
Dean of Graduate Studies: R. Gomez
Dean of College: J.W. Whalen

DEPARTMENT OF BIOLOGICAL SCIENCES
Phone: Chairman: 747-5164 Others: 747-5629

Chairman: A.G. Canaris
Professors: A.G. Canaris, C.E. Eklund, A.H. Harris, A.L. Metcalf, J.B. Reeves, G.W. Robertstad, R.G. Webb
Associate Professors: J.R. Bristol, P.S. Chrapliwy, E.L. Duke, J.D. Hunter, R.D. Worthington
Assistant Professors: J.T. Ellzey, C.E. Freeman, L.P. Jones, E.C. Manning, K.A. Redetzke

Degree Programs: Microbiology, Plant Sciences or Zoology, Medical Technology

Undergraduate Degrees: B.A., B.S.
Undergraduate Degree Requirements: Not Stated

Graduate Degree: M.S.
Graduate Degree Requirements: Master: 1. A thesis (6 semester hours) plus 24 semester hours of course work. Only 9 hours of 3300 or 3400 courses are permitted in a program, and no more than 6 hours may be included in either major or minor. 2. A major with a minimum of 18 semester hours including the thesis. Major fields for the M.S. include Biology, Chemistry, English, Geology, Health and Physical Education, Mathematics and Physics.

Graduate and Undergraduate Courses Available for Graduate Credit: Selected Topics, Developmental Cytology, Herpetology, Biology of the Pleistocene, Microbial Genetics, Recent Advances in Microbiology, Biogeography, Physiological Ecology, Analytical Cytology, Biosystematics, Plant Ecology, Identification and Ecology of Desert Plants, Ecology and Physiology of Animal Parasites, Neuroendocrine Physiology, Archaeobiology, Malacology, Ultrastructure, Mammalogy, Thesis

UNIVERSITY OF TEXAS AT HOUSTON
Houston, Texas 77025 Phone: (713) 792-6628

School of Public Health
Dean: R. Stallones

DEPARTMENT OF BIOMETRY
Phone: Convener: (713) 792-4421 Others: 792-2121

Convener: R.F. Frankowski
Professors: A.S. Littell, F.M.Hemphill
Associate Professors: T.D. Downs, R.F. Frankowski, J. H. Glasser, C.M. Hawkins, B.P. Hsi
Assistant Professors: G.Cutter, R.N. Forthofer, R.B. Harrist, E.S. Lee
Research Assistants: 1
Associate Research Biometrician: 1
Research Statistical Aides: 5
Traineeships: 6
Research Support: 4

Degree Programs: Biometry, Biostatistics

Graduate Degrees: M.S., Ph.D.
Graduate Degree Requirements: Master: Students entering the programs may hold an undergraduate degree in mathematics or one of the biological, physical or social sciences. Mathematics at least through the calculus is needed for entrance, and an overall B (3.0 on a 4 point system) average in prior academic work is ordinarily required. Doctoral: Satisfactory completion of a prescribed course of study of at least one academic year in duration in preparation for the qualifying examination, and satisfactory performance in a qualifying examination deemed by the faculty to test depth of knowledge in the chosen field of the candidate and at least two related fields, and a capacity to conceive and conduct independent research in the chosen field, and completion and presentation in written form of an original research project that makes a substantial contribution to knowledge.

Graduate and Undergraduate Courses Available for Graduate Credit: Biometry, Survey Sampling, Distribution-free methods, Bioassay, Demography, Operations Research, Biometric Methods, Linear Models, Quantitative Methods in Health Services, Multivariate Analysis, Stochastic Life Table Theory, Analysis of Variance and Factorial Experiments, Special Topics in Biometry, Individual Study in Biometry, Experimental Design, Advanced Statistical Inference.

UNIVERSITY OF TEXAS AT PERMIAN BASIN
Odessa, Texas 79762 Phone: (915) 367-2217
Dean of Graduate Studies: V.R. Cardezier
Dean of Science and Engineering: L. Harrisberger

FACULTY OF LIFE SCIENCE
Phone: Chairman: (915) 367-2217 Others: 367-2011

Chairman: E.B. Kurtz
Professor: E.B. Kurtz
Assistant Professors: C O. McKinney, C. Wisdom
Visiting Lecturers: 1

Field Stations/Laboratories: Crane County Field Site

Undergraduate Degree: B.S.
Undergraduate Degree Requirements: 30 hours of Life Science including Genetics, Suborganismic, and Organismic Courses; Chemistry through Organic Chemistry, 6 hours Mathematics or Computer Science

Graduate Degree: M.S.
Graduate Degree Requirements: Master: 30 hours including thesis or project

Graduate and Undergraduate Courses Available for Graduate Credit: Virology (U), Population Genetics (U), Evolution (U), Cell Biology (U)

UNIVERSITY OF TEXAS
HEALTH SCIENCE CENTER AT SAN ANTONIO
7703 Floyd Curl Drive Phone: (512) 696-6011
San Antonio, Texas 78284
Dean of Graduate School: A.I. Guarino
Dean of Medical School: S.E. Crawford
Dean of Dental School: P.J. Boyne

DEPARTMENT OF ANATOMY
Phone: Chairman: (512) 696-6535 Others: 969-6533

Chairman: E.C. Rennels
Professors: E.K. Adrian, Jr., A.L. Burton, I.L. Cameron, A.B. Cruz, Jr., J.J. Ghidoni, F. Harrison, W. H. Knisely, F.T. Lynd, C.W. McNutt, E.M. Nelson (Adjunct), R.J. Reiter, E.G. Rennels, J.L. Story
Associate Professors: G.L. Colborn, N. Hagino, M.L. Houston, V.F. Williams, W.B. Winborn
Assistant Professors: E.P. Bowie, H.C. Dung, D.C. Herbert, M.A. Kramen, W.W. Morgan, A.E. Sanders (Adjunct)
Instructors: M.G. Williams
Visiting Lecturers: 4
Teaching Fellows: 7
Research Assistants: 13
Visiting Associate Professor: 1

Degree Program: Anatomy

Graduate Degrees: M.A., Ph.D.
Graduate Degree Requirements: Master: For the Master of Arts degree in anatomy a minimum of 30 semester hours of graduate credit and a thesis are required. Doctoral: Students working toward the Ph.D. in Anatomy must take courses in Gross Anatomy, Microscopic Anatomy, Human Embryology, Neuroscience, Biochemistry, Physiology, and Statistics. In addition, at least two of the following advanced courses must be taken: an advanced course in Gross Anatomy, Cell and Quantitative Biology, Advanced Neuroscience Endocrinology of Reproduction, and Electron Microscopy. Before admission to candidacy a foreign language examination and departmental qualifying examinations must be passed. Each student must assist in the teaching of at least two different anatomy courses. A dissertation based on the results of an original research project is required.

Graduate and Undergraduate Courses Available for Graduate Credit: Gross Anatomy, Microscopic Anatomy, Human Embryology, Neuroscience, Cell and Quantitative Biology, Electron Microscopy, Problems in Neurochemistry, Endocrinology of Reproduction, Problems in Neuroendocrinology, Anatomy of the Newborn, Regional Anatomy of the Back, Head and Neck, Problems in Cell Biology, Medical Genetics, Scientific Cinematography, Applied to Basic Sciences, Histopathologic Studies of Mutant Mice, Supervised Teaching in Anatomical Sciences, History of Anatomy, Advanced Anatomy of the Trunk, Principles and Techniques of Ratioautography

DEPARTMENT OF BIOCHEMISTRY
Phone: Chairman: (512) 696-6201

Chairman: A.J. Guarino
Professors: D.A. Clark (Adjunct), A.D. Elbein, J.S. Nishimura, P.N. Rao (Adjunct), J.R. Rowlands (Adjunct), D.C. Wharton, R.A. Weisman
Associate Professors: R.W. Keenan, J.C. Lee
Assistant Professors: A.W. Rees, D.M. Shapiro
Teaching Fellows: 8
Research Assistants: 9
Graduate Students: 8

Degree Program: Biochemistry

Graduate Degrees: M.S., Ph.D.
Graduate Degree Requirements: Master: A thesis or an approved substitute. Requirements are less rigorous than those for the Ph.D. degree Doctoral: Comprehensive knowledge of biochemistry, as determined by a series of qualifying examinations. Proficiency in a cognate area (e.g., Chemistry, Microbiology, Physiology). A dissertation, which represents an original contribution to the field of Biochemistry and which is publishable in a reputable, scholarly journal.

Graduate and Undergraduate Courses Available for Graduate Credit: General Biochemistry, Biochemical Techniques, Nucleic Acid Metabolism, Biochemistry of Development, Enzymology, Biochemistry of the Amino Acids, Physical Biochemistry, Current Aspects of Protein Biosynthesis, Biochemistry of Carbohydrates, Lipid Biochemistry, Biochemistry Seminar, Supervised Teaching

DEPARTMENT OF BIOENGINEERING
(No reply received)

DEPARTMENT OF MICROBIOLOGY
Phone: Chairman: (512) 696-6501

Chairman: A. Shelokov
Professors: J.A. Bass, R.T. Jensen (Adjunct), S.S. Kalter (Adjunct), W.T. Kniker (Joint appointment), A. W. McCracken (Adjunct), E.L. Oginsky (Joint appointment), R.H. Persellin (Joint appointment), B.P. Sagik (Joint appointment), A. Shelokov, K.O. Smith, R.L. Taylor
Associate Professors: J. Eller (Joint appointment), C.J. Gauntt, D D. Madorsky (Adjunct), R.E. Paque, B.A. Sanford, D.E. Thor
Assistant Professors: R.E. Ellis (Joint appointment), J.T. Harrington, Jr. (Joint appointment), R.E. Harris (Adjunct), S.J. Mattingly, E.E. Moody, F.A. Rommel (Joint appointment)
Instructors: N.L. Funderburk (Joint appointment), J.H. Jorgensen (Joint appointment), V.L. Thomas, M.D. Trousdale
Visiting Lecturers: 4
Teaching Fellows: 1
Research Assistants: 9
Teaching Assistants: 5

Degree Program: Medical Microbiology

Graduate Degrees: M.S., Ph.D.
Graduate Degree Requirements: Master: 30 semester hours minimum, including: 6 hours thesis, 6-9 hours in minor areas Doctoral: No specific number of hours. No foreign language requirements. Qualifying written and oral examinations in Microbiology (Virology, Immunology, Mycology, Pathogenic Bacteriology, Bacterial Genetics and Physiology) and related minor area(s), usually to include Biochemistry. Dissertation based on independent research problem.

Graduate and Undergraduate Courses Available for Graduate Credit: Medical Microbiology, Microbial Physiology, Microbial Genetics, Pathogenic Bacteriology, Introduction to Virology, Introduction to Immunology, Parasitic Protozoology and Helminthology, Pathogenic Fungi, Clinical and Diagnostic Microbiology, Oncogenic and Slow and Latent Viruses, Molecular Basis of Animal Virology, Experimental Immunology and Immunopathology, Immunobiology, Immunochemistry, Advanced Microbial Genetics, Techniques inMicrobiology (To include Cell and Organ Culture, Purification Methodology, Electron Microscopy, Use of Radioisotopes, Special Immunologic Techniques)

DEPARTMENT OF PHARMACOLOGY
Phone: (512) 969-6411

Chairman: A.H. Briggs
Professors: V.S. Bishop, B.J. Blankenship, A.H. Briggs, W.B. Stavinoha
Associate Professors: O. Carrier, Jr., R. Huffman, M.A. Medina
Assistant Professors: D.F. Peterson, L. Felpel, C. G. Smith, J. Wallace, D. Jones, K. Blum, B. Hodgson
Visiting Lecturers: 15
Teaching Fellows: 11

Degree Programs: Pharmacology

Graduate Degree: Ph.D.
Graduate Degree Requirements: Doctoral: Pass required courses, laboratory qualifying examinations, write dissertation

Graduate and Undergraduate Courses Available for Graduate Credit: Advanced Pharmacology I, Advanced Pharmacology II, III, IV, Medical Pharmacology, Endocrine Pharmacology, Chemical Pharmacology, Analytical Toxicology, Clinical Pharmacology, Biophysical Pharmacology, Literature Evaluation, Techniques, Seminar, Special Topics in Pharmacology

DEPARTMENT OF PHYSIOLOGY
Phone: Chairman (512) 696-5618

Chairman: E.J. Masoro
Associate Professors: B.P. Yu, C. Levinson
Assistant Professors: P. Hegstad, R.J.M. McCarter, T.M. Mikiten, T.C. Smith
Instructor: D.C. Proppe
Research Assistants: 3

Degree Program: Physiology

Graduate Degrees: M.A., Ph.D.
Graduate Degree Requirements: Master: 30 semester hours of coursework, thesis Doctoral: no course requirements for Ph.D., dissertation

Graduate and Undergraduate Courses Available for Graduate Credit: Principles of Physiology, Endocrine Physiology, Renal Physiology and Electrolyte Metabolism, Respiratory Physiology, Cardiovascular Physiology, Gastrointestinal Physiology, Membrane Biophysics, Biochemical Aspects of Membrane Structure, Physiological Chemistry of Lipids, Regulation of Metabolism, History of Physiology, Mathematical Physiology, Physiology of Nerve and Muscle, Neurophysiology, Sensory Physiology, Physiology of Muscle, Molecular Basis of Hormone Action, Selected Topics in Physiology

WAYLAND COLLEGE

Plainview, Texas 79072 Phone: (806) 296-5521
Dean of College: K. Perrin

DEPARTMENT OF BIOLOGY
Phone: Chairman: Ext. 43

Chairman: W.H. Reese
Professors: W.H. Reese
Associate Professor: J.H. Bowers,
Assistant Professors: G. Thompson
Instructor: C. Steele
Visiting Lecturers: 1
Research Assistants: 3

Field Stations/Laboratories: Great Plains Research Station Sunray

Degree Program: Biological Sciences

Undergraduate Degrees: B.A., B.S.
Undergraduate Degree Requirements: 30 plus 24 in 2 teaching fields (30 plus 18 major plus minor), Genetics Physiology, Ecology, Invertebrate Zoology, 130 hours

WEST TEXAS STATE UNIVERSITY

Canyon, Texas 79015 Phone: (806) 656-0111
Dean of Graduate Studies: D. Wheeler

College of Arts and Sciences

Dean: T.D. Freidell

DEPARTMENT OF BIOLOGY
Phone: Chairman: (806) 656-3513 Others: 656-3681

Head: H.H. Bailey
Professors: H.H. Bailey, D.L. Brooks, R.C. Busteed, P.A. Caraway, W.A. Cooper, R.A. Wright
Associate Professors: D.P. Bingham, L.C. Higgins, D. A. LaBrie, C.E. Wright
Assistant Professors: N.P. Killian, C. Smith

Degree Program: Biology

Undergraduate Degrees: B.S., M.S.
Undergraduate Degree Requirements: 8 courses of which 4 must be advanced Mathematics, 3 semesters through Calculus, Chemistry 12 credit hours including Organic, Physics or Geology 2 semesters

Graduate Degree: M.S.
Graduate Degree Requirements: Master: 30 hours advanced and 6 hours thesis

Graduate and Undergraduate Courses Available for Graduate Credit: Genetics (U), Cytology (U), Human Genetics (U), Entomology (U), Medical Entomology (U), Plant Physiology (U), Invertebrate Zoology (U), Vertebrate Anatomy (U), Wildlife Management (U), Field Biology (U), Parasitology (U), Plant Classification (U), Plant Anatomy (U), History of Vertebrates (U), Histology (U), Embryology (U), Medical Microbiology (U), General Ecology (U), Plant Ecology (U), Biology of Freshwater (U), Ichthyology (U), Mammalogy (U), Animal Behavior (U), Advanced Physiology (U), Herpetology (U), Plant Pathology (U), Molecular Biology (U), Immunology and Serology (U), Thesis, Biological Investigations, Zoogeography, Fisheries Biology, Developmental Biology, Microbiology, Endocrinology, Speciation, Population Genetics, Methods in Physiology

College of Agriculture

Dean: C. Smellwood

DEPARTMENT OF ANIMAL SCIENCE
Phone: Chairman: (806) 656-3524

Head: J. McNeill
Professors: C. Smellwood
Associate Professors: J. McNeill, K. Wilson
Assistant Professors: D. Beerwinkle, R. Harbin
Instructors: J. McManigal
Research Assistants: 2
Part-time Faculty: 4

Field Stations/Laboratories: Nance Ranch

Degree Program: Animal Science

Undergraduate Degree: B.S.
Undergraduate Degree Requirements: 37 hours school of agriculture core (Including 8 Biology), 51 hours Animal Science, Biology, Chemistry and Electives

Graduate Degree: M. Agr.
Graduate Degree Requirements: Master: Master of Agriculture (Non-thesis - 36 hours (Including 6 hours Management Training and Professional Paper)

Graduate and Undergraduate Courses Available for Graduate Credit: Hormones in Animal Science (U), Sheep Production (U), Beef Cattle Production (U), Swine Production (U), Advanced Equitation (U), Market Classes and Grades of Livestock (U), Animal Nutrition (U), Feedlot Management (U), Special Topics (U), Individual Problems (U), Advanced Animal Nutrition (U), Artificial Insemination of Farm Animals (U), Animal Breeding (U), Specialized Horse Enterprises (U), Techniques in Animal Research, Advanced Animal Breeding, Animal Nutrition I, Advances in Meat Science, Advanced Livestock Management Systems

DEPARTMENT OF PLANT SCIENCE
Phone: Chairman: (806) 656-2100

Head: J. Green
Professor: J. Green
Associate Professors: R. Thomason, K. Wilson
Assistant Professor: L. Wilson
Visiting Lecturers: 2
Research Assistant: 1
Part-time. 1

Degree Programs: Plant Science, Wildlife

Undergraduate Degree: B.S.
Undergraduate Degree Requirements: 37 hours School of Agriculture Core (Including 8 Biology), 51 hours Plant Science - Biology - Chemistry and Electives

Graduate Degree: M. Agr.
Graduate Degree Requirements: Master: 36 hours non-thesis (Including 6 hours Management Training and Professional Paper)

Graduate and Undergraduate Courses Available for Graduate Credit: Genetics (U), Pesticides (U), Plant Breeding (U), Weeds and Weed Control (U), Crop Physiology (U), Special Topics (U), Surveying (U), Range Management (U), Individual Problems (U), Grain Crops (U), Forage Crops (U), Soil Conditions and Plant Growth (U), Soil Fertility (U), Soil Development Classification (U), Soil-Plant-Water Relations (U), Plant Pathology (U), Advanced Plant Breeding, Genetics, Forage and Grain Production, Analysis and Interpretation of Agricultural Data, Soil Fertility Relationships

WILEY COLLEGE

Marshall, Texas 75670 Phone: (214) 938-8341
Dean of College: D. Houston

DEPARTMENT OF BIOLOGY
Phone: Chairman: Ext. 77

Chairman: J.T. Baccus
Professor: K.K. Ganguli
Associate Professor: J.T. Baccus
Instructor: E. Andrey

Degree Program: Biology

Undergraduate Degree: B.S.
Undergraduate Degree Requirements: Zoology, Botany, Genetics, Invertebrate Zoology, Physiology, Directed Study, 18 hours Electives, 1 year Physics, General and Organic Chemistry

WILLIAM MARSH RICE UNIVERSITY

(see Rice University)

UTAH

BRIGHAM YOUNG UNIVERSITY

Provo, Utah 84601 Phone: (801) 374-1211
Dean of Graduate Studies: C.C. Riddle

College of Biological and Agricultural Sciences

Dean: A.L. Allen

DEPARTMENT OF AGRONOMY AND HORTICULTURE
(No reply received)

DEPARTMENT OF ANIMAL SCIENCE
Phone: Chairman: (801) 374-4294

Chairman: L.E. Orme
Professors: C.Y. Cannon (Emeritus), R.W. Gardner, K. H. Hoopes, L.E. Orme, R.L. Park, R.P. Shumway, M. V. Wallentine (Associate Dean)
Assistant Professors: N.P. Johnston, L.W. Smith, K. Andrus, R.T. Pace, W.T. Thompson
Research Assistants: 6-10

Field Stations/Laboratories: B.Y.U. Farm and Spanish Fork

Degree Programs: Animal Science, Dairy Science, Poultry Science

Undergraduate Degree: B.S. (Animal Science)
Undergraduate Degree Requirements: General Education Requirements-Biological Science, 170 American History and Government, English Composition 111, 316, Health 130, Humanities, Language or Mathematics-Statistics-Logic, Physical Education, Physical Science, Social Science, Religion 121. In addition to above 20-28 hours in major, 40 hours upper division credit and a total of 128 hours. The undergraduate can choose several study options, namely Science and Pre-Veterinary, Business and Industry or production.

Graduate Degree: M.S.
Graduate Degree Requirements: Master: Bachelor of Science degree or equivalent Minimum G.P.A. of 2.5 or a 4 point scale. Thirty semester hours total, 20 hours of graduate classes. Thesis. Must adhere to L.D.S. or University Standards.

Graduate and Undergraduate Courses Available for Graduate Credit: Soil Physics, Saline and Alkali Soils, Advanced Crop Production, Advanced Horticulture, Soil and Plant Analysis, Conferences and Reports, Chemistry of Soil-Plant Relationships, Soil Physical Conditions, Advanced Soil Microbiology, Advanced Plant Breeding, Seminar, Animal Nutrition, Animal Nutrition Laboratory, Advanced Animal Breeding, Management of Ranch Resources, Meat and Food Processing Plant Operations, Statistics, Advanced Dairy Production, Experimental Animal Techniques

DEPARTMENT OF BOTANY AND RANGE SCIENCE
Phone: Chairman: (801) 374-2582 Others: 374-1211

Chairman: K.T. Harper
Professors: W.R. Andersen, K.T. Harper, B.F. Harrison (Emeritus), W.M Hess, O. Julander (Emeritus), G. Moore, J.R. Murdock, H.C. Stutz, J.F. Vallentine, D.J. Weber, S.L. Welsh
Associate Professors: B.N. Smith, W.D. Tidwell, L. Whitton
Assistant Professors: J.D. Brotherson, S.R. Rushforth, B.W. Wood
Instructors: W.R. Leichty, J.W. Van Cott
Visiting Lecturers: 1
Teaching Fellows: 8
Research Assistants: 4

Degree Programs: Biological Education, Botany, Genetics, Range Science

Undergraduate Degree: B.S. (Botany, Range Science)
Undergraduate Degree Requirements: 128 semester hours of Credit in approved areas.

Graduate Degrees: M.S., Ph.D. (Botany, Range Science)
Graduate Degree Requirements: Master: Botany - 24 semester hours beyond bachelors plus thesis
Range Science - 24 semester hours plus thesis
Doctoral: Botany - 30 semester hours of acceptable course work beyond the masters degree plus dissertation

Graduate and Undergraduate Courses Available for Graduate Credit: Histological Technique (U), Advanced Taxonomy (U), Agrostology (U), Biological Instrumentation (U), Algology (U), Advanced Mycology (U), Paleobotany (U), Plant Geography (U), Quantitative Ecology (U), Experimental Ecology (U), Botanical Terminology and Nomenclature, Cell Biology, Electron Microscopy, Angiosperm Morphology, Morphogenesis, Organic Evolution, Genetics of the Fungi, Physiology of Fungi and Algae, Advanced Topics in Physiology, Plant Nutrition and Growth, Population Genetics, Advanced Topics in Ecology and Cytogenetics

GRADUATE SECTION OF BIOCHEMISTRY
Phone: Chairman: (801) 374-3667

Chairman: C.J. Gubler
Professors: C.J. Gubler, J.H. Mangum, A.D. Swensen, F.G. White
Associate Professors: M.A. Smith
Teaching Assistants: 6
Research Assistants: 10

Degree Program: Biochemistry

Graduate Degrees: M.S., Ph.D.
Graduate Degree Requirements: Master: 30 semester hours in major and related fields with research thesis and final oral examination Doctoral: Entrance examination in organic and physical chemistry, comprehensive written examination and research proposition. 6 semesters study beyond the B.S. Two languages, two tool or language and tool required, minimum of 12 semester hours in major written dissertation and oral final examination

Graduate and Undergraduate Courses Available for Graduate Credit: Biochemistry (U), Biochemistry Laboratory (U), Advanced Biochemistry, Advanced Biochemistry Laboratory, Biochemistry Seminar, Selected Topics in Biochemistry

ZOOLOGY DEPARTMENT
Phone: Chairman: (801) 374-2006

Chairman: C.D. Jorgensen (Acting)
Professors: J.R. Murphy, D.M. Allred, F.L. Anderson, A.O. Chapman, H.H. Frost, R.W. Heninger, A.W. Jaussi, C.D. Jorgensen, H.J. Nicholes, W.W. Tanner, V.J Tipton, S.L. Wood
Associate Professors: J.R. Barnes, G.M. Booth, L. Braithwaite, R. Heckmann, W.E. Miller, H.D. Smith, C.M. White
Assistant Professors: J.L. Farmer, D.E. Jeffery,

C.L. Pritchett, R.W. Rhees, R.E. Seegmiller, E.R. Simmons, D.A. White, A.T. Whitehead

Degree Programs: Zoology

Undergraduate Degree: B.S.
Undergraduate Degree Requirements: Not Stated

Graduate Degrees: M.S., Ph.D.
Graduate Degree Requirements: Not Stated

Graduate and Undergraduate Courses Available for Graduate Credit: Not Stated

SOUTHERN UTAH STATE COLLEGE
Cedar City, Utah 84720 Phone: (801) 586-4411
Dean of College: H.E. Judd

DEPARTMENT OF LIFE SCIENCE
Phone: Chairman: Ext. 244 Others: Ext. 251

Chairman: A.G. Wahlquist
Professors: R. Anderson, W. Larsen, V.R. Magleby, D. Matthews, A.G. Wahlquist
Associate Professors: J. Bowns, B. Palmer, A. Tait
Assistant Professors: J.A. Bowns, D. Braegger, P. Burgoyne, R. Dotsen
Research Assistants: 2

Degree Programs: Botany, Biology, Zoology

Undergraduate Degree: B.S.
Undergraduate Degree Requirements: Not Stated

UNIVERSITY OF UTAH
University Street and Second Avenue Phone: (801) 581-7200
Salt Lake City, Utah 84112
Dean of Graduate Studies: S.M. McMurrin

College of Science
Dean: F.E. Harris

DEPARTMENT OF BIOLOGY
Phone: Chairman: (801) 581-6517

Chairman: Not Stated
Professors: W.H. Behle, M R. Capecchi, G.F, Edmunds, F.R. Evans, P.D. Gardner, A.R. Gaufin, W.R. Gray, A.W. Grundmann, E.W. Hanley, H.F. Hirth, K.G. Lark, J.M. Legler, I.B. McNulty, N.C. Negus, L.T. Nielsen, B.M Olivera, J D. Spikes, M. Treshow, R .W. VanNorman, R.K. Vickery, D.R. Wolstenholme, D. Wiens
Associate Professors: J.H. Brown, R.R. Hathaway, A.C. Hill, W.L. Ingram (Clinical), M.S. James (Adjunct), L.G. Klikoff, C.A. Lark, J.L. Lords, L.M Okun, R.C. Pendleton, M.C. Rechsteiner, L.H. Wullstein
Assistant Professors: P.J. Berger (Research), O. Cuellar, W.J. Dickinson, R.G. Ellis (Clinical), T. Kogoma (Research), J.S. Parkinson, D.H. Parma, S.R. Poulter (Research), W.M. Schaffer, T.L. Shininger
Visiting Lecturers: 14
Teaching Fellows: 30
Research Assistants: 31

Degree Programs: Animal Behavior, Animal Disease, Animal Genetics, Bacteriology, Bacteriology and Immunology, Biological Chemistry, Biological Sciences, Biology, Biometry, Biostatistics, Botany, Ecology, Entomology, Entomology and Parasitology, Ethology, Forest Chemistry, Forest Entomology, Genetics, Histology, Human Genetics, Immunology, Life Sciences, Limnology, Microbiology, Parasitology, Radiation Biology, Radiobiology, Science, Science Education, Virology, Zoology, Zoology and Entomology

Undergraduate Degrees: B.A., B.S.
Undergraduate Degree Requirements: Mathematics through Calculus, Chemistry through Inorganic and Organic, Physics (Calculus Series), English (two quarters), General Education Electives, General Biology, General Botany, General Zoology, Core and six senior level option courses listed below

Graduate Degrees: M.S., M.Ph., Ph.D.
Graduate Degree Requirements: Master: Determined by supervisory committee and graduate committee
Doctoral: same as Master

Graduate and Undergraduate Courses Available for Graduate Credit: Instrumentation Theory (U), Biometry (U), Molecular Genetics (U), Photobiology (U), Cell Physiology (U), Plant Physiology (U), Comparative Animal Physiology (U), Mammalian Physiology (U), Neurobiology, Protein Structure and Function (U), Protein Evolution (U), Histology (U), Molecular and Microbiology Genetics Laboratory (U), Microbial Physiology (U), Microbial Genetics (U), Comparative Vertebrate Anatomy (U), Embryology (U), Mycology (U), Developmental Genetics (U), Plant Structure and Development (U), Chromosome Mechanics (U), Cytogenetics (U), Taxonomy (U), Comparative Morphology of Plants (U), Protozoology (U), Advanced Invertebrate Zoology (U), General Entomology (U), Ichthyology (U), Herpetology (U), Mammalogy (U), Adaptative Stratum Organisms (U), Theoretical Biogeography, (U), Population and Community Dynamics (U), Ethology (U), Plant Population and Environments (U), Population and Behavioral Ecology (U), Community Analysis (U), Medical Entomology (U), Parasitology (U), Limnology (U), Water Pollution Ecology (U), Fisheries Biology (U), Radiation Biology (U), Radiation Ecology (U), Plant Pathology (U), Air Pollution Ecology (U), Biological Techniques, Radiation Tracer Techniques, Microtechnique, Topics in Cell Biology, Topics in Developmental Biology, Topics in Genetics, Topics in Population Genetics, Topics in Molecular Genetics, Topics in Neurobiology, Topics in Plant Physiology, Topics in Photobiology, Topics in Plant Development, Topics in Insect Morphology, Insect Physiology, Mechanics of Evolution, Systematic Entomology, Mosquito Systematics, Aquatic Entomology, Community Integration and Structure, Ecological Physiology, Analysis of Ecosystems, Advanced Limnology Advanced Parasitology, Advanced Topics in Evolution, Advanced Topics in Zoology, Advanced Topics in Ecology

College of Medicine
Dean: J.A. Dixon

DEPARTMENT OF ANATOMY
Phone: Chairman: (801) 581-6728

Chairman: W.S.S. Jee (Acting)
Professors: D.L. Berliner (Adjunct - Research), E. I. Hashimoto, W.S.S. Jee, W. Stevens
Associate Professors: J.H. Dougherty (Research), C.W. Mays (Research), B.J. Stover (Adjunct - Research), G.N. Taylor (Research), A.R. Wennhold (Research), L.A. Woodbury (Research)
Assistant Professors: L.A. Dethlefsen (Research), R.D Lloyd (Research), O.E. Millhouse, C.J. Nabors, Jr., S.A. Schafer
Instructors: D.R. Atherton (Research), F.W. Bruenger (Research), G.L. Schneebeli (Research), L. Shabestari (Research), S.S. Stensaas, G.D. Westenskow (Research), J.L. Williams (Research)
Visiting Assistant Professor: 1

Degree Program: Anatomy

Graduate Degrees: M.S., Ph.D.
Graduate Degree Requirements: Master: Basic background in Biology, Chemistry and Mathematics, deficiencies may be made up by arrangements with staff Course requirements: Anatomy 601, 602, 603, 604, 605. Anatomy seminar and journal club, courses in

allied fields. Option in radiation biology. Course work arranged upon consultation with staff according to background and career plans. Doctoral: Basic background in Biology, Chemistry through Organic, and Mathematics, through Calculus. Desirable to have Physical Chemistry, Modern Physics or Biophysics and 3 quarters of following: 701, 702, 703, 706, 707, 708, 771, 773, 775, 776, 777, 778, 779. Anatomy seminars, journal clubs, courses in allied fields, preliminary examination, oral examination on dissertation and research proposals.

Graduate and Undergraduate Courses Available for Graduate Credit: Gross Anatomy of the Human, Histology, Advanced Graduate Histology, Neuroanatomy, Prosection, Graduate Instruction in Various Aspects of Anatomy, Fundamentals of Hematology, Reticuloendothelial System, Hematopoietic Organs and Connective Tissues Advanced Human Gross Anatomy, Radiation Physics Applied to Biology, Physical Chemistry Applied to Radiobiology, Techniques in Radiation and Molecular Biology, Seminar in Neuroanatomy, Techniques in Experimental Cancer Biology

DEPARTMENT OF BIOLOGICAL CHEMISTRY
Phone: Chairman: (801) 581-6795

Chairman: S.F. Velick
Professors: S.R. Dickman, S.A. Kuby, A. Linker (Research), H.C. Rilling, L.T. Samuels (Emeritus), S.F. Velick, C.D. West
Associate Professors: L. Kuehl, O.C. Richards
Assistant Professors: C.L. Atkin (Research), P.J. Lawrence, F.W. Sweat
Postdoctorates: 6
Research Assistants: 10

Degree Programs: Biochemistry, Biological Chemistry

Graduate Degrees: M.S., Ph.D.
Graduate Degree Requirements: Master: Not Stated
Doctoral: General Biochemistry (3 quarters), Topics in Biochemistry (2 quarters), other advanced courses and subjects as determined by thesis advisor and/or supervisory committee.

Graduate and Undergraduate Courses Available for Graduate Credit: Medical Biochemistry, General Biochemistry, Clinical Nutrition, Biochemical Endocrinology, Lipid Biochemistry and Membranes, Nucleic Acids and Protein Synthesis, Topics in Carbohydrate Biochemistry, Enzyme Chemistry, Topics in Biochemistry, Research Techniques in Biochemistry

DEPARTMENT OF MICROBIOLOGY
Phone: Chairman: (801) 581-8777 Others: 581-8778

Chairman: D.F. Summers
Professors: L.A. Glasgow, S. Marcus, P.S. Nicholes
Associate Professors: B.C. Cole (Research), D.W. Hill, J.C. Overall, Jr., C. Smith, J.L. Swanson, B.B. Wiley
Assistant Professors: G.A. Hill (Clinical), B. Janice, A.E. Larsen, P.S. Lombardi, F. Miya (Clinical), F.J. O'Neill, M. Rogolsky
Instructor: E.R. Kern
Research Assistants: 5

Degree Programs: Bacteriology, Bacteriology and Immunology, Immunology, Medical Microbiology, Microbiology, Virology, Bacterial and Viral Genetics

Graduate Degrees: M.S., Ph.D.
Graduate Degree Requirements: Master: 9 credit hours of thesis. 15 to 20 hours of laboratory, graduate research, 3 hours of graduate seminar, study in related area to complete 45 credit hours, and a thesis and final oral examination. Doctoral: Independent Laboratory research for not less than 3 years after baccalaureate degree, proficiency in foreign language, preliminary written (18 hours) and oral examination, and a written dissertation defended through an oral examination, Course work in departments related to the dissertation subject is suggested.

Graduate and Undergraduate Courses Available for Graduate Credit: General and Pathogenic Microbiology, Diagnostic Microbiology, Immunology and Serology, Microbiology Journals, Bacterial Metabolism and Physiology, Medical Microbiology, Advanced Microbial Genetics, Immunochemistry, Advanced Immunology Laboratory, Advanced Virology, Advanced Virology Laboratory, Veterinary Microbiology, Recent advances in Microbiological Techniques, Fluorescent Antibody Techniques, Blood Banking and Blood Component Preparation and Use

DEPARTMENT OF PHYSIOLOGY
(No reply received)

UTAH STATE UNIVERSITY
Logan, Utah 84322 Phone: (801) 752-4100
Dean of Graduate Studies: E. Gaidner

College of Agriculture
Dean: D.J. Matthews

DEPARTMENT OF DAIRY SCIENCE
Phone: Chairman: Ext. 7144

Head: Not Stated
Professors: C.D. Funk, G.E. Stoddard, M.J. Anderson
Associate Professors: J.J. Barnard (Extension), R.C. Lamb, D. Stokes (Extension)
Assistant Professors: C.W. Arave, W. Barnes (Extension), C.H. Mickelsen, D. Andrus (Extension)
Research Assistants: 1
Research Technicians: 1

Degree Program: Dairy Science

Undergraduate Degree: B.S.
Undergraduate Degree Requirements: 186 quarter credits three options each with some differences in requirements.

Graduate Degree: M.S.
Graduate Degree Requirements: Master: 45 quarter credits minimum with thesis of 9-15 credits. Specific course requirements approved by committee. Thesis and special report options.

Graduate and Undergraduate Courses Available for Graduate Credit: Dairy Breeding (U), Milk Secretion (U), Dairy Herd Evaluation (U), Dairy Herd Management (U), Dairy Herd Planning (U), Special Problems (U), Research

DEPARTMENT OF SOIL SCIENCE AND METEOROLOGY
Phone: Chairman: Ext. 7185 Others: Ext. 7186

Head: R.L. Smith
Professors: D.W. Carter (Adjunct), J.W. Cary (Adjunct) P.D. Christensen (Extension), I. Dirmhirn, R.J. Hanks, J.J. Jurinak, G.E. Leggett (Adjunct), R.W. Miller, J.H. Smith (Adjunct), D.W. Thorne (Director, Agricultural Experiment Station)
Associate Professors: G.A. Ashcroft, P.E. Daniels (Extension), D.W. James, H.F. Mayland (Adjunct), R.F. Nielsen, E.A. Richardson, J.S. Skujins, A.R. Southard, G.L. Wooldridge, J.L. Wright (Adjunct)
Assistant Professors: V.E. Hunsaker (Extension)
Research Assistants: 15

Degree Programs: Agronomy, Soil Science, Soils and Plant Nutrition, Soil Science and Biometerology, Biometerology

Undergraduate Degree: B.S.
Undergraduate Degree Requirements: 186 credit hours, 30 hours in major, Chemistry, Mathematics, Physics, Plant Science, Plant Physiology

Graduate Degrees: M.S., Ph.D.
Graduate Degree Requirements: Master: 45 hours of which 15 may be research and thesis Doctoral: 90 hours beyond M.S. of which 45 hours may be research and thesis

Graduate and Undergraduate Courses Available for Graduate Credit: Chemistry of Soil Water Systems, Soil Identification and Interpretation, Soil Microbiology, Soil and Plant Nutrition, Physical Properties of Soils, Soil Physics, Physical Chemistry of Soils, Saline and Alkali Soils, Soil Chemistry, Soil Biochemistry and Microbiobiology, Genesis, Morphology and Classification of Soils, Environmental Biogeochemistry, Introduction to Meterology, Climatology, Physical Meterology, Dynamic Meterology, Synoptic Meterology, Cloud and Preciptation Physics, Paleoclimatology, Environmental Field Experiments, Mountain Climatology, Environmental Remote Sensing, Biometeorology, Extratropical Cyclones, Physical Climatology, Instrumentation Laboratory

DEPARTMENT OF NUTRITION AND FOOD SCIENCES
Phone: Chairman: Ext. 7691

Head: C.A. Ernstrom
Professors: C.A. Ernstrom, C.I. Draper, M. Merkley, G.H. Richardson, D.K. Salunkhe, P. Snow
Associate Professors: J.C. Batty, F. Bardwell, D.G. Hendricks, A. Kearsley, P. Larsen, A.W. Mahoney
Assistant Professors: C. Brennand, T.A. Gillett, V.T. Mendenhall, F. Taylor, B. Wyse
Instructor: R. Reeder, F. Garrett
Research Assistants: 9

Degree Programs: Food Science, Nutrition

Undergraduate Degree: B.S.
Undergraduate Degree Requirements: 14 credits Biology, 11 credits Mathematics, 28 credits Chemistry, 46 credits Nutrition and Food Science

Graduate Degrees: M.S., Ph.D.
Graduate Degree Requirements: Master: 36 credits, qualifying oral examination, thesis Doctoral: 62 credits, one foreign language, preliminary comprehensive examination, thesis, final oral, teaching experience, examination

Graduate and Undergraduate Courses Available for Graduate Credit: Perspectives of Dietetics (U), Food Insepction and Regulations (U), Sensory Evaluation of Foods (U), Food Engineering (U), Food Analysis (U), Food Chemistry (U), Ice Cream, Concentrated Milks and Confections (U), Cheese (U), Processing and Storage of Fruits and Vegetables (U), Meat Processing (U), Science in Food Preparation (U), Nutrition (U), Nutrition Laboratory Technology (U), Community Nutrition (U), Clinical Dietetics (U), Medical Dietetics (U), Quantity Food Preparation (U), Institutional Organization and Management (U), Maternal and Child Nutrition (U), Food Science Seminar (U), Nutrition Laboratory, Food Toxicology, Human Nutrition, Continued Graduate Advisement, Nutrition and Growth

DEPARTMENT OF PLANT SCIENCE
(No reply received)

VETERINARY SCIENCE DEPARTMENT
Phone: Chairman: Ext. 7584

Head: J.L. Shupe
Professors: R.A. Bagley (Adjunct), J.T. Blake, D.S. Dennis (Adjunct), R.W. Jones, (Adjunct), N C. Leone, (Adjunct), M.L. Miner, A.E. Olson (Research), J.L. Shupe, D.W. Thomas, W.R. Thornley (Research), J.E. Tugaw (Adjunct)
Associate Professors: J.W. Call, R.A. Smart
Assistant Professors: R.P. Sharma

Degree Programs: Interdepartmental Curriculum in Toxicology

Graduate Degrees: M.S., Ph.D.
Graduate Degree Requirements: Master: Minimum 30 quarter hours of course work and an acceptable thesis Doctoral: Approximately 90 quarter hours of course work above B.S. level including a core curriculum of nearly 50 credit hours, a comprehensive written and/or oral examination after course work, an acceptable thesis (defended before the supervisory committee), and a foreign language

Graduate and Undergraduate Courses Available for Graduate Credit: Reproduction of Domestic Animals, General Pharmacology, Veterinary Parasitology, Reproductive Physiology, General Pathology, Special Pathology, Toxicological Animal Pathology, Principles of Toxicology

College of Natural Resources
Dean: T. Box

DEPARTMENT OF FORESTRY AND OUTDOOR RECREATION
Phone: Chairman: Ext. 7845 Others: (801) 752-4100

Head: L.S. Davis
Professors: T.W. Daniel (Emeritus), L.S. Davis, J.W. Floyd (Emeritus), J.D. Hunt, R.R. Moore
Associate Professors: G.E. Hart, R.H. Hawkins, C.M. Johnson, R.M. Lanner
Assistant Professors: J. Baden, J. Henderson, J.J. Kennedy, S.F. McCool, C. Romesburg, L. Royer
Visiting Lecturers: 1
Research Assistants: 10

Field Stations/Laboratories: Experimental Forest, Summer Camp

Degree Programs: Forest Botany, Forest Resources, Forestry, Forestry and Conservation, Natural Resources, Silviculture, Watershed Management, Outdoor Recreation

Undergraduate Degree: B.S.
Undergraduate Degree Requirements: 195 quarter credit hours (includes summer camp)

Graduate Degrees: M.S., Ph.D.
Graduate Degree Requirements: Not Stated

Graduate and Undergraduate Courses Available for Graduate Credit: Forestry - Principles of Conservation (U), Principles of Conservation Workshop (U), Silviculture (U), Seeding and Planting (U), Forest Administration and Policy (U), Forest Economics (U), Forest Valuation (U), Forest Management (U), Logging (U), Forest and Tundra Ecosystem (U), Trees of North America (U), Forest Protection I, II (U), Population and Resources Persepctives, Advanced Silviculture, Tree Improvement and Forest Genetics, Forest Management, Forest Economics, Natural Resources Policy and Administration, Forest Ecology, Forest Ecosystem Analysis, Forest Resources Seminar, Ecology Seminar, Environmental Remote Sensing, Biometerology, Forest Resources Seminar Outdoor Recreation - Regional Recreation Planning (U), Interpretive Planning (U), Forest Recreation Management (U), Recreation Facility Management (U), Outdoor Recreation Seminar Watershed Science - Forest and Range Hydrology (U), Wildland Water Quality (U), Watershed Science Problems (U), Watershed Operations (U), Snowpack Management, Snow Hydrology, Watershed Science Seminar, Watershed Science Problems, Forest Influences, Watershed Analysis

DEPARTMENT OF RANGE SCIENCE
Phone: Chairman: (801) 752-4100 Ext. 7621

Head: D.D. Dwyer
Professors: T. Box, D.D. Dwyer, C. McKell, O. Olpin (Adjunct), K. Parker, A. Smith (Emeritus)
Associate Professors: M. Caldwell, G. Coltharp, G. Gifford, J. Hooper (Adjunct), P. Urness, N. West
Assistant Professors: J. Bowns, F. Busby, J. Malechek, B. Norton, J. Workman
Research Assistants: 20
Teaching Assistants: 2

Degree Programs: Ecology, Range Science, Watershed Management

Undergraduate Degree: B.S.
Undergraduate Degree Requirements: 195 quarter hours of resident credit

Graduate Degrees: M.S., Ph.D.
Graduate Degree Requirements: Master: 45 quarter hours resident credit (including up to 15 hours for thesis) Doctoral: 90 quarter hours resident credit (including up to 45 hours for dissertation.

Graduate and Undergraduate Courses Available for Graduate Credit: Range Improvement and Environmental Quality (U), Range Resource Economics (U), Range Wildlife Relationships (U), Rangeland Appraisal (U), Range Inventory and Management Planning (U), Plant Autecology, Plant Synecology, Plant Geography, Plant Ecophysiology, Systems Ecology, Advanced Range Economics, Land Use, Range Animal Nutrition, Graduate Seminar, Research Methods, Ecology Seminar, Environmental Remote Sensing, Biometeorology

WILDLIFE SCIENCE DEPARTMENT
Phone: Chairman: (801) 752-4100 Ext. 7928

Head: J.A. Kadlec
Professors: D.A. Balph, J.M. Neuhold, A.W. Stokes, F.H. Wagner, J.E. Low, W.F. Sigler (Emeritus), R. Geode, O. Olpin (Adjunct)
Associate Professors: W. Helm, G.S. Innis, F.K. Knowlton, J.J. Spillett, G.W. Workman, R. Wydoski
Assistant Professors: C.B. Stalnaker, M.L. Wolfe
Instructor: G.K. O'Bryan
Teaching Fellows: 2
Research Assistants: 11

Field Stations/Laboratories: Bear Lake Biological Laboratory Green Canyon Ecology Laboratory

Degree Programs: Wildlife Science, Wildlife Ecology, Wildlife Biology, Fishery Biology

Undergraduate Degree: B.S. (Wildlife Science)
Undergraduate Degree Requirements: 188 credit hours in undergraduate courses including 55 credit hours of approved General Education courses. GE courses may be fulfilled through CLEP, Advanced Placement, or Special Examination.

Graduate Degrees: M.S., Ph.D.
Graduate Degree Requirements: Master: 45 credit hours in graduate level courses including 9-15 credit hours of research and thesis Doctoral: 90 credit hours above M.S. or 135 above B.S.

Graduate and Undergraduate Courses Available for Graduate Credit: Behavioral Ecology, Advanced Big Game Management, Ecology of Animal Populations, Pollution Ecology, Limnology II, Seminar in Animal Behavior, Aquatic Environmental Interactions, Ecology Seminar, Directed Study, Fish Population Theory, Seminar in Animal Populations, Dissertation and Thesis Research, Continuing Graduate Advisement, Ichthyology (U), Fishery Principles (U), Techniques in Fishery Management (U), Fishery Biology (U), Wildlife Seminar (U), Wildlife Problems (U), Systems Ecology (U), Diseases of Fish (U), Wildlife Techniques (U), Principles of Fish Culture (U), Terrestrial Wildlife Field Studies (U), Management Aspects of Wildlife Behavior (U), Management of Wildlife Populations (U), Management of Wildlife Habitat (U), Animal Behavior (U), Wildlife Law Enforcement (U)

College of Science

Dean: R.M. Johnson

DEPARTMENT OF BIOLOGY
Phone: Chairman: (801) 752-4100 Ext. 7771

Chairman: G.W. Miller
Professors: W.S. Boyle, T.L. Bahler, G.W. Cochran, D.W. Davis, K.L. Dixon, L.C. Ellis, E.J. Gardner, B.A. Haws, A.H. Holmgren, G. Knowlton, J.A. MacMahon, C.H. Richardson, R.J. Sanders, R.J. Shaw, J.P. Simmons, R.S. Spendlove, H.H. Wiebe
Associate Professors: J.T. Bowman, W.A. Brindley, P.B. Carter, L.N. Egbert, J.A. Gessaman, W.R.J. Hanson, T.H. Hsiao, R.L. Lynn, I.G. Palmblad, F. Parker (Adjunct), F.J. Post, R.S. Roberts, H.P. Stanley, J.J. Skujins, G.U. Welkie
Assistant Professors: E.R. Cronin, E.C. Oaks, Parker, N.N. Youssef
Instructor: Williams
Visiting Lecturer: 1 - 16
Teaching Fellows: 22
Research Assistants: 15
Traineeships in Genetics: 6

Field Stations/Laboratories: Bear Lake Biological Station

Degree Programs: Animal Behavior, Bacteriology, Bacteriology and Immunology, Biology, Botany, Ecology, Entomology, Genetics, Medical Microbiology, Microbiology, Physiology, Plant Pathology, Virology, Public Health Toxicology, Applied Biology

Undergraduate Degree: B.S. (Biology, Applied Biology, Medical Technology)
Undergraduate Degree Requirements: B.S. in Biology: 46 hours in biology, Mathematics through Calculus II 26-28 hours in Chemistry, 15 hours in Physics, B.S. in Applied Biology: 46 hours in Biology, College Algebra, 15-30 hours in Chemistry, 5-15 hours in Physics

Graduate Degrees: M.S., Ph.D. (Biology, Biological Ecology)
Graduate Degree Requirements: Master: Thesis in area of study and a course schedule outlined by the student's committee, 45 hours of credit or minimum requirement 9-21 may be research and thesis credits; 15 hours must be taken on Logan campus. Doctoral: Dissertation in area of study and course schedule outlined by student's committee. At least 3 years full-time graduate study beyond B.S. 135 hours beyond B.S. or 90 beyond M.S. 45 hours may be in research and thesis. One year of residence. Reading knowledge of one foreign language

Graduate and Undergraduate Courses Available for Graduate Credit: Microbiology - Pathogenic Microbiology (U), Immunology (U), Microbial Physiology (U), Soil Microbiology (U), Aquatic Microbiology (U), Virology (U), Food Microbiology (U), Food Fermentation (U), Special Topics in Microbiology (U), Bacterial Taxonomy, Microbial Biosynthesis, Advanced Soil Biochemistry and Microbiology, Environmental Biogeochemistry, Genetics of Lower Organisms, Seminar in Microbiology Botany - Microtechnique (U), Plant Anatomy (U), Morphology of Vascular Plants (U), Mycology (U), Freshwater Algae (U), Taxonomy of Wildland Plants (U), Evolution of Cultivated Plants (U), Principles of Plant Pathology (U), Forest Pathology (U), Cytogenetics, Advanced Plant Taxonomy, Evolutionary Ecology, Plant Geography, Ecology of

Soil Fungi, Ecology of Aquatic Fungi, Water Relations of Plants, Plant Virology, Plant Respiration and Metabolism, Photosynthesis in Higher Plants, Field Plant Pathology, Plant Physiology Seminar, Plant Pathology Seminar Physiology - Mammalian Physiology (U), Endocrinology (U), Cellular Physiology, Comparative Physiology, Ecological Vertebrate Physiology, Advanced Reproductive Physiology, Seminar in Physiology Zoology-Invertebrate Zoology (U), History and Literature of Biology (U), Parasitology (U), Principles of Development (U), Comparative Anatomy (U), Ornithology (U), Mammalogy (U), Elements of Histology (U), Histological Techniques (U), Evolution and Systematics (U), Herpetology (U), Animal Communities (U), Orientation for Graduate Students, Theoretical Biology, Biochemical Genetics, Current Topics in Genetics, Genetics of Drosophila and Maize, Biological Electron Microscopy, Electron Microscopy Laboratory, Current Topics in Developmental Biology, Zoogeography, Protozoology, Advanced Parasitology, Seminar in Vertebrate Zoology, Seminar in Genetics, Seminar in Parasitology, Seminar in Cellular and Developmental Biology Biology - Science Colloquium (U), Analytical Methods (U), Principles of Genetics (U), Cell Biology Entomology - Systemic Entomology (U), Principles of Entomology (U), Medical and Veterinary Entomology (U), Insect Pollination in Relation to Agriculture (U), Aquatic Entomology (U), Nematology (U), Economic Entomology (U), Agricultural Sprays and Dusts (U), Advanced Systematics Insect Ecology, Insect Ecology Laboratory, Insect Physiology, Insect Toxicology, Insect Toxicology Laboratory, Biological Control of Insect Pests, Seminar in Entomology

ECOLOGY CENTER
Phone: Chairman: Ext. 7411

Director: J.M. Neuhold
Professors: D.F. Balph, G.E. Bohart, T.W. Box, T.W. Daniel, I. Dirmhirn, K.L. Dixon, D.D. Dwyer, L.C. Ellis, D.W. Goodall, R.J. Hanks, J.J. Jurinak, W.A. Laycock, J.B. Low, C.M. McKell, J.M. Neuhold, F.B. Salisbury, R.J. Shaw, W.F. Sigler, A.W. Stokes, F.H. Wagner, H.H. Wiebe
Associate Professors: G.L. Ashcroft, M.M. Caldwell, E. H. Cronin, G.E. Hart, W.T. Helm, T.H. Hsiao, G.S. Innis, F.F. Knowlton, R.M. Lanner, R.I. Lynn, J. A. MacMahon, R.Q. Oaks, I.G. Palmblad, F.J. Post, J.J. Skujins, A.R. Southard, J J. Spillett, N.E. West, G.W. Workman, R.S. Wydoski
Assistant Professors: R.R. Alexander, J.A. Gessaman, J.A. Henderson, B.E. Norton, E.C. Oaks, C.B. Stalnaker, M.L. Wolfe
Visiting Lecturers: 7/year
Research Assistants: 32

Field Stations/Laboratories: Green Canyon Ecology Compound, Bear Lake Biological Station

Degree Program: Ecology

Graduate Degrees: M.S., Ph.D.
Graduate Degree Requirements: Master: Take a minimum of five courses from those listed below including at least one integrated ecology, plant ecology and animal ecology course. One credit hour of ecology seminar is required. Those interested in plant ecology, in addition, are required to show credit for soil survey and classification (or equivalent) and plant physiology (or equivalent). Students emphasizing animal ecology are required to show credit for animal physiology. Doctoral: Take a minimum of 25 quarter hours from the list below, including at least one integrated ecology course and two courses each in plant ecology and animal ecology; expected to earn credit for the interdepartmental ecology seminar each quarter he is in residence.

Graduate and Undergraduate Courses Available for Graduate Credit: General Ecology, Field Ecology, Evolution Ecology and Man, Evolutionary Ecology, Systems Ecology, Limnology, Plant Geography, Forest Ecology, Plant Autecology, Plant Synecology, Plant Ecophysiology, Insect Ecology, Ecology of Vertebrate Physiology, Animal Behavior, Behavioral Ecology, Ecology of Animal Populations, Animal Communities, Zoogeography, Soil Microbiology, Aquatic Microbiology, Climatology, Introduction to Meteorology, Biometerology, Biometeorology Instrumentation, Environmental Field Experiments, Environmental Remote Sensing, Biometeorology, Instrument Laboratory, Advanced Plant Taxonomy, Plant Growth and Development, Mineral Nutrition of Plants, Photosynthesis in Higher Plants, Insect Physiology, Surficial Geology, Paleoecology, Crop Ecology, Soil and Plant Nutrition, Physical Properties of Soils, Soil Biochemistry and Microbiology, Evolution and Systematics

WEBER STATE COLLEGE
3750 Harrison Boulevard Phone: (801) 399-5941
Ogden, Utah 84403

College of Arts, Letters and Science
Dean: G. Welch

DEPARTMENT OF BOTANY
Phone: Chairman: Ext. 508

Chairman: E.L. Hobbs
Professors: H. Buchanan, H.K. Harrison, E.L. Hobbs, R.W. Monk
Associate Professors: E.G. Bozniak, D.L. Chadwick
Assistant Professor: S.L. Clark
Research Assistant: 1

Degree Program: Botany

Undergraduate Degree: B.S.
Undergraduate Degree Requirements: 45 hours in Botany including specific courses designed to a broad basic understanding of Botany, Also 15-27 hours Chemistry, 5 hours Mathematics, 5-15 hours Physics and 10 hours Zoology as supporting classes

DEPARTMENT OF MICROBIOLOGY
Phone: Chairman: Ext. 275

Chairman: L.E. Jackson
Professor: S.P. Hayes
Associate Professors: L.S. Adams, L.E. Jackson
Assistant Professors: A.E. Stockland

Degree Program: Microbiology

Undergraduate Degree: B.S.
Undergraduate Degree Requirements: A candidate for graduation with a Bachelor of Arts or a Bachelor of Science Degree must have completed a minimum of 183 credits. A minimum of 60 credits of upper division work. A major in any one department will consist of not fewer than 40 and not more than 60 credits in the major field. A minor consists of not fewer than 20 credits.

DEPARTMENT OF ZOOLOGY
Phone: Chairman: Ext. 288

Chairman: E.A Jensen
Professors: D.J. Graff, D.S. Havertz, E.A. Jensen, E.W. Smart
Associate Professors: J.N. Jensen, C. Marti
Assistant Professors: E.A. Jenne, R. Richins, B.W. Winterton

Degree Program: Zoology

Undergraduate Degree: B.S.
Undergraduate Degree Requirements: (Depending on the area of emphasis), Mathematics through College

Algebra or first quarter of Calculus. Chemistry: 15 hours of Inorganic Chemistry or a comination of 15 hours of Inorganic and 15 hours of Organic Chemistry. Physics: 5 hours or 15 hours, Botany: about 10 hours of botany, Additional area requirements as required by Weber State College

WESTMINSTER COLLEGE

Salt Lake City, Utah 84105 Phone: (801) 484-7651
Dean of College: H. Hofmann

DEPARTMENT OF BIOLOGY
Phone: Chairman: Ext. 51

Chairman: B.G. Quinn
Professors: B.G. Quinn, R. Warnock
Undergraduate Teaching Assistants: 4

Degree Programs: Biology

Undergraduate Degree: B.S.
Undergraduate Degree Requirements: Minimum of 34 semester hours in Biology, including General Biology, General Botany, Invertebrate Zoology, Evolution, Chordate Anatomy, Genetics, and at least 12 additional upper division hours, one year of College Mathematics, one year of Physics, and one year of Chemistry, and a second year in any non-biological science, including earth sciences.

VERMONT

CASTLETON STATE COLLEGE

Castleton, Vermont 05735 Phone: (802) 468-5611
Dean of Graduate Studies: W.S. Reuling
Dean of College: D.M. Burns

DEPARTMENT OF NATURAL SCIENCES
Phone: Chairman: Ext. 238, 239

Chairman: S. Anderson
Professors: W.H. Feaster, J.V. Freeman
Assistant Professors: R.S. Egan, D.J. Gemmell

Degree Program: Biology

Undergraduate Degree: B.S.
Undergraduate Degree Requirements: Not Stated

Graduate Degree: M.A.Ed.
Graduate Degree Requirements: Master of Arts in Education: General Degree Requirements, Professional Self-Assessment, Educational Research, Graduate Seminar, Total 7 credits

JOHNSON STATE COLLEGE

Johnson, Vermont 05656 Phone: 635-2356

DIVISION OF SCIENCE AND MATHEMATICS
Phone: Chairman: Ext. 212

Division Chairman: J. Knapczyk
Associate Professors: P. ChiaraValle
Assistant Professors: G. Mercer, D. Salter

Degree Program: Life Sciences

Undergraduate Degree: B.S.
Undergraduate Degree Requirements: 40 credit hours in the Biological Sciences, 5 credits of Science Seminar, It is expected that the student will work closely with his/her advisor to plan the best possible 4 year program to accomodate his/her individual needs.

LYNDON STATE COLLEGE

Lyndonville, Vermont 05856 Phone: (802) 626-3335
Dean of College: F. McKay

SCIENCE DEPARTMENT
Phone: Chairman: (802) 626-3335

Chairman: D.H. Miller
Professors: N. Doberczack, D.H. Miller
Laboratory Assistant: 1

Degree Programs: Environmental Science Major - Biology or Geology emphasis

Undergraduate Degrees: B.A., B.S.
Undergraduate Degree Requirements: 122 credit hours

MARLBORO COLLEGE

(No reply received)

MIDDLEBURY COLLEGE

Middlebury, Vermont 05753 Phone: (802) 388-4948
Dean of Graduate Studies: R. Turner
Dean of College: G. Harnest

DEPARTMENT OF BIOLOGY
Phone: Chairman: (802) 388-2802

Chairman: G.B. Saul 2nd
Irene Heinz and John Laporte Given Professor of Pre-medical Sciences: G.B. Saul 2nd
Professors: D.J. McDonald, G.B. Saul 2nd, H.E. Woodin
Associate Professors: C. Watters
Assistant Professors: H. Fischer, M. Greenwood
Instructor: W.H. Bell
Visiting Lecturer: 1
Teaching Fellows: 2
Teaching Associate: 1

Field Stations/Laboratories: High Pond - Brandon, Battell Preserve,- Middlebury

Degree Program: Biology

Undergraduate Degree: B.S.
Undergraduate Degree Requirements: 10 courses in Biology; 3 cognates in Chemistry, Physics, and/or Mathe matics

Graduate Degree: M S.
Graduate Degree Requirements: Master: 9 courses of which at least 6 must be in Biology, and of which at least 3 must be research.

Graduate and Undergraduate Courses Available for Graduate Credit: Not Stated

NORWICH UNIVERSITY

Northfield, Vermont 05663 Phone: (802) 485-5011
Dean of Graduate Studies: G.H. Lane
Dean of College: S.H. McIntire

DEPARTMENT OF BIOLOGY
Phone: Chairman: Ext. 43

Head: R. Detwyler
Professors: W.D. Countryman (Adjunct), R. Detwyler
Assistant Professors: R.D. Bair, W.H. Barnard, E.M. Carney, L. Davison, J.W. McDaniel, M.G. Sinclair, R.C. White

Degree Programs: Biology, Medical Technology

Undergraduate Degrees: B.A., B.S.
Undergraduate Degree Requirements: Biology (BS) - 36 courses, Biology - 10, Chemistry 4, English - 4, Mathematics - 2, Physics - 2, Electives - 14 (BA) 36 courses, Biology - 8, Chemistry - 4, English - 4, Mathematics - 2, Physics - 2, Language - 4, Electives - 12

ST. MICHAEL'S COLLEGE

Winooski, Vermont 05604 Phone: (802) 655-2000
Dean of College: Not Stated

BIOLOGY DEPARTMENT
Phone: Chairman: (802) 655-2000

Chairman: J.C. Hartnett
Professors: J.C Hartnett, D.T. Klein, T.D. Sullivan
Associate Professors: D.J. Bean, R.N. DiLorenzo
Assistant Professors: A.C. Hessler, P. Thompson

Degree Program: Biology

Undergraduate Degree: B.A.
Undergraduate Degree Requirements: Total 40 courses, 13 courses in Biology Department, 4 in Chemistry, 2 in Physics (recommended), 2 in Mathematics. General Biology, Ecology, Cell Biochemistry, Genetics, Senior Seminar and Thesis required. Senior Research optional. 2 courses each in Humanities, Philosophy, Religious Studies and Social Sciences required.

TRINITY COLLEGE
(No reply received)

UNIVERSITY OF VERMONT
South Prospect Street Phone: (802) 656-3480
Burlington, Vermont 05401
Dean of Graduate Studies: W.H. Macmillan

College of Arts and Sciences
Dean: J.C. Weiger

DEPARTMENT OF ZOOLOGY
Phone: Chairman: (802) 656-2921 Others: 656-2922

Chairman: R.W. Glade
Professors: R.T. Bell, R.W. Glade, E.B. Henson, J.H. Lochhead, P.A. Moody (Emeritus), M. Potash, H. Rothstein
Associate Professors: J.A Davison, D.F. Stevens
Assistant Professors: J.D. Brammer, R.E. Keen, C.W. Kilpatrick, R.H. Landesman, C.A. Woods
Teaching Fellows: 16

Degree Programs: Cell Biology (Interdepartmental M.S., Ph.D.), Zoology

Undergraduate Degree: B.A.
Undergraduate Degree Requirements: Calculus, Chemistry, 2 semesters, General Physics, 2 semesters, Principles of Biology, 2 semesters; Genetics, Environmental Biology, Cell Structure and Function, Comparative Structure and Function, 7 additional hours of Zoology at an advanced level.

Graduate Degrees: M.S., Ph.D.
Graduate Degree Requirements: Master: 30 hours of which 2 must be Zoology Colloquium, 13-20 additional hours in Zoology courses or courses in a related field, and 8-15 hours in thesis research, teaching experience, written comprehensive examination in field of specialization; oral examination in defense of the thesis. Doctoral: Reading knowledge of 2 appropriate foreign languages or one language and an adjunct area of special competency, teaching experience in at least one undergraduate course, of 75 hours required at least 40 must be in course work including 4 hours of graduate colloquia and at least 13 hours in courses other than Zoology, at least 20 of the 75 but not more than 40 must be in thesis research, written comprehensive examination, oral thesis defense.

Graduate and Undergraduate Courses Available for Graduate Credit: Control of Growth and Differentiation, Population Ecology, Natural History of Birds and Mammals, Natural History of Lower Vertebrates, General Entomology, Field Zoology, Embryology, Comparative Histology, Human Genetics, Comparative Vertebrate Anatomy, Mechanisms of Cell Division, Experimental Embryology, Biochemical Embryology, Environmental Invertebrate Ecology of the Mountains, Invertebrate Zoology, Insect Structure and Function, Comparative Animal Physiology, Genetics of Development, Modern Evolutionary Theory, Advanced Limnology, Zoology Colloquia, Special Topics

College of Agriculture and Home Economics
Dean: T.W. Dowe

DEPARTMENT OF VOCATIONAL EDUCATION AND TECHNOLOGY (AGRICULTURAL ENGINEERING AND TECHNOLOGY PROGRAM)
Phone: Chairman: (802) 656-2000

Chairman: G.R. Fuller
Professor: E. Schneider
Associate Professor: J. Bornstein (Adjunct)
Assistant Professors: M.J. Moore (Adjunct), G.D. Wells (Adjunct, Research)
Research Technicians: 2

Degree Program: Agricultural Engineering
Undergraduate Degree Requirements: 120 semester hours

DEPARTMENT OF ANIMAL SCIENCES
Phone: (802) 656-2070 Others: 656-3131

Chairman: A.M. Smith
Professors: K.S. Gibson, W P. Leamy, A.H. Duthie, H.V. Atherton, D.J. Balch, J.G. Welch,
Associate Professors: L.B. Carew, D.C. Foss, K.M. Nilson, K.R. Simmons, C. Woelfel (Adjunct)
Assistant Professors: L. Mercia (Adjunct), J.J. Rutledge
Teaching Fellow: 1
Research Assistant: 8

Degree Program: Animal Science

Undergraduate Degree: B.S.
Undergraduate Degree Requirements: Eight semester courses in Animal Science, including at least five of advanced standing, minimum of 120 credit hours of course work plus credit in required courses.

Graduate Degrees: M.S., Ph.D.
Graduate Degree Requirements: Master: Must maintain a 3.0 quality point average and a thesis is required. Thirty semester hours of resident graduate credits required. Doctoral: A minimum of 75 credit hours of courses and thesis research is required with a minimum of 15 taken in residence. A comprehensive written examination in the major field of study must be passed. A thesis is required and must be defended in an oral examination by the graduate committee.

Graduate and Undergraduate Courses Available for Graduate Credit: Ice Cream and Frozen Dairy Products, Advanced Nutrition, Nutrition Seminar, Advanced Dairy Cattle Management, Dairy Plant Management, Endocrinology, Physiology of Reproduction and Lactation, Animal Sciences Seminar, Special Problems in Animal Sciences, History of Nutrition (U), Advanced Concepts in Nutrition, Experimental Techniques in Nutrition

DEPARTMENT OF BOTANY
Phone: Chairman: (802) 656-2932 Others: 656-2930

Chairman: B.B. Hyde
Professors: B.B. Hyde, R.M. Klein, J.W. Marvin (Emeritus), T. Sproston (Emeritus), F.H. Taylor, H.W. Vogelmann
Associate Professors: P.W. Cook, B. Etherton
Assistant Professors: I.A. Worley, R.C. Ullrich
Visiting Lecturers: 1
Teaching Fellows: 6
Research Assistants: 4
Research Associates: 3
Assistant Plant Pathologist: 1
Curator: 1

Field Stations/Laboratories: Proctor Maple Research Farm, Underhill, Vermont

Degree Programs: Botany, Cell Biology (Interdepartmental)

Undergraduate Degrees: B.S.
Undergraduate Degree Requirements: Calculus, 2 sem-

esters, General Chemistry, Organic Chemistry, General Physics, 2 semesters, Principles of Biology 2 semesters, Genetics, Plant Physiology, Plant Diversity, Ecology or Systematics and Phylogeny, Two additional semester courses in Botany.

Graduate Degrees: M.S., Ph.D.
Graduate Degree Requirements: Master: 15-21 hours in Botany and closely related fields; thesis research (9-15 hours), a written comprehensive examination in the field of specialization, an oral examination in defense of the thesis, teaching experience. Doctoral: Reading knowledge of an appropriate foreign language, 75 hours of thesis research and course work, at least 40 hours of course work of which at least 20 must be in Botany and at least 20 in other sciences, supervised teaching of at least 6 semester contact hours, comprehensive examination in the field of study, oral examination in defense of the thesis.

Graduate and Undergraduate Courses Available for Graduate Credit: Electron Microscopy, Mineral Nutrition of Plants, Water Relations of Plants, Plant Communities, Botany Field Trip, Microtechnique, Mycology, Molecular Genetics of Eucaryotes, Cytology, Physiology of the Plant Cell, Plant Growth and Development, Phycology, Plant Cytogenetics

DEPARTMENT OF MICROBIOLOGY AND BIOCHEMISTRY
Phone: Chairman: (802) 656-2640

Chairman: D.W. Racusen
Professors: D. Racusen, J.E. Little
Associate Professors: M. Foote, R. Sjogren, D.L. Weller
Instructor: G. Husted
Teaching Fellows: 2
Research Assistants: 3

Degree Programs: Biochemistry, Microbiology

Undergraduate Degree: B.S.
Undergraduate Degree Requirements: Not Stated

Graduate Degrees: M.S., Ph.D.
Graduate Degree Requirements: Not Stated

Graduate and Undergraduate Courses Available for Graduate Credit: Not Stated

DEPARTMENT OF FORESTRY
Phone: Chairman: (802) 636-2620

Chairman: W.W. Christensen
Professors: W.W. Christensen, R.A. Whitmore, Jr.
Associate Professors: P.R. Hannah, M.L. McCormack,Jr.
Assistant Professors: F.H. Armstrong, J.R. Donnelly, C. M. Newton
Instructors: R.T. Foulds, Jr., T.L. Turner
Research Assistants: 3

Field Stations/Laboratories: Jericho Research Forest - Jericho, Wolcott Research Forest - Wolcott

Degree Program: Forestry

Undergraduate Degree: B.S.
Undergraduate Degree Requirements: 138 semester credit hours of prescribed and elective courses.

Graduate Degree: M.S.
Graduate Degree Requirements: Master: Advanced forestry and related courses (15-24 hours); thesis research (6-15 hours)

Graduate and Undergraduate Courses Available for Graduate Credit: Mineral Nutrition of Plants (U), Water Relations of Plants (U), Site Relations and Production Dynamics (U), Advanced Silviculture (U), Forest Valuation (U), Forest Management Decision Theory (U), Forestry Seminar (U), Selected Problems in Forestry or Wildlife Sciences

DEPARTMENT OF PLANT AND SOIL SCIENCE
Phone: Chairman: (802) 656-2630

Chairman: S.C. Wiggans
Professors: R.J. Bartlett, C.L. Calahan (Adjunct), G.B. MacCollom, S.C. Wiggans
Associate Professors: G.R. Benoit (Adjunct), B.R. Boyce, D.R. Evert, T.R. Flanagan, J.L. McIntosh, B.L Parker, N.E. Pellett, G.M. Wood
Assistant Professors: F.R. Magdoff
Teaching Associate: D.W. Bruckel
Visiting Lecturers: 3
Teaching Fellow: 1
Research Assistants: 7
Graduate Technicians: 3

Field Stations/Laboratories: Horticultural Research Center - South Burlington, Vermont

Degree Programs: Agronomy, Crop Science, Entomology, Farm Crops, Horticultural Science, Horticulture, Landscape Horticulture, Ornamental Horticulture, Plant Sciences, Pomology, Soil Science, Soils and Plant Nutrition, Vegetable Crops

Undergraduate Degree: B.S.
Undergraduate Degree Requirements: 120 credit hours

Graduate Degrees: M.S., Ph.D.
Graduate Degree Requirements: Master: 30 credits of graduate work, including an acceptable thesis Doctoral: 75 credits of graduate work, including an acceptable thesis

Graduate and Undergraduate Courses Available for Graduate Credit: Micrometerology, Plant Research Techniques, Mineral Nutrition of Plants, Water Relations of Plants, Advanced Tree Fruit Culture, Soil Classification and Land Use, Soil Chemistry, Soil Physics, Soil and Water Pollution, Seminar, Graduate Special Topics

DEPARTMENT OF RESOURCE ECONOMICS
Phone: Chairman: (802) 656-3036

Chairman: R.O. Sinclair
Professors: V.R. Houghaboom (Adjunct), F.O. Sargent, R.O. Sinclair, R.H. Tremblay, F.C. Webster
Associate Professors: D.K. Eddy (Adjunct), C.L. Fife
Assistant Professors: M.I. Bevins (Adjunct), A.H. Gilbert
Research Assistants: 3

Degree Program: Agricultural Economics

Undergraduate Degree: B.S.
Undergraduate Degree Requirements: 30 hours in Social Science, 24 of which must be in resource (Agricultural) Economics or Economics. Also meet core requirements of college.

Graduate Degree: M.S.
Graduate Degree Requirements: Master: 30 hours minimum in graduate level courses, thesis

Graduate and Undergraduate Courses Available for Graduate Credit: Farm Business Management (U), Rural Communities in Modern Society (U), Markets, Food and Consumers (U), Agricultural Policy (U), Community Organization and Development (U), Natural Resource Evaluation (U), Economics of Outdoor Recreation (U), Community Design (U), Regional Planning (U), Legal Aspects of Planning and Zoning (U), Spatial Analysis (U), Advanced Agricultural Economics (U), Managerial Decisions (U), Advanced Resource Economics, Advanced Regional Planning

College of Medicine

Dean: W.H. Luginbuhl

DEPARTMENT OF ANATOMY
Phone: Chairman: (802) 656-2230

Chairman: W.J. Young
Thayer Professor of Anatomy: W.J. Young
Professor: W.J. Young
Associate Professors: S.L. Freedman, B.A. Ring (Radiologic Anatomy), J. Wells
Assistant Professors: D. Boushey, G.R. Herst, P.P. Krupp, D. McCandless, W.K. Paull

Degree Programs: Anatomy, Anatomy and Histology

Graduate Degrees: M.S., Ph.D.
Graduate Degree Requirements: Master: Anatomy 301, 302, 311, Biochemistry 301, 30 credits, dissertation, comprehensive examination. Doctoral: As above, plus electives, teaching assignments, thesis, 75 credits

Graduate and Undergraduate Courses Available for Graduate Credit: Gross Anatomy, Neuroscience, Medical Histology, Neuroendocrinology, Advanced Neuroanatomy, Cytogenetics, Special Dissections, Special Topics, Histology

DEPARTMENT OF BIOCHEMISTRY
Phone: Chairman: (802) 656-2226 Others: 656-2220

Chairman: D.B. Melville
Professors: M.P. Lamden, D.B. Melville
Associate Professors: W.L. Meyer, J.W. Thanassi, R.C. Woodworth, R.E. Wuthier
Assistant Professors: B.A. Hart, P. Schofield, J.M. Willard
Instructor: Y. Ishikawa
Research Assistants: 2

Degree Program: Biochemistry

Graduate Degrees: M.S., Ph.D.
Graduate Degree Requirements: Master: Fifteen hours from graduate courses offered by the Department of Biochemistry; fifteen hours of Master's Thesis Research Doctoral: Twenty hours from graduate courses offered by the Department of Biochemistry, nine hours from graduate courses offered by the Department of Chemistry, ten additional hours from courses in Physical or Biological Sciences, thirty hours of Doctoral Thesis, Research, a reading knowledge of German or other appropriate foreign language

Graduate and Undergraduate Courses Available for Graduate Credit: General Biochemistry, Biochemistry Laboratory, Medical Biochemistry, General Enzymology, Enzyme Kinetics and Mechanisms, Nucleic Acids, Organic Biochemistry, Radioisotope Laboratory, Physical Biochemistry, Seminar

DEPARTMENT OF MEDICAL MICROBIOLOGY
Phone: Chairman: (802) 656-2164 Others: 656-3480

Chairman: W.R. Stinebring
Professor: W.R. Stinebring
Associate Professors: R.J. Albertini, D.K. Boraker, D.W. Gump, T.J. Moehring, C.P. Novotny, C.A. Phillips, W.I. Schaeffer
Assistant Professors: P.M. Fives-Taylor, G.J. Jakab (Adjunct), J.M. Moehring
Teaching Fellows: 3

Degree Program: Medical Microbiology

Graduate Degrees: M.S., Ph.D.
Graduate Degree Requirements: Master: Medical Microbiology 381-384, Thesis Research, approved selected courses from among Medical Microbiology 203, 205, 211, 223, 302, 325, a course in Biochemistry from among Microbiology and Biochemistry 201-250, 254, or Bio-Chemistry 301-302, 303-304, with the approval of the Department, Passage of a comprehensive examination in Medical Microbiology and related subjects. Doctoral: Participation in seminars offered by the Department during residency of students, Biochemistry 301-302, 303; approved selected courses from programs in Medical Microbiology, Biochemistry, Microbiology and Biochemistry, Physiology and Biophysics, Botany and Zoology or others at the discretion of the Department. Proficient in computer language and programming, 35 hours maximum for thesis research.

Graduate and Undergraduate Courses Available for Graduate Credit: The Mammalian Cell as a Microorganism, Pathogenic Bacteriology, Genetics of Microorganisms, Immunology, Medical Microbiology, Special Problems in Medical Microbiology, Virology, Seminar

DEPARTMENT OF PHYSIOLOGY AND BIOPHYSICS
(No reply received)

WINDHAM COLLEGE
Putney, Vermont 05346 Phone: (802) 387-5511
Dean of College: J.R. Watt

DEPARTMENT OF BIOLOGY
Phone: (802) 387-5511

Chairman: A.H. Westing
Professors: W. Beautyman (Adjunct), A.H. Westing
Associate Professor: C.C. Tseng
Assistant Professor: A.M. Stewart
Visiting Lecturer: 1

Degree Programs: Biology, Biological Education, Conservation, Medical Technology

Undergraduate Degrees: B.A., B.S.
Undergraduate Degree Requirements: Varies with program (but include about a dozen courses in Biology, Chemistry, Mathematics and Physics)

VIRGIN ISLANDS

COLLEGE OF THE VIRGIN ISLANDS
St. Thomas, Virgin Islands 00801 Phone: (809) 774-1252
Dean of College: A.A. Richards

DIVISION OF SCIENCE AND MATHEMATICS
Phone: Chairman: (809) 774-1252 Ext. 311 or 312

Chairman of Division: J. Dougherty
Associate Professors: H.W. Gjessing
Assistant Professors: W.P. MacLean III, S. Holtzman
Instructor: C. Aregood
Visiting Lecturers: 1

Field Stations/Laboratories: Ecological Research Station - Lameshur Bay; Water Resources Research Center - Caribbean Research Institute

Degree Programs: (Three year transfer program in Medical Technology), Biology, Marine and Environmental Sciences

Undergraduate Degree: B.S.
Undergraduate Degree Requirements: (For Marine Science requirements): General Biology, Geology, Chemistry, Oceanography, College Algebra and Trignometry, At least five courses selected from Biology, Chemistry, Mathematics, Physics, Geology and at least two specialized field courses. Biology Major: General Biology, Vertebrate Zoology or Plant Biology, Ecology, Genetics, Physiology, Senior Science Seminar, College Algebra and Trigonometry, General Chemistry, Any two courses from list including Invertebrate Zoology, Cell Biology, Microbiology, Evolution, Population Biology, Marine Ecology

VIRGINIA

AVERETT COLLEGE

Danville, Virginia 24541 Phone: 793-7911
Dean of College: R.K. Godsey

DEPARTMENT OF BIOLOGY
Phone: Chairman: Ext. 243

Head: R.C. Brachman
Professors: R.C. Brachman
Assistant Professor: E.W. Fisher
Hospital Staff in Medical Technology Program: 8

Degree Programs: Biology, Medical Technology

Undergraduate Degrees: B.S.
Undergraduate Degree Requirements: Biology: 32 semester hours in cognate subjects, Medical Technology: Human Anatomy and Physiology, Microbiology, 3 courses Chemistry

BRIDGEWATER COLLEGE

Bridgewater, Virginia 22812 Phone: (703) 828-2501
Dean of College: D.V. Ulrich

DEPARTMENT OF BIOLOGY
Phone: Chairman: Ext. 55

Head: W.L. Mengebier
Professors: H.G.M Jopson, W.L. Mengebier
Associate Professor: L.M. Hill
Assistant Professor: E. Kyger

Degree Program: Biology

Undergraduate Degree: B.A.
Undergraduate Degree Requirements: 32 hours in Biology, Mathematics through Calculus, minimum of 8 hours Chemistry

CHRISTOPHER NEWPORT COLLEGE

(see College of William and Mary)

CLINCH VALLEY COLLEGE

(see University of Virginia)

COLLEGE OF WILLIAM AND MARY

Williamsburg, Virginia 23185 Phone: (703) 229-3000
Dean of Graduate Studies: J.E. Selby

Faculty of Arts and Sciences

Dean: J. Edwards

DEPARTMENT OF BIOLOGY
Phone: Chairman: Ext. 212

Chairman: M.A. Byrd
Professors: J.T. Baldwin, R.E. Black, M A.C. Byrd, G.R. Brooks, B.W. Coursen, C.R. Terman
Associate Professors: G.W. Hall, B S. Grant, C.P. Mangum, M.C. Mathes, H. Aceto, Jr., S.A. Ware, C.W. Vermaulen, B.M. Speese
Assistant Professors: R.A. Beck, N.L. Fashing, E.L. Bradley, J.L. Scott, L.L. Wiseman, G. Capelli
Teaching Assistants: 6-8

Field Stations/Laboratories: Population Laboratory

Degree Programs: Biology, Botany, Ecology, Zoology

Undergraduate Degree: B.A. (Biology
Undergraduate Degree Requirements: Not Stated

Graduate Degrees: M.A.
Graduate Degree Requirements: Master: Courses as approved by Committee, Comprehensive examination, one foreign language, thesis, final oral

Graduate and Undergraduate Courses Available for Graduate Credit: Cytogenetics, Cellular Physiology, Mammalian Physiology, Virology, Animal Behavior, Biology of Vascular Plants, Biochemistry, General Endocrinology, Introduction to Ornithology, Biostatistics, Experimental Biochemistry, Plant Physiology, Cell Structure and Function, Introduction to Radiation Biology, Aquatic Ecology, General Entomology, Physiological Ecology of Plants, Biogeography, Physiological Aspects of Plant Development, Interactive Processes in Development, Radiobiology Seminar, Genetics of Evolutionary Processes, Biosystematics, Ecological Dynamics of Natural and Experimental Populations, Concepts of Synecology, Experimental Electron Microscopy, Experimental Endocrinology, Herpetology, Environmental Physiology, Design of Experiments

Christopher Newport College

Newport News, Virginia 23606 Phone: (803) 596-7611
Dean: T. Musial

DEPARTMENT OF BIOLOGY AND ENVIRONMENTAL SCIENCE
Phone: Chairman: (803) 596-7611 Ext. 215

Chairman: J.E. Pugh
Professors: A.S. Markusen, J.E. Pugh
Associate Professors: D.A. Bankes, R.S. Edwards (Associate Dean), L.C. Olson, E.S. Wise
Assistant Professor: R.O. Simmons
Instructor: H.H. Hobbs, III, R. Hyle, Jr.

Degree Programs: Biological Sciences, Biological Education

Undergraduate Degree: B.S.

Undergraduate Degree Requirements: 42 hours Biology, 20 hours Chemistry, 8 hours Physics, 6 hours Mathematics

EASTERN MENNONITE COLLEGE

Harrisonburg, Virginia 22801 Phone: (703) 433-2771
Dean of College: D. Yutzy

DEPARTMENT OF LIFE SCIENCES
Phone: Chairman: Ext. 247 Others: Ext. 246

Chairman: D.B. Suter
Professors: K.K. Brubaker, D.R. Hostetter (Emeritus), D.B. Suter
Associate Professors: A.C. Mellinger, H.A. Mumaw
Assistant Professors: J.M. Brubaker, R.D. Yoder

Degree Programs: Biology, Medical Technology

Undergraduate Degrees: B.A., B.S.
Undergraduate Degree Requirements: Regular college requirements, 9 courses in Biology, General Chemistry, Highly recommended: General Physics, Organic Chemistry, Mathematics through Calculus

EMORY AND HENRY COLLEGE

Emory, Virginia 24327 Phone: (703) 944-3121
Dean of College: S.H.Lalita

DEPARTMENT OF BIOLOGY
Phone: Chairman: Ext. 231

Chairman: M.C. Wicht, Jr.
Assistant Professors: J.M. Jones, G.E. Treadwell
Student Assistants: 5

Field Stations/Laboratories: MACCI Research Station - Norris Lake, Tennessee

Degree Program: Biology

Undergraduate Degrees: B.A., B.S.
Undergraduate Degree Requirements: B.A. - 8 courses in Biology, 2 courses in Chemistry, Mathematics through College Algebra, Others to total 31 B.S. - 19 courses in Biology, Chemistry through Organic, Mathematics through Calculus, Others to total 31 courses

GEORGE MASON UNIVERSITY

4400 University Drive Phone: (703) 323-2181
Fairfax, Virginia 22030
Dean of Graduate Studies: W.S. Willis
Dean of College: L.E. Boothe

DEPARTMENT OF BIOLOGY
Phone: Chairman: (703) 323-2181

Chairman: S.R. Tabu
Professors: M.G. Emsley, M.S. Stanley, S.R. Taub, J.R. Wall
Associate Professors: T.R Bradley, C.H. Ernst, J.C. Shaffer
Assistant Professors: G.E. Andrykovitch, R. Chairez, D.L. Crawford, J.L. Hart, F.W. Hinton, J.K. Joslin, E.C. Joyce, D.P.Kelso, C.R.Landgren, C.E.Nix, C.H. Pike, C.L. Rockwood, N.R. Sinclair, J.E. Skog, J.W. Wilson, III
Temporary Appointments (Lecturers): 8

Degree Programs: Biology, Environmental Biology

Undergraduate Degrees: B.A., B.S.
Undergraduate Degree Requirements: B.A. 32 semester hours in Biology including Cell Biology, General Genetics, and Ecology, General Chemistry, Mathematics proficiency, B.S. 44 semester hours in Biology including Cell Biology, General Genetics and Ecology, Organic Chemistry, Mathematics Proficiency

Graduate Degree: M.S.
Graduate Degree Requirements: Master: M.S. in Biology. 30 semester hours of graduate work with not more than two approved upper-division undergraduate courses; two semester hours credit in seminar, comprehensive examination

Graduate and Undergraduate Courses Available for Graduate Credit: Human Genetics, Origin of the Cultivated Plants and Their Impact on Man and Society, Food, Energy and Insects, Animal Behavior, Advanced Plant Systematics, Freshwater Ecology, Marine Ecology, Selected Topics in Zoogeography, Selected Topics in Plant Biology, Problems in Development, Comparative Physiology, Human Evolution and Ecology, Evolutionary Taxonomy, Current Topics in Biology, Thesis, Environmental Biology I, Environmental Biology II, Seminar in Environmental Biology, Analysis of Terrestrial Ecosystems Population Ecology, Biological Resource Management, Environmental Hazards to Human Health, Experimental Design and Analysis for the LifeSciences, Independent Study in Biology, Plant Morphology(U), Plant Taxonomy (U), Advanced Genetics (U), Microbial Genetics (U), Selected Topics in Genetics (U), Plant Anatomy (U), Symbiology (U), Advanced Cellular Physiology (U), Plant Physiology (U), Histology (U), Evolution (U), Field Studies in Biology (U), Microbial Ecology (U), Ichthyology (U), Field Studies in Biology (U), Microbial Ecology (U), Ichthyology (U), Ornithology (U), Mammalogy (U), Vertebrate Paleontology (U), Herpetology (U)

HAMPDEN-SYDNEY COLLEGE

Hampden-Sydney, Virginia 23943 Phone: (804) 223-4381
Dean of College: F.J. Simes

DEPARTMENT OF BIOLOGY
Phone: Chairman: Ext. 82

Chairman: T.H. Turney
Professor: T.H. Turney
Associate Professor: E.A. Crawford, S.R. Gemborys
Instructor: B. Johnston

Degree Programs: Biology (Interdepartmental degree in Biochemistry and Biophysics)

Undergraduate Degrees: B.A., B.S.
Undergraduate Degree Requirements: 33 hours in a coherent program designed to meet individual student needs

HAMPTON INSTITUTE

Hampton, Virginia 23668 Phone: (804) 727-5000
Dean of Graduate Studies: J. Handy, Jr.
Dean of College: E.C. Kollmann

DEPARTMENT OF BIOLOGICAL SCIENCES
Phone: Chairman: 727-5295 Others: 727-5267

Chairman: R.D. Bonner
Professors: J.B. Abram, Jr., H.G. Bonner
Associate Professors: O.M. Bowman, A.B. Hall, V.G. Palmer, A.P. McQueen
Assistant Professors: E.T. Eatman
Instructor: R.A. Banks
Research Assistants: 2

Degree Programs: Biology, Biology - Education

Undergraduate Degrees: B.A. Biology, B.S. Biology(Education)
Undergraduate Degree Requirements: (Semester hours) Biology, 38-40, Chemistry 16-24, Physics 8, Mathematics 6-12, Humanities and Fine Arts - 21, Social Sciences - 6, Physical and Health Education 4, foreign language 6-12, electives 6-13.

Graduate Degrees: M.A. Biology (Secondary Education)
Graduate Degree Requirements: Master: 30 - 32 semester hours, Comprehensive Examination or Thesis Plan, Graduate Record Examination, English Proficiency

Graduate and Undergraduate Courses Available for Graduate Credit: General Ecology (U), History and Principles of Biology (U), Genetics (U), General and Cellular Physiology (U), Organic Evolution (U), Research Problems (U), Vertebrate Embryology (U), Invertebrate Zoology (U), Histology (U), Parasitology (U), Marine Ecology (U), Biological Techniques (U), Mycology (U), Taxonomy of Vascular Plants (U), Applied Microbiology (U), Experimental Plant Physiology (U)

HOLLINS COLLEGE

Roanoke, Virginia 24020 Phone: (703) 362-6000
Dean of College: J.P. Wheeler, Jr.

DEPARTMENT OF BIOLOGY
Phone: Chairman: (703) 362-6547 Others: 362-6543

Chairman: A.L. Bull
Associate Professors: A.L. Bull, C. Morlang, Jr.

Assistant Professor: H. Gray
Research Assistants: 10

Degree Program: Biology

Undergraduate Degree: B.S.
Undergraduate Degree Requirements: 128 credits of academic work and 16 short term credits with an average of C for all work. In the Biology major 32 credits required in Biology and 12 credits in related fields.

LONGWOOD COLLEGE

Farmville, Virginia 23901 Phone: (804) 392-9351
Dean of Graduate Studies: J. Gussett
Dean of College: H. Blackwell

DEPARTMENT OF NATURAL SCIENCES
Phone: Chairman: (804) 392-9351

Chairman: M.W. Scott
Professors: E.B. Jackson, A.M. Harvill, Jr., L.J. Holman, M.W. Scott, C. Wells
Associate Professors: J.M. Austin, B.S. Batts, D.A. Breil, S.O. Breil, T.H. Ely, L.R. Fawcett, Jr., R.L. Heinemann, R.H. Lehman, F.S. McCombs, W.H. Tinnell
Assistant Professors: J.W. Curley, J.A. Hardy III, D.A. Kirchgessner, J.J. Law, W.K. Meshijian, R.B. Thomas, Jr.
Visiting Lecturers: 10 per year

Degree Program: Biology

Undergraduate Degree: B.S.
Undergraduate Degree Requirements: 34 hours in Biology, 126 hours total

LYNCHBURG COLLEGE

Lynchburg, Virginia 24504 Phone: (703) 845-9071
Dean of College: J.A. Huston

BIOLOGY DEPARTMENT
Phone: Chairman: Ext. 388

Chairman: J.E. Carico
Professors: J.E. Carico, P.J. Osborne, G.W. Ramsey
Associate Professors: W.G. Rivers, W.A. Sherwood
Assistant Professors: O.O. Stenroos, S.K. Whitt

Degree Programs: Biology, Life Sciences

Undergraduate Degree: B.S.
Undergraduate Degree Requirements: Biology - 37 credits Biology, 16 credits Chemistry, 8 credits Physics, 6 credits Mathematics (also general degree requirements), Life Sciences - 40 credits from Biology (15 minimum , Psychology (15 minimum) and 10 minimum credits from Physics, Chemistry, or Mathematics (also general degree requirements.

MADISON COLLEGE

Harrisonburg, Virginia 22801 Phone: (203) 433-6211
Dean of Graduate Studies: C. Caldwell
Dean of College: J. Sweigart

BIOLOGY DEPARTMENT
Phone: Chairman: (203) 433-6225

Head: G.S. Trelawny
Professors: J.D. Davis, J.K. Grimm, M.M. Jenkins, G.S. Trelawny
Associate Professors: E. Fisher, W. Jones, P.T. Nielsen, J. Winstead
Assistant Professors: N. Bodkin, D. Cocking, N. Garrison, M. Gordon, R. Graves, J. Heading, C. Sellers, B. Silver

Degree Program: Biology

Undergraduate Degree: B.S.
Undergraduate Degree Requirements: 32 semester hours Chemistry 16 hours, Mathematics 6 hours, 128 semester hours total including 41 semester hours in basic studies

Graduate Degree: M.S., M.A.T.
Graduate Degree Requirements: Master: M.S. 30 semester hours of graduate courses, including 6 semester hours of thesis research. M.A.T. Degree: 21 hours Biology, 12 hours in professional education

Graduate and Undergraduate Courses Available for Graduate Credit: Animal Ecology (U), Parasitology (U), Developmental Biology (U), Public Health Microbiology (U), Cytology (U), Cellular Physiology (U), Ecosystem and Community Dynamics (U), Population Ecology (U), Advanced Plant Physiology (U), Morphology of Non-Vascular Plants (U), Vertebrates Histology (U), Population Genetics, Advanced Developmental Biology, Physiological Genetics, Medical Entomology Radiation Biology, Plant Growth and Development, Experimental Plant Ecology, Mycology, Developmental Anatomy of Higher Plants, Aquatic Ecology, Systematics of Vascular Plants, Insect Ecology, Independent Study

MARY BALDWIN COLLEGE

Staunton, Virginia 24401 Phone: (703) 855-0811
Dean of College: Not Stated

DEPARTMENT OF BIOLOGY
Phone: Chairman: (703) 855-0811

Chairman: J.F. Mehner
Professors: J.F. Mehner
Associate Professors: E.B. Conant, E.V. Vopicka
Assistant Professor: B.M. Hohn
Teaching Assistant: 1

Degree Programs: Biology, Medical Technology

Undergraduate Degrees: B.A., B.S.
Undergraduate Degree Requirements: Biology - 9 semester courses in biology, including seminar, 2 semester courses in Chemistry. Medical Technology - Courses required by American Society of Clinical Pathologists plus 2 additional courses in either biology or chemistry

MARY WASHINGTON COLLEGE

Fredericksburg, Virginia 22401 Phone: (703) 373-7250
Dean of College: J.H. Croushore

DEPARTMENT OF BIOLOGY
Phone: Chairman: Ext. 365

Chairman: Not Stated
Professors: A.H. Hoye, R.M. Johnson, T.L. Johnson, H.J. Parrish, W.C. Pinschmidt
Associate Professors: R.T. Friedman
Assistant Professors: M.L. Bass, S.W. Fuller, M.W. Pinschmidt

Field Stations/Laboratories: Cross Rip Camp - Deltaville

Degree Programs: Biology, Medical Technology

Undergraduate Degrees: B.A., B.S.
Undergraduate Degree Requirements: 36 credit hours in Biology

MEDICAL COLLEGE OF VIRGINIA

(see Virginia Commonwealth University)

OLD DOMINION UNIVERSITY

Norfolk, Virginia 23508 Phone: (804) 489-0000
Dean of Graduate Studies: S.E. Ruach
Dean of College: M.A. Pittman

DEPARTMENT OF BIOLOGY
Phone: Chairman: Ext. 410

Chairman: H.G. Marshall
Professors: D.E. Delzell, D.E. Sonenshine, H.G. Marshall
Associate Professors: V.S. Bagley, R.S. Birdsong, J.R. Holsinger, P.J. Homsher, P.W. Kirk, Jr., G.F. Levy, J.H. Richardson, E.F. Stillwell
Assistant Professors: J.C. Besharse, J.F. Matta, L.J. Musselman, K. Nesius, H.B. Stamper, Jr., N.L. Wade, A.J. Ward, B.A. Weeks
Instructors: B.J. Innes (Adjunct), A.T. Wan (Adjunct), H.S. Wise (Adjunct)
Visiting Lecturers: 3
Research Assistants: 8

Degree Programs: Biology, Medical Technology

Undergraduate Degree: B.S.
Undergraduate Degree Requirements: 41 semester hours Biology, 16 semester hours Chemistry, 8 semester hours Physics, 6 semester hours Mathematics (includes courses in Genetics and Physiology)

Graduate Degree: M.S.
Graduate Degree Requirements: Master: 31 hours with thesis, 34 hours without thesis

Graduate and Undergraduate Courses Available for Graduate Credit: Oral Medicine and Research, Consultation and Individual Study, Immunology, Plant Physiology Marine Ecology, Mycology, Limnology, Aquatic and Wetland Plants, Marine and Estuarine Plankton, Ichthyology, Ornithology, Field Studies in Ornithology, Cytology, Animal Physiology, Microbial Physiology, Physiological Ethology, Mammalian Physiology, Marine Science, Biochemical Genetics, Biochemical Genetics (Laboratory), Medical Statistics, Marine Science Field Studies, Epidemiology, Population Ecology, The Principles of Plant Ecology, Methods in Molecular Biology, Pollution Ecology, Topics in Biology, Systematic Biology, Acarology, Advanced Plant Taxonomy, Special Studies in Ecology, Advanced Microbiology, Graduate Seminar, Special Readings, Virology, Virology (Laboratory), Cytogenetics, Medical Entomology, Advanced Physiology, Radiobiology, Speciation, The Insects of Marshes and Estuaries, Advanced Ichthyology, Dendrology, Biochemistry, Special Topics in Biology, Special Topics in Allied Health, Environmental Health Planning

RADFORD COLLEGE

Radford, Virginia 24142 Phone: (703) 731-5404
Dean of Graduate Studies: P.L. Durrill
Dean of College: D.W. Stump

DEPARTMENT OF BIOLOGY
Phone: Chairman: (703) 731-5124

Chairman: S. Chalgren
Professors: J.W. Clark, R.L.Hoffman, F.E. Jarvis
Associate Professors: K.A. Benson, E.V. Gourley, C.M Lutes, R.L. Rittenhouse
Assistant Professors: R.E. Batie, P.H. Ireland, V.L. Mah, P.B. Mikesell
Instructor: G.S. McBeth

Field Stations/Laboratories: Radford College Field Station-Radford, Virginia

Degree Program: Biology

Undergraduate Degrees: B.S.
Undergraduate Degree Requirements: 186 quarter hours including General Biology, Cell Biology, Developmental Biology, Genetics, Ecology and 75 quarter hours of Biology electives and 20 hours of Chemistry

Graduate Degree: M.S.
Graduate Degree Requirements: Master: 45 quarter hours Biology plus Thesis

Graduate and Undergraduate Courses Available for Graduate Credit: Developmental Genetics (U) , Virology (U), General Ecology (U), Biochemistry (U), Seminar, Honors Courses, Biogeography, Endocrinology, Human Genetics, Bacterial Genetics

RANDOLPH MACON COLLEGE

Ashland, Virginia 23005 Phone: (703) 798-8372
Dean of College: H.E. Davis

DEPARTMENT OF BIOLOGY
Phone: Chairman: Ext. 274

Chairman: W.W. Martin
Professors: J.I. McClurkin
Associate Professors: W.W. Martin
Assistant Professors: J.E. Gates, W.L. Kirk

Degree Program: Biology

Undergraduate Degrees: B.A., B.S.
Undergraduate Degree Requirements: 36 credits, 4 of in colloquies, Satisfy the distribution and proficiency requirements and the major requirements

RANDOLPH-MACON WOMAN'S COLLEGE

Lynchburg, Virginia 24504 Phone: (804) 846-7392
Dean of College: H.D. Hudson

BIOLOGY DEPARTMENT
Phone: Chairman: Ext. 432 Others: Ext. 393

Chairman: F.F. Flint
Professors: D.C. Bliss, F.F. Flint, J.E. Perham

Field Stations/Laboratories: Russell Woods Nature Preserve

Degree Program: Biology

Undergraduate Degree: B.A.
Undergraduate Degree Requirements: 124 total hours, 24 hours in Biology, core program, and Undergraduate Record Examination in Biology

RICHMOND COLLEGE

(see University of Richmond)

ROANOKE COLLEGE

Salem, Virginia 24153 Phone: (703) 389-2351
Dean of College: E.W. Lamtenschlager

DEPARTMENT OF BIOLOGY
Phone: Chairman: Ext. 271

Chairman: J.C. Thompson, Jr.
Professor: J.C. Thompson, Jr.
Associate Professors: G. Grubitz III, P.C. Lee, Jr.
Assistant Professors: K.P. Adkisson, G.A. Clarke, R.E. Jenkins

Degree Program: Biology

Undergraduate Degree: B.S.
Undergraduate Degree Requirements: Not Stated

ST. PAUL'S COLLEGE

Lawrenceville, Virginia 23868 Phone: (703) 848-3111
Dean of College: T.H.E. Jones

DEPARTMENT OF SCIENCE AND MATHEMATICS
Phone: Chairman: Ext. 231 Others: (703) 848-3111

Chairman: A. Samuel
Associate Professors: S.S. Bale, D.O. Baumbach, A. Samuel, H.W. Witten, J. Yates

Degree Program: Biology

Undergraduate Degree: B.S.
Undergraduate Degree Requirements: Must complete successfully the curriculum requirements of his/her major. Last 30 semester hours for the degree must be completed in residence. Must take the Graduate Record Examination after they satisfy graduation requirements. B.S. in Education will take National Teacher Examination.

SWEET BRIAR COLLEGE

Sweet Briar, Virginia 24595 Phone: (804) 381-5614

DEPARTMENT OF BIOLOGY
Phone: Chairman: (804) 381-5614

Chairman: S.C. Belcher (Acting)
Chairman: E. Sprague (On sabbatical leave)
Dorys McConnell Duberg Professor of Ecology: J.C. Belcher
Professors: J.C. Belcher, E.P. Edwards, E.F. Sprague
Associate Professor: M. Simpson
Assistant: N. Huston

Degree Program: Biology

Undergraduate Degree: B.A.
Undergraduate Degree Requirements: 35 "units" including 1 English, Foreign Language (ancient or modern) proficiency, 1 laboratory science, plus 2 years Physical Education, major ir interdepartmental major requirements.

UNIVERSITY OF RICHMOND

Richmond, Virginia 23173 Phone: (804) 285-6000

Richmond College
Westhampton College

Dean: A.E. Gregg

DEPARTMENT OF BIOLOGY
Phone: Chairman: (804) 285-6275 Others: 285-6000

Chairman: Not Stated
Professors: W.M. Reams, J.C. Strickland, N.E. Rice, W.R. Tenney, W.R. West, W.S. Woolcott
Associate Professors: J.W. Bishop, R.D. Decker, F.B. Leftwich
Assistant Professors: W.L. Kirk, D. Witowle
Instructors: E.Q. Falls, H.M. Smith

Degree Program: Biology

Undergraduate Degrees: B.A., B.S.
Undergraduate Degree Requirements: 32 hours in Biology Courses, General Chemistry and 10 semester hours selected outside Biology Department.

Graduate Degree: M.S.
Graduate Degree Requirements: Master: 27 semester hours in Biology Courses, plus a thesis

Graduate and Undergraduate Courses Available for Graduate Credit: Biology of Bacteria (U), Comparative Morphology of Higher Plants (U), Systematic Botany (U), Entomology (U), Mycology (U), Microanatomy (U), Developmental Biology (U), Genetics (U), Cell Physiology (U), Systematic Vertebrate Zoology (U), Ecology (U), Molecular Biology (U), Limnology (U), Plant Physiology (U), General Endocrinology (U), Ichthyology, Arthropod Physiology, Protozoology, Experimental Embryology, Nutrition of Fungi, Advanced Cell Biology, Phycology, Advanced Physiology, Advanced Ecology

UNIVERSITY OF VIRGINIA

University Station 3726 Phone: (703) 924-3417
Charlottesville, Virginia 22903
Dean of Graduate Studies: W.D. Whitehead

Clinch Valley College

Wise Virginia 24239 Phone: (703) 328-2431
Dean: E.F. Low

BIOLOGY DEPARTMENT
Phone: Chairman: Ext. 225 Others: Ext. 236

Chairman: J.R. Baird
Professors: J.R. Baird, J.C. Smiddy
Associate Professors: L.B. Hutzler, P.C. Shelton
Assistant Professors: C.F. Denny
Laboratory Assistants: 5

Degree Programs: Biology, Medical Technology

Undergraduate Degrees: B.A., B.S.
Undergraduate Degree Requirements: B.A. - 32 semester hours Biology, 3 courses Chemistry, 2 courses Mathematics, 1 course Physics B.S. - 40 semester hours Biology, 4 courses Chemistry, one course Physics, Mathematics through Calculus and Statistics

College of Arts and Sciences

Dean: I.B. Cauthen

DEPARTMENT OF BIOLOGY
Phone: Chairman: (703) 924-7118 Others: 924-0311

Chairman: O.L. Miller, Jr.
Lewis and Clark Professor of Biology: D. Bodenstein
Miller Professor of Biology: B.F.D. Runk
Professors: D. Bodenstein, J.N. Dent, H.L. Hamilton, I.R. Konigsberg, O.L. Miller, Jr., J.J. Murray, L.I. Rebhun, B.F.D. Runk
Associate Professors: R.H. Bauerle, R.H. Benzinger, F.A. Diehl, R.H. Garrett, R.J. Huskey, R.H. Kretsinger, S.P. Maroney, R.D. McKinsey, D.F. Mellon Jr., J.J. Rappaport, J.L. Riopel, T.R.F. Wright
Assistant Professors: C.P. Emerson, H.M. Phillips, R.D. Rodewald, C.P. Spirito, R.F.Swanson
Instructors: R.F. Hatcher, L. Leonard
Teaching Fellows: 17
Research Assistants: 5
Post-doctorals: 11

Field Stations/Laboratories: Mountain Lake Biological Station - Pembroke, Virginia

Degree Programs: Animal Genetics, Animal Physiology, Biochemistry and Biophysics, Biological Sciences, Biology, Biophysics, Genetics, Histology, Microbiology, Physiology, Developmental Biology, Cell Biology

Graduate Degrees: M.A., M.S., M.A.T. (Biology), Ph.D.
Graduate Degree Requirements: Master: Twenty-four semester hours plus a thesis Doctoral: First year: courses in Biochemistry, Advanced Genetics, Cell Biology, and Developmental Biology, followed by a comprehensive examination. Second year: qualifying examination in area of research interest, courses as required by the major professor and committee. General requirements: four colloquia, one language administered by departmental examination, Final examination and dissertation defense

Graduate and Undergraduate Courses Available for Graduate Credit: Biochemistry, Developmental Biology, Advanced Genetics, Cell Biology, Algae, Bacteria, Fungi, Plant Morphogenesis, Invertebrate Biology, Comparative Neurophysiology, Histology, Animal Behavior, Plant Growth and Development, Morphogenetic Phenomena, Developmental Physiology of Insects, History of Biology, Animal Physiology, Immunology, Colloquium in Physiology, Colloquium in Genetics, Colloquium in Biology, Bacteriophage, Laboratory in Genetics, Genetics of Populations, Macromolecular Structure, Developmental Genetics, Invertebrate Endocrinology, Tissue Culture Technique, Molecular Biology of Eukaryotes, Vertebrate Endocrinology, Plant Physiology, Neurophysiology, Techniques in EM, Techniques and Instrumentation, Selected Topics in Genetics, Selected Topics in Evolution, Selected Topics in Developmental Biology, Selected topics in Plant Physiology, Selected Topics in Physiology, Selected Topics in Molecular Genetics, Selected Topics in Cell Biology, Selected Topics in Developmental Genetics, Selected Topics in Biochemistry

DEPARTMENT OF ENVIRONMENTAL SCIENCES
Phone: Chairman: (804) 924-3995 Others: 924-3761

Chairman: H.G. Goodell
Professors: R. Dolan, M. Garstang, H.G. Goodell, R.S. Mitchell, B. Nelson, J.M. Simpson, R.H. Simpson (Research), S.F. Singer
Associate Professors: C.I. Aspliden, D.W. Barnes, R.L. Ellison, E.H. Ern, A.D. Howard, B.W. Perry (Research), W.E. Reed, P.R. Tett (Research), W.A. Wallace
Assistant Professors: R. Dueser, J.S. Fisher, E. Gardiner, R.B. Hanawalt, B.P. Hayden, G.M. Hornberger, M.G. Kelly, W.E. Odum, R. Pielke, S.S. Skjei, D.E. Wilson, J.C. Zieman
Visiting Assistant Professor: 1

Degree Programs: Applied Ecology, Conservation, Ecology, Environmental Engineering, Forest Chemistry, Forest Chemistry, Land Resources, Life Sciences, Natural Resources, Oceanography, Population Planning, Soil Science, Watershed Management, Environmental Science

Undergraduate Degree: B.A. (Environmental Sciences)
Undergraduate Degree Requirements: B.A. requires 30 semester hours credit within the Department, including a basic course in each of the following: Ecology, Geology, Hydroclimatology, and Resource Analysis, Eight semester hours in Biology, Chemistry, or Physics and at least 3 hours of Calculus. Program of independent research for exceptional students leads to B.A. with Distinction

Graduate Degrees: M.S., Ph.D.
Graduade Degree Requirements: Master: Degree requirements include the following: (a) 24 hours of graduate coursework, (b) one course at the 500 level in each of four areas, (c) thesis Doctoral: Degree requirements include the following: (a) three years of full time graduate work, (b) one course at the 500 level in each of four areas, (c) two advanced seminars, (d) dissertation

Graduate and Undergraduate Courses Available for Graduate Credit: Environmentography, Systems Analysis in Environmental Sciences, Aquatic Ecology, Terrestrial Ecology, Environmental Thermodynamics, Planetary Fluid Motions, Solar and Terrestrial Radiation, Fundamentals of Planning Process and Decision Theory I, Natural Resource Utilization, Soil Science, Sediments and Sedimentary Fluid Dynamics, Quantitative Methods in Environmental Science, Normative and Descriptive Models for Environmental Assessment, Fresh Water Ecology, Estuarine Ecology, Conservation Ecology, Oceanic Ecology, Independent Study, Ecology, Special Topics - Ecology, Dynamic Climatology, Dynamic Hydrology, Estuarine Circulation, Independent Study - Fluid Systems, Special Topics - Fluid Systems, Land Use Policies, Population, Resources and Environment, Urban Development Models, Regional Development and Environmental Impact, Planning Processes and Decision Theory II, Independent Study - Resource Analysis, Special Topics - Resource Analysis, Soil Physics, Environmental Chemistry, Marine Geology, Coastal Hydrodynamics, Aquatic Chemistry, Independent Study - Surface Processes, Special Topics - Surface Processes, Seminar in Ecology, Seminar in Fluid Systems, Seminar in Resource Analysis, Seminar in Surface Processes, Advanced Topics - Ecology, Advanced Topics - Fluid Systems, Advanced Topics Resource Analysis, Advanced Topics - Surface Processes

College of Engineering and Applied Sciences
Dean: J.E. Gibson

DEPARTMENT OF BIOMEDICAL ENGINEERING
Phone: Chairman: (804) 924-5095 Others: 924-5101

Chairman: E.O. Altbinger
Professors: F.L. Damman, W.B. Loomey, D.W. Mervis, O.L. Updike
Associate Professors: A. Anne, G. Theodoridis, Y.S. Hee
Assistant Professors: M.L. McCartney, M.G. Wilkins, R.N. Johnson
Research Assistants: 4
Graduate Students: 35

Degree Program: Biomedical Engineering

Graduate Degree: Ph.D., M.E.
Graduate Degree Requirements: Master: minimum 30 credit hours plus apprenticeship Doctoral: minimum 48 credit hours, comprehensive examination, thesis defense

Graduate and Undergraduate Courses Available for Graduate Credit: Physiology, Biomathematics, Systems Analysis, Computer Applications, Biomedical Instrumentation, Health Care, Biomechanics, Biological Control Mechanisms, Analysis of Societal Systems

School of Medicine
Dean: W.R. Drucker

DEPARTMENT OF ANATOMY
Phone: Chairman: (703): 924-2731 Others: 924-2732

Chairman: J. Langman
Professors: R.R. Cardell, E.O. Ebbesson, L. Heimer, E.W. Pullen, J. Langman
Associate Professors: J.D. Deck, C.J. Flickinger, D.M. Kochhar
Assistant Professors: S. Edwards, J.R. Keefe, P.M. Rodier, J. Spyker
Research Associate (Postdoctoral): 8

Degree Program: Anatomy

Graduate Degrees: M.S., Ph.D.
Graduate Degree Requirements: Master: Two year study - one year course work in basic anatomical subject and one year research minimum of 24 semester hours of graduate courses. completion of a M.S. thesis Doctoral: one year course work in basic anatomical subjects - preparation of a qualifying examination, completion of an individual research project and teaching activities submit independent of a doctoral thesis

Graduate and Undergraduate Courses Available for Graduate Credit: Gross Anatomy and Development of Human Body, Cytology and Histology, Neuroanatomy, Experimental Morphology, General Embryology, Research, Advanced Neuroanatomy, Colloquium in Embryology, Colloquium in Cell Biology

DEPARTMENT OF BIOCHEMISTRY
Phone: Chairman: (804) 924-5130 Others: 924-5139

Chairman: T.E. Thompson
Professors: G.R. Ackers, D.W. Kupke, R.G. Langdon, T.E. Thompson
Associate Professors: R.L. Biltonen, C. Bradbeer, C. Huang, J.W. Ogilvie
Assistant Professors: A.R.L. Gear, P.W. Holloway, B.J. Litman, W.B. Panko, E.P. Paulsen, R.P. Taylor
Research Assistant: 1
Research Associate: 5

Degree Program: Biochemistry

Graduate Degree: Ph.D.
Graduate Degree Requirements: Doctoral: The student will take the following graduate courses: General Biochemistry, Laboratory Projects, Colloquium, Biochemical Literature, Organic Chemistry and other courses in areas of special interest or deemed necessary. The student will be advanced to candidacy for the Ph.D. degree upon satisfactory presentation of a Research Proposal. The student must successfully defend his Ph.D. dissertation in a formal oral examination.

Graduate and Undergraduate Courses Available for Graduate Credit: General Biochemistry, Biochemical Projects, Physical Chemistry of Biopolymers, Enzyme Mechanisms, Selected Topics, Lipid Metabolism, Regulation of Metabolism, Metabolic Control Processes in Higher Plants, Biochemical Literature

DEPARTMENT OF MICROBIOLOGY
Phone: Chairman: (804) 924-5111

Chairman: R.R. Wagner
Professors: C.A. Schnaitman, W.A. Volk, R.R. Wagner
Associate Professors: D.E. Normansell (Joint), H.H. Winkler
Assistant Professors: D.C. Benjamin, J.C. Brown, S.U. Emerson, R.J. Kadner, W.M. Kuehl, J.T. Parsons, W. Steinberg
Research Assistants: 2
Research Associates (Postdoctoral trainees): 10

Degree Programs: Virology, Microbial Genetics, Immunology, Bacterial Physiology, Membrane Biochemistry

Graduate Degrees: M.S., Ph.D. (Microbiology)
Graduate Degree Requirements: Master: Same as those of Graduate School of Arts and Science of University of Virginia Doctoral: Same as those of Graduate School of Arts and Sciences of University of Virginia

Graduate and Undergraduate Courses Available for Graduate Credit: Principles of Microbiology, Virology, Advanced Immunology, Advanced Bacteriology, Cell Biology, Research in Bacterial Physiology and Cell Wall Biosysthesis, Research in Biochemistry and Physiology of Active Transport, Research in Electron Microscopy and Biochemistry of Membranes and Cell Organelles, Research in Biochemical Virology, Research in Structure of Antibodies, Research in Bacteriophage, Research in Immunoglobulin Synthesis and Structure, Research in Immunology, Research in Molecular Genetics, Research in Molecular Biology of Development, Research in Membranes of Nerve Cells and Tumors, Research in Molecular Virology, Research in Tumor Virology, Research in Cellular Genetics

DEPARTMENT OF PHYSIOLOGY
Phone: Chairman: (804) 924-5108

Chairman: R.M. Berne
Charles Slaughter Professor of Physiology: R.M. Berne
Professors: R.M. Berne, D.H. Cohen, J.I. Kitay (Joint), N. Sperelakis
Associate Professors: B.R. Duling, T.W. Lamb (Joint), R.A. Murphy, G.C. Pitts, R. Rubio, S.M. Sherman
Assistant Professors: R.S. Fager, J.T. Hackett, H. Kutchai, L.C. Parsons (Joint), O. Steward (Joint)
Visiting Lecturers: 35
Teaching Fellows: 6
Research Associates: 1

Degree Program: Physiology

Graduate Degrees: M.S., Ph.D.
Graduate Degree Requirements: Master: 1 1/2 year course work in Physiological Sciences, Preliminary examination and Dissertation Doctoral: 2 1/2 year course work in Physiological Sciences, Preliminary examination and Dissertation

Graduate and Undergraduate Courses Available for Graduate Credit: Structure and Function of the Cell (U), General Physiology (U), Organ Systems Physiology (U), Cellular Physiology (U), Neurophysiology (U), Seminar in Physiology, Cardiovascular Physiology, Transport Across Biological Membranes, Endocrine Physiology, Electrophysiology of Nerve and Muscle, The Contractile Process, Central Neurophysiology, Literature Review , Directed Readings in Physiology

Virginia Institute of Marine Science
Gloucester Point, Virginia 23062 Phone: (804) 642-2111
Dean: W.J. Hargis, Jr.

Professors: W.J. Hargis, Jr., J.D. Andrews, R.E.L. Black, W.A. Van Engel, J.L. Wood
Associate Professors: M.E. Bender, R.H. Bieri, R.L. Byrne, W.J. Davis, P.A. Haefner, Jr., D.S. Haven, M.M. Nichols, J.J. Norcross, M.L. Wass, P.L. Zubkoff
Assistant Professors: D.R. Calder, J.L. Dupuy, C.S. Fang, V. Goldsmith, G.C. Grant, P.V. Hyer, F.Y. Kazama, A.Y. Kuo, J. Loesch, M.P. Lynch, W.G. MacIntyre, K.L. Marcellus, J.A. Musick, F.D. Ott, F.O. Perkins, E.P. Ruzecki, C.L. Smith, K.L. Webb, C.S. Welsh
Instructors: R.L. Bolus, J.V. Merriner, R.W. Moncure, J.E. Warinner, F.J. Wojcik, D.E. Zwerner
Associate Faculty: K.F. Bick, M.A. Byrd, R.L. Ellison, W.G. Hewatt, S.H. Hopkins, H.J. Humm, V.A. Liguori, C.P. Mangum, J.B. Morrill, B.W. Nelson, S.Y. Tyree, Jr.

Field Stations/Laboratories: VIMS, Wachapreague, Virginia

Degree Programs: Biological Oceanography, General Oceanography, Fisheries Oceanography

Graduate Degrees: M.A., M.S., Ph.D. (Marine Science)
Graduate Degree Requirements: Master: 30 semester credits advanced work, thesis, one foreign language, comprehensive examination Doctoral: 3 years graduate study beyond baccalaureate. Course requirements at discretion of school. Two foreign language thesis, comprehensive examination

Graduate and Undergraduate Courses Available for Graduate Credit: (Only courses of direct biological content are listed.) Problems in Marine Science, Introduction to Marine Science, Biometry I, Marine and Freshwater Invertebrates, Marine Botany, Marine Science Seminar, Advanced Biological Oceanography, Biology of Selected Marine Invertebrates, Radiobiology, Marine Microbiology, Ichthyology, Pollution Biology, Parasites of Marine Organisms, Marine Biogeography, Littoral Processes, Embryology of Marine Invertebrates, Marine Fishery Science, Biometry II, Environmental Physiology, Environmental Physiology Laboratory, Topics in Applied Marine Science, Physiology of Marine Organisms, Population Dynamics, Marine Mycology, Marine Phytoplankton, Marine Zooplankton, Marine Benthos, Marine Protozoology

VIRGINIA COMMONWEALTH UNIVERSITY
910 West Franklin Street Phone: (703) 770-6357
Richmond, Virginia 23220
Dean of Graduate Studies: P. Minton

School of Arts and Sciences
Dean: P. Minton

DEPARTMENT OF BIOLOGY
Phone: Chairman: (703) 770-7231

Chairman: R.V. Brown
Professors: R.V. Brown, L. Goldstein, R.R. Mills
Associate Professors: J.E. Jeffrey, M.F. Johnson, T.D. Kimbrough, M. May, J.R. Reed, Jr., J. Reynolds, W. Richards, Jr., J.M. Sharpley
Assistant Professors: C. Blem, W.V. Dashek, R. Pagels, R. Parker, A. Seidenberg, D. Cundall, J. Ruffolo, Jr., L.D. Montroy
Instructors: L. Blem, E. Waldrip
Research Assistants: 5

Degree Program: Biology

Undergraduate Degree: B.S.
Undergraduate Degree Requirements: 36 credits in Biology, in addition to general Arts and Sciences requirements.

Graduate Degree: M.S.
Graduate Degree Requirements: Master: 32 credits

Graduate and Undergraduate Courses Available for Graduate Credit: Not Stated

School of Graduate Studies, Health Sciences Division
Dean: D.T. Watts

DEPARTMENT OF ANATOMY
Phone: Chairman: (703) 770-4688 Others: 770-4689

Chairman: W.P. Jollie
Professors: J.D. Burke, W.P. Jollie, D.L. Odor, R.J. Weymouth
Associate Professors: J.A. Astruc, F.M. Bush, T.M. Harris, R.J. Jordan, J.E. Norvell, N.O. Owers, H.R. Seibel
Assistant Professors: S.J. Goldberg, J.L. Haar, J.A. Hightower, J.H. Johnson, G.R. Leichnetz, G.D. Meetz, A.S. Pakurar, W.L. Poteat, M.J. Snodgrass
Teaching Fellows: 6
Research Assistants: 1

Degree Program: Anatomy

Graduate Degrees: M.S., Ph.D.
Graduate Degree Requirements: Master: 1 year residency, 24 semester hours credit, core anatomy courses, (Histology, Gross, Neuroanatomy, master's level research, and thesis Doctoral: 2 years residency, 48 semester hours credit, 6 anatomy courses, 2 cognate courses, 1 foreign language, research and thesis

Graduate and Undergraduate Courses Available for Graduate Credit: Principles of Human Anatomy (U), Embryology (U), Microscopic Anatomy (Dentistry), Introduction to Facial Growth and Development (Dentistry), Anatomy of the Head and Neck (Dentistry), History of Anatomy, Gross Anatomy, Microscopic Anatomy, Neuroanatomy, Neuroanatomy Laboratory, Advanced Studies in Anatomy, Techniques in Electron Microscopy, Anatomy of the Eye, Anatomy Seminar, Fine Structure of Cells and Tissues, Embryology, Research in Anatomy

DEPARTMENT OF BIOCHEMISTRY
Phone: Chairman: (804) 770-4403

Chairman: L.D. Abbott, Jr.
Professors: L.D. Abbott, Jr., W.L. Banks, Jr., C.C. Clayton, E.S. Higgins, L. Swell (Research)
Associate Professors: J.S. Bond, R.B. Brandt, E.S. Kline, J.P. Liberti, K.S. Rogers
Assistant Professors: A.S. Bhatnagar, J.M. Collins, G.H. DeVries, R.F. Diegelmann, R.Z. Eanes, H.J. Evans, K.F. Guyer, Jr., K.R. Shelton, G.C. Van Tuyle
Instructors: S.S. Jennings
Teaching Fellows: 6
Research Assistants: 4

Degree Program: Biochemistry

Graduate Degrees: M.S., Ph.D.
Graduate Degree Requirements: Master: Based on Chemistry major or Biology major with equivalent in Chemistry and Mathematics courses. Course work, research, thesis, 1 language (usual time, 2 years) Doctoral: Comprehensive course work, written and oral examinations, research, thesis (usual time, 4 years)

Graduate and Undergraduate Courses Available for Graduate Credit: General Biochemistry, Experimental Biochemistry, Biochemical Preparations, Biochemistry Seminar, Lipids, Proteins, Enzymology, Nutritional Biochemistry, Controls of Metabolic Processes

DEPARTMENT OF BIOMETRY
Phone: Chairman: (804) 770-4065

Chairman: S.J. Kilpatrick, Jr.
Professors: S.J. Kilpatrick, Jr., P.F. Minton (Adjunct)
Associate Professors: W.H. Carter, Jr., R.E. Flora
Assistant Professors: W.B. Cummings, W.H. Fellner, E.F. Meydrech

Degree Program: Biometry

Graduate Degrees: M.S., Ph.D.
Graduate Degree Requirements: Master: One to two years residence, one foreign language, thesis, final oral Doctoral: Three years beyond baccalaureate, two years beyond M.S., M.D. or D.D.S. course work as specified by committee, two foreign languages written comprehensive, thesis, final oral

Graduate and Undergraduate Courses Available for Graduate Credit: Statistical Principles of Health Care Information, Operations Analysis: Biometry Seminar, Methods of Statistical Analysis, Special Topics in Biometry, Nonparametric Statistics, Advanced Statistical Inference, Operations Research, Methods of Multivariate Analysis, Sampling, Design and Analysis of Response Surface Experiments, Biometrical Methods of Population Research, Stochastic Processes, Statistical Methods, Nonparametric Statistics, Techniques of Linear Models, Design of Experiments, Theory of Probability and Statistical Inference, Research in Biometry, Linear Statistical Models

DEPARTMENT OF BIOPHYSICS
Phone: (804) 770-4041

Chairman: W.T. Hamg, Jr.,
Associate Professors: A. M. Clarke, S.F. Cleary, L.I. Epstein, C.H. O'Neal
Assistant Professors: W.E. Keefe, R.C. Williams
Instructors: E.C. Campbell, R.S. Ruffin

Degree Program: Biophysics

Graduate Degrees: M.S., Ph.D.
Graduate Degree Requirements: Master: 24 hours plus thesis Doctoral: Satisfaction of University - Graduate Council requirements

Graduate and Undergraduate Courses Available for Graduate Credit: Introduction to Biophysics, Radioisotopes, Applied Electronics, Circuit Design and Analysis, Photobiology, Information Theory, Optical Physics,

Molecular Biology, X-ray Crystallography, Macromolecules in Solution

DEPARTMENT OF MICROBIOLOGY
Phone: Chairman: (804) 770-4622 Others: 770-4632

Chairman: S.G. Bradley
Professors: P.H. Coleman, H.J. Welshimer
Associate Professors: J.V. Formica, H.S. Hsu, H.J. Shadomy
Assistant Professors: P.B. Hylemon, D.T. John, W.E. Keefe, R.M. Loria, F. Macrina, D.N. Mardon, P.S. Morahan, P.V. Phibbs, T. Tang, J.G. Tew
Visiting Lecturers: 1
Teaching Fellows: 6
Research Assistants: 5
Trainees: 6

Degree Programs: Bacteriology, Bacteriology and Immunology, Immunology, Medical Microbiology, Microbiology, Virology

Graduate Degrees: M.S., Ph.D.
Graduate Degree Requirements: Master: 20 semester hours including general Microbiology, Microbial Physiology and Immunology, one modern foreign language, thesis Doctoral: 40 semester hours including general microbiology, microbial physiology, immunology and an adjunct field of study, one modern foreign language, biometry, radioisotopes, original research leading to a dissertation

Graduate and Undergraduate Courses Available for Graduate Credit: Cell Physiology, Bacterial Physiology and Metabolism, Bacterial Physiology Laboratory, Immunobiology, Clinical Immunology, Special Topics in Immunology, Virology, Principles of Oncology, Medical Mycology, Medical Parasitology, Pathogenic Bacteriology

DEPARTMENT OF PHYSIOLOGY
Phone: Chairman: (804) 770-4113

Chairman: F.N. Briggs
Professors: E.B. Bodganove, F.N. Briggs, L.E. Edwards, E. Fischer (Emeritus), E.G. Huf
Associate Professors: D.C. Mikulecky, S. Price, A. Szumski
Assistant Professors: G.H. Bond, G.T. Campbell, H.P. Clamann, P.M. Conklin, K.C. Corley, J.A. DeSimone, G.D. Ford, E.L. Hardie, M.L. Hess, D.J. Mayer, R.N. Pittman, J.L.Poland, D.D. Price, E.B. Ridgway, R.J. Solaro, C. Trowbridge, R.J. Witorsch
Instructors: M.E. Soulsby
Research Assistants: 3

Degree Program: Physiology

Graduate Degrees: M.S., Ph.D.
Graduate Degree Requirements: Master: 1 year of residence, research thesis Doctoral: 3 years of residence (or 2 years with M.S.), 1 foreign language, research thesis

Graduate and Undergraduate Courses Available for Graduate Credit: General Physiology, Advanced Mammalian Physiology, Special Topics in Physiology, Physiology Seminar, Membrane Transport, Mathematical Physiology

VIRGINIA MILITARY INSTITUTE

Lexington, Virginia 24450 Phone: (703) 463-6247
Dean of College: J.M. Morgan, Jr.

DEPARTMENT OF BIOLOGY
Phone: Chairman: (703) 463-6247

Chairman: A.G.C. White
Professors: A.G.C. White, L.R. Humdley, O.W. Gupton, J.H. Reeves, Jr.
Associate Professors: E.B. Winefield, F.W. Swope

Degree Program: Biology

Undergraduate Degree: B.S.
Undergraduate Degree Requirements: Introductory Chemistry, Organic Chemistry 1 year, Analytic Chemistry 1 semester, Calculus 1 year, Biology 36 semester hours, including Developmental, Genetics, Cytology, Physiology, English 9 semesters, History 1 year, Physics 1 year, ROTC 4 years

VIRGINIA POLYTECHNIC INSTITUTE AND STATE UNIVERSITY

Blacksburgh, Virginia 24061 Phone: (703) 951-6000
Dean of Graduate Studies: F.W. Bull

College of Arts and Sciences

Dean: W.C. Havard

DEPARTMENT OF BIOLOGY
Phone: Chairman: (703) 951-6407 Others: 951-6000

Head: R.A. Paterson
University Professor: J. Cairns, Jr.
Professors: R.B. Holliman, P.C. Holt, N.R. Krieg, O.K. Miller, B.C. Parker, R.A. Paterson, R.D. Ross
Associate Professors: R.E. Benoit, A.L. Buikema, K.L. Dickson, A.G. Heath, G.M Simmons, H.R. Steeves, E.R. Stout, D.A. West, W.H. Yongue
Assistant Professors: C.S. Adkisson, R.C. Bates, E.F. Benfield, J.M. Byrne, W.F. Calhoun, G.W. Claus, P.L. Dalby, B.K. Davis, Falkingham, A.C. Hendricks, T.A. Jenssen, F.M.A. McNabb, R.A. McNabb, R.S. Mitchell, R.W. Rhoades, C.L. Rutherford, D.A. Stetler, J.G. Waines, A.A. Yousten
Teaching Fellows: 33
Research Assistants: 43
Research Associates: 3

Degree Programs: Biology, Botany, Genetics, Microbiology, Zoology

Undergraduate Degree: B.S. (Biology)
Undergraduate Degree Requirements: 50 quarter hours Biology (Genetics, Microbiology, Physiology required) 24 quarter hours Chemistry, 12 quarter hours Physics, 9 quarter hours Mathematics, 24 quarter hours Humanities, 24 quarter hours social sciences

Graduate Degrees: M.S., Ph.D.
Graduate Degree Requirements: Master: Research and thesis - 9 to 15 quarter hours course work 30 to 36 quarter hours Doctoral: Research and Dissertation 45 to 85 quarter hours, course work - 55 to 90 quarter hours, foreign languages - none, one or two - at committee option

Graduate and Undergraduate Courses Available for Graduate Credit: Limnology, Environmental Science Water, Water Resource Management, Evolution, Evolutionary Genetics, Selected Topics in Development, History of Biology, Developmental Biology, Developmental Biology Laboratory, Principles of Ecology, Advanced Genetics, Cytogenetics, Cytogenetics Laboratory, Cytology, Taxonomy of Vascular Plants, Plant Ecology, Biology of Fungi, Algal Ecology, Experimental Phycology, Aquatic Vascular Plants, Developmental Plant Anatomy, Morphology of Vascular Plants, Plant Geography, Phycology, Introductory Mycology, Topics in Virology, Systematic Bacteriology, Immunology, Viruses and Rickettsias, Soil Microbiology, Physiology of Microorganisms, Genetics of Microorganisms, Food Microbiology, Dairy Microbiology, Soil Bacteriology, Microbiology of Aquatic Systl ms, Advanced Microbiology, Pathogenic Bacteriology, Virology, Virology Laboratory, Animal Ecology, Advanced Topics in Comparative Animal Physiology, Histochemistry,

Experimental Parasitology, Ichthyology, Social Behavior of Birds and Mammals, Animal Histology, Comparative Physiology, Parasitology, Vertebrate Natural History, Invertebrate Zoology

College of Agriculture and Life Science

Dean: J.E. Martin

DEPARTMENT OF ANIMAL SCIENCE

Phone: Chairman: (703) 951-6311 Others: 951-5134

Head: M.B. Wise
Professors: R.C. Carter, J.P .Fontenot, E.T. Kornegay, T.J. Marlowe, M.B. Wise, G.A. Allen, C.C. Mast, K.C. Williamson
Associate Professors: K.P. Bovard, J.A. Gaines, G.G. Green, T.N. Meacham, G.L. Minish, J.S. Copenhaver, A.N. Huff
Assistant Professors: D.E. Webb, Jr., C.R. Cooper, A. L. Eller, Jr., H.J .Gerken, T.R. Jambuth
Instructor: J.D. Chadwell
Teaching Fellows: 4
Research Assistants: 8
Technicians: 36

Field Stations/Laboratories: Shenandoah Valley Research Station, Steeles Tavern, Southwest Virginia Research Station - Glade Spring, Tidewater Research and Continuing Education Center - Holland

Degree Programs: Genetics, Nutrition, Physiology, Management

Undergraduate Degrees: B.S. (Animal Science)
Undergraduate Degree Requirements: 19 credits Biology, 9 credits Mathematics, 18 credits Chemistry, 43 credits Animal Sciences

Graduate Degrees: M.S., Ph.D. (Animal Science)
Graduate Degree Requirements: Master: 45 quarter hours plus acceptable M.S. thesis. Non-thesis option is available. Doctoral: Sufficient course work to satisfy special graduate committee plus dissertation

Graduate and Undergraduate Courses Available for Graduate Credit: Advanced Beef Production, Animal Nutrition Experimentation, Animal Radiophysiology, Genetics of Population, Advanced Animal Nutrition, Comparative Reproduction, Modern Aspects of Ruminant Nutrition, Problems in Population Genetics, Animal Nutrition Seminar, Seminar, Advanced Anatomy and Physiology of Domestic Animals, Advanced Avian and Mammalian Nutrition, Independent Study, Special Study, Research and Thesis, Heterosis, Animal Experimentation, Research and Dissertation, Pork Production, Animal Nutrition, Beef Production, Sheep Production, Survey of Recent Development in Animal Science, Animal Genetics, Social Behavior of Birds and Mammals, Physiology of Reproduction in Domestic Animals, Horse Production and Management

DEPARTMENT OF BIOCHEMISTRY AND NUTRITION

Phone: Chairman: (703) 951-6315

Head: B.M. Anderson
Professors: C.J. Ackerman, B.M. Anderson, T.C. Campbell, R.R. Schmidt, J.R. Vercellotti
Associate Professors: L.B. Barnett, G.E. Bunce
Assistant Professors: R.D. Brown, Jr., J.P. Fox, J.L. Hess, M.H. Samli, R.W. Young, J.H. Yuan
Teaching Fellows: 4
Research Assistants: 9

Degree Program: Biochemistry

Undergraduate Degree: B.S.
Undergraduate Degree Requirements: Mathematics, 18 quarter hours, Physics, 12 quarter hours, Biology 20 quarter hours, Biochemistry 18 quarter hours, Chemistry 34 quarter hours, plus college requirements. Total of 190 hours

Graduate Degrees: M.S., Ph.D.
Graduate Degree Requirements: Master: 45 quarter hours, including 21 hours of Advanced Biochemistry courses plus Masters thesis Doctoral: 134 quarter hours, including 31 hours of Advanced Biochemistry courses plus Doctoral thesis.

Graduate and Undergraduate Courses Available for Graduate Credit: Advanced Methods of Biochemical Analysis, Theoretical Nutrition, Advanced Biochemistry, Biopolymers, Seminar in Biochemistry and Nutrition, Ezymology, Comparative Biochemistry, Independent Study, Special Study, Research and Thesis, Research and Dissertation, Bioinorganic Chemistry (U), General Biochemistry (U), Laboratory Problems in Biochemistry and Nutrition (U)

DEPARTMENT OF DAIRY SCIENCE

Phone: Chairman: (703) 951-6332

Chairman: R.G. Cragle
Professors: P.T. Chandler, R.G. Cragle, W.M. Etgen, W.R.Murley, C.E. Polan, R.G. Saacke, J.M. White
Associate Professors: G.M. Jones, J.A. Lineweaver, W.N. Patterson
Assistant Professors: R.E. Buffington, F.C. Gwazdauskas, C.W. Heald, M.L. McGilliard, W.E. Vinson
Instructors: C.N. Miller

Degree Programs: Animal Science, Dairy Science

Undergraduate Degree: B.S. (Dairy Science)
Undergraduate Degree Requirements: 190 quarter credits, 47 quarter credits in Mathematics and Natural Sciences, 40 quarter credits in Dairy Science, 33 quarter credits in Humanities and Social Sciences, 70 quarter credits of electives

Graduate Degrees: M.S. (Dairy Science), Ph.D. (Animal Science)
Graduate Degree Requirements: Master: 45 quarter credits minimum, 18 quarter credits minimum 500 level or higher, 18 quarter credits maximum 400 level or higher, 15 quarter credits maximum Research and thesis, 7 quarter credits maximum Independent Study Doctoral: 135 quarter credits minimum, 40 quarter credits minimum 500 level or higher, 45 quarter credits minimum Research and Thesis, 85 quarter credits maximum Research and thesis, 6 quarter credits maximum seminar, 6 quarter credits maximum 400 level

Graduate and Undergraduate Courses Available for Graduate Credit: Physiology of Reproduction in Domestic Animals (U), Cytology (U), Dairy Cattle Management (U), Animal Genetics (U), Applied Animal Genetics (U), Physiology of Lactation (U), Advanced Problems in Dairy Cattle Physiology, Animal Nutrition, Experimentation, Animal Radiophysiology, Genetics of Population, Advanced Topics in Selection, Comparative Reproduction, Problems in Population Genetics, Problems in Selection, Animal Nutrition Seminar, Seminar, Advanced Anatomy and Physiology of Domestic Animals, Advanced Avian and Mammalian Nutrition, Advanced Animal Nutrition, Modern Aspects of Ruminant Nutrition, Independent Study

DEPARTMENT OF ENTOMOLOGY

(No information available)

DEPARTMENT OF FISHERIES AND WILDLIFE SCIENCES

Phone: Chairman: (703) 951-5573 Others: 951-6000

Head: H.S. Mosby
Professors: R.H. Giles, Jr., B.S. McGinnes (Adjunct), H.S. Mosby, T.H. Ripley (Adjunct), H.E. Crawford (Adjunct)

Associate Professors: H.E. Burkhart (Joint), R.L. Kirkpatrick, R.T. Lackey, R.L. McElwee (Joint-Extension), R.F. Raleigh (Adjunct)
Assistant Professors: W. Flick (Joint), D. Grove (Joint, Extension), P.F. Scanlon, C.B. Schreck, R.B. Vasey (Joint), J.B. Whelan (Adjunct)
Teaching Fellows: 8
Visiting Lecturers: 8
Research Assistants: 6

Field Stations/Laboratories: Broad Run Wildlife Research Area (in cooperation with Jefferson National Forest)

Degree Programs: Fishery Wildlife, Wildlife Management, Wildlife Biology

Undergraduate Degree: B.S. (wildlife fisheries option)
Undergraduate Degree Requirements: 200 quarter hours (wildlife option must meet Civil Service Wildlife Biologist requirements of a minimum of 14 hours in Wildlife Biology, 18 hours in Zoology, 14 hours in Botany and 24 in Physical Sciences)

Graduate Degrees: M.S., Ph.D.
Graduate Degree Requirements: Master: 45 hours (quarter) including thesis. Minimum of 18 hours of graduate course work, maximum of 15 hours for research and thesis Doctoral: 135 quarter hours minimum, with dissertation, Minimum of 40 hours of graduate course work, research and dissertation minimum of 45 hours, maximum of 85 hours

Graduate and Undergraduate Courses Available for Graduate Credit: Fishery Biology, Fishery Theory, Population Dynamics, Wild Animal Physiology, Techniques in Wildlife Management, Outdoor Recreation Theory, Practices in Outdoor Recreation, Research in Outdoor Recreation, Graduate Seminar, Independent Study, Special Study, Fish Ecology, Fisheries Management (U), Ornithology (U), Game Mammals (U), Game Management (U)

DEPARTMENT OF HORTICULTURE
Phone: Chairman: (703) 951-5451

Head: C.L. McCombs
Professors: A.S. Beecher, W.P. Judkins, R.S. Lindstrom, G.E. Mattus, C.L. McCombs, P.L. Smeal, G.R. Williams
Associate Professors: J.A. Barden, F.R. Dreiling, J.A. Faiszt, E.L. Phillips
Assistant Professors: L.H. Aung, R.E. Byers, C. Elstrodt, J.C. Garrett, H.H. Hohlt, C.R. O'Dell, R.D. Wright, J.D. Martin, G.C. Whiting
Teaching Fellows: 8

Field Stations/Laboratories: Winchester Fruit Research Laboratory, Truck and Ornamental Research Station, Truck and Ornamental Research Station, Shenandoah Valley Research Station, Horticulture Farm

Degree Programs: Horticultural Science

Undergraduate Degree: B.S.
Undergraduate Degree Requirements: 190 hours of course work

Graduate Degrees: M.S., Ph.D.
Graduate Degree Requirements: Master: Thesis - 30 to 36 hours of course work and 9 - 15 research hours Non-Thesis - 45 hours of course work Doctoral: 135 hours of course work

Graduate and Undergraduate Courses Available for Graduate Credit: Recreation Planning and Development (U), Advanced Pomology (U), Planting Design (U), Nursery Management (U), Post Harvest Horticulture (U), Landscaping Contracting and Maintenance (U), Controlled Flowering of Floricultural Crops (U), Commercial Vegetable Production (U), Physiology of Horticulture Crops, Morphology of Horticulture Plants

DEPARTMENT OF HUMAN NUTRITION AND FOODS
Phone: Chairman: (703) 951-6783

Head: R.E. Webb
Professors: R.P. Abernathy, S.J. Ritchey, R.E. Webb
Associate Professors: M.K. Korslund, J.A. Phillips, M. E. Quam
Assistant Professors: J.C. Johnson, N.L. Marable, F. W. Thye, J. Wentworth
Instructors: T.M. Hollern, J.C. Wolgemuth
Teaching Fellows: 10
Research Assistants: 6
Trainees: 9

Degree Programs: Food Science, Nutrition

Undergraduate Degree: B.S.
Undergraduate Degree Requirements: 180 quarter hours

Graduate Degrees: M.S., Ph.D.
Graduate Degree Requirements: Master: 45 quarter credits, thesis Doctoral: 135 credits (beyond B.S.) dissertation

VIRGINIA STATE COLLEGE

Petersburg, Virginia 23834 Phone: (703) 526-5111

DEPARTMENT OF LIFE SCIENCES
Phone: Chairman: (703) 526-5111 Ext. 382

Chairman: P.S. Benepal
Professors: P.S. Benepal, N. Boggs, H.D. Hamlett, J. Upadhyay, B.R. Woodson
Associate Professors: E.W. Jemison, S.K. Sen, N.C. Brewington, E. Wilson
Assistant Professors: F.C. Divers, R. Newkirk, S.M. Siddigi, V. Taylor
Research Assistants: 13

Degree Programs: Biology, Microbiology

Undergraduate Degrees: B.S.
Undergraduate Degree Requirements: 128 semester hours

Graduate Degree: M.S.
Graduate Degree Requirements: Master: 30 semester hours

Graduate and Undergraduate Courses Available for Graduate Credit: Systematic Botany, Protozoology, Invertebrate Zoology, Vertebrate Biology (U), Principles of Microtechniques (U), Vertebrate Histology (U), Vertebrate Embryology (U), Animal Physiology (U), Plant Physiology (U), Cell Physiology (U), Principles of Genetics (U), Vertebrate Morphogenesis (U), General Biochemistry (U), Advanced Genetics, Topics in Modern Biology, Advanced Protozoology, Parasitology, Biology and Human Affairs, Endocrinology, Advanced Invertebrate Zoology, Phycology, Mycology, Advanced Plant Physiology, Experimental Embryology, Teaching an Advanced General Biology Course, Cytology, Plant Morphogenesis Microbiology - Virology (U), Microbial Physiology (U), Microbial Research Problem (U), Immunology and Serology (U), Pathogenic Microbiology (U), Microbial Biochemistry, Graduate Seminar, Advanced Immunobiology, Microbiology for High School Teachers of Science, Molecular Biology, Experimental Microbiology

VIRGINIA UNION UNIVERSITY

1500 N. Lombardy Street Phone: (804) 359-9331
Richmond, Virginia 23220
Dean of College: F.J. Gayles

DEPARTMENT OF BIOLOGY
Phone: Chairman: Ext. 313

Chairman: W.O. Bradley
Professors: W.O. Bradley, H.L. Strader, M.E. Toney, Jr.
Assistant Professor: R.D. Hargrove

Degree Program: Biology

Undergraduate Degree: B.S.
Undergraduate Degree Requirements: 35 semester hours Biology, 16 Chemistry, 9 Mathematics, 8 Physics, 6 German or French, 8 Humanities, 4 Physical Education, 9 Social Sciences, 3 Religion, 26 electives

VIRGINIA WESLEYAN COLLEGE

Norfolk, Virginia 23502 Phone: (804) 464-6291
Dean of College: W. Wilson

DEPARTMENT OF BIOLOGY

Phone: Chairman: Ext. 247

Chairman: V.M. Keefer
Assistant Professors: L. Ferreri, D. Gouoni

Degree Programs: Biology, Pre-Medicine, Field Studies

Undergraduate Degree: B.A.
Undergraduate Degree Requirements: Minimum of 32 hours in Biology, plus General Chemistry, Majors are urged to take Analytical Chemistry, Organic Chemistry, Biochemistry, Calculus, Statistics and Physics

WASHINGTON AND LEE UNIVERSITY

Lexington, Virginia 24450 Phone: (703) 463-9111
Dean of College: W.J. Watt

DEPARTMENT OF BIOLOGY

Phone: Chairman: Ext. 268 Others: Exts. 208 and 218

Head: J.H. Starling (Acting)
Professors: L.R. Emmons, C.P. Hickman, Jr., J.H. Starling
Associate Professors: T.G. Nye
Assistant Professors: W.E. Bryant

Degree Program: Biology

Undergraduate Degree: B.S.
Undergraduate Degree Requirements: 50 credits in Mathematics and Sciences, 36 in distribution requirements, 29-33 electives - total of 115 credits

WESTHAMPTON COLLEGE

(see University of Richmond)

WASHINGTON

CENTRAL WASHINGTON STATE COLLEGE

Ellensburg, Washington 98926 Phone: (509) 963-2731
Dean of Graduate Studies: D. Comstock
Dean of School of Natural Sciences and Mathematics: B. Martin

DEPARTMENT OF BIOLOGICAL SCIENCES
Phone: Chairman: (509) 963-2731

Chairman: P.C. Dumas
Professors: G.W. Clark, P.C. Dumas, E.J. Harrington, E.P. Klucking, J.M. Lowe, R.E. Pacha, J.S. Shrader
Associate Professors: R.J. Boles, W.W. Barker, R.H. Brown, S.R. Johnson, L.L. Kunz (Adjunct), E.C. McDonald (Adjunct), S.D. Smith, T.H. Thelen, C.A. Wiberg
Assistant Professors: J.E. Carr, D.R. Hosford, R.F. Lapen, R.S. Shook
Teaching Fellows: 10
Research Assistants: 5

Field Stations/Laboratories: Mobile Field Station

Degree Programs: Bacteriology and Immunology, Biology, Ecology, Zoology, Botany

Undergraduate Degree: B.A.
Undergraduate Degree Requirements: Biology - 60 credits, including general Biology, 48 credits by advisement, Botany - 48 credits, including General Biology, Plant Taxonomy, Anatomy, and Physiology, Zoology - 48 credits, including general Biology, Embryology, Vertebrate Zoology, Invertebrate Zoology, Genetics, Zoophysiology

Graduate Degree: M.S.
Graduate Degree Requirements: Master: 1) 45 credits in Biological Sciences and related subjects (including 1 credit seminar and 6 credit thesis) 2) 2 consecutive quarters in residence 3) an acceptable thesis 4) pass a final oral and/or written examination

Graduate and Undergraduate Courses Available for Graduate Credit: Environmental Microbiology, General Virology, Invertebrate Paleontology, Paleobotany, Selected Studies in Biology, Paleoclimatology, Advanced Genetics, Limnology, Modern Developments in Evolution, Biological Field Experience, Special Topics, Cellular Physiology, Microbial Physiology, Morphology, Physiology and Systematics of the Higher Bacteria, Graduate Research, Individual Study, Seminar, Thesis Plant Geography, Advanced Plant Ecology, Advanced Plant Physiology, Agrostology, Selected Studies in Botany, Plant Pathology, Comparative Morphology of Vascular Plants, Mycology, Phycology, Basidiomycetes, Individual Study, Thesis, Protozoology, Selected Studies in Zoology, Advanced Physiology, Histology, Animal Ecology, Zoogeography, Ethology, Advanced Mammalogy, Advanced Ornithology, Advanced Herpetology, Advanced Ichthyology, Vertebrate Paleontology, Aquatic Entomology, Environmental Physiology, Advanced Parasitology, Individual Study, Thesis

EASTERN WASHINGTON STATE COLLEGE

Cheney, Washington 99004 Phone: (509) 235-6221
Dean of Graduate Studies: R. Whilfield
Dean of College: H.Y. Steiner

CENTER FOR ENVIRONMENTAL STUDIES
Phone: Chairman: Ext. 2355

Chairman: F.D. Nicol
(No courses or degrees. A center for research only)
Visiting Lecturers: 5

DEPARTMENT OF BIOLOGY
Phone: Chairman: Ext. 2339

Chairman: J.L. Hanegan
Professors: M. Bacon, J.E. Johns, B.Z. Lang, F.D. Nicol, H.R. Simms, K.C. Swedberg
Associate Professors: N.V. Vigfusson, R.J. White
Assistant Professors: R.D. Andrews, R.L. Carr, F. Gibson, J.L. Hanegan, S.K. Kasuga, R.A. Soltero
Teaching Fellows: 7

Field Stations/Laboratories: Turnbull Laboratory - Turnbull Wildlife Refuge

Degree Program: Biology, Medical Technology

Undergraduate Degrees: B.S., B.A.
Undergraduate Degree Requirements: 151 Organismal Biology, Cellular Biology, Ecology and Evolution, Microbiology, Botany, Zoology, Physiology, Genetics, General Chemistry, Organic Chemistry, Precalculus or Statistics or Mathematical Modeling

Graduate Degree: M.S.
Graduate Degree Requirements: Master: Molecular Biology, Advanced Cellular Biology, Advanced Physiology, Biological Instrumentation, Research Design and Literature, Advanced Ecology, Advanced Evolution

Graduate and Undergraduate Courses Available for Graduate Credit: Plant Taxonomy, Ornithology, Entomology, Field Entomology, Invertebrate Zoology, Physiology, Limnology, Mycology and Plant Pathology, Medical Bacteriology, Organic Evolution, Immunology, Virology, Animal Behavior, Microtechnique, Mammalogy, Parasitology, Ichthyology, Plant Physiology, Vertebrate Physiology, Endocrinology, Endocrinology Laboratory, Advanced Genetics

GONZAGA UNIVERSITY

E. 502 Boone Avenue Phone: (509) 328-7243
Spokane, Washington 99202
Dean of Graduate Studies: J. Byrne
Dean of College: P.J. Ford

DEPARTMENT OF BIOLOGY
Phone: Chairman: Ext. 264 Others: (509) 328-7243

Chairman: R.C. Hurd
Professors: W.T. Barry, R.C. Hurd, H.B. Stough (Emeritus)
Associate Professor: M.M. Stanton
Assistant Professor: J.J. McGivern
Instructor: D.J. Guthrie
Research Assistant: 1

Degree Program: Biology, Medical Technology

Undergraduate Degrees: B.A., B.S.
Undergraduate Degree Requirements: B.S. in Biology - 36 semester hours in Biology, 16 hours Chemistry, 8 hours Physics, University Core Curriculum, including Mathematics, B.A. in Biology - 30 hours Biology, 8 hours Chemistry, University Core Curriculum, B.S. Medical Technology - 31 hours Biology, 16 hours Chemistry, 4 hours Physics, plus credits from 12

month internship in clinical laboratory and University core

PACIFIC LUTHERAN UNIVERSITY
Park Avenue Phone: (206) 537-6900
Tacoma, Washington 98447
Dean of Graduate Studies: R. Moe
Provost of the University: R. Jungkuntz
Chairman, Division of Natural Sciences: W. Giddings

DEPARTMENT OF BIOLOGY
Phone: Chairman: Ext. 382

Chairman: J.S. Jensen
Professors: J.S. Jensen, J.W. Knudsen, B.T. Ostenson
Associate Professors: R. Bohannon, A. Gee, J. Main
Assistant Professors: A. Alexander, M. Forster, D. Hansen, R. Johnson, J. Lerum, R. McGinnis

Degree Program: Biology

Undergraduate Degree: B.A., B.S.
Undergraduate Degree Requirements: Minimum: B.S. 40 semester credits Biology, Chemistry 18 credits (through Organic, including Laboratory), 10 credits Physics including laboratory, Mathematics through Calculus, B.A.: 28 credits Biology, 18 credits Chemistry (including Organic with Laboratory), Mathematics through College Algebra and Trigonometry, Calculus and Physics strongly recommended

ST. MARTIN'S COLLEGE
Olympia, Washington 98503 Phone: (206) 491-4700
Dean of College: R. Cebula

DEPARTMENT OF BIOLOGY
Phone: Chairman: (206) 491-4700

Head: P.G. Reischman
Professor: P.G. Reischman
Assistant Professor: C. Abair

Degree Program: Biology

Undergraduate Degree: B.S.
Undergraduate Degree Requirements: Lower Division: Two courses each introductory Biology, Chemistry, Mathematics and Physics, or the equivalent. Upper Division: (1) Twenty semester hours of upper division Biology courses. (2) one course in Genetics (3) Senior Seminar, two semesters, (4) One semester of Organic Chemistry

SEATTLE PACIFIC COLLEGE
3319 3rd Avenue W Phone: (206) 281-2165
Seattle, Washington 98119
Dean of College: L. Montzingo

BIOLOGY DEPARTMENT
Phone: Chairman: (206) 281-2203 Others: 281-2165

Chairman: R.C. Phillips
Professor: R.F. Shaw, H.T. Wiebe
Instructor: E. Rempel

Field Stations/Laboratories: Casey Campus - Whibbey Island

Degree Program: Biology

Undergraduate Degree: B.S.
Undergraduate Degree Requirements: Minimum of 40 quarter credits (Upper Division credits), Specific courses required are: Cell Biology, Ecology, Principles of Development, Genetics and Seminar. Also required are Organic Chemistry and Statistics

SEATTLE UNIVERSITY
Seattle, Washington 98122 Phone: (206) 626-6200
Dean of Graduate Studies: Rev. J. Cowgill
Dean of College: G. Zimmerman

DEPARTMENT OF BIOLOGY
Phone: Chairman: (206) 626-6200

Chairman: L.E. Aldrich, Jr.
Professors: E.A. Healy, L.A. Schmid
Associate Professors: L.E. Aldrich, Jr., P.P. Cook, Jr., G.D. Davis, G.A.Santisteban
Instructor: D.W. Boisseau

Degree Program: Biology

Undergraduate Degrees: B.A., B.S.
Undergraduate Degree Requirements: B.A. - 50 credits in Biology; 25 in Chemistry, Core to total 180
B.S. (Biology) - 60 credits in Biology, 25 in Chemistry, 15 in Physics, 10 Modern Language, Psychology, Mathematics, Core to 180

UNIVERSITY OF PUGET SOUND
(No reply received)

UNIVERSITY OF WASHINGTON
Seattle, Washington 98195 Phone: (206) 543-2100
Dean of Graduate Studies: J.L. McCarthy

Center for Bioengineering
Dean, School of Medicine: R.L. VanCitters
Dean, College of Engineering: W.R. Hill

CENTER FOR BIOENGINEERING
Phone: Chairman: (206) 543-0160 Others: 543-6124

Director: R.F. Rushmer
Professors: A.S. Hoffman, R.F. Rushmer
Associate Professors: L.L. Huntsman (Research), G. H. Pollack
Assistant Professors: J.E. Chimoskey, S.L. Johnson, W.E. Moritz (Research), W.G. Yates
Instructors: G.A. Holloway, A.B. LaVigne
Research Assistants: 20
Postdoctoral Fellows: 7

Degree Programs: About 75 collaborative projects between Health Scientists and Engineers exist at any given time.

Graduate and Undergraduate Courses Available for Graduate Credit: Introduction to Bioengineering (U), Fundamentals of Bioengineering (U), Engineering Prescriptions for Health Care Crises (U), Medical Instrumentation (U), Wave Effects in Biomaterials (U), Diagnostic Ultrasound (U), Engineering Materials for Biomedical Applications, Special Projects, Engineering Aspects of the Fluid Mechanics of the Human Body, Computer Applications in Medicine, Engineering Approaches to the Cardiovascular System

College of Arts and Sciences
Dean: G.M. Beckman

DEPARTMENT OF BOTANY
Phone: Chairman: (206) 543-1942

Chairman: A.R. Kruckeberg
Professors: H.W. Blaser, R.E. Cleland, C.L. Hitchcock (Emeritus), A.R. Kruckeberg, B.J.D. Meeuse, R.E. Norris, D.E. Stuntz, M. Tsukada, R.B. Walker, H.C. Whisler
Associate Professors: W. Halperin, E.F. Haskins
Assistant Professors: A. Bendich, R. del Moral, M.F. Denton, J.R. Wasland
Instructors: C.V. Muhlick (Lecturer Emeritus), E. Lawton (Professor Emeritus)

Teaching Fellows: 17
Research Assistants: 5
Research Assistant Professor: 1
Research Technicians: 2
EM Technicians: 2
Marine Technician: 1
Botanists: 2

Field Stations/Laboratories: Friday Harbor Laboratories - Friday Harbor

Degree Programs: Algology, Anatomy and Morphology of Bascular Plants, Cytology, Ecology, Mycology, Physiology, and Taxonomy

Undergraduate Degrees: B.S. (Botany)
Undergraduate Degree Requirements: Courses must include general chemistry including elementary organic, general Biology, molecular biology, plant classification, plant kingdom, plant physiology, cryptogamic botany, plus three elective upper division botany courses, for a minimum of 59 quarter credits. It is recommended that students planning to go on to graduate school develop one botanical subject indepth and include statistics and a foreign language in their program.

Graduate Degrees: M.S., Ph.D. (Botany)
Graduate Degree Requirements: Master: Non-thesis: 45 credits including a special problem (Botany 600), a minimum of 25 being in Botany, 18 credits in courses numbered 500 and above as required by Graduate School, reading knowledge of one foreign language. Thesis: 27 credits in approved course work, and 9 credits in Botany 700 Masters thesis, 18 of 36 credits must be in courses 500 or ablve, reading knowledge of one foreign language or reading knowledge of two, examinations in 2 subsidiary fields in areas other than student's speciality, pass creditably the general and final examinations as required by the Graduate School, Dissertation

Graduate and Undergraduate Courses Available for Graduate Credit: Advanced Systematics (U), Algology (U), Ascomycetes (U), Basidiomycetes (U), Bryology (U), Development in Lower Plants (U), Environmental Control of Plant Growth and Development, Field Ecology, Freshwater Algae (U), Graduate Research, Graduate Thesis, Marine Algal Ecology (U), Marine Algology, Marine Mycology, Mineral Nutrition (U), Palynology and Quaternary Phytogeography (U), Phycomycetes and Related Fungi (U), Phytoplankton Morphology and Taxonomy (U), Plant Anatomy (U), Plant Cytology (U), Plant Problems in Algal Physiology, Rusts, Smuts and Fungi Imperfecti (U), Selected Topics in Mycology, Seminar in Morphology and Taxonomy, Taxonomy (U), Terrestrial Plant Ecology (U), Topics in Algology, Topics in Palynology, Topics inPlant Ecology, Topics in Plant Physiology, Tutorial, Water Relations

DEPARTMENT OF ZOOLOGY
Phone: Chairman: (206) 543-1620

Chairman: D.S. Farner
Professors: R.A. Cloney, I. D. Olsen, W.T. Edmondson, J.S. Edwards, D.S. Farner, A. Gorbman, P. Illig, A. J. Kohn, E.N. Kozloff, A.W. Martin, G.H. Orians, R.T. Paine, R.C. Snyder, M. Stuiver, A.H. Whiteley
Associate Professors: M.W. Griffiths, S.P. Hauschka (Adjunct), C.D. Laird, K.L. Osterud, J.M. Palka, R.B. Pinter (Adjunct), D.L. Ray (on leave), L.M. Riddiford, T.W. Schoener, A.O. D. Willows
Assistant Professors: A. Bakken, D.P. Barash (Adjunct), Y. Palka, S. Rohwer, G. Schubiger, R.R. Strathmann, J.W. Truman
Visiting Scholars: 4
Teaching Fellows: 37
Research Assistants: 10
Trainees: 7
National Fellowships: 4
Research Associates: 43
Acting and Lecturer (Temporary) Appointments: 15

Field Stations/Laboratories: Friday Harbor Laboratories - Friday Harbor

Degree Programs: Zoology

Undergraduate Degree: B.S.
Undergraduate Degree Requirements: B.S.: 50 credits Zoology, approved electives from related fields, Physics, Chemistry, Mathematics, two years approved language. B.A.: 50 credits Zoology, approved electives from rel ated fields, Chemistry and Mathematics requirements less than for B.S.

Graduate Degrees: M.S., Ph.D.
Graduate Degree Requirements: Master: Thesis - 36 credits, of which 18 must be at the 500 level or above and 9 in thesis research, satisfy the departmental foreign-language and teaching requirements, thesis, final examination. Without Thesis: Substitute 9 credits of course work at the 500 level or above for thesis, satisfy the departmental foreign-language and teaching requirements, final examination
Doctoral: A minimum of three academic years of study, one quarter of which is spent at a biological field station, satisfy departmental foreign-language and teaching requirement, General Examination, Dissertation, Final Examination

Graduate and Undergraduate Courses Available for Graduate Credit: Topics in Experimental Embryology, Comparative Developmental Physiology, Advanced Topics in Physiology, Advanced Invertebrate Zoology, Topics in Advanced Invertebrate Zoology, Comparative Invertebrate Embryology, Advanced Vertebrate Morphology, Insect Development, Chemical Integration, Topics in Ecology, Ecology of Marine Communities, Environmental Marine Physiology, Advanced Ecology, Systematic Zoology, Advanced Techniques in Microscopy, Topics in Limnology, Topics in Physical and Chemical Limnology, Analysis of Development, Advanced Cytology, Cellular Physiology, Problems in Biological Instruction, Advanced Invertebrate Physiology

College of Fisheries
Dean: D.G. Chapman

COLLEGE OF FISHERIES
Phone: Chairman: (206) 543-4270 Others: 543-2100

Dean: D.G. Chapman
Professors: M.C. Bell, D.E. Bevan, K. Bonham (Research), R.L. Burgner, D.G. Chapman, K.K. Chew, A.C. DeLacy, L.R. Donaldson (Emeritus), R.D. Dugdale (Research), A.W. Erickson (Research), E.E. Held (Research), B.S. Jayne, J. Liston, J.E. Lynch (Emeriuts), B.W. Mar (Adjunct), O.A. Mathisen, R.C. Meier (Adjunct), R.E. Nakatani, W.T. Newell (Adjunct), E.J. Ordal (Adjunct), G.M. Pigott, E.O. Salo, A.H. Seymour, L.S. Smith, F.B. Taub, R. Van Cleve, A.D. Welander, R.R. Whitney, D.L. McKernan
Affiliates: D.L. Alverson, N. Bourne, L.E. Eberhardt, J.E. Halver, H.O. Hodgins, D.R. Johnson, J. Joseph, M. Katz, R.R. Rucker, A.K. Sparks, W. Templeton
Associate Professors: G.W. Brown, Jr., I. Fletcher (Acting), J.C. Kelley III, J.R. Matches, W.R. Schell (Research), Q.J. Stober (Research). Affiliates: G.I. Jones, T. Joyner, W. Pereyra, B.E. Skud, M. Southward, M.A. Steinberg, V.F. Stout, R.B. Thompson, G.A. Wedemeyer, C.E. Woelke, R.S. Wydoski, Research Associates: A. Nevissi, O. Agarawala, B. Satia, S. Felton, J. Richey, R.C. Wissmar
Assistant Professors: L.J. Bledsoe (Research), E.L. Brannon, V.F. Gallucci, W.K. Hershberger, D.A. McCaughran, S.B. Mathews, G. Pauley, D.E. Rogers (Research), J.J. Walsh (Research). Affiliates: P.K. Bergman, B.G. D'Aoust, S.M. Olsen, M. Tillman, F.M. Utter, D.D. Weber, Senior Research Asso-

ciates: G.E. Lord, B.S. Miller, R.E. Thorne
Instructors: G. Finne, R. Hansen (Lecturer)
Teaching Fellows: 10
Research Assistants: 40

Field Stations/Laboratories: Big Beef, Seabeck; Fernlake - Purdy, several small stations in Bristol Bay, Alaska

Degree Programs: Fisheries, Food Science, Wildlife Management

Undergraduate Degree: B.S.
Undergraduate Degree Requirements: Not Stated

Graduate Degrees: M.S., Ph.D. (Fisheries)
Graduate Degree Requirements: Master: (Fisheries) 27 course credits, 18 thesis credits, 9 credits at 500 level or above with 3 in Seminar. (Food Science) 27 course credits, 18 thesis credits, 9 credits at 500 level or above with 6 in Seminar. Doctoral: 54 course credits, 36 thesis credits. Usually M.S. required as prerequisite.

Graduate and Undergraduate Courses Available for Graduate Credit: Not Stated

College of Forest Resources
(No reply received)

School of Medicine
Dean: R.L. Van Citters

DEPARTMENT OF BIOCHEMISTRY
Phone: Chairman: (206) 543-1660 Others: 543-1600

Chairman: H. Neurath
Professors: P. Bornstein, E.W. Davie, E.H. Fischer, J.A. Glomset (Adjunct), M.P. Gordon, B.D. Hall (Adjunct) A. Kaplan (Joint), H. Neurath, K.A. Walsh
Associate Professors: S.D. Hauschka, D.R. Morris, W.W. Parson, B.M. Shapiro, D.C. Teller, K. Titani (Research)
Assistant Professors: N. Agabian, E.A. Boeker (Research), B.E. Byers (Adjunct), D.A. Deranleau, K. Fujikawa (Research), M. Hermodson (Joint - Research), J.R. Herriott, J.M. Keller, R.D. Palmiter, P.H. Petra (Joint), A. Pocker (Research), E.T. Young II
Visiting Lecturers: 2
Research Assistants: 33

Field Stations/Laboratories: Friday Harbor Laboratories - San Juan Island

Degree Program: Biochemistry

Graduate Degrees: M.S., Ph.D.
Graduate Degree Requirements: Master: 5 courses of Biochemistry in each of 3 successive trimesters, thesis Doctoral: M.S. or equivalent, courses as decided by advisory committee. Teaching experience, Preliminary qualifying examination, thesis, final oral

Graduate and Undergraduate Courses Available for Graduate Credit: Introduction to Biochemistry (U), Medical Students' Laboratory, Molecular Biology, Molecular Biology Laboratory, Seminar, Advanced Biochemistry, Literature Review, Physical Biochemistry, Biochemical Basis of Disease, Advanced Techniques in Biochemistry, Nucleic Acids in Biochemistry, Engyme Regulation, Seminar in Animal Cell Membranes, Current Topics in Molecular and Cellular Biology, Connective Tissue Macromolecules, Proteins and Enzymes Seminar, Seminar on Protein Structures, Topics in the Biochemistry of Regulation, Glycogen Metabolism Seminar, The Role of the Cell Surface in Cell Division and Development Seminar, Clinical Chemistry Seminar, Plant Viruses Seminar, Seminar in Developmental Biology, Seminar in Physical Chemistry of Polymers, Independent Study

DEPARTMENT OF BIOLOGICAL STRUCTURE
Phone: Chairman: (206) 543-1860

Chairman: N.B. Everett
Professors: R.J. Blandau, E.A. Boyden (Research), N. B. Everett, L.H. Jensen, J.H. Luft, G.F. Odland, E.C. Roosen-Runge, M.R. Schwarts, A. Tamarin (Adjunct)
Associate Professors: D.O. Graney, H.K. Kashiwa, J.K. Koehler, B.R. Landau, E.P. Lasher, R.D. Lund, J.W. Prothero, C. Rosse, R.E. Rumery (Research), J. W. Sundsten, L.E. Westrum
Assistant Professors: S.H. Broderson, P.W. Coates, E.M. Eddy, P.Gaddum-Rosse, S.A. Halbert (Acting), J.W. Lindsay (Acting), M.A. Nameroff, H.Z. Park (Research), W.D. Perkins, L.C. Robson, B. Szubinska-Luft (Research), P. Verdugo (Research), K.D. Watenpaugh (Research)
Instructors: B.B. Gallucci (Acting), K.A. Holbrook, F.W. Merchant, K.Graubard (Research Associate), J.M. Hodsdon (Research Associate), L.C. Sieker (Research Associate), P. Smith (Research Associate), M.B. Colman (Associate), E. Adman (Research Associate), A. Fitzgerald (Research Associate)
Visiting Lecturers: 4
Research Assistant: 1
Lecturer: 1

Field Stations/Laboratories: Friday Harbor Marine Laboratory - Friday Harbor

Degree Program: Biological Structure (Anatomy)

Graduate Degrees: M.S., Ph.D.
Graduate Degree Requirements: Master: 27 credits, 9 thesis credits, B.A. degree or equivalent, level 400 or better in major (Anatomy), level 300 for supporting classes Doctoral: 9 full time quarters in residence Credits 81 minimum of which 27 at 800 level, 18 at 500 level (graded), level 400 or over in major (anatomy) level 300 for supporting classes, B.A. degree or equivalent, Must maintain 3.00 GPA

Graduate and Undergraduate Courses Available for Graduate Credit: Gross Anatomy (Thorax, Abdomen, Pelvis and Perineum), Gross Anatomy (Extremities), Gross Anatomy (Head and Neck), Human Embryology and Development, Cytology, Cell Structure and Function, Human Microanatomy, Histological Basis of Biomechanics, Neuroanatomy, Cellular Differentiation, Hemopoesis, Comparative General Histology, Biological X-Ray Structure Analysis, Bioinstrumentation and Research Methods, Electron Microscopy, Functional Neuroanatomy

DEPARTMENT OF MICROBIOLOGY
Phone: Chairman: (206) 543-5824 Others: 543-2100

Chairman: J.C. Sherris
Professors: H.C. Douglas, C.A. Evans, S. Falkow, N. B. Groman, S. Hakomori, I. Hellstrom, G.E. Kenny, (Adjunct), S.B. Klebanoff (Adjunct), M. Mannik (Adjunct), E.W. Nester, E.J. Ordal, C.G. Ray, R.S. Weiser, H.R. Whiteley
Associate Professor: V.C. Chambers (Research)
Assistant Professors: J.J. Champoux, M.D. Chilton (Research), J.A. Clagett (Research), C.R. Clausen (Clinical), M.B. Coyle, E.D. Kiehn, J.C. Lara, N. N. Pearsall, S.B. Pollack (Research), F.D. Schoenknecht, J.T. Staley, U.B. Storb
Instructor: H.M. Pollock
Visiting Lecturer: 1
Research Associate: 1
Lecturers: 6
Associate: 1

Degree Programs: Immunology, Medical Microbiology, Microbiology

Undergraduate Degree: B.S. (Microbiology)

Undergraduate Degree Requirements: A minimum of 180 academic credits with 45 credits in biological sciences, including Biology 210, 211, 212)preferred) or an equivalent 10 or 15 credits in botany or zoology, or both; minimum of 30 credits in microbiology courses, including microbiology 400, 401, 430, 431, 441, 442, 443, and 496 (Microbiology 101, 301, 302, 351 cannot be used); Physics 114, 115, 116 or 121, 122, 123, Chemistry 140, 150, 151, 160, Chemistry 231, 232 or 231, 235, 236, or 335, 336, 337 (three-quarter sequence preferred), Chemistry 221, Mathematics 124 or 157 or Science 281 or 291, a grade-point average of 2.00 in microbiology courses.

Graduate Degrees: M.S., Ph.D.

Graduate Degree Requirements: Master: A minimum of three full-time quarters of residence. With thesis: includes 45 credits of course work and preparation of a thesis based on laboratory research. Without thesis: includes 45 credits of course work and an individually supervised laboratory report. No foreign language requirement Doctoral: A minimum of three academic years of resident study, two of them at the university of Washington. Microbiology Option: one course in three areas must be taken from among the general areas of virology, microbial physiology, advanced general microbiology, and immunology. Two courses must be taken from among the research methods courses offered in bacteriophage studies, enzymology, nucleic acid chemistry, immunochemistry, microbial genetics, and electron microscopy. Research. Laboratory teaching experience. General examination, dissertation, and final examination. No foreign language requirement. Immunology Option: same as Microbiology option except for specific additional course requirements

Graduate and Undergraduate Courses Available for Graduate Credit: Fundamentals of Bacteriology and Laboratory (U), Microbial Metabolism and Laboratory (U), Microbial Ecology and Laboratory (U), Medical Bacteriology, Virology and Immunology (U), Medical Microbiology Laboratory (U), Medical Mycology and Parasitology and Laboratory (U), Fundamentals of Immunology (U), Molecular Biology of Viruses (U), Research Techniques in Virology, Research Techniques in the Study of Microbial Genetics, Immunological Techniques, Techniques in Electron Microscopy of Microorganisms, Physiology of Bacteria, Cell Surface Membrane in Cell Sociology and Immunology, Advanced General Microbiology and Laboratory, Virology, Selected Topics in Immunology, Pathogenesis of Infectious Diseases of Man, Advanced Clinical Microbiology, Clinical Microbiology, Training and Research, Tumor Biology, Topics in Microbiology, Independent Study or Research

DEPARTMENT OF PHYSIOLOGY AND BIOPHYSICS

Phone: Chairman: (206) 543-0954 Others: 543-0950

Chairman: H.D. Patton

Professors: A.C. Brown, E.O. Feigl, B. Hille, T. Hornbein, T.T. Kennedy, H.D. Patton, L.B. Rowell, T.C. Ruch, A.M. Scher, O.A. Smith, C.F. Stevens, D.Y. Teller (Adjunct), A.L. Towe, R.L. Van Citters, C. Wiederhielm, A.C. Young (Emeritus)

Associate Professors: G.L. Brengelmann, J.T. Conrad, W.E. Crill, A.F. Fuchs, C.C. Gale, A.M. Gordon, J. Hildebrandt, T. Kehl, B. Landau, E.S. Luschei, J. Miller, W.L. Stahl, C.E. Stirling, H.J. Van Hassel, J.G. Skahen (Emeritus)

Assistant Professors: W. Almers, M. Anderson, S.K. Donaldson, E.E. Fetz, M. Hlastala, W.G.L. Kerrick, D.T. Koerker, P. Schwindt, D. Stromberg, B. Walike, M. Biedenbach (Research), V. Dionne (Acting), F. Harris (Research), P. Illner (Research), M. Shaw (Research)

Instructor: S. Lewis (Affiliate)

Postdoctoral Fellows: 15

Degree Programs: Physiology and Biophysics (Physiology-Psychology joint with Department of Psychology)

Graduate Degrees: M.S., Ph.D.

Graduate Degree Requirements: Master: course work and thesis (usually offered only to people already having an M.D.) Doctoral: course work and thesis

Graduate and Undergraduate Courses Available for Graduate Credit: Approximately 45 courses offered at the graduate level in all areas of physiology and biophysics

School of Public Health and Community Medicine

Dean: R. Day

DEPARTMENT OF BIOSTATISTICS

Phone: Chairman: (206) 281-1044

Chairman: D.J. Thompson

Professors: L. Fisher, D.J. Thompson

Associate Professors: N.K. Brown, P. Feigl, W.J. Kennedy, R.A. Kronmal, D.C. Martin, R. Prentice, G. van Belle (Visiting)

Assistant Professors: P. Diehr, V. Gallucci (Adjunct), D. Newman, L. Polissar, P. Wahl

Research Associates: 2

Visiting Scholar: 1

Visiting Professor: 1

Visiting Associate Professor: 1

Degree Program: Biomathematics

Graduate Degrees: M.S., Ph.D.

Graduate Degree Requirements: Master: 18 (quarter) credits in particular Mathematics courses, 19 credits in particular Biostatistics and Quantitative Science courses, 6-10 credits in approved electives, and at least 9 of the above-mentioned credits must be in courses numbered above 500. Also certain other general requirements of the University of Washington. Graduate School must be met, and a thesis must be written. Doctoral: Specific courses, successful completion of examinations and a dissertation, residence, total credit, and other general requirements specified by the University of Washington Graduate School.

Graduate and Undergraduate Courses Available for Graduate Credit: Biomathematics, Stochastic Processes in the Life Sciences, Seminar in Quantitative Ecology, Special Topics in Quantitative Ecology, Research in Quantitative Ecology, Independent Study or Research, Master's Thesis, Doctoral Dissertation, Biostatistics, Principles of Communicable Disease, Control and Biostatistics (U), Applied Statistics in Health Sciences (U), Sample Survey Techniques (U), Statistical Methods in Biological Assay (U), Application of Vital and Health Statistics (U), Biostatistics Special Elective (U), Undergraduate Thesis (U), Undergraduate Research (U), Medical Biometry, Operations Research for Health Services, Special Topics in Advanced Biostatistics, Seminar in Biostatistics, Seminar in Bio-Statistics Applied To Health Service Research, Epidemiology and Biostatistics Research Seminar, and Biostatistical Consulting

WALLA WALLA COLLEGE

College Place, Washington 99324 Phone: (509) 527-2602

Dean of Graduate Studies: D.W. Rigby

Dean of College: R.D. McCune

DEPARTMENT OF BIOLOGICAL SCIENCES

Phone: Chairman: (509) 527-2603

Chairman: D.W. Rigby

Associate Professors: D.Clayton, A.E. Grable, A.E. Perry

Assistant Professors: L.G. Dickson, L. Fisk, L.R. McCloskey

Research Assistant: 1

Teaching Assistants: 8

Field Stations/Laboratories: Walla Walla College Marine Biological Station - Anacortes

Degree Programs: Animal Behavior, Animal Physiology, Biological Sciences, Biology, Botany, Ecology, Entomology, Limnology, Microbiology, Parasitology, Pharmacy, Physiological Sciences, Plant Physiology, Zoology, Invertebrate Zoology, Marine Biology, Vertebrate Natural History and Ecology

Undergraduate Degree: B.S. (Biology)
Undergraduate Degree Requirements: General Biology, Genetics, Developmental Biology, Biostatistics, Research Methods I, II, III, Cell Physiology, General Ecology, Philosophy of Origins and Speciation, Electives: upper division in biology, one course in zoology and one in Botany required. Colloquium - required each year of junior and senior while in residence, Fundamentals of Mathematics, General Physics, Analytical Geometry and Calculus I, General Physics Laboratory, General Chemistry, Elementary Organic Chemistry

Graduate Degree: M.S.
Graduate Degree Requirements: Master: 45 quarter credits minimum total requirement; 25 credits of the total must be graduate numbered courses, the program is designed to fit the interests and goals of the individual student

Graduate and Undergraduate Courses Available for Graduate Credit: Plant Physiology (U), Ornithology (U), General Entomology (U), Plant Anatomy (U), Comparative Anatomy (U), Herpetology (U), Systematic Botany (U), Limnology (U), Microtechnique (U), Mammalogy (U), General Ecology (U), Parasitology (U), Vertebrate Histology (U), Invertebrate Zoology (U), Research Methods III (U), Ichthyology (U), Marine Botany (U), Animal Behavior (U), Bacteriology (U), Biological Oceanography (U), Comparative Physiology (U), Biophysics (U), Marine Invertebrates (U), Philosophy of Origins and Speciation (U), Techniques in Field Biology (U), Colloquium (U), Readings in Physiology, Research in Biology, Readings in Invertebrate Zoology, Genetics and Evolution, Morphology of Plants, Readings in Entomology, Physiology of Algae, Readings in Ecology, Graduate Seminar, Readings in Symbiosis, Biosystematics, Readings in Biosystematics, Symbiosis, Systematic Entomology, Principles of Economic Entomology, Cellular Biology, Thesis

WASHINGTON STATE UNIVERSITY

Pullman, Washington 99163 Phone: (509) 335-3564
Dean of Graduate Studies: C.C. Nyman

College of Liberal Arts

Dean: B.R. Ray

DEPARTMENT OF BACTERIOLOGY AND PUBLIC HEALTH

Phone: Chairman: (509) 335-3323

Chairman: H.M. Nakata
Professors: C.H. Drake, E.R. Hall, O.H. Johnson, H.M. Nakata, J.L. Stokes
Associate Professors: R.E. Hurlbert, L.P. Mallavia, K.D. Spence
Assistant Professors: D.J. Hinrichs, K.L. McIvor
Research Assistants: 1

Degree Program: Bacteriology

Undergraduate Degree: B.S.
Undergraduate Degree Requirements: Minimum: 120 semester hours, 1 year General Chemistry, Organic Chemistry, Quantitative Analysis, Biochemistry, Physics - 1 year, Biological Sciences - 1 year, Mathematics (precalculus), Foreign language 1 year or equivalent, English Composition, 21 semester hours social science and humanities, 28 semester hours Bacteriology, 2 supporting courses outside major advanced level.

Graduate Degrees: M.S., Ph.D.
Graduate Degree Requirements: Master: 21 semester hours formal graduate credit in courses, 2 seminar courses, 1 special project (repeat possible), 1 thesis courses (repeat possible), oral examination and acceptable thesis Doctoral: 34 semester hours formal course work, 2 seminars, 1 special project (repeat), 1 thesis course (repeat), written and oral preliminary, final oral examination (defense of thesis)

Graduate and Undergraduate Courses Available for Graduate Credit: Advanced Medical Microbiology (U), Immunology (U), General Virology and Laboratory (U), Microbiology of Foods (U), Epidemiology (U), Advanced Microbiology and Laboratory (U), Higher Bacteria and Fungi (U), Selected Topics, Epidemiology, Microbial Physiology, Seminar, Molecular Genetics, Advanced Immunology and Immunogenesis, Bacterial Techniques in Genetics, Special Projects or Independent Study, Master's Research, Thesis and/or examination, Doctoral Research, Dissertation and/or examination

DEPARTMENT OF GENERAL BIOLOGY

Phone: Chairman: (509) 335-8649

Chairman: J.L. Hindman
Professors: H.E. Brewer, A.C. Cohen, R.J. Jones, J.C. Hindman, W.H. Matchett, C.W. McNeil, V. Schultz
Associate Professors: E.S. Broch, F. Hungate (Adjunct), L.P. Mallavia, K.D. Spence, D.R. Strouck, H.A. Went, G.L.Young
Assistant Professors: J.W. Crane, H.L. Hosick, S.B. Moffett, W.R. Rayburn, G.W. Williams
Teaching Fellows: 25

Degree Program: Biology

Undergraduate Degree: B.S.
Undergraduate Degree Requirements: 1 course in Botany, Bacteriology, Zoology, Ecology, Genetics, 15 upper division semester hours, Organic Chemistry, Calculus Physics

Graduate Degree: M.S.
Graduate Degree Requirements: Master: 15 hours, Zoology, Botany or Bacteriology background, in all-32 total hours

Graduate and Undergraduate Courses Available for Graduate Credit: General Ecology (U), Electron Microscopy, Cell Biology (U), Statistical Ecology, Radiation Ecology, Teaching Techniques (both high school and college), Human Ecology (U)

DEPARTMENT OF ZOOLOGY

Phone: Chairman: (509) 335-3553 Others: 335-3564

Chairman: J.R. King
Professors: J.R. King, L.B. Kirschner, C.W. McNeil, J. Mizell (Adjunct), R. Moree, R.A. Parker, V. Schultz
Associate Professors: R.J. Adkins, E.S. Broch, R.J. Jonas, D. King, A. Koch, J. Larsen, P.C. Schroeder, H.A. Went
Assistant Professors: J.W. Crane, H.L. Hosick, R.E. Johnson, K.V. Kardong, D.E. Miller, S.R. Moffett, W.J. Turner, I.J. Ball
Teaching Fellows: 30
Research Assistants: 8
Postdoctoral Fellows: 3
NIH Trainees: 6

Degree Programs: Physiological Sciences, Zoology, Wildlife Management

Undergraduate Degree: B.S.
Undergraduate Degree Requirements: English Composi-

tion (sophomore level), Elementary Logic, 2 years of Chemistry (including Organic), 1 year of Physics, Mathematics through Differential Calculus (or statistics or computer science), 1 year of general Biology, Genetics, Vertebrate Biology, Invertebrate Biology, Vertebrate Morphogenesis, Physiology, Ecology, Undergraduate Seminar

Graduate Degrees: M.S., Ph.D.
Graduate Degree Requirements: Master: (non-thesis): A minimum of 30 semester hours of course work including 2 to 4 hours of special problems. (thesis): a minimum of 21 hours of course work, plus thesis and proficiency in one foreign language. Doctoral: Minimum of 34 semester hours of course work beyond the bachelor's degree, written and oral preliminary examinations, doctoral dissertation defended orally, proficiency in one or two modern foreign languages at the discretion of the supervisory committee

Graduate and Undergraduate Courses Available for Graduate Credit: Invertebrate Ecology, Parasitology, Microanatomy, Ornithology, Mammalogy, Biology of Amphibians and Reptiles, Principles of Wildlife Ecology, Animal Behavior, Radiation Ecology, Experimental Embryology, Principles of Systematic Biology, Limnology, Statistical Ecology, Population Dynamics, Comparative Physiology, Mathematical Modeling of Biological Systems, General and Cellular Neurophysiology, Advanced Vertebrate Physiology, Environmental Physiology, Principles of Animal Development, Laboratory in Animal Development, Advanced Topics in Zoology . (The following are available for graduate credit for extradepartmental majors): Principles of Organic Exolution, Introduction to Mathematical Biology, Gross and Microanatomy, Vertebrate Morphogenesis, Principles of Conservation, Principles of Zoophysiology

College of Agriculture
Dean: J.S. Robins

DEPARTMENT OF AGRONOMY AND SOILS
Phone: Chairman: (509) 335-3475 Others: 335-3564

Chairman: J.C. Engibous
Professors: R.E. Allan, S.N. Brooks, C.L. Canode, J.C. Engibous, W.H. Gardner, R.A. Gilkeson, A.R. Halvorson, R.L. Hausenbuiller, J.A. Kittrick, F.E. Koehler, C.F. Konzak, A.G. Law, K.J. Morrison, T.J. Muzik, R.A. Nilan, H.W. Smith, W.A. Starr, O.A. Vogel
Associate Professors: D.F. Bezdicek, H.H. Cheng, A. M. Davis, C.B. Harston, A. Kleinhofs, C.B. Kresge, A. Lejeune, B.L. McNeal, J.D. Maguire, D.G. Miller, R.I. Papendick, C.J. Peterson, W.C. Robocker, D.G. Swan
Assistant Professors: G.S. Campbell, E. Donaldson, C.F. Engle, D.W. George, L.P. Lilley, F.J. Muehlbauer, C. Sander, R.D. Schirman, J. Schwendiman, S. von Broembsen, R.L. Warner, V.E. Wilson, R.E. Witters
Instructors: V.L. Cochran, E. Field, T.D. Wagner, F. Webb
Visiting Lecturers: 1
Teaching Fellows: 2
Research Assistants: 11

Degree Programs: Interdepartmental programs in Genetics, Environmental Science, Biophysics, Land Use, Pest Management, departmental soils program in Soil Chemistry, Soil Physics, Soil Mineralogy, Soil Fertility, Soil Microbiology, Soil Genesis, Morphology and Classification, Agricultural Climatology

Undergraduate Degrees: B.S. (Agronomy, Soils)
Undergraduate Degree Requirements: Washington State University requirements for B.S. degree = 120 hours. Agronomy: At least 30 of the total hours required for this degree must be in upper-division courses. Soils: Minimum of 15 hours in soils, 24 hours in Chemistry and Physics, with 30 of total hours in upper-division courses

Graduate Degrees: M.S., Ph.D. (Agronomy, Soils)
Graduate Degree Requirements: Master: Thesis -- Agronomy, Soils -- 30 hours of credit, including at least 21 hours formal course work, with at least 11 hours of 400 and 500 level courses Non-thesis--Agronomy, 30 hours of credit, including at least 26 hours formal, 2-4 hours special problems or directed study with at least ly hours of 400 and 500 level courses. Doctoral: Agronomy: High level reading proficiency in one approved foreign language, or a moderate level reading proficiency in two approved foreign languages, or a moderate level proficiency in one language supplemented by approved course work. Soils: A total of 72 hours post-baccalaureate credit at the upper division and graduate level, including no more than 12 hours of 300 level courses approved for graduate credit; at least 34 hours of 400 and 500 level formal course work and fulfillment of foreign language requirements or of an approved substitute.

Graduate and Undergraduate Courses Available for Graduate Credit: Agronomy: Plant Breeding (U), Seed Production and Processing (U), Physiological Crop Ecology (U), Agronomic Research Techniques; Advanced Plant Breeding, Hormones and Herbicides, Seed Physiology, Seminar, Topics in Agronomy Soils: Soil Chemistry, Soil Analysis, Soil Morphology and Classification, Forest Soils, Forest Soils Field Trip, Physics and Hydrology of Soils, Air Photo Interpretation, Introduction to Environmental Biophysics, Environmental Biophysics Laboratory, Advanced Soil Chemistry, Soil Fertility, Soil Geography, Soil Mineralogy, Soil Organic Matter, Biochemistry of Soil-Water Environment, Soil Physics and Physical Chemistry, Seminar, Groundwater Chemistry.

DEPARTMENT OF ANIMAL SCIENCES
Phone: Chairman: (509) 335-5523 Others: 335-3564

Chairman: T.H. Blosser
Professors: W.A. Becker, T.H. Blosser, I.A. Dyer, M. H. Ehlers, V.L. Estergreen, I.L.Kosin, J.McGinnis, C.C.O'Mary
Associate Professors: J.R. Carlson, R.S. Hansen, J.K. Hillers, R.J. Johnson, M.H. Pubols,
Assistant Professors: D.C. Anderson, J.A. Froseth, A.S. Hodgson, I.M Hughes, E.L. Martin, J.J. Reeves, R.W. Wallenius
Research Associates: 3

Field Stations/Laboratories: Western Washington Research and Extension Center, Irrigated Agriculture Research and Extension Center

Degree Programs: Nutrition, Animal Science, Genetics

Undergraduate Degrees: B.S.
Undergraduate Degree Requirements: Varies according to option (Animal Production, Business Animal Production, Animal Biology, Animal Nutrition)

Graduate Degrees: M.S., Ph.D.
Graduate Degree Requirements: Master: 9 hours Biology 1 semester each of Biometry and Biochemistry
Doctoral: Not Stated

Graduate and Undergraduate Courses Available for Graduate Credit: Not Stated

DEPARTMENT OF ENTOMOLOGY
Phone: Chairman: (509) 335-5504

Chairman: R.F. Harwood
Professors: R.F. Harwood, M.T. James (Emeritus), C. A. Johansen, H S. Telford, C. Shanks (Research), T. Anthon (Research), E. Burts (Research), S. Hoyt (Research), W. Cone (Research), E.C. Klostermeyer (Research)

Associate Professors: R.D. Akre, A.A. Berryman, R.L. Campbell
Assistant Professors: R.L. Benson, W.J. Turner
Research Assistants: 7
Postdoctoral Research Assistants: 2

Field Stations/Laboratories: Irrigated Agricultural Research and Extension Center, Tree Fruit Research Center, Western Washington Research and Extension Center, Southwestern Washington Research Station, Northwestern Research and Extension Center

Degree Program: Entomology

Undergraduate Degree: B.S.
Undergraduate Degree Requirements: Minimum 18 hours Entomology courses, including general and agricultural Entomology for the B.S. degree in Entomology

Graduate Degrees: M.S., Ph.D.
Graduate Degree Requirements: Master: 21 hours minimum course work Doctoral: Minimum 34 hours coursework beyond Bachelors degree

Graduate and Undergraduate Courses Available for Graduate Credit: Agricultural Entomology, General Entomology, Field Entomology, Systematics Entomology, Entomology History and Literature, Insect Morphology, Medical Entomology, Biological and Internal Control, Principles of Applied Entomology, Insect Physiology, Pest and the Environment, Insect Morphogenesis, Pest Toxicology, Principles of Systematics Biology, Entomology Research Methods, Taxonomy of Immature Insects, Insect Ecology, Insect Behavior, Pest Management, Insect Toxicology, Plant Resistance to Insects, Advanced Forest Entomology, Insect Biochemistry, Aquatic Entomology, Aquatic Entomology Laboratory, Insect Physiology Ecology, Seminar

DEPARTMENT OF FORESTRY AND RANGE MANAGEMENT
Phone: Chairman: (509) 335-5584

Chairman: G.A. Harris
Professors: R.W. Dingle, C.J. Geobel, G.G. Marra, M.M. Mosher, B.F. Roche, J.D. Rogers, D.R. Satterland
Associate Professors: A.A. Berryman, P.E. Heilman, A.E. Noskowiak
Assistant Professors: R.W. Bruce, R.C. Chapman, J.R. Nelson, R.L. Shaw, B.A. Zamora
Teaching Fellows: 8
Research Assistants: 7
Experimental Aides: 5

Field Stations/Laboratories: Steffen Center, Colockum Multiple Use Center

Degree Programs: Forest Management, Range Management

Undergraduate Degree: B.S.
Undergraduate Degree Requirements: 128 semester hours, 95 hours in core curriculum, plus an option summer session professional integration course

Graduate Degree: M.S.
Graduate Degree Requirements: Master: Thesis: 30 hours minimum plus quantitative analysis ability. Non-thesis: 30 hours minimum plus special problems paper and quantitative decision making ability

Graduate and Undergraduate Courses Available for Graduate Credit: Conservation of Renewable Resources (U), Silviculture (U), Principles of Range Ecosystems Management (U), Range Plant Communities (U), Forestation (U), Forest Finance and Valuation (U), Forest and Range Policy and Administration (U), Timber Management (U), Wood Technology (U), Range Habitat Analysis (U), Range Management (U), Watershed Management (U), Wildland Recreation (U), Interpretive Techniques (U), Wildland Recreation Management (U), Land Use Seminars (U), Advanced Topics in Silviculture, Advanced Forest Economics, Multiple Use Management, Advanced Forest Pathology, Advanced Range Ecology, Range Regeneration, Advanced Range Plant Communities, Graduate Seminar, Forest and Range Special Projects and Independent Study, Master's Research and Thesis, Master's Special Problems

DEPARTMENT OF HORTICULTURE
Phone: Chairman: (509) 335-3616 Others: 335-3617

Chairman: W.B. Ackley
Professors: W.B. Ackley, D.R. Bienz, E.W. Kalin, R. Kunkel (Research), F.E. Larsen, M.E. Patterson, R.B. Tukey (Extension), C.G. Woodbridge
Associate Professors: W.I. Ashland, Jr., W.M. Iritani (Research), R.E. Thornton (Extension)
Assistant Professors: L.K. Hiller, L.W. Hudson (Research), T.E. Nelson, K.A. Schekel, K.A. Struckmeyer
Visiting Lecturers: 2
Teaching Fellows: 4
Research Assistants: 3

Field Stations/Laboratories: Irrigated Agriculture Research and Extension Center - Prosser, Tree Fruit Research Center - Wenatchee, Northwestern Washington Research and Extension Unit - Mount Vernon, Southwestern Washington Research Unit - Vancouver, Coastal Washington Research and Extension Unit - Long Beach, Western Washington Research and Extension Center - Puyallup

Degree Program: Horticulture

Undergraduate Degree: B.S.
Undergraduate Degree Requirements: A minimum of 120 semester hours

Graduate Degrees: M.S., Ph.D.
Graduate Degree Requirements: Master: A minimum of 30 semester hours beyond the baccalaureate
Doctoral: A minimum of 72 semester hours beyond the baccalaureate

Graduate and Undergraduate Courses Available for Graduate Credit: Commerical Flower Design and Retail Shop (U), Plant Breeding (U), Physiology of Horticultural Crop Plants (U), Plant Pest Control (U), Post-Harvest Physiology (U), Current Topics in Horticulture (U), Greenhouse Flower Crops (U), Horticultural Research Techniques, Analytical Methods in Horticultural Research, Advanced Pomology, Advanced Horticulture, Realizing Potato Production and Processing Potentials

DEPARTMENT OF PLANT PATHOLOGY
Phone: Chairman: (509) 335-3741 Others: 335-3737

Chairman: J.F. Schafer
Professors: J.L. Allison, G.W. Bruehl, D.W. Burke (Collaboratory, USDA), R.J. Cook (Collaboratory USDA) A.D. Davison (Extension), S.M. Dietz (Collaborator USDA), R. Duran, P.R. Fridlund, C.J. Gould, S.O. Graham, L.A. Hadwiger, W.A. Haglund, J.W. Hendrix, W.G. Hoyman (Collaboratory USDA), J.M. Kraft (Collaboratory USDA), R.F. Line, O.C. Maloy (Extension), G.I. Mink, C.L. Parish (Collaborator USDA), C.F. Pierson (Collaborator USDA), J.D. Rogers, J.F. Schafer, C.G. Shaw, M.J. Silbernagel (Collaborator USDA), C.B. Skotland, P.E. Thomas (Collaborator USDA)
Associate Professors: R.P. Covey, G.D. Easton, R.L. Gabrielson, D.S. Jackson (Collaborator USDA)
Research Assistants: 12

Field Stations/Laboratories: Irrigated Agriculture Research and Extension Center - Prosser, Western Washington Research and Extension Center - Puyallup, Northwestern Research and Extension Unit - Mt. Vernon,

Tree Fruit Research Center - Wenatchee

Degree Programs: Plant Pathology, Mycology

Undergraduate Degree: B.S.
Undergraduate Degree Requirements: English and Speech, 12 hours, Chemistry 15 hours, Mathematics and Biometry 10 hours, Biological Sciences 8 hours, Botany 9 hours, Economics 4 hours, Bacteriology 8 hours, Genetics 3 hours, Zoology 3 hours

Graduate Degrees: M.S., Ph.D.
Graduate Degree Requirements: The degree of Master of Science in Plant Pathology involves a formal courses in Plant Pathology and the supporting areas, a thesis, and a final oral examination. The degree provides preparation for public or commerical plant health service or for participation as a member of a research team. Doctoral: The degree of Doctor of Philosophy in Plant Pathology requires further formal courses in plant pathology and the background areas, a knowledge of a modern foreign language, a preliminary examination to establish the student's qualification for degree candidacy, a research dissertation, and a final examination including the defense of the research.

Graduate and Undergraduate Courses Available for Graduate Credit: Diseases of Plants, General Mycology, Plant Disease Diagnosis and Control, Principles and Practices of Plant Disease Control, Viruses and Virus Diseases of Plants, Nematodes and Nematode Diseases of Plants, Seminar, Basidiomycetes, Ascomycetes and Fungi Imperfecti, Lower Fungi, Field Plant Pathology, Methods and Techniques, Advanced Forest Pathology, Physiology and Genetics of Parasitism, Physiology and Genetics of Parasitism Laboratory, Field Mycology Genetics of Fungi

College of Veterinary Medicine

Dean: L.K. Bustad

DEPARTMENT OF VETERINARY ANATOMY
Phone: Chairman: (509) 335-3266

Chairman: Not Stated
Professors: J.A. McCurdy, R.P. Worthman
Associate Professors: D.J. Fuxton, R.W. Compton, B.C. Cummings
Assistant Professor: C.S. Zamord

Degree Program: Veterinary Science

Undergraduate Degree: B.S.
Undergraduate Degree Requirements: 2 years pre-veterinary curriculum, 2 years in professional curriculum

Graduate Degrees: M.S., Ph.D.
Graduate Degree Requirements: Not Stated

Graduate and Undergraduate Courses Available for Graduate Credit: Advanced Anatomy, Neuroanatomy

DEPARTMENT OF VETERINARY MICROBIOLOGY
Phone: Chairman: (509) 335-3923

Chairman: T. Moll (Acting)
Professors: S.G. Kenzy, L. Ringen, T. Moll
Associate Professors: D. Burger, B.R. Cho, W.C. Davis, C.S. McCain
Assistant Professor: E.D. Erickson
Visiting Lecturers: 1
Teaching Fellow: 1
Research Assistants: 1

Degree Program: Veterinary Microbiology

Graduate Degrees: M.S., Ph.D.
Graduate Degree Requirements: Masters: 30 hours, thesis Doctoral: 72 hours, thesis

Graduate and Undergraduate Courses Available for Graduate Credit: Not Stated

WESTERN WASHINGTON STATE COLLEGE

Bellingham, Washington 98225 Phone: (206) 676-3000
Dean of Graduate Studies: J.A. Ross
Dean of College: R.L. Monahan

DEPARTMENT OF BIOLOGY
Phone: Chairman: (206) 676-3628

Chairperson: C.M. Senger
Professors: A.C. Broad, J.S. Martin, J.S. Parakh, J.R.P. Ross, C.M. Senger, I.L. Slesnick, R.J. Taylor
Associate Professors: A.L. Balzer, H.A. Brown, M.A. Dube, J.F. Erickson, R.W. Fonda, H. Kohn, G.F. Kraft, A.L. Neckelson, M.M. Riffey, D.E. Schneider, D.J. Schwemmin, D.C. Williams
Teaching Fellows: 6

Field Stations/Laboratories: Shannon Point - Anacortes

Degree Program: Biology

Undergraduate Degree: B.A., B.S.
Undergraduate Degree Requirements: Bachelor of Arts - 50 credits plus supporting courses Bachelor of Science - 110 credits

Graduate Degree: M.S.
Graduate Degree Requirements: Master: 45 hours course work, thesis, language requirement

Graduate and Undergraduate Courses Available for Graduate Credit: Special Projects in Biology, Advanced Topics in Ecology, Biological Instrumentation, Aquatic Entomology, Topics in Developmental and Comparative Morphology, Advanced Topics in Genetics, Advanced Topics in Physiology, Protein Structure, Function and Evolution, Teaching Practicum, Seminar in Biology, Thesis Research, Enzymology, Laboratory, Radiation Biology, Advanced Topics in Biosystematics, Microbial Ecology (U), Limnology (U), Physiological Ecology (U), Plant Communities (U), Field Ecology (U), General Oceanography (U), Current Environment (U), Molecular Ecology (U), Entomology (U), Entomology Laboratory (U), Parasitology (U), Phylogeny of Flowering Plants (U), Montane Biology (U), Mycology (U), Algae (U), Bryophytes (U), Marine Invertebrate Zoology (U), Ornithology (U), Mammalogy (U), Physiological Genetics (U), Structure and Development of Vascular Plants (U), Cellular Physiology (U), Physiological Responses of Plants to Environment (U), Principles of Organic Evolution (U), Teaching Biological Science (U), Invertebrate Zoology (U), Marine Invertebrate Zoology (U), Development of Modern Ideas in Biology (U)

WHITMAN COLLEGE

Walla Walla, Washington 99362 Phone: (509) 529-5100
Dean of College: R.A. Skotheim

DEPARTMENT OF BIOLOGY
Phone: Chairman: Ext. 219

Chairman: E.F. Anderson
Associate Professors: E.F. Anderson, R.C. Brown
Assistant Professors: E.W. Fleck, W.E. Nodnan

Degree Program: Biology

Undergraduate Degree: B.S.
Undergraduate Degree Requirements: Zoology, Botany, or 2 biology and Botany, genetics, senior project or honors thesis, 20 additional credits in advanced biology courses, 2 years of Physical Science, including at least one year of Chemistry with Organic Chemistry

and Physics highly recommended, one year of foreign language at the college level or a departmentally-approved equivalent. It is strongly recommended that each student take at least one advanced course in most of the following areas within the department: 1) Physiology/Cell Biology, 2) Invertebrate Zoology, 3) Vertebrate Zoology, 4) Ecology/Natural History 5) Microbiology/Botany

WEST VIRGINIA

ALDERSON-BROADDUS COLLEGE
Philippi, West Virginia 26416 Phone: (304) 457-1700

DEPARTMENT OF BIOLOGY
Phone: Coordinator: (304) 457-1700, Ext. 252

Coordinator: B.L. Redd
Associate Professors: B.L. Redd, I.C. Aurelie
Assistant Professor: T.A. Redd

Field Stations/Laboratories: We are a member of the Marine Science Consortium, which has facilities in Wallops Island, Virginia and Lewes, Deleware

Degree Program: Biology

Undergraduate Degree: B.S.
Undergraduate Degree Requirements: Four courses in Chemistry, Three courses in Mathematics, two courses in Physics, Seven courses in Biology, and a basic command of a foreign language

BETHANY COLLEGE
Bethany, West Virginia 26032 Phone: (304) 829-7000
Dean of College: J. Callebs

DEPARTMENT OF BIOLOGY
Phone: Chairman: (304) 829-7642 Others: 829-7641

Chairman: G.E. Larson
Associate Professors: G.E. Larson, A.R. Buckelew
Assistant Professors: J.J.Sawtell, C.B. Carpenter
Instructor: V.S. Larson
Research Assistant: 1

Degree Program: Biology

Undergraduate Degree: B.S.
Undergraduate Degree Requirements: 32 semester hours including a senior Project, 12 semester hours in Chemistry at least 6 of which are Organic Chemistry, 6 semester hours in Physics, 1 semester of Calculus is strongly recommended

BLUEFIELD STATE COLLEGE
Bluefield, West Virginia 24701 Phone: (304) 325-7102
Dean of College: D.A. Puzzouli

DIVISION OF NATURAL SCIENCE
Phone: Chairman: Ext. 284

Chairman: D.A. Schoenefeld
Associate Professors: E.J. Bauer, Jr., S. Dodrill

Degree Program: Biology

Undergraduate Degree: B.S.
Undergraduate Degree Requirements: Not Stated

CONCORD COLLEGE
Athens, West Virginia 24712 Phone: (304) 384-3115
Dean of College: M.C. Edge

DEPARTMENT OF BIOLOGY
Phone: Chairman: Ext. 327

Chairman: K.D. Fezer
Professors: J.L. Blatt, K.D. Fezer
Associate Professors: L.E. Bayless, C.J. Chapman, D. S. Evans

Degree Programs: Biology, Biological Education, Medical Technology

Undergraduate Degrees: B.S., B.S.M.T.,B.S.Ed.
Undergraduate Degree Requirements: B.S. major in Biology: 28 hours core courses in Biology, 1 seminar, 6 research, 12 Chemistry, 8 physics, 8 Mathematics, 7 Science electives or minor field, general studies B.S.Ed.: 28 hours core courses in Biology, 1 seminar, 6 research, 8 Chemistry, 8 Physics, 4 Geology, 4 Astronomy, 20 Education, general studies, B.S.M.T. 16 hours Biology, 19 Chemistry, 8 Physics, 5 Mathematics, 3 Psychology, electives, general studies, 1 year at Appalachian Regional Hospital

DAVIS AND ELKINS COLLEGE
Elkins, West Virginia 26241 Phone: (304) 636-1900

DEPARTMENT OF BIOLOGY
Phone: Chairman: Ext. 43

Chairman: W.L. Tolstead
Professors: L. Elrod, A. Vaszquez

Degree Program: Biology

Undergraduate Degree: B.S.
Undergraduate Degree Requirements: 32 hours for major including 4 hours Plant Morphology, 4 hours Invertebrates, 8 hours General Biology

FAIRMONT STATE COLLEGE
Fairmont, West Virginia 26554 Phone: (304) 363-4000
Academic Dean: W.A. Boram

DEPARTMENT OF BIOLOGY
Phone: Chairman: Ext. 320 Others: Ext. 319

Chairman: W.D. Ruoff
Professors: R.E. Amos, Jr., W.D. Creasy, W.H. Pritchett
Associate Professors: R.K. Shan, C.W. Weems
Assistant Professors: A.F. Michna

Degree Program: Biology, Biology (Education)

Undergraduate Degrees: A.B., B.S.
Undergraduate Degree Requirements: 128 semester hours

GLENVILLE STATE COLLEGE
Glenville, West Virginia 26351 Phone: (304) 462-7361
Dean of College: C. Maze

DIVISION OF SCIENCE AND MATHEMATICS
Phone: Chairman: Ext. 272

Division Chairman: J.A. Chisler
Professor: J.A. Chisler
Associate Professors: R. Deal, B.L. Law
Assistant Professor: F. Jenio

Degree Programs: Biology, Biological Education

Undergraduate Degrees: B.Sc., A.B. (Teacher Prep.)
Undergraduate Degree Requirements: B.Sc. - 42 semester hours in Biology, A.B. - 24 semester hours in

Biology, 22 semester hours in Physical Science, 6 semester hours in Mathematics, 44 general studies

MARSHALL COLLEGE
(No reply received)

MORRIS HARVEY COLLEGE
Charleston, West Virginia 25304 Phone: (304) 346-9471
Dean of College: J.W. Rowley

BIOLOGY DEPARTMENT
Phone: Chairman: Ext. 235

Head: R.G. Nunley
Professors: R.G. Nunley, G.E. Smith
Associate Professors: C.T. Meadors, M.D. Meadors
Assistant Professor: R.B. Brandon
Instructor: C.B. Moss

Degree Program: Biology

Undergraduate Degree: B.S.
Undergraduate Degree Requirements: Biology - 31 credits hours in Biological Science (General Biology, 8, Botany 4, Zoology 4), 8 credit hours in General Chemistry, 7 credit hours in Mathematics, 15 electives depending on student's future plans

SALEM COLLEGE
Salem, West Virginia 26456 Phone: (304) 782-5011
Dean of College: W.H. England

DEPARTMENT OF BIOLOGY
Phone: Chairman: (304) 782-5257 Others: 782- 5011

Chairman: T.K. Pauley
Professor: J. McCoy
Associate Professor: T.K. Pauley
Instructor: E.L. Jarroll

Degree Programs: Biology - Medical Technology

Undergraduate Degree: B.S.
Undergraduate Degree Requirements: 32 hours in Biology, 16 of which in upper division courses - general and organic Chemistry - Algebra - Trigonometry

SHEPHERD COLLEGE
Shepherdstown, West Virginia 25443 Phone: (304) 876-2511
Dean of College: H. Schlossberg

BIOLOGY DEPARTMENT
Phone: Chairman: Ext. 227

Chairman: R.E. Harris
Professors: C.F. Bell, F.W. Sturges
Associate Professor: R.L. Latterell
Assistant Professors: J.C. Landolt, P.M. Saab, C.H. Woodward

Degree Programs: Biology, Biology Education

Undergraduate Degrees: B.S., B.S.Ed.
Undergraduate Degree Requirements: Core Courses: Plants as Organisms, Animals as Organisms, Cell Biology, Genetics, Developmental Biology, Aquatic Ecology, Directed Research, General Chemistry, General Physics, Mathematics, Seminar

WEST LIBERTY STATE COLLEGE
West Liberty, West Virginia 26074 Phone: (304) 336-5000
Dean of College: C.R. Whiting

DEPARTMENT OF BIOLOGY
Phone: Chairman: (304) 336-8020 Others: 336-5000

Chairman: H.A. Cook
Professors: R. Mitra , F.R. Swan
Associate Professors: S.L. Bressler, H.A. Cook, H.G. Sherrill
Assistant Professors: C.D. Cornell, W.M.Gordon, Jr.

Degree Program: Biology

Undergraduate Degree: B.S.
Undergraduate Degree Requirements: Not Stated

WEST VIRGINIA INSTITUTE OF TECHNOLOGY
Montgomery, West Virginia 25136 Phone: (304) 422-3071
Dean of College: J. Robertson

DEPARTMENT OF BIOLOGY
Phone: Chairman: (304) 422-3236

Coordinator: J.M. Parks
Professor: V.S. Phillips
Associate Professor: J.M. Parks
Assistant Professors: R.A. Gaertner, L. Esham, L. Abrash

Degree Program: Biology

Undergraduate Degree: B.S.
Undergraduate Degree Requirements: 32 hours in Biology, 8 hours in Chemistry, 22 hours in Education

WEST VIRGINIA STATE COLLEGE
Institute, West Virginia 25112 Phone: (304) 766-3143
Dean of College: D.E. Hoffman

BIOLOGY DEPARTMENT
Phone: (304) 766-3103

Chairperson: B.J. Oden
Associate Professors: M. Hawkins, Jr., T. Hutto B.F. Garrett, B.J. Oden, D.L.Edens
Assistant Professor: E. Franklin

Degree Program: Biology

Undergraduate Degree: B.S.
Undergraduate Degree Requirements: Principles of Biology, General Botany, General Zoology, General Ecology, General Physiology, Genetics, Senior Seminar, 12 hours Electives in either Botany or Zoology

WEST VIRGINIA UNIVERSITY
Morgantown, West Virginia 26506 Phone: (304) 293-0111
Dean of Graduate Studies: S. Wearden

College of Arts and Science
Dean: J.C. Wright

DEPARTMENT OF BIOLOGY
Phone: Chairman: (304) 293-5394 Others: 293-5201

Chairman: E.C. Keller, Jr.
Centennial Professor: M.W. Schein
Professors: C.H. Baer, H.D. Bennett, W.N. Bradshaw, R.B. Clarkson, J.F. Clovis, W.E. Collins, E.L. Core (Emeritus), E.C. Keller, Jr., C. Norman, R.P. Sutter
Associate Professors: L. Abrahamson, D.F. Blaydes, M.O. Coover, J.J. DeCosta, D.C. Dunning, R.H. Frist, R.L. Guthrie, W.H. Hertig, W.H. Hertig, H.W. Hurlbutt, J.A. Marshall, E.C. Montiegel, L.A. Williams
Assistant Professors: R.D. Allen, E.A. Bartholomew, A. Benson, R.L. Birch
Visiting Lecturers: 2
Research Assistants: 6
Teaching Assistants: 39

Part-Time Instructors: 6

Field Stations/Laboratories: Terra Alta Biological Station - Terra Alta, Member of Marine Science Consortium - Wallops Island, and Lewes, Delaware

Degree Programs: Behavioral Biology, Cellular and Molecular Biology, Developmental Biology and Physiology, Environmental Biology, Evolutionary and Systematic Biology

Undergraduate Degree: A.B. (Biology)
Undergraduate Degree Requirements: General Biology - 8 hours, Animals as Organisms 4 hours, Plants as Organisms 4 hours, Population Biology 4 hours, Cellular and Molecular Biology 4 hours

Graduate Degrees: M.A., M.S., Ph.D. (Biology)
Graduate Degree Requirements: Master: M.S. - 30 hours of graduate credit including 6 hours research and a thesis, M.A. - 30 hours of graduate credit including a special problem report Doctoral: Requirements are determined by the student's Advisory Committee

Graduate and Undergraduate Courses Available for Graduate Credit: Biology Seminar (U), Behavioral Ecology (U), Primary Production in Aquatic Environments (U), Plant Communities (U), Fisheries Science (U), Dynamics of Ecosystems (U), Biosystematics (U), Plant Morphology (U), Taxonomy of Vascular Plants (U), Fresh-Water Algae (U), Advanced Plant Systematics (U), Aquatic Seed Plants (U), Field Studies of Invertebrates (U), Field Studies of Vertebrates (U), Vascular Cryptogams (U), Developmental Biology (U), Advanced Plant Physiology (U), Plant Developmenta (U), Advanced Plant Ecology (U), Cytotaxonomy, Mammalogy, Developmental Genetics

College of Agriculture and Forestry

Dean: H.Evans

DIVISION OF ANIMAL SCIENCE
(No reply received)

DIVISION OF FORESTRY
Phone: Chairman: (304) 293-2941

Chairman: D.E. White
Professors: S.M. Brock, K.L. Carvell, F.C. Cech, J.R. Hamilton, C.B. Koch, R. L. Smith, E.H. Tryon, D.E. White, H.V. Wiant
Associate Professors: J.M. Hutchinson, N.D. Jackson (Assistant to Chairman), L.O. Keresztesy (Research), W.E. Kidd, Jr., E.D. Michael
Assistant Professors: G.H. Breiding, J.D. Gill, W.N. Grafton, C.S. Hall, E.P. Jenkins, W.R. Maxey, D.E. Samuel, H.R. Sanderson, B.A. Schick, H.P. Steinhagen, G.R. Trimble, B.W. Twight, W.L. Wylie, G.W. Zinn
Instructors: W.N. Healy, W.N. Santonas, S.C. Smith, A.C. Tomkowski, J.N. Yeager
Research Assistants: 14
Teaching Assistants: 8
Research Technicians: 2

Field Stations/Laboratories: Terra Alta Biological Station - Terra Alta, Camp Arthur Wood (Geology) - Alvon

Degree Programs: Forest Resources Management, Wood Industry, Wildlife Resources, Recreation and Parks, Wood Industry

Undergraduate Degree: B.S., B.S.F.
Undergraduate Degree Requirements: B.S.F. - Forest Resources Management and Wood Industry - 138 hours of approved study with a minimum grade point average 2.00. B.S. - Recreation - 134 hours of approved study with a minimum grade point average of 2.0 B.S. - Wildlife Resources - 136 hours of approved study with a minimum grade point average of 2.0

Graduate Degrees: M.S., M.S.F.
Graduate Degree Requirements: Master: M.S.F. - Must complete 30 hours of approved study, 6 of which shall constitute a thesis with a minimum grade point average of 3.00 M.S. - with the exception of those majoring in Recreation, must complete 30 hours of approved study with a minimum grade point average of 3.00. Students majoring in Recreation have the option of earning the degree on the basis of 30 hours with a thesis or 36 hours without a thesis

Graduate and Undergraduate Courses Available for Graduate Credit: Forestry - Forest WaterQuality, Forest Hydrology, Forest Policy and Administration, Remote Sensing of Environment, Principles of Industrial Forestry, Microclimatology Forest Management - Silvicultural Systems, Regional Silviculture, Principles of Artificial Forestation, Forest Genetics and Tree Improvement, Forest Mensuration, Principles of Forestry Economics, Forest Finance, Forest Management, Integrated Forest Resources Management, Advanced Principles of Forestry Economics, Environmental Relationships in Hardwood Forests, Silvicultural Practices for Hardwood Forest Types, Advanced Forest Regulation Wood Science - Wood Finishing, Theory and Practice of Wood Adhesion, Statistical Quality Control, Light-Frame Wood Construction, Wood Moisture Relationships, Forest Products Protection, Wood Microstructure, Seminar in Wood Utilization Wildlife Management - Wildlife Ecology, Field Ornithology, Forest Zoology, Wildlife Techniques, Principles of Wildlife Management, Wildlife Population Ecology, Wildlife Seminar, Ecology and Management of Upland Wildlife - Recreation Internship, Wildland Recreation Administration, Administration of Urban and Regional Services, Recreation Leadership, Program Planning, Functional Planning of Recreation and Park Facilities, Administration of Camping Services, Professional Synthesis, Philosophy of Recreation, Outdoor Recreation in Our Modern Society, Outdoor Education and School Camping, Practicum in Recreation, Leisure and Recreation, Human Interest Areas in Recreation Planning, Community Recreation

DIVISION OF PLANT SCIENCES
Phone: Chairman: (304) 293-4817 Others: 293-6023

Division Director: M.E. Gallegly
Professors: R.E. Adams, H.L. Barnett, J.G. Barrat, N.M. Baughman, E.S. Elliott, M.E. Gallegly, N.C. Hardin, L.M. Ingle, P.G. Moe, J. Nath, O.M. Neal, O.E. Schubert, E.G. Scott, R.M. Smith, V. Ulrich, W. van Eck
Associate Professors: B.C. Bearce, J.L. Brooks, L. Butler, A.P. Dye, E. Jencks, W.J. Kaczmarczyk, R.F. Keefer, D.O. Quinn, C.B. Sperow, D.A. Stelzig, B.B. Baker, J.A. Balaskó, J.F. Baniecki
Assistant Professors: J.J. Albert, O.L. Bennett, C.E. Hickman, W.L. MacDonald, R.N. Singh, L.P. Stevens, V.J. Valli, R.J. Young, R.S. Young
Instructors: F.L. Alt, S.H. Blizzard
Research Assistants: 5

Degree Programs: Agricultural Biochemistry, Agronomy, Bacteriology, Crop Science, Entomology, Genetics, Horticulture, Microbiology, Nematology, Ornamental Horticulture, Plant Pathology, Plant Science, Soil Science, Vegetable Crops, Environmental Production

Undergraduate Degree: B.S.
Undergraduate Degree Requirements: B.S. in Agriculture One major "Plant and Soil Sciences" with options checked on p. 4 . Requires 136 semester hours total with 45 hours of College of Agriculture and Forest courses as a minimum, plus 12 hours of University core in the Humanities and 12 hours in the Social Sciences, and 24 hours of Natural Sciences outside

the College

Graduate Degrees: M.S., Ph.D.
Graduate Degree Requirements: Master: 30 hours credit of graduate courses plus a thesis (6 hours credit allow for thesis toward 30 hours.) Committee established to specify other requirements - 1 year residency Doctoral: No set number of hours, Course requirements certified by advisory committee. Dissertation required. 2 years residency.

Graduate and Undergraduate Courses Available for Graduate Credit: Plant Sciences - Recognition and Diagnosis of Plant Disorders, Principles and Methods of Plant Pest Control, Crop Science - Turfgrass Management, Weed Control, Grain and Special Crops, Pasture and Forage Crops Soil Science - Soil Fertiligy, Soil Conservation and Management, Soil Physics, Geotechnic, Soil Genesis and Classification, Identification of Clay Minerals in Soil, Advanced Soil Fertility, Soil Chemistry, Chemistry of Soil Organic Matter, Agronomy, Bacteriology - Soil Microbiology, Foos Microbiology, Sanitary Bacteriology Entomology - Principles of Entomology Genetics - Crop Breeding, Basic Concepts of Modern Genetics, Human Genetics, Population Genetics, Cytogenetics, Advanced Biochemical Genetics, Seminar, Genetic Mechanisms of Evolution Horticulture - Plant Propagation, Landscape Design, Small-Fruits, Physiology of Vegetables, Handling and Storage of Horticultural Crops, Greenhouse Management, Post-Harvest Physiology Plant Pathology - General Plant Pathology, Diseases of Economic Plants, Principles of Plant Pathology, Mycology, Nematology, Physiology of Plant Diseases, Physiology of the Fungi, Taxonomy of the Fungi,

School of Medicine

Dean: J. Jones

DEPARTMENT OF ANATOMY

Phone: Chairman: (304) 293-6322 Others: 293-2212

Chairman: D.H. Enlow
Professors: D.H. Enlow, D.S. Jones, D.L. Kimmel (Emeritus), R.E. McCafferty, R.W. Reyer, T.W. Williams (Emeritus)
Associate Professors: W.A. Beresford, J.L. Culberson, M.H. Friedman, D.E. Haines, C.A. Pinkstaff
Assistant Professors: S.W. Carmichael, R.C. Frederickson, R.A. Hilloowala, D.O. Overman, R.S. Pope
Teaching Fellows: 6

Degree Program: Anatomy

Graduate Degrees: M.S., Ph.D.
Graduate Degree Requirements: Master: Gross Anatomy 10-11 hours, Microscopic Anatomy 6 hours, Introduction to Research 2 hours, Seminar 2 hours, Electives 3-4 hours, Thesis 6 hours, Total hours, 30 hours Doctoral: Gross Anatomy 10-11 hours, Microscopic Anatomy 6 hours, Introduction to Research 2 hours, Neurobiology 6 hours, Biochemistry 7 hours, Physiology 6 hours, Advanced Courses (electives) 10 hours, Seminar 6 hours, Total House 53 hours, Reading knowledge of one foreign language or approved substitute, dissertation

Graduate and Undergraduate Courses Available for Graduate Credit: Not Stated

DEPARTMENT OF BIOCHEMISTRY

Phone: Chairman: (304) 293-4649

Chairman: R.F. Krause
Professors: W.J. Canady, R. Koppelman, R.F. Krause, F.J. Lotspeich, G.W. Rafter, H. Resnick, G.H. Wirtz
Associate Professors: S. Katz, G.P. Tryfiates
Assistant Professors: J.B. Blair, J.S. Ellingson, C.L. Harris, D.K. Ponton
Instructor: S.K. Core
Teaching Fellow: 3
Visiting Lecturer: 1
Research Assistant: 7

Degree Program: Biochemistry

Graduate Degrees: M.S., Ph.D.
Graduate Degree Requirements: Master: 30 hours of course work and a thesis Doctoral: Requirements set-up by the students Advisory Committee.

Graduate and Undergraduate Courses Available for Graduate Credit: General Biochemistry, Advanced Study, Amino Acids, Peptides and Proteins, Enzymology, Immunochemistry, Nucleic Acids and Protein Synthesis, Lipids, Enzyme Kinetics, Seminar, Special Topics

DEPARTMENT OF MICROBIOLOGY

Phone: Chairman: (304) 293-2649

Chairman: I.S. Snyder
Professors: R. Burrell, S.J. Deal, J.E. Hall, J.M. Slack, I.S. Snyder, H.G. Voelz
Associate Professors: V.F. Gerencser, B.E. Kirk, R.S. Pore, R.W. Veltri
Assistant Professors: P.C. Major, H.F. Mengoli, D.B. Yelton
Instructors: H.M. Pavlech

Degree Program: Medical Microbiology

Graduate Degrees: M.S., Ph.D.
Graduate Degree Requirements: Master: 30 hours graduate work, thesis Doctoral: One foreign language (reading ability), dissertation

Graduate and Undergraduate Courses Available for Graduate Credit: Basic Microbiology, Pathogenic Microbiology, Diagnostic or Determinative Microbiology, Comparative Cytology, Electron Microscopy, Bacterial Physiology, Microbial Genetics, Immunology, Virology, Medical Mycology, Parasitology

DEPARTMENT OF PHYSIOLOGY AND BIOPHYSICS

Phone: Chairman: (304) 293-3814

Chairman: M.F. Wilson
Professors: L. Gutmann, R.J. Marshall, W.H. Moran, M.F. Wilson
Associate Professors: G.N. Franz, W.E. Gladfelter, P. Lee, T.W. McIntyre, R.J. Millecchia, K.C. Weber
Assistant Professors: P.B. Brown, H.D. Colby, P.R. Miles, L. Sherwood
Instructors: W.M. Caldwell, J. Hankinson, M.I. Morgan
Teaching Fellows: 6
Visiting Lecturers: 1
Research Assistants: 6

Degree Programs: Biophysics, Physiology

Graduate Degrees: M.S., Ph.D.
Graduate Degree Requirements: Master: 2 years program of didactic advanced courses (30 credit hours) plus thesis Doctoral: Normally a 4 year program with specified advanced courses, electives, qualifying examination and research dissertation

Graduate and Undergraduate Courses Available for Graduate Credit: Elementary Physiology (U), Introduction to Biophysics (U), Experimental Design, (U), Fundamentals of Physiology, Medical Physiology, Neurophysiology, Biophysical Analysis, Special Topics, Physiological Methods, Advanced Physiology, Cellular Biophysics, Systems Biophysics

WEST VIRGINIA WESLEYAN COLLEGE

Buckhannan, West Virginia 26201 Phone: (304) 473-8064
Dean of College: R. Cunningham (Acting)

DEPARTMENT OF BIOLOGY
Phone: Chairman: (304) 473-8064

Chairman: G.B. Rossbach
Professor: G.B. Rossbach
Associate Professors: C. Colson, W.P. Taylor
Assistant Professors: J. Glencoe, E. Weimer, A. Varkey

Degree Program: Biology

Undergraduate Degrees: A.B., B.S.
Undergraduate Degree Requirements: Not Stated

WHEELING COLLEGE
(No reply received)

WISCONSIN

ALVERNO COLLEGE

3401 S. 39th Street Phone: (414) 671-5400
Milwaukee, Wisconsin 53215

BIOLOGY DEPARTMENT
Phone: Chairman: Ext. 241

Chairman: L.C. Truchan
Associate Professor: M. Nemmor
Assistant Professors: R. Lewis, A. Been, S. Isakson
Instructors: A. Woelfel

Field Stations/Laboratories: Air Monitoring Site for State Department under Natural Resource Department

Degree Programs: Biology, Environmental Studies, Medical Technology

Undergraduate Degree: B.S.
Undergraduate Degree Requirements: Comptencies for Areas of Concentration specific in Biology differs for Environmental Studies major.

BELOIT COLLEGE

Beloit, Wisconsin 53511 Phone: (608) 365-3391
Dean of College: D.K. Adams

DEPARTMENT OF BIOLOGY
Phone: Chairman: Ext. 388

Chairman: J.E. Lutz
Professors: B.K. Kunny, E.J. Souter
Associate Professors: R.D. Newsome, L. Resseguie, D. R. Welch
Assistant Professors: D. Duff

Field Stations/Laboratories: Smith Limnology Laboratory - Rock River, Beloit , Wisconsin

Degree Programs: Biology, Psychobiology

Undergraduate Degrees: B.A., B.S.
Undergraduate Degree Requirements: 6.5 to 8.5 Biology courses, 5 to 7 Supporting Science Courses, 1 Mathematics course

CARDINAL STRITCH COLLEGE

6801 N. Yates Road Phone: (414) 352-5400
Milwaukee, Wisconsin 53217
Dean of Graduate Studies: Sr. C. Kleibhan (Acting)
Dean of College: Sr. B. M. Weithans

BIOLOGY DEPARTMENT
Phone: Chairman: Ext. 73

Chairman: Sr. M.A. Palasek
Associate Professor: Sr. M.A. Polasek
Assistant Professor: Sr. R. Costanzo

Degree Program: Biology

Undergraduate Degree: B.A.
Undergraduate Degree Requirements: 36 hours Biology (General I, II, Microbiology, Genetics, Cell Physiology, General Physiology, Developmental Biology, Histology, Seminar I, II), 8 hours Chemistry, 4 courses Mathematics

CARROLL COLLEGE

100 N. East Avenue Phone: (414) 547-1211
Waukesha, Wisconsin 53186
Dean of College: M.N. Spencer

DEPARTMENT OF BIOLOGY
Phone: Chairman: Ext. 282 Others: 547-1211

Chairman: R.J. Christoph
Professors: R.J. Christoph, T.C. Michaud
Associate Professors: J.V. Batha, B.A. Macintyre
Undergraduate Assistants: 7
Field Station Managers: 2

Field Stations/Laboratories: Howard T. Greene Conservatory and Scientific Study Area (40 acres), Genesee Depot, Wisconsin (9 miles west of campus), Genesee Creek Watershed Research Station - Genesee Depot, Carroll College Arboretum - Waukesha

Degree Programs: Biology, Medical Technology, Environmental Studies

Undergraduate Degree: B.S.
Undergraduate Degree Requirements: 10 courses (40 semester hours) in Biology, distributed in 5 areas of Biology, 2 courses Chemistry (8 semester hours) minimum, 1 course Mathematics (4 semester hours), secondary teachers of Biology take 6 courses in Education, Medical Technologists take fourth year at hospital affiliate, 3 course concentration in area different from major, 4 course distribution in two additional areas, 1 course religion, 2 general education courses, 32 courses total for graduation, UGRE in Biology Required

CARTHAGE COLLEGE

Kenosha, Wisconsin 53140 Phone: (414) 551-8500
Dean of College: E. Spangler

DEPARTMENT OF BIOLOGY
Phone: Chairman: Ext. 308 Others: (414) 551-8500

Chairman: R.M. Tiefel
Professors: H. Ogren, W. Suter, R.M. Tiefel
Associate Professor: E. Crump
Part-time Instructor: 1

Degree Program: Biology

Undergraduate Degree: A.B.
Undergraduate Degree Requirements: 44 credits Biology courses, 12 credits Physical Sciences courses, 16 credits Fundamental courses, 24 credits Electives for Diversity, 30 credits Free Electives

EDGEWOOD COLLEGE

855 Woodrow Street Phone: (608) 257-4861
Madison, Wisconsin 53711
Dean of College: Sr. M. Harty

BIOLOGY DEPARTMENT
Phone: Chairman: Ext. 234

Head: J. Feldballe
Instructors: 3

Degree Program: Biology

Undergraduate Degree: B.S.
Undergraduate Degree Requirements: 32 credits includ-

ing Genetics, Developmental Biology, Physiology and General Introductory course

HOLY FAMILY COLLEGE
(see Silverlake College)

LAWRENCE UNIVERSITY
Appleton, Wisconsin 54911 Phone: (414) 739-4846
Vice President: T. Headrick

DEPARTMENT OF BIOLOGY
Phone: Chairman: Ext. 472

Chairman: M.J. LaMarca
Alice J. Hulst - Professor of Life Sciences: S. Richman
Associate Professor: M.J. LaMarca
Assistant Professors: N.C. Maravolo, W. Perreault, G.A. Shibley, D.J. West, A.M. Young

Field Stations/Laboratories: Tropical Ecology Program - Costa Rica

Degree Program: Biology

Undergraduate Degree: B.A.
Undergraduate Degree Requirements: One year (3 terms) of Introductory Biology, including Cell Biology, General Zoology, General Botany, Introductory Chemistry (2 terms), 8 additional advanced courses in Biology, participation in weekly seminar series in senior year

MARIAN COLLEGE
Fond du Lac, Wisconsin 54935 Phone: (414) 921-3900
Dean of College: J.K. Kramer

DEPARTMENT OF NATURAL SCIENCES AND MATHEMATICS
Phone: Chairman: Ext. 233

Chairman: I. Palen
Professors: I. Palen, M.W. Saller
Assistant Professors: B. Prall, J. Bechner, J. Norbach, M. Conrey
Instructors: H.A. Bintz, C. Schwderer

Degree Programs: Biology, Environmental Studies

Undergraduate Degrees: B.S., B.A.
Undergraduate Degree Requirements: 32 hours in major, 22 hours in minor

MARQUETTE UNIVERSITY
615 North 11th Street Phone: (414) 224-7000
Milwaukee, Wisconsin 53233
Dean of Graduate School: J. Jache

College of Liberal Arts
Dean: Rev. R.G. Gassert, S.J.

DEPARTMENT OF BIOLOGY
Phone: Chairman: (414) 224-7355

Chairman: P. Abramoff
Professors: P. Abramoff, A. Krishnakumaran, E.S. Mc Donough (Emeritus), W.F. Millington, E.M. Rasch
Associate Professors: J.M. Barrett, W.W. Fredricks, S. Hennen, J.D. Irr, B.E. Piacsek, O.H. Smith, R.G. Thomson
Assistant Professors: J.B. Courtright, V.L. deVlaming, D.R. Hafemann, H.M. Miles, K.J. Tauvydas, B.R. Unsworth, E.L. Vigil
Teaching Fellows: 21
Research Assitants: 10

Degree Program: Biology

Undergraduate Degree: B.S.
Undergraduate Degree Requirements: 35 semester hours Biology, 16 semester hours Chemistry, 8 semester hours Physics and 6-8 semester hours Mathematics. Required courses: General Biology, Laboratory Investigations in Biology, Cell Biology, Genetics, Environmental Biology plus any two laboratory courses.

Graduate Degrees: M.Sc., Ph.D.
Graduate Degree Requirements: Master: Plan A: 24 hours Biology, 6 hours thesis, no language, teaching experience. Plan B: 30 hours Biology including courses in Molecular, Cell, Developmental and Regulatory Biology, no language, teaching experience Doctoral: 60 hours beyond baccalaureate, 12 hours doctoral dissertation, seminar participation, teaching experience

Graduate and Undergraduate Courses Available for Graduate Credit: Genetics (U), Cell Biology (U), Experimental Cell Biology (U), Biology and Evolution of Plants (U), Environmental Biology (U), Plant Physiology (U), Developmental Biology (U), Experimental Embryology (U), Vertebrate Physiology (U), Human Embryology and Biochemistry (U), Experimental Physiology (U), Microbiology (U), Immunobiology (U), Biology of Invertebrates (U), Experimental Invertebrate Physiology (U), Laboratory Studies in Plant Structure and Development (U), Molecular Genetics, Nucleic Acids, Molecular Control Mechanisms, Experimental Genetics, Advanced Cytology, Cell Fine Structure, Plant Growth and Development, Experimental Cell Biology, Laboratory Studies in Plant Morphogenesis, Molecular Biology, Cellular Functions and Interactions During Development, Methods in Research in Developmental Biology, Advanced Immunology, Hormonal Control of Growth and Development, Developmental Genetics, Biochemistry of Development, Introduction to Physiological Investigation, Physiological Ecology, Endocrinology, Neurobiology, Comparative Physiology

THE MEDICAL COLLEGE OF WISCONSIN
561 North 15th Street Phone: (414) 272-5450
Milwaukee, Wisconsin 53233
Dean of Graduate Studies: D. Grieff
Dean of Medical College: D. Kerrigan

DEPARTMENT OF ANATOMY
Phone: Chairman: Ext. 301

Chairman: F.D. Anderson
Professors: F.D. Anderson, H.B. Benjamin (Clinical), W.D. Gardner, M. Wagner (Clinical), W. Zeit (Emeritus)
Associate Professors: L.S. Cunningham, R.L. Curtis, S. Kaplan, L. Sether, K.A. Seigesmund
Assistant Professors: J.S. Haft, E. Kindley, S.Y. Long, W.P. Maher (Clinical)

Degree Program: Anatomy

Graduate Degrees: M.S., Ph.D.
Graduate Degree Requirements: Master: Approximately 2 years of graduate work including coursework and original laboratory research. A thesis is required which presents evidence of marked scholarly attainment. A comprehensive examination is also required. Doctoral: Approximately 4 years of graduate work beyond the bachelor's degree, including coursework and original laboratory research. A dissertation of publishable quality representing an original contribution to knowledte is required. A qualifying examination and dissertation defense is required.

Graduate and Undergraduate Courses Available for Graduate Credit: Human Gross Anatomy, Human Development, Human Microanatomy, Human Neuroanatomy, Basic Concepts of Gross Anatomy, Advanced Gross Anatomy, Introduction to Research, Special Cytology, Histological and Histochemical Methods, Interpretation of

Histochemical Results, Methods in Neuroanatomical Research, Electron Microscopic Techniques, The Interpretation of Ultrastructure, Teratology, Experimental Teratology, Endocrinology, Experimental Endocrinology, Special Senses I and II, Ascending Systems, Readings and Research, Supervised Teaching, Anatomy Seminar

DEPARTMENT OF BIOCHEMISTRY
Phone: Chairman: Ext. 265

Chairman: A.H. Mehler
Professors: B. Kassell, A.H. Mehler, S.A. Morell, P. Roll, F. Takete
Associate Professors: H. Ankel, P. Glick, R. Kemp
Assistant Professors: K. Chakraburtty, A. Girotti, J. Willis (Assistant Clinical Professor)
Instructor: 1
Research Assistants: 14

Degree Program: Biochemistry

Graduate Degrees: M.A., Ph.D.
Graduate Degree Requirements: Master: Courses determined by department, qualifying examination, thesis
Doctoral: Courses determined by department, qualifying and comprehensive examination, foreign language, thesis

DEPARTMENT OF MICROBIOLOGY
Phone: Chairman: Ext.282

Chairman: S.E. Grossberg
Professors: B.W. Catlin, J.N. Fink, S.E. Grossberg
Associate Professor: A. Bernstein
Assistant Professors: J.W. Bendler III, F.L A. Buckmire, F.E. Frerman, P. Jameson, J.J Sedmak, B.A. Brown, R.J. Duquesnoy, L. R. Heim, R. J. Zabransky,
Visiting Lectures: 15-20
Research Assistants: 2-4
Teaching Assistants: 2

Degree Program: Immunology, Microbiology

Graduate Degrees: M.S., Ph.D.
Graduate Degree Requirements: Master: Basic Courses; Microbiology, Biochemistry, Biostatistics, other courses as required, comprehensive examination; Thesis Doctoral: As for Masters degree but more extensive plus a defense of disseration.

Graduate and Undergraduate Courses Available for Graduate Study: General Microbiology, Advanced Immunology, Ultrastructure of Microorganisms, Special Project in Basic Research Methods and Concepts, Advanced Virology, Microbial Genetics, Laboratory in Microbial Genetics, Bacterial Chemistry and Physiology, Readings and Research in Microbiology

DEPARTMENT OF PHYSIOLOGY
Phone: Chairman: Ext. 270

Chairman: J. J. Smith
Professors: J. J. Smith, L. H. Hamilton, M. Klitgaard, G. B. Spurr, A. Baron, (Clinical), E. Eisman (Clinical) J. G. Llaurado (Clinical)
Associate Professors: H. F. Edelhauser, J. P. Kampine, M. G. Maksud (Clinical), R. W. Rasch, W. J. Stekiel, D. L. Van Horn, (Joint)
Assistant Professors: C. A. Dawson, H. V. Forster, (Joint), D. W. Glenister, R. W. Gotshall, D. J. McDermott, (Clinical), E. F. Banaszak, (Clinical) B. E. Piacsek, (Clinical) F. E. Tristani, (Clinical) K. E. Whisler (Clinical).

Degree Program: Physiology, Biophysics

Graduate Degrees: M.S., Ph.D.
Graduate Degree Requirements: B.S. or A.B. degree with a major in the sciences, who have demonstrated above average scholastic ability. GRE examination strongly encouraged.

Graduate and Undergraduate Courses Available for Graduate Study: Human Physiology, Cellular Functions in Mammalian Systems, Integrative Functions in Physiological Processes, Cardiovascular and Respiratory Physiology, Radioisotopes, Environmental Physiology, Biometry, Scientific Communication

MILTON COLLEGE
(No reply Received)

MOUNT MARY COLLEGE
Milwaukee, Wisconsin 53222 Phone: (414) 258 4810
Academic Dean: Sr. P. A. Preston

DEPARTMENT OF BIOLOGY
Phone: Chairman: (414) 258-4810

Chairman: Sr. P. Moehring
Assistant Professors: Sr. P. Moehring
Instructors: A. Sauro, S. Thom
Visiting Lectures: 1

Degree Program: Biology, Biological Education, Medical Techology

Undergraduate Degree: B.S.
Undergraduate Degree Requirements: (Biology Major): 24 credits including Cell Structure and Function, Vert. Physiology, Genetics and Evolution, and Ecology.; (Biology Teaching Major): 36 credits; requirements as set by school and State of Wisconsin; (Biology Medical Technology): 32 credits, plus Chemistry, Physics and Mathematics.

NORTHLAND COLLEGE
Ashland, Wisconsin 54806 Phone: (715) 682-4531
Dean of College: E. Stevens

DEPARTMENT OF BIOLOGY
Phone: Chairman: Ext. 208 Others: 682-4531

Chairman: R. L. Verch
Associate Professors: R. Mallampall, L. Stadnyk, R. Verch.
Assistant Professor: R. Maxwell

Degree Program: Biology

Undergraduate Degrees: B.S.
Undergraduate Degree Requirements: 10 courses in Biology (one unit each). Required courses: Ecology, Genetics, Physiology, Taxonomy. 1 yr. of general chemistry.

RIPON COLLEGE
Ripon, Wisconsin 54971 Phone: (414) 748-8122
Dean of College: Dean Pommer

DEPARTMENT OF BIOLOGY
Phone: Chairman: 748-8122

Chairman: C. Nichols
Professor: C. Nichols
Associate Professors: D. Brittain, W. Brooks, W. Bowen
Assistant Professor: R. F. Browning

Field Stations/Laboratories: ACM Wilderness Station - Ely

Degree Program: Biology

Undergraduate Degree: B.A.
Undergraduate Degree Requirements: 36 semester hours in Biology, 8 semester hours in Chemistry, plus the

Liberal Arts graduation requirements for all college students.

ST. NORBERT COLLEGE

DePere, Wisconsin 54115 Phone: (414) 336-3181
Dean of College: D.B. King

DIVISION OF NATURAL SCIENCES
Phone: Chairman: Ext. 468 Others: 336-3181

Chairman: H.J. Baeten
Professors: N.J. Flanigan
Associate Professors: H.J. Baeten, R.C. VandeHey

Degree Program: Biology

Undergraduate Degree: B.Sc.
Undergraduate Degree Requirements: 10 courses (40 semester credits) Biology, 3 courses (12 semester credits) Chemistry including one semester Organic. 32 courses (128 semester credits) total including General Education

SILVER LAKE COLLEGE

Manitowoc, Wisconsin 54220 Phone: (414) 684-6691
Dean of College: Sr. M. Van Ryzin

DIVISION OF NATURAL SCIENCES AND MATHEMATICS
Phone: Chairman: Ext. 28 Others: 684-6691

Chairperson: Sr. J. Van Denack
Professors: Sr. J. VanDenack, Sr. M. Van Ryzin
Assistant Professors: Sr. B. McCloseky, Sr. M.E. Schimpf
Instructor: Sr. M. Melko

Field Stations/Laboratories: Silver Lake Field Station

Degree Program: Biology, Natural Science

Undergraduate Degree: B.S.
Undergraduate Degree Requirements: Biology: 58 semester hours, Major: 36, Minor: 22, Natural Science: 56 semester hours, Upper biennium courses: 24 semester hours, Supporting courses in Mathematics and Natural Science: 32 semester hours

UNIVERSITY OF WISCONSIN - EAU CLAIRE

Eau Claire, Wisconsin 54701 Phone: (715) 826-0123
Dean of Graduate Studies: R.D. Dick
Dean of College: F. Haug

DEPARTMENT OF BIOLOGY
Phone: Chairman: (715) 836-4166 Others: 836-2637

Chairman: M.J. Fay
Professors: A. Bakken, D.B. Crowe, M.J. Fay, R.G. Fossland, J.B. Gerberich, J.K. Lim, O.S. Owen, C.S. Schildt, J.E. Woodruff
Associate Professors: V.A. Cvancara, J.C. Dixon, K.G. Foote, L.E. Ohl, T.C. Rouse, R.H. Saigo, B.H. Snudden
Assistant Professors: T.A. Balding, W. Barnes, Y.T. Ho, T.R. Jewell, J.C. Naughten, K.O'Connell
Instructor: J. Crowe
Graduate Teaching Assistants: 2

Field Stations/Laboratories: University of Wisconsin Pigeon Lake Field Station

Degree Program: Biology

Undergraduate Degree: B.S.
Undergraduate Degree Requirements: Thirty-six semester credits in Biology. Students may select a broad, general emphasis, a botanical emphasis, or a zoological emphasis. General Botany, General Zoology and Genetics are required

Graduate Degree: M.S.
Graduate Degree Requirements: Master: Thirty semester credits, a minimum of 24 must be in graduate-level biology courses

Graduate and Undergraduate Courses Available for Graduate Credit: Comparative Vertebrate Anatomy (U), Genetics (U), Human Heredity and Environment (U), General Entomology (U), Economic Entomology (U), Biological Techniques (U), Animal Physiology (U), Plant Anatomy (U), Taxonomy of Vascular Plants I, II (U), Economic Botany (U), Ecology (U), Dendrology (U), Field Botany (U), Plant Physiology (U), Mycology (U), Plant Ecology (U), Ornithology (U), General Bacteriology (U), Parasitology (U), Vertebrate Embryology (U), Natural History of Vertebrates (U), Radiobiology (U), Field Zoology (U), Animal Behavior (U), Phycology (U), Endocrinology (U), Population Genetics (U), Genetic Material (U), Mammalogy (U), Cytogenetics (U), Advanced General Bacteriology (U), Comparative Vertebrate Histology (U), Integrated Biological Concepts, History and Philosophy of Biological Science, Research Techniques, Advanced Laboratory Techniques, Organic Evolution, Advanced Plant Taxonomy, Advanced Ecology, Advanced Plant Physiology, Medical Mycology, Microbial Ecology, Endocrinology of Reproduction

UNIVERSITY OF WISCONSIN - GREEN BAY

Green Bay, Wisconsin 54302 Phone: (414) 465-2121
Dean of Graduate Studies: D. Jewett

College of Environmental Sciences

Dean: J. Beaten

This college does not have a departmentalized structure by topic [e.g. Biology] but by interdisciplinary units called "concentrations". The faculty shown below are biologists serving such units, two of which are given)

Chairman: L.J. Schwartz
Professors: H. Guilford, W. Kaufman, J. Reed, K. White
Associate Professors: R. Cook, A. Goldsby, H. Harris, N. Huber (on leave), E. McIntosh, M. Morgan, T. Mowbray, V.M.G. Nair, P. Sager, L. Schwartz
Assistant Professors: C. Ihrke, J. Maki, R. Stevens, D. Sager

CONCENTRATION IN ECOSYSTEMS ANALYSIS
Phone: Chairman: (414) 465-2370

Chairman: P.E. Sager
Professors: D.Jowett, J. Reed, K. White
Associate Professors: F. Fischbach, D. Girard, A. Goldsby, H. Harris, H. Huddleston, A. Loomer, M. Morgan, J. Norman, P. Sager, L. Schwartz, J. Wiersma
Assistant Professors: A. Mehra, C. Schwintzer, R. Simons, R. Starkey

Field Stations/Laboratories: Door County - Tofs Point, a Nature Conservatory Area.

Degree Programs: Ecosystems Analysis
Und
Undergraduate Degree: B.A., B.S.
Undergraduate Degree Requirements: 124 credits, 3 semesters Chemistry-Physics, Earth Science, Ecology, 2 semesters Biology 4 Mathematics courses, 30 credits (minimum) in major, 18 credits - Liberal Education Seminar, 15-18 credits as "Distribution credits"

Graduate Degree: Master of Environmental Arts and Sciences
Graduate Degree Requirements: 30 credits, 12-15 Graduate course work credits, 9-12 assigned study credits, 6 thesis credits

Graduate and Undergraduate Courses Available for Graduate Credit: Statistical Design and Analysis of Experiments, Seminar in Functional Ecology, Environmental Education Processes and Materials, Science, Structure and Money on Environmental Decision Making, Human Culture and Land Settlement Patterns, Coastal Zone Management, Scientific and Technical Communication, Population Impact, Physiological and Psychological Effects of Environmental Stress, Seminar in Structural Ecology, Survey and Systems Analysis, Soil Resour ce Management

CONCENTRATION IN POPULATION DYNAMICS
Phone: Chairman: (414) 465-2368

Chairman: T.B. Mowbray
Associate Professors: T.B. Mowbray, N.M. Huber
Assistant Professor: C.A. Ihrke, J.R. Maki, W.J. Mc Auley, D. Sager
Instructors: W.B. MacDonald, J. Westphal

Degree Programs: Population Dynamics, Biology, Botany, Zoology

Undergraduate Degree: B.S.
Undergraduate Degree Requirements: 124 credits, 36 credits , 300 in major, 18 credits of Liberal Education Seminars

Graduate Degree: M.S.
Graduate Degree Requirements: (This is an all university Graduate Program), 12-15 graduate course credits, 9-12 credits of Independent study (Assigned Study), 6 credits of thesis research

Graduate and Undergraduate Courses Available for Graduate Credit: Human Culture and Land Settlement Patterns, Transportation Systems, Human Communications and the Environment, Statistical Design and Analysis of Experiments, Behavioral Research Strategies, Seminar in Functional Ecology, Coastal Zone Management, Population Imput, Scientific and Technical Communication, Land Use in the Present and Future, Multivariate Statistical Analysis, Survey of Systems Analysis, Soil Resource Management, Seminar in Structural Ecology, Environmental Education Processes and Materials

UNIVERSITY OF WISCONSIN - LA CROSSE

La Crosse, Wisconsin 54601 Phone: (608) 784-6050
Dean of College: R.C. Voight

DEPARTMENT OF BIOLOGY
Phone: Chairman: Ext. 479 Others: Ext. 253

Chairman: R. Fletcher
Professors: T. Claflin (Director - River Study Center), A. Nelson, H. Young
Associate Professors: R.M. Burns, J. Davis, R. Fletcher, R. Nord, J. Parry, L. Schuh, R. Senff, P. Sparks, L. Unbehaun, T. Weeks, S. Sohmer
Assistant Professors: L. Buldhaupt, D.J. Grimes, J. Held, S. McIlraith, J. Warner, R. Mowbray
Instructors: E. Hughes, G. Smith
Research Associates: 2

Field Stations/Laboratories: 1 - 50 foot research vessel

Degree Program: Biology

Undergraduate Degree: B.S.
Undergraduate Degree Requirements: Major (letters and Science). 34 credits including Biology 100 (Principles of Biology, Plant Biology, Vertebrate Zoology or Invertebrate Zoology, Ecology or Plant Ecology, Genetics, Cell Physiology, Seminar, Problems in Biology. One year in Chemistry required.

Graduate Degree: M.S.
Graduate Degree Requirements: Master: 30 credits are required with 15 of these being at the Graduate level. Biology Seminar must be taken twice. Option of Plan A - Research and Thesis or Plan B - Research and Seminar

UNIVERSITY OF WISCONSIN - MADISON

500 Lincoln Drive Phone: (608) 262-1234
Madison, Wisconsin 53706
Dean of Graduate Studies: R.M. Bock

College of Letters and Science

Dean: E.C. Kleene

DEPARTMENT OF BOTANY
Phone: Chairman: (608) 262-1057

Chairman: R.F. Evert
William Trelease Professor of Bacteriology and Botany: K.B. Raper
C. Leonard Huskins Professor of Botany: F.K. Skoog
Professors: P.J. Allen, G. Cottam, R.F. Evert, G.C. Gerloff, H.H. Iltis, O.L. Loucks, E.H. Newcomb, K.B. Raper, H.A. Senn, F.K. Skogg, J.W. Thomson, W.F. Whittingham
Associate Professors: M.S. Adams, T.F. Allen, E.W. Beals, W.M. Becker, J.P. Helgeson,
Assistant Professors: H.M. Clarke, D.H. Franck, R.R. Kowall, W.J. Woelkerling
Research Assistants: 12
Teaching Assistants: 20
Graduate Fellows: 6

Field Stations/Laboratories: Kemp Station at Rhinelander

Degree Program: Botany

Undergraduate Degrees: B.A., B.S.
Undergraduate Degree Requirements: A grade-point average of 2.5 in all science courses, 30 credits of approved courses, with not more than one elementary course or cour se sequence, intermediate or advanced course in 5 of the 6 following areas: Genetics, Crypto gamic Botany, Anatomy or Morphology, Physiology and Ecology. Credits in Bacteriology can be applied toward the major, a one credit tutorial course is required both semesters of the junior year. Chemistry through Organic Chemistry

Graduate Degrees: M.S.,Ph.D.
Graduate Degree Requirements: Master: Courses beyond the elementary level in: Anatomy and Morphology, Field Botany, Plant Physiology, a minimum of 9 credits of graduate-level botany courses while in residence. A total of 50 credits (in combined undergraduate and graduate work) in natural science must be completed, seminar in Botany, 2 semesters, Organic Chemistry or Biochemistry, Calculus, Statistics (Biometry), or computer sciences recommended, final oral examination Doctoral: The course work must satisfy the requirements of the Certification Committee of the Graduate Biological Division, training in the 4 botanical areas enumerated in the master's degree program and a minimum of 4 semesters of seminar in Botany. Additional requirements may be set up at the discretion of the major professor, 1 foreign language, 1 semester of teaching experience, all requirements should be fulfilled within 5 years of residency

Graduate and Undergraduate Courses Available for Graduate Credit: Plant Anatomy (U), Comparative Morphology of Vascular Plants (U), Microtechnique (U), Algae (U), Fungi (U), Molds Yeasts and Actinomycetes (U), Bryophytes (U), Classification of Cultivated and Native Plants (U), Identification and Classification of Seed Plants (U), Dendrology (U), Trees and Shrubs (U), Field course in spring Flora of Wisconsin (U), Field Collections and Identification (U), Origins of Plant and Animal Species (U), Plant Geography (U), Recent Plant Migration (U), General Ecology (U),

Physiology and Ecology of Aquatic Plants for non-Biologists (U), Aquatic Plant Laboratory for Cytotaxonomy (U), Plant Biochemistry (U), General Genetics, (U), Introductory Cytogenetics (U), Cytotaxonomy (U), Plant Biochemistry (U), General Morphology and Classification (U), Marine Biology (U), Undergraduate Tutorial (U), Senior honors Thesis (U), Senior Thesis (U), Directed Studies (U), Advanced Plant Morphology, Advanced Mycology, Advanced Taxonomy, Taxonomy and Evolution of Grasses, Advanced Plant Ecology, Ecology, of the Tropics, Introduction to Graduate Research in Developmental Biology, Graduate Laboratory in Developmental Biology, Ecological Methods, Community Analysis, Physiological Research Techniques, Growth of Plants, Protein and Nucleic Acid Metabolism of Plants, Mineral Nutrition of Plants, Physiology of the Fungi, Special Topics in Plant Physiology, Plant Cell Biology, Fundamentals of Cytology, Advanced Cytology, Seminars inMorphology, Algology, Mycology, Plant Taxonomy, Plant Geography, Plant Ecology, Plant Physiology, Cell and Developmental Biology and Cytology, Oceanography and Limnology, Plant Ecology Journal Review, Topics in Plant Physiology, Research in Plant Morphology, Lichenology-Bryology, Mycology, Plant Taxonomy-Cytotaxonomy, Plant Ecology, Plant Physiology, Cytology-Cytogenetics, Plant Geography and Phycology

ZOOLOGY DEPARTMENT
Phone: Chairman: (608) 262-1051 Others: 262-1234

Chairman: W.C.Burns
Marshall Professor of Zoology: R.K. Meyer (Emeritus)
John Bascom Professor: J.F. Crow
Professors: S. Abrahamson, R. Auerbach, N.M. Bilstad (Emeritus), W.C. Burns, J.F. Crow, J.T. Emlen (Emeritus), L.A. Fraser, J.P. Hailman, A.D. Hasler, W.H.McShan (Emeritus), R.K.Meyer (Emeritus), J.C. Neess, L.M. Passano, W.S. Plaut, W.B. Quay, W.G. Reeder, H. Ris, J.T. Robinson, D.R. Sonneborn, G. Thorne (Emeritus)
Associate Professors: E.W. Beals, G.G. Borisy, M.D. Bownds, J.J. Magnuson, W.P. Porter, A.O.W. Stretton
Assistant Professors: S.I. Dodson, J. Lilien, M.P. Meyer, T.C. Moermond, M.R. Namenwirth, P.C. Whitehead (Emeritus)
Teaching Assistants: 37
Research Assistants: 25
Fellows: 6
Trainees: 7

Field Stations/Laboratories: Trout Lake Biological Station - Trout Lake, Maintains a cooperative arrangement with the University of Minnesota for summer course work at the Lake Itasca Biological Station in Minnesota

Degree Program: Zoology

Undergraduate Degree: B.S.
Undergraduate Degree Requirements: 30 credits in Zoology, with not more than one elementary course, one semester Mathematics, one year Chemistry, one year Physics, Recommended courses: Advanced Zoology, Mathematics, Physics, Organic Chemistry

Graduate Degrees: M.S., Ph.D.
Graduate Degree Requirements: Master: Applicants are expected to have completed year of Physics, two years of Chemistry including Organic Chemistry, Mathematics through Calculus, or equivalent, Botany and a semester of Advanced Botany or Bacteriology, have completed the following: (1) the required courses listed previously 2) the equivalent of the undergraduate major in Zoology at the University of Wisconsin, and 20 additional credits of graduate work of which 12 must be earned in formal courses (Not necessarily in Zoology), excluding research and seminar, of the 20 credits, at least 14 (research and seminars included must be taken in Zoology, written report or formal thesis based in part on research done by the candidate: and an oral examination which may be (1) a terminal examination for the M.A. or M.S. Degree or 2) a qualifying examination for admission to candidacy for the Ph.D. Doctoral: course requirements for the master's degree as given above, one year (20 contact hours or equivalent) of teaching, is required of all Ph.D. candidates, qualifying examination, thesis, final oral

Graduate and Undergraduate Courses Available for Graduate Credit: General Invertebrate Zoology (U), Insects in Our World (U), Introduction to Entomology (U), Animal Parasites of Man (U), Parasitology (U), Human Protozoology (U), Medical Entomology (U), Topics in Biology (U), Organic Evolution (U), Animal Taxonomy (U), Origins of Plant and Animal Species (U), Comparative Anatomy of Vertebrates (U), Paleontology of the Primates (U), Biology of the Vertebrates (U), Developmental Biology (U), Topics in Developmental Biology (U), Invertebrate Embryology (U), General Ecology (U), Advanced Developmental Biology (U), Biophysical Ecology, Field Zoology, Ecology of Fishes (U), Conservation of Aquatic Resources (U), Hydrobiology (U), Field Ornithology (U), Ethology (U), Ethology Ethology Laboratory (U), Environmental Toxicology (U), Experimental Genetics (U), General Genetics (U), Human Cytogenetics (U), Cellular Biology (U), Neurophysiology (U), Comparative Physiology (U), Biology of Aquatic Populations (U), Endocrinology (U), Neuroendocrinology, Histology and Histophysiology (U) (U), Ethological Theory and Analysis, Behavior Ecology, Problems in Oceanography, Ecology of the Tropic Indtroduction to Graduate Research in Developmental Biology, Graduate Laboratory in Developmental Biology, Community Analysis, Biometric Techniques in Ecological Research, Current Problems in Cell Biology, Advanced Cytology

College of Agricultural and Life Science
Dean: G.S. Pound

DEPARTMENT OF AGRONOMY
Phone: Chairman: (608) 262-1390

Chairman: J.W. Pendleton
Professors: R.H. Andrew, E.A.Brickbauer, R.E. Doersch P.N. Drolsom , R.A. Forsberg, R.F. Johannes, J.H. Lonnquist, J.W. Pendleton, D.R. Peterson, D.A. Rohweder, J.M. School, H.L. Shands, D. Smith, G.H. Tenpas, J.H. Torrie
Associate Professors: E.T. Bingham, E.T. Gritton, L.E. Schrader, R.R. Smith, J.M. Sund, V.L.Youngs
Assistant Professors: Z.M. Arawinko, R.S. Fawcett, R.G. Harvey, R. James, E.S. Oplinger, W.H. Paulson D.M. Peterson, J.N. Senturia
Research Assistants: 25
Research Associate: 1
Postdoctoral Fellow: 1

Field Stations/Laboratories: Experimental Farms at Arlington - Ashland, Spooner, Hancock, Lancaster

Degree Programs: Agronomy, Crop Science, Farm Crops, Natural Resources, Natural Science, Plant Breeding, Plant Sciences, Range Science, Seed Investigations, Soils and Plant Nutrition, Plant Physiology

Undergraduate Degree: B.Sc.
Undergraduate Degree Requirements: A total of 124 credits for the B.Sc. degree of which 45 credits must be in the College of Agriculture and 15 credits must be in Agronomy

Graduate Degrees: M.S., Ph.D.
Graduate Degree Requirements: Master: A total of 18 credits at A or B grade level with a formal thesis or 21 credits and a research report. No foreign language required Doctoral: Two years of residence beyond M.S. (or 3 beyond B.Sc.), course work agreed upon by candidate's committee, and a formal research thesis, no foreign language required

Graduate and Undergraduate Courses Available for Graduate Credit: Ecology of Agronomic Plants (U), Pastures and Pasture Problems (U), Grain Crops (U), Forage Problems (U), Crop Identification and Standards (U), Weed Control (U), The Breeding of Field Crops (U), Biometry (U), Physiology of Dry Matter Accumulation, Experimental Design, Biometrical Procedures, Advanced Plant Breeding, Chromosome Manipulations, Seminar in Crop Physiology, Seminar-Plant Breeding

DEPARTMENT OF BACTERIOLOGY
Phone: Chairman: (608) 262-2914

Chairman: R.S. Hanson
William Trelease Professor of Bacteriology and Botany: K.B. Raper
E.B. Fred, Professor of Natural Science: T.D. Brock
Professors: T.D. Brock, R.H. Deibel, J.C. Ensign, K.B. Raper, J.B. Wilson, R.S. Hanson, E.M. Foster, D.T. Berman (Joint), R.P. Hanson (Joint), E.H. Marth (Joint), H. Sugiyama (Joint)
Associate Professors: R.D. Hinsdill, J.L. Pate, W.J. Brill, D.O. Cliver (Joint), C.L.Duncan (Joint), J.M. Coepfert (Joint)
Assistant Professors: W.H. McClain, J.G. Zeikus
Research Assistants: 36
Fellows: 19
Teaching Specialists: 3

Field Stations/Laboratories: Laboratory of Thermal Biology: U. Yellowstone, Montana

Degree Program: Bacteriology

Undergraduate Degree: B.S.
Undergraduate Degree Requirements: At least 15 credits in the department including advanced general bacteriology and physiology of microorganisms. Fulfillment of college and university requirements.

Graduate Degrees: M.S., Ph.D.
Graduate Degree Requirements: Master: General regulations of the Graduate School, and 18 or more graduate course credits including Bacteriology 725, Bacteriology 726, Bacteriology 727 (see below). Presentation of M.S.thesis or report. Passage of written examination in bacteriology and and final oral examination
Doctoral: Courses: Bacteriology 725, 726, 727 (see below) plus 12 additional graduate course credits not uned to fulfill the minor requirements. Teaching: 2 semesters, part time. Passage of preliminary examination. Research resulting in an adequate thesis. General regulations of the Graduate School.

Graduate and Undergraduate Courses Available for Graduate Credit: General Survey of Bacteriology (U), General Microbiology (U), Advanced General Microbiology (U), Soil Bacteriology (U), Food Microbiology (U), Physiology of Microorganisms (U), Industrial Microbiology (U), Infection and Immunity in Animal Diseases (U), Molds, Yeasts, and Actinomyces (U), General Virology (U), Food-Borne Diseases (U),Genetics of Microorganisms (U), Microbial Ecology and Physiology (725), Bacterial Physiology (726), Microbial Ecology and Physiology Laboratory (727), Epizoology, Presentation of Bacteriological Reports, Advanced Seminar

DEPARTMENT OF BIOCHEMISTRY
Phone: Chairman: (608) 262-3026 Others: 262-3040

Chairman: H.F. DeLuca
Harry Steenbock Research Professor: H.F. DeLuca
Vilas Research Professor: H.A. Lardy
Conrad Elvehjem Professor: M. Nomura
E.B. Hart Professor: J. Adler
E.V. McCollum Professor of Nutritional Science: A.E. Harper
Professors: J. Adler, L. Anderson, C.A. Baumann, H. Beinert, R.M. Bock, R.H. Burris, W.W. Celland, J.E. Davies, H.F.DeLuca, J.C. Carver, J. Gorski, A.E. Harper, W.G. Hoekstra, R.B. Inman, P.J. Kaesberg, H.A. Lardy, M. Nomura, R. Rownd, R.R. Rueckert, M.A. Stahmann, M. Sundaralingam, J.W. Suttie, R. D. Wells
Associate Professors: W.H. Orme-Johnson, H.K. Schnoes
Assistant Professors: D.L. Nelson, S.T. Rao, W.S. Reznikoff
Visiting Lecturers: 62
Research Assistants: 36
Trainees: 57
Predoctoral Fellows: 12
Postdoctoral Fellows: 4
Research Associates: 36
Project Associates: 3
Project Assistants: 1
Honorary Fellows: 3

Degree Program: Biochemistry

Undergraduate Degree: B.S.
Undergraduate Degree Requirements: 28 credits Chemistry, Mathematics, through Calculus, 14 credits Biology, 8 credits Biochemistry, also general college requirements

Graduate Degrees: M.S., Ph.D.
Graduate Degree Requirements: Master: 7 credits in Organic chemistry, 3 credits in Physical Chemistry, 9 credits in Biochemistry, thesis, final examination
Doctoral: Biochemistry courses as specified by committee. One seminar each semester, one year tea teaching experience, foreign language recommended annual evaluation, preliminary written and oral examinations, thesis, oral defense

Graduate and Undergraduate Courses Available for Graduate Credit: Survey of Biochemistry (U), Introduction to Biochemistry (U), Biochemical Principles of Human and Animal Nutrition (U), Experimental Molecular Genetics (U), General Biochemistry (U), Plant Biochemistry, Vitamins and Metabolism, Metabolism and Function of Mineral Elements in Higher Animals, Metabolic Integration in Animals, The Behavior of Simple Organisms, Enzymatic and Metabolic Techniques, Isotopic Techniques, Chromatographic Techniques, Biochemistry of Microorganisms, Mechanisms of Enzyme Action, Molecular Genetics, Advanced Topics, Physical Chemistry of the Proteins and Nucleic Acids, Seminar in Nutrition and Metabolism, Seminar in Microbial Biochemistry, Seminar in Natural Products, Seminar in Molecular Genetics, Seminar in Enzymology

DEPARTMENT OF DAIRY SCIENCE
Phone: Chairman: (608) 263-3309 Others: 263-3308

Chairman: R.P. Niedermeier
Professors: J.W. Crowley, N.A. Jorgensen, R.P. Niedermeier, L.D. Satter, L.H. Schultz, E.E. Starkey W.J. Tyler, D.A. Wieckert
Associate Professors: A.N. Bringe, D.P. Dickson, W. T. Howard, E.L. Jensen, H.J. Larsen, C.C. Olson, A.G. Sendelbach, G.E. Shook
Assistant Professor: C.E. Zehner
Research Assistants: 22
Specialists: 7
Research Associates: 2

Degree Program: Dairy Science

Undergraduate Degree: B.S.
Undergraduate Degree Requirements: Varies with option selected. 1. Production and Technology, 2. Business and Industry, 3. Natural Science, 124 credits including 15 credits in major department and a total of 45 credits in the College of Agriculture and Life Sciences.

Graduate Degrees: M.S., Ph.D.
Graduate Degree Requirements: Master: Two plans

available. M.S. degree in course. 22 credits including a special problem in field of interest. M.S. degree in research. 18 credits including thesis, minimum of 2 semesters of residence credits plus passing a comprehensive oral examination Doctoral: A research degree with major emphasis on pursuit of a research project in a phase of dairy science. A joint major with related departments emphasizing basic sciences may be planned. A minor in such departments is required. The resident requirement is a minimum of 6 full semesters of resident credits.

Graduate and Undergraduate Courses Available for Graduate Credit: Principles of Animal Nutrition (U), Feed Ingredients and Ration Formulation (U), Dairy Herd Management (U), Reproductive Physiology (U), Dairy Herd Management (U), Reproductive Physiology (U), Dairy Cattle Nutrition (U), Physiology of Lactation (U), Dairy Cattle Breeding (U), Proseminar (U), Special Problems (U), Ruminology, Data Processing in Agricultural Research, Livestock Breeding (U)

DEPARTMENT OF ENTOMOLOGY
Phone: Chairman: (608) 262-3227

Chairman: G.R. DeFoliart
W.A. Henry Distinguished Professor: S.D. Beck
Professors: J.W. Apple, S.D. Beck, D.M. Benjamin, G.M. Boush, R.K. Chapman, H.C. Coppel, G.R. DeFoliart, R.J. Dicke, E.H. Fisher, W.L. Gojmerac, W.L. Hilsenhoff, C.F. Koval, J.L. Libby, E.P. Lichtenstein, F. Matsumura, J.T. Medler, D.M. Norris, R.D. Shenefelt
Associate Professors: W.E. Burkholder, F.E. Moeller
Assistant Professor: S.D. Carlson
Visiting Lecturers: 2
Research Assistants: 54
Research Associates: 12

Degree Program: Entomology

Undergraduate Degree: B.S. (Agriculture)
Undergraduate Degree Requirements: 124 credits required. including: 2 semesters Calculus or Analytical Chemistry, Statistics or Computer Science, Organic Chemistry with Laboratory, 2 semesters Physics, 3 credits in beginning courses in Plant or Animal Sciences, plus Invertebrate Zoology and students not in the Biology Core Curriculum Option must take 2 among the following: Bacteriology, Genetics, Biochemistry, Soils, and Plant Pathology. The Entomology requirement includes Introduction to Entomology, Insect Morphology, Proseminar and 3 additional entomology courses

Graduate Degrees: M.S., Ph.D. (Entomology)
Graduate Degree Requirements: Master: One year's residence; 18-24 graduate credits, a thesis or progress report in thesis form, pass final oral examination Doctoral: one course in each of the four following Biological Areas: a) Structure or Phylogeny of Organisms, b) Physiology or Cytology, c) Ecology or Population Dynamics, d) Genetics, e) Biochemistry; two semesters of General Chemistry, with Laboratory, and one semester of Organic Chemistry, with laboratory, a course in General Physics, College-level Mathematics, including Calculus and a course in Statistics. (a course, or non-credit series, in computer programming is advised), a minimum of one year of entomology colloquium, the student must present a dissertation and pass preliminary and final examinations.

Graduate and Undergraduate Courses Available for Graduate Credit: Insects in Our World (U), Introduction to Entomology (U), Insect Morphology, Histology and Physiology of Insects, Sensory Physiology and Behavior of Insects, Taxonomy of Mature Insects, Insect Ecology, Principles of Economic Entomology, Medical Entomology (U), Toxicology of Insecticides, Insects in Relation to Plant Diseases, Taxonomy of Immature Insects, Insects and Diseases in Forest Resource Management (U), Beekeeping (U), Aquatic Insects (U), Biological Control of Insects, Insect Population Ecology, Proseminar, Colloquium, Insect Pathology, Advanced Topics in Insecticide Toxicology

DEPARTMENT OF FOOD SCIENCE
Phone: Chairman: (608) 263-2010 Others: 262-3046

Chairman: H.E. Calbert
Professors: C.H. Amundson, M.S. Bergdoll, H.E. Calbert, B. David, M.P. Dean, O. Fennema, E.H. Marth, N.F. Olson, T. Richardson, A.M. Swanson, J.H. von Elbe, K.G. Weckel, W.C. Winder
Associate Professors: R.L. Bradley, Jr., F.S. Chu, C. M. Dunn, C.E. Johnson, R.C. Lindsay, D.B. Lund, M.E. Mennes, D. Stuiber
Assistant Professors: M.E. Matthews, M. Wahl
Visiting Lecturers: 3
Research Assistants: 64

Degree Programs: Graduate Options: Food Science, Food Chemistry, Food Microbiology

Undergraduate Degree: B.S. (Food Science)
Undergraduate Degree Requirements: Courses vary (usually 80-100 credits) according to option selected from Production and Technology, Business and Industry, Food Chemistry, Food Administration

Graduate Degrees: M.S., Ph.D. (Food Science)
Graduate Degree Requirements: Master: Courses vary (usually from 20 to 30 credits in all) according to option selected, thesis, oral defense Doctoral: M.S. is prerequisite, courses vary (usually from 30 to 50 credits in all) according to option selected, 1 year teaching experience, preliminary examination, thesis, final oral

Graduate and Undergraduate Courses Available for Graduate Credit: Survey of the Food Industries, Introduction to Meat Science and Technology, Consumer Aspects of Food Supply, Nature of Food, Quantity Food Purchasing and Preparation, Analysis of Food Products (U), Principles of Food Engineering (U), Regulatory and Quality Standards (U), Food Microbiology (U), Food Science I - Tissue Systems (U), Food Science II Fluid and Semisolid Systems (U), Food Processing Laboratory (U), Management Principles for Foods, Food-Borne Disease Hazards (U), Dairy Foods I-Market Milk, Butter, and Special Products, Food Fermentations (U), Dairy Foods II - Ice Cream, Concentrated and Dried Milks (U), Chemistry of the Food Lipids (U), Food Enzymes (U), Research Methods - Foods (U), Problems in Food Administration (U), Problems in Food Administration (U), Food Systems Operation (U), Principles of Food Chemistry (U), Principles of Food Chemistry Laboratory (U), Principles of Food Chemistry Laboratory (U), Physical Chemistry of Food Products (U), Principles of Food Chemistry (U), Dairy Chemistry (U), Principles of Packaging (U), Experimental Food Study (U), Organization and Management of Food Service (U), Analysis of Food Systems Operation (U)

DEPARTMENT OF FORESTRY
(see under School of Natural Resources)

LABORATORY OF GENETICS
(also in School of Medicine)
Phone: Chairman: (608) 262-3112

Chairman: M. Susman
E.B. Hart Professor of Biochemistry and Genetics: J. Adler
John Bascom Professor of Genetics, Medical Genetics, and Zoology: J.F. Crow
Conrad Elvehjem Professor of Genetics and Biochemistry: M. Nomura
Leon J. Cole Professor of Genetics and Medical Genetics O. Smithies

Professors: S. Abrahamson, J. Adler, F.H. Bach, R.A. Brink (Emeritus), L.E. Casida, A.B. Chapman, C.W. Cotterman, J.F. Crow, R.I. DeMars, A.S. Fox, R.W. Hougas, M.R. Irwin (Emeritus), N.P. Neal (Emeritus), O.E. Nelson, M. Nomura, J.M. Opitz, R.H. Osborne, K. Pateau, S.J. Peloquin, R.M. Shakelford, W.K. Smith (Emeritus), O. Smithies, W.H. Stone, M. Susman, W.J. Tyler, S. Wright (Emeritus)
Associate Professors: G.R. Craven, J.L. Kermicle, K.D. Munkres, J. Rapacz
Assistant Professors: F.R. Blattner, C. Denniston
Research Assistants: 60
Predoctoral Trainees: 28
Postdoctorals: 23

Degree Programs: Genetics, Plant Breeding, Medical Genetics

Undergraduate Degree: B.S.
Undergraduate Degree Requirements: A general Biological sciences major suitable for admission to graduate school, including: one year of Physics, Chemistry, through Organic, one year of Calculus, about 30 credits of Biology (6 of which must be in Genetics), and two years of foreign language

Graduate Degrees: M.S., Ph.D.
Graduate Degree Requirements: Master: A general genetics course, a semester of Advanced Genetics, nine additional nonseminar credits in graduate level genetics courses, two credits of seminar, a research thesis or review article, and such other courses as may be recommended by an advisory committee
Doctoral: The same as the master's degree plus three additional credits of Advanced Genetics, two additional seminar credits, two semesters of teaching experience, the preparation of a doctoral thesis on original research, and any other courses that might be required by an advisory committee.

Graduate and Undergraduate Courses Available for Graduate Credit: General Genetics (U), Experimental Genetics (U), Introductory Cytogenetics (U), Human Cytogenetics (U), Human Genetics (U), Advanced Genetics (U), Reproductive Physiology (U), Genetics of Microorganisms (U), Statistics in Design and Analysis of Animal Experiments (U), Immunogenetics (U), Population Genetics (U), Biochemical Genetics: Macromolecules and Evolution (U), Plant Genetics(U), Molecular and Cell Genetics, Genetics of Fungi, Genetic Counseling many seminars, and Research, and Independent Study (U)

DEPARTMENT OF HORTICULTURE
Phone: Chairman: (608) 262-1490

(Faculty are drawn from other departments)
Research Assistants: 20
Fellows: 3

Degree Programs: (B.S., M.S., Ph.D. joint major only Plant Breeding and Plant Genetics M.S., Ph.D.)

DEPARTMENT OF PLANT PATHOLOGY
Phone: Chairman: (608) 262-1436 Others: 262-1426

Chairman: A. Kelman
Professors: P.J. Allen (Joint), D.C. Arny, J.G. Berbee, D.M. Boone, H.M. Darling, R.D. Durbin, R.W. Fulton, D.J. Hagedorn, E.W. Hanson, A.C. Hildebrandt, A. Kelman, J.E. Kuntz, J.E. Mitchell, J.D. Moore, R.F. Patton, G.S. Pound (Dean), L. Sequeira, E.B. Smalley, E.K. Wade, P.H. Williams, G.L. Worf
Associate Professors: G.A. deZoeten, J.P. Helgeson, J. D. Kemp, D.P. Maxwell, C.D. Upper
Assistant Professors: T.K. Kirk
Research Assistants: 45
Postdoctoral Fellows: 6

Field Stations/Laboratories: Kemp Biological Station - Woodruff, Wisconsin

Degree Program: Plant Pathology

Undergraduate Degree: B.S.
Undergraduate Degree Requirements: Biological Sciences which are to include either Option A or Option B as follows: Option A - Zoology, Botany, or Biology and two courses from the following: Bacteriology, Genetics, Biochemistry, Soil Science, and Plant Pathology Option B (The Biology Core Curriculum), 15 Biocore credits

Graduate Degrees: M.S., Ph.D.
Graduate Degree Requirements: Master: (The M.S. degree frequently is required as an interim degree for students working toward the Ph.D. degree and occasionally is used as a terminal degree), course requirements for each candidate are arranged on an individual basis, 6 plant pathology courses are required, A minimum of 24 credits 9 of them in other departments, and no more than 3 of the 24 in Plant Pathology (Research), The first semester introductory Seminar and either a Proposition Seminar or a Special Topics course must be taken, a comprehensive oral examination, thesis, or equivalent research report
Doctoral: Option I: 9 core plant Pathology courses, in addition, eight course credits in Plant Pathology and related subjects are to be selected from a prescribed list of courses. A foreign language may or may not be required. Option II: A student's program will be designed for specific needs through consultation with the major professor and other members of the certification committee. Requirements are: Plant Pathology (4 credits), also required is a minimum of 25 additional credit hours including at least 14 hours in Biochemistry, the other 11 credit hours must be from courses in Bacteriology, Biochemistry, Biophysics, Chemistry, Botany, Genetics, Mathematics, Molecular Biology, Plant Pathology or Zoology, one language (French, German, or Russian), thesis

Graduate and Undergraduate Courses Available for Graduate Credit: Introduction to Plant Pathology, Plant Disease Control, Fungi, Coordinative Internship, Insects in Relation to Plant Disease, Insects and Diseases in Forest Resource Management, Forest Pathology, Fungus Deterioration of Forest Products, Plant Disease Resistance, Diseases of Economic Plants, Plant Pathogens and Pathogenesis, Ecology, Epidemiology and Control of Plant Diseases, Colloquium in Environmental Toxicology, Plant Parasitic Fungi and Bacteria, Cytology of Plant Pathogenesis, Ecology of Soil-borne Plant Pathogens, Plant Nematology Techniques, Variability of Plant Pathogens, Plant Pathology Reports and Illustrations, Senior Honors Thesis, Special Problems and Topics, Plant Virology, Plant Nematology, Plant Disease Physiology, Advanced Mycology, Physiology of the Fungi

POULTRY SCIENCE DEPARTMENT
Phone: Chairman: (608) 262-1243

Chairman: M.L. Sunde
Professors: H.R. Bird, W.H. McGibbon, J.L. Skinner, M.L. Sunde, B.C. Wentworth
Associate Professor: L.C. Arrington
Assistant Professors: R.W. Haller, A.J. Maurer
Instructors: M.J. Wineland
Research Assistants: 10

Field Stations/Laboratories: Poultry Science Laboratory - Observatory Drive, Arlington Turkey Farm and Arlington Chicken Farm both at Arlington, Wisconsin

Degree Program: Poultry Science

Undergraduate Degree: B.S.
Undergraduate Degree Requirements: 15 credits of Poultry Science from Department plus other college

requirements

Graduate Degree: M.S.
Graduate Degree Requirements: Master: 15 credits including 4 of 5 graduate level courses Doctoral: Ph.D. offered jointly with any other basic science department

Graduate and Undergraduate Courses Available for Graduate Credit: Poultry Feeds and Feeding, Avian Development, Advanced Poultry Management, Marketing Poultry Products, Poultry Breeding, Seminar in Poultry Management, Seminar in Poultry Research, Seminar on the Integrated Poultry Industry, Seminar in Avian Biology, Special Problems, Research

DEPARTMENT OF VETERINARY SCIENCE
Phone: Chairman: (608) 262-3177

Chairman: B.C. Easterday
Professors: D.T. Berman, R.F. Bristol, B.C. Easterday, R.E. Hall, R.P. Hanson, C. Olson, W.E. Ribelin, A.C. Todd, J.A. Will
Associate Professors: J.R. Anderson, G.E. Bisgard, O.J. Ginther, T.M. Yuill
Assistant Professor: L.D. Pearson
Visiting Lecturers: 2
Teaching Fellows: 2
Research Assistants: 36

Degree Programs: Immunology, Microbiology, Pathology, Parasitology, Veterinary Science, Wildlife Diseases, Cardio-pulmonary Physiology, Reproductive Physiology

Graduate Degrees: M.S., Ph.D.
Graduate Degree Requirements: Not Stated

Graduate and Undergraduate Courses Available for Graduate Credit: Physiology of Domesticated Animals (U), Veterinary Protozoology (U), Infection and Immunity in Animal Diseases (U), General Pathology (U), General Virology (U), General Virology Laboratory (U), Helminthology (U), Diseases of Wildlife (U), Pathology of Fishes (U), Pathology of Laboratory Animals and Fishes (U), Medical Entomology (U), Environmental Toxicology (U), Colloquium in Environmental Toxicology (U), Special Problems (U), Epizootiology, Comparative Systemic Pathology, Viral and Rickettsial Diseases of Vertebrates, Advanced Veterinary Pathology, Proseminar, General Seminar, Advanced Seminar, Research

DEPARTMENT OF WILDLIFE ECOLOGY
Phone: Chairman: (608) 262-2672

Chairman: R.A. McCabe
Professors: R.S. Ellarson, J.J. Hickey, L.B. Keith, R.A. McCabe
Associate Professors: O.J. Rongstad, T.M. Yuill
Assistant Professors: R.L. Ruff, D.H. Rusch
Research Assistants: 16

Field Stations/Laboratories: Rochester Wildlife Research Center - Rochester, Alberta, Canada

Degree Programs: Wildlife Ecology

Undergraduate Degree: B.S.
Undergraduate Degree Requirements: must attain a 2.5 GPA at end of sophomore year to continue, and must maintain a 3.0 each semester as upperclassmen

Graduate Degrees: M.S., Ph.D.
Graduate Degree Requirements: Master: 18 graduate credits, of which 9 must be in Wildlife Ecology, course work (or its equivalent) in Ornithology, Mammalogy, Taxonomic Botany, Plant Ecology and Biometry, and a publishable thesis Doctoral: four courses distributed among at least three of the following areas: Biochemistry, Genetics, Structure and or Function of Organisms, and Populations or Ethology, Chemistry through a semester of Organic, substantial second course in Physics, 2 semesters in Calculus and 1 semester in Statistics

Graduate and Undergraduate Courses Available for Graduate Credit: Wildlife Ecology (U), Diseases of Wild Animals (U), Animal Population Dynamics (U), Wildlife Management Techniques (U), Principles of Game Management (U), Avian Ecology (U), Mammalian Ecology (U), Ecologic Dimensions of Environmental Impact Assessment (U), Environmental Toxicology (U), Proseminar (U), Special Problems (U), Conservation Programs (U), Seminar, Field Problems in Wildlife Ecology

Medical School

Dean: L.G.Crowley

DEPARTMENT OF ANATOMY
Phone: Chairman: (608) 262-2888

Chairman: D.B. Slautterback
Professors: J.W. Anderson , R.W. Guillery, M.W. Orsini, D.B. Slautterback
Associate Professors: A.H. Martin, M.M. Miles, J.C. Pettersen
Assistant Professors: E.F. Allin, A.W. Clark, J.F. Fallon, J.K.Harting, R.E. Kalil, D.A. Langebartel, B.H. Lipton, J.K. Werner
NIH Trainees: 10

Degree Program: Anatomy

Graduate Degrees: M.S, Ph.D.
Graduate Degree Requirements: Master: Candidates must earn a minimum of two semesters of residence credit in which at least one semester (or equivalent) must have been earned at the University of Wisconsin Madison. Doctoral: The candidate must offer a minimum of six full semesters of residence credit for the Ph.D. degree. The residence requirement may not be satisfied by Summer Sessions or part-time attendance only. Qualifying examination for admission candidacy instead of oral preliminary consists of intensive review papers on subjects selected by student. Thesis and oral defense required.

Graduate and Undergraduate Courses Available for Graduate Credit: Anatomy for Nurses, Experimental Embryology (U), Experimental Neuroanatomy (U), Neuroanatomy Seminars (U), Contemporary Studies in Brain Structure (U), Independent Study (U), Mammalian Implantation (U), Histology and Organology (U), Gross Human Anatomy (U), Introduction to Neuroanatomy and Neurophysiology (U), Clinical Anatomy (U), Muscle Biology (U), The Morphology of Cell Membranes and Their Adnexae (U), Selected Topics in Dissection, Cell Anatomy (U), Research and Thesis (U), Advanced Independent Study (U)

LABORATORY IN GENETICS
(see under College of Agricultural and Life Sciences)

DEPARTMENT OF PHYSIOLOGICAL CHEMISTRY
Phone: Chairman: (608) 262-1348

Chairman: P.P. Cohen
H.C. Bradley Professor of Physiological Chemistry: P.P. Cohen
Professors: P.P. Cohen, A.E. Colas, H.F. Deutsch, W.M. Fitch, T. Gerritsen, M.R. Hokin, S.E. Kornguth, R.L. Metzenberg, J.W. Porter, H.J. Sallach
Associate Professors: J.E. Dahlberg, H.J. Karavolas, F.L. Siegel, D.R. Wilken
Assistant Professor: L. Kahan
Research Assistants: 2
Project Associates: 13
Ph.D. Trainees: 11

Degree Programs: Graduate study programs for the M.S. and Ph.D. degrees in the fundamentals of Biochemistry and Molecular Biology. The department has active research programs dealing with Enzymology, Intermediary Metabolism, Biophysical Aspects of Tissue and Plasma Proteins and Peptides, Immunochemistry, Biochemical Aspects of Morphogenesis and Development, Comparative Biochemistry, Biochemical Genetics, Biochemical Aspects of Viruses, Endocrinology and Neurochemistry

Graduate Degrees: M.S., Ph.D.
Graduate Degree Requirements: Not Stated

Graduate and Undergraduate Courses Available for Graduate Credit: Not Stated

DEPARTMENT OF MEDICAL MICROBIOLOGY
Phone: Chairman: (608) 262-3351

Chairman: D.L. Walker
Professors: R. Hong (Joint), D.W. Smith, D.L. Walker
Associate Professors: E. Balish, A.A. Blazkovec, H.C. Hinze, M.G. Lysenko, J.E. Osborn
Assistant Professors: G. Sundharadas, D. Manning

Degree Programs: Medical Microbiology

Undergraduate Degrees: B.S.
Undergraduate Degree Requirements: 6 semesters Chemistry, 1 semester Zoology, 1 semester Parasitology, 17 credits Bacteriology

Graduate Degrees: M.S., Ph.D.
Graduate Degree Requirements: Master: 10 course credits, thesis, oral examination Doctoral: qualifying examination, 18 course credits, thesis, oral defense

Graduate and Undergraduate Courses Available for Graduate Credit: Immunology (U), Animal Parasites of Man (U), Human Protozoology (U), Virus and Rickettsial Diseases of Man, Pathogenesis of Bacterial and Mycotic Infections, Journal Club

DEPARTMENT OF PHYSIOLOGY
Phone: Chairman: (608) 262-2939 Others: 262-2938

Chairman: R.C. Wolf
Professors: R.M. Benjamin, E.E. Bittar, Q.R. Murphy, W.E. Stone, R.C. Wolf, W.B. Youmans
Associate Professors: L.D. Davis, W.H. Dennis, D.D. Gilboe, J.E. Kendrick, F.J. Nagle
Assistant Professors: M.L. Birnbaum, S.S. Chen, H. Hift, P. Lipton, J.H.G. Rankin
Teaching Fellows: 11
Research Assistants: 5

Degree Program: Physiology

Graduate Degrees: M.S., Ph.D.
Graduate Degree Requirements: Master: Minimum of 18 credits (including Physiology 603 and one other Graduate School level Physiology course, or Physiology 715 and 716, and comprehensive examination for the M.S. degree Doctoral: Admitted to the Ph.D. program only when a staff member has agreed to serve as his major professor. Physiology 715, 716 and 901 are required. A course in statistics or biometry is required, either as an undergraduate or a graduate course. Additional courses are determined by a committee or the major professor. Required thesis, preliminary oral and final oral

Graduate and Undergraduate Courses Available for Graduate Credit: Environmental Physiology, Human Physiology and Biophysics, Physical Chemistry of Cell Systems, Human Physiology, Muscle Biology, Sensory Mechanisms, Renal Physiology in Ion Transport, Electrolytes and Acid-Base Balance, Cellular Bioenergetics, Physiology of the Fetus

School of Natural Resources
Dean: S.C. Smith

DEPARTMENT OF FORESTRY
Phone: Chairman: (608) 262-1780

Chairman: D.T. Lester (Acting)
Riker Professor of Forestry: T.T. Kozlowski
Professor: J.N. McGovern
Associate Professors: G.R. Cunningham, H. Kobler, D.T. Lester, T.A. Peterson
Assistant Professors: D.M. Adams, J.W. Balslger, J.D. Brodie, A.R. Ek, R.A. Oliverira
Research Assistants: 25

Field Stations/Laboratories: Kemp Biological Station - Woodruff, Wisconsin

Degree Program: Forestry

Undergraduate Degree: B.S.
Undergraduate Degree Requirements: 124 semester credits including 82 required and the remainder in distributed electives

Graduate Degrees: M.S., Ph.D.
Graduate Degree Requirements: Master: 18-36 semester credits depending on undergraduate background, cumulative GPA of 3.0 (on 4 point scale), thesis and non-thesis options Doctoral: 6 semesters of residence credit, major and minor fields, joint major or interdisciplinary emphasis, comprehensive preliminary examination, thesis, and final oral examination

Graduate and Undergraduate Courses Available for Graduate Credit: Forest Biometry (U), Forest Bionomics (U), Tree Physiology (U), Insects and Diseases in Forest Resource Management (U), Wood and A Building Material (U), Wood Science and Utilization (U), Forest Management Principles (U), Forest Resources Practicum (U), Forest Tree Improvement, Advanced Forest Management

UNIVERSITY OF WISCONSIN - MILWAUKEE
3201 North Maryland Phone: (414) 963-4586
Milwaukee, Wisconsin 53201
Dean of Graduate School: R. McQuistan

College of Letters and Sciences
Dean: W.F. Halloran

DEPARTMENT OF BOTANY
Phone: Chairman: (414) 963-4213

Chairman: F.W. Stearns
Professors: J. Baxter, J.L. Blum, R. Costello, D.W. Dunlop, J. Loewenberg, P.J. Salamun, F.W. Stearns P.B. Whitford, A. Throne (Emeritus)
Associate Professors: R. Grunewald, F.H. Kaufmann (Emeritus), D.K. Gehrz, T.C. Nelson
Assistant Professor: G. Kennedy
Research Assistants: 3
Teaching Assistants: 13
University Fellow: 1

Field Stations/Laboratories: University of Wisconsin Field Station - Saukville

Degree Program: Botany

Undergraduate Degree: B.S.
Undergraduate Degree Requirements: Not Stated

Graduate Degrees: M.S., Ph.D.
Graduate Degree Requirements: Master: 24 graduate credits, 18 in Botany, thesis, comprehensive examination Doctoral: M.S. or equivalent prerequisite, 54 credits beyond baccalaureate, one foreign language,

Dean of College: E. Norwood

LIFE SCIENCE DISCIPLINE, DIVISION OF SCIENCE
Phone: Chairman: (414) 553-2326

Head: A.M. Williams (Discipline Coordinator:
Professor: E.C. Gasiorkiewicz
Associate Professors: J.S. Balsano, C.M. Chen, S.P. Datta, B.E. Esser, E.M. Goodman, A.M. Williams
Assistant Professors: O.M. Amin, E.P. Wallen
Instructor: R.M. McKee, A.J. Renish

Degree Program: Life Science

Undergraduate Degree: B.S.
Undergraduate Degree Requirements: Required core of 20 credits life science courses, 20 credits life science electives, 1 semester college Algebra and Trigonometry, 2 semesters General Chemistry with Laboratory, 2 semesters Organic Chemistry

UNIVERSITY OF WISCONSIN - PLATTEVILLE
Platteville, Wisconsin 53818 Phone: (608) 342-1134
Dean of Graduate Studies: L. Gerside
Dean of College: Not Stated

BIOLOGY DEPARTMENT
Phone: Chairman: (608) 342-1134

Head: R. Wagner
Professors: R.H. Faulkes, J.L. Strohm, R.O. Wagner
Associate Professors: J.N. Dykstra, C.T. Dzukanowski, J.R. Holler, S.R. Tandon, M.J. Tufte, H.L. Willis
Assistant Professor: D.M. Molitor

Degree Programs: Biology, Botany, Zoology

Undergraduate Degree: B.S.
Undergraduate Degree Requirements: Not Stated

Graduate Degrees: M.S. (Ed.), M.A.T.
Graduate Degree Requirements: Not Stated

Graduate and Undergraduate Courses Available for Graduate Credit: Not Stated

UNIVERSITY OF WISCONSIN - RIVER FALLS
River Falls, Wisconsin 54022 Phone: (715) 425-3011
Dean of Graduate Studies: P.S. Anderson

College of Arts and Sciences
Dean: R.D. Swensen

DEPARTMENT OF BIOLOGY
Phone: Chairman: (715) 425-3591 Others: 425-3011

Chairman: J.M. Bostrack
Professors: V. Akins, J.M. Bostrack, R.L. Calentine, M.E. Michaelson
Associate Professors: S. Goddard, J. Richardson
Assistant Professors: J. Bjerke, C.D. Finstad, M. Harned, J. Hudson, T. Morrow

Field Stations/Laboratories: Pigeon Lake Field Station - Drummond

Degree Program: Biology

Undergraduate Degree: B.S.
Undergraduate Degree Requirements: Required Department Courses 34 credit hours, Free Electives in Biology 17 credit hours, Required Supporting Courses 23 credit hours

Graduate Degree: M.S.
Graduate Degree Requirements: Master: 9-13 credits in Education, 18-27 credits in the area of specialization, 9-13 credits outside the area of specialization

Graduate and Undergraduate Courses Available for Graduate Credit: Plant Anatomy (U), Plant Pathology (U), Animal Microtechnique (U), Wildlife Biology (U), Animal Histology (U), Plant Physiology (U), Plant Growth (U), Botanical Microtechnique (U), Morphology of Non-Vascular Plants (U), Ichthyology (U), Ornithology (U), Mammalogy (U), Vertebrate Embryology (U), Parasitology (U), Limnology (U), Seminar (U), Integrated Biological Concepts, Independent Research, Molecular Biology, Comparative Physiology, Environmental Biology, Thesis

College of Agriculture
Dean: J.C. Dollahan

DEPARTMENT OF ANIMAL AND FOOD SCIENCES
Phone: Chairman: (715) 425-3704 Others: 425-3011

Chairman: R.P. Johnston, Jr.
Distinguished Professor of Animal Science: M.E. Ensminger
Professors: J.C. Dollahon (Dean), R.C. Gray, L.S. Wittwer
Associate Professor: R.P. Johnston, Jr.
Assistant Professors: J.V. Chambers, D.W. Henderson, L.H. Kasten

Degree Program: Animal Science, Food Science

Undergraduate Degree: B.S.
Undergraduate Degree Requirements: 192 quarter credits

Graduate Degrees: The department offers courses for the M.S.T. and M.A.T. degree program in agricultural education and the M.S. degree in agricultural economics.

Graduate and Undergraduate Courses Available for Graduate Credit: Physiology of Lactation (U), Monogastric Nutrition (U), Ruminant Nutrition (U), Experimental Nutrition (U), Physiological Chemistry (U), Radiation Methods in Agriculture (U), Endocrinology (U), Animal Breeding (U), Advanced Dairy Management, Advanced Meat Science, Genetics of Animal Populations Swine Management, Beef Cattle Management, Livestock and Meat Evaluation (U), Physiology of Reproduction (U), Artificial Insemination (U)

DEPARTMENT OF PLANT AND EARTH SCIENCE
Phone: Chairman: (715) 425-3345 Others: 425-3011

Chairman: S.F. Huffman
Associate Professors: R.A. Swanson, L.J. Greub, D. Steinegger
Assistant Professor: J.L. Richardson

Degree Programs: Crop Science, Plant Sciences, Soil Science, Pre-Forestry

Undergraduate Degrees: B.S.
Undergraduate Degree Requirements: 192 quarter credits

UNIVERSITY OF WISCONSIN - STEVENS POINT
Stevens Point, Wisconsin 54481 Phone: (715) 346-2159
Dean of Graduate Studies: W.C. Difford
Dean of College: S.J. Woodka

BIOLOGY DEPARTMENT
Phone: Chairman: (715) 346-2159

Chairman: Not Stated
Professors: G.C. Becker, J.B. Harris, C. Long, E. Pierson, R.E. Simpson
Associate Professors: J. Barnes, F. Copes, G. Geeseman, G.Knopf, H. Smith, M.Temp, V. Thiesfeld, C. White, R. Whitmire, R. Wilde
Assistant Professors: R. Freckmann, K. Hall, D. Hay, V. Heig, D. Hillier, E. Pentecost, D. Post, S. Taft
Instructors: R. Quick, S.D. Van Horn

comprehensive preliminary, examination, thesis, oral defense

Graduate and Undergraduate Courses Available for Graduate Credit: Plant Anatomy (U), Botanical Microtechnique (U), Non-Flowering Vascular Plants (U), Morphology of Algae (U), Fungi (U), Applied Microbiology I (U), Applied Microbiology II (U), Bryophytes (U), Identification and Classification of Seed Plants (U), Dendrology (U), Field Collections and Identification (U), Plant Geography (U), Marine Algae (U), Principles of Plant Ecology (U), Environmental Measurements (U), Urban Ecology (U), Biological Principles in Environmental Measurements (U), Urban Ecology (U), Biological Principles in the Management of Natural Resources (U), Plant Physiology (U), Advanced Plant Physiology (U), Bionucleonics (U), Applied Bionucleonics (U), Bionucleonics Laboratory (U), General Genetics (U), Biological Science for Teachers (U), Botanical Diversity, Adaptation and Evolution, Evolutionary Morphology of Flowering Plants, Advanced Mycology, Problems in Applied Microbiology, Experimental Systematics and Evolution, Advanced Plant Ecology, Techniques of Research and Presentation for Biologists, Advances in Experimental and Applied Ecology, Advanced Genetics, Advanced Independent Studies in Botany, Biology Colloquium, Graduate Seminar, Graduate Seminar in Botany, Urban and Systems Ecology, Research, Research Anatomy and Morphology, Research, Phycology and Bryology, Research, Mycology, Research Plant Taxonomy, Research, Plant Ecology, Research Plant Physiology, Research, Genetics, Research, Plant Geography

DEPARTMENT OF ZOOLOGY
Phone: Chairman: (414) 963-4214

Chairman: C. Norden
Distinguished Professor: C. Mortimer
Michael F. Guyer Professor: J. Baier
Professors: J. Anthony, A. Beeton, C. Norden, E. Warner, R. Warren, C. Weise
Associate Professors: M. Ficken, J. Minnich, L. Pauly, N. Press
Assistant Professors: A. Brooks, R. Broyles, E. Lange, D. Mooren, R. Phillips, E. Sloane, M. Staab, D. Van Wynsberghe
Teaching Fellows: 26
Research Assistants: 10
Research Fellows: 1

Field Stations/Laboratories: Cedar - Sauk Field Station - Newburg

Degree Program: Zoology

Undergraduate Degree: B.A., B.S.
Undergraduate Degree Requirements: Thirty credits in Zoology which includes 21 credits in required courses

Graduate Degree: M.S.
Graduate Degree Requirements: Not Stated

Graduate and Undergraduate Courses Available for Graduate Credit: Biology of Animal Parasites (U), Animal Physiology (U), Animal Physiology Laboratory (U), Biometry (U), Ecology (U), Ichthyology (U), Limnology (U), Limnology Laboratory (U), Fishery Biology (U), Herpetology (U), Animal Behavior: Ethology (U), Comparative Animal Histology (U), Cell Biology (U), Physiological Ecology (U), Biological Instrumentation (U), Recent Advances in Limnology and Oceanography (U), Immunotaxonomy (U), Human Genetics (U), Endocrinology (U), Biological Theories (U), Sociobiology, Physiology of Reproduction (U), Lake and Ocean Dynamics: Particular Reference to the Great Lakes, Field Methods and Problems in Great Lakes Research, Avian Biology, Studies of Marine Fishes, Advanced Studies in Histology, Advanced Independent Studies in Zoology, Biology Colloquium, Seminars in Zoology, Molecular and Cell Biology, Animal Morphology and Development, Population and Community Ecology, Aquatic Biology, Behavioral Biology, Physiology and Endocrinology, Systematic Zoology and Evolution

UNIVERSITY OF WISCONSIN - OSHKOSH
Oshkosh, Wisconsin 54901 Phone: (414) 424-5111
Dean of Graduate Studies: Vacant
Dean of College: A. Darken

DEPARTMENT OF BIOLOGY
Phone: Chairman: (414) 424-5111

Chairman: M.A. Rouf
Professors: J.L. Kaspar, J.K. Klicka, I.Y. Mahmound, M.A. Rouf, E. Schwartz, M.H. Schwertfogor, J.W Unger
Associate Professors: A.S. Anand, H.G. Drecktrah, K. A. Feng, N.A. Harriman, M.N. Mahadeva, M. Rigney, W. Sloey, L. Tews, W.B. Willers, D.L. Wright
Assistant Professors: G. Bothner, R. Cyrus, P. Edmonds J. Hein, D. Parker, S. Randerson, F. Spangler, D. Strohmeyer, J. Zilinsky
Instructors: S. Beadle, E. Waddell, O. Willard
Research Assistants: 2
Graduate Assistants and Faculty Assistants: 8

Field Stations/Laboratories: Pigeon Lake and Clam Lake in cooperation with other University of Wisconsin campuses.

Degree Programs: Biology, Microbiology and Public Health

Undergraduate Degree: B.S.
Undergraduate Degree Requirements: B.S. Biology Major - minimum of 36 credits in Biology, plus one year of Chemistry and one year of Mathematics. B.S. Microbiology and Public Health Major - minimum of 43 credits in Microbiology and Public Health, plus two years of College Chemistry, one year of College Physics, and one year of College Mathematics

Graduate Degrees: M.S., M.S.T.
Graduate Degree Requirements: Master: M.S. and M.S.T. - 30 credits, specific requirements: Student must take a year of Organic Chemistry either as a part of the program or a deficiency and a course in calculus as a pre-requisite.

Graduate and Undergraduate Courses Available for Graduate Credit: Morphology of Vascular Plants (U), Food Microbiology (U), Immunology and Serology (U), Biological Microtechnique (U), Plant Physiology (U), General Ecology (U), Parasitology (U), Histology (U), Molecular Biology (U), Radiobiology (U), Microbial Physiology (U), and Public Health and Santitation (U), Integrated Biological Concepts, Systematics and Evolutionary Genetics, Introduction to Biological Research, Environmental Physiology, Plant Morphogenesis, Advanced Topics in Mycology, Botany Seminar, Zoology Seminar, Field Zoology, Biochemical Genetics, Advanced Topics in Microbiology, Microbiology Seminar, Introduction to Microbiological Research, Community Energetics, Biological Thesis, Genetics (U), Plant Taxonomy (U), Comparative Anatomy (U), Bacteriology (U), Animal Behavior (U), Medical Bacteriology and Epidemiology (U), Virology (U), Vertebrate Embryology (U), General Animal Physiology (U), Mycology (U), General Cytology (U), Human Genetics (U), Introductory Limnology (U), Microbial General Cytology (U), Human Genetics (U), Introductory Limnology (U), Microbial Ecology (U), Ornithology (U), Ichthyology (U), Entomology (U), Freshwater Algae (U)

UNIVERSITY OF WISCONSIN - PARKSIDE
Kenosha, Wisconsin 53440 Phone: (414) 553-2121

Research Assistants: 1

Field Stations/Laboratories: Aquatic Research Center - Stevens Point

Degree Program: Biology

Undergraduate Degrees: B.A., B.S.
Undergraduate Degree Requirements: Biology - 34 credits minimum, Botany, Zoology, Ecology, Genetics, Physiology, An advanced Animal Course, An advanced Plant Course, Seminar, Electives, 1 year Chemistry, Mathematics to Calculus, or Calculus or Statistics

Graduate Degrees: M.A.T., M.S.T.
Graduate Degree Requirements: Master: B.S. Distribution plus specific courses tailored to the individual, 12-18 Biology, 6-9 Education, 3-6 Liberal Arts

Graduate and Undergraduate Courses Available for Graduate Credit: Field Biology (U), Organic Evolution (U), Cell Biology (U), Biological Technique (U), Museum Methods (U), Plant Morphology (U), Plant Anatomy (U), General Bacteriology (U), Mycology (U), Plant Pathology (U), Vascular Plant Taxonomy (U), Agrostology (U), Aquatic Plants (U), Plant Physiology (U), Plant Ecology (U), Invertebrate Zoology (U), Animal Parasitilogy (U), Protozoology (U), General Entomology (U), Comparative Vertebrate Anatomy (U), Embryology (U), Natural History of Vertebrates (U), Histology (U), Ichthyology (U), Life Histories of Fishes (U), Herpetology (U), Mammalogy (U), Field Zoology (U), Comparative Animal Physiology (U), Human Reproduction (U), Animal Ecology (U), Animal Behavior (U), Selected Topics in Ecology (U), Problems in Quantitative Biology (U), Special Work (U), Seminar (U), Advanced Ecology, Advanced Genetics, Physiological Mechanisms, Problem Analysis, Integrated Biological Concepts

UNIVERSITY OF WISCONSIN - STOUT

Menomonie, Wisconsin 54751 Phone: (715) 232-0123
Dean of Graduate Studies: N. Runnalls
Dean of School of Liberal Studies: D. Agnew

BIOLOGY DEPARTMENT
Phone: Chairman: (715) 232-2268

Chairman: Not Stated
Professors: O. Anderson, O. Carlson, E. Lowry, L. Mahan
Associate Professors: H. Arneson, D. Dickmann, J. Kainski, G. Nelson, G. Olson
Assistant Professors: R. Wilson

Field Stations/Laboratories: Pigeon Lake Camp - Drummond

Degree Program: None

Undergraduate Degree: None

UNIVERSITY OF WISCONSIN - SUPERIOR

Superior, Wisconsin 54880 Phone: (715) 392-8101
Dean of Graduate Studies: J. Cronk
Dean of College: A. Forbes

DEPARTMENT OF BIOLOGY
Phone: Chairman: Ext. 263

Chairman: S.W. Oexemann
Professors: P. Lukens, S. Oexemann
Associate Professors: P. Arlansky, D. Davidson, E. Dennery, D. Kaufmann, R. Koch
Assistant Professors: R. Morden, W. Swenson
Instructors: L. Hawley
Research Assistant: 1

Field Stations/Laboratires: Wisconsin Point Field Station - Pigeon Lake Field Station

Degree Programs: Biology, Medical Technology

Undergraduate Degrees: B.S., A.B.
Undergraduate Degree Requirements: General Botany 4 credits, General Zoology 4 credits, Ecology 4 credits, Genetics 4 credits, Chemistry 8 credits, Mathematics 4 credits, Total Biology Credits 34

Graduate Degrees: M.S.T., M.A.T.
Graduate Degree Requirements: Master: 12-18 semester credits Biology, 6 semester credits minimum Education, 6 semester credits minimum Liberal Arts

Graduate and Undergraduate Courses Available for Graduate Credit: Economic Botany (U), Ecology (U), Limnology (U), Plant Taxonomy (U), Embryology of Vertebrates (U), Biogeography (U), Field Botany (U), Animal Behavior (U), Parasitology (U), Ornithology (U), Animal Physiology (U), Vertebrate Natural History (U), Ichthyology (U), Fisheries Management (U), Independent Study (U), Seminar in Biology (U)

UNIVERSITY OF WISCONSIN - WHITEWATER

800 West Main Street Phone: (414) 472-1234
Whitewater, Wisconsin 53190
Dean of Graduate Studies: E. Fulton
Dean of College: Not Stated

DEPARTMENT OF BIOLOGY
Phone: Chairman: (414) 472-1072

Chairman: C.W. Brady
Professors: C.W. Brady, J.A. Cummings, C.J. Dennis, H.N. Markham, R.G. Nash
Associate Professors: J.J. Chopp, L. Crone, J.S. Schlough, G.H. Seeburger, S.G. Smith
Assistant Professors: W. Brunckhorst, M. Follstad, W. L. Gross, M. Kalb, C.A. North, J.L. Briggs
Instructors: W.D. Sable, J. Young

Field Stations/Laboratories: Pigeon Lake Field Station - Drummond

Degree Programs: Animal Science, Biological Education, Biology, Plant Sciences

Undergraduate Degrees: B.S., B.A., B.S. in Ed.
Undergraduate Degree Requirements: Graduation requires 120 credits in addition to other requirements, the student must complete one major and minor or one Broadfield major. Students must have 2.0 average

Graduate Degree: M.S.T.
Graduate Degree Requirements: Master: 12-24 credits in subject, 6-9 in Education, 6-9 Liberal Arts, M.A.T. 1/2 subject, 1/2 Education plus state license requirements

Graduate and Undergraduate Courses Available for Graduate Credit: Bacteriology,(U), Immunology (U), Genetics (U), Endocrinology (U), Plant Anatomy (U), Plant Physiology (U), Ichthyology (U), Herpetology (U), Introduction to Ornithology (U), Introduction to Mammalogy (U), Animal Behavior (U), Microtechnique (U), Biological Techniques, History and Philosophy of Biology, Organic Evolution (U), Invertebrate Zoology (U), Entomology (U), Animal Histology (U), Vertebrate Field Biology (U), Parasitology, Radiation Biology (U), General Ecology, Taxonomy of Vascular Plants, Introductory Mycology, Cytogenetics, Molecular Biology, Invertebrate Field Biology, General Limnology, Pollution Biology, Biology Workshop, Seminar

VITERBO COLLEGE

La Crosse, Wisconsin 54601 Phone: (608) 785-3450
Dean of College: Sr. M. Gross

DEPARTMENT OF BIOLOGY

Phone: Chairman: (608) 785-3450

Chairman: J.A.Kawatski
Associate Professors: L.M. Senff
Assistant Professor: A.I. Pollpeter
Assistant Instructor (Temporary): 1

Degree Program: Biology

Undergraduate Degrees: B.S., B.A.

Undergraduate Degree Requirements: B.S. (in semester hours), 34 hours Biology including thesis research, 16 hours in Chemistry, 8 hours Physics, 12 hours Mathematics, B.A. - (in semester hours), 34 hours in Biology including thesis, research, 12 hours Chemistry, 3 hours Mathematics

WYOMING

UNIVERSITY OF WYOMING

Laramie, Wyoming 82070 Phone: (307) 766-1121
Dean of Graduate School: R.J. McColloch

College of Arts and Sciences

Dean: E.G. Meyer

DEPARTMENT OF BOTANY

Phone: Chairman: (307) 766-2236 Others: 766-2380

Head: J.R. Reeder
Professor: J.R. Reeder
Associate Professors: M. Christensen, D.J. Crawford, D.H. Knight, G.A. Pratt
Assistant Professors: A.T. Harrison, T.S. Moore
Visiting Lecturer: 1
Teaching Fellows: 10
Research Assistants: 3
Herbarium Assistant: 1

Field Stations/Laboratories: S.H. Knight Science Camp (Medicine Bow Mountains), Jackson Hole Research Station (Grand Teton National Park)

Degree Programs: Biology, Botany

Undergraduate Degree: B.S.
Undergraduate Degree Requirements: A minimum of 120 hours, 30 hours in Botany or Biology, 20 hours in Botany, including Genetics, or Ecology, Physiology and Genetics, plus ten more hours from Zoology, Microbiology

Graduate Degrees: M.S., Ph.D.
Graduate Degree Requirements: Master: 20 hours of Botany, or 16 hours of Botany plus six hours of Zoology, Chemistry, or Physics Doctoral: Reading knowledge of one foreign language, a minimum of 42 semester hours, with at least one-half of these in courses at 800 (graduate credit only) level.

Graduate and Undergraduate Courses Available for Graduate Credit: Taxonomy of Vascular Plants, Aquatic Botany, Plant Physiology, Algae and Bryophytes, Mycology, Genetics, Cell Structure and Function, Physiological Genetics, Ecology, Special Topics in Genetics, Plant Anatomy, Ecophysiology of Plants, Cytogenetics, Advanced Plant Ecology, Analysis of Primary Cycling and Production in Terrestrial Ecosystems, Evolution and Biosystematics, Research in Taxonomy of Vascular Plants, Phycology, Mycology, Bryology, Biochemical Systematics, Physiology, Morphology, Cytology, Cytogenetics, Ecology, Physiological Ecology

DEPARTMENT OF ZOOLOGY AND PHYSIOLOGY

Phone: Chairman: (307) 766-4207 Others: 766-1121

Head: O.H. Paris
Professors: G.T. Baxter, R.H. Denniston, K.L. Diem, R.W. Fautin (Emeritus), R.A.Jenkins, G.S. Kennington, O.H. Paris
Associate Professors: L.L. McDonald, W.L. Pancoe, M. Parker, J. Smith-Sonneborn
Assistant Professors: R.W. Atherton, R. Bowerman, P.G. Geisert, R.P. George, R.M. Kitchin, T.A. McKean, N. Staton, J.C. Turner, Jr.
Teaching Fellows: 18
Research Assistants: 10

Field Stations/Laboratories: Jackson Hole Biological Research Station - Moran, S.H. Knight Science Camp - (Medicine Bow Mountains)

Degree Program: Animal Physiology, Biology, Cell Biology, Ecology, Genetics, Physiology, Zoology, Zoology and Entomology

Undergraduate Degrees: B.S. (Biology, Zoology, Wildlife)
Undergraduate Degree Requirements: Introductory Calculus, Introduction to Statistics, General Chemistry (2 semesters) and two additional semesters of Chemistry,General Physics (2 semesters), and one course from each of the following areas: Cell and Developmental Biology, Environmental and Population Biology, Genetics, Organismal Biology and Physiology A total of 30 semester hours is required and may include (in addition to the above) certain courses in Biochemistry, Biology,Botany, Microbiology and Animal Science

Graduate Degrees: M.S., Ph.D. (Zoology and Physiology)
Graduate Degree Requirements: Master: Plan A: A minimum of 26 hours (B average) earned in coursework (up to 8 hours may be taken outside the department) and the completion of a research problem and its defense as a written thesis before an examining committee. Plan B: A minimum of 30 hours (B average) earned in course work with at least 14 hours in Zoology and the preparation of one or two papers which shall be denfeded as part of the comprehensive examination Doctoral: A minimum of 42 semester hours (B average) earned in course work with 21 semester hours at the 800 level (graduate only) and not more than 14 semester hours outside Zoology, A preliminary examination, a dissertation embodying the result of original research. The department requires that every Ph.D. student must teach during the period of his study, must have credit in genetics plus upper level credit in two of the three areas: Ecology, Physiology, Cell or Developmental Biology, must earn seminar credit each year, and must complete a minor acceptable to his advisory committee. Languages and other tools are designated by the advisory committee when deemed essential.

Graduate and Undergraduate Courses Available for Graduate Credit: Developmental Biology (U), Genetics (U), Genetics Laboratory (U), Biometry (U), Cytotechnique (U), Histology (U), General Physiology (U), General Vertebrate Physiology (U), Biological Effects of Radiation (U), Physiological Animal Ecology (U), Human Heredity (U), Evolution and Population Genetics (U), Limnology (U), Aquatic Zoology (U), Terrestrial Zoology (U), Animal Ecology (U), Protozoology (U), Comparative Vertebrate Physiology (U), Principles of Game Biology (U), Fisheries Management (U), Fish Culture (U), Ichthyology (U), Ornithology (U), Mammalogy (U), Research in Anatomy, Research in Physiology, Neurophysiology, Physiology of Behavior, Molecular and Cellular Aspects of Development, Research in Ecology, Testing Ecological Theory, Advanced Limnology, Research in Cell and Molecular Biology, Electron Microscopy, Big and Small Game Population Ecology, Game Bird Population Ecology, Advanced Fisheries Management, Graduate Seminar, These are departmental courses only, selected courses from other departments are accepted for graduate credit in many programs.

College of Agriculture

Dean: N.W. Hilston

DIVISION OF ANIMAL SCIENCE

Phone: Chairman: (307) 766-2224

Head: P.O. Stratton
Professors: M.P. Botkin, D.D. Deane, R.A. Field, N. W. Hilston, C.J. Kercher, G.Nelms, R.W. Rice, C.O. Schoonover, P.O. Stratton, T.R. Varnell
Associate Professors: T.G. Dunn , D. Hutto, C.C. Kaltenbach, J. Kunsman, H. Radloff, M. Riley
Assistant Professors: B. Ellis, C.K. Faulkner, C.L. Johnson, L. Paules, B. Steevens
Instructors: J. Kinnison, S. Larsen
Research Assistants: 8
Research Associates: 2

Field Stations/Laboratories: Field Stations at Torrington, Sheridan, Archer, Gillette, Afton and Powell, Wyoming

Degree Programs: Animal Genetics, Animal Husbandry, Animal Physiology, Animal Science, Dairy Science, Food Science, Meat Science, Nutrition, Poultry Husbandry, Wool

Undergraduate Degree: B.S. (Animal Science, Food Science)
Undergraduate Degree Requirements: Animal Science - Animal Science 20, Biological Science 10, Freshman English plus 3 hours in English Journalism, Language or Communications, Humanities and Social Science, Mathematics 5, Food Science - Animal Science 20, Biological Science 15, Freshman English plus 3 hours in English Journalism, Language or Communications, Humanities and Social Science, Mathematics 5

Graduate Degrees: M.S., Ph.D.
Graduate Degree Requirements: Master: Twenty-six semester hours of coursework with a minimum grade point of 3.0 or a b, with thesis for a minimum of 4 semester hours. Doctoral: Forty-two semester hours above the B.S. and an additional 18 semester course hours in allied area if the student prefers as alternative to language requirement. Minimum grade point of 3.0, successful completion of written and oral preliminary examination 6 months prior to defense of a thesis at the final oral.

Graduate and Undergraduate Courses Available for Graduate Credit: Animal Science - Physiology of Domestic Animals, Wool Growing, Marketing and Manufacturing Reproductive Physiology of Domestic Animals, Wool Structures and Properties, Poultry Problems, Advanced Dairy Production, Problems in Animal Science, Principles of Animal Breeding, Applied Animal Breeding, Seminar, Advances in Animal Breeding (U), Advanced Topics in Animal Nutrition (U), Advanced Animal Nutrition (U), Wool Measurement Methods (U), Animal Science (U), Wool Problem Analysis (U), Investigations in Poultry (U), Advanced Reproductive Physiology (U), Investigations in Animal Breeding (U), Advanced Seminar (U), Food Science - Meat Science, Food Plant Management, Food Fermentations, Frozen Products, Concentrated Products, Food Processing, Analysis, Seminar

DIVISION OF BIOCHEMISTRY
Phone: Chairman: (307) 766-5167 Others: 766-3300

Head: G.J. Miller
Professors: R.O. Asplund (Joint), Y.O.Chang,W.W. Ellis, J.W. Hamilton, R.J. McColloch (Dean of Graduate School), V.A. McMahon, G.J.Miller, C. L. Villemez, Jr.
Associate Professors: T.H. Ji, I.I. Kaiser
Assistant Professors: A.D. Anderson (Joint), R.E. Barden (Joint), B.E. Haley, W.R. Melander
Teaching Fellows: 2
Research Assistants: 3
Analyst: 1
Electronics Specialist: 1

Degree Program: Biochemistry

Undergraduate Degree: B.S.
Undergraduate Degree Requirements: Not Stated

Graduate Degrees: M.S., Ph.D.
Graduate Degree Requirements: Master: 26-30 semester hours of course work, thesis or research papers, final examination Doctoral: 42 semester hours of course work, not less than one-half in graduate level courses, preliminary examination, thesis, final examination

Graduate and Undergraduate Courses Available for Graduate Credit: General Biochemistry (U), Radioactive Tracer Techniques in Biochemistry (U), Biochemical Instrumentation and Preparation (U), Biophysical Instrumentation (U), Problems and Topics in Biochemistry (U), Plant Biochemistry, Biochemistry of Lipids, Biochemical Mechanism of Diseases, Biochemistry of Carbohydrates, Biochemistry of Amino Acids and Proteins, Biochemical Catalysis, Biochemistry of Pesticides and Antimetabolites, Biochemistry of Nucleic Acids, Structure and Function of Membranes, Advanced Problems and Topics in Biochemistry

PLANT SCIENCE DIVISION
Phone: Chairman: (307) 766-3103 Others: 766-1121

Chairman: L.I. Painter
Professors: H.P. Alley, C.C. Burkhardt, A.A. Beetle, G.H. Bridgmon, H.G. Fisser, B.J. Kolp, R.L. Lang, R.J. Lavigne, C.W. McAnelly, M. May, L.I. Painter, R.E. Pfadt
Associate Professors: H.W. Hough, K. Johnson, G.A. Lee, J.E. Lloyd, P.C. Singleton
Assistant Professors: K.E. Bohnenblust, R.H. Delaney, A.F. Gale, J.L. Moyer, W.J. Seamands, E.W. Spackman
Research Assistants: 14
Research Associates: 5

Field Stations/Laboratories: Agricultural Substations at Archer, Afton, Gillette, Powell, Sheridan, Torrington

Degree Programs: Agronomy, Crop Science, Entomology, Plant Pathology, Range Science, Soil Science

Undergraduate Degree: B.S.
Undergraduate Degree Requirements: Variable

Graduate Degrees: M.S., Ph.D.
Graduate Degree Requirements: Variable

Graduate and Undergraduate Courses Available for Graduate Credit: Crops - Weed Science and Technology (U), Plant Breeding (U), Crop Problems (U), Seminar in Crops (U), Auxins and Herbicides (U), Research in Crops, Advances in Plant Genetics, Advances in Crop Management, Laboratory Methods in Crop Research Entomology - Insects Affecting Livestock (U), External Morphology of Insects (U), Internal Morphology of Insects (U), Biology of the Honey Bee (U),, Insect Physiology (U), Insect Ecology (U), Classification of Insects (U), Advanced Economic Entomology (U), Problems in Entomology, Seminar in Entomology (U), Research in Entomology, Graduate Seminar, Insect Toxicology, Insect Population Biology, Theory of Insect Systematics, Insect Behavior, Insect Histology, Taxonomy of Immature Insects Plant Pathology - Host-Pathogen Interactions in Plant Disease (U), Plant Disease Problems, Virus Diseases of Plants, Microbiological Techniques Range Management - Field Applications of Range Management Techniques (U), Photo Interpretation for Range and Forest Studies (U), Range Resource Ecology (U), Range Biology (U), Seminar (U), Problems (U), Range Utilization and Improvement (U), Range Survey (U), Quantitative Methods of Range Analysis, Range Plant Distribution, Vegetative Influences, Graduate Seminar Soils - Soil Physics (U), Genesis, Morphology and Classification of Soils (U), Soil Chemistry and Plant Nutrition (U), Soils Problems

DIVISION OF MICROBIOLOGY AND VETERINARY MEDICINE

Phone: Chairman: (307) 766-2170

Head: J.O. Tucker
Professors: P.D. Bear, L.R. Maki, B.L. Swift, M.S. Trueblood, C.F. Wiesen
Associate Professors: R.C. Bergstrom, N. Kingston,
Assistant Professors: J.C. Adams, E.L. Belden, D.R. Caldwell

Field Stations/Laboratories: Wyo ming Veterinary Medical Research Center - Laramie

Degree Program: Microbiology

Undergraduate Degree: B.S.
Undergraduate Degree Requirements: Semester Hours: Microbiology (24), Chemistry, Biochemistry (14), Mathematics (5), Physics (8), Biological Science (15), Social Science, Humanities (15)

Graduate Degree: M.S.
Graduate Degree Requirements: Master: Semester hours: Microbiology (12), Related Courses (14), Research Thesis (4)

Graduate and Undergraduate Courses Available for Graduate Credit: Bacterial Metabolism (U), Problems in Microbiology (U), Advanced General Microbiology (U), Soil Microbiology (U), Immunology (U), Water Microbiology (U), Diagnostic Bacteriology (U), Microbial Genetics (U), Microbial Physiology (U), Food and Dairy Microbiology (U), Introductory Virology (U), Laboratory Techniques in Cell Culture (U), Seminar (U), Parasitology Seminar (U), Introduction to Veterinary Parasitology (U), Problems in Parasitology (U), Diagnostic Parasitology, Parasitic Protozoa (U), Helminthology (U), Diagnostic Parasitology (U), Laboratory Animal Care and Management (U), Veterinary Elements (U), Techniques in Molecular Microbiology, Investigations in Microbiology, Medical Mycology, Advanced Seminar

CANADA

A note by the editor

Canadian universities and colleges are not, with rare exceptions, organized along the same lines as their United States counterparts. It is the case that a few Canadian universities are subdivided into "schools" and "colleges" but the members of any department are assigned to the faculty of a basic discipline (e.g. "medicine", "science") rather than to the staff of an administrative division (e.g. School of Medicine, College of Agriculture). This results in two divergences from U.S. practice. First, replication of departments is almost unknown. The Department of Biochemistry, whether its members belong to the Faculty of Medicine or to the Faculty of Science, provides all the instruction and supervision of research in this topic that is conducted in the institution.

A word of explanation, for Canadian readers unacquainted with the United States system, is probably necessary at this point. It is not unusual for large United States universities to have five autonomous departments of, say, Microbiology variously assigned to the College of Arts and Sciences, the School of Medicine, the School of Dentistry, the College of Agriculture and the College of Vetinary Science.

A second variance between the two systems, in part due to the same circumstances, is that degrees in a Canadian university, in a far more real sense than in a U.S. university, are granted by the institution rather than by the department. Graduates of a Canadian university are accepted by a faculty before graduating from the university. Graduates of a United States university, though theoretically approved by a faculty, are in fact sponsored by a specific department.

These facts necessitate, in the pages that follow, a different method of presentation from that used for U.S. colleges and universities. Canadian institutions are similarly arranged alphabetically by provinces, also in alphabetic order. However degrees and degree topics pertinent to the biological sciences are listed immediately following the names of the deans of the relevant faculties and before the lists of departmental staffs and courses.

So far as is possible the names of departments and courses are listed in the language in which instruction is given.

Une note par le redacteur

Les universités et collèges canadiens demontrent guère le même plan d'organization que possedent leur complements aux États-Unis. Certainement, il existe quelques examplaires des institutiones que se subdivisent en "écoles" et "collèges" mais les savants d'un département sont rangés en facultés universitaires (e.g. médicine, sciences) et ne se trouvent pas isolés dans des divisions administratives particulières (e.g. école de medicine, collège d'agriculture). Il en resulte deux contrastes avec la mode d'operation aux États-Unis. Primo, la réplication des départments est presque inconnue. Le Départment de Biochimie, par example, soit adjoint au Faculté de Science soit au Faculté de Medicine, conduit le totalité d'instruction ou de récherche dans l'université.

Un mot d'explication parait necessaire à ce point pour ceux de mes lecteurs canadiens qui ne sont pas au courant avec la systѐme des États Unis où il est possible de trouver, dans une seule université, un assortiment de cinq départments autonomes de Biochimie departis, par example, au "Collège des Arts et Sciences", a l'"Ecole de Medicine" à l'"École de l'art dentaire" au "Collège d'agriculture" et au "Collège des science vétinaires".

Un deuxième désaccord se dérive, en grand parte, de cette réplication; c'est que les grades canadiens sont conferes en verité, comme ceux des Etats Unis en theorie, par l'université. En effet, les diplomes canadiens sont agrégés d'un faculté avant d'ètre agrégé du université: les diplomes aux Etats-Unis doivent être agregés d'un départment particulier avant d'etre agregés de l'université.

Il est donc necessaire d'offrir, dans les pages suivante, une mode de présentation qui differe de celle employée pour les États Unis. Les deux modes se ressemble en le rangement des institutions de chaque Province en ordre alphabetique. Les départments de chaque université sont aussi presentés en ordre alphabetique mais ils sont precedes par le noms des doyens des facultés pertinents, et des grades offerts, applicables aux sciences biologiques.

Dans la mesure de possible, les noms des départments, et des sujets, sont presentés dans la langue utilisée pour l'instruction.

ALBERTA

UNIVERSITY OF ALBERTA

Edmonton, Alberta T6G2E1

Deans of Faculties:
- Agriculture and Forestry: F.V. MacHardy
- Graduate Studies and Research: J.R. McGregor
- Science: D.M. Ross

Degrees offered:

B.Sc. Animal and Poultry Science, Biochemistry, Entomology, Horticulture, Microbiology, Plant Pathology, Plant Physiology, Soil Microbiology, Zoology

B.Sc. (Honors) Biochemistry, Botany, Entomology, Genetics, Microbiology, Zoology

M.S. Animal and Poultry Science, Bacteriology, Biochemistry, Botany, Entomology, Genetics, Horticulture, Microbiology, Physiology, Plant Pathology, Soil Microbiology, Zoology

Ph.D. Animal and Poultry Science, Biochemistry, Botany, Entomology, Genetics, Horticulture, Microbiology, Physiology, Plant Pathology, Soil Microbiology, Zoology

Faculty of Agriculture and Forestry

DEPARTMENT OF ANIMAL SCIENCE

Phone: Chairman: 432-3235

Chairman: L.P. Milligan
Professors: R.T. Berg, J.P. Bowland, D.R. Clandinin, A.R. Robblee
Associate Professors: C.M. Grieve, R.T. Hardin, G.W. Mathison, L.P. Milligan, B.A. Young
Assistant Professors: F.X. Aherne, R.J. Christopherson, J.R. Thompson
Sesssional Lecturers: 1
Temporary Appointments: 1

Field Stations/Laboratories: Edmonton Research Station - Edmonton, University of Alberta Ranch - Kinsella, Ellerslie Bull and Steer Testing Station - Ellerslie

Degree Programs: Animal Genetics, Animal Physiology, Poultry Science (Nutrition), Science, Animal Nutrition, Animal Biochemistry

Graduate and Undergraduate Courses Available for Graduate Credit: Fundamentals of Nutrition (U), Ruminant Nutrition (U), Applied Monogastic Nutrition (U), Nutritional Experimentation (U), Advanced Livestock Production and Management (U), Grazing Animal Management (U), Animal Population Genetics (U), Animal Improvement (U), Animal Biochemistry (U), Physiological Chemistry of Muscle and Meat (U), Research Project, Ruminant Physiology, Biometrical Techniques in Quantitative Animal Genetics, Seminar (Required each year of all graduate students), Advanced Animal Nutrition, Advanced Protein Nutrition, Applied Poultry Genetics, (U), Processing Poultry and Poultry Products (U), Poultry Diseases

DEPARTMENT OF ENTOMOLOGY

Phone: Chairman: 432-3237 Others: 432-3111

Chairman: G.E. Ball
Professors: G.E. Ball, W.G. Evans
Associate Professors: D.A.M. Craig, R.H. Gooding
Research Assistants: 1
Sessional Instructor: 1

Field Stations/Laboratories: George Lake Field Station - Busby, Alberta

Degree Program: Entomology

Graduate and Undergraduate Courses Available for Graduate Credit: Insect Taxonomy (U), Insect Biochemistry Project in Applied Entomology (U), Insect Toxicology (U), Forest Entomology (U), Medical and Veterinary Entomology (U), Insect Taxonomy, Insect Physiology, Insect Ecology, Advanced Insect Ecology, Advanced Morphology and Histology, Advanced Taxonomy, Project in Insect Physiology, Project in Insect Ecology, Project in Insect Toxicology, Project in Forest Entomology, Project in Medical Entomology

DEPARTMENT OF FOOD SCIENCE

Phone: Chairman: (403) 432-4591 Others: 432-3236

Chairman: H. Jackson
Professors: L.F.L. Clegg, H. Jackson
Associate Professors: D. Hadziyen, F.H. Wolfe
Assistant Professors: P. Jelen, M. LeMaguer
Research Assistants: 3

Degree Programs: Food Science, Food Chemistry, Food Microbiology, Food Processing, Food Engineering,

Graduate and Undergraduate Courses Available for Graduate Credit: Industrial Food Processing (U), Microbiology (U), Food Chemistry (U), Food Analysis and Quanity Control (U), Food Engineering, Sensory Analysis of Foods (U), Brewing and Enology, Industrial Fermentations, Advanced Food Microbiology, Industrial Microbiology, Food Engineering, Advanced Applied Microbiology, Advanced Food Chemistry, Chemistry and Technology of Edible Oils and Fats, Food Science Introduction to Food Science, Introductory Food Processing, Reading Project, Sensory Analysis of Foods, Brewing and Enology, Food Processing - Industrial Food Processing, Processing Dairy Products Food and Industrial Microbiology - Food Microbiology, Industrial Fermentations, Advanced Food Microbiology Industrial Microbiology, Advanced Applied Microbiology Food Chemistry - Food Chemistry I, II, Food Analysis and Quality Control, Advanced Food Chemistry, Chemistry and Technology of Edible Oils and Fats, Food Engineering - Food Engineering, Food Rheology

DEPARTMENT OF PLANT SCIENCE

Phone: Chairman: 432-3239 Others: 432-3111

Chairman: W.H. Vanden Born
Professors: W.T. Andrew, N. Colotelo, W.G. Corns, R.H. Knowles, W.P. Skoropad, M. Spencer, W.H. Vanden Born, P.D. Walton, S. Zacik
Associate Professors: A.W. Bailey, K.G. Briggs, C. Hiruki, Z.P. Kondra, C.T. Phan, E.W. Toop
Research Assistants: 4
Post Doctoral Fellows: 8

Field Stations/Laboratories: Edmonton Research Station - Parkland Farm, Ellerslie Research Station - Ellerslie, Kinsella Research Station - Kinsella

Degree Programs: Plant Breeding, Horticulture, Range Management, Plant Pathology, Weed Science, Plant Physiology and Biochemistry

Graduate and Undergraduate Courses Available for Graduate Credit: In addition to graduate level courses in the areas indicated above, nearly all senior undergraduate courses may be taken for graduate credit, particularly at the M.Sc. level.

Faculty of Science

DEPARTMENT OF BIOCHEMISTRY
Phone: Chairman: 432-3358

Chairman: J.S. Colter
Professors: J.S. Colter, C.M. Kay, N.B. Madsen, W. Paranchych, L.B. Smillie
Associate Professors: P.G. Barton, W. Bridger, S. Igaraski, M.N.G. James, R.N. McElhaney, A.R. Morgan, V.H. Paetkau, D.G. Scraba
Assistant Professors: R. Fletterick, R. Hodges
Instructor: L. Tyrrell
Visiting Lecturers: 2
Research Assistants: 8

Degree Programs: Biochemistry, Chemistry, Microbiology, Genetics

Graduate and Undergraduate Courses Available for Graduate Credit: Topics in Biochemistry (U), Advanced Laboratory (Biochemistry students only), Intermediary Metabolism, Enzymology, Metabolic and Cellular Control Mechanisms, Nucleic Acids, Biosynthesis and Function of Macromolecules, Chemistry of Proteins, Recent Advances (Biochemistry graduates only), Lipids and Membranes, Macromolecular Structure Analysis

DEPARTMENT OF BOTANY
Phone: Chairman: (403) 432-3247 Others: 432-3484

Chairman: P.R. Gorham
Professors: L.C. Bliss, E.A. Cossins, P.R. Gorham, L.L. Kennedy, S.K. Malhotra (Joint), J.G. Packer, W.N. Stewart
Associate Professors: D.D. Cass, G.H. LaRoi, G.D. Wes Weston
Assistant Professors: K.E. Denford, M. Hickman, J.M. Mayo, D.H. Vitt
Teaching Fellows: 11
Administrative/Professional Officers: 3

Field Stations/Laboratories: University of Alberta Botanic Garden and Field Laboratory - Devon

Degree Programs: Biochemistry, Biological Chemistry, Biological Sciences, Biology, Botany, Ecology, Bryology, Lichenology, Physiology, Limnology, Mycology, Paleobotany, Plant Physiology, Plant Anatomy and Morphology, Ecophysiology, Plant Taxonomy

Graduate and Undergraduate Courses Available for Graduate Credit: Flora of Alberta (U), Evolution Relationships in Vascular Plants (U), Plant Anatomy (U), Flowering Plants (U), Plant Physiology (U), Plant Ecology (U), Phycology (U), Bryology and Lichenology (U), Mycology I (U), Mycology II (U), Special Problems, Summer Field Course, Chemosystematics, Plant Geography, Intermediate Plant Physiology I, II, Paleobotany, Advanced Mycology, Vegetation Analysis, Advanced Plant Anatomy, Eco-Physiology, Advanced Plant Ecology, Vegetation of North America, Advanced Taxonomy, Advanced Phycology, Advanced Plant Biochemistry, Advanced Paleobotany

DEPARTMENT OF MICROBIOLOGY
(No reply received)

DEPARTMENT OF ZOOLOGY
Phone: Chairman: (403) 432-3464 Others: 432-3308

Chairman: J.R. Nursall
Professors: W.A. Fuller, J.C. Holmes, V. Lewin, S.K. Malhotra, J.R. Nursall, D.M. Ross, R.F. Ruth, A.L. Steiner
Associate Professors: D.D. Beatty, D.A. Boag, F.S. Chia, H.F. Clifford, R.C. Fox, J.K. Lauber, J.L. Mahrt, J.S. Nelson, R.E. Peter, L. Wang, S.E. Zalik, F.C. Zwickel
Assistant Professors: J.F. Addicott, P.L. Forey, D.N. Gallup, W.C. Mackay, J.O. Murie, W.M. Samuel
Visiting Professor: 1
Research Associates: 2
Associates: 2

Field Stations/Laboratories: R.B. Miller Biological Station - Gorge Creek near Turner Valley, Lac Ste. Anne Field Station - Lac Ste. Anne, Heart Lake Field Station - north of Enterprise, Northwest Territories, Western Canadian Universities Marine Biological Station - Bamfield, Vancouver Island, British Columbia

Degree Programs: Zoology and many combinations such as Zoology-Botany, Zoology-Geology, etc.

Graduate and Undergraduate Courses Available for Graduate Credit: The Chordates (U), The Invertebrates (U), Ichthyology (U), Ornithology (U), Herpetology (U), Mammalogy (U), Embryology (U), Experimental Embryology (U), Mechanisms of Evolution (U), Comparative Animal Physiology (U), Comparative Endocrinology (U), Comparative Animal Behavior (U), Aquatic Invertebrates of Alberta (U), Principles of Parasitism (U), Protozoology (U), Wildlife Parasitology (U), Protozoan Parasites (U), Marine Biology (U), Animal Ecology (U), Quantitative Ecology (U), Limnology (U), Boreal Ecology (U), Wildlife Conservation (U), Comparative Vertebrate Histology (U), Seminar in Fish Ecology and Systematics, Cellular Interactions in Morphogenesis, Immunogenesis, Seminar in Zoology, Special Problems in Zoology, Individual Study, Environmental Physiology, Seminar in Environmental Physiology, Seminar in Comparative Animal Physiology, Advanced Comparative Animal Physiology, Helminthology, Seminar in Advanced Animal Ecology, Population Ecology, Stream Ecology, Problems in Wildlife Conservation, Water Pollution Seminar, Advanced Animal Behavior, Ecological Aspects of Behavior

UNIVERSITY OF CALGARY

2920 - 24th Avenue Phone: (403) 284-5100
Calgary, Alberta
Deans of Faculties:
Arts and Science: R.G. Weyant
Graduate Studies: J.B. Hyne
Medicine: L.E. McLeod

Degrees offered:
B.Sc. Animal Biology, Biochemistry, Biology, Cellular and Microbial Biology, Environmental Biology, Plant Biology

M.Sc. Biology, Medical Science

Ph.D. Biology, Medical Science

Faculty of Arts and Sciences

DEPARTMENT OF BIOLOGY
Phone: Chairman: (403) 284-5260 Others: 284-5261

Head: D. Parkinson
Professors: P.K. Anderson, H.P. Arai, C.D. Bird, R.B. Church, J.D. Duerksen, R. Hartland-Rowe, N.E. Henderson, D. Parkinson, R.P. Pharis, K.E. Sanderson, E. Scheinberg, B.C. Sharman
Associate Professors: R.S. Anderson (Honorary), M. N. Arai, J.D. Bewley, L. Browder, J.W. Costerton, R.W. Davies, S. Herrero, M. Kappor, E.J. Laishley, B.R. McMahon, M.T. Myres, J.L. Wilkens, J.H. Williamson
Assistant Professors: G.A. Din, M.R. Lein
Instructors: G. Bourne, J.R. Holman, A.P. Russell
Part-time Sessionals: 2

Field Stations/Laboratories: Environmental Sciences Centre (Kananaskis) - Kananaskis

Degree Programs: Animal Biology, Cellular and Microbial Biology, Environmental Biology, Plant Biology, Biology.

Graduate and Undergraduate Courses Available for Graduate Credit: Principles of Biology, Organization of Diversity of Life, Principles of Cell Biology, Man as an Organism, Biology and Human Affairs, Principles of Genetics, Ecology of Individuals and Populations, Genetics of Man, Dynamics of Communities and Ecosystems, Cell in Development and Heredity, Biochemical Genetics, Quantitative Methods in Biology, Aquatic Ecosystems, Management of the Biosphere, Population Biology, Biology of Parasitism, Immunology, Environmental Physiology, Biogeography, The Genetics of Populations, Special Problems in Microbial Physiology, Special Problems in Human Genetics, Virology, Enzymes and Regulatory Mechanisms, Topics in Organismic Biology: Mycology, Topics in Organismic Biology: Ornithology, Topics in Organismic Biology: Invertebrate Zoology, Topics in Organismic Biology: Advanced Taxonomy of Vascular Plants, Advanced Systematic Entomology, Advanced Environmental Physiology, Techniques in Electron Microscopy, Biometrical Genetics, Soil Biology, Plant Biology, Plant Anatomy, Morphology and Taxonomy of Non-Vascular Plants, Morphology and Taxonomy of Vascular Plants, Physiology of Plant Growth and Development, Morphology and Taxonomy of Vascular Plants II, Plant Nutrition and Metabolism, Plant Synecology, Plant Autecology, Plant Morphogenesis, Experimental Plant Morphogenesis, Recent Advances in Plant Physiology, Advanced Plant Ecology Microbiology - The Life of Bacteria, Microbial Physiology, Microbial Genetics Zoology - Animal Biology, Introduction to Human Physiology, Introduction to Invertebrate Zoology, The Vertebrates, Mammalian Physiology, Systematics of the Vertebrates, General Entomology, Comparative Animal Physiology I, II, Ethology, Advanced Vertebrate Morphology, Experimental Parasitology, Advanced Invertebrate Physiology

UNIVERSITY OF LETHBRIDGE

4401 University Drive Phone: (403) 329-2111
Lethbridge, Alberta T1K3M1

Deans of Faculty:
Arts and Sciences: F.Q. Quo

Degrees offered:

B.A. Biological Sciences, Psychobiology

DEPARTMENT OF BIOLOGICAL SCIENCES
Phone: Chairman: (403) 329-2245

Chairman: J. Kuijt
Professors: J.B. Kuijt, E.B. Wagenaar
Associate Professors: R.E. Bullock, L.L. Stebbins
Assistant Professors: P.D. Lewis, K. Nakamura, W.H. Sharp
Instructor: T. Delane (Academic Assistant)
Postdoctoral Fellows: 1-2

Field Stations/Laboratories: West Castle, West Castle River, Rocky Mountains, Alberta

Degree Program: Biological Sciences

BRITISH COLUMBIA

NOTRE DAME UNIVERSITY

Nelson, British Columbia, Phone: (604) 352-2241

Deans of Faculties:

Academic Studies: D.V.E. George

Degrees offered:

B.A. Biology

B.Sc. Biology

DEPARTMENT OF BIOLOGICAL SCIENCES

Phone: Chairman: Ext. 46 Others: Ext. 35

Chairman: E.M. Randall
Professor: T.S. Bakshi
Associate Professor: E.M. Randall
Assistant Professors: J. Harling, J. Snyder
Laboratory Instructor: 1

Degree Programs: Biology, Botany, Zoology

SIMON FRASER UNIVERSITY

Burnaby 2, British Columbia Phone: (604) 291-3111

Deans of Faculties:

Graduate Studies: J. Wheatley
Science: A. Aronoff

Degrees offered:

B.Sc. (Honors) Biochemistry, Biological Sciences

M.Sc. Biological Sciences

DEPARTMENT OF BIOLOGICAL SCIENCES

Phone: Chairman: Ext. 3535 Others: 4475

Chairman: J.M. Webster
Professors: J.S. Barlow, B.P. Beirne, F.J.F. Fisher, J.P.M. Mackauer, K.K. Nair, L.M. Srivastava, A.L. Turnbull, A.P. van Overbeeke, W.E. Vidaver, J.M. Webster
Associate Professors: L.J. Albright, P. Belton, J.H. Borden, R.C. Brooke, L.D. Druehl, T. Finlayson, G.H. Geen, M. McClaren, R.M.F.S. Sadleir
Assistant Professors: A.H. Burr, L.M. Dill, P.V. Fankboner, I.R. Glen, C.L. Kemp, G.R. Lister, P.C. Oloffs, J.E. Rahe, H.L. Speer
Visiting Lecturers: 3
Teaching Fellows: 5
Research Assistants: 2

Field Stations/Laboratories: Pestology Field Station - Summerland

Degree Program: Biological Sciences

Graduate and Undergraduate Courses Available for Graduate Credit: Marine Biology, Cell Biology, Genetics, Plant Ecology, Comparative Endocrinology, Adaptation and Adaptability, Social Behavior, Biological Electron Microscopy, Student Seminar, Ionizing Radiation and Radioisotopes in Biology, Fungal Physiology and Development, Biochemistry of the Algae, Aquatic Ecology, Contemporary Problems in Plant Physiology, Biology of Forest Insects, Biology of Visual Photoreceptors, Selected Topics in Population Ecology, Forest Physiology, Advanced Electrophysiology, Photochemistry and Biology, Marine Microbiology, The Biology of Estuaries, Marine Plant Ecology, Photobiology, Vertebrate Reproductive Ecology, Economic Organisms, Internal Processes, Population Processes, Biology of Entomophagous Insects, Physical Controls, Insecticide Chemistry and Toxicology, Pest Prevention and Control Systems, Nematology, Selected Topics in Plant Development and Special Topics

UNIVERSITY OF BRITISH COLUMBIA

Vancouver 8, British Columbia Phone: (604) 228-2101

Deans of Faculties:

Agricultural Sciences: M. Shaw
Graduate Studies: D. McKie
Medicine: D.V. Bates
Science: G.M. Voekoff

Degrees offered:

B.Sc. Biochemistry, Biology, Botany, Microbiology, Physiology, Zoology

B.Sc. (Agriculture) Agricultural Microbiology, Agronomy, Animal Genetics, Animal Physiology, Dairy Science, Entomology, Horticulture, Plant Genetics, Plant Nutrition, Poultry Genetics, Poultry Physiology

M.Sc. Anatomy, Animal Science, Biochemistry, Biology, Botany, Fisheries, Forestry, Genetics, Microbiology, Oceanography, Physiology, Plant Science, Poultry Science, Zoology

Ph.D. Anatomy, Agricultural Microbiology, Animal and Poultry Science, Biochemistry, Botany, Fisheries, Forestry, Genetics, Microbiology, Oceanography, Physiology, Plant Science, Zoology

Faculty of Agricultural Sciences

DEPARTMENT OF ANIMAL SCIENCE

Phone: Chairman: (604) 228-2794

Chairman: W.D. Kitts
Professors: J. Hodges
Associate Professors: R.M. Beames, C.R. Krishnamurti, R.M. Tait
Assistant Professors: R.J. Hudson, H. Saben, J.A. Shelford, R.G. Peterson

Field Stations/Laboratories: Research Farm - Oyster River

Degree Programs: Nutrition, Genetics, Physiology, Management, Wildlife, Land Use

DEPARTMENT OF FOOD SCIENCE

Phone: Chairman: (604) 228-4411

Chairman: W.D. Powrie
Professors: J.F. Richards
Associate Professor: S. Wakei
Assistant Professors: P.M. Townsley, M.A. Tung, J. Vanderstoep
Lecturer: E. Watson
Research Assistants: 18
Teaching Fellows: 2

Degree Program: Food Science

Graduate and Undergraduate Courses Available for Graduate Credit: Food Lipids, Food Pigments and Colorimetry, Chemistry of Food Proteins, Molecular Basis of Chemoreception, Food Suspensions, Emulsions and Foams, Structure and Chemistry of Food Myosystems, Food Carbohydrates, Biorheology, Food Enzymes, Advanced Food Fermentations

DEPARTMENT OF PLANT SCIENCE
Phone: Chairman: (604) 228-3451 Others: 228-4384

Chairman: V.C. Runeckles
Professors: V.C. Brink, G.W. Eaton, A.J. Renney, V.C. Runeckles, M. Shaw, R.L. Taylor, W.G. Wellington, M. Weintraub (Honorary)
Associate Professors: C.A. Hornby, J.W. Neill
Assistant Professors: R.J. Copeman, P.A. Jolliffee, J. H. Myers, B.J.R. Philogène
Instructors: H. Andison (Honorary), J. Bandy (Honorary), D.G. Finlayson (Honorary), A.R. Forbes (Honorary), B. Frazer (Honorary), R.K. Hamilton (Honorary), G.G. Jacoli (Honorary), H.R. MacCarthy (Honorary), F.D. McElroy (Honorary), H.W.J. Ragetli, Jr. (Honorary), R. Stace-Smith (Honorary), J.H. Tremaine (Honorary), N.S. Wright (Honorary)
Visiting Lecturers: 1
Research Fellows: 3

Degree Programs: Agriculture, Range Science

Graduate and Undergraduate Courses Available for Graduate Credit: Applied Plant Ecology, Advances in Pomology, Advanced Plant Genetics and Breeding, Advanced Land-Architecture, Topics in Vegetable Crop Production, Control of Plant Growth, Crop Ecophysiology, Biological Control, Arthropod Vectors, Plant Virology, Topics in Plant Pathology, Topics in Weed Ecology, Responses of Plants to Air Pollutants plus all undergraduate courses at 300 and 400 levels

DEPARTMENT OF POULTRY SCIENCE
(No reply received)

DEPARTMENT OF SOIL SCIENCE
Phone: Chairman: (604) 228-2875

Chairman: C.A. Rowles
Professor: C.A. Rowles
Associate Professors: L.M. Lavkulich, L.E. Lowe
Assistant Professors: T.M. Ballard, T.A. Black, A.A. Bomke, J. de Vries, J.H. Wiens
Research Assistants: 6
Postdoctoral: 1

Degree Programs: Soil Science with concentrations in Soil Chemistry, Soil Physics, Soil Genesis, Cartography and Use, Biometeorology, Forest Soils, Soil Fertility

Graduate and Undergraduate Courses Available for Graduate Credit: Forest Soils, Soil and Water Conservation (U), Soil Fertility, Soil Biology, Chemical Properties of Soils, Physical Behavior of Soils, Biometeorology, Identification, Classification and Cartography of Soils, Interpretation and Use of Soil Survey Information, Methods of Soil Analysis (U)

Faculty of Science

DEPARTMENT OF BIOCHEMISTRY
Phone: Chairman: (604) 228-4487 Others: 228-3178

Head: W.J. Polglase (Acting)
Professors: C.T. Beer, M. Carrach, V.J. O'Donnell, W.J. Polglase, M. Smith, G.M.Tener, S. H. Zbarsky
Associate Professors: P.D. Bragg, A.F. Burton, J.F. Richards
Assistant Professors: E.P.M. Candido, D.E. Vance
Instructor: R. Barton
Teaching Fellows: 3
Research Assistants: 12

Degree Program: Biochemistry

Graduate and Undergraduate Courses Available for Graduate Credit: Not Stated

BIOLOGY PROGRAMME
Phone: Chairman: (604) 228-3366

Programme Chairman: C.V. Finnegan
(Faculty listed as members of Departments of Botany, Microbiology and Zoology)

Field Stations/Laboratories: Western Canadian Universities Marine Station - Bamfield

Degree Programs: Biology (Options - Cell Genetics, General Mathematical and Marine Biology)

Graduate and Undergraduate Courses Available for Graduate Credit: Biology - Biometrics, Biomathematics, Microscopy and Histology, Human Heredity and Evolution, Ecology and Man, Elementary Molecular Biology, Protistology, Population and Community Biology, Population and Community Biology, General Ecology, Cell Physiology, Fundamental Genetics, Principles of Genetics, Principles of Cytology, Marine Ecology, Microbial Ecology, Principles and Techniques in Electron Microscopy, Principles and Techniques in Electron Microscopy II, Comparative Biology, Principles of Radiotracer Methodology in Biological Research, Biological Applications of Radiotracers, Current Topics in Genetics, Advanced Biometrics Marine Science - Marine Invertebrate Zoology, Marine Phycology, Marine Ecology Botany - Non-Vascular Plants, Vascular Plants, Morphology and Taxonomy of Seed Plants

DEPARTMENT OF BOTANY
Phone: Chairman: (604) 228-3554 Others: 228-2133

Chairman: R.F. Scagel
Professors: R.J. Bandoni, B.A. Bohm, K.M. Cole, V.J. Krajina (Honorary), C.O. Person, G.E. Rouse, R.F. Scagel, W.B. Schofield, J.R. Stein, R.L. Taylor, G. H.N. Towers
Associate Professors: K.I. Beamish, T. Bisalputra, G. C. Hughes, J.R. Maze, F.J.R. Taylor, I.E.P. Taylor, E.B. Tregunna
Assistant Professors: C.E. Beil, R.E. Foreman, F.R. Ganders, B.R. Green, A.J.F. Griffiths, C.J. Marchant
Instructors: J.C. Andrews, D.S. Cameron, R.E. De Wreede, J. Luitjens, K.M. Patel
Visiting Research Associate: 1

Field Stations/Laboratories: Bamfield Marine Station - Bamfield

Degree Program: Botany

Graduate and Undergraduate Courses Available for Graduate Credit: Botany - Survey of Algae, Morphology and Taxonomy of Seed Plants, Structure and Evolution of The Bryophyta, Structure and Evolution of Ferns and Fern-allies, Structure and Reproduction of Fungi, Plants and Man, Plant Physiology, Plant Genetics, Plant Anatomy, Physiology and Ecology of Fungi, Biology of Marine Algae, Biology of Freshwater Algae, Phytogeography, Plant Ecology I, II, Plant Biochemistry Fundamentals of Cytogenetics, Paleobotany and Palynology, Botanical Research, Field Botany, Taxonomy of Vascular Plants, Cytogenetics of Natural Populations, Advanced Marine Phycology, Advanced Freshwater Phycology, Practical Marine Phytoplankton, Cytology of Marine Phytoplankton, Cytology of Marine Algae, Advanced Mycology, Advanced Phytogeography, Advanced Plant Autecology, Advanced Plant Synecology, Current Topics in Plant Biochemistry, Chemical Plant Taxonomy, Advanced Plant Physiology I, II, Advanced Paleontology and Palynology, Structure and Development of Pteridophytes and Gymnosperm, Recent Advances in the Biology of Plant Cells, Advanced Topics in Botany Biology - Biometrics, Biomathematics, Microscopy and Histology, Human Heredity and Evolution, Ecology and Man, Principles and Techniques in Electron Microscopy I, II, Comparative Biology, Principles of Radiotracer Methodology in Biological Research, Biological Applications of Radiotracers,

Current Topics in Genetics, Advanced Biometrics, Elementary Molecular Biology, Protistology, Protistology, Population and Community Biology I, II, General Ecology, Cell Physiology, Fundamental Genetics, Principles of Genetics, Principles of Cytology, Marine Ecology, Microbial Ecology, Directed Studies in Biology, Directed Biological Research Marine Phycology, Marine Ecology

DEPARTMENT OF GENETICS
Phone: Chairman: (403) 432-5381 Others: 432-3111

Chairman: R.C. von Borstel
Professors: R. Aksel, J. Kuspira, K.A. Lesins (Emeritus), W.E. Smith, R.C. von Borstel, G.W.R. Walker, J. Weijer
Associate Professors: A. Ahmed, P.J. Hastings, R.B. Hodgetts, K. Morgan, D. Nash
Assistant Professors: J.B. Bell, M.A. Russell

Field Stations/Laboratories: Ellerslie Field Laboratory - Edmonton

Degree Program: Genetics

Graduate and Undergraduate Courses Available for Graduate Credit: Cytogenetics (U), Quantitative Genetics (U), Recombination and Mutation (U), Molecular Genetics (U), Radiation Genetics (U), Genetic Improvement of Cultivated Plants (U), Human Genetics (U), Immunogenetics (U), Population Genetics (U), Developmental and Physiological Genetics (U), Reading Project (U), Research Project (U), Cytology, Current Topics in Molecular Genetics, Topics in Population Genetics, Genetics Seminar, Research Techniques and Molecular Genetics, Research Techniques and Molecular Genetics, Research Techniques in Classical and Molecular Cytogenetics

DEPARTMENT OF MICROBIOLOGY
Phone: Chairman: (604) 228-2501 Others: 228-2501

Professors: J.J.R. Campbell, J. Levy, D.M. McLean, J.J. Stock
Associate Professors: J. Bismanis, T.K. Blackburn, A.F. Gronlund, D.G. Kilburn, R.A.J. Warren
Assistant Professors: J.B. Hudson, E. Ishiguro, B.C. McBride, R.C. Miller, D. Syeklocha, G. Weeks
Instructor: L. Van Leeuwen

Field Stations/Laboratories: Marine Station - Bamfield

Degree Programs: Bacteriology and Immunology, Medical Microbiology, Microbiology

Graduate and Undergraduate Courses Available for Graduate Credit: Immunology (U), Immunology, Pathogenic Bacteriology (U), Virology, Molecular Biology, Bacterial Genetics (U), Metabolic Regulation of Growth (U)

DEPARTMENT OF ZOOLOGY
Phone: Chairman: (604) 228-3168 Others: 228-2131

Head: W.S. Hoar
Professors: A.B. Acton, J.R. Adams, N.A. Auersperg, D.H. Chitty, I. McT. Cowan, P.A. Dehnel, I.E. Efford, C.V. Finnegan, H.D. Fisher, W.S. Hoar, C.S. Holling, J. Kane, C.J. Krebs, P.A. Larkin, J.D. McPhail, T.G. Northcote, T.R. Parsons, A.M. Perks, J. E. Phillips, D.J. Randall, G.G.E. Scudder, H. Stich, D.T. Suzuki, J.M. Taylor, N.F. Wilimovsky
Associate Professors: T.H. Carefoot, P. Ford, P.W. Hochachka, D.R. Jones, A.G. Lewis, N.R. Liley, H. C. Nordan, C.F. Wehrhahn
Assistant Professors: J.D. Berger, C.L. Gass, J.M. Gosline, M.R. Hughes, M. Jackson, H. Kasinsky, W.E. Neill, O.R. Reeves, S.R. Shaw, J.N.M. Smith, C.J. Walters, N. Gilberg (Research)
Instructors: P.J. Ellickson, A. Redlich
Research Assistants: 25
Research Associates: 8

Field Stations/Laboratories: Bamfield Marine Station - Bamfield

Degree Programs: Particular Ph.D. programs are not offered. It has proven most flexible to provide whatever courses are necessary for whatever subject area is being pursued.

Graduate and Undergraduate Courses Available for Graduate Credit: Comparative Vertebrate Zoology, Comparative Invertebrate Zoology, Vertebrate Physiology, Developmental Biology, Biology of Vertebrates, Physiology, Comparative Biology of Reproduction, Introduction to Entomology, Introduction to Animal Behavior, Principles and Methodology in Zoological Research, Principles and History of Biology, Evolution, Terrestrial Animal Ecology, Aquatic Ecology, Molecular Adaptation of Animals to the Environment, Functional and Comparative Histology of the Vertebrates, Entomology, Zoogeography, Marine Invertebrate Zoology, Biology of Fishes, Terrestrial Vertebrate Zoology, Morphogenesis, Experimental Cytology, Histochemistry, Biology of Protozoa, Principles of Applied Ecology, Animal Behavior Laboratory, Comparative Histology and Histophysiology, Advanced Problems in Genetics, Comparative Physiology, Zoology Honours Thesis

Faculty of Medicine

DEPARTMENT OF ANATOMY
Phone: Chairman: Not Stated

Head: S.M. Friedman
Professors: S.M. Friedman, F.D. Garrett, J.A.M. Hinke, M.J. Hollenberg, W.A. Webber
Associate Professors: K.R. Donnelly, C.L. Friedman (Research), C.T. Friz, V. Palaty, C.E. Slonecker
Assistant Professors: W.K. Oudle, M.E. Todd
Teaching Fellows: 1

Degree Programs: Anatomy, Anatomy and Histology, Biophysics

Graduate and Undergraduate Courses Available for Graduate Credit: Gross Human Anatomy, Microscopic Human Anatomy, Microscopic Anatomy, Seminars in Ultrastructure, General Cytological Biophysics, Biophysics of Cell Membranes, Neuroanatomy, Neuroanatomy, Directed Studies in Anatomy, Master's Thesis, Ph.D. Thesis

DEPARTMENT OF PHYSIOLOGY
Phone: Chairman: (604) 228-2671 Others: 228-2494

Head: D.H. Copp
Professors: D.V. Bates, J.C. Brown, D.H. Copp, P. Keeler, J.R. Ledsome, R.H. McLennan, R.L. Noble,
Associate Professors: C.F. Cramer, F. Lioy, J.A. Pearson
Teaching Fellow: 1
Research Assistants: 3

Degree Program: Human Physiology

Graduate and Undergraduate Courses Available for Graduate Credit: Advanced Mammalian Physiology, Advanced Laboratory in Physiology, Elements of Neurophysiology, Advanced topics in Renal Physiology, Advanced Topics in Gastrointestinal Physiology, Advanced Topics in Cardiovascular Physiology, Advanced Topics in Endocrine Physiology, Advanced Topics in Neurophysiology, Seminar in Mammalian Physiology, Directed Studies

UNIVERSITY OF VICTORIA

P.O. Box 1700

Victoria, British Columbia
Deans of Faculties:
Arts and Science: J.P. Vinay
Graduate Study: S.A. Jennings

Degrees offered:
B.Sc. Bacteriology, Biochemistry, Biology

M.S. Biology

Ph.D. Biology

DEPARTMENT OF BACTERIOLOGY AND BIOCHEMISTRY
(No reply received)

DEPARTMENT OF BIOLOGY
Phone: Chairman: 477-6911 (737) Others: Ext. 736

Chairman: M.J. Ashwood-Smith
Professors: F.T. Algard, M.J. Ashwood-Smith, W.G. Fields, A.R. Fontaine, G.O. Mackie
Associate Professors: A.P. Austin, D.J. Ballantyne, M. A.M. Bell, D.V. Ellis, E.M. Hagmeier, J.S. Hayward, L.A. Hobson, J.L. Littlepage, J.E. McInerney, J.N. Owens, R.G.B. Reid, R.A. Ring, E.D. Styles
Assistant Professors: P.T. Gregory, J.W. Paden, M. Paul
Visiting Assistant Professors: 2
Research Assistants: 15
Postdoctoral Fellow: 1
Postdoctoral Assistant: 1
Part-time Lecturers: 2
Visiting Scientist: 1

Field Stations/Laboratories: The University of Victoria is a member of a consortium of western canadian universities operating Bamfield Marine Station on Vancouver Island

Degree Programs: Marine Biology, Terrestrial and Freshwater Ecology, Plant and Animal Physiology, Plant and Animal Morphology, Cellular and Developmental Biology, Systematic Biology

Graduate and Undergraduate Courses Available for Graduate Credit: This information cannot be condensed without being misleading. Interested individuals should send for the Calendar and other departmental information. Write: Chairman, Department of Biology, University of Victoria, P.O. Box 1700, Victoria B.C., Canada

MANITOBA

BRANDON UNIVERSITY

Brandon, Manitoba R7A6A9 Phone: (204) 727-5401
Deans of Faculties:
Science: D.R. Mour

Degrees offered:
B.Sc. Botany, Zoology

DEPARTMENT OF BOTANY
(No reply received)

DEPARTMENT OF ZOOLOGY
Phone: Chairman: Ext. 223

Head: D.B. Stewart (Acting)
Associate Professor: R.H. Hannah
Assistant Professor: C.L. Glenn, R.F.C. Smith

Field Stations/Laboratories: H. Stewart Perdue Field Station - Oak Lake

Degree Program: Botany

Graduate and Undergraduate Courses Available for Graduate Credit: Not Stated

UNIVERSITY OF MANITOBA

Winnipeg, Manitoba R3T 2N2 Phone:474-9204
Deans of Faculties:
Agriculture: L.H. Shebeski
Graduate Studies: S. Standil
Medicine: A. Naimark
Science: R.D. Connor

Degrees offered:
B.Sc. Botany, Microbiology, Zoology

B.Sc. (Honors) Botany, Microbiology, Zoology

B.S.A. Animal Science, Entomology, Food Science, Plant Science

B.H.Ec. Foods and Nutrition

M.Sc. Anatomy, Animal Science, Biochemistry, Botany, Entomology, Food Science, Medical Microbiology, Microbiology, Pharmaceutical Microbiology, Physiology, Plant Science, Zoology

Ph.D. Anatomy, Animal Science, Biochemistry, Botany, Entomology, Medical Microbiology, Physiology, Plant Science, Zoology

Faculty of Agriculture

DEPARTMENT OF ANIMAL SCIENCE
(No reply received)

DEPARTMENT OF ENTOMOLOGY
Phone: Chairman: 474-9257 Others: 474-9204

Head: A.J. Thorsteinson
Professors: P.S. Barker (Adjunct), R.A. Brust, J.E. Guthrie (Adjunct), A.L. Hamilton (Adjunct), S.C. Jay, A.G. Robinson, A.J. Thorsteinson, C.R.B. Webster, H. Westdal (Adjunct)
Associate Professor: G.M. Findlay
Assistant Professor: W. Hanec

Field Stations/Laboratories: Glenlea Research Station - Delta Research Station, Pesticide Testing Laboratory

Degree Program: Entomology

Graduate and Undergraduate Courses Available for Graduate Credit: Anatomy and Physiology of Insects, Insect Ethology, Principles of Insect Taxonomy, Principles of Insect Control, Insect Ecology, Social Insects, Pesticide Toxicology, Insects Affecting Man and Animals (U), Insect Taxonomy, Advanced Insect Physiology, Advanced Economic Entomology, Advanced Insect Ethology, Advanced Insect Ecology, Advanced Studies of Social Insects, Stored Products Entomology, Insect Population Management, Advances in Pesticide Toxicology, Graduate Courses

DEPARTMENT OF FOOD SCIENCE
(No reply received)

PLANT SCIENCE DEPARTMENT
Phone: Chairman: (204) 474-8221

Head: R.C. McGinnis
Rosner Research Chair: E.N. Larter
Professors: C.C. Bernier, W. Bushuk, J.D. Campbell, K.W. Clark, L.E. Evans, A.C. Ferguson, S.B. Helgason, L.J. LaCroix, E.N. Larter, R.C. McGinnis, B.R. Stefansson, A.K. Storgaard
Associate Professors: R.D. Hill, F.W. Hougen, P.J. Kalsikes, J.A. Menzies, E.H. Stobbe, G.M.Young
Assistant Professors: B.L. Dronzek, B.L. Jones, L.M. Lenz, W. Woodbury
Adjunct Professors: R.J. Baker, G.N. Irvine, C.F. Shaykewich, R. Tkachuk, G.R.B. Webster
Research Professor: J.P. Gustafson
Research Associates: 6

Field Stations/Laboratories: Glenlea Research Station - 12 miles south of Winnipeg City Limits

Degree Programs: Genetics - Plant Breeding, Horticulture, Cereal Chemistry, Physiology, Weed Science, Forage Crops

Graduate and Undergraduate Courses Available for Graduate Credit: Lipids, Laboratory Methods in Plant Breeding, Topics in Genetics, Topics in Plant Breeding, Genetics of Micro-organisms, Specialized Plant Pathology, Advanced Genetics, Advanced Plant Breeding, Aneuploidy, Cell Physiology, Nucleic Acids, Amino Acids and Proteins, Physical Biochemistry, Advanced Grain Science and Technology, Plant Science Seminar, Tracer Methodology in Biological Research, Control Mechanisms in Cellular Metabolism, Statistical Genetics and Plant Breeding, Breeding of Horticultural Plants, Nuclear Cytology, Cytogenetics, Carbohydrates, Advanced Weed Science, Special Problems in Plant Sciences I - Crops, II - Plant Protection, III - Biochemistry - Physiology, IV - Research Methodology Pesticide Analysis

Faculty of Medicine

DEPARTMENT OF ANATOMY
Phone: Chairman: 786-4333 Others: 786-3652

Head: K.L. Moore
Professors: F.D. Bertalanffy, K. Hoshino, K.L. Moore, E.J.H. Nathaniel, I.M. Thompson (Emeritus)
Associate Professors: C.R. Braekevelt, D.M. Cox, G. Froese, R.E. Grahame, J.L. Hamerton, J.C. Hay, J.B. Hyde, Y. Kameyama, K.P. Nagy, D.R. Nathaniel, T.V.N. Persaud, M.H.K. Shokeir, J. Stack-Haydon, C.E. Thomas

Assistant Professors: R.L. Cooke, J.E. Cooper, P.H. Decter, G. Hunzinger, Y.T. Kim (Research Associate), M. Ray, D.N.P. Singh, J.A. Thliveris
Instructors: B.R. Blakenberg, W.J. Dahlgren, A.J. Fernando, M.K. Kiernan, R.A. Mann, M. Peat, M.N. Pflueger
Visiting Lecturers: 1
Demonstrators: 5

Degree Programs: Anatomy, Genetics, Histology

Graduate and Undergraduate Courses Available for Graduate Credit: Human Microscopic Anatomy, Human Macroscopic (Gross) Anatomy, Neuroscience, Human Developmental Anatomy (Embryology), Advanced Human Macro scopic (Gross) Anatomy, Methodology of Research, Cell Biology, Introduction to Human Genetics, Clinical Genetics, Mammalian and Human Cytogenetics, Mammalian Somatic Cell Genetics, Introduction to Electron Microscopy, Advanced Course in Electron Microscopy, Human Biochemical Genetics, Seminars in Human Genetics, Cellular and Molecular Genetics, Human Cytology and General Histology, Experimental Teratology

DEPARTMENT OF BIOCHEMISTRY
Phone: Chairman: (204) 786-3593

Chairman: M.C. Blanchaer
Professors: M.C. Blanchaer, J.M. Bowness, K. Dakshinamurti, L.G. Israels (Adjunct)
Associate Professors: R.J. Hoeschen, W.M. Hryniuk (Adjunct), F.C. Stevens, J.H. Wang, K. Wrogemann, E.W. Yamada
Assistant Professors: P.R. Desjardins, H.K.L. Jacobs
Instructor: R.L.O. Boechx
Visiting Lecturers: 7
Research Assistants: 1
Teaching Assistants: 4

Degree Program: Biochemistry

Graduate and Undergraduate Courses Available for Graduate Credit: General and Comparative Biochemistry, Nucleic Acids, Proteins, Enzymology, Biochemical Techniques, Seminars in Biochemistry, Topics in Biochemistry, Lipid Metabolism, Regulation of Metabolism, Biochemistry (U)

DEPARTMENT OF MEDICAL MICROBIOLOGY
Phone: Chairman: 786-3522

Chairman: J.C. Wilt
Professors: J.C. Wilt, J.A. Romeyn, J.B.G. Kwapinski
Associate Professors: C.K. Hannan, N. Kordova, S. Parker, A. Ronald, H.I. Sayed, A.M. Wallbank, P. Warner, G.M. Wiseman
Assistant Professors: E.M. D. Cleveland, D.V. Cormack, G. Fox, M. Gurwith, A.F. Holloway, H.G. Stiver
Instructors: B. Harris, A.C. Maniar, W. Meyer, W. Stackiw (Lecturer)
Visiting Lecturers: 5
Teaching Fellows: 4
Research Assistants: 1
Research Associate: 1

Field Stations/Laboratories: Health Sciences Centre Microbiological Laboratories

Degree Program: Microbiology

Graduate and Undergraduate Courses Available for Graduate Credit: Virology, Medical Mycology, Clinical Bacteriology, Microbial Pathogenicity, Immunochemistry, Seminar Course

DEPARTMENT OF PATHOLOGY
(No information available)

DEPARTMENT OF PHYSIOLOGY
(No reply received)

DEPARTMENT OF IMMUNOLOGY
(No reply received)

Faculty of Science

DEPARTMENT OF BOTANY
Phone: Chairman: 474-9894 Others: 474-9368

Head: J. Reid (Acting)
Professors: P.K. Isaac, J. Reid, J.M. Stewart, E.R. Waygood, J. Dugle (Adjunct), D. Gillespie (Adjunct), K. Kim (Adjunct), J. Mills (Adjunct), R. Rohringer (Adjunct), D. Samborski (Adjunct)
Associate Professors: S. Badour, R.E. Longton, D. Punter, G.G.C. Robinson, J.M. Shay
Assistant Professors: T. Booth (Adjunct), B.R. Irvine, A. Olchowecki (Adjunct), C.E. Palmer (Adjunct), L. Van Caeseele

Field Stations/Laboratories: University Field Station - Delta Marsh, Manitoba

Degree Programs: Bryology, Ecology, Mycology and Plant Pathology, Phycology and Physiology

Graduate and Undergraduate Courses Available for Graduate Credit: Advanced Mycology (U), Advanced Plant Ecology, Advanced Plant Physiology, Advanced Taxonomy (U), Aquatic Mycology (U), Biology of Cells Biology of Organisms, Biology of Populations, Bryology (U), Experimental Phycology, Laboratory Methods, Microscopy and Microtechnique (U), Phycology (U), Phycology II (U), Physical Cytology, Physiology and Biochemistry of Algae (U), Physiology of Parasitism, Plant Ecology (U), Plant Growth and Development (U), Plant Metabolism (U), Plant Pathology, Pollution Biology (U), Wetland Ecology, Principles of Plant Pathology (U), Problems in Mycology, Special Topics in Botany

DEPARTMENT OF MICROBIOLOGY
Phone: Chairman: (204) 474-9372

Head: I. Suzuki
Professors: N.E.R. Campbell, H. Halvorson, R.Z. Hawirko, H. Lees, H.B. LeJohn, I. Suzuki
Associate Professors: D.N. Burton, R.M. Lyric, P.Y. Maeba
Assistant Professors: C.T. Chow, P.C. Loewen
Adjunct Professors: F.W.J. Davis, R.D. Hamilton, J.A. Wright
Instructors: G.R. Klassen
Research Assistants: 16
Teaching Assistants: 16

Degree Programs: Microbiology and Biochemistry

Graduate and Undergraduate Courses Available for Graduate Credit: Molecular Genetics (U), Immunology (U), Fermentations (U), Microbial Ecology, Biochemical Mechanisms, Biological Oxidations and Bioenergetics, Advanced Concepts in Molecular Biology, Advances in Microbial Genetics, Enzymology, Advanced Physiology of Bacteria, Advanced Pathogenic Microbiology and Immunology, Subcellular Microbial Physiology, Special Problems in Microbiology, Current Topics in Mammalian Cell Culture

DEPARTMENT OF ZOOLOGY
Phone: Chairman: (294) 474-9245

Head: H.E. Welch
Professors: H.E. Welch, J.G. Eales, C.C. Lindsey, W.O. Pruitt, Jr., F.J. Ward, G. Lubinsky (Emeritus)
Associate Professors: M. Aleksiuk, J.W.T. Dandy, R.M. Evans, J.H. Gee, R.H. Green, H.W. Laale, M. Samoiloff, K.W. Stewart
Assistant Professors: E. Heubner, J.C. Rauch, S.G. Sealy
Instructors: J.W. Clayton (Adjunct), T.A. Dick (Adjunct)

L.C. Graham (Adjunct), G.H. Lawler (Adjunct), K. Patalas (Adjunct), R.R. Riewe (Adjunct), D.W. Schindler (Adjunct)
Teaching Fellows: 3
Research Associates: 3
Honorary Lecturer: 1

Field Stations/Laboratories: West Blue Lake Field Station - Duck Mountains, Taiga Biological Station - Wallace Lake, plus a shared maintenance of the University of Manitoba Field Station - Delta Marsh

Degree Programs: Anatomy and Histology, Animal Behavior, Animal Physiology, Biometry, Ecology, Fisheries and Wildlife Biology, Histology, Limnology, Nematology, Parasitology, Physiology, Zoology, Wildlife Management, Cell Biology, Embryology

Graduate and Undergraduate Courses Available for Graduate Credit: Advanced Parasitology, Advanced Physiology, Animal Ecology, Advanced Topics in Zoology, Ichthyology, Problems in Evolution, Herpetology, Advanced Embryology, Selected Topics in Animal Behavior, Nematology, Boreal Ecology, Problems in Developmental Zoology I, Problems in Developmental Zoology II, Problems in Biological Statistics, Advanced Limnology, Advanced Mammalogy, Biological Resource Management I, Biological Resource Management II, Advanced Ornithology, plus 19 4th year undergraduate courses (U).

UNIVERSITY OF WINNIPEG

515 Portage Avenue Phone: (204) 786-7811
Winnipeg, Manitoba R3B2E9
Deans of Faculties:
Arts and Science: J. Clake

Degrees offered:
B.A. Botany, Environmental Studies, Molecular Biology

B.Sc. Biology

DEPARTMENT OF BIOLOGY
Phone: Chairman: Ext. 436 Others: 786-7811

Chairman: J.C. Conroy
Associate Professors: J.C. Conroy, F.W.J. Davis, W. D. Evans
Assistant Professors: L.A. Didon, J.A. Dowsett, G.E.E. Moodie, D.W. Galaugher, M.M. Novak, E.J. Ward
Instructors: R.C. Bollman, D.H. Bergman, D.J. Brown, M.A.H. Ismond

Degree Program: Biology

NEW BRUNSWICK

MOUNT ALLISON UNIVERSITY

Sackville, New Brunswick Phone: (506) 536-2040
Deans of Faculty: A.J. Motyer

Degrees offered:
B.A. Biology

B.Sc. Biologie

BIOLOGY DEPARTMENT
Phone: Chairman: (506) 536-2040 Ext. 494

Chairman: D.S. Fensom
Ruggles Gates Professor: W.B. Stallworthy
Associate Professors: D.C. Arnold, H. Harries
Assistant Professors: C.G. Paterson, R.G. Thompson, W.C. Trentini
Instructor: M. Whitla
Special Lecturer: H. Smith (Pathology)

Degree Programs: Biology, Botany, Ecology, Limnology, Medical Technology, Physiology, Plant Biophysics, Zoology

Graduate and Undergraduate Courses Available for Graduate Credit: Special Topics in Botany, Special Topics in Zoology, Special Topics in Microbiology

UNIVERSITÉ DE MONCTON

(Pas de réponse)

UNIVERSITY OF NEW BRUNSWICK

Fredericton, New Brunswick Phone: (506) 455-9471
Deans of Faculties:
Forestry: J.W. Ker
Graduate Studies: R.J. Kavanagh
Science: M. Franklin

Degrees offered:
B.A. Biology

B.Sc. Biology

B.Sc.F. Forestry

M.Sc. Biochemistry, Biology, Botany, Physiology, Zoology

M.Sc.F. Forestry

Ph.D. Biology, Microbiology

DEPARTMENT OF BIOLOGY
Phone: Chairman: 453-4582 Others: 453-4584

Chairman: M.D.B. Burt
Professors: C.W. Argue (Emeritus), M. Franklin (Dean of Science), M.D.B. Burt, O.T. Page, A.R.A. Taylor, N.W. Radforth, U. Paim, B.S. Wright, (Adjunct), B.G. Cumming
Associate Professors: M.D.B. Burt, M.O. Krause, L.A. Dionne, J.A. McKenzie, W.D. Seabrook, B.Y. Yoo, A.J. Wiggs
Assistant Professors: P.J. Cashion, W.H. Coulter, F.B. M. Cowan, T.G. Dilworth, D.W. Hagen, R.A. McAllister, R.T. Riding, R.W. Wein, N.J. Whitney
Instructors: P. Forbes, J. Meeking, B. Bacon, D. Crowe, A. Gloss

Field Stations/Laboratories: Curventon Field Station - Northwest Miramichi

Degree Program: Biology

Graduate and Undergraduate Courses Available for Graduate Credit: Genecology and Speciation (U), Cell Biology (U), Molecular Biology (U), Advanced General Ecology (U), Advanced Wildlife Ecology (U), Problems in Wildlife Biology (U), Phycology, Morphology, Taxonomy, Ecology and Physiology of Algae (U), Phycology II (U), Advanced Plant Genetics (U), Physiology and Ecology of Microorganisms (U), Recent Developments in Bacteriology (U), Advanced General Microbiology (U), Vascular Plant Systematics (U), Morphology of the Lower Vascular Plants (U), Morphology of the Gymnosperms (U), Advanced Plant Physiology (U), Principles of Phytopathology (U), Advanced Mychology (U), Advanced Parasitology (U), Vertebrate Morphology (U), Invertebrate Physiology (U), Fish Physiology (U), Neurophysiology (U), Sensory Physiology (U), Environmental Physiology (U), Comparative Animal Physiology (U), Neuroanatomy and Histology (U), General and Comparative Endocrinology (U), Biology of Animal Behaviour (U), Ethology (U)

NEWFOUNDLAND

MEMORIAL UNIVERSITY OF NEWFOUNDLAND

Elizabeth Avenue
St. Johns, Newfoundland
Deans of Faculties:
Graduate Studies: F.A. Aldrich
Science: W.D. Machin

Degrees offered:
B.Sc. (Honors) Biochemistry, Biology

M.Sc. Biochemistry, Biology

Ph.D. Biochemistry, Biology

Faculty of Science

DEPARTMENT OF BIOLOGY
Phone: Chairman: Ext. 2520

Head: J. Phipps
Professors: F.A. Aldrich, C.W. Andrews, G.F. Bennett, C.C. Davis
Associate Professors: M.M. Anderson, A.K. Bal, Y.A. S. Emara, J.W. Evans, J.M. Green, G. Moskovits, O. A. Olsen, G.R. South, D.H. Steele, V.J. Steele, W. Threlfall
Assistant Professors: V.C. Barber, G.R. Grassard, G.I. McT. Cowan, R. Gordon, R.A. Kahn, R.A. Nolan
Lecturers: P. Scott, A. Whittick, R. Morris (Honorary), F.C. Pollett (Honorary)

Field Stations/Laboratories: Bonne Bay Field Station - Norris Point

Degree Program: Biology

MARINE SCIENCE RESEARCH LABORATORY
Phone: Chairman: 726-6681

Director: D.R. Idler
Research Scientists: L. Crim, B. Truscott, R. Thompson, G. Fletcher, R. Buggeln
Joint Appointments: R. Khan, G. McT Cowan, J. Green, P. Harley
Research Assistants: 12

Degree Programs: Not Stated

Graduate and Undergraduate Courses Available for Graduate Credit: Not Stated

DEPARTMENT OF BIOCHEMISTRY
Phone: Chairman: (709) 753-1200 Ext. 2529

Chairman: C.C. Bigelow
Professors: C.C. Bigelow, L.A.W. Feltham, D.R. Idler (Director, Marine Sciences Research Laboratory), P.J. O'Brien
Associate Professor: E.A. Barnsley
Assistant Professors: J.T. Brosnan, M.E. Brosnan, C.L. Hew, K.M. Keough, P.E. Penner
Visiting Lecturers: 1
Postdoctoral Fellows: 3

Field Stations/Laboratories: Marine Sciences Research Laboratory - St. John's

Degree Program: Biochemistry

Graduate and Undergraduate Courses Available for Graduate Credit: Physical Biochemistry (U), Hormones and Their Mechanism of Action (U), Bioenergetics and Biological Oxidation (U), Membranes (U), Biochemical Research Techniques (U), Biochemical Function of Cellular Components (U), Biochemical Cytology, Advanced Intermediary Metabolism, Cellular Regulatory Mechanisms, Advanced Molecular Biology, Energy Transduction, Advanced Comparative Biochemistry, Biochemistry of Foreign Compounds

DEPARTMENT OF FORESTRY
(No reply received)

NOVA SCOTIA

ACADIA UNIVERSITY

Wolfville, Nova Scotia Phone: (902) 542-2201

Deans of Faculties:
Science: E.C. Smith

Degrees offered:
B.A. Biology

B.Sc. Biology

B.Sc. (Honors) Biology

M.Sc. Biology

DEPARTMENT OF BIOLOGY
Phone: Chairman: (902) 542-2201 Ext. 250

Head: G.M. Curry
Charles F. Myers Professor of Biology: E.C. Smith
Professors: J.S. Bleakney, G.M. Curry, D.G. Dodds, M.A. Gibson, E.C. Smith
Associate Professors: J. Basaraba, F.C. Bent, D.W. Grand
Assistant Professors: G.R. Daborn, D.P. Toews, S. Vander Kloet
Instructors: P.J. Austin-Smith, A.C. Smith, H.R. Taylor
Research Assistants: 3
Graduate Fellows: 9

Degree Program: Biology

Graduate and Undergraduate Courses Available for Graduate Credit: Cell Physiology (U), Plant Physiology (U), Plant Ecology (U), Mammalogy (U), Animal Ecology (U), Ornithology (U), Animal Behaviour (U), Cytogenetics (U), Radiobiology (U), Invertebrate Zoology (U), Comparative Animal Physiology (U), Plant Geography (U), Phycology (U), Advanced Mycology-Phycology (U), Limnology (U), WildlifeEcology, Fishery Biology, Graduate Research

DALHOUSIE UNIVERSITY

Halifax, Nova Scotia Phone: (902) 424-2211

Deans of Faculties:
Arts and Science: G.R. MacLean
Graduate Studies: K.T. Leffek
Medicine: L.B. Macpherson

Degrees offered:
B.A. (Honors) Biochemistry, Biology

B.Sc. (Honors) Biochemistry, Biology, Microbiology

M.Sc. Anatomy, Biochemistry, Biology, Microbiology, Oceanography, Physiology

Ph.D. Biochemistry, Biology, Microbiology, Oceanography Physiology

Institute of Oceanography

DEPARTMENT OF OCEANOGRAPHY
Phone: Chairman: 424-3557 or 3558

Chairman: L.M. Dickie
Professors: C.M. Boyd, A.J. Bowen, L.M. Dickie, E.L. Mills, G.A. Riley, P.J. Wangersky
Associate Professors: R.O. Fournier, C.J.R. Garrett, R.D. Hyndman
Assistant Professors: R.C. Cooke, D.A. Huntley, D.J.W. Piper
Visiting Lecturers: 27
Teaching Fellows: 3
Research Assistants: 2

Field Stations/Laboratories: Temporary Field Stations for work on theses by our students, and the Bedford Institute of Oceanography at Dartmouth lends government ships and gear to go out on oceanographic cruises of interest.

Degree Programs: Prigrams in various disciplines of Oceanography - Chemical, Biological, Physical and Geological.

Graduate and Undergraduate Courses Available for Graduate Credit: Introduction to Oceanography (U), Introduction to Biological Oceanography, Introduction to Chemical Oceanography, Introduction to Geological Oceanography, Introduction to Physical Oceanography, Advanced Biological Oceanography, Advanced Chemical Oceanography, Advanced Marine Geology and Geophysics, Advanced Physical Oceanography, Fluid Dynamics

Faculty of Arts and Sciences

DEPARTMENT OF BIOLOGY
Phone: Chairman: (902) 424-3514 Others: 424-3515

Chairman: K.H. Mann
Professors: C.M. Boyd (Oceanography), M.L. Cameron, F.R. Hayes, O.P. Kamra, W.C. Kimmins, K.E. von Maltzahn, K.H. Mann, I.A. McLaren, E.L. Mills (Oceanography), J.G. Ogden III, E.C. Pielou, G.A. Riley (Oceanography), L.C. Vining
Research Associates: D. Brewer, J.S. Craigie, T. Platt, D.P. Pielou, S. Russell, A. Taylor, J. Mortenson, M. Yoon
Associate Professors: E.W. Angelopoulos, R.G. Brown, R.W. Doyle, J. Farley, E.T. Garside, L.E. Haley, B.K. Hall, M.J. Harvey
Assistant Professors: A.R.O. Chapman, J.V. Collins, G.S. Hicks, P.A. Lane, R.W. Lee, R.P. McBride, R.K. O'Dor, E. Zouros
Instructors: C. Bays, W. Bohaychuk, J. Wilson, W. Joyce, B. Joyce, T. Mobbs, P. Malcolm, R. Pollock, D. Sarty, C. Schom, S. Singh, S. Silcox

Degree Program: Biology

Graduate and Undergraduate Courses Available for Graduate Credit: Advanced Molecular Biology (U), Concepts and Topics in Tissue Culture (U), Cytogenetics (U), Microbial Genetics (U), Biological Effects of Radiation (U), Population Genetics (U), Advanced Ecology-Seminar (U), Theoretical Ecology, Seminar in Population Biology, Pleistocene Biogeography (U), Microbial Ecology (U), Introduction to Oceanography (U), Advanced Biological Oceanography (U), Bacteriology (U), Virology, Immunology (U), Mycology (U), Advanced Topics in Algology (U), Advanced Entomology (U), Ichthyology (U), Ethology (U), Pharmacology: Influence of Chemical Agents on Living Organisms (U), Human Physiology (U), Functions and Structures of the Nervous System (U), Neurophysiology Laboratory (U), Special Topics (U), Honours Research and Thesis, Master Thesis, Ph. D. Thesis

Faculty of Medicine

DEPARTMENT OF ANATOMY
Phone: Chairman: 424-2051 Others: 424-2211

Head: D.G. Gwyn
D.G.J. Campbell Professor of Anatomy: D.G. Gwyn
Professors: D.M. Chapman, F.W. Fyfe, D.G. Gwyn
Associate Professors: R.C.J. Gonsalves, I. Holmes, D. Jimenez-Marin, J.G. Rutherford, S.B. Singh, V.G. Vethamany
Assistant Professors: J.R. Asuncion, Jr., D.H. Dickson, M.M. Hansell, A.C. Marshall
Lecturers: C.M. Morrison
Teaching Fellows: 1

Degree Programs: Anatomy, Anatomy and Histology, Anatomy and Physiology, Histology, Neuroanatomy

Graduate and Undergraduate Courses Available for Graduate Credit: Human Histology, Research-Microanatomical Methods, Seminar, Current Topics in Anatomy, Histochemistry, Advanced Histology, Gross Anatomy, Neuroanatomy, Comparative Animal Histology, Human Embryology, Anatomy for Paramedicals (U)

DEPARTMENT OF BIOCHEMISTRY
Phone: Chairman: (902) 424-2510 Others: 424-2211

Head: C.W. Helleiner
Professors: C.W. Helleiner, S.J. Patrick, D.W. Russell, S.D. Wainwright (Research)
Associate Professors: A.H. Blair, F.I. Maclean, C. Mezei, F.B Palmer, J.A. Verpoorte
Assistant Professors: W.F. Doolittle (Research), M.W. Gray (Research), C.B. Lazier (Research), F.M. Smith, M.W. Spence (Research), L.C. Stewart
Instructors: M.S. DeWolfe, E.S. McFarlane
Visiting Lecturer: 1
Teaching Fellows: 6
Research Assistants: 2

Degree Program: Biochemistry

Graduate and Undergraduate Courses Available for Graduate Credit: Intermediary Metabolism (U), Advanced Instrumentation Techniques (U), Experiments in Metabolism (U), Physical Biochemistry (U), Biochemical Regulatory Mechanisms, Structure and Function of Nucleic Acids, Special Topics in Biochemistry, Seminar in Biochemistry

DEPARTMENT OF MICROBIOLOGY
Phone: Chairman: 424-3587 Others: 424-3562

Chairman: K.R. Rozee
Professors: K.R. Rozee, K.B. Easterbrook
Associate Professors: L.S. Kind, S.H.S. Lee, J.A. Embil
Assistant Professors: E.S. McFarlane, D.E. Mahony, C. Stuttard
Instructors: L. Katz, E.V. Haldane, R.S. Martin, R.L. Ozere
Teaching Fellows: 2
Research Assistant: 1

Field Stations/Laboratories: Pathology Institute, Children's Hospital - Halifax

Degree Program: Microbiology

Graduate and Undergraduate Courses Available for Graduate Credit: Bacteriology (U), Immunology (U), Virology (U), Ultrastructure (U), Genetics (U)

DEPARTMENT OF PHYSIOLOGY AND BIOPHYSICS
(No reply received)

MOUNT SAINT VINCENT UNIVERSITE
Halifax, Nova Scotia Phone: (902) 453-4450
Deans of Faculty:
Academic Dean: Sr. M. Molloy

Degrees offered:
B.Sc. Biology

DEPARTMENT OF BIOLOGY
Phone: Chairman: 453-4450

Chairman: Sr. M.L. Gavin
Professor: L. Wainwright
Associate Professor: Sr. M.L. Gavin
Assistant Professors: Sr. M. Flinn, L. Spencer
Instructor: Sr. S. Martin

Degree Program: Biology

NOVA SCOTIA TECHNICAL COLLEGE
(No reply received)

ST. FRANCIS XAVIER UNIVERSITY
Antigonish, Nova Scotia Phone: (902) 863-3303
Dean of Faculty: J.T. Sears

Degrees offered:
B.Sc. Biology

M.Sc. Biology

DEPARTMENT OF BIOLOGY
Phone: Chairman: 867-2241 Others: 867-2137

Chairman: L.P. Chiasson
Professor: G.N.H. Greenidge
Associate Professors: P.A. Dill, P.G. Rovsell
Assistant Professors: R.H. Crawford, D.A. Drury, N. Seymour

Degree Program: Biological Sciences

ST. MARY'S UNIVERSITY
Robie Street Phone: (902) 422-7331
Halifax, Nova Scotia
Deans of Faculties:
Science: W.A. Bridgeo

Degrees offered:
B.Sc. Biology

BIOLOGY DEPARTMENT
Phone: Chairman: Ext. 165 Others: 422-7331

Chairman: B.M. Kapoor
Professor: A. Rojo
Associate Professors: B.M. Kapoor, K.K. Thomas, M. Wiles, E. Rojo
Assistant Professor: H. Bobr-Tylingo
Research Assistants: 3

Field Stations/Laboratories: Huntsman Marine Laboratory-St. Andrews

Degree Program: Biology

ONTARIO

BROCK UNIVERSITY

Merrittville Highway Phone: (416) 684-7201
St. Catherines, Ontario
Deans of Faculties:
Arts and Science: C. Plint

Degrees offered:
B.Sc. Biological Sciences

B.Sc. (Honors) Biological Sciences

M.Sc. Biological Sciences

DEPARTMENT OF BIOLOGICAL SCIENCES
Phone: Chairman: Ext. 200

Chairman: A.H. Houston
Professors: A.W.F. Banfield, A.H. Houston, B.M. Millman, R.P. Rand
Associate Professors: A.W. Bown, M. Dickman, J.C. Lewis, M.S. Manocha, R.D. Morris, M. Nwagwu
Assistant Professors: A.J.S. Ball, H.V. Danks, M. Tracey, S.M. Pearce
Visiting Lecturers: 1
Research Assistants: 6

Degree Programs: Biochemistry, Biochemistry and Biophysics, Biology, Biophysics, Ecology, Molecular Biophysics, Physiology, Radiation Ecology

Graduate and Undergraduate Courses Available for Graduate Credit: Developmental Biology (U), Vertebrate Histology (U), Biology of Fungi (U), Biology of Algae (U), Zoogeography (U), Plant Physiology (U), Cellular Control Mechanisms (U), Protein Synthesis (U), Environmental Physiology (U), Advanced Comparative Physiology (U), Population Dynamics (U), Advanced Histology (U), Light and Electron Microscopy, Cytology, Plant Physiology, X-ray Diffraction Techniques, Membrane Structure and Function, Muscle Structure and Function

CARLETON UNIVERSITY

Ottawa, Ontario K1S5B6 Phone: (613) 231-4321
Deans of Faculties:
Graduate Study: G. Paquet
Science: H.H.J. Nesbitt

Degrees offered:
B.A. Biology

B.Sc. Biology

B.Sc. (Honors) Biology, Biochemistry

M.Sc. Biology

Ph.D. Biology

DEPARTMENT OF BIOLOGY
Phone: Chairman: (613) 231-2792 Others: 231-3871

Chairman: J.A. Webb
Professors: C.A. Barlow, E.L. Bousfield (Adjunct), C.H. Buckner (Adjunct), G. Haggis (Adjunct), H.F. Howden, V.N. Iyer, K.W. Joy, L. Lefkovitch (Adjunct), E.E. Lindquist (Adjunct), A.T. Matheson (Adjunct), J. McNeill (Adjunct), J.M. Neelin, H.H.J. Nesbitt, D. Oliver (Adjunct), H.A. Robertson (Adjunct), G. Setterfield, J.A. Webb, F. Wightman, D.M Wood (Adjunct)
Associate Professors: I.L. Bayly, T.W. Betz, G.R. Carmody, M.B. Fenton, D.R. Gardner, W.I. Illman, P.E. Lee, M.E. McCully, H.G. Merriam, D.A. Smith, H. Yamazaki
Assistant Professors: S.L. Jacobson, J.D.H. Lambert, J. Sinclair
Instructors: R. Grey, A. Hutton
Visiting Lecturers: 2
Research Assistants: 7
Demonstrators: 8

Degree Programs: Biology, Combined Biology/Geology

Graduate and Undergraduate Courses Available for Graduate Credit: Molecular Genetics, Macromolecular Biosysthesis, Cell Biology, Plant Physiology, Plant Biochemistry, Animal Physiology, Endocrinology, Insect Morphology, Insect Taxonomy, Acarology, Evolution and Biogeography, Mammalogy, Mycology (U), Invertebrate Zoology (U), Plant Morphogenesis (U), Chordate Zoology (U), Molecular Genetics (U), Population Genetics (U), Cell Biology (U), Plant Physiology (U), Animal Physiology (U), Plant Taxonomy (U), Animal Development (U), Insect Morphology (U), Principles of Systematic Entomology (U), Quantitative Ecology (U), Evolution and Biogeography (U)

LAKEHEAD UNIVERSITY

Oliver Road Phone: (807) 345-2121
Thunder Bay, Ontario
Directors of Schools:
Forestry: K. Hearnden
Deans of Faculties:
Science: R. Ross

Degrees offered:
B.Sc. Biology, Natural Sciences

B.Sc. (Honors) Biology

B.Sc.F. Forestry

DEPARTMENT OF BIOLOGY
Phone: Chairman: (807) 345-2121 Ext. 289

Chairman: D.R. Lindsay
Associate Professors: D. Barclay, R. Freitag, W. Graham, S. Magwood, G. Ozburn, J. Ryder
Assistant Professors: G. Harvais, M. Lankester, A. MacDonald
Visiting Lecturer: 1

Field Stations/Laboratories: Black Sturgeon Research Centre - Thunder Bay

Degree Program: Biology

LAURENTRAN UNIVERSITY

Ramsey Lake Road Phone: (705) 675-1151
Sudbury, Ontario P3E2O6
Directors of Schools:
Graduate Studies: W.Y. Watson
Associate Deans of Faculties:
Science: D.H. Williamson

Degrees offered:
B.Sc. Biology

B.Sc. (Honors) Biology

M.S. Biology

DEPARTMENT OF BIOLOGY
Phone: Chairman: Ext. 501

Chairman: D.H.S. Richardson
Associate Professors: M.A. Alikhan, F.V. Clulow, G.M. Courtin, D.H.S. Richardson, A.B. Lakshman, Young
Assistant Professors: E.K. Winterhalder, J.R. Morris
Research Assistant: 1

Degree Program: Biology

Graduate and Undergraduate Courses Available for Graduate Credit: Not Stated

McMASTER UNIVERSITY
Hamilton 16, Ontario L854L8
Deans of Faculties:
Graduate Studies: L.J. King
Medicine: J.F. Mustard
Science: D.R. McCalla
Directors of Institutes:
Research Unit in Biochemistry, Biophysics and Molecular Biology: K.B. Freeman

Degrees offered:
B.Sc. Biology

B.Sc. (Honors) Biochemistry, Biology

M.Sc. Biochemistry, Biology, Biophysics, Molecular Biology

Ph.D. Biochemistry, Biology, Molecular Biology

DEPARTMENT OF BIOLOGY
Phone: Chairman: 525-9140 Ext. 4400

Chairman: D. Davidson
Professors: S.T. Bayley, D. Davidson, D.M. Davies, K.A. Kershaw, S. Mak, J.J. Miller, E.L. McCandless, B.A. Oaks, I. Takahaski, S.F.H. Threlkeld
Associate Professors: A.D. Dingle, D.E.N. Jensen, J.E. Mills-Westermann, R.A. Morton, L. Prevec, G.J. Sorger
Assistant Professors: G.P. Harris, J.N.A. Lott
Lecturers: L. Laking, B.C. Longstaff

Degree Programs: Biology, Biophysics

Graduate and Undergraduate Courses Available for Graduate Credit: Plant Physiology (U), Genetics (U), Plant Development (U), Biophysics (U), Mechanisms of Genetic Variation (U), Chromosome Evolution (U), Ecology of Inland Waters (U), Special Topics in Biology, Cell Biology, Topics in Molecular Biology I, II, Topics in Biophysics, Proteins, Nucleic Acids, Genetics of Fungi, Virology, Insect Development and Physiology, Metabolic Physiology, Plant Physiology, Comparative Animal Cytology and Histology, Topics in Developmental Biology Cytology, Ultrastructure of Plant Cells, Insect Ecology and Behaviour, Biology of Lichens, Systems Ecology, Environmental Microbiology

BIOCHEMISTRY DEPARTMENT
Phone: Chairman: Ext. 2454 Others: Ext. 2498

Chairman: K.B. Freeman
Professors: L.A. Branda, K.B. Freeman, R.H. Hall, B.L. Hillcoat, D.R. McCalla, T. Neilson
Associate Professors: W.W.C. Chan, R.M. Epand, B.M. Ferrier, H.P. Ghosh
Research Assistants: 4
Postdoctoral Fellows: 1

Degree Program: Biochemistry

Graduate and Undergraduate Courses Available for Graduate Credit: Advanced Biochemistry (U), Biochemistry in Contemporary Society (U), Energy Metabolism, General Biochemistry, Enzyme Kinetics, Mechanism of Enzyme Action, Mechanism of Action of Hormones, Biochemistry of Lipids, Special Topics in Biochemistry, Topics in Biosynthesis, Topics in Molecular Biology I, Proteins, Nucleic Acids

QUEEN'S UNIVERSITY
Kingston, Ontario Phone: (613) 547-5511
Deans of Faculties:
Arts and Science: R.L. Watts
Graduate Studies and Research: R.L. McIntosh
Medicine: D.D. Waugh

Degrees offered:
B.A. Biochemistry, Biology

B.A. (Honors) Biology

B.Sc. Biochemistry, Biology

B.Sc. (Honors) Biochemistry, Biology, Life Sciences, Microbiology and Immunology, Physiology

M.Sc. Anatomy, Biochemistry, Biology, Microbiology and Immunology, Physiology

Ph.D. Anatomy, Biochemistry, Biology, Microbiology and Immunology, Physiology

Faculty of Medicine

DEPARTMENT OF ANATOMY
(No reply received)

DEPARTMENT OF BIOCHEMISTRY
Phone: Chairman: (613) 547-5733 Others: 547-6174

Head: P.H. Jellinck
Professors: P.H. Jellinck, R.O. Hurst
Associate Professors: T. Anastassiades, A.F. Clark, J.S. Elce, T.G. Flynn, D.R. Forsdyke, R. Kisilvesky, F. Moller, T. Spencer, M. Sribney, D.J. Walton
Assistant Professors: B. Malchy
Instructor: A. Crowe, E. Walters

Degree Program: Biochemistry

Graduate and Undergraduate Courses Available for Graduate Credit: General Biochemistry, Protein Chemistry, Molecular Biology and Control Mechanisms, Advanced Topics in Molecular Biology, Biochemistry of Steroid Hormones, Advanced Enzymology, Advances in Protein Structure and Function, Biochemistry Seminar Program, Current Topics in Biochemistry

DEPARTMENT OF MICROBIOLOGY AND IMMUNOLOGY
Phone: Chairman: (613) 547-6623 Others: 547-6620

Head: R.B. Stewart
Professors: P. Chadwick, D. Eidinger, P. Faulkner, F. H. Milazzo, R.B. Stewart
Associate Professor: J.S. Chadwick
Assistant Professors: W.P. Aston, K.L. Chung, G.J. Delisle
Lecturers: A.M.B. Krowpinski, R.G. Lewis, M.V. O'Shaughnessy

Degree Programs: Bacteriology, Immunology, Medical Microbiology, Microbiology, Virology

Graduate and Undergraduate Courses Available for Graduate Credit: Microbiology and Immunology (U), Molecular Microbiology (U), Systematic Microbiology (U), Virology (U), Immunology-Immunochemistry (U), Bacterial Physiology (U), Medical Microbiology (U), Advances in Virology (U), Advances in Immunology (U), Advance in Bacterial (U), Advances in Bacterial Physiology (U), Advances in Medical Microbiology (U)

PHYSIOLOGY DEPARTMENT
Phone: Chairman: (613) 547-3022

Head: J.D. Hatcher
Professors: V.C. Abrahams, D.B. Jennings, J. Kraicer, R.E. Semple, D.G. Sinclair
Associate Professors: I.T. Beck, C.K. Chapler, M.A. Chiong, J.V. Milligan, D.H. York
Assistant Professors: G.M. Andrew (Joint), C.E. Bird (Joint), D.W. Lywood (Joint), S.H. Shin, Wales (Joint)
Instructor: P. Abrahams
Postdoctoral Fellow: 1

Degree Programs: Participating department in honors Life Science programme in conjunction with the Departments of Anatomy, Biochemistry, Microbiology, Immunology, Pharmacology and in collaboration with Departments of Biology and Psychology.

Graduate and Undergraduate Courses Available for Graduate Credit: Principles of General Physiology (U), Principles of Membrane Function (U), Advanced Physiology (U), Gastrointestinal Physiology (U), Cardiovascular Physiology (U), Mammalian Neurophysiology (U), Endocrinology (U), Medical Physiology (U)

Faculty of Arts and Sciences

DEPARTMENT OF BIOLOGY
Phone: Chairman: (613) 547-5905 Others: 547-5511

Head: G.R. Wyatt
Professors: R.G.S. Bidwell, S.R. Brown, D.T. Canvin, D.T. Dennis, A.E.R. Downe, H.M. Good, J.A. Keast, B.N. Smallman, A.S. West (Emeritus), G.R. Wyatt
Associate Professors: J.M. Bristow, K. Budd, A.B. Cairnie, F. Cooke, R. Harmsen, C.H. Hood, P.H. Johansen, G. Morris, N.E. Simpson
Assistant Professors: K.R. Brasch, P.W. Colgan, A.A. Crowder, D.A. Culver, C.R. Lyttle, R.J. Robertson, W.J. Roff, B.N. White, P.G. Young
Instructors: W. Garland, J. Hutchison, S. McAlpine
Teaching Fellows: 59

Field Stations/Laboratories: Biological Station - Lake Opinicon, Ontario

Degree Program: Biology

Graduate and Undergraduate Courses Available for Graduate Credit: Advanced Plant Physiology, Animal Physiology, and Biochemistry, Ecology and Evolutionary Biology, Advanced Cell and Molecular Biology, Isotopes in Biology, various courses in other Life Sciences Departments

TRENT UNIVERSITY
Peterborough, Ontario K9J7B8 Phone: (705) 748-1011
Deans of Faculties:
Arts and Science: W.G. Pitman

Degrees offered:
B.Sc. (Honors) Biology

DEPARTMENT OF BIOLOGY
Phone: Chairman: (705) 748-1011 Others: 748-1011

Chairman: I.M. Sandeman (Acting)
Professor: R.L. Edwards
Associate Professors: M. Berrill, J.E. Nighswander, P.M. Powles, I.M. Sandeman
Assistant Professors: R. Jones, D.C. Lasenby, T.R. Matthews
Lecturer: C.D. Johnson

Field Stations/Laboratories: Floating Laboratory

Degree Programs: Biology, Joint Programme Geography/Biology in "Watershed Ecosystem"

Graduate and Undergraduate Courses Available for Graduate Credit: Details may be obtained on application.

UNIVERSITY OF GUELPH
Guelph, Ontario N1G2W1 Phone: (519) 824-4120
Deans of Faculties and Colleges:
Biological Science: K. Ronald
Graduate Studies: H.S. Armstrong
Ontario Agricultural College: C.M. Switzer
Ontario Veterinary College: P.G. Howell
Research: W.E. Tossell

Degrees offered:
B.Sc. (Honors) Biophysics, Botany, Genetics, Fisheries and Wildlife, Marine Biology, Microbiology, Zoology

B.Sc. (Agriculture) Environmental and Agricultural Biology

B.Sc. (Engineering) Biological Engineering

M.Sc. Anatomy, Animal and Poultry Science, Biological Immunity, Crop Science, Environmental Biology, Genetics, Histology, Horticulture, Physiology, Veterinary Microbiology

Ph.D. Animal and Poultry Science, Biological Immunology, Cell Biology, Crop Science, Cytogenetics, Developmental Biology, Endocrinology, Environmental Biology, Genetics, Horticulture, Immunogenetics, Physiology, Veterinary Microbiology

College of Biological Sciences

DEPARTMENT OF BOTANY AND GENETICS
Phone: Chairman: (519) 824-4120 Ext. 2730

Chairman: W.G. Barker
Professors: W.G. Barker, G.L. Barron, D.M. Britton, H.M. Dale, R.T. Riddell
Associate Professors: R.F. Horton, B.C. Lu, R.L. Peterson, W.E. Rauser, D.W. Smith, R.E. Subden
Assistant Professors: J.F. Gerrath, H.L. Kim, J.P. Phillips, R.J. Reader
Research Associate: T.E. Dai
Teaching Fellows: 3

Degree Programs: Botany, Genetics

Graduate and Undergraduate Courses Available for Graduate Credit: Plant Physiology I, II, Plant Anatomy I, II, Morphogenesis, Autecology, Phytosociology, Genetics (Speciation), Genetics, (Regulatory Methods), Fungal Genetics, Cytology, Taxonomy I, II, Seminar, Current Topics, Principles of Quantitative Genetics, Statistical, Animal Breeding Plans, Advanced Cytogenetics, Cytogenetics in Plant Breeding, Forage Crop and Community Breeding, Crop Breeding, Colloquium in Genetics, Cytogenetics and Plant Breeding, Genetics and Cytogenetics in Horticultural Plants, Microbial Genetics

DEPARTMENT OF HUMAN KINETICS
Phone: Chairman: (519) 824-4120 Ext. 3869

Chairman: J.T. Powell
Professor: J.T. Powell
Associate Professors: L.A. Cooper, R.B. Walker
Assistant Professors: E. Bird, J.R. Bruce, J. Charteris, D. Heald, C.J. O'Brien, A. Peepre, R.K. Stallman
Lecturer: C.C. Kelly
Visiting Lecturer: 1

Degree Programs: Health and BiologicalSciences, Human

Kinetics

Graduate and Undergraduate Courses Available for Graduate Credit: Research Methods in Human Kinetics, Laboratory Instrumentation, Advanced Sports Medicine, History and Sociology of Human Kinetics, Physiokinetics, Kinesiotherapy, Biomechanics

DEPARTMENT OF MICROBIOLOGY
Phone: Chairman: (519) 824-4120 Ext. 3361

Chairman: D.C. Jordan
Professors: G.W. Anderson, C.T. Corke, R.J. Douglas, K.F. Gregory, D.C. Jordan, L.A. McDermott
Associate Professors: J.A. Carpenter, N.A. Epps, L.N. Gibbins, R.A. Johnston, R.E. Smith, P.R. Sweeny
Assistant Professors: P. Dobos, C.W. Forsberg, M.M. Hauser, D.K. Kidby, W.J. Vail
Research Assistants: 3

Degree Program: Microbiology

Graduate and Undergraduate Courses Available for Graduate Credit: Advanced Bacterial Physiology, Advanced Microbial Techniques, Advanced Instrumentation, Advanced Serology, Microbial Genetics, Electron Microscopy, Advanced Ultrastructure, Advanced Microbial Ecology

DEPARTMENT OF ZOOLOGY
Phone: Chairman: 824-4120 Ext. 3598 Others: Ext. 2710

Chairman: Not Stated
Professors: R.C. Anderson, F.W.H. Beamish, J.C. George, H.R. MacCrimmon, A.J. Musgrave, K. Ronald
Associate Professors: E.H. Anthony, E.D. Bailey, E.K. Balon, M. Beverely-Burton, S. Corey, D.E. Gaskin, A.L.A. Middleton, J.B. Sprague
Assistant Professors: R.J. Brooks, F.F. Gilbert, E.B. Hartwick, D.M. Lavigne, J.F. Leatherland, L. Lowe-Jinde, G.L. March, B.A. McKeown, D.L. Noakes, J.C. Roff, R.J. Wensler
Instructors: J.S. Francis, K.E. Mitchell, G.F. Ramprashad
Visiting Lecturers: 15

Field Stations/Laboratories: Cruikston Park Farm - Galt, Bradley Farms - Lake St. Clair

Degree Programs: Parasitology, Comparative Physiology Entomology, Freshwater and Marine Biology, Wildlife Biology, Ecology, Animal Behavior

Graduate and Undergraduate Courses Available for Graduate Credit: Ecological Foundation of Resource Use, Problems in Vertebrate Ecology, Advanced Wildlife Management, Aquaculture and Fish Production, Avian Physiology, Vertebrate Myology, Scientific Writing, Seminar, Topics in Advanced Zoology, According to students area of specialization the individual may be required to take undergraduate courses by their individual supervisors.

Ontario Agricultural College

DEPARTMENT OF ANIMAL AND POULTRY SCIENCE
Phone: Chairman: (519) 824-4120 Ext. 2251

Chairman: W.D. Morrison
Professors: G.H. Bowman, T.D. Burgess, G.W. Friars, G.K. MacLeod, J.W. Macpherson, J.H. Pettit, J.B. Stone, J.D. Summers
Associate Professors: R.G. Brown, E.B. Burnside, R.P. Forshaw, D.G. Grieve, E.C. Hunt, W.O. Kennedy, G. J. King, E.T. Moran, D.N. Mowat, H.L. Orr, W.R. Usborne, J.P. Walker, J.W. Wilton, L.G. Young
Assistant Professors: J.G. Buchanan-Smith, R.R. Hacker, J.F. Hurnik, G.M. Jenkinson, I. McMillan, B.S. Reinhart, H.J. Swatland

Degree Programs: Animal Science, Poultry Science

Graduate and Undergraduate Courses Available for Graduate Credit: A large number of undergraduate courses available with permission for graduate credit. Poultry Nutrition, Advanced Poultry Management, Special Project, Principles of Quantitative Genetics, Statistical Models in Quantitative Genetics, Animal Breeding Plans, Growth and Metabolism, Special Topics in Ruminant Nutrition, Meat Science, Animal Science Experimentation, Techniques in Large Animal Nutrition, Mammalian Reproduction I, II, Seminar

DEPARTMENT OF ENVIRONMENTAL BIOLOGY
Phone: Chairman: Ext. 3921

Chairman: F.L. McEwen
Professors: G.L. Barron, L.V. Busch, F.E. Chase, J.D. Cunningham, S.E. Dixon, L.V. Edgington, R.A. Fletcher, C.B. Kelly, B.H. MacNeill, D.H. Pengelly, J.B. Robinson, R.W. Shuel, C.M. Switzer (Dean), G.F. Townsend, F.L. McEwen
Associate Professors: J.F. Alex, W.C. Allan, P.W. Burke, S.G. Fushtey, R. Hall, G. Hofstra, M.V. Smith, G.R. Stephenson
Assistant Professors: C.R. Ellis, N.K. Kaushik, J.E. Laing, J.C. Sutton, R.E. Wright
Visiting Lecturer: 1
Research Assistants: 2
Postdoctorates: 5

Field Stations/Laboratories: Apiculture Field Station - Stone Road

Degree Programs: Bacteriology, Biology, Ecology, Entomology and Parasitology, Genetics, Microbiology, Nematology, Plant Pathology, Apiculture, Plant Physiology, Mycology, Taxonomy

Graduate and Undergraduate Courses Available for Graduate Credit: Advanced Apiculture Management (U), Advanced Economic Entomology (U), Applied Biology Colloquium (U), Bacterial Plant Diseases (U), Biological Activity of Pesticides (U), Biological Control, Biological Control (U), Biology of Aquatic Insects (U), Crop Pathology (U), Dairy Microbiology (U), Diagnostic Mycology I, II, Economic Entomology (U), Environmental Stress and Plant Response (U), Forestry (U), Food Microbiology, Food Microbiology (U), Genetics and Physiology of the Honey Bee (U), Industrial and Dairy Microbiology Laboratory, Insect Biology (U), Insect Ecology (U), Insect Taxonomy, Insect Transmission of Plant Pathogens, Invertebrate Physiology (U), Medical Entomology (U), Microbiological Quanity Control (U), Microbiology (U), Microbiology of Water (U), Natural Chemicals in the Environment (U), Natural History of Insects (U), Pesticides in the Environment (U), Physiology and Biochemistry of Herbicides, Physiology of Fungi (U), Physiology of Invertebrates, Phytopathology I, Phytopathology II, Plant Nematology, Plant Nematology (U), Plant Pathology (U), Plant Viruses I, Plant Viruses II, Poultry Microbiology (U), Principles of Insect Taxonomy, Problems in Applied Entomology, Problems in Environmental Biology (U), Scientific Method in Biological Research, Seminar, Seminar (U), Social Behavior of Insects (U), Soil Microbiology, Soil Microbiology (U), Techniques in Plant Pathology (U), Weed Science (U)

DEPARTMENT OF FOOD SCIENCE
Phone: Chairman: Ext. 3787 Others: 2281

Chairman: J.M.DeMan
Professors: D.R. Arnott, D.A. Biggs, D.H. Bullock, J. M. deMan, D.M. Irvine, A.M. Pearson
Associate Professors: A.N.Myhr, V. Rasper, D.W. Stanley
Assistant Professors: C.L. Duitschaever, J.I. Gray
Research Assistants: 6

Degree Programs: Food Science, Dairy Science

Graduate and Undergraduate Courses Available for Graduate Credit: Food Rheology, Food Flavour Chemistry, Quality Testing Methodology, Food Plant Management, Techniques, Food Additives, Chemistry of Food Lipids, Chemistry of Food Protein, Advances in Food Science, Food Carbohydrates, Seminar

DEPARTMENT OF HORTICULTURAL SCIENCE
Phone: Chairman: 824-4120 Ext. 2783

Courses taken from, and taught by, faculty of other Departments.

Field Stations/Laboratories: Cambridge Horticultural Station - Cambridge

Ontario Veterinary College

DEPARTMENT OF BIOMEDICAL SCIENCES
Phone: Chairman: Ext. 2668

Chairman: H.G. Downie
Professors: P.K. Basrur, P. Eyre, J.P.W. Gilman, A.R. Graham, R.M. Liptrap, J.I.Raeside, G.A. Robinson, R.D. Whiteford
Associate Professors: O.S. Atwal, J.K. Barclay, M.K. Bhatnagar, W.D. Black, W H. Goyd, H.W. Chapman, J. Gadhoke, H.D. Geissinger, M.H. Hardy-Fallding, F. Lotz, N.S. Platonow
Assistant Professors: V. deKleer,P.A. Gentry, I.B. Johnstone, P.M. Mann, R. Saison, S. Yamashiro

Degree Programs: Gross and Microscopic Anatomy, Physiology, Pharmacology, Genetics

Graduate and Undergraduate Courses Available for Graduate Credit: Advanced Gross Anatomy, Functional Neuroanatomy, Experimental Neuroanatomy, Advanced Anatomy of the Fowl, MicroAnatomy Seminar, Histological Techniques for Light and Electron Microscopy, Histochemical Techniques, Developmental Biology, Comparative Embryology, Advanced Comparative Vertebrate Embryology, Cell Biology, Advanced Cytogenetics, Analytical Cytol ogy, Tissue Culture Techniques, Neuropharmacology, Immunopharmacology, Systemic Pharmacology and Pharmacodynamics, Toxicology, Cardiovascular Physiology, Experimental Surgery and Physiology, Gastrointestinal Physiology, Hormonal Steroids, Reproductive Endocrinology, Biological Applications of Radioactive Isotopes, Numerous undergraduate courses as in undergraduate calendar.

DEPARTMENT OF VETERINARY MICROBIOLOGY AND IMMUNOLOGY
Phone: Chairman: Ext. 2576 Others: 824-4120

Chairman: D.A. Barnum
Professors: D.A. Barnum, J.B. Derbyshire, D.G. Ingram, W.R. Mitchell, F.H.S. Newbould
Associate Professors: N.A. Fish, C.L. Gyles, G. Lang, J.R. Long, M. Savan, J. Thorsen, I.R. Tizard, R.B. Truscott
Assistant Professors: S.W. Martin, B.N. Wilkie, K. Nielsen

Field Stations/Laboratories: Puslinch Research Station - Stone Road Research Station

Degree Programs: Veterinary Bacteriology, Veterinary Microbiology, Virology, Epidemiology

Graduate and Undergraduate Courses Available for Graduate Credit: Bacterial Diseases, Advanced Veterinary Bacteriology, Clinical Bacteriology, Mycotic Diseases, Medical Immunology, Advances in Immunology, Hygiene and Quality Control of Food Products I, II, Epidemiology, Epidemiology of Zoonoses, Viral Diseases, Diseases Exotic to North America, Advanced Animal Virology, Seminar

DEPARTMENT OF PATHOLOGY
Phone: Chairman: Ext. 3063 Others: 2592

Chairman: R.G. Thomson
Professors: T.J. Hulland, L.H.A. Karstad, B.M. McCraw, B.J. McSherry, J.D. Schroder
Associate Professors: J. Budd, H.C. Carlson, J.R. Geraci, P.B. Little, G.J. Losos, O. Slocombe, V.E.O. Valli
Assistant Professors: M.A. Fernando, J.H. Lumsden, R.B. Miller, M.G. Maxie
Lecturer: L. Sileo
Visiting Lecturers: 1
Research Assistants: 1

Field Stations/Laboratories: Gale Research Station, Puslinch Experimental Station

Degree Programs: Animal Pathology, Avian Disease, Parasitology, Veterinary Science, Wildlife Diseases

Graduate and Undergraduate Courses Available for Graduate Credit: Pathology, Diagnostic Pathology, Surgical Pathology, Experimental Pathology, Techniques in Pathology, Clinical Pathology, Diagnostic Hematology, Experimental Approaches to Parasitology, Diagnostic Parasitology, Wildlife Diseases - Special Problems, General Pathology (U), Systemic Pathology (U), Parasitology (U), Principles of Disease (U), Introduction to Laboratory Animal Care and Management (U)

UNIVERSITY OF OTTAWA

Ottawa, Ontario K1N6N5 Phone: (613) 231-3311
Deans of Faculties:
Medicine: J.J. Lussier
Arts and Sciences: A. D'Iorio

Degrees offered:
B.Sc. Biochemistry, Biology

M.Sc. Anatomy, Biochemistry, Biology, Histology and Embryology, Microbiology and Immunology, Physiology

Ph.D. Anatomy, Biochemistry, Biology, Histology and Embryology, Microbiology and Immunology, Physiology

Faculty of Medicine

DEPARTMENT OF ANATOMY
(No reply received)

BIOCHEMISTRY DEPARTMENT
Phone: Chairman: 231-2255 Others: 231-2256

Chairman: D.S. Layne
Professors: N.L. Benoiton, A. D'Iorio, P.S. Fitt, J. Himms-Hagen, M. Kates, D.S. Layne, P.R. Proulx, M.T. Ryan, D.R. Whitaker
Associate Professors: N. Begin-Heick, C. Maurides, J.C. Nixon
Assistant Professors: P.J. Anderson, H. Kaplan, D.G. Williamson
Instructors: H.M.C. Heick (Senior Lecturer), A. Quevedo (Senior Lecturer)

Degree Programs: Not Stated

Graduate and Undergraduate Courses Available for Graduate Credit: Not Stated

DEPARTMENT OF HISTOLOGY AND EMBRYOLOGY
Phone: Chairman: 231-4221

Chairman: L.F. Bélanger
Professors: L.F. Bélanger, R. Narbaitz, S. Tolnai
Associate Professors: S.S. Jande,

Visiting Lecturers: 2

Degree Programs: Histology, Embryology, Histochemistry

Graduate and Undergraduate Courses Available for Graduate Credit: Cytology, The Early Human Embryo, General Histology, Special Histo-Embryology, Preparation of Tissues of Light Microscopy, Microscopic Histochemistry, Applied Microscopy and Photomicrography, Radio-Autography, Methods of Tissue Culture, Introduction to Electron Microscopy, Advanced Electron Microscopy

DEPARTMENT OF MICROBIOLOGY AND IMMUNOLOGY
Phone: Chairman: 231-3394 Others: 231-3395

Chairman: J.C.N. Westwood
Professors: M. Beaulieu, R. Rossier
Associate Professors: E. Perry
Assistant Professors: R.V. Iyer, C.M. Johnson-Lussenberg, S.A. Sattar
Visiting Lecturers: 1
Research Assistants: 2

Degree Programs: Microbiology, Immunology

Graduate and Undergraduate Courses Available for Graduate Credit: Not Stated

DEPARTMENT OF PHYSIOLOGY
Phone: Chairman: (613) 231-2320

Chairman: G. Hetenyi, Jr.
Professors: B. Korecky, G.W. Mainwood, K. Rakusan
Associate Professors: K. Kako, G.A. Kinson
Assistant Professors: G.P. Biro, J.S. Cowan, C.A. Guzman (Joint), D.Z. Levine (Joint), K.C. Marshall, D. Parry
Postdoctoral Fellows: 2

Degree Programs: Physiology

Graduate and Undergraduate Courses Available for Graduate Credit: Physiology, Neuroanatomy, Neurophysiology

Faculty of Arts and Sciences

DEPARTMENT OF BIOLOGY
Phone: Chairman: (613) 231-2336 Others: 231-2338

Chairman: Q.N. LaHam
Professors: E.O. Dodson, G. Kaplan, D.J. Kushner, F. LeBlanc, V. Vladykov (Emeritus)
Associate Professors: C. Nozzolillo, S.U. Quadri, J. Vaillancourt, P. Weinberger
Assistant Professors: J.B. Armstrong, D.L. Brown, F. Briand, T. Moon, R.M. Reed
Instructor: N. McAllister
Visiting Lecturers: 1
Research Assistants: 6

Field Stations/Laboratories: Biology Station - Lake Henry - Gracefield, Quebec

Degree Programs: Cell and Molecular Biology, Ecology, Physiology

Graduate and Undergraduate Courses Available for Graduate Credit: Not Stated

UNIVERSITY OF TORONTO
Toronto, Ontario M5S1A1 Phone: (416) 928-2011
Deans of Faculties:
Arts and Science: R.A. Greene
Forestry: V.J. Nordin
Graduate Studies: A.E. Safarian
Medicine: R.B. Holmes

Degrees offered:

B.Sc.F. Forestry

B.Sc. Biochemistry, Biology, Botany, Microbiology, Parasitology, Physiology, Zoology

M.Sc. and Ph.D. Anatomy, Biochemistry, Botany, Forest Ecology, Forest Genetics, Forest Pathology, Immunology, Medical Biophysics, Natural Resources Management, Physiology, Tree Physiology, Wildlife Ecology, Zoology

Faculty of Medicine

DEPARTMENT OF ANATOMY
Phone: Chairman: 928-2692 Others: 928-2714

Chairman: J.S. Thompson
Professors: A.A. Axelrad, E.G. Bertram, J.W.A. Duckworth, D.L. McLeod
Associate Professor: K.O. McCuaig
Assistant Professors: M.E. Blackstein, W.M. Brown, D.H. Cormack, M. Harland, E. Hisaki, V.I. Kalnins, B. Liebgott, I.M. Taylor
Instructors: L. Harper, R.G. MacKenzie, A. Roberts, P. Rodney, C.G. Smith
Demonstrators: 30

Degree Programs: Anatomy, Genetics, Histology, Immunology

Graduate and Undergraduate Courses Available for Graduate Credit: Human Anatomy, including Embryology, Histology, Human Neuroanatomy, Human Growth, Constitution and Evolution, Immunogenetics, Advanced Neuroanatomy, Exploration through Anatomy

DEPARTMENT OF BIOCHEMISTRY
Phone: Chairman: (416) 928-2696 Others: 928-2700

Chairman: G.R. Williams
Professors: R.A. Anwar, G.E. Connell, J.M. Fisher, T. Hofmann, B.G. Lane, R.K. Murray, R.H. Painter, H. Schachter, W. Thompson
Associate Professors: A. Bennick, K.J. Dorrington, E.R.M. Kay, M.A. Packham, L. Pinteric, K.G. Scrimgeour, D.O. Tinker, J.T. Wong
Assistant Professor: N. Camerman
Instructors: L. Dove, J. Giles, D. Johnson, D. Painter

Degree Program: Biochemistry

Graduate and Undergraduate Courses Available for Graduate Credit: Introductory Course in Biochemistry (U), Structure and Function of Protein (U), Biochemistry of Membranes (U), Metabolic Enzymology and Control Mechanisms (U), Bio-organic Mechanisms (U), Nucleic Acids and Protein Synthesis (U), Physical Biochemistry (U), Advanced Biochemical Laboratory (U), Selected Topics in Biochemistry

DEPARTMENT OF EPIDEMIOLOGY AND BIOMETRICS
Phone: Chairman: (416) 928-2020

Head: W.H. le Riche
Professors: W.T. Anderson, D.B.W. Reid
Associate Professors: A. Csima, D. Hewitt (Research)
Assistant Professors: M.J. Ashley (Research), E.A. Clarke, P.N. Corey, J. Hsieh
Visiting Lecturers: 8

Degree Programs: Statistics, Biometry, Epidemiology

Graduate and Undergraduate Courses Available for Graduate Credit: Not Stated

DEPARTMENT OF MEDICAL BIOPHYSICS
Phone: Chairman: (416) 924-0671 Ext. 334

Chairman: G.F. Whitmore
Professors: C.L. Ash, W.R. Bruce, B. Cinader,

L.R. Christensen, C. Fuerst, J.W. Hunt, H.E. Johns, E.A. McCulloch, A. Rothstein, L. Siminovitch, J.E. Till, G.F. Whitmore

Associate Professors: W. Allt, R.S. Ruch, J.R. Cunningham, M. Gold, A.F. Howatson, L.A. MacHattie, R. G. Miller, F.P. Ottensmeyer, R.A. Phillips, A.M. Rauth, W.D. Rider, R. Sheinin, C.P. Stanners, M.A. W. Thompson

Assistant Professors: N. Aspin, A. Bernstein, M.J. Bronskill, M. Buchwald, J. Carver, G. DeBoer, R. Gorczynski, R.P. Hill, D.E. Housman, R.L. Juliano, P.A. Knauf, V. Ling, T.W. Mak, M.V. Peters, E.R. Stanley, K.W. Taylor, R. Worton

Visiting Scientist: 1

Postdoctoral Fellows: 7

Research Assistants: 1

Professional Assistant: 1

Degree Program: Medical Biophysics

Graduate and Undergraduate Courses Available for Graduate Credit: Radiation Biology and Chemotherapy, Genetics, Cell Surfaces of Normal and Neoplastic Cells, Physical Characterization of Macromolecules, Spermatogenesis, Mathematical Biophysics, Cellular Biophysics, Molecular and Medical Genetics , Medical Biophysics

DEPARTMENT OF MEDICAL MICROBIOLOGY
(No reply received)

DEPARTMENT OF PHYSIOLOGY
Phone: Chairman: (416) 928-2674

Chairman: R.E. Haist

Professors: D.W. Clarke, D.R. Crapper, I.B. Fritz, J. Grayson, J. Logothetopoulos, F.C. Monkhouse, J.T. Murphy, J.W. Pearce, J.W. Scott, R.J. Shephard, O.V. Sirek, G.A. Wrenshall, M. Vranic , H. Orrego (Visiting)

Associate Professors: M.A. Ashworth, F. Coceani, D. Fraser, R.C. Goode, P.E. Hallett, A.M. Hedlin, A. A. Horner, J. Hunter, W.H. Johnson, B.S.L. Kidd, B.J. Lin, J.M. Martin, K.E. Money, L.W. Organ, D. H. Osmond, M.J. Santalo, H. Sonnenberg, G. Steiner, A.T. Storey

Assistant Professors: R.D.G. Blair, G.M. Brown, N.E. Diamant, J. Duffin, N. Forbath, G.G. Forstner, J.M. Fredrickson, L.F. Greenwood, J. Kryspin, S.W. Kooh, K. Norwich, J.A. Satterberg, B.J. Sessle, G.R. Van Loon

Special Lecturer: J. Campbell

Lecturers: J.A. Coddling, B. Hines, H.C. Kwan, L. Lickley, A.M.F. Sun, G.J. Tomko

Degree Programs: Physiology

Graduate and Undergraduate Courses Available for Graduate Credit: An elementary course for Arts Students (U), A lecture course in Human Physiology (U), A general laboratory course to accompany course (U), A lecture course in Endocrinology (U), Electrophysiology (U), Tracer Methodology (U), Cardiovascular Regulation (U), Mathematical Aspects of Physiology (U), Mathematical Aspects of Physiology (U), Mathematical Theory of Turnover and Transport in Physiological Systems (U), Cardiovascular, Renal Physiology and Neurophysiology for Graduate Nurses, Advanced Neurophysiology (U), Advanced Respiratory Physiology (U)

Faculty of Arts and Sciences

DEPARTMENT OF BOTANY
(No reply received)

DEPARTMENT OF PARASITOLOGY
Phone: Chairman: (416) 928-2090 Others: 928-2759

Chairman: R.S. Freeman

Professor: A.M. Fallis

Associate Professor: S.S. Desser, S.B. McIver, K.A. Wright

Visiting Lecturers: 1

Teaching Fellows: 4

Research Assistants: 4

Postdoctoral: 2

Field Stations/Laboratories: Forestry Field Station - Dorset

Degree Program: Parasitology

Graduate and Undergraduate Courses Available for Graduate Credit: Fine Structure and Physiology of Parasites, Medical Parasitology, Medical Parasitology and Entomology, Special Topics in Parasitology, Arthropods as Parasites and Vectors, Helminthology, Physiology of Symbiotic and Parasitic Organisms

DEPARTMENT OF ZOOLOGY
Phone: Chairman: (416) 928-5567

Chairman: D.A. Chant

Professors: J. Berger, I. Brown, L. Butler, D.A.Chant, C.S. Churcher, G.M. Clark, J.B. Falls, W.G. Friend, F.E.J. Fry, H.H. Harvey, R.R. Langford, R.A. Liversage, D.F. Mettrick, T.S. Parsons, D.H. Pimlott, B.H. Pomeranz, T.E. Reed, H.A. Regier, F.H. Rigler, I. Tallan, F.A. Urquhart, A.M. Zimmerman

Associate Professors: H. Atwood, D.G. Butler, D.W. Dunham, G. Knerer, J.L. Fook, J. Machin, Y. Masui, J. Paloheimo, B.I.Roots, J.J.B. Smith, M. Telford,

Assistant Professors: M. Barrett, R. Elinson, C.K. Govind, R. Hansell, F. Holeton, R.C. Plowright, P.J. Pointing, J. Rising, Stokes, Yuyama

Instructors: J. Auerbach, C. Hughes

Visiting Lecturers: 2

Teaching Fellows: 3

Research Assistants: 5

Degree Programs: Animal Behavior, Animal Physiology, Applied Ecology,Biochemistry, Biology, Botany, Ecology, Entomology, Entomology and Parasitology, Ethology, Fisheries and Wildlife Biology, Forest Resources, Genetics, Health and Biological Sciences, Histology, Life Sciences, Limnology, Natural Resources, Parasitology, Radiation Biology, Zoology

Graduate and Undergraduate Courses Available for Graduate Credit: History of Biology, Advanced Seminar and Reading Course, Advanced Vertebrate Anatomy, Vertebrate Palaeontology, Advanced Invertebrate Zoology, Seminar on the Principles of Development, Experimental Embryology and Morphology, Advanced Genetics, Human Genetics, Cellular Physiology, Animal Ecology, Environmental Factors, Limnology, Ethology, Protozoology, Molecular Genetics, Ichthyology: Systematics and Distribution, Mammalogy: Taxonomy and Distribution, Ornithology: Taxonomy and Distribution, Comparative Endocrinology, Radiation Biology, Neurophysiology, Mathematical Ecology, Seminar in Invertebrate Physiology and Development, Statistical Classification in Biology, Entomology, Behavioural Ecology, Aquatic Entomology, Systematic Entomology, Seminar on the Principles of Systematic Zoology, Advanced Vertebrate Paleontology, Neurobiology: Physiology and Fine Structure, Selected Aspects of the Physiology of Symbiotic and Parasitic Organisms, Insect Behaviour, Seminar in Protozoology, Seminar inTheoretical Ecology, Seminar on Topics in Population Genetics and Evolution, Aquatic Macro-Ecology

UNIVERSITY OF WATERLOO

Waterloo, Ontario N2L3G1 Phone: (519) 885-1212

Deans of Faculties:

Environmental Studies: P.H. Nash

Graduate Studies: L.A.K. Watt

Science: W.B. Pearson

Degrees offered:

B.E.S. Man-Environment Studies

B.Sc.(Honors) Biology

M.Sc. Biology

Ph.D. Biology

DEPARTMENT OF BIOLOGY
Phone: Chairman: (519) 885-1211 Ext. 3943

Chairman: J.K. Morton
Professors: C.H. Fernando, A.D. Harrison, H.B.N. Hynes, W.B. Kendrick, J.K. Morton, G. Power
Associate Professors: C.R. Barnes, J.C.H. Carter, A.M. Charles, E.B. Dumbroff, H.C. Duthie, H.R.N Eydt, W.E. Inniss, J. Kruuv, P.E. Morrison, G.G. Mulamoottil, J.J. Pasternak, J.E. Thompson, K. Zachariah
Assistant Professors: R.D. Beauchamp, J.C. Carlson, R.G.H. Downer, H.W. Elmore, M. Globus, D.E. Hart, W.R. Hawthorn, C.I. Mayfield, S.M. Smith, J.M. Theberge, A.I. Dagg (Research), M.A. Morgan (Research), S. Vethamany-Globus (Research)
Teaching Fellows: 55
Research Assistants: 3

Degree Program: Biology

Graduate and Undergraduate Courses Available for Graduate Credit: Paleobotany, Advanced Plant Ecology and Plant Geography, Phycology, Advanced Microbiology, Advanced Genetics, Parasitology, Evolution and Biosystematics, Advanced Animal Physiology I, Mycology, Advanced Microbial Physiology, Advanced Cytology, Advanced Animal Physiology, Advanced Plant Physiology, Floral Morphology and Taxonomy, Advanced Limnology, Advanced Genetics, Statistical Procedures for Biologists, Advanced Entomology, Radiotracers in Biological Sciences, Recent Advances in Microbial Ecology, Comprehensive Examination, Molecular Biology

UNIVERSITY OF WESTERN ONTARIO

London, Ontario N7A3K7 Phone: (519) 679-2311

Deans of Faculties and Schools:
Graduate Studies: H.B. Stewart
School of Medical Sciences: D. Bocking
School of Science: A.E. Scott

Degrees offered:

B.A. Botany, Zoology

B.Sc. Biology

B.Sc. (Honors) Anatomy, Bacteriology and Immunology, Biochemistry, Botany, Biophysics, Pathology, Physiology, Zoology

M.Sc. and Ph.D. Anatomy, Bacteriology and Immunology, Biochemistry, Biophysics, Pathology, Physiology, Plant Sciences, Zoology

School of Medical Sciences

DEPARTMENT OF ANATOMY
Phone: Chairman: 679-3741 Others: 679-2111

Chairman: D.G. Montemurro
Professors: M.L. Barr, R.C. Buck, D.G. Montemurro
Associate Professors: G.G. Altmann, E.W. Donisch, E.B. Gammal, M.A. Hidayet, G.A. Leyshon, F.R. Sergovich, E.L. Shaver, R.P. Singh, H.C. Soltan
Assistant Professors: R.E. Clattenburg, B.A. Flumerfelt, K.A. Galil, J.A. Kiernan
Instructors: R.K. Coates, S.K. Warma, D.D. Wise

Degree Programs: Anatomy, Anatomy and Histology, Genetics, Histology, Human Genetics, Neurological Sciences

Graduate and Undergraduate Courses Available for Graduate Credit: Histology, Current Topics in Anatomy, Medical Genetics, Medical Histology, Neurological Sciences and Mind, Medical Neuroanatomy, Medical Gross Anatomy and Embryology, Dental Gross Anatomy and Embryology, Dental and Oral Histology, Gross Anatomy, Regional and Systemic, Systemic Gross Anatomy, Cytogenetics, Embryology, Neuroanatomy, Cell Structure

DEPARTMENT OF BACTERIOLOGY AND IMMUNOLOGY
Phone: Chairman: (519) 679-3752

Head: R.G.E. Murray
Professors: P.C. Fitz-James, R.G.E. Murray, C.F. Robinow, J. Robinson, J.L. Whitby, L.A. Hatch
Associate Professors: W. Chodirker, K. Ebisuzaki, E.L. Medzon, J. Rozanis, N.R.S. Sinclair, G.H. Strejan, J.E. Zajic
Assistant Professors: R. Behme, W.A. Black, W.P. Cheevers, D. Percy, S.K. Singhal, P.O. Wilkins, E.W.R. Campsall
Lecturer: S. Galsworthy
Visiting Lecturer: 1
Research Assistants: 14

Degree Programs: Bacteriology, Bacteriology and Immunology, Biochemistry, Biology, Genetics, Health and Biological Sciences, Immunology, Medical Microbiology, Virology

Graduate and Undergraduate Courses Available for Graduate Credit: Medical Microbiology, Oral and Dental Microbiology, Advanced Microbiology, Microbial Anatomy, Advanced Virology, Microbial Biochemistry, Microbial Genetics, Medical Bacteriology, Medical Mycology, Immunology, Immunochemistry, Immunology Seminar

DEPARTMENT OF BIOCHEMISTRY
Phone: Chairman: (519) 673-3764 Others: 673-6278

Chairman: B.D. Sanwal
Professors: Y.S. Brownstone, K.K. Carroll, P.C. Fitzjames, R. Hobkirk, J.A. McCarter, W.C.C. McMurray, R.J. Rossiter, B.D. Sanwal, D.B. Smith, H.B. Stewart, K.P. Strickland, I.G. Walker
Associate Professors: J.K. Ball, V. Donisch, J.C. Griffiths, F.C. Heagy, A.R. Henderson, W.L. Magee, J.R. Trevithick, E.R. Tustanoff, R.A. Cook
Assistant Professors: P.R. Galsworthy, P.D. Gatfield, B.A. Gordon, J.P. Guthrie, F.Y. Leung, G.A. Mackie, J.W.D. McDonald, F. Possmayer, A.L. Schincariol
Instructor: C.H. Lin
Lecturers: W.P. Cheevers, W. Chefurka, A.J. Hudson, R.M. Krupka, C.F.C. MacPherson, E.Y. Spencer, A. Vardanis, B.M. Wolfe
Temporary Assistant Professors: 2

Degree Program: Biochemistry

Graduate and Undergraduate Courses Available for Graduate Credit: General Biochemistry, Current Topics in Biochemistry, Topics in Lipids and Neurochemistry, Endocrinology, Metabolic Regulation, Protein Chemistry, Membrane Biochemistry, Clinical Biochemistry, Biochemistry (Medical), Basic Aspects, Applied Aspects, Integrative Biochemistry, Clinical Biochemistry, Intermediary Metabolism, Nucleic Acids and Protein Synthesis, Enzymology, Physical Biochemistry

DEPARTMENT OF BIOPHYSICS
Phone: Chairman: (519) 679-3801 Others: 679-3803

Chairman: M.R. Roach
Professors: A.C. Burton, A.C. Groom, M.R. Roach, C.P.S. Taylor
Associate Professors: P.C. Canhan, W.R. Inch,

M.H. Sherebrin, R.M. Sutherland
Assistant Professors: D.R. Boughner, R.I. Duncan, G.G. Ferguson, S.H. Song
Honorary Lecturers: J.B. Finlay, L.N. Johnson, D.M. Miller, T. Nagai

Degeee Program: Biophysics

Graduate and Undergraduate Courses Available for Graduate Credit: Not Stated

DEPARTMENT OF PATHOLOGY
Phone: Chairman: (519) 679-2782 Others: 679-2781

Chairman: A.C. Wallace
Professors: D.G. Gardner, M. Haust, J.C.E. Kaufmann, W.T.E. McCaughey, A.C. Wallace
Associate Professors: C. Anderson, J.V. Frei, J.P. Sapp, B.A. Warren
Assistant Professors: M.J. Ball, L.M. Beattie, F.N. Lewis, G.P. Wysocki
Clinical Professors: D.M. Mills, M.S. Smout
Clinical Associate Professors: E.M. Davies, I. Ramzy, R.P. Slinger, D.I. Turnbull
Clinical Assistant Professors: H.F. Chiu, A.A. Enriquez, J.J. Gilbert, M.M. Troster
Instructors: H.B. Adilman, R.F. Armstrong, W.A. Frishette

Degree Program: Pathology

Graduate and Undergraduate Courses Available for Graduate Credit: General Pathology, Special Pathology, Electives in Pathology, Medical Jurisprudence, Clinical Pathological Conferences, Pathology for Nursing, Medical Rehabilitation and Non-Medical Honors Students, Methods in Pathology,

DEPARTMENT OF PHYSIOLOGY
Phone: Chairman: (519) 679-3841 Others: 679-2111

Chairman: V.B. Brooks
Professors: D.T. Armstrong, V.B. Brooks, F.R. Calaresu, P.G. Dellow, G.J. Morgenson
Associate Professors: P.G. Harding, P.F. Mercer, J.J. Seguin, B.P. Squires, M. Wiesendanger
Assistant Professors: J.D. Cooke, D.A. Cunningham, T.G. Kennedy, R.L. Kline, S. Lodge, C. Prikle, D.K. Pomerantz, R.F. Weick
Instructors: E. Walker
Visiting Lecturers: 1
Teaching Fellows: 1
Sessional Lecturers: 6
Honorary Lecturers: 4
Post-doctoral Fellows: 4

Degree Program: Physiology

Graduate and Undergraduate Courses Available for Graduate Credit: Mammalian Physiology (U), Current Topics in Physiology (U), Neurophysiology (U), Systemic Mammalian Physiology (U), Body Water and Renal Physiology (U), Metabolic and Reproductive Physiology (U), Gastrointestinal Physiology (U), Systems Analysis in Physiology (U)

School of Science

DEPARTMENT OF PLANT SCIENCES
Phone: Chairman: (519) 679-3113 Others: 679-2111

Chairman: J.B. Bancroft
Professors: J.B. Bancroft, F.S. Cook, C.J. Hickman, W.E. McKeen, D.A. McLarty, L. Orloci, D.B. Walden, D.G. Wilson
Associate Professors: P.B. Cavers, A.W. Day, D. Fahselt, R.I. Greyson, W.G. Hopkins, R.C. Jancey, J.B. Phipps, R.B. van Huystee, A.M. Wellman
Assistant Professors: 4
Instructors: 3

Graduate and Undergraduate Courses Available for Graduate Credit: Advanced Morphogenesis and Differentiation, Advanced Plant Physiology and Biochemistry, Advanced Ecology, Advanced Genetics, Biochemical Systematics, Numerical Techniques in Taxonomy, Philosophy of Systematics, Freshwater Algae, Virus, Mycology, and Plant Pathology, Host-parasite Relations, Advanced Plant Physiology and Biochemistry, Advanced Ecology, Advanced Genetics, Advanced Radiobiology, M.Sc. Research and Thesis, Ph.D. Research and Thesis

DEPARTMENT OF ZOOLOGY
Phone: Chairman: (519) 679-3151 Others: 679-3155

Chairman: M. Locke
Professors: H.I. Battle (Emeritus), T.K.R. Bourns, W. W. Judd, M.H.A. Keenleyside, M. Locke, D.M. Scott
Associate Professors: J.A. George, D.B. McMillan, R. K. Misra, D.M. Ogilvie, J. Purko, J.E. Steele
Assistant Professors: C.D. Ankney, B.G. Atkinson, S. Caveney, A.C. Eggleston, D.E. Hay, G.M. Kidder, J.S. Millar, M.D. Owen, R.J. Planck, R.R. Roth, R.R. Shivers, W.H. Tam, J.P. Wiebe, B.H. Seghers, B.G. Wilson
Instructor: M.E. Moulton
Honorary Professor: 1
Honorary Lecturers: 8

Field Stations/Laboratories: McConnell River Field Station

Degree Programs: Biology, Zoology

Graduate and Undergraduate Courses Available for Graduate Credit: Advanced Histology (U), Advanced Insect Physiology (U), Nuerophysiology (U), Endocrinology (U), Immunity to Parasites (U), Advanced Population Biology (U), Physiological Ecology (U), Advanced Behaviour (U), Advanced Developmental Biology, Developmental Genetics (U), Evolution, Cell Biology (U), Advanced Cell Biology (U), Advanced Biometry (U), Population Genetics (U), Political Biology (U), Advanced Techniques (U), Advanced Microscopy (U), Current Topics in Experimental Biology, Current Topics in Reproductive Endocrinology, Insect Physiology (Seminar Course), Cell Biology, Developmental Biochemistry (Seminar Course), Cell Biology - Ultrastructure (Seminar course), Ecology (Seminar course), Behaviour (Seminar Course)

UNIVERSITY OF WINDSOR
(No reply received)

WILFRED LAURIER UNIVERSITY
Waterloo, Ontario N2L3C5 Phone: (519) 884-1970
Deans of Faculties:
Arts and Graduate Studies: G.R. Vallillee

Degrees offered:
B.Sc. Biology

DEPARTMENT OF BIOLOGY
Phone: Chairman: (519) 884-1970

Chairman: W.Y. Watson
Professors: A.A. Wellwood, D.A. MacLulich (Emeritus)
Associate Professors: E. Kott, R.A. MacCanley, K. Hayashida
Assistant Professor: D.A. Pierson
Instructor: F.F. Mallory

Degree Program: Biology

YORK UNIVERSITY
4700 Keele Street Phone: (416) 667-2100

Downsview, Ontario M3J1P3 Phone: (416) 667-2100

Deans of Faculties:
- Arts: S. Eisen
- Graduate Studies: G. Reed

Degrees offered:
- B.Sc. (Honors) Biology
- M.Sc. Biology
- Ph.D. Biology

DEPARTMENT OF BIOLOGY
Phone: Chairman: (416) 667-2335

Chairman: K.G. Davey
Professors: C.D. Fowle, J.D. Friesen, R.H. Haynes, P.B. Moens, D.M. Nicholls
Associate Professors: M.G. Boyer, D.M. Cameron, B. Colman, A. Forer, J.A.M. Heddle, D.M. Logan, B. B. Loughton, I.E. Pearce, A.S.M. Saleuddin, J.S. Tait
Assistant Professors: M.B. Coukell, L.M. Dill, I.B. Heath, L.E. Licht, D.J. McQueen
Instructors: E. Lander, C. Cheh, E. Gardonio
Visiting Lecturers: 2
Teaching Fellows: 5
Research Assistants: 1

Degree Program: Biology

Graduate and Undergraduate Courses Available for Graduate Credit: Molecular Biology, Cell Biology, Topics in Botany, Topics in Zoology, Molecular Radiation Biology, Biophysical Chemistry, Biochemistry of Photosynthesis, Insect Physiology, Evolution, Calcification in Biological Systems, Physiology of Fish, Physiology and Biochemistry of Plant Pathogens, Biochemistry, Regulation of Protein Biosynthesis in Mammalian Cells, Techniques in Biology

PRINCE EDWARD ISLAND

UNIVERSITY OF PRINCE EDWARD ISLAND

Charlottetown, Prince Edward Island Phone: (902) 892-4121

Deans of Faculties:

Science: J.R. Duffy

Degrees offered:

B.Sc. Biology

DEPARTMENT OF BIOLOGY

Phone: Chairman: Ext. 169 Others: Ext. 170

Chairman: E.L. Drake
Associate Professors: J.C. Cheverie, E.L. Drake, L.A. Hanic, N.N. LeBlanc, I.G. MacQuarrie
Assistant Professors: C.E. Johnston
Instructor: D.L. Guignion
Research Assistants: 2
Laboratory Technologist: 1

Degree Program: Biology

QUEBEC

BISHOP'S UNIVERSITY

Lennoxville, Quebec   Phone: (819) 569-9551
Dean of the Faculty: C.B. Haver
Division Chairman:
  Graduate Studies and Research: Rev. R.E. Reeve
  Natural Science: C.L. Arnot

Degrees offered:
  B.Sc. Biology

  B.Sc. (Honors) Biology

  M.S. Biology

DEPARTMENT OF BIOLOGICAL SCIENCES
  Phone: Chairman: Ext. 278 Others: 569-9551

  Chairman: A.N. Langford
  Professors: A. Langford
  Associate Professors: D.F. Brown, K. Moore
  Assistant Professors: D.F.C. Hilton

Degree Programs: Biology, Botany, Ecology, Genetics, Parasitology, Physiology, Zoology

McGILL UNIVERSITY

P.O. Box 6070   Phone: (514) 392-4311
Montreal 101, Quebec
Deans of Faculties:
  Agriculture: A.C. Blackwood
  Graduate Studies and Research: W.F. Hischfeld
  Medicine: R.F.P. Cronin
  Science: E.J. Stanbury

Degrees offered:
  B.Sc. Anatomical Sciences, Biology, Microbiology and Immunology, Physiology

  B.Sc. (Honors) Anatomical Sciences, Animal Behavior, Biochemistry, Cell Biology, Comparative Physiology, Ecology, Human Genetics, Marine Biology, Plant Science

  B.Sc. (Agriculture) Agricultural Sciences, Biological Sciences, Renewable Resources Development

  M.Sc. Agronomy, Anatomy, Animal Science, Biochemistry, Biology, Entomology, Horticulture, Marine Sciences, Microbiology, Microbiology and Immunology, Parasitology, Physiology, Plant Pathology, Wildlife Biology

  Ph.D. Agronomy, Anatomy, Animal Science, Biochemistry, Biology, Entomology, Marine Sciences, Microbiology, Microbiology and Immunology, Parasitology, Physiology, Plant Pathology

Faculty of Agriculture

DEPARTMENT OF AGRONOMY
(No reply received)

DEPARTMENT OF ANIMAL SCIENCE
(No information available)

DEPARTMENT OF HORTICULTURE
  Phone: Chairman: Not Stated

  Chairman: C.D. Taper
  Professors: J. David
  Associate Professor: C.D. Taper
  Assistant Professors: B.B. Bible
  Instructors: L. Perron, S. Wigdor, C. Chong
  Research Assistants: 5
  Research Associate: 1

Degree Program: Horticulture

Graduate and Undergraduate Courses Available for Graduate Credit: Pomology, Vegetable Crops, Fruit and Vegetable Preservation, Crop Physiology, Development, Crop Physiology, Processes, Special Topics, Seminar

DEPARTMENT OF PLANT PATHOLOGY
(see under MacDonald Campus)

Faculty of Medicine

DEPARTMENT OF ANATOMY
  Phone: Chairman: (514) 392-4931 Others: 392-4881

  Chairman: C.P. Leblond
  Professors: Y. Clermont, C.P. Leblond, D.G. Osmond
  Associate Professors: B.M. Kopriwa, P.K. Lala, N.J. Nadler, H. Warshawsky
  Assistant Professors: G.C. Bennett, E. Daniels, M.F. Lalli, E. Schultz, M. Weinstock, W.E. Wilson
  Lecturers: F. Brahim, L.D. Russell
  Teaching Fellow: 1
  Research Assistants: 5

Degree Program: Anatomical Sciences

Graduate and Undergraduate Courses Available for Graduate Credit: Dynamic Histology (U), Experimental Morphology (U), Histology, Gross Anatomy, Neuro-Anatomy

DEPARTMENT OF BIOCHEMISTRY
  Phone: Chairman: 392-4239

  Chairman: A.F. Graham
  Professors: M.J. Fraser, A.F. Graham, E.A. Hosein, D. Rubinstein, J.H. Spencer
  Associate Professors: P. Braun, D.T. Denhardt, R.M. Johnstone, A.R. Wasserman
  Assistant Professors: R.E. MacKenzie, E.A. Meighen, S. Millward, W. Mushynski
  Teaching Fellows: 21
  Technicians: 28
  Instructor: M.R. Reesal

Degree Program: Biochemistry

Graduate and Undergraduate Courses Available for Graduate Credit: Topics in Molecular Biology, Immunochemistry, Metabolism and Regulation (U), Protein Structure and Function (U), Biophysical Chemistry (U), Nucleic Acids (U), Neurochemistry (U)

DEPARTMENT OF MICROBIOLOGY AND IMMUNOLOGY
(No reply received)

DEPARTMENT OF PATHOLOGY
  Phone: Chairman: (842) 1251-594 Others: 1251-1781

  Chairman: R.H. More
  Strathcona Professor of Pathology: R.H. More
  Miranda Fraser Professor of Comparative Pathology: R.H. More
  Professors: R. Bolande, W.P. Duguid, S.N. Huang, D.S. Kahn, J.P.A. Latour, M.G. Lewis, G. Rona, H. Sheldon, G. Tremblay, F.W. Wiglesworth

Associate Professors: S. Carpenter, M.H. Finlayson, J. C. Hogg, E. Hosein, J. Knaack, K.G. Marshall, G. Mathieson, D. Murray, M. Reesal, N.S. Wang, B.I. Weigensberg
Assistant Professors: M.N. Ahmed, G. Berry, S. Brownstein, M.L. deChampain, A. Ferenczy, R.D.C. Forbes, W. Hauser, I. Huttner, S. Inour, M.E. Kirk, J.R.C. Lachance, J.O. Lough, M. Mandavia, Moinuddin, D. R. Murphy, J.B. Richardson, A. Rona, G. Rowden, P. Schopflocher, T.A. Seemayer, J.L. Webb
Lecturers: B. Artinian, J. Beland, D. Bishop, P. Bolduc, J.P. de Chaderavian, J. Emond, D. Kalousek, Y. Robitaille, H. Srolovitz, J. Viloria, B.M. Wolanskyj
Teaching Fellows: 2
Demonstrators: 4

Degree Program: Pathology

Graduate and Undergraduate Courses Available for Graduate Credit: General Biology of Disease, Special Biology of Disease, Principles of Disease, Clinical Pathological Conference, Clinical Pathology Courses, Biochemical and Ultrastructural Aspects of Blood, Cell Reaction to Microbiological Agents, Hematopathology

DEPARTMENT OF PHYSIOLOGY
Phone: Chairman: 392-4343 Others: 392-3079

Chairman: J. Milic-Emili
Drake Professor of Physiology: F.C. MacIntosh
Professors: R.I. Birks, T.M.S. Chang, B.A. Cooper, J.H. Dirks, P. Gold, K. Krnjevic, F.C. MacIntosh, G. Melvill-Jones, J. Milic-Emili
Associate Professors: M.W. Cohen, D.R. Firth, W.S. Lapp, M. Levy, G. Mandl, C. Polosa, P. Sekelj
Assistant Professors: M.M. Frojmovic, V. Klissouras, P. Kongshavn, M. Mackey, P. Noble, J.S. Outerbridge, A. Wechsler, P. Weldon, R.L. Williams
Instructor: M. Hazucha
Visiting Lecturers: 2
Teaching Fellows: 5
Research Assistants: 2

Degree Program: Physiology

Graduate and Undergraduate Courses Available for Graduate Credit: Physiology of Man: Organ and Tissue Physiology, Physiology of Man: Metabolic and Regulatory Physiology, Physiology, Mammalian Physiology with Emphasis on the Physiology of Man, Intermediate Physiology, Experimental Neurophysiology, Molecular Physiology, Experimental Physiology, Advanced Neurophysiology

Faculty of Science

BIOLOGY DEPARTMENT
Phone: Chairman: (514) 392-4649 Others: 392-4643

Chairman: G. Maclachlan
Professors: W.G. Boll, E.R. Boothroyd, P.S.B. Digby, S.P. Gibbs, W.F. Grant, P.R. Grant, G.A. Maclachlan, J. Marsden, J. Metrakos, B.B. Mukherjee, C.M. Wilson
Associate Professors: A.A. Auclair, A.H. Bussey, M.E. Goldstein, J. Kalff, W.C. Leggett, R.E. Lemon, G. Manley, V.M. Pasztor, R.J. Poole, R. Sattler, R. Sinclair, K. Sittmann, J.L. Southin, D. Trasler, H. Tyson, N. Wolfson
Assistant Professors: B. Brandhorst, R. Chase, D. Fromson, D. Madison, W.C. Oechel, R. Peters, D. P. Verma
Instructors: S. Cahn, J. Marvin, E. Shapiro
Part-time Demonstrators: 150

Field Stations/Laboratories: Gault Estate, St. Hilaire, Huntsman Marine Laboratory - St. Andrews, Bellairs Research Institute - Barbados, Morgan Arboretum - Ste. Anne de Bellvue

Degree Programs: Biology, Animal Behaviour, Cell, Molecular and Development, Ecology, Human Genetics, Marine Biology, Neurobiology and Comparative Physiology, Plant Science

Graduate and Undergraduate Courses Available for Graduate Credit: Developmental Mammalian Genetics, Advanced Plant Ecology, Seminar in Animal Ecology, Cytology, Cytogenetics, Microbial Genetics, Mutagenesis and Radiation Biology, Statistical Methods in Genetics, Techniques in Electron Microscopy, Developmental Mammalian Genetics, Topics in Genetic Regulation and Recombination, Ecosystems, Physiological Plant Ecology, Limnology, Marine Environment, Biological Oceanography, Advanced Plant Ecology, Seminar in Animal Ecology, Topics in Plant Biochemistry, Cytology, Cytogenetics, Microbial Genetics, Mutagenesis and Radiation Biology, Statistical Methods in Genetics, Techniques in Electron Microscopy, Comparative Metabolic Physiology, Advanced Neurobiology, Advances in Animal Behaviour

MARINE SCIENCES CENTRE
Phone: Chairman: (514) 392-5714 Others: 392-5733

Chairman: M.J. Dunbar
Professors: M.J. Dunbar, J.B. Lewis
Associate Professors: B.F. d'Anglejan, C.M. Lalli
Assistant Professors: R.G. Ingram, F. Sander, D.C. Maclellan
Research Assistants: 1

Field Stations/Laboratories: Bellairs Research Institute Barbados

Degree Programs: Oceanography, Marine Science

Graduate and Undergraduate Courses Available for Graduate Credit: Marine Environment (U), Biological Oceanography (U), Advanced Zoology of Molluscs and Crustacea, Tropical Marine Ecology, Advanced Marine Ecology, Descriptive Oceanography (U), Introduction to Dynamical Oceanography (U), Physics of the Ocean, Theoretical Aspects of the Oceanic Circulation, Marine Geology and Geochemistry, Recent Sediments and Marine Geology, Seminar in General Oceanography, Topics in Oceanography

DEPARTMENT OF ENTOMOLOGY
Phone: Chairman: (514) 457-6580

Chariman: F.O. Morrison
Professors: E.M. DuPorte (Emeritus), D.K. McE. Kevan, J.E. McFarlane, F.O. Morrison
Associate Professors: R.K. Stewart, V.R. Vickery
Assistant Professors: S.B. Hill
Instructors: F.E.A. Ali-Khan

Degree Programs: Agricultural Biology, Agriculture, Biology, Entomology, Zoology and Entomology

Graduate and Undergraduate Courses Available for Graduate Credit: Advanced Taxonomy and Zoogeography, General Insect Ecology, Soil Fauna, Soil Ecology, Aquatic Entomology, Advanced Economic Entomology, Seminar and Research Review, Ph.D. Comprehensive Examination, Advanced Insect Morphology, Advanced Insect Physiology, Insect Population Dynamics, General Insect Ecology and Population Dynamics, Soil Zoology, Veterinary and Medical Entomology, Nematology, Conservation Ecology, Wildlife Ecology, Applied Parasitology

MacDonald Campus

Ste. Anne de Bellevue Phone: (514) 457-6850

LYMAN ENTOMOLOGICAL MUSEUM AND RESEARCH LABORATORY
Phone: Director: Ext. 311 Others: Ext. 314

Director: D.K. Kevan
Professors: D.K. Kevan, V.R. Vickery
Assistant Professor: S.B. Hill
Instructors: F.E.A. Ali-Khan, A.C. Shepherd
Research Assistants: 5

Degree Programs: Systematic Entomology

Graduate and Undergraduate Courses Available for Graduate Credit: Courses are officially offered through the Department of Entomology and are not offered by this unit as such.

DEPARTMENT OF MICROBIOLOGY
Phone: Chairman: (514) 457-6580

Chairman: R. Knowles
Professors: A.C. Blackwood, R. Knowles, R.A. Macleod, C.E. Tanner
Associate Professor: E.S. Idziak
Assistant Professors: I.W. DeVoe, J.M. Ingram, P. Calcott (Research)
Research Assistants: 20

Degree Program: Microbiology

Graduate and Undergraduate Courses Available for Graduate Credit: Topics in Microbial Physiology, Microbial Biochemistry, Soil Microbiology, Invitation Lectures in Microbiology, Graduate Seminar, Microbial Physiology (U), Microbial Ecology (U), Biological Instrumentation (U), Isotopic Tracer Techniques (U), Immunology (U), Food Microbiology (U), Pathogenic Microbiology (U)

DEPARTMENT OF PLANT PATHOLOGY
Phone: Chairman: (415) 457-6580 Ext. 309

Chairman: R.H. Estey
Professors: R.H. Estey, R.L. Pelletier, W.E. Stackston
Assistant Professors: J.F. Peterson, M.A. Viswanathan, D.W. Woodland
Visiting Lecturers: 2
Research Assistants: 4

Field Stations/Laboratories: Macdonald College Farm - Morgan Arboretum, Ste. Anne de Bellevue

Degree Programs: Botanical Sciences, Plant Pathology, Mycology, Nematology, Plant Virology

Graduate and Undergraduate Courses Available for Graduate Credit: Diseases of Field and Forest Crops, Diseases of Horticultural Crops, Plant Pathology Research Methods, Plant Virus Diseases, Plant Virology, Methods, History of Plant Pathology, Advanced Mcyology, Seminar, Nematology, Nematode Diseases of Plants, Taxonomy and Ecology of Economic Plants, Principles of Plant Pathology, Advanced Plant Pathology

SIR GEORGE WILLIAMS UNIVERSITY

1455 de Maisonneuve Boulevard West Phone: (514) 879-5995
Montreal 107, Quebec
Deans of Faculties:
Graduate Studies: F. French
Science: R. Verschingel

Degrees offered:
B.Sc. Biochemistry, Biological Sciences

B.Sc. (Honors) Cell and Molecular Biology, Ecology, Physiology and Developmental Biology

M.Sc. Biological Sciences

Ph.D. Biological Sciences

DEPARTMENT OF BIOLOGICAL SCIENCES
Phone: Chairman: (514) 879-4452 Others: 879-2856

Chairman: Not Stated
Professor: D.L. Peets
Associate Professors: S.S. Ashtakala, H.E. Enesco, R.K. Ibrahim, G. Leduc, C.F. MacLeod, E.G. Newman
Assistant Professors: P.D. Anderson, R.L. Lowther, R.M. Roy, S.M. Ruby
Research Assistants: 3

Degree Program: Biology

Graduate and Undergraduate Courses Available for Graduate Credit: Special Topics in Ecology I, II, Studies in Organismal Biology, Special Topics in Organismal Biology I, II, Studies in Cell Biology, Special Topics in Cell Biology I, II, Special Topics in Biology, Recent Advances in Biology, Research and Thesis

UNIVERSITÉ LAVAL

Cité universitaire, Québec Phone: (418) 656-2131
Doyens des Facultés:
Agriculture et alimentation: V. Lavoie
Foresterie et géodesie: A. Lafond
Médicine: J-L. Beaudoin
Sciences: P. Grenier

Grades offerts:
B.Sc. Biochimie, Biologie, Microbiologie

B.Sc. app. Agro-économie, Bio-agronomie, Génie forestier

M.Sc. Agrobiologie, Anatomie, Biochimie, Biologie, Écologie forestière, Embryologie, Génétique, Histologie, Microbiologie, Physiologie, Phototechnie, Sylviculture, Zootechnie

D.Sc. (Mêmes options que pour les maîtrises)

Faculté des Sciences Biologiques

DÉPARTEMENT D'AGROBIOLOGIE
(Pas de réponse)

DÉPARTEMENT DE BIOCHIMIE
Phone: Chairman: (418) 656-2963 Others: 656-2131

Head: P. Tailleur
Associate Professors: R.H. Côté, C. Godin, M. Jean, P. Tailleur, G. Talbot
Assistant Professors: R.J. Bojanowski, J.L. Lavoie, M. Silver
Instructors: G. Bellemare, J. Lapointe
Research Assistants: 3

Degree Programs: Biochimie, Microbiologie

Graduate and Undergraduate Courses Available for Graduate Credit: Immunochimie, Inhibiteurs metaboliques, Metabolisme III, Biochimie des structures bactériennes Ribosomes I: Structure et fonction, Ribosomes II, action des antibiotique, Geomicrobiologie, Sporulation bactérienne

DÉPARTEMENT DE BIOLOGIE
Phone: Chairman: 656-2863

Chairman: L. Huot
Professors: W. Anderson, L. Ellison, L. Legendre, J.N. McNeil, G. Moreau, J. Remington (Adjoint), R. Bernard, G. Filteau, A. Gagnon, L. Huot, G. Lacroix, J.G. Lafontaine, R. Lagueux, J.M. Langlois, P. Trudel (titulaire), J. Bédard, J. Bovet, A. Cardinal, J. de La Noüe, P. Morisset, D. Pallotta, J.M. Perron (Agrégé)
Assistant Professors: A. Duval, J. Huot, J. Larochelle
Visiting Lecturers: 2
Teaching Fellows: 1
Research Assistants: 4

Field Stations/Laboratories: Forêt Montmorency, Faculté de Foresterie et Géodesie, Université Laval

Degree Program: Biology

Graduate and Undergraduate Courses Available for Graduate Credit: Cytologie biophysique, Génétique moleculaire, Physiologie générale (U), Physiologie comparée des insectes, Problèmes d'écologie marine, Problèmes de phycologie marine, Morphologie de l'insecte, Biologie des organismes de la zone intertidale, Problèmes d'éthologie, Dynamique des populations, La structure de l'éspèce et la speciation, Systématique et classification des insectes, Biologie du développement, Introduction à l'écologie des insectes, Neurophysiologie, Séminaires de biologie moleculaire I, II, Réproduction chez les oiseaux, Biologie cellulaire et moleculaire, Thermydynamique de l'écosysteme, Seminaires d'océanographie biologique, Mathematiques appliquées a l'écologie, Problèmes de biosystématique des plantes vasculaires, Écologie des poissons

DÉPARTEMENT DE MICROBIOLOGIE
(Pas de réponse)

DÉPARTEMENT DE PHYSIOLOGIE
(Pas de réponse)

DÉPARTEMENT DE PHYTOLOGIE
Phone: Chairman: 656-2165

Chairman: Not Stated
Professors: A. Alarie, R. Bedard, J.E. Chevrette, F.M. Gauthier, P. Gervais, R.A. Lachance, R.O. Lachance, G.M. Oláh, J.P. Pare, H.P. Therrien
Associate Professors: J.A. Fortin, J.M. Girard, F. Pauze, M.J. Trudel, R. Van Den Hende
Assistant Professors: P.A. Dube, S. Payette, P.M. Saint-Clair (Research)
Instructor: R. Gauthier, A. Vezina
Visiting Lecturer: 7
Teaching Fellow: 1
Research Assistants: 3

Field Stations/Laboratories: Station Agronomique St-Augustin, Station Agronomique St-Louis de Pintendre

Degree Programs: Plant Physiology and Management, Plant Ecology, Plant Breeding, Plant Protection (Mycology and Plant Pathology), Anatomy, Morphology and Taxonomy

Graduate and Undergraduate Courses Available for Graduate Credit: Amélioration génétique des plantes, Écophysiologie végétale, Herbages, Phytopathologie I, II, Physiologie du parasitisme I, II, Mycologie et diagnostics phytopathologiques, Histoire de la phytopathologie, Plantes horticoles, Dispositifs expérimentaux, Phytogéographie,Physiologie des champignons, Techniques d'inventaires et d'analyse écologique, Sujets speciaux, Systématique botanique, Systématique des plantes ornémentales, Génétique quantitative, Cytogénétique, Physiologie et nutrition des bactéries, Microbiologie des sols, Anatomie et morphologie, Séminaires

LOYOLA UNIVERSITY
(No reply received)

UNIVERSITÉ DE MONTRÉAL
Case postale 6128 Phone: (514) 343-6111
Montréal 101, Québec
Doyens des Facultés et Écoles:
Arts et Sciences: R. de Chantal
Études superieurs: H. Favre
Médicine: P. Bois

Grades offerts:
B.Sc. Biochimie, Biologie
M.Sc. Anatomie, Biochimie, Biologie, Microbiologie et Immunologie
Ph.D. Anatomie,Biochimie, Biologie, Microbiologie et Immunologie

Faculté de Médicine

DÉPARTEMENT D'ANATOMIE
Phone: Chairman: (514) 343-6289 Others: 343-6289

Chairman: P. Jean
Professors: J.-M. Blais, R. Daoust, R. Gagnon, P. Jean, G. Sainte-Marie, E.B. Sandborn
Associate Professors: Y. Bélanger, A.G. Gervais, B. Messier, P.-E. Messier, C.-L. Richer, T. Simard, R. Veilleux, S. Karasaki (Research), A. Simard (Research)
Assistant Professors: J. Déziel, J. Lamonica, M. Liskova,A. Pospisil, J. Roberte
Instructors: R. Bendjaballah
Teaching Fellows: 1
Research Assistants: 1

Degree Program: Anatomie

DÉPARTEMENT DE BIOCHIMIE
Phone: Chairman: (514) 343-6372 Others: 343-6374

Head: W. Verly
Professors: L.P. Bouthillier, G. de Lamirande, A. Gagnon, R. Gianetto, W. Verly
Associate Professors: J.M. Bourgault, M. Bratley-Mamet, R.J. Cedergren, G. Gingras, J.P. Lachance, R. Morais, K. Roberts, H. Simpkins
Assistant Professors: L. Brakier-Gingras

Field Stations/Laboratories: Institut du cancer and Hôpital Notre-Dame, Hôpital Hôtel-Dieu de Montréal, Hôpital Maisonneuve, Lady Davis Institute for Medical Research

Degree Programs: Biochimie

Graduate and Undergraduate Courses Available for Graduate Credit: Biochimie générale I et II, Chimie physiologique, Macromolecules biologiques, Enzymologie, Biophysique cellulaire, Biochimie expérimentale, Biochimie clinique, Acides nucléiques et biosynthèse des proteines(U), Techniques biochimiques, Proteines (U), Biochimie de la mitochondrie (U)

DÉPARTEMENT DE MICROBIOLOGIE ET IMMUNOLOGIE
Phone: Chairman: 343-6273 Others: 343-6274

Head: S. Sonea
Professors: V. Adamkiewicz, J. De Repentigny, V. Fredette, B. Martineau, L.G. Mathieu, S. Sonea, F. Turgeon
Associate Professors: J.C. Benoit, G. Drapeau, L. Gynes, J. Joncas, E. Kurstak, A. Leduc, R.A. Nelson Jr., V. Pavilanis, G. Richer, A. Sasarman
Assistant Professors: P. Auger, M. Brazeau, M. Domardski-Dobija, L. Lafleur, G. Lamoureaux, B. McLaughlin,R. Morisset, G. Nogrady, J. Robert, M. Saint-Martin, P.L. Turgeon, C. Vega-Jacome, C. Vezina, P. Viens, F. Zaccour
Visiting Lecturers: 9
Teaching Fellows: 3
Research Assistants: 6
Laboratory Instructors: 26

Degree Programs: Microbiologie et Immunologie

Graduate and Undergraduate Courses Available for Graduate Credit: Bactériologie générale, Physiologie microbienne avancée, Systematique bactérienne, Immunologie générale, Virologie et techniques virologiques, Colloques et séminaires, Parasitologie et protozoologie, Mycologie médicale, Génétique des bacteries et des virus, Virologie médicale, Micro-organismes: pathogénicité et immunité

DÉPARTEMENT DE PATHOLOGIE
Phone: Chairman: (514) 343-6308 Others: 343-6307

Chairman: G. Jasmin
Associate Professors: A. Dumont, N. Mandalenakis, B. Solymoss
Assistant Professor: F. Babai
Research Associates: J.G. Latour, S. Renaud
Visiting Lecturers: 5
Research Assistant: 1

Field Stations/Laboratories: Laboratoire de pathologie experimentale, Institut de cardiologie de Montréal

Degree Program: Pathologie

Graduate and Undergraduate Courses Available for Graduate Credit: Cancérologie - Histologie, pathologie générales et immunologie, Méthodologie et techniques de pathologie expérimentale - Acquisitions récentes en pathologie expérimentale et en anatomie pathologique, Séminaires d'anatomie pathologique, Pathologie cellulaire

Faculté des Arts et Sciences

DÉPARTEMENT DES SCIENCES BIOLOGIQUES
(Pas de réponse)

UNIVERSITÉ DE SHERBROOKE
boulevard de l'Université Phone: (819) 565-5970
Cité universitaire
Doyens des facultés:
Médicine: G. Pigeon
Sciences: L. O'Neil

Grades offerts:
B.Sc. Biologie

M.Sc. Anatomie, Biochimie, Biologie, Biologie Cellulaire, Biophysique, Microbiologie, Physiologie

Ph.D. Anatomie, Biochimie, Biologie, Biologie Cellulaire, Biophysique, Microbiologie, Physiologie

Faculté de Médicine

DÉPARTEMENT D'ANATOMIE
Phone: Chairman: 565-2021

Professors: T.F. Lévêque
Associate Professors: N. Brière, M. Nemirovsky, E. Ramon-Moliner
Assistant Professor: R. Calvert

Degree Program: Anatomie

Graduate and Undergraduate Courses Available for Graduate Credit: Neuroendocrinologie, La Base Anatomique des Troubles Neuropsyciatriques

DEPARTEMENT DE BIOCHIMIE
Phone: Chairman: (819) 505-2169 Others: 565-2092

Chairman: F. Lamy
Professors: F. Lamy, A. Wahba
Associate Professors: C. Petitclerk, L. Tan
Assistant Professors: E. De Medicis, G. DuPuis, D. Gibson, B. Preiss
Research Assistants: 10

Degree Program: Biochimie

Graduate and Undergraduate Courses Available for Graduate Credit: Biochimie Génétique I et II, Ginétique Enzymatique, Chimie des Proteines, Mechanismes des Réactions Biochimiques, Chimie des Membranes, Principles d'asymetrie Moléculaire en Biochimie

DÉPARTEMENT DE BIOPHYSIQUE
Phone: Chairman: (819) 565-2029 Others: 565-2031

Chairman: O.F. Schanne
Professors: M. Lavallee, O.F. Schanne
Associate Professors: E. de Ceretti, W. Seufert
Assistant Professors: F. Bessette, J.P. Caille
Visiting Lecturers: 1
Research Assistants: 1

Degree Programs: Biophysique

Graduate and Undergraduate Courses Available for Graduate Credit: Biophysique et Physiologie Cellulaire (U), Biophysique et Physiologie generale I, II, interactions proteines, ions, molécules neutres, phenomènes d'échange à travers la membrane biologique, structure moléculaire et spectroscopie, les propriétés electriques du muscle cardiaque, méthodes spectroscopiques en biophysique, contraction musculaire, les propriétés membranaires du muscle squelettique, technique des microélectrodes à bout ouvert, Membranes et leurs modèles, thermodynamique irreversible, biophysique du protoplasme, le transport ionique à travers des cellules epithéliales, mésures diélectriques sur des suspensions épithéliales, biocolloides

DEPARTEMENT DE MICROBIOLOGIE
Phone: Chairman: (819) 565-2067

Chairman: P. Bourgaux
Professor: P. Bourgaux
Associate Professor: D. Bourgaux
Assistant Professors: J.P. Thirion, J. Weber
Instructors: R. Capet, V. Fontaine
Visiting Lecturers: 2
Research Assistants: 5
Postdoctoral Fellows: 4

Degree Programs: Microbiologie, Microbiologie medicale

Graduate and Undergraduate Courses Available for Graduate Credit: Génétique virale, Les virus oncogènes, Génétique élementaire sur la lévure, Génétique avancée sur le phage, Séminaire de récherche, Séminaire de lecture, Séminaire d'orateurs invités

DEPARTEMENT DE PHYSIOLOGIE
(Pas de réponse)

Faculté des Sciences

DEPARTEMENT DE BIOLOGIE
Phone: Chairman: 565-3650 Others: 565-3652

Chairman: P. Matton
Professors: R. Desrochers, J. Dunnigan, J. Juillet, L.C. O'Neil, R. Saucier
Associate Professors: A. Legault, M.L. Sharma
Assistant Professors: A. Beaudoin, G. Beaumont, P. Bechard, J.M. Bergeron, P. Matton, J. Morisset, G. Veronneau, A. Villemaire
Instructors: J.L. Loubier, J. Robin
Research Assistants: 2

Degree Programs: Biologie, Entomologie, Zoologie

Graduate and Undergraduate Courses Available for Graduate Credit: Le pancréas exocrine: adaptation et régime alimentaire, Séminaires de récherches, Nomenclature botanique, Différenciation cellulaire, Écologie des insectes, Écologie microbienne, Écologie des mammifères II, Les pteridophytes, Les graminées, Les cyperacées, Le Pancréas exocrine et les glandes salivaires, Endocrinologie II, Mécanisme d'action hormonale, Les composées, Les coléoptères, Les hymenopteres, Les homopteres, Systématique zoologique, Physiologie de la réproduction, Les hormones gastro-intestinales et les enzymes du pancreas exocrine , Estomac: controle de la sécretion acide pepsine et mucus, Estomac: inhibition de la sécrétion acide, pepsine et mucus, Biochimie microbienne, Les membranes biologiques

UNIVERSITÉ DU QUÉBEC À CHICOUTIMI
930 est, rue Jacques-Cartier Phone: (418) 549-4354
Chicoutimi, Québec

Doyens:
Deuxièmes et troisièmes cycles: R. Bergeron
Directeurs de modules:
Biologie: A. Francoeur

Grades offerts:
B.Sp.Sc. (Baccalaureat specialisé en sciences) Biologie

MODULE DE BIOLOGIE
Phone: Chairman: 545-5414 Others: 545-5001

Faculty: Not Stated

Degree Program: Biologie

Graduate and Undergraduate Courses Available for Graduate Credit: Not Stated

UNIVERSITÉ DU QUÉBEC À MONTRÉAL
1180 rue Bleury C.P. 8888 Phone: (514) 876-5464
Montréal 101, Québec
Doyens:
Études avancée et récherche: M. Belanger
Vice-doyens de famille:
Sciences: G. Bolduc
Directeurs de module:
Biologie: C. Hamel

Grades offerts:
B.Sp.Sc. Biologie

M.Sc. Biologie, Ecologie

DÉPARTEMENT DES SCIENCES BIOLOGIQUES
Phone: Chairman: (514) 876-3102 Others: 876-5500

Chairman: L. Desnoyers
Professors: J.G. Alary, P. Bhéreur, S. Boileau, J.-P. Cheneval, G. Chevalier, M. D'Aoust, D. De Oliveria, L. Desnoyers, R. Elie, R. Fortin, J. Gingras, C. Hamel, R. Joyal, D. Mergler-Racine, B. Scherrer
(Il n'existe pas des catégories de faculté. Tous se nomment "Professeur"
Research Assistants: 2
Part-time Lecturers: 9

Field Stations/Laboratories: A Field Station is used for our summer trimester in Ecology

Degree Programs: Biologie, Ecologie, Sciences Biologique

Graduate and Undergraduate Courses Available for Graduate Credit: Conservation des réssources rénouvelables, Toxicologie, Séminaires, Écologie animale avancée, Traitement des données écologiques, Incidences physiologiques du milieu urbain, Limnologie avancée, Ichtyologie avancée, Taxonomie avancée des vasculaires, Écologie végétale avancée, Ornithologie

UNIVERSITÉ DU QUÉBEC À RIMOUSKI
(Pas de réponse)

UNIVERSITÉ DU QUÉBEC À TROIS-RIVIÈRES
CP500 Trois Rivières, Québec Phone: (819) 376-5454

Doyens:
Etudes avancées et récherche: J. Parent
Vice-doyens de familles:
Sciences pures et appliquées et de la santé: M. Lambert
Directeur de modules:
Biologie et Biochimie: G. Vaillancourt
Biologie Médicale: B. Feuille

Grades offerts:
B.Sp.Sc. Biochimie, Biologie, Biophysique

DÉPARTEMENT DE CHIMIE-BIOLOGIE
Phone: Chairman: (819) 376-5426 Others: 376-5011

Chairman: R.M. Leblanc
Associate Professors: A. Aubin, R. Gagnon, I. Gruda, J. Gruda, R.M. Leblanc, G. Vaillancourt
Assistant Professors: S. Belloncik, J.-P. Bourassa, C. Chapados, R. Couture, J.-P. Dumas, M. Fragata, E. Girard, E. Lacoursiere, L. Pazdernik
Instructors: J. Meunier, J.-P. Odile
Research Assistants: 5

Degree Programs: Biochimie, Biologie, Biophysique

Graduate and Undergraduate Courses Available for Graduate Credit: Not Stated

SASKATCHEWAN

UNIVERSITY OF REGINA

Regina, Saskatchewan Phone: (306) 584-4111

Deans of Faculties:

Arts and Science: (Vacant)
Graduate Studies and Research: A.B. Van Cleave

Degrees offered:

B.A. Biology

B.A. (Honors) Biology, Biochemistry

B.Sc. Biology

B.L.T. Laboratory Technology

M.Sc. Biology

Ph.D. Biology

DEPARTMENT OF BIOLOGY
Phone: Chairman: (306) 584-4224 Others: 584-4145

Chairman: P.W. Riegert
Professors: D.R. Cullimore, J.R. Hay (Adjunct), G.F. Ledingham, G.J. Mitchell, M.V.S. Raju, P.W. Riegert, R.Y. Zacharuk
Associate Professors: R.M. Agnew, H. Bertrand, W. Chapco, L.J. Mook, W.A. Quick, D.M. Secoy, D.H. Sheppard, A. Walther
Visiting Lecturer: 1
Teaching Assistants: 7

Field Stations/Laboratories: Cypress Hills Biology Field Station - Cypress Hills

Degree Programs: Biology, Botany, Ecology, Genetics, Microbiology, Virology, Wildlife Management, Zoology

Graduate and Undergraduate Courses Available for Graduate Credit: Molecular Genetics (U), Advanced Molecular Genetics (U), Microbiology (U), Current Problems in Microbiology, Principles of Plant Taxonomy (U), Introductory Biometrics (U), Advanced Biometrics, Population Genetics (U), Advanced Population Genetics, Plant Embryology, Plant Ecology (U), Applied Plant Physiology, Systematics of Vascular Plants, Animal Physiology (U), Community Ecology (U), Systematics of Vascular Plants, Animal Physiology (U), Community Ecology (U), Animal Population Dynamics (U), Advanced Population Dynamics, Behavioural Ecology, Insect Physiology (U), Insect Morphology, Advanced Insect Physiology, Ornithology (U), Wildlife Biology (U), Entomology (U)

UNIVERSITY OF SASKATCHEWAN

Saskatoon, Saskatchewan 57NOWO Phone: (306) 343-2100

Deans of Faculties and Colleges:

Agriculture: W.J. White
Arts and Science: D.R. Cherry
Graduate Studies: K.J. McCallum
Medicine: R.G. Murray
Veterinary Medicine: D.L.T. Smith

Degrees offered:

B.S.A. Agricultural Microbiology, Animal Science, Biology, Dairy and Food Sciences, Horticulture, Plant Ecology, Poultry Science

B.Sc. (M.L.T.) Medical Laboratory Technology

B.Sc. (Honors) Anatomy, Bacteriology, Biochemistry, Biology, Biomedical Engineering, Dairy Science, Horticulture, Physiology, Plant Ecology, Poultry Science

Ph.D. Anatomy, Animal Science, Bacteriology, Biochemistry, Biology, Biomedical Engineering, Dairy Science, Horticulture, Microbiology, Physiology, Plant Ecology, Poultry Science

College of Agriculture

DEPARTMENT OF ANIMAL SCIENCE
Phone: Chairman: (306) 343-2666

Chairman: J.M. Bell
Professors: J.M. Bell, W.E. Howell, B.D. Owen, C. M. Williams
Associate Professors: D.A. Christensen, R.D. Crawford H.H. Nicholson
Laboratory Demonstrators: 2
Professional Research Associate: 2

Field Stations/Laboratories: Animal Science Farm, Beef Cattle Research Station, Record of Performance Test Station

Degree Programs: Agriculture, Animal Genetics, Animal Husbandry, Animal Industries, Animal Physiology, Dairy Science, Meat Sciences, Nutrition, Physiology, Experimental Methodology

Graduate and Undergraduate Courses Available for Graduate Credit: Directed Studies in Protein Metabolism and Nutrition, Animal Breeding, Animal Nutrition, Animal Energetics, Seminar, Research in Animal Science

DEPARTMENT OF DAIRY AND FOOD SCIENCE
Phone: Chairman: 343-3530 Others: 343-3908

Chairman: D.L. Gibson
Professors: D.L. Gibson, E.S. Humbert, G.A. Jones, S.F. Chinn (Adjunct)
Associate Professors: W.M. Ingledew
Assistant Professors: G. Blankenagel
Instructor: N. Clancy
Research Assistants: 4
Technicians: 2

Degree Programs: Dairy Science, Food Science, Microbiology

Graduate and Undergraduate Courses Available for Graduate Credit: Fluid Products, Food Additives, Condensed Frozen Foods, Sanitation and Waste Disposal, High Fat Feeds, Dairy Chemistry, Food Microbiology, Utilization of Industrial Byproducts, Fermentations Analytical Microbiology, Rumen Microbiology

DEPARTMENT OF HORTICULTURE SCIENCES
Phone: Chairman: (306) 343-4241

Head: S.H. Nelson
Associate Professor: E.A. Maginnes, D.H. Dabbs
Visiting Lecturer: 1
Research Assistant: 1

Field Stations/Laboratories: Outlook, Saskatchewan

Degree Program: Horticulture

Graduate and Undergraduate Courses Available for Graduate

Credit: Ornamental Plants (U), Landscape Gardening (U), Floriculture (U), Advanced Fruit Growing, Advanced Vegetable Growing, Storage and Transportation (U), Propagation and Nursery Management, Turfgrass Culture (U)

DEPARTMENT OF PLANT ECOLOGY
Phone: Chairman: 343-3185

Head: R.T. Coupland
Professors: R.T. Coupland, J.S. Rowe
Associate Professors: V.L. Harms, R.E. Redmann, E.A. Ripley
Research Assistants: 4

Field Stations/Laboratories: Matador Field Station, Arctic Res~arch Training Centre

Degree Programs: Applied Ecology, Ecology, Plant Ecology

Graduate and Undergraduate Courses Available for Graduate Credit: Physiological Plant Ecology, Ecological Plant Geography, Advanced Plant Ecology, Wildland Ecology, plus additional classes in Ecology offered by the Biology Department

DEPARTMENT OF POULTRY SCIENCE
(No reply received)

College of Medicine

DEPARTMENT OF ANATOMY
Phone: Chairman: 343-2661

Head: S. Fedoroff
Professors: S. Fedoroff, H. Butler, B.S. Wenger, J.D. Newstead, I. Munkacsi, L. Hertz
Associate Professors: P. Innes, H.C. Wang, B.R. Brandell
Assistant Professors: G. Youssef, G. Burkholder, F. Oteruelo, F. Schoen, J. Bullaro
Teaching Fellows: 2
Research Assistants: 1
Department Assistants: 2

Degree Programs: Anatomy, Anatomy and Histology, Histology.

Graduate and Undergraduate Courses Available for Graduate Credit: Biophysics of Nerve and Muscle, Introduction to Electron Microscopy and Cellular Ultrastructure, Tissue Culture, Experimental Medicine, Biological Microscopy, Investigations in Locomotion, Histochemistry, Experimental Embryology, Neuroembryology, Biochemical Embryology, Mammalian Somatic Cell Genetics

DEPARTMENT OF BIOCHEMISTRY
Phone: Chairman: (306) K.J. McCallum

Chairman: J.D. Wood
Professors: C.S. McArthur, J.F. Morgan, F. Vella
Associate Professors: W.W. Kay, R.O. Martin, B.D. McLennan, P.D. Shargool, V.R. Woodford
Visiting Lecturers: 1
Research Assistants: 1

Degree Program: Biochemistry

Graduate and Undergraduate Courses Available for Graduate Credit: General Biochemistry, Enzymology, Lipids, Macromolecules, Neurochemistry, Cell Biochemistry, Medical Aspects of Biochemistry (U), Biochemical Techniques, Plant Biochemistry

DEPARTMENT OF PATHOLOGY
(No reply received)

DEPARTMENT OF PHYSIOLOGY
Phone: Chairman: (306) 343-3131

Head: J.W. Phillis
W.S. Lindsay Professor: L.B. Jaques
Professors: G.J. Millar, G.B. Sutherland
Associate Professors: N.M. McDuffie, S.S. Naidoo, K. Prasad
Assistant Professors: P.V. Sulakhe, G.G. Yarbrough
Teaching Fellows: 3
Research Assistants: 4

Degree Programs: Physiology, Physiological Sciences

Graduate and Undergraduate Courses Available for Graduate Credit: Biophysics of Nerve and Muscle, Vascular Physiology, Readings in Physiology, Quantitative Physiology, Cardiac Physiology, Blood, Mammalian Endocrinology, Molecular Biology of Connective Tissue, Blood Coagulation

College of Arts and Science

DEPARTMENT OF BACTERIOLOGY
Phone: Chairman: 343-3151

Head: J.F. Morgan (Acting)
Professor: G. Dempster
Associate Professors: M. Epp, E.I. Grodums, R.J.L. Paulton
Assistant Professors: F.A. Holden, R.G. Quist
Instructor: W.T. Martin

Degree Program: Bacteriology

Graduate and Undergraduate Courses Available for Graduate Credit: Immunology (U), Bacterial Physiology (U), Fermentation (U), Medical Mycology (U), Virology (U), The Genetics of Micro-organisms (U)

DEPARTMENT OF BIOLOGY
Phone: Chairman: 343-5980

Head: U.T. Hammer
Rawson Professor of Biology: J.M. Naylor
Professors: H.E. Gruen, U.T. Hammer, J.M. Naylor, T.A. Steeves
Associate Professors: C. Gillott, T.H.J. Gilmour, V. L. Harms, J. King, W.J. Maher, R.L. Randell, M. Rever-DuWors, J.W. Sheard, R.J.F. Smith, M.W. Zink
Assistant Professors: L.C. Fowke, D.M. Lehmkuhl, R.A.A. Morrall, B.D. Murphy, B.R. Neal, G.H. Rank, F.L.M. Turel
Instructors: R.C. Haynes, E. Wenger
Visiting Lecturers: 2
Research Assistants: 3

Field Stations/Laboratories: Emma Lake Field Station - Christopher Lake

Degree Program: Biology

Graduate and Undergraduate Courses Available for Graduate Credit: Undergraduate classes may be taken by graduate students if they fulfill a particular need. These would be 3rd or 4th year classes as a general rule.

College of Veterinary Medicine

DEPARTMENT OF VETERINARY ANATOMY
Phone: Chairman: (306) 343-3195

Head: W.D. Anderson
Professors: W.D. Anderson, F. Hrudka, F.J. MacCallum
Associate Professors: B.G. Anderson, W.K. Latshaw
Assistant Professors: L.W. Oliphant
Temporary Assistant Professor: 1

Field Stations/Laboratories: Fulton Laboratory

Degree Program: Veterinary Anatomy

Graduate and Undergraduate Courses Available for Graduate Credit: Gross Anatomy of Domestic Animals (U), Microscopic Anatomy of Domestic Animals (U), Neuroscience of Domestic Animals (U), Advanced Histology of Domestic Animals, Advanced Veterinary Embryology, Veterinary Radiographic Anatomy, Principles and Applications of Electron Microscopy, Cell Biology, Ultrastructural Cytology, Theoretical and Applied Histochemistry, Spermatology, Seminar in Veterinary Anatomy, Research, Special Problems in Anatomy

DEPARTMENT OF VETERINARY MICROBIOLOGY
Phone: Chairman: 343-5532 Others: 343-5531

Head: J.R. Saunders
Professors: J.R. Allen, C.H. Bigland, J.R. Saunders
Associate Professors: J.O. Iversen, B.T. Rouse, A.E. Sollod, G.E. Ward
Assistant Professors: L.A. Babiuk
Teaching Fellows: 3
Research Assistants: 2
Departmental Assistants: 1

Degree Program: Veterinary Microbiology and Veterinary Parasitology

Graduate and Undergraduate Courses Available for Graduate Credit: Immunology (U), Veterinary Virology (U), Epidemiology (U), Pathogenic Microbiology (U), Veterinary Parasitology (U), Recent Advances in Microbiology, Advanced Studies on Microorganisms and Parasites

DEPARTMENT OF VETERINARY PATHOLOGY
Phone: Chairman: (306) 343-4964 Others: 343-3388

Chairman: H.B. Schiefer
Professors: J.H.L. Mills, G.P. Searcy, D.L.T. Smith
Associate Professors: C.E. Doige, C. Riddell, L. Tryphonas, G. Wobeser
Assistant Professors: J.E.C. Bellamy
Instructor: R.J. Lewis
Teaching Fellow: 1
Professional Associates: 4

Degree Program: Pathology, Wildlife Diseases

Graduate and Undergraduate Courses Available for Graduate Credit: Comparative Histopathology, Advances in Veterinary Pathology, Advances in General Pathology, Advanced Systematic Pathology, Avian Pathology, Avian Necropsy, Mammalian Necropsy, Clinical Hematology, ClinicalChemistry, Surgical Pathology, Wildlife Disease, Experimental Pathology

DEPARTMENT OF VETERINARY PHYSIOLOGY SCIENCES
Phone: Chairman: (306) 343-2606

Chairman: W.E. Roe
Associate Professors: E.C. Crichlow, V.S. Gupta, F. M. Loew, J.G. Manns, C.S. Sisodia
Assistant Professors: G.W. Forsyth, R.K. Chaplin
Instructors: J. Katrusiak

Degree Programs: Veterinary Physiology

Graduate and Undergraduate Courses Available for Graduate Credit: Advanced Mammalian Physiology, Advanced Endocrinology, Advanced Neurophysiology, Advanced Pharmacology

INSTITUTIONAL INDEX

Acadia University,p .542
Adams State College, p.65
Adelbert College, p.355
Adelphi University, p.295
Adrian College, p.222
Agnew Scott College, p.99
Alabama A and M University, p.3
Alabama State University, p.3
Albany College of Pharmacy, p.295
Albany Medical Center, p.327
Albany Medical College, p.295
Albany State College, p.99
Albert Einstein College of Medicine, p.329
Albertus Magnus College, p.75
Albion College, p.222
Albright College, p.390
Alcorn State University, p.246
Alderson-Broaddus College, p.506
Alfred University, p.295
Allegheny College, p.390
Allentown College of Saint Francis de Sales, p.390
Alliance College, p.390
Alma College, p.222
Alvernia College, p.390
Alverno College, p.511
American International College, p.206
The American University, p.83
Amhurst College, p.206
Anderson College, p.142
Andrews University, p.222
Anna Maria College, p.206
Annhurst College, p.95
Antioch College, p.354
Appalachian State University, p.391
Aquinas College, p.222
Aristook State College, p.192
Arizona State University, p.15
Arkansas Polytechnic College, p.20
Arkansas State University, p.20
Armstrong State College, p.99
Asbury College, p.174
Ashland College, p.354
Ashville Biltmore College, p.331
Assumption College, p.206
Athens College, p.3
Atlanta University, p.99
Atlantic Union College, p.206
Auburn University, p.3
Augsburg College, p.237
Augusta College, p.100
Augustana College (IL), p.119
Augustana College (NC), p.431
Aurora College, p.119
Austin Peay State University, p.436
Averett College, p.484

B

Baker University, p.164
Baldwin-Wallace College, p.354
Baptist College at Charleston, p.423
Barat College, p.119
Barber-Scotia College, p.331
Bard College, p.295
Barnard College, p.300
Barrington College, p.419
Barry College, p.27
Bates College, p.192
Beaver College, p.390
Belhaven College, p.246
Bell State University, p.142
Bellarmine College, p.174
Belmont Abbey College, p.331
Belmont College, p.436
Beloit College, p.511
Bemidji State College, p.237
Benedict College, p.423
Benedictine College, p.164
Bennett College, p.331
Berea College, p.174
Bernard M. Baruch College, p.296
Berry College, p.100
Bethany College (KA), p.164
Bethany College (WV), p.506
Bethany Nazarene College, p.373
Bethel College (IN), p.142
Bethel College (KS), p.164
Bethel College (MN), p.237
Biola College, p.26
Birmingham-Southern College, p.6
Bishop's University, p.555
Blackburn College, p.119
Black Hills State College, p.431
Bloom field College, p.282
Bloomsburg State College, p.390
Bluefield State College, p.506
Bluffton College, p.354
Boise State University, p.116
Boston College, p.206
Boston State College, p.207
Boston University, p.207
Bowdoin College, p.192
Bowie State College, p.196
Bowman Gray School of Medicine, p.347
Bowling Green State University, p.354
Bradley University, p.119
Brandeis University, p.208
Brandon University, p.537
Brenan College, p.100
Brescia College, p.174
Briar Cliff College (IA), p.154
BriarCliff College (NY), p.295
Bridgewater College, p.484
Bridgewater State College, p.208
Brigham Young University (HI), p.111
Brigham Young University (UT), p.472
Brock University, p.544
Brooklyn College, p.296
Brooklyn College of Pharmacy, p.296
Brown University, p.419
Bryan College, p.436
Bryn Mawr College, p.391
Bucknell University, p.391
Butler University, p.142

C

Cabrini College, p.391
California Baptist College, p.26
California Lutheran College, p.26
California Institute of Technology, p.26
California Polytechnic State University, p.26
California State College (PA), p.391
California State College-Bakersfield, p.27
California State College-Dominguez Hill, p.27
California State College-Fresno, p.30
California State College-San Bernadino, p.28
California State College-Sacramento, p.31
California State College-Sonoma
California State College-Stanislaus, p.28
California State Polytechnic University, p.28
California State University-Chico, p.29

California State University-Fresno, p.29
California State University-Fullerton, p.30
California State University-Hayward, p.30
California State University-Humbolt, p.31
California State University-Long Beach, p.31
California State University-Los Angeles, p.31
California State University-Northridge, p.31
California State University-Sacramento, p.31
California State University-San Diego, p.32
Calvin College, p.223
Campbell College, p.331
Campbellsville College, p.174
Canisius College, p.296
Capital University, p.354
Cardinal Glennon College, p.253
Cardinal Stritch College, p.511
Carleton College, p.236
Carleton University, p.544
Carlow College, p.391
Carnegie Mellon University, p.391
Carroll College, p.511
Carson-Newman College, p.436
Carthage College, p.511
Case Western Reserve University, p.354
Castleton State College, p.479
Catawba College, p.331
The Catholic University of America, p.83
Catholic University of Puerto Rico, p.417
Cedar Crest College, p.392
Centenary College of Louisiana, p.182
Central State University, p.357
Central State University, p.374
Central Michigan University, p.223
Central Missouri State University, p.253
Central Methodist College, p.253
Central Washington State College, p.496
Central Wesleyan College, p.423
Centre College of Kentucky, p.174
Chadron State College, p.268
Chaminade College, p.111
Chapman College, p.32
Chatham College, p.392
Chestnut Hill College, p.392
Cheyney State College, p.392
Chicago State University, p.119
The City College of the City University of New York, p.296
City University of New York, p.269
Concord College, p.506
Concordia College, p.238
Concordia College-Saint Paul, p.238
Christian Brothers College, p.436
Christopher Newport College, p.484
The Citadel, p.423
City University of New York, p.297
Claflin College, p.423
Clarion State College, p.397
Clark University, p.209
Clarke College, p.154
Claremont Colleges, p.32
Clemson University, p.423
The Cleveland State University, p.357
Clinch Valley College, p.484
Coe College, p.154
Coker College, p.426
Colby College, p.192
College of Charleston, p.426
College Misericordia, p.392
College of the Holy Cross, p.209
The College of Idaho, p.116
College of Medicine and Dentistry of New Jersey, p.282
College of Mount St. Joseph, p.357
College of Mt. St. Vincent, p.299
College of New Rochelle, p.299
College of Notre Dame, p.33
College of Notre Dame of Maryland, p.196
College of Our Lady of the Elms, p.209
The College of the Ozarks, p.20
College of Saint Benedict, p.236
College of St. Catherine, p.236
College of Saint Elizabeth, p.284
College of St. Francis, p.120
College of Saint Mary, p.268
College of Saint Rose, p.299
College of Saint Scholastica, p.237
College of Saint Theresa, p.238
College of St. Thomas, p.238
College of Santa Fe, p.291
College of Steubenville, p.357
College of the Virgin Islands, p.483
College of William and Mary, p.484
College of White Plains, p.299
College of Wooster, p.358
The Colorado College, p.65
Colorado State University, p.65
Colorado Women's College, p.70
Colgate University, p.299
Columbia University, p.300
Columbia College, p.427
Columbia Union College, p.196
Columbian College of Arts and Sciences, p.24
Columbus College, p.100
Concordia College (MN), p.238
Concordia College (NY), p.301
Connecticut College, p.75
Converse College, p.427
Cook College, p.284
Coppin State College, p.196
Cornell College, p.154
Cornell Medical College, p.307
Cornell University, p.301
Covenant College, p.437
Creighton University, p.268
Culver-Stockton College, p.253
Curry College, p.209

D

Dakota State College, p.431
Dakota Wesleyan University, p.431
Dalhousie University, p.542
Dana College, p.269
Dartmouth College, p.277
David Lipscomb College, p.431
Davidson College, p.331
Davis and Elkins College, p.506
Defiance College, p.358
Delaware State College, p.82
Delta State University, p.246
Denison University, p.358
De Paul University, p.120
DePrauw University, p.143
Detroit Institute of Technology, p.223
Dickinson College, p.303
Delaware Valley College of Science and Agriculture, p.393
Dillard University, p.182
Doane College, p.269
Dominican College of San Rafael, p.33
Dordt College, p.154
Dorothy H. and Lewis Rosenthiel School of Marine and Atmospheric Science, p.96
Douglass College, p.284
Dowling College, p.308
Drake University, p.155
Drexel University, p.393
Drury College, p.253
Drew University, p.284
Duke University, p.332
Duquesne University, p.394
D'Youville College, p.308

E

Earlham College, p.143
East Carolina University, p.334
East Central State College, p.374
East Stroudsburg State College, p.394

East Tennessee State University, p.437
Eastern Baptist College, p.394
East Connecticut State College, p.79
Eastern College, p.394
Eastern Illinois University, p.120
Eastern Kentucky University, p.174
Eastern Mennonite College, p.484
Eastern Michigan University, p.223
Eastern Montana College, p.265
Eastern Nazarene College, p.209
Eastern New Mexico University, p.291
Eastern Washington State College, p.496
Eckerd College, p.87
Eastern Oregon State College, p.382
Edgecliff College, p.358
Edgewood College, p.511
Edinboro State College, p.394
Edward Waters College, p.87
Eisenhower College, p.308
Elizabethtown College, p.395
Elizabeth City State University, p.335
Elmira College, p.308
Elmurst College, p.120
Elon College, p.335
Emmanuel College, p.210
Emory and Henry College, p.485
Emory University, p.100
Emporia Kansas State College, p.164
Erskine College, p.427
Eureka College, p.121
Evangel College, p.253

F

Fairfield University, p.75
Fairleigh Dickinson University, p.284
Fairmont State College, p.506
Fayetteville State University, p.335
Federal City College, p.83
Ferris State College, p.223
Findlay College, p.358
Fisk University, p.431
Florence State University, p.6
Florida Atlantic University, p.87
Florida Institute of Technology, p.87
Florida Southern College, p.88
Florida State University, p.88
Florida Technological University, p.88
Fontbonne College, p.253
Fordham University, p.308
Fort Hays Kansas State College, p.164
Fort Lewis College, p.70
Framingham State College, p.210
Francis Marion College, p.427
Franconia College, p.278
Franklin College, p.143
Franklin College of Arts, p.105
Franklin and Marshall College, p.395
Franklin Pierce College, p.278
Friends University, p.165
Frostburg State College, p.196
Furman University, p.427

G

Galludet College, p.83
Gannon College, p.395
Gardner-Webb College, p.335
Geneva College, p.395
George Fox College, p.382
George Mason University, p.485
George Peabody College for Teachers, p.437
George Washington University, p.83
George Williams College, p.121
Sir George Williams University, p.557
Georgetown University, p.85
Georgia College, p.102
Georgia Institute of Technology, p.102
Georgia Southern College, p.102
Georgia Southwestern College, p.103
Georgian Court College, p.285
Georgetown College, p.175
Gettysburg College, p.395
Glassboro State College, p.285
Glenville State College, p.506
Gonzaga University, p.496
Gordon College, p.210
Goshen College, p.143
Goucher College, p.196
Graceland College, p.155
Grambling College, p.182
Grand Canyon College, p.15
Grand Valley State College, p.223
Greensboro College, p.335
Greenville College, p.121
Grinnell College, p.155
Grove City College, p.396
Guilford College, p.335
Gustavus Adolphus College, p.238
Gwynedd-Mercy College, p.396

H

Hamilton College, p.308
Hampden-Sydney College, p.485
Hampton Institute, p.485
Hanover College, p.144
Harding College, p.20
Hartwick College, p.309
Haverford College, p.396
Harvard University, p.210
Hastings College, p.269
Hawaii Loa College, p.111
Heidelberg College, p.358
Henderson State College, p.21
Hendrix College, p.21
Herbert H. Lehman College, p.309
High Point College, p.335
Hillsdale College, p.224
Hiram College, p.358
Hobart and William Smith Colleges, p.309
Hofstra University, p.309
Hollins College, p.485
Holy Family College (PA), p.396
Holy Family College (WI), p.512
Holy Names College, p.33
Hood College, p.196
Hope College, p.224
Houghton College, p.309
Howard University, p.86
Humbolt State University, p.33
Hunter College, p.298
Huntingdon College, p.7
Huron College, p.431
Huntington College, p.144

I

Idaho State University, p.116
Illinois Benedictine College, p.121
Illinois College, p.121
Illinois Institute of Technology, p.121
Illinois State University, p.122
Illinois Wesleyan University, p.122
Immaculata College, p.396
Immaculate Heart College, p.34
Indiana University at Bloomington, p.144
Indiana University-South Bend, p.147
Indiana University-Kokomo, p.145
Indiana University-Indianapolis, p.145
Indiana University Northwest, p.145
Indiana University Southeast, p.147
Indiana State University at Evansville, p.144
Indiana University of Pennsylvania, p.396
Inter American University of Puerto Rico, p.417
Iona College, p.309

Iowa State University, p.155
Ithaca College, p.309

J

Jackson State University, p.246
Jacksonville State University, p.7
Jacksonville University, p.89
Jamestown College, p.349
Jarvis Christian College, p.451
Jersey City State College, p.285
John Brown University, p.21
John Carroll University, p.359
Johns Hopkins University, p.197
Johnson C. Smith University, p.336
Johnson State College, p.479
Judson College (AL), p.7
Judson College (IL), p.122
Juniata College, p.397

K

Kalamazoo College, p.224
Kansas Newman College, p.165
Kansas State College of Pittsburgh, p.165
Kansas State University, p.166
Kansas Wesleyan, p.169
Kean College of New Jersey, p.286
Kearney State College, p.269
Keene State College, p.278
Kent State University, p.359
Kentucky State University, p.175
Kentucky Wesleyan College, p.175
Kenyon College, p.359
Keuka College, p.309
King College, p.437
King's College, p.397
The King's College, p.310
Kirkland College, p.310
Knox College, p.123
Knoxville College, p.439
Kutztown State College, p.397

L

La Salle College, p.397
La Verne College, p.35
Lafayette College, p.397
Lake Erie College, p.359
Lake Forest College, p.123
Lake Superior State College, p.224
Lakehead University, p.544
Lamar University, p.451
Lane College, p.439
Langston University, p.374
Lambuth College, p.439
Lander College, p.427
La Roche College, p.397
Laurentaran University, p.544
Lawrence University, p.512
Le Moyne College, p.310
Lebanon Valley College, p.398
Lee College, p.439
Lehigh University, p.398
Lehman College, p.310
Lemoyne-Owen College, p.439
Lenoir Rhyne College, p.336
Le Tourneau College, p.451
Lewis University, p.123
Lewis and Clark College, p.382
Limestone College, p.427
Lincoln Memorial University, p.439
Lincoln University, p.398
Lincoln University of Missouri, p.253
The Lindenwood College, p.254
Linfield College, p.382
Livingston College, p.286
Livingstone College, p.336
Livingston University, p.7
Lock Haven State College, p.398
Loma Linda University, p.35
Lone Mountain College, p.36
Long Island University, p.310
Longwood College, p.486
Loras College, p.159
Loretto Heights College, p.70
Los Angeles Baptist College, p.36
Louisiana College, p.182
Louisiana State University and Agricultural and Mechanical College, p.182
Louisiana State University in New Orleans, p.184
Louisiana Tech University, p.186
Loyola College, p.198
Loyola Marymount University, p.37
Loyola University (LA), p.187
Loyola University (QUE), p.558
Loyola University of Chicago, p.123
Lowell State College, p.212
Lowell Technological Institute, p.213
Lubbock Christian College, p.451
Luther College, p.159
Lycoming College, p.398
Lynchburg College, p.486
Lyndon State College, p.479

M

Macalester College, p.238
MacMurray College, p.125
Madison College, p.486
Malone College, p.359
Manchester College, p.147
Manhattan College, p.299
Mankato State College, p.238
Max C. Fleischman College, of Agriculture, p.275
Mansfield State College, p.399
Marian College, p.512
Marietta College, p.360
Marion College, p.147
Marist College, p.310
Marlboro College, p.479
Marquette University, p.512
Mars Hill College, p.336
Marshall College, p.507
Mary Baldwin College, p.486
Mary College, p.349
Mary Hardin-Baylor College, p.451
Mary Washington College, p.486
Marygrove College, p.224
Marymount College, p.311
Marymount College of Kansas, p.169
Maryville College, p.254
Marywood College, p.399
Massachusetts Institute of Technology, p:213
McGill University, p.555
McMaster University, p.545
McMurry College, p.451
McNeese State University, p.187
McPherson College, p.169
Medical College of Georgia, p.103
Medical College of Virginia, p.486
The Medical College of Wisconsin, p.512
Memorial University of Newfoundland, p.541
Memphis State University, p.459
Mercer University, p.104
Mayville State College, p.349
McMicken College of Arts and Sciences, p.369
The Medical College of Pennsylvania, p.399
Medical University of South Carolina, p.428
Mercy College, p.311
Mercy College of Detroit, p.224
Mercyhurst College, p.400
Meredith College, p.336
Merrimack College, p.213
Messiah College, p.400
Methodist College, p.336

Metropolitan State College, p.70
Miami University, p.360
Michigan State University, p.225
Michigan Technical University, p.228
Middle Tennessee State University, p.439
Middlebury College, p.479
Midland Lutheran College, p.270
Midwestern University, p.452
Miles College, p.7
Millersville State College, p.400
Milligan College, p.439
Millikin University, p.125
Mills College, p.37
Millsaps College, p.246
Milton College, p.513
Milton S. Hershey Medical Center, p.400
Minot State College, p.349
Mississippi College, p.247
Mississippi State University, p.247
Mississippi Valley State University, p.250
Mississippi University for Women, p.247
Missouri Southern State College, p.254
Missouri Valley College, p.254
Missouri Western State College, p.254
Mobile College, p.7
Molloy College, p.311
Monmouth College (NJ), p. 286
Monmouth College (ILL), p. 125
Montclair State College, p.286
Montana State University, p.265
Moorehouse College, p.99
Moorhead State College, p.239
Moravian College, p.400
Morehead State University, p.175
Morgan State College, p.199
Moringside College, p.159
Morris Brown College, p.100
Morris Harvey College, p.507
Mount Allison University, p.540
Mount Holyoke College, p.213
Mount Marty College, p.431
Mount Mary College, p.513
Mount Mercy College, p. 159
Mt. St. Mary College, p.311
Mount Saint Mary College, p.278
Mount St. Mary's College (CA), p.37
Mount St. Mary's College (MD), p.199
Mount St. Scholastica College, p.164
Mount Union College, p.361
Mount Saint Vincent University, p.543
Muhlenberg College, p.400
Mundelein College, p.129
Murray State University, p.175
Muskingum College, p.361

N

Nasson College, p.192
Nathaniel Hawthorne College, p.278
Nazareth College, p.229
Nazareth College of Rochester, p.312
Nebraska Wesleyan, p.210
New College, p.89
New England College, p.278
New Hampshire College, p.279
New Jersey Medical School, p.282
New Mexico Highlands University, p.291
New Mexico Institute of Mining and Technology, p.291
New Mexico State University, p.292
New York Institute of Technology, p.311
New York State Agricultural Experiment Station, p.311
New York State College of Agriculture and Sciences, p.311
New York State Veterinary College, p.311
New York University, p.312
Newark State College, p.286
Newberry College, p.429
Newcomb College, p.190
Niagara University, p.313
Nicholls State University, p.188
Norbert College, p.514
North Adams State College, p.214
North Carolina Agricultural and Technical State University, p. 336
North Carolina Central University, p.337
North Carolina State University, p.337
North Carolina Wesleyan College, p.337
North Dakota State University, p.349
North Georgia College, p.104
North Park College, p.125
North Texas State University, p.452
Northeast Louisiana University, p,188
Northeast Missouri State University, p.254
Northeastern State College, p.374, p.374
Northeastern University, p.214
Northern Arizona University, p.15
Northern Illinois University, p.126
Northern Kentucky State College, p.176
Northern Michigan University, p.228
Northern State College, p.431
Northland College, p.573
Northwest Missouri State University, p.254
Northwestern College, p.159
Northwestern State College, p.374
Northwestern State University, p.188
Northwestern University, p.126
Norwich University, p.479
Notre Dame College (MO), p.255
Notre Dame College (NH), p.279
Notre Dame College (OH), p.361
Notre Dame University, p.533
Nova Scotia Technical College, p.543

O

Oakland University, p.229
Oakwood College, p.7
Oberlin College, p.361
Occidental College, p.37
Oglethorpe University, p.104
Ohio Dominican College, p.361
Ohio Northern University, p.361
Ohio State University, p.362
Ohio University, p.368
Ohio Wesleyan University, p.369
Oklahoma Baptist University, p.374
Oklahoma Christian College, p.375
Oklahoma City University, p.375
Oklahoma Panhandle State University, p.375
Oklahoma State University, p.375
Old Dominion University, p.487
Olivet College, p.230
Olivet Nazarene College, p.127
Onachita College, p.21
Ontario Agricultural College, p.547
Ontario Vetinary College, p.548
Oral Roberts University, p.378
Oregon State University, p.382
Ottawa University, p.169
Otterbein College, p.369
Our Lady of Angels College, p.401
Our Lady of the Lake College, p.452

P

Pace University, p.313
Pacific College, p.37
Pacific Lutheran University, p.497
Pacific Union College, p.38
Pacific University, p.387
Palm Beach Atlantic College, p.89
Pan American University, p.452
Part College, p.255
Pembroke State University, p.337
Pennsylvania State University, p.401
Pepperdine University, p.38

Pfeiffer College, p.337
Philadelphia College of Pharmacy and Science, p.405
Phillips University, p.378
Piedmont College, p.104
Pitzer College, p.33
Plymouth State College, p.279
Point Loma College, p.38
Point Park College, p.406
Pomona College, p.32
Portland State University, p.387
Prairie View A and M University, p.453
Presbyterian College, p.429
Prescott College, p.16
Princeton University, p.286
Principia College, p.127
Pritzker School of Medicine, p.131
Providence College, p.419
Purdue University, p.147
Purdue University-Calumet Campus, p.150
Purdue University at Fort Wayne, p.150

Q

Queens College (NC), p.337
Queen's College (NY), p.298
Queens University, p.545
Quincy College, p.127
Quinnipiac College, p.76

R

Radford College, p.487
Randolph-Macon College, p.487
Randolph-Macon Woman's College, p.487
Reed College, p.387
Regis College (CO), p.70
Regis College (MA), p.215
Rensselaer Polytechnic Institute, p.314
Rhode Island College, p.419
Rice University, p.453
Richmond College, p.487
Ricker College, p.192
Rider College, p.287
Ripon College, p.513
Rio Grande College, p.369
Rivier College, p.279
Roanoke College, p.487
Roberts Wesleyan College, p.314
Rochester Institute of Technology, p.314
Rockford College, p.126
Rockhurst College, p.299
Roger Williams College, p.420
Rollins College, p.89
Roosevelt University, p.128
Rosary College, p.128
Rosary Hill College, p.315
Rosemont College, p.406
Russell Sage College, p.315
Rust College, p.250
Rutgers Medical School, p.283
Rutgers-The State University Camden, p.287
Rutgers-The State University New Brunswick, p.288
Rutgers-The State University Newark, p.288

S

Sacred Heart University, p.76
Saginaw Valley College, p.230
St. Ambrose College, p.159
St. Andrews Presbyterian College, p.337
St. Anselm's College, p.279
Saint Augustine's College, p.338
St. Bernard College, p.7
St. Bernadine of Siena College, p.315
St. Bonaventure University, p.315
St. Cloud State College, p.239
St. Edwards University, p.453
St. Francis College (IN), p.151
St. Francis College (ME), p.192
St. Francis College (NY), p.315
St. Francis College (PA), p.406
St. Francis Xavier University, p.543
St. John Fisher College, p.315
St. John's University (NY), p. 315
St. John's University (MN), p. 239
Saint Joseph College (CT), p.76
Saint Joseph's College (IN), p.151
St. Joseph College (NY), p.316
St. Joseph's Calumet College, p.151
Saint Joseph's College, p.406
St. Joseph's College, p.192
St. Lawrence University, p.316
St. Louis University, p.255
St. Martin's College, p.497
Saint Mary's College (IN), p.151
Saint Mary College (KS), p.169
St. Mary's College (MN), p.239
St. Mary's College of California, p.38
St. Mary's College of Maryland, p.199
St. Mary's Dominican College, p.189
St. Mary's University (CA), p.453
St. Mary's University (NS), p.543
St. Meinrad College, p.151
St. Michael's College, p.479
St. Olaf College, p.240
St. Paul's College, p.488
Saint Peter's College, p.209
St. Thomas Aquinas College, p.316
Saint Vincent College, p.406
St. Xavier College, p.128
Salem College, p.507
Salem State College, p.215
Salisbury State College, p.199
Salve Regina College, p.420
Sam Houston State University, p.454
San Diego State University, p.38
San Francisco State University, p.40
Samford University, p.8
Sarah Lawrence College, p.310
Salem College, p.338
Savannah State College, p.105
The School of the Ozarks, p.257
Scripps College, p.33
Seattle Pacific College, p.497
Seattle University, p.497
Seton Hall University, p.290
Seton Hill University, p.406
Shaw University, p.338
Shepherd College, p.507
Shimer College, p.128
Shippensburg State College, p.407
Shorter College, p.105
Siena College, p.316
Sienna Heights College, p.230
Silver Lake College, p.514
Simmons College, p.215
Simon Fraser University, p.533
Simpson College, p.159
Sioux Falls College, p.432
Skidmore College, p.316
Slippery Rock State College, p.407
Smith College, p.215
South Carolina State College, p.429
South Dakota State University, p.432
Southampton College, p.310
Southeast Missouri State University, p.257
Southeastern Louisiana University, p.189
Southeastern Massachusetts University, p.215
Southeastern State College, p.378
Southern California College, p.41
Southern Connecticut State College, p.76
Southern Colorado State College, p.70
Southern Illinois University at Carbondale, p.128
Southern Illinois University at Edwardsville, p.130
Southern Methodist University, p.454
Southern Missionary College, p.439
Southern Oregon College, p.387

Southern State College, p.21
Southern University, p.189
Southern Utah State College, p.473
Southwest Baptist College, p.257
Southwest Minnesota State College, p.249
Southwest Texas State University, p.454
Southwestern at Memphis, p.439
Southwestern College, p.169
Southwestern State College, p.378
Southwestern Union College, p.455
Southwestern University, p.455
Spring Hill College, p.8
Springfield College, p.216
Stanford University, p.41
State College of Arkansas, p.21
State University of New York at Albany, p.317
State University of New York at Binghampton, p.317
State University of New York at Buffalo, p.318
State University of New York College at Brockport, p.317
State University of New York College at Buffalo, p.317
State University of New York at Fredonia, p.323
State University of New York at Geneseo, p.323
State University of New York at Oneonta, p.323
State University of New York at Stony Brook, p.325
State University of New York College at Cortland, p.321
State University of New York College at New Paltz, p.323
State University of New York College at Oswego, p.324
State University of New York College at Plattsburgh, p.324
State University of New York College at Potsdam, p.324
State University of New York College of Environmental Sciences, p.321
State University of New York Downstate Medical Center, p.325
State University of New York Upstate Medical Center, p.325
Stephen F. Austin State University, p.455
Stephens College, p.258
Sterling College, p.170
Stetson University, p.89
Stillman College, p.8
Stonehill College, p.216
Stritch School of Medicine, p.124
Suffolk University, p.216
Sul Ross University, p.453
Susquehanna University, p.407
Swarthmore College, p.407
Sweet Briar College, p.488
Syracuse University, p.326

T

Tabor College, p.170
Talladega College, p.8
Tarkio College, p.258
Tarleton State University, p.456
Taylor University, p.151
Temple University, p.407
Tennessee State University, p.440
Tennessee Technological Institute, p.440
Tennessee Temple College, p.440
Tennessee Wesleyan College, p.440
Texas A and I University, p.456
Texas A and M University, p.456
Texas Christian University, p.460
Texas College, p.461
Texas Lutheran College, p.461
Texas Southern University, p.461
Texas Tech University, p.461
Texas Wesleyan College, p.463
Texas Woman's University, p.463
Thiel College, p.409
Thomas Jefferson College, p.230
Thomas More College, p.176
Tougaloo College, p.250
Towson State College, p.199
Transylvania University, p.176
Trent University, p.546
Trenton State College, p.290
Trinity Christian College, p.130
Trinity College (CT), p.76
Trinity College (DC), p.86
Trinity College (IL), p.131
Trinity College (VA), p.480
Trinity College of Arts and Sciences, p.332
Trinity University, p.463
Tri State College, p.151
Troy State University, p.8
Tufts University, p.216
Tulane University, p.190
Tusculum College, p.440
Tuskegee Institute, p.8

U

Union College, p.176
Union College and University, p.327
Union University, p.440
United States Air Force Academy, p.70
United States Naval Academy, p.199
Unity College, p.192
Université de Moncton, p.540
Université de Montréal, p.558
Université du Québec à Chicoutimi, p.560
Université du Québec à Montreal, p.560
Université du Québec à Rimouski, p.560
Université du Québec à Trois-Rivières, p.560
Université de Sherbrooke, p.559
Université Laval, p.557
University College, p.290
University of Alabama, p.10
University of Alabama in Birmingham, p.10
University of Alabama in Huntsville, p.12
University of Akron, p.369
University of Alaska, p.14
University of Alberta, p.530
University of Albuquerque, p.293
University of Arizona, p.16
University of Arkansas, p.22
University of Arkansas at Little Rock, p.25
University of Arkansas at Monticello, p.25
University of Bridgeport, p.76
University of British Columbia, p.533
University of Calgary, p.530
University of California-Davis, p.46
University of California-Berkeley, p.42
University of California-Irvine, p.55
University of California-Los Angeles, p.55
University of California-Riverside, p.57
University of California-San Francisco, p.60
University of California-San Diego, p.59
University of California-Santa Barbara, p.61
University of California-Santa Cruz, p.61
University of Chicago, p.131
University of Cincinnati, p.369
University of Colorado, p.71
University of Connecticut, p.77
University of Corpus Christie, p.463
University of Dallas, p.463
University of Dayton, p.371
University of Delaware, p.82
University of Denver, p.73
University of Detroit, p.230
University of Dubuque, p.159
University of Evansville, p.152
University of Florida, p.89
University of Georgia, p.105
University of Guam, p.110
University of Guelph, p.546
University of Hartford, p.79
University of Hawaii, p.111
University of Hawaii at Hilo, p.114
University of Houston, p.463
University of Iowa, p.160
The University of Health Science, p.133
University of Idaho, p.116
University of Illinois at Chicago Circle, p.134
University of Illinois at the Medical Center, p.134
University of Illinois at Urbana Champaign, p.135

University of Kansas, p.170
University of Kentucky, p.176
University of Lethbridge, p.532
University of Louisville, p.180
University of Maine at Fort Kent, p.193
University of Maine at Farmington, p.193
University of Maine at Orono, p.193
University of Maine at Portland-Gorham, p.195
University of Maine at Presque Isle, p.195
University of Manitoba, p.537
University of Maryland-Baltimore, p.203
University of Maryland-Baltimore County, p.205
University of Maryland-College Park, p.199
University of Massachusetts, p.217
University of Miami, p.95
University of Michigan, p.230
University of Michigan-Flint, p.234
University of Minnesota, p.240
University of Minnesota-Duluth, p.245
University of Minnesota-Morris, p.245
The University of Mississippi, p.251
University of Missouri-Columbia, p.258
University of Missouri-Kansas City, p.262
University of Missouri-St. Louis, p.262
University of Montana, p.266
University of Montevallo, p.12
University of Nevada-Las Vegas, p.275
University of Nevada-Reno, p.275
University of New Brunswick, p.540
University of New Hampshire, p.279
University of New Haven, p.79
University of New Mexico, p.293
University of North Carolina at Asheville, p.338
University of North Carolina at Chapel Hill, p.338
University of North Carolina at Charlotte, p.342
University of North Carolina at Greensboro, p.3 42
University of North Carolina State University at Raleigh, p.343
University of North Carolina at Wilmington, p.346
University of North Dakota, p.352
University of Northern Colorado, p.73
University of Notre Dame, p.152
University of Oklahoma, p.379
University of Oregon, p.387
University of Ottawa, p.548
University of Pennsylvania, p.409
University of Pittsburgh, p.412
University of Pittsburgh at Johnstown, p.414
University of Portland, p.389
University of Prince Edward Island, p.554
University of Puerto Rico-Mayaguez, p.417
University of Puerto Rico-Rio Pedras, p.418
University of Puget Sound, p.497
University of Redlands, p.62
University of Regina, p.561
University of Rhode Island, p.420
University of Richmond, p.488
University of Rochester, p.327
University of St. Thomas, p.464
University of San Diego, p.62
University of San Francisco, p.62
University of Santa Clara, p.62
University of Science and Arts of Oklahoma, p.380
University of Scranton, p.414
University of Saskatchewan, p.561
University of South Alabama, p.12
University of South California, p.63
University of South Carolina, p.429
University of South Dakota, p.434
University of South Dakota at Springfield, p.435
University of South Florida, p.97
University of Southern Mississippi, p.252
University of Southwestern Louisiana, p.191
University of Tampa, p.97
University of Tennessee at Chattanooga, p.440
University of Tennessee at Knoxville, p.441
University of Tennessee at Martin, p.444
University of Tennessee Medical Units, p.445
University of Texas at Arlington, p.464
University of Texas at Austin, p.465
University of Texas Health Science Center at Dallas, p.466
University of Texas at El Paso, p.468
University of Texas Health Science Center at San Antonio, p.469
University of Texas at Houston, p.469
University of Texas Medical Branch at Galveston, p.467
University of Texas at Permian Basin, p.469
University of the Pacific, p.61
University of the South, p.440
University of Toledo, p.371
University of Toronto, p.549
The University of Tulsa, p.381
University of Utah, p.473
University of Vermont, p.480
University of Victoria, p.535
University of Virginia, p.488
University of Washington, p.497
University of West Florida, p.98
University of Western Ontario, p.551
University of Winnipeg, p.539
University of Windsor, p.552
University of Wisconsin-Eau Claire, p.514
University of Wisconsin-Green Bay, p.514
University of Wisconsin-La Crosse, p.515
University of Wisconsin-Madison, p.515
University of Wisconsin-Milwaukee, p.521
University of Wisconsin-Oshkosh, p.523
University of Wiscon in-Parkside, p.523
University of Wisconsin-Platteville, p.522
University of Wisconsin-River Falls, p.522
University of Wisconsin-Stevens Point, p.522
University of Wisconsin-Stout, p.524
University of Wisconsin-Superior, p.524
University of Wisconsin-Whitewater, p.524
University of Wyoming, p.526
Upper Iowa College, p.162
Upsala College, p.290
Urbana College, p.371
Ursinus College, p.415
Ursuline College, p.371
Utah State University, p.474
Utica College, p.327

V

Valdosta State College, p.109
Valley City State College, p.353
Valparaiso University, p.153
Vassar College, p.329
Villa Maria College, p.415
Villanova University, p.415
Viterbo College, p.524
Virginia Commonwealth University, p.491
Virginia Institute of Marine Sciences, p.489
Virginia Military Institute, p.492
Virginia Polytechnic Institute and State University, p.492
Virginia State College, p.494
Virginia Union University, p.494
Virginia Wesleyan College, p.495
Voorhees College, p.429

W

Wabash College, p.153
Wagner College, p.329
Wake Forest College, p.347
Wake Forest University, p.347
Walla Walla College, p.500
Walsh College, p.371
Warner Pacific College, p.389
Warren Wilson College, p.348
Wartburg College, p.162
Washburn University, p.173
Washington College, p.205
Washington and Jefferson College, p.415
Washington and Lee University, p.495
Washington State University, p.501

FACULTY INDEX

Allen, L.H., 103
Allen, L.R., 423
Allen, M.K., 41
Allen, M.M., 220
Allen, P.J., 515, 519
Allen, P.Z., 328
Allen, R.D., 112, 507
Allen, R.K., 31
Allen, R.M., 18, 425
Allen, R.S., 184
Allen, R.T., 23
Allen, S.H.G., 327
Allen, S.L., 230, 231
Allen, T.C., 385
Allen, T.D., 110
Allen, T.F., 515
Allen, W.E., 334
Allen, W.R., 57
Allen, W.V., 34
Allenspach, A.L., 360
van Aller, R.T., 252
Allewell, N.M., 80
Alley, H.P., 527
Alley, W.P., 31
Allin, E.F., 520
Allinson, D.W., 79
Allison, B.M., 228
Allison, D.C., 125
Allison, E.B., 279
Allison, D.W., 78
Allison, J.E., 379
Allison, R.,4
Allison,J.L., 503
Allison, T., 452
Allison, V.P., 454
Allison, W., 393
Allimann, D.W., 146
Allred, D.M., 472
Allred, J.B., 365, 367
Allred, J.C., 464
Allt, W., 550
Alm, A.A., 243
Almers, W., 500
Al-Moslih, M.I., 273
Almquist, D.O., 401
Alo, W.A., 77
Aloridge, R., 256
Alper, R., 411
Alperin, J., 467
Alperin, M., 232
Alpers, D.H., 263
Alrutz, R.W., 358
Alsdorf, D.G., 364
Alsop, D.W., 298
Alsop, F.J., 437
Alspach, S., 205
Alston, R.P., 423
Alt, F.L., 508
Altbinger, E.O., 489
Altenbach, J.J., 293
Althouse, P.G., 365
Altig, R.G., 248
Altman, I., 414
Altman, J., 68, 150
Altman, K.I., 328
Altman, M.R., 81
Altman, S., 80
Altmann, G.G., 551
Altmann, S., 131, 132
Altmiller, Br.H., 453
Altshuller, A., 341
Altura, B.M., 330
Altwicker, E.R., 314
Alvarado, F., 418
Alvarado, R.H., 15, 386
Alvarez, A.M., 114
Alvarez, M., 96
Alvarez, M.R., 97
Alvarez-Rossario, E., 417
Alverson, D.L., 498
Amand, S.W., 251
Amann, R.P., 401
Amato, V.A., 454
Amborski, R.L., 185
Ambrose, C.T., 179
Ambrose, H.W.III, 136
Ambrus, C., 321
Ambrus, J.L., 320
Amdur, M.O., 212
Amen, R.D., 347
Amerine, M.A., 51
Amerson, M.G., 104
Ames, B.N., 42
Ames, D.R., 167
Ames, I., 325
Ames, R.W., 29
Amin, O.M., 522
Aminoff, D., 232
Amirana, M.T., 314
Amling, H.J., 5
Amliot, F.S., 372
Amme, R.C., 73
Ammerman, C.R., 249
Ammon, V.D., 249
Amos, D.B., 333
Amos, H., 211
Amos, R.E.Jr., 506
Ampey, R.M., 189
Amrein, U., 33
Amrein, Y.U., 333
Amsterdam, D., 310
Amstutz, H.C., 143
Amundson, C.C., 441
Amundson, C.H., 518
Amundson, J.C., 27
Amundson, R., 350
Amy, R.L., 439
Amzel, L.M., 197, 198
Anand, A.S., 523
Anastasia, A.A., 284
Anastassiades, T., 545
Anastos, G., 199
Anchel, M., 298
Ancclay, C., 417
Anderegg, D.E., 117
Anderes, E.A., 39
Anders, F.G., 223
Andersen, B., 222
Andersen, J.W., 129
Andersen, L.W., 271
Andersen, P., 308
Andersen, R.A., 177
Andersen, R.N., 240
Andersen, W., 254
Andersen, W.R., 472
Anderson, A.B., 44
Anderson, A.D., 521
Anderson, A.E., 382
Anderson, A.W., 351, 368, 386
Anderson, B., 127, 415
Anderson, B.G., 562
Anderson, B.M., 492
Anderson, C., 134, 155, 336, 552
Anderson, C.E., 343, 346
Anderson, C.K., 266
Anderson, D., 143 359
Anderson, D.C., 502
Anderson, D.E., 34
Anderson, D.G., 95
Anderson, D.K., 97, 228
Anderson, D.L., 219, 244
Anderson, D.W., 111
Anderson, E.C., 382
Anderson, E.F., 504
Anderson, F.D., 512
Anderson, F.L., 472
Anderson, F.O., 472
Anderson, G., 15, 37, 77
Anderson, G.B., 48
Anderson, G.L., 102
Anderson, G.W., 547
Anderson, H.G., 74
Anderson, J.D., 137
Anderson, J.E., 100
Anderson, J.F., 90, 219
Anderson, J.G., 389, 460
Anderson, J.H., 14
Anderson, J.M., 302, 374
Anderson, J.R., 44, 520
Anderson, J.S., 241
Anderson, J.W., 460, 520
Anderson, K., 31
Anderson, K.E., 315
Anderson, K.V., 101
Anderson, L., 517
Anderson, L.C., 166
Anderson, L.D., 58
Anderson, L.E., 258, 330
Anderson, L.P., 423
Anderson, M., 500
Anderson, M.J., 474
Anderson, M.K., 177
Anderson, M.M., 541
Anderson, M.W., 349
Anderson, N., 265, 393, 428
Anderson, N.H., 386
Anderson, N.O., 394
Anderson, O., 524
Anderson, O.E., 106
Anderson, P.A.W., 333
Anderson, P.D., 557
Anderson, P.J., 548
Anderson, P.K., 531
Anderson, P.L., 360
Anderson, P.S., 522
Anderson, R., 374, 473
Anderson, R.A., 73
Anderson, R.C., 22, 116, 547
Anderson, R.E., 304
Anderson, R.F., 334
Anderson, R.G.W., 466
Anderson, R.L., 225, 245
Anderson, R.O., 261
Anderson, R.R., 83, 259
Anderson, R.S., 531
Anderson, R.W., 333
Anderson, S., 297, 479
Anderson, S.H., 359
Anderson, S.R., 385
Anderson, T., 134, 392
Anderson, T.F., 410
Anderson, V.E., 242
Anderson, W., 155, 298, 557
Anderson, W.A., 131, 271
Anderson, W.B., 459
Anderson, W.D.Jr., 426
Anderson, W.D., 562
Anderson, W.E., 423
Anderson, W.L., 208
Anderson, W.P., 125, 401
Anderson, W.T., 549
Anderson, W.W., 106
Anderton, L.G., 342
Andison, H., 534
Andjelkovic, D., 342
Andoli, F.P., 27
Andres, G.A., 319
Andrew, G.M., 546
Andrew, N.J., 366
Andrew, R.H., 516
Andrew, W., 145
Andrew, W.T., 530
Andrews, A.K., 376
Andrews, C.H., 248
Andrews, C.L., 100
Andrews, C.W., 541
Andrews, D., 120
Andrews, E.J., 306
Andrews, F.N., 147, 150
Andrews, H., 77
Andrews, H.N., 77, 79
Andrews, H.T., 440
Andrews, J.C., 340, 534
Andrews, J.D., 490
Andrews, L.D., 23
Andrews, L.G., 100
Andrews, L.J., 51
Andrews, P.M., 466
Andrews, R.A., 280
Andrews, R.D., 496
Andrews, S.B., 80
Andrews, T.J., 218
Andrey, E., 471
Andriese, P.C., 32
Andris, K., 472
Andrius, O.K., 413
Andrus, D., 474
Andrus, G., 33
Andrus, W.W., 33
Andrykovitch, G.E., 485
Ang, J.K., 323
Angel, C., 24
Angel, K.Sr., 397
Angell, R.W., 73
Angelli, C.Sr., 315
Angelopoulos, E.W., 542
Angerio, A.J., 35
Anglemier, A.F., 383
Angleton, G.M., 69
Anglin, J.H., 380
Angstadt, R.B., 398
Ankel, H., 513
Anker, H.S., 131
Ankney, C.D., 552
Ankney, P.H.Jr., 102
Anlyan, W.G., 332
Ann, R.Sr., 396
Annan, M.E., 329
Anne, A., 489
Ansbacher, C., 258
Ansevin, K.D., 453
Anshel, D., 325
Anson, B.J., 126
Answini, A.F., 390
Anthon, T., 502
Anthoney, T., 129
Anthony, A., 403
Anthony, D.S., 90
Anthony, E.H., 547
Anthony, J., 523
Anthony, M.S., 29
Anthony, R., 246
Anthony, R.G., 402
Anthony, W.B., 3
Antley, R.M., 146
Antoine, A.D., 288
Antoine, P.P., 241
Antonovic, J., 332
Anwar, R.A., 549
Aogaichi, T., 408
Apirion, D., 265
Aplington, H.W.Jr., 206
App, A., 298
Appel, S.H., 333
Apple, J.L., 345
Apple, J.W., 518
Apple, S.B., 384
Applebury, M.L., 286
Appledorf, H., 92
Applegarth, A.G., 32
Applegate, A.L., 337
Applegate, J.E., 288
Applegate, R.L., 434
Appleman, M.D., 62
Appleman, M.M., 62
Applequist, J.B., 157
Appley, M., 317
Applie, H.P., 356
April, E.W., 300
Aprison, M.H., 146
Apt, W.J., 114
Aquinlivan, J.Sr., 33
Aragaki, M., 114
Arai, H.P., 531
Arai, K., 467
Arai, M.N., 531
Arakaki, D.T., 113
Araki, G.S., 40
Arant, F.S., 5
Araujo, P.E., 92
Arave, C.E., 474
Arawinko, Z.M., 516
Arbesman, C.E., 319
Arcadi, J., 64
Arce, G., 30
Arceneaus, J.L., 251
Arceneaux, J.L., 251
Arch, S.A., 387
Archdeacon, J.W., 180
Archer, M.C., 213
Archer, P.G., 72
Archer, S., 465
Archer, S.J., 15
Archer, V.G., 288
Archibald, H.L., 149
Archibald, P.A., 407
Arditti, J., 54
Ardoin, R., 189
Aregood, C., 483
Arendshorst, W.J., 340
Arendt, K.A., 36
Arey, L.B., 126
Arfin, S.M., 54
Arganbright, D.G., 44
Argenti, R.M., 8

Arersinger, W.J.Jr., 170
Argot, J., 398
Argue, C.W., 540
Argyris, B.F., 326
Arion, W.J., 306
Arkley, R.J., 46
Arlansky, P., 524
Arlian, L.G., 372
Armbruster, G., 306
Armentrout, D., 450
Armentrout, R.W., 370
Armer, J.M.Sr., 450
Armitage, K.B., 170, 171
Arsm, K., 302, 304
Armstead, N.W., 227
Armstrong, 207
Armstrong, D., 17
Armstrong, D.T., 552
Armstrong, F.B., 343, 345
Armstrong, F.H., 481
Armstrong, H.S., 546
Armstrong, J.A., 414
Armstrong, J.B., 549
Armstrong, P.B., 52
Armstrong, R., 60
Armstrong, R.B., 207
Armstrong, R.F., 552
Armstrong, R.L., 232
Armstrong, W., 146, 239
Arneman, H.F., 241
Arner, D.H., 250
Arneson, H., 524
Arneson, P.A., 304
Arnett, J.D., 425
Arnett, J.T., 431
Arnett, R.H.Jr., 316
Arnett, W.H., 275
Arnold, C.E., 92
Arnold, C.Y., 138
Arnold, D., 377, 540
Arnold, D.E., 403
Arnold, J.R., 28
Arnold, K.A., 459
Arnold, R.W., 302
Arnold, V., 337
Arnold, W.E., 433
Arnold, W.N., 172
Arnone, A., 161
Arnot, C.L., 555
Arnott, D.R., 547
Arnott, S., 150
Arntzen, C.J., 136
Arny, D.C., 519
Arny, M. 286
Aronoff, A., 533
Aronow, L. 42
Aronson, A.I., 150
Aronson, A.L., 306
Aronson, J.M., 15
Aronson, L., 296
Aronson, L.R., 297
Aronson, N.N.Jr., 404
Arora, H.L., 27
Arora, K.L., 127
Arquilla, E.R., 54
Arrington, L.C., 519
Arrington, L.R.
Arscott, G.H., 384
Arscott, T.G., 364
Arslavian, M.J., 448
Art, H.W., 220
Arteman, R.L., 122
Arthur, D., 119
Arthur, E.A., 315
Arthur, R.D., 248
Artinian, B., 556
Artist, R.C., 431
Arvey, M.D., 342
Arya, S.K., 320
Arzeni, C.B., 120
Asakura, T., 410
Asano, T., 152
Asato, Y., 215
Asay, K.H., 258
Ascanio, G., 409
Aschenbrener, C.A., 161
Ascher, P.D., 241
Ash, C.L., 549
Ash, W.J., 315
Ashburn, A.D., 251
Ashby, W.C., 128
Ashcroft, G.A., 474
Ashcroft, G.L., 477
Asher, J.H.Jr., 227
Ashland, W.I.Jr., 503
Ashley, D.A., 106
Ashley, M.D., 194
Ashley, M.J., 549
Ashley, R.A., 78, 79
Ashley, T., 73
Ashlock, P.D., 171, 172
Ashman, R.F., 56
Ashman, R.I ., 194
Ashmore, C.D., 427
Ashmore, C.R., 48
Ashtakala, S.S., 557
Ashton, F.M., 51
Ashwood-Smith, M.J., 536
Ashworth, L.J., 45
Ashworth, M.A., 550
Ashworth, R.P., 425
Asleson, J.A., 566
Asling, C.W., 60
Asokasrinivasan, G.S., 251
Aspin, N., 550
Aspinwall, N., 256
Aspliden, C.I., 489
Asplund, J.M., 259
Asplund, K., 16
Asplund, M., 16
Asplund, R.O., 527
Asquity, D., 402
Asterita, H.L., 286
Aston, W.P., 545
Astrachan, L., 356
Astruc, J.A., 491
Asuncion, J.R.Jr., 542
Atchley, W.R., 461
Atema, J., 207
Atencio, A.C., 294
Atherly, A.G., 157
Atherton, D.R., 473
Atherton, G.H., 384
Atherton, H.V., 480
Atherton, R.W., 526
Athow, K.L., 148
Atkin, C.L., 474
Atkins, C.G., 368
Atkins, D.L., 84
Atkins, E.G., 290
Atkins, M.D., 39
Atkins, R., 128
Atkinson, B.G., 552
Atkinson, R.L., 458
Atkinson, W., 348
Atlas, R.M., 180
Atnip, R.L., 445
Atsatt, P.R., 54
Attardi, G., 26
Attaway, J.A., 92
Attebery, J.T., 21
Attleberger, M.H.,6
Atwal, O.S., 548
Atwood, H., 550
Atwood, K.C., 301
Atyeo, W.T., 106
Aubin, A., 560
Auclair, A.A., 556
Auclair, W., 314
Audrick, E.T., 356
Auer, H.E., 328
Auerbach, J., 550
Auerbach, R., 516
Auerbach, S.I., 406
Auersperg, N.A., 535
Auffenberg, W., 90
Auger, P., 558
Augustine, J.R., 11
Auito, R.,222
Auleb, H.L., 40
Aulenbach, D.B., 314
Aulerich, R.J., 226
Aull, F., 313
Aulsebrook, L.H., 446
Ault, C.R., 382
Aumann, G.D., 463
Aune, J., 463
Aune, K.C., 448
Aung, L.H., 494
Aungst, C.W., 320
Aurand, L.W., 344
Aurelian, L., 198
Aurelie, I.C., 506
Aust, S.D., 225
Austic, R.E., 305
Austin, A.P., 536
Austin, B.P., 325
Austin, D.F., 87
Austin, E.B., 31
Austin, I., 416
Austin, J., 122
Austin, J.M., 486
Autian, J., 445
Autrey, K.M., 3
Autuori, M.J., 77, 79
Avalut, J.W., 183
Avenoso, A.C., 464
Avens, J.S., 65
Averett, J.E., 262
Averitt, J.N., 102
Averill, F., 122
Averre, C.W.III., 345
Avers, C.J., 288
Avery, D., 76
Avery, J., 125
Avet, P.R., 186
Avigad, G., 283
Avila, V.L., 40
Awad, W.M., 95
Awada, M., 111
Awbrey, F.T., 39
Axelrad, A.A., 549
Axelrad, S., 298
Axelrod, A.E., 413
Axelrod, B., 148
Axelrod, D.E., 288
Axelrod, D.I., 51
Axley, J.H., 200
Axman , C., 165
Axtell, J.D., 147
Axtell, R., 130
Axtell, R.C., 344
Ayala, F.J., 47
Aycock, M.K., 200
Aycock, R., 345
Aydelotte, M., 160
Ayers, C.I., 461
Ayers, J.E., 403
Ayling, J.E., 56
Aylward, D.C., 269
Aymath, Y., 417
Ayoub, E.M., 94
Ayres, J., 121
Ayres, J.C., 106
Azari, P., 68
Azarnia, R., 96
Azen, S.P., 63
Azer, P., 327
Azimi, P.H., 367
Azluha, A., 239
Azzam, N.A., 160

B

Baad, M.F., 31
Baalman, R.J., 30
Babai, F., 559
Babbel, G.R., 97
Babbs, C.F.Jr., 150
Babcock, K.L., 46
Babcock, M.J., 288
Babcock, W.J.V., 209
Babich, H., 312
Babin, D.R., 268
Babiuk, L.A., 563
Bablanian, R., 325
Babos, P., 27
Bacaner, M., 244
Bacchi, C., 313
Baccus, J.T., 471
Bach, F.H., 519
Bach, R.J., 333
Bacha, W.J.Jr., 287
Bachman, K.C., 91
Bachmann, M., 157
Bachmann, R., 157
Backens, V.W., 431
Backman, P.A., 4
Backus, W.R., 443
Bacon, A.L., 8
Bacon, B., 540
Bacon, I.R., 105
Bacon, J., 443
Bacon, J.A., 169
Bacon, L.D., 236
Bacon, M., 496
Bacon, O.G., 49
Bacon, R.L., 388
Bacon, W.L., 367
Badden, A., 514
Bade, M.L., 207
Badeer, H.S., 269
Baden, J., 475
Bader, P., 146
Bader, R.S., 262
Badger, A., 20
Badger, S., 246
Badger, S., 356
Badgett, A., 378
Badore, B., 206
Badour, S., 538
Baehr, L.F., 189
Baer, A.I., 37
Baer, C.H., 507
Baer, H., 94
Baer, R.Jr., 303
Baetcke, K., 24
Baeten, H.J., 514
Baez, S., 330
Bagby, J.R.Jr., 69
Bagdonas, K., 295
Bagley, R.A., 475
Bagley, V.S., 487
Bagley, W.T., 272
Bagnara, J.T., 18
Badr, S.A., 29
Bagby, R.M., 442
Bagshaw, J.C., 235
Bagshaw, R.J., 411
Baggett, B., 428
Bagget,, J.R., 384
Bagwell, J.N., 185
Bahadur, R., 250
Baham, A., 189
Bahl, O.P., 318
Bahler, T.L., 476
Bahlman, D.W.R., 220
Bahniuk, E., 356
Bahorik, J.W., 397
Bahr, T.G., 225
Baic, D., 231
Baich, A., 130
Baier, J., 523
Baier, R.E., 320
Baile, C.A., 401
Bailey, A.W., 530
Bailey, C.F., 22
Bailey, C.H., 288
Bailey, C.M., 275
Bailey, D.W., 181
Bailey, E.D., 547
Bailey, E.M., 460
Bailey, H., 8
Bailey, H.H., 471
Bailey, H.S., 432
Bailey, J.C., 249
Bailey, J.M., 84
Bailey, J.R., 332
Bailey, K.C., 237
Bailey, K.L., 435
Bailey, L.F., 22
Bailey, M.M., 391
Bailey, O.S., 99
Bailey, P.C., 6
Bailey, R.E., 223
Bailey, R.M., 231
Bailey, R.W., 82
Bailey, V., 223
Bailey, Z.E., 120
Bailie, M.D., 228
Bailin, M.H., 316
Bain, A., 316
Bain, J.A., 101
Bain, W.M., 194
Bainton, C.R., 60

Bainton, D.F., 60
Bair, R.D., 479
Bair, T.D., 31
Baird, C.J., 196
Baird, E., 467
Baird, I.L., 405
Baird, J.R., 488
Baird, J.L., 315
Baird, N.M., 125
Baird, R.C., 97
Baisden, V., 104
Baisted, D.J., 385
Bajema, C.J., 223
Bajer, A.S., 387
Bajpai, P.K., 371
Bajt, J., 120
Baker, A., 280
Baker, A.F., 29
Baker, B.B., 508
Baker, B.S., 339
Baker, C., 116
Baker, C., 351
Baker, C.H., 97
Baker, C., 323, 358
Baker, D.E., 401
Baker, D.G., 246
Baker, D.H., 138
Baker, D.K., 316
Baker, D.N., 248
Baker, E.E., 208
Baker, E.G.S., 284
Baker, F.T., 396
Baker, H.G., 42
Baker, J., 155, 243
Baker, J.C., 258
Baker, J.H., 219
Baker, J.M., 376
Baker, J.R., 157
Baker, J.R., 344
Baker, J.W., 427
Baker, K.F., 45
Baker, L.E., 185
Baker, L.O., 236
Baker, L.R., 226
Baker, M.A., 64
Baker, N.F., 53
Baker, P.C., 357
Baker, R., 68, 162, 236
Baker, R.A., 28
Baker, R.C., 305, 367
Baker, R.D., 468
Baker, R.F., 62, 64
Baker, R.H., 126, 225, 227
Baker, R.J., 432, 433, 461, 537
Baker, R.L., 202, 366
Baker, T.I., 294
Baker, T.S., 309
Baker, W., 132, 372
Baker-Cohen, F., 329
Bakke, J.E., 352
Bakken, A., 498
Bakko, E., 240
Bakshi, T.S., 533
Bakula, M., 256
Bakus, G., 62
Bakuzis, E.V., 242, 243
Bal, A.K., 541
Bal, H.S., 158
Balaban, M.M., 227
Balamuth, W., 43
Balasko, J.A., 508
Balbinder, E., 326
Balboni, E., 297, 298
Balch, D.J., 480
Bald, J.G., 58
Balda, R., 15
Balding, T.A., 514
Baldridge, J.D., 258
Baldsin, R.E., 423
Balduzzi, P.C., 328
Baldwin, B.A., 351
Baldwin, C.H., 260
Baldwin, D.M., 315
Baldwin, J.T., 484
Baldwin, K., 54
Baldwin, K.M., 86
Baldwin, K.V., 154
Baldwin, P.H., 68
Baldwin, R., 41
Baldwin, R.D., 402
Baldwin, R.E., 114
Baldwin, R.L.Jr., 48
Bale, S.S., 488
Bale, W.F., 328
Balek, R.W., 230
Balish, E., 521
Ball, A.J.S., 544
Ball, C.R., 251
Ball, E., 270
Ball, E.A., 54
Ball, E.M., 272
Ball, G.E., 530
Ball, G.L., 233
Ball, H.J., 271
Ball, H.R., 344
Ball, I.J., 501
Ball, J.K., 551
Ball, L., 69
Ball, M., 408
Ball, M.J., 552
Ball, R.C., 227
Ball, W.C., 197
Ball, W.S., 65
Balial, S.K., 440
Ballantyne, D.J., 536
Ballard, B., 378
Ballard, N., 238
Ballard, R., 372
Ballard, R.C., 32
Ballard, T.M., 534
Ballard, W.W., 277
Ballas, L.E., 36
Ballentine, R., 197
Ballesteros, D., 31
Balling, J.W., 391
Ballinger, R., 448
Ballinger, W.E., 345
Ballou, C.E., 42
Ballou, D., 232
Ballou, K.C., 222
Ballweg, R., 310
Balon, E.K., 547
Balph, D.A., 476
Balph, D.F., 477
Balraj, E.K., 356
Balsano, J.S., 522
Balsbaugh, E.U.Jr., 455
Balsiger, J.W., 521
Baluda, M.A., 56
Balzen, C.A., 438
Balzer, A.L., 504
Bamburg, J.R., 68
Bamforth, S.S., 190
Bamrick, J.F., 159
Banadyga, A.A., 345
Banasik, O.J., 351
Banaszak, E.F., 513
Banaszak, L.J., 262
Banchero, N., 72
Bancroft, C., 300
Bancroft, H.R., 438
Bancroft, J.B., 552
Band, N.R., 227
Bandel, V.A., 200
Bandelier, K.J., 267
Bandoni, R.J., 534
Bandy, B.E., 247
Bandy, J., 534
Banerjee, M., 270
Banerjee, S., 220, 406
Banerjee, U., 207
Banewicz, J.J., 466
Banfield, A.W.F., 544
Baniecki, J.F., 508
Bank, A., 301
Banker, B.Q., 356
Bankes, D.A., 484
Banks, D.J., 376
Banks, E.M., 136, 138
Banks, H.C., 246,
Banks, H.P., 302
Banks, R.O., 370
Banks, W.J., 68
Banks, W.L.Jr., 491
Banner, A.H., 112
Bannister, T.T., 327, 328
Banquet, R.S., 85
Banse, A.M., 321
Banta, W.C., 83
Banwart, G.J., 363
Baptista, L., 37
Barahas, L., 55
Barak, A.J., 273
Baranger, E., 412
Baranowski, R.M., 91
Barash, D.P., 498
Baratz, R.S., 126
Barau, R.G., 58
Barber, A.A., 55
Barber, G.A., 362
Barber, J., 69
Barber, J.T., 190
Barber, M.D., 286
Barber, M.L., 31
Barber, R.T., 332
Barber, S.A., 147
Barber, S.B., 398
Barber, V.C., 541
Barber, W.D., 444
Barbiers, A., 236
Barbosa, P., 218
Barbour, C.D., 8
Babour, M.G., 51
Barbour, R.W., 176
Barbour, S.D., 356
Barcellona, W.J., 461
Barchi, R.L., 410
Barclay, D., 544
Barclay, J.K., 548
Barclay, J.S., 376
Barclay, L.A., 27
Bard, G., 297, 298
Bardana, E.J., 388
Barden, A.A.Jr., 193
Barden, J.A., 494
Barden, N.T., 251
Barden, R.E., 527
Bardon, L., 273
Bardwell, F., 475
Barett, E.R., 188
Barfield, B.J., 179
Barfield, R.J., 288
Barger, A.C., 212
Barghoorn, E.S., 211
Barghusen, H., 134
Barghusen, H.R., 134
Baricos, W.H., 190
Barish, N., 30
Barkalow, F.S.Jr., 346
Barkat, S.J., 310
Barker, A.D., 363
Barker, A.V., 219
Barker, D., 239
Barker, D.J., 137
Barker, D.L., 387
Barker, G.J., 195
Barker, H., 239
Barker, H.A., 42
Barker, H.B., 186
Barker, J., 319
Barker, K.L., 273
Barker, K.R., 296, 345
Barker, L., 240
Barker, N.P., 437
Barker, P.S., 537
Barker, R., 225
Barker, S., 10
Barker, W.B., 546
Barker, W.T., 349
Barker, W.W., 496
Barkley, F.A., 214
Barkely, T.M., 166
Barklow, J.J., 356
Barkman, R.C., 216
Barksdale, A., 296
Barksdale, W.L., 313
Barla, P., 375
Barlet, R.L., 29
Barlow, B., 8
Barlow, C.A., 544
Barlow, D.J., 468
Barlow, G., 358
Barlow, G.W., 43
Barlow, J.L., 327
Barlow, J.P., 302, 303
Barlow, J.S., 533
Barlow, W.M., 166
Barman, E.H., 102
Barnabo, J., 215
Barnard, A.C.L., 11, 12
Barnard, E.A., 318
Barnard, J.E., 401, 402
Barnard, J.J., 747
Barnard, W.H., 479
Barnawell, E., 270
Barnes, A.U., 399
Barnes, B.V., 231
Barnes, C.R., 551
Barnes, D., 314
Barnes, D.K., 240
Barnes, D.W., 489
Barnes, E.M., 448
Barnes, G.L., 375
Barnes, J., 522
Barnes, J.R., 472
Barnes, K.L., 356
Barnes, M.M., 58
Barnes, R.D., 395
Barnes, R.F., 401
Barnes, R.H., 306
Barnes, R.L., 334
Barnes, R.N., 174
Barnes, W., 474, 514
Barnett, A., 285
Barnett, A.J., 199
Barnett, B.D., 424, 425
Barnett, C.A., 39
Barnett, D., 467, 474
Barnett, F.L., 166
Barnett, H.L., 508
Barnett, L.B., 415, 493
Barnett, N.M., 201
Barnett, O.W., 425, 426
Barnett, R.E., 241
Barney, C.W., 66
Barney, W.C., 199
Barnhard, R., 34
Barnhart, C.E., 177
Barnhart, D.D., 360
Barnhisel, R.I., 177
Barnitt, A., 323
Barnsley, E.A., 541
Barnum, D.A., 548
Barnwell, F.H., 242
Baron, A., 513
Baron, F., 394
Baron, R., 333
Barone, J.E., 285
Barone, M.C., 79
Barr, C., 317
Barr, E.M., 134
Barr, L., 157
Barr, M.L., 551
Barr, R.A., 407
Barr, T.C., 176
Barr, W.F., 117
Barraclough, C.A., 204
Barrance, S.III., 467
Barras, D., 8
Barrat, J.G., 508
Barrat,, R., 34
Barrentine, B.F., 248
Barrett, A.S., 235
Barrett, C.P., 203
Barrett, C., 73
Barrett, D.A., 161
Barrett, E., 160
Barrett, E.F., 96
Barrett, G.W., 360
Barrett, J.M., 512
Barrett, J.N., 96
Barrett, J.R., 280
Barrett, J.W., 445
Barrett, M., 550
Barrett, P., 217
Barrett, R., 436
Barrett, T.W., 445
Barrick, E.R., 343
Barridge, B.D., 187
Barriga, O.O., 412
Barrington, B.S., 428
Barrios, E.P., 184
Barron, A.L., 24
Barron, G.L., 546, 547
Barrow, E.M., 340

Barrow, J.H., 358
Barrow, R.E., 468
Barry, A., 53
Barry, A.C., 290
Barry, B.C., 363
Barry, C.D., 262
Barry, E., 191
Barry, E.G., 338
Barry, G., 297, 298
Barry, W.T., 496
Barsdate, R.J., 14
Barshad, I., 46
Barta, A.L., 364
Bartalos, M., 301
Bartel, A.H., 463, 464
Bartel, M.H., 239
Barteli, G.B., 268
Bartell, C.K., 185
Bartell, P.F., 282
Bartels, P.G., 18
Barth, K.M., 443
Barth, R.H., 465
Bartha, A., 125
Bartha, R., 288
Barthalmus, G.T., 346
Bartholic, J.F., 92
Bartholomew, E.A., 507
Bartholomew, G., 55
Bartholomew, J.W., 62
Bartholomew, W.R., 319
Bartle, E., 432
Bartlett, B.R., 58
Bartlett, D.Jr., 278
Bartlett, E.T., 67
Bartlett, J., 436
Bartlett, L.A., 249
Bartlett, L.M., 218
Bartlett, R.J., 481
Bartley, I.D., 258
Bartnicki-Garcia, S., 58
Bartio, P.B., 377
Barton, H.E., 20
Barton, J., 401
Barton, J.H., 169
Barton, L.L., 23
Barton, P.G., 531
Barton, R., 534
Bartos, J., 405
Bartsch, A.F., 385
Bartuska, D.G., 399
Bartz, J.A., 92
Baruch, S.B., 308
Barues, G., 246
Barzansky, B., 428
Basaraba, J., 542
Baseman, J.B., 339
Basford, R.E., 413
Basham, C.W., 66
Bashaw, E.C., 459
Bashirelahi, N., 204
Bashor, D., 342
Basile, D., 297, 298
Baskett, T.H., 261
Baskin, D., 33
Baskin, J.M., 176
Baskin, J.N., 29
Baskin, R.J., 52
Baskin, S.I., 399
Basler, E., 375
Basmanian, J.V., 101
Bason, C.R., 370
Basrur, P.K., 548
Bass, A.D., 446
Bass, B.G., 330
Bass, D.E., 421
Bass G.E.Jr., 445
Bass, J.A., 470
Bass, J.C., 165
Bass, M.H., 5
Bass, M.L., 486
Bassel, A.R., 326
Bassel, B.A.Jr., 326
Bassett, E.W., 282
Bassett, F.H., 332
Bassett, J.R., 231
Bassett, J.W., 458
Bassett, M.J., 93
Bassham, J.A., 42
Bassi, S.D., 164
Bassin, J.M., 295
Bast, R.Sr., 285
Baste, C.A., 101
Batchelder, G.L., 40
Bateman, C., 83
Bateman, D.R., 304
Bateman, E.C., 189
Bateman, G., 15
Bates, D.V., 533
Bates, F.W., 220
Bates, G.W., 458
Bates, H.B.Jr., 99
Bates, J., 24
Bates, M.N., 408
Bates, R.C., 492
Bates, R.P., 92
Bates, W.K., 342
Batey, J.W., 290
Bath, J.E., 226
Bath, J.L., 29
Batha, J.V., 511
Batie, R.E., 487
Batkin, S., 113
Batra, R.P., 372
Batsel, H.L., 57
Batson, J.D., 102
Batson, M.B., 252
Batson, W.E.Jr., 249
Batson, W.T., 429
Batt, E.R., 301
Batterman, S.C., 411
Battigelli, M.C., 341
Battin, W., 317
Battisto, J.R., 330
Battle, H.I., 552
Battle, W.R., 288
Battley, E., 325
Batts, B.S., 486
Batts, H.L., 224
Batty, J.C., 475
Batzing, B.L., 321
Batzli, G.O., 136
Bauer, D.C., 146
Bauer, E., 243
Bauer, E.J.Jr., 506
Bauer, F.K., 63
Bauer, H., 244
Bauer, M.H., 85
Bauer, R.E., 337
Bauerle, R.H., 488
Bauerle, W.L., 365
Baugh, C.L., 461
Baughman, N.M., 508
Baum, L.S., 188
Baum, S., 329
Baum, S.G., 330
Bauman, J.W., 283
Bauman, L.F., 147
Bauman, T.R., 10
Baumann, C.A., 517
Baumann, P., 51
Baumann, R.D., 452
Baumbach, D.O., 488
Baumbardner, K.D., 120
Baumel, J.J., 268
Baumgardner, J.H., 462
Baumgardner, M.F., 147
Baumgardner, R.A., 424
Baumgardner, R.K., 188
Baumgardt, B.R., 401, 402
Baumgartner, D., 129
Baumhoff, M.A., 146
Baumiller, R.C., 85
Baumstark, J.S., 268
Baur, J.M., 174
Baur, D., 467
Bauske, R.J., 156
Bausor, S.C., 391
Baust, J.G., 323
Bavel, V., 460
Baver, J., 387
Baver, L.D., 364
Baxter, A.J., 364
Baxter, A.W., 426
Baxter, C., 57
Baxter, C.H., 41
Baxter, G.T., 526
Baxter, J., 521
Baxter, L.W., 425
Baxter, R.W., 170
Baxter, W.D., 354
Baxter, W.L., 39
Bay, D.E., 458
Bay, E.C., 201
Bayer, D.E., 51
Bayer, F., 96
Bayless, L.E., 506
Bayley, S.T., 545
Baylis, J.R., 98
Bayliss, F.T., 395
Bayly, I.L., 544
Baynaski, M., 331
Bayne, C.J., 386
Bayne, R.D., 4
Bayona, J., 418
Bays, C., 542
Bayuzick, R.J., 446
Bazer, F.W., 91
Bazinet, G.F., 316
Bazlett, B.A., 231
Bazzano, G., 256
Bazzaz, F.A., 136
Be, A., 297
Be, A.W.H., 297
Beach, B., 176
Beach, J.L., 171
Beach, N.W., 395
Beadle, G., 132
Beadle, G.W., 132
Beadle, S., 523
Beadles, J.K., 20
Beal, E.O., 180, 181
Beal, J., 235
Beal, R.S., 15
Beall, F.C., 402
Beall, R.C., 27
Beals, E.W., 515, 516
Beals, H.O., 5
Beam, C.A., 296, 297
Beaman, T.C., 228
Beames, R.M., 533
Beamish, F.W.H., 547
Beamish, K.I., 534
Beams, H.W., 160
Bean, B., 398
Bean, D.J., 479
Bean, G.A., 201
Bean, J., 232
Beaney, W.D., 127
Bear, P.D., 528
Bearce, B.C., 508
Beard, B.H., 47
Beard, E.L., 187
Beard, J.D., 445
Beard, L.S., 331
Beard, R., 394
Beard, W., 25
Bearden, H.J., 249
Beardsley, J.W., 114
Beardsley, R.E., 299
Beardsley, W.G., 156
Beary, D.F., 455
Beasley, A.B., 399
Beasley, A.R., 96
Beasley, B., 451
Beasley, C.W., 451
Beasley, J.N., 23
Beasley, P.G., 12
Beason, S.L., 459
Beatden, B.J., 443
Beaten, J., 514
Beathard, G., 467
Beatley, J., 370
Beattie, A.J., 126
Beattie, D.S., 298
Beattie, J., 401
Beattie, L.M., 552
Beatty, D., 175
Beatty, D.D., 531
Beatty, J.A., 129
Beaty, E.R., 106
Beauchamp, R.D., 551
Beaudoin, A., 559
Beaudoin, J.L., 557
Beauge, L.A., 203
Beauliew, M., 549
Beaumont, G., 559
van Beaumont, K.W., 256
Beautyman, W., 482
Beaver, B.V., 460
Beaver, D., 373
Beaver, D.L., 227
Beaver, P.C., 191
Beaver, R.G., 265
Beaver, R.J., 59
Beavers, A.H., 137
Beavers, D.V., 383
Beavin, E.L., 175
Beceridge, D., 298
Bechard, P., 559
Bechner, J., 512
Bechtel, A.A., 186
Bechtel, E., 109
Bechtold, E., 356
Becicwith, J., 211
Beck, C.B., 230
Beck, I.T., 546
Beck, R., 316
Beck, R.A., 484
Beck, R.E., 352
Beck, R.F., 292
Beck, R.W., 442
Beck, S.D., 518
Beck, S.L., 220
Becker, B., 150
Becker, B., 311
Becker, C., 342
Becker, D.A., 22, 270
Becker, D.E., 138
Becker, E.L., 79
Becker, G., 70
Becker, G.C., 522
Becker, G.E., 161
Becker, H., 220, 298
Becker, J., 442
Becker, L.Y., 360
Becker, R.R., 385
Becker, S.A., 120
Becker, V.E., 347
Becker, W.A., 502
Becker, W.B., 218
Becker, W.M., 515
Beckert, W.H., 316
Beckett, J.B., 258
Beckett, S.D., 6
Becking, R.W., 34
Beckman, A.L., 411
Beckman, C.E., 421
Beckman, C.H., 422
Beckman, G.M., 497
Beckman, L.D., 327
Beckman, L.S., 147
Beckwith, J.R., 108
Bedard, J., 557
Bedard, R., 558
Bedell, G., 154
Bedell, M., 285
Bednarz, R., 92
Bedwell, J.E., 27
Bee, J.W., 172
Beebe, J., 223
Beebe, J.L., 308
Beech, J.A., 88
Beecher, A.S., 494
Beegle, C., 187
Beeks, J.C., 255
Beeler, D.A., 327
Beeler, M.F., 186
Beelman, R.B., 402
Beeman, E.A., 214
Beeman, R.D., 40
Been, A., 511
Beer, A.E., 466
Beer, C.E., 155
Beer, C.T., 534
Beer, M., 198
Beer, R.E., 171, 172
Beer, S.V., 304
Beering, S.C., 145
Beers, C.D., 339
Beers, T.W., 149
Beerwinkle, D., 471
Beery, G., 361
Beetle, A.A., 527
Beetley, B., 223
Beeton, A., 523
Beevers, H., 61
Beevers, L., 374
Begener, W.S., 146
Begg, E.L., 50
Begin, J.J., 177

Begin-Heick, N., 548
Behal, F.J., 461
Behan, M.J., 266
Behile, WH., 473
Behlke, C., 14
Behme, R., 551
Behmer, D.J., 224
Behnke, R.J., 66
Behnke, W.D., 370
Behr, L.C., 247
Behrend, D.F., 322
Behrens, R., 240
Behrens, W.C., 275
Behringer, M., 352
Behrisch, H., 14
Behrman, E.J., 362
Behrman, S.J., 234
Beibel, R.H., 517
Beidleman, R.G., 65
Beidler, L., 88
Beil, C.E., 534
Beilfuss, E.R., 34
Beilstein, H., 408
Beineke, W.F., 149
Beinert, H., 517
Beining, P.R., 415
Beinz, D.R., 503
Beirne, B.P., 533
Beitch, I., 76
Beitz, D.C., 157
Beitzel, R., 236
Bejnar, W., 291
Bekey, G.A., 63
Bekoff, M., 262
Belamrich, F.A., 207
Bélanger, L.F., 548
Bélanger, Y., 558
Belcik, F.P., 334
Belcher, J.C., 488
Belcher, S.C., 488
Belden, D.A.,Jr., 73
Belden, E.L., 528
Belding, R.C., 228
Belfour, W.M., 171
Belk, D., 15
Belk, F., 250
Belkin, J., 55
Belkin, R., 309
Belknap, H.W., 116
Belknap, R.W., 268
Bell, C.F., 507
Bell, C.R., 338
Bell, C.W., 32
Bell, D.M., 398
Bell, D.S., 365
Bell, E.L., 390
Bell, F.E., 340
Bell, F.F., 444
Bell, J., 29
Bell, J.B., 535
Bell, J.H.L., 271
Bell, J.M., 561
Bell, L., 3
Bell, M.A.M., 536
Bell, M.C., 443, 398
Bell, M.E., 254
Bell, O.E.Jr., 251
Bell, R.A., 351
Bell, R.D., 380
Bell, R.E., 246
Bell, R.M., 337
Bell, R.T., 480
Bell, S.L., 441
Bell, R.A., 344
Bell, W.H., 479
Bell, W.J., 171
Bellamy, J.E.C., 563
Bellan, G., 164
Bellemare, G., 557
Bellhorn, S., 330
Bellinger, P., 31
Bellion, E., 466
Bellis, E.D., 403
Bellis, V.J., 334
Bellmer, E.H., 86
Bello, C.T., 409
Bello, L.J., 412
Belloncik, S., 560
Bellone, C.J., 256
Belmonte, Rev., R.G., 289
Belse, W.L., 57
Belshe, J., 253
Belsky, M.M., 296
Belt, G.H., 118
Belt, W.D., 216
Belton, P., 533
Beltz, A.D., 99
Beltz, R.E., 35
Belzer, N.F., 397
Bemis, W.P., 17, 19
Benda, R.S., 223
Bender, F., 202
Bender, G.L., 15
Bender, H.A., 152
Bender, J., 100
Bender, L.D., 417
Bender, M.E., 490
Bendet, I.J., 413
Bendich, A., 497
Bendixen, J., 175
Bendixen, L.E., 364
Bendjaballah, R., 558
Bendler, J.W.III., 513
Beneddtti, D., 293
Benedict, A.A., 112
Beneke, E.S., 228
Benepal, P.S., 494
Benerito, R.R., 190
Benes, E.S., 31
Benesch, R., 300
Benesch, R.E., 300
Benevento, L.A., 134
Benfield, E.F., 492
Bengelink, H., 223
Benington, F., 11
Benisek, W.F., 53
Benjamin, C., 453
Benjamin, D.C., 490
Benjamin, D.M., 518
Benjamin, H.B., 512
Benjamin, M.M., 69
Benjamin, R.K., 32
Benjamin, R.M., 521
Benjamins, J., 340
Benjaminson, M.A., 297
Benke, A.C., 102
Benko, P.V., 28
Benner, D.B., 437
Bennett, A.C., 3
Bennett, C., 461
Bennett, D., 144
Bennett, G.C., 555
Bennett, G.F., 541
Bennett, G.W., 148
Bennett, H.D., 507
Bennett, H.J., 183
Bennett, I.L.Jr., 312
Bennett, J.C., 126
Bennett, J.L., 227
Bennett, J.W., 190
Bennett, L.L., 60
Bennett, M.F., 192
Bennett, O.L., 508
Bennett, R., 142
Bennett, R.B., 204
Bennett, R.R., 219
Bennett, S.N., 102
Bennett, T.P., 88
Bennett, W.F., 462
Bennick, A., 549
Benninghoff, W.S., 230
Benoit, E., 313
Benoit, G.R., 481
Benoit, J.C., 558
Benoit, R.E., 492
Benoit, S., 315
Benoiton, N.L., 548
Bensadoun, A., 305, 306
Bensen, L., 32
Bensend, D.W., 156
Benson, A., 507
Benson, B., 392
Benson, E.W., 99
Benson, G.L., 4
Benson, J.B., 105
Benson, K.A., 487
Benson, L., 32
Benson, N., 434
Benson, R.H., 77
Benson, R.L., 503
Benson, R.W., 180
Benson, W., 238
Benson, W.R., 340
Bensusan, H.B., 355
Bent, F.C., 542
Bentley, C.L., 25
Bentley, D.R., 43
Bentley, J.P., 388
Bentley, O.G., 137, 138
Bentley, R., 413
Bentley, W.R., 231
Benton, A.H., 323
Benton, A.W., 402
Benton, D.A., 423
Benton, R.S., 179
Benzer, B., 26
Benzing, D.H., 361
Benzinger, R.H., 488
Benzo, C., 325
Beran, G.W., 158
Beran, R.L., 367
Berau, J.M., 331
Berbee, J.G., 519
Berberet, R., 377
Berberian, D.A., 321
Berech, J., 297, 298
Berendsen, P., 282
Berendzen, R., 83
Beresford, W.A., 509
Berg, A.P., 399
Berg, B., 325
Berg, C., 77, 79, 297
Berg, C.C., 401
Berg, C.J.Jr., 297
Berg, C.O., 303
Berg, C.P., 161
Berg, D.H., 358
Berg, G.G., 328
Berg, H., 71
Berg, H.W., 51
Berg, I.A., 182
Berg, J.N., 262
Berg, P., 41
Berg, R.A., 283
Berg, R.T., 530
Berg, W.A., 65
Berg, W.E., 43
Berg, W.N., 159
Bergeaux, P.J., 106
Bergdoll, M.S., 518
Bergeland, M.E., 434
Bergen, S.S., 282
Bergen, W.G., 225
Berger, A.J., 112
Berger, H., 197
Berger, J., 320, 550
Berger, J.D., 535
Berger, L.R., 112
Berger, P.J., 473
Berger, R.J., 61
Berger, R.S., 5
Bergeron, J., 332
Bergeron, J.A., 332
Bergeron, J.M., 559
Bergerson, E., 66
Bergeson, G.B., 148
Berglund, E., 384
Berglund, J.V., 322
Bergman, D.H., 534
Bergman, E.N., 306
Bergman, P.K., 498
Bergman, R.A., 197
Bergman, R.N., 63
Bergquist, R.R., 114
Bergren, W.R., 64
Bergstresser, K.A., 400
Bergstrom, D.W., 360
Bergström, R.C., 528
Bergt, M., 397
Beringer, T., 216
Berisford, C.W., 106
Berk, R.S., 236
Berkovitz, A., 150
Berkowitz, D., 410
Berkson, B., 119
Berkut, M.K., 340
Berkut, M.L., 340
Berlin, C.I., 186
Berlin, J.D., 461
Berlind, A., 80
Berliner, D.L., 473
Berliner, M.D., 315
Berliner, R.W., 80
Berlinrood, M., 196
Berlowitz, L., 318
Berlyn, G.P., 81
Berman, D.T., 517, 520
Berman, H.S., 207
Berman, I., 95
Bermes, E.W., 124
Bern, C., 155
Bern, H.A., 43
Bernard, J.M., 309
Bernard, L., 182
Bernard, Sr.M., 396
Bernard, R., 557
Bernard, R.A., 228
Bernard, R.F., 76
Bernard, R.L., 137
Berne, R.M., 490
Bernard, W.A., 328
Berndt, W.L., 433
Berndt, W.O., 252, 277
Berner, L., 90
Berner, L.Jr., 459
Bernhard, R.A., 49
Bernhardt, J.C., 338
Bernhardt, R.W., 415
Bernheim, F., 333
Bernheim, M.L.C., 333
Bernheimer, A.E., 313
Bernheimer, A.W., 312
Bernice, E., 395
Bernick, S., 63
Bernier, C.C., 537
Bernier, P.E., 384
Berning, N., 279
Bernlohr, R.W., 404
Berns, D.S., 327
Berns, K.I., 198
Berns, M.W., 54
Bernsohn, J., 124
Bernstein, A., 513, 550
Bernstein, I.A., 232, 233
Bernstein, J.J., 94
Bernstein, K., 323
Bernstein, M., 292
Bernstein, M.E., 27
Bernstein, M.H., 235
Bernstein, R., 71
Beroit, Sr.M.J., 279
Berra, T., 364
Berrend, R.E., 40
Berrill, M., 546
Berrios, A., 417
Berry, B.L., 444
Berry, B.W., 65
Berry, C.E., 431
Berry, G., 556
Berry, J.E., 453
Berry, J.W., 16, 143
Berry, L., 258
Berry, L.J., 465
Berry, R., 126, 291
Berry, R.E., 386
Berry, S.J., 80
Berry, S.Z., 365
Berryhill, D.L., 351
Berryman, A.A., 503
Bertalanffy, F.D., 537
Berthold, R., 393
Bertke, E.M., 15
Bertram, E.G., 549
Bertrand, A.R., 462
Bertrand, H., 561
Bertrand, H.A., 399
Bertsch, W., 297, 298
Berwick, S.H., 81
Besch, E.L., 168
Besch, H., 146
Beschta, R., 384
Beschta, R.L., 18
Besharse, J.C., 487
Bessette, F., 559
Bessey, P.M., 17
Bessler, E.J.Jr., 357
Bessler, W.G., 238
Bessman, M., 197
Besso, J.A., 452
Best, G.K., 103

Best, J.B., 69
Best, S., 423
Beste, C.E., 202
Betencourt, C., 417
Betheil, J., 330
Bethune, J.E., 108
Betros, H., 313
Betters, D.R., 66
Betterton, H., 24
Bettis, R.B., 9
Betz, J.V., 97
Beta, T.W., 544
Betz, E., 72
Betzer, P.R., 97
Beuerlein, J.E., 364
Beulig, A., 299
Beute, M.K., 345
Beutner, E.H., 319
Beuving, L.J., 236
Bevan, D.E., 498
Beverely-Burton, M., 547
Bevington, J.M., 400
Bevins, M.I., 481
Bevmer, R.J., 215
Bewley, J.D., 531
Bey, C.F., 129
Beychok, S., 300
Beyer, A.F., 452
Beyer, R.E., 231
Beyers, R.J., 12
Bezdicek, D.F., 502
Bezy, R.L., 62
Bhagat, B.D., 256
Bhagavan, N.V., 112
Bhalla, R., 160
Bhalla, V.K., 103
Bhatnagar, A.S., 491
Bhatnagar, K.P., 180
Bhatnager, M.K., 548
Bhatt, B.D., 315
Bhattacharjee, J.K., 360
Bhattacharya, P.K., 145
Bhavanandan, V.P., 405
Bhéreur, P., 560
Bhullar, H.S., 261
Bhussry, B.R., 85
Bhuvaneswaran, C., 24
Biaglow, J., 355
Biale, J., 55
Bialey, H.H., 177
Bialey, J.A., 66
Bialey, R.L., 186
Bialey, S.F., 61
Bialy, H.S., 214
Bianchi, D.E., 31
Biava, C.G., 60
Bibeau, A.A., 279
Biber, T.U.L., 81
Biberstein, E.L., 53
Bible, B.B., 555
Biblis, E.J., 5
Bibring, J., 446
Bibring, T., 446
Bicak, L.J., 269
Bichel, M., 270
Bick, G., 151
Bick, K.F., 490
Bick, K.L., 85
Bickert, J., 100
Bicket, Z., 253
Bickley, H.C., 161
Bickley, W.E., 201
Bicknell, E.J., 18
Biddle, M.O., 64
Biddulph, D.M., 347
Bidner, T.D., 183
Bidwell, O.W., 166
Bidwell, R.G.S., 546
Biebel, P., 393
Bieber, G., 257
Bieber, L.L., 225
Bieberly, F.G., 166
Biedenbach, M., 500
Biehl, E.R., 466
Biehler, W., 29
Bienvenu, R.J., 188
Bier, M., 16
Bierer, B.W., 425
Bierhorst, D.W., 217
Bieri, R.H., 490
Bierly, M.Z., 400
Biernbaum, C.K., 426
Biemer, M.W., 441
Bieron, J., 296
Biersdorf, W.R., 362
Biesbrock, J.A., 104
Biesele, J.J., 465
Biesinger, K.E., 245
Biever, K.D., 259
Bigazzi, P.E., 319
Bigbee, D., 203
Bigbee, D.E., 202
Bigelow, C.C., 541
Bigelow, H.E., 217
Biggers, C.J., 438
Biggs, D.A., 547
Biggs, N.L., 449
Biggs, R.H., 92
Biggs, W.C., 347
Bigham, S.R., 374
Bigland, C.H., 563
Bigley, N.L., 133
Bigley, R.H., 388
Bigner, D.W., 333
Bilderback, T., 168
Bileau, Sr.C., 279
Bilheimer, M.G., 279
Billenstein, D.C., 68
Billheimer, F.E., 391
Billiar, R.B., 355, 356
Billingham, R.E., 466
Billings, W.D., 332
Bils, R.F., 62
Bilsborrow, R.E., 341
Bilstatd, N.M., 516
Biltonin, R.L., 490
Bing, A., 303
Binggeli, R.L., 63
Bingham, D.P., 471
Bingham, E.T., 516
Binhemmer, R.T., 370
Binkley, F., 101
Binns, E., 210
Bintliff, S.J., 113
Bintz, G.L., 265
Bintz, H.A., 512
Birch, M.C., 49
Birch, R.L., 507
Birch, W.G., 236
Bird, C.D., 531
Bird, C.E., 546
Bird, E., 546
Bird, E.H., 458
Bird, G.W., 226
Bird, H.R., 519
Bird, J.W.C., 288
Bird, R.M., 379
Bird, T.J., 133
Birdsell, D., 124
Birdsell, D.C., 94
Birdsong, R.S., 487
Birge, B.A., 180
Birge, E.A., 15
Birge, W.J., 176
Birk, J., 358
Birkeland, C.J., 138
Birkenholz, D.E., 122
Birks, R.I., 556
Birky, C.W.Jr., 363
Birnbaum, J.L., 521
Birnbaum, M., 206
Briney, E.C., 242
Biro, G.P., 549
Birshtein, B., 329
Bisalputra, T., 534
Bisazza, G., 285
Bisbee, J.W., 125
Bischoff, E.R., 265
Bischoff, H.W., 461
Bischoff, W.F.Jr., 371
Bisgard, G.E., 520
Bisggi, V., 417
Bish, J., 173
Bishop, B., 319
Bishop, C., 318
Bishop, C.F., 143
Bishop, D., 165, 556
Bishop, D.L., 288
Bishop, E.L., 10
Bishop, G.W., 117
Bishop, J., 450
Bishop, J.C., 51
Bishop, J.M., 60
Bishop, J.W., 488
Bishop, L.G., 62
Bishop, M.D., 168
Bishop, S.H., 448, 449
Bishop, V.S., 470
Bishop, Y.M.M., 212
Bismanis, J., 535
Bissett, O.R., 74
Bistis, G.N., 284
Biswas, P.K., 9
Biswell, H.H., 44
Bittar, E.E., 521
Bitters, W.P., 58
Bittikofer, J., 337
Bittman, R., 298
Bittner, G.D., 465
Bittum, M.T., 302
Bitzer, M.J., 177
Bixler, D., 146
Bizeau, E., 118
Bjeldanes, L., 45
Bjerke, J., 522
Bjork, C.M., 73
Bjorklund, R.G., 119
Bjornn, T.C., 118
Blac, B.L., 272
Black, A., 160
Black, A.L., 52
Black, B.C., 99
Black, C.C.Jr., 105
Black, C.C., 461
Black, D.L., 219
Black, H., 434
Black, H.D., 438
Black, J., 374
Black, J.A., 388
Black, J.B., 100, 187
Black, L., 203, 436
Black, L.M., 137
Black, M., 409
Black, P.E., 322
Black, R.E., 484
Black, R.E.L., 490
Black, S.H., 228
Black, T.A., 534
Black, V.H., 312
Black, W.A., 551
Black, W.D., 548
Black, W.G., 219
Black, W.L., 391
Blackburn, E.B., 451
Blackburn, J.G., 429
Blackburn, T.K., 535
Blackburn, T.R., 65
Blackburn, W., 175
Blackburn, W.H., 275
Blackhurst, H.T., 459
Blackler, A.W., 302
Blackman, C.W., 425
Blackmon, C.R., 393
Blackstein, M.E., 549
Blackstone, E.H., 12
Blackswell, J.V., 272
Blackweel, R.L., 266
Blackwelder, R.E., 129
Blackwell, C.S., 184
Blackwell, H., 486
Blackwell, H.R., 362
Blackwell, L.H., 445
Blackwell, W.H.Jr., 360
Blackwood, A.C., 555, 557
Blackwood, R., 451
Blaine, E.H., 414
Blair, A.H., 542
Blair, A.P., 381
Blair, B.D., 363
Blair, B.O., 147
Blair, C., 439
Blair, J.B., 509
Blair, M.L., 47
Blair, P.V., 146
Blair, R.D.G., 550
Blair, W.F., 465
Blais, J.M., 558
Blaisdell, J.L., 365
Blake, C.A., 332
Blake, C.T., 344
Blake, D.F., 250
Blake, G.H., 5
Blake, G.R., 240
Blake, J.T., 475
Blake, N.J., 97
Blake, R., 365
Blake, R.D., 193
Blake, W.D., 204
Blakely, L.M., 29
Blakeley, R.A., 149
Blakenberg, B.R., 538
Blakeney, E.W., 184
Blaker, G., 67
Blakley, R.L., 161
Blamire, J., 296, 297
Blanchaer, M.C., 538
Blanchard, O.J., 143
Blanchard, R., 280
Blanchard, R.O., 280
Blancher, R.W., 258
Blanco, J., 196
Bland, C.E., 334
Bland, J., 308
Bland, R., 238
Bland, R.G., 223
Blandau, R.J., 499
Blank, F., 408
Blank, M., 301
Blankespoor, G.B., 431
Blankenagel, G., 561
Blankenship, B.J., 470
Blankespoor, H.D., 231
Blankley, W.F., 332
Blanpied, G.D., 305
Blanton, F.S., 91
Blanton, H., 120
Blanton, P.L., 449
Blanton, W.G., 463
Blasdell, R.F., 296
Blaser, H.W., 497
Blasie, J.K., 410
Blaskey, J.W., 79
Blass, J.P., 56
Blatt, J.L., 506
Blatteis, C.M., 445
Blattner, F.R., 519
Blaumanis, O.R., 204
Blaustein, E.H., 207
Blaustein, M.P., 263
Blaydes, D.F., 507
Blazevic, D., 244
Blazkovec, A.A., 521
Bleakney, J.S., 542
Blecher, M., 85
Blechman, H., 312
Bledsoe, L.J., 498
Bleecker, A.L., 310
Blei, I., 297
Bleibtreu, H.K., 18
Bleiweis, A.S., 94
Blem, C., 491
Blem, L., 491
Blenden, D.C., 262
Blenkarn, D., 333
Bletner, J.K., 443
Blevins, C.E., 145
Blevins, R.L., 177
Blevins, W.T., 4
Bleyman, M., 340
Bleyman, M.A., 339
Blickensderfer, P., 270
Blickenstaff, R.T., 146
Blickle, A.H., 368
Blickle, R., 280
Bliese, J.C.W., 269
Blincoe, C.R., 275
Blinks, L.R., 61
Blinn, D., 15
Bliss, D., 296, 297
Bliss, L.C., 531
Blitz, R.R., 28
Blivaiss, B.B., 134
Blizzard, J., 253
Blizzard, S.H., 508
Bloch, A., 321
Bloch, D.P., 465
Bloch, E., 330
Bloch, E.H., 355
Bloch, K.E., 210
Bloch, P.H., 411

Block, B., 77, 79
Block, D.G., 267
Block W.D., 232
Blocker, H.D., 167
Bloebaum, A., 448
Bloedel, J., 244
Blogoslawski, W., 75
Blogoslowski, W.A., 77, 79
Bloodworth, M.E., 459
Bloom, B., 329
Bloom, B.R., 330
Bloom, G., 244
Bloom, J.R., 403
Bloom, S.E., 305
Bloom, W.L., 102, 170
Bloom, W.W., 153
Bloomfield, V.A., 241
Bloss, H.B., 18
Blosser, T.H., 502
Blount, J.L., 361
Blout, E.R., 211
Blu, J.J., 333
Blu, M.S., 106
Blue, W.G., 93
Blum, B., 61
Blum, J.L., 521
Blum, K., 470
Blum, K.E., 439
Blum, U., 343
Blume, F.D., 27
Blumenfield, D., 393
Blumenfeld, O., 330
Blumenson, L., 320
Blumenthal, H.J., 124
Blumenthal, T., 144
Blumer, F., 270
Blumer, T.N., 344
Blumeyer, Fr.A.J., 255
Bo, W.J., 347
Boag, D.A., 531
Boast, C.W., 138
Bobear, J., 317
Bobes, B., 278
Bobo, J.R., 185
Bobonis, J., 418
Bobr-Tylingo, H., 543
Bocage, A.J., 186
Boccabella, A.V., 282
Bocchieri, S.F., 327
Bocher, C., 309
Bock, C.E., 71
Bock, J.H., 71
Bock, R.M., 515, 517
Bock, W., 142
Bock, W.J., 300
Bockholt, A.J., 459
Bockhop, C.W., 155
Bocking, D., 551
Bockman, E.L., 186
Bockrath, R., 146
Boda, J.M., 47
Bode, H., 54
Bode, V.C., 166
Bode, W., 402
Bodenstein, D., 488
Bodian, D., 197
Bodie, J.L., 5
Bodkin, N., 486
Bodley, J.W., 244
Bodman, C., 125
Bodoh, J., 392
Bodyfelt, F.W., 383
Boe, A.A., 117
Boechx, R.L.O., 538
Boekenkamp, R.P., 416
Boeker, E.A., 499
Boekms, C.N., 436
Boelk, J., 78
Boelke, J., 79
Boell, E.J., 80
Boenig, R.W., 323
Boerman, H.R., 106
Boernke, W., 270
Boersman, L., 384
Boesel, M.W., 360
Boesman, M., 466
Boettner, E.A., 233
Boezi, J.A., 225
Bogacz, J., 397
Bogart, B.I., 312
Bogart, J.P., 187
Bogart, R., 382
Bogdanove, L., 121
Bogenschutz, R., 374
Boger, E.A., 221
Boger, L.L., 225
Bogert, C., 293
Boggan, W.O., 428
Boggs, N., 494
Bogin, C., 311
Bogle, A.L., 280
Bogorad, L., 210
Bohannon, J.L., 440
Bohannon, R., 497
Bohart, G.E., 477
Bohart, R.M., 49
Bohaychuk, W., 542
Bohen, B., 197
Bohl, E.H., 363, 366
Bohlke, J., 96
Bohm, B.A., 534
Bohman, V.R., 275
Bohmont, D.W., 275
Bohnenblust, K.E., 527
Bohning, R.H., 362
Bohnsack, K.K., 39
Bohr, D.F., 232
Bohrer, Sr.R.M., 415
Boicourt, A.W., 218
Boileau, S., 560
Boime, I., 265
Bois, J.J., 329
Bois, P., 558
Boisseau, D.W., 497
Bojanowski, R.J., 557
Boke, N.H., 379
Bolande, R., 555
Bolar, M.L., 31
Bold, H.C., 465
Boldberg, J.M., 132
Boldstein, S., 297
Bolduc, G., 560
Bolduc, P., 556
Boleman, L.L., 458
Boler, R.K., 97
Boles, R.J., 164, 496
Boley, R.B., 464
Bolin, M., 139
Boling, J.A., 177
Boll, W.G., 556
Bollag, J.M., 401
Bollinger, E.G., 128
Bollinger, O.P., 398
Bollis, N.J.Jr., 372
Bollman, R.C., 539
Bollon, A., 466
Bollum, F.J., 179
Bolognesi, D.P., 333
Bolooki, H., 95
Bolsen, K.K., 167
Bolus, R.L., 490
Boly, E.L., 286
Bomke, A.A., 534
Bonamo, P., 317
Bonaparte, T., 313
Bonar, D.A., 199
Bonaventura, J., 333
Bonavida, B., 56
Bond, A.N., 119
Bond, C.E., 383
Bond, G.C., 25
Bond, G.H., 492
Bond, J.H., 426
Bond, J.J., 331
Bond, J.S., 491
Bond, R., 195
Bond, T.P., 468
Bond, W.P., 189
Bondareff, W., 126
Bonde, E.K., 71
Bonds, D., 439
Bonds, E., 7
Bondurant, S., 327
Bone, D.W., 364
Boneham, R.F., 145
Bongiorno, S.F., 75
Bonham, C.D., 67
Bonham, K., 498
Bonhorst, C., 389
Bonilla, C.A., 69
Bonin, S.H., 74
Bonnemann, J.J., 433
Bonne, D.J., 167
Bonner, B.A., 51
Bonner, H.G., 485
Bonner, J.F., 26
Bonner, J.F.Jr., 146
Bonner, J.T., 287
Bonner, T., 317
Bonner, W.D., 410
Bonnet, A., 418
Bonnett, H.T.Jr., 387
Bonneville, M.A., 71
Bononi, S.R., 413
Bonsnes, R.W., 307
Bonte, F.J., 466
Bonventre, P.F., 370
Booden, T., 133
Boodley, J.W., 303
Boohar, R., 270
Book, H., 147
Bookhout, C.G., 332
Bookhout, T.A., 364
Books, M., 272
Boole, J.A., 102
Boon, J.K., 309
Boone, D.M., 519
Boone, G.C., 407
Boone, L.V., 138
Boone, M., 186
Boone, M.A., 425
Boord, R.L., 82
Boos, F.R., 223
Boosalis, M., 270
Boosalis, M.G., 272
Booth, G.M., 472
Booth, J., 439
Booth, J.A., 292
Booth, J.S., 27
Booth, N.H., 108
Booth, R.A., 74
Booth, T., 538
Boothe, L.E., 485
Boothroyd, C.W., 304
Boothroyd, E.R., 556
Boozen, R.B., 7
Bopp, G., 245
Boqusch, E.R., 456
Boraker, D.K., 482
Boram, W.A., 506
Borcbert, R., 171
Borchardt, R., 170
Borchers, H., 236
Borchers, R., 270
Borchers, R.L., 272
Borchert, R., 170
Bordelon, C.B., 447, 449
Borden, F.Y., 402
Borden, J.H., 533
Boree, E.C., 171
Borei, H.G., 409
Borek, E., 72
Borensztajn, J., 133
Borgatti, A.L., 215
Borgese, T.A., 297
Borgman, R.F., 424
Borie, A.B., 414
Borison, H.L., 277
Borisy, G.G., 516
Borkon, E., 129
Bormann, F.H., 51
Bornstein, J., 480
Bornstein, P., 499
von Borstel, R.C., 535
Borom, J., 7
Boros, D.L., 236
Borowsky, B., 299
Borowsky, R.L., 312
Borror, D.J., 363
Borst, D.C., 76
Borthelsen, J.D., 158
Bortner, C.E., 177
Borton, A., 219
Borton, R.J., 365
Bortone, S.A., 98
Bortree, A.L., 403
Borun, T., 408
Borysenko, J., 216
Bosch, A.W., 360
van den Bosch, R., 44
Boschung, H.T., 10
Bose, H.R., 465
Bose, S.K., 256
Boshko, E.B., 288
Boss, E., 122
Boss, K.J., 211
Boss, M.L., 87
Bossert, W.H., 211
Bossenga, L., 119
Bosshart, D.A., 468
Boston, A.C., 183
Boston, G.D., 367
Bostrack, J.M., 522
Boswell, F.C., 106
Boswell, J.L., 452
Bosworth, C., 76
Botelho, S.Y., 410
Botero, S., 186
Bothner, G., 523
Bothner, R.C., 315
Bothner-By, A.A., 391
Botkin, D.B., 81
Botkin, M.P., 527
Botsford, J., 292
Botstein, L., 278
Bott, K.F., 339
Bott, T.L., 409
Bottino, N.R., 458
Bottino, P.J., 201
Bottomley, R., 380
Bottrell, D.G., 458
Botts, D.J., 60
Botzler, R.G., 39
Bouchard, L.J., 374
Boucher, L.J., 285
Boughner, D.R., 552
Boulant, J.A., 97
Bouldin, D.R., 302
Boulpaep, E., 81
Boulware, R.F., 183
Bouma, A.H., 459
Bouman, F.L., 223
Bounds, H.C., 188
Bourassa, J.P., 560
Bourdo, E.A.Jr., 229
Bourgault, J.M., 558
Bourgaux, P., 559
Bourke, J.B., 307
Bourke, R., 320
Bourne, E.W., 293
Bourne, G., 157, 531
Bourne, J.R., 446
Bourne, M.C., 307
Bourne, N., 498
Bourns, T.K.R., 552
Bourque, D.P., 16
Bousfield, E.L., 544
Boush, G.M., 518
Boushey, D., 482
Bousman, L.C., 436
Bouthillier, L.P., 558
Boutwell, R., 254
Bouwkamp, J.C., 202
Bovard, K.P., 493
Bovbjerg, R.V., 16
Bovee, K., 411
Bovell, C.R., 57
Bovet, J., 557
Boving, B., 235
Bowden, B., 295
Bowden, J.A., 184
Bowden, R.L., 361
Bowden, R.L., 362
Bowden, S., 99
Bowen, A.J., 542
Bowen, C.C., 157
Bowen, E., 445
Bowen, H.H., 459
Bowen, J.E., 111
Bowen, S.T., 40
Bowen, W., 513
Bower, A., 37
Bower, J.M., 108
Bower, R.K., 22
Bowering, J., 306
Bowerman, R., 526
Bowers, D., 37
Bowers, J.H., 470
Bowers, J.L., 23
Bowers, L.E., 316

Bowers, M.C., 229
Bowers, R.R., 31
Bowers, W.S., 306
Bowes, G., 90
Bowes, H., 70
Bowes, J.E.II., 352
Bowes, M.J., 51
Bowie, E.P., 469
Bowie, W.C., 9
Bowker, L.S., 27
Bowland, J.P., 530
Bowles, D.A., 70
Bowles, J.B., 157
Bowling, J.J., 288
Bowling, R., 24
Bowling, R.A., 458
Bowman, B., 173
Bowman, B.H., 467
Bowman, B.L., 215
Bowman, B.U., 367
Bowman, D.E., 146
Bowman, D.L., 423
Bowman, E.A., 199
Bowman, G.H., 547
Bowman, H., 192
Bowman, J., 280
Bowman, J.E., 132, 133
Bowman, J.P., 68
Bowman, J.T., 476
Bowman, M., 125
Bowman, M.L., 366, 368
Bowman, O.M., 485
Bowman, R.I., 40
Bowman, T., 96
Bowman, T.C., 423
Bowmer, R.G., 116
Bownds, M.D., 516
Bowne, S., 395
Bowness, J.M., 538
Bowns, J., 473, 476
Bowns, J.A., 473
Bowring, J.R., 280
Box, E.D., 467
Box, C.O., 249
Box, H., 320
Box, Q.T., 467
Box, T., 475, 476
Box, T.W., 477
Boyarsky, L., 176
Boyarsky, L.L., 180
Boyarsky, S., 263
Boyce, A.M., 58
Boyce, B.R., 481
Boyce, J.S.Jr., 427
Boyce, R.P., 90
Boyd, A.H., 248
Boyd, C.E., 4
Boyd, C.M., 542
Boyd, C.T., 262
Boyd, F.M., 188
Boyd, G.B., 443
Boyd, J.C., 266
Boyd, L.J., 106
Boyd, M., 155, 331
Boyd, M.H., 34
Boyd, P., 191
Boyd, R., 336
Boyd, R.L., 6
Boyd, S.F., 438
Boyd, W.L., 69
Boyden, E.A., 499
Boyer, C.C., 142
Boyer, C.I.Jr., 306
Boyer, D.R., 173
Boyer, H.W., 60
Boyer, J.S., 136, 138
Boyer, M.G., 553
Boyer, R.A., 340
Boyer, S., 197
Boyer, S.D., 128
Boyer, W., 32
Boyese, T., 298
Boyette, J.G., 334
Boylan, J.W., 319
Boylan, D.R., 156
Boylan, E., 298
Boylan, R., 312
Boyle, Sr.A., 311
Boyle, C.H., 8
Boyle, J.III, 283
Boyle, J.R., 231
Boyle, J.S., 403
Boyle, W.S., 476
Boylen, C.W., 314
Boyles, C.R., 365
Boyles, J.M., 12
Boyne, A., 160
Boyne, P.J., 469
Boynton, E., 378
Boynton, J.E., 332
Boziam, R.C., 370
Bozniak, E.G., 477
Bozorth, R.G., 415
Braasch, N.L., 257
Brach, J., 95
Bracker, C.E., 148
Brachman, R.C., 484
Bradbeer, C., 490
Bradbury, M.G., 40
Bradbury, P.C., 346
Braddock, J.C., 227
Bradford, G.E., 48
Bradford, J.M., 258
Bradford, L., 128
Bradford, R.H., 380
Bradford, S.W., 219
Bradfute, O.E., 366
Bradley, B.P., 205
Bradley, C.M., 259
Bradley, E.L., 11
Bradley, E.L., 484
Bradley, G.A., 23
Bradley, G.L., 35
Bradley, J.R.Jr., 344
Bradley, J.W., 458
Bradley, K.A., 79
Bradley, N.W., 177
Bradley, R.C., 65
Bradley, R.E., 94
Bradley, R.L.Jr., 518
Bradley, S.G., 492
Bradley, T.R., 485
Bradley, W.O., 495
Bradshaw, A.S., 369
Bradshaw, J., 62
Bradshaw, L.H., 30
Bradshaw, M.W., 449
Bradshaw, R.A., 265
Bradshaw, W.E., 387
Bradshaw, W.N., 507
Brady, A., 224
Brady, A.H., 95
Brady, A.J., 57
Brady, C.W., 524
Brady, F.O., 434
Brady, R.J., 360
Brady, R.N., 446
Brady, R.O., 85
Brady, U.E.Jr., 106
Brage, B.L., 433
Bragg, J.D., 21
Bragg, P.D., 534
Bragonier, W.H., 65, 68
Brahim, F., 555
Braigger, D., 473
Brain, J.D., 217
Brainerd, J.W., 216
Braithwaite, L., 472
Braithwaite, R., 8
Brakier-Gingras, L., 558
Brakke, M., 270
Brakke, M.K., 272
Bram, A.J., 307
Bramante, P.O., 135
Brammer, J.D., 480
Brammer, M.I., 309
Bramucci, D., 205
Branam, G., 142
Branch, C.E., 6
Branch, J.C., 375
Brand, J.J., 465
Brand, K.G., 244
Brand, L., 197
Brand, L.R., 35
Brand, R.H., 139
Brand, R.J., 46
Branda, L.A., 545
Brandell, B.R., 562
Brandhorst, B., 556
Branding, A.E., 402
Brandom, W.F., 73
Brandon, R.A., 129
Brandon, R.B., 507
Brandt, C.L., 37
Brandt, E.E., 337
Brandt, E.N.Jr., 467, 468
Brandt, I.K., 146
Brandt, K.G., 148
Brandt, L.S., 374
Brandt, P.W., 300
Brandt, R.B., 491
Brandt, W.H., 385
Brank, J.S., 329
Brann, J.J.Jr., 303
Brannen, W.G., 109
Brannigan, D., 317
Branning, T., 246
Brannon, E.L., 498
Brannon, M.J.C., 7
Branson, T., 433
Brant, A.W., 40
Brant, D.H., 34
Brantigan, O.C., 203
Brasch, K.R., 546
Braselton, J.P., 368
Brasher, P.K., 247
Brashier, C.K., 431
Brasier, C., 432
Braswell, E.H., 77, 79
Bratcher, R.L., 326
Bratley-Mamet, M., 558
Bratt, A.L., 223
Bratton, G.R., 460
Brattstrom, B.H., 30
Bratz, R.D., 116
Brauerman, S., 408
Braum, E.J., 19
Braun, D.E., 37
Braun, G., 80
Braun, P., 555
Braune, M.O., 403
Braungart, D.C., 83
Braver, G., 379
Braverman, B., 125
Brawer, J., 216
Brawerman, G., 217
Brawley, G., 135
Brawn, R., 122
Braxter, S.E., 439
Bray, D.J., 138
Bray, D.M., 82
Bray, J., 150
Bray, J.M., 467
Braymer, H.D., 183
Brazda, F.G., 185
Brazeau, M., 558
Brazier, M.A.B., 57
Breakevelt, C.R., 537
Breakey, D.R., 389
Breakstone, K., 70
Breazile, J.E., 262
Brebbia, D.R., 284
Brecher, G.A., 380
Brecher, S., 214
Brecht, P.E., 305
BreDahl, R.L., 177
Breeden, J.E., 437
Breen, H., 69
Breen, M., 351
Breen, R., 88
Breer, J., 300
Breese, W.P., 383
Bregman, A., 323
Brehm, B., 387
Breidenbach, G., 360
Breidenbach, R.W., 47
Breiding, G.H., 508
Breil, D.A., 486
Breil, S.O., 486
Breitenbach, R.P., 260
Breland, H.L., 93
Breland, O.P., 465
Bremer, A., 320
Brenchley, J.E., 404
Breneman, W.R., 145
Brengle, K.G., 65
Brengelmann, G.L., 500
Brennan, E.G., 288
Brennan, E.L., 193
Brennan, J., 373
Brennan, J.J., 238
Brennan, J.R., 208
Brennand, C., 475
Brenneman, J.A., 152
Brenner, F.J., 396
Brenner, L.J., 357
Brenner, M., 211
Brenner, M.L., 241
Brenowitz, A.H., 295
Brent, B.E., 167
Brent, M.M., 354
Brescia, N., 123
Breslin, P.P., 414
Breslow, E.M., 307
Bresnick, E., 103
Bressler, G.O., 403
Bressler, S.L., 507
Brest, D., 27
Brett, C.H., 344
Bretz, H.W., 121
Brew, K., 95
Brewbaker, D., 358
Brewbaker, J.L., 114
Brewer, C.F., 330
Brewer, C.W., 183
Brewer, D., 542
Brewer, F.D., 247
Brewer, G., 64
Brewer, G.J., 232
Brewer, J.M., 105
Brewer, J.W., 68
Brewer, H.E., 501
Brewer, R., 236
Brewer, R.H., 76
Brewer, R.N., 5
Brewington, N.C., 494
Brewnig, De R., 300
Brewster, M., 24
Breyere, E.J., 83
Brezenoff, H.E., 283
Breznak, J.A., 228
Brezner, J., 322
Briand, F., 549
Brick, G., 299
Brick, I., 312
Brickbauer, E.A., 516
Brickler, S.K., 18
Briden, K., 284
Brides, W.H.Jr., 109
Bridger, W., 531
Bridger, W.A., 543
Bridges, K.W., 111
Bridgman, J., 20
Bridgmon, G.H., 527
Briehl, R., 336
Briehl, R.W., 336
Brierley, G.T., 79
Brierre, J.T., 191
Brière, N., 559
Brierley, J.A., 291
Briese, F.W., 72
Briggle, L.W., 240
Briggs, A.H., 470
Briggs, B.J., 256
Briggs, D.M., 346
Briggs, F.N., 492
Briggs, G.M., 45
Briggs, J.D., 363
Briggs, J.L., 524
Briggs, K.G., 530
Briggs, R.E., 16
Briggs, R.W., 145
Briggs, T., 380
Briggs, W.E., 71
Briggs, W.R., 41
Bright, D.B., 30
Bright, H.J., 410
Bright, P., 56
Bright, R.C., 242
Bright, T.J., 459
Brighton, C.T., 411
Briles, C.O., 8
Briles, W.E., 126
Brill, W.J., 517
Brillhart, W.E., 100
Brillinger, D.R., 46
Brim, C.A., 344
Brindley, T.A., 157
Brindley, W.A., 476
Bringe, A.N., 517

Brink, D.L., 44
Brink, J.J., 209
Brink, K.M., 66
Brink, R.A., 519
Brink, V.C., 534
Brinkerhoff, L.A., 375
Brinkhous, K.M., 340
Brinkley, B.R., 467
Brinkley, H.J., 199
Brinks, J.S., 65
Brinson, M.M., 334
Brinster, R.L., 411
Brinton, C.C.Jr., 413
Briody, B.A., 282
Brister, G.H., 108
Bristol, D.W., 352
Bristol, J.R., 468
Bristol, R.F., 520
Bristow, J.M., 546
Britt, E.M., 232
Britt, L.V., 359
Britt, N.W., 363
Britt, R.F., 337
Brittain, D., 513
Brittain, J.A., 424
Brittan, M.R., 31
Britten, B.T., 313
Britten, J.S., 301
Britton, D.M., 546
Britton, J.C., 461
Britton, W.M., 108
Broadbent, F.E., 50
Broad, A.D., 504
Broadbooks, H., 130
Brobeck, J.R., 410
Broch, E.S., 501
Brock, J.H., 112
Brock, R.O., 431
Brock, S.M., 508
Brock, T.D., 517
Brock, W.E., 377
Brocke, R.H., 322
Brockman, E.R., 223
Brockman, H.E., 122
Brockman, W.H., 158
Brocksen, R.W., 47
Brodbeck, M., 240
Brodehl, R.B., 199
Broderick, G.A., 458
Broderius, S., 240
Broderson, S.H., 499
Brodile, A.F., 64
Brodie, E.D.Jr., 295
Brodie, J.D., 521
Brodish, A., 370
Brodkey, J.S., 356
Brodkorb, P., 90
Brody, D.A., 445
Brody, M., 297, 298
Brody, S., 59
Brody, S.S., 312
Brody, T.M., 227
Broek, B.J., 223
Broersma, D.R., 148
Brogdon, J.E., 91
Broghamer, W.L.Jr., 181
Broitman, S.A., 208
Brokaw, C.J., 26
Bromel, M.C., 351
Bromley, S.C., 227
Broembsen, S., 502
Bronk, B.V., 426
Bronk, D.W., 410
Bronskill, M.J., 550
Bronson, F.H., 465
Bronson, H., 250
Bronson, R.D., 147
Brook, T.S., 249
Brookbank, J.W., 90
Brooke, R.C., 533
Brookes, V.J., 386
Brookhart, J.M., 389
Brooks, A., 523
Brooks, A.E., 153
Brooks, B.B., 25
Brooks, C.C., 113
Brooks, D.L., 471
Brooks, H.L., 167
Brooks, J.B., 179
Brooks, J.F., 345
Brooks, J.L., 508
Brooks, F.P., 410
Brooks, G.R., 484
Brooks, J.H., 187
Brooks, M.A., 240
Brooks, M.H., 375
Brooks, R.F., 91
Brooks, R.J., 547
Brooks, S.C.Jr., 235
Brooks, S.N., 502
Brooks, V.B., 552
Brooks, W., 513
Brooks, W.M., 364, 365
Broome, C.R., 201
Broquist, H.P., 446
Brosnan, J.T., 541
Brosnan, M.E., 541
Bross, I., 320
Brossette, K., 189
Brostoff, S., 428
Brot, N., 282
Brotherson, J.D., 472
Brotman, H.F., 305
Browder, L., 531
Browder, L.E., 168
Brower, A.D., 396
Brower, J., 126
Brower, L.P., 206
Brower, M.S., 99
Brown, A., 254, 442
Brown, A.C., 500
Brown, A.H., 409
Brown, A.L., 50
Brown, A.M., 468
Brown, A.R., 106
Brown, A.V., 203
Brown, A.W., 226, 227, 554
Brown, B.A., 243, 513
Brown, B.A.Jr., 459
Brown, B.I., 263
Brown, B.L., 12
Brown, C.A., 182
Brown, C.D., 438
Brown, C.H., 109
Brown, C.J., 23
Brown, C.L., 108
Brown, C.M., 137
Brown, C.S., 195
Brown, D., 189, 197
Brown, D.A., 22
Brown, D.F., 555
Brown, D.G., 233
Brown, D.H., 263
Brown, D.J., 539
Brown, D.L., 414, 549
Brown, D.M., 391
Brown, D.R., 204
Brown, E.B.Jr., 172
Brown, E.H., 137
Brown, E.M., 262
Brown, E.R., 133, 220, 374
Brown, F.A.Jr., 126
Brown, F.L., 436
Brown, G., 30
Brown, G.C., 234
Brown, G.G., 157
Brown, G.M., 550
Brown, G.N., 261
Brown, G.W., 106, 384, 440
Brown, G.W.Jr., 498
Brown, H.A., 504
Brown, H.B., 197
Brown, H.D., 438
Brown, H.E., 259
Brown, H.H., 327
Brown, H.J., 255
Brown, H.P., 379
Brown, H.S., 29
Brown, I., 550
Brwon, J., 327, 446
Brown, J.A., 28
Brown, J.C., 341, 490
Brown, J.E., 382, 446
Brown, J.F., 23
Brown, J.H., 366, 367, 473
Brown, J.H.Jr., 422
Brown, J.L., 72
Brown, J.M., 99
Brown, J.N., 251
Brown, J.R., 258
Brown, J.R.C., 199
Brown, J.S., 6
Brown, J.W., 11
Brown, K.E., 259
Brown, K.I., 366
Brown, K.M., 149, 277
Brown, K.W., 459
Brown, L., 122, 285, 335, 453
Brown, L.A., 176
Brown, L.N., 97
Brown, L.R., 58, 77, 247, 386
Brown, M., 205
Brown, M.A., 454
Brown, M.C., 261
Brown, M.F., 260
Brown, M.L., 196, 288
Brown, M.M., 439
Brown, M.P., 438
Brown, M.S., 459
Brown, M.T., 261
Brown, N.K., 500
Brown, N.S., 215
Brown, O.R., 262
Brown, P.B., 509
Brown, P.J., 452
Brown, P.S., 316
Brown, R., 21, 30, 182
Brown, R.B., 415
Brown, R.C., 504
Brown, R.D., 196
Brown, R.D.Jr., 493
Brown, R.F., 455
Brown, R.G., 542, 547
Brown, R.H., 106, 496
Brown, R.J., 27, 60, 62
Brown, R.K., 235
Brown, R.M., 338
Brown, R.T., 228
Brown, R.V., 260, 491
Brown, R.Z., 308
Brown, S.O., 460
Brown, S.R., 546
Brown, S.W., 45
Brown, T.W., 29
Brown, W.C., 59, 415
Brown, W.D., 49
Brown, W.E., 321, 379
Brown, W.H., 17, 122
Brown, W.J., 236
Brown, W.L.Jr., 303
Brown, W.M., 549
Brown, W.P., 360
Brown, W.S., 317
Brown, W.V., 465
Brown, W.W., 69
Browne, E.T., 438
Browne, M.G., 313
Brownell, G., 103
Brownell, J.R., 29
Brownell, W.E., 94
Brownie, A.C., 318
Browning, D.R., 138
Browning, H.C., 464
Browning, J.A., 157
Browning, R.A., 129
Browning, R.F., 513
Brownlee, C.G., 170
Brownold, D., 135
Brownson, R.H., 53
Brownstein, B.H., 179
Brownstein, S., 556
Brownstone, Y.S., 551
Broyles, R., 523
Brubaker, J.M., 484
Brubaker, K.K., 484
Brubaker, R.R., 228
Brubakery, G., 393
Bruce, A.K., 318
Bruce, D.S., 139
Bruce, J.I., 213
Bruce, J.R., 546
Bruce, R.C., 348
Bruce, R.W., 503
Bruce, W.R., 549
Bruck, D.L., 418
Bruck-Kan, R., 185
Bruckel, D.W., 481
Bruehl, G.W., 503
Bruenger, F.W., 473
Bruening, G.E., 48
Bruesch, S.R., 445
Brueske, C.H., 361
Brueske, W.A., 34
Brugh, T.,5
Bruhn, J.C., 49
Bruhn, J.G., 468
Brumbaugh, J., 270
Brumbaugh, J.H., 28
Brummett, A.R., 361
Brumsted, H.B., 303
Brun, W.A., 240
Brunckhorst, W., 524
Bruneau, G.P., 232
Bruneau, L.H., 375
Bruner, J.J., 293
Bruner, W.H., 365
Brunovskis, I., 213
Brunk, G., 55
Brunngraber, E., 300
Bruno, S., 299
Bruns, P.J., 302
Brunson, J.T., 324
Brunson, R.B., 267
Brusca, G.J., 34
Brush, A.H., 77, 79
Brushaber, J.A., 395
Brussard, P.F., 303
Brust, R.A., 537
Brusven, M.A., 117
Bryant, M.P., 137
Brutlag, D., 41
Bruton, C., 239
Bryan, C.R., 182
Bryan, G.A., 451
Bryan, H.H., 93
Bryan, J., 409
Bryan, J.H.D., 106
Bryan, S., 293
Bryan, S.E., 185
Bryan, W., 146
Bryan, W.P., 146
Bryank, J.K., 326
Bryans, J.T., 178
Bryant, E., 463
Bryant, E.S., 78, 79
Bryant, G.D., 112
Bryant, H.E., 178
Bryant, J.B.Jr., 189
Bryant, M.D., 293
Bryant, P.J., 54
Bryant, S.V., 54
Bryant, W.E., 495
Bryant, W.R., 459
Bryant, W.S., 176
Bryden, J.R., 176
Bryden, R.R., 335
Bryner, C.L., 415
Bryson, V., 288
Bubash, G.F., 403
Bubel, H.C., 370
Bubieniec, E., 173
Bublitz, C., 72
Bublitz, W.J., 384
Bucci, C.F., 203
Buchanan, A.M., 53
Buchanan, D.W., 92
Buchanan, G.A., 3
Buchanan, H., 477
Buchanan, J.R., 307
Buchanan, M.L., 350
Buchanan-Smith, J.G., 547
Buchele, W.F., 155
Buchell, B.R., 170
Buchenau, G., 433
Buchholz, R.H., 125
Buchring, G., 46
Buchwald, J.S., 57
Buchwald, M., 550
Buck, C.A., 166
Buck, C.E., 194
Buck, C.F., 177
Buck, G.J., 156
Buck, J.D., 77, 79
Buck, P., 381
Buck, R.C., 551
Buckalew, D.M., 401
Buckelew, A.R., 516
Bucklelew, T.P., 391
Buckholtz, N.S., 428
Buckler, A.F., 266

Buckle, B.R., 188
Buckley, C.E., 333
Buckley, H.R., 212
Buckley, N.M., 330
Buckley, R.H., 333
Buckley, S.D.Jr., 216
Buckley, W.J., 128
Buckman, R., 431
Buckmire, F.L.A., 513
Buckner, C.H., 544
Buckner, E.R., 443
Buckner, R.C., 177
Budd, J., 548
Budd, K., 546
Budd, T., 316
Budde, S.N.D., 176
Buddenhagen, I.W., 114
Buddingh, G.J., 185
Budelsky, C.A., 129
Budkin, A., 95
Buecher, E.J., 70
Buegel, J.E., 238
Buell, K.M., 269
Buell, W.H., 391
Buescher, R.W., 23
Buettner, Sr.J., 311
Buetow, D.E., 137
Buffaloe, N.D., 21
Buffett, R.F., 320
Buffington, R.E., 493
Bugbee, W., 352
Bugg, C., 11
Buggeln, R., 541
Buhan, P.J., 407
Buhs, C., 33
Buie, T.R., 454
Buikema, A.L., 492
Buist, N.R.M., 388
Bukovac, M.J., 225
Bukovsan, W., 323
Bula, R.J., 147
Buldhaupt, L., 515
Bulen, W.A., 362
Bulkin, B., 298
Bulkley, J.W., 232
Bulkley, R.V., 157
Bull, A.L., 485
Bull, F.W., 492
Bull, H., 161
Bull, L.S., 201
Bull, R., 117
Bulla, A.D., 66
Bullaro, J., 562
Buller, C.S., 171
Bullis, H.R., 96
Bullock, J., 283
Bullock, D.H., 547
Bullock, R.C., 88, 91
Bullock, R.E., 532
Bulmash, M.H., 203
Bulmer, G.S., 380
Bulow, F.J., 440
Bultmann, H., 453
Bultz, S.A., 217
Bumgardner, H.L., 346
Bumpus, M.F., 357
Bunce, G.E., 493
Bunde, D.E., 116
Bundenthal, T., 258
Bundy, M.M., 199
Bunge, R.P., 263
Bunger, P., 273
Bunn, C., 336
Bunn, J.M., 179
Bunn, R.F., 463
Bunt, J., 96
Bunting, D.L.II., 442
Bunting, J.R., 466
Buntley, G.J., 444
Buonaguro, P., 290
Burack-Cohn, E., 327
Burbach, J.A., 434
Burbanck, W.D., 100
Burbutis, P.P., 82
Burch, C., 329
Burch, D.G., 97
Burch, J.B., 231
Burchall, J.J., 333
Burcham, L.N., 292
Burchard, R.P., 205
Burchett, L.A., 167
Burchill, B.R., 171
Burdeshaw, J.A., 11
Burdette, L.A., 401
Burdick, A.B., 260
Burdick, B., 205
Burdick, C.J., 296, 297
Burdick, D., 30
Burdine, H.W., 93
Burfening, P.J., 266
Burgata, B., 240
Burger, A.W., 137
Burger, C.L., 326
Burger, C.P., 158
Burger, D., 504
Burger, D.R., 388
Burger, J., 288
Burger, J.W., 76
Burger, O.J., 29
Burger, R.E., 47, 48
Burgers, T., 128
Burgess, T.D., 547
Burgess, T.J., 21
Burgett, D.M., 386
Burgis, D.S., 93
Burgner, R.L., 498
Burgoyne, P., 473
Burhoe, S.O., 83
Buri, P., 89
Buric, J., 200, 202
Burk, C.J., 215
Burk, J.H., 30
Burk, L.G., 345
Burke, C.N., 78, 79
Burke, D.D., 137
Burke, D.W., 503
Burke, H.R., 458
Burke, J.D., 491
Burke, M., 204
Burke, M.J., 241
Burke, P.W., 547
Burke, T., 72
Burke, W., 310
Burke, W.J., 15
Burkett, B., 95
Burkhard, R.K., 166
Burkhardt, C.C., 527
Burkhart, H.E., 494
Burkhart, H.R., 382
Burkhart, L., 17
Burkholder, G., 562
Burkholder, J.H., 167
Burkholder, T.J., 151
Burkholder, W.E., 518
Burks, J., 285
Burks, S.L., 376
Burky, A.J., 371
Burley, J.W.A., 393
Burley, W., 394
Burlingham, B.T., 166
Burlington, R.F., 223
Burmester, B.R., 228
Burnard, R.K., 364
Burnes, D.J., 288
Burnett, A.L., 126
Burnett, J., 261
Burnett, J.P., 146
Burnham, K.D., 442
Burns, A.E., 84
Burns, D.M., 479
Burns, E.E., 459
Burns, E.M., 256
Burns, E.R., 24
Burns, G.W., 369
Burns, J., 164, 429
Burns, J.C., 344
Burns, J.M., 211, 461
Burns, M., 147
Burns, M.A., 416
Burns, M.F., 316
Burns, M.J., 6
Burns, P.E., 280
Burns, P.H., 445
Burns, P.Y., 183
Burns, R.D., 359
Burns, R.E., 106
Burns, R.E., 192
Burns, R.L., 74, 278
Burns, R.M., 515
Burns, R.O., 333
Burns, T.A., 188
Burns, V.W., 52
Burns, W.A., 315
Burns, W.C., 516
Burnside, E.B., 547
Burnside, M.B., 410
Burnside, O.C., 271
Burr, A.A., 314
Burr, A.H., 533
Burr, J.H.Jr., 455
Burrell, R., 509
Burris, C.A., 336
Burris, J.S., 157
Burris, R.H., 517
Burroughs, R., 436
Burseqicz, J., 253
Bursey, R.G., 424
Burson, B.L., 248
Burson, S., 342
Burstein, A., 356
Burstein, S., 449
Burstyn, H.L., 290
Burt, A.M.III., 446
Burt, G.W., 200
Burt, M.D.B., 540
Burtness, P.S., 126
Burton, A., 240
Burton, A.C., 551
Burton, A.F., 534
Burton, A.L., 469
Burton, D.N., 538
Burton, G.W., 106
Burton, H., 263
Burton, H.R., 177
Burton, P.R., 171
Burton, R., 377
Burton, V., 238
Burton, W.G., 152
Burts, E., 502
Bury, R.L., 458
Burzlaff, D.F., 463
Busbee, D.L., 452
Busbice, T.H., 344
Busby, F., 476
Buscemi, P.A., 291
Busch, D., 220
Busch, K.H.D., 272
Busch, L.V., 547
Busch, R., 34
Buschi, R.A., 77
Busci, R.A., 79
Buser, F.B., 394
Bush, F.M., 491
Bush, G.L., 465
Bush, L., 376
Bush, L.P., 177
Bush, P.B., 106
Bush, W.G., 398
Bushing, R.W., 49
Bushnell, J.H., 71
Bushnell, R.J., 28
Bushuk, W., 537
Buskirk, R., 304
Buskirk, W.H., 143
Busnel, R.G., 297
Busrel, R.G., 296
Buss, E.G., 403
Buss, J.T., 266
Buss, W., 73
Bussell, R.H., 171
Bussey, A.H., 556
Bussjaeger, C., 165
Busta, F., 244
Bustad, L.K., 504
Busteed, R.C., 471
Buston, D.R., 16
Buston, J.W., 178
Buszynski, A., 408
Butcher, D.G., 223
Butcher, H.C., 198
Butchko, G.M., 252
Buterbaugh, G.G., 204
Buterworth, T.M., 80
Buthala, D.A., 236
Butler, A., 84
Butler, A.P., 291
Butler, C.E., 374
Butler, D.G., 550
Butler, D.R., 362
Butler, E.E., 51
Butler, H., 562
Butler, J., 175
Butler, J.D., 66
Butler, J.E., 34, 161
Butler, J.F., 91
Butler, J.K., 464
Butler, J.L., 432
Butler, L., 295, 508, 550
Butler, L.G., 148
Butler, O.D., 458
Butler, R.L., 403
Butler, W., 427, 437
Butler, W.L., 59
Butler, W.R., 414
Butler, W.T., 449
Butow, R.A., 466
Butterworth, C.H.Jr., 11
Butterworth, F.M., 230
Buttlar, R., 359
Button, D.K., 14
Button, L.C., 36
Buttrey, B.W., 157
Butts, W., 323
Butwell, R., 323
Butz, A., 370
Buxton, J.A., 331
Byer, M.D., 418
Byerrum, R.U., 225
Byers, B.E., 499
Byers, B.R., 251
Byers, D.H., 233
Byers, G.L., 280
Byers, G.W., 171, 172
Byers, R.A., 402
Byers, R.E., 494
Byers, T.H., 105
Byers, T.J., 363
Byerumm, R.W., 226
Byfield, P.E., 56
Byram, J., 88
Byrd, D., 109
Byrd, J.R., 103
Byrd, M.A., 484, 490
Byrd, R.P., 181
Byrd, W.W., 20
Byrne, B., 329
Byrne, G.E.Jr., 133
Byrne, J., 496
Byrne, J.A., 238
Byrne, J.M., 492
Byrne, J.V., 385
Byrne, R.L., 490
Byrnes, E.W., 206
Byrnes, W.R., 149
Byron, Fr.W., 187

C

Cabaniss, M.L., 185
Cabelli, V.J., 421
Cable, P., 104
Cabrera, C., 95
Cacon, I.R., 105
Cade, T.J., 302, 303
Cadwell, L., 122
Cady, F.B., 302, 304
Cafferata, R., 327
Cagle, G.D., 363
Cahan, S., 556
Cahill, C.L., 346
Cahill, D., 95
Cahill, V.R., 365
Cahn, F., 277
Cahoon, G.A., 365
Cahoon, Sr.M.O., 237, 238
Caille, J.P., 559
Cain, B.W., 459
Cain, G.D., 160
Cain, J.R., 458
Cain, R.F., 383
Cain, W.A., 380
Caine, R.L., 25
Cairnie, A.B., 546
Cairns, G.M., 200, 201
Cairns, J., 320
Cairns, J.Jr., 492
Calabrese, A., 77, 79
Calahan, C.L., 481
Calarco, P., 60

Calaresu, F.R., 552
Calavan, E.C., 58
Calbert, H.E., 518
Calbert, H.W., 518
Calcott, P., 557
Caldecott, S.R., 241
Calder, D.R., 490
Calder, W.A., 18
Calderone, R.A., 415
Caldwell, A.C., 241
Caldwell, C., 250, 486
Caldwell, D.R., 528
Caldwell, E.K., 409
Caldwell, J., 458
Caldwell, J.L., 365, 428
Caldwell, L.D., 223
Caldwell, M., 476
Caldwell, M.M., 477
Caldwell, R.E., 93
Caldwell, R.L., 18, 43
Caldwell, R.S., 583
Caldwell, S.D., 102
Caldwell, W.M., 509
Calendar, R., 43
de Calesta, D.S., 346
Calhoon, R.A., 299
Calhoon, T.B., 181
Calhoun, F.G., 93
Calhoun, M.L., 227
Calhoun, R.R., 453
Calhoun, W.F., 492
Caliguiri, L.A., 327
Calkins, J.C., 177
Call, J.W., 475
Call, T.G., 27
Callaghan, J.J., 415
Callaham, M.A., 104
Callahan, J.C., 149
Callahan, K., 335
Callahan, L.M., 444
Callahan, R.F., 290
Callan, G., 467
Callard, I.P., 207
Callaway, H., 314
Callebs, J., 506
Callenbach, J.A., 351
Callihan, R.H., 117
Calloway, D.H., 45
Calloway, R.A., 249
Calnek, B.W., 305
Calpouzos, L., 117
Caluwe, P., 322
Calvert, O.H., 260
Calvert, R., 559
Calvin, H., 301
Calvin, M., 43
Calvo, J.M., 302
Camerman, N., 549
Cameron, D., 265
Cameron, D.M., 553
Cameron, D.R., 45
Cameron, E.A., 402
Cameron, G., 463
Cameron, H.R., 385
Cameron, I.L., 469
Cameron, J.A., 77, 79
Cameron, J.W., 58
Cameron, M.L., 542
Camhi, J.M., 302, 304
Camin, J.H., 171
Caminita, B.H., 104
Cammeron, D.S., 534
Cammeron, J.N., 14
Camp, B.J., 458, 460
Camp, E.D., 461
Camp, R.R., 210
Campana, R.J., 194
Campbell, A.M., 41
Campbell, A.R., 117
Campbell, B.J., 261
Campbell, C.M., 113
Campbell, C.W., 92
Campbell, D.G.J., 542
Campbell, E.C., 491
Campbell, F.R., 180
Campbell, G., 223
Campbell, G.S., 502
Campbell, G.T., 492
Campbell, H.J., 383
Campbell, I.M., 413
Campbell, J., 550
Campbell, J.B., 271
Campbell, J.D., 316, 537
Campbell, J.J.R., 535
Campbell, J.R., 259
Campbell, J.W., 453
Campbell, K.B., 411
Campbell, L., 88
Campbell, N., 444
Campbell, N.E.R., 538
Campbell, R., 215, 377
Campbell, R.D., 54
Campbel,, R.L., 503
Campbell, R.N., 51
Campbell, R.S., 261
Campbell, R.W., 168, 322
Campbell, S., 12
Campbell, T.C., 493
Campbell, T.H., 357
Campbell, W.R., 148
Campbell, W.V., 344
Camper, N.D., 425
Campsall, E.W.R., 551
Canaday, S., 428
Canady, W.J., 509
Canale-Parola, E., 218
Canaris, A.G., 468
Canry, J.J., 85
Cancilla, P.A., 161
Candelas, G., 418
Candido, E.P.M., 534
Canfield, E.L., 155
Canham, R.P., 267
Canham, S., 298
Canhan, P.C., 551
Cann, J.R., 72
Cannon, C.Y., 472
Cannon, D., 373
Cannon, F.T., 345
Cannon, G., 285
Cannon, M.S., 467
Cannon, R.E., 342
Cannon, R.Y., 3
Cano, A., 291
Cano, R.J., 27
Canode, C.L., 502
Canolty, N.L., 50
Canon, A.O., 337
Canright, J.E., 12
Cantliffe, D.J., 93
Cantor, A.H., 367
Cantor, C.R., 300
Cantor, M., 31
Cantrall, I.J., 231
Canvin, D.T., 546
Capecchi, M.R., 473
Capehart, J.R., 448
Capek, V., 135
Capella, R.O., 290
Capelli, G., 484
Capen, R.L., 65
Capers, J., 182
Capet, R., 559
Capin, R., 416
Capizzi, J., 380
Caplan, A., 355
Caplin, S.M., 31
Caplow, M., 340
Capodilupo, A.G., 224
Capon, B., 31
Caponetti, J.D., 441
Cappellini, R.A., 288
Capra, J.D., 466
Capranica, R.R., 302, 304
Capwell, R.L., 334
Caraway, P.A., 471
Carbaugh, B.T., 398
Carbó, G.R., 417
Carbon, J.A., 61
Carbonneau, M.C., 138
Card, C.S., 118
Card, G.L., 267
Cardasis, C.A., 325
Cardeilhac, P.T., 94
Cardell, R.R., 489
Cardenas, J., 385
Carder, C.A., 104
Carder, K.L., 97
Cardezier, V.R., 469
Cardinal, A., 557
Cardinet, G.H., 168
Cardonio, E., 553
Cardwell, J.T., 249
Cardwell, V.B., 240
Carefoot, T.H., 535
Carell, E.F., 412, 412
Carell, J.E., 260
Cares, C.W.Jr., 231
Carew, J., 225
Carew, L.B., 480
Carey, R., 32
Carey, W.E., 364
Carico, J.E., 486
Carison, J.H., 287
Carithers, J.R., 158
Carithers, R.W., 158
Carito, S., 310
Carl, P.L., 137
Carlander, K.D., 157
Carleton, B.H., 384
Carleton, S., 26
Carley, A.L., 128
Carlin, A.F., 155
Carlin, C.R., 150
Carlin, R., 285
Carlisle, C.L., 416
Carlisle, G.R., 138
Carlisle, V.W., 93
Carlo, I., 417
Carlquist, S., 32
Carls, R., 395
Carlson, A., 325
Carlson, C.A., 66
Carlson, C.W., 432
Carlson, D.L., 157
Carlson, D.M., 355
Carlson, E., 325
Carlson, F., 198
Carlson, G.L., 11
Carlson, G.M., 411
Carlson, H.C., 548
Carlson, J.B., 245
Carlson, J.G., 442
Carlson, J.R., 502
Carlson, O., 524
Carlson, R.B., 351
Carlson, R.F., 225
Carlson, S.D., 518
Carlson, W.H., 226
Carlson, W.R., 160
Carlton, B.C., 105
Carlton, R.A., 439
Carluccio, L.M., 74
Carman, G.E., 58
Carmer, S.G., 137
Carmichael, N.M., 39
Carmichael, S.W., 509
Carmin, R.L., 142
Carmody, G.R., 544
Carner, G.R., 424
Carner, J., 550
Carnery, E.M., 479
Carnes, R., 128
Carney, C.M., 340
Carolson, J.C., 551
Carolus, R.L., 225
Caron, D.M., 201
Carothers, Z.B., 156
Carow, J., 231
Carpelan, L.H., 57
Carpender, E.D.Jr., 78
Carpenter, A.M., 243
Carpenter, C.B., 506
Carpenter, C.C., 379
Carpenter, E.D., 79
Carpenter, F.H., 42
Carpenter, G., 151
Carpenter, G.P., 167
Carpenter, I.W.Jr., 331
Carpenter, J.A., 65, 106, 547
Carpenter, J.H, 176
Carpenter, J.M., 176
Carpenter, L.W., 54
Carpenter, M., 380
Carpenter, M.B., 300
Carpenter, N., 381
Carpenter, P.L., 149, 421
Carpenter, P.N., 195
Carpenter, R.E., 39
Carpenter, R.L., 216, 380
Carpenter, S., 556
Carpenter, S.J., 277
Carpenter, W., 225
Carpenter, Z.L., 458
Carpiles, V.A., 417
Carr, A.A., 104
Carr, A.F., 90
Carr, C.W., 244
Carr, D.O., 172
Carr, E.S., 337
Carr, J., 30
Carr, J.E., 496
Carr, R.D., 129
Carr, R.F., 186
Carr, R.L., 496
Carr, W.E.S., 90
Carrach, M., 534
Carraway, K.L., 377
Carrick, L.Jr., 236
Carrier, E.B., 189
Carrier, G.O., 104
Carrier, O.Jr., 470
Carriere, R.M., 325
Carrigus, U.S., 138
Carriker, M.E., 421
Carroll, A.G., 375
Carroll, C.R., 142
Carroll, D., 5
Carroll, D.E., 344
Carroll, E.J., 53, 69
Carroll, R.D., 48
Carroll, G.C., 367
Carroll, K.K., 551
Carroll, T., 265
Carroll, T.W., 265
Carrow, R.E., 227
Carrow, R.N., 219
Carruth, L.A., 17
Carruthers, C., 320
Carsiotis, M., 370
Carson, J.D., 61
Carson, P.L., 433
Carson, V., 32
Carson-Mayberry, K.J., 435
Carstens, H.B., 181
Cartee, R.E., 6
Carter, A.L., 191
Carter, C.H., 103
Carter, D.C., 461
Carter, D.W., 474
Carter, E., 272
Carter, G.E., 425
Carter, G.R., 228
Carter, H.P., 8
Carter, J.C., 103
Carter, J.C.H., 551
Carter, J.F., 350
Carter, J.L., 65
Carter, J.P., 446
Carter, J.R., 356
Carter, M.C., 149
Carter, P.B., 70, 476
Carter, R.C., 493
Carter, R.T., 387
Carter, T., 203
Carter, W.A., 320, 374
Carter, W.H.Jr., 491
Cartledge, M., 174
Cartmill, M., 332
Cartwright, G.W., 139
Cartwright, T.C., 458
Cartwright, T.E., 413
Carubelli, R., 380
Caruolo, E.V., 343
Carusi, E.A., 268
Caruso, J.L., 369, 370
Carvell, K.L., 508
Carver, D.H., 198
Carver, J., 335
Carver, J.C., 517
Carver, M.J., 273
Carver, V., 270
Cary, J.W., 474
Cary, N.E., 49
Casada, J.H., 179
Casamajor, P., 44
Casarett, G.W., 328
Cascarano, J., 55
Casciano, Sr.M.C., 299
Case, J.F., 61

Case, M., 106
Case, N.W., 35
Case, R.M., 272
Casey, K.L., 233
Cash, C., 335
Cash, E.H., 401
Cashion, P.J., 540
Cashner, R.C., 185
Casida, J.E., 44
Casida, L.E.Jr., 404
Casida, L.E., 519
Caskey, C.T., 448
Caspari, E., 327
Cass, D.D., 531
Cassaro, V.E., 289
Cassatt, J.C., 85
Cassel, J.C., 342
Cassel, J.F., 350
Cassidy, M.M., 84
Cassim, J.Y., 362
Cassin, S., 94
Cassista, J., 278
Cassman, M., 61
Castaner, D., 253
Castelfranco, P.A., 51
Castellucci, V., 313
Castenholz, R.W., 387
Castle, E.N., 382
Castle, G.B., 15
Castleberry, D.M., 40
Casto, -. 452
Caston, J.D., 355
Castric, P.A., 394
Castro, A.J., 185
Castro, C.E., 58
Castro, F.C., 86
Castro, P., 29
Catalano, R.A., 391
Catana, A.J.Jr., 435
Catchings, B.M., 185
Catchpole, R.H., 134
Catena, A., 40
Catenhusen, J., 224
Cater, C.M., 458, 459
Cather, J.N., 231
Cathou, R.E., 217
Catlett, R.H., 40
Catlin, B.W., 513
Catlin, J.E., 266
Catlin, R.W., 119
Catovic, S.C., 284
Catravas, G.N., 83
Catsimpoolas, N., 213
Catts, E.P., 82
Catz, J., 95
Caughey, W.S., 67
Causey, M.K., 5
Causey, N.B., 183
Cauthen, I.B., 488
Cavaletto, C.G., 113
Cavaliere, A.R., 395
Cavalieri, E., 273
Cavanah, L.E., 258
Cavanaugh, C.J., 182
Cavazos, L.F., 216
Cave, A., 371
Cave, M.D., 24
Cavender, J.C., 368
Cavender, T.M., 364
Caveney, S., 552
Cavers, P.B., 552
Cavert, H.M., 244
Caviness, C.E., 22
Cawidowica, E.A., 211
Cawley, E.T., 159
Cazier, M., 15
Cazin, J., 161
Cebra, J., 197
Cebula, R., 497
Ceccarini, C., 297, 298
Cech, F.C., 508
Cechner, R., 356
Cechner, R.L., 356
Cedergren, R.J., 558
Cedno-Maldonado, A., 417
Ceglowski, W., 408
Ceglowski, W.S., 404
Ceithaml, J., 131
Celine, S.A., 74
Celis, T.F.R., 313
Celland, W.W., 517
Center, E.M., 41
Center, M.S., 166
Cerroni, R.E., 357
Cerwonka, R.H., 324
Centifanto, Y.M., 94
Cernosek, R.M.Jr., 24
Cerny, J., 212
Cerutti, P.A., 90
Cetas, R.C., 304
Chabora, P.C., 297, 298
Chabot, B., 302, 303
Chabreck, R.H., 183
Chacko, G.K., 400
Chacko, K.J., 53
Chacko, S., 412
deChaderavian, J.P., 556
Chadwell, J.D., 493
Chadwick A.V., 35
Chadwick D.L., 477
Chadwick, J.S., 545
Chadwick, P., 545
Chae, C.B., 340
Chaet, A.B., 98
Chaffee, E.F., 342
Chaffin, W.L., 461
Chahal, K.S., 8
Chaiken, L.E., 334
Chairez, R., 485
Chakraburtty, K., 513
Chakrapani, B., 336
Chalcraft, R., 335
Chalgren, S., 487
Chalkley, R., 161
Challoner, D.R., 146
Chambelzand, D., 461
Chamberlain, C.C., 443
Chamberlain, D.W., 139
Chamberlain, J.L., 327
Chamberlain, J.P., 231
Chamberlain, N., 426
Chamberlain, W.D., 392
Chamberlin, M.J., 42
Chambers, E.L., 96
Chambers, H.W., 246
Chambers, J.V., 522
Chambers, K.L., 385
Chambers, R., 359
Chambers, R.E., 322
Chambers, V.C., 499
Chambers, W.F., 277
Chambers, W.W., 410
Chamblee, D.S., 344
Chambliss, C.G., 138
Chambliss, O.L., 5
Champ, M.A., 83
Champagne, G.B., 454
Champe, S.P., 288
Champion, R., 100
Champlin, A.K., 192
Champoin, L.R., 226
Champoux, J.J., 499
Champuey, W.S., 105
Chan, A.S., 86
Chan, F., 434
Chan, J.G., 114
Chan, L., 31
Chan, L.C.B., 449
Chan, S.H.P., 326
Chan, S.K., 179
Chan, S.W.C., 280
Chan, W.W.C., 545
Chance, B., 410
Chandler, A.M., 380
Chandler, C.M., 439
Chandler, J.A., 454
Chandler, L., 148
Chandler, P.T., 493
Chandler, W.K., 81
Chandra, P., 224
Chaney, A.H., 456
Chaney, C., 175
Chaney, S.G., 340
Chaney, W.R., 149
Chang, A.Y., 236
Chang, F., 114
Chang, G.C., 9
Chang, G.W., 45
Chang, J., 467
Chang, M.C., 207
Chang, P., 391
Chang, S.H., 184
Chang, T.M.S., 556
Chang, T.S., 226
Chang, Y.O., 527
Chang, Y.F., 204
Chanlett, E.T., 341
Channing, C.P., 204
Chansky, L.M., 428
Chant, D.A., 550
deChantal, R., 558
Chantell, C.J., 371
Chao, F.C., 207
Chao, T.T., 338
Chapados, C., 560
Chapco, W., 561
Chaperon, E.A., 269
Chapin, F.S., 14
Chapler, C.K., 546
Chaplin, C.E., 178
Chaplin, J.F., 344
Chaplin, M.H., 384
Chapman, D., 55
Chapman, A.B., 519
Chapman, A.O., 472
Chapman, A.R.O., 542
Chapman, C.J., 506
Chapman, D.G., 498
Chapman, D.M., 542
Chapman, E.S., 390
Chapman, G.B., 85
Chapman, H.W., 548
Chapman, J.A., 436
Chapman, J.P., 241
Chapman, L.F., 53, 260
Chapman, L.W., 64
Chapman, M.K., 120
Chapman, P., 244
Chapman, R.A., 178
Chapman, R.C., 503
Chapman, R.K., 518
Chapman, R.L., 182
Chapman, T.E., 168
Chapman, V.M., 319
Chapman, W.F., 429
Chappell, G.L.M., 177
Chappell, R., 297, 298
Chapple, W.D., 79
Charache, P.C., 198
Charalampous, F., 410
Charba, J.F., 88
Charchalis, G.W., 17
Charity, L., 155
Charkey, L.W., 67
Charman, E.C., 67
Charles, A.M., 551
Charnego, M., 396
Charteris, J., 546
Charters, Sr.E.M., 358
Chartock, M.A., 379
Charudattan, R., 92
Charvat, I.D., 241
Chase, F.E., 547
Chase, J., 288, 369
Chase, M., 57
Chase, R., 556
Chase, R.S., 397
Chase, T., 288
Chase, W., 230
Chasin, L.A., 300
Chasson, A.L., 344
Chasson, R.M., 122
Chatterjee, A.N., 318
Chaubal, K., 61
Chauvin, C.M., 221
Chauvin, R., 89
Chaves, M., 418
Chavin, W., 237
Chay, T.R., 415
Chaykin, S., 48
Checchi, A.C., 327
Cheda, G., 321
Cheeke, P.R., 382
Cheevers, W.P., 551
Chefurka, W., 551
Cheh, C., 553
Cheigh, J.S., 307
Chen, C., 127
Chen, C.H., 432
Chen, C.L., 168
Chen, C.L.H., 408
Chen, C.M., 522
Chen, I.W., 370
Chen, J., 467, 478
Chen, J.C.W., 288
Chen, J.L., 32
Chen, K.C., 234
Chen, L., 40
Chen, L.T., 197
Chen, P.K., 85
Chen, P.Y.S., 129
Chen, R.W., 356
Chen, S.M., 327
Chen, S.S., 521
Chen, T., 288
Chen, V., 355
Chency, J.M., 69
Cheneval, J.P., 560
Cheney, H.B., 384
Cheng, H.C., 370
Cheng, H.F., 162
Cheng, H.H., 502
Cheng, T.C., 398
Cheny, D.P., 114
Cherian, K.S., 349
Cherry, D.R., 561
Cherry, J.H., 149
Cherry, W.N., 175
Chertok, R., 246
Chesbro, W.R., 279
Chesemore, D., 30
Chesky, J., 96
Chesnin, L., 271
Chesnut, T.L., 102
Chesnutt, C., 423
Chessin, M., 266
Chester, E.H., 356
Chester, E.W., 436
Chestnut, A.F., 339
Cheuk, S.F., 30
Cheung, H.C., 11
Cheung, P.W., 356
Chevalier, G., 560
Cheverie, J.C., 554
Chevrette, J.E., 558
Chew, K.K., 498
Chew, R.M., 62
Chi, L.W., 27
Chia, F.S., 531
Chiaba, T , 160
Chiakulas, J.J., 133
Chiang, C.L., 46
Chiang, H., 242
Chiang, H.C., 240, 242
Chiang, K.S., 132
Chiang, T.S., 104
ChiaraValle, P., 479
Chiarenza, F., 79
Chiasson, L.P., 543
Chiasson, R.B., 18
Chibuzo, G.A., 9
Chichester, F.W., 364
Chichester, L.F., 74
Chick E.W., 179
Chidambaram, S., 296
Chien, P.K., 62
Chien, S., 301
Chik, L., 356
Chilcote, M.E., 318
Chilcote, W.W., 385
Child, F.M., 76
Childers, A.B., 460
Childers, N.F., 288
Childers, W.F., 138
Childress, E., 27
Childress, J.J., 61
Childs, B., 197
Childs, E.A., 444
Childs, O.A., 21
Chilgreen, D., 147
Chilman, C., 234
Chilman, K.C., 129
Chilson, O.P., 263
Chilton, M.D., 499
Chimoskey, J.E., 497
Chin, B., 233
Chin, C.M., 108
Chin, E.Jr., 32
Ching, M.C., 328
Chinn, S.F., 561

Chiong, M.A., 546
Chipley, J.R., 363, 365, 367
Chipman, P.B., 80
Chipman, R.K., 421
Chippendale, G.M., 259
Chiquoine, A.D., 308
Chiriboga, J., 418
Chirikjian, J.C., 85
Chiscon, J.A., 150
Chiscon, M.O., 150
Chisholm, A., 285
Chisler, J.A., 506
Chittenden, M.E., 459
Chittick, D.E., 382
Chitty, D.H., 535
Chiu, H.F., 552
Chiu, W., 5
Chizinsky, W., 296
Chmielewica, Z.F., 318
Cho, B.R., 504
Cho, J.J., 114
Cho, Y., 9
Choate, J.R., 164
Chobotar, B., 222
Chodirker, W., 551
Choe, B., 236
Choe, H.T., 238
Choi, S.Y., 290
Cholak, L., 372
Cholas, G., 69
Cholvin, N.R., 158
Chong, C., 555
Choong, E.T., 183
Chopp, J.J., 524
Chopra, B.K., 336
Chorpenning, F.W., 363
Chou, C., 228
Chovnick, A., 77, 79
Chow, C.T., 538
Chow, D., 139
Chow, T.L., 69
Chowdhury, L.N., 139
Chowhury, T.K., 380
Choy, S.H., 60
Chrapliwy, P.S., 468
Chrisanthopolous, G., 220
Chrispells, M.J., 59
Christen, H.E., 5
Christensen, A.K., 408
Christensen, C., 28
Christensen,D.A., 561
Christensen, H.N., 232
Christensen, J.R., 328
Christensen, K., 256
Christensen, L.R., 550
Christensen, M., 526
Christensen, N., 332
Christensen, P.D., 474
Chirstensen, R., 405
Christensen, W.W., 481
Christenson, C.M., 147
Christian, E.A.Jr., 31
Christian, F.A., 189
Christian, J., 38
Christian, J., 317
Christian, J.A., 186, 344
Christian, J.C., 146
Christian, P.J., 236
Christian, R.E., 117
Christian, T., 11
Christiansen, J.L., 155
Christiansen, K.A., 155
Christiansen, P., 154
Christianson, J.D., 329
Christianson, L.E., 61
Christie, F., 21
Christie, M., 284
Christie, R.J., 252
Christinsin, J.E., 423
Christman, J.F., 187
Christman, R.F., 341
Christofferson, J.P., 28
Christoph, R.J., 511
Christopher, Sr.M., 224
Christopher, Sr.M.H., 452
Christopherson, R.J., 530
Christopherson, W.M., 181
Christy, C.M., 258
Chrovser, W.H., 419
Chrusciel, Sr.I., 223
Chu, C.H.U., 353
Chu, C.Y.M., 138
Chu, E.H.Y., 232
Chu, F.S., 518
Chu, T.M., 320
Chu, Y-M, 84
Chuang, H. Y-K, 340
Chuang, R.Y., 333
Chuang, T.I., 122
Chucy, C., 373
Chun, P., 90
Chun-Hoon, H.C.F., 11
Chung, A.E., 413
Chung, C.S., 113
Chung, K.L., 545
Chung, K.W., 325
Chunosoff, L., 313
Churbuck, L., 162
Church, D.C., 382
Church, K.K., 15
Church, R.B., 531
Church, R.C., 77
Church, R.L., 79
Churcher, C.S., 550
Churchich, J.E., 441
Churchill, C.L., 431
Churchill, E.P., 434
Churchill, F.M., 448
Churchill, V., 286
Churney, L., 186
Chynoweth, D.P., 233
Chytil, F., 446
Ciaccio, L.A., 297
Ciani, S., 57
Cibula, A., 359
Cichuk, G., 286
Cicmanec, J.F., 370
Cifonelli, J.A., 131
Cilley, J., 408
Cimprica, H.A.R., 412
Cinader, B., 549
Cinnamon, C.G., 192
Cinquina, C.L., 415
Ciolkosz, E.J., 401
Cirino, E.F., 208
Cisneros, B., 452
Ciullo, R.H., 192
Civan, M.M., 411
Cizek, L.J., 301
Claborn, L.D., 460
Claflin, L.E., 168
Claflin, T., 515
Claflin, W.J., 349
Clagett, C.O., 404
Clagett, J.A., 499
Clague, D.W., 35
Claman, H., 72
Clamann, H.P., 492
Clambey, G.K., 349
Clancy, C.W., 387
Clancy, N., 561
Clancy, R.L., 172
Clancy-Hepburn, K., 306
Clandinin, D.R., 530
Clanetine, R.L., 222
Clapp, C.E., 241
Clapp, J.G.Jr., 344
Clapp, J.R., 333
Clapp, T., 239
Clare, Sr.R., 37
Clarenburg, R., 168
Clarin, B.J., 196
Clark, A.E., 545
Clark, A.J., 43
Clark, A.M., 82
Clark, A.W., 520
Clark, B.E., 307
Clark, B.P., 360
Clark, C., 162, 410
Clark, C.H., 6
Clark, C.W., 99
Clark, D.A., 469
Clark, D.L., 367
Clark, D.R., 460
Clark, D.T., 381
Clark, D.W., 334
Clark, E., 199
Clark, E., 336
Clark, E.A., 310
Clark, E.D., 38
Clark, E.M., 4
Clark, F.E., 89
Clark, F.J., 273
Clark, F.M., 100
Clark, G., 77, 175
Clark, G.A., 77, 79
Clark, G.E., 186
Clark, G.M., 550
Clark, G.T., 21
Clark, G.W., 496
Clark, H., 77, 79
Clark, H.E., 288
Clark, I., 340
Clark, J., 7, 35, 122, 192
Clark, J.A., 257
Clark, J.H., 449
Clark, J.L., 54, 259
Clark, J.W., 487
Clark, K., 327
Clark, K.B., 87
Clark, K.W., 537
Clark, M.A., 464
Clark, M.B., 452
Clark, M.E., 37
Clark, N.A., 200
Clark, N.B., 79
Clark, R., 95
Clark, R.B., 364
Clark, R.C., 427
Clark, R.J., 120, 416
Clark, R.M., 324
Clark, S.L., 477
Clark, V.A., 56
Clark, Rev.V.J., 8
Clark, W., 134
Clark, W.G., 467
Clark, W.H., 463
Clark, W.J., 459, 460
Clark, W.R., 55
Clarke, A.M., 491
Clarke, B.J., 83
Clarke, D.W., 550
Clarke, E.A., 549
Clarke, G., 371
Clarke, G.A., 487
Clarke, G.R., 374
Clarke, H.M., 515
Clarke, J., 539
Clarke, L.A., 431
Clarke, N., 189
Clarke, P., 355
Clarke, R., 316
Clarke, R.F., 164
Clarke, W.W., 402
Clarkson, J.D., 234
Clarkson, L.L., 461
Clarkson, R.B., 507
Clarkson, T.W., 328
Clattenburg, R.E., 551
Claus, G.W., 492
Clausen, B.L., 162
Clausen, C.D., 35
Clausen, C.P., 58
Clausen, C.R., 499
Clausen, R.T., 302
Claussen, D.L., 360
Clausz, J.C., 337
Clawson, A.J., 343
Clay, C., 30
Clay, J., 160
Clay, W.H., 180
Claybrook, D.L., 429
Claycomb, D.A., 393
Claypool, D.W., 382
Clayton, C.C., 491
Clayton, C.N., 345
Clayton, D., 41, 500
Clayton, F.E., 22
Clayton, G.T., 455
Clayton, J.W., 538
Clayton, K., 126
Clayton, P.C., 366
Clayton, R.K., 302
Cleary, Sr.M.B., 198
Cleary, P.P., 244
Cleary, S.F., 491
Cleaves, E.D., 272
Cleaves, G.R., 283
Clebsch, E.E.C., 441
Clegg, J., 95
Clegg, L.F.L., 530
Clegg, M.D., 271
Clegg, R.E., 166
Cleland, C.F., 211
Cleland, R.E., 497
Clem, J.R., 342
Clem, L.W., 94
Clemens, H.P., 379
Clemens, L., 227
Clement, A.C., 100
Clements, R.L., 364
Clemmer, G.H., 248, 250
Clench, W.J., 364
Clermont, Y., 555
Clesceri, L.S., 314
Clesceri, N.L., 314
Cleveland, A.G., 463
Cleveland, M.D., 538
Cleveland, J., 372
Cleveland, R.W., 401
Clewell, A., 88
Clewell, D.B., 232
Cliburn, J.W., 252
Cliff, F.S., 29
Clifford, A.J., 50
Clifford, F.B., 455
Clifford, H.F., 531
Clifford, R.O., 276
Clifron, C.M., 106
Cline, B., 10
Cline, G.B., 10
Cline, J.H., 365
Cline, M.G., 302, 362
Clise, R.L., 357
Cliver, D.O., 517
Clogston, F.L., 27
Cloney, R.A., 498
Clothier, G., 28
Clothier, R., 15
Cloutier, F., 317
Clovis, J.F., 507
Clower, B.R., 251
Cloyd, W.J., 436
Clugston, J.P., 108
Clulow, F.V., 545
Clutter, J.L., 108
Cmarik, G.F., 108
Coacher, J.B., 431
Coakley, J., 377
Coakley, S.M., 73
Coalson, R.E., 379
Coash, J., 27
Coates, P.W., 499
Coates, R.K., 551
Coats, G.E., 249
Coats, L.W., 250
Cobb, B.F., 458
Cobb, F., 333
Cobb, F.W., 45
Cobb, G.W., 187
Cobb, H., 434
Cobb, J., 75
Cobb, J.S., 421
Cobb, J.P., 75
Cobb, W.Y., 344
Cobbe, T.J., 360
Cobble, J.W., 38, 422
Coble, H.D., 344
Coburn, C.B., 440
Coburn, R.F., 411
Coceani, F., 550
Cochis, T., 7
Cochran, F.D., 345
Cochran, G.W., 476
Cochran, K.W., 234
Cochran, V.H., 254
Cochran, V.L., 502
Cochrane, J., 459
Cochrane, V.W., 80
Cockerham, L., 347
Cockerline, A.W., 463
Cocking, D., 486
Cockrell, R., 256
Cockrell, R.A., 44
Cockrum, E.L., 18
Cockrum, O.D., 452
Coddling, J.A., 550
Coduri, R.L., 422
Cody, M., 55
Cody, R.M., 4

Coe, E.H.Jr., 260
Coe, E.L., 127
Coe, N., 286
Coepfert, J.M., 517
Cofer, E., 344
Coffee, C.J., 413
Coffey, D.L., 444
Coffey, J.C., 337
Coffey, J.W., 285
Coffey, M., 387
Coffia, W., 375
Coffin, D.L., 333
Coffino, P., 60
Coffman, M.S., 229
Coffman, W.P., 412
Cofresis, F., 417
Cogar, V.F., 459
Cogen, R.B., 79
Cogen, T., 313
Coggeshall, R.E., 467
Coggin, J.H.Jr., 442
Coggins, C.W.Jr., 58
Coggins, L., 306
Coghlan, A.E., 215
Cogswell, H.L., 30
Cohen, A.C., 501
Cohen, A.E., 278
Cohen, A.I., 262
Cohen, A.M., 197
Cohen, B.J., 233
Cohen, D., 96
Cohen, D.H., 490
Cohen, E., 319
Cohen, E.H., 287
Cohen, E.P., 132
Cohen, G., 467
Cohen, G.M., 87
Cohen, G.H., 411
Cohen, I., 206
Cohen, J.E., 211
Cohen, J.R., 216
Cohen, K.I., 321
Cohen, L., 81, 410
Cohen, L.A., 228
Cohen, L.W., 33
Cohen, M., 308
Cohen, M.A., 345
Cohen, M.H., 132
Cohen, M.I., 330
Cohen, M.J., 80
Cohen, M.W., 556
Cohen, N., 328
Cohen, P.P., 520
Cohen, P.S., 421
Cohen, R., 70
Cohen, R.A., 214
Cohen, R.J., 90
Cohen, S., 79, 313, 446
Cohen, S.S., 72
Cohen, W., 190, 411
Cohen, W.D., 297, 298
Coher, E.I., 310
Cohn, D., 172
Cohn, M., 410
Cohn, N.S., 368
Cohn, S.A., 445
Cohn, T.J., 39
Cohoon, D.F., 425
Colahan, T., 323
Colas, A.E., 520
Colbert, D., 461
Colborn, G.L., 469
Colburn, R.B., 402
Colby, B., 219
Colby, C., 116
Colby, H.D., 509
Colby, T.W., 448
Cole, A.C., 442
Cole, A.W., 249
Cole, B.C., 474
Cole, B.D., 436
Cole, B.T., 429
Cole, C., 335
Cole, E.A., 206
Cole, E.J.Jr., 215
Cole, E.L., 182
Cole, E.M., 175
Cole, G.A., 15
Cole, G.T., 465
Cole, H., 403
Cole, H.E., 269
Cole, J.E., 390
Cole, K.M., 534
Cole, L.C., 302, 303, 304
Cole, M.A., 138
Cole, R., 128, 225
Cole, R.D., 42
Cole, R.H., 401
Cole, R.K., 305
Cole, R.S., 105
Cole, S.L., 8
Cole, T.A., 153
Cole, T.B.Jr., 434
Coleman, B., 134
Coleman, -., 391
Coleman, C., 392
Coleman, D.R., 80
Coleman, E.A., 462
Coleman, E.C., 461
Coleman, G.E.III, 5
Coleman, J.E., 121
Coleman, J.R., 328
Coleman, M.S., 179
Coleman, P.D., 328
Coleman, P.H., 492
Coleman, P.S., 312
Coleman, R., 380
Coleman, R.M., 213
Coleman, R.W., 162
Coleman, T.H., 226
Coleman, W., 79
Coleman, W.W., 402
Coleridge, H., 60
Coleridge, J.C., 60
Coles, R.C., 282
Coles, R.W., 263
Colgan, J.A., 134
Colgan, P.W., 546
Coli, W.H., 172
Colinvaux, P.A., 364
Colladay, T., 233
Collan, F., 95
Collard, S.B., 98
Collett, B.M., 44
Colley, D.G., 447
Collier, A., 88
Collier, B.D., 37
Collier, G., 40
Collier, J.R., 69
Collier, J.R., 297
Collier, J.W., 459
Collier, R.J., 55
Collies, J.R., 296
Colligan, R.D., 449
Collin, R.L., 328
Collings, E.D., 228
Collings, R.L., 183
Collins, A.D., 251
Collins, A.R., 319
Collins, B.J., 26
Collins, Sr.C.A., 255
Collins, C.E., 344
Collins, D., 101, 205
Collins, E.B., 49
Collins, E.H., 389
Collins, F.C., 22
Collins, G.B., 177
Collins, H.L., 245
Collins, H.M., 317
Collins, J.A., 23
Collins, J.L., 444
Collins, J.M., 491
Collins, J.V., 542
Collins, L.C., 336
Collins, M., 8
Collins, M.A., 124
Collins, O.R., 42
Collins, P.E., 433
Collins, R., 61
Collins, R.A., 21
Collins, R.M., 154
Collins, R.P., 77, 79
Collins, S., 76
Collins, T., 253
Collins, T.W., 239
Collins, W.E., 507
Collins, W.J., 363
Collins, W.K., 344
Collub, E., 300
Colman, B., 553
Colman, K.L., 260
Colman, Sr.M., 396
Colman, M.B., 499
Colome, J.S., 27
Colon, D., 417
Colón, J.I., 418
Colotelo, N., 530
Colowick, M., 447
Colowick, S.P., 447
Colson, C., 510
Colt, L.C.Jr., 207
Colter, J.S., 531
Coltharp, G., 476
Colville, W.L., 106
Colvin, D., 27
Colvin, H.W.Jr., 47
Colwell, R.K., 43
Colwell, R.N., 44
Colwell, R.R., 202
Colwin, A., 96
Comanor, P.L., 376
Combs, G.F.Jr., 5
Combs, M., 133
Combs, R.L., 249
Combs, T.J., 75
Comer, A.E., 214
Cominsky, N.C., 463
Comiskey, J.K., 87
Comita, G.W., 350
Commoner, B., 263
Compton, R.W., 504
Compton, W.A., 271
Comroe, J.H., 60
Comstock, D., 496
Comstock, J.A., 449
Comstock, R.E., 242
Comstock, V.E., 240
Conant, E.B., 486
Conant, N.F., 333
Conant, R., 293
Conard, E.C., 271
Conaway, C.H., 418
Conaway, H.H., 25
Concannon, J.N., 315
Condell, Y.C., 239
Condit, J.M., 364
Condouris, G.A., 283
Condren, S.M., 20
Cone, R., 198
Cone, W., 502
Conert, G.B., 292
Confer, J.L., 309
Congdon, P.U., 216
Coniglio, J.G., 446
Conklin, J.L., 227
Conklin, P.M., 492
Conley, W.H., 225
Conlon, H.P., 422
Conn, E.E., 48
Conneally, P.M., 146
Connell, C.E., 109
Connell, G.E., 549
Connell, J.H., 61
Connell, M.G., 299
Connell, Sr.R., 254
Connell, R.S., 388
Connell, W.A., 82
Connelly, J.L., 353
Conner, S., 354
Connin, R.V., 226
Connor, J.A., 131
Connor, L.I., 363
Connor, R.D., 537
Connor, R.S., 327
Connors, N.A., 256
Conquist, A., 300
Conrad, G.W., 166
Conrad, H.E., 135
Conrad, J.P., 47
Conrad, J.T., 500
Conrad, L., 223
Conrad, M.L., 254
Conrad, P., 206
Conran, P.B., 79
Conrey, M., 512
Conroy, G., 312
Conroy, J.C., 539
Conroy, J.S., 416
Considine, J., 159
Consigli, R.A., 166
Constance, L., 42
Constantin, R.J., 184
Conte, F.P., 386
Conti, S.F., 176
Contois, D.E., 111
Contois, L.A., 224
Contopoulos, A.N., 60
Contrera, J.F., 199
Converse, R.H., 385
Convery, F.J., 334
Convery, Sr.M.E., 393
Convey, E.M., 228
Conway, A.F., 157
Conway, C.M., 157
Conway, J.B., 372
Conway, T.E., 396
Conway, T.W., 161
Conzelman, G.Jr., 53
Coohill, T.P., 181
Cooil, B.J., 111
Cook, A.A., 92
Cook, A.H., 321
Cook, C.W., 67
Cook, D., 247, 251
Cook, D.E., 273
Cook, D.O., 390
Cook, D.R., 234
Cook, E., 459
Cook, E.A., 288
Cook, E.D., 451
Cook, E.E., 309
Cook, E.F., 240
Cook, F.S., 552
Cook, G.A., 455
Cook, G.B., 100
Cook, H., 131
Cook, H.A., 507
Cook, J.A., 51, 156
Cook, J.C., 293
Cook, J.R., 193, 194
Cook, K.M., 154
Cook, M.A., 80, 340
Cook, Sr.M.M., 285
Cook, P.P.Jr., 497
Cook, P.W., 480
Cook, R., 253, 514
Cook, R.A., 551
Cook, R.C., 542
Cook, R.E., 257, 346
Cook, R.J., 503
Cook, R.T., 356
Cook, S.A., 387
Cook, S.F., 43
Cook, T.M., 202
Cook, W.A., 333, 341
Cook, W.B., 67
Cook, W.L., 108
Cook, W.P., 424
Cooke, A.R., 60
Cooke, D., 359
Cooke, F., 546
Cooke, H.G., 335
Cooke, H.J., 162
Cooke, I.M., 112
Cooke, J.D., 552
Cooke, P.H., 171
Cooke, R.L., 538
Cooksey, K., 96
Cool, B.M., 425
Cool, W.S., 358
Coombs, M.C., 218
Coomes, R.C., 238
Coon, B.F., 402
Coon, C., 203
Coon, M.J., 232
Cooney, C.L., 213
Cooney, D.G., 276
Cooney, J.J., 371
Cooney, R.T., 14
Cooney, W.T., 382
Coons, M.P., 121
Cooper, A.F., 27
Cooper, A.W., 343, 346
Cooper, B.A., 556
Cooper, B.H., 408
Cooper, B.L., 371
Cooper, C., 355, 397
Cooper, C.R., 493
Cooper, D.M., 297
Cooper, E.A., 7

Cooper, E.L., 403
Cooper, G.A., 130
Cooper, G.E., 9
Cooper, G.W., 297
Cooper, H.A., 340
Cooper, J.B., 425
Cooper, J.E., 538
Cooper, J.H., 240, 351
Cooper, K., 299
Cooper, K.W., 52
Cooper, L.A., 546
Cooper, M., 235
Cooper, R.D., 177
Cooper, R.F., 183, 293
Cooper, R.L., 138
Cooper, R.W., 40
Cooper, S., 232
Cooper, T.G., 413
Cooper, W.A., 471
Cooper, W.E., 227
Cooperband, S.R., 208
Cooperrider, T., 359
Cooperstein, D., 295
Cooperstein, S.J., 79
Coorts, G.D., 130
Coover, M.O., 507
Cope, J.B., 143
Cope, J.T., 3
Cope, W.A., 344
Copeland, B.J., 346
Copeland, D.E., 190
Copeland, J.C., 363
Copeland, T.P., 437
Copeman, R.J., 534
Copenhaver, J.H., 273
Copenhaver, J.H.Jr., 277
Copenhaver, J.S., 493
Copenhaver, W., 95
Copes, F., 522
Copley, A.J., 254
Copp, D.H., 535
Coppel, H.C., 518
Coppenger, C.J., 40
Coppock, S., 377
Corah, L., 167
Corbett, J., 128
Corbett, M.K., 201
Corbin, F.T., 344
Corbin, Sr.G., 450
Corbin, J.E., 138
Corbin, K.W., 242
Corbin, L., 389
Corcino, J.J., 418
Corcoran, E., 96
Corcoran, J.W., 127
Corcoran, M., 31
Corcoran, M.L., 450
Corcoran, T.J., 194
Corden, M.E., 385
Corder, S.E., 384
Cordes, C.L., 191
Cordes, W.C., 123
Cordonnier, L.M., 257
Cordo, C.E., 294
Cords, H.P., 276
Cordts, A.F., 451
Cordy, D.R., 53
Core, E.L., 507
Core, H.A., 443
Core, S.K., 509
Corey, P.N., 549
Corey, R.R., 199
Corey, S., 547
Corgan, J.N., 293
Corke, C.T., 547
Corkum, K.C., 183
Corley, B., 293
Corley, K.C., 492
Corliss, C.E., 445
Corliss, J.O., 199
Cormack, D.V., 538
Cormack, D.H., 549
Cormier, M.J., 105
Cornatzer, W.E., 353
Cornelius, C.E., 94
Cornelius, P.L., 177
Cornell, C.D., 507
Cornell, C.N., 261
Cornell, R.G., 233
Cornell, R.P., 254
Corner, T.R., 228
Cornesky, R.A., 27
Cornfield, J., 414
Cornish, H.H., 233
Cornoni, J., 342
Corns, W.G., 530
Corpe, W.A., 300
Correa, P., 186
Correll, F.E., 345
Corrick, J.A., 443
Corrigan, J., 315
Corsa, L., 234
Corsaro, F., 396
Corson, G.E., 29
Corson, S.A., 362
Corss, R.L., 326
Corstvet, R.E., 377
Cort, H., 11
Cortelyou, J.R., 120
Cortelyou, Rev.M.T., 120
Corwin, G.M., 262
Corwin, H.O., 412
Coryell, M., 103
Coscia, C., 256
Cosenza, B., 76
Cosgriff, J.W.Jr., 234
Cosgrove, C.J., 422
Cosgrove, E.V., 207
Cosgrove, W.P., 106
Cossins, E.A., 531
Cossio, H.F., 62
Cost, K., 234
Costantin, L.L , 263
Costanzo, Sr.R., 511
Costello, D.P., 339
Costello, R., 251, 521
Costerton, J.W., 531
Costes, D.H., 8
Costilow, R.N., 228
Costisick, P., 308
Costlow, J.D.Jr., 332
Costoff, A., 103
Cota-Robles, E.H., 61
Cote, Br.A., 371
Côte, R.H., 557
Cothran, W.R., 49
Cothren, J.T., 22
Cotman, C.W., 55
Cotmore, J., 111
Cott, A.E., 156
Cottam, G., 515
Cottam, G.L., 466
Cottam, L.G., 466
Cotter, D.A., 215
Cotter, D.J., 102, 293
Cotter, Sr.M.V., 315
Cotter, W.B., 179
Cotterman, C.W., 519
Cottier, G.J., 5
Cotton, T.D., 461
Couch, E.F., 461
Couch, J.R., 458
Couch, R.B., 449
Couch, R.W., 378
Coughland, Rev.M.A., 213
Coughlin, J.H., 75
Coukell, M.B., 553
Coull, B., 429
Coulson, P.B., 442
Coulson, R., 185
Coulson, R.A., 185
Coulson, R.N., 458
Coulter, C.L., 131
Coulter, D., 243
Coulter, D.B., 108
Coulter, E.J., 341
Coulter, J., 239
Coulter, M.W., 194, 461
Coulter, N.A., 340
Coulter, W., 95
Coulter, W.H., 540
Counce, S., 332
Council, D., 196
Counter, S.A., 211
Countryman, W.D., 479
Coupland, R.T., 562
Coursen, B.W., 484
Coursey, P., 205
Courtade, S., 204
Courtenay, W.R.Jr., 87
Courter, J.C., 163
Courter, J.W., 138
Courtin, G.M., 545
Courtis, W.S., 145
Courtright, J.B., 512
Courtsal, F.R., 148
Couse, N.L., 73
Cousens, M.I., 98
Couser, R.D., 20
Cousins, R.J., 288
Coutsogeorgopoulos, C., 320
Couture, R., 560
Couvillan, G.A., 107
Covell, C.V., 180
Covert, S.V., 327
Covey, R.P., 503
Covich, A.P., 263
Covindjee, -., 137
Covington, H., 86
Cowan, A.M., 202
Cowan, G.I.McT., 541
Cowan, I.McT., 535
Cowan, J.S., 549
Cowan, J.T., 132
Cowan, M., 540
Cowan, M.E., 356
Cowan, R.L., 401
Cowan, W.A., 77
Cowan, W., 263
Coward, J.E., 301
Coward, S.J., 106
Cowden, R.R., 327
Cowdrey, C.R., 356
Cowell, B.C., 97
Cowen, B., 305
Cowen, R., 47
Cowen, R.D., 366
Cowen, W.F.Jr., 367
Cowgiel, J.M., 123
Cowgill, Rev,J., 497
Cowgill, R.W., 347
Cowgill, U.M., 412
Cowless, J., 463
Cowley, G.T., 424
Cowling, E.B., 345, 346
Cowlishaw, J.D., 230
Cox, A.C., 380
Cox, B., 374
Cox, D.J., 166
Cox, D., 313
Cox, C.D.Jr., 161, 218
Cox, D.C., 379
Cox, D.M., 537
Cox, E.C., 287
Cox, E.R., 450, 460
Cox, F.S., 7
Cox, G.S., 261
Cox, H.W., 228
Cox, J.L., 215
Cox, J.R., 263
Cox, M., 378
Cox, P.G., 247
Cox, R., 21
Cox, R.H., 411
Cox, V.S., 262
Coye, R.D., 235
Coyier, D.L., 385
Coyle, F.A., 348
Coyle, M.A., 70
Coyle, M.B., 499
Coyle, S.M.A., 76
Coyne, D.P., 272
Coyne, M.D., 220
Cozine, W.S., 16
Cozzarelli, N.R., 132
Crabill, E.V., 327
Crabtree, G., 384
Cracraft, J.L., 134
Craddock, G.R., 423
Crandall, E.D., 411
Craft, J.H., 65
Craft, T.J., 357
Crafts, A.S., 51
Crafts, C.B., 419
Crafts, R.C., 370
Cragle, R.G., 493
Craig, B., 173
Craig, D.A.M., 530
Craig, E.S., 445
Craig, G.B.Jr., 152
Craig, H.B., 344
Craig, J.M., 32
Craig, J.P., 325
Craig, M.W., 437
Craig, N.C., 205
Craig, S.P., 429
Craig, W.S., 259
Craighead, J.J., 267
Craigie, D.E., 34
Craigie, J.S., 542
Crain, J.L., 189
Crain, S.M., 330
Craine, L.E., 231
Cralley, J.C., 122
Cramblett, H.G., 367
Cramer, C.F., 535
Cramer, D.A., 65
Cramer, E.B., 325
Cramer, O.M., 204
Cramer, W.A., 150
Crampton, B., 47
Crampton, J., 269
Crandall, D.I., 370
Crandall, G.D., 210
Crandall, J.C., 38, 317
Crandall, M., 177
Crane, F.L., 150
Crane, J.W., 501
Crane, P.L., 147
Crane, R.K., 284
Crang, R.E., 354
Cranmer, M.F., 24
Crans, W.J., 288
Crapper, D.R., 550
Craul, P.J., 322
Craven, G.R., 519
Crawford, B.W., 91
Crawford, C.S., 293
Crawford, D.J., 526
Crawford, D.L., 383, 485
Crawford, E.A., 485
Crawford, E.C., 177
Crawford, G., 452
Crawford, H.E., 493
Crawford, L.M., 168
Crawford, R.B., 76
Crawford, R.D., 157, 561
Crawford, R.F., 26
Crawford, R.H., 543
Crawford, R.P., 460
Crawford, S.E., 469
Crawford, W., 63
Crawley, H.D., 20
Crayton, M.A., 238
Creamer, H.R., 388
Crean, D.E., 365
Creasy, Y.L., 305
Creasy, W.D., 506
Creech, R.G., 248
Creek, R.O., 269
Creel, G.C., 448
Creger, C.R., 458
Crenshaw, J.W.Jr., 102
Creps, I., 36
Crescitelli, F., 55
Creson, D.L., 468
Cress, D.C., 226, 268
Cresswell, P., 333
Creteau, Sr.R., 279
Creutz, C., 371
Crewe, A.V., 132
Cribben, L.D., 286
Cribbins, J., 310
Cribbs, W.H., 109
Crichlow, E.C., 563
Criddle, R.S., 46
Crigan, A.T., 66
Criley, B.B., 122
Criley, R.A., 114
Crill, W.E., 500
Crim, L., 541
Crim, S.H., 285
Crisan, E.V., 47, 49
Crisley, F.D., 214
Crisp, T., 85
Crispens, C.G., 10
Crissman, J.K.Jr., 153
Crissman, R., 126
Criswell, B.S., 449
Critchlow, V., 388

Crites, J.L., 364
Critz, J.B., 152
Crocker, D.W., 316
Crockett, C., 191
Crockett, C., 369
Crockett, L.J., 296, 297
Croft, B.A., 226
Crofutt, L.E., 120
Croley, T.E., 185
Crolla, L.J., 124
Cromartie, W.J., 339
Crombach, G.T., 315
Cromer, J.H., 427
Cromroy, H., 91
Crompton, A.W., 210
Cromwell, G.L., 177
Cromwell, H., 247
Cronan, M.R., 80
Crone, L., 524
Cronholm, L.S., 100
Cronin, E.H., 477
Cronin, E.R., 476
Cronin, R.F.P., 555
Cronk, J., 524
Cronkite, D., 61
Cronquist, A., 297, 298
Cronshaw, J., 61
Crook, J.R., 382
Crook, P.G., 299
Cropp, F.W., 358
Crosby, E.C., 11
Crosby, E.S., 423
Crosby, M., 127
Crosby, R., 426
Croshaw, L.M., 427
Cross, C., 210
Cross, D.L., 424, 425
Cross, E.A., 10
Cross, F.B., 172
Cross, H.Z., 350
Cross, R., 64
Cross, R.F., 367
Cross, S., 387
Cross, W.H., 249
Crossley, D.A.Jr., 106
Crothers, M.R., 80
Crotty, W.J., 312
Crouch, N.A., 161
Crouse, G.S., 408
Croushore, J.H., 486
Crovello, T.J., 152
Crow, A.B., 183
Crow, J.F., 516, 518, 519
Crow, J.H., 288
Crowder, A.A., 546
Crowder, H.M., 443
Crowder, L.A., 17
Crowder, L.V., 304
Crowe, A., 545
Crowe, D., 540
Crowe, D.B., 514
Crowe, J., 514
Crowe, J.H., 52
Crowe, M.W., 178
Crowe, R.S., 209
Crowell, H.H., 380
Crowell, K., 315
Crowell, P.S., 145
Crowell, R., 316
Crowl, R.H., 337
Crowle, A.J., 72
Crowle, P., 418
Crowley, J.W., 517
Crowley, L.G., 520
Crowley, R.H., 5
Crowther, C.R., 229
Croy, L.I., 376
Crozier, G.F., 10
Crozier, R.H., 106
Cruce, W., 86
Crum, H.A., 230
Crum, J.D., 28
Crumley, R., 254
Crump, E., 511
Crump, M.H., 158
Crumpacker, D.W., 71
Cruse, J.M., 252
Cruz, A., 71
de la Cruz, A.A., 248
Cruz, A.B.Jr., 469
Cruzan, J., 395
Cryer, G.L., 197
Csima, A., 549
Cuany, R.L., 65
Cudia, S.J., 323
Cudkowica, G., 319
Cuellar, O., 473
Cuevas-Ruiz, J., 417
Cuitjens, J.C., 276
Culberson, C.F., 322
Culberson, J.L., 509
Culberson, W.L., 332
Culbert, J.R., 138
Culbertson, W.R., 65
Culivan, P.C., 8
Culley, D.D., 183
Cullen, M.G., 415
Cullimore, D.R., 561
Cullison, A.E., 106
Cully, M.E., 544
Culp, F.B., 428
Culp, L.A., 356
Culp, W.J., 277
Culpepper, J.G., 25
Culver, A.J.Jr., 385
Culver, D.A., 546
Culver, D.C., 126
Culwell, D.E., 21
Cumbie, B.G., 260
Cummings, B.C., 504
Cummings, D.J., 72
Cummings, J., 359
Cummings, J.A., 524
Cummings, J.C., 415
Cummings, J.F., 304
Cummings, K.R., 249
Cummings, W.B., 491
Cummins, D.G., 106
Cummins, G., 18
Cummins, J.N., 307
Cummins, K.W., 225, 226, 227
Cundall, D., 491
Cundy, K., 408
Cunio, J., 95
Cunningham, B.A., 166
Cunningham, C.C., 347
Cunningham, C.H., 228
Cunningham, D.A., 552
Cunningham, D.D., 55
Cunningham, E.B., 428
Cunningham, G., 292
Cunningham, G.R., 521
Cunningham, H.B., 5
Cunningham, J., 76
Cunningham, J.D., 278, 547
Cunningham, J.G., 228
Cunningham, J.P., 3
Cunningham, J.R., 550
Cunningham, L.S., 512
Cunningham, L.W., 446
Cunningham, R., 509
Cunningham, R.K., 319
Cunningham, R.L., 401
Cunningham, W.P., 242
Cunningham-Rundles, M., 312
Cupp, E.W., 303
Cupps, P.T., 48
Curd, M.R., 376
Curet, J.S., 418
Curl, E.A., 4
Curl, H.Jr., 385
Curl, S.E., 462
Curles, W., 7
Curley, D.M., 310
Curley, J.W., 486
Curran, D.D., 386
Currie, C., 70
Currier, H.B., 51
Currier, R.A., 384
Currin, R.E., 423
Curry, A., 391, 401
Curry, D.L., 153
Curry, G.M., 542
Curry, J.R., 143
Curry, L.V.L., 223
Curry, W., 290
Curry, W.A., 200
Curthoys, N.P., 413
Curtin, C.B., 268
Curtin, D.H., 29
Curtin, T.J., 234
Curtin, T.W., 139
Curtis, C.R., 201
Curtis, G.L., 273
Curtis, H.L., 102
Curtis, J.C., 209
Curtis, J.S., 343
Curtis, O.F.Jr., 307
Curtis, P.C., 308
Curtis, R.L., 512
Curtis, R.W., 148
Curtis, S.E., 138
Curtis, S.K., 327
Curtis, W.E., 390
Curtiss, C.F., 157
Curtiss, J., 238
Curtz, W.B., 27
Cusanno, G.L., 403
Cushing, E.J., 241, 242
Cushing, J.E., 61
Cushman, L.K., 279
Cushman, S., 277
Custead, W.G., 111
Custer, E.W., 249
Custer, W., 144
Custodio-Gonzalez, A.A., 417
Cusumano, C.L., 94
Cutchins, E.C., 83
Cuthbert, N.L., 223
Cutkomp, L.K., 240
Cutler, E.B., 327
Culter, H., 203
Culter, Sr.M.L., 299
Cutler, S., 414
Cutroneo, K., 103
Cutshall, N., 365
Cutter, G., 469
Cutter, L.J., 342
Cvancara, A., 352
Cvancara, V.A., 514
Cynkin, M.A., 217
Cypess, R., 414
Cyrus, R., 523
Czajkowski, J., 78, 79
Czerlinski, G.H., 127
Czerwinski, A., 269
Czurles, Sr.C., 174

D

Dabbs, D.H., 561
Dabich, D., 235
Dabney, J.M., 228
Daborn, G.R., 542
Dachtler, S.L., 403
D'Adamo, A.F., 299
DaDavern, C.I., 61
Dagg, A.I., 551
Dagg, C.P., 10
Daggett, B., 467
Daggy, T., 332
Dagley, S., 241
D'Agrosa, L.S., 256
Dahl, A.O., 409
Dahl, B.E., 463
Dahl, E., 467
Dahl, N.A., 171
Dahlberg, J.E., 520
Dahlgren, R.B., 157
Dahlgren, W.J., 538
Dahling, G.V., 298
Dahlsten, D.L., 44
Dahm, P.A., 157
Dahmen, J., 117
Dahms, T.E., 256
Dahmus, M.E., 48
Dahneke, B.E., 329
Dahtstedt, W.A., 21
Dai, T.E., 546
Daiber, F.C., 82
Daigger, L.A., 271
Daigle, J.R., 184
Dail, W.H., 293
Dainello, F.J., 168
Daines, R.H., 288
Daitch, P.B., 314
Dakin, M.E., 191
Dakshinamurti, K., 538
Dalby, A., 148
Dalby, P.L., 492
Dale, E.E., 22
Dale, H.E., 262
Dale, H.M., 546
Dale, J.L., 24
Dale, R.F., 147
D'Alecy, L.G., 233
Dales, S., 313
Daley, C., 434
Dalgarn, M.C., 360
Dalgleish, A.T., 35
Dalke, P.D., 118
Dallam, R.D., 180
Dalldorf, F.G., 346
Dalley, A.F., 268
Dallimore, C., 118
Dallman, M.F., 60
Dalquest, W.W., 452
Dalrymple, G.V., 24
Dalton, H.C., 403
Dalton, P.D., 111
Dalvi, K., 9
Daly, H.V., 44
Daly, J., 24
Daly, J.M., 270
Daly, K., 31
Daly, M., 330
Daly, R., 3
Dalzell, Sr.R.C., 175
Dam, R., 270, 272
Damann, K., 317
Damatmat, M.M., 4
D'Ambruso, M., 315
Dameron, T.B., 346
Damian, R.T., 106
Damjanov, I., 79
Damman, A.W.H., 77, 79
Damman, F.L., 489
Dammon, R.Jr., 219
Damron, B.L., 93
Dana, J., 119
Dandit, H., 308
Dandowne, D., 96
Dandy, J.W.T., 138
Dane, B., 216
Daneo-Moore, L., 408
Danforth, W.F., 121
Dangle, R., 109
D'Anglejan, B.F., 556
Daniel, C.P., 102
Daniel, C.W., 61
Daniel, D.L., 390
Daniel, J., 126
Daniel, J.C.Jr., 442
Daniel, P.M., 360
Daniel, R.E., 175
Daniel, R.S., 29
Daniel, T.W., 475, 477
Daniel, W.H., 147
Daniel, W.L., 137
Daniell, D.L., 143
Danielli, J.F., 319
Daniels, A.C., 133
Daniels, E., 555
Daniels, G.L., 286
Daniels, L.B., 23
Daniels, J.D., 109
Daniels, P.E., 474
Daniels, R.S., 370
Daniels, W.H., 78, 79
Danielson, R., 350
Danielson, R.E., 65
Danks, H.V., 544
Danley, J., 374
Dantgler, H.T., 199
Dantzler, W.H., 19
Dao, T., 320
D'Aoust, B.G., 498
D'Aoust, M., 560
D'Aoust, R., 558
Dappen, G., 270
D'Appolonia, B.L., 351
Dapson, R.W., 234
Darby, R.E., 31
Darby, W.J., 446
Dardas, G.F., 228
Dardas, T.J., 228
Darden, W.H.Jr., 10
Darding, R.L., 120

Darley, E.F., 58
Darley, W.M., 105
Darling, H.M., 519
Darlington, J.T., 439
Darlington, R.W., 178
Darlington, W.W., 122
Darnell, J.E.Jr., 300
Darnell, R.M., 459
Darner, J., 342
Daron, H.H., 3
Darrah, H.H., 395
Darrah, W.C., 395
Darrow, T.D., 382
Darst, P.H., 251
Das, G.D., 150
Das, N.K., 179
Dasch, C.E., 361
Dashek, W.V., 491
Daspit, R.A., 188
Datta, P.K., 232
Datta, S.P., 522
Daubenspeck, J.A., 278
Daubs, E., 30
Daugherty, P., 334
Daugherty, R.M.Jr., 228
Daugherty, W.F., 39
Daughtrey, Z.W., 189
Daun, H., 299
Dauterman, W.C., 344
Dave, C., 321
Davenport, C.A., 30
Davenport, D., 61
Davenport, D.B., 343
Davenport, D.L., 175
Davenport, F.M., 234
Davenport, G.R., 446
Davenport, H.A., 126
Davenport, H.W., 232
Davenport, L.B.Jr., 99
Davenport, R., 137
Davey, A.B., 346
Davey, C.B., 346
Davey, K.G., 553
Davich, T.B., 249
David, B., 518
David, F.N., 57, 59
David, G.B., 73
David, J., 555
David, J.D., 260
Davidian, N.M., 340
Davidonis, G.H., 220
Davidson, B.M., 199
Davidson, D., 524, 545
Davidson, E., 26
Davidson, E.A., 405
Davidson, E.H., 25
Davidson, G.A., 265
Davidson, H., 225
Davidson, J., 270
Davidson, J.A., 201
Davidson, J.F., 407
Davidson, J.M., 376
Davidson, R.A., 83
Davidson, R.S., 364
Davidson, S.E., 253
Davidson, W.D., 383
Davie, E.W., 499
Davie, J.M., 262
Davies, D.M., 545
Davies, E., 270
Davies, E.M., 552
Davies, H., 410
Davies, J., 446
Davies, J.E., 517
Davies, P.J., 302
Davies, R., 408
Davies, R.E., 411
Davies, R.O., 411
Davies, R.V., 391
Davies, R.W., 531
Davies, T.C., 428
Davies, W., 150
Davis, A., 423
Davis, A.C., 306
Davis, A.H., 279
Davis, A.M., 502
Davis, B.D., 288
Davis, B.H., 184
Davis, B.J., 187
Davis, B.K., 492
Davis, C., 278
Davis, C.A., 186, 292
Davis, C.B., 157
Davis, C.C., 541
Davis, C.E., 189, 341
Davis, C.H., 39
Davis, D.D., 403
Davis, D.E., 4, 346
Davis, D.G., 10
Davis, D.L., 97, 177
Davis, D.M., 110
Davis, D.W., 241, 476
Davis, E., 463
Davis, E.A., 219
Davis, E.J., 146
Davis, E.L., 217
Davis, F.C., 90
Davis, F.F., 288
Davis, F.K., 393
Davis, F.M., 249
Davis, F.W.J., 538, 539
Davis, G., 458
Davis, G.D., 186, 497
Davis, G.E., 383
Davis, G.J., 334
Davis, H.E., 487
Davis, H.W., 388
Davis, J., 515
Davis, J.D., 247, 486
Davis, J.R., 19, 117, 124, 449
Davis, J.S., 90
Davis, K.B., 438
Davis, L.D., 521
Davis, L.E., 69, 326
Davis, L.S., 475
Davis, L.V., 438
Davis, M., 437
Davis, M.B., 80, 181
Davis, M.M., 166
Davis, M.P., 250
Davis, N., 130
Davis, N.C., 370
Davis, N.D., 4
Davis, N.T., 77, 79
Davis, R., 41, 87, 95
Davis, R.B., 85, 194, 456
Davis, R.F., 199, 201, 288
Davis, R.H., 230, 460
Davis, R.L., 74
Davis, R.P., 18, 355
Davis, R.R., 364
Davis, R.W., 68
Davis, S., 117
Davis, S.C., 423
Davis, S.H., 288
Davis, T.C., 4,5
Davis, W., 256, 257, 387
Davis, W.C., 504
Davis, W.D., 4
Davis, W.H., 176
Davis, W.J., 61, 439, 490
Davis, W.K., 455
Davis, W.L., 449
Davis, W.S., 180
Davison, A.D., 503
Davison, J.A., 480
Davison, L., 479
Davy, J., 349
Daw, J.C., 340
Daw, N., 263
Dawe, A.R., 125
Dawes, C.J., 97
Dawid, I., 197
Dawson, C.A., 513
Dawson, D.C., 162
Dawson, G., 131
Dawson, G.A., 18
Dawson, J., 333
Dawson, J.E., 362
Dawson, J.F., 454
Dawson, M.D., 384
Dawson, P.S., 386
Dawson, R.F., 288
Dawson, W.D., 429
Dawson, W.O., 58
Dawson, W.R., 231
Dawson, W.W., 94
Day, A.D., 16
Day, A.W., 552
Day, B.E., 45
Day, B.N., 259
Day, E.D., 333
Day, J., 224
Day, P.R., 46
Day, R., 500
Day, R.A., 370
Day, V., 270
Dayan, J., 300
Dayhoff, M.O., 85
Dayton, B.R., 323
Dayton, D.F., 138
De, R.K., 338
Deadwyler, S.A., 55
Deal, R., 506
Deal, S.J., 509
Deal, W.C.Jr., 225
Deamer, D.W., 52
Dean, B., 177
Dean, D.M., 12
Dean, D.S., 354
Dean, H.L., 160
Dean, J.M., 429
Dean, L.L., 117
Dean, Sr.M.B., 33
Dean, M.P., 518
Dean, R.D., 171
Deane, D.D., 527
Deans, R.J., 225
Dearborn, D., 355, 432
Dearborn, J.H., 193
Deason, T.R., 10
Deatherage, F.E., 362
DeBach, P.H., 58
Debacker, H.S., 428
Debarther, J.V., 200
De Bault, L.E., 160
DeBenedictis, P.A., 326
DeBoer, G., 550
DeBoer, K.F., 73
De Boer, R.H., 288
DeBor, I.W., 557
Debro, L., 250
DeBrunner, L.E., 5
DeBruyn, P.P.H., 131
DeBusk, A.G., 88
DeCaro, T.F., 416
DeCeretti, E., 559
DeChampain, M.L., 556
DeChatelet, L.R., 347
DeCicco, A., 312
DeCicco, B.T., 83
Deck, J.D., 489
Deck, J.E., 440
Deckard, E.L., 350
Decker, A.M., 200
Decker, C.F., 406
Decker, E., 66
Decker, J.M., 369
Decker, L.E., 406
Decker, P., 92
Decker, R.D., 488
Decker, R.S., 466
Deckert, M., 164
Decosta, J.J., 507
Decoursey, P.J., 429
Decter, P.H., 538
Decter, S.A., 288
Dedrick, M.C., 25
Deeds, D., 455
Deener, D.R., 190
Deep, I.W., 366
Deering, R., 393
Deering, R.A., 404
Deevey, E.S., 90
Defazio, S.R., 87
DeFelice, L.J., 101
DeFeo, V.J., 112
DeFigio, D., 395
Defoliart, G.R., 518
Defonti, N.L., 157
Deforest, A., 408
DeForw, D., 282
DeFrance, J., 235
DeGasperi, R.N., 96
Degenhardt, W.G., 293
DeGennaro, L.D., 310
DeGiusti, D.L., 234
DeGowin, R.L., 162
DeHaan, R.L., 101
Dehaase, H., 193
Dehertogh, A., 225
Dehnel, P.A., 535
DeHoff, P.H., 391
Dehner, E.W., 164
Dehority, B.A., 365
Deibel, R., 327
Deininger, R.A., 233
DeJong, A.A., 27
DeJong, D.W., 343
DeJong, P., 230
Dekker, C.A., 42
Dekker, E.E., 232
deKloet, S., 88
Dela, F., 359
Delahayes, J.F., 104
Delahunty, T., 466
Delaney, R., 380
DeLange, R.J., 56
DeLanney, L.E., 309
Delbruck, M., 26
Delco, E.A.Jr., 450
Delcomyn, F., 136
DeLemos, C.L., 312
Delevoryas, T., 465
Delgado, G.L., 83
DeLacy, A.C., 498
Delandy, R.H., 527
Delane, T., 532
Delisle, A.L., 31, 209
DeLisle, D.G., 159
Delisle, G.J., 545
Delivoria-Papadopoulos, M., 411
Dellmann, H.D., 262
Dellow, P.G., 552
Delmer, D., 226
Delmer, D.P., 225
De Long, K., 155
De Long, R.A., 155
De Long, S.K., 155
Delouche, J.C., 298
Delphia, J.M., 367
Dels, R.E., 66
DeLuca, C., 319
DeLuca, D.L., 24
Deluca, H.F., 517
DeLuca, P., 311
Delvecchio, V., 415
Delwiche, C.C., 50
Deizell, D.E., 487
De Maggio, A.E., 277
Demain, A.L., 213
DeMan, J.M., 547
DeMarinis, F., 357
Demars, R.I., 519
DeMartini, J.D., 34, 69
DeMartinis, F.D., 399
Dembitzer, H.M., 313
De Medicis, E., 559
DeMeester, W.A., 416
Demerree, R.S., 29
DeMoss, R.D., 137
Demott, B.J., 144
DeMott, H.E., 401
DeMott, T., 415
Dempsey, E.W., 300
Dempsey, M.E., 244
Dempsey, W.B., 466
Dempsey, W.H., 29
Dempster, E.R., 45, 46
Dempster, G., 562
Demski, L.S., 293
Den, H., 405
Denbo, J.R., 155
Dendinger, S.M., 236
Dendy, J.S., 4, 5
Deneke, F.J., 168
Denell, R.E., 166
De Neve, R., 324
Denford, K.E., 531
Deng, J., 92
Denhardt, D.T., 555
Denisen, E.L., 156
Denison, W.C., 385
Denman, C.E., 376
Denna, D.W., 66
Denner, A.G., 144
Denner, M., 144
Dennery, E., 524

Dennery, L.L., 258
Dennis, A., 244
Dennis, C.J., 524
Dennis, D.S., 475
Dennis, D.T., 546
Dennis, E.A., 288
Dennis, F.G., 225
Dennis, J.M., 203
Dennis, M.J., 60
Dennis, R.E., 16
Dennis, R.L., 385
Dennis, W.H., 521
Dennison, C., 439
Dennison, D.M., 142
Dennison, D.S., 277
Dennison, R.A., 91
Dennison, C., 519
Denniston, R.H., 526
Denny, C.F., 488
Denomme, Sr.Y., 224
Denoncourt, R.F., 416
Dent, J.M., 488
Dent, T., 210
Denton, J.F., 103
Denton, M.F., 497
Denton, T.E., 8
Denton, W.E., 91
DeNuccio, D.J., 74
Denyes, H.A., 76
Deodhar, S.S., 357
De Pew, L.J., 168
DePinto, J.A., 119
DePoe, C.E., 185
Deppe, C.S., 242
De Prospo, N.D., 290
Deputy, J.C., 111
Deranleau, D.A., 499
Derby, A., 262
Derbyshire, J.B., 548
Derechin, M., 318
Deren, J.J., 411
Derickson, E.H., 364
Derifield, K., 15
DerMarderosian, A., 405
Dermid, J.F., 347
Derrick, F.R., 331
Derrickson, C.M., 175
De Roos, R.M., 260
Derow, M.A., 208
Derr, W.F., 29
Derrenbacher, W.E., 47
Dersham, G.H., 218
DeSaix, L.S., 339
DeSantis, M., 85
Desboroligh, S.L., 244
DeSha, C., 317
Deshmukh, D.S., 408
Desiderio, D.M., 448
DeSiervo, A.J., 134
DeSimone, J.A., 492
Desjardins, C , 465
Desjardins, P.R., 58, 538
Desmond, A.H., 84
Desnoyers, L., 560
Desrochers, R., 559
Dessauer, H.C., 185
Desser, S.S., 550
DeStefano, J., 396
DeSy, R.J., 105
DeSylva, D., 96
Detar, R.L., 278
Deter, R.L., 449
Deters, M.E., 118
Dethier, V.G., 287
Dethlefsen, L.A., 473
Detling, J.K., 371
De Toma, F.J., 214
Detter, D.E., 260
DeTurck, J.E., 391
Detweiler, D.K., 411
Detwyler, R., 479
Deubler, E.E.Jr., 309
Deufel, R.D., 182
Deur, R.C., 236
Deuschle, F.M., 370
Deutsch, H.F., 520
Deutscher, G.H., 432
Deutschman, A.J., 16
Dev, V., 301
DeVail, W.B., 5
DeVault, D.C., 410
De Vay, J.E., 51
Devenport, E.C., 111
Dever, J.E., 70
DeVillez, E.J., 360
Devine, M., 306
Devine, Rev.R.J., 315
De Viney, G., 450
DeVires, G.H., 491
Devos, F., 138
DeVoy, J., 282
DevVartanian, D.V., 105
DeWalle, D.R., 402
DeWeer, P.J., 262
DeWein, L.F., 354
Dewel, R.A., 351
Dewel, W.C., 331
DeWet, J.M.J., 137
Dewey, D.D., 31
Dewey, D.H., 225
Dewey, J.E., 303
Dewey, W.C., 69
Dewhirst, L.W., 18
DeWitt, J., 33
Dewitt, J.C., 468
De Witt, J.R., 157
DeWitt, R.M., 126
DeWitt, W., 220
DeWolfe, B.B., 61
Dexter, A.G., 350
Dexter, D.M., 39
Dexter, R., 359
Deyal, H.H., 198
De Young, D.W., 158
Déziel, J., 558
Dezoeten, G.A., 519
Dhaliwal, A.S., 123, 124
Diab, I., 133
Diachun, S., 175
Diakow, C., 295
Dial, S.C., 337
Diamant, N.E., 550
Diamond, I., 333
Diamond, J.M., 57
Diamond, M., 112
Diamond, M.C., 43
Diana, J.N., 162
Diani, A.R., 236
Dias, J.K., 341
DiBattista, W.J., 400
Dibble, M.V., 326
Di Benedetto, F.E., 290
DiBerardino, M.A., 399
Dicerson, O.J., 168
Dichmann, M.E., 191
Dick, F.R.II., 161
Dick, J., 164
Dick, J.W., 425
Dick, N.P., 72
Dick, R., 378
Dick, R.D., 514
Dick T.A., 538
Dickason, E.A., 271
Dicke, R.J., 518
Dickens, J., 84
Dickens, L.E., 68
Dickens, R., 3
Dickenson, H., 436
Dickerman, L., 355
Dickerman, R.C., 357
Dickerman, R.W., 308
Dickerson, E.L., 452
Dickerson, L., 378
Dickey, M.A., 441
Dickey, R.S., 304, 403
Dickie, L.M., 542
Dickinson, A.B., 215
Dickinson, D.B., 136, 138
Dickinson, F.E., 44
Dickinson, J.C.Jr., 90
Dickinson, T.E., 47, 49
Dickinson, W.J., 473
Dickison, W.C., 338
Dickman, M., 544
Dickman, S.R., 474
Dickmann, D., 524
Dick-Peddie, W.A., 292
Dickson, A., 303
Dickson, D.H., 542
Dickson, D.P., 517
Dickson, D.R., 413
Dickson, D.W., 91
Dickson, K.L., 492
Dickson, L.G., 500
Dickson, M.H., 307
Dickson, R.C., 58
Dickson, S.S., 187
Dickson, W.J., 187
Dickson, W.M., 413
Didon, L.A., 539
Diebolt, J.R., 337
Diecke, F.P.J., 161
Dieckert, J.W., 458
Diefer, H.C., 127
Diegelmann, R.F., 491
Diehl, F.A., 488
Diehl, J., 398
Diehl, J.R., 138
Diehl, S.G., 257
Diehl, W.P., 39
Diehr, P., 500
Diem, K.L., 526
Diener, U.L., 4
Dienhart, C.M., 101
Diercks, F.H., 336
Dierks, R.D., 158, 266
Dierolf, B., 76
Dierolf, S., 392
Dietert, S.E., 293
Dietrich, L.S., 95
Dietrich, R., 14
Dietrich, R.V., 223
Dietrich, W.E.Jr., 396
Dietz, A., 106
Dietz, A.A.C., 124
Dietz, C., 408
Dietz, G.W.Jr., 307
Dietz, R., 8
Dietz, S.M., 503
Dietz, T.H., 183
Diferrante, N.M., 448
Difford, W.C., 522
Digby, P.S.B., 556
Diggens, Sr.M., 431
DiGiovanna, A.G., 199
DiGirolamo, R., 33
Dilcher, D.L., 144
Dilcher, R., 317
Diley, D., 29
Dilger, W.C., 302, 304
Dilks, E., 122
Dill, C.W., 458
Dill, L.M., 533, 553
Dill, P.A., 543
Dill, R.E., 449
Dillard, E.U., 343
Dillard, G.E., 180, 181
Dillard, W.L., 374
Dille, J.E., 430
Dillery, D.G., 222
Dilley, D.R., 225
Dilley, M.T., 293
Dilley, R.A., 150
Dilley, W.G., 300
Dillingham, E.O., 445
Dillon, C.R., 425
Dillon, J.A., 180
Dillon, L.S., 460
Dillon, R.D., 434
Dillon, W., 444
DiLorenzo, R.N., 479
Dilwort, T.G., 540
Dilworth, J.R.Jr., 385
Di Marco, G.R., 288
Di Meo, F., 394
Dimick, P.S., 401, 402
Dimit, J.E., 254
Dimitman, J.E., 29
Dimmick, J.F., 347
Dimmick, R.W., 443
Dimock, R.V., 347
Dimond, J.B., 194
Dimond, M.T., 86
Din, G.A., 531
Dina, S., 256
Dindal, D.L., 322
Dineen, C.F., 151
Dingle, A.D., 545
Dingle, H., 166
Dingle, R.W., 503
Dingman, R.E., 62
Dinkins, R.L., 165
Dinneen, Rev.J.A., 310
Dinsmore, B.H., 392
Dinsmore, C.E., 210
Dinsmore, J., 97
Dintzis, H., 198
Dintzis, H.M., 197
Dintzis, R., 198
Dintzie, R.Z., 197
Dinusson, W.E., 350
Dinwiddie, J.G.Jr., 100
Dionne, L.A., 540
Dionne, V., 500
Dippell, R.V., 145
Dirks, J.H., 556
Dirmhirn, I., 474, 477
Dirr, M.C., 138
Di Sabato, G., 446
DiSalvo, J., 370
Disinger, J.F., 366, 368
Disrud, D.T., 349
DiStefano, V.F., 328
Disterhoft, J., 126
Distler, D., 173
D'Itri, F.M., 225
Dittmer, H.J., 293
Dittmer, K., 387
Dittmer, J.E., 208
Divelbiss, J., 163
Dively, G.P., 201
Dively, J.L., 395
Diven, W.F., 413
Divers, F.C., 494
Diwan, J.J., 314
Dix, R.L., 68
Dix, V., 335
Dixit, P., 243
Dixon, C.F., 5
Dixon, C.L., 222
Dixon, E., 9
Dixon, H.N., 65
Dixon, J.A., 473
Dixon, J.B., 459
Dixon, J.C., 514
Dixon, J.E., 148
Dixon, J.R., 459
Dixon, K.L., 476, 477
Dixon, P.S., 54
Dixon, S.E., 547
Dixon, W.J., 56
Dixon, W.L., 175
D'Lorio, A., 548
Dlosterman, E.W., 365
Dobbins, D.A., 189
Dobbins, D.R., 220
Dobbs, H.D., 430
Doberczack, N., 479
Dobkin, S., 87
Dobos, P., 547
Dobrin, P.B., 125
Dobrogosz, W.J., 345
Dobbson, A., 306
Dobson, S.H., 344
Dobson, W.J., 460
Docherty, J.J., 404
Dochinger, L.S., 366
Docter, P.J., 120
Dodd, B.E., 37
Dodd, D.C., 411, 412
Dodd, J.D., 157
Dodd, M.C., 363
Dodds, D.G., 542
Dodge, W.F., 468
Dodgen, C.L., 251
Dodrill, S., 506
Dodson, E.O., 549
Dodson, R.C., 148
Dodson, S.I., 516
Doeg, K.A., 79
Doell, R.G., 40
Doemel, W.N., 153
Doepke, P.A., 229
Doerder, F.P., 412
Doerr, P.D., 346
Doersch, R.E., 516
Doetsch, R.N., 202
Doezema, C.P., 166
Doherty, J.M., 279
Doherty, R.A., 328, 329

Dohrenwend, R.E., 90
Doi, R.H., 48
Doige, C.E., 563
Doisy, E.A.Jr., 256
Doisy, R.J., 326
Dokma, R.M., 155
Doksum, K.A., 46
Dolan, R., 489
Dole, J., 31
Dole, R.M.Jr., 80
Doll, J.C., 258
Dollahite, J.W., 460
Dollahon, J.C., 522
Dollinger, E.J., 364
Dollwet, H., 369
Dolores, Sr., 192
Dolowy, W.C., 133
Dolph, G., 145
Dolphin, W.D., 157
Dolyak, F., 419
Dom, R., 428'
Doman, E., 290
Domardski-Dobija, M., 558
Domer, J.K., 190
Domingue, G.J., 190
Domizi, D., 133
Domnas, A.J., 338
Domoto, P.A., 156
Donachy, J.A., 392
Donachy, N.R., 392
Donady, J.J., 80
Donaher, D.J., 234
Donahue, E.H., 419
Donahue, Sr.F., 210
Donahue, W.H., 397
Donaldson, D.D., 27
Donaldson, D.J., 445
Donaldson, E., 502
Donaldson, -.Jr., 383
Donaldson, L.R., 498
Donaldson, S.K., 500
Donaldson, W.E., 346
Donati, E.J., 203
Donchin, E., 137
Donelson, D.H., 364
Donelson, J.E., 161
Donelson, J.R., 364
Doner, H.E., 46
Doney, R.C., 321
Donisch, E.W., 551
Donisch, V., 551
Donley, C.S., 428
Donnelley, W.F., 62
Donnelly, E.D., 3
Donnelly, J.R., 481
Donnelly, K.R., 535
Donnelly, R.D., 84
D'Onofrio, S., 298
Donoho, H.R., 113
Donohoo, J.T., 453
Donohue, E., 219
Donovan, A.B., 285
Donovan, G., 422
Donovan, C.A., 450
Donovan, Sr.M., 215
Donovan, O., 285
Doody, E.J., 436
Dooley, E.S., 440
Dooley, H., 385
Dooley, J.E., 348
Dooley, J.K., 295
Doolittle, W.F., 542
Doran, R.H., 20
Dorchester, J.E.C., 415
Dorer, F., 355
Dorfman, A., 130, 131, 132
Dorfman, D., 286
Dorg, K.A., 77
Dorgan, W.J., 265
Dorman, H.L., 450
Dorman, J., 159
Dorminey, R.W., 384
Dorn, C.R., 262
Dorn, J., 126
Dorner, R., 256
Dornfeld, E.J., 380
Dornfest, B.S., 325
Dornhoff, G.M., 271
Dorosz, L.G.Jr., 32
Dorough, H.W., 178
Dorrington, K.J., 549
Dorris, P.R., 21
Dorris, T.C., 375
Dorsch, J., 70
Dorsett, P.H., 445
Dorsey, O.L., 454
Doscher, M.S., 235
Doskotch, R.W., 362
Dossel, W.E., 268
Dostal, H.C., 149
Dotsen, R., 473
Doty, C.H., 93
Doty, M.S., 111
Doty, P., 210
Dotzenko, A.D., 65
Doubet, Sr.M.M., 415
Doubles, J., 336
Doubly, J.A., 351
Doucette, A.J., 183
Doudoroff, P., 383
Dougals, N.H., 188
Dougherty, E.III., 306
Dougherty, G.M., 245
Dougherty, J., 483
Dougherty, J.H., 473
Dougherty, M.A., 390
Dougherty, R.H., 149
Dougherty, R.M., 326
Dougherty, T.J., 150
Dougherty, W.J., 428
Douglas, A.G., 29, 248
Douglas, C.R., 93
Douglas, H.C., 499
Douglas, L.H., 279
Douglas, R.G., 328
Douglas, R.J., 547
Douglass, L.W., 201
Dounce, A.L., 328
Doupnik, B., 270
Douthit, H.A., 231
Dove, G.R., 165
Dove, G.N., 437
Dove, L., 549
Dove, L.D., 139
Dovre, P., 238
Dowben, R.M., 466
Dowd, F., 310
Dowd, M.D., 323
Dowdy, R., 67
Dowdy, R.H., 241
Dowe, T.W., 480
Dowell, C.E., 218
Dowell, V.E., 162
Dowlen, H.H., 443
Dowling, H.G., 312, 421
Dowling, J.E., 210, 211
Downe, A.E.R., 546
Downer, R.G.H., 551
Downes, J.D., 462
Downey, J.C., 162
Downey, J.M., 97
Downey, K., 95
Downey, R.J., 368
Downhower, J.F., 364
Downie, H.G., 548
Downing, D.L., 307
Downing, W.L., 238
Downs, F., 298
Downs, F.L., 358
Downs, R.J., 343
Downs, T.D., 469
Dowsett, J.A., 539
Doxtader, K.G., 65
Doyen, J.T., 44
Doyle, D., 319
Doyle, G.G., 258
Doyle, J.A., 231
Doyle, L.F., 97
Doyle, P.J., 439
Doyle, R.E., 262
Doyle, R.J., 180
Doyle, R.W., 542
Doyle, W.L., 131
Doyle, W.T., 61
Dozier, W.A., 5
Drabek, C., 374
Draft, K.J., 228
Drahmann, J.F., 62
Drake, C.H., 501
Drake, E.J., 395
Drake, E.L., 275, 554
Drake, M., 218
Drake, W.D., 231
Dramm, K.R., 228
Drapalik, D.J., 102
Drapeau, G., 558
Draper, C.I., 475
Draper, D.D., 158
Draper, L.R., 171
Drather, E.E., 4
Drawe, Sr.M.V., 191
Drayer, D., 409
Drecktrah, H.G., 523
Dregne, H.E., 462
Dreier, A.F., 271
Dreiling, C.E., 276
Dreiling, F.R., 494
Dreizen, P., 325
Dresden, C.F., 407
Dresden, M.H., 448
Dress, P.E., 108
Dressler, D., 210
Drew, L., 265
Drew, W.A., 377
Drew, W.H., 280
Drexler, H., 347
Drexler, R.V., 154
Dreyer, D., 333
Dreyer, W.A., 370
Dreyer, W.J., 26
Drickamer, L.C., 220
Drinkwater, W.O., 288
Driscoll, A.D., 220
Driver, P.J., 87
Drogmann, D.W., 148
Drolsom, P.N., 516
Dronzek, B.L., 537
Dropkin, V.H., 260
Dropp, J.J., 416
Drosdoff, M., 302
Drott, H.R., 410
Druce, R., 186
Drucker, W.R., 489
Drudge, J.H., 178
Druecker, J.D., 268
Druehl, L.D., 533
Druger, M., 326
Drumm, P., 397
Drummond, J., 266
Drury, D.A., 543
Drury, W.H., 211
Druse, A.C., 391
Dryden, G.L., 407
Dryer, R.L., 161
Drysdale, G.R., 263
Drysdale, J.W., 217
Dubbs, D.R., 449
Dube, M.A., 504
Dube, P.A., 558
Duberman, D., 310
Dubes, G.R., 273
Dubin, D.T., 283
Dubin, I.N., 399
Dubin, H.J., 194
Dubin, L.F., 30
Dubin, M.W., 71
Dubin, S., 394
Dubin, S.E., 393
Duble, R. L., 459
Dubnau, D.A., 313
DuBois, A.B., 410
DuBose, L.E., 448
Dubowski, K., 380
Du Brul, E.L., 134
Duby, R., 219
DuCharme, D.W., 236
Duckworth, D.H., 94
Duckworth, J.W.A., 549
Ducoff, H.S., 137
Duda, C.T., 234
Duda, E.J., 78, 79
Dudderar, G., 225
Duddy, P.J., 397
Dudeck, A., 92
Dudgeon, E., 257
Dudley, H.C., 135
Dudley, J.W., 137
Dudley, P.L., 300
Dudley, R.G., 108
Duell, E.A., 232
Duell, P., 389
Duell, R.W., 288
Duellmann, W.E., 172
Duerksen, J.D., 531
Duerr, F., 352
Duerrke, J.A., 353
Duesberg, P.H., 43
Dueser, R., 489
Dufe, D., 301
Dufeur, B.C., 279
Duff, D., 511
Duff, D.T., 422
Duff, F.L., 468
Duff, W., 79
Duffey, L.M., 435
Duffield, J.W., 346
Duffin, J., 550
Duffy, C.E., 24
Duffy, E., 311
Duffy, J.R., 554
Dugan, K.H., 380
Dugan, P.R., 363
Dugas, J.C., 461
Dugdale, C.B., 284
Dugdale, R.D., 498
Dugger, M.Jr., 57
Dugger, W.M., 57
Duggleby, W.F., 370
Dugle, J., 538
Duguid, W.P., 555
Duich, J.M., 401
Duitschaever, C.L., 547
Duke, E.L., 468
Duke, K.L., 332
Duke, R.D., 231
Duke, W.B., 302
Dukelow, W.R., 225, 228
Dukepoo, F., 39
Dulin, W.E., 236
Duling, B.R., 490
Duman, M.G., 406
Dumas, J.P., 560
Dumas, P.C., 496
Dumbroff, E.B., 551
Dumont, A., 559
Dumont, K., 298
Dumont, K.P., 297
Dunagan, T.T., 129
Dunbar, M.J., 556
Dunca, D.B., 198
Duncan, A.A., 241
Duncan, B., 165
Duncan, C.L., 517
Duncan, D., 467
Duncan, D.P., 261
Duncan, G.A., 179
Duncan, H.E., 345
Duncan, J.F., 138
Duncan, J.L., 127
Duncan, J.R., 175
Duncan, J.T., 40
Duncan, R.C., 428
Duncan, R.I., 552
Duncan, S., 207
Duncan, W.G., 177
Duncan, W.H., 105
Dundee, D.S., 185
Dundee, H.A., 190
Dunford, M.P., 292
Dung, H.C., 469
Dungworth, D.L., 53
Dunham, C.W., 82
Dunham, D.W., 550
Dunham, J., 157
Dunham, P.B., 326
Dunham, V., 323
Dunker, K., 316
Dunkle, L., 270
Dunkley, W.L., 49
Dunlap, D.G., 434
Dunlay, D., 431
Dunleavy, J.M., 157
Dunleavy, Sr.R., 151
Dunlop, B.L., 387
Dunlop, D.W., 521
Dunlop, S.G., 72
Dunn, A., 62
Dunn, A.J., 94
Dunn, B.A., 425

Dunn, C.M., 518
Dunn, D., 399, 436
Dunn, D.B., 260
Dunn, E.L., 105
Dunn, F., 137, 448
Dunn, J.H., 260
Dunn, G.M., 281
Dunn, J.D., 185
Dunn, M.C., 439
Dunn, M.F., 57
Dunn, N.K., 28
Dunn, O.J., 36
Dunn, P., 30
Dunn, R.J., 225
Dunn, S.J., 337
Dunn, T.G., 527
Dunne, H.W., 403
Dunne, M.J., 159
Dunnigan, J., 559
Dunning, D.C., 507
Dunnington, J.D., 422
Dunson, W.A., 403
Dunstan, T.C., 139
Duplantier, J., 182
Dupont, J., 67
DuPont, P., 121
DuPorte, E.M., 556
Dupree, D.E., 188
Du Puis, G., 559
Dupuy, J.L., 490
Duquella, J., 317
Duqesnoy, R.J., 513
Duran, R., 503
Duran, W., 333
Durand, D.P., 156
Durand, J.B., 287
DuRant, J.A., 424
Durbin, R.D., 519
Durbin, R.P., 60
Dure, L.S., 105
Durfee, W.K., 422
Durflinger, E., 143
Durgin, M., 431
Durgin, O.B., 280
Durham, A.M., 367
Durham, L., 120
Durham, N., 375
Durham, N.N., 375
Durham, R.M., 440, 462
Durham, W.F., 333
Durian, B., 323
Durig, J.R., 429
Durio, W.O., 191
Durkee, L.H., 155
Durkee, L.T., 155
Durkin, D.J., 288
Durnow, R.D., 364
Durrance, K.L., 91
Durrill, P.L., 487
Durso, D.F., 458
Durst, H., 164
Durvea, W., 406
Dus, K., 256
Dusenbery, D.B., 102
Dusi, J.L., 5
Dusing, A., 255
Dusseau, E.M., 234
Dusseau, J., 224
Dustman, J.H., 145
Duszinski, D., 299
Duszynski, D.W., 293
Duthie, A.H., 480
Duthie, H.C., 551
Dutrow, A.N., 169
Dutson, T.R., 458
Dutt, R.H., 177
Dutta, C.R., 235
Dutton, P.L., 410
Dutton, R.E.Jr., 314
Dutton, R.W., 59
Duval, A., 557
Duwe, A.E., 224
Duxbury, J.M., 302
Duysen, M.E., 349
Dwelle, R.L., 303
Dwordin, M., 244
Dwornik, J.J., 97
Dwyer, D.D., 476, 477
Dwyer, J., 256
Dyar, J.J., 174
Dybing, C.D., 433
Dyck, L.A., 425
Dycus, A.M., 15
Dye, A.J., 458
Dye, A.P., 508
Dye, E.L., 454
Dye, F.J., 80
Dye, W.E., 273
Dyer, A.A., 67
Dyer, A.J., 259
Dyer, G., 284
Dyer, I.A., 502
Dyer, M.I., 66
Dyer, R.F., 185
Dyer, W.G., 129
Dyke, J.W., 228
Dykman, R.A., 24
Dykstra, J.N., 522
Dynes, J.R., 266
Dyson, P.J., 108
Dziewiatkowski, D., 232
Dzuik, P.J., 138
Dzukanowski, C.T., 522

E

Eads, J.H.Jr., 336
Eager, J.F., 221
Eagle, H., 329
Eagles, D., 85
Eagles, J., 12
Eakin, R.M., 43
Eaks, I.L., 57
Eales, J.G., 538
Eanes, R.Z., 491
Earhart, C.F., 465
Earle, A.M., 273
Earley, E.B., 137
Easley, J.F., 91
Easter, S.S., 231
Easterbrook, K.B., 543
Easterday, B.C., 520
Easterla, D.A., 255
Easterly, N.W., 354
Easterly, W., 440
Eastin, J.D., 271
Easton, D., 88
Easton, G.D., 503
Easton, T.W., 192
Eaton, D.C., 468
Eatman, E.T., 485
Eaton, G.W., 534
Eaton, H.D., 78, 79
Eaton, J.A., 105
Eaton, M.D., 42
Eaton, N., 298
Eaton, N.R., 296
Eaton, S.W., 315
Eaton, T., 172
Ebben, J.A., 176
Ebbesson, E.O., 489
Ebeling, W., 58
Ebeltoft, D.C., 350
Eberhardt, L.E., 498
Eberhart, B.M., 342
Eberhart, R.J., 403
Eberiel, D., 212
Eberle, H.I., 329
Eberlin, J., 220
Eberly, W., 147
Ebersold, W.T., 55
Ebersole, L.A., 176
Ebersole, M.C., 407
Ebert, J., 197
Ebert, J.B., 337
Ebert, R.H., 211
Ebert, T.E., 39
Ebert, W.W., 28
Ebesson, S.O.E., 418
Ebinger, J.E., 120
Ebisuzaki, K., 551
Eble, A., 290
Ebner, K.E., 172, 377
Ebrey, T.G., 137
Eccles, J.C., 319
Echelle, A.A., 379
Echols, H., 43
Echols, J.W., 65
Echt, R., 227
Echternacht, A.C., 207
Eck, P., 288
Eckblad, J.W., 159
Eckelman, C.A., 149
Eckenhoff, J.E., 126
Eckenrode, C.J., 306
Eckert, E.A., 234
Eckert, J.W., 58
Eckert, Sr.M., 299
Eckert, R., 55
Eckert, R.E., 275
Eckert, T.E., 317
Eckess, E.D., 402
Eckhardt, E.T., 283
Eckhardt, R.A., 296
Eckhoff, G.A., 157
Eckner, F.A.O., 135
Eckstein, G., 370
Eckstein, J.W., 160
Eckstein, R., 175
Edberg, S.C., 300
Eddington, L.C., 26
Edds, G.T., 94
Edds, M.J.Jr., 217
Eddy, D.K., 481
Eddy, E.M., 499
Eddy, T., 164
Edelhauser, H.F., 513
Edelhoch, H., 85
Edelman, I., 60
Edelman, N., 284
Edelstein, S.J., 302
Eden, W.E., 394
Eden, W.G., 91
Edens, D.L., 507
Edens, F.W., 346
Ederer, G.M., 244
Edgar, A.L., 222
Edgar, R.K., 215
Edgar, R.S., 61
Edgar, S.A., 5
Edge, M.C., 506
Edgell, M.H., 339
Edgerly, C., 350
Edgerton, L.J., 305
Edgington, L.V., 547
Edgren, C.H., 125
Edidin, M., 197
Ediger, R.I., 29
Edinger, H., 283
Edmonds, M., 413
Edmonds, W.D., 29
Edmondson, W.T., 498
Edmounds, P., 523
Edmunds, G.F., 473
Edmunds, L., 325
Edmunds, L.K., 168
Edmunds, N., 274
Edmunds, R.A., 327
Edney, E., 55
Edney, N.A., 246
Edsall, D.W., 199
Edsall, J.T., 210
Edstrom, R.D., 244
Edward, A.G., 234
Edwards, B.F., 101
Edwards, B.R., 278
Edwards, C.R., 148
Edwards, D.C., 218
Edwards, D.I., 139
Edwards, D.J., 413
Edwards, E.P., 488
Edwards, H., 105, 438
Edwards, H.H., 139
Edwards, J., 155, 484
Edwards, J.A., 155
Edwards, J.G., 32
Edwards, J.S., 498
Edwards, J.W., 338
Edwards, L., 318
Edwards, L.E., 441, 492
Edwards, L.H., 376
Edwards, M.E., 440
Edwards, M.W., 411
Edwards, N., 342
Edwards, P., 288
Edwards, R.L., 177, 424, 546
Edwards, R.S., 484
Edwards, S., 455, 489
Edwards, W.M., 364
Edwin, G., 128
Eertmoed, G.E., 119
Efford, I.E., 535
Efthymiou, C., 316
Egan, R.S., 479
Egar, M., 355
Egbert, L.N., 476
Egge, A.S., 28
Eggen, D.A., 186
Eggert, F.G., 193
Eggler, W.A., 348
Eggleston, A.C., 552
Egli, C., 285
Egli, D.B., 177
Eglitis, I., 367
Eglitis, J.A., 367
Egloff, D.A., 361
Ehler, L.E., 49
Ehlers, M.H., 502
Ehren, F., 20
Ehrenfeld, E., 329
Ehrenreich, J.H., 118
Ehrhart, A.L., 357
Ehrhart, C.Y., 398
Ehrhart, L.M., 88
Ehrle, E., 238
Ehrlich, H.G., 394
Ehrlich, H.L., 314
Ehrlich, M., 190
Ehrlich, P.R., 41
Ehrmann, W., 154
Eiben, G.J., 162
Eichberg, J., 464
Eichen, A., 312
Eichenmuller, J.J., 199
Eichholz, A., 284
Eichler, V., 173
Eichor, M.E., 254
Eickelberg, B., 295
Eickwort, G.C., 303
Eidels, L., 466
Eidinger, D., 545
Eidinoff, M., 293
Eiduson, S., 56
Eighme, L.E., 38
Eigsti, O.J., 119
Eikenbary, R.D., 377
Eiler, J.J., 60
Eilers, F.I., 97
Eilers, L.J., 162
Eimers, L.E., 440
Einert, A.E., 23
Einhellig, F.A., 434
Einstein, N., 316
Eipper, A.W., 303
Eisel, M., 241
Eisen, E.J., 343
Eisen, S., 553
Eisenberg, F.Jr., 85
Eisenberg, R.M., 82
Eisenberg, J.F., 199
Eisenberg, M., 300
Eisenberg, R., 236
Eisenberg, R.J., 411
Eisenberg, R.S., 57
Eisenberg, W., 298
Eisendrath, E.R., 263
Eisenmann, G., 57
Eisenstark, A., 260
Eisenstat, L., 392
Eisenstein, T.K., 408
Eiserling, F.A., 55
Eisman, E., 513
Eisner, T., 302, 304
Eitenmiller, R.R., 107
Ek, A.R., 521
Ekblad, J., 119
Ekkens, D., 7
Eklund, C.E., 468
Ekstedt, R.D., 127
Elam, J., 88
Elam, W.W., 250
Elandt-Johnson, R.C., 341
El-Baroudi, H.M., 314
von Elbe, J.H., 518
Elvein, A.D., 469
Elberg, S.S., 46
El-Bermani, I., 208
Elce, J.S., 545
Eldefrawi, M., 302, 304

Elden, J., 95
Elder, J.T., 269
Elder, W.H., 261
Eldredge, F.A., 223
Eldredge, K.H., 31
Eldredge, L.G., 110
Eldredge, N., 297
Eldridge, F.L., 340
Eleuterio, M.K., 415
Eley, J.H., 177
Elford, H., 333
Elfving, D.C., 305
Elgin, S., 210
Elias, J.J., 60
Elick, G., 254
Elinson, R., 550
Elion, G.B., 333
Eliot, J.W., 234
Elkan, G.H., 345
El-Khatib, S., 418
Elkin, L.O., 30
Elkins, D., 224
Elkins, D.M., 130
El Koury, A.F., 418
Ellarson, R.S., 520
Elle, G.O., 462
Ellefson, P.V., 243
Ellegood, J.O., 103
Eller, A.L.Jr., 493
Eller, J., 470
Eller, J.G., 348
Eller, R.R., 464
Ellert, M.S., 257
Ellett, C.W., 396
Elligaard, E.G., 190
Ellias, L., 88
Ellickson, P.J., 535
Elliker, P.R., 383
Elling, L.J., 240
Ellington, J.J., 292
Ellinwood, E.H., 333
Elliot, A.M., 303
Elliot, A.M., 461
Elliott, A., 253
Elliott, A.Y., 244
Elliott, E., 274
Elliott, E.S., 508
Elliott, I.W., 431
Elliott, J., 192
Elliott, J.D., 454
Elliott, L.F., 271
Elliott, L.P., 181
Elliott, P., 88
Elliott, R., 319
Elliott, W., 256, 309
Elliott, W.A., 240
Elliott, W.B., 318
Ellis, B., 527
Ellis, C.H.Jr., 214
Ellis, C.J., 157
Ellis, C.R., 547
Ellis, D.V., 536
Ellis, F.A., 32
Ellis, G.F., 462
Ellis, G.H., 65
Ellis, G.W., 409
Ellis, J., 308
Ellis, J.R., 271
Ellis, L.C., 476, 477
Ellis, L.L., 88
Ellis, R., 166, 318, 390
Ellis, R.E., 470
Ellis, R.G., 473
Ellis, R.P., 433, 434
Ellis, T.T., 426
Ellis, W.C., 458
Ellis, W.H., 436
Ellis, W.W., 527
Ellison, J.H., 288
Ellison, J.R., 391
Ellison, L., 557
Ellison, M.L., 97
Ellison, R., 321
Ellison, R.L., 489, 490
Ellison, S.A., 399
Ellner, P.D., 301
Ells, J.E., 66
Ellwood, E.L., 346
Elizey, J.T., 468
Elmer, W.A., 100
Elmore, H.W., 551
Elmore, J.E., 143
Elmstrom, G.W., 93
El-Negoumy, A.M., 266
Elovson, J., 59
Elrod, L., 506
Elsas, L., 101
El-Sayed, S.Z., 459
Elsey, S.M.G., 70
Elsner, J.E., 106
Elsner, R., 14
Elsoth, G.D., 119
Elston, R.D., 341
Elstrodt, C., 494
Elwood, J.C., 326
Elwood, W.K., 179
Ely, B.E., 429
Ely, C.A., 164, 300
Ely, D.G., 177
Ely, J.K., 164
Ely, T.H., 486
Elzinga, R.J., 167
Embil, J.A., 543
Embleton, T.W., 58
Emboden, W., 31
Embree, R.W., 166
Embry, L.B., 432
Emeh, C.O., 105
Emerson, C.P., 488
Emerson, D.A., 196
Emerson, F.H., 149
Emerson, G.M., 11
Emerson, R., 42
Emerson, S.U., 490
Emerson, T.E.Jr., 228
Emery, D.A., 344
Emery, J.A., 458
Emery, T., 193
Emery, W., 455
Emino, E.R., 219
Emlen, J.M., 145
Emlen, J.T., 516
Emlen, S., 302, 304
Emlin, J., 15
Emmel, T.C., 126
Emmel, V.M., 328
Emmers, R., 301
Emmert, F.H., 78
Emmings, F.G., 319
Emmons, L.R., 495
Emond, J., 56, 556
Emrich, L.S., 290
Emsley, M.G., 485
Enders, A.C., 263
Enderson, J.H., 65
Endler, J.A., 287
Endo, R.M., 58
Endrizzi, J.E., 16, 119
Enesco, H.E., 557
Enfield, F.E., 242
Eng, R., 265
Engalichev, N., 280
Engel, H.N., 6
Engel, R., 298
Engel, R.A., 324
Engel, R.E., 288
van Engel, W.A., 490
Engelberg, J., 180
Engelhardt, D.L., 301
Engelhardt, F.R., 215
Engelman, -., 81
Engelmann, F., 55
Engels, W.L., 339
Engemann, J.G., 236
Engen, P., 35
Engen, R.L., 158
Enger, D.M., 194
Engh, H., 238
Engibous, J.C., 502
England, D.C., 382
England, D.R., 21
England, J.J., 69
England, M.W., 375
England, R.B., 107
England, W.H., 501
Englander, L., 422
Englander, S.W., 410
Englard, S., 330
Engle, C.F., 502
Engle, Sr.J., 33
Englert, D.C., 129
Englesberg, E., 61
Englin, D., 36
English, A.W., 101
English, D., 15
English, G.W., 152
English, J.A., 319
English, T.D., 341
English, W.H., 51
Engstrom, R., 285
Enna, C.D., 185
Enns, J., 270
Enns, W.R., 259
Enio, F., 506
Enlow, D.H., 509
Eno, C.F., 93
Enos, H.L., 65
Enriquez, A.A., 552
Enroth-Cugell, C., 126
Ensign, J.C., 517
Ensign, R.D., 117
Enslee, E.C., 290
Ensminger, A., 182
Ensminger, E., 29
Ensminger, L.E., 3
Ensminger, M.E., 522
Enterline, P.E., 414
Entingh, D.J., 340
Eollias, N.E., 55
Epand, R.M., 545
Epling, G.P., 68
Epp, L., 361
Epp, M., 562
Eppig, J., 297
Eppig, J.J., 296
Epple, G., 409
Epps, C.O., 439
Epps, N.A., 547
Epps, W.M., 425
Epstein, A.H., 157
Epstein, A.N., 409
Epstein, E., 50
Epstein, L.I., 491
Epstein, W., 131, 132
Epstien, C., 60
Epstien, H.F., 42
Equing, E.E., 305
Erasmus, B.deW., 319
Erb, K., 309
Erbe, L., 191
Erbisch, F.H., 228
Ercegovich, C.D., 402
Erdman, H., 463
Erdman, K.S., 407
Erdos, E.G., 467
Erecinska, M., 410
Erhardt, W.H., 195
Erickenberg, C.F., 88
Erickson, A.W., 498
Erickson, D.O., 350
Erickson, E.D., 504
Erickson, H.P., 332
Erickson, H.T., 149
Erickson, J., 399
Erickson, J.F., 504
Erickson, J.G., 70
Erickson, J.M., 32
Erickson, J.R., 350
Erickson, L.C., 57, 117
Erickson, L.E., 168
Erickson, L.G., 57
Erickson, R., 333
Erickson, R.J., 152
Erickson, R.L., 279
Erickson, R.O., 409
Ericson, E.E., 159
Ericson, G.C., 466
Eriksen, C., 33
Eriksen, J., 354
Erikson, L., 123, 124
Erion, G.W., 433
Erk, F., 325
Erlanger, G.F., 301
Erlenmeyer-Kimling, L., 301
Erman, D.C., 44
Ern, E.H., 489
Ernst, C.H., 485
Ernst, S.A., 408
Ernst, R.A., 48
Ernst, S.C., 408
Ernst-Fonberg, M.L., 80
Ernstrom, C.A., 475
Eroschenko, V.P., 117
Erpino, M.J., 29
Ershoff, B.H., 35
Erskine, I., 181
Erskine, T.L., 199
Erspamer, J.L., 29
Ervin, F., 418
Erway, L.C., 370
Erwin, D.C., 58
Erwin, J.A., 121
van Es, T., 288
Esau, K., 142
Esch, G.W., 347
Esch, H.E., 152
Eschenberg, K.M., 214
Eschner, A.R., 322
Escueta, A., 333
Esham, L., 507
Eshbaugh, E.L., 167
Eshbaugh, W.H., 360
Eskew, E.B., 423
Eskin, A., 451
Espey, S.L., 389
Espinosa, E., 181
Esplin, A.L., 65
Esposito, M., 132
Esposito, R., 132
Esposito, W.J., 80
Essenberg, R.C., 377
Esser, B.E., 522
Essex, M.E., 212
Essien, F.B., 285
Essig, A., 208, 211
Esslinger, J.H., 198
Essman, R.H., 363
Essman, W.B., 298
Estabrook, G.F., 231
Estabrook, R.W., 466
Estavillo, J., 277
Esteban, M., 283
Estell, G., 192
Estep, G., 451
Estergreen, V.L., 502
Esterly, J.R., 133
Estermann, E.F., 40
Estes, E.E., 199
Estes, G.O., 281
Estes, J.E., 314
Estes, J.R., 379
Estes, L.A., 249
Estes, M.S., 56
Estes, P.M., 5
Estes, R., 39
Estes, R.D., 440
Estey, R.H., 557
Etchells, J.L., 344, 345
Etgen, W.M., 493
Etges, F.J., 369
Etheridge, A.L., 25
Etheridge, R., 39
Etherton, B., 480
Etkin, W., 329
Etnier, D.A., 442
Ettinger, M.J., 318
Etzel, W.W., 184
Etzler, M.E., 48
Eubanks, C , 188
Eure, H.E., 347
Eustice, E., 127
Evan, A.P., 293
Evan, J.W., 541
Evans, A.H., 356
Evans, A.M., 441
Evans, B., 174
Evans, C.A., 499
Evans, C.E., 3
Evans, D., 95
Evans, D.A., 224
Evans, D.C., 254, 397
Evans, D.S., 506
Evans, E.A.Jr., 131
Evans, F.C., 231
Evans, F.R., 473
Evans, H., 508
Evans, H.E., 68, 305
Evans, H.H., 355
Evans, H.J., 209, 385, 491
Evans, H.L., 329

Evans, J., 463
Evans, J.B., 345
Evans, J.E., 464
Evans, J.L., 288
Evans, J.S., 445
Evans, J.T., 415
Evans, J.W., 48
Evans, L.E., 537
Evans, L.T., 278
Evans, M.J., 320
Evans, M.L., 362
Evans, R., 30
Evans, R.A., 275
Evans, R.J., 225
Evans, R.M., 538
Evans, R.T., 319
Evans, S.D., 241
Evans, T.C., 162
Evans, W.D., 539
Evans, W.G., 530
Evans, W.L., 22
Evard, R., 36
Eveland, L.K., 325
Eveland, W.C., 234
Eveleigh, D.E., 288
Evenson, P.D., 433
Everett, J.W., 332
Everett, K.R., 364
Everett, M.S., 215
Everett, N.B., 499
Everhart, W.H., 303
Eversman, S., 265
Eversmeyer, H., 175
Everson, L., 157
Everson, R.D., 310
Evert, D.R., 481
Evert, R.F., 515
Every, R.R., 189
Evoy, W., 95
Ewan, J.A., 190
Ewart, W.H., 58
Ewel, J., 90
Ewens, W.J., 409
Ewert, A., 467
Ewing, C.K., 103
Ewing, K., 359
Eydt, H.R.N., 551
Eyer, L.E., 222
Eylar, E.H., 428
Eylar, O.R., 203
Eyre, P., 548
Eyster, C., 7
Eyster, G.L., 191
Eyster, M.B., 191
Ezell, D.O., 424
Ezell, W., 239
Ezell, W.B., 423

F

Faber, D.S., 370
Faber, J.E., 202
Faber, J.J., 387
Faberge, A.C., 465
Fabian, M.W., 396
Fabricant, J., 305
Fadulu, S.O., 461
Fagen, R.M., 136
Fager, R.S., 391, 490
Fahey, J.L., 56
Fahey, Rev.J.R., 209
Fahlberg, W.J., 449
Fahlquist, D., 459
Fahlund, C., 329
Fahnestock, S.R., 404
Fahrney, D.E., 68
Fahs, E., 147
Fahselt, D., 552
Failey, C.F., 31
Fain, R.C., 456
Fairbairn, D., 218
Fairbanks, G.W., 427
Fairbrothers, D.E., 288
Fairchild, E.D., 165
Fairchild, M.L., 259
Fairchild, R.S., 22
Faircloth, W.R., 109
Fairclough, J.G.Jr., 307
Fairey, J.E.III., 425
Fairley, J.L., 225
Faiszt, J.A., 494
Faix, J.J., 138
Fajer, A.B., 204
Falcon, C.J., 188
Falek, A., 101
Fales, F.W., 101
Falgout, R.N., 188
Falk, M.A., 133
Falk, N.W., 400
Falk, R.H., 51
Falkingham, -., 492
Falker, W.A.Jr., 204
Falkow, S., 499
Faller, A., 357
Fallis, A.M., 550
Fallon, H.J., 340
Fallon, J.F., 520
Fallon, J.G., 38
Fallona, G.R., 466
Falls, E.Q., 488
Falls, J.B., 550
Falsom, M., 336
Falter, J.M., 344
Falvo, R.E., 129
Falter, C.M., 118
Fambrough, D., 197, 198
Famsworth, W.E., 318
Fanale, L.P., 290
Fand, I., 295
Fang, C.S., 490
Fangboner, R., 290
Fanger, M.W., 356
Fanguy, R.C., 458
Fankboner, P.V., 533
Fanning, D.S., 200
Fanning, K.A., 97
Fanning, M., 336
Fara, J., 325
Fara, A.J., 232
Farber, E., 408
Farber, J.P., 162
Farber, P.A., 390
Farber, R.H., 143
Farbman, A.L., 126
Farel, P.B., 340
Farelli, S., 301
Farhi, L.E., 319
Farish, D.J., 259, 260
Farley, F.E., 360
Farley, J., 542
Farley, J.D., 366
Farley, R.D., 57
Farley, W.H., 440
Farmanfarmaian, A., 288
Farmer, J.G., 24
Farmer, J.L., 472
Farmer, J.N., 260
Farner, D.S., 498
Farney, J.P., 267
Farnham, R.S., 241
Farnsworth, P., 233
Farny, D.O., 398
Farrar, D.R., 157
Farrar, E.S., 7, 15
Farrar, M.D., 424
Farrell, B., 186
Farrell, L.D., 116
Farrell, M.P., 360
Farrell, R., 123
Farrens, B., 62
Farrier, M.H., 344, 346
Farrington, J.E., 189
Farris, D.A., 37
Farris, G., 70
Farris, V., 317
Farrow, S., 449
Fasano, A., 282
Fasbender, V., 431, 432
Fashing, N.L., 484
Fasman, G.D., 208
Faso, P.J., 371
Fass, R.J., 367
Fassel, H., 157
Fast, T.N., 62
Fatour, J.G., 559
Fattal, G., 317
Fattig, W.D., 10
Fatula, R.P., 395
Fauchald, K., 62
Faulkes, R.H., 522
Faulkin, L.J.Jr., 52
Faulkner, C.K., 527
Faulkner, J.F., 232
Faulkner, K.K., 379
Faulkner, L.C., 69
Faulkner, L.R., 165
Faulkner, Sr.M.C.F., 151
Faulkner, P., 545
Faulkner, R.C., 455
Faulkner, W.R., 446
Fauntain, C.A., 337
Fausch, H.D., 28
Fausey, N.R., 364
Faust, E.C., 190
Faust, R.A., 267
Faust, R.G., 340
Faust, W.Z., 100
Fautin, R.W., 526
Faviola, G.E., 208
Favre, H., 558
Favretti, R.J., 78
Fawcett, J.D., 272
Fawcett, L.R.Jr., 486
Fawcett, R.S., 516
Fawley, J.P., 416
Fay, M.J., 514
Fazio, J.R., 118
Fazio, S., 17
Feagans, W., 318
Feary, W.T., 10
Feaster, C.V., 16
Feaster, W.H., 479
Feather, M.S., 261
Fedde, M.R., 168
Feder, H.M., 14
Federer, C.A., 280
Federer, W.T., 302, 304
Fedinec, A.A., 445
Fedoroff, S., 562
Feduccia, J.A., 339
Fee, J.A., 232
Feeney, R.E., 49
Feeny, P.P., 303
Feese, B.T., 174
Fehon, J.H., 337
Fehrenbacher, J.B., 138
Feick, J.R., 279
Feig, K., 135
Feigelson, M., 300
Feigelson, P., 300
Feigenson, G.W., 302
Feigl, E.O., 500
Feigl, P., 500
Feil, P., 405
Fein, A., 208
Feiner, R.R., 296, 297
Feingold, A., 95
Feingold, D.S., 413
Feinleib, M.E., 216
Feinsinger, P., 73
Feinstein, R.N., 131
Feir, D., 256
Feirstais, P., 316
Feisal, V.E., 438
Feiss, M.G., 161
Feist, C., 14
Feist, D.D., 14
Feit, I.N., 395
Feldballe, J., 511
Feldbush, T.L., 161
Felder, M.R., 429
Feldman, I., 328
Feldman, J.F., 61
Feldman, J.J., 212
Feldman, L.A., 282
Feldman, M., 41
Feldman, M.L., 208
Feldmeth, C.R., 33
Feldstein, A.M., 233
Feldstein, J., 393
Feldt, A.G., 231
Felix, M.D., 300
Felix, R., 256
Felkner, I.C., 461
Fell, C., 308
Fell, H.B., 211
Fell, J., 96
Fell, P., 75
Fell, R.G., 223
Feller, K.C., 356
Fellman, J.H., 388
Fellner, W.H., 491
Fellows, R.E., 333
Felpel, L., 470
Felten, D.L., 145
Feltham, L.A.W., 541
Felton, F.G., 380
Felton, J.L., 364
Felton, S., 498
Felts, J.M., 60
Felts, P.W., 446
Felts, W.J.L., 379
Fenbert, D.W., 316
Fencher, G.H., 66
Fender, D.H., 26
Feng, I.A., 523
Feng, S.Y., 77, 79
Fennema, O., 518
Fenner, H., 219
Fensom, D.S., 540
Fenstad, M., 243
Fenster, C.R., 271
Fenster, E.E., 241
Fenton, M.B., 544
Fentress, J.C., 387
Fenwick, H.S., 117
Fenwick, J.R., 147
Fenwick, M., 126
Ferchau, H.A., 73
Ferell, R.S., 130
Ferenczy, A., 556
Ferguson, A.C., 537
Fergus, C.L., 403
Ferguson, D., 387
Ferguson, D.V., 25
Ferguson, F.G., 403
Ferguson, G., 386
Ferguson, G.G., 552
Ferguson, G.W., 85, 461
Ferguson, J.E., 87
Ferguson, J.H., 117
Ferguson, J.J.Jr., 410
Ferguson, L.C., 363, 367
Ferguson, M., 359
Ferguson, M.B., 120
Ferguson, R.K., 227
Ferguson, T.M., 458
Ferguson, W., 32
Ferin, J., 328
Ferin, M., 301
Ferm, V.H., 277
Fernandez, A.S., 356
Fernández, N.A., 418
Fernando, A.J., 538
Fernando, C.H., 551
Fernando, M.A., 548
Ferner, J.W., 395
Fernstrom, J.D., 213
Ferrante, F.L., 283
Ferrario, C.M., 351
Ferraro, F.M., 269
Ferree, C.C., 380
Ferree, D.C., 365
Ferree, R.J., 424
Ferrell, W.J., 232
Ferrence, G., 396
Ferrendelli, J.A., 263
Ferreri, L., 495
Ferretti, J.J., 380
Ferretti, T., 246
Ferrier, B.M., 545
Ferrill, M.D., 78, 79
Ferris, B.G.Jr., 212
Ferris, H., 58
Ferris, J.M., 148
Ferris, V.R., 148
Ferris, W.R., 18
Ferron, W., 254
Fertiziger, A.P., 204
Fervasi, J., 273
Fessenden-Raden, J.M., 302
Fessler, J.H., 55
Fetner, R.H., 102
Fetz, E.E., 500
Feucht, J.R., 66
Feuille, B., 560
Fezer, K.D., 506
Ffolliott, P.F., 18
Fichter, E., 116

Fick, G.N., 350
Fick, G.W., 302
Ficken, M., 523
Ficorilli, G.A., 420
Ficsor, G., 236
Fiel, R., 320
Field, C.R., 459
Field, E., 502
Field, J., 57
Field, R.A., 527
Fieldhouse, D.J., 82
Fields, B., 329
Fields, B.N., 330
Fields, H., 60
Fields, H.M., 335
Fields, J.M., 234
Fields, M.J., 91
Fields, W.G., 536
Fienburg, R.H., 441
Fierro-Berwart, C.A., 417
Fierstine, H.L., 27
Fife, C.L., 481
Fife, T.H., 64
Fife, W.P., 460
Fike, W.T., 344
Filia, F.P., 62
Filkins, J.P., 125
Filmer, D.L., 150
Filner, B., 300
Filner, P., 225, 226
Filteau, G., 557
Fimian, W.J., 207
Fina, L.R., 166
Finch, C.E., 62
Finch, H.C., 27
Fincher, E.L., 102
Findley, D.L., 436
Findley, J.J., 293
Findley, W.R., 364
Fine, L.O., 433
Fine, M.H., 357
Fine, R., 208
Fineberg, R.A., 60
Finerty, J.C., 185
Fingal, W.A., 250
Finger, I., 396
Fingerhut, M.A., 289
Fingerman, M., 190
Fink, G.B., 456
Fink, G.R., 302
Fink, J.N., 513
Fink, R.M., 56
Finkelstein, A., 330
Finkelstein, R.A., 466
Finkenbinder, L.F., 373
Finkle, J.L., 234
Finkner, V.C., 177
Finlay, J.B., 552
Finlay, M.F., 423
Finlay, P.S., 295
Finlayson, D.G., 534
Finlayson, M.H., 556
Finlayson, T., 533
Finley, A.M., 117
Finley, K.T., 317
Finley, W.H., 11
Finn, F., 413
Finn, W.E., 460
Finne, G., 499
Finnegan, C.V., 534, 535
Finney, B.A., 70
Finney, H.R., 241
Finney, J., 450
Finney, K., 75
Finnegan, R., 135
Finni, G.R., 390
Finocchio, A.F., 315
Finstad, C.D., 522
Finstein, M., 285
Fiore, C., 74
Fiore, J., 216
Fiori, B.J., 306
Firehammer, B.D., 266
Firestone, E.E., 375
Firling, C.E., 245
Firpi, A., 418
Firriolo, N., 310
Firshein, W., 80
Firstman, B.L., 28
Firtel, R.A., 59
Firth, D.R., 556
Fischbach, F., 514
Fischbarg, J., 301
Fischer, A.G., 352
Fischer, C.Jr., 189
Fischer, C.C., 303
Fischer, C.J., 384
Fischer, E.H., 499
Fischer, G.M., 411
Fischer, H., 479
Fischer, J., 339
Fischer, L.E., 120
Fischer, M., 217
Fischer, R.B., 27
Fischer, R.G., 353
Fischer, R.L., 226
Fischer, V.W., 256
Fischer, W., 492
Fischette, V.A., 295
Fischman, D., 132
Fischman, D.A., 131
Fischthal, J., 317
Fiscus, A.G., 265
Fiset, P., 203
Fish, A.G., 252
Fish, L., 7
Fish, N.A., 548
Fish, W.A., 419
Fish, W.W., 428
Fishbeck, D., 373
Fisher, A.B., 411
Fisher, B., 374
Fisher, C.D., 455
Fisher, C.M., 337
Fisher, C.P., 27
Fisher, D.B., 105
Fisher, E.Jr., 387
Fisher, E., 486
Fisher, E.H., 518
Fisher, E.W., 484
Fisher, F., 395
Fisher, F.J.F., 533
Fisher, F.M.Jr., 453
Fisher, G.T., 280
Fisher, H., 172, 288
Fisher, H.D., 535
Fisher, H.I., 129
Fisher, H.W., 421
Fisher, J.M., 549
Fisher, J.S., 489
Fisher, K.W., 288
Fisher, L., 500
Fisher, L.H., 30
Fisher, P.L., 427
Fisher, R.L., 397
Fisher, S.G., 206
Fisher, S.K., 61
Fisher, T.N., 387
Fisher, T.R., 258, 354
Fisher, W.D., 16
Fishkin, A.P., 268
Fishman, H.M., 468
Fishman, I.Y., 155
Fishman, L., 95, 312
Fishman, M.M., 298
Fisk, F.W., 363
Fisk, L., 500
Fisk, W., 361
Fiske, V.M., 220
Fiskell, J.G.A., 93
Fissel, G.W., 401
Fisser, H.G., 527
Fister, G.J.Jr., 187
Fitch, C., 256
Fitch, F.W., 133
Fitch, H.S., 172
Fitch, K., 122
Fitch, W.C., 236
Fitch, W.M., 520
Fite, J., 335
Fites, R.C., 343
Fitt, P.S., 548
Fitzgeral, J., 73
Fitzgerald, A., 499
Fitzgerald, C.H., 108
Fitzgerald, L.R., 445
Fitzgerald, T.D., 321
Fitzgerald, W.T., 451
Fitzharrin, T.P., 428
Fitz-James, P.C., 551
Fitzpatrick, J.F.Jr., 12
Fitzpatrick, L.C., 452
Fitzsimons, J.M., 183
Fives-Taylor, P.M., 482
Fix, J.D., 145
Fix, M., 317
Fix, W.L., 149
Fjelde, A.F., 320
Flaim, F.R., 62
Flake, L.D., 434
Flaks, J.G., 410
Flanagan, P.W., 14
Flanagan, T.D., 319
Flanagan, T.R., 481
Flanders, S.E., 58
Flanigan, N.J., 514
Flannery, W.L., 191
Flashner, M., 321
Flaumenhaft, E., 369
Flavin, Rev.J.W., 209
Flechtner, V., 355
Fleck, E.W., 504
Fleeker, J.R., 352
Flegal, C.J., 226
Fleharty, E.D., 164
Fleischer, B., 446
Fleischer, S., 446
Fleischman, J., 263
Fleschner, C.A., 58
Fleishchmann, W.R.Jr., 116
Fleishman, D.A., 371
Fleming, A.A., 106
Fleming, B.R., 440
Fleming, H.P., 344
Fleming, M., 439
Fleming, R., 31
Fleming, T.H., 262
Fleming, W.R., 260
Fleming, W.J., 175
Flemister, L.J., 407
Flessed, C.P., 62
Fletchall, O.H., 258
Fletcher, D.W., 40
Fletcher, G., 541
Fletcher, I., 498
Fletcher, J.L., 439
Fletcher, J.R., 390
Fletcher, J.S., 374
Fletcher, P.W., 402
Fletcher, R., 515
Fletcher, R.A., 547
Fletcher, R.I., 143
Fletterick, R., 531
Flexer, A.S., 71
Flexner, L.B., 410
Flick, W., 494
Flickinger, C.J., 489
Flickinger, R.A., 318
Flickinger, S.A., 66
Flinders, J.T., 463
Flink, J.M., 213
Flinn, M., 543
Flint, F.F., 487
Flint, H.L., 149
Flipse, R.J., 401
Flocker, W.J., 51
Flook, J., 454
Flora, J.D.Jr., 233
Flora, R.E., 491
Flores, M.N., 86
Flores-Flores, L., 417
Florini, J.R., 326
Flory, W.S., 347
Flournoy, R.W., 187
Flower, M.J., 387
Flowerday, A.D., 271
Flowers, A., 332
Flowers, A.I., 460
Floyd, G.L., 288
Floyd, J.W., 475
Fluegel, W., 245
Fluharty, A.L., 64
Fluke, D.J., 332
Fluker, S., 91
Flumerfelt, B.A., 551
Flury, A., 448
Flynn, E.F., 283
Flynn, T.G., 545
Foder, H.M., 14
Foerster, L., 164
Foerster, J.W., 196
Fogel, M.M., 18
Fogel, S., 45
Fogg, K., 28
Fogg, P.J., 183
Foil, R.R., 250
Foin, T.C., 47, 49
Fold, G.E., 161
Fold, W.R., 232
Folden, D.B., 438
Folds, J.D., 339
Foldvari, T., 207
Foley, A., 372
Foley, A.L., 227
Foley, C.W., 262
Foley, D.C., 157
Foley, J.C., 279
Foley, M.A., 206
Folk, R.M., 367
Folkerts, G.W., 5
Follett, R.H., 166
Follett, R.H., 364
Follette, E., 431
Follstad, M., 524
Folse, D., 467
Folsome, C.E., 112
Foltz, C., 420
Foltz, E., 399
Foltz, V., 452
Fonda, M.L., 180
Fonda, R.W., 504
Fondy, Sr.B., 406
Fondy, T.P., 326
Fontaine, A.R., 536
Fontaine, N.E., 245
Fontaine, V., 559
Fontana, J.M., 12
Fontennelle, L.J., 116
Fontenot, J.F., 184
Fontenot, J.P., 493
Fontera, R., 215
Fontes, A.K., 400
Fontes, M.R., 17
Fooden, J., 119
Fook, J.L., 550
Foor, W.E., 234
Foos, K.M., 359
Foote, B.A., 359
Foote, F., 129
Foote, K.G., 514
Foote, M., 481
Foote, W.D., 275
Fopeano, J.V., 318
Foraker, R., 316
Forbath, N., 550
Forbes, A., 524
Forbes, A.R., 534
Forbes, G.B., 328
Forbes, M.H., 125
Forbes, O.C., 117
Forbes, P., 540
Forbes, R.D.C., 556
Forbes, R.M., 138
Forbes, W., 396
Forbes, W.C., 230
Forbes, Y.N., 154
Force, D.C., 29
Force, H.F., 254
Ford, C.E., 76
Ford, D., 222
Ford, D.H., 325
Ford, D.L., 135
Ford, F.M., 436
Ford, G.D., 492
Ford, H.W., 92
Ford, N.L., 239
Ford, P., 535
Ford, P.J., 496
Ford, R., 231
Ford, R.E., 139
Ford, R.F., 37
Ford, S., 181
Fordham, C.C.III., 339
Fordham, W., 87
Forehlich, H.A., 384
Foreman, D.L., 355
Foreman, R.E., 534
Foreny, R.B.Jr., 556
Forer, A., 553
Forest, H.S., 323

Forey, P.L., 531
Forgash, J.A., 288,
Forman, G.L., 128
Forman, J., 466
Forman, M., 150
Forman, R.T.T., 288
Formica, J.V., 492
Forney, F.W., 191
Forrest, E.J., 413
Forrest, H.S., 465
Forrest, W.D., 438
Forrester, D.J., 94
Forro, F.Jr., 242
Forseberg, C.W., 547
Forseberg, R.A., 516
Forsdyke, D.R., 545
Forshaw, R.P., 547
Forster, C.P., 419
Forster, H.V., 513
Forster, M., 497
Forster, R.E., 410
Forster, R.P., 277
Forster, W.O., 385
Forsthoefel, P.F., 230
Forshey, C.F., 307
Forstner, G.G., 550
Forsyth, B.J., 147
Forsyth, J.W., 461
Forsyth, R.H., 241
Forsyth, T.R., 147
Forsythe, A.B., 56
Forsythe, D.M., 423
Forsythe, H.Y.Jr., 194
Fort, B.H.Jr., 99
Forte, J.G., 43
Fortes, P.A.G., 59
Forthofer, R.N., 469
Fortin, J.A., 558
Fortin, R., 560
Fortman, J.R., 247
Fortson, J.C., 108
Forward, R.B., 332
Fosberg, M.A., 117
Fosgate, O.T., 106
Fosket, D.E., 54
Foss, D.C., 480
Foss, J.E., 200
Foss, J.G., 156, 157
Fossland, R.G., 514
Fossum, G.O., 352
Foster, A.E., 350
Foster, B.B., 280
Foster, B.G., 460
Foster, C.E., 324
Foster, D., 308
Foster, E.E., 62
Foster, E.M., 517
Foster, G.M., 138
Foster, J., 15
Foster, J.E., 148
Foster, M., 231, 288
Foster, M.S., 30
Foster, R., 132, 257
Foster, R.E., 17
Foster, W., 288
Foster, W.A., 363
Fotheringham, N., 463
Fotino, M., 71
Fougeron, M.G., 269
Foulds, R.T.Jr., 481
Fountain, J., 348
Fountain, K.R., 254
Fournier, M.J., 217
Fournier, R.O., 542
Fourquette, M.J., 15
Fourtner, C.R., 318
Fouts, J.R., 344
Fowell, E., 207
Fowke, L.C., 562
Fowle, C.D., 553
Fowler, D., 236
Fowler, D.R., 449
Fowler, G., 232
Fowler, I., 179
Fowler, M.E., 47
Fowler, R.G., 287
Fowles, B., 192
Fox, A.C., 108
Fox, A.S., 519
Fox, C.F., 55
Fox, C.L.Jr., 301
Fox, D.J., 442
Fox, C.A., 233
Fox, E.P., 450
Fox, D.L., 341
Fox, G., 538
Fox, H., 311
Fox, I.J., 244
Fox, J.D., 177
Fox, J.E., 170
Fox, J.L., 465
Fox, J.P., 493
Fox, K.A., 323
Fox, L.R., 88
Fox, R., 89
Fox, R.C., 424, 531
Fox, R.H., 83, 302
Fox, R.T., 303
Fox, T.W., 219
Foxworthy, J., 37
Fozdar, B.S., 279
Fozzard, H.A., 133
Fradkin, C.M., 32
Fraembs, F.A., 120
Fraenkel, D., 211
Fraenkel, G.K., 300
Fraenkel, G.S., 136
Fraenkel-Conrat, H.L., 43
Fragata, M., 560
Frahm, R.R., 376
Frair, W., 310
Fraizer, L., 450
Fraker, P., 225
Fraleigh, P., 371
Fraley, L.W., 69
Fralick, R.A., 279
Fralish, J.S., 130
Frampton, E., 126
Franch, D., 157
Francis, D.W., 82, 292
Francis, F., 256
Francis, J.C., 185
Francis, J.S., 547
Francis, M.E., 341
Francis, R.L., 129
Franck, D.H., 515
Franclemont, J.G., 303
Francoeur, A., 560
Frandsen, L.C., 5
Frandson, R.D., 68
Frane, J.W., 56
Frank, F.W., 118
Frank, H.A., 113
Frank, J.A., 194
Frank, K.D., 271
Frank, P.W., 387
Frank, R.E., 352
Frank, S., 89
Franke, D.E., 91
Franke, G.R., 254
Franke, H.W., 458
Franke, R.G., 157
Frankel, A., 317
Frankel, H.M., 288
Frankel, J., 160
Frankel, V., 356
Frankfater, A., 124
Frankie, G.W., 458
Franklin, B.C., 188
Franklin, D., 117
Franklin, E., 507
Franklin, L.E., 463
Franklin, M., 540
Franklin, R.E., 364
Franklin, R.T., 106, 108
Franklin, W.T., 65
Frankowski, R.F., 469
Franks, E.C., 139
Frans, R.E., 22
Fransworth, M.W., 318
Frantz, I.D.Jr., 244
Frantz, M.L., 169
Franz, D., 297
Franz, D.R., 296
Franz, E., 105
Franz, G.N., 509
Franz, J.M., 261
Franz, W.L., 228
Franzen, D.S., 122
Franzen, J.S., 413
Franzmeier, D.P., 147
Fraser, A.S., 369
Fraser, C.E.O., 212
Fraser, D., 144, 550
Fraser, D.A., 341
Fraser, D.F., 316
Fraser, L.A., 516
Fraser, M.J., 555
Fraser, R.C., 447
Frasher, W.F.Jr., 64
Frazier, R., 240
Frasier, S.D., 64
Fratianne, D.G., 362
Frazer, B., 534
Frazer, R.H., 440
Frazier, D.M., 340
Frazier, D.T., 180
Frazier, F., 374
Frazier, J.L., 249
Frazier, J.R., 5
Frazier, T.M., 212
Frazzetta, T.H., 136
Frea, J.I., 363
Freckmann, R., 522
Frederick, C.M., 371
Frederick, D.J., 229
Frederick, L., 99
Frederick, L., 189
Fredericksen, D.W., 157
Frederickson, E.L., 101
Frederickson, R.C., 509
Frederiksen, D.W., 446
Fredette, V., 558
Fredricks, W.W., 512
Fredrickson, J.M., 550
Fredrickson, L.H., 261
Fredrickson, R.W., 406
Fredrickson, T.N., 78, 79
Free, S.M., 408
Freebairn, H.T., 463
Freeberg, J.A., 219
Freed, J.M., 369
Freedland, R.A., 52
Freedman, J., 387
Freedman, S.L., 482
Freeland, M.Q., 254
Freeling, M.R., 45
Freeman, A.F., 409
Freeman, B.A., 445
Freeman, B.W., 357
Freeman, C.B., 395
Freeman, C.E., 468
Freeman, G., 465
Freeman, H.W., 426
Freeman, J.A., 186, 430, 446
Freeman, J.D., 4
Freeman, J.R., 440
Freeman, J.V., 479
Freeman, K.B., 545
Freeman, M., 88
Freeman, M.H., 286, 349
Freeman, R.S., 550
Freeman, T.E., 92
Freeman, T.P., 349
Freeman, W.J., 43
Fregly, M.J., 94
Frehn, J.L., 122
Frei, J.K., 87
Frei, J.V., 552
Freiburg, M.W., 125
Freiburg, R.E., 125
Freidell, T.D., 471
Freifelder, D.M., 208
Freinkel, N., 127
Freisheim, J.H., 370
Freitag, J., 44
Freitag, R., 544
Fremount, H.N., 394
French, D.W., 243
French, E., 7, 393
French, E.R., 345
French, F., 557
French, F.E., 102
French, W.L., 123
Frenkel, A.W., 241
Frenkel, C., 288
Frenkel, R., 466
Frenzel, D.L., 240
Frerman, F.E., 513
Fresco, J.R., 286
Freter, R., 232
Fretwell, D.S., 166
Fretz, T.A., 365
Frey, D.F., 27
Frey, D.G., 145
Frey, H., 50
Frey, J., 238
Frey, J.R., 358
Frey, M.L., 377
Frey, R.A., 168
Frey, R.D., 351
Frey, R.H., 279
Freytag, P.H., 178
Friar, R.E., 223
Friars, G.W., 547
Fribourg, H.A., 444
Frick, G.E., 280
Friday, D.T., 364
Fridlund, P.R., 503
Fridovich, I., 333
Fried, B., 397
Fried, G.H., 297, 298
Fried, J., 131
Fried, M., 90
Fried, R., 268
Friedberg, W., 379, 380
Friede, R.L., 356
Frieden, C., 263
Friederici, H.R., 126
Friedkin, M.E., 59
Friedl, F.E., 97
Friedlander, L.M., 60
Friedman, A.H., 124
Friedman, B., 356
Friedman, C.L., 535
Friedman, D.I., 232
Friedman, D.L., 446
Friedman, E., 208
Friedman, H., 59
Friedman, H., 408
Friedman, H.P., 262
Friedman, J., 146
Friedman, L., 190
Friedman, L.D., 262
Friedman, M.H., 509
Friedman, M.S., 298
Friedman, R.T., 486
Friedman, S., 136
Friedman, S.B., 236
Friedman, S.M., 297, 298, 535
Friedman, T.B., 230
Friedmann, E., 88
Friedmann, H.C., 131
Friend, D.J.C., 111
Friend, D.S., 60
Friend, E., 395
Friend, J.P., 180
Friend, W.G., 550
Friendley, D.I., 436
Fries, J.A., 431
Friesen, B.S., 171
Friesen, J.D., 553
Frings, H.W., 379
Frisby, B.E., 189
Frisch, A.W., 380
Frischknecht, W.D., 382
Frisell, W.R., 282
Frishette, W.A., 552
Frist, R.H., 507
Fristrom, J.W., 45
Fritchman, H.K., 116
Fritiz, H.I., 372
Fritton, D.D., 401
Fritts, D.H., 266
Fritz, A.W., 359
Fritz, E., 44
Fritz, G.R., 414
Fritz, I.B., 550
Friz, C.T., 535
Frizzell, R.A., 414
Froelich, J.S., 355
Froese, G., 537
Frohberg, R.C., 350
Frohlich, E.D., 280
Froiland, S.G., 431
Froiland, T.G., 229
Frojmovic, M.M., 556
Frola, J., 369
Frolander, H.F., 385

Froman, S., 56
Fromm, E., 393, 394
Fromm, H.J., 157
Fromm, P.O., 228
Frommer, J., 216
Fromson, D., 556
Froning, G.W., 272
Fronk, W.D., 68
Frontera, J., 418
Frosch, D., 374
Froseth, J.A., 502
Frost, G.W., 74
Frost, H.H., 472
Frost, P., 236
Frost, S.L., 368
Fruhman, G.J., 329
Fruidenier, F.J., 137
Fruton, J., 81
Fruton, J.S., 80
Fry, A.E., 369
Fry, B.W., 104
Fry, F.E.J., 550
Fry, J.L., 93
Fry, K.A., 463
Fry, R., 126
Fry, R.E., 17
Fry, W.E., 304
Frydendall, M., 238
Frye, B.E., 231
Frye, B.L., 464
Frye, W.W., 440
Fryer, J.L., 386
Frystak, R.W., 111
Fryer, M.E., 443
Fryxell, P.A., 459
Fu, S.C.J., 282
Fu, W., 74
Fuchigami, L.H., 384
Fuchigami, R.Y., 28
Fuchs, A.F., 500
Fuchs, F., 414
Fuchs, J.A., 241
Fuchs, J.C., 333
Fuchs, M.S., 152
Fudenberg, H.H., 428
Fudner, R., 297
Fuerst, C., 550
Fuerst, R., 463
Fugita, T.S., 388
Fugler, C.M., 347
Fuhrman, G., 301
Fuikawa, K., 499
Fujii, K., 150
Fujimoto, G., 350
Fukuto, T.R., 58
Fulcher, W., 335
Fulco, A.J., 56
Fulford, M., 370
Fuller, A., 33
Fuller, C.W., 121
Fuller, D.L., 89
Fuller, E., 116
Fuller, E.O., 101
Fuller, M.D., 143
Fuller, G., 467
Fuller, G.R., 480
Fuller, H.L., 108
Fuller, I., 96
Fuller, J., 467
Fuller, M., 247
Fuller, M.J., 175
Fuller, M.S., 105
Fuller, R., 355
Fuller, R.C., 217
Fuller, S.W., 486
Fuller, W.A., 531
Fullerton, C.M., 114
Fullerton, G., 32
Fulmer, J.P., 424
Fulton, D.E., 371
Fulton, E., 524
Fulton, G.P., 207
Fulton, H.F., 127
Fulton, J.P., 24
Fulton, N.D., 24
Fulton, R.W., 519
Fults, J.L., 68
Funderburk, H.H.Jr., 4
Funderburk, N.L., 470
Fung, D.Y.C., 404
Funk, C.D., 474
Funk, C.R., 288
Funk, D.T., 130
Funk, F., 405
Funk, H.B., 196
Funk, H.C., 440
Funk, J.E., 179
Funk, R.C., 120
Funk, R.N., 258
Funke, B.R., 351
Funkhouser, A., 61
Funt, R.C., 202
Fuquay, J.W., 249
Furlong, C.E., 57
Furman, D.P., 44
Furness, G., 282
Furnish, W.F., 180
Furnival, G.M., 81
Furr, R.D., 462
Furr, R.S., 224
Furrer, J.D., 271
Furumoto, W., 31
Fushtey, S.G., 547
Fustafson, J.F., 40
Futcher, A.G., 196
Futcher, C., 439
Futhe, K.F., 231
Futrell, M.C., 249
Fuxton, D.J., 504
Fyfe, F.W., 542

G

Gaafar, H., 327
Gabel, J.R., 40
Gable, M., 75
Gabliks, J.Z., 214
Gabor, A., 53
Gabor, W.E., 391
Gabridge, M.G., 137
Gabriel, D.M., 296
Gabriel, K., 399
Gabriel, M., 297
Gabriel, O., 85
Gabriele, T.C., 314
Gabrielson, F.C., 10
Gabrielson, R.L., 503
Gaddis, C.W., 18
Gaddum-Rosse, P., 499
Gadhoke, J., 548
Gadsden, R.H., 428
Gadt, L.O., 9
Gaertner, R.A., 507
Gaffney, E.V., 40
Gaffney, G.R., 130
Gaffon, H., 88
Gafford, L.G., 251
Gagliano, N.Jr., 136
Gagliano, V.J., 34
Gagnon, A., 557, 558
Gagnon, L.C., 266
Gagnon, R., 558, 560
Gahn, M.M., 364
Gaik, G.C., 413
Gailani, S., 321
Gailey, F.B., 174
Gaines, J.A., 493
Gaines, M.S., 172
Gainor, C., 214
Gaither, T.W., 407
Gal, E.M., 161
Galat, J., 285
Galuauger, D.W., 539
Galbert, G.B., 325
Galbraith, D.B., 76
Galbreath, E., 129
Galdfelter, W.E., 509
Gale, A.F., 527
Gale, C.C., 500
Gale, H.H., 269
Gale, T., 277, 292
Gale, T.F., 277
Galeotti, C., 376
Galey, W., 294
Galia, L.N., 285
Galil, K.A., 551
Galinsky, I., 309
Galitz, D.S., 349
Gall, G.A.E., 48
Gall, H., 132
Gall, J.G., 80
Gallagher, J.P., 401
Gallaher, A., 176
Gallaher, W.R., 185
Gallanar, J., 396
Gallander, J.F., 365
Gallati, W.W., 396
Gallegly, M.E., 508
Galletta, G.J., 345
Gallian, J., 276
Gallicchio, V., 359
Gallo, A., 355
Gallo, L.K., 128
Gallo, L.L., 84
Gallo, R., 60
Galloway, H.M., 147
Galloway, L., 254
Galloway, R.A., 201
Gallucci, B.B., 499
Gallucci, V., 500
Gallucci, V.F., 498
Gallun, R.L., 148
Gallup, A.H., 39
Gallup, D.N., 531
Galsky, A.G., 119
Galton, P.M., 77
Galston, A.W., 81
Galsworthy, P.R., 551
Galsworthy, S., 551
Galton, P.M., 79
Galton, V.A., 278
Galvas, P.E., 372
Gambal, D., 268
Gamble, W., 385
Gambrell, C.E.Jr., 424
Gambs, R.D., 27
Gammal, E.B., 551
Gammon, N.Jr., 93
Gammon, J.R., 143
Gammon, R.A., 395
Gan, J., 467
Ganders, F.R., 534
Gangarosa, L.P., 104
Gangstad, V., 342
Ganguli, K.K., 471
Gangwere, S.K., 234
Ganion, L.R., 142
Ganjam, V.K., 411
Gann, D.S., 197
Gann, J., 188
Gannon, J.J., 233
Gannong, W.F., 60
Gans, C., 231
Ganschow, R.E., 370
Gant, K., 324
Ganuea, W.J., 466
Gapter, J., 73
Garattini, S., 298
Garay, L.A., 211
Garber, B., 132
Garber, B.B., 131
Garber, M.J., 59
Garber, M.P., 156
Garcia, A.M., 325
Garica, C., 186
Garcia, D., 29
Garcia, J.D., 463
Gard, R., 66
Garden, R., 70
Gardiner, E., 489
Gardner, A.W., 109
Gardner, C.L., 271
Gardner, C.O., 271
Gardner, D., 308
Gardner, D.G., 552
Gardner, D.R., 544
Gardner, D.T., 250
Gardner, E., 53, 159, 313, 474
Gardner, E.H., 384
Gardner, E.J., 476
Gardner, E.W., 461
Gardner, F.A., 458
Gardner, M.D., 445
Gardner, P.D., 473
Gardner, P.J., 273
Gardner, R.W., 472
Gardner, W.D., 512
Gardner, W.H., 502
Gardner, W.S., 433
Garen, A., 81
Garfinkel, D., 411
Garg, B.D., 86
Garibanks, M.B., 223
Garica-Castro, Y., 418
Garin, G.I., 5
Garland, W., 546
Garlich, J.D., 346
Garner, G.B., 261
Garner, H.W., 456
Garner, J.G., 117
Garnett, P.J., 25
Garolian, G., 129
Garono, J.E., 285
Garoutte, B.C., 60
Garrard, W.T., 466
Garraway, M.O., 366
Garren, H., 106
Garren, R.Jr., 384
Garrett, B.F., 507
Garrett, C.J.R., 542
Garrett, F., 475
Garrett, F.D., 535
Garrett, J.C., 494
Garrett, M.H., 445
Garrett, P.W., 286
Garrett, R.H., 488
Garrett, W.N., 48, 108
Garrick, M.D., 318
Garrick, R.A., 290
Garriett, J.C., 467
Garrigus, W.P., 177
Garrison, D., 96
Garrison, N., 486
Garrison, R.L., 383
Garrison, S.A., 288
Garrity, R., 151
Garside, E.T., 542
Garstang, M., 489
Garth, J.S., 62
Garth, R.E., 440
Garthe, W., 126
Gartner, F.R., 432
Gartner, J.B., 138
Garver, F., 103
Garverick, H.A., 259
Garvey, J.S., 326
Garvey, W., 400
Garvin, J.E., 127
Gary, J.T., 99
Gary, R.T., 455
Gary, W.J., 271
Gascay, S.E., 206
Gasdorf, E.C., 119
Gashler, S.T., 382
Gasiokiewicz, E.C., 522
Gaskin, D.E., 547
Gaskin, J.M., 94
Gaskin, J.R., 338
Gaskins, C.T., 462
Gass, C.L., 535
Gass, G.H., 129
Gasser, R.F., 185
Gassert, Rev.R.G., 512
Gasslier, E.B., 399
Gast, R.G., 241
Gasteiger, E.L., 302, 304
Gaston, L.K., 58
Gastonguay, P.R., 216
Gastony, G.J., 144
Gates, A.H., 328, 329
Gates, D.E., 167
Gates, D.H., 382
Gates, D.M., 230
Gates, G.O., 61
Gates, J.E., 487
Gatfield, P.D., 551
Gatherum, G., 367
Gatherum, G.E., 366
Gaties, V.G., 328
Gatipon, G.B., 252
Gatlin, G.W., 451
Gatten, R.E.Jr., 358
Gaud, W., 15
Gaudin, A., 31
Gaudin, D., 11
Gaudy, E., 375
Gaufin, A.R., 473
Gauger, W., 270

Gaughran, G.R.L., 367
Gault, J., 368
Gault, J.H., 409
Gault, N.L.Jr., 243
Gaumer, A.E.H., 400
Gaumer, H.R., 241
Gaunt, A.S., 364
Gaunt, D.D., 163
Gaunt, J.J., 354
Gaunt, S., 219, 354
Gauntt, C.J., 470
Gaunya, W.S., 77
Gause, R.H., 439
Gaush, C.R., 435
Gausenbuiller, R.L., 502
Gautheir, D., 10
Gauthier, F.M., 558
Gauthier, J.J., 10
Gauthier, P., 199
Gauthier, R., 558
Gauthreaux, S.A.Jr., 426
Gavan, T.L., 357
Gavin, D., 359
Gavin, M.L., 543
Gavin, R.H., 296
Gavin, R.J., 33
Gawienowski, A.M., 217
Gawlik, S.R., 315
Gawyrs, D., 311
Gay, H., 231
Gay, N., 177
Gay, R.A., 111
Gay, V.L., 414
Gayle, L.G., 460
Gayles, F.J., 494
Gaylor, D.W., 24
Gaylor, J.L., 302, 306
Geadelmann, J.L., 246
Gear, A.R.L., 490
Gebben, A., 223
Gebber, G.L., 227
Geber, W.F., 104
Geder, L., 405
Geduldig, D.S., 203
Gee, A., 497
Gee, C., 439
Gee, D.H., 428
Gee, J.H., 538
Gee, L.L., 375
Gee, R.J., 357
Geen, G.H., 533
Geeseman, G., 522
Geeslin, C., 95
Gehres, J.L., 390
Gehrig, R.F., 315
Gehring, R.Z., 20
Gehris, G., 317
Gehrke, C.W., 261
Gehrmann, W.H., 456
Gehrz, D.K., 521
Geiduschek, E.P., 59
Geiger, D.R., 371
Geiger, R.S., 124
Geis, J.W., 321
Geise, H.A., 433
Geisert, P.G., 526
Geisler, Sr.G., 311
Geisman, J.R., 365
Geissinger, H.D., 548
Gelb, L.D., 262
Gelbart, S., 124
Gelderd, J.B., 185
Gelehrter, T.D., 232
Gelinas, D.A., 194
Geller, D.M., 263
Gellos, G.J., 390
Gelperin, A., 287
Gemborys, S.R., 485
Genaux, C., 14
Genco, R.J., 319
Genelly, R.E., 34
Gengenbach, B.G., 240
Genghof, D.S., 330
Gengozian, N., 347
Gengozian, N., 442
Gennaro, A.L., 291
Gennaro, J.F., 312
Gennaro, R.N., 88
Gennis, R., 135
Gentile, A., 379
Gentile, G., 311
Gentry, C.E., 174
Gentry, G.A., 251
Gentry, P.A., 548
Gentry, R.F., 403
Genung, W.G., 91
Geobel, C.J., 503
Geode, R., 476
George, D.V.E., 533
George, D.W., 502
George, J.A., 552
George, J.C., 547
George, J.E., 461
George, J.L., 402
George, P., 409
George, R.C., 458
George, R.P., 526
George, R.W., 225
George, S.A., 206
George, W.G., 129
George, W.L. Jr., 265
Georghiou, G.P., 58
Georgi, C., 270
Georgi, J.R., 306
Geortz, J.W., 187
Geoschl, J.D., 460
Geottemoeller, R., 364
Gephart, D.S., 277
Gepner, J., 286
Ger, R., 329
Geraci, J.R., 548
Gerald, J.W., 40
Geratz, J.D., 340
Gerber, B.R., 409
Gerber, D.B., 365
Gerber, E., 132
Gerber, J.F., 92
Gerber, J.N., 455
Gerber, N., 288
Gerber, W., 142
Gerberding, W., 37
Gerberich, J.B., 514
Gerchakov, S.M., 96
Gerdemann, J.W., 139
Gerencser, G.A., 173
Gerencser, V.F., 509
Gerhardt, H.C., 260
Gerhardt, N.B., 388
Gerhardt, P., 228
Gerhardt, P.D., 17
Gerhardt, R., 443
Gerhart, D.Z., 245
Gerhart, J.C., 43
Gerhold, H.D., 402
Gerity, P., 150
Gerken, H.J., 493
Gerlomg. S.D., 15
Gerlach, E. Hugh, 173
Gerlach, G.F., 149
Gerlings, E.D., 273
Gerloff, E.D., 433
Gerloff, G.C., 515
Germann, P.J., 238
Germer, H., 169
Germer, H.A., 468
Gerold, N.J., 308
Gerrard, D.J., 243
Gerrath, J.F., 546
Gerritsen, T., 520
Gershenowitz, G., 285
Gershman, H., 355
Gershowitz, H., 232
Gerside, L., 522
Gersper, P.L., 46
Gerst, J.W., 350
Gerstein, G.L., 410
Gerstel, D.U., 344
Gersten, J., 367
Gertner, S.B., 283
Gervais, A.G., 558
Gervais, P., 558
Geschwind, I.I., 48
Gese, E.C., 416
Gesell, S.G., 402
Gesler, J.T., 28
Gesmer, A.B., 207
Gessaman, J.A., 476, 477
Gessner, T., 321
Gest, H., 144
Gesteland, R.C., 126
Getchell, T., 81
Gethmann, R.C., 205
Getting, P., 41
Getz, G.S., 131, 133
Getz, L.L., 136
Geyer, R.A., 459
Geyer, W.A., 168
Ghabrial, S.A., 178
Ghayasuddin, M., 290
Ghent, A.W., 136
Ghidone, J.J., 464
Ghidoni, Sr.E., 299
Ghiron, C.A., 261
Ghiselin, M.T., 43
Gholson, J.T., 129
Gholson, R.K., 377
Ghosh, B.K., 284, 288
Ghosh, D., 461
Ghosh, H.P., 545
Ghosh, K.K., 338
Ghosh, S., 127
Ghoshal, N.G., 158
Giachetti, A., 467
Gianetto, R., 558
Gianfagna, A., 317
Gianferrari, E.A., 278
Giannasi, D., 298
Giannini, M.S., 299
Gianoulakis, C., 228
Gianturco, C., 137
Gibbins, L.N., 547
Gibbons, D.F., 356
Gibbons, I.R., 112
Gibbons, J.W., 347
Gibbs, F.P., 328
Gibbs, K.E., 194
Gibbs, R.H., 421
Gibbs, S.P., 556
Giblin, D.R., 330
Gibor, A., 61
Gibson, A.C., 18
Gibson, A.J., 302
Gibson, D., 559
Gibson, D.M., 146
Gibson, D.T., 465
Gibson, E.S., 254
Gibson, F., 496
Gibson, J.C., 26
Gibson, J.C., 268
Gibson, J.E., 227, 489
Gibson, K.S., 480
Gibson, M.A., 542
Gibson, M.D., 146
Gibson, N.H.L., 227
Gibson, Q., 302
Gibson, Q.H., 302
Gibson, R.E., 197
Gibson, S., 247, 259
Gibson, W.L., 3
Gibson, W.W., 455
Giddens, J.E., 106
Giddings, E.L., 194
Giddings, G.G., 344
Giddings, W., 497
Gidez, L., 330
Giebisch, G.H., 81
Gier, H.T., 166, 168
Giere, F.A., 123
Giersch, Sr.M.C., 169
Giese, H., 155
Giese, R.L., 148
Giesel, J.T., 90
Gieske, T., 147
Giesmann, L.A., 176
Giesy, R.M., 362
Gifford, E.D., 67
Gifford, G., 476
Gifford, G.E., 94
Gifford, P., 206
Gilani, S., 282
Gilberg, N., 535
Gilberston, R.L., 18
Gilbert, A.H., 481
Gilbert, B.E., 449
Gilbert, C.R., 126
Gilbert, D.L., 66
Gilbert, E.E., 109
Gilbert, F.F., 547
Gilbert, G.E., 362
Gilbert, I.K., 126
Gilbert, J., 150
Gilbert, J.C., 114
Gilbert, J.J., 277, 552
Gilbert, L., 465
Gilbert, M.L., 88
Gilbert, P.W., 302, 304
Gilbert, R., 393
Gilbert, S., 288
Gilbert, W., 210
Gilbert, W.B., 344
Gilbert, W.G., 192
Gilbert, W.J., 222
Gilbertson, J.R., 413
Gilboe, D.D., 521
Gilbreath, R.L., 288
Gilbreth, H.B., 429
Gilchrist, R.D., 34
Gilder, H., 307
Gile, L.H., 462
Giles, F.A., 138
Giles, F.E., 198
Giles, J., 549
Giles, N.H., 105, 106
Giles, R.H.Jr., 493
Gilfford, E.M.Jr., 51
Gilford, J.H., 196
Gilgut, C.J., 219
Gilham, P.T., 150
Gilkeson, R.A., 502
Gill, A.E., 55
Gill, D.E., 199
Gill, D.M., 211
Gill, F.B., 409
Gill, G.D., 229
Gill, J.D., 508
Gill, R.W., 57
Gillary, L., 113
Gillaspy, J.E., 456
Gille, G.L., 255
Gilles, K.A., 351
Gillespie, D., 538
Gillespie, J.H., 409
Gillespie, J.R., 52
Gillespie, A., 33
Gillespie, D.M., 102
Gillett, G.W., 57
Gillett, L., 239
Gillett, T.A., 475
Gillette, D.D., 158
Gillette, E.L., 69
Gillham, N.W., 332
Gillian, L.A., 179
Gillies, C.G., 79
Gilliland, F.R., 5
Gilliland, S.E., 344
Gillingham, J.T., 424
Gillings, D.B., 341
Gillis, E., 396
Gillis, G., 216
Gillis, R.E., 282
Gillman, C., 319
Gillott, C., 562
Gillum, R., 467
Gilman, J.P.W., 548
Gilman, L., 95
Gilmer, P., 467
Gilmer, R.M., 307
Gilmore, C.R., 192
Gilmore, E.C., 459
Gilmore, H.C., 401
Gilmore, H.R., 139
Gilmore, J.P., 273
Gilmore, S.A., 24
Gilmore, T.D., 224
Gilmour, T.H.J., 562
Gilpartick, J.D., 307
Gilpin, M., 59
Gilpin, R.H., 196
Gilpin, R.W., 399
Gilroy, J.J., 207
Giltz, M.L., 364
Gilvarg, C., 280
Gilvert, W., 211
Gingery, R.E., 366
Gingras, G., 558
Gingras, J., 560
Ginoza, W., 404
Ginsberg, H.S., 301
Ginsburg, J.M., 104
Ginski, J.M., 445

Ginther, O.J., 520
Giolli, R.A., 55
Gionet, C.L., 428
Girard, D., 514
Girard, E., 560
Girard, J.M., 558
Girardi, A.J., 412
Giri, S.N., 53
Girod, C., 418
Girotti, A., 513
Girvin, E.C., 455
Gisin, B.E., 333
Gisolfi, C., 162
Gist, D.H., 370
Gist, G.R.Jr., 364
Githens, S., 185
Gittelgohn, A.M., 198
Gittins, A.R., 117
Given, R.R., 62
Gjessing, H.W., 483
Gladden, F.G., 464
Glade, R.W., 480
Glagov, S., 133
Glaha, G.C., 370
Glantz, R.M., 453
Glantz, M., 298
Glantz, P.J., 403
Glaser, D.A., 43
Glaser, E.M., 204
Glaser, L., 263
Glaser, M., 135
Glaser, R.M., 405
Glasgon, L.L., 183
Glasgow, L.A., 474
Glass, A.O., 3
Glass, B.P., 375
Glass, E.D., 401, 402
Glass, E.H., 306
Glass, H.B., 325
Glass, L.E., 60
Glass, R.J., 269
Glass, R.L., 241
Glasser, M.E., 270
Glasser, J.H., 469
Glasser, R.L., 340
Glasser, R.M., 372
Glasser, S.G., 449
Glassman, E., 340
Glauser, E., 409
Glauser, S.C., 409
Glaviano, V.V., 134
Glazer, A.N., 56
Glazer, R.I., 101
Glazewski, Sr.J., 284
Glazier, L.R., 77
Gleason, G.J., 224
Gleason, G.R., 224
Gleason, L.N., 180, 181
Gleeson, Rev.T.F., 406
Gleich, C.S., 161
Glen, I.R., 533
Glencoe, J., 510
Glendening, M., 15
Glenister, D.W., 513
Glenn, C.L., 537
Glenn, J., 327
Glenn, J.L., 327
Glenn, R.B., 229
Glenn, R.C., 195
Glew, R.H., 413
Glick, P., 513
Glickson, J., 11
Glime, J., 228
Gliss, D.C., 487
Glisson, S., 124
Glitz, D.G., 56
Globus, M., 551
Glomset, J.A., 499
Gloss, A., 540
Glover, D.V., 147
Glover, S., 331
Glover, T.J., 309
Gloyd, H.K., 18
Gnewuch, W., 76
Glum, T.A., 349
Glusker, J., 410
Goan, H.C., 443
Gobbins, R.H., 189
Gobson, D.L., 561
Gochenaur, S.E., 295
Godard, J., 348
Godchaux, W., 206
Goddard, D.R., 409
Goddard, S., 522
Goddard, W.B., 49
Goder, H.A., 278
Godfrey, C.L., 459
Godfrey, E.B., 118
Godfrey, P.J., 217
Godfrey, R., 88
Godin, C., 557
Godish, T., 127
Godke, R.A., 183
Godleski, J.J., 399
Godman, G.C., 301
Godsey, R.K., 484
Godzyk, Sr.J., 278
Goebel, N.B., 425
Goeden, R.D., 58
Godhler, B.H., 29
Goellner, E.K., 154
Goering, K., 265
Goering, J.J., 14
Goering, R., 240
Goeringer, G.C., 85
Goerke, R.J., 60
Goerner, L.W., 453
Goerte-Miller, C.C.Jr., 429
Goetinck, P.F., 77, 78, 79
Goetsch, D.D., 108
Goetz, H., 349
Goff, A.M., 354
Goff, R.A., 379
Goff, R.C., 399
Goff, T.L., 236
Goforth, R., 350
Goggans, J.F., 5
Goh, K., 328
Goin, C., 15
Gojmerac, W.L., 518
Gokhale, D.V., 59
Gold, A., 300
Gold, A.H., 45
Gold, L., 71
Gold, L.L., 396
Gold, P., 556
Gold, P.E., 55
Gold, P.S., 318
Goldberg, D., 298
Goldberg, E., 126
Goldberg, E.B., 217
Goldberg, E.H., 294
Goldberg, J.M., 47, 124, 133
Goldberg, L.I., 133
Goldberg, M.L., 60
Goldberg, P.B., 399
Goldberg, R.B., 234
Goldberg, S.J., 491
Goldberg, S.R., 64
Goldberger, M.E., 399
Goldblatt, P.J., 79
Goldblith, S.A., 213
Golden, B.R., 317
Golden, E., 3
Golden, J.F., 11
Goldey, J.A., 360
Goldfarb, A.R., 235
Goldie, M., 123
Goldman, C.R., 47, 48
Goldman, D., 399
Goldman, D.E., 399
Goldman, L., 204
Goldman, M., 434
Goldman, P.C., 254
Goldman, R.D., 391
Goldman, S., 371
Goldner, H., 408
Goldner, K.J., 445
Goldring, S., 265
Goldsberry, K.L., 66
Goldsberry, S., 9
Goldsby, A., 514
Goldschneider, I., 79
Goldschmidt, E.P., 463, 464
Goldsmith, C., 95
Goldsmith, D.P.J., 273
Goldsmith, M.H., 80
Goldsmith, M.R., 54
Goldsmith, T.H., 80
Goldsmith, V., 490
Goldstein, A., 42, 467
Goldstein, B., 40
Goldstein, D.B., 42
Goldstein, E., 15
Goldstein, H.E., 365
Goldstein, I.J., 232
Goldstein, J., 307
Goldstein, J.S., 208
Goldstein, L., 71, 491
Goldstein, M.A., 328
Goldstein, M.E., 556
Goldstein, M.N., 262, 328
Goldstein, N., 30
Goldstein, P.J., 405
Goldstein, R., 211
Goldstein, S., 296
Goldstein, S.F., 242
Goldstick, T.K., 126
Goldstone, A., 127
Goldthwait, D.A., 355
Goldthwaite, J.J., 207, 211
Goldwasser, E., 131
Goldwater, L.J., 341
Goleman, L., 363
Golemon, C., 187
Goler, P., 212
Goll, D.E., 155, 157
Golley, F.B., 106
Goltz, S.M., 195
Golub, E.S., 150
Golub, S., 56
Golübic, S., 207
Golubow, J., 297, 298
Golus, H.M., 65
Gomez, R., 468
Gona, A., 282
Gona, O.D., 286
Gonor, J.J., 385
Gondos, B., 60
Gong, J.K., 319
Gonsalves, N., 420
Gonsalves, R.C.J., 541
Gonsoulin, G.J., 437
Gonzales, F., 126
Gonzales, R., 150
Gonzales, R.R., 36
Gonzalez, D., 58
Gonzalez, G., 95
Gonzalez, N., 239
Gonzalez, N.C., 172
Gooch, J.L., 397
Good, B.H., 225
Good, B.J., 185
Good, D.L., 167
Good, E.E., 366
Good, G.L., 303
Good, H., 35
Good, H.M., 546
Good, R., 287
Goodall, D.W., 477
Goodard, W.J.G.B., 219
Goodchild, C.G., 100
Goode, L., 343
Goode, M.D., 199
Goode, M.J., 24
Goode, R., 297
Goode, R.C., 550
Goode, R.P., 297
Goodell, H.G., 489
Gooden, J.H., 196
Goodenough, U., 211
Gooder, H., 339
Goodfellow, J., 205
Goodin, J.R., 461
Gooding, G.V.Jr., 345,
Gooding, R.H., 530
Goodman, A.B., 69
Goodman, B.L., 129
Goodman, D.C., 325
Goodman, E.M., 522
Goodman, F.R., 467
Goodman, G.J., 379
Goodman, H., 60
Goodman, I., 300
Goodman, J.D., 142
Goodman, J.I., 227
Goodman, J.W., 60
Goodman, K., 243
Goodman, M., 235
Goodman, M.F., 62
Goodman, R.E., 28
Goodman, R.M., 139
Goodman, R.N., 260
Goodman, V.H., 57
Goodnight, C.J., 236
Goodnow, R., 419
Goodrich, C.A., 357
Goodrich, K., 326
Goodrich, M.A., 128
Goodrick, F.E., 149
Goodson, C., 21
Goodwill, R.N., 177
Goodwin, E.E., 200
Goodwin, H., 135
Goodwin, K., 402, 403
Goodwin, R.E., 299
Goodwin, R.H., 75
Goodwin, T.L., 23
Goodyear, C.P., 440
Goonewardene, H.F., 148
Goos, R.D., 421
Goppel, S., 313
Gorbach, S., 56
Gorbman, A., 498
Gorczynski, R., 550
Gordeuk, S.Jr., 403
Gordon, A.M., 500
Gordon, A.R., 258
Gordon, A.S., 312
Gordon, B.A., 551
Gordon, C., 96
Gordon, C.C., 266
Gordon, C.N., 54
Gordon, C.V., 199
Gordon, D., 238
Gordon, D.B., 60
Gordon, D.F., 76
Gordon, D.T., 366
Gordon, I., 64
Gordon, J., 125, 190
Gordon, J.C., 156
Gordon, J.E., 213
Gordon, K.R., 409
Gordon, M., 55, 486
Gordon, M.A., 327
Gordon, M.P., 499
Gordon, R., 541
Gordon, R.C., 228
Gordon, R.E., 152, 288
Gordon, V., 24
Gordon, W.E., 453
Gordon, W.M.Jr., 507
Gordon, W.K., 287
Gore, R.W., 19
Gorecki, D., 128
Goree, H.K., 360
Gorfein, D., 89
Gorham, E., 241, 242
Gorham, P.R., 531
Gorini, L., 211
Gorka, R.R., 406
Gorman, A.I.F., 208
Gorman, G., 55
Gorrell, R.M., 276
Gorsic, J., 120
Gorski, J., 517
Gorski, L.J., 74
Gorthy, W.C., 68
Gortner, R.A., 80
Gortzig, C.F., 303
Gorvosky, M., 327
Gorz, H.J., 271
Gorzynski, E.A., 319
Goshen, C.E., 446
Goslin, W.E., 4
Gosline, J.M., 535
Goslow, G., 15
Goss, C.M., 84
Goss, J.A., 166
Goss, R.C., 162
Gosselin, R.E., 277
Gossett, B.J., 423
Gosz, J.R., 293
Gotelli, D.M., 28
Goth, A., 467'
Gotshall, R.W., 513
Gottlieb, A.A., 288
Gottlieb, D., 139
Gottlieb, F.J., 412

Gottlieb, S., 150
Gotto, A.M., 448
Gottschalk, C.W., 340
Gottschang, J.L., 369
Gotwald, W.H.Jr., 327
Goudsmit, E.M., 230
Gough, P., 158
Gouin, F.R., 202
Gould, A.B., 82
Gould, C., 503
Gould, E.M., 211
Gould, F.W., 458
Gould, M.D., 420
Gould, S., 382
Gould, S.J., 211
Gould, S.R., 373
Gould, W.A., 365
Gould, W.P., 422
Goulden, C.E., 409
Goulding, R.L., 386
Goulson, H.T., 342
Gouoni, D., 495
Gourley, E.V., 487
Gourley, L.M., 248
Goust, J.M., 428
Govind, C.K., 550
Govindjee, J.F., 136
Gowans, C.S., 260
Gowgiel, J.M., 123
Gowings, D.D., 142
Goyd, W.H., 548
Goyne, G., 38
Gozad, G.C., 379
Grabau, M.C., 255
Graber, C.D., 428
Grable, A.E., 500
Grabowski, C., 95
Grabowski, M., 95
Grabowski, P.H., 271
Grabowski, S.J., 390
Grabowski, W.B., 291
Gracen, V.E., 304
Grady, D.V., 27
Graef, P.E., 427
Graetz, D.A., 93
Graf, G., 352
Graf, W.T., 32
Graff, D.J., 477
Graff, P.W., 133
Graff, R.J., 263
Graffis, D.W., 138
Graffius, J.H., 368
Grafstein, B., 308
Grafton, B.F., 186
Grafton, T.S., 319
Grafton, W.N., 508
Grah, R.F., 44
Graham, A., 359
Graham, A.F., 555
Graham, A.R., 548
Graham, B.F., 155
Graham, C.R., 198
Graham, D.E., 150
Graham, E.R., 258
Graham, E.T., 444
Graham, J., 154
Graham, J.B., 338, 340
Graham, J.D., 354
Graham, L.C., 539
Graham, L.T., 146, 191
Graham, M.D., 356
Graham, R.C.Jr., 356
Graham, R.D., 384
Graham, S.O., 503
Graham, T., 9
Graham, T.E., 221
Graham, T.M., 10
Graham, W., 540
Graham, W.D., 423
Graham, W.J., 323
Grahame, R.E., 537
Grahn, D., 126
Granberry, D., 9
Grand, D.W., 542
Grand, L.F., 345, 346
Grandall, R.B., 94
Grande, J.A., 288
Granett, P., 288
Graney, D.O., 499
Granger, C.R., 262
Granger, G.A., 54
Granholm, N., 432
Granner, D.K., 161
Granstaff, E.L., 376
Grant, B., 7
Grant, B.S., 484
Grant, C.T., 332
Grant, D., 26
Grant, D.C., 332
Grant, D.W., 69
Grant, G.C., 490
Grant, M.A., 128
Grant, N.G., 297
Grant, P., 387
Grant, P.R., 556
Grant, R., 298
Grant, R.J., 297
Grant, V.E., 465
Grant, W.C., 96
Grant, W.C.Jr., 220
Grant, W.F., 330
Grantham, B.J., 255
Grappel, S., 408
Grass, F.S., 152
Grassard, G.R., 541
Grassmick, R.A., 360
Grasso, J.A., 208
Grat, P.F., 261
Grau, C.R., 48
Graubard, K., 499
Graul, W., 342
Grave, J., 241
Graver, H.T., 411
Graves, A., 354
Graves, A.P., 336
Graves, C.H.Jr., 249
Graves, D., 411
Graves, D.J., 157
Graves, H.B., 403
Graves, J., 175
Graves, L.L., 461
Graves, R., 486
Graves, R.C., 354
Graves, W.E., 321, 322
Gravett, H.L., 460
de Graw, W., 272
Gray, B.H., 244
Gray, B.W., 6
Gray, C.T., 277
Gray, E., 181
Gray, E.D., 244
Gray, F., 376
Gray, G.G., 120
Gray, G.R., 241
Gray, H., 192
Gray, H.B.Jr., 464
Gray, H.G., 422
Gray, I., 85
Gray, J., 387
Gray, J.I., 547
Gray, L.E., 139
Gray, M.W., 542
Gray, P., 380, 412,
Gray, R.C., 522
Gray, R.D., 180
Gray, R.G., 397, 460
Gray, R.H., 233
Gray, S.W., 101
Gray, T., 378
Gray, T.C., 177
Gray, W.D., 126
Gray, W.R., 473
Graybill, D.L., 390
Grayson, G., 3
Grayson, J., 550
Grayson, R.I., 552
Grayson, R.L., 27
Graziadei, P., 88
Graziadei, W.D., 324
Greason, A.L., 192
Greathouse, T.R., 65
Greaves, W.S., 396
Greco, D.G., 199
Grecz, N., 121
Greear, P.F-C., 105
Green, B.R., 534
Green, C., 285
Green, C.E., 240
Green, D., 455
Green, D.M., 280
Green, E.A., 99
Green, G.G., 493
Green, H.B., 249
Green, J., 293, 316, 471, 541
Green, J.G., 119
Green, J.H., 188
Green, J.M., 541
Green, J.T., 433
Green, J.W., 283, 288
Green, M., 256, 287, 298
Green, M.H., 59
Green, P.B., 41
Green, R., 76
Green, R.F., 59
Green, R.H., 538
Green, R.J., 148
Green, R.L., 429
Green, R.W., 152
Green, S., 402
Green, W., 382, 398
Green, W.A., 397
Green, W.W., 200
Greenberg, B.G., 341
Greenberg, D., 60
Greenberg, G.R., 232
Greenberg, M., 88
Greenberg, R., 135
Greenblatt, I.M., 77
Greenblatt, R.B., 103
Greene, A.A., 288
Greene, D.W., 219
Greene, E., 331
Greene, G.L., 91
Greene, G.S., 17
Greene, H.L., 446
Greene, J.E., 6
Greene, J.M., 215
Greene, J.T., 108
Greene, R.A., 549
Greene, R.C., 333
Greene, V.W., 244
Greenfield, D.W., 126
Greenfield, J.E., 333
Greenfield, L., 95
Greenfield, L.J., 95
Greenfield, S.S., 288
Greenhouse, G., 54
Greenidge, G.N.H., 543
Greenleaf, W.H., 5
Greenough, W.B.III., 198
Greenspan, K., 146
Greenwald, G., 180
Greenwald, L., 364
Greenwood, F.C., 112
Greenwood, H.L., 195
Greenwood, L.F., 550
Greenwood, M., 479
Greer, B.W., 123
Greer, H.A.L., 376
Greer, J.K., 379
Greer, M.H., 157
Greer, R.T., 158
Greer, S.B., 95, 96
Greffenius, R.J., 26
Grega, G.J., 228
Gregersen, H.M., 243
Gregg, A.E., 488
Gregg, C.M., 454
Gregg, E.W., 430
Gregg, J., 29
Gregg, J.H., 90
Gregg, J.R., 332
Gregg, R.E., 71
Gregg, R.V., 180
Gregg, T.G., 360
Gregoire, M.H., 213
Gregorek, Rev.J.C., 395
Gregorian, V., 409
Gregory, D.P., 278
Gregory, G.R., 231
Gregory, K.F., 547
Gregory, M.E., 344
Gregory, P.T., 536
Gregory, R.W., 193
Gregory, W.C., 344
Gregory, W.W.Jr., 178
Greocjis. A., 433
Greif, R.L., 308
Greig, J.K., 168
Greij, E.D., 224
Greisman, S.E., 204
Grell, Sr.M., 239
Grell, R.F., 442
Greller, A.M., 299
Gremillion, L.C., 185
Grenard, R.S., 149
Grenier, P., 557
Gressin, A., 316
Grether, D., 239
Greub, L.J., 522
Grewe, A., 239
Grey, R., 544
Grey, R.D., 52
Gribble, D.H., 53
Grieb, E., 431
Grieff, D., 512
Griel, L.C.Jr., 403
Grier, J.W., 350
Grierson, J., 317
Grierson, W., 92
Gries, C.L., 306
Gries, G., 375
Griesbach, R.A., 120
Griesel, W.O., 31
Grieve, C.M., 530
Grieve, D.G., 547
Griffen, S.A., 440
Griffin, B., 466
Griffin, B.R., 83, 283
Griffin, D., 369
Griffin, D.G.III., 90
Griffin, D.H., 321
Griffin, D.N., 187
Griffin, F., 308
Griffin, G.F., 78, 79
Griffin, H.E., 414
Griffin, M.J., 380
Griffin, Sr.M.R., 278
Griffin, R., 373
Griffin, R.H., 194
Griffin, R.P., 424
Griffin, W., 186
Griffing. B., 363
Griffith, B., 104
Griffith, D., 157
Griffith, M.M., 90
Griffith, R.B., 177
Griffith, R.E., 360
Griffiths, A.E., 422
Griffiths, A.J.F., 534
Griffiths, J.C., 551
Griffiths, J.M., 302
Griffiths, M.W., 498
Grifo, A.P.Jr., 365
Grigal, D.F., 241, 243
Grigarick, A.A., 49
Grigg, B.J., 201
Griggs, G., 21
Grigsby, E.M., 374
Grigsby, R.D., 458
Grilione, P.L., 32
Grillo, R.S., 295
Grillos, S.J., 28
Grim, E., 244
Grim, J.N., 15
Grimes, D.J., 515
Grimes, G., 309
Grimes, J.V., 452
Grimes, W.J., 19
Griminger, P., 288
Grimm, J.K., 486
Grimm, R.B., 87
Grimshaw, R.H., 365
Grindley, G., 321
Grindle, D.H., 216
Gring, D.M., 398
Grinnell, A., 55
Grinnell, A.D., 57
Grinnell, E.H., 269
Grinnell, F., 466
Grip, C.M., 121
Grisham, J.W., 340
Grisolia, S., 172
Grissom, P.E., 25
Griswold, B., 364
Griswold, J., 297
Griswold, J.G., 297
Gritton, E.T., 516
Grizzle, J.E., 341

Groat, C.S., 354
Grob, H.S., 295
Grobstein, C., 59
Grobstein, P., 133
Groder, R.H., 394
Grodins, F.S., 63, 64
Grodner, M.L., 183
Grodsky, G., 60
Grodums, E.I., 562
Groesbeck, E., 396
Grogan, C.O., 304
Grogan, D.E., 262
Grogan, J.B., 69, 251
Grogan, R.G., 50
Grollman, S., 199
Groman, N.B., 499
Gronau, D.M., 167
Gronlund, A.F., 535
Gronvall, J.A., 232
Gronwall, R.R., 168
Groody, T.C., 30
Groom, A.C., 551
deGroot, J., 60
Grosch, D.S., 344, 345
Groseclose, N.P., 99
Grosklags, J.H., 126
Gross, C.F., 401
Gross, E.L., 362
Gross, E.M., 79
Gross, F.B., 394
Gross, H.D., 344
Gross, J., 66
Gross, J.B., 463
Gross, J.H., 197
Gross, Sr.M., 524
Gross, P.P., 30
Gross, P.R., 320
Gross, R., 210
Gross, S.R., 333
Gross, W.L., 524
Grossberg, A.L., 319
Grossberg, R., 121
Grossberg, S.E., 513
Grossfield, J., 297
Grossfeld, R., 302
Grossman, J.A., 174
Grossman, L., 208
Grossman, L.I., 235
Grossman, M.I., 57
Grosso, L., 310
Grosvenor, C.E., 445
Groth, D.P., 101
Grove, A.R., 403
Grove, D., 494
Grove, D.G., 416
Grove, S., 97
Grover, J.H., 4
Grover, R.M., 219
Grover, T.A., 428
Groves, D.L., 347
Groves, P.A., 275
Groves, W.E., 428
Grubb, R., 380
Grubb, T., 364
Grubbs, D., 30
Grubbs, D.Z., 404
Grubbs, J.C., 445
Grube, G.E., 269
Gruber, P.J., 214
Gruber, S., 96
Gruber, S.E., 214
Grubitz, G.III., 487
Gruda, J., 560
Gruen, H.E., 562
Gruendling, G.K., 324
Gruener, R.P., 19
Gruenwedel, D.W., 49
Grula, E.A., 375
Grun, J., 288
Grun, P., 403
Grunau, J.A., 260
Grunberger, D., 300
Grundmann, A.W., 473
Grundset, E., 439
Grundy, E., 270
Grunes, D.L., 302
Grunwald, C., 177
Grunewald, J.O., 172
Grunewald, R., 521
Gruszczyk, Rev.J.H., 289
Gryder, R.M., 203
Guarino, A.I., 469
Guarino, A.J., 469
Guarlnick, S.A., 121
Guas, A.E., 260
Gubanich, A.A., 276
Gubler, C.J., 472
Guckian, J.C., 467
Gudauskas, R.T., 4
Gude, R.H., 97
Guenther, E., 432
Guentherm, J.J., 376
Guenthner, H.R., 276
Guest, M.F., 154
Guest, M.M., 468
Guest, W.C., 22
Gugler, C., 270
Guidotti, G., 210
Guidroz, G.P., 189
Guidry, D.J., 185
Guier, W.H., 197
Guiford, H., 514
Guiha, N., 399
Guiher, J.K., 139
Guild, P.A., 341
Guild, W.R., 333
Guillery, R.W., 520
Guillory, R.J., 112
Guillou, L.J.Jr., 99
Guimond, R., 207
Guisti, J., 420
Gull, D.D., 93
Gulley, R., 96
Gumbreck, L.G., 380
Gump, D.W., 482
Gumpf, D.J., 58
Gumport, R., 135
Gunasekaran, M., 250
Gunckel, J.E., 288
Gundersen, K.R., 112
Gundersen, R., 239
Gunderson, H., 270
Gundy, S.C., 397
Gunn, R.B., 333
Gunnea, P., 196
Gunning, G.E., 190
Gunsalus, I.C., 135
Gunter, A.V., 407
Gunter, K., 329
Gunter, T.E., 329
Gunther, F.A., 58
Gunther, F.G., 58
Gunther, W.C., 153
Gupta, A.P., 288
Gupta, K.C., 86
Gupta, N., 270
Gupta, V.S., 563
Gupton, O.W., 492
Guralnick, L., 341
Gurin, S., 90
Gurley, W.H., 106
Gurney, T., 43
Gurwith, M., 538
Gusseck, D.J., 36
Gussett, J., 486
Gussin, G.N., 160
Gustafson, A.H., 192
Gustafson, D., 415
Gustafson, J.A., 321
Gustafson, J.P., 537
Gustafson, R., 465
Gustafson, R.T., 399
Gustin, R.D., 433
Gustine, D.L., 401
Guthrie, C., 60
Guthrie, D., 33
Guthrie, D.J., 496
Guthrie, F.E., 344
Guthrie, J.E., 537
Guthrie, J.P., 551
Guthrie, J.W., 117
Guthrie, P., 374
Guthrie, R., 319
Guthrie, R.D., 14
Guthrie, R.K., 426
Guthrie, R.L., 507
Guthrie, W.D., 157
Guthwin, H., 297, 298
Gutknecht, J., 333
Gutmann, H.R., 244
Gutman, L.E., 333
Gutmann, L., 509
Gutook, H., 321
Guttay, A.J.R., 78
Guttay, J.R., 78
Guttentag, M., 82
Guttman, R., 296
Guttman, S.I., 366
Guyer, G.E., 226
Guyer, K.F.Jr., 491
Guymon, J.F., 51
Guyon, J.C., 128
Guyselman, J.B., 222
Guyton, R.F., 401
Guzman, C.A., 549
Guzman, V.L., 93
Gwazdauskas, F.C., 493
Gwilliam, G.F., 387
Gwinn, J., 369
Gwym, E., 205
Gwyn, D.G., 542
Gwynn, G.R., 344
Gwynn, R.H., 79
Gygi, F.R., 163
Gyles, C.L., 548
Gyles, N.R., 23
Gynes, L., 558
Gynn, G.J., 138
Gyrisco, G.G., 303
Gysel, L.W., 225

H

Ha, S.J., 400
Haack, D., 273
Haagens, M.J., 420
Haak, R., 146
Haaland, R.L., 3
Haan, C.T., 179
Haar, J.L., 491
Haard, N., 288
Haas, F., 150, 313
Haas, H.J., 129
Haas, K.B., 236
Haas, R., 30
Haas, R.H., 458
de la Haba, G., 410
Habbe, D., 434'
Habeck, D.H., 91
Habeck, J., 266
Habeeb, A., 11
Habel, R.E., 305
Haber, A., 317
Haber, A.H., 317
Haber, B., 467
Haber, L., 299
Haberfield, P., 298
Habermann, H.M., 196
Habig, R.L., 333
Hackbarth, W.P., 186
Hackenbrock, C.R., 466
Hacker, B., 327
Hacker, P., 463
Hacker, R.G., 68
Hacker, R.R., 547
Hackett, E.R., 186
Hackett, J.T., 490
Hackney, P.A., 366
Hackwell, G.A., 28
Hadden, N.C., 443
Hadder, J.C., 354
Haddock, J., 150
Haddy, F.J., 228
Hadfield, R.R., 199
Hadle, F.B., 168
Hadley, H.H., 138
Hadley, M.E., 18
Hadley, N., 15
Hadley, W.F., 318
Hadow, H.H., 70
Hadwiger, L.A., 503
Hadziyen, D., 530
Haefner, P.A.Jr., 490
Haen, P., 37
Haensly, W.E., 460
Haenzel, W., 414
Haertel, J.D., 433
Haertel, L., 432
Haesloop, J.G., 337
Hafeez, M.A., 177
Hafemann, D.R., 512
Hafen, L., 149
Hafer, P.E., 324
Haffley, P., 145
Hafley, W.L., 346
Hafs, H.D., 228
Haft, J.S., 512
Hagan, J., 124
Hagadorn, I.R., 339
Hagedorn, D.J., 519
Hagedorn, H.H., 218
Hagemann, L.E., 288
Hageman, R.H., 138
Hagemoser, W.A., 158
Hagen, A.F., 271
Hagen, C.W.Jr., 144
Hagen, D.W., 540
Hagen, H.K., 66
Hagen, K.S., 44
Hagen, P.O., 333
Hager, E., 290
Hager, L.P., 135
Hagerman, D., 72
Hagerman, F.C., 368
Hagerman, H.E., 227
Haggard, B., 21
Haggard, J.D., 165
Haggis, A.J., 230
Haggis, G., 544
Haghiri, F., 364
Hagino, N., 469
Hagiwara, S., 57
Haglan, B.W., 155
Haglund, B.H., 431
Haglund, W.A., 503
Hagmeier, E.M., 536
Hague, D.R., 387
Hahlman, D.L., 178
Hahn, A., 164
Hahn, M.E., 290
Hahn, B., 265
Hahn, R.J
Haibak, G.L., 327
Haight, C.P., 431
Haight, R.D., 32
Haight, T., 210
Hailman, J.P., 516
Haines, B., 122
Haines, D.E., 509
Haines, H.B., 379
Haines, T., 317
Haines, T.H., 298
Haines, T.R., 104
Hainsworth, F.R., 326
Hair, J.D., 424
Haire, C., 251
Hairston, N.C., 231
Haist, R.E., 550
Hajek, B.F., 3
Hajra, A.K., 232
Hakala, M., 320
Hakala, R., 224
Hakes, J.E., 131
Hakim, A.M., 314
Hakim, F., 177
Hakim, R.S., 177
Hakomori, S., 499
Halbert, S.A., 499
Haldane, E.V., 543
Halde, C., 60
Haldeman, J.H., 427
Haldeman, J.R., 88
Haldiman, J.T., 253
Hale, E.B., 403
Hale, F., 458
Hale, J., 411
Hale, K.K.Jr., 108
Hale, N.S., 77
Hale, R.A., 194
Hale, W.L., 319
Hales, D.C., 434
Haley, A.J., 199
Haley, B.E., 527
Haley, L.E., 542
Haley, S.R., 112
Halfacre, R.G., 424
Halgren, L., 240
Halgren, L.A., 240
Hall, A.B., 485

Hall, A.E., 58
Hall, B.D., 499
Hall, B.K., 542
Hall, B.V., 136
Hall, C.B., 93
Hall, C.C., 464
Hall, C.E., 468
Hall, C.S., 508
Hall, C.V., 156
Hall, D.H., 333
Hall, D.J., 227
Hall, D.L., 207
Hall, D.W., 218
Hall, E., 182
Hall, E.R., 501
Hall, F.R., 280, 363
Hall, G.F., 364
Hall, G.W., 484
Hall, I.M., 58
Hall, J., 8, 317
Hall, J.B., 112
Hall, J.D., 383
Hall, J.E., 333, 509
Hall, J.G., 40
Hall, J.K., 401
Hall, J.L., 288, 370
Hall, J.R., 200
Hall, J.S., 390
Hall, J.W., 241
Hall, K., 522
Hall, K.N., 78, 79
Hall, K.R., 279
Hall, L.M., 11
Hall, L.T., 378
Hall, M., 103, 357
Hall, M.T., 138
Hall, M.W.S., 5
Hall, N.M., 454
Hall, O.F., 280
Hall, O.G., 442
Hall, P., 69
Hall, P.F., 54
Hall, R., 547
Hall, R.B., 156
Hall, R.C., 148
Hall, R.E., 520
Hall, R.G., 36
Hall, R.H., 545
Hall, R.J., 399
Hall, T., 64
Hall, T.S., 262, 263
Hall, V.A., 431
Hall, V.E., 57
Hall, W., 317
Hall, W.C., 317, 332
Hall, W.H., 244
Hall, W.K., 103
Hall, W.P.III., 418
Hallberg, C.W., 354
Hallberg, R., 302
Haller, E., 145
Haller, J.R., 61
Haller, R.W., 519
Hallett, P.E., 550
Hallgren, A.R., 243
Hallin, R., 37
Hallinan, E.J., 390
Halling, S., 161
Halloran, W.F., 521
Hallum, J.V., 380
Halma, R., 392
Halmagyi, D., 301
Halperin, V., 186
Halperin, W., 497
Halpern, B., 127
Halpern, B.P., 302, 304
Halpern, M., 311, 325
Halpin, Sr.M., 311
Halsey, C.F., 241
Halsey, L.H., 93
Halver, J.E., 498
Halvorson, A.R., 502
Halvorson, H., 538
Halvorson, W.L., 421
Ham, G.E., 241
Ham, R.G., 71
Hamada, S.H., 27
Haman, A.C., 162
Hamann, D.D., 344
Hamann, C.B., 174
Hamburger, V., 263
Hamburgh, M., 297
Hamdy, M.K., 106
Hamel, C., 560
Hamel, G.Jr., 11
Hamelink, J.L., 149
Hamer, D.W., 69
Hamg, W.T.Jr., 491
Hamilton, A.L., 537
Hamilton, B., 85
Hamilton, B.A., 288
Hamilton, C., 378
Hamilton, C.L., 411
Hamilton, E.S., 354
Hamilton, H.E., 179
Hamilton, H.L., 488
Hamilton, J., 146
Hamilton, J.A., 146
Hamilton, J.B., 325
Hamilton, J.G., 285
Hamilton, J.M., 255
Hamilton, J.R., 508
Hamilton, J.W., 527
Hamilton, K.C., 16
Hamilton, L.H., 513
Hamilton, L.S., 303
Hamilton, M., 374
Hamilton, M.A., 65
Hamilton, M.G., 424
Hamilton, P.B., 345, 346
Hamilton, R., 408
Hamilton, R.A., 114
Hamilton, R.B., 183
Hamilton, R.D., 538
Hammerstedt, R.H., 403, 404
Hamilton, R.K., 534
Hamilton, R.L., 60
Hamilton, R.W., 123
Hamilton, T.H., 465
Hamilton, W.G., 324
Hamilton, W.G.III., 49
Hamilton, W.J., 47
Hamkalo, B., 54
Hamkalo, B.A., 54
Hamlen, R.A., 91
Hamlett, H.D., 494
Hamlin, C.R., 356
Hamm, D.C., 434
Hammen, C.S., 421
Hammer, K., 55
Hammer, P.A., 149
Hammer, U.T., 562
Hammerman, D., 310
Hammerstrom, H.E., 159
Hammes, G.G., 302
Hammett, H.L., 249
Hammett, L.K., 345
Hammond, A., 173
Hammond, B.F., 411
Hammond, B.R., 453
Hammond, C.T., 153
Hammond, D., 317
Hammond, E.G., 155, 157
Hammond, G.S., 61
Hammond, J.J., 350
Hammond, L.C., 93
Hammond, M.A.R., 450
Hammond, R.K., 174
Hamon, J.H., 176
Hamori, E., 190
Hamosh, P., 85
Hamparian, A., 235
Hamparian, V.V., 367
Hampel, A., 120
Hampton, J.K.Jr., 295
Hampton, J.R., 375
Hampton, R.O., 384
Hampton, S., 299
Hampton, R.E., 223
Hamre, H.T., 354
Hamrick, J.L.III., 170
Hanahan, D.J., 19
Hanan, J.J., 66
Hanawalt, P.C., 41
Hanawalt, R.B., 489
Hanaway, J., 263
Hanchey, P.J., 68
Hanchey, R.H., 183, 184
Hancock, D.R., 385
Hancock, H.M., 175
Hancock, J.G., 45
Hancock, J.M., 192
Hancock, K.F., 100
Hancock, K.P., 100
Hancock, M.B., 467
Hand, C., 61
Hand, C.H.Jr., 43
Hand, G.S., 10
Hand, W., 37
Handbenger, M., 188
Handel, M.A., 442
Handelman, G.H., 314
Handler, E., 298
Handler, E.E., 297
Handler, E.S., 297
Handler, P.H., 333
Handler, S., 450
Handley, E.J., 171
Handlin, D.L., 424
Handricks, F.S., 459
Hanebrink, E.L., 20
Hanegan, J.L., 496
Hanes, D.F., 28
Hanes, T.L., 30
Haney, A.W., 136
Hang, Y.D., 307
Hanic, L.A., 554
Hanig, R.C., 145
Haning, Q., 126
Hankinson, J., 509
Hanks, A.R., 458
Hanks, D.L., 254
Hanks, G.D., 145
Hanks, J., 75
Hanks, J.P., 297
Hanks, R.J., 474, 477
Hanks, S.L., 290
Hanle, R.L., 395
Hanley, E.W., 473
Hanlin, R.T., 105, 108
Hann, R., 418
Hann, W.D., 354
Hanna, R.L., 458
Hannah, L.C., 93
Hannah, P.R., 481
Hannah, R.H., 537
Hannah, R.S., 103
Hannan, C.K., 538
Hanne, J.M., 113
Hannum, D., 119
Hansel, A.K., 38
Hanselka, C.W., 456
Hansell, R., 550
Hansen, A.P., 344
Hansen, Sr.B., 222
Hansen, D., 497
Hansen, H., 258
Hansen, H.E., 155
Hansen, H.L., 243
Hansen, H.W., 240
Hansen, K.L., 89
Hansen, M.F., 166
Hansen, P., 159
Hansen, P.M.T., 365
Hansen, R., 499
Hansen, R.J., 53
Hansen, R.M., 67
Hansen, R.S., 502
Hansens, E.J., 288
Hanshe, J., 316
Hansing, E.D., 168
Hansman, C., 70
Hanson, A.W., 64
Hanson, E.D., 80
Hanson, E.W., 519
Hanson, F.E., 205
Hanson, H , 15
Hanson, J.B., 136, 138
Hanson, J.C., 28
Hanson, J.F., 218
Hanson, L.D., 241
Hanson, R., 10, 300, 408
Hanson, R.P., 520
Hanson, R.S., 517
Hanson, W.D., 345
Hanson, W.R., 31
Hanson, W.R.J., 476
Hanten, H.B., 245
Hantsbarger, W.M., 68
Hanway, D.G., 271
Hanzely, L., 126
Happ, G.M., 68
Happel, L.T., 186
Harakal, C., 409
Haramoti, F.H., 114
Harary, I., 56
Harayanan, S., 285
Harbaugh, D., 440
Harber, P.A., 239
Harbers, L.H., 167
Harbin, R., 471
Harbison, R.D., 446
Harbury, H.A., 277
Harclerode, J.E., 391
Harcombe, P., 453
Hard, C.G., 241
Hard, G.S., 11
Hard, W.L., 273
Hardee, D.D., 249
Harden, P.H., 314
Harder, J.D., 364
Harder, R.W., 117
Hardie, E.L., 492
Hardin, H., 24
Hardin, J.W., 343, 346
Hardin, N.C., 508
Hardin, R.T., 530
Hardin, S.H., 428
Harding, B., 64
Harding, C.V., 230
Harding, D.E., 157
Harding, J.A., 47
Harding, P.G., 552
Hardinge, M.G., 36
Hardison, J.R., 385
Hardman, K., 126
Hardy, D.E., 114
Hardy, G., 246
Hardy, G.W., 22
Hardy, H.B., 21
Hardy, J.A.III., 486
Hardy, J.L., 46
Hardy, J.W., 90
Hardy, P.H.Jr., 198
Hardy, S.M., 288
Hardy, W., 208
Hardy-Fallding, M.H., 548
Hare, J.D., 328
Hare, L.N., 222
Hare, M.L., 247
Hare, W.W., 249
Harein, P.K., 240
Haresign, T., 310
Harford, A.G., 318
Hargis, W.J.Jr., 490
Hargraves, P.E., 421
Hargreaves, L.A., 108
Hargrove, G.L., 402
Hargrove, R.D., 495
Hariharan, P.V., 90
Harkaway, R.W., 234
Harkema, R., 346
Harker, D., 320
Harkins, R.M., 380
Harkness, D.R., 95
Harlan, J.R., 138
Harland, M., 549
Harley, P., 541
Harlin, M.M., 421
Harling, J., 533
Harm, W., 173
Harman, A.S., 196
Harman, D., 273
Harman, G.E., 307
Harman, W.N, 323
Harman, W.J., 183
Harmet, K., 126
Harmon, B.G., 138
Harmon, G.A., 392
Harmon, K.W., 350
Harmon, L.D., 356
Harmon, W., 30, 73
Harms, C.E., 416
Harms, P.G., 458
Harms, R.H., 93
Harms, V.L., 562
Harms, W.S., 103
Harmsen, R., 546
Harned, M., 522
Harnest, G., 479

Harold, F., 72
Harp, G.L., 20
Harper, A.E., 517
Harper, E.T., 146
Harper, H., 60
Harper, J.A., 384
Harper, J.D., 5
Harper, J.E., 138
Harper, K.T., 472
Harper, L., 549
Harper, W.J., 362, 365
Harper, W.L., 78, 79
Harpst, H., 395
Harpst, J.A., 355
Harrap, K.A., 283
Harrell, B.E., 434
Harrell, B.R., 454
Harrelson, M., 335
Harrendorf, K., 259
Harrill, I., 67
Harries, H., 540
Harriman, P.H., 333
Harrington, B.D., 343
Harrington, D., 28
Harrington, E.M., 336
Harrington, G., 74
Harrington, J.D., 401
Harrington, J.F., 51
Harrington, Sr.M.M., 209
Harrington, R.D., 279
Harrington, W.F., 197
Harrington, W.H., 356
Harris, A., 188
Harris, A.H., 468
Harris, A.J., 3
Harris, A.K., 339
Harris, B., 91, 538
Harris, B.G., 452
Harris, B.H., 220
Harris, B.L., 384
Harris, C., 450
Harris, C.L., 324, 509
Harris, D.L., 158
Harris, D.O., 177
Harris, E.D., 458
Harris, F., 500
Harris, F.A., 249
Harris, F.E., 473
Harris, G.A., 503
Harris, G.C.Jr., 23
Harris, G.P., 545
Harris, H., 5, 253, 514
Harris, H.A., 32
Harris, H.B., 106
Harris, I., 182
Harris, J., 414
Harris, J.A., 137
Harris, J.B., 522
Harris, J.C., 249
Harris, J.E., 101
Harris, J.O., 166
Harris, J.P., 454
Harris, J.R., 346
Harris, J.S., 333
Harris, J.W., 440
Harris, K.R., 295
Harris, L., 236
Harris, M., 43
Harris, M.K., 458
Harris, N.S., 467
Harris, R.A., 146
Harris, R.E., 470, 507
Harris, R.L., 341
Harris, R.M., 18, 19
Harris, R.R., 3
Harris, S.E., 449
Harris, S.W., 34
Harris, T.L., 148
Harris, T.M., 491
Harris, T.T., 73
Harris, W.D., 164
Harris, W.K., 219
Harris, W.M., 22
Harresberger, L., 469
Harriman, N.A., 523
Harrington, E.J., 496
Harrington, J.T.Jr., 470
Harrises, A.E., 215
Harrison, A.D., 551
Harrison, A.P.Jr., 260
Harrison, A.T., 526
Harrison, B.F., 472
Harrison, B.H., 219
Harrison, C.W., 35
Harrison, E., 224
Harrison, F., 469
Harrison, F.P., 201
Harrison, F.W., 327
Harrison, G., 467
Harrison, G.K., 201
Harrison, G.W., 151
Harrison, H.K., 477
Harrison, J., 426
Harrison, J.L., 398
Harrison, J.O., 104
Harrison, J.R., 324
Harrison, M.B., 304
Harrison, M.D., 68
Harrison, N.A., 440
Harrison, P.C., 138
Harrison, R., 337
Harrison, R.D., 29
Harrison, R.P., 407
Harrison, R.W., 421
Harrison, S.C., 210
Harrison, T., 374, 391
Harrison, W., 224
Harrison, W.K., 197
Harrist, R.B., 469
Harrold, R., 350
Harry, H.H., 460
Harry, M., 463
Harsanyi, Z.P., 308
Harsch, M., 415
Harshfield, G.S., 434
Harshman, S., 447
Harstirn, W., 139
Harston, C.B., 502
Hart, B.A., 482
Hart, B.L., 52
Hart, D.A., 466
Hart, D.E., 551
Hart, E., 157
Hart, G.E., 475, 477
Hart, J.L., 213, 485
Hart, J.M., 271
Hart, N.H., 288
Hart, R.A., 255
Hart, R.G., 407
Hart, R.W., 362
Hart, T.W., 415
Hart, W.H., 50
Hartberg, W.K., 102
Hartenstein, R.C., 322
Harter, D.H., 301
Harter, R.D., 280
Hartesveldt, R.J., 32
Harthill, M.P., 29
Hartig, W.J., 316
Harting, J.K., 520
Hartke, G.T., 168
Hartl, D.L., 150
Hartland-Rowe, R., 531
Hartley, D.E., 260
Hartley, J.C., 177
Hartley, M.W., 10
Hartline, D.K., 59
Hartline, P.H., 137
Hartman, A.D., 135
Hartman, B.J., 276
Hartman, F.O., 365
Hartman, G.G., 396
Hartman, H.B., 195
Hartman, H.L., 446
Hartman, J., 87
Hartman, J.R., 178
Hartman, K.A., 421
Hartman, M.J., 31
Hartman, P., 197
Hartman, P.A., 155, 156
Hartman, R., 340
Hartman, R.D., 378
Hartman, R.E., 315
Hartman, R.T., 412
Hartman, W.D., 80
Hartman, W.L., 364
Hartmann, G., 420
Hartmann, G.C., 420
Hartmann, R.W., 114
Hartnett, J.C., 479
Hartroft, P.M., 263
Hartsfie, D.S.M., 460
Hartshorne, D.J., 391
Hartsook, E.W., 401
Hartsuck, J., 380
Hartung, R., 233
Hartwick, E.G., 547
Hartwig, N., 155
Hartwig, N.A., 433
Hartwig, N.L., 401
Hartwig, Q.L., 406
Harty, M., 410
Harry, Sr.N., 511
Hartzell, C.R., 404
Harvais, G., 544
Harvel, R.C.Jr., 451
Harvenstein, H., 95
Harvey, A., 258
Harvey, C., 335, 462
Harvey, E.J., 246
Harvey, H.H., 550
Harvey, H.T., 32
Harvey, J.E., 263
Harvey, L.H., 266, 423
Harvey, M.J., 438, 452
Harvey, P.H., 344
Harvey, R., 64, 439
Harvey, R.A., 283
Harvey, R.B., 244, 454
Harvey, R.G., 516
Harvey, R.W., 343
Harvey, S., 11
Harvey, S.C., 449
Harvey, T.L., 167
Harvey, W.A., 49
Harvey, W.H., 143
Harvey, W.R., 365, 396
Harvie, N.R., 232
Harvill, A.M.Jr., 486
Harward, M.E., 384
Harwood, R.F., 502
Hasbrouck, F., 15
Hascall, V.C., 232
Haschemeyer, A., 298
Haschemeyer, A.E., 297, 298
Haschemeyer, R.H., 307
Haselkorn, R., 131, 132
Haselton, S., 286
Hash, J.H., 447
Hasheen, A.M., 178
Hashimoto, E.I., 473
Hashimoto, K., 445
Hashimoto, T., 124
Hasiak, R.J., 156
Haskell, D.A., 215
Haskill, J.S., 428
Haskin, H.H., 288
Haskins, A., 349
Haskins, E.F., 497
Haskins, F.A., 271
Haslam, D.F., 437
Hasler, A.D., 516
Hass, L.F., 405
Hasse, B.L., 394
Hassell, M.D., 175
Hassett, J.J., 138
Hassid, A., 340
Hassler, T., 34
Hassler, W.W., 346
Hassur, R., 32
Hastings, J.W., 210, 211
Hastings, P.J., 535
Hastings, R., 287
Hatch, A.H., 66
Hatch, C.R., 118
Hatch, H., 192
Hatch, J.J., 219
Hatch, L.A., 551
Hatch, R.C., 108
Hatch, R.W., 193
Hatch, S., 458
Hatcher, J.D., 546
Hatcher, R.F., 488
Hatchett, J.H., 249
Hatfield, E.E., 138
Hatfield, G.W., 55
Hathaway, G.M., 57
Hathaway, R., 65
Hathaway, R.R., 473
Hattan, D.G., 204
Hatten, B.A., 452
Hattman, S., 327
Haubrich, R., 358
Hauck, F.E., 206
Haug, A., 226
Haug, F., 514
Haugh, J., 317
Haughton, G., 339
Haugse, C., 350
Hauke, R.L., 421
Haun, J.R., 424
Haun, C. K., 63
Haupt, R., 157
Hauschka, S.D., 498, 499
Hauschka, T., 319
Hauser, Sr.M.M., 397
Hauser, M.M., 547
Hauser, W., 556
Hausler, C.L., 129
Hausmann, E., 319
Haussler, M.R., 19
Haust, M., 552
Hauxwell, D.L., 34
Havard, W.C., 492
Havas, H.F., 408
Haven, D.S., 490
Haven, G.T., 273
Haver, C.B., 555
Haverkamp, H.J., 144
Havertz, D.S., 477
Haves, C.J., 316
Havgstad, M., 280
Haviland, J., 342
Havis, J.R., 218
Hawes, D.T., 200
Hawke, S.D., 389
Hawkins, C.M., 469
Hawkins, D.R., 225
Hawkins, G.E., 3
Hawkins, I.K., 103
Hawkins, M.Jr., 507
Hawkins, R., 256
Hawkins, R.H., 475
Hawkins, W.B., 6
Hawkinson, S.W., 441
Hawks, S.N.Jr., 344
Hawksley, O., 253
Hawley, E., 297
Hawley, E.S., 296
Hawley, J.W., 462
Hawley, L., 524
Hawley, P.L., 135
Hawley, R., 397
Haworth, S., 290
Haws, B.A., 476
Hawthorn, W.R., 551
Hawthorne, P.L., 184
Hay, D., 522
Hay, D.E., 552
Hay, J.C., 537
Hay, J.F.Jr., 451
Hay, J.R., 561
Hay, R.J., 372
Hay, R.L., 443
Hayashi, T., 121
Hayashida, T., 60
Hayashida, K., 552
Hayat, M.A., 286
Hayden, B.P., 489
Hayden, R.A., 149
Hayden, V.B., 246
Hayes, B.M., 454
Hayes, E.C.III., 345
Hayes, E.S., 451
Hayes, F.R., 542
Hayes, J.M., 70
Hayes, R., 177
Hayes, S.P., 477
Hayes, R.E., 360
Hayes, R.E., 127
Hayes, S.J., 54
Hayes, T.G., 367
Hayes, W., 416
Hayes, W.J., 446
Hayflick, L., 41
Hayne, D.W., 346
Haynes, D.L., 226
Haynes, F.L.Jr., 345
Haynes, J.D., 317
Haynes, J.F., 193

Haynes, R.C., 117, 562
Haynes, R.E., 367
Haynes, R.H., 553
Hays, D.H., 188, 189
Hays, D.L., 186
Hays, H.A., 165
Hays, H.E.Jr., 407
Hays, K.L., 5
Hays, R.L., 39, 426
Hays, R.M., 176
Hays, S.B., 424
Hays, V.W., 177
Hayse, F.A., 8
Haysman, P., 102
Hayward, J.S., 536
Haywood, B., 359
Haywood, C.R., 173
Hazard, E., 236
Hazel, D.W., 357
Hazel, J., 270
Hazelwood, D.H., 260
Hazelwood, R.L., 463
Hazen, A.G., 350
Hazen, T.E., 155
Hazen, W.E., 37
Hazlett, J., 235
Hazlett, L., 235
Hazucha, M., 556
Head, H.H., 91
Head, M.K., 344
Head, S.C., 382
Heading, J., 486
Headrick, T., 512
Heady, H.F., 44
Heagy, F.C., 551
Heald, C.W., 473
Heald, D., 546
Healey, P.K., 121
Heagle, A.S., 345
Health, M.S.Jr., 341
Healy, E.A., 497
Healy, W.N., 508
Healy, W.R., 209
Heard, W., 88
Hearson, L.L., 153
Hearst, J.A., 116
Heath, A.G., 492
Heath, E.C., 413
Heath, H.D., 30
Heath, I.B., 553
Heath, J.E., 90
Heath, J.L., 203
Heath, M.E., 147
Heath, M.S., 334
Heath, R., 359
Heath, R.L., 57
Heathcott, E., 175
Hebard, W.B., 147
Heberlein, G.T., 262
Hebert, T.T., 345
Hechenbleikner, H.H., 342
Hecht, F., 388
Hecht, M.K., 297, 298
Hecht, N.B., 216
Heck, H.A., 188
Heck, M.C., 23
Heck, O., 290
Heck, O.B., 368
Heck, W.W., 343
Hecker, L.H., 233
Heckman, J.L., 409
Heckman, J.R., 395
Heckmann, R., 472
Heckrotte, C., 334
Heddle, J.A.M., 553
Hedeen, R.A., 199
Hedgcoth, C.Jr., 166
Hedge, G.A., 19
Hedgecock, L.W., 256
Hedges, F.H., 120
Hedgpeth, J.W., 385
Hedine, P.A., 248
Hedlin, A.M., 550
Hedtke, J.L., 383
Hedlund, L., 259
Hedman, S.C., 245
Hedreen, J.C., 197
Hedrich, P.W., 172
Hedrick, D., 33
Hedrick, H.G., 186
Hedrick, J.L., 48
Hee, Y.S., 489
Heed, W.B., 18, 19
Heeney, M.W., 65
Heere, L.J., 428
Hefferren, J., 127
Hefner, R.A., 360
Heft, F.E., 364
Hegarty, T., 207
Hegeman, G., 144
Heggestad, C., 243
Heggie, A.D., 356
Hegmann, J., 160
Hegre, O., 243
Hegstad, G.D., 159
Hegstad, P., 470
Hehman, R., 370
Hehre, E., 310
Hehre, E.J., 330
Heick, H.M.C., 548
van der Heide, L., 79
Heidger, P.M., 160
Heidinger, R.C., 129
Heidrick, M.L., 273
Heidt, G.A., 25
Heidtke, H.E., 222
Heifetz, R.M., 39
Heig, V., 522
Heikes, P.E., 68
Heil, R.D., 65
Heiligmann, R.B., 156
Heilman, P.E., 503
Heim, L.R., 513
Heim, W.G., 65
Heimbigner, M.C., 387
Heimbrook, M., 73
Heimer, L., 489
Heimsch, C., 360
Heimstra, N.W., 434
Hein, C.H., 229
Hein, D., 66
Hein, J., 523
Hein, R.R., 290
Hein, W.W., 431
Heinemann, R.L., 486
Heiner, R.E., 240
Heinrich, B., 44
Heinrikson, R.L., 131
Heisler, G.M., 322
Heinz, J.M., 197
Heise, E.R., 347
Heise, J.J., 102
Heiser, C.B.Jr., 144
Heisey, S.R., 228
Heisler, C.R., 275
Heithaus, E.R., 126
Heitman, H.Jr., 48
Heitz, J.R., 248
Held, A., 297, 298
Held, E.E., 498
Held, J., 515
Helgason, S.B., 537
Helgeson, J.P., 515, 519
Helgesen, R.G., 303
Helinski, D.R., 59
Hellam, D.C., 263
Helleiner, C.W., 542
Heller, A., 133
Heller, H.D., 41
Heller, H.F., 390
Heller, J.J., 400
Heller, R.C., 118, 390
Hellerqvist, C.G., 446
Hellier, T.E., 464
Helling, R.B., 231
Hellman, J.L., 201
Hellmann, R., 317
Hellmers, H., 332, 334
Hellquist, C.G., 207
Hellstrom, I., 499
Helm, A.C., 137
Helm, J.P.III., 196
Helm, R.E., 454
Helm, W., 476
Helm, W.T., 477
Helman, S.I., 137
Helmreich, J.E., 390
Helms, C.W., 106
Helms, D.R., 426
Helms, T.J., 271
Helms, R.W., 341
Helmstetter, C., 319
Helseth, F., 331
Helsm, J.A., 44
Heltne, P.G., 197
Helton, A.W., 117
Helyer, B.J., 356
Hembrough, F.B., 158
Hembry, F.G., 183
Hemken, R.W., 177
Hemmes, R.B., 329
Hempel, F., 333
Hemphill, A.F., 8
Hemphill, D.D., 260
Hemphill, D.V., 38
Hemphill, F.M., 469
Hemphill, H.E., 326
Hempling, H.G., 429
Henderlong, P.R., 364
Henderson, A., 400
Henderson, A.R., 551
Henderson, B., 246
Henderson, B.Jr., 422
Henderson, D.M., 117
Henderson, D.W., 522
Henderson, J., 475
Henderson, J.A., 477
Henderson, J.H.M., 8
Henderson, J.M., 49
Henderson, K., 299
Henderson, L.E., 453
Henderson, L.L., 461
Henderson, L.M., 241
Henderson, M.A., 89
Henderson, N.E., 531
Henderson, N.S., 227, 345
Henderson, R.M., 435
Henderson, V., 182
Henderson, W.R., 345
Hendrich, C.E., 104
Hendricks, A.C., 492
Hendricks, D.G., 475
Hendricks, J.J., 32
Hendricks, J.R., 342
Hendrickson, H.T., 342
Hendrickson, J.R., 18
Hendrix, C.N.G., 199
Hendrix, F.F.Jr., 108
Hendrix, J.E., 68
Hendrix, J.W., 178, 503
Hendrix, R., 413
Hendrix, S.S., 395
Hendry, C.S., 97
Henegan, J.B., 186
Heninger, R.W., 472
Henken, H.W., 35
Henley, R., 92
Henneman, E., 212
Henneman, H.A., 225
Hennen, J.F., 148
Hennen, S., 512
Hennessy, A.V., 234
Henney, C.S., 198
Henney, H., 463
Henning, W.L., 436
Henninger, M.R., 288
Henninger-Trax, A., 398
Hennings, G., 286
Henny, G.C., 409
Henricks, D.M., 424
Henrickson, C.E., 177
Henrickson, J.S., 31
Henrikson, R.C., 300
Henrickson, R.L., 376
Henry, B.L., 67
Henry, C.S., 84
Henry, E., 298
Henry, E.W., 230
Henry, G.F., 336
Henry, H., 104
Henry, J.P., 64
Henry, M.D., 397
Henry, R., 239
Henry, R.C., 164
Henry, R.D., 139
Henry, W.F., 280
Hensill, J.S., 40
Hensley, M.M., 227
Hensleigh, P.A., 172
Hensley, S., 374
Henson, E.B., 480
Henson, H., 378
Henson, J.W., 444
Henson, R., 331
Henson, W.H.Jr., 179
Hentges, J.F.Jr., 91
Henzlik, R.E., 142
Heogler, C., 311
Heoksema, W., 223
Hepler, J.Q., 450
Hepler, P.K., 41
Hepler, P.R., 195
Hepler, R.W., 402
Hepner, L.W., 249
Heppel, L., 302
Heppel, L.A., 302
Heppner, F., 421
Hepting, G.H., 345, 346
Herald, S.L., 76
Herbek, J., 177
Herbel, C.H., 292
Herbener, G.E., 180
Herbert, D.C., 469
Herbert, J.D., 185
Herbert, J.H., 93
Herbert, M., 390
Herbert, T., 95
Herbst, E.J., 280
Herbst, G., 243
Herdendorf, C.E.III., 364
Herforth, R.S., 237
Hergenrader, G., 270
Herin, R.A., 69
Herman, D.C., 351
Herman, F.K., 242
Herman, N.J., 164
Herman, R., 288
Herman, S.S., 398
Hermann, H.R., 106
Hermans, C.O., 28
Hermans, J., 340
Hermanson, H.P., 337
Hermodson, M., 499
Hermreck, A.S., 172
Hernandez, A.C., 97
Hernandez, M., 418
Hernandez, T.P., 184
Hernandez-Perez, M.J., 405
Herndon, K., 358
Herndon, W.R., 441
Herner, R.C., 226
Herold, R.C., 411
Herr, D.E., 364
Herr, J.M., 429
Herr, L.J., 366
Herreid, C.F., 318
Herrer-Alva, A., 228
Herrera, N.M., 8
Herrero, S., 531
Herrick, A.M., 108
Herrick, R.B., 113
Herring, S.W., 134
Herrington, L.P., 322
Herriott, J.R., 499
Herrmann, H., 79
Herrman, J.L., 180
Herrmann, E.C., 34
Herrmann, H., 77
Herrmann, J.E., 212
Herrmann, K., 148
Herrmann, S., 70
Herrnkind, W., 88
Herron, J.W., 177
Herron, M.A., 460
Herschel, -., 26
Herschler, M., 369
Herschman, H.R., 56
Hersey, J.R., 20
Hersey, S.J., 101
Hersh, E.M., 449
Hersh, L.B., 466
Hersh, R.T., 170
Hershberger, C.L., 137
Hershberger, T.V., 401
Hershberger, W.K., 498
Hershey, A.L., 6
Hershey, J.W.B., 53
Herskowitz, I., 387
Herskowitz, I.H., 297
Herskowitz, I., 298

Hershkowitz, M., 340
Herson, D.S., 82
Herst, G.R., 482
Hertel, E.W., 162
Hertelendy, F., 256
Hertig, W.H., 507
Hertz, L., 562
Hertz, L.B., 241
Hervey, A., 298
Herzberg, D.G., 85
Herzberg, M., 112
Hesby, J.H., 458
Heslop, A., 33
Hespenheide, H., 55
Hess, A., 283
Hess, C.E., 288
Hess, C.J., 166
Hess, G.D., 400
Hess, G.P., 302
Hess, G.S., 177
Hess, J., 253
Hess, J.L., 493
Hess, M., 185
Hess, M.L., 492
Hess, W.M., 472
Hessinger, D.A., 97
Hessler, A.C., 479
Hetenyi, G.Jr., 549
Hester, D., 450
Hester, J.L., 186'
Hesterberg, G.A., 229
Hetling, L.J., 314
Hetzel, H.R., 122
Heubner, E., 538
Heuer, A.E., 30
Heuser, E.D., 398
Heusner, A.A., 52
Heusser, C.J., 31
Hevly, R., 15
Hew, C.L., 541
Hewatt, W.G., 490
Hewitson, W.M., 208
Hewitt, A.A., 29
Hewitt, D., 549
Hewitt, S., 253
Hewitt, W.B., 50
Hewston, J.G., 34
Hexter, W.M., 206
Hey, J., 154
Heyn, A.N.J., 185
Heyne, E.G., 166
Heywood, S.M., 77, 79
Hiatt, A.J., 177
Hiatt, H.H., 212
Hiatt, V., 354
Hibbard, A.D., 260
Hibbard, E., 403
Hibbs, E.T., 102, 147
Hibler, C.P., 69
Hichar, J.K., 317
Hickey, J.J., 520
Hickey-Weber, E.D., 315
Hickle, P.C., 302
Hickman, C.A., 458
Hickman, C.E., 508
Hickman, C.J., 552
Hickman, C.P.Jr., 495
Hickman, J.C., 407
Hickman, M., 531
Hickman, W., 308
Hicks, A., 105
Hicks, D.R., 5, 240
Hicks, E.A., 157
Hicks, F.W., 403
Hicks, G., 437
Hicks, G.S., 542
Hicks, J.R., 93
Hickson, F.T., 109
Hidayet, M.A., 551
Hiebert, E., 92
Hielman, A.S., 441
Hiesinger, J.F., 434
Hift,H., 521
Higerd, T.B., 428
Higginbotham, E.H., 189
Higginbotham, R.D., 180
Higgins, A., 380
Higgins, D., 21
Higgins, E.A., 201
Higgins, E.S., 491
Higgins, I.T.T., 234
Higgins, J., 411
Higgins, J.A., 81
Higgins, J.J., 410
Higgins, J.V., 227
Higgins, L.C., 471
Higgins, M.L., 408
Higgins, M.W., 234
Higgins, R.E., 117
Higgins, W.J., 199
High, W.I., 113
Highkin, H.R., 31
Highley, S.W., 9
Highton, R.T., 199
Hightower, D., 460
Hightower, J.A., 491
Higman, J., 96
Hikawyj, I., 186
Hikida, R.S., 368
Hilbert, M.S., 233
Hild, W.J., 467
Hildebrand, H.H., 456
Hildebrand, M., 52
Hildebrandt, A.C., 519
Hildebrandt, J., 500
Hildemann, W.H., 56
Hildreth, P., 342
Hileman, L.H., 22
Hilf, R., 328
Hilferty, F.J., 208
Hilgard, H.R., 61
Hilgeman, R.H., 17
Hilger, A.E., 429
Hill, A., 182, 336
Hill, A.C., 473
Hill, C., 393
Hill, C.H., 346
Hill, C.W., 405
Hill, D.R., 436
Hill, D.W., 474
Hill, E.P., 238
Hill, F.C., 147
Hill, F.W., 47, 50
Hill, G., 393
Hill, G.A., 474
Hill, G.B., 333
Hill, G.C., 69
Hill, G.J., 236
Hill, J.H., 157
Hill, J.L., 280, 297
Hill, J.M., 103
Hill, L.B., 379
Hill, L.M., 484
Hill, L.V., 374
Hill, L.W., 349, 352
Hill, M.K., 344
Hill, Sr.M.M., 268
Hill, R., 270
Hill, R.B., 421
Hill, R.D., 537
Hill, R.P., 550
Hill, R.E., 271
Hill, R.G.Jr., 365
Hill, R.L., 333
Hill, R.M., 272, 362
Hill, R.R.Jr., 401
Hill, R.W., 227
Hill, S.B., 556, 557
Hill, W.D., 265
Hill, W.F., 10
Hill, W.R., 497
Hillar, M., 461
Hillcoat, B.L., 545
Hille, B., 500
Hillemann, H.H., 386
Hiller, L.K., 503
Hillerbrand, H., 297
Hillers, J.K., 502
Hilliard, S.D., 354
Hillier, D., 522
Hillier, J.C., 376
Hillis-Colinvaux, L., 364
Hillman, D., 162
Hillmann, R.C., 344
Hilloowala, R.A., 509
Hillson, C.J., 403
Hillyer, G.V., 418
Hillyer, I.G., 130
Hilpert, M., 431
Hilsenhoff, W.L., 518
Hilston, N.W., 526, 527
Hiltbold, A.E., 3
Hiltibran, R.C., 138
Hilton, D.F.C., 555
Hilton, F.K., 180
Hilton, M.A., 180
Hilton, R., 207
Hilty, J., 443
Himel, C.M., 106
Himelick, E.B., 139
Himes, C.L., 390
Himes, F.L., 364
Himes, J.A., 94
Himes, M., 296
Himes, M.H., 297
Himes, R.H., 170
Himms-Hagen, J., 548
Himwich, W., 273
Hinck, L.W., 20
Hinckley, C., 213
Hinculey, T.M., 261
Hinde, H.P., 264
Hindel, R., 151
Hindle, R.J., 422
Hindman, J.C., 501
Hindman, J.L., 501
Hinds, D.S., 27
Hinds, F.C., 138
Hinds, J.W., 208
Hinds, M., 147
Hine, A., 96
Hine, R.B., 18
Hinegardner, R.T., 61
Hinerman, C., 253
Hines, B., 550
Hines, J., 463
Hines, M.H., 367
Hines, R.H., 167
Hinesly, T.D., 138
Hing, F.S., 113
Hink, W.F., 363
Hinke, J.A.M., 535
Hinkel, P.C., 302
Hinkle, D., 327
Hinkle, D.A., 22
Hinkley, R., 126
Hinkson, R.S.Jr., 422
Hinman, A.R., 314
Hinners, S.W., 129
Hinni, J.B., 257
Hinrichs, C.H., 382
Hinrichs, D.J., 501
Hinrichs, H.A., 377
Hinsdill, R.D., 517
Hinshaw, D.B., 35
Hinshaw, L.B., 380
Hinson, F.D., 348
Hinton, C.W., 358
Hinton, F.W., 485
Hinton, P.F., 252
Hintz, H., 358
Hintz, S.D., 157
Hintz, T.R., 149
Hinze, G.O., 65
Hinze, H.C., 521
Hippensteel, P., 152
Hipple, D., 128
Hiraizumi, Y., 465
Hirata, A.A., 171
Hirce, E., 254
Hironaka, M., 118
Hirsch, A.M., 241
Hirsch, C., 299
Hirsch, C.R., 356
Hirsch, H.R., 180
Hirschberg, C., 256
Hirschberg, E., 282
Hirschberg, J., 95
Hirschberg, R., 298
Hirschfeld, W.H., 119
Hirschfield, H.I., 312
Hirschman, A., 325
Hirschmann, H., 355
Hirsh, D.C., 53
Hirsh, D.I., 71
Hirshon, J., 310
Hirst, K.G., 313
Hirt, B.J., 274
Hirth, H.F., 473
Hiruki, C., 530
Hisaki, E., 549
Hisaw, F.L., 380
Hischfeld, W.F., 555
Hitchcock, C.L., 497
Hitchcock, D.J., 224
Hitchins, A.D., 177
Hitchings, G.H., 333
Hitchner, S.P., 305
Hite, J.M., 20
Hite, R.E., 403
Hites, R., 335
Hitt, J.B., 372
Hittle, C.N., 138
Hiu, D., 111
Hively, R.H., 95
Hixson, F., 29
Hlastala, M., 500
Ho, C., 413
Ho, H.H., 323
Ho, H-Z., 150
Ho, M., 414
Ho, R., 95
Ho, Y.L., 36
Ho, Y.T., 514
Hoadley, A.D., 286
Hoadley, A.W., 102
Hoak, D.C., 142
Hoar, R.M., 300
Hoar, W.S., 535
Hobbie, J.E., 346
Hobbs, C., 359
Hobbs, E.L., 477
Hobbs, H.H.III., 484
Hoberman, H., 330
Hobkirk, R., 551
Hobson, L.A., 536
Hoch, F.L., 232
Hoch, G., 321
Hochachka, P.W., 535
Hochman, B., 442
Hocker, R., 374
Hocker, W.H.Jr., 280
Hockman, C.H., 137
Hoctor, Sr.M., 311
Hodges, C.S., 334
Hodges, C.S.Jr., 345
Hodges, J., 533
Hodges, R., 531
Hodgdon, A., 280
Hodge, D., 222
Hodges, C.F., 156
Hodges, C.S., 346
Hodges, H.F., 147
Hodges, J.L., 46
Hodges, T.K., 148
Hodgetts, R.B., 535
Hodgins, D., 467
Hodgins, D.S., 380
Hodgins, H.O., 498
Hodgkins, E.J., 5
Hodgson, A.S., 502
Hodgson, B., 470
Hodgson, E., 344
Hodgson, E.S., 216
Hodgson, R.H., 350
Hodgson, R.W., 47
Hodgson, V., 210
Hodgson, W., 117
Hodson, A., 242
Hodson, A.C., 240
Hodson, H.H.Jr., 129
Hodsdon, J.M., 499
Hodson, M., 147
Hodson, R.C., 82
Hodson, W., 8
Hoefling, A., 455
Hoeft, R.G., 138
Hoehler, K.A., 340
Hoel, D.G., 341
Hoerlein, A.B., 69
Hoerner, T.A., 155
Hoeschen, R.J., 538
Hoekstra, W.G., 517
Hoese, H.D., 191
Hoessle, C., 256
van der Hoeven G.A., 168
Hof, L.B., 327
Hofer, K., 88
Hoff, D.J., 364
Hoff, J.E., 149

Hoff, J.G., 215
Hoff, K.M., 357
Hoff, V.J., 455
Hoffee, P.A., 413
Hoffer, R.M., 149
Hoffert, J.R., 228
Hoffler, K., 150
Hoffman, A.C., 400
Hoffman, A.S., 497
Hoffman, D., 391
Hoffman, D.A., 77, 79
Hoffman, D.E., 507
Hoffman, E.M., 94
Hoffman, F., 395
Hoffman, F.M., 407
Hoffman, G., 336
Hoffman, G.R., 434
Hoffman, H.A., 87
Hoffman, H.H., 11
Hoffman, J., 226
Hoffman, J.C., 112
Hoffman, J.D., 259
Hoffman, J.F., 81, 199
Hoffman, J.K., 91
Hoffman, J.L., 180
Hoffman, L.D., 401
Hoffman, L.G., 161
Hoffman, L.H., 446
Hoffman, L.W., 136, 234
Hoffman, N.A., 340
Hoffman, R.A., 299
Hoffman, R.L., 336, 487
Hoffman, R.S., 172
Hoffmann, E.O., 186
Hoffmann, H.P., 283
Hoffmann, P.C., 133
Hoffmeister, D.F., 136
Hofman, W.F., 104
Hofmann, H., 478
Hofmann, K., 413
Hofmann, I., 200
Hofmann, T., 549
Hofreter, J.H., 327
Hofsaess, F., 393
Hofslund, P.B., 245
Hofstad, M.S., 158
Hofstad, O.M., 291
Hofstetter, R., 95
Hofstra, G., 547
Hofstrand, H., 118
Hofwolt, C., 437
Hogan, E.L., 428
Hogan, L., 17
Hogan, P.M., 319
Hogan, R., 342
Hogan, Y., 461
Hogben, C.A.M., 162
Hogenkamp, H.P.C., 161
Hogg, E.C., 174
Hogg, J.C., 556
Hogg, J.F., 298
Hogg, J.M., 29
Hogg, R.W., 356
Hogle, R.M., 158
Hogness, D., 41
Hogue, C.C., 62
Hogue, C.R., 341
Hogue, R., 450
Hoham, R.W., 299
Hohenboken, W.D., 382
Hohlt, H.H., 494
Hohn, B.M., 486
Hohn, M.H., 223
Hokin, M.R., 520
Hoilund, L., 243
Hokanson, J.F., 403
Hokanson, K.E., 245
Holbrook, D.J., 340
Holbrook, J.P., 16
Holbrook, J.R., 445
Holbrook, K.A., 499
Holdaway, P.K., 201
Holden, F.A., 562
Holder, D.J., 432
Holdsworth, R.P., 363
Holeton, F., 550
Holladay, L.A., 446
Holladay, W.G., 446
Holland, I.I., 139
Holland, J.C., 228
Holland, J.F., 225
Holland, J.J., 59
Holland, J.P., 145
Holland, J.W.Jr., 6
Holland, L.A., 292
Holland, N.S., 351
Holland, R., 7
Holland, V.L., 27
Hollander, P., 299
Hollander, P.B., 362
Holldobler, B., 211
Holle, P.A., 221
Holleman, J., 128
Holleman, K.A., 425
Holleman, W.H., 124
Hollenbeck, R.R., 70
Hollenberg, M.J., 535
Hollenberg, P.F., 127
Hollenhorst, D., 119
Holler, J.R., 522
Hollern, T.M., 494
Holley, D.L., 346
Holley, F.B., 335
Holley, W.D., 66
Holliman, D.C., 6, 10
Holliman, R.B., 492
Holling, C.S., 535
Hollingworth, R.M., 148
Hollinshead, M., 282
Hollis, C.G., 438
Hollis, T.M., 403
Hollocher, T.G., 208
Holloway, A.F., 538
Holloway, C.L., 6
Holloway, G.A., 497
Holloway, H.L.Jr., 352
Holloway, J., 437
Holloway, J.R., 209
Holloway, P.W., 490
Holloway, R.L., 424
Holm, R.W., 41
Holman, C.N., 388
Holman, J.A., 227
Holman, J.R., 531
Holman, L.J., 486
Holman, R.T., 244
Holmes, A., 240
Holmes, E., 151
Holmes, E.B., 139
Holmes, F.W., 219
Holmes, I., 542
Holmes, J.C., 531
Holmes, K.V., 466
Holmes, P.K., 195
Holmes, R.B., 549
Holmes, R.T., 277
Holmes, W.F., 263
Holmes, W.N., 61
Holmgren, A.H., 476
Holmgren, N., 298
Holmgren, P., 298
Holmquist, D.R.G., 185
Holmquist, N.D., 186
Holmstedt, J.O.V., 185
Holmstrand, L.L., 245
Holoubek, V., 467
Holowaychuk, N., 364
Holowczak, J.A., 283
Holroyd, R., 397
Holsinger, J.R., 487
Holstein, T.J., 420
Holt, B.R., 262
Holt, C., 236
Holt, D.A., 147
Holt, E., 150
Holt, E.C., 459
Holt, H.A., 23
Holt, H.C., 444
Holt, I.V., 236
Holt, J., 424
Holt, J.G., 156
Holt, K., 214
Holt, P.C., 492
Holt, S.C., 218
Holte, K.E., 116
Holten, D.D., 57
Holtermann, O.A., 320
Holthaus, J.M., 268
Holthuis, L., 96
Holton, R.L., 385
Holton, R.W., 441
Holtz, R.B., 365
Holtz, R.E., 238
Holtzer, A.M., 263
Holtzer, H., 410
Holtzman, E., 300
Holtzman, S., 483
Holtzman, S.G., 101
Holtzmann, O.V., 114
Holoway, J.G., 323
Holyoke, E., 320
Holyoke, E.A., 273
Holyoke, V.H., 195
Holz, G.G.Jr., 326
Holzman, H.E., 416
Homach, L.J., 62
Homan, D.N., 27
Homann, P., 88
Homer, J.T., 378
Hommersand, M.H., 338
Homola, R.L., 194
Homsher, E., 57
Homsher, P.J., 487
Honda, S., 372
Honeycutt, T.J., 454
Honeyell, L.R., 399
Hong, E., 102
Hong, J.S., 208
Hong, R., 521
Hong, S.K., 113
Honig, F., 123
Honigberg, B.M., 218
Honkala, F., 199
Honma, S., 225
Honour, J.H., 120
Hoober, K.J., 408
Hood, C.H., 546
Hood, E.L., 147
Hood, J.B., 227
Hood, J.T., 3
Hood, L.E., 25
Hood, M.N., 88
Hood, R.D., 10
Hood, S., 391
Hoogakier, T., 195
Hook D.D., 425
Hook J.E., 401
Hook, P.W., 166
Hooker, A.L., 139
Hooker, W., 35
Hoomani, J., 12
Hooper, A.B., 242, 244
Hooper, E.T., 231
Hooper, F.F., 231
Hooper, G., 226
Hooper, G.B., 310
Hooper, J., 467, 476
Hoopes, K.H., 472
Hoopingarner, R.A., 226
Hoover, J.D., 167
Hoover, K.B., 400
Hoover, M.W., 344
Hoover, W.L., 459
Hopen, J.H., 138
Hopfer, D.A., 281
Hopkins, A., 464
Hopkins, A.L., 355
Hopkins, C.D., 242
Hopkins, F.S., 156
Hopkins, H., 239
Hopkins, J.W., 263
Hopkins, S.H., 490
Hopkins, T.F., 77
Hopkins, T.L., 97, 167
Hopkins, T.S., 98
Hopkins, W.G., 552
Hopkirk, J.D., 28
Hopla, C.F., 379
Hoppe, D., 68
Hopper, A.F., 288
Hopper, F.A., 291
Hopper, N.W., 364
Hoppin, F.G.Jr., 212
Hopson, J.A., 131
Hopwood, A., 239
Hopwood, M.L., 69
Horak, D.H., 71
Horak, K., 321
Horber, E.K., 167
Horecker, B.L., 307
Horel, F.A., 325
Horgan, D., 316
Horn, A.S., 117
Horn, G.W., 108
Horn, E.G., 315
Horn, D.J., 363
Horn, H.S., 287
Horn, L., 283
Horn, M.H., 30
Hornbein, T., 500
Hornberger, G.M., 489
Hornby, C.A., 534
Horner, A.A., 550
Horner, B.E., 215
Horner, C.E., 385
Horner, H.T., 157
Horner, J.A., 103
Horner, N.V., 452
Horner, R., 14
Hornick, S.A.B., 401
Horner, W., 85
Horner, W.H., 85
Horning, J.F., 277
Hornocker, M., 118
Hornstein, N., 285
Hornung, D., 316
Hornung, M.O., 190
Horoszewicz, J.S., 320
Horowitz, J., 157
Horowitz, J.M.Jr., 47
Horowitz, N.H., 26
Horowitz, P.M., 277
Horowitz, S.B., 234
Horrocks, T.F., 258
Horsfall, W.R., 136
Horst, R.K., 304
Hortink, H.A.J., 366
Hortenstine, C.C., 93
Horton, H., 348
Horton, H.F., 383
Horton, H.R., 343
Horton, K., 245
Horton, M.L., 433
Horton, R.F., 546
Horvat, B.L., 356
Horvath, E., 33
Horvath, R., 354
Horvitz, D.H., 341
Horwitt, M., 256
Horwitz, A., 410
Horwitz, B.A., 47
Horwitz, M., 329
Horwitz, M.S., 330
Horwitz, S., 329
Hosein, E., 556
Hosein, E.A., 555
Hosek, R.S., 95
Hoseknecht, C., 416
Hosford, D.R., 496
Hosford, R., 352
Hoshaw, R.W., 18
Hosick, H.L., 501
Hoshino, K., 537
Hoshizaki, T., 30
Hosier, D.W., 395
Hosier, P.E., 347
Hoskin, F.C.G., 121
Hoskin, G., 397
Hoskins, B., 252
Hossner, J.R., 460
Hostetler, J.R., 367
Hostetler, R.L., 458
Hostetter, D.L., 259
Hostetter, D.R., 484
Hostetter, H.P., 362
Hostetter, R.K., 367
Hotchin, J.E., 327
Hotchkiss, A.T., 180
Hotchkiss, D., 147
Hotchkiss, J., 414
Houck, D., 439
Houck, D.J., 321
Houck, J.C., 85
Houck, W.J., 34
Houde, E., 96
Hougas, R.W., 519
Hougen, F.W., 537
Hough, H.W., 527
Hough, L.F., 288
Hough, R.A., 234

Houghaboom, V.R., 481
Houghton, F.B., 255
Houghton, J.E., 118
Houk, R.D., 430
Houle, J.E., 313
Houp, K., 174
Houpt, T.R., 306
House, E.W., 116
House, H., 427
House, L.E., 282
House, V.H., 363
Houser, H.B., 355
Houser, T., 159
Housman, D.E., 550
Houston, A.H., 544
Houston, C.W., 421
Houston, D., 471
Houston, D.B., 366, 367
Houston, F., 467
Houston, J., 455
Houston, L.L., 170
Houston, M., 181
Houston, M.L., 469
Hoveland, C.S., 3
Hoversland, A.S., 29
Hovin, A.W., 240
Hovorka, J., 209
Howard, A.D., 489
Howard, B., 411
Howard, B.D., 56
Howard, D.H., 56
Howard, E., 103
Howard, F.D., 51
Howard, F.J.Jr., 424
Howard, H.H., 316, 394
Howard, H.M., 387
Howard, J.B., 244
Howard, J.C., 103
Howard, L.V.Jr., 212
Howard, R.A., 210, 211
Howard, R.S., 415
Howard, W.E., 47
Howard, W.T., 517
Howard, W.W.Jr., 292
Howard-Flanders, -., 81
Howarth, B., 108
Howatson, A.F., 550
Howden, H.F., 544
Howe, C., 185
Howe, G., 219
Howe, G.F., 36
Howe, J.P., 118
Howe, K.J., 208
Howe, M., 325
Howe, V.K., 139
Howell, B.J., 319
Howell, C.D., 61
Howell, D.E., 377
Howell, F.G., 252
Howell, G.F., 202
Howell, G.L., 438
Howell, G.S., 326
Howell, H.H., 174
Howell, J.C., 442
Howell, J.N., 414
Howell, L., 258
Howell, M., 188
Howell, M.J., 186
Howell, N.L., 247
Howell, P.G., 546
Howell, R.W., 137, 138
Howell, S., 59
Howell, T., 55
Howell, W.E., 561
Howell, W.M., 8, 302, 303
Howells, D.H., 341
Hower, A.A., 402
Howes, D., 117
Howes, E.L., 60
Howes, R.I., 380
Howes, V.L., 215
Howitt, A.J., 226
Howland, H., 302, 304
Howland, J.L., 192
Howland, I.W., 328
Howland, R.D., 283
Howmiller, R.P., 61
Howton, D.R., 56
Hoye, A.H., 486
Hoy, R., 302, 304
Hoyer, H., 298
Hoyert, J.H., 200
Hoyle, R., 437
Hoyler, G., 387
Hoyman, W.G., 503
Hoyt, D., 143
Hoyt, R.Jr., 208
Hoyt, R.D., 180
Hoyt, S., 502
Hrazdina, G., 307
Hrudka, F., 562
Hruban, Z., 133
Hryniuk, W.M., 538
Hsi, B.P., 469
Hsi, E., 240
Hsia, S., 333
Hsia, S.L., 95
Hsiao, T.H., 476, 477
Hsieh, J., 549
Hsu, H.S., 492
Hsu, P.H., 288
Hsu, R.Y., 326
Hsu, W.T., 131
Hu, A.S.L., 179
Hu, C.Y., 290
Hu, K.H., 356
Huang, A.H., 429
Huang, A.S., 211
Huang, C., 320, 490
Huang, K-Y., 84
Huang, R., 197
Huas, T.E., 65
Huang, S.N., 555
Hubbard, J.E., 180
Hubbard, J.S., 102
Hubbard, L.R., 150
Hubbard, P.L., 189
Hubbard, R.H., 211
Hubbard, W.B., 430
Hubbard, W.R., 331
Hubbell, D.H., 93
Hubbell, S., 160
Hubbs, C., 465
Hubby, J.L., 132
Huber, D.M., 148
Huber, N., 514
Huber, N.M., 515
Huber, T.L., 108
Hubert, W.J., 99
Hubhard, R.W., 36
Hubschman, J.H., 372
Huckabay, J.P., 257
Huckins, C., 449
Huddle, H.L., 323
Huddleston, H., 514
Huddleston, W., 257
Hudock, G.A., 145
Hudor, R., 175
Hudson, A.J., 551
Hudson, B.F., 99
Hudson, B.G., 172, 377
Hudson, E.A., 462
Hudson, E.E., 20
Hudson, F.A., 462
Hudson, H.D., 487
Hudson, J., 24, 522
Hudson, J.B., 535
Hudson, J.W., 302, 303
Hudson, L.W., 424, 503
Hudson, R.A., 235
Hudson, R.J., 533
Hudson, R.P., 340
Hue, L.L., 396
Huether, C.A., 370
Huey, J.W., 25
Huey, M.E., 463
Huf, E.G., 492
Huff, A.N., 493
Huff, C.B., 335
Huff, D.K., 31
Huff, G.C., 155
Huffaker, C.B., 44
Huffaker, R.C., 47
Huffine, W.W., 376
Huffines, W.D., 340
Huffman, D., 455
Huffman, D.L., 3
Huffman, D.M., 154
Huffman, E.W., 39
Huffman, N.M., 269
Huffman, R., 470
Huffman, S.F., 522
Hufford, T.L., 84
Hufnagel, L., 421
Hufschmidt, M.M., 341
Hug, C.C., 101
Hugelet, J., 352
Huggans, J.L., 259
Hugghins, E.J., 433
Huggins, S.E., 463
Hugh, R., 84
Hughes, A.F.W., 355, 356
Hughes, B.L., 425
Hughes, C., 550
Hughes, D., 4
Hughes, E., 515
Hughes, G.C., 534
Hughes, G.R., 345
Hughes, I.M., 502
Hughes, J.H., 367
Hughes, K.W., 441
Hughes, M.B., 424
Hughes, M.R., 535
Hughes, T.D., 138
Huijing, F., 95
Huish, M.T., 346
Huisingh, D., 345
Huisman, O.C., 45
Huisman, T.H.J., 103
Huizinga, H., 122
Huizenga, P.A., 223
Hukill, P.B., 79
Hulbary, R.L., 160
Hulbert, L.C., 166
Hulcher, F.H., 347
Hulett, G.K., 164
Hulka, B.S., 342
Hull, Sr.F.M., 24
Hull, G.Jr., 437
Hull, H.M., 18
Hull, I.M., 64
Hull, J.Jr., 225
Hull, M.W., 266
Hull, McA.H.Jr., 318
Hull, R.J., 422
Hulland, T.J., 548
Hulse, F.S., 19
Hultquist, D.E., 232
Humbert, E.S., 561
Humbertson, A.O., 367
Humble, B.J., 448
Humburg, N.E., 167
Humdley, L.R., 492
Hume, A.S., 252
Hume, J.C., 198
Humes, A.G., 207
Humes, P.E., 183
Humm, D.G., 339
Humm, H.J., 97, 490
Hummon, W.D., 368
Humora, P., 214
Humphrey, A.B., 19
Humphrey, A.E., 411
Humphrey, D.R., 101
Humphrey, G.L., 135
Humphrey, P.S., 172
Humphrey, R.D., 453
Humphrey, R.L., 198
Humphrey, R.R., 18
Humphrey, S.R., 90
Humphrey, W.D., 246
Humphreys, J., 396
Humphreys, T., 112
Humphreys, W.F., 176
Humphreys, W.J., 106
Humphries, A.A.Jr., 100
Humphries, J.C., 176
Humphryes, T.E., 90
Hunderfund, R.C., 40
Hung, G.W., 445
Hung, P.P., 124
Hungate, F., 501
Hungate, R.E., 51
Hungerford, G.F., 63
Hungerford, K., 118
Hunnicutt, M.B., 251
Hunsaker, D., 39
Hunsley, J., 256
Hunt, A., 104
Hunt, B., 95
Hunt, C.B., 437
Hunt, C.C., 263
Hunt, E.C., 547
Hunt, E.L., 100
Hunt, G.E., 441
Hunt, G.L., 54
Hunt, G.M.Jr., 35
Hunt, H.H., 428
Hunt, J., 398
Hunt, J.A., 113
Hunt, J.D., 475
Hunt, J.W., 550
Hunt, L.B., 120
Hunt, M.D., 451
Hunt, M.O., 149
Hunt, O.J., 276
Hunt, T.E., 11
Hunter, A., 61
Hunter, B.B., 391
Hunter, C., 169
Hunter, F.E., 263
Hunter, F.R., 61, 105
Hunter, G.E., 440
Hunter, J., 550
Hunter, J.D., 468
Hunter, J.E., 307
Hunter, J.R., 463
Hunter, M.J., 232
Hunter, P.E., 106
Hunter, R., 99
Hunter, R.D., 230
Hunter, R.L., 53, 133
Hunter, W.B., 75
Hunter, W.S., 257
Huntington, C.E., 192
Huntington, G.L., 50
Huntley, D.A., 542
Huntsman, L.L., 497
Hunzinger, G., 538
Huot, J., 557
Huot, L., 557
Hupp, E.W., 463
Huramitsu, H., 127
Hurd, L.E., 82
Hurd, R.C., 496
Hurlbert, R.E., 501
Hurlbert, S.H., 39
Hurlbutt, H.W., 507
Hurley, Rev.F.J., 216
Hurley, H., 331
Hurley, J.E., 374
Hurley, L.S., 52
Hurley, R.L., 34
Hurley, R.T., 151
Hurnik, J.F., 547
Hurnuff, L., 374
Hurst, D.C., 11
Hurst, D.D., 296, 297
Hurst, G.A., 250
Hurst, R.E., 11
Hurst, R.N., 150
Hurst, R.O., 545
Hurwitz, C., 314, 327
Husa, J., 125
Husain, A., 99
Husain, S., 217, 285
Husband, D.D., 429
Husband, R.W., 222
Huser, C.F., 257
Huskey, R.J., 488
Huss, T., 336
Hussain, S.T., 86
Hussey, R.S., 108
Husted, G., 481
Huston, N., 486, 488
Huston, T.M., 108
Hutchens, J.O., 133
Hutcheon, D.E., 283
Hutcheson, D.P., 262
Hutcheson, H.L., 432
Hutcheson, R.E., 324
Hutchings, B.L., 372
Hutchinson, D., 309
Hutchinson, F.E., 193
Hutchinson, G.P., 20
Hutchinson, J., 356
Hutchinson, J.A., 20
Hutchinson, J.M., 508
Hutchinson, R.L., 76
Hutchison, C.A., 339

Hutchison, G., 11
Hutchison, H.T., 467
Hutchison, J., 546
Hutchison, V.H., 379
Huttner, I., 556
Hutner, S.H., 313
Hutnik, R.J., 402
Hutson, D., 95
Hutt, F.B., 305
Hutt, W., 336
Hutto, D., 527
Hutto, T., 507
Hutton, A., 544
Hutton, C.A., 365
Hutton, J.J.Jr., 179
Hutton, K.E., 32
Hutzler, L.B., 488
van Huystee, R.B., 552
Hwang, J., 412
Hwang, U.K., 256
Hwang, Y.C., 86
Hwang, Y.L., 31
Hybertson, R., 238
Hybl, A., 203
Hyche, L.L., 5
Hyde, B.B., 480
Hyde, C.E., 189
Hyde, C.T., 102
Hyde, J.B., 537
Hyde, P.M., 185
Hyde, R.M., 67, 380
Hyde, R.W., 328
Hyder, D., 440
Hyder, W.C., 440
Hyer, P.V., 490
Hyink, D.M., 5
Hyland, K.E.Jr., 421
Hylander, W.L., 332
Hyle, R.Jr., 484
Hylemon, P.B., 492
Hyman, R., 405
Hymer, W.C., 403
Hymowitz, T., 138
Hyndman, R.D., 542
Hyne, J.B., 531
Hynes, C.D., 27
Hynes, H.B.N., 551
Hynes, J.B., 428
Hyons, P., 196

I

Iandolo, J.J., 166
Ianuzzo, D., 207
Ibanez, M.L., 185
Ibara, R., 215
Ibrahim, M.A., 342
Ibrahim, R.K., 557
Ibsen, K.H., 54
Ichida, A.A., 369
Ichinose, L., 128
Ichiye, T., 459
Ichniowski, C.T., 204
Idler, D.R., 541
Idziak, E.S., 557
Idzkowsky, H.J., 414
Ickes, R.A., 415
Ifft, D.D., 208
Igarashi, A., 283
Igaraski, S., 531
Igbokwe, E., 250
Igelsrud, D., 126
Iglewski, B.H., 388
Iglinsky, W.Jr., 187
Ignoffo, C.M., 255
Igwegbe, E.C.K., 3
Ihler, G.M., 413
Ihler, K., 413
Ihrke, C., 514
Ihrke, C.A., 515
Ikawa, M., 280
Ikenberry, G.J., 169
Ikenberry, R.D., 437
Ikenberry, R.W., 269
Ikuma, H., 231
Ikun, D.A., 341
Ilan, J., 355
Ilchman, W.J., 207
Illig, P., 498
Illman, W.I., 544
Illner, P., 500
Ilnicki, R.D., 288
Iltis, H.H., 515
Iltis, W.G., 32
Imaeda, I., 282
Imagawa, D., 56
Imberski, R.B., 199
Imbrahim, I.M., 424
Imig, C.J., 162
Immaculata, Sr.M., 396
Imrey, P.B., 341
Imsande, J.D., 157
Inagami, T., 446
Incardona, N., 445
Inch, W.R., 551
Inders, A., 77
Infanger, Sr.A., 406
Ingersoll, E.M., 360
Ingorsoll, R.J., 409
Ingham, C., 272
Ingle, L., 37
Ingle, L.M., 508
Ingledew, W.M., 561
Ingling, C.R., 362
Inglis, J.M., 459
Ingoglia, N., 283
Ingold, D.A., 450
Ingraham, J.S., 146
Ingraham, L.L., 48, 51
Ingraham, R.L., 32
Ingram, B.R., 426
Ingram, D.G., 548
Ingram, F.D., 162
Ingram, J.M., 557
Ingram, R.G., 556
Ingram, W.L., 473
Ingram, W.R., 160
Inman, R.B., 517
Innes, B.J., 487
Innes, P., 562
Innis, G.S., 476, 477
Inniss, W.E., 551
Inoue, S., 409
Inour, S., 556
Inouye, T., 127
Inselberg, E., 236
Inselburg, J.W., 277
Iovino, A., 310
Irby, H.D., 459
Ireland, P.H., 487
Irgens, R.L., 258
Iritani, W.M., 503
Irr, J.D., 512
Irvin, J.L., 340
Irvine, B.R., 538
Irvine, D.M., 547
Irvine, G.N., 537
Irvine, J., 455
Irving, F.D., 243
Irving, L., 14
Irwin, G.H.III., 173
Irwin, H., 298
Irwin, H.S.Jr., 300
Irwin, M.R., 519
Irwin, R.D., 85
Isaac, D.E., 28
Issac, P.K., 538
Isaac, R.A., 106
Isaacson, A., 290
Isaacson, R.L., 94
Isaak, D., 37
Isakson, S., 511
Isely, D., 157
Isenberg, F.M.R., 305
Isenberg, G.R., 324
Isenberg, I., 385
Ishaq, M., 145
Ishiguro, E., 535
Ishii, M., 114
Ishikawa, Y., 482
Ishimoto, T.T., 29
Ishizaka, K., 197, 198
Ishizaka, T., 198
Ishizaki, S.M., 113
Isler, G.A., 365
Ismail, A.A., 195
Ismond, M.A.H., 539
Isreal, J.V., 103
Israels, L.G., 538
Issenberg, P., 273
Isseroff, H., 317
Istock, C., 327
Itano, H.A., 59
Ito, M., 24
Ito, P.J., 114
Ito, T., 334
Itofsaess, F., 393
Itzkowitz, S., 224
Ivens, M.S., 182
Iversen, E., 96
Iversen, J.O., 563
Iverson, C.A., 155
Iverson, M.A., 295
Iverson, R.M., 87
Ives, D.H., 362
Ives, J.D., 139
Ives, P.T., 206
Ivey, W.D., 5
Ivler, D., 64
Iwamoto, R., 111
Iyer, R.V., 549
Iyer, V.N., 544

J

Jaap, R.G., 366
Jache, J., 512
Jachowski, L.A., 199
Jack, R.C., 316
Jacks, L., 175
Jackson, A.O., 148
Jackson, B.E., 183
Jackson, C.D., 24
Jackson, C.M., 263
Jackson, D., 369, 450
Jackson, D.A., 232
Jackson, D.S., 503
Jackson, E.B., 16, 486
Jackson, E.N., 232
Jackson, H., 530
Jackson, J., 119, 432
Jackson, J.A., 248, 378
Jackson, J.E., 106
Jackson, J.L.W., 380
Jackson, J.O., 29
Jackson, L.E., 477
Jackson, L.K., 92
Jackson, M., 535
Jackson, M.J., 84
Jackson, N., 422
Jackson, N.D., 508
Jackson, R.C., 461
Jackson, R.L., 449
Jackson, S.J., 445
Jackson, T.L., 384
Jackson, W.B., 354
Jackson, W.J., 104, 277
Jacob, H.S., 455
Jacobs, A.W., 227
Jacobs, F.A., 353
Jacobs, H.K.L., 538
Jacobs, H.S., 166
Jacobs, J., 117
Jacobs, J.J., 185
Jacobs, M.E., 143
Jacobs, N., 378
Jacobs, N.J., 277
Jacobs, P.A., 112
Jacobs, R., 61, 124
Jacobs, R.D., 326
Jacobs, S.C., 193
Jacobs, W.P., 287
Jacobsen, B.J., 139
Jacobson, A.G., 465
Jacobson, A.P., 233
Jacobson, D.R., 177
Jacobson, H., 298
Jacobson, H.I., 327
Jacobson, L., 310, 413
Jacobson, L.O., 131
Jacobson, M., 96
Jacobson, S., 216, 302
Jacobson, S.L., 544
Jacobus, W.E., 52
Jacoli, G.G., 534
Jacques, F.A., 315
Jacques, R.L., 284
Jacquez, J., 232
Jacquez, J.A., 233
Jaeger, M.J., 94
Jaeger, R.J., 212
Jaenke, R.S., 69
Jaffe, L.F., 150
Jaffe, R.B., 60
Jaffee, M.J., 368
Jaffee, O.C., 371
Jagger, R., 358
Jagendorf, A.T., 302
Jagiello, G., 301
Jagschitz, J.A., 422
Jahn, L.A., 139
Jahoda, J.C., 208
Jain, N.C., 53
Jain, S.K., 47
Jakab, G.J., 482
Jakobsson, E.G., 137
Jakway, G.E., 31
Jakway, J.S., 325
Jalal, S.M., 352
Jambuth, T.R., 493
James, C.W., 105
James, D.A., 22
James, D.W., 474
James, F.C., 429
James, H.A., 77, 79
James, L.E., 64
James, M.N.G., 531
James, M.S., 473
James, M.T., 502
James, P., 452
James, R., 119, 516
James, S.A., 342
James, T., 55
James, T.R., 444
James, W.S., 120
Jameson, D.A., 67
Jameson, D.L., 463
Jameson, E.W.Jr., 52
Jameson, P., 513
Jamieson, G.A., 85
Jamison, H.M., 443
Jamison, W., 121
Jamnback, H., 322
Jan, K., 301
Jancey, R.C., 552
Jande, S.S., 548
Jander, R., 171, 172
Janes, D.J., 5
Janes, M.J., 91
Janeway, R., 347
Janfaza, P., 216
Jani, V.M., 286
Janice, B., 474
Janick, J., 149
Jankay, P.T., 27
Janke, R.A., 228
Janky, D.M., 93
Jann, G.J., 55
Janovy, J.Jr., 270
Jansen, A.H., 204
Jansen, E.F., 280
Jansen, G.R., 67
Janson, B.F., 366
Jansons, V.K., 282
Janu, R.C., 336
Janzen, K.L., 238
Janzen, D.H., 231
Jaques, L.B., 562
Jardetzky, O., 42
Jarett, L., 263
Jarial, M.S., 412
Jaroslow, B.N., 126
Jarroll, E.L., 507
Jarvis, B.W., 186
Jarvis, F.E., 487
Jarvis, F.G., 116
Jarvis, J.L., 157
Jarvis, R.L., 383
Jaskiel-Hamilton, L., 219
Jaskoski, B.J., 123
Jasmin, G., 559
Jasper, D.E., 53
Jasper, D.K., 121
Jaussi, A.W., 472
Jawetz, E., 60
Jaworski, A., 105
Jay, A.E., 254
Jay, J.M., 234

Jay, S.C., 537
Jaycox, E.R., 136, 138
Jayne, B.S., 498
Jaynes, C.C., 462
Jaynes, H.O., 444
Jean, M., 557
Jean, Sr.M., 396
Jean, P., 558
Jeanne, R., 207
Jeansonne, B.G., 186
Jedlinski, H., 139
Jee, W.S.S., 473
Jeffery, D.E., 472
Jeffery, L.S., 444
Jeffery, P.R., 186
Jeffery, W.E., 464
Jeffrey, J., 263
Jeffrey, J.E., 491
Jeffrey, L., 459
Jeffreys, D.B., 334
Jeffers, D.L., 364
Jeffers, W., 397
Jefferson, C.A., 87
Jefferson, L.S., 405
Jefferson, R.N., 58
Jeffries, C.D., 236
Jeffries, H.E., 341
Jeffries, W.B., 393
Jegla, D., 359
Jegla, T.C., 359
Jelen, P., 530
Jelenkovic, G., 288
Jellinck, P.H., 545
Jellinek, M., 256
Jellum, M.D., 106
Jemison, E.W., 494
Jen, J.J., 424
Jencks, E., 508
Jencks, W.P., 208
Jenden, D.J., 56
Jendrasiak, G.L., 137
Jeng, S., 112
Jenifer, F.C., 288
Jenkin, H.M., 244
Jenkins, A.C., 3
Jenkins, A.L., 259
Jenkins, D.T., 10
Jenkins, E., 397
Jenkins, E.M., 9
Jenkins, E.P., 508
Jenkins, Rev.F., 37
Jenkins, F.A., 211
Jenkins, J.B., 407
Jenkins, J.H., 108, 181
Jenkins, J.N., 248
Jenkins, J.R., 335
Jenkins, L.E., 68
Jenkins, M.M., 486
Jenkins, R.A., 526
Jenkins, R.E., 365, 487
Jenkins, S.F.Jr., 345
Jenkins, T.W., 227
Jenkins, W.R., 288
Jenkinson, G.M., 547
Jenne, E.A., 477
Jenner, C.E., 334
Jenness, R., 241
Jenni, D.A., 267
Jennings, D., 293
Jennings, D.B., 546
Jennings, F.L., 467
Jennings, P.H., 219
Jennings, S.A., 536
Jennings, S.S., 491
Jennings, V.M., 157
Jennings, W.F., 49
Jennrich, R.I., 56
Jenny, B.F., 424
Jensen, A.H., 138
Jensen, D.E., 69
Jensen, D.E.N., 545
Jensen, D.R., 122
Jensen, E.A., 477
Jensen, E.H., 276
Jensen, E.J., 132
Jensen, E.L., 517
Jensen, G., 258
Jensen, G.L., 218
Jensen, H.J., 385
Jensen, H.W., 348
Jensen, J.A., 96
Jensen, J.N., 477
Jensen, J.S., 4, 497
Jensen, L.A., 158
Jensen, L.H., 499
Jensen, L.S., 108
Jensen, N.F., 304
Jensen, N.H., 125
Jensen, P., 236
Jensen, R.A., 449
Jensen, R.G., 79
Jensen, R.H., 273
Jensen, R.L., 69
Jensen, R.T., 470
Jensen, S.G., 433
Jensen, S.L., 258
Jensen, T., 298
Jensen, T.E., 297
Jensen, W.A., 42
Jenson, P.G., 192
Jenson, R.G., 78
Jenssen, T.A., 492
Jeon, K.W., 442
Jeppson, L.R., 58
Jepson, J., 399
Jernigan, C.A., 468
Jerram, D.C., 104
Jersild, R., 145, 146
Jervis, T., 278
Jeska, E.L., 157
Jeskey, H.A., 466
Jesse, G.W., 259
Jester, D., 174
Jetkiewicz, W., 392
Jett, S.C., 47
Jew, J.Y., 160
Jewel, D.D., 257
Jewell, T.R., 514
Jewett, D., 514
Jewett, D.L., 60
Jewett, J.G., 368
Jewett, R.E., 101
Jezerinac, R., 364
Ji, T.H., 527
Jiamondstone, T.I., 399
Jimerez, J., 150
Jimenez-Marin, V.G., 542
Jividen, G.M., 343
Jobbins, K.M., 288
Jobsis, F., 333
Jochim, K., 232
Jocobson, I., 46
Joerg, F.C., 334
Joern, E., 150
Johannes, R.E., 106
Johannes, R.F., 516
Johannsen, C.J., 258
Johansen, C.A., 502
Johansen, H.W., 209
Johansen, P.H., 546
Johansen, R.H., 351
Johanson, L., 456
Johansson, K.R., 452
Johansson, T.S.K., 298
John, D.T., 492
John, K.R., 395
John, T.H., 407
Johndrew, O.F.Jr., 305
Johns, F., 214
Johns, H.E., 550
Johns, J.E., 496
Johns, M., 208
Johns, R.J., 197
Johnsen, R.C., 295
Johnsen, R.E., 68
Johnsgard, P., 270
Johnson, A.A., 21
Johnson, -., 452
Johnson, A.D., 106, 434
Johnson, A.G., 232, 241
Johnson, A.R., 467
Johnson, A.T., 125
Johnson, A.W., 37, 352, 424
Johnson, B., 11
Johnson, B.C., 380
Johnson, B.D., 20
Johnson, B.H., 343
Johnson, B.J., 106
Johnson, B.L., 58
Johnson, B.R., 344
Johnson, C., 312
Johnson, C.D., 15, 231, 546
Johnson, C.E., 189, 518
Johnson, C.L., 527
Johnson, C.M., 27, 46, 475
Johnson, C.R., 92
Johnson, C.S., 366, 368
Johnson, D., 549
Johnson, D.C., 172
Johnson, D.D., 65
Johnson, D.E., 65, 162, 411
Johnson, D.L., 426
Johnson, D.M., 453
Johnson, D.R., 117, 258, 498
Johnson, D.W., 116, 350
Johnson, E., 72, 155, 399
Johnson, E.A., 333
Johnson, E.E., 445
Johnson, E.J., 190
Johnson, E.V., 27
Johnson, E.W., 5
Johnson, F.A., 91
Johnson, F.C., 90
Johnson, F.D., 118
Johnson, F.H., 287
Johnson, F.M., 345
Johnson, G.B., 263
Johnson, G.C., 405
Johnson, G.E., 352
Johnson, G.P., 229
Johnson, G.R., 365
Johnson, G.T., 22
Johnson, G.V., 293
Johnson, H., 57, 103
Johnson, H.A., 21
Johnson, H.B., 57
Johnson, H.D., 259
Johnson, H.E., 225
Johnson, H.L., 233
Johnson, H.P., 155
Johnson, H.S., 138
Johnson, H.W., 240
Johnson, I., 238
Johnson, J., 273
Johnson, J.A., 69, 244
Johnson, J.B., 24
Johnson, J.C., 165, 494
Johnson, J.D., 341
Johnson, J.E., 65, 69
Johnson, J.F., 451
Johnson, J.H., 491
Johnson, J.I., 227
Johnson, J.L., 435
Johnson, J.N., 350
Johnson, J.R., 269, 270, 433
Johnson, J.W., 138, 322, 384
Johnson, K., 527
Johnson, K.D., 39
Johnson, K.E., 332
Johnson, L., 70, 118, 442, 461
Johnson, L.B., 168
Johnson, L.C., 433
Johnson, L.D., 398
Johnson, L.E., 446
Johnson, L.G., 431
Johnson, L.J., 401
Johnson, L.L., 552
Johnson, L.P., 155
Johnson, L.R., 147
Johnson, L.W., 5
Johnson, M., 128
Johnson, M.A.Jr., 83, 357
Johnson, M.D., 143
Johnson, M.F., 491
Johnson, M.G., 424, 426
Johnson, M.H., 104
Johnson, M.I., 53
Johnson, M.K., 190
Johnson, M.P., 166
Johnson, M.R., 177
Johnson, M.W., 401
Johnson, N.E., 346
Johnson, N.K., 43
Johnson, N.S., 59
Johnson, O., 15
Johnson, O.H., 501
Johnson, O.W., 239
Johnson, P., 329, 342
Johnson, P.C., 19
Johnson, P.H., 239
Johnson, R., 88, 309, 350, 393, 440
Johnson, R.B., 91
Johnson, R.C., 428
Johnson, R.E., 123, 173, 501
Johnson, R.D., 118
Johnson, R.G., 175, 359
Johnson, R.J., 410, 502
Johnson, R.K., 376
Johnson, R.M., 15, 235, 476 486, 489
Johnson, R.R., 138, 443
Johnson, R.S., 11
Johnson, R.T., 198
Johnson, R.W., 236
Johnson, S., 145
Johnson, S.E., 209
Johnson, S.L., 193, 413, 497
Johnson, S.R., 496
Johnson, T.B., 238
Johnson, T.C., 127
Johnson, T.J., 362
Johnson, T.L., 486
Johnson, T.M., 437
Johnson, T.N., 84
Johnson, T.W.Jr., 332
Johnson, V., 239
Johnson, V.A., 271
Johnson, V.K., 350
Johnson, V.W., 229
Johnson, W., 161
Johnson, W.C., 3
Johnson, W.C., 385
Johnson, W.A., 5, 184
Johnson, W.E., 9, 236
Johnson, W.H., 153, 314, 550
Johnson, W.J., 231
Johnson, W.L., 345
Johnson, W.T., 303
Johnson, W.W., 293
Johnson-Lussenberg, C.M., 549
Johnston, B., 485
Johnston, C.E., 554
Johnston, D.E., 362, 363
Johnston, D.W., 126
Johnston, F.M., 223
Johnston, G.W., 247
Johnston, H.H., 338
Johnston, J.D., 385
Johnston, J.M., 466
Johnston, K., 127
Johnston, M., 256
Johnston, M.A., 209
Johnston, M.C., 465
Johnston, N., 134
Johnston, N.P., 472
Johnston, P.B., 180
Johnston, P.M., 22
Johnston, R., 172, 270, 450
Johnston, R.A., 47, 49, 547
Johnston, R.A., 547
Johnston, R.F., 228
Johnston, R.P.Jr., 522
Johnstone, F.E., 107
Johnstone, I.B., 548
Johnstone, R.M., 555
Johnstone, R.L., 174
Joiner, G.N., 460
Joiner, J.N., 92
Jokela, A.T., 39
Jokela, J.J., 139
Joklik, W.K., 333
Jolley, R.J.Jr., 388
Jollick, J.D., 236
Jollie, M.T., 126
Jollie, W.P., 491
Jolliffee, P.A., 534
Joly, R.A., 371
Jonas, H., 241
Jones, A.C., 96
Jones, A.L., 60
Jones, A.M., 166
Jonas, R.J., 501
Joncas, J., 558
Jones, A.W., 442
Jones, B.L., 537
Jones, B.M., 366
Jones, B.R., 439

Jones, B.W., 3
Jones, C., 72, 414, 448
Jones, C.E., 30
Jones, C.M., 149, 423
Jones, C.R., 298
Jones, C.S., 339
Jones, D., 175, 470
Jones, D.A., 151
Jones, D.D., 10
Jones, D.E., 10
Jones, D.H., 327
Jones, D.J., 3
Jones, D.L., 199
Jones, D.R., 535
Jones, D.S., 509
Jones, E.D., 304
Jones, E.E., 5, 12, 345
Jones, E.G., 263
Jones, E.V., 424
Jones, E.W., 212, 391
Jones, G.A., 178, 561
Jones, G.E., 57, 279
Jones, G.F., 62
Jones, G.H., 231
Jones, G.I., 498
Jones, G.L., 344
Jones, G.M., 493
Jones, H., 196, 287
Jones, H.B., 43
Jones, H.E., 166
Jones, H.G., 416
Jones, H.L., 142
Jones, I.D., 344
Jones, I.M., 155
Jones, J., 207, 273, 461, 509
Jones, Sr.J., 311
Jones, J.A., 392, 438
Jones, J.B., 106
Jones, J.B.Jr., 107
Jones, J.C., 201
Jones, J.E., 367, 424, 425
Jones, J.H., 130
Jones, J.K., 461
Jones, J.K.Jr., 461
Jones, J.M., 152, 485
Jones, J.P., 24, 117
Jones, K., 28
Jones, K.C., 31
Jones, K.L., 230
Jones, L., 116
Jones, L.A., 236, 371
Jones, L.G., 184
Jones, L.P., 468
Jones, L.V., 338
Jones, L.W., 441
Jones, M.B., 47
Jones, M.E., 64
Jones, M.E.Jr., 99
Jones, M.R., 338
Jones, N.A., 101
Jones, O.N., 246
Jones, P., 353, 368
Jones, R., 298, 382, 463, 546
Jones, R.D., 295
Jones, R.E., 71, 186
Jones, R.H., 87
Jones, R.J., 501
Jones, R.K., 345
Jones, R.L., 424, 138
Jones, R.R., 191
Jones, R.S., 110, 333
Jones, R.T., 388
Jones, R.W., 108, 475
Jones, S.B., 105
Jones, T.C., 214
Jones, T.H., 488
Jones, U.S., 423
Jones, V.A., 344
Jones, W., 175, 486
Jones, W.D., 17
Jones, W.F., 248
Jones, W.F.Jr., 102
Jones, W.L., 435
Jones, W.O., 9
Jones, W.R., 401, 402
Jones, W.W., 58
Jonsson, H.T., 428
Joppa, L.R., 350
Jopson, H.G.M., 484
Jordan, C., 182, 210
Jordan, C.S., 450
Jordan, D.C., 547
Jordan, G.L., 18
Jordan, J.P., 67
Jordan, H.C., 403
Jordan, H.E., 378
Jordan, L.B., 405
Jordan, L.S., 58
Jordan, O.R., 440
Jordan, P., 240
Jordan, R.J., 491
Jordan, R.M., 354
Jordan, W., 299
Jordan, W.P., 62
Jordan, W.S., 424
Jordan, W.S.Jr., 179
Jordan-Molero, F.L., 417
Jorgensen, C.D., 472
Jorgensen, C.J., 66
Jorgensen, G.N., 449
Jorgensen, J.H., 470
Jorgensen, N., 291
Jorgensen, N.A., 517
Jose, J., 398
Joseph, J., 498
Joseph, J.M., 204
Joseph, S.A., 328
Joseph, S.L., 76
Josephson, B.L., 295
Josephson, D.V., 401, 402
Josephson, L.M., 444
Josephson, R.K., 54, 55
Josephson, R.V., 365
Joshi, J.G., 441
Joshi, M.S., 325
Joshi, S.G., 327
Joshi, S.H., 405
Joshi, V.C., 448
Joslin, J.K., 485
Jost, D.N., 210
Jost, M., 226
Josten, J.J., 236
Jourdian, G.W., 232
Jowell, W.E., 454
Jowett, D., 514
Joy, K.W., 544
Joy, R.M., 53
Joy, W.W., 233
Joyal, R., 560
Joyce, B., 542
Joyce, E.C., 485
Joyce, W., 542
Joyner, T., 498
Joyner, W.L., 273
Joys, T.M., 404
Jubb, G., 402
Judd, B.H., 465
Judd, D., 262
Judd, F., 452
Judd, H.E., 473
Judd, R.W., 373
Judd, Sr.T., 236
Judd, W.W., 552
Judkins, R.L., 9
Judkins, W.P., 494
Judson, C.L., 49
Judy, J.K., 365
Judy, M.M., 466
Juergensmeyer, E.B., 122
Juge, D.M., 34
Juillet, J., 559
Julander, O., 472
Julian, L.M., 52
Juliano, R.L., 550
Jump, E., 388
Jump, J.A., 120
Jung, G.A., 401
Junge, D., 57
Jungkuntz, R., 497
Jungmann, R.A., 127
Jungreis, A.M., 442
Jungwirth, C.R., 420
Juni, E., 232
Juntunen, E.T., 383
Juo, P., 324
Juras, D.S., 120
Jurf, A.N., 204
Jurgensen, M.F., 229
Jurinak, J.J., 474, 477
Jurtshuk, P., 463
Just, J.J., 177
Justh, G., 399
Justice, K.E., 54
Justice, J.K., 448
Justin, J.R., 288
Justus, D.E., 180
Justus, J., 15
Jutila, J.W., 265
Jutras, M.W., 423
Jyung, W.H., 371

K

Kabat, D., 388
Kabat, E.A., 301
Kacian, D., 301
Kackler, L.R., 307
Kackobs, J.A., 138
Kaczmarczyk, W.J., 508
Kaczor, V.B., 284
Kadiec, J.A., 476
Kadner, C.G., 37
Kadner, R.J., 490
Kado, C.I., 50
Kadoum, A.H., 167
Kaeberle, M.L., 158
Kaeferstein, Sr.V., 154
Kaelber, W.W., 160
Kaempfer, R.O.R., 211
Kaempfer, W.W., 12
Kaesberg, P.J., 517
Kafatos, F.C., 210
Kafer, E.R., 340
Kagowski, J., 465
Kahan, E., 228
Kahan, L., 520
Kahle, J.B., 150
Kahler, G.A., 258
Kahn, A.J., 283
Kahn, D., 285
Kahn, D.S., 555
Kahn, J.R., 356
Kahn, J.S., 343
Kahn, N.A., 297
Kahn, R.A., 541
Kahn, R.L., 313
Kaile, J.D., 367
Kaill, W.M., 61
Kainer, R.A., 68
Kainski, J., 524
Kaiser, A.D., 41
Kaiser, C.J., 138
Kaiser, E.T., 131
Kaiser, I.I., 527
Kaiserman-Abramof, I.R., 355
Kaizer, H., 198
Kakamura, M.J., 267
Kakar, R.S., 189
Kako, K., 549
Kakolewski, J.W., 435
Kala, P.K., 555
Kalaba, R.E., 63
Kalb, M., 524
Kalbach, P.M., 75
Kalberer, C., 349
Kalch, L.W., 93
Kaldor, G.J., 399
Kalff, J., 556
Kalicki, H.G., 295
Kalil, R.E., 520
Kalin, E.W., 503
Kaliss, B.S., 199
Kalland, G., 27
Kalland, G.A., 27
Kallen, R.G., 410
Kallenbach, N.R., 409
Kallio, A., 241
Kallio, R.E., 137
Kalman, D.L., 79
Kalman, S., 42
Kalnins, V.I., 549
Kalnitsky, G., 161
Kalousek, D., 556
Kalra, V., 64
Kalser, M., 96
Kalt, M.R., 79
Kaltenbach, C.C., 527
Kalter, S.S., 470
Kalsikes, P.J., 537
Kaly, J.M., 272
Kamalu, T.N., 9
Kambysellis, M.P., 312
Kamel, I., 349
Kamemoto, F.I., 112
Kamemoto, H., 114
Kameyama, Y., 537
Kamien, E.N., 212
Kamin, H., 333
Kamin, L., 121
Kaminer, B., 208
Kaminski, Z.C., 282
Kaml, A.S., 68
Kamm, J.A., 386
Kammer, A.E., 166
Kammlade, W.G., 129
Kammula, R.G., 9
Kampine, J.P., 513
Kampky, J.R., 223
Kamra, O.P., 542
Kamran, M.A., 308
Kamstra, L.D., 432
Kanczak, N.M., 413
Kandel, E.R., 301
Kandel, J., 30
Kane, B.E.Jr., 337
Kane, E.S., 131
Kane, J., 535
Kane, J.F., 445
Kane, J.O., 404
Kane, R.T., 286
Kane, T.C., 370
Kaneko, J.J., 53
Kanemasu, E.T., 167
Kaneshiro, E., 370
Kanfert, F.H., 243
Kang, C-Y., 466
Kang, K.W., 146
Kang, S.S., 121
Kangas, D.A., 254
Kanich, R.E., 345
Kankel, D.R., 80
Kannowski, P.B., 352
Kano, K., 319
Kanpp, E.W., 178
Kanpp, W.R., 302
Kantack, V.H., 433
Kantor, G.J., 372
Kanungo, K., 80
Kantz, P.T., 31
Kantzes, J.G., 201
Kanzler, W.W., 329
Kao, F.T., 72
Kao, Y-S., 186
Kapica, S., 284
Kaplan, A., 256, 499
Kaplan, A.M., 422
Kaplan, A.S., 447
Kaplan, B., 330
Kaplan, B.H., 342
Kaplan, D.R., 42
Kaplan, E., 309
Kaplan, E.B., 341
Kaplan, G., 549
Kaplan, H., 548
Kaplan, H.M., 129
Kaplan, L., 219
Kaplan, M.L., 298
Kaplan, M.T., 55
Kaplan, R., 232
Kaplan, S., 137, 512
Kaplan, S.L., 417
Kapler, J.E., 159
Kapoor, B.M., 543
Kapp, R.O., 222
Kappler, J.W., 328
Kappor, M., 531
Kapral, F.A., 367
Kapraum, D.F., 347
Kaprielian, Z.A., 63
Kapur, S., 85
Kapusta, G., 130
Kar, S.B., 234
Karachorlu, K.V., 135
Karakawa, W.W., 404
Karam, J.D., 428
Karas, G., 373
Karasaki, S., 558
Karavolas, H.J., 520

Karczmar, A.G., 124
Kardong, K.V., 501
Kardos, L.T., 401
Karel, M., 213
Karetzky, M., 284
Karfunkel, P., 206
Karim, M.R., 431
Karimer, J.L., 465
Karis, P.C., 61
Karkas, J., 300
Karlander, E.P., 201
Karle, H.P., 29
Karlin, A., 301
Karmas, E., 288
Karlson, A.G., 244
Karlsson, U.L., 160
Karow, A.M.Jr., 104
Karp, E., 312
Karp, G.C., 90
Karpeles, L.M., 204
Karpoff, A.J., 180
Karr, A.L., 260
Karr, G.L., 372
Karr, J.R., 150
Karreman, G., 410
Karstad, L.H.A., 548
Karsten, M., 223
Karstetter, A.B., 192
Kartha, G., 320
Kasapligil, B., 37
Kasbekar, D.K., 85
Kasel, J.A., 449
Kashin, P., 297, 299
Kashiwa, H.K., 499
Kasinsky, H., 535
Kaspar, J.L., 523
Kasper, A.E., 288
Kasper, B., 311
Kasperbauer, M.J., 177
Kasperson, R.E., 209
Kassell, B., 513
Kastella, K., 294
Kasten, F., 185
Kasten, L.H., 522
Kaster, M.C., 39
Kasuga, S.K., 496
Kasweck, K.L., 87
Kataja, E.I., 62
Kater, S.B., 160
Kates, G., 83
Kates, K.C., 89
Kates, M., 548
Katholi, C., 11
Katona, P.G., 356
Katsch, S., 72
Kattan, A.A., 23
Katterman, F.R., 18
Katterman, F.R.H., 16
Katz, F.F., 290
Katz, G., 325
Katz, J., 62
Katz, L., 312, 543
Katz, M., 498
Katz, S., 429, 509
Katz, S.E., 288
Katz, S.S., 76
Katzberg, A.A., 145
Katze, J.R., 64
Katzenellenbogen, B.S., 137
Katzman, P., 256
Kauanaugh, J.F., 403
Kauffman, J.W., 126
Kauffman, S., 132
Kaufman, B., 333
Kaufman, C., 77
Kaufman, C., 79
Kaufman, D.G., 360
Kaufman, H., 300
Kaufman, P.B., 230
Kaufman, W., 514
Kaufmann, A.J., 453
Kaufmann, D., 524
Kaufmann, F.H., 521
Kaufmann, G.W., 159
Kaufmann, J.C.E., 552
Kaufmann, J.E., 303
Kaufmann, J.H., 90
Kaufmann, M., 354
Kaufmann, M.R., 58
Kaul, R., 270
Kaulenas, M.S., 218
Kaushik, N.K., 547
Kauss, T., 182
Kavaljian, L.G., 31
Kavanagh, R.J., 540
Kavanau, J.L., 55
Kawase, M., 365
Kawatski, J.A., 525
Kay, C.M., 531
Kay, E.R.M., 549
Kay, R., 332
Kay, W.W., 562
Kaya, C., 265
Kaye, J., 327
Kayhart, M., 392
Kayne, F., 410
Kayne, H., 208
Kays, J.M., 77
Kazama, F.Y., 490
Kazma, E.J., 354
Kean, E.L., 355
Kean, G.R., 401, 402
Keane, J.A., 316
Kearby, W.H., 259, 261
Kearns, D.H., 449
Kearsley, A., 475
Keast, J.A., 546
Keaster, A.J., 259
Keating, Sr., A., 222
Keating, E.K., 28
Keating, R., 130
Keck, E., 401
Keck, K., 18
Keck, R.W., 145
Kedich, N.M., 333
Kee, D.T., 188
Keebler, E., 7
Keefe, J.R., 489
Keefe, M., 419
Keefe, W.E., 491, 492
Keefer, R.F., 508
Keefer, V.M., 495
Keeler, P., 535
Keeley, L.L., 458
Keeling, R., 164
Keely, T.J.Jr., 198
Keen, N.T., 58
Keen, R.A., 168
Keen, R.E., 480
Keen, V.F., 65
Keenan, R.W., 469
Keene, J.J., 418
Keene, O.D., 403
Keener, C.S., 403
Keener, H., 280
Keener, N., 451
Keeney, A.H., 180
Keeney, C.E., 216
Keeney, P.G., 401, 402
Keeney, M., 201
Keenleyside, M.H.A., 552
Keeton, K.S., 53
Keeton, W.T., 302, 304
Kefalides, N., 410
Keffer, C.J., 238
Keffler, L., 174
Kefford, N.P., 111
Kegeles, G., 77, 79
Kehl, T., 500
Kehoe, B., 30
Kehr, W.R., 271
Keiffer, G.H., 136
Keim, P., 131
Keim, W.F., 147
Keiner, M., 311
Keiser, E.D., 191
Keiser, T.D., 362
Keisling, T.C., 106
Keister, T.D., 183
Keith, A., 404
Keith, D.E., 461
Keith, D.L., 271
Keith, L.B., 520
Keith, W., 251
Keith, W.A., 336
Keith, W.H., 199
Keithly, J., 298
Keleti, G., 413
Kelleher, J.J., 353
Keller, D.E., 313
Keller, E., 302
Keller, E.B., 302
Keller, E.C.Jr., 507
Keller, H.W., 372
Keller, J.R., 290
Keller, J.M., 499
Keller, J.W., 95
Keller, K.F., 180
Keller, L.E., 165
Keller, R., 369
Keller, S.J., 370
Kelley, A.G., 398
Kelley, B.J., 423
Kelley, B.L., 116
Kelley, G., 191, 373
Kelley, J.C.III., 498
Kelley, J.M., 252
Kelley, J.R., 459
Kelley, Sr.M.C., 299
Kelley, R.O., 293
Kelley, V.C., 4
Kelley, W.A., 138
Kelley, W.D., 4
Kelley, W.N., 333
Kelley, W.R., 407
Kelley, W.S., 391
Kelliher, G.J., 399
Kellison, R.C., 346
Kellogg, D.W., 292
Kellogg, R.H., 60
Kellogg, T.F., 248
Kellogg, W.P., 87
Kelly, A.M., 411, 412
Kelly, B.L., 39
Kelly, C.C., 546
Kelly, C.V., 547
Kelly, D.E., 63, 95
Kelly, D.G., 311
Kelly, E.C., 305
Kelly, F., 116
Kelly, J.B., 138
Kelly, J.D., 156
Kelly, J.F., 93
Kelley, J.W., 303
Kelly, K., 426
Kelly, M.G., 489
Kelly, R., 60
Kelly, R.E., 134
Kelly, R.P., 289
Kelly, R.W., 427
Kelly, T.C., 380
Kelman, A., 519
Kelmm, W.R., 460
Kelpper, L.A., 271
Kelsey, L.P., 82
Kelsey, R.C., 394
Kelso, D.P., 485
Kelting, R.W., 165
Kemen, M.J., 306
Kemler, A.G., 460
Kemmerer, A.R., 16
Kemmerer, W.W., 468
Kemp, C.L., 533
Kemp, J.D., 177, 519
Kemp, J.R., 439
Kemp, N.E., 231
Kemp, R., 513
Kemp, T.R., 178
Kemp, W.M., 448
Kempe, L.L., 232
Kemper, B.W., 137
Kemper, G.W., 175
Kemper, W.D., 65
Kenaga, C.B., 148
Kendal, A.P., 205
Kendall, M.W., 251
Kendall, W.A., 401
Kende, H.J., 226
Kendeigh, S.C., 136
Kender, D., 313
Kender, W.J., 307
Kendrick, E., 18
Kendrick, J.E., 521
Kendrick, W.B., 551
Kenefick, D.G., 433
Kenk, V.C., 32
Kennamer, J.E., 5
Kennaway, N., 388
Kennedy, Sr.B., 299
Kennedy, B.M., 45
Kennedy, D., 41, 71, 333
Kennedy, D.M., 445
Kennedy, E.J., 125
Kennedy, E.R., 83
Kennedy, G., 58, 521
Kennedy, J., 104
Kennedy, J.E., 361
Kennedy, J.J., 475
Kennedy, J.R., 442
Kennedy, J.W., 342
Kennedy, L.L., 531
Kennedy, M., 438
Kennedy, P.C., 53
Kennedy, T.G., 552
Kennedy, T.T., 500
Kennedy, W.J., 500
Kennedy, W.K., 301, 302
Kennedy, W.O., 547
Kennell, E., 263
Kennelly, Sr.K., 236
Kennerly, T.E., 464
Kenney, M.J., 395
Kenney, R.A., 84
Kenney, R.M., 411
Kennick, W.H., 382
Kennington, G.S., 526
Kenny, G.E., 499
Kent, B.J., 6
Kent, B.M., 402
Kent, G.C.Jr., 183
Kent, H.A., 106
Kent, J.F., 75
Kent, R.C., 405
Kent, T.H., 161
Kenworthy, A.L., 225
Kenyon, A.J., 78, 79
Kenyon, D.H., 40
Kenyon, J.P., 365
Kenzy, S.G., 504
Keogh, J.L., 22
Keogh, R.N., 419
Keolee, D., 314
Keoppe, D.E., 138
Keopsell, P.A., 385
Keough, K.M., 541
Kepner, G., 244
Kepner, R.A., 254
Keppler, W.J., 120
Keppner, E., 253
Ker, J.W., 540
Kercher, C.J., 527
Keresztes-Nagy, S., 124
Keresztesy, L.O., 508
Kerkay, J., 357
Kerkoff, P.R., 293
Kerland, J.T., 309
Kerley, D.E., 382
Kerman, A.B., 286
Kermicle, J.L., 519
Kern, A.K., 195
Kern, H., 330
Kernis, M., 134
Kerr, D., 355
Kerr, E., 270
Kerr, H.D., 258
Kerr, J., 287
Kerr, J.H., 422
Kerr, M.A., 326
Kerr, N., 242
Kerr, S.H., 91
Kerr, W.E., 407
Kerrick, W.G.L., 500
Kerrigan, D., 512
Kersavage, P.C., 402
Kerschner, J., 205
Kershae, K.A., 545
Kerstetter, R.E., 427
Kerstetter, T.H., 34
Kersting, E.J., 77
Kesler, E.M., 401
Kesner, C., 226
Kessel, E.K., 62
Kessel, R.G., 160
Kessel, R.W.I., 203
Kessler, D., 396
Kessler, F.W., 364
Kessler, G., 226
Kessler, G.D., 425
Kessler, R.E., 312
Kester, A.S., 452

Kestner, F.M., 192
Ketchledge, E.H., 321
Ketchum, D.C., 377
Ketchum, P.A., 230
Ketchum, R., 214
Ketellapper, H.J., 51
Kethley, T.W., 102
Ketterer, J.J., 125
Ketterer, F.D., 411
Kettering, J.D., 36
Kettman, J.R., 466
Keuhne, R.A., 177
Keumpel, P.L., 71
Kevan, D.K., 557
Kevan, D.K.M., 556
Kevern, N.R., 225
Key, J.L., 105
Keyes, J.L., 389
Keyl, J.M., 380
Keys, C.E., 6
Keys, D., 247
Keys, G.D., 209
Kezdy, F.J., 131
Kezer, J., 387
Khairallah, E.A., 77
Khairallah, P.A., 357, 359
Khalaf, K.T., 187
Khan, A.A., 307
Khan, N.A., 296
Khan, R., 541
Khan, S.A., 335
Khan, S-U-I., 196
Khazan, N., 204
Khoury, E.J., 187
Khoury, P.B., 412
Khudairi, A.K., 214
Kiang, Y.T., 281
Kibbey, D.E., 326
Kibel, C.J., 219
Kibler, R.F., 101
Kicliter, E.E.Jr., 137
Kidby, D.K., 547
Kidd, B.S.L., 550
Kidd, D.E., 293
Kidd, W.E.Jr., 508
Kidder, G.M., 552
Kidder, G.W.III., 204, 206
Kieckhefer, R.W., 433
Kiefer, B.I., 80
Kieffer, N.M., 458
Kiehl, E.R., 258, 261
Kiehn, E.D., 499
Kielan-Jaworowska, Z., 211
Kiely, L.J., 313
Kiely, M.L., 123
Kienholz, E.W., 69
Kierder, J.L., 183
Kiernan, J.A., 551
Kiernan, M.K., 538
Kiesling, R.L., 352
Kifer, P.E., 383
Kikeman, M.E., 167
Kikudome, G.Y., 260
Kilambi, R.V., 22
Kilburn, D.G., 535
Kilen, S.M., 135
Kiley, C., 329
Kilgen, M., 188
Kilgen, R., 188
Kilgore, D.L., 267
Kilham, L., 277
Kilham, P., 231
Kilian, D.J., 468
Killackey, H.P., 55
Killeen, L.A., 199
Killian, N.P., 471
Killingsworth, L.M., 340
Kilpatrick, C.W., 480
Kilpatrick, S.J.Jr., 491
Kilpper, R.W., 329
Kilroy, J.F., 446
Kim, H.D., 19
Kim, H.L., 460, 546
Kim, K., 538
Kim, K.C., 402
Kim, K.H., 148
Kim, K.S., 313
Kim, S.H., 333
Kim, S.N., 78, 79
Kim, Y-H., 217
Kim, Y.T., 538
Kimball, A.P., 464
Kimball, A.W., 198
Kimball, G., 416
Kimball, J., 216
Kimber, G., 258
Kimble, C.E., 213
Kimble, R., 401
Kimbrough, E.L., 344
Kimler, L., 327
Kimmel, C.B., 387
Kimmel, D.L.Jr., 332, 509
Kimmel, J., 172
Kimmel, R., 119
Kimmich, G.A., 329
Kimmins, W.C., 542
Kimsey, L.S., 180
Kinard, F.W., 428
Kinbacher, E.J., 272
Kinbrough, T.D., 491
Kinch, R.C., 433
Kind, C.A., 77, 79
Kind, L.S., 543
Kind, P., 84
Kindade, J.M.Jr., 101
Kindel, P., 225
Kinderman, N., 134
Kindler, S.D., 271
Kindley, E., 512
Kindschy, D.L., 117
Kinerson, R., 280
King, -., 316
King, B.F., 262
King, C.C., 3, 363
King, C.E., 97
King, C.L., 168
King, D., 247, 269, 501
King, D.A., 18
King, D.B., 395, 514
King, E.B., 60
King, E.N., 430
King, E.W., 424
King, F.A., 94
King, G.J., 547
King, G.T., 458
King, H., 226
King, J., 45, 562
King, J.A., 227
King, J.C., 313
King, J.D., 199
King, J.G., 118
King, J.L., 61
King, J.M., 306
King, J.R., 501
King, J.S., 367
King, J.W., 22, 357
King, L.E., 445
King, L.J., 323, 545
King, M., 21
King, P.H., 446
King, R.A., 340
King, R.C., 126, 299
King, R.L., 94, 201, 202
King, R.S., 29
King, W.A., 424
Kingdon, H., 340
Kingman, A.R., 424
Kingsbury, D.T., 55
Kingsbury, J.M., 302
Kingsbury, P.J., 155
Kingsbury, R.D., 352
Kingsland, G.C., 425
Kingsley, K., 16
Kingsley, Q.S., 433
Kingsley, R.E., 152
Kingsolver, C.H., 403
Kingston, N., 528
Kinkel, C.A., 432
Kinloch, R.A., 91
Kinnard, W.J., 204
Kinnington, M.H., 28
Kinnison, J., 527
Kinraide, T.B., 65
Kinser, E.R., 21
Kinsey, K., 420
Kinsey, V.E., 230
Kinsky, S.C., 263
Kinsman, D.M., 77
Kinsman, R.S., 55
Kintley, M.E., 203
Kintzis, R.Z., 197
Kinzer, H.G., 292
Kinzie, R.A.III., 112
Kiplinger, D.C., 365
Kiracofe, G.H., 167
Kiraly, R.J., 356
Kirby, E., 408
Kirby, J.T., 87
Kirchanski, S.J., 465
Kircher, H.W., 16
Kirchgessner, D.A., 486
Kirchman, R., 25
Kirchner, M.W., 286
Kirdani, R., 320
Kirk, B.E., 509
Kirk, B.T., 189
Kirk, D.E., 331
Kirk, D.L., 263
Kirk, G.R., 262
Kirk, K.A., 11
Kirk, M.E., 556
Kirk, M.V., 120
Kirk, P.W.Jr., 487
Kirk, R.E., 145
Kirk, R.G., 333
Kirk, T.K., 519
Kirk, V.M., 433
Kirk, W.L., 487, 488
Kirkbride, C.A., 433, 434
Kirkham, D., 254
Kirkland, G.L.Jr., 407
Kirkland, H., 378
Kirkpatrick, C.M., 149
Kirkpatrick, R.L., 494
Kirksey, C., 196
Kirkwood, J., 145
Kirkwood, R.T., 21
Kirkwood, S., 241
Kirleis, A.W., 147
Kirpatrick, J.F., 265
Kirsch, A., 128
Kirsch, J.F., 42
Kirsch, J.A.W., 80
Kirsch, M., 406
Kirschenmann, F., 209
Kirshner, L.B., 501
Kirshner,N., 333
Kirschner, R.H., 133
Kirst, R.C., 25
Kirsten, H., 132
Kirsten, W.H., 133
Kirton, K.T., 236
Kischev, C.W., 467
Kiser, R.S., 100
Kish, H., 285
Kisilvesky, R., 545
Kisliuk, R.L., 217
Kisner, R., 8
Kissam, J.B., 424
Kistner, D.H., 29
Kit, S., 449
Kitai, S.T., 234, 235
Kitay, J.I., 490
Kitchell, R.L., 52
Kitchen, D.W., 34
Kitchen, H., 225
Kitchin, I.C., 251
Kitchin, R.M., 526
Kite, J.H.Jr., 319
Kitiyakara, A., 186
Kitos, P.A., 170
Kittilson, H., 444
Kittman, R.M., 365
Kittredgy, J., 467
Kittrick, J.A., 502
Kitts, J.R., 27
Kitts, W.D., 533
Kitzmiller, J.B., 137
Kiviat, E., 295
Kjeldsen, C.K., 28
Klaaren, D.S., 360
Klaaren, H.E., 360
Klaassen, H.E., 166
Klabunde, K.J., 352
Klafter, R., 394
Klapper, C., 11
Klapper, D., 466
Karman, W.L., 201
Klatt, K.P., 358
Klaus, E.F., 450
Klebanoff, S.B., 499
Klebe, R., 467
Klebenow, D.A., 275
Kleeman, K., 30
Kleene, E.C., 515
deKleer, V., 548
Kleerekoper, H., 460
Klei, T.R., 400
Kleibhan, Sr.C., 511
Klein, A.W., 203
Klein, D., 244
Klein, D.A., 69
Klein, D.N., 67
Klein, D.T., 479
Klein, E., 320
Klein, H., 324
Klein, J., 466
Klein, M., 408
Klein, N.W., 77, 78, 79
Klein, R.L., 252
Klein, R.M., 480
Kleinbaum, D.G., 341
Kleiner, E.F., 276'
Kleinerman, J., 356
Kleinerman, J.I., 356
Kleinhofs, A., 502
Kleinfeld, R.G., 112
Kleinholz, L.H., 387
Kleinman, J.C., 212
Kleinschuster, S.J., 68
Kleinsmith, L.J., 231
Kleinzeller, A.K., 410
Klelowski, E.J.Jr., 217
Klement, V., 64
Klemm, R.D., 168
Klemmedson, J.O., 18
Klenner, J.J., 390
Klens, P.F., 398
Kleppenstein, G.L., 280
Klesius, P., 6
Klett, J.E., 433
Kleyn, D., 288
Klicka, J.K., 523
Kliebenstein, J.B., 255
Kliewer, W.M., 51
Klikoff, L.G., 473
Klimas, J.E., 75
Klimisch, Sr.J., 431
Klimstra, W.D., 129
Klinckmann, E., 36
Kline, B.C., 441
Kline, D.L., 370
Kline, E., 395
Kline, E.A., 155
Kline, E.S., 491
Kline, I., 355
Kline, J., 95
Kline, K., 317
Kline, R., 175
Kline, R.L., 552
Kling, G.F., 384
Kling, J.M., 104
Kling, M., 263
Kling, O.R., 380
Klingener, D.J., 218
Klingensmith, J., 315
Klingman, J.D., 318
Klinman, J., 410
Klip, D.A., 12
Klippstein, R., 306
Klissouras, V., 556
Klitgaard, M., 513
Klock, G., 175
Kloetzel, J.A., 205
Klontz, G.W., 118
Kloos, W.E., 345
Klooster, B., 223
Klopfer, P.H., 332
Klopfenstein, W.E., 166
Klosek, R.C., 284
Klosterman, H.J., 352
Klostermeyer, E.C., 502
Klotz, F., 335
Klotz, L., 210
Klotz, L.J., 58
Klucas, R., 270
Kluca, R.V., 272
Klucking, E.P., 496
Klug, M.J., 227, 228
Klug, S., 290

Klug, W.S., 290
Kluge, A.G., 231
Kluger, M., 233
Klun, J.A., 157
Klute, A., 65
Kmetec, E., 372
Kmetz, D.R., 181
Knaack, J., 556
Knable, A.E., 27
Knake, E.L., 138
Knapczyk, J., 479
Knaphus, G., 157
Knapich, S., 393
Knapp, B.K., 394
Knapp, F.M., 89
Knapp, F.W., 92
Knapp, P., 275
Knapp, W.W., 302
Knauel, R., 126
Knauf, P.A., 550
Knauff, R., 408
Knaus, R., 464
Knavel, D.E., 178
Knecht, G.N., 27
Kneebone, L.R., 403
Kneebone, W.R., 16
Knerer, G., 550
Kneuss, F.M., 428
Knierim, J.A., 226
Knievel, D.P., 401
Knigge, K.M., 328
Knight, A., 28
Knight, C.A., 42, 43
Knight, C.B., 334
Knight, C.W., 167
Knight, D.H., 526
Knight, F.B., 194
Knight, H.H., 157
Knight, J.W., 189
Knight, K.L., 344
Knight, L., 251
Knight, R., 399
Knight, V., 449
Knight, W.E., 248
Kniker, W.T., 470
Knill, L.M., 29
Knipp, L., 125
Knipp, T., 256
Knisely, M.H., 428
Knisely, W.H., 469
Kniskern, V.B., 120
Knisley, C.B., 143
Knobil, E., 414
Knoche, H., 270
Knoche, H.W., 272
Knoerr, K.R., 334
Knoke, J., 355
Knoke, J.K., 363
Knoll, J., 276
Knopf, G., 522
Knopp, J.A., 343
Knopp, T.B., 243
Knorr, P.N., 18
Knott, R., 325
Knovicka, J., 456
Knowles, C.O., 259
Knowles, P.F., 47
Knowles, R., 557
Knowles, R.H., 530
Knowlton, F.F., 477
Knowlton, F.K., 476
Knowlton, G., 476
Knowlton, R.E., 84
Knox, C., 244
Knox, D.M., 183
Knox, J., 77
Knox, J.M., 104
Knox, K.L., 78, 79
Knox, V.I., 199
Knuckles, J.L., 335
Knudsen, D., 271
Knudsen, J.W., 497
Knudsen, R.J., 255
Knudson, D.M., 149
Knudson, V., 224
Knudtson, W., 434
Knupp, D., 192
Knutson, D.M., 385
Knutson, H., 167
Knutson, K., 239
Knutson, K.W., 66
Knutson, R.M., 159
Knuttgen, H.G., 207
Ko, W.H., 114
Kobayashi, G.S., 263
Kobler, H., 521
Koburger, J.A., 92
Kocan, A.A., 378
Koch, A., 144, 501
Koch, B.A., 167
Koch, C.B., 508
Koch, D.W., 281
Koch, G., 29
Koch, G.G., 341
Koch, R., 195, 524
Koch, R.B., 248
Koch, W.J., 338
Kochakian, C.D., 11
Kochan, I., 360
Kochan, W., 117
Kochert, G.D., 105
Kochhar, D.M., 489
Kocka, F.E., 133
Koditschek, L.K., 286
Kodrich, W.R., 398
Koehler, C.S., 44
Koehler, F.E., 502
Koehler, J.K., 499
Koehler, K.A., 340
Koehler, L.D., 223
Koehler, P.E., 107
Koehn, R., 455
Koenig, V., 467
Koenigs, J.W., 345, 346
Koenker, R., 142
Koepke, J.A., 161
Koepp, S.J., 286
Koeppe, O.J., 261
Koeppe, R.E., 377
Koeppen, A.W., 327
Koering, M., 84
Koerker, D.T., 500
Koerner, J.F., 244
Koester, J.D., 301
Koevenig, J.L., 88
Koff, T.R., 27
Koft, B.W., 288
Kogoma, T., 473
Kohel, R.J., 459
Kohen, E., 96
Kohl, D.H., 263
Kohl, D.M., 61
Kohler, E.M., 367
Kohler, H., 131, 133
Kohler, P.H., 432
Kohler, P.O., 449
Kohlhaw, G.B., 148
Kohlmeyer, J.J., 338
Kohlmiller, E.F., 395
Kohls, R.L., 147
Kohn, A.J., 498
Kohn, H., 504
Kohn, R.A., 433
Kohn, R.R., 356
Koide, F.T., 113
Koike, T.I., 25
Koirtyohann, S.R., 261
Kokas, E.B., 340
Kokatnur, M., 186
Kokjer, K.J., 14
Koklo-Cunningham, A., 137
Kolenbrander, P.E., 404
Koler, R.D., 388
Kolar, J.J., 117
Kolbe, M.H., 345
Kolbezen, M.J., 58
Kollas, D.A., 78, 79
Koller, H.R., 147
Kollmann, E.C., 485
Kollros, J.J., 160
Kolmen, S.N., 468
Kolmer, L., 155
Kolodziej, B.J., 363
Kolp, B.J., 527
Kolstad, O.A., 269
Kondra, Z., 530
Konetzka, W., 144
Kong, Y.M., 236
Kongshavn, P., 556
Konecke, C., 393
Konigsberg, -., 81
Konigsberg, I.R., 488
Konikoff, M.A., 191
Konishi, M., 287
Koniske, J., 137
Konsler, T.R., 345
Kontio, A.L., 223
Kontogiannis, J., 31
Konzak, C.F., 502
Koo, R.C.J., 92
Kooh, S.W., 550
Koonce, J., 355
Koontz, H.V., 77
Koopowitz, H., 54
Koostra, W.L., 267
Kopenski, M.L., 229
Koplin, J.R., 34
Koppen, J.D., 288
Koppenheffer, T.L., 220
Kopper, P.H., 173
Koppelman, R., 509
Kopriwa, B.M., 555
Kordisch, M.S., 187
Kordova, N., 538
Korecky, B., 549
Koreman, S.G., 161
Korf, R.P., 304
Koritz, H., 299
Korn, R., 174
Kornacker, K., 362
Kornegay, E.T., 493
Kornfield, S., 263
Kornguth, S.E., 520
Korns, C., 257
Korostoff, E., 411
Korslund, M.K., 494
Kort, M.A., 257
Kortiz, S., 298
Kos, E.S., 255
Kosher, R.A., 79
Koshland, D.E.Jr., 42
Koshi, J.H., 113
Kosin, I.L., 502
Kosman, D.J., 318
Kossove, M., 311
Kostellow, A.B., 330
Koster, W.J., 293
Kostewicz, S.R., 93
Kostir, W.J., 364
Kostyo, J.L., 101
Kosuge, T., 51
Kot, P.A., 85
Kothmann, M.M., 458
Koths, J.S., 78
Kotopoulis, J., 279
Kotsonis, H., 285
Kott, E., 552
Kottman, R.M., 364
Kouba, J., 299
Koukkari, W.L., 241
Koulish, S., 297
Kouskolekas, C.A., 5
Kovach, A., 410
Kovachevich, R., 181
Koval, C.F., 518
Kovar, A.J., 403
Kowalczyk, B., 331
Kowalczyk, R.S., 232
Kowall, R.R., 515
Kowalski, D.T., 29
Kowalski, J., 317
Kowalsky, A., 330
Kowles, R., 239
Kowning, E., 319
Koyama, T., 298
Kozak, M., 284
Kozam, G., 282
Kozel, P.C., 365
Kozelnicky, G.M., 108
Kozlik, C.J., 384
Kozloff, E.N., 498
Kozloff, L.M., 72
Kozlowski, T.T., 521
Krabbenhoft, K., 238
Kradel, D.C., 403
Kraemer, D.C., 458
Kraemer, L.R., 22
Kraft, A.A., 155
Kraft, D., 236
Kraft, G.F., 504
Kraft, J.M., 503
Krahl, V.E., 203
Krahmer, R.L., 384
Kraicer, J., 546
Krajina, V.J., 534
Krakow, J., 298
Krakow, J.S., 297
Krakower, C.A., 134
Krall, A.R., 428
Kramen, M.A., 469
Kramer, A., 202
Kramer, C.L., 166, 168
Kramer, D., 239
Kramer, F., 301
Kramer, J.K., 512
Kramer, J.P., 303
Kramer, K.J., 166
Kramer, P.J., 332
Kramer, P.R., 458
Kramer, S.Z., 290
Kramer, T.T., 6
Krampitz, L.O., 356
Krane, S.G., 219
Krantz, G., 96
Krantz, G.W., 386
Krasna, A., 300
Krasner, R.I., 419
Krassner, S.M., 54
Kratky, B.A., 114
Kratzer, D.D., 177
Krauch, H.C., 149
Krause, E., 290
Krause, G.F., 258
Krause, J.B., 82
Krause, J.R., 399
Krause, M.O., 540
Krause, R.A., 403
Krause, R.F., 509
Krauss, R.W., 385
Krauthamer, G., 283
Krawiec, S., 398
Krear, R., 328
Kress, C.J., 535
Krebs, C.T., 199
Krebs, E.G., 53
Kreger, A.S., 347
Kreger, L.W., 407
Krehbiel, E.B., 120
Krehbiel, R., 134
Kreibich, G., 312
Kreider, J., 405
Kreider, M.B., 221
Kreier, J.P., 363
Kreig, D.R., 462
Kreimer-Birnbaum, M., 318
Kreisman, A., 387
Kreizinger, J.D., 80
Krekeler, C.H., 153
Krekorian, C.O., 40
Krenetsky, J.C., 70
Krenzer, E.G.Jr., 344
Kreps, D., 147
Kresch, E., 394
Kresge, C.B., 502
Kress, D.D., 266
Kretchman, D.W., 365
Kretsinger, R.H., 488
Kreutner, W., 285
Kreutzer, R., 373
Krevens, J., 60
Krey, L.C., 414
Krezdorn, A.H., 92
Kriby, J.S., 376
Kricher, J.C., 220
Krider, H.M., 77
Kriebel, H.B., 366, 367
Krieg, D., 323
Krieg, D.R., 55
Krieg, N.R., 492
Krieg, W.J.S., 126
Krienke, W.A., 91
Krigman, M.R., 340
Krikariam, A., 325
Kriksgens, A.E., 282
Kringen, W.B., 176
Krinsky, N.I., 217
Krisans, S., 39
Krise, G.M., 460
Krishna, K., 297
Krishnakumaran, A., 512

Krishnamurti, C.R., 533
Krishnamurti, P., 105
Krishnan, N., 355
Krista, L.M., 6
Kristoffersen, T., 365
Kritchevsky, D., 411
Kritikos, H.N., 411
Krizman, R., 110
Krnjevic, K., 556
Kroening, G.H., 129
Kroetz, M.L., 364
Kroger, M., 401, 402
Krogstad, B.O., 245
Krohn, S., 355
Krokosska, D., 173
Krolak, P.D., 446
Kroll, R.J., 406
Kronenberg, R., 244
Kronfeld, D.S., 411
Kronmal, R.A., 500
Kronman, M.J., 326
Krook, L., 306
Krooth, R.S., 301
Kropf, D.H., 167
Kropp, B., 376
Kroschewsky, J.R., 390
Kroth, E.M., 258
Krouse, T.B., 399
Krovetz, L.J., 197
Krowpinski, A.M.B., 545
Kruckeberg, A.R., 497
Kruczynski, W., 309
Krueger, C.R., 433
Krueger, D.W., 254
Krueger, E.C., 391
Krueger, H.R., 363
Krueger, M., 44
Krueger, R.C., 370
Krueger, W.A., 347, 444
Krueger, W.B., 363
Krueger, W.C., 382
Krueger, W.F., 458
Krug, J.L., 177
Krug, M.L., 39
Krukowski, M., 263
Krull, J.N., 223
Krum, A.A., 25
Krum, L., 29
Krumdieck, C., 11
Krumholz, L.A., 180
Krupa, P.L., 297
Krupka, R.M., 551
Krupp, R.H., 308
Krusberg, L.R., 201
Kruschwitz, L., 375
Kruse, C.E., 415
Kruse, D.N., 188
Krutchkoff, D., 79
Kruuv, J., 551
Krygier, G., 319
Krysan, J.L., 433
Kryspin, J., 550
Krywolap, G.N., 204
Ksiazek, R., 397
Kubersky, E.S., 290
Kubitschek, H., 126
Kubota, J., 302
Kuby, S.A., 474
Kubzansky, P.E., 207, 208
Kuc, J., 178
Kucera, C.L., 260
Kucera, L.S., 347
Kucharek, T.A., 92
Kuchler, R.J., 288
Kuchnow, K.P., 460
Kuck, J.F.R.Jr., 101
Kuebler, R.R., 341
Kuehl, L., 474
Kuehl, W.M., 490
Kuehnert, C.C., 326
Kuellmer, F.J., 291
Kuenzel, W., 203
Kuenzler, E.J., 338, 341
Kuerti, R., 355
Kugrens, P., 68
Kuhl, D.E., 411
Kuhlman, E.G., 345, 346
Kuhlman, H.H., 439
Kuhn, C.III., 262
Kuhn, C.W., 108
Kuhn, D., 13
Kuhn, D.T., 88
Kuhn, G.D., 402
Kuhn, L.W., 383
Kuhn, R.E., 347
Kuhnen, S.M., 286
Kuhnley, L.C., 461
Kuhr, R.J., 306
Kuijt, J., 532
Kuipers, P., 243
Kuitert, L.C., 91
Kujdych, I., 288
Kukezic, F.L., 403
Kukton, A., 298
Kulangara, A.C., 399
Kuld, P.H., 26
Kulfinski, F., 130
Kulhanek, J., 70
Kull, F., 317
Kullberg, R.G., 257
Kulman, H.M., 240, 242, 243
Kulpa, C.F., 152
Kumamoto, J., 58
Kumar, A., 298
Kumar, S., 282
Kumaroo, K.K., 340
Kumler, M., 130
Kummel, B., 211
Kummerow, J., 39
Kun, E., 60
Kundig, W., 197
Kundt, J.F., 202
Kunenker, J., 207
Kung, C., 61
Kung, F.H., 130
Kung, H., 282
Kung, S., 205
Kunisaki, J.T., 114
Kunkee, R.E., 51
Kunkel, H.O., 458
Kunkel, J.G., 218
Kunkel, R., 503
Kunkle, J.B., 199
Kunkle, L.D., 272
Kunkle, L.E., 365
Kunny, B.K., 511
Kuno, G., 417
Kuno, M., 340
Kunsman, J., 527
Kuntz, J.E., 519
Kuntzman, A.J., 372
Kunz, L.L., 496
Kunz, T.H., 207
Kunze, D.L., 468
Kunze, G.W., 459
Kuo, A.Y., 490
Kuo, D.K., 337
Kuo, J.F., 101
Kupa, J.J., 422
Kupke, D.W., 490
Kupor, S.R., 440
Kupper, L.L., 341
Kuramoto, R., 27
Kurczewski, F.E., 322
Kurmis, V., 243
Kurosky, A., 467
Kurstak, E., 558
Kurtz, A.C., 380
Kurtz, Sr.E., 278
Kurtz, E.B., 469
Kurtz, E.U., 433
Kurtz, H.M., 62
Kurtz, L.T., 138
Kusano, K., 121
Kusche, R., 174
Kushner, D.J., 549
Kushner, S.R., 105
Kuslan, L., 76
Kuspira, J., 535
Kuster, R.J., 257
Kustu, S., 51
Kuta, V.A., 123
Kutchai, H., 490
Kutscha, N.P., 194
Kuykendall, J.R., 17
Kuyper, A.C., 235
Kuzma, J.W., 36
Kwan, H.C., 550
Kwapinski, J.B.G., 538
Kwart, H., 411
Kwasikpui, D., 338
Kyger, E., 484
Kyhos, D.W., 51
Kylen, A.M., 67
Kynard, B.E., 18

L

Laale, H.W., 538
van der Laan, K., 220
Labanauskas, C.K., 58
Labar, M., 423
LaBatt, D., 223
LaBelle, R.L., 307
Laben, R.D., 48
Laber, L.J., 194
LaBerge, W.E., 136
Labosky, P., 425
Labrie, D.A., 471
LaColle, P.L., 328
Lacey, R.B., 247
Lacey, R.J., 295
Lachance, J.P., 558
Lachance, J.R.C., 556
LaChance, L.E., 351
Lachance, P., 288
Lachance, R.A., 558
Lachance, R.O., 558
Lachapelle, R.C., 371
Lachenbruch, P.A., 341
Lachman, W.H., 219
Lachmanan, N.K., 439
Lack, L., 333
Lackard, I., 428
Lackey, F., 446
Lackey, J.A., 324
Lackey, O.F., 120
Lackey, R.T., 494
Lacoursiere, E., 560
LaCroix, J.D., 230
La Croix, L.J., 537
Lacroix, G., 557
Lacy, A.M., 196
Lacy, J.C., 11
Lacy, P.E., 265
Ladd, T.L., 363
Ladman, A.J., 293
Laemle, L., 282
Laemmli, U.K., 286
Laetsch, W.M., 42
Lafaver, J., 6
Lafever, H.L., 364
Laffer, N.C., 202
Laffin, R.J., 327
La Fleur, K.S., 423
Lafleur, L., 558
Lafond, A., 557
Lafontaine, J.G., 557
LaForge, W.F., 246
LaFountain, J.R., 318
LaFrance, C.R., 145
Lagler, K.F., 231
Laglois, J.M., 557
La Grange, W.S., 155
Lagueux, R., 557
LaHam, Q.N., 549
LaHaye, P.A., 324
Lahiri, S., 411
LaHood, M., 215
de Lahunta, A., 305
Lai, M.M.C., 64
Laidlaw, H.H., 49
Laif, A., 246
Laing, C.C., 362
Laing, J.E., 547
Laird, C.D., 498
Laishley, E.J., 531
Lakars, T.C., 134
Lake, F.T., 103
Lake, J.A., 312
Lake, L.F., 263
Lake, R., 355
Lake, R.B., 356
Laking, L., 545
Lakshman, A.B., 545
Lakso, A.N., 307
Lal, P., 419
Lalita, S.H., 485
Lall, A., 297
Lalli, C.M., 556
Lalli, M.F., 555
Lally, D.A., 180
Lalonde, R., 322
Lam, C.F., 428
Lam, K.M., 408
Lam, K.W., 327
Lam, S.L., 149
La Marca, M.J., 512
Lamb, C., 455
Lamb, C.A., 364
Lamb, J.C.III., 341
Lamb, R., 387
Lamb, R.C., 307, 474
Lamb, T.W., 490
Lambert, A., 195
Lambert, C.C., 30
Lambert, D.B., 455
Lambert, H., 214
Lambert, J.D.H., 544
Lambert, J.W., 240
Lambert, R.G., 180
Lambert, R.J., 138, 303
Lambert, R.M., 319
Lambert, W.W., 301
Lamberti, A.J., 408
Lambeth, V.N., 260
Lambooy, J., 204
Lambooy, J.F., 203
Lambson, R.O., 179
Lamden, M.P., 482
Lammers, W.T., 332
Lammi, J.O., 346
deLamirande, G., 558
Lamonica, J., 558
Lamont, H.C., 216
LaMotte, C.C., 197
LaMotte, C.E., 157
Lamoureaux, G., 558
Lamoureux, C.H., 111
Lamoureux, G.L., 352
Lampe, K.F., 95
Lampen, J.O., 288
Lamperti, A.A., 370
Lamport, D.T.A., 226
Lampton, R.K., 109
Lamtenschlager, E.W., 487
Lamy, F., 559
Lana, E.P., 351
Lancaster, D.W., 271
Lancaster, J.D., 248
Lancaster, J.H., 379
Lancaster, J.L.Jr., 23
Lanchantin, G.E., 64
Lanciani, C.A., 90
Lanclos, K., 103
Land, S.B., 250
Landa, F., 260
Landahl, H., 60
Landan, D., 293
Landau, B., 500
Landau, B.R., 499
Landau, J.V., 314
Lande, M.A., 312
Landecker, E., 285
Lander, E., 553
Landers, E.J., 15
Landers, J.H.Jr., 382
Landers, K., 7
Landers, K.E., 7
Landers, R.B., 187
Landers, R.Q., 157
Landes, B.A., 120
Landesman, R.H., 480
Landgren, C.R., 485
Landman, O.E., 85
Landmann, W.A., 458
Landolt, J.C., 507
Landolt, P., 270
Landphair, H.C., 249
Landry, S., 317
Lands, W.E.M., 232
Lane, B.G., 549
Lane, C., 96
Lane, C.J., 392
Lane, C.L., 425
Lane, F.E., 22
Lane, G.T., 458
Lane, G.H., 479
Lane, H.C., 247, 347

Lane, J.D., 187
Lane, M.R., 196
Lane, P.A., 542
Lane, R.M., 393
Lane, R.S., 318
Lang, A., 226
Lang, B.Z., 496
Lang, C.A., 180
Lang, D.J., 333
Lang, D.R., 370
Lang, F., 207, 387
Lang, G., 548
Lang, K.L., 34
Lang, M., 208
Lang, N.J., 51
Lang, R.L., 527
Lang, R.W., 367
Lang, S., 263
Langbein, M.E., 121
Langdell, R.D., 340
Langdon, H.L., 413
Langdon,J.W, 12
Langdon, R.G., 490
Lange, C., 124
Lange, C.S., 329
Lange, E., 523
Lange, K., 56
Lange, L.H., 32
Lange, W.H., 49
Lange, Y., 211
Langenbach, R., 273
Langenberg, W., 270
Langenberg, W.G., 272
Langenheim, J.H., 61
Langer, G.A., 57
Langer, P., 415
Langerbartel, D.A., 520
Langerbeck, M.T., 336
Langerman, N.R., 217
Langford, A.N., 555
Langford, G.M., 219
Langford, R.R., 550
Langham, J.M., 31
Langhans, R.W., 303
Langhauser, C.A., 46
Langille, A.R., 195
Langin, E.J., 65
de Langlade, R.A., 372
Langley, C.H., 341
Langley, H.M., 100
Langlinais, J.W., 453
Langlinais, Br.R., 291
Langloius, B.E., 177
Langman,J, 489
Langridge, R., 286
Langston, D.T., 17
Langvardt, A.L., 269
Langwig, J.E., 377
Langworthy, T.A., 435
Lanier, G.N., 322
Lankester, M., 544
Lankford, C.E., 465
Lankford, H.C., 268
Lannan, J.E., 383
Lannan, Sr.M., 315
Lanner, R.M., 475, 477
LaNoüe, K., 410
Lansford, E.M., 455
Lantz, C.W., 377
Lantz, R.L., 383
Lapen, R.F., 496
LaPietra, R.A., 310
LaPlante, A.A., 114
LaPointe, J., 292, 557
Lapp, D., 103
Lapp, J.A., 143
Lapp, N.A., 345
Lapp, W.S., 556
LaPrade, J.C., 425
Lara, J.C., 499
Larcom, L.L., 426
Lardy, H.A., 517
Largent, D.L., 34
Larison, L.L., 219
Lark, C.A., 473
Lark, K.G., 473
Larkin, G., 359
Larkin, J.H., 139
Larkin, J.M., 183
Larkin, P.A., 535
Larmie, W.E., 422
Larochelle, J., 557
Laroi, G.H., 531
Larow, E.J., 316
Larrabee, M., 198
Larramendi, L.M.H., 131
Larrison, E.J., 117
Larsen, A.E., 474
Larsen, A.L., 68
Larsen, E.G., 380
Larsen, F.E., 503
Larsen, G., 33
Larsen, H.J., 517
Larsen, H.S., 5
Larsen, J., 432, 501
Larsen, J.B., 252
Larsen, J.E., 460
Larsen, J.R., 136, 137
Larsen, M., 270
Larsen, P., 475
Larsen, P.C., 349
Larsen, P.O., 366
Larsen, S., 527
Larsen, W., 473
Larsen, W.J., 96
Larsh, H.W., 379
Larsh, J.E., 333, 342
Larson, A.D., 183
Larson, C.L., 267
Larson, C.W.R., 359
Larson, D.A., 318
Larson, G.E., 506
Larson, G.L., 383
Larson, I., 119
Larson, K.A., 69
Larson, K.L., 258
Larson, L., 29, 455
Larson, L.A., 368
Larson, M.M., 366, 387
Larson, O.R., 352
Larson, P.F., 186
Larson, R., 261
Larson, R.C., 307
Larson, R.L., 258
Larson, V.S., 506
Larson, W.E., 241
Larter, E.N., 537
Lascelles, J., 55
Lasek, R.J., 355
Lasenby, D.C., 546
Laseter, J.L., 185
Lash, J., 410
Lasher, E.P., 499
Lasker, G.W., 235
Laskowski, M., 320
Lasley, J.F., 259
Lasser, N., 282
Lassiter, J.W., 106
Laster, L., 325
Laster, M.L., 249
Lastra, I., 418
Lata, G.F., 161
Latham, A.J., 4
Latham, M.C., 306
Lathers, C.M., 399
Lathwell, D.J., 302
Lathrop, E.W., 35
Laties, G., 55
Latif, A.A., 103
Latimer, H., 30
Latina, A.A., 97
Latorella, A.H., 323
Latour, J.P.A., 555
Latshaw, J.D., 367
Latshaw, W.K., 562
Latt, R.H., 403
Latterell, R.L., 507
Lattin, J.D., 386
Lau, B.H.S., 36
Lau, D.R., 34
Laub, L., 121
Lauber, J.K., 531
Lauck, D.R., 34
Laude, H.M., 47
Lauderdale, R.W., 275
Lauer, F.I., 241
Lauerman, L.H.Jr., 69
Lauf, P.K., 333
Laufer, H., 77, 79
Laufer, N.P., 397
Laufersweiler, J.D., 371
Lauff, G.H., 225, 227
Lauffer, M.A., 413
Laug, G.M., 317
Laughlin, C.W., 226
Laughlin, R.L., 395
Laughnan, J.R., 137, 138
Laughner, W.J., 358
Laun, C., 258
Laurence, Sr.M., 176
Lauria, D.T., 341
Laurie, J.S., 334
Laury, L., 454
Lausch, R.N., 405
Lauson, H.D., 330
Lauter, F.H., 429
Laux, D.C., 421
Laux, G., 8
Laux, L.J., 372
Lavallee, M., 559
Laveglia, J.G., 15
LaVelle, A., 134
LaVelle, J.W., 70
Laver, M.L., 384
LaVigne, A.B., 497
Lavigne, D.M., 547
Lavigne, R.J., 527
Lavine, R.A., 84
Laviolette, F.A., 148
Lavkulich, L.M., 534
Lavoie, J.L., 557
Lavoie, V., 557
Lavoipierre, M.M.J., 53
Lavy, T.L., 271
Law, A.G., 502
Law, B.L., 506
Law, G.R.J., 65
Law, J.H., 131
Law, J.J., 486
Lawler, G.H., 539
Lawlor, Sr.A.C., 284
Lawlor, L., 465
Lawlor, T.E., 34
Lawrence, A.L., 463
Lawrence, A.W., 73
Lawrence, C.W., 328
Lawrence, D.B., 241, 242
Lawrence, F.B., 5
Lawrence, F.P., 92
Lawrence, J., 97, 236
Lawrence, J.A., 117
Lawrence, J.B., 368
Lawrence, J.F., 211
Lawrence, J.M., 97
Lawrence, P.J., 474
Lawrence, R.W., 279
Lawrence, V.M., 415
Lawrence, W., 412
Lawrence, W.H., 445
Lawreng, J.M., 4
Lawson, F.R., 374
Lawson, H.R., 268
Lawson, J.E., 437
Lawson, J.W., 445
Lawson, L.A., 251
Lawson, V.R., 246
Lawton, E., 497
Lay, C.L., 433
Lay, R., 175
Laycock, W.A., 477
Layman, H., 181
Layne, D.S., 548
Layne, J.S., 438
Layton, W.M., 277
Lazaroff, N., 317
Lazarow, A., 243
Lazaruk, W., 75
Lazarus, M., 134
Lazarys, A.W., 186
Lazerson, E., 130
Lazier, C.B., 542
Lea, A.O., 106
Lea, M.A., 282
Lea, M.S., 387
Lea, R.V., 322
Leach, B.J., 253
Leach, C.M., 385
Leach, D.M., 116
Leach, E.D., 439
Leach, F.R., 377
Leach, L.D., 50
Leach, L.J., 329
Leach, R.M., 403
Leadbetter, E.R., 206
Leader, R.W., 78, 79
Leaf, A.L., 322
Leahy, Sr.M.G., 37
Leak, B.W., 280
Leak, L.V., 86
Leake, E.S., 347
Leake, W., 415
Leamy, W.P., 480
Leary, D., 4
Leary, D.E., 419, 420
Leary, J.V., 58
Lease, M.H., 245
Leasure, U.K., 130
Leath, K.T., 403
Leathem, J.H., 288
Leather, R.P., 314
Leathers, C.R., 15
Leatherland, J.F., 547
Leatherwood, M.J., 343
Leav, R.C., 96
Leavitt, B.B., 90
Leavitt, W.W., 370
Lebaron, F.N., 294
LeBaron, M.J., 117
LeBeau, L.J., 134
Lebel, J.L., 69
Leben, C., 366
LeBlanc, F., 549
Le Blanc, H.J., 186
LeBlanc, N.N., 554
LeBlanc, R., 210
LeBlanc, R.M., 560
LeBlond, C.P., 555
LeBoeuf, B.J., 61
LeBourhis, E., 297
Lebovitz, H.E., 333
Lebowitz, L., 326
Leboy, P.S., 411
LeCam, L., 46
Lecce, J.E., 345
Lecce, J.G., 343
Lechevalier, H.A., 288
Lechleitner, F.W., 68
Lechtenberg, V.L., 147
Leck, C.F., 288
Leck, M.A., 287
Ledbetter, J.W.Jr., 428
Leddy, J., 328
Ledeen, R., 330
Lederberg, E.M., 42
Lederberg, J., 41
Lederer, R.J., 29
Ledford, B.E., 428
Ledgerwood, L.B., 331
Ledig, F.T., 81
Ledingham, G.F., 561
Ledinko, N., 369
Lediy, R.B., 343
Ledley, R.S., 85
Ledsome, J.R., 535
Leduc, A., 558
Leduc, G., 557
Lee, C., 244, 410
Lee, C.B.T., 205
Lee, C.C., 96, 397
Lee, C.S., 465
Lee, D.Y., 307
Lee, D., 324, 380
Lee, D.D.Jr., 129
Lee, D.R., 450
Lee, E.S., 469
Lee, E.Y.C., 95
Lee, G.A., 527
Lee, H., 287, 371
Lee, J., 244, 317
Lee, J.A., 344
Lee, J.C., 469
Lee, J.G., 184
Lee, J.J., 297
Lee, J.W., 105
Lee, K., 290
Lee, K.Y., 273
Lee, L.E.Jr., 356
Lee, L.T., 185
Lee, M.M.L., 300
Lee, M.R., 136

Lee, N.L., 61
Lee, P., 509
Lee, P.C., 127
Lee, P.C.Jr., 487
Lee, P.E., 544
Lee, R.G., 44
Lee, R.W., 542
Lee, S., 95, 212
Lee, S.H.S., 543
Lee, S.Y., 34
Lee, T.F., 279
Lee, T.W., 74
Lee, V.A., 67
Lee, W.C., 21
Lee, W.R., 183
Lee, Y., 197
Lee, Y.P., 353
Leeder, J., 288
Leedy, J.L., 139
Leeling, N., 223
Leeper, G.F., 105
Lees, N.D., 145
Lees, H., 538
Lees, R.S., 213
LeFebvre, E., 129
Le Fever, H.M., 164
Lefevre, G.L., 31
LeFevre, P., 325
Leffek, K.T., 542
Leffel, E.C., 200
Lefkovitch, L., 544
Lefkowitz, R., 333
Lefkowitz, S.S., 461
Leftwich, F.B., 488
LeGasse, A.A., 414
Legault, A., 559
Legates, J.E., 343
Legendre, L., 557
Legg, P.D., 177
Legge, T., 395
Leggett, G.E., 474
Leggett, J.E., 177
Leggett, W.C., 556
Legler, J.M., 473
Legler, W.K., 172
Legner, E.F., 58
LeGrand, F.E., 376
Lehman, G.L., 373
Lehman, G.S., 18
Lehman, H.E., 239
Lehmann, H.P., 186
Lehman, I.R., 41
Lehman, R.H., 486
Lehman, W., 208
Lehmkuhl, D.M., 562
Lehner, P.N., 68
Lehnert, J., 223
Leibach, F., 103
Leibovitz, L., 305
Leiby, P.D., 349
Leichnetz, G.R., 491
Leichty, W.R., 472
Leidahl, G.A., 31
Leif, R., 95
Leigh, J.S., 410
Leigh, T.F., 49
Leigh, W.H., 95
Leighton, J., 299
Leighton, R.E., 458
Lein, M.R., 531
Leise, E., 85
Leinweber, C.L., 458
Leischner, T., 123
Leiser, A.T., 47
Leisman, G.A., 164
Leiter, E.H., 297
Leith, D.E., 212
LeJeune, A., 502
Le John, H.B., 538
Leland, E., 209
Lele, P.P., 213
Lelong, M.G., 12
Le Maguer, M., 530
Lembach, K.J., 446
Lembi, C.A., 148
Lemeshka, T., 420
Lemke, P.A., 391
Lemmon, B.E., 191
Lemon, E.R., 302
Lemon, R.E., 556
Lemons, J.E., 11
Lenard, J., 284
Leng, E.R., 138
Lenggel, -., 81
Lenhert, P.G., 446
Lenhoff, H.M., 54
Lenneberg, E.H., 302
Lennebert, E., 304
Lennon, C., 236
Leno, G., 349
LeNoir, W.C., 100
Lenon, H.L., 223
Lent, C., 325, 379
Lent, J.M., 78
Lentz, G.L., 17
Lentz, K.E., 355
Lentz, T.L., 81
Lenz, L.M., 537
Lenz, L.W., 32
Leon, M.A., 356
Leonard, A., 369
Leonard, C.B.Jr., 204
Leonard, C.D., 92
Leonard, Rev.C.J., 313
Leonard, D.E., 194
Leonard, J.W., 231
Leonard, K.J., 345
Leonard, L., 488
Leonard, O.A., 51
Leonard, S.J., 49
Leonard, R., 243
Leonard, R.C., 12
Leonard, R.T., 56
Leonard, W.R., 430
Leonards, J.R., 355
Leone, C., 354
Leone, I.A., 288
Leone, N.C., 475
Leone, S., 298
Leong, K.L., 27
Leonier, D.L., 360
Leonora, J., 36
Leopold, A.C., 149
Leopold, A.S., 43, 44
Leopold, E.B., 71
Leopold, R.A., 350
Leotscher, F.W.Jr., 174
Lepow, I.J., 220
Lepper, R.Jr., 420, 421
LeQuire, V.S., 446
LeRay, N.L., 280
Lerman, L.S., 446
Lerner, E., 84
Lerner, M.P., 380
Lerner, I.M., 45
Lerner, J., 193
Lerner, S., 132
Lernmark, A., 131
Leroy, E.P., 134
Lersten, N.R., 157
LeRud, R., 77, 79
Lerum, J., 497
Lesh, T.A., 142
Lesher, S., 412
Leshin, A., 123
Lesh-Laurie, G., 355
Lesins, K.A., 535
Leskowitz, I., 76
Leslie, G.A., 388
Lesnaw, J., 177
Lesperance, A.L., 275
Lessard, G.M., 36
Lessard, R.H., 291
Lesser, G.V., 306
Lessie, T.G., 218
Lessman, K.J., 147
Lester, C.T., 100
Lester, D.T., 521
Lester, H.A., 25
Lester, R.L., 179
Lester, W.L., 34
LeStourgeon, W.M., 446
Letbetter, W.D., 101
Letendre, J.B., 206
Lett, J.T., 69
Lettvin, J.Y., 466
Leu, R.W., 379
Leung, W.C., 449
Leung, F.Y., 551
Leupschen, N.S., 66
Leure-DuPree, A.E., 405
Leuthold, L.D., 168
Lev, M., 330
Levandovsky, M., 299
Levardsen, N.O., 223
Levene, H., 300, 301
Levengood, C.A., 381
Levenson, G.E., 411
Leveque, T.F., 559
Lever, J.A., 427
Levering, D.F.Jr., 214
Leverton, I., 84
Levetin, E., 381
Levi, H.W., 210
Levi, M.P., 345
Levi, S.F., 131
Levi-Montalcini, R., 263
Levin, B., 322
Levin, B.R., 218
Levin, D.A., 465
Levin, H.E., 301
Levin, N.L., 296, 297
Levin, R.A., 361
Levin, S.A., 302, 303, 304
Levine, A.J., 286
Levine, A.S., 146
Levine, D.A., 549
Levine, D.M., 290
Levine, H.G., 468
Levine, H.R., 76
Levine, L., 208, 234, 297
Levine, M., 232, 355
Levine, R., 96
Levine, R.J.C., 399
Levine, R.P., 210
Levine, S., 236
Levings, C.S., 345
Levins, R., 132
Levinson, C., 470
Levinson, F.J., 213
Levinson, W.E., 60
Levinthal, C., 300
Levinthal, M., 150
Levintow, L., 60
Levitan, H., 199
Levitsky, D.A., 306
Levitt, D., 244
Levitt, J., 260
LeVora, N., 369
Levy, A., 217
Levy, C.K., 207
Levy, D., 64
Levy, G.F., 487
Levy, H., 325
Levy, H.R., 326
Levy, J., 356, 535
Levy, M., 150, 556
Levy, M.N., 356
Levy, N.L., 333
Levy, R., 206
Levy, R.S., 180
Lew, G.M., 227
Lewert, R.M., 132
Lewin, D.C., 157
Lewin, V., 531
Lewis, A.G., 535
Lewis, A.J.III., 424
Lewis, A.R., 451
Lewis, B., 306
Lewis, C., 195
Lewis, D.T., 271
Lewis, E.B., 26
Lewis, F.H., 55, 403
Lewis, F.N., 552
Lewis, G.C., 117
Lewis, G.D., 288
Lewis, G.T., 95
Lewis, J.B., 556
Lewis, J.C., 376, 544
Lewis, J.C., 544
Lewis, J.K., 432
Lewis, J.M., 138
Lewis, K., 282
Lewis, L., 299
Lewis, L.C., 157
Lewis, L.N., 58
Lewis, M.G., 555
Lewis, M.J., 49
Lewis, P.A., 416
Lewis, P.D., 532
Lewis, P.K., 23
Lewis, R., 310, 511
Lewis, R.A., 275
Lewis, R.E., 157, 352
Lewis, R.G., 545
Lewis, R.J., 444, 563
Lewis, R.W., 458
Lewis, S., 239, 357, 500
Lewis, S.A., 425
Lewis, W.H., 263
Lewis, W.M., 129, 344
Lewontin, R.C., 210
Ley, K.D., 94
Leye, P.L., 257
Leyendecker, P.J., 292
Leyshon, G.A., 551
Leyton, M., 408
L'Heureux, M.V., 124
Lhotka, J.F., 379
Li, C.C., 414
Li, C.H., 60
Li, E.H., 253
Li, H.J., 298
Li, H.L., 409
Li, H.W., 47
Li, J.B., 405
Li, P.H., 241
Li, S.C., 190
Li, T.K., 146
Li, Y.T., 190
Liang, G.H., 166
Liao, S., 131
Libal, G.W., 432
Libby, D.W., 9
Libby, J., 195
Libby, J.L., 518
Libby, P.R., 320
Libby, W.J., 44, 45
Liberta, A.E., 122
Liberti, A.V., 316
Liberti, J.P., 491
Libet, B., 60
Libonati, J.P., 204
Lichstein, H.C., 370
Licht, L.E., 553
Licht, P., 43
Lichtenstein, E.P., 518
Lichtenstein, H., 301
Lichtenstein, L.M., 198
Lichtenwainer, R.E., 458
Lichtman, M.A., 328
Lichtwardt, R.W., 170
Lickley, L., 550
Liddle, L.B., 418
Lider, L.A., 51
Lidicker, W.Z., 43
Lidvall, E.R., 443
Lieb, J.R., 406
Lieb, M., 64
Lieb, S., 91
Lieberman, A.S., 303
Lieberman, M., 333
Lieberman, M.E., 109
Lieberman, S., 300
Liebgott, B., 549
Liebhaber, H., 263
Liebhauser, K., 284
Liebman, J., 356
Liebman, P.A., 410
Liebow, C., 308
Liegey, F.W., 396
Liem, K.F., 210, 211
Liener, I.E., 241
Lienk, S.E., 306
Lier, F.G., 300
Liesveld, D., 157
Lieth, H.H., 338
Lieux, M.H., 182
Lifson, N., 244
Light, G., 95
Light, M., 350
Lightbody, J.J.Jr., 235
Lightfoot, W.C., 383
Lignowski, E.M., 400
Ligon, J.D., 293
Liguori, V.A., 490
Likens, G.E., 302, 303
Likhite, V., 312
Liles, B.G.Jr., 193
Liles, S.L., 186

Liley, N.R., 535
Lilien, J., 516
Lilienfield, L.S., 85
Liljemark, W., 244
Lillard, D.A., 107
Lillegraven, J.A., 40
Lillevik, H.A., 225
Lilley, L.P., 502
Lillian, Sr.M., 396
Lillie, R.D., 186
Lilly, D.M., 316
Lilly, J.H., 218
Lilly, P.L., 358
Lillywhite, H.B., 171
Lim, J.K., 514
Lim, R., 131
Lim, S.M., 139
Lim, T.P.K., 273
Lim, Z., 159
Limnartz, N.E., 183
Lin, B.J., 550
Lin, C., 285
Lin, C.H., 551
Lin, C.S., 450
Lin, E.C., 211
Lin, H., 160
Lin, J.C., 188
Lin, S., 198
Lin, Y.C., 113
Linam, J., 70
Linck, A.J., 240
Lincoln, C., 23
Lind, A.R., 256
Lindamood, J.B., 365
Lindauer, I., 73
Lindeberg, L.H., 32
Lindeborg, R.G., 291
Lindell, D.L., 364
Lindell, E.A., 73
Lindemann, C.B., 230
Linden, D.B., 286
Linder, A.D., 116
Linder, F.E., 341
Linder, H.J., 199
Linder, M.C., 213
Linder, R.L., 434
Linderman, R.G., 385
Lindgren, A., 236
Lindgren, D.L., 58
Lindley, C.E., 248
Lindley, D., 308
Lindmark, R.D., 130
Lindorfer, R.K., 244
Lindquist, E.E., 544
Lindquist, R.K., 363
Lindquist, R.R., 79
Lindsay, D M., 175
Lindsay, D.R., 544
Lindsay,-D.T., 106
Lindsay, H.L.Jr., 381
Lindsay, J.W., 499
Lindsay, R.C., 518
Lindsay, V., 130
Lindsay, W., 308
Lindsay, W.L., 65
Lindsey, C.C., 538
Lindsey, D.L., 292
Lindsey, J., 467
Lindsley, D.B., 37
Lindsley, D.F., 64
Lindsley, D.L., 59
Lindstrom, E.S., 404
Lindstrom, L., 239
Lindstrom, R.S., 494
Linduska, J.P., 201
Line, R.F., 503
Linenhard, G.E., 277
Lineweaver, J.A., 493
Ling, D.L., 89
Ling, H., 80
Ling, V., 550
Lingappa. B.T., 209
Linger, B.L., 361
Lingner, J.W., 341
Linhart, Y., 71
Link, C.B., 202
Linkenheil, M., 435
Linker, A., 474
Linn, D.W., 387
Linn, M.B., 138
Linn, R.M., 228
Linn, S.M., 42
Linn, T.C., 466
Linna, T.J., 408
Linnell, J., 223
Linscott, D.L., 302
Linscott, W.D., 60
Linsley, E.G., 44
Lint, H.L., 29
Linton, C.D., 84
Linton, J.E., 97
Linton, K.J., 392
Linzey, D.W., 12
Lionetti, F.J., 207
Lioy, F., 535
Lipe, R.S., 349
Lipely, G.E., 359
Lipetz, J., 393
Lipetz, L.E., 362
Lipetz, M., 71
Lipke, H., 219
Lipke, W., 15
Lipkin, A., 348
Lipner, H., 88
Lippert, A.L., 82
Lippert, L.F., 58
Lippincott, J.A., 126
Lipps, E.L., 105
Lipps, J.H., 47
Lipscomb, W.N., 211
Lipscomb, W.T., 273
Lipton, A., 405
Lipton, B.H., 520
Lipton, G., 290
Lipton, M., 340
Lipton, P., 521
Liptrap, D., 177
Liptrap, R.M., 548
Lisanne, Sr.M., 361
Lisano, M.E., 5
Lisanti, A., 323
Lischner, H., 408
Lisk, R.D., 287
Liskova, M., 558
Liss, M., 207
Lissant, E., 253
List, A., 393
Lister, R.M., 148
Lister, W.R., 392
Liston, C., 225
Liston, J., 498
Listowsky, I., 330
Litchfield, C., 288
Litchfield, C.D., 288
Litchford, R.G., 440
Litman, B.J., 490
Litman, R.M., 446
Litt, M., 388, 411
Littell, A.S., 469
Litterfield, L., 352
Littge, W.G., 94
Little, A., 45
Little, A.B., 356
Little, C., 15
Little, H.F., 114
Little, H.N., 217
Little, J.A., 3
Little, J.B., 212
Little, J.C., 448
Little, J.E., 481
Little, J.R., 186, 263
Little, L.W., 341
Little, M.D., 190
Little, P.B., 548
Little, P.L., 454
Little, R.C., 104
Little, S.N., 456
Littlechild, J.A., 286
Littlefield, L., 195
Littlepage, J.L., 536
Littler, M.M., 54
Littwack, G., 408
Litzinger, B., 315
Liu, C.N., 410
Liu, C.L., 28
Liu, E.H., 429
Liu, H.Z., 324
Liu, M.S., 186
Liu, P.T., 394
Liu, P.V., 180
Liugrel, J.B., 370
Live, I., 412
Liversage, R.A., 550
Livezey, R.L., 31
Livingston, C.H., 68
Livingston, K.W., 5
Livingston, L.G., 407
Livingston, R., 88
Livingston, R.B., 217
Livingstone, D.A., 332
Ljungdahl, L.G., 105
Llaurado, J.G., 513
Llera, F.J., 418
Llinas, R., 162
Lloyd, E., 352
Lloyd, E.B., 83
Lloyd, E.P., 249
Lloyd, J.E., 527
Lloyd, L.E., 91
Lloyd, M., 132
Lloyd, R.D., 473
Lloyd, R.M., 368
Lloyd, R.S., 86
Lloyg, J.E., 90
Loadholt, C.B., 428
Loan, R.W., 262
Loats, K.V., 358
Lobenstein, C.W., 260
Lobl, R.T., 79
Loblick, D.C., 121
LoBue, J., 312
LoCascio, N.J., 318
Locascio, S.J., 93
Loch, W.E., 259
Lochhead, J.H., 480
Lockard, J.D., 201
Lockard, R.G., 178
Lockard, V.G., 252
Locke, J.F., 247
Locke, M., 552
Lockhart, C.H., 166
Lockhart, J.A., 217
Lockhart, L., 467
Lockhart, W.R., 156
Lockley, O.E., 99
Lockner, F.R., 28
Lockwood, L.B., 181
Lockwood, T.E., 136
Lockwood, W.R., 252
Loder, C., 195
Lodge, S., 552
Loe, P.R., 340
Loeblich, A.W., 211
Loegering, W.Q., 260
Loendorf, L.L., 352
Loercher, L., 287
Loesch, J., 490
Loesche, W.J., 232
Loew, C., 236
Loew, F.M., 563
Loewen, C., 389
Loewen, P.C., 538
Loewenberg, J., 521
Loewenstein, M.L., 106
Loewenstein, W.R., 96
Loewenthal, L.A., 231
Loewer, O.J., 179
Loewus, F.A., 318
Loewy, A.G., 396
Lofgreen, G.P., 48
Lofgren, J.A., 240
Loftfield, R.B., 294
Loftis, G., 257
Logan, D.M., 553
Logan, L.A., 21
Logan, M.E., 104
Logan, R.E., 396
Logan, T.J., 364
Logomarisino, J.V., 323
Logothetopoulos, J., 550
Loh, P.C., 112
Loher, W.J., 44
Lohse, C.L., 52
Loizzi, R.F., 135
Loken, S.C., 82
Lollis, D., 164, 379
Lomasney, J., 191
Lombard, R.E., 131
Lombardi, P.S., 474
Lommason, R., 270
Lonard, R., 452
Long, C., 522
Long, C.E., 168
Long, C.R., 458
Long, G.G., 260
Long, H., 188
Long, J., 285
Long, J.A., 60
Long, J.M., 8
Long, Rev.J.P., 357
Long, J.R., 548
Long, R.C., 344
Long, R.J., 451
Long, R.W., 97
Long, S., 81
Long, S.Y., 512
Long, T.A., 401
Long, T.C., 339
Long, T.J., 354
Long, W.K., 465
Longe, R.L., 104
Longest, P., 336
Longest, W.D., 251
Longfellow, S.N., 269
Longley, J.B., 180
Longley, W., 332
Longmore, W., 256
Longmuir, I.S., 343
Longo, F.J., 445
Longo, L.D., 36
Longpre, E.K, 73
Lonski, J., 39
Longstaff, B.C., 545
Longsworth, R.M., 361
Longton, R.E., 538
Lonnquist, J.H., 516
Lontz, J.F., 396
Loofbourrow, G.N., 172
Loomer, A., 514
Loomey, W.B., 489
Loomis, B., 135
Loomis, R.S., 47
Loomis, W.D., 385
Loomis, W.F.Jr., 59
Lootsey, J.M., 333
Loper, G.M., 16
Loper, J.C., 370
López, A.P., 417
López, B.R., 417
Lopez, D., 96
Lopez, R.J., 104
LoPinto, R.W., 284
Lorance, E.D., 41
Lorbeer, J.W., 304
Lorber, B., 408
Lorber, M., 85
Lorber, V., 244
Lorch, I.J., 296
Lord, G.E., 499
Lord, L.P., 439
Lord, W.J., 219
Lords, J.L., 471
Lorenz, F.W., 47
Lorenz, K., 67
Lorenz, O.A., 51
Lorenzo, A., 198
Loria, R.M., 492
Lorig, R.J., 356
Lorincz, A.A., 11
Lorio, W.J., 250
Lorkovic, H., 162
Lorz, A.P., 93
Losche, C.K., 129
Losey, G.S., 112
Losick, R., 210
Losick, R.M., 211
Loskutoff, D., 297
Losos, G.J., 548
Lospalluto, J., 466
Lothers, J.E.Jr., 437
Lotlikar, P., 408
Lotrich, V.A., 82
Lotse, E.G., 195
Lotspeich, F.J., 509
Lott, D.F., 47
Lott, E.J., 149
Lott, J.N.A., 545
Lott, J.R., 452
Lott, R.L., 446
Lotz, F., 548

Loublier, J.L., 559
Louch, C.D., 123
Loucks, O.L., 515
Lougee, R.W., 79
Lough, I., 420
Lough, J.B., 67
Lough, J.O., 556
Loughane, M.H., 409
Loughlin, Sr.M., 191
Loughlin, P.T., 253
Loughton, B.B., 553
Louie, R., 366
Louis, L.H., 232
Louis-Ferdinand, R.T., 204
Louius, R.E., 105
Loungee, R.W., 77
Loushin, L.L., 4
Love, B., 328
Love, C.M., 31
Love, D.D., 355
Love, H.S., 374
Love, J., 244
Love, J.E., 184
Love, L.D., 11
Love, R.M., 47
Love, S.H., 347
Love, W.E., 198
Lovejoy, B.P., 102
Lovejoy, D., 220
Lovelace, C.J., 34
Loveland, R.E., 288
Lovell, J.F., 378
Lovell, R., 290
Lovell, R.T., 4
Lovett, J.S., 150
Lovett, P.S., 205
Lovrien, E., 385
Lovrien, R.E., 241
Low, B., 300
Low, E.F., 488
Low, F.N., 352
Low, J.B., 477
Low, J.E., 476
Low, P.F., 147
Lowe, C.C., 304
Lowe, C.H., 18
Lowe, J.H.Jr., 267
Lowe, J.L., 321
Lowe, L.E., 534
Lowe, J.M., 496
Lowe, R.L., 354
Lowe, R.H., 177
Lowe-Jinde, L., 547
Lowell, R.D., 125
Lowenhaupt, B., 395
Lowenhaupt, R.W., 370
Lowenstein, H., 118
Lowenstein, J.M., 208
Lower, R.L., 345
Lowery, G.H., 183
Lowery, H., 106
Lowery, T.J., 397
Lowey, S., 208
Lownsbery, B.F., 50
Lowndes, H.E., 283
Lowrey, R.S., 106
Lowry, D.C., 48
Lowry, E., 524
Lowry, O.H., 263
Lowry, R.A., 407
Lowry, R.J., 230
Lowther, R.L., 557
Lowy, B., 182, 330
Loy, J.B., 281
Loyacano, H.A., 424, 426
Loyle, M.B., 281
Loynt, R.J., 328
Loyt, A.P.S., 64
Loyt, R., 181
Loza, U., 319
Lozano, E.A., 266
Lozeron, H., 256
Lozier, J., 431
Lozlowski, G.P., 68
Lroxel, M.M., 354
Lu, B.C., 546
Lubin, M., 277
Lubinsky, G., 538
Lubs, H., 72
Lucansky, T.W., 90
Lucas, E.A., 24
Lucas, G.A., 155
Lucas, G.B., 345
Lucas, J.D., 372
Lucas, L.T., 345
Lucas-Lenard, J., 77
Lucchesi, J.C., 339
Lucchino, R., 327
Luce, T.G., 150
Luce, W., 137
Lucey, R.F., 302
Lucido, P.J., 255
Luck, D.N., 361
Luck, R.F., 58
Lucken, K.A., 350
Luckenbill, L., 215
Luckett, W.P., 300
Luckey, T.D., 261
Luckman, C.E., 139
Luckmann, L.D., 62
Luckmann, W.H., 136
Ludlam, S.D., 218
Ludlam, T.S., 354
Ludtke, R.D., 352
Ludwick, A.E., 65
Ludwick, T.L., 365
Ludwig, C.E., 31
Ludwig, E.H., 404
Ludwig, J., 292
Ludwig, J.F., 290
Ludwig, M., 232
Luebbe, W.D., 65
Luecke, R.W., 225
Luedders, V.D., 258
Lueker, D.C., 69
Luft, J.H., 499
Luger, M., 404
Luginbuhl, R.E., 78, 79
Luginbuhl, W.H., 481
Lugo, A.E., 90
Lugo, H., 418
Lugthart, G.J., 310
Luh, B.S., 49
Luick, J.R., 14
Luisada, A., 134
Luitjens, J., 534
Lukas, G., 29
Luke, H.H., 92
Luke, W.J., 182
Lukens, L.N., 80
Lukens, P., 524
Luker, G.W., 456
Lukton, P., 498
Lumb, E.S., 329
Lumb, J.R., 99
Lumb, R.H., 348
Lumeng, L., 146
Lumsden, J.H., 548
Lumsden, R.D., 190
Lund, A., 238
Lund, D.B., 518
Lund, D.L., 269
Lund, H.O., 106
Lund, H.R., 350
Lund, R., 295
Lund, R.D., 499
Lund, S., 288
Lund, W.A., 77
Lundberg, J.G., 332
Lundblad, R.L., 340
Lunde, A.S., 341
Lundeen, C.V., 347
Lundell, C.L., 465
Lundgren, D., 293
Lundgren, D.G., 326
Lundquist, L., 427
Lundquist, M.G., 167
Lundzey, G., 465
Lunger, P.D., 82
Lungstrom, L., 164
Lunt, S.R., 272
Luper, M., 353
LuQui, I., 235
Lurie, D., 428
Luschei, E.S., 500
Lusk, J.W., 249
Lussier, J.J., 548
Lustick, S.I., 364
Lustig, H., 296
Lutes, C.M., 487
Lutes, D.D., 186
Luttrell, E.S., 105, 108
Lutz, J.E., 511
Lutz, P.E., 342
Luykx, P., 95
Lydon, C., 27
Lyerla, T.A., 209
Lyford, S., 219
Lyford, W.H., 211
Lyke, E.B., 30
Lyke, W.A., 443
Lyle, E.S., 5
Lyle, J.A., 4
Lyles, D., 452
Lyles, D.I., 9
Lyles, S.T., 461
Lylis, J., 79
Lyman, C.P., 211
Lyman, H., 325
Lyman, R.L., 45
Lynch, Br.D., 453
Lynch, C.B., 80
Lynch, D.L., 126
Lynch, E.C., 448
Lynch, E.J., 70
Lynch, G.S., 55
Lynch, Sr.J., 191
Lynch, J., 270
Lynch, J.E., 498
Lynch, K.D., 5
Lynch, M., 299
Lynch, M.P., 490
Lynch, P.R., 409
Lynch, R.D., 212
Lynch, R.G., 263
Lynch, R.L., 193
Lynd, F.T., 469
Lynd, J.Q., 376
Lynn, R., 375
Lynn, R.I., 477
Lynn, R.J., 434
Lynn, R.L., 476
Lynn, W.S., 333
Lyon, D.L., 154
Lyon, J., 212
Lyon, J.B.Jr., 101
Lyon, W.F., 363
Lyons, C.T., 290
Lyons, E.E., 99
Lyons, E.T., 178
Lyons, J.M., 51
Lyons, M.J., 308
Lyric, R.M., 538
Lysenko, M.G., 521
Lyser, K., 298
Lyser, K.M., 297
Lysko, J.Jr., 286
Lytle, C.F., 346
Lytle, I.M., 18
Lytle, J.B., 100
Lytle, L.D., 213
Lyttle, C.R., 546
Lyung, R.D., 150
Lywood, D.W., 546

M

Ma, T.H., 139
Maack, R.M., 308
Maan, S.S., 350
Maassab, H.F., 234
Maber, M.J., 171
Mabini, T.R., 356
MacCabe, J.A., 442
Macagno, E., 300
MacAllister, H., 286
Macaluso, Sr.M.C., 268
Macarak, E., 411
Macay, R.I., 43
MacCallum, D.B., 63
MacCallum, F.J., 562
MacCanley, R.A., 552
Mac Carthy, H.R., 534
Macchi, L.A., 207
MacCluer, J.W., 403
MacCollom, G.B., 481
MacCrimmon, H.R., 547
Mac Donald, A., 544
MacDonald, A.B., 212
Mac Donald, D.L., 385
MacDonald, E.L., 416
Macdonald, E.M., 467
MacDonald, G., 410
MacDonald, G.J., 283
MacDonald, H.A., 302
MacDonald, J.H., 349
MacDonald, P., 213
MacDonald, R., 302
MacDonald, R.C., 126
MacDonald, R.E., 302
MacDonald, R.H., 206
MacDonald, T., 213
Macdonald, W.B., 515
Macdonald, W.L., 508
MacDonnell, M.F., 288
MacDowell, R., 358
Mace, A.C.Jr., 243
Mace, K.D., 455
Mac Entee, F., 415
Mac Fadden, D.L., 437
MacGee, J., 370
MacGregor, J.M., 241
MacGregor, R., 10
Mac Hardy, F.V., 530
MacHardy, W., 280
MacHattie, L.A., 550
Machean, D., 373
Machia, B.M., 453
Machin, J., 550
Machin, W.D., 541
Machlis, L., 42
Machtiger, N., 220
MacInnis, A., 55
MacIntosh, F.C., 556
Macintyre, B.A., 511
MacIntyre, G.T., 297, 298
Macintyre, M.N., 355
MacIntyre, R.J., 302
MacIntyre, W.G., 490
Macior, L.W., 369
Mack, C.O., 139
Mack, H.J., 384
Mack, R., 359
Mack, W.N., 228
Mackal, R.P., 131
Mackauer, J.P.M., 533
Mackay, R.S., 207
MacKay, W., 355
Mackay, W.C., 531
MacKellar, I., 79
MacKeller, I., 78
Mackenzie, C.G., 72
Mac Kenzie, D., 214
MacKenzie, D.R., 403
Mackenzie, J.B., 72
Mackenzie, J.W., 283
MacKenzie, R.E., 555
MacKenzie, R.G., 549
Mackey, H., 414
Mackey, J.P., 40
Makcey, M., 556
Mackie, G.A., 551
Mackie, G.O., 536
Mackie, R., 265
Macklin, M., 356
Macksam, W.G., 66
Maclachlan, G., 556
Maclachlan, G.A., 556
MacLaughlin, L., 20
Mac Laury, D.W., 177
Maclean, F.I., 542
Mac Lean, G.R., 542
MacLean, S.F.Jr., 14
MacLean, W.P.III., 483
Maclellan, D.C., 556
Macleod, C.F., 557
MacLeod, G.K., 547
MacLeod, E., 137
MacLeod, E.G., 136
Macleod, J.C., 195
Macleod, R.A., 557
MacLulich, D.A., 552
Macmahon, J.A., 476m 477
MacMahon, R., 214
MacMaters, W.J., 360
Macmillan, J.D., 288
MacMillan, P.C., 144
Macmillan, W.H., 480
MacMillen, R.E., 54

MacNab, A.A., 403
Macnamara, J.P., 8
MacNeill, B.H., 547
MacNeil, J.H., 403
MacNichols, E., 208
MacPhee, C., 118
MacPherson, C.F.F., 551
Macpherson, C.R., 367
Macpherson, J.W., 547
Macpherson, L.B., 542
MacQuarrie, I.G., 554
MacQuillan, A.M., 202
Macri, J., 295
Macrina, F., 492
MacSwan, I.C., 385
Mac Vean, D.W., 378
Macy, J., 11, 12
Macy, J.Jr., 12
Macy, R., 387
Madden, B.S., 343
Madden, R., 311
Madden, Sr.T.A., 284
Mader, E.L., 166
Maderson, P.F.A., 295
Madge, J.E., 193
Madgwick, G.A., 222
Madhok, O.E., 349
Madin, S.H., 46
Madison, C.R., 286
Madison, D., 556
Madonia, J.V., 124
Madorsky, D.D., 470
Madsen, G., 88
Madsen, N.B., 531
Madson, M., 240
Maeba, P.Y., 538
Maenza, R.M., 79
Maes, J.L., 275
Maestas, S., 291
Maga, J.A., 67
Magalhaes, H., 391
Magdoff, F.R., 481
Magee, D.F., 269
Magee, J.C., 157
Magee, L.A., 251
Magee, W.L., 551
Magee, W.T., 225
Magelli, P.J., 173
Maggenti, A.R., 47, 50
Maggiora, G.M., 170
Magill, J.M., 458
Magilton, J.H., 158
Maginnes, E.A., 561
Magleby, K.L., 96
Magleby, V.R., 473
Magnuson, H.J., 233
Magnuson, J.J., 516
Magrath, L.K., 381
Magruder, W.J., 254
Maguder, T., 79
Maguire, B., 297, 300, 465
Maguire, M.H., 172
Magun, B.E., 445
Magwood, S., 544
Mah, V.L., 487
Mahadeva, M.N., 523
Mahadik, S., 300
Mahaffey, B.D., 168
Mahagan, S.C., 439
Mahan, D.C., 365
Mahan, H.D., 355
Mahan, J.M., 468
Mahan, L., 524
Mahan, P.E., 94
Mahannah, C.M., 276
Maher, W.J., 562
Maher, W.P., 512
Mahesh, V.B., 103
Mahlberg, P.G., 144
Mahler, W.F., 454
Mahloch, J.L., 247
Mahmound, I.Y., 523
Mahon, P.J., 206
Mahoney, A.W., 475
Mahoney, D.P., 31
Mahoney, J.J., 286
Mahoney, R.P., 316
Mahony, D.E., 543
Mahowald, A., 145
Mahowald, T.A., 273
Mahrt, J.L., 531
Mahvi, T., 428
Mai, W.F., 304
Maiello, J.M., 288
Maier, A., 11
Maier, C.R., 274
Maier, J.T., 198
Maier, S., 368
Maillie, H.D., 328
Mailman, D., 463
Mailman, D.S., 464
Main, A.R., 343
Main, C.E., 345
Main, J., 497
Main, R.A., 30
Main, S.P., 162
Mainland, C.M., 345
Mainland, R., 7
Mainwaring, H.R., 348
Mainwood, G.W., 549
Maio, J., 329
Maisel, H., 235
Maitra, U., 300
Majer, J., 127
Majerus, P.W., 263
Major, C.W., 193
Major, J., 51
Major, P.C., 509
Majumdar, S., 220, 397
Mak, S., 545
Mak, T.W., 550
Makarewicz, J., 317
Makarski, J., 390
Maki, J., 514
Maki, J.R., 515
Maki, L.R., 528
Maki, T.E., 334, 336
Makinen, M.W., 132
Makley, J.T., 356
Makman, M., 330
Maksud, M.G., 513
Malacinski, G., 145
Malamed, S., 283
Malamy, M.H., 217
Malaney, G.W., 446
Malchy, B., 545
Malcolm, D., 387
Malcolm, D.R., 387
Malcolm, P., 542
Malcom, G.T., 186
Male, C.J., 72
Malecha, S.R., 113
Malechek, J., 476
Maleike, R.R., 130
Malek, E.A., 190
Malek, R.B., 139
Males, J.R., 129
Maley, F., 327
Malhotra, H., 399
Malhotra, O.P., 356
Malhotra, S.K., 531
Malick, J., 285
Malicky, N., 164
Malik, D.D., 451
Malindzak, G.S.Jr., 340
Malkin, L.I., 235
Malkus, L.A., 77
Mallampall, R., 513
Mallavia, L.P., 501
Mallery, C., 95
Mallett, J.C., 213
Mallette, J., 446
Mallette, M.F., 404
Malley, A., 388
Malling, H.V., 345
Mallinson, G.G., 236
Mallis, A., 402
Mallmann, V.H., 228
Mallory, F.F., 552
Mallory, T., 30
Malloy, Sr.M., 452
Malnassy, P., 451
Malo, D.D., 433
Malo, S.E., 92
Malone, T.C., 297
Maloney, Sr.K., 311
Maloney, M.A., 9
Maloney, M.H.Jr., 9
Maloney, T.E., 385
Maloof, G., 116
Maloy, O.C., 503
Malsberger, R.G., 358
von Maltzahn, K.E., 542
Malvin, R., 232
Maly, E., 216
Mameesh, M., 380
Manahan, C.O., 109
Manaligod, J.R., 134, 135
Manalis, R.R., 370
Manasek, F., 131
Mancinelli, A.L., 300
Mancy, K.H., 233
Mandalenakis, N., 559
Mandavia, M., 556
Mandel, B., 313
Mandel, L., 333
Mandel, M., 112
Mandl, G., 556
Mandy, W.J., 465
Mangan, J., 30
Mangat, B.S., 3
Mange, A.P., 218
Manges, J.D., 227
Manglitz, G.R., 271
Mangum, C.P., 484, 490
Mangum, J.H., 472
Manheim, F.T., 97
Maniar, A.C., 538
Maniatis, G., 301
Maniloff, J., 328
Maning, J., 51
Manion, P.D., 321
Maniotis, J., 263
Manire, G.P., 339
Mankau, R., 58
Mankau, S.K., 28
Mankin, C.J., 433
Mankinen, C., 452
Manley, G., 556
Manley, T.R., 390
Manly, J.O., 337
Manly, K.F., 320
Mann, D., 317
Mann, G.V., 446
Mann, H.O., 65
Mann, J.B., 95
Mann, K.G., 241
Mann, K.H., 542
Mann, L.A., 371
Mann, M., 86
Mann, M.D., 273
Mann, P.M., 548
Mann, R.A., 538
Mann, T.J., 345
Manner, H.W., 123
Mannering, J.V., 147
Mannik, M., 499
Manning, D., 521
Manning, E.C., 468
Manning, J., 406
Manning, J.E., 54
Manning, J.W., 101
Manning, R., 96
Mannings, C.L., 9
Manns, J.G., 563
Manocha, M.S., 544
Mans, R.J., 90
Mansell, R.L., 97
Mansell, R.S., 93
Mansfield, J.M., 180
Manski, W., 301
Mansour, N.S., 384
Mansour, T., 42
Mansukhani, S., 399
Mantal, K., 323
Mantel, L.H., 297
Mantel, N., 414
Manteuffel, M.D., 124
Manthey, A.A., 445
Manwiller, A., 423
Manzer, F.E., 194
Manzi, J.J., 426
Maple, B.L., 295
Maple, G.R.Jr., 131
Maples, R., 22
Maples, R.S.Jr., 187
Maples, W.P., 109
Mapp, F.E., 99
Maquire, B., 298
Maquire, J.D., 502
Mar, B.W., 498
Mara, J., 278
Mara, V.J., 210
Marable, N.L., 494
Marangu, J.P., 127
Maranville, J.W., 271
Maravolo, N.C., 512
Marble, D.W., 275
Marbut, S.A., 425
Marby, T.J., 465
Marcellus, K.L., 490
March, G.L., 547
March, R.B., 58
Marchant, C.J., 534
Marchant, W.H., 106
Marchin, G.L., 166
Marchinton, R.L., 108
Marcum, J., 219
Marcus, D.M., 330
Marcus, K., 5
Marcus, L.F., 297
Marcus, L.M., 298
Marcus, P.I., 77, 79
Marcus, S., 474
Marcus, S.L., 313
Marcusiu, E.C., 138
Mardon, D.N., 492
Mare, C.J., 158
Mares, M., 412
Margaretten, W., 60
Marge, M., 326
Margen, S.M., 45
Margeson, R., 101
Margolin, P., 313
Margolis, F., 298
Margrave, J.L., 453
Margulies, L., 298
Margulis, L.A., 207
Maris, Sr.R., 396
Marie, Sr.E., 396
Marie, Sr.J., 396, 401
Mariella, R.P., 23
Marien, D., 297, 298
Marietta, Sr.M., 401
Marin, A.H., 278
Marines, H.M., 365
Marinetti, G.V., 328
Marini, M., 127
Marion, K.R., 10
Marion, W.W., 155
Marionneaux, M., 186
Mariscal, R., 88
Maritn, S.C., 18
Mark, R., 399
Mark, W.R., 27
Marker, D., 224
Markert, C., 80
Markert, C.L., 80
Markham, H.N., 524
Markland, F.S., 64
Markovitz, A., 132
Markovetz, A.J., 161
Marks, D., 408
Marks, E.P., 351
Marks, G.C., 153
Marks, L.S., 406
Marks, P., 302
Marks, P.A., 301
Marks, P.L., 303
Marks, R.H.L., 282
Markus, D.J., 118
Markus, G., 320
Markusen, A.S., 484
Markwald, R.R., 428
Marlatt, S., 225
Marley, S.J., 155
Marlin, C.B., 183
Marlow, P., 85
Marlowe, G.A.Jr., 93
Marlowe, T.J., 493
Marmur, J., 330
Marocco, D.A., 279
Maroney, S.P., 488
Marotta, S.F., 135
Marple, D.N., 3
Marquardt, W.C., 68
Marquez, B.J., 226
Marquez, E.D., 405
Marquis, R.E., 328
Marquis, S.D.Jr., 232

Marr, A.G., 46, 51
Marr, C.D., 323
Marr, C.W., 168
Marr, J.J., 263
Marr, J.W., 71
Marra, G.G., 503
Marriott, L.F., 401
Marrotte, P., 220
Marrow, E., 336
Marrs, B., 256
Marsden, J., 556
Marsh, D., 78
Marsh, D.G., 198
Marsh, D.J., 63
Marsh, D.L., 21
Marsh, G., 160
Marsh, G.A., 87, 386
Marsh, G.L., 49
Marsh, H.V.Jr., 219
Marsh, J.A., 110
Marsh, J.B., 411
Marsh, J.M., 95
Marshak, R.R., 411
Marshall, A.C., 542
Marshall, F.D., 434
Marshall, H.G., 401, 487
Marshall, G.R., 263
Marshall, J.A., 507
Marshall, J.D., 116
Marshall, J.J., 95
Marshall, J.T., 249
Marshall, K.C., 549
Marshall, K.G., 556
Marshall, N.L., 4
Marshall, R., 102, 251, 467
Marshall, R.J., 509
Marshall, S.P., 91
Marshall, W.H., 240, 242
Marsho, T.V., 205
Marston, N.L., 259
Martan, J., 129
Marten, G.C., 240
Marth, E.H., 517, 518
Marti, C., 477
Martin, A., 355
Martin, A.H., 520
Martin, A.P., 261
Martin, A.R., 72, 271
Martin, A.W., 498
Martin, B., 496
Martin, B.J., 252
Martin, C., 298
Martin, C.E., 247
Martin, C.M., 365
Martin, C.R., 297
Martin, D., 182, 364, 434
Martin, D.C., 178, 500
Martin, D.F., 249
Martin, D.L., 356
Martin, D.P., 364
Martin, D.W., 22
Martin, E., 270
Martin, E.C., 226, 249
Martin, E.L., 502
Martin, E.W., 354, 453
Martin, G.A., 346
Martin, G.F., 367
Martin, G.L., 259
Martin, H., 412
Martin, J., 175, 189, 375
Martin, Rev.J., 62
Martin, J.D., 494
Martin, J.E., 460, 493
Martin, J.H., 40, 432, 449
Martin, J.M., 550
Martin, J.S., 504
Martin, K.H., 324
Martin, L.D., 172
Martin, Sr.M.C., 151
Martin, M.M., 231
Martin, N.P., 240
Martin, P., 356
Martin, P.C., 211
Martin, R.E., 444
Martin, R.J., 401
Martin, R.L., 193
Martin, R.O., 562
Martin, R.P., 67
Martin, R.R., 449
Martin, R.S., 543
Martin, S., 543
Martin, S.K., 382
Martin, S.W., 548
Martin, T.E., 132
Martin, V.L., 337
Martin, W.C., 62, 293
Martin, W.C.Jr., 5
Martin, W.D., 179
Martin, W.P., 241
Martin, W.R., 132
Martin, W.T., 562
Martin, W.W., 487
Martine, R.L., 356
Martineau, B., 558
Martineau, J.R., 271
Martinek, C.A., 199
Martinek, G.W., 29
Martinek, J.J., 155
Martinez, A.P., 114
Martinez, H.M., 60
Martinez, I.R., 185
Martinez, J.C., 120
Martinez, L., 417
Martinez, M.Y., 415
Martinez, R.J., 55
Martinez, R.M., 76
Martinez-Maldonado, M., 418
Martinsen, D., 382
Martinson, C.A., 157
Martinson, I., 244
Martof, B.S., 346
Martonosi, A., 256
Martorano, J.J., 315
Maety, W.G., 163
Martz, F.A., 259
Marucci, A.A., 326
Marucci, P.E., 288
Marushige, K., 340
Marvel, M.E., 93
Marvin, J., 556
Marvin, J.W., 480
Marvin, H.N., 24
Marwin, R.M., 353
Marx, G.A., 307
Marx, W., 64
Mary, Sr.B., 396
Marzett, M., 182
Marzluf, G.A., 362
Marzolf, G.R., 166
Mas, A.G., 417
Masat, R.J., 159
Masatir, M., 371
Mascona, A., 132
Masha, R., 175
Mashimo, P.A., 319
Masiak, S., 325
Masincupp, F.B., 443
Masken, J.F., 69
Maslin, P.E., 29
Mason, A., 280
Mason, C.E., 17
Mason, C.T., 18
Mason, D.L., 372
Mason, E.B., 321
Mason, E.S., 352
Mason, G.R., 204
Mason, H.S., 388
Mason, K.E., 328
Mason, M., 232
Mason, M.E., 401, 402
Mason, R.G., 340
Mason, R.L., 337, 428
Mason, R.S., 254
Mason, T.L., 217
Mason, W.H., 4, 5, 409
Masoro, E.J., 470
Massaro, E.J., 318
Masse, R.F., 275
Massengale, M.A., 16
Massey, F.J., 56
Massey, H., 378
Massey, H.F., 177
Massey, J.T., 197
Massey, J.W., 259
Massey, L.H., 307
Massey, R.U., 79
Massey, V., 232
Massion, W.H., 380
Massopust, L.C., 256
Massuci, M., 199
Mast, C.C., 493
Mast, M.G., 402
Mastandrea, M., 95
Masters, B., 309
Masters, B.S., 466
Masters, J.M., 203
Masterson, J.P., 120
Masterson, W.K., 221
Masui, Y., 550
Mataric, S., 396
Matches, A.G., 258
Matches, J.R., 498
Matchett, W.H., 501
Matelski, R.P., 401
Matesich, Sr.M.A., 361
Mathemeier, P.F., 191
Mathelson, G., 556
Matheny, J.L., 104
Mather, F.B., 272
Mather, R.E., 288
Mathers, O., 126
Mathes, M.C., 484
Matheson, A.T., 544
Mathews, A.M., 182
Mathews, C.K., 19
Mathews, F.S., 263
Mathews, M.A., 131
Mathews, R., 112
Mathews, S.B., 498
Mathews, W.W., 234
Mathewson, J.A., 421
Mathias, M., 67
Mathias, M.H., 55
Mathies, M., 33
Mathieson, A., 280
Mathieu, L.G., 558
Mathis, B.J., 119
Mathis, P.M., 439
Mathisen, O.A., 498
Mathison, G.W., 530
Matioli, G., 64
Matlock, R.S., 376
Matrone, G., 343
Matschiner, J.T., 273
Matschinsky, F., 263
Matsen, J.M., 244
Matsuda, K., 18
Matsueda, G.R., 72
Matsumura, F., 518
Matsumoto, B., 215
Matsumoto, Y., 101
Matsushima, C.Y., 28
Matsushima, J.K., 65
Matta, J.F., 487
Mattas, R.E., 258
Matten, L.C., 128
Matteo, M.R., 219
Mattern, P.J., 271
Matteson, M.R., 136
Mattfeld, G.F., 322
Matthew, D.L., 148
Matthews, D., 473
Matthews, D.J., 474
Matthews, H., 24, 33
Matthews, J., 342
Matthews, J.L., 449
Matthews, M.A., 185
Matthews, M.E., 518
Matthews, R.A., 47
Matthews, R.F., 92
Matthews, R.W., 106
Matthews, T.R., 546
Matthews, W.R., 431
Matthias, R., 162
Matthies, D.L., 352
Matthysse, A.G., 146
Matthysse, J.G., 303
Mattice, W.L., 184
Mattick, J.F., 201, 202
Mattick, L.R., 307
Mattil, K.F., 459
Mattingly, M.E., 106
Mattingly, S.J., 470
Mattman, L.H., 234
Matton, P., 559
Mattox, K.R., 360
Mattson, F.H., 370
Mattson, R.H., 168
Mattus, G.E., 494
Matulionis, D.H., 179
Maturen, A., 135
Maturo, F.J.S., 90
Matz, L.L., 319
Matzinger, D.F., 345
Mauney, J.R., 16
Maur, K., 102
Maurer, A.J., 519
Maurer, R., 28
Maurides, C., 548
Mauriello, D.A., 39
Mauriello, G., 313
Mautner, H.G., 217
Mautz, W.W., 280
Maverakis, N.H., 234
Mavis, R.D., 127
Mawe, R.C., 297, 298
Maxey, C., 384
Maxey, W.R., 508
Maxfield, J.E., 276
Maxie, M.G., 548
Maxson, J., 377
Maxson, L.E., 137
Maxwell, B.F., 339
Maxwell, D.P., 519
Maxwell, G.R., 324
Maxwell, H.D., 337
Maxwell, F.G., 249
Maxwell, J., 31
Maxwell, J.D., 423
Maxwell, R., 513
Maxwell, R.A., 333
Maxwell, R.H., 147
May, D.S., 73
May, J.T., 108
May, M., 491
May, R.M., 287
May, S.W., 145
May, V.C.P., 12
Maya, J.A., 120
Mayashi, M., 59
Mayberry, B.D., 9
Mayberry, W.R., 435
Mayeda, K., 234
Mayer, D.J., 492
Mayer, M.M., 198
Mayer, T.C., 287
Mayer, W.D., 261
Mayer, W.V., 71
Mayeri, E., 60
Mayers, C., 120
Mayers, L., 398
Mayerson, P., 312
Mayes, J.S., 380
Mayfield, C.I., 551
Mayfield, J.E., 3, 391
Mayhew, E., 320
Mayhew, W.W., 57
Mayland, H.F., 474
Maynard, D.N., 219
Maynard, E.A., 387
Maynard, P.R., 286
Mayne, R., 11
Mayo, J.A., 185
Mayo, J.M., 531
Mayo, M.J., 142
Mayo, Z.B., 271
Mayol, P.S., 28
Mayor, H.D., 449
Mayr, E., 210
Mayrovitz, H.N., 409
Mays, C.E., 143
Mays, C.W., 473
Mays, L., 37
Mays, M., 374
Maysilles, J.H., 144
Maze, C., 506
Maze, J.R., 534
Mazelis, M., 49
Mazia, D., 43
Mazur, A., 298
Mazur, A.R., 424
Mazumdar, S., 414
Mazurak, A.P., 271
Mazurkiewicz, M., 195
Mazzer, S., 359
McAdaragh, J., 434
McAdaragh, J.P., 433
McAdee, D., 464
McAfee, J.A., 106
McAffee, D.A., 96

McAlester, A.L., 454
McAlipine, S., 546
McAlister, D.F., 16
McAlister, R.O., 454
McAlister, S.L., 216
McAllister, H.Y., 407
McAllister, N., 549
McAllister, R.A., 540
McAllister, W.T., 283
McAnelly, C.W., 527
McArdle, F.J., 402
McArdle, J.J., 283
McArthur, C.S., 562
McArthur, F., 6
McArthur, N.H., 460
McArthur, W., 439
McAuley, A.A., 308
McAuley, W.J., 515
McBay, A.J., 340
McBee, R.H., 265
McBee, R.M., 265
McBeth, G.S., 487
McBlair, W., 38
McBrayer, J.D., 253
McBride, A., 224
McBride, B.C., 535
McBride, G.E., 231
McBride, J.R., 44
McBride, R., 390
McBride, R.P., 542
McBride, W.J., 358
McBrien, T.H., 419
McBroom, M.J., 435
McBurney, M.L., 21
McCaa, C.S., 251
McCabe, R.A., 520
McCafferty, R.E., 509
McCafferty, W.P., 148
McCaffree, J., 393
McCague, T., 119
McCain, C.S., 504
McCall, J.P., 200
McCall, P.K., 331
McCalla, D.R., 545
McCalla, T.M., 271
McCalley, D.V., 162
McCallister, L.P., 405
McCallum, B.J., 337
McCallum, C.A.Jr., 11
McCallum, G.A., 32
McCallum, K.J., 561
McCallum, R.E., 380
McCally, M., 84
McCamish, J., 247
McCammon, J.R., 180
McCampbell, H.C., 106
McCandless, D., 482
McCandless, E.L., 545
McCann, D., 232
McCann, F.V., 278
McCann, L.J., 238
McCarl, R.L., 404
McCarter, J.A., 551
McCarter, R.J.M., 470
McCarter, S.M., 108
McCarthy, B.J., 60
McCarthy, C., 292
McCarthy, F., 27
McCarthy, J.L., 454, 497
McCarthy, M.D., 30
McCarthy, P.C., 416
McCarthy, R.D., 401, 402
McCarthy, R.E., 273
McCartney, M.G., 108
McCartney, M.L., 489
McCarty, J.W., 432
McCarty, K.S., 333
McCarty, M.K., 271
McCarty, R., 302
McCarty, R.E., 302
McCaskey, T.A., 3
McCashland, B., 239
McCaslin, G.A., 398
McCaughey, W.F., 16
McCaughey, W.T.E., 552
McCaughran, D.A., 498
McCauley, C.E., 85
McCauley, J.E., 385
McCauley, R.N., 207
McCauley, W., 392
McCay, P.B., 380
McChesney, A.E., 69
McChesney, J.D., 170
McClain, C., 400
McClain, E.F., 423
McClain, W.H., 517
McClanahan, L.L., 30
McClaren, M., 533
McClary, J.E., 244
McClay, D.R., 332
McClean, P., 239
McCleary, J.A., 126
McCleave, J., 193
McClellan, J.F., 68
McClellan, R., 293
McClelland, G.A.H., 49
McClenahen, J.R., 367
McClendon, J., 270
McCleone, M.P., 337
McCleskey, C.S., 183
McClintic, J.R., 30
McClinton, H., 186
McClish, R., 418
McCloseky, Sr.B., 514
McCloskey, L.R., 500
McCloskey, R., 391
McClugage, S.G., 185
McClung, J.R., 185
McClung, L., 144
McClung, N.M., 97
McClure, J., 311
McClure, J.W., 360
McClure, K.E., 365
McClure, M.A., 18
McClure, M.C., 449
McClure, R.C., 262
McClure, S., 412
McClure, T.D., 380
McClure, W., 210
McClure, W.O., 135
McClurg, C.A., 202
McClurg, J., 273
McClurkin, D., 251
McClurkin, J.I., 487
McCluskey, E.S., 35, 36
McCluskey, R.L., 455
McClymont, J.W., 76
McCollum, C.G., 162
McCollum, W.H., 178
McCombs, C.L., 494
McCombs, F.S., 486
McColloch, R.J., 526, 527
McCone, W.C., 432
McConkey, E.H., 71
McConnaughey, B.H., 387
McConnell, B., 112
McConnell, D., 225
McConnell, D.B., 92
McConnell, H., 41
McConnell, J.C., 424
McConnell, K.P., 180
McConnell, M., 421
McConnell, R.A., 413
McConnell, W.J., 66
McConville, D., 239
McCool, S.F., 475
McCord, C.P., 233
McCord, J., 333
McCord, N.F., 455
McCormack, C.E., 134
McCormack, M.L.Jr., 481
McCormick, D.B., 306
McCormick, D.E., 302
McCormick, J.F., 338
McCormick, J.H., 245
McCormick, J.M., 286
McCormick, L.H., 402
McCormick, R., 288
McCormick, W., 58, 205, 476
McCourt, R., 309
McCoy, C., 412
McCoy, C.W., 91
McCoy, D.E., 223
McCoy, G., 285
McCoy, G.C., 129
McCoy, J., 507
McCoy, J.F., 24
McCoy, J.J., 338
McCoy, J.W., 389
McCoy, R.H., 413
McCracken, A.W., 470
McCracken, D., 122
McCrady, E.III., 342
McCrady, J.D., 460
McCrady, W.B., 464
McCraken, M.D., 461
McCrary, A.B., 347
McCraw, B.M., 548
McCray, C., 230
McCray, J.A., 410
McCree, K.J., 459
McCreery, R.A., 106
McCreight, C.E., 347
McCroskey, J.M., 117
McCrum, R.C., 194
McCuaig, K.O., 549
McCue, J., 239
McCulloch, E.A., 550
McCulloch, J.A., 394
McCullough, J.D., 455
McCullough, D.R., 232
McCullough, H.A., 8
McCullough, N.B., 228
McCullough, W.E., 92
McCune, E.L., 262
McCune, R.D., 500
McCune, R.W., 116
McCune, T., 238
McCurdy, J.A., 504
McCurdy, D.R., 129
McCurry, J.R., 181
McCuskey, R.S., 370
McCutchan, M.C., 236
McCutcheon, E.P., 180
McCutcheon, H., 420
McDaniel, A.T., 23
McDaniel, B., 433
McDaniel, B.T., 343
McDaniel, G.L., 444
McDaniel, G.R., 5
McDaniel, I.N., 194
McDaniel, J.C., 138
McDaniel, J.S., 334
McDaniel, J.W., 479
McDaniel, L.E., 288
McDaniel, L.T., 382
McDaniel, M.E., 459
McDaniel, R.G., 16, 19
McDaniel, S.T., 247
McDaniel, T.R., 427
McDaniel, V.R., 20
McDaniel, W.M., 82
McDermott, D.J., 513
McDermott, J.J., 395
McDermott, L.A., 547
McDiarmid, R.W., 97
McDermid, R.W., 183
McDevitt, H., 396
McDiffett, W., 391
McDole, R.E., 118
McDonagh, R.P., 340
McDonald, B.B., 393
McDonald, C.E., 351
McDonald, D.J., 393
McDonald, D.M., 60
McDonald, E.C., 496
McDonald, D.J., 479
McDonald, H.J., 124
McDonald, H.S., 455
McDonald, J.C., 347
McDonald, J.D., 190
McDonald, J.W.D., 551
McDonald, L.E., 108
McDonald, L.L., 526
McDonald, T.F., 103
McDonald, W.C., 464
McDonnell, J.M., 415
McDonough, E.S., 512
McDonough, J.A., 110
McDougal, D.M., 262
McDougall, K.F., 371
McDowall, L.R., 91
McDowell, J.W., 100
McDowell, L.L., 321
McDowell, R.D., 78
McDuffie, G.T., 358
McDuffie, N.M., 562
McEarchran, J.D., 459
McElhaney, R.N., 531
McElligott, J.G., 409
McElree, H., 164
McElroy, F.D., 534
McElroy, W.D., 59
McElwee, R.L., 494
McEnery, M.E., 313
McEnery, W., 238
McEvoy, J., 49
McEvoy, J.III., 47
McEwen, F., 282
McEwen, W.R., 245
McFadden, H.W.Jr., 273
McFadden, J., 270
McFadden, J.M., 269
McFadden, J.T., 231
McFadden, S.E., 92
McFadden, T.E., 336
McFadden, W.D., 292
McFall, E., 313
McFarlan, E.S., 543
McFarland, C.R., 372
McFarland, W., 302
McFarland, W.N., 303
McFarlane, J.E., 556
McFee, W.W., 147
McFeeley, J., 450
McFeely, R.A., 411
McFerran, J., 23
McFeters, G.A., 265
McGaha, Y.J., 251
McGahan, M.W., 291
McGandy, R.B., 212
McGarry, J.D., 466
McGarry, M.P., 320
McGarry, P.A., 186
McGaugh, J.L., 55
McGaughey, R., 15
McGaughey, W.H., 167
McGeachin, R.L., 180
McGee, R., 461
McGee, W.H., 249
McGehee, H.W., 258
McGhee, C.R., 439
McGhee, R.B., 106
McGibbon, W.H., 519
McGill, D.P., 271
McGill, J.F.Jr., 106
McGill, J.J., 289
McGill, L.A., 383
McGill, W., 415
McGilliard, M.L., 493
McGinn, M.P., 284
McGinnes, B.S., 493
McGinnes, F.A., 261
McGinnis, J., 502
McGinnis, R., 497, 537
McGinnis, S.M., 30
McGirt, J.A., 337
McGivern, J., 391
McGiven, J.J., 496
McGivern, Sr.M.J., 311
McGlamery, M.D., 138
McGlinchy, S., 311
McGovern, F., 396
McGovern, J.N., 521
McGowan, R.E., 296, 297
McGowan, R.W., 438
McGrath, J., 359
McGrath, J.J., 152, 288
McGrath, J.T., 411, 412
McGrath, Sr.M.J., 209
McGraw, J.C., 324
McGraw, J.L., 451
McGreary, K.R., 76
McGregor, J.R., 530
McGregor, R.R., 170
McGrew, E.A., 134
McGuigan, J.E., 94
McGuinness, M.J., 311
McGuire, F., 198
McGuire, J.J., 422
McGuire, J.W., 233, 234
McGuire, M., 24
McGuire, R.F., 12
McHale, J.T., 187
McHale, P., 333
McHargue, L.T., 41
McHenry, M.G., 121
McIlraith, S., 515
McIlrath, W.J., 126

McIlwain, D.L., 340
McInerney, J.E., 536
McInnis, T.M.Jr., 425
McIntire, C.D., 385
McIntire, S.H., 479
McIntosh, E., 514
McIntosh, J.E., 449
McIntosh, J.L., 481
McIntosh, J.R., 71
McIntosh, R.L., 545
McIntosh, R.P., 152
McIntyre, G.A., 193
McIntyre, H.K.Jr., 109
McIntyre, J.D., 383
McIntyre, R.L., 331
McIntyre, T.W., 509
McIver, S.B., 550
McIvor, K.L., 501
McJunkin, F.E., 341
McKaughan, H.P., 111
McKauna, J., 325
McKay, D.G., 60
McKay, F., 479
McKay, H., 117
McKay, L., 244
McKay, R., 112
McKean, T.A., 526
McKee, C.G., 200
McKee, G.W., 401
McKee, J.C., 247
McKee, P., 333
McKee, R.M., 167, 522
McKee, R.W., 56
McKeegan, E.J., 313
McKeen, W.E., 552
McKeever, S., 102
McKeewn, J.P., 246
McKeil, R.R., 135
McKell, C., 476
McKell, C.M., 477
McKelvy, J.F., 79
McKelvey, M.E., 437
McKenna, J.M., 461
McKenry, M.V., 58
McKenzie, -., 9
McKenzie, D.S., 382
McKenzie, G.M., 333
McKenzie, J., 380
McKenzie, J.A., 540
McKenzie, J.W., 103
McKenzie, W.H., 345
McKeown, B.A., 547
McKercher, D.G., 53
McKernan, D.L., 498
McKibben, G.E., 138
McKibben, J.S., 6
McKibbin, -., 11
McKibbins, J.M., 11
McKie, D., 533
McKiel, C.G., 422
McKillop, W.L.M., 44
McKimmy, M.D., 384
McKinley, C., 150
McKinnell, R.G., 242
McKinney, C.O., 469
McKinney, D.F., 242
McKinney, J.C., 332
McKinney, P.S., 210
McKinnon, J.P., 241
McKinsey, R.D., 488
McKhann, C.F., 244
McKnight, T.J., 374
McKusick, V., 197
McLain, D., 450
McLaren, A.D., 46
McLaren, I.A., 542
McLaren, J.B., 443
McLaren, J.P., 142
McLaren, L.C., 294
McLarty, D.A., 552
McLaughlin, B., 558
McLaughlin, B.J., 445
McLaughlin, C.S., 54
McLaughlin, D.J., 241
McLaughlin, E., 8
McLaughlin, F.M., 344
McLaughlin, J.L., 238
McLaughlin, J.P., 213
McLaughlin, P., 96
McLaughlin, R.E., 249
McLaughlin, S., 325
McLean, A.F., 406
McLean, D.L., 49
McLean, D.M., 535
McLean, E.B., 359
McLean, E.O., 364
McLean, J.H., 62
McLean, M.P., 331
McLean, N., 39
McLean, R., 317
McLellan, C.R.Jr., 186
McLemore, J.A., 188
McLendon, W.W., 340
McLennan, B.S., 562
McLennan, R.H., 535
McLeod, D.E., 338
McLeod, D.L., 549
McLeod, G., 207
McLeod, L.E., 531
McLeod, M.G., 27
McLimans, W., 319
McMagan, E., 437
McMahan, E.A., 339
McMahon, B.R., 531
McMahon, D., 25
McMahon, K.J., 351
McMahon, R., 80
McMahon, R.F., 464
McMahon, R.O., 384
McMahon, V.A., 527
McManigal, J., 471
McManus, I.R., 413
McManus, L., 308
McManus, Sr.M.A., 159
McManus, M.L., 81
McManus, T.J., 333
McMaster, D.W., 21
McMenamy, R.H., 318
McMenanmin, J., 37
McMichaee, A., 342
McMillan, C., 465
McMillan, D.B., 552
McMillan, Sr.E., 391
McMillan, E.L., 247
McMillan, H.L., 20
McMillan, I., 547
McMillan, J., 265
McMillan, J.F., 238
McMillian, P., 35
McMillion, T.M., 395
McMorrow, A., 206
McMullen, Rev.A.C., 289
McMullen, C.R., 432
McMullen, J.L., 117
McMurphy, W.E., 376
McMurray, W.C.C., 551
McMurray, M., 67
McMurty, J.A., 58
McNab, B.K., 90
McNab, J., 387
McNabb, C.D., 225
McNabb, F.M.A., 492
McNabb, H.S.Jr., 157
McNabb, R.A., 492
McNair, J.L., 251
McNairn, R.B., 29
McNary, W.F.Jr., 208
McNaughton, S.J., 326
McNay, J.L., 101
McNeal, B.L., 502
McNeal, D.W., 61
McNeil, C.W., 501
McNeil, J.N., 557
McNeil, R.J., 303
McNeill, J., 471, 544
McNeill, J.J., 343, 345
McNeill, R., 103
McNeill, R., 223
McNulty, I.B., 473
McNutt, C.W., 469
McNutt, N.S., 60
McNutt, W.S., 217
McOwen, F.L., 547
McPhail, J.D., 535
McPherson, A., 405
McPherson, A.B., 182
McPherson, J.E., 29
McPherson, J.K., 375
McPhillips, Sr.M.M., 316
McQuate, J.M., 368
McQueen, A.P., 485
McQueen, D.J., 553
McQuistan, R., 521
McReynolds, H.D., 124
McRorie, R.A., 105
McRoy, C., 14
McSharry, J.J., 327
McShan, W.H., 516
McSweeney, J.H., 209
McSherry, B.J., 548
McVaugh, R., 230
McVickar, D.L., 56
McWhinnie, D.J., 120
McWhirter, E.L., 248
McWilliams, E.L., 459
McWilliams, K.L., 30
Meacham, T.N., 493
Meacham, W.R., 464
Mead, A.R., 18
Mead, J., 212
Mead, R., 414
Mead, R.W., 276
Meade, F.M., 23
Meade, J.H.Jr., 24
Meade, G., 62
Meade, R.A., 117
Meade, T.L., 422
Meader, R.D., 283
Meador, D.B., 138
Meadors, C.T., 507
Meadors, M.D., 507
Meadows, G.B., 3
Meagher, R., 285
Means, A.R., 449
Means, G.E., 362
Mears, J., 409
Mech, K.F., 203
Mecham, J.S., 461
Mechanic, G.L., 340
Medhlin, K.A., 372
Mechling, K.R., 392
Mecklenburg, H., 265
Mecklenburg, R., 226
Mecom, J.O., 454
Medearis, D.N.Jr., 413
Mederski, H.J., 364
Medici, P.T., 315
Medina, D., 449
Medlen, A.B., 456
Medler, J.T., 518
Medoff, G., 263
Medoff, J., 256
Medve, R.J., 407
Medway, W., 411
Medzihradsky, F., 232
Medzon, E.L., 551
Meeding, J., 540
Meehan, J.P., 64
Meek, C.L., 458
Meek, E.S., 161
Meeker, G.L., 31
Meetz, G.D., 491
Meeuse, B.J.D., 497
Meeuwig, R.O., 275
Megard, R.O., 242
Meglitsch, P.A., 155
Mego, J.L., 10
Mehaffey, L., 329
Mehall, A.G., 151
Mehler, A.H., 513
Mehler, W.R., 60
Mehlquist, G., 77
Mehlquist, G.A.L., 77, 79
Mehner, J.F., 486
Mehra, A., 514
Mehra, K.N., 427
Mehra, R., 147
Mehrotra, B., 251
Meier, A.H., 183
Meier, P., 133
Meier, P.G., 233
Meier, R.C., 498
Meier, R.J., 5
Meierotto, R., 238
Meighen, E.A., 555
Meigs, R.A., 355, 356
Meijer, W., 177
Meineke, H.A., 370
Meinershagen, F., 259
Meinkoth, N.A., 407
Meins, F.Jr., 287
Meints, R., 270
Meisch, M.V., 23
Meiselman, H.J., 64
Meisler, A., 328
Meisler, M.H., 318
Meissner, G., 446
Meissner, G.W.D., 340
Meister, A., 307
Meites, J., 228
Meizel, S., 53
Mela, L., 410
Melander, W.R., 527
Melbin, J., 411
Melcher, U., 466
Melchert, T.E., 160
Melchior, N.C., 124
Melchlor, R., 236
Mele, F., 285
Meleca, C.B., 362
Melechen, N.E., 256
Melius, M.E., 405
Melko, Sr.M., 514
Mellett, J.S., 312
Melley, M.A., 447
Mellinger, A.C., 484
Mellinger, T.J., 104
Mellinkoff, S., 56
Mellon, D.F.Jr., 488
Mellor, D.B., 458
Mellor, R.S., 18
Melnik, J.L., 448
Melnick, N., 176
Melquist, G.A.L., 78, 79
Melsted, S.W., 138
Melton, A.A., 458
Melton, C.C., 444
Melton, C.E., 379, 380
Melton, J.R., 459
Melton, S.L., 444
Meltzer, H., 300
Melvill-Jones, G., 556
Melville, D.B., 482
Melville, J., 311
Memmerly, T.E., 439
Memon, J., 232
Menaker, M., 465
Mendall, H.L., 194
Mende, T., 95
Mende, T.J., 95
Mendel, V.E., 47
Mendell, L., 333
Mendelson, M., 325
Mendelson, N.H., 19
Mendenhall, C.L., 370
Mendenhall, V.T., 475
de Mendez, C.B., 418
Mendicino, J.F., 105
Mendiola-Morgenthaler, L.R., 288
Mendoza, G., 155
Menefee, M.G., 370
Meng, H., 323
Menge, B.A., 219
Menge, J.A., 58
Mengebier, W.L., 484
Mengel, R.M., 172
Mengoli, H.F., 509
Menhinick, E., 342
Mennega, A., 154
Mennes, M.E., 518
Menninger, R.P., 97
Menon, M.D., 300
Menter, J.M., 11
Menton, D.N., 263
Mentzer, L.W., 122
Menzel, B.W., 157
Menzel, D., 333
Menzel, M., 88
Menzell, D.W., 102
Menzer, R.E., 201
Menzies, J.A., 537
Meon, M.A., 236
Mercer, D.N., 427
Mercer, E.H., 114
Mercer, E.K., 29
Mercer, G., 479
Mercer, P.F., 552
Mercer, T.T., 328
Merchant, F.W., 499

Merchant, H., 84
Mercia, L., 480
Mercurio, A., 312
Merczak, N., 121
Meredith, F.R., 34
Meredith, W.E., 354
Meredith, W.G., 199
Merell, D.J., 242
Mergen, F., 81
Mergler-Racine, D., 560
Merilan, C.P., 259
Merk, F., 216
Merkatz, I., 356
Merkel, R.A., 225
Merkle, M.G., 459
Merkle, O.G., 459
Merkley, M., 475
Merkley, W.B., 155
Merner, D.T., 97
Merriam, H.G., 544
Merriam, J., 55
Merriam, R., 325
Merriam, V., 37
Merrian, L.C., 243
Merrick, J.M., 319
Merrifield, R.G., 458
Merrill, L.G., 288
Merrill, W., 403
Merriman, D., 80
Merriman, G.M., 443
Merriner, J.V., 490
Merritt, A.D., 146
Merritt, J.F., 347
Merritt, K., 277
Merritt, R., 215
Merritt, R.E., 396
Merritt, R.H., 288
Merritt, R.W., 226
Merritt, S., 35
Merritt, T.L., 401
Merritte, C., 149
Merkl, M.E., 249
Merry, W.J., 229
Mershon, W.R., 420
Merson, R.L., 49
Mertz, D., 260
Mertz, E.T., 148
Mertz, J., 393
Mervis, D.W., 489
Merwine, N.C., 248
Merzenich, M.M., 60
Meschia, G., 72
Mesel, E., 12
Meselson, M.S., 210
Meserve, L.A., 354
Meseth, E.H., 121
Meshijian, W.K., 486
Meslow, E.C., 383
Messenger, P.S., 44
Messersmith, C.G., 350
Messersmith, D.H., 201
Messier, B., 558
Messier, P.E., 558
Messineo, L., 357
Messmer, D., 378
Meszoely, C.A., 214
Metcalf, A.L., 468
Metcalf, D.S., 16
Metcalf, I.S.H., 428
Metcalf, N., 273
Metcalf, R.L., 136
Metcalf, R.M., 31
Metcalf, T.G., 279
Metcalf, W.K., 273
Metcalf, Z.W.Jr., 8
Metcoff, J., 380
Metevia, L.A., 189
Metrakos, J., 556
Mettler, L.E., 345
Mettrick, D.F., 550
Metts, I.S., 428
Metz, L.J., 334, 346
Metzenberg, R.L., 520
Metzgar, L.H., 267
Metzgar, R.S., 333
Metzger, J., 57
Metzler, D.E., 157
Meunier, J., 560
Meux, J.W., 452
Mewaldt, L.R., 32
Meydrech, E.F., 491
Meyer, C.F., 288
Meyer, D.B., 235
Meyer, D.E., 354
Meyer, D.L., 323
Meyer, D.W., 350
Meyer, E.A., 388
Meyer, E.F., 458
Meyer, E.G., 526
Meyer, E.R., 153
Meyer, F.D., 326
Meyer, G., 238
Meyer, H., 234
Meyer, J., 322
Meyer, J.R., 181
Meyer, K.A., 399
Meyer, M., 244
Meyer, M.M., 138
Meyer, M.P., 516
Meyer, R., 30
Meyer, R.C., 137
Meyer, R.E., 462
Meyer, R.J., 34
Meyer, R.K., 516
Meyer, R.L., 22
Meyer, R.R., 370p
Meyer, R.W., 149
Meyer, S.L., 362
Meyer, W., 538
Meyer, W.E., 257
Meyer, W.L., 482
Meyerholz, G.W., 94
Meyerriecks, A.J., 97
Meyers, G., 455
Meyers, H.G., 92
Meyers, J.L., 464
Meyers, M.W., 394
Meyers, R.S., 317
Mezei, C., 542
Mi, M.P., 113
Michael, C., 81
Michael, D.N., 231
Michael, E.D., 508
Michael, J.G., 370
Michael, R.P., 101
Michaeli, D., 60
Michaelides, B., 97
Michaelson, M.E., 522
Michaelson, S.M., 328
Michals, B.E., 27
Michaud, T.C., 511
Michel, B.E., 105
Michel, H., 96
Michel, K.E., 407
Michelakis, A.M., 227
Michels, C., 299
Michels, C.V.A., 297
Michener, C.D., 171, 172
Michie, C.R., 409
Michna, A.F., 506
Mickel, J., 296, 297
Mickel, J., 298
Mickelberg, E.D., 237
Mickelsen, C.H., 474
Mickelson, J.C., 247
Mickey, M.R., 56
Mickle, J.B., 376
Micks, D.W., 468
Micola, M., 128
Mickus, J.C., 88
Middaugh, L.D., 428
Middaugh, P.R., 433
Middendorf, F.E., 126
Middlekauff, W.W., 44
Middleton, A.L.A., 547
Mielke, L.N., 271
Miethaner, W.L., 319
Miettinen, O.S., 212
Mieyal, J.J., 127
deMignard, V.A., 63
Migelow, M.E., 217
Mignone, A., 316
Mihich, E., 321
Mikesell, J.E., 396
Mikesell, P.B., 487
Mikiten, T.M., 470
Mikkelsen, D.S., 47
Mikolajcik, E.M., 365
Mikolon, A.G., 39
Mikula, B.C., 358
Mikulecky, D.C., 492
Mikulick, J., 175
Milan, F., 14
Milazzo, F.H., 545
Milbrath, G.M., 139
Milburn, N.S., 216
Milby, T.H., 379
Mildvan, A., 410
Miles, B., 182
Miles, B.R., 458
Miles, C.D., 260
Miles, E.T., 9
Miles, H.M., 512
Miles, J.T., 444
Miles, M.M., 520
Miles, N.W., 168
Miles, P.G., 318
Miles, P.R., 509
Miles, R.L., 106
Miles, W.R., 243
Milford, H.E., 293
Milford, M.H., 459
Milgrom, F., 319
Milholland, R.D., 345
Milic-Emili, J., 556
Militzer, W., 270
Miljkovic, M., 405
Milkman, R.D., 160
Millar, J.S., 552
Millar, R.I., 422
Millecchia, R.J., 509
Millemann, R.E., 383
Miller, A., 190
Miller, A.E., 425
Miller, A.J., 135
Miller, A.T.Jr., 340
Miller, B.F., 65
Miller, B.S., 499
Miller, C.B., 385
Miller, C.E., 368
Miller, C.G., 356
Miller, C.H., 146, 285, 345
Miller, C.I., 149
Miller, C.N., 493
Miller, C.O., 144
Miller, C.W., 69
Miller, D., 270
Miller, D.A., 104, 138, 301
Miller, D.C., 297
Miller, D.D., 292
Miller, D.E., 501
Miller, D.F., 364
Miller, D.G., 502
Miller, D.H., 361, 479
Miller, D.J., 384
Miller, D.M., 129, 552
Miller, D.R., 78, 79
Miller, E.C., 225
Miller, E.J., 11
Miller, E.L., 275
Miller, E.R., 225
Miller, F.P., 200
Miller, G., 79, 173
Miller, G.C., 346
Miller, G.J., 527
Miller, G.L., 308
Miller, G.N., 431
Miller, G.N.Jr., 266
Miller, G.R., 240
Miller, G.W., 476
Miller, H., 298
Miller, H.A., 85
Miller, H.C., 145, 228, 342
Miller, H.I., 186
Miller, H.N., 92
Miller, I.J.Jr., 347
Miller, J., 164, 350, 352, 393, 500
Miller, J.A., 324, 364
Miller, J.F., 106
Miller, J.G., 263
Miller, J.H., 326, 396
Miller, J.J., 545
Miller, J.M., 346
Miller, J.N., 56
Miller, J.R., 200
Miller, J.W., 354
Miller, K.G., 400
Miller, K.I., 89
Miller, L., 126, 354, 288
Miller, L.C., 424
Miller, L.K., 14
Miller, L.L., 328
Miller, M., 122, 246, 316, 370
Miller, M.A., 52
Miller, M.C.Sr., 169
Miller, M.C.III., 428
Miller, M.R., 60
Miller, M.W., 12, 49, 329
Miller, N., 142
Miller, N.A., 438
Miller, N.C., 344
Miller, N.G., 273, 338
Miller, O.J., 301
Miller, O.K., 492
Miller, O.L.Jr., 488
Miller, P., 206
Miller, P.A., 344
Miller, P.C., 38
Miller, R., 152
Miller, R.A., 327
Miller, R.B., 18, 214, 548
Miller, R.C., 535
Miller, R.D., 302
Miller, R.F., 319
Miller, R.G., 550
Miller, R.H., 363, 364
Miller, R.J., 376
Miller, R.K., 229
Miller, R.L., 304, 342, 363, 377, 428
Miller, R.M., 160
Miller, R.N., 263
Miller, R.R., 231
Miller, R.S., 80, 81
Miller, R.V.Jr., 454
Miller, R.W., 364, 425, 474
Miller, S., 120, 350
Miller, S.A., 213
Miller, S.E., 333
Miller, T.A., 58
Miller, W.A., 319
Miller, W.D., 346
Miller, W. E., 472
Miller, W.F., 250
Miller, W.J., 106
Miller, W.L., 432
Miller, W W., 276
Miller, W.W.III., 188
Milleville, H.P., 383
Millhouse, E.W., 133
Millhouse, O.E., 473
Millican, T., 246
Milligan, J.F., 205
Milligan, J.V., 546
Milligan, L.P., 530
Millikan, D.F., 260
Milliken, H.R., 35
Millington, W.F., 512
Millis, D.E., 138
Millman, B.M., 544
Millman, M.S., 135
Millner, S., 135
Mills, D., 301, 409
Mills, D.M., 552
Mills, E., 333
Mills, E.L., 542
Mills, G., 467
Mills, J., 538
Mills, J.B.III., 101
Mills, J.F., 459
Mills, J.H.L., 563
Mills, J.W., 408
Mills, K., 265
Mills, J.T., 453
Mills, R., 172
Mills, R.B., 167
Mills, R.C., 455
Mills, R.R., 491
Mills, S., 253
Mills, S.E., 59
Mills, T.M., 103
Mills, W.C.Jr., 346
Mills-Westermann, J.E., 545
Millward, S., 555
Milman, G., 42
Milowicki, E., 37
Milstead, W.L., 265
Mims, C.W., 455

Min, H.S., 426
Minard, E.L., 256
Minckler, L.S., 322
Minckley, W.L., 15
Miner, F.D., 23
Miner, M.L., 475
Mines, A.H., 60
Minges, P.A., 305
Mingrone, L.V., 390
Minick, D.R., 91
Minion, D., 175
Minis, H., 189
Minish, G.L., 493
Mink, G.I., 503
Mink, I.B., 320
Minkel, C.W., 225
Minnich, J., 523
Minninger, J.R., 160,
Minor, R., 410, 412
Minotti, P.L., 305
Minowada, J., 320
Minsavage, E.J., 397
Minser, W.G., 443
Minshall, G.W., 116
Minter, K.W., 255
Minton, P., 491
Minton, P.F., 491
Minton, S.A.Jr., 146
Minuse, E., 234
Minutoli, F., 299
Miovic, M.L., 407
Miraldi, F.D., 356
Mirand, E.A., 319
Miranti, J.P., 129
Mirkes, P.E., 429
Mirmow, E.L., 37
Mirouti, P., 282
Misch, D.W., 339
Mish, L.B., 208
Mishra, N.C., 429
Miskimen, M., 364
de Miskimen, G.R., 417
Miskimen, G.W., 417
Misner, D.E., 121
Misra, R.D., 552
Misra, R.P., 356
Mistric, W.J.Jr., 344
Mistry, S.P., 138
Misuraca, S.A., 189
Mitacek, E., 311
Mitchell, B., 219
Mitchell, C.A., 149
Mitchell, D., 139
Mitchell, D., 387
Mitchell, D.J., 92
Mitchell, E.D., 377
Mitchell, F.H., 12
Mitchell, G., 53, 340, 372
Mitchell, G.E., 177
Mitchell, G.J., 561
Mitchell, H.L., 166
Mitchell, J., 126, 325
Mitchell, J.A., 235
Mitchell, J.B., 400
Mitchell, J.E., 118, 519
Mitchell, J.H.Jr., 424
Mitchell, J.K., 288
Mitchell, J.P., 10, 338
Mitchell, J.R., 281
Mitchell, K., 26
Mitchell, K.E., 547
Mitchell, L.G., 157
Mitchell, Sr.M., 33
Mitchell, M.E., 250
Mitchell, N.L., 35
Mitchell, O.G., 312
Mitchell, P., 7
Mitchell, P.J., 377
Mitchell, R., 60, 323, 395
Mitchell, R.A., 235
Mitchell, R.B., 403
Mitchell, R.D., 362, 364
Mitchell, R.G., 288
Mitchell, R.M., 211
Mitchell, R.S., 489, 492
Mitchell, R.W., 461
Mitchell, T.O., 128
Mitchell, T.R., 366, 367
Mitchell, W., 336m 374
Mitchell, W.C., 114
Mitchell, W.H., 82
Mitchell, W.R., 548
Mitich, L.W., 350
Mitlin, N., 249
Mitra, J., 312
Mitra, R., 507
Mitsui, A., 96
Mittelman, A., 320
Mittenthal, J.E., 150
Mitterling, L.A., 78, 79
Mittermeyer, F.C., 121
Mittler, E.F., 254
Mittler, S., 126
Mityga, H.B., 202
Mixner, J.P., 288
Mixon, F.O., 341
Mixter, R.L., 139
Miya, F., 474
Miyahara, A.Y., 113
Miyamoto, V.K., 466
Mize, G.E., 466
Mizell, J., 501
Mizell, M., 190
Mizer, O., 122
Mizeres, N.J., 235
Mizukami, H., 234
Moak, J.E., 250
Moawad, A., 133
Mobbs, T., 542
Moberg, G.P., 48
Mobini, J., 399
Mobley, H.M., 73
Mochrie, R.D., 343
Mock, O.B., 254
Mockford, E.L., 122
Modabber, F.Z., 212
Moder, J., 95
Moe, P.G., 508
Moe, R., 497
Moehn, L., 121
Moehring, D.M., 458
Moehring, J.M., 482
Moehring, Sr.P., 513
Moehring, T.J., 482
Moeller, D., 159
Moeller, F.E., 518
Moeller, H.W., 308
Moen, A.N., 303
Moens, P.B., 553
Moffatt, D.J., 160
Moerchen, R.M., 453
Moermond, T.C., 516
Moffat, J.K., 302
Moffett, S.B., 501
Moffett, S.R., 501
Moinuddin, D., 556
Moffitt, B., 298
Moffitt, B.P., 297
Mogensen, L., 15
Mogren, E.W., 66
Mohanty, M.K., 439
Mohlenbrock, R.H., 128
Mohler, J.D., 160
Mohn, C.A., 243
Mohn, J.F., 319
Mohr, H.C., 178
Mohr, J.L., 62
Mohrenweiser, H., 24
Moisand, R.E., 317
Mokrasch, L.C., 185
Moldave, K., 54
Moldenke, A.R., 61
Mole, P.A., 186
Molero, P.L.J., 417
Molina, J.A., 241
Molitor, D.M., 522
Moll, E.O., 120
Moll, R.H., 345
Moll, T., 504
Moller, F., 545
Molliver, M.E., 197
Molnar, C.E., 263
Molnar, R.E., 350
Molloy, M., 543
Momberg, H.L., 253
Mommaerts, W.F.H.M., 57
Momot, W.T., 364, 366
Monaco, J.F., 345
Monaghan, Sr.M.L.F., 311
Monahan, T.J., 288
Monahan, R.L., 504
Monahan, T., 467
Monaloy, S.E., 98
Moncure, R.W., 490
Mondy, N., 306
Moner, J.G., 218
Money, K.E., 550
Money, W.L., 22
Mongold, D.W., 155
Monheimer, R.H., 234
Monie, I.W., 60
Monk, C.D., 105
Monk, R.W., 477
Monkhouse, F.C., 550
Monoson, H.L., 119
Monroe, B.G., 63
Monroe, B.L., 180
Monroe, R.E., 39
Monsen, H., 134
Monson, P.H., 245
Monson, R.G., 406
Monsour, V., 187
Montagna, P.A., 214
Montaldi, F., 331
Montanucci, R.R., 426
Montelaro, J., 93
Montemurro, D.B., 551
Montgomery, C.K., 60
Montgomery, D.H., 27
Montgomery, F., 174
Montgomery, L., 142
Montgomery, M.J., 443
Montgomery, M.W., 383
Montgomery, P.C., 411
Montgomery, R., 161
Montgomery, T.H., 375
Montgomery, W.R., 6
Montie, T.C., 442
Montiegel, E.C., 507
Montiel, F., 124
Monto, A.S., 234
Montroy, L.D., 491
Monty, K.J., 441
Montzingo, L., 497
Moodie, G.E.E., 539
Moody, D.P., 102
Moody, E E., 470
Moody, E.L., 260
Moody, J.K., 309
Moody, P.A., 480
Moody, R., 147
Moody, W.G., 177
Moog, F., 263
Mook, L.J., 561
Moomaw, R.S., 271
Moon, R.L., 228
Moon, T., 549
Moon, T.C., 391
Moore, -., 81
Moore, A.M., 348
Moore, B., 350
Moore, B.G., 10
Moore, B.J., 385
Moore, B.W., 263
Moore, C., 327, 330, 417
Moore, C.H., 5
Moore, D., 199
Moore, D.P., 384
Moore, D.V., 466
Moore, E.L., 249
Moore, E.N., 411
Moore, F.D., 66, 359
Moore, F.E., 233
Moore, G., 75, 389, 472
Moore, G.P., 63, 64
Moore, H., 122
Moore, H.B., 39, 344
Moore, H.D., 65
Moore, I.R., 393
Moore, J., 13, 394
Moore, J.A., 57
Moore, J.C., 181, 434
Moore, J.D., 437, 519
Moore, J.E., 21
Moore, J.H., 6
Moore, J.K., 215
Moore, J.M., 444
Moore, J.N., 23
Moore, J.R., 392
Moore, J.W., 333
Moore, K., 555
Moore, K.E., 227, 261
Moore, K.L., 537
Moore, L., 17
Moore, L.W., 385
Moore, M., 20
Moore, M.J., 480
Moore, N.A., 97
Moore, N.J., 362
Moore, P.J., 103
Moore, R., 265, 412
Moore, R.D., 324
Moore, R.N., 396
Moore, R.R., 475
Moore, R.O., 362
Moore, R.P., 344
Moore, S., 467
Moore, T., 394
Moore, T.C., 385
Moore, T.E., 231
Moore, T.O., 113
Moore, T.O., 506
Moore, W.A., 167
Moore, W.G., 187
Moore, W.S., 234
Mooren, D., 523
Moorhead, P.D., 367
Mooring, J.S., 62
Moorman, D.G., 454
Moorman, R.B., 157
Moosnick, M., 176
deMooy, C.J., 65
Mora, E.C., 5
Morad, M., 411
Morafka, D., 27
Morahan, P.S., 492
Morais, R., 558
del Moral, R., 497
Morales, M.F., 60
Moran, C.R., 145
Moran, D., 323
Moran, E.T., 547
Moran, J.F., 318
Moran, N.C., 101
Moran, W.C., 100
Moran, W.E., 234
Moran, W.H., 509
Morden, R., 524
More, R.H., 555
More, R.T., 286
Moreau, D.H., 341
Moreau, G., 195, 557
Moree, R., 501
Morehead, J., 216
Morejohn, G.V., 32
Morek, Sr.M., 174
Moreland, D.E., 344
Moreland, J.E., 129
Morell, P., 340
Morell, S.A., 513
Morelli, R., 40
Morello, J.A., 133
Moreton, E.J., 204
Morey, P.R., 461
Morgan, W.C., 432
Morgan, A.R., 531
Morgan, B., 236, 401
Morgan, C., 301
Morgan, D.T.Jr., 201
Morgan, D.W., 152
Morgan, E.C., 450
Morgan, E.L., 440
Morgan, F.D., 144
Morgan, G.C.Jr., 97
Morgan, H.E., 405
Morgan, H.R., 328
Morgan, J.F., 562
Morgan, J.M.Jr., 492
Morgan, K., 535
Morgan, L., 104
Morgan, Rev, M., 8
Morgan, M., 514
Morgan, M.A., 551
Morgan, M.E., 383
Morgan, M.I., 509
Morgan, O.D., 201
Morgan, P., 24
Morgan, R.A., 372
Morgan, R.H., 197
Morgan, R.J., 69

Morgan, R.S., 404
Morgan, T.B., 26
Morgan, V.F., 335
Morgan, W.C., 432
Morgan, W.W., 469
Morgane, P.J., 207
Morgan-Jones, G., 4
Morgans, L.F., 25
Morge, J.F., 417
Morgenson, G.J., 552
Morhardt, J.E., 263
Moriarty, C.M., 273
Moriarty, G.C., 273
Moriber, L.G., 297, 298
Moriber, L.G., 298
Morimoto, H., 391
Morimoto, T., 312
Morin, J.G., 55
Morin, W.A., 208
Morishige, W., 112
Morison, R.S., 304
Morisset, J., 559
Morisset, P., 557
Morisset, R., 558
Morita, M., 69
Morita, R.Y., 385, 386
Moritz, W.E., 497
Mork, D., 239
Morken, D.A., 328
Morlang, C.Jr., 485
Morley, T., 241
Morley, Sr.V., 311
Morningstar, D., 407
Morningstar, W., 356
Morocco, C., 215
Morowitz, -., 81
Morrall, R.A.A., 562
Morre, D.J., 148, 150
Morrill, G.A., 330
Morrill, J.B., 89, 490
Morril, K., 432
Morril, L.G., 376
Morrill, M.B., 348
Morris, A.J., 225
Morris, C., 64
Morris, C.A., 256
Morris, C.D., 322
Morris, D.B., 100
Morris, D.E., 280
Morris, D.R., 499
Morris, E., 15
Morris, E.F., 139
Morris, G., 546
Morris, H.D., 106
Morris, H.G., 444
Morris, J.D., 450
Morris, J.E., 386
Morris, J.G., 48, 53, 87
Morris, J.R., 23, 545
Morris, J.S., 299
Morris, L., 378
Morris, L.L., 51
Morris, M., 24, 204
Morris, M.R., 271
Morris, R.C., 249
Morris, R.J., 275
Morris, R.W., 387
Morris, T.B., 346
Morrison, C.M., 542
Morrison, E.O., 456
Morrison, F.D., 168
Morrison, F.O., 556
Morrison, J.A., 376
Morrison, J.H., 357
Morrison, K.J., 502
Morrison, L.S., 375
Morrison, M.A., 306
Morrison, M.E., 251
Morrison, P., 14
Morrison, P.E., 551
Morrison, R.M., 342
Morrison, S.L., 301
Morrison, S.M., 69
Morrison, W.D., 547
Morrison, W.J., 407
Morrison, W.S., 407
Morrissey, J.E., 169
Morrissey, R., 133
Morrow, J., 461
Morrow, J.E., 14
Morrow, L.A., 271
Morrow, J.T., 248
Morrow, P.E., 328
Morrow, R.E., 262
Morrow, R.R., 303
Morrow, T., 522
Morse, D.E., 84
Morse, D.H., 199
Morse, H.G., 72
Morse, J.T., 31
Morse, M.L., 72
Morse, M.P., 214
Morse, R.A., 303
Morse, R.E., 288
Morse, S.A., 388
Morse, S.I., 325
Mortensen, R.A., 35
Mortensen, R.D., 222
Mortenson, J., 542
Mortenson, L.E., 150
Mortenson, T.H., 32
Mortimer, C., 523
Mortimer, J.T., 356
Mortimore, B., 155
Mortimore, G.E., 405
Mortlock, R.P., 218
Morton, B., 112
Morton, E.S., 199
Morton, G.H., 288
Morton, H.L., 232
Morton, J.H., 328
Morton, J.K., 551
Morton, M., 37
Morton, R.A., 545
Morton, T.H., 313
Morzlock, F., 119
Mosby, H.S., 493
Moscatelli, E., 261
Moschler, W.M., 443
Moscona, A.A., 132
Moscovici, C., 94
Moseby, E., 316
Moseley, M.F., 61
Moseley, Y.C., 454
Moser, B.C., 288
Moser, C.R., 31
Moser, J.W., 149
Moser, J.W., 397
Moser, L.E., 271
Moses, E., 24
Moses, L., 41
Moses, M.J., 332
Moses, R.E., 448
Mosher, J., 317
Mosher, M.M., 503
Moshiri, G.A., 98
Mosig, G., 446
Moskovits, G., 541
Moskowitz, G., 394
Moskowitz, M., 150
Moslemi, A.A., 129
Mosley, A.R., 365
Moseley, M., 335
Moss, B.R., 266
Moss, C.B., 507
Moss, D.D., 4
Moss, D.N., 240
Moss, G., 314
Moss, G.E., 234
Moss, I.R., 313
Moss, M.L., 300
Moss, S.A., 215
Moss, W.W., 409
Mossman, A.S., 34
Mostardi, R., 369
Mosteller, R.D., 64
Mote, E.M., 363
Motes, J., 226
Moticka, E.J., 466
Mott, R.L., 343
de Motta, G.E., 418
Motta, J.J., 201
Mottinger, J.M., 421
Mottinger, J.P., 421
Motyer, A.J., 540
Moudrianakis, E., 197
Moudy, E., 451
Mouftah, S.P., 461
Moulder, B.C., 121
Moulder, J.W., 132
Moulton, B.C., 370
Moulton, D.G., 411
Moulton, J., 225
Moulton, J.E., 53
Moulton, J.M., 192
Moulton, M.E., 522
Mount, D.I., 245
Mount, D.T., 319
Mount, M.S., 219
Mount, R.H., 5
Mour, D.R., 537
Mouser, G.W., 225
Mouw, D., 233
Mowat, D.N., 547
Mowbray, R., 515
Mowbray, T., 514
Mowbray, T.B., 515
Mower, H.F., 112
Mower, R.G., 303
Mowry, J.B., 138
Mowshowitz, D.B., 300
Moxon, A.L., 362
Moxon, A.L., 365
Moy, J.H., 113
Moye, H.A., 92
Moyer, F.H., 262
Moyer, J.C., 307
Moyer, J.L., 527
Moyer, R., 300
Moyer, W., 299
Moyer, W.A., 167
Moyle, P.B., 47
Moyle, S.M., 174
Mozejko, A., 387
Mozingo, H.N., 276
Mrak, E.M., 49
Mroczynski, R.P., 149
Much, D.H., 287
Muchmore, H.G., 380
Muchmore, W., 327
Muckenthaler, F.A., 208
Muckil, R.D., 269
Mudd, J.B., 57
Mudge, G.H., 277
Mudkur, B., 77
Muehlbauer, F.J., 502
Mueller, A., 317
Mueller, A.J., 23
Mueller, C.F., 407
Mueller, D.D., 166
Mueller, D.M., 460
Mueller, E., 467
Mueller, H.C., 339
Mueller, I.M., 255
Mueller, J.F., 326
Mueller, W.C., 422
Mueller, W.J., 403
Mueller, W.P., 152
Mueller-Dombois, D., 111
Muench, E., 241
Muench, K.H., 95
Muhlick, C.V., 497
Muhrer, M., 261
Muilenberg, V., 159
Muir, R.M., 160
Muir, W.H., 236
Muka, A.A., 303
Mukai, T., 345
Mukherjee, B.B., 556
Mukkada, A.J., 370
Mulamoottil, G.G., 551
Mulcahy, D.L., 217
Mulcare, D., 215
Mulchi, C.L., 200
Muldoon, J.J., 397
Muldoon, T.G., 103
Muldrey, J.E., 190
Muldrow, H.H., 335
Mulford, D., 172
Mulhollan, P., 517
Mulkern, G.B., 351
Mull, L.E., 91
Mulla, M.S., 58
Mullahy, J.H., 187
Mullen, D.A., 62
Mullen, T.E., 347
Mullen, T.L., 388
Muller, C.H., 61, 465
Muller, J., 315
Muller, W., 311
Muller, W.H., 61
Muller-Schwarze, D., 322
Mullet, R., 443
Mulligan, J., 256
Mullin, R., 241
Mullin, R.S., 92
Mullins, A.M., 117
Mullins, J.T., 90
Mullins, L.J., 203
Mulloney, B., 52, 59
Mulrennan, Sr.C.A., 215
Mulroy, T., 32
Mulvaney, D.L., 138
Mulvey, P.F., 216
Mumaw, H.A., 484
Mumford, D.M., 449
Mumford, G., 216
Mumford, R.E., 149
Mumm, R.F., 271
Mumma, R.O., 402
Mun, A.M., 193
Munavalli, S.N., 336
Munch, S., 196
Munck, A.U., 278
Muncy, R.J., 157
Mundell, R.D., 413
Mundt, J.O., 442, 444
Munger, B.L., 405
Munger, H.M., 304, 305
Munk, V., 324
Munkacsi, I., 562
Munkres, K.D., 519
Munnecke, D.E., 58
Munns, D.N., 50
Munns, T., 256
Munro, D.W., 309
Munro, H.N., 213
Munson, B., 320
Munson, D., 196
Munson, F.C., 234
Munson, J.B., 94
Munson, R.E., 259
Munson, S.T., 241
Munyon, W.H., 320
Munz, F.W., 387
Murad, J.L., 187
Murakami, W.T., 208
Muramoto, H., 16
Murashige, T., 58
Murdoch, C.L., 114
Murdoch, W.W., 61
Murdock, A., 172
Murdock, G., 15
Murdock, J.K., 12
Murdock, J.R., 472
Murdock, L.W., 177
Murdy, W.H., 100
Murie, J.O., 531
Muriente, J., 418
Murley, W.R., 493
Murnane, J.P., 206
Murnik, M.R., 139
Murphey, F.J., 82
Murphey, M., 91
Murphey, R., 160
Murphey, W.K., 402
Murphree, R.L., 443
Murphy, B., 378
Murphy, B.C., 248
Murphy, B.C., 562
Murphy, C.F., 344, 461
Murphy, D.H., 120
Murphy, D.R., 556
Murphy, E., 197
Murphy, G.G., 439
Murphy, G.J., 387
Murphy, G.P., 320, 420
Murphy, H.D., 32
Murphy, H.J., 195
Murphy, J., 21, 164
Murphy, J.R., 72, 472
Murphy, J.T., 550
Murphy, J.W., 379
Murphy, L.S., 166
Murphy, M.B., 256
Murphy, M.N., 125
Murphy, Q.R., 521
Murphy, R.A., 490
Murphy, R.L.H.Jr., 212
Murphy, R.P., 304

Murphy, S.D., 212
Murphy, T., 27
Murphy, T.M., 51
Murphy, W.H., 232
Murphy, W.J., 258
Murr, J.L., 144
Murr, S., 317
Murray, D., 556
Murray, D.F., 14
Murray, E.S., 212
Murray, G.A., 117
Murray, H., 70, 160
Murray, I.M., 325
Murray, J., 207
Murray, J.C., 376
Murray, J.J., 488
Murray, M., 181, 399
Murray, M.A., 120
Murray, M.L., 185
Murray, R., 358
Murray, R.G., 561
Murray, R.G.E., 551
Murray, R.K., 549
Murray, S.G., 279
Murray, S.N., 30
Murray, W.D., 7
Murrell, J., 437
Murrell, L.R., 445
Murrish, D., 355
Murthy, P.V.N., 335
Murthy, V., 284
Murthy, V.K., 56
Musbach, R.A., 261
Muscatine, L., 55
Muse, B., 393
Musen, H.L., 423
Musgrave, A.J., 547
Musgrave, O.L., 364
Musgrave, R.B., 302
Mushak, P., 340
Musher, D.M., 449
Mushinsky, H.R., 183
Mushynski, W., 555
Musial, T., 484
Musick, G.J., 363
Musick, J.A., 490
Musselman, L.J., 487
Mustard, J.F., 545
Mustard, M., 95
Mutchell, J.H., 424
Mutchmor, J.A., 159
Muth, G.M., 38
Mutts, J., 342
Muzik, T.J., 502
Muzopappa, F.P., 397
Mycek, M.J., 283
Myer, D., 130
Myer, M.P., 243
Myerburg, R.J., 69, 95
Myers, B.W., 400
Myers, C.C., 129
Myers, D., 112, 176
Myers, D.K., 364
Myers, G.A., 432
Myers, G.G., 273
Myers, J., 128
Myers, J.B., 99
Myers, J.E., 465
Myers, J.H., 129, 534
Myers, L.L., 266
Myers, O., 128
Myers, O.Jr., 130
Myers, R.D., 150
Myers, R.F., 288
Myers, R.L., 258
Myers, R.M., 139, 343
Myers, T.B., 364
Myers, T.D., 82
Myers, W.F., 203
Myhr, A.N., 547
Myhre, D.L., 248
Myhrman, R., 122
Mylon, W.K., 428
Myquist, W.E., 147
Myrberg, A., 96
Myres, M.T., 531
Myrop, L.O., 49
Myrup, L.O., 47
Myrvik, Q.N., 347
Myser, W.C., 364
Mysong, D.S., 272
Myszewski, M., 155

N

Naber, E.C., 366, 367
Nabors, C.J.Jr., 473
Nabors, M.W., 68
Nace, G.W., 231
Nachbar, M.S., 313
Nachmias, V.T., 410
Nachreiner, R.F., 6
Nachtmann, W., 100
Nadakavukaren, M., 122
Nadler, N.J., 555
Nadolinski, V.J., 317
Naf, U., 299
Nafpaktitis, B.G., 62
Nagai, T., 552
Nagel, C.M., 433
Nagel, H.G., 270
Nagel, J.W., 437
Nagel, N.H., 269
Nagel, S.A., 38
Nagel, W.P., 386
Nagle, F.J., 521
Nagle, J.F., 391
Nagle, J.J., 284
Nagle, N., 264
Nagy, F., 327
Nagy, J.G., 66
Nagy, K., 53
Nagy, K.P., 537
Nagyvary, J., 458
Nahas, A., 328
Nahhas, F., 61
Naidoo, S.S., 562
Naimark, A., 537
Nair, C.N., 103
Nair, K.K., 533
Nair, R., 428
Nair, S., 130
Nair, V.M.G., 514
Najarian, H.H., 195
Najjar, V., 217
Nakada, D., 413
Nakada, H.I., 61
Nakagawa, S., 330
Nakajima, S., 150
Nakajima, Y., 150
Nakamoto, T., 131
Nakamura, K., 532
Nakamura, M.J., 267
Nakamura, R., 215
Nakamura, R.M., 113
Nakasone, H.Y., 114
Nakata, H.M., 501
Nakata, S., 111
Nakatani, R.E., 498
Nakatsugawa, T., 322
Nakayama, R.M., 293
Nakayama, T., 113
Nalbandov, A.V., 137, 138
Nalewaja, J.D., 350
Nalmi, N.Jr., 161
Nalmi, N.S., 160
Namba, R., 114
Nambiar, G.K., 105
Namenwirth, M.R., 516
Nameroff, M.A., 499
Namkoong, G., 345, 346
Namm, D.H., 333
Nancarrow, V., 10
Nance, F.C., 186
Nance, J.A., 454
Nance, J.F., 136
Nance, J.W., 24
Nance, K., 100
Nance, W.E., 146
Nancollas, C.H., 318
Nanda, R.S., 379
Nandi, S., 43
Nandy, K., 101
Nanney, D.L., 137
Nanniga, L.B., 468
Napier, E.A., 232
Napolitano, J.J., 295
Napolitano, L.M., 293
Naqvi, S., 246
Narahara, H.T., 327
Narahashi, T., 333
Narayanan, C.H., 185
Narbaitz, R., 548
Nardell, B., 204
Nardone, R.M., 83
Nash, A.J., 261
Nash, C.B., 406
Nash, D., 535
Nash, D.J., 68
Nash, P.H., 550
Nash, R.G., 524
Nash, R.R., 423
Nash, T.H., 15
Nash, V.E., 248
Nash, W.G., 84
Naskali, R.J., 117
Nason, A., 197
Nasrallah, M., 321
Nastuk, W.L., 301
Natalini, J., 136
Natarella, N.J., 107
Nath, J., 508
Nathan, H., 329
Nathan, P., 370
Nathaniel, D.R., 537
Nathaniel, E.J.H., 537
Nathans, D., 198
Nathanson, M., 299
Nathanson, M.E., 297
Nathenson, S., 329
Nathenson, S.G., 330
Nation, J.L., 41
Nations, C., 454
Naughten, J.C., 514
Naughton, M., 72
Nault, L.R., 363
Nauman, R.K., 204
Naumann, L., 10
Nava, P.B., 35
Navia, J., 11
Nayak, P., 36
Nayfeh, S.N., 340
Naylor, A.W., 332
Naylor, D.V., 117
Naylor, E.E., 162
Naylor, G.W., 436
Naylor, J.M., 562
Nayyar, R., 126
Neal, A.L., 302
Neal, B.R., 562
Neal, F.C., 94
Neal, N.P., 519
Neal, O.M., 508
Neal, R.A., 446
Neathery, M.B., 106
Neaves, W.B., 466
Neckelson, A.L., 504
Nedelkoff, G., 181
Needham, G., 244
Needleman, P., 263
Neel, J.vanG., 232
Neel, J.K., 352
Neel, J.W., 38
Neel, W.W., 249
Neeley, J.C., 389
Neelin, J.M., 544
Neelon, V., 340
Neely, F., 119
Neely, J.R., 405
Neely, P.M., 172
Neely, R.D., 139
Neess, J.C., 516
Neet, K.E., 355
Neff, J.M., 460
Neff, R.D., 458
Neff, R.J., 446
Neff, S., 180
Neff, W.H., 403
Negm, H.A., 182
Negulesco, J.A., 367
Negus, N.C., 473
Neher, R., 35
Nehlsen, S.L., 330
Neiderheiser, D., 355
Neiderpruem, D.J., 146
Neidhardt, A.P., 200
Neidhardt, F.C., 232
Neidle, E.A., 312
Neikam, W.C., 279
Neikle, T.H.Jr., 307
Neiland, B.J., 14
Neilands, J.B., 45
Neild, R.E., 272
Neill, J., 355
Neill, J.D., 101
Neill, J.W., 534
Neill, R.L., 464
Neill, U.M., 30
Neill, W.E., 535
Neill, W.H., 459
Neillis, J.M., 311
Neilsen, D.G., 363
Neilsen, I.R., 36
Neilson, J.T., 94
Neilson, S.W., 79
Neilson, T., 545
Neimark, H., 325
Nelbach, I.G., 284
Nelchers, L.E., 168
Nellermoe, R., 238
Nellis, L.F., 309
Nellis, S.H., 405
Nellor, J.E., 228
Nelms, G., 527
Nelsestuen, G.L., 241
Nelson, A., 515
Nelson, A.B., 292
Nelson, A.H., 10
Nelson, B., 489
Nelson, B.W., 490
Nelson, C.A., 24
Nelson, C.B., 26
Nelson, C.E., 145
Nelson, C.G., 216
Nelson, C.H., 440
Nelson, C.J., 258
Nelson, D.C., 351
Nelson, D.D., 29
Nelson, D.H., 222
Nelson, D.L., 517
Nelson, D.M., 29
Nelson, D.R., 352
Nelson, D.W., 147
Nelson, E.A., 28
Nelson, E.C., 377
Nelson, E.E., 385
Nelson, E.M., 469
Nelson, E.V., 362
Nelson, F.E., 97
Nelson, F.S., 246
Nelson, G., 336, 524
Nelson, G.E., 97
Nelson, H., 128
Nelson, I.J., 9
Nelson, J., 77, 79
Nelson, J.M., 378
Nelson, J.N., 17
Nelson, J.R., 503
Nelson, J.S., 531
Nelson, K.E., 51
Nelson, L.A., 271
Nelson, L.E., 248
Nelson, M.B., 112
Nelson, M.R., 18
Nelson, N.M., 265
Nelson, N.N., 251
Nelson, O.E., 519
Nelson, P.E., 149, 403
Nelson, P.K., 296
Nelson, P.V., 345
Nelson, R., 159
Nelson, R.A.Jr., 558
Nelson, R.E., 139, 503
Nelson, R.F., 27
Nelson, R.H., 225
Nelson, R.L., 138
Nelson, R.R., 403
Nelson, S.H., 561
Nelson, S.O.Jr., 324
Nelson, T.C., 521
Nelson, T.S., 23
Nelson, V., 76
Nelson, V.E., 164
Nelson, W., 9, 240, 302
Nelson, W.C., 341
Nelson, W.F., 444
Nelson, W.R., 138
Nemerofsky, A., 323
Nemeth, A.M., 410

Nemirovsky, M., 559
Nemmer, M., 511
Nentwig, R., 310
Nequin, L.G., 129
Nerenberg, S.T., 134
Neri, R., 285
Nes, W.R., 393
Nesbitt, H.H.J., 544
Nesbitt, M.N., 59
Nesbitt, W.B., 345
Nesheim, M.C., 306
Nesius, K., 487
NeSmith, J., 93
Ness, G.D., 234
Nester, E.W., 499
Nesthene, E., 288
Nestor, K.E., 367
Neter, E., 319
Netrick, F.M., 202
Nett, T.N., 69
Nettles, V.F., 93
Nettles, W.C., 424
Neubauer, B.F., 194
Neuburger, A.K., 169
Neufeld, E., 354
Neufeld, G., 164
Neufeld, G.R., 411
Neuffer, Y.G., 260
Neuhaus, O.W., 434
Neuhold, J.M., 476, 477
Neulieb, M.K., 357
Neuman, M.R., 356
Neuman, M.W., 329
Neuman, W.F., 328
Neumann, A.L., 138
Neumann, R., 308
Neunzig, H.H., 344
Neurath, C., 282
Neurath, H., 499
Neushul, N., 61
Nevalainen, D., 328
Neve, R., 14
Nevill, W.A., 145
Neville, M., 72
Neville, M., 122
Nevin, F.R., 324
Nevin, T.A., 87
Nevins, D., 157
Nevins, R.B., 246
Nevissi, A., 498
New, J.G., 323
Newberne, P.M., 213
Newberry, A.T., 61
Newbold, J.E., 339
Newbould, F.H.S., 548
Newburgh, R.W., 385
Newby, F.L., 118
Newcomer, J.L., 200
Newell, E.B., 331
Newell, I.M., 57, 58
Newell, L.C., 271
Newell, W.T., 498
Newfeld, B.R., 35
Newkirk, M.R., 354
Newkirk, R., 494
Newland, H.W., 365
Newland, L.W., 461
Newlin, G.E., 428
Newlon, C., 160, 264
Newman, C.W., 266
Newman, D., 500
Newman, D.W., 360
Newman, E.G., 557
Newman, F., 365
Newman, F.S., 266
Newman, J.E., 147
Newman, J.P., 228
Newman, R., 3
Newman, W.P.III., 186
Newman, W.R., 321
Newmark, M.Z., 170
Newsom, D.W., 184
Newsome, R.D., 511
Newson, H.D., 226, 228
Newstead, J.D., 562
Newton, C.M., 36, 481
Newton, D.C., 74
Newton, O., 290
Newton, R.J., 460
Newton, W.A., 287
Ney, E.P., 76
Ney, R.L., 340
Neyman, J., 46
Ng, A.B.P., 356
Niblett, C.L., 168
Nicely, K.A., 181
Nichlas, R.B., 332
Nichoalds, G., 446
Nichol, C.A., 333
Nicholai, J.H., 177
Nicholas, C.H., 223
Nicholas, H., 256
Nicholas, T.E., 113
Nicholes, H.J., 472
Nicholes, P.S., 474
Nicholls, D.M., 553
Nichols, C., 513
Nichols, E.W., 285
Nichols, H., 71
Nichols, H.E., 156
Nichols, H.W., 263
Nichols, J., 9
Nichols, J.L., 333
Nichols, J.M., 261
Nichols, J.R., 21
Nichols, J.T., 271
Nichols, K.E., 153
Nichols, L.P., 403
Nichols, M.M., 490
Nichols, R.L., 212
Nicholson, B.L., 194
Nicholson, D.P., 161
Nicholson, H.H., 561
Nicholson, J.C., 162
Nicholson, J.L., 203
Nicholson, N.L., 62
Nicholson, R.A., 164
Nicholson, R.L., 148
Nicholson, S., 323
Nickel, P.A., 26
Nickell, C.D., 167
Nickerson, J.L., 133, 134
Nickerson, N.H., 216
Nickerson, R.P., 215
Nickerson, T.A., 49
Nickerson, W.J., 288
Nickla, H., 268
Nickol, B., 270
Nickum, J.B., 303
Nicol, F.D., 496
Nicol, J.A.C., 465
Nicoli, M.Z., 211
Nicoll, C.S., 43
Nicoll, R.A., 319
Nicolls, K.E., 352
Nicolosi, G.R., 97
Nicosia, N.J., 185
Niederman, R.A., 288
Niedermeier, R.P., 517
Niedermeier, W., 11
Niehaus, M.H., 364
Niehaus, W.G., 404
Nield, V.K., 199
Nielsen, D., 239
Nielsen, D.H., 319
Nielsen, K., 548
Nielsen, L.W., 345
Nielsen, L.T., 473
Nielsen, P.J., 139
Nielsen, P.T., 486
Nielsen, R.F., 474
Nielsen, S.L., 60
Nielson, A.H., 441
Nielson, M.W., 17
Nielson, R.R., 360
Nielson, S.W., 78
Niemczyk, H., 363
Niemeck, R., 279
Nierderhuber, J.E., 232
Nierenberg, W.A., 59
Niering, W.A., 75
Nierlich, D.P., 55
Niewenhuis, R.J., 370
Niewoehner, D.E., 356
Niggemann, Fr.B., 239
Nigh, E.L., 18
Nighswander, J.E., 546
Nighswonger, P., 374
Nightingale, A.E., 459
Nilan, R.A., 502
Niles, G.A., 459
Nilsen, H.C., 120
Nilsen, K.N., 156
Nilson, E.B., 166
Nilson, K.M., 480
Nimi, M., 64
Nimitz, Sr.M., 33
Nippo, M., 422
Nir, S., 320
Nisbet, R., 359
Nisengard, R.J., 319
Nishi, S., 124
Nishida, T., 114
Nishimoto, R.K., 114
Nishimura, H., 438, 445
Nishimura, J.S., 469
Niswander, R.E., 147
Niswender, G.D., 69
Nitsoso, R., 387
Nittler, L.W., 307
Niver, J., 372
Niver, M.B., 359
Nix, C.E., 485
Nix, D.A., 349
Nix, L.E., 425
Nixon, A., 7
Nixon, E.S., 455
Nixon, H., 246
Nixon, J.C., 548
Noakes, D.L., 547
Noall, M.W., 448
Noback, C.R., 300
Nobel, P., 55
Noble, A.C., 51
Noble, D.J., 374
Noble, E.A., 352
Noble, E.P., 55
Noble, N.L., 95
Noble, P., 556
Noble, R.D., 354
Noble, R.E., 183
Noble, R.L., 316, 459, 535
Noble, R.W., 318
Noble-Harvey, J., 82
Noblet, G.P., 424, 426
Noblet, R., 424
Nocenti, M.R., 301
Nockels, C.F., 65
Noda, K., 114
Noda, L.H., 277
Noddergraaf, A., 411
Noden, D.M., 218
Nodnan, W.E., 504
Noe, B.D., 101
Noe, C.R., 121
Noelken, M.E., 172
Noell, W.K., 319
Noetzel, D.M., 246
Noggle, G.R., 343
Nogrady, G., 558
Nohlgren, J.E., 389
Nohlgren, S.R., 338
Nokes, R., 369
Nolan, C.N., 423
Nolan, J., 324
Nolan, J.C.Jr., 113
Nolan, R., 124
Nolan, R.A., 541
Nolan, V., 145
Noland, G.B., 371
Noland, P.R., 23
Nolasco, J.B., 283
Nollen, P.M., 139
Nollenberger, E.L., 407
Noller, H.F., 61
Noltmann, E.A., 57
Nomura, M., 517, 518, 519
Nooden, L.D., 231
Noon, T.H., 18
Noonan, K.D., 90
van der Noort, S., 55
Nopanitaya, W., 340
Norbach, J., 512
Norcia, L., 408
Norcross, J.J., 490
Nord, R., 515
Nordahl, C., 272
Nordan, H.C., 535
Nordby, G.L., 232
Norden, C., 523
Nordheim, E., 236
Nordin, G.L., 178
Nordin, J., 217
Nordin, P., 166
Nordin, V.J., 549
Nordland, F.N., 150
Nordlander, R., 355
Nordlie, F.G., 90
Nordlie, R.C., 353
Nordquist, P.T., 271
Nordstrom, P.E., 433
Nordyke, L., 464
Norkin, L.C., 218
Norland, C.E., 40
Norman, A.W., 57
Norman, B., 128
Norman, C., 507
Norman, G.A., 230
Norman, J., 514
Norman, M.A., 99
Norman, R., 388
Norman, R.S., 77
Norman, W., 85
Norman, W.H., 251
Normand, R.A., 188
Normandin, R.F., 279
Normansell, D.E., 490
Normen, E.M., 89
Norment, B.R., 249
Nornes, H.O., 68
Noronha, F., 306
Norrdin, R.W., 69
Norris, C.H., 71
Norris, D.E., 252
Norris, D.H., 34
Norris, D.M., 518
Norris, D.O., 71
Norris, F.H., 441
Norris, G.R., 358
Norris, J.E., 121
Norris, K.S., 61
Norris, L.C., 48
Norris, R.E., 497
Norris, R.F., 51
Norris, T.E., 99
Norris, W.W.Jr., 188
Norstog, K., 126
North, C.A., 524
North, D., 351
Northcote, T.G., 535
Northcutt, R.G., 231
Northern, P.T., 28
Northey, W.T., 15
Northington, D.K., 461
Norton, B., 476
Norton, B.E., 477
Norton, D.C., 157
Norton, H.W., 131
Norton, J., 146, 408
Norton, J.A., 146
Norton, J.D., 5
Nortrup, J., 151
Norvell, J.E., 491
Norwich, K., 550
Norwood, E., 522
Norwood, J.S., 450
Nose, Y., 356
Noskowiak, A.E., 503
Noteboom, W.D., 261
Nothdurft, R.R., 254
de la Noüe, J., 557
Novacky, A., 260
Novak, A., 258
Novak, J.A., 299
Novak, J.F., 287
Novak, J.R., 184
Novak, M.A., 218
Novak, M.M., 539
Novales, R.R., 126
Novick, A., 80, 387
Novick, R.P., 313
Novitski, E., 387
Novotny, C.P., 482
Novotny, R.T., 199
Nowak, P.F., 129
Nowlin, W.Jr., 459
Nowotny, A.H., 408
Nowsu, A., 331
de Noyelles, F.Jr., 379
Noyes, W.F., 320

Nozaki, Y., 333
Nozzolillo, C., 549
Null, R., 249
Nunez, L.J., 445
Nunley, R.G., 507
Nunez, W.J., 230
Nunn, D., 369
Nunneley, S.A., 319
Nunnemacher, R.F., 209
Nur, U., 327
Nursall, J.R., 531
Nussbaum, F.E., 427
Nussbaum, N.S., 372
Nussbaum, R.A., 231
Nutter, R.L., 36
Nutter, W.L., 108
Nutting, D.F., 445
Nutting, W.B., 218
Nutting, W.L., 17
Nwagwu, M., 544
Nybakken, J.W., 30
Nyc, J.R., 56
Nydegger, L., 323
Nye, R.E.Jr., 278
Nye, T.G., 495
Nye, W.E., 159
Nygaard, O.F., 355
Nyirady, S.A., 206
Nyland, G., 50
Nylund, R.E., 241
Nyman, C.C., 501
Nyquist, S., 391
Nystrom, R., 135
Nyvall, R.F., 157

O

Oace, S., 45
Oakes, J.B., 197
Oakley, B., 231
Oakley, D., 234
Oaks, B.A., 545
Oaks, E.C., 476, 477
Oaks, J.A., 160
Oaks, R.Q., 477
Oalmann, M.C., 186
Oaten, A., 61
Oatman, E.R., 58
O'Bannon, R., 439
Obee, D.J., 116
Obendorf, R.L., 302
Oberlander, G.T., 40
Oberly, G.H., 305
Obermire, R.F., 385
Oberpriller, J.O., 352
Oblinger, J.L., 92
O'Brien, C.A., 462
O'Brien, C.J., 546
O'Brien, F., 215
O'Brien, J., 299
O'Brien, J.A., 83
O'Brien, J.F., 310
O'Brien, R.D., 302, 304
O'Brien, R.E., 311
O'Brien, R.T., 242, 311
O'Brien, T.D., 275
O'Brien, T.W., 90
O'Brien, W.J., 172
O'Bryan, G.K., 476
O'Callaghan, D.J., 251
O'Callahan, R.J., 185
Ocasio-Cabanas, W., 417
Occino, J.C., 355
Ochs, B.O., 407
Ochs, S., 146
Ockerman, H.W., 365
O'Connell, C.J., 319
O'Connell, J.R., 315
O'Connell, K., 514
O'Connor, B.L., 145
O'Connor, G.M., 255
O'Connor, J., 55, 208
O'Connor, M.L., 161
O'Connor, T.M., 173
O'Connor, W.B., 218
Oddis, L., 287
Ode, P.E., 409
O'Dell, B.L., 261
O'Dell, C.R., 494
O'Dell, G.D., 424
Odell, D., 96
Odell, G.V., 377
Odell, W.D., 57
Oden, B.J., 507
Odenheimer, K., 186
Odile, J.P., 560
Odland, G.F., 499
Odlaug, T.O., 245
Odom, R.E., 168
Odom, W., 309
O'Donell, A.A., 468
O'Donnell, V.J., 534
O'Dor, R.K., 542
Odor, D.L., 491
Odum, E.P., 105, 106
Odum, W.E., 489
Oebker, N.F., 17
Oechel, W.C., 556
Oehling, R.A., 206
Oelke, E.A., 240
Oels, H.C., 408
Oehme, F.W., 168
Oeltgen, P.D., 124
Oester, Y.T., 124
Oestreich, C.H., 461
Oetjen, R.A., 369
Oexemann, S., 524
Oexemann, S.W., 524
Offutt, G.C., 297
Offutt, M.S., 22
O'Flaherty, L.M., 139
Ofosu, G.A., 99
Ogassawara, F.X., 48
Ogawa, J.M., 50
Ogburn, C.A., 399
Ogden, J.G.III., 542
Ogden, P.R., 18
Ogden, R.L., 272
Ogg, J.E., 69
Ogilvie, D.M., 552
Ogilvie, J.W., 490
Olilvie, P., 387
Ogilvie, R.W., 428
Oginsky, E.L., 470
Ogkerse, R., 224
Ogle, J.D., 370
Ogle, W.L., 424
Oglesby, L.C., 33
Oglesby, R.T., 303
Ogra, P.L., 319
Ogren, H., 511
Ogren, R., 416
Ogren, W.L., 138
Oh, Y.H., 448
O'Halloran, B., 450
O'Hare, K., 216
O'Hayre, A.P., 243
O'Heron, T.M., 150
Ohki, K., 106
Ohl, L.E., 514
Ohlrogge, A.J., 147
Ohlsson-Wilhelm, B., 328
Ohm, H.W., 147
Ohmart, R., 15
Ohmart, O., 257
Ohms, R.E., 117
Ohnishi, T., 410
Ohr, E.A., 319
Oiler, L.W., 153
Oishi, M., 313
Oitto, W.A., 365
O'Kane, D.J., 409
O'Keefe, R.B., 272
O'Keefe, T.A., 85
O'Keefe, L.E., 117
O'Kelley, J.C., 10
O'Kelley, B., 352
Okelo, O., 8
Okonkuro, A., 432
Okonkwo, A., 431
Okun, L.M., 473
Olafson, R.P., 101
Olah, G.M., 558
Olchowecki, A., 538
Olcott, H.S., 47, 49
Oldfield, D.G., 120
Oldfield, G.V., 171
Oldfield, J.E., 382
Oldham, M.G., 138
Oldham, S., 64
Olds, D., 177
Olds, J., 26
O'Leary, G.P., 420
O'Leary, J.W., 18
O'Leary, W.M., 308
Oleinick, N., 355
Olenick, S.R., 380
Oleson, A.E., 352
Olexia, P.D., 224
Olin, P.J., 241
Oliphant, L.W., 562
Oliphant, M.W., 111
Olive, A.T., 347
Olive, J., 369
Olive, L.S., 338
Olive, C.P., 465
Oliver, D., 544
Oliver, D.R., 161
Oliver, H.M., 278
Oliver, J., 79
Oliver, J.H.Jr., 102
Oliver, K.H.Jr., 21
Oliver, L., 348
Oliver, V.L., 21
Olivera, B.M., 473
de Oliveria, D., 560
Oliverira, R.A., 521
Oliviere, J., 3
Olivo, M.A., 215
Olivo, R.F., 215
Ollerich, D.A., 352
Olmo, H.P., 51
Olmsted, C.A., 185
Olmsted, J., 327
Olmsted, S.P., 372
Olney, A.A., 29
Oloffs, P.C., 533
Olpin, O., 476
Olsen, B.R., 283
Olsen, C.E.Jr., 231
Olsen, F.J., 130
Olsen, I.D., 498
Olsen, O.A., 541
Olsen, R.H., 232
Olsen, R.W., 57, 236
Olsen, S.M., 498
Olson, A.C., 40
Olson, A.E., 475
Olson, C., 520
Olson, C.C., 517
Olson, D.P., 280
Olson, E., 55
Olson, E.S., 432
Olson, G., 155, 524
Olson, G.A., 90
Olson, G.W., 302
Olson, H.H., 129
Olson, J.B., 150, 430
Olson, J.K., 458
Olson, J.O., 193
Olson, J.S., 441
Olson, K.C., 257
Olson, K.N., 243
Olson, K.R., 288
Olson, L.A., 20
Olson, L.C., 146, 484
Olson, L.J., 467
Olson, M., 242
Olson, M.D., 84
Olson, M.S., 19
Olson, N.F., 518
Olson, R.A., 271
Olson, R.E., 194, 256
Olson, V.A., 236
Olson, W., 139
Olson, W.H., 446
Olsory, A.M., 207
Olsson, G.P., 234
Olszowka, A.J., 319
Olton, G.S., 17
Olum, P., 465
Omachi, A., 135
O'Malley, B.W., 449
O'Malley, J., 152
O'Malley, J.A., 320
O'Malley, S.L., 289
O'Malley, T.P., 206
Oman, P., 386
Omara, J.C., 403
O'Mary, C.C., 502
Omdahl, J.L., 294
O'Melia, C.R., 341
Omran, A., 342
Oncley, J.L., 232
Ondo, J.G., 429
O'Neal, C.H., 491
O'Neal, T.D., 314
O'Neil, L., 559
O'Neil, L.C., 559
O'Neill, C.E., 285
O'Neill, F.J., 474
O'Neill, M.C., 205
Ontko, J., 380
Onyia, K.A., 392
Oomens, F.W., 255
Oostenink, W.J., 299
Opdyke, D.F., 283
Opitz, J.M., 519
Oplinger, C.S., 400
Oplinger, E.S., 516
Oppenheim, B., 330
Oppenheim, J.D., 313
Oppenheimer, S.B., 31
Oppensiek, G.C., 305
Opplt, J.J., 356
O'Quinn, S., 185
O'Rahilly, R., 53
O'Rahilly, R., 235
Oram, T.F.D., 327
Oratz, M., 312
Orcutt, S., 369
Ordal, E.J., 498, 499
Ordal, Z.J., 137
Ordin, L., 57
Ordway, E., 245
O'Reagan, W.G., 44
O'Rear, C.W., 334
Orenstein, N.S., 212
Oreskes, I., 298
Organ, J.A., 296, 297
Organ, L.W., 550
Organisciak, D., 319
Orians, G.H., 498
Orihel, T.C., 190
Oring, L.W., 352
Orkand, R., 55
Orkney, G.D., 382
Orlando, J.A., 207
Orloci, L., 552
Orlove, B., 49
Orlove, B.S., 47
Orme, L.E., 472
Orme-Johnson, W.H., 517
Ormsby, A., 467
Ornduff, R., 42
Ornston, L.N., 80
Oro, J., 464
O'Rourke, E.N., 184
Orpurt, P.A., 147
Orr, A.R., 162
Orr, H.D., 240
Orr, H.L., 547
Orr, H.P., 5
Orr, L.P., 359
Orr, M.F., 126
Orr, O.E., 254
Orrego, H., 550
Orsenigo, J.R., 93
Orsi, E.V., 290
Orsini, M.W., 520
Ortega, J., 452
Orten, J.M., 235
de Ortia, L.O., 418
Ortiz, C.L., 61
Ortíz, E., 418
de Ortiz, E.M., 418
Ortiz, G., 96
Ortiz, M.E., 27
Ortman, E.E., 148
Ortman, L.L., 165
Ortman, R., 297
Ortman, R.A., 297
Orton, E.R., 288
Ortwerth, B., 261
Orzechowski, R.F., 405
Osborn, C.M., 318
Osborn, H.H., 238
Osborn, J.E., 521
Osborn, M.J., 79

Osborn, N., 70
Osborne, D.R., 360
Osborne, F.H., 286
Osborne, J.A., 88
Osborne, J.L., 54
Osborne, J.W., 162
Osborne, P.J., 486
Osburg, H., 323
Osburn, B.I., 53
Oschman, J., 126
Oschwald, W.R., 138
Osebold, J.W., 53
Osgood, D.W., 143
O'Shaughnessy, M.V., 545
Oshima, G., 253
Oshima, N., 68
Osinchak, J., 297
Oshino, N., 410
Osler, A.G., 313
Osman, G., 320
Osmond, D.G., 555
Osmond, D.H., 550
Osmun, J.V., 148
Osness, W., 171
van Oss, C.J., 319
Ost, D.H., 27
Ostarello, G., 33
Ostdick, T., 151
Ostdiek, F.J.L., 127
O'Steen, W.K., 101
Ostenson, B.T., 497
Oster, G., 44, 298
Oster, I.I., 354
Oster, K.A., 75
Osterberg, D.M., 324
Osterhout, S., 333
Osterman, G.O., 361
Ostermeier, D.M., 443
Osterud, K.L., 498
Osteryoung, J.G., 69
Ostle, O., 88
Ostovar, K., 402
Ostrander, C.E., 305
Ostrander, D.R., 74
Ostrander, K., 169
Ostrander, L.E., 356
Ostrovsky, D., 400
Ostrowski, R., 342
Ostroy, S.E., 150
Ostwald, R., 45
Osuji, F., 250
Oswald, C., 268
Oswald, E.O., 340
Oswald, J.W., 403
Oswald, V.H., 29
Oswalt, E.S., 276
Osweiler, G.D., 262
O'Tanyi, T.J., 416
Otero, J.G., 234
Otero, L.R., 418
Oteruelo, F., 562
Otis, A.B., 94
O'Toole, Rev.A., 395
Ott, D., 369
Ott, E.A., 91
Ott, F.D., 490
Ott, K.J., 152
Otta, J.D., 433
Ottbacher, A.G., 138
Ottensmeyer, F.P., 550
Otto, D., 258
Otto, G.F., 199
Otto, H.J., 240
Ottolenghi, A., 333
Ottolenghi, A.C., 367
Oudle, W.K., 535
Ough, C.S., 51
Ourecky, D.K., 307
Outerbridge, J.S., 556
Outka, D.E., 157
Outland, R.H., 188
Outten, L.M., 336
Ouzts, J.D., 246
Overall, J.C.Jr., 474
Overall, J.E., 468
Overbeck, H.W., 228
Overbeek, V., 460
van,Overbeeke, A.P., 533
Overcash, J.P., 249
Overcast, W.W., 444
Overdahl, C.J., 241
Overhults, D.G., 179
Overlease, W.R., 415
Overley, C.B., 166
Overman, A.J., 91
Overman, D.O., 509
Overman, J.L., 148
Overman, R.R., 445
Overn, O.B., 238
Overstreet, R., 251
Overton, J., 132
Overton, M.L., 392
Overweg, N., 312
Owasoyo, J.O., 9
Owczarzak, A., 386
Owen, B.B., 398
Owen, B.D., 561
Owen, B.L., 169
Owen, G.E., 197
Owen, H.E., 70
Owen, J.B., 352
Owen, M.D., 552
Owen, O.S., 514
Owen, R.B., 194
Owen, R.D., 26
Owen, T., 82
Owen, W., 21
Owen, W.G., 161
Owens, C.A., 437, 439
Owens, E.E., 117
Owens, F.N., 138, 376
Owens, J.C., 157
Owens, J.N., 536
Owens, M., 433
Owens, W.C., 437
Owensby, C.E., 167
Owers, N.O., 491
Owings, A.D., 189
Ownbey, G.B., 241
Ownby, J.S., 377
Owre, O., 95
Owre, O.T., 95
Owsley, W., 230
Owyer, J.F., 139
Oxborne, R.H., 519
Oxender, D.L., 232
Oxendine, L.B., 337
Oxley, J.W., 65
Oxnard, C.E., 131
Oyakawa, E.K., 31
Ozaki, H., 227
Ozaki, H.Y., 93
Ozbun, J.L., 305
Ozburn, G., 544
Ozere, R.L., 543

P

Paas, C., 317
Paavola, L.G., 408
Pace, A., 299
Pace, B., 70
Pace, C.N., 458
Pace, G.L., 234
Pace, N., 43, 72
Pace, R.J.E., 91
Pace, R.T., 472
Pacha, R.E., 496
Pack, F., 38
Packard, D., 325
Packard, G.C., 68
Packard, P.L., 116
Packard, R.H., 28
Packard, R.L., 461
Packchanian, A., 467
Packer, J.G., 531
Packer, L., 43
Packer, R.K., 84
Packer, R.R., 158
Packham, M.A., 549
Packman, E.W., 405
Padawer, J., 329
Paddock, E.F., 362
Paden, J.W., 536
Padgett, C.A., 394
Padgitt, D.D., 255
Padilla, G., 333
Padovarii, F., 417
Paetkau, V.H., 531
Paganelli, C.V., 319
Pagano, J.S., 339
Page, C., 288
Page, C.H., 368
Page, C.R., 190
Page, E., 133
Page, O.T., 540
Pagels, R., 491
Pagliaro, H.E., 407
Pai, A.C., 286
Paigen, B., 319
Paigen, K., 319
Paik, W.K., 408
Paik, Y.K., 113
Paine, C.M., 215
Paine, D.A., 302
Paine, R.T., 498
Painter, C.G., 47
Painter, D., 549
Painter, G.H., 331
Painter, L.I., 527
Painter, P., 57
Painter, R., 60
Painter, R.H., 549
Paim, U., 540
Pair, J.C., 168
Paisley, R., 310
Paisley, R.B., 435
Pak, M.J., 230
Pak, W.L., 150
Pakkock, E.F., 363
Pakurar, A.S., 491
Palafox, A.L., 113
Palasek, Sr.M.A., 511
Palaty, V., 535
Palczuk, N.C., 288
Palen, I., 512
Palese, V.J., 315
Palincsar, E.E., 123
Palka, J.M., 498
Palka, Y., 498
Palko, T.M., 20
Pallone, N., 288
Palm, D.J., 240
Palm, E.W., 260
Palm, J., 412
Palmatier, E.A., 421
Palmblad, I.G., 476, 477
Palmer, B., 473
Palmer, C.E., 538
Palmer, C.G., 146
Palmer, E.G., 158
Palmer, F.B., 542
Palmer, J.H., 423
Palmer, L., 270
Palmer, L.T., 272
Palmer, R., 95
Palmer, V.G., 485
Palmersheim, J.J., 341
Palmiter, R.D., 499
Palmore, W.P., 94
Palmour, R.M., 45
Palms, J.M., 100
Paloheimo, J., 550
Pallotta, D., 557
Palser, B.F., 288
Paltan, J.D., 413
Panagiotis, N., 60
Pancoe, W.L., 526
Pandya, D.N., 208
Pang, P.K.C., 296
Pang, S.F., 160
Pangborn, R.M., 49
Pangle, C.C., 440
Panko, W.B., 490
Pankratz, H.V., 274
Pannabecker, R., 354
Pantle, C., 293
Pantle, E., 293
Panuska, J.A., 85
Paolella, P., 216
Paoletti, R.A., 397
Paolillo, D., 302
Paolini, P.J., 39
Papa, K.E., 108
Papaconstantinou, J., 106
Papacostas, C.A., 409
Papahadjopoulos, D., 318
Papahadjopoulos, P.D., 320
Papendick, R.I., 502
Papermaster, B., 467
Papir, Y.S., 298
Papka, R.E., 179
Papenfuss, H., 116
Pappas, G.D., 329
Pappas, P.W., 364
Pappelis, A.J., 128
Pappenheimer, A.M., 210
Pappenheimer, J.R., 212
Paque, R.E., 470
Paquet, G., 544
Paracer, S.M., 221
Paradise, N., 244
Paradise, R., 146
Parakh, J.S., 504
Paranchych, W., 531
Parchman, G.W., 438
Parchman, L.G., 82
Parchment, J.G., 439
Pardee, A.B., 286
Pardee, W.D., 302, 304
Pardini, R.S., 275
Pardue, F.E., 424
Pardue, G.B., 346
Pare, J.P., 558
Parejko, R.A., 229
Parent, J., 560
Parente, W., 414
Parenti, R.L., 164
Paretsky, D., 171
Parham, B.T., 440
Parikh, G.C., 433
Parimanath, A., 70
Paris, O.H., 526
Parish, C.L., 503
Parizek, W.J., 78, 79
Park, H.Z., 499
Park, J.T., 217
Park, M.C., 356
Park, R., 42
Park, R.B., 42
Park, R.L., 472
Park, R.M., 143
Park, T., 132
Parke, R.V., 68
Parker, B.C., 492
Parker, B.F., 179
Parker, B.L., 481
Parker, B.M., 122
Parker, C., 465
Parker, C.F., 365
Parker, C.J.Jr., 235
Parker, C.W., 263
Parker, D., 523
Parker, D.R., 57
Parker, E.E., 292
Parker, F., 476
Parker, G.E., 154
Parker, G.R., 149
Parker, H.E., 148
Parker, H.L., 246
Parker, H.R., 53
Parker, J., 79, 335
Parker, J.C., 340
Parker, J.E., 181, 384
Parker, J.R., 108
Parker, K., 44, 84, 476
Parker, L., 348
Parker, L.R., 27
Parker, M., 7, 526
Parker, M.B., 106
Parker, N., 130
Parker, R., 130, 491
Parker, R.A., 501
Parker, R.B., 388
Parker, S., 538
Parker, W.S., 247
Parkes, K.C., 412
Parkhurst, C.R., 346
Parkhurst, L., 270
Parkinson, D., 531
Parkinson, J.S., 473
Parks, B., 252
Parks, C.E., 249
Parks, C.L., 423
Parks, C.R., 338
Parks, H.F., 179
Parks, J.C., 400
Parks, J.M., 507
Parks, L.W., 386

Parks, P., 3
Parks, P.F., 3
Parks, W., 444
Parks, W.W., 444
Parma, D.H., 473
Parmelee, D.F., 242
Parmeter, J.R., 45
Parnell, D.R., 30
Parnell, J.F., 347
Parochetti, J.V., 200
Parodiz, J., 412
Parr, J.C., 178
Parrett, N.A., 365
Parrish, D.B., 166
Parrish, F.C., 155
Parrish, H.J., 486
Parrish, W.B., 362, 364
Parrondo, R.T., 182
Parrott, W.L., 249
Parry, D., 540
Parry, J., 515
Parson, W.W., 499
Parsons, J.A., 39
Parsons, J.G., 432
Parsons, J.L., 16, 239, 384
Parsons, J.T., 490
Parsons, L.C., 490
Parsons, L.R., 241
Parsons, P., 207
Parsons, R.H., 314
Parsons, T.R., 535
Parsons, T.S., 550
Partanen, C.R., 412
Partch, M., 239
Parthasarathy, M., 302
Parthasarathy, R., 320
Partida, G.J., 167
Partish, S., 335
Partridge, A.D., 118
Partridge, L.D., 445
Parvin, P.E., 114
Paschall, H.D., 142
Pasley, J.N., 25
Pass, B.C., 178
Pass, H., 376
Passano, L.M., 516
Passey, R., 467
Passmore, H.C., 288
Passmore, J.C., 181
Passon, R., 268
Past, W.L., 181
Pasternak, J.J., 551
Pastoret, J.P., 261
Pasztor, V.M., 556
Patach, D., 272
Patalas, K., 539
Pate, J.L., 517
Pateau, K., 519
Patek, P.R., 63
Patel, G.L., 106
Patel, K.M., 534
Patel, M., 408
Patel, N.G., 82
Patent, G., 267
Paterson, C.G., 540
Paterson, J.W., 288
Paterson, P.Y., 127
Paterson, R.A., 492
Pates, A., 88
Patil, S.S., 114
Patno, M.E., 233
Patnode, R.A., 380
Pato, M., 72
Patrenois, J., 7
Patric, E.F., 422
Patrick, J.W., 277
Patrick, O.R., 343
Patrick, R., 342, 409
Patrick, R.A., 228
Patrick, S.J., 542
Patt, D.I., 207
Patt, G.R., 207
Patt, H.M., 60
Pattee, H.E., 343
Pattee, P.A., 156
Patten, B.C., 106
Patten, D.R., 62
Patten, D.T., 15
Patten, J.A., 439
Patterson, A.E., 108
Patterson, B., 211
Patterson, D., 72
Patterson, E.B., 138
Patterson, D.F., 411
Patterson, F.L., 147
Patterson, G., 285
Patterson, G.W., 201
Patterson, H.R., 32
Patterson, I.G., 461
Patterson, J.I.H., 405
Patterson, L.T., 23
Patterson, M., 468
Patterson, M.E., 503
Patterson, R.A., 15
Patterson, R.G., 439
Patterson, R.J., 228
Patterson, R.L., 231
Patterson, R.M., 4
Patterson, R.P., 344
Patterson, S., 427
Patterson, T.B., 3
Patterson, W.M., 493
Patton, D., 299
Patton, E.G., 430
Patton, H.D., 500
Patton, J.L., 43
Patton, R.F., 519
Patton, R.L., 303
Patton, S., 401
Patton, T.H., 90
Patton, W.H., 403
Patton, W.K., 369
Pau, D.N., 437
Paucker, K., 399
Paul, C.A., 220
Paul, H.A., 29
Paul, J.L., 44
Paul, K.B., 9
Paul, K.G., 410
Paul, M., 536
Paul, Sr.M., 396
Paule, M.R., 68
Paule, W.J., 63
Paules, L., 527
Pauley, G., 498
Pauley, T.K., 507
Paulin, J.J., 106
Pauling, C., 57
Paulissen, L.J., 22
Paull, W.K., 482
Pauls, J., 408
Paulsell, L.K., 261
Paulsen, D.F., 352
Paulsen, E.P., 490
Paulsen, G.M., 166
Paulson, -., 81
Paulson, C., 238
Paulson, G.D., 352
Paulson, K.N., 51
Paulson, W.H., 516
Paulton, R.J.L., 562
Paulus, A.O., 58
Pauly, J.E., 24
Pauly, L., 523
Paustian, F., 273
Pautler, E.L., 69
Pauze, F., 558
Pavan, C., 465
Pavilanis, V., 558
Pavlak, -., 391
Pavlech, H.M., 509
Pavlek, J., 238
Pawlowski, R., 355
Paxson, J., 395
Payette, S., 558
Payne, A., 232
Payne, A.F., 467
Payne, B.R., 275
Payne, C., 7
Payne, F.E., 234
Payne, G.F., 266
Payne, H.H., 322
Payne, J.F., 438
Payne, J.J., 189
Payne, R.B., 231
Payne, R.N., 377
Payne, R.R., 26
Payne, T.L., 458
Payne, W.A., 142
Payne, W.W., 90
Paynter, M.J.B., 426
Paynter, R.A., 211
Pax, R.A., 227
Paxton, J.D., 139
Pazdernik, L., 560
Pazur, J.H., 404
Peabody, F.R., 228
Peachey, L., 410
Peachey, L.D., 71, 409
Peacock, J.T., 456
Peadon, A.M., 83
Peakman, D., 72
Peanasky, R.J., 434
Pearce, I.E., 553
Pearce, J.W., 550
Pearce, S.M., 544
Pearce, T.L., 299
Pearcy, W.G., 385
Pearlman, S., 263
Pearsall, N.N., 499
Pearse, J.E., 61
Pearson, A.A., 388
Pearson, A.M., 547
Pearson, D., 391
Pearson, E.S., 247
Pearson, F., 109, 420
Pearson, G.D., 385
Pearson, H., 64
Pearson, J.A., 535
Pearson, J.B., 452
Pearson, J.L., 422
Pearson, L.D., 520
Pearson, P., 181, 419
Pearson, P.B., 16
Pearson, P.F., 181
Pearson, P.G., 288
Pearson, R., 126
Pearson, W.B., 550
Pearson, W.D., 452
Peaslee, D.E., 177
Peaslee, M.H., 434
Peat, M., 538
Peavler, R.J., 254
Pechuman, L.L., 303
Peck, B., 253
Peck, D., 179
Peck, E.J., 449
Peck, H.P.Jr., 105
Peck, J., 239
Peck, N.H., 307
Peck, T.R., 138
Peckham, M.C., 305
Peckham, R.S., 311
Pecknold, P.C., 148
Pecora, R.A., 191
Pedersoli, W.M., 6
Pederson, V., 352
Pederson, V.C., 139
Pedigo, L.P., 157
Pedrini, V.A., 161
Peech, M., 302
Peek, F.W., 311
Peek, J.M., 118
Peek, S.Z., 193
Peele, T., 332
Peele, T.C., 423
Peeler, E.N., 246
Peeples, E., 73
Peepre, A., 546
Peets, D.L., 557
Pegg, W.J., 196
Peightel, W.E., 407
Peirce, L.C., 281
Peiss, C.N., 125
Pekarthy, J.M., 185
Pelias, M., 190
Pelikan, J., 81
Pelizza, J.J., 142
Pell, E.J., 403
Pelle, T., 393
Peller, L.M., 60
Pellerin, G.J., 223
Pelletier, R.L., 557
Pellett, H.M., 241
Pellett, N.E., 481
Pelligrino, P., 76
Pelon, W., 185
Peloquin, R.L., 150
Peloquin, S.J., 519
Pelthier, D., 399
Pelton, J.F., 142
Pelton, M.R., 443
Pelz, D.R., 139
Pelzer, C., 230
Pemble, R.H., 239
Penas, E.J., 271
Pendergast, D.R., 319
Pendleton, J.W., 516
Pendleton, R.C., 473
Pendse, P.C., 27
Pene, J.J., 82
Penfound, W.T., 348
Peng, A.C., 365
Peng, J.Y., 234
Pengelley, E.T., 57
Pengelly, D.H., 547
Pengra, R.M., 433
Penhoet, E.E., 45
Penhos, J.C., 85
Penick, G.D., 161
Pennak, R.W., 71
Penner, L., 77
Penner, L.R., 77, 79
Penner, P.E., 541
Penney, D.P., 328
Penniall, R., 340
Pennington, F.C., 29
Pennington, M.T., 426, 428
Pennington, R.C., 428
Pennock, R.Jr., 401
Penny, J.S., 397
Penny, L., 310
Penny, L.H., 433
Pennypacker, S.P., 403
Penquite, W., 372
Pentecost, E., 522
Peoples, S.A., 52
Pepe, F.A., 410
Pepper, R.E., 395
Peppler, R.D., 185
Pequegnat, W., 459
Peraino, C., 126
Percy, D., 551
Pereira, G.P., 300
Peretz, B., 180
Pereyra, W., 498
Perez, J.C., 452, 456
de Pérez, N.R., 418
Perez, R., 465
Perez-Chiesa, Y., 418
Perez-Lopez, A., 417
Perez-Sosa, R.F., 417
Perfetti, P.A., 440
Perham, J.E., 487
Perincott, J.V., 214
Perkel, D., 41
Perkins, B.L., 191
Perkins, D.D., 41
Perkins, D.L., 379
Perkins, D.Y., 5
Perkins, E.M., 62
Perkins, F.O., 490
Perkins, H.F., 106
Perkins, J.L., 23
Perkins, L.C., 145
Perkins, R., 126
Perkins, R.H., 149
Perkins, R.J., 172
Perkins, R.L., 367
Perkins, W.D., 499
Perks, A.M., 535
Perl, E.R., 340
Perl, W., 283
Perley, J.E., 358
Perlin, M., 179
Perlman, P.S., 363
Perlmutt, J.H., 340
Perlmutter, A., 312
Perlmutter, H.I., 355
Perlmutter, J., 338
Perlmutter, O.W., 239
Perper, T.P., 288
Perreault, W., 512
Perrill, S.A., 143
Perrin, E.V., 356
Perrin, K., 470
Perron, J.M., 557
Perron, L., 555
Perry, A.E., 500
Perry, B.Jr., 5

Perry, B.W., 489
Perry, D., 127
Perry, E., 549
Perry, E.A., 123
Perry, J.A., 249
Perry, J.D., 338
Perry, J.E., 468
Perry, J.H., 449
Perry, J.J., 345
Perry, J.W., 360
Perry, L.J., 271, 348
Perry, R.K., 410
Perry, T.O., 346
Perry, V.G., 91
Perryman, E.K., 27
Perryman, J., 464
Persaud, T.V.N., 537
Persellin, R.H., 470
Person, C.O., 534
Person, R.J., 380
Person, S., 404
Pertuit, A.J., 424
Pesce, A.J., 370
Pesetsky, I., 329
Pessoney, G.F., 252
Peter, R.E., 531
Peter, T., 88
Petering, H.G., 370
Peterjohn, G.W., 354
Peterle, T.J., 364
Peters, A., 208
Peters, D.B., 138
Peters, D.C., 377
Peters, E.J., 159, 258
Peters, H., 153
Peters, H.T., 236
Peters, J.M., 212
Peters, L.E., 229
Peters, L.L., 271
Peters, M., 253
Peters, M.V., 550
Peters, R., 556
Peters, R.A., 78, 79
Peters, T., 300
Peters, T.M., 218
Peters, W., 88
Peters, W.P., 123
Petersen, D., 290
Petersen, D.H., 403
Petersen, J.C., 520
Petersen, K.W., 203
Petersen, R.H., 441
Peterson, A.G., 240
Peterson, A.J., 240
Peterson, C., 301
Peterson, C.E., 392
Peterson, C.F., 117
Peterson, C.H., 205
Peterson, C.J., 125, 502
Peterson, C.M., 4
Peterson, D., 239, 244
Peterson, D.F., 470
Peterson, D.M., 516
Peterson, D.R., 516
Peterson, D.W., 48
Peterson, E., 95
Peterson, E.H., 23
Peterson, F.F., 276
Peterson, G., 270, 432, 438
Peterson, G.A., 271
Peterson, G.E., 438
Peterson, G.W., 272, 401
Peterson, H., 230
Peterson, H.P., 23
Peterson, J.A., 466
Peterson, J.E., 164
Peterson, J.F., 557
Peterson, J.L., 288
Peterson, J.W., 467
Peterson, L.H., 410
Peterson, M., 157
Peterson, M.L., 47
Peterson, N.K., 280
Peterson, P., 373
Peterson, R., 130, 387
Peterson, R.G., 533
Peterson, R.L., 546
Peterson, R.M., 433
Peterson, R.R., 263
Peterson, S., 360
Peterson, S.P., 118
Peterson, T.A., 521
Peterson, V.H., 166
Peterson, W.D.Jr., 236
Peterson, W.J., 343, 368
Petit, M.G., 69
Petitclerk, C., 559
Petra, P.H., 499
Petrali, J.P., 203
Petrelli, M., 356
Petri, L.H., 162
Petrides, G.A., 225, 227
Petriello, R.P., 288
Petrucci, R.H., 28
Petrucci, V.E., 29
Petruska, J.A., 62
Petrychenko, T., 393
Pettigrew, J.D., 25
Pettegrew, R.K., 358
Pettijohn, D.E., 72
Pettinger, W.A., 467
Pettit, B.J., 285
Pettit, J.H., 547
Pettit, R.D., 463
Pettus, D., 68
Petty, A., 344
Petty, R.O., 153
Peverly, J.H., 302
Pew, W.D., 17
Peyton, S., 250
Pfadt, R.E., 527
Pfaender, F.K., 341
Pfaffman, M.A., 252
Pfander, W.H., 259
Pfau, C.J., 314
Pfefferkorn, E.R.Jr., 277
Pfefier, H.W., 77
Pfeifer, C.H., 293
Pfeifer, H., 77
Pfeifer, R.P., 401
Pfeiffer, C.A., 418
Pfeiffer, E.W., 267
Pfeiffer, J.B., 114
Pfeiffer, R.R., 263
Pfeiffer, S., 79
Pfeiffer, W.F., 327
Pfiester, L., 379
Pfister, R.M., 362, 363
Pflanzer, R.G., 145
Pfohl, R.J., 360
Pflueger, M.N., 538
Phaff, H.J., 49, 51
Phalora, O.S., 142
Phan, C.T., 530
Phares, C.K., 273
Phares, K., 273
Phares, R.E., 129
Pharis, R.P., 531
Pharr, D.M., 345
Phelps, C.H., 79
Phelps, L.N., 39
Phemister, R.D., 69
Phibbs, P.V., 492
Philibert, R.L., 258
Philip, B.A., 14
Philip, E.L., 128
Phillips, A.H., 77
Phillips, A.T., 404
Phillips, A.W., 326
Phillips, B.A., 413
Phillips, B.U., 468
Phillips, C., 309
Phillips, C.A., 482
Phillips, D., 265
Phillips, D.E., 257
Phillips, E., 535
Phillips, E.A., 32
Phillips, E.L., 494
Phillips, H.C., 436
Phillips, H.J., 269
Phillips, H.M., 12, 488
Phillips, I.A., 32
Phillips, J., 220
Phillips, J.A., 494
Phillips, J.B., 284
Phillips, J.H.Jr., 41
Phillips, J.P., 546
Phillips, J.R., 23
Phillips, J.T., 468
Phillips, L., 27
Phillips, L.L., 344
Phillips, M.I., 162
Phillips, M.W., 147
Phillips, P., 277
Phillips, R., 523
Phillips, R.A., 550
Phillips, R.C., 497
Phillips, R.E., 177
Phillips, R.E., 242
Phillips, R.L., 36, 240, 382
Phillips, R.R., 323
Phillips, R.W., 69
Phillips, S.J., 408
Phillips, S.L., 413
Phillips, T.L., 136
Phillips, W.D., 410
Phillips, W.R., 445
Phillis, J.W., 562
Philogene, B.J.R., 534
Philpot, V.B., 190
Philpott, C.W., 453
Philpott, J., 332
Phinney, B.O., 55
Phinney, G., 369
Phinney, H.K., 385
Phipps, J., 541
Phipps, J.B., 552
Phizackerley, R.P., 197, 198
Piacsek, B.E., 512, 513
Pianka, E.R., 465
Piantadosi, G., 340
Piatka, B., 396
Piatt, J., 410
Pice, J.S., 423
Pick, J.R., 340
Pickard, B.G., 263
Pickens, P.E., 18
Pickering, J.D., 395
Pickering, J.L., 396
Pickering, R.J., 327
Pickett, B.W., 69
Pickett, E.E., 261
Pickett, J.M., 265
Pickett, M.J., 55
Pickett, R.C., 147
Pickett-Heaps, J., 71
Picklum, W.E., 162
Pickren, J., 320
Picton, H., 265
Pidcoe, V., 408
Piddington, R., 411
Pie, E., 151
Piedmont, E., 217
Piehl, M.A., 182
Pielke, R., 489
Pielou, D.P., 542
Pielou, E.C., 542
Pielou, W.P., 427
Pienaar, L.V., 108
Pieper, R.D., 292
Pieper, R.E., 62
Pier, A.C., 158
Pierce, D., 220
Pierce, D.A., 439
Pierce, J.G., 56
Pierce, J.M., 366
Pierce, R., 434
Pierce, R.H., 59
Pierce, R.S., 286
Pierce, T., 291
Pierce, S., 298
Pierce, S.K., 199
Pierce, W.A., 190
Pierce, W.S., 28
Pieringer, R., 408
Pierro, L.J., 77, 78, 79
Pierson, C.F., 503
Pierson, D.A., 552
Pierson, D.W., 164
Pierson, E., 522
Pierucci, O., 319
Pietruszko, R., 288
Piette, L.H., 112
Pigeon, G., 559
Pigg, C.J., 30
Pigiet, V., 197
Pigon, A., 69
Pigott, G.M., 498
Pike, C.H., 485
Pike, C.S., 395
Pike, K., 15
Pike, L.M., 459
Pike, R., 466
Pilar, G.R.J., 77, 79
Pilcher, B.L., 451
Pilcher, K.S., 386
Pillsbury, N.H., 27
Pilmanis, A.A., 64
Pimentel, D., 303
Pimentel, M., 306
Pimentel, R.A., 27
Pimlott, D.H., 550
Pimm, S.L., 426
Pincus, I.J., 64
Pincus, J.H., 127
Pine, M., 320
Pineus, M., 399
Pingel, A.L., 99
Pings, C.J., 26
Pinjani, M., 3
Pinkava, D.J., 15
Pinkerton, A.C., 464
Pinkerton, J.M., 146
Pinkstaff, C.A., 509
Pinnell, R., 33
Pinon, R., 58
Pinschmidt, M.W., 486
Pinschmidt, W.C., 486
Pinter, A.J., 185
Pinter, G.G., 204
Pinter, R.B., 498
Pinteric, L., 549
Pinto, J.D., 58
Pinto, L.H., 150
Pipa, R.L., 44
Piper, D.J.W., 542
Piper, E.L., 23
Piper, R.W., 200
Pipes, W.O., 126
Pipes, W.O., 393
Pippen, R.W., 236
Pippitt, D., 128
Pirofsky, B., 388
Pirone, T.P., 178
Pirozok, R.P., 78
Pisacano, N.J., 176
Pisano, M.A., 316
Pisano, R.G., 32
Piscitelli, J., 397
Pistey, W.E., 77, 79
Pistole, T.G., 279
Pisziewicz, D., 54
Pitelka, D.R., 43
Pitkin, F., 118
Pitman, W.J., 175
Pitner, J.B., 423
Pitre, H.N., 249
Pitre, V., 188
Pitt, D.G., 202
Pittendrigh, C.S., 41
Pittenger, T.H., 166
Pittillo, J.D., 348
Pittman, A.B., 347
Pittman, D., 456
Pittman, J.A.Jr., 11
Pittman, J.M., 228
Pittman, M.A., 487
Pittman, R.N., 492
Pittman, W.G., 546
Pitts, C.W., 167
Pitts, G.C., 490
Pitts, M.B., 335
Pitts, R.D., 151
Pitts, R.F., 308
Pitts, T.D., 37, 444
Piver, S., 320
Pivonka, W.C., 255
Pivorun, E.B., 426
Pixley, E.E., 161
Pizzolato, P., 186
Plagemann, P.G.W., 244
Plager, J.E., 320
Plagge, J., 134
Plaisted, R.L., 304
Plakke, R.K., 73
Planck, R.J., 552
Plant, J., 235
Plantenberg, Sr.D., 239
Plapp, B.V., 161
Plapp, F.W.Jr., 458

Plato, P.A., 233
Platonow, N.S., 548
Platt, A.P., 209
Platt, D.R., 164
Platt, J.E., 73
Platt, R.B., 100
Platt, R.S., 362
Platt, T., 224, 540
Platt, W.J., 160
Platz, C.E., 133
Platz, J., 268
Platzer, E.G., 57, 58
Plaut, G.W.E., 408
Plaut, W.S., 516
Pleasants, J.R., 152
Plescia, O.J., 288
Pleshkewych, A., 315
Pless, C., 443
Plimpton, R.F.Jr., 365
Plint, C., 544
Pliske, T., 93
Plocke, D.J., 207
Plonsey, R., 356
Plough, H.H., 206
Ploux, Sr.D.M., 37
Plowright, R.C., 550
Pluenneke, R.H., 248
Plumart, P.E., 432
Plumb, J.A., 4
Plummer, G., 124
Plummer, G.L., 105
Plummer, H.A., 194
Plummer, J., 393
Plummer, J.R., 393
Plunkett, R.W., 319
Plyler, D.B., 347
Plymale, H.H., 40
Paog, C.W., 459
Pocetto, Rev.A.T., 390
Pocker, A., 499
Podelski, T., 302, 304
Poe, S.L., 91
Poehlman, J.M., 258
Poff, K., 226
Poffenbarger, P., 467
Poffenberger, T., 234
Pogany, G., 15
Pogell, B.M., 256
Pogge, A., 127
Poglayen, I., 180
Pohl, R.W., 157
Pohlo, R., 13
Poin, L.V., 406
Pointer, J.L., 444
Pointing, P.J., 550
Poirier, G.R., 10
Poirrier, M.L., 185
Pokorny, F.A., 107
Polacek, B., 323
Polakowski, K.J., 231
Polan, C.E., 493
Poland, J.L., 492
Polansky, S., 311
Polasek, Sr.M.A., 511
Polechko, M.A., 405
Polensek, A., 384
Polglase, W.J., 534
Polic, R.B., 261
Policansky, D.J., 219
Polidora, V.J., 53
Polijak, R., 198
Polin, D., 226
Polissar, L., 500
Polites, L., 393
Polivanov, S., 83
Poljak, R.J., 197
Pollack, B.L., 288
Pollack, G.H., 497
Pollack, J.D., 367
Pollack, S.A., 145
Pollack, S.B., 499
Pollack, S.R., 411
Pollack, W., 408
Pollak, D.G., 465
Pollard, D., 222
Pollard, J.E., 281
Pollard, M., 152
Pollet, D.K., 424
Pollett, F.C., 541
Polley, E., 134
Pollitt, E., 213
Pollock, E., 13
Pollock, H.M., 499
Pollock, L.W., 284
Pollock, R., 542
Pollock, R.J., 124, 125
Pollock, R.T., 352
Pollpeter, A.I., 525
Polly, S.M., 269
Politoff, A., 208
Polmar, S., 355
Polosa, C., 556
Poluhowich, J.J., 77, 79
Pomerantz, D.K., 552
Pomeranz, B.H., 550
Pomeranz, J.R., 556
Pomerontz, S., 203
Pomeroy, L.R., 106
Pommer, D., 513
Pon, N.G., 57
Pond, S., 79
Poneleit, C.G., 177
Ponthier, R.L., 185
Ponticorvo, L., 300
Pontle, Sr.C., 33
Ponton, D.K., 509
Poodry, C.A., 61
Pool, R.R., 53
Poole, R.J.. 556
Pooler, J.P., 101
Poorman, A.E., 269
Pootjes, C.F., 404
Pope, D.H., 314
Pope, D.T., 345
Pope, H., 103
Pope, R.S., 509
Pope, W.K., 117
Popenoe, H.H., 90
Popescu, E., 135
Popham, R.A., 362
Popjak, G.J., 56
Poponoe, H.L., 93
Popovic, V., 101
Poppele, R., 244
Popper, A.N., 112
Pore, R.S., 509
Poretz, R.D., 288
Poritsky, R., 355
Porot, T.B., 447
Porter, A., 358
Porter, C.W., 32
Porter, D., 105
Porter, D.A., 5
Porter, H.L., 366
Porter, J.R., 161, 186
Porter, J.W., 520
Porter, K.G., 231
Porter, K.R., 71
Porter, M.M., 399
Porter, R.J., 234
Porter, T.A., 74
Porter, T.E., 230
Porter, T.R., 28
Porter, T.W., 227
Porter, W., 440
Porter, W.P., 516
Porterfield, I.D., 343
Porterfield, S.P., 102
Portnoff, C.L., 63
Portz, H.L., 130
Posey, H.G., 5
Posner, A.S., 307
Posner, G.S., 297, 298
Posner, H., 317
Posner, H.S., 333
Posner, P., 94
Pospisil, A., 558
Posser, C.L., 137
Possey, A.F., 22
Possmayer, F., 551
Post, D., 522
Post, D.M., 40
Post, F.J., 476, 477
Post, G., 66, 69
Post, J.E., 306
Post, R.L., 276, 351
Postlethwait, J.H., 387
Postlethwait, S.N., 150
Postwald, H.E., 218
Potash, M., 480
Poteat, W.L., 491
Potgieter, L.N.D., 158
Pothier, L., 320
Potter, D., 396
Potter, G.D., 458
Potter, J.H., 199
Potter, L., 96, 308
Potter, L.D., 293
Potter, N.A., 420
Potter, R., 31
Potts, H.C., 248
Potts, R.C., 459
Pou, C., 239
Pough, F.H., 302, 303
Poulik, M.D., 236
Poulson, D.F., 80
Poulter, S.R., 473
Pouncey, P., 300
Pound, G.S., 516, 519
Pourcho, R., 235
Povell, J.A., 44
Powders, V.N., 100
Powell, A.E., 355
Powell, A.M., 455
Powell, C., 83
Powell, C.C.Jr., 366
Powell, E.C., 157
Powell, E.W., 24
Powell, G., 467
Powell, J., 216
Powell, J.A., 215
Powell, J.E., 9
Powell, J.H., 28
Powell, J.R., 80
Powell, J.T., 546
Powell, L.E., 305
Powell, N.T., 345
Powell, R., 246
Powell, R.W.Jr., 427
Powell, T.M., 47, 49
Powell, V.M., 153
Powell, W.M., 108
Powelson, E.E., 372
Powelson, R.L., 385
Power, G., 551
Power, G.G., 36
Powers, D., 197
Powers, E.E., 123
Powers, E.L., 465
Powers, H.O., 324
Powers, J., 389
Powers, J.J., 106
Powers, L.J., 445
Powers, J.R., 238
Powers, S.R., 314
Powers, W., 228
Powers, W.L., 166
Powrie, W.D., 533
Poydock, Sr.E., 400
Poznansky, M.J., 211
Pragay, D.A., 318
Prager, J.C., 421
Prager, M.D., 466
Prahlad, K.V., 126
Prakash, L., 329
Prakash, S., 327
Prall, B., 512
Pramer, D., 288
Prance, G., 297, 298
Prange, H.D., 90
Prasad, K., 562
Prasad, M., 331
Prashar, P., 433
Pratley, J., 32
Pratt, D., 51
Pratt, D.B., 194
Pratt, D.C., 241
Pratt, D.R., 456
Pratt, G.A., 526
Pratt, H.K., 51
Pratt, J.D., 180
Pratt, N.E., 408
Prau, E.G., 144
Pray, F., 154
Pray, T.R., 62
Preece, S.J.Jr., 266
Preedy, J.R.K., 101
Preer, J., 145
Preiss, B., 559
Preiss, J., 48
Prentice, R., 500
Prentice, V., 254
Presch, W.F., 30
Prescott, D.M., 71
Prescott, G., 388
Prescott, G.W., 266
Prescott, J.M., 458
Prescitt, L.M., 431
Preska, M., 35
Presley, B.J., 459
Press, N., 523
Presser, B.D., 407
Pressick, M.L , 43
Pressman, B.C., 96
Pressman, D., 319
Prest, D.B., 310
Preston, A.L., 258
Preston, A.M., 418
Preston, D.A., 39
Preston, E., 418
Preston, E.C., 278
Preston, H., 300
Preston, L., 418
Preston, Sr.P.A., 513
Preston, R.J., 346
Preston, R.L., 365
Preston, S.B., 231
Preston, W.C., 278
Pretlow, T.G.II., 11
Prevec, L., 545
Previte, J.P., 210
Prevost, N.H., 185
Preyer, P., 155
Pribor, D., 371
Price, A.R., 232
Price, C.A., 288
Price, D.D., 492
Price, E.O., 322
Price, F.W., 317
Price, H., 226
Price, H.J., 88
Price, J., 310
Price, J.A., 247
Price, J.L., 263
Price, J.W., 364
Price, M.A., 458
Price, P.A., 59
Price, P.B., 433
Price, P.W., 136
Price, R.D., 240
Price, R.G., 311
Price, R.L., 17
Price, W., 64
Price, S., 492
Prichard, H.N., 398
Prichard, P., 317
Prichard, P.M., 103
Prichett, J.F., 5
Priddy, R., 144
Priddy, R.B., 329
Prieb, W., 170
Prier, J., 408
Priest, D.G., 428
Priest, S., 207
Prieto, A., 285
Prigmore, G.T., 193
Prikle, C., 552
Prikle, J., 440
Primiano, F.P.Jr., 356
Prince, A.B., 280
Prince, B.E., 188
Prince, H.H., 225
Prince, J., 95
Prince, J.T., 244
Prince, R., 119
Prince, W.R., 346
Princiotto, J.V., 85
Pring, D.R., 92
Pring, M., 411
Prins, R., 181
Prinz, P., 333
Priola, D., 294
Prior, D.J., 177
Prior, P.V., 272
Prioreschi, P., 269
Pritchard, A.W., 386
Pritchard, J.B., 429
Pritchard, M.L., 272
Pritchard, W.R., 52
Pritchett, C.L., 473

Pritchett, W.H., 506
Pritchett, W.L., 93
Privitera, C.A., 318
Probst, C.J.Jr., 185
Procaccini, D.J., 210
Prockop, D.J., 283
Procopio, D., 212
Proctor, B.S., 392
Proctor, N., 76
Prokesch, J., 75
Promnitz, L.C., 156
Prophet, C., 164
Prophet, J., 51
Propst, H.D., 99
Prothero, J.W., 499
Protopapas, P., 212
Protor, V.W., 461
Prough, R.A., 466
Prouix, P.R., 548
Prout, T., 57
Prouty, W.F., 413
Prouvidenti, R., 307
Provasoli, L., 80
Prover, C., 309
Provie, M.M., 436
Provonsha, A.V., 149
Provost, E.E., 106, 108
Prowse, G.A., 111
Prosles, P.M., 546
Pruess, K.P., 271
Pruitt, D.W., 101
Pruitt, K., 11
Pruitt, W.O.Jr., 538
Prusso, D.C., 276
Prutkin, L., 312
Pruzansky, J.J., 127
Prychodoko, W., 234
Pryor, J.E., 20
Pryor, M.Z., 214
Prystowsky, H., 405
Przybylski, R.J., 355
Ptashne, M., 210
Pubols, B.H., 405
Pubols, M.H., 502
Puck, T.T., 72
Puckett, D.H., 181
Pudelkiewicz, W.J., 78, 79
Puett, J.D., 446
Pugh, J.E., 484
Puglia, C.R., 278, 279
Pugliese, F.A., 407
Puleo, J., 319
Pullen, A.M., 450
Pullen, E.W., 489
Pullen, T.M., 5
Pulley, L.T., 53
Pulliam, H.R., 18
Pullin, W.E., 182
Pullon, T.M., 251
Pulse, R.E., 382
Pun, P.P., 139
Punter, D., 538
Purcell, A.E., 344
Purcell, G.M., 228
Purcell, W.P., 445
Purcifull, D.E., 92
Purdy, L.N., 92
Purinton, P.T., 168
Purko, J., 552
Purple, R., 244
Pursell, R.A., 403
Purves, D., 263
Purves, W.K., 77, 79
Purvis, B.S., 109
Putman, E.W., 111
Putnam, A., 226
Putnam, D.N., 401
Putnam, F.W., 145
Putnam, G.B., 383
Putnam, J.L., 332
Putnam, L.S., 364
Putttler, B., 259
Puzzouli, D.A., 506
Pyburn, W.F., 464
Pye, E.K., 410
Pyle, T.H., 97
Pylman, R.W.Jr., 433
Pyrah, G.L., 250
Pysh, J.J., 126

Q

Quackenbush, F.W., 148
Quadagno, D.M., 171
Quade, A., 341
Quade, H.W., 238
Quadri, S.U., 549
Qualset, C.O., 47
Quam, M.E., 494
Quarles, C.L., 65
Quarles, T.S., 6
Quatrano, R.S., 385
Quattropani, S.L., 405
Quay, T.L., 346
Quay, W.B., 516
Quertekmus, C.J., 109
Quevedo, A., 548
Quibell, C.F., 28
Quick, F.W., 336
Quick, J.S., 350
Quick, R., 522
Quick W.A., 561
Quigley, J., 325
Quimby, D., 265
Quin, Br.P., 291
Quinn, B.G., 478
Quinn, D., 361
Quinn, D.O., 508
Quinn, Br.E., 299
Quinn, Sr.G., 278
Quinn, J., 288
Quinn, J.A., 411
Quinn, L.Y., 156
Quinn, R.D., 29
Quinn, R.E., 459
Quinn, W.G., 287
Quintal, C., 206
Quintana, J.M., 417
Quinton, D.A., 463
Quinton-Cox, R., 388
Quissell, D., 261
Quist, R.G., 562
Quo, F.Q., 532
Quraishi, M.S., 351

R

Raabe, R.D., 45
Raache, I.D., 207
Rabalais, F.C., 354
Rabb, D.D., 390
Rabb, R.L., 344
Rabbege, J.R., 356
Rabe, F.W., 117
Rabinowitz, M., 131
Rabeni, C., 192
Raber, R.L., 320
Rabinovitch, B., 380
Rabinowitz, J.C., 42
Rabinowitz, J.L., 411
Rabinovitch, M.P., 312
Race, J., 461
Race, S.R., 288
Rachele, J.R., 307
Rachlin, J., 298
Rachlin, J.W., 297
Rachmeler, M., 127
Rachon, T.E., 254
Racker, E., 302
Racusen, D., 481
Racusen, D.W., 481
Radabaugh, D.C., 369
Radandt, F.K.I., 123
Radcliffe, C., 70
Radcliffe, E.B., 240
Radding, -., 81
Rademaker, A.W., 341
Raden, J.F., 302
Rader, B.W., 257
Radewald, J.D., 58
Radford, A.E., 338
Radforth, N.W., 540
Radichevich, I., 300
Radin, N., 232
Radinovsky, S., 400
Radinsky, L., 131
Radivoyevitch, M.A., 356
Radke, F.H., 193
Radloff, H., 527
Radloff, R.J., 294
Radmall, M.B., 275
Raeside, J.K., 548
Raff, R., 145
Raffel, S., 41
Raffensperger, E.M., 303
Rafferty, N.S., 126
Rafols, J., 235
Rafter, G.W., 509
Rafuse, D.D., 36
Ragan, J., 188
Ragetli, H.W.J.Jr., 534
Raghavan, V., 362
Raghuvir, N., 77, 79
Ragin, J.F., 68
Ragland, J.L., 177
Ragsdale, H.L., 100
Raguse, C.A., 47
Raha, C., 273
Rahe, J.E., 533
Rahi, G.S., 9
Rahil, K.S., 86
Rahman, M., 175
Rahman, S., 356
Rahman, Y., 126
Rahn, H., 319
Rai, K.S., 152
Rai, P.M.M., 80
Raidt, D.J., 55
Raidt, H., 146
Raijman, L., 64
Raikow, R.J., 412
Raimist, R., 285
Rain, A.P., 108
Rainbolt, L., 121
Rainer, J.K., 301
Raines, P.L., 20
Rains, D.D., 66
Rains, D.W., 47
Rainwater, F.L., 455
Raisbeck, B., 214
Raitt, R.J., 292
Raizen, E.C., 394
Raj, B., 246
Rajagopalan, K.V., 333
Rajalakshmi, S., 408
Rajasekhara, K., 331
Raju, M.V.S., 561
Rakes, A.H., 343
Rakes, J.M., 23
Rakowitz, F., 312
Rakusan, K., 549
Raleigh, R.F., 494
Raleigh, R.J., 302
Ralin, D.B., 175
Ralph, C.J., 393
Ralph, C.L., 68
Ralston, A.T., 382
Ralston, C.W., 334
Ralston, H.J., 60
Ralston, R.D., 349
Ram, I., 243
Ramachandran, G.N., 132
Ramachandran, J., 60
Ramage, C.H., 288
Ramage, D.R., 436
Ramage, R.T., 16, 19
Ramaley, R., 273
Ramazzotto, L.J., 285
Rambo, O.N., 60, 62
Rambo, T.C., 176
Ramey, E.R., 85
Ramger, R.C., 439
Ramirez, V.D., 137
Ramlau, R., 235
Ramm, A.E., 357
Ramm, G.M., 199
Ramon, F., 333
Ramon-Moliner, E., 559
Ramos, J.A., 417
Ramos, R., 418
Rampacek, G., 106
Rampone, A.J., 389
Ramprashad, G.F., 547
Ramsay, F.J., 203
Ramsdell, F.W., 29
Ramsey, C.B., 462
Ramsey, D.S., 249
Ramsey, G.W., 486
Ramsey, H.A., 345
Ramsey, J.J., 451
Ramsey, J.M., 371
Ramsey, J.S., 4, 5
Ramsey, P.R., 429
Ramus, J.S., 80
Ramwell, P.W., 85
Ramzy, I., 552
Rana, M.W., 256
Ranck, J.B.Jr., 232
Rand, A.G., 422
Rand, C., 310
Rand, L., 373
Rand, P., 270
Rand, R.P., 544
Rand, W.M., 213
Randale, A.G., 194
Randall, C., 318
Randall, C.C., 251
Randall, D.D., 261
Randall, D.J., 535
Randall, E., 318
Randall, E.H., 73
Randall, E.M., 533
Randall, H.M., 186
Randall, J.F., 331
Randall, R., 110
Randall, W.C., 125
Randel, P.F., 417
Randell, R.L., 562
Randerson, S., 523
Randic, M., 217
Randles, C.I., 363
Randolph, H.E., 458
Randolf, K., 449
Randolph, N.M., 458
Randolph, P., 145
Raney, H.G., 178
Raney, R.J., 167
Rank, G.H., 562
Rank, G.L., 70
Rankin, B.J., 292
Rankin, J.H.G., 521
Rankin, J.S., 77, 79
Rankin, M.A., 465
Rannels, D.E., 405
Ransbottom, J.R., 199
Ransom, J., 164
Ranzoni, F.V., 329
Rao, B.K., 394
Rao, B.O.N., 410
Rao, B.R., 414
Rao, H.V.R., 416
Rao, K.R., 98
Rao, M.A., 307
Rao, N.V., 203
Rao, P.N., 469
Rao, P.S., 256
Rao, R., 335
Rao, S., 85
Rao, S.T., 517
Rapacz, J., 519
Raper, K.B., 515, 517
Rapp, F., 405
Rapp, G.W., 124
Rappaport, H., 133
Rappaport, J.J., 488
Rappaport, L., 51
Rappleye, R.D., 201
Rapport, M., 300
Raps, S., 298
Rasch, C.A., 356
Rasch, E.M., 512, 513
Raschke, K., 226
Rashad, A.L., 388
Rashad, M.N., 113
Raska, K.Jr., 283
Raskas, H.J., 263
Raski, D.J., 50
Rasmusen, B.A., 138
Rasmussen, A.F., 56
Rasmussen, D.I., 15
Rasmussen, H., 410
Rasmussen, H.P., 226
Rasmussen, J., 431
Rasmussen, R.A., 34
Rasmusson, D.C., 240
Rasor, W.W., 80
Rasper, V., 547

Raspet, M., 246
Raspter, J., 210
Rastorfer, J., 119
Rasweiler, J.J., 300
Rathkamp, W., 230
Ratner, A., 294
Rattner, M., 77, 79
Ratty, F.J., 39
Ratzlaff, K., 130
Ratzlaff, M.H., 227
Ratzlaff, W., 400
Rauch, F.D., 114
Rauch, H., 218
Raudzens, L., 217
Rauganathan, V.S., 8
Raulerson, L., 110
Raun, C.G., 448
Raun, E.S., 271
Raunio, E.K., 116
Rausch, R., 14
Rausch, R.L., 14
Rauser, W.E., 546
Rauth, A.M., 550
Raveling, D.G., 47
Raven, P., 256
Raven, P.H., 263
Ravin, A., 132
Ravin, R.W., 132
Rawls, H.C., 120
Rawls, J.M., 12
Rawson, J., 105
Rawson, K.S., 407
Ray, B., 344
Ray, B.R., 501
Ray, B.W., 138
Ray, C.Jr., 100
Ray, C.G., 499
Ray, D.A., 364
Ray, D.L., 498
Ray, E.E., 292
Ray, J.D.Jr., 97
Ray, M., 538
Ray, M.L., 23
Ray, P.D., 353
Ray, P.H., 345
Ray, P.M., 41
Ray, W., 270
Ray, W.J., 150
Rayburn, J.C., 456
Rayburn, W.R., 501
Rayford, P., 467
Rayle, D.L., 39
Rayle, R.E., 360
Raymo, R., 312
Raymond, A.M., 21
Rayner, M.D., 113
Rea, J.C., 259
Read, J.K., 342
Read, K.B., 267
Read, K.R.H., 207
Read, P.E., 241
Read, V.H., 251
Read, W.O., 435
Reader, R.J., 546
Readon, R.M., 122
Reagan, C.R., 101
Reagan, J.O., 107, 382
Reagan, J.W., 356
Reagan, R.E., 250
Reams, D.C., 79
Reams, W.M., 488
Reaney, B., 239
Reardon, J.J., 215
Reaves, J.A., 7
Rebach, J.O., 199
Rebeiz, C.A., 138
Rebeiz, C.C., 138
Reber, L.A., 405
Rebers, P.A., 157
Rebhun, L.I., 488
Rebuck, A., 358
Rech, R.H., 227
Rechsteiner, M.C., 473
Reck, D.G., 62
Reckard, E.C., 174
Redd, B.L., 506
Redd, T.A., 506
Reddan, J.R., 230
Redden, D.R., 452
Redding, E., 163
Redding, R.W., 6
Redding, W., 223
Reddish, R.L., 91
Reddish, P.S., 335
Reddy, C.A., 228
Reddy, K.N.N., 370
Reddy, S.K., 398
Reddy, V.K., 9
Reddy, V.N., 230
Redei, G.P., 258
Refearn, P.L.Jr., 258
Redfield, A.G., 208
Redgate, E.S., 414
Redick, T.F., 196
Redington, C.B., 216
Redlich, A., 535
Redman, D.R., 367
Redmann, R.E., 562
Redmond, B.L., 323
Redmond, J.R., 157
Redmond, C.K., 414
Redmond, R.B., 39
Redwood, W.R., 282
Reece, R.P., 288
Reece, W.O., 158
Reeck, G.R., 166
Reed, D.E., 433, 434
Reed, D.J., 385
Reed, D.L., 365
Reed, G., 553
Reed, G.B.Jr., 356
Reed, G.M., 445
Reed, G.R., 410
Reed, H., 443
Reed, H.B., 439
Reed, J., 394, 514
Reed, J.K., 423
Reed, J.R., 288
Reed, J.R.Jr., 491
Reed, L.W., 376
Reed, M., 205
Reed, M.J., 319
Reed, N., 7
Reed, N.D., 265
Reed, R.B., 212
Reed, R.D., 407
Reed, R.E., 18
Reed, R.M., 376, 549
Reed, R.R., 365
Reed, S.A., 112
Reed, S.C., 242
Reed, T., 146
Reed, T.E., 550
Reed, W.E., 489
Reed, W.M., 108
Reeder, J.R., 526
Reeder, R., 197, 475
Reeder, W.G., 516
Reedy, J.J., 313
Reedy, M.K., 332
Reedy, W., 11
Rees, A.W., 469
Rees, C.P., 199
Rees, J.R., 142
Reesal, M., 556
Reesal, M.R., 555
Reese, C.R., 364
Reese, E.S., 112
Reese, R.W., 453
Reese, W.D., 191
Reese, W.H., 470
Reetz, G.R., 303
Reeve, M., 96
Reeve, R.E., 555
Reeves, A.F., 22
Reeves, B.E., 452
Reeves, D.L., 433
Reeves, F., 68
Reeves, G., 429
Reeves, H.C., 15
Reeves, J.B., 468
Reeves, J.D., 374
Reeves, J.H.Jr., 492
Reeves, J.J., 502
Reeves, M.M., 280
Reeves, O.R., 535
Reeves, R.B., 319
Reeves, R.E., 185
Reeves, W.C., 46
Regal, P.J., 242
Regan, G.T., 8
Regenstein, J.M., 305
Regier, H.A., 550
Reggio, R.B., 299
Reggion, R.B., 299
Register, U.D., 35
Regnery, D.C., 41
Regnier, F.E., 148
Rehm, G.W., 271
Rehman, I., 63
Rehwaldt, C., 239
Reice, S.R., 339
Reich, M., 84
Reich, V.H., 444
Reichart, C.V., 419
Reichelderfer, C.F., 201
Reichert, L.E.Jr., 101
Reichle, R.F., 31
Reichlin, M., 318
Reichmann, M.E., 137
Reid, -., 81
Reid, A., 323
Reid, A.F., 323
Reid, B.R., 57
Reid, D.B.W., 549
Reid, G.K., 87
Reid, H.B.Jr., 286'
Reid, J., 538
Reid, J.D., 356
Reid, J.R., 352
Reid, K.H., 181
Reid, M.R., 246
Reid, P.D., 215
Reid, R.D., 98
Reid, R.G.B., 536
Reid, R.O., 459
Reid, W.S., 302
Reid, W.W., 345
Reidenberg, M.M., 409
Reif, C.B., 416
Reif, J.S., 411
Reifsynder, W.F., 81
Reigel, G.T., 120
Reilly, H.C., 288
Reilly, J.R., 352
Reilly, K.D., 11, 12
Reilly, M., 266
Reilly, R.E., 224
Reilly, R.W., 323
Reimer, A.A., 403
Reimer, D., 37
Reimer, K.D., 113
Reimer, R.D., 459
Reimold, R.J., 106
Rein, R., 320
Reinecke, R., 418
Reineke, E.P., 228
Reiner, A.M., 218
Reiner, J.M., 327
Reiners, W., 277
Reinhart, B.S., 547
Reinert, J.A., 91
Reinert, R.A., 345
Reines, M., 108
Reinhardt, W.O., 60
Reinhart, R., 284
Reinhold, G.W., 155
Reinke, D.A., 227
Reinke, R.L., 435
Reinmuth, O.M., 95
Reisa, S.M., 125
Reischman, P.G., 497
Reisenauer, H.M., 50
Reiser, R., 458
Reising, J.W., 144
Reiskind, J., 90
Reisner, E., 333
Reisner, G.S., 390
Reiss, O.K., 72
Reiss, W.D., 147
Reissing, W.H., 306
Reist, P.C., 341
Reist, W., 361
Reitan, P.J., 159
Reiter, R.J., 469
Reitmeyer, J.C., 467
Reitz, H.J., 92
Reitz, R.C., 340
Relyea, K.G., 89
Rembert, D.H., 429
Remington, C.L., 80
Remington, J., 557
Remington, R., 233
Remmers, J.E., 278
Remondini, D.J., 228
Remp, M., 522
Rempel, E., 497
Remsen, J., 90
Remy, C.N., 347
Renaud, F.L., 418
Renaud, S., 559
Rencricca, N.J., 212
Rendig, V.V., 50
Rendon, O.R., 418
Renfro, W.C., 385
Renish, A.J., 522
Renlund, R.N., 285
Rennels, E.C., 469
Rennels, E.G., 469
Rennels, M.L., 203
Renner, E.R., 351
Rennert, O.M., 94
Renney, A.J., 534
Rennie, D.W., 319
Rennie, J.C., 443
Rennie, T., 119
Renwick, G.A., 147
Repak, A.J., 76
de Repentigny, J., 558
Resch, H., 384
Rescigno, A., 244
Reseick, J.B., 63
Reseonich, E., 406
Resh, D., 120
Resko, J.A., 379
Resnick, H., 509
Resnick, M.A., 329
Resnikoff, G., 30
Resseguie, L., 511
Ressler, C., 307
Ressler, N., 135
Rettenmeyer, C., 77
Rettenmeyer, C.W., 77, 79
Reuben, J.P., 301
Reuling, W.S., 479
Reuning, W., 407
Reuppel, L., 431
Reuss, J.O., 65
Reuss, R.M., 317
Reuter, F., 460
Reuter, L.A., 287
Reuter, Sr.M.A., 224
Reuther, W., 58
Reveal, J.L., 201
Revel, J.P., 25
Rever-DuWors, M., 562
Rex, M.A., 219
Rexroad, P., 261
Reyburn, N.J., 221
Reynelds, W.A., 134
Reyer, R.W., 509
Reyes, P., 294
Reynhout, J.K., 209
Reynolds, A., 73
Reynolds, A.E., 143
Reynolds, C.W., 202
Reynolds, D.A., 275
Reynolds, D.M., 51
Reynolds, D.R., 88
Reynolds, H.C., 164
Reynolds, H.K., 321
Reynolds, H.T., 58
Reynolds, I., 219
Reynolds, J., 491
Reynolds, J.A., 333
Reynolds, J.B., 261
Reynolds, J.H., 444
Reynolds, J.T., 209
Reynolds, L.B., 445
Reynolds, M., 411
Reynolds, N.B., 321
Reynolds, O.B., 229
Reynolds, R.S., 20
Reynolds, T.M., 6
Rezak, R., 459
Reznikoff, W.S., 517
Rha, C.K., 213
Rhea, M.B., 34
Rhee, C.H., 121
Rhees, R.W., 473

Rhein, R.R., 397
Rheins, M.S., 363
Rhines, R., 131
Rhoades, E.R., 380
Rhoades, H.L., 31
Rhoades, M., 197
Rhoades, M.M., 144
Rhoades, R., 354
Rhoads, H., 26
Rhoads, R.E., 179
Rhode, R.D., 456
Rhodes, D.G., 186
Rhodes, F.H.T., 230
Rhodes, H.D., 16, 19
Rhodes, J.B., 172
Rhodes, R., 359
Rhodes, R.R., 458
Rhodes, R.W., 492
Rhodes, S.A., 390
Rhodes, W.H., 411
Rhoten, W.B., 137
Rhoton, V.D., 416
Rhum, G.J., 162
Rhykerd, C.L., 147
Rhymer, I., 122
Ribbens, D.C., 154
Ribbons, D.W., 95
Ribelin, W.E., 520
Ribinson, N.E., 228
Riblet, D., 285
Riccardi, L., 132
Riccardi, V., 72
Rice, A., 311
Rice, E.L., 379
Rice, F.J., 75
Rice, J., 164
Rice, J.C., 344
Rice, J.T., 108
Rice, L., 116
Rice, L.V., 222
Rice, M., 439
Rice, M.M., 159
Rice, N.E., 488
Rice, P., 79
Rice, R.E., 49
Rice, R.W., 527
Rice, R.V., 391
Rice, S.A., 132
Rich, A., 280
Rich, C., 41
Rich, E., 95
Rich, M.A., 236
Rich, P., 77
Rich, R., 73
Rich, R.P., 197
Rich, R.R., 449
Richard, D.I., 89
Richard, J., 96
Richards, A.A., 483
Richards, A.G., 240, 242
Richards, A.H., 327
Richards, B.D., 122
Richards, C.D., 194
Richards, C.L., 176
Richards, E., 73
Richards, E.G., 466
Richards, E.L., 20
Richards, F.M., 81
Richards, J.F., 533, 534
Richards, M.L., 186'
Richards, N.R., 322
Richards, O.C., 474
Richards, R.E., 74
Richards, T.C., 388
Richards, T.L., 27
Richards, W., 96
Richards, W.Jr., 491
Richardson, A., 122
Richardson, A.P., 101
Richardson, A.W., 129
Richardson, B., 369
Richardson, C.H., 476
Richardson, D., 167
Richardson, D.C., 333
Richardson, D.G., 384
Richardson, D.H.S., 545
Richardson, D.O., 443
Richardson, D.R., 180
Richardson, E.A., 474
Richardson, F.C., 145
Richardson, G.H., 475
Richardson, H.R., 174
Richardson, J., 522
Richardson, J.B., 556
Richardson, J.H., 487
Richardson, J.L., 395, 522
Richardson, J.W., 128
Richardson, L., 375
Richardson, N., 71
Richardson, O.L.Jr., 175
Richardson, P.E., 375
Richardson, R.H., 465
Richardson, R.L., 161
Richardson, S., 385
Richardson, S.H., 347
Richardson, T., 437, 518
Richardson, V.B., 8
Richardson, W.L., 376
leRiche, W.H., 549
Richens, V.B., 194
Richer, C.L., 558
Richer, G., 558
Richerson, P.J., 47, 49
Riches, R.H., 165
Richess, D.R., 264
Richey, J., 498
Richins, C.A., 256
Richins, R., 477
Richkus, W., 416
Richmond, J., 45
Richmond, M., 439
Richmond, M.E., 303
Richmond, R., 145
Richmond, W.H., 64
Richter, K.M., 379
Richter, R.E., 32
Richter, R.L., 91
Richter, W.R., 133
Riciputi, R.H., 195
Rick, C.M., 51
Rickard, C.G., 306
Rickborn, B., 61
Rickenberg, H.V., 72
Rickert, D.E., 227
Rickert, R.K., 415
Rickett, J.R., 25
Ricketts, G.E., 138
Ricketts, R.E., 259
Ricklefs, R.E., 409
Ricks, B.L., 188
Rickson, F.R., 385
Rico-Ballester, M., 417
Riddell, C., 563
Riddell, R.T., 546
Riddick, C., 348
Riddiford, L.M., 498
Riddle, C.C., 472
Rider, W.D., 550
Ridgel, G.C., 175
Ridgeway, B.T., 120
Ridgway, E.B., 492
Ridgway, J.E., 262
Ridgway, P.N., 438
Riding, R.T., 540
Ridlen, S.F., 138
Ridley, B.L., 440
Ridley, J.D., 424
Riechert, S.E., 442
Rieck, C.E., 177
Rieck, N., 224
Riedel, R.M., 366
Riedesel, D.H., 158
Riedesel, M.L., 293
Riegel, G.T., 120
Rieger, R.M., 339
Riegert, P.W., 561
Riegle, G.D., 225, 228
Riehl, L.A., 58
Riehm, J.P., 98
Riewe, R.R., 539
Rieke, W.D., 172
Riemenschneider, V., 147
Rierson, H.A., 76
Ries, S.K., 226
Ries, S.M., 139
Riesen, J.W., 77
von Riesen, V.L., 273
Rieser, P., 207
Rietsma, C., 323
Riffel, F., 173
Riffey, M.M., 504
Riffle, G., 270
Riffle, J.W., 272
Rifkin, J., 299
Rifkind, R.A., 301
Rigamonti, D., 85
Rigas, D.A., 386, 388
Rigby, D.W., 500
Rigby, P., 273
Rigdon, R., 467
Riggan, W.B., 341
Riggins-Pimentel, R., 27
Riggio, R.R., 307
Riggs, A.F., 465
Riggs, D.L., 162
Riggs, J.K., 458
Riggs, M.P., 439
Riggs, R.A., 338
Riggs, R.D., 24
Riggs, T.R., 232
Riggsby, W.S., 442
Righthand, V.F., 236
Rights, F.L., 234, 236
Rigler, F.H., 550
Rigney, M., 523
Rihan, T., 250
Rikel, J.R., 64
Rikuris, E., 83
Rilett, R.O., 122
Riley, C., 359
Riley, C.V., 359
Riley, D., 60
Riley, E.F., 162
Riley, G.A., 542
Riley, J.G., 167
Riley, M., 527
Riley, M.V., 230
Riley, M.W., 104
Riley, P.H., 279
Riley, R., 171
Riley, R.K., 196
Rilling, H.C., 474
Rinard, G.A., 101
Rinaudo, P., 79
Rinch, R.A., 347
Rinehart, F.A., 379
Rinehart, R.R., 39
Rines, H., 105
Ring, B.A., 482
Ring, R.A., 536
Ringen, L., 504
Ringenberg, L.A., 120
Ringer, R.K., 226, 228
Ringkob, T.P., 275
Rings, R.W., 363
Ringwald, D.W., 376
Rini, J., 206
Rink, R.D., 180
Rink, S., 125
Rinker, G.C., 434
Rinne, R.W., 138
Rinson, G.A., 549
del Río, F., 417
Rio, G., 316
Riopel, J.L., 488
de Rios, N.E., 417
Ripley, E.A., 562
Ripley, T.H., 493
Rippe, D., 96
Ripperton, L.A., 341
Ris, H., 516
Riser, R.W., 214
Riser, W.H., 412
Risher, C.F., 425
Rising, J., 550
Risinger, G.E., 184
Risius, M.L., 401
Riss, W., 325
Risser, A.C., 276
Risser, P.G., 379
Ristau, A.E., 162
Ristow, B., 323
Ristuben, P., 26
Ritcher, P.O., 386
Ritchey, S.J., 494
Ritchie, B.B., 76
Ritchie, D.D., 300
Ritchie, H.D., 225
Ritenour, G., 29
Ritland, R.M., 222
Rittenberg, M.B., 388
Rittenberg, S.C., 55
Rittenhouse, L.R., 382
Rittenhouse, R.L., 487
Ritter, E., 323
Ritter, H.T.M.Jr., 106
Ritter, R.C., 187
Ritter, R.D., 359
Ritterson, A.L., 328
Ritschel, G., 176
Rivard, C.E., 200
Rivas, M.L., 288
de Rivera, B.C., 418
Rivera, E.M., 227
de Rivera, W.T., 418
Rivero, J., 417
Rivers, I.L., 276
Rivers, J.M., 306
Rivers, N.P., 442
Rivers, W.G., 486
Rizki, T.M., 231
Riznyk, R.Z., 29
Rizzo, D.C., 224
Roach, A.W., 452
Roach, E., 146
Roach, F.C., 75
Roach, M.R., 551
Roaden, A.L., 362
Roantree, R.J., 42
Roark, D.E., 405
Roark, J.L., 270
Robbins, A.J., 289
Robbins, C.E., 123
Robbins, D., 254
Robbins, E., 329
Robbins, E.B., 330
Robbins, E.L., 356
Robbins, E.S., 312
Robbins, F.C., 355
Robbins, L.G., 227
Robbins, M.L., 424
Robbins, N., 355
Robbins, W.C., 194
Robblee, A.R., 530
Robel, R.J., 166
Roberson, W.B., 20
Robert, A., 236
Robert, J., 558
Roberte, J., 558
Roberts, A., 549
Roberts, A.N., 384, 445
Roberts, B.R., 369
Roberts, C.R., 178
Roberts, D., 410
Roberts, D.A., 92
Roberts, D.R., 142
Roberts, E., 64, 432
Roberts, E.C., 422
Roberts, E.G., 250
Roberts, E.P., 450
Roberts, F.L., 193
Roberts, G., 288
Roberts, H.R., 340
Roberts, J., 268, 399
Roberts, J.H., 83, 185
Roberts, J.I., 302
Roberts, J.J., 147
Roberts, J.L., 218
Roberts, J.R., 346
Roberts, K., 558
Roberts, L.S., 218
Roberts, L.W., 117
Roberts, P.A., 380, 386
Roberts, R.H., 249
Roberts, R.M., 90
Roberts, R.S., 476
Roberts, S., 56
Roberts, T.L., 221
Roberts, T.R., 211
Roberts, W., 72
Roberts, W.H.B., 35
Roberts, W.M., 344
Robertson, A.D.J., 132
Robertson, A.L.Jr., 356
Robertson, B.S., 202
Robertson, B.T., 6
Robertson, C.A., 59
Robertson, C.H., 337
Robertson, D., 243, 325
Robertson, D.C., 171

Robertson, G.G.,445
Robertson, G.L., 183
Robertson, H.A., 544
Robertson, H.J., 199
Robertson, J., 507
Robertson, J.A., 358
Robertson, J.D., 332
Robertson, J.H., 275
Robertson, J.S., 89
Robertson, P.A., 68
Robertson, P.B., 449
Robertson, R.G., 328
Robertson, R.J., 546
Robertson, R.L., 344
Robertson, W.K., 93
Robertson, W.V., 455
Robertstad, G.W., 468
Robey, R.C., 288
Robillard, D., 79
Robin, J., 559
Robinette, H.R., 256
Robinow, C.F., 551
Robins, C., 96
Robins, C.R., 96
Robins, J.D., 169
Robins, J.S., 502
Robins, M.L., 84
Robins, R.C., 92
Robinson, A., 72
Robinson, A.D., 324
Robinson, A.G., 537
Robinson, C., 395
Robinson, C.W., 65
Robinson, D.A., 197
Robinson, D.C., 99
Robinson, D.M., 85
Robinson, D.W., 48, 377
Robinson, E.J.Jr., 238
Robinson, F.A., 91
Robinson, G., 97
Robinson, G.C., 121
Robinson, G.A., 548
Robinson, G.G.C., 538
Robinson, G.N., 458
Robinson, G.W., 179
Robinson, H.L., 277
Robinson, H.W., 32
Robinson, J., 135, 447, 551
Robinson, J.B., 547
Robinson, J.L., 392
Robinson, J.R., 432
Robinson, J.T., 516
Robinson, K.R., 331
Robinson, M.C., 105
Robinson, N., 89
Robinson, O.W., 343
Robinson, R., 189, 335
Robinson, R.G.Jr., 127, 240
Robinson, R.H., 292
Robinson, R.R., 386
Robinson, R.W., 193, 307
Robinson, T., 217
Robinson, T.S., 180
Robinson, W.B., 307
Robinson, W.L., 229
Robison, H.W., 21
Robitaille, H.A., 149
Robitaille, Y., 556
Robocker, W.C., 502
Robson, D.S., 302, 304
Robson, L.C., 499
Robson, R.M., 157
Robyt, J.F., 157
Rocha, V.M., 61
Roche, B.F., 503
Roche, E.T., 29
Roche, M., 312
Roche, T.E., 166
Rochman, H., 133
Rock, B.N., 295
Rock, G.C., 344
Rock, K.C., 72
Rock, M., 196
Rock, W.A., 186
Rockel, E., 290
Rockett, C.L., 354
Rockette, H.E., 414
Rockrath, R.C., 146
Rockstein, M., 96
Rockwell, K.H., 397
Rockwood, C.L., 485
Rockwood, S.W., 360
Rockwood, W.P., 315
Roddy, L.R., 189
Rode, L.J., 465
Roden, L., 11
Rodewald, R.D., 488
Rodgers, C.L., 427
Rodgers, O.A., 246
Rodgers, T.A., 3
Rodier, P.M., 489
Rodin, A., 467
Rodin, M., 235
Rodin, R.J., 27
Rodkey, L.S., 166
Rodman, J.E., 80
Rodney, D.R., 17
Rodney, J.M., 416
Rodney, P., 549
Rodney, R., 265
Rodrigues-Kabana, R., 4
Rodriguez, E., 89
Rodriguez, J.E., 161
Rodriguez, J.G., 178
Rodriguez, K., 418
Rodriguez, L., 417
Rodriguez, L.D., 418
de Rodriguez, O.B., 418
Rodriguez-Lopez, B., 417
Rodriguez-Peralta, L.G., 408
Rodriguez de Perez, N., 418
Rodwell, V.W., 148
Roe, D.A., 306
Roe, P., 28
Roe, R.Jr., 466
Roe, W.E., 563
Roecker, R.M., 323
Roeder, K.D., 216
Roeder, M., 88
Roeder, R.G., 265
Roelofs, E.W., 225
Roelofs, T., 34
Roelofs, W.L., 306
Roels, O.A., 297, 298
Roemhild, G., 265
Roeper, R.A., 222
Roesel, C., 103
Roeske, R.W., 146
Roess, W.B., 87
Roessler, M.A., 96
Roessmann, U., 356
Roest, A.E., 27
Roeth, F.W., 148
Roff, J.C., 547
Roff, W.J., 546
Roger, J.D., 445
Rogers, B.L., 202
Rogers, C.M., 234
Rogers, D.E., 498
Rogers, D.J., 71, 431
Rogers, D.P., 136
Rogers, D.T., 10
Rogers, E., 154
Rogers, E.A., 66
Rogers, F.A., 155
Rogers, H.J., 342
Rogers, J.D., 503
Rogers, J.S., 185
Rogers, K.S., 491
Rogers, M.A., 174
Rogers, M.N., 260
Rogers, O.M., 281
Rogers, P., 241, 244
Rogers, Q.R., 53
Rogers, R., 157, 182
Rogers, R.A., 155
Rogers, R.W., 135
Rogers, S.H., 381
Rogers, T.A., 112
Rogers, W., 95
Rogers, W.A., 4
Rogers, W.E., 407
Rogers, W.J., 150
Rogerson, C., 297, 298
Rogolsky, M., 474
Rogowski, A.S., 401
Rohde, C.A., 198
Rohde, C.J., 126
Rohde, R.A., 219
Roheim, P.S., 330
Rohlfing, D.L., 429
Rohrbach, K.G., 114
Rohringer, R., 538
Rohrs, H.C., 284
Rohweder, D.A., 516
Rohwer, S., 498
Roia, F.C., 405
Roias, E., 61
Roizman, B., 132
Rojo, A., 543
Rojo, E., 543
Roland, C.R., 106
Roland, D.A., 93
Rolfe, G.L., 139
Rolfson, G., 239
Roll, P., 513
Rollason, G.S., 218
Rollason, H.D., 218
Roller, H.A., 458, 460
Roller, M.H., 433
Rollins, G.H., 3
Rollins, H.A.Jr., 365
Rollins, J.A., 192
Rollins, M., 7
Rollins, R., 7
Rollins, R.C., 210, 211
Rollins, W.C., 48
Rolston, D.E., 50
Rom, R.C., 23
Romach, E.H., 216
Romanko, R.R., 117
Romano, A.H., 77, 79
Romanoff, A.L., 305
Romanowski, R.R., 149
Romans, J.R., 138
Romans, R.C., 354
Romans, W.R., 466
Romeo, J.T., 230
Romesburg, C., 475
Romeyn, J.A., 538
Romig, W.R., 55
Romine, M., 291
Roming, R.F., 415
Rominger, J., 15
Rommann, L.M., 376
Rommel, F.A., 470
Romoser, W.S., 368
Rona, A., 556
Rona, G., 555
Ronald, A., 538
Ronald, K., 546, 547
Ronan, S.G., 135
Roncadori, R.W., 108
Rongone, E.L., 268
Rongstad, O.J., 520
Ronka, E.K.F., 208
Ronnenkamp, R.E., 147
Ronning, M., 48
Ronzio, R.A., 225
Rooney, L.W., 459
Rook, O.W., 21
Roon, R.J., 244
Rooney, D., 256
Rooney, F.V., 255
Rooney, J.R., 412
Rooney, S., 192
Roop, T., 427
Roos, A., 263
Roos, E.E., 66
Roos, H., 75
Roos, T.B., 277
Roosen-Runge, E.C., 499
Root, R.B., 303
Roots, B.I., 550
Roovos, M., 280
Rorabaugh, J., 400
Rornberg, A., 41
Rosamilia, H.G., 319
Rosan, B., 411
Rosario, C., 35
Rosario, J.T., 417
Rose, B., 96
Rose, D., 290
Rose, D.W., 156, 243
Rose, E.F., 161
Rose, F.L., 116, 461
Rose, G.W., 233
Rose, I., 410
Rose, J.C., 85
Rose, J.D., 101
Rose, N.R., 236
Rose, R., 176
Rose, R.C., 405
Rose, S.M., 190
Rose, T.L., 51
Roseanne, Sr.M., 396
Rosebery, D.A., 254
Roselle, R.E., 271
Roseman, S., 197
Rosen, B.P., 203
Rosen, D.E., 297
Rosen, E., 309
Rosen, F., 320
Rosen, G., 333
Rosen, H., 31
Rosen, J., 288
Rosen, J.M., 449
Rosen, L., 300
Rosen, M., 95, 356
Rosen, R., 128, 130
Rosen, S., 363
Rosen, W., 219
Rosenau, W., 60
Rosenau, W.A., 219
Rosenbaum, J.L., 80
Rosenberg, A., 249, 405
Rosenberg, B., 75
Rosenberg, D., 288
Rosenberg, F.A., 214
Rosenberg, E.K., 285
Rosenberg, J.L., 412
Rosenberg, L.E., 52
Rosenberg, L.L., 43
Rosenberg, L.T., 42
Rosenberg, M., 240, 355
Rosenberg, M.D., 242
Rosenberg, M.J., 30, 355
Rosenberg, M.M., 286
Rosenberg, N.J., 272
Rosenberger, W.S., 155
Rosenberry, T., 300
Rosenblatt, J., 355
Rosenbloom, J., 411
Rosenblum, E., 466
Rosenbluth, J., 313
Rosengren, J.H., 290
Rosenkranz, E.E., 249
Rosenkranz, H., 301
Rosenmann, M., 14
Rosenquist, A.C., 410
Rosenquist, B.D., 262
Rosenquist, T.H., 103
Rosenstein, L., 95
Rosenstein, R., 277
Rosenstiel, -., 386
Rosenthal, A.F., 298
Rosenthal, A.F., 418
Rosenthal, G.A., 177
Rosenthal, J., 286
Rosenthal, M., 333
Rosenwinlsel, E., 387
Rosenzweig, M., 408
Rosenzweig, M.L., 293
Roshal, J.Y., 245
Rosing, L.M., 12
Rosinski, J., 220
Rosinski, M.A., 142
Roskey, C.T., 210
Roskowski, R.Jr., 161
Roslien, D.J., 159
Rosohke, S., 301
Rosovsky, H., 210
Rosowski, J., 270
Ross, A., 198
Ross, B.B., 389
Ross, C.H., 185
Ross, D.J., 75
Ross, D.M., 530, 531
Ross, C.V., 259
Ross, C.W., 68
Ross, E., 113
Ross, G., 57
Ross, G.L., 138
Ross, G.N., 189
Ross, H., 106
Ross, I.J., 179
Ross, I.K., 61
Ross, J.E., 259
Ross, J.A., 504
Ross, J.F., 216

Ross, J.G., 433
Ross, J.P., 345
Ross, J.R.P., 504
Ross, L., 151
Ross, L.L., 399
Ross, L.M., 227
Ross, M.A., 148
Ross, M.L., 468
Ross, Q., 152
Ross, R., 117 , 452, 544
Ross, R.D., 250, 492
Ross, R.F., 158
Ross, R.T., 362
Ross, S.T., 252
Ross, W.M., 271
Rossario, E.A., 417
Rossbach, G.B., 510
Rosse, C., 499
Rosse, W.F., 333
Rossen, R.D., 449
Rosser, R., 222
Rossi, C.R., 6
Rossi, E.P., 356
Rossi, G.V., 405
Rossier, R., 549
Rossiter, R.J., 551
Rossman, D.A., 183
Rossmann, M.G., 150
Rossmiller, J.D., 372
Rossmoore, H.W., 234
Rosso, S.W., 252
van Rossu, G.D.V., 409
Rost, M.C., 435
Roszman, T.R., 179
Rotenberg, S.L., 393
Rotermund, A.J., 123
deRoth, G.C., 358
Roth, A.A., 35
Roth, C.W., 386
Roth, C.B., 147
Roth, E., 169
Roth, F.J.Jr., 96
Roth, J.A., 258
Roth, J.N., 143
Roth, J.R., 43
Roth, J.S., 77
Roth, L.E., 166
Roth, L.F., 385
Roth, L.J., 133
Roth, O.H., 406
Roth, P.L., 129
Roth, R.E., 366, 368
Roth, R.M., 121
Roth, R.P., 26
Roth, R.R., 82, 552
Roth, S., 197
Roth, T.F., 205
Rothchild, I., 356
Rothe, C., 146
Rotheim, M.B., 326
Rothenbacher, N., 403
Rothenberg, H., 135
Rothenberger, R.R., 260
Rothenbuhler, W.C., 363, 364
Rothfield, L.I., 94
Rothman, A.H., 30
Rothman, S.S., 60
Rothman-Denes, L.B., 132
Rothschild, H., 185
Rothstein, A., 550
Rothstein, H., 480
Rothstein, J., 362
Rothstein, M., 318
Rothstein, S., 61
Rothwell, D.F., 93
Rothwell, N., 310
Rottman, F.M., 225
Roubicek, C.B., 19
Rouf, M.A., 523
Roufa, D.J., 448
Rough, G.E., 295
Roughgarden, J., 41
Rougvie, M.A., 157
Rouke, A.W., 117
Rounds, D.E., 63
Rounds, H., 173
Rounds, H.D., 173
Rounsefell, G.A., 10
Rourke, R.V., 195
Rousch, J.P., 295
Rouse, B.T., 563
Rouse, G.E., 534
Rouse, H., 283
Rouse, R., 286
Rouse, R.D., 3
Rouse, T.C., 514
Rousek, E., 29
Roush, A.H., 121
Rouslin, W., 288
Rousseau, J.E., 78, 79
Roussell, N., 182
Routh, J.I., 161
Routley, D.G., 280, 281
Routt, R.L., 44
Routtenberg, A., 126
Roux, K., 190
Roux, S.J., 412
Rovainen, C.M., 263
Rover, I., 205
Rovie, L.A., 417
Rovner, J.S., 368
Rovozzo, G., 75
Rovsell, P.G., 543
Rowan, J.M., 315
Rowbotham, N.D., 20
Rownd, R., 517
Rowden, D.W., 468
Rowden, G., 556
Rowe, E.C., 164
Rowe, J.S., 562
Rowe, L.D., 460
Rowe, W.B., 307
Rowell, C.H.F., 43
Rowell, C.M.Jr., 448
Rowell, L.B., 500
Rowell, T.E., 43
Rowland, E., 12
Rowland, W.J., 145
Rowlands, J.R., 469
Rowlatt, U.F., 135
Rowles, C.A., 534
Rowley, D.A., 133
Rowley, J., 132
Rowley, J.W., 507
Rowley, P.T., 328
Rowley, W., 157
Roy, A.K., 230
Roy, R.M., 557
Royall, R.M., 198
Roy-Burman, P., 64
Royden, H., 41
Roye, D.B., 347
Royer, G.P., 362
Royer, L., 475
Roylance, H.B., 117
Royse, N.D., 20
Royster, W.C., 176
Roxby, R., 408
Rozanis, J., 551
Roze, J., 297
Roze, U., 299
Rozee, K.R., 543
Rozen, J., 297
Rozen, J.G., 297
Ruach, S.E., 487
Ruark, A., 329
Ruben, L.N., 387
Rubenstein, J., 242
Rubenstein, N.M., 220
Ruber, E., 214
Rubin, A.L., 307
Rubin, D.C., 357
Rubin, H., 43
Rubin, L.F., 411
Rubin, M., 85
Rubin, R., 69, 89
Rubin, W., 399
Rubin, S.I., 307
Rubins, E.J., 78, 79
Rubinson, K., 312
Rubinstein, B., 217
Rubinstein, D., 555
Rubinstein, E.H., 57
Rubio, R., 490
Rubio, V., 283
Rubis, D.D., 16
Ruby, J.R., 185
Ruby, S.M., 557
Ruch, R.S., 550
Ruch, T.C., 500
Ruchkin, D.S., 204
Rucinsky, T.E., 404
Rucker, E.S., 439
Rucker, R.B., 50
Rucker, R.H., 459
Rucker, R.R., 498
Rucknagel, D.L., 232
Rudbach, J.A., 267
Rudd, R.L., 47, 52
Ruddat, M., 132
Ruddle, -., 81
Ruddle, F.H., 80
Rudenberg, F.H., 468
Rudenstine, N.L., 286
Rudick, M., 463
Rudick, V., 452
Rudinsky, J.A., 386
Rudko, A., 300
Rudman, D., 101
Rudner, M.J., 80
Rudner, R., 298
Rudney, H., 370
Rudolph, A.M., 60
Rudolph, E.D., 362
Rudolph, J., 175
Rudolph, L.C., 299
Rudy, P.P., 387
Rueckert, R.R., 517
Ruedisueli, Sr.A., 224
Ruegamer, W.R., 273
Ruehr, T.A., 65
Ruesink, A.W., 144
Ruf, R.H., 276
Ruff, M.D., 108
Ruff, R.L., 520
Ruffa, I., 337
Ruffin, J., 99
Ruffin, R.S., 491
Ruffolo, J.Jr., 491
Rufh, C., 373
Ruh, F.M., 257
Ruibal, R., 57
Ruiz, J.C., 417
Rule, A.H., 207
Ruliffson, W.S., 166
Rulon, R.R., 159
Rumbaugh, M.D., 433
Rumble, T.C., 234
Rumburg, C.B., 65
Rumely, J., 265
Rumery, R.E., 499
Rumph, P.F., 6
Rumpp, W.R., 339
Runbinstein, A., 330
Rundel, P.W., 54
Rundell, H.L., 159
Runeckles, V.C., 534
Runey, G.L., 423
Runey, M.W., 426
Runge, R.R., 123
Rungle, E.C.A., 258
Runk, B.F.D., 488
Runkles, J.R., 459
Runnalls, N., 524
Runnels, W.C., 451
Runner, M.N., 71
Runyan, M., 427
Ruoff, W.D., 506
Rupp, -., 81
Ruppel, R.F., 226
Rusch, D.H., 520
Rush, H., 15
Rush, J.E., 12
Rushforth, N., 355
Rushforth, S.R., 472
Rushmer, R.F., 497
Rushton, P.S., 438
Rushton, W.F., 20
Russ, O.G., 166
Russell, A., 246
Russell, A.P., 531
Russell, B.D., 221
Russell, C., 298
Russell, C.C., 375
Russell, C.E., 143
Russell, D., 251
Russell, D.W., 542
Russell, F.E., 64
Russell, G., 135
Russell, G.F., 49
Russell, G.V., 467
Russell, H.T.Jr., 455
Russel, J., 3
Russell, J.M., 468
Russell, J.W., 103
Russell, K., 66
Russell, K.R., 157
Russell, L.D., 555
Russell, L.H., 460
Russell, L.M., 344
Russell, M.A., 535
Russell, M.B., 138
Russell, M.P., 31
Russell, M.W., 181
Russell, N., 374
Russell, P.J., 59, 387
Russell, P.T., 370
Russell, R.A., 186
Russell, R.L., 25
Russell, S., 542
Russell, S.M., 18
Russell, T.,
Russell, T.E., 18
Russell, W.A., 156
Russell-Hunter, W.D., 326
Russelmann, H.B., 233
Rust, J., 133
Rust, R.H., 241
Rust, R.W., 82
Rustad, R., 355
Rustioni, A., 340
Rusy, B.F., 409
Rutger, J.N., 47
Ruth, E.A., 197
Ruth, G.R., 434
Ruth, R.F., 531
Rutherford, C.L., 492
Rutherford, J.G., 542
Rutherford, P., 236
Rutland, P.M., 382
Rutledge, E.M., 22
Rutledge, J.A., 126
Rutledge, J.J., 480
Rutledge, L.T., 232
Rutley, M.S., 324
Rutschky, C.W., 402
Rutter, W.J., 60
Ruttle, J.L., 292
Ruzecki, E.P., 490
Ryals, G.L.Jr., 335
Ryan, C.B., 458
Ryan, E.P., 334
Ryan, M.T., 548
Ryan, R., 358
Ryan, R.A., 309
Ryan, S.O., 157
Ryan, T.I., 215
Ryan, W., 382
Ryan, W.L., 273
Ryckman, R.E., 36
Ryder, C.A., 323
Ryder, G.J., 364
Ryder, J., 544
Ryder, R.A., 66
Rydl, G.M., 454
Ryen, K., 177
Ryerson, D.E., 266
Rylander, M.K., 461
Rymal, K.S., 5
Rymon, L.M., 394
Rynbrandt, D., 355
Rypke, G., 293
Ryser, F., 276

S

Saab, P.M., 507
Saacke, R.G., 493
Saavada, E., 417
Sabagh, G., 341
Sabatier, P.A., 47, 49
Sabatini, D.D., 312
Sabbe, W.E., 22
Saben, H., 533
Sabey, B.R., 65
Sabharwal, P.S., 177
Sabin, A., 428
Sabin, G.E., 425
Sable, H.Z., 355

Sable, W.D., 524
Sacalis, J.N., 288
Sacamand, C.M., 17
Saccoman, F.M., 236
Sacher, J.A., 31
Sachs, J., 81
Sachs, M.B., 197
Sack, C., 315
Sack, R.B., 198
Sack, W.O., 305
Sackett, R.G., 16
Sackett, W., 459
Sackner, M., 95, 96
Sacks, M., 297
Sacks, M.L., 215
Sadag, P., 327
Sadek, F., 348
Sadleir, R.M.F.S., 533
Sadler, J.R., 72
Sadoff, H.L., 228
Safarian, A.E., 549
Safwat, F.M., 219
Sagar, R.G., 390
Sagawa, K., 197
Sagawa, Y., 114
Sage, H.J., 333
Sage, M., 262
Sager, D., 514, 515
Sager, P., 514
Sager, P.E., 514
Sager, R., 297, 298
Sagik, B.P., 470
Saha, A., 282
Said, S.I., 467
Saidel, G., 355
Saidel, G.M., 356
Saier, M., 59
Saif, Y.M., 367
Saigo, R.H., 514
Saila, S.B., 421
Saini, P.K., 372
Saini, R.S., 8
St. Amand, W., 251
Saint-Clair, P.M., 558
St. Damian, Sr.M., 361
de St. Jeor, S.C., 405
St. John, F.L., 364
St. John, P.A., 143
St. John, W.M., 25
St. Lawrence, P., 45
Saint-Martin, M., 558
St. Omer, V.V., 262
St. Omer, V.V.E., 168
St. Pierre, R.L., 363
St. Pierre, R.S., 367
Sainte-Marie, G., 558
Saison, R., 548
Sajdak, R.L., 229
Sak, M.F., 215
Sakami, W., 355
Saks, N.M., 297
Saksena, V., 361
Sakuras, D., 128
Salac, S.S., 272
Saladino, C., 309
Salamun, P.J., 521
Salch, R.K., 394
Salde, N.A., 172
Saletijn, L.M., 300
Saleuddin, A.S.M., 553
Salganicoff, L., 409
Salisbury, F.B., 477
Sallach, H.J., 520
Saller, M.W., 512
Salley, J.J., 204
Sallman, B., 96
Salmon, M., 136
Salmons, G.B., 279
Salo, E.O., 498
Salo, T.P., 441
Salpeter, M., 302, 304
Salser, W., 55
Salt, G.W., 47, 52
Saltarelli, C., 319
Salter, D., 479
Salter, E.G.Jr., 97
Salter, F.C., 251
Salter, L.S., 123
Salthe, S.N., 287, 296, 298
Saltman, P.D., 59
Salton, M.R.J., 313
Salunkhe, D.K., 475
Salvin, S.B., 413
Salzano, J., 333
Samborski, D., 538
Sambury, S., 282
Sames, G.L., 216
Samet, P., 95
Samli, M.H., 493
Samoiloff, M., 538
Sampath, A.C., 301
Sampert, H.C., 44
Samson, F.E., 171, 172
Sampson, H., 253
Sampson, H.W., 449
Sampson, S.R., 60
Samuel, A., 488
Samuel, C., 61
Samuel, D.E., 508
Samuel, G., 392
Samuel, W.M., 531
Samuels, L.T., 474
Samuels, R., 145
San Clemente, C.L., 228
San Pietro, A., 144
Sanborn, E.B., 558
Sanborn, R.C., 145
Sanchez, E., 418
Sanchez, G., 291
de Sánchez, S., 418
de Sánchez, S.S., 418
Sand, R.S., 91
Sandal, P.C., 350
Sandberg, A., 320
Sandberg, D.V., 66
Sandeman, I.M., 546
Sander, C., 502
Sander, D.A., 276
Sander, D.H., 271
Sander, E.G., 90
Sander, F., 556
Sanderlin, K. E., 186
Sanders, A.E., 469
Sanders, A.P., 333
Sanders, B.E., 318
Sanders, B.G., 465
Sanders, C.C., 269
Sanders, D.C., 345
Sanders, D.F., 148
Sanders, D.R., 188
Sanders, E., 228
Sanders, G.P., 39
Sanders, R. B., 170
Sanders, R.J., 476
Sanders, T.G., 287
Sanders, W.E.Jr. 269
Sanderson, H. R., 508
Sanderson, K. C., 5
Sanderson, K.E., 531
Sanford, F.R., 154
Sandford, J.O., 248
Sandhu, R.S., 283
Sandifer, C.K., 21
Sando, R.W., 243
Sandox, M., 233
Sands, F. B., 305
Sands, S., 343
Sandsted, R.F., 305
Sandstrom, C., 89
Sanfiorenzo, J.H., 417
Sanford, B.A., 470
Sanford, G., 220
Sanford, L. G., 7,365
Sanford, W. C.,376
Sanger, J., 410
Sanger, J. E., 369
Sanik, J., 223
Sanjur, D., 306
Sanland, E. K., 467
Sanschagrin, Sr. H. J., 236
Santalo, M.J., 550
Santamarina, E., 288
Santelmann, P. W., 376
Santer, E. A., 117
Santer, M., 396
Santi, D.V., 60
Santilli, V., 318
Santisteban, G.A., 497
Santlucito, J., 95
Santolucito, J.A., 343
Santonas, W.N., 508
Santoro, T., 323
Santos, C. G., 86
Santos, M., 299
Sanwal, B.D., 551
Sapp, J.P., 552
Sapp, W.J., 8
Sappenfield, W.P., 258
Saracco, C.G., 413
Sarachek, A.,173
Sarcione, E., 320
Sargent, F.O., 481
Sargent, M.L., 137
Sargent, T.D., 218
Sargent, W.Q., 445
Sarkar, P.K., 121
Sarkissian, I.V., 460
Sarko, A., 322
Sarmiento, A., 95
Sarot, D., 299
Sartain, J.B., 93
Sartor, G.F., 249
Sartori, L., 206
Sarty, D., 542
Sasarman, A., 558
Sasava, D., 33
Sasin, R., 400
Saslaw, S., 363, 367
Sass, B., 193
Sassaman, J.F., 192
Sasse, C.E., 462
Sassenrath, E., 53
Sasser, G., 117
Sasser, J.N., 345
Sasseville, N., 215
Sata, H., 409
Satchell, D. J., 130
Satchell, D.P., 129
Sather, J. H., 139
Satia, B., 498
Satir, P., 43
Sato, G., 59
Sattar, S.A., 549
Satter, L.D., 517
Satterberg, J. A., 550
Satterfield, J., 224
Satterland, D.R., 503
Sattler, R., 556
Saucier, R., 559
Saudine, W.E., 386
Sauer, D.B., 168
Sauer, D.L., 437
Sauer, J.C., 437
Sauer, J.R., 377
Sauer, R.J., 226
Sauerbier, W., 72
Saul, G.B, 479
Saul, W.E., 116
Saunders, J.H., 246
Saunders, J.K., 431
Saunders, J. L., 303
Saunders, J.P., 467
Saunders, J.R., 563
Saunders, L., 306
Saunders, P.R., 64
Saunders, R., 258
Saupe, D.C., 138
de Saurez, V.T., 418
Saurino, V.R., 87
Sauro, A., 513
Sauro, J., 215
Sausing, M.G., 165
Savage, C.R., 327
Savage, D. C., 137
Savage, E.J., 147
Savage, E.P., 69
Savage, J.M., 62
Savage, N.L., 452
Savage, R. E., 407
Savage, W., 32
Savan, M., 548
Savelle, G.D., 382
Saver, P., 162
Savery, H., 253
Savery, H.P., 253
Savitz, J., 123
Savoldi, M.T., 28
Savory, J., 340
Savos, M.G., 78,79
Sawhill, D., 399
Sawits, M.S., 286
Sawson, S.V., 212
Sawtell, J. J., 506
Sawver, M., 285
Sawyer, G.P., Jr., 426
Sawyer, J.O., 34
Sawyer, R., 48
Sawyer, R.L., 345
Sawyer, W.D., 146
Saxena, K.M.S., 182
Sayed, H. I., 538
Sayers, E.R., 10
Sayles, J.H., 351
Saylor, L.C., 346
Saylor, R., 418
Sayner, D.B., 18
Sayre, J.D., 364
Saz, H.J., 152
Scagel, R.F., 534
Scalet, C.G., 434
Scaletti, J.V., 294
Scalia, F.R., 325
Scallen, T.J., 294
Scanlon, P.F., 494
Scanlan, R. A., 383
Scannell, R. J., 303
Scanu, A.M., 131
Scapino, R. P., 134
Scarborough, A., 14
Scarborough, G.A., 72
Scarpa, A., 410
Scarpelli, E.M., 33
Scarsbrook, C.E., 3
Scarsbrook, E.W., 4
Scarth, R.D., 106
Schaar, W.L., 466
Schabilion, J.T., 160
Schach, H., 178
Schachet, M., 112
Schachman, H. K., 42,43
Schacht, J.H., 232
Shachtman, R. H., 341
Schachter, D., 301
Schachter, H., 549
Schad, G.A., 412
Schaechter, M., 217
Schaedle, M., 321
Schaefer, A., 273
Schaefer, C., 77
Schaefer, C.W., 77
Schaefer, H.P., 205
Schaefer, K. E., 421
Schaefer, R.O. F., 87
Schaeffer, J.F., 216
Schaeffer, Sr. M. J., 151
Schaeffer, R. L., 400
Schaeffer, W. I., 482
Schaefers, G. A., 306
Schafer, D. E., 167
Schafer, J. F., 503
Schafer, J., 453
Schafer, R., 37
Schafer, R. R., 231
Schafer, S. A., 473
Schaffer, H. E., 345
Schaffer, W. M., 473
Schaffner, C.P., 288
Schafner, J. C., 458
Schaiberger, G. E., 96
Schaible, R. H., 145,146
Schaier, C. E., 150
Schake, L. M., 458
Schales, F. D., 202
Schall, B., 463
Schall, E. D., 148
Schaller, W. E., 142
Schalles, R. R., 167
Schalm, O. W., 53
Schanberg, S., 333
Schanne, O. F., 559
Schano, E. A., 305
Schapiro, H. C., 39
Scharff, D., 265
Scharff, M.D., 329
Scharrer, B. V., 328
Schatte, C. L., 69
Schatz, G., 302
Schauer, R. C., 343
Schaufler, E. F., 303
Schaup, H. W., 385

Schechter, J.E., 63
Schechter, R. J., 28
Scheel, K.W., 445
Scheer, B.T., 387
Scheer, E.W., 39
Scheetz, R.W., 252
Scheff, G.J., 133
Scheffler, I., 59
Schefler, W.C., 317
Schieibner, R.A., 178
Schein, M.W., 507
Schein, R.D., 403
Scheinberg, E., 531
Scheinblum, T.S., 74
Schekel, K.A., 503
Schell, C.A., 223
Schell, K.F., 443
Schell, W.R., 498
Schemnitz, S.D., 194
Schenck, N.C., 92
Schengrund, C.L., 405
Schenk, J.A., 117,118
Schepers, J.S., 106
Scher, A.M., 500
Scherba, G.M., 28
Scherer, R., 398
Scherer, W.F., 308
Scherrer, B., 560
Schertz, K.F., 459
Scherzer, N., 298
Scheuer, J., 330
Schevill, B.L., 211
Scheving, L.E., 24
Schexnayder, C.A., 182
Schick, B.A., 508
Schick, E.B., 315
Schick, H., 366
Schiefer, H.B., 563
Schiff, E., 95
Schiff, J., 312
Schiff, S.O., 84
Schiffenbauer, M., 313
Schiffman, G., 325
Schiffman, R. H., 262
Schilb, T.P., 186
Schildkraut, C., 329
Schildt, C.S., 514
Schimpf, Sr. M.E., 51
Schimke, R., 42
Schimke, R.T., 41
Schimmel, R. J., 414
Schin, K. S., 324
Schincariol, A.L., 551
Schindler, D.W., 539
Schindler, J. E., 106
Schingoethe, D. J., 433
Schink, D. R., 459
Schinski, R. A., 31
Schirman, R. D., 502
Schisler, L. C., 403
Schjeide, O. A., 126
Schkade, L. L., 464
Schlaeger, Sr. M. D., 151
Schlafke, S. J., 263
Schlager, G., 172
Schlegel, D. E., 45
Schleich, R., 397
Schleicher, S.J.d'A., 70
Schleidt, W.M., 199
Schleif, R. F., 208
Schlein, J.M., 299
Schlenk, F., 120,131
Schlenk, H., 24
Schlesinger, A.B., 268
Schlesinger, M.J., 263
Schlesinger, R.W., 283
Schlesinger, S., 262
Schlessinger, D., 262
Schlinger, E. I., 44
Schlising, R. A., 29
Schlossberg, H., 507
Schlough, J. S., 524
Schlueter, E. A., 52
Schmedtje, J. F., 145
Schmehl, W. R., 65
Schmetz, D., 151
Schmickel, R. D., 232
Schmid, A. R., 240
Schmid, L. A., 497
Schmid, R., 42
Schmid, W. D., 242
Schmid, W. E., 128
Schmidly, D. J., 459
Schmidt, A., 77, 79
Schmidt, A. H., 134
Schmidt, B., 119
Schmidt, B. L., 364
Schmidt, C. L., 35
Schmidt, D. C., 446
Schmidt, E. L., 241,242,244
Schmidt, G., 73, 217
Schmidt, G. R., 138
Schmidt, G. T., 259
Schmidt, H. W. 399
Schmidt, J., 11
Schmidt, J. L., 66
Schmidt, J.M., 15
Schmidt, J.P. 452
Schmidt, J. W., 271
Schmidt, N. D., 150
Schmidt, P., 135
Schmidt, R. H., 92
Schmidt, R. P., 325
Schmidt, R. R., 493
Schmidt, R. S., 124
Schmidt, R. W., 164
Schmidt, W. H., 364
Schmidt-Koenig, K., 332
Schmidt-Nielsen, B., 355
Scymidt-Neilson, K., 332
Schmidtke, J., 244
Schmieder, L. A., 309
Schmieding, W. R., 380
Schmir, L., 8
Schmitt, J. A., 362
Schmitt, J. B., 288
Schmitthenner, A. F., 366
Schmittner, S.M., 109
Schmitton, H. R., 4
Schmitz, E. H., 22
Schmulbach, J. C., 434
Schmutz, E.M., 18
Schnaitman, C. A., 490
Schnatz, P.T., 356
Schneck, C.D., 408
Schneebeli, G.L., 473
Schneeder, L., 373
Schneeweis, S., 309
Schneidau, J.D.Jr., 190
Schneider, D., 217, 402
Schneider, D.E., 504
Schneider, E., 455, 480
Schneider, E.G., 445
Schneider, F.E., 352
Schneider, F.H., 101
Schneider, G., 225
Schneider, G.H., 127
Schneider, H., 58
Schneider, H.A., 340, 343
Schneider, M.D., 290
Schneider, O.R., 354
Schneider, R.G., 467
Schneider, W., 235
Schneiderman, H.A., 54
Schneiderman, M.A., 414
Schneiderman, N., 95
Schneiter, A.A., 350
Schneiweiss, J.W., 309
Schnell, G.D., 379
Schneller, Sr.M.B., 316
Schniewind, A.P., 44
Schnitzlein, H.N., 97
Schochet, S., 467
Schock, A.J., 274
Schoen, F., 562
Schoener, T.W., 211, 498
Schoener, W., 243
Schoenefeld, D.A., 506
Schoenfeld, E., 293
Schoenhard, D.E., 228
Schoenheider, W., 123
Schoenholz, W.K., 30
Schoenike, R.E., 425
Schoenknecht, F.D., 499
Schoenthal, N.D., 265
Schoenweiss, D.F., 139
Schofield, C.M., 76
Schofield, P., 482
Schofield, W.B., 534
Scholes, N.W., 269
Scholten, H., 243
Scholz, E.W., 351
Scholz, R.W., 403
Schom, C., 542
Schomaker, C.E., 194
Schomberg, D.W., 333
Schön, M.A., 197
Schonbrod, R.D., 386
Schonhorst, M.H., 16
School, J.M., 516
School, R.L., 363
Schooler, A.B., 350
Schooler, J.M., 333
Schooley, J.B., 215
Schooley, R.A., 62
Schoonover, C.O., 527
Schopflocher, P., 556
Schor, S., 408
Schork, M.A., 233
Schotté, O.C., 206
Schottelius, B.A., 162
Schoultz, T.W., 24
Schrader, J.W., 344
Schrader, L.E., 516
Schrader, W.T., 447, 449
Schraer, H., 404
Schraer, R.S., 404
Schram, F.R., 120
Schramm, G., 232
Schramm, L.P., 197
Schramm, P., 123
Schramm, R.J.Jr., 78, 79
Schramm, V.L., 408
Schrank, A.R., 465
Schreck, C.B., 494
Schreckenberg, G., 284
Schreiber, M.M., 148
Schreiber, R.L., 435
Schreibman, M.P., 296, 297
Schrieber, F., 30
Schrieber, R., 280
Schrock, A.E., 196
Schrock, G.F., 396
Schroder, E.K., 164
Schrodt, G.R., 181
Schroeder, A., 24
Schroeder, A.C., 396
Schroeder, Sr.D., 238
Schroeder, D.B., 79
Schroeder, E.E., 152
Schroeder, H.A., 278
Schroeder, P.C., 501
Schroeder, R.A., 260
Schrohenloher, R., 11
Schroth, M.N., 45
Schs, H.G., 134
Schubert, B.G., 211
Schubert, E.T., 307
Schubert, O.E., 508
Schubiger, G., 498
Schuerch, C., 322
Schuh, J.D., 17
Schuh, Rev.J.E., 289
Schuh, L., 515
Schuhardt, V.T., 465
Schuhmacher, R.W., 286
Schulter, R.P., 203
Schultes, R.E., 210, 211
Schultz, A.M., 44
Schultz, B., 236
Schultz, E., 555
Schultz, G.A., 285
Schultz, H.W., 383
Schultz, J., 54, 95
Schultz, J.H., 219
Schultz, J.R., 236
Schultz, L.H., 517
Schultz, O.E., 304
Schultz, R.C., 108
Schultz, R.G., 357
Schultz, R.J., 77, 79
Schultz, R.L., 35
Schultz, R.M., 124
Schultz, S.G., 414
Schultz, V., 501
Schultz, W.M., 271
Schulz, A.R., 146
Schulz, H., 298
Schulz, J.T., 351
Schulz, P., 128
Schulz, R.K., 46
Schulze, I.T., 256
Schulze, R.O., 73
Schumacher, G., 317
Schuman, G.E., 271
Schumann, D., 236
Schumann, H., 454
Schumann, W., 215
Schurr, K.M., 354
Schushmann, J.S., 83
Schuster, C., 133
Schuster, E.G., 118
Schuster, F. L., 296, 297
Schuster, M., 270
Schuster, M.L., 272
Schuster, R.M., 217
Schuster, T., 77
Schwab, J.H., 339
Schwab, R.G., 47
Schwabe, C., 428
Schwabe, O., 288
Schwalb, M.N., 282
Schwalbe, P., 398
Schwalm, F., 122
Schwan, H.P., 411
Schwark, W.S., 306
Schwarts, J.H., 313
Schwarts, L.W., 55
Schwarts, M.R., 499
Schwartz, A., 33, 408
Schwartz, D., 144
Schwartz, E., 523
Schwartz, E.A., 423
Schwartz, J.H., 301
Schwartz, L., 514
Schwartz, L.J., 514
Schwartz, M., 205
Schwartz, M.A., 318
Schwartz, N.M., 297
Schwartz, O.J., 441
Schwartz, R., 306
Schwartz, S., 223
Schwartz, S.I., 47, 99
Schwartz, T.L., 77
Schwartzman, R.M., 411
Schwassmann, H.O., 90
Schwderer, C., 512
Schweber, M., 220
Schweigert, B.S., 49
Schwelitz, F.D., 371
Schwemmin, D.J., 504
Schwendiman, J., 502
Schwengel, J., 144
Schwenk, F.W., 168
Schweppe, J., 127
Schwerdt, C.E., 41
Schwert, G.W., 179
Schwertfeger, M.H., 523
Schwinck, I., 77
Schwindt, P., 500
Schwinek, L., 195
Schwinghamer, J.M., 228
Schwintzer, C., 514
Scifres, C.J., 458
Sciorra, L., 288
Sclabassi, R.J., 56
Scobey, R., 53
Scoby, D.R., 349
Scocolar, S.J., 96
Scoggins, C.E., 249'
Scora, R.W., 58
Scogin, R., 32
Scornik, O.A., 277
Scorpio, R.M., 196
Scott, A., 175
Scott, A.E., 551
Scott, A.F., 176
Scott, B.D., 255
Scott, C., 10
Scott, C.J., 285
Scott, D., 103, 314
Scott, D.Jr., 411
Scott, D.B., 355
Scott, D.C., 106
Scott, D.E., 328
Scott, D.H., 148
Scott, D.M., 552
Scott, D.R., 117
Scott, D.W., 333
Scott, E.B., 434

Scott, E.G., 508
Scott, E.L., 46
Scott, G.T., 361
Scott, H.A., 24
Scott, H.D., 22
Scott, H.E., 344
Scott, J., 38, 423
Scott, J.B., 228
Scott, J.H., 258
Scott, J.L., 77, 484
Scott, J.W., 36, 101, 550
Scott, L., 189
Scott, L.V., 380
Scott, M.L., 305
Scott, M.W., 486
Scott, N., 452
Scott, P., 541
Scott, P.J., 231
Scott, R., 392
Scott, R.A.III., 362
Scott, R.W., 222
Scott, T.K., 338
Scott, T.W., 302
Scott, U., 239
Scott, W.E., 96
Scott, W.O., 138
Scott, W.W., 120
Scottino, J., 395
Scraba, D.G., 531
Scranton, J.R., 160
Scribner, T.K., 458
Scrimgeour, K.G., 549
Scrimshaw, N.S., 213
Scrivner, C.L., 258
Scrutton, M.C., 408
Scudday, J., 455
Scudder, C.L., 124
Scudder, G.G.E., 535
Scudder, H.I., 30
Scudder, W.T., 93
Scudo, F.M., 218
Scullen, H.A., 386
Scurvey, H.L., 175
Seabrook, W.D., 540
Seabury, F.S., 423
Seabloom, R.W., 352
Seagrave, R.C., 158
Seakes, S.J., 169
Seal, U.S., 241
Sealander, J.A., J.A., 22
Seale, R.H., 118
Seale, T., 88
Sealy, S.G., 538
Seaman, D.E., 47
Seaman, E., 219
Seaman, G.R., 128
Seamands, W.J., 527
Seaney, R.R., 302, 304
Seapy, R.R., 30
Searcey, M., 256
Searcy, D.G., 218
Searcy, G.P., 563
Searle, G., 60
Searle, G.W., 162
Searle, S.R., 302
Searles, R.B., 332
Searls, J.C., 160
Sears, E.R., 258
Sears, J.T., 543
Sears, J.W., 20
Seatz, L.F., 444
Seavey, S.R., 27
Seay, T., 175
Sebastiani, A., 373
Sechler, D.T., 258
Sechtem, E.A., 270
Secker-Walker, R.H., 256
Secor, J.B., 291
Secoy, D.M., 561
Sedell, J.R., 383
Sederoff, R.R., 300
Sedmak, J.J., 513
Sedman, Y.S., 139
Sedovic, W.M., 202
Seeburger, G.H., 524
Seeds, A.E., 244
Seeds, N.W., 72
Seegmiller, R.E., 473
Seeley, J.G., 303
Seeley, R.R., 116
Seely, C.L., 117
Seely, D.J., 201
Seely, J.R., 380
Seemayer, T.A., 556
Seerley, R.W., 106
Seery, V., 101
Seewald, R.H., 397
Sefick, H.J., 424
Segal, A., 11
Segal, A.J., 356
Segal, E., 13
Segal, H.L., 318
Segal, I.H., 48
Segal, R., 378
Segal, W., 71
Segall, S., 393
Segars, W.I., 106
Segelman, M., 207
Seghers, B.H., 552
Segina, M.R., 309
Seguin, J.J., 552
Sehe, C., 238
Sehgal, O., 260
Sehgal, P.P., 334
Seibel, H.R., 491
Seibert, H.J., 368
Seibert, R., 378
Seidel, G.E.Jr., 69
Seidel, M.E., 293
Seiden, D., 283
Seiden, L., 133
Seidenberg, A., 491
Seidler, R.J., 386
Seifter, E., 330
Seifter, S., 330
Seigel, M.R., 178
Seiger, M.B., 372
Seigesmund, K.A., 512
Seigler, D.S., 136
Seigler, H.F., 333
Seilheimer, J., 70
Seim, W.K., 383
Seinwill, M.R., 119
Seischab, F., 315
Sekelj, P., 556
Sekhorn, B.S., 246
Sekioka, T.T., 114
Selander, R., 327
Selander, R.B., 137
Selby, J.E., 484
Selby, L.A., 262
Self, J.T., 379, 380
Seliger, H., 197
Selinger, C., 295
Seliskar, C.E., 369
Selkirk, R., 29
Sell, H.M., 225
Sell, J., 350
Sell, W.H., 106
Selleck, B.H., 228
Sellemire, C.T., 192
Sellers, A.F., 306
Sellers, C., 486
Sellers, L.G., 157
Sellers, L.G., 187
Sellers, M.I., 56
Sellmann, L.R., 5
Sellmer, G.P., 290
Sells, G.D., 254
de Selm, H.R., 441
Selman, R.B., 72
Selser, W.L., 254
Selsky, M.I., 296, 297
Seltmann, H., 343
Seltzer, E., 288
Selverston, A.I., 59
Selvin, S., 46
Semancik, J.S., 58
Semel, M., 303
Semeniuk, G., 433
Semple, R.E., 546
Sen, P.K., 341
Sen, S.K., 494
Senay, L.S.Jr., 256
Sendelbach, A.G., 517
Seneca, E.D., 343
Sener, R., 62
Senff, L.M., 525
Senff, R., 515
Senft, J.F., 149
Senft, J.P., 397
Seng, J., 173
Senger, C.M., 504
Senior, A.E., 328
Senn, H.A., 515
Senn, T.L., 424
Senterfit, L.B., 308
Senturia, J.B., 357
Senturia, J.N., 516
Sequeira, L., 519
Sercarz, E.E., 55
Sergovich, F.R., 551
Serianni, R.W., 199
Serif, G.S., 362
Serinko, R.J., 391
de Serrano, H.L., 418
Seshachalam, D., 269
Session, J.J., 461
Sessle, B.J., 550
Sether, L., 512
Sethi, S.L., 250
Seto, F., 379
Seto, J.T., 31
Setterfield, G., 544
Settlage, J.H., 254
Settle, W.J., 323
Settlemyer, K., 398
Seufert, W., 559
Sevacherian, V., 58, 59
Severin, Br.C., 239
Severin, M.J., 269
Severson, D.K., 62
Severson, J., 256
Sevet, R.W., 408
Sevoian, M., 219
Seward, L., 21
Sewell, H.B., 259
Sexton, O.J., 263
Seymour, A.H., 498
Seymour, E.W., 216
Seymour, N., 543
Seymour, R.L., 362
Sguros, P.L., 87
Shabestari, L., 473
Shacklett, R.S., 161
Shaddy, J.H., 254
Shade, R.E., 148, 445
Shade, R.W., 314
Shadomy, H.J., 492
Shaeowen, H.E., 181
Shafer, J.A., 232
Shafer, S.J., 308
Shaffer, C.F.Jr., 372
Shaffer, H.C., 371
Shaffer, H.E., 401
Shaffer, J.C., 485
Shaffer, M.F., 282
Shaffer, R.L., 230
Shaffner, C.S., 203
Shafia, F., 29
Shafland, J.L., 312
Shafritz, D., 330
Shah, B.V., 341
Shah, D.B., 179
Shahidi, A., 301
Shahn, E., 297, 298
Shain, L., 178
Shain, W.A., 425
Shake, R., 448
Shakelford, R.M., 519
Shalla, T.A., 50
Shallenberger, R.S., 307
Shalucha, B., 144
Shambaugh, G.F., 363
Shamoo, A., 329
Shan, D., 207
Shan, R.K., 506
Shanahan, Rev.R., 268
Shandland, D.L., 148
Shands, H.L., 516
Shands, J.W.Jr., 94
Shands, W.A.Jr., 4
Shane, W., 117
Shankel, D.M., 171
Shaner, G.E., 148
Shank, D.B., 433
Shank, R.C., 213
Shanks, C., 502
Shanks, J.B., 202
Shanks, R.A., 275
Shannon, C.K., 286
Shannon, F., 309
Shannon, J.G., 258
Shannon, L.M., 57
Shannon, R., 97, 294
Shannon, S., 307
Shanor, L., 90
Shao, D., 61
Shapira, R., 101
Shapiro, A.M., 47, 52
Shapiro, B.I., 210
Shapiro, B.M., 499
Shapiro, B.S., 220
Shapiro, D.M., 469
Shapiro, D.W., 69
Shapiro, E., 556
Shapiro, H.S., 282
Shapiro, I.M., 411
Shapiro, J., 442
Shapiro, J.A., 132
Shapiro, N., 75
Shapiro, S., 217
Shappard, L.C., 12
Shappirio, D.G., 231
Sharawy, M., 103
Share, L., 445
Sharer, A.W., 337
Shargool, P.D., 562
Sharitz, R., 105
Sharma, C., 444
Sharma, M.L., 559
Sharma, M.P., 246
Sharma, O.P., 178
Sharma, R., 408
Sharma, R.P., 475
Sharma, U.D., 3
Sharman, B.C., 531
Sharman, D.D., 331
Sharman, M.L., 323
Sharp, A.J., 441
Sharp, D.G., 339
Sharp, J.G., 273
Sharp, J.J., 277
Sharp, L.A., 118
Sharp, M.S., 464
Sharp, R.E., 439
Sharp, W.H., 532
Sharp, W.R., 363
Sharpe, R., 272
Sharpe, R.H., 92
Sharples, G.C., 17
Sharpley, J.M., 491
Sharry, J.J., 428
Shaughnessy, J.J., 66
Shaulis, L.G., 401
Shaulis, N.J., 307
Shave, H., 434
Shave, R.S., 433
Shaver, E.L., 551
Shaver, J.R., 227
Shaw, A., 416
Shaw, C.G., 503
Shaw, D.R.D., 208
Shaw, E.T., 171
Shaw, G.G., 346
Shaw, J.G., 178
Shaw, J.H., 376
Shaw, J.S., 168
Shaw, K.C., 157
Shaw, M., 500, 533, 534
Shaw, M.V., 144
Shaw, P.D., 139
Shaw, R.F., 497
Shaw, R.H., 147
Shaw, R.J., 420, 476, 477
Shaw, R.L., 503
Shaw, S.R., 535
Shaw, W.V., 95
Shawhan, F.M., 155
Shay, D.E., 204
Shay, F.J., 393
Shay, G.D., 371
Shay, J.M., 538
Shay, J.R., 385
Shaykewich, C.F., 537
Shea, A.V., 215
Shea, J.F., 256
Shea, M.G., 34
Shear, C.A., 101
Shear, D.B., 261

Sheard, J.W., 562
Shearer, C.A., 136
Shearin, A.T., 425
Shearn, A., 197
Shebeski, L.H., 537
Shechter, Y., 298
Shecter, Y., 297
Shedlarski, J.G.Jr., 185
Sheehan, J.E., 422
Sheehan, M.R., 92
Sheehy, R.J., 99
Sheeler, P., 13
Sheeley, E.E., 109
Sheen, S., 178
Sheen, S.J., 177
Shhets, T.J., 344
Sheinin, R., 550
Shelar, E., 76
Sheldon, A.L., 267
Sheldon, H., 555
Sheldon, J.K., 394
Sheldrake, R.Jr., 305
Shelford, J.A., 533
Shell, E.W., 4
Shell, L.C., 125
Shell, R.R., 454
Shellenberger, G.W., 20
Shellenberger, P.R., 401
Shellgren, M.A., 407
Shellhamer, R.H., 145
Shellhammer, H.S., 32
Shelokov, A., 470
Shelton, G.C., 460
Shelton, K.R., 491
Shelton, L.R., 250
Shelton, M.E., 158
Shelton, P.C., 488
Shelton, V.R., 175
Shelton, W.L., 4
Shemesh, A., 243
Shen, L., 340
Shen, N.H.C., 285
Shen, S.C., 300
Shenefelt, R.D., 518
Shenk, J.S., 401
Shepard, B.M., 424
Shepard, D.C., 39
Shepard, L., 343
Shepard, R., 214
Shepard, R.H., 197
Shephard, R.J., 550
Shepherd, A.C., 557
Shepherd, B.A., 129
Shepherd, D.P., 189
Shepherd, G., 81
Shepherd, H.G., 124
Shepherd, L.N., 364
Shepherd, P., 212
Shepherd, R.J.,-50
Sheppard, C.C., 226
Sheppard, C.W., 445
Sheppard, D.E., 82
Sheppard, D.H., 561
Sheepard, J.R., 242
Sheppard, V.F., 439
Sheppe, W., 369
Shepperson, J.R., 348
Shepro, D., 207
Sher, S.A., 58
Sherbine, K.B., 398
Sherebrin, M.H., 552
Sherf, A.F., 304
Sheridan, J.D., 242
Sheridan, M.N., 328
Sheridan, R.P., 260
Sheridan, W.F., 260
Sherman, F., 328
Sherman, B.S., 325
Sherman, F.G., 326
Sherman, H.L., 247
Sherman, I.W., 57
Sherman, J.K., 24
Sherman, L.A., 260
Sherman, J.H., 233
Sherman, M., 114
Sherman, R.G., 209
Sherman, R.J., 28
Sherman, S.M., 490
Sherman, T.F., 361
Sherman, W.B., 92
Sherman, W.R., 262
Shermer, R.W., 340
Sherp, D.C., 91
Sherr, S.I., 282
Sherrill, H.G., 507
Sherris, J.C., 499
Sherritt, G.W., 401
Sherrod, L., 462
Sherrod, L.L., 451
Sherry, A.D., 466
Sherwani, J.K., 341
Sherwin, R.H., 144
Sherwood, L., 509
Sherwood, O.D., 137
Sherwood, R.T., 403
Sherwood, W.A., 486
Shiavi, R.G., 446
Shibley, G.A., 512
Shidlovsky, B., 286
Shieh, J., 112
Shields, A.R., 439
Shields, L.M., 291
Shields, R.J., 297
Shiffman, M.A., 341
Shiflet, G.W., 427
Shiflett,,R.B., 174
Shifrine, M., 53
Shifriss, O., 288
Shigeura, G.T., 114
Shigley, J.W., 404
Shigo, A., 280
Shigo, A.L., 194
Shillcock, J.A., 286
Shimabukuro, M.A., 239
Shimabukuro, R.H., 350
Shiman, R., 405
Shimaoka, K., 320
Shin, A.F., 290
Shin, H.S., 198
Shin, S.H., 546
Shinehouse, E.J., 415
Shiner, V.J., 144
Shininger, T.L., 473
Shinn, R.F.Jr., 419
Shiota, T., 10, 11
Shipman, C.Jr., 232
Shipman, R.D., 402
Shipp, C.M., 264
Shipp, R.L., 12
Shippe, R.H., 150
Shipton, H., 162
Shirer, H.W., 171
Shirley, B., 381
Shirley, H.V., 443
Shirley, R.L., 91
Shirling, E.B., 369
Shirreffs, J.H., 142
Shively, C., 295
Shively, J.N., 306
Shivers, R.R., 552
Shivvers, D.W., 421
Shizuya, H., 448
Shleser, R.A., 49
Shleser, R.H., 47
Shockman, G.D., 408
Shoemaker, H.H., 136
Shoemaker, P.B., 345
Shoemaker, V.H., 57
Shoger, R.L., 236
Shokeir, M.H.K., 537
Shook, G.E., 517
Shook, R.S., 496
Shoop, C.R., 421
Shore, P.A., 467
Shooter, E., 41
Shore, J.D., 235
Shore, R., 371
Shorey, H., 58
Shorey, H.H., 57
Shorey, W.K., 24
Short, D.E., 91
Short, J.E., 267
Short, L., 297
Short, R., 88
Short, S.H., 326
Shortess, D.K., 291
Shostak, S., 412
Shott, H.I.II., 104
Shottafer, J.E., 194
Shoukas, A.A., 197
Shoup, J.R., 150
Showalter, R.K., 93
Showe, M., 396
Showers, W.B., 157
Shows, T., 319
Shrader, J.S., 496
Shreffler, D.C., 232
Shrewsburg, M.M.Jr., 32
Shrift, A., 317
Shrigley, E.W., 146
Shrivastav, B., 333
Shrode, R.R., 443
Shropshire, W.A.Jr., 84
Shryock, G.D., 431
Shryock, H., 35
Shubeck, F.E., 433
Shubect, P.P., 286
Shubert, L.E., 352
Shubert, M.L., 73
Shuel, R.W., 547
Shuey, W.C., 351
Shuffett, R., 437
Shugarman, P.M., 62
Shugars, J.P., 174
Shugart, H.H., 441
Shukri, S., 214
Shuler, C.E., 194
Shull, J.K., 187
Shull, L.R., 382
Shulls, W.A., 71
Shum, A., 31
Shumaker, J.R., 93
Shuman, M.S., 341
Shuman, R., 399
Shuman, S., 122
Shumway, R.P., 472
Shupe, J.L., 475
Shure, D.J., 100
Shurtleff, M.C., 139
Shushan, S., 71
Shuster, C.W., 356
Shuster, L., 217
Shuster, R.C., 101
Shuster, W.W., 314
Shutak, V.G., 422
Shwe, H., 394
Sibbitt, L.D., 351
Sibley, C.S., 80
Siccama, T.C., 81
Sick, L.V., 102
Siddigi, S.M., 494
Siddique, I.H., 9
Sideman, M., 297
Sideropoulos, A.S., 399
Sides, S.I., 452
Sidhu, B., 348
Sidle, R.C., 401
Siebeling, R.J., 183
Sieben, V., 239
Sieburth, J.M., 421
Siefken, M., 121
Siegel, A., 234, 282
Siegel, B.Z., 112
Siegel, C., 312
Siegel, E.C., 216
Siegel, F.L., 520
Siegel, H.S., 108
Siegel, L., 333
Siegel, S.M., 111
Siegenthaler, E.J., 233
Siegrist, Rev.U.J., 151
Sieker, L.C., 499
Siemer, E.G., 65
Siemer, R., 154
Sieren, D.J., 346, 347
Sierk, H.A., 440
Sigafoos, R.S., 84
Sigafus, R.E., 177
Sigel, M.M., 96
Siger, A., 62
Sigler, P.B., 132
Sigler, W.F., 476, 477
Sigman, D.S., 56
Sikes, J.D., 259
Sikes, L.N., 454
Sikorowski, P.P., 249
Silber, M., 312
Silberglied, R.F., 211
Silbernagel, M.J., 503
Silbert, D.F., 263
Silcox, S., 542
Sileo, L., 548
Siler, W., 11
Silker, T.H., 377
Sill, W.H., 434
Sill, W.H.Jr., 434
Silletto, T., 155
Sillman, A., 412
Sillman, E.I., 394
Silson, C.P., 113
Silver, A., 420
Silver, B., 486
Silver, B.B., 181
Silver, D.M., 142
Silver, M., 557
Silver, M.E., 61
Silver, S.D., 263
Silver, W.S., 97
Silverborg, S.B., 321
Silverman, D.J., 203
Silverman, M.S., 339
Silverman, P.H., 293
Silverman, W.B., 238
Silverstein, M., 322
Silverstein, R., 172
Silverstein, S.J., 301
Silvey, J.K.G., 452
Silvy, N.J., 459
Simard, A., 558
Simard, T., 558
Simberloff, D., 88
Simborg, D.W., 197
Simco, B.A., 438
Simeone, J.B., 322
Simes, F.J., 485
Siminovitch, L., 550
Simkins, C.A., 241
Simmons, -., 81
Simmons, A., 161
Simmons, D., 57
Simmons, D.M., 70
Simmons, E.R., 473
Simmons, G., 234
Simmons, G.A., 194
Simmons, G.M., 492
Simmons, J.E., 43, 76
Simmons, J.P., 476
Simmons, K.R., 480
Simmons, M., 444
Simmons, R.L., 244
Simmons, R.O., 484
Simms, E., 263
Simms, H.R., 496
Simon, E.H., 150
Simon, J., 97
Simon, J.M., 406
Simon, M.I., 59
Simon, R., 327
Simon, S., 333
Simone, L.D., 324
Simoni, R.D., 41
Simons, M.D., 157
Simons, R., 514
Simons, R.K., 138
Simonson, G.H., 384
Simonton, P.R., 438
Simpkins, H., 558
Simpson, B., 88
Simpson, C.D., 459
Simpson, C.F., 94
Simpson, E.C., 334
Simpson, E.J., 454
Simpson, G.W., 194
Simpson, I., 96
Simpson, J.M., 489
Simpson, L., 55
Simpson, M., 488
Simpson, N.E., 546
Simpson, R.E., 522
Simpson, R.G., 68
Simpson, R.H., 489
Simpson, R.J., 162
Simpson, R.L., 287
Simpson, R.W., 288
Simpson, S.B.Jr., 126
Simpson, T., 79, 99
Simpson, W., 102
Simpson, W.R., 117
Sims, E.T.Jr., 424
Sims, J.J., 58

Sims, J.L., 177
Sims, M.H., 6
Sims, P.L., 67
Sims, R.A., 463
Sinclair, C.B., 25
Sinclair, D.G., 546
Sinclair, J., 544
Sinclair, J.B., 139
Sinclair, J.H., 145
Sinclair, M.G., 479
Sinclair, N.R., 485
Sinclair, N.R.S., 551
Sinclair, R., 556
Sinclair, R.O., 481
Sinclair, R.J., 380
Sinclair, W., 126
Sinclair, W.A., 304
Sinclair, W.B., 57
Sing, C.F., 232
Singal, S.A., 103
Singer, I., 244
Singer, M., 355
Singer, M.J., 50
Singer, P.C., 341
Singer, R., 131, 132
Singer, S., 139
Singer, S.F., 489
Singer, S.J., 59
Singer, T.P., 60
Singh, B., 414
Singh, C.H., 312
Singh, D., 119
Singh, D.N.P., 538
Singh, I.J., 312
Singh, J., 256
Singh, M., 189, 323, 348
Singh, R., 423
Singh, R.N., 508
Singh, R.P., 551
Singh, S., 542
Singh, S.P., 165, 166
Singhai, S.K., 551
Singletary, C.C., 249
Singletary, R.A., 77
Singleton, L.L., 249
Singleton, P.C., 527
Singleton, V.L., 51
Singsen, E.P., 78, 79
Singley, M.E., 288
Sinha, A., 242
Sinha, A.K., 284
Sinha, D., 320
Siniff, D.B., 242
Siniscalchi, J.R., 285
Sink, J.D., 401, 402
Sink, K.C., 226
Sinks, L., 320
Sinnhuber, R.O., 383
Sinnock, P., 193
Sinsheimer, R.L., 25, 26
Sinskey, A.J., 213
Siocombe, O., 548
Sipe, F.P., 387
Sipe, J.E., 142
Siplet, H., 408
Siplock, N., 122
Siraki, C., 124
Sirek, O.V., 550
Sirvish, S., 273
Sis, R.F., 460
Sisk, M.E., 175
Sisken, J.E., 179
Sisler, E.C., 343
Sisler, H.D., 201
Sisler, H.H., 89
Sissom, S., 455
Sistrom, W.R., 387
Sistrunk, D.R., 199
Sistrunk, W.A., 23
Sites, J.E., 100
Sites, J.W., 92
Sitterly, W.R., 425
Sittmann, K., 556
Siverly, R.E., 142
Six, E.W., 161
Sizer, I.W., 213
Sjodin, R.A., 203
Sjogren, R., 481
Sjolund, R.D., 160
Slaar, P.D., 265
Skahen, J.G., 500
Skalko, R.G., 327
Skandalakis, J.E., 101
Skau, C.M., 275
Skavaril, R.V., 363
Skean, J.D., 181
Skeggs, L.T., 355
Skelley, G.C., 424
Skelton, B.J., 424
Skelton, T.E., 424
Skinner, J.L., 519
Skinner, R.D., 24
Skjegstad, K.R., 239
Skjei, S.S., 489
Skoch, E.J., 359
Skog, J.E., 485
Skogg, F.K., 515
Skogley, C.R., 422
Skok, R.A., 243
Skok, S., 126
Skogerson, L., 300
Skold, B.H., 158
Skold, L.N., 444
Skoultchi, A., 330
Skotheim, R.A., 309, 504
Skopik, S.D., 82
Skoropad, W.P., 530
Skotland, C.B., 503
Skrepetos, C., 387
Skroch, W.A., 345
Skud, B.E., 498
Skujins, J.J., 476, 477
Skujins, J.S., 474
Skvarla, J.J., 379
Slabaugh, W.R., 151
Slack, C.E., 444
Slack, D.A., 24
Slack, J.M., 5, 509
Slack, N., 320
Slack, N.G., 315
Slade, H.D., 127
Slade, L.M., 65
Sladek, J.R., 328
Slagle, W.G., 455
Slakey, L., 217
Slanetz, L.W., 279
Slapikoff, S.A., 216
Slater, J., 77, 323
Slater, J.A., 77, 79
Slater, N., 70
Slater, R., 399
Slatis, H.M., 227
Slatkin, M., 132
Slatter, W.L., 365
Slattery, C.W., 36
Slattery, Sr.J., 311
Slaunwhite, W.R.Jr., 318
Slautterback, D.B., 520
Slavin, A.J., 180
Slavin, B.G., 63
Slayman, C., 81
Slayter, H., 211
Sleator, W.W., 137
Slechta, R.F., 207
Sledge, E.B., 12
Sledge, J.L., 354
Sleeper, B.P., 351
Sleeper, D.A., 309
Sleesman, J.P., 363
Slemmer, R.L., 370
Slentz, R., 104
Slepecky, R.A., 326
Slesinski, R.S., 166, 168
Slesnick, I.L., 504
Slife, F.W., 138
Slinger, R.P., 552
Slinger, W.A., 444
Sloan, B.L., 436
Sloan, N.F., 229
Sloan, R.F., 167
Sloan, R.P., 280
Sloan, W.C., 39
Sloane, E., 523
Slobodchikoff, C.N., 15
Sloey, W., 523
Slome, C., 342
Slonecker, C.E., 535
Slonka, G.F., 53
Slosky, W.E., 391
Slotkin, T., 333
Slotkoff, L.M., 85
Slotta, K.H., 95
Slusher, J.P., 261
Slysh, A.R., 407
Slyter, A.L., 432
Smail, J.P., 238
Small, E.B., 199
Small, G.D., 434
Small, J.W., 89
Small, L.F., 385
Small, P.A.Jr., 94
Small, P.F., 415
Small, R.D., 428
Smalley, A.E., 190
Smalley, E.B., 519
Smalley, K., 164
Smalling, J.D., 443
Smallman, B.N., 546
Smallwood, J.E., 460
Smart, E.W., 477
Smart, G.C.Jr., 91
Smart, K.M., 295
Smart, L.I., 183
Smart, R.A., 475
Smayda, T.J., 421
Smeal, P.L., 494
Smeby, R.R., 357
Smeck, N.E., 364
Smeins, F.E., 458
Smellwood, C., 471
Smichowski, V.P., 399
Smid, J., 322
Smiddy, J.C., 488
Smiley, J., 178, 426
Smiley, J.H., 177
Smiley, K.L., 445
Smillie, L.B., 531
Smilowitz, Z., 402
Smit, C.J.B., 106
Smith, A., 95, 217, 476
Smith, A.A., 416
Smith, A.C., 30, 77, 217
Smith, A.E., 317
Smith, A.H., 47, 230
Smith, A.I., 439
Smith, A.J., 139
Smith, A.L., 286
Smith, A.M., 480
Smith, A.P., 409
Smith, A.S., 166
Smith, B., 121
Smith, B.B., 416
Smith, B.C., 252
Smith, B.E., 224
Smith, B.N., 472
Smith, B.R., 423
Smith, B.S., 84
Smith, B.W., 345
Smith, C., 37, 210, 428, 471, 474
Smith, C.A., 354
Smith, C.C., 166, 370, 374
Smith, C.E., 10, 318
Smith, C.F., 344
Smith, C.G., 175, 214, 470, 549
Smith, C.J., 371
Smith, C.K., 367
Smith, C.L., 297, 490
Smith, C.M., 408
Smith, C.N., 445
Smith, C.R., 288
Smith, C.W., 111, 227
Smith, D., 76, 317, 436, 516
Smith, D.A., 544
Smith, D.B., 551
Smith, D.C., 424
Smith, D.D., 21, 162
Smith, D.E., 89, 346
Smith, D.G., 325
Smith, D.J., 285
Smith, D.L., 97, 451
Smith, D.L.T., 561, 563
Smith, D.M., 61, 81, 255
Smith, D.R., 66, 110
Smith, D.T., 459
Smith, D.W., 59, 452, 521, 546
Smith, E., 20, 273
Smith, E.A., 87
Smith, E.B., 22
Smith, E.C., 542
Smith, E.E., 95
Smith, E.F., 167
Smith, E.H., 303
Smith, E.L., 11, 18, 56, 277, 376
Smith, E.M., 179, 365
Smith, E.W., 223, 254
Smith, E.Y., 440
Smith, F., 387
Smith, F.A., 328
Smith, F.E., 211
Smith, F.G., 157
Smith, F.H., 425
Smith, F.L., 47
Smith, F.M., 542
Smith, G., 298, 515
Smith, G.C., 427, 458
Smith, G.E., 258, 507
Smith, G.H., 468
Smith, G.R., 231
Smith, G.S., 92, 292, 330
Smith, G.W., 415
Smith, H., 35, 522, 540
Smith, H.C., 444
Smith, H.D., 472
Smith, H.G., 275
Smith, H.J., 345
Smith, H.L., 434
Smith, H.M., 71, 488
Smith, H.O., 198
Smith, H.W., 117, 445, 502
Smith, I.K., 368
Smith, J., 182, 188, 253
Smith, J.A., 134
Smith, J.D., 30, 461
Smith, J.E.Jr., 260, 326
Smith, J.H., 474
Smith, J.J.B., 550
Smith, J.M., 174
Smith, J.N.M., 535
Smith, J.P., 34
Smith, J.R., 262
Smith, J.W., 185
Smith, J.W.Jr., 458
Smith, K.J., 322
Smith, K.K., 150
Smith, K.L., 91, 367
Smith, K.O., 470
Smith, K.R., 364
Smith, L., 294, 467
Smith, L.C., 315, 448
Smith, L.D., 150, 169
Smith, L.H., 147, 240
Smith, L.J., 297, 298
Smith, L.K., 62
Smith, L.L., 103, 240, 513
Smith, L.M., 49
Smith, L.S., 498
Smith, L.T., 422
Smith, L.W., 472
Smith, Sr.M., 400
Smith, M., 534
Smith, M.A., 472
Smith, M.E., 182, 243
Smith, M.H., 106
Smith, M.R., 198
Smith, M.S., 101
Smith, M.V., 547
Smith, N., 205
Smith, N.A., 339
Smith, N.E., 48
Smith, O.A., 500
Smith, O.D., 459
Smith, O.E., 58, 438
Smith, O.H., 512
Smith, P., 30, 499
Smith, P.D., 100
Smith, P.E., 364, 380
Smith, P.F., 434
Smith, P.G., 51
Smith, Q.T., 244
Smith, R., 155, 309
Smith, R.A., 405
Smith, R.C., 3, 7, 90, 261, 291, 365
Smith, R.E., 52, 219, 333, 547
Smith, R.F., 44

Smith, R.F.C., 537
Smith, R.G., 233, 449
Smith, R.H., 246
Smith, R.I., 43
Smith, R.J., 230
Smith, R.J.F., 562
Smith, R.L., 14, 120, 474, 508
Smith, R.M., 102, 113, 508
Smith, R.O., 365
Smith, R.P., 277, 278
Smith, Sr.R.P., 308
Smith, R.R., 516
Smith, R.W., 150
Smith, S., 251
Smith, S.A., 453
Smith, S.C., 280, 508, 521
Smith, S.D., 179, 496
Smith, S.G., 524
Smith, S.H., 403
Smith, S.M., 551
Smith, S.W., 362
Smith, T.C., 470
Smith, T.E., 466
Smith, V.S., 380
Smith, W., 310, 374
Smith, W.C., 341
Smith, W.E., 535
Smith, W.G., 24, 288, 361
Smith, W.H., 81, 167
Smith, W.J., 80, 334, 409
Smith, W.K., 333, 519
Smith, W.L., 98, 444
Smith, W.R., 438, 451
Smith-Gill, S., 84
Smith-Sonneborn, J., 526
Smithberg, M., 243
Smithberg, M.H., 241
Smitherman, R.O., 4
Smithies, O., 518, 519
Smithson, J.A., 125
Smittle, R.B., 286
Smoake, J.A., 291
Smock, R.M., 305
Smolders, A.P., 37
Smolens, J., 408
Smosky, J., 214
Smothers, J.L., 180
Smouse, P.E., 232
Smout, M.S., 552
Smulson, M.E., 85
Smyser, C., 219
Smyth, J.R.Jr., 219
Smyth, T.Jr., 402, 404
Smythies, J.R., 11
Snazelle, T., 446
Snedecor, J.G., 218
Snegoski, Sr.M.C., 268
Snelgrove, J.J., 454
Snell, E.E., 42
Snell, J., 362
Snell, J.F., 362, 363
Snell, R.S., 84
Snelson, F.F., 88
Snetsinger, R.J., 402
Snider, J.A., 370
Snider, P., 463
Snider, P.G., 464
Snider, R.S., 328
Snipes, W., 404
Snitgen, N.A., 229
Snodgrass, M.J., 491
Snoek, E., 309
Snoeyenbox, G., 219
Snoke, J.E., 56
Snook, T., 352
Snow, B.L., 216
Snow, C.D., 383
Snow, G.E., 222
Snow, J.P., 109
Snow, Sr.L., 37
Snow, M., 96
Snow, M.D., 389
Snow, P., 475
Snowbarger, W.E., 127
Snudden, B.H., 514
Snyder, B.W., 407
Snyder, D., 147, 395
Snyder, D.H., 436
Snyder, G.K., 57
Snyder, H.E., 155, 157, 419
Snyder, H.Z., 151
Snyder, I.S., 509
Snyder, J., 533
Snyder, J.A., 427
Snyder, J.C., 212
Snyder, J.R., 196
Snyder, J.W., 121
Snyder, L.A., 242
Snyder, L.C., 241, 260
Snyder, L.R., 228
Snyder, M.J., 204
Snyder, R.A., 129
Snyder, R.C., 498
Snyder, R.D., 102
Snyder, S.P., 69
Snyder, W.E., 288
Snyder, W.Z., 313
So, A.G., 95
Soares, J., 203
Soave, O.A., 42
Sobek, J.M., 191
Sobell, H.M., 328
Sobieski, R., 173
Sobin, S.S., 64
Sobonya, J., 356
Sobota, A., 373
Sochard, M., 83
Socher, S.H., 449
Socolofsky, M.D., 183
Socquet, I., 206
Soderquist, M.R., 383
Soderwall, A.L., 387
Sodicoff, M., 408
Soeiro, R., 329, 330
Soergel, K.P., 202
Sofer, W., 197
Sogandares, F., 267
Sohacki, L.P., 323
Sohal, R.S., 454
Sohmer, S., 515
Soike, K.F., 327
Sojka, G., 144
Sokal, J., 320
Sokatch, J., 380
Sokatch, J.R., 380
Sokol, H.W., 278
Sokol, R.J., 356
Sokoloff, A., 28
Sokoloff, L., 295
Sokolove, P., 205
Sokolove, P.A., 41
Sokolove, P.G., 205
Solari, J., 331
Solaro, R.J., 492
Solberg, M., 288
Solberg, R.A., 260
Solberg, L.A., 257
Solbrig, O.T., 211
Soldo, A.T., 95
Soletsky, S., 356
Solie, T.N., 69
Soliman, K.F., 9
Soll, -., 81
Soll, D.R., 160
Soll, L., 71
Sollberger, R.A., 129
Solliday, S., 449
Sollod, A.E., 563
Solnitzky, O., 85
Solomon, A.K., 211
Solomon, D.L., 302, 304
Solomon, G.B., 412
Solomon, G.C., 68
Solomon, H., 84
Solomon, J.B., 207
Solomon, J.D., 249
Solomon, S., 294
Solorano, R.F., 262
Solotorovsky, M., 288
Soltan, H.C., 551
Soltanpour, P.N., 65
Soltero, R.A., 496
Soltz, D.L., 31
Solursh, M., 160
Solymoss, B., 559
Soma, L.R., 411
Somer, L.S., 95
Somers, G.F., 82
Somers, K., 449
Somers, M.E., 77, 79
Somerson, N.L., 367
Somerville, J., 338
Somerville, M.H., 352
Somerville, R.L., 148
Somes, R.G., 78, 79
Somjen, G., 333
Somlyo, A.P., 411
Sommerfeld, M.R., 15
Sommers, L.E., 147
Sonea, S., 558
Sonenshein, A.L., 217
Sonenshine, D.E., 487
Song, S.H., 552
Songster, C.L., 181
Sonleitner, F.J., 379
Sonn, J., 251
Sonneborn, D.R., 516
Sonneborn, T.M., 145
Sonnenberg, H., 550
Sonnenschein, R.R., 57
Sonnenwirth, A.C., 263
Sonstein, S.A., 100
Soodak, M., 208
Soost, R.K., 58
Soots, R.F.Jr., 331
Sophianopoulos, A.J., 101
Sopper, W.E., 402
Sorant, A.M., 341
Sordahl, L., 467
Sorensen, A.M., 458
Sorensen, K.A., 344
Sorensen, L.O., 452
Sorensen, P.D., 126
Sorensen, R.A., 120
Sorensen, R.C., 271
Sorenson, M.W., 260
Sorenson, R., 243
Sorenson, R.M., 268
Sorenson, W.G., 379
Sorger, G.J., 545
Sorokin, S.P., 212
Sorsoli, W., 387
Sorrentino, S., 328
Sosebee, R.E., 463
Soter, R.P., 324
Sotereanos, G., 413
Soule, J., 92
Soule, M.E., 59
Soulen, T.K., 241
Soules, J.A., 357
Soulsby, E.L., 412
Soulsby, E.J.L., 411
Soulsby, M.E., 492
Souter, E.J., 511
South, F.E., 262
South, G.R., 541
Southall, R.M., 345
Southard, A.R., 474, 477
Southards, C.J., 442
Southern, W., 126
Southin, J.L., 556
Southward, M., 498
Southwick, F.W., 218
Sowinski, R., 315
Sowles, K.M., 118
Slach, M., 333
Spackman, E.W., 527
Spackman, W., 403
Spaeth, Rev.F.W., 390
Spaethling, R., 219
Spain, J.D., 228
Spalding, C.A., 200
Spang, R.L., 226
Spangler, E., 511
Spangler, F., 523
Spanis, C.W., 62
Spann, C., 246
Spanswick, R., 302
Spar, I.L., 328
Spargo, B.H., 133
Sparks, A.K., 498
Sparks, D., 107, 199
Sparks, E., 8
Sparks, H.V., 233
Sparks, L.M., 424
Sparks, P., 285, 515
Sparks, W.C., 117
Sparling, P.F., 339
Sparling, S.R., 27
Sparrowe, R.D., 261
Sparwasser, C., 127
Spaulding, E.H., 408
Spaulding, J., 395
Spaur, C., 273
Spaziani, E., 160
Speake, D.W., 5
Spear, I., 465
Spear, P.G., 132
Spear, W.D., 382
Specht, L.W., 401
Speck, J.C.Jr., 225
Speck, M.L., 344, 345
Speck, R.S., 62
Speck, S.J., 106
Spector, A.A., 161
Spector, B., 95, 311
Spedel, E., 245
Speer, H.L., 533
Speer, J.M., 120
Speese, B.M., 484
Speicher, B.R., 193
Speidel, D.H., 298
Speirs, R.S., 325
Spellenberg, R., 292
Spence, A.P., 321
Spence, H.A., 185
Spence, K.D., 501
Spence, M.A., 56
Spence, W.L., 210
Spence, M.W., 542
Spencer, A.W., 70
Spencer, D.L., 164
Spencer, E.Y., 551
Spencer, F., 189
Spencer, J.H., 555
Spencer, J.W., 291
Spencer, L., 543
Spencer, L.T., 279
Spencer, M., 530
Spencer, M.N., 511
Spencer, P.W., 138
Spencer, S.B., 402
Spencer, S.J., 323
Spencer, S.M., 187
Spencer, T., 545
Spencer, W.A., 301
Spendlove, R.S., 476
Sperelakis, N., 490
Sperling, J.A., 299
Sperow, C.B., 508
Sperry, J.H., 460
Sperry, R.W., 25, 26
Sperry, T.M., 165
Speth, C.F., 275
Spiegel, L.E., 286
Spiegel, M., 277
Spiegelman, S., 301
Spieler, R., 30
Spiegler, P.E., 84
Spielman, A.A., 218
Spies, C.D., 147
Spies, H., 388
Spiess, L.D., 123
Spieth, H.T., 52
Spieth, P.T., 45
Spikes, J.D., 473
Spillett, J.J., 476, 477
Spiltoir, C.F., 77, 79
Spinks, V., 246
Spirito, C.P., 488
Spiroff, B.E.N., 123
Spittler, R.P., 41
Spitz, A., 279
Spitzer, J.J., 186
Spitzer, N.C., 59
Spitznagel, J.K., 339
Spivak, M.L., 290
Spivey, H.O., 377
Spivey, R.A., 88
Splittstoesser, C.M., 306
Splittstoesser, O.F., 307
Splittstoesser, W.E., 138
Spofford, J., 132
Spomer, G.G., 117
Spomer, L.A., 138
Spoon, D.M., 85
Spooner, A.E., 22
Spooner, B.S., 166
Spotila, J.R., 318

Spotts, C.R., 13
Spradling, M., 236
Sprague, E., 488
Sprague, E.F., 488
Sprague, I.B., 214
Sprague, J.B., 547
Sprague, J.M., 410
Sprague, M.A., 288
Spremulli, G.H., 308
Sprenger, T.S., 373
Spriestersbach, D.C., 160
Spring, T.G., 464
Springer, C., 269
Springer, G.F., 127
Springer, M.E., 389, 444
Sprinson, D., 300
Sproston, T., 480
Sproul, E., 320
Sprouse, R.F., 262
Spruill, A., 336
Spryer, J.F., 77, 79
Spuller, R.L., 336
Spulnik, J.B., 116
Spurgeon, K.R., 432
Spurgpon, H., 294
Spurr, A.R., 51
Spurr, G.B., 513
Spurr, H.W.Jr., 345
Spurr, S., 465
Spyker, J., 489
Spyropoulos, C.S., 132
Squares, B.P., 552
Squire, P.G., 67
Squires, C.D., 3
Srb, A.M., 302
Srebnik, H.H., 43
Srebro, R., 319
Srere, P.A., 466
Sribney, M., 545
Srinivasan, D., 300
Srinivasan, P.R., 300
Srinivasan, V.R., 183
Srivastava, L.M., 533
Srivastava, P.N., 105
S'Souza, S.M., 198
Srolovita, H., 556
Staab, M., 523
Stableford, L.T., 397
Stabler, T.A., 145
Staby, G.L., 365
Stace-Smith, R., 534
Stacey, J.K., 66
Stachow, C., 207
Stack, S.M., 68
Stack-Haydon, J., 537
Stackiw, W., 538
Stackston, W.E., 557
Stacy, P., 335
Stacy, R.W., 129
Stadelmann, E.J., 241
Stadnyk, L., 513
Stadtherr, R.J., 184
Staebler, A., 30
Staehelin, L.A., 71
Staffeldt, E., 292
Stafford, D.W., 339, 340
Stafford, H.A., 387
Stage, G., 77
Stage, G.I., 77
Stager, K.E., 62
Stagg, R.M., 309
Stahl, B.J., 279
Stahl, F.W., 387
Stahl, J.B., 129
Stahl, P.D., 262
Stahl, W.L., 500
Stahly, D.P., 161
Stahmann, M.A., 517
Staiger, J., 96
Stains, H.J., 129
Stainsby, W.N., 94
Stairs, G.R., 16, 363
Staley, J.T., 499
Staley, T.F., 381
Staling, L., 204
Stalker, H.D., 263
Stall, R.E., 92
Stallard, J.R., 359
Stallard, R.E., 208
Stallcup, O.T., 23
Stallcup, P.L., 382
Stallcup, W.B., 454
Stalling, D.T., 188
Stallknecht, G., 117
Stallman, R.K., 546
Stallones, R., 469
Stallworth, H., 436
Stallworthy, W.B., 540
Stalnaker, C.B., 476, 477
Stalter, R., 316
Stam, J.C., 254
Stambaugh, W.J., 334
Stambrook, P., 355
Stamer, J.R., 307
Stamey, W.L., 166
Stamier, F.W., 161
Stamper, H.B.Jr., 487
Stamps, J.A., 52
Stanbridge, E.J., 42
Stanbury, E.J., 555
Stanbury, J.B., 213
Standaert, W.F., 199
Stander, J.A., 371
Standifer, L.C., 184
Standil, S., 537
Standing, J., 393
Standing, K., 30
Stanfield, A.B., 16
Stanfield, J.K., 190
Stanford, E.H., 47
Stanford, J., 452
Stanford, J.W., 450
Stanford, L.M., 116
Stang, E.J., 365
Stangeland, R.J., 433
Stanghellini, M.E., 18
Stanglein, T., 400
Staniforth, D.W., 157
Stanley, A.J., 380
Stanley, D.W., 547
Stanley, E.R., 550
Stanley, H.P., 476
Stanley, M.S., 485
Stanley, W.F., 323
Stanley, W.M.Jr., 54
Stannard, J.N., 328
Stannard, L.J., 136
Stanners, C.P., 550
Stansbery, D.H., 364
Stansfield, W.D., 27
Stanton, G., 134
Stanton, G.E., 100
Stanton, G.J., 467
Stanton, M.M., 496
Stanton, P., 210
Stanziale, W.G., 76
Staple, P.H., 319
Staples, R., 271
Staples, W.D.Jr., 7
Stapleton, M.L., 396
Stapp, W.B., 231
Star, A., 290
Stark, E., 315
Stark, E.W., 149
Stark, F.C., 202
Stark, G., 41
Stark, R., 116
Stark, R.W., 117, 118
Starkey, E.E., 517
Starkey, R., 514
Starks, K.F., 377
Starlard, V., 25
Starling, J.H., 495
Starling, J.L., 401
Starr, A., 55
Starr, J.W., 250
Starr, M.A., 389
Starr, M.P., 51
Starr, P.R., 137
Starr, R.C., 144
Starr, T.J., 317
Starr, W.A., 502
Starrett, A., 31
Starrett, H., 62
Stary, H.C., 186
Starzyk, M.J., 126
Stasek, C., 88
Stasiak, R., 272
Staszak, D.J., 102
Staten, R.D., 459
Staton, N., 526
States, J., 15
Statler, G., 352
Staub, N.C., 60
Staub, R.J., 436
Stauber, E.H., 118
Stauble, A., 290
Staudinger, W.L., 270
Staudt, E., 373
Stauffer, J.P., 35
Staugaard, B.T., 79
Stavinoha, W.B., 470
Stavitsky, A.B., 356
Stavn, R.H., 343
Stavrianopoulos, J., 300
Stay, B.A., 160
Steadman, J., 270
Steadman, J.R., 272
Stearns, F.W., 521
Stearns, L.W., 31
Stearns, M., 234
Stearns, W.F., 521
Stebbins, L.L., 532
Stebbins, R.C., 43
Stebbins, R.L., 384
Stebbins, W., 254
Steebes, T.A., 562
Stechschulte, A.L., 87
Steck, T.L., 131
Steckel, J.E., 288
Stecker, R.E., 32
Stedman, E., 123
Steele, C., 70, 470
Steele, D., 29
Steele, D.H., 541
Steele, J.R., 552
Steele, R.H., 190
Steele, V.J., 541
Steelman, H.C., 25
Steen, J.S., 246
Steen, T., 387
Steenbergen, J.F., 39
Steensen, D.H.J., 346
Steere, W., 298
Steere, W.C., 297
Steergren, C.D., 261
Steevens, B., 527
Steeves, H.R., 492
Stefano, H.D., 325
Stefansson, B.R., 537
Steffensen, D.N., 137
Stegink, L.C., 161
Stegner, R.W., 82
Stehr, F.W., 226
Stein, A.A., 314
Stein, A.M., 95
Stein, G., 90
Stein, H.J., 223
Stein, J.R., 534
Stein, O.L., 217
Stein, P., 323
Stein, P.S.G., 263
Stein, R.C., 317
Steinbeck, K., 108
Steinberg, A., 355
Steinberg, A.G., 355
Steinberg, A.H., 356
Steinberg, M.A., 498
Steinberg, M.S., 287
Steinberg, R., 60, 432
Steinberg, S.A., 411
Steinberg, W., 490
Steinbridge, G.E., 424
Steinegger, D., 522
Steiner, A.L., 531
Steiner, D.F., 131
Steiner, E.E., 230
Steiner, G., 550
Steiner, H.Y., 496
Steiner, W.W.M., 137
Steinhagen, H.P., 508
Steinhardt, G.L., 147
Steinhardt, R.A., 43
Steinhauer, A.L., 201
Steinhilb, H.M., 229
Steinhoff, H.W., 66
Steinkraus, K.K., 307
Steinle, J.A., 375
Steinman, H., 333
Steinmetz, C., 76
Steinrauf, L., 146
Steinrauf, L.K., 146
Steir, E.F., 288
Steitz, -., 81
Stekiel, W.H., 513
Steldt, F., 145
Stellwagen, E., 161
Stellwagen, R.H., 64
Stelzig, D.A., 508
Stelzner, D., 325
Stemmier, E., 410
Stempak, J.G., 325
Stempen, H., 287
Stenroos, O.O., 486
Stensaas, S.S., 473
Stensland, G.J., 314
Stent, G.S., 43
Stenzel, K.H., 307
Stephan, J., 400
Stephen, F.M., 23
Stephen, W.P., 386
Stephens, B.G., 430
Stephens, G.C., 54
Stephens, J., 37
Stephens, J.C., 105
Stephens, J.F., 367
Stephens, J.M., 93
Stephens, M.W., 359
Stephens, N., 174
Stephens, R.L., 26, 423
Stephens, R.R., 119
Stephens, S.G., 345
Stephens, W.L., 29
Stephens, W.R., 121
Stephenson, D.L., 450
Stephenson, E.L., 23
Stephenson, G.R., 547
Stephenson, W.K., 143
Steponkus, P.L., 303
Sterin, W.K., 356
Sterling, C., 49
Sterling, P., 410
Sterling, R., 64
Sterling, W.L., 458
Stern, A.C., 341
Stern, A.I., 217
Stern, C., 45
Stern, D., 219
Stern, G., 369
Stern, H., 59
Stern, I., 124
Stern, J.S., 50
Stern, K.R., 29
Stern, O., 327
Stern, V.M., 58
Stern, W.L., 201
Sternberger, L.A., 198
Sternburg, J.G., 136
Sterrett, A., 358
Stetler, D.A., 492
Stetson, C.A., 94
Stetson, M.H., 82
Steucek, G.L., 400
Stevens, C.E., 306
Stevens, C.F., 500
Stevens, C.L., 413
Stevens, C.M., 64
Stevens, D.F., 480
Stevens, E., 513
Stevens, E.D., 112
Stevens, F.C., 538
Stevens, G.L., 62
Stevens, J.G., 56
Stevens, L.P., 508
Stevens, M.A., 51
Stevens, M.H., 189
Stevens, R., 514
Stevens, Sr.R.C., 316
Stevens, R.G., 462
Stevens, R.H., 56
Stevens, V.C., 364
Stevens, W., 159, 473
Stevens, W., 473
Stevens, W.C., 448
Stevenson, E.L., 99
Stevenson, F.J., 138
Stevenson, H., 88
Stevenson, H.L., 429
Stevenson, H.Q., 76
Stevenson, J., 223

Stevenson, J.C., 201
Stevenson, J.H., 20
Stevenson, N., 284
Stevenson, R., 96, 223
Stevenson, R.T., 258
Stevenson, W.R., 148
Steward, C.F., 49
Steward, C.R., 157
Steward, J.P., 42
Steward, O., 490
Stewart, A.M., 482
Stewart, C.R., 453
Stewart, D.B., 537
Stewart, G.L., 219
Stewart, G.R., 29, 186
Stewart, H.B., 551
Stewart, J., 278
Stewart, J.A., 280
Stewart, J.M., 72, 266, 538
Stewart, J.P., 341
Stewart, J.W., 329
Stewart, K.D., 360
Stewart, K.M., 318
Stewart, K.W., 452, 538
Stewart, Br.L., 151
Stewart, L.C., 542
Stewart, L.W., 92
Stewart, R., 246
Stewart, R.B., 545
Stewart, R.K., 556
Stewart, S.E., 427
Stewart, T., 186
Stewart, V., 454
Stewart, W.G., 65
Stewart, W.N., 531
Steyer, -., 81
Stibitz, G.R., 278
Stich, H., 535
Stickle, W.B., 183
Stickney, J.L., 227
Stickney, R.R., 102
Stidd, B.M., 139
Stidham, J.D., 429
Stiedemann, M., 238
Stiegler, J.H., 376
Stiehl, W.L., 418
Stiffler, D., 376
Stiles, H.D., 202
Stiles, R.. 3
Stiles, R.N., 445
Stiles, W.C., 195
Still, C.C., 288
Still, S.M., 168
Stiller, M., 150
Stillway, L.W., 428
Stillwell, E.F., 487
Stillwell, R., 309
Stilson, D.W., 72
Stimpson, D., 195
Stimson, J.S., 112
Stimson, W.R., 285
Stinebring, W.R., 482
Stinner, R.E., 344
Stinski, M.F., 161
Stinson, A.W., 227
Stinson, H., 302
Stinson, M.W., 319
Stinson, W.W., 273
Stirewalt, H.L., 100
Stirewalt, W.S., 284
Stirling, C.E., 500
Stirm, W.L., 147
Stish, R., 244
Stith, L.S., 16
Stith, R.D., 380
Stiven, A.E., 339
Stiver, H.G., 538
Stivers, R.K., 147
Stjernholm, R.L., 190
Stobbe, E.H., 537
Stober, Q.J., 498
Stock, D.A., 89
Stock, J.J., 535
Stock, S., 121
Stockdale, D., 348
Stockdale, F.E., 41
Stockdale, H.J., 157
Stockdale, T.M., 366
Stockelman, Sr.M.E., 358
Stocker, B.A.D., 41
Stockhammer, K.A., 171
Stockhouse, R., 37
Stocking, C.R., 51
Stockland, A.E., 477
Stockton, B., 170
Stocum, D.L., 137
Stoddard, G.E., 474
Stodolsky, M., 124
Stoeckenius, W., 60
Stoewsand, G.S., 307
Stoffolano, J.G., 218
Stoin, H.R., 22
Stojanovic, B.C., 249
Stojanovic, B.J., 247, 248
Stojda, R., 311
Stojda, R.J., 311
Stokes, -., 550
Stokes, A.W., 476, 477
Stokes, D., 474
Stokes, D.E., 230
Stokes, D.R., 100
Stokes, G.W., 177, 178
Stokes, J.L., 501
Stokes, R., 359
Stokstad, E.L.R., 45
Stolk, J.M., 277
Stollar, B.D., 217
Stollar, V., 283
Stoller, W.E., 138
Stolman, S., 227
Stolpe, S.G., 137
Stolte, W.F., 174
Stoltenberg, C., 384
Stoltz, L.P., 178
Stoltzfus, C.M., 447
Stoltzfus, W.B., 157
Stombaugh, T.A., 258
Stone, B.P., 436
Stone, D.B., 60
Stone, D.E., 332
Stone, E.C., 44
Stone, E.L., 301, 302
Stone, G.E., 73
Stone, H.O.Jr., 171
Stone, J.B., 547
Stone, J.D., 167
Stone, J.F., 376
Stone, L.E., 269
Stone, L.R., 167
Stone, M., 270
Stone, M.J., 105
Stone, R., 219
Stone, R.W., 404
Stone, S.S., 157
Stone, W.E., 521
Stone, W.H., 519
Stoner, M.F., 29
Stoner, W.N., 433
Stones, R.C., 228
Stoney, S.D., 104
Stonier, T., 299
Stoolmiller, A.C., 131
Stoops, J.K., 448
Storb, U.B., 499
Storch, R.H., 194
Storck, R., 453
Storer, R.W., 231
Storey, A.T., 550
Storey, B.T., 410, 411
Storey, J.B., 459
Storey, W.B., 58
Storgaard, A.K., 537
Storm, D.R., 135
Storm, G.L., 402
Storm, R.M., 386
Stormer, F.A., 229
Stormshak, F., 382
Storr, J.F., 318
Story, J., 126
Story, J.L., 469
Story, L.E., 252
Storz, J., 69
Stothart, J.R., 188
Stotler, B.C., 128
Stotler, R.E., 128
Stotler, W.A., 388
Stott, G.G., 460
Stott, G.H., 17
Stott, P.B., 236
Stotz, E.H., 328
Stoudt, H., 285
Stouffer, J.E., 448
Stouffer, R.F., 403
Stough, H.B., 496
Stout, B.B., 288
Stout, E.R., 492
Stout, F.M., 382
Stout, I.J., 88
Stout, J.F., 222
Stout, L.C., 467
Stout, P.R., 50
Stout, R., 398
Stout, V.F., 498
Stout, W.L., 401
Stoutamire, W.P., 369
Stouter, V.P., 296
Stow, R.W., 362
Stowe, B.B., 80, 81
Stowell, E.A., 222
Stover, B.J., 473
Stozky, G., 312
Strack, L.E., 129
Strader, H.L., 495
Strain, B.R., 332
Straley, D., 160
Strand, E.M., 12
Strand, F., 312
Strand, O.E., 240
Strandtmann, R.W., 461
Strange, J.R., 102
Strathmann, R.R., 498
Stratton, D.B., 155
Stratton, L.P., 427
Stratton, P.O., 527
Straub, K.D., 24
Straub, R.S., 306
Straughan, D., 62
Straughan, I.R., 62
Straughn, W.R., 339
Straus, F.H., 133
Straus, L.P., 131, 136
Strauss, B.S., 132
Strauss, C.H., 402
Strauss, D.J., 59
Strauss, H.C., 333
Strauss, J.A., 26
Strauss, N., 318
Strauss, P.R., 214
Strauss, R.H., 113
Straw, T., 245
Strawcutter, R., 396
Strawn, R.K., 459
Strayer, J.L., 91
Streams, F.A., 77
Strecker, H., 330
Streeter, J.G., 364
Streett, J.C., 463
Streett, J.W., 76
Strehler, A.F., 381
Strehler, B.L., 62
Streilein, J.W., 466
Streips, U.N., 180
Streisinger, G., 387
Strejan, G.H., 551
Streitfeld, M.M., 96
Strength, D.R., 3
Stretch, A.W., 288
Stretton, A.O.W., 516
Streu, H.T., 288
Strickberger, M.W., 262
Stricker, E.M., 412
Strickland, H., 414
Strickland, J.C., 488
Strickland, K.P., 551
Strickland, W., 331
Strickler, D.J., 127
Strickler, T., 332
Strickler, J.C., 277
Strickling, E., 200
Strider, D.L., 345
Strittmatter, C.F., 347
Stritzke, J.F., 376
Strobel, E.F., 439
Strobel, J.W., 92, 345
Strohl, W.A., 283
Strohm, J.L., 522
Strohman, R.C., 43
Strohmeyer, D., 523
Stroike, J.E., 271
Strolle, R.S., 229
Stromberg, B.E., 412
Stromberg, D., 500
Stromer, M.H., 156
Strominger, J.L., 210
Strominger, N.L., 327
Strong, D., 88
Strong, E.B.Jr., 381
Strong, J.P., 186
Strong, J.S., 16
Strong, R.G., 58
Strother, G.K., 404
Stroube, E.W., 364
Strouck, D.R., 501
Stroud, L.M., 343
Stroupe, D.B., 360
Stroupe, H.S., 347
Struchtemeyer, R.A., 195
Struckmeyer, K.A., 503
Strum, T., 335
Strumeyer, D.H., 288
Strumwasser, F., 25
Struve, W.M., 29
Stuart, A.M., 218
Stuart, C.M., 34
Stuart, D.G., 19, 265
Stuart, G., 31
Stuart, J., 331
Stuart, K., 97
Stubbs, E.L., 412
Stubbs, J.D., 40
Stubbs, D.W., 468
Stuber, C.W., 345
Stucker, R.E., 240
Stuckey, I.H., 422
Stuckey, R.L., 362
Stuckwisch, C.G., 95
Stucky, D.J., 130
Studier, E.H., 234
Stueben, E.B., 191
Stuerckow, B.I.E., 214
Stuessy, T.F., 362
Stuewer, C., 236
Stuiber, D., 518
Stuit, D.B., 160
Stuiver, M., 498
Stukus, P., 358
Stulberg, C.C., 236
Stull, E.A., 18
Stull, J.W., 17
Stull, J.T.Jr., 53
Stull, W.D., 369
Stullken, R.E., 100
Stumpf, P.K., 48
Sturch, E., 378
Sturgeon, E.E., 377
Sturgeon, R.V.Jr., 375
Sturges, F.W., 507
Stump, D.W., 487
Stuntz, D.E., 497
Sturkie, P.D., 288
Sturm, N., 373
Sturman, L.S., 327
Sturrock, T.T., 87
Stushnoff, C., 241
Stuteville, D.L., 168
Stuthman, D.D., 240
Stuttard, C., 543
Stutte, C.A., 22
Stutz, H.C., 472
Stuuon, C., 95
Stuy, J., 88
Stutzman, L., 320
Styles, E.D., 536
Styron, C.E., 337
Suarez, A., 186
Suarez, F., 85, 417
Suarez, M.A., 418
Suarez-Hoyos, J., 356
Subden, R.E., 546
Subhas, T., 394
Sublette, J.E., 291
Subtelny, S., 453
Sucheston, M.E., 367
Suchindran, C.H., 341
Suchy, T.D., 121
Sucoff, E.I., 243
Sud, G.C., 236
Suddarth, S.K., 149
Sudds, R.D., 324
Sudilovsky, O., 356

Sudo, S., 244
Suelter, C.H., 225
Sueoka, N., 71
Suerth, C., 299
Suga, N., 263
Sugihara, J., 349
Sugiyama, H., 517
Suhadolnik, R., 408
Suie, T., 363
Sukowski, E.J., 134
Sulakhe, P.V., 562
Sulerud, R.L., 237
Sulkin, N.M., 347
Sulkowski, E., 320
Sullivan, A.D., 250
Sullivan, A.L., 334
Sullivan, A.M., 279
Sullivan, C.Y., 272
Sullivan, D.T., 293, 326
Sullivan, F.L., 215
Sullivan, G., 70
Sullivan, G.A., 344
Sullivan, G.H., 149
Sullivan, J., 387
Sullivan, J.B., 333
Sullivan, J.F., 307
Sullivan, J.T., 464
Sullivan, L.P., 172
Sullivan, M.J., 424
Sullivan, N., 397
Sullivan, P., 356
Sullivan, P.A., 329
Sullivan, R.L., 347
Sullivan, T.D., 479
Sullivan, T.P., 415
Sullivan, T.W., 272
Sullivan, V.I., 191
Sullivan, W.D., 207
Sullivan, W.J., 313
Sultzer, B.M., 325
Sulya, L.L., 251
Suman, R.F., 423
Summer, G.K., 340
Summerfelt, R.J., 376
Summerour, C.W.III., 7
Summers, -., 81
Summers, D., 329
Summers, D.F., 330
Summers, D.F., 474
Summers, J.D., 547
Summers, J.S., 264
Summers, M.D., 465
Summers, R.G.Jr., 193
Summers, W.A., 146
Summitt, R.L., 445
Sumpter, P., 356
Sun, A.M.F., 550
Sun, B.C., 229
Sun, H.H., 394
Sund, J.M., 516
Sundaralingam, M., 517
Sundberg, D., 243
Sunde, M.L., 519
Sunder, J., 406
Sunderwirth, S., 70
Sundet, N., 238
Sundharadas, G., 521
Sundick, R.S., 236
Sundsten, J.W., 499
Sunshine, I., 356
Sunstad, D.P., 242
Supdich, J., 60
Surak, J.G., 92
Surdy, T., 240
Surgalla, M.J., 320
Surgenor, D.M., 318
Suriano, R.A., 123
Surratt, A.L., 175
Surrette, G.H., 257
Survant, W.G., 177
Surzycki, S.J., 160
Susalla, A., 151
Suskind, S.R., 197
Susman, M., 518, 519
Sussdorf, D.H., 308
Sussex, I.M., 80
Sussman, A.S., 231
Sutcliffe, R., 392
Suter, D.B., 484
Suter, R., 354
Suter, W., 511
Sutherland, B.M., 54
Sutherland, D., 272
Sutherland, D.J., 288
Sutherland, G.B., 562
Sutherland, J., 54
Sutherland, J.H.R., 104
Sutherland, J.P., 332
Sutherland, R.C., 274
Sutherland, R.M., 552
Sutherland, T.M., 65
Sutin, J., 101
Sutter, G.R., 433
Sutter, R.P., 507
Sutterfield, W., 431
Suttie, J.W., 517
Suttkus, R.D., 190
Suttle, A., 467
Sutton, D.A., 29
Sutton, D.D., 30
Sutton, H.E., 465
Sutton, J.G., 547
Sutton, P., 364
Sutton, T.S., 365
Sutton, W.W., 182
Suyama, Y., 409
Suzuki, D.T., 535, 538
Svajgr, L., 271
Svendsen, J., 368
Swade, R., 31
Swader, F.N., 302
Swader, J.A., 58
Swafford, J.R., 15
Swails, L.F., 427
Swain, E.R., 79
Swain, S.H., 439
Swaisgood, H.E., 344
Swallow, C., 167
Swallow, R.L., 426
Swan, D.G., 502
Swan, F.R., 507
Swan, J.P., 241
Swan, L.W., 40
Swanborg, R.H., 236
Swaney, J., 330
Swango, L.J., 6
Swank, R.T., 319
Swann, C.W., 106
Swanson, A.A., 428
Swanson, A.M., 518
Swanson, B.T.Jr., 66
Swanson, C.A., 362
Swanson, C.J., 234
Swanson, C.P., 217
Swanson, E.W., 443
Swanson, G.A., 66
Swanson, H.D., 155
Swanson, J., 31, 268
Swanson, J.L., 474
Swanson, L.V., 382
Swanson, R.A., 522
Swanson, R.E., 389
Swanson, R.F., 488
Swanson, R.N., 432
Swanson, V.B., 65
Swanton, M.C., 340
Sward, S.M., 329
Swartz, F.J., 180
Swartz, G.E., 318
Swartz, L.G., 14
Swartzendruber, D., 147
Swasey, J.E., 195
Swatek, J.A., 296
Swatland, H.J., 547
Swearingen, K., 37
Swearingin, M.L., 147
Sweat, F.W., 474
Sweat, Sr.J., 169
Swedberg, K.C., 496
Sweeley, C.C., 225
Sweeney, B.M., 61
Sweeney, C.R., 317
Sweeney, D., 136
Sweeney, D.C., 137
Sweeney, E.F., 215
Sweeney, J.R., 40
Sweeney, M.J., 88
Sweeney, P.R., 547
Sweeney, R.A., 317
Sweet, E.E.Jr., 415
Sweet, G., 173
Sweet, H.C., 88
Sweet, H.R., 216
Sweet, M.H., 460
Sweet, R.D., 305
Sweigart, J., 486
Sweissing, F.D., 66
Swell, L., 491
Swenseid, M.E., 56
Swensen, A.D., 472
Swensen, R.D., 522
Swenson, C.A., 161
Swenson, G., 309
Swenson, K.G., 386
Swenson, R., 408
Swenson, L., 65
Swenson, M.J., 158
Swenson, W., 524
Swerczek, T.W., 178
Swift, B.L., 528
Swift, C.C., 62
Swift, C.S., 158
Swift, E., 421
Swift, F.C., 288
Swift, H., 132
Swigart, R.H., 180
Swiger, L.A., 365
Swihart, S.L., 97
Swim, H.E., 179
Swindel, B.F., 346
Swingle, H.D., 444
Swink, J.F., 66
Swinyer, B.P., 39
Switzer, B.R., 340
Switzer, C.M., 546, 547
Switzer, G.L., 250
Switzer, R., 135
Switzer, W.P., 158
Swoboda, A.R., 459
Sydiskis, R.J., 204
Syeklocha, D., 535
Sykes, E.E., 99
Sydnor, T.D., 365
Sylvester, E.S., 44
Sylvester, M.A., 391
Sylvestre, M.T., 279
Sylwester, E.P., 157
Symmons, R.A., 30
Symons, M.J., 341
Synder, D.P., 218
Synn, W.K.Jr., 108
Sype, W.E., 386
Sypert, G.W., 94
Sypherd, P.S., 55
Syrocki, J., 317
Szabo, A., 37, 86
Szabo, G., 57
Szabo, P., 298
Szal, R., 215
Szalay, J., 297, 299
Szaniszlo, P.J., 465
Szebenyi, A., 310
Szebenyi, E., 284
Szepsenwol, J., 418
Szijj, L.J., 29
Szilagyi, G., 330
Szkolnik, M., 307
Szubinska-Luft, B., 499
Szumski, A., 492
Szwarc, M., 322

T

Taaffe, J.G., 354
Tabb, D., 96
Taber, B.G., 258
Taber, C.A., 258
Taber, H.G., 156
Taber, H.W., 328
Taber, W.A., 460
Tabita, R.F., 465
Taboada, O., 226
Tabor, E., 428
Taborsky, G., 61
Tabu, S.R., 485
Tachibana, H., 157
Tack, P.I., 225, 227
Tackett, A., 175
Tafanelli, R.J., 456
Taffe, W.J., 279
Taft, C.E., 362
Taft, H., 80
Taft, S., 522
Tager, H.S., 131
Taggart, J.F., 301
Taggart, J.V., 301
Tai, C., 178
Tailleur, P., 557
Tainter, F.H., 24
Tait, A., 473
Tait, J.S., 553
Tait, R.M., 533
Taiz, L., 61
Takacs, W., 406
Takahashi, M., 284
Takahashi, T., 330
Takahaski, I., 545
Takashima, S., 411
Takeshita, J.Y., 234
Takete, F., 513
Talamantes, F.J., 61
Talbert, R.E., 22
Talbot, G., 557
Talbott, R.E., 389
Talburt, D.E., 22
Taliaferro, C.M., 376
Tallan, I., 550
Tallarida, R.J., 409
Talmage, D.W., 72, 77
Talmadge, M.B., 287
Tam, W.H., 552
Tamarin, A., 499
Tamarin, R., 207
Tamas, I.A., 309
Tamashiro, M., 114
Tamblyn, E., 31
Tamir, H., 300
Tamm, S.L., 145
Tammariello, R., 395
Tammen, J., 403
Tamper, W., 436
Tamppari, R., 15
Tan, C., 256
Tan, I.C.H., 153
Tan, K.H., 106
Tan, L., 559
Tanabe, T.Y., 402
Tanada, Y., 44
Tanaka, D., 86
Tanaka, J.S., 114
Tanaka, M., 112
Tanaka, T., 113
Tandler, B., 355
Tandon, S.R., 522
Tang, J., 380
Tang, T., 492
Tang, P.C., 134
Tang, P.L., 356
Tanksley, T.D., 454
Tannenbaum, S.R., 213
Tanenbaum, S.W., 321
Tanford, C., 333
Tanner, C.E., 557
Tanner, N., 210
Tanner, W.W., 472
Tansey, M.R., 144
Taper, C.D., 555
Tapia, L., 307
Tapley, D.F., 300
Tapley, E.M., 152
Tappa, D., 416
Tappan, W.B., 91
Tappel, A.L., 49
Tapper, D., 304
Targowski, S.P., 319
Tarjan, A.C., 91
Tarjan, E., 298
Tarjan, P., 95
Tarpley, W.A., 437
Tarr, C.M., 172
Tarrants, L., 7
Tarter, M.E., 46
Tarver, F.R., 344
Tarver, M.G., 121
Tarzwell, C.M., 422
Tasca, R.J., 82
Taschdjian, C., 315
Taschenberg, E.P., 306
Tash, D., 267

Tashian, R.E., 232
Tashiro, H., 306
Taslitz, N., 355
Taso, P.H., 58
Tassava, R.A., 364
Tate, L.G., 12
Tate, S.S., 307
Tattar, T., 280
Tattar, T.A., 219
Tatum, B.L., 20
Tatum, M.E., 460
Taub, F.B., 498
Taub, S., 297
Taube, S., 79
Tauber, M.J., 303
Tauber, O.E., 157. 159
Taubler, J.H., 406
Taurog, A., 467
Tauvydas, K.J., 512
Tavares, J.E., 318
Taven, R.E., 260
Taves, D., 150
Taves, D.R., 328
Tavolga, W.N., 297
Tawakami, T.G., 53
Tayama, H.K., 365
Taylor, A., 542
Taylor, A.B., 137
Taylor, A.L., 72
Taylor, A.R., 236
Taylor, A.R.A., 540
Taylor, B., 96
Taylor, B.B., 16
Taylor, C.A., 432
Taylor, C.L., 249
Taylor, C.P.S., 551
Taylor, C.R., 211
Taylor, D., 96
Taylor, D.A., 371
Taylor, D.H., 360
Taylor, D.L., 170
Taylor, D.W., 80
Taylor, E.L., 28
Taylor, E.W., 132
Taylor, F., 475
Taylor, F.H., 480
Taylor, F.J.R., 534
Taylor, G., 377
Taylor, G.A., 288
Taylor, G.D., 444
Taylor, G.F., 104
Taylor, G.S., 364
Taylor, G.N., 473
Taylor, G.T., 129
Taylor, H.L., 70
Taylor, I.E.P., 534
Taylor, I.M., 549
Taylor, J., 16, 108, 234, 295, 405, 434
Taylor, J.B., 374
Taylor, J.D., 234
Taylor, J.E., 266
Taylor, J.F., 227
Taylor, J.H., 88
Taylor, J.J., 256, 267
Taylor, J.K., 220
Taylor, J.L., 226
Taylor, J.M., 535
Taylor, J.R., 180
Taylor, K.B., 11
Taylor, K.M., 39
Taylor, L.W., 550
Taylor, M., 144
Taylor, M.H., 82
Taylor, M.L., 387
Taylor, N.L., 177
Taylor, O.C., 58
Taylor, O.R., 171, 172
Taylor, R.C., 106, 210
Taylor, R.E., 65, 407
Taylor, R.E.Jr., 445
Taylor, R.J., 242, 504
Taylor, R.G., 291
Taylor, R.L., 275, 470, 534
Taylor, R.P., 490
Taylor, S.L., 75
Taylor, T., 439
Taylor, T.H., 177
Taylor, T.N., 362
Taylor, V., 494
Taylor, W., 427
Taylor, W.D., 404
Taylor, W.H., 35
Taylor, W.K., 88
Taylor, W.P., 510
Taylor, Y., 372
Tchao, R., 399
Teare, I.D., 166
Teas, H., 95
Teate, J.L., 377
Teater, R.W., 364, 366, 367
Teeguarden, D.E., 44
Teel, M.R., 82
Teel, R., 36
Teer, J.G., 459
Teeri, A.E., 280
Tegenkamp, T., 369
Teichman, R.J., 112
Teigen, J.B., 147
Teigler, H.I., 137
Te Krony, D.M., 177
Telfer, W.H., 409
Telford, H.S., 502
Telford, I., 85
Telford, M., 550
Teller, D.C., 499
Teller, D.Y., 500
Telser, A., 126
Temp, M., 522
Tempelis, C.H., 46
Temple, K.L., 265
Temple, L.C., 247
Templet, A., 188
Templet, P., 188
Templeton, A.R., 465
Templeton, G.E., 24
Templeton, J., 388
Templeton, W., 498
Templeton, W.C.Jr., 177
Tend, C.T., 449
Tener, G.M., 534
Tenerowicz, P., 76
Teng, C., 449
Tengerdy, R.P., 69
Tenney, S.M., 278
Tenney, W.R., 488
Tennyson, L.C., 261
Tenpas, G.H., 516
Teodoro, R.R., 234
Tepaske, E.R., 162
Tepfer, S.S., 387
Tepper, H.B., 321
Terada, M., 301
Teraguchi, M., 355
Terborgh, J.W., 287
Terebey, N., 312
Terepka, A.R., 328
Terjung, R.L., 137
Terman, C.R., 484
Terman, M.P., 170
Terner, C., 207
Terrel, T.L., 5
Terrell, A., 348
Terrell, C.R., 215
Terri, J., 132
Terriere, L.C., 386
Terry, R.J., 461
Terry, N., 46
Terry, R.A., 324
Terry, R.L., 192
Terry, T.M., 77
Tershak, D.R., 404
Tershakovec, G.A., 95
Terwilliger, R.C., 387
Terwilliger, C., 67
Terzuolo, C., 244
Tesseneer, R.A., 176
Tessman, I., 150
Test, F.H., 231
Tester, A.L., 112
Tester, C.F., 344
Tester, J.R., 242
Tetrault, R.C., 402
Tett, P.R., 489
Teusink, J.T., 223
Tew, J.G., 492
Tewari, K.K., 54
Tewari, R.P., 288
Tews, L., 523
Thach, R.E., 263
Thach, T., 81
Thacker, G.H., 305
Thacker, S.W., 108
Thacore, H.R., 319
Thacore, H.R., 413
Thaeler, C.S., 292
Thaemert, J.C., 133
Thalmar, R.H., 449
Thalter, D., 117
Thames, J.L., 18
Thames, S.F., 252
Thanassi, J.W., 482
Tharp, G., 270
Thatcher, W.W., 91
Thaxton, J.P., 346
Thayer, D.W., 461
Theberge, J.M., 551
Theil, E.C., 343
Theis, J.H., 53
Theisen, N., 349
Thelen, T.H., 496
Theobald, W.L., 111
Theodoridis, G., 489
Therrien, C.D., 403
Therrien, H.P., 558
Thibodeau, G.A., 433
Thiele, V.F., 326
Thielges, B.A., 183
Thien, L.B., 190
Thieret, J.W., 176
Thies, R.E., 380
Thirey, H., 393
Thirey, H.E., 393
Thiers, H.D., 40
Thiesfeld, V., 522
Thilly, W.G., 213
Thimann, K.V., 61
Thirion, J.P., 559
Thiruvathukal, K.V., 123
Thliveris, J.A., 538
Thockmorton, L., 132
Thode, F.W., 424
Thodes, A.M., 138
Thoele, M.W., 402
Thom, S., 513
Thoma, Sr.M., 397
Thomas, A.S., 213
Thomas, B., 73
Thomas, C., 317
Thomas, C.A., 424
Thomas, C.E., 537
Thomas, C.H., 133
Thomas, C.W., 356
Thomas, D.C., 367
Thomas, D.D., 266
Thomas, D.E., 335
Thomas, D.L., 185
Thomas, D.W., 31, 475
Thomas, E.R., 400
Thomas, G.D., 259
Thomas, G.J., 106
Thomas, G.W., 177, 259
Thomas, J., 147
Thomas, J.A., 157, 434
Thomas, J.E., 8, 375, 450
Thomas, J.J., 87
Thomas, J.L., 10
Thomas, J.P., 89
Thomas, J.W., 27
Thomas, K.K., 543
Thomas, L., 70, 96
Thomas, L.J., 263
Thomas, M.B., 347
Thomas, O.P., 203
Thomas, P., 395
Thomas, P.E., 503
Thomas, R.B.Jr., 486
Thomas, R.D., 188
Thomas, R.E., 29
Thomas, R.J., 268
Thomas, S., 394
Thomas, V.L., 470
Thomason, I.J., 58
Thomason, R., 471
Thomerson, J., 130
Thommes, R.C., 120
Thompkins, T.B., 347
Thompson, A.C., 248
Thompson, A.E., 17
Thompson, A.H., 202
Thompson, A.M., 172
Thompson, B., 284
Thompson, B.D., 93
Thompson, C., 24
Thompson, C.C., 258
Thompson, C.F., 360
Thompson, D., 189
Thompson, D.C., 182
Thompson, D.J., 500
Thompson, D.L., 344
Thompson, D.Q., 303
Thompson, E., 31
Thompson, E.F., 250, 429
Thompson, E.M., 28
Thompson, F.J., 94
Thompson, F.N., 108
Thompson, G., 126, 470
Thompson, G.A.Jr., 465
Thompson, G.B., 259
Thompson, Sr.H., 154
Thompson, H.E., 167
Thompson, I.M., 537
Thompson, J., 11, 408
Thompson, J.C.Jr., 487
Thompson, J.D., 162
Thompson, J.E., 551
Thompson, J.J., 185
Thompson, J.M., 97
Thompson, J.N., 11
Thompson, J.R., 530
Thompson, J.S., 549
Thompson, K., 408
Thompson, L.F., 22
Thompson, L.H., 462
Thompson, M.A.W., 550
Thompson, M.B., 349
Thompson, M.C., 169
Thompson, M.H., 314
Thompson, M.M., 384
Thompson, M.P., 401, 402
Thompson, N.P., 92
Thompson, O.C., 4
Thompson, P., 479
Thompson, P.E., 106
Thompson, R., 317, 541
Thompson, R.B., 498
Thompson, R.F., 55
Thompson, R.G., 540
Thompson, R.H., 170
Thompson, R.L., 240
Thompson, T.E., 490
Thompson, S.N., 58
Thompson, S.R., 309
Thompson, T., 32, 270
Thompson, W., 549
Thompson, W.F., 217
Thompson, W.L., 234
Thompron, W.M., 468
Thompson, W.R., 164
Thompson, W.T., 472
Thomsen, S.L., 133
Thomson, B., 264
Thomson, B.F., 75
Thomson, C.L., 219
Thomson, D.A., 18
Thomson, D.H., 27
Thomson, D.S., 359
Thomson, G.W., 156
Thomson, J.W., 515
Thomson, K.S., 80
Thomson, M., 292
Thomson, R.G., 512, 548
Thomson, S.T., 45
Thomson, W.W., 57
Thomulka, K.W., 405
Thor, D.E., 470
Thoe, W., 443
Thoren, R.K.E.,2 29
Thoresen, A.C., 222
Thoresett, G.O., 389
Thorhaug, A., 96
Thornber, J.P., 55
Thornburg, J.E., 227
Thorne, C.B., 218
Thorne, D.W., 474
Thorne, F.M.III., 99
Thorne, G., 516
Thorne, G.M., 428
Thorne, M.D., 138
Thorne, O., 71

Thorne, R.E., 499
Thorne, R.F., 32
Thornley, W.R., 475
Thornton, C.S., 227
Thornton, J.W., 376
Thornton, K.W., 354
Thornton, M.L., 266
Thornton, P.A., 177
Thornton, R.E., 503
Thornton, R.M., 51
Thornton, W.A., 448
Thorp, R.W., 49
Thorpe, B., 73
Thorpe, N.O., 237
Thorsen, J., 548
Thorson, R.E., 152
Thorson, T., 270
Thorsteinson, A.J., 537
Thorud, D.B., 18
Thrasher, D.M., 183
Thrasher, F.F., 26
Thrasher, S.G., 327
Threlfall, W., 541
Threlkeld, S.F.H., 545
Thrift, F.A., 177
Thro, Sr.P., 254
Throckmorton, G.S., 134
Thron, C.D., 277
Throne, A., 521
Thorneberry, G.O., 292
Thorneberry, J.B., 21
Thureson-Klein, A., 252
Thurlow, D.L., 3
Thurman, L.D., 378
Thurman, R.G., 410
Thurman, W.J., 190
Thurmond, W., 27
Thurow, G.R., 139
Thurston, A.W., 221
Thurston, E.L., 460
Thurston, H.D., 304
Thurston, R., 178
Thwaites, W.M., 39
Thye, F.W., 494
Thysell, J.R., 433
Thysen, B., 330
Tibbitts, F.D., 276
Tibbs, G.T., 102
Tibbs, J.F., 261
Tibby, R.B., 62
Tice, L.F., 405
Tidball, C.S., 84
Tidball, M.E., 84
Tidwell, H.C., 466
Tidwell, W.D., 472
Tidwell, W.L., 32
Tieben, G.L., 151
Tiedje, J.M., 228
Tiefel, R.M., 511
Tiemeier, O.W., 166
van Tienhoven, A., 305
Tiernan, C.F., 275
Tierney, M., 207
Tieszen, L.L., 431
Tietjen, J., 297
Tietjen, J.H., 297
Tietjen, W.L., 103
Tietz, W.J., 68
Tiffany, L.H., 157
Tiffney, W., 219
Tigchelaar, E.C., 149
Tigges, J.W., 101
Tihen, J.A., 152
Tijia, B., 178
Till, G.G., 352
Till, J.E., 550
Tillack, T.W., 263
Tilley, S.G., 215
Tillman, M., 498
Tillman, R.W., 168
Tilney, L.G., 409
Tilton, P.I., 461
Timasheff, S.N., 208
Timell, T., 322
Timerlake, B., 84
Timian, R., 352
Timken, R.L., 267
Timm, H., 51
Timmermann, D., 20
Timmons, E.H., 129
Timothy, D.H., 344
Tindall, D., 128
Ting, F.T.C., 352
Ting, I.P., 57
Ting, Y.C., 207
Tinga, J.H., 107
Ting-Beall, H.P., 333
Tingey, W.M., 303
Tinker, D.O., 549
Tinkle, D.W., 231
Tinnell, W.H., 486
Tinnin, R., 387
Tipple, R.E., 362
Tippo, O., 217
Tipton, C.L., 157
Tipton, C.M., 162
Tipton, H.C., 97
Tipton, R.L., 400
Tipton, S.R., 442
Tipton, V.J., 472
Tischer, R.G., 247
Tischfield, J., 355
Tisdale, E.W., 118
Tisniras, P.S., 43
Titani, K., 499
Titman, P.W., 119
Tittler, I.A., 296
Titus, T.C., 424
Tizard, I.R., 548
Tkacz, J., 405
Tobias, C., 46
Tobach, E., 297
Toben, H.R., 466
Tobin, A.J., 211
Tobin, R.B., 273
Tobin, T., 227
Tobin, T.J., 397
Tobolski, J., 150
Toby, Rev.N., 159
Tocher, S.R., 231
Toczek, D., 224
Todar, K., 465
Todd, A.C., 520
Todd, E.H., 448
Todd, F.A., 345
Todd, G.L., 268
Todd, G.W., 375
Todd, M.E., 535
Todd, N.B., 207
Todd, P.W., 404
Todd, R., 272
Todd, R.L., 106
Todd, S.H., 202
Todd, W.M., 445
Todo, O.M., 409
Toepfer, J., 373
Toetz, D.W., 376
Toews, D.P., 542
Tofte, R., 125
Togassaki, R.K., 144
Tojstem, J.L., 159
Tokay, E., 329
Tokes, Z., 64
Tokuda, S., 294
Tokuhata, G.K., 414
Tokumaru, T., 301
Tolbert, N.E., 225
Tolbert, R.J., 239
Toledo, R.T., 107
Toler, Sr.C., 406
Tolins, I.S., 428
Toliver, A.P., 48
Tollefson, J.J., 157
Tolmach, L.J., 263
Tolnai, S., 548
Tolstead, W.L., 506
Toman, F.R., 181
Tomanek, G.W., 164
Tomanek, R.J., 160
Tomasczewski, J., 123
Tomasulo, J.A., 296
Tomasz, M., 298
Tombes, A.S., 426
Tomes, M.L., 148
Tometsko, A.M., 328
Tomkins, G.M., 60
Tomkins, J.P., 305
Tomko, G.J., 550
Tomkowski, A.C., 508
Tomlinson, G., 437
Tomlinson, G.A., 62
Tomlinson, J.T., 40
Tomlinson, P.B., 211
Tomoff, C., 16
Tompkins, D.R., 23
Tompkins, R., 287
Tonascia, J.A., 198
Tone, J., 122
Toney, M.E.Jr., 495
Tong, E.Y., 220
Tong, W., 414
Tonzetich, J., 391
Toomey, W.G., 344
Toop, E.W., 530
Top, W., 7
Topel, D.C., 155
Topham, W.S., 356
Topoleski, L.D., 305
Topping, M.S., 258
Torack, R., 263
Torch, R., 229
Tordoff, H.B., 242
Tordoff, W.III., 28
Torell, C.R., 275
Torell, P., 117
Torgerson, R.L., 461
Toribara, T.Y., 328
Tormey, J.McD., 57
Tornabene, H.S., 196
Tornabene, T.G., 69
Tornheim, P., 370
de Toro, C.M., 418
del Toro, R.S., 417
Toro-Goyco, E., 418
Toro-Rosario, J., 417
Torres, A.M., 170
Torres, del S.M., 417
Torres-Pinedo, R., 418
Torrey, H., 287
Torrey, J.G., 211
Torrie, J.H., 516
Torstveit, E., 238
Torstveit, O., 238
Tossell, W.E., 546
Tosteson, D.C., 333
Tosteson, M.T., 333
Totel, G.L., 257
Toth, L.A., 186
Totusek, R., 376
Tou, J.S., 190
Toue, S.R., 338
Toulouse, R.B., 452
Tourtellotte, M.E., 78, 79
Touse, R.D., 366, 367
Toussoun, T.A., 403
Touster, O., 446
Tove, S.B., 343
Towe, A.L., 500
Towers, G.H.N., 534
Towle, A.L., 40
Towner, H., 37
Townes, P.L., 328
Townley, R.G., 269
Townsend, G.F., 547
Townsend, H.E., 189
Townsend, J.C., 214
Townsend, N.J., 251
Townsend, S.F., 370
Townsend, T.W., 366
Townsley, P.M., 533
Townsley, S.J., 112
Tozloski, A.H., 74
Traber, D.L., 468
Tracey, Sr.K., 299
Tracey, M., 544
Tracy, E., 239
Tracy, R., 113
Tracy, R.E., 186
Train, C.T., 257
Traina, R., 395
Trainer, J.E., 89
Trainer, J.E., 400
Trainor, F.R., 77, 79
Trama, F.B., 288
Trame, Rev.R., 37
Tramel, B.R., 250
Tramer, E.J., 371
Trammel, K., 306
Trank, J.W., 172
Trankle, R.J., 143
Trapp, G.R., 31
Trappe, J.M., 385
Trasler, D., 556
Traub, R., 203
Traugh, J.A., 57
Traurig, H.H., 179
Traut, R.R., 53
Trautman, M.A., 364
Trautman, M.B., 364
Traverse, A., 403
Travis, D.M., 360
Travis, J., 105
Travis, J., 342
Travis, R.H., 142
Travis, R.V., 416
Trawinski, B.J., 11
Trawinski, I., 11
Traxler, R.W., 421, 422
Traynham, J.C., 182
Treadwell, C.R., 84
Treadwell, G.E., 485
Treagan, L., 62
Treanor, K.P., 296
Treble, D.H., 327
Tredway, T., 119
Treece, R.E., 363
Treese, Sr.E.D., 147
Treese, W.W., 255
Tregunna, E.B., 534
Trei, J.E., 28
Treick, R.W., 360
Trela, J., 370
Trelawny, G.S., 486
Trelease, R.N., 15
Trelease, T., 275
Trelka, D.G., 415
Tremaine, J.H., 534
Tremblay, G., 555
Tremblay, R.H., 481
Tremaine, M.M., 273
Tremmel, B., 164
Trench, R.K., 50
Trentini, W.C., 540
Tresch, P.S., 303
Treshow, M., 473
Tretiak, O., 394
Treumann, W.B., 239
Treuting, H., 290
Trevino, D.L., 340
Trevithick, J.R., 551
Trezise, W.J., 415
Triantaphyllou, A.C., 345
Triantaphyllou, H.H., 345
Tribbey, B., 30
Tribbey, B.A., 30
Tribble, J., 270
Tribble, L.F., 462
Trieff, N.M., 468
Trierweiler, J.F., 364
Trihey, P., 236
Trimble, G.R., 508
Trimble, J.J.III., 183
Trinkaus, J.P., 80
Trione, E.J., 385
Triplehorn, C.A., 363
Triplett, E.L., 61
Triplett, G.B., 364
Tripodi, D., 408
Tripp, J.R., 88
Tripp, L.D., 376
Tripp, M.R., 80
Triska, F.J., 383
Tristani, F.E., 513
Tritsch, G., 320
Trivers, R.L., 211
Trivett, T., 38
Trlica, M.J., 67
Troeger, T., 152
Troll, J., 219
Troll, R., 119
Troncale, L.R., 29
Tropp, B.E., 298
Trost, C., 116
Troster, M.M., 552
Trotter, D.M., 168
Trotter, M., 263
Trountner, W.T., 26
Trousdale, M.D., 470
Troutman, E.L., 364
Troutman, J.L., 18

Trowbridge, C., 492
Troxel, A.W., 366
Troxell, H.E., 66
Troy, F.A., 53
Troyer, H., 103
Troyer, J.R., 343, 408
Truchan, L.C., 511
Trudel, M.J., 558
Trudel, P., 557
True, C.E., 269, 270
Trueblood, K., 55
Trueblood, M.S., 528
Truelove, B., 4
Truesdale, F.M., 183
Truesdale, R., 372
Truex, R.C., 408
Trujillo, E.E., 114
Trula, D., 449
Truman, J.W., 498
Trumbore, R., 317
Trumpower, B.L., 277
Trupin, G.L., 283
Truscott, B., 541
Truscott, R.B., 548
Truxal, F.S., 62
Trybom, J., 376
Tryfiates, G.P., 509
Tryon, E.H., 508
Tryon, R.M., 211
Tryphonas, L., 563
Tsai, C.Y., 148
Tsai, J.A., 91
Tsai, M.J., 449
Tsai, S., 24
Tsang, J., 122
Tsao, C.H., 106
Tsapogas, M.J., 314
Tschinkel, W., 88
Tsein, R., 81
Tseng, C.C., 482
Tseng, M., 325, 338
Tsernoglou, D., 235
Tsibris, J.C.M., 90
Tsotsis, B., 126
Tsuchiya, H.M., 244
Tsuchiya, T., 65
Tsuda, R.T., 110
Tsukada, M., 497
Tsutsui, E.A., 458
Tu, A.T., 67
Tubb, R.A., 364
Tucker, B.B., 376
Tucker, C., 72
Tucker, C.E., 7
Tucker, F.S., 247
Tucker, G.E., 20
Tucker, H.A., 228
Tucker, H.F., 3
Tucker, H.H.Jr., 295
Tucker, J.M., 51
Tucker, J.O., 528
Tucker, Sr.M.E., 253
Tucker, R., 8
Tucker, R.E., 177
Tucker, S.C., 182
Tucker, V.A., 332
Tuckey, J.S., 150
Tudor, D.C., 288
Tuech, J.K., 436
Tueller, P.T., 275
Tuff, D., 455
Tufte, M.J., 522
Tugaw, J.E., 475
Tugwell, N.P., 23
Tugwell, R.L., 443
Tuinstra, K.E., 154
Tuite, J.F., 148
Tukell, R.A., 184
Tukey, H.B.Jr., 303
Tukey, R.B., 503
Tuleen, N.A., 459
Tulenko, T.N., 409
Tulis, J.J., 345
Tullis, J.E., 116
Tullis, R.E., 30
Tuma, D., 273
Tuma, H.J., 167
Tumbleson, M.E., 262
Tummalal, R.L., 226
Tung, A.K., 25
Tung, M.A., 533
Tunik, B., 325
Tunis, M., 318
Tunturi, A.R., 388
Tupper, C.J., 53
Tupper, J.T., 326
Turano, J., 65
Turbes, C.C., 268
Turcotte, W.H., 247
Turel, F.L.M., 562
Turetsky, S., 312
Turgeon, A.J., 138
Turgeon, F., 558
Turgeon, J.L., 204
Turgeon, P.L., 558
Turk, D.E., 424, 425
Turk, R.F., 340
Turkki, P., 326
Turko, A., 76
Turley, H.P., 310
Turley, R., 369
Turman, E.J., 376
Turnage, C.D., 335
Turnbull, A.L., 533
Turnbull, C.D., 341
Turnbull, D.I., 552
Turner, A.G.Jr., 341
Turner, B.H., 86, 247
Turner, B.J., 402
Turner, B.L., 465
Turner, D., 342
Turner, E.E., 456
Turner, G.M., 179
Turner, H.F., 12
Turner, J.A., 56
Turner, J.C.Jr., 526
Turner, J.E., 347
Turner, J.H., 414
Turner, J.K., 139
Turner, J.L., 5
Turner, L.G., 344
Turner, M.Jr., 11
Turner, M.E.Jr., 11
Turner, M.X., 285
Turner, P.R., 292
Turner, R., 479
Turner, R.B., 292
Turner, R.D., 211
Turner, T.B., 198
Turner, T.L., 481
Turner, W.J., 501, 503
Turney, C., 331
Turney, T.H., 485
Turnipseed, G., 20
Turnipseed, S.G., 424
Turoczi, L., 416
Turpin, F.T., 148
Turnquist, O.C., 241
Turnquist, R., 119
Turnquist, T., 361
Tustanoff, E.R., 551
Tuttle, D.M., 17
Tuttle, E.L., 215
Tuttle, J.W., 177
Tuveson, R.W., 136, 137
Tvrdik, G.M., 27
Twardock, A.R., 137
Twarog, B.M., 216
Twarog, R., 339
Tweedel, C.D., 227
Tweedell, K.S., 152
Tweedy, J.A., 130
Twiest, G.L., 392
Twight, B.W., 508
Twitchell, C.G., 257
Twente, J.W., 260
Twigg, B.A., 202
Twohy, D.W., 228
Twomey, J.A., 66
Tyan, M.L., 339
Tyberg, J.V., 60
Tyler, W.J., 517. 519
Tyler, W.S., 52
Tylutki, E.E., 117
Tyner, E.H., 138
Tyree, S.Y.Jr., 490
Tyrey, L., 332
Tyrl, R.J., 375
Tyroler, H.A., 342
Tyrrell, E.A., 215
Tysl, G., 121
Tyson, H., 556
Tyznik, W.J., 365

U

Ubben, M.S., 392
Ubelaker, J.E., 454
Udall, R.H., 69
Udenfriend, S., 298, 301
Udvardy, M.D.F., 31
Ueckert, D.N., 463
Uffen, R.L., 228
Ugent, D., 128
Uglem, G., 177
Uhl, C., 302
Uhlenbeck, O., 135
Uhler, L., 302
Uhler, L.D., 303
Uhlrich, H.M., 391
Uhr, J.W., 466
Uitto, J.J., 283
Ulam, S.M., 72
Ulberg, L.C., 343
Ulbrich, R.W., 123
Ulinski, P.S., 123
Ullery, C.H., 384
Ulliman, J.J., 118
Ullman, B.M., 233
Ullrey, D.E., 225
Ullrich, R.C., 211, 480
Ullrick, W., 208
Ulmer, D.H.B.Jr., 323
Ulmer, M.J., 157
Ulrich, A., 46
Ulrich, A.L., 164
Ulrich, D.V., 484
Ulrich, J.A., 291
Ulrich, J.T., 116
Ulrich, M.G., 163
Ulrich, V., 508
Ultsch, G.R., 10
Umbarger, H.E., 150
Umbreit, W.W., 288
Umminger, B.L., 369, 370
Umphlett, C.J., 425
Unakar, N.J., 230
Unbehaun, L., 515
Underhill, D.W., 217
Underjerg, G.K.L., 168
Underwood, J.T., 364
Underwood, V.H., 345
Ungar, F., 244
Ungar, I.A., 368
Unger, J.W., 523
Unger, L., 380
Unglaube, J.M., 336
Unrath, C.R., 345
Unsworth, B.R., 512
Unz, R.F., 404
Upadhyay, J., 494
Upadhyay, J.M., 187
Upchurch, W.J., 258
Updike, O.L., 489
Upham, S.D., 248
Upper, C.D., 519
Upson, D.W., 168
Upton, A., 325
Urban, J., 317
Urban, J.E., 166
Urbaitis, B.K., 204
Urbscheit, N.L., 319
Uretz, R.B., 132
Urey, J.C., 220
Uricchio, W.A., 391
Urness, P., 476
Urquhart, F.A., 550
Urry, D.W., 11
Urry, F., 467
Usborne, W.R., 547
Ushijima, R.N., 267
Utermohlen-Lovelace, V., 302
Utter, F.M., 498
Utter, M.F., 355
Utting, J., 312
Utz, J.P., 85
Utzinger, J.D., 365
Uyeda, K., 466
Uyeshiro, S.J., 150
Uzemoto, J.K., 307
Uzodinma, J.E., 246
Uzzell, T., 409

V

Vacca, L.L., 313
Vacey, V., 352
Vacquier, V.D., 52
Vadas, R.L., 194
Vagelos, P.R., 263
Vahouny, G.V., 84
Vaile, J.E., 23
Vaillancourt, G., 560
Vaillancourt, J., 549
Vaituzis, Z., 202
Valderrama, C.R., 418
Valdovinos, J., 298
Valdovinos, J.G., 297
Valek, D.A., 223
Valentine, B.D., 364
Valentine, F.A., 321
Valentine, J.W., 47
Valentour, J.C., 356
Valeriote, F.A., 263
Valiela, I., 207
Vallentine, J.F., 472
del Valle, M.R., 418
Valli, V.E.O., 548
Valli, V.J., 508
Vallillee, G.R., 552
van der Valk, A., 157
Valk, H.S., 102
Valleau, W.G., 193
Vallowe, H.H., 396
Van, N.T., 449
Vanable, J.W., 150
Van Alten, P., 134
Vanaman, T.C., 333
Van Amburg, G., 238
Vanarsdel, E.P., 458
Van Asdall, W., 18
Van Bavel, C.H.M., 459
Van Belle, G., 500
Van Blaricom, L.O., 424
VanBreeman, C., 96
Vanbreemen, V.L., 199
Van Bruggen, J.T., 388
Van Bruggen, T., 434
Van Bruiningen, H.J.P., 78
Van Buijtenen, J.P., 458
Van Buren, J.P., 307
Van Caeseele, L., 538
Vance, B.D., 452
Vance, D.E., 534
Vance, R., 55
Vance, V.J., 31
Van Citters, R.L., 497, 499, 500
Van Cleave, A.B., 561
Van Cleave, H.W., 458
Van Cleve, K., 14
Van Cleve, R., 498
Van Cott, J.W., 472
VandeBerg, J.S., 234
Vande Berg, W.J., 229
Vandemark, J.S., 138
Van Denack, Sr.J., 514
Vandenberg, J., 223
Vanden Born, W.H., 530
Vanden Branden, R.J., 155
Van Den Brink, C., 226
Van Den Hende, R., 558
Vander, A.J., 232
Vanderahe, A.R., 370
Vander, Beek, L.C., 236
Vanderford, H.B., 248
Van der Heide, L., 78
Vanderhoef, L.N., 136
Vanderhoek, J., 232
Vander Hoeven, T., 232
Vander Jagt, D.L., 294
Vander Kloet, S., 542
Van Der Kloot, W.G., 325
Vanderlip, R.L., 166
Vandermeer, J.H., 231
Vander Molen, G.E., 88
Vander Noot, G.W., 288
Vandersall, J.H., 201

Van Der Schalie, H., 231
Vanderstoep, J., 533
Vander Werf, C.A., 90
Vanderwoude, W.J., 57
Vanderzant, C., 458
Vander Zee, D., 154
VanDeventer, W.C., 236
Vandiver, R., 37
Van Doren, D.M., 364
Vandover, B., 455
Van Dyke, C.G., 343, 345
VanDyke, G., 131
Van Dyne, G., 67
Van Eck, E.A., 159
Van Eck, W., 508
Van Elswyk, M., 29
Van Etten, H.D., 304
Van Etten, J., 270, 272
VanFaasen, P., 224
Van Fleet, D.S., 105
Van Gelder, G.A., 262
VanGorder, P., 124
Van Gundy, J., 77, 79
Van Gundy, S.D., 58
Van Hartesveldt, C.J., 94
Van Hassel, H.J., 500
Van Hemert, D., 76
Van Holde, K.E., 385
Van Horn, D.L., 513
Van Horn, G., 223
Van Horn, G.S., 441
Van Horn, H.H., 91
Van Horn, J.L., 266
Van Horn, S.D., 522
Vanicek, C.D., 31
Vankat, J.L., 360
Van Keuren, R.W., 364
Vankin, G.L., 220
Van Kirk, R., 34
Van Kley, H., 256
Vanko, M., 327
Van Lear, D.H., 425
Van Lenten, L., 85
Van Leeuwen, L., 535
Van Liew, H.D., 319
Van Loon, G.R., 550
Van Loon, P., 418
Van Meter, W.G., 158
Van Middlesworth, L., 445
Vann, D.C., 113
Vannorman, R.W., 473
VanOrmer, D.W., 243
Van Pelt, A.F., 335
Van Pelt, R., 14
Van Pilsum, J.F., 244
Van Ryzin, Sr. M., 514
VanScoyoc, G.E., 147
Vanselow, N.A., 19
Van Stavern, B.D., 365
Van Stone, J.M., 76
Van Tuyle, G.C., 491
Van Ummerson, C.A., 219
VanValen, L., 132
Van Valkenburg, S.D., 201
Van Vliet, A.C., 381
Van Vunakis, H., 208
Van Winkle, Q., 362
Vanwright, A.Jr., 461
Van Wynsberghe, D., 523
Van Zandt, P.D., 373
Vaples, J.R., 433
Vardanis, A., 551
Varghese, S., 226
Vargo, A., 65
Vargo, R.A., 104
Variakokis, D., 133
Varkey, A., 510
Varmus, H.E., 60
Varnell, T.R., 527
Varner, J., 226
Varner, J.E., 263
Varney, E.H., 288
Varney, W.Y., 177
Varon, H., 450
Varon, S.S., 59
Varsa, E.C., 130
Vasek, F.C., 57
Vasey, C.E., 323
Vasey, R.B., 494
Varseveld, G.W., 383
Vasconcelos, A.C., 288
Vasil, I.K., 90
Vasington, F.D., 77, 79
Vassallo, G., 389
Vastano, A.C., 459
Vasu, B., 24
Vasvary, L.M., 288
Vaszquez, A., 506
Vaughan, C.E., 248
Vaughan, D.W., 208
Vaughan, F.L., 233
Vaughan, G.A., 253
Vaughan, G.L., 442
Vaughan, J., 374
Vaughan, J.P., 390
Vaughan, J.R., 400
Vaughan, M.H., 413
Vaughan, T., 15
Vaughn, C.M., 360
Vaughn, H.W., 118
Vaughn, J.C., 360
Vaughn, P., 55
Vaughn, R.H., 49
Vaux, H.J., 44
Vavich, M.G., 16
Vavra, M., 382
de Vazquez, M.I.C., 418
Veach, C.W., 154
Veale, F., 207
Vedder, D.K., 327
Vedros, N.A., 46
de la Vega, G., 396
Vega-Jacome, C., 558
Vehling, B., 420
Veilleux, R., 558
Veis, A., 127
Vela, G.R., 452
Velez, M.J., 418
Velicer, L.F., 228
Velick, S.F., 474
Vella, A.R., 186
Vella, F., 562
Veltri, R.W., 509
Vena, J., 290
Vener, K.J., 123
Vengris, J., 219
Venketeswaran, S., 463
Vennes, J.W., 352
Venzke, W.G., 365
Verbal, B.J., 336
Vercellotti, J.R., 493
Verch, R., 513
Verch, R.L., 513
Verdile, R.N., 391
Verdugo, P., 499
Verduin, J., 128
Verean, L., 427
Verhalen, L.M., 376
Verlangieri, A.J., 288
Verley, F.A., 229
Verly, W., 558
Verma, O.D., 9
Verman, D.P., 556
Vermaulen, C.W., 484
Vermeij, G.J., 199
Vernall, D.G., 367
Vernberg, W.B., 429
Verne, D., 355
Verner, J., 122
Veronneau, G., 559
Verpoorte, J.A., 542
Verrett, J.M., 182
Verschingel, R., 557
Verses, C., 76
Vertrees, G.L., 463
Verts, B.J., 383
Verwey, W.F., 467
Verzeano, M., 55
Vessey, S.H., 354
Vest, G., 226
Vestal, J.R., 370
Vester, J.W., 370
Vestling, C.S., 161
Vethamany, V.G., 542
Vethemany-Globus, S., 551
Veum, T.L., 259
Vezina, A., 558
Vezina, C., 558
Vial, C.C., 186
Vial, L., 186
Viale, R.O., 410
Viamonte, M.Jr., 95
Viator, C., 188
Viator, S.J., 186
Vice, J., 124
Vick, A.R., 336
Vickers, D.H., 88
Vickery, R.K., 473
Vickery, V.R., 556, 557
Vidaver, A., 270
Vidaver, A.K., 272
Vidaver, G., 270
Vidaver, W.E., 533
Vidgoff, J., 388
Vidic, B., 85
Vidoli, V., 30
Viens, P., 558
Vierck, C.J.Jr., 94
Viers, C.E., 188
Vig, B.K., 276
Vigfusson, N.V., 496
Vigil, E.L., 512
Viglierchio, D.R., 50
Vigue, C.L., 76
Vijayan, V.K., 53
Vilcek, J., 313
Viles, J., 157
deVillafranca, G.W., 215
Villa, V., 292
Villalard-Bohnsack, M., 420
Villanueva, G.B., 299
Villella, J.B., 417
Villemaire, A., 559
Villemez, C.L.Jr., 527
Viloria, J., 556
Vimmerstedt, J.P., 366
Vimmerstedt, J.P., 367
Vimmerstedt, J.R., 364
Vinay, J.P., 536
Vincent, N.J., 393
Vincent, W.S., 82
Vinegar, A., 372
Vines, H.M., 107
Vining, L.C., 542
Vinje, M.M., 159
Vinograd, J., 25
Vinogradov, S.N., 235
Vinopal, R.T., 77
Vinson, J.W., 212
Vinson, S.B., 458
Vinson, W.E., 493
Vinyard, W.C., 34
Virella, G., 428
Virkar, R.A., 286
Visnyei, K., 306
Visscher, M.B., 244
Visscher, S., 265
Visser, D.W., 64
Viswanathan, M.A., 557
Vitale, R., 449
Vitetta, E., 466
Vitt, D.H., 531
Vittor, B.A., 10
Vittum, M.T., 307
Vivian, V., 285
Vivrette, N.J., 42
Vizzini, E.A., 414
Vladykov, V., 549
deVlaming, V.L., 512
Vlamis, J., 46
Vodkin, M.H., 429
Voegeli, H.E., 79
Voekoff, G.M., 533
Voelz, H.G., 509
Voge, M., 56
Vogel, H., 438
Vogel, H.E., 425
Vogel, H.H., 445
Vogel, H.J., 301
Vogel, M.B., 339
Vogel, N.W., 415
Vogel, O.A., 502
Vogel, S., 332
Vogelmann, H.W., 480
Vogl, R.J., 31
Vogt, A.R., 366, 367
Vogt, D.W., 113
Vogt, P.K., 64
Vohra, P.N., 48
Vohs, P.A.Jr., 383, 434
Voigt, G.K., 81
Voight, J.W., 128
Voight, R.C., 515
Voight, R.L., 16
Voilker, H.H., 432
Voitle, R.A., 93
Volini, M., 131
Volk, B.G., 93
Volk, G.M., 93
Volk, G.W., 364
Volk, S.L., 33
Volk, V.V., 384
Volk, W.A., 490
Volker, V., 10
Volker, W.H., 176
Volkert, D., 286
Voll, M.J., 202
Vollerston, E.M., 278
Vollmer, D.A., 195
Volpe, E.P., 190
Volz, E.C., 156
Vomocil, J.A., 384
Von Foerster, H., 137
Voneida, T., 355
Voneida, T.J., 355
Von Gunten, R., 376
Von Hagen, S., 283
Von Zellen, B., 126
Voogt, J.L., 181
Voorhees, F.R., 123
Voos, J.R., 290
Voorhies, E.S., 439
Vopicka, E.V., 486
Vorbeck, M., 261
Vorhies, M.W., 434
Vorst, J.J., 147
Vose, R., 239
Voss, E.G., 231
Voss, E.W., 137
Voss, G., 96
Voss, N., 96
Voth, D., 70
Voth, E., 382
Vournakis, J.N., 326
Vranic, M., 550
Vratsanos, S.M., 301
Vredrichs, A.V., 189
Vredeveld, G.N., 440
de Vries, J., 534
Vukasin, P., 317, 323
Vukmonich, F., 237
Vukovich, F.M., 334
Vyse, E., 265

W

Wabeck, C., 203
Wabeck, C.J., 202
Wacha, R.S., 155
Wachowski, Rev.H.E., 34
Wachsmann, J.T., 137
Wachtel, A., 77, 79
Wack, P.E., 389
Waddell, E., 523
Waddell, H., 451
Wade, B.A., 189
Wade, B.F., 88
Wade, D., 66
Wade, E.K., 519
Wade, K., 175
Wade, N.L., 487
Wade, W., 437
Waddington, D.V., 401
Waddle, B.A., 22
Waddle, F., 335
Wadkins, C.L., 24
Wadson, V.H., 248
Wadsworth, D.F., 375
Wadsworth, G.E., 203
Wadsworth, S.E., 384
Waechter, C., 203
Waechter, R.F., 396
Wagenaar, E.B., 532
Wagenbach, G.E., 236
Wagenknecht, B.L., 264
Waggoner, G.R., 170
Waggoner, J., 62
Waggoner, P., 235
Waggoner, P.E., 81

Waggoner, W.J., 223
Wagh, P.V., 24
Wagle, R.F., 18
Wagman, I.H., 47
Wagner, C., 446
Wagner, C.E., 180
Wagner, C.K., 439
Wagner, D.G., 376
Wagner, E.D., 36
Wagner, E.K., 54
Wagner, F., 270, 272
Wagner, F.H., 476, 477
Wagner, G.H., 258
Wagner, H.E., 382
Wagner, K., 223
Wagner, J.L., 333
Wagner, K., 399
Wagner, M., 152
Wagner, N., 390
Wagner, R., 316, 522
Wagner, R.C., 82
Wagner, R.H., 340
Wagner, R.O., 522
Wagner, R.P., 465
Wagner, R.R., 490
Wagner, T.D., 502
Wagner, W., 62
Wagner, W.C., 158
Wagner, W.H., 231
Wagoner, D.E., 351
Wahba, A., 559
Wahl, M., 518
Wahl, P., 500
Wahl, R.W., 407
Wahlstrom, R.C., 433
Wahlquist, A.G., 473
Wailly, L.F., 70
Waincoat, T., 445
Waines, J.G., 492
Wainio, W.W., 288
Wainwright, L., 543
Wainwright, S.A., 332
Wainwright, S.D., 542
Waite, M.R.F., 465
Waite, B.M., 347
Waite, J.G., 98
Waites, R.E., 91
Waits, E.D., 6
Wake, D.B., 43
Wake, M.H., 43
Wakefield, R.C., 422
Wakei, S., 533
Wakeman, D.L., 91
Wakil, S.J., 448
Walbot, V., 263
Walch, H.A., 39
Walcott, C., 325
Wald, G., 210
Waldbauer, E.C., 321
Waldbauer, G.P., 136
Walden, D.B., 552
Waldman, B., 152
Waldman, R.H., 94
Waldorf, E.S., 183
Waldren, C., 72
Waldrip, E., 491
Waldron, A.C., 363, 364
Waldron, I.L., 409
Waldron, L.J., 46
Waldroup, P.W., 23
Walensky, N.A., 288
Wales, -., 546
Wales, J.H., 383
Walgenbach, D.D., 433
Wali, M.K., 352
Walike, B., 500
Walker, B., 55
Walker, B.E., 227
Walker, C.A., 9, 269
Walker, D., 105, 417
Walker, D.E., 254
Walker, D.G., 86, 197
Walker, D.H.Jr., 161
Walker, D.K., 34
Walker, D.L., 521
Walker, D.W., 94
Walker, E., 552
Walker, G.A., 84
Walker, G.W.R., 535
Walker, H.O., 50
Walker, H.W., 155
Walker, I.G., 551
Walker, J.A., 269
Walker, J.G., 65
Walker, J.K., 458
Walker, J.M., 22
Walker, J.N., 179
Walker, J.P., 547
Walker, J.R., 465, 468
Walker, J.W., 217
Walker, K., 196
Walker, L., 336
Walker, M.E., 106
Walker, N., 377
Walker, N.A., 164
Walker, N.E., 255
Walker, P.C., 324
Walker, R., 170
Walker, R.B., 497, 546
Walker, R.E., 123
Walker, R.F., 426
Walker, R.H., 456, 463
Walker, R.V., 368
Walker, R.W., 191
Walker, S.H., 72
Walker, T.J., 91
Walker, T.D., 149
Walker, W., 155
Walker, W.F.Jr., 361
Walker, W.H., 406
Walker, W.M., 138
Walker, W.S., 155, 425
Walkington, D.L., 30
Wall, J.R., 485
Wall, R., 56
Wall, W.J., 208
Wallace, A.C., 552
Wallace, A.G., 333
Wallace, B., 302
Wallace, D.H., 304, 305
Wallace, E., 290
Wallace, F.L., 423
Wallace, G., 439
Wallace, H.D., 91
Wallace, H.S., 188
Wallace, H.W., 411
Wallace, J., 470
Wallace, J.B., 106
Wallace, J.D., 292
Wallace, J.H., 180
Wallace, J.M., 58
Wallace, J.T., 413
Wallace, O.P., 280
Wallace, R.J., 392
Wallace, R.L., 117
Wallace, S., 298
Wallace, S.L., 100
Wallace, S.S., 297
Wallace, T.P., 315
Wallace, W.A., 489
Wallace, W.C.Jr., 201
Wallace, W.R., 189
Wallbank, A.M., 538
Wallbrunn, H.M., 90
Walle, E.M., 301
Wallen, E.P., 522
Wallen, L.E., 374
Wallenius, R.W., 502
Wallentine, M.V., 472
Waller, D., 359
Waller, G.D., 17
Waller, G.R., 377
Waller, J.H., 414
Waller, J.R., 353
Walley, W.W., 246
Wallin, J.R., 157
Wallin, R., 317
Wallner, W.E., 226
Walls, J.A., 9
Walls, N.W., 102
Walls, R.C., 24
Walne, P.L., 441
Walp, R.L., 360
Walsh, C., 412
Walsh, D.A., 53
Walsh, D.E., 351
Walsh, Br.J., 291
Walsh, J.J., 498
Walsh, K.A., 499
Walsh, M.A., 412
Walsh, P.J., 341
Walsh, R.R., 256
Walstrom, R.J., 433
Walt, J.W., 159
Walter, D.O., 57
Walter, R.G., 68
Walter, R.H., 307
Walter, T.L., 167
Walter, W.G., 265
Walter, W.M., 139, 344
Walters, C.J., 535
Walters, C.S., 139, 186
Walters, D.R., 27, 207
Walters, E., 262, 545
Walters, H.J., 24
Walters, J.B.Jr., 12
Walters, J.L., 61
Walters, L., 376
Walters, R.N., 246
Walters, T., 36
Walters, T.W., 230
Walther, F.R., 459
Waltner, A.W., 341
Walton, B.A., 441
Walton, D., 322
Walton, D.J., 545
Walton, P.D., 530
Walton, G.F., 288
Walton, Br.J., 299
Walz, A.J., 117
Walzak, W., 124
Wampler, J.E., 105
Wan, A.T., 487
Wanamaker, J.F., 127
Wandesforde-Smith, G.A., 49
Wanek, W., 236
Wang, A.C., 428
Wang, A.C.S., 315
Wang, A.W., 247
Wang, C.J.K., 321
Wang, C.S., 88
Wang, C.W., 118
Wang, D.I.C., 213
Wang, H.C., 562
Wang, H.H., 61
Wang, I., 428
Wang, J.H., 318, 538
Wang, J.S., 320
Wang, L., 531
Wang, L.K., 314
Wang, M.B., 409
Wang, N.S., 556
Wang, R.J., 260
Wang, T.Y., 318
Wangaard, F.F., 66
Wangensteen, O.D., 244
Wanger, D.F., 444
Wangersky, P.J., 542
Wangsness, P.J., 401
Wann, D.F., 203
Wannamaker, L.W., 244
Wanner, A., 95
Warburton, D., 301
Warburton, F.E., 300
Warchalowski, G., 285
Ward, -., 81
Ward, A.J., 487
Ward, C.F., 175
Ward, C.H., 453
Ward, C.L., 332
Ward, C.M., 25
Ward, C.S., 106
Ward, C.Y., 248
Ward, D.B., 90
Ward, D.K., 286
Ward, E.J., 539
Ward, F.E., 333
Ward, F.J., 538
Ward, G.E., 563
Ward, G.H., 123
Ward, G.L., 143
Ward, G.M., 65
Ward, H., 72
Ward, H.S., 7
Ward, J., 122, 192
Ward, J.A., 349
Ward, J.B., 346
Ward, J.E.Jr., 335
Ward, J.V., 68
Ward, J.W., 97, 446
Ward, Sr.M., 33
Ward, M.A., 111
Ward, O.G., 19
Ward, P., 437
Ward, P.A., 79
Ward, R.T., 68, 325
Ward, R.W., 223
Ward, W.F., 395
Ward, W.W., 402
Wardell, W.L., 205
Warden, J.C., 437
Ware, C.B., 325
Ware, F., 273
Ware, G.H., 139
Ware, G.L., 104
Ware, G.W., 17
Ware, K., 108
Ware, S.A., 484
Ware, W., 452
Warinner, J.E., 490
Waring, G., 129
Waring, G.H., 129
Waring, T., 234
Warma, S.K., 551
Warmbrod, J.G., 443
Warme, P.K., 404
Warner, C.M., 157
Warner, E., 523
Warner, F.D., 326
Warner, H.R., 241
Warner, J., 329, 330, 515
Warner, J.R., 425
Warner, L., 85
Warner, M.H., 364
Warner, P., 538
Warner, R.C., 54
Warner, R.J., 24
Warner, R.L., 502
Warner, R.M., 114
Warnes, C.E., 338
Warnhoff, E.H., 182
Warnick, A.C., 91
Warnke, E., 238
Warnock, B.H., 455
Warnock, J.E., 139
Warnock, L.G., 446
Warnock, M.L., 133
Warnock, R., 478
Warr, W.B., 208
Warram, J.H., 212
Warren, B.A., 552
Warren, C.E., 383
Warren, C.O., 439
Warren, D., 127
Warren, D.W., 64
Warren, F.W., 104
Warren, G.F., 149
Warren, H.L., 148
Warren, J., 224
Warren, J.C., 263
Warren, M., 437
Warren, M.E., 451
Warren, R., 96, 523
Warren, R.A.J., 535
Warren, R.S., 75
Warren, S.R., 411
Warren, W., 295
Warren, W.M., 3
Warshawsky, H., 555
Wartell, R.M., 102
Wartik, T., 403
Warzeski, W., 397
Wascom, E.R., 189
Washako, W.W., 78
Washborn, M., 306
Washburn, K.W., 108
Washburn, L.L., 31
Washington, A., 453
Washington, B.J., 99
Washington, E., 119
Washington, G., 246
Washington, V., 336
Washington, W.J., 357
Washino, R.K., 49
Washko, J.B., 401
Washko, W.W., 79
Waskom, J.D., 189
Waskoskie, W., 396
Wasland, J.R., 497
Wass, M.L., 490

Wasser, C., 66, 67
Wasserman, A.O., 297
Wasserman, A.R., 555
Wasserman, M., 297, 298
Wasson, C.E., 166
Wasti, S.S., 420
Waston, D.P., 114
Watabe, N., 429
Watanabe, S., 60
Watenpaugh, K.D., 499
Waterbury, A.M., 27
Waterhouse, J.S., 324
Waterman, F.E., 306
Waterman, M.R., 466
Waterman, R.E., 293
Waterman, T.H., 80
Waters, C., 427
Waters, J.F., 34
Waters, N.D., 117
Waters, T.F., 240
Watertor, J., 173
Watkins, D.T., 79
Watkins, D.W., 84
Watkins, E.A., 39
Watkins, E.M., 171
Watkins, H., 177
Watkins, J.L., 401, 402
Watling, H., 265
Watrous, G.H., 401, 402
Watrous, J.J., 406
Watschke, T.L., 401
Watson, B.B., 9
Watson, B.F., 460
Watson, D.W., 244
Watson, E., 533
Watson, J.A., 60
Watson, J.C., 187
Watson, J.D., 210
Watson, J.E., 5
Watson, J.L., 164
Watson, J.R., 247
Watson, L., 369
Watson, M.L., 159
Watson, R.A., 192
Watson, R.B., 342
Watson, R.D., 117
Watson, R.L., 25
Watson, R.R., 146, 252
Watson, R.T., 94
Watson, T.F., 17
Watson, T.J., 266
Watson, W.F., 250
Watson, W.Y., 552, 554
Watt, D.D., 268
Watt, J.R., 482
Watt, K.E.F., 47, 49
Watt, L.A.K., 550
Watt, W.B., 41
Watt, W.J., 495
Watters, C., 479
Watterson, R.L., 137
Watts, A.B., 184
Watts, D.T., 491
Watts, J.G., 292
Watts, P.W., 29
Watts, R.B., 365
Watts, R.E., 18
Watts, R.L., 545
Watts, V.M., 23
Waugh, D.D., 545
Wave, H.E., 195
Wax, L.M., 138
Waxman, S., 78
Way, J.S., 408
Way, R.D., 307
Waygood, E.R., 538
Wayman, O., 113
Waymire, J.C., 55
Wayne, P., 378
Wead, W.B., 181
Weakly, J.N., 340
Wear, J.I., 3
Wearden, S., 507
Weathers, G., 58
Weathers, L.G., 58
Weathers, W.W., 288
Weathersby, A.B., 106
Weathersby, H.T., 466
Weathersby, S.M., 187
Weaver, A.A., 358
Weaver, B., 265
Weaver, E.C., 32
Weaver, G.T., 130
Weaver, J.B., 106
Weaver, L.O., 201
Weaver, N., 219
Weaver, P.W.H.Jr., 400
Weaver, R.F., 170
Weaver, R.J., 51
Weaver, R.M., 302
Weaver, R.W., 459
Webb, A., 336
Webb, A.D., 51
Webb, B., 126
Webb, B.B., 376
Webb, B.C., 337
Webb, D.E.Jr., 493
Webb, D.W., 91
Webb, F., 502
Webb, G.N., 197
Webb, H.M., 191
Webb, J.A., 544
Webb, J.L., 556
Webb, K.L., 490
Webb, L.G., 424
Webb, Sr.M., 33
Webb, N.B., 344
Webb, N.E., 445
Webb, P.M., 152
Webb, R., 372
Webb, R.E., 494
Webb, R.G., 468
Webb, S.C., 102
Webb, S.D., 90
Webb, W.L., 322
Webber, E.E., 310
Webber, H.H., 96
Webber, P.J., 71
Webber, W.A., 535
Webent, H., 184
Weber, A., 410
Weber, A.T., 272
Weber, Sr.C., 33
Weber, C.S., 205
Weber, D., 122
Weber, D.A., 329
Weber, D.D., 498
Weber, D.J., 472
Weber, E.J., 138
Weber, G., 135, 137
Weber, G.R., 142, 397
Weber, H.P., 87
Weber, J., 559
Weber, J.B., 344
Weber, K.C., 509
Weber, L., 419
Weber, L.J., 383
Weber, M.J., 137
Weber, M.M., 256
Weber, P.B., 327
Weber, R.J., 85
Weber, W.R., 258, 411
Weber, W.T., 412
Webster, B.D., 47, 51
Webster, C.R.B., 537
Webster, D., 185
Webster, D.A., 121, 303
Webster, F.C., 481
Webster, G.C., 87
Webster, G.L., 51
Webster, G.R.B., 537
Webster, G.T., 177
Webster, H.H., 156
Webster, J., 226, 320
Webster, J.D., 144
Webster, J.E., 377
Webster, J.F., 288
Webster, J.M., 533
Webster, J.S., 271
Webster, P., 217
Webster, R., 145, 333
Webster, R.K., 50
Webster, R.M., 185
Webster, R.N., 142
Webster, S.K., 32
Webster, S.W., 415
Webster, T., 77
Webster, T.Jr., 77
Webster, W., 453
Webster, W.P., 340
Wechsler, A., 556
Wechsler, J.A., 300
Weckel, K.G., 518
Wecker, S., 297
Wedberg, H.L., 39
Wedding, R.T., 57
Wedemeyer, G.A., 498
Weed, C., 399
Weed, R.I., 328
Weedon, R., 268
Weekman, G.T., 344
Weeks, B.A., 487
Weeks, G., 535
Weeks, K., 159
Weeks, L., 335
Weeks, M.E., 219
Weeks, S.B., 280
Weeks, T., 515
Weeks, W.W., 344
van Weel, P.B., 112
Weems, C.W., 506
Weete, J.D., 4
Weeth, H.J., 275
Weg, R.B., 62
Wegman, D.H., 212
Wegmann, T.C., 210
Wegner, K.H., 434
Wegner, T.N., 17
Weiant, E.A., 215
Weibel, D.E., 376
Weibel, M.K., 410
Weibust, R.S., 239
Weichenthal, B.A., 138
Weick, R.F., 552
Weidensaul, T.C., 366
Weidman, S.W., 263
Weidner, E.H., 183
Weidner, T.M., 120
Weier, T.E., 51
Weierich, A., 336
Weigel, R.D., 122
Weigensberg, B.I., 556
Weiger, J.C., 480
Weigl, A., 348
Weigl, P.D., 347
Weigle, J.L., 156
Weihing, J., 270
Weijer, J., 535
Weik, K.L., 123
Weil, J.D., 446
Weil, M.H., 63
Weiler, J., 30
Weiler, T.C., 149
Weiler, W.A., 120
Weimer, E., 510
Weimer, H.E., 56
Wein, R.W., 540
Weinack, O., 219
Weinbaum, C.M., 111
Weinbaum, G., 408
Weinberg, E., 144, 197
Weinberger, N.M., 55
Weinberger, P., 549
Weiner, H., 148
Weiner, L.M., 236
Weiner, R., 296
Weiner, R.M., 202
Weinfeld, H., 320
Weingartner, D.L., 199
Weingartner, R.H., 126
Weinhouse, S., 408
Weinhold, A.R., 45
Weinhold, P.A., 232
Weinmann, C.J., 44
Weinsieder, A., 235
Weinstein, P.P., 152
Weinstock, M., 555
Weintraub, H.M., 286
Weintraub, M., 534
Weintraub, R.L., 84
Weintrub, J.D., 30
Weir, J.A., 171
Weir, W.C., 50
Weires, R.W.Jr., 306
Weirich, H.R., 400
Weis, A., 67
Weis, D.S., 357
Weis, J.S., 166
Weis, P., 282
Weisbart, M., 234
Weisbrot, D., 290
Weise, C., 523
Weisel, G.F., 267
Weiser, R.S., 499
Weisenbeck, J., 174
Weiser, C.J., 383, 384
Weiser, P., 71
Weisinger, H., 325
Weisman, R.A., 469
Weismeyer, H., 446
Weisner, R., 300
Weiss, B., 198, 300, 399
Weiss, B.F., 210
Weiss, C.M., 341
Weiss, E.M., 234
Weiss, G.B., 467
Weiss, H., 284
Weiss, I.T., 312
Weiss, J., 294
Weiss, L., 197, 320
Weiss, S.B., 131, 132
Weissbach, A., 301
Weissbach, H., 301
Weissman, G.S., 287
Weithans, Sr.B.M., 511
Weitkamp, L.P., 328
Weitlauf, H., 388
Weitsen, H.A., 185
Weitzel, T.O., 89
Weitzman, M.C., 329
Wekstein, D.R., 180
Welander, A.D., 498
Welbourne, F.F., 427
Welch, A., 270
Welch, C.A., 238
Welch, D.R., 511
Welch, G., 448, 477
Welch, H.E., 538
Welch, J.E., 51
Welch, J.G., 480
Welch, J.T., 199
Welch, L.F., 138
Welch, R.M., 109, 333
Welch, W.H., 275
Welden, A.L., 190
Weldon, P., 556
Welker, E.M., 11
Welker, N.E., 126
Welker, R., 310, 455
Welker, W.V., 288
Welkie, G.U., 476
Well, M.L., 159
Wellband, W.A., 103
Wellborn, T.L., 250
Weller, D.L., 481
Weller, E., 10
Weller, E.M., 11
Weller, H., 360
Weller, M.W., 157
Weller, R.I., 394
Wellington, J.S., 60
Wellington, W.G., 534
Wellner, D., 307
Wellons, J.D., 384
Wells, B., 329
Wells, C., 486
Wells, C.G., 346
Wells, C.H., 468
Wells, C.T., 8
Wells, C.V., 336
Wells, D.G., 433
Wells, G.D., 480
Wells, G.N., 87
Wells, G.R., 443
Wells, H.B., 341
Wells, I.C., 268
Wells, J., 482
Wells, J.A., 124
Wells, J.C., 345
Wells, J.D., 364
Wells, J.H., 254
Wells, K., 51
Wells, K.L., 177
Wells, L.G., 179
Wells, M., 396, 465
Wells, M.A., 19
Wells, M.R., 439
Wells, O.S., 281
Wells, P.H., 37
Wells, P.V., 170, 172

Wells, R.D., 517
Wells, R.F., 316
Wells, V., 193
Wells, W., 127
Wells, W.W., 225
Wellso, S., 226
Wellwood, A.A., 552
Welsch, C.W., 227
Welsch, F., 227
Welsh, C.S., 490
Welsh, B.L., 77
Welsh, J.F., 34
Welsh, J.P., 33
Welsh, J.R., 65
Welsh, R.A., 186
Welsh, S.L., 472
Welshimer, H.J., 492
Welter, D.A., 103
Welter, J.F., 425
Welton, R., 387
Welty, J.D., 435
Welty, R.C., 165
Welty, R.E., 345
Wemyss, C.T., 309
Wende, R.D., 449
Wendel, E.O., 329
Wendlandt, R.M., 177
Wendt, G., 310
Wengbusch, H.G., 317
Wengel, R.W., 78, 79
Wenger, B.S., 562
Wenger, E., 562
Wenger, J., 437
Wenger, J.I., 405
Wenger, R.W., 149
Weniger, D., 452
Wenke, T.L., 164
Wenner, C., 320
Wenner, K.A., 366
Wennhold, A.R., 473
Wensel, L.C., 44
Wensler, R.J., 547
Went, H.A., 501
Wentland, M.J., 315
Wentworth, B., 228
Wentworth, B.C., 519
Wentworth, D., 340
Wentworth, J., 494
Wentz, W.B., 356
Wenzel, B.M., 57
van de Werken, H., 444
Wermers, G.W., 213
Wermuth, J.E., 150
Werner, E.E., 227
Werner, F.G., 17
Werner, H.J., 183
Werner, J.K., 229, 520
Werner, P.A., 227
Werner, R., 95
Werner, R.G., 322
Werner, W.E.Jr., 119
Wernsman, E.A., 344
Werntz, H.O., 214
Wersham, M., 186
Wert, J.J., 446
Werth, J.M., 290
Werth, R.J., 150
Wertz, G., 339
Weseli, D.G., 458
Wesley, D.E., 250
Wesley, I., 70
Wesner, G., 127
Wessells, N.K., 41
Wessenberg, H.S., 40
Wessling, W.H., 344
West, A.J.II., 216
West, A.S., 546
West, C.D., 474
West, D.A., 492
West, D.J., 512
West, G.C., 14
West, H.H., 109
West, J.A., 42
West, J.L., 348
West, J.R., 346
West, K.P., 393
West, L.S., 229
West, M.L., 31
West, N., 476
West, N.E., 477
West, R.F., 288
West, R.L., 91
West, S.S., 11
West, W.R., 488
West, W.T., 325
Westby, C.A., 433
Westdal, H., 537
Westenskow, G.D., 473
Wester, W.C., 358
Westerberg, H.E., 121
Westerfield, W.W., 326
Westergaard, J.M., 6
Westerling, J., 316
Westervelt, C.A.Jr., 32
Westfall, J.A., 168
Westfall, J.J., 105
Westfall, M.J., 90
Westfall, R.D., 322
Westhead, E.W., 217
Westhiwer, G., 43
Westhoff, D.C., 201, 202
Westin, F.C., 433
Westin, J., 315
Westing, A.H., 482
Westing, T.W., 28
Westlake, W., 408
Westley, J., 131
Westmeyer, H.W., 167
Weston, C., 31
Weston, G.D., 531
Weston, H.G., 32
Weston, J.A., 387
Weston, J.C., 400
Westphal, J., 515
Westphal, U.F., 180
Westra, R., 103
Westrum, L.E., 499
Westwood, J.C.N., 549
Westwood, M.N., 384
deWet, J.M.J., 136
Wetherell, D.F., 77
Wetherell, D.F., 79
Wethington, Sr.M.J., 358
Wetlaufer, D.B., 244
Wetmore, C.M., 241
Wetta, Sr.T., 165
Wettemann, R., 376
Wettstein, F., 56
Wetzel, R., 77
Wetzel, K., 389
Wetzel, R., 439
Wetzel, R.G., 227
Wetzel, R.M., 77, 79
Weyant, R.G., 531
Weybrew, J.A., 344
Weyerhaeuser, J.P.Jr., 81
Weyerts, P.R., 456
Weygandt, C., 411
Weyl, F.J., 298
Weyland, H.B., 455
Weymouth, R.J., 491
Weyrick, R.R., 280
Weyter, F.W., 299
Whalen, J.W., 468
Whalen, R.E., 55
Whalen, R.H., 432
Whalen, T.A., 316
Whaley, J., 29
Whaley, W.G., 465
Whalls, M.J., 26
Whang, H.Y., 319
Wharton, D.C., 469
Wharton, G.W., 362, 363
Wharton, M.E., 175
Wharton, W.W., 365
Whatley, B.T., 9, 305
Whatley, E.C., 188
Whatley, J.A., 376
Whayne, T., 380
Wheat, J.D., 167
Wheat, R.W., 333
Wheatley, J., 533
Wheatley, J.F., 368
Wheatley, J.H., 366
Wheaton, H.N., 258
Wheeler, A.G., 402
Wheeler, B., 75
Wheeler, D., 395, 471
Wheeler, D.D., 429
Wheeler, E.J., 103
Wheeler, G., 298
Wheeler, G.E., 296, 297
Wheeler, H., 178
Wheeler, J.P.Jr., 485
Wheeler, L.C., 62
Wheeler, M.R., 465
Wheeler, M.W., 79
Wheeler, R.F., 424
Wheeler, W.B., 92
Whelan, J.B., 494
Whelan, W.J., 95
Whellis, M.L., 51
Whicker, F.W., 69
Whigham, D.F., 287
Whigham, D.K., 138
Whipp, B., 57
Whipple, G.H., 233
Whipple, H.L., 102
Whipple, S.D., 5
Whisenhunt, D., 291
Whisler, F.D., 248
Whisler, H.C., 497
Whisler, K.E., 513
Whistler, R.L., 148
Whitaker, D.R., 548
Whitaker, E., 215
Whitaker, J.R., 49
Whitby, J.L., 551
Whitcomb, C.E., 377
Whitcomb, W.H., 91
White, A., 41
White, A.A., 261
White, A.G.C., 492
White, A.M., 359
White, B., 157
White, B.J., 218
White, B.N., 546
White, C., 522
White, C.A., 122
White, C.E., 445
White, C.H., 106
White, C.M., 472'
White, D., 144
White, D.A., 473
White, D.B., 241
White, D.C., 88
White, D.E., 508
White, D.G., 139
White, D.J., 206
White, E.H., 243
White, E.L., 220
White, E.M., 8, 433
White, E.P., 70
White, F., 336
White, F.G., 472
White, F.H., 94
White, F.J., 334
White, F.M., 405
White, F.N., 57
White, G.M., 179
White, G.V.S., 187
White, H.B.Jr., 251
White, J., 147, 336, 339
White, J.Jr., 375
White, J.A., 116
White, J.C., 186
White, J.E., 310
White, J.F., 101, 211
White, J.I., 204
White, J.L., 147
White, J.M., 93, 126, 493
White, J.P., 315
White, J.R., 340
White, J.S., 246
White, K., 514
White, L.L., 189
White, M., 44
White, M.P., 12
White, R., 118
White, R.A., 332
White, R.C., 479
White, R.G., 14
White, R.H., 62, 219
White, R.J., 225, 273, 496
White, R.S., 88
White, S.H., 54
White, W.J., 561
White, W.S., 257
Whited, D.A., 350
Whiteford, R.D., 548
Whitehead, A.T., 473
Whitehead, P.C., 516
Whitehead, R.A., 238
Whitehead, W.D., 488
Whitehill, A.R., 194
Whitehouse, F., 232
Whitehouse, U.G., 460
Whiteker, M.D., 177
Whiteley, A.H., 498
Whiteley, H.R., 499
Whitely, E.L., 459
Whiteman, C.E., 69
Whiteman, J.V., 376
Whitenberg, D., 455
Whiteny, M.I., 223
Whitesell, J.H., 25
Whiteside, B.G., 455
Whiteside, M.C., 442
Whiteside, W.C., 120
Whitfield, C.F., 405
Whitfield, G.B., 429
Whitfield, H.J., 232
Whitford, P.B., 521
Whitford, W., 292
Whiting, A.M., 309
Whiting, C.R., 507
Whiting, F.M., 17
Whiting, G.C., 494
Whitis, D., 15
Whitla, M., 540
Whitley, K., 455
Whitley, L.S., 120
Whitlock, J.H., 306
Whitman, W.C., 349
Whitmire, R., 522
Whitmon, S.R., 437
Whitmore, D.H., 464
Whitmore, F.W., 366, 367
Whitmore, G.F., 550
Whitmore, G.V., 549
Whitmore, M.R., 379
Whitmore, R.A.Jr., 481
Whitney, C., 370
Whitney, D.A., 166
Whitney, J.B.Jr., 425
Whitney, J.E., 24, 25
Whitney, N.J., 540
Whitney, R., 308
Whitney, R.R., 498
Whitney, R.S., 65
Whitsel, B.L., 340
Whitsett, J.M., 346
Whitson, P.D., 162
Whitson, G.L., 442
Whitson, R.E., 458
Whitt, G.S., 137
Whitt, S.K., 486
Whittaker, F.H., 180
Whittaker, J.C., 194
Whittaker, R., 302
Whittaker, R.H., 303
Whitten, B.K., 228
Whitten, D.N., 10
Whittenberger, J.L., 212
Whittick, A., 541
Whittier, H.O., 88
Whittig, L.D., 50
Whittingham, W.F., 515
Whittinghill, M., 339
Whitton, L., 472
Whittow, G.C., 113
Whitworth, C., 105
Whitworth, N., 445
Whitworth, W.R., 78, 79
Whorton, E.B.Jr., 468
Whyatt, P.L., 445
Whybrew, W., 439
Wiant, H.V., 508
Wiberg, J.S., 328
Wiberley, S.E., 314
Wicher, K., 319
Wicht, M.C.Jr., 485
Wick, E.L., 214
Wick, J., 15
Wick, Sr.J.A., 154
Wick, J.R., 15
Wickelgren, W.O., 72
Wickersham, E.W., 403
Wickland, W.B., 318
Wickliff, J.L., 22

Wicklow, D.T., 412
Wickner, W.T., 56
Wickoff, H.A., 102
Wicks, G.A., 271
Wickstrom, C.E., 100
Wickware, R., 75
Widdows, R.E., 437
Widdowson, D.C., 8
Wideman, C.H., 359
Widholm, J.M., 138
Widmayer, D.J., 220
Widmer, R.E., 241
Widmoyer, F.B., 293
Wiebe, H.H., 476, 477
Wiebe, H.T., 497
Wiebe, J.P., 552
Wiebe, M.E., 308
Wiebenga, W.M., 460
Wiebers, J.E., 150
Wieckert, D.A., 517
Wiedeman, M.P., 409
Wiedeman, V.E., 180
Wiederhielm, C., 500
Wiedman, H.W., 31
Wiedmeier, V.T., 104
Wiegers, H.L., 272
Wiegert, R.G., 106
Wiener, E.L., 95
Wiens, A.W., 224
Wiens, D., 473
Wiens, J.A., 386
Wiens, J.H., 534
Wiersig, D.O., 460
Wiersma, C.A.G., 25
Wiersma, D., 147
Wiersma, J., 514
Wiese, L., 88
Wiese, R.A., 271
Wiesen, C.F., 528
Wiesendanger, M., 552
Wigdor, S., 555
Wiggans, D.S., 466
Wiggans, S.C., 481
Wiggins, E.L., 3
Wiggins, W., 198
Wiggs, A.J., 540
Wightman, F., 544
Wiglesworth, F.W., 555
Wiitanen, W.A., 387
Wikström, M.K.F., 410
Wilber, C.G., 60
Wilberg, C.A., 496
Wilbur, H.M., 332
Wilbur, K.M., 332
Wilbur, R.L., 332
Wilce, R.T., 217
Wilcox, C.J., 91
Wilcox, D., 395
Wilcox, G.E., 149
Wilcox, H.E., 321
Wilcox, H.H., 445
Wilcox, R.B., 35
Wilcox, S., 463
Wilcox, W.C., 412
Wilcox, W.W., 44
Wilcoz, J.R., 147
Wild, F.B., 445
Wild, G.C., 294
Wilde, C.E.Jr., 411
Wilde, G.E., 167
Wilde, R., 522
Wilde, W.S., 232
Wilder, B., 309
Wilder, B.J., 94
Wilder, C.D., 175
Wilder, M.S., 218
Wilding, L.D., 364
Wildins, H.D., 350
Wildman, R.B., 157
Wildman, S.G., 55
Wildman, W.C., 157
Wiles, A.B., 249
Wiley, B.B., 474
Wiley, D.C., 210
Wiley, L., 30
Wiley, R.C., 202
Wiley, R.H., 339
Wiley, R.L., 360
Wiley, R.W., 25
Wiley, W.H., 425
Wilfert, C.M., 333
Wilfong, A.J., 358
Wilhelm, A.R., 29
Wilhelm, H.G.Jr., 145
Wilhelm, J.M., 328
Wilhelm, S., 45
Wilhelm, W.E., 438
Wilhelmi, A.E., 101
Wilhm, J., 375
Wilhm, J.L., 376
Wilimovsky, N.F., 535
Wilinson, R.E., 106
Wilk, J.C., 343
Wilke, A., 262
Wilken, D.H., 68
Wilken, D.R., 520
Wilkening, M.H., 291
Wilkens, J.L., 531
Wilkes, H.G., 219
Wilkes, J.C., 8
Wilkes, F.W., 8
Wilkes, S., 15
Wilkes, W., 408
Wilkie, B.N., 548
Wilkie, D., 297
Wilkins, B.T., 303
Wilkins, H.F., 241
Wilkins, M.G., 489
Wilkins, O.P., 182
Wilkins, P.O., 551
Wilkinson, C.F., 303
Wilkinson, R.C., 91
Wilkinson, R.E., 304
Wilkinson, R.F., 258
Wilkinson, R.R., 346
Wilkinson, T.R., 433
Will, D.H., 69
Will, J.A., 520
Willard, B., 71
Willard, C.J., 364
Willard, H.B., 355
Willard, J.M., 482
Willard, M., 263
Willard, O., 523
Willemsen, R.W., 288
Willers, W.B., 523
Willes, M., 543
Willett, H.P., 333
Willey, T.J., 36
Willhoit, D.G., 341
Williams, A., 45, 144
Williams, A.M., 522
Williams, A.S., 178
Williams, B., 15
Williams, Sr.B., 285
Williams, C., 210, 331
Williams, C.H., 232
Williams, C.M., 210, 561
Williams, C.R., 378
Williams, D., 152
Williams, D.B., 329, 444
Williams, D.C., 504
Williams, D.E., 46
Williams, D.F., 28
Williams, D.J., 138
Williams, E., 210, 395
Williams, E.B., 148
Williams, E.C., 153
Williams, F., 96, 450
Williams, F.D., 156
Williams, F.M., 403
Williams, G., 372, 431
Williams, G.Jr., 189
Williams, G.R., 144, 494, 549
Williams, G.W., 233, 501
Williams, H., 443
Williams, H.B., 184
Williams, H.W., 264
Williams, J., 165, 199, 384
Williams, J.A., 60
Williams, J.A.Jr., 336
Williams, J.C.Jr., 4
Williams, J.E., 392
Williams, J.F., 228
Williams, J.G., 378
Williams, J.H., 271
Williams, J.L., 148, 473
William, J.N.II., 424
Williams, J.R., 212
Williams, J.W., 8, 70
Williams, K.B., 448
Williams, K.L., 188
Williams, L.A., 507
Williams, L.F., 78, 79
Williams, L.G., 10, 166
Williams, L.P., 69
Williams, L.S., 150
Williams, Sr.M., 37
Williams, M.E., 437
Williams, M.G., 469
Williams, M.H., 330
Williams, M.L., 5, 269
Williams, N., 88
Williams, N.B., 411
Williams, N.D., 350
Williams, N.E., 160
Williams, O., 71
Williams, O.D., 341
Williams, P.H., 519
Williams, P.K., 371
Williams, P.P., 351
Williams, R., 80, 95
Williams, R.C., 9, 43, 80, 318, 491
Williams, R.K., 450
Williams, R.L., 401, 556
Williams, R.N., 363
Williams, R.P., 449
Williams, R.R., 415
Williams, R.W., 6, 176, 317
Williams, S., 398
Williams, S.C., 40
Williams, T., 88, 192
Williams, T.F., 328
Williams, T.H., 160
Williams, T.S., 9
Williams, T.W., 509
Williams, V.F., 469
Williams, W.A., 17
Williams, W.F., 201
Williams, W.L., 105, 251
Williams, W.P., 424
Williams-Ashman, H.G., 131, 133
Williamson, C.E., 304
Williamson, C.K., 360
Williamson, D.G., 548
Williamson, D.H., 544
Williamson, H., 398
Williamson, J.H., 531
Williamson, J.R., 46, 263
Williamson, K.C., 493
Williamson, O.C., 8
Williamson, S., 215
Williges, G.G., 456
Willingham, A.K., 273
Willis, E.R., 122
Willis, G.L., 371
Willis, H.L., 522
Willis, J., 369, 513
Willis, J.H., 136, 137
Willis, J.S., 137
Willis, N., 5
Willis, W.D., 467
Willis, W.G., 168
Willis, W.S., 485
Willman, A.M., 552
Willoughby, E.J., 199
Willoughby, P.J., 199
Willows, A.O.D., 498
Wills, C.J., 59
Wills, I., 21
Willson, G.P.III., 363
Willson, M.F., 136
Wilmes, N., 224
Wilmoth, J., 317
Wilson, A.C., 42
Wilson, B., 188
Wilson, B.E., 450
Wilson, B.G., 552
Wilson, B.J., 446
Wilson, B.R., 288
Wilson, B.W., 48
Wilson, C.A., 249
Wilson, C.L., 366
Wilson, C.M., 138, 199, 556
Wilson, C.R., 321
Wilson, D., 302
Wilson, D.A., 133
Wilson, D.B., 53, 302
Wilson, D.E., 154, 489
Wilson, D.E.Jr., 314
Wilson, D.F., 360, 410
Wilson, D.G., 552
Wilson, D.L., 96
Wilson, E., 410, 494
Wilson, E.E., 50
Wilson, E.M., 9, 12
Wilson, E.O., 211
Wilson, F., 283
Wilson, F.E., 166
Wilson, G., 428
Wilson, G.A., 328
Wilson, G.M., 189
Wilson, G.R., 365
Wilson, H., 132
Wilson, H.C., 224
Wilson, H.J., 12
Wilson, H.R., 93
Wilson, I.E., 340
Wilson, J., 542
Wilson, J.A., 368, 390
Wilson, J.B., 100, 384, 517
Wilson, J.C., 178
Wilson, J.E., 225, 448
Wilson, J.F., 342
Wilson, J.H., 222, 345, 384
Wilson, J.L., 370, 445, 446
Wilson, J.S., 164
Wilson, J.W.III., 485
Wilson, K., 31, 471
Wilson, K.G., 360
Wilson, K.S., 150
Wilson, L., 42, 226, 443, 471
Wilson, L.A., 185
Wilson, L.L., 401
Wilson, M.C., 148
Wilson, M.F., 509
Wilson, M.L., 317
Wilson, N.A., 162
Wilson, P.H., 422
Wilson, R., 524
Wilson, R.C., 28
Wilson, R.E., 199, 255, 360
Wilson, R.F., 365
Wilson, R.H., 329
Wilson, R.P., 246
Wilson, S.L., 155
Wilson, S.Z., 449
Wilson, T., 74
Wilson, T.E., 191
Wilson, T.H., 7, 212
Wilson, T.K., 360
Wilson, T.L., 286
Wilson, V.E., 502
Wilson, W., 333, 495
Wilson, W.D., 323
Wilson, W.E., 360, 555
Wilson, W.E.Jr., 341
Wilson, W.J., 40
Wilson, W.L., 230
Wilson, W.O., 48
Wilson, W.T., 81
Wilt, J.C., 538
Wilt, F.H., 43
Wilt, G.R., 4
Wiltbank, W.J., 92
Wilton, A.C., 401
Wilton, J.W., 547
Wiltshire, C.T., 253
Wilz, K., 395
Wimber, D.E., 387
Wimer, L.T., 429
Wimmer, R.B., 169
Wimsatt, W., 302
Winborn, W.B., 469
Winch, F.E.Jr., 303
Windell, J.T., 71
Winder, R.B., 312
Winder, W.C., 518
Windhager, E.E., 308
Windom, H.L., 102
Winegrad, S., 411
Wineland, M.J., 519
Winer, A.D., 179
Winer, M.N., 179
Winfree, A.T., 150
Wing, J.M., 91
Wing, L.D., 157
Wing, M., 382

Wing, M.W., 395
Wingenfeld, J., 356
Winger, M., 352
Winget, G.D., 370
Wingo, C.O., 199
Wingo, C.W., 259
Wingo, W.J., 11
Winicur, S., 147
Winier, L.P., 162
Winkelmann, J.R., 396
Winkelstein, W.W., 46
Winking, E.M., 255
Winkler, A.J., 51
Winkler, B.S., 230
Winkler, H.H., 490
Winn, H.E., 421
Winner, A.M., 61
Winner, R.W., 360
Winnett, G., 288
Winsness, K.A., 243
Winstead, C., 186
Winstead, C.W., 186
Winstead, J., 181, 486
Winstead, N.N., 345
Winsten, S., 408
Winston, P.W., 71
Winston, R.A., 192
Winston, S., 60
Winter, C.E., 36
Winter, C.G., 24
Winter, F.H., 49
Winter, H.C., 404
Winter, J.E., 313
Winter, P.A., 98
Winterhalder, E.K., 545
Winternheimer, P.L., 152
Winters, R., 95
Winters, W.D., 56
Winterton, B.W., 477
Winward, A.H., 382
Winzer, J., 168
Wirt, N.S., 380
Wirth, M., 279
Wirtz, G.H., 509
Wirtz, W.O.II., 33
Wisdon, C., 469
Wise, A.Jr., 261
Wise, B., 438
Wise, B.N., 359
Wise, D.D., 551
Wise, D.L., 358
Wise, E.G., 324
Wise, E.S., 484
Wise, G., 96
Wise, G.H., 343
Wise, H.S., 487
Wise, J.P., 96
Wise, M.B., 493
Wise, W.C., 429
Wiseman, G.M., 538
Wiseman, H.A.B., 95
Wiseman, L.L., 484
Wiseman, R.F., 176
Wiser, C.W., 439
Wislinsky, B., 36
Wismar, B.L., 367
Wisner, H.P., 381
Wison, R.G., 282
Wisseman, C.L.Jr., 203
Wissig, S.L., 60
Wissing, T.E., 360
Wissler, R.W., 133
Wissmar, R.C., 498
Wistendahl, W.A., 368
Wistrand, H., 99
Witcher, W., 425
Witham, F.H., 403
Withee, L.V., 166
Witherspoon, A.M., 343
Witherspoon, E.C., 331
Witherspoon, J.D., 439
Withner, C.L., 296, 297
Witkin, E.M., 288
Witkowski, J., 157
Witkovsky, P., 301
Witlin, B., 405
Witman, G.B., 287
Witorsch, R.J., 492
Witowle, D., 488
Witt, W., 177
Witte, D.L., 161
Witten, H.W., 488
Wittenbach, Sr.A., 223
Wittenberg, B.A., 330
Wittenberg, J.B., 330
Witters, R.E., 502
Witters, W., 368
Wittig, G., 130
Wittig, K.P., 316
Wittlake, E.G., 20
Wittle, L.W., 222
Wittliff, J.L., 328
Wittmeyer, E.C., 365
Wittrup, R., 126
Wittry, Sr.E., 236
Wittwer, L.S., 522
Wittwer, S., 226
Witzell, O., 393
Wivagg, D.E., 123
Wixom, R.L., 261
Wixon, S.E., 12
Wobeser, G., 563
Wochok, Z.S., 10
Wodzinski, R.J., 88
Wodzinski, R.S., 309
Woelfel, A., 511
Woelke, C.E., 498
Woelkerling, W.J., 515
Woelfel, C., 480
Woese, C.R., 137
Woessner, J.F., 95
Woessner, R.A., 458
Wogan, C.N., 213
Wohler, J.R., 390
Wohlgemuth, K., 434
Wohlpart, A., 359
Wohlschlag, D.E., 465
Wojciechowski, N., 445
Wojcik, F.J., 490
Wolanskyj, B.M., 556
Wolbarsht, M., 333
Wolcott, T.G., 346
Wold, F., 241
Woldow, N.L., 254
Woletz, E.W., 118
Wolf, A.V., 134
Wolf, B., 411, 412
Wolf, D.C., 200
Wolf, E.A., 93
Wolf, G., 213
Wolf, K.E., 304
Wolf, L.L., 326
Wolf, P.L., 398
Wolf, R.C., 521
Wolf, T.M., 173
Wolfdenen, R.V., 340
de Wolfe, M.S., 542
Wolfe, A.F., 398
Wolfe, B.M., 551
Wolfe, F.H., 530
Wolfe, H.G., 171
Wolfe, J.L., 248, 250
Wolfe, J.S., 80
Wolfe, M.L., 476, 477
Wolfe, R., 175
Wolfe, R.R., 288
Wolfe, R.S., 137
Wolfe, S., 312
Wolfe, S.L., 52
Wolfenbarger, D.O., 91
Wolff, A.E., 11
Wolff, D.A., 363
Wolff, D.M., 162
Wolff, E.T., 309
Wolff, G., 24
Wolff, S., 60
Wolford, F.T., 393
Wolfson, A., 126
Wolfson, N., 556
Wolgamott, G., 378
Wolgast, L.J., 288
Wolgemuth, J.C., 494
Wolinsky, E., 356
Wolk, C.P., 226
Wolley, J.T., 138
Wollin, E., 408
Wollman, H., 431
Wollman, N., 293
Wollmuth, R., 350
Wollum, A.G., 345
Wolstenholme, D.R., 473
Wolterink, L.F., 228
Womack, F.C., 447
Womack, H., 199
Womack, J.B., 448
Womack, S.J., 336
Womble, C.H., 426
Wong, J.R., 549
Wong, R.L., 134
Wong, T.W., 133
Wong-Riley, M.T.T., 60
Woo, A.L.W., 431
Wood, B.W., 472
Wood, C.E., 211, 456
Wood, C.L., 302, 304
Wood, D.L., 44
Wood, D.M., 544
Wood, D.R., 65
Wood, F.E., 201
Wood, G.C., 445
Wood, G.M., 481
Wood, H.G., 355
Wood, J.D., 172, 562
Wood, J.G., 445
Wood, J.L., 490
Wood, J.M., 241, 242, 260
Wood, J.S., 236
Wood, K.G., 323
Wood, N.P., 421
Wood, P.B., 423
Wood, P.J., 230
Wood, R., 95, 175, 261
Wood, R.C., 467
Wood, R.D., 421
Wood, S.L., 472
Wood, T.K., 372
Wood, T.S., 372
Wood, W.A., 225
Wood, W.B., 25
Woodard, A.E., 48
Woodard, W., 113
Woodbridge, C.G., 503
Woodbury, L.A., 473
Woodbury, R., 418
Woodbury, W., 537
Woodcock, A.E.R., 220
Woodcock, C.L.F., 218
Woodfin, B.M., 294
Woodford, V.R., 562
Woodhouse, B.L., 105
Woodin, H.E., 479
Woodin, S.A., 199
Woodka, S.J., 522
Woodland, D.W., 557
Woodland, J., 207
Woodley, C.L., 251
Woodliff, F., 7
Woodring, J.P., 183
Woodring, R., 394
Woodruff, D.S., 150
Woodruff, G.M., 258
Woodruff, G.W., 20
Woodruff, J.E., 514
Woodruff, J.J., 325
Woodruff, J.R., 423
Woodruff, K.H., 60
Woodruff, R.I., 415
Woods, C.A., 480
Woods, F.H., 253
Woods, F.W., 445
Woods, J.E., 120
Woods, J.J., 76
Woods, J.W., 380
Woods, K.R., 307
Woods, P., 298
Woods, P.S., 297
Woods, W.J., 338
Woodside, K.H., 405
Woodson, B.R., 494
Woodward, A.E., 298
Woodward, C.H., 507
Woodward, C.K., 241
Woodward, D.O., 41
Woodward, J.M., 442
Woodward, V.W., 242
Woodwell, G.M., 81
Woodwick, K., 30
Woodworth, R.C., 482
Woody, C.D., 57
Woody, C.O., 77
Woodyard, W.T., 70
Wool, I.G., 131
Woolcott, W.S., 488
Wooldridge, G.L., 474
Woolf, C.H., 15
Woolf, C.M., 15
Woolfenden, G.E., 97
Woolfold, P.G., 177
Woolfolk, C.A., 54
Woolfolk, E.O., 437
Woollacott, R.M., 211
Wooley, D.E., 47
Woolley, T.A., 68
Woolpy, J.H., 143
Woolsey, M., 381
Woolsey, T.A., 262
Wooster, W.W., 96
Wooten, J.W., 252
Wooten, T.E., 425
Wootton, D.M., 29
Worcel, A., 286
Worcester, J., 212
Wordinger, R.J., 315
Worf, G.L., 519
Work, R., 96
Work, T.H., 56
Workman, G.W., 476, 477
Workman, J., 476
Workman, M., 66
Workman, R.G., 91
World, P., 434
Worley, D., 265, 369
Worley, D.E., 266
Worley, E., 297
Worley, I.A., 480
Wormuth, J., 459
Worrell, A.C., 81
Worsham, A.D., 344
Wortham, K., 349
Worthington, C.R., 391
Worthington, R.D., 468
Worthington, W.C., 428
Worthman, R.P., 504
Worton, R., 550
Wostman, B.S., 152
Wott, J.A., 149
Wrathall, C.R., 83
Wray, G.W., 449
Wray, V.P., 449
de Wreede, R.E., 534
Wrenn, W.J., 352
Wrenshall, G.A., 550
Wrestler, F.A., 142
Wright, A., 217
Wright, B.S., 540
Wright, C.E., 471
Wright, C.G., 344
Wright, C.P., 348
Wright, D.K.Jr., 356
Wright, D.L., 523
Wright, E., 177
Wright, E.M., 57
Wright, E.R., 150
Wright, F., 81
Wright, Br.G., 291
Wright, H.A., 463
Wright, H.E., 242
Wright, H.F., 79
Wright, H.M., 383
Wright, J., 265
Wright, J.A., 186, 538
Wright, J.C., 21, 507
Wright, J.E., 403
Wright, J.L., 474
Wright, J.W., 62
Wright, K.A., 550
Wright, L.D., 302, 306
Wright, L.N., 16
Wright, M., 329
Wright, M.J., 301, 302
Wright, Sr.M.L., 209
Wright, N.S., 534
Wright, P.L., 267
Wright, R., 61
Wright, R.A., 471
Wright, R.D., 62, 494
Wright, R.E., 547
Wright, R.R., 60
Wright, R.T., 210
Wright, R.W., 127

Wright, S., 519
Wright, T.D., 228
Wright, T.R.F., 488
Wright, W., 3
Wright, W.R., 422
Wrightstone, R., 103
Wrogemann, K., 538
Wrolstad, R.E., 383
Wu, A.S.H., 382
Wu, C., 124, 232
Wu, C.H., 333
Wu, C.K., 222
Wu, C.W., 330
Wu, H.C., 79
Wu, J.H., 29
Wu, R., 302
Wu, W.G., 40
Wu, Y.H., 330
Wuest, P.J., 403
Wuevsohor, J.F., 334
Wujek, D.E., 223
Wulff, B., 75
Wulff, D.L., 54
Wullstein, L.H., 473
Wunder, B.A., 68
Wunder, C., 162
Wurf, H.K., 299
Wurst, R.P., 74
Wurster, R.D., 125
Wurtman, R.J., 213
Wust, C.J., 442
Wuthier, P., 72
Wuthier, R.E., 482
Wyand, D.S., 78
Wyandt, H., 388
Wyatt, E.J., 238
Wyatt, G.R., 546
Wyatt, R.L., 347
Wyche, J.H., 261
Wyckoff, -., 81
Wycoff, H., 103
Wydoski, R., 476
Wydoski, R.S., 477, 498
Wyland, J.V., 209
Wylie, W.D., 23
Wylie, W.L., 508
Wyllie, G., 116
Wyllie, T.D., 260
Wyman, J.F., 247
Wyman, R.H., 80
Wynn, T.E., 343
Wynne, J.C., 344
Wynne, M.J., 465
Wynne, P.F., 255
Wypych, J.I., 319
Wyrick, P., 339
Wyse, B., 475
Wyse, D.L., 240
Wyse, G.A., 218
Wysocki, A.A., 28
Wysocki, G.P., 552
Wysong, D., 270
Wyss, O., 465
Wythe, L.D., 458
Wyttenbach, C.W., 171

X

Xavier, K.S., 222
Xong, N., 59

Y

Yadav, R., 189
Yaden, S., 451
Yaeger, R.G., 190
Yager, J.G., 445
Yahiku, P.Y., 36
Yahner, J.E., 147
Yahr, F.I., 55
Yakatis-Surbis, A., 96
Yale, S.H., 134
Yam, P., 273
Yamada, E.W., 538
Yamaguchi, H., 144
Yamaguchi, H.G., 145
Yamaguchi, M., 51, 110
Yamaguchi, S., 8
Yamamoto, H.Y., 113
Yamamoto, M., 445
Yamamoto, N., 408
Yamamoto, R.T., 344
Yamamoto, W.S., 56
Yamashiro, S., 548
Yamashiro, S.M., 63
Yamashiroya, H.M., 135
Yamazaki, H., 544
Yamazaki, W.T., 364
Yambert, P.A., 129
Yanagimachi, R., 112
Yand, C.C., 12
Yand, S.P., 461
Yanders, A.F., 260
Yandle, D.O., 334
Yang, C.H., 445
Yahg, C.S., 282
Yang, H.Y., 383
Yang, J.T., 60
Yang, S.F., 51
Yang, W., 96
Yankeelov, J.A.Jr., 180
Yankow, M., 313
Yannone, M.E., 428
Yanofsky, C., 41
Yarbrough, C.G., 331
Yarbrough, G.G., 562
Yarbrough, J.A., 336
Yarbrough, J.D., 248
Yarger, W., 333
Yarnall, J.L., 34
Yarns, D., 329
Yarrison, C.W., 397
Yarus, M.J., 71
Yarwood, C.E., 45
Yasunobu, K.T., 112
Yates, F.E., 62
Yates, H.O., 437
Yates, J., 488
Yates, L.A., 337
Yates, W.F.Jr., 142
Yates, W.G., 497
Yaw, K., 205
Yaw, W.M., 129
Ycas, M., 326
Yeager, J.N., 508
Yeager, V.L., 256
Yeargers, E.K., 102
Yearian, W.C., 23
Yeaton, R.I., 409
Yeats, F., 335
Yee, W., 114
Yee, Y.C., 53
Yeh, J.Z., 333
Yeh, Y.C., 24
Yeh, Y.Y., 445
Yelenc, J., 406
Yelling, E.L., 330
Yelton, D.B., 509
Yemma, J., 373
Yendol, W.G., 402
Yengoyan, A.A., 234
Yensen, A.E., 246
Yerg, D., 228
Yermanos, D.M., 58
Yguerabide, J., 59
Yielding, K.L., 11
Yingling, H.C.Jr., 309
Yipintsoi, T., 330
Yochim, J.M., 171
Yocom, C.F., 34
Yocum, E.A., 389
Yocom, T.R., 139
Yocum, C.F., 231
Yocum, C.S., 231
Yoder, O.C., 304
Yoder, R.D., 484
Yoder, R.E., 364
Yoder, W.A., 196
Yoes, M.E., 454
Yoesting, D.R., 156
Yohn, D.S., 363
Yokel, B., 96
Yokey, P., 6
Yonce, H.D., 423
Yonce, R.L., 340
Yonenaka, H.H., 40
Yonetani, T., 410
Yongue, W.H., 492
Yonke, T.R., 259
Yoo, B.Y., 510
Yoon, C.H., 207
Yoon, M., 542
Yopp, J.H., 128
York, A.C., 149
York, D.H., 546
York, G.K., 47, 49
York, J.L., 24
York, J.O., 22
York, R., 195
Yos, D.A., 291
Yoshida, T., 79
Yoshikawa, F., 344
Yost, H.T., 206
Yotis, W., 124
Youmans, A.S., 127
Youmans, G.P., 127
Youmans, W.B., 521
Younathan, E.S., 184
Young, -., 545
Young, A.C., 500
Young, A.M., 512
Young, B., 448
Young, B.A., 530
Young, B.G., 202
Young, B.L., 3
Young, C.S.H., 301
Young, D.A., 328, 344
Young, D.F., 249, 360
Young, D.L., 209
Young, D.W., 5
Young, E., 202
Young, E.P., 200
Young, E.T.II., 499
Young, F.E., 328
Young, F.N., 145
Young, G., 207
Young, G.B., 399
Young, G.E., 382
Young, G.L., 501
Young, G.M., 537
Young, H., 515
Young, H.C.Jr., 375
Young, H.E., 194
Young, J., 377, 524
Young, J.A., 275
Young, J.H., 32
Young, J.L., 384
Young, J.O., 272
Young, L.B., 26
Young, L.G., 101, 547
Young, M.C., 467
Young, P.A., 256
Young, P.G., 546
Young, P.L., 182
Young, R., 420
Young, R.A., 139, 276
Young, R.E., 58, 250
Young, R.G., 303
Young, R.J., 305, 508
Young, R.S., 508
Young, R.W., 493
Young, S., 69
Young, S.C., 373
Young, S.S.Y., 362, 363
Young, S.Y.III., 23
Young, V.R., 213
Young, W.A., 184
Young, W.C., 455
Young, W.J., 482
Youngberg, C.T., 384
Younggren, N.A., 18
Youngman, A., 173
Youngman, V.E., 65
Youngner, J.S., 413
Youngner, V.B., 58
Youngs, P., 15
Youngs, V.L., 516
Youngs, W.D., 303, 332
Younkin, D.E., 401
Younoszai, R., 243
Yount, W.J., 339
Youse, H.R., 143
Youssef, G., 562
Youssef, N.N., 476
Yousten, A.A., 492
Yow, F.W., 359
Yozwiak, B.J., 373
Yphantis, D.A., 77, 79
Yu, B.P., 470
Yu, M., 282
Yu, P.L., 146
Yu, R.B., 414
Yu, R.C., 467
Yu, S., 386
Yu, T.C., 383
Yuan, J.H., 493
Yuan, L.L., 124
Yuan, M., 326
Yuan, T.L., 93
Yuhas, J.G., 358
Yuill, T.M., 520
Yund, M.A., 234
Yungbluth, T.A., 181
Yunghans, W., 323
Yunis, A.A., 95
Yurkiewicz, W.J., 400
YuSun, C.C.C., 293
Yutzy, D., 484
Yuyama, -., 550

Z

Zabel, R.A., 321
Zabik, M., 226
Zabin, I., 56
Zable, G.L., 3
Zablotney, S., 223
Zabransky, R.J., 513
Zacakowski, N.K., 239
Zaccaria, R.A., 398
Zaccour, F., 558
Zachariah, K., 551
Zacharuk, R.Y., 561
Zachary, A.L., 147
Zacik, S., 530
Zadorozny, E., 391
Zadunaisaky, J., 313
Zady, M., 181
Zaffarano, D.J., 155
Zagata, M.D., 194
Zagroski, S.J., 395
Zahalsky, A., 130
Zahalsky, A.C., 130
Zahara, M.B., 51
Zaharis, J., 214
Zaharopoulos, P., 467
Zahler, S., 302
Zahler, W.L., 261
Zahn, L., 59
Zaitlin, M., 304
Zajac, B.A., 399
Zajic, J.E., 551
Zak, B., 385
Zak, J.M., 219
Zakrzewski, S.F., 321
Zaleski, M.B., 227
Zalik, S.E., 531
Zalkin, H., 148
Zam, S.G., 90
Zamenhof, P.J., 56
Zamenhof, S., 56
Zamora, B.A., 503
Zamord, C.S., 504
Zamrernard, J.L., 327
Zand, R., 232
Zapisek, W.F., 296
Zar, J.H., 126
Zaranek, J., 415
Zarcaro, R.M., 419
Zarenko, P., 395
Zarkower, A., 403
Zartman, D.L., 292
Zaugg, W., 36
Zavarin, E., 44
Zbarsky, S.H., 534
Zeakes, S.J., 169
Zechman, F.W., 180
Zedler, J.B., 39
Zee, P., 445
Zeevaart, J.A.D., 226
Zehm, W., 222
Zehnder, L.R., 393
Zehner, C.E., 517
Zehr, E.I., 425
Zehr, J.E., 137
Zeigel, R.F., 320
Zeiger, D.C., 345

Zeikus, J.G., 517
Zeineh, R.A., 133
Zeiner, F.N., 73
Zeit, W., 512
Zeitler, E., 132
Zelazny, L.W., 93
Zelenski, K., 319
Zelitch, I., 81
Zell, L.W., 238
Zelle, M.R., 69'
Zeller, E.A., 127
Zeller, F., 145
Zellis, R.F., 405
Zemach, R.B., 233
Zeman, F.J., 50
Zeman, W., 146
Zemp, J.W., 428
Zengel-Messer, J., 96
Zenisek, C., 396
Zenner, E., 33
Zentmyer, G.A., 58
Zeppa, R., 95
Zerbe, G.O., 72
Zerbion, V.R., 312
Zettergren, L.D., 6
Zettler, F.W., 92
Zeuthen, M., 37
Ziboh, V., 95
Ziegler, E.L., 138
Ziegler, H.R., 328
Ziegler, J.H., 401, 402
Ziegler, J.R., 218
Ziegler, L.W., 92
Ziegler, R.F., 228
Zieman, J.C., 489
Ziemke, D., 253
Zigmond, M., 412
Zilinsky, J., 523
Ziller, H.H., 189
Zillioux, E., 96
Zilversmit, D.B., 302, 306
Zimbrick, J.D., 171
Zimdahl, R.L., 68
Zimmer, D., 352
Zimmer, G.P., 323
Zimmer, R.L., 62
Zimmerer, R.P., 397
Zimmerman, A.M., 550
Zimmerman, B., 197, 198
Zimmerman, D.C., 352
Zimmerman, E.G., 452
Zimmerman, G., 497
Zimmerman, G.S., 372
Zimmerman, J., 216, 292
Zimmerman, J.L., 166
Zimmerman, L.H., 16
Zimmerman, T.F., 135
Zimmerman, U.D., 120
Zimmerman, W.F., 206
Zimmermann, I.D., 400
Zimmermann, L.N., 404
Zimmermann, M.H., 210
Zimmermann, R., 217
Zimmy, M.L., 185
Zimpfer, P.E., 354
Zimprich, R., 350
Zindel, H.C., 226
Zingmark, R.G., 429
Zink, D.L., 458
Zink, F.W., 51
Zink, M.W., 562
Zinke, P.J., 44
Zinn, D.J., 421
Zinn, D.W., 462
Zinn, G.W., 508
Zinnerman, H., 244
Zinsmeister, P.P., 104
Zipf, Sr.M.E., 299
Zischke, J., 240
Zissler, J.F., 244
Zivnuska, J.A., 44
Zobel, B.J., 346
Zobel, D.B., 385
Zoellner, K.O., 167
Zolman, J.F., 180
Zorach, T., 329
Zorn, E.E., 365
Zornetzer, S.F., 94
Zorzoli, A., 329
Zouros, E., 542
Zscheile, E.P., 47
Zsigray, R.M., 279
Zubay, G., 300
Zuber, M.S., 258
Zubkoff, P.L., 490
Zubrzycki, L., 408
Zuck, F.M., 284
Zuck, R.K., 284
Zuck, R.R., 38
Zucker, I.H., 273
Zucker, N., 292
Zucker, W.V., 18
Zull, J., 355
Zusy, D.R., 154
Zuzolo, R., 297
Zwadyk, P.J., 333
Zweerink, H.J., 333
Zwerman, P.J., 302
Zwerner, D.E., 490
Zwick, M., 132
Zwickel, F.C., 531
Zwilling, B.S., 363
Zwolinski, M.J., 18
Zych, C.C., 138